Conversion Factors

DIMENSION	METRIC	METRIC/ENGLISH
Acceleration	$1 \text{ m/s}^2 = 100 \text{ cm/s}^2$	$1 \text{ m/s}^2 = 3.2808 \text{ ft/s}^2$ $1 \text{ ft/s}^2 = 0.3048^* \text{ m/s}^2$
Area	$1 \text{ m}^2 = 10^4 \text{ cm}^2 = 10^6 \text{ mm}^2 = 10^{-6} \text{ km}^2$	$1 \text{ m}^2 = 1550 \text{ in}^2 = 10.764 \text{ ft}^2$ $1 \text{ ft}^2 = 144 \text{ in}^2 = 0.09290304^* \text{ m}^2$
Density	$1 \text{ g/cm}^3 = 1 \text{ kg/L} = 1000 \text{ kg/m}^3$	$1 \text{ g/cm}^3 = 62.428 \text{ lbm/ft}^3 = 0.036127 \text{ lbm/in}^3$ $1 \text{ lbm/in}^3 = 1728 \text{ lbm/ft}^3$ $1 \text{ kg/m}^3 = 0.062428 \text{ lbm/ft}^3$
Energy, heat, work, and specific energy	$1 \text{ kJ} = 1000 \text{ J} = 1000 \text{ N} \cdot \text{m} = 1 \text{ kPa} \cdot \text{m}^3$ $1 \text{ kJ/kg} = 1000 \text{ m}^2/\text{s}^2$ $1 \text{ kWh} = 3600 \text{ kJ}$	$1 \text{ kJ} = 0.94782 \text{ Btu}$ $1 \text{ Btu} = 1.055056 \text{ kJ}$ $\quad = 5.40395 \text{ psia} \cdot \text{ft}^3 = 778.169 \text{ lbf} \cdot \text{ft}$ $1 \text{ Btu/lbm} = 25{,}037 \text{ ft}^2/\text{s}^2 = 2.326^* \text{ kJ/kg}$ $1 \text{ kWh} = 3412.14 \text{ Btu}$
Force	$1 \text{ N} = 1 \text{ kg} \cdot \text{m/s}^2 = 10^5 \text{ dyne}$ $1 \text{ kgf} = 9.80665 \text{ N}$	$1 \text{ N} = 0.22481 \text{ lbf}$ $1 \text{ lbf} = 32.174 \text{ lbm} \cdot \text{ft/s}^2 = 4.44822 \text{ N}$ $1 \text{ lbf} = 1 \text{ slug} \cdot \text{ft/s}^2$
Length	$1 \text{ m} = 100 \text{ cm} = 1000 \text{ mm} = 10^6 \text{ }\mu\text{m}$ $1 \text{ km} = 1000 \text{ m}$	$1 \text{ m} = 39.370 \text{ in} = 3.2808 \text{ ft} = 1.0926 \text{ yd}$ $1 \text{ ft} = 12 \text{ in} = 0.3048^* \text{ m}$ $1 \text{ mile} = 5280 \text{ ft} = 1.6093 \text{ km}$ $1 \text{ in} = 2.54^* \text{ cm}$
Mass	$1 \text{ kg} = 1000 \text{ g}$ $1 \text{ metric ton} = 1000 \text{ kg}$	$1 \text{ kg} = 2.2046226 \text{ lbm}$ $1 \text{ lbm} = 0.45359237^* \text{ kg}$ $1 \text{ ounce} = 28.3495 \text{ g}$ $1 \text{ slug} = 32.174 \text{ lbm} = 14.5939 \text{ kg}$ $1 \text{ short ton} = 2000 \text{ lbm} = 907.1847 \text{ kg}$
Power	$1 \text{ W} = 1 \text{ J/s}$ $1 \text{ kW} = 1000 \text{ W} = 1 \text{ kJ/s}$ $1 \text{ hp}^\ddagger = 745.7 \text{ W}$	$1 \text{ kW} = 3412.14 \text{ Btu/h} = 1.341 \text{ hp}$ $\quad = 737.56 \text{ lbf} \cdot \text{ft/s}$ $1 \text{ hp} = 550 \text{ lbf} \cdot \text{ft/s} = 0.7068 \text{ Btu/s}$ $\quad = 42.41 \text{ Btu/min} = 2544.5 \text{ Btu/h}$ $\quad = 0.74570 \text{ kW}$ $1 \text{ Btu/h} = 1.055056 \text{ kJ/h}$
Pressure or stress, and pressure expressed as a head	$1 \text{ Pa} = 1 \text{ N/m}^2$ $1 \text{ kPa} = 10^3 \text{ Pa} = 10^{-3} \text{ MPa}$ $1 \text{ atm} = 101.325 \text{ kPa} = 1.01325 \text{ bar}$ $\quad = 760 \text{ mm Hg at } 0°\text{C}$ $\quad = 1.03323 \text{ kgf/cm}^2$ $1 \text{ mm Hg} = 0.1333 \text{ kPa}$	$1 \text{ Pa} = 1.4504 \times 10^{-4} \text{ psi}$ $\quad = 0.020886 \text{ lbf/ft}^2$ $1 \text{ psi} = 144 \text{ lbf/ft}^2 = 6.894757 \text{ kPa}$ $1 \text{ atm} = 14.696 \text{ psi}$ $\quad = 29.92 \text{ inches Hg at } 30°\text{F}$ $1 \text{ inch Hg} = 13.60 \text{ inches H}_2\text{O} = 3.387 \text{ kPa}$
Specific heat	$1 \text{ kJ/kg} \cdot °\text{C} = 1 \text{ kJ/kg} \cdot \text{K}$ $\quad = 1 \text{ J/g} \cdot °\text{C}$	$1 \text{ Btu/lbm} \cdot °\text{F} = 4.1868 \text{ kJ/kg} \cdot °\text{C}$ $1 \text{ Btu/lbmol} \cdot \text{R} = 4.1868 \text{ kJ/kmol} \cdot \text{K}$ $1 \text{ kJ/kg} \cdot °\text{C} = 0.23885 \text{ Btu/lbm} \cdot °\text{F}$ $\quad = 0.23885 \text{ Btu/lbm} \cdot \text{R}$
Specific volume	$1 \text{ m}^3/\text{kg} = 1000 \text{ L/kg}$ $\quad = 1000 \text{ cm}^3/\text{g}$	$1 \text{ m}^3/\text{kg} = 16.02 \text{ ft}^3/\text{lbm}$ $1 \text{ ft}^3/\text{lbm} = 0.062428 \text{ m}^3/\text{kg}$
Temperature	$T(\text{K}) = T(°\text{C}) = 273.15$ $\Delta T(\text{K}) = \Delta T(°\text{C})$	$T(\text{R}) = T(°\text{F}) + 459.67 = 1.8 T(\text{K})$ $T(°\text{F}) = 1.8 \, T(°\text{C}) + 32$ $\Delta T(°\text{F}) = \Delta T(\text{R}) = 1.8^* \, \Delta T(\text{K})$
Velocity	$1 \text{ m/s} = 3.60 \text{ km/h}$	$1 \text{ m/s} = 3.2808 \text{ ft/s} = 2.237 \text{ mi/h}$ $1 \text{ mi/h} = 1.46667 \text{ ft/s}$ $1 \text{ mi/h} = 1.6093 \text{ km/h}$
Viscosity, dynamic	$1 \text{ kg/m} \cdot \text{s} = 1 \text{ N} \cdot \text{s/m}^2 = 1 \text{ Pa} \cdot \text{s} = 10 \text{ poise}$	$1 \text{ kg/m} \cdot \text{s} = 2419.1 \text{ lbm/ft} \cdot \text{h}$ $\quad = 0.020886 \text{ lbf} \cdot \text{s/ft}^2$ $\quad = 0.67197 \text{ lbm/ft} \cdot \text{s}$

*Exact conversion factor between metric and English units.

‡Mechanical horsepower. The electrical horsepower is taken to be exactly 746 W.

DIMENSION	METRIC	METRIC/ENGLISH
Viscosity, kinematic	1 m²/s = 10⁴ cm²/s 1 stoke = 1 cm²/s = 10⁻⁴ m²/s	1 m²/s = 10.764 ft²/s = 3.875 × 10⁴ ft²/h 1 m²/s = 10.764 ft²/s
Volume	1 m³ = 1000 L = 10⁶ cm³ (cc)	1 m³ = 6.1024 × 10⁴ in³ = 35.315 ft³ = 264.17 gal (U.S.) 1 U.S. gallon = 231 in³ = 3.7854 L 1 fl ounce = 29.5735 cm³ = 0.0295735 L 1 U.S. gallon = 128 fl ounces
Volume flow rate	1 m³/s = 60,000 L/min = 10⁶ cm³/s	1 m³/s = 15,850 gal/min = 35.315 ft³/s = 2118.9 ft³/min (CFM)

*Exact conversion factor between metric and English units.

Some Physical Constants

PHYSICAL CONSTANT	METRIC	ENGLISH
Standard acceleration of gravity	g = 9.80665 m/s²	g = 32.174 ft/s²

Standard atmospheric pressure	P_{atm} = 1 atm = 101.325 kPa
	    = 1.01325 bar
	    = 760 mm Hg (0°C)
	    = 10.3323 m H₂O (4°C)
		P_{atm} = 1 atm = 14.696 psia
		    = 2116.2 lbf/ft²
		    = 29.9213 inches Hg (32°F)
		    = 406.78 inches H₂O (39.2°F)

Universal gas constant	R_u = 8.31447 kJ/kmol · K
	    = 8.31447 kN · m/kmol · K
		R_u = 1.9859 Btu/lbmol · R
		    = 1545.37 ft · lbf/lbmol · R

Commonly Used Properties

PROPERTY	METRIC	ENGLISH
Air at 20°C (68°F) and 1 atm		

Specific gas constant*	R_{air} = 0.2870 kJ/kg · K
	    = 287.0 m²/s² · K
		R_{air} = 0.06855 Btu/lbm · R
		    = 53.34 ft · lbf/lbm · R
		    = 1716 ft²/s² · R

Specific heat ratio	$k = c_p/c_v$ = 1.40	$k = c_p/c_v$ = 1.40

Specific heats	c_p = 1.007 kJ/kg · K
	    = 1007 m²/s² · K
	c_v = 0.7200 kJ/kg · K
	    = 720.0 m²/s² · K
		c_p = 0.2404 Btu/lbm · R
		    = 187.1 ft · lbf/lbm · R
		    = 6019 ft²/s² · R
		c_v = 0.1719 Btu/lbm · R
		    = 133.8 ft · lbf/lbm · R
		    = 4304 ft²/s² · R

Speed of sound	c = 343.2 m/s = 1236 km/h	c = 1126 ft/s = 767.7 mi/h
Density	ρ = 1.204 kg/m³	ρ = 0.07518 lbm/ft³
Viscosity	μ = 1.825 × 10⁻⁵ kg/m · s	μ = 1.227 × 10⁻⁵ lbm/ft · s
Kinematic viscosity	ν = 1.516 × 10⁻⁵ m²/s	ν = 1.632 × 10⁻⁴ ft²/s
Liquid water at 20°C (68°F) and 1 atm		

Specific heat ($c = c_p = c_v$)	c = 4.182 kJ/kg · K
	    = 4182 m²/s² · K
		c = 0.9989 Btu/lbm · R
		    = 777.3 ft · lbf/lbm · R
		    = 25,009 ft²/s² · R

Density	ρ = 998.0 kg/m³	ρ = 62.30 lbm/ft³
Viscosity	μ = 1.002 × 10⁻³ kg/m · s	μ = 6.733 × 10⁻⁴ lbm/ft · s
Kinematic viscosity	ν = 1.004 × 10⁻⁶ m²/s	ν = 1.081 × 10⁻⁵ ft²/s

* Independent of pressure or temperature

FLUID MECHANICS

FUNDAMENTALS AND APPLICATIONS

Third Edition in SI Units

FLUID MECHANICS

FUNDAMENTALS AND APPLICATIONS

THIRD EDITION IN SI UNITS

YUNUS A. ÇENGEL
*Department of Mechanical Engineering
University of Nevada, Reno*

JOHN M. CIMBALA
*Department of Mechanical and Nuclear Engineering
The Pennsylvania State University*

SI Conversion by
MEHMET KANOĞLU
*Department of Mechanical Engineering
University of Gaziantep*

FLUID MECHANICS: FUNDAMENTALS AND APPLICATIONS
THIRD EDITION IN SI UNITS

Copyright © 2014 by McGraw-Hill Education. All rights reserved. Previous editions © 2006 and 2010. No part of this publication may be reproduced or distributed in any form or by any means, or stored in a database or retrieval system, without the prior written consent of the publisher, including, but not limited to, in any network or other electronic storage or transmission, or broadcast for distance learning.

Some ancillaries, including electronic and print components, may not be available to customers outside the United States.

Exclusive rights by McGraw-Hill Education (Asia) for adaptation, manufacture and export. This book cannot be re-exported from the country to which it is sold by McGraw-Hill.

Cover image © Shutterstock.com

10 9 8 7 6 5 4 3 2 1
CTP MPM
20 16 15 14 13

When ordering this title, use **ISBN 978-1-259-01122-1 or MHID 1-259-01122-4**.

Printed in Singapore

Dedication

To all students, with the hope of stimulating their desire to explore our marvelous world, of which fluid mechanics is a small but fascinating part. And to our wives Zehra and Suzy for their unending support.

About the Authors

Yunus A. Çengel is Professor Emeritus of Mechanical Engineering at the University of Nevada, Reno. He received his B.S. in mechanical engineering from Istanbul Technical University and his M.S. and Ph.D. in mechanical engineering from North Carolina State University. His research areas are renewable energy, desalination, energy analysis, heat transfer enhancement, radiation heat transfer, and energy conservation. He served as the director of the Industrial Assessment Center (IAC) at the University of Nevada, Reno, from 1996 to 2000. He has led teams of engineering students to numerous manufacturing facilities in Northern Nevada and California to do industrial assessments, and has prepared energy conservation, waste minimization, and productivity enhancement reports for them.

Dr. Çengel is the coauthor of the widely adopted textbook *Thermodynamics: An Engineering Approach*, 7th edition (2011), published by McGraw-Hill. He is also the co-author of the textbook *Heat and Mass Transfer: Fundamentals & Applications*, 4th Edition (2011), and the coauthor of the textbook *Fundamentals of Thermal-Fluid Sciences*, 4th edition (2012), both published by McGraw-Hill. Some of his textbooks have been translated to Chinese, Japanese, Korean, Spanish, Turkish, Italian, and Greek.

Dr. Çengel is the recipient of several outstanding teacher awards, and he has received the ASEE Meriam/Wiley Distinguished Author Award for excellence in authorship in 1992 and again in 2000.

Dr. Çengel is a registered Professional Engineer in the State of Nevada, and is a member of the American Society of Mechanical Engineers (ASME) and the American Society for Engineering Education (ASEE).

John M. Cimbala is Professor of Mechanical Engineering at The Pennsylvania State University, University Park. He received his B.S. in Aerospace Engineering from Penn State and his M.S. in Aeronautics from the California Institute of Technology (CalTech). He received his Ph.D. in Aeronautics from CalTech in 1984 under the supervision of Professor Anatol Roshko, to whom he will be forever grateful. His research areas include experimental and computational fluid mechanics and heat transfer, turbulence, turbulence modeling, turbomachinery, indoor air quality, and air pollution control. Professor Cimbala completed sabbatical leaves at NASA Langley Research Center (1993-94), where he advanced his knowledge of computational fluid dynamics (CFD), and at Weir American Hydo (2010-11), where he performed CFD analyses to assist in the design of hydroturbines.

Dr. Cimbala is the coauthor of three other textbooks: *Indoor Air Quality Engineering: Environmental Health and Control of Indoor Pollutants* (2003), published by Marcel-Dekker, Inc.; *Essentials of Fluid Mechanics: Fundamentals and Applications* (2008); and *Fundamentals of Thermal-Fluid Sciences*, 4th edition (2012), both published by McGraw-Hill. He has also contributed to parts of other books, and is the author or co-author of dozens of journal and conference papers. More information can be found at www.mne.psu.edu/cimbala.

Professor Cimbala is the recipient of several outstanding teaching awards and views his book writing as an extension of his love of teaching. He is a member of the American Institute of Aeronautics and Astronautics (AIAA), the American Society of Mechanical Engineers (ASME), the American Society for Engineering Education (ASEE), and the American Physical Society (APS).

BRIEF CONTENTS

CHAPTER ONE
INTRODUCTION AND BASIC CONCEPTS 1

CHAPTER TWO
PROPERTIES OF FLUIDS 37

CHAPTER THREE
PRESSURE AND FLUID STATICS 75

CHAPTER FOUR
FLUID KINEMATICS 133

CHAPTER FIVE
BERNOULLI AND ENERGY EQUATIONS 185

CHAPTER SIX
MOMENTUM ANALYSIS OF FLOW SYSTEMS 243

CHAPTER SEVEN
DIMENSIONAL ANALYSIS AND MODELING 291

CHAPTER EIGHT
INTERNAL FLOW 347

CHAPTER NINE
DIFFERENTIAL ANALYSIS OF FLUID FLOW 437

CHAPTER TEN
APPROXIMATE SOLUTIONS OF THE NAVIER–STOKES EQUATION 515

CHAPTER ELEVEN
EXTERNAL FLOW: DRAG AND LIFT 607

CHAPTER TWELVE
COMPRESSIBLE FLOW 659

CHAPTER THIRTEEN
OPEN-CHANNEL FLOW 725

CHAPTER FOURTEEN
TURBOMACHINERY 787

CHAPTER FIFTEEN
INTRODUCTION TO COMPUTATIONAL FLUID DYNAMICS 879

Contents

Preface   xvii

CHAPTER ONE
INTRODUCTION AND BASIC CONCEPTS   1

1–1 Introduction   2
 What Is a Fluid?   2
 Application Areas of Fluid Mechanics   4

1–2 A Brief History of Fluid Mechanics   6

1–3 The No-Slip Condition   8

1–4 Classification of Fluid Flows   9
 Viscous versus Inviscid Regions of Flow   10
 Internal versus External Flow   10
 Compressible versus Incompressible Flow   10
 Laminar versus Turbulent Flow   11
 Natural (or Unforced) versus Forced Flow   11
 Steady versus Unsteady Flow   12
 One-, Two-, and Three-Dimensional Flows   13

1–5 System and Control Volume   14

1–6 Importance of Dimensions and Units   15
 Some SI and English Units   17
 Dimensional Homogeneity   19
 Unity Conversion Ratios   20

1–7 Modeling in Engineering   21

1–8 Problem-Solving Technique   23
 Step 1: Problem Statement   24
 Step 2: Schematic   24
 Step 3: Assumptions and Approximations   24
 Step 4: Physical Laws   24
 Step 5: Properties   24
 Step 6: Calculations   24
 Step 7: Reasoning, Verification, and Discussion   25

1–9 Engineering Software Packages   25
 Engineering Equation Solver (EES)   26
 CFD Software   27

1–10 Accuracy, Precision, and Significant Digits   28
 Summary   31
 References and Suggested Reading   31
 Application Spotlight: What Nuclear Blasts and Raindrops Have in Common   32
 Problems   33

CHAPTER TWO
PROPERTIES OF FLUIDS   37

2–1 Introduction   38
 Continuum   38

2–2 Density and Specific Gravity   39
 Density of Ideal Gases   40

2–3 Vapor Pressure and Cavitation   41

2–4 Energy and Specific Heats   43

2–5 Compressibility and Speed of Sound   44
 Coefficient of Compressibility   44
 Coefficient of Volume Expansion   46
 Speed of Sound and Mach Number   48

2–6 Viscosity   50

2–7 Surface Tension and Capillary Effect   55
 Capillary Effect   58
 Summary   61
 Application Spotlight: Cavitation   62
 References and Suggested Reading   63
 Problems   63

CHAPTER THREE
PRESSURE AND FLUID STATICS   75

3–1 Pressure   76
 Pressure at a Point   77
 Variation of Pressure with Depth   78

3–2 Pressure Measurement Devices   81
 The Barometer   81
 The Manometer   84
 Other Pressure Measurement Devices   88

3–3 Introduction to Fluid Statics   89

3–4 Hydrostatic Forces on Submerged Plane Surfaces   89
 Special Case: Submerged Rectangular Plate   92

3–5 Hydrostatic Forces on Submerged Curved Surfaces   95

3–6 Buoyancy and Stability 98
    Stability of Immersed and Floating Bodies 101
3–7 Fluids in Rigid-Body Motion 103
    Special Case 1: Fluids at Rest 105
    Special Case 2: Free Fall of a Fluid Body 105
    Acceleration on a Straight Path 106
    Rotation in a Cylindrical Container 107

    Summary 111
    References and Suggested Reading 112
    Problems 112

CHAPTER FOUR
FLUID KINEMATICS 133

4–1 Lagrangian and Eulerian Descriptions 134
    Acceleration Field 136
    Material Derivative 139
4–2 Flow Patterns and Flow Visualization 141
    Streamlines and Streamtubes 141
    Pathlines 142
    Streaklines 144
    Timelines 146
    Refractive Flow Visualization Techniques 147
    Surface Flow Visualization Techniques 148
4–3 Plots of Fluid Flow Data 148
    Profile Plots 149
    Vector Plots 149
    Contour Plots 150
4–4 Other Kinematic Descriptions 151
    Types of Motion or Deformation of Fluid Elements 151
4–5 Vorticity and Rotationality 156
    Comparison of Two Circular Flows 159
4–6 The Reynolds Transport Theorem 160
    Alternate Derivation of the Reynolds Transport Theorem 165
    Relationship between Material Derivative and RTT 167

    Summary 168
    Application Spotlight: Fluidic Actuators 169
    References and Suggested Reading 170
    Problems 170

CHAPTER FIVE
BERNOULLI AND ENERGY EQUATIONS 185

5–1 Introduction 186
    Conservation of Mass 186
    The Linear Momentum Equation 186
    Conservation of Energy 186
5–2 Conservation of Mass 187
    Mass and Volume Flow Rates 187
    Conservation of Mass Principle 189
    Moving or Deforming Control Volumes 191
    Mass Balance for Steady-Flow Processes 191
    Special Case: Incompressible Flow 192
5–3 Mechanical Energy and Efficiency 194
5–4 The Bernoulli Equation 199
    Acceleration of a Fluid Particle 199
    Derivation of the Bernoulli Equation 200
    Force Balance across Streamlines 202
    Unsteady, Compressible Flow 202
    Static, Dynamic, and Stagnation Pressures 202
    Limitations on the Use of the Bernoulli Equation 204
    Hydraulic Grade Line (HGL) and Energy Grade Line (EGL) 205
    Applications of the Bernoulli Equation 207
5–5 General Energy Equation 214
    Energy Transfer by Heat, Q 215
    Energy Transfer by Work, W 215
5–6 Energy Analysis of Steady Flows 219
    Special Case: Incompressible Flow with No Mechanical Work Devices and Negligible Friction 221
    Kinetic Energy Correction Factor, α 221

    Summary 228
    References and Suggested Reading 229
    Problems 230

CHAPTER SIX
MOMENTUM ANALYSIS OF FLOW SYSTEMS 243

6–1 Newton's Laws 244
6–2 Choosing a Control Volume 245
6–3 Forces Acting on a Control Volume 246
6–4 The Linear Momentum Equation 249
    Special Cases 251
    Momentum-Flux Correction Factor, β 251
    Steady Flow 253
    Flow with No External Forces 254
6–5 Review of Rotational Motion and Angular Momentum 263

6–6 The Angular Momentum Equation   265
   Special Cases   267
   Flow with No External Moments   268
   Radial-Flow Devices   269

   Application Spotlight: Manta Ray Swimming   273

   Summary   275
   References and Suggested Reading   275
   Problems   276

CHAPTER SEVEN
DIMENSIONAL ANALYSIS AND MODELING   291

7–1 Dimensions and Units   292

7–2 Dimensional Homogeneity   293
   Nondimensionalization of Equations   294

7–3 Dimensional Analysis and Similarity   299

7–4 The Method of Repeating Variables and The Buckingham Pi Theorem   303

   Historical Spotlight: Persons Honored by Nondimensional Parameters   311

7–5 Experimental Testing, Modeling, and Incomplete Similarity   319
   Setup of an Experiment and Correlation of Experimental Data   319
   Incomplete Similarity   320
   Wind Tunnel Testing   320
   Flows with Free Surfaces   323

   Application Spotlight: How a Fly Flies   326

   Summary   327
   References and Suggested Reading   327
   Problems   327

CHAPTER EIGHT
INTERNAL FLOW   347

8–1 Introduction   348

8–2 Laminar and Turbulent Flows   349
   Reynolds Number   350

8–3 The Entrance Region   351
   Entry Lengths   352

8–4 Laminar Flow in Pipes   353
   Pressure Drop and Head Loss   355
   Effect of Gravity on Velocity and Flow Rate in Laminar Flow   357
   Laminar Flow in Noncircular Pipes   358

8–5 Turbulent Flow in Pipes   361
   Turbulent Shear Stress   363
   Turbulent Velocity Profile   364
   The Moody Chart and the Colebrook Equation   367
   Types of Fluid Flow Problems   369

8–6 Minor Losses   374

8–7 Piping Networks and Pump Selection   381
   Series and Parallel Pipes   381
   Piping Systems with Pumps and Turbines   383

8–8 Flow Rate and Velocity Measurement   391
   Pitot and Pitot-Static Probes   391
   Obstruction Flowmeters: Orifice, Venturi, and Nozzle Meters   392
   Positive Displacement Flowmeters   396
   Turbine Flowmeters   397
   Variable-Area Flowmeters (Rotameters)   398
   Ultrasonic Flowmeters   399
   Electromagnetic Flowmeters   401
   Vortex Flowmeters   402
   Thermal (Hot-Wire and Hot-Film) Anemometers   402
   Laser Doppler Velocimetry   404
   Particle Image Velocimetry   406
   Introduction to Biofluid Mechanics   408

   Application Spotlight: PIV Applied to Cardiac Flow   416

   Summary   417
   References and Suggested Reading   418
   Problems   419

CHAPTER NINE
DIFFERENTIAL ANALYSIS OF FLUID FLOW   437

9–1 Introduction   438

9–2 Conservation of Mass—The Continuity Equation   438
   Derivation Using the Divergence Theorem   439
   Derivation Using an Infinitesimal Control Volume   440
   Alternative Form of the Continuity Equation   443
   Continuity Equation in Cylindrical Coordinates   444
   Special Cases of the Continuity Equation   444

9–3 The Stream Function   450
   The Stream Function in Cartesian Coordinates   450
   The Stream Function in Cylindrical Coordinates   457
   The Compressible Stream Function   458

9–4 The Differential Linear Momentum Equation—Cauchy's Equation 459
  Derivation Using the Divergence Theorem 459
  Derivation Using an Infinitesimal Control Volume 460
  Alternative Form of Cauchy's Equation 463
  Derivation Using Newton's Second Law 463

9–5 The Navier–Stokes Equation 464
  Introduction 464
  Newtonian versus Non-Newtonian Fluids 465
  Derivation of the Navier–Stokes Equation for Incompressible, Isothermal Flow 466
  Continuity and Navier–Stokes Equations in Cartesian Coordinates 468
  Continuity and Navier–Stokes Equations in Cylindrical Coordinates 469

9–6 Differential Analysis of Fluid Flow Problems 470
  Calculation of the Pressure Field for a Known Velocity Field 470
  Exact Solutions of the Continuity and Navier–Stokes Equations 475
  Differential Analysis of Biofluid Mechanics Flows 493

  Application Spotlight: The No-Slip Boundary Condition 498

  Summary 499
  References and Suggested Reading 499
  Problems 499

CHAPTER TEN
APPROXIMATE SOLUTIONS OF THE NAVIER–STOKES EQUATION 515

10–1 Introduction 516
10–2 Nondimensionalized Equations of Motion 517
10–3 The Creeping Flow Approximation 520
  Drag on a Sphere in Creeping Flow 523
10–4 Approximation for Inviscid Regions of Flow 525
  Derivation of the Bernoulli Equation in Inviscid Regions of Flow 526
10–5 The Irrotational Flow Approximation 529
  Continuity Equation 529
  Momentum Equation 531
  Derivation of the Bernoulli Equation in Irrotational Regions of Flow 531
  Two-Dimensional Irrotational Regions of Flow 534
  Superposition in Irrotational Regions of Flow 538
  Elementary Planar Irrotational Flows 538
  Irrotational Flows Formed by Superposition 545

10–6 The Boundary Layer Approximation 554
  The Boundary Layer Equations 559
  The Boundary Layer Procedure 564
  Displacement Thickness 568
  Momentum Thickness 571
  Turbulent Flat Plate Boundary Layer 572
  Boundary Layers with Pressure Gradients 578
  The Momentum Integral Technique for Boundary Layers 583

  Summary 591
  References and Suggested Reading 592

  Application Spotlight: Droplet Formation 593
  Problems 594

CHAPTER ELEVEN
EXTERNAL FLOW: DRAG AND LIFT 607

11–1 Introduction 608
11–2 Drag and Lift 610
11–3 Friction and Pressure Drag 614
  Reducing Drag by Streamlining 615
  Flow Separation 616
11–4 Drag Coefficients of Common Geometries 617
  Biological Systems and Drag 618
  Drag Coefficients of Vehicles 621
  Superposition 623
11–5 Parallel Flow Over Flat Plates 625
  Friction Coefficient 627
11–6 Flow Over Cylinders And Spheres 629
  Effect of Surface Roughness 632
11–7 Lift 634
  Finite-Span Wings and Induced Drag 638
  Lift Generated by Spinning 639

  Summary 643
  References and Suggested Reading 644

  Application Spotlight: Drag Reduction 645
  Problems 646

CHAPTER TWELVE
COMPRESSIBLE FLOW 659

12–1 Stagnation Properties 660
12–2 One-Dimensional Isentropic Flow 663
  Variation of Fluid Velocity with Flow Area 665
  Property Relations for Isentropic Flow of Ideal Gases 667

12–3 Isentropic Flow Through Nozzles 669
   Converging Nozzles 670
   Converging–Diverging Nozzles 674

12–4 Shock Waves and Expansion Waves 678
   Normal Shocks 678
   Oblique Shocks 684
   Prandtl–Meyer Expansion Waves 688

12–5 Duct Flow With Heat Transfer and Negligible Friction (Rayleigh Flow) 693
   Property Relations for Rayleigh Flow 699
   Choked Rayleigh Flow 700

12–6 Adiabatic Duct Flow With Friction (Fanno Flow) 702
   Property Relations for Fanno Flow 705
   Choked Fanno Flow 708

   Application Spotlight: Shock-Wave/Boundary-Layer Interactions 712

   Summary 713
   References and Suggested Reading 714
   Problems 714

CHAPTER THIRTEEN
OPEN-CHANNEL FLOW 725

13–1 Classification of Open-Channel Flows 726
   Uniform and Varied Flows 726
   Laminar and Turbulent Flows in Channels 727

13–2 Froude Number and Wave Speed 729
   Speed of Surface Waves 731

13–3 Specific Energy 733

13–4 Conservation of Mass and Energy Equations 736

13–5 Uniform Flow in Channels 737
   Critical Uniform Flow 739
   Superposition Method for Nonuniform Perimeters 740

13–6 Best Hydraulic Cross Sections 743
   Rectangular Channels 745
   Trapezoidal Channels 745

13–7 Gradually Varied Flow 747
   Liquid Surface Profiles in Open Channels, $y(x)$ 749
   Some Representative Surface Profiles 752
   Numerical Solution of Surface Profile 754

13–8 Rapidly Varied Flow and The Hydraulic Jump 757

13–9 Flow Control and Measurement 761
   Underflow Gates 762
   Overflow Gates 764

   Application Spotlight: Bridge Scour 771
   Summary 772
   References and Suggested Reading 773
   Problems 773

CHAPTER FOURTEEN
TURBOMACHINERY 787

14–1 Classifications and Terminology 788

14–2 Pumps 790
   Pump Performance Curves and Matching a Pump to a Piping System 791
   Pump Cavitation and Net Positive Suction Head 797
   Pumps in Series and Parallel 800
   Positive-Displacement Pumps 803
   Dynamic Pumps 806
   Centrifugal Pumps 806
   Axial Pumps 816

14–3 Pump Scaling Laws 824
   Dimensional Analysis 824
   Pump Specific Speed 827
   Affinity Laws 829

14–4 Turbines 833
   Positive-Displacement Turbines 834
   Dynamic Turbines 834
   Impulse Turbines 835
   Reaction Turbines 837
   Gas and Steam Turbines 847
   Wind Turbines 847

14–5 Turbine Scaling Laws 855
   Dimensionless Turbine Parameters 855
   Turbine Specific Speed 857

   Application Spotlight: Rotary Fuel Atomizers 861
   Summary 862
   References and Suggested Reading 862
   Problems 863

CHAPTER FIFTEEN
INTRODUCTION TO COMPUTATIONAL FLUID DYNAMICS 879

15–1 Introduction and Fundamentals 880
   Motivation 880
   Equations of Motion 880

Solution Procedure   881
Additional Equations of Motion   883
Grid Generation and Grid Independence   883
Boundary Conditions   888
Practice Makes Perfect   893

15–2 Laminar CFD Calculations   893
Pipe Flow Entrance Region at Re = 500   893
Flow around a Circular Cylinder at Re = 150   897

15–3 Turbulent CFD Calculations   902
Flow around a Circular Cylinder at Re = 10,000   905
Flow around a Circular Cylinder at Re = 10^7 907
Design of the Stator for a Vane-Axial Flow Fan   907

15–4 CFD With Heat Transfer   915
Temperature Rise through a Cross-Flow Heat Exchanger   915
Cooling of an Array of Integrated Circuit Chips   917

15–5 Compressible Flow CFD Calculations   922
Compressible Flow through a Converging–Diverging Nozzle   923
Oblique Shocks over a Wedge   927

15–6 Open-Channel Flow CFD Calculations   928
Flow over a Bump on the Bottom of a Channel   929
Flow through a Sluice Gate (Hydraulic Jump)   930

Application Spotlight: A Virtual Stomach   931
Summary   932
References and Suggested Reading   932
Problems   933

APPENDIX
PROPERTY TABLES AND CHARTS   939

TABLE A–1	Molar Mass, Gas Constant, and Ideal-Gas Specfic Heats of Some Substances 940	
TABLE A–2	Boiling and Freezing Point Properties 941	
TABLE A–3	Properties of Saturated Water 942	
TABLE A–4	Properties of Saturated Refrigerant-134a 943	
TABLE A–5	Properties of Saturated Ammonia 944	
TABLE A–6	Properties of Saturated Propane 945	
TABLE A–7	Properties of Liquids 946	
TABLE A–8	Properties of Liquid Metals 947	
TABLE A–9	Properties of Air at 1 atm Pressure 948	
TABLE A–10	Properties of Gases at 1 atm Pressure 949	
TABLE A–11	Properties of the Atmosphere at High Altitude 951	
FIGURE A–12	The Moody Chart for the Friction Factor for Fully Developed Flow in Circular Pipes 952	
TABLE A–13	One-Dimensional Isentropic Compressible Flow Functions for an Ideal Gas with $k = 1.4$ 953	
TABLE A–14	One-Dimensional Normal Shock Functions for an Ideal Gas with $k = 1.4$ 954	
TABLE A–15	Rayleigh Flow Functions for an Ideal Gas with $k = 1.4$ 955	
TABLE A–16	Fanno Flow Functions for an Ideal Gas with $k = 1.4$ 956	

Glossary   957
Index   971

PREFACE

BACKGROUND

Fluid mechanics is an exciting and fascinating subject with unlimited practical applications ranging from microscopic biological systems to automobiles, airplanes, and spacecraft propulsion. Fluid mechanics has also historically been one of the most challenging subjects for undergraduate students because proper analysis of fluid mechanics problems requires not only knowledge of the concepts but also physical intuition and experience. Our hope is that this book, through its careful explanations of concepts and its use of numerous practical examples, sketches, figures, and photographs, bridges the gap between knowledge and the proper application of that knowledge.

Fluid mechanics is a mature subject; the basic equations and approximations are well established and can be found in any introductory textbook. Our book is distinguished from other introductory books because we present the subject in a *progressive order* from simple to more difficult, building each chapter upon foundations laid down in earlier chapters. We provide more diagrams and photographs that other books because fluid mechanics, is by its nature, a highly visual subject. Only by illustrating the concepts discussed, can students fully appreciate the mathematical significance of the material.

OBJECTIVES

This book has been written for the first fluid mechanics course for undergraduate engineering students. There is sufficient material for a two-course sequence, if desired. We assume that readers will have an adequate background in calculus, physics, engineering mechanics, and thermodynamics. The objectives of this text are

- To present the *basic principles and equations* of fluid mechanics.
- To show numerous and diverse real-world *engineering examples* to give the student the intuition necessary for correct application of fluid mechanics principles in engineering applications.
- To develop an *intuitive understanding* of fluid mechanics by emphasizing the physics, and reinforcing that understanding through illustrative figures and photographs.

The book contains enough material to allow considerable flexibility in teaching the course. Aeronautics and aerospace engineers might emphasize potential flow, drag and lift, compressible flow, turbomachinery, and CFD, while mechanical or civil engineering instructors might choose to emphasize pipe flows and open-channel flows, respectively.

NEW TO THE THIRD EDITION

In this edition, the overall content and order of presentation has not changed significantly except for the following: the visual impact of all figures and photographs has been enhanced by a full color treatment. We also added new photographs throughout the book, often replacing existing diagrams with photographs in order to convey the practical real-life applications of the material. Several new Application Spotlights have been added to the end of selected chapters. These introduce students to industrial applications and exciting research projects being conducted by leaders in the field about material presented in the chapter. We hope these motivate students to see the relevance and application of the materials they are studying. New sections on Biofluids have been added to Chapters 8 and 9, written by guest author Keefe Manning of The Pennsylvania State University, along with bio-related examples and homework problems in those chapters.

New solved example problems were added to some chapters and several new end-of-chapter problems or modifications to existing problems were made to make them more versatile and practical. Most significant is the addition of Fundamentals of Engineering (FE) exam-type problems to help students prepare to take their Professional Engineering exams. Finally, the end-of-chapter problems that require Computational Fluid Dynamics (CFD) have been moved to the text website (www.mheducation.asia/olc/cengel) where updates based on software or operating system changes can be better managed.

PHILOSOPHY AND GOAL

The Third Edition of *Fluid Mechanics: Fundamentals and Applications* has the same goals and philosophy as the other texts by lead author Yunus Çengel.

- Communicates directly with tomorrow's engineers in a *simple yet precise* manner
- Leads students toward a clear understanding and firm grasp of the *basic principles* of fluid mechanics
- Encourages creative thinking and development of a *deeper understanding* and *intuitive feel* for fluid mechanics
- Is read by students with *interest* and *enthusiasm* rather than merely as a guide to solve homework problems

The best way to learn is by practice. Special effort is made throughout the book to reinforce the material that was presented earlier (in each chapter as well as in material from previous chapters). Many of the illustrated example problems and end-of-chapter problems are comprehensive and encourage students to review and revisit concepts and intuitions gained previously.

Throughout the book, we show examples generated by computational fluid dynamics (CFD). We also provide an introductory chapter on the subject. Our goal is not to teach the details about numerical algorithms associated with CFD—this is more properly presented in a separate course. Rather, our intent is to introduce undergraduate students to the capabilities and limitations of CFD as an *engineering tool*. We use CFD solutions in much the same way as experimental results are used from wind tunnel tests (i.e., to reinforce understanding of the physics of fluid flows and to provide quality flow visualizations that help explain fluid behavior). With dozens of CFD end-of-chapter problems posted on the website, instructors have ample opportunity to introduce the basics of CFD throughout the course.

CONTENT AND ORGANIZATION

This book is organized into 15 chapters beginning with fundamental concepts of fluids, fluid properties, and fluid flows and ending with an introduction to computational fluid dynamics.

- Chapter 1 provides a basic introduction to fluids, classifications of fluid flow, control volume versus system formulations, dimensions, units, significant digits, and problem-solving techniques.

- Chapter 2 is devoted to fluid properties such as density, vapor pressure, specific heats, speed of sound, viscosity, and surface tension.

- Chapter 3 deals with fluid statics and pressure, including manometers and barometers, hydrostatic forces on submerged surfaces, buoyancy and stability, and fluids in rigid-body motion.

- Chapter 4 covers topics related to fluid kinematics, such as the differences between Lagrangian and Eulerian descriptions of fluid flows, flow patterns, flow visualization, vorticity and rotationality, and the Reynolds transport theorem.

- Chapter 5 introduces the fundamental conservation laws of mass, momentum, and energy, with emphasis on the proper use of the mass, Bernoulli, and energy equations and the engineering applications of these equations.

- Chapter 6 applies the Reynolds transport theorem to linear momentum and angular momentum and emphasizes practical engineering applications of finite control volume momentum analysis.

- Chapter 7 reinforces the concept of dimensional homogeneity and introduces the Buckingham Pi theorem of dimensional analysis, dynamic similarity, and the method of repeating variables—material that is useful throughout the rest of the book and in many disciplines in science and engineering.

- Chapter 8 is devoted to flow in pipes and ducts. We discuss the differences between laminar and turbulent flow, friction losses in pipes and ducts, and minor losses in piping networks. We also explain how to properly select a pump or fan to match a piping network. Finally, we discuss various experimental devices that are used to measure flow rate and velocity, and provide a brief introduction to biofluid mechanics.

- Chapter 9 deals with differential analysis of fluid flow and includes derivation and application of the continuity equation, the Cauchy equation, and the Navier-Stokes equation. We also introduce the stream function and describe its usefulness in analysis of fluid flows, and we provide a brief introduction to biofluids. Finally, we point out some of the unique aspects of differential analysis related to biofluid mechanics.

- Chapter 10 discusses several *approximations* of the Navier–Stokes equation and provides example solutions for each approximation, including creeping flow, inviscid flow, irrotational (potential) flow, and boundary layers.

- Chapter 11 covers forces on bodies (drag and lift), explaining the distinction between friction and pressure drag, and providing drag coefficients for many common geometries. This chapter emphasizes the practical application of wind tunnel measurements coupled with dynamic similarity and dimensional analysis concepts introduced earlier in Chapter 7.

- Chapter 12 extends fluid flow analysis to compressible flow, where the behavior of gases is greatly affected by the Mach number. In this chapter, the concepts of expansion waves, normal and oblique shock waves, and choked flow are introduced.
- Chapter 13 deals with open-channel flow and some of the unique features associated with the flow of liquids with a free surface, such as surface waves and hydraulic jumps.
- Chapter 14 examines turbomachinery in more detail, including pumps, fans, and turbines. An emphasis is placed on how pumps and turbines work, rather than on their detailed design. We also discuss overall pump and turbine design, based on dynamic similarity laws and simplified velocity vector analyses.
- Chapter 15 describes the fundamental concepts of computational fluid dyamics (CFD) and shows students how to use commercial CFD codes as tools to solve complex fluid mechanics problems. We emphasize the *application* of CFD rather than the algorithms used in CFD codes.

Each chapter contains a wealth of end-of-chapter homework problems. A comprehensive set of appendix is provided, giving the thermodynamic and fluid properties of several materials, in addition to air and water, along with some useful plots and tables. Many of the end-of-chapter problems require the use of material properties from the appendices to enhance the realism of the problems.

LEARNING TOOLS

EMPHASIS ON PHYSICS

A distinctive feature of this book is its emphasis on the physical aspects of the subject matter in addition to mathematical representations and manipulations. The authors believe that the emphasis in undergraduate education should remain on *developing a sense of underlying physical mechanisms* and a *mastery of solving practical problems* that an engineer is likely to face in the real world. Developing an intuitive understanding should also make the course a more motivating and worthwhile experience for the students.

EFFECTIVE USE OF ASSOCIATION

An observant mind should have no difficulty understanding engineering sciences. After all, the principles of engineering sciences are based on our *everyday experiences* and *experimental observations*. Therefore, a physical, intuitive approach is used throughout this text. Frequently, *parallels are drawn* between the subject matter and students' everyday experiences so that they can relate the subject matter to what they already know.

SELF-INSTRUCTING

The material in the text is introduced at a level that an average student can follow comfortably. It speaks *to* students, not *over* students. In fact, it is *self-instructive*. Noting that the principles of science are based on experimental observations, most of the derivations in this text are largely based on physical arguments, and thus they are easy to follow and understand.

EXTENSIVE USE OF ARTWORK AND PHOTOGRAPHS

Figures are important learning tools that help the students "get the picture," and the text makes effective use of graphics. It contains more figures, photographs, and illustrations than any other book in this category. Figures attract attention and stimulate curiosity and interest. Most of the figures in this text are intended to serve as a means of emphasizing some key concepts that would otherwise go unnoticed; some serve as page summaries.

CONSISTENT COLOR SCHEME FOR FIGURES

The figures have a consistent color scheme applied for all arrows.

- Blue: (→) motion related, like velocity vectors
- Green: (→) force and pressure related, and torque
- Black: (→) distance related arrows and dimensions
- Red: (→) energy related, like heat and work
- Purple: (→) acceleration and gravity vectors, vorticity, and miscellaneous

NUMEROUS WORKED-OUT EXAMPLES

All chapters contain numerous worked-out *examples* that both clarify the material and illustrate the use of basic principles in a context that helps develops the student's intuition. An *intuitive* and *systematic* approach is used in the solution of all example problems. The solution methodology starts with a statement of the problem, and all objectives are identified. The assumptions and approximations are then stated together with their justifications. Any properties needed to solve the problem are listed separately. Numerical values are used together with numbers to emphasize that without units, numbers are meaningless. The significance of each example's result is discussed following the solution. This methodical approach is also followed and provided in the solutions to the end-of-chapter problems, available to instructors.

A WEALTH OF REALISTIC END-OF-CHAPTER PROBLEMS

The end-of-chapter problems are grouped under specific topics to make problem selection easier for both instructors and students. Within each group of problems are *Concept Questions,* indicated by "C," to check the students' level of understanding of basic concepts. Problems under *Fundamentals of Engineering (FE) Exam Problems* are designed to help students prepare for the *Fundamentals of Engineering* exam, as they prepare for their Professional Engineering license. The problems under *Review Problems* are more comprehensive in nature and are not directly tied to any specific section of a chapter—in some cases they require review of material learned in previous chapters. Problems designated as *Design and Essay* are intended to encourage students to make engineering judgments, to conduct independent exploration of topics of interest, and to communicate their findings in a professional manner. Problems with the icon are solved using EES, and complete solutions together with parametric studies are included the text website. Problems with the icon are comprehensive in nature and are intended to be solved with a computer, preferably using the EES software. Several economics- and safety-related problems are incorporated throughout to enhance cost and safety awareness among engineering students. Answers to selected problems are listed immediately following the problem for convenience to students.

USE OF COMMON NOTATION

The use of different notation for the same quantities in different engineering courses has long been a source of discontent and confusion. A student taking both fluid mechanics and heat transfer, for example, has to use the notation Q for volume flow rate in one course, and for heat transfer in the other. The need to unify notation in engineering education has often been raised, even in some reports of conferences sponsored by the National Science Foundation through Foundation Coalitions, but little effort has been made to date in this regard. For example, refer to the final report of the *Mini-Conference on Energy Stem Innovations*, May 28 and 29, 2003, University of Wisconsin. In this text we made a conscious effort to minimize this conflict by adopting the familiar thermodynamic notation $\dot{V}$ for volume flow rate, thus reserving the notation Q for heat transfer. Also, we consistently use an overdot to denote time rate. We think that both students and instructors will appreciate this effort to promote a common notation.

COMBINED COVERAGE OF BERNOULLI AND ENERGY EQUATIONS

The Bernoulli equation is one of the most frequently used equations in fluid mechanics, but it is also one of the most misused. Therefore, it is important to emphasize the limitations on the use of this idealized equation and to show how to properly account for imperfections and irreversible losses. In Chapter 5, we do this by introducing the energy equation right after the Bernoulli equation and demonstrating how the solutions of many practical engineering problems differ from those obtained using the Bernoulli equation. This helps students develop a realistic view of the Bernoulli equation.

A SEPARATE CHAPTER ON CFD

Commercial *Computational Fluid Dynamics* (CFD) codes are widely used in engineering practice in the design and analysis of flow systems, and it has become exceedingly important for engineers to have a solid understanding of the fundamental aspects, capabilities, and limitations of CFD. Recognizing that most undergraduate engineering curriculums do not have room for a full course on CFD, a separate chapter is included here to make up for this deficiency and to equip students with an adequate background on the strengths and weaknesses of CFD.

APPLICATION SPOTLIGHTS

Throughout the book are highlighted examples called *Application Spotlights* where a real-world application of fluid mechanics is shown. A unique feature of these special examples is that they are written by *guest authors*. The Application Spotlights are designed to show students how fluid mechanics has diverse applications in a wide variety of fields. They also include eye-catching photographs from the guest authors' research.

GLOSSARY OF FLUID MECHANICS TERMS

Throughout the chapters, when an important key term or concept is introduced and defined, it appears in **black** boldface type. Fundamental fluid mechanics terms and concepts appear in **red** boldface type, and these fundamental terms also appear in a comprehensive end-of-book glossary developed by Professor James Brasseur of The Pennsylvania State University. This unique glossary is an excellent learning and review tool for students as they move forward in their study of fluid mechanics. In addition, students can test their knowledge of these fundamental terms by using the interactive flash cards and other resources located on our accompanying website (www.mheducation.asia/olc/cengel).

CONVERSION FACTORS

Frequently used conversion factors, physical constants, and properties of air and water at 20°C and atmospheric pressure are listed on the front inner cover pages of the text for easy reference.

NOMENCLATURE

A list of the major symbols, subscripts, and superscripts used in the text are listed on the inside back cover pages of the text for easy reference.

SUPPLEMENTS

These supplements are available to adopters of the book:

Text Website

Web support is provided for the book on the text specific website at www.mheducation.asia/olc/cengel. Visit this robust site for book and supplement information, errata, author information, and further resources for instructors and students.

Engineering Equation Solver (EES)

Developed by Sanford Klein and William Beckman from the University of Wisconsin–Madison, this software combines equation-solving capability and engineering property data. EES can do optimization, parametric analysis, and linear and nonlinear regression, and provides publication-quality plotting capabilities. Thermodynamics and transport properties for air, water, and many other fluids are built-in and EES allows the user to enter property data or functional relationships.

ACKNOWLEDGMENTS

The authors would like to acknowledge with appreciation the numerous and valuable comments, suggestions, constructive criticisms, and praise from the following evaluators and reviewers of the third edition:

Bass Abushakra
Milwaukee School of Engineering

John G. Cherng
University of Michigan—Dearborn

Peter Fox
Arizona State University

Sathya Gangadbaran
Embry Riddle Aeronautical University

Jonathan Istok
Oregon State University

Tim Lee
McGill University

Nagy Nosseir
San Diego State University

Robert Spall
Utah State University

We also thank those who were acknowledged in the first and second editions of this book, but are too numerous to mention again here. Special thanks go to Gary S. Settles and his associates at Penn State (Lori Dodson-Dreibelbis, J. D. Miller, and Gabrielle Tremblay) for creating the exciting narrated video clips that are found on the book's website. The authors also thank James Brasseur of Penn State for creating the precise glossary of fluid mechanics terms, Glenn Brown of Oklahoma State for providing many items of historical interest throughout the text, guest authors David F. Hill (parts of Chapter 13) and Keefe Manning (sections on biofluids), Mehmet Kanoğlu of University of Gaziantep for preparing FE Exam problems and the solutions of EES problems, and Tahsin Engin of Sakarya University for contributing several end-of-chapter problems.

We also acknowledge the Korean translation team, who in the translation process, pointed out several errors and inconsistencies in the first and second editions that have now been corrected. The team includes Yun-ho Choi, Ajou University; Nae-Hyun Kim, University of Incheon; Woonjean Park, Korea University of Technology & Education; Wonnam Lee, Dankook University; Sang-Won Cha, Suwon University; Man Yeong Ha, Pusan National University; and Yeol Lee, Korea Aerospace University.

Finally, special thanks must go to our families, especially our wives, Zehra Çengel and Suzanne Cimbala, for their continued patience, understanding, and support throughout the preparation of this book, which involved many long hours when they had to handle family concerns on their own because their husbands' faces were glued to a computer screen.

Yunus A. Çengel
John M. Cimbala

Online Resources for Students and Instructors

Online Resources available at www.mheducation.asia/olc/cengel

Your home page for teaching and studying fluid mechanics, the *Fluid Mechanics: Fundamentals and Applications* text-specific website offers resources for both instructors and students.

For the student, this website offers various resources, including:

- **FE Exam Interactive Review Quizzes**—chapter-based self-quizzes provide hints for solutions and correct solution methods, and help students prepare for the NCEES Fundamentals of Engineering Examination.
- **Glossary of Key Terms in Fluid Mechanics**—full text and chapter-based glossaries.
- **Weblinks**—helpful weblinks to relevant fluid mechanics sites.

For the instructor, this password-protected website offers various resources, including:

- **Electronic Solutions Manual**—provides PDF files with detailed solutions to all text homework problems.
- **PowerPoint® Slides**—provide presentation slides for all chapters in the text for use in lectures.
- **Sample Syllabi**—make it easier for you to map out your course using this text for different course durations (one quarter, one semester, etc.) and for different disciplines (ME approach, Civil approach, etc.).
- **Transition Guides**—compare coverage to other popular introductory fluid mechanics books at the section level to aid transition to teaching from our text.
- **Links to ANSYS Workbench®, FLUENT FLOWLAB®, and EES (Engineering Equation Solver) download sites**—the academic versions of these powerful software programs are available free to departments of educational institutions who adopt this text.
- CFD homework problems and solutions designed for use with various CFD packages.

McGraw-Hill Connect® Engineering provides online presentation, assignment, and assessment solutions. It connects your students with the tools and resources they'll need to achieve success. With Connect Engineering, you can deliver assignments, quizzes, and tests online. A robust set of questions and activities are presented and aligned with the textbook's learning outcomes. As an instructor, you can edit existing questions and author entirely new problems. Track individual student performance—by question, assignment, or in relation to the class overall—with detailed grade reports. Integrate grade reports easily with Learning Management Systems (LMS), such as WebCT and Blackboard—and much more. ConnectPlus Engineering provides students with all the advantages of Connect Engineering, plus 24/7 online access to an eBook. This media-rich version of the book is available through the McGraw-Hill Connect platform and allows seamless integration of text, media, and assessments. To learn more, visit **www.mcgrawhillconnect.com.**

CHAPTER 1

INTRODUCTION AND BASIC CONCEPTS

In this introductory chapter, we present the basic concepts commonly used in the analysis of fluid flow. We start this chapter with a discussion of the phases of matter and the numerous ways of classification of fluid flow, such as *viscous versus inviscid regions of flow*, *internal versus external flow*, *compressible versus incompressible flow*, *laminar versus turbulent flow*, *natural versus forced flow*, and *steady versus unsteady flow*. We also discuss the *no-slip condition* at solid–fluid interfaces and present a brief history of the development of fluid mechanics.

After presenting the concepts of *system* and *control volume*, we review the *unit systems* that will be used. We then discuss how mathematical models for engineering problems are prepared and how to interpret the results obtained from the analysis of such models. This is followed by a presentation of an intuitive systematic *problem-solving technique* that can be used as a model in solving engineering problems. Finally, we discuss accuracy, precision, and significant digits in engineering measurements and calculations.

OBJECTIVES

When you finish reading this chapter, you should be able to

- Understand the basic concepts of fluid mechanics
- Recognize the various types of fluid flow problems encountered in practice
- Model engineering problems and solve them in a systematic manner
- Have a working knowledge of accuracy, precision, and significant digits, and recognize the importance of dimensional homogeneity in engineering calculations

Schlieren image showing the thermal plume produced by Professor Cimbala as he welcomes you to the fascinating world of fluid mechanics.
Michael J. Hargather and Brent A. Craven, Penn State Gas Dynamics Lab. Used by Permission.

INTRODUCTION AND BASIC CONCEPTS

FIGURE 1–1
Fluid mechanics deals with liquids and gases in motion or at rest.
© D. Falconer/PhotoLink/Getty RF

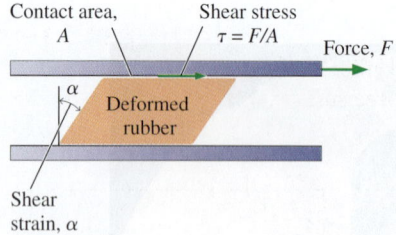

FIGURE 1–2
Deformation of a rubber block placed between two parallel plates under the influence of a shear force. The shear stress shown is that on the rubber—an equal but opposite shear stress acts on the upper plate.

1–1 ■ INTRODUCTION

Mechanics is the oldest physical science that deals with both stationary and moving bodies under the influence of forces. The branch of mechanics that deals with bodies at rest is called **statics,** while the branch that deals with bodies in motion is called **dynamics.** The subcategory **fluid mechanics** is defined as the science that deals with the behavior of fluids at rest (*fluid statics*) or in motion (*fluid dynamics*), and the interaction of fluids with solids or other fluids at the boundaries. Fluid mechanics is also referred to as **fluid dynamics** by considering fluids at rest as a special case of motion with zero velocity (Fig. 1–1).

Fluid mechanics itself is also divided into several categories. The study of the motion of fluids that can be approximated as incompressible (such as liquids, especially water, and gases at low speeds) is usually referred to as **hydrodynamics.** A subcategory of hydrodynamics is **hydraulics,** which deals with liquid flows in pipes and open channels. **Gas dynamics** deals with the flow of fluids that undergo significant density changes, such as the flow of gases through nozzles at high speeds. The category **aerodynamics** deals with the flow of gases (especially air) over bodies such as aircraft, rockets, and automobiles at high or low speeds. Some other specialized categories such as **meteorology, oceanography,** and **hydrology** deal with naturally occurring flows.

What Is a Fluid?

You will recall from physics that a substance exists in three primary phases: solid, liquid, and gas. (At very high temperatures, it also exists as plasma.) A substance in the liquid or gas phase is referred to as a **fluid.** Distinction between a solid and a fluid is made on the basis of the substance's ability to resist an applied shear (or tangential) stress that tends to change its shape. A solid can resist an applied shear stress by deforming, whereas *a fluid deforms continuously under the influence of a shear stress*, no matter how small. In solids, stress is proportional to *strain*, but in fluids, stress is proportional to *strain rate*. When a constant shear force is applied, a solid eventually stops deforming at some fixed strain angle, whereas a fluid never stops deforming and approaches a constant *rate* of strain.

Consider a rectangular rubber block tightly placed between two plates. As the upper plate is pulled with a force F while the lower plate is held fixed, the rubber block deforms, as shown in Fig. 1–2. The angle of deformation α (called the *shear strain* or *angular displacement*) increases in proportion to the applied force F. Assuming there is no slip between the rubber and the plates, the upper surface of the rubber is displaced by an amount equal to the displacement of the upper plate while the lower surface remains stationary. In equilibrium, the net force acting on the upper plate in the horizontal direction must be zero, and thus a force equal and opposite to F must be acting on the plate. This opposing force that develops at the plate–rubber interface due to friction is expressed as $F = \tau A$, where τ is the shear stress and A is the contact area between the upper plate and the rubber. When the force is removed, the rubber returns to its original position. This phenomenon would also be observed with other solids such as a steel block provided that the applied force does not exceed the elastic range. If this experiment were repeated with a fluid (with two large parallel plates placed in a large body of water, for example), the fluid layer in contact with the upper plate would move with the plate continuously at the velocity of the plate no matter how small the force F. The fluid velocity would decrease with depth because of friction between fluid layers, reaching zero at the lower plate.

You will recall from statics that **stress** is defined as force per unit area and is determined by dividing the force by the area upon which it acts. The normal component of a force acting on a surface per unit area is called the **normal stress,** and the tangential component of a force acting on a surface per unit area is called **shear stress** (Fig. 1–3). In a fluid at rest, the normal stress is called **pressure.** A fluid at rest is at a state of zero shear stress. When the walls are removed or a liquid container is tilted, a shear develops as the liquid moves to re-establish a horizontal free surface.

In a liquid, groups of molecules can move relative to each other, but the volume remains relatively constant because of the strong cohesive forces between the molecules. As a result, a liquid takes the shape of the container it is in, and it forms a free surface in a larger container in a gravitational field. A gas, on the other hand, expands until it encounters the walls of the container and fills the entire available space. This is because the gas molecules are widely spaced, and the cohesive forces between them are very small. Unlike liquids, a gas in an open container cannot form a free surface (Fig. 1–4).

Although solids and fluids are easily distinguished in most cases, this distinction is not so clear in some borderline cases. For example, *asphalt* appears and behaves as a solid since it resists shear stress for short periods of time. When these forces are exerted over extended periods of time, however, the asphalt deforms slowly, behaving as a fluid. Some plastics, lead, and slurry mixtures exhibit similar behavior. Such borderline cases are beyond the scope of this text. The fluids we deal with in this text will be clearly recognizable as fluids.

Intermolecular bonds are strongest in solids and weakest in gases. One reason is that molecules in solids are closely packed together, whereas in gases they are separated by relatively large distances (Fig. 1–5). The molecules in a solid are arranged in a pattern that is repeated throughout. Because of the small distances between molecules in a solid, the attractive forces of molecules on each other are large and keep the molecules at fixed positions. The molecular spacing in the liquid phase is not much different from that of

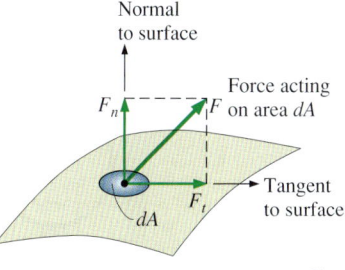

Normal stress: $\sigma = \dfrac{F_n}{dA}$

Shear stress: $\tau = \dfrac{F_t}{dA}$

FIGURE 1–3
The normal stress and shear stress at the surface of a fluid element. For fluids at rest, the shear stress is zero and pressure is the only normal stress.

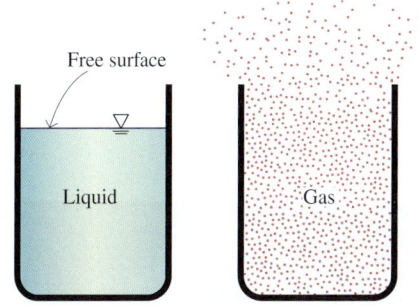

FIGURE 1–4
Unlike a liquid, a gas does not form a free surface, and it expands to fill the entire available space.

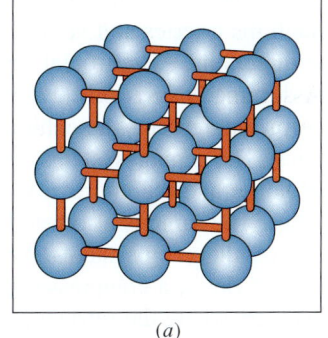
(a)

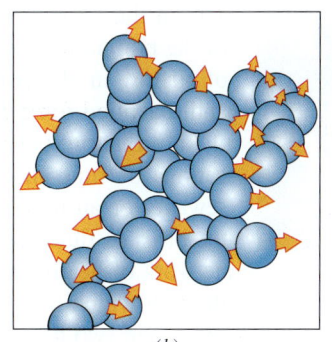
(b)

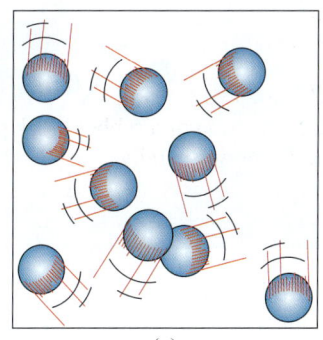
(c)

FIGURE 1–5
The arrangement of atoms in different phases: (a) molecules are at relatively fixed positions in a solid, (b) groups of molecules move about each other in the liquid phase, and (c) individual molecules move about at random in the gas phase.

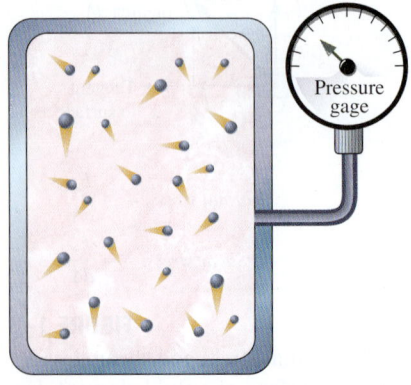

FIGURE 1–6
On a microscopic scale, pressure is determined by the interaction of individual gas molecules. However, we can measure the pressure on a macroscopic scale with a pressure gage.

the solid phase, except the molecules are no longer at fixed positions relative to each other and they can rotate and translate freely. In a liquid, the intermolecular forces are weaker relative to solids, but still strong compared with gases. The distances between molecules generally increase slightly as a solid turns liquid, with water being a notable exception.

In the gas phase, the molecules are far apart from each other, and molecular ordering is nonexistent. Gas molecules move about at random, continually colliding with each other and the walls of the container in which they are confined. Particularly at low densities, the intermolecular forces are very small, and collisions are the only mode of interaction between the molecules. Molecules in the gas phase are at a considerably higher energy level than they are in the liquid or solid phase. Therefore, the gas must release a large amount of its energy before it can condense or freeze.

Gas and *vapor* are often used as synonymous words. The vapor phase of a substance is customarily called a *gas* when it is above the critical temperature. *Vapor* usually implies that the current phase is not far from a state of condensation.

Any practical fluid system consists of a large number of molecules, and the properties of the system naturally depend on the behavior of these molecules. For example, the pressure of a gas in a container is the result of momentum transfer between the molecules and the walls of the container. However, one does not need to know the behavior of the gas molecules to determine the pressure in the container. It is sufficient to attach a pressure gage to the container (Fig. 1–6). This macroscopic or *classical* approach does not require a knowledge of the behavior of individual molecules and provides a direct and easy way to analyze engineering problems. The more elaborate microscopic or *statistical* approach, based on the average behavior of large groups of individual molecules, is rather involved and is used in this text only in a supporting role.

Application Areas of Fluid Mechanics

It is important to develop a good understanding of the basic principles of fluid mechanics, since fluid mechanics is widely used both in everyday activities and in the design of modern engineering systems from vacuum cleaners to supersonic aircraft. For example, fluid mechanics plays a vital role in the human body. The heart is constantly pumping blood to all parts of the human body through the arteries and veins, and the lungs are the sites of airflow in alternating directions. All artificial hearts, breathing machines, and dialysis systems are designed using fluid dynamics (Fig. 1–7).

An ordinary house is, in some respects, an exhibition hall filled with applications of fluid mechanics. The piping systems for water, natural gas, and sewage for an individual house and the entire city are designed primarily on the basis of fluid mechanics. The same is also true for the piping and ducting network of heating and air-conditioning systems. A refrigerator involves tubes through which the refrigerant flows, a compressor that pressurizes the refrigerant, and two heat exchangers where the refrigerant absorbs and rejects heat. Fluid mechanics plays a major role in the design of all these components. Even the operation of ordinary faucets is based on fluid mechanics.

We can also see numerous applications of fluid mechanics in an automobile. All components associated with the transportation of the fuel from the fuel tank to the cylinders—the fuel line, fuel pump, and fuel injectors or

FIGURE 1–7
Fluid dynamics is used extensively in the design of artificial hearts. Shown here is the Penn State Electric Total Artificial Heart.

Photo courtesy of the Biomedical Photography Lab, Penn State Biomedical Engineering Institute. Used by Permission.

carburetors—as well as the mixing of the fuel and the air in the cylinders and the purging of combustion gases in exhaust pipes—are analyzed using fluid mechanics. Fluid mechanics is also used in the design of the heating and air-conditioning system, the hydraulic brakes, the power steering, the automatic transmission, the lubrication systems, the cooling system of the engine block including the radiator and the water pump, and even the tires. The sleek streamlined shape of recent model cars is the result of efforts to minimize drag by using extensive analysis of flow over surfaces.

On a broader scale, fluid mechanics plays a major part in the design and analysis of aircraft, boats, submarines, rockets, jet engines, wind turbines, biomedical devices, cooling systems for electronic components, and transportation systems for moving water, crude oil, and natural gas. It is also considered in the design of buildings, bridges, and even billboards to make sure that the structures can withstand wind loading. Numerous natural phenomena such as the rain cycle, weather patterns, the rise of ground water to the tops of trees, winds, ocean waves, and currents in large water bodies are also governed by the principles of fluid mechanics (Fig. 1–8).

Natural flows and weather
© Glen Allison/Betty RF

Boats
© Doug Menuez/Getty RF

Aircraft and spacecraft
© Photo Link/Getty RF

Power plants
© Malcom Fife/Getty RF

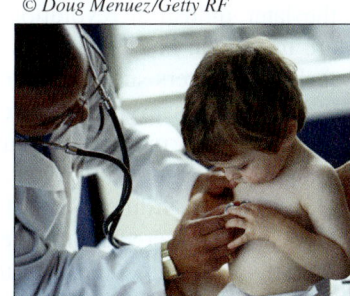
Human body
© Ryan McVay/Getty RF

Cars
© Mark Evans/Getty RF

Wind turbines
© F. Schussler/PhotoLink/Getty RF

Piping and plumbing systems
Photo by John M. Cimbala.

Industrial applications
Digital Vision/PunchStock

FIGURE 1–8
Some application areas of fluid mechanics.

1–2 ▪ A BRIEF HISTORY OF FLUID MECHANICS[1]

One of the first engineering problems humankind faced as cities were developed was the supply of water for domestic use and irrigation of crops. Our urban lifestyles can be retained only with abundant water, and it is clear from archeology that every successful civilization of prehistory invested in the construction and maintenance of water systems. The Roman aqueducts, some of which are still in use, are the best known examples. However, perhaps the most impressive engineering from a technical viewpoint was done at the Hellenistic city of Pergamon in present-day Turkey. There, from 283 to 133 BC, they built a series of pressurized lead and clay pipelines (Fig. 1–9), up to 45 km long that operated at pressures exceeding 1.7 MPa (180 m of head). Unfortunately, the names of almost all these early builders are lost to history.

FIGURE 1–9
Segment of Pergamon pipeline. Each clay pipe section was 13 to 18 cm in diameter.
Courtesy Gunther Garbrecht. Used by permission.

The earliest recognized contribution to fluid mechanics theory was made by the Greek mathematician Archimedes (285–212 BC). He formulated and applied the buoyancy principle in history's first nondestructive test to determine the gold content of the crown of King Hiero I. The Romans built great aqueducts and educated many conquered people on the benefits of clean water, but overall had a poor understanding of fluids theory. (Perhaps they shouldn't have killed Archimedes when they sacked Syracuse.)

During the Middle Ages, the application of fluid machinery slowly but steadily expanded. Elegant piston pumps were developed for dewatering mines, and the watermill and windmill were perfected to grind grain, forge metal, and for other tasks. For the first time in recorded human history, significant work was being done without the power of a muscle supplied by a person or animal, and these inventions are generally credited with enabling the later industrial revolution. Again the creators of most of the progress are unknown, but the devices themselves were well documented by several technical writers such as Georgius Agricola (Fig. 1–10).

The Renaissance brought continued development of fluid systems and machines, but more importantly, the scientific method was perfected and adopted throughout Europe. Simon Stevin (1548–1617), Galileo Galilei (1564–1642), Edme Mariotte (1620–1684), and Evangelista Torricelli (1608–1647) were among the first to apply the method to fluids as they investigated hydrostatic pressure distributions and vacuums. That work was integrated and refined by the brilliant mathematician and philosopher, Blaise Pascal (1623–1662). The Italian monk, Benedetto Castelli (1577–1644) was the first person to publish a statement of the continuity principle for fluids. Besides formulating his equations of motion for solids, Sir Isaac Newton (1643–1727) applied his laws to fluids and explored fluid inertia and resistance, free jets, and viscosity. That effort was built upon by Daniel Bernoulli (1700–1782), a Swiss, and his associate Leonard Euler (1707–1783). Together, their work defined the energy and momentum equations. Bernoulli's 1738 classic treatise *Hydrodynamica* may be considered the first fluid mechanics text. Finally, Jean d'Alembert (1717–1789) developed the idea of velocity and acceleration components, a differential expression of

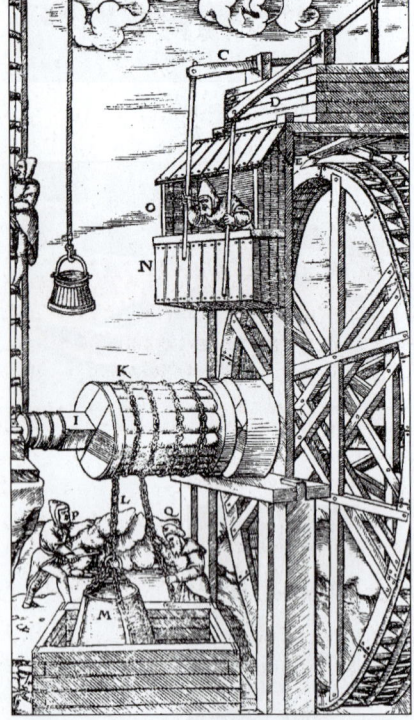

FIGURE 1–10
A mine hoist powered by a reversible water wheel.
G. Agricola, De Re Metalica, *Basel, 1556.*

---

[1] This section is contributed by Professor Glenn Brown of Oklahoma State University.

continuity, and his "paradox" of zero resistance to steady uniform motion over a body.

The development of fluid mechanics theory through the end of the eighteenth century had little impact on engineering since fluid properties and parameters were poorly quantified, and most theories were abstractions that could not be quantified for design purposes. That was to change with the development of the French school of engineering led by Riche de Prony (1755–1839). Prony (still known for his brake to measure shaft power) and his associates in Paris at the École Polytechnique and the École des Ponts et Chaussées were the first to integrate calculus and scientific theory into the engineering curriculum, which became the model for the rest of the world. (So now you know whom to blame for your painful freshman year.) Antonie Chezy (1718–1798), Louis Navier (1785–1836), Gaspard Coriolis (1792–1843), Henry Darcy (1803–1858), and many other contributors to fluid engineering and theory were students and/or instructors at the schools.

By the mid nineteenth century, fundamental advances were coming on several fronts. The physician Jean Poiseuille (1799–1869) had accurately measured flow in capillary tubes for multiple fluids, while in Germany Gotthilf Hagen (1797–1884) had differentiated between laminar and turbulent flow in pipes. In England, Lord Osborne Reynolds (1842–1912) continued that work (Fig. 1–11) and developed the dimensionless number that bears his name. Similarly, in parallel to the early work of Navier, George Stokes (1819–1903) completed the general equation of fluid motion (with friction) that takes their names. William Froude (1810–1879) almost single-handedly developed the procedures and proved the value of physical model testing. American expertise had become equal to the Europeans as demonstrated by James Francis' (1815–1892) and Lester Pelton's (1829–1908) pioneering work in turbines and Clemens Herschel's (1842–1930) invention of the Venturi meter.

In addition to Reynolds and Stokes, many notable contributions were made to fluid theory in the late nineteenth century by Irish and English scientists, including William Thomson, Lord Kelvin (1824–1907), William Strutt, Lord Rayleigh (1842–1919), and Sir Horace Lamb (1849–1934). These individuals investigated a large number of problems, including dimensional analysis, irrotational flow, vortex motion, cavitation, and waves. In a broader sense,

FIGURE 1–11
Osborne Reynolds' original apparatus for demonstrating the onset of turbulence in pipes, being operated by John Lienhard at the University of Manchester in 1975.

Photo courtesy of John Lienhard, University of Houston. Used by permission.

FIGURE 1–12
The Wright brothers take flight at Kitty Hawk.
Library of Congress Prints & Photographs Division [LC-DIG-ppprs-00626]

FIGURE 1–13
Old and new wind turbine technologies north of Woodward, OK. The modern turbines have 1.6 MW capacities.
Photo courtesy of the Oklahoma Wind Power Initiative. Used by permission.

their work also explored the links between fluid mechanics, thermodynamics, and heat transfer.

The dawn of the twentieth century brought two monumental developments. First, in 1903, the self-taught Wright brothers (Wilbur, 1867–1912; Orville, 1871–1948) invented the airplane through application of theory and determined experimentation. Their primitive invention was complete and contained all the major aspects of modern aircraft (Fig. 1–12). The Navier–Stokes equations were of little use up to this time because they were too difficult to solve. In a pioneering paper in 1904, the German Ludwig Prandtl (1875–1953) showed that fluid flows can be divided into a layer near the walls, the *boundary layer,* where the friction effects are significant, and an outer layer where such effects are negligible and the simplified Euler and Bernoulli equations are applicable. His students, Theodor von Kármán (1881–1963), Paul Blasius (1883–1970), Johann Nikuradse (1894–1979), and others, built on that theory in both hydraulic and aerodynamic applications. (During World War II, both sides benefited from the theory as Prandtl remained in Germany while his best student, the Hungarian-born von Kármán, worked in America.)

The mid twentieth century could be considered a golden age of fluid mechanics applications. Existing theories were adequate for the tasks at hand, and fluid properties and parameters were well defined. These supported a huge expansion of the aeronautical, chemical, industrial, and water resources sectors; each of which pushed fluid mechanics in new directions. Fluid mechanics research and work in the late twentieth century were dominated by the development of the digital computer in America. The ability to solve large complex problems, such as global climate modeling or the optimization of a turbine blade, has provided a benefit to our society that the eighteenth-century developers of fluid mechanics could never have imagined (Fig. 1–13). The principles presented in the following pages have been applied to flows ranging from a moment at the microscopic scale to 50 years of simulation for an entire river basin. It is truly mind-boggling.

Where will fluid mechanics go in the twenty-first century and beyond? Frankly, even a limited extrapolation beyond the present would be sheer folly. However, if history tells us anything, it is that engineers will be applying what they know to benefit society, researching what they don't know, and having a great time in the process.

1–3 ■ THE NO-SLIP CONDITION

Fluid flow is often confined by solid surfaces, and it is important to understand how the presence of solid surfaces affects fluid flow. We know that water in a river cannot flow through large rocks, and must go around them. That is, the water velocity normal to the rock surface must be zero, and water approaching the surface normally comes to a complete stop at the surface. What is not as obvious is that water approaching the rock at any angle also comes to a complete stop at the rock surface, and thus the tangential velocity of water at the surface is also zero.

Consider the flow of a fluid in a stationary pipe or over a solid surface that is nonporous (i.e., impermeable to the fluid). All experimental observations indicate that a fluid in motion comes to a complete stop at the surface and assumes a zero velocity relative to the surface. That is, a fluid in direct contact with a solid "sticks" to the surface, and there is no slip. This is known as the **no-slip condition.** The fluid property responsible for the no-slip condition and the development of the boundary layer is *viscosity* and is discussed in Chap. 2.

The photograph in Fig. 1–14 clearly shows the evolution of a velocity gradient as a result of the fluid sticking to the surface of a blunt nose. The layer that sticks to the surface slows the adjacent fluid layer because of viscous forces between the fluid layers, which slows the next layer, and so on. A consequence of the no-slip condition is that all velocity profiles must have zero values with respect to the surface at the points of contact between a fluid and a solid surface (Fig. 1–15). Therefore, the no-slip condition is responsible for the development of the velocity profile. The flow region adjacent to the wall in which the viscous effects (and thus the velocity gradients) are significant is called the **boundary layer.** Another consequence of the no-slip condition is the *surface drag*, or *skin friction drag*, which is the force a fluid exerts on a surface in the flow direction.

When a fluid is forced to flow over a curved surface, such as the back side of a cylinder, the boundary layer may no longer remain attached to the surface and separates from the surface—a process called **flow separation** (Fig. 1–16). We emphasize that the no-slip condition applies *everywhere* along the surface, even downstream of the separation point. Flow separation is discussed in greater detail in Chap. 9.

A phenomenon similar to the no-slip condition occurs in heat transfer. When two bodies at different temperatures are brought into contact, heat transfer occurs such that both bodies assume the same temperature at the points of contact. Therefore, a fluid and a solid surface have the same temperature at the points of contact. This is known as **no-temperature-jump condition.**

1–4 · CLASSIFICATION OF FLUID FLOWS

Earlier we defined *fluid mechanics* as the science that deals with the behavior of fluids at rest or in motion, and the interaction of fluids with solids or other fluids at the boundaries. There is a wide variety of fluid flow problems encountered in practice, and it is usually convenient to classify them on the basis of some common characteristics to make it feasible to study them in groups. There are many ways to classify fluid flow problems, and here we present some general categories.

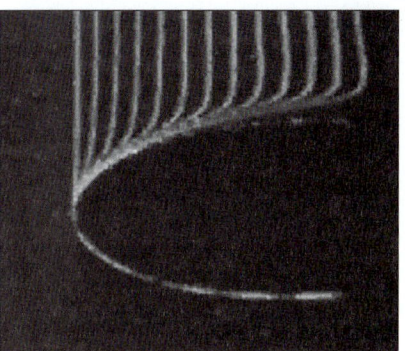

FIGURE 1–14
The development of a velocity profile due to the no-slip condition as a fluid flows over a blunt nose.

"Hunter Rouse: Laminar and Turbulent Flow Film." Copyright IIHR-Hydroscience & Engineering, The University of Iowa. Used by permission.

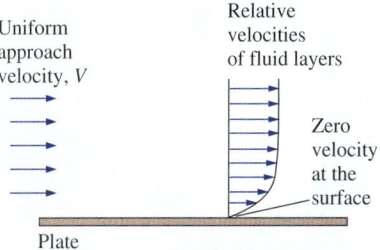

FIGURE 1–15
A fluid flowing over a stationary surface comes to a complete stop at the surface because of the no-slip condition.

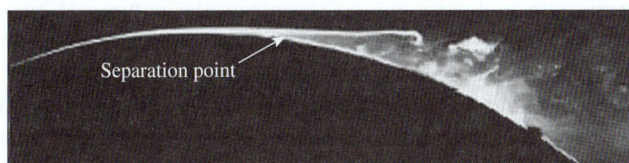

FIGURE 1–16
Flow separation during flow over a curved surface.

From G. M. Homsy et al, "Multi-Media Fluid Mechanics," Cambridge Univ. Press (2001). ISBN 0-521-78748-3. Reprinted by permission.

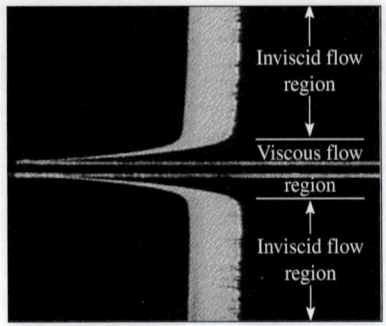

FIGURE 1–17
The flow of an originally uniform fluid stream over a flat plate, and the regions of viscous flow (next to the plate on both sides) and inviscid flow (away from the plate).

*Fundamentals of Boundary Layers,
National Committee from Fluid Mechanics Films,
© Education Development Center.*

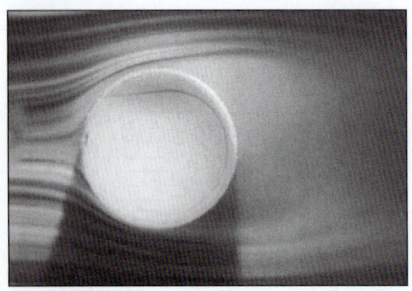

FIGURE 1–18
External flow over a tennis ball, and the turbulent wake region behind.

Courtesy NASA and Cislunar Aerospace, Inc.

Viscous versus Inviscid Regions of Flow

When two fluid layers move relative to each other, a friction force develops between them and the slower layer tries to slow down the faster layer. This internal resistance to flow is quantified by the fluid property *viscosity*, which is a measure of internal stickiness of the fluid. Viscosity is caused by cohesive forces between the molecules in liquids and by molecular collisions in gases. There is no fluid with zero viscosity, and thus all fluid flows involve viscous effects to some degree. Flows in which the frictional effects are significant are called **viscous flows.** However, in many flows of practical interest, there are *regions* (typically regions not close to solid surfaces) where viscous forces are negligibly small compared to inertial or pressure forces. Neglecting the viscous terms in such **inviscid flow regions** greatly simplifies the analysis without much loss in accuracy.

The development of viscous and inviscid regions of flow as a result of inserting a flat plate parallel into a fluid stream of uniform velocity is shown in Fig. 1–17. The fluid sticks to the plate on both sides because of the no-slip condition, and the thin boundary layer in which the viscous effects are significant near the plate surface is the *viscous flow region*. The region of flow on both sides away from the plate and largely unaffected by the presence of the plate is the *inviscid flow region*.

Internal versus External Flow

A fluid flow is classified as being internal or external, depending on whether the fluid flows in a confined space or over a surface. The flow of an unbounded fluid over a surface such as a plate, a wire, or a pipe is **external flow.** The flow in a pipe or duct is **internal flow** if the fluid is completely bounded by solid surfaces. Water flow in a pipe, for example, is internal flow, and airflow over a ball or over an exposed pipe during a windy day is external flow (Fig. 1–18). The flow of liquids in a duct is called *open-channel flow* if the duct is only partially filled with the liquid and there is a free surface. The flows of water in rivers and irrigation ditches are examples of such flows.

Internal flows are dominated by the influence of viscosity throughout the flow field. In external flows the viscous effects are limited to boundary layers near solid surfaces and to wake regions downstream of bodies.

Compressible versus Incompressible Flow

A flow is classified as being *compressible* or *incompressible*, depending on the level of variation of density during flow. Incompressibility is an approximation, in which the flow is said to be **incompressible** if the density remains nearly constant throughout. Therefore, the volume of every portion of fluid remains unchanged over the course of its motion when the flow is approximated as incompressible.

The densities of liquids are essentially constant, and thus the flow of liquids is typically incompressible. Therefore, liquids are usually referred to as *incompressible substances.* A pressure of 210 atm, for example, causes the density of liquid water at 1 atm to change by just 1 percent. Gases, on the other hand, are highly compressible. A pressure change of just 0.01 atm, for example, causes a change of 1 percent in the density of atmospheric air.

When analyzing rockets, spacecraft, and other systems that involve high-speed gas flows (Fig. 1–19), the flow speed is often expressed in terms of the dimensionless **Mach number** defined as

$$\text{Ma} = \frac{V}{c} = \frac{\text{Speed of flow}}{\text{Speed of sound}}$$

where c is the **speed of sound** whose value is 346 m/s in air at room temperature at sea level. A flow is called **sonic** when Ma = 1, **subsonic** when Ma < 1, **supersonic** when Ma > 1, and **hypersonic** when Ma ≫ 1. Dimensionless parameters are discussed in detail in Chapter 7.

Liquid flows are incompressible to a high level of accuracy, but the level of variation of density in gas flows and the consequent level of approximation made when modeling gas flows as incompressible depends on the Mach number. Gas flows can often be approximated as incompressible if the density changes are under about 5 percent, which is usually the case when Ma < 0.3. Therefore, the compressibility effects of air at room temperature can be neglected at speeds under about 100 m/s.

Small density changes of liquids corresponding to large pressure changes can still have important consequences. The irritating "water hammer" in a water pipe, for example, is caused by the vibrations of the pipe generated by the reflection of pressure waves following the sudden closing of the valves.

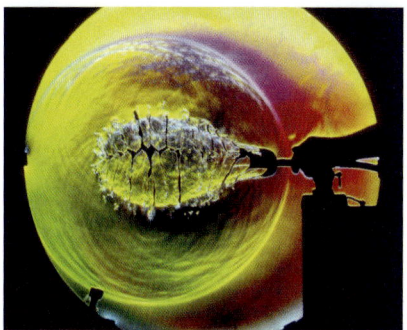

FIGURE 1–19
Schlieren image of the spherical shock wave produced by a bursting ballon at the Penn State Gas Dynamics Lab. Several secondary shocks are seen in the air surrounding the ballon.
Photo by G. S. Settles, Penn State University. Used by permission.

Laminar versus Turbulent Flow

Some flows are smooth and orderly while others are rather chaotic. The highly ordered fluid motion characterized by smooth layers of fluid is called **laminar.** The word *laminar* comes from the movement of adjacent fluid particles together in "laminae." The flow of high-viscosity fluids such as oils at low velocities is typically laminar. The highly disordered fluid motion that typically occurs at high velocities and is characterized by velocity fluctuations is called **turbulent** (Fig. 1–20). The flow of low-viscosity fluids such as air at high velocities is typically turbulent. A flow that alternates between being laminar and turbulent is called **transitional.** The experiments conducted by Osborne Reynolds in the 1880s resulted in the establishment of the dimensionless **Reynolds number, Re,** as the key parameter for the determination of the flow regime in pipes (Chap. 8).

Natural (or Unforced) versus Forced Flow

A fluid flow is said to be natural or forced, depending on how the fluid motion is initiated. In **forced flow,** a fluid is forced to flow over a surface or in a pipe by external means such as a pump or a fan. In **natural flows,** fluid motion is due to natural means such as the buoyancy effect, which manifests itself as the rise of warmer (and thus lighter) fluid and the fall of cooler (and thus denser) fluid (Fig. 1–21). In solar hot-water systems, for example, the thermosiphoning effect is commonly used to replace pumps by placing the water tank sufficiently above the solar collectors.

Laminar

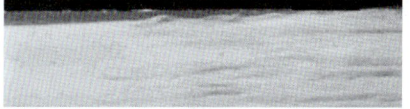
Transitional

Turbulent

FIGURE 1–20
Laminar, transitional, and turbulent flows over a flat plate.
Courtesy ONERA, photograph by Werlé.

12
INTRODUCTION AND BASIC CONCEPTS

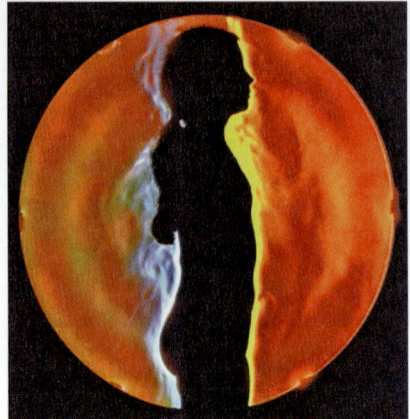

FIGURE 1–21
In this schlieren image of a girl in a swimming suit, the rise of lighter, warmer air adjacent to her body indicates that humans and warm-blooded animals are surrounded by thermal plumes of rising warm air.

G. S. Settles, Gas Dynamics Lab, Penn State University. Used by permission.

Steady versus Unsteady Flow

The terms *steady* and *uniform* are used frequently in engineering, and thus it is important to have a clear understanding of their meanings. The term **steady** implies *no change of properties, velocity, temperature, etc., at a point with time.* The opposite of steady is **unsteady.** The term **uniform** implies *no change with location* over a specified region. These meanings are consistent with their everyday use (steady girlfriend, uniform distribution, etc.).

The terms *unsteady* and *transient* are often used interchangeably, but these terms are not synonyms. In fluid mechanics, *unsteady* is the most general term that applies to any flow that is not steady, but **transient** is typically used for developing flows. When a rocket engine is fired up, for example, there are transient effects (the pressure builds up inside the rocket engine, the flow accelerates, etc.) until the engine settles down and operates steadily. The term **periodic** refers to the kind of unsteady flow in which the flow oscillates about a steady mean.

Many devices such as turbines, compressors, boilers, condensers, and heat exchangers operate for long periods of time under the same conditions, and they are classified as *steady-flow devices*. (Note that the flow field near the rotating blades of a turbomachine is of course unsteady, but we consider the overall flow field rather than the details at some localities when we classify devices.) During steady flow, the fluid properties can change from point to point within a device, but at any fixed point they remain constant. Therefore, the volume, the mass, and the total energy content of a steady-flow device or flow section remain constant in steady operation. A simple analogy is shown in Fig. 1–22.

Steady-flow conditions can be closely approximated by devices that are intended for continuous operation such as turbines, pumps, boilers, condensers, and heat exchangers of power plants or refrigeration systems. Some cyclic devices, such as reciprocating engines or compressors, do not satisfy the steady-flow conditions since the flow at the inlets and the exits is

FIGURE 1–22
Comparison of (*a*) instantaneous snapshot of an unsteady flow, and (*b*) long exposure picture of the same flow.

Photos by Eric A. Paterson. Used by permission.

(*a*)

(*b*)

pulsating and not steady. However, the fluid properties vary with time in a periodic manner, and the flow through these devices can still be analyzed as a steady-flow process by using time-averaged values for the properties.

Some fascinating visualizations of fluid flow are provided in the book *An Album of Fluid Motion* by Milton Van Dyke (1982). A nice illustration of an unsteady-flow field is shown in Fig. 1–23, taken from Van Dyke's book. Figure 1–23*a* is an instantaneous snapshot from a high-speed motion picture; it reveals large, alternating, swirling, turbulent eddies that are shed into the periodically oscillating wake from the blunt base of the object. The eddies produce shock waves that move upstream alternately over the top and bottom surfaces of the airfoil in an unsteady fashion. Figure 1–23*b* shows the *same* flow field, but the film is exposed for a longer time so that the image is time averaged over 12 cycles. The resulting time-averaged flow field appears "steady" since the details of the unsteady oscillations have been lost in the long exposure.

One of the most important jobs of an engineer is to determine whether it is sufficient to study only the time-averaged "steady" flow features of a problem, or whether a more detailed study of the unsteady features is required. If the engineer were interested only in the overall properties of the flow field (such as the time-averaged drag coefficient, the mean velocity, and pressure fields), a time-averaged description like that of Fig. 1–23*b*, time-averaged experimental measurements, or an analytical or numerical calculation of the time-averaged flow field would be sufficient. However, if the engineer were interested in details about the unsteady-flow field, such as flow-induced vibrations, unsteady pressure fluctuations, or the sound waves emitted from the turbulent eddies or the shock waves, a time-averaged description of the flow field would be insufficient.

Most of the analytical and computational examples provided in this textbook deal with steady or time-averaged flows, although we occasionally point out some relevant unsteady-flow features as well when appropriate.

One-, Two-, and Three-Dimensional Flows

A flow field is best characterized by its velocity distribution, and thus a flow is said to be one-, two-, or three-dimensional if the flow velocity varies in one, two, or three primary dimensions, respectively. A typical fluid flow involves a three-dimensional geometry, and the velocity may vary in all three dimensions, rendering the flow three-dimensional [$\vec{V}(x, y, z)$ in rectangular or $\vec{V}(r, \theta, z)$ in cylindrical coordinates]. However, the variation of velocity in certain directions can be small relative to the variation in other directions and can be ignored with negligible error. In such cases, the flow can be modeled conveniently as being one- or two-dimensional, which is easier to analyze.

Consider steady flow of a fluid entering from a large tank into a circular pipe. The fluid velocity everywhere on the pipe surface is zero because of the no-slip condition, and the flow is two-dimensional in the entrance region of the pipe since the velocity changes in both the *r*- and *z*-directions, but not in the θ-direction. The velocity profile develops fully and remains unchanged after some distance from the inlet (about 10 pipe diameters in turbulent flow, and less in laminar pipe flow, as in Fig. 1–24), and the flow in this region is said to be *fully developed*. The fully developed flow in a circular pipe is *one-dimensional* since the velocity varies in the radial *r*-direction but not in the angular θ- or axial *z*-directions, as shown in Fig. 1–24. That is, the velocity profile is the same at any axial *z*-location, and it is symmetric about the axis of the pipe.

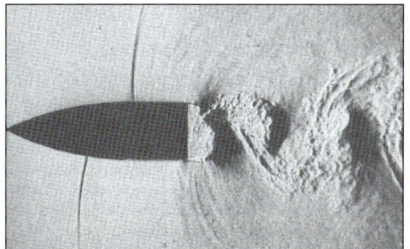

(*a*)

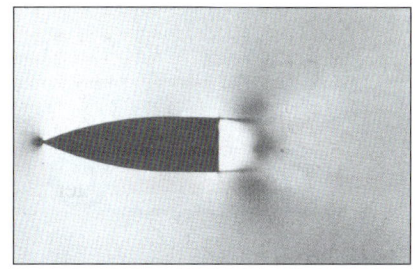

(*b*)

FIGURE 1–23
Oscillating wake of a blunt-based airfoil at Mach number 0.6. Photo (*a*) is an instantaneous image, while photo (*b*) is a long-exposure (time-averaged) image.

(*a*) Dyment, A., Flodrops, J. P. & Gryson, P. 1982 in Flow Visualization II, W. Merzkirch, ed., 331–336. Washington: Hemisphere. *Used by permission of Arthur Dyment.*

(*b*) Dyment, A. & Gryson, P. 1978 in Inst. Mèc. Fluides Lille, No. 78-5. *Used by permission of Arthur Dyment.*

14
INTRODUCTION AND BASIC CONCEPTS

FIGURE 1–24
The development of the velocity profile in a circular pipe. $V = V(r, z)$ and thus the flow is two-dimensional in the entrance region, and becomes one-dimensional downstream when the velocity profile fully develops and remains unchanged in the flow direction, $V = V(r)$.

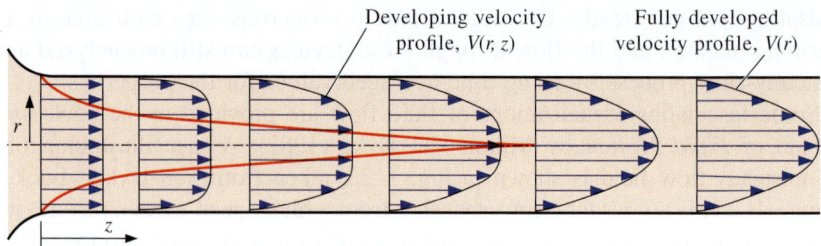

Note that the dimensionality of the flow also depends on the choice of coordinate system and its orientation. The pipe flow discussed, for example, is one-dimensional in cylindrical coordinates, but two-dimensional in Cartesian coordinates—illustrating the importance of choosing the most appropriate coordinate system. Also note that even in this simple flow, the velocity cannot be uniform across the cross section of the pipe because of the no-slip condition. However, at a well-rounded entrance to the pipe, the velocity profile may be approximated as being nearly uniform across the pipe, since the velocity is nearly constant at all radii except very close to the pipe wall.

A flow may be approximated as *two-dimensional* when the aspect ratio is large and the flow does not change appreciably along the longer dimension. For example, the flow of air over a car antenna can be considered two-dimensional except near its ends since the antenna's length is much greater than its diameter, and the airflow hitting the antenna is fairly uniform (Fig. 1–25).

FIGURE 1–25
Flow over a car antenna is approximately two-dimensional except near the top and bottom of the antenna.

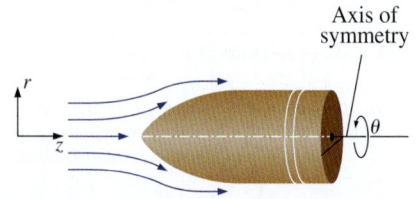

FIGURE 1–26
Axisymmetric flow over a bullet.

EXAMPLE 1–1   Axisymmetric Flow over a Bullet

Consider a bullet piercing through calm air during a short time interval in which the bullet's speed is nearly constant. Determine if the time-averaged airflow over the bullet during its flight is one-, two-, or three-dimensional (Fig. 1–26).

SOLUTION   It is to be determined whether airflow over a bullet is one-, two-, or three-dimensional.
Assumptions   There are no significant winds and the bullet is not spinning.
Analysis   The bullet possesses an axis of symmetry and is therefore an axisymmetric body. The airflow upstream of the bullet is parallel to this axis, and we expect the time-averaged airflow to be rotationally symmetric about the axis—such flows are said to be axisymmetric. The velocity in this case varies with axial distance z and radial distance r, but not with angle θ. Therefore, the time-averaged airflow over the bullet is **two-dimensional.**
Discussion   While the time-averaged airflow is axisymmetric, the *instantaneous* airflow is not, as illustrated in Fig. 1–23. In Cartesian coordinates, the flow would be three-dimensional. Finally, many bullets also spin.

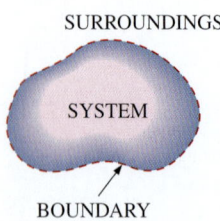

FIGURE 1–27
System, surroundings, and boundary.

1–5 ■ SYSTEM AND CONTROL VOLUME

A **system** is defined as a *quantity of matter or a region in space chosen for study*. The mass or region outside the system is called the **surroundings**. The real or imaginary surface that separates the system from its surroundings is called the **boundary** (Fig. 1–27). The boundary of a system can be *fixed* or *movable*. Note that the boundary is the contact surface shared by both the system and the surroundings. Mathematically speaking, the boundary has zero thickness, and thus it can neither contain any mass nor occupy any volume in space.

Systems may be considered to be *closed* or *open*, depending on whether a fixed mass or a volume in space is chosen for study. A **closed system** (also known as a **control mass** or simply a *system* when the context makes it clear) consists of a fixed amount of mass, and no mass can cross its boundary. But energy, in the form of heat or work, can cross the boundary, and the volume of a closed system does not have to be fixed. If, as a special case, even energy is not allowed to cross the boundary, that system is called an **isolated system.**

Consider the piston–cylinder device shown in Fig. 1–28. Let us say that we would like to find out what happens to the enclosed gas when it is heated. Since we are focusing our attention on the gas, it is our system. The inner surfaces of the piston and the cylinder form the boundary, and since no mass is crossing this boundary, it is a closed system. Notice that energy may cross the boundary, and part of the boundary (the inner surface of the piston, in this case) may move. Everything outside the gas, including the piston and the cylinder, is the surroundings.

An **open system,** or a **control volume,** as it is often called, is a *selected region in space*. It usually encloses a device that involves mass flow such as a compressor, turbine, or nozzle. Flow through these devices is best studied by selecting the region within the device as the control volume. Both mass and energy can cross the boundary (the *control surface*) of a control volume.

A large number of engineering problems involve mass flow in and out of an open system and, therefore, are modeled as *control volumes*. A water heater, a car radiator, a turbine, and a compressor all involve mass flow and should be analyzed as control volumes (open systems) instead of as control masses (closed systems). In general, *any arbitrary region in space* can be selected as a control volume. There are no concrete rules for the selection of control volumes, but a wise choice certainly makes the analysis much easier. If we were to analyze the flow of air through a nozzle, for example, a good choice for the control volume would be the region within the nozzle, or perhaps surrounding the entire nozzle.

A control volume can be fixed in size and shape, as in the case of a nozzle, or it may involve a moving boundary, as shown in Fig. 1–29. Most control volumes, however, have fixed boundaries and thus do not involve any moving boundaries. A control volume may also involve heat and work interactions just as a closed system, in addition to mass interaction.

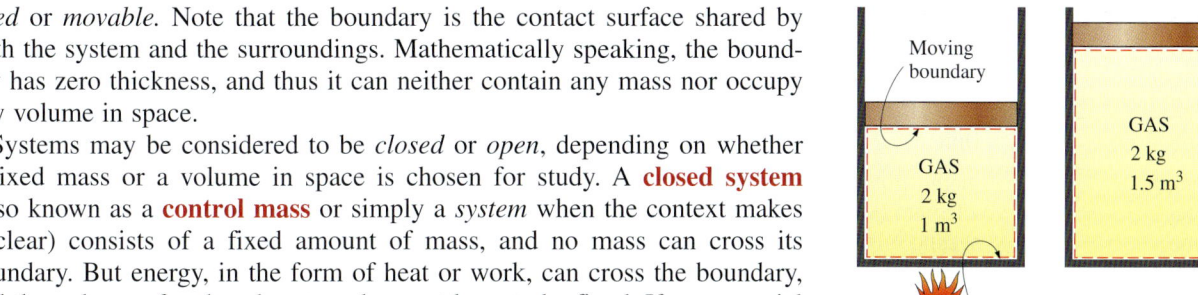

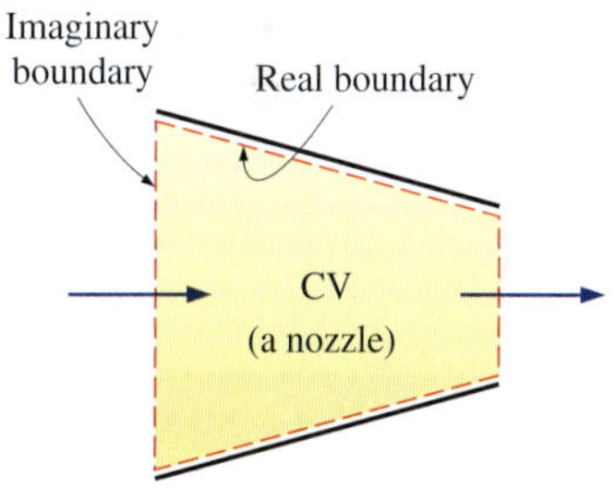

FIGURE 1–28
A closed system with a moving boundary.

(*a*) A control volume (CV) with real and imaginary boundaries

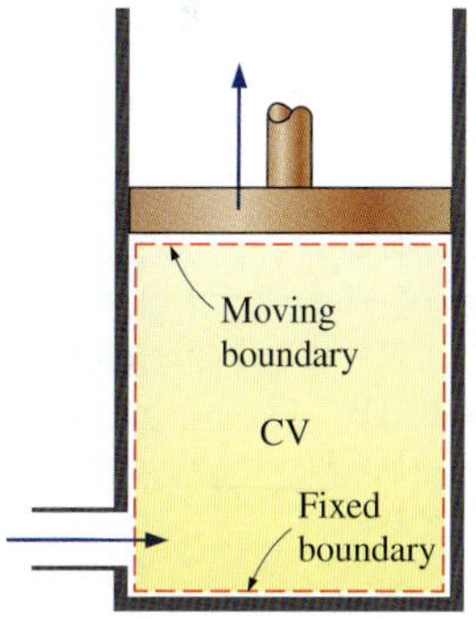

(*b*) A control volume (CV) with fixed and moving boundaries as well as real and imaginary boundaries

FIGURE 1–29
A control volume may involve fixed, moving, real, and imaginary boundaries.

1–6 ■ IMPORTANCE OF DIMENSIONS AND UNITS

Any physical quantity can be characterized by **dimensions.** The magnitudes assigned to the dimensions are called **units.** Some basic dimensions such as mass m, length L, time t, and temperature T are selected as **primary** or **fundamental dimensions,** while others such as velocity V, energy E, and volume V are expressed in terms of the primary dimensions and are called **secondary dimensions,** or **derived dimensions.**

TABLE 1–1

The seven fundamental (or primary) dimensions and their units in SI

Dimension	Unit
Length	meter (m)
Mass	kilogram (kg)
Time	second (s)
Temperature	kelvin (K)
Electric current	ampere (A)
Amount of light	candela (cd)
Amount of matter	mole (mol)

TABLE 1–2

Standard prefixes in SI units

Multiple	Prefix
10^{24}	yotta, Y
10^{21}	zetta, Z
10^{18}	exa, E
10^{15}	peta, P
10^{12}	tera, T
10^{9}	giga, G
10^{6}	mega, M
10^{3}	kilo, k
10^{2}	hecto, h
10^{1}	deka, da
10^{-1}	deci, d
10^{-2}	centi, c
10^{-3}	milli, m
10^{-6}	micro, μ
10^{-9}	nano, n
10^{-12}	pico, p
10^{-15}	femto, f
10^{-18}	atto, a
10^{-21}	zepto, z
10^{-24}	yocto, y

A number of unit systems have been developed over the years. Despite strong efforts in the scientific and engineering community to unify the world with a single unit system, two sets of units are still in common use today: the **English system,** which is also known as the *United States Customary System* (USCS), and the metric **SI** (from *Le Système International d' Unités*), which is also known as the *International System*. The SI is a simple and logical system based on a decimal relationship between the various units, and it is being used for scientific and engineering work in most of the industrialized nations, including England. The English system, however, has no apparent systematic numerical base, and various units in this system are related to each other rather arbitrarily (12 in = 1 ft, 1 mile = 5280 ft, 4 qt = 1 gal, etc.), which makes it confusing and difficult to learn. The United States is the only industrialized country that has not yet fully converted to the metric system.

The systematic efforts to develop a universally acceptable system of units dates back to 1790 when the French National Assembly charged the French Academy of Sciences to come up with such a unit system. An early version of the metric system was soon developed in France, but it did not find universal acceptance until 1875 when *The Metric Convention Treaty* was prepared and signed by 17 nations, including the United States. In this international treaty, meter and gram were established as the metric units for length and mass, respectively, and a *General Conference of Weights and Measures* (CGPM) was established that was to meet every six years. In 1960, the CGPM produced the SI, which was based on six fundamental quantities, and their units were adopted in 1954 at the Tenth General Conference of Weights and Measures: *meter* (m) for length, *kilogram* (kg) for mass, *second* (s) for time, *ampere* (A) for electric current, *degree Kelvin* (°K) for temperature, and *candela* (cd) for luminous intensity (amount of light). In 1971, the CGPM added a seventh fundamental quantity and unit: *mole* (mol) for the amount of matter.

Based on the notational scheme introduced in 1967, the degree symbol was officially dropped from the absolute temperature unit, and all unit names were to be written without capitalization even if they were derived from proper names (Table 1–1). However, the abbreviation of a unit was to be capitalized if the unit was derived from a proper name. For example, the SI unit of force, which is named after Sir Isaac Newton (1647–1723), is *newton* (not Newton), and it is abbreviated as N. Also, the full name of a unit may be pluralized, but its abbreviation cannot. For example, the length of an object can be 5 m or 5 meters, *not* 5 ms or 5 meter. Finally, no period is to be used in unit abbreviations unless they appear at the end of a sentence. For example, the proper abbreviation of meter is m (not m.).

The recent move toward the metric system in the United States seems to have started in 1968 when Congress, in response to what was happening in the rest of the world, passed a Metric Study Act. Congress continued to promote a voluntary switch to the metric system by passing the Metric Conversion Act in 1975. A trade bill passed by Congress in 1988 set a September 1992 deadline for all federal agencies to convert to the metric system. However, the deadlines were relaxed later with no clear plans for the future.

As pointed out, the SI is based on a decimal relationship between units. The prefixes used to express the multiples of the various units are listed in Table 1–2.

They are standard for all units, and the student is encouraged to memorize some of them because of their widespread use (Fig. 1–30).

Some SI and English Units

In SI, the units of mass, length, and time are the kilogram (kg), meter (m), and second (s), respectively. The respective units in the English system are the pound-mass (lbm), foot (ft), and second (s). The pound symbol *lb* is actually the abbreviation of *libra*, which was the ancient Roman unit of weight. The English retained this symbol even after the end of the Roman occupation of Britain in 410. The mass and length units in the two systems are related to each other by

$$1 \text{ lbm} = 0.45359 \text{ kg}$$
$$1 \text{ ft} = 0.3048 \text{ m}$$

In the English system, force is often considered to be one of the primary dimensions and is assigned a nonderived unit. This is a source of confusion and error that necessitates the use of a dimensional constant (g_c) in many formulas. To avoid this nuisance, we consider force to be a secondary dimension whose unit is derived from Newton's second law, i.e.,

$$\text{Force} = (\text{Mass})(\text{Acceleration})$$

or
$$F = ma \quad (1\text{–}1)$$

In SI, the force unit is the newton (N), and it is defined as the *force required to accelerate a mass of 1 kg at a rate of 1 m/s²*. In the English system, the force unit is the **pound-force** (lbf) and is defined as the *force required to accelerate a mass of 32.174 lbm (1 slug) at a rate of 1 ft/s²* (Fig. 1–31). That is,

$$1 \text{ N} = 1 \text{ kg·m/s}^2$$
$$1 \text{ lbf} = 32.174 \text{ lbm·ft/s}^2$$

A force of 1 N is roughly equivalent to the weight of a small apple ($m = 102$ g), whereas a force of 1 lbf is roughly equivalent to the weight of four medium apples ($m_{\text{total}} = 454$ g), as shown in Fig. 1–32. Another force unit in common use in many European countries is the *kilogram-force* (kgf), which is the weight of 1 kg mass at sea level (1 kgf = 9.807 N).

The term **weight** is often incorrectly used to express mass, particularly by the "weight watchers." Unlike mass, weight *W* is a *force*. It is the gravitational force applied to a body, and its magnitude is determined from an equation based on Newton's second law,

$$W = mg \quad (\text{N}) \quad (1\text{–}2)$$

where *m* is the mass of the body, and *g* is the local gravitational acceleration (*g* is 9.807 m/s² or 32.174 ft/s² at sea level and 45° latitude). An ordinary bathroom scale measures the gravitational force acting on a body. The weight per unit volume of a substance is called the **specific weight** γ and is determined from $\gamma = \rho g$, where ρ is density.

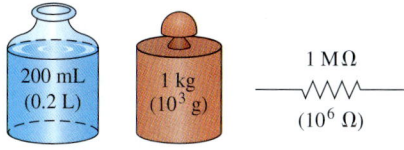

FIGURE 1–30
The SI unit prefixes are used in all branches of engineering.

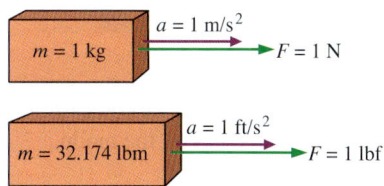

FIGURE 1–31
The definition of the force units.

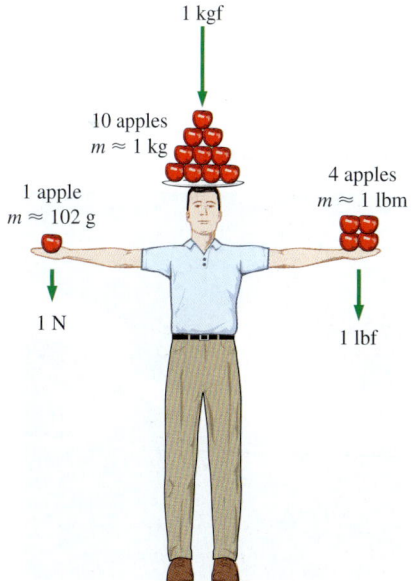

FIGURE 1–32
The relative magnitudes of the force units newton (N), kilogram-force (kgf), and pound-force (lbf).

FIGURE 1–33
A body weighing 72 kgf on earth will weigh only 12 kgf on the moon.

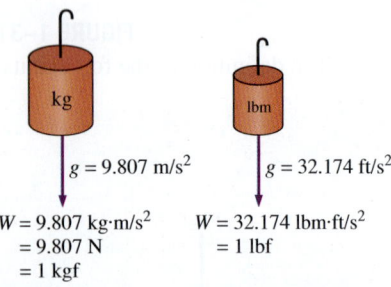

FIGURE 1–34
The weight of a unit mass at sea level.

FIGURE 1–35
A typical match yields about one Btu (or one kJ) of energy if completely burned.

Photo by John M. Cimbala.

The mass of a body remains the same regardless of its location in the universe. Its weight, however, changes with a change in gravitational acceleration. A body weighs less on top of a mountain since g decreases (by a small amount) with altitude. On the surface of the moon, an astronaut weighs about one-sixth of what she or he normally weighs on earth (Fig. 1–33).

At sea level a mass of 1 kg weighs 9.807 N, as illustrated in Fig. 1–34. A mass of 1 lbm, however, weighs 1 lbf, which misleads people to believe that pound-mass and pound-force can be used interchangeably as pound (lb), which is a major source of error in the English system.

It should be noted that the *gravity force* acting on a mass is due to the *attraction* between the masses, and thus it is proportional to the magnitudes of the masses and inversely proportional to the square of the distance between them. Therefore, the gravitational acceleration g at a location depends on the *local density* of the earth's crust, the *distance* to the center of the earth, and to a lesser extent, the positions of the moon and the sun. The value of g varies with location from 9.8295 m/s^2 at 4500 m below sea level to 7.3218 m/s^2 at 100,000 m above sea level. However, at altitudes up to 30,000 m, the variation of g from the sea-level value of 9.807 m/s^2, is less than 1 percent. Therefore, for most practical purposes, the gravitational acceleration can be assumed to be *constant* at 9.807 m/s^2, often rounded to 9.81 m/s^2. It is interesting to note that the value of g increases with distance below sea level, reaches a maximum at about 4500 m below sea level, and then starts decreasing. (What do you think the value of g is at the center of the earth?)

The primary cause of confusion between mass and weight is that mass is usually measured *indirectly* by measuring the *gravity force* it exerts. This approach also assumes that the forces exerted by other effects such as air buoyancy and fluid motion are negligible. This is like measuring the distance to a star by measuring its red shift, or measuring the altitude of an airplane by measuring barometric pressure. Both of these are also indirect measurements. The correct *direct* way of measuring mass is to compare it to a known mass. This is cumbersome, however, and it is mostly used for calibration and measuring precious metals.

Work, which is a form of energy, can simply be defined as force times distance; therefore, it has the unit "newton-meter (N·m)," which is called a **joule** (J). That is,

$$1 \text{ J} = 1 \text{ N·m} \tag{1–3}$$

A more common unit for energy in SI is the kilojoule (1 kJ = 10^3 J). In the English system, the energy unit is the **Btu** (British thermal unit), which is defined as the energy required to raise the temperature of 1 lbm of water at 68°F by 1°F. In the metric system, the amount of energy needed to raise the temperature of 1 g of water at 14.5°C by 1°C is defined as 1 **calorie** (cal), and 1 cal = 4.1868 J. The magnitudes of the kilojoule and Btu are very nearly the same (1 Btu = 1.0551 kJ). Here is a good way to get a feel for these units: If you light a typical match and let it burn itself out, it yields approximately one Btu (or one kJ) of energy (Fig. 1–35).

The unit for time rate of energy is joule per second (J/s), which is called a **watt** (W). In the case of work, the time rate of energy is called *power*. A commonly used unit of power is horsepower (hp), which is equivalent to 745.7 W. Electrical energy typically is expressed in the unit kilowatt-hour (kWh), which is equivalent to 3600 kJ. An electric appliance with a rated power of 1 kW consumes 1 kWh of electricity when running continuously for one hour. When dealing with electric power generation, the units kW and kWh are often confused. Note that kW or kJ/s is a unit of power, whereas kWh is a unit of energy. Therefore, statements like "the new wind turbine will generate 50 kW of electricity per year" are meaningless and incorrect. A correct statement should be something like "the new wind turbine with a rated power of 50 kW will generate 120,000 kWh of electricity per year."

Dimensional Homogeneity

We all know that you cannot add apples and oranges. But we somehow manage to do it (by mistake, of course). In engineering, all equations must be *dimensionally homogeneous*. That is, every term in an equation must have the same dimensions. If, at some stage of an analysis, we find ourselves in a position to add two quantities that have different dimensions or units, it is a clear indication that we have made an error at an earlier stage. So checking dimensions (or units) can serve as a valuable tool to spot errors.

EXAMPLE 1–2 Electric Power Generation by a Wind Turbine

A school is paying $0.09/kWh for electric power. To reduce its power bill, the school installs a wind turbine (Fig 1–36) with a rated power of 30 kW. If the turbine operates 2200 hours per year at the rated power, determine the amount of electric power generated by the wind turbine and the money saved by the school per year.

SOLUTION A wind turbine is installed to generate electricity. The amount of electric energy generated and the money saved per year are to be determined.
Analysis The wind turbine generates electric energy at a rate of 30 kW or 30 kJ/s. Then the total amount of electric energy generated per year becomes

$$\text{Total energy} = (\text{Energy per unit time})(\text{Time interval})$$
$$= (30 \text{ kW})(2200 \text{ h})$$
$$= \mathbf{66{,}000 \text{ kWh}}$$

The money saved per year is the monetary value of this energy determined as

$$\text{Money saved} = (\text{Total energy})(\text{Unit cost of energy})$$
$$= (66{,}000 \text{ kWh})(\$0.09/\text{kWh})$$
$$= \mathbf{\$5940}$$

Discussion The annual electric energy production also could be determined in kJ by unit manipulations as

$$\text{Total energy} = (30 \text{ kW})(2200 \text{ h})\left(\frac{3600 \text{ s}}{1 \text{ h}}\right)\left(\frac{1 \text{ kJ/s}}{1 \text{ kW}}\right) = 2.38 \times 10^8 \text{ kJ}$$

which is equivalent to 66,000 kWh (1 kWh = 3600 kJ).

FIGURE 1–36
A wind turbine, as discussed in Example 1–2.
Photo by Andy Cimbala.

We all know from experience that units can give terrible headaches if they are not used carefully in solving a problem. However, with some attention and skill, units can be used to our advantage. They can be used to check formulas; sometimes they can even be used to *derive* formulas, as explained in the following example.

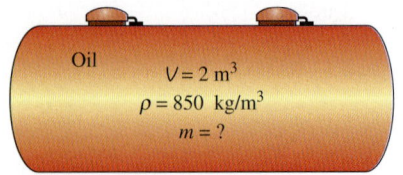

FIGURE 1–37
Schematic for Example 1–3.

EXAMPLE 1–3   Obtaining Formulas from Unit Considerations

A tank is filled with oil whose density is $\rho = 850$ kg/m^3. If the volume of the tank is $V = 2$ m^3, determine the amount of mass m in the tank.

SOLUTION   The volume of an oil tank is given. The mass of oil is to be determined.
Assumptions   Oil is a nearly incompressible substance and thus its density is constant.
Analysis   A sketch of the system just described is given in Fig. 1–37. Suppose we forgot the formula that relates mass to density and volume. However, we know that mass has the unit of kilograms. That is, whatever calculations we do, we should end up with the unit of kilograms. Putting the given information into perspective, we have

$$\rho = 850 \text{ kg/m}^3 \quad \text{and} \quad V = 2 \text{ m}^3$$

It is obvious that we can eliminate m^3 and end up with kg by multiplying these two quantities. Therefore, the formula we are looking for should be

$$m = \rho V$$

Thus,

$$m = (850 \text{ kg/m}^3)(2 \text{ m}^3) = \mathbf{1700 \text{ kg}}$$

Discussion   Note that this approach may not work for more complicated formulas. Nondimensional constants also may be present in the formulas, and these cannot be derived from unit considerations alone.

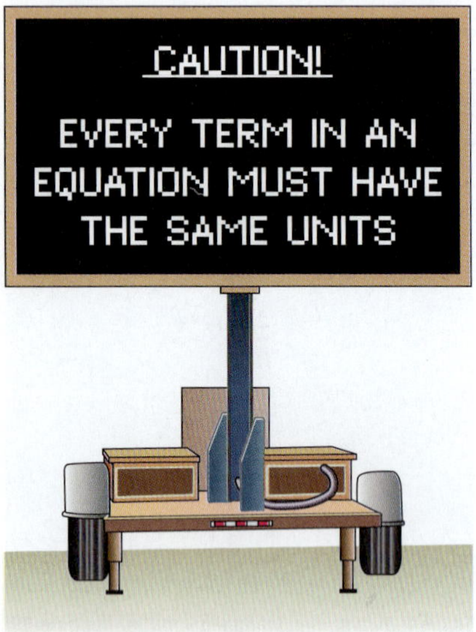

You should keep in mind that a formula that is not dimensionally homogeneous is definitely wrong (Fig. 1–38), but a dimensionally homogeneous formula is not necessarily right.

Unity Conversion Ratios

Just as all nonprimary dimensions can be formed by suitable combinations of primary dimensions, *all nonprimary units (**secondary units**) can be formed by combinations of primary units.* Force units, for example, can be expressed as

$$N = kg \frac{m}{s^2} \quad \text{and} \quad lbf = 32.174 \, lbm \frac{ft}{s^2}$$

They can also be expressed more conveniently as **unity conversion ratios** as

$$\frac{N}{kg \cdot m/s^2} = 1 \quad \text{and} \quad \frac{lbf}{32.174 \, lbm \cdot ft/s^2} = 1$$

FIGURE 1–38
Always check the units in your calculations.

Unity conversion ratios are identically equal to 1 and are unitless, and thus such ratios (or their inverses) can be inserted conveniently into any calculation to properly convert units (Fig 1–39). You are encouraged to always use unity conversion ratios such as those given here when converting units. Some textbooks insert the archaic gravitational constant g_c defined as $g_c = 32.174$ lbm·ft/lbf·s^2 = kg·m/N·s^2 = 1 into equations in order to force units to match. This practice leads to unnecessary confusion and is strongly discouraged by the present authors. We recommend that you instead use unity conversion ratios.

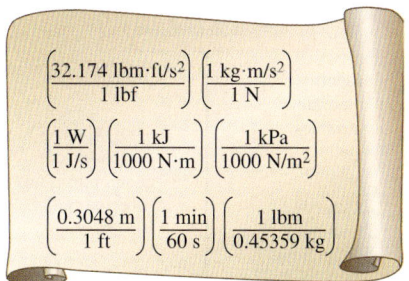

FIGURE 1–39
Every unity conversion ratio (as well as its inverse) is exactly equal to one. Shown here are a few commonly used unity conversion ratios.

EXAMPLE 1–4 The Weight of One Pound-Mass

Using unity conversion ratios, show that 1.00 lbm weighs 1.00 lbf on earth (Fig. 1–40).

Solution A mass of 1.00 lbm is subjected to standard earth gravity. Its weight in lbf is to be determined.
Assumptions Standard sea-level conditions are assumed.
Properties The gravitational constant is $g = 32.174$ ft/s^2.
Analysis We apply Newton's second law to calculate the weight (force) that corresponds to the known mass and acceleration. The weight of any object is equal to its mass times the local value of gravitational acceleration. Thus,

$$W = mg = (1.00 \text{ lbm})(32.174 \text{ ft/s}^2)\left(\frac{1 \text{ lbf}}{32.174 \text{ lbm·ft/s}^2}\right) = \mathbf{1.00 \text{ lbf}}$$

Discussion The quantity in large parentheses in this equation is a unity conversion ratio. Mass is the same regardless of its location. However, on some other planet with a different value of gravitational acceleration, the weight of 1 lbm would differ from that calculated here.

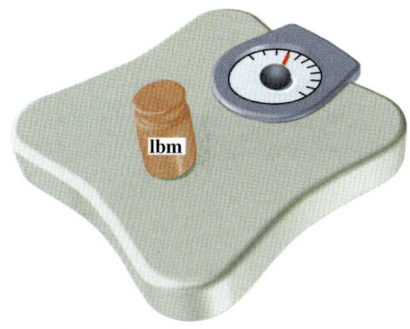

FIGURE 1–40
A mass of 1 lbm weighs 1 lbf on earth.

When you buy a box of breakfast cereal, the printing may say "Net weight: One pound (454 grams)." (See Fig. 1–41.) Technically, this means that the cereal inside the box weighs 1.00 lbf on earth and has a *mass* of 453.6 g (0.4536 kg). Using Newton's second law, the actual weight of the cereal on earth is

$$W = mg = (453.6 \text{ g})(9.81 \text{ m/s}^2)\left(\frac{1 \text{ N}}{1 \text{ kg·m/s}^2}\right)\left(\frac{1 \text{ kg}}{1000 \text{ g}}\right) = 4.49 \text{ N}$$

1–7 · MODELING IN ENGINEERING

An engineering device or process can be studied either *experimentally* (testing and taking measurements) or *analytically* (by analysis or calculations). The experimental approach has the advantage that we deal with the actual physical system, and the desired quantity is determined by measurement,

FIGURE 1–41
A quirk in the metric system of units.

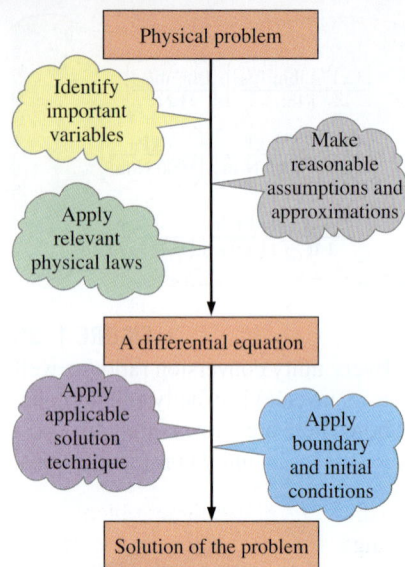

FIGURE 1–42
Mathematical modeling of physical problems.

within the limits of experimental error. However, this approach is expensive, time-consuming, and often impractical. Besides, the system we are studying may not even exist. For example, the entire heating and plumbing systems of a building must usually be sized *before* the building is actually built on the basis of the specifications given. The analytical approach (including the numerical approach) has the advantage that it is fast and inexpensive, but the results obtained are subject to the accuracy of the assumptions, approximations, and idealizations made in the analysis. In engineering studies, often a good compromise is reached by reducing the choices to just a few by analysis, and then verifying the findings experimentally.

The descriptions of most scientific problems involve equations that relate the changes in some key variables to each other. Usually the smaller the increment chosen in the changing variables, the more general and accurate the description. In the limiting case of infinitesimal or differential changes in variables, we obtain *differential equations* that provide precise mathematical formulations for the physical principles and laws by representing the rates of change as *derivatives*. Therefore, differential equations are used to investigate a wide variety of problems in sciences and engineering (Fig. 1–42). However, many problems encountered in practice can be solved without resorting to differential equations and the complications associated with them.

The study of physical phenomena involves two important steps. In the first step, all the variables that affect the phenomena are identified, reasonable assumptions and approximations are made, and the interdependence of these variables is studied. The relevant physical laws and principles are invoked, and the problem is formulated mathematically. The equation itself is very instructive as it shows the degree of dependence of some variables on others, and the relative importance of various terms. In the second step, the problem is solved using an appropriate approach, and the results are interpreted.

Many processes that seem to occur in nature randomly and without any order are, in fact, being governed by some visible or not-so-visible physical laws. Whether we notice them or not, these laws are there, governing consistently and predictably over what seem to be ordinary events. Most of these laws are well defined and well understood by scientists. This makes it possible to predict the course of an event before it actually occurs or to study various aspects of an event mathematically without actually running expensive and time-consuming experiments. This is where the power of analysis lies. Very accurate results to meaningful practical problems can be obtained with relatively little effort by using a suitable and realistic mathematical model. The preparation of such models requires an adequate knowledge of the natural phenomena involved and the relevant laws, as well as sound judgment. An unrealistic model will obviously give inaccurate and thus unacceptable results.

An analyst working on an engineering problem often finds himself or herself in a position to make a choice between a very accurate but complex model, and a simple but not-so-accurate model. The right choice depends on the situation at hand. The right choice is usually the simplest model that

 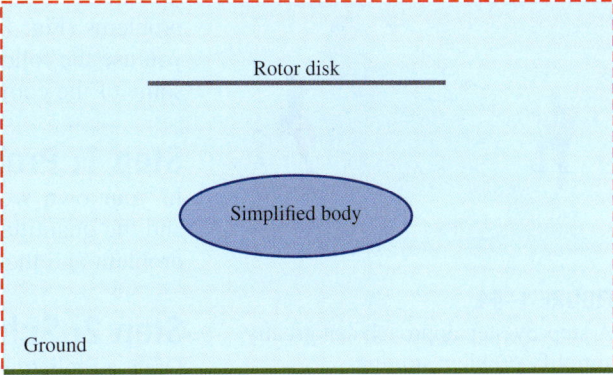

(a) Actual engineering problem　　　(b) Minimum essential model of the engineering problem

FIGURE 1–43
Simplified models are often used in fluid mechanics to obtain approximate solutions to difficult engineering problems. Here, the helicopter's rotor is modeled by a disk, across which is imposed a sudden change in pressure. The helicopter's body is modeled by a simple ellipsoid. This simplified model yields the essential features of the overall air flow field in the vicinity of the ground.
Photo by John M. Cimbala.

yields satisfactory results (Fig 1–43). Also, it is important to consider the actual operating conditions when selecting equipment.

Preparing very accurate but complex models is usually not so difficult. But such models are not much use to an analyst if they are very difficult and time-consuming to solve. At the minimum, the model should reflect the essential features of the physical problem it represents. There are many significant real-world problems that can be analyzed with a simple model. But it should always be kept in mind that the results obtained from an analysis are at best as accurate as the assumptions made in simplifying the problem. Therefore, the solution obtained should not be applied to situations for which the original assumptions do not hold.

A solution that is not quite consistent with the observed nature of the problem indicates that the mathematical model used is too crude. In that case, a more realistic model should be prepared by eliminating one or more of the questionable assumptions. This will result in a more complex problem that, of course, is more difficult to solve. Thus any solution to a problem should be interpreted within the context of its formulation.

1–8 ■ PROBLEM-SOLVING TECHNIQUE

The first step in learning any science is to grasp the fundamentals and to gain a sound knowledge of it. The next step is to master the fundamentals by testing this knowledge. This is done by solving significant real-world problems. Solving such problems, especially complicated ones, requires a systematic approach. By using a step-by-step approach, an engineer can reduce the

INTRODUCTION AND BASIC CONCEPTS

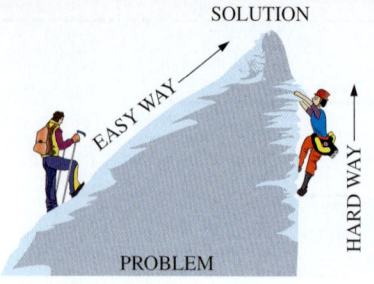

FIGURE 1–44
A step-by-step approach can greatly simplify problem solving.

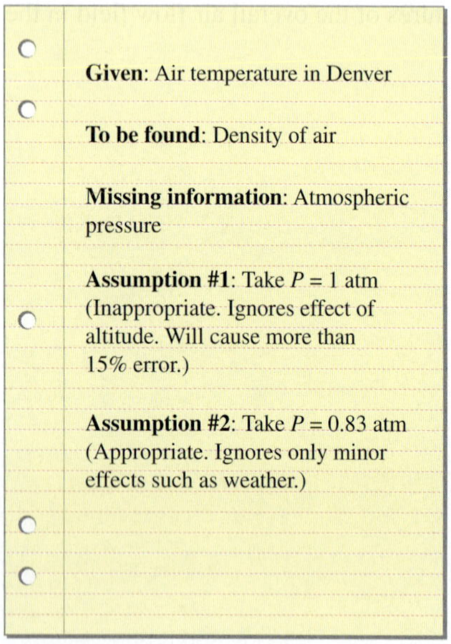

solution of a complicated problem into the solution of a series of simple problems (Fig. 1–44). When you are solving a problem, we recommend that you use the following steps zealously as applicable. This will help you avoid some of the common pitfalls associated with problem solving.

Step 1: Problem Statement
In your own words, briefly state the problem, the key information given, and the quantities to be found. This is to make sure that you understand the problem and the objectives before you attempt to solve the problem.

Step 2: Schematic
Draw a realistic sketch of the physical system involved, and list the relevant information on the figure. The sketch does not have to be something elaborate, but it should resemble the actual system and show the key features. Indicate any energy and mass interactions with the surroundings. Listing the given information on the sketch helps one to see the entire problem at once. Also, check for properties that remain constant during a process (such as temperature during an isothermal process), and indicate them on the sketch.

Step 3: Assumptions and Approximations
State any appropriate assumptions and approximations made to simplify the problem to make it possible to obtain a solution. Justify the questionable assumptions. Assume reasonable values for missing quantities that are necessary. For example, in the absence of specific data for atmospheric pressure, it can be taken to be 1 atm. However, it should be noted in the analysis that the atmospheric pressure decreases with increasing elevation. For example, it drops to 0.83 atm in Denver (elevation 1610 m) (Fig. 1–45).

Given: Air temperature in Denver

To be found: Density of air

Missing information: Atmospheric pressure

Assumption #1: Take $P = 1$ atm (Inappropriate. Ignores effect of altitude. Will cause more than 15% error.)

Assumption #2: Take $P = 0.83$ atm (Appropriate. Ignores only minor effects such as weather.)

FIGURE 1–45
The assumptions made while solving an engineering problem must be reasonable and justifiable.

Step 4: Physical Laws
Apply all the relevant basic physical laws and principles (such as the conservation of mass), and reduce them to their simplest form by utilizing the assumptions made. However, the region to which a physical law is applied must be clearly identified first. For example, the increase in speed of water flowing through a nozzle is analyzed by applying conservation of mass between the inlet and outlet of the nozzle.

Step 5: Properties
Determine the unknown properties at known states necessary to solve the problem from property relations or tables. List the properties separately, and indicate their source, if applicable.

Step 6: Calculations
Substitute the known quantities into the simplified relations and perform the calculations to determine the unknowns. Pay particular attention to the units and unit cancellations, and remember that a dimensional quantity without a unit is meaningless. Also, don't give a false implication of high precision

by copying all the digits from the screen of the calculator—round the final results to an appropriate number of significant digits (Section 1–10).

Step 7: Reasoning, Verification, and Discussion

Check to make sure that the results obtained are reasonable and intuitive, and verify the validity of the questionable assumptions. Repeat the calculations that resulted in unreasonable values. For example, under the same test conditions the aerodynamic drag acting on a car should *not* increase after streamlining the shape of the car (Fig. 1–46).

Also, point out the significance of the results, and discuss their implications. State the conclusions that can be drawn from the results, and any recommendations that can be made from them. Emphasize the limitations under which the results are applicable, and caution against any possible misunderstandings and using the results in situations where the underlying assumptions do not apply. For example, if you determined that using a larger-diameter pipe in a proposed pipeline will cost an additional $5000 in materials, but it will reduce the annual pumping costs by $3000, indicate that the larger-diameter pipeline will pay for its cost differential from the electricity it saves in less than two years. However, also state that only additional material costs associated with the larger-diameter pipeline are considered in the analysis.

Keep in mind that the solutions you present to your instructors, and any engineering analysis presented to others, is a form of communication. Therefore neatness, organization, completeness, and visual appearance are of utmost importance for maximum effectiveness (Fig 1–47). Besides, neatness also serves as a great checking tool since it is very easy to spot errors and inconsistencies in neat work. Carelessness and skipping steps to save time often end up costing more time and unnecessary anxiety.

The approach described here is used in the solved example problems without explicitly stating each step, as well as in the Solutions Manual of this text. For some problems, some of the steps may not be applicable or necessary. For example, often it is not practical to list the properties separately. However, we cannot overemphasize the importance of a logical and orderly approach to problem solving. Most difficulties encountered while solving a problem are not due to a lack of knowledge; rather, they are due to a lack of organization. You are strongly encouraged to follow these steps in problem solving until you develop your own approach that works best for you.

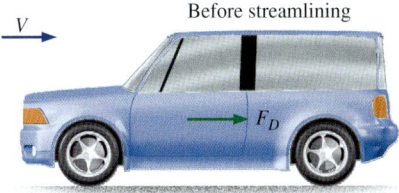
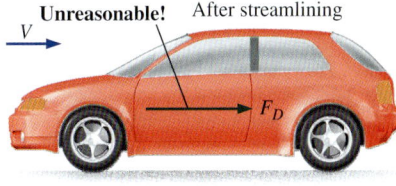

FIGURE 1–46
The results obtained from an engineering analysis must be checked for reasonableness.

FIGURE 1–47
Neatness and organization are highly valued by employers.

1–9 ▪ ENGINEERING SOFTWARE PACKAGES

You may be wondering why we are about to undertake an in-depth study of the fundamentals of another engineering science. After all, almost all such problems we are likely to encounter in practice can be solved using one of several sophisticated software packages readily available in the market today. These software packages not only give the desired numerical results, but also supply the outputs in colorful graphical form for impressive presentations. It is unthinkable to practice engineering today without using some of these packages. This tremendous computing power available to us at the touch of a button is both a blessing and a curse. It certainly enables engineers to solve problems easily and quickly, but it also opens the door for abuses and misinformation. In the hands of poorly educated people, these software packages are as dangerous as sophisticated powerful weapons in the hands of poorly trained soldiers.

Thinking that a person who can use the engineering software packages without proper training in the fundamentals can practice engineering is like thinking that a person who can use a wrench can work as a car mechanic. If it were true that the engineering students do not need all these fundamental courses they are taking because practically everything can be done by computers quickly and easily, then it would also be true that the employers would no longer need high-salaried engineers since any person who knows how to use a word-processing program can also learn how to use those software packages. However, the statistics show that the need for engineers is on the rise, not on the decline, despite the availability of these powerful packages.

We should always remember that all the computing power and the engineering software packages available today are just *tools*, and tools have meaning only in the hands of masters. Having the best word-processing program does not make a person a good writer, but it certainly makes the job of a good writer much easier and makes the writer more productive (Fig. 1–48). Hand calculators did not eliminate the need to teach our children how to add or subtract, and sophisticated medical software packages did not take the place of medical school training. Neither will engineering software packages replace the traditional engineering education. They will simply cause a shift in emphasis in the courses from mathematics to physics. That is, more time will be spent in the classroom discussing the physical aspects of the problems in greater detail, and less time on the mechanics of solution procedures.

All these marvelous and powerful tools available today put an extra burden on today's engineers. They must still have a thorough understanding of the fundamentals, develop a "feel" of the physical phenomena, be able to put the data into proper perspective, and make sound engineering judgments, just like their predecessors. However, they must do it much better, and much faster, using more realistic models because of the powerful tools available today. The engineers in the past had to rely on hand calculations, slide rules, and later hand calculators and computers. Today they rely on software packages. The easy access to such power and the possibility of a simple misunderstanding or misinterpretation causing great damage make it more important today than ever to have solid training in the fundamentals of engineering. In this text we make an extra effort to put the emphasis on developing an intuitive and physical understanding of natural phenomena instead of on the mathematical details of solution procedures.

FIGURE 1–48
An excellent word-processing program does not make a person a good writer; it simply makes a good writer a more efficient writer.

© Ingram Publishing RF

Engineering Equation Solver (EES)

EES is a program that solves systems of linear or nonlinear algebraic or differential equations numerically. It has a large library of built-in thermodynamic property functions as well as mathematical functions, and allows the user to supply additional property data. Unlike some software packages, EES does not solve engineering problems; it only solves the equations supplied by the user. Therefore, the user must understand the problem and formulate it by applying any relevant physical laws and relations. EES saves the user considerable time and effort by simply solving the resulting mathematical equations. This makes it possible to attempt significant engineering problems not suitable for hand calculations and to conduct parametric studies quickly and conveniently. EES is a very powerful yet intuitive program that is very easy to use, as shown in Example 1–5. The use and capabilities of EES are explained in Appendix 3 on the text website.

EXAMPLE 1–5   Solving a System of Equations with EES

The difference of two numbers is 4, and the sum of the squares of these two numbers is equal to the sum of the numbers plus 20. Determine these two numbers.

SOLUTION Relations are given for the difference and the sum of the squares of two numbers. The two numbers are to be determined.
Analysis We start the EES program by double-clicking on its icon, open a new file, and type the following on the blank screen that appears:

$$x-y=4$$
$$x\char`^2+y\char`^2=x+y+20$$

which is an exact mathematical expression of the problem statement with *x* and *y* denoting the unknown numbers. The solution to this system of two nonlinear equations with two unknowns is obtained by a single click on the "calculator" icon on the taskbar. It gives (Fig. 1–49)

$$\mathbf{x = 5} \quad \text{and} \quad \mathbf{y = 1}$$

Discussion Note that all we did is formulate the problem as we would on paper; EES took care of all the mathematical details of solution. Also note that equations can be linear or nonlinear, and they can be entered in any order with unknowns on either side. Friendly equation solvers such as EES allow the user to concentrate on the physics of the problem without worrying about the mathematical complexities associated with the solution of the resulting system of equations.

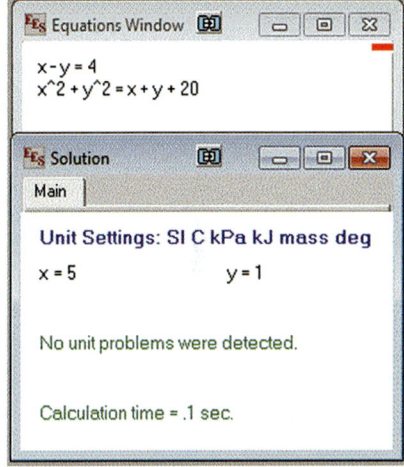

FIGURE 1–49
EES screen images for Example 1–5.

CFD Software

Computational fluid dynamics (CFD) is used extensively in engineering and research, and we discuss CFD in detail in Chapter 15. We also show example solutions from CFD throughout the textbook since CFD graphics are great for illustrating flow streamlines, velocity, and pressure distributions, etc.– beyond what we are able to visualize in the laboratory. However, because there are several different commercial CFD packages available for users, and student access to these codes is highly dependent on departmental licenses, we do not provide end-of-chapter CFD problems that are tied to any particular CFD package. Instead, we provide some general CFD problems in Chapter 15, and we also maintain a website (see link at www.mheducation.asia/olc/cengel) containing CFD problems that can be solved with a number of different CFD programs. Students are encouraged to work through some of these problems to become familiar with CFD.

1–10 • ACCURACY, PRECISION, AND SIGNIFICANT DIGITS

In engineering calculations, the supplied information is not known to more than a certain number of significant digits, usually three digits. Consequently, the results obtained cannot possibly be precise to more significant digits. Reporting results in more significant digits implies greater precision than exists, and it should be avoided.

Regardless of the system of units employed, engineers must be aware of three principles that govern the proper use of numbers: accuracy, precision, and significant digits. For engineering measurements, they are defined as follows:

- **Accuracy error** (*inaccuracy*) is the value of one reading minus the true value. In general, accuracy of a set of measurements refers to the closeness of the average reading to the true value. Accuracy is generally associated with repeatable, fixed errors.
- **Precision error** is the value of one reading minus the average of readings. In general, precision of a set of measurements refers to the fineness of the resolution and the repeatability of the instrument. Precision is generally associated with unrepeatable, random errors.
- **Significant digits** are digits that are relevant and meaningful.

A measurement or calculation can be very precise without being very accurate, and vice versa. For example, suppose the true value of wind speed is 25.00 m/s. Two anemometers A and B take five wind speed readings each:

Anemometer A: 25.50, 25.69, 25.52, 25.58, and 25.61 m/s. Average of all readings = 25.58 m/s.

Anemometer B: 26.3, 24.5, 23.9, 26.8, and 23.6 m/s. Average of all readings = 25.02 m/s.

Clearly, anemometer A is more precise, since none of the readings differs by more than 0.11 m/s from the average. However, the average is 25.58 m/s, 0.58 m/s greater than the true wind speed; this indicates significant **bias error,** also called **constant error** or **systematic error.** On the other hand, anemometer B is not very precise, since its readings swing wildly from the average; but its overall average is much closer to the true value. Hence, anemometer B is more accurate than anemometer A, at least for this set of readings, even though it is less precise. The difference between accuracy and precision can be illustrated effectively by analogy to shooting arrows at a target, as sketched in Fig. 1–50. Shooter A is very precise, but not very accurate, while shooter B has better overall accuracy, but less precision.

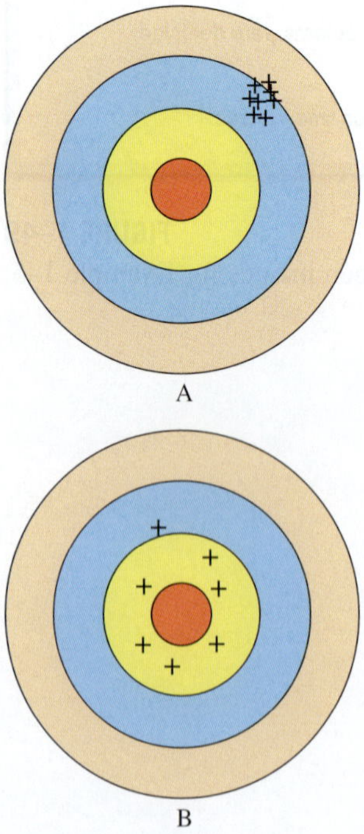

Many engineers do not pay proper attention to the number of significant digits in their calculations. The least significant numeral in a number implies the precision of the measurement or calculation. For example, a result written as 1.23 (three significant digits) *implies* that the result is precise to within one digit in the second decimal place; i.e., the number is somewhere between 1.22 and 1.24. Expressing this number with any more digits would be misleading. The number of significant digits is most easily evaluated when the number is written in exponential notation; the number of significant digits can then simply be counted, including zeroes. Alternatively, the

FIGURE 1–50
Illustration of accuracy versus precision. Shooter A is more precise, but less accurate, while shooter B is more accurate, but less precise.

least significant digit can be underlined to indicate the author's intent. Some examples are shown in Table 1–3.

When performing calculations or manipulations of several parameters, the final result is generally only as precise as the least precise parameter in the problem. For example, suppose A and B are multiplied to obtain C. If $A = 2.3601$ (five significant digits), and $B = 0.34$ (two significant digits), then $C = 0.80$ (only two digits are significant in the final result). Note that most students are tempted to write $C = 0.802434$, with six significant digits, since that is what is displayed on a calculator after multiplying these two numbers.

Let's analyze this simple example carefully. Suppose the exact value of B is 0.33501, which is read by the instrument as 0.34. Also suppose A is exactly 2.3601, as measured by a more accurate and precise instrument. In this case, $C = A \times B = 0.79066$ to five significant digits. Note that our first answer, $C = 0.80$ is off by one digit in the second decimal place. Likewise, if B is 0.34499, and is read by the instrument as 0.34, the product of A and B would be 0.81421 to five significant digits. Our original answer of 0.80 is again off by one digit in the second decimal place. The main point here is that 0.80 (to two significant digits) is the best one can expect from this multiplication since, to begin with, one of the values had only two significant digits. Another way of looking at this is to say that beyond the first two digits in the answer, the rest of the digits are meaningless or not significant. For example, if one reports what the calculator displays, 2.3601 times 0.34 equals 0.802434, the last four digits are *meaningless*. As shown, the final result may lie between 0.79 and 0.81—any digits beyond the two significant digits are not only meaningless, but *misleading*, since they imply to the reader more precision than is really there.

As another example, consider a 3.75-L container filled with gasoline whose density is 0.845 kg/L, and determine its mass. Probably the first thought that comes to your mind is to multiply the volume and density to obtain 3.16875 kg for the mass, which falsely implies that the mass so determined is precise to six significant digits. In reality, however, the mass cannot be more precise than three significant digits since both the volume and the density are precise to three significant digits only. Therefore, the result should be rounded to three significant digits, and the mass should be reported to be 3.17 kg instead of what the calculator displays (Fig. 1–51). The result 3.16875 kg would be correct only if the volume and density were given to be 3.75000 L and 0.845000 kg/L, respectively. The value 3.75 L implies that we are fairly confident that the volume is precise within ± 0.01 L, and it cannot be 3.74 or 3.76 L. However, the volume can be 3.746, 3.750, 3.753, etc., since they all round to 3.75 L.

You should also be aware that sometimes we knowingly introduce small errors in order to avoid the trouble of searching for more accurate data. For example, when dealing with liquid water, we often use the value of 1000 kg/m^3 for density, which is the density value of pure water at 0°C. Using this value at 75°C will result in an error of 2.5 percent since the density at this temperature is 975 kg/m^3. The minerals and impurities in the water will introduce additional error. This being the case, you should have no reservation in rounding the final results to a reasonable number of significant digits. Besides, having a few percent uncertainty in the results of engineering analysis is usually the norm, not the exception.

TABLE 1–3

Significant digits

Number	Exponential Notation	Number of Significant Digits
12.3	1.23×10^1	3
12<u>3</u>,000	1.23×10^5	3
0.00123	1.23×10^{-3}	3
40,<u>3</u>00	4.03×10^4	3
40,300	4.0300×10^4	5
0.005600	5.600×10^{-3}	4
0.0056	5.6×10^{-3}	2
0.006	$6. \times 10^{-3}$	1

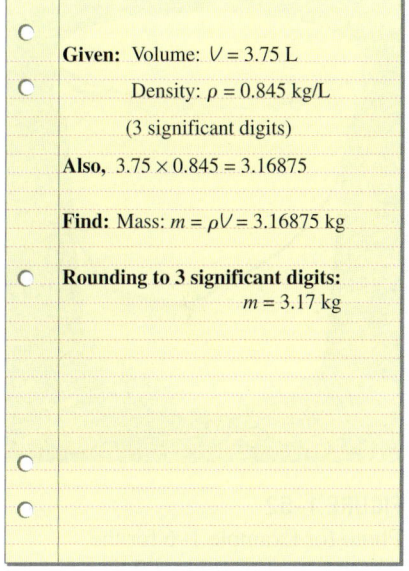

FIGURE 1–51

A result with more significant digits than that of given data falsely implies more precision.

When writing intermediate results in a computation, it is advisable to keep several "extra" digits to avoid round-off errors; however, the final result should be written with the number of significant digits taken into consideration. You must also keep in mind that a certain number of significant digits of precision in the result does not necessarily imply the same number of digits of overall *accuracy*. Bias error in one of the readings may, for example, significantly reduce the overall accuracy of the result, perhaps even rendering the last significant digit meaningless, and reducing the overall number of reliable digits by one. Experimentally determined values are subject to measurement errors, and such errors are reflected in the results obtained. For example, if the density of a substance has an uncertainty of 2 percent, then the mass determined using this density value will also have an uncertainty of 2 percent.

Finally, when the number of significant digits is unknown, the accepted engineering standard is three significant digits. Therefore, if the length of a pipe is given to be 40 m, we will assume it to be 40.0 m in order to justify using three significant digits in the final results.

FIGURE 1–52
Photo for Example 1–6 for the measurement of volume flow rate.
Photo by John M. Cimbala.

EXAMPLE 1–6   Significant Digits and Volume Flow Rate

Jennifer is conducting an experiment that uses cooling water from a garden hose. In order to calculate the volume flow rate of water through the hose, she times how long it takes to fill a container (Fig. 1–52). The volume of water collected is $V = 4.2$ L in time period $\Delta t = 45.62$ s, as measured with a stopwatch. Calculate the volume flow rate of water through the hose in units of cubic meters per minute.

SOLUTION Volume flow rate is to be determined from measurements of volume and time period.
Assumptions **1** Jennifer recorded her measurements properly, such that the volume measurement is precise to two significant digits while the time period is precise to four significant digits. **2** No water is lost due to splashing out of the container.
Analysis Volume flow rate $\dot{V}$ is volume displaced per unit time and is expressed as

Volume flow rate: $$\dot{V} = \frac{\Delta V}{\Delta t}$$

Substituting the measured values, the volume flow rate is determined to be

$$\dot{V} = \frac{4.2 \text{ L}}{45.62 \text{ s}} \left(\frac{1 \text{ m}^3}{1000 \text{ L}}\right)\left(\frac{60 \text{ s}}{1 \text{ min}}\right) = \mathbf{5.5 \times 10^{-3} \text{ m}^3/\text{min}}$$

Discussion The final result is listed to two significant digits since we cannot be confident of any more precision than that. If this were an intermediate step in subsequent calculations, a few extra digits would be carried along to avoid accumulated round-off error. In such a case, the volume flow rate would be written as $\dot{V} = 5.5239 \times 10^{-3}$ m^3/min. Based on the given information, we cannot say anything about the *accuracy* of our result, since we have no information about systematic errors in either the volume measurement or the time measurement.

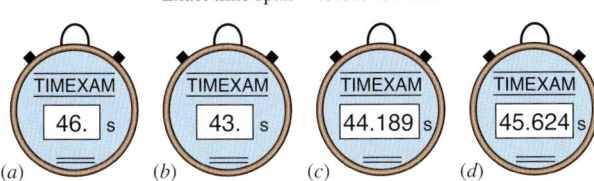

FIGURE 1–53
An instrument with many digits of resolution (stopwatch *c*) may be less accurate than an instrument with few digits of resolution (stopwatch *a*). What can you say about stopwatches *b* and *d*?

Also keep in mind that good precision does not guarantee good accuracy. For example, if the batteries in the stopwatch were weak, its accuracy could be quite poor, yet the readout would still be displayed to four significant digits of precision.

In common practice, precision is often associated with *resolution*, which is a measure of how finely the instrument can report the measurement. For example, a digital voltmeter with five digits on its display is said to be more precise than a digital voltmeter with only three digits. However, the number of displayed digits has nothing to do with the overall *accuracy* of the measurement. An instrument can be very precise without being very accurate when there are significant bias errors. Likewise, an instrument with very few displayed digits can be more accurate than one with many digits (Fig. 1–53).

SUMMARY

In this chapter some basic concepts of fluid mechanics are introduced and discussed. A substance in the liquid or gas phase is referred to as a *fluid*. *Fluid mechanics* is the science that deals with the behavior of fluids at rest or in motion and the interaction of fluids with solids or other fluids at the boundaries.

The flow of an unbounded fluid over a surface is *external flow*, and the flow in a pipe or duct is *internal flow* if the fluid is completely bounded by solid surfaces. A fluid flow is classified as being *compressible* or *incompressible*, depending on the density variation of the fluid during flow. The densities of liquids are essentially constant, and thus the flow of liquids is typically incompressible. The term *steady* implies *no change with time*. The opposite of steady is *unsteady*. The term *uniform* implies *no change with location* over a specified region. A flow is said to be *one-dimensional* when the properties or variables change in one dimension only. A fluid in direct contact with a solid surface sticks to the surface and there is no slip. This is known as the *no-slip condition*, which leads to the formation of *boundary layers* along solid surfaces. In this book we concentrate on steady incompressible viscous flows—both internal and external.

A system of fixed mass is called a *closed system*, and a system that involves mass transfer across its boundaries is called an *open system* or *control volume*. A large number of engineering problems involve mass flow in and out of a system and are therefore modeled as control volumes.

In engineering calculations, it is important to pay particular attention to the units of the quantities to avoid errors caused by inconsistent units, and to follow a systematic approach. It is also important to recognize that the information given is not known to more than a certain number of significant digits, and the results obtained cannot possibly be accurate to more significant digits. The information given on dimensions and units; problem-solving technique; and accuracy, precision, and significant digits will be used throughout the entire text.

REFERENCES AND SUGGESTED READING

1. American Society for Testing and Materials. *Standards for Metric Practice*. ASTM E 380-79, January 1980.
2. G. M. Homsy, H. Aref, K. S. Breuer, S. Hochgreb, J. R. Koseff, B. R. Munson, K. G. Powell, C. R. Robertson, and S. T. Thoroddsen. *Multi-Media Fluid Mechanics* (CD). Cambridge: Cambridge University Press, 2000.
3. M. Van Dyke. *An Album of Fluid Motion*. Stanford, CA: The Parabolic Press, 1982.

INTRODUCTION AND BASIC CONCEPTS

APPLICATION SPOTLIGHT ■ What Nuclear Blasts and Raindrops Have in Common

Guest Author: Lorenz Sigurdson, Vortex Fluid Dynamics Lab, University of Alberta

(a)   (b)

FIGURE 1–54
Comparison of the vortex structure created by: (*a*) a water drop after impacting a pool of water (inverted, from Peck and Sigurdson, 1994), and (*b*) an above-ground nuclear test in Nevada in 1957 (U.S. Department of Energy). The 2.6 mm drop was dyed with fluorescent tracer and illuminated by a strobe flash 50 ms after it had fallen 35 mm and impacted the clear pool. The drop was approximately spherical at the time of impact with the clear pool of water. Interruption of a laser beam by the falling drop was used to trigger a timer that controlled the time of the strobe flash after impact of the drop. Details of the careful experimental procedure necessary to create the drop photograph are given by Peck and Sigurdson (1994) and Peck et al. (1995). The tracers added to the flow in the bomb case were primarily heat and dust. The heat is from the original fireball which for this particular test (the "Priscilla" event of Operation Plumbob) was large enough to reach the ground from where the bomb was initially suspended. Therefore, the tracer's initial geometric condition was a sphere intersecting the ground.

(*a*) *From Peck, B., and Sigurdson, L. W., Phys. Fluids, 6(2)(Part 1), 564, 1994. Used by permission of the author.*

(*b*) *United States Department of Energy. Photo from Lorenz Sigurdson.*

Why do the two images in Fig. 1–54 look alike? Figure 1–54*b* shows an above-ground nuclear test performed by the U.S. Department of Energy in 1957. An atomic blast created a fireball on the order of 100 m in diameter. Expansion is so quick that a compressible flow feature occurs: an expanding spherical shock wave. The image shown in Fig. 1–54*a* is an everyday innocuous event: an *inverted* image of a dye-stained water drop after it has fallen into a pool of water, looking from below the pool surface. It could have fallen from your spoon into a cup of coffee, or been a secondary splash after a raindrop hit a lake. Why is there such a strong similarity between these two vastly different events? The application of fundamental principles of fluid mechanics learned in this book will help you understand much of the answer, although one can go much deeper.

The water has higher *density* (Chap. 2) than air, so the drop has experienced negative *buoyancy* (Chap. 3) as it has fallen through the air before impact. The fireball of hot gas is less dense than the cool air surrounding it, so it has positive buoyancy and rises. The *shock wave* (Chap. 12) reflecting from the ground also imparts a positive upward force to the fireball. The primary structure at the top of each image is called a *vortex ring*. This ring is a mini-tornado of concentrated *vorticity* (Chap. 4) with the ends of the tornado looping around to close on itself. The laws of *kinematics* (Chap. 4) tell us that this vortex ring will carry the fluid in a direction toward the top of the page. This is expected in both cases from the forces applied and the law of conservation of momentum applied through a *control volume analysis* (Chap. 5). One could also analyze this problem with *differential analysis* (Chaps. 9 and 10) or with *computational fluid dynamics* (Chap. 15). But why does the *shape* of the tracer material look so similar? This occurs if there is approximate *geometric* and *kinematic similarity* (Chap. 7), and if the *flow visualization* (Chap. 4) technique is similar. The passive tracers of heat and dust for the bomb, and fluorescent dye for the drop, were introduced in a similar manner as noted in the figure caption.

Further knowledge of kinematics and vortex dynamics can help explain the similarity of the vortex structure in the images to much greater detail, as discussed by Sigurdson (1997) and Peck and Sigurdson (1994). Look at the lobes dangling beneath the primary vortex ring, the striations in the "stalk," and the ring at the base of each structure. There is also topological similarity of this structure to other vortex structures occurring in turbulence. Comparison of the drop and bomb has given us a better understanding of how turbulent structures are created and evolve. What other secrets of fluid mechanics are left to be revealed in explaining the similarity between these two flows?

References

Peck, B., and Sigurdson, L. W., "The Three-Dimensional Vortex Structure of an Impacting Water Drop," *Phys. Fluids*, 6(2) (Part 1), p. 564, 1994.

Peck, B., Sigurdson, L. W., Faulkner, B., and Buttar, I., "An Apparatus to Study Drop-Formed Vortex Rings," *Meas. Sci. Tech.*, 6, p. 1538, 1995.

Sigurdson, L. W., "Flow Visualization in Turbulent Large-Scale Structure Research," Chapter 6 in *Atlas of Visualization*, Vol. III, Flow Visualization Society of Japan, eds., CRC Press, pp. 99–113, 1997.

PROBLEMS*

Introduction, Classification, and System

1–1C Consider the flow of air over the wings of an aircraft. Is this flow internal or external? How about the flow of gases through a jet engine?

1–2C Define incompressible flow and incompressible fluid. Must the flow of a compressible fluid necessarily be treated as compressible?

1–3C Define internal, external, and open-channel flows.

1–4C How is the Mach number of a flow defined? What does a Mach number of 2 indicate?

1–5C When an airplane is flying at a constant speed relative to the ground, is it correct to say that the Mach number of this airplane is also constant?

1–6C Consider the flow of air at a Mach number of 0.12. Should this flow be approximated as being incompressible?

1–7C What is the no-slip condition? What causes it?

1–8C What is forced flow? How does it differ from natural flow? Is flow caused by winds forced or natural flow?

1–9C What is a boundary layer? What causes a boundary layer to develop?

1–10C What is the difference between the classical and the statistical approaches?

1–11C What is a steady-flow process?

1–12C Define stress, normal stress, shear stress, and pressure.

1–13C When analyzing the acceleration of gases as they flow through a nozzle, what would you choose as your system? What type of system is this?

1–14C When is a closed system, and when is it a control volume?

1–15C You are trying to understand how a reciprocating air compressor (a piston-cylinder device) works. What system would you use? What type of system is this?

1–16C What are system, surroundings, and boundary?

Mass, Force, and Units

1–17C Explain why the light-year has the dimension of length.

1–18C What is the difference between kg-mass and kg-force?

1–19C What is the difference between pound-mass and pound-force?

1–20C In a news article, it is stated that a recently developed geared turbofan engine produces 15,000 pounds of thrust to propel the aircraft forward. Is "pound" mentioned here lbm or lbf? Explain.

1–21C What is the net force acting on a car cruising at a constant velocity of 70 km/h (*a*) on a level road and (*b*) on an uphill road?

1–22 A 6-kg plastic tank that has a volume of 0.18 m^3 is filled with liquid water. Assuming the density of water is 1000 kg/m^3, determine the weight of the combined system.

1–23 What is the weight, in N, of an object with a mass of 200 kg at a location where $g = 9.6$ m/s^2?

1–24 What is the weight of a 1-kg substance in N, kN, kg·m/s^2, kgf, lbm·ft/s^2, and lbf?

1–25 Determine the mass and the weight of the air contained in a room whose dimensions are 6 m × 6 m × 8 m. Assume the density of the air is 1.16 kg/m^3. *Answers:* 334.1 kg, 3277 N

1–26 While solving a problem, a person ends up with the equation $E = 16$ kJ $+ 7$ kJ/kg at some stage. Here *E* is the total energy and has the unit of kilojoules. Determine how to correct the error and discuss what may have caused it.

1–27 The acceleration of high-speed aircraft is sometimes expressed in *g*'s (in multiples of the standard acceleration of gravity). Determine the net force, in N, that a 90-kg man would experience in an aircraft whose acceleration is 6 *g*'s.

1–28 A 5-kg rock is thrown upward with a force of 150 N at a location where the local gravitational acceleration is 9.79 m/s^2. Determine the acceleration of the rock, in m/s^2.

1–29 Solve Prob. 1–28 using EES (or other) software. Print out the entire solution, including the numerical results with proper units.

1–30 The value of the gravitational acceleration *g* decreases with elevation from 9.807 m/s^2 at sea level to 9.767 m/s^2 at an altitude of 13,000 m, where large passenger planes cruise. Determine the percent reduction in the weight of an airplane cruising at 13,000 m relative to its weight at sea level.

1–31 At 45° latitude, the gravitational acceleration as a function of elevation *z* above sea level is given by $g = a - bz$, where $a = 9.807$ m/s^2 and $b = 3.32 \times 10^{-6}$ s^{-2}. Determine the height above sea level where the weight of an object will decrease by 1 percent. *Answer:* 29,500 m

---

* Problems designated by a "C" are concept questions, and students are encouraged to answer them all. Problems with the icon are solved using EES, and complete solutions together with parametric studies are included on the text website. Problems with the icon are comprehensive in nature and are intended to be solved with an equation solver such as EES.

1–32 A 4-kW resistance heater in a water heater runs for 2 hours to raise the water temperature to the desired level. Determine the amount of electric energy used in both kWh and kJ.

1–33 The gas tank of a car is filled with a nozzle that discharges gasoline at a constant flow rate. Based on unit considerations of quantities, obtain a relation for the filling time in terms of the volume V of the tank (in L) and the discharge rate of gasoline ($\dot{V}$, in L/s).

1–34 A pool of volume V (in m³) is to be filled with water using a hose of diameter D (in m). If the average discharge velocity is V (in m/s) and the filling time is t (in s), obtain a relation for the volume of the pool based on unit considerations of quantities involved.

1–35 Based on unit considerations alone, show that the power needed to accelerate a car of mass m (in kg) from rest to velocity V (in m/s) in time interval t (in s) is proportional to mass and the square of the velocity of the car and inversely proportional to the time interval.

1–36 An airplane flies horizontally at 70 m/s. Its propeller delivers 1500 N of thrust (forward force) to overcome aerodynamic drag (backward force). Using dimensional reasoning and unity converstion ratios, calculate the useful power delivered by the propeller in units of kW and horsepower.

1–37 If the airplane of Problem 1–36 weighs 1450 lbf, estimate the lift force produced by the airplane's wings (in lbf and newtons) when flying at 70.0 m/s.

1–38 The boom of a fire truck raises a fireman (and his equipment—total weight 1250 N) 18 m into the air to fight a building fire. (*a*) Showing all your work and using unity conversion ratios, calculate the work done by the boom on the fireman in units of kJ. (*b*) If the useful power supplied by the boom to lift the fireman is 2.60 kW, estimate how long it takes to lift the fireman.

1–39 A man goes to a traditional market to buy a steak for dinner. He finds a 12-oz steak (1 lbm = 16 oz) for $3.15. He then goes to the adjacent international market and finds a 320-g steak of identical quality for $3.30. Which steak is the better buy?

1–40 Water at 20°C from a garden hose fills a 2.0 L container in 2.85 s. Using unity converstion ratios and showing all your work, calculate the volume flow rate in liters per minute (Lpm) and the mass flow rate in kg/s.

1–41 A forklift raises a 90.5 kg crate 1.80 m. (*a*) Showing all your work and using unity conversion ratios, calculate the work done by the forklift on the crane, in units of kJ. (*b*) If it takes 12.3 seconds to lift the crate, calculate the useful power supplied to the crate in kilowatts.

Modeling and Solving Engineering Problems

1–42C When modeling an engineering process, how is the right choice made between a simple but crude and a complex but accurate model? Is the complex model necessarily a better choice since it is more accurate?

1–43C What is the difference between the analytical and experimental approach to engineering problems? Discuss the advantages and disadvantages of each approach.

1–44C What is the importance of modeling in engineering? How are the mathematical models for engineering processes prepared?

1–45C What is the difference between precision and accuracy? Can a measurement be very precise but inaccurate? Explain.

1–46C How do the differential equations in the study of a physical problem arise?

1–47C What is the value of the engineering software packages in (*a*) engineering education and (*b*) engineering practice?

1–48 Solve this system of three equations with three unknowns using EES:

$$2x - y + z = 9$$
$$3x^2 + 2y = z + 2$$
$$xy + 2z = 14$$

1–49 Solve this system of two equations with two unknowns using EES:

$$x^3 - y^2 = 10.5$$
$$3xy + y = 4.6$$

1–50 Determine a positive real root of this equation using EES:

$$3.5x^3 - 10x^{0.5} - 3x = -4$$

1–51 Solve this system of three equations with three unknowns using EES:

$$x^2y - z = 1.5$$
$$x - 3y^{0.5} + xz = -2$$
$$x + y - z = 4.2$$

Review Problems

1–52 The reactive force developed by a jet engine to push an airplane forward is called thrust, and the thrust developed by the engine of a Boeing 777 is about 85,000 lbf. Express this thrust in N and kgf.

1–53 The weight of bodies may change somewhat from one location to another as a result of the variation of the gravitational acceleration g with elevation. Accounting for this variation using the relation in Prob. 1–33, determine the weight of an 80.0-kg person at sea level ($z = 0$), in Denver ($z = 1610$ m), and on the top of Mount Everest ($z = 8848$ m).

1–54 For liquids, the dynamic viscosity μ, which is a measure of resistance against flow is approximated as $\mu = a10^{b/(T-c)}$, where T is the absolute temperature, and a, b and c are experimental constants. Using the data listed in Table A–7 for methanol at 20°C, 40°C and 60°C, determine the constant a, b and c.

1–55 An important design consideration in two-phase pipe flow of solid-liquid mixtures is the terminal settling velocity below, which the flow becomes unstable and eventually the pipe becomes clogged. On the basis of extended transportation tests, the terminal settling velocity of a solid particle in the rest water given by $V_L = F_L \sqrt{2gD(S-1)}$, where F_L is an experimental coefficient, g the gravitational acceleration, D the pipe diameter, and S the specific gravity of solid particle. What is the dimension of F_L? Is this equation dimensionally homogeneous?

1–56 Consider the flow of air through a wind turbine whose blades sweep an area of diameter D (in m). The average air velocity through the swept area is V (in m/s). On the bases of the units of the quantities involved, show that the mass flow rate of air (in kg/s) through the swept area is proportional to air density, the wind velocity, and the square of the diameter of the swept area.

1–57 The drag force exerted on a car by air depends on a dimensionless drag coefficient, the density of air, the car velocity, and the frontal area of the car. That is, F_D = function (C_{Drag}, A_{front}, ρ, V). Based on unit considerations alone, obtain a relation for the drag force.

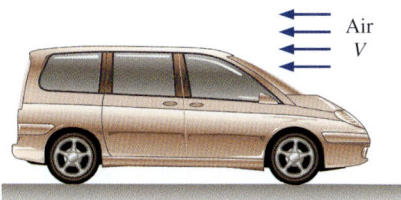

FIGURE P1–57

Fundamentals of Engineering (FE) Exam Problems

1–58 The speed of an aircraft is given to be 260 m/s in air. If the speed of sound at that location is 330 m/s, the flight of aircraft is
(a) Sonic   (b) Subsonic   (c) Supersonic   (d) Hypersonic

1–59 The speed of an aircraft is given to be 1250 km/h. If the speed of sound at that location is 315 m/s, the Mach number is
(a) 0.5   (b) 0.85   (c) 1.0   (d) 1.10   (e) 1.20

1–60 If mass, heat, and work are not allowed to cross the boundaries of a system, the system is called
(a) Isolated   (b) Isothermal   (c) Adiabatic   (d) Control mass   (e) Control volume

1–61 The weight of a 10-kg mass at sea level is
(a) 9.81 N   (b) 32.2 kgf   (c) 98.1 N   (d) 10 N   (e) 100 N

1–62 The weight of a 1-lbm mass is
(a) 1 lbm·ft/s^2 (b) 9.81 lbf   (c) 9.81 N   (d) 32.2 lbf   (e) 1 lbf

1–63 One kJ is *not* equal to
(a) 1 kPa·m^3 (b) 1 kN·m   (c) 0.001 MJ   (d) 1000 J   (e) 1 kg·m^2/s^2

1–64 Which is a unit for the amount of energy?
(a) Btu/h   (b) kWh   (c) kcal/h   (d) hp   (e) kW

1–65 A hydroelectric power plant operates at its rated power of 7 MW. If the plant has produced 26 million kWh of electricity in a specified year, the number of hours the plant has operated that year is
(a) 1125 h   (b) 2460 h   (c) 2893 h   (d) 3714 h   (e) 8760 h

Design and Essay Problems

1–66 Write an essay on the various mass- and volume-measurement devices used throughout history. Also, explain the development of the modern units for mass and volume.

1–67 Search the Internet to find out how to properly add or subtract numbers while taking into consideration the number of significant digits. Write a summary of the proper technique, then use the technique to solve the following cases: (a) 1.006 + 23.47, (b) 703,200 − 80.4, and (c) 4.6903 − 14.58. Be careful to express your final answer to the appropriate number of significant digits.

CHAPTER 2

PROPERTIES OF FLUIDS

In this chapter, we discuss properties that are encountered in the analysis of fluid flow. First we discuss *intensive* and *extensive properties* and define *density* and *specific gravity*. This is followed by a discussion of the properties *vapor pressure*, *energy* and its various forms, the *specific heats* of ideal gases and incompressible substances, the *coefficient of compressibility*, and the *speed of sound*. Then we discuss the property *viscosity*, which plays a dominant role in most aspects of fluid flow. Finally, we present the property *surface tension* and determine the *capillary rise* from static equilibrium conditions. The property *pressure* is discussed in Chap. 3 together with fluid statics.

OBJECTIVES

When you finish reading this chapter, you should be able to

- Have a working knowledge of the basic properties of fluids and understand the continuum approximation
- Have a working knowledge of viscosity and the consequences of the frictional effects it causes in fluid flow
- Calculate the capillary rise (or drop) in tubes due to the surface tension effect

A drop forms when liquid is forced out of a small tube. The shape of the drop is determined by a balance of pressure, gravity, and surface tension forces.
Royalty-Free/CORBIS

PROPERTIES OF FLUIDS

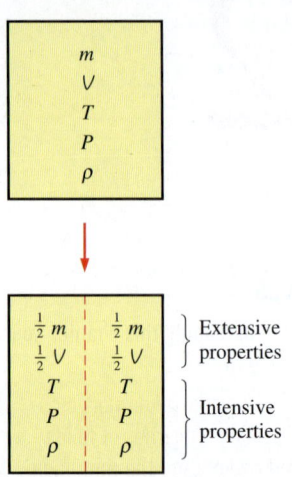

FIGURE 2–1
Criterion to differentiate intensive and extensive properties.

FIGURE 2–2
The length scale associated with most flows, such as seagulls in flight, is orders of magnitude larger than the mean free path of the air molecules. Therefore, here, and for all fluid flows considered in this book, the continuum idealization is appropriate.
PhotoLink/Getty RF

2–1 ▪ INTRODUCTION

Any characteristic of a system is called a **property**. Some familiar properties are pressure P, temperature T, volume V, and mass m. The list can be extended to include less familiar ones such as viscosity, thermal conductivity, modulus of elasticity, thermal expansion coefficient, electric resistivity, and even velocity and elevation.

Properties are considered to be either *intensive* or *extensive*. **Intensive properties** are those that are independent of the mass of the system, such as temperature, pressure, and density. **Extensive properties** are those whose values depend on the size—or extent—of the system. Total mass, total volume V, and total momentum are some examples of extensive properties. An easy way to determine whether a property is intensive or extensive is to divide the system into two equal parts with an imaginary partition, as shown in Fig. 2–1. Each part will have the same value of intensive properties as the original system, but half the value of the extensive properties.

Generally, uppercase letters are used to denote extensive properties (with mass m being a major exception), and lowercase letters are used for intensive properties (with pressure P and temperature T being the obvious exceptions).

Extensive properties per unit mass are called **specific properties.** Some examples of specific properties are specific volume ($v = V/m$) and specific total energy ($e = E/m$).

The state of a system is described by its properties. But we know from experience that we do not need to specify all the properties in order to fix a state. Once the values of a sufficient number of properties are specified, the rest of the properties assume certain values. That is, specifying a certain number of properties is sufficient to fix a state. The number of properties required to fix the state of a system is given by the **state postulate:** *The state of a simple compressible system is completely specified by two independent, intensive properties.*

Two properties are independent if one property can be varied while the other one is held constant. Not all properties are independent, and some are defined in terms of others, as explained in Section 2–2.

Continuum

A fluid is composed of molecules which may be widely spaced apart, especially in the gas phase. Yet it is convenient to disregard the atomic nature of the fluid and view it as continuous, homogeneous matter with no holes, that is, a **continuum.** The continuum idealization allows us to treat properties as point functions and to assume that the properties vary continually in space with no jump discontinuities. This idealization is valid as long as the size of the system we deal with is large relative to the space between the molecules (Fig. 2–2). This is the case in practically all problems, except some specialized ones. The continuum idealization is implicit in many statements we make, such as "the density of water in a glass is the same at any point."

To have a sense of the distances involved at the molecular level, consider a container filled with oxygen at atmospheric conditions. The diameter of an oxygen molecule is about 3×10^{-10} m and its mass is 5.3×10^{-26} kg. Also, the *mean free path* of oxygen at 1 atm pressure and 20°C is 6.3×10^{-8} m. That is, an oxygen molecule travels, on average, a distance of 6.3×10^{-8} m (about 200 times its diameter) before it collides with another molecule.

Also, there are about 3×10^{16} molecules of oxygen in the tiny volume of 1 mm³ at 1 atm pressure and 20°C (Fig. 2–3). The continuum model is applicable as long as the characteristic length of the system (such as its diameter) is much larger than the mean free path of the molecules. At very low pressure, e.g., at very high elevations, the mean free path may become large (for example, it is about 0.1 m for atmospheric air at an elevation of 100 km). For such cases the **rarefied gas flow theory** should be used, and the impact of individual molecules should be considered. In this text we limit our consideration to substances that can be modeled as a continuum.

2–2 ▪ DENSITY AND SPECIFIC GRAVITY

Density is defined as *mass per unit volume* (Fig. 2–4). That is,

Density: $$\rho = \frac{m}{V} \quad (\text{kg/m}^3) \quad (2\text{–}1)$$

The reciprocal of density is the **specific volume v**, which is defined as *volume per unit mass*. That is, $v = V/m = 1/\rho$. For a differential volume element of mass δm and volume δV, density can be expressed as $\rho = \delta m/\delta V$.

The density of a substance, in general, depends on temperature and pressure. The density of most gases is proportional to pressure and inversely proportional to temperature. Liquids and solids, on the other hand, are essentially incompressible substances, and the variation of their density with pressure is usually negligible. At 20°C, for example, the density of water changes from 998 kg/m³ at 1 atm to 1003 kg/m³ at 100 atm, a change of just 0.5 percent. The density of liquids and solids depends more strongly on temperature than it does on pressure. At 1 atm, for example, the density of water changes from 998 kg/m³ at 20°C to 975 kg/m³ at 75°C, a change of 2.3 percent, which can still be neglected in many engineering analyses.

Sometimes the density of a substance is given relative to the density of a well-known substance. Then it is called **specific gravity,** or **relative density,** and is defined as *the ratio of the density of a substance to the density of some standard substance at a specified temperature* (usually water at 4°C, for which $\rho_{H_2O} = 1000$ kg/m³). That is,

Specific gravity: $$SG = \frac{\rho}{\rho_{H_2O}} \quad (2\text{–}2)$$

Note that the specific gravity of a substance is a dimensionless quantity. However, in SI units, the numerical value of the specific gravity of a substance is exactly equal to its density in g/cm³ or kg/L (or 0.001 times the density in kg/m³) since the density of water at 4°C is 1 g/cm³ = 1 kg/L = 1000 kg/m³. The specific gravity of mercury at 20°C, for example, is 13.6. Therefore, its density at 20°C is 13.6 g/cm³ = 13.6 kg/L = 13,600 kg/m³. The specific gravities of some substances at 20°C are given in Table 2–1. Note that substances with specific gravities less than 1 are lighter than water, and thus they would float on water (if immiscible).

The weight of a unit volume of a substance is called **specific weight** or **weight density** and is expressed as

Specific weight: $$\gamma_s = \rho g \quad (\text{N/m}^3) \quad (2\text{–}3)$$

where g is the gravitational acceleration.

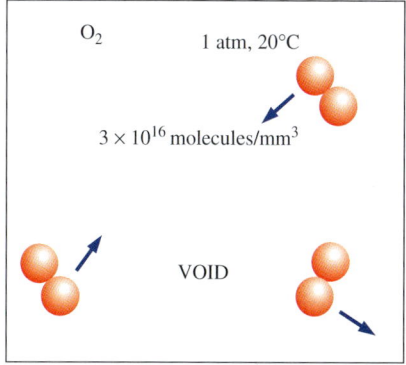

FIGURE 2–3
Despite the relatively large gaps between molecules, a gas can usually be treated as a continuum because of the very large number of molecules even in an extremely small volume.

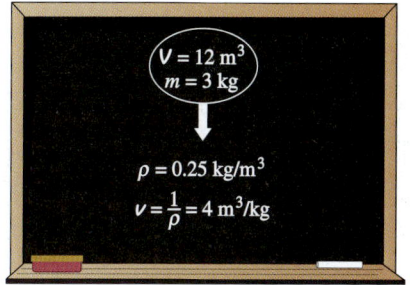

FIGURE 2–4
Density is mass per unit volume; specific volume is volume per unit mass.

TABLE 2–1
The specific gravity of some substances at 20°C and 1 atm unless stated otherwise

Substance	SG
Water	1.0
Blood (at 37°C)	1.06
Seawater	1.025
Gasoline	0.68
Ethyl alcohol	0.790
Mercury	13.6
Balsa wood	0.17
Dense oak wood	0.93
Gold	19.3
Bones	1.7–2.0
Ice (at 0°C)	0.916
Air	0.001204

Recall from Chap. 1 that the densities of liquids are essentially constant, and thus they can often be approximated as being incompressible substances during most processes without sacrificing much in accuracy.

Density of Ideal Gases

Property tables provide very accurate and precise information about the properties, but sometimes it is convenient to have some simple relations among the properties that are sufficiently general and reasonably accurate. Any equation that relates the pressure, temperature, and density (or specific volume) of a substance is called an **equation of state.** The simplest and best-known equation of state for substances in the gas phase is the **ideal-gas equation of state,** expressed as

$$P v = RT \quad \text{or} \quad P = \rho RT \tag{2-4}$$

where P is the absolute pressure, v is the specific volume, T is the thermodynamic (absolute) temperature, ρ is the density, and R is the gas constant. The gas constant R is different for each gas and is determined from $R = R_u/M$, where R_u is the **universal gas constant** whose value is $R_u = 8.314$ kJ/kmol·K, and M is the *molar mass* (also called *molecular weight*) of the gas. The values of R and M for several substances are given in Table A–1.

The thermodynamic temperature scale in the SI is the **Kelvin scale,** and the temperature unit on this scale is the **kelvin,** designated by K. In the English system, it is the **Rankine scale,** and the temperature unit on this scale is the **rankine,** R. Various temperature scales are related to each other by

$$T(\text{K}) = T(°\text{C}) + 273.15 = T(\text{R})/1.8 \tag{2-5}$$

$$T(\text{R}) = T(°\text{F}) + 459.67 = 1.8\, T(\text{K}) \tag{2-6}$$

It is common practice to round the constants 273.15 and 459.67 to 273 and 460, respectively, but we do not encourage this practice.

Equation 2–4, the ideal-gas equation of state, is also called simply the **ideal-gas relation,** and a gas that obeys this relation is called an **ideal gas.** For an ideal gas of volume V, mass m, and number of moles $N = m/M$, the ideal-gas equation of state can also be written as $PV = mRT$ or $PV = NR_uT$. For a fixed mass m, writing the ideal-gas relation twice and simplifying, the properties of an ideal gas at two different states are related to each other by $P_1V_1/T_1 = P_2V_2/T_2$.

An ideal gas is a hypothetical substance that obeys the relation $Pv = RT$. It has been experimentally observed that the ideal-gas relation closely approximates the P-v-T behavior of real gases at low densities. At low pressures and high temperatures, the density of a gas decreases and the gas behaves like an ideal gas (Fig. 2–5). In the range of practical interest, many familiar gases such as air, nitrogen, oxygen, hydrogen, helium, argon, neon, and krypton and even heavier gases such as carbon dioxide can be treated as ideal gases with negligible error (often less than 1 percent). Dense gases such as water vapor in steam power plants and refrigerant vapor in refrigerators, air conditioners, and heat pumps, however, should not be treated as ideal gases since they usually exist at a state near saturation.

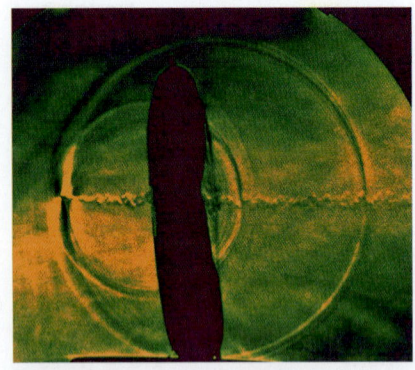

FIGURE 2–5

Air behaves as an ideal gas, even at very high speeds. In this schlieren image, a bullet traveling at about the speed of sound bursts through both sides of a balloon, forming two expanding shock waves. The turbulent wake of the bullet is also visible.

Photograph by Gary S. Settles, Penn State Gas Dynamics Lab. Used by permission.

EXAMPLE 2–1   Density, Specific Gravity, and Mass of Air in a Room

Determine the density, specific gravity, and mass of the air in a room whose dimensions are 4 m × 5 m × 6 m at 100 kPa and 25°C (Fig. 2–6).

SOLUTION The density, specific gravity, and mass of the air in a room are to be determined.
Assumptions At specified conditions, air can be treated as an ideal gas.
Properties The gas constant of air is $R = 0.287$ kPa·m³/kg·K.
Analysis The density of the air is determined from the ideal-gas relation $P = \rho RT$ to be

$$\rho = \frac{P}{RT} = \frac{100 \text{ kPa}}{(0.287 \text{ kPa·m}^3/\text{kg·K})(25 + 273.15) \text{ K}} = \mathbf{1.17 \text{ kg/m}^3}$$

Then the specific gravity of the air becomes

$$SG = \frac{\rho}{\rho_{H_2O}} = \frac{1.17 \text{ kg/m}^3}{1000 \text{ kg/m}^3} = \mathbf{0.00117}$$

Finally, the volume and the mass of the air in the room are

$$V = (4 \text{ m})(5 \text{ m})(6 \text{ m}) = 120 \text{ m}^3$$

$$m = \rho V = (1.17 \text{ kg/m}^3)(120 \text{ m}^3) = \mathbf{140 \text{ kg}}$$

Discussion Note that we converted the temperature to (absolute) unit K from (relative) unit °C before using it in the ideal-gas relation.

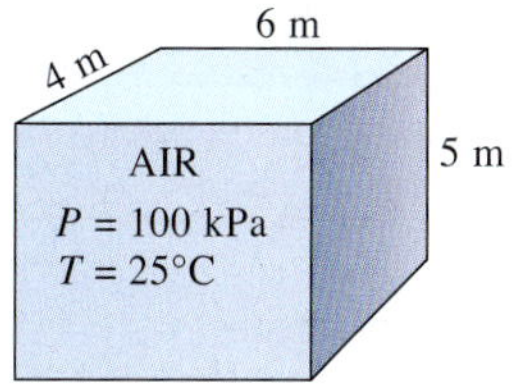

FIGURE 2–6
Schematic for Example 2–1.

2–3 ▪ VAPOR PRESSURE AND CAVITATION

It is well-established that temperature and pressure are dependent properties for pure substances during phase-change processes, and there is one-to-one correspondence between temperature and pressure. At a given pressure, the temperature at which a pure substance changes phase is called the **saturation temperature** T_{sat}. Likewise, at a given temperature, the pressure at which a pure substance changes phase is called the **saturation pressure** P_{sat}. At an absolute pressure of 1 standard atmosphere (1 atm or 101.325 kPa), for example, the saturation temperature of water is 100°C. Conversely, at a temperature of 100°C, the saturation pressure of water is 1 atm.

The **vapor pressure** P_v of a pure substance is defined as *the pressure exerted by its vapor in phase equilibrium with its liquid at a given temperature* (Fig. 2–7). P_v is a property of the pure substance, and turns out to be identical to the saturation pressure P_{sat} of the liquid ($P_v = P_{sat}$). We must be careful not to confuse vapor pressure with *partial pressure*. **Partial pressure** is defined as *the pressure of a gas or vapor in a mixture with other gases*. For example, atmospheric air is a mixture of dry air and water vapor, and atmospheric pressure is the sum of the partial pressure of dry air and the partial pressure of water vapor. The partial pressure of water vapor constitutes a small fraction (usually under 3 percent) of the atmospheric pressure since air is mostly nitrogen and oxygen. The partial pressure of a vapor must be less than or equal to the vapor pressure if there is no liquid present. However, when both vapor and liquid are present and the system is in phase equilibrium, the partial pressure of the vapor must equal the vapor pressure, and the system is said to be *saturated*. The rate of evaporation from open water

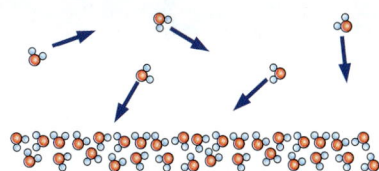

FIGURE 2–7
The vapor pressure (saturation pressure) of a pure substance (e.g., water) is the pressure exerted by its vapor molecules when the system is in phase equilibrium with its liquid molecules at a given temperature.

TABLE 2–2

Saturation (or vapor) pressure of water at various temperatures

Temperature T, °C	Saturation Pressure P_{sat}, kPa
−10	0.260
−5	0.403
0	0.611
5	0.872
10	1.23
15	1.71
20	2.34
25	3.17
30	4.25
40	7.38
50	12.35
100	101.3 (1 atm)
150	475.8
200	1554
250	3973
300	8581

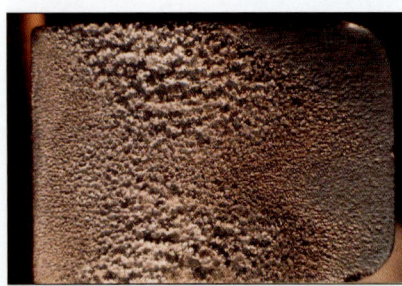

FIGURE 2–8

Cavitation damage on a 16-mm by 23-mm aluminum sample tested at 60 m/s for 2.5 hours. The sample was located at the cavity collapse region downstream of a cavity generator specifically designed to produce high damage potential.

Photo by David Stinebring, ARL/Pennsylvania State University. Used by permission.

bodies such as lakes is controlled by the difference between the vapor pressure and the partial pressure. For example, the vapor pressure of water at 20°C is 2.34 kPa. Therefore, a bucket of water at 20°C left in a room with dry air at 1 atm will continue evaporating until one of two things happens: the water evaporates away (there is not enough water to establish phase equilibrium in the room), or the evaporation stops when the partial pressure of the water vapor in the room rises to 2.34 kPa at which point phase equilibrium is established.

For phase-change processes between the liquid and vapor phases of a pure substance, the saturation pressure and the vapor pressure are equivalent since the vapor is pure. Note that the pressure value would be the same whether it is measured in the vapor or liquid phase (provided that it is measured at a location close to the liquid–vapor interface to avoid any hydrostatic effects). Vapor pressure increases with temperature. Thus, a substance at higher pressure boils at higher temperature. For example, water boils at 134°C in a pressure cooker operating at 3 atm absolute pressure, but it boils at 93°C in an ordinary pan at a 2000-m elevation, where the atmospheric pressure is 0.8 atm. The saturation (or vapor) pressures are given in Appendices 1 and 2 for various substances. An abridged table for water is given in Table 2–2 for easy reference.

The reason for our interest in vapor pressure is the possibility of the liquid pressure in liquid-flow systems dropping below the vapor pressure at some locations, and the resulting unplanned vaporization. For example, water at 10°C may vaporize and form bubbles at locations (such as the tip regions of impellers or suction sides of pumps) where the pressure drops below 1.23 kPa. The vapor bubbles (called **cavitation bubbles** since they form "cavities" in the liquid) collapse as they are swept away from the low-pressure regions, generating highly destructive, extremely high-pressure waves. This phenomenon, which is a common cause for drop in performance and even the erosion of impeller blades, is called **cavitation,** and it is an important consideration in the design of hydraulic turbines and pumps.

Cavitation must be avoided (or at least minimized) in most flow systems since it reduces performance, generates annoying vibrations and noise, and causes damage to equipment. We note that some flow systems use cavitation to their *advantage*, e.g., high-speed "supercavitating" torpedoes. The pressure spikes resulting from the large number of bubbles collapsing near a solid surface over a long period of time may cause erosion, surface pitting, fatigue failure, and the eventual destruction of the components or machinery (Fig. 2–8). The presence of cavitation in a flow system can be sensed by its characteristic tumbling sound.

EXAMPLE 2–2   Minimum Pressure to Avoid Cavitation

In a water distribution system, the temperature of water is observed to be as high as 30°C. Determine the minimum pressure allowed in the system to avoid cavitation.

SOLUTION The minimum pressure in a water distribution system to avoid cavitation is to be determined.
Properties The vapor pressure of water at 30°C is 4.25 kPa (Table 2–2).

Analysis To avoid cavitation, the pressure anywhere in the flow should not be allowed to drop below the vapor (or saturation) pressure at the given temperature. That is,

$$P_{min} = P_{sat@30°C} = \mathbf{4.25\ kPa}$$

Therefore, the pressure should be maintained above 4.25 kPa everywhere in the flow.

Discussion Note that the vapor pressure increases with increasing temperature, and thus the risk of cavitation is greater at higher fluid temperatures.

2–4 · ENERGY AND SPECIFIC HEATS

Energy can exist in numerous forms such as thermal, mechanical, kinetic, potential, electrical, magnetic, chemical, and nuclear (Fig. 2–9) and their sum constitutes the **total energy** E (or e on a unit mass basis) of a system. The forms of energy related to the molecular structure of a system and the degree of the molecular activity are referred to as the *microscopic energy*. The sum of all microscopic forms of energy is called the **internal energy** of a system, and is denoted by U (or u on a unit mass basis).

The *macroscopic* energy of a system is related to motion and the influence of some external effects such as gravity, magnetism, electricity, and surface tension. The energy that a system possesses as a result of its motion is called **kinetic energy.** When all parts of a system move with the same velocity, the kinetic energy per unit mass is expressed as $ke = V^2/2$ where V denotes the velocity of the system relative to some fixed reference frame. The energy that a system possesses as a result of its elevation in a gravitational field is called **potential energy** and is expressed on a per-unit mass basis as $pe = gz$ where g is the gravitational acceleration and z is the elevation of the center of gravity of the system relative to some arbitrarily selected reference plane.

In daily life, we frequently refer to the sensible and latent forms of internal energy as **heat,** and we talk about the heat content of bodies. In engineering, however, those forms of energy are usually referred to as **thermal energy** to prevent any confusion with *heat transfer*.

The international unit of energy is the *joule* (J) or *kilojoule* (1 kJ = 1000 J). A joule is 1 N times 1 m. In the English system, the unit of energy is the *British thermal unit* (Btu), which is defined as the energy needed to raise the temperature of 1 lbm of water at 68°F by 1°F. The magnitudes of kJ and Btu are almost identical (1 Btu = 1.0551 kJ). Another well-known unit of energy is the *calorie* (1 cal = 4.1868 J), which is defined as the energy needed to raise the temperature of 1 g of water at 14.5°C by 1°C.

In the analysis of systems that involve fluid flow, we frequently encounter the combination of properties u and Pv. For convenience, this combination is called **enthalpy** h. That is,

Enthalpy:
$$h = u + Pv = u + \frac{P}{\rho} \qquad (2\text{–}7)$$

where P/ρ is the *flow energy*, also called the *flow work*, which is the energy per unit mass needed to move the fluid and maintain flow. In the energy analysis of flowing fluids, it is convenient to treat the flow energy as part of the energy of the fluid and to represent the microscopic energy of a fluid

(a)

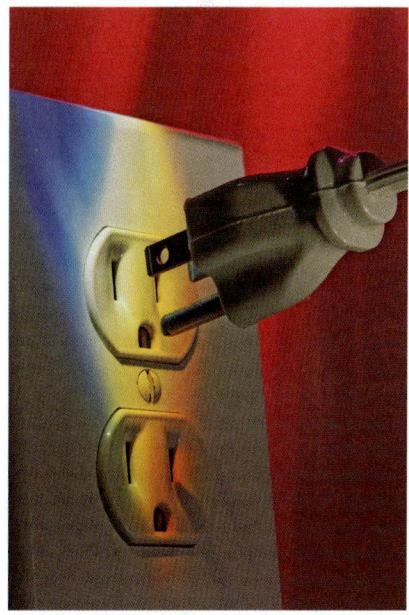

(b)

FIGURE 2–9
At least six different forms of energy are encountered in bringing power from a nuclear plant to your home, nuclear, thermal, mechanical, kinetic, magnetic, and electrical.

(a) © Creatas/PunchStock RF
(b) Comstock Images/Jupiterimages RF

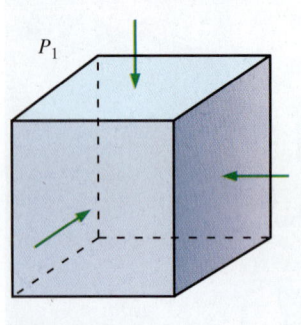

FIGURE 2–10
The *internal energy u* represents the microscopic energy of a nonflowing fluid per unit mass, whereas *enthalpy h* represents the microscopic energy of a flowing fluid per unit mass.

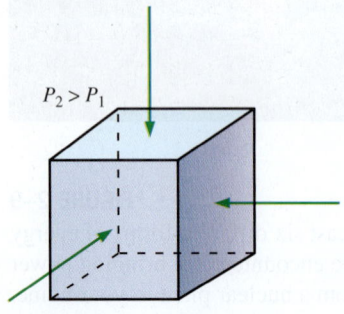

FIGURE 2–11
Fluids, like solids, compress when the applied pressure is increased from P_1 to P_2.

stream by enthalpy h (Fig. 2–10). Note that enthalpy is a quantity per unit mass, and thus it is a *specific* property.

In the absence of such effects as magnetic, electric, and surface tension, a system is called a simple compressible system. The total energy of a simple compressible system consists of three parts: internal, kinetic, and potential energies. On a unit-mass basis, it is expressed as $e = u + \text{ke} + \text{pe}$. The fluid entering or leaving a control volume possesses an additional form of energy—the *flow energy* P/ρ. Then the total energy of a **flowing fluid** on a unit-mass basis becomes

$$e_{\text{flowing}} = P/\rho + e = h + \text{ke} + \text{pe} = h + \frac{V^2}{2} + gz \quad \text{(kJ/kg)} \quad (2\text{–}8)$$

where $h = P/\rho + u$ is the enthalpy, V is the magnitude of velocity, and z is the elevation of the system relative to some external reference point.

By using the enthalpy instead of the internal energy to represent the energy of a flowing fluid, we do not need to be concerned about the flow work. The energy associated with pushing the fluid is automatically taken care of by enthalpy. In fact, this is the main reason for defining the property enthalpy.

The differential and finite changes in the internal energy and enthalpy of an *ideal gas* can be expressed in terms of the specific heats as

$$du = c_v\, dT \quad \text{and} \quad dh = c_p\, dT \quad (2\text{–}9)$$

where c_v and c_p are the constant-volume and constant-pressure specific heats of the ideal gas. Using specific heat values at the average temperature, the finite changes in internal energy and enthalpy can be expressed approximately as

$$\Delta u \cong c_{v,\text{avg}} \Delta T \quad \text{and} \quad \Delta h \cong c_{p,\text{avg}} \Delta T \quad (2\text{–}10)$$

For *incompressible substances*, the constant-volume and constant-pressure specific heats are identical. Therefore, $c_p \cong c_v \cong c$ for liquids, and the change in the internal energy of liquids can be expressed as $\Delta u \cong c_{\text{avg}} \Delta T$.

Noting that $\rho = $ constant for incompressible substances, the differentiation of enthalpy $h = u + P/\rho$ gives $dh = du + dP/\rho$. Integrating, the enthalpy change becomes

$$\Delta h = \Delta u + \Delta P/\rho \cong c_{\text{avg}} \Delta T + \Delta P/\rho \quad (2\text{–}11)$$

Therefore, $\Delta h = \Delta u \cong c_{\text{avg}} \Delta T$ for constant-pressure processes, and $\Delta h = \Delta P/\rho$ for constant-temperature processes in liquids.

2–5 ▪ COMPRESSIBILITY AND SPEED OF SOUND

Coefficient of Compressibility

We know from experience that the volume (or density) of a fluid changes with a change in its temperature or pressure. Fluids usually expand as they are heated or depressurized and contract as they are cooled or pressurized. But the amount of volume change is different for different fluids, and we need to define properties that relate volume changes to the changes in pressure and temperature. Two such properties are the bulk modulus of elasticity κ and the coefficient of volume expansion β.

It is a common observation that a fluid contracts when more pressure is applied on it and expands when the pressure acting on it is reduced (Fig. 2–11). That is, fluids act like elastic solids with respect to pressure. Therefore, in an analogous manner to Young's modulus of elasticity for solids, it is appropriate to define a **coefficient of compressibility** κ (also called the **bulk modulus of compressibility** or **bulk modulus of elasticity**) for fluids as

$$\kappa = -v\left(\frac{\partial P}{\partial v}\right)_T = \rho\left(\frac{\partial P}{\partial \rho}\right)_T \qquad \text{(Pa)} \qquad (2\text{-}12)$$

It can also be expressed approximately in terms of finite changes as

$$\kappa \cong -\frac{\Delta P}{\Delta v/v} \cong \frac{\Delta P}{\Delta \rho/\rho} \qquad (T = \text{constant}) \qquad (2\text{-}13)$$

Noting that $\Delta v/v$ or $\Delta\rho/\rho$ is dimensionless, κ must have the dimension of pressure (Pa). Also, the coefficient of compressibility represents the change in pressure corresponding to a fractional change in volume or density of the fluid while the temperature remains constant. Then it follows that the coefficient of compressibility of a truly incompressible substance (v = constant) is infinity.

A large value of κ indicates that a large change in pressure is needed to cause a small fractional change in volume, and thus a fluid with a large κ is essentially incompressible. This is typical for liquids, and explains why liquids are usually considered to be *incompressible*. For example, the pressure of water at normal atmospheric conditions must be raised to 210 atm to compress it 1 percent, corresponding to a coefficient of compressibility value of κ = 21,000 atm.

Small density changes in liquids can still cause interesting phenomena in piping systems such as the *water hammer*—characterized by a sound that resembles the sound produced when a pipe is "hammered." This occurs when a liquid in a piping network encounters an abrupt flow restriction (such as a closing valve) and is locally compressed. The acoustic waves that are produced strike the pipe surfaces, bends, and valves as they propagate and reflect along the pipe, causing the pipe to vibrate and produce the familiar sound. In addition to the irritating sound, water hammering can be quite destructive, leading to leaks or even structural damage. The effect can be suppressed with a *water hammer arrestor* (Fig. 2–12), which is a volumetric chamber containing either a bellows or piston to absorb the shock. For large pipes, a vertical tube called a *surge tower* often is used. A surge tower has a free air surface at the top and is virtually maintenance free.

Note that volume and pressure are inversely proportional (volume decreases as pressure is increased and thus $\partial P/\partial v$ is a negative quantity), and the negative sign in the definition (Eq. 2–12) ensures that κ is a positive quantity. Also, differentiating $\rho = 1/v$ gives $d\rho = -dv/v^2$, which can be rearranged as

$$\frac{d\rho}{\rho} = -\frac{dv}{v} \qquad (2\text{-}14)$$

That is, the fractional changes in the specific volume and the density of a fluid are equal in magnitude but opposite in sign.

For an ideal gas, $P = \rho RT$ and $(\partial P/\partial \rho)_T = RT = P/\rho$, and thus

$$\kappa_{\text{ideal gas}} = P \qquad \text{(Pa)} \qquad (2\text{-}15)$$

Therefore, the coefficient of compressibility of an ideal gas is equal to its absolute pressure, and the coefficient of compressibility of the gas increases

(a)

(b)

FIGURE 2–12
Water hammer arrestors:
(a) A large surge tower built to protect the pipeline against water hammer damage.
Photo by Arris S. Tijsseling, visitor of the University of Adelaide, Australia. Used by permission.
(b) Much smaller arrestors used for supplying water to a household washing machine.
Photo provided courtesy of Oatey Co.

with increasing pressure. Substituting $\kappa = P$ into the definition of the coefficient of compressibility and rearranging gives

Ideal gas: $$\frac{\Delta \rho}{\rho} = \frac{\Delta P}{P} \quad (T = \text{constant}) \quad (2\text{--}16)$$

Therefore, the percent increase of density of an ideal gas during isothermal compression is equal to the percent increase in pressure.

For air at 1 atm pressure, $\kappa = P = 1$ atm and a decrease of 1 percent in volume ($\Delta V/V = -0.01$) corresponds to an increase of $\Delta P = 0.01$ atm in pressure. But for air at 1000 atm, $\kappa = 1000$ atm and a decrease of 1 percent in volume corresponds to an increase of $\Delta P = 10$ atm in pressure. Therefore, a small fractional change in the volume of a gas can cause a large change in pressure at very high pressures.

The inverse of the coefficient of compressibility is called the **isothermal compressibility** α and is expressed as

$$\alpha = \frac{1}{\kappa} = -\frac{1}{v}\left(\frac{\partial v}{\partial P}\right)_T = \frac{1}{\rho}\left(\frac{\partial \rho}{\partial P}\right)_T \quad (1/\text{Pa}) \quad (2\text{--}17)$$

The isothermal compressibility of a fluid represents the fractional change in volume or density corresponding to a unit change in pressure.

Coefficient of Volume Expansion

The density of a fluid, in general, depends more strongly on temperature than it does on pressure, and the variation of density with temperature is responsible for numerous natural phenomena such as winds, currents in oceans, rise of plumes in chimneys, the operation of hot-air balloons, heat transfer by natural convection, and even the rise of hot air and thus the phrase "heat rises" (Fig. 2–13). To quantify these effects, we need a property that represents the *variation of the density of a fluid with temperature at constant pressure*.

The property that provides that information is the **coefficient of volume expansion** (or *volume expansivity*) β, defined as (Fig. 2–14)

$$\beta = \frac{1}{v}\left(\frac{\partial v}{\partial T}\right)_P = -\frac{1}{\rho}\left(\frac{\partial \rho}{\partial T}\right)_P \quad (1/\text{K}) \quad (2\text{--}18)$$

It can also be expressed approximately in terms of finite changes as

$$\beta \approx \frac{\Delta v/v}{\Delta T} = -\frac{\Delta \rho/\rho}{\Delta T} \quad (\text{at constant } P) \quad (2\text{--}19)$$

A large value of β for a fluid means a large change in density with temperature, and the product $\beta \, \Delta T$ represents the fraction of volume change of a fluid that corresponds to a temperature change of ΔT at constant pressure.

It can be shown that the volume expansion coefficient of an *ideal gas* ($P = \rho RT$) at a temperature T is equivalent to the inverse of the temperature:

$$\beta_{\text{ideal gas}} = \frac{1}{T} \quad (1/\text{K}) \quad (2\text{--}20)$$

where T is the *absolute* temperature.

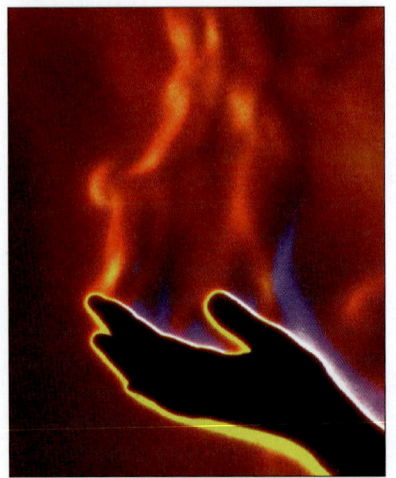

FIGURE 2–13
Natural convection over a woman's hand.
Photograph by Gary S. Settles, Penn State Gas Dynamics Lab. Used by permission.

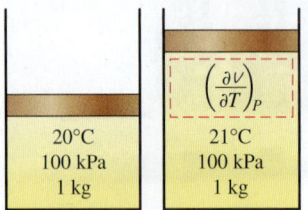

(*a*) A substance with a large β

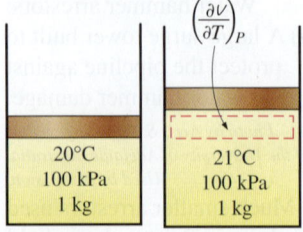

(*b*) A substance with a small β

FIGURE 2–14
The coefficient of volume expansion is a measure of the change in volume of a substance with temperature at constant pressure.

In the study of natural convection currents, the condition of the main fluid body that surrounds the finite hot or cold regions is indicated by the subscript "infinity" to serve as a reminder that this is the value at a distance where the presence of the hot or cold region is not felt. In such cases, the volume expansion coefficient can be expressed approximately as

$$\beta \approx -\frac{(\rho_\infty - \rho)/\rho}{T_\infty - T} \quad \text{or} \quad \rho_\infty - \rho = \rho\beta(T - T_\infty) \quad (2\text{-}21)$$

where ρ_∞ is the density and T_∞ is the temperature of the quiescent fluid away from the confined hot or cold fluid pocket.

We will see in Chap. 3 that natural convection currents are initiated by the *buoyancy force*, which is proportional to the *density difference*, which is in turn proportional to the *temperature difference* at constant pressure. Therefore, the larger the temperature difference between the hot or cold fluid pocket and the surrounding main fluid body, the *larger* the buoyancy force and thus the *stronger* the natural convection currents. A related phenomenon sometimes occurs when an aircraft flies near the speed of sound. The sudden drop in temperature produces condensation of water vapor on a visible vapor cloud (Fig. 2–15).

The combined effects of pressure and temperature changes on the volume change of a fluid can be determined by taking the specific volume to be a function of T and P. Differentiating $v = v(T, P)$ and using the definitions of the compression and expansion coefficients α and β give

$$dv = \left(\frac{\partial v}{\partial T}\right)_P dT + \left(\frac{\partial v}{\partial P}\right)_T dP = (\beta\, dT - \alpha\, dP)v \quad (2\text{-}22)$$

Then the fractional change in volume (or density) due to changes in pressure and temperature can be expressed approximately as

$$\frac{\Delta v}{v} = -\frac{\Delta \rho}{\rho} \cong \beta\, \Delta T - \alpha\, \Delta P \quad (2\text{-}23)$$

FIGURE 2–15
Vapor cloud around an F/A-18F Super Hornet as it flies near the speed of sound.
U.S. Navy photo by Photographer's Mate 3rd Class Jonathan Chandler.

■ EXAMPLE 2–3   Variation of Density with Temperature and Pressure

Consider water initially at 20°C and 1 atm. Determine the final density of the water (*a*) if it is heated to 50°C at a constant pressure of 1 atm, and (*b*) if it is compressed to 100-atm pressure at a constant temperature of 20°C. Take the isothermal compressibility of water to be $\alpha = 4.80 \times 10^{-5}$ atm^{-1}.

SOLUTION Water at a given temperature and pressure is considered. The densities of water after it is heated and after it is compressed are to be determined.
Assumptions **1** The coefficient of volume expansion and the isothermal compressibility of water are constant in the given temperature range. **2** An approximate analysis is performed by replacing differential changes in quantities by finite changes.
Properties The density of water at 20°C and 1 atm pressure is $\rho_1 = 998.0$ kg/m^3. The coefficient of volume expansion at the average temperature of $(20 + 50)/2 = 35°C$ is $\beta = 0.337 \times 10^{-3}$ K^{-1}. The isothermal compressibility of water is given to be $\alpha = 4.80 \times 10^{-5}$ atm^{-1}.
Analysis When differential quantities are replaced by differences and the properties α and β are assumed to be constant, the change in density in

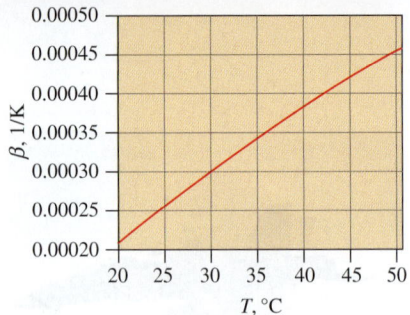

FIGURE 2–16
The variation of the coefficient of volume expansion β of water with temperature in the range of 20°C to 50°C.
Data were generated and plotted using EES.

terms of the changes in pressure and temperature is expressed approximately as (Eq. 2–23)

$$\Delta\rho = \alpha\rho\,\Delta P - \beta\rho\,\Delta T$$

(a) The change in density due to the change of temperature from 20°C to 50°C at constant pressure is

$$\Delta\rho = -\beta\rho\,\Delta T = -(0.337 \times 10^{-3}\text{ K}^{-1})(998\text{ kg/m}^3)(50 - 20)\text{ K}$$
$$= -10.0\text{ kg/m}^3$$

Noting that $\Delta\rho = \rho_2 - \rho_1$, the density of water at 50°C and 1 atm is

$$\rho_2 = \rho_1 + \Delta\rho = 998.0 + (-10.0) = \mathbf{988.0\text{ kg/m}^3}$$

which is almost identical to the listed value of 988.1 kg/m³ at 50°C in Table A–3. This is mostly due to β varying with temperature almost linearly, as shown in Fig. 2–16.

(b) The change in density due to a change of pressure from 1 atm to 100 atm at constant temperature is

$$\Delta\rho = \alpha\rho\,\Delta P = (4.80 \times 10^{-5}\text{ atm}^{-1})(998\text{ kg/m}^3)(100 - 1)\text{ atm} = 4.7\text{ kg/m}^3$$

Then the density of water at 100 atm and 20°C becomes

$$\rho_2 = \rho_1 + \Delta\rho = 998.0 + 4.7 = \mathbf{1002.7\text{ kg/m}^3}$$

Discussion Note that the density of water decreases while being heated and increases while being compressed, as expected. This problem can be solved more accurately using differential analysis when functional forms of properties are available.

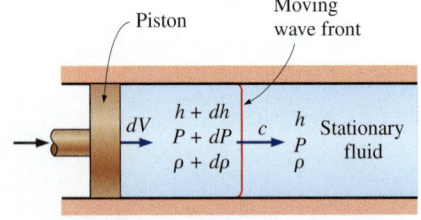

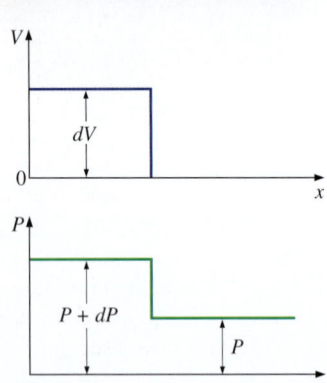

FIGURE 2–17
Propagation of a small pressure wave along a duct.

Speed of Sound and Mach Number

An important parameter in the study of compressible flow is the **speed of sound** (or the **sonic speed**), defined as the speed at which an infinitesimally small pressure wave travels through a medium. The pressure wave may be caused by a small disturbance, which creates a slight rise in local pressure.

To obtain a relation for the speed of sound in a medium, consider a duct that is filled with a fluid at rest, as shown in Fig. 2–17. A piston fitted in the duct is now moved to the right with a constant incremental velocity dV, creating a sonic wave. The wave front moves to the right through the fluid at the speed of sound c and separates the moving fluid adjacent to the piston from the fluid still at rest. The fluid to the left of the wave front experiences an incremental change in its thermodynamic properties, while the fluid on the right of the wave front maintains its original thermodynamic properties, as shown in Fig. 2–17.

To simplify the analysis, consider a control volume that encloses the wave front and moves with it, as shown in Fig. 2–18. To an observer traveling with the wave front, the fluid to the right appears to be moving toward the wave front with a speed of c and the fluid to the left to be moving away from the wave front with a speed of $c - dV$. Of course, the observer sees the control volume that encloses the wave front (and herself or himself) as stationary, and the observer is witnessing a steady-flow process. The mass balance for this single-stream, steady-flow process is expressed as

$$\dot{m}_{\text{right}} = \dot{m}_{\text{left}}$$

or

$$\rho A c = (\rho + d\rho) A (c - dV)$$

By canceling the cross-sectional (or flow) area A and neglecting the higher-order terms, this equation reduces to

$$c\, d\rho - \rho\, dV = 0$$

No heat or work crosses the boundaries of the control volume during this steady-flow process, and the potential energy change can be neglected. Then the steady-flow energy balance $e_{in} = e_{out}$ becomes

$$h + \frac{c^2}{2} = h + dh + \frac{(c - dV)^2}{2}$$

which yields

$$dh - c\, dV = 0$$

where we have neglected the second-order term dV^2. The amplitude of the ordinary sonic wave is very small and does not cause any appreciable change in the pressure and temperature of the fluid. Therefore, the propagation of a sonic wave is not only adiabatic but also very nearly isentropic. Then the thermodynamic relation $T\, ds = dh - dP/\rho$ (see Çengel and Boles, 2011) reduces to

$$T\, ds\overset{0}{=} dh - \frac{dP}{\rho}$$

or

$$dh = \frac{dP}{\rho}$$

Combining the above equations yields the desired expression for the speed of sound as

$$c^2 = \frac{dP}{d\rho} \quad \text{at } s = \text{constant}$$

or

$$c^2 = \left(\frac{\partial P}{\partial \rho}\right)_s \qquad (2\text{–}24)$$

It is left as an exercise for the reader to show, by using thermodynamic property relations, that Eq. 2–24 can also be written as

$$c^2 = k\left(\frac{\partial P}{\partial \rho}\right)_T \qquad (2\text{–}25)$$

where $k = c_p/c_v$ is the specific heat ratio of the fluid. Note that the speed of sound in a fluid is a function of the thermodynamic properties of that fluid (Fig. 2–19).

When the fluid is an ideal gas ($P = \rho RT$), the differentiation in Eq. 2–25 can be performed to yield

$$c^2 = k\left(\frac{\partial P}{\partial \rho}\right)_T = k\left[\frac{\partial(\rho RT)}{\partial \rho}\right]_T = kRT$$

or

$$c = \sqrt{kRT} \qquad (2\text{–}26)$$

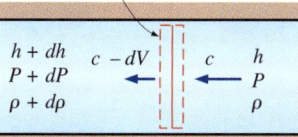

FIGURE 2–18
Control volume moving with the small pressure wave along a duct.

FIGURE 2–19
The speed of sound in air increases with temperature. At typical outside temperatures, c is about 340 m/s. In round numbers, therefore, the sound of thunder from a lightning strike travels about 1 km in 3 seconds. If you see the lightning and then hear the thunder less than 3 seconds later, you know that the lightning is close, and it is time to go indoors!

© Bear Dancer Studios/Mark Dierker

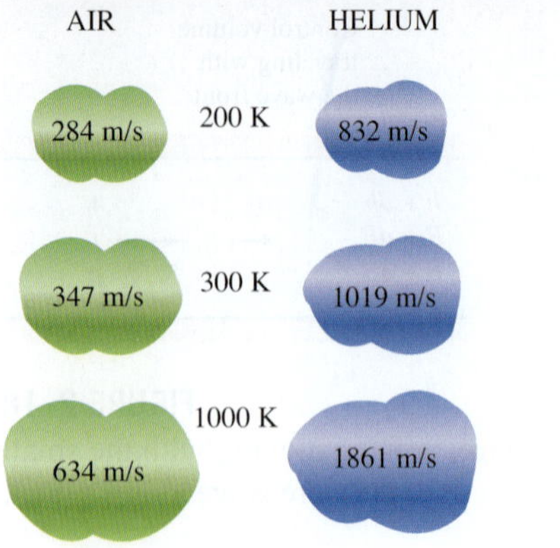

FIGURE 2–20
The speed of sound changes with temperature and varies with the fluid.

Noting that the gas constant R has a fixed value for a specified ideal gas and the specific heat ratio k of an ideal gas is, at most, a function of temperature, we see that the speed of sound in a specified ideal gas is a function of temperature alone (Fig. 2–20).

A second important parameter in the analysis of compressible fluid flow is the **Mach number Ma**, named after the Austrian physicist Ernst Mach (1838–1916). It is the ratio of the actual speed of the fluid (or an object in still fluid) to the speed of sound in the same fluid at the same state:

$$\text{Ma} = \frac{V}{c} \qquad (2\text{–}27)$$

Note that the Mach number depends on the speed of sound, which depends on the state of the fluid. Therefore, the Mach number of an aircraft cruising at constant velocity in still air may be different at different locations (Fig. 2–21).

Fluid flow regimes are often described in terms of the flow Mach number. The flow is called **sonic** when Ma = 1, **subsonic** when Ma < 1, **supersonic** when Ma > 1, **hypersonic** when Ma ≫ 1, and **transonic** when Ma ≅ 1.

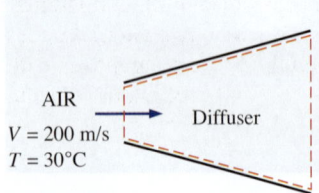

FIGURE 2–21
The Mach number can be different at different temperatures even if the flight speed is the same.
© Alamy RF

EXAMPLE 2–4   Mach Number of Air Entering a Diffuser

Air enters a diffuser shown in Fig. 2–22 with a speed of 200 m/s. Determine (a) the speed of sound and (b) the Mach number at the diffuser inlet when the air temperature is 30°C.

SOLUTION Air enters a diffuser at high speed. The speed of sound and the Mach number are to be determined at the diffuser inlet.
Assumption Air at the specified conditions behaves as an ideal gas.
Properties The gas constant of air is $R = 0.287$ kJ/kg·K, and its specific heat ratio at 30°C is 1.4.
Analysis We note that the speed of sound in a gas varies with temperature, which is given to be 30°C.

(a) The speed of sound in air at 30°C is determined from Eq. 2–26 to be

$$c = \sqrt{kRT} = \sqrt{(1.4)(0.287 \text{ kJ/kg·K})(303 \text{ K})\left(\frac{1000 \text{ m}^2/\text{s}^2}{1 \text{ kJ/kg}}\right)} = \mathbf{349 \text{ m/s}}$$

(b) Then the Mach number becomes

$$\text{Ma} = \frac{V}{c} = \frac{200 \text{ m/s}}{349 \text{ m/s}} = \mathbf{0.573}$$

Discussion The flow at the diffuser inlet is subsonic since Ma < 1.

FIGURE 2–22
Schematic for Example 12–4.

2–6 ■ VISCOSITY

When two solid bodies in contact move relative to each other, a friction force develops at the contact surface in the direction opposite to motion. To move a table on the floor, for example, we have to apply a force to the table in the horizontal direction large enough to overcome the friction force.

The magnitude of the force needed to move the table depends on the *friction coefficient* between the table legs and the floor.

The situation is similar when a fluid moves relative to a solid or when two fluids move relative to each other. We move with relative ease in air, but not so in water. Moving in oil would be even more difficult, as can be observed by the slower downward motion of a glass ball dropped in a tube filled with oil. It appears that there is a property that represents the internal resistance of a fluid to motion or the "fluidity," and that property is the **viscosity.** The force a flowing fluid exerts on a body in the flow direction is called the **drag force,** and the magnitude of this force depends, in part, on viscosity (Fig. 2–23).

To obtain a relation for viscosity, consider a fluid layer between two very large parallel plates (or equivalently, two parallel plates immersed in a large body of a fluid) separated by a distance ℓ (Fig. 2–24). Now a constant parallel force F is applied to the upper plate while the lower plate is held fixed. After the initial transients, it is observed that the upper plate moves continuously under the influence of this force at a constant speed V. The fluid in contact with the upper plate sticks to the plate surface and moves with it at the same speed, and the shear stress τ acting on this fluid layer is

$$\tau = \frac{F}{A} \qquad (2\text{-}28)$$

where A is the contact area between the plate and the fluid. Note that the fluid layer deforms continuously under the influence of shear stress.

The fluid in contact with the lower plate assumes the velocity of that plate, which is zero (because of the no-slip condition—see Section 1–2). In steady laminar flow, the fluid velocity between the plates varies linearly between 0 and V, and thus the *velocity profile* and the *velocity gradient* are

$$u(y) = \frac{y}{\ell} V \quad \text{and} \quad \frac{du}{dy} = \frac{V}{\ell} \qquad (2\text{-}29)$$

where y is the vertical distance from the lower plate.

During a differential time interval dt, the sides of fluid particles along a vertical line MN rotate through a differential angle $d\beta$ while the upper plate moves a differential distance $da = V\,dt$. The angular displacement or deformation (or shear strain) can be expressed as

$$d\beta \approx \tan d\beta = \frac{da}{\ell} = \frac{V\,dt}{\ell} = \frac{du}{dy} dt \qquad (2\text{-}30)$$

Rearranging, the rate of deformation under the influence of shear stress τ becomes

$$\frac{d\beta}{dt} = \frac{du}{dy} \qquad (2\text{-}31)$$

Thus we conclude that the rate of deformation of a fluid element is equivalent to the velocity gradient du/dy. Further, it can be verified experimentally that for most fluids the rate of deformation (and thus the velocity gradient) is directly proportional to the shear stress τ,

$$\tau \propto \frac{d\beta}{dt} \quad \text{or} \quad \tau \propto \frac{du}{dy} \qquad (2\text{-}32)$$

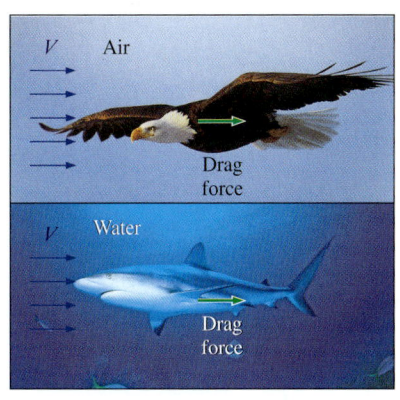

FIGURE 2–23
A fluid moving relative to a body exerts a drag force on the body, partly because of friction caused by viscosity.
© Digital Vision/Getty RF

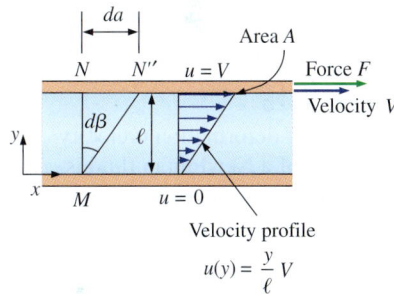

FIGURE 2–24
The behavior of a fluid in laminar flow between two parallel plates when the upper plate moves with a constant velocity.

PROPERTIES OF FLUIDS

Fluids for which the rate of deformation is linearly proportional to the shear stress are called **Newtonian fluids** after Sir Isaac Newton, who expressed it first in 1687. Most common fluids such as water, air, gasoline, and oils are Newtonian fluids. Blood and liquid plastics are examples of non-Newtonian fluids.

In one-dimensional shear flow of Newtonian fluids, shear stress can be expressed by the linear relationship

Shear stress: $$\tau = \mu \frac{du}{dy} \quad (N/m^2) \tag{2-33}$$

where the constant of proportionality μ is called the **coefficient of viscosity** or the **dynamic** (or **absolute**) **viscosity** of the fluid, whose unit is kg/m·s, or equivalently, N·s/m² (or Pa·s where Pa is the pressure unit pascal). A common viscosity unit is **poise,** which is equivalent to 0.1 Pa·s (or *centipoise*, which is one-hundredth of a poise). The viscosity of water at 20°C is 1.002 centipoise, and thus the unit centipoise serves as a useful reference. A plot of shear stress versus the rate of deformation (velocity gradient) for a Newtonian fluid is a straight line whose slope is the viscosity of the fluid, as shown in Fig. 2–25. Note that viscosity is independent of the rate of deformation for Newtonian fluids. Since the rate of deformation is proportional to the strain rate, Fig. 2–25 reveals that viscosity is actually a coefficient in a stress–strain relationship.

The **shear force** acting on a Newtonian fluid layer (or, by Newton's third law, the force acting on the plate) is

Shear force: $$F = \tau A = \mu A \frac{du}{dy} \quad (N) \tag{2-34}$$

where again A is the contact area between the plate and the fluid. Then the force F required to move the upper plate in Fig. 2–24 at a constant speed of V while the lower plate remains stationary is

$$F = \mu A \frac{V}{\ell} \quad (N) \tag{2-35}$$

This relation can alternately be used to calculate μ when the force F is measured. Therefore, the experimental setup just described can be used to measure the viscosity of fluids. Note that under identical conditions, the force F would be very different for different fluids.

For non-Newtonian fluids, the relationship between shear stress and rate of deformation is not linear, as shown in Fig. 2–26. The slope of the curve on the τ versus du/dy chart is referred to as the *apparent viscosity* of the fluid. Fluids for which the apparent viscosity increases with the rate of deformation (such as solutions with suspended starch or sand) are referred to as *dilatant* or *shear thickening fluids*, and those that exhibit the opposite behavior (the fluid becoming less viscous as it is sheared harder, such as some paints, polymer solutions, and fluids with suspended particles) are referred to as *pseudoplastic* or *shear thinning fluids*. Some materials such as toothpaste can resist a finite shear stress and thus behave as a solid, but deform continuously when the shear stress exceeds the yield stress and behave as a fluid. Such materials are referred to as Bingham plastics after Eugene C. Bingham (1878–1945), who did pioneering work on fluid viscosity for the U.S. National Bureau of Standards in the early twentieth century.

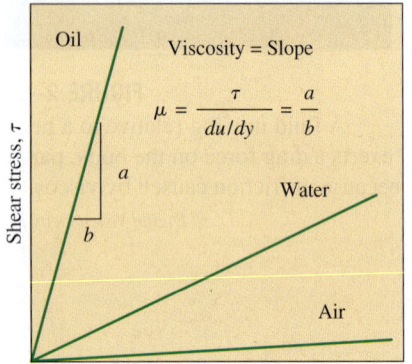

FIGURE 2–25
The rate of deformation (velocity gradient) of a Newtonian fluid is proportional to shear stress, and the constant of proportionality is the viscosity.

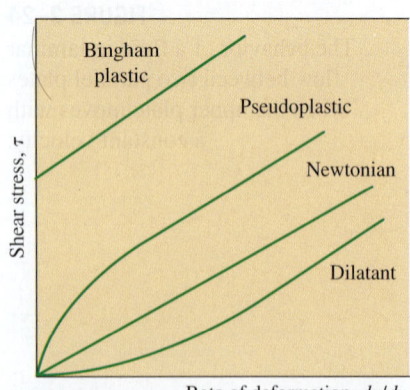

FIGURE 2–26
Variation of shear stress with the rate of deformation for Newtonian and non-Newtonian fluids (the slope of a curve at a point is the apparent viscosity of the fluid at that point).

In fluid mechanics and heat transfer, the ratio of dynamic viscosity to density appears frequently. For convenience, this ratio is given the name **kinematic viscosity** ν and is expressed as $\nu = \mu/\rho$. Two common units of kinematic viscosity are m^2/s and **stoke** (1 stoke = 1 cm^2/s = 0.0001 m^2/s).

In general, the viscosity of a fluid depends on both temperature and pressure, although the dependence on pressure is rather weak. For *liquids*, both the dynamic and kinematic viscosities are practically independent of pressure, and any small variation with pressure is usually disregarded, except at extremely high pressures. For *gases*, this is also the case for dynamic viscosity (at low to moderate pressures), but not for kinematic viscosity since the density of a gas is proportional to its pressure (Fig. 2–27).

The viscosity of a fluid is a measure of its "resistance to deformation." Viscosity is due to the internal frictional force that develops between different layers of fluids as they are forced to move relative to each other.

The viscosity of a fluid is directly related to the pumping power needed to transport a fluid in a pipe or to move a body (such as a car in air or a submarine in the sea) through a fluid. Viscosity is caused by the cohesive forces between the molecules in liquids and by the molecular collisions in gases, and it varies greatly with temperature. The viscosity of liquids decreases with temperature, whereas the viscosity of gases increases with temperature (Fig. 2–28). This is because in a liquid the molecules possess more energy at higher temperatures, and they can oppose the large cohesive intermolecular forces more strongly. As a result, the energized liquid molecules can move more freely.

In a gas, on the other hand, the intermolecular forces are negligible, and the gas molecules at high temperatures move randomly at higher velocities. This results in more molecular collisions per unit volume per unit time and therefore in greater resistance to flow. The kinetic theory of gases predicts the viscosity of gases to be proportional to the square root of temperature. That is, $\mu_{gas} \propto \sqrt{T}$. This prediction is confirmed by practical observations, but deviations for different gases need to be accounted for by incorporating some correction factors. The viscosity of gases is expressed as a function of temperature by the Sutherland correlation (from The U.S. Standard Atmosphere) as

Gases:
$$\mu = \frac{aT^{1/2}}{1 + b/T} \quad (2\text{–}36)$$

where T is absolute temperature and a and b are experimentally determined constants. Note that measuring viscosity at two different temperatures is sufficient to determine these constants. For air at atmospheric conditions, the values of these constants are $a = 1.458 \times 10^{-6}$ kg/(m·s·$K^{1/2}$) and $b = 110.4$ K. The viscosity of gases is independent of pressure at low to moderate pressures (from a few percent of 1 atm to several atm). But viscosity increases at high pressures due to the increase in density.

For liquids, the viscosity is approximated as

Liquids:
$$\mu = a10^{b/(T-c)} \quad (2\text{–}37)$$

where again T is absolute temperature and a, b, and c are experimentally determined constants. For water, using the values $a = 2.414 \times 10^{-5}$ N·s/m^2, $b = 247.8$ K, and $c = 140$ K results in less than 2.5 percent error in viscosity in the temperature range of 0°C to 370°C (Touloukian et al., 1975).

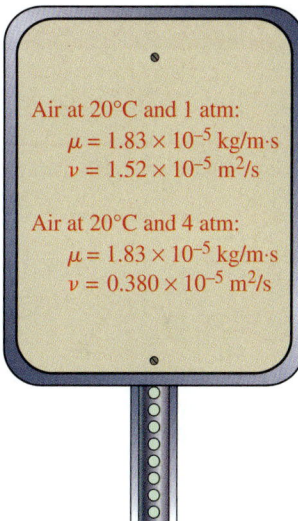

FIGURE 2–27
Dynamic viscosity, in general, does not depend on pressure, but kinematic viscosity does.

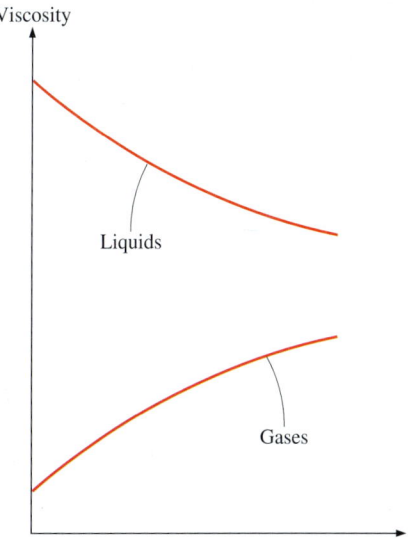

FIGURE 2–28
The viscosity of liquids decreases and the viscosity of gases increases with temperature.

TABLE 2–3

Dynamic viscosity of some fluids at 1 atm and 20°C (unless otherwise stated)

Fluid	Dynamic Viscosity μ, kg/m·s
Glycerin:	
−20°C	134.0
0°C	10.5
20°C	1.52
40°C	0.31
Engine oil:	
SAE 10W	0.10
SAE 10W30	0.17
SAE 30	0.29
SAE 50	0.86
Mercury	0.0015
Ethyl alcohol	0.0012
Water:	
0°C	0.0018
20°C	0.0010
100°C (liquid)	0.00028
100°C (vapor)	0.000012
Blood, 37°C	0.00040
Gasoline	0.00029
Ammonia	0.00015
Air	0.000018
Hydrogen, 0°C	0.0000088

The viscosities of some fluids at room temperature are listed in Table 2–3. They are plotted against temperature in Fig. 2–29. Note that the viscosities of different fluids differ by several orders of magnitude. Also note that it is more difficult to move an object in a higher-viscosity fluid such as engine oil than it is in a lower-viscosity fluid such as water. Liquids, in general, are much more viscous than gases.

Consider a fluid layer of thickness ℓ within a small gap between two concentric cylinders, such as the thin layer of oil in a journal bearing. The gap between the cylinders can be modeled as two parallel flat plates separated by the fluid. Noting that torque is $T = FR$ (force times the moment arm, which is the radius R of the inner cylinder in this case), the tangential velocity is $V = \omega R$ (angular velocity times the radius), and taking the wetted surface area of the inner cylinder to be $A = 2\pi RL$ by disregarding the shear stress acting on the two ends of the inner cylinder, torque can be expressed as

$$T = FR = \mu \frac{2\pi R^3 \omega L}{\ell} = \mu \frac{4\pi^2 R^3 \dot{n} L}{\ell} \qquad (2\text{--}38)$$

where L is the length of the cylinder and $\dot{n}$ is the number of revolutions per unit time, which is usually expressed in rpm (revolutions per minute). Note that the angular distance traveled during one rotation is 2π rad, and thus the

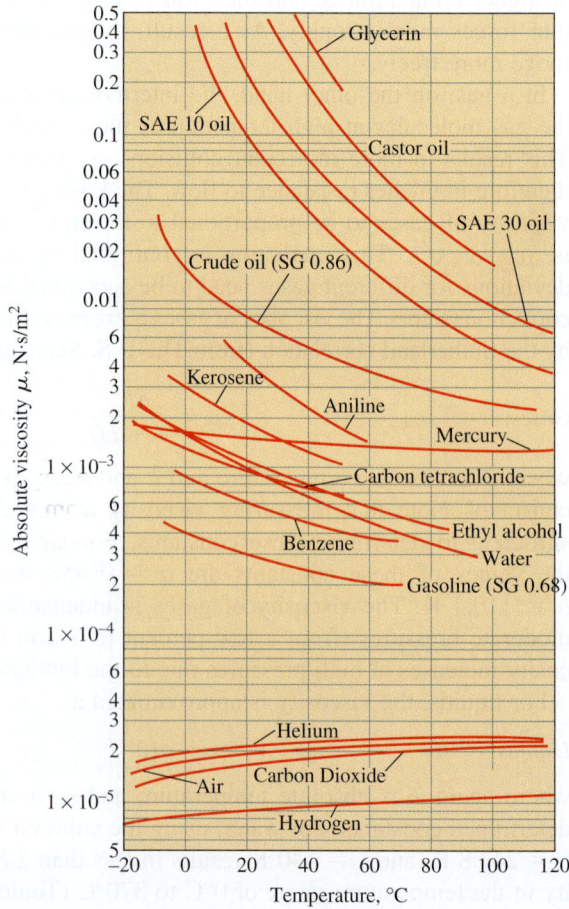

FIGURE 2–29

The variation of dynamic (absolute) viscosity of common fluids with temperature at 1 atm (1 N·s/m² = 1 kg/m·s).

Data from EES and F. M. White, Fluid Mechanics 7e. Copyright © 2011 The McGraw-Hill Companies, Inc. Used by permission.

relation between the angular velocity in rad/min and the rpm is $\omega = 2\pi\dot{n}$. Equation 2–38 can be used to calculate the viscosity of a fluid by measuring torque at a specified angular velocity. Therefore, two concentric cylinders can be used as a *viscometer*, a device that measures viscosity.

EXAMPLE 2–5   Determining the Viscosity of a Fluid

The viscosity of a fluid is to be measured by a viscometer constructed of two 40-cm-long concentric cylinders (Fig. 2–30). The outer diameter of the inner cylinder is 12 cm, and the gap between the two cylinders is 0.15 cm. The inner cylinder is rotated at 300 rpm, and the torque is measured to be 1.8 N·m. Determine the viscosity of the fluid.

SOLUTION The torque and the rpm of a double cylinder viscometer are given. The viscosity of the fluid is to be determined.
Assumptions 1 The inner cylinder is completely submerged in the fluid. 2 The viscous effects on the two ends of the inner cylinder are negligible.
Analysis The velocity profile is linear only when the curvature effects are negligible, and the profile can be approximated as being linear in this case since $\ell/R = 0.025 \ll 1$. Solving Eq. 2–38 for viscosity and substituting the given values, the viscosity of the fluid is determined to be

$$\mu = \frac{T\ell}{4\pi^2 R^3 \dot{n} L} = \frac{(1.8 \text{ N·m})(0.0015 \text{ m})}{4\pi^2 (0.06 \text{ m})^3 \left(300 \frac{1}{\text{min}}\right)\left(\frac{1 \text{ min}}{60 \text{ s}}\right)(0.4 \text{ m})} = \mathbf{0.158 \text{ N·s/m}^2}$$

Discussion Viscosity is a strong function of temperature, and a viscosity value without a corresponding temperature is of little usefulness. Therefore, the temperature of the fluid should have also been measured during this experiment, and reported with this calculation.

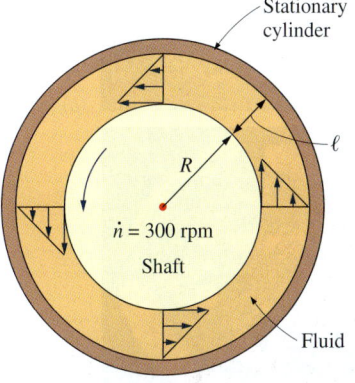

FIGURE 2–30
Schematic for Example 2–5 (not to scale).

2–7 · SURFACE TENSION AND CAPILLARY EFFECT

It is often observed that a drop of blood forms a hump on a horizontal glass; a drop of mercury forms a near-perfect sphere and can be rolled just like a steel ball over a smooth surface; water droplets from rain or dew hang from branches or leaves of trees; a liquid fuel injected into an engine forms a mist of spherical droplets; water dripping from a leaky faucet falls as nearly spherical droplets; a soap bubble released into the air forms a nearly spherical shape; and water beads up into small drops on flower petals (Fig. 2–31a).

In these and other observances, liquid droplets behave like small balloons filled with the liquid, and the surface of the liquid acts like a stretched elastic membrane under tension. The pulling force that causes this tension acts parallel to the surface and is due to the attractive forces between the molecules of the liquid. The magnitude of this force per unit length is called **surface tension** or *coefficient of surface tension* σ_s and is usually expressed in the unit N/m. This effect is also called *surface energy* (per unit area) and is expressed in the equivalent unit of N·m/m² or J/m². In this case, σ_s represents the stretching work that needs to be done to increase the surface area of the liquid by a unit amount.

56
PROPERTIES OF FLUIDS

(a)

(b)

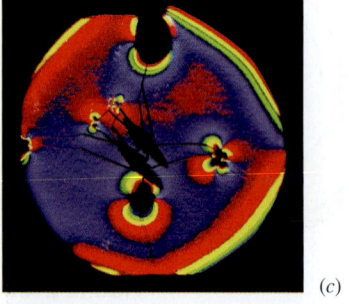
(c)

FIGURE 2–31
Some consequences of surface tension: (*a*) drops of water beading up on a leaf, (*b*) a water strider sitting on top of the surface of water, and (*c*) a color schlieren image of the water strider revealing how the water surface dips down where its feet contact the water (it looks like two insects but the second one is just a shadow).
(*a*) © Don Paulson Photography/Purestock/SuperStock RF
(*b*) NPS Photo by Rosalie LaRue.
(*c*) Photo courtesy of G. S. Settles, Gas Dynamics Lab, Penn State University, used by permission.

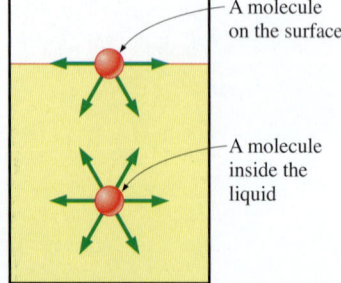

FIGURE 2–32
Attractive forces acting on a liquid molecule at the surface and deep inside the liquid.

To visualize how surface tension arises, we present a microscopic view in Fig. 2–32 by considering two liquid molecules, one at the surface and one deep within the liquid body. The attractive forces applied on the interior molecule by the surrounding molecules balance each other because of symmetry. But the attractive forces acting on the surface molecule are not symmetric, and the attractive forces applied by the gas molecules above are usually very small. Therefore, there is a net attractive force acting on the molecule at the surface of the liquid, which tends to pull the molecules on the surface toward the interior of the liquid. This force is balanced by the repulsive forces from the molecules below the surface that are trying to be compressed. The result is that the liquid minimizes its surface area. This is the reason for the tendency of liquid droplets to attain a spherical shape, which has the minimum surface area for a given volume.

You also may have observed, with amusement, that some insects can land on water or even walk on water (Fig. 2–31*b*) and that small steel needles can float on water. These phenomena are made possible by surface tension which balances the weights of these objects.

To understand the surface tension effect better, consider a liquid film (such as the film of a soap bubble) suspended on a U-shaped wire frame with a movable side (Fig. 2–33). Normally, the liquid film tends to pull the movable wire inward in order to minimize its surface area. A force F needs to be applied on the movable wire in the opposite direction to balance this pulling effect. Both sides of the thin film are surfaces exposed to air, and thus the length along which the surface tension acts in this case is $2b$. Then a force balance on the movable wire gives $F = 2b\sigma_s$, and thus the surface tension can be expressed as

$$\sigma_s = \frac{F}{2b} \quad (2\text{–}39)$$

Note that for $b = 0.5$ m, the measured force F (in N) is simply the surface tension in N/m. An apparatus of this kind with sufficient precision can be used to measure the surface tension of various liquids.

In the U-shaped wire frame apparatus, the movable wire is pulled to stretch the film and increase its surface area. When the movable wire is pulled a distance Δx, the surface area increases by $\Delta A = 2b\,\Delta x$, and the work W done during this stretching process is

$$W = \text{Force} \times \text{Distance} = F\,\Delta x = 2b\sigma_s\,\Delta x = \sigma_s\,\Delta A$$

where we have assumed that the force remains constant over the small distance. This result can also be interpreted as *the surface energy of the film is increased by an amount $\sigma_s\,\Delta A$ during this stretching process*, which is consistent with the alternative interpretation of σ_s as surface energy per unit area. This is similar to a rubber band having more potential (elastic) energy after it is stretched further. In the case of liquid film, the work is used to move liquid molecules from the interior parts to the surface against the attraction forces of other molecules. Therefore, surface tension also can be defined as *the work done per unit increase in the surface area of the liquid*.

The surface tension varies greatly from substance to substance, and with temperature for a given substance, as shown in Table 2–4. At 20°C, for example, the surface tension is 0.073 N/m for water and 0.440 N/m for mercury surrounded by atmospheric air. The surface tension of mercury is large enough that mercury droplets form nearly spherical balls that can be rolled like a solid ball on a smooth surface. The surface tension of a liquid, in general, decreases with temperature and becomes zero at the critical point (and thus there is no distinct liquid–vapor interface at temperatures above the critical point). The effect of pressure on surface tension is usually negligible.

The surface tension of a substance can be changed considerably by *impurities*. Therefore, certain chemicals, called *surfactants*, can be added to a liquid to decrease its surface tension. For example, soaps and detergents lower the surface tension of water and enable it to penetrate the small openings between fibers for more effective washing. But this also means that devices whose operation depends on surface tension (such as heat pipes) can be destroyed by the presence of impurities due to poor workmanship.

We speak of surface tension for liquids only at liquid–liquid or liquid–gas interfaces. Therefore, it is imperative that the adjacent liquid or gas be specified when specifying surface tension. Surface tension determines the size of the liquid droplets that form, and so a droplet that keeps growing by the addition of more mass breaks down when the surface tension can no longer hold it together. This is like a balloon that bursts while being inflated when the pressure inside rises above the strength of the balloon material.

A curved interface indicates a pressure difference (or "pressure jump") across the interface with pressure being higher on the concave side. Consider, for example, a droplet of liquid in air, an air (or other gas) bubble in water, or a soap bubble in air. The excess pressure ΔP above atmospheric pressure can be determined by considering a free-body diagram of half the droplet or bubble (Fig. 2–34). Noting that surface tension acts along the circumference and the pressure acts on the area, horizontal force balances for the droplet or air bubble and the soap bubble give

$$\text{Droplet or air bubble:} \quad (2\pi R)\sigma_s = (\pi R^2)\Delta P_{\text{droplet}} \rightarrow \Delta P_{\text{droplet}} = P_i - P_o = \frac{2\sigma_s}{R} \quad (2\text{–}40)$$

$$\text{Soap bubble:} \quad 2(2\pi R)\sigma_s = (\pi R^2)\Delta P_{\text{bubble}} \rightarrow \Delta P_{\text{bubble}} = P_i - P_o = \frac{4\sigma_s}{R} \quad (2\text{–}41)$$

where P_i and P_o are the pressures inside and outside the droplet or bubble, respectively. When the droplet or bubble is in the atmosphere, P_o is simply atmospheric pressure. The extra factor of 2 in the force balance for the soap bubble is due to the existence of a soap film with *two* surfaces (inner and outer surfaces) and thus two circumferences in the cross section.

The excess pressure in a droplet of liquid in a gas (or a bubble of gas in a liquid) can also be determined by considering a differential increase in the radius of the droplet due to the addition of a differential amount of mass and interpreting the surface tension as the increase in the surface energy per unit area. Then the increase in the surface energy of the droplet during this differential expansion process becomes

$$\delta W_{\text{surface}} = \sigma_s \, dA = \sigma_s \, d(4\pi R^2) = 8\pi R \sigma_s \, dR$$

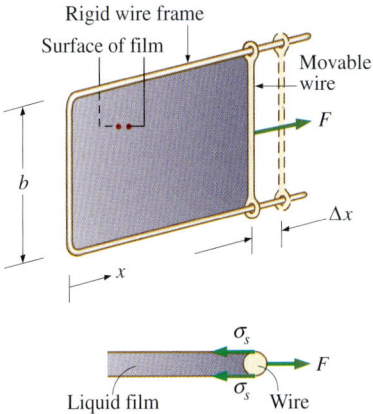

FIGURE 2–33
Stretching a liquid film with a U-shaped wire, and the forces acting on the movable wire of length b.

TABLE 2–4

Surface tension of some fluids in air at 1 atm and 20°C (unless otherwise stated)

Fluid	Surface Tension σ_s, N/m
†Water:	
0°C	0.076
20°C	0.073
100°C	0.059
300°C	0.014
Glycerin	0.063
SAE 30 oil	0.035
Mercury	0.440
Ethyl alcohol	0.023
Blood, 37°C	0.058
Gasoline	0.022
Ammonia	0.021
Soap solution	0.025
Kerosene	0.028

† See Appendices for more precise data for water.

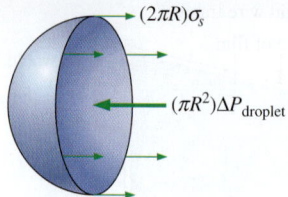

(a) Half of a droplet or air bubble

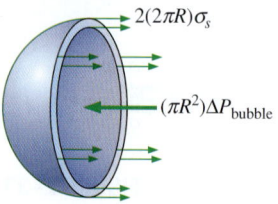

(b) Half of a soap bubble

FIGURE 2–34
The free-body diagram of half of a droplet or air bubble and half of a soap bubble.

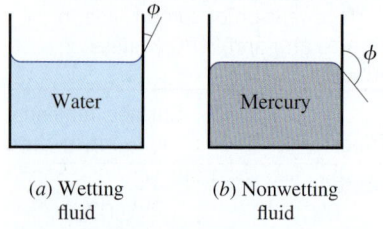

(a) Wetting fluid

(b) Nonwetting fluid

FIGURE 2–35
The contact angle for wetting and nonwetting fluids.

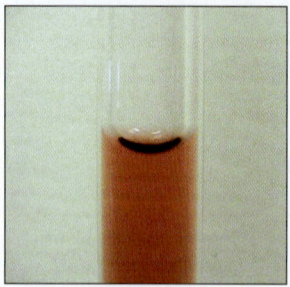

FIGURE 2–36
The meniscus of colored water in a 4-mm-inner-diameter glass tube. Note that the edge of the meniscus meets the wall of the capillary tube at a very small contact angle.

Photo by Gabrielle Tremblay. Used by permission.

The expansion work done during this differential process is determined by multiplying the force by distance to obtain

$$\delta W_{expansion} = \text{Force} \times \text{Distance} = F\, dR = (\Delta P A)\, dR = 4\pi R^2\, \Delta P\, dR$$

Equating the two expressions above gives $\Delta P_{droplet} = 2\sigma_s/R$, which is the same relation obtained before and given in Eq. 2–40. Note that the excess pressure in a droplet or bubble is inversely proportional to the radius.

Capillary Effect

Another interesting consequence of surface tension is the **capillary effect,** which is the rise or fall of a liquid in a small-diameter tube inserted into the liquid. Such narrow tubes or confined flow channels are called **capillaries.** The rise of kerosene through a cotton wick inserted into the reservoir of a kerosene lamp is due to this effect. The capillary effect is also partially responsible for the rise of water to the top of tall trees. The curved free surface of a liquid in a capillary tube is called the **meniscus.**

It is commonly observed that water in a glass container curves up slightly at the edges where it touches the glass surface; but the opposite occurs for mercury: it curves down at the edges (Fig. 2–35). This effect is usually expressed by saying that water *wets* the glass (by sticking to it) while mercury does not. The strength of the capillary effect is quantified by the **contact** (or *wetting*) **angle** ϕ, defined as *the angle that the tangent to the liquid surface makes with the solid surface at the point of contact*. The surface tension force acts along this tangent line toward the solid surface. A liquid is said to wet the surface when $\phi < 90°$ and not to wet the surface when $\phi > 90°$. In atmospheric air, the contact angle of water (and most other organic liquids) with glass is nearly zero, $\phi \approx 0°$ (Fig. 2–36). Therefore, the surface tension force acts upward on water in a glass tube along the circumference, tending to pull the water up. As a result, water rises in the tube until the weight of the liquid in the tube above the liquid level of the reservoir balances the surface tension force. The contact angle is 130° for mercury–glass and 26° for kerosene–glass in air. Note that the contact angle, in general, is different in different environments (such as another gas or liquid in place of air).

The phenomenon of the capillary effect can be explained microscopically by considering *cohesive forces* (the forces between like molecules, such as water and water) and *adhesive forces* (the forces between unlike molecules, such as water and glass). The liquid molecules at the solid–liquid interface are subjected to both cohesive forces by other liquid molecules and adhesive forces by the molecules of the solid. The relative magnitudes of these forces determine whether a liquid wets a solid surface or not. Obviously, the water molecules are more strongly attracted to the glass molecules than they are to other water molecules, and thus water tends to rise along the glass surface. The opposite occurs for mercury, which causes the liquid surface near the glass wall to be suppressed (Fig. 2–37).

The magnitude of the capillary rise in a circular tube can be determined from a force balance on the cylindrical liquid column of height h in the tube (Fig. 2–38). The bottom of the liquid column is at the same level as the free surface of the reservoir, and thus the pressure there must be atmospheric pressure. This balances the atmospheric pressure acting at the top surface of the liquid column, and thus these two effects cancel each other. The weight of the liquid column is approximately

$$W = mg = \rho Vg = \rho g(\pi R^2 h)$$

Equating the vertical component of the surface tension force to the weight gives

$$W = F_{\text{surface}} \rightarrow \rho g(\pi R^2 h) = 2\pi R \sigma_s \cos \phi$$

Solving for h gives the capillary rise to be

Capillary rise: $\quad h = \dfrac{2\sigma_s}{\rho g R} \cos \phi \quad (R = \text{constant})$ (2–42)

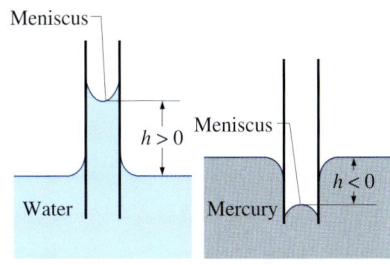

FIGURE 2–37
The capillary rise of water and the capillary fall of mercury in a small-diameter glass tube.

This relation is also valid for nonwetting liquids (such as mercury in glass) and gives the capillary drop. In this case $\phi > 90°$ and thus $\cos \phi < 0$, which makes h negative. Therefore, a negative value of capillary rise corresponds to a capillary drop (Fig. 2–37).

Note that the capillary rise is inversely proportional to the radius of the tube. Therefore, the thinner the tube is, the greater the rise (or fall) of the liquid in the tube. In practice, the capillary effect for water is usually negligible in tubes whose diameter is greater than 1 cm. When pressure measurements are made using manometers and barometers, it is important to use sufficiently large tubes to minimize the capillary effect. The capillary rise is also inversely proportional to the density of the liquid, as expected. Therefore, in general, lighter liquids experience greater capillary rises. Finally, it should be kept in mind that Eq. 2–42 is derived for constant-diameter tubes and should not be used for tubes of variable cross section.

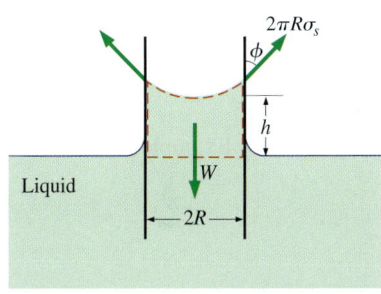

FIGURE 2–38
The forces acting on a liquid column that has risen in a tube due to the capillary effect.

■ EXAMPLE 2–6   The Capillary Rise of Water in a Tube

A 0.6-mm-diameter glass tube is inserted into water at 20°C in a cup. Determine the capillary rise of water in the tube (Fig. 2–39).

SOLUTION The rise of water in a slender tube as a result of the capillary effect is to be determined.
Assumptions 1 There are no impurities in the water and no contamination on the surfaces of the glass tube. 2 The experiment is conducted in atmospheric air.
Properties The surface tension of water at 20°C is 0.073 N/m (Table 2–4). The contact angle of water with glass is approximately 0° (from preceding text). We take the density of liquid water to be 1000 kg/m³.
Analysis The capillary rise is determined directly from Eq. 2–42 by substituting the given values, yielding

$$h = \frac{2\sigma_s}{\rho g R} \cos \phi = \frac{2(0.073 \text{ N/m})}{(1000 \text{ kg/m}^3)(9.81 \text{ m/s}^2)(0.3 \times 10^{-3} \text{m})} (\cos 0°) \left(\frac{1 \text{kg} \cdot \text{m/s}^2}{1 \text{ N}} \right)$$

$$= 0.050 \text{ m} = \mathbf{5.0 \text{ cm}}$$

Therefore, water rises in the tube 5 cm above the liquid level in the cup.
Discussion Note that if the tube diameter were 1 cm, the capillary rise would be 0.3 mm, which is hardly noticeable to the eye. Actually, the capillary rise in a large-diameter tube occurs only at the rim. The center does not rise at all. Therefore, the capillary effect can be ignored for large-diameter tubes.

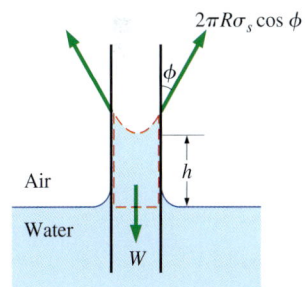

FIGURE 2–39
Schematic for Example 2–6.

EXAMPLE 2–7   Using Capillary Rise to Generate Power in a Hydraulic Turbine

Reconsider Example 2–6. Realizing that water rises by 5 cm under the influence of surface tension without requiring any energy input from an external source, a person conceives the idea that power can be generated by drilling a hole in the tube just below the water level and feeding the water spilling out of the tube into a turbine (Fig. 2–40). The person takes this idea even further by suggesting that a series of tube banks can be used for this purpose and cascading can be incorporated to achieve practically feasible flow rates and elevation differences. Determine if this idea has any merit.

SOLUTION Water that rises in tubes under the influence of the capillary effect is to be used to generate power by feeding it into a turbine. The validity of this suggestion is to be evaluated.

Analysis The proposed system may appear like a stroke of genius, since the commonly used hydroelectric power plants generate electric power by simply capturing the potential energy of elevated water, and the capillary rise provides the mechanism to raise the water to any desired height without requiring any energy input.

When viewed from a thermodynamic point of view, the proposed system immediately can be labeled as a perpetual motion machine (PMM) since it continuously generates electric power without requiring any energy input. That is, the proposed system creates energy, which is a clear violation of the first law of thermodynamics or the conservation of energy principle, and it does not warrant any further consideration. But the fundamental principle of conservation of energy did not stop many from dreaming about being the first to prove nature wrong, and to come up with a trick to permanently solve the world's energy problems. Therefore, the impossibility of the proposed system should be demonstrated.

As you may recall from your physics courses (also to be discussed in the next chapter), the pressure in a static fluid varies in the vertical direction only and increases with increasing depth linearly. Then the pressure difference across the 5-cm-high water column in the tube becomes

$$\Delta P_{\text{water column in tube}} = P_2 - P_1 = \rho_{\text{water}}gh$$

$$= (1000 \text{ kg/m}^2)(9.81 \text{ m/s}^2)(0.05 \text{ m})\left(\frac{1 \text{ kN}}{1000 \text{ kg·m/s}^2}\right)$$

$$= 0.49 \text{ kN/m}^2 \, (\approx 0.005 \text{ atm})$$

That is, the pressure at the top of the water column in the tube is 0.005 atm *less* than the pressure at the bottom. Noting that the pressure at the bottom of the water column is atmospheric pressure (since it is at the same horizontal line as the water surface in the cup) the pressure anywhere in the tube is below atmospheric pressure with the difference reaching 0.005 atm at the top. Therefore, if a hole is drilled in the tube, air will leak into the tube rather than water leaking out.

Discussion The water column in the tube is motionless, and thus, there cannot be any unbalanced force acting on it (zero net force). The force due to the pressure difference across the meniscus between the atmospheric air and the water at the top of water column is balanced by the surface tension. If this surface-tension force were to disappear, the water in the tube would drop down under the influence of atmospheric pressure to the level of the free surface in the tube.

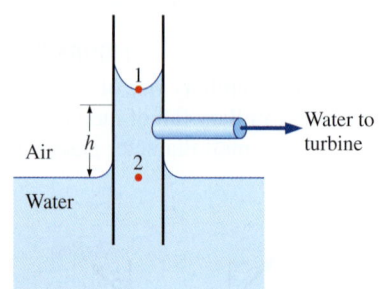

FIGURE 2–40
Schematic for Example 2-7.

SUMMARY

In this chapter various properties commonly used in fluid mechanics are discussed. The mass-dependent properties of a system are called *extensive properties* and the others, *intensive properties*. *Density* is mass per unit volume, and *specific volume* is volume per unit mass. The *specific gravity* is defined as the ratio of the density of a substance to the density of water at 4°C,

$$SG = \frac{\rho}{\rho_{H_2O}}$$

The ideal-gas equation of state is expressed as

$$P = \rho RT$$

where P is the absolute pressure, T is the thermodynamic temperature, ρ is the density, and R is the gas constant.

At a given temperature, the pressure at which a pure substance changes phase is called the *saturation pressure*. For phase-change processes between the liquid and vapor phases of a pure substance, the saturation pressure is commonly called the *vapor pressure* P_v. Vapor bubbles that form in the low-pressure regions in a liquid (a phenomenon called *cavitation*) collapse as they are swept away from the low-pressure regions, generating highly destructive, extremely high-pressure waves.

Energy can exist in numerous forms, and their sum constitutes the *total energy* E (or e on a unit-mass basis) of a system. The sum of all microscopic forms of energy is called the *internal energy* U of a system. The energy that a system possesses as a result of its motion relative to some reference frame is called *kinetic energy* expressed per unit mass as $ke = V^2/2$, and the energy that a system possesses as a result of its elevation in a gravitational field is called *potential energy* expressed per unit mass as $pe = gz$.

The compressibility effects in a fluid are represented by the *coefficient of compressibility* κ (also called the *bulk modulus of elasticity*) defined as

$$\kappa = -v\left(\frac{\partial P}{\partial v}\right)_T = \rho\left(\frac{\partial P}{\partial \rho}\right)_T \cong -\frac{\Delta P}{\Delta v/v}$$

The property that represents the variation of the density of a fluid with temperature at constant pressure is the *volume expansion coefficient* (or volume expansivity) β, defined as

$$\beta = \frac{1}{v}\left(\frac{\partial v}{\partial T}\right)_P = -\frac{1}{\rho}\left(\frac{\partial \rho}{\partial T}\right)_P \cong -\frac{\Delta\rho/\rho}{\Delta T}$$

The velocity at which an infinitesimally small pressure wave travels through a medium is the *speed of sound*. For an ideal gas it is expressed as

$$c = \sqrt{\left(\frac{\partial P}{\partial \rho}\right)_s} = \sqrt{kRT}$$

The *Mach number* is the ratio of the actual speed of the fluid to the speed of sound at the same state:

$$\text{Ma} = \frac{V}{c}$$

The flow is called *sonic* when $\text{Ma} = 1$, *subsonic* when $\text{Ma} < 1$, *supersonic* when $\text{Ma} > 1$, *hypersonic* when $\text{Ma} >> 1$, and *transonic* when $\text{Ma} \cong 1$.

The *viscosity* of a fluid is a measure of its resistance to deformation. The tangential force per unit area is called *shear stress* and is expressed for simple shear flow between plates (one-dimensional flow) as

$$\tau = \mu \frac{du}{dy}$$

where μ is the coefficient of viscosity or the *dynamic* (or *absolute*) *viscosity* of the fluid, u is the velocity component in the flow direction, and y is the direction normal to the flow direction. Fluids that obey this linear relationship are called *Newtonian fluids*. The ratio of dynamic viscosity to density is called the *kinematic viscosity* ν.

The pulling effect on the liquid molecules at an interface caused by the attractive forces of molecules per unit length is called *surface tension* σ_s. The excess pressure ΔP inside a spherical droplet or soap bubble, respectively, is given by

$$\Delta P_{\text{droplet}} = P_i - P_o = \frac{2\sigma_s}{R} \quad \text{and} \quad \Delta P_{\text{soap bubble}} = P_i - P_o = \frac{4\sigma_s}{R}$$

where P_i and P_o are the pressures inside and outside the droplet or soap bubble. The rise or fall of a liquid in a small-diameter tube inserted into the liquid due to surface tension is called the *capillary effect*. The capillary rise or drop is given by

$$h = \frac{2\sigma_s}{\rho g R}\cos\phi$$

where ϕ is the *contact angle*. The capillary rise is inversely proportional to the radius of the tube; for water, it is negligible for tubes whose diameter is larger than about 1 cm.

Density and viscosity are two of the most fundamental properties of fluids, and they are used extensively in the chapters that follow. In Chap. 3, the effect of density on the variation of pressure in a fluid is considered, and the hydrostatic forces acting on surfaces are determined. In Chap. 8, the pressure drop caused by viscous effects during flow is calculated and used in the determination of the pumping power requirements. Viscosity is also used as a key property in the formulation and solutions of the equations of fluid motion in Chaps. 9 and 10.

APPLICATION SPOTLIGHT ■ Cavitation

Guest Authors: G. C. Lauchle and M. L. Billet, Penn State University

Cavitation is the rupture of a liquid, or of a fluid–solid interface, caused by a reduction of the local static pressure produced by the dynamic action of the fluid in the interior and/or boundaries of a liquid system. The rupture is the formation of a visible bubble. Liquids, such as water, contain many microscopic voids that act as *cavitation nuclei*. Cavitation occurs when these nuclei grow to a significant, visible size. Although boiling is also the formation of voids in a liquid, we usually separate this phenomenon from cavitation because it is caused by an increase in temperature, rather than by a reduction in pressure. Cavitation can be used in beneficial ways, such as in ultrasonic cleaners, etchers, and cutters. But more often than not, cavitation is to be avoided in fluid flow applications because it spoils hydrodynamic performance, it causes extremely loud noise and high vibration levels, and it damages (erodes) the surfaces that support it. When cavitation bubbles enter regions of high pressure and collapse, the underwater shock waves sometimes create minute amounts of light. This phenomenon is called *sonoluminescence*.

Body cavitation is illustrated in Fig. 2–41. The body is a model of the underwater bulbulous bow region of a surface ship. It is shaped this way because located within it is a *so*und *na*vigation and *r*anging (sonar) system that is spherical in shape. This part of the surface ship is thus called a *sonar dome*. As ship speeds get faster and faster some of these domes start to cavitate and the noise created by the cavitation renders the sonar system useless. Naval architects and fluid dynamicists attempt to design these domes so that they will not cavitate. Model-scale testing allows the engineer to see first hand whether a given design provides improved cavitation performance. Because such tests are conducted in water tunnels, the conditions of the test water should have sufficient nuclei to model those conditions in which the prototype operates. This assures that the effect of liquid tension (nuclei distribution) is minimized. Important variables are the gas content level (nuclei distribution) of the water, the temperature, and the hydrostatic pressure at which the body operates. Cavitation first appears—as either the speed V is increased, or as the submergence depth h is decreased—at the minimum pressure point $C_{p\text{min}}$ of the body. Thus, good hydrodynamic design requires $2(P_\infty - P_v)/\rho V^2 > C_{p\text{min}}$, where ρ is density, $P_\infty = \rho g h$ is the reference to static pressure, C_p is the pressure coefficient (Chap. 7), and P_v is the vapor pressure of water.

References

Lauchle, G. C., Billet, M. L., and Deutsch, S., "High-Reynolds Number Liquid Flow Measurements," in *Lecture Notes in Engineering,* Vol. 46, *Frontiers in Experimental Fluid Mechanics,* Springer-Verlag, Berlin, edited by M. Gad-el-Hak, Chap. 3, pp. 95–158, 1989.

Ross, D., *Mechanics of Underwater Noise,* Peninsula Publ., Los Altos, CA, 1987.

Barber, B. P., Hiller, R. A., Löfstedt, R., Putterman, S. J., and Weninger, K. R., "Defining the Unknowns of Sonoluminescence," *Physics Reports*, Vol. 281, pp. 65–143, 1997.

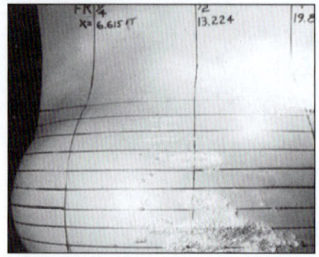

(a)

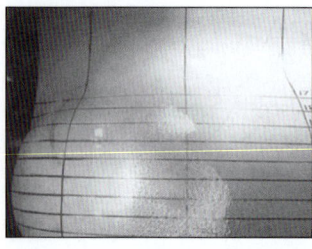

(b)

FIGURE 2–41

(a) *Vaporous cavitation* occurs in water that has very little entrained gas, such as that found very deep in a body of water. Cavitation bubbles are formed when the speed of the body—in this case the bulbulous bow region of a surface ship sonar dome—increases to the point where the local static pressure falls below the vapor pressure of the water. The cavitation bubbles are filled essentially with water vapor. This type of cavitation is very violent and noisy. (b) On the other hand, in shallow water, there is much more entrained gas in the water to act as cavitation nuclei. That's because of the proximity of the dome to the atmosphere at the free surface. The cavitation bubbles first appear at a slower speed, and hence at a higher local static pressure. They are predominantly filled with the gases that are entrained in the water, so this is known as *gaseous cavitation*.

Reprinted by permission of G. C. Lauchle and M. L. Billet, Penn State University.

REFERENCES AND SUGGESTED READING

1. J. D. Anderson *Modern Compressible Flow with Historical Perspective*, 3rd ed. New York: McGraw-Hill, 2003.
2. E. C. Bingham. "An Investigation of the Laws of Plastic Flow," *U.S. Bureau of Standards Bulletin*, 13, pp. 309–353, 1916.
3. Y. A. Cengel and M. A. Boles. *Thermodynamics: An Engineering Approach*, 7th ed. New York: McGraw-Hill, 2011.
4. D. C. Giancoli. *Physics*, 6th ed. Upper Saddle River, NJ: Pearson, 2004.
5. Y. S. Touloukian, S. C. Saxena, and P. Hestermans. *Thermophysical Properties of Matter, The TPRC Data Series*, Vol. 11, *Viscosity*. New York: Plenum, 1975.
6. L. Trefethen. "Surface Tension in Fluid Mechanics." In *Illustrated Experiments in Fluid Mechanics*. Cambridge, MA: MIT Press, 1972.
7. *The U.S. Standard Atmosphere*. Washington, DC: U.S. Government Printing Office, 1976.
8. M. Van Dyke. *An Album of Fluid Motion*. Stanford, CA: Parabolic Press, 1982.
9. C. L. Yaws, X. Lin, and L. Bu. "Calculate Viscosities for 355 Compounds. An Equation Can Be Used to Calculate Liquid Viscosity as a Function of Temperature," *Chemical Engineering*, 101, no. 4, pp. 1110–1128, April 1994.
10. C. L. Yaws. *Handbook of Viscosity*. 3 Vols. Houston, TX: Gulf Publishing, 1994.

PROBLEMS*

Density and Specific Gravity

2–1C What is the difference between intensive and extensive properties?

2–2C What is specific gravity? How is it related to density?

2–3C The specific weight of a system is defined as the weight per unit volume (note that this definition violates the normal specific property-naming convention). Is the specific weight an extensive or intensive property?

2–4C What is the state postulate?

2–5C Under what conditions is the ideal-gas assumption suitable for real gases?

2–6C What is the difference between R and R_u? How are these two related?

2–7 A fluid that occupies a volume of 24 L weighs 225 N at a location where the gravitational acceleration is 9.80 m/s^2. Determine the mass of this fluid and its density.

2–8 A 100-L container is filled with 1 kg of air at a temperature of 27°C. What is the pressure in the container?

2–9 What is the specific volume of oxygen at 300 kPa and 27°C?

2–10 The air in an automobile tire with a volume of 0.0740 m^3 is at 30°C and 140 kPa gage. Determine the amount of air that must be added to raise the pressure to the recommended value of 210 kPa gage. Assume the atmospheric pressure to be 100 kPa and the temperature and the volume to remain constant. *Answer:* 0.0596 kg

2–11 The pressure in an automobile tire depends on the temperature of the air in the tire. When the air temperature is 25°C, the pressure gage reads 210 kPa. If the volume of the tire is 0.025 m^3, determine the pressure rise in the tire when the air temperature in the tire rises to 50°C. Also, determine the amount of air that must be bled off to restore pressure to its original value at this temperature. Assume the atmospheric pressure to be 100 kPa.

FIGURE P2–11

Stockbyte/GettyImages

2–12 A spherical balloon with a diameter of 9 m is filled with helium at 20°C and 200 kPa. Determine the mole number and the mass of the helium in the balloon. *Answers:* 31.3 kmol, 125 kg

---

* Problems designated by a "C" are concept questions, and students are encouraged to answer them all. Problems with the ⊙ icon are solved using EES, and complete solutions together with parametric studies are included on the text website. Problems with the [EES] icon are comprehensive in nature and are intended to be solved with an equation solver such as EES.

2–13 Reconsider Prob. 2–12. Using EES (or other) software, investigate the effect of the balloon diameter on the mass of helium contained in the balloon for the pressures of (*a*) 100 kPa and (*b*) 200 kPa. Let the diameter vary from 5 m to 15 m. Plot the mass of helium against the diameter for both cases.

2–14 A cylindrical tank of methanol has a mass of 40 kg and a volume of 51 L. Determine the methanol's weight, density, and specific gravity. Take the gravitational acceleration to be 9.81 m/s^2. Also, estimate how much force is needed to accelerate this tank linearly at 0.25 m/s^2.

2–15 The density of saturated liquid refrigerant–134a for $-20°C \leq T \leq 100°C$ is given in Table A–4. Using this value develop an expression in the form $\rho = aT^2 + bT + c$ for the density of refrigerant–134a as a function of absolute temperature, and determine relative error for each data set.

2–16 The density of atmospheric air varies with elevation, decreasing with increasing altitude. (*a*) Using the data given in the table, obtain a relation for the variation of density with elevation, and calculate the density at an elevation of 7000 m. (*b*) Calculate the mass of the atmosphere using the correlation you obtained. Assume the earth to be a perfect sphere with a radius of 6377 km, and take the thickness of the atmosphere to be 25 km.

r, km	ρ, kg/m^3
6377	1.225
6378	1.112
6379	1.007
6380	0.9093
6381	0.8194
6382	0.7364
6383	0.6601
6385	0.5258
6387	0.4135
6392	0.1948
6397	0.08891
6402	0.04008

Vapor Pressure and Cavitation

2–17C What is cavitation? What causes it?

2–18C Does water boil at higher temperatures at higher pressures? Explain.

2–19C If the pressure of a substance is increased during a boiling process, will the temperature also increase or will it remain constant? Why?

2–20C What is vapor pressure? How is it related to saturation pressure?

2–21 The analysis of a propeller that operates in water at 20°C shows that the pressure at the tips of the propeller drops to 1 kPa at high speeds. Determine if there is a danger of cavitation for this propeller.

2–22 A pump is used to transport water to a higher reservoir. If the water temperature is 20°C, determine the lowest pressure that can exist in the pump without cavitation.

2–23 In a piping system, the water temperature remains under 30°C. Determine the minimum pressure allowed in the system to avoid cavitation.

2–24 The analysis of a propeller that operates in water at 20°C shows that the pressure at the tips of the propeller drops to 2 kPa at high speeds. Determine if there is a danger of cavitation for this propeller.

Energy and Specific Heats

2–25C What is flow energy? Do fluids at rest possess any flow energy?

2–26C How do the energies of a flowing fluid and a fluid at rest compare? Name the specific forms of energy associated with each case.

2–27C What is the difference between the macroscopic and microscopic forms of energy?

2–28C What is total energy? Identify the different forms of energy that constitute the total energy.

2–29C List the forms of energy that contribute to the internal energy of a system.

2–30C How are heat, internal energy, and thermal energy related to each other?

2–31C Using average specific heats, explain how internal energy changes of ideal gases and incompressible substances can be determined.

2–32C Using average specific heats, explain how enthalpy changes of ideal gases and incompressible substances can be determined.

2–33 Saturated water vapor at 150°C (enthalpy $h = 2745.9$ kJ/kg) flows in a pipe at 50 m/s at an elevation of $z = 10$ m. Determine the total energy of vapor in J/kg relative to the ground level.

Compressibility

2–34C What does the coefficient of compressibility of a fluid represent? How does it differ from isothermal compressibility?

2–35C What does the coefficient of volume expansion of a fluid represent? How does it differ from the coefficient of compressibility?

2–36C Can the coefficient of compressibility of a fluid be negative? How about the coefficient of volume expansion?

2–37 Water at 15°C and 1 atm pressure is heated to 100°C at constant pressure. Using coefficient of volume expansion data, determine the change in the density of water. *Answer:* −38.7 kg/m³

2–38 It is observed that the density of an ideal gas increases by 10 percent when compressed isothermally from 10 atm to 11 atm. Determine the percent increase in density of the gas if it is compressed isothermally from 1000 atm to 1001 atm.

2–39 Using the definition of the coefficient of volume expansion and the expression $\beta_{\text{ideal gas}} = 1/T$, show that the percent increase in the specific volume of an ideal gas during isobaric expansion is equal to the percent increase in absolute temperature.

2–40 Water at 1 atm pressure is compressed to 400 atm pressure isothermally. Determine the increase in the density of water. Take the isothermal compressibility of water to be 4.80×10^{-5} atm⁻¹.

2–41 The volume of an ideal gas is to be reduced by half by compressing it isothermally. Determine the required change in pressure.

2–42 Saturated refrigerant-134a liquid at 10°C is cooled to 0°C at constant pressure. Using coefficient of volume expansion data, determine the change in the density of the refrigerant.

2–43 A water tank is completely filled with liquid water at 20°C. The tank material is such that it can withstand tension caused by a volume expansion of 0.8 percent. Determine the maximum temperature rise allowed without jeopardizing safety. For simplicity, assume β = constant = β at 40°C.

2–44 Repeat Prob. 2–43 for a volume expansion of 1.5 percent for water.

2–45 The density of seawater at a free surface where the pressure is 98 kPa is approximately 1030 kg/m³. Taking the bulk modulus of elasticity of seawater to be 2.34×10^9 N/m² and expressing variation of pressure with depth z as $dP = \rho g\, dz$ determine the density and pressure at a depth of 2500 m. Disregard the effect of temperature.

2–46 Taking the coefficient of compressibility of water to be 5×10^6 kPa, determine the pressure increase required to reduce the volume of water by (a) 1 percent and (b) 2 percent.

2–47 Prove that the coefficient of volume expansion for an ideal gas is $\beta_{\text{ideal gas}} = 1/T$.

2–48 The ideal gas equation of state is very simple, but its range of applicability is limited. A more accurate but complicated equation is the Van der Waals equation of state given by

$$P = \frac{RT}{v - b} - \frac{a}{v^2}$$

where a and b are constants depending on critical pressure and temperatures of the gas. Predict the coefficient of compressibility of nitrogen gas at $T = 175$ K and $v = 0.00375$ m³/kg, assuming the nitrogen to obey the Van der Waals equation of state. Compare your result with the ideal gas value. Take $a = 0.175$ m⁶·kPa/kg² and $b = 0.00138$ m³/kg for the given conditions. The experimentally measured pressure of nitrogen is 10,000 kPa.

2–49 A frictionless piston-cylinder device contains 10 kg of water at 20°C at atmospheric pressure. An external force F is then applied on the piston until the pressure inside the cylinder increases to 100 atm. Assuming the coefficient of compressibility of water remains unchanged during the compression; estimate the energy needed to compress the water isothermally. *Answer:* 29.4 J

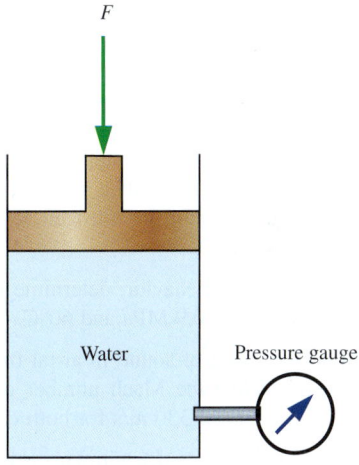

FIGURE P2–49

2–50 Reconsider Prob. 2–49. Assuming a linear pressure increase during the compression, estimate the energy needed to compress the water isothermally.

Speed of Sound

2–51C What is sound? How is it generated? How does it travel? Can sound waves travel in a vacuum?

2–52C In which medium does a sound wave travel faster: in cool air or in warm air?

2–53C In which medium will sound travel fastest for a given temperature: air, helium, or argon?

2–54C In which medium does a sound wave travel faster: in air at 20°C and 1 atm or in air at 20°C and 5 atm?

2–55C Does the Mach number of a gas flowing at a constant velocity remain constant? Explain.

2–56C Is it realistic to approximate that the propagation of sound waves is an isentropic process? Explain.

2–57C Is the sonic velocity in a specified medium a fixed quantity, or does it change as the properties of the medium change? Explain.

2–58 The Airbus A-340 passenger plane has a maximum takeoff weight of about 260,000 kg, a length of 64 m, a wing span of 60 m, a maximum cruising speed of 945 km/h, a seating capacity of 271 passengers, a maximum cruising altitude of 14,000 m, and a maximum range of 12,000 km. The air temperature at the crusing altitude is about $-60°C$. Determine the Mach number of this plane for the stated limiting conditions.

2–59 Carbon dioxide enters an adiabatic nozzle at 1200 K with a velocity of 50 m/s and leaves at 400 K. Assuming constant specific heats at room temperature, determine the Mach number (a) at the inlet and (b) at the exit of the nozzle. Assess the accuracy of the constant specific heat approximation. *Answers:* (a) 0.0925, (b) 3.73

2–60 Nitrogen enters a steady-flow heat exchanger at 150 kPa, 10°C, and 100 m/s, and it receives heat in the amount of 120 kJ/kg as it flows through it. Nitrogen leaves the heat exchanger at 100 kPa with a velocity of 200 m/s. Determine the Mach number of the nitrogen at the inlet and the exit of the heat exchanger.

2–61 Assuming ideal gas behavior, determine the speed of sound in refrigerant-134a at 0.9 MPa and 60°C.

2–62 Determine the speed of sound in air at (a) 300 K and (b) 800 K. Also determine the Mach number of an aircraft moving in air at a velocity of 330 m/s for both cases.

2–63 Steam flows through a device with a pressure of 825 kPa, a temperature of 400°C, and a velocity of 275 m/s. Determine the Mach number of the steam at this state by assuming ideal-gas behavior with $k = 1.3$. *Answer:* 0.433

2–64 Reconsider Prob. 2–63. Using EES (or other) software, compare the Mach number of steam flow over the temperature range 200 to 400°C. Plot the Mach number as a function of temperature.

2–65 Air expands isentropically from 2.2 MPa and 77°C to 0.4 MPa. Calculate the ratio of the initial to the final speed of sound. *Answer:* 1.28

2–66 Repeat Prob. 2–65 for helium gas.

2–67 The isentropic process for an ideal gas is expressed as Pv^k = constant. Using this process equation and the definition of the speed of sound (Eq. 2–24), obtain the expression for the speed of sound for an ideal gas (Eq. 2–26).

Viscosity

2–68C What is viscosity? What is the cause of it in liquids and in gases? Do liquids or gases have higher dynamic viscosities?

2–69C What is a Newtonian fluid? Is water a Newtonian fluid?

2–70C How does the kinematic viscosity of (a) liquids and (b) gases vary with temperature?

2–71C How does the dynamic viscosity of (a) liquids and (b) gases vary with temperature?

2–72C Consider two identical small glass balls dropped into two identical containers, one filled with water and the other with oil. Which ball will reach the bottom of the container first? Why?

2–73 The viscosity of a fluid is to be measured by a viscometer constructed of two 1.5-m-long concentric cylinders. The inner diameter of the outer cylinder is 16 cm, and the gap between the two cylinders is 0.09 cm. The outer cylinder is rotated at 250 rpm, and the torque is measured to be 1.4 N.m. Determine the viscosity of the fluid. *Answer:* 0.00997 N.s/m²

2–74 A 50-cm × 30-cm × 20-cm block weighing 150 N is to be moved at a constant velocity of 0.80 m/s on an inclined surface with a friction coefficient of 0.27. (a) Determine the force F that needs to be applied in the horizontal direction. (b) If a 0.40-mm-thick oil film with a dynamic viscosity of 0.012 Pa·s is applied between the block and inclined surface, determine the percent reduction in the required force.

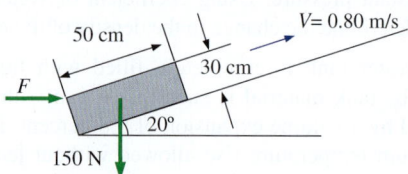

FIGURE P2–74

2–75 Consider the flow of a fluid with viscosity μ through a circular pipe. The velocity profile in the pipe is given as $u(r) = u_{max}(1 - r^n/R^n)$, where u_{max} is the maximum flow velocity, which occurs at the centerline; r is the radial distance from the centerline; and $u(r)$ is the flow velocity at any position r. Develop a relation for the drag force exerted on the pipe wall by the fluid in the flow direction per unit length of the pipe.

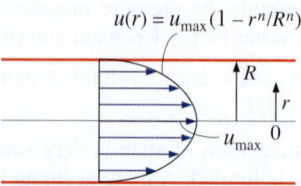

FIGURE P2–75

2–76 A thin 30-cm × 30-cm flat plate is pulled at 3 m/s horizontally through a 3.6-mm-thick oil layer sandwiched between two plates, one stationary and the other moving at a constant velocity of 0.3 m/s, as shown in Fig. P2–76. The dynamic viscosity of the oil is 0.027 Pa·s. Assuming the velocity in each oil layer to vary linearly, (a) plot the velocity profile and find the location where the oil velocity is zero and (b) determine the force that needs to be applied on the plate to maintain this motion.

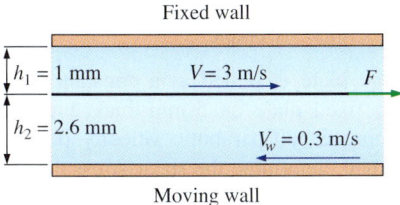

FIGURE P2–76

2–77 A rotating viscometer consists of two concentric cylinders – an inner cylinder of radius R_i rotating at angular velocity (rotation rate) ω_i, and a stationary outer cylinder of inside radius R_o. In the tiny gap between the two cylinders is the fluid of viscosity μ. The length of the cylinders (into the page in Fig. P2–77) is L. L is large such that end effects are negligible (we can treat this as a two-dimensional problem). Torque (T) is required to rotate the inner cylinder at constant speed. (a) Showing all of your work and algebra, generate an approximate expression for T as a function of the other variables. (b) Explain why your solution is only an *approximation*. In particular, do you expect the velocity profile in the gap to remain linear as the gap becomes larger and larger (i.e., if the outer radius R_o were to increase, all else staying the same)?

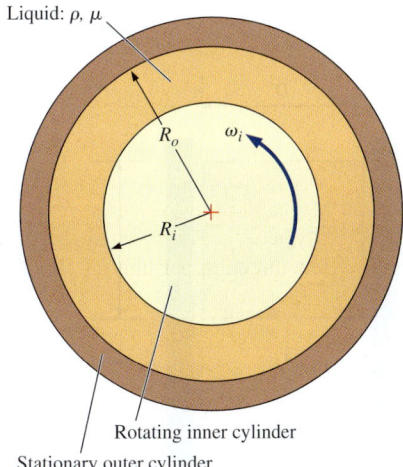

FIGURE P2–77

2–78 The clutch system shown in Fig. P2–78 is used to transmit torque through a 2-mm-thick oil film with μ = 0.38 N·s/m² between two identical 30-cm-diameter disks. When the driving shaft rotates at a speed of 1450 rpm, the driven shaft is observed to rotate at 1398 rpm. Assuming a linear velocity profile for the oil film, determine the transmitted torque.

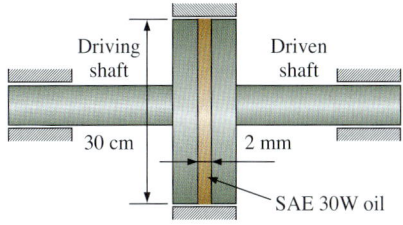

FIGURE P2–78

2–79 Reconsider Prob. 2–78. Using EES (or other) software, investigate the effect of oil film thickness on the torque transmitted. Let the film thickness vary from 0.1 mm to 10 mm. Plot your results, and state your conclusions.

2–80 The dynamic viscosity of carbon dioxide at 50°C and 200°C are 1.612×10^{-5} Pa·s and 2.276×10^{-5} Pa·s, respectively. Determine the constants a and b of Sutherland correlation for carbon dioxide at atmospheric pressure. Then predict the viscosity of carbon dioxide at 100°C and compare your result against the value given in Table A–10.

2–81 One of the widely used correlations to describe the variation of the viscosity of gases is the power-law equation given by $\mu/\mu_0 = (T/T_0)^n$, where μ_0 and T_0 are the reference viscosity and temperature, respectively. Using the power and Sutherland laws, examine the variation of the air viscosity for the temperature range 100°C (373 K) to 1000°C (1273 K). Plot your results to compare with values listed in Table A-9. Take the reference temperature as 0°C and $n = 0.666$ for the atmospheric air.

2–82 For flow over a plate, the variation of velocity with vertical distance y from the plate is given as $u(y) = ay - by^2$ where a and b are constants. Obtain a relation for the wall shear stress in terms of a, b, and μ.

2–83 In regions far from the entrance, fluid flow through a circular pipe is one dimensional, and the velocity profile for laminar flow is given by $u(r) = u_{max}(1 - r^2/R^2)$, where R is the radius of the pipe, r is the radial distance from the center of the pipe, and u_{max} is the maximum flow velocity, which occurs at the center. Obtain (a) a relation for the drag force applied by the fluid on a section of the pipe of length L and (b) the value of the drag force for water flow at 20°C with $R = 0.08$ m, $L = 30$ m, $u_{max} = 3$ m/s, and $\mu = 0.0010$ kg/m·s.

68
PROPERTIES OF FLUIDS

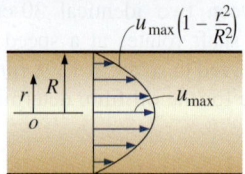

FIGURE P2–83

2–84 Repeat Prob. 2–83 for $u_{max} = 7$ m/s.   *Answer:* (b) 2.64 N

2–85 A frustum-shaped body is rotating at a constant angular speed of 200 rad/s in a container filled with SAE 10W oil at 20°C ($\mu = 0.100$ Pa·s), as shown in Fig. P2–85. If the thickness of the oil film on all sides is 1.2 mm, determine the power required to maintain this motion. Also determine the reduction in the required power input when the oil temperature rises to 80°C ($\mu = 0.0078$ Pa·s).

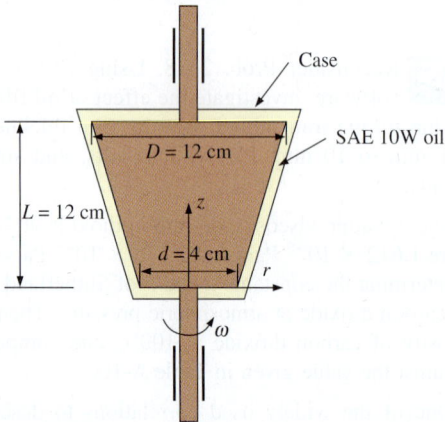

FIGURE P2–85

2–86 A rotating viscometer consists of two concentric cylinders—a stationary inner cyliner of radius R_i and an outer cylinder of inside radius R_o rotating at angular velocity (rotation rate) ω_o. In the tiny gap between the two cylinders is the fluid whose viscosity (μ) is to be measured. The length of the cylinders (into the page in Fig. P2–86) is L. L is large such that end effects are negligible (we can treat this as a two-dimensional problem). Torque (T) is required to rotate the inner cylinder at constant speed. Showing all your work and algebra, generate an approximate expression of T as a function of the other variables.

2–87 A large plate is pulled at a constant speed of $U = 4$ m/s over a fixed plate on 5-mm-thick engine oil film at 20°C. Assuming a half-parabolic velocity profile in the oil film, as sketched, determine the shear stress developed on the upper plate and its direction. What would happen if a linear velocity profile were assumed?

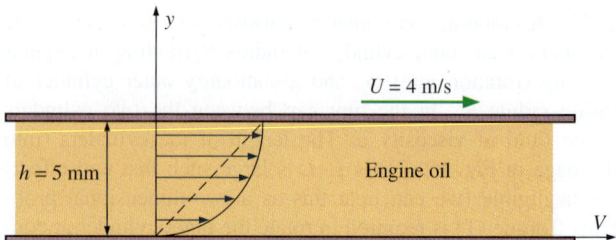

FIGURE P2–87

2–88 A cylinder of mass m slides down from rest in a vertical tube whose inner surface is covered by a viscous oil of film thickness h. If the diameter and height of the cylinder are D and L, respectively, derive an expression for the velocity of the cylinder as a function of time, t. Discuss what will happen as $t \to \infty$. Can this device serve as a viscometer?

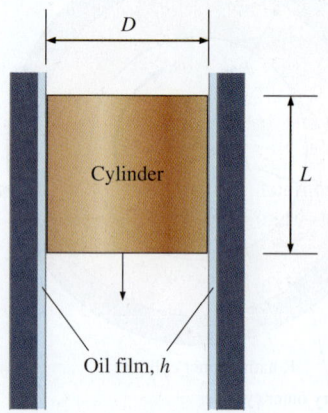

FIGURE P2–88

2–89 A thin plate moves between two parallel, horizontal, stationary flat surfaces at a constant velocity of 5 m/s. The two stationary surfaces are spaced 4 cm apart, and the medium between them is filled with oil whose viscosity is 0.9 N·s/m².

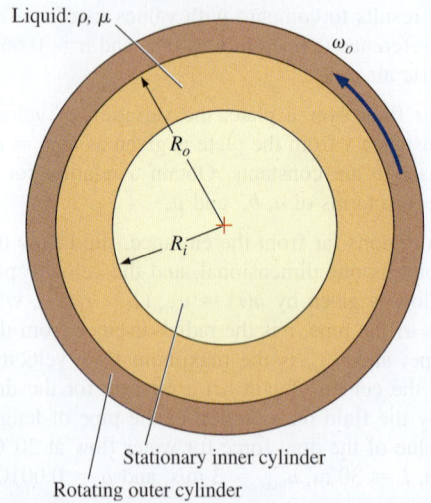

FIGURE P2–86

The part of the plate immersed in oil at any given time is 2-m long and 0.5-m wide. If the plate moves through the mid-plane between the surfaces, determine the force required to maintain this motion. What would your response be if the plate was 1 cm from the bottom surface (h_2) and 3 cm from the top surface (h_1)?

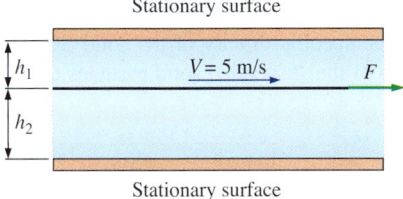

FIGURE P2–89

2–90 Reconsider Prob. 2–89. If the viscosity of the oil above the moving plate is 4 times that of the oil below the plate, determine the distance of the plate from the bottom surface (h_2) that will minimize the force needed to pull the plate between the two oils at constant velocity.

Surface Tension and Capillary Effect

2–91C What is surface tension? What is its cause? Why is the surface tension also called surface energy?

2–92C A small-diameter tube is inserted into a liquid whose contact angle is 110°. Will the level of liquid in the tube be higher or lower than the level of the rest of the liquid? Explain.

2–93C What is the capillary effect? What is its cause? How is it affected by the contact angle?

2–94C Consider a soap bubble. Is the pressure inside the bubble higher or lower than the pressure outside?

2–95C Is the capillary rise greater in small- or large-diameter tubes?

2–96 Consider a 0.15-mm diameter air bubble in a liquid. Determine the pressure difference between the inside and outside of the air bubble if the surface tension at the air-liquid interface is (*a*) 0.080 N/m and (*b*) 0.12 N/m.

2–97 A 6-cm-diameter soap bubble is to be enlarged by blowing air into it. Taking the surface tension of soap solution to be 0.039 N/m, determine the work input required to inflate the bubble to a diameter of 6.9 cm.

2–98 A 1.2-mm-diameter tube is inserted into an unknown liquid whose density is 960 kg/m³, and it is observed that the liquid rises 5 mm in the tube, making a contact angle of 15°. Determine the surface tension of the liquid.

2–99 Determine the gage pressure inside a soap bubble of diameter (*a*) 0.2 cm and (*b*) 5 cm at 20°C.

2–100 A 0.8-mm-diameter glass tube is inserted into kerosene at 20°C. The contact angle of kerosene with a glass surface is 26°. Determine the capillary rise of kerosene in the tube. *Answer:* 16 mm

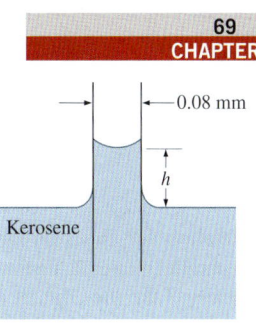

FIGURE P2–100

2–101 The surface tension of a liquid is to be measured using a liquid film suspended on a U-shaped wire frame with an 8-cm-long movable side. If the force needed to move the wire is 0.024 N, determine the surface tension of this liquid in air.

2–102 A capillary tube of 1.2 mm diameter is immersed vertically in water exposed to the atmosphere. Determine how high water will rise in the tube. Take the contact angle at the inner wall of the tube to be 6° and the surface tension to be 1.00 N/m. *Answer:* 0.338 m

2–103 A capillary tube is immersed vertically in a water container. Knowing that water starts to evaporate when the pressure drops below 2 kPa, determine the maximum capillary rise and tube diameter for this maximum-rise case. Take the contact angle at the inner wall of the tube to be 6° and the surface tension to be 1.00 N/m.

2–104 Contrary to what you might expect, a solid steel ball can float on water due to the surface tension effect. Determine the maximum diameter of a steel ball that would float on water at 20°C. What would your answer be for an aluminum ball? Take the densities of steel and aluminum balls to be 7800 kg/m³ and 2700 kg/m³, respectively.

2–105 Nutrients dissolved in water are carried to upper parts of plants by tiny tubes partly because of the capillary effect. Determine how high the water solution will rise in a

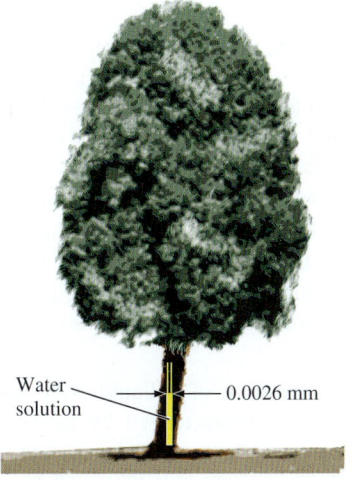

FIGURE P2–105

tree in a 0.0026-mm-diameter tube as a result of the capillary effect. Treat the solution as water at 20°C with a contact angle of 15°. *Answer:* 11.1 m

Review Problems

2–106 Derive a relation for the capillary rise of a liquid between two large parallel plates a distance t apart inserted into the liquid vertically. Take the contact angle to be ϕ.

2–107 Consider a 55-cm-long journal bearing that is lubricated with oil whose viscosity is 0.1 kg/m·s at 20°C at the beginning of operation and 0.008 kg/m·s at the anticipated steady operating temperature of 80°C. The diameter of the shaft is 8 cm, and the average gap between the shaft and the journal is 0.08 cm. Determine the torque needed to overcome the bearing friction initially and during steady operation when the shaft is rotated at 1500 rpm.

2–108 The diameter of one arm of a U-tube is 5 mm while the other arm is large. If the U-tube contains some water, and both surfaces are exposed to atmospheric pressure, determine the difference between the water levels in the two arms.

2–109 The combustion in a gasoline engine may be approximated by a constant volume heat addition process, and the contents of the combustion chamber both before and after combustion as air. The conditions are 1.80 MPa and 450°C before the combustion and 1300°C after it. Determine the pressure at the end of the combustion process. *Answer:* 3916 kPa

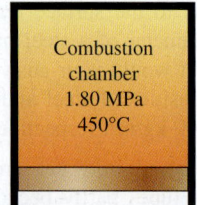

FIGURE P2–109

2–110 A rigid tank contains an ideal gas at 300 kPa and 600 K. Half of the gas is withdrawn from the tank and the gas is at 100 kPa at the end of the process. Determine (*a*) the final temperature of the gas and (*b*) the final pressure if no mass were withdrawn from the tank and the same final temperature were reached at the end of the process.

2–111 The absolute pressure of an automobile tire is measured to be 320 kPa before a trip and 335 kPa after the trip. Assuming the volume of the tire remains constant at 0.022 m³, determine the percent increase in the absolute temperature of the air in the tire.

2–112 The composition of a liquid with suspended solid particles is generally characterized by the fraction of solid particles either by weight or mass, $C_{s,\,mass} = m_s/m_m$ or by volume, $C_{s,\,vol} = V_s/V_m$ where m is mass and V is volume. The subscripts s and m indicate solid and mixture, respectively. Develop an expression for the specific gravity of a water-based suspension in terms of $C_{s,\,mass}$ and $C_{s,\,vol}$.

2–113 The specific gravities of solids and carrier fluids of a slurry are usually known, but the specific gravity of the slurry depends on the concentration of the solid particles. Show that the specific gravity of a water-based slurry can be expressed in terms of the specific gravity of the solid SG_s and the mass concentration of the suspended solid particles $C_{s,\,mass}$ as

$$SG_m = \frac{1}{1 + C_{s,\,mass}(1/SG_s - 1)}$$

2–114 A 10-m³ tank contains nitrogen at 25°C and 800 kPa. Some nitrogen is allowed to escape until the pressure in the tank drops to 600 kPa. If the temperature at this point is 20°C, determine the amount of nitrogen that has escaped. *Answer:* 21.5 kg

2–115 A closed tank is partially filled with water at 60°C. If the air above the water is completely evacuated, determine the absolute pressure in the evacuated space. Assume the temperature to remain constant.

2–116 The variation of the dynamic viscosity of water with absolute temperature is given as

T, K	μ, Pa·s
273.15	1.787×10^{-3}
278.15	1.519×10^{-3}
283.15	1.307×10^{-3}
293.15	1.002×10^{-3}
303.15	7.975×10^{-4}
313.15	6.529×10^{-4}
333.15	4.665×10^{-4}
353.15	3.547×10^{-4}
373.15	2.828×10^{-4}

Using these tabulated data, develop a relation for viscosity in the form of $\mu = \mu(T) = A + BT + CT^2 + DT^3 + ET^4$. Using the relation developed, predict the dynamic viscosity of water at 50°C at which the reported value is 5.468×10^{-4} Pa·s. Compare your result with the results of Andrade's equation, which is given in the form of $\mu = D \cdot e^{B/T}$, where D and B are constants whose values are to be determined using the viscosity data given.

2–117 A newly produced pipe with diameter of 2 m and length 15 m is to be tested at 10 MPa using water at 15°C. After sealing both ends, the pipe is first filled with water and then the pressure is increased by pumping additional water into the test pipe until the test pressure is reached. Assuming no deformation in the pipe, determine how much additional water needs to be pumped into the pipe. Take the coefficient of compressibility to be 2.10×10^9 Pa. *Answer:* 224 kg

2–118 Although liquids, in general, are hard to compress, the compressibility effect (variation in the density) may become unavoidable at the great depths in the oceans due to enormous pressure increase. At a certain depth the pressure is reported to be 100 MPa and the average coefficient of compressibility is about 2350 MPa.

(*a*) Taking the liquid density at the free surface to be $\rho_0 = 1030$ kg/m³, obtain an analytical relation between density and pressure, and determine the density at the specified pressure. *Answer:* 1074 kg/m³

(*b*) Use Eq. 2–13 to estimate the density for the specified pressure and compare your result with that of part (*a*).

2–119 Consider laminar flow of a Newtonian fluid of viscosity μ between two parallel plates. The flow is one-dimensional, and the velocity profile is given as $u(y) = 4u_{max}[y/h - (y/h)^2]$, where y is the vertical coordinate from the bottom surface, h is the distance between the two plates, and u_{max} is the maximum flow velocity that occurs at midplane. Develop a relation for the drag force exerted on both plates by the fluid in the flow direction per unit area of the plates.

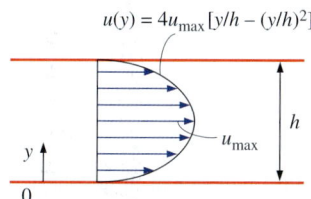

FIGURE P2–119

2–120 Two immiscible Newtonian liquids flow steadily between two large parallel plates under the influence of an applied pressure gradient. The lower plate is fixed while the upper one is pulled with a constant velocity of $U = 10$ m/s. The thickness, h, of each layer of fluid is 0.5 m. The velocity profile for each layer is given by

$V_1 = 6 + ay - 3y^2$, $\quad -0.5 \leq y \leq 0$
$V_2 = b + cy - 9y^2$, $\quad 0 \leq y \leq -0.5$

where a, b, and c are constants.

(*a*) Determine the values of constants a, b, and c.

(*b*) Develop an expression for the viscosity ratio, e.g., $\mu_1/\mu_2 = ?$

(*c*) Determine the forces and their directions exerted by the liquids on both plates if $\mu_1 = 10^{-3}$ Pa·s and each plate has a surface area of 4 m².

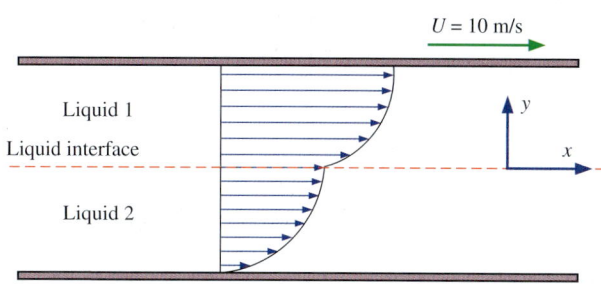

FIGURE P2–120

2–121 A shaft with a diameter of $D = 80$ mm and a length of $L = 400$ mm, shown in Fig. P2–121 is pulled with a constant velocity of $U = 5$ m/s through a bearing with variable diameter. The clearance between shaft and bearing, which varies from $h_1 = 1.2$ mm to $h_2 = 0.4$ mm, is filled with a Newtonian lubricant whose dynamic viscosity is 0.10 Pa·s. Determine the force required to maintain the axial movement of the shaft. *Answer:* 69 N

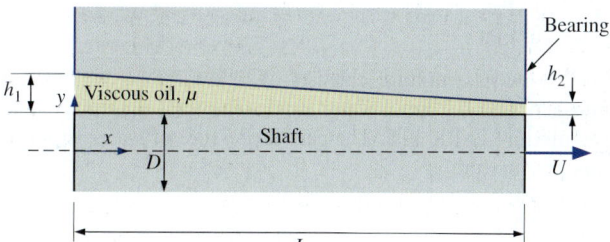

FIGURE P2–121

2–122 Reconsider Prob. 2–121. The shaft now rotates with a constant angular speed of $n = 1450$ rpm in a bearing with variable diameter. The clearance between shaft and bearing, which varies from $h_1 = 1.2$ mm to $h_2 = 0.4$ mm, is filled with a Newtonian lubricant whose dynamic viscosity is 0.1 Pa·s. Determine the torque required to maintain the motion.

2–123 A 10-cm-diameter cylindrical shaft rotates inside a 40-cm-long 10.3-cm diameter bearing. The space between the shaft and the bearing is completely filled with oil whose viscosity at anticipated operating temperature is 0.300 N·s/m². Determine the power required to overcome friction when the shaft rotates at a speed of (*a*) 600 rpm and (*b*) 1200 rpm.

2–124 Some rocks or bricks contain small air pockets in them and have a spongy structure. Assuming the air spaces form columns of an average diameter of 0.006 mm, determine how high water can rise in such a material. Take the surface tension of the air–water interface in that material to be 0.085 N/m.

PROPERTIES OF FLUIDS

Fundamentals of Engineering (FE) Exam Problems

2–125 The specific gravity of a fluid is specified to be 0.82. The specific volume of this fluid is
(a) 0.00100 m³/kg
(b) 0.00122 m³/kg
(c) 0.0082 m³/kg
(d) 82 m³/kg
(e) 820 m³/kg

2–126 The specific gravity of mercury is 13.6. The specific weight of mercury is
(a) 1.36 kN/m³
(b) 9.81 kN/m³
(c) 106 kN/m³
(d) 133 kN/m³
(e) 13,600 kN/m³

2–127 An ideal gas flows in a pipe at 20°C. The density of the gas is 1.9 kg/m³ and its molar mass is 44 kg/kmol. The pressure of the gas is
(a) 7 kPa
(b) 72 kPa
(c) 105 kPa
(d) 460 kPa
(e) 4630 kPa

2–128 A gas mixture consists of 3 kmol oxygen, 2 kmol nitrogen, and 0.5 kmol water vapor. The total pressure of the gas mixture is 100 kPa. The partial pressure of water vapor in this gas mixture is
(a) 5 kPa
(b) 9.1 kPa
(c) 10 kPa
(d) 22.7 kPa
(e) 100 kPa

2–129 Liquid water vaporizes into water vapor as it flows in the piping of a boiler. If the temperature of water in the pipe is 180°C, the vapor pressure of the water in the pipe is
(a) 1002 kPa
(b) 180 kPa
(c) 101.3 kPa
(d) 18 kPa
(e) 100 kPa

2–130 In a water distribution system, the pressure of water can be as low as 1.4 psia. The maximum temperature of water allowed in the piping to avoid cavitation is
(a) 50°F
(b) 77°F
(c) 100°F
(d) 113°F
(e) 140°F

2–131 The thermal energy of a system refers to
(a) Sensible energy
(b) Latent energy
(c) Sensible + latent energies
(d) Enthalpy
(e) Internal energy

2–132 The difference between the energies of a flowing and stationary fluid per unit mass of the fluid is equal to
(a) Enthalpy
(b) Flow energy
(c) Sensible energy
(d) Kinetic energy
(e) Internal energy

2–133 The pressure of water is increased from 100 kPa to 1200 kPa by a pump. The temperature of water also increases by 0.15°C. The density of water is 1 kg/L and its specific heat is c_p = 4.18 kJ/kg·°C. The enthalpy change of the water during this process is
(a) 1100 kJ/kg
(b) 0.63 kJ/kg
(c) 1.1 kJ/kg
(d) 1.73 kJ/kg
(e) 4.2 kJ/kg

2–134 The coefficient of compressibility of a truly incompressible substance is
(a) 0
(b) 0.5
(c) 1
(d) 100
(e) Infinity

2–135 The pressure of water at atmospheric pressure must be raised to 210 atm to compress it by 1 percent. Then, the coefficient of compressibility value of water is
(a) 209 atm
(b) 20,900 atm
(c) 21 atm
(d) 0.21 atm
(e) 210,000 atm

2–136 When a liquid in a piping network encounters an abrupt flow restriction (such as a closing valve), it is locally compressed. The resulting acoustic waves that are produced strike the pipe surfaces, bends, and valves as they propagate and reflect along the pipe, causing the pipe to vibrate and produce a familiar sound. This is known as
(a) Condensation
(b) Cavitation
(c) Water hammer
(d) Compression
(e) Water arrest

2–137 The density of a fluid decreases by 5 percent at constant pressure when its temperature increases by 10°C. The coefficient of volume expansion of this fluid is
(a) 0.01 K^{-1}
(b) 0.005 K^{-1}
(c) 0.1 K^{-1}
(d) 0.5 K^{-1}
(e) 5 K^{-1}

2–138 Water is compressed from 100 kPa to 5000 kPa at constant temperature. The initial density of water is 1000 kg/m³ and the isothermal compressibility of water is $\alpha = 4.8 \times 10^{-5}$ atm^{-1}. The final density of the water is
(a) 1000 kg/m³
(b) 1001.1 kg/m³
(c) 1002.3 kg/m³
(d) 1003.5 kg/m³
(e) 997.4 kg/m³

2–139 The speed of a spacecraft is given to be 1250 km/h in atmospheric air at −40°C. The Mach number of this flow is
(a) 35.9
(b) 0.85
(c) 1.0
(d) 1.13
(e) 2.74

2–140 The dynamic viscosity of air at 20°C and 200 kPa is 1.83×10^{-5} kg/m·s. The kinematic viscosity of air at this state is
(a) 0.525×10^{-5} m²/s
(b) 0.77×10^{-5} m²/s
(c) 1.47×10^{-5} m²/s
(d) 1.83×10^{-5} m²/s
(e) 0.380×10^{-5} m²/s

2–141 A viscometer constructed of two 30-cm-long concentric cylinders is used to measure the viscosity of a fluid. The outer diameter of the inner cylinder is 9 cm, and the gap between the two cylinders is 0.18 cm. The inner cylinder is rotated at 250 rpm, and the torque is measured to be 1.4 N·m. The viscosity of the fluid is
(a) 0.0084 N·s/m²
(b) 0.017 N·s/m²
(c) 0.062 N·s/m²
(d) 0.0049 N·s/m²
(e) 0.56 N·s/m²

2–142 Which one is *not* a surface tension or surface energy (per unit area) unit?
(a) lbf/ft
(b) N·m/m²
(c) lbf/ft²
(d) J/m²
(e) Btu/ft²

2–143 The surface tension of soap water at 20°C is $\sigma_s = 0.025$ N/m. The gage pressure inside a soap bubble of diameter 2 cm at 20°C is
(a) 10 Pa
(b) 5 Pa
(c) 20 Pa
(d) 40 Pa
(e) 0.5 Pa

2–144 A 0.4-mm-diameter glass tube is inserted into water at 20°C in a cup. The surface tension of water at 20°C is $\sigma_s = 0.073$ N/m. The contact angle can be taken as zero degrees. The capillary rise of water in the tube is
(a) 2.9 cm
(b) 7.4 cm
(c) 5.1 cm
(d) 9.3 cm
(e) 14.0 cm

Design and Essay Problems

2–145 Design an experiment to measure the viscosity of liquids using a vertical funnel with a cylindrical reservoir of height h and a narrow flow section of diameter D and length L. Making appropriate assumptions, obtain a relation for viscosity in terms of easily measurable quantities such as density and volume flow rate.

2–146 Write an essay on the rise of the fluid to the top of trees by capillary and other effects.

2–147 Write an essay on the oils used in car engines in different seasons and their viscosities.

2–148 Consider the flow of water through a clear tube. It is sometimes possible to observe cavitation in the throat created by pinching off the tube to a very small diameter as sketched. We assume incompressible flow with negligible gravitational effects and negligible irreversibilities. You will learn later (Chap. 5) that as the duct cross-sectional area decreases, the velocity increases and the pressure decreases according to

$$V_1 A_1 = V_2 A_2 \quad \text{and} \quad P_1 + \rho \frac{V_1^2}{2} = P_2 + \rho \frac{V_2^2}{2}$$

respectively, where V_1 and V_2 are the average velocities through cross-sectional areas A_1 and A_2. Thus, both the maximum velocity and minimum pressure occur at the throat. (a) If the water is at 20°C, the inlet pressure is 20.803 kPa, and the throat diameter is one-twentieth of the inlet diameter, estimate the minimum average inlet velocity at which cavitation is likely to occur in the throat. (b) Repeat at a water temperature of 50°C.

Explain why the required inlet velocity is higher or lower than that of part (*a*).

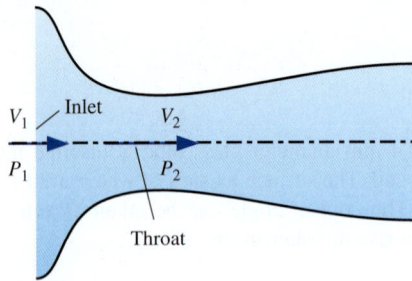

FIGURE P2–148

2–149 Even though steel is about 7 to 8 times denser than water, a steel paper clip or razor blade can be made to float on water! Explain and discuss. Predict what would happen if you mix some soap with the water.

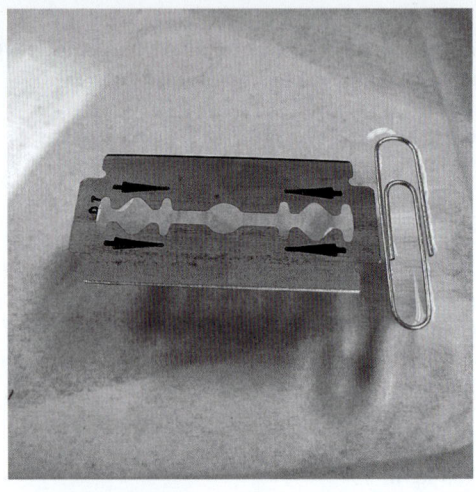

FIGURE P2–149
Photo by John M. Cimbala.

CHAPTER 3

PRESSURE AND FLUID STATICS

This chapter deals with forces applied by fluids at rest or in rigid-body motion. The fluid property responsible for those forces is *pressure*, which is a normal force exerted by a fluid per unit area. We start this chapter with a detailed discussion of pressure, including *absolute* and *gage pressures*, the pressure at a *point*, the *variation of pressure with depth* in a gravitational field, the *barometer*, the *manometer*, and other pressure measurement devices. This is followed by a discussion of the *hydrostatic forces* applied on submerged bodies with plane or curved surfaces. We then consider the *buoyant force* applied by fluids on submerged or floating bodies, and discuss the *stability* of such bodies. Finally, we apply Newton's second law of motion to a body of fluid in motion that acts as a rigid body and analyze the variation of pressure in fluids that undergo linear acceleration and in rotating containers. This chapter makes extensive use of force balances for bodies in static equilibrium, and it would be helpful if the relevant topics from statics are first reviewed.

OBJECTIVES

When you finish reading this chapter, you should be able to:

- Determine the variation of pressure in a fluid at rest
- Calculate pressure using various kinds of manometers
- Calculate the forces and moments exerted by a fluid at rest on plane or curved submerged surfaces
- Analyze the stability of floating and submerged bodies
- Analyze the rigid-body motion of fluids in containers during linear acceleration or rotation

John Ninomiya flying a cluster of 72 helium-filled balloons over Temecula, California in April of 2003. The helium balloons displace approximately 230 m³ of air, providing the necessary buoyant force. Don't try this at home!

Photograph by Susan Dawson. Used by permission.

3–1 · PRESSURE

Pressure is defined as *a normal force exerted by a fluid per unit area.* We speak of pressure only when we deal with a gas or a liquid. The counterpart of pressure in solids is *normal stress.* Since pressure is defined as force per unit area, it has the unit of newtons per square meter (N/m²), which is called a **pascal** (Pa). That is,

$$1\,\text{Pa} = 1\,\text{N/m}^2$$

The pressure unit pascal is too small for most pressures encountered in practice. Therefore, its multiples *kilopascal* (1 kPa = 10^3 Pa) and *megapascal* (1 MPa = 10^6 Pa) are commonly used. Three other pressure units commonly used in practice, especially in Europe, are *bar, standard atmosphere,* and *kilogram-force per square centimeter:*

$$1\,\text{bar} = 10^5\,\text{Pa} = 0.1\,\text{MPa} = 100\,\text{kPa}$$

$$1\,\text{atm} = 101{,}325\,\text{Pa} = 101.325\,\text{kPa} = 1.01325\,\text{bars}$$

$$1\,\text{kgf/cm}^2 = 9.807\,\text{N/cm}^2 = 9.807 \times 10^4\,\text{N/m}^2 = 9.807 \times 10^4\,\text{Pa}$$
$$= 0.9807\,\text{bar}$$
$$= 0.9679\,\text{atm}$$

Note the pressure units bar, atm, and kgf/cm² are almost equivalent to each other. In the English system, the pressure unit is *pound-force per square inch* (lbf/in², or psi), and 1 atm = 14.696 psi. The pressure units kgf/cm² and lbf/in² are also denoted by kg/cm² and lb/in², respectively, and they are commonly used in tire gages. It can be shown that 1 kgf/cm² = 14.223 psi.

Pressure is also used on solid surfaces as synonymous to *normal stress,* which is the force acting perpendicular to the surface per unit area. For example, a 70-kg person with a total foot imprint area of 343 cm² exerts a pressure of (70 × 9.81/1000)kN/0.0343 m² = 20 kPa on the floor (Fig. 3–1). If the person stands on one foot, the pressure doubles. If the person gains excessive weight, he or she is likely to encounter foot discomfort because of the increased pressure on the foot (the size of the bottom of the foot does not change with weight gain). This also explains how a person can walk on fresh snow without sinking by wearing large snowshoes, and how a person cuts with little effort when using a sharp knife.

The actual pressure at a given position is called the **absolute pressure,** and it is measured relative to absolute vacuum (i.e., absolute zero pressure). Most pressure-measuring devices, however, are calibrated to read zero in the atmosphere (Fig. 3–2), and so they indicate the difference between the absolute pressure and the local atmospheric pressure. This difference is called the **gage pressure.** P_gage can be positive or negative, but pressures below atmospheric pressure are sometimes called **vacuum pressures** and are measured by vacuum gages that indicate the difference between the atmospheric pressure and the absolute pressure. Absolute, gage, and vacuum pressures are related to each other by

$$P_\text{gage} = P_\text{abs} - P_\text{atm} \tag{3-1}$$

$$P_\text{vac} = P_\text{atm} - P_\text{abs} \tag{3-2}$$

This is illustrated in Fig. 3–3.

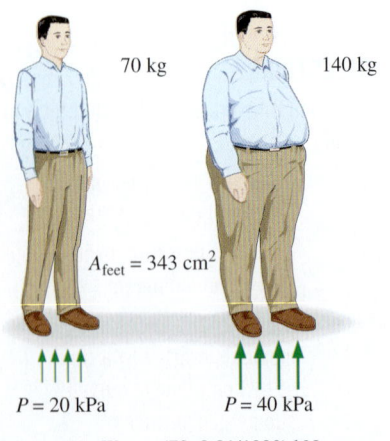

FIGURE 3–1
The normal stress (or "pressure") on the feet of a chubby person is much greater than on the feet of a slim person.

FIGURE 3–2
Some basic pressure gages.
Dresser Instruments, Dresser, Inc. Used by permission.

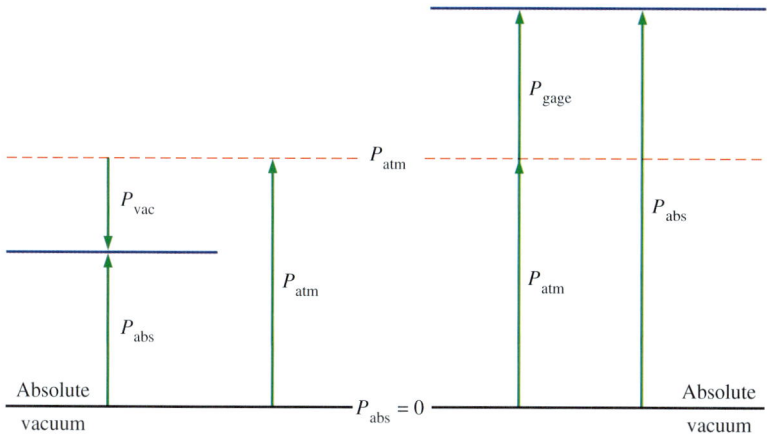

FIGURE 3–3
Absolute, gage, and vacuum pressures.

Like other pressure gages, the gage used to measure the air pressure in an automobile tire reads the gage pressure. Therefore, the common reading of 32.0 psi (2.25 kgf/cm²) indicates a pressure of 32.0 psi above the atmospheric pressure. At a location where the atmospheric pressure is 14.3 psi, for example, the absolute pressure in the tire is 32.0 + 14.3 = 46.3 psi.

In thermodynamic relations and tables, absolute pressure is almost always used. Throughout this text, the pressure P will denote *absolute pressure* unless specified otherwise. Often the letters "a" (for absolute pressure) and "g" (for gage pressure) are added to pressure units (such as psia and psig) to clarify what is meant.

EXAMPLE 3–1 Absolute Pressure of a Vacuum Chamber

A vacuum gage connected to a chamber reads 40 kPa at a location where the atmospheric pressure is 100 kPa. Determine the absolute pressure in the chamber.

SOLUTION The gage pressure of a vacuum chamber is given. The absolute pressure in the chamber is to be determined.
Analysis The absolute pressure is easily determined from Eq. 3–2 to be

$$P_{abs} = P_{atm} - P_{vac} = 100 - 40 = \mathbf{60\ kPa}$$

Discussion Note that the *local* value of the atmospheric pressure is used when determining the absolute pressure.

Pressure at a Point

Pressure is the *compressive force* per unit area, and it gives the impression of being a vector. However, pressure at any point in a fluid is the same in all directions (Fig. 3–4). That is, it has magnitude but not a specific direction, and thus it is a scalar quantity. This can be demonstrated by considering a small wedge-shaped fluid element of unit length ($\Delta y = 1$ into the page) in equilibrium, as shown in Fig. 3–5. The mean pressures at the three surfaces are P_1, P_2, and P_3, and the force acting on a surface is the product of mean

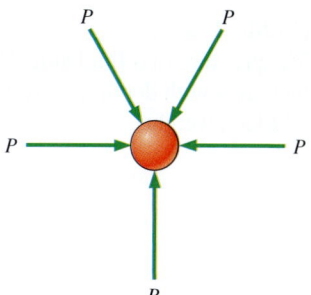

FIGURE 3–4
Pressure is a *scalar* quantity, not a vector; the pressure at a point in a fluid is the same in all directions.

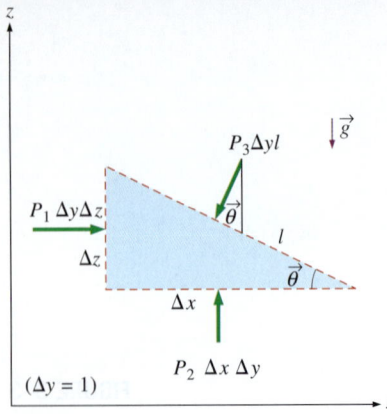

FIGURE 3–5
Forces acting on a wedge-shaped fluid element in equilibrium.

pressure and the surface area. From Newton's second law, a force balance in the x- and z-directions gives

$$\sum F_x = ma_x = 0: \quad P_1 \Delta y \Delta z - P_3 \Delta y l \sin\theta = 0 \quad (3\text{-}3a)$$

$$\sum F_z = ma_z = 0: \quad P_2 \Delta y \Delta x - P_3 \Delta y l \cos\theta - \frac{1}{2}\rho g \Delta x \Delta y \Delta z = 0 \quad (3\text{-}3b)$$

where ρ is the density and $W = mg = \rho g \Delta x \Delta y \Delta z/2$ is the weight of the fluid element. Noting that the wedge is a right triangle, we have $\Delta x = l\cos\theta$ and $\Delta z = l\sin\theta$. Substituting these geometric relations and dividing Eq. 3–3a by $\Delta y\, \Delta z$ and Eq. 3–3b by $\Delta x\, \Delta y$ gives

$$P_1 - P_3 = 0 \quad (3\text{-}4a)$$

$$P_2 - P_3 - \frac{1}{2}\rho g \Delta z = 0 \quad (3\text{-}4b)$$

The last term in Eq. 3–4b drops out as $\Delta z \to 0$ and the wedge becomes infinitesimal, and thus the fluid element shrinks to a point. Then combining the results of these two relations gives

$$P_1 = P_2 = P_3 = P \quad (3\text{-}5)$$

regardless of the angle θ. We can repeat the analysis for an element in the yz-plane and obtain a similar result. Thus we conclude that *the pressure at a point in a fluid has the same magnitude in all directions*. This result is applicable to fluids in motion as well as fluids at rest since pressure is a *scalar*, not a vector.

Variation of Pressure with Depth

It will come as no surprise to you that pressure in a fluid at rest does not change in the horizontal direction. This can be shown easily by considering a thin horizontal layer of fluid and doing a force balance in any horizontal direction. However, this is not the case in the vertical direction in a gravity field. Pressure in a fluid increases with depth because more fluid rests on deeper layers, and the effect of this "extra weight" on a deeper layer is balanced by an increase in pressure (Fig. 3–6).

To obtain a relation for the variation of pressure with depth, consider a rectangular fluid element of height Δz, length Δx, and unit depth ($\Delta y = 1$ into the page) in equilibrium, as shown in Fig. 3–7. Assuming the density of the fluid ρ to be constant, a force balance in the vertical z-direction gives

$$\sum F_z = ma_z = 0: \quad P_1 \Delta x \Delta y - P_2 \Delta x \Delta y - \rho g \Delta x \Delta y \Delta z = 0$$

where $W = mg = \rho g \Delta x \Delta y \Delta z$ is the weight of the fluid element and $\Delta z = z_2 - z_1$. Dividing by $\Delta x\, \Delta y$ and rearranging gives

$$\Delta P = P_2 - P_1 = -\rho g \Delta z = -\gamma_s \Delta z \quad (3\text{-}6)$$

where $\gamma_s = \rho g$ is the *specific weight* of the fluid. Thus, we conclude that the pressure difference between two points in a constant density fluid is proportional to the vertical distance Δz between the points and the density ρ of the fluid. Noting the negative sign, *pressure in a static fluid increases linearly with depth*. This is what a diver experiences when diving deeper in a lake.

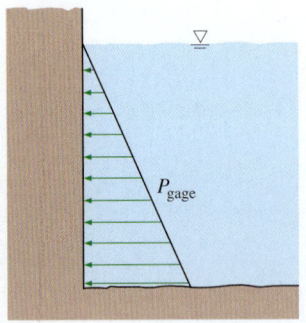

FIGURE 3–6
The pressure of a fluid at rest increases with depth (as a result of added weight).

An easier equation to remember and apply between any two points in the same fluid under hydrostatic conditions is

$$P_{below} = P_{above} + \rho g |\Delta z| = P_{above} + \gamma_s |\Delta z| \quad (3\text{–}7)$$

where "below" refers to the point at lower elevation (deeper in the fluid) and "above" refers to the point at higher elevation. If you use this equation consistently, you should avoid sign errors.

For a given fluid, the vertical distance Δz is sometimes used as a measure of pressure, and it is called the *pressure head*.

We also conclude from Eq. 3–6 that for small to moderate distances, the variation of pressure with height is negligible for gases because of their low density. The pressure in a tank containing a gas, for example, can be considered to be uniform since the weight of the gas is too small to make a significant difference. Also, the pressure in a room filled with air can be approximated as a constant (Fig. 3–8).

If we take the "above" point to be at the free surface of a liquid open to the atmosphere (Fig. 3–9), where the pressure is the atmospheric pressure P_{atm}, then from Eq. 3–7 the pressure at a depth h below the free surface becomes

$$P = P_{atm} + \rho g h \quad \text{or} \quad P_{gage} = \rho g h \quad (3\text{–}8)$$

Liquids are essentially incompressible substances, and thus the variation of density with depth is negligible. This is also the case for gases when the elevation change is not very large. The variation of density of liquids or gases with temperature can be significant, however, and may need to be considered when high accuracy is desired. Also, at great depths such as those encountered in oceans, the change in the density of a liquid can be significant because of the compression by the tremendous amount of liquid weight above.

The gravitational acceleration g varies from 9.807 m/s² at sea level to 9.764 m/s² at an elevation of 14,000 m where large passenger planes cruise. This is a change of just 0.4 percent in this extreme case. Therefore, g can be approximated as a constant with negligible error.

For fluids whose density changes significantly with elevation, a relation for the variation of pressure with elevation can be obtained by dividing Eq. 3–6 by Δz, and taking the limit as $\Delta z \to 0$. This yields

$$\frac{dP}{dz} = -\rho g \quad (3\text{–}9)$$

Note that dP is negative when dz is positive since pressure decreases in an upward direction. When the variation of density with elevation is known, the pressure difference between any two points 1 and 2 can be determined by integration to be

$$\Delta P = P_2 - P_1 = -\int_1^2 \rho g \, dz \quad (3\text{–}10)$$

For constant density and constant gravitational acceleration, this relation reduces to Eq. 3–6, as expected.

Pressure in a fluid at rest is independent of the shape or cross section of the container. It changes with the vertical distance, but remains constant

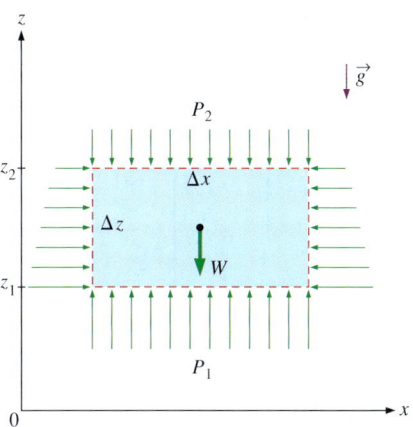

FIGURE 3–7
Free-body diagram of a rectangular fluid element in equilibrium.

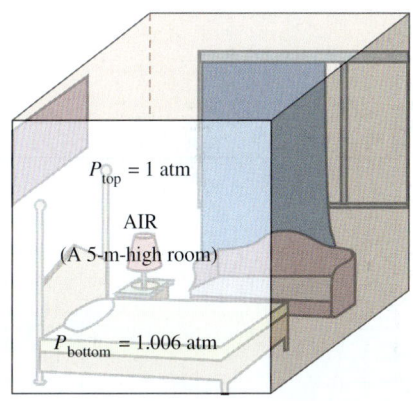

FIGURE 3–8
In a room filled with a gas, the variation of pressure with height is negligible.

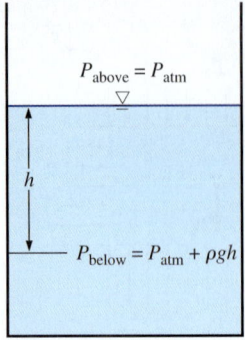

FIGURE 3–9
Pressure in a liquid at rest increases linearly with distance from the free surface.

in other directions. Therefore, the pressure is the same at all points on a horizontal plane in a given fluid. The Dutch mathematician Simon Stevin (1548–1620) published in 1586 the principle illustrated in Fig. 3–10. Note that the pressures at points A, B, C, D, E, F, and G are the same since they are at the same depth, and they are interconnected by the same static fluid. However, the pressures at points H and I are not the same since these two points cannot be interconnected by the same fluid (i.e., we cannot draw a curve from point I to point H while remaining in the same fluid at all times), although they are at the same depth. (Can you tell at which point the pressure is higher?) Also notice that the pressure force exerted by the fluid is always normal to the surface at the specified points.

A consequence of the pressure in a fluid remaining constant in the horizontal direction is that *the pressure applied to a confined fluid increases the pressure throughout by the same amount*. This is called **Pascal's law**, after Blaise Pascal (1623–1662). Pascal also knew that the force applied by a fluid is proportional to the surface area. He realized that two hydraulic cylinders of different areas could be connected, and the larger could be used to exert a proportionally greater force than that applied to the smaller. "Pascal's machine" has been the source of many inventions that are a part of our daily lives such as hydraulic brakes and lifts. This is what enables us to lift a car easily by one arm, as shown in Fig. 3–11. Noting that $P_1 = P_2$ since both pistons are at the same level (the effect of small height differences is negligible, especially at high pressures), the ratio of output force to input force is determined to be

$$P_1 = P_2 \quad \rightarrow \quad \frac{F_1}{A_1} = \frac{F_2}{A_2} \quad \rightarrow \quad \frac{F_2}{F_1} = \frac{A_2}{A_1} \qquad (3\text{–}11)$$

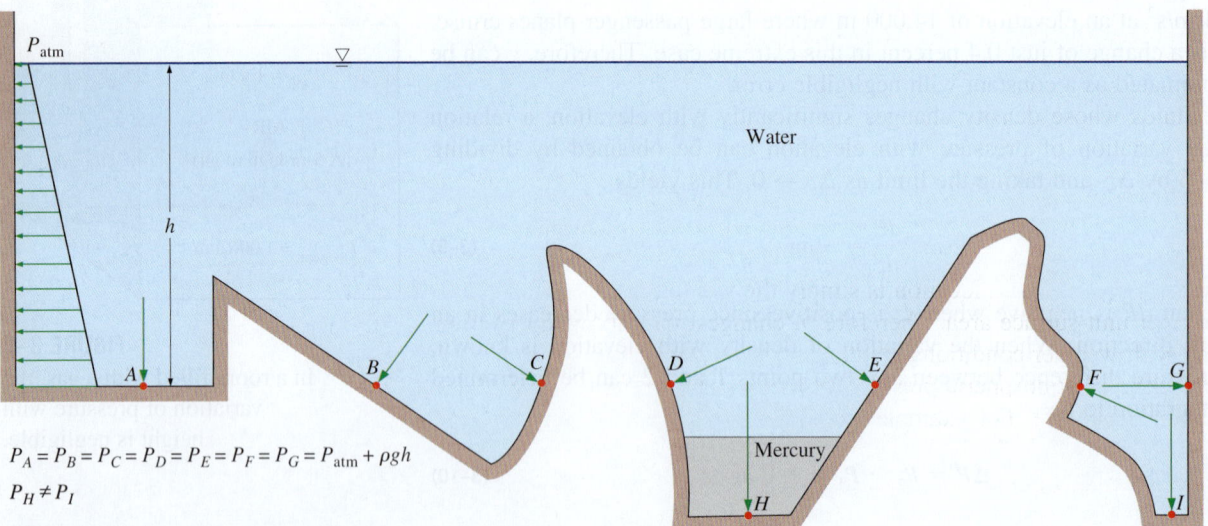

FIGURE 3–10
Under hydrostatic conditions, the pressure is the same at all points on a horizontal plane in a given fluid regardless of geometry, provided that the points are interconnected by the same fluid.

The area ratio A_2/A_1 is called the *ideal mechanical advantage* of the hydraulic lift. Using a hydraulic car jack with a piston area ratio of $A_2/A_1 = 100$, for example, a person can lift a 1000-kg car by applying a force of just 10 kgf (= 90.8 N).

3–2 ▪ PRESSURE MEASUREMENT DEVICES

The Barometer

Atmospheric pressure is measured by a device called a **barometer;** thus, the atmospheric pressure is often referred to as the *barometric pressure*.

The Italian Evangelista Torricelli (1608–1647) was the first to conclusively prove that the atmospheric pressure can be measured by inverting a mercury-filled tube into a mercury container that is open to the atmosphere, as shown in Fig. 3–12. The pressure at point B is equal to the atmospheric pressure, and the pressure at point C can be taken to be zero since there is only mercury vapor above point C and the pressure is very low relative to P_{atm} and can be neglected to an excellent approximation. Writing a force balance in the vertical direction gives

$$P_{atm} = \rho g h \quad (3\text{--}12)$$

where ρ is the density of mercury, g is the local gravitational acceleration, and h is the height of the mercury column above the free surface. Note that the length and the cross-sectional area of the tube have no effect on the height of the fluid column of a barometer (Fig. 3–13).

A frequently used pressure unit is the *standard atmosphere*, which is defined as the pressure produced by a column of mercury 760 mm in height at 0°C (ρ_{Hg} = 13,595 kg/m³) under standard gravitational acceleration (g = 9.807 m/s²). If water instead of mercury were used to measure the standard atmospheric pressure, a water column of about 10.3 m would be needed. Pressure is sometimes expressed (especially by weather forecasters) in terms of the height of the mercury column. The standard atmospheric pressure, for example, is 760 mmHg at 0°C. The unit mmHg is also called the **torr** in honor of Torricelli. Therefore, 1 atm = 760 torr and 1 torr = 133.3 Pa.

Atmospheric pressure P_{atm} changes from 101.325 kPa at sea level to 89.88, 79.50, 54.05, 26.5, and 5.53 kPa at altitudes of 1000, 2000, 5000, 10,000, and 20,000 meters, respectively. The typical atmospheric pressure in Denver (elevation = 1610 m), for example, is 83.4 kPa. Remember that the atmospheric pressure at a location is simply the weight of the air above that location per unit surface area. Therefore, it changes not only with elevation but also with weather conditions.

The decline of atmospheric pressure with elevation has far-reaching ramifications in daily life. For example, cooking takes longer at high altitudes since water boils at a lower temperature at lower atmospheric pressures. Nose bleeding is a common experience at high altitudes since the difference between the blood pressure and the atmospheric pressure is larger in this case, and the delicate walls of veins in the nose are often unable to withstand this extra stress.

For a given temperature, the density of air is lower at high altitudes, and thus a given volume contains less air and less oxygen. So it is no surprise

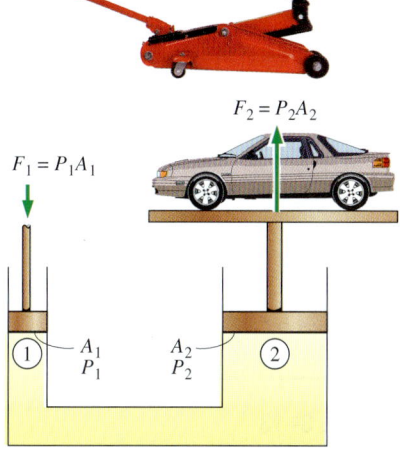

FIGURE 3–11
Lifting of a large weight by a small force by the application of Pascal's law. A common example is a hydraulic jack.
(Top) © Stockbyte/Getty RF

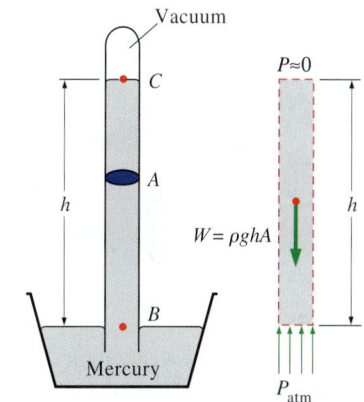

FIGURE 3–12
The basic barometer.

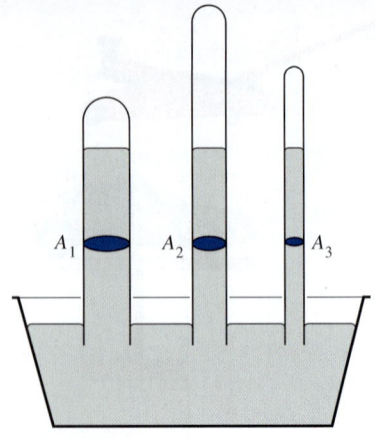

FIGURE 3–13
The length or the cross-sectional area of the tube has no effect on the height of the fluid column of a barometer, provided that the tube diameter is large enough to avoid surface tension (capillary) effects.

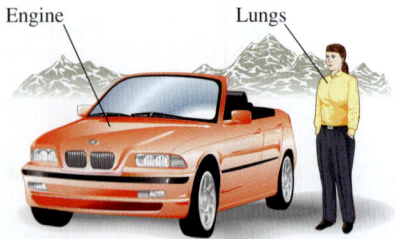

FIGURE 3–14
At high altitudes, a car engine generates less power and a person gets less oxygen because of the lower density of air.

that we tire more easily and experience breathing problems at high altitudes. To compensate for this effect, people living at higher altitudes develop more efficient lungs. Similarly, a 2.0-L car engine will act like a 1.7-L car engine at 1500 m altitude (unless it is turbocharged) because of the 15 percent drop in pressure and thus 15 percent drop in the density of air (Fig. 3–14). A fan or compressor will displace 15 percent less air at that altitude for the same volume displacement rate. Therefore, larger cooling fans may need to be selected for operation at high altitudes to ensure the specified mass flow rate. The lower pressure and thus lower density also affects lift and drag: airplanes need a longer runway at high altitudes to develop the required lift, and they climb to very high altitudes for cruising in order to reduce drag and thus achieve better fuel efficiency.

EXAMPLE 3–2   Measuring Atmospheric Pressure with a Barometer

Determine the atmospheric pressure at a location where the barometric reading is 740 mm Hg and the gravitational acceleration is $g = 9.805$ m/s^2. Assume the temperature of mercury to be 10°C, at which its density is 13,570 kg/m^3.

SOLUTION   The barometric reading at a location in height of mercury column is given. The atmospheric pressure is to be determined.
Assumptions   The temperature of mercury is assumed to be 10°C.
Properties   The density of mercury is given to be 13,570 kg/m^3.
Analysis   From Eq. 3–12, the atmospheric pressure is determined to be

$$P_{atm} = \rho g h$$
$$= (13{,}570 \text{ kg/m}^3)(9.805 \text{ m/s}^2)(0.740 \text{ m})\left(\frac{1 \text{ N}}{1 \text{ kg} \cdot \text{m/s}^2}\right)\left(\frac{1 \text{ kPa}}{1000 \text{ N/m}^2}\right)$$
$$= \mathbf{98.5 \text{ kPa}}$$

Discussion   Note that density changes with temperature, and thus this effect should be considered in calculations.

EXAMPLE 3–3   Gravity Driven Flow from an IV Bottle

Intravenous infusions usually are driven by gravity by hanging the fluid bottle at sufficient height to counteract the blood pressure in the vein and to force the fluid into the body (Fig. 3–15). The higher the bottle is raised, the higher the flow rate of the fluid will be. (*a*) If it is observed that the fluid and the blood pressures balance each other when the bottle is 1.2 m above the arm level, determine the gage pressure of the blood. (*b*) If the gage pressure of the fluid at the arm level needs to be 20 kPa for sufficient flow rate, determine how high the bottle must be placed. Take the density of the fluid to be 1020 kg/m^3.

SOLUTION   It is given that an IV fluid and the blood pressures balance each other when the bottle is at a certain height. The gage pressure of the blood and elevation of the bottle required to maintain flow at the desired rate are to be determined.

Assumptions 1 The IV fluid is incompressible. 2 The IV bottle is open to the atmosphere.
Properties The density of the IV fluid is given to be $\rho = 1020$ kg/m³.
Analysis (*a*) Noting that the IV fluid and the blood pressures balance each other when the bottle is 1.2 m above the arm level, the gage pressure of the blood in the arm is simply equal to the gage pressure of the IV fluid at a depth of 1.2 m,

$$P_{gage, arm} = P_{abs} - P_{atm} = \rho g h_{arm-bottle}$$
$$= (1020 \text{ kg/m}^3)(9.81 \text{ m/s}^2)(1.20 \text{ m})\left(\frac{1 \text{ kN}}{1000 \text{ kg} \cdot \text{m/s}^2}\right)\left(\frac{1 \text{ kPa}}{1 \text{ kN/m}^2}\right)$$
$$= \mathbf{12.0 \text{ kPa}}$$

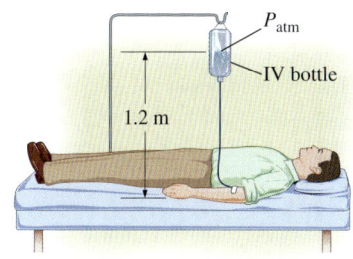

FIGURE 3–15
Schematic for Example 3–3.

(*b*) To provide a gage pressure of 20 kPa at the arm level, the height of the surface of the IV fluid in the bottle from the arm level is again determined from $P_{gage, arm} = \rho g h_{arm-bottle}$ to be

$$h_{arm-botttle} = \frac{P_{gage, arm}}{\rho g}$$
$$= \frac{20 \text{ kPa}}{(1020 \text{ kg/m}^3)(9.81 \text{ m/s}^2)}\left(\frac{1000 \text{ kg} \cdot \text{m/s}^2}{1 \text{ kN}}\right)\left(\frac{1 \text{ kN/m}^2}{1 \text{ kPa}}\right)$$
$$= \mathbf{2.00 \text{ m}}$$

Discussion Note that the height of the reservoir can be used to control flow rates in gravity-driven flows. When there is flow, the pressure drop in the tube due to frictional effects also should be considered. For a specified flow rate, this requires raising the bottle a little higher to overcome the pressure drop.

■ **EXAMPLE 3–4**   **Hydrostatic Pressure in a Solar Pond with Variable Density**

Solar ponds are small artificial lakes of a few meters deep that are used to store solar energy. The rise of heated (and thus less dense) water to the surface is prevented by adding salt at the pond bottom. In a typical salt gradient solar pond, the density of water increases in the gradient zone, as shown in Fig. 3–16, and the density can be expressed as

$$\rho = \rho_0 \sqrt{1 + \tan^2\left(\frac{\pi}{4} \frac{s}{H}\right)}$$

where ρ_0 is the density on the water surface, s is the vertical distance measured downward from the top of the gradient zone ($s = -z$), and H is the thickness of the gradient zone. For $H = 4$ m, $\rho_0 = 1040$ kg/m³, and a thickness of 0.8 m for the surface zone, calculate the gage pressure at the bottom of the gradient zone.

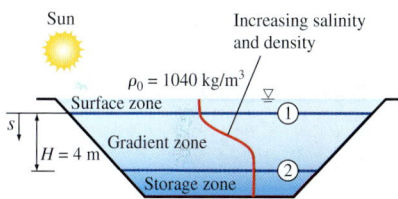

FIGURE 3–16
Schematic for Example 3–4.

SOLUTION The variation of density of saline water in the gradient zone of a solar pond with depth is given. The gage pressure at the bottom of the gradient zone is to be determined.
Assumptions The density in the surface zone of the pond is constant.
Properties The density of brine on the surface is given to be 1040 kg/m³.
Analysis We label the top and the bottom of the gradient zone as 1 and 2, respectively. Noting that the density of the surface zone is constant, the

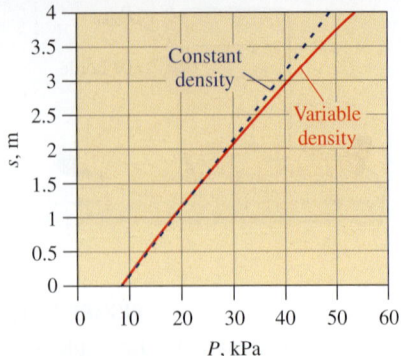

FIGURE 3–17
The variation of gage pressure with depth in the gradient zone of the solar pond.

FIGURE 3–18
A simple U-tube manometer, with high pressure applied to the right side.
Photo by John M. Cimbala.

gage pressure at the bottom of the surface zone (which is the top of the gradient zone) is

$$P_1 = \rho g h_1 = (1040 \text{ kg/m}^3)(9.81 \text{ m/s}^2)(0.8 \text{ m})\left(\frac{1 \text{ kN}}{1000 \text{ kg}\cdot\text{m/s}^2}\right) = 8.16 \text{ kPa}$$

since $1 \text{ kN/m}^2 = 1 \text{ kPa}$. Since $s = -z$, the differential change in hydrostatic pressure across a vertical distance of ds is given by

$$dP = \rho g \, ds$$

Integrating from the top of the gradient zone (point 1 where $s = 0$) to any location s in the gradient zone (no subscript) gives

$$P - P_1 = \int_0^s \rho g \, ds \quad \rightarrow \quad P = P_1 + \int_0^s \rho_0 \sqrt{1 + \tan^2\left(\frac{\pi}{4}\frac{s}{H}\right)} g \, ds$$

Performing the integration gives the variation of gage pressure in the gradient zone to be

$$P = P_1 + \rho_0 g \frac{4H}{\pi} \sinh^{-1}\left(\tan\frac{\pi}{4}\frac{s}{H}\right)$$

Then the pressure at the bottom of the gradient zone ($s = H = 4$ m) becomes

$$P_2 = 8.16 \text{ kPa} + (1040 \text{ kg/m}^3)(9.81 \text{ m/s}^2)\frac{4(4 \text{ m})}{\pi} \sinh^{-1}\left(\tan\frac{\pi}{4}\frac{4}{4}\right)\left(\frac{1 \text{ kN}}{1000 \text{ kg}\cdot\text{m/s}^2}\right)$$

$$= 54.0 \text{ kPa (gage)}$$

Discussion The variation of gage pressure in the gradient zone with depth is plotted in Fig. 3–17. The dashed line indicates the hydrostatic pressure for the case of constant density at 1040 kg/m³ and is given for reference. Note that the variation of pressure with depth is not linear when density varies with depth. That is why integration was required.

The Manometer

We notice from Eq. 3–6 that an elevation change of $-\Delta z$ in a fluid at rest corresponds to $\Delta P/\rho g$, which suggests that a fluid column can be used to measure pressure differences. A device based on this principle is called a **manometer**, and it is commonly used to measure small and moderate pressure differences. A manometer consists of a glass or plastic U-tube containing one or more fluids such as mercury, water, alcohol, or oil (Fig. 3–18). To keep the size of the manometer to a manageable level, heavy fluids such as mercury are used if large pressure differences are anticipated.

Consider the manometer shown in Fig. 3–19 that is used to measure the pressure in the tank. Since the gravitational effects of gases are negligible, the pressure anywhere in the tank and at position 1 has the same value. Furthermore, since pressure in a fluid does not vary in the horizontal direction within a fluid, the pressure at point 2 is the same as the pressure at point 1, $P_2 = P_1$.

The differential fluid column of height h is in static equilibrium, and it is open to the atmosphere. Then the pressure at point 2 is determined directly from Eq. 3–7 to be

$$P_2 = P_{\text{atm}} + \rho g h \quad \text{(3–13)}$$

where ρ is the density of the manometer fluid in the tube. Note that the cross-sectional area of the tube has no effect on the differential height h, and thus the pressure exerted by the fluid. However, the diameter of the tube should be large enough (more than several millimeters) to ensure that the surface tension effect and thus the capillary rise is negligible.

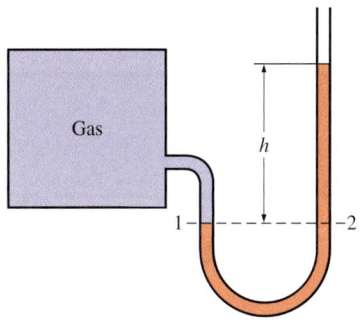

FIGURE 3–19
The basic manometer.

EXAMPLE 3–5 Measuring Pressure with a Manometer

A manometer is used to measure the pressure of a gas in a tank. The fluid used has a specific gravity of 0.85, and the manometer column height is 55 cm, as shown in Fig. 3–20. If the local atmospheric pressure is 96 kPa, determine the absolute pressure within the tank.

SOLUTION The reading of a manometer attached to a tank and the atmospheric pressure are given. The absolute pressure in the tank is to be determined.
Assumptions The density of the gas in the tank is much lower than the density of the manometer fluid.
Properties The specific gravity of the manometer fluid is given to be 0.85. We take the standard density of water to be 1000 kg/m³.
Analysis The density of the fluid is obtained by multiplying its specific gravity by the density of water,

$$\rho = SG\,(\rho_{H_2O}) = (0.85)(1000\text{ kg/m}^3) = 850\text{ kg/m}^3$$

Then from Eq. 3–13,

$$P = P_{atm} + \rho g h$$

$$= 96\text{ kPa} + (850\text{ kg/m}^3)(9.81\text{ m/s}^2)(0.55\text{ m})\left(\frac{1\text{ N}}{1\text{ kg·m/s}^2}\right)\left(\frac{1\text{ kPa}}{1000\text{ N/m}^2}\right)$$

$$= \mathbf{100.6\text{ kPa}}$$

Discussion Note that the gage pressure in the tank is 4.6 kPa.

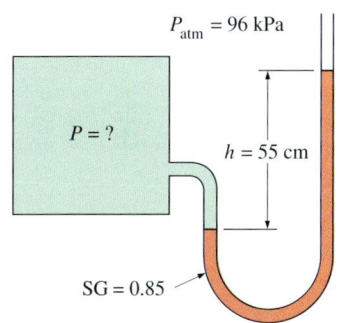

FIGURE 3–20
Schematic for Example 3–5.

Some manometers use a slanted or inclined tube in order to increase the resolution (precision) when reading the fluid height. Such devices are called **inclined manometers.**

Many engineering problems and some manometers involve multiple immiscible fluids of different densities stacked on top of each other. Such systems can be analyzed easily by remembering that (1) the pressure change across a fluid column of height h is $\Delta P = \rho g h$, (2) pressure increases downward in a given fluid and decreases upward (i.e., $P_{bottom} > P_{top}$), and (3) two points at the same elevation in a continuous fluid at rest are at the same pressure.

The last principle, which is a result of *Pascal's law*, allows us to "jump" from one fluid column to the next in manometers without worrying about pressure change as long as we stay in the same continuous fluid and the fluid is at rest. Then the pressure at any point can be determined by starting with a point of known pressure and adding or subtracting $\rho g h$ terms as we advance toward the point of interest. For example, the pressure at the bottom of the tank in Fig. 3–21 can be determined by starting at the free

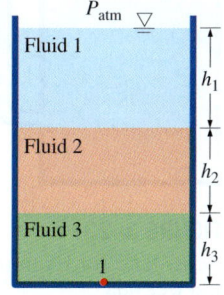

FIGURE 3–21
In stacked-up fluid layers at rest, the pressure change across each fluid layer of density ρ and height h is $\rho g h$.

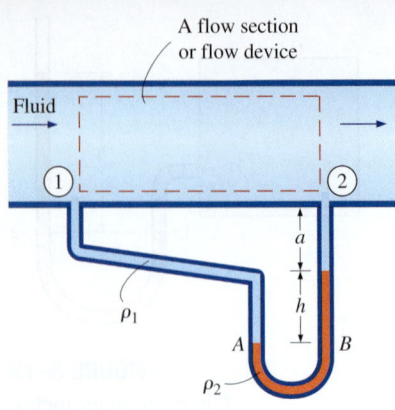

FIGURE 3–22
Measuring the pressure drop across a flow section or a flow device by a differential manometer.

surface where the pressure is P_{atm}, moving downward until we reach point 1 at the bottom, and setting the result equal to P_1. It gives

$$P_{atm} + \rho_1 g h_1 + \rho_2 g h_2 + \rho_3 g h_3 = P_1$$

In the special case of all fluids having the same density, this relation reduces to $P_{atm} + \rho g(h_1 + h_2 + h_3) = P_1$.

Manometers are particularly well-suited to measure pressure drops across a horizontal flow section between two specified points due to the presence of a device such as a valve or heat exchanger or any resistance to flow. This is done by connecting the two legs of the manometer to these two points, as shown in Fig. 3–22. The working fluid can be either a gas or a liquid whose density is ρ_1. The density of the manometer fluid is ρ_2, and the differential fluid height is h. The two fluids must be immiscible, and ρ_2 must be greater than ρ_1.

A relation for the pressure difference $P_1 - P_2$ can be obtained by starting at point 1 with P_1, moving along the tube by adding or subtracting the $\rho g h$ terms until we reach point 2, and setting the result equal to P_2:

$$P_1 + \rho_1 g(a + h) - \rho_2 g h - \rho_1 g a = P_2 \qquad (3\text{–}14)$$

Note that we jumped from point A horizontally to point B and ignored the part underneath since the pressure at both points is the same. Simplifying,

$$P_1 - P_2 = (\rho_2 - \rho_1)gh \qquad (3\text{–}15)$$

Note that the distance a must be included in the analysis even though it has no effect on the result. Also, when the fluid flowing in the pipe is a gas, then $\rho_1 \ll \rho_2$ and the relation in Eq. 3–15 simplifies to $P_1 - P_2 \cong \rho_2 g h$.

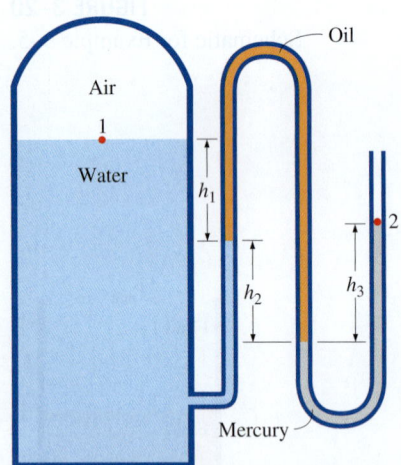

FIGURE 3–23
Schematic for Example 3–3; drawing not to scale.

EXAMPLE 3–6   Measuring Pressure with a Multifluid Manometer

The water in a tank is pressurized by air, and the pressure is measured by a multifluid manometer as shown in Fig. 3–23. The tank is located on a mountain at an altitude of 1400 m where the atmospheric pressure is 85.6 kPa. Determine the air pressure in the tank if $h_1 = 0.1$ m, $h_2 = 0.2$ m, and $h_3 = 0.35$ m. Take the densities of water, oil, and mercury to be 1000 kg/m³, 850 kg/m³, and 13,600 kg/m³, respectively.

SOLUTION The pressure in a pressurized water tank is measured by a multifluid manometer. The air pressure in the tank is to be determined.
Assumption The air pressure in the tank is uniform (i.e., its variation with elevation is negligible due to its low density), and thus we can determine the pressure at the air–water interface.
Properties The densities of water, oil, and mercury are given to be 1000 kg/m³, 850 kg/m³, and 13,600 kg/m³, respectively.
Analysis Starting with the pressure at point 1 at the air–water interface, moving along the tube by adding or subtracting the $\rho g h$ terms until we reach point 2, and setting the result equal to P_{atm} since the tube is open to the atmosphere gives

$$P_1 + \rho_{water} g h_1 + \rho_{oil} g h_2 - \rho_{mercury} g h_3 = P_2 = P_{atm}$$

Solving for P_1 and substituting,

$$P_1 = P_{atm} - \rho_{water}gh_1 - \rho_{oil}gh_2 + \rho_{mercury}gh_3$$
$$= P_{atm} + g(\rho_{mercury}h_3 - \rho_{water}h_1 - \rho_{oil}h_2)$$
$$= 85.6 \text{ kPa} + (9.81 \text{ m/s}^2)[(13{,}600 \text{ kg/m}^3)(0.35 \text{ m}) - (1000 \text{ kg/m}^3)(0.1 \text{ m})$$
$$- (850 \text{ kg/m}^3)(0.2 \text{ m})]\left(\frac{1 \text{ N}}{1 \text{ kg·m/s}^2}\right)\left(\frac{1 \text{ kPa}}{1000 \text{ N/m}^2}\right)$$
$$= \mathbf{130 \text{ kPa}}$$

Discussion Note that jumping horizontally from one tube to the next and realizing that pressure remains the same in the same fluid simplifies the analysis considerably. Also note that mercury is a toxic fluid, and mercury manometers and thermometers are being replaced by ones with safer fluids because of the risk of exposure to mercury vapor during an accident.

EXAMPLE 3–7   Analyzing a Multifluid Manometer with EES

Reconsider the multifluid manometer discussed in Example 3–6. Determine the air pressure in the tank using EES. Also determine what the differential fluid height h_3 would be for the same air pressure if the mercury in the last column were replaced by seawater with a density of 1030 kg/m³.

SOLUTION The pressure in a water tank is measured by a multifluid manometer. The air pressure in the tank and the differential fluid height h_3 if mercury is replaced by seawater are to be determined using EES.
Analysis We start the EES program, open a new file, and type the following on the blank screen that appears (we express the atmospheric pressure in Pa for unit consistency):

```
g=9.81
Patm=85600
h1=0.1; h2=0.2; h3=0.35
rw=1000; roil=850; rm=13600
P1+rw*g*h1+roil*g*h2−rm*g*h3=Patm
```

Here P1 is the only unknown, and it is determined by EES to be

$$P_1 = 129647 \text{ Pa} \cong \mathbf{130 \text{ kPa}}$$

which is identical to the result obtained in Example 3–6. The height of the fluid column h_3 when mercury is replaced by seawater is determined easily by replacing "h3=0.35" by "P1=129647" and "rm=13600" by "rm=1030," and clicking on the calculator symbol. It gives

$$h_3 = \mathbf{4.62 \text{ m}}$$

Discussion Note that we used the screen like a paper pad and wrote down the relevant information together with the applicable relations in an organized manner. EES did the rest. Equations can be written on separate lines or on the same line by separating them by semicolons, and blank or comment lines can be inserted for readability. EES makes it very easy to ask "what if" questions and to perform parametric studies, as explained in Appendix 3 on the text website.

PRESSURE AND FLUID STATICS

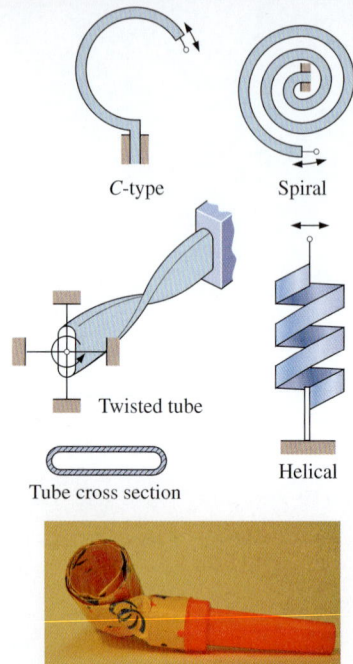

FIGURE 3–24
Various types of Bourdon tubes used to measure pressure. They work on the same principle as party noise-makers (bottom photo) due to the flat tube cross section.
(Bottom) Photo by John M. Cimbala.

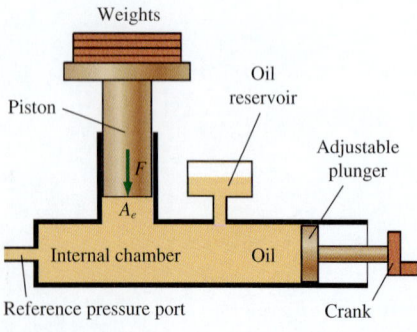

FIGURE 3–25
A deadweight tester is able to measure extremely high pressures (up to 70 MPa in some applications).

Other Pressure Measurement Devices

Another type of commonly used mechanical pressure measurement device is the **Bourdon tube,** named after the French engineer and inventor Eugene Bourdon (1808–1884), which consists of a bent, coiled, or twisted hollow metal tube whose end is closed and connected to a dial indicator needle (Fig. 3–24). When the tube is open to the atmosphere, the tube is undeflected, and the needle on the dial at this state is calibrated to read zero (gage pressure). When the fluid inside the tube is pressurized, the tube stretches and moves the needle in proportion to the applied pressure.

Electronics have made their way into every aspect of life, including pressure measurement devices. Modern pressure sensors, called **pressure transducers,** use various techniques to convert the pressure effect to an electrical effect such as a change in voltage, resistance, or capacitance. Pressure transducers are smaller and faster, and they can be more sensitive, reliable, and precise than their mechanical counterparts. They can measure pressures from less than a millionth of 1 atm to several thousands of atm.

A wide variety of pressure transducers is available to measure gage, absolute, and differential pressures in a wide range of applications. *Gage pressure transducers* use the atmospheric pressure as a reference by venting the back side of the pressure-sensing diaphragm to the atmosphere, and they give a zero signal output at atmospheric pressure regardless of altitude. *Absolute pressure transducers* are calibrated to have a zero signal output at full vacuum. *Differential pressure transducers* measure the pressure difference between two locations directly instead of using two pressure transducers and taking their difference.

Strain-gage pressure transducers work by having a diaphragm deflect between two chambers open to the pressure inputs. As the diaphragm stretches in response to a change in pressure difference across it, the strain gage stretches and a Wheatstone bridge circuit amplifies the output. A capacitance transducer works similarly, but capacitance change is measured instead of resistance change as the diaphragm stretches.

Piezoelectric transducers, also called solid-state pressure transducers, work on the principle that an electric potential is generated in a crystalline substance when it is subjected to mechanical pressure. This phenomenon, first discovered by brothers Pierre and Jacques Curie in 1880, is called the piezoelectric (or press-electric) effect. Piezoelectric pressure transducers have a much faster frequency response compared to diaphragm units and are very suitable for high-pressure applications, but they are generally not as sensitive as diaphragm-type transducers, especially at low pressures.

Another type of mechanical pressure gage called a **deadweight tester** is used primarily for *calibration* and can measure extremely high pressures (Fig. 3–25). As its name implies, a deadweight tester measures pressure *directly* through application of a weight that provides a force per unit area—the fundamental definition of pressure. It is constructed with an internal chamber filled with a fluid (usually oil), along with a tight-fitting piston, cylinder, and plunger. Weights are applied to the top of the piston, which exerts a force on the oil in the chamber. The total force F acting on the oil at the piston–oil interface is the sum of the weight of the piston plus the applied weights. Since the piston cross-sectional area A_e is known, the pressure is calculated as $P = F/A_e$. The only significant source of error is that due to static friction along the interface between the piston and cylinder, but even this error is usually negligibly small. The reference pressure port is connected to either an unknown pressure that is to be measured or to a pressure sensor that is to be calibrated.

3–3 · INTRODUCTION TO FLUID STATICS

Fluid statics deals with problems associated with fluids at rest. The fluid can be either gaseous or liquid. Fluid statics is generally referred to as *hydrostatics* when the fluid is a liquid and as *aerostatics* when the fluid is a gas. In fluid statics, there is no relative motion between adjacent fluid layers, and thus there are no shear (tangential) stresses in the fluid trying to deform it. The only stress we deal with in fluid statics is the *normal stress*, which is the pressure, and the variation of pressure is due only to the weight of the fluid. Therefore, the topic of fluid statics has significance only in gravity fields, and the force relations developed naturally involve the gravitational acceleration g. The force exerted on a surface by a fluid at rest is normal to the surface at the point of contact since there is no relative motion between the fluid and the solid surface, and thus there are no shear forces acting parallel to the surface.

Fluid statics is used to determine the forces acting on floating or submerged bodies and the forces developed by devices like hydraulic presses and car jacks. The design of many engineering systems such as water dams and liquid storage tanks requires the determination of the forces acting on their surfaces using fluid statics. The complete description of the resultant hydrostatic force acting on a submerged surface requires the determination of the magnitude, the direction, and the line of action of the force. In the following two sections, we consider the forces acting on both plane and curved surfaces of submerged bodies due to pressure.

FIGURE 3–26
Hoover Dam.
Courtesy United States Department of the Interior, Bureau of Reclamation-Lower Colorado Region.

3–4 · HYDROSTATIC FORCES ON SUBMERGED PLANE SURFACES

A plate (such as a gate valve in a dam, the wall of a liquid storage tank, or the hull of a ship at rest) is subjected to fluid pressure distributed over its surface when exposed to a liquid (Fig. 3–26). On a *plane* surface, the hydrostatic forces form a system of parallel forces, and we often need to determine the *magnitude* of the force and its *point of application*, which is called the **center of pressure.** In most cases, the other side of the plate is open to the atmosphere (such as the dry side of a gate), and thus atmospheric pressure acts on both sides of the plate, yielding a zero resultant. In such cases, it is convenient to subtract atmospheric pressure and work with the gage pressure only (Fig. 3–27). For example, $P_{gage} = \rho g h$ at the bottom of the lake.

Consider the top surface of a flat plate of arbitrary shape completely submerged in a liquid, as shown in Fig. 3–28 together with its normal view. The plane of this surface (normal to the page) intersects the horizontal free surface at angle θ, and we take the line of intersection to be the x-axis (out of the page). The absolute pressure above the liquid is P_0, which is the local atmospheric pressure P_{atm} if the liquid is open to the atmosphere (but P_0

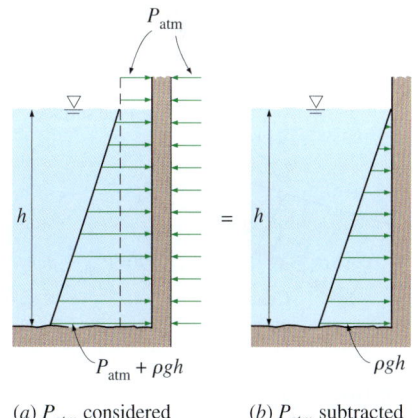

FIGURE 3–27
When analyzing hydrostatic forces on submerged surfaces, the atmospheric pressure can be subtracted for simplicity when it acts on both sides of the structure.

PRESSURE AND FLUID STATICS

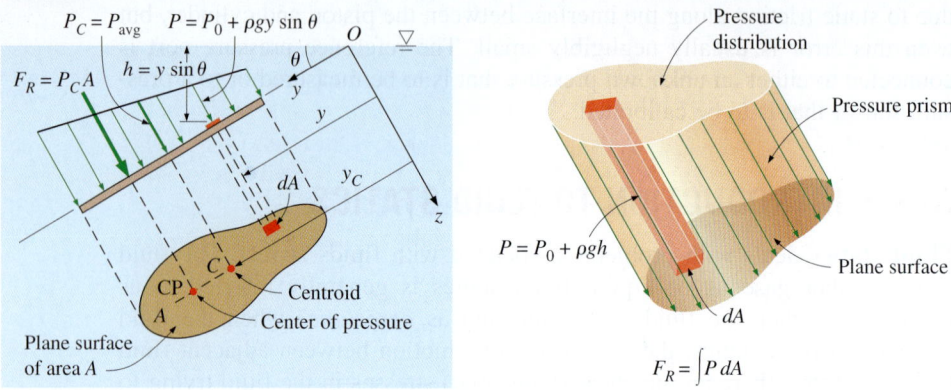

FIGURE 3–28
Hydrostatic force on an inclined plane surface completely submerged in a liquid.

may be different than P_{atm} if the space above the liquid is evacuated or pressurized). Then the absolute pressure at any point on the plate is

$$P = P_0 + \rho g h = P_0 + \rho g y \sin\theta \tag{3-16}$$

where h is the vertical distance of the point from the free surface and y is the distance of the point from the x-axis (from point O in Fig. 3–28). The resultant hydrostatic force F_R acting on the surface is determined by integrating the force $P\,dA$ acting on a differential area dA over the entire surface area,

$$F_R = \int_A P\,dA = \int_A (P_0 + \rho g y \sin\theta)\,dA = P_0 A + \rho g \sin\theta \int_A y\,dA \tag{3-17}$$

But the *first moment of area* $\int_A y\,dA$ is related to the y-coordinate of the centroid (or center) of the surface by

$$y_C = \frac{1}{A}\int_A y\,dA \tag{3-18}$$

Substituting,

$$F_R = (P_0 + \rho g y_C \sin\theta)A = (P_0 + \rho g h_C)A = P_C A = P_{avg} A \tag{3-19}$$

where $P_C = P_0 + \rho g h_C$ is the pressure at the centroid of the surface, which is equivalent to the *average* pressure P_{avg} on the surface, and $h_C = y_C \sin\theta$ is the *vertical distance* of the centroid from the free surface of the liquid (Fig. 3–29). Thus we conclude that:

> The magnitude of the resultant force acting on a plane surface of a completely submerged plate in a homogeneous (constant density) fluid is equal to the product of the pressure P_C at the centroid of the surface and the area A of the surface (Fig. 3–30).

The pressure P_0 is usually atmospheric pressure, which can be ignored in most force calculations since it acts on both sides of the plate. When this is not the case, a practical way of accounting for the contribution of P_0 to

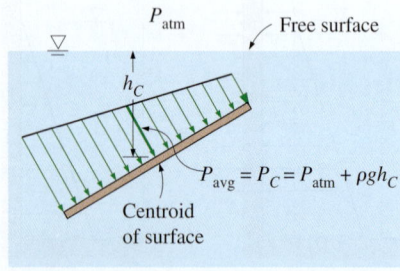

FIGURE 3–29
The pressure at the centroid of a plane surface is equivalent to the *average* pressure on the surface.

the resultant force is simply to add an equivalent depth $h_{\text{equiv}} = P_0/\rho g$ to h_C; that is, to assume the presence of an additional liquid layer of thickness h_{equiv} on top of the liquid with absolute vacuum above.

Next we need to determine the line of action of the resultant force F_R. Two parallel force systems are equivalent if they have the same magnitude and the same moment about any point. The line of action of the resultant hydrostatic force, in general, does not pass through the centroid of the surface—it lies underneath where the pressure is higher. The point of intersection of the line of action of the resultant force and the surface is the **center of pressure.** The vertical location of the line of action is determined by equating the moment of the resultant force to the moment of the distributed pressure force about the x-axis:

$$y_P F_R = \int_A yP\, dA = \int_A y(P_0 + \rho g y \sin\theta)\, dA = P_0 \int_A y\, dA + \rho g \sin\theta \int_A y^2\, dA$$

or

$$y_P F_R = P_0 y_C A + \rho g \sin\theta\, I_{xx,O} \quad (3\text{-}20)$$

where y_P is the distance of the center of pressure from the x-axis (point O in Fig. 3–30) and $I_{xx,O} = \int_A y^2\, dA$ is the *second moment of area* (also called the *area moment of inertia*) about the x-axis. The second moments of area are widely available for common shapes in engineering handbooks, but they are usually given about the axes passing through the centroid of the area. Fortunately, the second moments of area about two parallel axes are related to each other by the *parallel axis theorem,* which in this case is expressed as

$$I_{xx,O} = I_{xx,C} + y_C^2 A \quad (3\text{-}21)$$

where $I_{xx,C}$ is the second moment of area about the x-axis passing through the centroid of the area and y_C (the y-coordinate of the centroid) is the distance between the two parallel axes. Substituting the F_R relation from Eq. 3–19 and the $I_{xx,O}$ relation from Eq. 3–21 into Eq. 3–20 and solving for y_P yields

$$y_P = y_C + \frac{I_{xx,C}}{[y_C + P_0/(\rho g \sin\theta)]A} \quad (3\text{-}22a)$$

For $P_0 = 0$, which is usually the case when the atmospheric pressure is ignored, it simplifies to

$$y_P = y_C + \frac{I_{xx,C}}{y_C A} \quad (3\text{-}22b)$$

Knowing y_P, the vertical distance of the center of pressure from the free surface is determined from $h_P = y_P \sin\theta$.

The $I_{xx,C}$ values for some common areas are given in Fig. 3–31. For areas that possess symmetry about the y-axis, the center of pressure lies on the y-axis directly below the centroid. The location of the center of pressure in such cases is simply the point on the surface of the vertical plane of symmetry at a distance h_P from the free surface.

Pressure acts normal to the surface, and the hydrostatic forces acting on a flat plate of any shape form a volume whose base is the plate area and

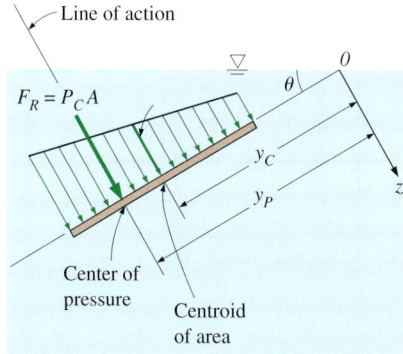

FIGURE 3–30
The resultant force acting on a plane surface is equal to the product of the pressure at the centroid of the surface and the surface area, and its line of action passes through the center of pressure.

92
PRESSURE AND FLUID STATICS

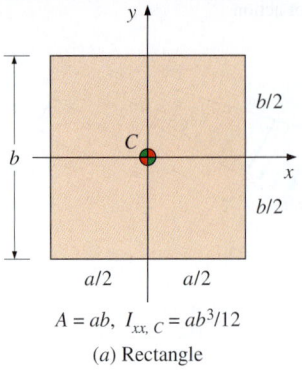

$A = ab$, $I_{xx,\,C} = ab^3/12$

(a) Rectangle

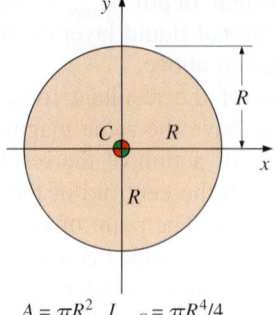

$A = \pi R^2$, $I_{xx,\,C} = \pi R^4/4$

(b) Circle

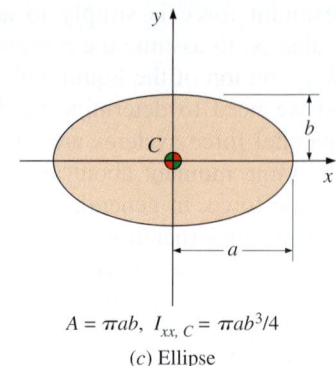

$A = \pi ab$, $I_{xx,\,C} = \pi ab^3/4$

(c) Ellipse

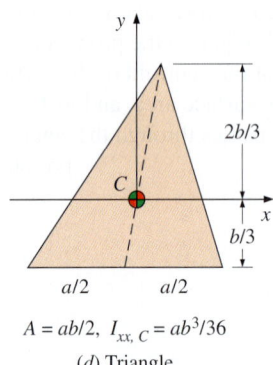

$A = ab/2$, $I_{xx,\,C} = ab^3/36$

(d) Triangle

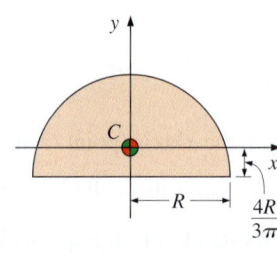

$A = \pi R^2/2$, $I_{xx,\,C} = 0.109757 R^4$

(e) Semicircle

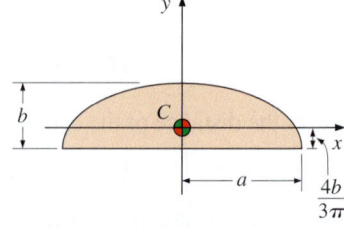

$A = \pi ab/2$, $I_{xx,\,C} = 0.109757 ab^3$

(f) Semiellipse

FIGURE 3–31
The centroid and the centroidal moments of inertia for some common geometries.

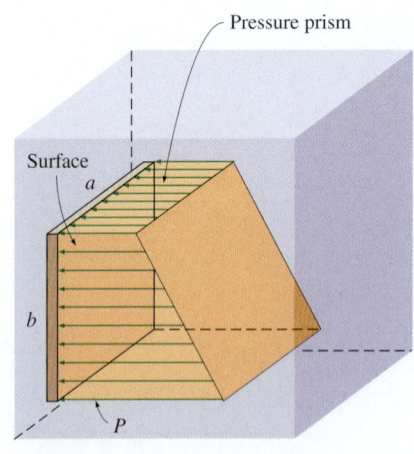

FIGURE 3–32
The hydrostatic forces acting on a plane surface form a pressure prism whose base (left face) is the surface and whose length is the pressure.

whose length is the linearly varying pressure, as shown in Fig. 3–32. This virtual **pressure prism** has an interesting physical interpretation: its *volume* is equal to the *magnitude* of the resultant hydrostatic force acting on the plate since $F_R = \int P\,dA$, and the line of action of this force passes through the *centroid* of this homogeneous prism. The projection of the centroid on the plate is the *pressure center*. Therefore, with the concept of pressure prism, the problem of describing the resultant hydrostatic force on a plane surface reduces to finding the volume and the two coordinates of the centroid of this pressure prism.

Special Case: Submerged Rectangular Plate
Consider a completely submerged rectangular flat plate of height b and width a tilted at an angle θ from the horizontal and whose top edge is horizontal and is at a distance s from the free surface along the plane of the plate, as shown in Fig. 3–33a. The resultant hydrostatic force on the upper surface is equal to the average pressure, which is the pressure at the midpoint of the surface, times the surface area A. That is,

Tilted rectangular plate: $\qquad F_R = P_C A = [P_0 + \rho g(s + b/2)\sin\theta]ab \qquad$ (3–23)

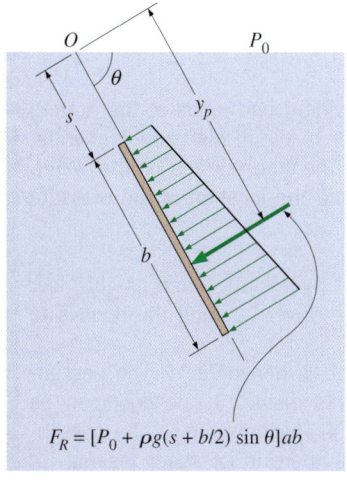
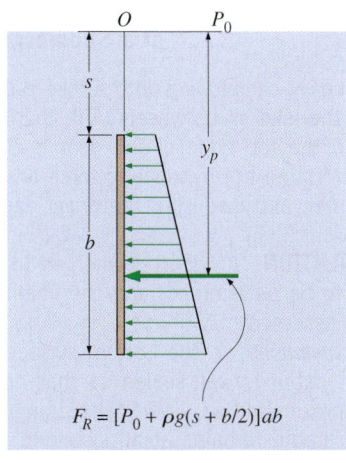
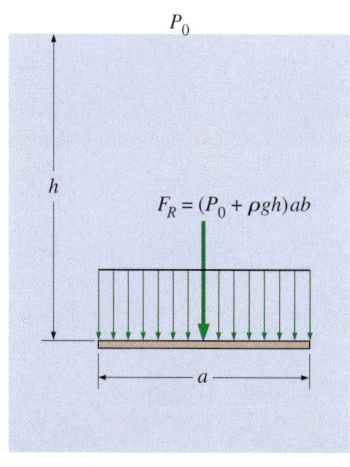

(a) Tilted plate   (b) Vertical plate   (c) Horizontal plate

FIGURE 3–33
Hydrostatic force acting on the top surface of a submerged rectangular plate for tilted, vertical, and horizontal cases.

The force acts at a vertical distance of $h_P = y_P \sin \theta$ from the free surface directly beneath the centroid of the plate where, from Eq. 3–22a,

$$y_P = s + \frac{b}{2} + \frac{ab^3/12}{[s + b/2 + P_0/(\rho g \sin \theta)]ab}$$

$$= s + \frac{b}{2} + \frac{b^2}{12[s + b/2 + P_0/(\rho g \sin \theta)]} \quad (3\text{–}24)$$

When the upper edge of the plate is at the free surface and thus $s = 0$, Eq. 3–23 reduces to

Tilted rectangular plate ($s = 0$): $\quad F_R = [P_0 + \rho g(b \sin \theta)/2]ab \quad (3\text{–}25)$

For a completely submerged *vertical* plate ($\theta = 90°$) whose top edge is horizontal, the hydrostatic force can be obtained by setting $\sin \theta = 1$ (Fig. 3–33b)

Vertical rectangular plate: $\quad F_R = [P_0 + \rho g(s + b/2)]ab \quad (3\text{–}26)$

Vertical rectangular plate ($s = 0$): $\quad F_R = (P_0 + \rho g b/2)ab \quad (3\text{–}27)$

When the effect of P_0 is ignored since it acts on both sides of the plate, the hydrostatic force on a vertical rectangular surface of height b whose top edge is horizontal and at the free surface is $F_R = \rho g a b^2/2$ acting at a distance of $2b/3$ from the free surface directly beneath the centroid of the plate.

The pressure distribution on a submerged *horizontal* surface is uniform, and its magnitude is $P = P_0 + \rho g h$, where h is the distance of the surface from the free surface. Therefore, the hydrostatic force acting on a horizontal rectangular surface is

Horizontal rectangular plate: $\quad F_R = (P_0 + \rho g h)ab \quad (3\text{–}28)$

and it acts through the midpoint of the plate (Fig. 3–32c).

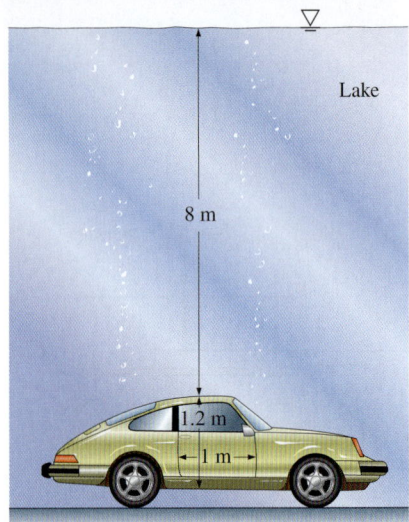

FIGURE 3–34
Schematic for Example 3–8.

EXAMPLE 3–8   Hydrostatic Force Acting on the Door of a Submerged Car

A heavy car plunges into a lake during an accident and lands at the bottom of the lake on its wheels (Fig. 3–34). The door is 1.2 m high and 1 m wide, and the top edge of the door is 8 m below the free surface of the water. Determine the hydrostatic force on the door and the location of the pressure center, and discuss if the driver can open the door.

SOLUTION A car is submerged in water. The hydrostatic force on the door is to be determined, and the likelihood of the driver opening the door is to be assessed.
Assumptions **1** The bottom surface of the lake is horizontal. **2** The passenger cabin is well-sealed so that no water leaks inside. **3** The door can be approximated as a vertical rectangular plate. **4** The pressure in the passenger cabin remains at atmospheric value since there is no water leaking in, and thus no compression of the air inside. Therefore, atmospheric pressure cancels out in the calculations since it acts on both sides of the door. **5** The weight of the car is larger than the buoyant force acting on it.
Properties We take the density of lake water to be 1000 kg/m³ throughout.
Analysis The average (gage) pressure on the door is the pressure value at the centroid (midpoint) of the door and is determined to be

$$P_{avg} = P_C = \rho g h_C = \rho g(s + b/2)$$
$$= (1000 \text{ kg/m}^3)(9.81 \text{ m/s}^2)(8 + 1.2/2 \text{ m})\left(\frac{1 \text{ kN}}{1000 \text{ kg}\cdot\text{m/s}^2}\right)$$
$$= \mathbf{84.4 \text{ kN/m}^2}$$

Then the resultant hydrostatic force on the door becomes

$$F_R = P_{avg} A = (84.4 \text{ kN/m}^2)(1 \text{ m} \times 1.2 \text{ m}) = \mathbf{101.3 \text{ kN}}$$

The pressure center is directly under the midpoint of the door, and its distance from the surface of the lake is determined from Eq. 3–24 by setting $P_0 = 0$, yielding

$$y_P = s + \frac{b}{2} + \frac{b^2}{12(s + b/2)} = 8 + \frac{1.2}{2} + \frac{1.2^2}{12(8 + 1.2/2)} = \mathbf{8.61 \text{ m}}$$

Discussion A strong person can lift 100 kg, which is a weight of 981 N or about 1 kN. Also, the person can apply the force at a point farthest from the hinges (1 m farther) for maximum effect and generate a moment of 1 kN·m. The resultant hydrostatic force acts under the midpoint of the door, and thus a distance of 0.5 m from the hinges. This generates a moment of 50.6 kN·m, which is about 50 times the moment the driver can possibly generate. Therefore, it is impossible for the driver to open the door of the car. The driver's best bet is to let some water in (by rolling the window down a little, for example) and to keep his or her head close to the ceiling. The driver should be able to open the door shortly before the car is filled with water since at that point the pressures on both sides of the door are nearly the same and opening the door in water is almost as easy as opening it in air.

3–5 · HYDROSTATIC FORCES ON SUBMERGED CURVED SURFACES

In many practical applications, submerged surfaces are not flat (Fig. 3–35). For a submerged curved surface, the determination of the resultant hydrostatic force is more involved since it typically requires integration of the pressure forces that change direction along the curved surface. The concept of the pressure prism in this case is not much help either because of the complicated shapes involved.

The easiest way to determine the resultant hydrostatic force F_R acting on a two-dimensional curved surface is to determine the horizontal and vertical components F_H and F_V separately. This is done by considering the free-body diagram of the liquid block enclosed by the curved surface and the two plane surfaces (one horizontal and one vertical) passing through the two ends of the curved surface, as shown in Fig. 3–36. Note that the vertical surface of the liquid block considered is simply the projection of the curved surface on a *vertical plane*, and the horizontal surface is the projection of the curved surface on a *horizontal plane*. The resultant force acting on the curved solid surface is then equal and opposite to the force acting on the curved liquid surface (Newton's third law).

The force acting on the imaginary horizontal or vertical plane surface and its line of action can be determined as discussed in Section 3–4. The weight of the enclosed liquid block of volume V is simply $W = \rho g V$, and it acts downward through the centroid of this volume. Noting that the fluid block is in static equilibrium, the force balances in the horizontal and vertical directions give

Horizontal force component on curved surface: $\qquad F_H = F_x \qquad$ (3–29)

Vertical force component on curved surface: $\qquad F_V = F_y \pm W \qquad$ (3–30)

FIGURE 3–35
In many structures of practical application, the submerged surfaces are not flat, but curved as here at Glen Canyon Dam in Utah and Arizona.
© Corbis RF

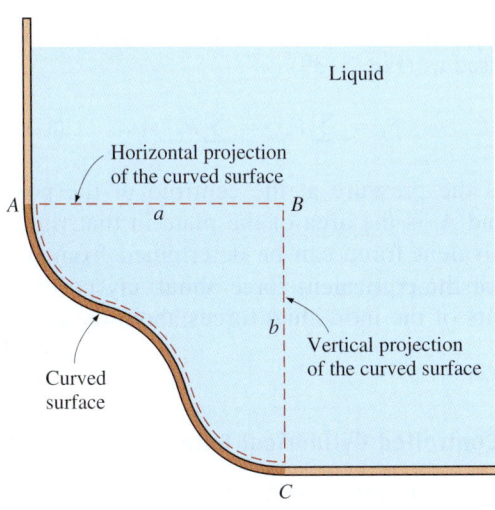
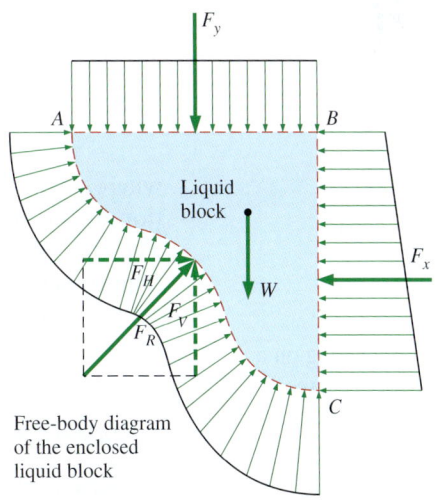

FIGURE 3–36
Determination of the hydrostatic force acting on a submerged curved surface.

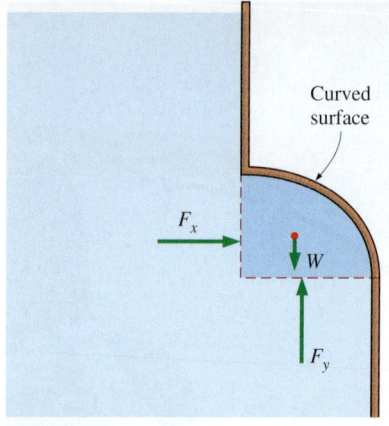

FIGURE 3–37
When a curved surface is above the liquid, the weight of the liquid and the vertical component of the hydrostatic force act in the opposite directions.

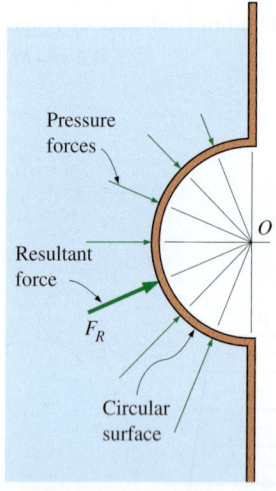

FIGURE 3–38
The hydrostatic force acting on a circular surface always passes through the center of the circle since the pressure forces are normal to the surface and they all pass through the center.

where the summation $F_y \pm W$ is a vector addition (i.e., add magnitudes if both act in the same direction and subtract if they act in opposite directions). Thus, we conclude that

1. The horizontal component of the hydrostatic force acting on a curved surface is equal (in both magnitude and the line of action) to the hydrostatic force acting on the vertical projection of the curved surface.
2. The vertical component of the hydrostatic force acting on a curved surface is equal to the hydrostatic force acting on the horizontal projection of the curved surface, plus (minus, if acting in the opposite direction) the weight of the fluid block.

The magnitude of the resultant hydrostatic force acting on the curved surface is $F_R = \sqrt{F_H^2 + F_V^2}$, and the tangent of the angle it makes with the horizontal is $\tan \alpha = F_V/F_H$. The exact location of the line of action of the resultant force (e.g., its distance from one of the end points of the curved surface) can be determined by taking a moment about an appropriate point. These discussions are valid for all curved surfaces regardless of whether they are above or below the liquid. Note that in the case of a *curved surface above a liquid*, the weight of the liquid is *subtracted* from the vertical component of the hydrostatic force since they act in opposite directions (Fig. 3–37).

When the curved surface is a *circular arc* (full circle or any part of it), the resultant hydrostatic force acting on the surface always passes through the center of the circle. This is because the pressure forces are normal to the surface, and all lines normal to the surface of a circle pass through the center of the circle. Thus, the pressure forces form a concurrent force system at the center, which can be reduced to a single equivalent force at that point (Fig. 3–38).

Finally, the hydrostatic force acting on a plane or curved surface submerged in a **multilayered fluid** of different densities can be determined by considering different parts of surfaces in different fluids as different surfaces, finding the force on each part, and then adding them using vector addition. For a plane surface, it can be expressed as (Fig. 3–39)

Plane surface in a multilayered fluid: $\qquad F_R = \sum F_{R,i} = \sum P_{C,i} A_i \qquad$ (3–31)

where $P_{C,i} = P_0 + \rho_i g h_{C,i}$ is the pressure at the centroid of the portion of the surface in fluid i and A_i is the area of the plate in that fluid. The line of action of this equivalent force can be determined from the requirement that the moment of the equivalent force about any point is equal to the sum of the moments of the individual forces about the same point.

EXAMPLE 3–9   **A Gravity-Controlled Cylindrical Gate**

A long solid cylinder of radius 0.8 m hinged at point A is used as an automatic gate, as shown in Fig. 3–40. When the water level reaches 5 m, the gate opens by turning about the hinge at point A. Determine (*a*) the hydrostatic force acting on the cylinder and its line of action when the gate opens and (*b*) the weight of the cylinder per m length of the cylinder.

SOLUTION The height of a water reservoir is controlled by a cylindrical gate hinged to the reservoir. The hydrostatic force on the cylinder and the weight of the cylinder per m length are to be determined.
Assumptions 1 Friction at the hinge is negligible. 2 Atmospheric pressure acts on both sides of the gate, and thus it cancels out.
Properties We take the density of water to be 1000 kg/m³ throughout.
Analysis (*a*) We consider the free-body diagram of the liquid block enclosed by the circular surface of the cylinder and its vertical and horizontal projections. The hydrostatic forces acting on the vertical and horizontal plane surfaces as well as the weight of the liquid block are determined as

Horizontal force on vertical surface:

$$F_H = F_x = P_{avg} A = \rho g h_C A = \rho g (s + R/2) A$$
$$= (1000 \text{ kg/m}^3)(9.81 \text{ m/s}^2)(4.2 + 0.8/2 \text{ m})(0.8 \text{ m} \times 1 \text{ m}) \left(\frac{1 \text{ kN}}{1000 \text{ kg} \cdot \text{m/s}^2} \right)$$
$$= 36.1 \text{ kN}$$

Vertical force on horizontal surface (upward):

$$F_y = P_{avg} A = \rho g h_C A = \rho g h_{bottom} A$$
$$= (1000 \text{ kg/m}^3)(9.81 \text{ m/s}^2)(5 \text{ m})(0.8 \text{ m} \times 1 \text{ m}) \left(\frac{1 \text{ kN}}{1000 \text{ kg} \cdot \text{m/s}^2} \right)$$
$$= 39.2 \text{ kN}$$

Weight (downward) of fluid block for one m width into the page:

$$W = mg = \rho g V = \rho g (R^2 - \pi R^2/4)(1 \text{ m})$$
$$= (1000 \text{ kg/m}^3)(9.81 \text{ m/s}^2)(0.8 \text{ m})^2 (1 - \pi/4)(1 \text{ m}) \left(\frac{1 \text{ kN}}{1000 \text{ kg} \cdot \text{m/s}^2} \right)$$
$$= 1.3 \text{ kN}$$

Therefore, the net upward vertical force is

$$F_V = F_y - W = 39.2 - 1.3 = 37.9 \text{ kN}$$

Then the magnitude and direction of the hydrostatic force acting on the cylindrical surface become

$$F_R = \sqrt{F_H^2 + F_V^2} = \sqrt{36.1^2 + 37.9^2} = \mathbf{52.3 \text{ kN}}$$
$$\tan \theta = F_V/F_H = 37.9/36.1 = 1.05 \rightarrow \theta = 46.4°$$

Therefore, the magnitude of the hydrostatic force acting on the cylinder is 52.3 kN per m length of the cylinder, and its line of action passes through the center of the cylinder making an angle 46.4° with the horizontal.

(*b*) When the water level is 5 m high, the gate is about to open and thus the reaction force at the bottom of the cylinder is zero. Then the forces other than those at the hinge acting on the cylinder are its weight, acting through the center, and the hydrostatic force exerted by water. Taking a moment about point *A* at the location of the hinge and equating it to zero gives

$$F_R R \sin \theta - W_{cyl} R = 0 \rightarrow W_{cyl} = F_R \sin \theta = (52.3 \text{ kN}) \sin 46.4° = \mathbf{37.9 \text{ kN}}$$

Discussion The weight of the cylinder per m length is determined to be 37.9 kN. It can be shown that this corresponds to a mass of 3863 kg per m length and to a density of 1921 kg/m³ for the material of the cylinder.

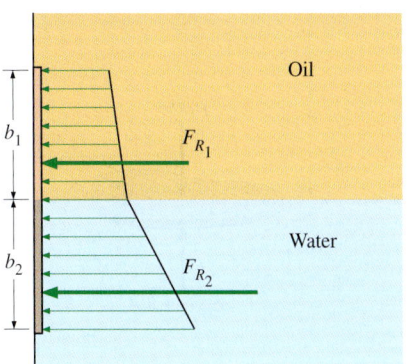

FIGURE 3–39
The hydrostatic force on a surface submerged in a multilayered fluid can be determined by considering parts of the surface in different fluids as different surfaces.

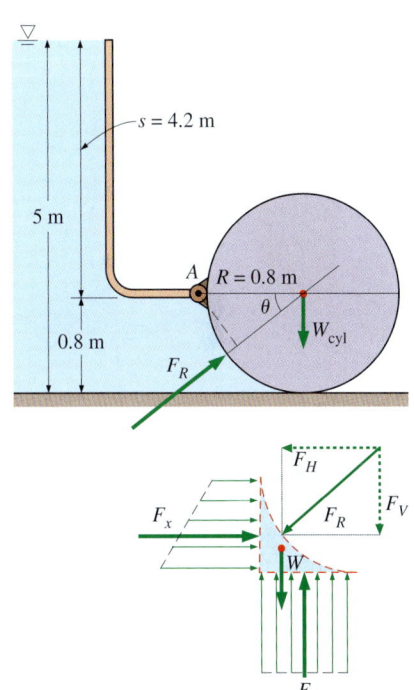

FIGURE 3–40
Schematic for Example 3–9 and the free-body diagram of the liquid underneath the cylinder.

PRESSURE AND FLUID STATICS

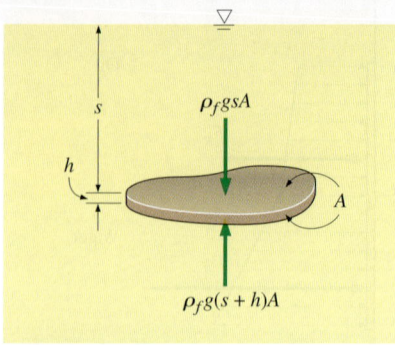

FIGURE 3–41
A flat plate of uniform thickness h submerged in a liquid parallel to the free surface.

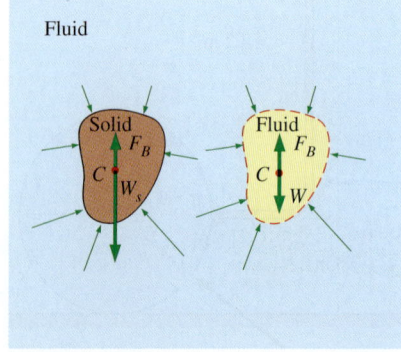

FIGURE 3–42
The buoyant forces acting on a solid body submerged in a fluid and on a fluid body of the same shape at the same depth are identical. The buoyant force F_B acts upward through the centroid C of the displaced volume and is equal in magnitude to the weight W of the displaced fluid, but is opposite in direction. For a solid of uniform density, its weight W_s also acts through the centroid, but its magnitude is not necessarily equal to that of the fluid it displaces. (Here $W_s > W$ and thus $W_s > F_B$; this solid body would sink.)

3–6 · BUOYANCY AND STABILITY

It is a common experience that an object feels lighter and weighs less in a liquid than it does in air. This can be demonstrated easily by weighing a heavy object in water by a waterproof spring scale. Also, objects made of wood or other light materials float on water. These and other observations suggest that a fluid exerts an upward force on a body immersed in it. This force that tends to lift the body is called the **buoyant force** and is denoted by F_B.

The buoyant force is caused by the increase of pressure with depth in a fluid. Consider, for example, a flat plate of thickness h submerged in a liquid of density ρ_f parallel to the free surface, as shown in Fig. 3–41. The area of the top (and also bottom) surface of the plate is A, and its distance to the free surface is s. The gage pressures at the top and bottom surfaces of the plate are $\rho_f g s$ and $\rho_f g(s + h)$, respectively. Then the hydrostatic force $F_{top} = \rho_f g s A$ acts downward on the top surface, and the larger force $F_{bottom} = \rho_f g(s + h)A$ acts upward on the bottom surface of the plate. The difference between these two forces is a net upward force, which is the *buoyant force*,

$$F_B = F_{bottom} - F_{top} = \rho_f g(s + h)A - \rho_f g s A = \rho_f g h A = \rho_f g V \quad (3\text{–}32)$$

where $V = hA$ is the volume of the plate. But the relation $\rho_f g V$ is simply the weight of the liquid whose volume is equal to the volume of the plate. Thus, we conclude that *the buoyant force acting on the plate is equal to the weight of the liquid displaced by the plate.* For a fluid with constant density, the buoyant force is independent of the distance of the body from the free surface. It is also independent of the density of the solid body.

The relation in Eq. 3–32 is developed for a simple geometry, but it is valid for any body regardless of its shape. This can be shown mathematically by a force balance, or simply by this argument: Consider an arbitrarily shaped solid body submerged in a fluid at rest and compare it to a body of fluid of the same shape indicated by dashed lines at the same vertical location (Fig. 3–42). The buoyant forces acting on these two bodies are the same since the pressure distributions, which depend only on elevation, are the same at the boundaries of both. The imaginary fluid body is in static equilibrium, and thus the net force and net moment acting on it are zero. Therefore, the upward buoyant force must be equal to the weight of the imaginary fluid body whose volume is equal to the volume of the solid body. Further, the weight and the buoyant force must have the same line of action to have a zero moment. This is known as **Archimedes' principle,** after the Greek mathematician Archimedes (287–212 BC), and is expressed as

> The buoyant force acting on a body of uniform density immersed in a fluid is equal to the weight of the fluid displaced by the body, and it acts upward through the centroid of the displaced volume.

For *floating* bodies, the weight of the entire body must be equal to the buoyant force, which is the weight of the fluid whose volume is equal to the volume of the *submerged portion* of the floating body. That is,

$$F_B = W \rightarrow \rho_f g V_{sub} = \rho_{avg, body} g V_{total} \rightarrow \frac{V_{sub}}{V_{total}} = \frac{\rho_{avg, body}}{\rho_f} \quad (3\text{–}33)$$

Therefore, the submerged volume fraction of a floating body is equal to the ratio of the average density of the body to the density of the fluid. Note that when the density ratio is equal to or greater than one, the floating body becomes completely submerged.

It follows from these discussions that a body immersed in a fluid (1) remains at rest at any location in the fluid where its average density is equal to the density of the fluid, (2) sinks to the bottom when its average density is greater than the density of the fluid, and (3) rises to the surface of the fluid and floats when the average density of the body is less than the density of the fluid (Fig. 3–43).

The buoyant force is proportional to the density of the fluid, and thus we might think that the buoyant force exerted by gases such as air is negligible. This is certainly the case in general, but there are significant exceptions. For example, the volume of a person is about 0.1 m^3, and taking the density of air to be 1.2 kg/m^3, the buoyant force exerted by air on the person is

$$F_B = \rho_f g V = (1.2 \text{ kg/m}^3)(9.81 \text{ m/s}^2)(0.1 \text{ m}^3) \cong 1.2 \text{ N}$$

The weight of an 80-kg person is $80 \times 9.81 = 788$ N. Therefore, ignoring the buoyancy in this case results in an error in weight of just 0.15 percent, which is negligible. But the buoyancy effects in gases dominate some important natural phenomena such as the rise of warm air in a cooler environment and thus the onset of natural convection currents, the rise of hot-air or helium balloons, and air movements in the atmosphere. A helium balloon, for example, rises as a result of the buoyancy effect until it reaches an altitude where the density of air (which decreases with altitude) equals the density of helium in the balloon—assuming the balloon does not burst by then, and ignoring the weight of the balloon's skin. Hot air balloons (Fig. 3–44) work by similar principles.

Archimedes' principle is also used in geology by considering the continents to be floating on a sea of magma.

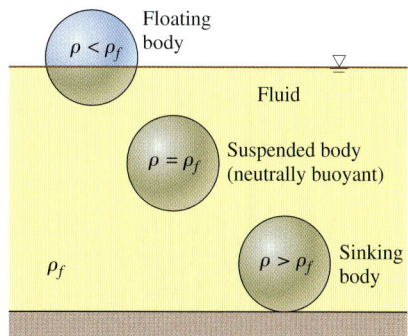

FIGURE 3–43
A solid body dropped into a fluid will sink, float, or remain at rest at any point in the fluid, depending on its average density relative to the density of the fluid.

EXAMPLE 3–10   Measuring Specific Gravity by a Hydrometer

If you have a seawater aquarium, you have probably used a small cylindrical glass tube with a lead-weight at its bottom to measure the salinity of the water by simply watching how deep the tube sinks. Such a device that floats in a vertical position and is used to measure the specific gravity of a liquid is called a *hydrometer* (Fig. 3–45). The top part of the hydrometer extends above the liquid surface, and the divisions on it allow one to read the specific gravity directly. The hydrometer is calibrated such that in pure water it reads exactly 1.0 at the air–water interface. (*a*) Obtain a relation for the specific gravity of a liquid as a function of distance Δz from the mark corresponding to pure water and (*b*) determine the mass of lead that must be poured into a 1-cm-diameter, 20-cm-long hydrometer if it is to float halfway (the 10-cm mark) in pure water.

SOLUTION The specific gravity of a liquid is to be measured by a hydrometer. A relation between specific gravity and the vertical distance from the reference level is to be obtained, and the amount of lead that needs to be added into the tube for a certain hydrometer is to be determined.
Assumptions 1 The weight of the glass tube is negligible relative to the weight of the lead added. 2 The curvature of the tube bottom is disregarded.

FIGURE 3–44
The altitude of a hot air balloon is controlled by the temperature difference between the air inside and outside the balloon, since warm air is less dense than cold air. When the balloon is neither rising nor falling, the upward buoyant force exactly balances the downward weight.
© PhotoLink/Getty RF

100
PRESSURE AND FLUID STATICS

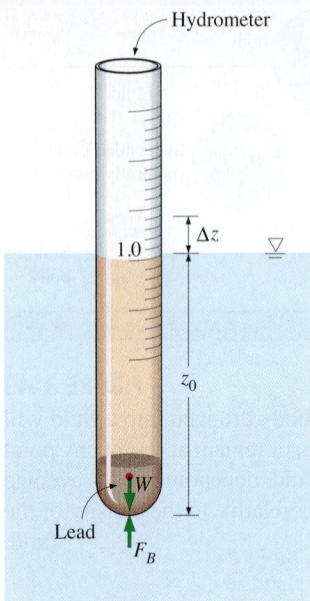

FIGURE 3–45
Schematic for Example 3–10.

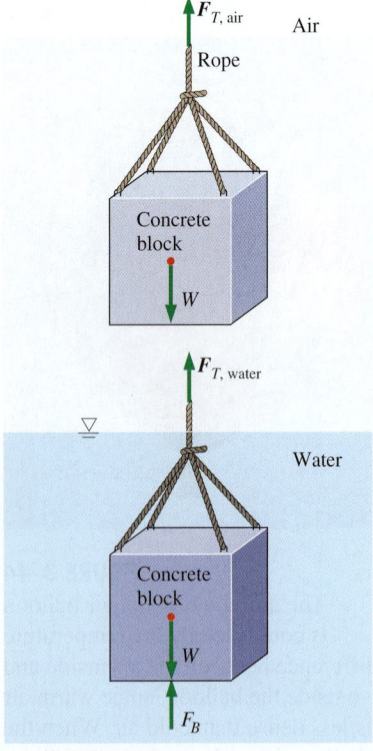

FIGURE 3–46
Schematic for Example 3–11.

Properties We take the density of pure water to be 1000 kg/m³.
Analysis (a) Noting that the hydrometer is in static equilibrium, the buoyant force F_B exerted by the liquid must always be equal to the weight W of the hydrometer. In pure water (subscript w), we let the vertical distance between the bottom of the hydrometer and the free surface of water be z_0. Setting $F_{B,w} = W$ in this case gives

$$W_{hydro} = F_{B,w} = \rho_w g V_{sub} = \rho_w g A z_0 \quad (1)$$

where A is the cross-sectional area of the tube, and ρ_w is the density of pure water.

In a fluid lighter than water ($\rho_f < \rho_w$), the hydrometer will sink deeper, and the liquid level will be a distance of Δz above z_0. Again setting $F_B = W$ gives

$$W_{hydro} = F_{B,f} = \rho_f g V_{sub} = \rho_f g A (z_0 + \Delta z) \quad (2)$$

This relation is also valid for fluids heavier than water by taking Δz to be a negative quantity. Setting Eqs. (1) and (2) here equal to each other since the weight of the hydrometer is constant and rearranging gives

$$\rho_w g A z_0 = \rho_f g A (z_0 + \Delta z) \quad \rightarrow \quad SG_f = \frac{\rho_f}{\rho_w} = \frac{z_0}{z_0 + \Delta z}$$

which is the relation between the specific gravity of the fluid and Δz. Note that z_0 is constant for a given hydrometer and Δz is negative for fluids heavier than pure water.

(b) Disregarding the weight of the glass tube, the amount of lead that needs to be added to the tube is determined from the requirement that the weight of the lead be equal to the buoyant force. When the hydrometer is floating with half of it submerged in water, the buoyant force acting on it is

$$F_B = \rho_w g V_{sub}$$

Equating F_B to the weight of lead gives

$$W = mg = \rho_w g V_{sub}$$

Solving for m and substituting, the mass of lead is determined to be

$$m = \rho_w V_{sub} = \rho_w (\pi R^2 h_{sub}) = (1000 \text{ kg/m}^3)[\pi (0.005 \text{ m})^2 (0.1 \text{ m})] = \mathbf{0.00785 \text{ kg}}$$

Discussion Note that if the hydrometer were required to sink only 5 cm in water, the required mass of lead would be one-half of this amount. Also, the assumption that the weight of the glass tube is negligible is questionable since the mass of lead is only 7.85 g.

EXAMPLE 3–11   Weight Loss of an Object in Seawater

A crane is used to lower weights into the sea (density = 1025 kg/m³) for an underwater construction project (Fig. 3–46). Determine the tension in the rope of the crane due to a rectangular 0.4-m × 0.4-m × 3-m concrete block (density = 2300 kg/m³) when it is (a) suspended in the air and (b) completely immersed in water.

SOLUTION A concrete block is lowered into the sea. The tension in the rope is to be determined before and after the block is in water.

Assumptions 1 The buoyant force in air is negligible. 2 The weight of the ropes is negligible.
Properties The densities are given to be 1025 kg/m³ for seawater and 2300 kg/m³ for concrete.
Analysis (*a*) Consider a free-body diagram of the concrete block. The forces acting on the concrete block in air are its weight and the upward pull action (tension) by the rope. These two forces must balance each other, and thus the tension in the rope must be equal to the weight of the block:

$$V = (0.4 \text{ m})(0.4 \text{ m})(3 \text{ m}) = 0.48 \text{ m}^3$$

$$F_{T,\text{air}} = W = \rho_{\text{concrete}} g V$$

$$= (2300 \text{ kg/m}^3)(9.81 \text{ m/s}^2)(0.48 \text{ m}^3)\left(\frac{1 \text{ kN}}{1000 \text{ kg}\cdot\text{m/s}^2}\right) = \mathbf{10.8 \text{ kN}}$$

(*b*) When the block is immersed in water, there is the additional force of buoyancy acting upward. The force balance in this case gives

$$F_B = \rho_f g V = (1025 \text{ kg/m}^3)(9.81 \text{ m/s}^2)(0.48 \text{ m}^3)\left(\frac{1 \text{ kN}}{1000 \text{ kg}\cdot\text{m/s}^2}\right) = 4.8 \text{ kN}$$

$$F_{T,\text{water}} = W - F_B = 10.8 - 4.8 = \mathbf{6.0 \text{ kN}}$$

Discussion Note that the weight of the concrete block, and thus the tension of the rope, decreases by (10.8 − 6.0)/10.8 = 55 percent in water.

FIGURE 3–47
For floating bodies such as ships, stability is an important consideration for safety.
© Corbis RF

Stability of Immersed and Floating Bodies

An important application of the buoyancy concept is the assessment of the stability of immersed and floating bodies with no external attachments. This topic is of great importance in the design of ships and submarines (Fig. 3–47). Here we provide some general qualitative discussions on vertical and rotational stability.

We use the classic "ball on the floor" analogy to explain the fundamental concepts of stability and instability. Shown in Fig. 3–48 are three balls at rest on the floor. Case (*a*) is **stable** since any small disturbance (someone moves the ball to the right or left) generates a restoring force (due to gravity) that returns it to its initial position. Case (*b*) is **neutrally stable** because if someone moves the ball to the right or left, it would stay put at its new location. It has no tendency to move back to its original location, nor does it continue to move away. Case (*c*) is a situation in which the ball may be at rest at the moment, but any disturbance, even an infinitesimal one, causes the ball to roll off the hill—it does not return to its original position; rather it *diverges* from it. This situation is **unstable.** What about a case where the ball is on an *inclined* floor? It is not appropriate to discuss stability for this case since the ball is not in a state of equilibrium. In other words, it cannot be at rest and would roll down the hill even without any disturbance.

For an immersed or floating body in static equilibrium, the weight and the buoyant force acting on the body balance each other, and such bodies are

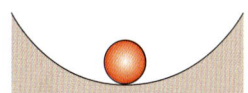
(*a*) Stable

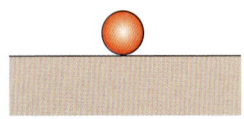
(*b*) Neutrally stable

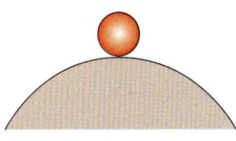
(*c*) Unstable

FIGURE 3–48
Stability is easily understood by analyzing a ball on the floor.

102
PRESSURE AND FLUID STATICS

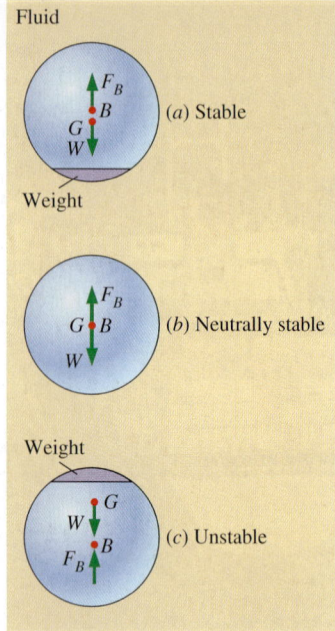

FIGURE 3–49
An immersed neutrally buoyant body is (*a*) stable if the center of gravity *G* is directly below the center of buoyancy *B* of the body, (*b*) neutrally stable if *G* and *B* are coincident, and (*c*) unstable if *G* is directly above *B*.

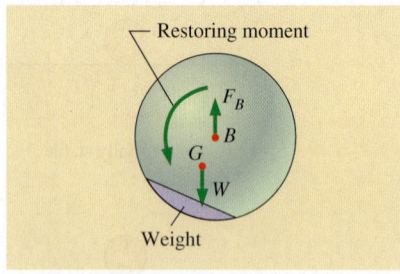

FIGURE 3–50
When the center of gravity *G* of an immersed neutrally buoyant body is not vertically aligned with the center of buoyancy *B* of the body, it is not in an equilibrium state and would rotate to its stable state, even without any disturbance.

inherently stable in the *vertical direction*. If an immersed neutrally buoyant body is raised or lowered to a different depth in an incompressible fluid, the body will remain in equilibrium at that location. If a floating body is raised or lowered somewhat by a vertical force, the body will return to its original position as soon as the external effect is removed. Therefore, a floating body possesses vertical stability, while an immersed neutrally buoyant body is neutrally stable since it does not return to its original position after a disturbance.

The *rotational stability* of an *immersed body* depends on the relative locations of the *center of gravity G* of the body and the *center of buoyancy B*, which is the centroid of the displaced volume. An immersed body is *stable* if the body is bottom-heavy and thus point *G* is directly below point *B* (Fig. 3–49*a*). A rotational disturbance of the body in such cases produces a *restoring moment* to return the body to its original stable position. Thus, a stable design for a submarine calls for the engines and the cabins for the crew to be located at the lower half in order to shift the weight to the bottom as much as possible. Hot-air or helium balloons (which can be viewed as being immersed in air) are also stable since the heavy cage that carries the load is at the bottom. An immersed body whose center of gravity *G* is directly above point *B* is *unstable*, and any disturbance will cause this body to turn upside down (Fig 3–49*c*). A body for which *G* and *B* coincide is *neutrally stable* (Fig 3–49*b*). This is the case for bodies whose density is constant throughout. For such bodies, there is no tendency to overturn or right themselves.

What about a case where the center of gravity is not vertically aligned with the center of buoyancy, as in Fig. 3–50? It is not appropriate to discuss stability for this case since the body is not in a state of equilibrium. In other words, it cannot be at rest and would rotate toward its stable state even without any disturbance. The restoring moment in the case shown in Fig. 3–50 is counterclockwise and causes the body to rotate counterclockwise so as to align point *G* vertically with point *B*. Note that there may be some oscillation, but eventually the body settles down at its stable equilibrium state [case (*a*) of Fig. 3–49]. The initial stability of the body of Fig. 3–50 is analogous to that of the ball on an inclined floor. Can you predict what would happen if the weight in the body of Fig. 3–50 were on the opposite side of the body?

The rotational stability criteria are similar for *floating bodies*. Again, if the floating body is bottom-heavy and thus the center of gravity *G* is directly below the center of buoyancy *B*, the body is always stable. But unlike immersed bodies, a floating body may still be stable when *G* is directly above *B* (Fig. 3–51). This is because the centroid of the displaced volume shifts to the side to a point *B'* during a rotational disturbance while the center of gravity *G* of the body remains unchanged. If point *B'* is sufficiently far, these two forces create a restoring moment and return the body to the original position. A measure of stability for floating bodies is the **metacentric height** *GM*, which is the distance between the center of gravity *G* and the metacenter *M*—the intersection point of the lines of action of the buoyant force through the body before and after rotation. The metacenter may be considered to be a fixed point for most hull shapes for small rolling angles up to about 20°. A floating body is stable if point *M* is above point *G*, and thus *GM* is positive, and unstable if point *M* is below point *G*, and thus *GM* is negative. In the

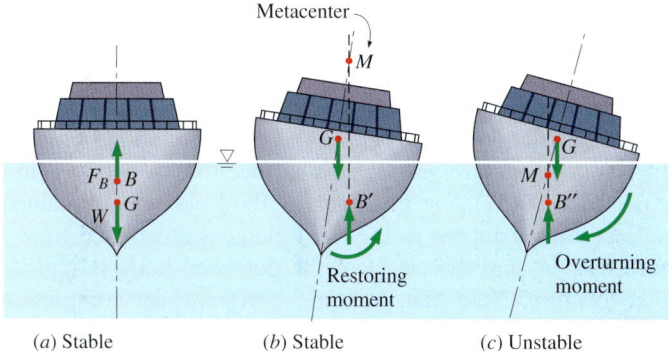

FIGURE 3–51
A floating body is *stable* if the body is (*a*) bottom-heavy and thus the center of gravity *G* is below the centroid *B* of the body, or (*b*) if the metacenter *M* is above point *G*. However, the body is (*c*) *unstable* if point *M* is below point *G*.

latter case, the weight and the buoyant force acting on the tilted body generate an overturning moment instead of a restoring moment, causing the body to capsize. The length of the metacentric height *GM* above *G* is a measure of the stability: the larger it is, the more stable is the floating body.

As already discussed, a boat can tilt to some maximum angle without capsizing, but beyond that angle it overturns (and sinks). We make a final analogy between the stability of floating objects and the stability of a ball rolling along the floor. Namely, imagine the ball in a trough between two hills (Fig. 3–52). The ball returns to its stable equilibrium position after being perturbed—up to a limit. If the perturbation amplitude is too great, the ball rolls down the opposite side of the hill and does not return to its equilibrium position. This situation is described as stable up to some limiting level of disturbance, but unstable beyond.

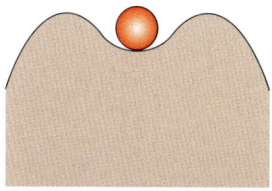

FIGURE 3–52
A ball in a trough between two hills is stable for small disturbances, but unstable for large disturbances.

3–7 ■ FLUIDS IN RIGID-BODY MOTION

We showed in Section 3–1 that pressure at a given point has the same magnitude in all directions, and thus it is a *scalar* function. In this section we obtain relations for the variation of pressure in fluids moving like a solid body with or without acceleration in the absence of any shear stresses (i.e., no motion between fluid layers relative to each other).

Many fluids such as milk and gasoline are transported in tankers. In an accelerating tanker, the fluid rushes to the back, and some initial splashing occurs. But then a new free surface (usually nonhorizontal) is formed, each fluid particle assumes the same acceleration, and the entire fluid moves like a rigid body. No shear stresses exist within the fluid body since there is no deformation and thus no change in shape. Rigid-body motion of a fluid also occurs when the fluid is contained in a tank that rotates about an axis.

Consider a differential rectangular fluid element of side lengths *dx*, *dy*, and *dz* in the *x*-, *y*-, and *z*-directions, respectively, with the *z*-axis being upward in the vertical direction (Fig. 3–53). Noting that the differential fluid element behaves like a *rigid body, Newton's second law of motion* for this element can be expressed as

$$\delta \vec{F} = \delta m \cdot \vec{a} \tag{3-34}$$

where $\delta m = \rho \, dV = \rho \, dx \, dy \, dz$ is the mass of the fluid element, $\vec{a}$ is the acceleration, and $\delta \vec{F}$ is the net force acting on the element.

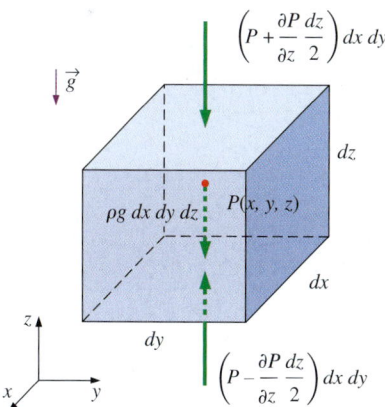

FIGURE 3–53
The surface and body forces acting on a differential fluid element in the vertical direction.

The forces acting on the fluid element consist of *body forces* such as gravity that act throughout the entire body of the element and are proportional to the volume of the body (and also electrical and magnetic forces, which will not be considered in this text), and *surface forces* such as the pressure forces that act on the surface of the element and are proportional to the surface area (shear stresses are also surface forces, but they do not apply in this case since the relative positions of fluid elements remain unchanged). The surface forces appear as the fluid element is isolated from its surroundings for analysis, and the effect of the detached body is replaced by a force at that location. Note that pressure represents the compressive force applied on the fluid element by the surrounding fluid and is always normal to the surface and inward toward the surface.

Taking the pressure at the center of the element to be P, the pressures at the top and bottom surfaces of the element can be expressed as $P + (\partial P/\partial z)\,dz/2$ and $P - (\partial P/\partial z)\,dz/2$, respectively. Noting that the pressure force acting on a surface is equal to the average pressure multiplied by the surface area, the net surface force acting on the element in the z-direction is the difference between the pressure forces acting on the bottom and top faces,

$$\delta F_{S,z} = \left(P - \frac{\partial P}{\partial z}\frac{dz}{2}\right)dx\,dy - \left(P + \frac{\partial P}{\partial z}\frac{dz}{2}\right)dx\,dy = -\frac{\partial P}{\partial z}dx\,dy\,dz \qquad (3\text{–}35)$$

Similarly, the net surface forces in the x- and y-directions are

$$\delta F_{S,x} = -\frac{\partial P}{\partial x}dx\,dy\,dz \quad \text{and} \quad \delta F_{S,y} = -\frac{\partial P}{\partial y}dx\,dy\,dz \qquad (3\text{–}36)$$

Then the surface force (which is simply the pressure force) acting on the entire element can be expressed in vector form as

$$\delta \vec{F}_S = \delta F_{S,x}\vec{i} + \delta F_{S,y}\vec{j} + \delta F_{S,z}\vec{k}$$

$$= -\left(\frac{\partial P}{\partial x}\vec{i} + \frac{\partial P}{\partial y}\vec{j} + \frac{\partial P}{\partial z}\vec{k}\right)dx\,dy\,dz = -\vec{\nabla}P\,dx\,dy\,dz \qquad (3\text{–}37)$$

where $\vec{i}, \vec{j}$, and $\vec{k}$ are the unit vectors in the x-, y-, and z-directions, respectively, and

$$\vec{\nabla}P = \frac{\partial P}{\partial x}\vec{i} + \frac{\partial P}{\partial y}\vec{j} + \frac{\partial P}{\partial z}\vec{k} \qquad (3\text{–}38)$$

is the *pressure gradient*. Note that $\vec{\nabla}$ or "del" is a vector operator that is used to express the gradients of a scalar function compactly in vector form. Also, the *gradient* of a scalar function is expressed in a given *direction* and thus it is a *vector* quantity.

The only body force acting on the fluid element is the weight of the element acting in the negative z-direction, and it is expressed as $\delta F_{B,z} = -g\delta m = -\rho g\,dx\,dy\,dz$ or in vector form as

$$\delta \vec{F}_{B,z} = -g\delta m\vec{k} = -\rho g\,dx\,dy\,dz\vec{k} \qquad (3\text{–}39)$$

Then the total force acting on the element becomes

$$\delta \vec{F} = \delta \vec{F}_S + \delta \vec{F}_B = -(\vec{\nabla}P + \rho g\vec{k})\,dx\,dy\,dz \qquad (3\text{–}40)$$

Substituting into Newton's second law of motion $\delta \vec{F} = \delta m \cdot \vec{a} = \rho \, dx \, dy \, dz \cdot \vec{a}$ and canceling $dx \, dy \, dz$, the general **equation of motion** for a fluid that acts as a rigid body (no shear stresses) is determined to be

Rigid-body motion of fluids: $$\vec{\nabla} P + \rho g \vec{k} = -\rho \vec{a} \quad (3\text{–}41)$$

Resolving the vectors into their components, this relation can be expressed more explicitly as

$$\frac{\partial P}{\partial x}\vec{i} + \frac{\partial P}{\partial y}\vec{j} + \frac{\partial P}{\partial z}\vec{k} + \rho g \vec{k} = -\rho(a_x\vec{i} + a_y\vec{j} + a_z\vec{k}) \quad (3\text{–}42)$$

or, in scalar form in the three orthogonal directions as

Accelerating fluids: $\quad \dfrac{\partial P}{\partial x} = -\rho a_x, \quad \dfrac{\partial P}{\partial y} = -\rho a_y, \quad \text{and} \quad \dfrac{\partial P}{\partial z} = -\rho(g + a_z) \quad (3\text{–}43)$

where a_x, a_y, and a_z are accelerations in the x-, y-, and z-directions, respectively.

Special Case 1: Fluids at Rest

For fluids at rest or moving on a straight path at constant velocity, all components of acceleration are zero, and the relations in Eqs. 3–43 reduce to

Fluids at rest: $\quad \dfrac{\partial P}{\partial x} = 0, \quad \dfrac{\partial P}{\partial y} = 0, \quad \text{and} \quad \dfrac{dP}{dz} = -\rho g \quad (3\text{–}44)$

which confirm that, in fluids at rest, the pressure remains constant in any horizontal direction (P is independent of x and y) and varies only in the vertical direction as a result of gravity [and thus $P = P(z)$]. These relations are applicable for both compressible and incompressible fluids (Fig. 3–54).

Special Case 2: Free Fall of a Fluid Body

A freely falling body accelerates under the influence of gravity. When the air resistance is negligible, the acceleration of the body equals the gravitational acceleration, and acceleration in any horizontal direction is zero. Therefore, $a_x = a_y = 0$ and $a_z = -g$. Then the equations of motion for accelerating fluids (Eqs. 3–43) reduce to

Free-falling fluids: $\quad \dfrac{\partial P}{\partial x} = \dfrac{\partial P}{\partial y} = \dfrac{\partial P}{\partial z} = 0 \quad \rightarrow \quad P = \text{constant} \quad (3\text{–}45)$

Therefore, in a frame of reference moving with the fluid, it behaves like it is in an environment with zero gravity. (This is the situation in an orbiting spacecraft, by the way. Gravity is *not* zero up there, despite what many people think!) Also, the gage pressure in a drop of liquid in free fall is zero throughout. (Actually, the gage pressure is slightly above zero due to surface tension, which holds the drop intact.)

When the direction of motion is reversed and the fluid is forced to accelerate vertically with $a_z = +g$ by placing the fluid container in an elevator or a space vehicle propelled upward by a rocket engine, the pressure gradient in the z-direction is $\partial P/\partial z = -2\rho g$. Therefore, the pressure difference across a fluid layer now doubles relative to the stationary fluid case (Fig. 3–55).

FIGURE 3–54
A glass of water at rest is a special case of a fluid in rigid-body motion. If the glass of water were moving at constant velocity in any direction, the hydrostatic equations would still apply.
© Imagestate Media (John Foxx)/Imagestate RF

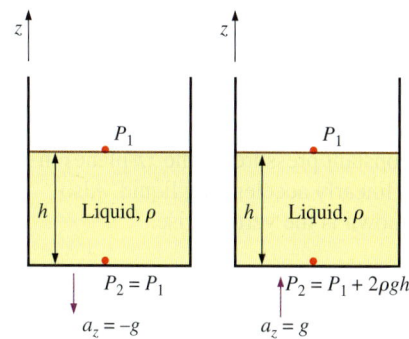

(a) Free fall of a liquid

(b) Upward acceleration of a liquid with $a_z = +g$

FIGURE 3–55
The effect of acceleration on the pressure of a liquid during free fall and upward acceleration.

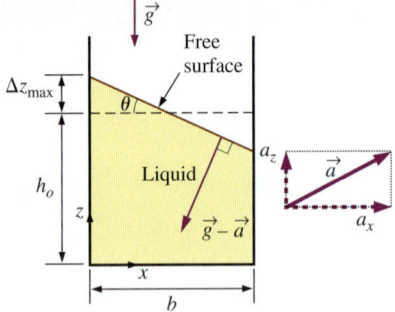

FIGURE 3–56
Rigid-body motion of a liquid in a linearly accelerating tank. The system behaves like a fluid at rest except that $\vec{g} - \vec{a}$ replaces $\vec{g}$ in the hydrostatic equations.

Acceleration on a Straight Path

Consider a container partially filled with a liquid. The container is moving on a straight path with a constant acceleration. We take the projection of the path of motion on the horizontal plane to be the x-axis, and the projection on the vertical plane to be the z-axis, as shown in Fig. 3–56. The x- and z-components of acceleration are a_x and a_z. There is no movement in the y-direction, and thus the acceleration in that direction is zero, $a_y = 0$. Then the equations of motion for accelerating fluids (Eqs. 3–43) reduce to

$$\frac{\partial P}{\partial x} = -\rho a_x, \quad \frac{\partial P}{\partial y} = 0, \quad \text{and} \quad \frac{\partial P}{\partial z} = -\rho(g + a_z) \quad (3\text{–}46)$$

Therefore, pressure is independent of y. Then the total differential of $P = P(x, z)$, which is $(\partial P/\partial x)dx + (\partial P/\partial z)dz$, becomes

$$dP = -\rho a_x \, dx - \rho(g + a_z) \, dz \quad (3\text{–}47)$$

For $\rho =$ constant, the pressure difference between two points 1 and 2 in the fluid is determined by integration to be

$$P_2 - P_1 = -\rho a_x(x_2 - x_1) - \rho(g + a_z)(z_2 - z_1) \quad (3\text{–}48)$$

Taking point 1 to be the origin ($x = 0$, $z = 0$) where the pressure is P_0 and point 2 to be any point in the fluid (no subscript), the pressure distribution is expressed as

Pressure variation: $\quad P = P_0 - \rho a_x x - \rho(g + a_z)z \quad (3\text{–}49)$

The vertical rise (or drop) of the free surface at point 2 relative to point 1 is determined by choosing both 1 and 2 on the free surface (so that $P_1 = P_2$), and solving Eq. 3–48 for $z_2 - z_1$ (Fig. 3–57),

Vertical rise of surface: $\quad \Delta z_s = z_{s2} - z_{s1} = -\dfrac{a_x}{g + a_z}(x_2 - x_1) \quad (3\text{–}50)$

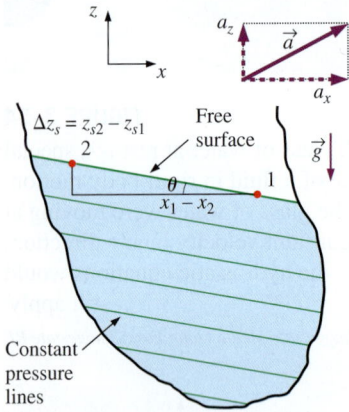

FIGURE 3–57
Lines of constant pressure (which are the projections of the surfaces of constant pressure on the xz-plane) in a linearly accelerating liquid. Also shown is the vertical rise.

where z_s is the z-coordinate of the liquid's free surface. The equation for surfaces of constant pressure, called **isobars**, is obtained from Eq. 3–47 by setting $dP = 0$ and replacing z by z_{isobar}, which is the z-coordinate (the vertical distance) of the surface as a function of x. It gives

Surfaces of constant pressure: $\quad \dfrac{dz_{\text{isobar}}}{dx} = -\dfrac{a_x}{g + a_z} = \text{constant} \quad (3\text{–}51)$

Thus we conclude that the isobars (including the free surface) in an incompressible fluid with constant acceleration in linear motion are parallel surfaces whose slope in the xz-plane is

Slope of isobars: $\quad \text{Slope} = \dfrac{dz_{\text{isobar}}}{dx} = -\dfrac{a_x}{g + a_z} = -\tan\theta \quad (3\text{–}52)$

Obviously, the free surface of such a fluid is a *plane* surface, and it is inclined unless $a_x = 0$ (the acceleration is in the vertical direction only). Also, conservation of mass, together with the assumption of incompressibility ($\rho =$ *constant*), requires that the volume of the fluid remain constant before and during acceleration. Therefore, the rise of fluid level on one side must be balanced by a drop of fluid level on the other side.

■ **EXAMPLE 3–12** Overflow from a Water Tank During Acceleration

An 80-cm-high fish tank of cross section 2 m × 0.6 m that is partially filled with water is to be transported on the back of a truck (Fig. 3–58). The truck accelerates from 0 to 90 km/h in 10 s. If it is desired that no water spills during acceleration, determine the allowable initial water height in the tank. Would you recommend the tank to be aligned with the long or short side parallel to the direction of motion?

SOLUTION A fish tank is to be transported on a truck. The allowable water height to avoid spill of water during acceleration and the proper orientation are to be determined.
Assumptions 1 The road is horizontal during acceleration so that acceleration has no vertical component ($a_z = 0$). 2 Effects of splashing, braking, shifting gears, driving over bumps, climbing hills, etc., are assumed to be secondary and are not considered. 3 The acceleration remains constant.
Analysis We take the x-axis to be the direction of motion, the z-axis to be the upward vertical direction, and the origin to be the lower left corner of the tank. Noting that the truck goes from 0 to 90 km/h in 10 s, the acceleration of the truck is

$$a_x = \frac{\Delta V}{\Delta t} = \frac{(90 - 0) \text{ km/h}}{10 \text{ s}} \left(\frac{1 \text{ m/s}}{3.6 \text{ km/h}}\right) = 2.5 \text{ m/s}^2$$

The tangent of the angle the free surface makes with the horizontal is

$$\tan \theta = \frac{a_x}{g + a_z} = \frac{2.5}{9.81 + 0} = 0.255 \quad \text{(and thus } \theta = 14.3°\text{)}$$

The maximum vertical rise of the free surface occurs at the back of the tank, and the vertical midplane experiences no rise or drop during acceleration since it is a plane of symmetry. Then the vertical rise at the back of the tank relative to the midplane for the two possible orientations becomes

Case 1: The long side is parallel to the direction of motion:

$$\Delta z_{s1} = (b_1/2) \tan \theta = [(2 \text{ m})/2] \times 0.255 = 0.255 \text{ m} = \mathbf{25.5 \text{ cm}}$$

Case 2: The short side is parallel to the direction of motion:

$$\Delta z_{s2} = (b_2/2) \tan \theta = [(0.6 \text{ m})/2] \times 0.255 = 0.076 \text{ m} = \mathbf{7.6 \text{ cm}}$$

Therefore, assuming tipping is not a problem, **the tank should definitely be oriented such that its short side is parallel to the direction of motion.** Emptying the tank such that its free surface level drops just 7.6 cm in this case will be adequate to avoid spilling during acceleration.
Discussion Note that the orientation of the tank is important in controlling the vertical rise. Also, the analysis is valid for any fluid with constant density, not just water, since we used no information that pertains to water in the solution.

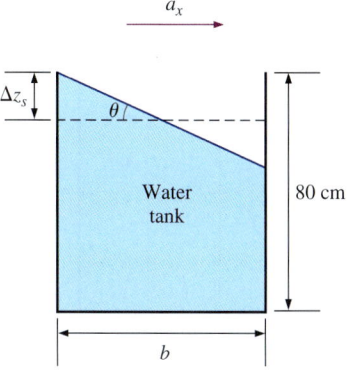

FIGURE 3–58
Schematic for Example 3–12.

Rotation in a Cylindrical Container

We know from experience that when a glass filled with water is rotated about its axis, the fluid is forced outward as a result of the so-called centrifugal

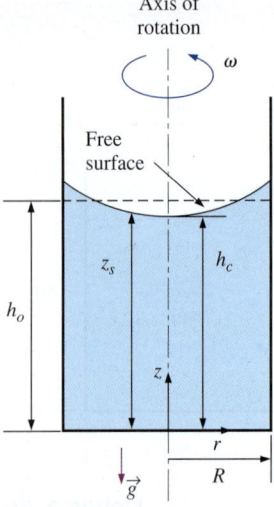

FIGURE 3–59
Rigid-body motion of a liquid in a rotating vertical cylindrical container.

force (but more properly explained in terms of centripetal acceleration), and the free surface of the liquid becomes concave. This is known as the *forced vortex motion*.

Consider a vertical cylindrical container partially filled with a liquid. The container is now rotated about its axis at a constant angular velocity of ω, as shown in Fig. 3–59. After initial transients, the liquid will move as a rigid body together with the container. There is no deformation, and thus there can be no shear stress, and every fluid particle in the container moves with the same angular velocity.

This problem is best analyzed in cylindrical coordinates (r, θ, z), with z taken along the centerline of the container directed from the bottom toward the free surface, since the shape of the container is a cylinder, and the fluid particles undergo a circular motion. The centripetal acceleration of a fluid particle rotating with a constant angular velocity of ω at a distance r from the axis of rotation is $r\omega^2$ and is directed radially toward the axis of rotation (negative r-direction). That is, $a_r = -r\omega^2$. There is symmetry about the z-axis, which is the axis of rotation, and thus there is no θ dependence. Then $P = P(r, z)$ and $a_\theta = 0$. Also, $a_z = 0$ since there is no motion in the z-direction.

Then the equation of motion for accelerating fluids (Eq. 3–41) reduces to

$$\frac{\partial P}{\partial r} = \rho r \omega^2, \quad \frac{\partial P}{\partial \theta} = 0, \quad \text{and} \quad \frac{\partial P}{\partial z} = -\rho g \qquad (3\text{–}53)$$

Then the total differential of $P = P(r, z)$, which is $dP = (\partial P/\partial r)dr + (\partial P/\partial z)dz$, becomes

$$dP = \rho r \omega^2 \, dr - \rho g \, dz \qquad (3\text{–}54)$$

The equation for surfaces of constant pressure is obtained by setting $dP = 0$ and replacing z by z_{isobar}, which is the z-value (the vertical distance) of the surface as a function of r. It gives

$$\frac{dz_{\text{isobar}}}{dr} = \frac{r\omega^2}{g} \qquad (3\text{–}55)$$

Integrating, the equation for the surfaces of constant pressure is determined to be

Surfaces of constant pressure: $\qquad z_{\text{isobar}} = \dfrac{\omega^2}{2g} r^2 + C_1 \qquad (3\text{–}56)$

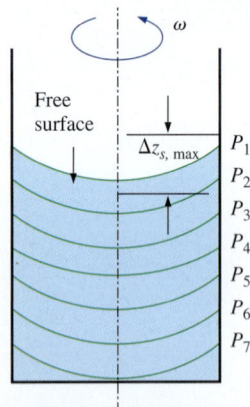

FIGURE 3–60
Surfaces of constant pressure in a rotating liquid.

which is the equation of a *parabola*. Thus we conclude that the surfaces of constant pressure, including the free surface, are *paraboloids of revolution* (Fig. 3–60).

The value of the integration constant C_1 is different for different paraboloids of constant pressure (i.e., for different isobars). For the free surface, setting $r = 0$ in Eq. 3–56 gives $z_{\text{isobar}}(0) = C_1 = h_c$, where h_c is the distance of the free surface from the bottom of the container along the axis of rotation (Fig. 3–59). Then the equation for the free surface becomes

$$z_s = \frac{\omega^2}{2g} r^2 + h_c \qquad (3\text{–}57)$$

where z_s is the distance of the free surface from the bottom of the container at radius r. The underlying assumption in this analysis is that there is sufficient liquid in the container so that the entire bottom surface remains covered with liquid.

The volume of a cylindrical shell element of radius r, height z_s, and thickness dr is $dV = 2\pi r z_s\, dr$. Then the volume of the paraboloid formed by the free surface is

$$V = \int_{r=0}^{R} 2\pi z_s r\, dr = 2\pi \int_{r=0}^{R} \left(\frac{\omega^2}{2g} r^2 + h_c\right) r\, dr = \pi R^2 \left(\frac{\omega^2 R^2}{4g} + h_c\right) \quad (3\text{-}58)$$

Since mass is conserved and density is constant, this volume must be equal to the original volume of the fluid in the container, which is

$$V = \pi R^2 h_0 \quad (3\text{-}59)$$

where h_0 is the original height of the fluid in the container with no rotation. Setting these two volumes equal to each other, the height of the fluid along the centerline of the cylindrical container becomes

$$h_c = h_0 - \frac{\omega^2 R^2}{4g} \quad (3\text{-}60)$$

Then the equation of the free surface becomes

Free surface: $$z_s = h_0 - \frac{\omega^2}{4g}(R^2 - 2r^2) \quad (3\text{-}61)$$

The paraboloid shape is independent of fluid properties, so the same free surface equation applies to *any* liquid. For example, spinning liquid mercury forms a parabolic mirror that is useful in astronomy (Fig. 3–61).

The maximum vertical height occurs at the edge where $r = R$, and the *maximum height difference* between the edge and the center of the free surface is determined by evaluating z_s at $r = R$ and also at $r = 0$, and taking their difference,

Maximum height difference: $$\Delta z_{s,\,max} = z_s(R) - z_s(0) = \frac{\omega^2}{2g} R^2 \quad (3\text{-}62)$$

When ρ = constant, the pressure difference between two points 1 and 2 in the fluid is determined by integrating $dP = \rho r \omega^2\, dr - \rho g\, dz$. This yields

$$P_2 - P_1 = \frac{\rho \omega^2}{2}(r_2^2 - r_1^2) - \rho g(z_2 - z_1) \quad (3\text{-}63)$$

Taking point 1 to be the origin ($r = 0$, $z = 0$) where the pressure is P_0 and point 2 to be any point in the fluid (no subscript), the pressure distribution is expressed as

Pressure variation: $$P = P_0 + \frac{\rho \omega^2}{2} r^2 - \rho g z \quad (3\text{-}64)$$

Note that at a fixed radius, the pressure varies hydrostatically in the vertical direction, as in a fluid at rest. For a fixed vertical distance z, the pressure varies with the square of the radial distance r, increasing from the centerline toward the outer edge. In any horizontal plane, the pressure difference between the center and edge of the container of radius R is $\Delta P = \rho \omega^2 R^2 / 2$.

FIGURE 3–61
The 6-meter spinning liquid-mercury mirror of the Large Zenith Telescope located near Vancouver, British Columbia.

Photo courtesy of Paul Hickson, The University of British Columbia. Used by permission.

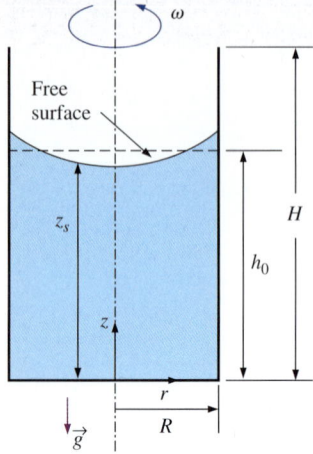

FIGURE 3–62
Schematic for Example 3–13.

EXAMPLE 3–13 Rising of a Liquid During Rotation

A 20-cm-diameter, 60-cm-high vertical cylindrical container, shown in Fig. 3–62, is partially filled with 50-cm-high liquid whose density is 850 kg/m^3. Now the cylinder is rotated at a constant speed. Determine the rotational speed at which the liquid will start spilling from the edges of the container.

SOLUTION A vertical cylindrical container partially filled with a liquid is rotated. The angular speed at which the liquid will start spilling is to be determined.

Assumptions 1 The increase in the rotational speed is very slow so that the liquid in the container always acts as a rigid body. 2 The bottom surface of the container remains covered with liquid during rotation (no dry spots).

Analysis Taking the center of the bottom surface of the rotating vertical cylinder as the origin ($r = 0$, $z = 0$), the equation for the free surface of the liquid is given as

$$z_s = h_0 - \frac{\omega^2}{4g}(R^2 - 2r^2)$$

Then the vertical height of the liquid at the edge of the container where $r = R$ becomes

$$z_s(R) = h_0 + \frac{\omega^2 R^2}{4g}$$

where $h_0 = 0.5$ m is the original height of the liquid before rotation. Just before the liquid starts spilling, the height of the liquid at the edge of the container equals the height of the container, and thus $z_s(R) = H = 0.6$ m. Solving the last equation for ω and substituting, the maximum rotational speed of the container is determined to be

$$\omega = \sqrt{\frac{4g(H - h_0)}{R^2}} = \sqrt{\frac{4(9.81 \text{ m/s}^2)[(0.6 - 0.5) \text{ m}]}{(0.1 \text{ m})^2}} = \mathbf{19.8 \text{ rad/s}}$$

Noting that one complete revolution corresponds to 2π rad, the rotational speed of the container can also be expressed in terms of revolutions per minute (rpm) as

$$\dot{n} = \frac{\omega}{2\pi} = \frac{19.8 \text{ rad/s}}{2\pi \text{ rad/rev}}\left(\frac{60 \text{ s}}{1 \text{ min}}\right) = \mathbf{189 \text{ rpm}}$$

Therefore, the rotational speed of this container should be limited to 189 rpm to avoid any spill of liquid as a result of the centrifugal effect.

Discussion Note that the analysis is valid for any liquid since the result is independent of density or any other fluid property. We should also verify that our assumption of no dry spots is valid. The liquid height at the center is

$$z_s(0) = h_0 - \frac{\omega^2 R^2}{4g} = 0.4 \text{ m}$$

Since $z_s(0)$ is positive, our assumption is validated.

SUMMARY

The normal force exerted by a fluid per unit area is called *pressure*, and its SI unit is the *pascal*, 1 Pa ≡ 1 N/m². The pressure relative to absolute vacuum is called the *absolute pressure*, and the difference between the absolute pressure and the local atmospheric pressure is called the *gage pressure*. Pressures below atmospheric pressure are sometimes called *vacuum pressures*. The absolute, gage, and vacuum pressures are related by

$$P_{gage} = P_{abs} - P_{atm}$$
$$P_{vac} = P_{atm} - P_{abs} = -P_{gage}$$

The pressure at a point in a fluid has the same magnitude in all directions. The variation of pressure with elevation in a fluid at rest is given by

$$\frac{dP}{dz} = -\rho g$$

where the positive z-direction is taken to be upward by convention. When the density of the fluid is constant, the pressure difference across a fluid layer of thickness Δz is

$$P_{below} = P_{above} + \rho g |\Delta z| = P_{above} + \gamma_s |\Delta z|$$

The absolute and gage pressures in a static liquid open to the atmosphere at a depth h from the free surface are

$$P = P_{atm} + \rho g h \quad \text{and} \quad P_{gage} = \rho g h$$

The pressure in a fluid at rest does not vary in the horizontal direction. *Pascal's law* states that the pressure applied to a confined fluid increases the pressure throughout by the same amount. The atmospheric pressure can be measured by a *barometer* and is given by

$$P_{atm} = \rho g h$$

where h is the height of the liquid column.

Fluid statics deals with problems associated with fluids at rest, and it is called *hydrostatics* when the fluid is a liquid. The magnitude of the resultant force acting on a plane surface of a completely submerged plate in a homogeneous fluid is equal to the product of the pressure P_C at the centroid of the surface and the area A of the surface and is expressed as

$$F_R = (P_0 + \rho g h_C)A = P_C A = P_{avg} A$$

where $h_C = y_C \sin \theta$ is the *vertical distance* of the centroid from the free surface of the liquid. The pressure P_0 is usually atmospheric pressure, which cancels out in most cases since it acts on both sides of the plate. The point of intersection of the line of action of the resultant force and the surface is the *center of pressure*. The vertical location of the line of action of the resultant force is given by

$$y_P = y_C + \frac{I_{xx,C}}{[y_C + P_0/(\rho g \sin \theta)]A}$$

where $I_{xx,C}$ is the second moment of area about the x-axis passing through the centroid of the area.

A fluid exerts an upward force on a body immersed in it. This force is called the *buoyant force* and is expressed as

$$F_B = \rho_f g V$$

where V is the volume of the body. This is known as *Archimedes' principle* and is expressed as: the buoyant force acting on a body immersed in a fluid is equal to the weight of the fluid displaced by the body; it acts upward through the centroid of the displaced volume. In a fluid with constant density, the buoyant force is independent of the distance of the body from the free surface. For *floating* bodies, the submerged volume fraction of the body is equal to the ratio of the average density of the body to the density of the fluid.

The general *equation of motion* for a fluid that acts as a rigid body is

$$\vec{\nabla} P + \rho g \vec{k} = -\rho \vec{a}$$

When gravity is aligned in the $-z$-direction, it is expressed in scalar form as

$$\frac{\partial P}{\partial x} = -\rho a_x, \quad \frac{\partial P}{\partial y} = -\rho a_y, \quad \text{and} \quad \frac{\partial P}{\partial z} = -\rho(g + a_z)$$

where a_x, a_y, and a_z are accelerations in the x-, y-, and z-directions, respectively. During *linearly accelerating motion* in the xz-plane, the pressure distribution is expressed as

$$P = P_0 - \rho a_x x - \rho(g + a_z)z$$

The surfaces of constant pressure (including the free surface) in a liquid with constant acceleration in linear motion are parallel surfaces whose slope in some xz-plane is

$$\text{Slope} = \frac{dz_{isobar}}{dx} = -\frac{a_x}{g + a_z} = -\tan \theta$$

During rigid-body motion of a liquid in a *rotating cylinder*, the surfaces of constant pressure are *paraboloids of revolution*. The equation for the free surface is

$$z_s = h_0 - \frac{\omega^2}{4g}(R^2 - 2r^2)$$

where z_s is the distance of the free surface from the bottom of the container at radius r and h_0 is the original height of the fluid in the container with no rotation. The variation of pressure in the liquid is expressed as

$$P = P_0 + \frac{\rho \omega^2}{2} r^2 - \rho g z$$

where P_0 is the pressure at the origin ($r = 0$, $z = 0$).

Pressure is a fundamental property, and it is hard to imagine a significant fluid flow problem that does not involve pressure. Therefore, you will see this property in all chapters in the rest of this book. The consideration of hydrostatic forces acting on plane or curved surfaces, however, is mostly limited to this chapter.

REFERENCES AND SUGGESTED READING

1. F. P. Beer, E. R. Johnston, Jr., E. R. Eisenberg, and G. H. Staab. *Vector Mechanics for Engineers, Statics,* 10th ed. New York: McGraw-Hill, 2012.

2. D. C. Giancoli. *Physics,* 6th ed. Upper Saddle River, NJ: Prentice Hall, 2012.

PROBLEMS*

Pressure, Manometer, and Barometer

3–1C A tiny steel cube is suspended in water by a string. If the lengths of the sides of the cube are very small, how would you compare the magnitudes of the pressures on the top, bottom, and side surfaces of the cube?

3–2C Express Pascal's law, and give a real-world example of it.

3–3C Consider two identical fans, one at sea level and the other on top of a high mountain, running at identical speeds. How would you compare (*a*) the volume flow rates and (*b*) the mass flow rates of these two fans?

3–4C What is the difference between gage pressure and absolute pressure?

3–5C Explain why some people experience nose bleeding and some others experience shortness of breath at high elevations.

3–6 The piston of a vertical piston-cylinder device containing a gas has a mass of 40 kg and a cross-sectional area of 0.012 m² (Fig P3–6). The local atmospheric pressure is 95 kPa, and the gravitational acceleration is 9.81 m/s². (*a*) Determine the pressure inside the cylinder. (*b*) If some heat is transferred to the gas and its volume is doubled, do you expect the pressure inside the cylinder to change?

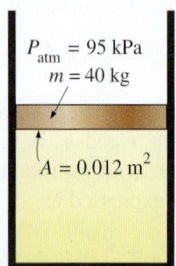

FIGURE P3–6

3–7 A vacuum gage connected to a chamber reads 36 kPa at a location where the atmospheric pressure is 92 kPa. Determine the absolute pressure in the chamber.

3–8 A manometer is used to measure the air pressure in a tank. The fluid used has a specific gravity of 1.25, and the differential height between the two arms of the manometer is 70 cm. If the local atmospheric pressure is 88 kPa, determine the absolute pressure in the tank for the cases of the manometer arm with the (*a*) higher and (*b*) lower fluid level being attached to the tank.

3–9 The water in a tank is pressurized by air, and the pressure is measured by a multifluid manometer as shown in Fig. P3–9. Determine the gage pressure of air in the tank if $h_1 = 0.4$ m, $h_2 = 0.6$ m, and $h_3 = 0.8$ m. Take the densities of water, oil, and mercury to be 1000 kg/m³, 850 kg/m³, and 13,600 kg/m³, respectively.

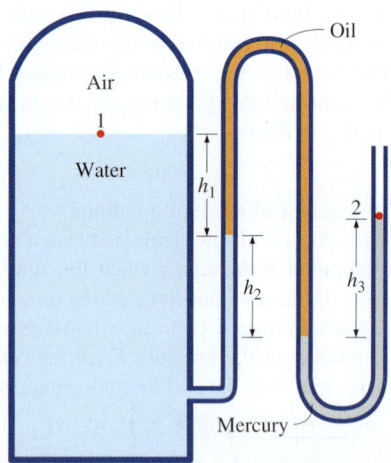

FIGURE P3–9

3–10 Determine the atmospheric pressure at a location where the barometric reading is 735 mmHg. Take the density of mercury to be 13,600 kg/m³.

3–11 The gage pressure in a liquid at a depth of 3 m is read to be 28 kPa. Determine the gage pressure in the same liquid at a depth of 12 m.

3–12 The absolute pressure in water at a depth of 8 m is read to be 175 kPa. Determine (*a*) the local atmospheric pressure, and (*b*) the absolute pressure at a depth of 8 m in a liquid whose specific gravity is 0.78 at the same location.

---

* Problems designated by a "C" are concept questions, and students are encouraged to answer them all. Problems with the icon are solved using EES, and complete solutions together with parametric studies are included on the text website. Problems with the icon are comprehensive in nature and are intended to be solved with an equation solver such as EES.

3–13 A 90-kg man has a total foot imprint area of 450 cm². Determine the pressure this man exerts on the ground if (a) he stands on both feet and (b) he stands on one foot.

3–14 Consider a 55-kg woman who has a total foot imprint area of 400 cm². She wishes to walk on the snow, but the snow cannot withstand pressures greater than 0.5 kPa. Determine the minimum size of the snowshoes needed (imprint area per shoe) to enable her to walk on the snow without sinking.

3–15 A vacuum gage connected to a tank reads 45 kPa at a location where the barometric reading is 755 mmHg. Determine the absolute pressure in the tank. Take $\rho_{Hg} = 13{,}590$ kg/m³. *Answer:* 55.6 kPa

3–16 A pressure gage connected to a tank reads 350 kPa at a location where the barometric reading is 740 mmHg. Determine the absolute pressure in the tank. Take $\rho_{Hg} = 13{,}590$ kg/m³. *Answer:* 449 kPa

3–17 A pressure gage connected to a tank reads 500 kPa at a location where the atmospheric pressure is 94 kPa. Determine the absolute pressure in the tank.

3–18 If the pressure inside a rubber balloon is 1500 mmHg, what is this pressure in bar? *Answer:* 2.00 bar

3–19 The vacuum pressure of a condenser is given to be 80 kPa. If the atmospheric pressure is 98 kPa, what is the gage pressure and absolute pressure in kPa, kN/m², lbf/in², psi, and mmHg.

3–20 Water from a reservoir is raised in a vertical tube of internal diameter $D = 30$ cm under the influence of the pulling force F of a piston. Determine the force needed to raise the water to a height of $h = 1.5$ m above the free surface. What would your response be for $h = 3$ m? Also, taking the atmospheric pressure to be 96 kPa, plot the absolute water pressure at the piston face as h varies from 0 to 3 m.

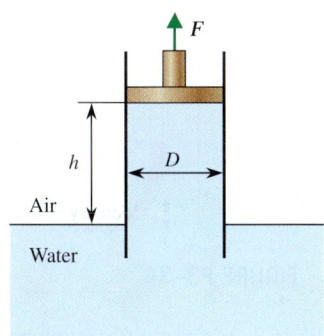

FIGURE P3–20

3–21 The barometer of a mountain hiker reads 980 mbars at the beginning of a hiking trip and 790 mbars at the end. Neglecting the effect of altitude on local gravitational acceleration, determine the vertical distance climbed. Assume an average air density of 1.20 kg/m³. *Answer:* 1614 m

3–22 The basic barometer can be used to measure the height of a building. If the barometric readings at the top and at the bottom of a building are 730 and 755 mmHg, respectively, determine the height of the building. Assume an average air density of 1.18 kg/m³.

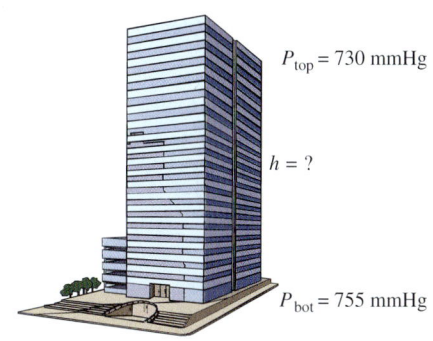

FIGURE P3–22

3–23 Solve Prob. 3–22 using EES (or other) software. Print out the entire solution, including the numerical results with proper units, and take the density of mercury to be 13,600 kg/m³.

3–24 Determine the pressure exerted on a diver at 20 m below the free surface of the sea. Assume a barometric pressure of 101 kPa and a specific gravity of 1.03 for seawater. *Answer:* 303 kPa

3–25 Determine the pressure exerted on the surface of a submarine cruising 70 m below the free surface of the sea. Assume that the barometric pressure is 101 kPa and the specific gravity of seawater is 1.03.

3–26 A gas is contained in a vertical, frictionless piston–cylinder device. The piston has a mass of 4 kg and a cross-sectional area of 35 cm². A compressed spring above the piston exerts a force of 60 N on the piston. If the atmospheric pressure is 95 kPa, determine the pressure inside the cylinder. *Answer:* 123.4 kPa

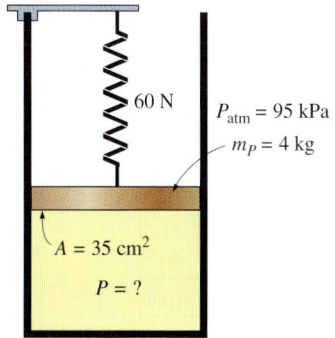

FIGURE P3–26

3–27 Reconsider Prob. 3–26. Using EES (or other) software, investigate the effect of the spring force in the range of 0 to 500 N on the pressure inside the cylinder. Plot the pressure against the spring force, and discuss the results.

3–28 Both a gage and a manometer are attached to a gas tank to measure its pressure. If the reading on the pressure gage is 65 kPa, determine the distance between the two fluid levels of the manometer if the fluid is (a) mercury ($\rho = 13,600$ kg/m³) or (b) water ($\rho = 1000$ kg/m³).

3–32 The manometer shown in the figure is designed to measure pressures of up to a maximum of 100 Pa. If the reading error is estimated to be ±0.5 mm, what should the ratio of d/D be in order for the error associated with pressure measurement not to exceed 2.5% of the full scale.

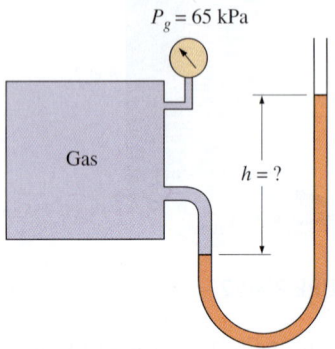

FIGURE P3–28

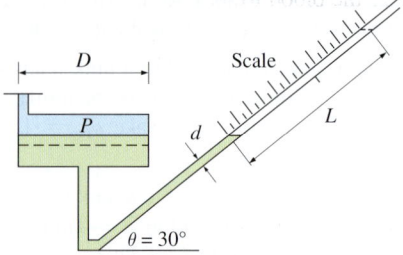

FIGURE P3–32

3–29 Reconsider Prob. 3–28. Using EES (or other) software, investigate the effect of the manometer fluid density in the range of 800 to 13,000 kg/m³ on the differential fluid height of the manometer. Plot the differential fluid height against the density, and discuss the results.

3–30 The variation of pressure P in a gas with density ρ is is given by $P = C\rho^n$ where C and n and are constants with $P = P_0$ and $\rho = \rho_0$ at elevation $z = 0$. Obtain a relation for the variaton of P with elevation in terms of z, g, n, P_0 and ρ_0.

3–31 The system shown in the figure is used to accurately measure the pressure changes when the pressure is increased by ΔP in the water pipe. When $\Delta h = 70$ mm, what is the change in the pipe pressure?

3–33 A manometer containing oil ($\rho = 850$ kg/m³) is attached to a tank filled with air. If the oil-level difference between the two columns is 150 cm and the atmospheric pressure is 98 kPa, determine the absolute pressure of the air in the tank. *Answer:* 111 kPa

3–34 A mercury manometer ($\rho = 13,600$ kg/m³) is connected to an air duct to measure the pressure inside. The difference in the manometer levels is 10 mm, and the atmospheric pressure is 100 kPa. (a) Judging from Fig. P3–34, determine if the pressure in the duct is above or below the atmospheric pressure. (b) Determine the absolute pressure in the duct.

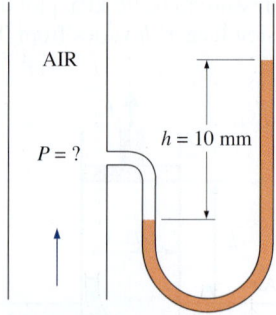

FIGURE P3–34

3–35 Repeat Prob. 3–34 for a differential mercury height of 30 mm.

3–36 Blood pressure is usually measured by wrapping a closed air-filled jacket equipped with a pressure gage around the upper arm of a person at the level of the heart. Using a mercury manometer and a stethoscope, the systolic pressure (the maximum pressure when the heart is pumping) and the diastolic pressure (the minimum pressure when the heart is resting) are measured in mmHg. The systolic and diastolic pressures of a healthy person are about 120 mmHg and 80 mmHg,

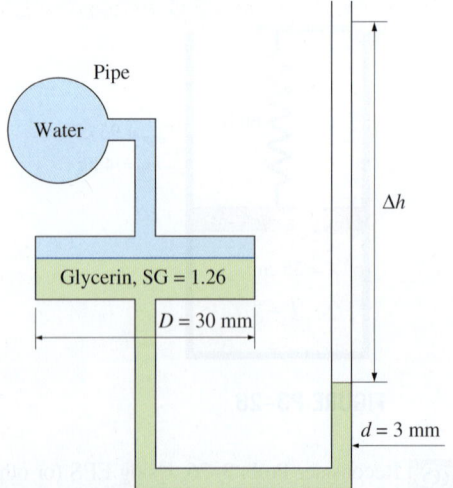

FIGURE P3–31

respectively, and are indicated as 120/80. Express both of these gage pressures in kPa, psi, and meter water column.

3–37 The maximum blood pressure in the upper arm of a healthy person is about 120 mmHg. If a vertical tube open to the atmosphere is connected to the vein in the arm of the person, determine how high the blood will rise in the tube. Take the density of the blood to be 1040 kg/m^3.

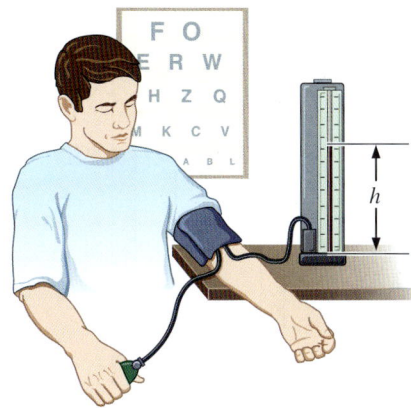

FIGURE P3–37

3–38 Consider a 1.73-m-tall man standing vertically in water and completely submerged in a pool. Determine the difference between the pressure acting on the head and on the toes of this man, in kPa.

3–39 Consider a U-tube whose arms are open to the atmosphere. Now water is poured into the U-tube from one arm, and light oil (ρ = 790 kg/m^3) from the other. One arm contains 70-cm-high water, while the other arm contains both fluids with an oil-to-water height ratio of 6. Determine the height of each fluid in that arm.

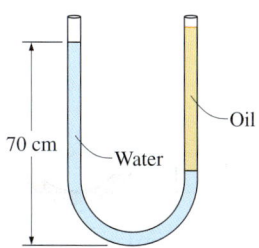

FIGURE P3–39

3–40 The hydraulic lift in a car repair shop has an output diameter of 40 cm and is to lift cars up to 1800 kg. Determine the fluid gage pressure that must be maintained in the reservoir.

3–41 Freshwater and seawater flowing in parallel horizontal pipelines are connected to each other by a double U-tube manometer, as shown in Fig. P3–41. Determine the pressure difference between the two pipelines. Take the density of seawater at that location to be ρ = 1035 kg/m^3. Can the air column be ignored in the analysis?

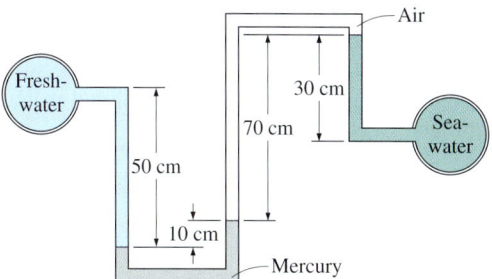

FIGURE P3–41

3–42 Repeat Prob. 3–41 by replacing the air with oil whose specific gravity is 0.72.

3–43 The gage pressure of the air in the tank shown in Fig. P3–43 is measured to be 65 kPa. Determine the differential height h of the mercury column.

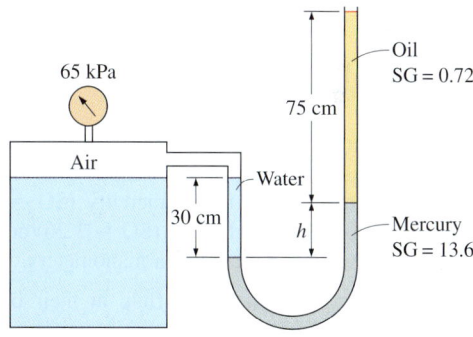

FIGURE P3–43

3–44 Repeat Prob. 3–43 for a gage pressure of 45 kPa.

3–45 The 500-kg load on the hydraulic lift shown in Fig. P3–45 is to be raised by pouring oil (ρ = 780 kg/m^3) into a thin tube. Determine how high h should be in order to begin to raise the weight.

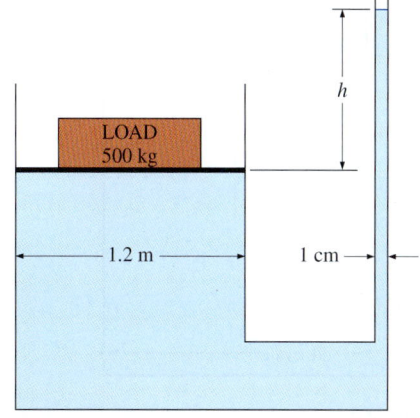

FIGURE P3–45

3–46 Two oil tanks are connected to each other through a manometer. If the difference between the mercury levels in the two arms is 80 cm, determine the pressure difference between the two tanks. The densities of oil and mercury are 721 kg/m³ and 13,600 kg/m³, respectively.

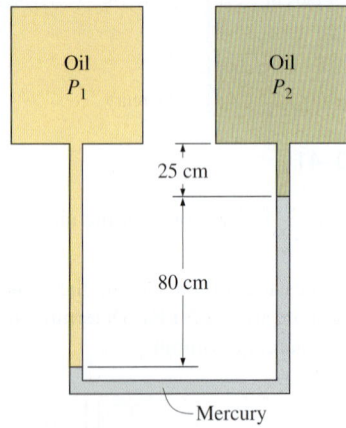

FIGURE P3–46

3–47 Pressure is often given in terms of a liquid column and is expressed as "pressure head." Express the standard atmospheric pressure in terms of (a) mercury (SG = 13.6), (b) water (SG = 1.0), and (c) glycerin (SG = 1.26) columns. Explain why we usually use mercury in manometers.

3–48 Two chambers with the same fluid at their base are separated by a 30-cm-diameter piston whose weight is 25 N, as shown in Fig. P3–48. Calculate the gage pressures in chambers A and B.

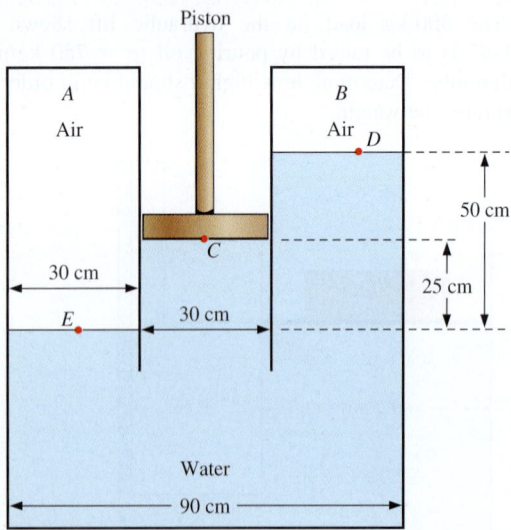

FIGURE P3–48

3–49 Consider a double-fluid manometer attached to an air pipe shown in Fig. P3–49. If the specific gravity of one fluid is 13.55, determine the specific gravity of the other fluid for the indicated absolute pressure of air. Take the atmospheric pressure to be 100 kPa. *Answer:* 1.34

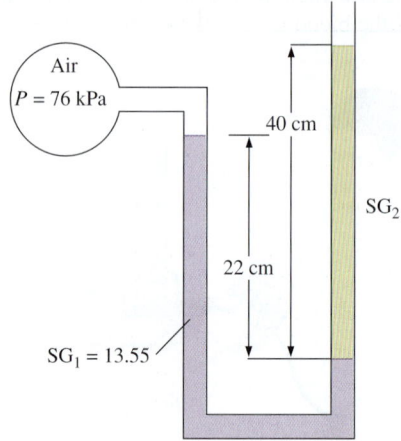

FIGURE P3–49

3–50 The pressure difference between an oil pipe and water pipe is measured by a double-fluid manometer, as shown in Fig. P3–50. For the given fluid heights and specific gravities, calculate the pressure difference $\Delta P = P_B - P_A$.

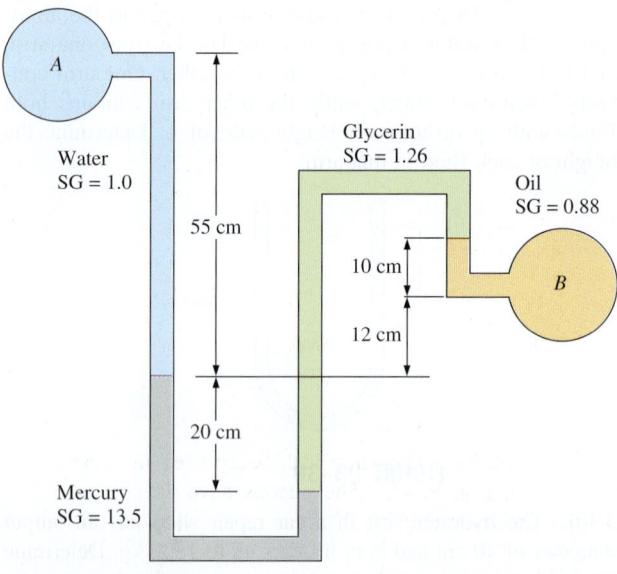

FIGURE P3–50

3–51 Consider the system shown in Fig. P3–51. If a change of 0.9 kPa in the pressure of air causes the brine-mercury interface in the right column to drop by 5 mm in the brine level in the right column while the pressure in the brine pipe remains constant, determine the ratio of A_2/A_1.

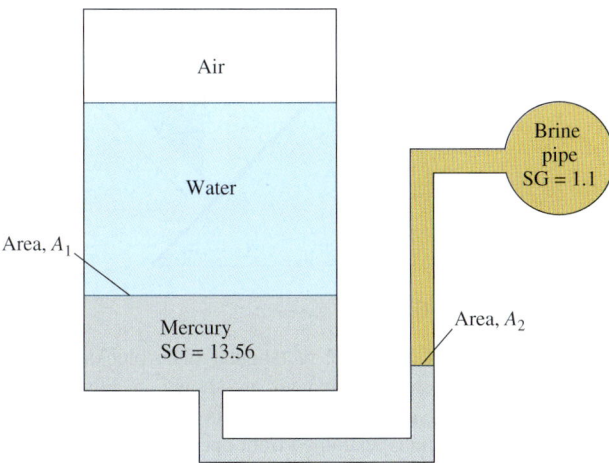

FIGURE P3–51

3–52 Two water tanks are connected to each other through a mercury manometer with inclined tubes, as shown in Fig. P3–52. If the pressure difference between the two tanks is 20 kPa, calculate a and θ.

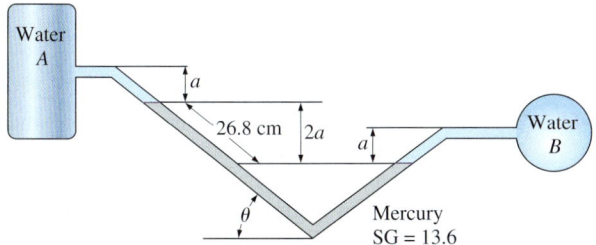

FIGURE P3–52

3–53 Consider a hydraulic jack being used in a car repair shop, as in Fig. P3–53. The pistons have an area of $A_1 = 0.8$ cm² and $A_2 = 0.04$ m². Hydraulic oil with a specific gravity of 0.870 is pumped in as the small piston on the left side is pushed up and down, slowly raising the larger piston on the right side. A car that weighs 13,000 N is to be jacked up. (a) At the beginning, when both pistons are at the same elevation ($h = 0$), calculate the force F_1 in newtons required to hold the weight of the car. (b) Repeat the calculation after the car has been lifted two meters ($h = 2$ m). Compare and discuss.

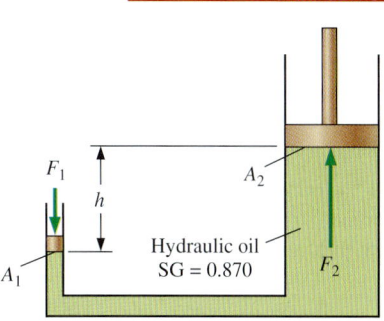

FIGURE P3–53

Fluid Statics: Hydrostatic Forces on Plane and Curved Surfaces

3–54C Define the resultant hydrostatic force acting on a submerged surface, and the center of pressure.

3–55C Someone claims that she can determine the magnitude of the hydrostatic force acting on a plane surface submerged in water regardless of its shape and orientation if she knew the vertical distance of the centroid of the surface from the free surface and the area of the surface. Is this a valid claim? Explain.

3–56C A submerged horizontal flat plate is suspended in water by a string attached at the centroid of its upper surface. Now the plate is rotated 45° about an axis that passes through its centroid. Discuss the change in the hydrostatic force acting on the top surface of this plate as a result of this rotation. Assume the plate remains submerged at all times.

3–57C You may have noticed that dams are much thicker at the bottom. Explain why dams are built that way.

3–58C Consider a submerged curved surface. Explain how you would determine the horizontal component of the hydrostatic force acting on this surface.

3–59C Consider a submerged curved surface. Explain how you would determine the vertical component of the hydrostatic force acting on this surface.

3–60C Consider a circular surface subjected to hydrostatic forces by a constant density liquid. If the magnitudes of the horizontal and vertical components of the resultant hydrostatic force are determined, explain how you would find the line of action of this force.

3–61 Consider a heavy car submerged in water in a lake with a flat bottom. The driver's side door of the car is 1.1 m high and 0.9 m wide, and the top edge of the door is 10 m below the water surface. Determine the net force acting on the door (normal to its surface) and the location of the pressure center if (a) the car is well-sealed and it contains air at atmospheric pressure and (b) the car is filled with water.

3–62 Consider a 8-m-long, 8-m-wide, and 2-m-high aboveground swimming pool that is filled with water to the rim. (*a*) Determine the hydrostatic force on each wall and the distance of the line of action of this force from the ground. (*b*) If the height of the walls of the pool is doubled and the pool is filled, will the hydrostatic force on each wall double or quadruple? Why? *Answer:* (*a*) 157 kN

3–63 Consider a 60-m-high, 360-m-wide dam filled to capacity. Determine (*a*) the hydrostatic force on the dam and (*b*) the force per unit area of the dam near the top and near the bottom.

3–64 A room in the lower level of a cruise ship has a 30-cm-diameter circular window. If the midpoint of the window is 4 m below the water surface, determine the hydrostatic force acting on the window, and the pressure center. Take the specific gravity of seawater to be 1.025. *Answers:* 2840 N, 4.001 m

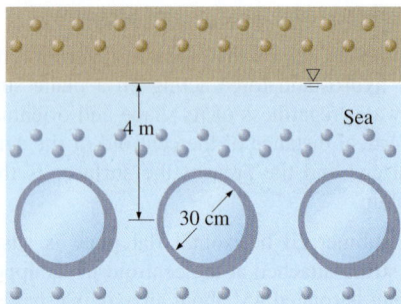

FIGURE P3–64

3–65 The water side of the wall of a 70-m-long dam is a quarter circle with a radius of 7 m. Determine the hydrostatic force on the dam and its line of action when the dam is filled to the rim.

3–66 For a gate width of 2 m into the paper (Fig. P3–66), determine the force required to hold the gate *ABC* at its location. *Answer:* 17.8 kN

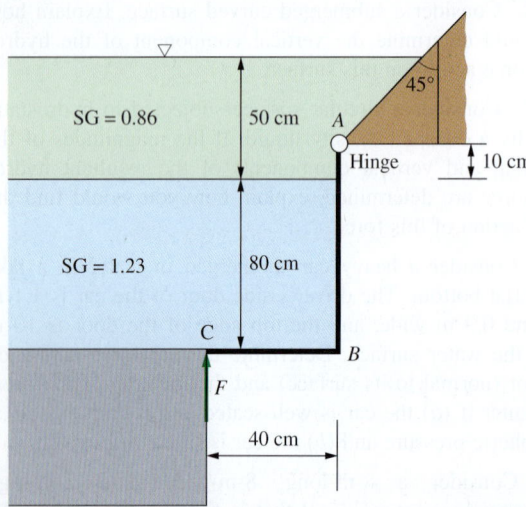

FIGURE P3–66

3–67 Determine the resultant force acting on the 0.7-m-high and 0.7-m-wide triangular gate shown in Fig. P3–67 and its line of action.

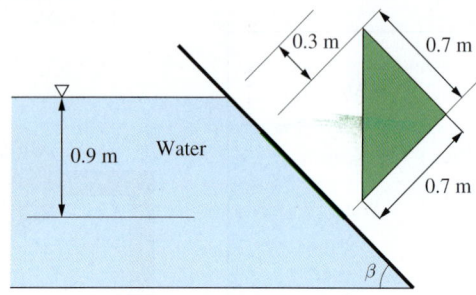

FIGURE P3–67

3–68 A 6-m-high, 5-m-wide rectangular plate blocks the end of a 5-m-deep freshwater channel, as shown in Fig. P3–68. The plate is hinged about a horizontal axis along its upper edge through a point *A* and is restrained from opening by a fixed ridge at point *B*. Determine the force exerted on the plate by the ridge.

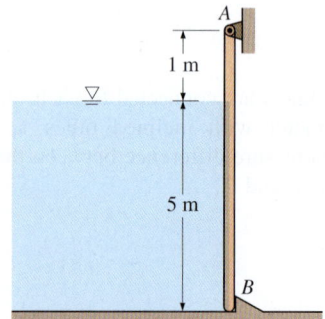

FIGURE P3–68

3–69 Reconsider Prob. 3–68. Using EES (or other) software, investigate the effect of water depth on the force exerted on the plate by the ridge. Let the water depth vary from 0 to 5 m in increments of 0.5 m. Tabulate and plot your results.

3–70 The flow of water from a reservoir is controlled by a 1.5-m-wide L-shaped gate hinged at point *A*, as shown in

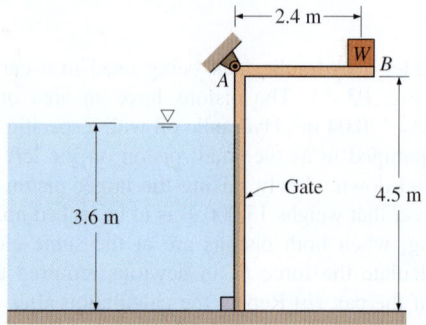

FIGURE P3–70

Fig. P3–70. If it is desired that the gate open when the water height is 3.6 m, determine the mass of the required weight W. *Answer:* 13,400 kg

3–71 Repeat Prob. 3–70 for a water height of 2.4 m.

3–72 A water trough of semicircular cross section of radius 0.6 m consists of two symmetric parts hinged to each other at the bottom, as shown in Fig. P3–72. The two parts are held together by a cable and turnbuckle placed every 3 m along the length of the trough. Calculate the tension in each cable when the trough is filled to the rim.

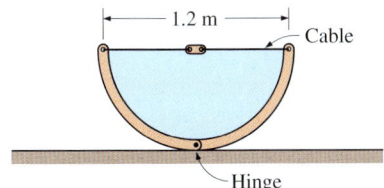

FIGURE P3–72

3–73 A cylindrical tank is fully filled with water (Fig. P3–73). In order to increase the flow from the tank, an additional pressure is applied to the water surface by a compressor. For $P_0 = 0$, $P_0 = 3$ bar, and $P_0 = 10$ bar, calculate the hydrostatic force on the surface A exerted by water.

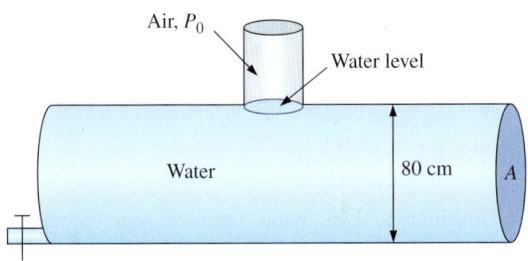

FIGURE P3–73

3–74 An open settling tank shown in the figure contains a liquid suspension. Determine the resultant force acting on the gate and its line of action if the liquid density is 850 kg/m³.
Answers: 140 kN, 1.64 m from bottom

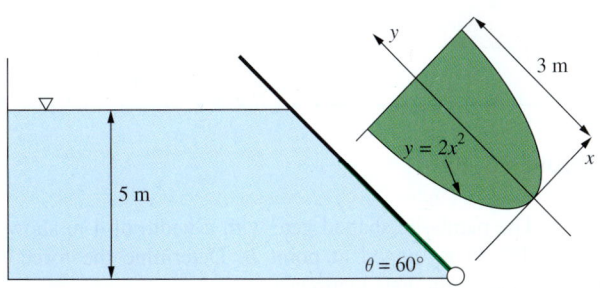

FIGURE P3–74

3–75 From Prob. 3-74, knowing that the density of the suspension depends on liquid depth and changes linearly from 800 kg/m³ to 900 kg/m³ in the vertical direction, determine the resultant force acting on the gate ABC, and its line of action.

3–76 The 2.5 m × 8.1 m × 6 m tank shown below is filled by oil of SG = 0.88. Determine (a) the magnitude and the location of the line of action of the resultant force acting on surface AB and (b) the pressure force acting on surface BD. Will the force acting on surface BD equal the weight of the oil in the tank? Explain.

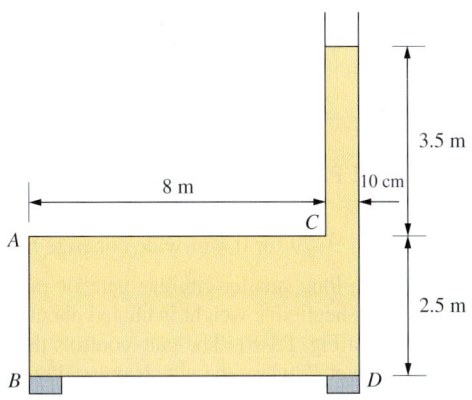

FIGURE P3–76

3–77 The two sides of a V-shaped water trough are hinged to each other at the bottom where they meet, as shown in Fig. P3–77, making an angle of 45° with the ground from both sides. Each side is 0.75 m wide, and the two parts are held together by a cable and turnbuckle placed every 6 m along the length of the trough. Calculate the tension in each cable when the trough is filled to the rim. *Answer:* 5510 N

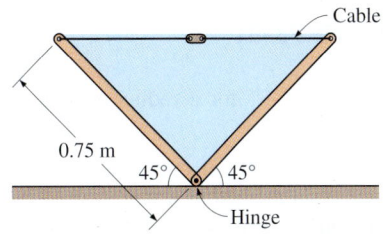

FIGURE P3–77

3–78 Repeat Prob. 3–77 for the case of a partially filled trough with a water height of 0.4 m directly above the hinge.

3–79 A retaining wall against a mud slide is to be constructed by placing 1.2-m-high and 0.25-m-wide rectangular concrete blocks ($\rho = 2700$ kg/m³) side by side, as shown in

Fig. P3–79. The friction coefficient between the ground and the concrete blocks is $f = 0.4$, and the density of the mud is about 1400 kg/m³. There is concern that the concrete blocks may slide or tip over the lower left edge as the mud level rises. Determine the mud height at which (a) the blocks will overcome friction and start sliding and (b) the blocks will tip over.

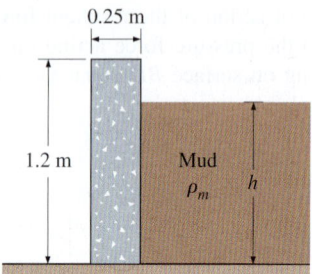

FIGURE P3–79

3–80 Repeat Prob. 3–79 for 0.4-m-wide concrete blocks.

3–81 A 4-m-long quarter-circular gate of radius 3 m and of negligible weight is hinged about its upper edge A, as shown in Fig. P3–81. The gate controls the flow of water over the ledge at B, where the gate is pressed by a spring. Determine the minimum spring force required to keep the gate closed when the water level rises to A at the upper edge of the gate.

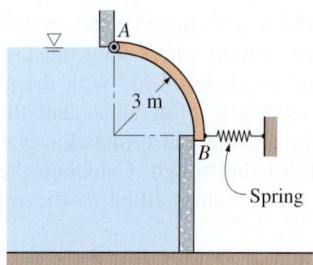

FIGURE P3–81

3–82 Repeat Prob. 3–81 for a radius of 4 m for the gate.
Answer: 314 kN

3–83 Consider a flat plate of thickness t, width w into the page, and length b submerged in water, as in Fig. P3–83. The depth of water from the surface to the center of the plate is H, and angle θ is defined relative to the center of the plate. (a) Generate an equation for the force F on the upper face of the plate as a function of (at most) H, b, t, w, g, ρ, and θ. Ignore atmospheric pressure. In other words, calculate the force that is *in addition to* the force due to atmospheric pressure. (b) As a test of your equation, let $H = 1.25$ m, $b = 1$ m, $t = 0.2$ m, $w = 1$ m, $g = 9.807$ m/s², $\rho = 998.3$ kg/m³, and $\theta = 30°$. If your equation is correct, you should get a force of 11.4 kN.

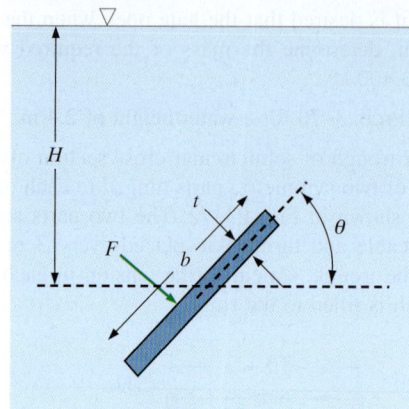

FIGURE P3–83

3–84 The weight of the gate separating the two fluids is such that the system shown in Fig. P3–84 is at static equilibrium. If it is known that $F_1/F_2 = 1.70$, determine h/H.

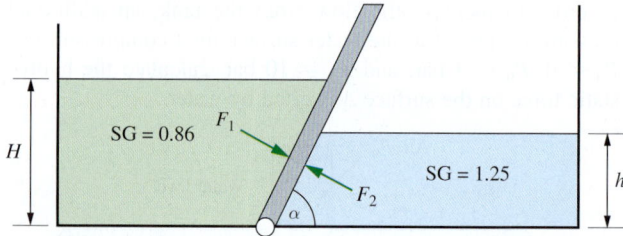

FIGURE P3–84

3–85 Consider a 1-m wide inclined gate of negligible weight that separates water from another fluid. What would be the volume of the concrete block (SG = 2.4) immersed in water to keep the gate at the position shown? Disregard any frictional effects.

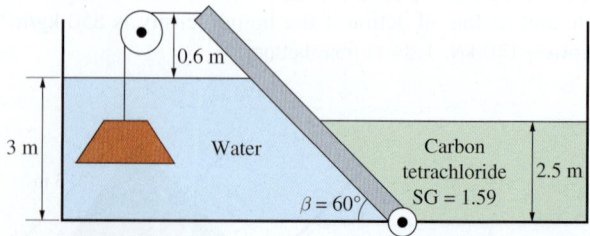

FIGURE P3–85

3–86 The parabolic shaped gate with a width of 4 m shown in Fig. P3–86 is hinged at point B. Determine the force F needed to keep the gate stationary.

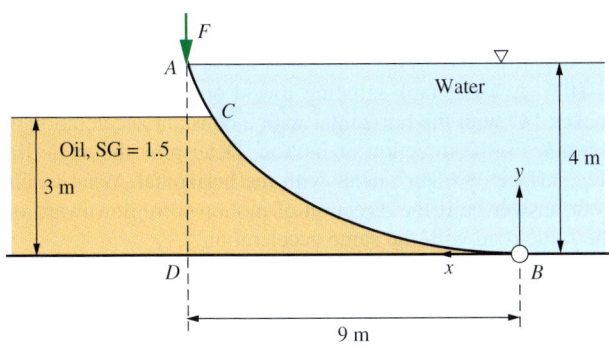

FIGURE P3–86

Buoyancy

3–87C What is buoyant force? What causes it? What is the magnitude of the buoyant force acting on a submerged body whose volume is V? What are the direction and the line of action of the buoyant force?

3–88C Consider two identical spherical balls submerged in water at different depths. Will the buoyant forces acting on these two balls be the same or different? Explain.

3–89C Consider two 5-cm-diameter spherical balls—one made of aluminum, the other of iron—submerged in water. Will the buoyant forces acting on these two balls be the same or different? Explain.

3–90C Consider a 3-kg copper cube and a 3-kg copper ball submerged in a liquid. Will the buoyant forces acting on these two bodies be the same or different? Explain.

3–91C Discuss the stability of (a) a submerged and (b) a floating body whose center of gravity is above the center of buoyancy.

3–92 The density of a liquid is to be determined by an old 1-cm-diameter cylindrical hydrometer whose division marks are completely wiped out. The hydrometer is first dropped in water, and the water level is marked. The hydrometer is then dropped into the other liquid, and it is observed that the mark for water has risen 0.3 cm above the liquid–air interface (Fig. P3–92). If the height of the original water mark is 12.3 cm, determine the density of the liquid.

3–93 The volume and the average density of an irregularly shaped body are to be determined by using a spring scale. The body weighs 7200 N in air and 4790 N in water. Determine the volume and the density of the body. State your assumptions.

3–94 Consider a large cubic ice block floating in seawater. The specific gravities of ice and seawater are 0.92 and 1.025, respectively. If a 25-cm-high portion of the ice block extends above the surface of the water, determine the height of the ice block below the surface.   *Answer:* 2.19 m

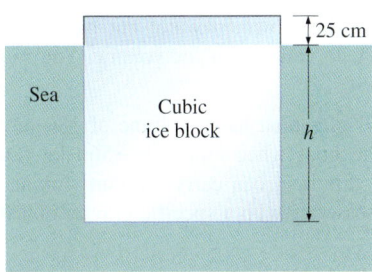

FIGURE P3–94

3–95 A spherical shell made of a material with a density of 1600 kg/m^3 is placed in water. If the inner and outer radii of the shell are $R_1 = 5$ cm, $R_2 = 6$ cm, determine the percentage of the shell's total volume that would be submerged.

3–96 An inverted cone is placed in a water tank as shown. If the weight of the cone is 16.5 N, what is the tensile force in the cord connecting the cone to the bottom of the tank?

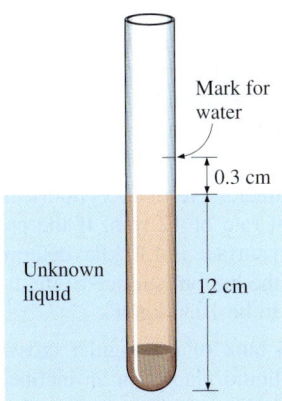

FIGURE P3–92

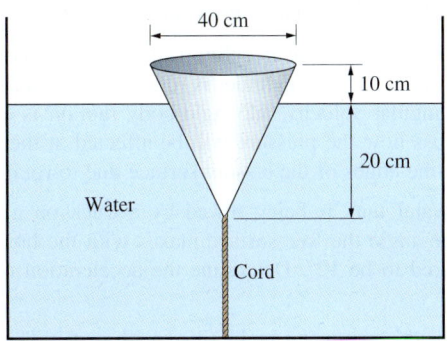

FIGURE P3–96

PRESSURE AND FLUID STATICS

3–97 The weight of a body is usually measured by disregarding buoyancy force applied by the air. Consider a 20-cm-diameter spherical body of density 7800 kg/m³. What is the percentage error associated with the neglecting of air buoyancy?

3–98 A 170-kg granite rock (ρ = 2700 kg/m³) is dropped into a lake. A man dives in and tries to lift the rock. Determine how much force the man needs to apply to lift it from the bottom of the lake. Do you think he can do it?

3–99 It is said that Archimedes discovered his principle during a bath while thinking about how he could determine if King Hiero's crown was actually made of pure gold. While in the bathtub, he conceived the idea that he could determine the average density of an irregularly shaped object by weighing it in air and also in water. If the crown weighed 3.55 kgf (= 34.8 N) in air and 3.25 kgf (= 31.9 N) in water, determine if the crown is made of pure gold. The density of gold is 19,300 kg/m³. Discuss how you can solve this problem without weighing the crown in water but by using an ordinary bucket with no calibration for volume. You may weigh anything in air.

3–100 The hull of a boat has a volume of 180 m³, and the total mass of the boat when empty is 8560 kg. Determine how much load this boat can carry without sinking (*a*) in a lake and (*b*) in seawater with a specific gravity of 1.03.

Fluids in Rigid-Body Motion

3–101C Under what conditions can a moving body of fluid be treated as a rigid body?

3–102C Consider a glass of water. Compare the water pressures at the bottom surface for the following cases: the glass is (*a*) stationary, (*b*) moving up at constant velocity, (*c*) moving down at constant velocity, and (*d*) moving horizontally at constant velocity.

3–103C Consider two identical glasses of water, one stationary and the other moving on a horizontal plane with constant acceleration. Assuming no splashing or spilling occurs, which glass will have a higher pressure at the (*a*) front, (*b*) midpoint, and (*c*) back of the bottom surface?

3–104C Consider a vertical cylindrical container partially filled with water. Now the cylinder is rotated about its axis at a specified angular velocity, and rigid-body motion is established. Discuss how the pressure will be affected at the midpoint and at the edges of the bottom surface due to rotation.

3–105 A water tank is being towed by a truck on a level road, and the angle the free surface makes with the horizontal is measured to be 12°. Determine the acceleration of the truck.

3–106 Consider two water tanks filled with water. The first tank is 8 m high and is stationary, while the second tank is 2 m high and is moving upward with an acceleration of 5 m/s². Which tank will have a higher pressure at the bottom?

3–107 A water tank is being towed on an uphill road that makes 14° with the horizontal with a constant acceleration of 3.5 m/s² in the direction of motion. Determine the angle the free surface of water makes with the horizontal. What would your answer be if the direction of motion were downward on the same road with the same acceleration?

3–108 A 0.9-m-diameter vertical cylindrical tank open to the atmosphere contains 0.3-m-high water. The tank is now rotated about the centerline, and the water level drops at the center while it rises at the edges. Determine the angular velocity at which the bottom of the tank will first be exposed. Also determine the maximum water height at this moment.

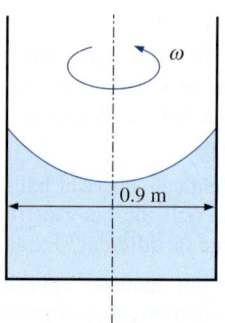

FIGURE P3–108

3–109 A 60-cm-high, 40-cm-diameter cylindrical water tank is being transported on a level road. The highest acceleration anticipated is 4 m/s². Determine the allowable initial water height in the tank if no water is to spill out during acceleration. *Answer:* 51.8 cm

3–110 A 30-cm-diameter, 90-cm-high vertical cylindrical container is partially filled with 60-cm-high water. Now the cylinder is rotated at a constant angular speed of 180 rpm. Determine how much the liquid level at the center of the cylinder will drop as a result of this rotational motion.

3–111 A fish tank that contains 60-cm-high water is moved in the cabin of an elevator. Determine the pressure at the bottom of the tank when the elevator is (*a*) stationary, (*b*) moving up with an upward acceleration of 3 m/s², and (*c*) moving down with a downward acceleration of 3 m/s².

3–112 A 3-m-diameter vertical cylindrical milk tank rotates at a constant rate of 12 rpm. If the pressure at the center of the bottom surface is 130 kPa, determine the pressure at the edge of the bottom surface of the tank. Take the density of the milk to be 1030 kg/m³.

3–113 Consider a tank of rectangular cross-section partially filled with a liquid placed on an inclined surface, as shown in the figure. When frictional effects are negligible, show that the slope of the liquid surface will be the same as the slope of the inclined surface when the tank is released. What can you say about the slope of the free surface when the friction is significant?

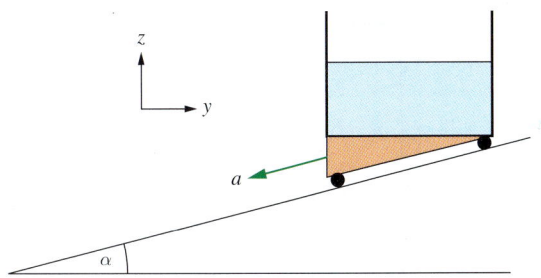

FIGURE P3–113

3–114 The bottom quarter of a vertical cylindrical tank of total height 0.4 m and diameter 0.3 m is filled with a liquid (SG > 1, like glycerin) and the rest with water, as shown in the figure. The tank is now rotated about its vertical axis at a constant angular speed of ω. Determine (*a*) the value of the angular speed when the point *P* on the axis at the liquid-liquid interface touches the bottom of the tank and (*b*) the amount of water that would be spilled out at this angular speed.

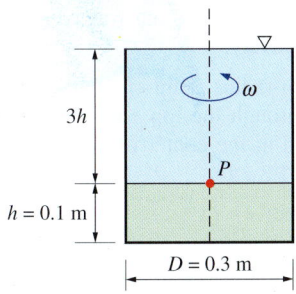

FIGURE P3–114

3–115 Milk with a density of 1020 kg/m³ is transported on a level road in a 9-m-long, 3-m-diameter cylindrical tanker. The tanker is completely filled with milk (no air space), and it accelerates at 4 m/s². If the minimum pressure in the tanker is 100 kPa, determine the maximum pressure difference and the location of the maximum pressure. *Answer:* 66.7 kPa

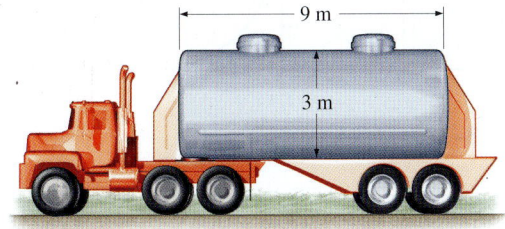

FIGURE P3–115

3–116 Repeat Prob. 3–115 for a deceleration of 2.5 m/s².

3–117 The distance between the centers of the two arms of a U-tube open to the atmosphere is 30 cm, and the U-tube contains 20-cm-high alcohol in both arms. Now the U-tube is rotated about the left arm at 4.2 rad/s. Determine the elevation difference between the fluid surfaces in the two arms.

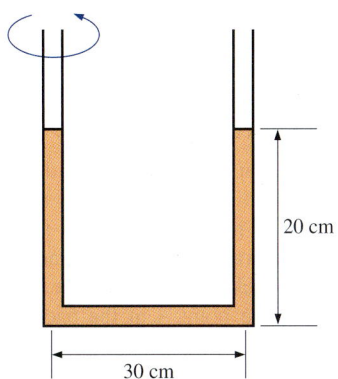

FIGURE P3–117

3–118 A 1.2-m-diameter, 3-m-high sealed vertical cylinder is completely filled with gasoline whose density is 740 kg/m³. The tank is now rotated about its vertical axis at a rate of 70 rpm. Determine (*a*) the difference between the pressures at the centers of the bottom and top surfaces and (*b*) the difference between the pressures at the center and the edge of the bottom surface.

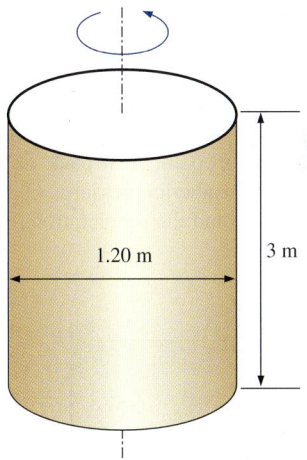

FIGURE P3–118

3–119 Reconsider Prob. 3–118. Using EES (or other) software, investigate the effect of rotational speed on the pressure difference between the center and the edge of the bottom surface of the cylinder. Let the rotational speed vary from 0 rpm to 500 rpm in increments of 50 rpm. Tabulate and plot your results.

124
PRESSURE AND FLUID STATICS

3–120 A 5-m-long, 1.8-m-high rectangular tank open to the atmosphere is towed by a truck on a level road. The tank is filled with water to a depth of 1.5 m. Determine the maximum acceleration or deceleration allowed if no water is to spill during towing.

3–121 A 3-m-diameter, 7-m-long cylindrical tank is completely filled with water. The tank is pulled by a truck on a level road with the 7-m-long axis being horizontal. Determine the pressure difference between the front and back ends of the tank along a horizontal line when the truck (*a*) accelerates at 3 m/s² and (*b*) decelerates at 4 m/s².

3–122 The rectangular tank is filled with heavy oil (like glycerin) at the bottom and water at the top, as shown in the figure. The tank is now moved to the right horizontally with a constant acceleration and ¼ of water is spilled out as a result from the back. Using geometrical considerations, determine how high the point *A* at the back of the tank on the oil-water interface will rise under this acceleration.
Answer: 0.25 m

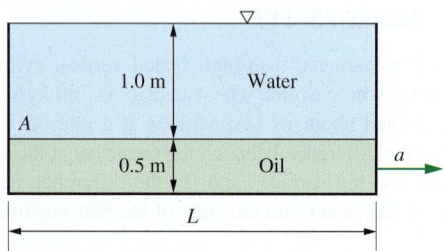

FIGURE P3–122

3–123 A sealed box filled with a liquid shown in the figure can be used to measure the acceleration of vehicles by measuring the pressure at top point *A* at back of the box while point *B* is kept at atmospheric pressure. Obtain a relation between the pressure P_A and the acceleration *a*.

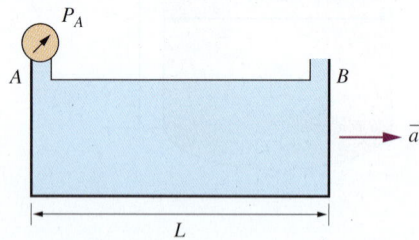

FIGURE P3–123

3–124 A centrifugal pump consists simply of a shaft and a few blades attached normally to the shaft. If the shaft is rotated at a constant rate of 2400 rpm, what would the theoretical pump head due to this rotation be? Take the impeller diameter to be 35 cm and neglect the blade tip effects.
Answer: 98.5 m

3–125 A U-tube is rotating at a constant angular velocity of ω. The liquid (glycerin) rises to the levels shown in Fig. P3–125. Obtain a relation for ω in terms of *g*, *h*, and *L*.

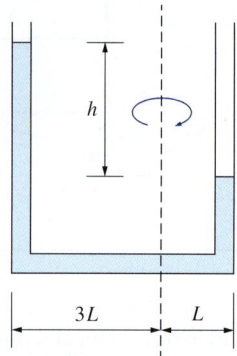

FIGURE P3–125

Review Problems

3–126 An air-conditioning system requires a 34-m-long section of 12-cm-diameter ductwork to be laid underwater. Determine the upward force the water will exert on the duct. Take the densities of air and water to be 1.3 kg/m³ and 1000 kg/m³, respectively.

3–127 The 0.5-m-radius semi-circular gate shown in the figure is hinged through the top edge *AB*. Find the required force to be applied at the center of gravity to keep the gate closed. *Answer:* 11.3 kN

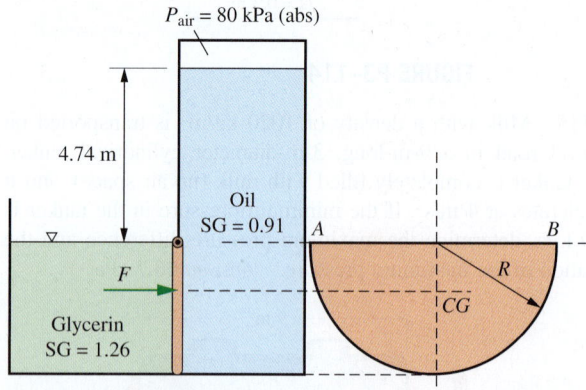

FIGURE P3–127

3–128 If the rate of rotational speed of the 3-tube system shown in Fig. P3–128 is $\omega = 10$ rad/s, determine the water heights in each tube leg. At what rotational speed will the middle tube be completely empty?

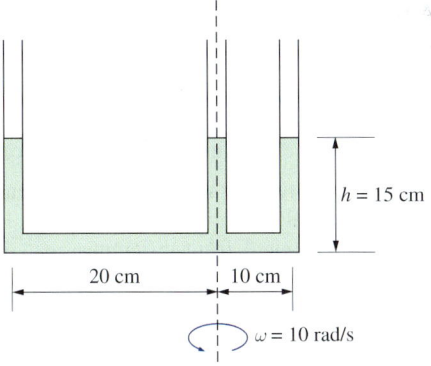

FIGURE P3–128

3–129 A 30-cm-diameter vertical cylindrical vessel is rotated about its vertical axis at a constant angular velocity of 100 rad/s. If the pressure at the midpoint of the inner top surface is atmospheric pressure like the outer surface, determine the total upward force acting upon the entire top surface inside the cylinder.

3–130 Balloons are often filled with helium gas because it weighs only about one-seventh of what air weighs under identical conditions. The buoyancy force, which can be expressed as $F_b = \rho_{air} g V_{balloon}$, will push the balloon upward. If the balloon has a diameter of 12 m and carries two people, 70 kg each, determine the acceleration of the balloon when it is first released. Assume the density of air is $\rho = 1.16$ kg/m³, and neglect the weight of the ropes and the cage.   *Answer:* 25.7 m/s²

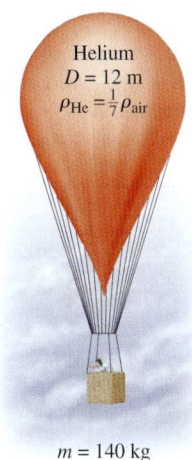

FIGURE P3–130

3–131 Reconsider Prob. 3–130. Using EES (or other) software, investigate the effect of the number of people carried in the balloon on acceleration. Plot the acceleration against the number of people, and discuss the results.

3–132 Determine the maximum amount of load, in kg, the balloon described in Prob. 3–130 can carry.   *Answer:* 521 kg

3–133 The pressure in a steam boiler is given to be 90 kgf/cm². Express this pressure in psi, kPa, atm, and bars.

3–134 The basic barometer can be used as an altitude-measuring device in airplanes. The ground control reports a barometric reading of 760 mmHg while the pilot's reading is 420 mmHg. Estimate the altitude of the plane from ground level if the average air density is 1.20 kg/m³.   *Answer:* 3853 m

3–135 The lower half of a 12-m-high cylindrical container is filled with water ($\rho = 1000$ kg/m³) and the upper half with oil that has a specific gravity of 0.85. Determine the pressure difference between the top and bottom of the cylinder.   *Answer:* 109 kPa

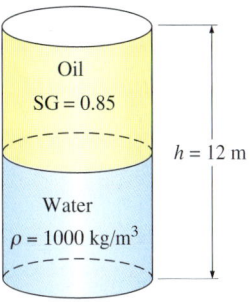

FIGURE P3–135

3–136 A vertical, frictionless piston–cylinder device contains a gas at 500 kPa. The atmospheric pressure outside is 100 kPa, and the piston area is 30 cm². Determine the mass of the piston.

3–137 A pressure cooker cooks a lot faster than an ordinary pan by maintaining a higher pressure and temperature inside. The lid of a pressure cooker is well sealed, and steam can escape only through an opening in the middle of the lid. A separate metal piece, the petcock, sits on top of this opening and

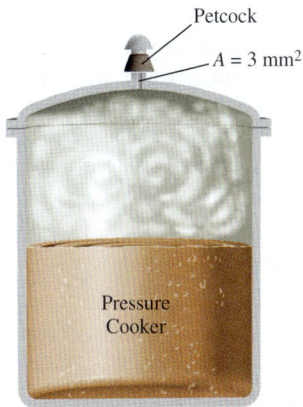

FIGURE P3–137

prevents steam from escaping until the pressure force overcomes the weight of the petcock. The periodic escape of the steam in this manner prevents any potentially dangerous pressure buildup and keeps the pressure inside at a constant value. Determine the mass of the petcock of a pressure cooker whose operation pressure is 120 kPa gage and has an opening cross-sectional area of 3 mm². Assume an atmospheric pressure of 101 kPa, and draw the free-body diagram of the petcock. *Answer:* 36.7 g

3–138 A glass tube is attached to a water pipe, as shown in Fig. P3–138. If the water pressure at the bottom of the tube is 115 kPa and the local atmospheric pressure is 98 kPa, determine how high the water will rise in the tube, in m. Assume $g = 9.8$ m/s² at that location and take the density of water to be 1000 kg/m³.

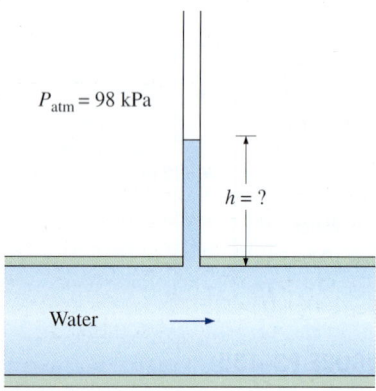

FIGURE P3–138

3–139 The average atmospheric pressure on earth is approximated as a function of altitude by the relation $P_{atm} = 101.325 (1 - 0.02256z)^{5.256}$, where P_{atm} is the atmospheric pressure in kPa and z is the altitude in km with $z = 0$ at sea level. Determine the approximate atmospheric pressures at Atlanta ($z = 306$ m), Denver ($z = 1610$ m), Mexico City ($z = 2309$ m), and the top of Mount Everest ($z = 8848$ m).

3–140 When measuring small pressure differences with a manometer, often one arm of the manometer is inclined to improve the accuracy of the reading. (The pressure difference is still proportional to the *vertical* distance and not the actual length of the fluid along the tube.) The air pressure in a circular duct is to be measured using a manometer whose open arm is inclined 25° from the horizontal, as shown in Fig. P3–140. The density of the liquid in the manometer is 0.81 kg/L, and the vertical distance between the fluid levels in the two arms of the manometer is 8 cm. Determine the gage pressure of air in the duct and the length of the fluid column in the inclined arm above the fluid level in the vertical arm.

3–141 Consider a U-tube whose arms are open to the atmosphere. Now equal volumes of water and light oil ($\rho = 790$ kg/m³) are poured from different arms. A person blows from the oil side of the U-tube until the contact surface of the two fluids moves to the bottom of the U-tube, and thus the liquid levels in the two arms are the same. If the fluid height in each arm is 102 cm, determine the gage pressure the person exerts on the oil by blowing.

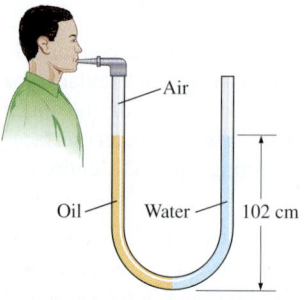

FIGURE P3–141

3–142 An elastic air balloon having a diameter of 30 cm is attached to the base of a container partially filled with water at +4°C, as shown in Fig. P3–142. If the pressure of the air above the water is gradually increased from 100 kPa to 1.6 MPa, will the force on the cable change? If so, what is the percent change in the force? Assume the pressure on the free surface and the diameter of the balloon are related by $P = CD^n$, where C is a constant and $n = -2$. The weight of the balloon and the air in it is negligible. *Answer:* 98.4 percent

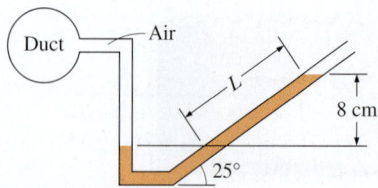

FIGURE P3–140

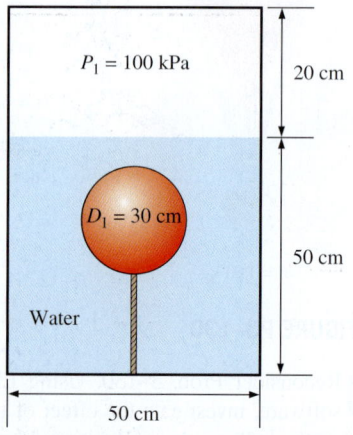

FIGURE P3–142

3–143 Reconsider Prob. 3–142. Using EES (or other) software, investigate the effect of air pressure above water on the cable force. Let this pressure vary from 0.5 MPa to 15 MPa. Plot the cable force versus the air pressure.

3–144 A gasoline line is connected to a pressure gage through a double-U manometer, as shown in Fig. P3–144. If the reading of the pressure gage is 260 kPa, determine the gage pressure of the gasoline line.

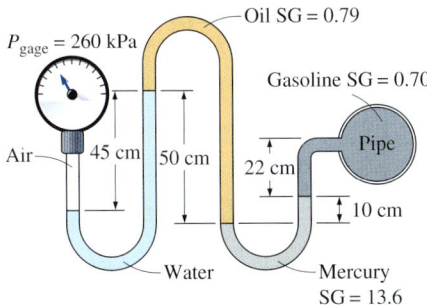

FIGURE P3–144

3–145 Repeat Prob. 3–144 for a pressure gage reading of 330 kPa.

3–146 The pressure of water flowing through a pipe is measured by the arrangement shown in Fig. P3–146. For the values given, calculate the pressure in the pipe.

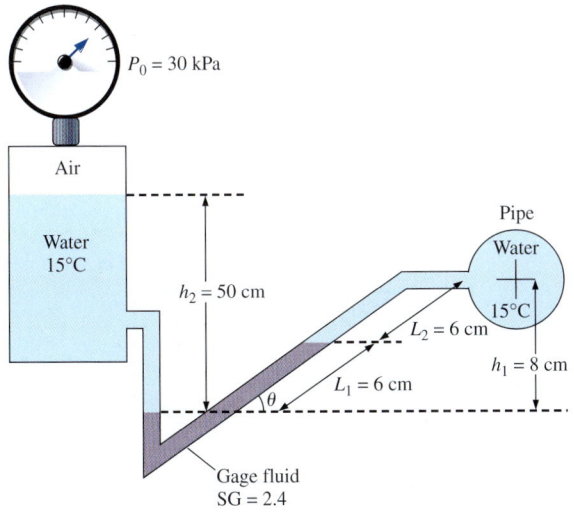

FIGURE P3–146

3–147 Consider a U-tube filled with mercury as shown in Fig. P3–147. The diameter of the right arm of the U-tube is $D = 1.5$ cm, and the diameter of the left arm is twice that. Heavy oil with a specific gravity of 2.72 is poured into the left arm, forcing some mercury from the left arm into the right one. Determine the maximum amount of oil that can be added into the left arm. *Answer:* 0.0884 L

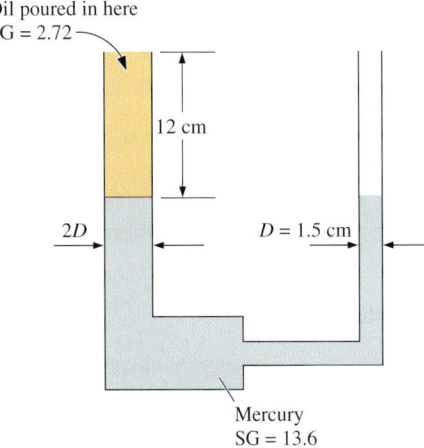

FIGURE P3–147

3–148 It is well known that the temperature of the atmosphere varies with altitude. In the troposphere, which extends to an altitude of 11 km, for example, the variation of temperature can be approximated by $T = T_0 - \beta z$, where T_0 is the temperature at sea level, which can be taken to be 288.15 K, and $\beta = 0.0065$ K/m. The gravitational acceleration also changes with altitude as $g(z) = g_0/(1 + z/6,370,320)^2$ where $g_0 = 9.807$ m/s² and z is the elevation from sea level in m. Obtain a relation for the variation of pressure in the troposphere (*a*) by ignoring and (*b*) by considering the variation of g with altitude.

3–149 The variation of pressure with density in a thick gas layer is given by $P = C\rho^n$, where C and n are constants. Noting that the pressure change across a differential fluid layer of thickness dz in the vertical z-direction is given as $dP = -\rho g\, dz$, obtain a relation for pressure as a function of elevation z. Take the pressure and density at $z = 0$ to be P_0 and ρ_0, respectively.

3–150 A 3-m-high, 6-m-wide rectangular gate is hinged at the top edge at A and is restrained by a fixed ridge at B. Determine the hydrostatic force exerted on the gate by the 5-m-high water and the location of the pressure center.

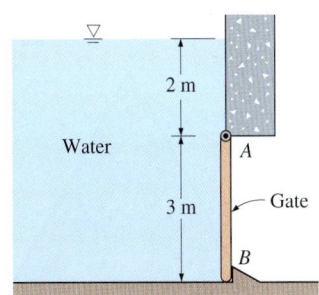

FIGURE P3–150

3–151 Repeat Prob. 3–150 for a total water height of 2 m.

3–152 A semicircular 12-m-diameter tunnel is to be built under a 45-m-deep, 240-m-long lake, as shown in Fig. P3–152. Determine the total hydrostatic force acting on the roof of the tunnel.

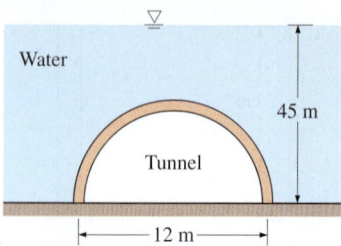

FIGURE P3–152

3–153 A 30-ton, 4-m-diameter hemispherical dome on a level surface is filled with water, as shown in Fig. P3–153. Someone claims that he can lift this dome by making use of Pascal's law by attaching a long tube to the top and filling it with water. Determine the required height of water in the tube to lift the dome. Disregard the weight of the tube and the water in it. *Answer:* 1.72 m

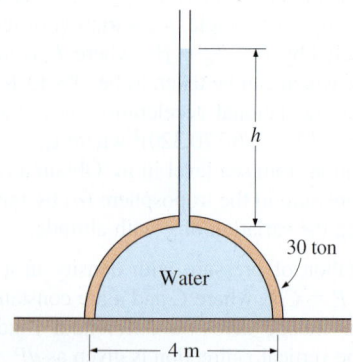

FIGURE P3–153

3–154 The water in a 25-m-deep reservoir is kept inside by a 150-m-wide wall whose cross section is an equilateral triangle, as shown in Fig. P3–154. Determine (*a*) the total force (hydrostatic + atmospheric) acting on the inner surface of the wall and its line of action and (*b*) the magnitude of the horizontal component of this force. Take P_{atm} = 100 kPa.

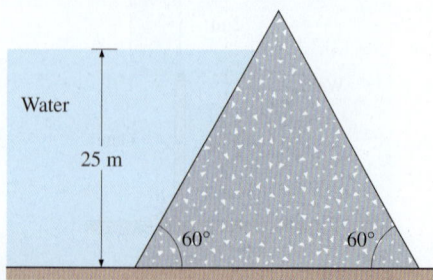

FIGURE P3–154

3–155 A U-tube contains water in the right arm, and another liquid in the left arm. It is observed that when the U-tube rotates at 50 rpm about an axis that is 15 cm from the right arm and 5 cm from the left arm, the liquid levels in both arms become the same, and the fluids meet at the axis of rotation. Determine the density of the fluid in the left arm.

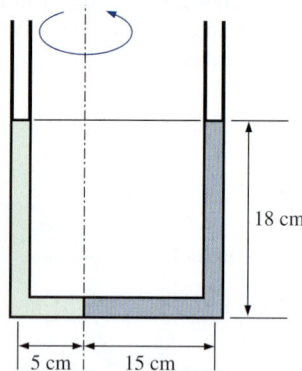

FIGURE P3–155

3–156 A 1-m-diameter, 2-m-high vertical cylinder is completely filled with gasoline whose density is 740 kg/m³. The tank is now rotated about its vertical axis at a rate of 130 rpm, while being accelerated upward at 5 m/s². Determine (*a*) the difference between the pressures at the centers of the bottom and top surfaces and (*b*) the difference between the pressures at the center and the edge of the bottom surface.

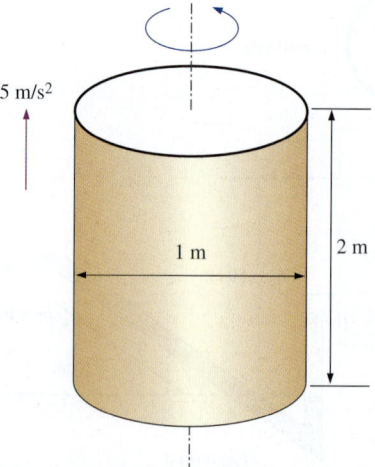

FIGURE P3–156

3–157 A 5-m-long, 4-m-high tank contains 2.5-m-deep water when not in motion and is open to the atmosphere through a vent in the middle. The tank is now accelerated to the right on a level surface at 2 m/s². Determine the maximum pressure in the tank relative to the atmospheric pressure. *Answer:* 29.5 kPa

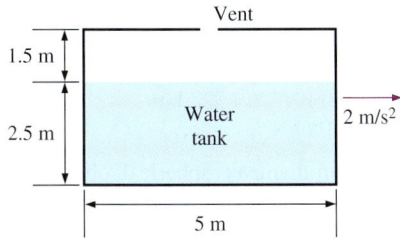

FIGURE P3–157

3–158 Reconsider Prob. 3–157. Using EES (or other) software, investigate the effect of acceleration on the slope of the free surface of water in the tank. Let the acceleration vary from 0 m/s² to 15 m/s² in increments of 1 m/s². Tabulate and plot your results.

3–159 A cylindrical container whose weight is 65 N is inverted and pressed into the water, as shown in Fig. P3–159. Determine the differential height h of the manometer and the force F needed to hold the container at the position shown.

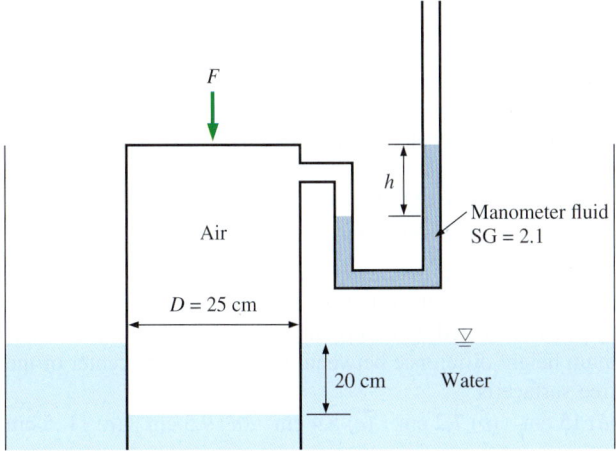

FIGURE P3–159

3–160 The average density of icebergs is about 917 kg/m³. (a) Determine the percentage of the total volume of an iceberg submerged in seawater of density 1042 kg/m³. (b) Although icebergs are mostly submerged, they are observed to turn over. Explain how this can happen. (*Hint*: Consider the temperatures of icebergs and seawater.)

3–161 The density of a floating body can be determined by tying weights to the body until both the body and the weights are completely submerged, and then weighing them separately in air. Consider a wood log that weighs 1540 N in air. If it takes 34 kg of lead (ρ = 11,300 kg/m³) to completely sink the log and the lead in water, determine the average density of the log. *Answer:* 835 kg/m³

3–162 The 280-kg, 6-m-wide rectangular gate shown in Fig. P3–162 is hinged at B and leans against the floor at A making an angle of 45° with the horizontal. The gate is to be opened from its lower edge by applying a normal force at its center. Determine the minimum force F required to open the water gate. *Answer:* 626 kN

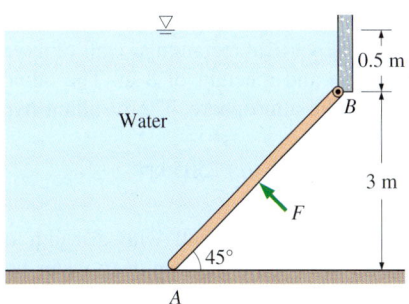

FIGURE P3–162

3–163 Repeat Prob. 3–162 for a water height of 0.8 m above the hinge at B.

Fundamentals of Engineering (FE) Exam Problems

3–164 The absolute pressure in a tank is measured to be 35 kPa. If the atmospheric pressure is 100 kPa, the vacuum pressure in the tank is
(a) 35 kPa (b) 100 kPa (c) 135 psi
(d) 0 kPa (e) 65 kPa

3–165 The pressure difference between the top and bottom of a water body with a depth of 10 m is (Take the density of water to be 1000 kg/m³.)
(a) 98,100 kPa (b) 98.1 kPa (c) 100 kPa
(d) 10 kPa (e) 1.9 kPa

3–166 The gage pressure in a pipe is measured by a manometer containing mercury (ρ = 13,600 kg/m³). The top of the mercury is open to the atmosphere and the atmospheric pressure is 100 kPa. If the mercury column height is 24 cm, the gage pressure in the pipe is
(a) 32 kPa (b) 24 kPa (c) 76 kPa
(d) 124 kPa (e) 68 kPa

3–167 Consider a hydraulic car jack with a piston diameter ratio of 9. A person can lift a 2000-kg car by applying a force of
(a) 2000 N (b) 200 N (c) 19,620 N
(d) 19.6 N (e) 18,000 N

3–168 The atmospheric pressure in a location is measured by a mercury (ρ = 13,600 kg/m³) barometer. If the height of the mercury column is 715 mm, the atmospheric pressure at that location is
(a) 85.6 kPa (b) 93.7 kPa (c) 95.4 kPa
(d) 100 kPa (e) 101 kPa

3–169 A manometer is used to measure the pressure of a gas in a tank. The manometer fluid is water (ρ = 1000 kg/m³)

and the manometer column height is 1.8 m. If the local atmospheric pressure is 100 kPa, the absolute pressure within the tank is

(a) 17,760 kPa   (b) 100 kPa   (c) 180 kPa
(d) 101 kPa   (e) 118 kPa

3–170 Consider the vertical rectangular wall of a water tank with a width of 5 m and a height of 8 m. The other side of the wall is open to the atmosphere. The resultant hydrostatic force on this wall is

(a) 1570 kN   (b) 2380 kN   (c) 2505 kN
(d) 1410 kN   (e) 404 kN

3–171 A vertical rectangular wall with a width of 20 m and a height of 12 m is holding a 7-m-deep water body. The resultant hydrostatic force acting on this wall is

(a) 1370 kN   (b) 4807 kN   (c) 8240 kN
(d) 9740 kN   (e) 11,670 kN

3–172 A vertical rectangular wall with a width of 20 m and a height of 12 m is holding a 7-m-deep water body. The line of action y_p for the resultant hydrostatic force on this wall is (disregard the atmospheric pressure)

(a) 5 m   (b) 4.0 m   (c) 4.67 m   (d) 9.67 m   (e) 2.33 m

3–173 A rectangular plate with a width of 16 m and a height of 12 m is located 4 m below a water surface. The plate is tilted and makes a 35° angle with the horizontal. The resultant hydrostatic force acting on the top surface of this plate is

(a) 10,800 kN   (b) 9745 kN   (c) 8470 kN
(d) 6400 kN   (e) 5190 kN

3–174 A 2-m-long and 3-m-wide horizontal rectangular plate is submerged in water. The distance of the top surface from the free surface is 5 m. The atmospheric pressure is 95 kPa. Considering atmospheric pressure, the hydrostatic force acting on the top surface of this plate is

(a) 307 kN   (b) 688 kN   (c) 747 kN
(d) 864 kN   (e) 2950 kN

3–175 A 1.8-m-diameter and 3.6-m-long cylindrical container contains a fluid with a specific gravity of 0.73. The container is positioned vertically and is full of the fluid. Disregarding atmospheric pressure, the hydrostatic force acting on the top and bottom surfaces of this container, respectively, are

(a) 0 kN, 65.6 kN   (b) 65.6 kN, 0 kN   (c) 65.6 kN, 65.6 kN
(d) 25.5 kN, 0 kN   (e) 0 kN, 25.5 kN

3–176 Consider a 6-m-diameter spherical gate holding a body of water whose height is equal to the diameter of the gate. Atmospheric pressure acts on both sides of the gate. The horizontal component of the hydrostatic force acting on this curved surface is

(a) 709 kN   (b) 832 kN   (c) 848 kN
(d) 972 kN   (e) 1124 kN

3–177 Consider a 6-m-diameter spherical gate holding a body of water whose height is equal to the diameter of the gate. Atmospheric pressure acts on both sides of the gate. The vertical component of the hydrostatic force acting on this curved surface is

(a) 89 kN   (b) 270 kN   (c) 327 kN
(d) 416 kN   (e) 505 kN

3–178 A 0.75-cm-diameter spherical object is completely submerged in water. The buoyant force acting on this object is

(a) 13,000 N   (b) 9835 N   (c) 5460 N
(d) 2167 N   (e) 1267 N

3–179 A 3-kg object with a density of 7500 kg/m³ is placed in water. The weight of this object in water is

(a) 29.4 N   (b) 25.5 N   (c) 14.7 N   (d) 30 N   (e) 3 N

3–180 A 7-m-diameter hot air balloon is neither rising nor falling. The density of atmospheric air is 1.3 kg/m³. The total mass of the balloon including the people on board is

(a) 234 kg   (b) 207 kg   (c) 180 kg   (d) 163 kg   (e) 134 kg

3–181 A 10-kg object with a density of 900 kg/m³ is placed in a fluid with a density of 1100 kg/m³. The fraction of the volume of the object submerged in water is

(a) 0.637   (b) 0.716   (c) 0.818   (d) 0.90   (e) 1

3–182 Consider a cubical water tank with a side length of 3 m. The tank is half filled with water, and is open to the atmosphere with a pressure of 100 kPa. Now, a truck carrying this tank is accelerated at a rate of 5 m/s². The maximum pressure in the water is

(a) 115 kPa   (b) 122 kPa   (c) 129 kPa
(d) 137 kPa   (e) 153 kPa

3–183 A 15-cm-diameter, 40-cm-high vertical cylindrical container is partially filled with 25-cm-high water. Now the cylinder is rotated at a constant speed of 20 rad/s. The maximum height difference between the edge and the center of the free surface is

(a) 15 cm   (b) 7.2 cm   (c) 5.4 cm   (d) 9.5 cm   (e) 11.5 cm

3–184 A 20-cm-diameter, 40-cm-high vertical cylindrical container is partially filled with 25-cm-high water. Now the cylinder is rotated at a constant speed of 15 rad/s. The height of water at the center of the cylinder is

(a) 25 cm   (b) 19.5 cm   (c) 22.7 cm
(d) 17.7 cm   (e) 15 cm

3–185 A 15-cm-diameter, 50-cm-high vertical cylindrical container is partially filled with 30-cm-high water. Now the cylinder is rotated at a constant speed of 20 rad/s. The pressure difference between the center and edge of the container at the base surface is

(a) 7327 Pa   (b) 8750 Pa   (c) 9930 Pa
(d) 1045 Pa   (e) 1125 Pa

Design and Essay Problems

3–186 Shoes are to be designed to enable people of up to 80 kg to walk on freshwater or seawater. The shoes are to be made of blown plastic in the shape of a sphere, a (American) football, or a loaf of French bread. Determine the equivalent diameter of each shoe and comment on the proposed shapes from the stability point of view. What is your assessment of the marketability of these shoes?

3–187 The volume of a rock is to be determined without using any volume measurement devices. Explain how you would do this with a waterproof spring scale.

3–188 The density of stainless steel is about 8000 kg/m³ (eight times denser than water), but a razor blade can float on water, even with some added weights. The water is at 20°C. The blade shown in the photograph is 4.3 cm long and 2.2 cm wide. For simplicity, the center cut-out area of the razor blade has been taped so that only the outer edges of the blade contribute to surface tension effects. Because the razor blade has sharp corners, the contact angle is not relevant. Rather, the limiting case is when the water contacts the blade vertically as sketched (effective contact angle along the edge of the blade is 180°). (*a*) Considering surface tension *alone*, estimate (in grams) how much total mass (razor blade + weights placed on top of it) can be supported. (*b*) Refine your analysis by considering that the razor blade pushes the water down, and thus hydrostatic pressure effects are also present. *Hint:* You will also need to know that due to the curvature of the meniscus, the maximum possible depth is $h = \sqrt{\dfrac{2\sigma_s}{\rho g}}$.

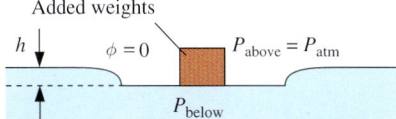

FIGURE P3–188
(Bottom) Photo by John M. Cimbala.

CHAPTER 4

FLUID KINEMATICS

Fluid kinematics deals with describing the motion of fluids without necessarily considering the forces and moments that *cause* the motion. In this chapter, we introduce several kinematic concepts related to flowing fluids. We discuss the *material derivative* and its role in transforming the conservation equations from the *Lagrangian description of fluid flow* (following a *fluid particle*) to the *Eulerian description of fluid flow* (pertaining to a *flow field*). We then discuss various ways to visualize flow fields—*streamlines*, *streaklines*, *pathlines*, *timelines*, optical methods *schlieren* and *shadowgraph*, and *surface* methods; and we describe three ways to plot flow data—*profile plots*, *vector plots*, and *contour plots*. We explain the four fundamental kinematic properties of fluid motion and deformation—*rate of translation*, *rate of rotation*, *linear strain rate*, and *shear strain rate*. The concepts of *vorticity*, *rotationality*, and *irrotationality* in fluid flows are then discussed. Finally, we discuss the *Reynolds transport theorem* (*RTT*), emphasizing its role in transforming the equations of motion from those following a *system* to those pertaining to fluid flow into and out of a *control volume*. The analogy between material derivative for infinitesimal fluid elements and RTT for finite control volumes is explained.

OBJECTIVES

When you finish reading this chapter, you should be able to

- Understand the role of the material derivative in transforming between Lagrangian and Eulerian descriptions
- Distinguish between various types of flow visualizations and methods of plotting the characteristics of a fluid flow
- Appreciate the many ways that fluids move and deform
- Distinguish between rotational and irrotational regions of flow based on the flow property vorticity
- Understand the usefulness of the Reynolds transport theorem

Satellite image of a hurricane near the Florida coast; water droplets move with the air, enabling us to visualize the counterclockwise swirling motion. However, the major portion of the hurricane is actually *irrotational*, while only the core (the eye of the storm) is *rotational*.
© StockTrek/Getty RF

FLUID KINEMATICS

FIGURE 4–1
With a small number of objects, such as billiard balls on a pool table, individual objects can be tracked.

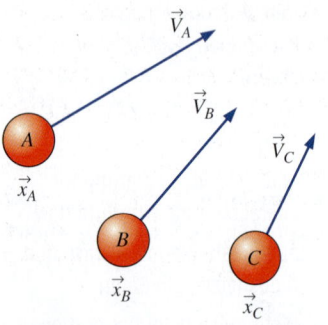

FIGURE 4–2
In the Lagrangian description, we must keep track of the position and velocity of individual particles.

4–1 ■ LAGRANGIAN AND EULERIAN DESCRIPTIONS

The subject called **kinematics** concerns the study of *motion*. In fluid dynamics, *fluid kinematics* is the study of how fluids flow and how to describe fluid motion. From a fundamental point of view, there are two distinct ways to describe motion. The first and most familiar method is the one you learned in high school physics—to follow the path of individual objects. For example, we have all seen physics experiments in which a ball on a pool table or a puck on an air hockey table collides with another ball or puck or with the wall (Fig. 4–1). Newton's laws are used to describe the motion of such objects, and we can accurately predict where they go and how momentum and kinetic energy are exchanged from one object to another. The kinematics of such experiments involves keeping track of the **position vector** of each object, $\vec{x}_A$, $\vec{x}_B$, . . . , and the **velocity vector** of each object, $\vec{V}_A$, $\vec{V}_B$, . . . , as functions of time (Fig. 4–2). When this method is applied to a flowing fluid, we call it the **Lagrangian description** of fluid motion after the Italian mathematician Joseph Louis Lagrange (1736–1813). Lagrangian analysis is analogous to the (closed) **system analysis** that you learned in thermodynamics; namely, we follow a mass of fixed identity. The Lagrangian description requires us to track the position and velocity of each individual fluid parcel, which we refer to as a **fluid particle,** and take to be a parcel of fixed identity.

As you can imagine, this method of describing motion is much more difficult for fluids than for billiard balls! First of all we cannot easily define and identify fluid particles as they move around. Secondly, a fluid is a **continuum** (from a macroscopic point of view), so interactions between fluid particles are not as easy to describe as are interactions between distinct objects like billiard balls or air hockey pucks. Furthermore, the fluid particles continually *deform* as they move in the flow.

From a *microscopic* point of view, a fluid is composed of *billions* of molecules that are continuously banging into one another, somewhat like billiard balls; but the task of following even a subset of these molecules is quite difficult, even for our fastest and largest computers. Nevertheless, there are many practical applications of the Lagrangian description, such as the tracking of passive scalars in a flow to model contaminant transport, rarefied gas dynamics calculations concerning reentry of a spaceship into the earth's atmosphere, and the development of flow visualization and measurement systems based on particle tracking (as discussed in Section 4–2).

A more common method of describing fluid flow is the **Eulerian description** of fluid motion, named after the Swiss mathematician Leonhard Euler (1707–1783). In the Eulerian description of fluid flow, a finite volume called a **flow domain** or **control volume** is defined, through which fluid flows in and out. Instead of tracking individual fluid particles, we define **field variables,** functions of space and time, within the control volume. The field variable at a particular location at a particular time is the value of the variable for whichever fluid particle happens to occupy that location at that time. For example, the **pressure field** is a **scalar field variable**; for general unsteady three-dimensional fluid flow in Cartesian coordinates,

Pressure field: $$P = P(x, y, z, t) \tag{4–1}$$

We define the **velocity field** as a **vector field variable** in similar fashion,

Velocity field: $$\vec{V} = \vec{V}(x, y, z, t) \tag{4–2}$$

Likewise, the **acceleration field** is also a vector field variable,

Acceleration field: $\quad\vec{a} = \vec{a}(x, y, z, t)$ (4–3)

Collectively, these (and other) field variables define the **flow field.** The velocity field of Eq. 4–2 is expanded in Cartesian coordinates (x, y, z), $(\vec{i}, \vec{j}, \vec{k})$ as

$$\vec{V} = (u, v, w) = u(x, y, z, t)\vec{i} + v(x, y, z, t)\vec{j} + w(x, y, z, t)\vec{k} \quad (4-4)$$

A similar expansion can be performed for the acceleration field of Eq. 4–3. In the Eulerian description, all such field variables are defined at any location (x, y, z) in the control volume and at any instant in time t (Fig. 4–3). In the Eulerian description we don't really care what happens to individual fluid particles; rather we are concerned with the pressure, velocity, acceleration, etc., of whichever fluid particle happens to be at the location of interest at the time of interest.

The difference between these two descriptions is made clearer by imagining a person standing beside a river, measuring its properties. In the Lagrangian approach, he throws in a probe that moves downstream with the water. In the Eulerian approach, he anchors the probe at a fixed location in the water.

While there are many occasions in which the Lagrangian description is useful, the Eulerian description is often more convenient for fluid mechanics applications. Furthermore, experimental measurements are generally more suited to the Eulerian description. In a wind tunnel, for example, velocity or pressure probes are usually placed at a fixed location in the flow, measuring $\vec{V}(x, y, z, t)$ or $P(x, y, z, t)$. However, whereas the equations of motion in the Lagrangian description following individual fluid particles are well known (e.g., Newton's second law), the equations of motion of fluid flow are not so readily apparent in the Eulerian description and must be carefully derived. We do this for control volume (integral) analysis via the Reynolds transport theorem at the end of this chapter. We derive the differential equations of motion in Chap. 9.

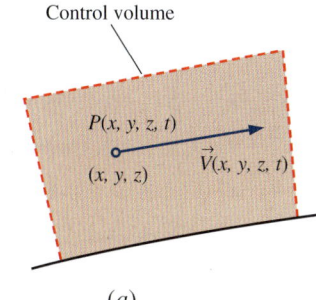

(a)

(b)

FIGURE 4–3
(*a*) In the Eulerian description, we define field variables, such as the pressure field and the velocity field, at any location and instant in time. (*b*) For example, the air speed probe mounted under the wing of an airplane measures the air speed at that location.
(Bottom) Photo by John M. Cimbala.

■ **EXAMPLE 4–1**   **A Steady Two-Dimensional Velocity Field**

A steady, incompressible, two-dimensional velocity field is given by

$$\vec{V} = (u, v) = (0.5 + 0.8x)\vec{i} + (1.5 - 0.8y)\vec{j} \quad (1)$$

where the *x*- and *y*-coordinates are in meters and the magnitude of velocity is in m/s. A **stagnation point** is defined as *a point in the flow field where the velocity is zero.* (*a*) Determine if there are any stagnation points in this flow field and, if so, where? (*b*) Sketch velocity vectors at several locations in the domain between $x = -2$ m to 2 m and $y = 0$ m to 5 m; qualitatively describe the flow field.

SOLUTION   For the given velocity field, the location(s) of stagnation point(s) are to be determined. Several velocity vectors are to be sketched and the velocity field is to be described.
Assumptions   **1** The flow is steady and incompressible. **2** The flow is two-dimensional, implying no *z*-component of velocity and no variation of *u* or *v* with *z*.
Analysis   (*a*) Since $\vec{V}$ is a vector, *all* its components must equal zero in order for $\vec{V}$ itself to be zero. Using Eq. 4–4 and setting Eq. 1 equal to zero,

Stagnation point:
$$u = 0.5 + 0.8x = 0 \quad \rightarrow \quad x = -0.625 \text{ m}$$
$$v = 1.5 - 0.8y = 0 \quad \rightarrow \quad y = 1.875 \text{ m}$$

Yes. There is one stagnation point located at $x = -0.625$ **m,** $y = 1.875$ **m.**

136
FLUID KINEMATICS

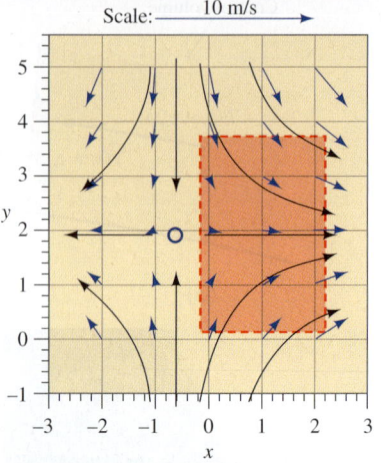

FIGURE 4–4
Velocity vectors (blue arrows) for the velocity field of Example 4–1. The scale is shown by the top arrow, and the solid black curves represent the approximate shapes of some streamlines, based on the calculated velocity vectors. The stagnation point is indicated by the blue circle. The shaded region represents a portion of the flow field that can approximate flow into an inlet (Fig. 4–5).

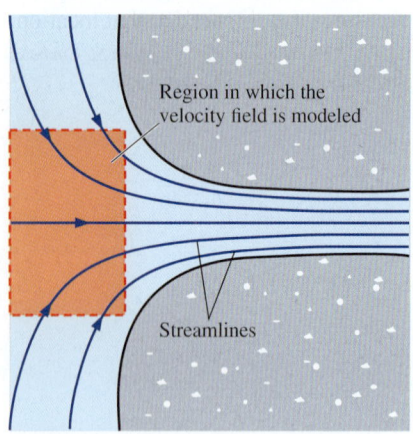

FIGURE 4–5
Flow field near the bell mouth inlet of a hydroelectric dam; a portion of the velocity field of Example 4–1 may be used as a first-order approximation of this physical flow field. The shaded region corresponds to that of Fig. 4–4.

(*b*) The *x*- and *y*-components of velocity are calculated from Eq. 1 for several (*x*, *y*) locations in the specified range. For example, at the point ($x = 2$ m, $y = 3$ m), $u = 2.10$ m/s and $v = -0.900$ m/s. The magnitude of velocity (the *speed*) at that point is 2.28 m/s. At this and at an array of other locations, the velocity vector is constructed from its two components, the results of which are shown in Fig. 4–4. The flow can be described as stagnation point flow in which flow enters from the top and bottom and spreads out to the right and left about a horizontal line of symmetry at $y = 1.875$ m. The stagnation point of part (*a*) is indicated by the blue circle in Fig. 4–4.

If we look only at the shaded portion of Fig. 4–4, this flow field models a converging, accelerating flow from the left to the right. Such a flow might be encountered, for example, near the submerged bell mouth inlet of a hydroelectric dam (Fig. 4–5). The useful portion of the given velocity field may be thought of as a first-order approximation of the shaded portion of the physical flow field of Fig. 4–5.

Discussion   It can be verified from the material in Chap. 9 that this flow field is physically valid because it satisfies the differential equation for conservation of mass.

Acceleration Field

As you should recall from your study of thermodynamics, the fundamental conservation laws (such as conservation of mass and the first law of thermodynamics) are expressed for a *system* of fixed identity (also called a *closed system*). In cases where analysis of a *control volume* (also called an *open system*) is more convenient than system analysis, it is necessary to rewrite these fundamental laws into forms applicable to the control volume. The same principle applies here. In fact, there is a direct analogy between systems versus control volumes in thermodynamics and Lagrangian versus Eulerian descriptions in fluid dynamics. The equations of motion for fluid flow (such as Newton's second law) are written for a fluid particle, which we also call a **material particle.** If we were to follow a particular fluid particle as it moves around in the flow, we would be employing the Lagrangian description, and the equations of motion would be directly applicable. For example, we would define the particle's location in space in terms of a **material position vector** ($x_{\text{particle}}(t)$, $y_{\text{particle}}(t)$, $z_{\text{particle}}(t)$). However, some mathematical manipulation is then necessary to convert the equations of motion into forms applicable to the Eulerian description.

Consider, for example, Newton's second law applied to our fluid particle,

Newton's second law:
$$\vec{F}_{\text{particle}} = m_{\text{particle}} \vec{a}_{\text{particle}} \tag{4-5}$$

where $\vec{F}_{\text{particle}}$ is the net force acting on the fluid particle, m_{particle} is its mass, and $\vec{a}_{\text{particle}}$ is its acceleration (Fig. 4–6). By definition, the acceleration of the fluid particle is the time derivative of the particle's velocity,

Acceleration of a fluid particle:
$$\vec{a}_{\text{particle}} = \frac{d\vec{V}_{\text{particle}}}{dt} \tag{4-6}$$

However, at any instant in time t, the velocity of the particle is the same as the local value of the velocity *field* at the location ($x_{\text{particle}}(t)$, $y_{\text{particle}}(t)$, $z_{\text{particle}}(t)$) of the particle, since the fluid particle moves with the fluid by definition. In other words, $\vec{V}_{particle}(t) \equiv \vec{V}(x_{particle}(t), y_{particle}(t), z_{particle}(t), t)$. To take the time derivative in Eq. 4–6, we must therefore use the *chain rule*, since the dependent variable ($\vec{V}$) is a function of *four* independent variables ($x_{particle}, y_{particle}, z_{particle}$, and t),

$$\vec{a}_{particle} = \frac{d\vec{V}_{particle}}{dt} = \frac{d\vec{V}}{dt} = \frac{d\vec{V}(x_{particle}, y_{particle}, z_{particle}, t)}{dt}$$

$$= \frac{\partial \vec{V}}{\partial t}\frac{dt}{dt} + \frac{\partial \vec{V}}{\partial x_{particle}}\frac{dx_{particle}}{dt} + \frac{\partial \vec{V}}{\partial y_{particle}}\frac{dy_{particle}}{dt} + \frac{\partial \vec{V}}{\partial z_{particle}}\frac{dz_{particle}}{dt} \quad (4\text{–}7)$$

In Eq. 4–7, ∂ is the **partial derivative operator** and d is the **total derivative operator.** Consider the second term on the right-hand side of Eq. 4–7. Since the acceleration is defined as that *following a fluid particle* (Lagrangian description), the rate of change of the particle's x-position with respect to time is $dx_{particle}/dt = u$ (Fig. 4–7), where u is the x-component of the velocity vector defined by Eq. 4–4. Similarly, $dy_{particle}/dt = v$ and $dz_{particle}/dt = w$. Furthermore, at any instant in time under consideration, the material position vector ($x_{particle}, y_{particle}, z_{particle}$) of the fluid particle in the Lagrangian frame is equal to the position vector (x, y, z) in the Eulerian frame. Equation 4–7 thus becomes

$$\vec{a}_{particle}(x, y, z, t) = \frac{d\vec{V}}{dt} = \frac{\partial \vec{V}}{\partial t} + u\frac{\partial \vec{V}}{\partial x} + v\frac{\partial \vec{V}}{\partial y} + w\frac{\partial \vec{V}}{\partial z} \quad (4\text{–}8)$$

where we have also used the (obvious) fact that $dt/dt = 1$. Finally, at any instant in time t, the acceleration field of Eq. 4–3 must equal the acceleration of the fluid particle that happens to occupy the location (x, y, z) at that time t. Why? Because the fluid particle is by definition accelerating with the fluid flow. Hence, *we may replace $\vec{a}_{particle}$ with $\vec{a}(x, y, z, t)$ in Eqs. 4–7 and 4–8 to transform from the Lagrangian to the Eulerian frame of reference.* In vector form, Eq. 4–8 is written as

Acceleration of a fluid particle expressed as a field variable:

$$\vec{a}(x, y, z, t) = \frac{d\vec{V}}{dt} = \frac{\partial \vec{V}}{\partial t} + (\vec{V} \cdot \vec{\nabla})\vec{V} \quad (4\text{–}9)$$

where $\vec{\nabla}$ is the **gradient operator** or **del operator,** a vector operator that is defined in Cartesian coordinates as

Gradient or del operator: $\quad \vec{\nabla} = \left(\frac{\partial}{\partial x}, \frac{\partial}{\partial y}, \frac{\partial}{\partial z}\right) = \vec{i}\frac{\partial}{\partial x} + \vec{j}\frac{\partial}{\partial y} + \vec{k}\frac{\partial}{\partial z} \quad$ (4–10)

In Cartesian coordinates then, the components of the acceleration vector are

Cartesian coordinates:
$$a_x = \frac{\partial u}{\partial t} + u\frac{\partial u}{\partial x} + v\frac{\partial u}{\partial y} + w\frac{\partial u}{\partial z}$$
$$a_y = \frac{\partial v}{\partial t} + u\frac{\partial v}{\partial x} + v\frac{\partial v}{\partial y} + w\frac{\partial v}{\partial z} \quad (4\text{–}11)$$
$$a_z = \frac{\partial w}{\partial t} + u\frac{\partial w}{\partial x} + v\frac{\partial w}{\partial y} + w\frac{\partial w}{\partial z}$$

The first term on the right-hand side of Eq. 4–9, $\partial \vec{V}/\partial t$, is called the **local acceleration** and is nonzero only for unsteady flows. The second term, $(\vec{V} \cdot \vec{\nabla})\vec{V}$, is called the **advective acceleration** (sometimes the **convective**

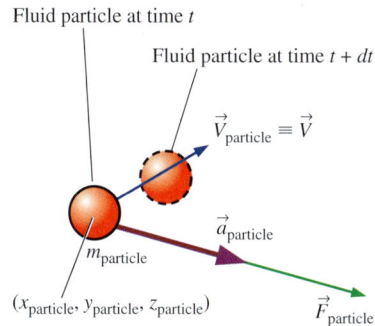

FIGURE 4–6
Newton's second law applied to a fluid particle; the acceleration vector (purple arrow) is in the same direction as the force vector (green arrow), but the velocity vector (blue arrow) may act in a different direction.

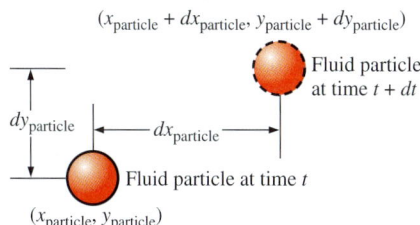

FIGURE 4–7
When following a fluid particle, the x-component of velocity, u, is defined as $dx_{particle}/dt$. Similarly, $v = dy_{particle}/dt$ and $w = dz_{particle}/dt$. Movement is shown here only in two dimensions for simplicity.

138
FLUID KINEMATICS

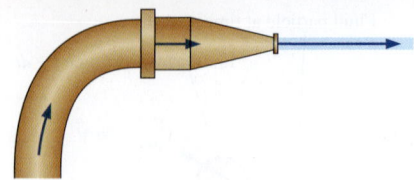

FIGURE 4–8
Flow of water through the nozzle of a garden hose illustrates that fluid particles may accelerate, even in a steady flow. In this example, the exit speed of the water is much higher than the water speed in the hose, implying that fluid particles have accelerated even though the flow is steady.

acceleration); *this term can be nonzero even for steady flows*. It accounts for the effect of the fluid particle moving (advecting or convecting) to a new location in the flow, where the velocity field is different. For example, consider steady flow of water through a garden hose nozzle (Fig. 4–8). We define *steady* in the Eulerian frame of reference to be when properties at any point in the flow field do not change with respect to time. Since the velocity at the exit of the nozzle is larger than that at the nozzle entrance, fluid particles clearly accelerate, even though the flow is steady. The acceleration is nonzero because of the advective acceleration terms in Eq. 4–9. Note that while the flow is steady from the point of view of a fixed observer in the Eulerian reference frame, it is *not* steady from the Lagrangian reference frame moving with a fluid particle that enters the nozzle and accelerates as it passes through the nozzle.

EXAMPLE 4–2   Acceleration of a Fluid Particle through a Nozzle

Nadeen is washing her car, using a nozzle similar to the one sketched in Fig. 4–8. The nozzle is 9.91 cm long, with an inlet diameter of 1.07 cm and an outlet diameter of 0.460 cm (Fig. 4–9). The volume flow rate through the garden hose (and through the nozzle) is $\dot{V} = 0.0530$ L/S, and the flow is steady. Estimate the magnitude of the acceleration of a fluid particle moving down the centerline of the nozzle.

SOLUTION   The acceleration following a fluid particle down the center of a nozzle is to be estimated.
Assumptions **1** The flow is steady and incompressible. **2** The *x*-direction is taken along the centerline of the nozzle. **3** By symmetry, $v = w = 0$ along the centerline, but u increases through the nozzle.
Analysis The flow is steady, so you may be tempted to say that the acceleration is zero. However, even though the local acceleration $\partial \vec{V}/\partial t$ is identically zero for this steady flow field, the advective acceleration $(\vec{V} \cdot \vec{\nabla})\vec{V}$ is *not* zero. We first calculate the average *x*-component of velocity at the inlet and outlet of the nozzle by dividing volume flow rate by cross-sectional area:

Inlet speed:

$$u_{inlet} \cong \frac{\dot{V}}{A_{inlet}} = \frac{4\dot{V}}{\pi D_{inlet}^2} = \frac{4(5.30 \times 10^{-5}\,\text{m}^3/\text{s})}{\pi(0.0107\,\text{m})^2} = 0.589 \text{ m/s}$$

Similarly, the average outlet speed is $u_{outlet} = 3.19$ m/s. We now calculate the acceleration two ways, with equivalent results. First, a simple average value of acceleration in the *x*-direction is calculated based on the change in speed divided by an estimate of the **residence time** of a fluid particle in the nozzle, $\Delta t = \Delta x / u_{avg}$ (Fig. 4–10). By the fundamental definition of acceleration as the rate of change of velocity,

Method A: $a_x \cong \dfrac{\Delta u}{\Delta t} = \dfrac{u_{outlet} - u_{inlet}}{\Delta x / u_{avg}} = \dfrac{u_{outlet} - u_{inlet}}{2\,\Delta x / (u_{outlet} + u_{inlet})} = \dfrac{u_{outlet}^2 - u_{inlet}^2}{2\,\Delta x}$

The second method uses the equation for acceleration field components in Cartesian coordinates, Eq. 4–11,

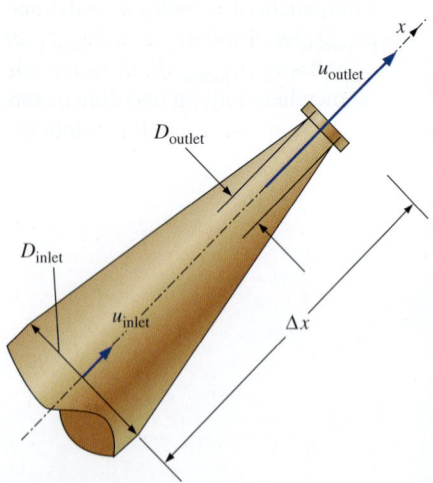

FIGURE 4–9
Flow of water through the nozzle of Example 4–2.

Method B: $\quad a_x = \underbrace{\cancel{\dfrac{\partial u}{\partial t}}}_{\text{Steady}} + u\dfrac{\partial u}{\partial x} + \underbrace{v\cancel{\dfrac{\partial u}{\partial y}}}_{v=0 \text{ along centerline}} + \underbrace{w\cancel{\dfrac{\partial u}{\partial z}}}_{w=0 \text{ along centerline}} \cong u_{\text{avg}}\dfrac{\Delta u}{\Delta x}$

Here we see that only one advective term is nonzero. We approximate the average speed through the nozzle as the average of the inlet and outlet speeds, and we use a **first-order finite difference approximation** (Fig. 4–11) for the average value of derivative $\partial u/\partial x$ through the centerline of the nozzle:

$$a_x \cong \dfrac{u_{\text{outlet}} + u_{\text{inlet}}}{2}\,\dfrac{u_{\text{outlet}} - u_{\text{inlet}}}{\Delta x} = \dfrac{u_{\text{outlet}}^2 - u_{\text{inlet}}^2}{2\,\Delta x}$$

The result of method B is identical to that of method A. Substitution of the given values yields

Axial acceleration:

$$a_x \cong \dfrac{u_{\text{outlet}}^2 - u_{\text{inlet}}^2}{2\,\Delta x} = \dfrac{(3.19 \text{ m/s})^2 - (0.589 \text{ m/s})^2}{2(0.0991 \text{ m})} = \mathbf{49.6 \text{ m/s}^2}$$

Discussion Fluid particles are accelerated through the nozzle at nearly five times the acceleration of gravity (almost five *g*'s)! This simple example clearly illustrates that the acceleration of a fluid particle can be nonzero, even in steady flow. Note that the acceleration is actually a **point function**, whereas we have estimated a simple average acceleration through the entire nozzle.

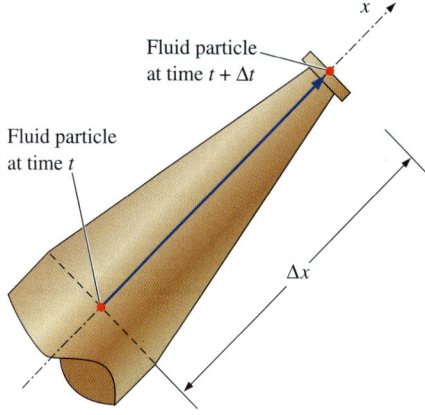

FIGURE 4–10
Residence time Δt is defined as the time it takes for a fluid particle to travel through the nozzle from inlet to outlet (distance Δx).

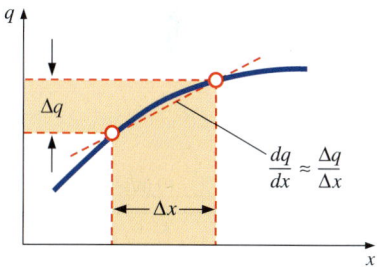

FIGURE 4–11
A *first-order finite difference approximation* for derivative dq/dx is simply the change in dependent variable (q) divided by the change in independent variable (x).

Material Derivative

The total derivative operator d/dt in Eq. 4–9 is given a special name, the **material derivative**; it is assigned a special notation, D/Dt, in order to emphasize that it is formed by *following a fluid particle as it moves through the flow field* (Fig. 4–12). Other names for the material derivative include **total, particle, Lagrangian, Eulerian,** and **substantial derivative.**

Material derivative: $\quad \dfrac{D}{Dt} = \dfrac{d}{dt} = \dfrac{\partial}{\partial t} + (\vec{V}\cdot\vec{\nabla})$ (4–12)

When we apply the material derivative of Eq. 4–12 to the velocity field, the result is the acceleration field as expressed by Eq. 4–9, which is thus sometimes called the **material acceleration,**

Material acceleration: $\quad \vec{a}(x,y,z,t) = \dfrac{D\vec{V}}{Dt} = \dfrac{d\vec{V}}{dt} = \dfrac{\partial \vec{V}}{\partial t} + (\vec{V}\cdot\vec{\nabla})\vec{V}$ (4–13)

Equation 4–12 can also be applied to other fluid properties besides velocity, both scalars and vectors. For example, the material derivative of pressure is written as

Material derivative of pressure: $\quad \dfrac{DP}{Dt} = \dfrac{dP}{dt} = \dfrac{\partial P}{\partial t} + (\vec{V}\cdot\vec{\nabla})P$ (4–14)

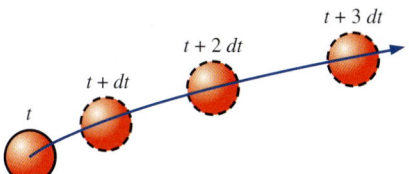

FIGURE 4–12
The material derivative D/Dt is defined by following a fluid particle as it moves throughout the flow field. In this illustration, the fluid particle is accelerating to the right as it moves up and to the right.

FLUID KINEMATICS

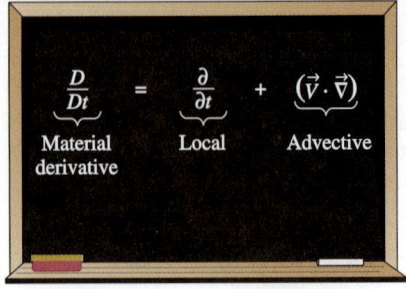

FIGURE 4–13
The material derivative D/Dt is composed of a *local* or *unsteady* part and a *convective* or *advective* part.

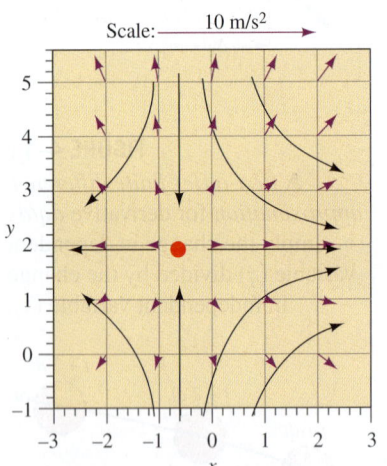

FIGURE 4–14
Acceleration vectors (purple arrows) for the velocity field of Examples 4–1 and 4–3. The scale is shown by the top arrow, and the solid black curves represent the approximate shapes of some streamlines, based on the calculated velocity vectors (see Fig. 4–4). The stagnation point is indicated by the red circle.

Equation 4–14 represents the time rate of change of pressure following a fluid particle as it moves through the flow and contains both local (unsteady) and advective components (Fig. 4–13).

EXAMPLE 4–3 Material Acceleration of a Steady Velocity Field

Consider the steady, incompressible, two-dimensional velocity field of Example 4–1. (*a*) Calculate the material acceleration at the point ($x = 2$ m, $y = 3$ m). (*b*) Sketch the material acceleration vectors at the same array of x- and y-values as in Example 4–1.

SOLUTION For the given velocity field, the material acceleration vector is to be calculated at a particular point and plotted at an array of locations in the flow field.
Assumptions **1** The flow is steady and incompressible. **2** The flow is two-dimensional, implying no z-component of velocity and no variation of u or v with z.
Analysis (*a*) Using the velocity field of Eq. 1 of Example 4–1 and the equation for material acceleration components in Cartesian coordinates (Eq. 4–11), we write expressions for the two nonzero components of the acceleration vector:

$$a_x = \frac{\partial u}{\partial t} + u\frac{\partial u}{\partial x} + v\frac{\partial u}{\partial y} + w\frac{\partial u}{\partial z}$$
$$= 0 + (0.5 + 0.8x)(0.8) + (1.5 - 0.8y)(0) + 0 = (0.4 + 0.64x) \text{ m/s}^2$$

and

$$a_y = \frac{\partial v}{\partial t} + u\frac{\partial v}{\partial x} + v\frac{\partial v}{\partial y} + w\frac{\partial v}{\partial z}$$
$$= 0 + (0.5 + 0.8x)(0) + (1.5 - 0.8y)(-0.8) + 0 = (-1.2 + 0.64y) \text{ m/s}^2$$

At the point ($x = 2$ m, $y = 3$ m), $a_x = 1.68$ m/s^2 and $a_y = 0.720$ m/s^2.
(*b*) The equations in part (*a*) are applied to an array of x- and y-values in the flow domain within the given limits, and the acceleration vectors are plotted in Fig. 4–14.
Discussion The acceleration field is nonzero, even though the flow is *steady*. Above the stagnation point (above $y = 1.875$ m), the acceleration vectors plotted in Fig. 4–14 point upward, increasing in magnitude away from the stagnation point. To the right of the stagnation point (to the right of $x = -0.625$ m), the acceleration vectors point to the right, again increasing in magnitude away from the stagnation point. This agrees qualitatively with the velocity vectors of Fig. 4–4 and the streamlines sketched in Fig. 4–14; namely, in the upper-right portion of the flow field, fluid particles are accelerated in the upper-right direction and therefore veer in the counterclockwise direction due to **centripetal acceleration** toward the upper right. The flow below $y = 1.875$ m is a mirror image of the flow above this symmetry line, and the flow to the left of $x = -0.625$ m is a mirror image of the flow to the right of this symmetry line.

4–2 ▪ FLOW PATTERNS AND FLOW VISUALIZATION

While quantitative study of fluid dynamics requires advanced mathematics, much can be learned from **flow visualization**—the visual examination of flow field features. Flow visualization is useful not only in physical experiments (Fig. 4–15), but in *numerical* solutions as well [**computational fluid dynamics (CFD)**]. In fact, the very first thing an engineer using CFD does after obtaining a numerical solution is simulate some form of flow visualization, so that he or she can see the "whole picture" rather than merely a list of numbers and quantitative data. Why? Because the human mind is designed to rapidly process an incredible amount of visual information; as they say, a picture is worth a thousand words. There are many types of flow patterns that can be visualized, both physically (experimentally) and/or computationally.

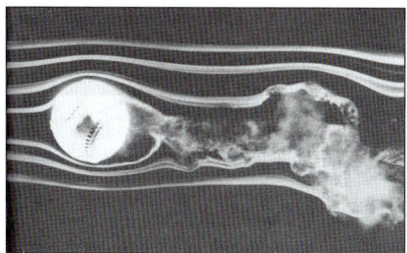

FIGURE 4–15
Spinning baseball. The late F. N. M. Brown devoted many years to developing and using smoke visualization in wind tunnels at the University of Notre Dame. Here the flow speed is about 23 m/s and the ball is rotated at 630 rpm.

Photograph courtesy of T. J. Mueller.

Streamlines and Streamtubes

A **streamline** is a curve that is everywhere tangent to the instantaneous local velocity vector.

Streamlines are useful as indicators of the instantaneous direction of fluid motion throughout the flow field. For example, regions of recirculating flow and separation of a fluid off of a solid wall are easily identified by the streamline pattern. Streamlines cannot be directly observed experimentally except in steady flow fields, in which they are coincident with pathlines and streaklines, to be discussed next. Mathematically, however, we can write a simple expression for a streamline based on its definition.

Consider an infinitesimal arc length $d\vec{r} = dx\vec{i} + dy\vec{j} + dz\vec{k}$ along a streamline; $d\vec{r}$ must be parallel to the local velocity vector $\vec{V} = u\vec{i} + v\vec{j} + w\vec{k}$ by definition of the streamline. By simple geometric arguments using similar triangles, we know that the components of $d\vec{r}$ must be proportional to those of $\vec{V}$ (Fig. 4–16). Hence,

Equation for a streamline:
$$\frac{dr}{V} = \frac{dx}{u} = \frac{dy}{v} = \frac{dz}{w} \quad (4\text{–}15)$$

where dr is the magnitude of $d\vec{r}$ and V is the speed, the magnitude of velocity vector $\vec{V}$. Equation 4–15 is illustrated in two dimensions for simplicity in Fig. 4–16. For a known velocity field, we integrate Eq. 4–15 to obtain equations for the streamlines. In two dimensions, (x, y), (u, v), the following differential equation is obtained:

Streamline in the xy-plane:
$$\left(\frac{dy}{dx}\right)_{\text{along a streamline}} = \frac{v}{u} \quad (4\text{–}16)$$

In some simple cases, Eq. 4–16 may be solvable analytically; in the general case, it must be solved numerically. In either case, an arbitrary constant of integration appears. Each chosen value of the constant represents a different streamline. The *family* of curves that satisfy Eq. 4–16 therefore represents streamlines of the flow field.

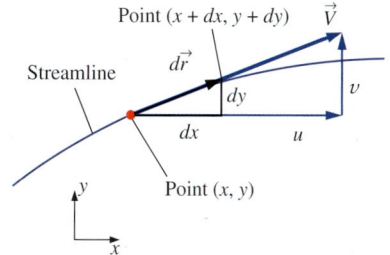

FIGURE 4–16
For two-dimensional flow in the *xy*-plane, arc length $d\vec{r} = (dx, dy)$ along a *streamline* is everywhere tangent to the local instantaneous velocity vector $\vec{V} = (u, v)$.

FLUID KINEMATICS

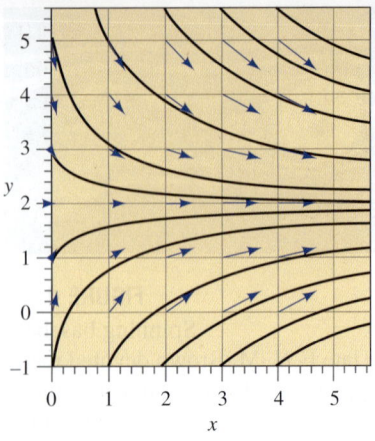

FIGURE 4–17
Streamlines (solid black curves) for the velocity field of Example 4–4; velocity vectors of Fig. 4–4 (blue arrows) are superimposed for comparison.

EXAMPLE 4–4   Streamlines in the *xy*-Plane—An Analytical Solution

For the steady, incompressible, two-dimensional velocity field of Example 4–1, plot several streamlines in the right half of the flow ($x > 0$) and compare to the velocity vectors plotted in Fig. 4–4.

SOLUTION An analytical expression for streamlines is to be generated and plotted in the upper-right quadrant.
Assumptions **1** The flow is steady and incompressible. **2** The flow is two-dimensional, implying no *z*-component of velocity and no variation of u or v with z.
Analysis Equation 4–16 is applicable here; thus, along a streamline,

$$\frac{dy}{dx} = \frac{v}{u} = \frac{1.5 - 0.8y}{0.5 + 0.8x}$$

We solve this differential equation by separation of variables:

$$\frac{dy}{1.5 - 0.8y} = \frac{dx}{0.5 + 0.8x} \quad \rightarrow \quad \int \frac{dy}{1.5 - 0.8y} = \int \frac{dx}{0.5 + 0.8x}$$

After some algebra, we solve for y as a function of x along a streamline,

$$y = \frac{C}{0.8(0.5 + 0.8x)} + 1.875$$

where C is a constant of integration that can be set to various values in order to plot the streamlines. Several streamlines of the given flow field are shown in Fig. 4–17.
Discussion The velocity vectors of Fig. 4–4 are superimposed on the streamlines of Fig. 4–17; the agreement is excellent in the sense that the velocity vectors point everywhere tangent to the streamlines. Note that speed cannot be determined directly from the streamlines alone.

A **streamtube** consists of a bundle of streamlines (Fig. 4–18), much like a communications cable consists of a bundle of fiber-optic cables. Since streamlines are everywhere parallel to the local velocity, fluid cannot cross a streamline by definition. By extension, *fluid within a streamtube must remain there and cannot cross the boundary of the streamtube.* You must keep in mind that both streamlines and streamtubes are instantaneous quantities, defined at a particular instant in time according to the velocity field at that instant. In an *unsteady* flow, the streamline pattern may change significantly with time. Nevertheless, at any instant in time, the mass flow rate passing through any cross-sectional slice of a given streamtube must remain the same. For example, in a converging portion of an incompressible flow field, the diameter of the streamtube must decrease as the velocity increases in order to conserve mass (Fig. 4–19a). Likewise, the streamtube diameter increases in diverging portions of an incompressible flow (Fig. 4–19b).

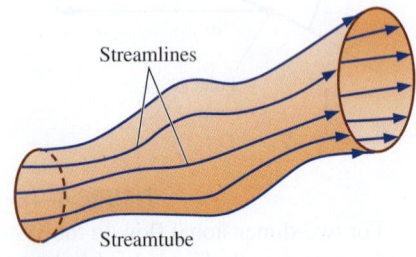

FIGURE 4–18
A *streamtube* consists of a bundle of individual streamlines.

Pathlines

A **pathline** is the actual path traveled by an individual fluid particle over some time period.

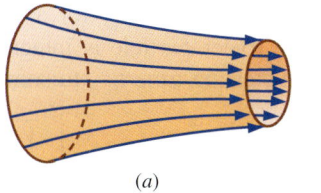
(a)

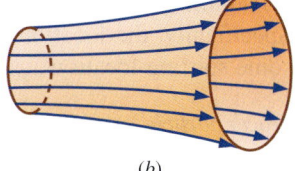
(b)

FIGURE 4–19
In an incompressible flow field, a streamtube (*a*) decreases in diameter as the flow accelerates or converges and (*b*) increases in diameter as the flow decelerates or diverges.

Pathlines are the easiest of the flow patterns to understand. A pathline is a Lagrangian concept in that we simply follow the path of an individual fluid particle as it moves around in the flow field (Fig. 4–20). Thus, a pathline is the same as the fluid particle's material position vector ($x_{particle}(t)$, $y_{particle}(t)$, $z_{particle}(t)$), discussed in Section 4–1, traced out over some finite time interval. In a physical experiment, you can imagine a tracer fluid particle that is marked somehow—either by color or brightness—such that it is easily distinguishable from surrounding fluid particles. Now imagine a camera with the shutter open for a certain time period, $t_{start} < t < t_{end}$, in which the particle's path is recorded; the resulting curve is called a pathline. An intriguing example is shown in Fig. 4–21 for the case of waves moving along the surface of water in a tank. Neutrally buoyant white **tracer particles** are suspended in the water, and a time-exposure photograph is taken for one complete wave period. The result is pathlines that are elliptical in shape, showing that fluid particles bob up and down and forward and backward, but return to their original position upon completion of one wave period; there is no net forward motion. You may have experienced something similar while bobbing up and down on ocean waves at the beach.

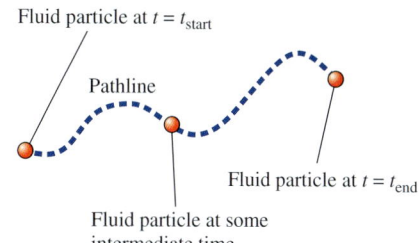

FIGURE 4–20
A *pathline* is formed by following the actual path of a fluid particle.

FIGURE 4–21
Pathlines produced by white tracer particles suspended in water and captured by time-exposure photography; as waves pass horizontally, each particle moves in an elliptical path during one wave period.

Wallet, A. & Ruellan, F. 1950, La Houille Blanche 5:483–489. Used by permission.

A modern experimental technique called **particle image velocimetry (PIV)** utilizes short segments of particle pathlines to measure the velocity field over an entire plane in a flow (Adrian, 1991). (Recent advances also extend the technique to three dimensions.) In PIV, tiny tracer particles are suspended in the fluid, much like in Fig. 4–21. However, the flow is illuminated by two flashes of light (usually a light sheet from a laser as in Fig. 4–22) to produce two bright spots (recorded by a camera) for each moving particle. Then, both the magnitude and direction of the velocity vector at each particle location can be inferred, assuming that the tracer particles are small enough that they move with the fluid. Modern digital photography and fast computers have enabled PIV to be performed rapidly enough so that *unsteady* features of a flow field can also be measured. PIV is discussed in more detail in Chap. 8.

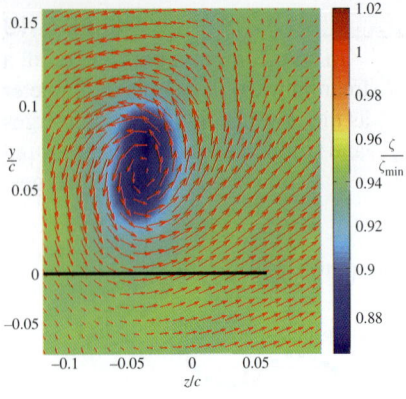

FIGURE 4–22

Stereo PIV measurements of the wing tip vortex in the wake of a NACA-66 airfoil at angle of attack. Color contours denote the local vorticity, normalized by the minimum value, as indicated in the color map. Vectors denote fluid motion in the plane of measurement. The black line denotes the location of the upstream wing trailing edge. Coordinates are normalized by the airfoil chord, and the origin is the wing root.

Photo by Michael H. Krane, ARL-Penn State.

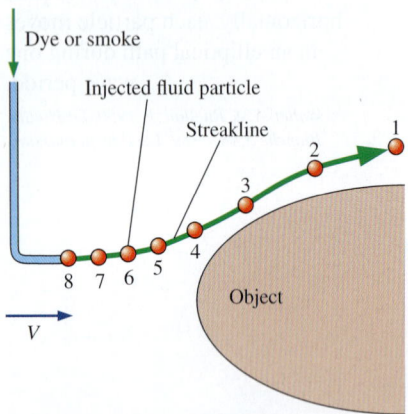

FIGURE 4–23

A *streakline* is formed by continuous introduction of dye or smoke from a point in the flow. Labeled tracer particles (1 through 8) were introduced sequentially.

Pathlines can also be calculated numerically for a known velocity field. Specifically, the location of the tracer particle is integrated over time from some starting location $\vec{x}_{start}$ and starting time t_{start} to some later time t.

Tracer particle location at time t:
$$\vec{x} = \vec{x}_{start} + \int_{t_{start}}^{t} \vec{V}\,dt \qquad (4\text{–}17)$$

When Eq. 4–17 is calculated for t between t_{start} and t_{end}, a plot of $\vec{x}(t)$ is the pathline of the fluid particle during that time interval, as illustrated in Fig. 4–20. For some simple flow fields, Eq. 4–17 can be integrated analytically. For more complex flows, we must perform a numerical integration.

If the velocity field is steady, individual fluid particles follow streamlines. Thus, *for steady flow, pathlines are identical to streamlines.*

Streaklines

A **streakline** is the locus of fluid particles that have passed sequentially through a prescribed point in the flow.

Streaklines are the most common flow pattern generated in a physical experiment. If you insert a small tube into a flow and introduce a continuous stream of tracer fluid (dye in a water flow or smoke in an airflow), the observed pattern is a streakline. Figure 4–23 shows a tracer being injected into a free-stream flow containing an object, such as a wing. The circles represent individual injected tracer fluid particles, released at a uniform time interval. As the particles are forced out of the way by the object, they accelerate around the shoulder of the object, as indicated by the increased distance between individual tracer particles in that region. The streakline is formed by connecting all the circles into a smooth curve. In physical experiments in a wind or water tunnel, the smoke or dye is injected *continuously*, not as individual particles, and the resulting flow pattern is by definition a streakline. In Fig. 4–23, tracer particle 1 was released at an earlier time than tracer particle 2, and so on. The location of an individual tracer particle is determined by the surrounding velocity field from the moment of its injection into the flow until the present time. If the flow is unsteady, the surrounding velocity field changes, and we cannot expect the resulting streakline to resemble a streamline or pathline at any given instant in time. However, *if the flow is steady, streamlines, pathlines, and streaklines are identical* (Fig. 4–24).

Streaklines are often confused with streamlines or pathlines. While the three flow patterns are identical in steady flow, they can be quite different in unsteady flow. The main difference is that a streamline represents an *instantaneous* flow pattern at a given instant in time, while a streakline and a pathline are flow patterns that have some *age* and thus a *time history* associated with them. A streakline is an instantaneous snapshot of a *time-integrated* flow pattern. A pathline, on the other hand, is the *time-exposed* flow path of an individual particle over some time period.

The time-integrative property of streaklines is vividly illustrated in an experiment by Cimbala et al. (1988), reproduced here as Fig. 4–25. The authors used a **smoke wire** for flow visualization in a wind tunnel. In operation, the smoke wire is a thin vertical wire that is coated with mineral oil. The oil breaks up into beads along the length of the wire due to surface tension effects. When an electric current heats the wire, each little bead of oil produces a streakline of smoke. In Fig. 4–25a, streaklines are introduced from a smoke wire located just downstream of a circular cylinder of diameter D aligned normal to the plane of view. (When multiple streaklines are introduced along a line, as in Fig. 4–25, we refer to this as a **rake** of streaklines.) The Reynolds number of the flow is Re = $\rho VD/\mu$ = 93. Because of unsteady **vortices** shed in an alternating pattern from the cylinder, the smoke collects into a clearly defined periodic pattern called a **Kármán vortex street.** A similar pattern can be seen at much larger scale in the air flow in the wake of an island (Fig. 4–26).

From Fig. 4–25a alone, you may think that the shed vortices continue to exist to several hundred diameters downstream of the cylinder. However, the streakline pattern of this figure is misleading! In Fig. 4–25b, the smoke wire is placed 150 diameters downstream of the cylinder. The resulting streaklines are straight, indicating that the shed vortices have in reality disappeared by this downstream distance. The flow is steady and parallel at this location, and there are no more vortices; viscous diffusion has caused adjacent vortices of opposite sign to cancel each other out by around 100 cylinder diameters. The patterns of Fig. 4–25a near x/D = 150 are merely *remnants* of the vortex street that existed upstream. The streaklines of Fig. 4–25b, however, show the correct features of the flow at that location. The streaklines generated at x/D = 150 are identical to streamlines or pathlines in that region of the flow—straight, nearly horizontal lines—since the flow is steady there.

For a known velocity field, a streakline can be generated numerically. We need to follow the paths of a continuous stream of tracer particles from the time of their injection into the flow until the present time, using Eq. 4–17. Mathematically, the location of a tracer particle is integrated over time from the time of its injection t_{inject} to the present time t_{present}. Equation 4–17 becomes

Integrated tracer particle location: $\qquad \vec{x} = \vec{x}_{\text{injection}} + \int_{t_{\text{inject}}}^{t_{\text{present}}} \vec{V} \, dt \qquad$ (4–18)

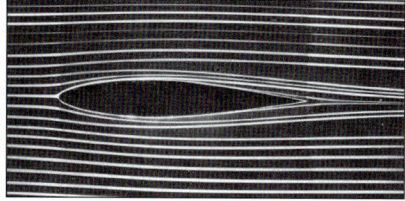

FIGURE 4–24
Streaklines produced by colored fluid introduced upstream; since the flow is steady, these streaklines are the same as streamlines and pathlines.

Courtesy ONERA. Photograph by Werlé.

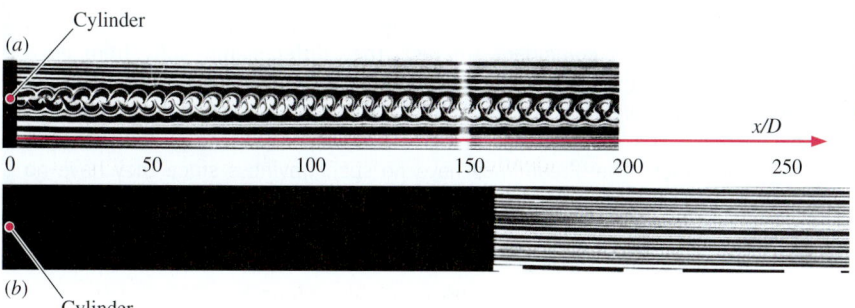

FIGURE 4–25
Smoke streaklines introduced by a smoke wire at two different locations in the wake of a circular cylinder: (*a*) smoke wire just downstream of the cylinder and (*b*) smoke wire located at x/D = 150. The time-integrative nature of streaklines is clearly seen by comparing the two photographs.

Photos by John M. Cimbala.

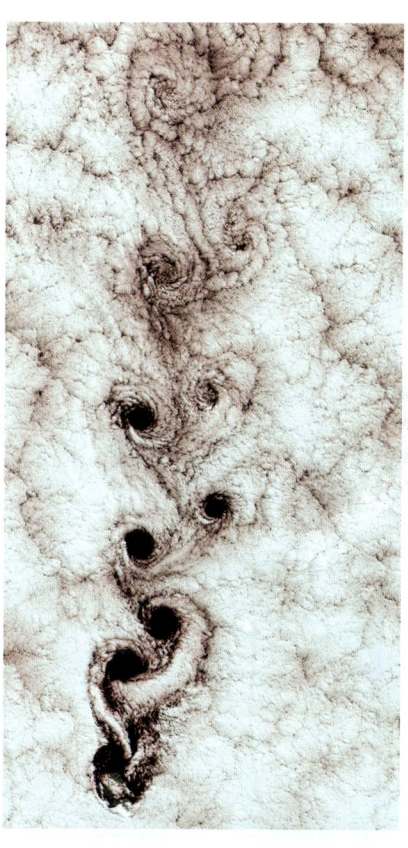

FIGURE 4–26
Kármán vortices visible in the clouds in the wake of Alexander Selkirk Island in the southern Pacific Ocean.

Photo from Landsat 7 WRS Path 6 Row 83, center: -33.18, -79.99, 9/15/1999, earthobservatory.nasa.gov. Courtesy of NASA.

146
FLUID KINEMATICS

In a complex unsteady flow, the time integration must be performed numerically as the velocity field changes with time. When the locus of tracer particle locations at $t = t_{present}$ is connected by a smooth curve, the result is the desired streakline.

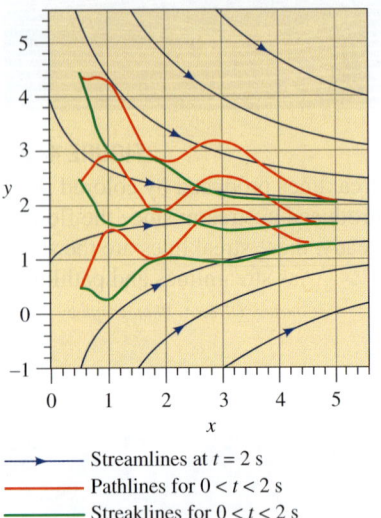

— Streamlines at $t = 2$ s
— Pathlines for $0 < t < 2$ s
— Streaklines for $0 < t < 2$ s

FIGURE 4–27
Streamlines, pathlines, and streaklines for the oscillating velocity field of Example 4–5. The streaklines and pathlines are wavy because of their integrated time history, but the streamlines are not wavy since they represent an instantaneous snapshot of the velocity field.

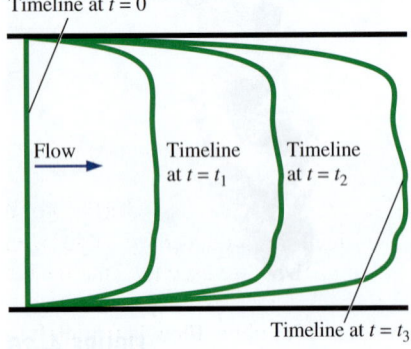

FIGURE 4–28
Timelines are formed by marking a line of fluid particles, and then watching that line move (and deform) through the flow field; timelines are shown at $t = 0$, t_1, t_2, and t_3.

EXAMPLE 4–5 Comparison of Flow Patterns in an Unsteady Flow

An *unsteady*, incompressible, two-dimensional velocity field is given by

$$\vec{V} = (u, v) = (0.5 + 0.8x)\vec{i} + (1.5 + 2.5 \sin(\omega t) - 0.8y)\vec{j} \quad (1)$$

where the angular frequency ω is equal to 2π rad/s (a physical frequency of 1 Hz). This velocity field is identical to that of Eq. 1 of Example 4–1 except for the additional periodic term in the v-component of velocity. In fact, since the period of oscillation is 1 s, when time t is any integral multiple of $\frac{1}{2}$ s ($t = 0, \frac{1}{2}, 1, \frac{3}{2}, 2, \ldots$ s), the sine term in Eq. 1 is zero and the velocity field is instantaneously identical to that of Example 4–1. Physically, we imagine flow into a large bell mouth inlet that is oscillating up and down at a frequency of 1 Hz. Consider two complete cycles of flow from $t = 0$ s to $t = 2$ s. Compare instantaneous streamlines at $t = 2$ s to pathlines and streaklines generated during the time period from $t = 0$ s to $t = 2$ s.

SOLUTION Streamlines, pathlines, and streaklines are to be generated and compared for the given unsteady velocity field.
Assumptions 1 The flow is incompressible. 2 The flow is two-dimensional, implying no z-component of velocity and no variation of u or v with z.
Analysis The instantaneous streamlines at $t = 2$ s are identical to those of Fig. 4–17, and several of them are replotted in Fig. 4–27. To simulate pathlines, we use the Runge–Kutta numerical integration technique to march in time from $t = 0$ s to $t = 2$ s, tracing the path of fluid particles released at three locations: ($x = 0.5$ m, $y = 0.5$ m), ($x = 0.5$ m, $y = 2.5$ m), and ($x = 0.5$ m, $y = 4.5$ m). These pathlines are shown in Fig. 4–27, along with the streamlines. Finally, streaklines are simulated by following the paths of *many* fluid tracer particles released at the given three locations at times between $t = 0$ s and $t = 2$ s, and connecting the locus of their positions at $t = 2$ s. These streaklines are also plotted in Fig. 4–27.
Discussion Since the flow is unsteady, the streamlines, pathlines, and streaklines are *not* coincident. In fact, they differ significantly from each other. Note that the streaklines and pathlines are wavy due to the undulating v-component of velocity. Two complete periods of oscillation have occurred between $t = 0$ s and $t = 2$ s, as verified by a careful look at the pathlines and streaklines. The streamlines have no such waviness since they have no time history; they represent an instantaneous snapshot of the velocity field at $t = 2$ s.

Timelines

A **timeline** is a set of adjacent fluid particles that were marked at the same (earlier) instant in time.

Timelines are particularly useful in situations where the uniformity of a flow (or lack thereof) is to be examined. Figure 4–28 illustrates timelines in a channel flow between two parallel walls. Because of friction at the walls, the fluid velocity there is zero (the no-slip condition), and the top and bottom of the timeline are anchored at their starting locations. In regions of the flow away from the walls, the marked fluid particles move at the local fluid velocity, deforming the timeline. In the example of Fig. 4–28, the speed near the center of the channel is fairly uniform, but small deviations tend to amplify with time as the timeline stretches. Timelines can be generated experimentally in a water channel through use of a **hydrogen bubble wire.** When a short burst of electric current is sent through the cathode wire, electrolysis of the water occurs and tiny hydrogen gas bubbles form at the wire. Since the bubbles are so small, their buoyancy is nearly negligible, and the bubbles follow the water flow nicely (Fig. 4–29).

Refractive Flow Visualization Techniques

Another category of flow visualization is based on the **refractive property** of light waves. As you recall from your study of physics, the speed of light through one material may differ somewhat from that in another material, or even in the *same* material if its density changes. As light travels through one fluid into a fluid with a different index of refraction, the light rays bend (they are **refracted**).

There are two primary flow visualization techniques that utilize the fact that the index of refraction in air (or other gases) varies with density. They are the **shadowgraph technique** and the **schlieren technique** (Settles, 2001). **Interferometry** is a visualization technique that utilizes the related *phase change* of light as it passes through air of varying densities as the basis for flow visualization and is not discussed here (see Merzkirch, 1987). All these techniques are useful for flow visualization in flow fields where density changes from one location in the flow to another, such as natural convection flows (temperature differences cause the density variations), mixing flows (fluid species cause the density variations), and supersonic flows (shock waves and expansion waves cause the density variations).

Unlike flow visualizations involving streaklines, pathlines, and timelines, the shadowgraph and schlieren methods do not require injection of a visible

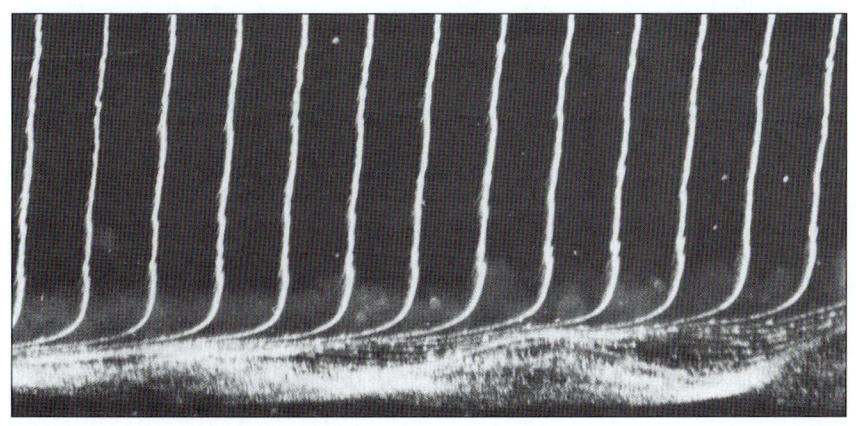

FIGURE 4–29

Timelines produced by a hydrogen bubble wire are used to visualize the boundary layer velocity profile shape along a flat plate. Flow is from left to right, and the hydrogen bubble wire is located to the left of the field of view. Bubbles near the wall reveal a flow instability that leads to turbulence.

Bippes, H. 1972 Sitzungsber, Heidelb. Akad. Wiss. Math. Naturwiss. Kl., no. 3, 103–180; NASA TM-75243, 1978.

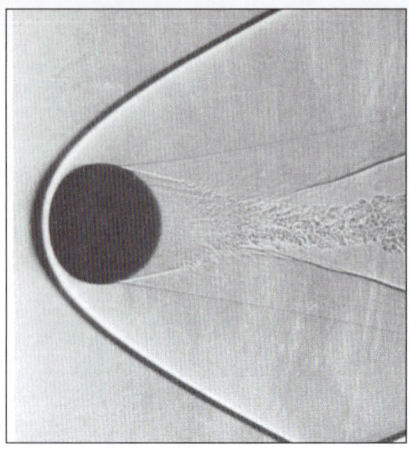

FIGURE 4–30
Shadowgram of a 14.3 mm sphere in free flight through air at Ma = 3.0. A shock wave is clearly visible in the shadow as a dark band that curves around the sphere and is called a *bow wave* (see Chap. 12).

A. C. Charters, Air Flow Branch, U.S. Army Ballistic Research Laboratory.

tracer (smoke or dye). Rather, density differences and the refractive property of light provide the necessary means for visualizing regions of activity in the flow field, allowing us to "see the invisible." The image (a **shadowgram**) produced by the shadowgraph method is formed when the refracted rays of light rearrange the shadow cast onto a viewing screen or camera focal plane, causing bright or dark patterns to appear in the shadow. The dark patterns indicate the location where the refracted rays *originate*, while the bright patterns mark where these rays *end up*, and can be misleading. As a result, the dark regions are less distorted than the bright regions and are more useful in the interpretation of the shadowgram. In the shadowgram of Fig. 4–30, for example, we can be confident of the shape and position of the bow shock wave (the dark band), but the refracted bright light has distorted the front of the sphere's shadow.

A shadowgram is not a true optical image; it is, after all, merely a shadow. A **schlieren image,** however, involves lenses (or mirrors) and a knife edge or other cutoff device to block the refracted light and is a true focused optical image. Schlieren imaging is more complicated to set up than is shadowgraphy (see Settles, 2001 for details) but has a number of advantages. For example, a schlieren image does not suffer from optical distortion by the refracted light rays. Schlieren imaging is also more sensitive to weak density gradients such as those caused by natural convection (Fig. 4–31) or by gradual phenomena like expansion fans in supersonic flow. Color schlieren imaging techniques have also been developed. Finally, one can adjust more components in a schlieren setup, such as the location, orientation, and type of the cutoff device, in order to produce an image that is most useful for the problem at hand.

Surface Flow Visualization Techniques

Finally, we briefly mention some flow visualization techniques that are useful along solid surfaces. The direction of fluid flow immediately above a solid surface can be visualized with **tufts**—short, flexible strings glued to the surface at one end that point in the flow direction. Tufts are especially useful for locating regions of flow separation, where the flow direction reverses.

A technique called **surface oil visualization** can be used for the same purpose—oil placed on the surface forms streaks called **friction lines** that indicate the direction of flow. If it rains lightly when your car is dirty (especially in the winter when salt is on the roads), you may have noticed streaks along the hood and sides of the car, or even on the windshield. This is similar to what is observed with surface oil visualization.

Lastly, there are pressure-sensitive and temperature-sensitive paints that enable researchers to observe the pressure or temperature distribution along solid surfaces.

4–3 ▪ PLOTS OF FLUID FLOW DATA

Regardless of how the results are obtained (analytically, experimentally, or computationally), it is usually necessary to *plot* flow data in ways that enable the reader to get a feel for how the flow properties vary in time and/or space. You are already familiar with *time plots*, which are especially useful in turbulent flows (e.g., a velocity component plotted as a function

FIGURE 4–31
Schlieren image of natural convection due to a barbeque grill.

G. S. Settles, Gas Dynamics Lab, Penn State University. Used by permission.

of time), and *xy*-plots (e.g., pressure as a function of radius). In this section, we discuss three additional types of plots that are useful in fluid mechanics—profile plots, vector plots, and contour plots.

Profile Plots

> A **profile plot** indicates how the value of a scalar property varies along some desired direction in the flow field.

Profile plots are the simplest of the three to understand because they are like the common *xy*-plots that you have generated since grade school. Namely, you plot how one variable *y* varies as a function of a second variable *x*. In fluid mechanics, profile plots of *any* scalar variable (pressure, temperature, density, etc.) can be created, but the most common one used in this book is the *velocity profile plot*. We note that since velocity is a vector quantity, we usually plot either the magnitude of velocity or one of the components of the velocity vector as a function of distance in some desired direction.

For example, one of the timelines in the boundary layer flow of Fig. 4–29 is converted into a velocity profile plot by recognizing that at a given instant in time, the horizontal distance traveled by a hydrogen bubble at vertical location *y* is proportional to the local *x*-component of velocity *u*. We plot *u* as a function of *y* in Fig. 4–32. The values of *u* for the plot can also be obtained analytically (see Chaps. 9 and 10), experimentally using PIV or some kind of local velocity measurement device (see Chap. 8), or computationally (see Chap. 15). Note that it is more physically meaningful in this example to plot *u* on the *abscissa* (horizontal axis) rather than on the *ordinate* (vertical axis) even though it is the dependent variable, since position *y* is then in its proper orientation (up) rather than across.

Finally, it is common to add arrows to velocity profile plots to make them more visually appealing, although no additional information is provided by the arrows. If more than one component of velocity is plotted by the arrow, the *direction* of the local velocity vector is indicated and the velocity profile plot becomes a velocity *vector* plot.

Vector Plots

> A **vector plot** is an array of arrows indicating the magnitude and direction of a vector property at an instant in time.

While streamlines indicate the *direction* of the instantaneous velocity field, they do not directly indicate the *magnitude* of the velocity (i.e., the speed). A useful flow pattern for both experimental and computational fluid flows is thus the vector plot, which consists of an array of arrows that indicate both magnitude *and* direction of an instantaneous vector property. We have already seen an example of a velocity vector plot in Fig. 4–4 and an acceleration vector plot in Fig. 4–14. These were generated analytically. Vector plots can also be generated from experimentally obtained data (e.g., from PIV measurements) or numerically from CFD calculations.

To further illustrate vector plots, we generate a two-dimensional flow field consisting of free-stream flow impinging on a block of rectangular cross section. We perform CFD calculations, and the results are shown in

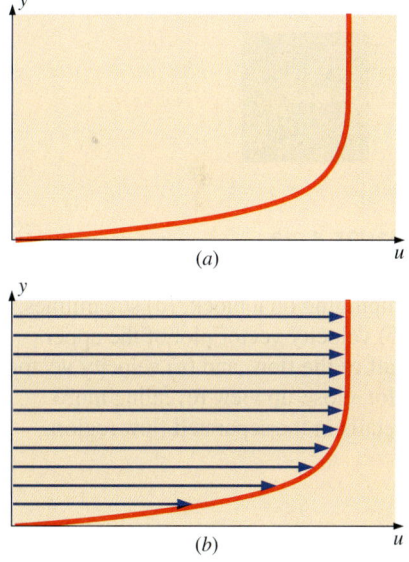

FIGURE 4–32
Profile plots of the horizontal component of velocity as a function of vertical distance; flow in the boundary layer growing along a horizontal flat plate: (*a*) standard profile plot and (*b*) profile plot with arrows.

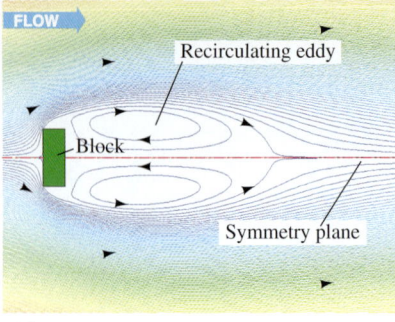

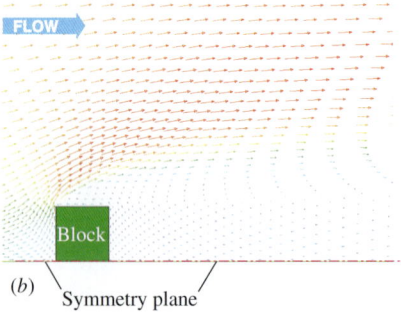

FIGURE 4–33
Results of CFD calculations of flow impinging on a block; (*a*) streamlines, (*b*) velocity vector plot of the upper half of the flow, and (*c*) velocity vector plot, close-up view revealing more details in the separated flow region.

Fig. 4–33. Note that this flow is by nature turbulent and unsteady, but only the long-time averaged results are calculated and displayed here. Streamlines are plotted in Fig. 4–33*a*; a view of the entire block and a large portion of its wake is shown. The closed streamlines above and below the symmetry plane indicate large recirculating eddies, one above and one below the line of symmetry. A velocity vector plot is shown in Fig. 4–33*b*. (Only the upper half of the flow is shown because of symmetry.) It is clear from this plot that the flow accelerates around the upstream corner of the block, so much so in fact that the boundary layer cannot negotiate the sharp corner and separates off the block, producing the large recirculating eddies downstream of the block. (Note that these velocity vectors are time-averaged values; the instantaneous vectors change in both magnitude and direction with time as vortices are shed from the body, similar to those of Fig. 4–25*a*.) A close-up view of the separated flow region is plotted in Fig. 4–33*c*, where we verify the reverse flow in the lower half of the large recirculating eddy.

The vectors of Fig. 4–33 are colored by velocity magnitude, but with modern CFD codes and postprocessors, the vectors can be colored according to some other flow property such as pressure (red for high pressure and blue for low pressure) or temperature (red for hot and blue for cold). In this manner, one can easily visualize not only the magnitude and direction of the flow, but other properties as well, simultaneously.

Contour Plots

A **contour plot** shows curves of constant values of a scalar property (or magnitude of a vector property) at an instant in time.

If you do any hiking, you are familiar with contour maps of mountain trails. The maps consist of a series of closed curves, each indicating a constant elevation or altitude. Near the center of a group of such curves is the mountain peak or valley; the actual peak or valley is a *point* on the map showing the highest or lowest elevation. Such maps are useful in that not only do you get a bird's-eye view of the streams and trails, etc., but you can also easily see your elevation and where the trail is flat or steep. In fluid mechanics, the same principle is applied to various scalar flow properties; contour plots (also called **isocontour plots**) are generated of pressure, temperature, velocity magnitude, species concentration, properties of turbulence, etc. A contour plot can quickly reveal regions of high (or low) values of the flow property being studied.

A contour plot may consist simply of curves indicating various levels of the property; this is called a **contour line plot**. Alternatively, the contours can be filled in with either colors or shades of gray; this is called a **filled contour plot**. An example of pressure contours is shown in Fig. 4–34 for the same flow as in Fig. 4–33. In Fig. 4–34*a*, filled contours are shown using color to identify regions of different pressure levels—blue regions indicate low pressure and red regions indicate high pressure. It is clear from this figure that the pressure is highest at the front face of the block and lowest along the top of the block in the separated zone. The pressure is also low in the wake of the block, as expected. In Fig. 4–34*b*, the same pressure contours are shown, but as a contour line plot with labeled levels of gage pressure in units of pascal.

In CFD, contour plots are often displayed in vivid colors with red usually indicating the highest value of the scalar and blue the lowest. A healthy human eye can easily spot a red or blue region and thus locate regions of high or low value of the flow property. Because of the pretty pictures produced by CFD, computational fluid dynamics is sometimes given the nickname "colorful fluid dynamics."

4–4 ■ OTHER KINEMATIC DESCRIPTIONS

Types of Motion or Deformation of Fluid Elements

In fluid mechanics, as in solid mechanics, an element may undergo four fundamental types of motion or deformation, as illustrated in two dimensions in Fig. 4–35: (*a*) **translation,** (*b*) **rotation,** (*c*) **linear strain** (sometimes called **extensional strain**), and (*d*) **shear strain.** The study of fluid dynamics is further complicated by the fact that all four types of motion or deformation usually occur simultaneously. Because fluid elements may be in constant motion, it is preferable in fluid dynamics to describe the motion and deformation of fluid elements in terms of *rates*. In particular, we discuss *velocity* (rate of translation), *angular velocity* (rate of rotation), *linear strain rate* (rate of linear strain), and *shear strain rate* (rate of shear strain). In order for these **deformation rates** to be useful in the calculation of fluid flows, we must express them in terms of velocity and derivatives of velocity.

Translation and rotation are easily understood since they are commonly observed in the motion of solid particles such as billiard balls (Fig. 4–1). A vector is required in order to fully describe the rate of translation in three dimensions. The **rate of translation vector** is described mathematically as the **velocity vector.** In Cartesian coordinates,

Rate of translation vector in Cartesian coordinates:

$$\vec{V} = u\vec{i} + v\vec{j} + w\vec{k} \quad (4\text{–}19)$$

In Fig. 4–35*a*, the fluid element has moved in the positive horizontal (*x*) direction; thus *u* is positive, while *v* (and *w*) are zero.

Rate of rotation (angular velocity) at a point is defined as *the average rotation rate of two initially perpendicular lines that intersect at that point*. In Fig. 4–35*b*, for example, consider the point at the bottom-left corner of the initially square fluid element. The left edge and the bottom edge of the element intersect at that point and are initially perpendicular. Both of these lines rotate counterclockwise, which is the mathematically positive direction. The angle between these two lines (or between *any* two initially perpendicular lines on this fluid element) remains at 90° since solid body rotation is illustrated in the figure. Therefore, both lines rotate at the same rate, and the rate of rotation in the plane is simply the component of angular velocity in that plane.

In the more general, but still two-dimensional case (Fig. 4–36), the fluid particle translates and deforms as it rotates, and the rate of rotation is calculated according to the definition given in the previous paragraph. Namely, we begin at time t_1 with two initially perpendicular lines (lines *a* and *b* in Fig. 4–36) that intersect at point *P* in the *xy*-plane. We follow these lines as they move and rotate in an infinitesimal increment of time $dt = t_2 - t_1$.

FIGURE 4–34
Contour plots of the pressure field due to flow impinging on a block, as produced by CFD calculations; only the upper half is shown due to symmetry; (*a*) filled color contour plot and (*b*) contour line plot where pressure values are displayed in units of Pa (pascals) gage pressure.

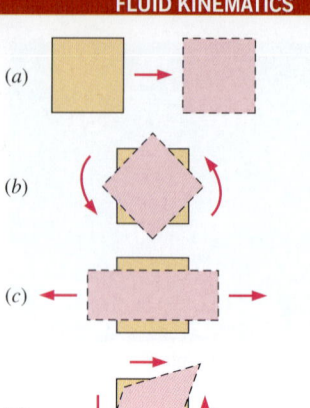

FIGURE 4–35
Fundamental types of fluid element motion or deformation: (*a*) translation, (*b*) rotation, (*c*) linear strain, and (*d*) shear strain.

At time t_2, line a has rotated by angle α_a, and line b has rotated by angle α_b, and both lines have moved with the flow as sketched (both angle values are given in radians and are shown mathematically positive in the sketch). The average rotation angle is thus $(\alpha_a + \alpha_b)/2$, and the *rate of rotation* or angular velocity in the xy-plane is equal to the time derivative of this average rotation angle,

Rate of rotation of fluid element about point P in Fig. 4–36:

$$\omega = \frac{d}{dt}\left(\frac{\alpha_a + \alpha_b}{2}\right) = \frac{1}{2}\left(\frac{\partial v}{\partial x} - \frac{\partial u}{\partial y}\right) \quad (4\text{–}20)$$

It is left as an exercise to prove the right side of Eq. 4–20 where we have written ω in terms of velocity components u and v in place of angles α_a and α_b.

In three dimensions, we must define a *vector* for the rate of rotation at a point in the flow since its magnitude may differ in each of the three dimensions. Derivation of the rate of rotation vector in three dimensions can be found in many fluid mechanics books such as Kundu and Cohen (2011) and White (2005). The **rate of rotation vector** is equal to the **angular velocity vector** and is expressed in Cartesian coordinates as

Rate of rotation vector in Cartesian coordinates:

$$\vec{\omega} = \frac{1}{2}\left(\frac{\partial w}{\partial y} - \frac{\partial v}{\partial z}\right)\vec{i} + \frac{1}{2}\left(\frac{\partial u}{\partial z} - \frac{\partial w}{\partial x}\right)\vec{j} + \frac{1}{2}\left(\frac{\partial v}{\partial x} - \frac{\partial u}{\partial y}\right)\vec{k} \quad (4\text{–}21)$$

Linear strain rate is defined as *the rate of increase in length per unit length*. Mathematically, the linear strain rate of a fluid element depends on the initial orientation or direction of the line segment upon which we measure the linear strain. Thus, it cannot be expressed as a scalar or vector quantity. Instead, we define linear strain rate in some arbitrary direction, which we denote as the x_α-direction. For example, line segment PQ in Fig. 4–37 has an initial length of dx_α, and it grows to line segment $P'Q'$ as shown. From the given definition and using the lengths marked in Fig. 4–37, the linear strain rate in the x_α-direction is

$$\varepsilon_{\alpha\alpha} = \frac{d}{dt}\left(\frac{P'Q' - PQ}{PQ}\right)$$

$$\cong \frac{d}{dt}\left(\frac{\overbrace{\left(u_\alpha + \frac{\partial u_\alpha}{\partial x_\alpha}dx_\alpha\right)dt + dx_\alpha - u_\alpha dt}^{\text{Length of } P'Q' \text{ in the } x_\alpha\text{-direction}} - \overbrace{dx_\alpha}^{\text{Length of } PQ \text{ in the } x_\alpha\text{-direction}}}{\underbrace{dx_\alpha}_{\text{Length of } PQ \text{ in the } x_\alpha\text{-direction}}}\right) = \frac{\partial u_\alpha}{\partial x_\alpha} \quad (4\text{–}22)$$

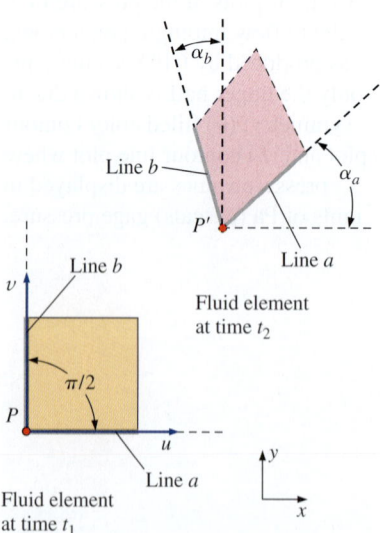

FIGURE 4–36
For a fluid element that translates and deforms as sketched, the *rate of rotation* at point P is defined as the average rotation rate of two initially perpendicular lines (lines a and b).

In Cartesian coordinates, we normally take the x_α-direction as that of each of the three coordinate axes, although we are not restricted to these directions.

Linear strain rate in Cartesian coordinates:

$$\varepsilon_{xx} = \frac{\partial u}{\partial x} \quad \varepsilon_{yy} = \frac{\partial v}{\partial y} \quad \varepsilon_{zz} = \frac{\partial w}{\partial z} \quad (4\text{–}23)$$

For the more general case, the fluid element moves and deforms as sketched in Fig. 4–36. It is left as an exercise to show that Eq. 4–23 is still valid for the general case.

Solid objects such as wires, rods, and beams stretch when pulled. You should recall from your study of engineering mechanics that when such an object stretches in one direction, it usually shrinks in direction(s) normal to that direction. The same is true of fluid elements. In Fig. 4–35c, the originally square fluid element stretches in the horizontal direction and shrinks in the vertical direction. The linear strain rate is thus positive horizontally and negative vertically.

If the flow is *incompressible*, the net volume of the fluid element must remain constant; thus if the element stretches in one direction, it must shrink by an appropriate amount in other direction(s) to compensate. The volume of a *compressible* fluid element, however, may increase or decrease as its density decreases or increases, respectively. (The mass of a fluid element must remain constant, but since $\rho = m/V$, density and volume are inversely proportional.) Consider for example a parcel of air in a cylinder being compressed by a piston (Fig. 4–38); the volume of the fluid element decreases while its density increases such that the fluid element's mass is conserved. The rate of increase of volume of a fluid element per unit volume is called its **volumetric strain rate** or **bulk strain rate**. This kinematic property is defined as *positive* when the volume *increases*. Another synonym of volumetric strain rate is **rate of volumetric dilatation**, which is easy to remember if you think about how the iris of your eye dilates (enlarges) when exposed to dim light. It turns out that the volumetric strain rate is the sum of the linear strain rates in three mutually orthogonal directions. In Cartesian coordinates (Eq. 4–23), the volumetric strain rate is thus

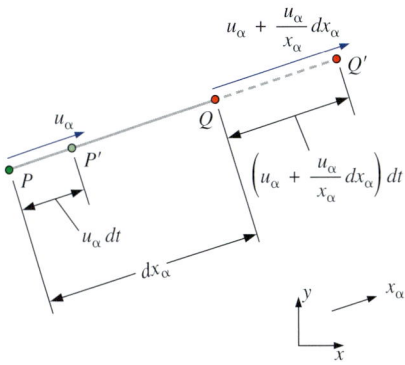

FIGURE 4–37
Linear strain rate in some arbitrary direction x_α is defined as the rate of increase in length per unit length in that direction. Linear strain rate would be *negative* if the line segment length were to *decrease*. Here we follow the increase in length of line segment PQ into line segment $P'Q'$, which yields a positive linear strain rate. Velocity components and distances are truncated to first-order since dx_α and dt are infinitesimally small.

Volumetric strain rate in Cartesian coordinates:

$$\frac{1}{V}\frac{DV}{Dt} = \frac{1}{V}\frac{dV}{dt} = \varepsilon_{xx} + \varepsilon_{yy} + \varepsilon_{zz} = \frac{\partial u}{\partial x} + \frac{\partial v}{\partial y} + \frac{\partial w}{\partial z} \qquad (4\text{–}24)$$

In Eq. 4–24, the uppercase D notation is used to stress that we are talking about the volume *following a fluid element*, that is to say, the *material volume* of the fluid element, as in Eq. 4–12.

The volumetric strain rate is zero in an incompressible flow.

Shear strain rate is a more difficult deformation rate to describe and understand. **Shear strain rate** at a point is defined as *half of the rate of decrease of the angle between two initially perpendicular lines that intersect at the point.* (The reason for the half will become clear later when we combine shear strain rate and linear strain rate into one tensor.) In Fig. 4–35d, for example, the initially 90° angles at the lower-left corner and upper-right corner of the square fluid element decrease; this is by definition a *positive* shear strain. However, the angles at the upper-left and lower-right corners of the square fluid element increase as the initially square fluid element deforms; this is a *negative* shear strain. Obviously we cannot describe the shear strain rate in terms of only one scalar quantity or even in terms of one *vector* quantity for that matter. Rather, a full mathematical description of shear strain rate requires its specification in any *two mutually perpendicular directions*. In Cartesian coordinates, the axes themselves are the most obvious choice, although we are not restricted to these. Consider a fluid element in two dimensions in the *xy*-plane. The element translates and deforms with time as sketched in Fig. 4–39. Two initially mutually perpendicular lines

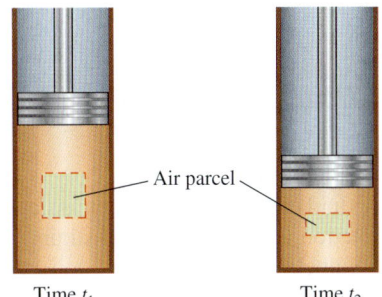

FIGURE 4–38
Air being compressed by a piston in a cylinder; the volume of a fluid element in the cylinder decreases, corresponding to a negative rate of volumetric dilatation.

154
FLUID KINEMATICS

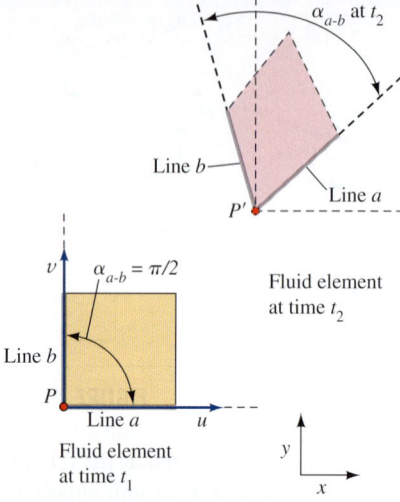

FIGURE 4–39
For a fluid element that translates and deforms as sketched, the *shear strain rate* at point P is defined as half of the rate of decrease of the angle between two initially perpendicular lines (lines *a* and *b*).

(lines *a* and *b* in the *x*- and *y*-directions, respectively) are followed. The angle between these two lines decreases from $\pi/2$ (90°) to the angle marked $\alpha_{a\text{-}b}$ at t_2 in the sketch. It is left as an exercise to show that the shear strain rate at point *P* for initially perpendicular lines in the *x*- and *y*-directions is given by

Shear strain rate, initially perpendicular lines in the x- and y-directions:

$$\varepsilon_{xy} = -\frac{1}{2}\frac{d}{dt}\alpha_{a\text{-}b} = \frac{1}{2}\left(\frac{\partial u}{\partial y} + \frac{\partial v}{\partial x}\right) \quad (4\text{–}25)$$

Equation 4–25 can be easily extended to three dimensions. The shear strain rate is thus

Shear strain rate in Cartesian coordinates:

$$\varepsilon_{xy} = \frac{1}{2}\left(\frac{\partial u}{\partial y} + \frac{\partial v}{\partial x}\right) \quad \varepsilon_{zx} = \frac{1}{2}\left(\frac{\partial w}{\partial x} + \frac{\partial u}{\partial z}\right) \quad \varepsilon_{yz} = \frac{1}{2}\left(\frac{\partial v}{\partial z} + \frac{\partial w}{\partial y}\right) \quad (4\text{–}26)$$

Finally, it turns out that we can mathematically combine linear strain rate and shear strain rate into one symmetric second-order tensor called the **strain rate tensor,** which is a combination of Eqs. 4–23 and 4–26:

Strain rate tensor in Cartesian coordinates:

$$\varepsilon_{ij} = \begin{pmatrix} \varepsilon_{xx} & \varepsilon_{xy} & \varepsilon_{xz} \\ \varepsilon_{yx} & \varepsilon_{yy} & \varepsilon_{yz} \\ \varepsilon_{zx} & \varepsilon_{zy} & \varepsilon_{zz} \end{pmatrix} = \begin{pmatrix} \dfrac{\partial u}{\partial x} & \dfrac{1}{2}\left(\dfrac{\partial u}{\partial y}+\dfrac{\partial v}{\partial x}\right) & \dfrac{1}{2}\left(\dfrac{\partial u}{\partial z}+\dfrac{\partial w}{\partial x}\right) \\ \dfrac{1}{2}\left(\dfrac{\partial v}{\partial x}+\dfrac{\partial u}{\partial y}\right) & \dfrac{\partial v}{\partial y} & \dfrac{1}{2}\left(\dfrac{\partial v}{\partial z}+\dfrac{\partial w}{\partial y}\right) \\ \dfrac{1}{2}\left(\dfrac{\partial w}{\partial x}+\dfrac{\partial u}{\partial z}\right) & \dfrac{1}{2}\left(\dfrac{\partial w}{\partial y}+\dfrac{\partial v}{\partial z}\right) & \dfrac{\partial w}{\partial z} \end{pmatrix} \quad (4\text{–}27)$$

The strain rate tensor obeys all the laws of mathematical tensors, such as tensor invariants, transformation laws, and principal axes.

Figure 4–40 shows a general (although two-dimensional) situation in a compressible fluid flow in which all possible motions and deformations are present simultaneously. In particular, there is translation, rotation, linear strain, and shear strain. Because of the compressible nature of the fluid flow, there is also volumetric strain (dilatation). You should now have a better appreciation of the inherent complexity of fluid dynamics, and the mathematical sophistication required to fully describe fluid motion.

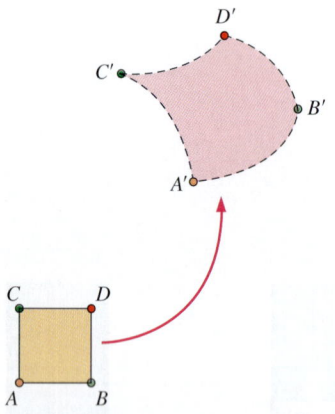

FIGURE 4–40
A fluid element illustrating translation, rotation, linear strain, shear strain, and volumetric strain.

EXAMPLE 4–6   Calculation of Kinematic Properties in a Two-Dimensional Flow

Consider the steady, two-dimensional velocity field of Example 4–1:

$$\vec{V} = (u, v) = (0.5 + 0.8\,x)\vec{i} + (1.5 - 0.8\,y)\vec{j} \quad (1)$$

where lengths are in units of m, time in s, and velocities in m/s. There is a stagnation point at (−0.625, 1.875) as shown in Fig. 4–41. Streamlines of the flow are also plotted in Fig. 4–41. Calculate the various kinematic properties, namely, the rate of translation, rate of rotation, linear strain rate, shear strain rate, and volumetric strain rate. Verify that this flow is incompressible.

SOLUTION We are to calculate several kinematic properties of a given velocity field and verify that the flow is incompressible.

Assumptions 1 The flow is steady. 2 The flow is two-dimensional, implying no z-component of velocity and no variation of u or v with z.

Analysis By Eq. 4–19, the rate of translation is simply the velocity vector itself, given by Eq. 1. Thus,

Rate of translation: $\quad u = \mathbf{0.5 + 0.8}x \quad v = \mathbf{1.5 - 0.8}y \quad w = \mathbf{0}$ (2)

The rate of rotation is found from Eq. 4–21. In this case, since $w = 0$ everywhere, and since neither u nor v vary with z, the only nonzero component of rotation rate is in the z-direction. Thus,

Rate of rotation: $\quad \vec{\omega} = \frac{1}{2}\left(\frac{\partial v}{\partial x} - \frac{\partial u}{\partial y}\right)\vec{k} = \frac{1}{2}(0-0)\vec{k} = \mathbf{0}$ (3)

In this case, we see that there is no net rotation of fluid particles as they move about. (This is a significant piece of information, to be discussed in more detail later in this chapter and also in Chap. 10.)

Linear strain rates can be calculated in any arbitrary direction using Eq. 4–23. In the x-, y-, and z-directions, the linear strain rates are

$$\varepsilon_{xx} = \frac{\partial u}{\partial x} = \mathbf{0.8\ s^{-1}} \quad \varepsilon_{yy} = \frac{\partial v}{\partial y} = \mathbf{-0.8\ s^{-1}} \quad \varepsilon_{zz} = \mathbf{0} \quad (4)$$

Thus, we predict that fluid particles *stretch* in the x-direction (positive linear strain rate) and *shrink* in the y-direction (negative linear strain rate). This is illustrated in Fig. 4–42, where we have marked an initially square parcel of fluid centered at (0.25, 4.25). By integrating Eqs. 2 with time, we calculate the location of the four corners of the marked fluid after an elapsed time of 1.5 s. Indeed this fluid parcel has stretched in the x-direction and has shrunk in the y-direction as predicted.

Shear strain rate is determined from Eq. 4–26. Because of the two-dimensionality, nonzero shear strain rates can occur only in the xy-plane. Using lines parallel to the x- and y-axes as our initially perpendicular lines, we calculate ε_{xy},

$$\varepsilon_{xy} = \frac{1}{2}\left(\frac{\partial u}{\partial y} + \frac{\partial v}{\partial x}\right) = \frac{1}{2}(0+0) = \mathbf{0} \quad (5)$$

Thus, there is no shear strain in this flow, as also indicated by Fig. 4–42. Although the sample fluid particle deforms, it remains rectangular; its initially 90° corner angles remain at 90° throughout the time period of the calculation.

Finally, the volumetric strain rate is calculated from Eq. 4–24:

$$\frac{1}{V}\frac{DV}{Dt} = \varepsilon_{xx} + \varepsilon_{yy} + \varepsilon_{zz} = (0.8 - 0.8 + 0)\ \text{s}^{-1} = \mathbf{0} \quad (6)$$

Since the volumetric strain rate is zero everywhere, we can say definitively that fluid particles are neither dilating (expanding) nor shrinking (compressing) in volume. Thus, **we verify that this flow is indeed incompressible.** In Fig. 4–42, the area of the shaded fluid particle (and thus its volume since it is a 2-D flow) remains constant as it moves and deforms in the flow field.

Discussion In this example it turns out that the linear strain rates (ε_{xx} and ε_{yy}) are nonzero, while the shear strain rates (ε_{xy} and its symmetric partner ε_{yx})

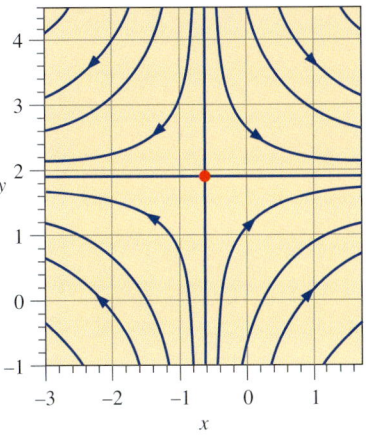

FIGURE 4–41
Streamlines for the velocity field of Example 4–6. The stagnation point is indicated by the red circle at $x = -0.625$ m and $y = 1.875$ m.

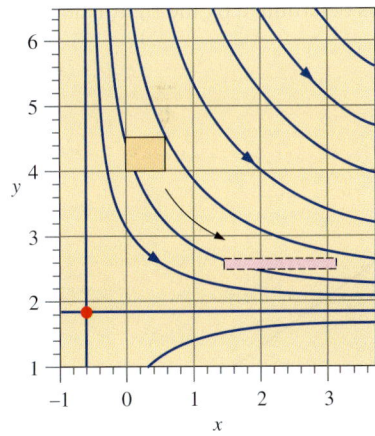

FIGURE 4–42
Deformation of an initially square parcel of marked fluid subjected to the velocity field of Example 4–6 for a time period of 1.5 s. The stagnation point is indicated by the red circle at $x = -0.625$ m and $y = 1.875$ m, and several streamlines are plotted.

are zero. This means that *the x- and y-axes of this flow field are the principal axes.* The (two-dimensional) strain rate tensor in this orientation is thus

$$\varepsilon_{ij} = \begin{pmatrix} \varepsilon_{xx} & \varepsilon_{xy} \\ \varepsilon_{yx} & \varepsilon_{yy} \end{pmatrix} = \begin{pmatrix} 0.8 & 0 \\ 0 & -0.8 \end{pmatrix} \text{s}^{-1} \quad (7)$$

If we were to rotate the axes by some arbitrary angle, the new axes would *not* be principal axes, and all four elements of the strain rate tensor would be nonzero. You may recall rotating axes in your engineering mechanics classes through use of Mohr's circles to determine principal axes, maximum shear strains, etc. Similar analyses are performed in fluid mechanics.

4–5 ▪ VORTICITY AND ROTATIONALITY

We have already defined the rate of rotation vector of a fluid element (see Eq. 4–21). A closely related kinematic property of great importance to the analysis of fluid flows is the **vorticity vector,** defined mathematically as the curl of the velocity vector $\vec{V}$,

Vorticity vector: $\qquad \vec{\zeta} = \vec{\nabla} \times \vec{V} = \text{curl}(\vec{V}) \qquad$ (4–28)

Physically, you can tell the direction of the vorticity vector by using the right-hand rule for cross product (Fig. 4–43). The symbol ζ used for vorticity is the Greek letter *zeta*. You should note that this symbol for vorticity is *not* universal among fluid mechanics textbooks; some authors use the Greek letter *omega* (ω) while still others use uppercase *omega* (Ω). In this book, $\vec{\omega}$ is used to denote the rate of rotation vector (angular velocity vector) of a fluid element. It turns out that the rate of rotation vector is equal to half of the vorticity vector,

Rate of rotation vector: $\qquad \vec{\omega} = \dfrac{1}{2}\vec{\nabla} \times \vec{V} = \dfrac{1}{2}\text{curl}(\vec{V}) = \dfrac{\vec{\zeta}}{2} \qquad$ (4–29)

Thus, *vorticity is a measure of rotation of a fluid particle.* Specifically,

Vorticity is equal to twice the angular velocity of a fluid particle (Fig. 4–44).

If the vorticity at a point in a flow field is nonzero, the fluid particle that happens to occupy that point in space is rotating; the flow in that region is called **rotational.** Likewise, if the vorticity in a region of the flow is zero (or negligibly small), fluid particles there are not rotating; the flow in that region is called **irrotational.** Physically, fluid particles in a rotational region of flow rotate end over end as they move along in the flow. For example, fluid particles within the viscous boundary layer near a solid wall are rotational (and thus have nonzero vorticity), while fluid particles outside the boundary layer are irrotational (and their vorticity is zero). Both of these cases are illustrated in Fig. 4–45.

Rotation of fluid elements is associated with wakes, boundary layers, flow through turbomachinery (fans, turbines, compressors, etc.), and flow with heat transfer. The vorticity of a fluid element cannot change except through the action of viscosity, nonuniform heating (temperature gradients), or other nonuniform phenomena. Thus if a flow originates in an irrotational region, it remains irrotational until some nonuniform process alters it. For example,

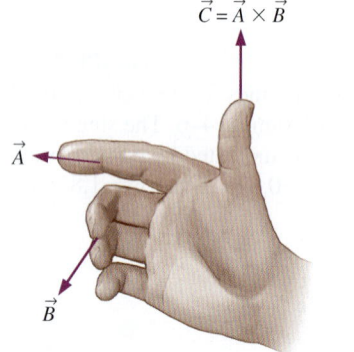

FIGURE 4–43
The direction of a vector cross product is determined by the right-hand rule.

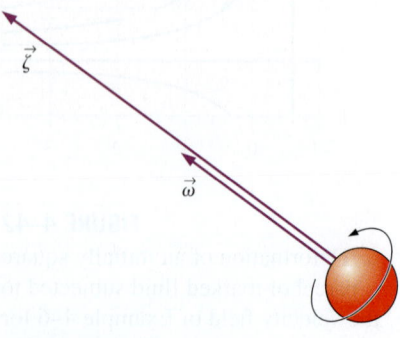

FIGURE 4–44
The *vorticity vector* is equal to twice the angular velocity vector of a rotating fluid particle.

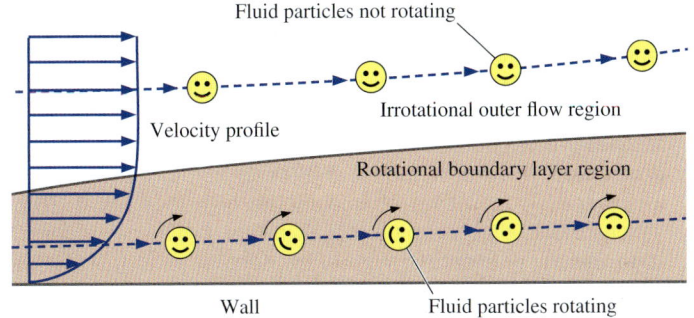

FIGURE 4–45
The difference between rotational and irrotational flow: fluid elements in a rotational region of the flow rotate, but those in an irrotational region of the flow do not.

air entering an inlet from quiescent (still) surroundings is irrotational and remains so unless it encounters an object in its path or is subjected to non-uniform heating. If a region of flow can be approximated as irrotational, the equations of motion are greatly simplified, as you will see in Chap. 10.

In Cartesian coordinates, $(\vec{i}, \vec{j}, \vec{k})$, (x, y, z), and (u, v, w), Eq. 4–28 is expanded as follows:

Vorticity vector in Cartesian coordinates:

$$\vec{\zeta} = \left(\frac{\partial w}{\partial y} - \frac{\partial v}{\partial z}\right)\vec{i} + \left(\frac{\partial u}{\partial z} - \frac{\partial w}{\partial x}\right)\vec{j} + \left(\frac{\partial v}{\partial x} - \frac{\partial u}{\partial y}\right)\vec{k} \quad (4\text{–}30)$$

If the flow is two-dimensional in the xy-plane, the z-component of velocity (w) is zero and neither u nor v varies with z. Thus the first two components of Eq. 4–30 are identically zero and the vorticity reduces to

Two-dimensional flow in Cartesian coordinates:

$$\vec{\zeta} = \left(\frac{\partial v}{\partial x} - \frac{\partial u}{\partial y}\right)\vec{k} \quad (4\text{–}31)$$

Note that if a flow is two-dimensional in the xy-plane, the vorticity vector must point in either the z- or $-z$-direction (Fig. 4–46).

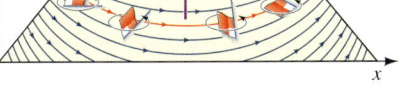

FIGURE 4–46
For two-dimensional flow in the xy-plane, the vorticity vector always points in the z- or $-z$-direction. In this illustration, the flag-shaped fluid particle rotates in the counterclockwise direction as it moves in the xy-plane; its vorticity points in the positive z-direction as shown.

EXAMPLE 4–7 Vorticity Contours in a Two-Dimensional Flow

Consider the CFD calculation of two-dimensional free-stream flow impinging on a block of rectangular cross section, as shown in Figs. 4–33 and 4–34. Plot vorticity contours and discuss.

SOLUTION We are to calculate the vorticity field for a given velocity field produced by CFD and then generate a contour plot of vorticity.
Analysis Since the flow is two-dimensional, the only nonzero component of vorticity is in the z-direction, normal to the page in Figs. 4–33 and 4–34. A contour plot of the z-component of vorticity for this flow field is shown in Fig. 4–47. The blue region near the upper-left corner of the block indicates large negative values of vorticity, implying *clockwise* rotation of fluid particles in that region. This is due to the large velocity gradients encountered in this portion of the flow field; the boundary layer separates off the wall at the corner

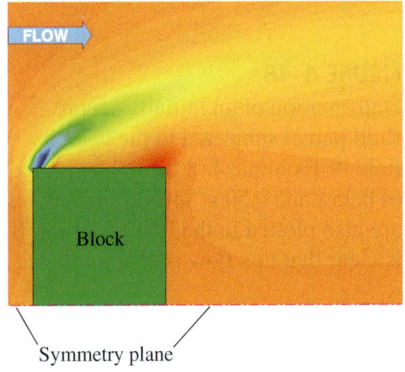

FIGURE 4–47
Contour plot of the vorticity field ζ_z due to flow impinging on a block, as produced by CFD calculations; only the upper half is shown due to symmetry. Blue regions represent large negative vorticity, and red regions represent large positive vorticity.

of the body and forms a thin **shear layer** across which the velocity changes rapidly. The concentration of vorticity in the shear layer diminishes as vorticity diffuses downstream. The small red region near the top right corner of the block represents a region of *positive* vorticity (counterclockwise rotation)—a secondary flow pattern caused by the flow separation.

Discussion We expect the magnitude of vorticity to be highest in regions where spatial derivatives of velocity are high (see Eq. 4–30). Close examination reveals that the blue region in Fig. 4–47 does indeed correspond to large velocity gradients in Fig. 4–33. Keep in mind that the vorticity field of Fig. 4–47 is time-averaged. The instantaneous flow field is in reality turbulent and unsteady, and vortices are shed from the bluff body.

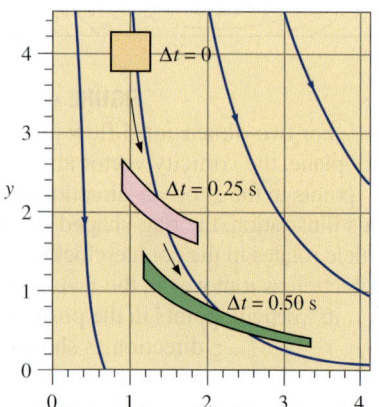

FIGURE 4–48
Deformation of an initially square fluid parcel subjected to the velocity field of Example 4–8 for a time period of 0.25 s and 0.50 s. Several streamlines are also plotted in the first quadrant. It is clear that this flow is *rotational*.

EXAMPLE 4–8 Determination of Rotationality in a Two-Dimensional Flow

Consider the following steady, incompressible, two-dimensional velocity field:

$$\vec{V} = (u, v) = x^2\vec{i} + (-2xy - 1)\vec{j} \quad (1)$$

Is this flow rotational or irrotational? Sketch some streamlines in the first quadrant and discuss.

SOLUTION We are to determine whether a flow with a given velocity field is rotational or irrotational, and we are to draw some streamlines in the first quadrant.

Analysis Since the flow is two-dimensional, Eq. 4–31 is applicable. Thus,

Vorticity: $\quad \vec{\zeta} = \left(\dfrac{\partial v}{\partial x} - \dfrac{\partial u}{\partial y}\right)\vec{k} = (-2y - 0)\vec{k} = -2y\vec{k} \quad (2)$

Since the vorticity is nonzero, this flow is **rotational.** In Fig. 4–48 we plot several streamlines of the flow in the first quadrant; we see that fluid moves downward and to the right. The translation and deformation of a fluid parcel is also shown: at $\Delta t = 0$, the fluid parcel is square, at $\Delta t = 0.25$ s, it has moved and deformed, and at $\Delta t = 0.50$ s, the parcel has moved farther and is further deformed. In particular, the right-most portion of the fluid parcel moves faster to the right and faster downward compared to the left-most portion, stretching the parcel in the *x*-direction and squashing it in the vertical direction. It is clear that there is also a net *clockwise* rotation of the fluid parcel, which agrees with the result of Eq. 2.

Discussion From Eq. 4–29, individual fluid particles rotate at an angular velocity equal to $\vec{\omega} = -y\vec{k}$, half of the vorticity vector. Since $\vec{\omega}$ is not constant, this flow is *not* solid-body rotation. Rather, $\vec{\omega}$ is a linear function of *y*. Further analysis reveals that this flow field is incompressible; the area (and volume) of the shaded regions representing the fluid parcel in Fig. 4–48 remains constant at all three instants in time.

In cylindrical coordinates, $(\vec{e}_r, \vec{e}_\theta, \vec{e}_z)$, (r, θ, z), and (u_r, u_θ, u_z), Eq. 4–28 is expanded as

Vorticity vector in cylindrical coordinates:

$$\vec{\zeta} = \left(\dfrac{1}{r}\dfrac{\partial u_z}{\partial \theta} - \dfrac{\partial u_\theta}{\partial z}\right)\vec{e}_r + \left(\dfrac{\partial u_r}{\partial z} - \dfrac{\partial u_z}{\partial r}\right)\vec{e}_\theta + \dfrac{1}{r}\left(\dfrac{\partial(ru_\theta)}{\partial r} - \dfrac{\partial u_r}{\partial \theta}\right)\vec{e}_z \quad (4\text{–}32)$$

For two-dimensional flow in the $r\theta$-plane, Eq. 4–32 reduces to

Two-dimensional flow in cylindrical coordinates:

$$\vec{\zeta} = \frac{1}{r}\left(\frac{\partial(ru_\theta)}{\partial r} - \frac{\partial u_r}{\partial \theta}\right)\vec{k} \qquad (4\text{--}33)$$

where $\vec{k}$ is used as the unit vector in the z-direction in place of $\vec{e}_z$. Note that if a flow is two-dimensional in the $r\theta$-plane, the vorticity vector must point in either the z- or $-z$-direction (Fig. 4–49).

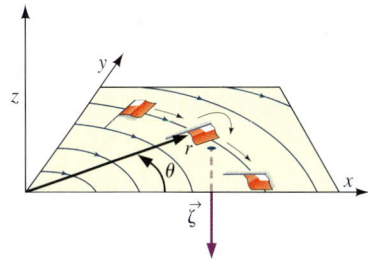

FIGURE 4–49

For a two-dimensional flow in the $r\theta$-plane, the vorticity vector always points in the z (or $-z$) direction. In this illustration, the flag-shaped fluid particle rotates in the clockwise direction as it moves in the $r\theta$-plane; its vorticity points in the $-z$-direction as shown.

Comparison of Two Circular Flows

Not all flows with circular streamlines are rotational. To illustrate this point, we consider two incompressible, steady, two-dimensional flows, both of which have circular streamlines in the $r\theta$-plane:

Flow A—solid-body rotation: $\qquad u_r = 0 \quad$ and $\quad u_\theta = \omega r \qquad (4\text{--}34)$

Flow B—line vortex: $\qquad u_r = 0 \quad$ and $\quad u_\theta = \dfrac{K}{r} \qquad (4\text{--}35)$

where ω and K are constants. (Alert readers will note that u_θ in Eq. 4–35 is infinite at $r = 0$, which is of course physically impossible; we ignore the region close to the origin to avoid this problem.) Since the radial component of velocity is zero in both cases, the streamlines are circles about the origin. The velocity profiles for the two flows, along with their streamlines, are sketched in Fig. 4–50. We now calculate and compare the vorticity field for each of these flows, using Eq. 4–33.

Flow A—solid-body rotation: $\qquad \vec{\zeta} = \dfrac{1}{r}\left(\dfrac{\partial(\omega r^2)}{\partial r} - 0\right)\vec{k} = 2\omega\vec{k} \qquad (4\text{--}36)$

Flow B—line vortex: $\qquad \vec{\zeta} = \dfrac{1}{r}\left(\dfrac{\partial(K)}{\partial r} - 0\right)\vec{k} = 0 \qquad (4\text{--}37)$

Not surprisingly, the vorticity for solid-body rotation is nonzero. In fact, it is a constant of magnitude twice the angular velocity and pointing in the same direction. (This agrees with Eq. 4–29.) *Flow A is rotational.* Physically, this means that individual fluid particles rotate as they revolve around the origin (Fig. 4–50a). By contrast, the vorticity of the line vortex is zero everywhere (except right at the origin, which is a mathematical singularity). *Flow B is irrotational.* Physically, fluid particles do *not* rotate as they revolve in circles about the origin (Fig. 4–50b).

A simple analogy can be made between flow A and a merry-go-round or roundabout, and flow B and a Ferris wheel (Fig. 4–51). As children revolve around a roundabout, they also rotate at the same angular velocity as that of the ride itself. This is analogous to a rotational flow. In contrast, children on a Ferris wheel always remain oriented in an upright position as they trace out their circular path. This is analogous to an irrotational flow.

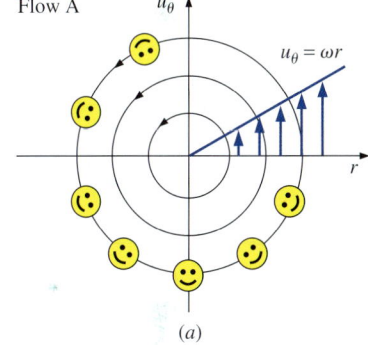

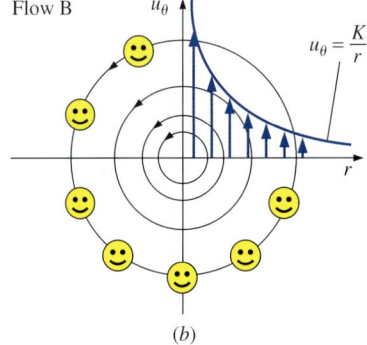

FIGURE 4–50

Streamlines and velocity profiles for (a) flow A, solid-body rotation and (b) flow B, a line vortex. Flow A is rotational, but flow B is irrotational everywhere except at the origin.

160
FLUID KINEMATICS

(a)

(b)

FIGURE 4–51
A simple analogy: (a) *rotational* circular flow is analogous to a roundabout, while (b) *irrotational* circular flow is analogous to a Ferris wheel.
(a) Mc Graw-Hill Companies, Inc. Mark Dierker, photographer (b) © DAJ/Getty RF

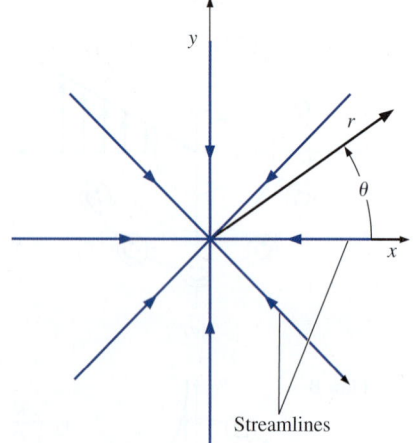

FIGURE 4–52
Streamlines in the $r\theta$-plane for the case of a line sink.

EXAMPLE 4–9   **Determination of Rotationality of a Line Sink**

A simple two-dimensional velocity field called a **line sink** is often used to simulate fluid being sucked into a line along the z-axis. Suppose the volume flow rate per unit length along the z-axis, $\dot{V}/L$, is known, where $\dot{V}$ is a negative quantity. In two dimensions in the $r\theta$-plane,

Line sink: $\qquad u_r = \dfrac{\dot{V}}{2\pi L}\dfrac{1}{r}$ and $u_\theta = 0$ (1)

Draw several streamlines of the flow and calculate the vorticity. Is this flow rotational or irrotational?

SOLUTION Streamlines of the given flow field are to be sketched and the rotationality of the flow is to be determined.
Analysis Since there is only radial flow and no tangential flow, we know immediately that all streamlines must be rays into the origin. Several streamlines are sketched in Fig. 4–52. The vorticity is calculated from Eq. 4–33:

$$\vec{\zeta} = \dfrac{1}{r}\left(\dfrac{\partial(ru_\theta)}{\partial r} - \dfrac{\partial}{\partial \theta}u_r\right)\vec{k} = \dfrac{1}{r}\left(0 - \dfrac{\partial}{\partial \theta}\left(\dfrac{\dot{V}}{2\pi L}\dfrac{1}{r}\right)\right)\vec{k} = 0 \qquad (2)$$

Since the vorticity vector is everywhere zero, this flow field is **irrotational**.
Discussion Many practical flow fields involving suction, such as flow into inlets and hoods, can be approximated quite accurately by assuming irrotational flow (Heinsohn and Cimbala, 2003).

4–6 ■ THE REYNOLDS TRANSPORT THEOREM

In thermodynamics and solid mechanics we often work with a *system* (also called a *closed system*), defined as a *quantity of matter of fixed identity*. In fluid dynamics, it is more common to work with a *control volume* (also called an *open system*), defined as a *region in space chosen for study*. The size and shape of a system may change during a process, but no mass crosses its boundaries. A control volume, on the other hand, allows mass to flow in or out across its boundaries, which are called the **control surface.** A control volume may also move and deform during a process, but many real-world applications involve fixed, nondeformable control volumes.

Figure 4–53 illustrates both a system and a control volume for the case of deodorant being sprayed from a spray can. When analyzing the spraying process, a natural choice for our analysis is either the moving, deforming fluid (a system) or the volume bounded by the inner surfaces of the can (a control volume). These two choices are identical before the deodorant is sprayed. When some contents of the can are discharged, the system approach considers the discharged mass as part of the system and tracks it (a difficult job indeed); thus the mass of the system remains constant. Conceptually, this is equivalent to attaching a flat balloon to the nozzle of the can and letting the spray inflate the balloon. The inner surface of the balloon now becomes part of the boundary of the system. The control volume approach, however, is not concerned at all with the deodorant that has escaped the can (other than its properties at the exit), and thus the mass of the control volume decreases during this process while its volume remains constant. Therefore, the system approach treats the spraying process as an expansion of the system's volume, whereas the control volume approach considers it as a fluid discharge through the control surface of the fixed control volume.

Most principles of fluid mechanics are adopted from solid mechanics, where the physical laws dealing with the time rates of change of extensive properties are expressed for systems. In fluid mechanics, it is usually more convenient to work with control volumes, and thus there is a need to relate the changes in a control volume to the changes in a system. The relationship between the time rates of change of an extensive property for a system and for a control volume is expressed by the **Reynolds transport theorem (RTT),** which provides the link between the system and control volume approaches (Fig. 4–54). RTT is named after the English engineer, Osborne Reynolds (1842–1912), who did much to advance its application in fluid mechanics.

The general form of the Reynolds transport theorem can be derived by considering a system with an arbitrary shape and arbitrary interactions, but the derivation is rather involved. To help you grasp the fundamental meaning of the theorem, we derive it first in a straightforward manner using a simple geometry and then generalize the results.

Consider flow from left to right through a diverging (expanding) portion of a flow field as sketched in Fig. 4–55. The upper and lower bounds of the fluid under consideration are *streamlines* of the flow, and we assume uniform flow through any cross section between these two streamlines. We choose the control volume to be fixed between sections (1) and (2) of the flow field. Both (1) and (2) are normal to the direction of flow. At some initial time t, the system coincides with the control volume, and thus the system and control volume are identical (the greenish-shaded region in Fig. 4–55). During time interval Δt, the system moves in the flow direction at uniform speeds V_1 at section (1) and V_2 at section (2). The system at this later time is indicated by the hatched region. The region uncovered by the system during this motion is designated as section I (part of the CV), and the new region

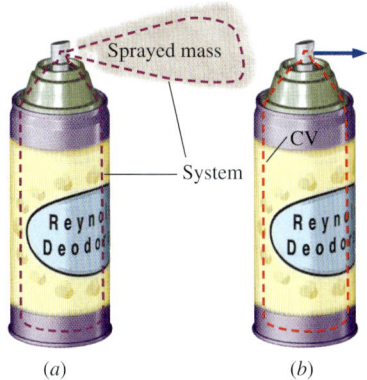

FIGURE 4–53
Two methods of analyzing the spraying of deodorant from a spray can: (*a*) We follow the fluid as it moves and deforms. This is the *system approach*—no mass crosses the boundary, and the total mass of the system remains fixed. (*b*) We consider a fixed interior volume of the can. This is the *control volume approach*—mass crosses the boundary.

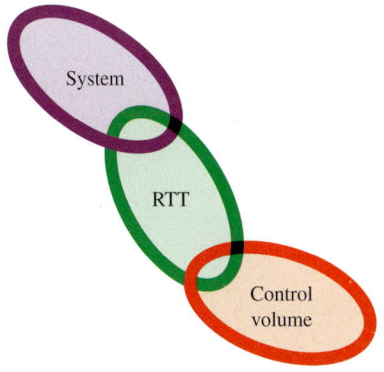

FIGURE 4–54
The *Reynolds transport theorem* (RTT) provides a link between the system approach and the control volume approach.

162
FLUID KINEMATICS

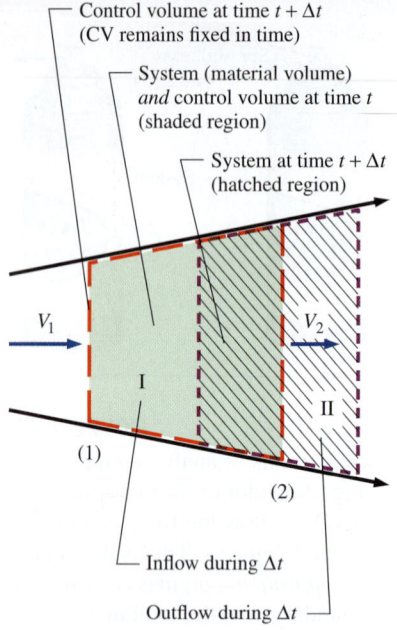

FIGURE 4–55
A moving *system* (hatched region) and a fixed *control volume* (shaded region) in a diverging portion of a flow field at times t and $t + \Delta t$. The upper and lower bounds are streamlines of the flow.

covered by the system is designated as section II (not part of the CV). Therefore, at time $t + \Delta t$, the system consists of the same fluid, but it occupies the region CV − I + II. The control volume is fixed in space, and thus it remains as the shaded region marked CV at all times.

Let B represent any **extensive property** (such as mass, energy, or momentum), and let $b = B/m$ represent the corresponding **intensive property.** Noting that extensive properties are additive, the extensive property B of the system at times t and $t + \Delta t$ is expressed as

$$B_{\text{sys},\,t} = B_{\text{CV},\,t} \quad \text{(the system and CV coincide at time } t\text{)}$$

$$B_{\text{sys},\,t+\Delta t} = B_{\text{CV},\,t+\Delta t} - B_{\text{I},\,t+\Delta t} + B_{\text{II},\,t+\Delta t}$$

Subtracting the first equation from the second one and dividing by Δt gives

$$\frac{B_{\text{sys},\,t+\Delta t} - B_{\text{sys},\,t}}{\Delta t} = \frac{B_{\text{CV},\,t+\Delta t} - B_{\text{CV},\,t}}{\Delta t} - \frac{B_{\text{I},\,t+\Delta t}}{\Delta t} + \frac{B_{\text{II},\,t+\Delta t}}{\Delta t}$$

Taking the limit as $\Delta t \to 0$, and using the definition of derivative, we get

$$\frac{dB_{\text{sys}}}{dt} = \frac{dB_{\text{CV}}}{dt} - \dot{B}_{\text{in}} + \dot{B}_{\text{out}} \quad (4\text{–}38)$$

or

$$\frac{dB_{\text{sys}}}{dt} = \frac{dB_{\text{CV}}}{dt} - b_1 \rho_1 V_1 A_1 + b_2 \rho_2 V_2 A_2$$

since

$$B_{\text{I},\,t+\Delta t} = b_1 m_{\text{I},\,t+\Delta t} = b_1 \rho_1 \mathcal{V}_{\text{I},\,t+\Delta t} = b_1 \rho_1 V_1 \Delta t\, A_1$$

$$B_{\text{II},\,t+\Delta t} = b_2 m_{\text{II},\,t+\Delta t} = b_2 \rho_2 \mathcal{V}_{\text{II},\,t+\Delta t} = b_2 \rho_2 V_2 \Delta t\, A_2$$

and

$$\dot{B}_{\text{in}} = \dot{B}_{\text{I}} = \lim_{\Delta t \to 0} \frac{B_{\text{I},\,t+\Delta t}}{\Delta t} = \lim_{\Delta t \to 0} \frac{b_1 \rho_1 V_1 \Delta t\, A_1}{\Delta t} = b_1 \rho_1 V_1 A_1$$

$$\dot{B}_{\text{out}} = \dot{B}_{\text{II}} = \lim_{\Delta t \to 0} \frac{B_{\text{II},\,t+\Delta t}}{\Delta t} = \lim_{\Delta t \to 0} \frac{b_2 \rho_2 V_2 \Delta t\, A_2}{\Delta t} = b_2 \rho_2 V_2 A_2$$

where A_1 and A_2 are the cross-sectional areas at locations 1 and 2. Equation 4–38 states that *the time rate of change of the property B of the system is equal to the time rate of change of B of the control volume plus the net flux of B out of the control volume by mass crossing the control surface*. This is the desired relation since it relates the change of a property of a system to the change of that property for a control volume. Note that Eq. 4–38 applies at any instant in time, where it is assumed that the system and the control volume occupy the same space at that particular instant in time.

The influx $\dot{B}_{\text{in}}$ and outflux $\dot{B}_{\text{out}}$ of the property B in this case are easy to determine since there is only one inlet and one outlet, and the velocities are approximately normal to the surfaces at sections (1) and (2). In general, however, we may have several inlet and outlet ports, and the velocity may not be normal to the control surface at the point of entry. Also, the velocity may not be uniform. To generalize the process, we consider a differential surface area dA on the control surface and denote its **unit outer normal** by $\vec{n}$. The flow rate of property b through dA is $\rho b \vec{V} \cdot \vec{n}\, dA$ since the dot product $\vec{V} \cdot \vec{n}$ gives the normal component of the velocity. Then the net rate of outflow through the entire control surface is determined by integration to be (Fig. 4–56)

$$\dot{B}_{net} = \dot{B}_{out} - \dot{B}_{in} = \int_{CS} \rho b \vec{V} \cdot \vec{n} \, dA \quad \text{(inflow if negative)} \quad (4\text{–}39)$$

An important aspect of this relation is that it automatically subtracts the inflow from the outflow, as explained next. The dot product of the velocity vector at a point on the control surface and the outer normal at that point is $\vec{V} \cdot \vec{n} = |\vec{V}||\vec{n}| \cos\theta = |\vec{V}| \cos\theta$, where θ is the angle between the velocity vector and the outer normal, as shown in Fig. 4–57. For $\theta < 90°$, $\cos\theta > 0$ and thus $\vec{V} \cdot \vec{n} > 0$ for outflow of mass from the control volume, and for $\theta > 90°$, $\cos\theta < 0$ and thus $\vec{V} \cdot \vec{n} < 0$ for inflow of mass into the control volume. Therefore, the differential quantity $\rho b \vec{V} \cdot \vec{n} \, dA$ is positive for mass flowing out of the control volume, and negative for mass flowing into the control volume, and its integral over the entire control surface gives the rate of net outflow of the property B by mass.

The properties within the control volume may vary with position, in general. In such a case, the total amount of property B within the control volume must be determined by integration:

$$B_{CV} = \int_{CV} \rho b \, dV \quad (4\text{–}40)$$

The term dB_{CV}/dt in Eq. 4–38 is thus equal to $\dfrac{d}{dt}\int_{CV} \rho b \, dV$, and represents the time rate of change of the property B content of the control volume. A positive value for dB_{CV}/dt indicates an increase in the B content, and a negative value indicates a decrease. Substituting Eqs. 4–39 and 4–40 into Eq. 4–38 yields the Reynolds transport theorem, also known as the *system-to-control-volume transformation* for a fixed control volume:

RTT, fixed CV:
$$\frac{dB_{sys}}{dt} = \frac{d}{dt}\int_{CV} \rho b \, dV + \int_{CS} \rho b \vec{V} \cdot \vec{n} \, dA \quad (4\text{–}41)$$

Since the control volume is not moving or deforming with time, the time derivative on the right-hand side can be moved inside the integral, since the domain of integration does not change with time. (In other words, it is irrelevant whether we differentiate or integrate first.) But the time derivative in that case must be expressed as a *partial* derivative ($\partial/\partial t$) since density and the quantity b may depend not only on time, but also on the position within the control volume. Thus, an alternate form of the Reynolds transport theorem for a fixed control volume is

Alternate RTT, fixed CV:
$$\frac{dB_{sys}}{dt} = \int_{CV} \frac{\partial}{\partial t}(\rho b) \, dV + \int_{CS} \rho b \vec{V} \cdot \vec{n} \, dA \quad (4\text{–}42)$$

It turns out that Eq. 4–42 is also valid for the most general case of a moving and/or deforming control volume, provided that velocity vector $\vec{V}$ is an *absolute* velocity (as viewed from a fixed reference frame).

Next we consider yet *another* alternative form of the RTT. Equation 4–41 was derived for a *fixed* control volume. However, many practical systems such as turbine and propeller blades involve nonfixed control volumes. Fortunately, Eq. 4–41 is also valid for *moving* and/or *deforming* control volumes provided that the absolute fluid velocity $\vec{V}$ in the last term is replaced by the **relative velocity** $\vec{V}_r$,

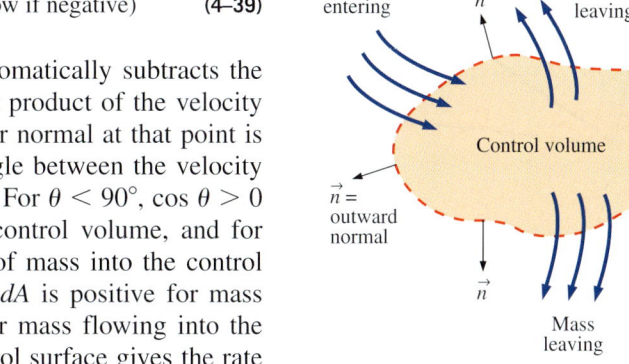

FIGURE 4–56
The integral of $b\rho \vec{V} \cdot \vec{n} \, dA$ over the control surface gives the net amount of the property B flowing out of the control volume (into the control volume if it is negative) per unit time.

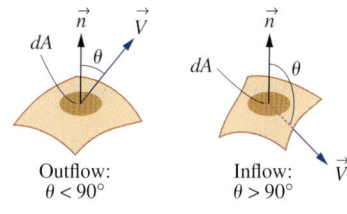

$\vec{V} \cdot \vec{n} = |\vec{V}||\vec{n}| \cos\theta = V \cos\theta$
If $\theta < 90°$, then $\cos\theta > 0$ (outflow).
If $\theta > 90°$, then $\cos\theta < 0$ (inflow).
If $\theta = 90°$, then $\cos\theta = 0$ (no flow).

FIGURE 4–57
Outflow and inflow of mass across the differential area of a control surface.

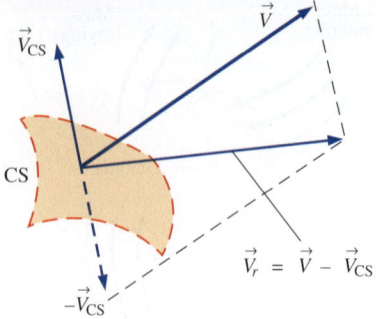

FIGURE 4–58
Relative velocity crossing a control surface is found by vector addition of the absolute velocity of the fluid and the negative of the local velocity of the control surface.

Relative velocity: $$\vec{V}_r = \vec{V} - \vec{V}_{CS} \tag{4-43}$$

where $\vec{V}_{CS}$ is the local velocity of the control surface (Fig. 4–58). The most general form of the Reynolds transport theorem is thus

RTT, nonfixed CV: $$\frac{dB_{sys}}{dt} = \frac{d}{dt}\int_{CV} \rho b\, dV + \int_{CS} \rho b \vec{V}_r \cdot \vec{n}\, dA \tag{4-44}$$

Note that for a control volume that moves and/or deforms with time, the time derivative is applied *after* integration in Eq. 4–44. As a simple example of a moving control volume, consider a toy car moving at a constant absolute velocity $\vec{V}_{car} = 10$ km/h to the right. A high-speed jet of water (absolute velocity = $\vec{V}_{jet} = 25$ km/h to the right) strikes the back of the car and propels it (Fig. 4–59). If we draw a control volume around the car, the relative velocity is $\vec{V}_r = 25 - 10 = 15$ km/h to the right. This represents the velocity at which an observer moving with the control volume (moving with the car) would observe the fluid crossing the control surface. In other words, $\vec{V}_r$ is the fluid velocity expressed relative to a coordinate system moving *with* the control volume.

Finally, by application of the Leibniz theorem, it can be shown that the Reynolds transport theorem for a general moving and/or deforming control volume (Eq. 4–44) is equivalent to the form given by Eq. 4–42, which is repeated here:

Alternate RTT, nonfixed CV: $$\frac{dB_{sys}}{dt} = \int_{CV} \frac{\partial}{\partial t}(\rho b)\, dV + \int_{CS} \rho b \vec{V} \cdot \vec{n}\, dA \tag{4-45}$$

In contrast to Eq. 4–44, the velocity vector $\vec{V}$ in Eq. 4–45 must be taken as the *absolute* velocity (as viewed from a fixed reference frame) in order to apply to a nonfixed control volume.

During steady flow, the amount of the property B within the control volume remains constant in time, and thus the time derivative in Eq. 4–44 becomes zero. Then the Reynolds transport theorem reduces to

RTT, steady flow: $$\frac{dB_{sys}}{dt} = \int_{CS} \rho b \vec{V}_r \cdot \vec{n}\, dA \tag{4-46}$$

Note that unlike the control volume, the property B content of the system may still change with time during a steady process. But in this case the change must be equal to the net property transported by mass across the control surface (an advective rather than an unsteady effect).

In most practical engineering applications of the RTT, fluid crosses the boundary of the control volume at a finite number of well-defined inlets and outlets (Fig. 4–60). In such cases, it is convenient to cut the control surface directly across each inlet and outlet and replace the surface integral in Eq. 4–44 with approximate algebraic expressions at each inlet and outlet based on the *average* values of fluid properties crossing the boundary. We define ρ_{avg}, b_{avg}, and $V_{r,avg}$ as the average values of ρ, b, and V_r, respectively, across an inlet or outlet of cross-sectional area A [e.g., $b_{avg} = \frac{1}{A}\int_A b\, dA$]. The surface integrals in the RTT (Eq. 4–44), when applied over an inlet or outlet

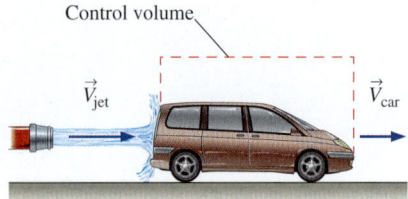

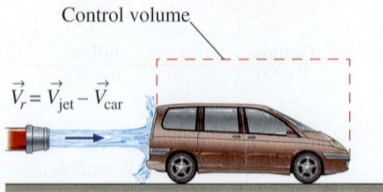

FIGURE 4–59
Reynolds transport theorem applied to a control volume moving at constant velocity.

of cross-sectional area A, are then *approximated* by pulling property b out of the surface integral and replacing it with its average. This yields

$$\int_A \rho b \vec{V}_r \cdot \vec{n} \, dA \cong b_{avg} \int_A \rho \vec{V}_r \cdot \vec{n} \, dA = b_{avg} \dot{m}_r$$

where $\dot{m}_r$ is the mass flow rate through the inlet or outlet relative to the (moving) control surface. The approximation in this equation is exact when property b is uniform over cross-sectional area A. Equation 4–44 thus becomes

$$\frac{dB_{sys}}{dt} = \frac{d}{dt}\int_{CV} \rho b \, dV + \sum_{out} \underbrace{\dot{m}_r b_{avg}}_{\text{for each outlet}} - \sum_{in} \underbrace{\dot{m}_r b_{avg}}_{\text{for each inlet}} \quad (4\text{–}47)$$

In some applications, we may wish to rewrite Eq. 4–47 in terms of volume (rather than mass) flow rate. In such cases, we make a further approximation that $\dot{m}_r \approx \rho_{avg} \dot{V}_r = \rho_{avg} V_{r,avg} A$. This approximation is exact when fluid density ρ is uniform over A. Equation 4–47 then reduces to

Approximate RTT for well-defined inlets and outlets:

$$\frac{dB_{sys}}{dt} = \frac{d}{dt}\int_{CV} \rho b \, dV + \sum_{out} \underbrace{\rho_{avg} b_{avg} V_{r,avg} A}_{\text{for each outlet}} - \sum_{in} \underbrace{\rho_{avg} b_{avg} V_{r,avg} A}_{\text{for each inlet}} \quad (4\text{–}48)$$

Note that these approximations simplify the analysis greatly but may not always be accurate, especially in cases where the velocity distribution across the inlet or outlet is not very uniform (e.g., pipe flows; Fig. 4–60). In particular, the control surface integral of Eq. 4–45 becomes *nonlinear* when property b contains a velocity term (e.g., when applying RTT to the linear momentum equation, $b = \vec{V}$), and the approximation of Eq. 4–48 leads to errors. Fortunately we can eliminate the errors by including *correction factors* in Eq. 4–48, as discussed in Chaps. 5 and 6.

Equations 4–47 and 4–48 apply to fixed *or* moving control volumes, but as discussed previously, the *relative velocity* must be used for the case of a nonfixed control volume. In Eq. 4–47 for example, the mass flow rate $\dot{m}_r$ is relative to the (moving) control surface, hence the r subscript.

*Alternate Derivation of the Reynolds Transport Theorem

A more elegant mathematical derivation of the Reynolds transport theorem is possible through use of the **Leibniz theorem** (see Kundu and Cohen, 2011). You may be familiar with the one-dimensional version of this theorem, which allows you to differentiate an integral whose limits of integration are functions of the variable with which you need to differentiate (Fig. 4–61):

One-dimensional Leibniz theorem:

$$\frac{d}{dt}\int_{x=a(t)}^{x=b(t)} G(x,t) \, dx = \int_a^b \frac{\partial G}{\partial t} \, dx + \frac{db}{dt} G(b,t) - \frac{da}{dt} G(a,t) \quad (4\text{–}49)$$

---

* This section may be omitted without loss of continuity.

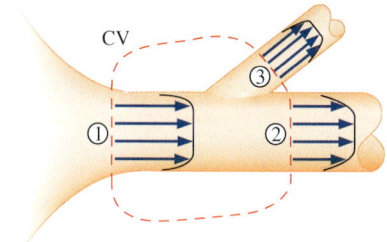

FIGURE 4–60

An example control volume in which there is one well-defined inlet (1) and two well-defined outlets (2 and 3). In such cases, the control surface integral in the RTT can be more conveniently written in terms of the average values of fluid properties crossing each inlet and outlet.

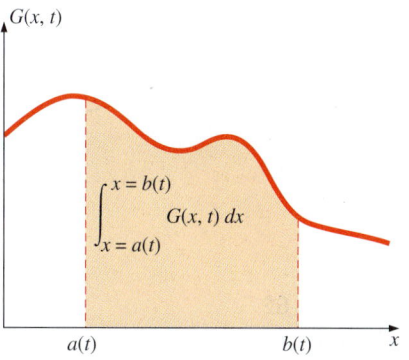

FIGURE 4–61

The *one-dimensional Leibniz theorem* is required when calculating the time derivative of an integral (with respect to x) for which the limits of the integral are functions of time.

The Leibniz theorem takes into account the change of limits $a(t)$ and $b(t)$ with respect to time, as well as the unsteady changes of integrand $G(x, t)$ with time.

EXAMPLE 4–10   One-Dimensional Leibniz Integration

Reduce the following expression as far as possible:

$$F(t) = \frac{d}{dt} \int_{x=0}^{x=Ct} e^{-x^2} \, dx \qquad (1)$$

SOLUTION $F(t)$ is to be evaluated from the given expression.
Analysis We could try integrating first and then differentiating, but since Eq. 1 is of the form of Eq. 4–49, we use the one-dimensional Leibniz theorem. Here, $G(x, t) = e^{-x^2}$ (G is not a function of time in this simple example). The limits of integration are $a(t) = 0$ and $b(t) = Ct$. Thus,

$$F(t) = \underbrace{\int_a^b \frac{\partial G}{\partial t} dx}_{0} + \underbrace{\frac{db}{dt}}_{C} \underbrace{G(b, t)}_{e^{-b^2}} - \underbrace{\frac{da}{dt}}_{0} G(a, t) \quad \to \quad F(t) = Ce^{-C^2 t^2} \qquad (2)$$

Discussion You are welcome to try to obtain the same solution without using the Leibniz theorem.

In three dimensions, the Leibniz theorem for a *volume* integral is

Three-dimensional Leibniz theorem:

$$\frac{d}{dt} \int_{V(t)} G(x, y, z, t) \, dV = \int_{V(t)} \frac{\partial G}{\partial t} dV + \int_{A(t)} G \vec{V}_A \cdot \vec{n} \, dA \qquad (4\text{–}50)$$

where $V(t)$ is a moving and/or deforming volume (a function of time), $A(t)$ is its surface (boundary), and $\vec{V}_A$ is the absolute velocity of this (moving) surface (Fig. 4–62). Equation 4–50 is valid for *any* volume, moving and/or deforming arbitrarily in space and time. For consistency with the previous analyses, we set integrand G to ρb for application to fluid flow,

Three-dimensional Leibniz theorem applied to fluid flow:

$$\frac{d}{dt} \int_{V(t)} \rho b \, dV = \int_{V(t)} \frac{\partial}{\partial t} (\rho b) \, dV + \int_{A(t)} \rho b \vec{V}_A \cdot \vec{n} \, dA \qquad (4\text{–}51)$$

If we apply the Leibniz theorem to the special case of a **material volume** (a system of fixed identity moving with the fluid flow), then $\vec{V}_A = \vec{V}$ everywhere on the material surface since it moves *with* the fluid. Here $\vec{V}$ is the local fluid velocity, and Eq. 4–51 becomes

Leibniz theorem applied to a material volume:

$$\frac{d}{dt} \int_{V(t)} \rho b \, dV = \frac{dB_{sys}}{dt} = \int_{V(t)} \frac{\partial}{\partial t} (\rho b) \, dV + \int_{A(t)} \rho b \vec{V} \cdot \vec{n} \, dA \qquad (4\text{–}52)$$

Equation 4–52 is valid at any instant in time t. We define our control volume such that at this time t, the control volume and the system occupy the same space; in other words, they are *coincident*. At some later time $t + \Delta t$, the system has moved and deformed with the flow, but the control volume

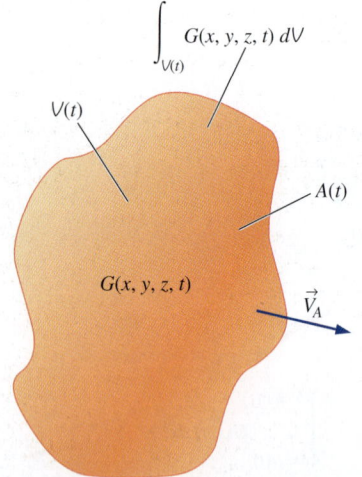

FIGURE 4–62

The *three-dimensional Leibniz theorem* is required when calculating the time derivative of a volume integral for which the volume itself moves and/or deforms with time. It turns out that the three-dimensional form of the Leibniz theorem can be used in an alternative derivation of the Reynolds transport theorem.

may have moved and deformed differently (Fig. 4–63). The key, however, is that *at time t, the system (material volume) and control volume are one and the same*. Thus, the volume integral on the right-hand side of Eq. 4–52 can be evaluated over the *control volume* at time *t*, and the surface integral can be evaluated over the *control surface* at time *t*. Hence,

General RTT, nonfixed CV:
$$\frac{dB_{sys}}{dt} = \int_{CV} \frac{\partial}{\partial t}(\rho b)\, dV + \int_{CS} \rho b \vec{V} \cdot \vec{n}\, dA \quad (4\text{–}53)$$

This expression is identical to that of Eq. 4–42 and is valid for an arbitrarily shaped, moving, and/or deforming control volume at time *t*. Keep in mind that $\vec{V}$ in Eq. 4–53 is the *absolute* fluid velocity.

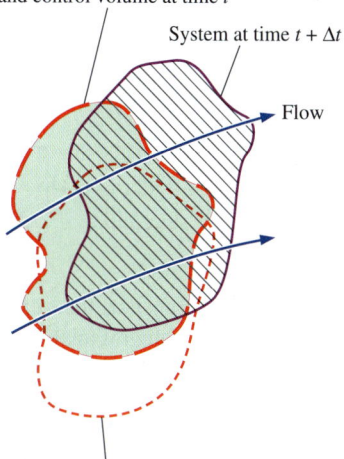

FIGURE 4–63
The material volume (system) and control volume occupy the same space at time *t* (the greenish shaded area), but move and deform differently. At a later time they are *not* coincident.

■ **EXAMPLE 4–11 Reynolds Transport Theorem in Terms of Relative Velocity**

Beginning with the Leibniz theorem and the general Reynolds transport theorem for an arbitrarily moving and deforming control volume, Eq. 4–53, prove that Eq. 4–44 is valid.

SOLUTION Equation 4–44 is to be proven.
Analysis The general three-dimensional version of the Leibniz theorem, Eq. 4–50, applies to *any* volume. We choose to apply it to the control volume of interest, which can be moving and/or deforming differently than the material volume (Fig. 4–63). Setting G to ρb, Eq. 4–50 becomes

$$\frac{d}{dt}\int_{CV} \rho b\, dV = \int_{CV} \frac{\partial}{\partial t}(\rho b)\, dV + \int_{CS} \rho b \vec{V}_{CS} \cdot \vec{n}\, dA \quad (1)$$

We solve Eq. 4–53 for the control volume integral,

$$\int_{CV} \frac{\partial}{\partial t}(\rho b)\, dV = \frac{dB_{sys}}{dt} - \int_{CS} \rho b \vec{V} \cdot \vec{n}\, dA \quad (2)$$

Substituting Eq. 2 into Eq. 1, we get

$$\frac{d}{dt}\int_{CV} \rho b\, dV = \frac{dB_{sys}}{dt} - \int_{CS} \rho b \vec{V} \cdot \vec{n}\, dA + \int_{CS} \rho b \vec{V}_{CS} \cdot \vec{n}\, dA \quad (3)$$

Combining the last two terms and rearranging,

$$\frac{dB_{sys}}{dt} = \frac{d}{dt}\int_{CV} \rho b\, dV + \int_{CS} \rho b (\vec{V} - \vec{V}_{CS}) \cdot \vec{n}\, dA \quad (4)$$

But recall that the relative velocity is defined by Eq. 4–43. Thus,

RTT in terms of relative velocity:
$$\frac{dB_{sys}}{dt} = \frac{d}{dt}\int_{CV} \rho b\, dV + \int_{CS} \rho b \vec{V}_r \cdot \vec{n}\, dA \quad (5)$$

Discussion Equation 5 is indeed identical to Eq. 4–44, and the power and elegance of the Leibniz theorem are demonstrated.

Relationship between Material Derivative and RTT

You may have noticed a similarity or analogy between the material derivative discussed in Section 4–1 and the Reynolds transport theorem discussed here. In fact, both analyses represent methods to transform from

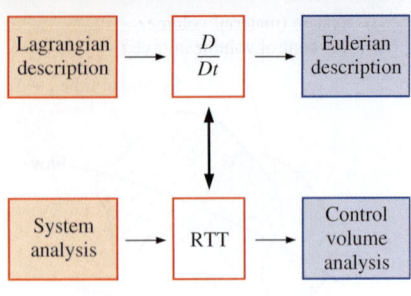

FIGURE 4–64
The Reynolds transport theorem for finite volumes (integral analysis) is analogous to the material derivative for infinitesimal volumes (differential analysis). In both cases, we transform from a Lagrangian or system viewpoint to an Eulerian or control volume viewpoint.

fundamentally Lagrangian concepts to Eulerian interpretations of those concepts. While the Reynolds transport theorem deals with finite-size control volumes and the material derivative deals with infinitesimal fluid particles, the same fundamental physical interpretation applies to both (Fig. 4–64). In fact, the Reynolds transport theorem can be thought of as the integral counterpart of the material derivative. In either case, the total rate of change of some property following an identified portion of fluid consists of two parts: There is a local or unsteady part that accounts for changes in the flow field with time (compare the first term on the right-hand side of Eq. 4–12 to that of Eq. 4–45). There is also an advective part that accounts for the movement of fluid from one region of the flow to another (compare the second term on the right-hand sides of Eqs. 4–12 and 4–45).

Just as the material derivative can be applied to any fluid property, scalar or vector, the Reynolds transport theorem can be applied to any scalar or vector property as well. In Chaps. 5 and 6, we apply the Reynolds transport theorem to conservation of mass, energy, momentum, and angular momentum by choosing parameter B to be mass, energy, momentum, and angular momentum, respectively. In this fashion we can easily convert from the fundamental system conservation laws (Lagrangian viewpoint) to forms that are valid and useful in a control volume analysis (Eulerian viewpoint).

SUMMARY

Fluid kinematics is concerned with describing fluid motion, without necessarily analyzing the forces responsible for such motion. There are two fundamental descriptions of fluid motion—*Lagrangian* and *Eulerian*. In a Lagrangian description, we follow individual fluid particles or collections of fluid particles, while in the Eulerian description, we define a *control volume* through which fluid flows in and out. We transform equations of motion from Lagrangian to Eulerian through use of the *material derivative* for infinitesimal fluid particles and through use of the *Reynolds transport theorem* (*RTT*) for systems of finite volume. For some extensive property B or its corresponding intensive property b,

Material derivative:
$$\frac{Db}{Dt} = \frac{\partial b}{\partial t} + (\vec{V} \cdot \vec{\nabla})b$$

General RTT, nonfixed CV:
$$\frac{dB_{sys}}{dt} = \int_{CV} \frac{\partial}{\partial t}(\rho b)\, dV + \int_{CS} \rho b \vec{V} \cdot \vec{n}\, dA$$

In both equations, the total change of the property following a fluid particle or following a system is composed of two parts: a *local* (unsteady) part and an *advective* (movement) part.

There are various ways to visualize and analyze flow fields—*streamlines, streaklines, pathlines, timelines, surface imaging, shadowgraphy, schlieren imaging, profile plots, vector plots,* and *contour plots*. We define each of these and provide examples in this chapter. In general unsteady flow, streamlines, streaklines, and pathlines differ, but *in steady flow, streamlines, streaklines, and pathlines are coincident.*

Four fundamental rates of motion (*deformation rates*) are required to fully describe the kinematics of a fluid flow: *velocity* (rate of translation), *angular velocity* (rate of rotation), *linear strain rate*, and *shear strain rate*. *Vorticity* is a property of fluid flows that indicates the *rotationality* of fluid particles.

Vorticity vector: $\vec{\zeta} = \vec{\nabla} \times \vec{V} = \text{curl}(\vec{V}) = 2\vec{\omega}$

A region of flow is *irrotational* if the vorticity is zero in that region.

The concepts learned in this chapter are used repeatedly throughout the rest of the book. We use the RTT to transform the conservation laws from closed systems to control volumes in Chaps. 5 and 6, and again in Chap. 9 in the derivation of the differential equations of fluid motion. The role of vorticity and irrotationality is revisited in greater detail in Chap. 10 where we show that the irrotationality approximation leads to greatly reduced complexity in the solution of fluid flows. Finally, we use various types of flow visualization and data plots to describe the kinematics of example flow fields in nearly every chapter of this book.

APPLICATION SPOTLIGHT ■ Fluidic Actuators

Guest Author: Ganesh Raman,
Illinois Institute of Technology

Fluidic actuators are devices that use fluid logic circuits to produce oscillatory velocity or pressure perturbations in jets and shear layers for delaying separation, enhancing mixing, and suppressing noise. Fluidic actuators are potentially useful for shear flow control applications for many reasons: they have no moving parts; they can produce perturbations that are controllable in frequency, amplitude, and phase; they can operate in harsh thermal environments and are not susceptible to electromagnetic interference; and they are easy to integrate into a functioning device. Although fluidics technology has been around for many years, recent advances in miniaturization and microfabrication have made them very attractive candidates for practical use. The fluidic actuator produces a self-sustaining oscillatory flow using the principles of wall attachment and backflow that occur within miniature passages of the device.

Figure 4–65 demonstrates the application of a fluidic actuator for jet thrust vectoring. Fluidic thrust vectoring is important for future aircraft designs, since they can improve maneuverability without the complexity of additional surfaces near the nozzle exhaust. In the three images of Fig. 4–65, the primary jet exhausts from right to left and a single fluidic actuator is located at the top. Figure 4–65*a* shows the unperturbed jet. Figures 4–65*b* and *c* show the vectoring effect at two fluidic actuation levels. Changes to the primary jet are characterized using particle image velocimetry (PIV). A simplified explanation is as follows: In this technique tracer particles are introduced into the flow and illuminated by a thin laser light sheet that is pulsed to freeze particle motion. Laser light scattered by the particles is recorded at two instances in time using a digital camera. Using a spatial cross correlation, the local displacement vector is obtained. The results indicate that there exists the potential for integrating multiple fluidic sub-elements into aircraft components for improved performance.

Figure 4–65 is actually a combination vector plot and contour plot. Velocity vectors are superimposed on contour plots of velocity magnitude (speed). The red regions represent high speeds, and the blue regions represent low speeds.

References
Raman, G., Packiarajan, S., Papadopoulos, G., Weissman, C., and Raghu, S., "Jet Thrust Vectoring Using a Miniature Fluidic Oscillator," ASME FEDSM 2001-18057, 2001.
Raman, G., Raghu, S., and Bencic, T. J., "Cavity Resonance Suppression Using Miniature Fluidic Oscillators," AIAA Paper 99-1900, 1999.

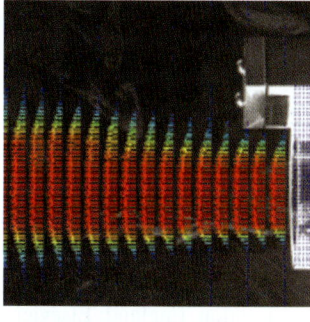

(*a*)

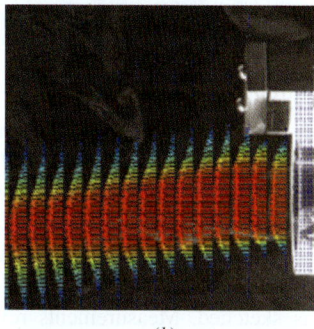

(*b*)

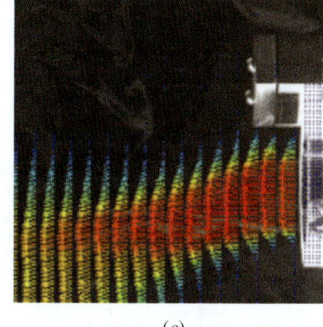

(*c*)

FIGURE 4–65
Time-averaged mean velocity field of a fluidic actuator jet. Results are from 150 PIV realizations, overlaid on an image of the seeded flow. Every seventh and second velocity vector is shown in the horizontal and vertical directions, respectively. The color levels denote the magnitude of the velocity field. (*a*) No actuation; (*b*) single actuator operating at 20 kPa (gage); (*c*) single actuator operating at 60 kPa (gage).

Courtesy Ganesh Raman, Illinois Institute of Technology. Used by permission.

FLUID KINEMATICS

REFERENCES AND SUGGESTED READING

1. R. J. Adrian. "Particle-Imaging Technique for Experimental Fluid Mechanics," *Annual Reviews in Fluid Mechanics*, 23, pp. 261–304, 1991.
2. J. M. Cimbala, H. Nagib, and A. Roshko. "Large Structure in the Far Wakes of Two-Dimensional Bluff Bodies," *Journal of Fluid Mechanics,* 190, pp. 265–298, 1988.
3. R. J. Heinsohn and J. M. Cimbala. *Indoor Air Quality Engineering*. New York: Marcel-Dekker, 2003.
4. P. K. Kundu and I. M. Cohen *Fluid Mechanics*. Ed. 5, London, England: Elsevier Inc. 2011.
5. W. Merzkirch. *Flow Visualization*, 2nd ed. Orlando, FL: Academic Press, 1987.
6. G. S. Settles. *Schlieren and Shadowgraph Techniques: Visualizing Phenomena in Transparent Media*. Heidelberg: Springer-Verlag, 2001.
7. M. Van Dyke. *An Album of Fluid Motion*. Stanford, CA: The Parabolic Press, 1982.
8. F. M. White. *Viscous Fluid Flow*, 3rd ed. New York: McGraw-Hill, 2005.

PROBLEMS*

Introductory Problems

4–1C Briefly discuss the difference between derivative operators d and ∂. If the derivative $\partial u/\partial x$ appears in an equation, what does this imply about variable u?

4–2 Consider steady flow of water through an axisymmetric garden hose nozzle (Fig. P4–2). Along the centerline of the nozzle, the water speed increases from $u_{entrance}$ to u_{exit} as sketched. Measurements reveal that the centerline water speed increases parabolically through the nozzle. Write an equation for centerline speed $u(x)$, based on the parameters given here, from $x = 0$ to $x = L$.

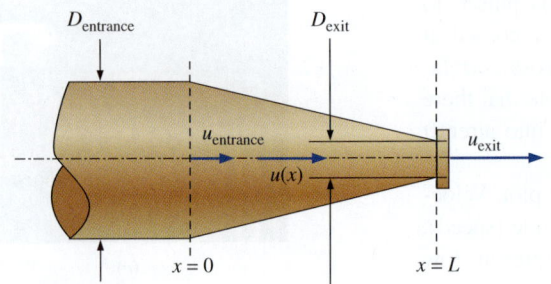

FIGURE P4–2

4–3 Consider the following steady, two-dimensional velocity field:
$$\vec{V} = (u, v) = (a^2 - (b - cx)^2)\vec{i} + (-2cby + 2c^2xy)\vec{j}$$
Is there a stagnation point in this flow field? If so, where is it?

4–4 A steady, two-dimensional velocity field is given by
$$\vec{V} = (u, v) = (-0.781 - 4.67x)\vec{i} + (-3.54 + 4.67y)\vec{j}$$
Calculate the location of the stagnation point.

4–5 Consider the following steady, two-dimensional velocity field:
$$\vec{V} = (u, v) = (0.66 + 2.1x)\vec{i} + (-2.7 - 2.1y)\vec{j}$$
Is there a stagnation point in this flow field? If so, where is it?
Answer: Yes; $x = -0.314$, $y = -1.29$

Lagrangian and Eulerian Descriptions

4–6C What is the *Eulerian description* of fluid motion? How does it differ from the Lagrangian description?

4–7C Is the Lagrangian method of fluid flow analysis more similar to study of a system or a control volume? Explain.

4–8C What is the *Lagrangian description* of fluid motion?

4–9C A stationary probe is placed in a fluid flow and measures pressure and temperature as functions of time at one

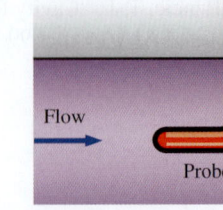

FIGURE P4–9C

---

* Problems designated by a "C" are concept questions, and students are encouraged to answer them all. Problems with the 🌀 icon are solved using EES, and complete solutions together with parametric studies are included on the text website. Problems with the 📖 icon are comprehensive in nature and are intended to be solved with an equation solver such as EES.

location in the flow (Fig. P4–9C). Is this a Lagrangian or an Eulerian measurement? Explain.

4–10C A tiny neutrally buoyant electronic pressure probe is released into the inlet pipe of a water pump and transmits 2000 pressure readings per second as it passes through the pump. Is this a Lagrangian or an Eulerian measurement? Explain.

4–11C Define a *steady flow field* in the Eulerian reference frame. In such a steady flow, is it possible for a fluid particle to experience a nonzero acceleration?

4–12C List at least three other names for the material derivative, and write a brief explanation about why each name is appropriate.

4–13C A weather balloon is launched into the atmosphere by meteorologists. When the balloon reaches an altitude where it is neutrally buoyant, it transmits information about weather conditions to monitoring stations on the ground (Fig. P4–13C). Is this a Lagrangian or an Eulerian measurement? Explain.

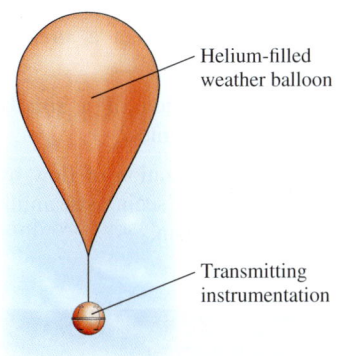

FIGURE P4–13C

4–14C A Pitot-static probe can often be seen protruding from the underside of an airplane (Fig. P4–14C). As the airplane flies, the probe measures relative wind speed. Is this a Lagrangian or an Eulerian measurement? Explain.

4–15C Is the Eulerian method of fluid flow analysis more similar to study of a system or a control volume? Explain.

4–16 Consider steady, incompressible, two-dimensional flow through a converging duct (Fig. P4–16). A simple approximate velocity field for this flow is

$$\vec{V} = (u, v) = (U_0 + bx)\vec{i} - by\vec{j}$$

where U_0 is the horizontal speed at $x = 0$. Note that this equation ignores viscous effects along the walls but is a reasonable approximation throughout the majority of the flow field. Calculate the material acceleration for fluid particles passing through this duct. Give your answer in two ways: (1) as acceleration components a_x and a_y and (2) as acceleration vector $\vec{a}$.

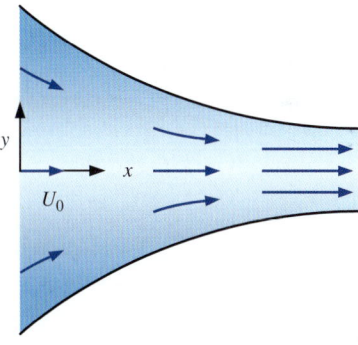

FIGURE P4–16

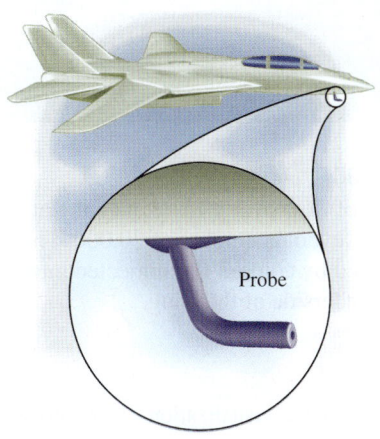

FIGURE P4–14C

4–17 Converging duct flow is modeled by the steady, two-dimensional velocity field of Prob. 4–16. The pressure field is given by

$$P = P_0 - \frac{\rho}{2}\left[2U_0bx + b^2(x^2 + y^2)\right]$$

where P_0 is the pressure at $x = 0$. Generate an expression for the rate of change of pressure *following a fluid particle*.

4–18 A steady, incompressible, two-dimensional velocity field is given by the following components in the xy-plane:

$$u = 1.85 + 2.33x + 0.656y$$

$$v = 0.754 - 2.18x - 2.33y$$

Calculate the acceleration field (find expressions for acceleration components a_x and a_y), and calculate the acceleration at the point $(x, y) = (-1, 2)$. *Answers:* $a_x = 0.806$, $a_y = 2.21$

4–19 A steady, incompressible, two-dimensional velocity field is given by the following components in the xy-plane:

$$u = 0.205 + 0.97x + 0.851y$$
$$v = -0.509 + 0.953x - 0.97y$$

Calculate the acceleration field (find expressions for acceleration components a_x and a_y) and calculate the acceleration at the point $(x, y) = (2, 1.5)$.

4–20 The velocity field for a flow is given by $\vec{V} = u\vec{i} + v\vec{j} + w\vec{k}$ where $u = 3x$, $v = -2y$, $w = 2z$. Find the streamline that will pass through the point $(1, 1, 0)$.

4–21 Consider steady flow of air through the diffuser portion of a wind tunnel (Fig. P4–21). Along the centerline of the diffuser, the air speed decreases from u_{entrance} to u_{exit} as sketched. Measurements reveal that the centerline air speed decreases parabolically through the diffuser. Write an equation for centerline speed $u(x)$, based on the parameters given here, from $x = 0$ to $x = L$.

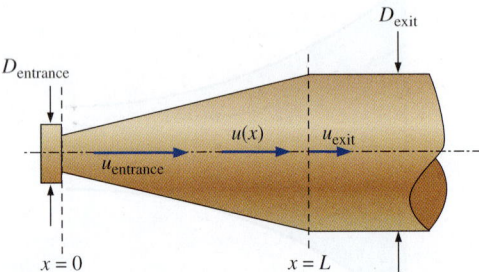

FIGURE P4–21

4–22 For the velocity field of Prob. 4–21, calculate the fluid acceleration along the diffuser centerline as a function of x and the given parameters. For $L = 1.56$ m, $u_{\text{entrance}} = 24.3$ m/s, and $u_{\text{exit}} = 16.8$ m/s, calculate the acceleration at $x = 0$ and $x = 1.0$ m. *Answers:* 0, -131 m/s^2

4–23 A steady, incompressible, two-dimensional (in the xy-plane) velocity field is given by

$$\vec{V} = (0.523 - 1.88x + 3.94y)\vec{i} + (-2.44 + 1.26x + 1.88y)\vec{j}$$

Calculate the acceleration at the point $(x, y) = (-1.55, 2.07)$.

4–24 For the velocity field of Prob. 4–2, calculate the fluid acceleration along the nozzle centerline as a function of x and the given parameters.

Flow Patterns and Flow Visualization

4–25C What is the definition of a *pathline*? What do pathlines indicate?

4–26C Consider the visualization of flow over a 12° cone in Fig. P4–26C. Are we seeing streamlines, streaklines, pathlines, or timelines? Explain.

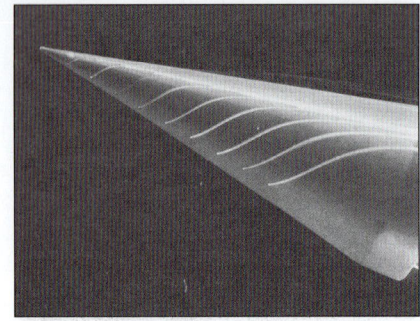

FIGURE P4–26C

Visualization of flow over a 12° cone at a 16° angle of attack at a Reynolds number of 15,000. The visualization is produced by colored fluid injected into water from ports in the body.

Courtesy ONERA. Photograph by Werlé.

4–27C What is the definition of a *streamline*? What do streamlines indicate?

4–28C What is the definition of a *streakline*? How do streaklines differ from streamlines?

4–29C Consider the visualization of flow over a 15° delta wing in Fig. P4–29C. Are we seeing streamlines, streaklines, pathlines, or timelines? Explain.

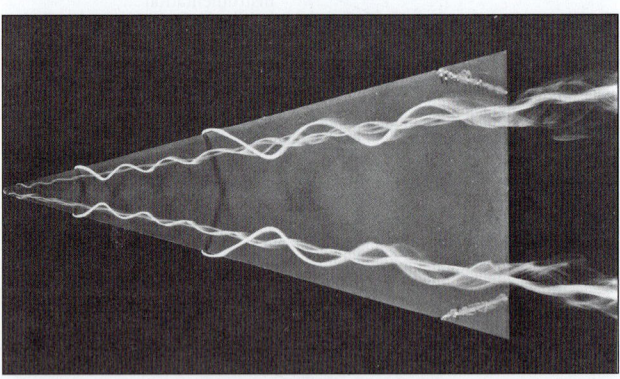

FIGURE P4–29C

Visualization of flow over a 15° delta wing at a 20° angle of attack at a Reynolds number of 20,000. The visualization is produced by colored fluid injected into water from ports on the underside of the wing.

Courtesy ONERA. Photograph by Werlé.

4–30C Consider the visualization of ground vortex flow in Fig. P4–30C. Are we seeing streamlines, streaklines, pathlines, or timelines? Explain.

FIGURE P4–30C
Visualization of ground vortex flow. A high-speed round air jet impinges on the ground in the presence of a free-stream flow of air from left to right. (The ground is at the bottom of the picture.) The portion of the jet that travels upstream forms a recirculating flow known as a **ground vortex.** The visualization is produced by a smoke wire mounted vertically to the left of the field of view.
Photo by John M. Cimbala.

4–31C Consider the visualization of flow over a sphere in Fig. P4–31C. Are we seeing streamlines, streaklines, pathlines, or timelines? Explain.

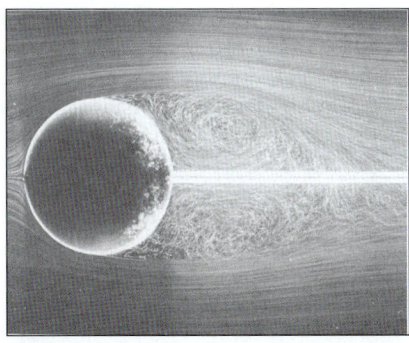

FIGURE P4–31C
Visualization of flow over a sphere at a Reynolds number of 15,000. The visualization is produced by a time exposure of air bubbles in water.
Courtesy ONERA. Photograph by Werlé.

4–32C What is the definition of a *timeline*? How can timelines be produced in a water channel? Name an application where timelines are more useful than streaklines.

4–33C Consider a cross-sectional slice through an array of heat exchanger tubes (Fig. P4–33C). For each desired piece of information, choose which kind of flow visualization plot (vector plot or contour plot) would be most appropriate, and explain why.

(a) The location of maximum fluid speed is to be visualized.
(b) Flow separation at the rear of the tubes is to be visualized.
(c) The temperature field throughout the plane is to be visualized.
(d) The distribution of the vorticity component normal to the plane is to be visualized.

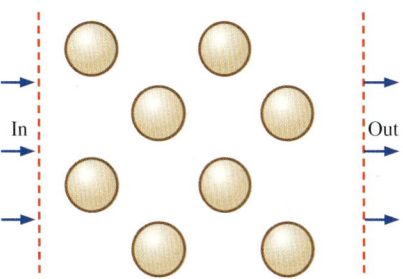

FIGURE P4–33C

4–34 Converging duct flow (Fig. P4–16) is modeled by the steady, two-dimensional velocity field of Prob. 4–16. Generate an analytical expression for the flow streamlines.
Answer: $y = C/(U_0 + bx)$

4–35 The velocity field of a flow is described by $\vec{V} = (4x)\vec{i} + (5y + 3)\vec{j} + (3t^2)\vec{k}$. What is the pathline of a particle at a location (1 m, 2 m, 4 m) at time $t = 1$ s?

4–36 Consider the following steady, incompressible, two-dimensional velocity field:

$$\vec{V} = (u, v) = (4.35 + 0.656x)\vec{i} + (-1.22 - 0.656y)\vec{j}$$

Generate an analytical expression for the flow streamlines and draw several streamlines in the upper-right quadrant from $x = 0$ to 5 and $y = 0$ to 6.

4–37 Consider the steady, incompressible, two-dimensional velocity field of Prob. 4–36. Generate a velocity vector plot in the upper-right quadrant from $x = 0$ to 5 and $y = 0$ to 6.

4–38 Consider the steady, incompressible, two-dimensional velocity field of Prob. 4–36. Generate a vector plot of the acceleration field in the upper-right quadrant from $x = 0$ to 5 and $y = 0$ to 6.

4–39 A steady, incompressible, two-dimensional velocity field is given by

$$\vec{V} = (u, v) = (1 + 2.5x + y)\vec{i} + (-0.5 - 3x - 2.5y)\vec{j}$$

where the x- and y-coordinates are in m and the magnitude of velocity is in m/s.

(a) Determine if there are any stagnation points in this flow field, and if so, where they are.

(b) Sketch velocity vectors at several locations in the upper-right quadrant for $x = 0$ m to 4 m and $y = 0$ m to 4 m; qualitatively describe the flow field.

4–40 Consider the steady, incompressible, two-dimensional velocity field of Prob. 4–39.
(a) Calculate the material acceleration at the point ($x = 2$ m, $y = 3$ m). *Answers:* $a_x = 8.50$ m/s^2, $a_y = 8.00$ m/s^2
(b) Sketch the material acceleration vectors at the same array of x- and y-values as in Prob. 4–39.

4–41 The velocity field for *solid-body rotation* in the $r\theta$-plane (Fig. P4–41) is given by

$$u_r = 0 \qquad u_\theta = \omega r$$

where ω is the magnitude of the angular velocity ($\vec{\omega}$ points in the z-direction). For the case with $\omega = 1.5$ s^{-1}, plot a contour plot of velocity magnitude (speed). Specifically, draw curves of constant speed $V = 0.5, 1.0, 1.5, 2.0,$ and 2.5 m/s. Be sure to label these speeds on your plot.

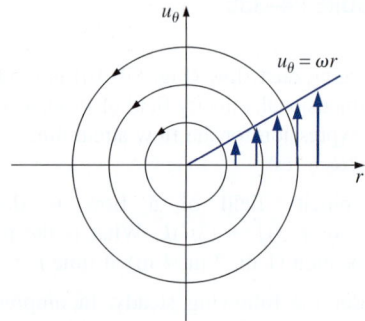

FIGURE P4–41

4–42 The velocity field for a *line vortex* in the $r\theta$-plane (Fig. P4–42) is given by

$$u_r = 0 \qquad u_\theta = \frac{K}{r}$$

where K is the *line vortex strength*. For the case with $K = 1.5$ m/s^2, plot a contour plot of velocity magnitude (speed). Specifically, draw curves of constant speed $V = 0.5, 1.0, 1.5, 2.0,$ and 2.5 m/s. Be sure to label these speeds on your plot.

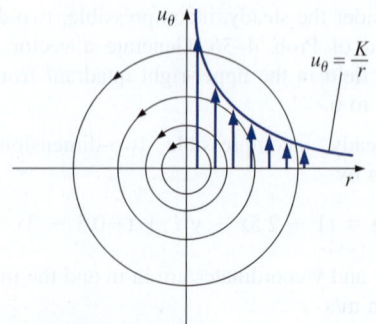

FIGURE P4–42

4–43 The velocity field for a *line source* in the $r\theta$-plane (Fig. P4–43) is given by

$$u_r = \frac{m}{2\pi r} \qquad u_\theta = 0$$

where m is the line source strength. For the case with $m/(2\pi) = 1.5$ m^2/s, plot a contour plot of velocity magnitude (speed). Specifically, draw curves of constant speed $V = 0.5, 1.0, 1.5, 2.0,$ and 2.5 m/s. Be sure to label these speeds on your plot.

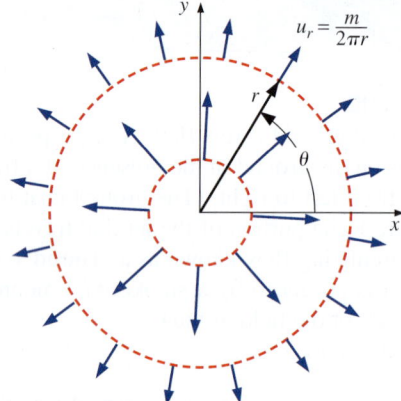

FIGURE P4–43

4–44 A very small circular cylinder of radius R_i is rotating at angular velocity ω_i inside a much larger concentric cylinder of radius R_o that is rotating at angular velocity ω_o. A liquid of density ρ and viscosity μ is confined between the two cylinders, as in Fig. P4–44. Gravitational and end effects can be neglected (the flow is two-dimensional into the page).

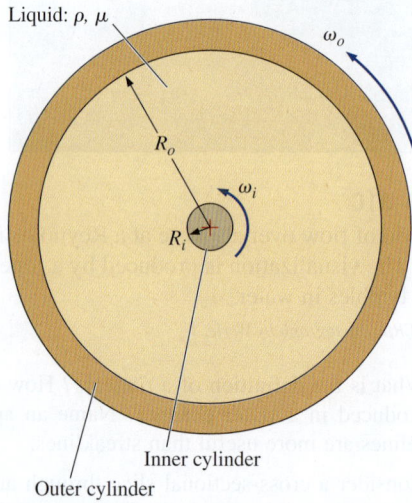

FIGURE P4–44

If $\omega_i = \omega_o$ and a long time has passed, generate an expression for the tangential velocity profile, u_θ as a function of (at most) r, ω, R_i, R_o, ρ, and μ, where $\omega = \omega_i = \omega_o$. Also, calculate the torque exerted by the fluid on the inner cylinder and on the outer cylinder.

4–45 Consider the same two concentric cylinders of Prob. 4–44. This time, however, the inner cylinder is rotating, but the outer cylinder is stationary. In the limit, as the outer cylinder is very large compared to the inner cylinder (imagine the inner cylinder spinning very fast while its radius gets very small), what kind of flow does this approximate? Explain. After a long time has passed, generate an expression for the tangential velocity profile, namely u_θ as a function of (at most) r, ω_i, R_i, R_o, ρ, and μ. *Hint*: Your answer may contain an (unknown) constant, which can be obtained by specifying a boundary condition at the inner cylinder surface.

Motion and Deformation of Fluid Elements; Vorticity and Rotationality

4–46C Explain the relationship between vorticity and rotationality.

4–47C Name and briefly describe the four fundamental types of motion or deformation of fluid particles.

4–48 Converging duct flow (Fig. P4–16) is modeled by the steady, two-dimensional velocity field of Prob. 4–16. Is this flow field rotational or irrotational? Show all your work. *Answer:* irrotational

4–49 Converging duct flow is modeled by the steady, two-dimensional velocity field of Prob. 4–16. A fluid particle (A) is located on the x-axis at $x = x_A$ at time $t = 0$ (Fig. P4–49). At some later time t, the fluid particle has moved downstream with the flow to some new location $x = x_{A'}$, as shown in the figure. Since the flow is symmetric about the x-axis, the fluid particle remains on the x-axis at all times. Generate an analytical expression for the x-location of the fluid particle at some arbitrary time t in terms of its initial location x_A and constants U_0 and b. In other words, develop an expression for $x_{A'}$. (*Hint:* We know that $u = dx_{\text{particle}}/dt$ following a fluid particle. Plug in u, separate variables, and integrate.)

4–50 Converging duct flow is modeled by the steady, two-dimensional velocity field of Prob. 4–16. Since the flow is symmetric about the x-axis, line segment AB along the x-axis remains on the axis, but stretches from length ξ to length $\xi + \Delta\xi$ as it flows along the channel centerline (Fig. P4–50). Generate an analytical expression for the change in length of the line segment, $\Delta\xi$. (*Hint:* Use the result of Prob. 4–49.) *Answer:* $(x_B - x_A)(e^{bt} - 1)$

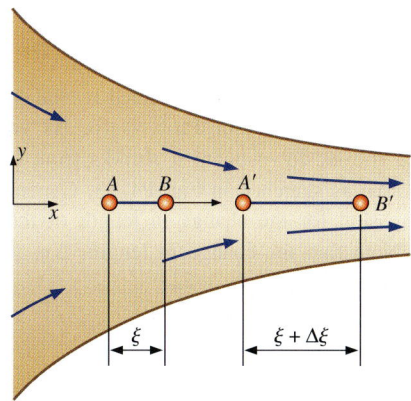

FIGURE P4–50

4–51 Using the results from Prob. 4–50 and the fundamental definition of linear strain rate (the rate of increase in length per unit length), develop an expression for the linear strain rate in the x-direction (ε_{xx}) of fluid particles located on the centerline of the channel. Compare your result to the general expression for ε_{xx} in terms of the velocity field, i.e., $\varepsilon_{xx} = \partial u/\partial x$. (*Hint:* Take the limit as time $t \to 0$. You may need to apply a truncated series expansion for e^{bt}.) *Answer:* b

4–52 Converging duct flow is modeled by the steady, two-dimensional velocity field of Prob. 4–16. A fluid particle (A) is located at $x = x_A$ and $y = y_A$ at time $t = 0$ (Fig. P4–52). At some later time t, the fluid particle has moved downstream with the flow to some new location $x = x_{A'}$, $y = y_{A'}$, as shown in the figure. Generate an analytical expression for the y-location of the fluid particle at arbitrary time t in terms of its initial y-location y_A and constant b. In other words, develop an expression for $y_{A'}$. (*Hint:* We know that $v = dy_{\text{particle}}/dt$ following a fluid particle. Substitute the equation for v, separate variables, and integrate.) *Answer:* $y_A e^{-bt}$

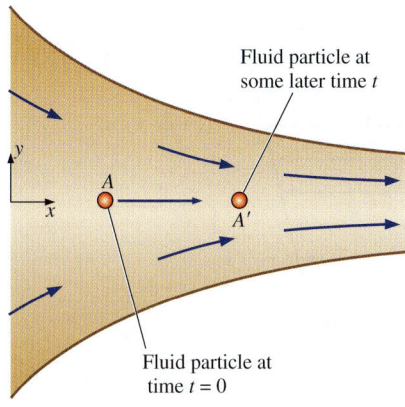

FIGURE P4–49

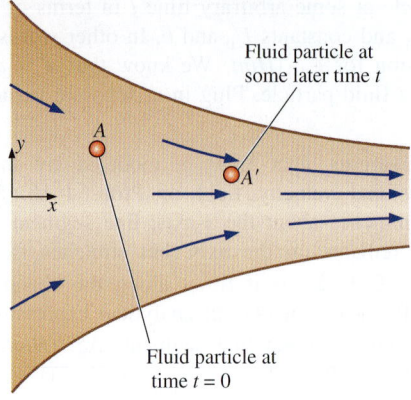

FIGURE P4–52

for volumetric strain rate to verify that this flow field is incompressible.

4–56 A general equation for a steady, two-dimensional velocity field that is linear in both spatial directions (x and y) is

$$\vec{V} = (u, v) = (U + a_1 x + b_1 y)\vec{i} + (V + a_2 x + b_2 y)\vec{j}$$

where U and V and the coefficients are constants. Their dimensions are assumed to be appropriately defined. Calculate the x- and y-components of the acceleration field.

4–57 For the velocity field of Prob. 4–56, what relationship must exist between the coefficients to ensure that the flow field is incompressible?   *Answer:* $a_1 + b_2 = 0$

4–58 For the velocity field of Prob. 4–56, calculate the linear strain rates in the x- and y-directions.   *Answers:* a_1, b_2

4–59 For the velocity field of Prob. 4–56, calculate the shear strain rate in the xy-plane.

4–60 Combine your results from Probs. 4–58 and 4–59 to form the two-dimensional strain rate tensor ε_{ij} in the xy-plane,

$$\varepsilon_{ij} = \begin{pmatrix} \varepsilon_{xx} & \varepsilon_{xy} \\ \varepsilon_{yx} & \varepsilon_{yy} \end{pmatrix}$$

Under what conditions would the x- and y-axes be principal axes?   *Answer:* $b_1 + a_2 = 0$

4–61 For the velocity field of Prob. 4–56, calculate the vorticity vector. In which direction does the vorticity vector point?   *Answer:* $(a_2 - b_1)\vec{k}$ in z (or $-z$) direction

4–62 Consider steady, incompressible, two-dimensional **shear flow** for which the velocity field is

$$\vec{V} = (u, v) = (a + by)\vec{i} + 0\vec{j}$$

where a and b are constants. Sketched in Fig. P4–62 is a small rectangular fluid particle of dimensions dx and dy at time t. The fluid particle moves and deforms with the flow such that at a later time ($t + dt$), the particle is no longer rectangular,

4–53 Converging duct flow is modeled by the steady, two-dimensional velocity field of Prob. 4–16. As vertical line segment AB moves downstream it shrinks from length η to length $\eta + \Delta\eta$ as sketched in Fig. P4–53. Generate an analytical expression for the change in length of the line segment, $\Delta\eta$. Note that the change in length, $\Delta\eta$, is *negative*. (*Hint*: Use the result of Prob. 4–52.)

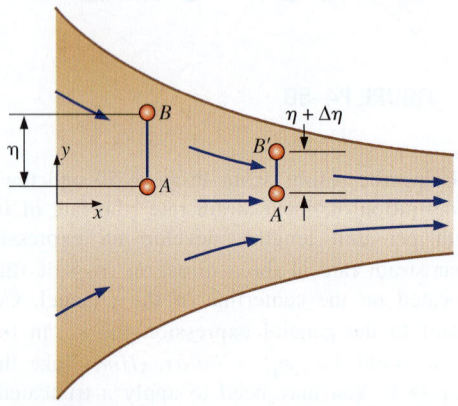

FIGURE P4–53

4–54 Using the results of Prob. 4–53 and the fundamental definition of linear strain rate (the rate of increase in length per unit length), develop an expression for the linear strain rate in the y-direction (ε_{yy}) of fluid particles moving down the channel. Compare your result to the general expression for ε_{yy} in terms of the velocity field, i.e., $\varepsilon_{yy} = \partial v/\partial y$. (*Hint*: Take the limit as time $t \to 0$. You may need to apply a truncated series expansion for e^{-bt}.)

4–55 Converging duct flow is modeled by the steady, two-dimensional velocity field of Prob. 4–16. Use the equation

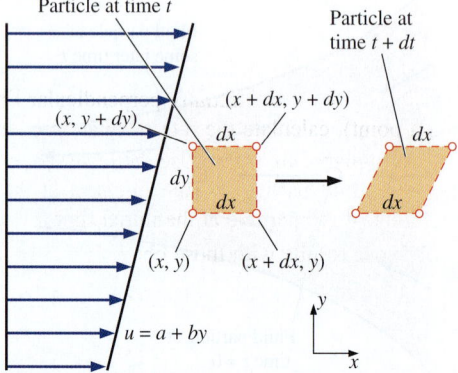

FIGURE P4–62

as also shown in the figure. The initial location of each corner of the fluid particle is labeled in Fig. P4–62. The lower-left corner is at (x, y) at time t, where the x-component of velocity is $u = a + by$. At the later time, this corner moves to $(x + u\,dt, y)$, or

$$(x + (a + by)\,dt, y)$$

(a) In similar fashion, calculate the location of each of the other three corners of the fluid particle at time $t + dt$.
(b) From the fundamental definition of *linear strain rate* (the rate of increase in length per unit length), calculate linear strain rates ε_{xx} and ε_{yy}. *Answers:* 0, 0
(c) Compare your results with those obtained from the equations for ε_{xx} and ε_{yy} in Cartesian coordinates, i.e.,

$$\varepsilon_{xx} = \frac{\partial u}{\partial x} \quad \varepsilon_{yy} = \frac{\partial v}{\partial y}$$

4–63 Use two methods to verify that the flow of Prob. 4–62 is incompressible: (a) by calculating the volume of the fluid particle at both times, and (b) by calculating the volumetric strain rate. Note that Prob. 4–62 should be completed before this problem.

4–64 Consider the steady, incompressible, two-dimensional flow field of Prob. 4–62. Using the results of Prob. 4–62(a), do the following:
(a) From the fundamental definition of *shear strain rate* (half of the rate of decrease of the angle between two initially perpendicular lines that intersect at a point), calculate shear strain rate ε_{xy} in the xy-plane. (*Hint*: Use the lower edge and the left edge of the fluid particle, which intersect at 90° at the lower-left corner of the particle at the initial time.)
(b) Compare your results with those obtained from the equation for ε_{xy} in Cartesian coordinates, i.e.,

$$\varepsilon_{xy} = \frac{1}{2}\left(\frac{\partial u}{\partial y} + \frac{\partial v}{\partial x}\right)$$

Answers: (a) $b/2$, (b) $b/2$

4–65 Consider the steady, incompressible, two-dimensional flow field of Prob. 4–62. Using the results of Prob. 4–62(a), do the following:
(a) From the fundamental definition of the *rate of rotation* (average rotation rate of two initially perpendicular lines that intersect at a point), calculate the rate of rotation of the fluid particle in the xy-plane, ω_z. (*Hint*: Use the lower edge and the left edge of the fluid particle, which intersect at 90° at the lower-left corner of the particle at the initial time.)
(b) Compare your results with those obtained from the equation for ω_z in Cartesian coordinates, i.e.,

$$\omega_z = \frac{1}{2}\left(\frac{\partial v}{\partial x} - \frac{\partial u}{\partial y}\right)$$

Answers: (a) $-b/2$, (b) $-b/2$

4–66 From the results of Prob. 4–65,
(a) Is this flow rotational or irrotational?
(b) Calculate the z-component of vorticity for this flow field.

4–67 A two-dimensional fluid element of dimensions dx and dy translates and distorts as shown in Fig. P4–67 during the infinitesimal time period $dt = t_2 - t_1$. The velocity components at point P at the initial time are u and v in the x- and y-directions, respectively. Show that the magnitude of the rate of rotation (angular velocity) about point P in the xy-plane is

$$\omega_z = \frac{1}{2}\left(\frac{\partial v}{\partial x} - \frac{\partial u}{\partial y}\right)$$

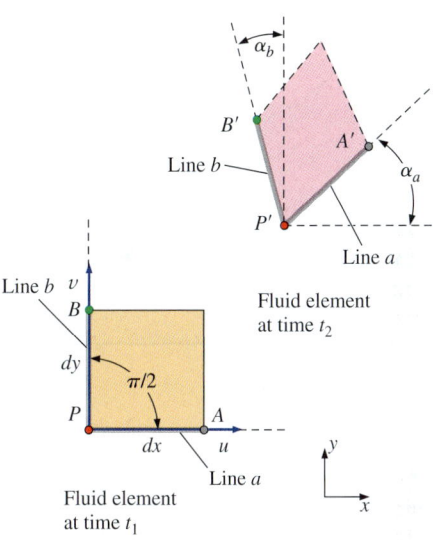

FIGURE P4–67

4–68 A two-dimensional fluid element of dimensions dx and dy translates and distorts as shown in Fig. P4–67 during the infinitesimal time period $dt = t_2 - t_1$. The velocity components at point P at the initial time are u and v in the x- and y-directions, respectively. Consider the line segment PA in Fig. P4–67, and show that the magnitude of the linear strain rate in the x-direction is

$$\varepsilon_{xx} = \frac{\partial u}{\partial x}$$

4–69 A two-dimensional fluid element of dimensions dx and dy translates and distorts as shown in Fig. P4–67 during the infinitesimal time period $dt = t_2 - t_1$. The velocity components at point P at the initial time are u and v in the x- and y-directions, respectively. Show that the magnitude of the shear strain rate about point P in the xy-plane is

$$\varepsilon_{xy} = \frac{1}{2}\left(\frac{\partial u}{\partial y} + \frac{\partial v}{\partial x}\right)$$

4–70 Consider a steady, two-dimensional, incompressible flow field in the *xy*-plane. The linear strain rate in the *x*-direction is 2.5 s^{-1}. Calculate the linear strain rate in the *y*-direction.

4–71 A cylindrical tank of water rotates in solid-body rotation, counterclockwise about its vertical axis (Fig. P4–71) at angular speed $\dot{n}$ = 175 rpm. Calculate the vorticity of fluid particles in the tank. *Answer:* 36.7 $\vec{k}$ rad/s

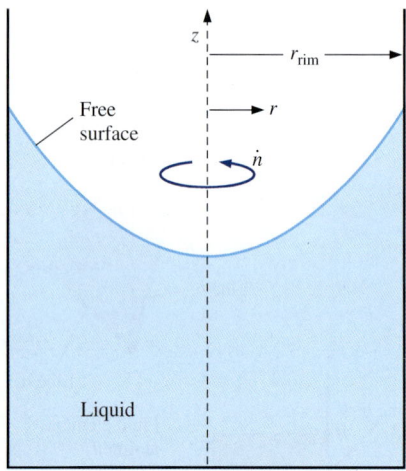

FIGURE P4–71

4–72 A cylindrical tank of water rotates about its vertical axis (Fig. P4–71). A PIV system is used to measure the vorticity field of the flow. The measured value of vorticity in the *z*-direction is −45.4 rad/s and is constant to within ±0.5 percent everywhere that it is measured. Calculate the angular speed of rotation of the tank in rpm. Is the tank rotating clockwise or counterclockwise about the vertical axis?

4–73 A cylindrical tank of radius r_{rim} = 0.354 m rotates about its vertical axis (Fig. P4–71). The tank is partially filled with oil. The speed of the rim is 3.61 m/s in the counterclockwise direction (looking from the top), and the tank has been spinning long enough to be in solid-body rotation. For any fluid particle in the tank, calculate the magnitude of the component of vorticity in the vertical *z*-direction. *Answer:* 20.4 rad/s

4–74 Consider a two-dimensional, incompressible flow field in which an initially square fluid particle moves and deforms. The fluid particle dimension is *a* at time *t* and is aligned with the *x*- and *y*-axes as sketched in Fig. P4–74. At some later time, the particle is still aligned with the *x*- and *y*-axes, but has deformed into a rectangle of horizontal length 2*a*. What is the vertical length of the rectangular fluid particle at this later time?

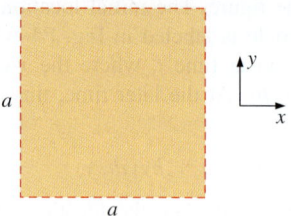

FIGURE P4–74

4–75 Consider a two-dimensional, *compressible* flow field in which an initially square fluid particle moves and deforms. The fluid particle dimension is *a* at time *t* and is aligned with the *x*- and *y*-axes as sketched in Fig. P4–74. At some later time, the particle is still aligned with the *x*- and *y*-axes but has deformed into a rectangle of horizontal length 1.08*a* and vertical length 0.903*a*. (The particle's dimension in the *z*-direction does not change since the flow is two-dimensional.) By what percentage has the density of the fluid particle increased or decreased?

4–76 Consider the following steady, three-dimensional velocity field:

$$\vec{V} = (u, v, w)$$
$$= (3.0 + 2.0x - y)\vec{i} + (2.0x - 2.0y)\vec{j} + (0.5xy)\vec{k}$$

Calculate the vorticity vector as a function of space (*x, y, z*).

4–77 Consider fully developed **Couette flow**—flow between two infinite parallel plates separated by distance *h*, with the top plate moving and the bottom plate stationary as illustrated in Fig. P4–77. The flow is steady, incompressible, and two-dimensional in the *xy*-plane. The velocity field is given by

$$\vec{V} = (u, v) = V\frac{y}{h}\vec{i} + 0\vec{j}$$

Is this flow rotational or irrotational? If it is rotational, calculate the vorticity component in the *z*-direction. Do fluid particles in this flow rotate clockwise or counterclockwise? *Answers:* yes, −*V/h*, clockwise

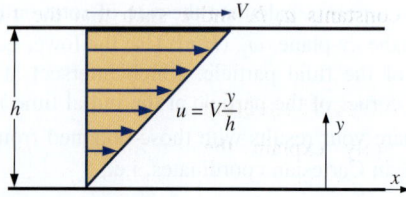

FIGURE P4–77

4–78 For the Couette flow of Fig. P4–77, calculate the linear strain rates in the x- and y-directions, and calculate the shear strain rate ε_{xy}.

4–79 Combine your results from Prob. 4–78 to form the two-dimensional strain rate tensor ε_{ij},

$$\varepsilon_{ij} = \begin{pmatrix} \varepsilon_{xx} & \varepsilon_{xy} \\ \varepsilon_{yx} & \varepsilon_{yy} \end{pmatrix}$$

Are the x- and y-axes principal axes?

4–80 A steady, three-dimensional velocity field is given by

$$\vec{V} = (u, v, w)$$
$$= (2.49 + 1.36x - 0.867y)\vec{i}$$
$$+ (1.95x - 1.36y)\vec{j} + (-0.458xy)\vec{k}$$

Calculate the vorticity vector as a function of space variables (x, y, z).

4–81 A steady, two-dimensional velocity field is given by

$$\vec{V} = (u, v)$$
$$= (2.85 + 1.26x - 0.896y)\vec{i}$$
$$+ (3.45x + cx - 1.26y)\vec{j}$$

Calculate constant c such that the flow field is irrotational.

4–82 A steady, three-dimensional velocity field is given by

$$\vec{V} = (1.35 + 2.78x + 0.754y + 4.21z)\vec{i}$$
$$+ (3.45 + cx - 2.78y + bz)\vec{j}$$
$$+ (-4.21x - 1.89y)\vec{k}$$

Calculate constants b and c such that the flow field is irrotational.

4–83 A steady, three-dimensional velocity field is given by

$$\vec{V} = (0.657 + 1.73x + 0.948y + az)\vec{i}$$
$$+ (2.61 + cx + 1.91y + bz)\vec{j}$$
$$+ (-2.73x - 3.66y - 3.64z)\vec{k}$$

Calculate constants a, b, and c such that the flow field is irrotational.

Reynolds Transport Theorem

4–84C Briefly explain the similarities and differences between the material derivative and the Reynolds transport theorem.

4–85C Briefly explain the purpose of the Reynolds transport theorem (RTT). Write the RTT for extensive property B as a "word equation," explaining each term in your own words.

4–86C True or false: For each statement, choose whether the statement is true or false and discuss your answer briefly.

(a) The Reynolds transport theorem is useful for transforming conservation equations from their naturally occurring control volume forms to their system forms.

(b) The Reynolds transport theorem is applicable only to nondeforming control volumes.

(c) The Reynolds transport theorem can be applied to both steady and unsteady flow fields.

(d) The Reynolds transport theorem can be applied to both scalar and vector quantities.

4–87 Consider the integral $\dfrac{d}{dt}\displaystyle\int_{t}^{2t} x^{-2}dx$. Solve it two ways:

(a) Take the integral first and then the time derivative.
(b) Use Leibniz theorem. Compare your results.

4–88 Solve the integral $\dfrac{d}{dt}\displaystyle\int_{t}^{2t} x^{x}dx$ as far as you are able.

4–89 Consider the general form of the Reynolds transport theorem (RTT) given by

$$\frac{dB_{sys}}{dt} = \frac{d}{dt}\int_{CV} \rho b\, dV + \int_{CS} \rho b \vec{V}_r \cdot \vec{n}\, dA$$

where $\vec{V}_r$ is the velocity of the fluid relative to the control surface. Let B_{sys} be the mass m of a closed system of fluid particles. We know that for a system, $dm/dt = 0$ since no mass can enter or leave the system by definition. Use the given equation to derive the equation of conservation of mass for a control volume.

4–90 Consider the general form of the Reynolds transport theorem (RTT) as stated in Prob. 4–89. Let B_{sys} be the linear momentum $m\vec{V}$ of a system of fluid particles. We know that for a system, Newton's second law is

$$\sum \vec{F} = m\vec{a} = m\frac{d\vec{V}}{dt} = \frac{d}{dt}(m\vec{V})_{sys}$$

Use the RTT and Newton's second law to derive the linear momentum equation for a control volume.

4–91 Consider the general form of the Reynolds transport theorem (RTT) as stated in Prob. 4–89. Let B_{sys} be the angular momentum $\vec{H} = \vec{r} \times m\vec{V}$ of a system of fluid particles, where $\vec{r}$ is the moment arm. We know that for a system, conservation of angular momentum is

$$\sum \vec{M} = \frac{d}{dt}\vec{H}_{sys}$$

where $\sum \vec{M}$ is the net moment applied to the system. Use the RTT and the above equation to derive the equation of conservation of angular momentum for a control volume.

4–92 Reduce the following expression as far as possible:

$$F(t) = \frac{d}{dt}\int_{x=At}^{x=Bt} e^{-2x^2} dx$$

(*Hint*: Use the one-dimensional Leibniz theorem.) Answer: $Be^{-B^2 t^2} - Ae^{-A^2 t^2}$

Review Problems

4–93 Consider a steady, two-dimensional flow field in the *xy*-plane whose *x*-component of velocity is given by

$$u = a + b(x - c)^2$$

where a, b, and c are constants with appropriate dimensions. Of what form does the *y*-component of velocity need to be in order for the flow field to be incompressible? In other words, generate an expression for v as a function of x, y, and the constants of the given equation such that the flow is incompressible. Answer: $-2b(x - c)y + f(x)$

4–94 In a steady, two-dimensional flow field in the *xy*-plane, the *x*-component of velocity is

$$u = ax + by + cx^2$$

where a, b, and c are constants with appropriate dimensions. Generate a general expression for velocity component v such that the flow field is incompressible.

4–95 Consider fully developed two-dimensional **Poiseuille flow**—flow between two infinite parallel plates separated by distance h, with both the top plate and bottom plate stationary, and a forced pressure gradient dP/dx driving the flow as illustrated in Fig. P4–95. (dP/dx is constant and negative.) The flow is steady, incompressible, and two-dimensional in the *xy*-plane. The velocity components are given by

$$u = \frac{1}{2\mu}\frac{dP}{dx}(y^2 - hy) \qquad v = 0$$

where μ is the fluid's viscosity. Is this flow rotational or irrotational? If it is rotational, calculate the vorticity component in the *z*-direction. Do fluid particles in this flow rotate clockwise or counterclockwise?

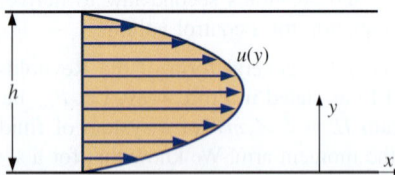

FIGURE P4–95

4–96 For the two-dimensional Poiseuille flow of Prob. 4–95, calculate the linear strain rates in the *x*- and *y*-directions, and calculate the shear strain rate ε_{xy}.

4–97 Combine your results from Prob. 4–96 to form the two-dimensional strain rate tensor ε_{ij} in the *xy*-plane,

$$\varepsilon_{ij} = \begin{pmatrix} \varepsilon_{xx} & \varepsilon_{xy} \\ \varepsilon_{yx} & \varepsilon_{yy} \end{pmatrix}$$

Are the *x*- and *y*-axes principal axes?

4–98 Consider the two-dimensional Poiseuille flow of Prob. 4–95. The fluid between the plates is water at 40°C. Let the gap height $h = 1.6$ mm and the pressure gradient $dP/dx = -230$ N/m³. Calculate and plot seven *pathlines* from $t = 0$ to $t = 10$ s. The fluid particles are released at $x = 0$ and at $y = 0.2, 0.4, 0.6, 0.8, 1.0, 1.2,$ and 1.4 mm.

4–99 Consider the two-dimensional Poiseuille flow of Prob. 4–95. The fluid between the plates is water at 40°C. Let the gap height $h = 1.6$ mm and the pressure gradient $dP/dx = -230$ N/m³. Calculate and plot seven *streaklines* generated from a dye rake that introduces dye streaks at $x = 0$ and at $y = 0.2, 0.4, 0.6, 0.8, 1.0, 1.2,$ and 1.4 mm (Fig. P4–99). The dye is introduced from $t = 0$ to $t = 10$ s, and the streaklines are to be plotted at $t = 10$ s.

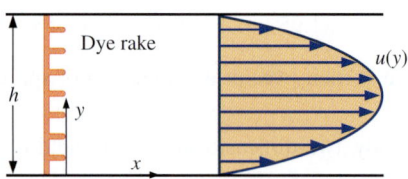

FIGURE P4–99

4–100 Repeat Prob. 4–99 except that the dye is introduced from $t = 0$ to $t = 10$ s, and the streaklines are to be plotted at $t = 12$ s instead of 10 s.

4–101 Compare the results of Probs. 4–99 and 4–100 and comment about the linear strain rate in the *x*-direction.

4–102 Consider the two-dimensional Poiseuille flow of Prob. 4–95. The fluid between the plates is water at 40°C. Let the gap height $h = 1.6$ mm and the pressure gradient $dP/dx = -230$ N/m³. Imagine a hydrogen bubble wire stretched vertically through the channel at $x = 0$ (Fig. P4–102). The wire is pulsed on and off such that bubbles are produced periodically to create *timelines*. Five distinct timelines are generated at $t = 0, 2.5, 5.0, 7.5,$ and 10.0 s. Calculate and plot what these five timelines look like at time $t = 12.5$ s.

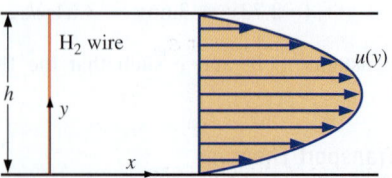

FIGURE P4–102

4–103 The velocity field of a flow is given by $\vec{V} = k(x^2 - y^2)\vec{i} - 2kxy\vec{j}$ where k is a constant. If the radius of curvature of a streamline is $R = [1 + y'^2]^{3/2}/|y''|$, determine the normal acceleration of a particle (which is normal to the streamline) passing through the position $x = 1, y = 2$.

4–104 The velocity field for an incompressible flow is given as $\vec{V} = 5x^2\vec{i} - 20xy\vec{j} + 100t\vec{k}$. Determine if this flow is steady. Also determine the velocity and acceleration of a particle at (1, 3, 3) at $t = 0.2$ s.

4–105 Consider fully developed axisymmetric Poiseuille flow—flow in a round pipe of radius R (diameter $D = 2R$), with a forced pressure gradient dP/dx driving the flow as illustrated in Fig. P4–105. (dP/dx is constant and negative.) The flow is steady, incompressible, and axisymmetric about the x-axis. The velocity components are given by

$$u = \frac{1}{4\mu}\frac{dP}{dx}(r^2 - R^2) \qquad u_r = 0 \qquad u_\theta = 0$$

where μ is the fluid's viscosity. Is this flow rotational or irrotational? If it is rotational, calculate the vorticity component in the circumferential (θ) direction and discuss the sign of the rotation.

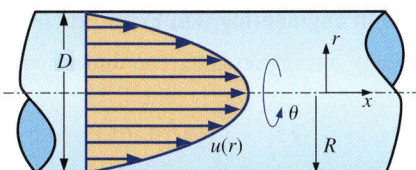

FIGURE P4–105

4–106 For the axisymmetric Poiseuille flow of Prob. 4–105, calculate the linear strain rates in the x- and r-directions, and calculate the shear strain rate ε_{xr}. The strain rate tensor in cylindrical coordinates (r, θ, x) and (u_r, u_θ, u_x), is

$$\varepsilon_{ij} = \begin{pmatrix} \varepsilon_{rr} & \varepsilon_{r\theta} & \varepsilon_{rx} \\ \varepsilon_{\theta r} & \varepsilon_{\theta\theta} & \varepsilon_{\theta x} \\ \varepsilon_{xr} & \varepsilon_{x\theta} & \varepsilon_{xx} \end{pmatrix}$$

$$= \begin{pmatrix} \dfrac{\partial u_r}{\partial r} & \dfrac{1}{2}\left(r\dfrac{\partial}{\partial r}\left(\dfrac{u_\theta}{r}\right) + \dfrac{1}{r}\dfrac{\partial u_r}{\partial \theta}\right) & \dfrac{1}{2}\left(\dfrac{\partial u_r}{\partial x} + \dfrac{\partial u_x}{\partial r}\right) \\ \dfrac{1}{2}\left(r\dfrac{\partial}{\partial r}\left(\dfrac{u_\theta}{r}\right) + \dfrac{1}{r}\dfrac{\partial u_r}{\partial \theta}\right) & \dfrac{1}{r}\dfrac{\partial u_\theta}{\partial \theta} + \dfrac{u_r}{r} & \dfrac{1}{2}\left(\dfrac{1}{r}\dfrac{\partial u_x}{\partial \theta} + \dfrac{\partial u_\theta}{\partial x}\right) \\ \dfrac{1}{2}\left(\dfrac{\partial u_r}{\partial x} + \dfrac{\partial u_x}{\partial r}\right) & \dfrac{1}{2}\left(\dfrac{1}{r}\dfrac{\partial u_x}{\partial \theta} + \dfrac{\partial u_\theta}{\partial x}\right) & \dfrac{\partial u_x}{\partial x} \end{pmatrix}$$

4–107 Combine your results from Prob. 4–106 to form the axisymmetric strain rate tensor ε_{ij},

$$\varepsilon_{ij} = \begin{pmatrix} \varepsilon_{rr} & \varepsilon_{rx} \\ \varepsilon_{xr} & \varepsilon_{xx} \end{pmatrix}$$

Are the x- and r-axes principal axes?

4–108 We approximate the flow of air into a vacuum cleaner attachment by the following velocity components in the centerplane (the xy-plane):

$$u = \frac{-\dot{V}x}{\pi L}\frac{x^2 + y^2 + b^2}{x^4 + 2x^2y^2 + 2x^2b^2 + y^4 - 2y^2b^2 + b^4}$$

and

$$v = \frac{-\dot{V}y}{\pi L}\frac{x^2 + y^2 - b^2}{x^4 + 2x^2y^2 + 2x^2b^2 + y^4 - 2y^2b^2 + b^4}$$

where b is the distance of the attachment above the floor, L is the length of the attachment, and $\dot{V}$ is the volume flow rate of air being sucked up into the hose (Fig. P4–108). Determine the location of any stagnation point(s) in this flow field.
Answer: at the origin

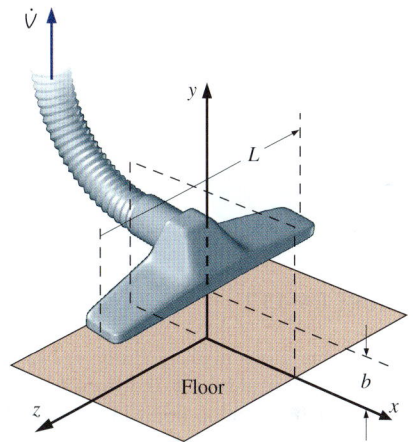

FIGURE P4–108

4–109 Consider the vacuum cleaner of Prob. 4–108. For the case where $b = 2.0$ cm, $L = 35$ cm, and $\dot{V} = 0.1098$ m³/s, create a velocity vector plot in the upper half of the xy-plane from $x = -3$ cm to 3 cm and from $y = 0$ cm to 2.5 cm. Draw as many vectors as you need to get a good feel of the flow field. *Note*: The velocity is infinite at the point $(x, y) = (0, 2.0$ cm$)$, so do not attempt to draw a velocity vector at that point.

4–110 Consider the approximate velocity field given for the vacuum cleaner of Prob. 4–108. Calculate the flow speed along the floor. Dust particles on the floor are most likely to be sucked up by the vacuum cleaner at the location of maximum speed. Where is that location? Do you think the vacuum cleaner will do a good job at sucking up dust directly below the inlet (at the origin)? Why or why not?

4–111 In a steady, two-dimensional flow field in the xy-plane, the x-component of velocity is

$$u = ax + by + cx^2 - dxy$$

where a, b, c, and d are constants with appropriate dimensions. Generate a general expression for velocity component v such that the flow field is incompressible.

4–112 There are numerous occasions in which a fairly uniform free-stream flow encounters a long circular cylinder aligned normal to the flow (Fig. P4–112). Examples include air flowing around a car antenna, wind blowing against a flag pole or telephone pole, wind hitting electrical wires, and ocean currents impinging on the submerged round beams that support oil platforms. In all these cases, the flow at the rear of the cylinder is separated and unsteady, and usually turbulent. However, the flow in the front half of the cylinder is much more steady and predictable. In fact, except for a very thin boundary layer near the cylinder surface, the flow field may be approximated by the following steady, two-dimensional velocity components in the xy- or $r\theta$-plane:

$$u_r = V\cos\theta\left(1 - \frac{a^2}{r^2}\right) \quad u_\theta = -V\sin\theta\left(1 + \frac{a^2}{r^2}\right)$$

Is this flow field rotational or irrotational? Explain.

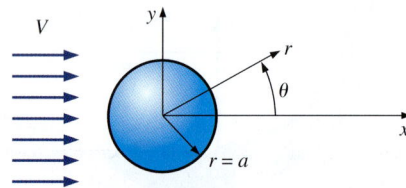

FIGURE P4–112

4–113 Consider the flow field of Prob. 4–112 (flow over a circular cylinder). Consider only the front half of the flow ($x < 0$). There is one stagnation point in the front half of the flow field. Where is it? Give your answer in both cylindrical (r, θ) coordinates and Cartesian (x, y) coordinates.

4–114 Consider the upstream half ($x < 0$) of the flow field of Prob. 4–112 (flow over a circular cylinder). We introduce a parameter called the **stream function** ψ, which is *constant along streamlines* in two-dimensional flows such as the one being considered here (Fig. P4–114). The velocity field of Prob. 4–112 corresponds to a stream function given by

$$\psi = V\sin\theta\left(r - \frac{a^2}{r}\right)$$

(*a*) Setting ψ to a constant, generate an equation for a streamline. (*Hint*: Use the quadratic rule to solve for r as a function of θ.)

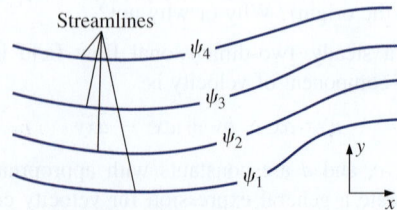

FIGURE P4–114

(*b*) For the particular case in which $V = 1.00$ m/s and cylinder radius $a = 10.0$ cm, plot several streamlines in the upstream half of the flow ($90° < \theta < 270°$). For consistency, plot in the range -0.4 m $< x < 0$ m, -0.2 m $< y < 0.2$ m, with stream function values evenly spaced between -0.16 m^2/s and 0.16 m^2/s.

4–115 Consider the flow field of Prob. 4–112 (flow over a circular cylinder). Calculate the two linear strain rates in the $r\theta$-plane; i.e., calculate ε_{rr} and $\varepsilon_{\theta\theta}$. Discuss whether fluid line segments stretch (or shrink) in this flow field. (*Hint*: The strain rate tensor in cylindrical coordinates is given in Prob. 4–106.)

4–116 Based on your results of Prob. 4–115, discuss the compressibility (or incompressibility) of this flow. *Answer:* flow is incompressible

4–117 Consider the flow field of Prob. 4–112 (flow over a circular cylinder). Calculate $\varepsilon_{r\theta}$, the shear strain rate in the $r\theta$-plane. Discuss whether fluid particles in this flow deform with shear or not. (*Hint*: The strain rate tensor in cylindrical coordinates is given in Prob. 4–106.)

Fundamentals of Engineering (FE) Exam Problems

4–118 A steady, incompressible, two-dimensional velocity field is given by

$$\vec{V} = (u, v) = (2.5 - 1.6x)\vec{i} + (0.7 + 1.6y)\vec{j}$$

where the x- and y-coordinates are in meters and the magnitude of velocity is in m/s. The values of x and y at the stagnation point, respectively, are

(*a*) 0.9375 m, 0.375 m   (*b*) 1.563 m, -0.4375 m
(*c*) 2.5 m, 0.7 m   (*d*) 0.731 m, 1.236 m   (*e*) -1.6 m, 0.8 m

4–119 Water is flowing in a 3-cm-diameter garden hose at a rate of 30 L/min. A 20-cm nozzle is attached to the hose which decreases the diameter to 1.2 cm. The magnitude of the acceleration of a fluid particle moving down the centerline of the nozzle is

(*a*) 9.81 m/s^2 (*b*) 14.5 m/s^2 (*c*) 25.4 m/s^2 (*d*) 39.1 m/s^2
(*e*) 47.6 m/s^2

4–120 A steady, incompressible, two-dimensional velocity field is given by

$$\vec{V} = (u, v) = (2.5 - 1.6x)\vec{i} + (0.7 + 1.6y)\vec{j}$$

where the x- and y-coordinates are in meters and the magnitude of velocity is in m/s. The x-component of the acceleration vector a_x is

(*a*) $0.8y$ (*b*) $-1.6x$ (*c*) $2.5x - 1.6$ (*d*) $2.56x - 4$
(*e*) $2.56x + 0.8y$

4–121 A steady, incompressible, two-dimensional velocity field is given by

$$\vec{V} = (u, v) = (2.5 - 1.6x)\vec{i} + (0.7 + 1.6y)\vec{j}$$

where the x- and y-coordinates are in meters and the magnitude of velocity is in m/s. The x- and y-component of material acceleration a_x and a_y at the point ($x = 1$ m, $y = 1$ m), respectively, in m/s^2, are

(*a*) -1.44, 3.68   (*b*) -1.6, 1.5   (*c*) 3.1, -1.32
(*d*) 2.56, -4 (*e*) -0.8, 1.6

4–122 A steady, incompressible, two-dimensional velocity field is given by

$$\vec{V} = (u, v) = (0.65 + 1.7x)\vec{i} + (1.3 - 1.7y)\vec{j}$$

where the x- and y-coordinates are in meters and the magnitude of velocity is in m/s. The y-component of the acceleration vector a_y is
(a) 1.7y   (b) −1.7y   (c) 2.89y − 2.21   (d) 3.0x − 2.73
(e) 0.84y + 1.42

4–123 A steady, incompressible, two-dimensional velocity field is given by

$$\vec{V} = (u, v) = (0.65 + 1.7x)\vec{i} + (1.3 - 1.7y)\vec{j}$$

where the x- and y-coordinates are in meters and the magnitude of velocity is in m/s. The x- and y-component of material acceleration a_x and a_y at the point ($x = 0$ m, $y = 0$ m), respectively, in m/s², are
(a) 0.37, −1.85   (b) −1.7, 1.7   (c) 1.105, −2.21
(d) 1.7, −1.7   (e) 0.65, 1.3

4–124 A steady, incompressible, two-dimensional velocity field is given by

$$\vec{V} = (u, v) = (0.65 + 1.7x)\vec{i} + (1.3 - 1.7y)\vec{j}$$

where the x- and y-coordinates are in meters and the magnitude of velocity is in m/s. The x- and y-component of velocity u and v at the point ($x = 1$ m, $y = 2$ m), respectively, in m/s, are
(a) 0.54, −2.31   (b) −1.9, 0.75   (c) 0.598, −2.21
(d) 2.35, −2.1   (e) 0.65, 1.3

4–125 The actual path traveled by an individual fluid particle over some period is called a
(a) Pathline   (b) Streamtube   (c) Streamline
(d) Streakline   (e) Timeline

4–126 The locus of fluid particles that have passed sequentially through a prescribed point in the flow is called a
(a) Pathline   (b) Streamtube   (c) Streamline
(d) Streakline   (e) Timeline

4–127 A curve that is everywhere tangent to the instantaneous local velocity vector is called a
(a) Pathline   (b) Streamtube   (c) Streamline
(d) Streakline   (e) Timeline

4–128 An array of arrows indicating the magnitude and direction of a vector property at an instant in time is called a
(a) Profiler plot   (b) Vector plot   (c) Contour plot
(d) Velocity plot   (e) Time plot

4–129 The CFD stands for
(a) Compressible fluid dynamics
(b) Compressed flow domain
(c) Circular flow dynamics
(d) Convective fluid dynamics
(e) Computational fluid dynamics

4–130 Which one is not a fundamental type of motion or deformation an element may undergo in fluid mechanics?

(a) Rotation   (b) Converging   (c) Translation
(d) Linear strain   (e) Shear strain

4–131 A steady, incompressible, two-dimensional velocity field is given by

$$\vec{V} = (u, v) = (2.5 - 1.6x)\vec{i} + (0.7 + 1.6y)\vec{j}$$

where the x- and y-coordinates are in meters and the magnitude of velocity is in m/s. The linear strain rate in the x-direction in s^{-1} is
(a) −1.6   (b) 0.8   (c) 1.6   (d) 2.5   (e) −0.875

4-132 A steady, incompressible, two-dimensional velocity field is given by

$$\vec{V} = (u, v) = (2.5 - 1.6x)\vec{i} + (0.7 + 1.6y)\vec{j}$$

where the x- and y-coordinates are in meters and the magnitude of velocity is in m/s. The shear strain rate in s^{-1} is
(a) −1.6   (b) 1.6   (c) 2.5   (d) 0.7   (e) 0

4–133 A steady, two-dimensional velocity field is given by

$$\vec{V} = (u, v) = (2.5 - 1.6x)\vec{i} + (0.7 + 0.8y)\vec{j}$$

where the x- and y-coordinates are in meters and the magnitude of velocity is in m/s. The volumetric strain rate in s^{-1} is
(a) 0   (b) 3.2   (c) −0.8   (d) 0.8   (e) −1.6

4–134 If the vorticity in a region of the flow is zero, the flow is
(a) Motionless   (b) Incompressible   (c) Compressible
(d) Irrotational   (e) Rotational

4–135 The angular velocity of a fluid particle is 20 rad/s. The vorticity of this fluid particle is
(a) 20 rad/s   (b) 40 rad/s   (c) 80 rad/s   (d) 10 rad/s
(e) 5 rad/s

4–136 A steady, incompressible, two-dimensional velocity field is given by

$$\vec{V} = (u, v) = (0.75 + 1.2x)\vec{i} + (2.25 - 1.2y)\vec{j}$$

where the x- and y-coordinates are in meters and the magnitude of velocity is in m/s. The vorticity of this flow is
(a) 0   (b) $1.2y\vec{k}$ (c) $-1.2y\vec{k}$ (d) $y\vec{k}$ (e) $-1.2xy\vec{k}$

4–137 A steady, incompressible, two-dimensional velocity field is given by

$$\vec{V} = (u, v) = (2xy + 1)\vec{i} + (-y^2 - 0.6)\vec{j}$$

where the x- and y-coordinates are in meters and the magnitude of velocity is in m/s. The angular velocity of this flow is
(a) 0   (b) $-2y\vec{k}$ (c) $2y\vec{k}$ (d) $-2x\vec{k}$ (e) $-x\vec{k}$

4–138 A cart is moving at a constant absolute velocity $\vec{V}_{cart} = 5$ km/h to the right. A high-speed jet of water at an absolute velocity of $\vec{V}_{jet} = 15$ km/h to the right strikes the back of the car. The relative velocity of the water is
(a) 0 km/h   (b) 5 km/h   (c) 10 km/h   (d) 15 km/h   (e) 20 km/h

4–122 A steady, incompressible, two-dimensional velocity field is given by

$$\vec{V} = (u, v) = (0.65 + 1.7x)\vec{i} + (1.3 - 1.7y)\vec{j}$$

where the x- and y-coordinates are in meters and the magnitude of velocity is in m/s. The y-component of the acceleration vector a_y is
(a) 1.7y   (b) $-1.7y$ (c) $2.89y - 2.21$ (d) $3.0x - 2.73$
(e) $0.84y + 1.42$

4–123 A steady, incompressible, two-dimensional velocity field is given by

$$\vec{V} = (u, v) = (0.65 + 1.7x)\vec{i} + (1.3 - 1.7y)\vec{j}$$

where the x- and y-coordinates are in meters and the magnitude of velocity is in m/s. The x- and y-component of material acceleration a_x and a_y at the point ($x = 0$ m, $y = 0$ m), respectively, in m/s², are
(a) $0.37, -1.85$ (b) $-1.7, 1.7$ (c) $1.105, -2.21$
(d) $1.7, -1.7$ (e) $0.65, 1.3$

4–124 A steady, incompressible, two-dimensional velocity field is given by

$$\vec{V} = (u, v) = (0.65 + 1.7x)\vec{i} + (1.3 - 1.7y)\vec{j}$$

where the x- and y-coordinates are in meters and the magnitude of velocity is in m/s. The x- and y-component of velocity u and v at the point ($x = 1$ m, $y = 2$ m), respectively, in m/s, are
(a) $0.54, -2.31$ (b) $-1.9, 0.75$ (c) $0.598, -2.21$
(d) $2.35, -2.1$ (e) $0.65, 1.3$

4–125 The actual path traveled by an individual fluid particle over some period is called a
(a) Pathline   (b) Streamtube   (c) Streamline
(d) Streakline   (e) Timeline

4–126 The locus of fluid particles that have passed sequentially through a prescribed point in the flow is called a
(a) Pathline   (b) Streamtube   (c) Streamline
(d) Streakline   (e) Timeline

4–127 A curve that is everywhere tangent to the instantaneous local velocity vector is called a
(a) Pathline   (b) Streamtube   (c) Streamline
(d) Streakline   (e) Timeline

4–128 An array of arrows indicating the magnitude and direction of a vector property at an instant in time is called a
(a) Profiler plot   (b) Vector plot   (c) Contour plot
(d) Velocity plot   (e) Time plot

4–129 The CFD stands for
(a) Compressible fluid dynamics
(b) Compressed flow domain
(c) Circular flow dynamics
(d) Convective fluid dynamics
(e) Computational fluid dynamics

4–130 Which one is not a fundamental type of motion or deformation an element may undergo in fluid mechanics?

(a) Rotation   (b) Converging   (c) Translation
(d) Linear strain   (e) Shear strain

4–131 A steady, incompressible, two-dimensional velocity field is given by

$$\vec{V} = (u, v) = (2.5 - 1.6x)\vec{i} + (0.7 + 1.6y)\vec{j}$$

where the x- and y-coordinates are in meters and the magnitude of velocity is in m/s. The linear strain rate in the x-direction in s^{-1} is
(a) -1.6 (b) 0.8   (c) 1.6   (d) 2.5   (e) -0.875

4-132 A steady, incompressible, two-dimensional velocity field is given by

$$\vec{V} = (u, v) = (2.5 - 1.6x)\vec{i} + (0.7 + 1.6y)\vec{j}$$

where the x- and y-coordinates are in meters and the magnitude of velocity is in m/s. The shear strain rate in s^{-1} is
(a) -1.6 (b) 1.6   (c) 2.5   (d) 0.7   (e) 0

4–133 A steady, two-dimensional velocity field is given by

$$\vec{V} = (u, v) = (2.5 - 1.6x)\vec{i} + (0.7 + 0.8y)\vec{j}$$

where the x- and y-coordinates are in meters and the magnitude of velocity is in m/s. The volumetric strain rate in s^{-1} is
(a) 0   (b) 3.2   (c) -0.8 (d) 0.8   (e) -1.6

4–134 If the vorticity in a region of the flow is zero, the flow is
(a) Motionless   (b) Incompressible   (c) Compressible
(d) Irrotational   (e) Rotational

4–135 The angular velocity of a fluid particle is 20 rad/s. The vorticity of this fluid particle is
(a) 20 rad/s   (b) 40 rad/s   (c) 80 rad/s   (d) 10 rad/s
(e) 5 rad/s

4–136 A steady, incompressible, two-dimensional velocity field is given by

$$\vec{V} = (u, v) = (0.75 + 1.2x)\vec{i} + (2.25 - 1.2y)\vec{j}$$

where the x- and y-coordinates are in meters and the magnitude of velocity is in m/s. The vorticity of this flow is
(a) 0   (b) $1.2y\vec{k}$ (c) $-1.2y\vec{k}$ (d) $y\vec{k}$ (e) $-1.2xy\vec{k}$

4–137 A steady, incompressible, two-dimensional velocity field is given by

$$\vec{V} = (u, v) = (2xy + 1)\vec{i} + (-y^2 - 0.6)\vec{j}$$

where the x- and y-coordinates are in meters and the magnitude of velocity is in m/s. The angular velocity of this flow is
(a) 0   (b) $-2y\vec{k}$ (c) $2y\vec{k}$ (d) $-2x\vec{k}$ (e) $-x\vec{k}$

4–138 A cart is moving at a constant absolute velocity $\vec{V}_{cart} = 5$ km/h to the right. A high-speed jet of water at an absolute velocity of $\vec{V}_{jet} = 15$ km/h to the right strikes the back of the car. The relative velocity of the water is
(a) 0 km/h   (b) 5 km/h   (c) 10 km/h   (d) 15 km/h   (e) 20 km/h

CHAPTER 5

BERNOULLI AND ENERGY EQUATIONS

This chapter deals with three equations commonly used in fluid mechanics: the mass, Bernoulli, and energy equations. The *mass equation* is an expression of the conservation of mass principle. The *Bernoulli equation* is concerned with the conservation of kinetic, potential, and flow energies of a fluid stream and their conversion to each other in regions of flow where net viscous forces are negligible and where other restrictive conditions apply. The *energy equation* is a statement of the conservation of energy principle. In fluid mechanics, it is convenient to separate *mechanical energy* from *thermal energy* and to consider the conversion of mechanical energy to thermal energy as a result of frictional effects as *mechanical energy loss*. Then the energy equation becomes the *mechanical energy balance*.

We start this chapter with an overview of conservation principles and the conservation of mass relation. This is followed by a discussion of various forms of mechanical energy and the efficiency of mechanical work devices such as pumps and turbines. Then we derive the Bernoulli equation by applying Newton's second law to a fluid element along a streamline and demonstrate its use in a variety of applications. We continue with the development of the energy equation in a form suitable for use in fluid mechanics and introduce the concept of *head loss*. Finally, we apply the energy equation to various engineering systems.

OBJECTIVES

When you finish reading this chapter, you should be able to

- Apply the conservation of mass equation to balance the incoming and outgoing flow rates in a flow system
- Recognize various forms of mechanical energy, and work with energy conversion efficiencies
- Understand the use and limitations of the Bernoulli equation, and apply it to solve a variety of fluid flow problems
- Work with the energy equation expressed in terms of heads, and use it to determine turbine power output and pumping power requirements

Wind turbine "farms" are being constructed all over the world to extract kinetic energy from the wind and convert it to electrical energy. The mass, energy, momentum, and angular momentum balances are utilized in the design of a wind turbine. The Bernoulli equation is also useful in the preliminary design stage.
© J. Luke/PhotoLink/Getty RF

185

BERNOULLI AND ENERGY EQUATIONS

FIGURE 5–1
Many fluid flow devices such as this Pelton wheel hydraulic turbine are analyzed by applying the conservation of mass and energy principles, along with the linear momentum equation.
Courtesy of Hydro Tasmania, www.hydro.com.au. Used by permission.

5–1 ▪ INTRODUCTION

You are already familiar with numerous **conservation laws** such as the laws of conservation of mass, conservation of energy, and conservation of momentum. Historically, the conservation laws are first applied to a fixed quantity of matter called a *closed system* or just a *system*, and then extended to regions in space called *control volumes*. The conservation relations are also called *balance equations* since any conserved quantity must balance during a process. We now give a brief description of the conservation of mass and energy relations, and the linear momentum equation (Fig. 5–1).

Conservation of Mass

The conservation of mass relation for a closed system undergoing a change is expressed as m_{sys} = constant or $dm_{sys}/dt = 0$, which is the statement that the mass of the system remains constant during a process. For a control volume (CV), mass balance is expressed in rate form as

$$\text{Conservation of mass:} \qquad \dot{m}_{in} - \dot{m}_{out} = \frac{dm_{CV}}{dt} \qquad (5\text{--}1)$$

where $\dot{m}_{in}$ and $\dot{m}_{out}$ are the total rates of mass flow into and out of the control volume, respectively, and dm_{CV}/dt is the rate of change of mass within the control volume boundaries. In fluid mechanics, the conservation of mass relation written for a differential control volume is usually called the *continuity equation*. Conservation of mass is discussed in Section 5–2.

The Linear Momentum Equation

The product of the mass and the velocity of a body is called the *linear momentum* or just the *momentum* of the body, and the momentum of a rigid body of mass m moving with a velocity $\vec{V}$ is $m\vec{V}$. Newton's second law states that the acceleration of a body is proportional to the net force acting on it and is inversely proportional to its mass, and that the rate of change of the momentum of a body is equal to the net force acting on the body. Therefore, the momentum of a system remains constant only when the net force acting on it is zero, and thus the momentum of such systems is conserved. This is known as the *conservation of momentum principle*. In fluid mechanics, Newton's second law is usually referred to as the *linear momentum equation*, which is discussed in Chap. 6 together with the *angular momentum equation*.

Conservation of Energy

Energy can be transferred to or from a closed system by heat or work, and the conservation of energy principle requires that the net energy transfer to or from a system during a process be equal to the change in the energy content of the system. Control volumes involve energy transfer via mass flow also, and the *conservation of energy principle*, also called the *energy balance*, is expressed as

$$\text{Conservation of energy:} \qquad \dot{E}_{in} - \dot{E}_{out} = \frac{dE_{CV}}{dt} \qquad (5\text{--}2)$$

where $\dot{E}_{in}$ and $\dot{E}_{out}$ are the total rates of energy transfer into and out of the control volume, respectively, and dE_{CV}/dt is the rate of change of energy within the control volume boundaries. In fluid mechanics, we usually limit our consideration to mechanical forms of energy only. Conservation of energy is discussed in Section 5–6.

5–2 · CONSERVATION OF MASS

The conservation of mass principle is one of the most fundamental principles in nature. We are all familiar with this principle, and it is not difficult to understand. A person does not have to be a rocket scientist to figure out how much vinegar-and-oil dressing will be obtained by mixing 100 g of oil with 25 g of vinegar. Even chemical equations are balanced on the basis of the conservation of mass principle. When 16 kg of oxygen reacts with 2 kg of hydrogen, 18 kg of water is formed (Fig. 5–2). In an electrolysis process, the water separates back to 2 kg of hydrogen and 16 kg of oxygen.

Technically, mass is not exactly conserved. It turns out that mass m and energy E can be converted to each other according to the well-known formula proposed by Albert Einstein (1879–1955):

$$E = mc^2 \qquad (5\text{-}3)$$

where c is the speed of light in a vacuum, which is $c = 2.9979 \times 10^8$ m/s. This equation suggests that there is equivalence between mass and energy. All physical and chemical systems exhibit energy interactions with their surroundings, but the amount of energy involved is equivalent to an extremely small mass compared to the system's total mass. For example, when 1 kg of liquid water is formed from oxygen and hydrogen at normal atmospheric conditions, the amount of energy released is 15.8 MJ, which corresponds to a mass of only 1.76×10^{-10} kg. However, in nuclear reactions, the mass equivalence of the amount of energy interacted is a significant fraction of the total mass involved. Therefore, in most engineering analyses, we consider both mass and energy as conserved quantities.

For *closed systems,* the conservation of mass principle is implicitly used by requiring that the mass of the system remain constant during a process. For *control volumes,* however, mass can cross the boundaries, and so we must keep track of the amount of mass entering and leaving the control volume.

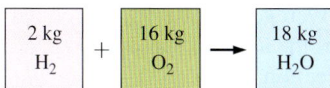

FIGURE 5–2
Mass is conserved even during chemical reactions.

Mass and Volume Flow Rates

The amount of mass flowing through a cross section per unit time is called the **mass flow rate** and is denoted by $\dot{m}$. The dot over a symbol is used to indicate *time rate of change.*

A fluid flows into or out of a control volume, usually through pipes or ducts. The differential mass flow rate of fluid flowing across a small area element dA_c in a cross section of a pipe is proportional to dA_c itself, the fluid density ρ, and the component of the flow velocity normal to dA_c, which we denote as V_n, and is expressed as (Fig. 5–3)

$$\delta \dot{m} = \rho V_n \, dA_c \qquad (5\text{-}4)$$

Note that both δ and d are used to indicate differential quantities, but δ is typically used for quantities (such as heat, work, and mass transfer) that are *path functions* and have *inexact differentials,* while d is used for quantities

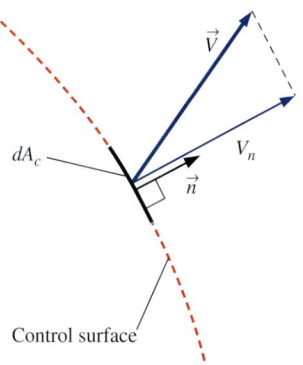

FIGURE 5–3
The normal velocity V_n for a surface is the component of velocity perpendicular to the surface.

188
BERNOULLI AND ENERGY EQUATIONS

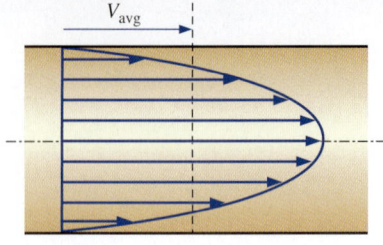

FIGURE 5–4
Average velocity V_{avg} is defined as the average speed through a cross section.

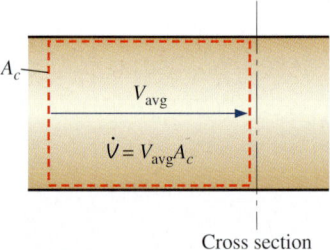

FIGURE 5–5
The volume flow rate is the volume of fluid flowing through a cross section per unit time.

(such as properties) that are *point functions* and have *exact differentials*. For flow through an annulus of inner radius r_1 and outer radius r_2, for example,

$$\int_1^2 dA_c = A_{c2} - A_{c1} = \pi(r_2^2 - r_1^2) \text{ but } \int_1^2 \delta\dot{m} = \dot{m}_{total} \text{ (total mass flow rate}$$

through the annulus), not $\dot{m}_2 - \dot{m}_1$. For specified values of r_1 and r_2, the value of the integral of dA_c is fixed (thus the names point function and exact differential), but this is not the case for the integral of $\delta\dot{m}$ (thus the names path function and inexact differential).

The mass flow rate through the entire cross-sectional area of a pipe or duct is obtained by integration:

$$\dot{m} = \int_{A_c} \delta\dot{m} = \int_{A_c} \rho V_n \, dA_c \quad \text{(kg/s)} \tag{5–5}$$

While Eq. 5–5 is always valid (in fact it is *exact*), it is not always practical for engineering analyses because of the integral. We would like instead to express mass flow rate in terms of average values over a cross section of the pipe. In a general compressible flow, both ρ and V_n vary across the pipe. In many practical applications, however, the density is essentially uniform over the pipe cross section, and we can take ρ outside the integral of Eq. 5–5. Velocity, however, is *never* uniform over a cross section of a pipe because of the no-slip condition at the walls. Rather, the velocity varies from zero at the walls to some maximum value at or near the centerline of the pipe. We define the **average velocity** V_{avg} as the average value of V_n across the entire cross section of the pipe (Fig. 5–4),

Average velocity:
$$V_{avg} = \frac{1}{A_c}\int_{A_c} V_n \, dA_c \tag{5–6}$$

where A_c is the area of the cross section normal to the flow direction. Note that if the speed were V_{avg} all through the cross section, the mass flow rate would be identical to that obtained by integrating the actual velocity profile. Thus for incompressible flow or even for compressible flow where ρ is approximated as uniform across A_c, Eq. 5–5 becomes

$$\dot{m} = \rho V_{avg} A_c \quad \text{(kg/s)} \tag{5–7}$$

For compressible flow, we can think of ρ as the bulk average density over the cross section, and then Eq. 5–7 can be used as a reasonable approximation. For simplicity, we drop the subscript on the average velocity. Unless otherwise stated, V denotes the average velocity in the flow direction. Also, A_c denotes the cross-sectional area normal to the flow direction.

The volume of the fluid flowing through a cross section per unit time is called the **volume flow rate** $\dot{V}$ (Fig. 5–5) and is given by

$$\dot{V} = \int_{A_c} V_n \, dA_c = V_{avg} A_c = V A_c \quad \text{(m}^3\text{/s)} \tag{5–8}$$

An early form of Eq. 5–8 was published in 1628 by the Italian monk Benedetto Castelli (circa 1577–1644). Note that many fluid mechanics textbooks use Q instead of $\dot{V}$ for volume flow rate. We use $\dot{V}$ to avoid confusion with heat transfer.

The mass and volume flow rates are related by

$$\dot{m} = \rho\dot{V} = \frac{\dot{V}}{v} \tag{5–9}$$

where v is the specific volume. This relation is analogous to $m = \rho V = V/v$, which is the relation between the mass and the volume of a fluid in a container.

Conservation of Mass Principle

The **conservation of mass principle** for a control volume can be expressed as: *The net mass transfer to or from a control volume during a time interval Δt is equal to the net change (increase or decrease) of the total mass within the control volume during Δt.* That is,

$$\begin{pmatrix} \text{Total mass entering} \\ \text{the CV during } \Delta t \end{pmatrix} - \begin{pmatrix} \text{Total mass leaving} \\ \text{the CV during } \Delta t \end{pmatrix} = \begin{pmatrix} \text{Net change of mass} \\ \text{within the CV during } \Delta t \end{pmatrix}$$

or

$$m_{in} - m_{out} = \Delta m_{CV} \quad (\text{kg}) \quad (5\text{-}10)$$

where $\Delta m_{CV} = m_{final} - m_{initial}$ is the change in the mass of the control volume during the process (Fig. 5–6). It can also be expressed in *rate form* as

$$\dot{m}_{in} - \dot{m}_{out} = dm_{CV}/dt \quad (\text{kg/s}) \quad (5\text{-}11)$$

where $\dot{m}_{in}$ and $\dot{m}_{out}$ are the total rates of mass flow into and out of the control volume, and dm_{CV}/dt is the rate of change of mass within the control volume boundaries. Equations 5–10 and 5–11 are often referred to as the **mass balance** and are applicable to any control volume undergoing any kind of process.

Consider a control volume of arbitrary shape, as shown in Fig. 5–7. The mass of a differential volume dV within the control volume is $dm = \rho \, dV$. The total mass within the control volume at any instant in time t is determined by integration to be

Total mass within the CV:
$$m_{CV} = \int_{CV} \rho \, dV \quad (5\text{-}12)$$

Then the time rate of change of the amount of mass within the control volume is expressed as

Rate of change of mass within the CV:
$$\frac{dm_{CV}}{dt} = \frac{d}{dt} \int_{CV} \rho \, dV \quad (5\text{-}13)$$

For the special case of no mass crossing the control surface (i.e., the control volume is a closed system), the conservation of mass principle reduces to $dm_{CV}/dt = 0$. This relation is valid whether the control volume is fixed, moving, or deforming.

Now consider mass flow into or out of the control volume through a differential area dA on the control surface of a fixed control volume. Let $\vec{n}$ be the outward unit vector of dA normal to dA and $\vec{V}$ be the flow velocity at dA relative to a fixed coordinate system, as shown in Fig. 5–7. In general, the velocity may cross dA at an angle θ off the normal of dA, and the mass flow rate is proportional to the normal component of velocity $\vec{V}_n = \vec{V} \cos \theta$ ranging from a maximum outflow of $\vec{V}$ for $\theta = 0$ (flow is normal to dA) to a minimum of zero for $\theta = 90°$ (flow is tangent to dA) to a maximum *inflow* of $\vec{V}$ for $\theta = 180°$ (flow is normal to dA but in the opposite direction).

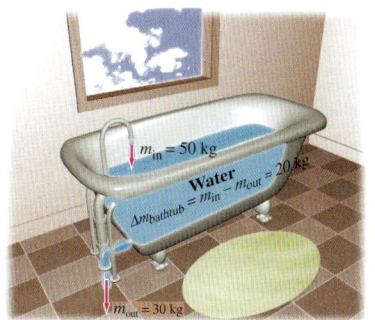

FIGURE 5–6
Conservation of mass principle for an ordinary bathtub.

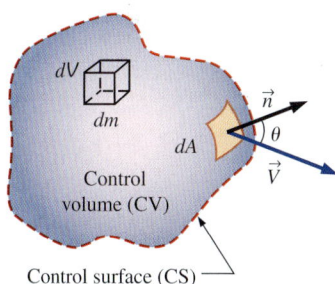

FIGURE 5–7
The differential control volume dV and the differential control surface dA used in the derivation of the conservation of mass relation.

Making use of the concept of dot product of two vectors, the magnitude of the normal component of velocity is

Normal component of velocity: $\qquad V_n = V \cos \theta = \vec{V} \cdot \vec{n}$ (5–14)

The mass flow rate through dA is proportional to the fluid density ρ, normal velocity V_n, and the flow area dA, and is expressed as

Differential mass flow rate: $\quad \delta \dot{m} = \rho V_n \, dA = \rho (V \cos \theta) \, dA = \rho (\vec{V} \cdot \vec{n}) \, dA$ (5–15)

The net flow rate into or out of the control volume through the entire control surface is obtained by integrating $\delta \dot{m}$ over the entire control surface,

Net mass flow rate: $\qquad \dot{m}_{net} = \int_{CS} \delta \dot{m} = \int_{CS} \rho V_n \, dA = \int_{CS} \rho (\vec{V} \cdot \vec{n}) \, dA$ (5–16)

Note that $V_n = \vec{V} \cdot \vec{n} = V \cos \theta$ is positive for $\theta < 90°$ (outflow) and negative for $\theta > 90°$ (inflow). Therefore, the direction of flow is automatically accounted for, and the surface integral in Eq. 5–16 directly gives the *net mass flow rate*. A positive value for $\dot{m}_{net}$ indicates a net outflow of mass and a negative value indicates a net inflow of mass.

Rearranging Eq. 5–11 as $dm_{CV}/dt + \dot{m}_{out} - \dot{m}_{in} = 0$, the conservation of mass relation for a fixed control volume is then expressed as

General conservation of mass: $\qquad \dfrac{d}{dt} \int_{CV} \rho \, dV + \int_{CS} \rho (\vec{V} \cdot \vec{n}) \, dA = 0$ (5–17)

It states that *the time rate of change of mass within the control volume plus the net mass flow rate through the control surface is equal to zero.*

The general conservation of mass relation for a control volume can also be derived using the Reynolds transport theorem (RTT) by taking the property B to be the mass m (Chap. 4). Then we have $b = 1$ since dividing mass by mass to get the property per unit mass gives unity. Also, the mass of a closed system is constant, and thus its time derivative is zero. That is, $dm_{sys}/dt = 0$. Then the Reynolds transport equation reduces immediately to Eq. 5–17, as shown in Fig. 5–8, and thus illustrates that the Reynolds transport theorem is a very powerful tool indeed.

Splitting the surface integral in Eq. 5–17 into two parts—one for the outgoing flow streams (positive) and one for the incoming flow streams (negative)—the general conservation of mass relation can also be expressed as

$$\dfrac{d}{dt} \int_{CV} \rho \, dV + \sum_{out} \rho |V_n| A - \sum_{in} \rho |V_n| A = 0 \qquad (5\text{–}18)$$

where A represents the area for an inlet or outlet, and the summation signs are used to emphasize that *all* the inlets and outlets are to be considered. Using the definition of mass flow rate, Eq. 5–18 can also be expressed as

$$\dfrac{d}{dt} \int_{CV} \rho \, dV = \sum_{in} \dot{m} - \sum_{out} \dot{m} \quad \text{or} \quad \dfrac{dm_{CV}}{dt} = \sum_{in} \dot{m} - \sum_{out} \dot{m} \qquad (5\text{–}19)$$

There is considerable flexibility in the selection of a control volume when solving a problem. Many control volume choices are available, but some are more convenient to work with. A control volume should not introduce any unnecessary complications. A wise choice of a control volume can make the solution of a seemingly complicated problem rather easy. A simple rule in selecting a control volume is to make the control surface *normal to the*

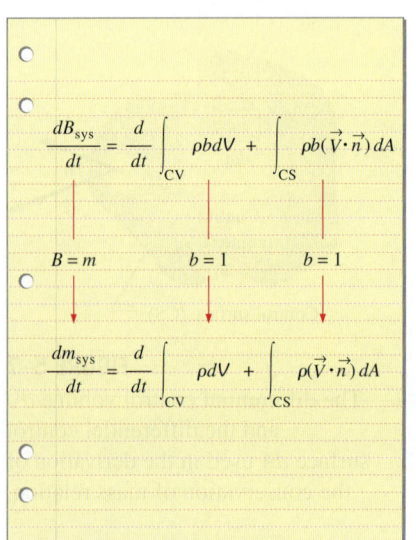

FIGURE 5–8
The conservation of mass equation is obtained by replacing B in the Reynolds transport theorem by mass m, and b by 1 (m per unit mass = $m/m = 1$).

flow at all locations where it crosses the fluid flow, whenever possible. This way the dot product $\vec{V}\cdot\vec{n}$ simply becomes the magnitude of the velocity, and the integral $\int_A \rho(\vec{V}\cdot\vec{n})\,dA$ becomes simply ρVA (Fig. 5–9).

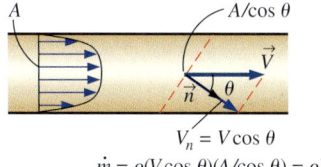

$V_n = V\cos\theta$
$\dot{m} = \rho(V\cos\theta)(A/\cos\theta) = \rho VA$

(a) Control surface *at an angle* to the flow

Moving or Deforming Control Volumes

Equations 5–17 and 5–19 are also valid for moving control volumes provided that the *absolute velocity* $\vec{V}$ is replaced by the *relative velocity* $\vec{V}_r$, which is the fluid velocity relative to the control surface (Chap. 4). In the case of a moving but nondeforming control volume, relative velocity is the fluid velocity observed by a person moving with the control volume and is expressed as $\vec{V}_r = \vec{V} - \vec{V}_{CS}$, where $\vec{V}$ is the fluid velocity and $\vec{V}_{CS}$ is the velocity of the control surface, both relative to a fixed point outside. Note that this is a *vector* subtraction.

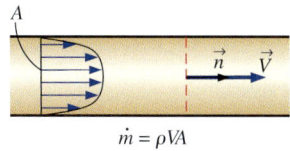

$\dot{m} = \rho VA$

(b) Control surface *normal* to the flow

FIGURE 5–9
A control surface should always be selected *normal to the flow* at all locations where it crosses the fluid flow to avoid complications, even though the result is the same.

Some practical problems (such as the injection of medication through the needle of a syringe by the forced motion of the plunger) involve *deforming* control volumes. The conservation of mass relations developed can still be used for such deforming control volumes provided that the velocity of the fluid crossing a deforming part of the control surface is expressed relative to the control surface (that is, the fluid velocity should be expressed relative to a reference frame attached to the deforming part of the control surface). The relative velocity in this case at any point on the control surface is expressed again as $\vec{V}_r = \vec{V} - \vec{V}_{CS}$, where $\vec{V}_{CS}$ is the local velocity of the control surface at that point relative to a fixed point outside the control volume.

Mass Balance for Steady-Flow Processes

During a steady-flow process, the total amount of mass contained within a control volume does not change with time (m_{CV} = constant). Then the conservation of mass principle requires that the total amount of mass entering a control volume equal the total amount of mass leaving it. For a garden hose nozzle in steady operation, for example, the amount of water entering the nozzle per unit time is equal to the amount of water leaving it per unit time.

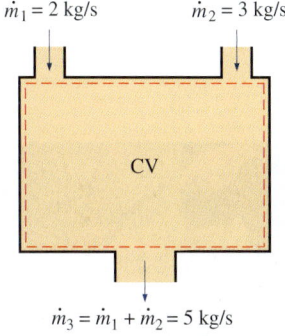

FIGURE 5–10
Conservation of mass principle for a two-inlet–one-outlet steady-flow system.

When dealing with steady-flow processes, we are not interested in the amount of mass that flows in or out of a device over time; instead, we are interested in the amount of mass flowing per unit time, that is, *the mass flow rate* $\dot{m}$. The conservation of mass principle for a general steady-flow system with multiple inlets and outlets is expressed in rate form as (Fig. 5–10)

Steady flow: $$\sum_{in}\dot{m} = \sum_{out}\dot{m} \quad \text{(kg/s)} \tag{5-20}$$

It states that *the total rate of mass entering a control volume is equal to the total rate of mass leaving it.*

Many engineering devices such as nozzles, diffusers, turbines, compressors, and pumps involve a single stream (only one inlet and one outlet). For these cases, we typically denote the inlet state by the subscript 1 and the outlet state by the subscript 2, and drop the summation signs. Then Eq. 5–20 reduces, for *single-stream steady-flow systems,* to

Steady flow (single stream): $$\dot{m}_1 = \dot{m}_2 \quad \rightarrow \quad \rho_1 V_1 A_1 = \rho_2 V_2 A_2 \tag{5-21}$$

192
BERNOULLI AND ENERGY EQUATIONS

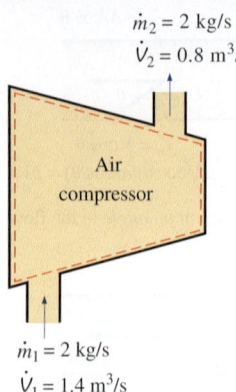

FIGURE 5–11
During a steady-flow process, volume flow rates are not necessarily conserved although mass flow rates are.

Special Case: Incompressible Flow

The conservation of mass relations can be simplified even further when the fluid is incompressible, which is usually the case for liquids. Canceling the density from both sides of the general steady-flow relation gives

Steady, incompressible flow: $\quad \sum_{in} \dot{V} = \sum_{out} \dot{V} \quad$ (m³/s) (5–22)

For single-stream steady-flow systems Eq. 5–22 becomes

Steady, incompressible flow (single stream): $\quad \dot{V}_1 = \dot{V}_2 \rightarrow V_1 A_1 = V_2 A_2$ (5–23)

It should always be kept in mind that there is no such thing as a "conservation of volume" principle. Therefore, the volume flow rates into and out of a steady-flow device may be different. The volume flow rate at the outlet of an air compressor is much less than that at the inlet even though the mass flow rate of air through the compressor is constant (Fig. 5–11). This is due to the higher density of air at the compressor exit. For steady flow of liquids, however, the volume flow rates remain nearly constant since liquids are essentially incompressible (constant-density) substances. Water flow through the nozzle of a garden hose is an example of the latter case.

The conservation of mass principle requires every bit of mass to be accounted for during a process. If you can balance your checkbook (by keeping track of deposits and withdrawals, or by simply observing the "conservation of money" principle), you should have no difficulty applying the conservation of mass principle to engineering systems.

FIGURE 5–12
Schematic for Example 5–1.
Photo by John M. Cimbala.

EXAMPLE 5–1 Water Flow through a Garden Hose Nozzle

A garden hose attached with a nozzle is used to fill a 40-L bucket. The inner diameter of the hose is 2 cm, and it reduces to 0.8 cm at the nozzle exit (Fig. 5–12). If it takes 50 s to fill the bucket with water, determine (*a*) the volume and mass flow rates of water through the hose, and (*b*) the average velocity of water at the nozzle exit.

SOLUTION A garden hose is used to fill a water bucket. The volume and mass flow rates of water and the exit velocity are to be determined.
Assumptions 1 Water is a nearly incompressible substance. 2 Flow through the hose is steady. 3 There is no waste of water by splashing.
Properties We take the density of water to be 1000 kg/m³ = 1 kg/L.
Analysis (*a*) Noting that 40 L of water are discharged in 50 s, the volume and mass flow rates of water are

$$\dot{V} = \frac{V}{\Delta t} = \frac{40 \text{ L}}{50 \text{ s}} = \mathbf{0.800 \text{ L/s}}$$

$$\dot{m} = \rho \dot{V} = (1 \text{ kg/L})(0.800 \text{ L/s}) = \mathbf{0.800 \text{ kg/s}}$$

(*b*) The cross-sectional area of the nozzle exit is

$$A_e = \pi r_e^2 = \pi (0.4 \text{ cm})^2 = 0.5027 \text{ cm}^2 = 0.5027 \times 10^{-4} \text{ m}^2$$

The volume flow rate through the hose and the nozzle is constant. Then the average velocity of water at the nozzle exit becomes

$$V_e = \frac{\dot{V}}{A_e} = \frac{0.800 \text{ L/s}}{0.5027 \times 10^{-4} \text{ m}^2}\left(\frac{1 \text{ m}^3}{1000 \text{ L}}\right) = \mathbf{15.9 \text{ m/s}}$$

Discussion It can be shown that the average velocity in the hose is 2.5 m/s. Therefore, the nozzle increases the water velocity by over six times.

■ EXAMPLE 5–2   Discharge of Water from a Tank

A 1.2-m-high, 0.9-m-diameter cylindrical water tank whose top is open to the atmosphere is initially filled with water. Now the discharge plug near the bottom of the tank is pulled out, and a water jet whose diameter is 1.3 cm streams out (Fig. 5–13). The average velocity of the jet is approximated as $V = \sqrt{2gh}$, where h is the height of water in the tank measured from the center of the hole (a variable) and g is the gravitational acceleration. Determine how long it takes for the water level in the tank to drop to 0.6 m from the bottom.

SOLUTION The plug near the bottom of a water tank is pulled out. The time it takes for half of the water in the tank to empty is to be determined.
Assumptions **1** Water is a nearly incompressible substance. **2** The distance between the bottom of the tank and the center of the hole is negligible compared to the total water height. **3** The gravitational acceleration is 9.81 m/s².
Analysis We take the volume occupied by water as the control volume. The size of the control volume decreases in this case as the water level drops, and thus this is a variable control volume. (We could also treat this as a fixed control volume that consists of the interior volume of the tank by disregarding the air that replaces the space vacated by the water.) This is obviously an unsteady-flow problem since the properties (such as the amount of mass) within the control volume change with time.

The conservation of mass relation for a control volume undergoing any process is given in rate form as

$$\dot{m}_{in} - \dot{m}_{out} = \frac{dm_{CV}}{dt} \quad (1)$$

During this process no mass enters the control volume ($\dot{m}_{in} = 0$), and the mass flow rate of discharged water is

$$\dot{m}_{out} = (\rho V A)_{out} = \rho\sqrt{2gh}A_{jet} \quad (2)$$

where $A_{jet} = \pi D_{jet}^2/4$ is the cross-sectional area of the jet, which is constant. Noting that the density of water is constant, the mass of water in the tank at any time is

$$m_{CV} = \rho V = \rho A_{tank} h \quad (3)$$

where $A_{tank} = \pi D_{tank}^2/4$ is the base area of the cylindrical tank. Substituting Eqs. 2 and 3 into the mass balance relation (Eq. 1) gives

$$-\rho\sqrt{2gh}A_{jet} = \frac{d(\rho A_{tank} h)}{dt} \rightarrow -\rho\sqrt{2gh}(\pi D_{jet}^2/4) = \frac{\rho(\pi D_{tank}^2/4)dh}{dt}$$

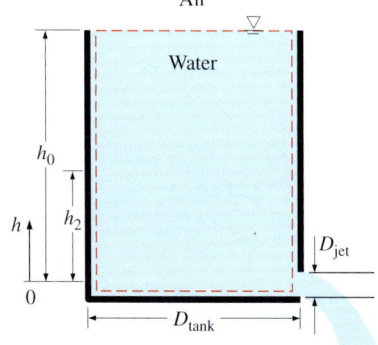

FIGURE 5–13
Schematic for Example 5–2.

BERNOULLI AND ENERGY EQUATIONS

Canceling the densities and other common terms and separating the variables give

$$dt = -\frac{D_{tank}^2}{D_{jet}^2} \frac{dh}{\sqrt{2gh}}$$

Integrating from $t = 0$ at which $h = h_0$ to $t = t$ at which $h = h_2$ gives

$$\int_0^t dt = -\frac{D_{tank}^2}{D_{jet}^2 \sqrt{2g}} \int_{h_0}^{h_2} \frac{dh}{\sqrt{h}} \rightarrow t = \frac{\sqrt{h_0} - \sqrt{h_2}}{\sqrt{g/2}} \left(\frac{D_{tank}}{D_{jet}}\right)^2$$

Substituting, the time of discharge is determined to be

$$t = \frac{\sqrt{1.2 \text{ m}} - \sqrt{0.6 \text{ m}}}{\sqrt{9.81/2 \text{ m/s}^2}} \left(\frac{0.9 \text{ m}}{0.013 \text{ m}}\right)^2 = 694 \text{ s} = \textbf{11.6 min}$$

Therefore, it takes 11.6 min after the discharge hole is unplugged for half of the tank to be emptied.

Discussion Using the same relation with $h_2 = 0$ gives $t = 39.5$ min for the discharge of the entire amount of water in the tank. Therefore, emptying the bottom half of the tank takes much longer than emptying the top half. This is due to the decrease in the average discharge velocity of water with decreasing h.

5–3 ▪ MECHANICAL ENERGY AND EFFICIENCY

Many fluid systems are designed to transport a fluid from one location to another at a specified flow rate, velocity, and elevation difference, and the system may generate mechanical work in a turbine or it may consume mechanical work in a pump or fan during this process (Fig. 5–14). These systems do not involve the conversion of nuclear, chemical, or thermal energy to mechanical energy. Also, they do not involve heat transfer in any significant amount, and they operate essentially at constant temperature. Such systems can be analyzed conveniently by considering only the *mechanical forms of energy* and the frictional effects that cause the mechanical energy to be lost (i.e., to be converted to thermal energy that usually cannot be used for any useful purpose).

The **mechanical energy** is defined as *the form of energy that can be converted to mechanical work completely and directly by an ideal mechanical device such as an ideal turbine*. Kinetic and potential energies are the familiar forms of mechanical energy. Thermal energy is not mechanical energy, however, since it cannot be converted to work directly and completely (the second law of thermodynamics).

A pump transfers mechanical energy to a fluid by raising its pressure, and a turbine extracts mechanical energy from a fluid by dropping its pressure. Therefore, the pressure of a flowing fluid is also associated with its mechanical energy. In fact, the pressure unit Pa is equivalent to Pa = N/m² = N·m/m³ = J/m³, which is energy per unit volume, and the product Pv or its equivalent P/ρ has the unit J/kg, which is energy per unit mass. Note that pressure itself is not a form of energy. But a pressure force acting on a fluid through a distance produces work, called *flow work*, in the amount of P/ρ per unit mass. Flow work is expressed in terms of fluid properties, and it is convenient to view it as part of the energy of a flowing fluid and call

FIGURE 5–14
Mechanical energy is a useful concept for flows that do not involve significant heat transfer or energy conversion, such as the flow of gasoline from an underground tank into a car.
Royalty-Free/CORBIS

it *flow energy*. Therefore, the mechanical energy of a flowing fluid can be expressed on a unit-mass basis as

$$e_{mech} = \frac{P}{\rho} + \frac{V^2}{2} + gz$$

where P/ρ is the *flow energy*, $V^2/2$ is the *kinetic energy*, and gz is the *potential energy* of the fluid, all per unit mass. Then the mechanical energy change of a fluid during incompressible flow becomes

$$\Delta e_{mech} = \frac{P_2 - P_1}{\rho} + \frac{V_2^2 - V_1^2}{2} + g(z_2 - z_1) \quad (kJ/kg) \quad (5\text{--}24)$$

Therefore, the mechanical energy of a fluid does not change during flow if its pressure, density, velocity, and elevation remain constant. In the absence of any irreversible losses, the mechanical energy change represents the mechanical work supplied to the fluid (if $\Delta e_{mech} > 0$) or extracted from the fluid (if $\Delta e_{mech} < 0$). The maximum (ideal) power generated by a turbine, for example, is $\dot{W}_{max} = \dot{m} \Delta e_{mech}$, as shown in Fig. 5–15.

Consider a container of height h filled with water, as shown in Fig. 5–16, with the reference level selected at the bottom surface. The gage pressure and the potential energy per unit mass are, respectively, $P_{gage,\,A} = 0$ and $pe_A = gh$ at point A at the free surface, and $P_{gage,\,B} = \rho gh$ and $pe_B = 0$ at point B at the bottom of the container. An ideal hydraulic turbine at the bottom elevation would produce the same work per unit mass $w_{turbine} = gh$ whether it receives water (or any other fluid with constant density) from the top or from the bottom of the container. Note that we are assuming ideal flow (no irreversible losses) through the pipe leading from the tank to the turbine and negligible kinetic energy at the turbine outlet. Therefore, the total available mechanical energy of water at the bottom is equivalent to that at the top.

The transfer of mechanical energy is usually accomplished by a rotating shaft, and thus mechanical work is often referred to as *shaft work*. A pump or a fan receives shaft work (usually from an electric motor) and transfers it to the fluid as mechanical energy (less frictional losses). A turbine, on the other hand, converts the mechanical energy of a fluid to shaft work. Because of irreversibilities such as friction, mechanical energy cannot be converted entirely from one mechanical form to another, and the **mechanical efficiency** of a device or process is defined as

$$\eta_{mech} = \frac{\text{Mechanical energy output}}{\text{Mechanical energy input}} = \frac{E_{mech,\,out}}{E_{mech,\,in}} = 1 - \frac{E_{mech,\,loss}}{E_{mech,\,in}} \quad (5\text{--}25)$$

A conversion efficiency of less than 100 percent indicates that conversion is less than perfect and some losses have occurred during conversion. A mechanical efficiency of 74 percent indicates that 26 percent of the mechanical energy input is converted to thermal energy as a result of frictional heating (Fig 5–17), and this manifests itself as a slight rise in the temperature of the fluid.

In fluid systems, we are usually interested in increasing the pressure, velocity, and/or elevation of a fluid. This is done by *supplying mechanical energy* to the fluid by a pump, a fan, or a compressor (we refer to all of them as pumps). Or we are interested in the reverse process of *extracting mechanical*

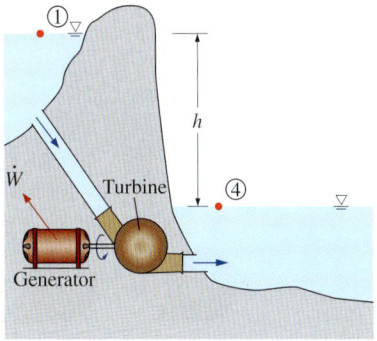

$\dot{W}_{max} = \dot{m}\Delta e_{mech} = \dot{m}g(z_1 - z_4) = \dot{m}gh$
since $P_1 \approx P_4 = P_{atm}$ and $V_1 = V_4 \approx 0$
(a)

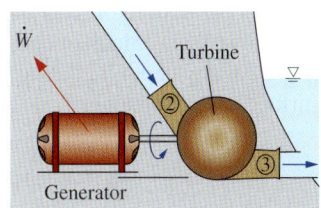

$\dot{W}_{max} = \dot{m}\Delta e_{mech} = \dot{m}\dfrac{P_2 - P_3}{\rho} = \dot{m}\dfrac{\Delta P}{\rho}$
since $V_2 \approx V_3$ and $z_2 \approx z_3$
(b)

FIGURE 5–15
Mechanical energy is illustrated by an ideal hydraulic turbine coupled with an ideal generator. In the absence of irreversible losses, the maximum produced power is proportional to (a) the change in water surface elevation from the upstream to the downstream reservoir or (b) (close-up view) the drop in water pressure from just upstream to just downstream of the turbine.

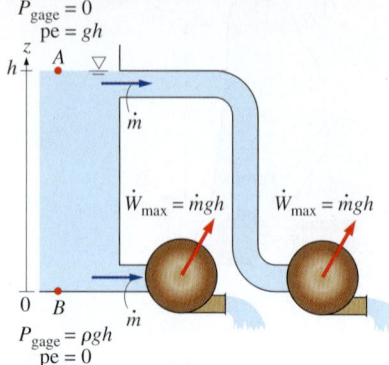

FIGURE 5–16
The available mechanical energy of water at the bottom of a container is equal to the available mechanical energy at any depth including the free surface of the container.

energy from a fluid by a turbine and producing mechanical power in the form of a rotating shaft that can drive a generator or any other rotary device. The degree of perfection of the conversion process between the mechanical work supplied or extracted and the mechanical energy of the fluid is expressed by the **pump efficiency** and **turbine efficiency.** In rate form, these are defined as

$$\eta_{pump} = \frac{\text{Mechanical power increase of the fluid}}{\text{Mechanical power input}} = \frac{\Delta \dot{E}_{mech, fluid}}{\dot{W}_{shaft, in}} = \frac{\dot{W}_{pump, u}}{\dot{W}_{pump}} \quad (5\text{--}26)$$

where $\Delta \dot{E}_{mech, fluid} = \dot{E}_{mech, out} - \dot{E}_{mech, in}$ is the rate of increase in the mechanical energy of the fluid, which is equivalent to the **useful pumping power** $\dot{W}_{pump, u}$ supplied to the fluid, and

$$\eta_{turbine} = \frac{\text{Mechanical power output}}{\text{Mechanical power decrease of the fluid}} = \frac{\dot{W}_{shaft, out}}{|\Delta \dot{E}_{mech, fluid}|} = \frac{\dot{W}_{turbine}}{\dot{W}_{turbine, e}} \quad (5\text{--}27)$$

where $|\Delta \dot{E}_{mech, fluid}| = \dot{E}_{mech, in} - \dot{E}_{mech, out}$ is the rate of decrease in the mechanical energy of the fluid, which is equivalent to the mechanical power extracted from the fluid by the turbine $\dot{W}_{turbine, e}$, and we use the absolute value sign to avoid negative values for efficiencies. A pump or turbine efficiency of 100 percent indicates perfect conversion between the shaft work and the mechanical energy of the fluid, and this value can be approached (but never attained) as the frictional effects are minimized.

The mechanical efficiency should not be confused with the **motor efficiency** and the **generator efficiency**, which are defined as

Motor:
$$\eta_{motor} = \frac{\text{Mechanical power output}}{\text{Electric power input}} = \frac{\dot{W}_{shaft, out}}{\dot{W}_{elect, in}} \quad (5\text{--}28)$$

and

Generator:
$$\eta_{generator} = \frac{\text{Electric power output}}{\text{Mechanical power input}} = \frac{\dot{W}_{elect, out}}{\dot{W}_{shaft, in}} \quad (5\text{--}29)$$

A pump is usually packaged together with its motor, and a turbine with its generator. Therefore, we are usually interested in the **combined** or **overall efficiency** of pump–motor and turbine–generator combinations (Fig. 5–18), which are defined as

$$\eta_{pump\text{-}motor} = \eta_{pump}\,\eta_{motor} = \frac{\dot{W}_{pump, u}}{\dot{W}_{elect, in}} = \frac{\Delta \dot{E}_{mech, fluid}}{\dot{W}_{elect, in}} \quad (5\text{--}30)$$

and

$$\eta_{turbine\text{-}gen} = \eta_{turbine}\,\eta_{generator} = \frac{\dot{W}_{elect, out}}{\dot{W}_{turbine, e}} = \frac{\dot{W}_{elect, out}}{|\Delta \dot{E}_{mech, fluid}|} \quad (5\text{--}31)$$

All the efficiencies just defined range between 0 and 100 percent. The lower limit of 0 percent corresponds to the conversion of the entire mechanical or electric energy input to thermal energy, and the device in this case functions like a resistance heater. The upper limit of 100 percent corresponds to the case of perfect conversion with no friction or other irreversibilities, and thus no conversion of mechanical or electric energy to thermal energy (no losses).

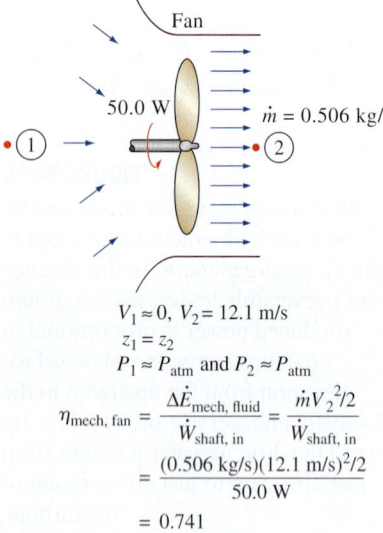

FIGURE 5–17
The mechanical efficiency of a fan is the ratio of the rate of increase of the mechanical energy of the air to the mechanical power input.

EXAMPLE 5–3 Performance of a Hydraulic Turbine–Generator

The water in a large lake is to be used to generate electricity by the installation of a hydraulic turbine–generator. The elevation difference between the free surfaces upstream and downstream of the dam is 50 m (Fig. 5–19). Water is to be supplied at a rate of 5000 kg/s. If the electric power generated is measured to be 1862 kW and the generator efficiency is 95 percent, determine (a) the overall efficiency of the turbine–generator, (b) the mechanical efficiency of the turbine, and (c) the shaft power supplied by the turbine to the generator.

SOLUTION A hydraulic turbine–generator is to generate electricity from the water of a lake. The overall efficiency, the turbine efficiency, and the shaft power are to be determined.
Assumptions 1 The elevation of the lake and that of the discharge site remain constant. 2 Irreversible losses in the pipes are negligible.
Properties The density of water is taken to be $\rho = 1000$ kg/m^3.
Analysis (a) We perform our analysis from inlet (1) at the free surface of the lake to outlet (2) at the free surface of the downstream discharge site. At both free surfaces the pressure is atmospheric and the velocity is negligibly small. The change in the water's mechanical energy per unit mass is then

$$e_{\text{mech, in}} - e_{\text{mech, out}} = \underbrace{\frac{P_{\text{in}} - P_{\text{out}}}{\rho}}_{0} + \underbrace{\frac{V_{\text{in}}^2 - V_{\text{out}}^2}{2}}_{0} + g(z_{\text{in}} - z_{\text{out}})$$

$$= gh$$

$$= (9.81 \text{ m/s}^2)(50 \text{ m})\left(\frac{1 \text{ kJ/kg}}{1000 \text{ m}^2/\text{s}^2}\right) = 0.491 \frac{\text{kJ}}{\text{kg}}$$

Then the rate at which mechanical energy is supplied to the turbine by the fluid and the overall efficiency become

$$|\Delta \dot{E}_{\text{mech, fluid}}| = \dot{m}(e_{\text{mech, in}} - e_{\text{mech, out}}) = (5000 \text{ kg/s})(0.491 \text{ kJ/kg}) = 2455 \text{ kW}$$

$$\eta_{\text{overall}} = \eta_{\text{turbine-gen}} = \frac{\dot{W}_{\text{elect, out}}}{|\Delta \dot{E}_{\text{mech, fluid}}|} = \frac{1862 \text{ kW}}{2455 \text{ kW}} = \mathbf{0.760}$$

(b) Knowing the overall and generator efficiencies, the mechanical efficiency of the turbine is determined from

$$\eta_{\text{turbine-gen}} = \eta_{\text{turbine}} \eta_{\text{generator}} \rightarrow \eta_{\text{turbine}} = \frac{\eta_{\text{turbine-gen}}}{\eta_{\text{generator}}} = \frac{0.76}{0.95} = \mathbf{0.800}$$

(c) The shaft power output is determined from the definition of mechanical efficiency,

$$\dot{W}_{\text{shaft, out}} = \eta_{\text{turbine}} |\Delta \dot{E}_{\text{mech, fluid}}| = (0.800)(2455 \text{ kW}) = 1964 \text{ kW} \approx \mathbf{1960 \text{ kW}}$$

Discussion Note that the lake supplies 2455 kW of mechanical power to the turbine, which converts 1964 kW of it to shaft power that drives the generator, which generates 1862 kW of electric power. There are irreversible losses through each component. Irreversible losses in the pipes are ignored here; you will learn how to account for these in Chap. 8.

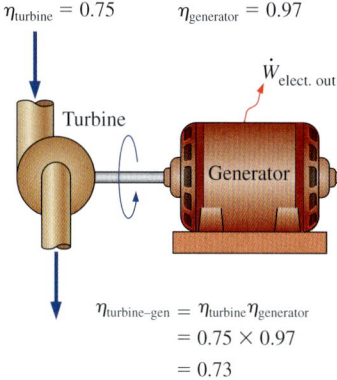

FIGURE 5–18
The overall efficiency of a turbine–generator is the product of the efficiency of the turbine and the efficiency of the generator, and represents the fraction of the mechanical power of the fluid converted to electrical power.

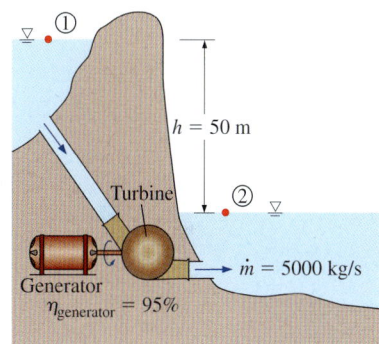

FIGURE 5–19
Schematic for Example 5–3.

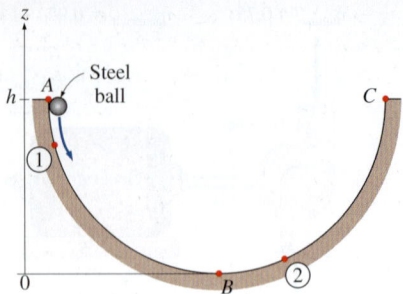

FIGURE 5–20
Schematic for Example 5–4.

EXAMPLE 5–4 Conservation of Energy for an Oscillating Steel Ball

The motion of a steel ball in a hemispherical bowl of radius h shown in Fig. 5–20 is to be analyzed. The ball is initially held at the highest location at point A, and then it is released. Obtain relations for the conservation of energy of the ball for the cases of frictionless and actual motions.

SOLUTION A steel ball is released in a bowl. Relations for the energy balance are to be obtained.
Assumptions For the frictionless case, friction between the ball, the bowl, and the air is negligible.
Analysis When the ball is released, it accelerates under the influence of gravity, reaches a maximum velocity (and minimum elevation) at point B at the bottom of the bowl, and moves up toward point C on the opposite side. In the ideal case of frictionless motion, the ball will oscillate between points A and C. The actual motion involves the conversion of the kinetic and potential energies of the ball to each other, together with overcoming resistance to motion due to friction (doing frictional work). The general energy balance for any system undergoing any process is

$$\underbrace{E_{in} - E_{out}}_{\text{Net energy transfer by heat, work, and mass}} = \underbrace{\Delta E_{system}}_{\text{Change in internal, kinetic, potential, etc., energies}}$$

Then the energy balance (per unit mass) for the ball for a process from point 1 to point 2 becomes

$$-w_{friction} = (ke_2 + pe_2) - (ke_1 + pe_1)$$

or

$$\frac{V_1^2}{2} + gz_1 = \frac{V_2^2}{2} + gz_2 + w_{friction}$$

since there is no energy transfer by heat or mass and no change in the internal energy of the ball (the heat generated by frictional heating is dissipated to the surrounding air). The frictional work term $w_{friction}$ is often expressed as e_{loss} to represent the loss (conversion) of mechanical energy into thermal energy.

For the idealized case of frictionless motion, the last relation reduces to

$$\frac{V_1^2}{2} + gz_1 = \frac{V_2^2}{2} + gz_2 \quad \text{or} \quad \frac{V^2}{2} + gz = C = \text{constant}$$

where the value of the constant is $C = gh$. That is, *when the frictional effects are negligible, the sum of the kinetic and potential energies of the ball remains constant.*
Discussion This is certainly a more intuitive and convenient form of the conservation of energy equation for this and other similar processes such as the swinging motion of a pendulum. The relation obtained is analogous to the Bernoulli equation derived in Section 5–4.

Most processes encountered in practice involve only certain forms of energy, and in such cases it is more convenient to work with the simplified versions of the energy balance. For systems that involve only *mechanical forms of energy* and its transfer as *shaft work*, the conservation of energy principle can be expressed conveniently as

$$E_{\text{mech, in}} - E_{\text{mech, out}} = \Delta E_{\text{mech, system}} + E_{\text{mech, loss}} \quad (5\text{–}32)$$

where $E_{\text{mech, loss}}$ represents the conversion of mechanical energy to thermal energy due to irreversibilities such as friction. For a system in steady operation, the rate of mechanical energy balance becomes $\dot{E}_{\text{mech, in}} = \dot{E}_{\text{mech, out}} + \dot{E}_{\text{mech, loss}}$ (Fig. 5–21).

5–4 ■ THE BERNOULLI EQUATION

The **Bernoulli equation** is *an approximate relation between pressure, velocity, and elevation,* and is valid in *regions of steady, incompressible flow where net frictional forces are negligible* (Fig. 5–22). Despite its simplicity, it has proven to be a very powerful tool in fluid mechanics. In this section, we derive the Bernoulli equation by applying the *conservation of linear momentum principle*, and we demonstrate both its usefulness and its limitations.

The key approximation in the derivation of the Bernoulli equation is that *viscous effects are negligibly small compared to inertial, gravitational, and pressure effects.* Since all fluids have viscosity (there is no such thing as an "inviscid fluid"), this approximation cannot be valid for an entire flow field of practical interest. In other words, we cannot apply the Bernoulli equation *everywhere* in a flow, no matter how small the fluid's viscosity. However, it turns out that the approximation *is* reasonable in certain *regions* of many practical flows. We refer to such regions as *inviscid regions of flow*, and we stress that they are *not* regions where the fluid itself is inviscid or frictionless, but rather they are regions where net viscous or frictional forces are negligibly small compared to other forces acting on fluid particles.

Care must be exercised when applying the Bernoulli equation since it is an approximation that applies only to inviscid regions of flow. In general, frictional effects are always important very close to solid walls (*boundary layers*) and directly downstream of bodies (*wakes*). Thus, the Bernoulli approximation is typically useful in flow regions outside of boundary layers and wakes, where the fluid motion is governed by the combined effects of pressure and gravity forces.

Acceleration of a Fluid Particle

The motion of a particle and the path it follows are described by the *velocity vector* as a function of time and space coordinates and the initial position of the particle. When the flow is *steady* (no change with time at a specified location), all particles that pass through the same point follow the same path (which is the *streamline*), and the velocity vectors remain tangent to the path at every point.

Often it is convenient to describe the motion of a particle in terms of its distance *s* along a streamline together with the radius of curvature along the streamline. The speed of the particle is related to the distance by $V = ds/dt$, which may vary along the streamline. In two-dimensional flow, the acceleration can be decomposed into two components: *streamwise acceleration* a_s along the streamline and *normal acceleration* a_n in the direction normal to the streamline, which is given as $a_n = V^2/R$. Note that streamwise

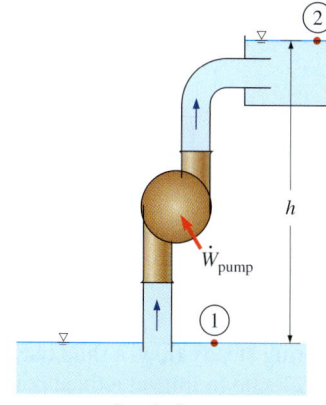

Steady flow
$V_1 = V_2 \approx 0$
$z_2 = z_1 + h$
$P_1 = P_2 = P_{\text{atm}}$

$\dot{E}_{\text{mech, in}} = \dot{E}_{\text{mech, out}} + \dot{E}_{\text{mech, loss}}$
$\dot{W}_{\text{pump}} + \dot{m}gz_1 = \dot{m}gz_2 + \dot{E}_{\text{mech, loss}}$
$\dot{W}_{\text{pump}} = \dot{m}gh + \dot{E}_{\text{mech, loss}}$

FIGURE 5–21
Many fluid flow problems involve mechanical forms of energy only, and such problems are conveniently solved by using a rate of *mechanical energy* balance.

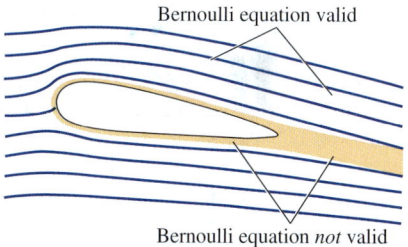

FIGURE 5–22
The *Bernoulli equation* is an *approximate* equation that is valid only in *inviscid regions of flow* where net viscous forces are negligibly small compared to inertial, gravitational, or pressure forces. Such regions occur outside of *boundary layers* and *wakes*.

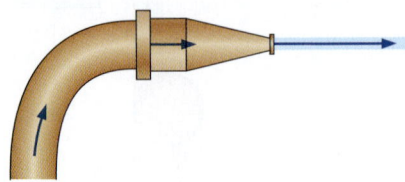

FIGURE 5–23
During steady flow, a fluid may not accelerate in time at a fixed point, but it may accelerate in space.

acceleration is due to a change in speed along a streamline, and normal acceleration is due to a change in direction. For particles that move along a *straight path*, $a_n = 0$ since the radius of curvature is infinity and thus there is no change in direction. The Bernoulli equation results from a force balance along a streamline.

One may be tempted to think that acceleration is zero in steady flow since acceleration is the rate of change of velocity with time, and in steady flow there is no change with time. Well, a garden hose nozzle tells us that this understanding is not correct. Even in steady flow and thus constant mass flow rate, water accelerates through the nozzle (Fig. 5–23 as discussed in Chap. 4). *Steady* simply means *no change with time at a specified location*, but the value of a quantity may change from one location to another. In the case of a nozzle, the velocity of water remains constant at a specified point, but it changes from the inlet to the exit (water accelerates along the nozzle).

Mathematically, this can be expressed as follows: We take the velocity V of a fluid particle to be a function of s and t. Taking the total differential of $V(s, t)$ and dividing both sides by dt yield

$$dV = \frac{\partial V}{\partial s} ds + \frac{\partial V}{\partial t} dt \quad \text{and} \quad \frac{dV}{dt} = \frac{\partial V}{\partial s} \frac{ds}{dt} + \frac{\partial V}{\partial t} \quad (5\text{-}33)$$

In steady flow $\partial V/\partial t = 0$ and thus $V = V(s)$, and the acceleration in the *s*-direction becomes

$$a_s = \frac{dV}{dt} = \frac{\partial V}{\partial s} \frac{ds}{dt} = \frac{\partial V}{\partial s} V = V \frac{dV}{ds} \quad (5\text{-}34)$$

where $V = ds/dt$ if we are following a fluid particle as it moves along a streamline. Therefore, acceleration in steady flow is due to the change of velocity with position.

Derivation of the Bernoulli Equation

Consider the motion of a fluid particle in a flow field in steady flow. Applying Newton's second law (which is referred to as the *linear momentum equation* in fluid mechanics) in the *s*-direction on a particle moving along a streamline gives

$$\sum F_s = ma_s \quad (5\text{-}35)$$

In regions of flow where net frictional forces are negligible, there is no pump or turbine, and there is no heat transfer along the streamline, the significant forces acting in the *s*-direction are the pressure (acting on both sides) and the component of the weight of the particle in the *s*-direction (Fig. 5–24). Therefore, Eq. 5–35 becomes

$$P\, dA - (P + dP)\, dA - W \sin\theta = mV \frac{dV}{ds} \quad (5\text{-}36)$$

where θ is the angle between the normal of the streamline and the vertical *z*-axis at that point, $m = \rho \mathcal{V} = \rho\, dA\, ds$ is the mass, $W = mg = \rho g\, dA\, ds$ is the weight of the fluid particle, and $\sin\theta = dz/ds$. Substituting,

$$-dP\, dA - \rho g\, dA\, ds \frac{dz}{ds} = \rho\, dA\, ds\, V \frac{dV}{ds} \quad (5\text{-}37)$$

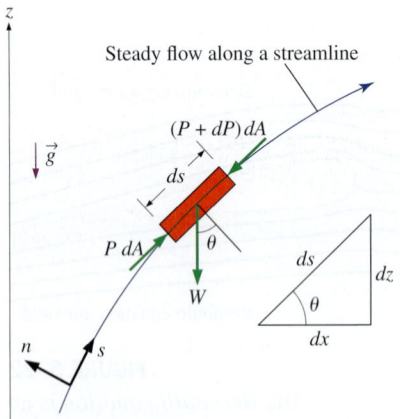

FIGURE 5–24
The forces acting on a fluid particle along a streamline.

Canceling dA from each term and simplifying,

$$-dP - \rho g\, dz = \rho V\, dV \qquad (5\text{--}38)$$

Noting that $V\, dV = \tfrac{1}{2} d(V^2)$ and dividing each term by ρ gives

$$\frac{dP}{\rho} + \frac{1}{2} d(V^2) + g\, dz = 0 \qquad (5\text{--}39)$$

Integrating,

Steady flow: $\qquad \displaystyle\int \frac{dP}{\rho} + \frac{V^2}{2} + gz = \text{constant (along a streamline)} \qquad (5\text{--}40)$

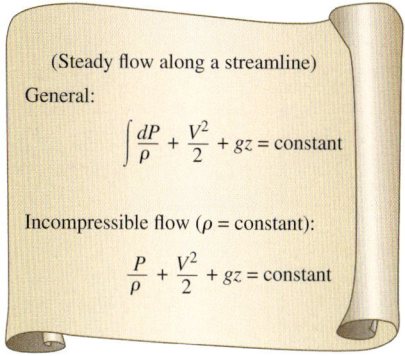

FIGURE 5–25
The incompressible Bernoulli equation is derived assuming incompressible flow, and thus it should not be used for flows with significant compressibility effects.

since the last two terms are exact differentials. In the case of incompressible flow, the first term also becomes an exact differential, and integration gives

Steady, incompressible flow: $\quad \dfrac{P}{\rho} + \dfrac{V^2}{2} + gz = \text{constant (along a streamline)} \quad (5\text{--}41)$

This is the famous **Bernoulli equation** (Fig. 5–25), which is commonly used in fluid mechanics for steady, incompressible flow along a streamline in inviscid regions of flow. The Bernoulli equation was first stated in words by the Swiss mathematician Daniel Bernoulli (1700–1782) in a text written in 1738 when he was working in St. Petersburg, Russia. It was later derived in equation form by his associate Leonhard Euler (1707–1783) in 1755.

The value of the constant in Eq. 5–41 can be evaluated at any point on the streamline where the pressure, density, velocity, and elevation are known. The Bernoulli equation can also be written between any two points on the same streamline as

Steady, incompressible flow: $\quad \dfrac{P_1}{\rho} + \dfrac{V_1^2}{2} + gz_1 = \dfrac{P_2}{\rho} + \dfrac{V_2^2}{2} + gz_2 \quad (5\text{--}42)$

We recognize $V^2/2$ as *kinetic energy*, gz as *potential energy*, and P/ρ as *flow energy*, all per unit mass. Therefore, the Bernoulli equation can be viewed as an expression of *mechanical energy balance* and can be stated as follows (Fig. 5–26):

> The sum of the kinetic, potential, and flow energies of a fluid particle is constant along a streamline during steady flow when compressibility and frictional effects are negligible.

The kinetic, potential, and flow energies are the mechanical forms of energy, as discussed in Section 5–3, and the Bernoulli equation can be viewed as the "conservation of mechanical energy principle." This is equivalent to the general conservation of energy principle for systems that do not involve any conversion of mechanical energy and thermal energy to each other, and thus the mechanical energy and thermal energy are conserved separately. The Bernoulli equation states that during steady, incompressible flow with negligible friction, the various forms of mechanical energy are converted to each other, but their sum remains constant. In other words, there is no dissipation of mechanical energy during such flows since there is no friction that converts mechanical energy to sensible thermal (internal) energy.

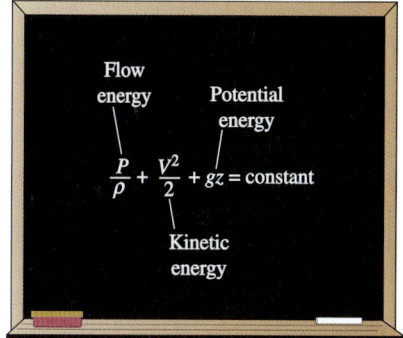

FIGURE 5–26
The Bernoulli equation states that the sum of the kinetic, potential, and flow energies (all per unit mass) of a fluid particle is constant along a streamline during steady flow.

Recall that energy is transferred to a system as work when a force is applied to the system through a distance. In the light of Newton's second law of motion, the Bernoulli equation can also be viewed as: *The work done by the pressure and gravity forces on the fluid particle is equal to the increase in the kinetic energy of the particle.*

The Bernoulli equation is obtained from Newton's second law for a fluid particle moving along a streamline. It can also be obtained from the *first law of thermodynamics* applied to a steady-flow system, as shown in Section 5–6.

Despite the highly restrictive approximations used in its derivation, the Bernoulli equation is commonly used in practice since a variety of practical fluid flow problems can be analyzed to reasonable accuracy with it. This is because many flows of practical engineering interest are steady (or at least steady in the mean), compressibility effects are relatively small, and net frictional forces are negligible in some regions of interest in the flow.

Force Balance across Streamlines

It is left as an exercise to show that a force balance in the direction n normal to the streamline yields the following relation applicable *across* the streamlines for steady, incompressible flow:

$$\frac{P}{\rho} + \int \frac{V^2}{R} dn + gz = \text{constant} \quad \text{(across streamlines)} \quad (5\text{--}43)$$

where R is the local radius of curvature of the streamline. For flow along curved streamlines (Fig 5–27a), the pressure *decreases* towards the center of curvature, and fluid particles experience a corresponding centripetal force and centripetal acceleration due to this pressure gradient.

For flow along a straight line, $R \to \infty$ and Eq. 5–43 reduces to $P/\rho + gz = $ constant or $P = -\rho gz + $ constant, which is an expression for the variation of hydrostatic pressure with vertical distance for a stationary fluid body. Therefore, the variation of pressure with elevation in steady, incompressible flow along a straight line in an inviscid region of flow is the same as that in the stationary fluid (Fig. 5–27b).

Unsteady, Compressible Flow

Similarly, using both terms in the acceleration expression (Eq. 5–33), it can be shown that the Bernoulli equation for *unsteady, compressible flow* is

Unsteady, compressible flow: $\quad \int \frac{dP}{\rho} + \int \frac{\partial V}{\partial t} ds + \frac{V^2}{2} + gz = \text{constant} \quad (5\text{--}44)$

Static, Dynamic, and Stagnation Pressures

The Bernoulli equation states that the sum of the flow, kinetic, and potential energies of a fluid particle along a streamline is constant. Therefore, the kinetic and potential energies of the fluid can be converted to flow energy (and vice versa) during flow, causing the pressure to change. This phenomenon can be made more visible by multiplying the Bernoulli equation by the density ρ,

$$P + \rho \frac{V^2}{2} + \rho gz = \text{constant (along a streamline)} \quad (5\text{--}45)$$

Each term in this equation has pressure units, and thus each term represents some kind of pressure:

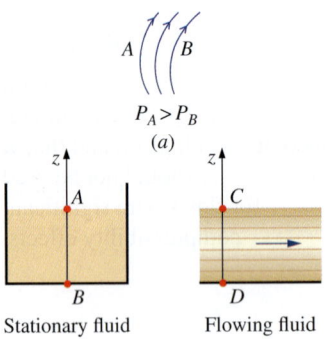

FIGURE 5–27
Pressure decreases towards the center of curvature when streamlines are curved (*a*), but the variation of pressure with elevation in steady, incompressible flow along a straight line (*b*) is the same as that in stationary fluid.

- P is the **static pressure** (it does not incorporate any dynamic effects); it represents the actual thermodynamic pressure of the fluid. This is the same as the pressure used in thermodynamics and property tables.
- $\rho V^2/2$ is the **dynamic pressure**; it represents the pressure rise when the fluid in motion is brought to a stop isentropically.
- $\rho g z$ is the **hydrostatic pressure** term, which is not pressure in a real sense since its value depends on the reference level selected; it accounts for the elevation effects, i.e., fluid weight on pressure. (Be careful of the sign—unlike hydrostatic pressure $\rho g h$ which *increases* with fluid depth h, the hydrostatic pressure term $\rho g z$ *decreases* with fluid depth.)

The sum of the static, dynamic, and hydrostatic pressures is called the **total pressure**. Therefore, the Bernoulli equation states that *the total pressure along a streamline is constant.*

The sum of the static and dynamic pressures is called the **stagnation pressure**, and it is expressed as

$$P_{\text{stag}} = P + \rho \frac{V^2}{2} \quad \text{(kPa)} \tag{5–46}$$

The stagnation pressure represents the pressure at a point where the fluid is brought to a complete stop isentropically. The static, dynamic, and stagnation pressures are shown in Fig. 5–28. When static and stagnation pressures are measured at a specified location, the fluid velocity at that location is calculated from

$$V = \sqrt{\frac{2(P_{\text{stag}} - P)}{\rho}} \tag{5–47}$$

Equation 5–47 is useful in the measurement of flow velocity when a combination of a static pressure tap and a Pitot tube is used, as illustrated in Fig. 5–28. A **static pressure tap** is simply a small hole drilled into a wall such that the plane of the hole is parallel to the flow direction. It measures the static pressure. A **Pitot tube** is a small tube with its open end aligned *into* the flow so as to sense the full impact pressure of the flowing fluid. It measures the stagnation pressure. In situations in which the static and stagnation pressure of a flowing *liquid* are greater than atmospheric pressure, a vertical transparent tube called a **piezometer tube** (or simply a **piezometer**) can be attached to the pressure tap and to the Pitot tube, as sketched in Fig. 5–28. The liquid rises in the piezometer tube to a column height (*head*) that is proportional to the pressure being measured. If the pressures to be measured are below atmospheric, or if measuring pressures in *gases*, piezometer tubes do not work. However, the static pressure tap and Pitot tube can still be used, but they must be connected to some other kind of pressure measurement device such as a U-tube manometer or a pressure transducer (Chap. 3). Sometimes it is convenient to integrate static pressure holes on a Pitot probe. The result is a **Pitot-static probe** (also called a **Pitot-Darcy probe**), as shown in Fig. 5–29 and discussed in more detail in Chap. 8. A Pitot-static probe connected to a pressure transducer or a manometer measures the dynamic pressure (and thus infers the fluid velocity) directly.

When the static pressure is measured by drilling a hole in the tube wall, care must be exercised to ensure that the opening of the hole is flush with the wall surface, with no extrusions before or after the hole (Fig. 5–30). Otherwise the reading would incorporate some dynamic effects, and thus it would be in error.

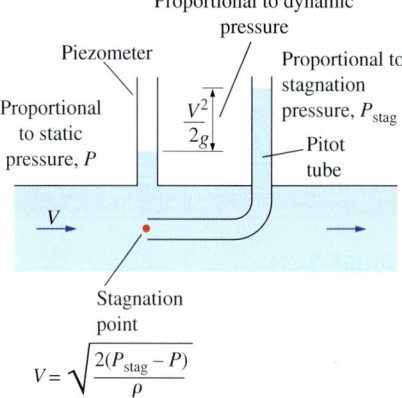

FIGURE 5–28
The static, dynamic, and stagnation pressures measured using piezometer tubes.

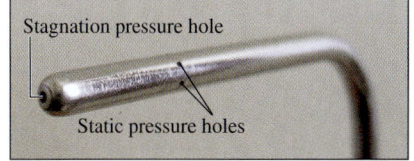

FIGURE 5–29
Close-up of a Pitot-static probe, showing the stagnation pressure hole and two of the five static circumferential pressure holes.
Photo by Po-Ya Abel Chuang. Used by permission.

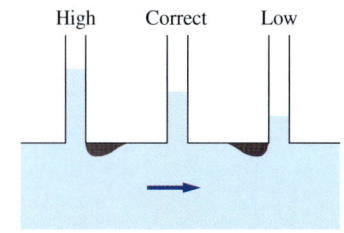

FIGURE 5–30
Careless drilling of the static pressure tap may result in an erroneous reading of the static pressure head.

When a stationary body is immersed in a flowing stream, the fluid is brought to a stop at the nose of the body (the **stagnation point**). The flow streamline that extends from far upstream to the stagnation point is called the **stagnation streamline** (Fig. 5–31). For a two-dimensional flow in the *xy*-plane, the stagnation point is actually a *line* parallel to the *z*-axis, and the stagnation streamline is actually a *surface* that separates fluid that flows *over* the body from fluid that flows *under* the body. In an incompressible flow, the fluid decelerates nearly isentropically from its free-stream velocity to zero at the stagnation point, and the pressure at the stagnation point is thus the stagnation pressure.

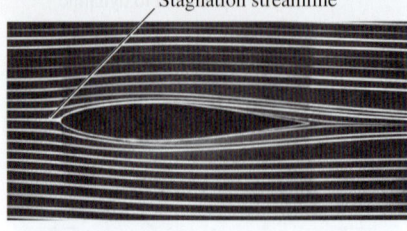

FIGURE 5–31
Streaklines produced by colored fluid introduced upstream of an airfoil; since the flow is steady, the streaklines are the same as streamlines and pathlines. The stagnation streamline is marked.

Courtesy ONERA. Photograph by Werlé.

Limitations on the Use of the Bernoulli Equation

The Bernoulli equation (Eq. 5–41) is one of the most frequently used and *misused* equations in fluid mechanics. Its versatility, simplicity, and ease of use make it a very valuable tool for use in analysis, but the same attributes also make it very tempting to misuse. Therefore, it is important to understand the restrictions on its applicability and observe the limitations on its use, as explained here:

1. **Steady flow**   The first limitation on the Bernoulli equation is that it is applicable to *steady flow*. Therefore, it should not be used during the transient start-up and shut-down periods, or during periods of change in the flow conditions. Note that there is an unsteady form of the Bernoulli equation (Eq. 5–44), discussion of which is beyond the scope of the present text (see Panton, 2005).

2. **Negligible viscous effects**   Every flow involves some friction, no matter how small, and *frictional effects* may or may not be negligible. The situation is complicated even more by the amount of error that can be tolerated. In general, frictional effects are negligible for short flow sections with large cross sections, especially at low flow velocities. Frictional effects are usually significant in long and narrow flow passages, in the wake region downstream of an object, and in *diverging flow sections* such as diffusers because of the increased possibility of the fluid separating from the walls in such geometries. Frictional effects are also significant near solid surfaces, and thus the Bernoulli equation is usually applicable along a streamline in the core region of the flow, but not along a streamline close to the surface (Fig. 5–32).

   A component that disturbs the streamlined structure of flow and thus causes considerable mixing and backflow such as a sharp entrance of a tube or a partially closed valve in a flow section can make the Bernoulli equation inapplicable.

3. **No shaft work**   The Bernoulli equation was derived from a force balance on a particle moving along a streamline. Therefore, the Bernoulli equation is not applicable in a flow section that involves a pump, turbine, fan, or any other machine or impeller since such devices disrupt the streamlines and carry out energy interactions with the fluid particles. When the flow section considered involves any of these devices, the energy equation should be used instead to account for the shaft work input or output. However, the Bernoulli equation can still be applied to a flow section prior to or past a machine (assuming, of course, that the other

restrictions on its use are satisfied). In such cases, the Bernoulli constant changes from upstream to downstream of the device.

4. **Incompressible flow** One of the approximations used in the derivation of the Bernoulli equation is that ρ = constant and thus the flow is incompressible. This condition is satisfied by liquids and also by gases at Mach numbers less than about 0.3 since compressibility effects and thus density variations of gases are negligible at such relatively low velocities. Note that there is a compressible form of the Bernoulli equation (Eqs. 5–40 and 5–44).

5. **Negligible heat transfer** The density of a gas is inversely proportional to temperature, and thus the Bernoulli equation should not be used for flow sections that involve significant temperature change such as heating or cooling sections.

6. **Flow along a streamline** Strictly speaking, the Bernoulli equation $P/\rho + V^2/2 + gz = C$ is applicable along a streamline, and the value of the constant C is generally different for different streamlines. However, when a region of the flow is *irrotational* and there is no *vorticity* in the flow field, the value of the constant C remains the same for all streamlines, and the Bernoulli equation becomes applicable *across* streamlines as well (Fig. 5–33). Therefore, we do not need to be concerned about the streamlines when the flow is irrotational, and we can apply the Bernoulli equation between any two points in the irrotational region of the flow (Chap. 10).

We derived the Bernoulli equation by considering two-dimensional flow in the *xz*-plane for simplicity, but the equation is valid for general three-dimensional flow as well, as long as it is applied along the same streamline. We should always keep in mind the approximations used in the derivation of the Bernoulli equation and make sure that they are valid before applying it.

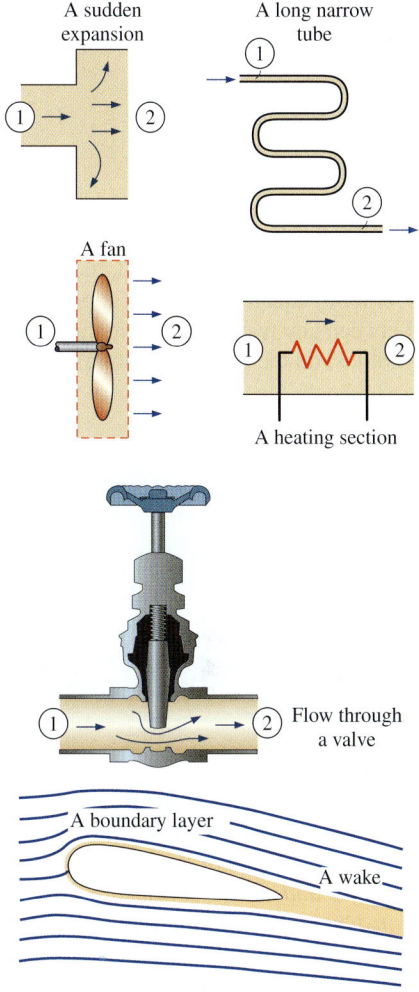

FIGURE 5–32
Frictional effects, heat transfer, and components that disturb the streamlined structure of flow make the Bernoulli equation invalid. It should *not* be used in any of the flows shown here.

Hydraulic Grade Line (HGL) and Energy Grade Line (EGL)

It is often convenient to represent the level of mechanical energy graphically using *heights* to facilitate visualization of the various terms of the Bernoulli equation. This is done by dividing each term of the Bernoulli equation by g to give

$$\frac{P}{\rho g} + \frac{V^2}{2g} + z = H = \text{constant} \quad \text{(along a streamline)} \quad (5\text{–}48)$$

Each term in this equation has the dimension of length and represents some kind of "head" of a flowing fluid as follows:

- $P/\rho g$ is the **pressure head**; it represents the height of a fluid column that produces the static pressure P.
- $V^2/2g$ is the **velocity head**; it represents the elevation needed for a fluid to reach the velocity V during frictionless free fall.
- z is the **elevation head**; it represents the potential energy of the fluid.

Also, H is the **total head** for the flow. Therefore, the Bernoulli equation is expressed in terms of heads as: *The sum of the pressure, velocity, and elevation heads along a streamline is constant during steady flow when compressibility and frictional effects are negligible* (Fig. 5–34).

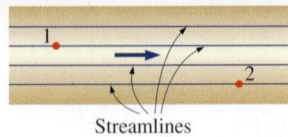

$$\frac{P_1}{\rho} + \frac{V_1^2}{2} + gz_1 = \frac{P_2}{\rho} + \frac{V_2^2}{2} + gz_2$$

FIGURE 5–33
When the flow is irrotational, the Bernoulli equation becomes applicable between any two points along the flow (not just on the same streamline).

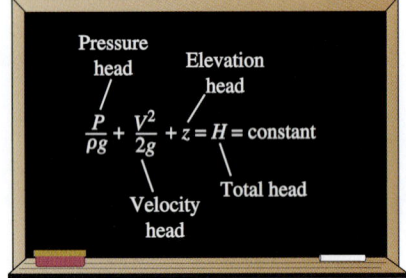

FIGURE 5–34
An alternative form of the Bernoulli equation is expressed in terms of heads as: *The sum of the pressure, velocity, and elevation heads is constant along a streamline.*

If a piezometer (which measures static pressure) is tapped into a pressurized pipe, as shown in Fig. 5–35, the liquid would rise to a height of $P/\rho g$ above the pipe center. The *hydraulic grade line* (HGL) is obtained by doing this at several locations along the pipe and drawing a curve through the liquid levels in the piezometers. The vertical distance above the pipe center is a measure of pressure within the pipe. Similarly, if a Pitot tube (measures static + dynamic pressure) is tapped into a pipe, the liquid would rise to a height of $P/\rho g + V^2/2g$ above the pipe center, or a distance of $V^2/2g$ above the HGL. The *energy grade line* (EGL) is obtained by doing this at several locations along the pipe and drawing a curve through the liquid levels in the Pitot tubes.

Noting that the fluid also has elevation head z (unless the reference level is taken to be the centerline of the pipe), the HGL and EGL are defined as follows: The line that represents the sum of the static pressure and the elevation heads, $P/\rho g + z$, is called the **hydraulic grade line.** The line that represents the total head of the fluid, $P/\rho g + V^2/2g + z$, is called the **energy grade line.** The difference between the heights of EGL and HGL is equal to the dynamic head, $V^2/2g$. We note the following about the HGL and EGL:

- For *stationary bodies* such as reservoirs or lakes, the EGL and HGL coincide with the free surface of the liquid. The elevation of the free surface z in such cases represents both the EGL and the HGL since the velocity is zero and the static (gage) pressure is zero.

- The EGL is always a distance $V^2/2g$ above the HGL. These two curves approach each other as the velocity decreases, and they diverge as the velocity increases. The height of the HGL decreases as the velocity increases, and vice versa.

- In an *idealized Bernoulli-type flow,* EGL is horizontal and its height remains constant. This would also be the case for HGL when the flow velocity is constant (Fig. 5–36).

- For *open-channel flow,* the HGL coincides with the free surface of the liquid, and the EGL is a distance $V^2/2g$ above the free surface.

- At a *pipe exit,* the pressure head is zero (atmospheric pressure) and thus the HGL coincides with the pipe outlet (location 3 on Fig. 5–35).

- The *mechanical energy loss* due to frictional effects (conversion to thermal energy) causes the EGL and HGL to slope downward in the direction of flow. The slope is a measure of the head loss in the pipe

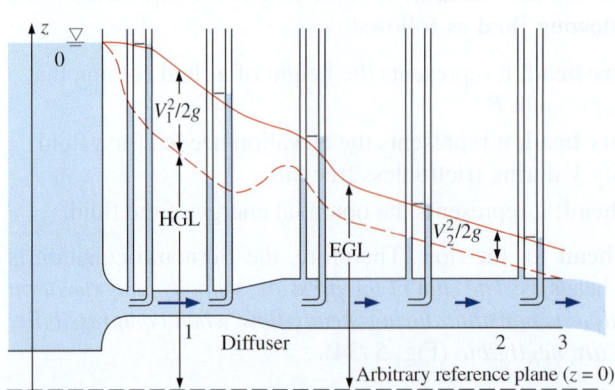

FIGURE 5–35
The *hydraulic grade line* (HGL) and the *energy grade line* (EGL) for free discharge from a reservoir through a horizontal pipe with a diffuser.

(discussed in detail in Chap. 8). A component that generates significant frictional effects such as a valve causes a sudden drop in both EGL and HGL at that location.

- A *steep jump* occurs in EGL and HGL whenever mechanical energy is added to the fluid (by a pump, for example). Likewise, a *steep drop* occurs in EGL and HGL whenever mechanical energy is removed from the fluid (by a turbine, for example), as shown in Fig. 5–37.
- The gage pressure of a fluid is zero at locations where the HGL *intersects* the fluid. The pressure in a flow section that lies above the HGL is negative, and the pressure in a section that lies below the HGL is positive (Fig. 5–38). Therefore, an accurate drawing of a piping system overlaid with the HGL can be used to determine the regions where the gage pressure in the pipe is negative (below atmospheric pressure).

The last remark enables us to avoid situations in which the pressure drops below the vapor pressure of the liquid (which may cause *cavitation*, as discussed in Chap. 2). Proper consideration is necessary in the placement of a liquid pump to ensure that the suction side pressure does not fall too low, especially at elevated temperatures where vapor pressure is higher than it is at low temperatures.

Now we examine Fig. 5–35 more closely. At point 0 (at the liquid surface), EGL and HGL are even with the liquid surface since there is no flow there. HGL decreases rapidly as the liquid accelerates into the pipe; however, EGL decreases very slowly through the well-rounded pipe inlet. EGL declines continually along the flow direction due to friction and other irreversible losses in the flow. EGL cannot increase in the flow direction unless energy is supplied to the fluid. HGL can rise or fall in the flow direction, but can never exceed EGL. HGL rises in the diffuser section as the velocity decreases, and the static pressure recovers somewhat; the total pressure does *not* recover, however, and EGL decreases through the diffuser. The difference between EGL and HGL is $V_1^2/2g$ at point 1, and $V_2^2/2g$ at point 2. Since $V_1 > V_2$, the difference between the two grade lines is larger at point 1 than at point 2. The downward slope of both grade lines is larger for the smaller diameter section of pipe since the frictional head loss is greater. Finally, HGL decays to the liquid surface at the outlet since the pressure there is atmospheric. However, EGL is still higher than HGL by the amount $V_2^2/2g$ since $V_3 = V_2$ at the outlet.

Applications of the Bernoulli Equation

So far, we have discussed the fundamental aspects of the Bernoulli equation. Now, we demonstrate its use in a wide range of applications through examples.

EXAMPLE 5–5   Spraying Water into the Air

Water is flowing from a garden hose (Fig. 5–39). A child places his thumb to cover most of the hose outlet, causing a thin jet of high-speed water to emerge. The pressure in the hose just upstream of his thumb is 400 kPa. If the hose is held upward, what is the maximum height that the jet could achieve?

SOLUTION   Water from a hose attached to the water main is sprayed into the air. The maximum height the water jet can rise is to be determined.

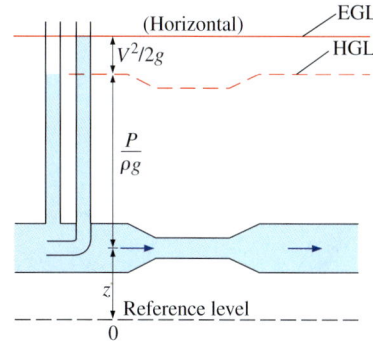

FIGURE 5–36
In an idealized Bernoulli-type flow, EGL is horizontal and its height remains constant. But this is not the case for HGL when the flow velocity varies along the flow.

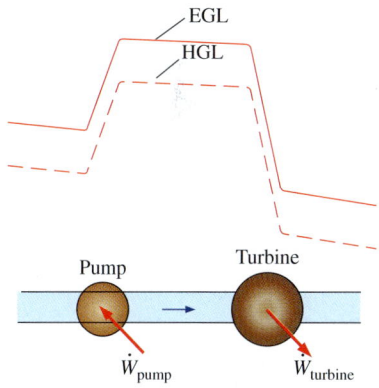

FIGURE 5–37
A *steep jump* occurs in EGL and HGL whenever mechanical energy is added to the fluid by a pump, and a *steep drop* occurs whenever mechanical energy is removed from the fluid by a turbine.

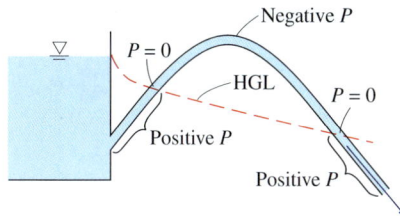

FIGURE 5–38
The gage pressure of a fluid is zero at locations where the HGL *intersects* the fluid, and the gage pressure is negative (vacuum) in a flow section that lies above the HGL.

208
BERNOULLI AND ENERGY EQUATIONS

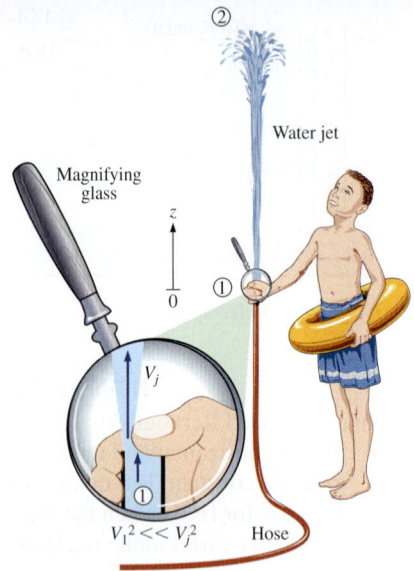

FIGURE 5–39
Schematic for Example 5–5. Inset shows a magnified view of the hose outlet region.

Assumptions 1 The flow exiting into the air is steady, incompressible, and irrotational (so that the Bernoulli equation is applicable). 2 The surface tension effects are negligible. 3 The friction between the water and air is negligible. 4 The irreversibilities that occur at the outlet of the hose due to abrupt contraction are not taken into account.
Properties We take the density of water to be 1000 kg/m³.
Analysis This problem involves the conversion of flow, kinetic, and potential energies to each other without involving any pumps, turbines, and wasteful components with large frictional losses, and thus it is suitable for the use of the Bernoulli equation. The water height will be maximum under the stated assumptions. The velocity inside the hose is relatively low ($V_1^2 \ll V_j^2$, and thus $V_1 \cong 0$ compared to V_j) and we take the elevation just below the hose outlet as the reference level ($z_1 = 0$). At the top of the water trajectory $V_2 = 0$, and atmospheric pressure pertains. Then the Bernoulli equation along a streamline from 1 to 2 simplifies to

$$\frac{P_1}{\rho g} + \underbrace{\frac{V_1^2}{2g}}_{\approx 0} + \underbrace{z_1}_{0} = \frac{P_2}{\rho g} + \underbrace{\frac{V_2^2}{2g}}_{0} + z_2 \quad \rightarrow \quad \frac{P_1}{\rho g} = \frac{P_{\text{atm}}}{\rho g} + z_2$$

Solving for z_2 and substituting,

$$z_2 = \frac{P_1 - P_{\text{atm}}}{\rho g} = \frac{P_{1,\text{gage}}}{\rho g} = \frac{400 \text{ kPa}}{(1000 \text{ kg/m}^3)(9.81 \text{ m/s}^2)} \left(\frac{1000 \text{ N/m}^2}{1 \text{ kPa}}\right)\left(\frac{1 \text{ kg} \cdot \text{m/s}^2}{1 \text{ N}}\right)$$

$$= \mathbf{40.8 \text{ m}}$$

Therefore, the water jet can rise as high as 40.8 m into the sky in this case.
Discussion The result obtained by the Bernoulli equation represents the upper limit and should be interpreted accordingly. It tells us that the water cannot possibly rise more than 40.8 m, and, in all likelihood, the rise will be much less than 40.8 m due to irreversible losses that we neglected.

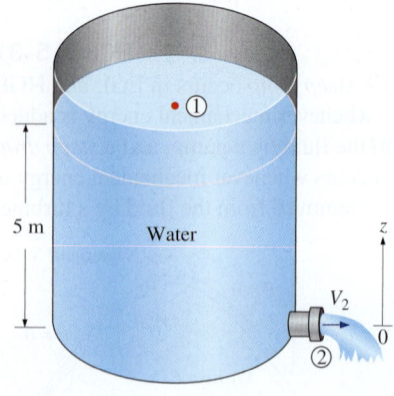

FIGURE 5–40
Schematic for Example 5–6.

EXAMPLE 5–6 Water Discharge from a Large Tank

A large tank open to the atmosphere is filled with water to a height of 5 m from the outlet tap (Fig. 5–40). A tap near the bottom of the tank is now opened, and water flows out from the smooth and rounded outlet. Determine the maximum water velocity at the outlet.

SOLUTION A tap near the bottom of a tank is opened. The maximum exit velocity of water from the tank is to be determined.
Assumptions 1 The flow is incompressible and irrotational (except very close to the walls). 2 The water drains slowly enough that the flow can be approximated as steady (actually quasi-steady when the tank begins to drain). 3 Irreversible losses in the tap region are neglected.
Analysis This problem involves the conversion of flow, kinetic, and potential energies to each other without involving any pumps, turbines, and wasteful components with large frictional losses, and thus it is suitable for the use of the Bernoulli equation. We take point 1 to be at the free surface of water so that $P_1 = P_{\text{atm}}$ (open to the atmosphere), $V_1^2 \ll V_2^2$ and thus $V_1 \cong 0$ compared to V_2 (the tank is very large relative to the outlet), $z_1 = 5$ m and $z_2 = 0$ (we take the reference level at the center of the outlet). Also, $P_2 = P_{\text{atm}}$ (water discharges into the atmosphere). For flow along a streamline from 1 to 2, the Bernoulli equation simplifies to

$$\frac{\cancel{P_1}}{\rho g} + \cancel{\frac{V_1^2}{2g}}^{\approx 0} + z_1 = \frac{\cancel{P_2}}{\rho g} + \frac{V_2^2}{2g} + \cancel{z_2}^{0} \quad \rightarrow \quad z_1 = \frac{V_2^2}{2g}$$

Solving for V_2 and substituting,

$$V_2 = \sqrt{2gz_1} = \sqrt{2(9.81 \text{ m/s}^2)(5 \text{ m})} = \mathbf{9.9 \text{ m/s}}$$

The relation $V = \sqrt{2gz}$ is called the **Torricelli equation.**

Therefore, the water leaves the tank with an initial maximum velocity of 9.9 m/s. This is the same velocity that would manifest if a solid were dropped a distance of 5 m in the absence of air friction drag. (What would the velocity be if the tap were at the bottom of the tank instead of on the side?)

Discussion If the orifice were sharp-edged instead of rounded, then the flow would be disturbed, and the average exit velocity would be less than 9.9 m/s. Care must be exercised when attempting to apply the Bernoulli equation to situations where abrupt expansions or contractions occur since the friction and flow disturbance in such cases may not be negligible. From conversion of mass, $(V_1/V_2)^2 = (D_2/D_1)^4$. So, for example, if $D_2/D_1 = 0.1$, then $(V_1/V_2)^2 = 0.0001$, and our approximation that $V_1^2 << V_2^2$ is justified.

EXAMPLE 5–7   Siphoning Out Gasoline from a Fuel Tank

During a trip to the beach ($P_{atm} = 1$ atm $= 101.3$ kPa), a car runs out of gasoline, and it becomes necessary to siphon gas out of the car of a Good Samaritan (Fig. 5–41). The siphon is a small-diameter hose, and to start the siphon it is necessary to insert one siphon end in the full gas tank, fill the hose with gasoline via suction, and then place the other end in a gas can below the level of the gas tank. The difference in pressure between point 1 (at the free surface of the gasoline in the tank) and point 2 (at the outlet of the tube) causes the liquid to flow from the higher to the lower elevation. Point 2 is located 0.75 m below point 1 in this case, and point 3 is located 2 m above point 1. The siphon diameter is 5 mm, and frictional losses in the siphon are to be disregarded. Determine (*a*) the minimum time to withdraw 4 L of gasoline from the tank to the can and (*b*) the pressure at point 3. The density of gasoline is 750 kg/m³.

SOLUTION Gasoline is to be siphoned from a tank. The minimum time it takes to withdraw 4 L of gasoline and the pressure at the highest point in the system are to be determined.

Assumptions **1** The flow is steady and incompressible. **2** Even though the Bernoulli equation is not valid through the pipe because of frictional losses, we employ the Bernoulli equation anyway in order to obtain a *best-case estimate*. **3** The change in the gasoline surface level inside the tank is negligible compared to elevations z_1 and z_2 during the siphoning period.

Properties The density of gasoline is given to be 750 kg/m³.

Analysis (*a*) We take point 1 to be at the free surface of gasoline in the tank so that $P_1 = P_{atm}$ (open to the atmosphere), $V_1 \cong 0$ (the tank is large relative to the tube diameter), and $z_2 = 0$ (point 2 is taken as the reference level). Also, $P_2 = P_{atm}$ (gasoline discharges into the atmosphere). Then the Bernoulli equation simplifies to

$$\frac{\cancel{P_1}}{\rho g} + \cancel{\frac{V_1^2}{2g}}^{\approx 0} + z_1 = \frac{\cancel{P_2}}{\rho g} + \frac{V_2^2}{2g} + \cancel{z_2}^{0} \quad \rightarrow \quad z_1 = \frac{V_2^2}{2g}$$

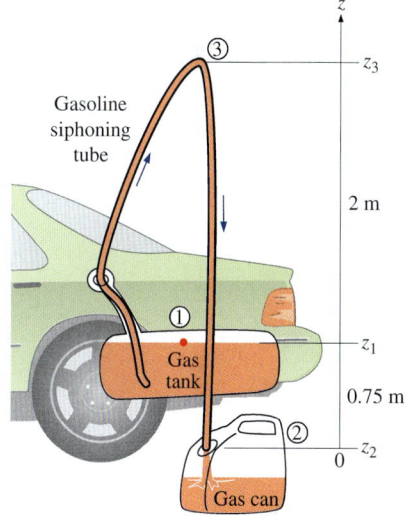

FIGURE 5–41
Schematic for Example 5–7.

Solving for V_2 and substituting,

$$V_2 = \sqrt{2gz_1} = \sqrt{2(9.81 \text{ m/s}^2)(0.75 \text{ m})} = 3.84 \text{ m/s}$$

The cross-sectional area of the tube and the flow rate of gasoline are

$$A = \pi D^2/4 = \pi(5 \times 10^{-3} \text{ m})^2/4 = 1.96 \times 10^{-5} \text{ m}^2$$

$$\dot{V} = V_2 A = (3.84 \text{ m/s})(1.96 \times 10^{-5} \text{ m}^2) = 7.53 \times 10^{-5} \text{ m}^3/\text{s} = 0.0753 \text{ L/s}$$

Then the time needed to siphon 4 L of gasoline becomes

$$\Delta t = \frac{V}{\dot{V}} = \frac{4 \text{ L}}{0.0753 \text{ L/s}} = \mathbf{53.1 \text{ s}}$$

(b) The pressure at point 3 is determined by writing the Bernoulli equation along a streamline between points 3 and 2. Noting that $V_2 = V_3$ (conservation of mass), $z_2 = 0$, and $P_2 = P_{atm}$,

$$\frac{P_2}{\rho g} + \frac{\cancel{V_2^2}}{\cancel{2g}} + \cancel{z_2}^{\;0} = \frac{P_3}{\rho g} + \frac{\cancel{V_3^2}}{\cancel{2g}} + z_3 \quad \rightarrow \quad \frac{P_{atm}}{\rho g} = \frac{P_3}{\rho g} + z_3$$

Solving for P_3 and substituting,

$$P_3 = P_{atm} - \rho g z_3$$

$$= 101.3 \text{ kPa} - (750 \text{ kg/m}^3)(9.81 \text{ m/s}^2)(2.75 \text{ m})\left(\frac{1 \text{ N}}{1 \text{ kg·m/s}^2}\right)\left(\frac{1 \text{ kPa}}{1000 \text{ N/m}^2}\right)$$

$$= \mathbf{81.1 \text{ kPa}}$$

Discussion The siphoning time is determined by neglecting frictional effects, and thus this is the *minimum time* required. In reality, the time will be longer than 53.1 s because of friction between the gasoline and the tube surface, along with other irreversible losses, as discussed in Chap. 8. Also, the pressure at point 3 is below the atmospheric pressure. If the elevation difference between points 1 and 3 is too high, the pressure at point 3 may drop below the vapor pressure of gasoline at the gasoline temperature, and some gasoline may evaporate (cavitate). The vapor then may form a pocket at the top and halt the flow of gasoline.

EXAMPLE 5–8 Velocity Measurement by a Pitot Tube

A piezometer and a Pitot tube are tapped into a horizontal water pipe, as shown in Fig. 5–42, to measure static and stagnation (static + dynamic) pressures. For the indicated water column heights, determine the velocity at the center of the pipe.

SOLUTION The static and stagnation pressures in a horizontal pipe are measured. The velocity at the center of the pipe is to be determined.
Assumptions 1 The flow is steady and incompressible. 2 Points 1 and 2 are close enough together that the irreversible energy loss between these two points is negligible, and thus we can use the Bernoulli equation.
Analysis We take points 1 and 2 along the streamline at the centerline of the pipe, with point 1 directly under the piezometer and point 2 at the tip of

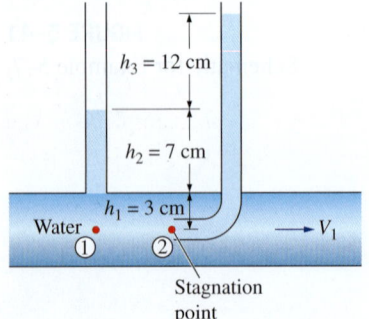

FIGURE 5–42
Schematic for Example 5–8.

the Pitot tube. This is a steady flow with straight and parallel streamlines, and the gage pressures at points 1 and 2 can be expressed as

$$P_1 = \rho g(h_1 + h_2)$$

$$P_2 = \rho g(h_1 + h_2 + h_3)$$

Noting that $z_1 = z_2$, and point 2 is a stagnation point and thus $V_2 = 0$, the application of the Bernoulli equation between points 1 and 2 gives

$$\frac{P_1}{\rho g} + \frac{V_1^2}{2g} + \cancel{z_1} = \frac{P_2}{\rho g} + \cancel{\frac{V_2^2}{2g}}^{0} + \cancel{z_2} \quad \rightarrow \quad \frac{V_1^2}{2g} = \frac{P_2 - P_1}{\rho g}$$

Substituting the P_1 and P_2 expressions gives

$$\frac{V_1^2}{2g} = \frac{P_2 - P_1}{\rho g} = \frac{\rho g(h_1 + h_2 + h_3) - \rho g(h_1 + h_2)}{\rho g} = h_3$$

Solving for V_1 and substituting,

$$V_1 = \sqrt{2gh_3} = \sqrt{2(9.81 \text{ m/s}^2)(0.12 \text{ m})} = \mathbf{1.53 \text{ m/s}}$$

Discussion Note that to determine the flow velocity, all we need is to measure the height of the excess fluid column in the Pitot tube compared to that in the piezometer tube.

EXAMPLE 5–9 The Rise of the Ocean Due to a Hurricane

A hurricane is a tropical storm formed over the ocean by low atmospheric pressures. As a hurricane approaches land, inordinate ocean swells (very high tides) accompany the hurricane. A Class-5 hurricane features winds in excess of 250 km/h, although the wind velocity at the center "eye" is very low.

Figure 5–43 depicts a hurricane hovering over the ocean swell below. The atmospheric pressure 320 km from the eye is 762 mmHg (at point 1, generally normal for the ocean) and the winds are calm. The atmospheric pressure at the eye of the storm is 560 mmHg. Estimate the ocean swell at (*a*) the eye of the hurricane at point 3 and (*b*) point 2, where the wind velocity is 250 km/h. Take the density of seawater and mercury to be 1025 kg/m³ and 13,600 kg/m³, respectively, and the density of air at normal sea-level temperature and pressure to be 1.2 kg/m³.

SOLUTION A hurricane is moving over the ocean. The amount of ocean swell at the eye and at active regions of the hurricane are to be determined.
Assumptions **1** The airflow within the hurricane is steady, incompressible, and irrotational (so that the Bernoulli equation is applicable). (This is certainly a very questionable assumption for a highly turbulent flow, but it is justified in the discussion.) **2** The effect of water sucked into the air is negligible.
Properties The densities of air at normal conditions, seawater, and mercury are given to be 1.2 kg/m³, 1025 kg/m³, and 13,600 kg/m³, respectively.
Analysis (*a*) Reduced atmospheric pressure over the water causes the water to rise. Thus, decreased pressure at point 2 relative to point 1 causes the ocean water to rise at point 2. The same is true at point 3, where the storm air

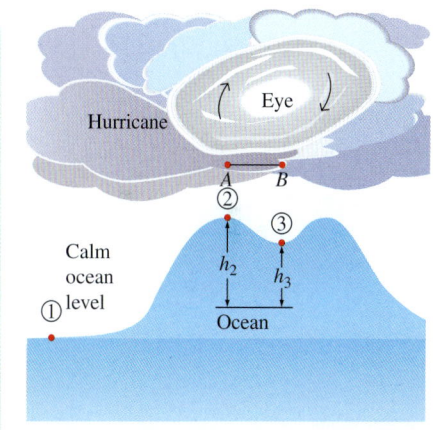

FIGURE 5–43
Schematic for Example 5–9. The vertical scale is greatly exaggerated.

FIGURE 5–44
The eye of hurricane Linda (1997 in the Pacific Ocean near Baja California) is clearly visible in this satellite photo.
© Brand X Pictures/PunchStock RF

velocity is negligible. The pressure difference given in terms of the mercury column height is expressed in terms of the seawater column height by

$$\Delta P = (\rho g h)_{Hg} = (\rho g h)_{sw} \rightarrow h_{sw} = \frac{\rho_{Hg}}{\rho_{sw}} h_{Hg}$$

Then the pressure difference between points 1 and 3 in terms of the seawater column height becomes

$$h_3 = \frac{\rho_{Hg}}{\rho_{sw}} h_{Hg} = \left(\frac{13{,}600 \text{ kg/m}^3}{1025 \text{ kg/m}^3}\right)[(762 - 560) \text{ mmHg}]\left(\frac{1 \text{ m}}{1000 \text{ mm}}\right) = \mathbf{2.68 \text{ m}}$$

which is equivalent to the storm surge at the *eye of the hurricane* (Fig. 5–44) since the wind velocity there is negligible and there are no dynamic effects.

(*b*) To determine the additional rise of ocean water at point 2 due to the high winds at that point, we write the Bernoulli equation between points *A* and *B*, which are on top of points 2 and 3, respectively. Noting that $V_B \cong 0$ (the eye region of the hurricane is relatively calm) and $z_A = z_B$ (both points are on the same horizontal line), the Bernoulli equation simplifies to

$$\frac{P_A}{\rho g} + \frac{V_A^2}{2g} + \cancel{z_A} = \frac{P_B}{\rho g} + \cancel{\frac{V_B^2}{2g}}^{0} + \cancel{z_B} \rightarrow \frac{P_B - P_A}{\rho g} = \frac{V_A^2}{2g}$$

Substituting,

$$\frac{P_B - P_A}{\rho g} = \frac{V_A^2}{2g} = \frac{(250 \text{ km/h})^2}{2(9.81 \text{ m/s}^2)}\left(\frac{1 \text{ m/s}}{3.6 \text{ km/h}}\right)^2 = 246 \text{ m}$$

where ρ is the density of air in the hurricane. Noting that the density of an ideal gas at constant temperature is proportional to absolute pressure and the density of air at the normal atmospheric pressure of 101 kPa $\cong$ 762 mmHg is 1.2 kg/m^3, the density of air in the hurricane is

$$\rho_{air} = \frac{P_{air}}{P_{atm\ air}} \rho_{atm\ air} = \left(\frac{560 \text{ mmHg}}{762 \text{ mmHg}}\right)(1.2 \text{ kg/m}^3) = 0.882 \text{ kg/m}^3$$

Using the relation developed above in part (*a*), the seawater column height equivalent to 246 m of air column height is determined to be

$$h_{dynamic} = \frac{\rho_{air}}{\rho_{sw}} h_{air} = \left(\frac{0.882 \text{ kg/m}^3}{1025 \text{ kg/m}^3}\right)(246 \text{ m}) = 0.21 \text{ m}$$

Therefore, the pressure at point 2 is 0.21 m seawater column lower than the pressure at point 3 due to the high wind velocities, causing the ocean to rise an additional 0.21 m. Then the total storm surge at point 2 becomes

$$h_2 = h_3 + h_{dynamic} = 2.68 + 0.21 = \mathbf{2.89 \text{ m}}$$

Discussion This problem involves highly turbulent flow and the intense breakdown of the streamlines, and thus the applicability of the Bernoulli equation in part (*b*) is questionable. Furthermore, the flow in the eye of the storm is *not* irrotational, and the Bernoulli equation constant changes across streamlines (see Chap. 10). The Bernoulli analysis can be thought of as the limiting, ideal case, and shows that the rise of seawater due to high-velocity winds cannot be more than 0.21 m.

The wind power of hurricanes is not the only cause of damage to coastal areas. Ocean flooding and erosion from excessive tides is just as serious, as are high waves generated by the storm turbulence and energy.

EXAMPLE 5–10 Bernoulli Equation for Compressible Flow

Derive the Bernoulli equation when the compressibility effects are not negligible for an ideal gas undergoing (*a*) an isothermal process and (*b*) an isentropic process.

SOLUTION The Bernoulli equation for compressible flow is to be obtained for an ideal gas for isothermal and isentropic processes.
Assumptions **1** The flow is steady and frictional effects are negligible. **2** The fluid is an ideal gas, so the relation $P = \rho RT$ is applicable. **3** The specific heats are constant so that $P/\rho^k = $ constant during an isentropic process.
Analysis (*a*) When the compressibility effects are significant and the flow cannot be assumed to be incompressible, the Bernoulli equation is given by Eq. 5–40 as

$$\int \frac{dP}{\rho} + \frac{V^2}{2} + gz = \text{constant} \quad \text{(along a streamline)} \quad (1)$$

The compressibility effects can be properly accounted for by performing the integration $\int dP/\rho$ in Eq. 1. But this requires a relation between P and ρ for the process. For the *isothermal* expansion or compression of an ideal gas, the integral in Eq. 1 is performed easily by noting that $T = $ constant and substituting $\rho = P/RT$,

$$\int \frac{dP}{\rho} = \int \frac{dP}{P/RT} = RT \ln P$$

Substituting into Eq. 1 gives the desired relation,

Isothermal process: $\quad RT \ln P + \dfrac{V^2}{2} + gz = \text{constant} \quad (2)$

(*b*) A more practical case of compressible flow is the *isentropic flow of ideal gases* through equipment that involves high-speed fluid flow such as nozzles, diffusers, and the passages between turbine blades (Fig. 5–45). Isentropic (i.e., reversible and adiabatic) flow is closely approximated by these devices, and it is characterized by the relation $P/\rho^k = C = $ constant, where k is the specific heat ratio of the gas. Solving for ρ from $P/\rho^k = C$ gives $\rho = C^{-1/k} P^{1/k}$. Performing the integration,

$$\int \frac{dP}{\rho} = \int C^{1/k} P^{-1/k}\, dP = C^{1/k}\frac{P^{-1/k+1}}{-1/k+1} = \frac{P^{1/k}}{\rho}\frac{P^{-1/k+1}}{-1/k+1} = \left(\frac{k}{k-1}\right)\frac{P}{\rho} \quad (3)$$

Substituting, the Bernoulli equation for steady, isentropic, compressible flow of an ideal gas becomes

Isentropic flow: $\quad \left(\dfrac{k}{k-1}\right)\dfrac{P}{\rho} + \dfrac{V^2}{2} + gz = \text{constant} \quad (4a)$

FIGURE 5–45
Compressible flow of a gas through turbine blades is often modeled as isentropic, and the compressible form of the Bernoulli equation is a reasonable approximation.
Royalty-Free/CORBIS

or

$$\left(\frac{k}{k-1}\right)\frac{P_1}{\rho_1} + \frac{V_1^2}{2} + gz_1 = \left(\frac{k}{k-1}\right)\frac{P_2}{\rho_2} + \frac{V_2^2}{2} + gz_2 \quad (4b)$$

A common practical situation involves the acceleration of a gas from rest (stagnation conditions at state 1) with negligible change in elevation. In that case we have $z_1 = z_2$ and $V_1 = 0$. Noting that $\rho = P/RT$ for ideal gases, P/ρ^k = constant for isentropic flow, and the Mach number is defined as Ma = V/c where $c = \sqrt{kRT}$ is the local speed of sound for ideal gases, Eq. 4b simplifies to

$$\frac{P_1}{P_2} = \left[1 + \left(\frac{k-1}{2}\right)\text{Ma}_2^2\right]^{k/(k-1)} \quad (4c)$$

where state 1 is the stagnation state and state 2 is any state along the flow.
Discussion It can be shown that the results obtained using the compressible and incompressible equations deviate no more than 2 percent when the Mach number is less than 0.3. Therefore, the flow of an ideal gas can be considered to be incompressible when Ma ≤ 0.3. For atmospheric air at normal conditions, this corresponds to a flow speed of about 100 m/s or 360 km/h.

5–5 · GENERAL ENERGY EQUATION

One of the most fundamental laws in nature is the **first law of thermodynamics,** also known as the **conservation of energy principle,** which provides a sound basis for studying the relationships among the various forms of energy and energy interactions. It states that *energy can be neither created nor destroyed during a process; it can only change forms.* Therefore, every bit of energy must be accounted for during a process.

A rock falling off a cliff, for example, picks up speed as a result of its potential energy being converted to kinetic energy (Fig. 5–46). Experimental data show that the decrease in potential energy equals the increase in kinetic energy when the air resistance is negligible, thus confirming the conservation of energy principle. The conservation of energy principle also forms the backbone of the diet industry: a person who has a greater energy input (food) than energy output (exercise) will gain weight (store energy in the form of fat), and a person who has a smaller energy input than output will lose weight. The change in the energy content of a system is equal to the difference between the energy input and the energy output, and the conservation of energy principle for any system can be expressed simply as $E_{in} - E_{out} = \Delta E$.

The transfer of any quantity (such as mass, momentum, and energy) is recognized *at the boundary* as the quantity *crosses the boundary*. A quantity is said to *enter* a system (or control volume) if it crosses the boundary from the outside to the inside, and to *exit* the system if it moves in the reverse direction. A quantity that moves from one location to another within a system is not considered as a transferred quantity in an analysis since it does not enter or exit the system. Therefore, it is important to specify the system and thus clearly identify its boundaries before an engineering analysis is performed.

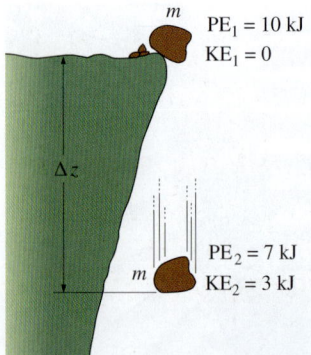

FIGURE 5–46
Energy cannot be created or destroyed during a process; it can only change forms.

The energy content of a fixed quantity of mass (a closed system) can be changed by two mechanisms: *heat transfer Q* and *work transfer W*. Then the conservation of energy for a fixed quantity of mass can be expressed in rate form as (Fig. 5–47)

$$\dot{Q}_{net\ in} + \dot{W}_{net\ in} = \frac{dE_{sys}}{dt} \quad \text{or} \quad \dot{Q}_{net\ in} + \dot{W}_{net\ in} = \frac{d}{dt}\int_{sys} \rho e\, dV \quad (5\text{–}49)$$

where the overdot stands for time rate of change, and $\dot{Q}_{net\ in} = \dot{Q}_{in} - \dot{Q}_{out}$ is the net rate of heat transfer to the system (negative, if from the system), $\dot{W}_{net\ in} = \dot{W}_{in} - \dot{W}_{out}$ is the net power input to the system in all forms (negative, if power output), and dE_{sys}/dt is the rate of change of the total energy content of the system. For simple compressible systems, total energy consists of internal, kinetic, and potential energies, and it is expressed on a unit-mass basis as (see Chap. 2)

$$e = u + \text{ke} + \text{pe} = u + \frac{V^2}{2} + gz \quad (5\text{–}50)$$

Note that total energy is a property, and its value does not change unless the state of the system changes.

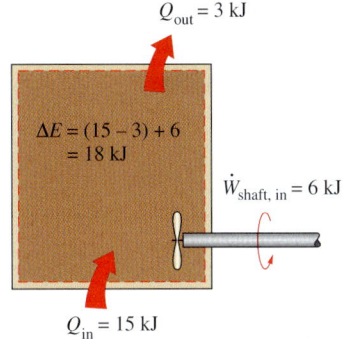

FIGURE 5–47
The energy change of a system during a process is equal to the *net* work and heat transfer between the system and its surroundings.

Energy Transfer by Heat, *Q*

In daily life, we frequently refer to the sensible and latent forms of internal energy as *heat*, and talk about the heat content of bodies. Scientifically the more correct name for these forms of energy is *thermal energy*. For single-phase substances, a change in the thermal energy of a given mass results in a change in temperature, and thus temperature is a good representative of thermal energy. Thermal energy tends to move naturally in the direction of decreasing temperature. The transfer of energy from one system to another as a result of a temperature difference is called **heat transfer.** The warming up of a canned drink in a warmer room, for example, is due to heat transfer (Fig. 5–48). The time rate of heat transfer is called **heat transfer rate** and is denoted by $\dot{Q}$.

The direction of heat transfer is always from the higher-temperature body to the lower-temperature one. Once temperature equality is established, heat transfer stops. There cannot be any net heat transfer between two systems (or a system and its surroundings) that are at the same temperature.

A process during which there is no heat transfer is called an **adiabatic process.** There are two ways a process can be adiabatic: Either the system is well insulated so that only a negligible amount of heat can pass through the system boundary, or both the system and the surroundings are at the same temperature and therefore there is no driving force (temperature difference) for net heat transfer. An adiabatic process should not be confused with an isothermal process. Even though there is no net heat transfer during an adiabatic process, the energy content and thus the temperature of a system can still be changed by other means such as work transfer.

FIGURE 5–48
Temperature difference is the driving force for heat transfer. The larger the temperature difference, the higher is the rate of heat transfer.

Energy Transfer by Work, *W*

An energy interaction is **work** if it is associated with a force acting through a distance. A rising piston, a rotating shaft, and an electric wire crossing the system boundary are all associated with work interactions. The time rate of doing work is called **power** and is denoted by $\dot{W}$. Car engines and hydraulic,

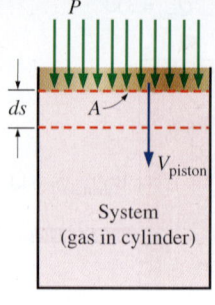

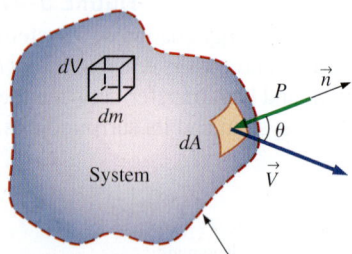

FIGURE 5–49
The pressure force acting on (a) the moving boundary of a system in a piston-cylinder device, and (b) the differential surface area of a system of arbitrary shape.

steam, and gas turbines produce power ($\dot{W}_{shaft,\,in} < 0$); compressors, pumps, fans, and mixers consume power ($\dot{W}_{shaft,\,in} > 0$).

Work-consuming devices transfer energy to the fluid, and thus increase the energy of the fluid. A fan in a room, for example, mobilizes the air and increases its kinetic energy. The electric energy a fan consumes is first converted to mechanical energy by its motor that forces the shaft of the blades to rotate. This mechanical energy is then transferred to the air, as evidenced by the increase in air velocity. This energy transfer to air has nothing to do with a temperature difference, so it cannot be heat transfer. Therefore, it must be work. Air discharged by the fan eventually comes to a stop and thus loses its mechanical energy as a result of friction between air particles of different velocities. But this is not a "loss" in the real sense; it is simply the conversion of mechanical energy to an equivalent amount of thermal energy (which is of limited value, and thus the term *loss*) in accordance with the conservation of energy principle. If a fan runs a long time in a sealed room, we can sense the buildup of this thermal energy by a rise in air temperature.

A system may involve numerous forms of work, and the total work can be expressed as

$$W_{total} = W_{shaft} + W_{pressure} + W_{viscous} + W_{other} \qquad (5\text{-}51)$$

where W_{shaft} is the work transmitted by a rotating shaft, $W_{pressure}$ is the work done by the pressure forces on the control surface, $W_{viscous}$ is the work done by the normal and shear components of viscous forces on the control surface, and W_{other} is the work done by other forces such as electric, magnetic, and surface tension, which are insignificant for simple compressible systems and are not considered in this text. We do not consider $W_{viscous}$ either, since moving walls (such as fan blades or turbine runners) are usually *inside* the control volume and are not part of the control surface. But it should be kept in mind that the work done by shear forces as the blades shear through the fluid may need to be considered in a refined analysis of turbomachinery.

Shaft Work

Many flow systems involve a machine such as a pump, a turbine, a fan, or a compressor whose shaft protrudes through the control surface, and the work transfer associated with all such devices is simply referred to as *shaft work* W_{shaft}. The power transmitted via a rotating shaft is proportional to the shaft torque T_{shaft} and is expressed as

$$\dot{W}_{shaft} = \omega T_{shaft} = 2\pi \dot{n} T_{shaft} \qquad (5\text{-}52)$$

where ω is the angular speed of the shaft in rad/s and $\dot{n}$ is the number of revolutions of the shaft per unit time, often expressed in rev/min or rpm.

Work Done by Pressure Forces

Consider a gas being compressed in the piston-cylinder device shown in Fig. 5–49a. When the piston moves down a differential distance ds under the influence of the pressure force PA, where A is the cross-sectional area of the piston, the boundary work done *on* the system is $\delta W_{boundary} = PA\,ds$. Dividing both sides of this relation by the differential time interval dt gives the time rate of boundary work (i.e., *power*),

$$\delta \dot{W}_{pressure} = \delta \dot{W}_{boundary} = PAV_{piston}$$

where $V_{piston} = ds/dt$ is the piston speed, which is the speed of the moving boundary at the piston face.

Now consider a material chunk of fluid (a system) of arbitrary shape that moves with the flow and is free to deform under the influence of pressure, as shown in Fig. 5–49b. Pressure always acts inward and normal to the surface, and the pressure force acting on a differential area dA is PdA. Again noting that work is force times distance and distance traveled per unit time is velocity, the time rate at which work is done by pressure forces on this differential part of the system is

$$\delta\dot{W}_{pressure} = -P\,dA\,V_n = -P\,dA(\vec{V}\cdot\vec{n}) \qquad (5\text{-}53)$$

since the normal component of velocity through the differential area dA is $V_n = V\cos\theta = \vec{V}\cdot\vec{n}$. Note that $\vec{n}$ is the outward normal of dA, and thus the quantity $\vec{V}\cdot\vec{n}$ is positive for expansion and negative for compression. The negative sign in Eq. 5–53 ensures that work done by pressure forces is positive when it is done *on* the system, and negative when it is done *by* the system, which agrees with our sign convention. The total rate of work done by pressure forces is obtained by integrating $\delta\dot{W}_{pressure}$ over the entire surface A,

$$\dot{W}_{pressure,\,net\,in} = -\int_A P(\vec{V}\cdot\vec{n})\,dA = -\int_A \frac{P}{\rho}\rho(\vec{V}\cdot\vec{n})\,dA \qquad (5\text{-}54)$$

In light of these discussions, the net power transfer can be expressed as

$$\dot{W}_{net\,in} = \dot{W}_{shaft,\,net\,in} + \dot{W}_{pressure,\,net\,in} = \dot{W}_{shaft,\,net\,in} - \int_A P(\vec{V}\cdot\vec{n})\,dA \qquad (5\text{-}55)$$

Then the rate form of the conservation of energy relation for a closed system becomes

$$\dot{Q}_{net\,in} + \dot{W}_{shaft,\,net\,in} + \dot{W}_{pressure,\,net\,in} = \frac{dE_{sys}}{dt} \qquad (5\text{-}56)$$

To obtain a relation for the conservation of energy for a *control volume*, we apply the Reynolds transport theorem by replacing B with total energy E, and b with total energy per unit mass e, which is $e = u + ke + pe = u + V^2/2 + gz$ (Fig. 5–50). This yields

$$\frac{dE_{sys}}{dt} = \frac{d}{dt}\int_{CV} e\rho\,dV + \int_{CS} e\rho(\vec{V_r}\cdot\vec{n})A \qquad (5\text{-}57)$$

Substituting the left-hand side of Eq. 5–56 into Eq. 5–57, the general form of the energy equation that applies to fixed, moving, or deforming control volumes becomes

$$\dot{Q}_{net\,in} + \dot{W}_{shaft,\,net\,in} + \dot{W}_{pressure,\,net\,in} = \frac{d}{dt}\int_{CV} e\rho\,dV + \int_{CS} e\rho(\vec{V_r}\cdot\vec{n})\,dA \qquad (5\text{-}58)$$

which is stated in words as

$$\begin{pmatrix}\text{The net rate of energy}\\\text{transfer into a CV by}\\\text{heat and work transfer}\end{pmatrix} = \begin{pmatrix}\text{The time rate of}\\\text{change of the energy}\\\text{content of the CV}\end{pmatrix} + \begin{pmatrix}\text{The net flow rate of}\\\text{energy out of the control}\\\text{surface by mass flow}\end{pmatrix}$$

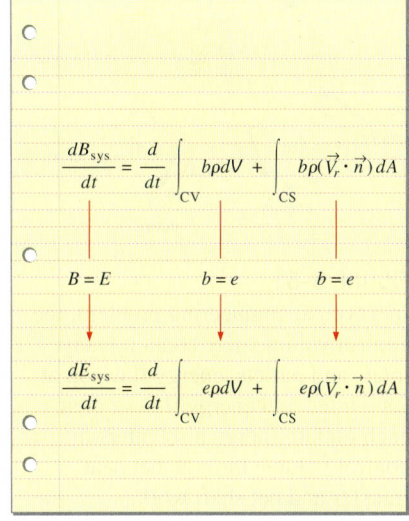

FIGURE 5–50
The conservation of energy equation is obtained by replacing B in the Reynolds transport theorem by energy E and b by e.

BERNOULLI AND ENERGY EQUATIONS

Here $\vec{V}_r = \vec{V} - \vec{V}_{CS}$ is the fluid velocity relative to the control surface, and the product $\rho(\vec{V}_r \cdot \vec{n})\, dA$ represents the mass flow rate through area element dA into or out of the control volume. Again noting that $\vec{n}$ is the outward normal of dA, the quantity $\vec{V}_r \cdot \vec{n}$ and thus mass flow is positive for outflow and negative for inflow.

Substituting the surface integral for the rate of pressure work from Eq. 5–54 into Eq. 5–58 and combining it with the surface integral on the right give

$$\dot{Q}_{\text{net in}} + \dot{W}_{\text{shaft, net in}} = \frac{d}{dt}\int_{CV} e\rho\, dV + \int_{CS}\left(\frac{P}{\rho} + e\right)\rho(\vec{V}_r \cdot \vec{n})\, dA \quad (5\text{–}59)$$

This is a convenient form for the energy equation since pressure work is now combined with the energy of the fluid crossing the control surface and we no longer have to deal with pressure work.

The term $P/\rho = Pv = w_{\text{flow}}$ is the **flow work,** which is the work per unit mass associated with pushing a fluid into or out of a control volume. Note that the fluid velocity at a solid surface is equal to the velocity of the solid surface because of the no-slip condition. As a result, the pressure work along the portions of the control surface that coincide with nonmoving solid surfaces is zero. Therefore, pressure work for fixed control volumes can exist only along the imaginary part of the control surface where the fluid enters and leaves the control volume, i.e., inlets and outlets.

For a fixed control volume (no motion or deformation of the control volume), $\vec{V}_r = \vec{V}$ and the energy equation Eq. 5–59 becomes

Fixed CV: $\quad \dot{Q}_{\text{net in}} + \dot{W}_{\text{shaft, net in}} = \dfrac{d}{dt}\int_{CV} e\rho\, dV + \int_{CS}\left(\dfrac{P}{\rho} + e\right)\rho(\vec{V}\cdot\vec{n})\, dA \quad (5\text{–}60)$

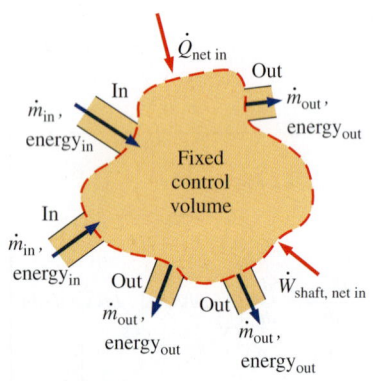

FIGURE 5–51
In a typical engineering problem, the control volume may contain many inlets and outlets; energy flows in at each inlet, and energy flows out at each outlet. Energy also enters the control volume through net heat transfer and net shaft work.

This equation is not in a convenient form for solving practical engineering problems because of the integrals, and thus it is desirable to rewrite it in terms of average velocities and mass flow rates through inlets and outlets. If $P/\rho + e$ is nearly uniform across an inlet or outlet, we can simply take it outside the integral. Noting that $\dot{m} = \int_{A_c} \rho(\vec{V}\cdot\vec{n})\, dA_c$ is the mass flow rate across an inlet or outlet, the rate of inflow or outflow of energy through the inlet or outlet can be approximated as $\dot{m}(P/\rho + e)$. Then the energy equation becomes (Fig. 5–51)

$$\dot{Q}_{\text{net in}} + \dot{W}_{\text{shaft, net in}} = \frac{d}{dt}\int_{CV} e\rho\, dV + \sum_{\text{out}} \dot{m}\left(\frac{P}{\rho} + e\right) - \sum_{\text{in}} \dot{m}\left(\frac{P}{\rho} + e\right) \quad (5\text{–}61)$$

where $e = u + V^2/2 + gz$ (Eq. 5–50) is the total energy per unit mass for both the control volume and flow streams. Then,

$$\dot{Q}_{\text{net in}} + \dot{W}_{\text{shaft, net in}} = \frac{d}{dt}\int_{CV} e\rho\, dV + \sum_{\text{out}} \dot{m}\left(\frac{P}{\rho} + u + \frac{V^2}{2} + gz\right) - \sum_{\text{in}} \dot{m}\left(\frac{P}{\rho} + u + \frac{V^2}{2} + gz\right)$$

(5–62)

or

$$\dot{Q}_{\text{net in}} + \dot{W}_{\text{shaft, net in}} = \frac{d}{dt}\int_{CV} e\rho\, dV + \sum_{\text{out}} \dot{m}\left(h + \frac{V^2}{2} + gz\right) - \sum_{\text{in}} \dot{m}\left(h + \frac{V^2}{2} + gz\right)$$

(5–63)

where we used the definition of specific enthalpy $h = u + Pv = u + P/\rho$. The last two equations are fairly general expressions of conservation of energy, but their use is still limited to fixed control volumes, uniform flow at inlets and outlets, and negligible work due to viscous forces and other effects. Also, the subscript "net in" stands for "net input," and thus any heat or work transfer is positive if *to* the system and negative if *from* the system.

5–6 · ENERGY ANALYSIS OF STEADY FLOWS

For steady flows, the time rate of change of the energy content of the control volume is zero, and Eq. 5–63 simplifies to

$$\dot{Q}_{\text{net in}} + \dot{W}_{\text{shaft, net in}} = \sum_{\text{out}} \dot{m}\left(h + \frac{V^2}{2} + gz\right) - \sum_{\text{in}} \dot{m}\left(h + \frac{V^2}{2} + gz\right) \quad (5\text{–}64)$$

It states that *during steady flow the net rate of energy transfer to a control volume by heat and work transfers is equal to the difference between the rates of outgoing and incoming energy flows by mass flow.*

Many practical problems involve just one inlet and one outlet (Fig. 5–52). The mass flow rate for such **single-stream devices** is the same at the inlet and outlet, and Eq. 5–64 reduces to

$$\dot{Q}_{\text{net in}} + \dot{W}_{\text{shaft, net in}} = \dot{m}\left(h_2 - h_1 + \frac{V_2^2 - V_1^2}{2} + g(z_2 - z_1)\right) \quad (5\text{–}65)$$

where subscripts 1 and 2 refer to the inlet and outlet, respectively. The steady-flow energy equation on a unit-mass basis is obtained by dividing Eq. 5–65 by the mass flow rate $\dot{m}$,

$$q_{\text{net in}} + w_{\text{shaft, net in}} = h_2 - h_1 + \frac{V_2^2 - V_1^2}{2} + g(z_2 - z_1) \quad (5\text{–}66)$$

where $q_{\text{net in}} = \dot{Q}_{\text{net in}}/\dot{m}$ is the net heat transfer to the fluid per unit mass and $w_{\text{shaft, net in}} = \dot{W}_{\text{shaft, net in}}/\dot{m}$ is the net shaft work input to the fluid per unit mass. Using the definition of enthalpy $h = u + P/\rho$ and rearranging, the steady-flow energy equation can also be expressed as

$$w_{\text{shaft, net in}} + \frac{P_1}{\rho_1} + \frac{V_1^2}{2} + gz_1 = \frac{P_2}{\rho_2} + \frac{V_2^2}{2} + gz_2 + (u_2 - u_1 - q_{\text{net in}}) \quad (5\text{–}67)$$

where u is the *internal energy*, P/ρ is the *flow energy*, $V^2/2$ is the *kinetic energy*, and gz is the *potential energy* of the fluid, all per unit mass. These relations are valid for both compressible and incompressible flows.

The left side of Eq. 5–67 represents the mechanical energy input, while the first three terms on the right side represent the mechanical energy output. If the flow is ideal with no irreversibilities such as friction, the total mechanical energy must be conserved, and the term in parentheses $(u_2 - u_1 - q_{\text{net in}})$ must equal zero. That is,

Ideal flow (no mechanical energy loss): $\quad q_{\text{net in}} = u_2 - u_1 \quad (5\text{–}68)$

Any increase in $u_2 - u_1$ above $q_{\text{net in}}$ is due to the irreversible conversion of mechanical energy to thermal energy, and thus $u_2 - u_1 - q_{\text{net in}}$ represents the mechanical energy loss per unit mass (Fig. 5–53). That is,

Real flow (with mechanical energy loss): $\quad e_{\text{mech, loss}} = u_2 - u_1 - q_{\text{net in}} \quad (5\text{–}69)$

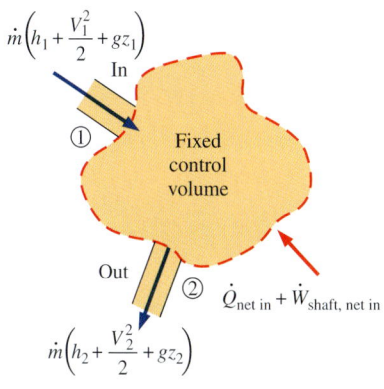

FIGURE 5–52
A control volume with only one inlet and one outlet and energy interactions.

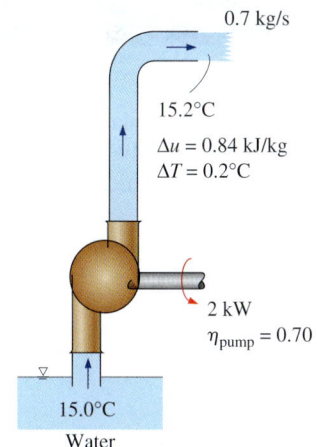

FIGURE 5–53
The lost mechanical energy in a fluid flow system results in an increase in the internal energy of the fluid and thus in a rise of fluid temperature.

For single-phase fluids (a gas or a liquid), $u_2 - u_1 = c_v(T_2 - T_1)$ where c_v is the constant-volume specific heat.

The steady-flow energy equation on a unit-mass basis can be written conveniently as a **mechanical energy** balance,

$$e_{\text{mech, in}} = e_{\text{mech, out}} + e_{\text{mech, loss}} \tag{5-70}$$

or

$$w_{\text{shaft, net in}} + \frac{P_1}{\rho_1} + \frac{V_1^2}{2} + gz_1 = \frac{P_2}{\rho_2} + \frac{V_2^2}{2} + gz_2 + e_{\text{mech, loss}} \tag{5-71}$$

Noting that $w_{\text{shaft, net in}} = w_{\text{pump}} - w_{\text{turbine}}$, the mechanical energy balance can be written more explicitly as

$$\frac{P_1}{\rho_1} + \frac{V_1^2}{2} + gz_1 + w_{\text{pump}} = \frac{P_2}{\rho_2} + \frac{V_2^2}{2} + gz_2 + w_{\text{turbine}} + e_{\text{mech, loss}} \tag{5-72}$$

where w_{pump} is the mechanical work input (due to the presence of a pump, fan, compressor, etc.) and w_{turbine} is the mechanical work output (due to a turbine). When the flow is incompressible, either absolute or gage pressure can be used for P since P_{atm}/ρ would appear on both sides and would cancel out.

Multiplying Eq. 5–72 by the mass flow rate $\dot{m}$ gives

$$\dot{m}\left(\frac{P_1}{\rho_1} + \frac{V_1^2}{2} + gz_1\right) + \dot{W}_{\text{pump}} = \dot{m}\left(\frac{P_2}{\rho_2} + \frac{V_2^2}{2} + gz_2\right) + \dot{W}_{\text{turbine}} + \dot{E}_{\text{mech, loss}} \tag{5-73}$$

where $\dot{W}_{\text{pump}}$ is the shaft power input through the pump's shaft, $\dot{W}_{\text{turbine}}$ is the shaft power output through the turbine's shaft, and $\dot{E}_{\text{mech, loss}}$ is the *total* mechanical power loss, which consists of pump and turbine losses as well as the frictional losses in the piping network. That is,

$$\dot{E}_{\text{mech, loss}} = \dot{E}_{\text{mech loss, pump}} + \dot{E}_{\text{mech loss, turbine}} + \dot{E}_{\text{mech loss, piping}}$$

By convention, irreversible pump and turbine losses are treated separately from irreversible losses due to other components of the piping system (Fig. 5–54). Thus, the energy equation is expressed in its most common form in terms of *heads* by dividing each term in Eq. 5–73 by $\dot{m}g$. The result is

$$\frac{P_1}{\rho_1 g} + \frac{V_1^2}{2g} + z_1 + h_{\text{pump, }u} = \frac{P_2}{\rho_2 g} + \frac{V_2^2}{2g} + z_2 + h_{\text{turbine, }e} + h_L \tag{5-74}$$

where

- $h_{\text{pump, }u} = \dfrac{w_{\text{pump, }u}}{g} = \dfrac{\dot{W}_{\text{pump, }u}}{\dot{m}g} = \dfrac{\eta_{\text{pump}}\dot{W}_{\text{pump}}}{\dot{m}g}$ is the *useful head delivered to the fluid by the pump.* Because of irreversible losses in the pump, $h_{\text{pump, }u}$ is *less* than $\dot{W}_{\text{pump}}/\dot{m}g$ by the factor η_{pump}.

- $h_{\text{turbine, }e} = \dfrac{w_{\text{turbine, }e}}{g} = \dfrac{\dot{W}_{\text{turbine, }e}}{\dot{m}g} = \dfrac{\dot{W}_{\text{turbine}}}{\eta_{\text{turbine}}\dot{m}g}$ is the *extracted head removed from the fluid by the turbine.* Because of irreversible losses in the turbine, $h_{\text{turbine, }e}$ is *greater* than $\dot{W}_{\text{turbine}}/\dot{m}g$ by the factor η_{turbine}.

- $h_L = \dfrac{e_{\text{mech loss, piping}}}{g} = \dfrac{\dot{E}_{\text{mech loss, piping}}}{\dot{m}g}$ is the *irreversible head loss* between 1 and 2 due to all components of the piping system other than the pump or turbine.

FIGURE 5–54
A typical power plant has numerous pipes, elbows, valves, pumps, and turbines, all of which have irreversible losses.
© Brand X Pictures PunchStock RF

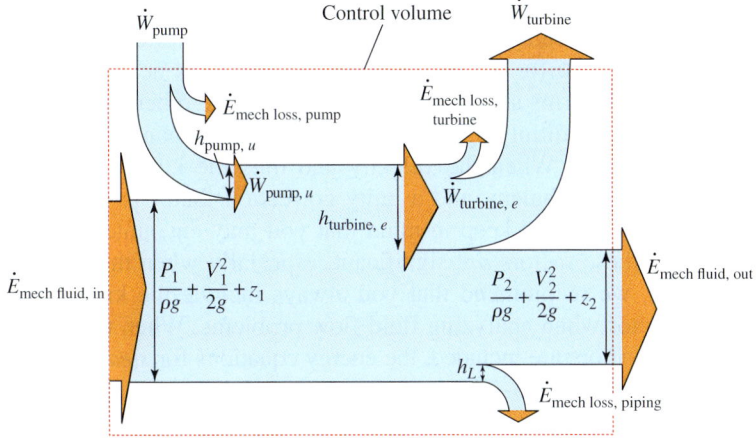

FIGURE 5–55
Mechanical energy flow chart for a fluid flow system that involves a pump and a turbine. Vertical dimensions show each energy term expressed as an equivalent column height of fluid, i.e., *head*, corresponding to each term of Eq. 5–74.

Note that the head loss h_L represents the frictional losses associated with fluid flow in piping, and it does not include the losses that occur within the pump or turbine due to the inefficiencies of these devices—these losses are taken into account by η_{pump} and $\eta_{turbine}$. Equation 5–74 is illustrated schematically in Fig. 5–55.

The *pump head* is zero if the piping system does not involve a pump, a fan, or a compressor, and the *turbine head* is zero if the system does not involve a turbine.

Special Case: Incompressible Flow with No Mechanical Work Devices and Negligible Friction

When piping losses are negligible, there is negligible dissipation of mechanical energy into thermal energy, and thus $h_L = e_{mech\ loss,\ piping}/g \cong 0$, as shown later in Example 5–11. Also, $h_{pump,\ u} = h_{turbine,\ e} = 0$ when there are no mechanical work devices such as fans, pumps, or turbines. Then Eq. 5–74 reduces to

$$\frac{P_1}{\rho g} + \frac{V_1^2}{2g} + z_1 = \frac{P_2}{\rho g} + \frac{V_2^2}{2g} + z_2 \quad \text{or} \quad \frac{P}{\rho g} + \frac{V^2}{2g} + z = \text{constant} \quad (5\text{–}75)$$

which is the **Bernoulli equation** derived earlier using Newton's second law of motion. Thus, the Bernoulli equation can be thought of as a degenerate form of the energy equation.

Kinetic Energy Correction Factor, α

The average flow velocity V_{avg} was defined such that the relation $\rho V_{avg} A$ gives the actual mass flow rate. Therefore, there is no such thing as a correction factor for mass flow rate. However, as Gaspard Coriolis (1792–1843) showed, the kinetic energy of a fluid stream obtained from $V^2/2$ is not the same as the actual kinetic energy of the fluid stream since the square of a sum is not equal to the sum of the squares of its components (Fig. 5–56). This error can be corrected by replacing the kinetic energy terms $V^2/2$ in the energy equation by $\alpha V_{avg}^2/2$, where α is the **kinetic energy correction factor.** By using equations for the variation of velocity with the radial distance, it can be shown that the correction factor is 2.0 for fully developed laminar pipe flow, and it ranges between 1.04 and 1.11 for fully developed turbulent flow in a round pipe.

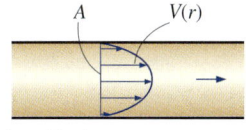

$$\dot{m} = \rho V_{avg} A, \quad \rho = \text{constant}$$

$$\dot{KE}_{act} = \int ke\, \delta \dot{m} = \int_A \frac{1}{2} [V(r)]^2 [\rho V(r)\, dA]$$

$$= \frac{1}{2}\rho \int_A [V(r)]^3\, dA$$

$$\dot{KE}_{avg} = \frac{1}{2} \dot{m} V_{avg}^2 = \frac{1}{2} \rho A V_{avg}^3$$

$$\alpha = \frac{\dot{KE}_{act}}{\dot{KE}_{avg}} = \frac{1}{A} \int_A \left(\frac{V(r)}{V_{avg}}\right)^3 dA$$

FIGURE 5–56
The determination of the *kinetic energy correction factor* using the actual velocity distribution $V(r)$ and the average velocity V_{avg} at a cross section.

The kinetic energy correction factors are often ignored (i.e., α is set equal to 1) in an elementary analysis since (1) most flows encountered in practice are turbulent, for which the correction factor is near unity, and (2) the kinetic energy terms are often small relative to the other terms in the energy equation, and multiplying them by a factor less than 2.0 does not make much difference. When the velocity and thus the kinetic energy are high, the flow turns turbulent, and a unity correction factor is more appropriate. However, you should keep in mind that you may encounter some situations for which these factors *are* significant, especially when the flow is laminar. Therefore, we recommend that you always include the kinetic energy correction factor when analyzing fluid flow problems. When the kinetic energy correction factors are included, the energy equations for *steady incompressible flow* (Eqs. 5–73 and 5–74) become

$$\dot{m}\left(\frac{P_1}{\rho} + \alpha_1 \frac{V_1^2}{2} + gz_1\right) + \dot{W}_{pump} = \dot{m}\left(\frac{P_2}{\rho} + \alpha_2 \frac{V_2^2}{2} + gz_2\right) + \dot{W}_{turbine} + \dot{E}_{mech,\,loss}$$

(5–76)

$$\frac{P_1}{\rho g} + \alpha_1 \frac{V_1^2}{2g} + z_1 + h_{pump,\,u} = \frac{P_2}{\rho g} + \alpha_2 \frac{V_2^2}{2g} + z_2 + h_{turbine,\,e} + h_L \quad (5\text{–}77)$$

If the flow at an inlet or outlet is fully developed turbulent pipe flow, we recommend using $\alpha = 1.05$ as a reasonable estimate of the correction factor. This leads to a more conservative estimate of head loss, and it does not take much additional effort to include α in the equations.

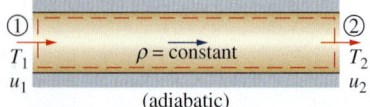

FIGURE 5–57
Schematic for Example 5–11.

EXAMPLE 5–11   Effect of Friction on Fluid Temperature and Head Loss

Show that during steady and incompressible flow of a fluid in an adiabatic flow section (*a*) the temperature remains constant and there is no head loss when friction is ignored and (*b*) the temperature increases and some head loss occurs when frictional effects are considered. Discuss if it is possible for the fluid temperature to decrease during such flow (Fig. 5–57).

SOLUTION Steady and incompressible flow through an adiabatic section is considered. The effects of friction on the temperature and the heat loss are to be determined.
Assumptions 1 The flow is steady and incompressible. 2 The flow section is adiabatic and thus there is no heat transfer, $q_{net\,in} = 0$.
Analysis The density of a fluid remains constant during incompressible flow and the entropy change is

$$\Delta s = c_v \ln \frac{T_2}{T_1}$$

This relation represents the entropy change of the fluid per unit mass as it flows through the flow section from state 1 at the inlet to state 2 at the outlet. Entropy change is caused by two effects: (1) heat transfer and (2) irreversibilities. Therefore, in the absence of heat transfer, entropy change is due to irreversibilities only, whose effect is always to increase entropy.

(*a*) The entropy change of the fluid in an adiabatic flow section ($q_{\text{net in}} = 0$) is zero when the process does not involve any irreversibilities such as friction and swirling, and thus for *reversible flow* we have

Temperature change: $\Delta s = c_v \ln \dfrac{T_2}{T_1} = 0 \;\;\rightarrow\;\; T_2 = T_1$

Mechanical energy loss:

$$e_{\text{mech loss, piping}} = u_2 - u_1 - q_{\text{net in}} = c_v(T_2 - T_1) - q_{\text{net in}} = 0$$

Head loss: $h_L = e_{\text{mech loss, piping}}/g = 0$

Thus we conclude that when heat transfer and frictional effects are negligible, (1) the temperature of the fluid remains constant, (2) no mechanical energy is converted to thermal energy, and (3) there is no irreversible head loss.

(*b*) When irreversibilities such as friction are taken into account, the entropy change is positive and thus we have:

Temperature change: $\Delta s = c_v \ln \dfrac{T_2}{T_1} > 0 \;\rightarrow\; T_2 > T_1$

Mechanical energy loss: $e_{\text{mech loss, piping}} = u_2 - u_1 - q_{\text{net in}} = c_v(T_2 - T_1) > 0$

Head loss: $h_L = e_{\text{mech loss, piping}}/g > 0$

Thus we conclude that when the flow is adiabatic and irreversible, (1) the temperature of the fluid increases, (2) some mechanical energy is converted to thermal energy, and (3) some irreversible head loss occurs.

Discussion It is impossible for the fluid temperature to decrease during steady, incompressible, adiabatic flow since this would require the entropy of an adiabatic system to decrease, which would be a violation of the second law of thermodynamics.

■ EXAMPLE 5–12   Pumping Power and Frictional Heating in a Pump

The pump of a water distribution system is powered by a 15-kW electric motor whose efficiency is 90 percent (Fig. 5–58). The water flow rate through the pump is 50 L/s. The diameters of the inlet and outlet pipes are the same, and the elevation difference across the pump is negligible. If the absolute pressures at the inlet and outlet of the pump are measured to be 100 kPa and 300 kPa, respectively, determine (*a*) the mechanical efficiency of the pump and (*b*) the temperature rise of water as it flows through the pump due to mechanical inefficiencies.

SOLUTION The pressures across a pump are measured. The mechanical efficiency of the pump and the temperature rise of water are to be determined.

Assumptions **1** The flow is steady and incompressible. **2** The pump is driven by an external motor so that the heat generated by the motor is dissipated to the atmosphere. **3** The elevation difference between the inlet and outlet of the pump is negligible, $z_1 \cong z_2$. **4** The inlet and outlet diameters are the

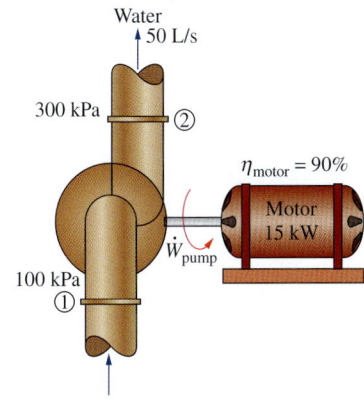

FIGURE 5–58
Schematic for Example 5–12.

same and thus the average inlet and outlet velocities are equal, $V_1 = V_2$.
5 The kinetic energy correction factors are equal, $\alpha_1 = \alpha_2$.
Properties We take the density of water to be 1 kg/L = 1000 kg/m³ and its specific heat to be 4.18 kJ/kg·°C.
Analysis (*a*) The mass flow rate of water through the pump is

$$\dot{m} = \rho \dot{V} = (1 \text{ kg/L})(50 \text{ L/s}) = 50 \text{ kg/s}$$

The motor draws 15 kW of power and is 90 percent efficient. Thus the mechanical (shaft) power it delivers to the pump is

$$\dot{W}_{\text{pump, shaft}} = \eta_{\text{motor}} \dot{W}_{\text{electric}} = (0.90)(15 \text{ kW}) = 13.5 \text{ kW}$$

To determine the mechanical efficiency of the pump, we need to know the increase in the mechanical energy of the fluid as it flows through the pump, which is

$$\Delta \dot{E}_{\text{mech, fluid}} = \dot{E}_{\text{mech, out}} - \dot{E}_{\text{mech, in}} = \dot{m}\left(\frac{P_2}{\rho} + \alpha_2 \frac{V_2^2}{2} + gz_2\right) - \dot{m}\left(\frac{P_1}{\rho} + \alpha_1 \frac{V_1^2}{2} + gz_1\right)$$

Simplifying it for this case and substituting the given values,

$$\Delta \dot{E}_{\text{mech, fluid}} = \dot{m}\left(\frac{P_2 - P_1}{\rho}\right) = (50 \text{ kg/s})\left(\frac{(300 - 100) \text{ kPa}}{1000 \text{ kg/m}^3}\right)\left(\frac{1 \text{ kJ}}{1 \text{ kPa} \cdot \text{m}^3}\right) = 10.0 \text{ kW}$$

Then the mechanical efficiency of the pump becomes

$$\eta_{\text{pump}} = \frac{\dot{W}_{\text{pump, }u}}{\dot{W}_{\text{pump, shaft}}} = \frac{\Delta \dot{E}_{\text{mech, fluid}}}{\dot{W}_{\text{pump, shaft}}} = \frac{10.0 \text{ kW}}{13.5 \text{ kW}} = 0.741 \text{ or } \mathbf{74.1\%}$$

(*b*) Of the 13.5-kW mechanical power supplied by the pump, only 10.0 kW is imparted to the fluid as mechanical energy. The remaining 3.5 kW is converted to thermal energy due to frictional effects, and this "lost" mechanical energy manifests itself as a heating effect in the fluid,

$$\dot{E}_{\text{mech, loss}} = \dot{W}_{\text{pump,shaft}} - \Delta \dot{E}_{\text{mech, fluid}} = 13.5 - 10.0 = 3.5 \text{kW}$$

The temperature rise of water due to this mechanical inefficiency is determined from the thermal energy balance, $\dot{E}_{\text{mech, loss}} = \dot{m}(u_2 - u_1) = \dot{m}c\Delta T$. Solving for ΔT,

$$\Delta T = \frac{\dot{E}_{\text{mech, loss}}}{\dot{m}c} = \frac{3.5 \text{ kW}}{(50 \text{ kg/s})(4.18 \text{ kJ/kg} \cdot {}^\circ\text{C})} = \mathbf{0.017{}^\circ\text{C}}$$

Therefore, the water experiences a temperature rise of 0.017°C which is very small, due to mechanical inefficiency, as it flows through the pump.
Discussion In an actual application, the temperature rise of water would probably be less since part of the heat generated would be transferred to the casing of the pump and from the casing to the surrounding air. If the entire pump and motor were submerged in water, then the 1.5 kW dissipated due to motor inefficiency would also be transferred to the surrounding water as heat.

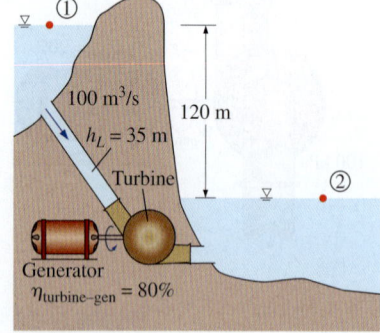

FIGURE 5–59
Schematic for Example 5–13.

EXAMPLE 5–13   Hydroelectric Power Generation from a Dam

In a hydroelectric power plant, 100 m³/s of water flows from an elevation of 120 m to a turbine, where electric power is generated (Fig. 5–59). The total irreversible head loss in the piping system from point 1 to point 2 (excluding the turbine unit) is determined to be 35 m. If the overall efficiency of the turbine–generator is 80 percent, estimate the electric power output.

SOLUTION The available head, flow rate, head loss, and efficiency of a hydroelectric turbine are given. The electric power output is to be determined.
Assumptions 1 The flow is steady and incompressible. 2 Water levels at the reservoir and the discharge site remain constant.
Properties We take the density of water to be 1000 kg/m³.
Analysis The mass flow rate of water through the turbine is

$$\dot{m} = \rho \dot{V} = (1000 \text{ kg/m}^3)(100 \text{ m}^3/\text{s}) = 10^5 \text{ kg/s}$$

We take point 2 as the reference level, and thus $z_2 = 0$. Also, both points 1 and 2 are open to the atmosphere ($P_1 = P_2 = P_{atm}$) and the flow velocities are negligible at both points ($V_1 = V_2 = 0$). Then the energy equation for steady, incompressible flow reduces to

$$\cancel{\frac{P_1}{\rho g}} + \cancel{\alpha_1 \frac{V_1^2}{2g}}^0 + z_1 + \cancel{h_{pump,u}}^0 = \cancel{\frac{P_2}{\rho g}} + \cancel{\alpha_2 \frac{V_2^2}{2g}} + \cancel{z_2}^0 + h_{turbine,e} + h_L$$

or

$$h_{turbine,e} = z_1 - h_L$$

Substituting, the extracted turbine head and the corresponding turbine power are

$$h_{turbine,e} = z_1 - h_L = 120 - 35 = 85 \text{ m}$$

$$\dot{W}_{turbine,e} = \dot{m}gh_{turbine,e} = (10^5 \text{ kg/s})(9.81 \text{ m/s}^2)(85 \text{ m})\left(\frac{1 \text{ kJ/kg}}{1000 \text{ m}^2/\text{s}^2}\right) = 83,400 \text{ kW}$$

Therefore, a perfect turbine–generator would generate 83,400 kW of electricity from this resource. The electric power generated by the actual unit is

$$\dot{W}_{electric} = \eta_{turbine-gen} \dot{W}_{turbine,e} = (0.80)(83.4 \text{ MW}) = \mathbf{66.7 \text{ MW}}$$

Discussion Note that the power generation would increase by almost 1 MW for each percentage point improvement in the efficiency of the turbine–generator unit. You will learn how to determine h_L in Chap. 8.

EXAMPLE 5–14 Fan Selection for Air Cooling of a Computer

A fan is to be selected to cool a computer case whose dimensions are 12 cm × 40 cm × 40 cm (Fig. 5–60). Half of the volume in the case is expected to be filled with components and the other half to be air space. A 5-cm-diameter hole is available at the back of the case for the installation of the fan that is to replace the air in the void spaces of the case once every second. Small low-power fan–motor combined units are available in the market and their efficiency is estimated to be 30 percent. Determine (a) the wattage of the fan–motor unit to be purchased and (b) the pressure difference across the fan. Take the air density to be 1.20 kg/m³.

Solution A fan is to cool a computer case by completely replacing the air inside once every second. The power of the fan and the pressure difference across it are to be determined.
Assumptions 1 The flow is steady and incompressible. 2 Losses other than those due to the inefficiency of the fan–motor unit are negligible. 3 The flow

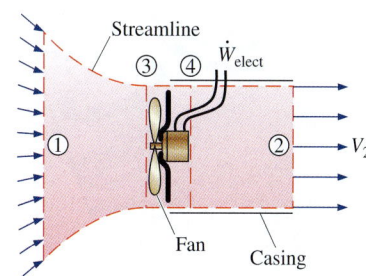

FIGURE 5–60
Schematic for Example 5–14.

at the outlet is fairly uniform except near the center (due to the wake of the fan motor), and the kinetic energy correction factor at the outlet is 1.10.
Properties The density of air is given to be 1.20 kg/m³.
Analysis (*a*) Noting that half of the volume of the case is occupied by the components, the air volume in the computer case is

$$V = (\text{Void fraction})(\text{Total case volume})$$
$$= 0.5(12 \text{ cm} \times 40 \text{ cm} \times 40 \text{ cm}) = 9600 \text{ cm}^3$$

Therefore, the volume and mass flow rates of air through the case are

$$\dot{V} = \frac{V}{\Delta t} = \frac{9600 \text{ cm}^3}{1 \text{ s}} = 9600 \text{ cm}^3/\text{s} = 9.6 \times 10^{-3} \text{ m}^3/\text{s}$$

$$\dot{m} = \rho \dot{V} = (1.20 \text{ kg/m}^3)(9.6 \times 10^{-3} \text{ m}^3/\text{s}) = 0.0115 \text{ kg/s}$$

The cross-sectional area of the opening in the case and the average air velocity through the outlet are

$$A = \frac{\pi D^2}{4} = \frac{\pi (0.05 \text{ m})^2}{4} = 1.96 \times 10^{-3} \text{ m}^2$$

$$V = \frac{\dot{V}}{A} = \frac{9.6 \times 10^{-3} \text{ m}^3/\text{s}}{1.96 \times 10^{-3} \text{ m}^2} = 4.90 \text{ m/s}$$

We draw the control volume around the fan such that both the inlet and the outlet are at atmospheric pressure ($P_1 = P_2 = P_{atm}$), as shown in Fig. 5–60, where the inlet section 1 is large and far from the fan so that the flow velocity at the inlet section is negligible ($V_1 \cong 0$). Noting that $z_1 = z_2$ and frictional losses in the flow are disregarded, the mechanical losses consist of fan losses only and the energy equation (Eq. 5–76) simplifies to

$$\dot{m}\left(\cancel{\frac{P_1}{\rho}} + \alpha_1\cancel{\frac{V_1^2}{2}}^{0} + \cancel{gz_1}\right) + \dot{W}_{\text{fan}} = \dot{m}\left(\cancel{\frac{P_2}{\rho}} + \alpha_2\frac{V_2^2}{2} + \cancel{gz_2}\right) + \cancel{\dot{W}_{\text{turbine}}}^{0} + \dot{E}_{\text{mech loss, fan}}$$

Solving for $\dot{W}_{\text{fan}} - \dot{E}_{\text{mech loss, fan}} = \dot{W}_{\text{fan}, u}$ and substituting,

$$\dot{W}_{\text{fan}, u} = \dot{m}\alpha_2\frac{V_2^2}{2} = (0.0115 \text{ kg/s})(1.10)\frac{(4.90 \text{ m/s})^2}{2}\left(\frac{1 \text{ N}}{1 \text{ kg} \cdot \text{m/s}^2}\right) = 0.152 \text{ W}$$

Then the required electric power input to the fan is determined to be

$$\dot{W}_{\text{elect}} = \frac{\dot{W}_{\text{fan}, u}}{\eta_{\text{fan-motor}}} = \frac{0.152 \text{ W}}{0.3} = \mathbf{0.506 \text{ W}}$$

Therefore, a fan–motor rated at about a half watt is adequate for this job (Fig. 5–61). (*b*) To determine the pressure difference across the fan unit, we take points 3 and 4 to be on the two sides of the fan on a horizontal line. This time $z_3 = z_4$ again and $V_3 = V_4$ since the fan is a narrow cross section, and the energy equation reduces to

$$\dot{m}\frac{P_3}{\rho} + \dot{W}_{\text{fan}} = \dot{m}\frac{P_4}{\rho} + \dot{E}_{\text{mech loss, fan}} \quad \rightarrow \quad \dot{W}_{\text{fan}, u} = \dot{m}\frac{P_4 - P_3}{\rho}$$

FIGURE 5–61
The cooling fans used in computers and computer power supplies are typically small and consume only a few watts of electrical power.
© PhotoDisc/Getty RF

Solving for $P_4 - P_3$ and substituting,

$$P_4 - P_3 = \frac{\rho \dot{W}_{fan,u}}{\dot{m}} = \frac{(1.2 \text{ kg/m}^3)(0.152 \text{ W})}{0.0115 \text{ kg/s}} \left(\frac{1\text{Pa}\cdot\text{m}^3}{1 \text{ Ws}}\right) = \mathbf{15.8 \text{ Pa}}$$

Therefore, the pressure rise across the fan is 15.8 Pa.

Discussion The efficiency of the fan–motor unit is given to be 30 percent, which means 30 percent of the electric power $\dot{W}_{electric}$ consumed by the unit is converted to useful mechanical energy while the rest (70 percent) is "lost" and converted to thermal energy. Also, a more powerful fan is required in an actual system to overcome frictional losses inside the computer case. Note that if we had ignored the kinetic energy correction factor at the outlet, the required electrical power and pressure rise would have been 10 percent lower in this case (0.460 W and 14.4 Pa, respectively).

EXAMPLE 5–15 Pumping Water from a Lake to a Pool

A submersible pump with a shaft power of 5 kW and an efficiency of 72 percent is used to pump water from a lake to a pool through a constant diameter pipe (Fig. 5–62). The free surface of the pool is 25 m above the free surface of the lake. If the irreversible head loss in the piping system is 4 m, determine the discharge rate of water and the pressure difference across the pump.

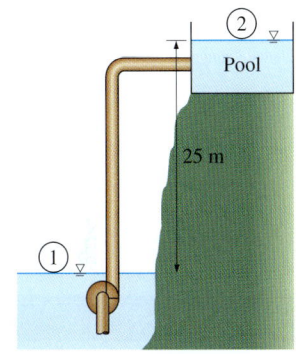

FIGURE 5–62
Schematic for Example 5–15.

SOLUTION Water from a lake is pumped to a pool at a given elevation. For a given head loss, the flow rate and the pressure difference across the pump are to be determined.
Assumptions 1 The flow is steady and incompressible. 2 Both the lake and pool are large enough that their surface elevations remain fixed.
Properties We take the density of water to be 1 kg/L = 1000 kg/m³.
Analysis The pump delivers 5 kW of shaft power and is 72 percent efficient. The useful mechanical power it imparts to the water is

$$\dot{W}_{pump,u} = \eta_{pump} \dot{W}_{shaft} = (0.72)(5 \text{ kW}) = 3.6 \text{ kW}$$

We take point 1 at the free surface of the lake, which is also taken as the reference level ($z_1 = 0$), and point 2 at the free surface of the pool. Also, both points 1 and 2 are open to the atmosphere ($P_1 = P_2 = P_{atm}$), and the velocities are negligible there ($V_1 \cong V_2 \cong 0$). Then the energy equation for steady, incompressible flow through a control volume between these two surfaces that includes the pump is expressed as

$$\dot{m}\left(\frac{P_1}{\rho} + \alpha_1 \frac{V_1^2}{2} + gz_1\right) + \dot{W}_{pump,u} = \dot{m}\left(\frac{P_2}{\rho} + \alpha_2 \frac{V_2^2}{2} + gz_2\right)$$
$$+ \dot{W}_{turbine,e} + \dot{E}_{mech\,loss,\,piping}$$

Under the stated assumptions, the energy equation reduces to

$$\dot{W}_{pump,u} = \dot{m}gz_2 + \dot{E}_{mech\,loss,\,piping}$$

Noting that $\dot{E}_{\text{mech loss, piping}} = \dot{m}gh_L$, the mass and volume flow rates of water become

$$\dot{m} = \frac{\dot{W}_{\text{pump},u}}{gz_2 + gh_L} = \frac{\dot{W}_{\text{pump},u}}{g(z_2 + h_L)} = \frac{3.6 \text{ kJ/s}}{(9.81 \text{m/s}^2)(25 + 4 \text{ m})}\left(\frac{1000 \text{ m}^2/\text{s}^2}{1 \text{ kJ}}\right) = 12.7 \text{ kg/s}$$

$$\dot{V} = \frac{\dot{m}}{\rho} = \frac{12.7 \text{ kg/s}}{1000 \text{ kg/m}^3} = 12.7 \times 10^{-3} \text{ m}^3/\text{s} = 12.7 \text{ L/s}$$

We now take the pump as the control volume. Assuming that the elevation difference and the kinetic energy change across the pump are negligible, the energy equation for this control volume yields

$$\Delta P = P_{\text{out}} - P_{\text{in}} = \frac{\dot{W}_{\text{pump},u}}{\dot{V}} = \frac{3.6 \text{ kJ/s}}{12.7 \times 10^{-3} \text{ m}^3/\text{s}}\left(\frac{1 \text{ kN·m}}{1 \text{ kJ}}\right)\left(\frac{1 \text{ kPa}}{1 \text{ kN/m}^2}\right)$$

$$= 283 \text{ kPa}$$

Discussion It can be shown that in the absence of head loss ($h_L = 0$) the flow rate of water would be 14.7 L/s, which is an increase of 16 percent. Therefore, frictional losses in pipes should be minimized since they always cause the flow rate to decrease.

SUMMARY

This chapter deals with the mass, Bernoulli, and energy equations and their applications. The amount of mass flowing through a cross section per unit time is called the *mass flow rate* and is expressed as

$$\dot{m} = \rho V A_c = \rho \dot{V}$$

where ρ is the density, V is the average velocity, $\dot{V}$ is the volume flow rate of the fluid, and A_c is the cross-sectional area normal to the flow direction. The conservation of mass relation for a control volume is expressed as

$$\frac{d}{dt}\int_{CV} \rho \, dV + \int_{CS} \rho(\vec{V}\cdot\vec{n}) \, dA = 0$$

It states that *the time rate of change of the mass within the control volume plus the net mass flow rate out of the control surface is equal to zero*.

In simpler terms,

$$\frac{dm_{CV}}{dt} = \sum_{\text{in}} \dot{m} - \sum_{\text{out}} \dot{m}$$

For steady-flow devices, the conservation of mass principle is expressed as

Steady flow:
$$\sum_{\text{in}} \dot{m} = \sum_{\text{out}} \dot{m}$$

Steady flow (single stream):
$$\dot{m}_1 = \dot{m}_2 \rightarrow \rho_1 V_1 A_1 = \rho_2 V_2 A_2$$

Steady, incompressible flow:
$$\sum_{\text{in}} \dot{V} = \sum_{\text{out}} \dot{V}$$

Steady, incompressible flow (single stream):
$$\dot{V}_1 = \dot{V}_2 \rightarrow V_1 A_1 = V_2 A_2$$

Mechanical energy is the form of energy associated with the velocity, elevation, and pressure of the fluid, and it can be converted to mechanical work completely and directly by an ideal mechanical device. The efficiencies of various *real* devices are defined as

$$\eta_{\text{pump}} = \frac{\Delta \dot{E}_{\text{mech, fluid}}}{\dot{W}_{\text{shaft, in}}} = \frac{\dot{W}_{\text{pump},u}}{\dot{W}_{\text{pump}}}$$

$$\eta_{\text{turbine}} = \frac{\dot{W}_{\text{shaft, out}}}{|\Delta \dot{E}_{\text{mech, fluid}}|} = \frac{\dot{W}_{\text{turbine}}}{\dot{W}_{\text{turbine},e}}$$

$$\eta_{\text{motor}} = \frac{\text{Mechanical power output}}{\text{Electric power input}} = \frac{\dot{W}_{\text{shaft, out}}}{\dot{W}_{\text{elect, in}}}$$

$$\eta_{generator} = \frac{\text{Electric power output}}{\text{Mechanical power input}} = \frac{\dot{W}_{elect,\,out}}{\dot{W}_{shaft,\,in}}$$

$$\eta_{pump\text{-}motor} = \eta_{pump}\eta_{motor} = \frac{\Delta \dot{E}_{mech,\,fluid}}{\dot{W}_{elect,\,in}} = \frac{\dot{W}_{pump,\,u}}{\dot{W}_{elect,\,in}}$$

$$\eta_{turbine\text{-}gen} = \eta_{turbine}\eta_{generator} = \frac{\dot{W}_{elect,\,out}}{|\Delta \dot{E}_{mech,\,fluid}|} = \frac{\dot{W}_{elect,\,out}}{\dot{W}_{turbine,\,e}}$$

The *Bernoulli equation* is a relation between pressure, velocity, and elevation in steady, incompressible flow, and is expressed along a streamline and in regions where net viscous forces are negligible as

$$\frac{P}{\rho} + \frac{V^2}{2} + gz = \text{constant}$$

It can also be expressed between any two points on a streamline as

$$\frac{P_1}{\rho} + \frac{V_1^2}{2} + gz_1 = \frac{P_2}{\rho} + \frac{V_2^2}{2} + gz_2$$

The Bernoulli equation is an expression of mechanical energy balance and can be stated as: *The sum of the kinetic, potential, and flow energies of a fluid particle is constant along a streamline during steady flow when the compressibility and frictional effects are negligible.* Multiplying the Bernoulli equation by density gives

$$P + \rho\frac{V^2}{2} + \rho gz = \text{constant}$$

where P is the *static pressure*, which represents the actual pressure of the fluid; $\rho V^2/2$ is the *dynamic pressure*, which represents the pressure rise when the fluid in motion is brought to a stop; and ρgz is the *hydrostatic pressure*, which accounts for the effects of fluid weight on pressure. The sum of the static, dynamic, and hydrostatic pressures is called the *total pressure*. The Bernoulli equation states that *the total pressure along a streamline is constant*. The sum of the static and dynamic pressures is called the *stagnation pressure*, which represents the pressure at a point where the fluid is brought to a complete stop in an isentropic manner.

The Bernoulli equation can also be represented in terms of "heads" by dividing each term by g,

$$\frac{P}{\rho g} + \frac{V^2}{2g} + z = H = \text{constant}$$

where $P/\rho g$ is the *pressure head*, which represents the height of a fluid column that produces the static pressure P; $V^2/2g$ is the *velocity head*, which represents the elevation needed for a fluid to reach the velocity V during frictionless free fall; and z is the *elevation head*, which represents the potential energy of the fluid. Also, H is the *total head* for the flow. The curve that represents the sum of the static pressure and the elevation heads, $P/\rho g + z$, is called the *hydraulic grade line* (HGL), and the curve that represents the total head of the fluid, $P/\rho g + V^2/2g + z$, is called the *energy grade line* (EGL).

The *energy equation* for steady, incompressible flow is

$$\frac{P_1}{\rho g} + \alpha_1\frac{V_1^2}{2g} + z_1 + h_{pump,\,u}$$
$$= \frac{P_2}{\rho g} + \alpha_2\frac{V_2^2}{2g} + z_2 + h_{turbine,\,e} + h_L$$

where

$$h_{pump,\,u} = \frac{w_{pump,\,u}}{g} = \frac{\dot{W}_{pump,\,u}}{\dot{m}g} = \frac{\eta_{pump}\dot{W}_{pump}}{\dot{m}g}$$

$$h_{turbine,\,e} = \frac{w_{turbine,\,e}}{g} = \frac{\dot{W}_{turbine,\,e}}{\dot{m}g} = \frac{\dot{W}_{turbine}}{\eta_{turbine}\dot{m}g}$$

$$h_L = \frac{e_{mech\,loss,\,piping}}{g} = \frac{\dot{E}_{mech\,loss,\,piping}}{\dot{m}g}$$

$$e_{mech,\,loss} = u_2 - u_1 - q_{net\,in}$$

The mass, Bernoulli, and energy equations are three of the most fundamental relations in fluid mechanics, and they are used extensively in the chapters that follow. In Chap. 6, either the Bernoulli equation or the energy equation is used together with the mass and momentum equations to determine the forces and torques acting on fluid systems. In Chaps. 8 and 14, the mass and energy equations are used to determine the pumping power requirements in fluid systems and in the design and analysis of turbomachinery. In Chaps. 12 and 13, the energy equation is also used to some extent in the analysis of compressible flow and open-channel flow.

REFERENCES AND SUGGESTED READING

1. R. C. Dorf, ed. in chief. *The Engineering Handbook*, 2nd ed. Boca Raton, FL: CRC Press, 2004.
2. R. L. Panton. *Incompressible Flow*, 3rd ed. New York: Wiley, 2005.
3. M. Van Dyke. *An Album of Fluid Motion*. Stanford, CA: The Parabolic Press, 1982.

BERNOULLI AND ENERGY EQUATIONS

PROBLEMS*

Conservation of Mass

5–1C Define mass and volume flow rates. How are they related to each other?

5–2C Does the amount of mass entering a control volume have to be equal to the amount of mass leaving during an unsteady-flow process?

5–3C When is the flow through a control volume steady?

5–4C Consider a device with one inlet and one outlet. If the volume flow rates at the inlet and at the outlet are the same, is the flow through this device necessarily steady? Why?

5–5 In climates with low night-time temperatures, an energy-efficient way of cooling a house is to install a fan in the ceiling that draws air from the interior of the house and discharges it to a ventilated attic space. Consider a house whose interior air volume is 720 m³. If air in the house is to be exchanged once every 20 minutes, determine (a) the required flow rate of the fan and (b) the average discharge speed of air if the fan diameter is 0.5 m.

5–6 Air whose density is 1.3 kg/m³ enters the duct of an air-conditioning system at a volume flow rate of 12.7 m³/min. If the diameter of the duct is 40 cm, determine the velocity of the air at the duct inlet and the mass flow rate of air.

5–7 A 0.75-m³ rigid tank initially contains air whose density is 1.18 kg/m³. The tank is connected to a high-pressure supply line through a valve. The valve is opened, and air is allowed to enter the tank until the density in the tank rises to 4.95 kg/m³. Determine the mass of air that has entered the tank. *Answer:* 2.83 kg

5–8 Consider the flow of an incompressible Newtonian fluid between two parallel plates. If the upper plate moves to right with $u_1 = 3$ m/s while the bottom one moves to the left with $u_2 = 0.75$ m/s, what would be the net flow rate at a cross-section between two plates? Take the plate width to be $b = 5$ cm.

5–9 Consider a fully filled tank of semi-circular cross section tank with radius R and width of b into the page, as shown in Fig. P5-9. If the water is pumped out of the tank at flow rate of $\dot{V} = Kh^2$, where K is a positive constant and h is the water depth at time t. Determine the time needed to drop the water level to a specified h value of h_o in terms of R, K, and h_o.

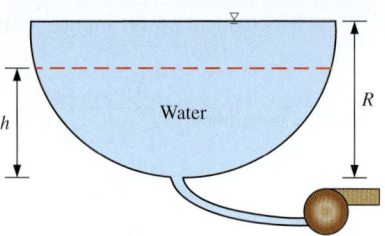

FIGURE P5–9

5–10 A desktop computer is to be cooled by a fan whose flow rate is 0.40 m³/min. Determine the mass flow rate of air through the fan at an elevation of 3400 m where the air density is 0.7 kg/m³. Also, if the average velocity of air is not to exceed 110 m/min, determine the minimum diameter of the casing of the fan. *Answers:* 0.00467 kg/s, 0.0569 m

5–11 A smoking lounge is to accommodate 40 heavy smokers. The minimum fresh air requirement for smoking lounges is specified to be 30 L/s per person (ASHRAE, Standard 62, 1989). Determine the minimum required flow rate of fresh air that needs to be supplied to the lounge, and the minimum diameter of the duct if the air velocity is not to exceed 8 m/s.

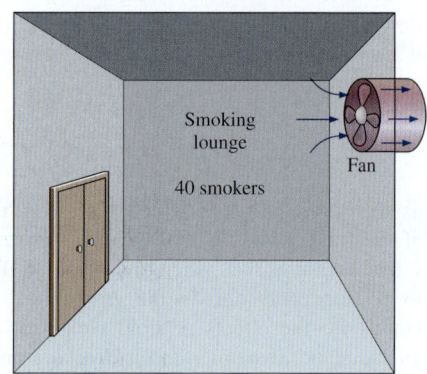

FIGURE P5–11

5–12 The minimum fresh air requirement of a residential building is specified to be 0.35 air changes per hour (ASHRAE, Standard 62, 1989). That is, 35 percent of the entire air contained in a residence should be replaced by fresh outdoor air every hour. If the ventilation requirement of a 2.7-m-high, 200-m² residence is to be met entirely by a fan, determine the flow capacity in L/min of the fan that needs to be installed. Also determine the minimum diameter of the duct if the average air velocity is not to exceed 5 m/s.

5–13 Air enters a nozzle steadily at 2.21 kg/m³ and 20 m/s and leaves at 0.762 kg/m³ and 150 m/s. If the inlet area of the nozzle is 60 cm², determine (a) the mass flow rate through

---

* Problems designated by a "C" are concept questions, and students are encouraged to answer them all. Problems with the icon are solved using EES, and complete solutions together with parametric studies are included on the text website. Problems with the icon are comprehensive in nature and are intended to be solved with an equation solver such as EES.

the nozzle, and (*b*) the exit area of the nozzle. *Answers:* (*a*) 0.265 kg/s, (*b*) 23.2 cm²

5–14 Air at 40°C flow steadily through the pipe shown in Fig. P5–14. If P_1 = 50 kPa (gage), P_2 = 10 kPa (gage), $D = 3d$, $P_{atm} \cong 100$ kPa, the average velocity at section 2 is V_2 = 30 m/s, and air temperature remains nearly constant, determine the average speed at section 1.

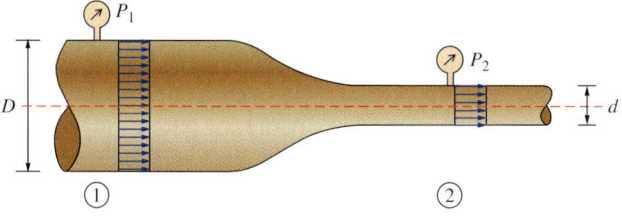

FIGURE P5–14

5–15 A hair dryer is basically a duct of constant diameter in which a few layers of electric resistors are placed. A small fan pulls the air in and forces it through the resistors where it is heated. If the density of air is 1.20 kg/m³ at the inlet and 1.05 kg/m³ at the exit, determine the percent increase in the velocity of air as it flows through the hair dryer.

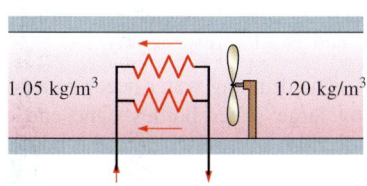

FIGURE P5–15

Mechanical Energy and Efficiency

5–16C Define turbine efficiency, generator efficiency, and combined turbine–generator efficiency.

5–17C What is mechanical efficiency? What does a mechanical efficiency of 100 percent mean for a hydraulic turbine?

5–18C How is the combined pump–motor efficiency of a pump and motor system defined? Can the combined pump–motor efficiency be greater than either the pump or the motor efficiency?

5–19C What is mechanical energy? How does it differ from thermal energy? What are the forms of mechanical energy of a fluid stream?

5–20 At a certain location, wind is blowing steadily at 8 m/s. Determine the mechanical energy of air per unit mass and the power generation potential of a wind turbine with 50-m-diameter blades at that location. Also determine the actual electric power generation assuming an overall efficiency of 30 percent. Take the air density to be 1.25 kg/m³.

5–21 Reconsider Prob. 5–20. Using EES (or other) software, investigate the effect of wind velocity and the blade span diameter on wind power generation. Let the velocity vary from 5 to 20 m/s in increments of 5 m/s, and the diameter to vary from 20 to 80 m in increments of 20 m. Tabulate the results, and discuss their significance.

5–22 Electric power is to be generated by installing a hydraulic turbine–generator at a site 110 m below the free surface of a large water reservoir that can supply water steadily at a rate of 900 kg/s. If the mechanical power output of the turbine is 800 kW and the electric power generation is 750 kW, determine the turbine efficiency and the combined turbine–generator efficiency of this plant. Neglect losses in the pipes.

5–23 Consider a river flowing toward a lake at an average speed of 4 m/s at a rate of 500 m³/s at a location 70 m above the lake surface. Determine the total mechanical energy of the river water per unit mass and the power generation potential of the entire river at that location. *Answer:* 347 MW

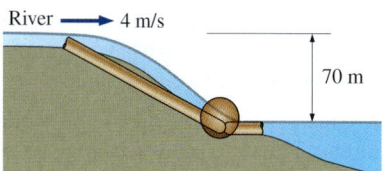

FIGURE P5–23

5–24 Water is pumped from a lake to a storage tank 18 m above at a rate of 70 L/s while consuming 20.4 kW of electric power. Disregarding any frictional losses in the pipes and any changes in kinetic energy, determine (*a*) the overall efficiency of the pump–motor unit and (*b*) the pressure difference between the inlet and the exit of the pump.

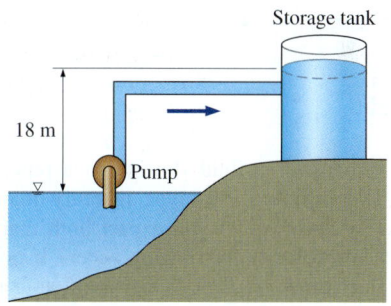

FIGURE P5–24

Bernoulli Equation

5–25C What is stagnation pressure? Explain how it can be measured.

5–26C Express the Bernoulli equation in three different ways using (*a*) energies, (*b*) pressures, and (*c*) heads.

5–27C What are the three major assumptions used in the derivation of the Bernoulli equation?

5–28C Define static, dynamic, and hydrostatic pressure. Under what conditions is their sum constant for a flow stream?

BERNOULLI AND ENERGY EQUATIONS

5–29C What is streamwise acceleration? How does it differ from normal acceleration? Can a fluid particle accelerate in steady flow?

5–30C Define pressure head, velocity head, and elevation head for a fluid stream and express them for a fluid stream whose pressure is P, velocity is V, and elevation is z.

5–31C Explain how and why a siphon works. Someone proposes siphoning cold water over a 7-m-high wall. Is this feasible? Explain.

5–32C How is the location of the hydraulic grade line determined for open-channel flow? How is it determined at the outlet of a pipe discharging to the atmosphere?

5–33C In a certain application, a siphon must go over a high wall. Can water or oil with a specific gravity of 0.8 go over a higher wall? Why?

5–34C What is the hydraulic grade line? How does it differ from the energy grade line? Under what conditions do both lines coincide with the free surface of a liquid?

5–35C A glass manometer with oil as the working fluid is connected to an air duct as shown in Fig. P5–35C. Will the oil levels in the manometer be as in Fig. P5–35C*a* or *b*? Explain. What would your response be if the flow direction is reversed?

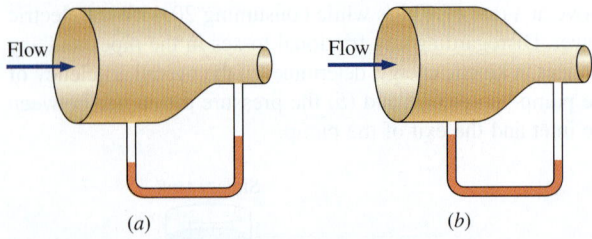

FIGURE P5–35C

5–36C The velocity of a fluid flowing in a pipe is to be measured by two different Pitot-type mercury manometers shown in Fig. P5–36C. Would you expect both manometers to predict the same velocity for flowing water? If not, which would be more accurate? Explain. What would your response be if air were flowing in the pipe instead of water?

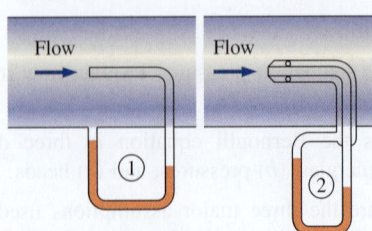

FIGURE P5–36C

5–37C The water level of a tank on a building roof is 20 m above the ground. A hose leads from the tank bottom to the ground. The end of the hose has a nozzle, which is pointed straight up. What is the maximum height to which the water could rise? What factors would reduce this height?

5–38C A student siphons water over a 8.5-m-high wall at sea level. She then climbs to the summit of Mount Shasta (elevation 4390 m, P_{atm} = 58.5 kPa) and attempts the same experiment. Comment on her prospects for success.

5–39 In a hydroelectric power plant, water enters the turbine nozzles at 800 kPa absolute with a low velocity. If the nozzle outlets are exposed to atmospheric pressure of 100 kPa, determine the maximum velocity to which water can be accelerated by the nozzles before striking the turbine blades.

5–40 A Pitot-static probe is used to measure the speed of an aircraft flying at 3000 m. If the differential pressure reading is 3 kPa, determine the speed of the aircraft.

5–41 The air velocity in the duct of a heating system is to be measured by a Pitot-static probe inserted into the duct parallel to the flow. If the differential height between the water columns connected to the two outlets of the probe is 2.4 cm, determine (*a*) the flow velocity and (*b*) the pressure rise at the tip of the probe. The air temperature and pressure in the duct are 45°C and 98 kPa, respectively.

5–42 The drinking water needs of an office are met by large water bottles. One end of a 0.6-cm-diameter plastic hose is inserted into the bottle placed on a high stand, while the other end with an on/off valve is maintained 0.6 m below the bottom of the bottle. If the water level in the bottle is 0.45 m when it is full, determine how long it will take at the minimum to fill a 0.25-L glass (*a*) when the bottle is first opened and (*b*) when the bottle is almost empty. Neglect frictional losses.

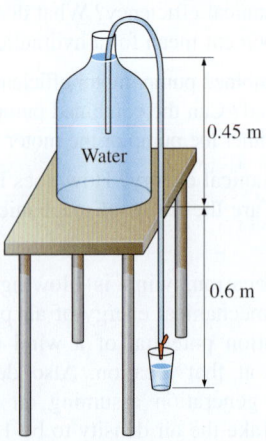

FIGURE P5–42

5–43 A piezometer and a Pitot tube are tapped into a 4-cm-diameter horizontal water pipe, and the height of the water columns are measured to be 26 cm in the piezometer and 35 cm in the Pitot tube (both measured from the top surface of the pipe). Determine the velocity at the center of the pipe.

5–44 The diameter of a cylindrical water tank is D_o and its height is H. The tank is filled with water, which is open to the atmosphere. An orifice of diameter D with a smooth entrance (i.e., negligible losses) is open at the bottom. Develop a relation for the time required for the tank (*a*) to empty halfway and (*b*) to empty completely.

5–45 A siphon pumps water from a large reservoir to a lower tank that is initially empty. The tank also has a rounded orifice 6 m below the reservoir surface where the water leaves the tank. Both the siphon and the orifice diameters are 5 cm. Ignoring frictional losses, determine to what height the water will rise in the tank at equilibrium.

5–46 Water enters a tank of diameter D_T steadily at a mass flow rate of $\dot{m}_{in}$. An orifice at the bottom with diameter D_o allows water to escape. The orifice has a rounded entrance, so the frictional losses are negligible. If the tank is initially empty, (*a*) determine the maximum height that the water will reach in the tank and (*b*) obtain a relation for water height z as a function of time.

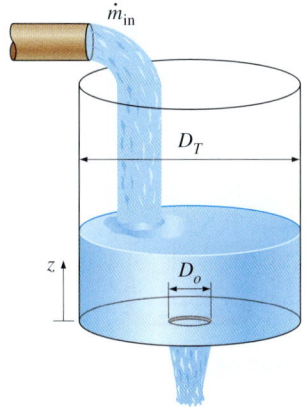

FIGURE P5–46

5–47 An airplane is flying at an altitude of 12,000 m. Determine the gage pressure at the stagnation point on the nose of the plane if the speed of the plane is 300 km/h. How would you solve this problem if the speed were 1050 km/h? Explain.

5–48 While traveling on a dirt road, the bottom of a car hits a sharp rock and a small hole develops at the bottom of its gas tank. If the height of the gasoline in the tank is 30 cm, determine the initial velocity of the gasoline at the hole. Discuss how the velocity will change with time and how the flow will be affected if the lid of the tank is closed tightly. *Answer:* 2.43 m/s

5–49 The water in an 8-m-diameter, 3-m-high above-ground swimming pool is to be emptied by unplugging a 3-cm-diameter, 25-m-long horizontal pipe attached to the bottom of the pool. Determine the maximum discharge rate of water through the pipe. Also, explain why the actual flow rate will be less.

5–50 Reconsider Prob. 5–49. Determine how long it will take to empty the swimming pool completely. *Answer:* 15.4 h

5–51 Reconsider Prob. 5–50. Using EES (or other) software, investigate the effect of the discharge pipe diameter on the time required to empty the pool completely. Let the diameter vary from 1 to 10 cm in increments of 1 cm. Tabulate and plot the results.

5–52 Air at 105 kPa and 37°C flows upward through a 6-cm-diameter inclined duct at a rate of 65 L/s. The duct diameter is then reduced to 4 cm through a reducer. The pressure change across the reducer is measured by a water manometer. The elevation difference between the two points on the pipe where the two arms of the manometer are attached is 0.20 m. Determine the differential height between the fluid levels of the two arms of the manometer.

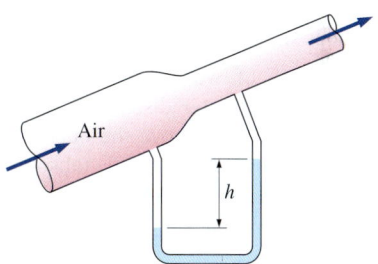

FIGURE P5–52

5–53 A handheld bicycle pump can be used as an atomizer to generate a fine mist of paint or pesticide by forcing air at a high velocity through a small hole and placing a short tube between the liquid reservoir and the high-speed air jet. The pressure across a subsonic jet exposed to the atmosphere is nearly atmospheric, and the surface of the liquid in the reservoir is also open to atmospheric pressure. In light of this, explain how the liquid is sucked up the tube. *Hint:* Read Sec. 5-4 carefully.

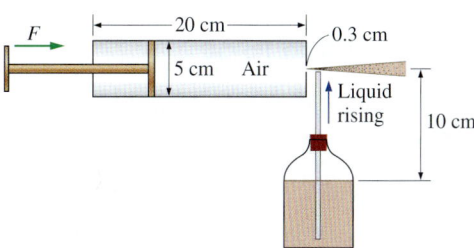

FIGURE P5–53

5–54 Water at 20°C is siphoned from a reservoir as shown in Fig. P5–54. For $d = 10$ cm and $D = 16$ cm, determine (a) the minimum flow rate that can be achieved without cavitation occurring in the piping system and (b) the maximum elevation of the highest point of the piping system to avoid cavitation.

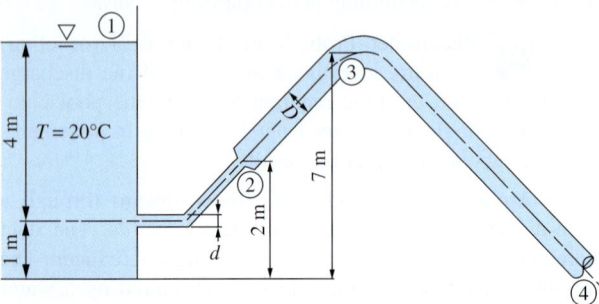

FIGURE P5–54

5–55 The water pressure in the mains of a city at a particular location is 270 kPa gage. Determine if this main can serve water to neighborhoods that are 25 m above this location.

5–56 A pressurized tank of water has a 10-cm-diameter orifice at the bottom, where water discharges to the atmosphere. The water level is 2.5 m above the outlet. The tank air pressure above the water level is 250 kPa (absolute) while the atmospheric pressure is 100 kPa. Neglecting frictional effects, determine the initial discharge rate of water from the tank. *Answer:* 0.147 m³/s

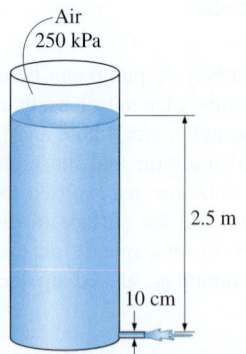

FIGURE P5–56

5–57 Reconsider Prob. 5–56. Using EES (or other) software, investigate the effect of water height in the tank on the discharge velocity. Let the water height vary from 0 to 5 m in increments of 0.5 m. Tabulate and plot the results.

5–58 Air is flowing through a venturi meter whose diameter is 6.6 cm at the entrance part (location 1) and 4.6 cm at the throat (location 2). The gage pressure is measured to be 84 kPa at the entrance and 81 kPa at the throat. Neglecting frictional effects, show that the volume flow rate can be expressed as

$$\dot{V} = A_2 \sqrt{\frac{2(P_1 - P_2)}{\rho(1 - A_2^2/A_1^2)}}$$

and determine the flow rate of air. Take the air density to be 1.2 kg/m³.

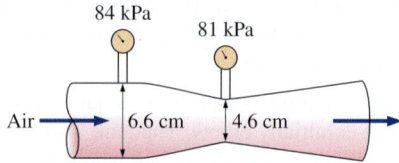

FIGURE P5–58

5–59 The water level in a tank is 15 m above the ground. A hose is connected to the bottom of the tank, and the nozzle at the end of the hose is pointed straight up. The tank cover is airtight, and the air pressure above the water surface is 3 atm gage. The system is at sea level. Determine the maximum height to which the water stream could rise. *Answer:* 46.0 m

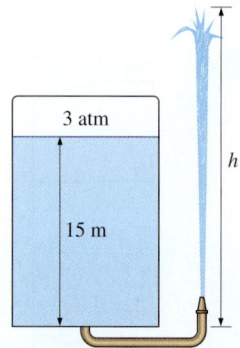

FIGURE P5–59

5–60 A Pitot-static probe connected to a water manometer is used to measure the velocity of air. If the deflection (the vertical distance between the fluid levels in the two arms) is 5.5 cm, determine the air velocity. Take the density of air to be 1.16 kg/m³.

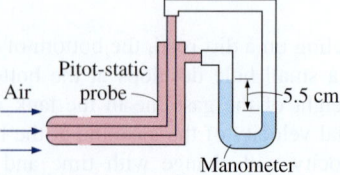

FIGURE P5–60

5–61 The air velocity in a duct is measured by a Pitot-static probe connected to a differential pressure gage. If the air is at 92 kPa absolute and 20°C and the reading of the differential pressure gage is 1.0 kPa, determine the air velocity. *Answer:* 42.8 m/s

5–62 In cold climates, water pipes may freeze and burst if proper precautions are not taken. In such an occurrence, the exposed part of a pipe on the ground ruptures, and water shoots up to 42 m. Estimate the gage pressure of water in the pipe. State your assumptions and discuss if the actual pressure is more or less than the value you predicted.

5–63 A well-fitting piston with 4 small holes in a sealed water-filled cylinder, shown in Fig. P5–63, is pushed to the right at a constant speed of 4 mm/s while the pressure in the right compartment remains constant at 50 kPa gage. Disregarding the frictional effects, determine the force F that needs to be applied to the piston to maintain this motion.

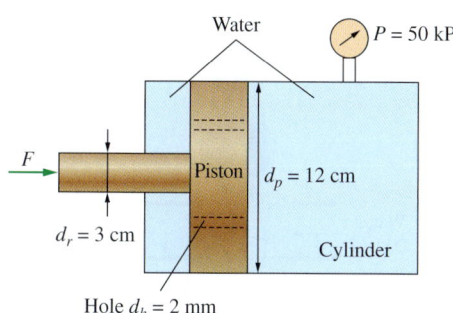

FIGURE P5–63

5–64 A fluid of density ρ and viscosity μ flows through a section of horizontal converging–diverging duct. The duct cross-sectional areas A_{inlet}, A_{throat}, and A_{outlet} are known at the inlet, throat (minimum area), and outlet, respectively. Average pressure P_{outlet} is measured at the outlet, and average velocity V_{inlet} is measured at the inlet. (*a*) Neglecting any irreversibilities such as friction, generate expressions for the average velocity and average pressure at the inlet and the throat in terms of the given variables. (*b*) In a real flow (with irreversibilities), do you expect the actual pressure at the inlet to be higher or lower than the prediction? Explain.

Energy Equation

5–65C What is useful pump head? How is it related to the power input to the pump?

5–66C Consider the steady adiabatic flow of an incompressible fluid. Can the temperature of the fluid decrease during flow? Explain.

5–67C What is irreversible head loss? How is it related to the mechanical energy loss?

5–68C Consider the steady adiabatic flow of an incompressible fluid. If the temperature of the fluid remains constant during flow, is it accurate to say that the frictional effects are negligible?

5–69C What is the kinetic energy correction factor? Is it significant?

5–70C The water level in a tank is 20 m above the ground. A hose is connected to the bottom of the tank, and the nozzle at the end of the hose is pointed straight up. The water stream from the nozzle is observed to rise 25 m above the ground. Explain what may cause the water from the hose to rise above the tank level.

5–71C A person is filling a knee-high bucket with water using a garden hose and holding it such that water discharges from the hose at the level of his waist. Someone suggests that the bucket will fill faster if the hose is lowered such that water discharges from the hose at the knee level. Do you agree with this suggestion? Explain. Disregard any frictional effects.

5–72C A 3-m-high tank filled with water has a discharge valve near the bottom and another near the top. (*a*) If these two valves are opened, will there be any difference between the discharge velocities of the two water streams? (*b*) If a hose whose discharge end is left open on the ground is first connected to the lower valve and then to the higher valve, will there be any difference between the discharge rates of water for the two cases? Disregard any frictional effects.

5–73 An oil pump is drawing 25 kW of electric power while pumping oil with $\rho = 860$ kg/m³ at a rate of 0.1 m³/s. The inlet and outlet diameters of the pipe are 8 cm and 12 cm, respectively. If the pressure rise of oil in the pump is measured to be 250 kPa and the motor efficiency is 90 percent, determine the mechanical efficiency of the pump. Take the kinetic energy correction factor to be 1.05.

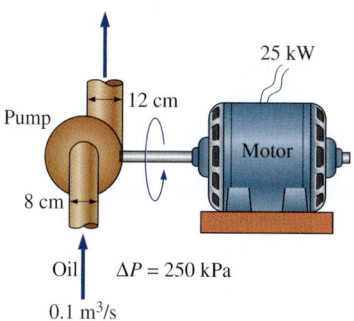

FIGURE P5–73

5–74 Water is being pumped from a large lake to a reservoir 25 m above at a rate of 25 L/s by a 10-kW (shaft) pump. If the irreversible head loss of the piping system is 5 m, determine the mechanical efficiency of the pump. *Answer:* 73.6 percent

236
BERNOULLI AND ENERGY EQUATIONS

5–75 Reconsider Prob. 5–74. Using EES (or other) software, investigate the effect of irreversible head loss on the mechanical efficiency of the pump. Let the head loss vary from 0 to 15 m in increments of 1 m. Plot the results, and discuss them.

5–76 A 15-hp (shaft) pump is used to raise water to a 45-m higher elevation. If the mechanical efficiency of the pump is 82 percent, determine the maximum volume flow rate of water.

5–77 Water flows at a rate of 0.035 m³/s in a horizontal pipe whose diameter is reduced from 15 cm to 8 cm by a reducer. If the pressure at the centerline is measured to be 480 kPa and 445 kPa before and after the reducer, respectively, determine the irreversible head loss in the reducer. Take the kinetic energy correction factors to be 1.05. *Answer:* 1.18 m

5–78 The water level in a tank is 20 m above the ground. A hose is connected to the bottom of the tank, and the nozzle at the end of the hose is pointed straight up. The tank is at sea level, and the water surface is open to the atmosphere. In the line leading from the tank to the nozzle is a pump, which increases the pressure of water. If the water jet rises to a height of 27 m from the ground, determine the minimum pressure rise supplied by the pump to the water line.

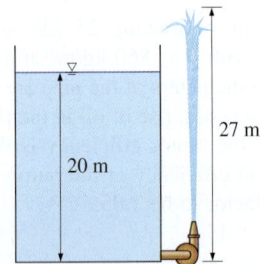

FIGURE P5–78

5–79 A hydraulic turbine has 50 m of head available at a flow rate of 1.30 m³/s, and its overall turbine–generator efficiency is 78 percent. Determine the electric power output of this turbine.

5–80 A fan is to be selected to ventilate a bathroom whose dimensions are 2 m × 3 m × 3 m. The air velocity is not to exceed 8 m/s to minimize vibration and noise. The combined efficiency of the fan–motor unit to be used can be taken to be 50 percent. If the fan is to replace the entire volume of air in 10 min, determine (*a*) the wattage of the fan–motor unit to be purchased, (*b*) the diameter of the fan casing, and (*c*) the pressure difference across the fan. Take the air density to be 1.25 kg/m³ and disregard the effect of the kinetic energy correction factors.

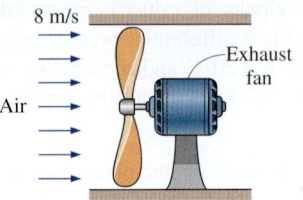

FIGURE P5–80

5–81 Water flows at a rate of 20 L/s through a horizontal pipe whose diameter is constant at 3 cm. The pressure drop across a valve in the pipe is measured to be 2 kPa, as shown in Fig P5–81. Determine the irreversible head loss of the valve, and the useful pumping power needed to overcome the resulting pressure drop. *Answers:* 0.204 m, 40 W

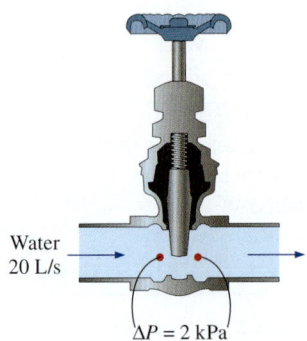

FIGURE P5–81

5–82 The water level in a tank is 10 m above the ground. A hose is connected to the bottom of the tank at the ground level and the nozzle at the end of the hose is pointed straight up. The tank cover is airtight, but the pressure over the water surface is unknown. Determine the minimum tank air pressure (gage) that will cause a water stream from the nozzle to rise 22 m from the ground.

5–83 A large tank is initially filled with water 5 m above the center of a sharp-edged 10-cm-diameter orifice. The tank water surface is open to the atmosphere, and the orifice drains to the atmosphere. If the total irreversible head loss in the system is 0.3 m, determine the initial discharge velocity of water from the tank. Take the kinetic energy correction factor at the orifice to be 1.2.

5–84 Water enters a hydraulic turbine through a 30-cm-diameter pipe at a rate of 0.6 m³/s and exits through a 25-cm-diameter pipe. The pressure drop in the turbine is measured by a mercury manometer to be 1.2 m. For a combined turbine–generator efficiency of 83 percent, determine the net electric power output. Disregard the effect of the kinetic energy correction factors.

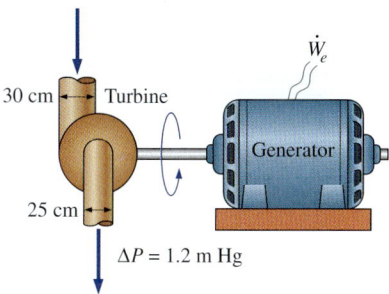

FIGURE P5–84

5–85 The velocity profile for turbulent flow in a circular pipe is approximated as $u(r) = u_{max}(1 - r/R)^{1/n}$, where $n = 9$. Determine the kinetic energy correction factor for this flow. *Answer:* 1.04

5–86 Water is pumped from a lower reservoir to a higher reservoir by a pump that provides 20 kW of useful mechanical power to the water. The free surface of the upper reservoir is 45 m higher than the surface of the lower reservoir. If the flow rate of water is measured to be 0.03 m³/s, determine the irreversible head loss of the system and the lost mechanical power during this process.

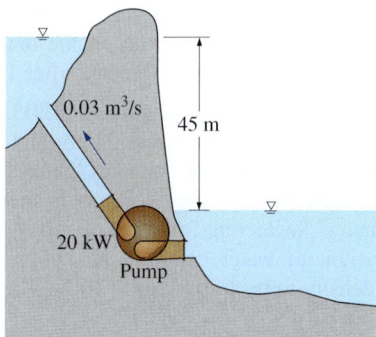

FIGURE P5–86

5–87 Water in a partially filled large tank is to be supplied to the roof top, which is 8 m above the water level in the tank, through a 2.5-cm-internal-diameter pipe by maintaining a constant air pressure of 300 kPa (gage) in the tank. If the head loss in the piping is 2 m of water, determine the discharge rate of the supply of water to the roof top.

5–88 Underground water is to be pumped by a 78 percent efficient 5-kW submerged pump to a pool whose free surface is 30 m above the underground water level. The diameter of the pipe is 7 cm on the intake side and 5 cm on the discharge

side. Determine (*a*) the maximum flow rate of water and (*b*) the pressure difference across the pump. Assume the elevation difference between the pump inlet and the outlet and the effect of the kinetic energy correction factors to be negligible.

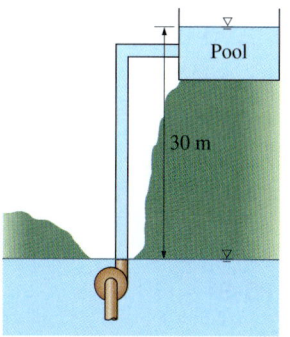

FIGURE P5–88

5–89 Reconsider Prob. 5–88. Determine the flow rate of water and the pressure difference across the pump if the irreversible head loss of the piping system is 4 m.

5–90 A 73-percent efficient 8.9-kW pump is pumping water from a lake to a nearby pool at a rate of 0.035 m³/s through a constant-diameter pipe. The free surface of the pool is 11 m above that of the lake. Determine the irreversible head loss of the piping system, in ft, and the mechanical power used to overcome it.

5–91 The demand for electric power is usually much higher during the day than it is at night, and utility companies often sell power at night at much lower prices to encourage consumers to use the available power generation capacity and to avoid building new expensive power plants that will be used only a short time during peak periods. Utilities are also willing to purchase power produced during the day from private parties at a high price.

Suppose a utility company is selling electric power for $0.06/kWh at night and is willing to pay $0.13/kWh for power produced during the day. To take advantage of this opportunity, an entrepreneur is considering building a large reservoir 50 m above the lake level, pumping water from the lake to the reservoir at night using cheap power, and letting the water flow from the reservoir back to the lake during the day, producing power as the pump–motor operates as a turbine–generator during reverse flow. Preliminary analysis shows that a water flow rate of 2 m³/s can be used in either direction, and the irreversible head loss of the piping system is 4 m. The combined pump–motor and turbine–generator efficiencies are expected to be 75 percent each. Assuming the system operates for 10 h each in the pump and turbine modes during a typical day, determine the potential revenue this pump–turbine system can generate per year.

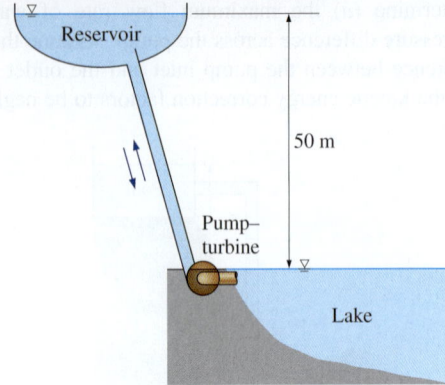

FIGURE P5–91

FIGURE P5–93

Review Problems

5–94 The velocity of a liquid flowing in a circular pipe of radius R varies from zero at the wall to a maximum at the pipe center. The velocity distribution in the pipe can be represented as $V(r)$, where r is the radial distance from the pipe center. Based on the definition of mass flow rate $\dot{m}$, obtain a relation for the average velocity in terms of $V(r)$, R, and r.

5–95 Air at 2.50 kg/m³ enters a nozzle that has an inlet-to-exit area ratio of 2:1 with a velocity of 120 m/s and leaves with a velocity of 330 m/s. Determine the density of air at the exit. *Answer:* 1.82 kg/m³

5–96 A pressurized 2-m-diameter tank of water has a 10-cm-diameter orifice at the bottom, where water discharges to the atmosphere. The water level initially is 3 m above the outlet. The tank air pressure above the water level is maintained at 450 kPa absolute and the atmospheric pressure is 100 kPa. Neglecting frictional effects, determine (*a*) how long it will take for half of the water in the tank to be discharged and (*b*) the water level in the tank after 10 s.

5–97 Air flows through a pipe at a rate of 120 L/s. The pipe consists of two sections of diameters 22 cm and 10 cm with a smooth reducing section that connects them. The pressure difference between the two pipe sections is measured by a water manometer. Neglecting frictional effects, determine the differential height of water between the two pipe sections. Take the air density to be 1.20 kg/m³. *Answer:* 1.37 cm

5–92 When a system is subjected to a linear rigid body motion with constant linear acceleration a along a distance L, the modified Bernoulli Equation takes the form

$$\left(\frac{P_1}{\rho} + \frac{V_1^2}{2} + gz_1\right) - \left(\frac{P_2}{\rho} + \frac{V_2^2}{2} + gz_2\right) = aL + \text{Losses}$$

where V_1 and V_2 are velocities relative to a fixed point and 'Losses' which represents frictional losses is zero when the frictional effects are negligible. The tank with two discharge pipes shown in Fig. P5–92 accelerates to the left at a constant linear acceleration of 3 m/s². If volumetric flow rates from both pipes are to be identical, determine the diameter D of the inclined pipe. Disregard any frictional effects.

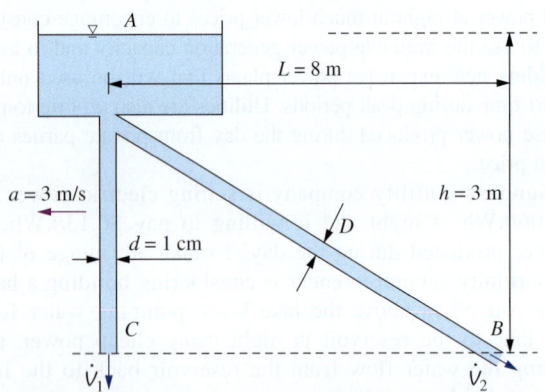

FIGURE P5–92

5–93 A fireboat is to fight fires at coastal areas by drawing seawater with a density of 1030 kg/m³ through a 10-cm-diameter pipe at a rate of 0.04 m³/s and discharging it through a hose nozzle with an exit diameter of 5 cm. The total irreversible head loss of the system is 3 m, and the position of the nozzle is 3 m above sea level. For a pump efficiency of 70 percent, determine the required shaft power input to the pump and the water discharge velocity. *Answers:* 39.2 kW, 20.4 m/s

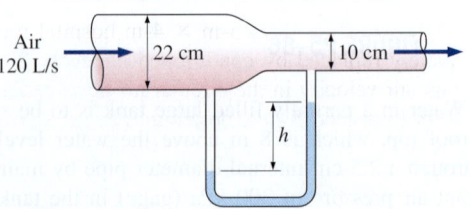

FIGURE P5–97

5–98 Air at 100 kPa and 25°C flows in a horizontal duct of variable cross section. The water column in the manometer that measures the difference between two sections has a vertical displacement of 8 cm. If the velocity in the first section is low and the friction is negligible, determine the velocity at the second section. Also, if the manometer reading has a possible error of ±2 mm, conduct an error analysis to estimate the range of validity for the velocity found.

5–99 A very large tank contains air at 102 kPa at a location where the atmospheric air is at 100 kPa and 20°C. Now a 2-cm-diameter tap is opened. Determine the maximum flow rate of air through the hole. What would your response be if air is discharged through a 2-m-long, 4-cm-diameter tube with a 2-cm-diameter nozzle? Would you solve the problem the same way if the pressure in the storage tank were 300 kPa?

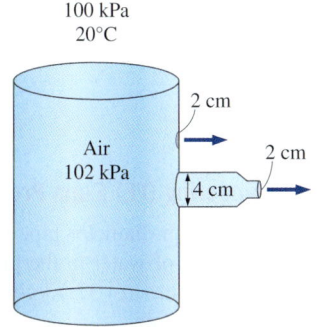

FIGURE P5–99

5–100 Water is flowing through a Venturi meter whose diameter is 7 cm at the entrance part and 4 cm at the throat. The pressure is measured to be 380 kPa at the entrance and 150 kPa at the throat. Neglecting frictional effects, determine the flow rate of water. *Answer:* 0.0285 m³/s

5–101 Water flows at a rate of 0.011 m³/s in a horizontal pipe whose diameter increases from 6 to 11 cm by an enlargement section. If the head loss across the enlargement section is 0.65 m and the kinetic energy correction factor at both the inlet and the outlet is 1.05, determine the pressure change.

5–102 The air in a 6-m × 5-m × 4-m hospital room is to be completely replaced by conditioned air every 20 min. If the average air velocity in the circular air duct leading to the room is not to exceed 5 m/s, determine the minimum diameter of the duct.

5–103 Underground water is being pumped into a pool whose cross section is 3 m × 4 m while water is discharged through a 5-cm-diameter orifice at a constant average velocity of 5 m/s. If the water level in the pool rises at a rate of 1.5 cm/min, determine the rate at which water is supplied to the pool, in m³/s.

5–104 A 3-m-high large tank is initially filled with water. The tank water surface is open to the atmosphere, and a sharp-edged 10-cm-diameter orifice at the bottom drains to the atmosphere through a horizontal 80-m-long pipe. If the total irreversible head loss of the system is determined to be 1.5 m, determine the initial velocity of the water from the tank. Disregard the effect of the kinetic energy correction factors. *Answer:* 5.42 m/s

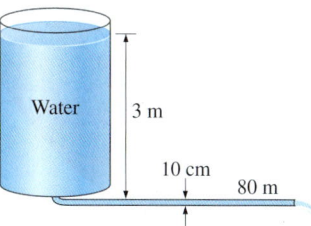

FIGURE P5–104

5–105 Reconsider Prob. 5–104. Using EES (or other) software, investigate the effect of the tank height on the initial discharge velocity of water from the completely filled tank. Let the tank height vary from 2 to 15 m in increments of 1 m, and assume the irreversible head loss to remain constant. Tabulate and plot the results.

5–106 Reconsider Prob. 5–104. In order to drain the tank faster, a pump is installed near the tank exit. Determine the pump head input necessary to establish an average water velocity of 6.5 m/s when the tank is full.

5–107 A D_0 = 8-m-diameter tank is initially filled with water 2 m above the center of a D = 10-cm-diameter valve near the bottom. The tank surface is open to the atmosphere, and the tank drains through a L = 80-m-long pipe connected to the valve. The friction factor of the pipe is given to be f = 0.015, and the discharge velocity is expressed as

$$V = \sqrt{\frac{2gz}{1.5 + fL/D}}$$

where z is the water height above the center of the valve. Determine (a) the initial discharge velocity from the tank and (b) the time required to empty the tank. The tank can be considered to be empty when the water level drops to the center of the valve.

5–108 In some applications, elbow-type flow meters like the one shown in Fig. P5–108 are used to measure flow rates. The pipe radius is R, the radius of curvature of the elbow is λ, and the pressure difference ΔP across the curvature inside the pipe is measured. From the potential flow theory, it is known that $Vr = C$, where V is the fluid velocity at a distance r from the center of curvature O, and C is a constant. Assuming frictionless steady-state flow and thus the Bernoulli equation across streamlines to be applicable, obtain a relation for the flow rate as a function of ρ, g, ΔP, λ, and R.

Answer: $\dot{V} = \pi\sqrt{\dfrac{2\Delta P}{\rho g \lambda R}}\,(\lambda^2 - R^2)(\lambda - \sqrt{\lambda^2 - R^2})$

BERNOULLI AND ENERGY EQUATIONS

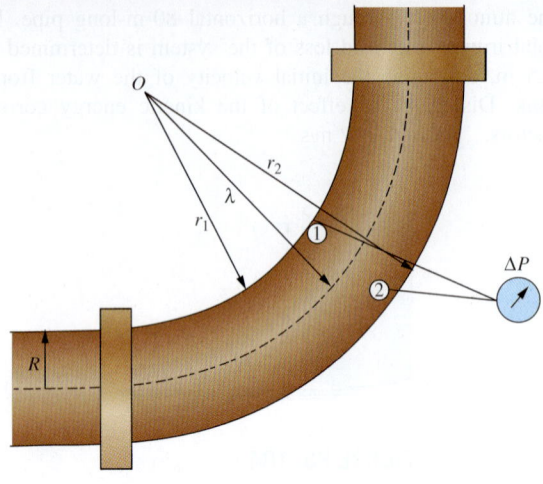

FIGURE P5–108

5–109 The cylindrical water tank with a valve at the bottom shown in Fig. P5–109 contains air at the top part at the local atmospheric pressure of 100 kPa and water as shown. Is it possible to completely empty this tank by fully opening the valve? If not, determine the water height in the tank when water stops flowing out of the fully open valve. Assume the temperature of the air inside the cylinder to remain constant during the discharging process.

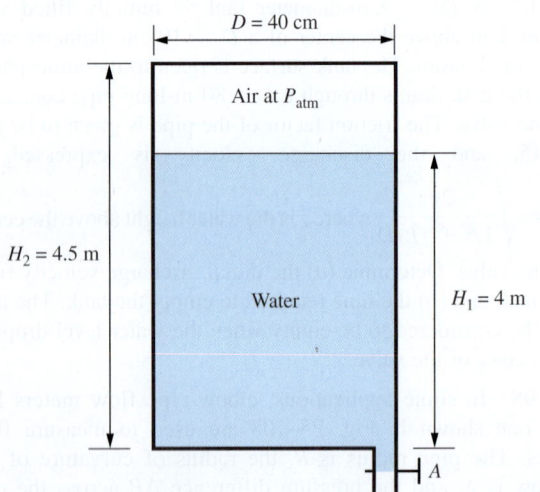

FIGURE P5–109

5–110 A rigid tank of volume 1.5 m³ initially contains atmospheric air at 20°C and 150 kPa. Now a compressor is turned on, and atmospheric air at a constant rate of 0.05 m³/s is supplied to the tank. If the pressure and density in the tank varies as $P/\rho^{1.4}$ = constant during charging, (a) obtain a relation for the variation of pressure in the tank with time and (b) calculate how long it will take for the absolute pressure in the tank to triple.

5–111 A wind tunnel draws atmospheric air at 20°C and 101.3 kPa by a large fan located near the exit of the tunnel. If the air velocity in the tunnel is 80 m/s, determine the pressure in the tunnel.

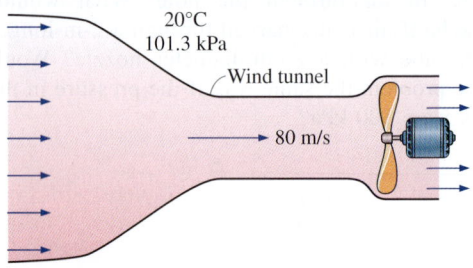

FIGURE P5–111

Fundamentals of Engineering (FE) Exam Problems

5–112 Water flows in a 5-cm-diameter pipe at a velocity of 0.75 m/s. The mass flow rate of water in the pipe is
(a) 353 kg/min   (b) 75 kg/min   (c) 37.5 kg/min
(d) 1.47 kg/min   (e) 88.4 kg/min

5–113 Air at 100 kPa and 20°C flows in a 12-cm-diameter pipe at a rate of 9.5 kg/min. The velocity of air in the pipe is
(a) 1.4 m/s   (b) 6.0 m/s   (c) 9.5 m/s   (d) 11.8 m/s
(e) 14.0 m/s

5–114 A water tank initially contains 140 L of water. Now, equal rates of cold and hot water enter the tank for a period of 30 minutes while warm water is discharged from the tank at a rate of 25 L/min. The amount of water in the tank at the end of this 30-min period is 50 L. The rate of hot water entering the tank is
(a) 33 L/min   (b) 25 L/min   (c) 11 L/min   (d) 7 L/min
(e) 5 L/min

5–115 Water enters a 4-cm-diameter pipe at a velocity of 1 m/s. The diameter of the pipe is reduced to 3 cm at the exit. The velocity of the water at the exit is
(a) 1.78 m/s   (b) 1.25 m/s   (c) 1 m/s   (d) 0.75 m/s
(e) 0.50 m/s

5–116 The pressure of water is increased from 100 kPa to 900 kPa by a pump. The mechanical energy increase of water is
(a) 0.9 kJ/kg   (b) 0.5 kJ/kg   (c) 500 kJ/kg   (d) 0.8 kJ/kg
(e) 800 kJ/kg

5–117 A 75-m-high water body that is open to the atmosphere is available. Water is run through a turbine at a rate of 200 L/s at the bottom of the water body. The pressure difference across the turbine is
(a) 736 kPa   (b) 0.736 kPa   (c) 1.47 kPa   (d) 1470 kPa
(e) 368 kPa

5–118 A pump is used to increase the pressure of water from 100 kPa to 900 kPa at a rate of 160 L/min. If the shaft power input to the pump is 3 kW, the efficiency of the pump is
(a) 0.532 (b) 0.660 (c) 0.711 (d) 0.747 (e) 0.855

5–119 A hydraulic turbine is used to generate power by using the water in a dam. The elevation difference between the free surfaces upstream and downstream of the dam is 120 m. The water is supplied to the turbine at a rate of 150 kg/s. If the shaft power output from the turbine is 155 kW, the efficiency of the turbine is
(a) 0.77 (b) 0.80 (c) 0.82 (d) 0.85 (e) 0.88

5–120 The motor of a pump consumes 1.05 hp of electricity. The pump increases the pressure of water from 120 kPa to 1100 kPa at a rate of 35 L/min. If the motor efficiency is 94 percent, the pump efficiency is
(a) 0.75 (b) 0.78 (c) 0.82 (d) 0.85 (e) 0.88

5–121 The efficiency of a hydraulic turbine-generator unit is specified to be 85 percent. If the generator efficiency is 96 percent, the turbine efficiency is
(a) 0.816 (b) 0.850 (c) 0.862 (d) 0.885 (e) 0.960

5–122 Which parameter is *not* related in the Bernoulli equation?
(a) Density (b) Velocity (c) Time (d) Pressure
(e) Elevation

5–123 Consider incompressible, frictionless flow of a fluid in a horizontal piping. The pressure and velocity of a fluid is measured to be 150 kPa and 1.25 m/s at a specified point. The density of the fluid is 700 kg/m³. If the pressure is 140 kPa at another point, the velocity of the fluid at that point is
(a) 1.26 m/s (b) 1.34 m/s (c) 3.75 m/s (d) 5.49 m/s
(e) 7.30 m/s

5–124 Consider incompressible, frictionless flow of water in a vertical piping. The pressure is 240 kPa at 2 m from the ground level. The velocity of water does not change during this flow. The pressure at 15 m from the ground level is
(a) 227 kPa (b) 174 kPa (c) 127 kPa (d) 120 kPa
(e) 113 kPa

5–125 Consider water flow in a piping network. The pressure, velocity, and elevation at a specified point (point 1) of the flow are 150 kPa, 1.8 m/s, and 14 m. The pressure and velocity at point 2 are 165 kPa and 2.4 m/s. Neglecting frictional effects, the elevation at point 2 is
(a) 12.4 m (b) 9.3 m (c) 14.2 m (d) 10.3 m (e) 7.6 m

5–126 The static and stagnation pressures of a fluid in a pipe are measured by a piezometer and a pitot tube to be 200 kPa and 210 kPa, respectively. If the density of the fluid is 550 kg/m³, the velocity of the fluid is
(a) 10 m/s (b) 6.03 m/s (c) 5.55 m/s (d) 3.67 m/s
(e) 0.19 m/s

5–127 The static and stagnation pressures of a fluid in a pipe are measured by a piezometer and a pitot tube. The heights of the fluid in the piozemeter and pitot tube are measured to be 2.2 m and 2.0 m, respectively. If the density of the fluid is 5000 kg/m³, the velocity of the fluid in the pipe is
(a) 0.92 m/s (b) 1.43 m/s (c) 1.65 m/s (d) 1.98 m/s
(e) 2.39 m/s

5–128 The difference between the heights of energy grade line (EGL) and hydraulic grade line (HGL) is equal to
(a) z (b) $P/\rho g$ (c) $V^2/2g$ (d) $z + P/\rho g$ (e) $z + V^2/2g$

5–129 Water at 120 kPa (gage) is flowing in a horizontal pipe at a velocity of 1.15 m/s. The pipe makes a 90° angle at the exit and the water exits the pipe vertically into the air. The maximum height the water jet can rise is
(a) 6.9 m (b) 7.8 m (c) 9.4 m (d) 11.5 m (e) 12.3 m

5–130 Water is withdrawn at the bottom of a large tank open to the atmosphere. The water velocity is 6.6 m/s. The minimum height of the water in the tank is
(a) 2.22 m (b) 3.04 m (c) 4.33 m (d) 5.75 m (e) 6.60 m

5–131 Water at 80 kPa (gage) enters a horizontal pipe at a velocity of 1.7 m/s. The pipe makes a 90° angle at the exit and the water exits the pipe vertically into the air. Take the correction factor to be 1. If the irreversible head loss between the inlet and exit of the pipe is 3 m, the height the water jet can rise is
(a) 3.4 m (b) 5.3 m (c) 8.2 m (d) 10.5 m (e) 12.3 m

5–132 Seawater is to be pumped into a large tank at a rate of 165 kg/min. The tank is open to the atmosphere and the water enters the tank from a 80-m-height. The overall efficiency of the motor-pump unit is 75 percent and the motor consumes electricity at a rate of 3.2 kW. Take the correction factor to be 1. If the irreversible head loss in the piping is 7 m, the velocity of the water at the tank inlet is
(a) 2.34 m/s (b) 4.05 m/s (c) 6.21 m/s (d) 8.33 m/s
(e) 10.7 m/s

5–133 Water enters a pump at 350 kPa at a rate of 1 kg/s. The water leaving the pump enters a turbine in which the pressure is reduced and electricity is produced. The shaft power input to the pump is 1 kW and the shaft power output from the turbine is 1 kW. Both the pump and turbine are 90 percent efficient. If the elevation and velocity of the water remain constant throughout the flow and the irreversible head loss is 1 m, the pressure of water at the turbine exit is
(a) 350 kPa (b) 100 kPa (c) 173 kPa (d) 218 kPa
(e) 129 kPa

5–134 An adiabatic pump is used to increase the pressure of water from 100 kPa to 500 kPa at a rate of 400 L/min. If the efficiency of the pump is 75 percent, the maximum temperature rise of the water across the pump is
(a) 0.096°C (b) 0.058°C (c) 0.035°C (d) 1.52°C
(e) 1.27°C

5–135 The shaft power from a 90 percent-efficient turbine is 500 kW. If the mass flow rate through the turbine is 575 kg/s, the extracted head removed from the fluid by the turbine is
(a) 48.7 m (b) 57.5 m (c) 147 m (d) 139 m (e) 98.5 m

Design and Essay Problems

5–136 Using a large bucket whose volume is known and measuring the time it takes to fill the bucket with water from a garden hose, determine the mass flow rate and the average velocity of water through the hose.

5–137 Your company is setting up an experiment that involves the measurement of airflow rate in a duct, and you are to come up with proper instrumentation. Research the available techniques and devices for airflow rate measurement, discuss the advantages and disadvantages of each technique, and make a recommendation.

5–138 Computer-aided designs, the use of better materials, and better manufacturing techniques have resulted in a tremendous increase in the efficiency of pumps, turbines, and electric motors. Contact one or more pump, turbine, and motor manufacturers and obtain information about the efficiency of their products. In general, how does efficiency vary with rated power of these devices?

5–139 Using a handheld bicycle pump to generate an air jet, a soda can as the water reservoir, and a straw as the tube, design and build an atomizer. Study the effects of various parameters such as the tube length, the diameter of the exit hole, and the pumping speed on performance.

5–140 Using a flexible drinking straw and a ruler, explain how you would measure the water flow velocity in a river.

5–141 The power generated by a wind turbine is proportional to the cube of the wind velocity. Inspired by the acceleration of a fluid in a nozzle, someone proposes to install a reducer casing to capture the wind energy from a larger area and accelerate it before the wind strikes the turbine blades, as shown in Fig. P5–141. Evaluate if the proposed modification should be given a consideration in the design of new wind turbines.

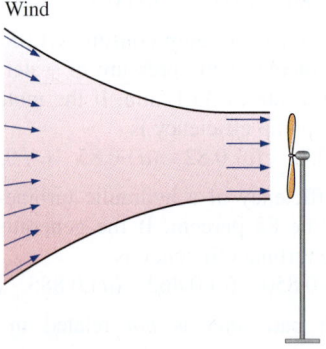

FIGURE P5–141

CHAPTER 6

MOMENTUM ANALYSIS OF FLOW SYSTEMS

When dealing with engineering problems, it is desirable to obtain fast and accurate solutions at minimal cost. Most engineering problems, including those associated with fluid flow, can be analyzed using one of three basic approaches: differential, experimental, and control volume. In *differential approaches*, the problem is formulated accurately using differential quantities, but the solution of the resulting differential equations is difficult, usually requiring the use of numerical methods with extensive computer codes. *Experimental approaches* complemented with dimensional analysis are highly accurate, but they are typically time consuming and expensive. The *finite control volume approach* described in this chapter is remarkably fast and simple and usually gives answers that are sufficiently accurate for most engineering purposes. Therefore, despite the approximations involved, the basic finite control volume analysis performed with paper and pencil has always been an indispensable tool for engineers.

In Chap. 5, the control volume mass and energy analysis of fluid flow systems was presented. In this chapter, we present the finite control volume momentum analysis of fluid flow problems. First we give an overview of Newton's laws and the conservation relations for linear and angular momentum. Then using the Reynolds transport theorem, we develop the linear momentum and angular momentum equations for control volumes and use them to determine the forces and torques associated with fluid flow.

OBJECTIVES

When you finish reading this chapter, you should be able to

- Identify the various kinds of forces and moments acting on a control volume
- Use control volume analysis to determine the forces associated with fluid flow
- Use control volume analysis to determine the moments caused by fluid flow and the torque transmitted

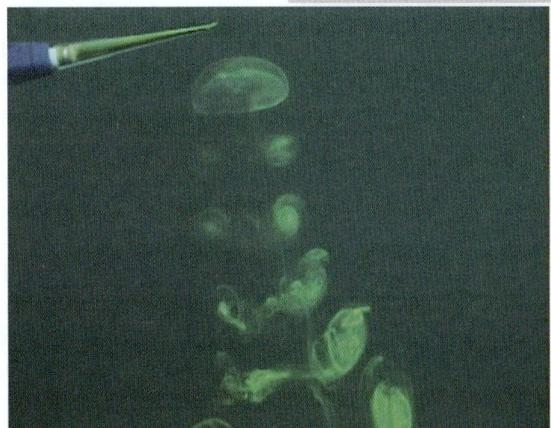

Steady swimming of the jellyfish Aurelia aurita. Fluorescent dye placed directly upstream of the animal is drawn underneath the bell as the body relaxes and forms vortex rings below the animal as the body contracts and ejects fluid. The vortex rings simultaneously induce flows for both feeding and propulsion.
*Adapted from Dabiri et al., J. Exp. Biol. 208: 1257–1265.
Photo credit: Sean P. Colin and John H. Costello.*

6–1 · NEWTON'S LAWS

Newton's laws are relations between motions of bodies and the forces acting on them. Newton's first law states that *a body at rest remains at rest, and a body in motion remains in motion at the same velocity in a straight path when the net force acting on it is zero.* Therefore, a body tends to preserve its state of inertia. Newton's second law states that *the acceleration of a body is proportional to the net force acting on it and is inversely proportional to its mass.* Newton's third law states that *when a body exerts a force on a second body, the second body exerts an equal and opposite force on the first.* Therefore, the direction of an exposed reaction force depends on the body taken as the system.

For a rigid body of mass m, Newton's second law is expressed as

Newton's second law: $$\vec{F} = m\vec{a} = m\frac{d\vec{V}}{dt} = \frac{d(m\vec{V})}{dt} \tag{6-1}$$

where $\vec{F}$ is the net force acting on the body and $\vec{a}$ is the acceleration of the body under the influence of $\vec{F}$.

The product of the mass and the velocity of a body is called the *linear momentum* or just the *momentum* of the body. The momentum of a rigid body of mass m moving with velocity $\vec{V}$ is $m\vec{V}$ (Fig. 6–1). Then Newton's second law expressed in Eq. 6–1 can also be stated as *the rate of change of the momentum of a body is equal to the net force acting on the body* (Fig. 6–2). This statement is more in line with Newton's original statement of the second law, and it is more appropriate for use in fluid mechanics when studying the forces generated as a result of velocity changes of fluid streams. Therefore, in fluid mechanics, Newton's second law is usually referred to as the *linear momentum equation*.

The momentum of a system remains constant only when the net force acting on it is zero, and thus the momentum of such a system is conserved. This is known as the *conservation of momentum principle*. This principle has proven to be a very useful tool when analyzing collisions such as those between balls; between balls and rackets, bats, or clubs; and between atoms or subatomic particles; and explosions such as those that occur in rockets, missiles, and guns. In fluid mechanics, however, the net force acting on a system is typically *not* zero, and we prefer to work with the linear momentum equation rather than the conservation of momentum principle.

Note that force, acceleration, velocity, and momentum are vector quantities, and as such they have direction as well as magnitude. Also, momentum is a constant multiple of velocity, and thus the direction of momentum is the direction of velocity as shown in Fig 6–1. Any vector equation can be written in scalar form for a specified direction using magnitudes, e.g., $F_x = ma_x = d(mV_x)/dt$ in the x-direction.

The counterpart of Newton's second law for rotating rigid bodies is expressed as $\vec{M} = I\vec{\alpha}$, where $\vec{M}$ is the net moment or torque applied on the body, I is the moment of inertia of the body about the axis of rotation, and $\vec{\alpha}$ is the angular acceleration. It can also be expressed in terms of the rate of change of angular momentum $d\vec{H}/dt$ as

Angular momentum equation: $$\vec{M} = I\vec{\alpha} = I\frac{d\vec{\omega}}{dt} = \frac{d(I\vec{\omega})}{dt} = \frac{d\vec{H}}{dt} \tag{6-2}$$

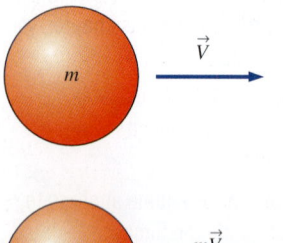

FIGURE 6–1
Linear momentum is the product of mass and velocity, and its direction is the direction of velocity.

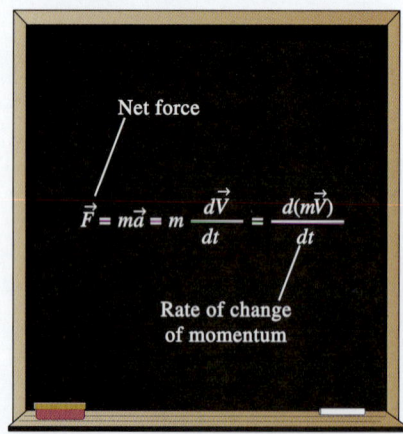

FIGURE 6–2
Newton's second law is also expressed as *the rate of change of the momentum of a body is equal to the net force acting on it.*

where $\vec{\omega}$ is the angular velocity. For a rigid body rotating about a fixed x-axis, the angular momentum equation is written in scalar form as

Angular momentum about x-axis: $$M_x = I_x \frac{d\omega_x}{dt} = \frac{dH_x}{dt} \quad (6\text{-}3)$$

The angular momentum equation can be stated as *the rate of change of the angular momentum of a body is equal to the net torque acting on it* (Fig. 6–3).

The total angular momentum of a rotating body remains constant when the net torque acting on it is zero, and thus the angular momentum of such systems is conserved. This is known as the *conservation of angular momentum principle* and is expressed as $I\omega$ = constant. Many interesting phenomena such as ice skaters spinning faster when they bring their arms close to their bodies and divers rotating faster when they curl after the jump can be explained easily with the help of the conservation of angular momentum principle (in both cases, the moment of inertia *I* is decreased and thus the angular velocity ω is increased as the outer parts of the body are brought closer to the axis of rotation).

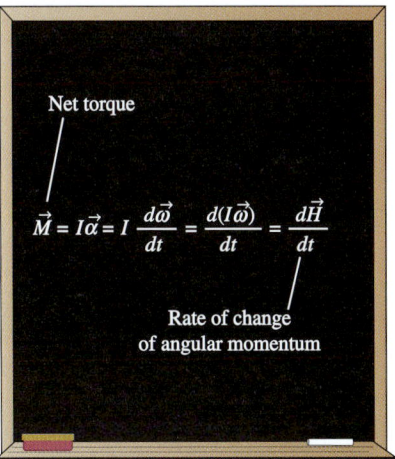

FIGURE 6–3
The rate of change of the angular momentum of a body is equal to the net torque acting on it.

6–2 ▪ CHOOSING A CONTROL VOLUME

We now briefly discuss how to *wisely* select a control volume. A control volume can be selected as any arbitrary region in space through which fluid flows, and its bounding control surface can be fixed, moving, and even deforming during flow. The application of a basic conservation law is a systematic procedure for bookkeeping or accounting of the quantity under consideration, and thus it is extremely important that the boundaries of the control volume are well defined during an analysis. Also, the flow rate of any quantity into or out of a control volume depends on the flow velocity *relative to the control surface*, and thus it is essential to know if the control volume remains at rest during flow or if it moves.

Many flow systems involve stationary hardware firmly fixed to a stationary surface, and such systems are best analyzed using *fixed* control volumes. When determining the reaction force acting on a tripod holding the nozzle of a hose, for example, a natural choice for the control volume is one that passes perpendicularly through the nozzle exit flow and through the bottom of the tripod legs (Fig. 6–4*a*). This is a fixed control volume, and the water velocity relative to a fixed point on the ground is the same as the water velocity relative to the nozzle exit plane.

When analyzing flow systems that are moving or deforming, it is usually more convenient to allow the control volume to *move* or *deform*. When determining the thrust developed by the jet engine of an airplane cruising at constant velocity, for example, a wise choice of control volume is one that encloses the airplane and cuts through the nozzle exit plane (Fig. 6–4*b*). The control volume in this case moves with velocity $\vec{V}_{CV}$, which is identical to the cruising velocity of the airplane relative to a fixed point on earth. When determining the flow rate of exhaust gases leaving the nozzle, the proper velocity to use is the velocity of the exhaust gases relative to the nozzle exit plane, that is, the *relative velocity* $\vec{V}_r$. Since the entire control volume moves at velocity $\vec{V}_{CV}$, the relative velocity becomes $\vec{V}_r = \vec{V} - \vec{V}_{CV}$, where $\vec{V}$ is the *absolute velocity* of the exhaust gases, i.e., the velocity relative to a fixed

246
MOMENTUM ANALYSIS OF FLOW SYSTEMS

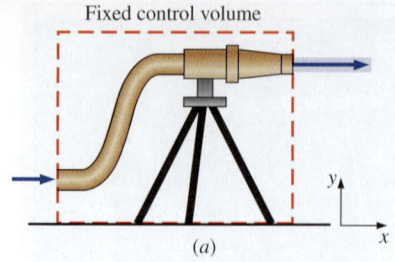

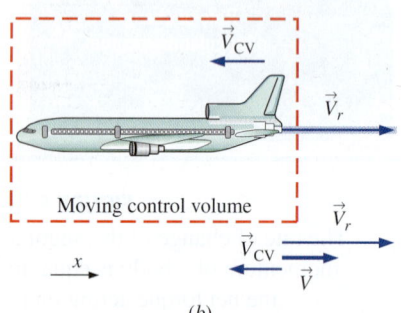

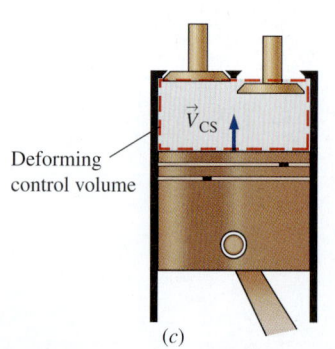

FIGURE 6–4
Examples of (a) fixed, (b) moving, and (c) deforming control volumes.

point on earth. Note that $\vec{V}_r$ is the fluid velocity expressed relative to a coordinate system moving *with* the control volume. Also, this is a vector equation, and velocities in opposite directions have opposite signs. For example, if the airplane is cruising at 500 km/h to the left, and the velocity of the exhaust gases is 800 km/h to the right relative to the ground, the velocity of the exhaust gases relative to the nozzle exit is

$$\vec{V}_r = \vec{V} - \vec{V}_{CV} = 800\vec{i} - (-500\vec{i}) = 1300\vec{i} \text{ km/h}$$

That is, the exhaust gases leave the nozzle at 1300 km/h to the right relative to the nozzle exit (in the direction opposite to that of the airplane); this is the velocity that should be used when evaluating the outflow of exhaust gases through the control surface (Fig. 6–4b). Note that the exhaust gases would appear motionless to an observer on the ground if the relative velocity were equal in magnitude to the airplane velocity.

When analyzing the purging of exhaust gases from a reciprocating internal combustion engine, a wise choice for the control volume is one that comprises the space between the top of the piston and the cylinder head (Fig. 6–4c). This is a *deforming* control volume, since part of the control surface moves relative to other parts. The relative velocity for an inlet or outlet on the deforming part of a control surface (there are no such inlets or outlets in Fig. 6–4c) is then given by $\vec{V}_r = \vec{V} - \vec{V}_{CS}$ where $\vec{V}$ is the absolute fluid velocity and $\vec{V}_{CS}$ is the control surface velocity, both relative to a fixed point outside the control volume. Note that $\vec{V}_{CS} = \vec{V}_{CV}$ for moving but nondeforming control volumes, and $\vec{V}_{CS} = \vec{V}_{CV} = 0$ for fixed ones.

6–3 ■ FORCES ACTING ON A CONTROL VOLUME

The forces acting on a control volume consist of **body forces** that act throughout the entire body of the control volume (such as gravity, electric, and magnetic forces) and **surface forces** that act on the control surface (such as pressure and viscous forces and reaction forces at points of contact). Only external forces are considered in the analysis. Internal forces (such as the pressure force between a fluid and the inner surfaces of the flow section) are not considered in a control volume analysis unless they are exposed by passing the control surface through that area.

In control volume analysis, the sum of all forces acting on the control volume at a particular instant in time is represented by $\Sigma \vec{F}$ and is expressed as

Total force acting on control volume: $\quad \Sigma \vec{F} = \Sigma \vec{F}_{\text{body}} + \Sigma \vec{F}_{\text{surface}} \quad$ (6–4)

Body forces act on each volumetric portion of the control volume. The body force acting on a differential element of fluid of volume dV within the control volume is shown in Fig. 6–5, and we must perform a volume integral to account for the net body force on the entire control volume. *Surface forces* act on each portion of the control surface. A differential surface element of area dA and unit outward normal $\vec{n}$ on the control surface is shown in Fig. 6–5, along with the surface force acting on it. We must perform an area integral to obtain the net surface force acting on the entire control surface. As sketched, the surface force may act in a direction independent of that of the outward normal vector.

The most common body force is that of **gravity,** which exerts a downward force on every differential element of the control volume. While other body forces, such as electric and magnetic forces, may be important in some analyses, we consider only gravitational forces here.

The differential body force $d\vec{F}_{body} = d\vec{F}_{gravity}$ acting on the small fluid element shown in Fig. 6–6 is simply its weight,

Gravitational force acting on a fluid element: $\quad d\vec{F}_{gravity} = \rho \vec{g}\, dV \quad$ (6–5)

where ρ is the average density of the element and $\vec{g}$ is the gravitational vector. In Cartesian coordinates we adopt the convention that $\vec{g}$ acts in the negative z-direction, as in Fig. 6–6, so that

Gravitational vector in Cartesian coordinates: $\quad \vec{g} = -g\vec{k} \quad$ (6–6)

Note that the coordinate axes in Fig. 6–6 are oriented so that the gravity vector acts *downward* in the $-z$-direction. On earth at sea level, the gravitational constant g is equal to 9.807 m/s². Since gravity is the only body force being considered, integration of Eq. 6–5 yields

Total body force acting on control volume: $\quad \sum \vec{F}_{body} = \int_{CV} \rho \vec{g}\, dV = m_{CV}\vec{g} \quad$ (6–7)

Surface forces are not as simple to analyze since they consist of both *normal* and *tangential* components. Furthermore, while the physical force acting on a surface is independent of orientation of the coordinate axes, the *description* of the force in terms of its coordinate components changes with orientation (Fig. 6–7). In addition, we are rarely fortunate enough to have each of the control surfaces aligned with one of the coordinate axes. While not desiring to delve too deeply into tensor algebra, we are forced to define a **second-order tensor** called the **stress tensor** σ_{ij} in order to adequately describe the surface stresses at a point in the flow,

Stress tensor in Cartesian coordinates: $\quad \sigma_{ij} = \begin{pmatrix} \sigma_{xx} & \sigma_{xy} & \sigma_{xz} \\ \sigma_{yx} & \sigma_{yy} & \sigma_{yz} \\ \sigma_{zx} & \sigma_{zy} & \sigma_{zz} \end{pmatrix} \quad$ (6–8)

The diagonal components of the stress tensor, σ_{xx}, σ_{yy}, and σ_{zz}, are called **normal stresses;** they are composed of pressure (which always acts inwardly normal) and viscous stresses. Viscous stresses are discussed in more detail in Chap. 9. The off-diagonal components, σ_{xy}, σ_{zx}, etc., are called **shear stresses;** since pressure can act only normal to a surface, shear stresses are composed entirely of viscous stresses.

When the face is not parallel to one of the coordinate axes, mathematical laws for axes rotation and tensors can be used to calculate the normal and tangential components acting at the face. In addition, an alternate notation called **tensor notation** is convenient when working with tensors but is usually reserved for graduate studies. (For a more in-depth analysis of tensors and tensor notation see, for example, Kundu and Cohen, 2011.)

In Eq. 6–8, σ_{ij} is defined as the stress (force per unit area) in the *j*-direction acting on a face whose normal is in the *i*-direction. Note that *i* and *j* are merely *indices* of the tensor and are not the same as unit vectors $\vec{i}$ and $\vec{j}$. For example, σ_{xy} is defined as positive for the stress pointing in the *y*-direction on a face whose outward normal is in the *x*-direction. This component of the

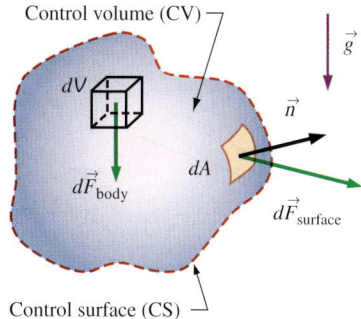

FIGURE 6–5
The total force acting on a control volume is composed of body forces and surface forces; body force is shown on a differential volume element, and surface force is shown on a differential surface element.

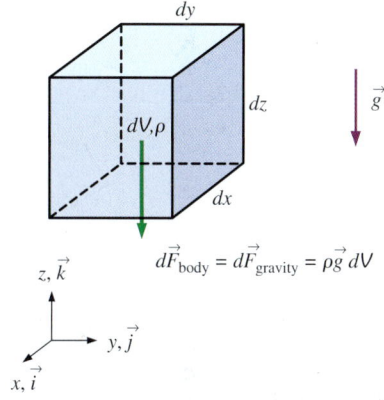

FIGURE 6–6
The gravitational force acting on a differential volume element of fluid is equal to its weight; the axes are oriented so that the gravity vector acts *downward* in the negative z-direction.

MOMENTUM ANALYSIS OF FLOW SYSTEMS

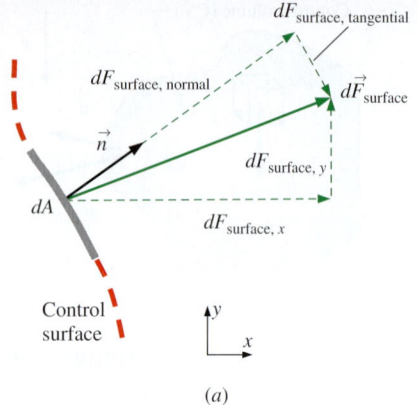

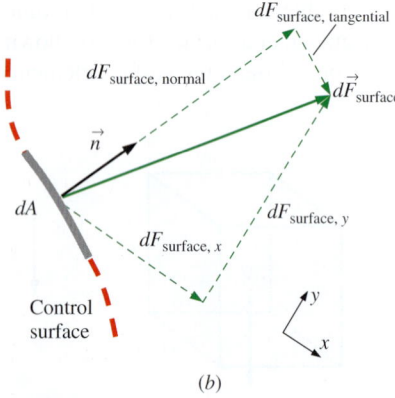

FIGURE 6–7

When coordinate axes are rotated (*a*) to (*b*), the components of the surface force change, even though the force itself remains the same; only two dimensions are shown here.

stress tensor, along with the other eight components, is shown in Fig. 6–8 for the case of a differential fluid element aligned with the axes in Cartesian coordinates. All the components in Fig. 6–8 are shown on positive faces (right, top, and front) and in their positive orientation by definition. Positive stress components on the *opposing* faces of the fluid element (not shown) point in exactly opposite directions.

The dot product of a second-order tensor and a vector yields a second vector; this operation is often called the **contracted product** or the **inner product** of a tensor and a vector. In our case, it turns out that the inner product of the stress tensor σ_{ij} and the unit outward normal vector $\vec{n}$ of a differential surface element yields a vector whose magnitude is the force per unit area acting on the surface element and whose direction is the direction of the surface force itself. Mathematically we write

Surface force acting on a differential surface element: $\quad d\vec{F}_{\text{surface}} = \sigma_{ij} \cdot \vec{n}\, dA \quad$ (6–9)

Finally, we integrate Eq. 6–9 over the entire control surface,

Total surface force acting on control surface: $\quad \sum \vec{F}_{\text{surface}} = \int_{CS} \sigma_{ij} \cdot \vec{n}\, dA \quad$ (6–10)

Substitution of Eqs. 6–7 and 6–10 into Eq. 6–4 yields

$$\sum \vec{F} = \sum \vec{F}_{\text{body}} + \sum \vec{F}_{\text{surface}} = \int_{CV} \rho \vec{g}\, dV + \int_{CS} \sigma_{ij} \cdot \vec{n}\, dA \quad (6\text{–}11)$$

This equation turns out to be quite useful in the derivation of the differential form of conservation of linear momentum, as discussed in Chap. 9. For practical control volume analysis, however, it is rare that we need to use Eq. 6–11, especially the cumbersome surface integral that it contains.

A careful selection of the control volume enables us to write the total force acting on the control volume, $\sum \vec{F}$, as the sum of more readily available quantities like weight, pressure, and reaction forces. We recommend the following for control volume analysis:

Total force: $\quad \underbrace{\sum \vec{F}}_{\text{total force}} = \underbrace{\sum \vec{F}_{\text{gravity}}}_{\text{body force}} + \underbrace{\sum \vec{F}_{\text{pressure}} + \sum \vec{F}_{\text{viscous}} + \sum \vec{F}_{\text{other}}}_{\text{surface forces}} \quad$ (6–12)

The first term on the right-hand side of Eq. 6–12 is the body force *weight*, since gravity is the only body force we are considering. The other three terms combine to form the net surface force; they are pressure forces, viscous forces, and "other" forces acting on the control surface. $\Sigma \vec{F}_{\text{other}}$ is composed of reaction forces required to turn the flow; forces at bolts, cables, struts, or walls through which the control surface cuts; etc.

All these surface forces arise as the control volume is isolated from its surroundings for analysis, and the effect of any detached object is accounted for by a force at that location. This is similar to drawing a free-body diagram in your statics and dynamics classes. We should choose the control volume such that forces that are not of interest remain internal, and thus they do not complicate the analysis. A well-chosen control volume exposes only the forces that are to be determined (such as reaction forces) and a minimum number of other forces.

A common simplification in the application of Newton's laws of motion is to subtract the *atmospheric pressure* and work with gage pressures. This is because atmospheric pressure acts in all directions, and its effect cancels out in every direction (Fig. 6–9). This means we can also ignore the pressure forces at outlet sections where the fluid is discharged at subsonic velocities to the atmosphere since the discharge pressures in such cases are very near atmospheric pressure.

As an example of how to wisely choose a control volume, consider control volume analysis of water flowing steadily through a faucet with a partially closed gate valve spigot (Fig. 6–10). It is desired to calculate the net force on the flange to ensure that the flange bolts are strong enough. There are many possible choices for the control volume. Some engineers restrict their control volumes to the fluid itself, as indicated by CV A (the purple control volume) in Fig 6–10. With this control volume, there are pressure forces that vary along the control surface, there are viscous forces along the pipe wall and at locations inside the valve, and there is a body force, namely, the weight of the water in the control volume. Fortunately, to calculate the net force on the flange, we do *not* need to integrate the pressure and viscous stresses all along the control surface. Instead, we can lump the unknown pressure and viscous forces together into one reaction force, representing the net force of the walls on the water. This force, plus the weight of the faucet and the water, is equal to the net force on the flange. (We must be very careful with our signs, of course.)

When choosing a control volume, you are not limited to the fluid alone. Often it is more convenient to slice the control surface *through* solid objects such as walls, struts, or bolts as illustrated by CV B (the red control volume) in Fig. 6–10. A control volume may even surround an entire object, like the one shown here. Control volume B is a wise choice because we are not concerned with any details of the flow or even the geometry inside the control volume. For the case of CV B, we assign a net reaction force acting at the portions of the control surface that slice through the flange bolts. Then, the only other things we need to know are the gage pressure of the water at the flange (the inlet to the control volume) and the weights of the water and the faucet assembly. The pressure everywhere else along the control surface is atmospheric (zero gage pressure) and cancels out. This problem is revisited in Section 6–4, Example 6–7.

6–4 ■ THE LINEAR MOMENTUM EQUATION

Newton's second law for a system of mass m subjected to net force $\Sigma \vec{F}$ is expressed as

$$\Sigma \vec{F} = m\vec{a} = m\frac{d\vec{V}}{dt} = \frac{d}{dt}(m\vec{V}) \quad (6\text{–}13)$$

where $m\vec{V}$ is the **linear momentum** of the system. Noting that both the density and velocity may change from point to point within the system, Newton's second law can be expressed more generally as

$$\Sigma \vec{F} = \frac{d}{dt}\int_{\text{sys}} \rho \vec{V} \, d\mathcal{V} \quad (6\text{–}14)$$

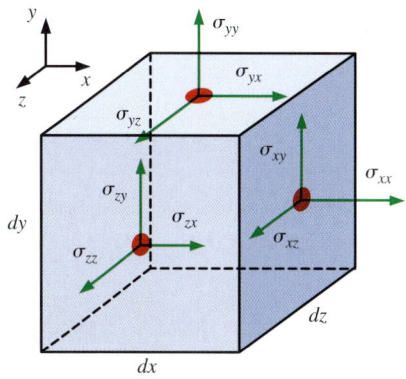

FIGURE 6–8
Components of the stress tensor in Cartesian coordinates on the right, top, and front faces.

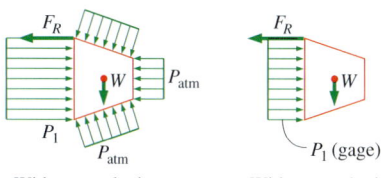

FIGURE 6–9
Atmospheric pressure acts in all directions, and thus it can be ignored when performing force balances since its effect cancels out in every direction.

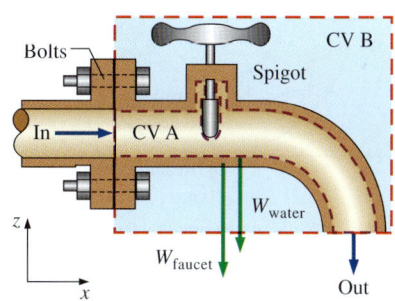

FIGURE 6–10
Cross section through a faucet assembly, illustrating the importance of choosing a control volume wisely; CV B is much easier to work with than CV A.

MOMENTUM ANALYSIS OF FLOW SYSTEMS

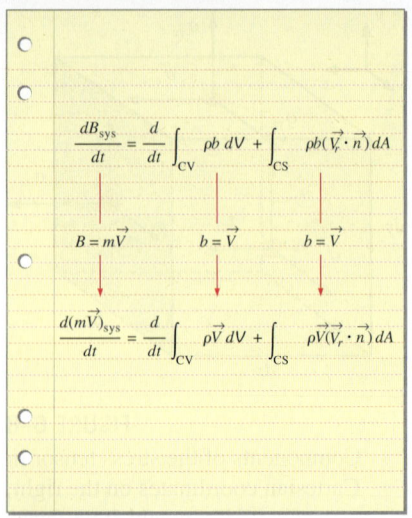

FIGURE 6–11
The linear momentum equation is obtained by replacing B in the Reynolds transport theorem by the momentum $m\vec{V}$, and b by the momentum per unit mass $\vec{V}$.

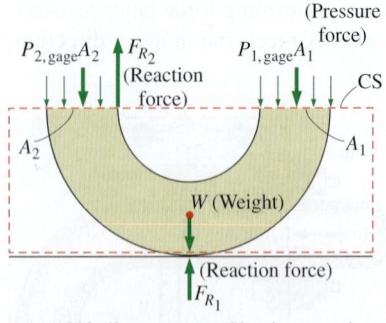

An 180° elbow supported by the ground

FIGURE 6–12
In most flow systems, the sum of forces $\Sigma \vec{F}$ consists of weights, pressure forces, and reaction forces. Gage pressures are used here since atmospheric pressure cancels out on all sides of the control surface.

where $\rho \vec{V}\, dV$ is the momentum of a differential element dV, which has mass $\delta m = \rho\, dV$. Therefore, Newton's second law can be stated as *the sum of all external forces acting on a system is equal to the time rate of change of linear momentum of the system*. This statement is valid for a coordinate system that is at rest or moves with a constant velocity, called an *inertial coordinate system* or *inertial reference frame*. Accelerating systems such as aircraft during takeoff are best analyzed using noninertial (or accelerating) coordinate systems fixed to the aircraft. Note that Eq. 6–14 is a vector relation, and thus the quantities $\vec{F}$ and $\vec{V}$ have direction as well as magnitude.

Equation 6–14 is for a given mass of a solid or fluid and is of limited use in fluid mechanics since most flow systems are analyzed using control volumes. The *Reynolds transport theorem* developed in Section 4–6 provides the necessary tools to shift from the system formulation to the control volume formulation. Setting $b = \vec{V}$ and thus $B = m\vec{V}$, the Reynolds transport theorem is expressed for linear momentum as (Fig. 6–11)

$$\frac{d(m\vec{V})_{sys}}{dt} = \frac{d}{dt}\int_{CV} \rho\vec{V}\, dV + \int_{CS} \rho\vec{V}(\vec{V}_r \cdot \vec{n})\, dA \qquad (6\text{--}15)$$

The left-hand side of this equation is, from Eq. 6–13, equal to $\Sigma \vec{F}$. Substituting, the general form of the linear momentum equation that applies to fixed, moving, or deforming control volumes is

General: $$\Sigma \vec{F} = \frac{d}{dt}\int_{CV} \rho\vec{V}\, dV + \int_{CS} \rho\vec{V}(\vec{V}_r \cdot \vec{n})\, dA \qquad (6\text{--}16)$$

which is stated in words as

$$\begin{pmatrix}\text{The sum of all}\\ \text{external forces}\\ \text{acting on a CV}\end{pmatrix} = \begin{pmatrix}\text{The time rate of change}\\ \text{of the linear momentum}\\ \text{of the contents of the CV}\end{pmatrix} + \begin{pmatrix}\text{The net flow rate of}\\ \text{linear momentum out of the}\\ \text{control surface by mass flow}\end{pmatrix}$$

Here $\vec{V}_r = \vec{V} - \vec{V}_{CS}$ is the fluid velocity relative to the control surface (for use in mass flow rate calculations at all locations where the fluid crosses the control surface), and $\vec{V}$ is the fluid velocity as viewed from an inertial reference frame. The product $\rho(\vec{V}_r \cdot \vec{n})\, dA$ represents the mass flow rate through area element dA into or out of the control volume.

For a fixed control volume (no motion or deformation of the control volume), $\vec{V}_r = \vec{V}$ and the linear momentum equation becomes

Fixed CV: $$\Sigma \vec{F} = \frac{d}{dt}\int_{CV} \rho\vec{V}\, dV + \int_{CS} \rho\vec{V}(\vec{V} \cdot \vec{n})\, dA \qquad (6\text{--}17)$$

Note that the momentum equation is a *vector equation*, and thus each term should be treated as a vector. Also, the components of this equation can be resolved along orthogonal coordinates (such as x, y, and z in the Cartesian coordinate system) for convenience. The sum of forces $\Sigma \vec{F}$ in most cases consists of weights, pressure forces, and reaction forces (Fig. 6–12). The momentum equation is commonly used to calculate the forces (usually on support systems or connectors) induced by the flow.

Special Cases

Most momentum problems considered in this text are steady. During *steady flow*, the amount of momentum within the control volume remains constant, and thus the time rate of change of linear momentum of the contents of the control volume (the second term of Eq. 6–16) is zero. Thus,

Steady flow:
$$\sum \vec{F} = \int_{CS} \rho \vec{V} (\vec{V}_r \cdot \vec{n}) \, dA \qquad (6\text{–}18)$$

For a case in which a non-deforming control volume moves at constant velocity (an inertial reference frame), the *first* $\vec{V}$ in Eq. 6-18 may *also* be taken relative to the moving control surface.

While Eq. 6–17 is exact for fixed control volumes, it is not always convenient when solving practical engineering problems because of the integrals. Instead, as we did for conservation of mass, we would like to rewrite Eq. 6–17 in terms of average velocities and mass flow rates through inlets and outlets. In other words, our desire is to rewrite the equation in *algebraic* rather than *integral* form. In many practical applications, fluid crosses the boundaries of the control volume at one or more inlets and one or more outlets, and carries with it some momentum into or out of the control volume. For simplicity, we always draw our control surface such that it slices normal to the inflow or outflow velocity at each such inlet or outlet (Fig. 6–13).

The mass flow rate $\dot{m}$ into or out of the control volume across an inlet or outlet at which ρ is nearly constant is

Mass flow rate across an inlet or outlet:
$$\dot{m} = \int_{A_c} \rho(\vec{V}\cdot\vec{n}) \, dA_c = \rho V_{avg} A_c \qquad (6\text{–}19)$$

Comparing Eq. 6–19 to Eq. 6–17, we notice an extra velocity in the control surface integral of Eq. 6–17. If $\vec{V}$ were uniform ($\vec{V} = \vec{V}_{avg}$) across the inlet or outlet, we could simply take it outside the integral. Then we could write the rate of inflow or outflow of momentum through the inlet or outlet in simple algebraic form,

Momentum flow rate across a uniform inlet or outlet:
$$\int_{A_c} \rho \vec{V}(\vec{V}\cdot\vec{n}) \, dA_c = \rho V_{avg} A_c \vec{V}_{avg} = \dot{m} \vec{V}_{avg} \qquad (6\text{–}20)$$

The uniform flow approximation is reasonable at some inlets and outlets, e.g., the well-rounded entrance to a pipe, the flow at the entrance to a wind tunnel test section, and a slice through a water jet moving at nearly uniform speed through air (Fig. 6–14). At each such inlet or outlet, Eq. 6–20 can be applied directly.

Momentum-Flux Correction Factor, β

Unfortunately, the velocity across most inlets and outlets of practical engineering interest is *not* uniform. Nevertheless, it turns out that we can still convert the control surface integral of Eq. 6–17 into algebraic form, but a dimensionless correction factor β, called the **momentum-flux correction factor,** is required, as first shown by the French scientist Joseph Boussinesq

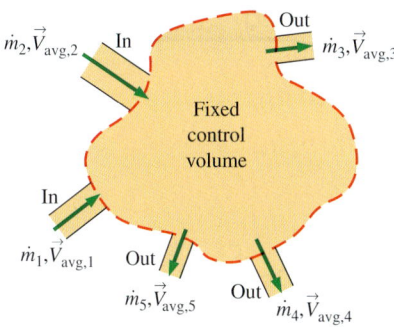

FIGURE 6–13
In a typical engineering problem, the control volume may contain multiple inlets and outlets; at each inlet or outlet we define the mass flow rate $\dot{m}$ and the average velocity $\vec{V}_{avg}$.

MOMENTUM ANALYSIS OF FLOW SYSTEMS

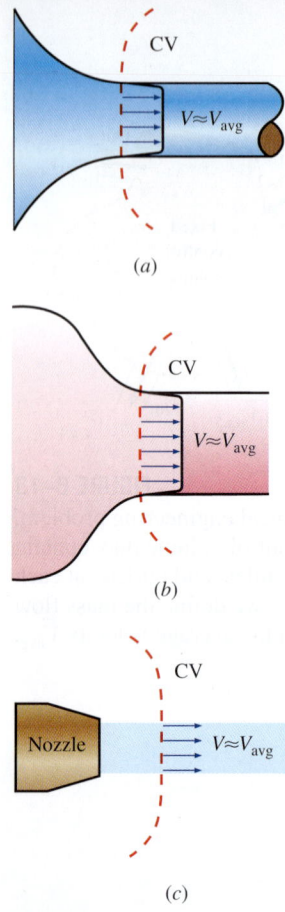

FIGURE 6–14
Examples of inlets or outlets in which the uniform flow approximation is reasonable: (*a*) the well-rounded entrance to a pipe, (*b*) the entrance to a wind tunnel test section, and (*c*) a slice through a free water jet in air.

(1842–1929). The algebraic form of Eq. 6–17 for a fixed control volume is then written as

$$\sum \vec{F} = \frac{d}{dt}\int_{CV} \rho \vec{V}\, dV + \sum_{out} \beta \dot{m} \vec{V}_{avg} - \sum_{in} \beta \dot{m} \vec{V}_{avg} \quad (6\text{–}21)$$

where a unique value of momentum-flux correction factor is applied to each inlet and outlet in the control surface. Note that $\beta = 1$ *for the case of uniform flow* over an inlet or outlet, as in Fig. 6–14. For the general case, we define β such that the integral form of the momentum flux into or out of the control surface at an inlet or outlet of cross-sectional area A_c can be expressed in terms of mass flow rate $\dot{m}$ through the inlet or outlet and average velocity $\vec{V}_{avg}$ through the inlet or outlet,

Momentum flux across an inlet or outlet: $\quad \int_{A_c} \rho \vec{V}(\vec{V}\cdot\vec{n})\, dA_c = \beta \dot{m} \vec{V}_{avg} \quad (6\text{–}22)$

For the case in which density is uniform over the inlet or outlet and $\vec{V}$ is in the same direction as $\vec{V}_{avg}$ over the inlet or outlet, we solve Eq. 6–22 for β,

$$\beta = \frac{\int_{A_c} \rho V(\vec{V}\cdot\vec{n})\, dA_c}{\dot{m} V_{avg}} = \frac{\int_{A_c} \rho V(\vec{V}\cdot\vec{n})\, dA_c}{\rho V_{avg} A_c V_{avg}} \quad (6\text{–}23)$$

where we have substituted $\rho V_{avg} A_c$ for $\dot{m}$ in the denominator. The densities cancel and since V_{avg} is constant, it can be brought inside the integral. Furthermore, if the control surface slices normal to the inlet or outlet area, $(\vec{V}\cdot\vec{n})\, dA_c = V\, dA_c$. Then, Eq. 6–23 simplifies to

Momentum-flux correction factor: $\quad \beta = \frac{1}{A_c} \int_{A_c} \left(\frac{V}{V_{avg}}\right)^2 dA_c \quad (6\text{–}24)$

It may be shown that β is always greater than or equal to unity.

EXAMPLE 6–1   **Momentum-Flux Correction Factor for Laminar Pipe Flow**

Consider laminar flow through a very long straight section of round pipe. It is shown in Chap. 8 that the velocity profile through a cross-sectional area of the pipe is parabolic (Fig. 6–15), with the axial velocity component given by

$$V = 2V_{avg}\left(1 - \frac{r^2}{R^2}\right) \quad (1)$$

where R is the radius of the inner wall of the pipe and V_{avg} is the average velocity. Calculate the momentum-flux correction factor through a cross section of the pipe for the case in which the pipe flow represents an outlet of the control volume, as sketched in Fig. 6–15.

SOLUTION For a given velocity distribution we are to calculate the momentum-flux correction factor.

Assumptions 1 The flow is incompressible and steady. 2 The control volume slices through the pipe normal to the pipe axis, as sketched in Fig. 6–15.
Analysis We substitute the given velocity profile for V in Eq. 6–24 and integrate, noting that $dA_c = 2\pi r\, dr$,

$$\beta = \frac{1}{A_c}\int_{A_c}\left(\frac{V}{V_{avg}}\right)^2 dA_c = \frac{4}{\pi R^2}\int_0^R \left(1 - \frac{r^2}{R^2}\right)^2 2\pi r\, dr \qquad (2)$$

Defining a new integration variable $y = 1 - r^2/R^2$ and thus $dy = -2r\, dr/R^2$ (also, $y = 1$ at $r = 0$, and $y = 0$ at $r = R$) and performing the integration, the momentum-flux correction factor for fully developed laminar flow becomes

Laminar flow: $\qquad \beta = -4\int_1^0 y^2\, dy = -4\left[\frac{y^3}{3}\right]_1^0 = \frac{4}{3} \qquad (3)$

Discussion We have calculated β for an outlet, but the same result would have been obtained if we had considered the cross section of the pipe as an inlet to the control volume.

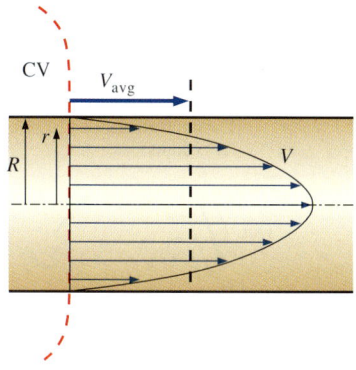

FIGURE 6–15
Velocity profile over a cross section of a pipe in which the flow is fully developed and laminar.

From Example 6–1 we see that β is not very close to unity for fully developed laminar pipe flow, and ignoring β could potentially lead to significant error. If we were to perform the same kind of integration as in Example 6–1 but for fully developed *turbulent* rather than laminar pipe flow, we would find that β ranges from about 1.01 to 1.04. Since these values are so close to unity, many practicing engineers completely disregard the momentum-flux correction factor. While the neglect of β in turbulent flow calculations may have an insignificant effect on the final results, it is wise to keep it in our equations. Doing so not only improves the accuracy of our calculations, but reminds us to include the momentum-flux correction factor when solving laminar flow control volume problems.

> For turbulent flow β may have an insignificant effect at inlets and outlets, but for laminar flow β may be important and should not be neglected. It is wise to include β in all momentum control volume problems.

Steady Flow

If the flow is also *steady*, the time derivative term in Eq. 6–21 vanishes and we are left with

Steady linear momentum equation: $\qquad \sum \vec{F} = \sum_{out} \beta \dot{m} \vec{V} - \sum_{in} \beta \dot{m} \vec{V} \qquad (6\text{–}25)$

where we have dropped the subscript "avg" from average velocity. Equation 6–25 states that *the net force acting on the control volume during steady flow is equal to the difference between the rates of outgoing and incoming momentum flows.* This statement is illustrated in Fig. 6–16. It can also be expressed for any direction, since Eq. 6–25 is a vector equation.

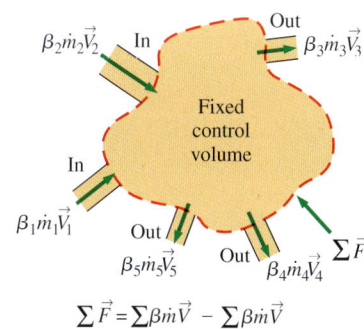

FIGURE 6–16
The net force acting on the control volume during steady flow is equal to the difference between the outgoing and the incoming momentum fluxes.

MOMENTUM ANALYSIS OF FLOW SYSTEMS

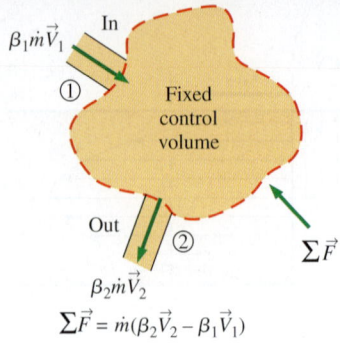

FIGURE 6–17
A control volume with only one inlet and one outlet.

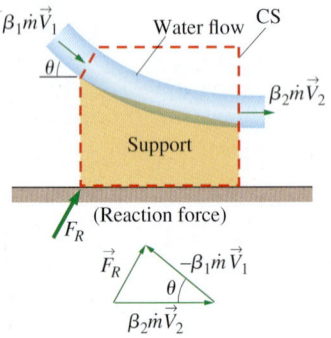

FIGURE 6–18
The determination by vector addition of the reaction force on the support caused by a change of direction of water.

Steady Flow with One Inlet and One Outlet
Many practical engineering problems involve just one inlet and one outlet (Fig. 6–17). The mass flow rate for such **single-stream systems** remains constant, and Eq. 6–25 reduces to

One inlet and one outlet: $$\sum \vec{F} = \dot{m}(\beta_2 \vec{V}_2 - \beta_1 \vec{V}_1) \quad (6\text{–}26)$$

where we have adopted the usual convention that subscript 1 implies the inlet and subscript 2 the outlet, and $\vec{V}_1$ and $\vec{V}_2$ denote the *average* velocities across the inlet and outlet, respectively.

We emphasize again that all the preceding relations are *vector* equations, and thus all the additions and subtractions are *vector* additions and subtractions. Recall that subtracting a vector is equivalent to adding it after reversing its direction (Fig. 6–18). When writing the momentum equation for a specified coordinate direction (such as the *x*-axis), we use the projections of the vectors on that axis. For example, Eq. 6–26 is written along the *x*-coordinate as

Along x-coordinate: $$\sum F_x = \dot{m}(\beta_2 V_{2,x} - \beta_1 V_{1,x}) \quad (6\text{–}27)$$

where $\sum F_x$ is the vector sum of the *x*-components of the forces, and $V_{2,x}$ and $V_{1,x}$ are the *x*-components of the outlet and inlet velocities of the fluid stream, respectively. The force or velocity components in the positive *x*-direction are positive quantities, and those in the negative *x*-direction are negative quantities. Also, it is good practice to take the direction of unknown forces in the positive directions (unless the problem is very straightforward). A negative value obtained for an unknown force indicates that the assumed direction is wrong and should be reversed.

Flow with No External Forces
An interesting situation arises when there are no external forces (such as weight, pressure, and reaction forces) acting on the body in the direction of motion—a common situation for space vehicles and satellites. For a control volume with multiple inlets and outlets, Eq. 6–21 reduces in this case to

No external forces: $$0 = \frac{d(m\vec{V})_{\text{CV}}}{dt} + \sum_{\text{out}} \beta \dot{m} \vec{V} - \sum_{\text{in}} \beta \dot{m} \vec{V} \quad (6\text{–}28)$$

This is an expression of the conservation of momentum principle, which is stated in words as *in the absence of external forces, the rate of change of the momentum of a control volume is equal to the difference between the rates of incoming and outgoing momentum flow rates.*

When the mass *m* of the control volume remains nearly constant, the first term of Eq. 6–28 becomes simply mass times acceleration, since

$$\frac{d(m\vec{V})_{\text{CV}}}{dt} = m_{\text{CV}} \frac{d\vec{V}_{\text{CV}}}{dt} = (m\vec{a})_{\text{CV}} = m_{\text{CV}} \vec{a}$$

Therefore, the control volume in this case can be treated as a solid body (a fixed-mass system) with a net thrusting force (or just **thrust**) of

Thrust: $$\vec{F}_{\text{thrust}} = m_{\text{body}} \vec{a} = \sum_{\text{in}} \beta \dot{m} \vec{V} - \sum_{\text{out}} \beta \dot{m} \vec{V} \quad (6\text{–}29)$$

acting on the body. In Eq 6–29, fluid velocities are relative to an inertial reference frame—that is, a coordinate system that is fixed in space or is moving uniformly at constant velocity on a straight path. When analyzing the motion of bodies moving at constant velocity on a straight path, it is convenient to choose an inertial reference frame that moves with the body at the same velocity on the same path. In this case the velocities of fluid streams relative to the inertial reference frame are identical to the velocities relative to the moving body, which are much easier to apply. This approach, while not strictly valid for noninertial reference frames, can also be used to calculate the *initial* acceleration of a space vehicle when its rocket is fired (Fig. 6–19).

Recall that thrust is a mechanical force typically generated through the reaction of an accelerating fluid. In the jet engine of an aircraft, for example, hot exhaust gases are accelerated by the action of expansion and outflow of gases through the back of the engine, and a thrusting force is produced by a reaction in the opposite direction. The generation of thrust is based on Newton's third law of motion, which states that *for every action at a point there is an equal and opposite reaction*. In the case of a jet engine, if the engine exerts a force on exhaust gases, then the exhaust gases exert an equal force on the engine in the opposite direction. That is, the pushing force exerted on the departing gases by the engine is equal to the thrusting force the departing gases exert on the remaining mass of the aircraft in the opposite direction $\vec{F}_{thrust} = -\vec{F}_{push}$. On the free-body diagram of an aircraft, the effect of outgoing exhaust gases is accounted for by the insertion of a force in the opposite direction of motion of the exhaust gases.

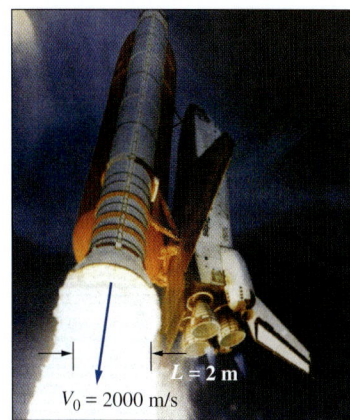

FIGURE 6–19
The thrust needed to lift the space shuttle is generated by the rocket engines as a result of momentum change of the fuel as it is accelerated from about zero to an exit speed of about 2000 m/s after combustion.

NASA

EXAMPLE 6–2 The Force to Hold a Deflector Elbow in Place

A reducing elbow is used to deflect water flow at a rate of 14 kg/s in a horizontal pipe upward 30° while accelerating it (Fig. 6–20). The elbow discharges water into the atmosphere. The cross-sectional area of the elbow is 113 cm² at the inlet and 7 cm² at the outlet. The elevation difference between the centers of the outlet and the inlet is 30 cm. The weight of the elbow and the water in it is considered to be negligible. Determine (*a*) the gage pressure at the center of the inlet of the elbow and (*b*) the anchoring force needed to hold the elbow in place.

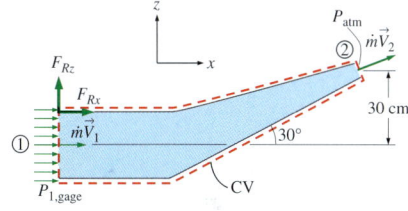

FIGURE 6–20
Schematic for Example 6–2.

SOLUTION A reducing elbow deflects water upward and discharges it to the atmosphere. The pressure at the inlet of the elbow and the force needed to hold the elbow in place are to be determined.
Assumptions 1 The flow is steady, and the frictional effects are negligible. 2 The weight of the elbow and the water in it is negligible. 3 The water is discharged to the atmosphere, and thus the gage pressure at the outlet is zero. 4 The flow is turbulent and fully developed at both the inlet and outlet of the control volume, and we take the momentum-flux correction factor to be $\beta = 1.03$ (as a conservative estimate) at both the inlet and the outlet.
Properties We take the density of water to be 1000 kg/m³.
Analysis (*a*) We take the elbow as the control volume and designate the inlet by 1 and the outlet by 2. We also take the *x*- and *z*-coordinates as shown. The continuity equation for this one-inlet, one-outlet, steady-flow system is $\dot{m}_1 = \dot{m}_2 = \dot{m} = 14$ kg/s. Noting that $\dot{m} = \rho AV$, the inlet and outlet velocities of water are

$$V_1 = \frac{\dot{m}}{\rho A_1} = \frac{14 \text{ kg/s}}{(1000 \text{ kg/m}^3)(0.0113 \text{ m}^2)} = 1.24 \text{ m/s}$$

$$V_2 = \frac{\dot{m}}{\rho A_2} = \frac{14 \text{ kg/s}}{(1000 \text{ kg/m}^3)(7 \times 10^{-4} \text{ m}^2)} = 20.0 \text{ m/s}$$

We use the Bernoulli equation (Chap. 5) as a first approximation to calculate the pressure. In Chap. 8 we will learn how to account for frictional losses along the walls. Taking the center of the inlet cross section as the reference level ($z_1 = 0$) and noting that $P_2 = P_{atm}$, the Bernoulli equation for a streamline going through the center of the elbow is expressed as

$$\frac{P_1}{\rho g} + \frac{V_1^2}{2g} + z_1 = \frac{P_2}{\rho g} + \frac{V_2^2}{2g} + z_2$$

$$P_1 - P_2 = \rho g \left(\frac{V_2^2 - V_1^2}{2g} + z_2 - z_1 \right)$$

$$P_1 - P_{atm} = (1000 \text{ kg/m}^3)(9.81 \text{ m/s}^2)$$

$$\times \left(\frac{(20 \text{ m/s})^2 - (1.24 \text{ m/s})^2}{2(9.81 \text{ m/s}^2)} + 0.3 - 0 \right) \left(\frac{1 \text{ kN}}{1000 \text{ kg·m/s}^2} \right)$$

$$P_{1,\text{gage}} = 202.2 \text{ kN/m}^2 = \mathbf{202.2 \text{ kPa}} \quad \text{(gage)}$$

(*b*) The momentum equation for steady flow is

$$\sum \vec{F} = \sum_{\text{out}} \beta \dot{m} \vec{V} - \sum_{\text{in}} \beta \dot{m} \vec{V}$$

We let the *x*- and *z*-components of the anchoring force of the elbow be F_{Rx} and F_{Rz}, and assume them to be in the positive direction. We also use gage pressure since the atmospheric pressure acts on the entire control surface. Then the momentum equations along the *x*- and *z*-axes become

$$F_{Rx} + P_{1,\text{gage}} A_1 = \beta \dot{m} V_2 \cos\theta - \beta \dot{m} V_1$$

$$F_{Rz} = \beta \dot{m} V_2 \sin\theta$$

where we have set $\beta = \beta_1 = \beta_2$. Solving for F_{Rx} and F_{Rz}, and substituting the given values,

$$F_{Rx} = \beta \dot{m}(V_2 \cos\theta - V_1) - P_{1,\text{gage}} A_1$$

$$= 1.03(14 \text{ kg/s})[(20 \cos 30° - 1.24) \text{ m/s}] \left(\frac{1 \text{ N}}{1 \text{ kg·m/s}^2} \right)$$

$$- (202{,}200 \text{ N/m}^2)(0.0113 \text{ m}^2)$$

$$= 232 - 2285 = \mathbf{-2053 \text{ N}}$$

$$F_{Rz} = \beta \dot{m} V_2 \sin\theta = (1.03)(14 \text{ kg/s})(20 \sin 30° \text{ m/s}) \left(\frac{1 \text{ N}}{1 \text{ kg·m/s}^2} \right) = \mathbf{144 \text{ N}}$$

The negative result for F_{Rx} indicates that the assumed direction is wrong, and it should be reversed. Therefore, F_{Rx} acts in the negative *x*-direction.

Discussion There is a nonzero pressure distribution along the inside walls of the elbow, but since the control volume is outside the elbow, these pressures do not appear in our analysis. The weight of the elbow and the water in it could be added to the vertical force for better accuracy. The actual value of $P_{1,\text{gage}}$ will be higher than that calculated here because of frictional and other irreversible losses in the elbow.

EXAMPLE 6–3 The Force to Hold a Reversing Elbow in Place

The deflector elbow in Example 6–2 is replaced by a reversing elbow such that the fluid makes a 180° U-turn before it is discharged, as shown in Fig. 6–21. The elevation difference between the centers of the inlet and the exit sections is still 0.3 m. Determine the anchoring force needed to hold the elbow in place.

SOLUTION The inlet and the outlet velocities and the pressure at the inlet of the elbow remain the same, but the vertical component of the anchoring force at the connection of the elbow to the pipe is zero in this case ($F_{Rz} = 0$) since there is no other force or momentum flux in the vertical direction (we are neglecting the weight of the elbow and the water). The horizontal component of the anchoring force is determined from the momentum equation written in the x-direction. Noting that the outlet velocity is negative since it is in the negative x-direction, we have

$$F_{Rx} + P_{1,\text{gage}} A_1 = \beta_2 \dot{m}(-V_2) - \beta_1 \dot{m} V_1 = -\beta \dot{m}(V_2 + V_1)$$

Solving for F_{Rx} and substituting the known values,

$$F_{Rx} = -\beta \dot{m}(V_2 + V_1) - P_{1,\text{gage}} A_1$$

$$= -(1.03)(14 \text{ kg/s})[(20 + 1.24) \text{ m/s}]\left(\frac{1 \text{ N}}{1 \text{ kg} \cdot \text{m/s}^2}\right) - (202{,}200 \text{ N/m}^2)(0.0113 \text{ m}^2)$$

$$= -306 - 2285 = \mathbf{-2591 \text{ N}}$$

Therefore, the horizontal force on the flange is 2591 N acting in the negative x-direction (the elbow is trying to separate from the pipe). This force is equivalent to the weight of about 260 kg mass, and thus the connectors (such as bolts) used must be strong enough to withstand this force.

Discussion The reaction force in the x-direction is larger than that of Example 6–2 since the walls turn the water over a much greater angle. If the reversing elbow is replaced by a straight nozzle (like one used by firefighters) such that water is discharged in the positive x-direction, the momentum equation in the x-direction becomes

$$F_{Rx} + P_{1,\text{gage}} A_1 = \beta \dot{m} V_2 - \beta \dot{m} V_1 \rightarrow F_{Rx} = \beta \dot{m}(V_2 - V_1) - P_{1,\text{gage}} A_1$$

since both V_1 and V_2 are in the positive x-direction. This shows the importance of using the correct sign (positive if in the positive direction and negative if in the opposite direction) for velocities and forces.

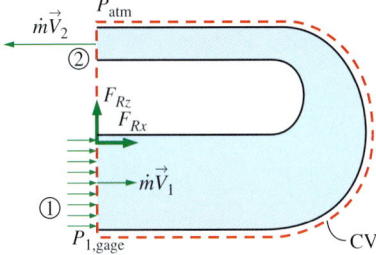

FIGURE 6–21
Schematic for Example 6–3.

EXAMPLE 6–4 Water Jet Striking a Stationary Plate

Water is accelerated by a nozzle to an average speed of 20 m/s, and strikes a stationary vertical plate at a rate of 10 kg/s with a normal velocity of 20 m/s (Fig. 6–22). After the strike, the water stream splatters off in all directions in the plane of the plate. Determine the force needed to prevent the plate from moving horizontally due to the water stream.

SOLUTION A water jet strikes a vertical stationary plate normally. The force needed to hold the plate in place is to be determined.
Assumptions 1 The flow of water at the nozzle outlet is steady. 2 The water splatters in directions normal to the approach direction of the water jet.

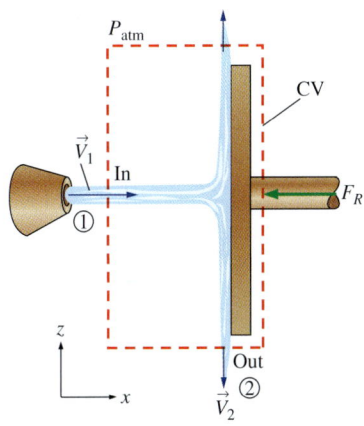

FIGURE 6–22
Schematic for Example 6–4.

3 The water jet is exposed to the atmosphere, and thus the pressure of the water jet and the splattered water leaving the control volume is atmospheric pressure, which is disregarded since it acts on the entire system. **4** The vertical forces and momentum fluxes are not considered since they have no effect on the horizontal reaction force. **5** The effect of the momentum-flux correction factor is negligible, and thus $\beta \cong 1$ at the inlet.

Analysis We draw the control volume for this problem such that it contains the entire plate and cuts through the water jet and the support bar normally. The momentum equation for steady flow is given as

$$\sum \vec{F} = \sum_{out} \beta \dot{m} \vec{V} - \sum_{in} \beta \dot{m} \vec{V} \qquad (1)$$

Writing Eq. 1 for this problem along the *x*-direction (without forgetting the negative sign for forces and velocities in the negative *x*-direction) and noting that $V_{1,x} = V_1$ and $V_{2,x} = 0$ gives

$$-F_R = 0 - \beta \dot{m} V_1$$

Substituting the given values,

$$F_R = \beta \dot{m} V_1 = (1)(10 \text{ kg/s})(20 \text{ m/s})\left(\frac{1 \text{ N}}{1 \text{ kg·m/s}^2}\right) = \mathbf{200 \text{ N}}$$

Therefore, the support must apply a 200-N horizontal force (equivalent to the weight of about a 20-kg mass) in the negative *x*-direction (the opposite direction of the water jet) to hold the plate in place. A similar situation occurs in the downwash of a helicopter (Fig. 6–23).

Discussion The plate absorbs the full brunt of the momentum of the water jet since the *x*-direction momentum at the outlet of the control volume is zero. If the control volume were drawn instead along the interface between the water and the plate, there would be additional (unknown) pressure forces in the analysis. By cutting the control volume through the support, we avoid having to deal with this additional complexity. This is an example of a "wise" choice of control volume.

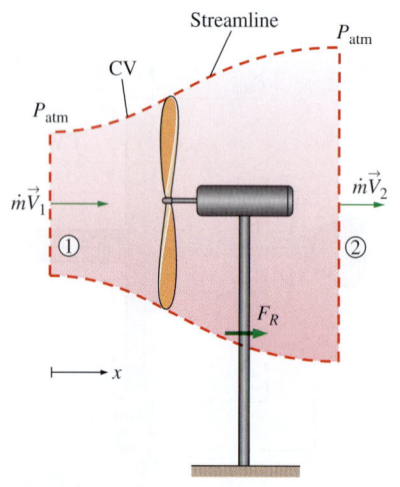

FIGURE 6–23
The downwash of a helicopter is similar to the jet discussed in Example 6–4. The jet impinges on the surface of the water in this case, causing circular waves as seen here.
© Purestock/SuperStock RF

FIGURE 6–24
Schematic for Example 6–5.

EXAMPLE 6–5 Power Generation and Wind Loading of a Wind Turbine

A wind generator with a 9-m-diameter blade span has a cut-in wind speed (minimum speed for power generation) of 11 km/h, at which velocity the turbine generates 0.4 kW of electric power (Fig. 6–24). Determine (*a*) the efficiency of the wind turbine–generator unit and (*b*) the horizontal force exerted by the wind on the supporting mast of the wind turbine. What is the effect of doubling the wind velocity to 22 km/h on power generation and the force exerted? Assume the efficiency remains the same, and take the density of air to be 1.22 kg/m³.

SOLUTION The power generation and loading of a wind turbine are to be analyzed. The efficiency and the force exerted on the mast are to be determined, and the effects of doubling the wind velocity are to be investigated.
Assumptions **1** The wind flow is steady and incompressible. **2** The efficiency of the turbine–generator is independent of wind speed. **3** The frictional effects are negligible, and thus none of the incoming kinetic energy is converted to

thermal energy. **4** The average velocity of air through the wind turbine is the same as the wind velocity (actually, it is considerably less—see Chap. 14). **5** The wind flow is nearly uniform upstream and downstream of the wind turbine and thus the momentum-flux correction factor is $\beta = \beta_1 = \beta_2 \cong 1$.
Properties The density of air is given to be 1.22 kg/m³.
Analysis Kinetic energy is a mechanical form of energy, and thus it can be converted to work entirely. Therefore, the power potential of the wind is proportional to its kinetic energy, which is $V^2/2$ per unit mass, and thus the maximum power is $\dot{m}V^2/2$ for a given mass flow rate:

$$V_1 = (11 \text{ km/h})\left(\frac{1 \text{ m/s}}{3.6 \text{ km/h}}\right) = 3.056 \text{ m/s}$$

$$\dot{m} = \rho_1 V_1 A_1 = \rho_1 V_1 \frac{\pi D^2}{4} = (1.22 \text{ kg/m}^3)(3.056 \text{ m/s})\frac{\pi (9 \text{ m})^2}{4} = 237.2 \text{ kg/s}$$

$$\dot{W}_{\text{max}} = \dot{m}\text{ke}_1 = \dot{m}\frac{V_1^2}{2}$$

$$= (237.2 \text{ kg/s})\frac{(3.056 \text{ m/s})^2}{2}\left(\frac{1 \text{ kN}}{1000 \text{ kg·m/s}^2}\right)\left(\frac{1 \text{ kW}}{1 \text{ kN·m/s}}\right)$$

$$= 1.108 \text{ kW}$$

Therefore, the available power to the wind turbine is 1.108 kW at the wind velocity of 11 km/h. Then the turbine–generator efficiency becomes

$$\eta_{\text{wind turbine}} = \frac{\dot{W}_{\text{act}}}{\dot{W}_{\text{max}}} = \frac{0.4 \text{ kW}}{1.108 \text{ kW}} = \mathbf{0.361} \quad (\text{or } \mathbf{36.1\%})$$

(*b*) The frictional effects are assumed to be negligible, and thus the portion of incoming kinetic energy not converted to electric power leaves the wind turbine as outgoing kinetic energy. Noting that the mass flow rate remains constant, the exit velocity is determined to be

$$\dot{m}\text{ke}_2 = \dot{m}\text{ke}_1(1 - \eta_{\text{wind turbine}}) \rightarrow \dot{m}\frac{V_2^2}{2} = \dot{m}\frac{V_1^2}{2}(1 - \eta_{\text{wind turbine}}) \quad (1)$$

or

$$V_2 = V_1\sqrt{1 - \eta_{\text{wind turbine}}} = (3.056 \text{ m/s})\sqrt{1 - 0.361} = 2.443 \text{ m/s}$$

To determine the force on the mast (Fig. 6–25), we draw a control volume around the wind turbine such that the wind is normal to the control surface at the inlet and the outlet and the entire control surface is at atmospheric pressure (Fig. 6–23). The momentum equation for steady flow is given as

$$\sum \vec{F} = \sum_{\text{out}} \beta \dot{m}\vec{V} - \sum_{\text{in}} \beta \dot{m}\vec{V} \quad (2)$$

Writing Eq. 2 along the *x*-direction and noting that $\beta = 1$, $V_{1,x} = V_1$, and $V_{2,x} = V_2$ give

$$F_R = \dot{m}V_2 - \dot{m}V_1 = \dot{m}(V_2 - V_1) \quad (3)$$

Substituting the known values into Eq. 3 gives

$$F_R = \dot{m}(V_2 - V_1) = (237.2 \text{ kg/s})(2.443 - 3.056 \text{ m/s})\left(\frac{1 \text{ N}}{1 \text{ kg·m/s}^2}\right)$$

$$= -145 \text{ N}$$

FIGURE 6–25
Forces and moments on the supporting mast of a modern wind turbine can be substantial, and increase like V^2; thus the mast is typically quite large and strong.
© Ingram Publishing/SuperStock RF

MOMENTUM ANALYSIS OF FLOW SYSTEMS

The negative sign indicates that the reaction force acts in the negative x-direction, as expected. Then the force exerted by the wind on the mast becomes $F_{\text{mast}} = -F_R =$ **145 N**.

The power generated is proportional to V^3 since the mass flow rate is proportional to V and the kinetic energy to V^2. Therefore, doubling the wind velocity to 22 km/h will increase the power generation by a factor of $2^3 = 8$ to $0.4 \times 8 = 3.2$ kW. The force exerted by the wind on the support mast is proportional to V^2. Therefore, doubling the wind velocity to 22 km/h will increase the wind force by a factor of $2^2 = 4$ to $145 \times 4 = 580$ N.

Discussion Wind turbines are treated in more detail in Chap. 14.

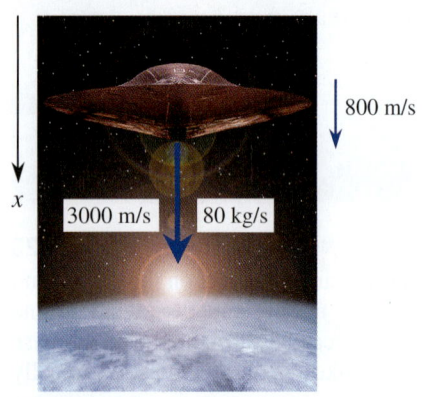

FIGURE 6–26
Schematic for Example 6–6.
© Brand X Pictures/PunchStock

EXAMPLE 6–6   Deceleration of a Spacecraft

A spacecraft with a mass of 12,000 kg is dropping vertically towards a planet at a constant speed of 800 m/s (Fig. 6–26). To slow down the spacecraft, a solid-fuel rocket at the bottom is fired, and combustion gases leave the rocket at a constant rate of 80 kg/s and at a velocity of 3000 m/s relative to the spacecraft in the direction of motion of the spacecraft for a period of 5 s. Disregarding the small change in the mass of the spacecraft, determine (*a*) the deceleration of the spacecraft during this period, (*b*) the change of velocity of the spacecraft, and (*c*) the thrust exerted on the spacecraft.

SOLUTION The rocket of a spacecraft is fired in the direction of motion. The deceleration, the velocity change, and the thrust are to be determined.
Assumptions **1** The flow of combustion gases is steady and one-dimensional during the firing period, but the flight of the spacecraft is unsteady. **2** There are no external forces acting on the spacecraft, and the effect of pressure force at the nozzle outlet is negligible. **3** The mass of discharged fuel is negligible relative to the mass of the spacecraft, and thus, the spacecraft may be treated as a solid body with a constant mass. **4** The nozzle is well designed such that the effect of the momentum-flux correction factor is negligible, and thus, $\beta \cong 1$.
Analysis (*a*) For convenience, we choose an inertial reference frame that moves with the spacecraft at the same initial velocity. Then the velocities of the fluid stream relative to an inertial reference frame become simply the velocities relative to the spacecraft. We take the direction of motion of the spacecraft as the positive direction along the x-axis. There are no external forces acting on the spacecraft, and its mass is essentially constant. Therefore, the spacecraft can be treated as a solid body with constant mass, and the momentum equation in this case is, from Eq. 6–29,

$$\vec{F}_{\text{thrust}} = m_{\text{spacecraft}} \vec{a}_{\text{spacecraft}} = \sum_{\text{in}} \beta \dot{m} \vec{V} - \sum_{\text{out}} \beta \dot{m} \vec{V}$$

where the fluid stream velocities relative to the inertial reference frame in this case are identical to the velocities relative to the spacecraft. Noting that the motion is on a straight line and the discharged gases move in the positive x-direction, we write the momentum equation using magnitudes as

$$m_{\text{spacecraft}} a_{\text{spacecraft}} = m_{\text{spacecraft}} \frac{dV_{\text{spacecraft}}}{dt} = -\dot{m}_{\text{gas}} V_{\text{gas}}$$

Noting that gases leave in the positive x-direction and substituting, the acceleration of the spacecraft during the first 5 seconds is determined to be

$$a_{spacecraft} = \frac{dV_{spacecraft}}{dt} = -\frac{\dot{m}_{gas}}{m_{spacecraft}}V_{gas} = -\frac{80 \text{ kg/s}}{12{,}000 \text{ kg}}(+3000 \text{ m/s}) = -20 \text{ m/s}^2$$

The negative value confirms that the spacecraft is decelerating in the positive x direction at a rate of 20 m/s².

(b) Knowing the deceleration, which is constant, the velocity change of the spacecraft during the first 5 seconds is determined from the definition of acceleration to be

$$dV_{spacecraft} = a_{spacecraft}dt \rightarrow \Delta V_{spacecraft} = a_{spacecraft}\Delta t = (-20 \text{ m/s}^2)(5 \text{ s})$$
$$= -100 \text{ m/s}$$

(c) The thrusting force exerted on the space aircraft is, from Eq. 6-29,

$$F_{thrust} = 0 - \dot{m}_{gas}V_{gas} = 0 - (80 \text{ kg/s})(+3000 \text{ m/s})\left(\frac{1 \text{ kN}}{1000 \text{ kg·m/s}^2}\right) = -240 \text{ kN}$$

The negative sign indicates that the trusting force due to firing of the rocket acts on the aircraft in the negative x-direction.

Discussion Note that if this fired rocket were attached somewhere on a test stand, it would exert a force of 240 kN (equivalent to the weight of about 24 tons of mass) to its support in the opposite direction of the discharged gases.

EXAMPLE 6–7 Net Force on a Flange

Water flows at a rate of 70 L/min through a flanged faucet with a partially closed gate valve spigot (Fig. 6–27). The inner diameter of the pipe at the location of the flange is 2 cm, and the pressure at that location is measured to be 90 kPa (gage). The total weight of the faucet assembly plus the water within it is 57 N. Calculate the net force on the flange.

SOLUTION Water flow through a flanged faucet is considered. The net force acting on the flange is to be calculated.
Assumptions 1 The flow is steady and incompressible. 2 The flow at the inlet and at the outlet is turbulent and fully developed so that the momentum-flux correction factor is about 1.03. 3 The pipe diameter at the outlet of the faucet is the same as that at the flange.
Properties The density of water at room temperature is 997 kg/m³.
Analysis We choose the faucet and its immediate surroundings as the control volume, as shown in Fig. 6–27 along with all the forces acting on it. These forces include the weight of the water and the weight of the faucet assembly, the gage pressure force at the inlet to the control volume, and the net force of the flange on the control volume, which we call $\vec{F}_R$. We use gage pressure for convenience since the gage pressure on the rest of the control surface is zero (atmospheric pressure). Note that the pressure through the outlet of the control volume is also atmospheric since we are assuming incompressible flow; hence, the gage pressure is also zero through the outlet.

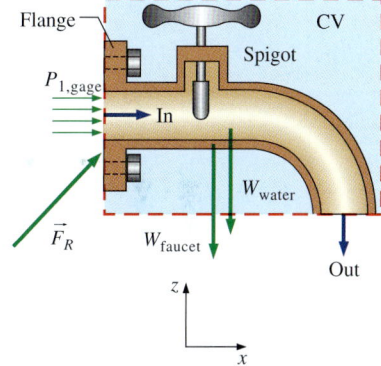

FIGURE 6–27
Control volume for Example 6–7 with all forces shown; gage pressure is used for convenience.

We now apply the control volume conservation laws. Conservation of mass is trivial here since there is only one inlet and one outlet; namely, the mass flow rate into the control volume is equal to the mass flow rate out of the control volume. Also, the outflow and inflow average velocities are identical since the inner diameter is constant and the water is incompressible, and are determined to be

$$V_2 = V_1 = V = \frac{\dot{V}}{A_c} = \frac{\dot{V}}{\pi D^2/4} = \frac{70 \text{ L/min}}{\pi (0.02 \text{ m})^2/4} \left(\frac{1 \text{ m}^3}{1000 \text{ L}}\right)\left(\frac{1 \text{ min}}{60 \text{ s}}\right) = 3.714 \text{ m/s}$$

Also,

$$\dot{m} = \rho \dot{V} = (997 \text{ kg/m}^3)(70 \text{ L/min})\left(\frac{1 \text{ m}^3}{1000 \text{ L}}\right)\left(\frac{1 \text{ min}}{60 \text{ s}}\right) = 1.163 \text{ kg/s}$$

Next we apply the momentum equation for steady flow,

$$\sum \vec{F} = \sum_{\text{out}} \beta \dot{m} \vec{V} - \sum_{\text{in}} \beta \dot{m} \vec{V} \tag{1}$$

We let the x- and z-components of the force acting on the flange be F_{Rx} and F_{Rz}, and assume them to be in the positive directions. The magnitude of the velocity in the x-direction is $+V_1$ at the inlet, but zero at the outlet. The magnitude of the velocity in the z-direction is zero at the inlet, but $-V_2$ at the outlet. Also, the weight of the faucet assembly and the water within it acts in the $-z$-direction as a body force. No pressure or viscous forces act on the chosen (wise) control volume in the z-direction.

The components of Eq. 1 along the x- and z-directions become

$$F_{Rx} + P_{1, \text{gage}} A_1 = 0 - \dot{m}(+V_1)$$

$$F_{Rz} - W_{\text{faucet}} - W_{\text{water}} = \dot{m}(-V_2) - 0$$

Solving for F_{Rx} and F_{Rz}, and substituting the given values,

$$F_{Rx} = -\dot{m} V_1 - P_{1, \text{gage}} A_1$$

$$= -(1.163 \text{ kg/s})(3.714 \text{ m/s})\left(\frac{1 \text{ N}}{1 \text{ kg·m/s}^2}\right) - (90{,}000 \text{ N/m}^2)\frac{\pi (0.02 \text{ m})^2}{4}$$

$$= -32.6 \text{ N}$$

$$F_{Rz} = -\dot{m} V_2 + W_{\text{faucet + water}}$$

$$= -(1.163 \text{ kg/s})(3.714 \text{ m/s})\left(\frac{1 \text{ N}}{1 \text{ kg·m/s}^2}\right) + 57 \text{ N} = 52.7 \text{ N}$$

Then the net force of the flange on the control volume is expressed in vector form as

$$\vec{F}_R = F_{Rx} \vec{i} + F_{Rz} \vec{k} = -32.6 \vec{i} + 52.7 \vec{k} \text{ N}$$

From Newton's third law, the force the faucet assembly exerts on the flange is the negative of $\vec{F}_R$,

$$\vec{F}_{\text{faucet on flange}} = -\vec{F}_R = 32.6 \vec{i} - 52.7 \vec{k} \text{ N}$$

Discussion The faucet assembly pulls to the right and down; this agrees with our intuition. Namely, the water exerts a high pressure at the inlet, but the outlet pressure is atmospheric. In addition, the momentum of the water at the inlet in the *x*-direction is lost in the turn, causing an additional force to the right on the pipe walls. The faucet assembly weighs much more than the momentum effect of the water, so we expect the force to be downward. Note that labeling forces such as "faucet on flange" clarifies the direction of the force.

6–5 ▪ REVIEW OF ROTATIONAL MOTION AND ANGULAR MOMENTUM

The motion of a rigid body can be considered to be the combination of translational motion of its center of mass and rotational motion about its center of mass. The translational motion is analyzed using the linear momentum equation, Eq. 6–1. Now we discuss the rotational motion—a motion during which all points in the body move in circles about the axis of rotation. Rotational motion is described with angular quantities such as angular distance θ, angular velocity $\vec{\omega}$, and angular acceleration $\vec{\alpha}$.

The amount of rotation of a point in a body is expressed in terms of the angle θ swept by a line of length r that connects the point to the axis of rotation and is perpendicular to the axis. The angle θ is expressed in radians (rad), which is the arc length corresponding to θ on a circle of unit radius. Noting that the circumference of a circle of radius r is $2\pi r$, the angular distance traveled by any point in a rigid body during a complete rotation is 2π rad. The physical distance traveled by a point along its circular path is $l = \theta r$, where r is the normal distance of the point from the axis of rotation and θ is the angular distance in rad. Note that 1 rad corresponds to $360/(2\pi) \cong 57.3°$.

The magnitude of angular velocity ω is the angular distance traveled per unit time, and the magnitude of angular acceleration α is the rate of change of angular velocity. They are expressed as (Fig. 6–28),

$$\omega = \frac{d\theta}{dt} = \frac{d(l/r)}{dt} = \frac{1}{r}\frac{dl}{dt} = \frac{V}{r} \quad \text{and} \quad \alpha = \frac{d\omega}{dt} = \frac{d^2\theta}{dt^2} = \frac{1}{r}\frac{dV}{dt} = \frac{a_t}{r} \quad (6\text{–}30)$$

or

$$V = r\omega \quad \text{and} \quad a_t = r\alpha \quad (6\text{–}31)$$

where V is the linear velocity and a_t is the linear acceleration in the tangential direction for a point located at a distance r from the axis of rotation. Note that ω and α are the same for all points of a rotating rigid body, but V and a_t are not (they are proportional to r).

Newton's second law requires that there must be a force acting in the tangential direction to cause angular acceleration. The strength of the rotating effect, called the *moment* or *torque*, is proportional to the magnitude of the force and its distance from the axis of rotation. The perpendicular distance from the axis of rotation to the line of action of the force is called the *moment arm*, and the magnitude of torque M acting on a point mass m at normal distance r from the axis of rotation is expressed as

$$M = rF_t = rma_t = mr^2\alpha \quad (6\text{–}32)$$

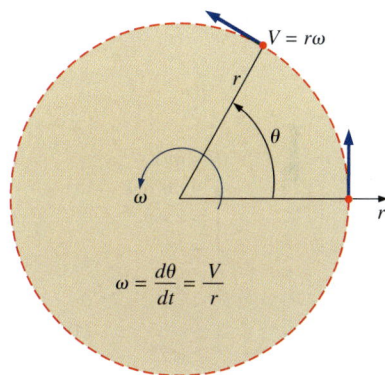

FIGURE 6–28
The relations between angular distance θ, angular velocity ω, and linear velocity V in a plane.

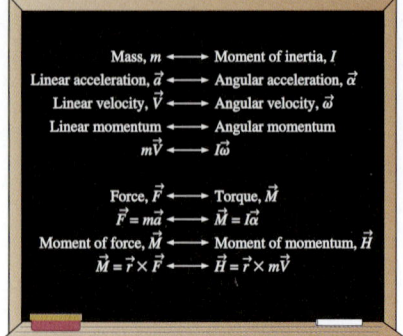

FIGURE 6–29
Analogy between corresponding linear and angular quantities.

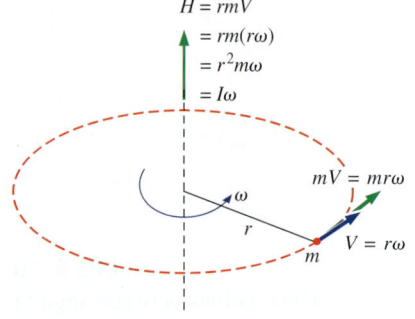

FIGURE 6–30
Angular momentum of point mass m rotating at angular velocity ω at distance r from the axis of rotation.

The total torque acting on a rotating rigid body about an axis is determined by integrating the torque acting on differential mass δm over the entire body to give

Magnitude of torque: $\quad M = \int_{mass} r^2 \alpha \, \delta m = \left[\int_{mass} r^2 \, \delta m \right] \alpha = I\alpha \quad$ (6–33)

where I is the *moment of inertia* of the body about the axis of rotation, which is a measure of the inertia of a body against rotation. The relation $M = I\alpha$ is the counterpart of Newton's second law, with torque replacing force, moment of inertia replacing mass, and angular acceleration replacing linear acceleration (Fig. 6–29). Note that unlike mass, the rotational inertia of a body also depends on the distribution of the mass of the body with respect to the axis of rotation. Therefore, a body whose mass is closely packed about its axis of rotation has a small resistance against angular acceleration, while a body whose mass is concentrated at its periphery has a large resistance against angular acceleration. A flywheel is a good example of the latter.

The linear momentum of a body of mass m having velocity $\vec{V}$ is $m\vec{V}$, and the direction of linear momentum is identical to the direction of velocity. Noting that the moment of a force is equal to the product of the force and the normal distance, the magnitude of the moment of momentum, called the **angular momentum,** of a point mass m about an axis is expressed as $H = rmV = r^2 m\omega$, where r is the normal distance from the axis of rotation to the line of action of the momentum vector (Fig. 6–30). Then the total angular momentum of a rotating rigid body is determined by integration to be

Magnitude of angular momentum: $\quad H = \int_{mass} r^2 \omega \, \delta m = \left[\int_{mass} r^2 \, \delta m \right] \omega = I\omega \quad$ (6–34)

where again I is the *moment of inertia* of the body about the axis of rotation. It can also be expressed more generally in vector form as

$$\vec{H} = I\vec{\omega} \quad (6\text{–}35)$$

Note that the angular velocity $\vec{\omega}$ is the same at every point of a rigid body.

Newton's second law $\vec{F} = m\vec{a}$ was expressed in terms of the rate of change of linear momentum in Eq. 6–1 as $\vec{F} = d(m\vec{V})/dt$. Likewise, the counterpart of Newton's second law for rotating bodies $\vec{M} = I\vec{\alpha}$ is expressed in Eq. 6–2 in terms of the rate of change of angular momentum as

Angular momentum equation: $\quad \vec{M} = I\vec{\alpha} = I\dfrac{d\vec{\omega}}{dt} = \dfrac{d(I\vec{\omega})}{dt} = \dfrac{d\vec{H}}{dt} \quad$ (6–36)

where $\vec{M}$ is the net torque applied on the body about the axis of rotation.

The angular velocity of rotating machinery is typically expressed in rpm (number of revolutions per minute) and denoted by $\dot{n}$. Noting that velocity is distance traveled per unit time and the angular distance traveled during each revolution is 2π, the angular velocity of rotating machinery is $\omega = 2\pi\dot{n}$ rad/min or

Angular velocity versus rpm: $\quad \omega = 2\pi\dot{n} \text{ (rad/min)} = \dfrac{2\pi\dot{n}}{60} \text{ (rad/s)} \quad$ (6–37)

Consider a constant force F acting in the tangential direction on the outer surface of a shaft of radius r rotating at an rpm of $\dot{n}$. Noting that work W is force times distance, and power $\dot{W}$ is work done per unit time and thus force times velocity, we have $\dot{W}_{shaft} = FV = Fr\omega = M\omega$. Therefore, the power transmitted by a shaft rotating at an rpm of $\dot{n}$ under the influence of an applied torque M is (Fig. 6–31)

Shaft power: $$\dot{W}_{shaft} = \omega M = 2\pi \dot{n} M \tag{6-38}$$

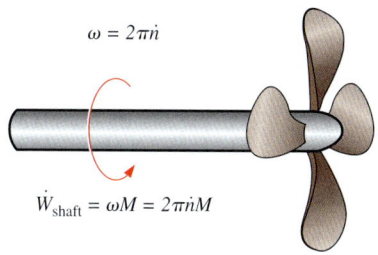

FIGURE 6–31
The relations between angular velocity, rpm, and the power transmitted through a rotating shaft.

The kinetic energy of a body of mass m during translational motion is $KE = \frac{1}{2}mV^2$. Noting that $V = r\omega$, the rotational kinetic energy of a body of mass m at a distance r from the axis of rotation is $KE = \frac{1}{2}mr^2\omega^2$. The total rotational kinetic energy of a rotating rigid body about an axis is determined by integrating the rotational kinetic energies of differential masses dm over the entire body to give

Rotational kinetic energy: $$KE_r = \frac{1}{2}I\omega^2 \tag{6-39}$$

where again I is the moment of inertia of the body and ω is the angular velocity.

During rotational motion, the direction of velocity changes even when its magnitude remains constant. Velocity is a vector quantity, and thus a change in direction constitutes a change in velocity with time, and thus acceleration. This is called **centripetal acceleration.** Its magnitude is

$$a_r = \frac{V^2}{r} = r\omega^2$$

Centripetal acceleration is directed toward the axis of rotation (opposite direction of radial acceleration), and thus the radial acceleration is negative. Noting that acceleration is a constant multiple of force, centripetal acceleration is the result of a force acting on the body toward the axis of rotation, known as the **centripetal force,** whose magnitude is $F_r = mV^2/r$. Tangential and radial accelerations are perpendicular to each other (since the radial and tangential directions are perpendicular), and the total linear acceleration is determined by their vector sum, $\vec{a} = \vec{a}_t + \vec{a}_r$. For a body rotating at constant angular velocity, the only acceleration is the centripetal acceleration. The centripetal force does not produce torque since its line of action intersects the axis of rotation.

6–6 ▪ THE ANGULAR MOMENTUM EQUATION

The linear momentum equation discussed in Section 6–4 is useful for determining the relationship between the linear momentum of flow streams and the resultant forces. Many engineering problems involve the moment of the linear momentum of flow streams, and the rotational effects caused by them. Such problems are best analyzed by the *angular momentum equation*, also called the *moment of momentum equation*. An important class of fluid devices, called *turbomachines,* which include centrifugal pumps, turbines, and fans, is analyzed by the angular momentum equation.

The *moment of a force* $\vec{F}$ about a point O is the vector (or cross) product (Fig. 6–32)

Moment of a force: $$\vec{M} = \vec{r} \times \vec{F} \tag{6-40}$$

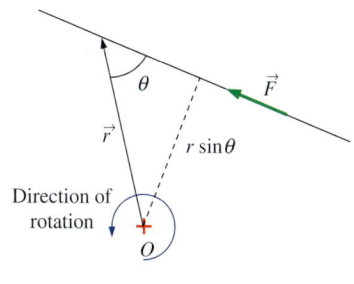

FIGURE 6–32
The moment of a force $\vec{F}$ about a point O is the vector product of the position vector $\vec{r}$ and $\vec{F}$.

MOMENTUM ANALYSIS OF FLOW SYSTEMS

where $\vec{r}$ is the position vector from point O to any point on the line of action of $\vec{F}$. The vector product of two vectors is a vector whose line of action is normal to the plane that contains the crossed vectors ($\vec{r}$ and $\vec{F}$ in this case) and whose magnitude is

Magnitude of the moment of a force: $\qquad M = Fr \sin\theta \qquad$ (6–41)

where θ is the angle between the lines of action of the vectors $\vec{r}$ and $\vec{F}$. Therefore, the magnitude of the moment about point O is equal to the magnitude of the force multiplied by the normal distance of the line of action of the force from the point O. The sense of the moment vector $\vec{M}$ is determined by the right-hand rule: when the fingers of the right hand are curled in the direction that the force tends to cause rotation, the thumb points the direction of the moment vector (Fig. 6–33). Note that a force whose line of action passes through point O produces zero moment about point O.

The vector product of $\vec{r}$ and the momentum vector $m\vec{V}$ gives the *moment of momentum*, also called the *angular momentum*, about a point O as

Moment of momentum: $\qquad \vec{H} = \vec{r} \times m\vec{V} \qquad$ (6–42)

Therefore, $\vec{r} \times \vec{V}$ represents the angular momentum per unit mass, and the angular momentum of a differential mass $\delta m = \rho\, dV$ is $d\vec{H} = (\vec{r} \times \vec{V})\rho\, dV$. Then the angular momentum of a system is determined by integration to be

Moment of momentum (system): $\qquad \vec{H}_{sys} = \int_{sys} (\vec{r} \times \vec{V})\rho\, dV \qquad$ (6–43)

The rate of change of the moment of momentum is

Rate of change of moment of momentum: $\qquad \dfrac{d\vec{H}_{sys}}{dt} = \dfrac{d}{dt}\int_{sys} (\vec{r} \times \vec{V})\rho\, dV \qquad$ (6–44)

The angular momentum equation for a system was expressed in Eq. 6–2 as

$$\sum \vec{M} = \dfrac{d\vec{H}_{sys}}{dt} \qquad (6\text{–}45)$$

where $\sum \vec{M} = \sum(\vec{r} \times \vec{F})$ is the net torque or moment applied on the system, which is the vector sum of the moments of all forces acting on the system, and $d\vec{H}_{sys}/dt$ is the rate of change of the angular momentum of the system. Equation 6–45 is stated as *the rate of change of angular momentum of a system is equal to the net torque acting on the system*. This equation is valid for a fixed quantity of mass and an inertial reference frame, i.e., a reference frame that is fixed or moves with a constant velocity in a straight path.

The general control volume formulation of the angular momentum equation is obtained by setting $b = \vec{r} \times \vec{V}$ and thus $B = \vec{H}$ in the general Reynolds transport theorem. It gives (Fig. 6–34)

$$\dfrac{d\vec{H}_{sys}}{dt} = \dfrac{d}{dt}\int_{CV} (\vec{r} \times \vec{V})\rho\, dV + \int_{CS} (\vec{r} \times \vec{V})\rho(\vec{V}_r \cdot \vec{n})\, dA \qquad (6\text{–}46)$$

The left-hand side of this equation is, from Eq. 6–45, equal to $\sum \vec{M}$. Substituting, the angular momentum equation for a general control volume (stationary or moving, fixed shape or distorting) is

General: $\qquad \sum \vec{M} = \dfrac{d}{dt}\int_{CV} (\vec{r} \times \vec{V})\rho\, dV + \int_{CS} (\vec{r} \times \vec{V})\rho(\vec{V}_r \cdot \vec{n})\, dA \qquad$ (6–47)

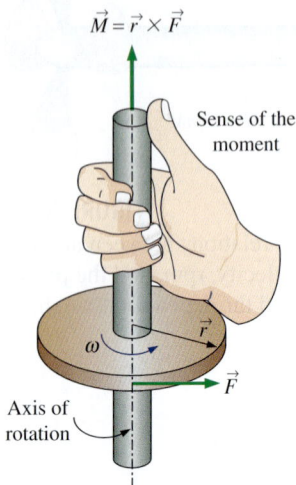

FIGURE 6–33
The determination of the direction of the moment by the right-hand rule.

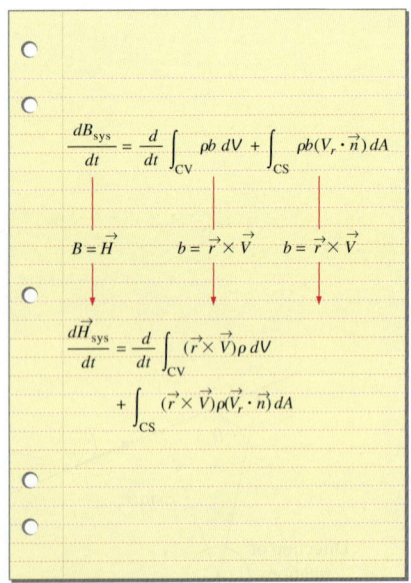

FIGURE 6–34
The angular momentum equation is obtained by replacing B in the Reynolds transport theorem by the angular momentum $\vec{H}$, and b by the angular momentum per unit mass $\vec{r} \times \vec{V}$.

which is stated in words as

$$\begin{pmatrix} \text{The sum of all} \\ \text{external moments} \\ \text{acting on a CV} \end{pmatrix} = \begin{pmatrix} \text{The time rate of change} \\ \text{of the angular momentum} \\ \text{of the contents of the CV} \end{pmatrix} + \begin{pmatrix} \text{The net flow rate of} \\ \text{angular momentum} \\ \text{out of the control} \\ \text{surface by mass flow} \end{pmatrix}$$

Again, $\vec{V}_r = \vec{V} - \vec{V}_{CS}$ is the fluid velocity relative to the control surface (for use in mass flow rate calculations at all locations where the fluid crosses the control surface), and $\vec{V}$ is the fluid velocity as viewed from a fixed reference frame. The product $\rho(\vec{V}_r \cdot \vec{n})\, dA$ represents the mass flow rate through dA into or out of the control volume, depending on the sign.

For a fixed control volume (no motion or deformation of the control volume), $\vec{V}_r = \vec{V}$ and the angular momentum equation becomes

Fixed CV: $\quad \sum \vec{M} = \dfrac{d}{dt} \displaystyle\int_{CV} (\vec{r} \times \vec{V}) \rho\, dV + \displaystyle\int_{CS} (\vec{r} \times \vec{V}) \rho (\vec{V} \cdot \vec{n})\, dA$ (6–48)

Also, note that the forces acting on the control volume consist of *body forces* that act throughout the entire body of the control volume such as gravity, and *surface forces* that act on the control surface such as the pressure and reaction forces at points of contact. The net torque consists of the moments of these forces as well as the torques applied on the control volume.

Special Cases

During *steady flow*, the amount of angular momentum within the control volume remains constant, and thus the time rate of change of angular momentum of the contents of the control volume is zero. Then,

Steady flow: $\quad \sum \vec{M} = \displaystyle\int_{CS} (\vec{r} \times \vec{V}) \rho (\vec{V}_r \cdot \vec{n})\, dA$ (6–49)

In many practical applications, the fluid crosses the boundaries of the control volume at a certain number of inlets and outlets, and it is convenient to replace the area integral by an algebraic expression written in terms of the average properties over the cross-sectional areas where the fluid enters or leaves the control volume. In such cases, the angular momentum flow rate can be expressed as the difference in the angular momentum of outgoing and incoming streams. Furthermore, in many cases the moment arm $\vec{r}$ is either constant along the inlet or outlet (as in radial flow turbomachines) or is large compared to the diameter of the inlet or outlet pipe (as in rotating lawn sprinklers, Fig. 6–35). In such cases, the *average* value of $\vec{r}$ is used throughout the cross-sectional area of the inlet or outlet. Then, an approximate form of the angular momentum equation in terms of average properties at inlets and outlets becomes

$$\sum \vec{M} \cong \frac{d}{dt} \int_{CV} (\vec{r} \times \vec{V}) \rho\, dV + \sum_{\text{out}} (\vec{r} \times \dot{m}\vec{V}) - \sum_{\text{in}} (\vec{r} \times \dot{m}\vec{V}) \quad (6\text{–}50)$$

FIGURE 6–35
A rotating lawn sprinkler is a good example of application of the angular momentum equation.
© *John A. Rizzo/Getty RF*

You may be wondering why we don't introduce a correction factor into Eq. 6–50, like we did for conservation of energy (Chap. 5) and for conservation of linear momentum (Section 6–4). The reason is that the cross product of $\vec{r}$ and $\dot{m}\vec{V}$ is dependent on problem geometry, and thus, such a correction

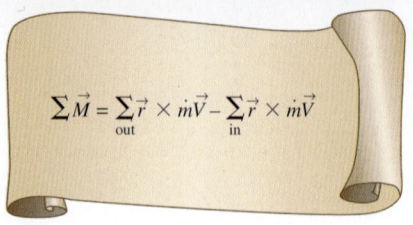

FIGURE 6–36
The net torque acting on a control volume during steady flow is equal to the difference between the outgoing and incoming angular momentum flow rates.

factor would vary from problem to problem. Therefore, whereas we can readily calculate a kinetic energy flux correction factor and a momentum flux correction factor for fully developed pipe flow that can be applied to various problems, we cannot do so for angular momentum. Fortunately, in many problems of practical engineering interest, the error associated with using average values of radius and velocity is small, and the approximation of Eq. 6–50 is reasonable.

If the flow is *steady*, Eq. 6–50 further reduces to (Fig. 6–36)

Steady flow: $$\sum \vec{M} = \sum_{\text{out}} (\vec{r} \times \dot{m}\vec{V}) - \sum_{\text{in}} (\vec{r} \times \dot{m}\vec{V}) \qquad (6\text{–}51)$$

Equation 6–51 states that *the net torque acting on the control volume during steady flow is equal to the difference between the outgoing and incoming angular momentum flow rates.* This statement can also be expressed for any specified direction. Note that velocity $\vec{V}$ in Eq. 6–51 is the velocity relative to an inertial coordinate system.

In many problems, all the significant forces and momentum flows are in the same plane, and thus all give rise to moments in the same plane and about the same axis. For such cases, Eq. 6–51 can be expressed in scalar form as

$$\sum M = \sum_{\text{out}} r\dot{m}V - \sum_{\text{in}} r\dot{m}V \qquad (6\text{–}52)$$

where r represents the average normal distance between the point about which moments are taken and the line of action of the force or velocity, provided that the sign convention for the moments is observed. That is, all moments in the counterclockwise direction are positive, and all moments in the clockwise direction are negative.

Flow with No External Moments

When there are no external moments applied, the angular momentum equation Eq. 6–50 reduces to

No external moments: $$0 = \frac{d\vec{H}_{\text{CV}}}{dt} + \sum_{\text{out}} (\vec{r} \times \dot{m}\vec{V}) - \sum_{\text{in}} (\vec{r} \times \dot{m}\vec{V}) \qquad (6\text{–}53)$$

This is an expression of the conservation of angular momentum principle, which can be stated as *in the absence of external moments, the rate of change of the angular momentum of a control volume is equal to the difference between the incoming and outgoing angular momentum fluxes.*

When the moment of inertia I of the control volume remains constant, the first term on the right side of Eq. 6–53 becomes simply moment of inertia times angular acceleration, $I\vec{\alpha}$. Therefore, the control volume in this case can be treated as a solid body, with a net torque of

$$\vec{M}_{\text{body}} = I_{\text{body}} \vec{\alpha} = \sum_{\text{in}} (\vec{r} \times \dot{m}\vec{V}) - \sum_{\text{out}} (\vec{r} \times \dot{m}\vec{V}) \qquad (6\text{–}54)$$

(due to a change of angular momentum) acting on it. This approach can be used to determine the angular acceleration of space vehicles and aircraft when a rocket is fired in a direction different than the direction of motion.

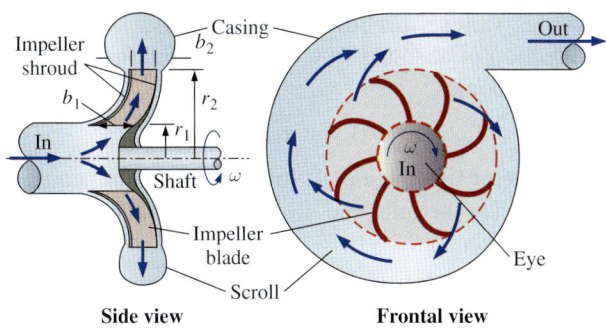

FIGURE 6–37
Side and frontal views of a typical centrifugal pump.

Radial-Flow Devices

Many rotary-flow devices such as centrifugal pumps and fans involve flow in the radial direction normal to the axis of rotation and are called *radial-flow devices* (Chap. 14). In a centrifugal pump, for example, the fluid enters the device in the axial direction through the eye of the impeller, turns outward as it flows through the passages between the blades of the impeller, collects in the scroll, and is discharged in the tangential direction, as shown in Fig. 6–37. Axial-flow devices are easily analyzed using the linear momentum equation. But radial-flow devices involve large changes in angular momentum of the fluid and are best analyzed with the help of the angular momentum equation.

To analyze a centrifugal pump, we choose the annular region that encloses the impeller section as the control volume, as shown in Fig. 6–38. Note that the average flow velocity, in general, has normal and tangential components at both the inlet and the outlet of the impeller section. Also, when the shaft rotates at angular velocity ω, the impeller blades have tangential velocity ωr_1 at the inlet and ωr_2 at the outlet. For steady, incompressible flow, the conservation of mass equation is written as

$$\dot{V}_1 = \dot{V}_2 = \dot{V} \quad \rightarrow \quad (2\pi r_1 b_1)V_{1,n} = (2\pi r_2 b_2)V_{2,n} \quad \text{(6-55)}$$

where b_1 and b_2 are the flow widths at the inlet where $r = r_1$ and at the outlet where $r = r_2$, respectively. (Note that the actual circumferential cross-sectional area is somewhat less than $2\pi rb$ since the blade thickness is not zero.) Then the average normal components $V_{1,n}$ and $V_{2,n}$ of absolute velocity can be expressed in terms of the volumetric flow rate $\dot{V}$ as

$$V_{1,n} = \frac{\dot{V}}{2\pi r_1 b_1} \quad \text{and} \quad V_{2,n} = \frac{\dot{V}}{2\pi r_2 b_2} \quad \text{(6-56)}$$

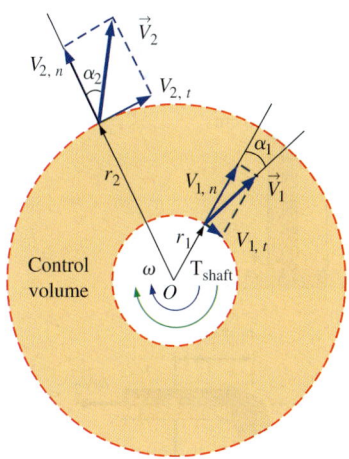

FIGURE 6–38
An annular control volume that encloses the impeller section of a centrifugal pump.

The normal velocity components $V_{1,n}$ and $V_{2,n}$ as well as pressure acting on the inner and outer circumferential areas pass through the shaft center, and thus they do not contribute to torque about the origin. Then only the tangential velocity components contribute to torque, and the application of the angular momentum equation $\sum M = \sum_{\text{out}} r\dot{m}V - \sum_{\text{in}} r\dot{m}V$ to the control volume gives

Euler's turbine equation: $\quad T_{\text{shaft}} = \dot{m}(r_2 V_{2,t} - r_1 V_{1,t}) \quad \text{(6-57)}$

270
MOMENTUM ANALYSIS OF FLOW SYSTEMS

which is known as **Euler's turbine equation.** When the angles α_1 and α_2 between the direction of absolute flow velocities and the radial direction are known, Eq. 6–57 becomes

$$T_{shaft} = \dot{m}(r_2 V_2 \sin \alpha_2 - r_1 V_1 \sin \alpha_1) \qquad (6\text{–}58)$$

In the idealized case of the tangential fluid velocity being equal to the blade angular velocity both at the inlet and the exit, we have $V_{1,t} = \omega r_1$ and $V_{2,t} = \omega r_2$, and the torque becomes

$$T_{shaft,\,ideal} = \dot{m}\omega(r_2^2 - r_1^2) \qquad (6\text{–}59)$$

where $\omega = 2\pi \dot{n}$ is the angular velocity of the blades. When the torque is known, the shaft power is determined from $\dot{W}_{shaft} = \omega T_{shaft} = 2\pi \dot{n} T_{shaft}$.

EXAMPLE 6–8   Bending Moment Acting at the Base of a Water Pipe

Underground water is pumped through a 10-cm-diameter pipe that consists of a 2-m-long vertical and 1-m-long horizontal section, as shown in Fig. 6–39. Water discharges to atmospheric air at an average velocity of 3 m/s, and the mass of the horizontal pipe section when filled with water is 12 kg per meter length. The pipe is anchored on the ground by a concrete base. Determine the bending moment acting at the base of the pipe (point A) and the required length of the horizontal section that would make the moment at point A zero.

SOLUTION Water is pumped through a piping section. The moment acting at the base and the required length of the horizontal section to make this moment zero is to be determined.
Assumptions **1** The flow is steady. **2** The water is discharged to the atmosphere, and thus the gage pressure at the outlet is zero. **3** The pipe diameter is small compared to the moment arm, and thus we use average values of radius and velocity at the outlet.
Properties We take the density of water to be 1000 kg/m³.
Analysis We take the entire L-shaped pipe as the control volume, and designate the inlet by 1 and the outlet by 2. We also take the x- and z-coordinates as shown. The control volume and the reference frame are fixed.
  The conservation of mass equation for this one-inlet, one-outlet, steady-flow system is $\dot{m}_1 = \dot{m}_2 = \dot{m}$, and $V_1 = V_2 = V$ since $A_c = $ constant. The mass flow rate and the weight of the horizontal section of the pipe are

$$\dot{m} = \rho A_c V = (1000 \text{ kg/m}^3)[\pi(0.10 \text{ m})^2/4](3 \text{ m/s}) = 23.56 \text{ kg/s}$$

$$W = mg = (12 \text{ kg/m})(1 \text{ m})(9.81 \text{ m/s}^2)\left(\frac{1 \text{ N}}{1 \text{ kg·m/s}^2}\right) = 117.7 \text{ N}$$

To determine the moment acting on the pipe at point A, we need to take the moment of all forces and momentum flows about that point. This is a steady-flow problem, and all forces and momentum flows are in the same plane. Therefore, the angular momentum equation in this case is expressed as

$$\sum M = \sum_{out} r\dot{m}V - \sum_{in} r\dot{m}V$$

where r is the average moment arm, V is the average speed, all moments in the counterclockwise direction are positive, and all moments in the clockwise direction are negative.

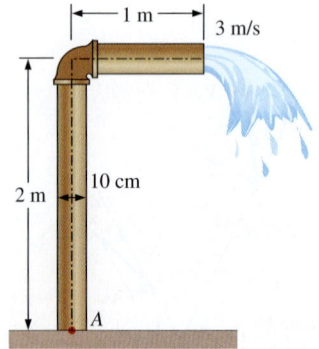

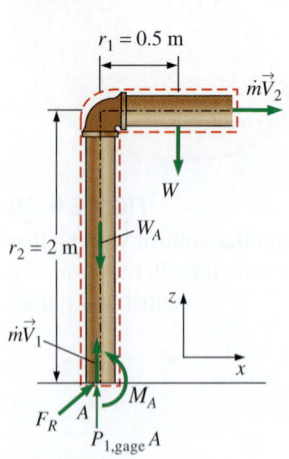

FIGURE 6–39
Schematic for Example 6–8 and the free-body diagram.

The free-body diagram of the L-shaped pipe is given in Fig. 6–39. Noting that the moments of all forces and momentum flows passing through point A are zero, the only force that yields a moment about point A is the weight W of the horizontal pipe section, and the only momentum flow that yields a moment is the outlet stream (both are negative since both moments are in the clockwise direction). Then the angular momentum equation about point A becomes

$$M_A - r_1 W = -r_2 \dot{m} V_2$$

Solving for M_A and substituting give

$$M_A = r_1 W - r_2 \dot{m} V_2$$
$$= (0.5 \text{ m})(118 \text{ N}) - (2 \text{ m})(23.56 \text{ kg/s})(3 \text{ m/s})\left(\frac{1 \text{ N}}{1 \text{ kg}\cdot\text{m/s}^2}\right)$$
$$= -82.5 \text{ N}\cdot\text{m}$$

The negative sign indicates that the assumed direction for M_A is wrong and should be reversed. Therefore, a moment of 82.5 N·m acts at the stem of the pipe in the clockwise direction. That is, the concrete base must apply a 82.5 N·m moment on the pipe stem in the clockwise direction to counteract the excess moment caused by the exit stream.

The weight of the horizontal pipe is $w = W/L = 117.7$ N per m length. Therefore, the weight for a length of Lm is Lw with a moment arm of $r_1 = L/2$. Setting $M_A = 0$ and substituting, the length L of the horizontal pipe that would cause the moment at the pipe stem to vanish is determined to be

$$0 = r_1 W - r_2 \dot{m} V_2 \rightarrow 0 = (L/2)Lw - r_2 \dot{m} V_2$$

or

$$L = \sqrt{\frac{2 r_2 \dot{m} V_2}{w}} = \sqrt{\frac{2(2 \text{ m})(23.56 \text{ kg/s})(3 \text{ m/s})}{117.7 \text{ N/m}}\left(\frac{\text{N}}{\text{kg}\cdot\text{m/s}^2}\right)} = \mathbf{1.55 \text{ m}}$$

Discussion Note that the pipe weight and the momentum of the exit stream cause opposing moments at point A. This example shows the importance of accounting for the moments of momentums of flow streams when performing a dynamic analysis and evaluating the stresses in pipe materials at critical cross sections.

FIGURE 6–40
Lawn sprinklers often have rotating heads to spread the water over a large area.

© Andy Sotiriou/Getty RF

■ **EXAMPLE 6–9** **Power Generation from a Sprinkler System**

A large lawn sprinkler (Fig. 6–40) with four identical arms is to be converted into a turbine to generate electric power by attaching a generator to its rotating head, as shown in Fig. 6–41. Water enters the sprinkler from the base along the axis of rotation at a rate of 20 L/s and leaves the nozzles in the tangential direction. The sprinkler rotates at a rate of 300 rpm in a horizontal plane. The diameter of each jet is 1 cm, and the normal distance between the axis of rotation and the center of each nozzle is 0.6 m. Estimate the electric power produced.

SOLUTION A four-armed sprinkler is used to generate electric power. For a specified flow rate and rotational speed, the power produced is to be determined.

272
MOMENTUM ANALYSIS OF FLOW SYSTEMS

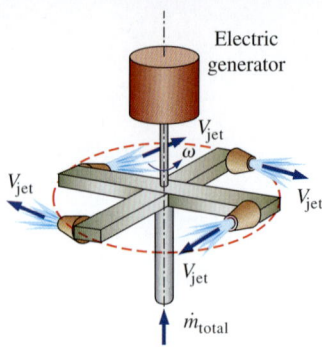

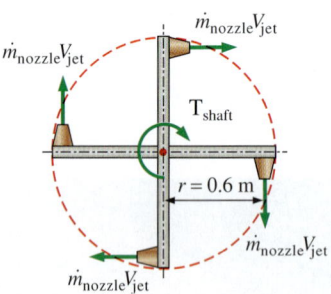

FIGURE 6–41
Schematic for Example 6–9 and the free-body diagram.

Assumptions 1 The flow is cyclically steady (i.e., steady from a frame of reference rotating with the sprinkler head). 2 The water is discharged to the atmosphere, and thus the gage pressure at the nozzle exit is zero. 3 Generator losses and air drag of rotating components are neglected. 4 The nozzle diameter is small compared to the moment arm, and thus we use average values of radius and velocity at the outlet.
Properties We take the density of water to be 1000 kg/m³ = 1 kg/L.
Analysis We take the disk that encloses the sprinkler arms as the control volume, which is a stationary control volume.

The conservation of mass equation for this steady-flow system is $\dot{m}_1 = \dot{m}_2 = \dot{m}_{total}$. Noting that the four nozzles are identical, we have $\dot{m}_{nozzle} = \dot{m}_{total}/4$ or $\dot{V}_{nozzle} = \dot{V}_{total}/4$ since the density of water is constant. The average jet exit velocity relative to the rotating nozzle is

$$V_{jet,r} = \frac{\dot{V}_{nozzle}}{A_{jet}} = \frac{5 \text{ L/s}}{[\pi(0.01 \text{ m})^2/4]}\left(\frac{1 \text{ m}^3}{1000 \text{ L}}\right) = 63.66 \text{ m/s}$$

The angular and tangential velocities of the nozzles are

$$\omega = 2\pi \dot{n} = 2\pi(300 \text{ rev/min})\left(\frac{1 \text{ min}}{60 \text{ s}}\right) = 31.42 \text{ rad/s}$$

$$V_{nozzle} = r\omega = (0.6 \text{ m})(31.42 \text{ rad/s}) = 18.85 \text{ m/s}$$

Note that water in the nozzle is also moving at an average velocity of 18.85 m/s in the opposite direction when it is discharged. The average absolute velocity of the water jet (velocity relative to a fixed location on earth) is the vector sum of its relative velocity (jet velocity relative to the nozzle) and the absolute nozzle velocity,

$$\vec{V}_{jet} = \vec{V}_{jet,r} + \vec{V}_{nozzle}$$

All of these three velocities are in the tangential direction, and taking the direction of jet flow as positive, the vector equation can be written in scalar form using magnitudes as

$$V_{jet} = V_{jet,r} - V_{nozzle} = 63.66 - 18.85 = 44.81 \text{ m/s}$$

Noting that this is a cyclically steady-flow problem, and all forces and momentum flows are in the same plane, the angular momentum equation is approximated as $\sum M = \sum_{out} r\dot{m}V - \sum_{in} r\dot{m}V$, where r is the moment arm, all moments in the counterclockwise direction are positive, and all moments in the clockwise direction are negative.

The free-body diagram of the disk that contains the sprinkler arms is given in Fig. 6–41. Note that the moments of all forces and momentum flows passing through the axis of rotation are zero. The momentum flows via the water jets leaving the nozzles yield a moment in the clockwise direction and the effect of the generator on the control volume is a moment also in the clockwise direction (thus both are negative). Then the angular momentum equation about the axis of rotation becomes

$$-T_{shaft} = -4r\dot{m}_{nozzle}V_{jet} \quad \text{or} \quad T_{shaft} = r\dot{m}_{total}V_{jet}$$

Substituting, the torque transmitted through the shaft is

$$T_{shaft} = r\dot{m}_{total}V_{jet} = (0.6 \text{ m})(20 \text{ kg/s})(44.81 \text{ m/s})\left(\frac{1 \text{ N}}{1 \text{ kg·m/s}^2}\right) = 537.7 \text{ N·m}$$

since $\dot{m}_{total} = \rho\dot{V}_{total} = (1\text{ kg/L})(20\text{ L/s}) = 20\text{ kg/s}$.

Then the power generated becomes

$$\dot{W} = \omega T_{shaft} = (31.42\text{ rad/s})(537.7\text{ N·m})\left(\frac{1\text{ kW}}{1000\text{ N·m/s}}\right) = \mathbf{16.9\text{ kW}}$$

Therefore, this sprinkler-type turbine has the potential to produce 16.9 kW of power.

Discussion To put the result obtained in perspective, we consider two limiting cases. In the first limiting case, the sprinkler is stuck, and thus, the angular velocity is zero. The torque developed is maximum in this case, since $V_{nozzle} = 0$. Thus $V_{jet} = V_{jet,\,r} = 63.66$ m/s, giving $T_{shaft,\,max} = 764$ N·m. The power generated is zero since the generator shaft does not rotate.

In the second limiting case, the sprinkler shaft is disconnected from the generator (and thus both the useful torque and power generation are zero), and the shaft accelerates until it reaches an equilibrium velocity. Setting $T_{shaft} = 0$ in the angular momentum equation gives the absolute water-jet velocity (jet velocity relative to an observer on earth) to be zero, $V_{jet} = 0$. Therefore, the relative velocity $V_{jet,\,r}$ and absolute velocity V_{nozzle} are equal but in opposite direction. So, the absolute tangential velocity of the jet (and thus torque) is zero, and the water mass drops straight down like a waterfall under gravity with zero angular momentum (around the axis of rotation). The angular speed of the sprinkler in this case is

$$\dot{n} = \frac{\omega}{2\pi} = \frac{V_{nozzle}}{2\pi r} = \frac{63.66\text{ m/s}}{2\pi(0.6\text{ m})}\left(\frac{60\text{ s}}{1\text{ min}}\right) = 1013\text{ rpm}$$

Of course, the $T_{shaft} = 0$ case is possible only for an ideal, frictionless nozzle (i.e., 100 percent nozzle efficiency, as a no-load ideal turbine). Otherwise, there would be a resisting torque due to friction of the water, shaft, and surrounding air.

The variation of power produced with angular speed is plotted in Fig. 6–42. Note that the power produced increases with increasing rpm, reaches a maximum (at about 500 rpm in this case), and then decreases. The actual power produced would be less than this due to generator inefficiency (Chap. 5) and other irreversible losses such as fluid friction within the nozzle (Chap. 8), shaft friction, and aerodynamic drag (Chap. 11).

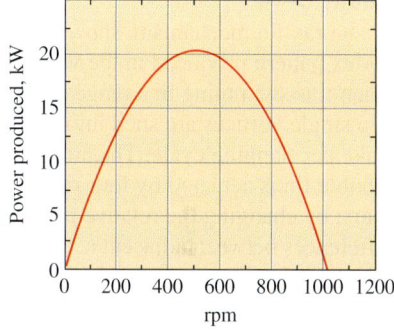

FIGURE 6–42
The variation of power produced with angular speed for the turbine of Example 6–9.

APPLICATION SPOTLIGHT ■ Manta Ray Swimming

Guest Authors: Alexander Smits, Keith Moored and Peter Dewey, Princeton University

Aquatic animals propel themselves using a wide variety of mechanisms. Most fish flap their tail to produce thrust, and in doing so they shed two single vortices per flapping cycle, creating a wake that resembles a reverse von Kármán vortex street. The non-dimensional number that describes this vortex shedding is the Strouhal number St, where $St = fA/U_\infty$, where f is the frequency of actuation, A is the peak-to-peak amplitude of the trailing edge motion at the half-span, and U_∞ is the steady swimming velocity. Remarkably, a wide variety of fish and mammals swim in the range $0.2 < St < 0.35$.

In manta rays (Fig. 6–43), propulsion is achieved by combining oscillatory and undulatory motions of flexible pectoral fins. That is, as the manta ray

FIGURE 6–43
The manta ray is the largest of the rays, reaching up to 8 m in span. They swim with a motion that is a combination of flapping and undulation of their large pectoral fins.

© Frank & Joyce Burek/Getty RF

274
MOMENTUM ANALYSIS OF FLOW SYSTEMS

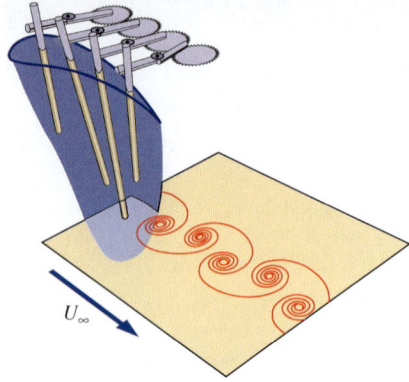

FIGURE 6–44
Manta ray fin mechanism, showing the vortex pattern produced in the wake when it is swimming in a range where two single vortices are shed into the wake per flapping cycle. The artificial flexible fin is actuated by four rigid spars; by changing the relative phase differences between adjacent actuators, undulations of varying wavelength can be produced.

flaps its fins, it is also generating a traveling wave motion along the chord, opposite to the direction of its motion. This wave motion is not readily apparent because the wavelength is 6 to 10 times greater than the chord length. A similar undulation is observed in sting rays, but there it is more obvious because the wavelength is less than the chord length. Field observations indicate that many species of manta ray are migratory, and that they are very efficient swimmers. They are difficult to study in the laboratory because they are a protected and somewhat fragile creature. However, it is possible to study many aspects of their swimming behavior by mimicking their propulsive techniques using robots or mechanical devices such as that shown in Fig. 6–44. The flow field generated by such a fin displays the vortex shedding seen in other fish studies, and when time-averaged displays a high momentum jet that contributes to the thrust (Fig 6–45). The thrust and efficiencies can also be measured directly, and it appears that the undulatory motion due to the traveling wave is most important to thrust production at high efficiency in the manta ray.

References

G. S. Triantafyllou, M. S. Triantafyllou, and M. A. Grosenbaugh. Optimal thrust development in oscillating foils with application to fish propulsion. *J. Fluid. Struct.*, 7:205–224, 1993.

Clark, R.P. and Smits, A.J., Thrust production and wake structure of a batoid-inspired oscillating fin. *Journal of Fluid Mechanics*, 562, 415–429, 2006.

Moored, K. W., Dewey, P. A., Leftwich, M. C., Bart-Smith, H. and Smits, A. J., "Bio-inspired propulsion mechanisms based on lamprey and manta ray locomotion." *The Marine Technology Society Journal*, Vol. 45(4), pp. 110–118, 2011.

Dewey, P. A., Carriou, A. and Smits, A. J. "On the relationship between efficiency and wake structure of a batoid-inspired oscillating fin." *Journal of Fluid Mechanics*, Vol. 691, pp. 245–266, 2011.

FIGURE 6–45
Measurements of the wake of the manta ray fin mechanism, with the flow going from bottom to top. On the left, we see the vortices shed in the wake, alternating between positive vorticity (red) and negative vorticity (blue). The induced velocities are shown by the black arrows, and in this case we see that thrust is being produced. On the right, we see the time-averaged velocity field. The unsteady velocity field induced by the vortices produces a high velocity jet in the time-averaged field. The momentum flux associated with this jet contributes to the total thrust on the fin.

Image courtesy of Peter Dewey, Keith Moored and Alexander Smits. Used by permission.

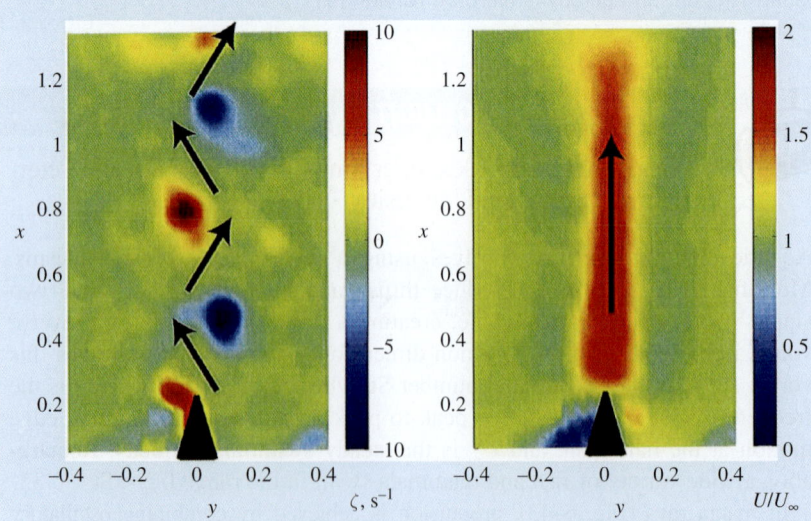

SUMMARY

This chapter deals mainly with the conservation of momentum for finite control volumes. The forces acting on the control volume consist of *body forces* that act throughout the entire body of the control volume (such as gravity, electric, and magnetic forces) and *surface forces* that act on the control surface (such as the pressure forces and reaction forces at points of contact). The sum of all forces acting on the control volume at a particular instant in time is represented by $\Sigma \vec{F}$ and is expressed as

$$\underbrace{\sum \vec{F}}_{\text{total force}} = \underbrace{\sum \vec{F}_{\text{gravity}}}_{\text{body force}} + \underbrace{\sum \vec{F}_{\text{pressure}} + \sum \vec{F}_{\text{viscous}} + \sum \vec{F}_{\text{other}}}_{\text{surface forces}}$$

Newton's second law can be stated as *the sum of all external forces acting on a system is equal to the time rate of change of linear momentum of the system*. Setting $b = \vec{V}$ and thus $B = m\vec{V}$ in the Reynolds transport theorem and utilizing Newton's second law gives the *linear momentum equation* for a control volume as

$$\sum \vec{F} = \frac{d}{dt} \int_{CV} \rho \vec{V} \, dV + \int_{CS} \rho \vec{V}(\vec{V}_r \cdot \vec{n}) \, dA$$

which reduces to the following special cases:

Steady flow: $\qquad \sum \vec{F} = \int_{CS} \rho \vec{V}(\vec{V}_r \cdot \vec{n}) \, dA$

Unsteady flow (algebraic form):

$$\sum \vec{F} = \frac{d}{dt} \int_{CV} \rho \vec{V} \, dV + \sum_{\text{out}} \beta \dot{m} \vec{V} - \sum_{\text{in}} \beta \dot{m} \vec{V}$$

Steady flow (algebraic form): $\quad \sum \vec{F} = \sum_{\text{out}} \beta \dot{m} \vec{V} - \sum_{\text{in}} \beta \dot{m} \vec{V}$

No external forces: $\quad 0 = \dfrac{d(m\vec{V})_{CV}}{dt} + \sum_{\text{out}} \beta \dot{m} \vec{V} - \sum_{\text{in}} \beta \dot{m} \vec{V}$

where β is the momentum-flux correction factor. A control volume whose mass m remains constant can be treated as a solid body (a fixed-mass system) with a *net thrusting force* (also called simply the *thrust*) of

$$\vec{F}_{\text{thrust}} = m_{CV} \vec{a} = \sum_{\text{in}} \beta \dot{m} \vec{V} - \sum_{\text{out}} \beta \dot{m} \vec{V}$$

acting on the body.

Newton's second law can also be stated as *the rate of change of angular momentum of a system is equal to the net torque acting on the system*. Setting $b = \vec{r} \times \vec{V}$ and thus $B = \vec{H}$ in the general Reynolds transport theorem gives the *angular momentum equation* as

$$\sum \vec{M} = \frac{d}{dt} \int_{CV} (\vec{r} \times \vec{V}) \rho \, dV + \int_{CS} (\vec{r} \times \vec{V}) \rho (\vec{V}_r \cdot \vec{n}) \, dA$$

which reduces to the following special cases:

Steady flow: $\qquad \sum \vec{M} = \int_{CS} (\vec{r} \times \vec{V}) \rho (\vec{V}_r \cdot \vec{n}) \, dA$

Unsteady flow (algebraic form):

$$\sum \vec{M} = \frac{d}{dt} \int_{CV} (\vec{r} \times \vec{V}) \rho \, dV + \sum_{\text{out}} \vec{r} \times \dot{m} \vec{V} - \sum_{\text{in}} \vec{r} \times \dot{m} \vec{V}$$

Steady and uniform flow:

$$\sum \vec{M} = \sum_{\text{out}} \vec{r} \times \dot{m} \vec{V} - \sum_{\text{in}} \vec{r} \times \dot{m} \vec{V}$$

Scalar form for one direction:

$$\sum M = \sum_{\text{out}} r \dot{m} V - \sum_{\text{in}} r \dot{m} V$$

No external moments:

$$0 = \frac{d\vec{H}_{CV}}{dt} + \sum_{\text{out}} \vec{r} \times \dot{m} \vec{V} - \sum_{\text{in}} \vec{r} \times \dot{m} \vec{V}$$

A control volume whose moment of inertia I remains constant can be treated as a solid body (a fixed-mass system), with a net torque of

$$\vec{M}_{CV} = I_{CV} \vec{\alpha} = \sum_{\text{in}} \vec{r} \times \dot{m} \vec{V} - \sum_{\text{out}} \vec{r} \times \dot{m} \vec{V}$$

acting on the body. This relation is used to determine the angular acceleration of a spacecraft when a rocket is fired.

The linear and angular momentum equations are of fundamental importance in the analysis of turbomachinery and are used extensively in Chap. 14.

REFERENCES AND SUGGESTED READING

1. P. K. Kundu, I. M. Cohen, and D. R. Dowling. *Fluid Mechanics,* ed. 5. San Diego, CA: Academic Press, 2011.

2. Terry Wright, *Fluid Machinery: Performance, Analysis, and Design*, Boca Raton, FL: CRC Press, 1999.

MOMENTUM ANALYSIS OF FLOW SYSTEMS

PROBLEMS*

Newton's Laws and Conservation of Momentum

6–1C Express Newton's second law of motion for rotating bodies. What can you say about the angular velocity and angular momentum of a rotating nonrigid body of constant mass if the net torque acting on it is zero?

6–2C Is momentum a vector? If so, in what direction does it point?

6–3C Express the conservation of momentum principle. What can you say about the momentum of a body if the net force acting on it is zero?

Linear Momentum Equation

6–4C Two firefighters are fighting a fire with identical water hoses and nozzles, except that one is holding the hose straight so that the water leaves the nozzle in the same direction it comes, while the other holds it backward so that the water makes a U-turn before being discharged. Which firefighter will experience a greater reaction force?

6–5C How do surface forces arise in the momentum analysis of a control volume? How can we minimize the number of surface forces exposed during analysis?

6–6C Explain the importance of the Reynolds transport theorem in fluid mechanics, and describe how the linear momentum equation is obtained from it.

6–7C What is the importance of the momentum-flux correction factor in the momentum analysis of flow systems? For which type(s) of flow is it significant and must it be considered in analysis: laminar flow, turbulent flow, or jet flow?

6–8C Write the momentum equation for steady one-dimensional flow for the case of no external forces and explain the physical significance of its terms.

6–9C In the application of the momentum equation, explain why we can usually disregard the atmospheric pressure and work with gage pressures only.

6–10C A rocket in space (no friction or resistance to motion) can expel gases relative to itself at some high velocity V. Is V the upper limit to the rocket's ultimate velocity?

6–11C Describe in terms of momentum and airflow how a helicopter is able to hover.

FIGURE P6–11C
© JupiterImages/Thinkstock/Alamy RF

6–12C Does it take more, equal, or less power for a helicopter to hover at the top of a high mountain than it does at sea level? Explain.

6–13C In a given location, would a helicopter require more energy in summer or winter to achieve a specified performance? Explain.

6–14C A horizontal water jet from a nozzle of constant exit cross section impinges normally on a stationary vertical flat plate. A certain force F is required to hold the plate against the water stream. If the water velocity is doubled, will the necessary holding force also be doubled? Explain.

6–15C Describe body forces and surface forces, and explain how the net force acting on a control volume is determined. Is fluid weight a body force or surface force? How about pressure?

6–16C A constant-velocity horizontal water jet from a stationary nozzle impinges normally on a vertical flat plate that rides on a nearly frictionless track. As the water jet hits the plate, it begins to move due to the water force. Will the acceleration of the plate remain constant or change? Explain.

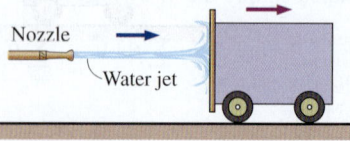

FIGURE P6–16C

6–17C A horizontal water jet of constant velocity V from a stationary nozzle impinges normally on a vertical flat plate that rides on a nearly frictionless track. As the water jet hits

---
* Problems designated by a "C" are concept questions, and students are encouraged to answer them all. Problems with the icon are solved using EES, and complete solutions together with parametric studies are included on the text website. Problems with the [EES] icon are comprehensive in nature and are intended to be solved with an equation solver such as EES.

the plate, it begins to move due to the water force. What is the highest velocity the plate can attain? Explain.

6–18 Water enters a 10-cm-diameter pipe steadily with a uniform velocity of 3 m/s and exits with the turbulent flow velocity distribution given by $u = u_{max}(1 - r/R)^{1/7}$. If the pressure drop along the pipe is 10 kPa, determine the drag force exerted on the pipe by water flow.

6–19 A 2.5-cm-diameter horizontal water jet with a speed of $V_j = 40$ m/s relative to the ground is deflected by a 60° stationary cone whose base diameter is 25 cm. Water velocity along the cone varies linearly from zero at the cone surface to the incoming jet speed of 40 m/s at the free surface. Disregarding the effect of gravity and the shear forces, determine the horizontal force F needed to hold the cone stationary.

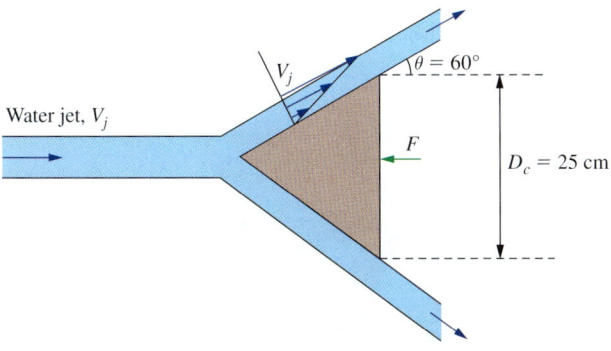

FIGURE P6–19

6–20 A horizontal water jet of constant velocity V impinges normally on a vertical flat plate and splashes off the sides in the vertical plane. The plate is moving toward the oncoming water jet with velocity $\frac{1}{2}V$. If a force F is required to maintain the plate stationary, how much force is required to move the plate toward the water jet?

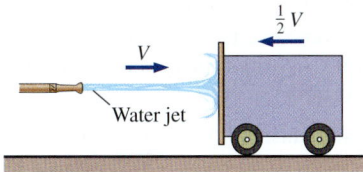

FIGURE P6–20

6–21 A 90° elbow in a horizontal pipe is used to direct water flow upward at a rate of 40 kg/s. The diameter of the entire elbow is 10 cm. The elbow discharges water into the atmosphere, and thus the pressure at the exit is the local atmospheric pressure. The elevation difference between the centers of the exit and the inlet of the elbow is 50 cm. The weight of the elbow and the water in it is considered to be negligible. Determine (a) the gage pressure at the center of the inlet of the elbow and (b) the anchoring force needed to hold the elbow in place. Take the momentum-flux correction factor to be 1.03 at both the inlet and the outlet.

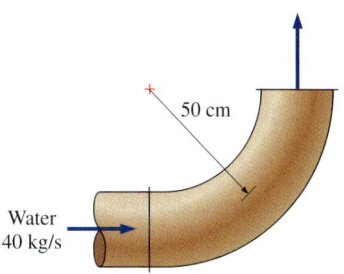

FIGURE P6–21

6–22 Repeat Prob. 6–21 for the case of another (identical) elbow attached to the existing elbow so that the fluid makes a U-turn. *Answers:* (a) 9.81 kPa, (b) −497 N

6–23 A reducing elbow in a horizontal pipe is used to deflect water flow by an angle $\theta = 45°$ from the flow direction while accelerating it. The elbow discharges water into the atmosphere. The cross-sectional area of the elbow is 150 cm² at the inlet and 25 cm² at the exit. The elevation difference between the centers of the exit and the inlet is 40 cm. The mass of the elbow and the water in it is 50 kg. Determine the anchoring force needed to hold the elbow in place. Take the momentum-flux correction factor to be 1.03 at both the inlet and outlet.

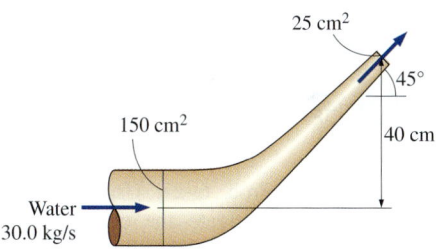

FIGURE P6–23

6–24 Repeat Prob. 6–23 for the case of $\theta = 110°$.

6–25 Water accelerated by a nozzle to 35 m/s strikes the vertical back surface of a cart moving horizontally at a constant velocity of 10 m/s in the flow direction. The mass flow rate of water through the stationary nozzle is 30 kg/s. After the strike, the water stream splatters off in all directions in the plane of the back surface. (a) Determine the force that needs to be applied by the brakes of the cart to prevent it from accelerating. (b) If this force were used to generate power instead of wasting it on the brakes, determine the maximum amount of power that could ideally be generated. *Answers:* (a) −536 N, (b) 5.36 kW

278
MOMENTUM ANALYSIS OF FLOW SYSTEMS

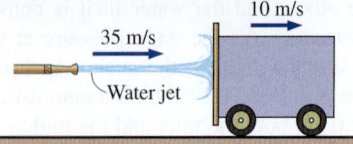

FIGURE P6–25

6–26 Reconsider Prob. 6–25. If the mass of the cart is 400 kg and the brakes fail, determine the acceleration of the cart when the water first strikes it. Assume the mass of water that wets the back surface is negligible.

6–27 A 2.8 m³/s water jet is moving in the positive x-direction at 5.5 m/s. The stream hits a stationary splitter, such that half of the flow is diverted upward at 45° and the other half is directed downward, and both streams have a final average speed of 5.5 m/s. Disregarding gravitational effects, determine the x- and z-components of the force required to hold the splitter in place against the water force.

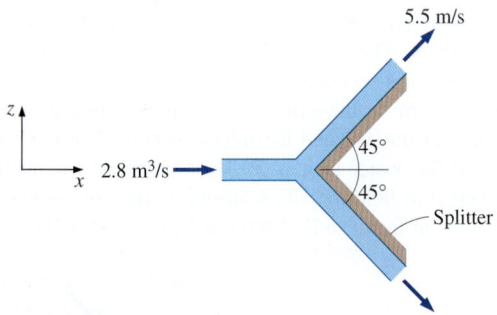

FIGURE P6–27

6–28 Reconsider Prob. 6–27. Using EES (or other) software, investigate the effect of the splitter angle on the force exerted on the splitter in the incoming flow direction. Let the half splitter angle vary from 0° to 180° in increments of 10°. Tabulate and plot your results, and draw some conclusions.

6–29 A horizontal 5-cm-diameter water jet with a velocity of 18 m/s impinges normally upon a vertical plate of mass 1000 kg. The plate rides on a nearly frictionless track and is initially stationary. When the jet strikes the plate, the plate begins to move in the direction of the jet. The water always splatters in the plane of the retreating plate. Determine (a) the acceleration of the plate when the jet first strikes it (time = 0), (b) the time it takes for the plate to reach a velocity of 9 m/s, and (c) the plate velocity 20 s after the jet first strikes the plate. For simplicity, assume the velocity of the jet is increased as the cart moves such that the impulse force exerted by the water jet on the plate remains constant.

6–30 A fan with 61-cm-diameter blades moves 0.95 m³/s of air at 20°C at sea level. Determine (a) the force required to hold the fan and (b) the minimum power input required for the fan. Choose a control volume sufficiently large to contain the fan, with the inlet sufficiently far upstream so that the gage pressure at the inlet is nearly zero. Assume air approaches the fan through a large area with negligible velocity and air exits the fan with a uniform velocity at atmospheric pressure through an imaginary cylinder whose diameter is the fan blade diameter. *Answers:* (a) 3.72 N, (b) 6.05 W

6–31 Firefighters are holding a nozzle at the end of a hose while trying to extinguish a fire. If the nozzle exit diameter is 8 cm and the water flow rate is 12 m³/min, determine (a) the average water exit velocity and (b) the horizontal resistance force required of the firefighters to hold the nozzle. *Answers:* (a) 39.8 m/s, (b) 7958 N

FIGURE P6–31

6–32 A 5-cm-diameter horizontal jet of water with a velocity of 40 m/s relative to the ground strikes a flat plate that is moving in the same direction as the jet at a velocity of 10 m/s. The water splatters in all directions in the plane of the plate. How much force does the water stream exert on the plate?

6–33 Reconsider Prob. 6–32. Using EES (or other) software, investigate the effect of the plate velocity on the force exerted on the plate. Let the plate velocity vary from 0 to 30 m/s, in increments of 3 m/s. Tabulate and plot your results.

6–34 A 7.5-cm-diameter horizontal water jet having a velocity of 28 m/s strikes a curved plate, which deflects the water 180° at the same speed. Ignoring the frictional effects, determine the force required to hold the plate against the water stream.

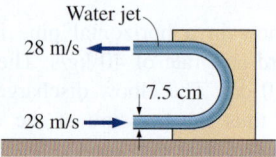

FIGURE P6–34

6–35 An unloaded helicopter of mass 12,000 kg hovers at sea level while it is being loaded. In the unloaded hover mode, the blades rotate at 550 rpm. The horizontal blades above the helicopter cause a 18-m-diameter air mass to move downward at an average velocity proportional to the overhead blade rotational velocity (rpm). A load of 14,000 kg is loaded onto the helicopter, and the helicopter slowly rises. Determine (a) the volumetric airflow rate downdraft that the helicopter generates during unloaded hover and the required power input and (b) the rpm of the helicopter blades to hover with the 14,000-kg load and the required power input. Take the density of atmospheric air to be 1.18 kg/m^3. Assume air approaches the blades from the top through a large area with negligible velocity and air is forced by the blades to move down with a uniform velocity through an imaginary cylinder whose base is the blade span area.

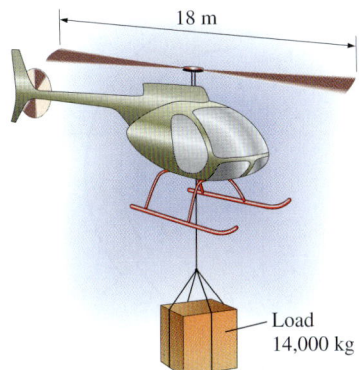

FIGURE P6–35

6–36 Reconsider the helicopter in Prob. 6–35, except that it is hovering on top of a 2800-m-high mountain where the air density is 0.928 kg/m^3. Noting that the unloaded helicopter blades must rotate at 550 rpm to hover at sea level, determine the blade rotational velocity to hover at the higher altitude. Also determine the percent increase in the required power input to hover at 3000-m altitude relative to that at sea level. *Answers:* 620 rpm, 12.8 percent

6–37 Water is flowing through a 10-cm-diameter water pipe at a rate of 0.1 m^3/s. Now a diffuser with an outlet diameter of 20 cm is bolted to the pipe in order to slow down water, as shown in Fig. P6–37. Disregarding frictional effects, determine the force exerted on the bolts due to the water flow.

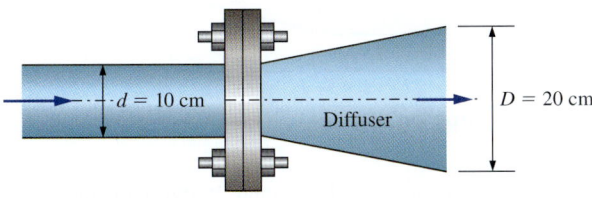

FIGURE P6–37

6–38 The weight of a water tank open to the atmosphere is balanced by a counterweight, as shown in Fig. P6–38. There is a 4-cm hole at the bottom of the tank with a discharge coefficient of 0.90, and water level in the tank is maintained constant at 50 cm by water entering the tank horizontally. Determine how much mass must be added to or removed from the counterweight to maintain balance when the hole at the bottom is opened.

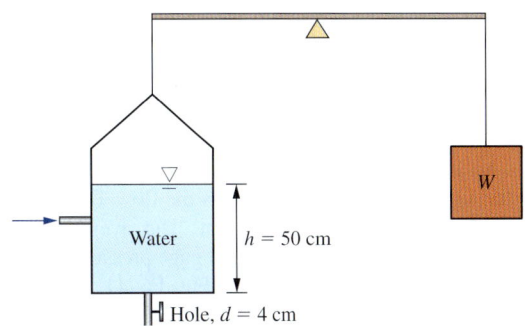

FIGURE P6–38

6–39 Commercially available large wind turbines have blade span diameters larger than 100 m and generate over 3 MW of electric power at peak design conditions. Consider a wind turbine with a 60-m blade span subjected to 30-km/h steady winds. If the combined turbine–generator efficiency of the wind turbine is 32 percent, determine (a) the power generated by the turbine and (b) the horizontal force exerted by the wind on the supporting mast of the turbine. Take the density of air to be 1.25 kg/m^3, and disregard frictional effects on mast.

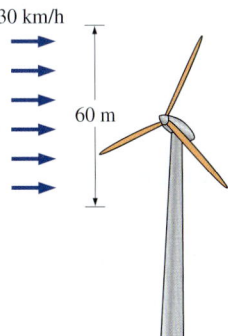

FIGURE P6–39

6–40 Water enters a centrifugal pump axially at atmospheric pressure at a rate of 0.09 m^3/s and at a velocity of 5 m/s, and leaves in the normal direction along the pump casing, as shown in Fig. P6–40. Determine the force acting on the shaft (which is also the force acting on the bearing of the shaft) in the axial direction.

MOMENTUM ANALYSIS OF FLOW SYSTEMS

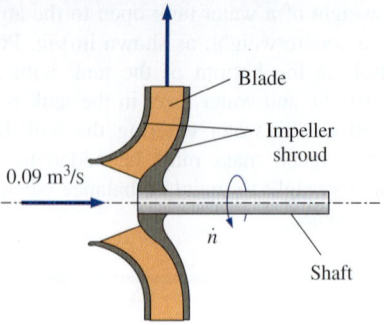

FIGURE P6–40

6–41 An incompressible fluid of density ρ and viscosity μ flows through a curved duct that turns the flow 180°. The duct cross-sectional area remains constant. The average velocity, momentum flux correction factor, and gage pressure are known at the inlet (1) and outlet (2), as in Fig. P6–41. (*a*) Write an expression for the horizontal force F_x of the fluid on the walls of the duct in terms of the given variables. (*b*) Verify your expression by plugging in the following values: $\rho = 998.2$ kg/m³, $\mu = 1.003 \times 10^{-3}$ kg/m·s, $A_1 = A_2 = 0.025$ m², $\beta_1 = 1.01$, $\beta_2 = 1.03$, $V_1 = 10$ m/s, $P_{1,\text{gage}} = 78.47$ kPa, and $P_{2,\text{gage}} = 65.23$ kPa. *Answer:* (*b*) $F_x = 8680$ N to the right

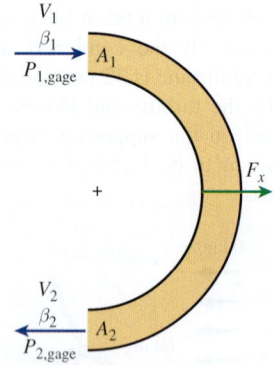

FIGURE P6–41

6–42 Consider the curved duct of Prob. 6–41, except allow the cross-sectional area to vary along the duct ($A_1 \neq A_2$). (*a*) Write an expression for the horizontal force F_x of the fluid on the walls of the duct in terms of the given variables. (*b*) Verify your expression by plugging in the following values: $\rho = 998.2$ kg/m³, $A_1 = 0.025$ m², $A_2 = 0.015$ m², $\beta_1 = 1.02$, $\beta_2 = 1.04$, $V_1 = 20$ m/s, $P_{1,\text{gage}} = 88.34$ kPa, and $P_{2,\text{gage}} = 67.48$ kPa. *Answer:* (*b*) $F_x = 30{,}700$ N to the right

6–43 As a follow-up to Prob. 6–41, it turns out that for a large enough area ratio A_2/A_1, the inlet pressure is actually *smaller* than the outlet pressure! Explain how this can be true in light of the fact that there is friction and other irreversibilities due to turbulence, and pressure must be lost along the axis of the duct to overcome these irreversibilities.

6–44 An incompressible fluid of density ρ and viscosity μ flows through a curved duct that turns the flow through angle θ. The cross-sectional area also changes. The average velocity, momentum flux correction factor, gage pressure, and area are known at the inlet (1) and outlet (2), as in Fig. P6–44. (*a*) Write an expression for the horizontal force F_x of the fluid on the walls of the duct in terms of the given variables. (*b*) Verify your expression by plugging in the following values: $\theta = 135°$, $\rho = 998.2$ kg/m³, $\mu = 1.003 \times 10^{-3}$ kg/m·s, $A_1 = 0.025$ m², $A_2 = 0.050$ m², $\beta_1 = 1.01$, $\beta_2 = 1.03$, $V_1 = 6$ m/s, $P_{1,\text{gage}} = 78.47$ kPa, and $P_{2,\text{gage}} = 65.23$ kPa. (*Hint:* You will first need to solve for V_2.) (*c*) At what turning angle is the force maximized? *Answers:* (*b*) $F_x = 5500$ N to the right, (*c*) 180°

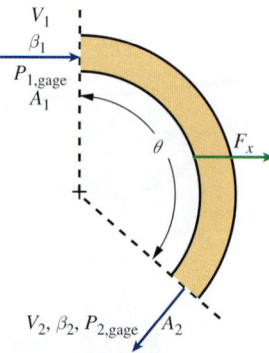

FIGURE P6–44

6–45 Water of density $\rho = 998.2$ kg/m³ flows through a fireman's nozzle—a converging section of pipe that accelerates the flow. The inlet diameter is $d_1 = 0.100$ m, and the outlet diameter is $d_2 = 0.050$ m. The average velocity, momentum flux correction factor, and gage pressure are known at the inlet (1) and outlet (2), as in Fig. P6–45. (*a*) Write an expression for the horizontal force F_x of the fluid on the walls of the nozzle in terms of the given variables. (*b*) Verify your expression by plugging in the following values: $\beta_1 = 1.03$, $\beta_2 = 1.02$, $V_1 = 4$ m/s, $P_{1,\text{gage}} = 123{,}000$ Pa, and $P_{2,\text{gage}} = 0$ Pa. *Answer:* (*b*) $F_x = 583$ N to the right

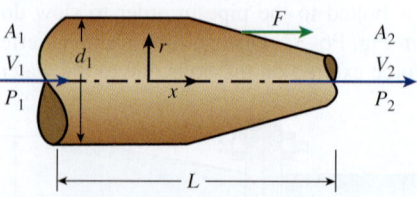

FIGURE P6–45

6–46 Water flowing in a horizontal 25-cm-diameter pipe at 8 m/s and 300 kPa gage enters a 90° bend reducing section, which connects to a 15-cm-diameter vertical pipe. The inlet of the bend is 50 cm above the exit. Neglecting any frictional and gravitational effects, determine the net resultant force exerted on the reducer by the water. Take the momentum-flux correction factor to be 1.04.

6–47 A sluice gate, which controls flow rate in a channel by simply raising or lowering a vertical plate, is commonly used in irrigation systems. A force is exerted on the gate due to the difference between the water heights y_1 and y_2 and the flow velocities V_1 and V_2 upstream and downstream from the gate, respectively. Take the width of the sluice gate (into the page) to be w. Wall shear stresses along the channel walls may be ignored, and for simplicity, we assume steady, uniform flow at locations 1 and 2. Develop a relationship for the force F_R acting on the sluice gate as a function of depths y_1 and y_2, mass flow rate $\dot{m}$, gravitational constant g, gate width w, and water density ρ.

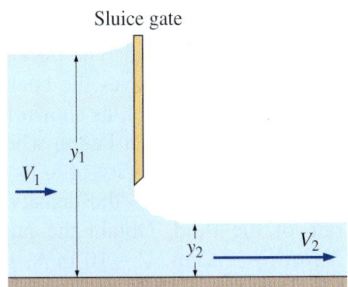

FIGURE P6–47

Angular Momentum Equation

6–48C How is the angular momentum equation obtained from Reynolds transport equations?

6–49C Express the angular momentum equation in scalar form about a specified axis of rotation for a fixed control volume for steady and uniform flow.

6–50C Express the unsteady angular momentum equation in vector form for a control volume that has a constant moment of inertia I, no external moments applied, one outgoing uniform flow stream of velocity $\vec{V}$, and mass flow rate $\dot{m}$.

6–51C Consider two rigid bodies having the same mass and angular speed. Do you think these two bodies must have the same angular momentum? Explain.

6–52 Water is flowing through a 15-cm-diameter pipe that consists of a 3-m-long vertical and 2-m-long horizontal section with a 90° elbow at the exit to force the water to be discharged downward, as shown in Fig. P6–52, in the vertical direction. Water discharges to atmospheric air at a velocity of 7 m/s, and the mass of the pipe section when filled with water is 15 kg per meter length. Determine the moment acting at the intersection of the vertical and horizontal sections

of the pipe (point A). What would your answer be if the flow were discharged upward instead of downward?

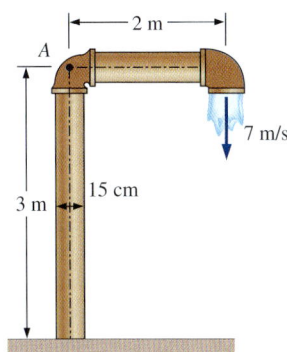

FIGURE P6–52

6–53 A large lawn sprinkler with two identical arms is used to generate electric power by attaching a generator to its rotating head. Water enters the sprinkler from the base along the axis of rotation at a rate of 20 L/s and leaves the nozzles in the tangential direction. The sprinkler rotates at a rate of 180 rpm in a horizontal plane. The diameter of each jet is 1.3 cm, and the normal distance between the axis of rotation and the center of each nozzle is 0.6 m. Determine the maximum possible electrical power produced.

6–54 Reconsider the lawn sprinkler in Prob. 6–53. If the rotating head is somehow stuck, determine the moment acting on the head.

6–55 The impeller of a centrifugal pump has inner and outer diameters of 13 and 30 cm, respectively, and a flow rate of 0.15 m³/s at a rotational speed of 1200 rpm. The blade width of the impeller is 8 cm at the inlet and 3.5 cm at the outlet. If water enters the impeller in the radial direction and exits at an angle of 60° from the radial direction, determine the minimum power requirement for the pump.

6–56 The impeller of a centrifugal blower has a radius of 18 cm and a blade width of 6.1 cm at the inlet, and a radius of

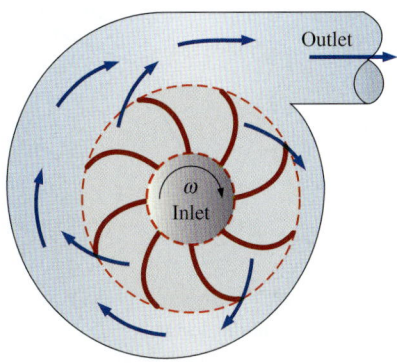

FIGURE P6–56

30 cm and a blade width of 3.4 cm at the outlet. The blower delivers atmospheric air at 20°C and 95 kPa. Disregarding any losses and assuming the tangential components of air velocity at the inlet and the outlet to be equal to the impeller velocity at respective locations, determine the volumetric flow rate of air when the rotational speed of the shaft is 900 rpm and the power consumption of the blower is 120 W. Also determine the normal components of velocity at the inlet and outlet of the impeller.

6–57 Water enters vertically and steadily at a rate of 35 L/s into the sprinkler shown in Fig. P6–57 with unequal arms and unequal discharge areas. The smaller jet has a discharge area of 3 cm² and a normal distance of 50 cm from the axis of rotation. The larger jet has a discharge area of 5 cm² and a normal distance of 35 cm from the axis of rotation. Disregarding any frictional effects, determine (*a*) the rotational speed of the sprinkler in rpm and (*b*) the torque required to prevent the sprinkler from rotating.

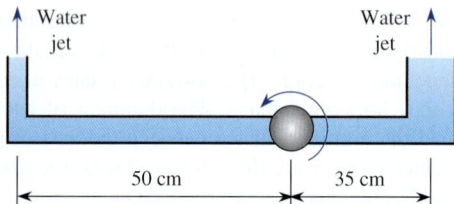

FIGURE P6–57

6–58 Repeat Prob. 6–57 for a water flow rate of 50 L/s.

6–59 Consider a centrifugal blower that has a radius of 20 cm and a blade width of 8.2 cm at the impeller inlet, and a radius of 45 cm and a blade width of 5.6 cm at the outlet. The blower delivers air at a rate of 0.70 m³/s at a rotational speed of 700 rpm. Assuming the air to enter the impeller in the radial direction and to exit at an angle of 50° from the radial direction, determine the minimum power consumption of the blower. Take the density of air to be 1.25 kg/m³.

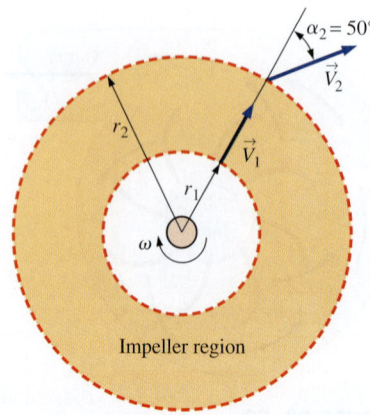

FIGURE P6–59

6–60 Reconsider Prob. 6–59. For the specified flow rate, investigate the effect of discharge angle α_2 on the minimum power input requirements. Assume the air to enter the impeller in the radial direction ($\alpha_1 = 0°$), and vary α_2 from 0° to 85° in increments of 5°. Plot the variation of power input versus α_2, and discuss your results.

6–61 A lawn sprinkler with three identical arms is used to water a garden by rotating in a horizontal plane by the impulse caused by water flow. Water enters the sprinkler along the axis of rotation at a rate of 60 L/s and leaves the 1.5-cm-diameter nozzles in the tangential direction. The bearing applies a retarding torque of $T_0 = 50$ N·m due to friction at the anticipated operating speeds. For a normal distance of 40 cm between the axis of rotation and the center of the nozzles, determine the angular velocity of the sprinkler shaft.

6–62 Pelton wheel turbines are commonly used in hydroelectric power plants to generate electric power. In these turbines, a high-speed jet at a velocity of V_j impinges on buckets, forcing the wheel to rotate. The buckets reverse the direction of the jet, and the jet leaves the bucket making an angle β with the direction of the jet, as shown in Fig. P6–62. Show that the power produced by a Pelton wheel of radius r rotating steadily at an angular velocity of ω is $\dot{W}_{shaft} = \rho \omega r \dot{V} (V_j - \omega r)(1 - \cos\beta)$, where ρ is the density and $\dot{V}$ is the volume flow rate of the fluid. Obtain the numerical value for $\rho = 1000$ kg/m³, $r = 2$ m, $\dot{V} = 10$ m³/s, $\dot{n} = 150$ rpm, $\beta = 160°$, and $V_j = 50$ m/s.

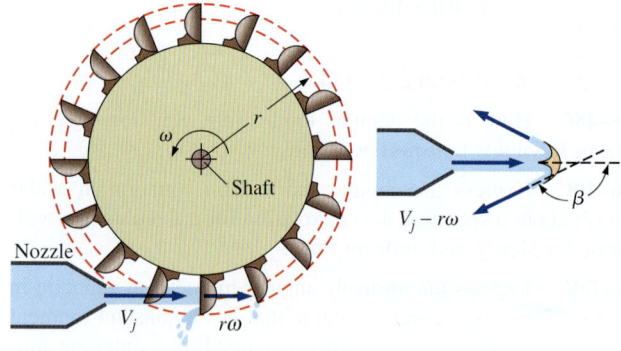

FIGURE P6–62

6–63 Reconsider Prob. 6–62. The maximum efficiency of the turbine occurs when $\beta = 180°$, but this is not practical. Investigate the effect of β on the power generation by allowing it to vary from 0° to 180°. Do you think we are wasting a large fraction of power by using buckets with a β of 160°?

Review Problems

6–64 Water flowing steadily at a rate of 0.16 m³/s is deflected downward by an angled elbow as shown in Fig. P6–64. For $D = 30$ cm, $d = 10$ cm, and $h = 50$ cm, determine the force acting on the flanges of the elbow and the angle its line of action makes with the horizontal. Take the internal volume of the elbow to be 0.03 m³ and disregard the weight of the elbow material and the frictional effects.

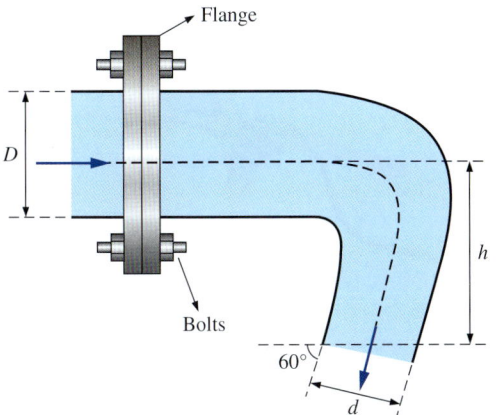

FIGURE P6–64

6–65 Repeat Prob. 6–64 by taking into consideration the weight of the elbow whose mass is 5 kg.

6–66 A 12-cm diameter horizontal water jet with a speed of $V_j = 25$ m/s relative to the ground is deflected by a 40° cone moving to the left at $V_c = 10$ m/s. Determine the external force, F, needed to maintain the motion of the cone. Disregard the gravity and surface shear effects and assume the cross-sectional area of water jet normal to the direction of motion remains constant throughout the flow. *Answer:* 3240 N to left

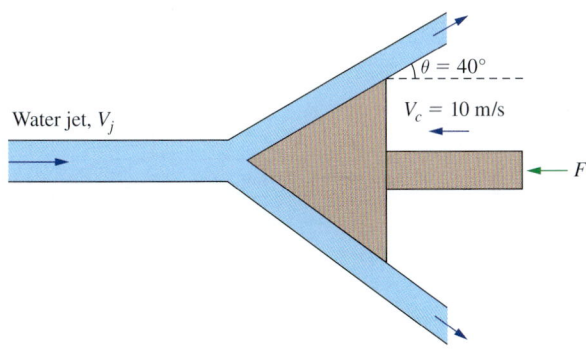

FIGURE P6–66

6–67 Water enters vertically and steadily at a rate of 10 L/s into the sprinkler shown in Fig. P6–67. Both water jets have a diameter of 1.2 cm. Disregarding any frictional effects, determine (*a*) the rotational speed of the sprinkler in rpm and (*b*) the torque required to prevent the sprinkler from rotating.

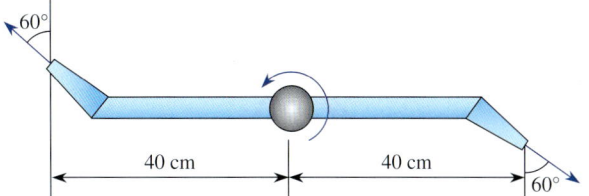

FIGURE P6–67

6–68 Repeat Prob. 6–67 for the case of unequal arms—the left one being 60 cm and the right one 20 cm from the axis of rotation.

6–69 A 6-cm-diameter horizontal water jet having a velocity of 25 m/s strikes a vertical stationary flat plate. The water splatters in all directions in the plane of the plate. How much force is required to hold the plate against the water stream? *Answer:* 1770 N

6–70 Consider steady developing laminar flow of water in a constant-diameter horizontal discharge pipe attached to a tank. The fluid enters the pipe with nearly uniform velocity V and pressure P_1. The velocity profile becomes parabolic after a certain distance with a momentum correction factor of 2 while the pressure drops to P_2. Obtain a relation for the horizontal force acting on the bolts that hold the pipe attached to the tank.

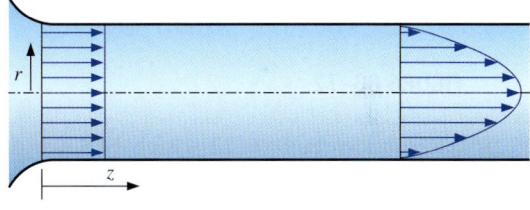

FIGURE P6–70

6–71 A tripod holding a nozzle, which directs a 5-cm-diameter stream of water from a hose, is shown in Fig. P6–71. The nozzle mass is 10 kg when filled with water. The tripod is rated to provide 1800 N of holding force. A firefighter was standing 60 cm behind the nozzle and was hit by the nozzle when the tripod suddenly failed and released the nozzle. You have been hired as an accident reconstructionist and, after testing the tripod, have determined that as water flow rate increased, it did collapse at 1800 N. In your final report you must state the water velocity and the flow rate consistent with the failure and the nozzle velocity when it hit the firefighter. For simplicity, ignore pressure and momentum effects in the upstream portion of the hose. *Answers:* 30.3 m/s, 0.0595 m³/s, 14.7 m/s

284
MOMENTUM ANALYSIS OF FLOW SYSTEMS

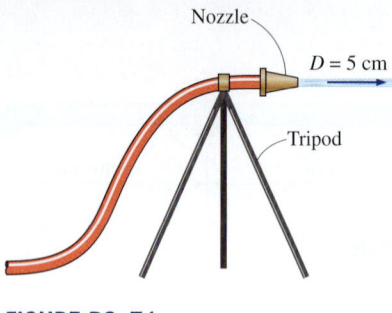

FIGURE P6–71

6–72 Consider an airplane with a jet engine attached to the tail section that expels combustion gases at a rate of 18 kg/s with a velocity of $V = 300$ m/s relative to the plane. During landing, a thrust reverser (which serves as a brake for the aircraft and facilitates landing on a short runway) is lowered in the path of the exhaust jet, which deflects the exhaust from rearward to 150°. Determine (*a*) the thrust (forward force) that the engine produces prior to the insertion of the thrust reverser and (*b*) the braking force produced after the thrust reverser is deployed.

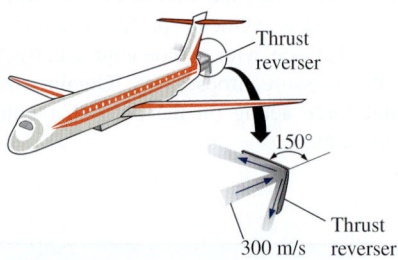

FIGURE P6–72

6–73 Reconsider Prob. 6–72. Using EES (or other) software, investigate the effect of thrust reverser angle on the braking force exerted on the airplane. Let the reverser angle vary from 0° (no reversing) to 180° (full reversing) in increments of 10°. Tabulate and plot your results and draw conclusions.

6–74 A spacecraft cruising in space at a constant velocity of 600 m/s has a mass of 11,000 kg. To slow down the spacecraft, a solid fuel rocket is fired, and the combustion gases leave the rocket at a constant rate of 70 kg/s at a velocity of 1500 m/s in the same direction as the spacecraft for a period of 5 s. Assuming the mass of the spacecraft remains constant, determine (*a*) the deceleration of the spacecraft during this 5-s period, (*b*) the change of velocity of the spacecraft during this time period, and (*c*) the thrust exerted on the spacecraft.

6–75 A 60-kg ice skater is standing on ice with ice skates (negligible friction). She is holding a flexible hose (essentially weightless) that directs a 2-cm-diameter stream of water horizontally parallel to her skates. The water velocity at the hose outlet is 10 m/s relative to the skater. If she is initially standing still, determine (*a*) the velocity of the skater and the distance she travels in 5 s and (*b*) how long it will take to move 5 m and the velocity at that moment. *Answers:* (*a*) 2.62 m/s, 6.54 m, (*b*) 4.4 s, 2.3 m/s

FIGURE P6–75

6–76 A 5-cm-diameter horizontal jet of water, with velocity 30 m/s, strikes the tip of a horizontal cone, which deflects the water by 45° from its original direction. How much force is required to hold the cone against the water stream?

6–77 Water is flowing into and discharging from a pipe U-section as shown in Fig. P6–77. At flange (1), the total absolute pressure is 200 kPa, and 55 kg/s flows into the pipe. At flange (2), the total pressure is 150 kPa. At location (3), 15 kg/s of water discharges to the atmosphere, which is at 100 kPa. Determine the total *x*- and *z*-forces at the two flanges connecting the pipe. Discuss the significance of gravity force for this problem. Take the momentum-flux correction factor to be 1.03 throughout the pipes.

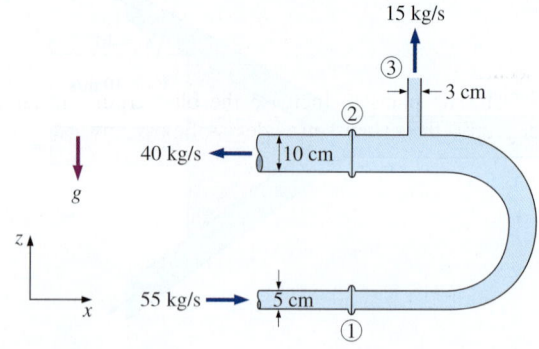

FIGURE P6–77

6–78 Indiana Jones needs to ascend a 10-m-high building. There is a large hose filled with pressurized water hanging down from the building top. He builds a square platform and mounts four 4-cm-diameter nozzles pointing down at each corner. By connecting hose branches, a water jet with

15-m/s velocity can be produced from each nozzle. Jones, the platform, and the nozzles have a combined mass of 150 kg. Determine (*a*) the minimum water jet velocity needed to raise the system, (*b*) how long it takes for the system to rise 10 m when the water jet velocity is 18 m/s and the velocity of the platform at that moment, and (*c*) how much higher will the momentum raise Jones if he shuts off the water at the moment the platform reaches 10 m above the ground. How much time does he have to jump from the platform to the roof? *Answers:* (*a*) 17.1 m/s, (*b*) 4.37 s, 4.57 m/s, (*c*) 1.07 m, 0.933 s

FIGURE P6–78

6–79 An engineering student considers using a fan as a levitation demonstration. She plans to face the box-enclosed fan so the air blast is directed face down through a 0.9-m-diameter blade span area. The system weighs 22 N, and the student will secure the system from rotating. By increasing the power to the fan, she plans to increase the blade rpm and air exit velocity until the exhaust provides sufficient upward force to cause the box fan to hover in the air. Determine (*a*) the air exit velocity to produce 22 N, (*b*) the volumetric flow rate needed, and (*c*) the minimum mechanical power that must be supplied to the airstream. Take the air density to be 1.25 kg/m³

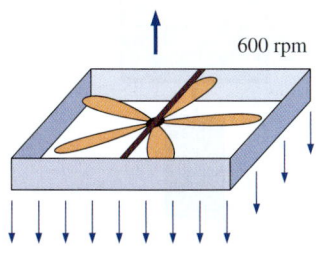

FIGURE P6–79

6–80 Nearly frictionless vertical guide rails maintain a plate of mass m_p in a horizontal position, such that it can slide freely in the vertical direction. A nozzle directs a water stream of area A against the plate underside. The water jet splatters in the plate plane, applying an upward force against the plate. The water flow rate $\dot{m}$ (kg/s) can be controlled. Assume that distances are short, so the velocity of the rising jet can be considered constant with height. (*a*) Determine the minimum mass flow rate $\dot{m}_{min}$ necessary to just levitate the plate and obtain a relation for the steady-state velocity of the upward moving plate for $\dot{m} > \dot{m}_{min}$. (*b*) At time $t = 0$, the plate is at rest, and the water jet with $\dot{m} > \dot{m}_{min}$ is suddenly turned on. Apply a force balance to the plate and obtain the integral that relates velocity to time (do not solve).

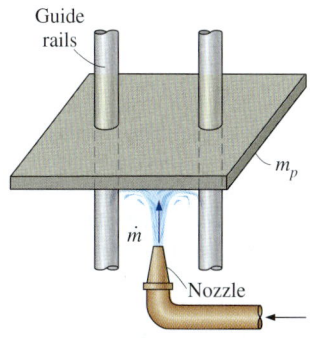

FIGURE P6–80

6–81 A walnut with a mass of 50 g requires a force of 200 N applied continuously for 0.002 s to be cracked. If walnuts are to be cracked by dropping them from a high place onto a hard surface, determine the minimum height required. Disregard air friction.

6–82 A 7-cm diameter vertical water jet is injected upwards by a nozzle at a speed of 15 m/s. Determine the maximum weight of a flat plate that can be supported by this water jet at a height of 2 m from the nozzle.

6–83 Repeat Prob. 6–82 for a height of 8 m from the nozzle.

6–84 Show that the force exerted by a liquid jet on a stationary nozzle as it leaves with a velocity V is proportional to V^2 or, alternatively, to $\dot{m}^2$. Assume the jet stream is perpendicular to the incoming liquid flow line.

6–85 A soldier jumps from a plane and opens his parachute when his velocity reaches the terminal velocity V_T. The parachute slows him down to his landing velocity of V_F. After the parachute is deployed, the air resistance is proportional to the velocity squared (i.e., $F = kV^2$). The soldier, his parachute, and his gear have a total mass of m. Show that $k = mg/V_F^2$ and develop a relation for the soldier's velocity after he opens the parachute at time $t = 0$.

Answer: $V = V_F \dfrac{V_T + V_F + (V_T - V_F)e^{-2gt/V_F}}{V_T + V_F - (V_T - V_F)e^{-2gt/V_F}}$

MOMENTUM ANALYSIS OF FLOW SYSTEMS

FIGURE P6–85
© Corbis RF

6–86 A horizontal water jet with a flow rate of $\dot{V}$ and cross-sectional area of A drives a covered cart of mass m_c along a level and nearly frictionless path. The jet enters a hole at the rear of the cart and all water that enters the cart is retained, increasing the system mass. The relative velocity between the jet of constant velocity V_J and the cart of variable velocity V is $V_J - V$. If the cart is initially empty and stationary when the jet action is initiated, develop a relation (integral form is acceptable) for cart velocity versus time.

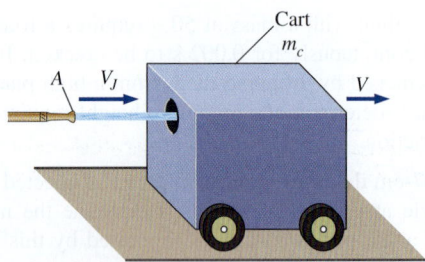

FIGURE P6–86

6–87 Water accelerated by a nozzle enters the impeller of a turbine through its outer edge of diameter D with a velocity of V making an angle α with the radial direction at a mass flow rate of $\dot{m}$. Water leaves the impeller in the radial direction. If the angular speed of the turbine shaft is $\dot{n}$, show that the maximum power that can be generated by this radial turbine is $\dot{W}_{shaft} = \pi \dot{n} \dot{m} D V \sin \alpha$.

6–88 Water enters a two-armed lawn sprinkler along the vertical axis at a rate of 75 L/s, and leaves the sprinkler nozzles as 2-cm diameter jets at an angle of θ from the tangential direction, as shown in Fig. P6–88. The length of each sprinkler arm is 0.52 m. Disregarding any frictional effects, determine the rate of rotation $\dot{n}$ of the sprinkler in rev/min for (a) $\theta = 0°$, (b) $\theta = 30°$, and (c) $\theta = 60°$.

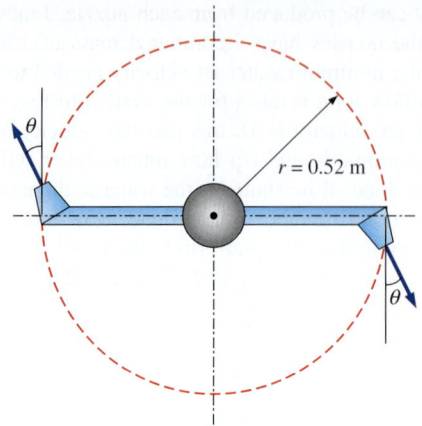

FIGURE P6–88

6–89 Reconsider Prob. 6–88. For the specified flow rate, investigate the effect of discharge angle θ on the rate of rotation $\dot{n}$ by varying θ from 0° to 90° in increments of 10°. Plot the rate of rotation versus θ, and discuss your results.

6–90 A stationary water tank of diameter D is mounted on wheels and is placed on a nearly frictionless level surface. A smooth hole of diameter D_o near the bottom of the tank allows water to jet horizontally and rearward and the water jet force propels the system forward. The water in the tank is much heavier than the tank-and-wheel assembly, so only the mass of water remaining in the tank needs to be considered in this problem. Considering the decrease in the mass of water with time, develop relations for (a) the acceleration, (b) the velocity, and (c) the distance traveled by the system as a function of time.

6–91 An orbiting satellite has a mass of 3400 kg and is traveling at a constant velocity of V_0. To alter its orbit, an attached rocket discharges 100 kg of gases from the reaction of solid fuel at a speed of 3000 m/s relative to the satellite

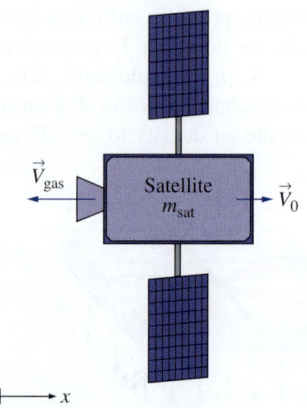

FIGURE P6–91

in a direction opposite V_0. The fuel discharge rate is constant for 3s. Determine (a) the thrust exerted on the satellite, (b) the acceleration of the satellite during this 3-s period, and (c) the change of velocity of the satellite during this time period.

6–92 Water enters a mixed flow pump axially at a rate of 0.3 m³/s and at a velocity of 7 m/s, and is discharged to the atmosphere at an angle of 75° from the horizontal, as shown in Fig. P6–92. If the discharge flow area is half the inlet area, determine the force acting on the shaft in the axial direction.

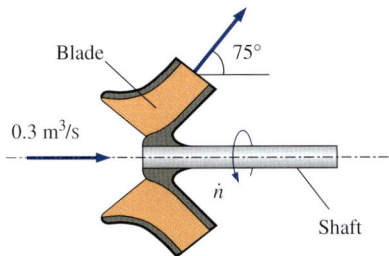

FIGURE P6–92

6–93 Water flows steadily through a splitter as shown in Fig. P6–93 with $\dot{V}_1 = 0.08$ m³/s, $\dot{V}_2 = 0.05$ m³/s, $D_1 = D_2 = 12$ cm, $D_3 = 10$ cm. If the pressure readings at the inlet and outlets of the splitter are $P_1 = 100$ kPa, $P_2 = 90$ kPa and $P_3 = 80$ kPa, determine external force needed to hold the device fixed. Disregard the weight effects.

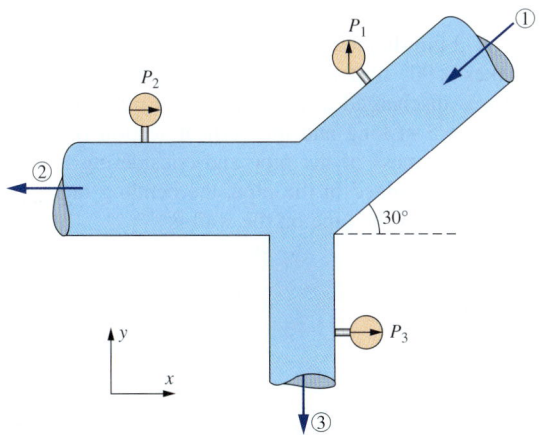

FIGURE P6–93

6–94 Water is discharged from a pipe through a 1.2-m long 5-mm wide rectangular slit underneath of the pipe. Water discharge velocity profile is parabolic, varying from 3 m/s on one end of the slit to 7 m/s on the other, as shown in Fig. P6–94. Determine (a) the rate of discharge through the slit and (b) the vertical force acting on the pipe due to this discharge process.

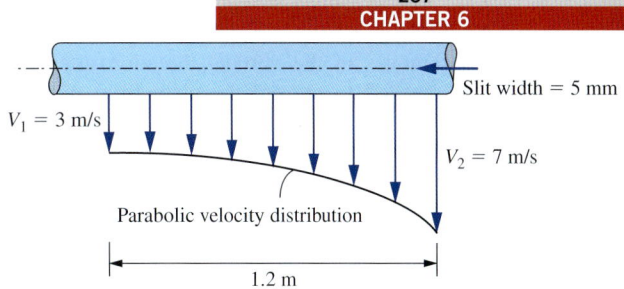

FIGURE P6–94

Fundamentals of Engineering (FE) Exam Problems

6–95 When determining the thrust developed by a jet engine, a wise choice of control volume is
(a) Fixed control volume
(b) Moving control volume
(c) Deforming control volume
(d) Moving or deforming control volume
(e) None of these

6–96 Consider an airplane cruising at 850 km/h to the right. If the velocity of exhaust gases is 700 km/h to the left relative to the ground, the velocity of the exhaust gases relative to the nozzle exit is
(a) 1550 km/h
(b) 850 km/h
(c) 700 km/h
(d) 350 km/h
(e) 150 km/h

6–97 Consider water flow through a horizontal, short garden hose at a rate of 30 kg/min. The velocity at the inlet is 1.5 m/s and that at the outlet is 14.5 m/s. Disregard the weight of the hose and water. Taking the momentum-flux correction factor to be 1.04 at both the inlet and the outlet, the anchoring force required to hold the hose in place is
(a) 2.8 N
(b) 8.6 N
(c) 17.5 N
(d) 27.9 N
(e) 43.3 N

6–98 Consider water flow through a horizontal, short garden hose at a rate of 30 kg/min. The velocity at the inlet is 1.5 m/s and that at the outlet is 11.5 m/s. The hose makes a 180° turn before the water is discharged. Disregard the weight of the hose and water. Taking the momentum-flux correction factor to be 1.04 at both the inlet and the outlet, the anchoring force required to hold the hose in place is
(a) 7.6 N
(b) 28.4 N
(c) 16.6 N
(d) 34.1 N
(e) 11.9 N

6–99 A water jet strikes a stationary vertical plate horizontally at a rate of 5 kg/s with a velocity of 35 km/h. Assume

the water stream moves in the vertical direction after the strike. The force needed to prevent the plate from moving horizontally is
(a) 15.5 N
(b) 26.3 N
(c) 19.7 N
(d) 34.2 N
(e) 48.6 N

6–100 Consider water flow through a horizontal, short garden hose at a rate of 40 kg/min. The velocity at the inlet is 1.5 m/s and that at the outlet is 16 m/s. The hose makes a 90° turn to a vertical direction before the water is discharged. Disregard the weight of the hose and water. Taking the momentum-flux correction factor to be 1.04 at both the inlet and the outlet, the reaction force in the vertical direction required to hold the hose in place is
(a) 11.1 N
(b) 10.1 N
(c) 9.3 N
(d) 27.2 N
(e) 28.9 N

6–101 Consider water flow through a horizontal, short pipe at a rate of 80 kg/min. The velocity at the inlet is 1.5 m/s and that at the outlet is 16.5 m/s. The pipe makes a 90° turn to a vertical direction before the water is discharged. Disregard the weight of the pipe and water. Taking the momentum-flux correction factor to be 1.04 at both the inlet and the outlet, the reaction force in the horizontal direction required to hold the pipe in place is
(a) 73.7 N
(b) 97.1 N
(c) 99.2 N
(d) 122 N
(e) 153 N

6–102 A water jet strikes a stationary horizontal plate vertically at a rate of 18 kg/s with a velocity of 24 m/s. The mass of the plate is 10 kg. Assume the water stream moves in the horizontal direction after the strike. The force needed to prevent the plate from moving vertically is
(a) 192 N
(b) 240 N
(c) 334 N
(d) 432 N
(e) 530 N

6–103 The velocity of wind at a wind turbine is measured to be 6 m/s. The blade span diameter is 24 m and the efficiency of the wind turbine is 29 percent. The density of air is 1.22 kg/m^3. The horizontal force exerted by the wind on the supporting mast of the wind turbine is
(a) 2524 N
(b) 3127 N
(c) 3475 N
(d) 4138 N
(e) 4313 N

6–104 The velocity of wind at a wind turbine is measured to be 8 m/s. The blade span diameter is 12 m. The density of air is 1.2 kg/m^3. If the horizontal force exerted by the wind on the supporting mast of the wind turbine is 1620 N, the efficiency of the wind turbine is
(a) 27.5%
(b) 31.7%
(c) 29.5%
(d) 35.1%
(e) 33.8%

6–105 The shaft of a turbine rotates at a speed of 800 rpm. If the torque of the shaft is 350 N·m, the shaft power is
(a) 112 kW
(b) 176 kW
(c) 293 kW
(d) 350 kW
(e) 405 kW

6–106 A 3-cm-diameter horizontal pipe attached to a surface makes a 90° turn to a vertical upward direction before the water is discharged at a velocity of 9 m/s. The horizontal section is 5 m long and the vertical section is 4 m long. Neglecting the mass of the water contained in the pipe, the bending moment acting on the base of the pipe on the wall is
(a) 286 N·m
(b) 229 N·m
(c) 207 N·m
(d) 175 N·m
(e) 124 N·m

6–107 A 3-cm-diameter horizontal pipe attached to a surface makes a 90° turn to a vertical upward direction before the water is discharged at a velocity of 6 m/s. The horizontal section is 5 m long and the vertical section is 4 m long. Neglecting the mass of the pipe and considering the weight of the water contained in the pipe, the bending moment acting on the base of the pipe on the wall is
(a) 11.9 N·m
(b) 46.7 N·m
(c) 127 N·m
(d) 104 N·m
(e) 74.8 N·m

6–108 A large lawn sprinkler with four identical arms is to be converted into a turbine to generate electric power by attaching a generator to its rotating head. Water enters the sprinkler from the base along the axis of rotation at a rate of 15 kg/s and leaves the nozzles in the tangential direction at a velocity of 50 m/s relative to the rotating nozzle. The sprinkler rotates at a rate of 400 rpm in a horizontal plane. The normal distance between the axis of rotation and the center of each nozzle is 30 cm. Estimate the electric power produced.

(a) 5430 W
(b) 6288 W
(c) 6634 W
(d) 7056 W
(e) 7875 W

6–109 Consider the impeller of a centrifugal pump with a rotational speed of 900 rpm and a flow rate of 95 kg/min. The impeller radii at the inlet and outlet are 7 cm and 16 cm, respectively. Assuming that the tangential fluid velocity is equal to the blade angular velocity both at the inlet and the exit, the power requirement of the pump is
(a) 83 W
(b) 291 W
(c) 409 W
(d) 756 W
(e) 1125 W

6–110 Water enters the impeller of a centrifugal pump radially at a rate of 450 L/min when the shaft is rotating at 400 rpm. The tangential component of absolute velocity of water at the exit of the 70-cm outer diameter impeller is 55 m/s. The torque applied to the impeller is
(a) 144 N·m
(b) 93.6 N·m
(c) 187 N·m
(d) 112 N·m
(e) 235 N·m

Design and Essay Problem

6–111 Visit a fire station and obtain information about flow rates through hoses and discharge diameters. Using this information, calculate the impulse force to which the firefighters are subjected when holding a fire hose.

CHAPTER 7

DIMENSIONAL ANALYSIS AND MODELING

In this chapter, we first review the concepts of *dimensions* and *units*. We then review the fundamental principle of *dimensional homogeneity*, and show how it is applied to equations in order to *nondimensionalize* them and to identify *dimensionless groups*. We discuss the concept of *similarity* between a *model* and a *prototype*. We also describe a powerful tool for engineers and scientists called *dimensional analysis*, in which the combination of dimensional variables, nondimensional variables, and dimensional constants into *nondimensional parameters* reduces the number of necessary independent parameters in a problem. We present a step-by-step method for obtaining these nondimensional parameters, called the *method of repeating variables*, which is based solely on the dimensions of the variables and constants. Finally, we apply this technique to several practical problems to illustrate both its utility and its limitations.

OBJECTIVES

When you finish reading this chapter, you should be able to

- Develop a better understanding of dimensions, units, and dimensional homogeneity of equations
- Understand the numerous benefits of dimensional analysis
- Know how to use the method of repeating variables to identify nondimensional parameters
- Understand the concept of dynamic similarity and how to apply it to experimental modeling

A 1:46.6 scale model of an Arleigh Burke class U.S. Navy fleet destroyer being tested in the 100-m-long towing tank at the University of Iowa. The model is 3.048 m long. In tests like this, the Froude number is the most important nondimensional parameter.

Photograph courtesy of IIHR-Hydroscience & Engineering, University of Iowa. Used by permission.

DIMENSIONAL ANALYSIS AND MODELING

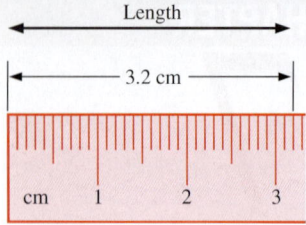

FIGURE 7–1
A *dimension* is a measure of a physical quantity without numerical values, while a *unit* is a way to assign a number to the dimension. For example, length is a dimension, but centimeter is a unit.

7–1 ▪ DIMENSIONS AND UNITS

A **dimension** is a measure of a physical quantity (without numerical values), while a **unit** is a way to assign a *number* to that dimension. For example, length is a dimension that is measured in units such as microns (μm), feet (ft), centimeters (cm), meters (m), kilometers (km), etc. (Fig. 7–1). There are seven **primary dimensions** (also called **fundamental** or **basic dimensions**)—mass, length, time, temperature, electric current, amount of light, and amount of matter.

> All nonprimary dimensions can be formed by some combination of the seven primary dimensions.

For example, force has the same dimensions as mass times acceleration (by Newton's second law). Thus, in terms of primary dimensions,

Dimensions of force: $\quad \{\text{Force}\} = \left\{ \text{Mass} \dfrac{\text{Length}}{\text{Time}^2} \right\} = \{mL/t^2\}$ (7–1)

where the brackets indicate "the dimensions of" and the abbreviations are taken from Table 7–1. You should be aware that some authors prefer force instead of mass as a primary dimension—we do not follow that practice.

TABLE 7–1
Primary dimensions and their associated primary SI and English units

Dimension	Symbol*	SI Unit	English Unit
Mass	m	kg (kilogram)	lbm (pound-mass)
Length	L	m (meter)	ft (foot)
Time†	t	s (second)	s (second)
Temperature	T	K (kelvin)	R (rankine)
Electric current	I	A (ampere)	A (ampere)
Amount of light	C	cd (candela)	cd (candela)
Amount of matter	N	mol (mole)	mol (mole)

* We italicize symbols for variables, but not symbols for dimensions.
† Note that some authors use the symbol T for the time dimension and the symbol θ for the temperature dimension. We do not follow this convention to avoid confusion between time and temperature.

FIGURE 7–2
The water strider is an insect that can walk on water due to surface tension.
NPS Photo by Rosalie LaRue.

EXAMPLE 7–1   Primary Dimensions of Surface Tension

An engineer is studying how some insects are able to walk on water (Fig. 7–2). A fluid property of importance in this problem is surface tension (σ_s), which has dimensions of force per unit length. Write the dimensions of surface tension in terms of primary dimensions.

SOLUTION   The primary dimensions of surface tension are to be determined.
Analysis   From Eq. 7–1, force has dimensions of mass times acceleration, or $\{mL/t^2\}$. Thus,

Dimensions of surface tension: $\quad \{\sigma_s\} = \left\{ \dfrac{\text{Force}}{\text{Length}} \right\} = \left\{ \dfrac{m \cdot L/t^2}{L} \right\} = \{m/t^2\}$ (1)

Discussion   The usefulness of expressing the dimensions of a variable or constant in terms of primary dimensions will become clearer in the discussion of the method of repeating variables in Section 7–4.

7–2 ■ DIMENSIONAL HOMOGENEITY

We've all heard the old saying, You can't add apples and oranges (Fig. 7–3). This is actually a simplified expression of a far more global and fundamental mathematical law for equations, the **law of dimensional homogeneity,** stated as

FIGURE 7–3
You can't add apples and oranges!

> Every additive term in an equation must have the same dimensions.

Consider, for example, the change in total energy of a simple compressible closed system from one state and/or time (1) to another (2), as illustrated in Fig. 7–4. The change in total energy of the system (ΔE) is given by

Change of total energy of a system: $\Delta E = \Delta U + \Delta KE + \Delta PE$ (7–2)

where E has three components: internal energy (U), kinetic energy (KE), and potential energy (PE). These components can be written in terms of the system mass (m); measurable quantities and thermodynamic properties at each of the two states, such as speed (V), elevation (z), and specific internal energy (u); and the gravitational acceleration constant (g),

$$\Delta U = m(u_2 - u_1) \qquad \Delta KE = \frac{1}{2}m(V_2^2 - V_1^2) \qquad \Delta PE = mg(z_2 - z_1) \quad (7\text{–}3)$$

It is straightforward to verify that the left side of Eq. 7–2 and all three additive terms on the right side of Eq. 7–2 have the same dimensions—energy. Using the definitions of Eq. 7–3, we write the primary dimensions of each term,

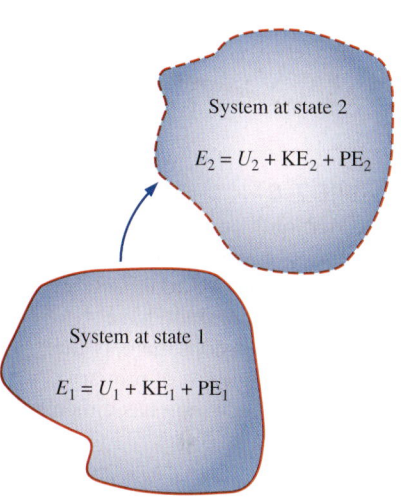

FIGURE 7–4
Total energy of a system at state 1 and at state 2.

$$\{\Delta E\} = \{\text{Energy}\} = \{\text{Force·Length}\} \rightarrow \{\Delta E\} = \{mL^2/t^2\}$$

$$\{\Delta U\} = \left\{\text{Mass}\frac{\text{Energy}}{\text{Mass}}\right\} = \{\text{Energy}\} \rightarrow \{\Delta U\} = \{mL^2/t^2\}$$

$$\{\Delta KE\} = \left\{\text{Mass}\frac{\text{Length}^2}{\text{Time}^2}\right\} \rightarrow \{\Delta KE\} = \{mL^2/t^2\}$$

$$\{\Delta PE\} = \left\{\text{Mass}\frac{\text{Length}}{\text{Time}^2}\text{Length}\right\} \rightarrow \{\Delta PE\} = \{mL^2/t^2\}$$

If at some stage of an analysis we find ourselves in a position in which two additive terms in an equation have *different* dimensions, this would be a clear indication that we have made an error at some earlier stage in the analysis (Fig. 7–5). In addition to dimensional homogeneity, calculations are valid only when the *units* are also homogeneous in each additive term. For example, units of energy in the above terms may be J, N·m, or kg·m²/s², all of which are equivalent. Suppose, however, that kJ were used in place of J for one of the terms. This term would be off by a factor of 1000 compared to the other terms. It is wise to write out *all* units when performing mathematical calculations in order to avoid such errors.

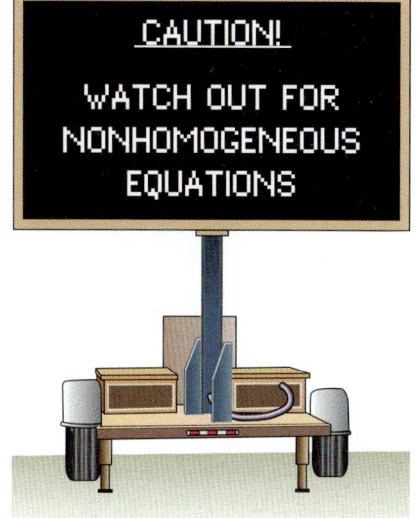

FIGURE 7–5
An equation that is not dimensionally homogeneous is a sure sign of an error.

294
DIMENSIONAL ANALYSIS AND MODELING

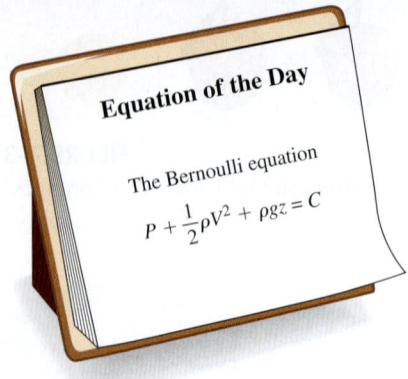

FIGURE 7–6
The Bernoulli equation is a good example of a *dimensionally homogeneous* equation. All additive terms, including the constant, have the *same* dimensions, namely that of pressure. In terms of primary dimensions, each term has dimensions $\{m/(t^2L)\}$.

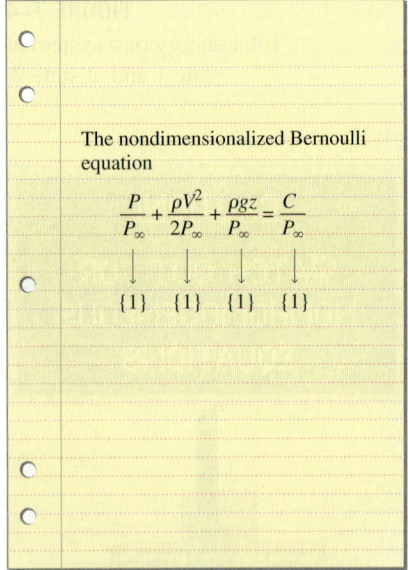

FIGURE 7–7
A *nondimensionalized* form of the Bernoulli equation is formed by dividing each additive term by a pressure (here we use P_∞). Each resulting term is *dimensionless* (dimensions of $\{1\}$).

EXAMPLE 7–2 Dimensional Homogeneity of the Bernoulli Equation

Probably the most well-known (and most misused) equation in fluid mechanics is the Bernoulli equation (Fig. 7–6), discussed in Chap. 5. One standard form of the Bernoulli equation for incompressible irrotational fluid flow is

Bernoulli equation: $$P + \frac{1}{2}\rho V^2 + \rho gz = C \qquad (1)$$

(*a*) Verify that each additive term in the Bernoulli equation has the same dimensions. (*b*) What are the dimensions of the constant *C*?

SOLUTION We are to verify that the primary dimensions of each additive term in Eq. 1 are the same, and we are to determine the dimensions of constant *C*.
Analysis (*a*) Each term is written in terms of primary dimensions,

$$\{P\} = \{\text{Pressure}\} = \left\{\frac{\text{Force}}{\text{Area}}\right\} = \left\{\text{Mass}\,\frac{\text{Length}}{\text{Time}^2}\,\frac{1}{\text{Length}^2}\right\} = \left\{\frac{m}{t^2 L}\right\}$$

$$\left\{\frac{1}{2}\rho V^2\right\} = \left\{\frac{\text{Mass}}{\text{Volume}}\left(\frac{\text{Length}}{\text{Time}}\right)^2\right\} = \left\{\frac{\text{Mass} \times \text{Length}^2}{\text{Length}^3 \times \text{Time}^2}\right\} = \left\{\frac{m}{t^2 L}\right\}$$

$$\{\rho gz\} = \left\{\frac{\text{Mass}}{\text{Volume}}\,\frac{\text{Length}}{\text{Time}^2}\,\text{Length}\right\} = \left\{\frac{\text{Mass} \times \text{Length}^2}{\text{Length}^3 \times \text{Time}^2}\right\} = \left\{\frac{m}{t^2 L}\right\}$$

Indeed, **all three additive terms have the same dimensions.**

(*b*) From the law of dimensional homogeneity, the constant must have the same dimensions as the other additive terms in the equation. Thus,

Primary dimensions of the Bernoulli constant: $$\{C\} = \left\{\frac{m}{t^2 L}\right\}$$

Discussion If the dimensions of any of the terms were different from the others, it would indicate that an error was made somewhere in the analysis.

Nondimensionalization of Equations

The law of dimensional homogeneity guarantees that every additive term in an equation has the same dimensions. It follows that if we divide each term in the equation by a collection of variables and constants whose product has those same dimensions, the equation is rendered **nondimensional** (Fig. 7–7). If, in addition, the nondimensional terms in the equation are of order unity, the equation is called **normalized.** Normalization is thus more restrictive than nondimensionalization, even though the two terms are sometimes (incorrectly) used interchangeably.

Each term in a nondimensional equation is dimensionless.

In the process of nondimensionalizing an equation of motion, **nondimensional parameters** often appear—most of which are named after a notable scientist or engineer (e.g., the Reynolds number and the Froude number). This process is referred to by some authors as **inspectional analysis.**

As a simple example, consider the equation of motion describing the elevation z of an object falling by gravity through a vacuum (no air drag), as in Fig. 7–8. The initial location of the object is z_0 and its initial velocity is w_0 in the z-direction. From high school physics,

Equation of motion: $$\frac{d^2z}{dt^2} = -g \qquad (7\text{--}4)$$

Dimensional variables are defined as dimensional quantities that change or vary in the problem. For the simple differential equation given in Eq. 7–4, there are two dimensional variables: z (dimension of length) and t (dimension of time). **Nondimensional** (or **dimensionless**) **variables** are defined as quantities that change or vary in the problem, but have no dimensions; an example is angle of rotation, measured in degrees or radians which are dimensionless units. Gravitational constant g, while dimensional, remains constant and is called a **dimensional constant**. Two additional dimensional constants are relevant to this particular problem, initial location z_0 and initial vertical speed w_0. While dimensional constants may change from problem to problem, they are fixed for a particular problem and are thus distinguished from dimensional variables. We use the term **parameters** for the combined set of dimensional variables, nondimensional variables, and dimensional constants in the problem.

Equation 7–4 is easily solved by integrating twice and applying the initial conditions. The result is an expression for elevation z at any time t:

Dimensional result: $$z = z_0 + w_0 t - \frac{1}{2}gt^2 \qquad (7\text{--}5)$$

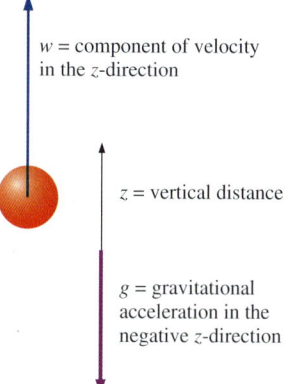

FIGURE 7–8
Object falling in a vacuum. Vertical velocity is drawn positively, so $w < 0$ for a falling object.

The constant $\frac{1}{2}$ and the exponent 2 in Eq. 7–5 are dimensionless results of the integration. Such constants are called **pure constants**. Other common examples of pure constants are π and e.

To nondimensionalize Eq. 7–4, we need to select **scaling parameters,** based on the primary dimensions contained in the original equation. In fluid flow problems there are typically at least *three* scaling parameters, e.g., L, V, and $P_0 - P_\infty$ (Fig. 7–9), since there are at least three primary dimensions in the general problem (e.g., mass, length, and time). In the case of the falling object being discussed here, there are only two primary dimensions, length and time, and thus we are limited to selecting only *two* scaling parameters. We have some options in the selection of the scaling parameters since we have three available dimensional constants g, z_0, and w_0. We choose z_0 and w_0. You are invited to repeat the analysis with g and z_0 and/or with g and w_0. With these two chosen scaling parameters we nondimensionalize the dimensional variables z and t. The first step is to list the primary dimensions of all dimensional variables and dimensional constants in the problem,

Primary dimensions of all parameters:

$\{z\} = \{L\} \quad \{t\} = \{t\} \quad \{z_0\} = \{L\} \quad \{w_0\} = \{L/t\} \quad \{g\} = \{L/t^2\}$

The second step is to use our two scaling parameters to nondimensionalize z and t (by inspection) into nondimensional variables z^* and t^*,

Nondimensionalized variables: $$z^* = \frac{z}{z_0} \qquad t^* = \frac{w_0 t}{z_0} \qquad (7\text{--}6)$$

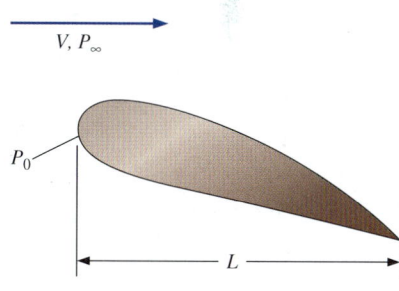

FIGURE 7–9
In a typical fluid flow problem, the scaling parameters usually include a characteristic length L, a characteristic velocity V, and a reference pressure difference $P_0 - P_\infty$. Other parameters and fluid properties such as density, viscosity, and gravitational acceleration enter the problem as well.

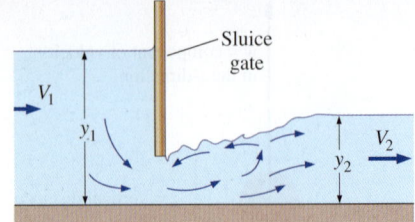

FIGURE 7–10
The *Froude number* is important in free-surface flows such as flow in open channels. Shown here is flow through a sluice gate. The Froude number upstream of the sluice gate is $\text{Fr}_1 = V_1/\sqrt{gy_1}$, and it is $\text{Fr}_2 = V_2/\sqrt{gy_2}$ downstream of the sluice gate.

Substitution of Eq. 7–6 into Eq. 7–4 gives

$$\frac{d^2z}{dt^2} = \frac{d^2(z_0 z^*)}{d(z_0 t^*/w_0)^2} = \frac{w_0^2}{z_0}\frac{d^2 z^*}{dt^{*2}} = -g \quad \rightarrow \quad \frac{w_0^2}{gz_0}\frac{d^2 z^*}{dt^{*2}} = -1 \quad (7\text{–}7)$$

which is the desired nondimensional equation. The grouping of dimensional constants in Eq. 7–7 is the square of a well-known **nondimensional parameter** or **dimensionless group** called the **Froude number**,

Froude number: $$\text{Fr} = \frac{w_0}{\sqrt{gz_0}} \quad (7\text{–}8)$$

The Froude (pronounced "Frude") number also appears as a nondimensional parameter in free-surface flows (Chap. 13), and can be thought of as the ratio of inertial force to gravitational force (Fig. 7–10). You should note that in some older textbooks, Fr is defined as the *square* of the parameter shown in Eq. 7–8. Substitution of Eq. 7–8 into Eq. 7–7 yields

Nondimensionalized equation of motion: $$\frac{d^2 z^*}{dt^{*2}} = -\frac{1}{\text{Fr}^2} \quad (7\text{–}9)$$

In dimensionless form, only one parameter remains, namely the Froude number. Equation 7–9 is easily solved by integrating twice and applying the initial conditions. The result is an expression for dimensionless elevation z^* as a function of dimensionless time t^*:

Nondimensional result: $$z^* = 1 + t^* - \frac{1}{2\text{Fr}^2} t^{*2} \quad (7\text{–}10)$$

Comparison of Eqs. 7–5 and 7–10 reveals that they are equivalent. In fact, for practice, substitute Eqs. 7–6 and 7–8 into Eq. 7–5 to verify Eq. 7–10.

It seems that we went through a lot of extra algebra to generate the same final result. *What then is the advantage of nondimensionalizing the equation?* Before answering this question, we note that the advantages are not so clear in this simple example because we were able to analytically integrate the differential equation of motion. In more complicated problems, the differential equation (or more generally the coupled *set* of differential equations) *cannot* be integrated analytically, and engineers must either integrate the equations numerically, or design and conduct physical experiments to obtain the needed results, both of which can incur considerable time and expense. In such cases, the nondimensional parameters generated by nondimensionalizing the equations are extremely useful and can save much effort and expense in the long run.

There are two key advantages of nondimensionalization (Fig. 7–11). First, *it increases our insight about the relationships between key parameters.* Equation 7–8 reveals, for example, that doubling w_0 has the same effect as decreasing z_0 by a factor of 4. Second, *it reduces the number of parameters in the problem.* For example, the original problem contains one dependent variable, z; one independent variable, t; and *three* additional dimensional constants, g, w_0, and z_0. The nondimensionalized problem contains one dependent parameter, z^*; one independent parameter, t^*; and only *one* additional parameter, namely the dimensionless Froude number, Fr. The number of additional parameters has been reduced from three to one! Example 7–3 further illustrates the advantages of nondimensionalization.

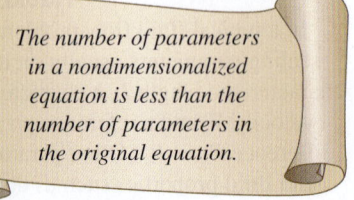

FIGURE 7–11
The two key advantages of nondimensionalization of an equation.

EXAMPLE 7–3   Illustration of the Advantages of Nondimensionalization

Your little brother's high school physics class conducts experiments in a large vertical pipe whose inside is kept under vacuum conditions. The students are able to remotely release a steel ball at initial height z_0 between 0 and 15 m (measured from the bottom of the pipe), and with initial vertical speed w_0 between 0 and 10 m/s. A computer coupled to a network of photosensors along the pipe enables students to plot the trajectory of the steel ball (height z plotted as a function of time t) for each test. The students are unfamiliar with dimensional analysis or nondimensionalization techniques, and therefore conduct several "brute force" experiments to determine how the trajectory is affected by initial conditions z_0 and w_0. First they hold w_0 fixed at 4 m/s and conduct experiments at five different values of z_0: 3, 6, 9, 12, and 15 m. The experimental results are shown in Fig. 7–12a. Next, they hold z_0 fixed at 10 m and conduct experiments at five different values of w_0: 2, 4, 6, 8, and 10 m/s. These results are shown in Fig. 7–12b. Later that evening, your brother shows you the data and the trajectory plots and tells you that they plan to conduct more experiments at different values of z_0 and w_0. You explain to him that by first nondimensionalizing the data, the problem can be reduced to just *one* parameter, and no further experiments are required. Prepare a nondimensional plot to prove your point and discuss.

SOLUTION A nondimensional plot is to be generated from all the available trajectory data. Specifically, we are to plot z^* as a function of t^*.
Assumptions The inside of the pipe is subjected to strong enough vacuum pressure that aerodynamic drag on the ball is negligible.
Properties The gravitational constant is 9.81 m/s².
Analysis Equation 7–4 is valid for this problem, as is the nondimensionalization that resulted in Eq. 7–9. As previously discussed, this problem combines three of the original dimensional parameters (g, z_0, and w_0) into *one* nondimensional parameter, the Froude number. After converting to the dimensionless variables of Eq. 7–6, the 10 trajectories of Fig. 7–12a and b are replotted in dimensionless format in Fig. 7–13. It is clear that all the trajectories are of the same family, with the Froude number as the only remaining parameter. Fr² varies from about 0.041 to about 1.0 in these experiments. If any more experiments are to be conducted, they should include combinations of z_0 and w_0 that produce Froude numbers outside of this range. A large number of additional experiments would be unnecessary, since all the trajectories would be of the same family as those plotted in Fig. 7–13.
Discussion At low Froude numbers, gravitational forces are much larger than inertial forces, and the ball falls to the floor in a relatively short time. At large values of Fr on the other hand, inertial forces dominate initially, and the ball rises a significant distance before falling; it takes much longer for the ball to hit the ground. The students are obviously not able to adjust the gravitational constant, but if they could, the brute force method would require many more experiments to document the effect of g. If they nondimensionalize first, however, the dimensionless trajectory plots already obtained and shown in Fig. 7–13 would be valid for *any* value of g; no further experiments would be required unless Fr were outside the range of tested values.

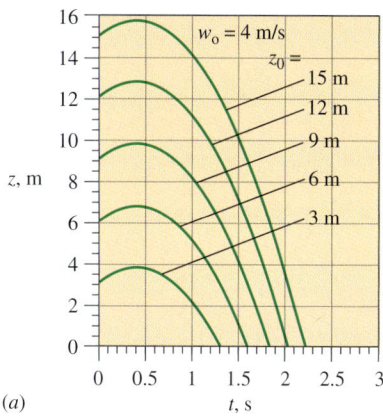
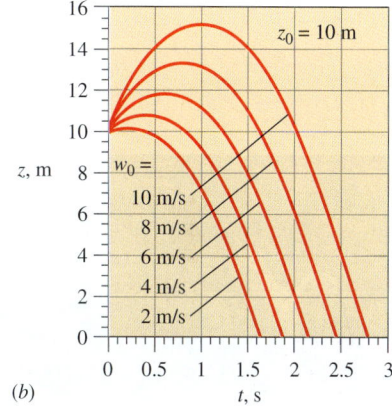

FIGURE 7–12
Trajectories of a steel ball falling in a vacuum: (a) w_0 fixed at 4 m/s, and (b) z_0 fixed at 10 m (Example 7–3).

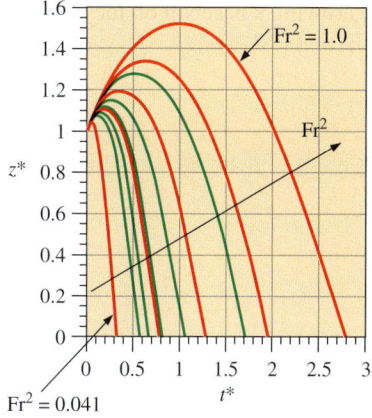

FIGURE 7–13
Trajectories of a steel ball falling in a vacuum. Data of Fig. 7–12a and b are nondimensionalized and combined onto one plot.

If you are still not convinced that nondimensionalizing the equations and the parameters has many advantages, consider this: In order to reasonably document the trajectories of Example 7–3 for a range of all three of the dimensional parameters g, z_0, and w_0, the brute force method would require several (say a minimum of four) additional plots like Fig. 7–12a at various values (levels) of w_0, plus several additional *sets* of such plots for a range of g. A complete data set for three parameters with five levels of each parameter would require $5^3 = 125$ experiments! Nondimensionalization reduces the number of parameters from three to one—a total of only $5^1 = 5$ experiments are required for the same resolution. (For five levels, only five dimensionless trajectories like those of Fig. 7–13 are required, at carefully chosen values of Fr.)

Another advantage of nondimensionalization is that extrapolation to untested values of one or more of the dimensional parameters is possible. For example, the data of Example 7–3 were taken at only one value of gravitational acceleration. Suppose you wanted to extrapolate these data to a different value of g. Example 7–4 shows how this is easily accomplished via the dimensionless data.

FIGURE 7–14
Throwing a baseball on the moon (Example 7–4).

EXAMPLE 7–4   **Extrapolation of Nondimensionalized Data**

The gravitational constant at the surface of the moon is only about one-sixth of that on earth. An astronaut on the moon throws a baseball at an initial speed of 21.0 m/s at a 5° angle above the horizon and at 2.0 m above the moon's surface (Fig. 7–14). (a) Using the dimensionless data of Example 7–3 shown in Fig. 7–13, predict how long it takes for the baseball to fall to the ground. (b) Do an *exact* calculation and compare the result to that of part (a).

SOLUTION Experimental data obtained on earth are to be used to predict the time required for a baseball to fall to the ground on the moon.
Assumptions 1 The horizontal velocity of the baseball is irrelevant. 2 The surface of the moon is perfectly flat near the astronaut. 3 There is no aerodynamic drag on the ball since there is no atmosphere on the moon. 4 Moon gravity is one-sixth that of earth.
Properties The gravitational constant on the moon is $g_{moon} \cong 9.81/6 = 1.63$ m/s^2.
Analysis (a) The Froude number is calculated based on the value of g_{moon} and the vertical component of initial speed,

$$w_0 = (21.0 \text{ m/s}) \sin(5°) = 1.830 \text{ m/s}$$

from which

$$\text{Fr}^2 = \frac{w_0^2}{g_{moon} z_0} = \frac{(1.830 \text{ m/s})^2}{(1.63 \text{ m/s}^2)(2.0 \text{ m})} = 1.03$$

This value of Fr^2 is nearly the same as the largest value plotted in Fig. 7–13. Thus, in terms of dimensionless variables, the baseball strikes the ground at $t^* \cong 2.75$, as determined from Fig. 7–13. Converting back to dimensional variables using Eq. 7–6,

Estimated time to strike the ground: $t = \dfrac{t^* z_0}{w_0} = \dfrac{2.75 (2.0 \text{ m})}{1.830 \text{ m/s}} = \mathbf{3.01 \text{ s}}$

(b) An exact calculation is obtained by setting z equal to zero in Eq. 7–5 and solving for time t (using the quadratic formula),

Exact time to strike the ground:

$$t = \frac{w_0 + \sqrt{w_0^2 + 2z_0 g}}{g}$$

$$= \frac{1.830 \text{ m/s} + \sqrt{(1.830 \text{ m/s})^2 + 2(2.0 \text{ m})(1.63 \text{ m/s}^2)}}{1.63 \text{ m/s}^2} = \mathbf{3.05 \text{ s}}$$

Discussion If the Froude number had landed between two of the trajectories of Fig. 7–13, interpolation would have been required. Since some of the numbers are precise to only two significant digits, the small difference between the results of part (*a*) and part (*b*) is of no concern. The final result is $t = 3.0$ s to two significant digits.

The differential equations of motion for fluid flow are derived and discussed in Chap. 9. In Chap. 10 you will find an analysis similar to that presented here, but applied to the differential equations for fluid flow. It turns out that the Froude number also appears in that analysis, as do three other important dimensionless parameters—the Reynolds number, Euler number, and Strouhal number (Fig. 7–15).

7–3 ■ DIMENSIONAL ANALYSIS AND SIMILARITY

Nondimensionalization of an equation by inspection is useful only when we know the equation to begin with. However, in many cases in real-life engineering, the equations are either not known or too difficult to solve; oftentimes *experimentation* is the only method of obtaining reliable information. In most experiments, to save time and money, tests are performed on a geometrically scaled **model,** rather than on the full-scale **prototype.** In such cases, care must be taken to properly scale the results. We introduce here a powerful technique called **dimensional analysis.** While typically taught in fluid mechanics, dimensional analysis is useful in *all* disciplines, especially when it is necessary to design and conduct experiments. You are encouraged to use this powerful tool in other subjects as well, not just in fluid mechanics. The three primary purposes of dimensional analysis are

- To generate nondimensional parameters that help in the design of experiments (physical and/or numerical) and in the reporting of experimental results
- To obtain scaling laws so that prototype performance can be predicted from model performance
- To (sometimes) predict trends in the relationship between parameters

Before discussing the *technique* of dimensional analysis, we first explain the underlying *concept* of dimensional analysis—the principle of **similarity.** There are three necessary conditions for complete similarity between a model and a prototype. The first condition is **geometric similarity**—the model must be the same shape as the prototype, but may be scaled by some constant scale factor. The second condition is **kinematic similarity,** which means that the velocity at any point in the model flow must be proportional

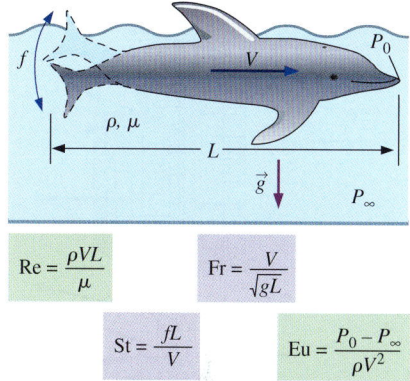

FIGURE 7–15
In a general unsteady fluid flow problem with a free surface, the scaling parameters include a characteristic length L, a characteristic velocity V, a characteristic frequency f, and a reference pressure difference $P_0 - P_\infty$. Nondimensionalization of the differential equations of fluid flow produces four dimensionless parameters: the Reynolds number, Froude number, Strouhal number, and Euler number (see Chap. 10).

DIMENSIONAL ANALYSIS AND MODELING

(by a constant scale factor) to the velocity at the corresponding point in the prototype flow (Fig. 7–16). Specifically, for kinematic similarity the velocity at corresponding points must scale in magnitude and must point in the same relative direction. You may think of geometric similarity as *length-scale* equivalence and kinematic similarity as *time-scale* equivalence. *Geometric similarity is a prerequisite for kinematic similarity.* Just as the geometric scale factor can be less than, equal to, or greater than one, so can the velocity scale factor. In Fig. 7–16, for example, the geometric scale factor is less than one (model smaller than prototype), but the velocity scale is greater than one (velocities around the model are greater than those around the prototype). You may recall from Chap. 4 that streamlines are kinematic phenomena; hence, the streamline pattern in the model flow is a geometrically scaled copy of that in the prototype flow when kinematic similarity is achieved.

The third and most restrictive similarity condition is that of **dynamic similarity.** Dynamic similarity is achieved when all *forces* in the model flow scale by a constant factor to corresponding forces in the prototype flow (*force-scale* equivalence). As with geometric and kinematic similarity, the scale factor for forces can be less than, equal to, or greater than one. In Fig. 7–16 for example, the force-scale factor is less than one since the force on the model building is less than that on the prototype. *Kinematic similarity is a necessary but insufficient condition for dynamic similarity.* It is thus possible for a model flow and a prototype flow to achieve both geometric and kinematic similarity, yet not dynamic similarity. All three similarity conditions must exist for complete similarity to be ensured.

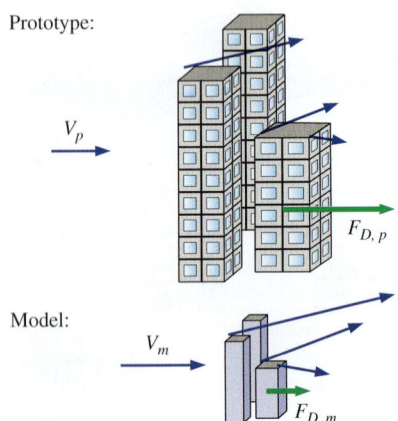

FIGURE 7–16
Kinematic similarity is achieved when, at all locations, the speed in the model flow is proportional to that at corresponding locations in the prototype flow, and points in the same direction.

> In a general flow field, complete similarity between a model and prototype is achieved only when there is geometric, kinematic, and dynamic similarity.

We let uppercase Greek letter Pi (Π) denote a nondimensional parameter. In Sec. 7–2, we have already discussed one Π, namely the Froude number, Fr. In a general dimensional analysis problem, there is one Π that we call the **dependent** Π, giving it the notation Π_1. The parameter Π_1 is in general a function of several other Π's, which we call **independent** Π's. The functional relationship is

Functional relationship between Π's: $\qquad \Pi_1 = f(\Pi_2, \Pi_3, \ldots, \Pi_k) \qquad$ (7–11)

where k is the total number of Π's.

Consider an experiment in which a scale model is tested to simulate a prototype flow. To ensure complete similarity between the model and the prototype, each independent Π of the model (subscript m) must be identical to the corresponding independent Π of the prototype (subscript p), i.e., $\Pi_{2,m} = \Pi_{2,p}$, $\Pi_{3,m} = \Pi_{3,p}$, ..., $\Pi_{k,m} = \Pi_{k,p}$.

> To ensure complete similarity, the model and prototype must be geometrically similar, and all independent Π groups must match between model and prototype.

Under these conditions the *dependent* Π of the model ($\Pi_{1,m}$) is guaranteed to also equal the dependent Π of the prototype ($\Pi_{1,p}$). Mathematically, we write a conditional statement for achieving similarity,

If $\quad \Pi_{2,m} = \Pi_{2,p} \quad$ and $\quad \Pi_{3,m} = \Pi_{3,p} \ldots \quad$ and $\quad \Pi_{k,m} = \Pi_{k,p},$

then $\quad \Pi_{1,m} = \Pi_{1,p} \qquad\qquad$ (7–12)

Consider, for example, the design of a new sports car, the aerodynamics of which is to be tested in a wind tunnel. To save money, it is desirable to test a small, geometrically scaled model of the car rather than a full-scale prototype of the car (Fig. 7–17). In the case of aerodynamic drag on an automobile, it turns out that if the flow is approximated as incompressible, there are only two Π's in the problem,

$$\Pi_1 = f(\Pi_2) \quad \text{where} \quad \Pi_1 = \frac{F_D}{\rho V^2 L^2} \quad \text{and} \quad \Pi_2 = \frac{\rho V L}{\mu} \quad (7\text{--}13)$$

The procedure used to generate these Π's is discussed in Section 7–4. In Eq. 7–13, F_D is the magnitude of the aerodynamic drag on the car, ρ is the air density, V is the car's speed (or the speed of the air in the wind tunnel), L is the length of the car, and μ is the viscosity of the air. Π_1 is a nonstandard form of the drag coefficient, and Π_2 is the **Reynolds number,** Re. You will find that many problems in fluid mechanics involve a Reynolds number (Fig. 7–18).

> The Reynolds number is the most well known and useful dimensionless parameter in all of fluid mechanics.

In the problem at hand there is only one independent Π, and Eq. 7–12 ensures that if the independent Π's match (the Reynolds numbers match: $\Pi_{2,m} = \Pi_{2,p}$), then the dependent Π's also match ($\Pi_{1,m} = \Pi_{1,p}$). This enables engineers to measure the aerodynamic drag on the model car and then use this value to predict the aerodynamic drag on the prototype car.

■ **EXAMPLE 7–5** **Similarity between Model and Prototype Cars**

The aerodynamic drag of a new sports car is to be predicted at a speed of 80.0 km/h at an air temperature of 25°C. Automotive engineers build a one-fifth scale model of the car to test in a wind tunnel. It is winter and the wind tunnel is located in an unheated building; the temperature of the wind tunnel air is only about 5°C. Determine how fast the engineers should run the wind tunnel in order to achieve similarity between the model and the prototype.

SOLUTION We are to utilize the concept of similarity to determine the speed of the wind tunnel.
Assumptions **1** Compressibility of the air is negligible (the validity of this approximation is discussed later). **2** The wind tunnel walls are far enough away so as to not interfere with the aerodynamic drag on the model car. **3** The model is geometrically similar to the prototype. **4** The wind tunnel has a moving belt to simulate the ground under the car, as in Fig. 7–19. (The moving belt is necessary in order to achieve kinematic similarity everywhere in the flow, in particular underneath the car.)
Properties For air at atmospheric pressure and at $T = 25°C$, $\rho = 1.184$ kg/m³ and $\mu = 1.849 \times 10^{-5}$ kg/m·s. Similarly, at $T = 5°C$, $\rho = 1.269$ kg/m³ and $\mu = 1.754 \times 10^{-5}$ kg/m·s.
Analysis Since there is only one independent Π in this problem, the similarity equation (Eq. 7–12) holds if $\Pi_{2,m} = \Pi_{2,p}$, where Π_2 is given by Eq. 7–13, and we call it the Reynolds number. Thus, we write

$$\Pi_{2,m} = \text{Re}_m = \frac{\rho_m V_m L_m}{\mu_m} = \Pi_{2,p} = \text{Re}_p = \frac{\rho_p V_p L_p}{\mu_p}$$

Prototype car

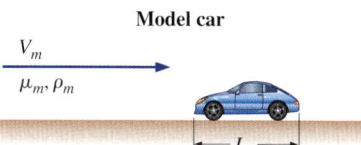

Model car

FIGURE 7–17
Geometric similarity between a prototype car of length L_p and a model car of length L_m.

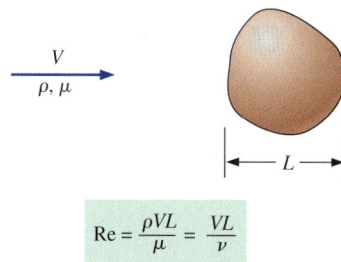

$$\text{Re} = \frac{\rho V L}{\mu} = \frac{V L}{\nu}$$

FIGURE 7–18
The *Reynolds number* Re is formed by the ratio of density, characteristic speed, and characteristic length to viscosity. Alternatively, it is the ratio of characteristic speed and length to *kinematic viscosity*, defined as $\nu = \mu/\rho$.

DIMENSIONAL ANALYSIS AND MODELING

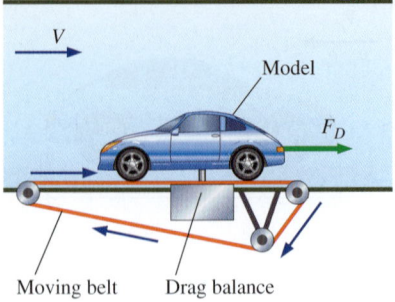

FIGURE 7–19
A *drag balance* is a device used in a wind tunnel to measure the aerodynamic drag of a body. When testing automobile models, a *moving belt* is often added to the floor of the wind tunnel to simulate the moving ground (from the car's frame of reference).

which we solve for the unknown wind tunnel speed for the model tests, V_m,

$$V_m = V_p \left(\frac{\mu_m}{\mu_p}\right)\left(\frac{\rho_p}{\rho_m}\right)\left(\frac{L_p}{L_m}\right)$$

$$= (80.0 \text{ km/h}) \left(\frac{1.754 \times 10^{-5} \text{ kg/m·s}}{1.849 \times 10^{-5} \text{ kg/m·s}}\right)\left(\frac{1.184 \text{ kg/m}^3}{1.269 \text{ kg/m}^3}\right)(5) = \mathbf{354 \text{ km/h}}$$

Thus, to ensure similarity, the wind tunnel should be run at 354 km/h (to three significant digits). Note that we were never given the actual length of either car, but the ratio of L_p to L_m is known because the prototype is five times larger than the scale model. When the dimensional parameters are rearranged as nondimensional ratios (as done here), the unit system is irrelevant. Since the units in each numerator cancel those in each denominator, no unit conversions are necessary.

Discussion This speed is quite high (about 100 m/s), and the wind tunnel may not be able to run at that speed. Furthermore, the incompressible approximation may come into question at this high speed (we discuss this in more detail in Example 7–8).

Once we are convinced that complete similarity has been achieved between the model tests and the prototype flow, Eq. 7–12 can be used again to predict the performance of the prototype based on measurements of the performance of the model. This is illustrated in Example 7–6.

EXAMPLE 7–6 Prediction of Aerodynamic Drag Force on a Prototype Car

This example is a follow-up to Example 7–5. Suppose the engineers run the wind tunnel at 354 km/h to achieve similarity between the model and the prototype. The aerodynamic drag force on the model car is measured with a **drag balance** (Fig. 7–19). Several drag readings are recorded, and the average drag force on the model is 94.3 N. Predict the aerodynamic drag force on the prototype (at 80 km/h and 25°C).

SOLUTION Because of similarity, the model results are to be scaled up to predict the aerodynamic drag force on the prototype.
Analysis The similarity equation (Eq. 7–12) shows that since $\Pi_{2,m} = \Pi_{2,p}$, $\Pi_{1,m} = \Pi_{1,p}$, where Π_1 is given for this problem by Eq. 7–13. Thus, we write

$$\Pi_{1,m} = \frac{F_{D,m}}{\rho_m V_m^2 L_m^2} = \Pi_{1,p} = \frac{F_{D,p}}{\rho_p V_p^2 L_p^2}$$

which we solve for the unknown aerodynamic drag force on the prototype car, $F_{D,p}$,

$$F_{D,p} = F_{D,m} \left(\frac{\rho_p}{\rho_m}\right)\left(\frac{V_p}{V_m}\right)^2 \left(\frac{L_p}{L_m}\right)^2$$

$$= (94.3 \text{ N}) \left(\frac{1.184 \text{ kg/m}^3}{1.269 \text{ kg/m}^3}\right)\left(\frac{80.0 \text{ km/h}}{354 \text{ km/h}}\right)^2 (5)^2 = \mathbf{112 \text{ N}}$$

Discussion By arranging the dimensional parameters as nondimensional ratios, the units cancel nicely. Because both velocity and length are squared in the equation for Π_1, the higher speed in the wind tunnel nearly compensates for the model's smaller size, and the drag force on the model is nearly the same as that on the prototype. In fact, if the density and viscosity of the air in the wind tunnel were *identical* to those of the air flowing over the prototype, the two drag forces would be identical as well (Fig. 7–20).

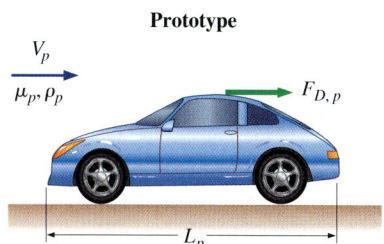

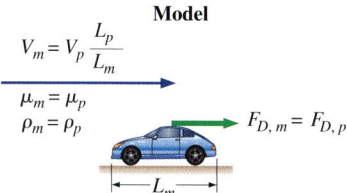

The power of using dimensional analysis and similarity to supplement experimental analysis is further illustrated by the fact that the actual values of the dimensional parameters (density, velocity, etc.) are irrelevant. As long as the corresponding independent Π's are set equal to each other, similarity is achieved—*even if different fluids are used*. This explains why automobile or aircraft performance can be simulated in a water tunnel, and the performance of a submarine can be simulated in a wind tunnel (Fig. 7–21). Suppose, for example, that the engineers in Examples 7–5 and 7–6 use a water tunnel instead of a wind tunnel to test their one-fifth scale model. Using the properties of water at room temperature (20°C is assumed), the water tunnel speed required to achieve similarity is easily calculated as

$$V_m = V_p \left(\frac{\mu_m}{\mu_p}\right)\left(\frac{\rho_p}{\rho_m}\right)\left(\frac{L_p}{L_m}\right)$$
$$= (80.0 \text{ km/h})\left(\frac{1.002 \times 10^{-3} \text{ kg/m·s}}{1.849 \times 10^{-5} \text{ kg/m·s}}\right)\left(\frac{1.184 \text{ kg/m}^3}{998.0 \text{ kg/m}^3}\right)(5) = 25.7 \text{ km/h}$$

As can be seen, one advantage of a water tunnel is that the required water tunnel speed is much lower than that required for a wind tunnel using the same size model.

FIGURE 7–20
For the special case in which the wind tunnel air and the air flowing over the prototype have the same properties ($\rho_m = \rho_p$, $\mu_m = \mu_p$), and under similarity conditions ($V_m = V_p L_p / L_m$), the aerodynamic drag force on the prototype is equal to that on the scale model. If the two fluids do *not* have the same properties, the two drag forces are *not* necessarily the same, even under dynamically similar conditions.

7–4 ■ THE METHOD OF REPEATING VARIABLES AND THE BUCKINGHAM PI THEOREM

We have seen several examples of the usefulness and power of dimensional analysis. Now we are ready to learn how to *generate* the nondimensional parameters, i.e., the Π's. There are several methods that have been developed for this purpose, but the most popular (and simplest) method is the **method of repeating variables,** popularized by Edgar Buckingham (1867–1940). The method was first published by the Russian scientist Dimitri Riabouchinsky (1882–1962) in 1911. We can think of this method as a step-by-step procedure or "recipe" for obtaining nondimensional parameters. There are six steps, listed concisely in Fig. 7–22, and in more detail in Table 7–2. These steps are explained in further detail as we work through a number of example problems.

As with most new procedures, the best way to learn is by example and practice. As a simple first example, consider a ball falling in a vacuum as discussed in Section 7–2. Let us pretend that we do not know that Eq. 7–4 is appropriate for this problem, nor do we know much physics concerning falling objects. In fact, suppose that all we know is that the instantaneous

FIGURE 7–21
Similarity can be achieved even when the model fluid is different than the prototype fluid. Here a submarine model is tested in a wind tunnel.
Courtesy NASA Langley Research Center.

DIMENSIONAL ANALYSIS AND MODELING

The Method of Repeating Variables

Step 1: List the parameters in the problem and count their total number n.

Step 2: List the primary dimensions of each of the n parameters.

Step 3: Set the *reduction* j as the number of primary dimensions. Calculate k, the expected number of Π's, $k = n - j$

Step 4: Choose j *repeating parameters*.

Step 5: Construct the k Π's, and manipulate as necessary.

Step 6: Write the final functional relationship and check your algebra.

FIGURE 7–22
A concise summary of the six steps that comprise the *method of repeating variables*.

TABLE 7–2

Detailed description of the six steps that comprise the *method of repeating variables*[*]

Step 1	List the parameters (dimensional variables, nondimensional variables, and dimensional constants) and count them. Let n be the total number of parameters in the problem, including the dependent variable. Make sure that any listed independent parameter is indeed independent of the others, i.e., it cannot be expressed in terms of them. (E.g., don't include radius r and area $A = \pi r^2$, since r and A are *not* independent.)
Step 2	List the primary dimensions for each of the n parameters.
Step 3	Guess the **reduction** j. As a first guess, set j equal to the number of primary dimensions represented in the problem. The expected number of Π's (k) is equal to n minus j, according to the **Buckingham Pi theorem,**

$$\text{The Buckingham Pi theorem:} \qquad k = n - j \qquad (7\text{–}14)$$

	If at this step or during any subsequent step, the analysis does not work out, verify that you have included enough parameters in step 1. Otherwise, go back and *reduce j by one* and try again.
Step 4	Choose j **repeating parameters** that will be used to construct each Π. Since the repeating parameters have the potential to appear in each Π, be sure to choose them *wisely* (Table 7–3).
Step 5	Generate the Π's one at a time by grouping the j repeating parameters with one of the remaining parameters, forcing the product to be dimensionless. In this way, construct all k Π's. By convention the first Π, designated as Π_1, is the *dependent* Π (the one on the left side of the list). Manipulate the Π's as necessary to achieve established dimensionless groups (Table 7–5).
Step 6	Check that all the Π's are indeed dimensionless. Write the final functional relationship in the form of Eq. 7–11.

[*] This is a step-by-step method for finding the dimensionless Π groups when performing a dimensional analysis.

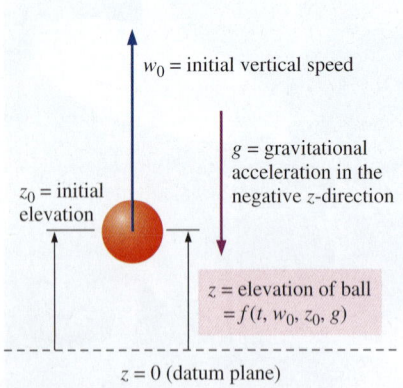

FIGURE 7–23
Setup for dimensional analysis of a ball falling in a vacuum. Elevation z is a function of time t, initial vertical speed w_0, initial elevation z_0, and gravitational constant g.

elevation z of the ball must be a function of time t, initial vertical speed w_0, initial elevation z_0, and gravitational constant g (Fig. 7–23). The beauty of dimensional analysis is that the only other thing we need to know is the primary dimensions of each of these quantities. As we go through each step of the method of repeating variables, we explain some of the subtleties of the technique in more detail using the falling ball as an example.

Step 1

There are five parameters (dimensional variables, nondimensional variables, and dimensional constants) in this problem; $n = 5$. They are listed in functional form, with the dependent variable listed as a function of the independent variables and constants:

List of relevant parameters: $\qquad z = f(t, w_0, z_0, g) \qquad n = 5$

Step 2
The primary dimensions of each parameter are listed here. We recommend writing each dimension with exponents since this helps with later algebra.

z	t	w_0	z_0	g
$\{L^1\}$	$\{t^1\}$	$\{L^1 t^{-1}\}$	$\{L^1\}$	$\{L^1 t^{-2}\}$

Step 3
As a first guess, j is set equal to 2, the number of primary dimensions represented in the problem (L and t).

Reduction: $\qquad j = 2$

If this value of j is correct, the number of Π's predicted by the Buckingham Pi theorem is

Number of expected Π's: $\qquad k = n - j = 5 - 2 = 3$

Step 4
We need to choose two repeating parameters since $j = 2$. Since this is often the hardest (or at least the most mysterious) part of the method of repeating variables, several guidelines about choosing repeating parameters are listed in Table 7–3.

Following the guidelines of Table 7–3 on the next page, the wisest choice of two repeating parameters is w_0 and z_0.

Repeating parameters: $\qquad w_0$ and z_0

Step 5
Now we combine these repeating parameters into products with each of the remaining parameters, one at a time, to create the Π's. The first Π is always the *dependent* Π and is formed with the dependent variable z.

Dependent Π: $\qquad \Pi_1 = z w_0^{a_1} z_0^{b_1} \qquad$ (7–15)

where a_1 and b_1 are constant exponents that need to be determined. We apply the primary dimensions of step 2 into Eq. 7–15 and *force* the Π to be dimensionless by setting the exponent of each primary dimension to zero:

Dimensions of Π_1: $\qquad \{\Pi_1\} = \{L^0 t^0\} = \{z w_0^{a_1} z_0^{b_1}\} = \{L^1 (L^1 t^{-1})^{a_1} L^{b_1}\}$

Since primary dimensions are by definition independent of each other, we equate the exponents of each primary dimension independently to solve for exponents a_1 and b_1 (Fig. 7–24).

Time: $\qquad \{t^0\} = \{t^{-a_1}\} \qquad 0 = -a_1 \qquad a_1 = 0$

Length: $\quad \{L^0\} = \{L^1 L^{a_1} L^{b_1}\} \quad 0 = 1 + a_1 + b_1 \quad b_1 = -1 - a_1 \quad b_1 = -1$

Equation 7–15 thus becomes

$$\Pi_1 = \frac{z}{z_0} \qquad (7\text{–}16)$$

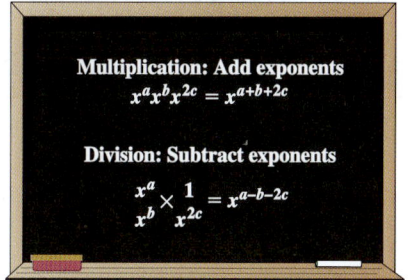

FIGURE 7–24
The mathematical rules for adding and subtracting exponents during multiplication and division, respectively.

TABLE 7–3

Guidelines for choosing *repeating parameters* in step 4 of the method of repeating variables[*]

Guideline	Comments and Application to Present Problem
1. Never pick the *dependent* variable. Otherwise, it may appear in all the Π's, which is undesirable.	In the present problem we cannot choose z, but we must choose from among the remaining four parameters. Therefore, we must choose two of the following parameters: t, w_0, z_0, and g.
2. The chosen repeating parameters must not *by themselves* be able to form a dimensionless group. Otherwise, it would be impossible to generate the rest of the Π's.	In the present problem, any two of the independent parameters would be valid according to this guideline. For illustrative purposes, however, suppose we have to pick three instead of two repeating parameters. We could not, for example, choose t, w_0, and z_0, because these can form a Π all by themselves (tw_0/z_0).
3. The chosen repeating parameters must represent *all* the primary dimensions in the problem.	Suppose for example that there were *three* primary dimensions (m, L, and t) and *two* repeating parameters were to be chosen. You could not choose, say, a length and a time, since primary dimension mass would not be represented in the dimensions of the repeating parameters. An appropriate choice would be a density and a time, which together represent all three primary dimensions in the problem.
4. Never pick parameters that are already dimensionless. These are Π's already, all by themselves.	Suppose an angle θ were one of the independent parameters. We could not choose θ as a repeating parameter since angles have no dimensions (radian and degree are dimensionless units). In such a case, one of the Π's is already known, namely θ.
5. Never pick two parameters with the *same* dimensions or with dimensions that differ by only an exponent.	In the present problem, two of the parameters, z and z_0, have the same dimensions (length). We cannot choose both of these parameters. (Note that dependent variable z has already been eliminated by guideline 1.) Suppose one parameter has dimensions of length and another parameter has dimensions of volume. In dimensional analysis, volume contains only one primary dimension (length) and *is not dimensionally distinct from length*—we cannot choose both of these parameters.
6. Whenever possible, choose dimensional constants over dimensional variables so that only *one* Π contains the dimensional variable.	If we choose time t as a repeating parameter in the present problem, it would appear in all three Π's. While this would not be *wrong*, it would not be *wise* since we know that ultimately we want some nondimensional height as a function of some nondimensional time and other nondimensional parameter(s). From the original four independent parameters, this restricts us to w_0, z_0, and g.
7. Pick common parameters since they may appear in each of the Π's.	In fluid flow problems we generally pick a length, a velocity, and a mass or density (Fig. 7–25). It is unwise to pick less common parameters like viscosity μ or surface tension σ_s, since we would in general not want μ or σ_s to appear in each of the Π's. In the present problem, w_0 and z_0 are wiser choices than g.
8. Pick simple parameters over complex parameters whenever possible.	It is better to pick parameters with only one or two basic dimensions (e.g., a length, a time, a mass, or a velocity) instead of parameters that are composed of several basic dimensions (e.g., an energy or a pressure).

[*] These guidelines, while not infallible, help you to pick repeating parameters that usually lead to established nondimensional Π groups with minimal effort.

In similar fashion we create the first independent Π (Π_2) by combining the repeating parameters with independent variable t.

First independent Π: $\qquad \Pi_2 = tw_0^{a_2}z_0^{b_2}$

Dimensions of Π_2: $\qquad \{\Pi_2\} = \{L^0 t^0\} = \{tw_0^{a_2}z_0^{b_2}\} = \{t(L^1 t^{-1})^{a_2} L^{b_2}\}$

Equating exponents,

Time: $\quad\{t^0\} = \{t^1 t^{-a_2}\} \quad 0 = 1 - a_2 \quad a_2 = 1$

Length: $\quad\{L^0\} = \{L^{a_2} L^{b_2}\} \quad 0 = a_2 + b_2 \quad b_2 = -a_2 \quad b_2 = -1$

Π_2 is thus

$$\Pi_2 = \frac{w_0 t}{z_0} \qquad (7\text{-}17)$$

Finally we create the second independent Π (Π_3) by combining the repeating parameters with g and *forcing* the Π to be dimensionless (Fig. 7–26).

Second independent Π: $\quad \Pi_3 = g w_0^{a_3} z_0^{b_3}$

Dimensions of Π_3: $\quad \{\Pi_3\} = \{L^0 t^0\} = \{g w_0^{a_3} z_0^{b_3}\} = \{L^1 t^{-2} (L^1 t^{-1})^{a_3} L^{b_3}\}$

Equating exponents,

Time: $\quad \{t^0\} = \{t^{-2} t^{-a_3}\} \quad 0 = -2 - a_3 \quad a_3 = -2$

Length: $\quad \{L^0\} = \{L^1 L^{a_3} L^{b_3}\} \quad 0 = 1 + a_3 + b_3 \quad b_3 = -1 - a_3 \quad b_3 = 1$

Π_3 is thus

$$\Pi_3 = \frac{g z_0}{w_0^2} \qquad (7\text{-}18)$$

All three Π's have been found, but at this point it is prudent to examine them to see if any manipulation is required. We see immediately that Π_1 and Π_2 are the same as the nondimensionalized variables z^* and t^* defined by Eq. 7–6—no manipulation is necessary for these. However, we recognize that the third Π must be raised to the power of $-\tfrac{1}{2}$ to be of the same form as an established dimensionless parameter, namely the Froude number of Eq. 7–8:

Modified Π_3: $\quad \Pi_{3,\text{modified}} = \left(\dfrac{g z_0}{w_0^2}\right)^{-1/2} = \dfrac{w_0}{\sqrt{g z_0}} = \text{Fr} \qquad (7\text{-}19)$

Such manipulation is often necessary to put the Π's into proper established form. The Π of Eq. 7–18 is not *wrong*, and there is certainly no mathematical advantage of Eq. 7–19 over Eq. 7–18. Instead, we like to say that Eq. 7–19 is more "socially acceptable" than Eq. 7–18, since it is a named, established nondimensional parameter that is commonly used in the literature. In Table 7–4 are listed some guidelines for manipulation of nondimensional Π groups into established nondimensional parameters.

Table 7–5 lists some established nondimensional parameters, most of which are named after a notable scientist or engineer (see Fig. 7–27 and the Historical Spotlight on p. 311). This list is by no means exhaustive. Whenever possible, you should manipulate your Π's as necessary in order to convert them into established nondimensional parameters.

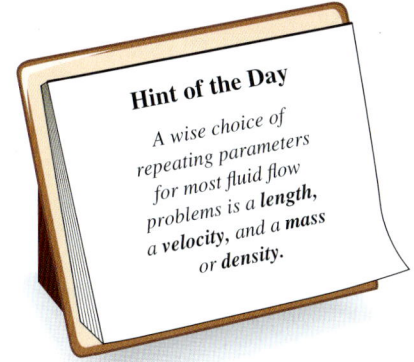

FIGURE 7–25
It is wise to choose *common* parameters as repeating parameters since they may appear in each of your dimensionless Π groups.

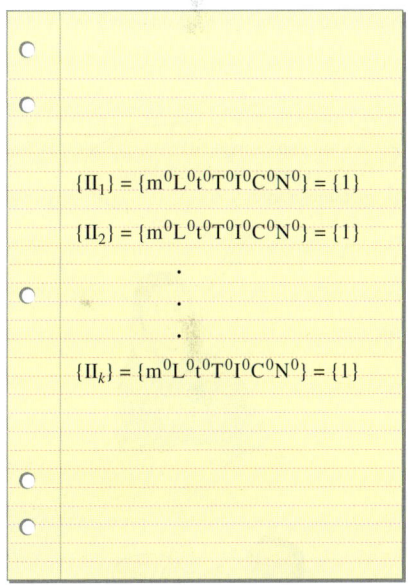

FIGURE 7–26
The Π groups that result from the method of repeating variables are guaranteed to be dimensionless because we *force* the overall exponent of all seven primary dimensions to be zero.

TABLE 7–4

Guidelines for manipulation of the Π's resulting from the method of repeating variables*

Guideline	Comments and Application to Present Problem
1. We may impose a constant (dimensionless) exponent on a Π or perform a functional operation on a Π.	We can raise a Π to any exponent n (changing it to Π^n) without changing the dimensionless stature of the Π. For example, in the present problem, we imposed an exponent of $-1/2$ on Π_3. Similarly we can perform the functional operation $\sin(\Pi)$, $\exp(\Pi)$, etc., without influencing the dimensions of the Π.
2. We may multiply a Π by a pure (dimensionless) constant.	Sometimes dimensionless factors of π, 1/2, 2, 4, etc., are included in a Π for convenience. This is perfectly okay since such factors do not influence the dimensions of the Π.
3. We may form a product (or quotient) of any Π with any other Π in the problem to replace one of the Π's.	We could replace Π_3 by $\Pi_3\Pi_1$, Π_3/Π_2, etc. Sometimes such manipulation is necessary to convert our Π into an established Π. In many cases, the established Π would have been produced if we would have chosen different repeating parameters.
4. We may use any of guidelines 1 to 3 in combination.	In general, we can replace any Π with some new Π such as $A\Pi_3^B \sin(\Pi_1^C)$, where A, B, and C are pure constants.
5. We may substitute a dimensional parameter in the Π with other parameter(s) of the same dimensions.	For example, the Π may contain the square of a length or the cube of a length, for which we may substitute a known area or volume, respectively, in order to make the Π agree with established conventions.

*These guidelines are useful in step 5 of the method of repeating variables and are listed to help you convert your nondimensional Π groups into standard, *established* nondimensional parameters, many of which are listed in Table 7–5.

FIGURE 7–27
Established nondimensional parameters are usually named after a notable scientist or engineer.

Step 6

We should double-check that the Π's are indeed dimensionless (Fig. 7–28). You can verify this on your own for the present example. We are finally ready to write the functional relationship between the nondimensional parameters. Combining Eqs. 7–16, 7–17, and 7–19 into the form of Eq. 7–11,

Relationship between Π's: $\quad \Pi_1 = f(\Pi_2, \Pi_3) \quad \rightarrow \quad \dfrac{z}{z_0} = f\left(\dfrac{w_0 t}{z_0}, \dfrac{w_0}{\sqrt{gz_0}}\right)$

Or, in terms of the nondimensional variables z^* and t^* defined previously by Eq. 7–6 and the definition of the Froude number,

Final result of dimensional analysis: $\quad z^* = f(t^*, \text{Fr}) \quad$ (7–20)

It is useful to compare the result of dimensional analysis, Eq. 7–20, to the exact analytical result, Eq. 7–10. The method of repeating variables properly predicts the functional relationship between dimensionless groups. However,

> The method of repeating variables cannot predict the exact mathematical form of the equation.

This is a fundamental limitation of dimensional analysis and the method of repeating variables. For some simple problems, however, the form of the equation *can* be predicted to within an unknown constant, as is illustrated in Example 7–7.

TABLE 7–5

Some common established nondimensional parameters or Π's encountered in fluid mechanics and heat transfer*

Name	Definition	Ratio of Significance
Archimedes number	$Ar = \dfrac{\rho_s g L^3}{\mu^2}(\rho_s - \rho)$	$\dfrac{\text{Gravitational force}}{\text{Viscous force}}$
Aspect ratio	$AR = \dfrac{L}{W}$ or $\dfrac{L}{D}$	$\dfrac{\text{Length}}{\text{Width}}$ or $\dfrac{\text{Length}}{\text{Diameter}}$
Biot number	$Bi = \dfrac{hL}{k}$	$\dfrac{\text{Surface thermal resistance}}{\text{Internal thermal resistance}}$
Bond number	$Bo = \dfrac{g(\rho_f - \rho_v)L^2}{\sigma_s}$	$\dfrac{\text{Gravitational force}}{\text{Surface tension force}}$
Cavitation number	$Ca \text{ (sometimes } \sigma_c) = \dfrac{P - P_v}{\rho V^2}$ $\left(\text{sometimes } \dfrac{2(P - P_v)}{\rho V^2}\right)$	$\dfrac{\text{Pressure} - \text{Vapor pressure}}{\text{Inertial pressure}}$
Darcy friction factor	$f = \dfrac{8\tau_w}{\rho V^2}$	$\dfrac{\text{Wall friction force}}{\text{Inertial force}}$
Drag coefficient	$C_D = \dfrac{F_D}{\frac{1}{2}\rho V^2 A}$	$\dfrac{\text{Drag force}}{\text{Dynamic force}}$
Eckert number	$Ec = \dfrac{V^2}{c_p T}$	$\dfrac{\text{Kinetic energy}}{\text{Enthalpy}}$
Euler number	$Eu = \dfrac{\Delta P}{\rho V^2}$ $\left(\text{sometimes } \dfrac{\Delta P}{\frac{1}{2}\rho V^2}\right)$	$\dfrac{\text{Pressure difference}}{\text{Dynamic pressure}}$
Fanning friction factor	$C_f = \dfrac{2\tau_w}{\rho V^2}$	$\dfrac{\text{Wall friction force}}{\text{Inertial force}}$
Fourier number	$Fo \text{ (sometimes } \tau) = \dfrac{\alpha t}{L^2}$	$\dfrac{\text{Physical time}}{\text{Thermal diffusion time}}$
Froude number	$Fr = \dfrac{V}{\sqrt{gL}}$ $\left(\text{sometimes } \dfrac{V^2}{gL}\right)$	$\dfrac{\text{Inertial force}}{\text{Gravitational force}}$
Grashof number	$Gr = \dfrac{g\beta|\Delta T|L^3\rho^2}{\mu^2}$	$\dfrac{\text{Buoyancy force}}{\text{Viscous force}}$
Jakob number	$Ja = \dfrac{c_p(T - T_{sat})}{h_{fg}}$	$\dfrac{\text{Sensible energy}}{\text{Latent energy}}$
Knudsen number	$Kn = \dfrac{\lambda}{L}$	$\dfrac{\text{Mean free path length}}{\text{Characteristic length}}$
Lewis number	$Le = \dfrac{k}{\rho c_p D_{AB}} = \dfrac{\alpha}{D_{AB}}$	$\dfrac{\text{Thermal diffusion}}{\text{Species diffusion}}$
Lift coefficient	$C_L = \dfrac{F_L}{\frac{1}{2}\rho V^2 A}$	$\dfrac{\text{Lift force}}{\text{Dynamic force}}$

(Continued)

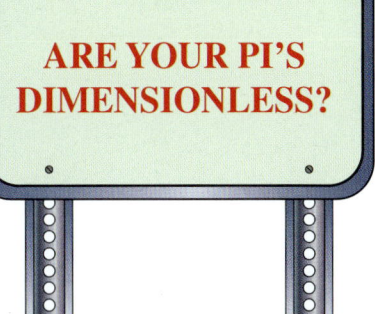

FIGURE 7–28
A quick check of your algebra is always wise.

DIMENSIONAL ANALYSIS AND MODELING

TABLE 7–5 (Continued)

Name	Definition	Ratio of Significance
Mach number	Ma (sometimes M) $= \dfrac{V}{c}$	$\dfrac{\text{Flow speed}}{\text{Speed of sound}}$
Nusselt number	$\text{Nu} = \dfrac{Lh}{k}$	$\dfrac{\text{Convection heat transfer}}{\text{Conduction heat transfer}}$
Peclet number	$\text{Pe} = \dfrac{\rho L V c_p}{k} = \dfrac{LV}{\alpha}$	$\dfrac{\text{Bulk heat transfer}}{\text{Conduction heat transfer}}$
Power number	$N_P = \dfrac{\dot{W}}{\rho D^5 \omega^3}$	$\dfrac{\text{Power}}{\text{Rotational inertia}}$
Prandtl number	$\text{Pr} = \dfrac{\nu}{\alpha} = \dfrac{\mu c_p}{k}$	$\dfrac{\text{Viscous diffusion}}{\text{Thermal diffusion}}$
Pressure coefficient	$C_p = \dfrac{P - P_\infty}{\frac{1}{2}\rho V^2}$	$\dfrac{\text{Static pressure difference}}{\text{Dynamic pressure}}$
Rayleigh number	$\text{Ra} = \dfrac{g\beta |\Delta T| L^3 \rho^2 c_p}{k\mu}$	$\dfrac{\text{Buoyancy force}}{\text{Viscous force}}$
Reynolds number	$\text{Re} = \dfrac{\rho V L}{\mu} = \dfrac{VL}{\nu}$	$\dfrac{\text{Inertial force}}{\text{Viscous force}}$
Richardson number	$\text{Ri} = \dfrac{L^5 g \Delta \rho}{\rho \dot{V}^2}$	$\dfrac{\text{Buoyancy force}}{\text{Inertial force}}$
Schmidt number	$\text{Sc} = \dfrac{\mu}{\rho D_{AB}} = \dfrac{\nu}{D_{AB}}$	$\dfrac{\text{Viscous diffusion}}{\text{Species diffusion}}$
Sherwood number	$\text{Sh} = \dfrac{VL}{D_{AB}}$	$\dfrac{\text{Overall mass diffusion}}{\text{Species diffusion}}$
Specific heat ratio	k (sometimes γ) $= \dfrac{c_p}{c_V}$	$\dfrac{\text{Enthalpy}}{\text{Internal energy}}$
Stanton number	$\text{St} = \dfrac{h}{\rho c_p V}$	$\dfrac{\text{Heat transfer}}{\text{Thermal capacity}}$
Stokes number	Stk (sometimes St) $= \dfrac{\rho_p D_p^2 V}{18 \mu L}$	$\dfrac{\text{Particle relaxation time}}{\text{Characteristic flow time}}$
Strouhal number	St (sometimes S or Sr) $= \dfrac{fL}{V}$	$\dfrac{\text{Characteristic flow time}}{\text{Period of oscillation}}$
Weber number	$\text{We} = \dfrac{\rho V^2 L}{\sigma_s}$	$\dfrac{\text{Inertial force}}{\text{Surface tension force}}$

* A is a characteristic area, D is a characteristic diameter, f is a characteristic frequency (Hz), L is a characteristic length, t is a characteristic time, T is a characteristic (absolute) temperature, V is a characteristic velocity, W is a characteristic width, $\dot{W}$ is a characteristic power, ω is a characteristic angular velocity (rad/s). Other parameters and fluid properties in these Π's include: c = speed of sound, c_p, c_v = specific heats, D_p = particle diameter, D_{AB} = species diffusion coefficient, h = convective heat transfer coefficient, h_{fg} = latent heat of evaporation, k = thermal conductivity, P = pressure, T_{sat} = saturation temperature, $\dot{V}$ = volume flow rate, α = thermal diffusivity, β = coefficient of thermal expansion, λ = mean free path length, μ = viscosity, ν = kinematic viscosity, ρ = fluid density, ρ_f = liquid density, ρ_p = particle density, ρ_s = solid density, ρ_v = vapor density, σ_s = surface tension, and τ_w = shear stress along a wall.

HISTORICAL SPOTLIGHT ■ Persons Honored by Nondimensional Parameters

Guest Author: Glenn Brown, Oklahoma State University

Commonly used, established dimensionless numbers have been given names for convenience, and to honor persons who have contributed in the development of science and engineering. In many cases, the namesake was not the first to define the number, but usually he/she used it or a similar parameter in his/her work. The following is a list of some, but not all, such persons. Also keep in mind that some numbers may have more than one name.

Archimedes (287–212 BC) Greek mathematician who defined buoyant forces.

Biot, Jean-Baptiste (1774–1862) French mathematician who did pioneering work in heat, electricity, and elasticity. He also helped measure the arc of the meridian as part of the metric system development.

Darcy, Henry P. G. (1803–1858) French engineer who performed extensive experiments on pipe flow and the first quantifiable filtration tests.

Eckert, Ernst R. G. (1904–2004) German–American engineer and student of Schmidt who did early work in boundary layer heat transfer.

Euler, Leonhard (1707–1783) Swiss mathematician and associate of Daniel Bernoulli who formulated equations of fluid motion and introduced the concept of centrifugal machinery.

Fanning, John T. (1837–1911) American engineer and textbook author who published in 1877 a modified form of Weisbach's equation with a table of resistance values computed from Darcy's data.

Fourier, Jean B. J. (1768–1830) French mathematician who did pioneering work in heat transfer and several other topics.

Froude, William (1810–1879) English engineer who developed naval modeling methods and the transfer of wave and boundary resistance from model to prototype.

Grashof, Franz (1826–1893) German engineer and educator known as a prolific author, editor, corrector, and dispatcher of publications.

Jakob, Max (1879–1955) German–American physicist, engineer, and textbook author who did pioneering work in heat transfer.

Knudsen, Martin (1871–1949) Danish physicist who helped develop the kinetic theory of gases.

Lewis, Warren K. (1882–1975) American engineer who researched distillation, extraction, and fluidized bed reactions.

Mach, Ernst (1838–1916) Austrian physicist who was first to realize that bodies traveling faster than the speed of sound would drastically alter the properties of the fluid. His ideas had great influence on twentieth-century thought, both in physics and in philosophy, and influenced Einstein's development of the theory of relativity.

Nusselt, Wilhelm (1882–1957) German engineer who was the first to apply similarity theory to heat transfer.

Peclet, Jean C. E. (1793–1857) French educator, physicist, and industrial researcher.

Prandtl, Ludwig (1875–1953) German engineer and developer of boundary layer theory who is considered the founder of modern fluid mechanics.

Lord Raleigh, John W. Strutt (1842–1919) English scientist who investigated dynamic similarity, cavitation, and bubble collapse.

Reynolds, Osborne (1842–1912) English engineer who investigated flow in pipes and developed viscous flow equations based on mean velocities.

Richardson, Lewis F. (1881–1953) English mathematician, physicist, and psychologist who was a pioneer in the application of fluid mechanics to the modeling of atmospheric turbulence.

Schmidt, Ernst (1892–1975) German scientist and pioneer in the field of heat and mass transfer. He was the first to measure the velocity and temperature field in a free convection boundary layer.

Sherwood, Thomas K. (1903–1976) American engineer and educator. He researched mass transfer and its interaction with flow, chemical reactions, and industrial process operations.

Stanton, Thomas E. (1865–1931) English engineer and student of Reynolds who contributed to a number of areas of fluid flow.

Stokes, George G. (1819–1903) Irish scientist who developed equations of viscous motion and diffusion.

Strouhal, Vincenz (1850–1922) Czech physicist who showed that the period of oscillations shed by a wire are related to the velocity of the air passing over it.

Weber, Moritz (1871–1951) German professor who applied similarity analysis to capillary flows.

312
DIMENSIONAL ANALYSIS AND MODELING

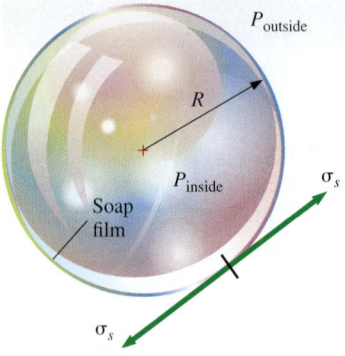

FIGURE 7–29
The pressure inside a soap bubble is greater than that surrounding the soap bubble due to surface tension in the soap film.

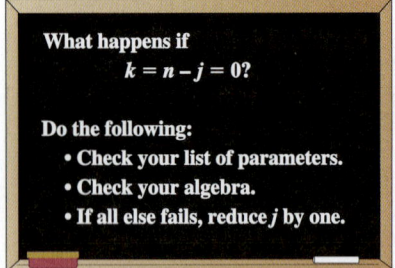

FIGURE 7–30
If the method of repeating variables indicates zero Π's, we have either made an error, or we need to reduce j by one and start over.

EXAMPLE 7–7 Pressure in a Soap Bubble

Some children are playing with soap bubbles, and you become curious as to the relationship between soap bubble radius and the pressure inside the soap bubble (Fig. 7–29). You reason that the pressure inside the soap bubble must be greater than atmospheric pressure, and that the shell of the soap bubble is under tension, much like the skin of a balloon. You also know that the property surface tension must be important in this problem. Not knowing any other physics, you decide to approach the problem using dimensional analysis. Establish a relationship between pressure difference $\Delta P = P_{inside} - P_{outside}$, soap bubble radius R, and the surface tension σ_s of the soap film.

SOLUTION The pressure difference between the inside of a soap bubble and the outside air is to be analyzed by the method of repeating variables.
Assumptions **1** The soap bubble is neutrally buoyant in the air, and gravity is not relevant. **2** No other variables or constants are important in this problem.
Analysis The step-by-step method of repeating variables is employed.

Step 1 There are three variables and constants in this problem; $n = 3$. They are listed in functional form, with the dependent variable listed as a function of the independent variables and constants:

List of relevant parameters: $\quad\quad \Delta P = f(R, \sigma_s) \quad\quad n = 3$

Step 2 The primary dimensions of each parameter are listed. The dimensions of surface tension are obtained from Example 7–1, and those of pressure from Example 7–2.

$$\begin{array}{ccc} \Delta P & R & \sigma_s \\ \{m^1 L^{-1} t^{-2}\} & \{L^1\} & \{m^1 t^{-2}\} \end{array}$$

Step 3 As a first guess, j is set equal to 3, the number of primary dimensions represented in the problem (m, L, and t).

Reduction (first guess): $\quad\quad j = 3$

If this value of j is correct, the expected number of Π's is $k = n - j = 3 - 3 = 0$. But how can we have zero Π's? Something is obviously not right (Fig. 7–30). At times like this, we need to first go back and make sure that we are not neglecting some important variable or constant in the problem. Since we are confident that the pressure difference should depend only on soap bubble radius and surface tension, we reduce the value of j by one,

Reduction (second guess): $\quad\quad j = 2$

If this value of j is correct, $k = n - j = 3 - 2 = 1$. Thus we expect *one* Π, which is more physically realistic than zero Π's.

Step 4 We need to choose two repeating parameters since $j = 2$. Following the guidelines of Table 7–3, our only choices are R and σ_s, since ΔP is the dependent variable.

Step 5 We combine these repeating parameters into a product with the dependent variable ΔP to create the dependent Π,

Dependent Π: $\quad\quad \Pi_1 = \Delta P R^{a_1} \sigma_s^{b_1}$ $\quad\quad$ (1)

We apply the primary dimensions of step 2 into Eq. 1 and force the Π to be dimensionless.

Dimensions of Π_1:

$$\{\Pi_1\} = \{m^0 L^0 t^0\} = \{\Delta P R^{a_1} \sigma_s^{b_1}\} = \{(m^1 L^{-1} t^{-2}) L^{a_1} (m^1 t^{-2})^{b_1}\}$$

We equate the exponents of each primary dimension to solve for a_1 and b_1:

Time: $\{t^0\} = \{t^{-2} t^{-2b_1}\}$ $0 = -2 - 2b_1$ $b_1 = -1$

Mass: $\{m^0\} = \{m^1 m^{b_1}\}$ $0 = 1 + b_1$ $b_1 = -1$

Length: $\{L^0\} = \{L^{-1} L^{a_1}\}$ $0 = -1 + a_1$ $a_1 = 1$

Fortunately, the first two results agree with each other, and Eq. 1 thus becomes

$$\Pi_1 = \frac{\Delta P R}{\sigma_s} \qquad (2)$$

From Table 7–5, the established nondimensional parameter most similar to Eq. 2 is the **Weber number,** defined as a pressure (ρV^2) times a length divided by surface tension. There is no need to further manipulate this Π.

Step 6 We write the final functional relationship. In the case at hand, there is only one Π, which is a function of *nothing*. This is possible only if the Π is constant. Putting Eq. 2 into the functional form of Eq. 7–11,

Relationship between Π's:

$$\Pi_1 = \frac{\Delta P R}{\sigma_s} = f(\text{nothing}) = \text{constant} \quad \rightarrow \quad \Delta P = \text{constant} \, \frac{\sigma_s}{R} \qquad (3)$$

Discussion This is an example of how we can sometimes predict *trends* with dimensional analysis, even without knowing much of the physics of the problem. For example, we know from our result that if the radius of the soap bubble doubles, the pressure difference decreases by a factor of 2. Similarly, if the value of surface tension doubles, ΔP increases by a factor of 2. Dimensional analysis cannot predict the value of the constant in Eq. 3; further analysis (or *one* experiment) reveals that the constant is equal to 4 (Chap. 2).

EXAMPLE 7–8 Lift on a Wing

Some aeronautical engineers are designing an airplane and wish to predict the lift produced by their new wing design (Fig. 7–31). The chord length L_c of the wing is 1.12 m, and its **planform area** A (area viewed from the top when the wing is at zero angle of attack) is 10.7 m². The prototype is to fly at $V = 52.0$ m/s close to the ground where $T = 25°C$. They build a one-tenth scale model of the wing to test in a pressurized wind tunnel. The wind tunnel can be pressurized to a maximum of 5 atm. At what speed and pressure should they run the wind tunnel in order to achieve dynamic similarity?

SOLUTION We are to determine the speed and pressure at which to run the wind tunnel in order to achieve dynamic similarity.

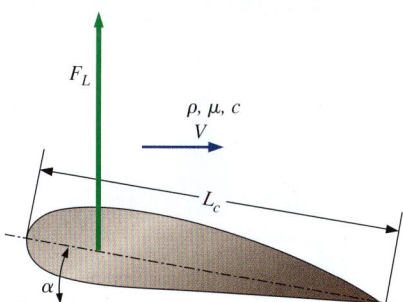

FIGURE 7–31

Lift F_L on a wing of chord length L_c at angle of attack α in a flow of free-stream speed V with density ρ, viscosity μ, and speed of sound c. The angle of attack α is measured relative to the free-stream flow direction.

DIMENSIONAL ANALYSIS AND MODELING

Assumptions **1** The prototype wing flies through the air at standard atmospheric pressure. **2** The model is geometrically similar to the prototype.

Analysis First, the step-by-step method of repeating variables is employed to obtain the nondimensional parameters. Then, the dependent Π's are matched between prototype and model.

Step 1 There are seven parameters (variables and constants) in this problem; $n = 7$. They are listed in functional form, with the dependent variable listed as a function of the independent parameters:

List of relevant parameters: $\quad F_L = f(V, L_c, \rho, \mu, c, \alpha) \quad n = 7$

where F_L is the lift force on the wing, V is the fluid speed, L_c is the chord length, ρ is the fluid density, μ is the fluid viscosity, c is the speed of sound in the fluid, and α is the angle of attack of the wing.

Step 2 The primary dimensions of each parameter are listed; angle α is dimensionless:

$$
\begin{array}{ccccccc}
F_L & V & L_c & \rho & \mu & c & \alpha \\
\{m^1 L^1 t^{-2}\} & \{L^1 t^{-1}\} & \{L^1\} & \{m^1 L^{-3}\} & \{m^1 L^{-1} t^{-1}\} & \{L^1 t^{-1}\} & \{1\}
\end{array}
$$

Step 3 As a first guess, j is set equal to 3, the number of primary dimensions represented in the problem (m, L, and t).

Reduction: $\quad\quad\quad\quad\quad\quad\quad\quad\quad j = 3$

If this value of j is correct, the expected number of Π's is $k = n - j = 7 - 3 = 4$.

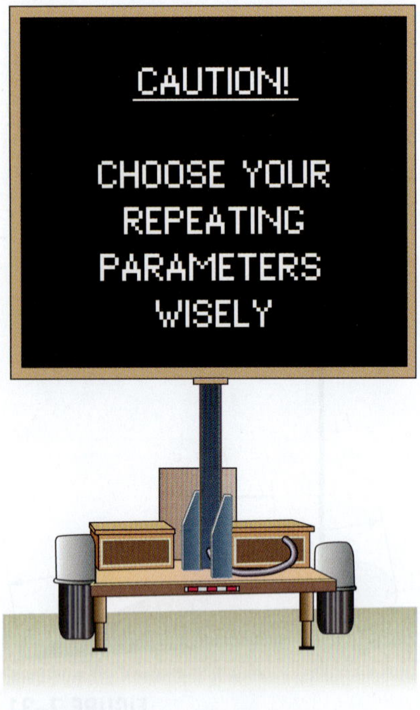

Step 4 We need to choose three repeating parameters since $j = 3$. Following the guidelines listed in Table 7–3, we cannot pick the dependent variable F_L. Nor can we pick α since it is already dimensionless. We cannot choose both V and c since their dimensions are identical. It would not be desirable to have μ appear in all the Π's. The best choice of repeating parameters is thus either V, L_c, and ρ or c, L_c, and ρ. Of these, the former is the better choice since the speed of sound appears in only one of the established nondimensional parameters of Table 7–5, whereas the velocity scale is more "common" and appears in several of the parameters (Fig. 7–32).

Repeating parameters: $\quad\quad V, L_c,$ and ρ

Step 5 The dependent Π is generated:

$$\Pi_1 = F_L V^{a_1} L_c^{b_1} \rho^{c_1} \rightarrow \{\Pi_1\} = \{(m^1 L^1 t^{-2})(L^1 t^{-1})^{a_1}(L^1)^{b_1}(m^1 L^{-3})^{c_1}\}$$

The exponents are calculated by forcing the Π to be dimensionless (algebra not shown). We get $a_1 = -2$, $b_1 = -2$, and $c_1 = -1$. The dependent Π is thus

$$\Pi_1 = \frac{F_L}{\rho V^2 L_c^2}$$

From Table 7–5, the established nondimensional parameter most similar to our Π_1 is the **lift coefficient**, defined in terms of planform area A rather than the square of chord length, and with a factor of 1/2 in the denominator. Thus, we may manipulate this Π according to the guidelines listed in Table 7–4 as follows:

Modified Π_1: $\quad\quad \Pi_{1,\text{modified}} = \dfrac{F_L}{\frac{1}{2}\rho V^2 A} =$ Lift coefficient $= C_L$

FIGURE 7–32
Oftentimes when performing the method of repeating variables, the most difficult part of the procedure is choosing the repeating parameters. With practice, however, you will learn to choose these parameters wisely.

Similarly, the first independent Π is generated:

$$\Pi_2 = \mu V^{a_2} L_c^{b_2} \rho^{c_2} \quad \rightarrow \quad \{\Pi_2\} = \{(m^1 L^{-1} t^{-1})(L^1 t^{-1})^{a_2}(L^1)^{b_2}(m^1 L^{-3})^{c_2}\}$$

from which $a_2 = -1$, $b_2 = -1$, and $c_2 = -1$, and thus

$$\Pi_2 = \frac{\mu}{\rho V L_c}$$

We recognize this Π as the inverse of the Reynolds number. So, after inverting,

Modified Π_2:
$$\Pi_{2,\,\text{modified}} = \frac{\rho V L_c}{\mu} = \text{Reynolds number} = \text{Re}$$

The third Π is formed with the speed of sound, the details of which are left for you to generate on your own. The result is

$$\Pi_3 = \frac{V}{c} = \text{Mach number} = \text{Ma}$$

Finally, since the angle of attack α is already dimensionless, it is a dimensionless Π group all by itself (Fig. 7–33). You are invited to go through the algebra; you will find that all the exponents turn out to be zero, and thus

$$\Pi_4 = \alpha = \text{Angle of attack}$$

Step 6 We write the final functional relationship as

$$C_L = \frac{F_L}{\frac{1}{2}\rho V^2 A} = f(\text{Re}, \text{Ma}, \alpha) \tag{1}$$

To achieve dynamic similarity, Eq. 7–12 requires that all three of the dependent nondimensional parameters in Eq. 1 match between the model and the prototype. While it is trivial to match the angle of attack, it is not so simple to simultaneously match the Reynolds number and the Mach number. For example, if the wind tunnel were run at the same temperature and pressure as those of the prototype, such that ρ, μ, and c of the air flowing over the model were the same as ρ, μ, and c of the air flowing over the prototype, Reynolds number similarity would be achieved by setting the wind tunnel air speed to 10 times that of the prototype (since the model is one-tenth scale). But then the Mach numbers would differ by a factor of 10. At 25°C, c is approximately 346 m/s, and the Mach number of the prototype airplane wing is $\text{Ma}_p = 52.0/346 = 0.150$—subsonic. At the required wind tunnel speed, Ma_m would be 1.50—supersonic! This is clearly unacceptable since the physics of the flow changes dramatically from subsonic to supersonic conditions. At the other extreme, if we were to match Mach numbers, the Reynolds number of the model would be 10 times too small.

What should we do? A common rule of thumb is that for Mach numbers less than about 0.3, as is the fortunate case here, compressibility effects are practically negligible. Thus, it is not necessary to exactly match the Mach number; rather, as long as Ma_m is kept below about 0.3, approximate dynamic similarity can be achieved by matching the Reynolds number. Now the problem shifts to one of how to match Re while maintaining a low Mach number. This is where the pressurization feature of the wind tunnel comes in. At constant temperature, density is proportional to pressure, while viscosity and speed of sound are very weak functions of pressure. If the wind tunnel pressure could be pumped to 10 atm, we could run the model test at the

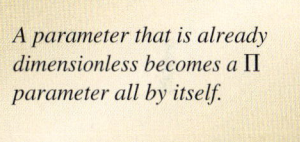

A parameter that is already dimensionless becomes a Π parameter all by itself.

FIGURE 7–33

A parameter that is dimensionless (like an angle) is already a nondimensional Π all by itself— we know this Π without doing any further algebra.

same speed as the prototype and achieve a nearly perfect match in both Re and Ma. However, at the maximum wind tunnel pressure of 5 atm, the required wind tunnel speed would be twice that of the prototype, or 104 m/s. The Mach number of the wind tunnel model would thus be Ma_m = 104/346 = 0.301—approximately at the limit of incompressibility according to our rule of thumb. In summary, the wind tunnel should be run at approximately **100 m/s, 5 atm**, and **25°C**.

Discussion This example illustrates one of the (frustrating) limitations of dimensional analysis; namely, *You may not always be able to match all the dependent* Π'*s simultaneously in a model test.* Compromises must be made in which only the most important Π's are matched. In many practical situations in fluid mechanics, the Reynolds number is not critical for dynamic similarity, provided that Re is high enough. If the Mach number of the prototype were significantly larger than about 0.3, we would be wise to precisely match the Mach number rather than the Reynolds number in order to ensure reasonable results. Furthermore, if a different gas were used to test the model, we would also need to match the specific heat ratio (*k*), since compressible flow behavior is strongly dependent on *k* (Chap. 12). We discuss such model testing problems in more detail in Section 7–5.

Recall that in Examples 7–5 and 7–6 the air speed of the prototype car is 80.0 km/h, and that of the wind tunnel is 354 km/h. At 25°C, this corresponds to a prototype Mach number of Ma_p = 0.065, and at 5°C, the Mach number of the wind tunnel is 0.29—on the borderline of the incompressible limit. In hindsight, we should have included the speed of sound in our dimensional analysis, which would have generated the Mach number as an additional Π. Another way to match the Reynolds number while keeping the Mach number low is to use a *liquid* such as water, since liquids are nearly incompressible, even at fairly high speeds.

EXAMPLE 7–9 Friction in a Pipe

Consider flow of an incompressible fluid of density ρ and viscosity μ through a long, horizontal section of round pipe of diameter *D*. The velocity profile is sketched in Fig. 7–34; *V* is the average speed across the pipe cross section, which by conservation of mass remains constant down the pipe. For a very long pipe, the flow eventually becomes hydrodynamically **fully developed,** which means that the velocity profile also remains uniform down the pipe. Because of frictional forces between the fluid and the pipe wall, there exists a shear stress τ_w on the inside pipe wall as sketched. The shear stress is also constant down the pipe in the fully developed region. We assume some constant average roughness height ε along the inside wall of the pipe. In fact, the only parameter that is *not* constant down the length of pipe is the pressure, which must decrease (linearly) down the pipe in order to "push" the fluid through the pipe to overcome friction. Develop a nondimensional relationship between shear stress τ_w and the other parameters in the problem.

SOLUTION We are to generate a nondimensional relationship between shear stress and other parameters.

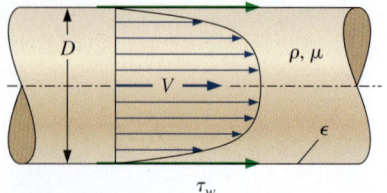

FIGURE 7–34
Friction on the inside wall of a pipe. The shear stress τ_w on the pipe walls is a function of average fluid speed *V*, average wall roughness height ε, fluid density ρ, fluid viscosity μ, and inside pipe diameter *D*.

Assumptions **1** The flow is hydrodynamically fully developed. **2** The fluid is incompressible. **3** No other parameters are significant in the problem.
Analysis The step-by-step method of repeating variables is employed to obtain the nondimensional parameters.

Step 1 There are six variables and constants in this problem; $n = 6$. They are listed in functional form, with the dependent variable listed as a function of the independent variables and constants:

List of relevant parameters: $\quad \tau_w = f(V, \varepsilon, \rho, \mu, D) \quad n = 6$

Step 2 The primary dimensions of each parameter are listed. Note that shear stress is a force per unit area, and thus has the same dimensions as pressure.

$$\begin{array}{cccccc} \tau_w & V & \varepsilon & \rho & \mu & D \\ \{m^1 L^{-1} t^{-2}\} & \{L^1 t^{-1}\} & \{L^1\} & \{m^1 L^{-3}\} & \{m^1 L^{-1} t^{-1}\} & \{L^1\} \end{array}$$

Step 3 As a first guess, j is set equal to 3, the number of primary dimensions represented in the problem (m, L, and t).

Reduction: $\quad\quad\quad\quad\quad\quad j = 3$

If this value of j is correct, the expected number of Π's is $k = n - j = 6 - 3 = 3$.

Step 4 We choose three repeating parameters since $j = 3$. Following the guidelines of Table 7–3, we cannot pick the dependent variable τ_w. We cannot choose both ε and D since their dimensions are identical, and it would not be desirable to have μ or ε appear in all the Π's. The best choice of repeating parameters is thus V, D, and ρ.

Repeating parameters: $\quad\quad V, D,$ and ρ

Step 5 The dependent Π is generated:

$$\Pi_1 = \tau_w V^{a_1} D^{b_1} \rho^{c_1} \quad\rightarrow\quad \{\Pi_1\} = \{(m^1 L^{-1} t^{-2})(L^1 t^{-1})^{a_1}(L^1)^{b_1}(m^1 L^{-3})^{c_1}\}$$

from which $a_1 = -2$, $b_1 = 0$, and $c_1 = -1$, and thus the dependent Π is

$$\Pi_1 = \frac{\tau_w}{\rho V^2}$$

From Table 7–5, the established nondimensional parameter most similar to this Π_1 is the **Darcy friction factor**, defined with a factor of 8 in the numerator (Fig. 7–35). Thus, we manipulate this Π according to the guidelines listed in Table 7–4 as follows:

Modified Π_1: $\quad\quad \Pi_{1,\text{modified}} = \dfrac{8\tau_w}{\rho V^2} =$ Darcy friction factor $= f$

Similarly, the two independent Π's are generated, the details of which are left for you to do on your own:

$$\Pi_2 = \mu V^{a_2} D^{b_2} \rho^{c_2} \quad\rightarrow\quad \Pi_2 = \frac{\rho V D}{\mu} = \text{Reynolds number} = \text{Re}$$

$$\Pi_3 = \varepsilon V^{a_3} D^{b_3} \rho^{c_3} \quad\rightarrow\quad \Pi_3 = \frac{\varepsilon}{D} = \text{Roughness ratio}$$

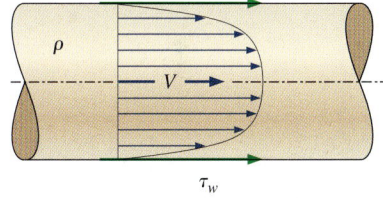

Darcy friction factor: $\quad f = \dfrac{8\tau_w}{\rho V^2}$

Fanning friction factor: $\quad C_f = \dfrac{2\tau_w}{\rho V^2}$

FIGURE 7–35
Although the *Darcy friction factor* for pipe flows is most common, you should be aware of an alternative, less common friction factor called the *Fanning friction factor*. The relationship between the two is $f = 4 C_f$.

Step 6 We write the final functional relationship as

$$f = \frac{8\tau_w}{\rho V^2} = f\left(Re, \frac{\varepsilon}{D}\right) \tag{1}$$

Discussion The result applies to both laminar and turbulent fully developed pipe flow; it turns out, however, that the second independent Π (roughness ratio ε/D) is not nearly as important in laminar pipe flow as in turbulent pipe flow. This problem presents an interesting connection between geometric similarity and dimensional analysis. Namely, it is necessary to match ε/D since it is an independent Π in the problem. From a different perspective, thinking of roughness as a geometric property, it is necessary to match ε/D to ensure *geometric similarity* between two pipes.

To verify the validity of Eq. 1 of Example 7–9, we use **computational fluid dynamics (CFD)** to predict the velocity profiles and the values of wall shear stress for two physically different but dynamically similar pipe flows:

- *Air* at 300 K flowing at an average speed of 4.42 m/s through a pipe of inner diameter 0.305 m and average roughness height 0.305 mm.
- *Water* at 300 K flowing at an average speed of 3.09 m/s through a pipe of inner diameter 0.0300 m and average roughness height 0.030 mm.

The two pipes are clearly geometrically similar since they are both round pipes. They have the same average roughness ratio ($\varepsilon/D = 0.0010$ in both cases). We have carefully chosen the values of average speed and diameter such that the two flows are also *dynamically* similar. Specifically, the other independent Π (the Reynolds number) also matches between the two flows.

$$Re_{air} = \frac{\rho_{air} V_{air} D_{air}}{\mu_{air}} = \frac{(1.225 \text{ kg/m}^3)(4.42 \text{ m/s})(0.305 \text{ m})}{1.789 \times 10^{-5} \text{ kg/m·s}} = 9.23 \times 10^4$$

where the fluid properties are those built into the CFD code, and

$$Re_{water} = \frac{\rho_{water} V_{water} D_{water}}{\mu_{water}} = \frac{(998.2 \text{ kg/m}^3)(3.09 \text{ m/s})(0.0300 \text{ m})}{0.001003 \text{ kg/m·s}} = 9.22 \times 10^4$$

Hence by Eq. 7–12, we expect that the *dependent* Π's should match between the two flows as well. We generate a computational mesh for each of the two flows, and use a commercial CFD code to generate the velocity profile, from which the shear stress is calculated. Fully developed, time-averaged, turbulent velocity profiles near the far end of both pipes are compared. Although the pipes are of different diameters and the fluids are vastly different, the velocity profile shapes look quite similar. In fact, when we plot *normalized* axial velocity (u/V) as a function of *normalized* radius (r/R), we find that the two profiles fall on top of each other (Fig. 7–36).

Wall shear stress is also calculated from the CFD results for each flow, a comparison of which is shown in Table 7–6. There are several reasons why the wall shear stress in the water pipe is orders of magnitude larger than that in the air pipe. Namely, water is over 800 times as dense as air and over 50 times as viscous. Furthermore, shear stress is proportional to the *gradient* of velocity, and the water pipe diameter is less than one-tenth that of the air

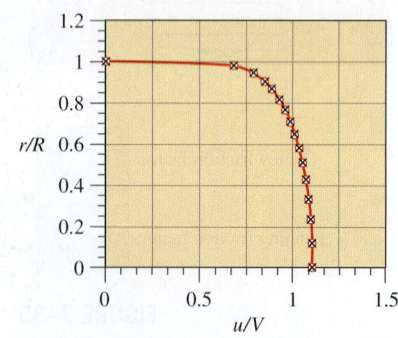

FIGURE 7–36
Normalized axial velocity profiles for fully developed flow through a pipe as predicted by CFD; profiles of air (circles) and water (crosses) are shown on the same plot.

TABLE 7–6

Comparison of wall shear stress and nondimensionalized wall shear stress for fully developed flow through an air pipe and a water pipe as predicted by CFD*

Parameter	Air Flow	Water Flow
Wall shear stress	$\tau_{w,\,air} = 0.0557$ N/m^2	$\tau_{w,\,water} = 22.2$ N/m^2
Dimensionless wall shear stress (Darcy friction factor)	$f_{air} = \dfrac{8\tau_{w,\,air}}{\rho_{air} V_{air}^2} = 0.0186$	$f_{water} = \dfrac{8\tau_{w,\,water}}{\rho_{water} V_{water}^2} = 0.0186$

* Data obtained with ANSYS-FLUENT using the standard k-ε turbulence model with wall functions.

pipe, leading to steeper velocity gradients. In terms of the *nondimensionalized* wall shear stress, f, however, Table 7–6 shows that the results are identical due to dynamic similarity between the two flows. Note that although the values are reported to three significant digits, the reliability of turbulence models in CFD is accurate to at most two significant digits (Chap. 15).

7–5 ■ EXPERIMENTAL TESTING, MODELING, AND INCOMPLETE SIMILARITY

One of the most useful applications of dimensional analysis is in designing physical and/or numerical experiments, and in reporting the results of such experiments. In this section we discuss both of these applications, and point out situations in which complete dynamic similarity is not achievable.

Setup of an Experiment and Correlation of Experimental Data

As a generic example, consider a problem in which there are five original parameters (one of which is the *dependent* parameter). A complete set of experiments (called a **full factorial** test matrix) is conducted by testing every possible combination of several levels of each of the four independent parameters. A full factorial test with five levels of each of the four independent parameters would require $5^4 = 625$ experiments. While experimental design techniques (**fractional factorial** test matrices; see Montgomery, 2013) can significantly reduce the size of the test matrix, the number of required experiments would still be large. However, assuming that three primary dimensions are represented in the problem, we can reduce the number of parameters from five to two ($k = 5 - 3 = 2$ nondimensional Π groups), and the number of *independent* parameters from four to one. Thus, for the same resolution (five tested levels of each independent parameter) we would then need to conduct a total of only $5^1 = 5$ experiments. You don't have to be a genius to realize that replacing 625 experiments by 5 experiments is cost effective. You can see why it is wise to perform a dimensional analysis *before* conducting an experiment.

Continuing our discussion of this generic example (a two-Π problem), once the experiments are complete, we plot the dependent dimensionless parameter (Π$_1$) as a function of the independent dimensionless parameter (Π$_2$), as in Fig. 7–37. We then determine the functional form of the relationship by

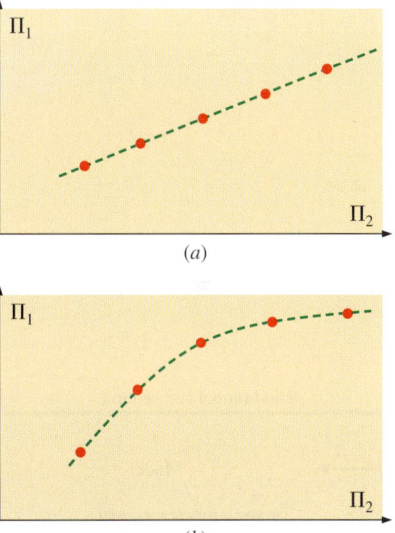

FIGURE 7–37
For a two-Π problem, we plot dependent dimensionless parameter (Π$_1$) as a function of independent dimensionless parameter (Π$_2$). The resulting plot can be (*a*) linear or (*b*) nonlinear. In either case, regression and curve-fitting techniques are available to determine the relationship between the Π's.

performing a **regression analysis** on the data. If we are lucky, the data may correlate linearly. If not, we can try linear regression on log–linear or log–log coordinates, polynomial curve fitting, etc., to establish an approximate relationship between the two Π's. See Holman (2001) for details about these curve-fitting techniques.

If there are more than two Π's in the problem (e.g., a three-Π problem or a four-Π problem), we need to set up a test matrix to determine the relationship between the dependent Π and the independent Π's. In many cases we discover that one or more of the dependent Π's has negligible effect and can be removed from the list of necessary dimensionless parameters.

As we have seen (Example 7–7), dimensional analysis sometimes yields only *one* Π. In a one-Π problem, we know the form of the relationship between the original parameters to within some unknown constant. In such a case, only *one* experiment is needed to determine that constant.

Incomplete Similarity

We have shown several examples in which the nondimensional Π groups are easily obtained with paper and pencil through straightforward use of the method of repeating variables. In fact, after sufficient practice, you should be able to obtain the Π's with ease—sometimes in your head or on the "back of an envelope." Unfortunately, it is often a much different story when we go to apply the results of our dimensional analysis to experimental data. The problem is that it is not always possible to match *all* the Π's of a model to the corresponding Π's of the prototype, even if we are careful to achieve geometric similarity. This situation is called **incomplete similarity**. Fortunately, in some cases of incomplete similarity, we are still able to extrapolate model test data to obtain reasonable full-scale predictions.

Wind Tunnel Testing

We illustrate incomplete similarity with the problem of measuring the aerodynamic drag force on a model truck in a wind tunnel (Fig. 7–38). Suppose we purchase a one-sixteenth scale die-cast model of a tractor-trailer rig (18-wheeler). The model is geometrically similar to the prototype—even in the details such as side mirrors, mud flaps, etc. The model truck is 0.991 m long, corresponding to a full-scale prototype length of 15.9 m. The model truck is to be tested in a wind tunnel that has a maximum speed of 70 m/s. The wind tunnel test section is 1.0 m tall and 1.2 m wide—big enough to accommodate the model without needing to worry about wall interference or blockage effects. The air in the wind tunnel is at the same temperature and pressure as the air flowing around the prototype. We want to simulate flow at $V_p = 96.5$ km/h (26.8 m/s) over the full-scale prototype truck.

The first thing we do is match the Reynolds numbers,

$$\text{Re}_m = \frac{\rho_m V_m L_m}{\mu_m} = \text{Re}_p = \frac{\rho_p V_p L_p}{\mu_p}$$

which can be solved for the required wind tunnel speed for the model tests V_m,

$$V_m = V_p \left(\frac{\mu_m}{\mu_p}\right)\left(\frac{\rho_p}{\rho_m}\right)\left(\frac{L_p}{L_m}\right) = (26.8 \text{ m/s})(1)(1)\left(\frac{16}{1}\right) = 429 \text{ m/s}$$

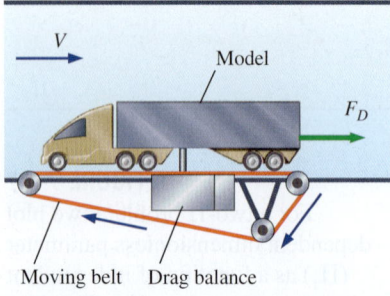

FIGURE 7–38
Measurement of aerodynamic drag on a model truck in a wind tunnel equipped with a *drag balance* and a *moving belt* ground plane.

Thus, to match the Reynolds number between model and prototype, the wind tunnel should be run at 429 m/s (to three significant digits). We obviously have a problem here, since this speed is more than six times greater than the maximum achievable wind tunnel speed. Moreover, even if we *could* run the wind tunnel that fast, the flow would be *supersonic*, since the speed of sound in air at room temperature is about 346 m/s. While the Mach number of the prototype truck moving through the air is 26.8/335 = 0.080, that of the wind tunnel air moving over the model would be 429/335 = 1.28 (if the wind tunnel could go that fast).

It is clearly not possible to match the model Reynolds number to that of the prototype with this model and wind tunnel facility. What do we do? There are several options:

- If we had a bigger wind tunnel, we could test with a larger model. Automobile manufacturers typically test with three-eighths scale model cars and with one-eighth scale model trucks and buses in very large wind tunnels. Some wind tunnels are even large enough for full-scale automobile tests (Fig. 7–39a). As you can imagine, however, the bigger the wind tunnel and the model the more expensive the tests. We must also be careful that the model is not too big for the wind tunnel. A useful rule of thumb is that the **blockage** (ratio of the model frontal area to the cross-sectional area of the test section) should be less than 7.5 percent. Otherwise, the wind tunnel walls adversely affect both geometric and kinematic similarity.

- We could use a different fluid for the model tests. For example, water tunnels can achieve higher Reynolds numbers than can wind tunnels of the same size, but they are much more expensive to build and operate (Fig. 7–39b).

- We could pressurize the wind tunnel and/or adjust the air temperature to increase the maximum Reynolds number capability. While these techniques can help, the increase in the Reynolds number is limited.

- If all else fails, we could run the wind tunnel at several speeds near the maximum speed, and then extrapolate our results to the full-scale Reynolds number.

Fortunately, it turns out that for many wind tunnel tests the last option is quite viable. While drag coefficient C_D is a strong function of the Reynolds number at low values of Re, C_D often levels off for Re above some value. In other words, for flow over many objects, especially "bluff" objects like trucks, buildings, etc., the flow is **Reynolds number independent** above some threshold value of Re (Fig. 7–40), typically when the boundary layer and the wake are both fully turbulent.

■ **EXAMPLE 7–10** Model Truck Wind Tunnel Measurements

A one-sixteenth scale model tractor-trailer truck (18-wheeler) is tested in a wind tunnel as sketched in Fig. 7–38. The model truck is 0.991 m long, 0.257 m tall, and 0.159 m wide. During the tests, the moving ground belt speed is adjusted so as to always match the speed of the air moving through the test section. Aerodynamic drag force F_D is measured as a function of

(a)

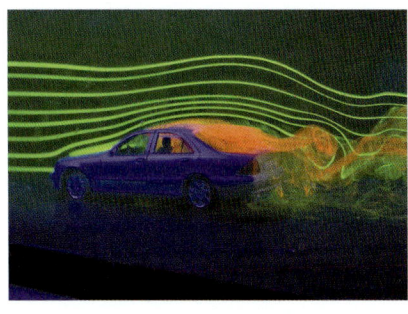

(b)

FIGURE 7–39
(a) The Langley full-scale wind tunnel (LFST) is large enough that full-scale vehicles can be tested. (b) For the same scale model and speed, water tunnels achieve higher Reynolds numbers than wind tunnels.
(b) NASA/Eric James

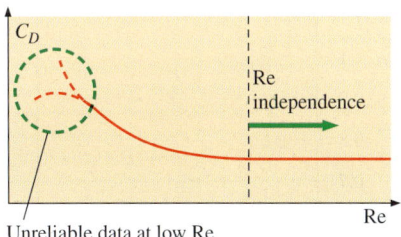

FIGURE 7–40
For many objects, the drag coefficient levels off at Reynolds numbers above some threshold value. This fortunate situation is called *Reynolds number independence*. It enables us to extrapolate to prototype Reynolds numbers that are outside of the range of our experimental facility.

DIMENSIONAL ANALYSIS AND MODELING

TABLE 7–7

Wind tunnel data: aerodynamic drag force on a model truck as a function of wind tunnel speed

V, m/s	F_D, N
20	12.4
25	19.0
30	22.1
35	29.0
40	34.3
45	39.9
50	47.2
55	55.5
60	66.0
65	77.6
70	89.9

wind tunnel speed; the experimental results are listed in Table 7–7. Plot the drag coefficient C_D as a function of the Reynolds number Re, where the area used for the calculation of C_D is the frontal area of the model truck (the area you see when you look at the model from upstream), and the length scale used for calculation of Re is truck width W. Have we achieved dynamic similarity? Have we achieved Reynolds number independence in our wind tunnel test? Estimate the aerodynamic drag force on the prototype truck traveling on the highway at 26.8 m/s. Assume that both the wind tunnel air and the air flowing over the prototype car are at 25°C and standard atmospheric pressure.

SOLUTION We are to calculate and plot C_D as a function of Re for a given set of wind tunnel measurements and determine if dynamic similarity and/or Reynolds number independence have been achieved. Finally, we are to estimate the aerodynamic drag force acting on the prototype truck.

Assumptions 1 The model truck is geometrically similar to the prototype truck. 2 The aerodynamic drag on the strut(s) holding the model truck is negligible.

Properties For air at atmospheric pressure and at $T = 25°C$, $\rho = 1.184$ kg/m³ and $\mu = 1.849 \times 10^{-5}$ kg/m·s.

Analysis We calculate C_D and Re for the last data point listed in Table 7–7 (at the fastest wind tunnel speed),

$$C_{D,m} = \frac{F_{D,m}}{\frac{1}{2}\rho_m V_m^2 A_m} = \frac{89.9 \text{ N}}{\frac{1}{2}(1.184 \text{ kg/m}^3)(70 \text{ m/s})^2(0.159 \text{ m})(0.257 \text{ m})}\left(\frac{1 \text{ kg·m/s}^2}{1 \text{ N}}\right)$$

$$= 0.758$$

and

$$\text{Re}_m = \frac{\rho_m V_m W_m}{\mu_m} = \frac{(1.184 \text{ kg/m}^3)(70 \text{ m/s})(0.159 \text{ m})}{1.849 \times 10^{-5} \text{ kg/m·s}} = 7.13 \times 10^5 \quad (1)$$

We repeat these calculations for all the data points in Table 7–7, and we plot C_D versus Re in Fig. 7–41.

Have we achieved dynamic similarity? Well, we have *geometric* similarity between model and prototype, but the Reynolds number of the prototype truck is

$$\text{Re}_p = \frac{\rho_p V_p W_p}{\mu_p} = \frac{(1.184 \text{ kg/m}^3)(26.8 \text{ m/s})[16(0.159 \text{ m})]}{1.849 \times 10^{-5} \text{ kg/m·s}} = 4.37 \times 10^6 \quad (2)$$

where the width of the prototype is specified as 16 times that of the model. Comparison of Eqs. 1 and 2 reveals that the prototype Reynolds number is more than six times larger than that of the model. Since we cannot match the independent Π's in the problem, **dynamic similarity has not been achieved**.

Have we achieved Reynolds number independence? From Fig. 7–41 we see that **Reynolds number independence has indeed been achieved**—at Re greater than about 5×10^5, C_D has leveled off to a value of about 0.76 (to two significant digits).

Since we have achieved Reynolds number independence, we can extrapolate to the full-scale prototype, assuming that C_D remains constant as Re is increased to that of the full-scale prototype.

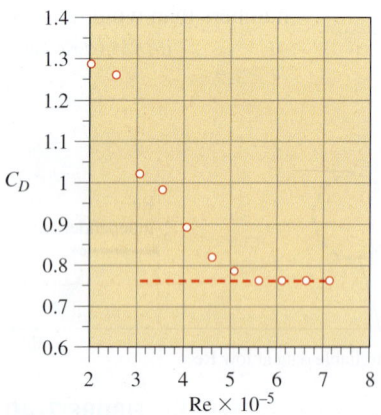

FIGURE 7–41
Aerodynamic drag coefficient as a function of the Reynolds number. The values are calculated from wind tunnel test data on a model truck (Table 7–7).

Predicted aerodynamic drag on the prototype:

$$F_{D,p} = \tfrac{1}{2}\rho_p V_p^2 A_p C_{D,p}$$

$$= \tfrac{1}{2}(1.184 \text{ kg/m}^3)(26.8 \text{ m/s})^2[16^2(0.159 \text{ m})(0.257 \text{ m})](0.76)\left(\frac{1 \text{ N}}{1 \text{ kg·m/s}^2}\right)$$

$$= \mathbf{3400 \text{ N}}$$

Discussion We give our final result to two significant digits. More than that cannot be justified. As always, we must exercise caution when performing an extrapolation, since we have no guarantee that the extrapolated results are correct.

Flows with Free Surfaces

For the case of model testing of flows with free surfaces (boats and ships, floods, river flows, aqueducts, hydroelectric dam spillways, interaction of waves with piers, soil erosion, etc.), complications arise that preclude complete similarity between model and prototype. For example, if a model river is built to study flooding, the model is often several hundred times smaller than the prototype due to limited lab space. If the vertical dimensions of the model were scaled proportionately, the depth of the model river would be so small that surface tension effects (and the Weber number) would become important, and would perhaps even dominate the model flow, even though surface tension effects are negligible in the prototype flow. In addition, although the flow in the actual river may be turbulent, the flow in the model river may be laminar, especially if the slope of the riverbed is geometrically similar to that of the prototype. To avoid these problems, researchers often use a **distorted model** in which the vertical scale of the model (e.g., river depth) is exaggerated in comparison to the horizontal scale of the model (e.g., river width). In addition, the model riverbed slope is often made proportionally steeper than that of the prototype. These modifications result in incomplete similarity due to lack of geometric similarity. Model tests are still useful under these circumstances, but other tricks (like deliberately roughening the model surfaces) and empirical corrections and correlations are required to properly scale up the model data.

In many practical problems involving free surfaces, both the Reynolds number and Froude number appear as relevant independent Π groups in the dimensional analysis (Fig. 7–42). It is difficult (often impossible) to match both of these dimensionless parameters simultaneously. For a free-surface flow with length scale L, velocity scale V, and kinematic viscosity ν, the Reynolds number is matched between model and prototype when

$$\text{Re}_p = \frac{V_p L_p}{\nu_p} = \text{Re}_m = \frac{V_m L_m}{\nu_m} \qquad (7\text{–}21)$$

The Froude number is matched between model and prototype when

$$\text{Fr}_p = \frac{V_p}{\sqrt{gL_p}} = \text{Fr}_m = \frac{V_m}{\sqrt{gL_m}} \qquad (7\text{–}22)$$

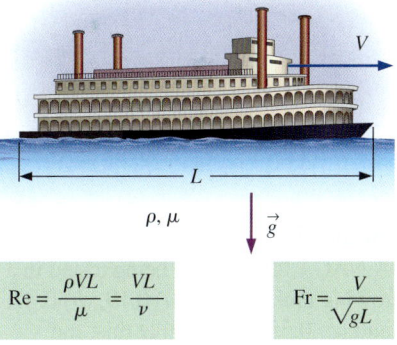

FIGURE 7–42
In many flows involving a liquid with a free surface, both the Reynolds number and Froude number are relevant nondimensional parameters. Since it is not always possible to match both Re and Fr between model and prototype, we are sometimes forced to settle for incomplete similarity.

324
DIMENSIONAL ANALYSIS AND MODELING

To match both Re and Fr, we solve Eqs. 7–21 and 7–22 simultaneously for the required length scale factor L_m/L_p,

$$\frac{L_m}{L_p} = \frac{\nu_m}{\nu_p}\frac{V_p}{V_m} = \left(\frac{V_m}{V_p}\right)^2 \quad (7\text{–}23)$$

Eliminating the ratio V_m/V_p from Eq. 7–23, we see that

Required ratio of kinematic viscosities to match both Re and Fr:

$$\frac{\nu_m}{\nu_p} = \left(\frac{L_m}{L_p}\right)^{3/2} \quad (7\text{–}24)$$

Thus, to ensure complete similarity (assuming geometric similarity is achievable without unwanted surface tension effects as discussed previously), we would need to use a liquid whose kinematic viscosity satisfies Eq. 7–24. Although it is sometimes possible to find an appropriate liquid for use with the model, in most cases it is either impractical or impossible, as Example 7–11 illustrates. In such cases, it is more important to match Froude number than Reynolds number (Fig. 7–43).

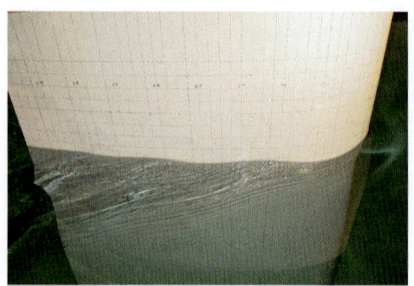
(a)

(b)

(c)

FIGURE 7–43
A NACA 0024 airfoil being tested in a towing tank at Fr = (a) 0.19, (b) 0.37, and (c) 0.55. In tests like this, the Froude number is the most important nondimensional parameter.

Photograph courtesy of IIHR-Hydroscience & Engineering, University of Iowa. Used by permission.

EXAMPLE 7–11   Model Lock and River

In the late 1990s the U.S. Army Corps of Engineers designed an experiment to model the flow of the Tennessee River downstream of the Kentucky Lock and Dam (Fig. 7–44). Because of laboratory space restrictions, they built a scale model with a length scale factor of $L_m/L_p = 1/100$. Suggest a liquid that would be appropriate for the experiment.

SOLUTION We are to suggest a liquid to use in an experiment involving a one-hundredth scale model of a lock, dam, and river.
Assumptions 1 The model is geometrically similar to the prototype. 2 The model river is deep enough that surface tension effects are not significant.
Properties For water at atmospheric pressure and at $T = 20°C$, the prototype kinematic viscosity is $\nu_p = 1.002 \times 10^{-6}$ m²/s.
Analysis From Eq. 7–24,

Required kinematic viscosity of model liquid:

$$\nu_m = \nu_p\left(\frac{L_m}{L_p}\right)^{3/2} = (1.002 \times 10^{-6}\,\text{m}^2/\text{s})\left(\frac{1}{100}\right)^{3/2} = \mathbf{1.00 \times 10^{-9}\,m^2/s} \quad (1)$$

Thus, we need to find a liquid that has a viscosity of 1.00×10^{-9} m²/s. A quick glance through the appendices yields no such liquid. Hot water has a lower kinematic viscosity than cold water, but only by a factor of about 3. Liquid mercury has a very small kinematic viscosity, but it is of order 10^{-7} m²/s—still two orders of magnitude too large to satisfy Eq. 1. Even if liquid mercury would work, it would be too expensive and too hazardous to use in such a test. What do we do? The bottom line is that *we cannot match both the Froude number and the Reynolds number in this model test.*

FIGURE 7–44
A 1:100 scale model constructed to investigate navigation conditions in the lower lock approach for a distance of 3.2 km downstream of the dam. The model includes a scaled version of the spillway, powerhouse, and existing lock. In addition to navigation, the model was used to evaluate environmental issues associated with the new lock and required railroad and highway bridge relocations. The view here is looking upstream toward the lock and dam. At this scale, 16 m on the model represents 1.6 km on the prototype. A (real, full-scale) pickup truck in the background gives you a feel for the model scale.

Photo courtesy of the U.S. Army Corps of Engineers, Nashville.

In other words, it is impossible to achieve complete similarity between model and prototype in this case. Instead, we do the best job we can under conditions of incomplete similarity. Water is typically used in such tests for convenience.

Discussion It turns out that for this kind of experiment, Froude number matching is more critical than Reynolds number matching. As discussed previously for wind tunnel testing, Reynolds number independence is achieved at high enough values of Re. Even if we are unable to achieve Reynolds number independence, we can often extrapolate our low Reynolds number model data to predict full-scale Reynolds number behavior (Fig. 7–45). A high level of confidence in using this kind of extrapolation comes only after much laboratory experience with similar problems.

In closing this section on experiments and incomplete similarity, we mention the importance of similarity in the production of Hollywood movies in which model boats, trains, airplanes, buildings, monsters, etc., are blown up or burned. Movie producers must pay attention to dynamic similarity in order to make the small-scale fires and explosions appear as realistic as possible. You may recall some low-budget movies where the special effects are unconvincing. In most cases this is due to lack of dynamic similarity between the small model and the full-scale prototype. If the model's Froude number and/or Reynolds number differ too much from those of the prototype, the special effects don't look right, even to the untrained eye. The next time you watch a movie, be on the alert for incomplete similarity!

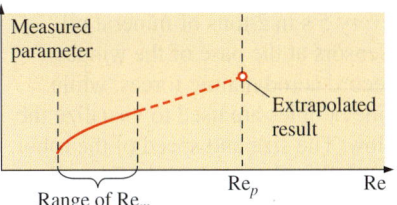

FIGURE 7–45
In many experiments involving free surfaces, we cannot match both the Froude number and the Reynolds number. However, we can often *extrapolate* low Re model test data to predict high Re prototype behavior.

DIMENSIONAL ANALYSIS AND MODELING

APPLICATION SPOTLIGHT ■ How a Fly Flies

Guest Author: Michael Dickinson, California Institute of Technology

An interesting application of dimensional analysis is in the study of how insects fly. The small size and fast wing speed of an insect, such as a tiny fruit fly, make it difficult to directly measure the forces or visualize the air motion created by the fly's wings. However, using principles of dimensional analysis, it is possible to study insect aerodynamics on a larger-scale, slowly moving model—a mechanical robot. The forces created by a hovering fly and flapping robot are dynamically similar if the Reynolds number is the same for each case. For a flapping wing, Re is calculated as $2\Phi RL_c\omega/\nu$, where Φ is the angular amplitude of the wing stroke, R is the wing length, L_c is the average wing width (chord length), ω is the angular frequency of the stroke, and ν is the kinematic viscosity of the surrounding fluid. A fruit fly flaps its 2.5-mm-long, 0.7-mm-wide wings 200 times per second over a 2.8-rad stroke in air with a kinematic viscosity of 1.5×10^{-5} m^2/s. The resulting Reynolds number is approximately 130. By choosing mineral oil with a kinematic viscosity of 1.15×10^{-4} m^2/s, it is possible to match this Reynolds number on a robotic fly that is 100 times larger, flapping its wings over 1000 times more slowly! If the fly is not stationary, but rather moving through the air, it is necessary to match another dimensionless parameter to ensure dynamic similarity, the reduced frequency, $\sigma = 2\Phi R\omega/V$, which measures the ratio of the flapping velocity of the wing tip ($2\Phi R\omega$) to the forward velocity of the body (V). To simulate forward flight, a set of motors tows *Robofly* through its oil tank at an appropriately scaled speed.

Dynamically scaled robots have helped show that insects use a variety of different mechanisms to produce forces as they fly. During each back-and-forth stroke, insect wings travel at high angles of attack, generating a prominent leading-edge vortex. The low pressure of this large vortex pulls the wings upward. Insects can further augment the strength of the leading-edge vortex by rotating their wings at the end of each stroke. After the wing changes direction, it can also generate forces by quickly running through the wake of the previous stroke.

Figure 7–46a shows a real fly flapping its wings, and Fig. 7–46b shows *Robofly* flapping its wings. Because of the larger length scale and shorter time scale of the model, measurements and flow visualizations are possible. Experiments with dynamically scaled model insects continue to teach researchers how insects manipulate wing motion to steer and maneuver.

(a)

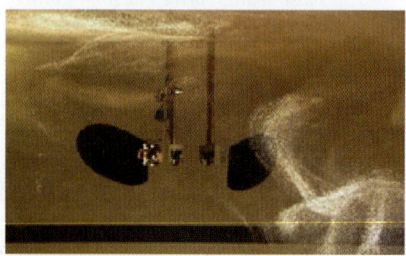

(b)

FIGURE 7–46
(a) The fruit fly, *Drosophila melanogaster*, flaps its tiny wings back and forth 200 times a second, creating a blurred image of the stroke plane. (b) The dynamically scaled model, *Robofly*, flaps its wings once every 5 s in 2 tons of mineral oil. Sensors at the base of the wings record aerodynamic forces, while fine bubbles are used to visualize the flow. The size and speed of the robot, as well as the properties of the oil, were carefully chosen to match the Reynolds number of a real fly.
Photos © Courtesy of Michael Dickinson, CALTECH.

References
Dickinson, M. H., Lehmann, F.-O., and Sane, S., "Wing rotation and the aerodynamic basis of insect flight," *Science*, 284, p. 1954, 1999.
Dickinson, M. H., "Solving the mystery of insect flight," *Scientific American*, 284, No. 6, pp. 35–41, June 2001.
Fry, S. N., Sayaman, R., and Dickinson, M. H., "The aerodynamics of free-flight maneuvers in *Drosophila*," *Science*, 300, pp. 495–498, 2003.

SUMMARY

There is a difference between *dimensions* and *units*; a *dimension* is a measure of a physical quantity (without numerical values), while a *unit* is a way to assign a number to that dimension. There are seven *primary dimensions*—not just in fluid mechanics, but in all fields of science and engineering. They are mass, length, time, temperature, electric current, amount of light, and amount of matter. *All other dimensions can be formed by combination of these seven primary dimensions.*

All mathematical equations must be *dimensionally homogeneous*; this fundamental principle can be applied to equations in order to nondimensionalize them and to identify *dimensionless groups*, also called *nondimensional parameters*. A powerful tool to reduce the number of necessary independent parameters in a problem is called *dimensional analysis*. The *method of repeating variables* is a step-by-step procedure for finding the nondimensional parameters, or Π's, based simply on the dimensions of the variables and constants in the problem. The six steps in the method of repeating variables are summarized here.

Step 1 List the n parameters (variables and constants) in the problem.

Step 2 List the primary dimensions of each parameter.

Step 3 Guess the *reduction j*, usually equal to the number of primary dimensions in the problem. If the analysis does not work out, reduce j by one and try again. The expected number of Π's (k) is equal to n minus j.

Step 4 Wisely choose j *repeating parameters* for construction of the Π's.

Step 5 Generate the k Π's one at a time by grouping the j repeating parameters with each of the remaining variables or constants, forcing the product to be dimensionless, and manipulating the Π's as necessary to achieve established nondimensional parameters.

Step 6 Check your work and write the final functional relationship.

When all the dimensionless groups match between a model and a prototype, *dynamic similarity* is achieved, and we are able to directly predict prototype performance based on model experiments. However, it is not always possible to match *all* the Π groups when trying to achieve similarity between a model and a prototype. In such cases, we run the model tests under conditions of *incomplete similarity*, matching the most important Π groups as best we can, and then extrapolating the model test results to prototype conditions.

We use the concepts presented in this chapter throughout the remainder of the book. For example, dimensional analysis is applied to fully developed pipe flows in Chap. 8 (friction factors, loss coefficients, etc.). In Chap. 10, we normalize the differential equations of fluid flow derived in Chap. 9, producing several dimensionless parameters. Drag and lift coefficients are used extensively in Chap. 11, and dimensionless parameters also appear in the chapters on compressible flow and open-channel flow (Chaps. 12 and 13). We learn in Chap. 14 that dynamic similarity is often the basis for design and testing of pumps and turbines. Finally, dimensionless parameters are also used in computations of fluid flows (Chap. 15).

REFERENCES AND SUGGESTED READING

1. D. C. Montgomery. *Design and Analysis of Experiments*, 8th ed. New York: Wiley, 2013.

2. J. P. Holman. *Experimental Methods for Engineers*, 7th ed. New York: McGraw-Hill, 2001.

PROBLEMS*

Dimensions and Units, Primary Dimensions

7–1C What is the difference between a *dimension* and a *unit*? Give three examples of each.

---

* Problems designated by a "C" are concept questions, and students are encouraged to answer them all. Problems with the [icon] icon are solved using EES, and complete solutions together with parametric studies are included on the text website. Problems with the [icon] icon are comprehensive in nature and are intended to be solved with an equation solver such as EES.

7–2 Write the primary dimensions of the *universal ideal gas constant* R_u. (*Hint:* Use the *ideal gas law*, $PV = nR_uT$ where P is pressure, V is volume, T is absolute temperature, and n is the number of moles of the gas.) *Answer:* $\{m^1L^2t^{-2}T^{-1}N^{-1}\}$

7–3 Write the primary dimensions of each of the following variables from the field of thermodynamics, showing all your work: (*a*) energy E; (*b*) specific energy $e = E/m$; (*c*) power W. *Answers:* (*a*) $\{m^1L^2t^{-2}\}$, (*b*) $\{L^2t^{-2}\}$, (*c*) $\{m^1L^2t^{-3}\}$

7–4 When performing a dimensional analysis, one of the first steps is to list the primary dimensions of each relevant parameter. It is handy to have a table of parameters and their

DIMENSIONAL ANALYSIS AND MODELING

primary dimensions. We have started such a table for you (Table P7–4), in which we have included some of the basic parameters commonly encountered in fluid mechanics. As you work through homework problems in this chapter, add to this table. You should be able to build up a table with dozens of parameters.

TABLE P7–4

Parameter Name	Parameter Symbol	Primary Dimensions
Acceleration	a	$L^1 t^{-2}$
Angle	θ, ϕ, etc.	1 (none)
Density	ρ	$m^1 L^{-3}$
Force	F	$m^1 L^1 t^{-2}$
Frequency	f	t^{-1}
Pressure	P	$m^1 L^{-1} t^{-2}$
Surface tension	σ_s	$m^1 t^{-2}$
Velocity	V	$L^1 t^{-1}$
Viscosity	μ	$m^1 L^{-1} t^{-1}$
Volume flow rate	$\dot{V}$	$L^3 t^{-1}$

7–5 Consider the table of Prob. 7–4 where the primary dimensions of several variables are listed in the mass–length–time system. Some engineers prefer the force–length–time system (force replaces mass as one of the primary dimensions). Write the primary dimensions of three of these (density, surface tension, and viscosity) in the force–length–time system.

7–6 On a periodic chart of the elements, molar mass (M), also called *atomic weight,* is often listed as though it were a dimensionless quantity (Fig. P7–6). In reality, atomic weight is the mass of 1 mol of the element. For example, the atomic weight of nitrogen $M_{\text{nitrogen}} = 14.0067$. We interpret this as 14.0067 g/mol of elemental nitrogen. What are the primary dimensions of atomic weight?

6	7	8
C	**N**	**O**
12.011	14.0067	15.9994
14	15	16
Si	**P**	**S**
28.086	30.9738	32.060

FIGURE P7–6

7–7 Some authors prefer to use *force* as a primary dimension in place of mass. In a typical fluid mechanics problem, then, the four represented primary dimensions m, L, t, and T are replaced by F, L, t, and T. The primary dimension of force in this system is {force} = {F}. Using the results of Prob. 7–2, rewrite the primary dimensions of the universal gas constant in this alternate system of primary dimensions.

7–8 We define the *specific ideal gas constant* R_{gas} for a particular gas as the ratio of the universal gas constant and the molar mass (also called *molecular weight*) of the gas, $R_{\text{gas}} = R_u/M$. For a particular gas, then, the ideal gas law is written as follows:

$$PV = mR_{\text{gas}}T \quad \text{or} \quad P = \rho R_{\text{gas}}T$$

where P is pressure, V is volume, m is mass, T is absolute temperature, and ρ is the density of the particular gas. What are the primary dimensions of R_{gas}? For air, $R_{\text{air}} = 287.0$ J/kg·K in standard SI units. Verify that these units agree with your result.

7–9 The *moment of force* ($\vec{M}$) is formed by the cross product of a moment arm ($\vec{r}$) and an applied force ($\vec{F}$), as sketched in Fig. P7–9. What are the primary dimensions of moment of force? List its unit in primary SI units.

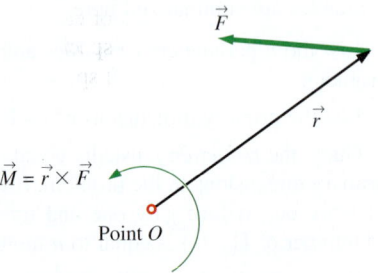

FIGURE P7–9

7–10 What are the primary dimensions of electric voltage (E)? (*Hint*: Make use of the fact that electric power is equal to voltage times current.)

7–11 You are probably familiar with *Ohm's law* for electric circuits (Fig. P7–11), where ΔE is the voltage difference or *potential* across the resistor, I is the electric current passing through the resistor, and R is the electrical resistance. What are the primary dimensions of electrical resistance? *Answer:* $\{m^1 L^2 t^{-3} I^{-2}\}$

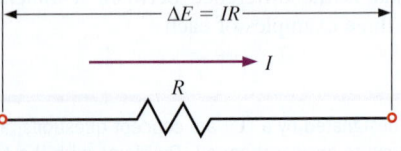

FIGURE P7–11

7–12 Write the primary dimensions of each of the following variables, showing all your work: (*a*) acceleration a; (*b*) angular velocity ω; (*c*) angular acceleration α.

7–13 *Angular momentum*, also called *moment of momentum* ($\vec{H}$), is formed by the cross product of a moment arm ($\vec{r}$) and the linear momentum ($m\vec{V}$) of a fluid particle, as sketched in Fig. P7–13. What are the primary dimensions of angular momentum? List the units of angular momentum in primary SI units and in primary English units. *Answers:* {$m^1L^2t^{-1}$}, kg·m²/s, lbm·ft²/s

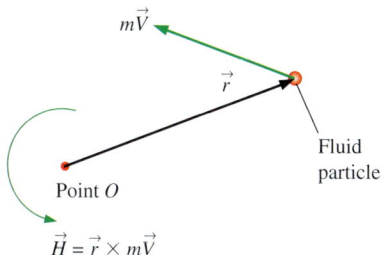

FIGURE P7–13

7–14 Write the primary dimensions of each of the following variables, showing all your work: (*a*) specific heat at constant pressure c_p; (*b*) specific weight ρg; (*c*) specific enthalpy h.

7–15 **Thermal conductivity** k is a measure of the ability of a material to conduct heat (Fig. P7–15). For conduction heat transfer in the *x*-direction through a surface normal to the *x*-direction, **Fourier's law of heat conduction** is expressed as

$$\dot{Q}_{conduction} = -kA\frac{dT}{dx}$$

where $\dot{Q}_{conduction}$ is the rate of heat transfer and A is the area normal to the direction of heat transfer. Determine the primary dimensions of thermal conductivity (k). Look up a value of k in the appendices and verify that its SI units are consistent with your result. In particular, write the primary SI units of k.

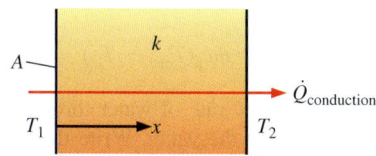

FIGURE P7–15

7–16 Write the primary dimensions of each of the following variables from the study of convection heat transfer (Fig. P7–16), showing all your work: (*a*) heat generation rate $\dot{g}$ (*Hint*: rate of conversion of thermal energy per unit volume); (*b*) heat flux $\dot{q}$ (*Hint*: rate of heat transfer per unit area); (*c*) heat transfer coefficient h (*Hint*: heat flux per unit temperature difference).

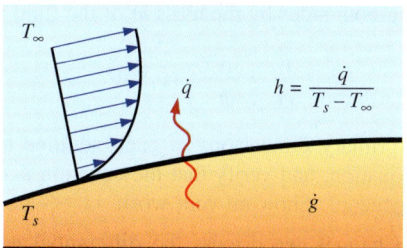

FIGURE P7–16

7–17 Thumb through the appendices of your thermodynamics book, and find three properties or constants not mentioned in Probs. 7–1 to 7–16. List the name of each property or constant and its SI units. Then write out the primary dimensions of each property or constant.

Dimensional Homogeneity

7–18C Explain the *law of dimensional homogeneity* in simple terms.

7–19 In Chap. 4, we defined the *material acceleration*, which is the acceleration following a fluid particle,

$$\vec{a}(x, y, z, t) = \frac{\partial \vec{V}}{\partial t} + (\vec{V}\cdot\vec{\nabla})\vec{V}$$

(*a*) What are the primary dimensions of the gradient operator $\vec{\nabla}$? (*b*) Verify that each additive term in the equation has the same dimensions. *Answers:* (*a*) {L^{-1}}; (*b*) {L^1t^{-2}}

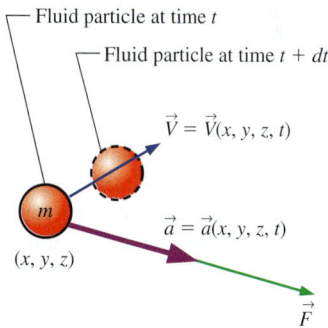

FIGURE P7–19

7–20 Newton's second law is the foundation for the differential equation of conservation of linear momentum (to be discussed in Chap. 9). In terms of the material acceleration following a fluid particle (Fig. P7–19), we write Newton's second law as follows:

$$\vec{F} = m\vec{a} = m\left(\frac{\partial \vec{V}}{\partial t} + (\vec{V}\cdot\vec{\nabla})\vec{V}\right)$$

Or, dividing both sides by the mass m of the fluid particle,

$$\frac{\vec{F}}{m} = \frac{\partial \vec{V}}{\partial t} + (\vec{V} \cdot \vec{\nabla})\vec{V}$$

Write the primary dimensions of each additive term in the (second) equation, and verify that the equation is dimensionally homogeneous. Show all your work.

7–21 In Chap. 9, we discuss the differential equation for conservation of mass, the *continuity equation*. In cylindrical coordinates, and for steady flow,

$$\frac{1}{r}\frac{\partial(ru_r)}{\partial r} + \frac{1}{r}\frac{\partial u_\theta}{\partial \theta} + \frac{\partial u_z}{\partial z} = 0$$

Write the primary dimensions of each additive term in the equation, and verify that the equation is dimensionally homogeneous. Show all your work.

7–22 The *Reynolds transport theorem* (RTT) is discussed in Chap. 4. For the general case of a moving and/or deforming control volume, we write the RTT as follows:

$$\frac{dB_{sys}}{dt} = \frac{d}{dt}\int_{CV} \rho b \, dV + \int_{CS} \rho b \vec{V}_r \cdot \vec{n} \, dA$$

where $\vec{V}_r$ is the *relative velocity*, i.e., the velocity of the fluid relative to the control surface. Write the primary dimensions of each additive term in the equation, and verify that the equation is dimensionally homogeneous. Show all your work. (*Hint*: Since B can be any property of the flow—scalar, vector, or even tensor—it can have a variety of dimensions. So, just let the dimensions of B be those of B itself, $\{B\}$. Also, b is defined as B per unit mass.)

7–23 An important application of fluid mechanics is the study of room ventilation. In particular, suppose there is a **source** S (mass per unit time) of air pollution in a room of volume V (Fig. P7–23). Examples include carbon monoxide from cigarette smoke or an unvented kerosene heater, gases like ammonia from household cleaning products, and vapors given off by evaporation of **volatile organic compounds** (VOCs) from an open container. We let c represent the **mass concentration** (mass of contaminant per unit volume of air). $\dot{V}$ is the volume flow rate of fresh air entering the room. If the room air is well mixed so that the mass concentration c is uniform throughout the room, but varies with time, the differential equation for mass concentration in the room as a function of time is

$$V\frac{dc}{dt} = S - \dot{V}c - cA_s k_w$$

where k_w is an **adsorption coefficient** and A_s is the surface area of walls, floors, furniture, etc., that adsorb some of the contaminant. Write the primary dimensions of the first three terms in the equation (including the term on the left side), and verify that those terms are dimensionally homogeneous. Then determine the dimensions of k_w. Show all your work.

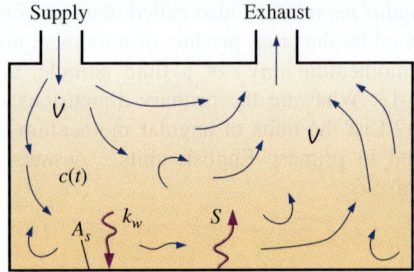

FIGURE P7–23

7–24 In Chap. 4 we defined *volumetric strain rate* as the rate of increase of volume of a fluid element per unit volume (Fig. P7–24). In Cartesian coordinates we write the volumetric strain rate as

$$\frac{1}{V}\frac{DV}{Dt} = \frac{\partial u}{\partial x} + \frac{\partial v}{\partial y} + \frac{\partial w}{\partial z}$$

Write the primary dimensions of each additive term, and verify that the equation is dimensionally homogeneous. Show all your work.

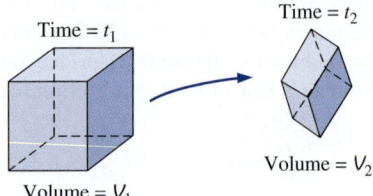

FIGURE P7–24

7–25 Cold water enters a pipe, where it is heated by an external heat source (Fig. P7–25). The inlet and outlet water temperatures are T_{in} and T_{out}, respectively. The total rate of heat transfer $\dot{Q}$ from the surroundings into the water in the pipe is

$$\dot{Q} = \dot{m}c_p(T_{out} - T_{in})$$

where $\dot{m}$ is the mass flow rate of water through the pipe, and c_p is the specific heat of the water. Write the primary dimensions of each additive term in the equation, and verify that the equation is dimensionally homogeneous. Show all your work.

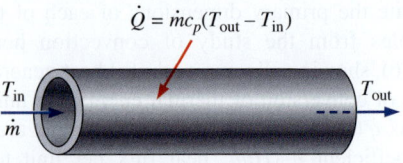

FIGURE P7–25

Nondimensionalization of Equations

7–26C What is the primary reason for *nondimensionalizing* an equation?

7–27 Recall from Chap. 4 that the volumetric strain rate is zero for a steady incompressible flow. In Cartesian coordinates we express this as

$$\frac{\partial u}{\partial x} + \frac{\partial v}{\partial y} + \frac{\partial w}{\partial z} = 0$$

Suppose the characteristic speed and characteristic length for a given flow field are V and L, respectively (Fig. P7–27). Define the following dimensionless variables,

$$x^* = \frac{x}{L}, \quad y^* = \frac{y}{L}, \quad z^* = \frac{z}{L},$$

$$u^* = \frac{u}{V}, \quad v^* = \frac{v}{V}, \quad \text{and} \quad w^* = \frac{w}{V}$$

Nondimensionalize the equation, and identify any established (named) dimensionless parameters that may appear. Discuss.

FIGURE P7–27

7–28 In an oscillating compressible flow field the volumetric strain rate is *not* zero, but varies with time following a fluid particle. In Cartesian coordinates we express this as

$$\frac{1}{V}\frac{DV}{Dt} = \frac{\partial u}{\partial x} + \frac{\partial v}{\partial y} + \frac{\partial w}{\partial z}$$

Suppose the characteristic speed and characteristic length for a given flow field are V and L, respectively. Also suppose that f is a characteristic frequency of the oscillation (Fig. P7–28). Define the following dimensionless variables,

$$t^* = ft, \quad V^* = \frac{V}{L^3}, \quad x^* = \frac{x}{L}, \quad y^* = \frac{y}{L},$$

$$z^* = \frac{z}{L}, \quad u^* = \frac{u}{V}, \quad v^* = \frac{v}{V}, \quad \text{and} \quad w^* = \frac{w}{V}$$

Nondimensionalize the equation and identify any established (named) dimensionless parameters that may appear.

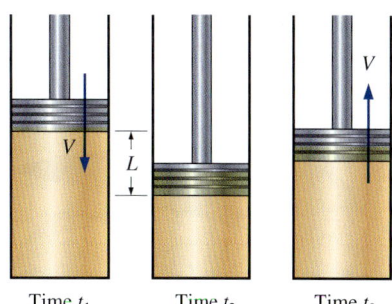

f = frequency of oscillation

FIGURE P7–28

7–29 In Chap. 9, we define the **stream function** ψ for two-dimensional incompressible flow in the xy-plane,

$$u = \frac{\partial \psi}{\partial y} \quad v = -\frac{\partial \psi}{\partial x}$$

where u and v are the velocity components in the x- and y-directions, respectively. (*a*) What are the primary dimensions of ψ? (*b*) Suppose a certain two-dimensional flow has a characteristic length scale L and a characteristic time scale t. Define dimensionless forms of variables x, y, u, v, and ψ. (*c*) Rewrite the equations in nondimensional form, and identify any established dimensionless parameters that may appear.

7–30 In an oscillating incompressible flow field the force per unit mass acting on a fluid particle is obtained from Newton's second law in intensive form (see Prob. 7–20),

$$\frac{\vec{F}}{m} = \frac{\partial \vec{V}}{\partial t} + (\vec{V}\cdot\vec{\nabla})\vec{V}$$

Suppose the characteristic speed and characteristic length for a given flow field are V_∞ and L, respectively. Also suppose that ω is a characteristic angular frequency (rad/s) of the oscillation (Fig. P7–30). Define the following nondimensionalized variables,

$$t^* = \omega t, \quad \vec{x}^* = \frac{\vec{x}}{L}, \quad \vec{\nabla}^* = L\vec{\nabla}, \quad \text{and} \quad \vec{V}^* = \frac{\vec{V}}{V_\infty}$$

Since there is no given characteristic scale for the force per unit mass acting on a fluid particle, we assign one, noting that $\{\vec{F}/m\} = \{L/t^2\}$. Namely, we let

$$(\vec{F}/m)^* = \frac{1}{\omega^2 L}\vec{F}/m$$

Nondimensionalize the equation of motion and identify any established (named) dimensionless parameters that may appear.

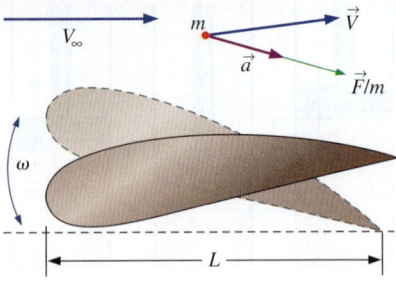

FIGURE P7–30

7–31 A wind tunnel is used to measure the pressure distribution in the airflow over an airplane model (Fig. P7–31). The air speed in the wind tunnel is low enough that compressible effects are negligible. As discussed in Chap. 5, the Bernoulli equation approximation is valid in such a flow situation everywhere except very close to the body surface or wind tunnel wall surfaces and in the wake region behind the model. Far away from the model, the air flows at speed V_∞ and pressure P_∞, and the air density ρ is approximately constant. Gravitational effects are generally negligible in airflows, so we write the Bernoulli equation as

$$P + \frac{1}{2}\rho V^2 = P_\infty + \frac{1}{2}\rho V_\infty^2$$

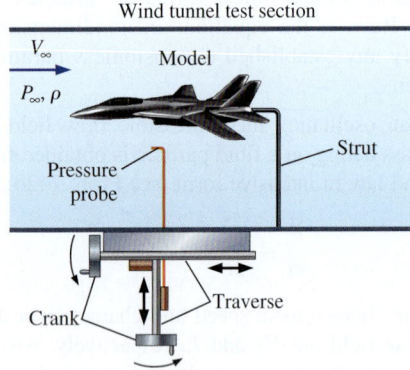

FIGURE P7–31

Nondimensionalize the equation, and generate an expression for the **pressure coefficient** C_p at any point in the flow where the Bernoulli equation is valid. C_p is defined as

$$C_p = \frac{P - P_\infty}{\frac{1}{2}\rho V_\infty^2}$$

Answer: $C_p = 1 - V^2/V_\infty^2$

7–32 Consider ventilation of a well-mixed room as in Fig. P7–23. The differential equation for mass concentration in the room as a function of time is given in Prob. 7–23 and is repeated here for convenience,

$$V\frac{dc}{dt} = S - \dot{V}c - cA_s k_w$$

There are three characteristic parameters in such a situation: L, a characteristic length scale of the room (assume $L = V^{1/3}$); $\dot{V}$, the volume flow rate of fresh air into the room, and c_{limit}, the maximum mass concentration that is not harmful. (*a*) Using these three characteristic parameters, define dimensionless forms of all the variables in the equation. (*Hint*: For example, define $c^* = c/c_{\text{limit}}$.) (*b*) Rewrite the equation in dimensionless form, and identify any established dimensionless groups that may appear.

Dimensional Analysis and Similarity

7–33C List the three primary purposes of dimensional analysis.

7–34C List and describe the three necessary conditions for complete similarity between a model and a prototype.

7–35 A student team is to design a human-powered submarine for a design competition. The overall length of the prototype submarine is 4.85 m, and its student designers hope that it can travel fully submerged through water at 0.440 m/s. The water is freshwater (a lake) at $T = 15°C$. The design team builds a one-fifth scale model to test in their university's wind tunnel (Fig. P7–35). A shield surrounds the drag balance strut so that the aerodynamic drag of the strut itself does not influence the measured drag. The air in the wind tunnel is at 25°C and at one standard atmosphere pressure. At what air speed do they need to run the wind tunnel in order to achieve similarity? *Answer:* 30.2 m/s

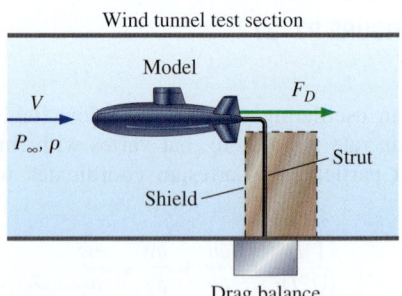

FIGURE P7–35

7–36 Repeat Prob. 7–35 with all the same conditions except that the only facility available to the students is a much smaller wind tunnel. Their model submarine is a one-twenty-fourth scale model instead of a one-fifth scale model. At what air speed do they need to run the wind tunnel in order to achieve similarity? Do you notice anything disturbing or suspicious about your result? Discuss your results.

7–37 This is a follow-up to Prob. 7–35. The students measure the aerodynamic drag on their model submarine in the wind tunnel (Fig. P7–35). They are careful to run the wind tunnel at conditions that ensure similarity with the prototype submarine. Their measured drag force is 5.70 N. Estimate the drag force on the prototype submarine at the conditions given in Prob. 7–35. *Answer:* 25.5 N

7–38 A lightweight parachute is being designed for military use (Fig. P7–38). Its diameter D is 7 m and the total weight W of the falling payload, parachute, and equipment is 1020 N. The design *terminal settling speed* V_t of the parachute at this weight is 5.5 m/s. A one-twelfth scale model of the parachute is tested in a wind tunnel. The wind tunnel temperature and pressure are the same as those of the prototype, namely 15°C and standard atmospheric pressure. (*a*) Calculate the drag coefficient of the prototype. (*Hint*: At terminal settling speed, weight is balanced by aerodynamic drag.) (*b*) At what wind tunnel speed should the wind tunnel be run in order to achieve dynamic similarity? (*c*) Estimate the aerodynamic drag of the model parachute in the wind tunnel (in N).

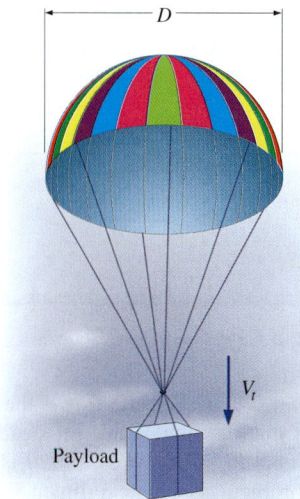

FIGURE P7–38

7–39 Some wind tunnels are *pressurized*. Discuss why a research facility would go through all the extra trouble and expense to pressurize a wind tunnel. If the air pressure in the tunnel increases by a factor of 1.8, all else being equal (same wind speed, same model, etc.), by what factor will the Reynolds number increase?

7–40 The aerodynamic drag of a new sports car is to be predicted at a speed of 95 km/h at an air temperature of 25°C. Automotive engineers build a one-third scale model of the car (Fig. P7–40) to test in a wind tunnel. The temperature of the wind tunnel air is also 25°C. The drag force is measured with a drag balance, and the moving belt is used to simulate the moving ground (from the car's frame of reference). Determine how fast the engineers should run the wind tunnel to achieve similarity between the model and the prototype.

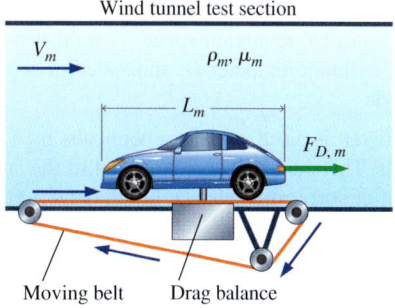

FIGURE P7–40

7–41 This is a follow-up to Prob. 7–40. The aerodynamic drag on the model in the wind tunnel (Fig. P7–40) is measured to be 150 N when the wind tunnel is operated at the speed that ensures similarity with the prototype car. Estimate the drag force on the prototype car at the conditions given in Prob. 7–40.

7–42 Consider the common situation in which a researcher is trying to match the Reynolds number of a large prototype vehicle with that of a small-scale model in a wind tunnel. Is it better for the air in the wind tunnel to be cold or hot? Why? Support your argument by comparing wind tunnel air at 10°C and at 40°C, all else being equal.

Dimensionless Parameters and the Method of Repeating Variables

7–43 Using primary dimensions, verify that the Archimedes number (Table 7–5) is indeed dimensionless.

7–44 Using primary dimensions, verify that the Grashof number (Table 7–5) is indeed dimensionless.

7–45 Using primary dimensions, verify that the Rayleigh number (Table 7–5) is indeed dimensionless. What other established nondimensional parameter is formed by the ratio of Ra and Gr? *Answer:* the Prandtl number

7–46 A periodic *Kármán vortex street* is formed when a uniform stream flows over a circular cylinder (Fig. P7–46). Use the method of repeating variables to generate a dimensionless relationship for Kármán vortex shedding frequency f_k as a function of free-stream speed V, fluid density ρ, fluid viscosity μ, and cylinder diameter D. Show all your work. *Answer:* St = f(Re)

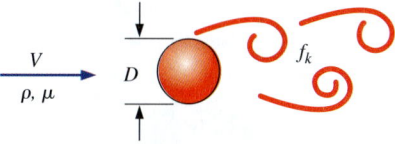

FIGURE P7–46

7–47 Repeat Prob. 7–46, but with an additional independent parameter included, namely, the speed of sound c in the fluid. Use the method of repeating variables to generate a dimensionless relationship for Kármán vortex shedding frequency f_k as a function of free-stream speed V, fluid density ρ, fluid viscosity μ, cylinder diameter D, and speed of sound c. Show all your work.

7–48 A stirrer is used to mix chemicals in a large tank (Fig. P7–48). The shaft power $\dot{W}$ supplied to the stirrer blades is a function of stirrer diameter D, liquid density ρ, liquid viscosity μ, and the angular velocity ω of the spinning blades. Use the method of repeating variables to generate a dimensionless relationship between these parameters. Show all your work and be sure to identify your Π groups, modifying them as necessary. *Answer:* $N_p = f(\text{Re})$

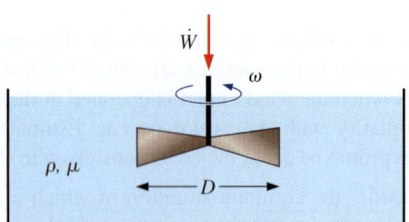

FIGURE P7–48

7–49 Repeat Prob. 7–48 except do not assume that the tank is large. Instead, let tank diameter D_{tank} and average liquid depth h_{tank} be additional relevant parameters.

7–50 Albert Einstein is pondering how to write his (soon-to-be-famous) equation. He knows that energy E is a function of mass m and the speed of light c, but he doesn't know the functional relationship ($E = m^2c$? $E = mc^4$?). Pretend that Albert knows nothing about dimensional analysis, but since you are taking a fluid mechanics class, you help Albert come up with his equation. Use the step-by-step method of repeating variables to generate a dimensionless relationship between these parameters, showing all of your work. Compare this to Einstein's famous equation—does dimensional analysis give you the correct form of the equation?

FIGURE P7–50

7–51 The *Richardson number* is defined as

$$\text{Ri} = \frac{L^5 g \, \Delta\rho}{\rho \dot{V}^2}$$

Miguel is working on a problem that has a characteristic length scale L, a characteristic velocity V, a characteristic density difference $\Delta\rho$, a characteristic (average) density ρ, and of course the gravitational constant g, which is always available. He wants to define a Richardson number, but does not have a characteristic volume flow rate. Help Miguel define a characteristic volume flow rate based on the parameters available to him, and then define an appropriate Richardson number in terms of the given parameters.

7–52 Consider fully developed **Couette flow**—flow between two infinite parallel plates separated by distance h, with the top plate moving and the bottom plate stationary as illustrated in Fig. P7–52. The flow is steady, incompressible, and two-dimensional in the xy-plane. Use the method of repeating variables to generate a dimensionless relationship for the x-component of fluid velocity u as a function of fluid viscosity μ, top plate speed V, distance h, fluid density ρ, and distance y. Show all your work. *Answer:* $u/V = f(\text{Re}, y/h)$

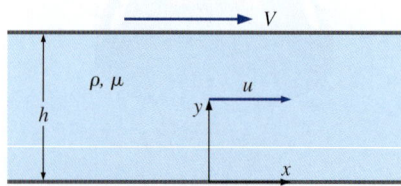

FIGURE P7–52

7–53 Consider *developing* Couette flow—the same flow as Prob. 7–52 except that the flow is not yet steady-state, but is developing with time. In other words, time t is an additional parameter in the problem. Generate a dimensionless relationship between all the variables.

7–54 The speed of sound c in an ideal gas is known to be a function of the ratio of specific heats k, absolute temperature T, and specific ideal gas constant R_{gas} (Fig. P7–54). Showing all your work, use dimensional analysis to find the functional relationship between these parameters.

FIGURE P7–54

7–55 Repeat Prob. 7–54, except let the speed of sound c in an ideal gas be a function of absolute temperature T, universal ideal gas constant R_u, molar mass (molecular weight) M of the gas, and ratio of specific heats k. Showing all your work, use dimensional analysis to find the functional relationship between these parameters.

7–56 Repeat Prob. 7–54, except let the speed of sound c in an ideal gas be a function only of absolute temperature T and specific ideal gas constant R_{gas}. Showing all your work, use dimensional analysis to find the functional relationship between these parameters. *Answer:* $c/\sqrt{R_{gas} T}$ = constant

7–57 Repeat Prob. 7–54, except let speed of sound c in an ideal gas be a function only of pressure P and gas density ρ. Showing all your work, use dimensional analysis to find the functional relationship between these parameters. Verify that your results are consistent with the equation for speed of sound in an ideal gas, $c = \sqrt{kR_{gas} T}$.

7–58 When small aerosol particles or microorganisms move through air or water, the Reynolds number is very small (Re ≪ 1). Such flows are called **creeping flows.** The aerodynamic drag on an object in creeping flow is a function only of its speed V, some characteristic length scale L of the object, and fluid viscosity μ (Fig. P7–58). Use dimensional analysis to generate a relationship for F_D as a function of the independent variables.

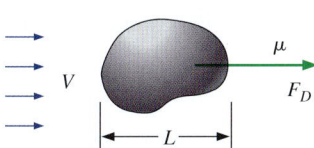

FIGURE P7–58

7–59 A tiny aerosol particle of density ρ_p and characteristic diameter D_p falls in air of density ρ and viscosity μ (Fig. P7–59). If the particle is small enough, the creeping flow approximation is valid, and the terminal settling speed of the particle V depends only on D_p, μ, gravitational constant g, and the density difference $(\rho_p - \rho)$. Use dimensional analysis to generate a relationship for V as a function of the independent variables. Name any established dimensionless parameters that appear in your analysis.

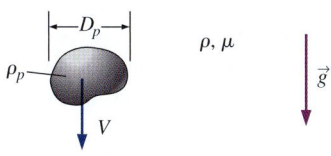

FIGURE P7–59

7–60 Combine the results of Probs. 7–58 and 7–59 to generate an equation for the settling speed V of an aerosol particle falling in air (Fig. P7–59). Verify that your result is consistent with the functional relationship obtained in Prob. 7–59. For consistency, use the notation of Prob. 7–59. (*Hint:* For a particle falling at constant settling speed, the particle's net weight must equal its aerodynamic drag. Your final result should be an equation for V that is valid to within some unknown constant.)

7–61 You will need the results of Prob. 7–60 to do this problem. A tiny aerosol particle falls at steady settling speed V. The Reynolds number is small enough that the creeping flow approximation is valid. If the particle size is doubled, all else being equal, by what factor will the settling speed go up? If the density difference $(\rho_p - \rho)$ is doubled, all else being equal, by what factor will the settling speed go up?

7–62 An incompressible fluid of density ρ and viscosity μ flows at average speed V through a long, horizontal section of round pipe of length L, inner diameter D, and inner wall roughness height ε (Fig. P7–62). The pipe is long enough that the flow is fully developed, meaning that the velocity profile does not change down the pipe. Pressure decreases (linearly) down the pipe in order to "push" the fluid through the pipe to overcome friction. Using the method of repeating variables, develop a nondimensional relationship between pressure drop $\Delta P = P_1 - P_2$ and the other parameters in the problem. Be sure to modify your Π groups as necessary to achieve established nondimensional parameters, and name them. (*Hint:* For consistency, choose D rather than L or ε as one of your repeating parameters.) *Answer:* Eu = f (Re, ε/D, L/D)

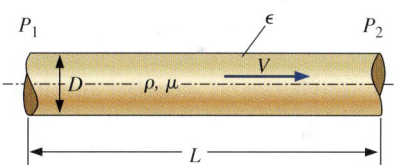

FIGURE P7–62

7–63 Consider *laminar* flow through a long section of pipe, as in Fig. P7–62. For laminar flow it turns out that wall roughness is not a relevant parameter unless ε is very large. The volume flow rate $\dot{V}$ through the pipe is a function of pipe diameter D, fluid viscosity μ, and axial pressure gradient dP/dx. If pipe diameter is doubled, all else being equal, by what factor will volume flow rate increase? Use dimensional analysis.

7–64 One of the first things you learn in physics class is the law of universal gravitation, $F = G\dfrac{m_1 m_2}{r^2}$, where F is the attractive force between two bodies, m_1 and m_2 are the masses of the two bodies, r is the distance between the two bodies, and G is the universal gravitational constant equal to $(6.67428 \pm 0.00067) \times 10^{-11}$ [the units of G are not given here]. (*a*) Calculate the SI units of G. For consistency, give your answer in terms of kg, m, and s. (*b*) Suppose you don't remember the law of universal gravitation, but you are clever enough to know that F is a function of G, m_1, m_2, and r. Use dimensional analysis and the method of repeating variables (show all your work) to generate a nondimensional expression for $F = F(G, m_1, m_2, r)$. Give your answer as $\Pi_1 =$ function of $(\Pi_2, \Pi_3, \ldots)$. (c) Dimensional analysis cannot yield the exact form of the function. However, compare your result to the law of universal gravitation to find the form of the function (e.g., $\Pi_1 = \Pi_2{}^2$ or some other functional form).

7–65 Jen is working on a spring–mass–damper system, as shown in Fig. P7–65. She remembers from her dynamic systems class that the damping ratio ζ is a nondimensional property of such systems and that ζ is a function of spring constant k, mass m, and damping coefficient c. Unfortunately, she does not recall the exact form of the equation for ζ. However, she is taking a fluid mechanics class and decides to use her newly acquired knowledge about dimensional analysis to recall the form of the equation. Help Jen develop the equation for ζ using the method of repeating variables, showing all of your work. (*Hint*: Typical units for k are N/m and those for c are N·s/m.)

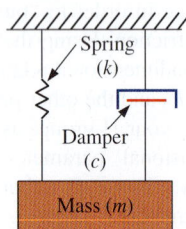

FIGURE P7–65

7–66 Bill is working on an electrical circuit problem. He remembers from his electrical engineering class that voltage drop ΔE is a function of electrical current I and electrical resistance R. Unfortunately, he does not recall the exact form of the equation for ΔE. However, he is taking a fluid mechanics class and decides to use his newly acquired knowledge about dimensional analysis to recall the form of the equation. Help Bill develop the equation for ΔE using the method of repeating variables, showing all of your work. Compare this to Ohm's law—does dimensional analysis give you the correct form of the equation?

7–67 A boundary layer is a thin region (usually along a wall) in which viscous forces are significant and within which the flow is rotational. Consider a boundary layer growing along a thin flat plate (Fig. P7–67). The flow is steady. The boundary layer thickness δ at any downstream distance x is a function of x, free-stream velocity V_∞, and fluid properties ρ (density) and μ (viscosity). Use the method of repeating variables to generate a dimensionless relationship for δ as a function of the other parameters. Show all your work.

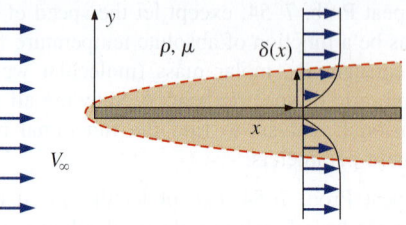

FIGURE P7–67

7–68 A liquid of density ρ and viscosity μ is pumped at volume flow rate $\dot{V}$ through a pump of diameter D. The blades of the pump rotate at angular velocity ω. The pump supplies a pressure rise ΔP to the liquid. Using dimensional analysis, generate a dimensionless relationship for ΔP as a function of the other parameters in the problem. Identify any established nondimensional parameters that appear in your result. *Hint*: For consistency (and whenever possible), it is wise to choose a length, a density, and a velocity (or angular velocity) as repeating variables.

7–69 A propeller of diameter D rotates at angular velocity ω in a liquid of density ρ and viscosity μ. The required torque T is determined to be a function of D, ω, ρ, and μ. Using dimensional analysis, generate a dimensionless relationship. Identify any established nondimensional parameters that appear in your result. *Hint*: For consistency (and whenever possible), it is wise to choose a length, a density, and a velocity (or angular velocity) as repeating variables.

7–70 Repeat Prob. 7–69 for the case in which the propeller operates in a compressible gas instead of a liquid.

7–71 In the study of turbulent flow, turbulent viscous dissipation rate ε (rate of energy loss per unit mass) is known to be a function of length scale l and velocity scale u' of the large-scale turbulent eddies. Using dimensional analysis (Buckingham pi and the method of repeating variables) and showing all of your work, generate an expression for ε as a function of l and u'.

7–72 The rate of heat transfer to water flowing in a pipe was analyzed in Prob. 7–25. Let us approach that same problem, but now with dimensional analysis. Cold water enters a pipe, where it is heated by an external heat source (Fig. P7–72). The inlet and outlet water temperatures are T_{in} and T_{out}, respectively. The total rate of heat transfer $\dot{Q}$ from the surroundings into the water in the pipe is known to be a function of mass flow rate $\dot{m}$, the specific heat c_p of the water, and the temperature difference between the incoming and outgoing water. Showing all your work, use dimensional analysis to find the functional relationship between these parameters, and compare to the analytical equation given in Prob. 7–25. (*Note*: We are pretending that we do not know the analytical equation.)

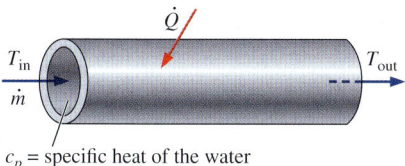

FIGURE P7–72

7–73 Consider a liquid in a cylindrical container in which both the container and the liquid are rotating as a rigid body (solid-body rotation). The elevation difference h between the center of the liquid surface and the rim of the liquid surface is a function of angular velocity ω, fluid density ρ, gravitational acceleration g, and radius R (Fig. P7–73). Use the method of repeating variables to find a dimensionless relationship between the parameters. Show all your work. *Answer:* $h/R = f(\text{Fr})$

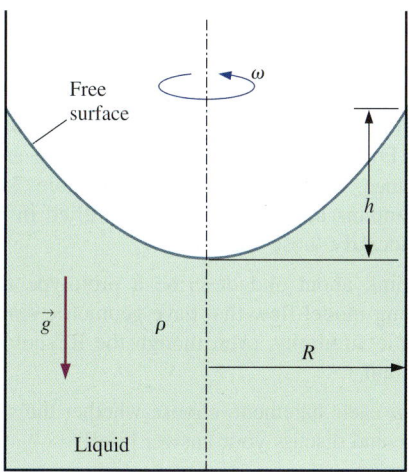

FIGURE P7–73

7–74 Consider the case in which the container and liquid of Prob. 7–73 are initially at rest. At $t = 0$ the container begins to rotate. It takes some time for the liquid to rotate as a rigid body, and we expect that the liquid's viscosity is an additional relevant parameter in the unsteady problem. Repeat Prob. 7–73, but with two additional independent parameters included, namely, fluid viscosity μ and time t. (We are interested in the development of height h as a function of time and the other parameters.)

Experimental Testing and Incomplete Similarity

7–75C Although we usually think of a model as being smaller than the prototype, describe at least three situations in which it is better for the model to be *larger* than the prototype.

7–76C Discuss the purpose of a moving ground belt in wind tunnel tests of flow over model automobiles. Think of an alternative if a moving ground belt is unavailable.

7–77C Consider again the model truck example discussed in Section 7–5, except that the maximum speed of the wind tunnel is only 50 m/s. Aerodynamic force data are taken for wind tunnel speeds between $V = 20$ and 50 m/s—assume the same data for these speeds as those listed in Table 7–7. Based on these data *alone,* can the researchers be confident that they have reached Reynolds number independence?

7–78C Define *wind tunnel blockage*. What is the rule of thumb about the maximum acceptable blockage for a wind tunnel test? Explain why there would be measurement errors if the blockage were significantly higher than this value.

7–79C What is the rule of thumb about the Mach number limit in order that the incompressible flow approximation is reasonable? Explain why wind tunnel results would be incorrect if this rule of thumb were violated.

7–80 A one-sixteenth scale model of a new sports car is tested in a wind tunnel. The prototype car is 4.37 m long, 1.30 m tall, and 1.69 m wide. During the tests, the moving ground belt speed is adjusted so as to always match the speed of the air moving through the test section. Aerodynamic drag force F_D is measured as a function of wind tunnel speed; the experimental results are listed in Table P7–80. Plot drag coefficient C_D as a function of the Reynolds number Re, where the area used for calculation of C_D is the frontal area of the model car (assume $A = $ width $\times$ height), and the length scale used for calculation of Re is car width W. Have we achieved dynamic similarity? Have we achieved Reynolds number independence in our wind tunnel test? Estimate the aerodynamic drag force on the prototype car traveling on the highway at 31.3 m/s. Assume that both the wind tunnel air and the air flowing over the prototype car are at 25°C and atmospheric pressure. *Answers:* no, yes, 408 N

TABLE P7–80

V, m/s	F_D, N
10	0.29
15	0.64
20	0.96
25	1.41
30	1.55
35	2.10
40	2.65
45	3.28
50	4.07
55	4.91

7–81 Water at 20°C flows through a long, straight pipe. The pressure drop is measured along a section of the pipe of length $L = 1.3$ m as a function of average velocity V through the pipe (Table P7–81). The inner diameter of the pipe is $D = 10.4$ cm. (*a*) Nondimensionalize the data and plot the Euler number as a function of the Reynolds number. Has the experiment been run at high enough speeds to achieve Reynolds number independence? (*b*) Extrapolate the experimental data to predict the pressure drop at an average speed of 80 m/s. *Answer:* 1,940,000 N/m²

TABLE P7–81

V, m/s	ΔP, N/m²
0.5	77.0
1	306
2	1218
4	4865
6	10,920
8	19,440
10	30,340
15	68,330
20	121,400
25	189,800
30	273,200
35	372,100
40	485,300
45	614,900
50	758,700

7–82 In the model truck example discussed in Section 7–5, the wind tunnel test section is 3.5 m long, 0.85 m tall, and 0.90 m wide. The one-sixteenth scale model truck is 0.991 m long, 0.257 m tall, and 0.159 m wide. What is the wind tunnel blockage of this model truck? Is it within acceptable limits according to the standard rule of thumb?

7–83 A small wind tunnel in a university's undergraduate fluid flow laboratory has a test section that is 50 by 50 cm in cross section and is 1.2 m long. Its maximum speed is 44 m/s. Some students wish to build a model 18-wheeler to study how aerodynamic drag is affected by rounding off the back of the trailer. A full-size (prototype) tractor-trailer rig is 16 m long, 2.5 m wide, and 3.7 m high. Both the air in the wind tunnel and the air flowing over the prototype are at 25°C and atmospheric pressure. (*a*) What is the largest scale model they can build to stay within the rule-of-thumb guidelines for blockage? What are the dimensions of the model truck in inches? (*b*) What is the maximum model truck Reynolds number achievable by the students? (*c*) Are the students able to achieve Reynolds number independence? Discuss.

7–84 Use dimensional analysis to show that in a problem involving shallow water waves (Fig. P7–84), both the Froude number and the Reynolds number are relevant dimensionless parameters. The wave speed c of waves on the surface of a liquid is a function of depth h, gravitational acceleration g, fluid density ρ, and fluid viscosity μ. Manipulate your Π's to get the parameters into the following form:

$$\text{Fr} = \frac{c}{\sqrt{gh}} = f(\text{Re}) \quad \text{where Re} = \frac{\rho c h}{\mu}$$

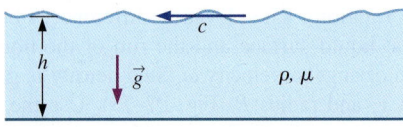

FIGURE P7–84

Review Problems

7–85C There are many established nondimensional parameters besides those listed in Table 7–5. Do a literature search or an Internet search and find at least three established, named nondimensional parameters that are *not* listed in Table 7–5. For each one, provide its definition and its ratio of significance, following the format of Table 7–5. If your equation contains any variables not identified in Table 7–5, be sure to identify those variables.

7–86C Think about and describe a prototype flow and a corresponding model flow that have geometric similarity, but not kinematic similarity, even though the Reynolds numbers match. Explain.

7–87C For each statement, choose whether the statement is true or false and discuss your answer briefly.

(*a*) Kinematic similarity is a necessary and sufficient condition for dynamic similarity.

(*b*) Geometric similarity is a necessary condition for dynamic similarity.

(*c*) Geometric similarity is a necessary condition for kinematic similarity.

(*d*) Dynamic similarity is a necessary condition for kinematic similarity.

7–88 Write the primary dimensions of each of the following variables from the field of solid mechanics, showing all your work: (*a*) moment of inertia I; (*b*) modulus of elasticity E, also called Young's modulus; (*c*) strain ε; (*d*) stress σ. (*e*) Finally, show that the relationship between stress and strain (Hooke's law) is a dimensionally homogeneous equation.

7–89 Force F is applied at the tip of a cantilever beam of length L and moment of inertia I (Fig. P7–89). The modulus of elasticity of the beam material is E. When the force is applied, the tip deflection of the beam is z_d. Use dimensional analysis to generate a relationship for z_d as a function of the independent variables. Name any established dimensionless parameters that appear in your analysis.

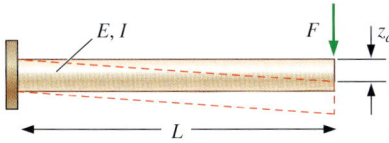

FIGURE P7–89

7–90 An explosion occurs in the atmosphere when an antiaircraft missile meets its target (Fig. P7–90). A **shock wave** (also called a **blast wave**) spreads out radially from the explosion. The pressure difference across the blast wave ΔP and its radial distance r from the center are functions of time t, speed of sound c, and the total amount of energy E released by the explosion. (*a*) Generate dimensionless relationships between ΔP and the other parameters and between r and the other parameters. (*b*) For a given explosion, if the time t since the explosion doubles, all else being equal, by what factor will ΔP decrease?

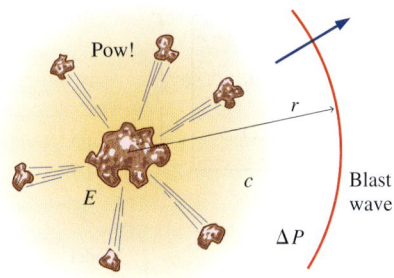

FIGURE P7–90

7–91 The Archimedes number listed in Table 7–5 is appropriate for buoyant *particles* in a fluid. Do a literature search or an Internet search and find an alternative definition of the Archimedes number that is appropriate for buoyant *fluids* (e.g., buoyant jets and buoyant plumes, heating and airconditioning applications). Provide its definition and its ratio of significance, following the format of Table 7–5. If your equation contains any variables not identified in Table 7–5, be sure to identify those variables. Finally, look through the established dimensionless parameters listed in Table 7–5 and find one that is similar to this alternate form of the Archimedes number.

7–92 Consider steady, laminar, fully developed, two-dimensional **Poiseuille flow**—flow between two infinite parallel plates separated by distance h, with both the top plate and bottom plate stationary, and a forced pressure gradient dP/dx driving the flow as illustrated in Fig. P7–92. (dP/dx is constant and negative.) The flow is steady, incompressible, and two-dimensional in the xy-plane. The flow is also *fully developed*, meaning that the velocity profile does not change with downstream distance x. Because of the fully developed nature of the flow, there are no inertial effects and density does not enter the problem. It turns out that u, the velocity component in the x-direction, is a function of distance h, pressure gradient dP/dx, fluid viscosity μ, and vertical coordinate y. Perform a dimensional analysis (showing all your work), and generate a dimensionless relationship between the given variables.

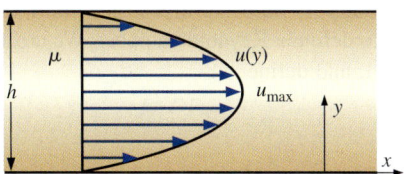

FIGURE P7–92

7–93 Consider the steady, laminar, fully developed, two-dimensional Poiseuille flow of Prob. 7–92. The maximum velocity u_{max} occurs at the center of the channel. (*a*) Generate a dimensionless relationship for u_{max} as a function of distance between plates h, pressure gradient dP/dx, and fluid viscosity μ. (*b*) If the plate separation distance h is doubled, all else being equal, by what factor will u_{max} change? (*c*) If the pressure gradient dP/dx is doubled, all else being equal, by what factor will u_{max} change? (*d*) How many experiments are required to describe the complete relationship between u_{max} and the other parameters in the problem?

7–94 The pressure drop $\Delta P = P_1 - P_2$ through a long section of round pipe can be written in terms of the shear stress τ_w along the wall. Shown in Fig. P7–94 is the shear stress acting by the wall on the fluid. The shaded region is a control volume composed of the fluid in the pipe between axial locations 1 and 2. There are two dimensionless parameters related to the pressure drop: the Euler number Eu and the Darcy friction factor f. (*a*) Using the control volume sketched in Fig. P7–94, generate a relationship for f in terms of Eu (and

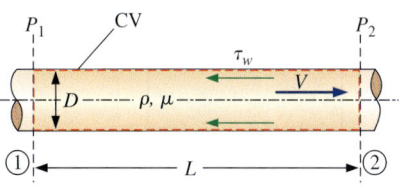

FIGURE P7–94

any other properties or parameters in the problem as needed). (b) Using the experimental data and conditions of Prob. 7–81 (Table P7–81), plot the Darcy friction factor as a function of Re. Does f show Reynolds number independence at large values of Re? If so, what is the value of f at very high Re?

Answers: (a) $f = 2\dfrac{D}{L}\mathrm{Eu}$; (b) yes, 0.0487

7–95 Oftentimes it is desirable to work with an established dimensionless parameter, but the characteristic scales available do not match those used to define the parameter. In such cases, we *create* the needed characteristic scales based on dimensional reasoning (usually by inspection). Suppose for example that we have a characteristic velocity scale V, characteristic area A, fluid density ρ, and fluid viscosity μ, and we wish to define a Reynolds number. We create a length scale $L = \sqrt{A}$, and define

$$\mathrm{Re} = \dfrac{\rho V \sqrt{A}}{\mu}$$

In similar fashion, define the desired established dimensionless parameter for each case: (a) Define a Froude number, given $\dot{V}'$ = volume flow rate per unit depth, length scale L, and gravitational constant g. (b) Define a Reynolds number, given $\dot{V}'$ = volume flow rate per unit depth and kinematic viscosity ν. (c) Define a Richardson number (see Table 7–5), given $\dot{V}'$ = volume flow rate per unit depth, length scale L, characteristic density difference $\Delta\rho$, characteristic density ρ, and gravitational constant g.

7–96 A liquid of density ρ and viscosity μ flows by gravity through a hole of diameter d in the bottom of a tank of diameter D (Fig. P7–96). At the start of the experiment, the liquid surface is at height h above the bottom of the tank, as sketched. The liquid exits the tank as a jet with average velocity V straight down as also sketched. Using dimensional analysis, generate a dimensionless relationship for V as a function of the other parameters in the problem. Identify any established nondimensional parameters that appear in your result. (*Hint*: There are three length scales in this problem. For consistency, choose h as your length scale.)

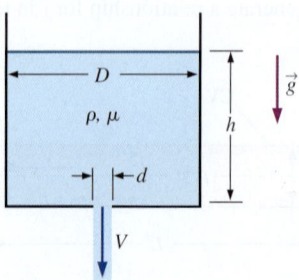

FIGURE P7–96

7–97 Repeat Prob. 7–96 except for a different dependent parameter, namely, the time required to empty the tank t_{empty}. Generate a dimensionless relationship for t_{empty} as a function of the following independent parameters: hole diameter d, tank diameter D, density ρ, viscosity μ, initial liquid surface height h, and gravitational acceleration g.

7–98 A liquid delivery system is being designed such that ethylene glycol flows out of a hole in the bottom of a large tank, as in Fig. P7–96. The designers need to predict how long it will take for the ethylene glycol to completely drain. Since it would be very expensive to run tests with a full-scale prototype using ethylene glycol, they decide to build a one-quarter scale model for experimental testing, and they plan to use *water* as their test liquid. The model is geometrically similar to the prototype (Fig. P7–98). (a) The temperature of the ethylene glycol in the prototype tank is 60°C, at which $\nu = 4.75 \times 10^{-6}$ m²/s. At what temperature should the water in the model experiment be set in order to ensure complete similarity between model and prototype? (b) The experiment is run with water at the proper temperature as calculated in part (a). It takes 3.27 min to drain the model tank. Predict how long it will take to drain the ethylene glycol from the prototype tank. Answers: (a) 45.8°C, (b) 6.54 min

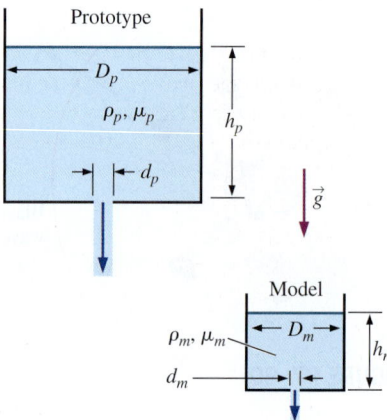

FIGURE P7–98

7–99 Liquid flows out of a hole in the bottom of a tank as in Fig. P7–96. Consider the case in which the hole is very small compared to the tank ($d \ll D$). Experiments reveal that average jet velocity V is nearly independent of d, D, ρ, or μ. In fact, for a wide range of these parameters, it turns out that V depends only on liquid surface height h and gravitational acceleration g. If the liquid surface height is doubled, all else being equal, by what factor will the average jet velocity increase? Answer: $\sqrt{2}$

7–100 An aerosol particle of characteristic size D_p moves in an airflow of characteristic length L and characteristic velocity V. The characteristic time required for the particle to adjust to a sudden change in air speed is called the **particle relaxation time** τ_p,

$$\tau_p = \frac{\rho_p D_p^2}{18\mu}$$

Verify that the primary dimensions of τ_p are time. Then create a dimensionless form of τ_p, based on some characteristic velocity V and some characteristic length L of the airflow (Fig. P7–100). What established dimensionless parameter do you create?

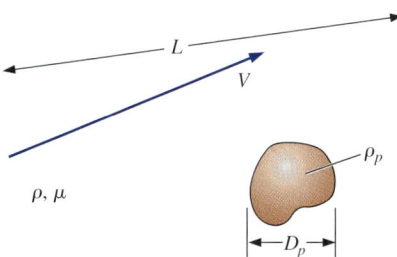

FIGURE P7–100

7–101 Compare the primary dimensions of each of the following properties in the *mass-based* primary dimension system (m, L, t, T, I, C, N) to those in the *force-based* primary dimension system (F, L, t, T, I, C, N): (*a*) pressure or stress; (*b*) moment or torque; (*c*) work or energy. Based on your results, explain when and why some authors prefer to use force as a primary dimension in place of mass.

7–102 The Stanton number is listed as a named, established nondimensional parameter in Table 7–5. However, careful analysis reveals that it can actually be formed by a combination of the Reynolds number, Nusselt number, and Prandtl number. Find the relationship between these four dimensionless groups, showing all your work. Can you also form the Stanton number by some combination of only *two* other established dimensionless parameters?

7–103 Consider a variation of the fully developed Couette flow problem of Prob. 7–52—flow between two infinite parallel plates separated by distance h, with the top plate moving at speed V_{top} and the bottom plate moving at speed V_{bottom} as illustrated in Fig. P7–103. The flow is steady, incompressible, and two-dimensional in the *xy*-plane. Generate a dimensionless relationship for the *x*-component of fluid velocity u as a function of fluid viscosity μ, plate speeds V_{top} and V_{bottom}, distance h, fluid density ρ, and distance y. (*Hint*: Think carefully about the list of parameters before rushing into the algebra.)

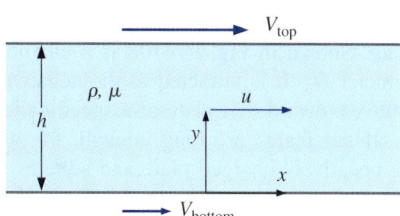

FIGURE P7–103

7–104 What are the primary dimensions of electric charge q, the units of which are **coulombs** (C)? (*Hint*: Look up the fundamental definition of electric current.)

7–105 What are the primary dimensions of electrical capacitance C, the units of which are **farads**? (*Hint*: Look up the fundamental definition of electrical capacitance.)

7–106 In many electronic circuits in which some kind of time scale is involved, such as filters and time-delay circuits (Fig. P7–106—*a low-pass filter*), you often see a resistor (R) and a capacitor (C) in series. In fact, the product of R and C is called the *electrical time constant*, RC. Showing all your work, what are the primary dimensions of RC? Using dimensional reasoning alone, explain why a resistor and capacitor are often found together in timing circuits.

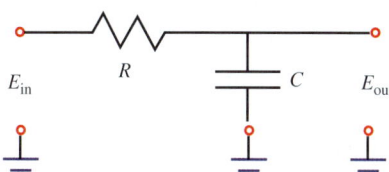

FIGURE P7–106

7–107 From fundamental electronics, the current flowing through a capacitor at any instant of time is equal to the capacitance times the rate of change of voltage (electromotive force) across the capacitor,

$$I = C \frac{dE}{dt}$$

Write the primary dimensions of both sides of this equation, and verify that the equation is dimensionally homogeneous. Show all your work.

7–108 An **electrostatic precipitator** (ESP) is a device used in various applications to clean particle-laden air. First, the dusty air passes through the *charging stage* of the ESP, where dust particles are given a positive charge q_p (coulombs) by charged ionizer wires (Fig. P7–108). The dusty air then enters the *collector stage* of the device, where it flows between two oppositely charged plates. The applied *electric* field strength between the plates is E_f (voltage difference per unit distance). Shown in Fig. P7–108 is a charged dust particle of diameter D_p. It is attracted to the negatively charged plate and moves toward that plate at a speed called the **drift velocity** w. If the plates are long enough, the dust particle impacts the negatively charged plate and adheres to it. Clean air exits the device. It turns out that for very small particles the drift velocity depends only on q_p, E_f, D_p, and air viscosity μ. (*a*) Generate a dimensionless relationship between the drift velocity through the collector stage of the ESP and the given parameters. Show all your work. (*b*) If the electric field strength is doubled, all else being equal, by what factor will the drift velocity change? (*c*) For a given ESP, if the particle diameter is doubled, all else being equal, by what factor will the drift velocity change?

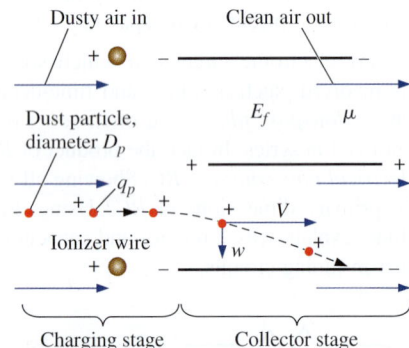

FIGURE P7–108

7–109 Experiments are being designed to measure the horizontal force F on a fireman's nozzle, as shown in Fig. P7–109. Force F is a function of velocity V_1, pressure drop $\Delta P = P_1 - P_2$, density ρ, viscosity μ, inlet area A_1, outlet area A_2, and length L. Perform a dimensional analysis for $F = f(V_1, \Delta P, \rho, \mu, A_1, A_2, L)$. For consistency, use V_1, A_1, and ρ as the repeating parameters and generate a dimensionless relationship. Identify any established nondimensional parameters that appear in your result.

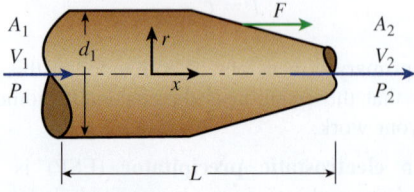

FIGURE P7–109

7–110 When a capillary tube of small diameter D is inserted into a container of liquid, the liquid rises to height h inside the tube (Fig. P7–110). h is a function of liquid density ρ, tube diameter D, gravitational constant g, contact angle ϕ, and the surface tension σ_s of the liquid. (*a*) Generate a dimensionless relationship for h as a function of the given parameters. (*b*) Compare your result to the exact analytical equation for h given in Chap. 2. Are your dimensional analysis results consistent with the exact equation? Discuss.

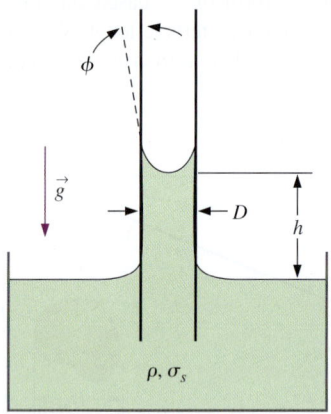

FIGURE P7–110

7–111 Repeat part (*a*) of Prob. 7–110, except instead of height h, find a functional relationship for the time scale t_{rise} needed for the liquid to climb up to its final height in the capillary tube. (*Hint*: Check the list of independent parameters in Prob. 7–110. Are there any additional relevant parameters?)

7–112 Sound intensity I is defined as the acoustic power per unit area emanating from a sound source. We know that I is a function of sound pressure level P (dimensions of pressure) and fluid properties ρ (density) and speed of sound c. (*a*) Use the method of repeating variables in mass-based primary dimensions to generate a dimensionless relationship for I as a function of the other parameters. Show all your work. What happens if you choose three repeating variables? Discuss. (*b*) Repeat part (*a*), but use the force-based primary dimension system. Discuss.

7–113 Repeat Prob. 7–112, but with the distance r from the sound source as an additional independent parameter.

7–114 Engineers at MIT have developed a mechanical model of a tuna fish to study its locomotion. The "Robotuna" shown in Fig. P7–114 is 1.0 m long and swims at speeds up to 2.0 m/s. Real bluefin tuna can exceed 3.0 m in length and have been clocked at speeds greater than 13 m/s. How fast would the 1.0-m Robotuna need to swim in order to match the Reynolds number of a real tuna that is 2.0 m long and swims at 10 m/s?

FIGURE P7–114

Photo by David Barrett of MIT, used by permission.

7–115 In Example 7–7, the mass-based system of primary dimensions was used to establish a relationship for the pressure difference $\Delta P = P_{inside} - P_{outside}$ between the inside and outside of a soap bubble as a function of soap bubble radius R and surface tension σ_s of the soap film (Fig. P7–115). Repeat the dimensional analysis using the method of repeating variables, but use the *force-based* system of primary dimensions instead. Show all your work. Do you get the same result?

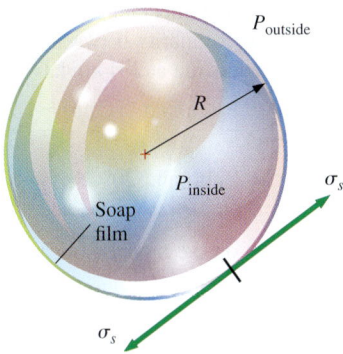

FIGURE P7–115

7–116 Many of the established nondimensional parameters listed in Table 7–5 can be formed by the product or ratio of two other established nondimensional parameters. For each pair of nondimensional parameters listed, find a *third* established nondimensional parameter that is formed by some manipulation of the two given parameters: (*a*) Reynolds number and Prandtl number; (*b*) Schmidt number and Prandtl number; (*c*) Reynolds number and Schmidt number.

7–117 A common device used in various applications to clean particle-laden air is the **reverse-flow cyclone** (Fig. P7–117). Dusty air (volume flow rate $\dot{V}$ and density ρ) enters tangentially through an opening in the side of the cyclone and swirls around in the tank. Dust particles are flung outward and fall out the bottom, while clean air is drawn out the top. The reverse-flow cyclones being studied are all geometrically similar; hence, diameter D represents the only length scale required to fully specify the entire cyclone geometry. Engineers are concerned about the pressure drop δP through the cyclone. (*a*) Generate a dimensionless relationship between the pressure drop through the cyclone and the given parameters. Show all your work. (*b*) If the cyclone size is doubled, all else being equal, by what factor will the pressure drop change? (*c*) If the volume flow rate is doubled, all else being equal, by what factor will the pressure drop change? *Answers:* (*a*) $D^4 \delta P / \rho \dot{V}^2$ = constant, (*b*) 1/16, (*c*) 4

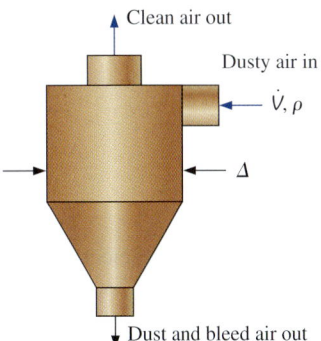

FIGURE P7–117

Fundamentals of Engineering (FE) Exam Problems

7–118 Which one is *not* a primary dimension?
(*a*) Velocity
(*b*) Time
(*c*) Electric current
(*d*) Temperature
(*e*) Mass

7–119 The primary dimensions of kinematic viscosity are
(*a*) $m \cdot L/t^2$
(*b*) $m/L \cdot t$
(*c*) L^2/t
(*d*) $L^2/m \cdot t$
(*e*) $L/m \cdot t^2$

7–120 The thermal conductivity of a substance may be defined as the rate of heat transfer per unit length per unit temperature difference. The primary dimensions of thermal conductivity are
(*a*) $m^2 \cdot L/t^2 \cdot T$
(*b*) $m^2 \cdot L^2/t \cdot T$
(*c*) $L^2/m \cdot t^2 \cdot T$
(*d*) $m \cdot L/t^3 \cdot T$
(*e*) $m \cdot L^2/t^3 \cdot T$

DIMENSIONAL ANALYSIS AND MODELING

7–121 The primary dimensions of the gas constant over the universal gas constant R/R_u are
(a) $L^2/t^2 \cdot T$
(b) $m \cdot L/N$
(c) $m/t \cdot N \cdot T$
(d) m/L^3
(e) N/m

7–122 The primary dimensions of the universal gas constant R_u are
(a) $m \cdot L/t^2 \cdot T$
(b) $m^2 \cdot L/N$
(c) $m \cdot L^2/t^2 \cdot N \cdot T$
(d) $L^2/t^2 \cdot T$
(e) $N/m \cdot t$

7–123 There are four additive terms in an equation, and their units are given below. Which one is not consistent with this equation?
(a) J
(b) W/m
(c) $kg \cdot m^2/s^2$
(d) $Pa \cdot m^3$
(e) $N \cdot m$

7–124 The heat transfer coefficient is a nondimensional parameter which is a function of viscosity μ, specific heat c_p (kJ/kg·K), and thermal conductivity k (W/m·K). This nondimensional parameter is expressed as
(a) $c_p/\mu k$
(b) $k/\mu c_p$
(c) $\mu/c_p k$
(d) $\mu c_p/k$
(e) $c_p k/\mu$

7–125 The nondimensional heat transfer coefficient is a function of convection coefficient h (W/m²·K), thermal conductivity k (W/m·K), and characteristic length L. This nondimensional parameter is expressed as
(a) hL/k
(b) h/kL
(c) L/hk
(d) hk/L
(e) kL/h

7–126 The drag coefficient C_D is a nondimensional parameter and is a function of drag force F_D, density ρ, velocity V, and area A. The drag coefficient is expressed as
(a) $\dfrac{F_D V^2}{2\rho A}$
(b) $\dfrac{2F_D}{\rho V A}$
(c) $\dfrac{\rho V A^2}{F_D}$
(d) $\dfrac{F_D A}{\rho V}$
(e) $\dfrac{2F_D}{\rho V^2 A}$

7–127 Which similarity condition is related to force-scale equivalence?
(a) Geometric
(b) Kinematic
(c) Dynamic
(d) Kinematic and dynamic
(e) Geometric and kinematic

7–128 A one-third scale model of a car is to be tested in a wind tunnel. The conditions of the actual car are $V = 75$ km/h and $T = 0°C$ and the air temperature in the wind tunnel is 20°C.

The properties of air at 1 atm and 0°C: $\rho = 1.292$ kg/m³, $v = 1.338 \times 10^{-5}$ m²/s.
The properties of air at 1 atm and 20°C: $\rho = 1.204$ kg/m³, $v = 1.516 \times 10^{-5}$ m²/s.

In order to achieve similarity between the model and the prototype, the wind tunnel velocity should be
(a) 255 km/h
(b) 225 km/h
(c) 147 km/h
(d) 75 km/h
(e) 25 km/h

7–129 A one-fourth scale model of a car is to be tested in a wind tunnel. The conditions of the actual car are $V = 45$ km/h and $T = 0°C$ and the air temperature in the wind tunnel is 20°C. In order to achieve similarity between the model and the prototype, the wind tunnel is run at 204 km/h.

The properties of air at 1 atm and 0°C: $\rho = 1.292$ kg/m³, $v = 1.338 \times 10^{-5}$ m²/s.

The properties of air at 1 atm and 20°C: $\rho = 1.204$ kg/m³, $v = 1.516 \times 10^{-5}$ m²/s.

If the average drag force on the model is measured to be 70 N, the drag force on the prototype is
(a) 17.5 N
(b) 58.5 N
(c) 70 N
(d) 93.2 N
(e) 280 N

7–130 A one-third scale model of an airplane is to be tested in water. The airplane has a velocity of 900 km/h in air at $-50°C$. The water temperature in the test section is 10°C.
The properties of air at 1 atm and $-50°C$: $\mu = 1.582$ kg/m³, $\mu = 1.474 \times 10^{-5}$ kg/m·s.
The properties of water at 1 atm and 10°C: $\mu = 999.7$ kg/m³, $\mu = 1.307 \times 10^{-3}$ kg/m·s.

In order to achieve similarity between the model and the prototype, the water velocity on the model should be
(a) 97 km/h
(b) 186 km/h
(c) 263 km/h
(d) 379 km/h
(e) 450 km/h

7–131 A one-fourth scale model of an airplane is to be tested in water. The airplane has a velocity of 700 km/h in air at −50°C. The water temperature in the test section is 10°C. In order to achieve similarity between the model and the prototype, the test is done at a water velocity of 393 km/h.

The properties of air at 1 atm and −50°C: $\rho = 1.582$ kg/m^3, $\mu = 1.474 \times 10^{-5}$ kg/m·s.

The properties of water at 1 atm and 10°C: $\rho = 999.7$ kg/m^3, $\mu = 1.307 \times 10^{-3}$ kg/m·s.

If the average drag force on the model is measured to be 13,800 N, the drag force on the prototype is
(a) 590 N
(b) 862 N
(c) 1109 N
(d) 4655 N
(e) 3450 N

7–132 Consider a boundary layer growing along a thin flat plate. This problem involves the following parameters: boundary layer thickness μ, downstream distance x, free-stream velocity V, fluid density μ, and fluid viscosity μ. The number of expected nondimensional parameters Πs for this problem is
(a) 5
(b) 4
(c) 3
(d) 2
(e) 1

7–133 Consider unsteady fully developed Couette flow-flow between two infinite parallel plates. This problem involves the following parameters: velocity component u, distance between the plates h, vertical distance y, top plate speed V, fluid density ρ, fluid viscosity μ, and time t. The number of expected nondimensional parameters μs for this problem is
(a) 6
(b) 5
(c) 4
(d) 3
(e) 2

7–134 Consider a boundary layer growing along a thin flat plate. This problem involves the following parameters: boundary layer thickness δ, downstream distance x, free-stream velocity V, fluid density ρ, and fluid viscosity μ. The number of primary dimensions represented in this problem is
(a) 1
(b) 2
(c) 3
(d) 4
(e) 5

7–135 Consider a boundary layer growing along a thin flat plate. This problem involves the following parameters: boundary layer thickness δ, downstream distance x, free-stream velocity V, fluid density ρ, and fluid viscosity μ. The dependent parameter is δ. If we choose three repeating parameters as x, ρ, and V, the dependent Π is
(a) $\delta x^2/V$
(b) $\delta V^2/x\rho$
(c) $\delta \rho/xV$
(d) $x/\delta V$
(e) δ/x

CHAPTER 8

INTERNAL FLOW

Fluid flow is classified as *external* or *internal*, depending on whether the fluid is forced to flow over a surface or in a conduit. Internal and external flows exhibit very different characteristics. In this chapter we consider *internal flow* where the conduit is completely filled with the fluid, and the flow is driven primarily by a pressure difference. This should not be confused with *open-channel flow* (Chap. 13) where the conduit is partially filled by the fluid and thus the flow is partially bounded by solid surfaces, as in an irrigation ditch, and the flow is driven by gravity alone.

We start this chapter with a general physical description of internal flow through pipes and ducts including the *entrance region* and the *fully developed* region. We continue with a discussion of the dimensionless *Reynolds number* and its physical significance. We then introduce the *pressure drop* correlations associated with pipe flow for both laminar and turbulent flows. Then, we discuss minor losses and determine the pressure drop and pumping power requirements for real-world piping systems. Finally, we present a brief overview of flow measurement devices.

OBJECTIVES

When you finish reading this chapter, you should be able to

- Have a deeper understanding of laminar and turbulent flow in pipes and the analysis of fully developed flow
- Calculate the major and minor losses associated with pipe flow in piping networks and determine the pumping power requirements
- Understand various velocity and flow rate measurement techniques and learn their advantages and disadvantages

Internal flows through pipes, elbows, tees, valves, etc., as in this oil refinery, are found in nearly every industry.
Royalty Free/CORBIS

347

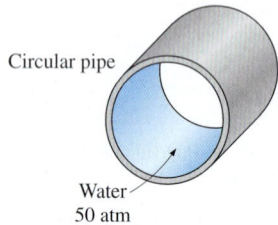

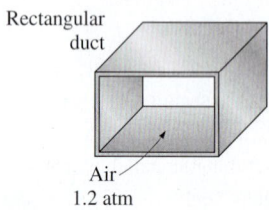

FIGURE 8–1
Circular pipes can withstand large pressure differences between the inside and the outside without undergoing any significant distortion, but noncircular pipes cannot.

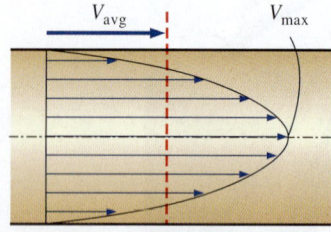

FIGURE 8–2
Average velocity V_{avg} is defined as the average speed through a cross section. For fully developed laminar pipe flow, V_{avg} is half of the maximum velocity.

8–1 · INTRODUCTION

Liquid or gas flow through *pipes* or *ducts* is commonly used in heating and cooling applications and fluid distribution networks. The fluid in such applications is usually forced to flow by a fan or pump through a flow section. We pay particular attention to *friction*, which is directly related to the *pressure drop* and *head loss* during flow through pipes and ducts. The pressure drop is then used to determine the pumping power requirement. A typical piping system involves pipes of different diameters connected to each other by various fittings or elbows to route the fluid, valves to control the flow rate, and pumps to pressurize the fluid.

The terms *pipe*, *duct*, and *conduit* are usually used interchangeably for flow sections. In general, flow sections of circular cross section are referred to as *pipes* (especially when the fluid is a liquid), and flow sections of noncircular cross section as *ducts* (especially when the fluid is a gas). Small-diameter pipes are usually referred to as *tubes*. Given this uncertainty, we will use more descriptive phrases (such as *a circular pipe* or *a rectangular duct*) whenever necessary to avoid any misunderstandings.

You have probably noticed that most fluids, especially liquids, are transported in *circular pipes*. This is because pipes with a circular cross section can withstand large pressure differences between the inside and the outside without undergoing significant distortion. *Noncircular pipes* are usually used in applications such as the heating and cooling systems of buildings where the pressure difference is relatively small, the manufacturing and installation costs are lower, and the available space is limited for ductwork (Fig. 8–1).

Although the theory of fluid flow is reasonably well understood, theoretical solutions are obtained only for a few simple cases such as fully developed laminar flow in a circular pipe. Therefore, we must rely on experimental results and empirical relations for most fluid flow problems rather than closed-form analytical solutions. Noting that the experimental results are obtained under carefully controlled laboratory conditions and that no two systems are exactly alike, we must not be so naive as to view the results obtained as "exact." An error of 10 percent (or more) in friction factors calculated using the relations in this chapter is the "norm" rather than the "exception."

The fluid velocity in a pipe changes from *zero* at the wall because of the no-slip condition to a maximum at the pipe center. In fluid flow, it is convenient to work with an *average* velocity V_{avg}, which remains constant in incompressible flow when the cross-sectional area of the pipe is constant (Fig. 8–2). The average velocity in heating and cooling applications may change somewhat because of changes in density with temperature. But, in practice, we evaluate the fluid properties at some average temperature and treat them as constants. The convenience of working with constant properties usually more than justifies the slight loss in accuracy.

Also, the friction between the fluid particles in a pipe does cause a slight rise in fluid temperature as a result of the mechanical energy being converted to sensible thermal energy. But this temperature rise due to *frictional heating* is usually too small to warrant any consideration in calculations and thus is disregarded. For example, in the absence of any heat transfer, no noticeable difference can be detected between the inlet and outlet temperatures of water flowing in a pipe. The primary consequence of friction in fluid flow is pressure drop, and thus any significant temperature change in the fluid is due to heat transfer.

The value of the average velocity V_{avg} at some streamwise cross-section is determined from the requirement that the *conservation of mass* principle be satisfied (Fig. 8–2). That is,

$$\dot{m} = \rho V_{avg} A_c = \int_{A_c} \rho u(r) \, dA_c \tag{8-1}$$

where $\dot{m}$ is the mass flow rate, ρ is the density, A_c is the cross-sectional area, and $u(r)$ is the velocity profile. Then the average velocity for incompressible flow in a circular pipe of radius R is expressed as

$$V_{avg} = \frac{\int_{A_c} \rho u(r) \, dA_c}{\rho A_c} = \frac{\int_0^R \rho u(r) 2\pi r \, dr}{\rho \pi R^2} = \frac{2}{R^2} \int_0^R u(r) r \, dr \tag{8-2}$$

Therefore, when we know the flow rate or the velocity profile, the average velocity can be determined easily.

8–2 · LAMINAR AND TURBULENT FLOWS

If you have been around smokers, you probably noticed that the cigarette smoke rises in a smooth plume for the first few centimeters and then starts fluctuating randomly in all directions as it continues its rise. Other plumes behave similarly (Fig. 8–3). Likewise, a careful inspection of flow in a pipe reveals that the fluid flow is streamlined at low velocities but turns chaotic as the velocity is increased above a critical value, as shown in Fig. 8–4. The flow regime in the first case is said to be **laminar**, characterized by *smooth streamlines* and *highly ordered motion,* and **turbulent** in the second case, where it is characterized by *velocity fluctuations* and *highly disordered motion.* The **transition** from laminar to turbulent flow does not occur suddenly; rather, it occurs over some region in which the flow fluctuates between laminar and turbulent flows before it becomes fully turbulent. Most flows encountered in practice are turbulent. Laminar flow is encountered when highly viscous fluids such as oils flow in small pipes or narrow passages.

We can verify the existence of these laminar, transitional, and turbulent flow regimes by injecting some dye streaks into the flow in a glass pipe, as the British engineer Osborne Reynolds (1842–1912) did over a century ago. We observe that the dye streak forms a *straight and smooth line* at low velocities when the flow is laminar (we may see some blurring because of molecular diffusion), has *bursts of fluctuations* in the transitional regime, and *zigzags rapidly and disorderly* when the flow becomes fully turbulent. These zigzags and the dispersion of the dye are indicative of the fluctuations in the main flow and the rapid mixing of fluid particles from adjacent layers.

The *intense mixing* of the fluid in turbulent flow as a result of rapid fluctuations enhances momentum transfer between fluid particles, which increases the friction force on the pipe wall and thus the required pumping power. The friction factor reaches a maximum when the flow becomes fully turbulent.

FIGURE 8–3
Laminar and turbulent flow regimes of a candle smoke plume.

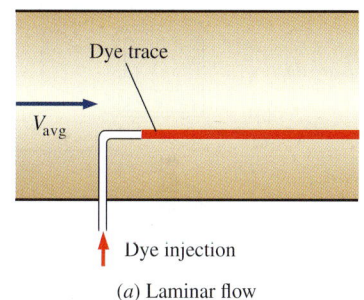

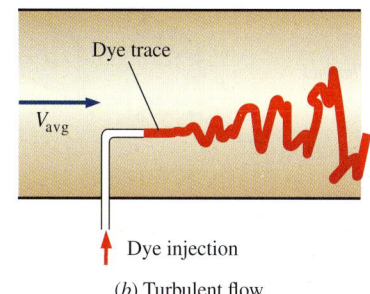

FIGURE 8–4
The behavior of colored fluid injected into the flow in (*a*) laminar and (*b*) turbulent flow in a pipe.

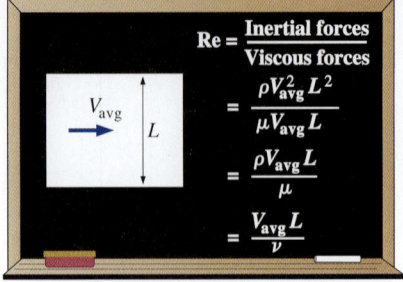

FIGURE 8–5
The Reynolds number can be viewed as the ratio of inertial forces to viscous forces acting on a fluid element.

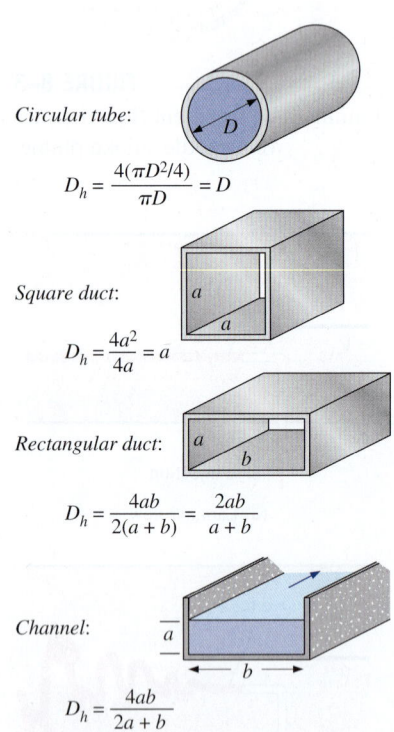

FIGURE 8–6
The hydraulic diameter $D_h = 4A_c/p$ is defined such that it reduces to ordinary diameter for circular tubes. When there is a free surface, such as in open-channel flow, the wetted perimeter includes only the walls in contact with the fluid.

Reynolds Number

The transition from laminar to turbulent flow depends on the *geometry, surface roughness, flow velocity, surface temperature,* and *type of fluid,* among other things. After exhaustive experiments in the 1880s, Osborne Reynolds discovered that the flow regime depends mainly on the ratio of *inertial forces* to *viscous forces* in the fluid (Fig. 8–5). This ratio is called the **Reynolds number** and is expressed for internal flow in a circular pipe as

$$\text{Re} = \frac{\text{Inertial forces}}{\text{Viscous forces}} = \frac{V_{avg} D}{\nu} = \frac{\rho V_{avg} D}{\mu} \qquad (8\text{–}3)$$

where V_{avg} = average flow velocity (m/s), D = characteristic length of the geometry (diameter in this case, in m), and $\nu = \mu/\rho$ = kinematic viscosity of the fluid (m^2/s). Note that the Reynolds number is a *dimensionless* quantity (Chap. 7). Also, kinematic viscosity has units m^2/s, and can be viewed as *viscous diffusivity* or *diffusivity for momentum.*

At large Reynolds numbers, the inertial forces, which are proportional to the fluid density and the square of the fluid velocity, are large relative to the viscous forces, and thus the viscous forces cannot prevent the random and rapid fluctuations of the fluid. At *small* or *moderate* Reynolds numbers, however, the viscous forces are large enough to suppress these fluctuations and to keep the fluid "in line." Thus the flow is *turbulent* in the first case and *laminar* in the second.

The Reynolds number at which the flow becomes turbulent is called the **critical Reynolds number, Re$_{cr}$.** The value of the critical Reynolds number is different for different geometries and flow conditions. For internal flow in a circular pipe, the generally accepted value of the critical Reynolds number is Re$_{cr}$ = 2300.

For flow through noncircular pipes, the Reynolds number is based on the **hydraulic diameter** D_h defined as (Fig. 8–6)

Hydraulic diameter: $$D_h = \frac{4A_c}{p} \qquad (8\text{–}4)$$

where A_c is the cross-sectional area of the pipe and p is its wetted perimeter. The hydraulic diameter is defined such that it reduces to ordinary diameter D for circular pipes,

Circular pipes: $$D_h = \frac{4A_c}{p} = \frac{4(\pi D^2/4)}{\pi D} = D$$

It certainly is desirable to have precise values of Reynolds numbers for laminar, transitional, and turbulent flows, but this is not the case in practice. It turns out that the transition from laminar to turbulent flow also depends on the degree of disturbance of the flow by *surface roughness, pipe vibrations,* and *fluctuations in the upstream flow.* Under most practical conditions, the flow in a circular pipe is laminar for Re ≲ 2300, turbulent for Re ≳ 4000, and transitional in between. That is,

$$\text{Re} \lesssim 2300 \quad \text{laminar flow}$$
$$2300 \lesssim \text{Re} \lesssim 4000 \quad \text{transitional flow}$$
$$\text{Re} \gtrsim 4000 \quad \text{turbulent flow}$$

In transitional flow, the flow switches between laminar and turbulent in a disorderly fashion (Fig. 8–7). It should be kept in mind that laminar flow can be maintained at much higher Reynolds numbers in very smooth pipes by avoiding flow disturbances and pipe vibrations. In such carefully controlled laboratory experiments, laminar flow has been maintained at Reynolds numbers of up to 100,000.

8–3 · THE ENTRANCE REGION

Consider a fluid entering a circular pipe at a uniform velocity. Because of the no-slip condition, the fluid particles in the layer in contact with the wall of the pipe come to a complete stop. This layer also causes the fluid particles in the adjacent layers to slow down gradually as a result of friction. To make up for this velocity reduction, the velocity of the fluid at the midsection of the pipe has to increase to keep the mass flow rate through the pipe constant. As a result, a velocity gradient develops along the pipe.

The region of the flow in which the effects of the viscous shearing forces caused by fluid viscosity are felt is called the **velocity boundary layer** or just the **boundary layer.** The hypothetical boundary surface divides the flow in a pipe into two regions: the **boundary layer region,** in which the viscous effects and the velocity changes are significant, and the **irrotational (core) flow region,** in which the frictional effects are negligible and the velocity remains essentially constant in the radial direction.

The thickness of this boundary layer increases in the flow direction until the boundary layer reaches the pipe center and thus fills the entire pipe, as shown in Fig. 8–8, and the velocity becomes fully developed a little farther downstream. The region from the pipe inlet to the point at which the velocity profile is fully developed is called the **hydrodynamic entrance region,** and the length of this region is called the **hydrodynamic entry length** L_h. Flow in the entrance region is called *hydrodynamically developing flow* since this is the region where the velocity profile develops. The region beyond the entrance region in which the velocity profile is fully developed and remains unchanged is called the **hydrodynamically fully developed region.** The flow is said to be **fully developed** when the normalized temperature profile remains unchanged as well. Hydrodynamically fully developed flow is equivalent to fully developed flow when the fluid in the pipe is not heated or cooled since the fluid temperature in this case remains

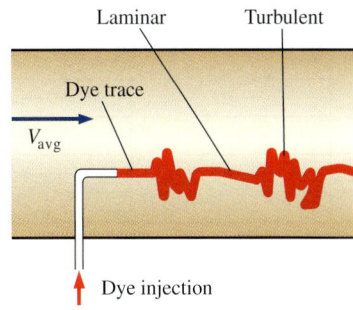

FIGURE 8–7
In the transitional flow region of $2300 \leq \text{Re} \leq 4000$, the flow switches between laminar and turbulent somewhat randomly.

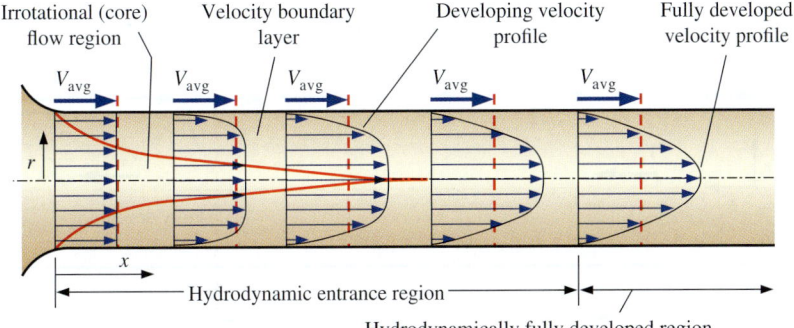

FIGURE 8–8
The development of the velocity boundary layer in a pipe. (The developed average velocity profile is parabolic in laminar flow, as shown, but much flatter or fuller in turbulent flow.)

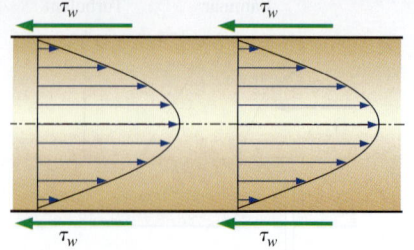

FIGURE 8–9
In the fully developed flow region of a pipe, the velocity profile does not change downstream, and thus the wall shear stress remains constant as well.

essentially constant throughout. The velocity profile in the fully developed region is *parabolic* in laminar flow and much *flatter* (or *fuller*) in turbulent flow due to eddy motion and more vigorous mixing in the radial direction. The time-averaged velocity profile remains unchanged when the flow is fully developed, and thus

Hydrodynamically fully developed: $\quad \dfrac{\partial u(r, x)}{\partial x} = 0 \quad \rightarrow \quad u = u(r)$ (8–5)

The shear stress at the pipe wall τ_w is related to the slope of the velocity profile at the surface. Noting that the velocity profile remains unchanged in the hydrodynamically fully developed region, the wall shear stress also remains constant in that region (Fig. 8–9).

Consider fluid flow in the hydrodynamic entrance region of a pipe. The wall shear stress is the *highest* at the pipe inlet where the thickness of the boundary layer is smallest, and decreases gradually to the fully developed value, as shown in Fig. 8–10. Therefore, the pressure drop is *higher* in the entrance regions of a pipe, and the effect of the entrance region is always to *increase* the average friction factor for the entire pipe. This increase may be significant for short pipes but is negligible for long ones.

Entry Lengths

The hydrodynamic entry length is usually taken to be the distance from the pipe entrance to where the wall shear stress (and thus the friction factor) reaches within about 2 percent of the fully developed value. In *laminar flow*, the nondimensional hydrodynamic entry length is given approximately as [see Kays and Crawford (2004) and Shah and Bhatti (1987)]

$$\dfrac{L_{h,\,\text{laminar}}}{D} \cong 0.05 \text{Re} \qquad (8\text{–}6)$$

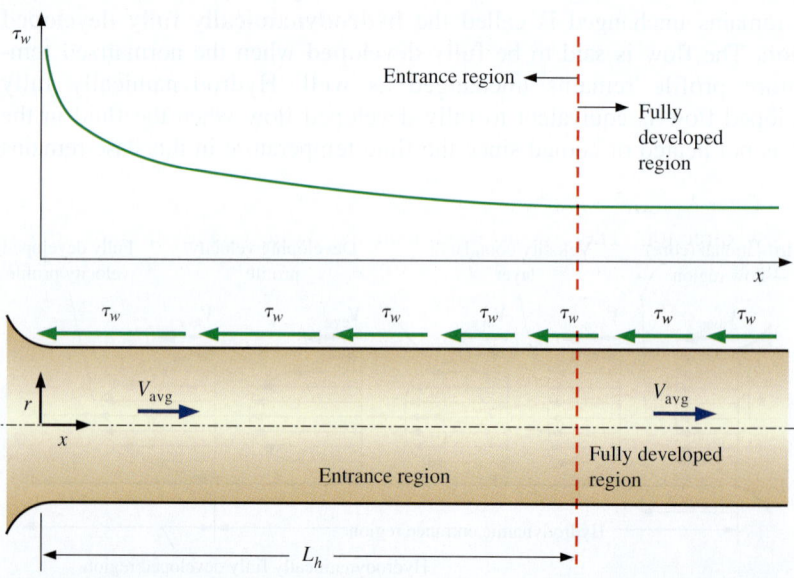

FIGURE 8–10
The variation of wall shear stress in the flow direction for flow in a pipe from the entrance region into the fully developed region.

For Re = 20, the hydrodynamic entry length is about the size of the diameter, but increases linearly with velocity. In the limiting laminar case of Re = 2300, the hydrodynamic entry length is 115D.

In *turbulent flow*, the intense mixing during random fluctuations usually overshadows the effects of molecular diffusion. The nondimensional hydrodynamic entry length for turbulent flow is approximated as [see Bhatti and Shah (1987) and Zhi-qing (1982)]

$$\frac{L_{h,\text{ turbulent}}}{D} = 1.359 \text{Re}^{1/4} \quad (8\text{–}7)$$

The entry length is much shorter in turbulent flow, as expected, and its dependence on the Reynolds number is weaker. In many pipe flows of practical engineering interest, the entrance effects become insignificant beyond a pipe length of about 10 diameters, and the nondimensional hydrodynamic entry length is approximated as

$$\frac{L_{h,\text{ turbulent}}}{D} \approx 10 \quad (8\text{–}8)$$

Precise correlations for calculating the frictional head losses in entrance regions are available in the literature. However, the pipes used in practice are usually several times the length of the entrance region, and thus the flow through the pipes is often assumed to be fully developed for the entire length of the pipe. This simplistic approach gives *reasonable* results for long pipes but sometimes poor results for short ones since it underpredicts the wall shear stress and thus the friction factor.

8–4 ▪ LAMINAR FLOW IN PIPES

We mentioned in Section 8–2 that flow in pipes is laminar for Re ≤ 2300, and that the flow is fully developed if the pipe is sufficiently long (relative to the entry length) so that the entrance effects are negligible. In this section, we consider the steady, laminar, incompressible flow of fluid with constant properties in the fully developed region of a straight circular pipe. We obtain the momentum equation by applying a momentum balance to a differential volume element, and we obtain the velocity profile by solving it. Then we use it to obtain a relation for the friction factor. An important aspect of the analysis here is that it is one of the few available for viscous flow.

In fully developed laminar flow, each fluid particle moves at a constant axial velocity along a streamline and the velocity profile $u(r)$ remains unchanged in the flow direction. There is no motion in the radial direction, and thus the velocity component in the direction normal to the pipe axis is everywhere zero. There is no acceleration since the flow is steady and fully developed.

Now consider a ring-shaped differential volume element of radius r, thickness dr, and length dx oriented coaxially with the pipe, as shown in Fig. 8–11. The volume element involves only pressure and viscous effects and thus the pressure and shear forces must balance each other. The pressure force acting on a submerged plane surface is the product of the pressure at the centroid of the surface and the surface area. A force balance on the volume element in the flow direction gives

$$(2\pi r\, dr\, P)_x - (2\pi r\, dr\, P)_{x+dx} + (2\pi r\, dx\, \tau)_r - (2\pi r\, dx\, \tau)_{r+dr} = 0 \quad (8\text{–}9)$$

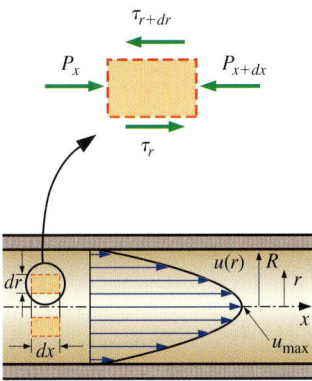

FIGURE 8–11
Free-body diagram of a ring-shaped differential fluid element of radius r, thickness dr, and length dx oriented coaxially with a horizontal pipe in fully developed laminar flow. (The size of the fluid element is greatly exaggerated for clarity.)

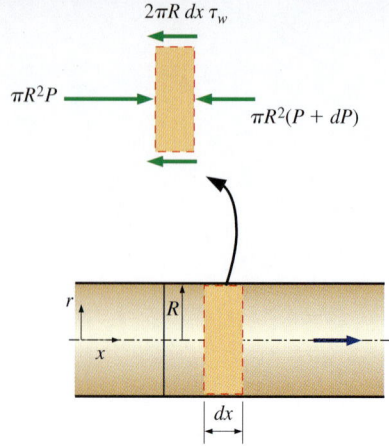

FIGURE 8–12
Free-body diagram of a fluid disk element of radius R and length dx in fully developed laminar flow in a horizontal pipe.

which indicates that in fully developed flow in a horizontal pipe, the viscous and pressure forces balance each other. Dividing by $2\pi drdx$ and rearranging,

$$r\frac{P_{x+dx}-P_x}{dx}+\frac{(r\tau)_{r+dr}-(r\tau)_r}{dr}=0 \quad (8\text{–}10)$$

Taking the limit as $dr, dx \to 0$ gives

$$r\frac{dP}{dx}+\frac{d(r\tau)}{dr}=0 \quad (8\text{–}11)$$

Substituting $\tau = -\mu(du/dr)$, dividing by r, and taking μ = constant gives the desired equation,

$$\frac{\mu}{r}\frac{d}{dr}\left(r\frac{du}{dr}\right)=\frac{dP}{dx} \quad (8\text{–}12)$$

The quantity du/dr is negative in pipe flow, and the negative sign is included to obtain positive values for τ. (Or, $du/dr = -du/dy$ if we define $y = R - r$.) The left side of Eq. 8–12 is a function of r, and the right side is a function of x. The equality must hold for any value of r and x, and an equality of the form $f(r) = g(x)$ can be satisfied only if both $f(r)$ and $g(x)$ are equal to the same constant. Thus we conclude that dP/dx = constant. This is verified by writing a force balance on a volume element of radius R and thickness dx (a slice of the pipe as in Fig. 8–12), which gives

$$\frac{dP}{dx}=-\frac{2\tau_w}{R} \quad (8\text{–}13)$$

Here τ_w is constant since the viscosity and the velocity profile are constants in the fully developed region. Therefore, dP/dx = constant.

Equation 8–12 is solved by rearranging and integrating it twice to give

$$u(r)=\frac{r^2}{4\mu}\left(\frac{dP}{dx}\right)+C_1 \ln r + C_2 \quad (8\text{–}14)$$

The velocity profile $u(r)$ is obtained by applying the boundary conditions $\partial u/\partial r = 0$ at $r = 0$ (because of symmetry about the centerline) and $u = 0$ at $r = R$ (the no-slip condition at the pipe wall),

$$u(r)=-\frac{R^2}{4\mu}\left(\frac{dP}{dx}\right)\left(1-\frac{r^2}{R^2}\right) \quad (8\text{–}15)$$

Therefore, the velocity profile in fully developed laminar flow in a pipe is *parabolic* with a maximum at the centerline and a minimum (zero) at the pipe wall. Also, the axial velocity u is positive for any r, and thus the axial pressure gradient dP/dx must be negative (i.e., pressure must decrease in the flow direction because of viscous effects—it takes pressure to push the fluid through the pipe).

The average velocity is determined from its definition by substituting Eq. 8–15 into Eq. 8–2, and performing the integration, yielding

$$V_{avg}=\frac{2}{R^2}\int_0^R u(r)r\,dr = \frac{-2}{R^2}\int_0^R \frac{R^2}{4\mu}\left(\frac{dP}{dx}\right)\left(1-\frac{r^2}{R^2}\right)r\,dr = -\frac{R^2}{8\mu}\left(\frac{dP}{dx}\right) \quad (8\text{–}16)$$

Combining the last two equations, the velocity profile is rewritten as

$$u(r)=2V_{avg}\left(1-\frac{r^2}{R^2}\right) \quad (8\text{–}17)$$

This is a convenient form for the velocity profile since V_{avg} can be determined easily from the flow rate information.

The maximum velocity occurs at the centerline and is determined from Eq. 8–17 by substituting $r = 0$,

$$u_{max} = 2V_{avg} \tag{8-18}$$

Therefore, *the average velocity in fully developed laminar pipe flow is one-half of the maximum velocity.*

Pressure Drop and Head Loss

A quantity of interest in the analysis of pipe flow is the *pressure drop* ΔP since it is directly related to the power requirements of the fan or pump to maintain flow. We note that $dP/dx =$ constant, and integrating from $x = x_1$ where the pressure is P_1 to $x = x_1 + L$ where the pressure is P_2 gives

$$\frac{dP}{dx} = \frac{P_2 - P_1}{L} \tag{8-19}$$

Substituting Eq. 8–19 into the V_{avg} expression in Eq. 8–16, the pressure drop is expressed as

Laminar flow: $$\Delta P = P_1 - P_2 = \frac{8\mu L V_{avg}}{R^2} = \frac{32\mu L V_{avg}}{D^2} \tag{8-20}$$

The symbol Δ is typically used to indicate the difference between the final and initial values, like $\Delta y = y_2 - y_1$. But in fluid flow, ΔP is used to designate pressure drop, and thus it is $P_1 - P_2$. A pressure drop due to viscous effects represents an irreversible pressure loss, and it is sometimes called **pressure loss** ΔP_L to emphasize that it is a *loss* (just like the head loss h_L, which as we shall see is proportional to ΔP_L.)

Note from Eq. 8–20 that the pressure drop is proportional to the viscosity μ of the fluid, and ΔP would be zero if there were no friction. Therefore, the drop of pressure from P_1 to P_2 in this case is due entirely to viscous effects, and Eq. 8–20 represents the pressure loss ΔP_L when a fluid of viscosity μ flows through a pipe of constant diameter D and length L at average velocity V_{avg}.

In practice, it is convenient to express the pressure loss for all types of fully developed internal flows (laminar or turbulent flows, circular or noncircular pipes, smooth or rough surfaces, horizontal or inclined pipes) as (Fig. 8–13)

Pressure loss: $$\Delta P_L = f \frac{L}{D} \frac{\rho V_{avg}^2}{2} \tag{8-21}$$

where $\rho V_{avg}^2/2$ is the *dynamic pressure* and f is the **Darcy friction factor,**

$$f = \frac{8\tau_w}{\rho V_{avg}^2} \tag{8-22}$$

It is also called the **Darcy–Weisbach friction factor,** named after the Frenchman Henry Darcy (1803–1858) and the German Julius Weisbach (1806–1871), the two engineers who provided the greatest contribution to its development. It should not be confused with the *friction coefficient* C_f

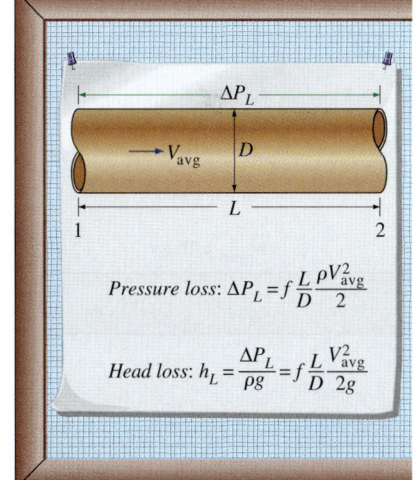

FIGURE 8–13
The relation for pressure loss (and head loss) is one of the most general relations in fluid mechanics, and it is valid for laminar or turbulent flows, circular or noncircular pipes, and pipes with smooth or rough surfaces.

[also called the *Fanning friction factor*, named after the American engineer John Fanning (1837–1911)], which is defined as $C_f = 2\tau_w/(\rho V_{avg}^2) = f/4$.

Setting Eqs. 8–20 and 8–21 equal to each other and solving for f gives the friction factor for fully developed laminar flow in a circular pipe,

Circular pipe, laminar: $$f = \frac{64\mu}{\rho D V_{avg}} = \frac{64}{\text{Re}} \qquad (8\text{–}23)$$

This equation shows that *in laminar flow, the friction factor is a function of the Reynolds number only and is independent of the roughness of the pipe surface* (assuming, of course, that the roughness is not extreme).

In the analysis of piping systems, pressure losses are commonly expressed in terms of the *equivalent fluid column height*, called the **head loss** h_L. Noting from fluid statics that $\Delta P = \rho g h$ and thus a pressure difference of ΔP corresponds to a fluid height of $h = \Delta P/\rho g$, the *pipe head loss* is obtained by dividing ΔP_L by ρg to give

Head loss: $$h_L = \frac{\Delta P_L}{\rho g} = f \frac{L}{D} \frac{V_{avg}^2}{2g} \qquad (8\text{–}24)$$

The head loss h_L represents *the additional height that the fluid needs to be raised by a pump in order to overcome the frictional losses in the pipe.* The head loss is caused by viscosity, and it is directly related to the wall shear stress. Equations 8–21 and 8–24 are valid for both laminar and turbulent flows in both circular and noncircular pipes, but Eq. 8–23 is valid only for fully developed laminar flow in circular pipes.

Once the pressure loss (or head loss) is known, the required pumping power *to overcome the pressure loss* is determined from

$$\dot{W}_{\text{pump}, L} = \dot{V} \Delta P_L = \dot{V}\rho g h_L = \dot{m} g h_L \qquad (8\text{–}25)$$

where $\dot{V}$ is the volume flow rate and $\dot{m}$ is the mass flow rate.

The average velocity for laminar flow in a horizontal pipe is, from Eq. 8–20,

Horizontal pipe: $$V_{avg} = \frac{(P_1 - P_2)R^2}{8\mu L} = \frac{(P_1 - P_2)D^2}{32\mu L} = \frac{\Delta P D^2}{32\mu L} \qquad (8\text{–}26)$$

Then the volume flow rate for laminar flow through a horizontal pipe of diameter D and length L becomes

$$\dot{V} = V_{avg} A_c = \frac{(P_1 - P_2)R^2}{8\mu L}\pi R^2 = \frac{(P_1 - P_2)\pi D^4}{128\mu L} = \frac{\Delta P \pi D^4}{128\mu L} \qquad (8\text{–}27)$$

This equation is known as **Poiseuille's law,** and this flow is called *Hagen–Poiseuille flow* in honor of the works of G. Hagen (1797–1884) and J. Poiseuille (1799–1869) on the subject. Note from Eq. 8–27 that *for a specified flow rate, the pressure drops and thus the required pumping power is proportional to the length of the pipe and the viscosity of the fluid, but it is inversely proportional to the fourth power of the radius (or diameter) of the pipe.* Therefore, the pumping power requirement for a laminar-flow piping system can be reduced by a factor of 16 by doubling the pipe diameter (Fig. 8–14). Of course the benefits of the reduction in the energy costs must be weighed against the increased cost of construction due to using a larger-diameter pipe.

The pressure drop ΔP equals the pressure loss ΔP_L in the case of a horizontal pipe, but this is not the case for inclined pipes or pipes with variable cross-sectional area. This can be demonstrated by writing the energy

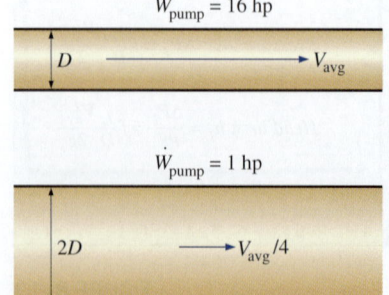

FIGURE 8–14

The pumping power requirement for a laminar-flow piping system can be reduced by a factor of 16 by doubling the pipe diameter.

equation for steady, incompressible one-dimensional flow in terms of heads as (see Chap. 5)

$$\frac{P_1}{\rho g} + \alpha_1 \frac{V_1^2}{2g} + z_1 + h_{\text{pump}, u} = \frac{P_2}{\rho g} + \alpha_2 \frac{V_2^2}{2g} + z_2 + h_{\text{turbine}, e} + h_L \quad (8\text{--}28)$$

where $h_{\text{pump}, u}$ is the useful pump head delivered to the fluid, $h_{\text{turbine}, e}$ is the turbine head extracted from the fluid, h_L is the irreversible head loss between sections 1 and 2, V_1 and V_2 are the average velocities at sections 1 and 2, respectively, and α_1 and α_2 are the *kinetic energy correction factors* at sections 1 and 2 (it can be shown that $\alpha = 2$ for fully developed laminar flow and about 1.05 for fully developed turbulent flow). Equation 8–28 can be rearranged as

$$P_1 - P_2 = \rho(\alpha_2 V_2^2 - \alpha_1 V_1^2)/2 + \rho g[(z_2 - z_1) + h_{\text{turbine}, e} - h_{\text{pump}, u} + h_L] \quad (8\text{--}29)$$

Therefore, the pressure drop $\Delta P = P_1 - P_2$ and pressure loss $\Delta P_L = \rho g h_L$ for a given flow section are equivalent if (1) the flow section is horizontal so that there are no hydrostatic or gravity effects ($z_1 = z_2$), (2) the flow section does not involve any work devices such as a pump or a turbine since they change the fluid pressure ($h_{\text{pump}, u} = h_{\text{turbine}, e} = 0$), (3) the cross-sectional area of the flow section is constant and thus the average flow velocity is constant ($V_1 = V_2$), and (4) the velocity profiles at sections 1 and 2 are the same shape ($\alpha_1 = \alpha_2$).

Effect of Gravity on Velocity and Flow Rate in Laminar Flow

Gravity has no affect on flow in horizontal pipes, but it has a significant effect on both the velocity and the flow rate in uphill or downhill pipes. Relations for inclined pipes can be obtained in a similar manner from a force balance in the direction of flow. The only additional force in this case is the component of the fluid weight in the flow direction, whose magnitude is

$$W_x = W \sin\theta = \rho g V_{\text{element}} \sin\theta = \rho g (2\pi r \, dr \, dx) \sin\theta \quad (8\text{--}30)$$

where θ is the angle between the horizontal and the flow direction (Fig. 8–15). The force balance in Eq. 8–9 now becomes

$$(2\pi r \, dr \, P)_x - (2\pi r \, dr \, P)_{x+dx} + (2\pi r \, dx \, \tau)_r$$
$$- (2\pi r \, dx \, \tau)_{r+dr} - \rho g (2\pi r \, dr \, dx) \sin\theta = 0 \quad (8\text{--}31)$$

which results in the differential equation

$$\frac{\mu}{r} \frac{d}{dr}\left(r \frac{du}{dr}\right) = \frac{dP}{dx} + \rho g \sin\theta \quad (8\text{--}32)$$

Following the same solution procedure as previously, the velocity profile is

$$u(r) = -\frac{R^2}{4\mu}\left(\frac{dP}{dx} + \rho g \sin\theta\right)\left(1 - \frac{r^2}{R^2}\right) \quad (8\text{--}33)$$

From Eq. 8–33, the *average velocity* and the *volume flow rate* relations for laminar flow through inclined pipes are, respectively,

$$V_{\text{avg}} = \frac{(\Delta P - \rho g L \sin\theta) D^2}{32 \mu L} \quad \text{and} \quad \dot{V} = \frac{(\Delta P - \rho g L \sin\theta) \pi D^4}{128 \mu L} \quad (8\text{--}34)$$

which are identical to the corresponding relations for horizontal pipes, except that ΔP is replaced by $\Delta P - \rho g L \sin\theta$. Therefore, the results already obtained for horizontal pipes can also be used for inclined pipes provided

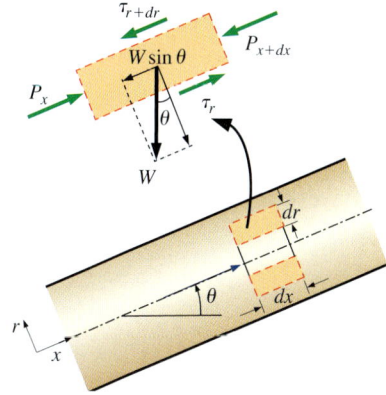

FIGURE 8–15
Free-body diagram of a ring-shaped differential fluid element of radius r, thickness dr, and length dx oriented coaxially with an inclined pipe in fully developed laminar flow.

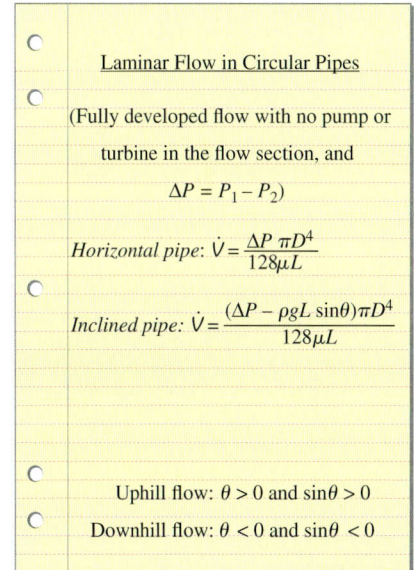

FIGURE 8–16
The relations developed for fully developed laminar flow through horizontal pipes can also be used for inclined pipes by replacing ΔP with $\Delta P - \rho g L \sin\theta$.

that ΔP is replaced by $\Delta P - \rho g L \sin \theta$ (Fig. 8–16). Note that $\theta > 0$ and thus $\sin \theta > 0$ for uphill flow, and $\theta < 0$ and thus $\sin \theta < 0$ for downhill flow.

In inclined pipes, the combined effect of pressure difference and gravity drives the flow. Gravity helps downhill flow but opposes uphill flow. Therefore, much greater pressure differences need to be applied to maintain a specified flow rate in uphill flow although this becomes important only for liquids, because the density of gases is generally low. In the special case of *no flow* ($\dot{V} = 0$), Eq. 8–34 yields $\Delta P = \rho g L \sin \theta$, which is what we would obtain from fluid statics (Chap. 3).

Laminar Flow in Noncircular Pipes

The friction factor f relations are given in Table 8–1 for *fully developed laminar flow* in pipes of various cross sections. The Reynolds number for flow in these pipes is based on the hydraulic diameter $D_h = 4A_c/p$, where A_c is the cross-sectional area of the pipe and p is its wetted perimeter.

TABLE 8–1

Friction factor for fully developed *laminar flow* in pipes of various cross sections ($D_h = 4A_c/p$ and Re $= V_{avg} D_h/\nu$)

Tube Geometry	a/b or $\theta°$	Friction Factor f
Circle	—	64.00/Re
Rectangle	a/b	
	1	56.92/Re
	2	62.20/Re
	3	68.36/Re
	4	72.92/Re
	6	78.80/Re
	8	82.32/Re
	∞	96.00/Re
Ellipse	a/b	
	1	64.00/Re
	2	67.28/Re
	4	72.96/Re
	8	76.60/Re
	16	78.16/Re
Isosceles triangle	θ	
	10°	50.80/Re
	30°	52.28/Re
	60°	53.32/Re
	90°	52.60/Re
	120°	50.96/Re

■ **EXAMPLE 8–1**   **Laminar Flow in Horizontal and Inclined Pipes**

Consider the fully developed flow of glycerin at 40°C through a 70-m-long, 4-cm-diameter, horizontal, circular pipe. If the flow velocity at the centerline is measured to be 6 m/s, determine the velocity profile and the pressure difference across this 70-m-long section of the pipe, and the useful pumping power required to maintain this flow. For the same useful pumping power input, determine the percent increase of the flow rate if the pipe is inclined 15° downward and the percent decrease if it is inclined 15° upward. The pump is located outside this pipe section.

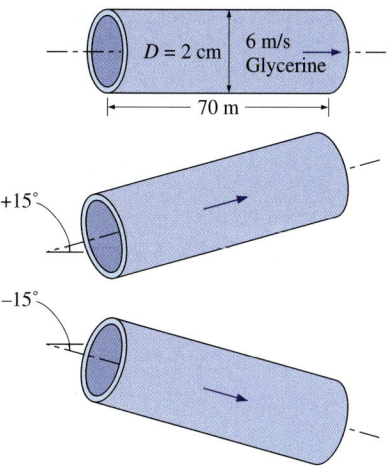

FIGURE 8–17
Schematic for Example 8–1.

SOLUTION The centerline velocity in a horizontal pipe in fully developed flow is measured. The velocity profile, the pressure difference across the pipe, and the pumping power required are to be determined. The effects of downward and upward tilting of the pipe on the flow rate is to be investigated.
Assumptions **1** The flow is steady, laminar, incompressible, and fully developed. **2** There are no pumps or turbines in the flow section. **3** There are no valves, elbows, or other devices that may cause local losses.
Properties The density and dynamic viscosity of glycerin at 40°C are $\rho = 1252$ kg/m^3 and $\mu = 0.3073$ kg/m·s, respectively.
Analysis The velocity profile in fully developed laminar flow in a circular pipe is expressed as

$$u(r) = u_{max}\left(1 - \frac{r^2}{R^2}\right)$$

Substituting, the velocity profile is determined to be

$$u(r) = (6 \text{ m/s})\left(1 - \frac{r^2}{(0.02 \text{ m})^2}\right) = \mathbf{6(1 - 2500r^2)}$$

where u is in m/s and r is in m. The average velocity, the flow rate, and the Reynolds number are

$$V = V_{avg} = \frac{u_{max}}{2} = \frac{6 \text{ m/s}}{2} = 3 \text{ m/s}$$

$$\dot{V} = V_{avg}A_c = V(\pi D^2/4) = (3 \text{ m/s})[\pi(0.04 \text{ m})^2/4] = 3.77 \times 10^{-3} \text{ m}^3/\text{s}$$

$$\text{Re} = \frac{\rho V D}{\mu} = \frac{(1252 \text{ kg/m}^3)(3 \text{ m/s})(0.04 \text{ m})}{0.3073 \text{ kg/m·s}} = 488.9$$

which is less than 2300. Therefore, the flow is indeed laminar. Then the friction factor and the head loss become

$$f = \frac{64}{\text{Re}} = \frac{64}{488.9} = 0.1309$$

$$h_L = f\frac{L}{D}\frac{V^2}{2g} = 0.1309\,\frac{(70 \text{ m})}{(0.04 \text{ m})}\,\frac{(3 \text{ m/s})^2}{2(9.81 \text{ m/s}^2)} = 105.1 \text{m}$$

The energy balance for steady, incompressible one-dimensional flow is given by Eq. 8–28 as

$$\frac{P_1}{\rho g} + \alpha_1\frac{V_1^2}{2g} + z_1 + h_{pump,u} = \frac{P_2}{\rho g} + \alpha_2\frac{V_2^2}{2g} + z_2 + h_{turbine,e} + h_L$$

For fully developed flow in a constant diameter pipe with no pumps or turbines, it reduces to

$$\Delta P = P_1 - P_2 = \rho g(z_2 - z_1 + h_L)$$

Then the pressure difference and the required useful pumping power for the horizontal case become

$$\Delta P = \rho g(z_2 - z_1 + h_L)$$

$$= (1252 \text{ kg/m}^3)(9.81 \text{ m/s}^2)(0 + 105.1 \text{ m})\left(\frac{1 \text{ kPa}}{1000 \text{ kg/m·s}^2}\right)$$

$$= \mathbf{1291 \text{ kPa}}$$

$$\dot{W}_{\text{pump, }u} = \dot{V}\Delta P = (3.77 \times 10^3 \text{ m}^3\text{/s})(1291 \text{ kPa})\left(\frac{1 \text{ kW}}{\text{kPa·m}^3\text{/s}}\right) = \mathbf{4.87 \text{ kW}}$$

The elevation difference and the pressure difference for a pipe inclined upwards 15° is

$$\Delta z = z_2 - z_1 = L\sin 15° = (70 \text{ m})\sin 15° = 18.1 \text{ m}$$

$$\Delta P_{\text{upward}} = (1252 \text{ kg/m}^3)(9.81 \text{ m/s}^2)(18.1 \text{ m} + 105.1 \text{ m})\left(\frac{1 \text{ kPa}}{1000 \text{ kg/m·s}^2}\right)$$

$$= 1366 \text{ kPa}$$

Then the flow rate through the upward inclined pipe becomes

$$\dot{V}_{\text{upward}} = \frac{\dot{W}_{\text{pump, }u}}{\Delta P_{\text{upward}}} = \frac{4.87 \text{ kW}}{1366 \text{ kPa}}\left(\frac{1 \text{ kPa·m}^3\text{/s}}{1 \text{ kW}}\right) = 3.57 \times 10^{-3} \text{ m}^3\text{/s}$$

which is a decrease of **5.6 percent** in flow rate. It can be shown similarly that when the pipe is inclined 15° downward from the horizontal, the flow rate will increase by **5.6 percent.**

Discussion Note that the flow is driven by the combined effect of pumping power and gravity. As expected, gravity opposes uphill flow, enhances downhill flow, and has no effect on horizontal flow. Downhill flow can occur even in the absence of a pressure difference applied by a pump. For the case of $P_1 = P_2$ (i.e., no applied pressure difference), the pressure throughout the entire pipe would remain constant, and the fluid would flow through the pipe under the influence of gravity at a rate that depends on the angle of inclination, reaching its maximum value when the pipe is vertical. When solving pipe flow problems, it is always a good idea to calculate the Reynolds number to verify the flow regime—laminar or turbulent.

EXAMPLE 8–2   Pressure Drop and Head Loss in a Pipe

Water at 5°C (ρ = 1000 kg/m³ and μ = 1.519 × 10⁻³ kg/m·s) is flowing steadily through a 0.3-cm diameter 9-m-long horizontal pipe at an average velocity of 0.9 m/s (Fig. 8–18). Determine (*a*) the head loss, (*b*) the pressure drop, and (*c*) the pumping power requirement to overcome this pressure drop.

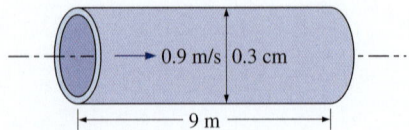

FIGURE 8–18
Schematic for Example 8–2.

SOLUTION The average flow velocity in a pipe is given. The head loss, the pressure drop, and the pumping power are to be determined.
Assumptions **1** The flow is steady and incompressible. **2** The entrance effects are negligible, and thus the flow is fully developed. **3** The pipe involves no components such as bends, valves, and connectors.
Properties The density and dynamic viscosity of water are given to be $\rho = 1000$ kg/m^3 and $\mu = 1.519 \times 10^{-3}$ kg/m·s, respectively.
Analysis (*a*) First we need to determine the flow regime. The Reynolds number is

$$\text{Re} = \frac{\rho V_{\text{avg}} D}{\mu} = \frac{(1000 \text{ kg/m}^3)(0.9 \text{ m/s})(0.003 \text{ m})}{1.519 \times 10^{-3} \text{ kg/m·s}} = 1777$$

which is less than 2300. Therefore, the flow is laminar. Then the friction factor and the head loss become

$$f = \frac{64}{\text{Re}} = \frac{64}{1777} = 0.0360$$

$$h_L = f \frac{L}{D} \frac{V_{\text{avg}}^2}{2g} = 0.0360 \frac{9 \text{ m}}{0.003 \text{ m}} \frac{(0.9 \text{ m/s})^2}{2(9.81 \text{ m/s}^2)} = \mathbf{4.46 \text{ m}}$$

(*b*) Noting that the pipe is horizontal and its diameter is constant, the pressure drop in the pipe is due entirely to the frictional losses and is equivalent to the pressure loss,

$$\Delta P = \Delta P_L = f \frac{L}{D} \frac{\rho V_{\text{avg}}^2}{2} = 0.0360 \frac{9 \text{ m}}{0.003 \text{ m}} \frac{(1000 \text{ kg/m}^3)(0.9 \text{ m/s})^2}{2} \left(\frac{1 \text{ N}}{1 \text{ kg·m/s}^2}\right)$$

$$= 43{,}740 \text{ N/m}^2 = \mathbf{43.7 \text{ kPa}}$$

(*c*) The volume flow rate and the pumping power requirements are

$$\dot{V} = V_{\text{avg}} A_c = V_{\text{avg}}(\pi D^2/4) = (0.9 \text{ m/s})[\pi(0.003 \text{ m})^2/4] = 6.36 \times 10^{-6} \text{ m}^3/\text{s}$$

$$\dot{W}_{\text{pump}} = \dot{V} \Delta P = (6.36 \times 10^{-6} \text{ m}^3/\text{s})(43{,}740 \text{ N/m}^2)\left(\frac{1 \text{ W}}{1 \text{ N·m/s}}\right) = \mathbf{0.28 \text{ W}}$$

Therefore, power input in the amount of 0.28 W is needed to overcome the frictional losses in the flow due to viscosity.
Discussion The pressure rise provided by a pump is often listed by a pump manufacturer in units of head (Chap. 14). Thus, the pump in this flow needs to provide 4.46 m of water head in order to overcome the irreversible head loss.

(*a*)

(*b*)

(*c*)

FIGURE 8–19
Water exiting a tube: (*a*) laminar flow at low flow rate, (*b*) turbulent flow at high flow rate, and (*c*) same as (*b*) but with a short shutter exposure to capture individual eddies.
Photos by Alex Wouden.

8–5 · TURBULENT FLOW IN PIPES

Most flows encountered in engineering practice are turbulent, and thus it is important to understand how turbulence affects wall shear stress. However, turbulent flow is a complex mechanism dominated by fluctuations, and despite tremendous amounts of work done in this area by researchers, turbulent flow still is not fully understood. Therefore, we must rely on experiments and the empirical or semi-empirical correlations developed for various situations.

Turbulent flow is characterized by disorderly and rapid fluctuations of swirling regions of fluid, called **eddies,** throughout the flow (Fig. 8–19). These fluctuations provide an additional mechanism for momentum and energy

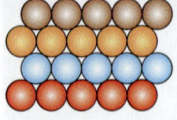
(a) Before turbulence

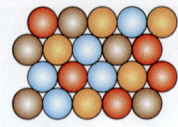
(b) After turbulence

FIGURE 8–20
The intense mixing in turbulent flow brings fluid particles at different momentums into close contact and thus enhances momentum transfer.

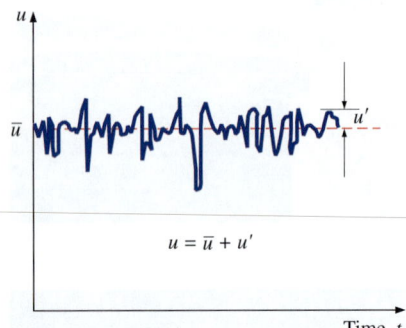

FIGURE 8–21
Fluctuations of the velocity component u with time at a specified location in turbulent flow.

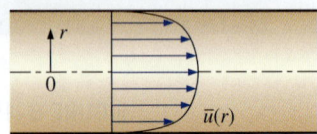

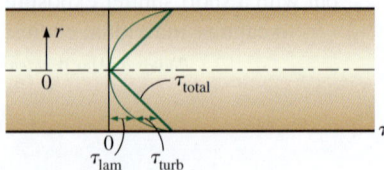

FIGURE 8–22
The velocity profile and the variation of shear stress with radial distance for turbulent flow in a pipe.

transfer. In laminar flow, fluid particles flow in an orderly manner along pathlines, and momentum and energy are transferred across streamlines by molecular diffusion. In turbulent flow, the swirling eddies transport mass, momentum, and energy to other regions of flow much more rapidly than molecular diffusion, greatly enhancing mass, momentum, and heat transfer. As a result, turbulent flow is associated with much higher values of friction, heat transfer, and mass transfer coefficients (Fig. 8–20).

Even when the average flow is steady, the eddy motion in turbulent flow causes significant fluctuations in the values of velocity, temperature, pressure, and even density (in compressible flow). Figure 8–21 shows the variation of the instantaneous velocity component u with time at a specified location, as can be measured with a hot-wire anemometer probe or other sensitive device. We observe that the instantaneous values of the velocity fluctuate about an average value, which suggests that the velocity can be expressed as the sum of an *average value* $\bar{u}$ and a *fluctuating component* u',

$$u = \bar{u} + u' \qquad (8\text{–}35)$$

This is also the case for other properties such as the velocity component v in the y-direction, and thus $v = \bar{v} + v'$, $P = \bar{P} + P'$, and $T = \bar{T} + T'$. The average value of a property at some location is determined by averaging it over a time interval that is sufficiently large so that the time average levels off to a constant. Therefore, the time average of fluctuating components is zero, e.g., $\bar{u}' = 0$. The magnitude of u' is usually just a few percent of $\bar{u}$, but the high frequencies of eddies (on the order of a thousand per second) make them very effective for the transport of momentum, thermal energy, and mass. In time-averaged *stationary* turbulent flow, the average values of properties (indicated by an overbar) are independent of time. The chaotic fluctuations of fluid particles play a dominant role in pressure drop, and these random motions must be considered in analyses together with the average velocity.

Perhaps the first thought that comes to mind is to determine the shear stress in an analogous manner to laminar flow from $\tau = -\mu \, d\bar{u}/dr$, where $\bar{u}(r)$ is the average velocity profile for turbulent flow. But the experimental studies show that this is not the case, and the effective shear stress is much larger due to the turbulent fluctuations. Therefore, it is convenient to think of the turbulent shear stress as consisting of two parts: the *laminar component,* which accounts for the friction between layers in the flow direction (expressed as $\tau_{\text{lam}} = -\mu \, d\bar{u}/dr$), and the *turbulent component,* which accounts for the friction between the fluctuating fluid particles and the fluid body (denoted as τ_{turb} and is related to the fluctuation components of velocity). Then the *total shear stress* in turbulent flow can be expressed as

$$\tau_{\text{total}} = \tau_{\text{lam}} + \tau_{\text{turb}} \qquad (8\text{–}36)$$

The typical average velocity profile and relative magnitudes of laminar and turbulent components of shear stress for turbulent flow in a pipe are given in Fig. 8–22. Note that although the velocity profile is approximately parabolic in laminar flow, it becomes flatter or "fuller" in turbulent flow, with a sharp drop near the pipe wall. The fullness increases with the Reynolds number, and the velocity profile becomes more nearly uniform, lending support to the commonly utilized uniform velocity profile approximation for fully developed turbulent pipe flow. Keep in mind, however, that the flow speed at the wall of a stationary pipe is always zero (no-slip condition).

Turbulent Shear Stress

Consider turbulent flow in a horizontal pipe, and the upward eddy motion of a fluid particle from a layer of lower velocity to an adjacent layer of higher velocity through a differential area dA as a result of the velocity fluctuation v', as shown in Fig. 8–23. The mass flow rate of the fluid particle rising through dA is $\rho v' dA$, and its net effect on the layer above dA is a reduction in its average flow velocity because of momentum transfer to the fluid particle with lower average flow velocity. This momentum transfer causes the horizontal velocity of the fluid particle to increase by u', and thus its momentum in the horizontal direction to increase at a rate of $(\rho v' dA)u'$, which must be equal to the decrease in the momentum of the upper fluid layer. Noting that force in a given direction is equal to the rate of change of momentum in that direction, the horizontal force acting on a fluid element above dA due to the passing of fluid particles through dA is $\delta F = (\rho v' dA)(-u') = -\rho u' v' dA$. Therefore, the shear force per unit area due to the eddy motion of fluid particles $\delta F/dA = -\rho u'v'$ can be viewed as the instantaneous turbulent shear stress. Then the **turbulent shear stress** can be expressed as

$$\tau_{\text{turb}} = -\rho \overline{u'v'} \tag{8–37}$$

where $\overline{u'v'}$ is the time average of the product of the fluctuating velocity components u' and v'. Note that $\overline{u'v'} \neq 0$ even though $\overline{u'} = 0$ and $\overline{v'} = 0$ (and thus $\overline{u'}\,\overline{v'} = 0$), and experimental results show that $\overline{u'v'}$ is usually a negative quantity. Terms such as $-\rho \overline{u'v'}$ or $-\rho \overline{u'^2}$ are called **Reynolds stresses** or **turbulent stresses.**

Many semi-empirical formulations have been developed that model the Reynolds stress in terms of average velocity gradients in order to provide mathematical *closure* to the equations of motion. Such models are called **turbulence models** and are discussed in more detail in Chap. 15.

The random eddy motion of groups of particles resembles the random motion of molecules in a gas—colliding with each other after traveling a certain distance and exchanging momentum in the process. Therefore, momentum transport by eddies in turbulent flows is analogous to the molecular momentum diffusion. In many of the simpler turbulence models, turbulent shear stress is expressed in an analogous manner as suggested by the French mathematician Joseph Boussinesq (1842–1929) in 1877 as

$$\tau_{\text{turb}} = -\rho \overline{u'v'} = \mu_t \frac{\partial \overline{u}}{\partial y} \tag{8–38}$$

where μ_t is the **eddy viscosity** or **turbulent viscosity,** which accounts for momentum transport by turbulent eddies. Then the total shear stress can be expressed conveniently as

$$\tau_{\text{total}} = (\mu + \mu_t)\frac{\partial u}{\partial y} = \rho(\nu + \nu_t)\frac{\partial u}{\partial y} \tag{8–39}$$

where $\nu_t = \mu_t/\rho$ is the **kinematic eddy viscosity** or **kinematic turbulent viscosity** (also called the *eddy diffusivity of momentum*). The concept of eddy viscosity is very appealing, but it is of no practical use unless its value can be determined. In other words, eddy viscosity must be modeled as a function of the average flow variables; we call this *eddy viscosity closure*. For example, in the early 1900s, the German engineer L. Prandtl introduced the concept of

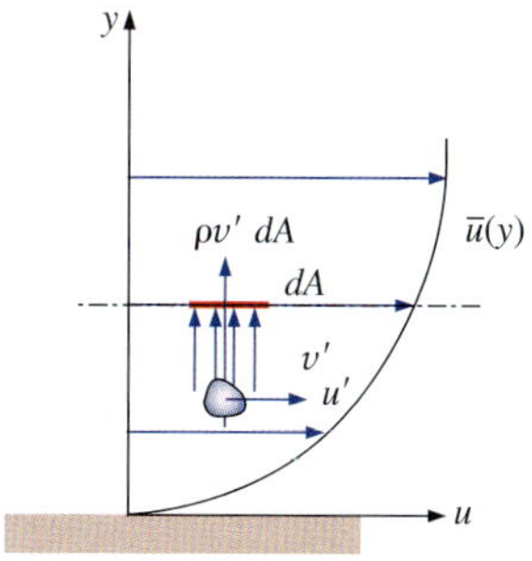

FIGURE 8–23
Fluid particle moving upward through a differential area dA as a result of the velocity fluctuation v'.

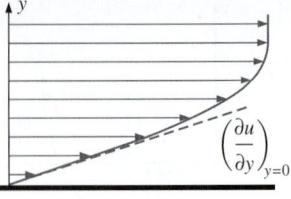

Laminar flow

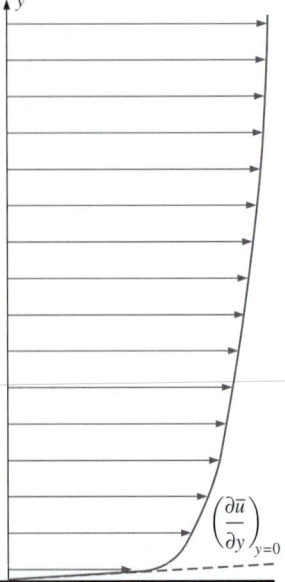

Turbulent flow

FIGURE 8–24

The velocity gradients at the wall, and thus the wall shear stress, are much larger for turbulent flow than they are for laminar flow, even though the turbulent boundary layer is thicker than the laminar one for the same value of free-stream velocity.

mixing length l_m, which is related to the average size of the eddies that are primarily responsible for mixing, and expressed the turbulent shear stress as

$$\tau_{\text{turb}} = \mu_t \frac{\partial \bar{u}}{\partial y} = \rho l_m^2 \left(\frac{\partial \bar{u}}{\partial y}\right)^2 \quad (8\text{–}40)$$

But this concept is also of limited use since l_m is not a constant for a given flow (in the vicinity of the wall, for example, l_m is nearly proportional to the distance from the wall) and its determination is not easy. Final mathematical closure is obtained only when l_m is written as a function of average flow variables, distance from the wall, etc.

Eddy motion and thus eddy diffusivities are much larger than their molecular counterparts in the core region of a turbulent boundary layer. The eddy motion loses its intensity close to the wall and diminishes at the wall because of the no-slip condition (u' and v' are identically zero at a stationary wall). Therefore, the velocity profile is very slowly changing in the core region of a turbulent boundary layer, but very steep in the thin layer adjacent to the wall, resulting in large velocity gradients at the wall surface. So it is no surprise that the wall shear stress is much larger in turbulent flow than it is in laminar flow (Fig. 8–24).

Note that the molecular diffusivity of momentum ν (as well as μ) is a fluid property, and its value is listed in fluid handbooks. Eddy diffusivity ν_t (as well as μ_t), however, is *not* a fluid property, and its value depends on flow conditions. Eddy diffusivity ν_t decreases toward the wall, becoming zero at the wall. Its value ranges from zero at the wall to several thousand times the value of the molecular diffusivity in the core region.

Turbulent Velocity Profile

Unlike laminar flow, the expressions for the velocity profile in a turbulent flow are based on both analysis and measurements, and thus they are semi-empirical in nature with constants determined from experimental data. Consider fully developed turbulent flow in a pipe, and let u denote the time-averaged velocity in the axial direction (and thus drop the overbar from $\bar{u}$ for simplicity).

Typical velocity profiles for fully developed laminar and turbulent flows are given in Fig. 8–25. Note that the velocity profile is parabolic in laminar flow but is much fuller in turbulent flow, with a sharp drop near the pipe wall. Turbulent flow along a wall can be considered to consist of four regions, characterized by the distance from the wall (Fig. 8–25). The very thin layer next to the wall where viscous effects are dominant is the **viscous** (or **laminar** or **linear** or **wall**) sublayer. The velocity profile in this layer is very nearly *linear*, and the flow is streamlined. Next to the viscous sublayer is the **buffer layer,** in which turbulent effects are becoming significant, but the flow is still dominated by viscous effects. Above the buffer layer is the **overlap** (or **transition**) **layer,** also called the **inertial sublayer,** in which the turbulent effects are much more significant, but still not dominant. Above that is the **outer** (or **turbulent**) **layer** in the remaining part of the flow in which turbulent effects dominate over molecular diffusion (viscous) effects.

Flow characteristics are quite different in different regions, and thus it is difficult to come up with an analytic relation for the velocity profile for the entire flow as we did for laminar flow. The best approach in the turbulent case turns out to be to identify the key variables and functional forms using dimensional analysis, and then to use experimental data to determine the numerical values of any constants.

The thickness of the viscous sublayer is very small (typically, much less than 1 percent of the pipe diameter), but this thin layer next to the wall plays a dominant role on flow characteristics because of the large velocity gradients it involves. The wall dampens any eddy motion, and thus the flow in this layer is essentially laminar and the shear stress consists of laminar shear stress which is proportional to the fluid viscosity. Considering that velocity changes from zero to nearly the core region value across a layer that is sometimes no thicker than a hair (almost like a step function), we would expect the velocity profile in this layer to be very nearly linear, and experiments confirm that. Then the velocity gradient in the viscous sublayer remains nearly constant at $du/dy = u/y$, and the wall shear stress can be expressed as

$$\tau_w = \mu \frac{u}{y} = \rho \nu \frac{u}{y} \quad \text{or} \quad \frac{\tau_w}{\rho} = \frac{\nu u}{y} \tag{8-41}$$

where y is the distance from the wall (note that $y = R - r$ for a circular pipe). The quantity τ_w/ρ is frequently encountered in the analysis of turbulent velocity profiles. The square root of τ_w/ρ has the dimensions of velocity, and thus it is convenient to view it as a fictitious velocity called the **friction velocity** expressed as $u_* = \sqrt{\tau_w/\rho}$. Substituting this into Eq. 8–41, the velocity profile in the viscous sublayer is expressed in dimensionless form as

Viscous sublayer:
$$\frac{u}{u_*} = \frac{y u_*}{\nu} \tag{8-42}$$

This equation is known as the **law of the wall**, and it is found to satisfactorily correlate with experimental data for smooth surfaces for $0 \le y u_*/\nu \le 5$. Therefore, the thickness of the viscous sublayer is roughly

Thickness of viscous sublayer:
$$y = \delta_{\text{sublayer}} = \frac{5\nu}{u_*} = \frac{25\nu}{u_\delta} \tag{8-43}$$

where u_δ is the flow velocity at the edge of the viscous sublayer (where $u_\delta \approx 5u_*$), which is closely related to the average velocity in a pipe. Thus we conclude that *the thickness of the viscous sublayer is proportional to the kinematic viscosity and inversely proportional to the average flow velocity.* In other words, the viscous sublayer is suppressed and it gets thinner as the velocity (and thus the Reynolds number) increases. Consequently, the velocity profile becomes nearly flat and thus the velocity distribution becomes more uniform at very high Reynolds numbers.

The quantity ν/u_* has dimensions of length and is called the **viscous length**; it is used to nondimensionalize the distance y from the surface. In boundary layer analysis, it is convenient to work with nondimensionalized distance and nondimensionalized velocity defined as

Nondimensionalized variables:
$$y^+ = \frac{y u_*}{\nu} \quad \text{and} \quad u^+ = \frac{u}{u_*} \tag{8-44}$$

Then the law of the wall (Eq. 8–42) becomes simply

Normalized law of the wall:
$$u^+ = y^+ \tag{8-45}$$

Note that the friction velocity u_* is used to nondimensionalize both y and u, and y^+ resembles the Reynolds number expression.

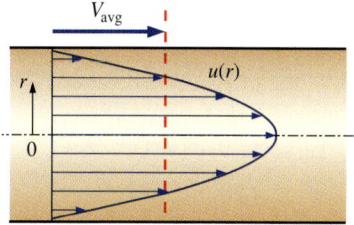

Laminar flow

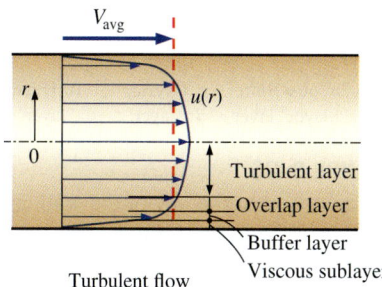

Turbulent flow

FIGURE 8–25
The velocity profile in fully developed pipe flow is parabolic in laminar flow, but much fuller in turbulent flow. Note that $u(r)$ in the turbulent case is the *time-averaged* velocity component in the axial direction (the overbar on u has been dropped for simplicity).

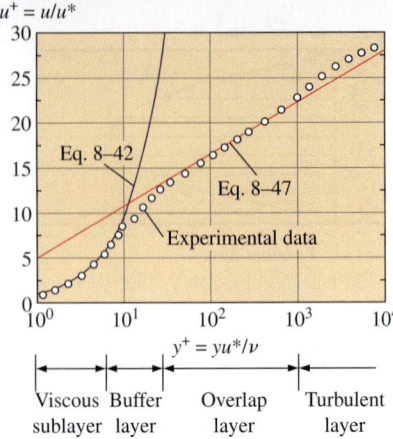

FIGURE 8–26
Comparison of the law of the wall and the logarithmic-law velocity profiles with experimental data for fully developed turbulent flow in a pipe.

In the overlap layer, the experimental data for velocity are observed to line up on a straight line when plotted against the logarithm of distance from the wall. Dimensional analysis indicates and the experiments confirm that the velocity in the overlap layer is proportional to the logarithm of distance, and the velocity profile can be expressed as

The logarithmic law:
$$\frac{u}{u_*} = \frac{1}{\kappa} \ln \frac{yu_*}{\nu} + B \qquad (8\text{-}46)$$

where κ and B are constants whose values are determined experimentally to be about 0.40 and 5.0, respectively. Equation 8–46 is known as the **logarithmic law**. Substituting the values of the constants, the velocity profile is determined to be

Overlap layer:
$$\frac{u}{u_*} = 2.5 \ln \frac{yu_*}{\nu} + 5.0 \quad \text{or} \quad u^+ = 2.5 \ln y^+ + 5.0 \qquad (8\text{-}47)$$

It turns out that the logarithmic law in Eq. 8–47 satisfactorily represents experimental data for the entire flow region except for the regions very close to the wall and near the pipe center, as shown in Fig. 8–26, and thus it is viewed as a *universal velocity profile* for turbulent flow in pipes or over surfaces. Note from the figure that the logarithmic-law velocity profile is quite accurate for $y^+ > 30$, but neither velocity profile is accurate in the buffer layer, i.e., the region $5 < y^+ < 30$. Also, the viscous sublayer appears much larger in the figure than it is since we used a logarithmic scale for distance from the wall.

A good approximation for the outer turbulent layer of pipe flow can be obtained by evaluating the constant B in Eq. 8–46 from the requirement that maximum velocity in a pipe occurs at the centerline where $r = 0$. Solving for B from Eq. 8–46 by setting $y = R - r = R$ and $u = u_{max}$, and substituting it back into Eq. 8–46 together with $\kappa = 0.4$ gives

Outer turbulent layer:
$$\frac{u_{max} - u}{u_*} = 2.5 \ln \frac{R}{R - r} \qquad (8\text{-}48)$$

The deviation of velocity from the centerline value $u_{max} - u$ is called the **velocity defect**, and Eq. 8–48 is called the **velocity defect law**. This relation shows that the normalized velocity profile in the core region of turbulent flow in a pipe depends on the distance from the centerline and is independent of the viscosity of the fluid. This is not surprising since the eddy motion is dominant in this region, and the effect of fluid viscosity is negligible.

Numerous other empirical velocity profiles exist for turbulent pipe flow. Among those, the simplest and the best known is the **power-law velocity profile** expressed as

Power-law velocity profile:
$$\frac{u}{u_{max}} = \left(\frac{y}{R}\right)^{1/n} \quad \text{or} \quad \frac{u}{u_{max}} = \left(1 - \frac{r}{R}\right)^{1/n} \qquad (8\text{-}49)$$

where the exponent n is a constant whose value depends on the Reynolds number. The value of n increases with increasing Reynolds number. The value $n = 7$ generally approximates many flows in practice, giving rise to the term *one-seventh power-law velocity profile*.

Various power-law velocity profiles are shown in Fig. 8–27 for $n = 6, 8$, and 10 together with the velocity profile for fully developed laminar flow for comparison. Note that the turbulent velocity profile is fuller than the laminar one, and it becomes more flat as n (and thus the Reynolds number)

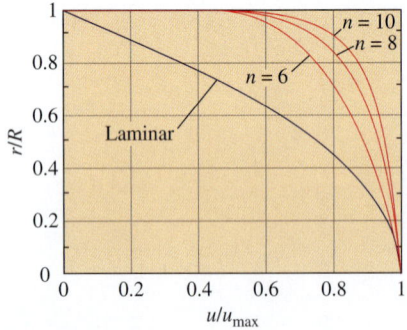

FIGURE 8–27
Power-law velocity profiles for fully developed turbulent flow in a pipe for different exponents, and its comparison with the laminar velocity profile.

increases. Also note that the power-law profile cannot be used to calculate wall shear stress since it gives a velocity gradient of infinity there, and it fails to give zero slope at the centerline. But these regions of discrepancy constitute a small portion of the overall flow, and the power-law profile gives highly accurate results for turbulent flow through a pipe.

Despite the small thickness of the viscous sublayer (usually much less than 1 percent of the pipe diameter), the characteristics of the flow in this layer are very important since they set the stage for flow in the rest of the pipe. Any irregularity or roughness on the surface disturbs this layer and affects the flow. Therefore, unlike laminar flow, the friction factor in turbulent flow is a strong function of surface roughness.

It should be kept in mind that roughness is a relative concept, and it has significance when its height ϵ is comparable to the thickness of the viscous sublayer (which is a function of the Reynolds number). All materials appear "rough" under a microscope with sufficient magnification. In fluid mechanics, a surface is characterized as being rough when the hills of roughness protrude out of the viscous sublayer. A surface is said to be *hydrodynamically smooth* when the sublayer submerges the roughness elements. Glass and plastic surfaces are generally considered to be hydrodynamically smooth.

The Moody Chart and the Colebrook Equation

The friction factor in fully developed turbulent pipe flow depends on the Reynolds number and the **relative roughness** ϵ/D, which is the ratio of the mean height of roughness of the pipe to the pipe diameter. The functional form of this dependence cannot be obtained from a theoretical analysis, and all available results are obtained from painstaking experiments using artificially roughened surfaces (usually by gluing sand grains of a known size on the inner surfaces of the pipes). Most such experiments were conducted by Prandtl's student J. Nikuradse in 1933, followed by the works of others. The friction factor was calculated from measurements of the flow rate and the pressure drop.

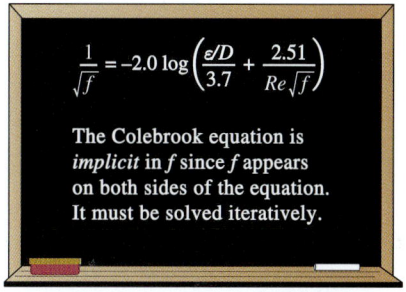

FIGURE 8–28
The Colebrook equation.

The experimental results are presented in tabular, graphical, and functional forms obtained by curve-fitting experimental data. In 1939, Cyril F. Colebrook (1910–1997) combined the available data for transition and turbulent flow in smooth as well as rough pipes into the following implicit relation (Fig. 8–28) known as the **Colebrook equation:**

$$\frac{1}{\sqrt{f}} = -2.0 \log\left(\frac{\varepsilon/D}{3.7} + \frac{2.51}{\text{Re}\sqrt{f}}\right) \quad \text{(turbulent flow)} \quad (8\text{-}50)$$

We note that the logarithm in Eq. 8–50 is a base 10 rather than a natural logarithm. In 1942, the American engineer Hunter Rouse (1906–1996) verified Colebrook's equation and produced a graphical plot of f as a function of Re and the product $\text{Re}\sqrt{f}$. He also presented the laminar flow relation and a table of commercial pipe roughness. Two years later, Lewis F. Moody (1880–1953) redrew Rouse's diagram into the form commonly used today. The now famous **Moody chart** is given in the appendix as Fig. A–12. It presents the Darcy friction factor for pipe flow as a function of Reynolds number and ϵ/D over a wide range. It is probably one of the most widely accepted and used charts in engineering. Although it is developed for circular pipes, it can also be used for noncircular pipes by replacing the diameter with the hydraulic diameter.

TABLE 8–2

Equivalent roughness values for new commercial pipes*

Material	Roughness, ε	
	ft	mm
Glass, plastic	0 (smooth)	
Concrete	0.003–0.03	0.9–9
Wood stave	0.0016	0.5
Rubber, smoothed	0.000033	0.01
Copper or brass tubing	0.000005	0.0015
Cast iron	0.00085	0.26
Galvanized iron	0.0005	0.15
Wrought iron	0.00015	0.046
Stainless steel	0.000007	0.002
Commercial steel	0.00015	0.045

*The uncertainty in these values can be as much as ±60 percent.

Relative Roughness, ε/D	Friction Factor, f
0.0*	0.0119
0.00001	0.0119
0.0001	0.0134
0.0005	0.0172
0.001	0.0199
0.005	0.0305
0.01	0.0380
0.05	0.0716

*Smooth surface. All values are for Re = 10^6 and are calculated from the Colebrook equation.

FIGURE 8–29
The friction factor is minimum for a smooth pipe and increases with roughness.

Commercially available pipes differ from those used in the experiments in that the roughness of pipes in the market is not uniform and it is difficult to give a precise description of it. Equivalent roughness values for some commercial pipes are given in Table 8–2 as well as on the Moody chart. But it should be kept in mind that these values are for new pipes, and the relative roughness of pipes may increase with use as a result of corrosion, scale buildup, and precipitation. As a result, the friction factor may increase by a factor of 5 to 10. Actual operating conditions must be considered in the design of piping systems. Also, the Moody chart and its equivalent Colebrook equation involve several uncertainties (the roughness size, experimental error, curve fitting of data, etc.), and thus the results obtained should not be treated as "exact." They are is usually considered to be accurate to ±15 percent over the entire range in the figure.

The Colebrook equation is implicit in f, and thus the determination of the friction factor requires iteration. An approximate *explicit* relation for f was given by S. E. Haaland in 1983 as

$$\frac{1}{\sqrt{f}} \cong -1.8 \log\left[\frac{6.9}{\text{Re}} + \left(\frac{\varepsilon/D}{3.7}\right)^{1.11}\right] \quad (8\text{–}51)$$

The results obtained from this relation are within 2 percent of those obtained from the Colebrook equation. If more accurate results are desired, Eq. 8–51 can be used as a good *first guess* in a Newton iteration when using a programmable calculator or a spreadsheet to solve for f with Eq. 8–50.

We make the following observations from the Moody chart:

- For laminar flow, the friction factor decreases with increasing Reynolds number, and it is independent of surface roughness.

- The friction factor is a minimum for a smooth pipe (but still not zero because of the no-slip condition) and increases with roughness (Fig. 8–29). The Colebrook equation in this case ($\epsilon = 0$) reduces to the **Prandtl equation** expressed as $1/\sqrt{f} = 2.0 \log(\text{Re}\sqrt{f}) - 0.8$.

- The transition region from the laminar to turbulent regime (2300 < Re < 4000) is indicated by the shaded area in the Moody chart (Figs. 8–30 and A–12). The flow in this region may be laminar or turbulent, depending on flow disturbances, or it may alternate between laminar and turbulent, and thus the friction factor may also alternate between the values for laminar and turbulent flow. The data in this range are the least reliable. At small relative roughnesses, the friction factor increases in the transition region and approaches the value for smooth pipes.

- At very large Reynolds numbers (to the right of the dashed line on the Moody chart) the friction factor curves corresponding to specified relative roughness curves are nearly horizontal, and thus the friction factors are independent of the Reynolds number (Fig. 8–30). The flow in that region is called *fully rough turbulent flow* or just *fully rough flow* because the thickness of the viscous sublayer decreases with increasing Reynolds number, and it becomes so thin that it is negligibly small compared to the surface roughness height. The viscous effects in this case are produced in the main flow primarily by the protruding roughness elements, and the contribution of the viscous sublayer is negligible. The Colebrook equation in the *fully rough* zone (Re → ∞) reduces to the **von Kármán equation**

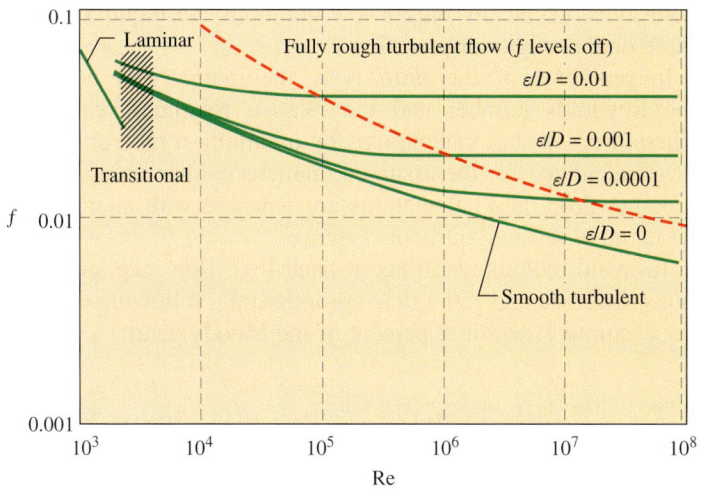

FIGURE 8–30
At very large Reynolds numbers, the friction factor curves on the Moody chart are nearly horizontal, and thus the friction factors are independent of the Reynolds number. See Fig. A–12 for a full-page Moody chart.

expressed as $1/\sqrt{f} = -2.0 \log[(\varepsilon/D)/3.7]$, which is explicit in f. Some authors call this zone *completely* (or *fully*) *turbulent flow*, but this is misleading since the flow to the left of the dashed blue line in Fig. 8–30 is also fully turbulent.

In calculations, we should make sure that we use the actual internal diameter of the pipe, which may be different than the nominal diameter. For example, the internal diameter of a steel pipe whose nominal diameter is 1 in is 1.049 in (Table 8–3).

Types of Fluid Flow Problems

In the design and analysis of piping systems that involve the use of the Moody chart (or the Colebrook equation), we usually encounter three types of problems (the fluid and the roughness of the pipe are assumed to be specified in all cases) (Fig. 8–31):

1. Determining the **pressure drop** (or head loss) when the pipe length and diameter are given for a specified flow rate (or velocity)
2. Determining the **flow rate** when the pipe length and diameter are given for a specified pressure drop (or head loss)
3. Determining the **pipe diameter** when the pipe length and flow rate are given for a specified pressure drop (or head loss)

Problems of the *first type* are straightforward and can be solved directly by using the Moody chart. Problems of the *second type* and *third type* are commonly encountered in engineering design (in the selection of pipe diameter, for example, that minimizes the sum of the construction and pumping costs), but the use of the Moody chart with such problems requires an iterative approach—an equation solver (such as EES) is recommended.

In problems of the *second type*, the diameter is given but the flow rate is unknown. A good guess for the friction factor in that case is obtained from the completely turbulent flow region for the given roughness. This is true for large Reynolds numbers, which is often the case in practice. Once the flow rate is obtained, the friction factor is corrected using the Moody chart or the Colebrook equation, and the process is repeated until the solution

TABLE 8–3

Standard sizes for Schedule 40 steel pipes

Nominal Size, in	Actual Inside Diameter, in
$\frac{1}{8}$	0.269
$\frac{1}{4}$	0.364
$\frac{3}{8}$	0.493
$\frac{1}{2}$	0.622
$\frac{3}{4}$	0.824
1	1.049
$1\frac{1}{2}$	1.610
2	2.067
$2\frac{1}{2}$	2.469
3	3.068
5	5.047
10	10.02

Problem type	Given	Find
1	$L, D, \dot{V}$	ΔP (or h_L)
2	$L, D, \Delta P$	$\dot{V}$
3	$L, \Delta P, \dot{V}$	D

FIGURE 8–31
The three types of problems encountered in pipe flow.

converges. (Typically only a few iterations are required for convergence to three or four digits of precision.)

In problems of the *third type*, the diameter is not known and thus the Reynolds number and the relative roughness cannot be calculated. Therefore, we start calculations by assuming a pipe diameter. The pressure drop calculated for the assumed diameter is then compared to the specified pressure drop, and calculations are repeated with another pipe diameter in an iterative fashion until convergence.

To avoid tedious iterations in head loss, flow rate, and diameter calculations, Swamee and Jain (1976) proposed the following explicit relations that are accurate to within 2 percent of the Moody chart:

$$h_L = 1.07 \frac{\dot{V}^2 L}{gD^5} \left\{ \ln\left[\frac{\varepsilon}{3.7D} + 4.62\left(\frac{\nu D}{\dot{V}}\right)^{0.9}\right] \right\}^{-2} \quad \begin{array}{c} 10^{-6} < \varepsilon/D < 10^{-2} \\ 3000 < \text{Re} < 3 \times 10^8 \end{array} \quad (8\text{–}52)$$

$$\dot{V} = -0.965\left(\frac{gD^5 h_L}{L}\right)^{0.5} \ln\left[\frac{\varepsilon}{3.7D} + \left(\frac{3.17\nu^2 L}{gD^3 h_L}\right)^{0.5}\right] \quad \text{Re} > 2000 \quad (8\text{–}53)$$

$$D = 0.66\left[\varepsilon^{1.25}\left(\frac{L\dot{V}^2}{gh_L}\right)^{4.75} + \nu\dot{V}^{9.4}\left(\frac{L}{gh_L}\right)^{5.2}\right]^{0.04} \quad \begin{array}{c} 10^{-6} < \varepsilon/D < 10^{-2} \\ 5000 < \text{Re} < 3 \times 10^8 \end{array} \quad (8\text{–}54)$$

Note that all quantities are dimensional and the units simplify to the desired unit (for example, to m or ft in the last relation) when consistent units are used. Noting that the Moody chart is accurate to within 15 percent of experimental data, we should have no reservation in using these approximate relations in the design of piping systems.

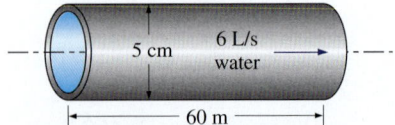

FIGURE 8–32
Schematic for Example 8–3.

EXAMPLE 8–3   Determining the Head Loss in a Water Pipe

Water at 15°C (ρ = 999 kg/m³ and μ = 1.138 × 10⁻³ kg/m·s) is flowing steadily in a 5-cm-diameter horizontal pipe made of stainless steel at a rate of 6 L/s (Fig. 8–32). Determine the pressure drop, the head loss, and the required pumping power input for flow over a 60-m-long section of the pipe.

SOLUTION The flow rate through a specified water pipe is given. The pressure drop, the head loss, and the pumping power requirements are to be determined.
Assumptions **1** The flow is steady and incompressible. **2** The entrance effects are negligible, and thus the flow is fully developed. **3** The pipe involves no components such as bends, valves, and connectors. **4** The piping section involves no work devices such as a pump or a turbine.
Properties The density and dynamic viscosity of water are given to be ρ = 999 kg/m³ and μ = 1.138 × 10⁻³ kg/m·s, respectively.
Analysis We recognize this as a problem of the first type, since flow rate, pipe length, and pipe diameter are known. First we calculate the average velocity and the Reynolds number to determine the flow regime:

$$V = \frac{\dot{V}}{A_c} = \frac{\dot{V}}{\pi D^2/4} = \frac{0.006 \text{ m}^3}{\pi(0.05 \text{ m})^2/4} = 3.06 \text{ m/s}$$

$$\text{Re} = \frac{\rho V D}{\mu} = \frac{(999 \text{ kg/m}^3)(3.06 \text{ m/s})(0.05 \text{ m})}{1.138 \times 10^{-3} \text{ kg/m·s}} = 134{,}300$$

Since Re is greater than 4000, the flow is turbulent. The relative roughness of the pipe is estimated using Table 8–2

$$\varepsilon/D = \frac{0.002 \text{ mm}}{50 \text{ mm}} = 0.000040$$

The friction factor corresponding to this relative roughness and Reynolds number is determined from the Moody chart. To avoid any reading error, we determine f from the Colebrook equation on which the Moody chart is based:

$$\frac{1}{\sqrt{f}} = -2.0 \log\left(\frac{\varepsilon/D}{3.7} + \frac{2.51}{\text{Re}\sqrt{f}}\right) \rightarrow \frac{1}{\sqrt{f}} = -2.0 \log\left(\frac{0.000040}{3.7} + \frac{2.51}{134{,}300\sqrt{f}}\right)$$

Using an equation solver or an iterative scheme, the friction factor is determined to be $f = 0.0172$. Then the pressure drop (which is equivalent to pressure loss in this case), head loss, and the required power input become

$$\Delta P = \Delta P_L = f\frac{L}{D}\frac{\rho V^2}{2} = 0.0172\frac{60 \text{ m}}{0.05 \text{ m}}\frac{(999 \text{ kg/m}^3)(3.06 \text{ m/s})^2}{2}\left(\frac{1 \text{ N}}{1 \text{ kg·m/s}^2}\right)$$
$$= 96{,}540 \text{ N/m}^2 = \mathbf{96.5 \text{ kPa}}$$

$$h_L = \frac{\Delta P_L}{\rho g} = f\frac{L}{D}\frac{V^2}{2g} = 0.0172\frac{60 \text{ m}}{0.05 \text{ m}}\frac{(3.06 \text{ m/s})^2}{2(9.81 \text{ m/s}^2)} = \mathbf{9.85 \text{ m}}$$

$$\dot{W}_{\text{pump}} = \dot{V}\,\Delta P = (0.006 \text{ m}^3/\text{s})(96{,}540 \text{ N/m}^2)\left(\frac{1 \text{ W}}{1 \text{ N·m/s}}\right) = \mathbf{579 \text{ W}}$$

Therefore, power input in the amount of 579 W is needed to overcome the frictional losses in the pipe.

Discussion It is common practice to write our final answers to three significant digits, even though we know that the results are accurate to at most two significant digits because of inherent inaccuracies in the Colebrook equation, as discussed previously. The friction factor could also be determined easily from the explicit Haaland relation (Eq. 8–51). It would give $f = 0.0170$, which is sufficiently close to 0.0172. Also, the friction factor corresponding to $\epsilon = 0$ in this case is 0.0169, which indicates that this stainless-steel pipe can be approximated as smooth with minimal error.

■ **EXAMPLE 8–4**   **Determining the Diameter of an Air Duct**

Heated air at 1 atm and 35°C is to be transported in a 150-m-long circular plastic duct at a rate of 0.35 m³/s (Fig. 8–33). If the head loss in the pipe is not to exceed 20 m, determine the minimum diameter of the duct.

SOLUTION The flow rate and the head loss in an air duct are given. The diameter of the duct is to be determined.
Assumptions **1** The flow is steady and incompressible. **2** The entrance effects are negligible, and thus the flow is fully developed. **3** The duct involves no components such as bends, valves, and connectors. **4** Air is an ideal gas. **5** The duct is smooth since it is made of plastic. **6** The flow is turbulent (to be verified).
Properties The density, dynamic viscosity, and kinematic viscosity of air at 35°C are $\rho = 1.145$ kg/m³, $\mu = 1.895 \times 10^{-5}$ kg/m·s, and $\nu = 1.655 \times 10^{-5}$ m²/s.

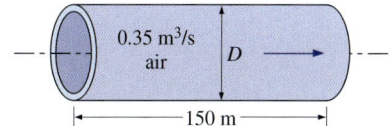

FIGURE 8–33
Schematic for Example 8–4.

Analysis This is a problem of the third type since it involves the determination of diameter for specified flow rate and head loss. We can solve this problem using three different approaches: (1) an iterative approach by assuming a pipe diameter, calculating the head loss, comparing the result to the specified head loss, and repeating calculations until the calculated head loss matches the specified value; (2) writing all the relevant equations (leaving the diameter as an unknown) and solving them simultaneously using an equation solver; and (3) using the third Swamee–Jain formula. We will demonstrate the use of the last two approaches.

The average velocity, the Reynolds number, the friction factor, and the head loss relations are expressed as (D is in m, V is in m/s, and Re and f are dimensionless)

$$V = \frac{\dot{V}}{A_c} = \frac{\dot{V}}{\pi D^2/4} = \frac{0.35 \text{ m}^3/\text{s}}{\pi D^2/4}$$

$$\text{Re} = \frac{VD}{\nu} = \frac{VD}{1.655 \times 10^{-5} \text{ m}^2/\text{s}}$$

$$\frac{1}{\sqrt{f}} = -2.0 \log\left(\frac{\varepsilon/D}{3.7} + \frac{2.51}{\text{Re}\sqrt{f}}\right) = -2.0 \log\left(\frac{2.51}{\text{Re}\sqrt{f}}\right)$$

$$h_L = f\frac{L}{D}\frac{V^2}{2g} \rightarrow 20 \text{ m} = f\frac{150 \text{ m}}{D}\frac{V^2}{2(9.81 \text{ m/s}^2)}$$

The roughness is approximately zero for a plastic pipe (Table 8–2). Therefore, this is a set of four equations and four unknowns, and solving them with an equation solver such as EES gives

$$D = \mathbf{0.267 \text{ m}}, \quad f = 0.0180, \quad V = 6.24 \text{ m/s}, \quad \text{and} \quad \text{Re} = 100{,}800$$

Therefore, the diameter of the duct should be more than 26.7 cm if the head loss is not to exceed 20 m. Note that Re > 4000, and thus the turbulent flow assumption is verified.

The diameter can also be determined directly from the third Swamee–Jain formula to be

$$D = 0.66\left[\varepsilon^{1.25}\left(\frac{L\dot{V}^2}{gh_L}\right)^{4.75} + \nu\dot{V}^{9.4}\left(\frac{L}{gh_L}\right)^{5.2}\right]^{0.04}$$

$$= 0.66\left[0 + (1.655 \times 10^{-5} \text{ m}^2/\text{s})(0.35 \text{ m}^3/\text{s})^{9.4}\left(\frac{150 \text{ m}}{(9.81 \text{ m/s}^2)(20 \text{ m})}\right)^{5.2}\right]^{0.04}$$

$$= \mathbf{0.271 \text{ m}}$$

Discussion Note that the difference between the two results is less than 2 percent. Therefore, the simple Swamee–Jain relation can be used with confidence. Finally, the first (iterative) approach requires an initial guess for D. If we use the Swamee–Jain result as our initial guess, the diameter converges to $D = 0.267$ m in short order.

EXAMPLE 8–5 Determining the Flow Rate of Air in a Duct

Reconsider Example 8–4. Now the duct length is doubled while its diameter is maintained constant. If the total head loss is to remain constant, determine the drop in the flow rate through the duct.

SOLUTION The diameter and the head loss in an air duct are given. The drop in the flow rate is to be determined.

Analysis This is a problem of the second type since it involves the determination of the flow rate for a specified pipe diameter and head loss. The solution involves an iterative approach since the flow rate (and thus the flow velocity) is not known.

The average velocity, Reynolds number, friction factor, and the head loss relations are expressed as (D is in m, V is in m/s, and Re and f are dimensionless)

$$V = \frac{\dot{V}}{A_c} = \frac{\dot{V}}{\pi D^2/4} \quad \rightarrow \quad V = \frac{\dot{V}}{\pi (0.267 \text{ m})^2/4}$$

$$\text{Re} = \frac{VD}{\nu} \quad \rightarrow \quad \text{Re} = \frac{V(0.267 \text{ m})}{1.655 \times 10^{-5} \text{ m}^2/\text{s}}$$

$$\frac{1}{\sqrt{f}} = -2.0 \log\left(\frac{\varepsilon/D}{3.7} + \frac{2.51}{\text{Re}\sqrt{f}}\right) \quad \rightarrow \quad \frac{1}{\sqrt{f}} = -2.0 \log\left(\frac{2.51}{\text{Re}\sqrt{f}}\right)$$

$$h_L = f \frac{L}{D} \frac{V^2}{2g} \quad \rightarrow \quad 20 \text{ m} = f \frac{300 \text{ m}}{0.267 \text{ m}} \frac{V^2}{2(9.81 \text{ m/s}^2)}$$

This is a set of four equations in four unknowns and solving them with an equation solver such as EES (Fig. 8–34) gives

$$\dot{V} = 0.24 \text{ m}^3/\text{s}, \quad f = 0.0195, \quad V = 4.23 \text{ m/s}, \quad \text{and} \quad \text{Re} = 68,300$$

Then the drop in the flow rate becomes

$$\dot{V}_{\text{drop}} = \dot{V}_{\text{old}} - \dot{V}_{\text{new}} = 0.35 - 0.24 = \mathbf{0.11 \text{ m}^3/\text{s}} \quad \text{(a drop of 31 percent)}$$

Therefore, for a specified head loss (or available head or fan pumping power), the flow rate drops by about 31 percent from 0.35 to 0.24 m³/s when the duct length doubles.

Alternative Solution If a computer is not available (as in an exam situation), another option is to set up a *manual iteration loop*. We have found that the best convergence is usually realized by first guessing the friction factor f, and then solving for the velocity V. The equation for V as a function of f is

Average velocity through the pipe: $\quad V = \sqrt{\dfrac{2gh_L}{fL/D}}$

Once V is calculated, the Reynolds number can be calculated, from which a *corrected* friction factor is obtained from the Moody chart or the Colebrook equation. We repeat the calculations with the corrected value of f until convergence. We guess $f = 0.04$ for illustration:

Iteration	f (guess)	V, m/s	Re	Corrected f
1	0.04	2.955	4.724×10^4	0.0212
2	0.0212	4.059	6.489×10^4	0.01973
3	0.01973	4.207	6.727×10^4	0.01957
4	0.01957	4.224	6.754×10^4	0.01956
5	0.01956	4.225	6.756×10^4	0.01956

Notice that the iteration has converged to three digits in only three iterations and to four digits in only four iterations. The final results are identical to those obtained with EES, yet do not require a computer.

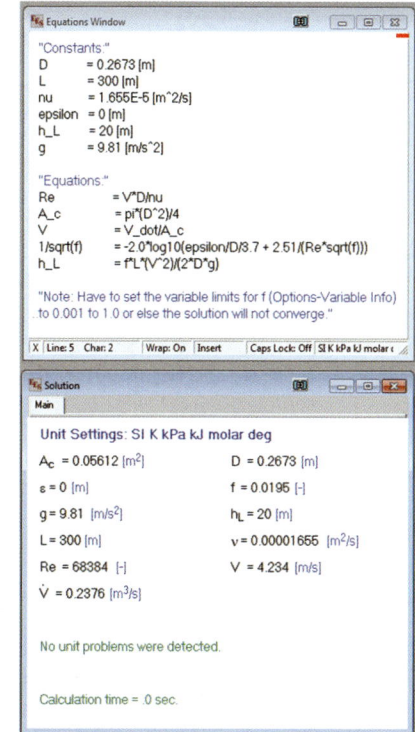

FIGURE 8–34
EES solution for Example 8–5.

Discussion The new flow rate can also be determined directly from the second Swamee–Jain formula to be

$$\dot{V} = -0.965\left(\frac{gD^5h_L}{L}\right)^{0.5} \ln\left[\frac{\varepsilon}{3.7D} + \left(\frac{3.17\nu^2 L}{gD^3 h_L}\right)^{0.5}\right]$$

$$= -0.965\left(\frac{(9.81 \text{ m/s}^2)(0.267 \text{ m})^5(20 \text{ m})}{300 \text{ m}}\right)^{0.5}$$

$$\times \ln\left[0 + \left(\frac{3.17(1.655 \times 10^{-5} \text{ m}^2/\text{s})^2(300 \text{ m})}{(9.81 \text{ m/s}^2)(0.267 \text{ m})^3(20 \text{ m})}\right)^{0.5}\right]$$

$$= 0.24 \text{ m}^3/\text{s}$$

Note that the result from the Swamee–Jain relation is the same (to two significant digits) as that obtained with the Colebrook equation using EES or using our manual iteration technique. Therefore, the simple Swamee–Jain relation can be used with confidence.

8–6 · MINOR LOSSES

The fluid in a typical piping system passes through various fittings, valves, bends, elbows, tees, inlets, exits, expansions, and contractions in addition to the straight sections of piping. These components interrupt the smooth flow of the fluid and cause additional losses because of the flow separation and mixing they induce. In a typical system with long pipes, these losses are minor compared to the head loss in the straight sections (the **major losses**) and are called **minor losses.** Although this is generally true, in some cases the minor losses may be greater than the major losses. This is the case, for example, in systems with several turns and valves in a short distance. The head loss introduced by a completely open valve, for example, may be negligible. But a partially closed valve may cause the largest head loss in the system, as evidenced by the drop in the flow rate. Flow through valves and fittings is very complex, and a theoretical analysis is generally not plausible. Therefore, minor losses are determined experimentally, usually by the manufacturers of the components.

Minor losses are usually expressed in terms of the **loss coefficient** K_L (also called the **resistance coefficient**), defined as (Fig. 8–35)

Loss coefficient:
$$K_L = \frac{h_L}{V^2/(2g)} \quad (8\text{–}55)$$

where h_L is the *additional* irreversible head loss in the piping system caused by insertion of the component, and is defined as $h_L = \Delta P_L/\rho g$. For example, imagine replacing the valve in Fig. 8–35 with a section of constant diameter pipe from location 1 to location 2. ΔP_L is defined as the pressure drop from 1 to 2 for the case *with* the valve, $(P_1 - P_2)_\text{valve}$, *minus* the pressure drop that would occur in the imaginary straight pipe section from 1 to 2 *without* the valve, $(P_1 - P_2)_\text{pipe}$ at the same flow rate. While the majority of the irreversible head loss occurs locally near the valve, some of it occurs downstream of the valve due to induced swirling turbulent eddies that are produced in the valve and continue downstream. These eddies "waste" mechanical energy because they are ultimately dissipated into heat while the flow in the

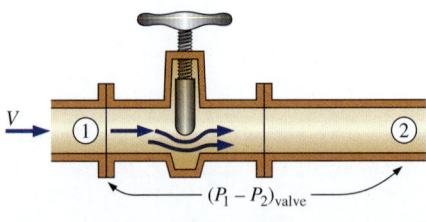

Pipe section with valve:

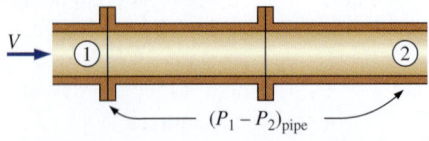

Pipe section without valve:

$\Delta P_L = (P_1 - P_2)_\text{valve} - (P_1 - P_2)_\text{pipe}$

FIGURE 8–35
For a constant-diameter section of a pipe with a minor loss component, the loss coefficient of the component (such as the gate valve shown) is determined by measuring the additional pressure loss it causes and dividing it by the dynamic pressure in the pipe.

downstream section of pipe eventually returns to fully developed conditions. When measuring minor losses in some minor loss components, such as *elbows*, for example, location 2 must be considerably far downstream (tens of pipe diameters) in order to fully account for the additional irreversible losses due to these decaying eddies.

When the pipe diameter downstream of the component *changes*, determination of the minor loss is even more complicated. In all cases, however, it is based on the *additional* irreversible loss of mechanical energy that would otherwise not exist if the minor loss component were not there. For simplicity, you may think of the minor loss as occurring *locally* across the minor loss component, but keep in mind that the component influences the flow for several pipe diameters downstream. By the way, this is the reason why most flow meter manufacturers recommend installing their flow meter at least 10 to 20 pipe diameters downstream of any elbows or valves—this allows the swirling turbulent eddies generated by the elbow or valve to largely disappear and the velocity profile to become fully developed before entering the flow meter. (Most flow meters are calibrated with a fully developed velocity profile at the flow meter inlet, and yield the best accuracy when such conditions also exist in the actual application.)

When the inlet diameter equals the outlet diameter, the loss coefficient of a component can also be determined by measuring the pressure loss across the component and dividing it by the dynamic pressure, $K_L = \Delta P_L/(\frac{1}{2}\rho V^2)$. When the loss coefficient for a component is available, the head loss for that component is determined from

Minor loss: $$h_L = K_L \frac{V^2}{2g} \quad (8\text{–}56)$$

The loss coefficient, in general, depends on the geometry of the component and the Reynolds number, just like the friction factor. However, it is usually assumed to be independent of the Reynolds number. This is a reasonable approximation since most flows in practice have large Reynolds numbers and the loss coefficients (including the friction factor) tend to be independent of the Reynolds number at large Reynolds numbers.

Minor losses are also expressed in terms of the **equivalent length** L_{equiv}, defined as (Fig. 8–36)

Equivalent length: $$h_L = K_L \frac{V^2}{2g} = f \frac{L_{equiv}}{D} \frac{V^2}{2g} \quad \rightarrow \quad L_{equiv} = \frac{D}{f} K_L \quad (8\text{–}57)$$

where f is the friction factor and D is the diameter of the pipe that contains the component. The head loss caused by the component is equivalent to the head loss caused by a section of the pipe whose length is L_{equiv}. Therefore, the contribution of a component to the head loss is accounted for by simply adding L_{equiv} to the total pipe length.

Both approaches are used in practice, but the use of loss coefficients is more common. Therefore, we also use that approach in this book. Once all the loss coefficients are available, the total head loss in a piping system is determined from

Total head loss (general):
$$h_{L,\,total} = h_{L,\,major} + h_{L,\,minor}$$
$$= \sum_i f_i \frac{L_i}{D_i} \frac{V_i^2}{2g} + \sum_j K_{L,j} \frac{V_j^2}{2g} \quad (8\text{–}58)$$

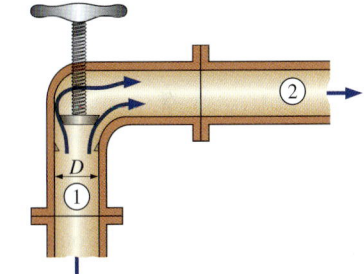

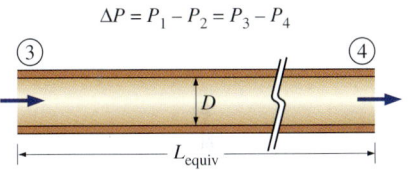

$\Delta P = P_1 - P_2 = P_3 - P_4$

FIGURE 8–36
The head loss caused by a component (such as the angle valve shown) is equivalent to the head loss caused by a section of the pipe whose length is the equivalent length.

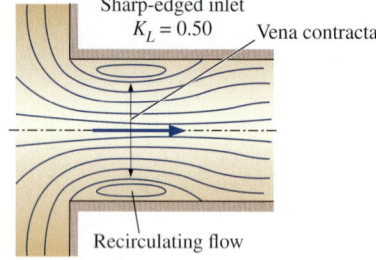
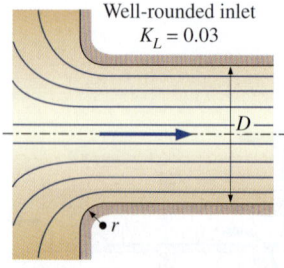

FIGURE 8–37
The head loss at the inlet of a pipe is almost negligible for well-rounded inlets ($K_L = 0.03$ for $r/D > 0.2$) but increases to about 0.50 for sharp-edged inlets.

where i represents each pipe section with constant diameter and j represents each component that causes a minor loss. If the entire piping system being analyzed has a constant diameter, Eq. 8–58 reduces to

Total head loss (D = constant): $\qquad h_{L,\,\text{total}} = \left(f\frac{L}{D} + \sum K_L \right) \frac{V^2}{2g}$ (8–59)

where V is the average flow velocity through the entire system (note that V = constant since D = constant).

Representative loss coefficients K_L are given in Table 8–4 for inlets, exits, bends, sudden and gradual area changes, and valves. There is considerable uncertainty in these values since the loss coefficients, in general, vary with the pipe diameter, the surface roughness, the Reynolds number, and the details of the design. The loss coefficients of two seemingly identical valves by two different manufacturers, for example, can differ by a factor of 2 or more. Therefore, the particular manufacturer's data should be consulted in the final design of piping systems rather than relying on the representative values in handbooks.

The head loss at the inlet of a pipe is a strong function of geometry. It is almost negligible for well-rounded inlets ($K_L = 0.03$ for $r/D > 0.2$), but increases to about 0.50 for sharp-edged inlets (Fig. 8–37). That is, a sharp-edged inlet causes half of the velocity head to be lost as the fluid enters the pipe. This is because the fluid cannot make sharp 90° turns easily, especially at high velocities. As a result, the flow separates at the corners, and the flow is constricted into the **vena contracta** region formed in the midsection of the pipe (Fig. 8–38). Therefore, a sharp-edged inlet acts like a flow constriction. The velocity increases in the vena contracta region (and the pressure decreases) because of the reduced effective flow area and then decreases as the flow fills the entire cross section of the pipe. There would be negligible loss if the pressure were increased in accordance with Bernoulli's equation (the velocity head would simply be converted into pressure head). However, this deceleration process is far from ideal and the viscous dissipation caused by intense mixing and the turbulent eddies converts part of the kinetic energy into frictional heating, as evidenced by a slight rise in fluid temperature. The end result is a drop in velocity without much pressure recovery, and the inlet loss is a measure of this irreversible pressure drop.

Even slight rounding of the edges can result in significant reduction of K_L, as shown in Fig. 8–39. The loss coefficient rises sharply (to about $K_L = 0.8$) when the pipe protrudes into the reservoir since some fluid near the edge in this case is forced to make a 180° turn.

The loss coefficient for a submerged pipe exit is often listed in handbooks as $K_L = 1$. More precisely, however, K_L is equal to the kinetic energy correction factor α at the exit of the pipe. Although α is indeed close to 1 for fully developed *turbulent* pipe flow, it is equal to 2 for fully developed *laminar* pipe flow. To avoid possible errors when analyzing laminar pipe flow, then, it is best to always set $K_L = \alpha$ at a submerged pipe exit. At any such exit, whether laminar or turbulent, the fluid leaving the pipe loses *all* of its kinetic energy as it mixes with the reservoir fluid and eventually comes to rest through the irreversible action of viscosity. This is true regardless of the shape of the exit (Table 8–4 and Fig. 8–40). Therefore, there is no advantage to rounding off the sharp edges of pipe exits.

TABLE 8–4

Loss coefficients K_L of various pipe components for turbulent flow (for use in the relation $h_L = K_L V^2/(2g)$, where V is the average velocity in the pipe that contains the component)*

Pipe Inlet

Reentrant: $K_L = 0.80$
($t \ll D$ and $l \approx 0.1D$)

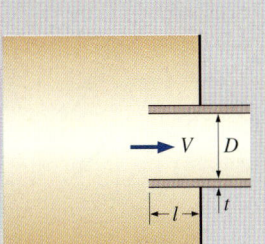

Sharp-edged: $K_L = 0.50$

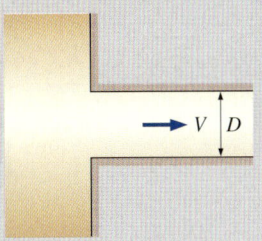

Well-rounded ($r/D > 0.2$): $K_L = 0.03$
Slightly rounded ($r/D = 0.1$): $K_L = 0.12$
(see Fig. 8–39)

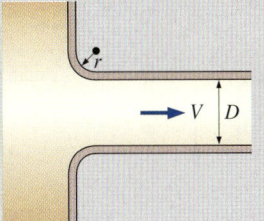

Pipe Exit

Reentrant: $K_L = \alpha$

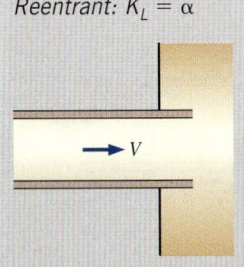

Sharp-edged: $K_L = \alpha$

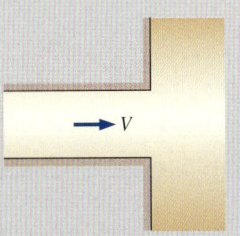

Rounded: $K_L = \alpha$

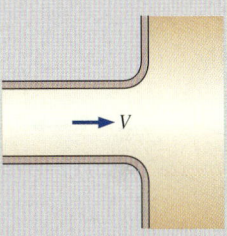

Note: The kinetic energy correction factor is $\alpha = 2$ for fully developed laminar flow, and $\alpha \approx 1.05$ for fully developed turbulent flow.

Sudden Expansion and Contraction (based on the velocity in the smaller-diameter pipe)

Sudden expansion: $K_L = \alpha \left(1 - \dfrac{d^2}{D^2}\right)^2$

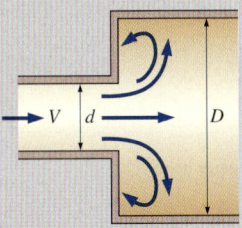

Sudden contraction: See chart.

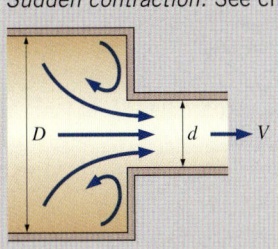

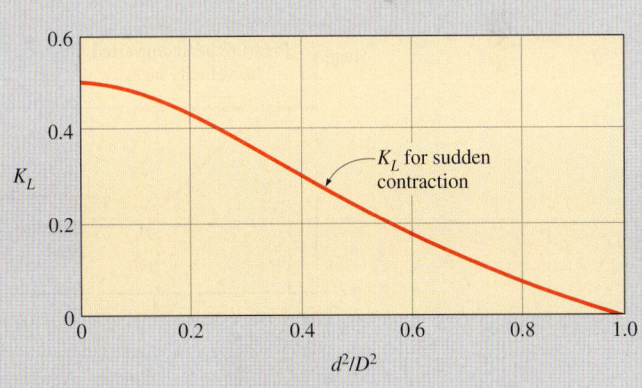

Gradual Expansion and Contraction (based on the velocity in the smaller-diameter pipe)

Expansion (for $\theta = 20°$):
$K_L = 0.30$ for $d/D = 0.2$
$K_L = 0.25$ for $d/D = 0.4$
$K_L = 0.15$ for $d/D = 0.6$
$K_L = 0.10$ for $d/D = 0.8$

Contraction:
$K_L = 0.02$ for $\theta = 30°$
$K_L = 0.04$ for $\theta = 45°$
$K_L = 0.07$ for $\theta = 60°$

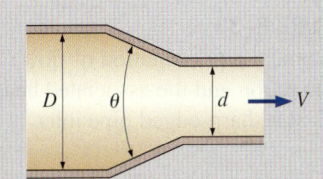

TABLE 8–4 (CONCLUDED)

Bends and Branches

90° smooth bend: Flanged: $K_L = 0.3$ Threaded: $K_L = 0.9$	90° miter bend (without vanes): $K_L = 1.1$	90° miter bend (with vanes): $K_L = 0.2$	45° threaded elbow: $K_L = 0.4$
180° return bend: Flanged: $K_L = 0.2$ Threaded: $K_L = 1.5$	Tee (branch flow): Flanged: $K_L = 1.0$ Threaded: $K_L = 2.0$	Tee (line flow): Flanged: $K_L = 0.2$ Threaded: $K_L = 0.9$	Threaded union: $K_L = 0.08$

Valves

Globe valve, fully open: $K_L = 10$
Angle valve, fully open: $K_L = 5$
Ball valve, fully open: $K_L = 0.05$
Swing check valve: $K_L = 2$

Gate valve, fully open: $K_L = 0.2$
$\tfrac{1}{4}$ closed: $K_L = 0.3$
$\tfrac{1}{2}$ closed: $K_L = 2.1$
$\tfrac{3}{4}$ closed: $K_L = 17$

* These are representative values for loss coefficients. Actual values strongly depend on the design and manufacture of the components and may differ from the given values considerably (especially for valves). Actual manufacturer's data should be used in the final design.

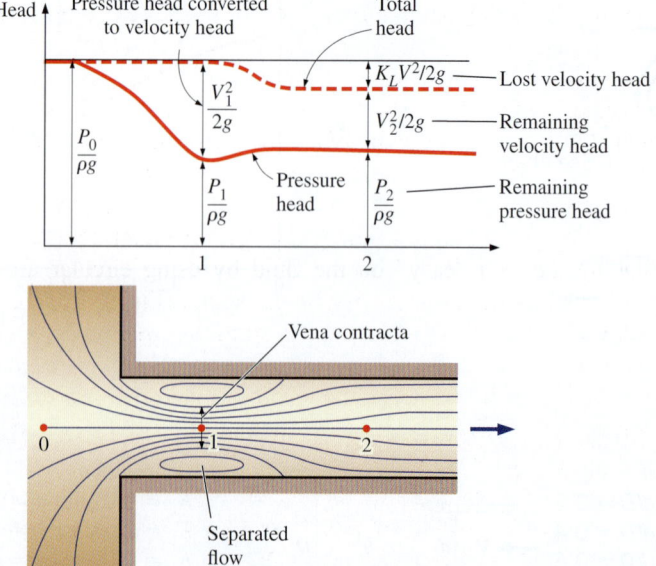

FIGURE 8–38
Graphical representation of flow contraction and the associated head loss at a sharp-edged pipe inlet.

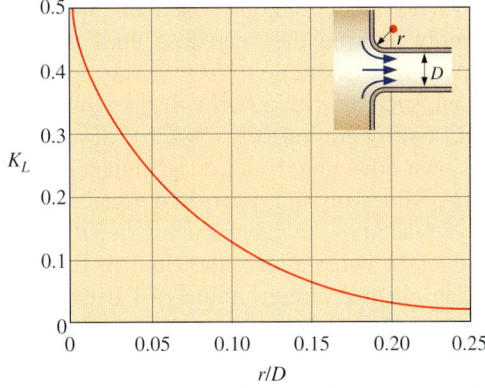

FIGURE 8–39
The effect of rounding of a pipe inlet on the loss coefficient.
Data from ASHRAE Handbook of Fundamentals.

Piping systems often involve *sudden* or *gradual* expansion or contraction sections to accommodate changes in flow rates or properties such as density and velocity. The losses are usually much greater in the case of *sudden* expansion and contraction (or wide-angle expansion) because of flow separation. By combining the equations of mass, momentum, and energy balance, the loss coefficient for the case of a **sudden expansion** is approximated as

$$K_L = \alpha\left(1 - \frac{A_{\text{small}}}{A_{\text{large}}}\right)^2 \quad \text{(sudden expansion)} \quad (8\text{-}60)$$

where A_{small} and A_{large} are the cross-sectional areas of the small and large pipes, respectively. Note that $K_L = 0$ when there is no area change ($A_{\text{small}} = A_{\text{large}}$) and $K_L = \alpha$ when a pipe discharges into a reservoir ($A_{\text{large}} \gg A_{\text{small}}$). No such relation exists for a sudden contraction, and the K_L values in that case must be read from a chart or table (e.g., Table 8–4). The losses due to expansions and contractions can be reduced significantly by installing conical gradual area changers (nozzles and diffusers) between the small and large pipes. The K_L values for representative cases of gradual expansion and contraction are given in Table 8–4. Note that in head loss calculations, the velocity in the *small pipe* is to be used as the reference velocity in Eq. 8–56. Losses during expansion are usually much higher than the losses during contraction because of flow separation.

Piping systems also involve changes in direction without a change in diameter, and such flow sections are called *bends* or *elbows*. The losses in these devices are due to flow separation (just like a car being thrown off the road when it enters a turn too fast) on the inner side and the swirling secondary flows that result. The losses during changes of direction can be minimized by making the turn "easy" on the fluid by using circular arcs (like 90° elbows) instead of sharp turns (like miter bends) (Fig. 8–41). But the use of sharp turns (and thus suffering a penalty in loss coefficient) may be necessary when the turning space is limited. In such cases, the losses can be minimized by utilizing properly placed guide vanes to help the flow turn in an orderly manner without being thrown off the course. The loss coefficients for some elbows and miter bends as well as tees are given in Table 8–4. These coefficients do not include the frictional losses along the pipe bend. Such losses should be calculated as in straight pipes (using the length of the centerline as the pipe length) and added to other losses.

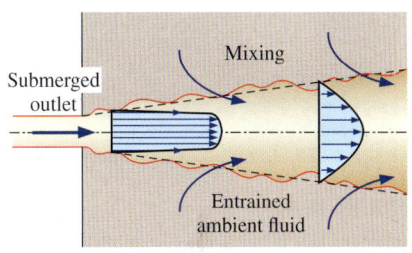

FIGURE 8–40
All the kinetic energy of the flow is "lost" (turned into thermal energy) through friction as the jet decelerates and mixes with ambient fluid downstream of a submerged outlet.

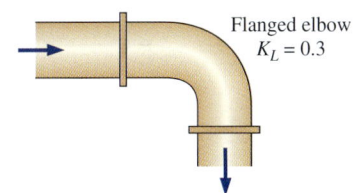

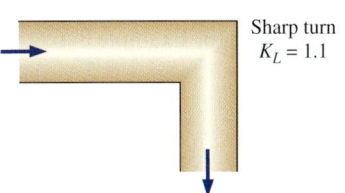

FIGURE 8–41
The losses during changes of direction can be minimized by making the turn "easy" on the fluid by using circular arcs instead of sharp turns.

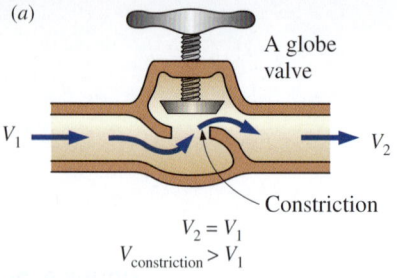

(a)

(b)

FIGURE 8–42
(a) The large head loss in a partially closed globe valve is due to irreversible deceleration, flow separation, and mixing of high-velocity fluid coming from the narrow valve passage. (b) The head loss through a fully-open ball valve, on the other hand, is quite small.
Photo by John M. Cimbala.

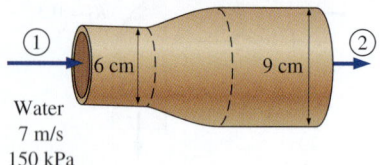

FIGURE 8–43
Schematic for Example 8–6.

Valves are commonly used in piping systems to control flow rates by simply altering the head loss until the desired flow rate is achieved. For valves it is desirable to have a very low loss coefficient when they are fully open, such as with a *ball valve*, so that they cause minimal head loss during full-load operation (Fig. 8–42b). Several different valve designs, each with its own advantages and disadvantages, are in common use today. The *gate valve* slides up and down like a gate, the *globe valve* (Fig. 8–42a) closes a hole placed in the valve, the *angle valve* is a globe valve with a 90° turn, and the *check valve* allows the fluid to flow only in one direction like a diode in an electric circuit. Table 8–4 lists the representative loss coefficients of the popular designs. Note that the loss coefficient increases drastically as a valve is closed. Also, the deviation in the loss coefficients for different manufacturers is greatest for valves because of their complex geometries.

EXAMPLE 8–6 Head Loss and Pressure Rise during Gradual Expansion

A 6-cm-diameter horizontal water pipe expands gradually to a 9-cm-diameter pipe (Fig. 8–43). The walls of the expansion section are angled 10° from the axis. The average velocity and pressure of water before the expansion section are 7 m/s and 150 kPa, respectively. Determine the head loss in the expansion section and the pressure in the larger-diameter pipe.

SOLUTION A horizontal water pipe expands gradually into a larger-diameter pipe. The head loss and pressure after the expansion are to be determined.
Assumptions 1 The flow is steady and incompressible. 2 The flow at sections 1 and 2 is fully developed and turbulent with $\alpha_1 = \alpha_2 \cong 1.06$.
Properties We take the density of water to be $\rho = 1000$ kg/m^3. The loss coefficient for a gradual expansion of total included angle $\theta = 20°$ and diameter ratio $d/D = 6/9$ is $K_L = 0.133$ (by interpolation using Table 8–4).
Analysis Noting that the density of water remains constant, the downstream velocity of water is determined from conservation of mass to be

$$\dot{m}_1 = \dot{m}_2 \rightarrow \rho V_1 A_1 = \rho V_2 A_2 \rightarrow V_2 = \frac{A_1}{A_2} V_1 = \frac{D_1^2}{D_2^2} V_1$$

$$V_2 = \frac{(0.06 \text{ m})^2}{(0.09 \text{ m})^2} (7 \text{ m/s}) = 3.11 \text{ m/s}$$

Then the irreversible head loss in the expansion section becomes

$$h_L = K_L \frac{V_1^2}{2g} = (0.133) \frac{(7 \text{ m/s})^2}{2(9.81 \text{ m/s}^2)} = \textbf{0.333 m}$$

Noting that $z_1 = z_2$ and there are no pumps or turbines involved, the energy equation for the expansion section is expressed in terms of heads as

$$\frac{P_1}{\rho g} + \alpha_1 \frac{V_1^2}{2g} + \cancel{z_1} + \cancel{h_{\text{pump},u}}^{0} = \frac{P_2}{\rho g} + \alpha_2 \frac{V_2^2}{2g} + \cancel{z_2} + \cancel{h_{\text{turbine},e}}^{0} + h_L$$

or

$$\frac{P_1}{\rho g} + \alpha_1 \frac{V_1^2}{2g} = \frac{P_2}{\rho g} + \alpha_2 \frac{V_2^2}{2g} + h_L$$

Solving for P_2 and substituting,

$$P_2 = P_1 + \rho\left\{\frac{\alpha_1 V_1^2 - \alpha_2 V_2^2}{2} - gh_L\right\} = (150 \text{ kPa}) + (1000 \text{ kg/m}^3)$$

$$\times \left\{\frac{1.06(7 \text{ m/s})^2 - 1.06(3.11 \text{ m/s})^2}{2} - (9.81 \text{ m/s}^2)(0.333 \text{ m})\right\}$$

$$\times \left(\frac{1 \text{ kN}}{1000 \text{ kg·m/s}^2}\right)\left(\frac{1 \text{ kPa}}{1 \text{ kN/m}^2}\right)$$

$$= \mathbf{168 \text{ kPa}}$$

Therefore, despite the head (and pressure) loss, the pressure *increases* from 150 to 168 kPa after the expansion. This is due to the conversion of dynamic pressure to static pressure when the average flow velocity is decreased in the larger pipe.

Discussion It is common knowledge that higher pressure upstream is necessary to cause flow, and it may come as a surprise to you that the downstream pressure has *increased* after the expansion, despite the loss. This is because the flow is driven by the sum of the three heads that comprise the total head (namely, pressure head, velocity head, and elevation head). During flow expansion, the higher velocity head upstream is converted to pressure head downstream, and this increase outweighs the nonrecoverable head loss. Also, you may be tempted to solve this problem using the Bernoulli equation. Such a solution would ignore the head loss (and the associated pressure loss) and result in an incorrect higher pressure for the fluid downstream.

8–7 ▪ PIPING NETWORKS AND PUMP SELECTION

Series and Parallel Pipes

Most piping systems encountered in practice such as the water distribution systems in cities or commercial or residential establishments involve numerous parallel and series connections as well as several sources (supply of fluid into the system) and loads (discharges of fluid from the system) (Fig. 8–44). A piping project may involve the design of a new system or the expansion of an existing system. The engineering objective in such projects is to design a piping system that will reliably deliver the specified flow rates at specified pressures at minimum total (initial plus operating and maintenance) cost. Once the layout of the system is prepared, the determination of the pipe diameters and the pressures throughout the system, while remaining within the budget constraints, typically requires solving the system repeatedly until the optimal solution is reached. Computer modeling and analysis of such systems make this tedious task a simple chore.

Piping systems typically involve several pipes connected to each other in series and/or in parallel, as shown in Figs. 8–45 and 8–46. When the pipes are connected **in series,** the flow rate through the entire system remains constant regardless of the diameters of the individual pipes in the system. This is a natural consequence of the conservation of mass principle for steady incompressible flow. The total head loss in this case is equal to the sum of the head losses in individual pipes in the system, including the minor losses. The

FIGURE 8–44
A piping network in an industrial facility.
Courtesy UMDE Engineering, Contracting, and Trading. Used by permission.

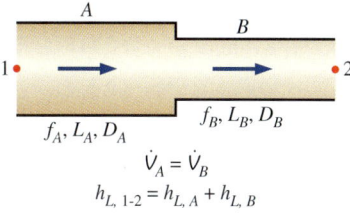

FIGURE 8–45
For pipes *in series*, the flow rate is the same in each pipe, and the total head loss is the sum of the head losses in the individual pipes.

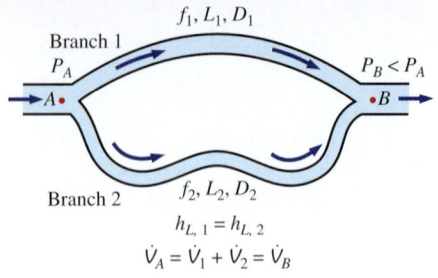

FIGURE 8–46
For pipes *in parallel*, the head loss is the same in each pipe, and the total flow rate is the sum of the flow rates in individual pipes.

expansion or contraction losses at connections are considered to belong to the smaller-diameter pipe since the expansion and contraction loss coefficients are defined on the basis of the average velocity in the smaller-diameter pipe.

For a pipe that branches out into two (or more) **parallel pipes** and then rejoins at a junction downstream, the total flow rate is the sum of the flow rates in the individual pipes. The pressure drop (or head loss) in each individual pipe connected in parallel must be the same since $\Delta P = P_A - P_B$ and the junction pressures P_A and P_B are the same for all the individual pipes. For a system of two parallel pipes 1 and 2 between junctions A and B with negligible minor losses, this is expressed as

$$h_{L,1} = h_{L,2} \quad \rightarrow \quad f_1 \frac{L_1}{D_1} \frac{V_1^2}{2g} = f_2 \frac{L_2}{D_2} \frac{V_2^2}{2g}$$

Then the ratio of the average velocities and the flow rates in the two parallel pipes become

$$\frac{V_1}{V_2} = \left(\frac{f_2 L_2 D_1}{f_1 L_1 D_2}\right)^{1/2} \quad \text{and} \quad \frac{\dot{V}_1}{\dot{V}_2} = \frac{A_{c,1} V_1}{A_{c,2} V_2} = \frac{D_1^2}{D_2^2}\left(\frac{f_2 L_2 D_1}{f_1 L_1 D_2}\right)^{1/2}$$

Therefore, the relative flow rates in parallel pipes are established from the requirement that the head loss in each pipe be the same. This result can be extended to any number of pipes connected in parallel. The result is also valid for pipes for which the minor losses are significant if the equivalent lengths for components that contribute to minor losses are added to the pipe length. Note that the flow rate in one of the parallel branches is proportional to its diameter to the power 5/2 and is inversely proportional to the square root of its length and friction factor.

The analysis of piping networks, no matter how complex they are, is based on two simple principles:

1. *Conservation of mass throughout the system must be satisfied.* This is done by requiring the total flow into a junction to be equal to the total flow out of the junction for all junctions in the system. Also, the flow rate must remain constant in pipes connected in series regardless of the changes in diameters.
2. *Pressure drop (and thus head loss) between two junctions must be the same for all paths between the two junctions.* This is because pressure is a point function and it cannot have two values at a specified point. In practice this rule is used by requiring that the algebraic sum of head losses in a loop (for all loops) be equal to zero. (A head loss is taken to be positive for flow in the clockwise direction and negative for flow in the counterclockwise direction.)

Therefore, the analysis of piping networks is very similar to the analysis of electric circuits (Kirchhoff's laws), with flow rate corresponding to electric current and pressure corresponding to electric potential. However, the situation is much more complex here since, unlike the electrical resistance, the "flow resistance" is a highly nonlinear function. Therefore, the analysis of piping networks requires the simultaneous solution of a system of nonlinear equations, which requires software such as EES, Mathcad, Matlab, etc., or commercially available software designed specifically for such applications.

Piping Systems with Pumps and Turbines

When a piping system involves a pump and/or turbine, the steady-flow energy equation on a unit-mass basis is expressed as (see Section 5–6)

$$\frac{P_1}{\rho} + \alpha_1 \frac{V_1^2}{2} + gz_1 + w_{pump,u} = \frac{P_2}{\rho} + \alpha_2 \frac{V_2^2}{2} + gz_2 + w_{turbine,e} + gh_L \quad (8\text{-}61)$$

or in terms of heads as

$$\frac{P_1}{\rho g} + \alpha_1 \frac{V_1^2}{2g} + z_1 + h_{pump,u} = \frac{P_2}{\rho g} + \alpha_2 \frac{V_2^2}{2g} + z_2 + h_{turbine,e} + h_L \quad (8\text{-}62)$$

where $h_{pump,u} = w_{pump,u}/g$ is the useful pump head delivered to the fluid, $h_{turbine,e} = w_{turbine,e}/g$ is the turbine head extracted from the fluid, α is the kinetic energy correction factor whose value is about 1.05 for most (turbulent) flows encountered in practice, and h_L is the total head loss in the piping (including the minor losses if they are significant) between points 1 and 2. The pump head is zero if the piping system does not involve a pump or a fan, the turbine head is zero if the system does not involve a turbine, and both are zero if the system does not involve any mechanical work-producing or work-consuming devices.

Many practical piping systems involve a pump to move a fluid from one reservoir to another. Taking points 1 and 2 to be at the *free surfaces* of the reservoirs (Fig. 8–47), the energy equation is solved for the required useful pump head, yielding

$$h_{pump,u} = (z_2 - z_1) + h_L \quad (8\text{-}63)$$

since the velocities at free surfaces are negligible for large reservoirs and the pressures are at atmospheric pressure. Therefore, the useful pump head is equal to the elevation difference between the two reservoirs plus the head loss. If the head loss is negligible compared to $z_2 - z_1$, the useful pump head is equal to the elevation difference between the two reservoirs. In the case of $z_1 > z_2$ (the first reservoir being at a higher elevation than the second one) with no pump, the flow is driven by gravity at a flow rate that causes a head loss equal to the elevation difference. A similar argument can be given for the turbine head for a hydroelectric power plant by replacing $h_{pump,u}$ in Eq. 8–63 by $-h_{turbine,e}$.

Once the useful pump head is known, the *mechanical power that needs to be delivered by the pump to the fluid* and the *electric power consumed by the motor of the pump* for a specified flow rate are determined from

$$\dot{W}_{pump,shaft} = \frac{\rho \dot{V} g h_{pump,u}}{\eta_{pump}} \quad \text{and} \quad \dot{W}_{elect} = \frac{\rho \dot{V} g h_{pump,u}}{\eta_{pump-motor}} \quad (8\text{-}64)$$

where $\eta_{pump-motor}$ is the *efficiency of the pump–motor combination*, which is the product of the pump and the motor efficiencies (Fig. 8–48). The pump–motor efficiency is defined as the ratio of the net mechanical energy delivered to the fluid by the pump to the electric energy consumed by the motor of the pump, and it typically ranges between 50 and 85 percent.

The head loss of a piping system increases (usually quadratically) with the flow rate. A plot of required useful pump head $h_{pump,u}$ as a function of flow rate is called the **system** (or **demand**) **curve**. The head produced by a pump is not a constant either. Both the pump head and the pump efficiency vary

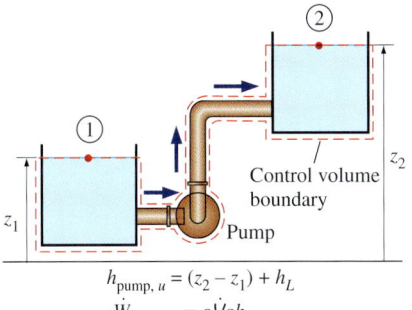

$h_{pump,u} = (z_2 - z_1) + h_L$
$\dot{W}_{pump,u} = \rho \dot{V} g h_{pump,u}$

FIGURE 8–47
When a pump moves a fluid from one reservoir to another, the useful pump head requirement is equal to the elevation difference between the two reservoirs plus the head loss.

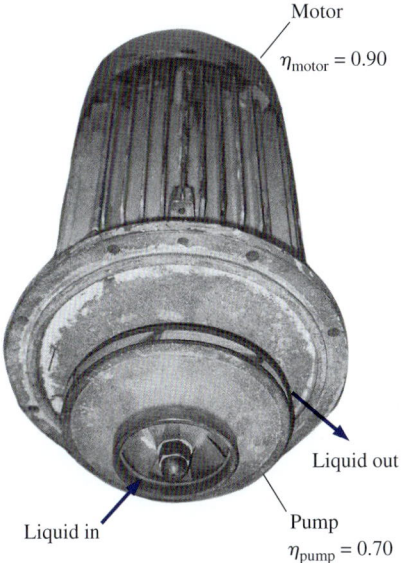

$\eta_{pump-motor} = \eta_{pump} \eta_{motor}$
$= 0.70 \times 0.90 = 0.63$

FIGURE 8–48
The efficiency of the pump–motor combination is the product of the pump and the motor efficiencies.

Photo by Yunus Çengel.

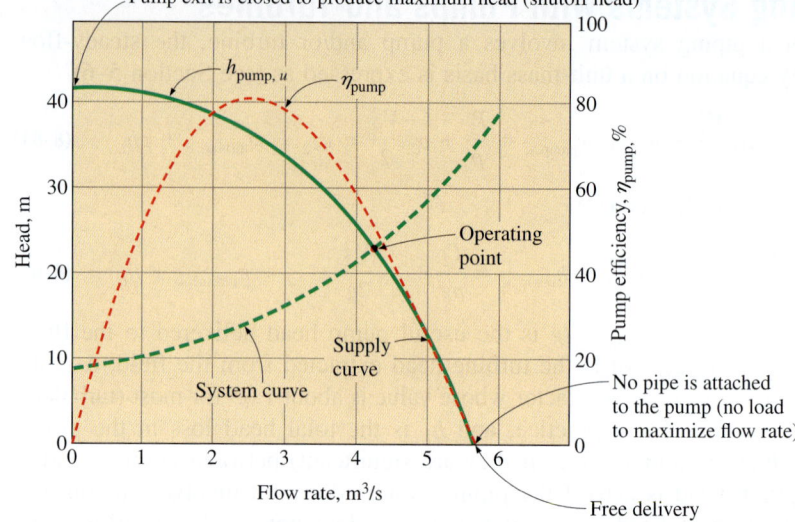

FIGURE 8–49
Characteristic pump curves for centrifugal pumps, the system curve for a piping system, and the operating point.

with the flow rate, and pump manufacturers supply this variation in tabular or graphical form, as shown in Fig. 8–49. These experimentally determined $h_{pump,\,u}$ and $\eta_{pump,\,u}$ versus $\dot{V}$ curves are called **characteristic** (or **supply** or **performance**) **curves.** Note that the flow rate of a pump increases as the required head decreases. The intersection point of the pump head curve with the vertical axis typically represents the *maximum head* (called the **shutoff head**) the pump can provide, while the intersection point with the horizontal axis indicates the *maximum flow rate* (called the **free delivery**) that the pump can supply.

The *efficiency* of a pump is highest at a certain combination of head and flow rate. Therefore, a pump that can supply the required head and flow rate is not necessarily a good choice for a piping system unless the efficiency of the pump at those conditions is sufficiently high. The pump installed in a piping system will operate at the point where the *system curve* and the *characteristic curve* intersect. This point of intersection is called the **operating point,** as shown in Fig. 8–46. The useful head produced by the pump at this point matches the head requirements of the system at that flow rate. Also, the efficiency of the pump during operation is the value corresponding to that flow rate.

> **EXAMPLE 8–7   Pumping Water through Two Parallel Pipes**
>
> Water at 20°C is to be pumped from a reservoir (z_A = 5 m) to another reservoir at a higher elevation (z_B = 13 m) through two 36-m-long pipes connected in parallel, as shown in Fig. 8–50. The pipes are made of commercial steel, and the diameters of the two pipes are 4 and 8 cm. Water is to be pumped by a 70 percent efficient motor–pump combination that draws 8 kW of electric power during operation. The minor losses and the head loss in pipes that connect the parallel pipes to the two reservoirs are considered to be negligible. Determine the total flow rate between the reservoirs and the flow rate through each of the parallel pipes.

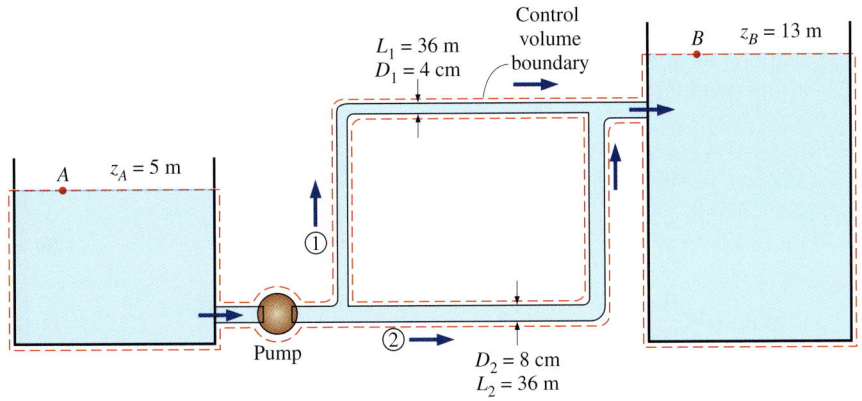

FIGURE 8–50
The piping system discussed in Example 8–7.

SOLUTION The pumping power input to a piping system with two parallel pipes is given. The flow rates are to be determined.
Assumptions **1** The flow is steady (since the reservoirs are large) and incompressible. **2** The entrance effects are negligible, and thus the flow is fully developed. **3** The elevations of the reservoirs remain constant. **4** The minor losses and the head loss in pipes other than the parallel pipes are negligible. **5** Flows through both pipes are turbulent (to be verified).
Properties The density and dynamic viscosity of water at 20°C are $\rho = 998$ kg/m^3 and $\mu = 1.002 \times 10^{-3}$ kg/m·s. The roughness of commercial steel pipe is $\epsilon = 0.000045$ m (Table 8–2).
Analysis This problem cannot be solved directly since the velocities (or flow rates) in the pipes are not known. Therefore, we would normally use a trial-and-error approach here. However, equation solvers such as EES are widely available, and thus, we simply set up the equations to be solved by an equation solver. The useful head supplied by the pump to the fluid is determined from

$$\dot{W}_{\text{elect}} = \frac{\rho \dot{V} g h_{\text{pump, }u}}{\eta_{\text{pump-motor}}} \rightarrow 8000 \text{ W} = \frac{(998 \text{ kg/m}^3)\dot{V}(9.81 \text{ m/s}^2) h_{\text{pump, }u}}{0.70} \quad (1)$$

We choose points *A* and *B* at the free surfaces of the two reservoirs. Noting that the fluid at both points is open to the atmosphere (and thus $P_A = P_B = P_{\text{atm}}$) and that the fluid velocities at both points are nearly zero ($V_A \approx V_B \approx 0$) since the reservoirs are large, the energy equation for a control volume between these two points simplifies to

$$\cancel{\frac{P_A}{\rho g}} + \alpha_A \cancel{\frac{V_A^2}{2g}}^{\,0} + z_A + h_{\text{pump, }u} = \cancel{\frac{P_B}{\rho g}} + \alpha_B \cancel{\frac{V_B^2}{2g}}^{\,0} + z_B + h_L$$

or

$$h_{\text{pump, }u} = (z_B - z_A) + h_L$$

or

$$h_{\text{pump, }u} = (13 \text{ m} - 5 \text{ m}) + h_L \quad (2)$$

where

$$h_L = h_{L,1} = h_{L,2} \quad (3)(4)$$

We designate the 4-cm-diameter pipe by 1 and the 8-cm-diameter pipe by 2. Equations for the average velocity, the Reynolds number, the friction factor, and the head loss in each pipe are

$$V_1 = \frac{\dot{V}_1}{A_{c,1}} = \frac{\dot{V}_1}{\pi D_1^2/4} \quad \rightarrow \quad V_1 = \frac{\dot{V}_1}{\pi (0.04 \text{ m})^2/4} \tag{5}$$

$$V_2 = \frac{\dot{V}_2}{A_{c,2}} = \frac{\dot{V}_2}{\pi D_2^2/4} \quad \rightarrow \quad V_2 = \frac{\dot{V}_2}{\pi (0.08 \text{ m})^2/4} \tag{6}$$

$$\text{Re}_1 = \frac{\rho V_1 D_1}{\mu} \quad \rightarrow \quad \text{Re}_1 = \frac{(998 \text{ kg/m}^3) V_1 (0.04 \text{ m})}{1.002 \times 10^{-3} \text{ kg/m·s}} \tag{7}$$

$$\text{Re}_2 = \frac{\rho V_2 D_2}{\mu} \quad \rightarrow \quad \text{Re}_2 = \frac{(998 \text{ kg/m}^3) V_2 (0.08 \text{ m})}{1.002 \times 10^{-3} \text{ kg/m·s}} \tag{8}$$

$$\frac{1}{\sqrt{f_1}} = -2.0 \log\left(\frac{\varepsilon/D_1}{3.7} + \frac{2.51}{\text{Re}_1 \sqrt{f_1}}\right)$$

$$\rightarrow \quad \frac{1}{\sqrt{f_1}} = -2.0 \log\left(\frac{0.000045}{3.7 \times 0.04} + \frac{2.51}{\text{Re}_1 \sqrt{f_1}}\right) \tag{9}$$

$$\frac{1}{\sqrt{f_2}} = -2.0 \log\left(\frac{\varepsilon/D_2}{3.7} + \frac{2.51}{\text{Re}_2 \sqrt{f_2}}\right)$$

$$\rightarrow \quad \frac{1}{\sqrt{f_2}} = -2.0 \log\left(\frac{0.000045}{3.7 \times 0.08} + \frac{2.51}{\text{Re}_2 \sqrt{f_2}}\right) \tag{10}$$

$$h_{L,1} = f_1 \frac{L_1}{D_1} \frac{V_1^2}{2g} \quad \rightarrow \quad h_{L,1} = f_1 \frac{36 \text{ m}}{0.04 \text{ m}} \frac{V_1^2}{2(9.81 \text{ m/s}^2)} \tag{11}$$

$$h_{L,2} = f_2 \frac{L_2}{D_2} \frac{V_2^2}{2g} \quad \rightarrow \quad h_{L,2} = f_2 \frac{36 \text{ m}}{0.08 \text{ m}} \frac{V_2^2}{2(9.81 \text{ m/s}^2)} \tag{12}$$

$$\dot{V} = \dot{V}_1 + \dot{V}_2 \tag{13}$$

This is a system of 13 equations in 13 unknowns, and their simultaneous solution by an equation solver gives

$\dot{V} = \mathbf{0.0300 \text{ m}^3/\text{s}}, \quad \dot{V}_1 = \mathbf{0.00415 \text{ m}^3/\text{s}}, \quad \dot{V}_2 = \mathbf{0.0259 \text{ m}^3/\text{s}}$

$V_1 = 3.30 \text{ m/s}, \quad V_2 = 5.15 \text{ m/s}, \quad h_L = h_{L,1} = h_{L,2} = 11.1 \text{ m}, \quad h_{\text{pump}} = 19.1 \text{ m}$

$\text{Re}_1 = 131{,}600, \quad \text{Re}_2 = 410{,}000, \quad f_1 = 0.0221, \quad f_2 = 0.0182$

Note that Re > 4000 for both pipes, and thus the assumption of turbulent flow is verified.

Discussion The two parallel pipes have the same length and roughness, but the diameter of the first pipe is half the diameter of the second one. Yet only 14 percent of the water flows through the first pipe. This shows the strong dependence of the flow rate on diameter. Also, it can be shown that if the free surfaces of the two reservoirs were at the same elevation (and thus $z_A = z_B$), the flow rate would increase by 20 percent from 0.0300 to 0.0361 m³/s. Alternately, if the reservoirs were as given but the irreversible head losses were negligible, the flow rate would become 0.0715 m³/s (an increase of 138 percent).

CHAPTER 8

EXAMPLE 8–8   Gravity-Driven Water Flow in a Pipe

Water at 10°C flows from a large reservoir to a smaller one through a 5-cm-diameter cast iron piping system, as shown in Fig. 8–51. Determine the elevation z_1 for a flow rate of 6 L/s.

SOLUTION The flow rate through a piping system connecting two reservoirs is given. The elevation of the source is to be determined.
Assumptions 1 The flow is steady and incompressible. 2 The elevations of the reservoirs remain constant. 3 There are no pumps or turbines in the line.
Properties The density and dynamic viscosity of water at 10°C are $\rho = 999.7$ kg/m³ and $\mu = 1.307 \times 10^{-3}$ kg/m·s. The roughness of cast iron pipe is $\epsilon = 0.00026$ m (Table 8–2).
Analysis The piping system involves 89 m of piping, a sharp-edged entrance ($K_L = 0.5$), two standard flanged elbows ($K_L = 0.3$ each), a fully open gate valve ($K_L = 0.2$), and a submerged exit ($K_L = 1.06$). We choose points 1 and 2 at the free surfaces of the two reservoirs. Noting that the fluid at both points is open to the atmosphere (and thus $P_1 = P_2 = P_{atm}$) and that the fluid velocities at both points are nearly zero ($V_1 \approx V_2 \approx 0$), the energy equation for a control volume between these two points simplifies to

$$\cancel{\frac{P_1}{\rho g}} + \alpha_1 \cancel{\frac{V_1^2}{2g}}^0 + z_1 = \cancel{\frac{P_2}{\rho g}} + \alpha_2 \cancel{\frac{V_2^2}{2g}}^0 + z_2 + h_L \quad \rightarrow \quad z_1 = z_2 + h_L$$

where

$$h_L = h_{L,\,total} = h_{L,\,major} + h_{L,\,minor} = \left(f\frac{L}{D} + \sum K_L\right)\frac{V^2}{2g}$$

since the diameter of the piping system is constant. The average velocity in the pipe and the Reynolds number are

$$V = \frac{\dot{V}}{A_c} = \frac{\dot{V}}{\pi D^2/4} = \frac{0.006 \text{ m}^3/\text{s}}{\pi(0.05 \text{ m})^2/4} = 3.06 \text{ m/s}$$

$$\text{Re} = \frac{\rho V D}{\mu} = \frac{(999.7 \text{ kg/m}^3)(3.06 \text{ m/s})(0.05 \text{ m})}{1.307 \times 10^{-3} \text{ kg/m·s}} = 117,000$$

The flow is turbulent since Re > 4000. Noting that $\epsilon/D = 0.00026/0.05 = 0.0052$, the friction factor is determined from the Colebrook equation (or the Moody chart),

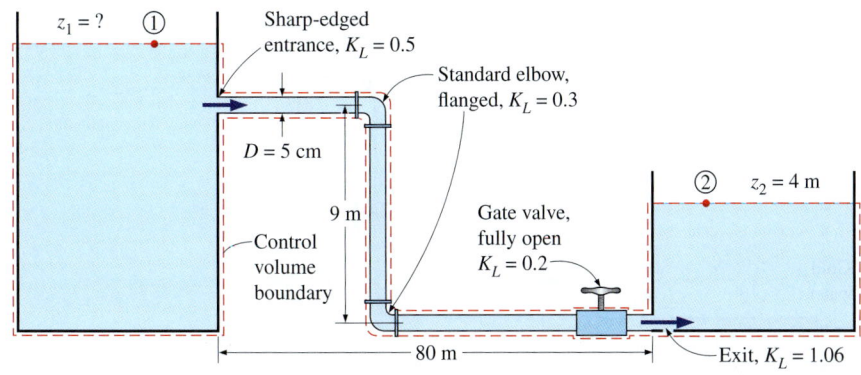

FIGURE 8–51
The piping system discussed in Example 8–8.

$$\frac{1}{\sqrt{f}} = -2.0 \log\left(\frac{\varepsilon/D}{3.7} + \frac{2.51}{Re\sqrt{f}}\right) \rightarrow \frac{1}{\sqrt{f}} = -2.0 \log\left(\frac{0.0052}{3.7} + \frac{2.51}{117{,}000\sqrt{f}}\right)$$

It gives $f = 0.0315$. The sum of the loss coefficients is

$$\sum K_L = K_{L,\text{entrance}} + 2K_{L,\text{elbow}} + K_{L,\text{valve}} + K_{L,\text{exit}}$$
$$= 0.5 + 2 \times 0.3 + 0.2 + 1.06 = 2.36$$

Then the total head loss and the elevation of the source become

$$h_L = \left(f\frac{L}{D} + \sum K_L\right)\frac{V^2}{2g} = \left(0.0315\frac{89 \text{ m}}{0.05 \text{ m}} + 2.36\right)\frac{(3.06 \text{ m/s})^2}{2(9.81 \text{ m/s}^2)} = 27.9 \text{ m}$$

$$z_1 = z_2 + h_L = 4 + 27.9 = \mathbf{31.9 \text{ m}}$$

Therefore, the free surface of the first reservoir must be 31.9 m above the ground level to ensure water flow between the two reservoirs at the specified rate.

Discussion Note that $fL/D = 56.1$ in this case, which is about 24 times the total minor loss coefficient. Therefore, ignoring the sources of minor losses in this case would result in about 4 percent error. It can be shown that at the same flow rate, the total head loss would be 35.9 m (instead of 27.9 m) if the valve were three-fourths closed, and it would drop to 24.8 m if the pipe between the two reservoirs were straight at the ground level (thus eliminating the elbows and the vertical section of the pipe). The head loss could be reduced further (from 24.8 to 24.6 m) by rounding the entrance. The head loss can be reduced significantly (from 27.9 to 16.0 m) by replacing the cast iron pipes by smooth pipes such as those made of plastic.

EXAMPLE 8–9   Effect of Flushing on Flow Rate from a Shower

The bathroom plumbing of a building consists of 1.5-cm-diameter copper pipes with threaded connectors, as shown in Fig. 8–52. (*a*) If the gage pressure at the inlet of the system is 200 kPa during a shower and the toilet reservoir is full (no flow in that branch), determine the flow rate of water through the shower head. (*b*) Determine the effect of flushing of the toilet on the flow rate through the shower head. Take the loss coefficients of the shower head and the reservoir to be 12 and 14, respectively.

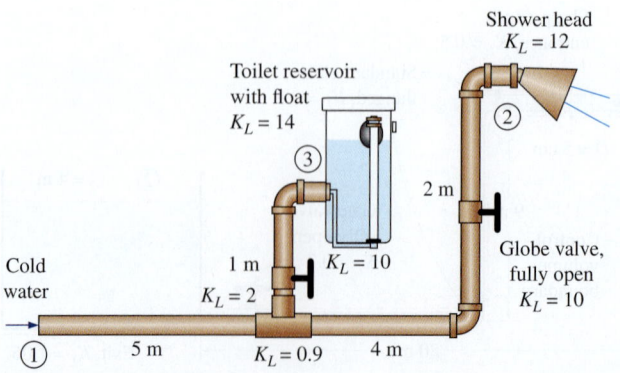

FIGURE 8–52
Schematic for Example 8–9.

SOLUTION The cold-water plumbing system of a bathroom is given. The flow rate through the shower and the effect of flushing the toilet on the flow rate are to be determined.

Assumptions **1** The flow is steady and incompressible. **2** The flow is turbulent and fully developed. **3** The reservoir is open to the atmosphere. **4** The velocity heads are negligible.

Properties The properties of water at 20°C are $\rho = 998$ kg/m^3, $\mu = 1.002 \times 10^{-3}$ kg/m·s, and $\nu = \mu/\rho = 1.004 \times 10^{-6}$ m^2/s. The roughness of copper pipes is $\epsilon = 1.5 \times 10^{-6}$ m.

Analysis This is a problem of the second type since it involves the determination of the flow rate for a specified pipe diameter and pressure drop. The solution involves an iterative approach since the flow rate (and thus the flow velocity) is not known.

(a) The piping system of the shower alone involves 11 m of piping, a tee with line flow ($K_L = 0.9$), two standard elbows ($K_L = 0.9$ each), a fully open globe valve ($K_L = 10$), and a shower head ($K_L = 12$). Therefore, $\sum K_L = 0.9 + 2 \times 0.9 + 10 + 12 = 24.7$. Noting that the shower head is open to the atmosphere, and the velocity heads are negligible, the energy equation for a control volume between points 1 and 2 simplifies to

$$\frac{P_1}{\rho g} + \alpha_1 \frac{V_1^2}{2g} + z_1 + h_{\text{pump},u} = \frac{P_2}{\rho g} + \alpha_2 \frac{V_2^2}{2g} + z_2 + h_{\text{turbine},e} + h_L$$

$$\rightarrow \quad \frac{P_{1,\text{gage}}}{\rho g} = (z_2 - z_1) + h_L$$

Therefore, the head loss is

$$h_L = \frac{200{,}000 \text{ N/m}^2}{(998 \text{ kg/m}^3)(9.81 \text{ m/s}^2)} - 2 \text{ m} = 18.4 \text{ m}$$

Also,

$$h_L = \left(f \frac{L}{D} + \sum K_L\right) \frac{V^2}{2g} \quad \rightarrow \quad 18.4 = \left(f \frac{11 \text{ m}}{0.015 \text{ m}} + 24.7\right) \frac{V^2}{2(9.81 \text{ m/s}^2)}$$

since the diameter of the piping system is constant. Equations for the average velocity in the pipe, the Reynolds number, and the friction factor are

$$V = \frac{\dot{V}}{A_c} = \frac{\dot{V}}{\pi D^2/4} \quad \rightarrow \quad V = \frac{\dot{V}}{\pi (0.015 \text{ m})^2/4}$$

$$\text{Re} = \frac{VD}{\nu} \quad \rightarrow \quad \text{Re} = \frac{V(0.015 \text{ m})}{1.004 \times 10^{-6} \text{ m}^2/\text{s}}$$

$$\frac{1}{\sqrt{f}} = -2.0 \log\left(\frac{\epsilon/D}{3.7} + \frac{2.51}{\text{Re}\sqrt{f}}\right)$$

$$\rightarrow \quad \frac{1}{\sqrt{f}} = -2.0 \log\left(\frac{1.5 \times 10^{-6} \text{ m}}{3.7(0.015 \text{ m})} + \frac{2.51}{\text{Re}\sqrt{f}}\right)$$

This is a set of four equations with four unknowns, and solving them with an equation solver such as EES gives

$\dot{V} = 0.00053$ m³/s, $f = 0.0218$, $V = 2.98$ m/s, and Re = 44,550

Therefore, the flow rate of water through the shower head is **0.53 L/s**.

(b) When the toilet is flushed, the float moves and opens the valve. The discharged water starts to refill the reservoir, resulting in parallel flow after the tee connection. The head loss and minor loss coefficients for the shower branch were determined in (a) to be $h_{L,2} = 18.4$ m and $\sum K_{L,2} = 24.7$, respectively. The corresponding quantities for the reservoir branch can be determined similarly to be

$$h_{L,3} = \frac{200{,}000 \text{ N/m}^2}{(998 \text{ kg/m}^3)(9.81 \text{ m/s}^2)} - 1 \text{ m} = 19.4 \text{ m}$$

$$\sum K_{L,3} = 2 + 10 + 0.9 + 14 = 26.9$$

The relevant equations in this case are

$$\dot{V}_1 = \dot{V}_2 + \dot{V}_3$$

$$h_{L,2} = f_1 \frac{5 \text{ m}}{0.015 \text{ m}} \frac{V_1^2}{2(9.81 \text{ m/s}^2)} + \left(f_2 \frac{6 \text{ m}}{0.015 \text{ m}} + 24.7\right) \frac{V_2^2}{2(9.81 \text{ m/s}^2)} = 18.4$$

$$h_{L,3} = f_1 \frac{5 \text{ m}}{0.015 \text{ m}} \frac{V_1^2}{2(9.81 \text{ m/s}^2)} + \left(f_3 \frac{1 \text{ m}}{0.015 \text{ m}} + 26.9\right) \frac{V_3^2}{2(9.81 \text{ m/s}^2)} = 19.4$$

$$V_1 = \frac{\dot{V}_1}{\pi(0.015 \text{ m})^2/4}, \quad V_2 = \frac{\dot{V}_2}{\pi(0.015 \text{ m})^2/4}, \quad V_3 = \frac{\dot{V}_3}{\pi(0.015 \text{ m})^2/4}$$

$$\text{Re}_1 = \frac{V_1(0.015 \text{ m})}{1.004 \times 10^{-6} \text{m}^2/\text{s}}, \quad \text{Re}_2 = \frac{V_2(0.015 \text{ m})}{1.004 \times 10^{-6} \text{m}^2/\text{s}}, \quad \text{Re}_3 = \frac{V_3(0.015 \text{ m})}{1.004 \times 10^{-6} \text{m}^2/\text{s}}$$

$$\frac{1}{\sqrt{f_1}} = -2.0 \log\left(\frac{1.5 \times 10^{-6} \text{ m}}{3.7(0.015 \text{ m})} + \frac{2.51}{\text{Re}_1 \sqrt{f_1}}\right)$$

$$\frac{1}{\sqrt{f_2}} = -2.0 \log\left(\frac{1.5 \times 10^{-6} \text{ m}}{3.7(0.015 \text{ m})} + \frac{2.51}{\text{Re}_2 \sqrt{f_2}}\right)$$

$$\frac{1}{\sqrt{f_3}} = -2.0 \log\left(\frac{1.5 \times 10^{-6} \text{ m}}{3.7(0.015 \text{ m})} + \frac{2.51}{\text{Re}_3 \sqrt{f_3}}\right)$$

Solving these 12 equations in 12 unknowns simultaneously using an equation solver, the flow rates are determined to be

$$\dot{V}_1 = 0.00090 \text{ m}^3/\text{s}, \quad \dot{V}_2 = 0.00042 \text{ m}^3/\text{s}, \text{ and } \dot{V}_3 = 0.00048 \text{ m}^3/\text{s}$$

Therefore, the flushing of the toilet **reduces the flow rate of cold water through the shower by 21 percent** from 0.53 to 0.42 L/s, causing the shower water to suddenly get very hot (Fig. 8–53).

Discussion If the velocity heads were considered, the flow rate through the shower would be 0.43 instead of 0.42 L/s. Therefore, the assumption of negligible velocity heads is reasonable in this case. Note that a leak in a piping system would cause the same effect, and thus an unexplained drop in flow rate at an end point may signal a leak in the system.

FIGURE 8–53
Flow rate of cold water through a shower may be affected significantly by the flushing of a nearby toilet.

8–8 ▪ FLOW RATE AND VELOCITY MEASUREMENT

A major application area of fluid mechanics is the determination of the flow rate of fluids, and numerous devices have been developed over the years for the purpose of flow metering. Flowmeters range widely in their level of sophistication, size, cost, accuracy, versatility, capacity, pressure drop, and the operating principle. We give an overview of the meters commonly used to measure the flow rate of liquids and gases flowing through pipes or ducts. We limit our consideration to incompressible flow.

Some flowmeters measure the flow rate directly by discharging and recharging a measuring chamber of known volume continuously and keeping track of the number of discharges per unit time. But most flowmeters measure the flow rate indirectly—they measure the average velocity V or a quantity that is related to average velocity such as pressure and drag, and determine the volume flow rate $\dot{V}$ from

$$\dot{V} = VA_c \qquad (8\text{–}65)$$

where A_c is the cross-sectional area of flow. Therefore, measuring the flow rate is usually done by measuring flow velocity, and many flowmeters are simply velocimeters used for the purpose of metering flow.

The velocity in a pipe varies from zero at the wall to a maximum at the center, and it is important to keep this in mind when taking velocity measurements. For laminar flow, for example, the average velocity is half the centerline velocity. But this is not the case in turbulent flow, and it may be necessary to take the weighted average or an integral of several local velocity measurements to determine the average velocity.

The flow rate measurement techniques range from very crude to very elegant. The flow rate of water through a garden hose, for example, can be measured simply by collecting the water in a bucket of known volume and dividing the amount collected by the collection time (Fig. 8–54). A crude way of estimating the flow velocity of a river is to drop a float on the river and measure the drift time between two specified locations. At the other extreme, some flowmeters use the propagation of sound in flowing fluids while others use the electromotive force generated when a fluid passes through a magnetic field. In this section we discuss devices that are commonly used to measure velocity and flow rate, starting with the Pitot-static probe introduced in Chap. 5.

FIGURE 8–54
A primitive (but fairly accurate) way of measuring the flow rate of water through a garden hose involves collecting water in a bucket and recording the collection time.

Pitot and Pitot-Static Probes

Pitot probes (also called *Pitot tubes*) and **Pitot-static probes**, named after the French engineer Henri de Pitot (1695–1771), are widely used for flow speed measurement. A Pitot probe is just a tube with a pressure tap at the stagnation point that measures stagnation pressure, while a Pitot-static probe has both a stagnation pressure tap and several circumferential static pressure taps and it measures both stagnation and static pressures (Figs. 8–55 and 8–56). Pitot was the first person to measure velocity with the upstream pointed tube, while French engineer Henry Darcy (1803–1858) developed most of the features of the instruments we use today, including the use of small openings and the placement of the static tube on the same assembly. Therefore, it is more appropriate to call the Pitot-static probes **Pitot–Darcy probes.**

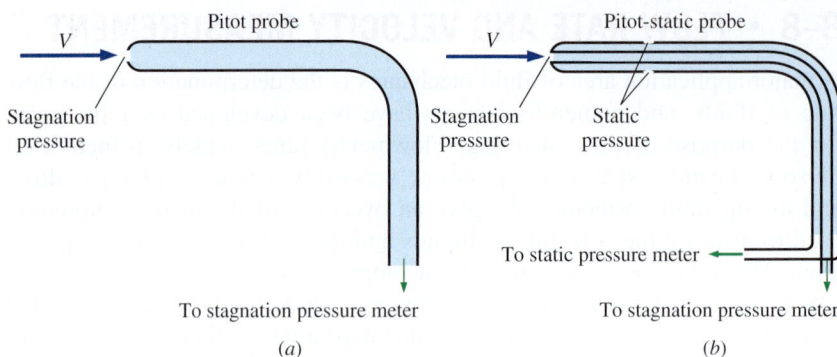

FIGURE 8–55
(a) A Pitot probe measures stagnation pressure at the nose of the probe, while (b) a Pitot-static probe measures both stagnation pressure and static pressure, from which the flow speed is calculated.

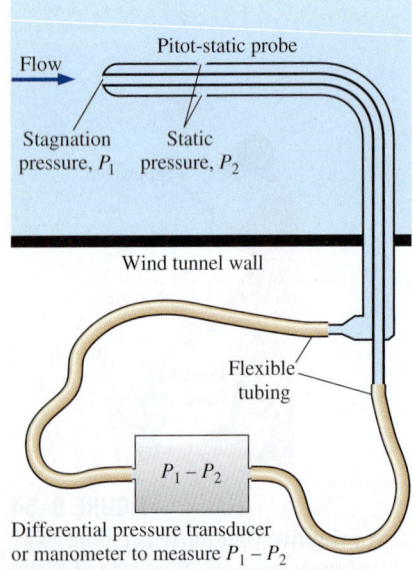

FIGURE 8–56
Measuring flow velocity with a Pitot-static probe. (A manometer may be used in place of the differential pressure transducer.)

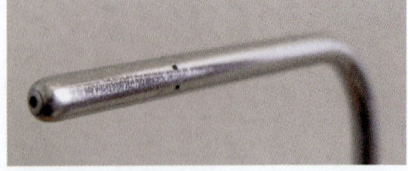

FIGURE 8–57
Close-up of a Pitot-static probe, showing the stagnation pressure hole and two of the five static circumferential pressure holes.
Photo by Po-Ya Abel Chuang.

The Pitot-static probe measures local velocity by measuring the pressure difference in conjunction with the Bernoulli equation. It consists of a slender double-tube aligned with the flow and connected to a differential pressure meter. The inner tube is fully open to flow at the nose, and thus it measures the stagnation pressure at that location (point 1). The outer tube is sealed at the nose, but it has holes on the side of the outer wall (point 2) and thus it measures the static pressure. For incompressible flow with sufficiently high velocities (so that the frictional effects between points 1 and 2 are negligible), the Bernoulli equation is applicable and is expressed as

$$\frac{P_1}{\rho g} + \frac{V_1^2}{2g} + z_1 = \frac{P_2}{\rho g} + \frac{V_2^2}{2g} + z_2 \quad (8\text{--}66)$$

Noting that $z_1 \cong z_2$ since the static pressure holes of the Pitot-static probe are arranged circumferentially around the tube and $V_1 = 0$ because of the stagnation conditions, the flow velocity $V = V_2$ becomes

Pitot formula: $\qquad V = \sqrt{\dfrac{2(P_1 - P_2)}{\rho}} \quad (8\text{--}67)$

which is known as the **Pitot formula.** If the velocity is measured at a location where the local velocity is equal to the average flow velocity, the volume flow rate can be determined from $\dot{V} = VA_c$.

The Pitot-static probe is a simple, inexpensive, and highly reliable device since it has no moving parts (Fig. 8–57). It also causes very small pressure drop and usually does not disturb the flow appreciably. However, it is important that it be properly aligned with the flow to avoid significant errors that may be caused by misalignment. Also, the difference between the static and stagnation pressures (which is the dynamic pressure) is proportional to the density of the fluid and the square of the flow velocity. It is used to measure velocity in both liquids and gases. Noting that gases have low densities, the flow velocity should be sufficiently high when the Pitot-static probe is used for gas flow such that a measurable dynamic pressure develops.

Obstruction Flowmeters: Orifice, Venturi, and Nozzle Meters

Consider incompressible steady flow of a fluid in a horizontal pipe of diameter D that is constricted to a flow area of diameter d, as shown in Fig. 8–58. The mass balance and the Bernoulli equations between a location before the constriction (point 1) and the location where constriction occurs (point 2) are written as

Mass balance: $\quad \dot{V} = A_1 V_1 = A_2 V_2 \rightarrow V_1 = (A_2/A_1)V_2 = (d/D)^2 V_2 \quad$ (8-68)

Bernoulli equation ($z_1 = z_2$): $\quad \dfrac{P_1}{\rho g} + \dfrac{V_1^2}{2g} = \dfrac{P_2}{\rho g} + \dfrac{V_2^2}{2g} \quad$ (8-69)

Combining Eqs. 8–68 and 8–69 and solving for velocity V_2 gives

Obstruction (with no loss): $\quad V_2 = \sqrt{\dfrac{2(P_1 - P_2)}{\rho(1 - \beta^4)}} \quad$ (8-70)

where $\beta = d/D$ is the diameter ratio. Once V_2 is known, the flow rate can be determined from $\dot{V} = A_2 V_2 = (\pi d^2/4) V_2$.

This simple analysis shows that the flow rate through a pipe can be determined by constricting the flow and measuring the decrease in pressure due to the increase in velocity at the constriction site. Noting that the pressure drop between two points along the flow is measured easily by a differential pressure transducer or manometer, it appears that a simple flow rate measurement device can be built by obstructing the flow. Flowmeters based on this principle are called **obstruction flowmeters** and are widely used to measure flow rates of gases and liquids.

The velocity in Eq. 8–70 is obtained by assuming no loss, and thus it is the maximum velocity that can occur at the constriction site. In reality, some pressure losses due to frictional effects are inevitable, and thus the actual velocity is less. Also, the fluid stream continues to contract past the obstruction, and the vena contracta area is less than the flow area of the obstruction. Both losses can be accounted for by incorporating a correction factor called the **discharge coefficient** C_d whose value (which is less than 1) is determined experimentally. Then the flow rate for obstruction flowmeters is expressed as

Obstruction flowmeters: $\quad \dot{V} = A_0 C_d \sqrt{\dfrac{2(P_1 - P_2)}{\rho(1 - \beta^4)}} \quad$ (8-71)

where $A_0 = A_2 = \pi d^2/4$ is the cross-sectional area of the throat or orifice and $\beta = d/D$ is the ratio of throat diameter to pipe diameter. The value of C_d depends on both β and the Reynolds number $\text{Re} = V_1 D/\nu$, and charts and curve-fit correlations for C_d are available for various types of obstruction meters.

Of the numerous types of obstruction meters available, those most widely used are orifice meters, flow nozzles, and Venturi meters (Fig. 8–59). For standardized geometries, the experimentally determined data for discharge coefficients are expressed as (Miller, 1997)

Orifice meters: $\quad C_d = 0.5959 + 0.0312\beta^{2.1} - 0.184\beta^8 + \dfrac{91.71\beta^{2.5}}{\text{Re}^{0.75}} \quad$ (8-72)

Nozzle meters: $\quad C_d = 0.9975 - \dfrac{6.53\beta^{0.5}}{\text{Re}^{0.5}} \quad$ (8-73)

These relations are valid for $0.25 < \beta < 0.75$ and $10^4 < \text{Re} < 10^7$. Precise values of C_d depend on the particular design of the obstruction, and thus the

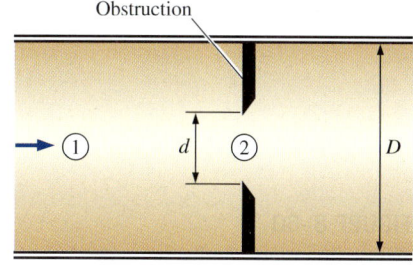

FIGURE 8–58
Flow through a constriction in a pipe.

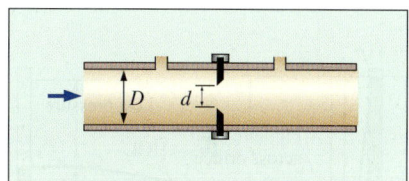

(a) Orifice meter

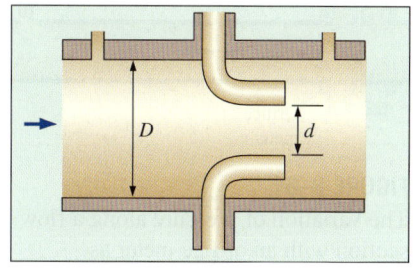

(b) Flow nozzle

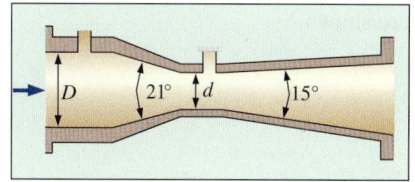

(c) Venturi meter

FIGURE 8–59
Common types of obstruction meters.

394
INTERNAL FLOW

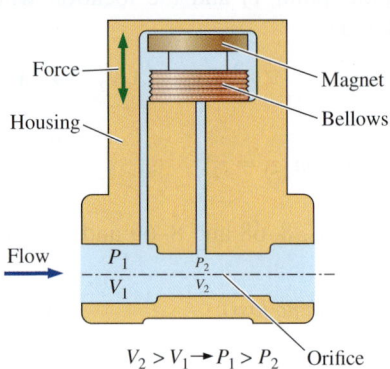

FIGURE 8–60
An orifice meter and schematic showing its built-in pressure transducer and digital readout.
Courtesy KOBOLD Instruments, Pittsburgh, PA. www.koboldusa.com. Used by permission.

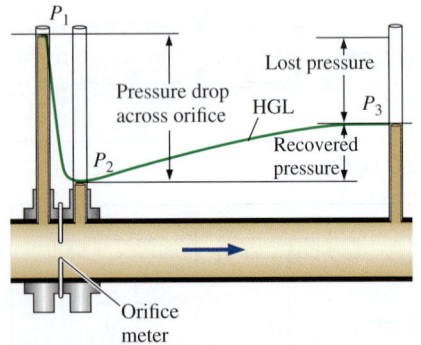

FIGURE 8–61
The variation of pressure along a flow section with an orifice meter as measured with piezometer tubes; the lost pressure and the pressure recovery are shown.

manufacturer's data should be consulted when available. Also, the Reynolds number depends on the flow velocity, which is not known a priori. Therefore, the solution is iterative in nature when curve-fit correlations are used for C_d. For flows with high Reynolds numbers (Re > 30,000), the value of C_d can be taken to be 0.96 for flow nozzles and 0.61 for orifices.

Owing to its streamlined design, the discharge coefficients of Venturi meters are very high, ranging between 0.95 and 0.99 (the higher values are for the higher Reynolds numbers) for most flows. In the absence of specific data, we can take $C_d = 0.98$ for Venturi meters.

The orifice meter has the simplest design and it occupies minimal space as it consists of a plate with a hole in the middle, but there are considerable variations in design (Fig. 8–60). Some orifice meters are sharp-edged, while others are beveled or rounded. The sudden change in the flow area in orifice meters causes considerable swirl and thus significant head loss or permanent pressure loss, as shown in Fig. 8–61. In nozzle meters, the plate is replaced by a nozzle, and thus the flow in the nozzle is streamlined. As a result, the vena contracta is practically eliminated and the head loss is smaller. However, flow nozzle meters are more expensive than orifice meters.

The Venturi meter, invented by the American engineer Clemens Herschel (1842–1930) and named by him after the Italian Giovanni Venturi (1746–1822) for his pioneering work on conical flow sections, is the most accurate flowmeter in this group, but it is also the most expensive. Its gradual contraction and expansion prevent flow separation and swirling, and it suffers only frictional losses on the inner wall surfaces. Venturi meters cause very low head losses, and thus, they should be preferred for applications that cannot allow large pressure drops.

When an obstruction flowmeter is placed in a piping system, its net effect on the flow system is like that of a minor loss. The minor loss coefficient of the flowmeter is available from the manufacturer, and should be included when summing minor losses in the system. In general, orifice meters have the highest minor loss coefficients, while Venturi meters have the lowest. Note that the pressure drop $P_1 - P_2$ measured to calculate the flow rate is *not* the same as the total pressure drop caused by the obstruction flowmeter because of the locations of the pressure taps.

Finally, obstruction flowmeters are also used to measure compressible-gas flow rates, but an additional correction factor must be inserted into Eq. 8–71 to account for compressibility effects. In such cases, the equation is written for *mass flow* rate instead of volume flow rate, and the compressible correction factor is typically an empirically curve-fitted equation (like the one for C_d) and is available from the flowmeter manufacturer.

■ EXAMPLE 8–10   Measuring Flow Rate with an Orifice Meter

The flow rate of methanol at 20°C (ρ = 788.4 kg/m³ and μ = 5.857 × 10^{-4} kg/m·s) through a 4-cm-diameter pipe is to be measured with a 3-cm-diameter orifice meter equipped with a mercury manometer across the orifice plate, as shown in Fig. 8–62. If the differential height of the manometer is 11 cm, determine the flow rate of methanol through the pipe and the average flow velocity.

SOLUTION The flow rate of methanol is to be measured with an orifice meter. For a given pressure drop across the orifice plate, the flow rate and the average flow velocity are to be determined.
Assumptions 1 The flow is steady and incompressible. 2 Our first guess for the discharge coefficient of the orifice meter is $C_d = 0.61$.
Properties The density and dynamic viscosity of methanol are given to be ρ = 788.4 kg/m³ and μ = 5.857 × 10^{-4} kg/m·s, respectively. We take the density of mercury to be 13,600 kg/m³.
Analysis The diameter ratio and the throat area of the orifice are

$$\beta = \frac{d}{D} = \frac{3}{4} = 0.75$$

$$A_0 = \frac{\pi d^2}{4} = \frac{\pi (0.03 \text{ m})^2}{4} = 7.069 \times 10^{-4} \text{ m}^2$$

The pressure drop across the orifice plate is

$$\Delta P = P_1 - P_2 = (\rho_{Hg} - \rho_{met})gh$$

Then the flow rate relation for obstruction meters becomes

$$\dot{V} = A_0 C_d \sqrt{\frac{2(P_1 - P_2)}{\rho(1 - \beta^4)}} = A_0 C_d \sqrt{\frac{2(\rho_{Hg} - \rho_{met})gh}{\rho_{met}(1 - \beta^4)}} = A_0 C_d \sqrt{\frac{2(\rho_{Hg}/\rho_{met} - 1)gh}{1 - \beta^4}}$$

Substituting, the flow rate is determined to be

$$\dot{V} = (7.069 \times 10^{-4} \text{ m}^2)(0.61)\sqrt{\frac{2(13,600/788.4 - 1)(9.81 \text{ m/s}^2)(0.11 \text{ m})}{1 - 0.75^4}}$$
$$= 3.09 \times 10^{-3} \text{ m}^3/\text{s}$$

which is equivalent to 3.09 L/s. The average flow velocity in the pipe is determined by dividing the flow rate by the cross-sectional area of the pipe,

$$V = \frac{\dot{V}}{A_c} = \frac{\dot{V}}{\pi D^2/4} = \frac{3.09 \times 10^{-3} \text{ m}^3/\text{s}}{\pi (0.04 \text{ m})^2/4} = 2.46 \text{ m/s}$$

The Reynolds number of flow through the pipe is

$$\text{Re} = \frac{\rho V D}{\mu} = \frac{(788.4 \text{ kg/m}^3)(2.46 \text{ m/s})(0.04 \text{ m})}{5.857 \times 10^{-4} \text{ kg/m·s}} = 1.32 \times 10^5$$

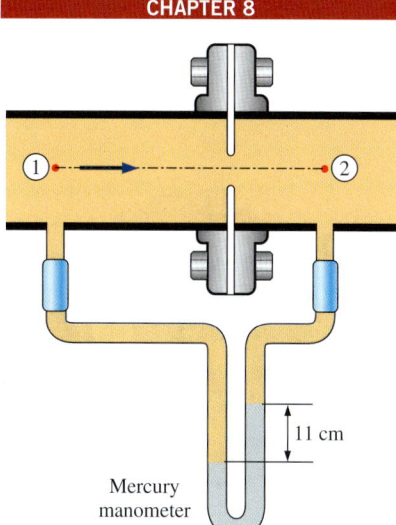

FIGURE 8–62
Schematic for the orifice meter considered in Example 8–10.

FIGURE 8–63
A positive displacement flowmeter with double helical three-lobe impeller design.

Courtesy Flow Technology, Inc.
Source: www.ftimeters.com.

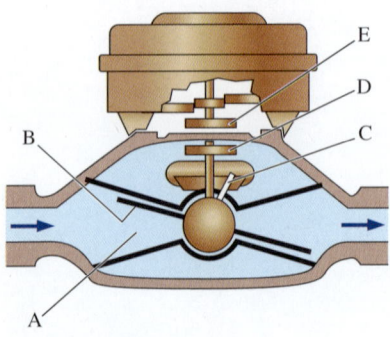

FIGURE 8–64
A nutating disk flowmeter.
(*Top*) *Courtesy Badger Meter, Inc. Used by Permission.*

Substituting $\beta = 0.75$ and $Re = 1.32 \times 10^5$ into the orifice discharge coefficient relation

$$C_d = 0.5959 + 0.0312\beta^{2.1} - 0.184\beta^8 + \frac{91.71\beta^{2.5}}{Re^{0.75}}$$

gives $C_d = 0.601$, which differs from the original guessed value of 0.61. Using this refined value of C_d, the flow rate becomes 3.04 L/s, which differs from our original result by 1.6 percent. After a couple iterations, the final converged flow rate is **3.04 L/s,** and the average velocity is **2.42 m/s** (to three significant digits).

Discussion If the problem is solved using an equation solver such as EES, then it can be formulated using the curve-fit formula for C_d (which depends on the Reynolds number), and all equations can be solved simultaneously by letting the equation solver perform the iterations as necessary.

Positive Displacement Flowmeters

When we buy gasoline for our car, we are interested in the total amount of gasoline that flows through the nozzle during the period we fill the tank rather than the flow rate of gasoline. Likewise, we care about the total amount of water or natural gas we use in our homes during a billing period. In these and many other applications, the quantity of interest is the total amount of mass or volume of a fluid that passes through a cross section of a pipe over a certain period of time rather than the instantaneous value of flow rate, and **positive displacement flowmeters** are well suited for such applications. There are numerous types of displacement meters, and they are based on continuous filling and discharging of the measuring chamber. They operate by trapping a certain amount of incoming fluid, displacing it to the discharge side of the meter, and counting the number of such discharge–recharge cycles to determine the total amount of fluid displaced.

Figure 8–63 shows a positive displacement flowmeter with two rotating impellers driven by the flowing liquid. Each impeller has three gear lobes, and a pulsed output signal is generated each time a lobe passes by a nonintrusive sensor. Each pulse represents a known volume of liquid that is captured in between the lobes of the impellers, and an electronic controller converts the pulses to volume units. The clearance between the impeller and its casing must be controlled carefully to prevent leakage and thus to avoid error. This particular meter has a quoted accuracy of 0.1 percent, has a low pressure drop, and can be used with high- or low-viscosity liquids at temperatures up to 230°C and pressures up to 7 MPa for flow rates of up to 50 L/s.

The most widely used flowmeters to measure liquid volumes are **nutating disk flowmeters,** shown in Fig. 8–64. They are commonly used as water and gasoline meters. The liquid enters the nutating disk meter through the chamber (A). This causes the disk (B) to nutate or wobble and results in the rotation of a spindle (C) and the excitation of a magnet (D). This signal is transmitted through the casing of the meter to a second magnet (E). The total volume is obtained by counting the number of these signals during a discharge process.

Quantities of gas flows, such as the amount of natural gas used in buildings, are commonly metered by using **bellows flowmeters** that displace a certain amount of gas volume (or mass) during each revolution.

(a)

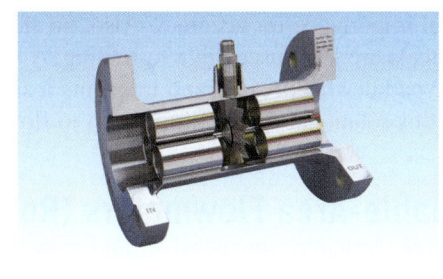
(b)

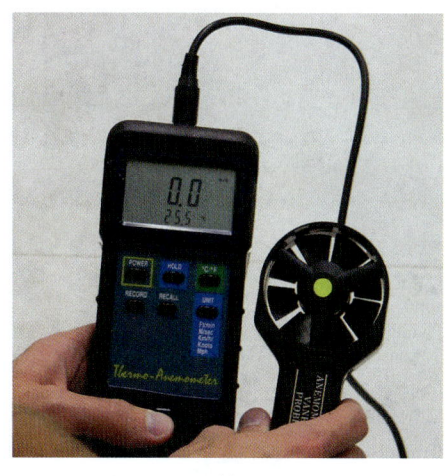
(c)

Turbine Flowmeters

We all know from experience that a propeller held against the wind rotates, and the rate of rotation increases as the wind velocity increases. You may also have seen that the turbine blades of wind turbines rotate rather slowly at low winds, but quite fast at high winds. These observations suggest that the flow velocity in a pipe can be measured by placing a freely rotating propeller inside a pipe section and doing the necessary calibration. Flow measurement devices that work on this principle are called **turbine flowmeters** or sometimes **propeller flowmeters**, although the latter is a misnomer since, by definition, propellers add energy to a fluid, while turbines extract energy from a fluid.

A turbine flowmeter consists of a cylindrical flow section that houses a turbine (a vaned rotor) that is free to rotate, additional stationary vanes at the inlet to straighten the flow, and a sensor that generates a pulse each time a marked point on the turbine passes by to determine the rate of rotation. The rotational speed of the turbine is nearly proportional to the flow rate of the fluid. Turbine flowmeters give highly accurate results (as accurate as 0.25 percent) over a wide range of flow rates when calibrated properly for the anticipated flow conditions. Turbine flowmeters have very few blades (sometimes just two blades) when used to measure liquid flow, but several blades when used to measure gas flow to ensure adequate torque generation. The head loss caused by the turbine is very small.

Turbine flowmeters have been used extensively for flow measurement since the 1940s because of their simplicity, low cost, and accuracy over a wide range of flow conditions. They are commercially available for both liquids and gases and for pipes of practically all sizes. Turbine flowmeters are also commonly used to measure flow velocities in unconfined flows such as winds, rivers, and ocean currents. The handheld device shown in Fig. 8–65c is used to measure wind velocity.

Paddlewheel Flowmeters

Paddlewheel flowmeters are low-cost alternatives to turbine flowmeters for flows where very high accuracy is not required. In paddlewheel flowmeters,

FIGURE 8–65
(a) An in-line turbine flowmeter to measure liquid flow, with flow from left to right, (b) a cutaway view of the turbine blades inside the flowmeter, and (c) a handheld turbine flowmeter to measure wind speed, measuring no flow at the time the photo was taken so that the turbine blades are visible. The flowmeter in (c) also measures the air temperature for convenience.

*Photos (a) and (c) by John M. Cimbala.
Photo (b) Courtesy Hoffer Flow Controls.*

398
INTERNAL FLOW

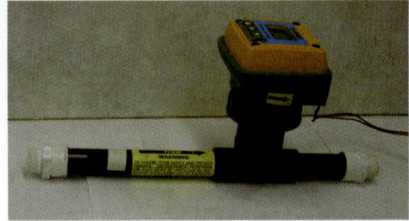

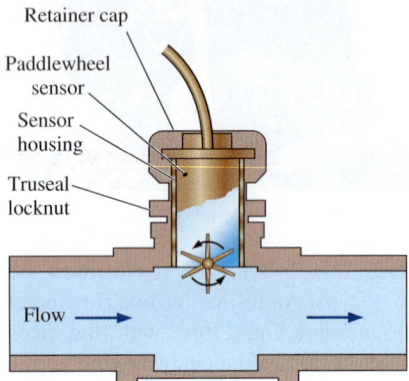

FIGURE 8–66
Paddlewheel flowmeter to measure liquid flow, with flow from left to right, and a schematic diagram of its operation.
Photo by John M. Cimbala.

the paddlewheel (the rotor and the blades) is perpendicular to the flow, as shown in Fig. 8–66, rather than parallel as was the case with turbine flowmeters. The paddles cover only a portion of the flow cross section (typically less than half), and thus the head loss is smaller compared to that of turbine flowmeters, but the depth of insertion of the paddlewheel into the flow is of critical importance for accuracy. Also, no strainers are required since the paddlewheels are less susceptible to fouling. A sensor detects the passage of each of the paddlewheel blades and transmits a signal. A microprocessor then converts this rotational speed information to flow rate or integrated flow quantity.

Variable-Area Flowmeters (Rotameters)

A simple, reliable, inexpensive, and easy-to-install flowmeter with reasonably low pressure drop and no electrical connections that gives a direct reading of flow rate for a wide range of liquids and gases is the **variable-area flowmeter,** also called a **rotameter** or **floatmeter.** A variable-area flowmeter consists of a vertical tapered conical transparent tube made of glass or plastic with a float inside that is free to move, as shown in Fig. 8–67. As fluid flows through the tapered tube, the float rises within the tube to a location where the float weight, drag force, and buoyancy force balance each other and the net force acting on the float is zero. The flow rate is determined by simply matching the position of the float against the graduated flow scale outside the tapered transparent tube. The float itself is typically either a sphere or a loose-fitting piston-like cylinder (as in Fig. 8–67a).

We know from experience that high winds knock down trees, break power lines, and blow away hats or umbrellas. This is because the drag force increases with flow velocity. The weight and the buoyancy force acting on the float are constant, but the drag force changes with flow velocity. Also, the velocity along the tapered tube decreases in the flow direction because of the increase in the cross-sectional area. There is a certain velocity that generates enough drag to balance the float weight and the buoyancy force, and the location at which this velocity occurs around the float is the location where the float settles. The degree of tapering of the tube can be made such that the vertical rise changes linearly with flow rate, and thus the tube can be calibrated linearly for flow rates. The transparent tube also allows the fluid to be seen during flow.

There are several kinds of variable-area flowmeters. The gravity-based flowmeter, as shown in Fig. 8–67a must be positioned vertically, with fluid entering from the bottom and leaving from the top. In spring-opposed flowmeters (Fig. 8–67b), the drag force is balanced by the spring force, and such flowmeters can be installed horizontally.

The accuracy of variable-area flowmeters is typically ±5 percent. Therefore, these flowmeters are not appropriate for applications that require precision measurements. However, some manufacturers quote accuracies of the order of 1 percent. Also, these meters depend on visual checking of the location of the float, and thus they cannot be used to measure the flow rate of fluids that are opaque or dirty, or fluids that coat the float since such fluids block visual access. Finally, glass tubes are prone to breakage and thus they pose a safety hazard if toxic fluids are handled. In such applications, variable-area flowmeters should be installed at locations with minimum traffic.

Ultrasonic Flowmeters

It is a common observation that when a stone is dropped into calm water, the waves that are generated spread out as concentric circles uniformly in all directions. But when a stone is thrown into flowing water such as a river, the waves move much faster in the flow direction (the wave and flow velocities are added since they are in the same direction) compared to the waves moving in the upstream direction (the wave and flow velocities are subtracted since they are in opposite directions). As a result, the waves appear spread out downstream while they appear tightly packed upstream. The difference between the number of waves in the upstream and downstream parts of the flow per unit length is proportional to the flow velocity, and this suggests that flow velocity can be measured by comparing the propagation of waves in the forward and backward directions with respect to the flow. **Ultrasonic flowmeters** operate on this principle, using sound waves in the ultrasonic range (beyond human hearing ability, typically at a frequency of 1 MHz).

Ultrasonic (or acoustic) flowmeters operate by generating sound waves with a transducer and measuring the propagation of those waves through a flowing fluid. There are two basic kinds of ultrasonic flowmeters: *transit time* and *Doppler-effect* (or *frequency shift*) flowmeters. The transit time flowmeter transmits sound waves in the upstream and downstream directions and measures the difference in travel time. A typical transit time ultrasonic meter is shown schematically in Fig. 8–68. It involves two transducers that alternately transmit and receive ultrasonic waves, one in the direction of flow and the other in the opposite direction. The travel time for each direction can be measured accurately, and the difference in the travel time is calculated. The average flow velocity V in the pipe is proportional to this travel time difference Δt, and is determined from

$$V = KL\,\Delta t \qquad (8\text{–}74)$$

where L is the distance between the transducers and K is a constant.

Doppler-Effect Ultrasonic Flowmeters

You have probably noticed that when a fast-moving car approaches with its horn blowing, the tone of the high-pitched sound of the horn drops to a lower pitch as the car passes by. This is due to the sonic waves being compressed in front of the car and being spread out behind it. This shift in frequency is called the **Doppler effect,** and it forms the basis for the operation of most ultrasonic flowmeters.

Doppler-effect ultrasonic flowmeters measure the average flow velocity along the sonic path. This is done by clamping a piezoelectric transducer on the outside surface of a pipe (or pressing the transducer against the pipe for handheld units). The transducer transmits a sound wave at a fixed frequency through the pipe wall and into the flowing liquid. The waves reflected by impurities, such as suspended solid particles or entrained gas bubbles, are relayed to a receiving transducer. The change in the frequency of the reflected waves is proportional to the flow velocity, and a microprocessor determines the flow velocity by comparing the frequency shift between the transmitted and reflected signals (Figs. 8–69 and 8–70). The flow rate and the total amount of flow can also be determined using the measured velocity by properly configuring the flowmeter for the given pipe and flow conditions.

FIGURE 8–67
Two types of variable-area flowmeters: (*a*) an ordinary gravity-based meter and (*b*) a spring-opposed meter.

(*a*) *Photo by Luke A. Cimbala and* (*b*) *Courtesy Insite, Universal Flow Monitors, Inc. Used by permission.*

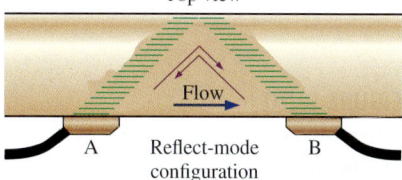

FIGURE 8–68
The operation of a transit time ultrasonic flowmeter equipped with two transducers.

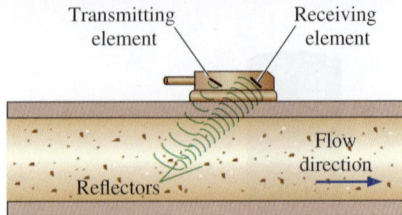

FIGURE 8–69
The operation of a Doppler-effect ultrasonic flowmeter equipped with a transducer pressed on the outer surface of a pipe.

The operation of ultrasonic flowmeters depends on the ultrasound waves being reflected off discontinuities in density. Ordinary ultrasonic flowmeters require the liquid to contain impurities in concentrations greater than 25 parts per million (ppm) in sizes greater than at least 30 μm. But advanced ultrasonic units can also measure the velocity of clean liquids by sensing the waves reflected off turbulent swirls and eddies in the flow stream, provided that they are installed at locations where such disturbances are nonsymmetrical and at a high level, such as a flow section just downstream of a 90° elbow.

Ultrasonic flowmeters have the following advantages:

- They are easy and quick to install by clamping them on the outside of pipes of 0.6 cm to over 3 m in diameter (Fig. 8–70), and even on open channels.
- They are nonintrusive. Since the meters clamp on, there is no need to stop operation and drill holes into piping, and no production downtime.
- There is no pressure drop since the meters do not interfere with the flow.
- Since there is no direct contact with the fluid, there is no danger of corrosion or clogging.
- They are suitable for a wide range of fluids from toxic chemicals to slurries to clean liquids, for permanent or temporary flow measurement.
- There are no moving parts, and thus the meters provide reliable and maintenance-free operation.
- They can also measure flow quantities in reverse flow.
- The quoted accuracies are 1 to 2 percent.

Ultrasonic flowmeters are noninvasive devices, and the ultrasonic transducers can effectively transmit signals through polyvinyl chloride (PVC), steel, iron,

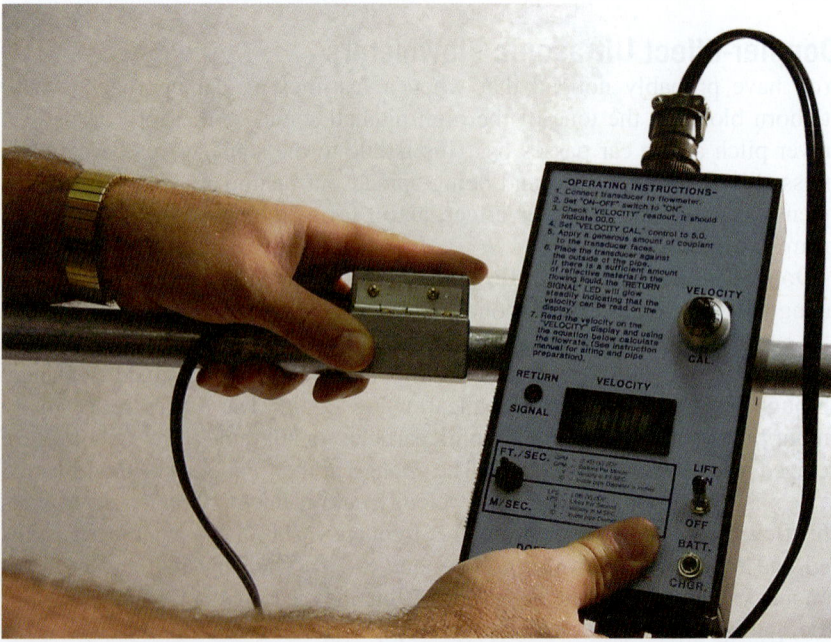

FIGURE 8–70
Ultrasonic clamp-on flowmeters enable one to measure flow velocity without even contacting (or disturbing) the fluid by simply pressing a transducer on the outer surface of the pipe.

Photo by J. Matthew Deepe.

and glass pipe walls. However, coated pipes and concrete pipes are not suitable for this measurement technique since they absorb ultrasonic waves.

Electromagnetic Flowmeters

It has been known since Faraday's experiments in the 1830s that when a conductor is moved in a magnetic field, an electromotive force develops across that conductor as a result of magnetic induction. Faraday's law states that the voltage induced across any conductor as it moves at right angles through a magnetic field is proportional to the velocity of that conductor. This suggests that we may be able to determine flow velocity by replacing the solid conductor by a conducting fluid, and **electromagnetic flowmeters** do just that. Electromagnetic flowmeters have been in use since the mid-1950s, and they come in various designs such as full-flow and insertion types.

A *full-flow electromagnetic flowmeter* is a nonintrusive device that consists of a magnetic coil that encircles the pipe, and two electrodes drilled into the pipe along a diameter flush with the inner surface of the pipe so that the electrodes are in contact with the fluid but do not interfere with the flow and thus do not cause any head loss (Fig. 8–71a). The electrodes are connected to a voltmeter. The coils generate a magnetic field when subjected to electric current, and the voltmeter measures the electric potential difference between the electrodes. This potential difference is proportional to the flow velocity of the conducting fluid, and thus the flow velocity can be calculated by relating it to the voltage generated.

Insertion electromagnetic flowmeters operate similarly, but the magnetic field is confined within a flow channel at the tip of a rod inserted into the flow, as shown in Fig. 8–71b.

Electromagnetic flowmeters are well-suited for measuring flow velocities of liquid metals such as mercury, sodium, and potassium that are used in some nuclear reactors. They can also be used for liquids that are poor conductors, such as water, provided that they contain an adequate amount of charged particles. Blood and seawater, for example, contain sufficient amounts of ions, and thus electromagnetic flowmeters can be used to measure their flow rates. Electromagnetic flowmeters can also be used to measure the flow rates of chemicals, pharmaceuticals, cosmetics, corrosive

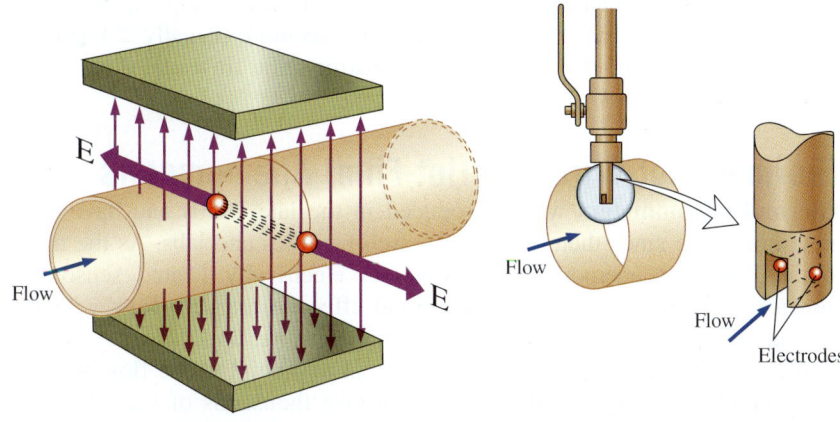

(a) Full-flow electromagnetic flowmeter

(b) Insertion electromagnetic flowmeter

FIGURE 8–71
(a) Full-flow and (b) insertion electromagnetic flowmeters, www.flocat.com.

liquids, beverages, fertilizers, and numerous slurries and sludges, provided that the substances have high enough electrical conductivities. Electromagnetic flowmeters are not suitable for use with distilled or deionized water.

Electromagnetic flowmeters measure flow velocity indirectly, and thus careful calibration is important during installation. Their use is limited by their relatively high cost, power consumption, and the restrictions on the types of suitable fluids with which they can be used.

Vortex Flowmeters

You have probably noticed that when a flow stream such as a river encounters an obstruction such as a rock, the fluid separates and moves around the rock. But the presence of the rock is felt for some distance downstream via the swirls generated by it.

Most flows encountered in practice are turbulent, and a disk or a short cylinder placed in the flow coaxially sheds vortices (see also Chap. 4). It is observed that these vortices are shed periodically, and the shedding frequency is proportional to the average flow velocity. This suggests that the flow rate can be determined by generating vortices in the flow by placing an obstruction in the flow and measuring the shedding frequency. The flow measurement devices that work on this principle are called **vortex flowmeters.** The *Strouhal number*, defined as $St = fd/V$, where f is the vortex shedding frequency, d is the characteristic diameter or width of the obstruction, and V is the velocity of the flow impinging on the obstruction, also remains constant in this case, provided that the flow velocity is high enough.

A vortex flowmeter consists of a sharp-edged bluff body (strut) placed in the flow that serves as the vortex generator, and a detector (such as a pressure transducer that records the oscillation in pressure) placed a short distance downstream on the inner surface of the casing to measure the shedding frequency. The detector can be an ultrasonic, electronic, or fiber-optic sensor that monitors the changes in the vortex pattern and transmits a pulsating output signal (Fig. 8–72). A microprocessor then uses the frequency information to calculate and display the flow velocity or flow rate. The frequency of vortex shedding is proportional to the average velocity over a wide range of Reynolds numbers, and vortex flowmeters operate reliably and accurately at Reynolds numbers from 10^4 to 10^7.

The vortex flowmeter has the advantage that it has no moving parts and thus is inherently reliable, versatile, and very accurate (usually ±1 percent over a wide range of flow rates), but it obstructs the flow and thus causes considerable head loss.

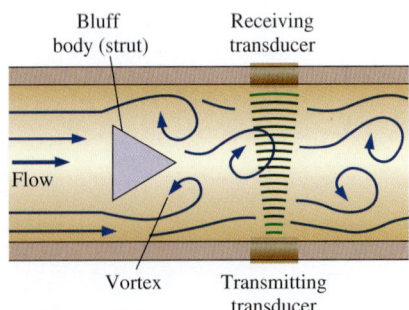

FIGURE 8–72
The operation of a vortex flowmeter.

Thermal (Hot-Wire and Hot-Film) Anemometers

Thermal anemometers were introduced in the late 1950s and have been in common use since then in fluid research facilities and labs. As the name implies, thermal anemometers involve an electrically heated sensor, as shown in Fig. 8–73, and utilize a thermal effect to measure flow velocity. Thermal anemometers have extremely small sensors, and thus they can be used to measure the instantaneous velocity at any point in the flow without appreciably disturbing the flow. They can take thousands of velocity measurements per second with excellent spatial and temporal resolution, and

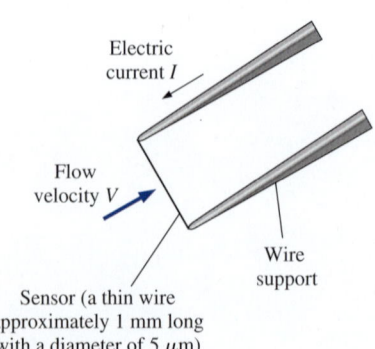

FIGURE 8–73
The electrically heated sensor and its support, which are the components of a hot-wire probe.

thus they can be used to study the details of fluctuations in turbulent flow. They can measure velocities in liquids and gases accurately over a wide range—from a few centimeters to over a hundred meters per second.

A thermal anemometer is called a **hot-wire anemometer** if the sensing element is a wire, and a **hot-film anemometer** if the sensor is a thin metallic film (less than 0.1 μm thick) mounted usually on a relatively thick ceramic support having a diameter of about 50 μm. The hot-wire anemometer is characterized by its very small sensor wire—usually a few microns in diameter and a couple of millimeters in length. The sensor is usually made of platinum, tungsten, or platinum–iridium alloys, and it is attached to the probe through needle-like holders. The fine wire sensor of a hot-wire anemometer is very fragile because of its small size and can easily break if the liquid or gas contains excessive amounts of contaminants or particulate matter. This is especially of consequence at high velocities. In such cases, the more rugged hot-film probes should be used. But the sensor of the hot-film probe is larger, has significantly lower frequency response, and interferes more with the flow; thus it is not always suitable for studying the fine details of turbulent flow.

The operating principle of a constant-temperature anemometer (CTA), which is the most common type and is shown schematically in Fig. 8–74, is as follows: the sensor is electrically heated to a specified temperature (typically about 200°C). The sensor tends to cool as it loses heat to the surrounding flowing fluid, but electronic controls maintain the sensor at a constant temperature by varying the electric current (which is done by varying the voltage) as needed. The higher the flow velocity, the higher the rate of heat transfer from the sensor, and thus the larger the voltage that needs to be applied across the sensor to maintain it at constant temperature. There is a close correlation between the flow velocity and voltage, and the flow velocity is determined by measuring the voltage applied by an amplifier or the electric current passing through the sensor.

The sensor is maintained at a constant temperature during operation, and thus its thermal energy content remains constant. The conservation of energy principle requires that the electrical Joule heating $\dot{W}_{elect} = I^2 R_w = E^2/R_w$ of the sensor must be equal to the total rate of heat loss from the sensor $\dot{Q}_{total}$, which consists of convection heat transfer since conduction to the wire supports and radiation to the surrounding surfaces are small and can be disregarded. Using proper relations for forced convection, the energy balance is expressed by **King's law** as

$$E^2 = a + bV^n \tag{8-75}$$

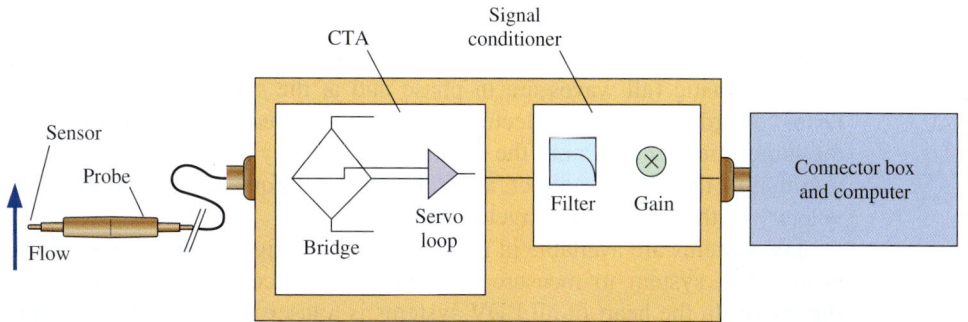

FIGURE 8–74
Schematic of a thermal anemometer system.

FIGURE 8–75
Thermal anemometer probes with single, double, and triple sensors to measure (*a*) one-, (*b*) two-, and (*c*) three-dimensional velocity components simultaneously.

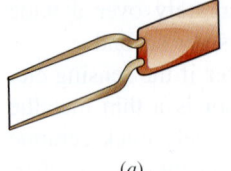
(*a*)

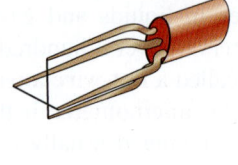
(*b*)

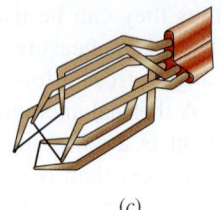
(*c*)

where E is the voltage, and the values of the constants a, b, and n are calibrated for a given probe. Once the voltage is measured, this relation gives the flow velocity V directly.

Most hot-wire sensors have a diameter of 5 μm and a length of approximately 1 mm and are made of tungsten. The wire is spot-welded to needle-shaped prongs embedded in a probe body, which is connected to the anemometer electronics. Thermal anemometers can be used to measure two- or three-dimensional velocity components simultaneously by using probes with two or three sensors, respectively (Fig. 8–75). When selecting probes, consideration should be given to the type and the contamination level of the fluid, the number of velocity components to be measured, the required spatial and temporal resolution, and the location of measurement.

Laser Doppler Velocimetry

Laser Doppler velocimetry (LDV), also called **laser velocimetry (LV)** or **laser Doppler anemometry (LDA),** is an optical technique to measure flow velocity at any desired point without disturbing the flow. Unlike thermal anemometry, LDV involves no probes or wires inserted into the flow, and thus it is a nonintrusive method. Like thermal anemometry, it can accurately measure velocity at a very small volume, and thus it can also be used to study the details of flow at a locality, including turbulent fluctuations, and it can be traversed through the entire flow field without intrusion.

The LDV technique was developed in the mid-1960s and has found widespread acceptance because of the high accuracy it provides for both gas and liquid flows; the high spatial resolution it offers; and, in recent years, its ability to measure all three velocity components. Its drawbacks are the relatively high cost; the requirement for sufficient transparency between the laser source, the target location in the flow, and the photodetector; and the requirement for careful alignment of emitted and reflected beams for accuracy. The latter drawback is eliminated for the case of a fiber-optic LDV system, since it is aligned at the factory.

The operating principle of LDV is based on sending a highly coherent monochromatic (all waves are in phase and at the same wavelength) light beam toward the target, collecting the light reflected by small particles in the target area, determining the change in frequency of the reflected radiation due to the Doppler effect, and relating this frequency shift to the flow velocity of the fluid at the target area.

LDV systems are available in many different configurations. A basic dual-beam LDV system to measure a single velocity component is shown in Fig. 8–76. In the heart of all LDV systems is a laser power source, which is

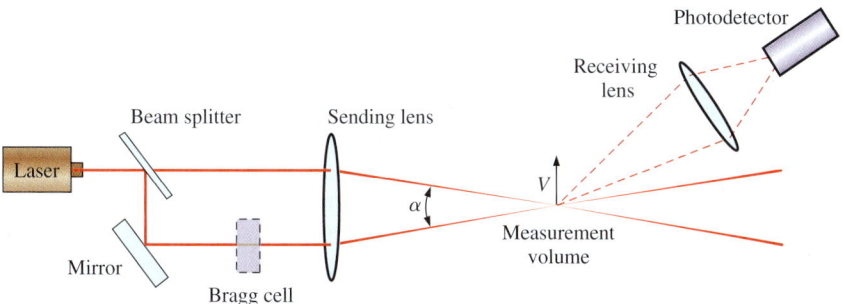

FIGURE 8–76
A dual-beam LDV system in forward scatter mode.

usually a helium–neon or argon-ion laser with a power output of 10 mW to 20 W. Lasers are preferred over other light sources since laser beams are highly coherent and highly focused. The helium–neon laser, for example, emits radiation at a wavelength of 0.6328 μm, which is in the reddish-orange color range. The laser beam is first split into two parallel beams of equal intensity by a half-silvered mirror called a *beam splitter*. Both beams then pass through a converging lens that focuses the beams at a point in the flow (the *target*). The small fluid volume where the two beams intersect is the region where the velocity is measured and is called the *measurement volume* or the *focal volume*. The measurement volume resembles an ellipsoid, typically of 0.1 mm diameter and 0.5 mm in length. The laser light is scattered by particles passing through this measurement volume, and the light scattered in a certain direction is collected by a receiving lens and is passed through a photodetector that converts the fluctuations in light intensity into fluctuations in a voltage signal. Finally, a signal processor determines the frequency of the voltage signal and thus the velocity of the flow.

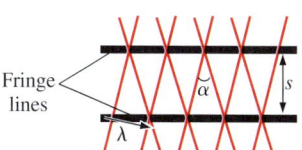

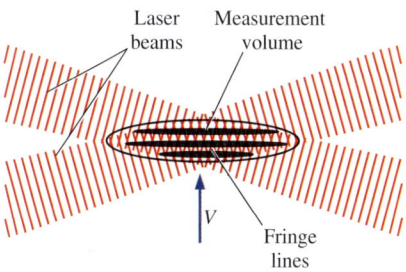

FIGURE 8–77
Fringes that form as a result of the interference at the intersection of two laser beams of an LDV system (lines represent peaks of waves). The top diagram is a close-up view of two fringes.

The waves of the two laser beams that cross in the measurement volume are shown schematically in Fig. 8–77. The waves of the two beams interfere in the measurement volume, creating a bright fringe where they are in phase and thus support each other, and creating a dark fringe where they are out of phase and thus cancel each other. The bright and dark fringes form lines parallel to the midplane between the two incident laser beams. Using trigonometry, the spacing s between the fringe lines, which can be viewed as the wavelength of fringes, can be shown to be $s = \lambda/[2\sin(\alpha/2)]$, where λ is the wavelength of the laser beam and α is the angle between the two laser beams. When a particle traverses these fringe lines at velocity V, the frequency of the scattered fringe lines is

$$f = \frac{V}{s} = \frac{2V\sin(\alpha/2)}{\lambda} \qquad (8\text{–}76)$$

This fundamental relation shows the flow velocity to be proportional to the frequency and is known as the **LDV equation**. As a particle passes through the measurement volume, the reflected light is bright, then dark, then bright, etc., because of the fringe pattern, and the flow velocity is determined by measuring the frequency of the reflected light. The velocity profile at a cross section of a pipe, for example, can be obtained by mapping the flow across the pipe (Fig. 8–78).

FIGURE 8–78
A time-averaged velocity profile in turbulent pipe flow obtained by an LDV system.

Courtesy Dantec Dynamics, www.dantecdynamics.com. Used by permission.

The LDV method obviously depends on the presence of scattered fringe lines, and thus the flow must contain a sufficient amount of small particles called *seeds* or *seeding particles*. These particles must be small enough to follow the flow closely so that the particle velocity is equal to the flow velocity, but large enough (relative to the wavelength of the laser light) to scatter an adequate amount of light. Particles with a diameter of 1 μm usually serve the purpose well. Some fluids such as tap water naturally contain an adequate amount of such particles, and no seeding is necessary. Gases such as air are commonly seeded with smoke or with particles made of latex, oil, or other materials. By using three laser beam pairs at different wavelengths, the LDV system is also used to obtain all three velocity components at any point in the flow.

Particle Image Velocimetry

Particle image velocimetry (PIV) is a double-pulsed laser technique used to measure the instantaneous velocity distribution in a plane of flow by photographically determining the displacement of particles in the plane during a very short time interval. Unlike methods like hot-wire anemometry and LDV that measure velocity at a point, PIV provides velocity values simultaneously throughout an entire cross section, and thus it is a whole-field technique. PIV combines the accuracy of LDV with the capability of flow visualization and provides instantaneous flow field mapping. The entire instantaneous velocity profile at a cross section of pipe, for example, can be obtained with a single PIV measurement. A PIV system can be viewed as a camera that can take a snapshot of velocity distribution at any desired plane in a flow. Ordinary flow visualization gives a qualitative picture of the details of flow. PIV also provides an accurate *quantitative* description of various flow quantities such as the velocity field, and thus the capability to analyze the flow numerically using the velocity data provided. Because of its whole-field capability, PIV is also used to validate computational fluid dynamics (CFD) codes (Chap. 15).

The PIV technique has been used since the mid-1980s, and its use and capabilities have grown in recent years with improvements in frame grabber and charge-coupled device (CCD) camera technologies. The accuracy, flexibility, and versatility of PIV systems with their ability to capture whole-field images with submicrosecond exposure time have made them extremely valuable tools in the study of supersonic flows, explosions, flame propagation, bubble growth and collapse, turbulence, and unsteady flow.

The PIV technique for velocity measurement consists of two main steps: visualization and image processing. The first step is to seed the flow with suitable particles in order to trace the fluid motion. Then a pulse of laser light sheet illuminates a thin slice of the flow field at the desired plane, and the positions of particles in that plane are determined by detecting the light scattered by particles on a digital video or photographic camera positioned at right angles to the light sheet (Fig. 8–79). After a very short time period Δt (typically in μs), the particles are illuminated again by a second pulse of laser light sheet, and their new positions are recorded. Using the information on these two superimposed camera images, the particle displacements Δs are determined for all particles, and the magnitude of velocity of each particle in the plane of the laser light sheet is determined from $\Delta s/\Delta t$. The direction of motion of the particles is also determined from the two

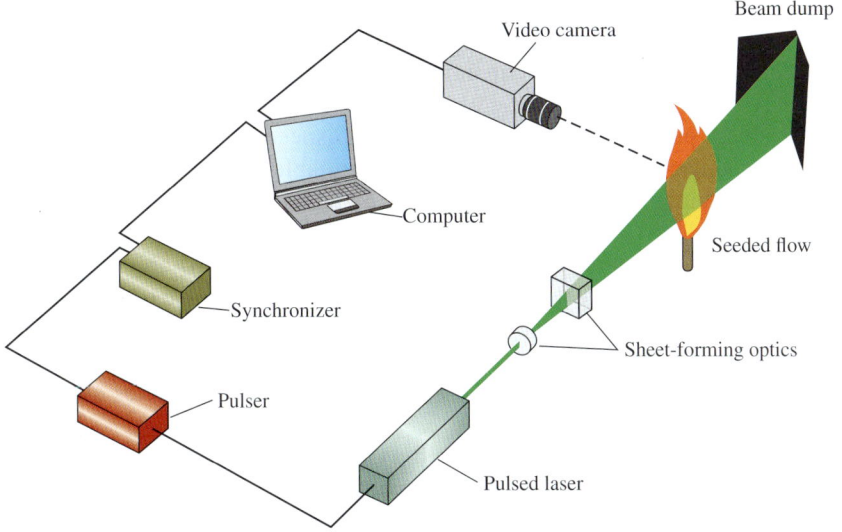

FIGURE 8–79
A PIV system to study flame stabilization.

positions, so that two components of velocity in the plane are calculated. The built-in algorithms of PIV systems determine the velocities at hundreds or thousands of area elements called *interrogation regions* throughout the entire plane and display the velocity field on the computer monitor in any desired form (Fig. 8–80).

The PIV technique relies on the laser light scattered by particles, and thus the flow must be seeded if necessary with particles, also called *markers*, in order to obtain an adequate reflected signal. Seed particles must be able to follow the pathlines in the flow for their motion to be representative of

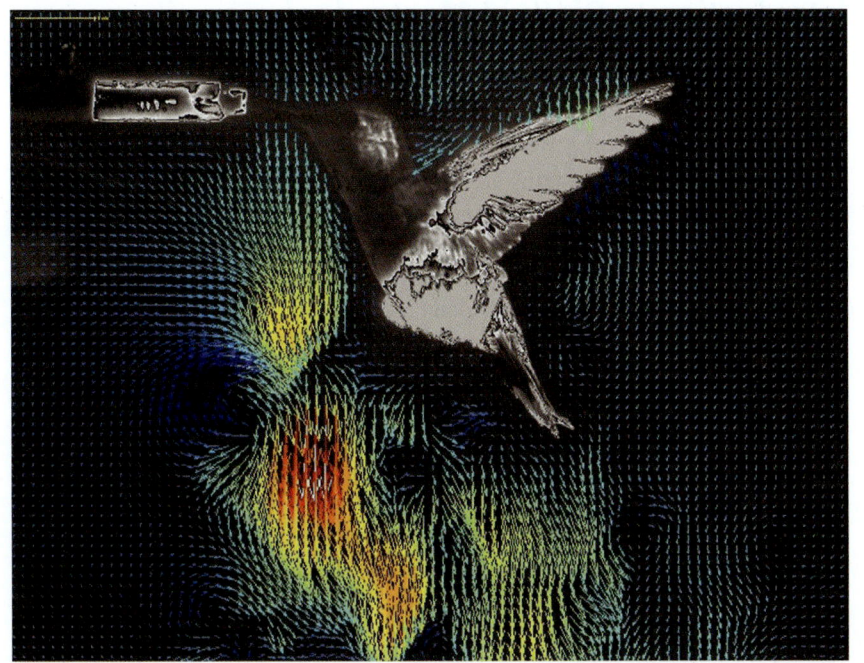

FIGURE 8–80
Instantaneous PIV velocity vectors superimposed on a hummingbird in hover. Color scale is from low velocity (blue) to high velocity (red).
Photo by Douglas Warrick. Used by permission.

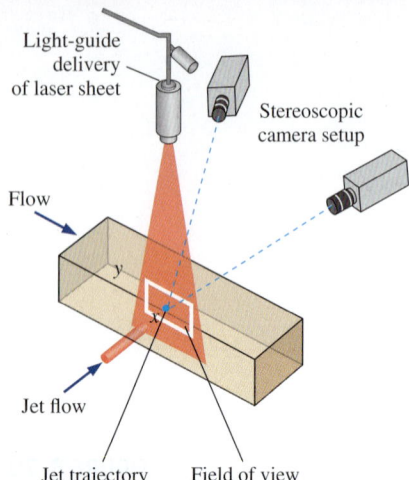

FIGURE 8–81
A three-dimensional PIV system set up to study the mixing of an air jet with cross duct flow.

the flow, and this requires the particle density to be equal to the fluid density (so that they are neutrally buoyant) or the particles to be so small (typically μm-sized) that their movement relative to the fluid is insignificant. A variety of such particles is available to seed gas or liquid flow. Very small particles must be used in high-speed flows. Silicon carbide particles (mean diameter of 1.5 μm) are suitable for both liquid and gas flow, titanium dioxide particles (mean diameter of 0.2 μm) are usually used for gas flow and are suitable for high-temperature applications, and polystyrene latex particles (nominal diameter of 1.0 μm) are suitable for low-temperature applications. Metallic-coated particles (mean diameter of 9.0 μm) are also used to seed water flows for LDV measurements because of their high reflectivity. Gas bubbles as well as droplets of some liquids such as olive oil or silicon oil are also used as seeding particles after they are atomized to μm-sized spheres.

A variety of laser light sources such as argon, copper vapor, and Nd:YAG can be used with PIV systems, depending on the requirements for pulse duration, power, and time between pulses. Nd:YAG lasers are commonly used in PIV systems over a wide range of applications. A beam delivery system such as a light arm or a fiber-optic system is used to generate and deliver a high-energy pulsed laser sheet at a specified thickness.

With PIV, other flow properties such as vorticity and strain rates can also be obtained, and the details of turbulence can be studied. Recent advances in PIV technology have made it possible to obtain three-dimensional velocity profiles at a cross section of a flow using two cameras (Fig. 8–81). This is done by recording the images of the target plane simultaneously by both cameras at different angles, processing the information to produce two separate two-dimensional velocity maps, and combining these two maps to generate the instantaneous three-dimensional velocity field.

Introduction to Biofluid Mechanics[1]

Biofluid mechanics can cover a number of physiological systems in the human body but the term also applies to all animal species as there are a number of basic fluid systems that are essentially a series of piping networks to transport a fluid (be it liquid or gas or perhaps both). If we focus on humans, these fluid systems are the cardiovascular, respiratory, lymphatic, ocular, and gastrointestinal to name several. We should keep in mind that all these systems are similar to other mechanical piping networks in that the fundamental constituents for the network include a pump, pipes, valves, and a fluid. For our purposes, we will focus more on the cardiovascular system to demonstrate the basic concepts of a piping network within a human.

Figure 8–82 illustrates the cardiovascular system, more specifically, the systemic circulation or the vessels (pipes) that carry the blood (fluid) from the heart, specifically the left ventricle (pump), to the rest of the body. Keep in mind there is a separate network of vessels from the right ventricle to the lungs to oxygenate the blood again. What is unique about the series of pipes in the systemic circulation is that the geometry or cross section is not circular but rather elliptical and in fact, unlike the typical mechanical systems for piping networks that have fittings to transition from one size pipe to

---

[1] This section was contributed by Professor Keefe Manning of Penn State University.

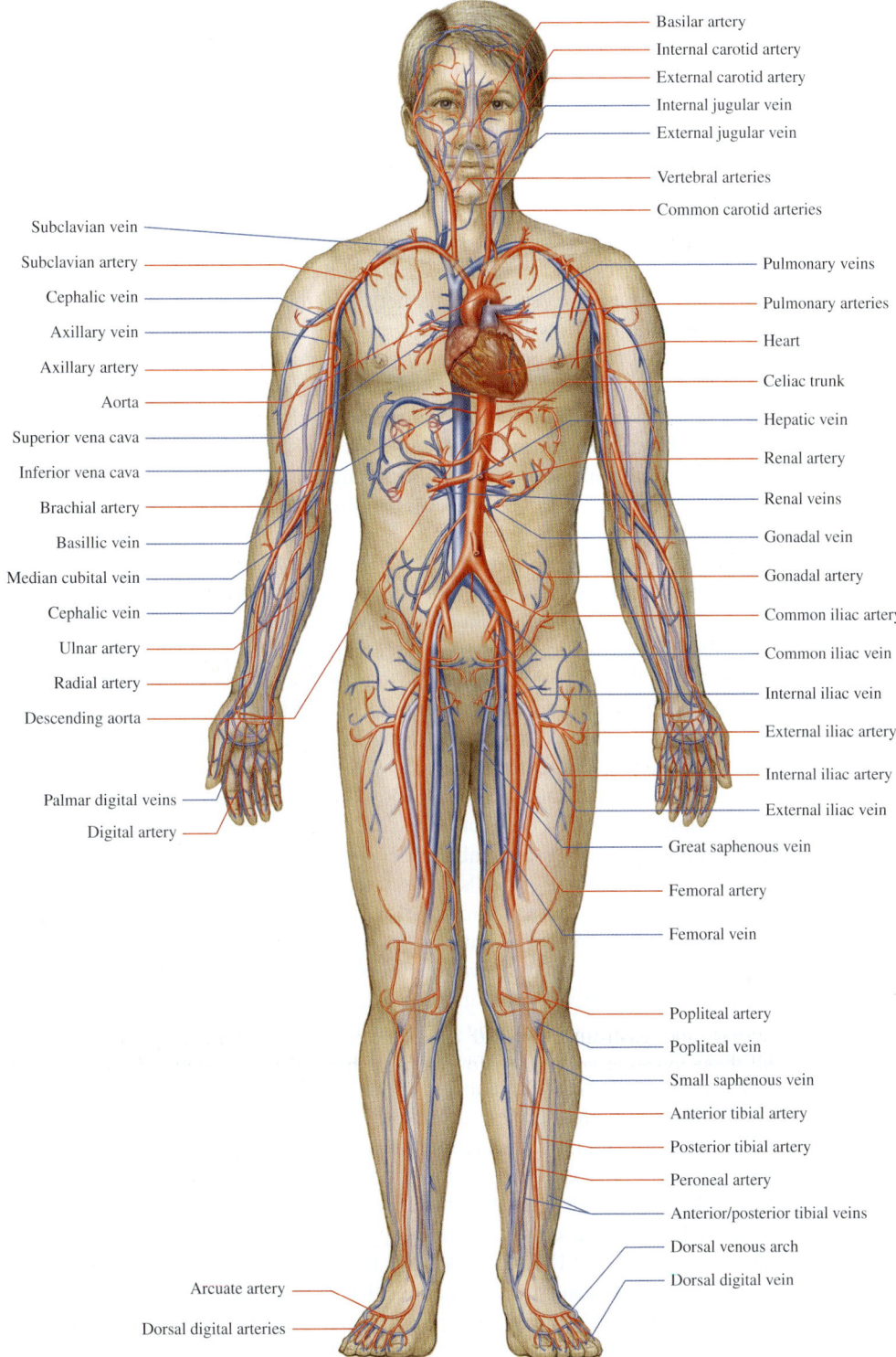

FIGURE 8–82
The cardiovascular system.
McGraw-Hill Companies, Inc.

another size pipe, the cardiovascular system starting with the aorta (the first vessel from the left ventricle) continually tapers from approximately 25 mm in diameter to 5 microns in diameter at the capillary level and then gradually increases in diameter to approximately 25 mm at the vena cava, which is the vessel connected to the right ventricle. Another important element of the circulation and specifically the vessels is that they are compliant and can expand to accommodate blood volume as needed to regulate pressure changes to maintain homeostasis.

The cardiovascular system is a complex network of pipes that themselves are living and respond to stresses as do blood elements that react when the norm has changed. Even with this network, the system is even more intricate given that the flow is continually moving based on pulses initiated from the heart to drive blood through the network. This pulsatility propagates through the blood and the vessel wall creating an interaction of waves and reflections within the system. Because of the discontinuities associated with the branching, bifurcations, and curvature as seen in Fig. 8-82 initial and boundary conditions are not straightforward. Understanding blood flow is a challenging endeavor given the complexities of the vessel network and the components themselves.

Flow measurement techniques like PIV and LDV are extremely useful in characterizing the flow within and about medical devices, particularly those implanted in the cardiovascular system. Much can be ascertained, and design changes can be made, using these techniques with respect to how blood might flow through or about these cardiovascular devices. Furthermore, we can even use these measurements to then estimate levels of blood damage and the potential for clotting to occur. To ensure we have an accurate representation of the cardiovascular system on the bench, engineers have designed mock circulatory loops or flow loops that allow the experimentalist to simulate cardiac flow and pressure waveforms for bench top studies. For example, Dr. Gus Rosenberg developed the Penn State mock circulatory loop in the early 1970s (Rosenberg *et al.*, 1981). We also need to simulate blood for these particular flow measurement techniques to ensure that the fluid is transparent but also mimics the behavior of blood as a non-Newtonian fluid. We have developed a blood analog that does that and also matches the refractive index of the acrylic models that represent the cardiovascular devices, thus allowing the laser light to pass through the acrylic into the flow field without any refraction. The simulated loop and fluid are critical to ensure that the measurements are acquired under controllable physiological conditions and with sufficient accuracy.

The Pennsylvania State University has been developing mechanical circulatory support devices (blood pumps) since the 1970s, which are devices that help patients stay alive as they await a heart transplant (former Vice President Dick Cheney used such technology while awaiting a heart transplant). Through the years, PIV and LDV have been used quite successfully to measure the flow and make design changes that reduce clotting. Our recent focus has been the development of a pulsatile pediatric ventricular assist device (PVAD) that helps children stay alive until they can receive a donor heart. The device operates pneumatically with air pulsing into a chamber which then causes a diaphragm to inflate against a polyurethane urea sac (the blood contacting surface within the PVAD). The flow is directed into the device from a tube attached to the left ventricle, passes through a mechanical

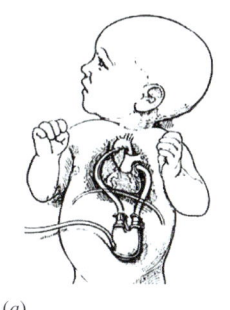
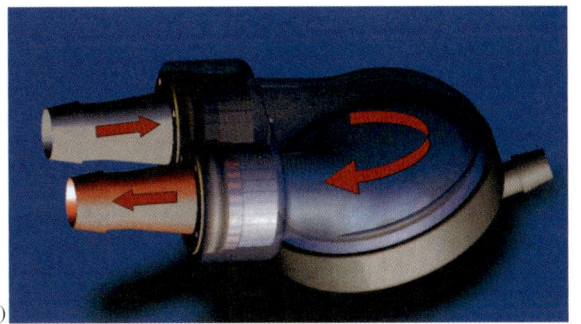

(a) (b)

FIGURE 8–83
(*a*) An artist rendering of the 12-cc pulsatile Penn State pediatric ventricular assist device with the inlet attached to the left atrium and the outlet attached to the ascending aorta (*b*) The direction of blood through the PVAD.

Photo (b) Permission granted from ASME, Cooper et al. JBME, 2008.

heart valve into the PVAD, and then flows through the outlet of the device through another mechanical heart valve and into a tube which is attached to the ascending aorta as shown in Fig. 8–83*a*. Fig. 8–83*b* shows the flow path through the PVAD, and it should be noted that it can be placed within the palm of an adult's hand. One of the first PIV PVAD studies was to determine which type of mechanical heart valve (tilting disc or bileaflet) would be used with the device. Fig. 8–84 illustrates part of the PIV study results

FIGURE 8–84
Particle traces for the BSM valve configuration at 250 ms (left column) and for the CM valve configuration at 350 ms (right column) for the 7 mm (top row), 8.2 mm (middle row), and 11 mm (bottom row) planes. These images highlight the first time step that the rotational flow pattern is fully developed.

Permission granted from ASME, Cooper et al. JBME, 2008.

(Cooper *et al.*, 2008). Here, we used particle traces as a way to examine how the vortical structure would develop inside the device, which for this technology is a way to ensure adequate wall washing (sufficient wall shear) to prevent clotting on the blood contacting surfaces within the device. The tighter rotation would lead to more momentum over the entire cardiac cycle and create a larger vortical structure.

Our research group has also looked at characterizing the flow through mechanical heart valves. In one study (Manning *et al.*, 2008), we focused on the flow within the housing of a Bjork-Shiley Monostrut mechanical heart valve (tilting disc valve) as shown in Fig. 8–85*b*. We removed part of the housing and inserted an optical window to allow access for the LDV system. Instead of using a flow-through loop for this study, we used a single-shot chamber (Fig. 8–85*a*) that mimicked the mitral valve position since we are more interested in the closure fluid dynamics. The mitral valve sits between the left atrium and left ventricle. The native heart valves, like the mitral valve, are passive, similar to a check valve, and

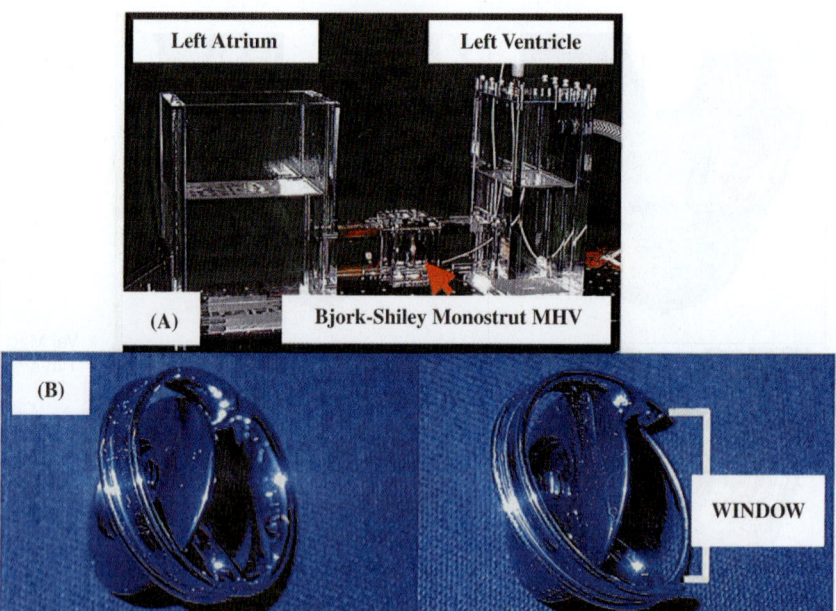

FIGURE 8–85
(*a*) The single shot chamber mimics the closure dynamics of the Bjork-Shiley Monostrut valve.
(*b*) On the lefthand side is a view of the intact Bjork-Shiley Monostrut mechanical heart valve. To the right, the modification to the valve housing is displayed. The window was later filled in with acrylic to maintain similar fluid dynamic patterns and rigidity.

Permission granted from ASME, Manning et al. JBME, 2008.

respond to the pressure changes within the heart's different structures. In this study, we measured how fast the fluid flowed through the small gap between the tilting disc and valve housing, and also how large was the vortex that is created as the tilting disc closes. Figure 8–86 is a schematic illustration of the flow, and Fig. 8–87 is a time sequence of the flow that was measured using LDV within a couple of milliseconds around impact of the valve housing during closure. The intense vortex can be measured right at impact. These data were collected over hundreds of simulated heart beats. We then used these velocity measurements to estimate the amount of potential blood damage by relating time duration and shear magnitude.

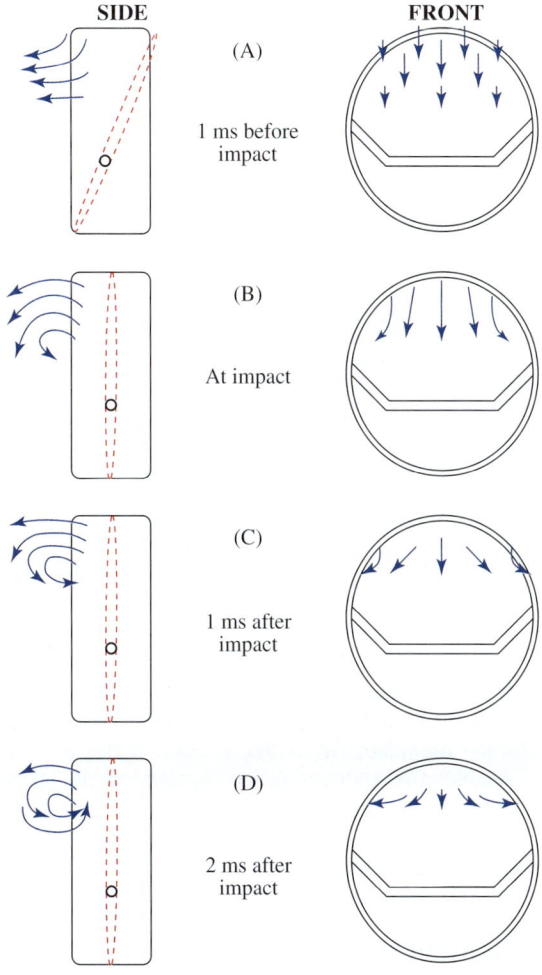

FIGURE 8–86
These schematics depict side and front views of the overall flow structure generated by the closing occluder for four successive times.

Permission granted from ASME, Manning et al. JBME, 2008.

414
INTERNAL FLOW

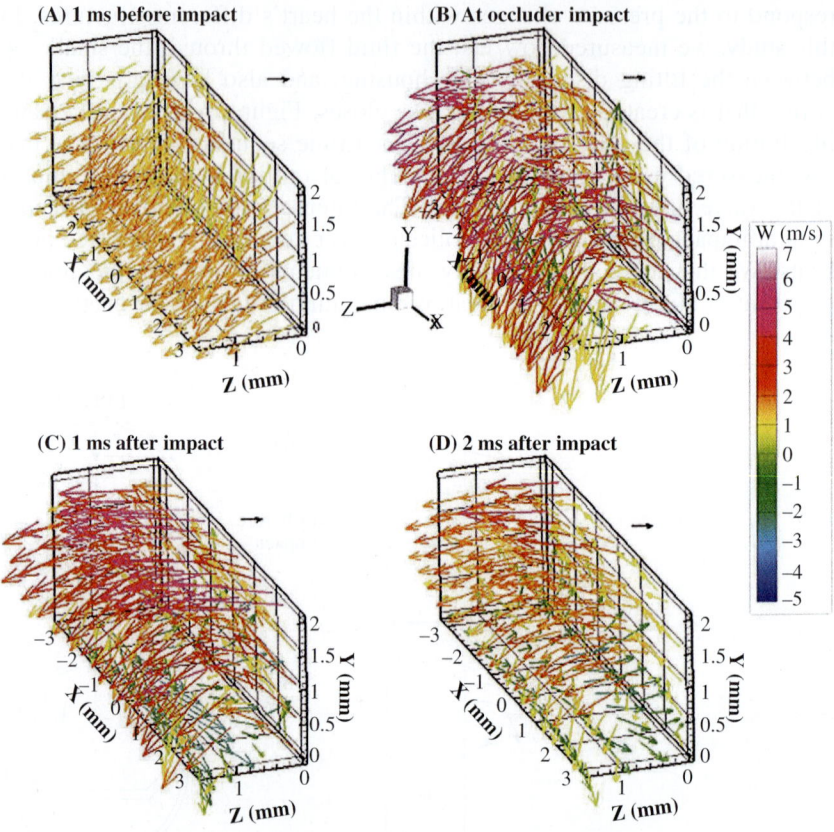

FIGURE 8–87
Three-dimensional flow structures are constructed with the vectors indicating direction and the color signifying axial velocity strength. The valve closes right to left, with $x = 0$ representing the centerline of the leaflet. The four plots show the flow (*a*) 1 ms before impact, (*b*) at impact, (*c*) 1 ms following closure, and (*d*) 2 ms after closure.

Permission granted from ASME, Manning et al. JBME, 2008.

EXAMPLE 8–11   Blood flow through the Aortic Bifurcation

Blood flows from the heart (specifically, the left ventricle) into the aorta to feed the rest of the body oxygen. As blood flow moves from the ascending aorta and downward to the abdominal aorta, some of the volume is directed through a branching network. As the blood reaches the pelvic region, there is a bifurcation (see Fig. 8–88) into the left and right common iliac arteries. This bifurcation is symmetrical but the common iliac vessels are not the same diameter. Given that the kinematic viscosity of blood is 4 cSt (centistokes), the abdominal aorta's diameter is 15 mm, the right common iliac artery's diameter is 10 mm, and the left common iliac artery's diameter is 8 mm, determine the mean flow rate through the right common iliac artery if the abdominal aorta's mean velocity is 30 cm/s and the left common iliac artery's mean velocity is 40 cm/s.

SOLUTION   The mean velocities for two of the three vessels is provided along with the diameters of all three vessels. Approximate the vessels as rigid pipes.
Assumptions   **1** The flow is steady even though the heart contracts and relaxes approximately 75 beats per minute creating a pulsatile flow. **2** The entrance effects are negligible and the flow is considered fully developed. **3** Blood acts as a Newtonian fluid.
Properties   The kinematic viscosity at 37°C is 4 cSt.

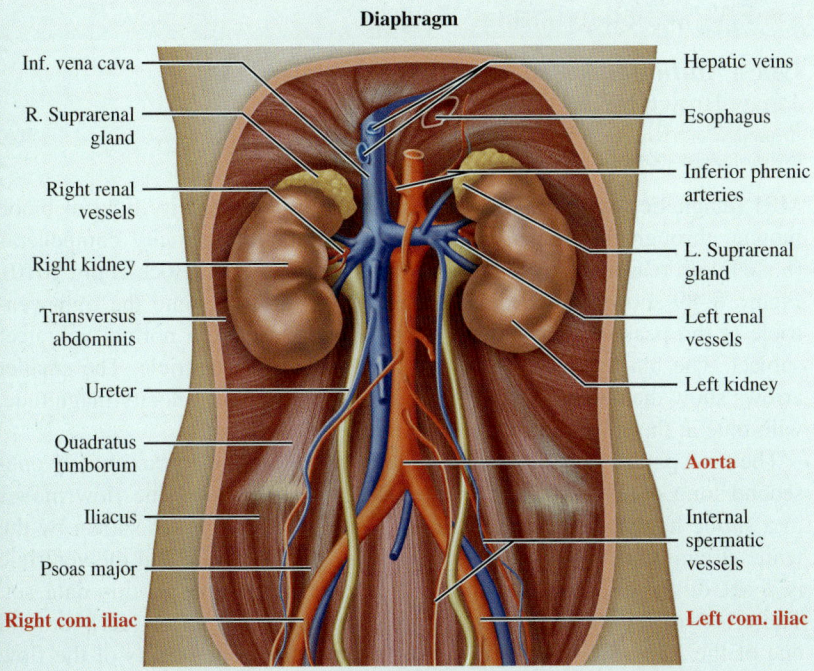

FIGURE 8-88
Anatomy of the human body. Note the aorta and left and right common iliac arteries.

Analysis Using conservation of mass, we can say the flow rate of the abdominal aorta ($\dot{V}_1$) equals the sum of both common iliac arteries ($\dot{V}_2$ for left and $\dot{V}_3$ for right). Thus,

$$\dot{V}_1 = \dot{V}_2 + \dot{V}_3$$

Since we are using the mean velocities, we know the diameters, and the density of blood is the same throughout this section of the circulatory system, we can rewrite the equation to be

$V_1 A_1 = V_2 A_2 + V_3 A_3$ where V are the average velocities and A are the areas.

By rearranging and solving for V_3, the equation becomes,

$$V_3 = (V_1 A_1 - V_2 A_2)/A_3$$

Inserting the values we know,

$V_3 = (30 \text{ cm/s} \times (1.5 \text{ cm})^2 - 40 \text{ cm/s} \times (0.8 \text{ cm})^2)/(1.0 \text{ cm})^2$

$V_3 = 41.9 \text{ cm/s}$

Discussion Since we assume a steady flow, the mean velocities are appropriate, but in reality there will be a maximum positive velocity and also some retrograde (or reverse) flow towards the heart as the left ventricle fills during diastole. The velocity profiles through these vessels and many large arteries will vary over a cardiac cycle. It is also assumed that blood will behave as a Newtonian fluid even though it is viscoelastic. Many researchers use this assumption since at this particular location, the shear rate is sufficient to reach the asymptotic value for blood viscosity.

INTERNAL FLOW

APPLICATION SPOTLIGHT ■ PIV Applied to Cardiac Flow

Guest Authors: Jean Hertzberg[1], Brett Fenster[2], Jamey Browning[1] and Joyce Schroeder[2]

[1]Department of Mechanical Engineering, University of Colorado, Boulder, CO.
[2]National Jewish Health Center, Denver, CO.

MRI (magnetic resonance imaging) can measure the velocity field of blood moving through the human heart, including all three velocity components (u,v,w) with reasonable resolution in 3-D space and time (Bock et al., 2010). Figure 8–89 shows blood moving from the right atrium into the right ventricle at the peak of diastole (the heart-filling phase) of a normal volunteer subject. The black arrow shows the long axis of the ventricle. The smaller arrows show the velocity vector field and are colored by velocity magnitude, with blue at the slow end of the scale, up to red at 0.5 m/s.

The flow patterns change rapidly with time during the approximately one-second long cardiac cycle, and show complex geometry. The flow moves in a subtle helical path from the atrium into the ventricle, as shown by the white stream tube. The tricuspid valve between the atrium and the ventricle is a set of three thin tissue flaps, which are not visible in this data set. The effect of the valve on flow patterns can be seen as flow curls around one of the flaps, shown by the yellow stream tube. The details of the flow (including *vorticity*, Chapter 4) are expected to reveal information about the underlying physics of the interaction between the heart and lungs, and lead to improved diagnostics for pathologic conditions like pulmonary hypertension (Fenster et al., 2012).

After the right ventricle is filled, the tricuspid valve closes, the ventricle contracts, and blood is ejected into the pulmonary arteries which lead to the lungs, where the blood is oxygenated. After that, the blood goes to the left side of the heart, where the pressure is raised by the contraction of the left ventricle. The oxygenated blood is then ejected into the aorta and is distributed to the body. In this way, the heart functions as two separate positive displacement pumps.

Since calibration of these data is difficult, it's important to check the data for consistency. One useful test, conservation of mass in the ventricle throughout one cardiac cycle, is applied by computing the volume flow of blood entering the ventricle during diastole, and comparing it to the volume that leaves during systole. Similarly, the net flow through the right side of the heart must match the net flow through the left side of the heart in each cycle.

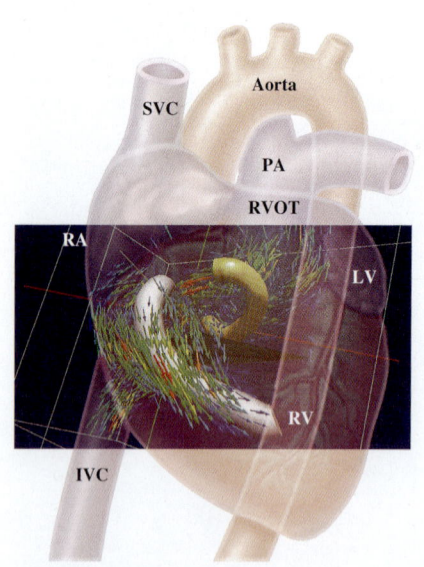

FIGURE 8–89
MRI-PIV measurements of flow through a human heart.
Photo courtesy of Jean Hertzberg.

References

Bock J, Frydrychowicz A, Stalder AF, Bley TA, Burkhardt H, Hennig J, and Markl M. 2010. 40 phase contrast MRI at 3 T: Effect of standard and blood-pool contrast agents on SNR, PC-MRA, and blood flow visualization. *Magnetic Resonance in Medicine* 63(2):330-338.

Fenster BE, Schroeder JD, Hertzberg JR, and Chung JH. 2012. 4-Dimensional Cardiac Magnetic Resonance in a Patient With Bicuspid Pulmonic Valve: Characterization of Post-Stenotic Flow. *J Am Coll Cardiol* 59(25):e49.

SUMMARY

In *internal flow*, a pipe is completely filled with a fluid. *Laminar flow* is characterized by smooth streamlines and highly ordered motion, and *turbulent flow* is characterized by unsteady disorderly velocity fluctuations and highly disordered motion. The *Reynolds number* is defined as

$$\text{Re} = \frac{\text{Inertial forces}}{\text{Viscous forces}} = \frac{V_{avg}D}{\nu} = \frac{\rho V_{avg}D}{\mu}$$

Under most practical conditions, the flow in a pipe is laminar at $\text{Re} < 2300$, turbulent at $\text{Re} > 4000$, and transitional in between.

The region of the flow in which the effects of the viscous shearing forces are felt is called the *velocity boundary layer*. The region from the pipe inlet to the point at which the flow becomes fully developed is called the *hydrodynamic entrance region*, and the length of this region is called the *hydrodynamic entry length* L_h. It is given by

$$\frac{L_{h,\,laminar}}{D} \cong 0.05\,\text{Re} \quad \text{and} \quad \frac{L_{h,\,turbulent}}{D} \cong 10$$

The friction coefficient in the fully developed flow region is constant. The *maximum* and *average* velocities in fully developed laminar flow in a circular pipe are

$$u_{max} = 2V_{avg} \quad \text{and} \quad V_{avg} = \frac{\Delta P D^2}{32\mu L}$$

The *volume flow rate* and the *pressure drop* for laminar flow in a horizontal pipe are

$$\dot{V} = V_{avg} A_c = \frac{\Delta P \pi D^4}{128\mu L} \quad \text{and} \quad \Delta P = \frac{32\mu L V_{avg}}{D^2}$$

The *pressure loss* and *head loss* for all types of internal flows (laminar or turbulent, in circular or noncircular pipes, smooth or rough surfaces) are expressed as

$$\Delta P_L = f\frac{L}{D}\frac{\rho V^2}{2} \quad \text{and} \quad h_L = \frac{\Delta P_L}{\rho g} = f\frac{L}{D}\frac{V^2}{2g}$$

where $\rho V^2/2$ is the *dynamic pressure* and the dimensionless quantity f is the *friction factor*. For fully developed laminar flow in a round pipe, the friction factor is $f = 64/\text{Re}$.

For noncircular pipes, the diameter in the previous relations is replaced by the *hydraulic diameter* defined as $D_h = 4A_c/p$, where A_c is the cross-sectional area of the pipe and p is its wetted perimeter.

In fully developed turbulent flow, the friction factor depends on the Reynolds number and the *relative roughness* ε/D. The friction factor in turbulent flow is given by the *Colebrook equation*, expressed as

$$\frac{1}{\sqrt{f}} = -2.0 \log\left(\frac{\varepsilon/D}{3.7} + \frac{2.51}{\text{Re}\sqrt{f}}\right)$$

The plot of this formula is known as the *Moody chart*. The design and analysis of piping systems involve the determination of the head loss, flow rate, or the pipe diameter. Tedious iterations in these calculations can be avoided by the approximate Swamee–Jain formulas expressed as

$$h_L = 1.07\frac{\dot{V}^2 L}{gD^5}\left\{\ln\left[\frac{\varepsilon}{3.7D} + 4.62\left(\frac{\nu D}{\dot{V}}\right)^{0.9}\right]\right\}^{-2}$$

$$10^{-6} < \varepsilon/D < 10^{-2}$$
$$3000 < \text{Re} < 3 \times 10^8$$

$$\dot{V} = -0.965\left(\frac{gD^5 h_L}{L}\right)^{0.5}\ln\left[\frac{\varepsilon}{3.7D} + \left(\frac{3.17\nu^2 L}{gD^3 h_L}\right)^{0.5}\right]$$

$$\text{Re} > 2000$$

$$D = 0.66\left[\varepsilon^{1.25}\left(\frac{L\dot{V}^2}{gh_L}\right)^{4.75} + \nu\dot{V}^{9.4}\left(\frac{L}{gh_L}\right)^{5.2}\right]^{0.04}$$

$$10^{-6} < \varepsilon/D < 10^{-2}$$
$$5000 < \text{Re} < 3 \times 10^8$$

The losses that occur in piping components such as fittings, valves, bends, elbows, tees, inlets, exits, expansions, and contractions are called *minor losses*. The minor losses are usually expressed in terms of the *loss coefficient* K_L. The head loss for a component is determined from

$$h_L = K_L \frac{V^2}{2g}$$

When all the loss coefficients are available, the total head loss in a piping system is

$$h_{L,\,total} = h_{L,\,major} + h_{L,\,minor} = \sum_i f_i \frac{L_i}{D_i}\frac{V_i^2}{2g} + \sum_j K_{L,j}\frac{V_j^2}{2g}$$

If the entire piping system is of constant diameter, the total head loss reduces to

$$h_{L,\,total} = \left(f\frac{L}{D} + \sum K_L\right)\frac{V^2}{2g}$$

The analysis of a piping system is based on two simple principles: (1) The conservation of mass throughout the system must be satisfied and (2) the pressure drop between two points must be the same for all paths between the two points. When the pipes are connected *in series*, the flow rate through the entire system remains constant regardless of the diameters of the individual pipes. For a pipe that branches out into two (or more) *parallel pipes* and then rejoins at a junction downstream, the total flow rate is the sum of the flow rates in the individual pipes but the head loss in each branch is the same.

When a piping system involves a pump and/or turbine, the steady-flow energy equation is expressed as

$$\frac{P_1}{\rho g} + \alpha_1 \frac{V_1^2}{2g} + z_1 + h_{pump,u} = \frac{P_2}{\rho g} + \alpha_2 \frac{V_2^2}{2g} + z_2 + h_{turbine,e} + h_L$$

When the useful pump head $h_{pump,u}$ is known, the mechanical power that needs to be supplied by the pump to the fluid and the electric power consumed by the motor of the pump for a specified flow rate are

$$\dot{W}_{pump,shaft} = \frac{\rho \dot{V} g h_{pump,u}}{\eta_{pump}} \quad \text{and} \quad \dot{W}_{elect} = \frac{\rho \dot{V} g h_{pump,u}}{\eta_{pump\text{-}motor}}$$

where $\eta_{pump\text{-}motor}$ is the *efficiency of the pump–motor combination,* which is the product of the pump and the motor efficiencies.

The plot of the head loss versus the flow rate $\dot{V}$ is called the *system curve.* The head produced by a pump is not a constant, and the curves of $h_{pump,u}$ and η_{pump} versus $\dot{V}$ are called the *characteristic curves.* A pump installed in a piping system operates at the *operating point,* which is the point of intersection of the system curve and the characteristic curve.

Flow measurement techniques and devices can be considered in three major categories: (1) volume (or mass) flow rate measurement techniques and devices such as obstruction flowmeters, turbine meters, positive displacement flowmeters, rotameters, and ultrasonic meters; (2) point velocity measurement techniques such as the Pitot-static probes, hot-wires, and LDV; and (3) whole-field velocity measurement techniques such as PIV.

The emphasis in this chapter has been on flow through pipes, including blood vessels. A detailed treatment of numerous types of pumps and turbines, including their operation principles and performance parameters, is given in Chap. 14.

REFERENCES AND SUGGESTED READING

1. H. S. Bean (ed.). *Fluid Meters: Their Theory and Applications,* 6th ed. New York: American Society of Mechanical Engineers, 1971.

2. M. S. Bhatti and R. K. Shah. "Turbulent and Transition Flow Convective Heat Transfer in Ducts." In *Handbook of Single-Phase Convective Heat Transfer,* ed. S. Kakaç, R. K. Shah, and W. Aung. New York: Wiley Interscience, 1987.

3 B. T. Cooper, B. N. Roszelle, T. C. Long, S. Deutsch, and K. B. Manning. "The 12 cc Penn State pulsatile pediatric ventricular assist device: fluid dynamics associated with valve selection." *J. of Biomechonicol Engineering.* 130 (2008) pp. 041019.

4. C. F. Colebrook. "Turbulent Flow in Pipes, with Particular Reference to the Transition between the Smooth and Rough Pipe Laws," *Journal of the Institute of Civil Engineers London.* 11 (1939), pp. 133–156.

5. F. Durst, A. Melling, and J. H. Whitelaw. *Principles and Practice of Laser-Doppler Anemometry,* 2nd ed. New York: Academic, 1981.

6. *Fundamentals of Orifice Meter Measurement.* Houston, TX: Daniel Measurement and Control, 1997.

7. S. E. Haaland. "Simple and Explicit Formulas for the Friction Factor in Turbulent Pipe Flow," *Journal of Fluids Engineering,* March 1983, pp. 89–90.

8. I. E. Idelchik. *Handbook of Hydraulic Resistance,* 3rd ed. Boca Raton, FL: CRC Press, 1993.

9. W. M. Kays, M. E. Crawford, B. Weigand. *Convective Heat and Mass Transfer,* 4th ed. New York: McGraw-Hill, 2004.

10. K. B. Manning, L. H. Herbertson, A. A. Fontaine, and S. S. Deutsch. "A detailed fluid mechanics study of tilting disk mechanical heart valve closure and the implications to blood damage." *J. Biomech. Eng.* 130(4) (2008), pp. 041001-1-4.

11. R. W. Miller. *Flow Measurement Engineering Handbook,* 3rd ed. New York: McGraw-Hill, 1997.

12. L. F. Moody. "Friction Factors for Pipe Flows," *Transactions of the ASME* 66 (1944), pp. 671–684.

13. G. Rosenberg, W. M. Phillips, D. L. Landis, and W. S. Pierce. "Design and evaluation of the Pennsylvania State University mock circulatory system." *ASAIO J.* 4 (1981) pp. 41–49.

14. O. Reynolds. "On the Experimental Investigation of the Circumstances Which Determine Whether the Motion of Water Shall Be Direct or Sinuous, and the Law of Resistance in Parallel Channels." *Philosophical Transactions of the Royal Society of London*, 174 (1883), pp. 935–982.

15. H. Schlichting. *Boundary Layer Theory*, 7th ed. New York: Springer, 2000.

16. R. K. Shah and M. S. Bhatti. "Laminar Convective Heat Transfer in Ducts." In *Handbook of Single-Phase Convective Heat Transfer*, ed. S. Kakaç, R. K. Shah, and W. Aung. New York: Wiley Interscience, 1987.

17. P. L. Skousen. *Valve Handbook*. New York: McGraw-Hill, 1998.

18. P. K. Swamee and A. K. Jain. "Explicit Equations for Pipe-Flow Problems," *Journal of the Hydraulics Division. ASCE* 102, no. HY5 (May 1976), pp. 657–664.

19. G. Vass. "Ultrasonic Flowmeter Basics," *Sensors*, 14, no. 10 (1997).

20. A. J. Wheeler and A. R. Ganji. *Introduction to Engineering Experimentation*. Englewood Cliffs, NJ: Prentice-Hall, 1996.

21. W. Zhi-qing. "Study on Correction Coefficients of Laminar and Turbulent Entrance Region Effects in Round Pipes," *Applied Mathematical Mechanics*, 3 (1982), p. 433.

PROBLEMS*

Laminar and Turbulent Flow

8–1C Consider laminar flow in a circular pipe. Is the wall shear stress τ_w higher near the inlet of the pipe or near the exit? Why? What would your response be if the flow were turbulent?

8–2C What is hydraulic diameter? How is it defined? What is it equal to for a circular pipe of diameter D?

8–3C How is the hydrodynamic entry length defined for flow in a pipe? Is the entry length longer in laminar or turbulent flow?

8–4C Why are liquids usually transported in circular pipes?

8–5C What is the physical significance of the Reynolds number? How is it defined for (a) flow in a circular pipe of inner diameter D and (b) flow in a rectangular duct of cross section $a \times b$?

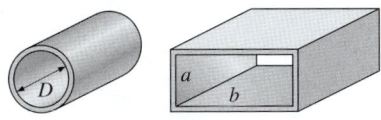

FIGURE P8–5C

---

* Problems designated by a "C" are concept questions, and students are encouraged to answer them all. Problems with the icon are solved using EES, and complete solutions together with parametric studies are included on the text website. Problems with the icon are comprehensive in nature and are intended to be solved with an equation solver such as EES.

8–6C Consider a person walking first in air and then in water at the same speed. For which motion will the Reynolds number be higher?

8–7C Show that the Reynolds number for flow in a circular pipe of diameter D can be expressed as $\text{Re} = 4\dot{m}/(\pi D \mu)$.

8–8C Which fluid at room temperature requires a larger pump to flow at a specified velocity in a given pipe: water or engine oil? Why?

8–9C What is the generally accepted value of the Reynolds number above which the flow in smooth pipes is turbulent?

8–10C How does surface roughness affect the pressure drop in a pipe if the flow is turbulent? What would your response be if the flow were laminar?

Fully Developed Flow in Pipes

8–11C Someone claims that the volume flow rate in a circular pipe with laminar flow can be determined by measuring the velocity at the centerline in the fully developed region, multiplying it by the cross-sectional area, and dividing the result by 2. Do you agree? Explain.

8–12C Someone claims that the average velocity in a circular pipe in fully developed laminar flow can be determined by simply measuring the velocity at $R/2$ (midway between the wall surface and the centerline). Do you agree? Explain.

8–13C Someone claims that the shear stress at the center of a circular pipe during fully developed laminar flow is zero. Do you agree with this claim? Explain.

8–14C Someone claims that in fully developed turbulent flow in a pipe, the shear stress is a maximum at the pipe wall. Do you agree with this claim? Explain.

8–15C How does the wall shear stress τ_w vary along the flow direction in the fully developed region in (a) laminar flow and (b) turbulent flow?

8–16C What fluid property is responsible for the development of the velocity boundary layer? For what kinds of fluids will there be no velocity boundary layer in a pipe?

8–17C In the fully developed region of flow in a circular pipe, does the velocity profile change in the flow direction?

8–18C How is the friction factor for flow in a pipe related to the pressure loss? How is the pressure loss related to the pumping power requirement for a given mass flow rate?

8–19C Discuss whether fully developed pipe flow is one-, two-, or three-dimensional.

8–20C Consider fully developed flow in a circular pipe with negligible entrance effects. If the length of the pipe is doubled, the head loss will (a) double, (b) more than double, (c) less than double, (d) reduce by half, or (e) remain constant.

8–21C Consider fully developed laminar flow in a circular pipe. If the diameter of the pipe is reduced by half while the flow rate and the pipe length are held constant, the head loss will (a) double, (b) triple, (c) quadruple, (d) increase by a factor of 8, or (e) increase by a factor of 16.

8–22C Explain why the friction factor is independent of the Reynolds number at very large Reynolds numbers.

8–23C What is turbulent viscosity? What causes it?

8–24C The head loss for a certain circular pipe is given by $h_L = 0.0826 fL(\dot{V}^2/D^5)$, where f is the friction factor (dimensionless), L is the pipe length, $\dot{V}$ is the volumetric flow rate, and D is the pipe diameter. Determine if the 0.0826 is a dimensional or dimensionless constant. Is this equation dimensionally homogeneous as it stands?

8–25C Consider fully developed laminar flow in a circular pipe. If the viscosity of the fluid is reduced by half by heating while the flow rate is held constant, how does the head loss change?

8–26C How is head loss related to pressure loss? For a given fluid, explain how you would convert head loss to pressure loss.

8–27C Consider laminar flow of air in a circular pipe with perfectly smooth surfaces. Do you think the friction factor for this flow is zero? Explain.

8–28C What is the physical mechanism that causes the friction factor to be higher in turbulent flow?

8–29 The velocity profile for the fully developed laminar flow of a Newtonian fluid between two large parallel plates is given by

$$u(y) = \frac{3u_0}{2}\left[1 - \left(\frac{y}{h}\right)^2\right]$$

where $2h$ is the distance between the two plates, u_0 is the velocity at the center plane, and y is the vertical coordinate from the center plane. For a plate width of b, obtain a relation for the flow rate through the plates.

8–30 Water flows steadily through a reducing pipe section. The flow upstream with a radius of R_1 is laminar with a velocity profile of $u_1(r) = u_{01}(1 - r^2/R_1^2)$ while the flow downstream is turbulent with a velocity profile of $u_2(r) = u_{02}(1 - r/R_2)^{1/7}$. For incompressible flow with $R_2/R_1 = 4/7$, determine the ratio of centerline velocities u_{01}/u_{02}.

8–31 Water at 10°C ($\rho = 999.7$ kg/m^3 and $\mu = 1.307 \times 10^{-3}$ kg/m·s) is flowing steadily in a 0.12-cm-diameter, 15-m-long pipe at an average velocity of 0.9 m/s. Determine (a) the pressure drop, (b) the head loss, and (c) the pumping power requirement to overcome this pressure drop. *Answers:* (a) 392 kPa, (b) 40.0 m, (c) 0.399 W

8–32 Consider an air solar collector that is 1 m wide and 5 m long and has a constant spacing of 3 cm between the glass cover and the collector plate. Air flows at an average temperature of 45°C at a rate of 0.15 m^3/s through the 1-m-wide edge of the collector along the 5-m-long passageway. Disregarding the entrance and roughness effects and the 90° bend, determine the pressure drop in the collector. *Answer:* 32.3 Pa

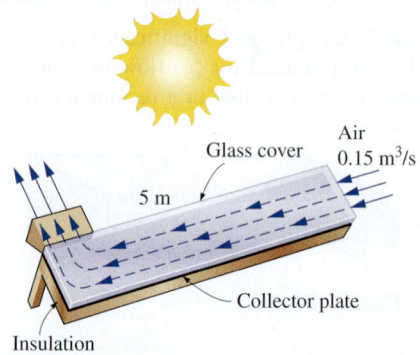

FIGURE P8–32

8–33 Heated air at 1 atm and 40°C is to be transported in a 120-m-long circular plastic duct at a rate of 0.35 m^3/s. If the head loss in the pipe is not to exceed 15 m, determine the minimum diameter of the duct.

8–34 In fully developed laminar flow in a circular pipe, the velocity at $R/2$ (midway between the wall surface and the centerline) is measured to be 11 m/s. Determine the velocity at the center of the pipe. *Answer:* 14.7 m/s

8–35 The velocity profile in fully developed laminar flow in a circular pipe of inner radius $R = 2$ cm, in m/s, is given by $u(r) = 4(1 - r^2/R^2)$. Determine the average and maximum velocities in the pipe and the volume flow rate.

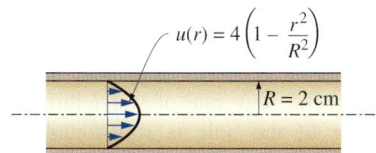

FIGURE P8–35

8–36 Repeat Prob. 8–35 for a pipe of inner radius 7 cm.

8–37 Water at 15°C ($\rho = 999.1$ kg/m³ and $\mu = 1.138 \times 10^{-3}$ kg/m·s) is flowing steadily in a 30-m-long and 5-cm-diameter horizontal pipe made of stainless steel at a rate of 9 L/s. Determine (*a*) the pressure drop, (*b*) the head loss, and (*c*) the pumping power requirement to overcome this pressure drop.

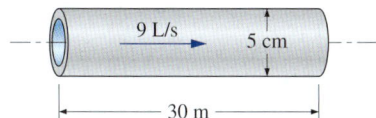

FIGURE P8–37

8–38 Consider the flow of oil with $\rho = 894$ kg/m³ and $\mu = 2.33$ kg/m·s in a 28-cm-diameter pipeline at an average velocity of 0.5 m/s. A 330-m-long section of the pipeline passes through the icy waters of a lake. Disregarding the entrance effects, determine the pumping power required to overcome the pressure losses and to maintain the flow of oil in the pipe.

8–39 Consider laminar flow of a fluid through a square channel with smooth surfaces. Now the average velocity of the fluid is doubled. Determine the change in the head loss of the fluid. Assume the flow regime remains unchanged.

8–40 Repeat Prob. 8–39 for turbulent flow in smooth pipes for which the friction factor is given as $f = 0.184\text{Re}^{-0.2}$. What would your answer be for fully turbulent flow in a rough pipe?

8–41 Air enters a 10-m-long section of a rectangular duct of cross section 15 cm × 20 cm made of commercial steel at 1 atm and 35°C at an average velocity of 7 m/s. Disregarding the entrance effects, determine the fan power needed to overcome the pressure losses in this section of the duct. *Answer:* 7.00 W

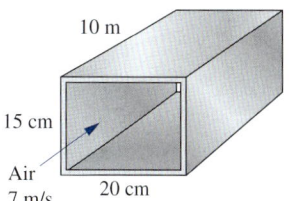

FIGURE P8–41

8–42 Water at 20°C passes through 2.0-cm-internal-diameter copper tubes at a rate of 0.23 kg/s. Determine the pumping power per m of pipe length required to maintain this flow at the specified rate.

8–43 Oil with $\rho = 876$ kg/m³ and $\mu = 0.24$ kg/m·s is flowing through a 1.5-cm-diameter pipe that discharges into the atmosphere at 88 kPa. The absolute pressure 15 m before the exit is measured to be 135 kPa. Determine the flow rate of oil through the pipe if the pipe is (*a*) horizontal, (*b*) inclined 8° upward from the horizontal, and (*c*) inclined 8° downward from the horizontal.

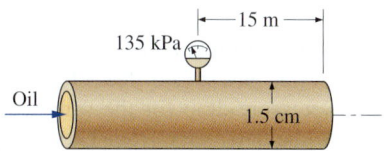

FIGURE P8–43

8–44 Glycerin at 40°C with $\rho = 1252$ kg/m³ and $\mu = 0.27$ kg/m·s is flowing through a 2-cm-diameter, 25-m-long pipe that discharges into the atmosphere at 100 kPa. The flow rate through the pipe is 0.048 L/s. (*a*) Determine the absolute pressure 25 m before the pipe exit. (*b*) At what angle θ must the pipe be inclined downward from the horizontal for the pressure in the entire pipe to be atmospheric pressure and the flow rate to be maintained the same?

8–45 Water enters into a cone of height H and base radius R through a small hole of cross-sectional area A_h and the discharge coefficient is C_d at the base with a constant uniform velocity of V. Obtain a relation for the variation of water height h from the cone base with time. Air escapes the cone through the tip at the top as water enters the cone from the bottom.

8–46 The velocity profile for incompressible turbulent flow in a pipe of radius R is given by $u(r) = u_{\text{max}}(1 - r/R_2)^{1/7}$. Obtain an expression for the average velocity in the pipe.

8–47 Oil with a density of 850 kg/m³ and kinematic viscosity of 0.00062 m²/s is being discharged by a 8-mm-diameter, 40-m-long horizontal pipe from a storage tank open to the atmosphere. The height of the liquid level above the center of the pipe is 4 m. Disregarding the minor losses, determine the flow rate of oil through the pipe.

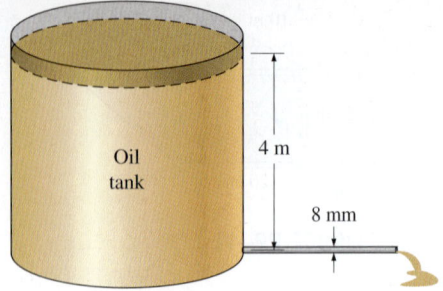

FIGURE P8–47

8–48 In an air heating system, heated air at 40°C and 105 kPa absolute is distributed through a 0.2 m × 0.3 m rectangular duct made of commercial steel at a rate of 0.5 m³/s. Determine the pressure drop and head loss through a 40-m-long section of the duct. *Answers:* 124 Pa, 10.8 m

8–49 Glycerin at 40°C with $\rho = 1252$ kg/m³ and $\mu = 0.27$ kg/m·s is flowing through a 4-cm-diameter horizontal smooth pipe with an average velocity of 3.5 m/s. Determine the pressure drop per 10 m of the pipe.

8–50 Reconsider Prob. 8–49. Using EES (or other) software, investigate the effect of the pipe diameter on the pressure drop for the same constant flow rate. Let the pipe diameter vary from 1 to 10 cm in increments of 1 cm. Tabulate and plot the results, and draw conclusions.

8–51 Liquid ammonia at −20°C is flowing through a 20-m-long section of a 5-mm-diameter copper tube at a rate of 0.09 kg/s. Determine the pressure drop, the head loss, and the pumping power required to overcome the frictional losses in the tube. *Answers:* 1240 kPa, 189 m, 0.167 kW

Minor Losses

8–52C During a retrofitting project of a fluid flow system to reduce the pumping power, it is proposed to install vanes into the miter elbows or to replace the sharp turns in 90° miter elbows by smooth curved bends. Which approach will result in a greater reduction in pumping power requirements?

8–53C Define equivalent length for minor loss in pipe flow. How is it related to the minor loss coefficient?

8–54C The effect of rounding of a pipe inlet on the loss coefficient is (a) negligible, (b) somewhat significant, or (c) very significant.

8–55C The effect of rounding of a pipe exit on the loss coefficient is (a) negligible, (b) somewhat significant, or (c) very significant.

8–56C Which has a greater minor loss coefficient during pipe flow: gradual expansion or gradual contraction? Why?

8–57C A piping system involves sharp turns, and thus large minor head losses. One way of reducing the head loss is to replace the sharp turns by circular elbows. What is another way?

8–58C What is minor loss in pipe flow? How is the minor loss coefficient K_L defined?

8–59 Water is to be withdrawn from an 8-m-high water reservoir by drilling a 2.2-cm-diameter hole at the bottom surface. Disregarding the effect of the kinetic energy correction factor, determine the flow rate of water through the hole if (a) the entrance of the hole is well-rounded and (b) the entrance is sharp-edged.

8–60 Consider flow from a water reservoir through a circular hole of diameter D at the side wall at a vertical distance H from the free surface. The flow rate through an actual hole with a sharp-edged entrance ($K_L = 0.5$) is considerably less than the flow rate calculated assuming "frictionless" flow and thus zero loss for the hole. Disregarding the effect of the kinetic energy correction factor, obtain a relation for the "equivalent diameter" of the sharp-edged hole for use in frictionless flow relations.

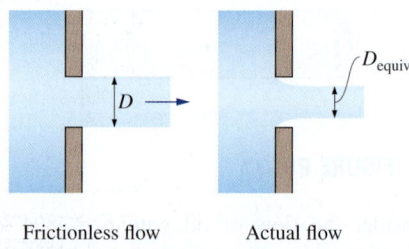

FIGURE P8–60

8–61 Repeat Prob. 8–60 for a slightly rounded entrance ($K_L = 0.12$).

8–62 A horizontal pipe has an abrupt expansion from $D_1 = 8$ cm to $D_2 = 16$ cm. The water velocity in the smaller section is 10 m/s and the flow is turbulent. The pressure in the smaller section is $P_1 = 410$ kPa. Taking the kinetic energy correction factor to be 1.06 at both the inlet and the

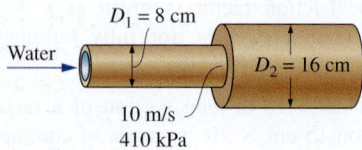

FIGURE P8–62

outlet, determine the downstream pressure P_2, and estimate the error that would have occurred if Bernoulli's equation had been used. *Answers: 432 kPa, 25.4 kPa*

Piping Systems and Pump Selection

8–63C Water is pumped from a large lower reservoir to a higher reservoir. Someone claims that if the head loss is negligible, the required pump head is equal to the elevation difference between the free surfaces of the two reservoirs. Do you agree?

8–64C A piping system equipped with a pump is operating steadily. Explain how the operating point (the flow rate and the head loss) is established.

8–65C A person filling a bucket with water using a garden hose suddenly remembers that attaching a nozzle to the hose increases the discharge velocity of water and wonders if this increased velocity would decrease the filling time of the bucket. What would happen to the filling time if a nozzle were attached to the hose: increase it, decrease it, or have no effect? Why?

8–66C Consider two identical 2-m-high open tanks filled with water on top of a 1-m-high table. The discharge valve of one of the tanks is connected to a hose whose other end is left open on the ground while the other tank does not have a hose connected to its discharge valve. Now the discharge valves of both tanks are opened. Disregarding any frictional loses in the hose, which tank do you think empties completely first? Why?

8–67C A piping system involves two pipes of different diameters (but of identical length, material, and roughness) connected in series. How would you compare the (*a*) flow rates and (*b*) pressure drops in these two pipes?

8–68C A piping system involves two pipes of different diameters (but of identical length, material, and roughness) connected in parallel. How would you compare the (*a*) flow rates and (*b*) pressure drops in these two pipes?

8–69C A piping system involves two pipes of identical diameters but of different lengths connected in parallel. How would you compare the pressure drops in these two pipes?

8–70C For a piping system, define the system curve, the characteristic curve, and the operating point on a head versus flow rate chart.

8–71 A 4-m-high cylindrical tank having a cross-sectional area of $A_T = 1.5$ m^2 is filled with equal volumes of water and oil whose specific gravity is SG = 0.75. Now a 1-cm-diameter hole at the bottom of the tank is opened, and water starts to flow out. If the discharge coefficient of the hole is $C_d = 0.85$, determine how long it will take for the water in the tank, which is open to the atmosphere to empty completely.

8–72 A semi-spherical tank of radius R is completely filled with water. Now a hole of cross sectional area A_h and discharge coefficient C_d at the bottom of the tank is fully opened and water starts to flow out. Develop an expression for the time needed to empty the tank completely.

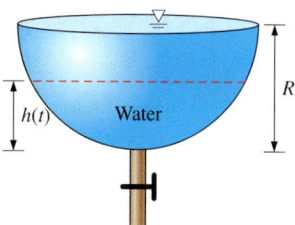

FIGURE P8–72

8–73 The water needs of a small farm are to be met by pumping water from a well that can supply water continuously at a rate of 4 L/s. The water level in the well is 20 m below the ground level, and water is to be pumped to a large tank on a hill, which is 58 m above the ground level of the well, using 5-cm internal diameter plastic pipes. The required length of piping is measured to be 420 m, and the total minor loss coefficient due to the use of elbows, vanes, etc. is estimated to be 12. Taking the efficiency of the pump to be 75 percent, determine the rated power of the pump that needs to be purchased, in kW. The density and viscosity of water at anticipated operation conditions are taken to be 1000 kg/m^3 and 0.00131 kg/m·s, respectively. Is it wise to purchase a suitable pump that meets the total power requirements, or is it necessary to also pay particular attention to the large elevation head in this case? Explain. *Answer: 6.0 kW*

8–74 Water at 20°C flows by gravity from a large reservoir at a high elevation to a smaller one through a 18-m-long, 5-cm-diameter cast iron piping system that includes four standard flanged elbows, a well-rounded entrance, a sharp-edged exit, and a fully open gate valve. Taking the free surface of the lower reservoir as the reference level, determine the elevation z_1 of the higher reservoir for a flow rate of 0.3 m^3/min. *Answer: 4.55 m*

8–75 A 2.4-m-diameter tank is initially filled with water 4 m above the center of a sharp-edged 10-cm-diameter orifice. The tank water surface is open to the atmosphere, and the orifice drains to the atmosphere. Neglecting the effect of the kinetic energy correction factor, calculate (*a*) the initial velocity from the tank and (*b*) the time required to empty the tank. Does the loss coefficient of the orifice cause a significant increase in the draining time of the tank?

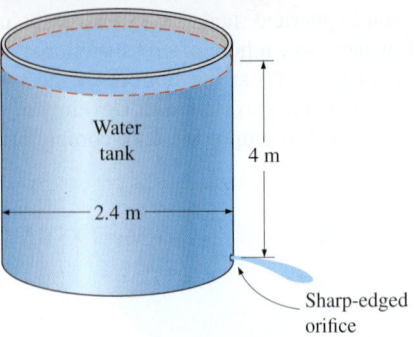

FIGURE P8–75

8–76 A 3-m-diameter tank is initially filled with water 2 m above the center of a sharp-edged 10-cm-diameter orifice. The tank water surface is open to the atmosphere, and the orifice drains to the atmosphere through a 100-m-long pipe. The friction coefficient of the pipe is taken to be 0.015 and the effect of the kinetic energy correction factor can be neglected. Determine (*a*) the initial velocity from the tank and (*b*) the time required to empty the tank.

8–77 Reconsider Prob. 8–76. In order to drain the tank faster, a pump is installed near the tank exit as in Fig. P8–81. Determine how much pump power input is necessary to establish an average water velocity of 4 m/s when the tank is full at $z = 2$ m. Also, assuming the discharge velocity to remain constant, estimate the time required to drain the tank.

Someone suggests that it makes no difference whether the pump is located at the beginning or at the end of the pipe, and that the performance will be the same in either case, but another person argues that placing the pump near the end of the pipe may cause cavitation. The water temperature is 30°C, so the water vapor pressure is $P_v = 4.246$ kPa $= 0.43$ m-H$_2$O, and the system is located at sea level. Investigate if there is the possibility of cavitation and if we should be concerned about the location of the pump.

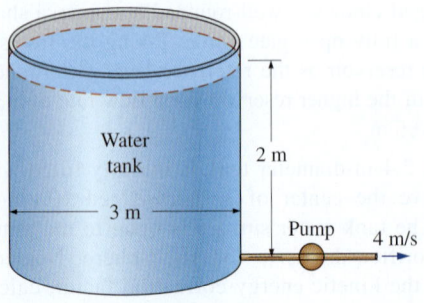

FIGURE P8–77

8–78 Water to a residential area is transported at a rate of 1.5 m³/s via 70-cm-internal-diameter concrete pipes with a surface roughness of 3 mm and a total length of 1500 m. In order to reduce pumping power requirements, it is proposed to line the interior surfaces of the concrete pipe with 2-cm-thick petroleum-based lining that has a surface roughness thickness of 0.04 mm. There is a concern that the reduction of pipe diameter to 66 cm and the increase in average velocity may offset any gains. Taking $\rho = 1000$ kg/m³ and $\nu = 1 \times 10^{-6}$ m²/s for water, determine the percent increase or decrease in the pumping power requirements due to pipe frictional losses as a result of lining the concrete pipes.

8–79 Oil at 20°C is flowing through a vertical glass funnel that consists of a 20-cm-high cylindrical reservoir and a 1-cm-diameter, 40-cm-high pipe. The funnel is always maintained full by the addition of oil from a tank. Assuming the entrance effects to be negligible, determine the flow rate of oil through the funnel and calculate the "funnel effectiveness," which is defined as the ratio of the actual flow rate through the funnel to the maximum flow rate for the "frictionless" case. *Answers:* 3.83×10^{-6} m³/s, 1.4 percent

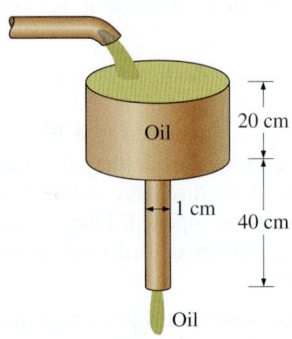

FIGURE P8–79

8–80 Repeat Prob. 8–79 assuming (*a*) the diameter of the pipe is tripled and (*b*) the length of the pipe is tripled while the diameter is maintained the same.

8–81 Water at 15°C is drained from a large reservoir using two horizontal plastic pipes connected in series. The first pipe is 20 m long and has a 10-cm diameter, while the second pipe is 35 m long and has a 4-cm diameter. The water level in the reservoir is 18 m above the centerline of the pipe. The pipe entrance is sharp-edged, and the contraction between the two pipes is sudden. Neglecting the effect of the kinetic energy correction factor, determine the discharge rate of water from the reservoir.

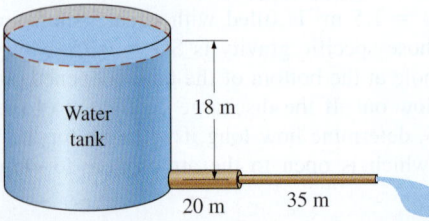

FIGURE P8–81

8–82 A farmer is to pump water at 20°C from a river to a water storage tank nearby using a 40-m-long, 12-cm-diameter plastic pipe with three flanged 90° smooth bends. The water velocity near the river surface is 1.8 m/s and the pipe inlet is placed in the river normal to the flow direction of water to take advantage of the dynamic pressure. The elevation difference between the river and the free surface of the tank is 3.5 m. For a flow rate of 0.042 m³/s and an overall pump efficiency of 70 percent, determine the required electric power input to the pump.

8–83 Reconsider Prob. 8–82. Using EES (or other) software, investigate the effect of the pipe diameter on the required electric power input to the pump. Let the pipe diameter vary from 2 to 20 cm, in increments of 2 cm. Tabulate and plot the results, and draw conclusions.

8–84 A water tank filled with solar-heated water at 40°C is to be used for showers in a field using gravity-driven flow. The system includes 35 m of 1.5-cm-diameter galvanized iron piping with four miter bends (90°) without vanes and a wide-open globe valve. If water is to flow at a rate of 1.2 L/s through the shower head, determine how high the water level in the tank must be from the exit level of the shower. Disregard the losses at the entrance and at the shower head, and neglect the effect of the kinetic energy correction factor.

8–85 Two water reservoirs A and B are connected to each other through a 40-m-long, 2-cm-diameter cast iron pipe with a sharp-edged entrance. The pipe also involves a swing check valve and a fully open gate valve. The water level in both reservoirs is the same, but reservoir A is pressurized by compressed air while reservoir B is open to the atmosphere at 88 kPa. If the initial flow rate through the pipe is 1.2 L/s, determine the absolute air pressure on top of reservoir A. Take the water temperature to be 10°C. *Answer:* 733 kPa

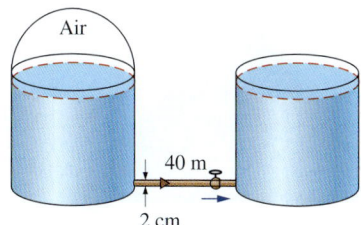

FIGURE P8–85

8–86 A vented tanker is to be filled with fuel oil with $\rho = 920$ kg/m³ and $\mu = 0.045$ kg/m·s from an underground reservoir using a 25-m-long, 4-cm-diameter plastic hose with a slightly rounded entrance and two 90° smooth bends. The elevation difference between the oil level in the reservoir and the top of the tanker where the hose is discharged is 5 m. The capacity of the tanker is 18 m³ and the filling time is 30 min. Taking the kinetic energy correction factor at the hose discharge to be 1.05 and assuming an overall pump efficiency of 82 percent, determine the required power input to the pump.

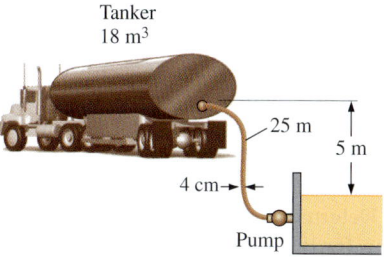

FIGURE P8–86

8–87 Two pipes of identical length and material are connected in parallel. The diameter of pipe A is twice the diameter of pipe B. Assuming the friction factor to be the same in both cases and disregarding minor losses, determine the ratio of the flow rates in the two pipes.

8–88 A certain part of cast iron piping of a water distribution system involves a parallel section. Both parallel pipes have a diameter of 30 cm, and the flow is fully turbulent. One of the branches (pipe A) is 1500 m long while the other branch (pipe B) is 2500 m long. If the flow rate through pipe A is 0.4 m³/s, determine the flow rate through pipe B. Disregard minor losses and assume the water temperature to be 15°C. Show that the flow is fully rough, and thus the friction factor is independent of Reynolds number. *Answer:* 0.310 m³/s

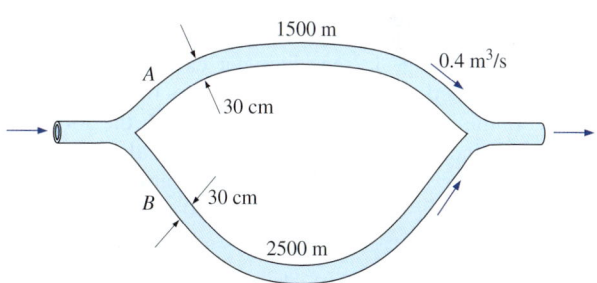

FIGURE P8–88

8–89 Repeat Prob. 8–88 assuming pipe A has a halfway-closed gate valve ($K_L = 2.1$) while pipe B has a fully open globe valve ($K_L = 10$), and the other minor losses are negligible.

8–90 A geothermal district heating system involves the transport of geothermal water at 110°C from a geothermal well to a city at about the same elevation for a distance of 12 km at a rate of 1.5 m³/s in 60-cm-diameter stainless-steel pipes. The fluid pressures at the wellhead and the arrival point in the city are to be the same.

The minor losses are negligible because of the large length-to-diameter ratio and the relatively small number of components that cause minor losses. (*a*) Assuming the pump–motor efficiency to be 80 percent, determine the electric power consumption of the system for pumping. Would you recommend the use of a single large pump or several smaller pumps of the same total pumping power scattered along the pipeline? Explain. (*b*) Determine the daily cost of power consumption of the system if the unit cost of electricity is $0.06/kWh. (*c*) The temperature of geothermal water is estimated to drop 0.5°C during this long flow. Determine if the frictional heating during flow can make up for this drop in temperature.

8–91 Repeat Prob. 8–90 for cast iron pipes of the same diameter.

8–92 Water is transported by gravity through a 12-cm-diameter 800-m-long plastic pipe with an elevation gradient of 0.01 (i.e., an elevation drop of 1 m per 100 m of pipe length). Taking $\rho = 1000$ kg/m³ and $\nu = 1 \times 10^{-6}$ m²/s for water, determine the flow rate of water through the pipe. If the pipe were horizontal, what would the power requirements be to maintain the same flow rate?

8–93 Gasoline ($\rho = 680$ kg/m³ and $\nu = 4.29 \times 10^{-7}$ m²/s) is transported at a rate of 240 L/s for a distance of 2 km. The surface roughness of the piping is 0.03 mm. If the head loss due to pipe friction is not to exceed 10 m, determine the minimum diameter of the pipe.

8–94 In large buildings, hot water in a water tank is circulated through a loop so that the user doesn't have to wait for all the water in long piping to drain before hot water starts coming out. A certain recirculating loop involves 40-m-long, 1.2-cm-diameter cast iron pipes with six 90° threaded smooth bends and two fully open gate valves. If the average flow velocity through the loop is 2 m/s, determine the required power input for the recirculating pump. Take the average water temperature to be 60°C and the efficiency of the pump to be 70 percent. *Answer:* 0.111 kW

8–95 Reconsider Prob. 8–94. Using EES (or other) software, investigate the effect of the average flow velocity on the power input to the recirculating pump. Let the velocity vary from 0 to 3 m/s in increments of 0.3 m/s. Tabulate and plot the results.

8–96 Repeat Prob. 8–94 for plastic (smooth) pipes.

8–97 Water at 20°C is to be pumped from a reservoir ($z_A = 2$ m) to another reservoir at a higher elevation ($z_B = 9$ m) through two 25-m-long plastic pipes connected in parallel. The diameters of the two pipes are 3 cm and 5 cm. Water is to be pumped by a 68 percent efficient motor–pump unit that draws 7 kW of electric power during operation. The minor losses and the head loss in the pipes that connect the parallel pipes to the two reservoirs are considered to be negligible. Determine the total flow rate between the reservoirs and the flow rates through each of the parallel pipes.

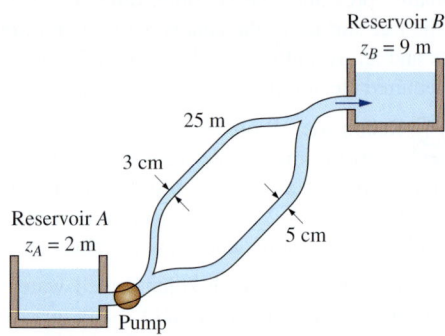

FIGURE P8–97

8–98 A 6-m-tall chimney shown in Fig. P8–98 is to be designed to discharge hot gases from a fireplace at 180°C at a constant rate of 0.15 m³/s when the atmospheric air temperature is 20°C. Assuming no heat transfer from the chimney and taking the chimney entrance loss coefficient to be 1.5 and the friction coefficient of the chimney to be 0.020, determine the chimney diameter that would discharge the hot gases at the desired rate. Note that $P_3 = P_4 = P_{atm}$ and $P_2 = P_1 = P_{atm} + \rho_{atm\ air}\ gh$, and assume the hot gases in the entire chimney are at 180°C.

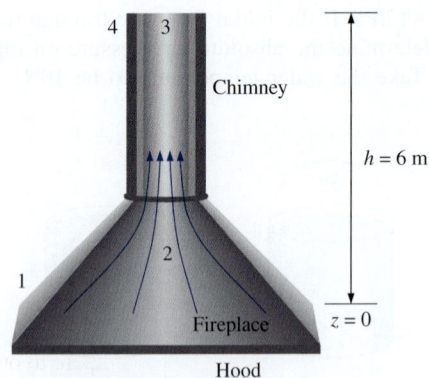

FIGURE P8–98

8–99 An inverted 3-m-high conical container shown in Fig. P8–99 is initially filled with 2-m-high water. At time $t = 0$, a faucet is opened to supply water into the container at a rate of 3 L/s. At the same time, a 4-cm-diameter hole with a discharge coefficient of 0.90 at the bottom of the container is opened. Determine how long it will take for the water level in the tank to drop to 1-m.

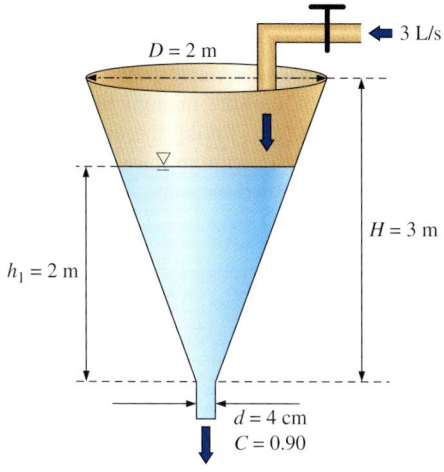

FIGURE P8–99

Flow Rate and Velocity Measurements

8–100C What is the difference between the operating principles of thermal and laser Doppler anemometers?

8–101C What is the difference between laser Doppler velocimetry (LDV) and particle image velocimetry (PIV)?

8–102C What are the primary considerations when selecting a flowmeter to measure the flow rate of a fluid?

8–103C Explain how flow rate is measured with a Pitot-static tube, and discuss its advantages and disadvantages with respect to cost, pressure drop, reliability, and accuracy.

8–104C Explain how flow rate is measured with obstruction-type flowmeters. Compare orifice meters, flow nozzles, and Venturi meters with respect to cost, size, head loss, and accuracy.

8–105C How do positive displacement flowmeters operate? Why are they commonly used to meter gasoline, water, and natural gas?

8–106C Explain how flow rate is measured with a turbine flowmeter, and discuss how they compare to other types of flowmeters with respect to cost, head loss, and accuracy.

8–107C What is the operating principle of variable-area flowmeters (rotameters)? How do they compare to other types of flowmeters with respect to cost, head loss, and reliability?

8–108 The flow rate of water at 20°C ($\rho = 998$ kg/m³ and $\mu = 1.002 \times 10^{-3}$ kg/m·s) through a 60-cm-diameter pipe is measured with an orifice meter with a 30-cm-diameter opening to be 400 L/s. Determine the pressure difference indicated by the orifice meter and the head loss.

8–109 A Pitot-static probe is mounted in a 2.5-cm-inner diameter pipe at a location where the local velocity is approximately equal to the average velocity. The oil in the pipe has density $\rho = 860$ kg/m³ and viscosity $\mu = 0.0103$ kg/m·s. The pressure difference is measured to be 95.8 Pa. Calculate the volume flow rate through the pipe in cubic meters per second.

8–110 Calculate the Reynolds number of the flow of Prob. 8–109. Is it laminar or turbulent?

8–111 A flow nozzle equipped with a differential pressure gage is used to measure the flow rate of water at 10°C ($\rho = 999.7$ kg/m³ and $\mu = 1.307 \times 10^{-3}$ kg/m·s) through a 3-cm-diameter horizontal pipe. The nozzle exit diameter is 1.5 cm, and the measured pressure drop is 3 kPa. Determine the volume flow rate of water, the average velocity through the pipe, and the head loss.

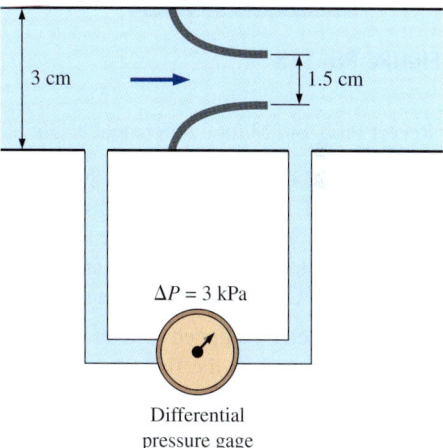

FIGURE P8–111

8–112 The flow rate of water through a 10-cm-diameter pipe is to be determined by measuring the water velocity at several locations along a cross section. For the set of measurements given in the table, determine the flow rate.

r, cm	V, m/s
0	6.4
1	6.1
2	5.2
3	4.4
4	2.0
5	0.0

8–113 An orifice with a 4.6-cm-diameter opening is used to measure the mass flow rate of water at 15°C ($\rho = 999.1$ kg/m³ and $\mu = 1.138 \times 10^{-3}$ kg/m·s) through a horizontal 10-cm-diameter pipe. A mercury manometer is used to measure the pressure difference across the orifice. If the differential height of the manometer is 18 cm, determine the volume flow rate of water through the pipe, the average velocity, and the head loss caused by the orifice meter.

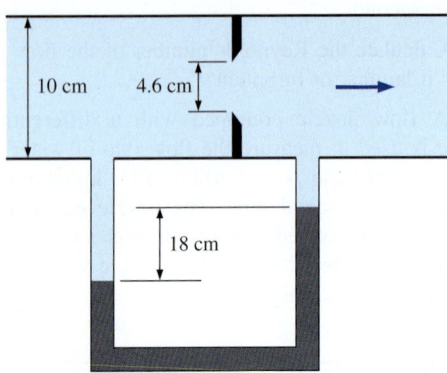

FIGURE P8–113

8–114 Repeat Prob. 8–113 for a differential height of 25 cm.

8–115 Air ($\rho = 1.225$ kg/m^3 and $\mu = 1.789 \times 10^{-5}$ kg/m·s) flows in a wind tunnel, and the wind tunnel speed is measured with a Pitot-static probe. For a certain run, the stagnation pressure is measured to be 472.6 Pa gage and the static pressure is 15.43 Pa gage. Calculate the wind-tunnel speed.

8–116 A Venturi meter equipped with a differential pressure gage is used to measure the flow rate of water at 15°C ($\rho = 999.1$ kg/m^3) through a 5-cm-diameter horizontal pipe. The diameter of the Venturi neck is 3 cm, and the measured pressure drop is 5 kPa. Taking the discharge coefficient to be 0.98, determine the volume flow rate of water and the average velocity through the pipe. *Answers:* 2.35 L/s and 1.20 m/s

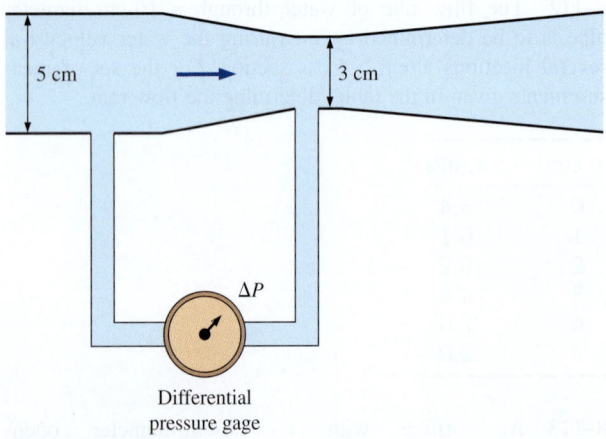

FIGURE P8–116

8–117 Reconsider Prob. 8–116. Letting the pressure drop vary from 1 kPa to 10 kPa, evaluate the flow rate at intervals of 1 kPa, and plot it against the pressure drop.

8–118 The mass flow rate of air at 20°C ($\rho = 1.204$ kg/m^3) through a 18-cm-diameter duct is measured with a Venturi meter equipped with a water manometer. The Venturi neck has a diameter of 5 cm, and the manometer has a maximum differential height of 40 cm. Taking the discharge coefficient to be 0.98, determine the maximum mass flow rate of air this Venturi meter/manometer can measure. *Answer:* 0.188 kg/s

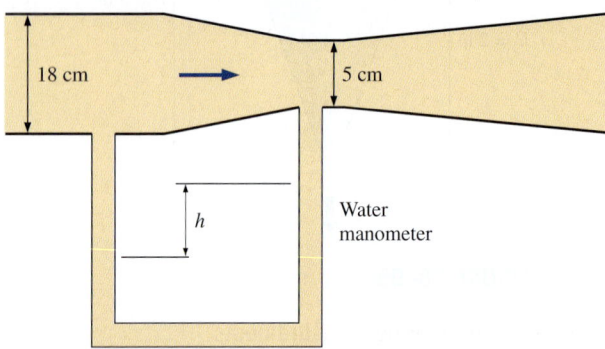

FIGURE P8–118

8–119 Repeat Prob. 8–118 for a Venturi neck diameter of 6 cm.

8–120 A vertical Venturi meter equipped with a differential pressure gage shown in Fig. P8–120 is used to measure the flow rate of liquid propane at 10°C ($\rho = 514.7$ kg/m^3) through an 10-cm-diameter vertical pipe. For a discharge coefficient of 0.98, determine the volume flow rate of propane through the pipe.

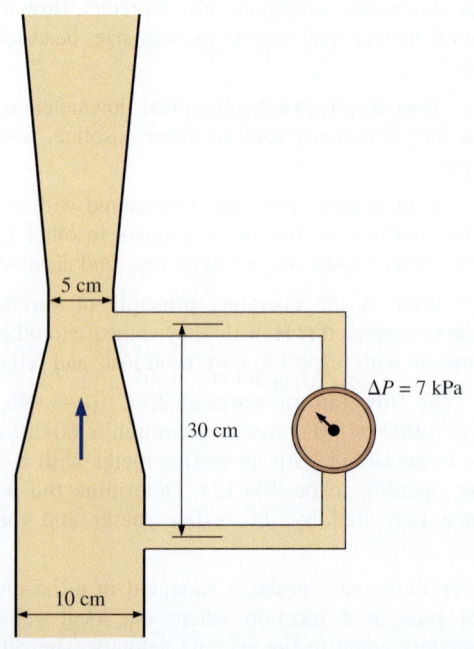

FIGURE P8–120

8–121 The volume flow rate of liquid refrigerant-134a at −10°C (ρ = 1327 kg/m³) is to be measured with a horizontal Venturi meter with a diameter of 12 cm at the inlet and 5 cm at the throat. If a differential pressure meter indicates a pressure drop of 44 kPa, determine the flow rate of the refrigerant. Take the discharge coefficient of the Venturi meter to be 0.98.

8–122 A 22-L kerosene tank (ρ = 820 kg/m³) is filled with a 2-cm-diameter hose equipped with a 1.5-cm-diameter nozzle meter. If it takes 20 s to fill the tank, determine the pressure difference indicated by the nozzle meter.

8–123 The flow rate of water at 20°C (ρ = 998 kg/m³ and μ = 1.002 × 10⁻³ kg/m·s) through a 4-cm-diameter pipe is measured with a 2-cm-diameter nozzle meter equipped with an inverted air–water manometer. If the manometer indicates a differential water height of 44 cm, determine the volume flow rate of water and the head loss caused by the nozzle meter.

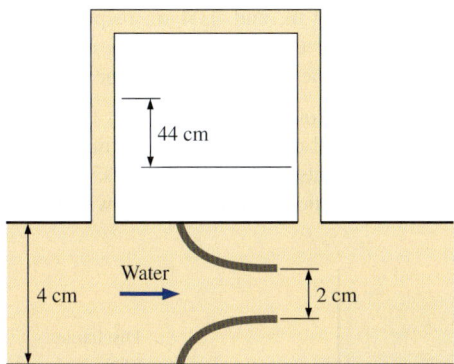

FIGURE P8–123

8–124 The flow rate of ammonia at 10°C (ρ = 624.6 kg/m³ and μ = 1.697 × 10⁻⁴ kg/m·s) through a 2-cm-diameter pipe is to be measured with a 1.5-cm-diameter flow nozzle equipped with a differential pressure gage. If the gage reads a pressure differential of 4 kPa, determine the flow rate of ammonia through the pipe, and the average flow velocity.

Review Problems

8–125 In a laminar flow through a circular tube of radius of R, the velocity and temperature profiles at a cross-section are given by $u = u_0(1 - r^2/R^2)$ and $T(r) = A + Br^2 - Cr^4$ where A, B and C are positive constants. Obtain a relation for the bulk fluid temperature at that cross section.

8–126 The conical container with a thin horizontal tube attached at the bottom, shown in Fig. P8–126, is to be used to measure the viscosity of an oil. The flow through the tube is laminar. The discharge time needed for the oil level to drop from h_1 to h_2 is to be measured by a stopwatch. Develop an expression for the viscosity of oil in the container as a function of the discharge time t.

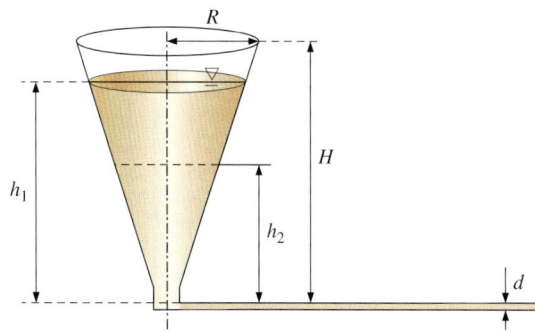

FIGURE P8–126

8–127 Shell-and-tube heat exchangers with hundreds of tubes housed in a shell are commonly used in practice for heat transfer between two fluids. Such a heat exchanger used in an active solar hot-water system transfers heat from a water-antifreeze solution flowing through the shell and the solar collector to fresh water flowing through the tubes at an average temperature of 60°C at a rate of 15 L/s. The heat exchanger contains 80 brass tubes 1 cm in inner diameter and 1.5 m in length. Disregarding inlet, exit, and header losses, determine the pressure drop across a single tube and the pumping power required by the tube-side fluid of the heat exchanger.

After operating a long time, 1-mm-thick scale builds up on the inner surfaces with an equivalent roughness of 0.4 mm. For the same pumping power input, determine the percent reduction in the flow rate of water through the tubes.

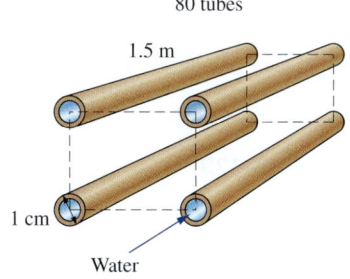

FIGURE P8–127

8–128 The compressed air requirements of a manufacturing facility are met by a 120-hp compressor that draws in air from the outside through an 9-m-long, 22-cm-diameter duct made of thin galvanized iron sheets. The compressor takes in air at a rate of 0.27 m³/s at the outdoor conditions of 15°C and 95 kPa. Disregarding any minor losses, determine the useful power used by the compressor to overcome the frictional losses in this duct. *Answer:* 6.74 W

430
INTERNAL FLOW

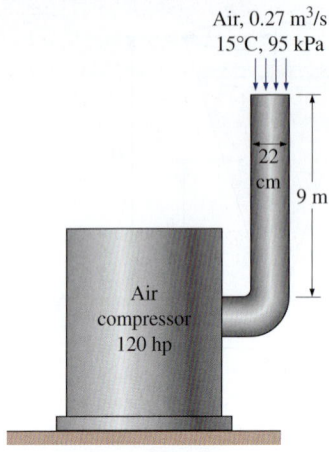

FIGURE P8–128

8–129 A house built on a riverside is to be cooled in summer by utilizing the cool water of the river. A 15-m-long section of a circular stainless-steel duct of 20-cm diameter passes through the water. Air flows through the underwater section of the duct at 3 m/s at an average temperature of 15°C. For an overall fan efficiency of 62 percent, determine the fan power needed to overcome the flow resistance in this section of the duct.

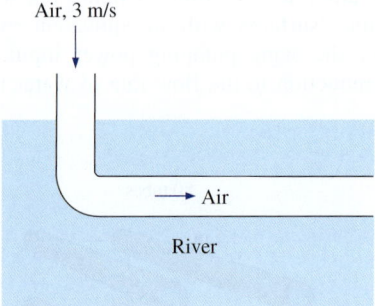

FIGURE P8–129

8–130 The velocity profile in fully developed laminar flow in a circular pipe, in m/s, is given by $u(r) = 6(1 - 100r^2)$, where r is the radial distance from the centerline of the pipe in m. Determine (a) the radius of the pipe, (b) the average velocity through the pipe, and (c) the maximum velocity in the pipe.

8–131 The velocity profile in a fully developed laminar flow of water at 5°C in a 75-m-long horizontal circular pipe, in m/s, is given by $u(r) = 0.24(1 - 6945r^2)$, where r is the radial distance from the centerline of the pipe in m. Determine (a) the volume flow rate of water through the pipe, (b) the pressure drop across the pipe, and (c) the useful pumping power required to overcome this pressure drop.

8–132 Repeat Prob. 8–131 assuming the pipe is inclined 12° from the horizontal and the flow is uphill.

8–133 Oil at 20°C is flowing steadily through a 5-cm-diameter 40-m-long pipe. The pressures at the pipe inlet and outlet are measured to be 745 and 97.0 kPa, respectively, and the flow is expected to be laminar. Determine the flow rate of oil through the pipe, assuming fully developed flow and that the pipe is (a) horizontal, (b) inclined 15° upward, and (c) inclined 15° downward. Also, verify that the flow through the pipe is laminar.

8–134 Consider flow from a reservoir through a horizontal pipe of length L and diameter D that penetrates into the side wall at a vertical distance H from the free surface. The flow rate through an actual pipe with a reentrant section ($K_L = 0.8$) is considerably less than the flow rate through the hole calculated assuming "frictionless" flow and thus zero loss. Obtain a relation for the "equivalent diameter" of the reentrant pipe for use in relations for frictionless flow through a hole and determine its value for a pipe friction factor, length, and diameter of 0.018, 10 m, and 0.04 m, respectively. Assume the friction factor of the pipe to remain constant and the effect of the kinetic energy correction factor to be negligible.

8–135 A highly viscous liquid discharges from a large container through a small-diameter tube in laminar flow. Disregarding entrance effects and velocity heads, obtain a relation for the variation of fluid depth in the tank with time.

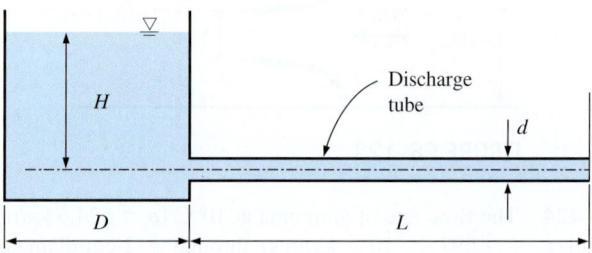

FIGURE P8–135

8–136 A student is to determine the kinematic viscosity of an oil using the system shown in Prob. 8–135. The initial fluid height in the tank is $H = 40$ cm, the tube diameter is $d = 6$ mm, the tube length is $L = 0.65$ m, and the tank diameter is $D = 0.63$ m. The student observes that it takes 1400 s for the fluid level in the tank to drop to 34 cm. Find the fluid viscosity.

8–137 A circular water pipe has an abrupt expansion from diameter $D_1 = 8$ cm to $D_2 = 24$ cm. The pressure and the average water velocity in the smaller pipe are $P_1 = 135$ kPa and 10 m/s, respectively, and the flow is turbulent. By applying the continuity, momentum, and energy equations and disregarding the effects of the kinetic energy and momentum-flux correction factors, show that the loss coefficient for sudden expansion is $K_L = (1 - D_1^2/D_2^2)^2$, and calculate K_L and P_2 for the given case.

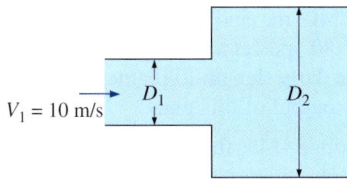

FIGURE P8–137

8–138 In a geothermal district heating system, 10,000 kg/s of hot water must be delivered a distance of 10 km in a horizontal pipe. The minor losses are negligible, and the only significant energy loss arises from pipe friction. The friction factor is taken to be 0.015. Specifying a larger-diameter pipe would reduce water velocity, velocity head, pipe friction, and thus power consumption. But a larger pipe would also cost more money initially to purchase and install. Otherwise stated, there is an optimum pipe diameter that will minimize the sum of pipe cost and future electric power cost.

Assume the system will run 24 h/day, every day, for 30 years. During this time the cost of electricity remains constant at $0.06/kWh. Assume system performance stays constant over the decades (this may not be true, especially if highly mineralized water is passed through the pipeline—scale may form). The pump has an overall efficiency of 80 percent. The cost to purchase, install, and insulate a 10-km pipe depends on the diameter D and is given by $Cost = \$10^6 \, D^2$, where D is in m. Assuming zero inflation and interest rate for simplicity and zero salvage value and zero maintenance cost, determine the optimum pipe diameter.

8–139 Water at 15°C is to be discharged from a reservoir at a rate of 18 L/s using two horizontal cast iron pipes connected in series and a pump between them. The first pipe is 20 m long and has a 6-cm diameter, while the second pipe is 35 m long and has a 4-cm diameter. The water level in the reservoir is 30 m above the centerline of the pipe. The pipe entrance is sharp-edged, and losses associated with the connection of the pump are negligible. Neglecting the effect of the kinetic energy correction factor, determine the required pumping head and the minimum pumping power to maintain the indicated flow rate.

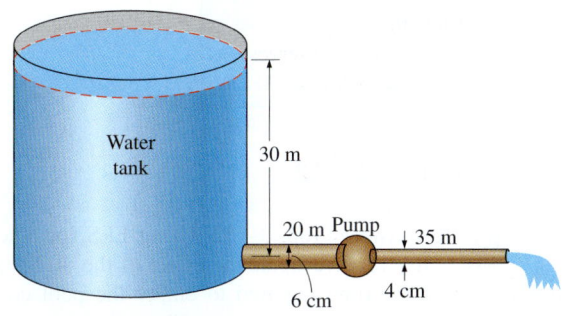

FIGURE P8–139

8–140 Reconsider Prob. 8–139. Using EES (or other) software, investigate the effect of the second pipe diameter on the required pumping head to maintain the indicated flow rate. Let the diameter vary from 1 to 10 cm in increments of 1 cm. Tabulate and plot the results.

8–141 Two pipes of identical diameter and material are connected in parallel. The length of pipe A is five times the length of pipe B. Assuming the flow is fully turbulent in both pipes and thus the friction factor is independent of the Reynolds number and disregarding minor losses, determine the ratio of the flow rates in the two pipes. *Answer:* 0.447

8–142 A pipeline that transports oil at 40°C at a rate of 3 m³/s branches out into two parallel pipes made of commercial steel that reconnect downstream. Pipe A is 500 m long and has a diameter of 30 cm while pipe B is 800 m long and has a diameter of 45 cm. The minor losses are considered to be negligible. Determine the flow rate through each of the parallel pipes.

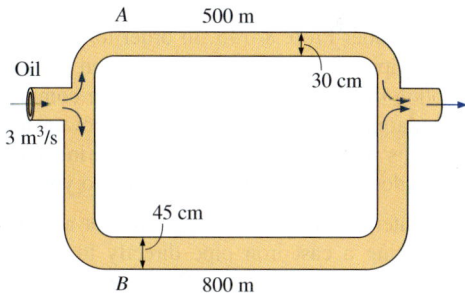

FIGURE P8–142

8–143 Repeat Prob. 8–142 for hot-water flow of a district heating system at 100°C.

8–144 A system that consists of two interconnected cylindrical tanks with $D_1 = 30$ cm and $D_2 = 12$ cm is to be used

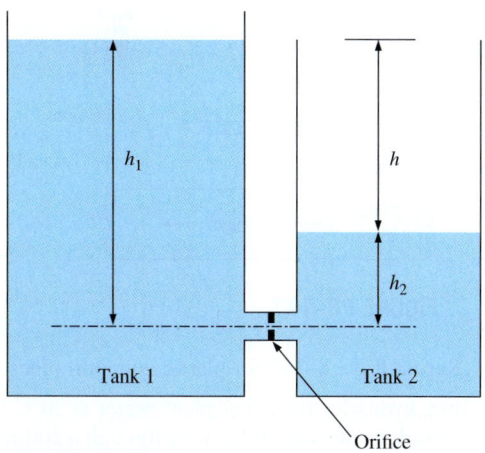

FIGURE P8–144

to determine the discharge coefficient of a short $D_0 = 5$ mm diameter orifice. At the beginning ($t = 0$ s), the fluid heights in the tanks are $h_1 = 50$ cm and $h_2 = 15$ cm, as shown in Fig. P8–144. If it takes 170 s for the fluid levels in the two tanks to equalize and the flow to stop, determine the discharge coefficient of the orifice. Disregard any other losses associated with this flow.

8–145 The compressed air requirements of a textile factory are met by a large compressor that draws in 0.6 m³/s air at atmospheric conditions of 20°C and 1 bar (100 kPa) and consumes 300 kW electric power when operating. Air is compressed to a gage pressure of 8 bar (absolute pressure of 900 kPa), and compressed air is transported to the production area through a 15-cm-internal-diameter, 83-m-long, galvanized steel pipe with a surface roughness of 0.15 mm. The average temperature of compressed air in the pipe is 60°C. The compressed air line has 8 elbows with a loss coefficient of 0.6 each. If the compressor efficiency is 85 percent, determine the pressure drop and the power wasted in the transportation line. *Answers:* 1.40 kPa, 0.125 kW

8–146 Reconsider Prob. 8–145. In order to reduce the head losses in the piping and thus the power wasted, someone suggests doubling the diameter of the 83-m-long compressed air pipes. Calculating the reduction in wasted power, and determine if this is a worthwhile idea. Considering the cost of replacement, does this proposal make sense to you?

8–147 A water fountain is to be installed at a remote location by attaching a cast iron pipe directly to a water main through which water is flowing at 20°C and 400 kPa (gage). The entrance to the pipe is sharp-edged, and the 15-m-long piping system involves three 90° miter bends without vanes, a fully open gate valve, and an angle valve with a loss coefficient of 5 when fully open. If the system is to provide water at a rate of 75 L/min and the elevation difference between the pipe and the fountain is negligible, determine the minimum diameter of the piping system. *Answer:* 1.92 cm

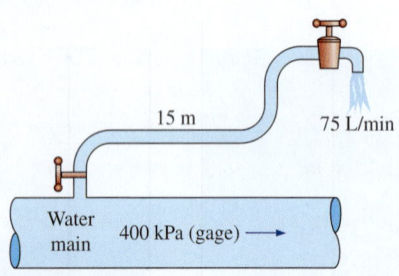

FIGURE P8–147

8–148 Repeat Prob. 8–147 for plastic (smooth) pipes.

8–149 In a hydroelectric power plant, water at 20°C is supplied to the turbine at a rate of 0.6 m³/s through a 200-m-long, 0.35-m-diameter cast iron pipe. The elevation difference between the free surface of the reservoir and the turbine discharge is 140 m, and the combined turbine–generator efficiency is 80 percent. Disregarding the minor losses because of the large length-to-diameter ratio, determine the electric power output of this plant.

8–150 In Prob. 8–149, the pipe diameter is tripled in order to reduce the pipe losses. Determine the percent increase in the net power output as a result of this modification.

8–151 Water is to be withdrawn from a 7-m-high water reservoir by drilling a well-rounded 4-cm-diameter hole with negligible loss near the bottom and attaching a horizontal 90° bend of negligible length. Taking the kinetic energy correction factor to be 1.05, determine the flow rate of water through the bend if (*a*) the bend is a flanged smooth bend and (*b*) the bend is a miter bend without vanes. *Answers:* (*a*) 12.7 L/s, (*b*) 10.0 L/s

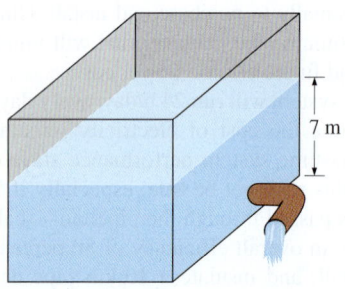

FIGURE P8–151

8–152 The water at 20°C in a 10-m-diameter, 2-m-high aboveground swimming pool is to be emptied by unplugging a 5-cm-diameter, 25-m-long horizontal plastic pipe attached to the bottom of the pool. Determine the initial rate of discharge of water through the pipe and the time (hours) it would take to empty the swimming pool completely assuming the entrance to the pipe is well-rounded with negligible loss. Take the friction factor of the pipe to be 0.022. Using the initial discharge velocity, check if this is a reasonable value for the friction factor. *Answers:* 3.55 L/s, 24.6 h

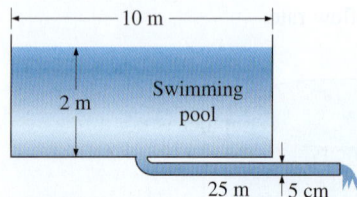

FIGURE P8–152

8–153 Reconsider Prob. 8–152. Using EES (or other) software, investigate the effect of the discharge pipe diameter on the time required to empty the pool completely. Let the diameter vary from 1 to 10 cm, in increments of 1 cm. Tabulate and plot the results.

8–154 Repeat Prob. 8–152 for a sharp-edged entrance to the pipe with $K_L = 0.5$. Is this "minor loss" truly "minor" or not?

8–155 An elderly woman is rushed to the hospital because she is having a heart attack. The emergency room doctor informs her that she needs immediate coronary artery (a vessel that wraps around the heart) bypass surgery because one coronary artery has 75 percent blockage (caused by atherosclerotic plaque). This surgery involves using an artificial graft (typically made of Dacron) to divert blood from the coronary artery around the blockage and reattach to the coronary artery beyond the blockage site as illustrated in Figure P8-155. The coronary artery diameter is 5.0 mm and its length is 15.0 mm. The bypass graft diameter is 4.0 mm and its length is 20.0 mm. The flow rate within the bypass graft is 0.45 liters per minute (recall 1 ml equals 1 cm^3). Blood has a density of 1060 kg/m^3 and a dynamic viscosity of 3.5 centipoise. Assume that the Dacron and coronary artery have the same material properties and ignore any minor losses. Assume the friction factor is the same in both tubes. Ignoring the plaque in determining the head loss for the coronary artery, calculate the velocity through the small gap between the plaque and the coronary artery.

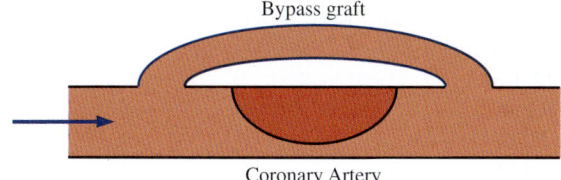

FIGURE P8–155

Fundamentals of Engineering (FE) Exam Problems

8–156 The average velocity for fully developed laminar pipe flow is
(a) $V_{max}/2$
(b) $V_{max}/3$
(c) V_{max}
(d) $2V_{max}/3$
(e) $3V_{max}/4$

8–157 The Reynolds number is not a function of
(a) Fluid velocity
(b) Fluid density
(c) Characteristic length
(d) Surface roughness
(e) Fluid viscosity

8–158 Air flows in a 5 cm by 8 cm cross section rectangular duct at a velocity of 4 m/s at 1 atm and 15°C. The Reynolds number for this flow is
(a) 13,605
(b) 16,745
(c) 17,690
(d) 21,770
(e) 23,235

8–159 Air at 1 atm and 20°C flows in a 4-cm-diameter tube. The maximum velocity of air to keep the flow laminar is
(a) 0.872 m/s
(b) 1.52 m/s
(c) 2.14 m/s
(d) 3.11 m/s
(e) 3.79 m/s

8–160 Consider laminar flow of water in a 0.8-cm-diameter pipe at a rate of 1.15 L/min. The velocity of water halfway between the surface and the center of the pipe is
(a) 0.381 m/s
(b) 0.762 m/s
(c) 1.15 m/s
(d) 0.874 m/s
(e) 0.572 m/s

8–161 Consider laminar flow of water at 15°C in a 0.7-cm-diameter pipe at a velocity of 0.4 m/s. The pressure drop of water for a pipe length of 50 m is
(a) 6.8 kPa
(b) 8.7 kPa
(c) 11.5 kPa
(d) 14.9 kPa
(e) 17.3 kPa

8–162 Engine oil at 40°C ($\rho = 876$ kg/m^3, $\mu = 0.2177$ kg/m·s) flows in a 20-cm-diameter pipe at a velocity of 1.2 m/s. The pressure drop of oil for a pipe length of 20 m is
(a) 4180 Pa
(b) 5044 Pa
(c) 6236 Pa
(d) 7419 Pa
(e) 8615 Pa

8–163 A fluid flows in a 25-cm-diameter pipe at a velocity of 4.5 m/s. If the pressure drop along the pipe is estimated to be 6400 Pa, the required pumping power to overcome this pressure drop is
(a) 452 W
(b) 640 W
(c) 923 W
(d) 1235 W
(e) 1508 W

8–164 Water flows in a 15-cm-diameter pipe at a velocity of 1.8 m/s. If the head loss along the pipe is estimated to be 16 m, the required pumping power to overcome this head loss is
(a) 3.22 kW
(b) 3.77 kW
(c) 4.45 kW
(d) 4.99 kW
(e) 5.54 kW

8–165 The pressure drop for a given flow is determined to be 100 Pa. For the same flow rate, if we reduce the diameter of the pipe by half, the pressure drop will be

(a) 25 Pa
(b) 50 Pa
(c) 200 Pa
(d) 400 Pa
(e) 1600 Pa

8–166 Air at 1 atm and 25°C ($v = 1.562 \times 10^{-5}$ m²/s) flows in a 9-cm-diameter cast iron pipe at a velocity of 5 m/s. The roughness of the pipe is 0.26 mm. The head loss for a pipe length of 24 m is

(a) 8.1 m
(b) 10.2 m
(c) 12.9 m
(d) 15.5 m
(e) 23.7 m

8–167 Consider air flow in a 10-cm-diameter pipe at a high velocity so that the Reynolds number is very large. The roughness of the pipe is 0.002 mm. The friction factor for this flow is

(a) 0.0311
(b) 0.0290
(c) 0.0247
(d) 0.0206
(e) 0.0163

8–168 Air at 1 atm and 40°C flows in a 8-cm-diameter pipe at a rate of 2500 L/min. The friction factor is determined from the Moody chart to be 0.027. The required power input to overcome the pressure drop for a pipe length of 150 m is

(a) 310 W
(b) 188 W
(c) 132 W
(d) 81.7 W
(e) 35.9 W

8–169 Water at 10°C ($\rho = 999.7$ kg/m³, $\mu = 1.307 \times 10^{-3}$ kg/m·s) is to be transported in a 5-cm-diamater, 30-m-long circular pipe. The roughness of the pipe is 0.22 mm. If the pressure drop in the pipe is not to exceed 19 kPa, the maximum flow rate of water is

(a) 324 L/min
(b) 281 L/min
(c) 243 L/min
(d) 195 L/min
(e) 168 L/min

8–170 The valve in a piping system causes a 3.1 m head loss. If the velocity of the flow is 6 m/s, the loss coefficient of this valve is

(a) 0.87
(b) 1.69
(c) 1.25
(d) 0.54
(e) 2.03

8–171 Consider a sharp-edged pipe exit for fully developed laminar flow of a fluid. The velocity of the flow is 4 m/s. This minor loss is equivalent to a head loss of

(a) 0.72 m
(b) 1.16 m
(c) 1.63 m
(d) 2.0 m
(e) 4.0 m

8–172 A water flow system involves a 180° return bend (threaded) and a 90° miter bend (without vanes). The velocity of water is 1.2 m/s. The minor losses due to these bends are equivalent to a pressure loss of

(a) 648 Pa
(b) 933 Pa
(c) 1255 Pa
(d) 1872 Pa
(e) 2600 Pa

8–173 A constant-diameter piping system involves multiple flow restrictions with a total loss coefficient of 4.4. The friction factor of piping is 0.025 and the diameter of the pipe is 7 cm. These minor losses are equivalent to the losses in a pipe of length

(a) 12.3 m
(b) 9.1 m
(c) 7.0 m
(d) 4.4 m
(e) 2.5 m

8–174 Air flows in an 8-cm-diameter, 33-m-long pipe at a velocity of 5.5 m/s. The piping system involves multiple flow restrictions with a total minor loss coefficient of 2.6. The friction factor of pipe is obtained from the Moody chart to be 0.025. The total head loss of this piping system is

(a) 13.5 m
(b) 7.6 m
(c) 19.9 m
(d) 24.5 m
(e) 4.2 m

8–175 Consider a pipe that branches out into two parallel pipes and then rejoins at a junction downstream. The two parallel pipes have the same lengths and friction factors. The diameters of the pipes are 2 cm and 4 cm. If the flow rate in one pipe is 10 L/min, the flow rate in the other pipe is

(a) 10 L/min
(b) 3.3 L/min
(c) 100 L/min
(d) 40 L/min
(e) 56.6 L/min

8–176 Consider a pipe that branches out into two parallel pipes and then rejoins at a junction downstream. The two parallel pipes have the same lengths and friction factors. The diameters of the pipes are 2 cm and 4 cm. If the head loss in one pipe is 0.5 m, the head loss in the other pipe is
(a) 0.5 m
(b) 1 m
(c) 0.25 m
(d) 2 m
(e) 0.125 m

8–177 A pump moves water from a reservoir to another reservoir through a piping system at a rate of 0.15 m³/min. Both reservoirs are open to the atmosphere. The elevation difference between the two reservoirs is 35 m and the total head loss is estimated to be 4 m. If the efficiency of the motor-pump unit is 65 percent, the electrical power input to the motor of the pump is
(a) 1664 W
(b) 1472 W
(c) 1238 W
(d) 983 W
(e) 805 W

8–178 Consider a pipe that branches out into three parallel pipes and then rejoins at a junction downstream. All three pipes have the same diameters ($D = 3$ cm) and friction factors ($f = 0.018$). The lengths of pipe 1 and pipe 2 are 5 m and 8 m, respectively while the velocities of the fluid in pipe 2 and pipe 3 are 2 m/s and 4 m/s, respectively. The length of pipe 3 is
(a) 8 m
(b) 5 m
(c) 4 m
(d) 2 m
(e) 1 m

Design and Essay Problems

8–179 Electronic boxes such as computers are commonly cooled by a fan. Write an essay on forced air cooling of electronic boxes and on the selection of the fan for electronic devices.

8–180 Design an experiment to measure the viscosity of liquids using a vertical funnel with a cylindrical reservoir of height h and a narrow flow section of diameter D and length L. Making appropriate assumptions, obtain a relation for viscosity in terms of easily measurable quantities such as density and volume flow rate. Is there a need for the use of a correction factor?

8–181 A pump is to be selected for a waterfall in a garden. The water collects in a pond at the bottom, and the elevation difference between the free surface of the pond and the location where the water is discharged is 3 m. The flow rate of water is to be at least 8 L/s. Select an appropriate motor–pump unit for this job and identify three manufacturers with product model numbers and prices. Make a selection and explain why you selected that particular product. Also estimate the cost of annual power consumption of this unit assuming continuous operation.

8–182 During a camping trip you notice that water is discharged from a high reservoir to a stream in the valley through a 30-cm-diameter plastic pipe. The elevation difference between the free surface of the reservoir and the stream is 70 m. You conceive the idea of generating power from this water. Design a power plant that will produce the most power from this resource. Also, investigate the effect of power generation on the discharge rate of water. What discharge rate maximizes the power production?

CHAPTER 9

DIFFERENTIAL ANALYSIS OF FLUID FLOW

In this chapter we derive the differential equations of fluid motion, namely, conservation of mass (the *continuity equation*) and Newton's second law (the *Navier–Stokes equation*). These equations apply to every point in the flow field and thus enable us to solve for all details of the flow everywhere in the *flow domain*. Unfortunately, most differential equations encountered in fluid mechanics are very difficult to solve and often require the aid of a computer. Also, these equations must be combined when necessary with additional equations, such as an equation of state and an equation for energy and/or species transport. We provide a step-by-step procedure for solving this set of differential equations of fluid motion and obtain analytical solutions for several simple examples. We also introduce the concept of the *stream function*; curves of constant stream function turn out to be *streamlines* in two-dimensional flow fields.

OBJECTIVES

When you finish reading this chapter, you should be able to

- Understand how the differential equation of conservation of mass and the differential linear momentum equation are derived and applied
- Calculate the stream function and pressure field, and plot streamlines for a known velocity field
- Obtain analytical solutions of the equations of motion for simple flow fields

The fundamental differential equations of fluid motion are derived in this chapter, and we show how to solve them analytically for some simple flows. More complicated flows, such as the air flow induced by a tornado shown here, cannot be solved exactly.
Royalty-Free/CORBIS

DIFFERENTIAL ANALYSIS OF FLUID FLOW

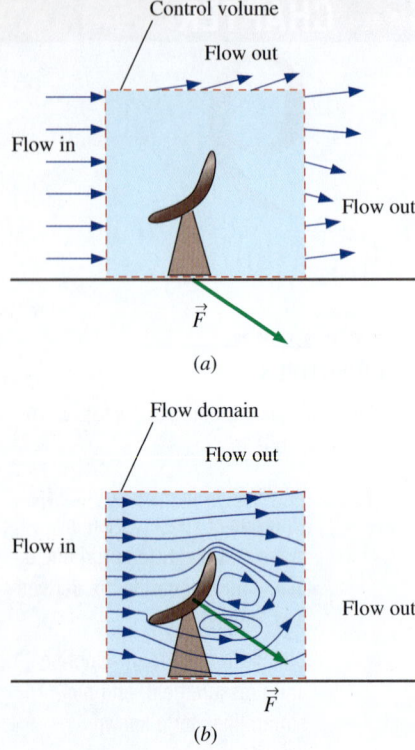

FIGURE 9–1
(*a*) In control volume analysis, the interior of the control volume is treated like a black box, but (*b*) in differential analysis, *all* the details of the flow are solved at *every* point within the flow domain.

9–1 • INTRODUCTION

In Chapter 5, we derived control volume versions of the laws of conservation of mass and energy, and in Chap. 6 we did the same for momentum. The control volume technique is useful when we are interested in the overall features of a flow, such as mass flow rate into and out of the control volume or net forces applied to bodies. An example is sketched in Fig. 9–1*a* for the case of wind flowing around a satellite dish. A rectangular control volume is taken around the vicinity of the satellite dish, as sketched. If we know the air velocity along the entire control surface, we can calculate the net reaction force on the stand without ever knowing any details about the geometry of the satellite dish. The interior of the control volume is in fact treated like a "black box" in control volume analysis—we *cannot* obtain detailed knowledge about flow properties such as velocity or pressure at points *inside* the control volume.

Differential analysis, on the other hand, involves application of differential equations of fluid motion to *any* and *every* point in the flow field over a region called the **flow domain.** You can think of the differential technique as the analysis of millions of tiny control volumes stacked end to end and on top of each other all throughout the flow field. In the limit as the number of tiny control volumes goes to infinity, and the size of each control volume shrinks to a point, the conservation equations simplify to a set of partial differential equations that are valid at any point in the flow. When solved, these differential equations yield details about the velocity, density, pressure, etc., at *every* point throughout the *entire* flow domain. In Fig. 9–1*b*, for example, differential analysis of airflow around the satellite dish yields streamline shapes, a detailed pressure distribution around the dish, etc. From these details, we can integrate to find gross features of the flow such as the net force on the satellite dish.

In a fluid flow problem such as the one illustrated in Fig. 9–1 in which air density and temperature changes are insignificant, it is sufficient to solve two differential equations of motion—conservation of mass and Newton's second law (the linear momentum equation). For three-dimensional incompressible flow, there are *four unknowns* (velocity components u, v, w, and pressure P) and *four equations* (one from conservation of mass, which is a scalar equation, and three from Newton's second law, which is a vector equation). As we shall see, the equations are **coupled,** meaning that some of the variables appear in all four equations; the set of differential equations must therefore be solved simultaneously for all four unknowns. In addition, **boundary conditions** for the variables must be specified at *all boundaries of the flow domain*, including inlets, outlets, and walls. Finally, if the flow is unsteady, we must march our solution along in time as the flow field changes. You can see how differential analysis of fluid flow can become quite complicated and difficult. Computers are a tremendous help here, as discussed in Chap. 15. Nevertheless, there is much we can do analytically, and we start by deriving the differential equation for conservation of mass.

9–2 • CONSERVATION OF MASS— THE CONTINUITY EQUATION

Through application of the Reynolds transport theorem (Chap. 4), we have the following general expression for conservation of mass as applied to a control volume:

Conservation of mass for a CV:

$$0 = \int_{CV} \frac{\partial \rho}{\partial t} dV + \int_{CS} \rho \vec{V} \cdot \vec{n} \, dA \quad (9\text{--}1)$$

Recall that Eq. 9–1 is valid for both fixed and moving control volumes, provided that the velocity vector is the *absolute* velocity (as seen by a fixed observer). When there are well-defined inlets and outlets, Eq. 9–1 is rewritten as

$$\int_{CV} \frac{\partial \rho}{\partial t} dV = \sum_{in} \dot{m} - \sum_{out} \dot{m} \quad (9\text{--}2)$$

In words, the net rate of change of mass within the control volume is equal to the rate at which mass flows into the control volume minus the rate at which mass flows out of the control volume. Equation 9–2 applies to *any* control volume, regardless of its size. To generate a differential equation for conservation of mass, we imagine the control volume shrinking to infinitesimal size, with dimensions dx, dy, and dz (Fig. 9–2). In the limit, the entire control volume shrinks to a *point* in the flow.

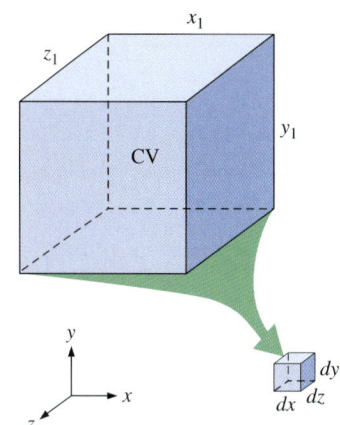

FIGURE 9–2
To derive a differential conservation equation, we imagine shrinking a control volume to infinitesimal size.

Derivation Using the Divergence Theorem

The quickest and most straightforward way to derive the differential form of conservation of mass is to apply the **divergence theorem** to Eq. 9–1. The divergence theorem is also called **Gauss's theorem,** named after the German mathematician Johann Carl Friedrich Gauss (1777–1855). The divergence theorem allows us to transform a volume integral of the divergence of a vector into an area integral over the surface that defines the volume. For any vector $\vec{G}$, the **divergence** of $\vec{G}$ is defined as $\vec{\nabla} \cdot \vec{G}$, and the divergence theorem is written as

Divergence theorem:
$$\int_V \vec{\nabla} \cdot \vec{G} \, dV = \oint_A \vec{G} \cdot \vec{n} \, dA \quad (9\text{--}3)$$

The circle on the area integral is used to emphasize that the integral must be evaluated around the *entire closed area A* that surrounds volume V. Note that the control surface of Eq. 9–1 *is* a closed area, even though we do not always add the circle to the integral symbol. Equation 9–3 applies to *any* volume, so we choose the control volume of Eq. 9–1. We also let $\vec{G} = \rho\vec{V}$ since $\vec{G}$ can be any vector. Substitution of Eq. 9–3 into Eq. 9–1 converts the area integral into a volume integral,

$$0 = \int_{CV} \frac{\partial \rho}{\partial t} dV + \int_{CV} \vec{\nabla} \cdot (\rho \vec{V}) \, dV$$

We now combine the two volume integrals into one,

$$\int_{CV} \left[\frac{\partial \rho}{\partial t} + \vec{\nabla} \cdot (\rho \vec{V}) \right] dV = 0 \quad (9\text{--}4)$$

Finally, we argue that Eq. 9–4 must hold for *any* control volume regardless of its size or shape. This is possible only if the integrand (the terms within square brackets) is identically zero. Hence, we have a general differential equation for conservation of mass, better known as the **continuity equation**:

Continuity equation:
$$\frac{\partial \rho}{\partial t} + \vec{\nabla}\cdot(\rho\vec{V}) = 0 \tag{9-5}$$

Equation 9–5 is the compressible form of the continuity equation since we have not assumed incompressible flow. It is valid at any point in the flow domain.

Derivation Using an Infinitesimal Control Volume

We derive the continuity equation in a different way, by starting with a control volume on which we apply conservation of mass. Consider an infinitesimal box-shaped control volume aligned with the axes in Cartesian coordinates (Fig. 9–3). The dimensions of the box are dx, dy, and dz, and the center of the box is shown at some arbitrary point P from the origin (the box can be located anywhere in the flow field). At the center of the box we define the density as ρ and the velocity components as u, v, and w, as shown. At locations away from the center of the box, we use a **Taylor series expansion** about the center of the box (point P). [The series expansion is named in honor of its creator, the English mathematician Brook Taylor (1685–1731).] For example, the center of the right-most face of the box is located a distance $dx/2$ from the middle of the box in the x-direction; the value of ρu at that point is

$$(\rho u)_{\text{center of right face}} = \rho u + \frac{\partial(\rho u)}{\partial x}\frac{dx}{2} + \frac{1}{2!}\frac{\partial^2(\rho u)}{\partial x^2}\left(\frac{dx}{2}\right)^2 + \cdots \tag{9-6}$$

As the box representing the control volume shrinks to a point, however, second-order and higher terms become negligible. For example, suppose $dx/L = 10^{-3}$, where L is some characteristic length scale of the flow domain. Then $(dx/L)^2 = 10^{-6}$, a factor of a thousand less than dx/L. In fact, the smaller dx, the better the assumption that second-order terms are negligible. Applying this truncated Taylor series expansion to the density times the normal velocity component at the center point of each of the six faces of the box, we have

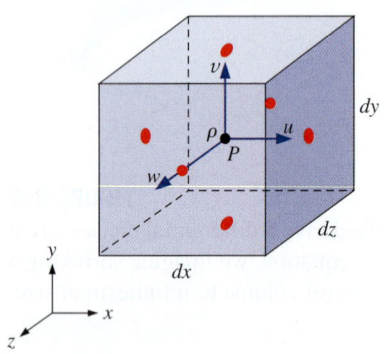

FIGURE 9–3

A small box-shaped control volume centered at point P is used for derivation of the differential equation for conservation of mass in Cartesian coordinates; the red dots indicate the center of each face.

Center of right face: $\quad (\rho u)_{\text{center of right face}} \cong \rho u + \dfrac{\partial(\rho u)}{\partial x}\dfrac{dx}{2}$

Center of left face: $\quad (\rho u)_{\text{center of left face}} \cong \rho u - \dfrac{\partial(\rho u)}{\partial x}\dfrac{dx}{2}$

Center of front face: $\quad (\rho w)_{\text{center of front face}} \cong \rho w + \dfrac{\partial(\rho w)}{\partial z}\dfrac{dz}{2}$

Center of rear face: $\quad (\rho w)_{\text{center of rear face}} \cong \rho w - \dfrac{\partial(\rho w)}{\partial z}\dfrac{dz}{2}$

Center of top face: $\quad (\rho v)_{\text{center of top face}} \cong \rho v + \dfrac{\partial(\rho v)}{\partial y}\dfrac{dy}{2}$

Center of bottom face: $\quad (\rho v)_{\text{center of bottom face}} \cong \rho v - \dfrac{\partial(\rho v)}{\partial y}\dfrac{dy}{2}$

The mass flow rate into or out of one of the faces is equal to the density times the normal velocity component at the center point of the face times the surface area of the face. In other words, $\dot{m} = \rho V_n A$ at each face, where V_n is the magnitude of the normal velocity through the face and A is the surface area of the face (Fig. 9–4). The mass flow rate through each face of our infinitesimal control volume is illustrated in Fig. 9–5. We could construct truncated Taylor series expansions at the center of each face for the remaining (nonnormal) velocity components as well, but this is unnecessary since these components are *tangential* to the face under consideration. For example, the value of ρv at the center of the right face can be estimated by a similar expansion, but since v is tangential to the right face of the box, it contributes nothing to the mass flow rate into or out of that face.

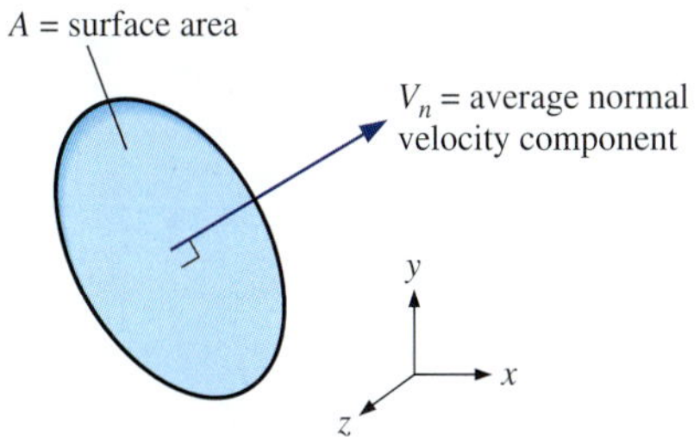

As the control volume shrinks to a point, the value of the volume integral on the left-hand side of Eq. 9–2 becomes

Rate of change of mass within CV:

$$\int_{CV} \frac{\partial \rho}{\partial t} dV \cong \frac{\partial \rho}{\partial t} dx\, dy\, dz \qquad (9\text{–}7)$$

since the volume of the box is $dx\, dy\, dz$. We now apply the approximations of Fig. 9–5 to the right-hand side of Eq. 9–2. We add up all the mass flow rates into and out of the control volume through the faces. The left, bottom, and back faces contribute to mass *inflow*, and the first term on the right-hand side of Eq. 9–2 becomes

Net mass flow rate into CV:

$$\sum_{in} \dot{m} \cong \underbrace{\left(\rho u - \frac{\partial(\rho u)}{\partial x}\frac{dx}{2}\right)dy\, dz}_{\text{left face}} + \underbrace{\left(\rho v - \frac{\partial(\rho v)}{\partial y}\frac{dy}{2}\right)dx\, dz}_{\text{bottom face}} + \underbrace{\left(\rho w - \frac{\partial(\rho w)}{\partial z}\frac{dz}{2}\right)dx\, dy}_{\text{rear face}}$$

FIGURE 9–4
The mass flow rate through a surface is equal to $\rho V_n A$.

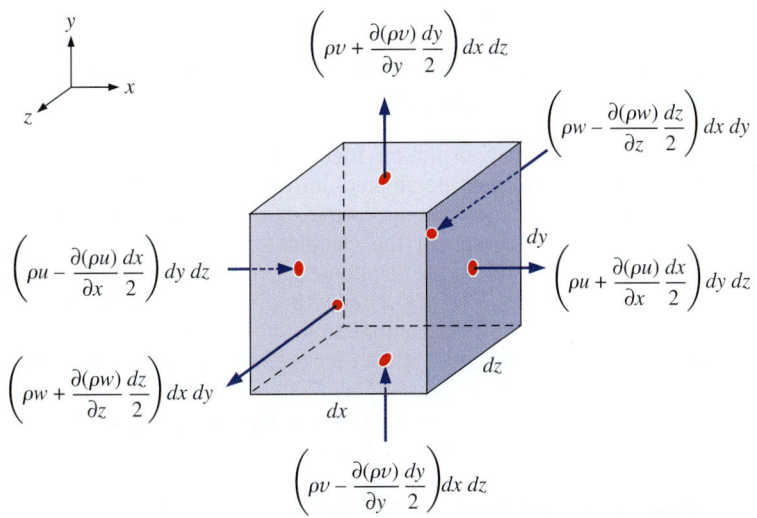

FIGURE 9–5
The inflow or outflow of mass through each face of the differential control volume; the red dots indicate the center of each face.

DIFFERENTIAL ANALYSIS OF FLUID FLOW

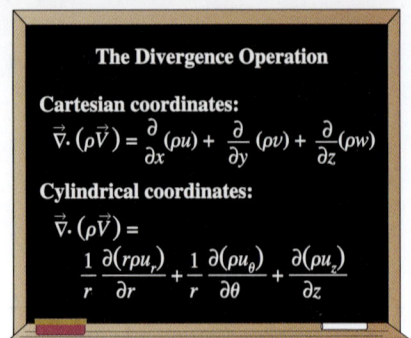

FIGURE 9–6
The divergence operation in Cartesian and cylindrical coordinates.

Similarly, the right, top, and front faces contribute to mass *outflow*, and the second term on the right-hand side of Eq. 9–2 becomes

Net mass flow rate out of CV:

$$\sum_{out} \dot{m} \cong \underbrace{\left(\rho u + \frac{\partial(\rho u)}{\partial x}\frac{dx}{2}\right)dy\,dz}_{\text{right face}} + \underbrace{\left(\rho v + \frac{\partial(\rho v)}{\partial y}\frac{dy}{2}\right)dx\,dz}_{\text{top face}} + \underbrace{\left(\rho w + \frac{\partial(\rho w)}{\partial z}\frac{dz}{2}\right)dx\,dy}_{\text{front face}}$$

We substitute Eq. 9–7 and these two equations for mass flow rate into Eq. 9–2. Many of the terms cancel each other out; after combining and simplifying the remaining terms, we are left with

$$\frac{\partial \rho}{\partial t}dx\,dy\,dz = -\frac{\partial(\rho u)}{\partial x}dx\,dy\,dz - \frac{\partial(\rho v)}{\partial y}dx\,dy\,dz - \frac{\partial(\rho w)}{\partial z}dx\,dy\,dz$$

The volume of the box, $dx\,dy\,dz$, appears in each term and can be eliminated. After rearrangement we end up with the following differential equation for conservation of mass in Cartesian coordinates:

Continuity equation in Cartesian coordinates:

$$\frac{\partial \rho}{\partial t} + \frac{\partial(\rho u)}{\partial x} + \frac{\partial(\rho v)}{\partial y} + \frac{\partial(\rho w)}{\partial z} = 0 \qquad (9\text{–}8)$$

Equation 9–8 is the compressible form of the continuity equation in Cartesian coordinates. It is written in more compact form by recognizing the divergence operation (Fig. 9–6), yielding the same equation as Eq. 9–5.

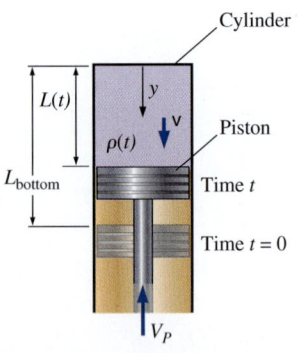

FIGURE 9–7
Fuel and air being compressed by a piston in a cylinder of an internal combustion engine.

EXAMPLE 9–1   Compression of an Air–Fuel Mixture

An air–fuel mixture is compressed by a piston in a cylinder of an internal combustion engine (Fig. 9–7). The origin of coordinate y is at the top of the cylinder, and y points straight down as shown. The piston is assumed to move up at constant speed V_P. The distance L between the top of the cylinder and the piston decreases with time according to the linear approximation $L = L_{bottom} - V_P t$, where L_{bottom} is the location of the piston when it is at the bottom of its cycle at time $t = 0$, as sketched in Fig. 9–7. At $t = 0$, the density of the air–fuel mixture in the cylinder is everywhere equal to $\rho(0)$. Estimate the density of the air–fuel mixture as a function of time and the given parameters during the piston's up stroke.

SOLUTION   The density of the air–fuel mixture is to be estimated as a function of time and the given parameters in the problem statement.
Assumptions   **1** Density varies with time, but not space; in other words, the density is uniform throughout the cylinder at any given time, but changes with time: $\rho = \rho(t)$. **2** Velocity component v varies with y and t, but not with x or z; in other words $v = v(y, t)$ only. **3** $u = w = 0$. **4** No mass escapes from the cylinder during the compression.
Analysis   First we need to establish an expression for velocity component v as a function of y and t. Clearly $v = 0$ at $y = 0$ (the top of the cylinder), and $v = -V_P$ at $y = L$. For simplicity, we approximate that v varies linearly between these two boundary conditions,

Vertical velocity component: $v = -V_P \dfrac{y}{L}$ (1)

where L is a function of time, as given. The compressible continuity equation in Cartesian coordinates (Eq. 9–8) is appropriate for solution of this problem.

$$\frac{\partial \rho}{\partial t} + \underbrace{\frac{\partial(\rho u)}{\partial x}}_{0 \text{ since } u = 0} + \frac{\partial(\rho v)}{\partial y} + \underbrace{\frac{\partial(\rho w)}{\partial z}}_{0 \text{ since } w = 0} = 0 \quad \rightarrow \quad \frac{\partial \rho}{\partial t} + \frac{\partial(\rho v)}{\partial y} = 0$$

By assumption 1, however, density is not a function of y and can therefore come out of the y-derivative. Substituting Eq. 1 for v and the given expression for L, differentiating, and simplifying, we obtain

$$\frac{\partial \rho}{\partial t} = -\rho \frac{\partial v}{\partial y} = -\rho \frac{\partial}{\partial y}\left(-V_P \frac{y}{L}\right) = \rho \frac{V_P}{L} = \rho \frac{V_P}{L_{\text{bottom}} - V_P t} \quad (2)$$

By assumption 1 again, we replace $\partial\rho/\partial t$ by $d\rho/dt$ in Eq. 2. After separating variables we obtain an expression that can be integrated analytically,

$$\int_{\rho = \rho(0)}^{\rho} \frac{d\rho}{\rho} = \int_{t=0}^{t} \frac{V_P}{L_{\text{bottom}} - V_P t}\, dt \quad \rightarrow \quad \ln \frac{\rho}{\rho(0)} = \ln \frac{L_{\text{bottom}}}{L_{\text{bottom}} - V_P t} \quad (3)$$

Finally then, we have the desired expression for ρ as a function of time,

$$\rho = \rho(0) \frac{L_{\text{bottom}}}{L_{\text{bottom}} - V_P t} \quad (4)$$

In keeping with the convention of nondimensionalizing results, Eq. 4 is rewritten as

$$\frac{\rho}{\rho(0)} = \frac{1}{1 - V_P t/L_{\text{bottom}}} \quad \rightarrow \quad \rho^* = \frac{1}{1 - t^*} \quad (5)$$

where $\rho^* = \rho/\rho(0)$ and $t^* = V_P t/L_{\text{bottom}}$. Equation 5 is plotted in Fig. 9–8.

Discussion At $t^* = 1$, the piston hits the top of the cylinder and ρ goes to infinity. In an actual internal combustion engine, the piston stops before reaching the top of the cylinder, forming what is called the *clearance volume*, which typically constitutes 4 to 12 percent of the maximum cylinder volume. The assumption of uniform density within the cylinder is the weakest link in this simplified analysis. In reality, ρ may be a function of both space and time.

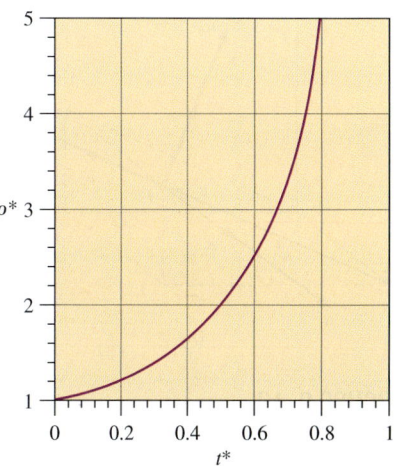

FIGURE 9–8
Nondimensional density as a function of nondimensional time for Example 9–1.

Alternative Form of the Continuity Equation

We expand Eq. 9–5 by using the product rule on the divergence term,

$$\underbrace{\frac{\partial \rho}{\partial t} + \vec{\nabla} \cdot (\rho \vec{V}) = \frac{\partial \rho}{\partial t} + \vec{V} \cdot \vec{\nabla} \rho}_{\text{Material derivative of } \rho} + \rho \vec{\nabla} \cdot \vec{V} = 0 \quad (9\text{–}9)$$

Recognizing the *material derivative* in Eq. 9–9 (see Chap. 4), and dividing by ρ, we write the compressible continuity equation in an alternative form,

Alternative form of the continuity equation:

$$\frac{1}{\rho}\frac{D\rho}{Dt} + \vec{\nabla} \cdot \vec{V} = 0 \quad (9\text{–}10)$$

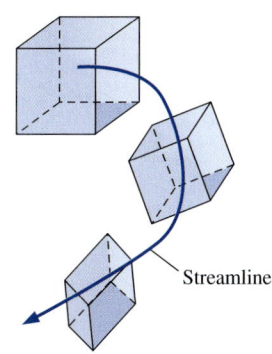

FIGURE 9–9
As a material element moves through a flow field, its density changes according to Eq. 9–10.

DIFFERENTIAL ANALYSIS OF FLUID FLOW

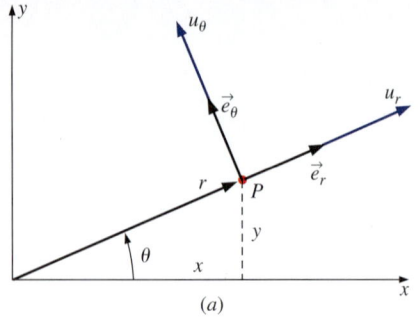

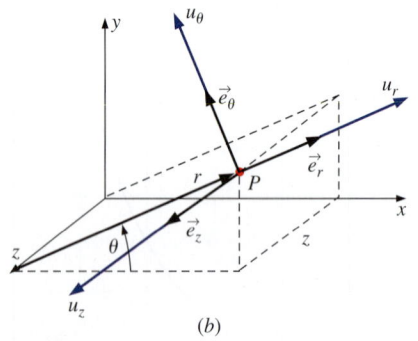

FIGURE 9–10
Velocity components and unit vectors in cylindrical coordinates: (*a*) two-dimensional flow in the *xy*- or *rθ*-plane, (*b*) three-dimensional flow.

Equation 9–10 shows that as we follow a fluid element through the flow field (we call this a **material element**), its density changes as $\vec{\nabla}\cdot\vec{V}$ changes (Fig. 9–9). On the other hand, if changes in the density of the material element are negligibly small compared to the magnitudes of the velocity gradients in $\vec{\nabla}\cdot\vec{V}$ as the element moves around, $\rho^{-1}D\rho/Dt \cong 0$, and the flow is approximated as **incompressible.**

Continuity Equation in Cylindrical Coordinates

Many problems in fluid mechanics are more conveniently solved in **cylindrical coordinates** (r, θ, z) (often called **cylindrical polar coordinates**), rather than in Cartesian coordinates. For simplicity, we introduce cylindrical coordinates in two dimensions first (Fig. 9–10*a*). By convention, r is the radial distance from the origin to some point (P), and θ is the angle measured from the *x*-axis (θ is always defined as mathematically positive in the counterclockwise direction). Velocity components, u_r and u_θ, and unit vectors, $\vec{e}_r$ and $\vec{e}_\theta$, are also shown in Fig. 9–10*a*. In three dimensions, imagine sliding everything in Fig. 9–10*a* out of the page along the *z*-axis (normal to the *xy*-plane) by some distance *z*. We have attempted to draw this in Fig. 9–10*b*. In three dimensions, we have a third velocity component, u_z, and a third unit vector, $\vec{e}_z$, also sketched in Fig. 9–10*b*.

The following coordinate transformations are obtained from Fig. 9–10:

Coordinate transformations:

$$r = \sqrt{x^2 + y^2} \quad x = r\cos\theta \quad y = r\sin\theta \quad \theta = \tan^{-1}\frac{y}{x} \quad (9\text{–}11)$$

Coordinate *z* is the same in cylindrical and Cartesian coordinates.

To obtain an expression for the continuity equation in cylindrical coordinates, we have two choices. First, we can use Eq. 9–5 directly, since it was derived without regard to our choice of coordinate system. We simply look up the expression for the divergence operator in cylindrical coordinates in a vector calculus book (e.g., Spiegel, 1968; see also Fig. 9–6). Second, we can draw a three-dimensional infinitesimal fluid element in cylindrical coordinates and analyze mass flow rates into and out of the element, similar to what we did before in Cartesian coordinates. Either way, we end up with

Continuity equation in cylindrical coordinates:

$$\frac{\partial \rho}{\partial t} + \frac{1}{r}\frac{\partial(r\rho u_r)}{\partial r} + \frac{1}{r}\frac{\partial(\rho u_\theta)}{\partial \theta} + \frac{\partial(\rho u_z)}{\partial z} = 0 \quad (9\text{–}12)$$

Details of the second method can be found in Fox and McDonald (1998).

Special Cases of the Continuity Equation

We now look at two special cases, or simplifications, of the continuity equation. In particular, we first consider steady compressible flow, and then incompressible flow.

Special Case 1: Steady Compressible Flow

If the flow is compressible but steady, $\partial/\partial t$ of any variable is equal to zero. Thus, Eq. 9–5 reduces to

Steady continuity equation: $\quad\quad \vec{\nabla}\cdot(\rho\vec{V}) = 0 \quad (9\text{–}13)$

In Cartesian coordinates, Eq. 9–13 reduces to

$$\frac{\partial(\rho u)}{\partial x} + \frac{\partial(\rho v)}{\partial y} + \frac{\partial(\rho w)}{\partial z} = 0 \qquad (9\text{–}14)$$

In cylindrical coordinates, Eq. 9–13 reduces to

$$\frac{1}{r}\frac{\partial(r\rho u_r)}{\partial r} + \frac{1}{r}\frac{\partial(\rho u_\theta)}{\partial \theta} + \frac{\partial(\rho u_z)}{\partial z} = 0 \qquad (9\text{–}15)$$

Special Case 2: Incompressible Flow

If the flow is approximated as incompressible, density is not a function of time or space. Thus $\partial\rho/\partial t \cong 0$ in Eq. 9–5, and ρ can be taken outside of the divergence operator. Equation 9–5 therefore reduces to

Incompressible continuity equation: $\qquad \vec{\nabla}\cdot\vec{V} = 0 \qquad (9\text{–}16)$

The same result is obtained if we start with Eq. 9–10 and recognize that for an incompressible flow, density does not change appreciably following a fluid particle, as pointed out previously. Thus the material derivative of ρ is approximately zero, and Eq. 9–10 reduces immediately to Eq. 9–16.

You may have noticed that *no time derivatives remain in Eq. 9–16*. We conclude from this that *even if the flow is unsteady, Eq. 9–16 applies at any instant in time*. Physically, this means that as the velocity field changes in one part of an incompressible flow field, the entire rest of the flow field immediately adjusts to the change such that Eq. 9–16 is satisfied at all times. For compressible flow this is not the case. In fact, a disturbance in one part of the flow is not even felt by fluid particles some distance away until the sound wave from the disturbance reaches that distance. Very loud noises, such as that from a gun or explosion, generate a **shock wave** that actually travels *faster* than the speed of sound. (The shock wave produced by an explosion is illustrated in Fig. 9–11.) Shock waves and other manifestations of compressible flow are discussed in Chap. 12.

In Cartesian coordinates, Eq. 9–16 is

Incompressible continuity equation in Cartesian coordinates:

$$\frac{\partial u}{\partial x} + \frac{\partial v}{\partial y} + \frac{\partial w}{\partial z} = 0 \qquad (9\text{–}17)$$

Equation 9–17 is the form of the continuity equation you will probably encounter most often. It applies to steady or unsteady, incompressible, three-dimensional flow, and you would do well to memorize it.

In cylindrical coordinates, Eq. 9–16 is

Incompressible continuity equation in cylindrical coordinates:

$$\frac{1}{r}\frac{\partial(ru_r)}{\partial r} + \frac{1}{r}\frac{\partial(u_\theta)}{\partial \theta} + \frac{\partial(u_z)}{\partial z} = 0 \qquad (9\text{–}18)$$

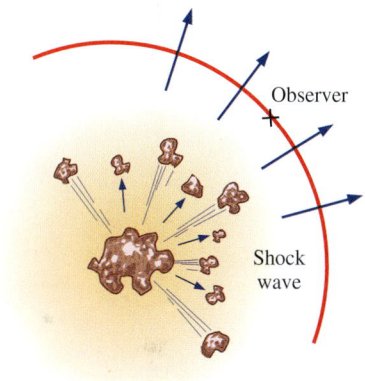

FIGURE 9–11
The disturbance from an explosion is not felt until the shock wave reaches the observer.

■ **EXAMPLE 9–2** Design of a Compressible Converging Duct

A two-dimensional converging duct is being designed for a high-speed wind tunnel. The bottom wall of the duct is to be flat and horizontal, and the top wall is to be curved in such a way that the axial wind speed *u* increases

446
DIFFERENTIAL ANALYSIS OF FLUID FLOW

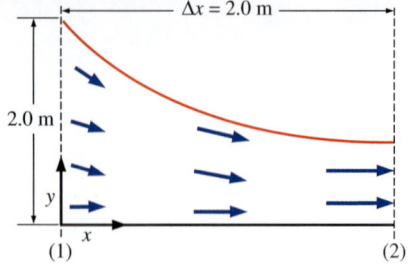

FIGURE 9–12
Converging duct, designed for a high-speed wind tunnel (not to scale).

approximately linearly from $u_1 = 100$ m/s at section (1) to $u_2 = 300$ m/s at section (2) (Fig. 9–12). Meanwhile, the air density ρ is to decrease approximately linearly from $\rho_1 = 1.2$ kg/m³ at section (1) to $\rho_2 = 0.85$ kg/m³ at section (2). The converging duct is 2.0 m long and is 2.0 m high at section (1). (a) Predict the y-component of velocity, $v(x, y)$, in the duct. (b) Plot the approximate shape of the duct, ignoring friction on the walls. (c) How high should the duct be at section (2), the exit of the duct?

SOLUTION For given velocity component u and density ρ, we are to predict velocity component v, plot an approximate shape of the duct, and predict its height at the duct exit.
Assumptions **1** The flow is steady and two-dimensional in the *xy*-plane. **2** Friction on the walls is ignored. **3** Axial velocity u increases linearly with x, and density ρ decreases linearly with x.
Properties The fluid is air at room temperature (25°C). The speed of sound is about 346 m/s, so the flow is subsonic, but compressible.
Analysis (a) We write expressions for u and ρ, forcing them to be linear in x,

$$u = u_1 + C_u x \quad \text{where} \quad C_u = \frac{u_2 - u_1}{\Delta x} = \frac{(300 - 100) \text{ m/s}}{2.0 \text{ m}} = 100 \text{ s}^{-1} \quad (1)$$

and

$$\rho = \rho_1 + C_\rho x \quad \text{where} \quad C_\rho = \frac{\rho_2 - \rho_1}{\Delta x} = \frac{(0.85 - 1.2) \text{ kg/m}^3}{2.0 \text{ m}} \quad (2)$$
$$= -0.175 \text{ kg/m}^4$$

The steady continuity equation (Eq. 9–14) for this two-dimensional compressible flow simplifies to

$$\frac{\partial(\rho u)}{\partial x} + \frac{\partial(\rho v)}{\partial y} + \underbrace{\frac{\partial(\rho w)}{\partial z}}_{0 \text{ (2-D)}} = 0 \quad \rightarrow \quad \frac{\partial(\rho v)}{\partial y} = -\frac{\partial(\rho u)}{\partial x} \quad (3)$$

Substituting Eqs. 1 and 2 into Eq. 3 and noting that C_u and C_ρ are constants,

$$\frac{\partial(\rho v)}{\partial y} = -\frac{\partial[(\rho_1 + C_\rho x)(u_1 + C_u x)]}{\partial x} = -(\rho_1 C_u + u_1 C_\rho) - 2C_u C_\rho x$$

Integration with respect to y gives

$$\rho v = -(\rho_1 C_u + u_1 C_\rho)y - 2C_u C_\rho xy + f(x) \quad (4)$$

Note that since the integration is a *partial* integration, we have added an arbitrary function of x instead of simply a constant of integration. Next, we apply boundary conditions. We argue that since the bottom wall is flat and horizontal, v must equal zero at $y = 0$ for any x. This is possible only if $f(x) = 0$. Solving Eq. 4 for v gives

$$v = \frac{-(\rho_1 C_u + u_1 C_\rho)y - 2C_u C_\rho xy}{\rho} \quad \rightarrow \quad v = \frac{-(\rho_1 C_u + u_1 C_\rho)y - 2C_u C_\rho xy}{\rho_1 + C_\rho x} \quad (5)$$

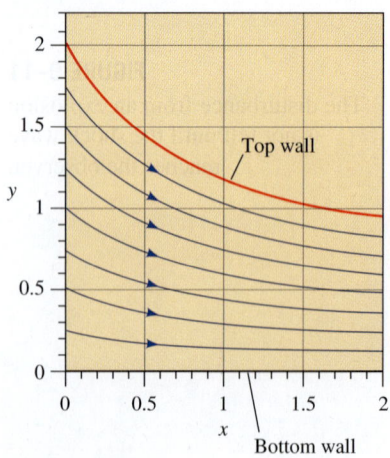

FIGURE 9–13
Streamlines for the converging duct of Example 9–2.

(b) Using Eqs. 1 and 5 and the technique described in Chap. 4, we plot several streamlines between $x = 0$ and $x = 2.0$ m in Fig. 9–13. The streamline starting at $x = 0$, $y = 2.0$ m approximates the top wall of the duct.

(c) At section (2), the top streamline crosses $y = 0.941$ m at $x = 2.0$ m. Thus, the predicted height of the duct at section (2) is **0.941 m.**

Discussion You can verify that the combination of Eqs. 1, 2, and 5 satisfies the continuity equation. However, this alone does not guarantee that the density and velocity components will actually *follow* these equations if the duct were to be built as designed here. The actual flow depends on the *pressure drop* between sections (1) and (2); only one unique pressure drop can yield the desired flow acceleration. Temperature may also change considerably in this kind of compressible flow in which the air accelerates toward sonic speeds.

EXAMPLE 9–3   Incompressibility of an Unsteady Two-Dimensional Flow

Consider the velocity field of Example 4–5—an unsteady, two-dimensional velocity field given by $\vec{V} = (u, v) = (0.5 + 0.8x)\vec{i} + [1.5 + 2.5 \sin(\omega t) - 0.8y]\vec{j}$, where angular frequency ω is equal to 2π rad/s (a physical frequency of 1 Hz). Verify that this flow field can be approximated as incompressible.

SOLUTION We are to verify that a given velocity field is incompressible.
Assumptions **1** The flow is two-dimensional, implying no z-component of velocity and no variation of u or v with z.
Analysis The components of velocity in the x- and y-directions, respectively, are

$$u = 0.5 + 0.8x \quad \text{and} \quad v = 1.5 + 2.5 \sin(\omega t) - 0.8y$$

If the flow is incompressible, Eq. 9–16 must apply. More specifically, in Cartesian coordinates Eq. 9–17 must apply. Let's check:

$$\underbrace{\frac{\partial u}{\partial x}}_{0.8} + \underbrace{\frac{\partial v}{\partial y}}_{-0.8} + \underbrace{\frac{\partial w}{\partial z}}_{0 \text{ since 2-D}} = 0 \quad \rightarrow \quad 0.8 - 0.8 = 0$$

So we see that the incompressible continuity equation is indeed satisfied at any instant in time, and **this flow field may be approximated as incompressible**.
Discussion Although there is an unsteady term in v, it has no y-derivative and drops out of the continuity equation.

EXAMPLE 9–4   Finding a Missing Velocity Component

Two velocity components of a steady, incompressible, three-dimensional flow field are known, namely, $u = ax^2 + by^2 + cz^2$ and $w = axz + byz^2$, where a, b, and c are constants. The y velocity component is missing (Fig. 9–14). Generate an expression for v as a function of x, y, and z.

SOLUTION We are to find the y-component of velocity, v, using given expressions for u and w.
Assumptions **1** The flow is steady. **2** The flow is incompressible.
Analysis Since the flow is steady and incompressible, and since we are working in Cartesian coordinates, we apply Eq. 9–17 to the flow field,

FIGURE 9–14
The continuity equation can be used to find a missing velocity component.

Condition for incompressibility:

$$\frac{\partial v}{\partial y} = -\underbrace{\frac{\partial u}{\partial x}}_{2ax} - \underbrace{\frac{\partial w}{\partial z}}_{ax + 2byz} \quad \rightarrow \quad \frac{\partial v}{\partial y} = -3ax - 2byz$$

Next we integrate with respect to y. Since the integration is a *partial* integration, we add some arbitrary function of x and z instead of a simple constant of integration.

Solution: $\quad\quad\quad\quad\quad\quad v = -3axy - by^2z + f(x,z)$

Discussion Any function $f(x,z)$ yields a v that satisfies the incompressible continuity equation, since there are no derivatives of v with respect to x or z in the continuity equation.

EXAMPLE 9–5 Two-Dimensional, Incompressible, Vortical Flow

Consider a two-dimensional, incompressible flow in cylindrical coordinates; the tangential velocity component is $u_\theta = K/r$, where K is a constant. This represents a class of vortical flows. Generate an expression for the other velocity component, u_r.

SOLUTION For a given tangential velocity component, we are to generate an expression for the radial velocity component.
Assumptions 1 The flow is two-dimensional in the xy- ($r\theta$-) plane (velocity is not a function of z, and $u_z = 0$ everywhere). 2 The flow is incompressible.
Analysis The incompressible continuity equation (Eq. 9–18) for this two-dimensional case simplifies to

$$\frac{1}{r}\frac{\partial(ru_r)}{\partial r} + \frac{1}{r}\frac{\partial u_\theta}{\partial \theta} + \underbrace{\frac{\partial u_z}{\partial z}}_{0\ (2\text{-D})} = 0 \quad \rightarrow \quad \frac{\partial(ru_r)}{\partial r} = -\frac{\partial u_\theta}{\partial \theta} \quad (1)$$

The given expression for u_θ is not a function of θ, and therefore Eq. 1 reduces to

$$\frac{\partial(ru_r)}{\partial r} = 0 \quad \rightarrow \quad ru_r = f(\theta, t) \quad (2)$$

where we have introduced an arbitrary function of θ and t instead of a constant of integration, since we performed a *partial* integration with respect to r. Solving for u_r,

$$u_r = \frac{f(\theta, t)}{r} \quad (3)$$

Thus, *any radial velocity component of the form given by Eq. 3 yields a two-dimensional, incompressible velocity field that satisfies the continuity equation.*

We discuss some specific cases. The simplest case is when $f(\theta, t) = 0$ ($u_r = 0$, $u_\theta = K/r$). This yields the **line vortex** discussed in Chap. 4, as sketched in Fig. 9–15a. Another simple case is when $f(\theta, t) = C$, where C is a constant. This yields a radial velocity whose magnitude decays as $1/r$. For negative C, imagine a spiraling line vortex/sink flow, in which fluid elements not only revolve around the origin, but get sucked into a sink at the origin (actually a line sink along the z-axis). This is illustrated in Fig. 9–15b.

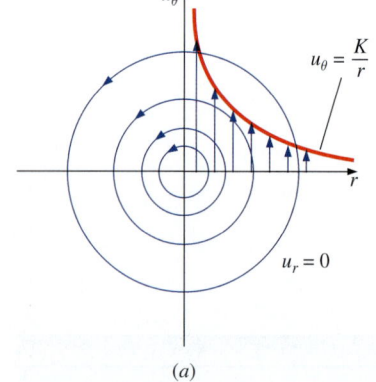
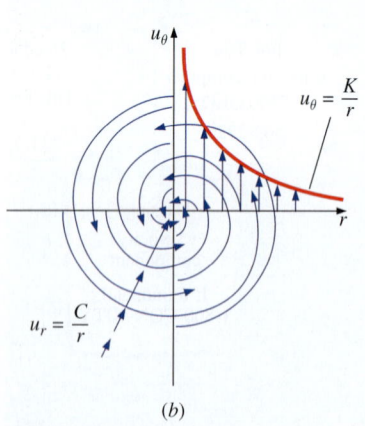

FIGURE 9–15
Streamlines and velocity profiles for (a) a line vortex flow and (b) a spiraling line vortex/sink flow.

Discussion Other more complicated flows can be obtained by setting $f(\theta, t)$ to some other function. For any function $f(\theta, t)$, the flow satisfies the two-dimensional, incompressible continuity equation at a given instant in time.

EXAMPLE 9–6 Comparison of Continuity and Volumetric Strain Rate

Recall the *volumetric strain rate*, defined in Chap. 4. In Cartesian coordinates,

$$\frac{1}{V}\frac{DV}{Dt} = \varepsilon_{xx} + \varepsilon_{yy} + \varepsilon_{zz} = \frac{\partial u}{\partial x} + \frac{\partial v}{\partial y} + \frac{\partial w}{\partial z} \quad (1)$$

Show that volumetric strain rate is zero for incompressible flow. Discuss the physical interpretation of volumetric strain rate for incompressible and compressible flows.

SOLUTION We are to show that volumetric strain rate is zero in an incompressible flow, and discuss its physical significance in incompressible and compressible flow.
Analysis If the flow is incompressible, Eq. 9–16 applies. More specifically, Eq. 9–17, in Cartesian coordinates, applies. Comparing Eq. 9–17 to Eq. 1,

$$\frac{1}{V}\frac{DV}{Dt} = 0 \quad \text{for incompressible flow}$$

Thus, *volumetric strain rate is zero in an incompressible flow field*. In fact, you can *define* incompressibility by $DV/Dt = 0$. Physically, as we follow a fluid element, parts of it may stretch while other parts shrink, and the element may translate, distort, and rotate, but its volume remains constant along its entire path through the flow field (Fig. 9–16a). This is true whether the flow is steady or unsteady, as long as it is incompressible. If the flow were compressible, the volumetric strain rate would not be zero, implying that fluid elements may expand in volume (dilate) or shrink in volume as they move around in the flow field (Fig. 9–16b). Specifically, consider Eq. 9–10, an alternative form of the continuity equation for compressible flow. By definition, $\rho = m/V$, where m is the mass of a fluid element. For a material element (following the fluid element as it moves through the flow field), m must be constant. Applying some algebra to Eq. 9–10 yields

$$\frac{1}{\rho}\frac{D\rho}{Dt} = \frac{V}{m}\frac{D(m/V)}{Dt} = -\frac{V}{m}\frac{m}{V^2}\frac{DV}{Dt} = -\frac{1}{V}\frac{DV}{Dt} = -\vec{\nabla}\cdot\vec{V} \;\;\rightarrow\;\; \frac{1}{V}\frac{DV}{Dt} = \vec{\nabla}\cdot\vec{V}$$

Discussion The final result is general—not limited to Cartesian coordinates. It applies to unsteady as well as steady flows.

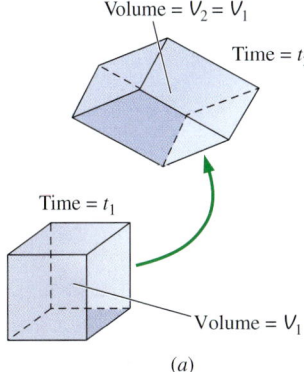
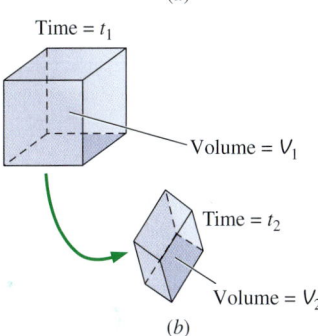

FIGURE 9–16
(*a*) In an incompressible flow field, fluid elements may translate, distort, and rotate, but they do not grow or shrink in volume; (*b*) in a compressible flow field, fluid elements may grow or shrink in volume as they translate, distort, and rotate.

EXAMPLE 9–7 Conditions for Incompressible Flow

Consider a steady velocity field given by $\vec{V} = (u, v, w) = a(x^2y + y^2)\vec{i} + bxy^2\vec{j} + cx\vec{k}$, where a, b, and c are constants. Under what conditions is this flow field incompressible?

450
DIFFERENTIAL ANALYSIS OF FLUID FLOW

SOLUTION We are to determine a relationship between constants *a*, *b*, and *c* that ensures incompressibility.

Assumptions 1 The flow is steady. 2 The flow is incompressible (under certain constraints to be determined).

Analysis We apply Eq. 9–17 to the given velocity field,

$$\underbrace{\frac{\partial u}{\partial x}}_{2axy} + \underbrace{\frac{\partial v}{\partial y}}_{2bxy} + \underbrace{\frac{\partial w}{\partial z}}_{0} = 0 \quad \rightarrow \quad 2axy + 2bxy = 0$$

Thus to guarantee incompressibility, constants *a* and *b* must be equal in magnitude but opposite in sign.

Condition for incompressibility: $\qquad a = -b$

Discussion If *a* were not equal to $-b$, this might still be a valid flow field, but density would have to vary with location in the flow field. In other words, the flow would be *compressible*, and Eq. 9–14 would need to be satisfied in place of Eq. 9–17.

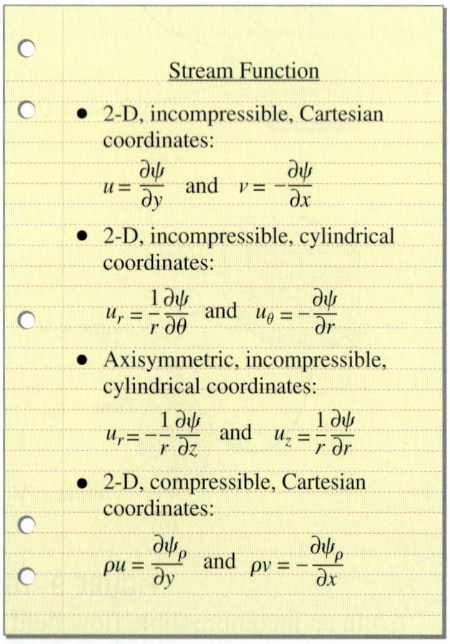

9–3 • THE STREAM FUNCTION

The Stream Function in Cartesian Coordinates

Consider the simple case of incompressible, two-dimensional flow in the *xy*-plane. The continuity equation (Eq. 9–17) in Cartesian coordinates reduces to

$$\frac{\partial u}{\partial x} + \frac{\partial v}{\partial y} = 0 \qquad (9\text{–}19)$$

A clever variable transformation enables us to rewrite Eq. 9–19 in terms of *one* dependent variable (ψ) instead of *two* dependent variables (*u* and *v*). We define the **stream function** ψ as (Fig. 9–17)

Incompressible, two-dimensional stream function in Cartesian coordinates:

$$u = \frac{\partial \psi}{\partial y} \quad \text{and} \quad v = -\frac{\partial \psi}{\partial x} \qquad (9\text{–}20)$$

The stream function and the corresponding velocity potential function (Chap. 10) were first introduced by the Italian mathematician Joseph Louis Lagrange (1736–1813). Substitution of Eq. 9–20 into Eq. 9–19 yields

$$\frac{\partial}{\partial x}\left(\frac{\partial \psi}{\partial y}\right) + \frac{\partial}{\partial y}\left(-\frac{\partial \psi}{\partial x}\right) = \frac{\partial^2 \psi}{\partial x\, \partial y} - \frac{\partial^2 \psi}{\partial y\, \partial x} = 0$$

which is identically satisfied for any smooth function $\psi(x, y)$, because the order of differentiation (*y* then *x* versus *x* then *y*) is irrelevant.

You may ask why we chose to put the negative sign on *v* rather than on *u*. (We could have defined the stream function with the signs reversed, and continuity would still have been identically satisfied.) The answer is that although the sign is arbitrary, the definition of Eq. 9–20 leads to flow from left to right as ψ increases in the *y*-direction, which is usually preferred. Most fluid mechanics books define ψ in this way, although sometimes ψ is

---

Stream Function

- 2-D, incompressible, Cartesian coordinates:

  $u = \dfrac{\partial \psi}{\partial y}$ and $v = -\dfrac{\partial \psi}{\partial x}$

- 2-D, incompressible, cylindrical coordinates:

  $u_r = \dfrac{1}{r}\dfrac{\partial \psi}{\partial \theta}$ and $u_\theta = -\dfrac{\partial \psi}{\partial r}$

- Axisymmetric, incompressible, cylindrical coordinates:

  $u_r = -\dfrac{1}{r}\dfrac{\partial \psi}{\partial z}$ and $u_z = \dfrac{1}{r}\dfrac{\partial \psi}{\partial r}$

- 2-D, compressible, Cartesian coordinates:

  $\rho u = \dfrac{\partial \psi_\rho}{\partial y}$ and $\rho v = -\dfrac{\partial \psi_\rho}{\partial x}$

FIGURE 9–17
There are several definitions of the stream function, depending on the type of flow under consideration as well as the coordinate system being used.

defined with the opposite signs (e.g., in some British text books and in the indoor air quality field, Heinsohn and Cimbala, 2003).

What have we gained by this transformation? First, as already mentioned, a single variable (ψ) replaces *two* variables (u and v)—once ψ is known, we can generate both u and v via Eq. 9–20, and we are guaranteed that the solution satisfies continuity, Eq. 9–19. Second, it turns out that the stream function has useful physical significance (Fig. 9–18). Namely,

Curves of constant ψ are **streamlines** of the flow.

This is easily proven by considering a streamline in the *xy*-plane, as sketched in Fig. 9–19. Recall from Chap. 4 that along such a streamline,

Along a streamline: $\quad \dfrac{dy}{dx} = \dfrac{v}{u} \quad \rightarrow \quad \underbrace{-v\,dx}_{\partial\psi/\partial x} + \underbrace{u\,dy}_{\partial\psi/\partial y} = 0$

where we have applied Eq. 9–20, the definition of ψ. Thus,

Along a streamline: $\quad \dfrac{\partial\psi}{\partial x} dx + \dfrac{\partial\psi}{\partial y} dy = 0 \quad$ (9–21)

But for any smooth function ψ of two variables x and y, we know by the chain rule of mathematics that the total change of ψ from point (x, y) to another point $(x + dx, y + dy)$ some infinitesimal distance away is

Total change of ψ: $\quad d\psi = \dfrac{\partial\psi}{\partial x} dx + \dfrac{\partial\psi}{\partial y} dy \quad$ (9–22)

By comparing Eq. 9–21 to Eq. 9–22 we see that $d\psi = 0$ along a streamline; thus we have proven the statement that ψ is constant along streamlines.

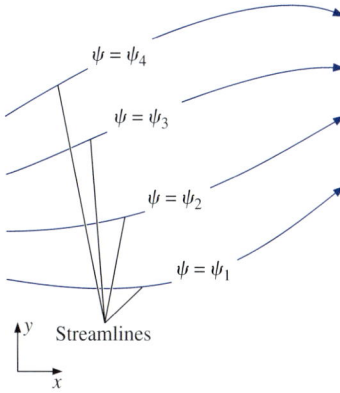

FIGURE 9–18
Curves of constant stream function represent streamlines of the flow.

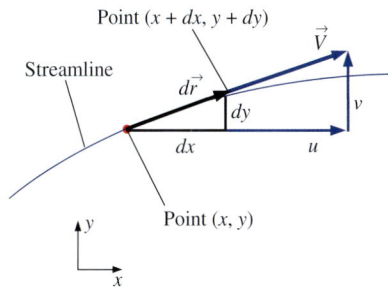

FIGURE 9–19
Arc length $d\vec{r} = (dx, dy)$ and local velocity vector $\vec{V} = (u, v)$ along a two-dimensional streamline in the *xy*-plane.

EXAMPLE 9–8   Calculation of the Velocity Field from the Stream Function

A steady, two-dimensional, incompressible flow field in the *xy*-plane has a stream function given by $\psi = ax^3 + by + cx$, where a, b, and c are constants: $a = 0.50$ (m·s)$^{-1}$, $b = -2.0$ m/s, and $c = -1.5$ m/s. (a) Obtain expressions for velocity components u and v. (b) Verify that the flow field satisfies the incompressible continuity equation. (c) Plot several streamlines of the flow in the upper-right quadrant.

SOLUTION For a given stream function, we are to calculate the velocity components, verify incompressibility, and plot flow streamlines.
Assumptions 1 The flow is steady. 2 The flow is incompressible (this assumption is to be verified). 3 The flow is two-dimensional in the *xy*-plane, implying that $w = 0$ and neither u nor v depend on z.
Analysis (a) We use Eq. 9–20 to obtain expressions for u and v by differentiating the stream function,

$$u = \dfrac{\partial\psi}{\partial y} = b \quad \text{and} \quad v = -\dfrac{\partial\psi}{\partial x} = -3ax^2 - c$$

(b) Since u is not a function of x, and v is not a function of y, we see immediately that the two-dimensional, incompressible continuity equation (Eq. 9–19) is satisfied. In fact, since ψ is smooth in x and y, the two-dimensional,

452
DIFFERENTIAL ANALYSIS OF FLUID FLOW

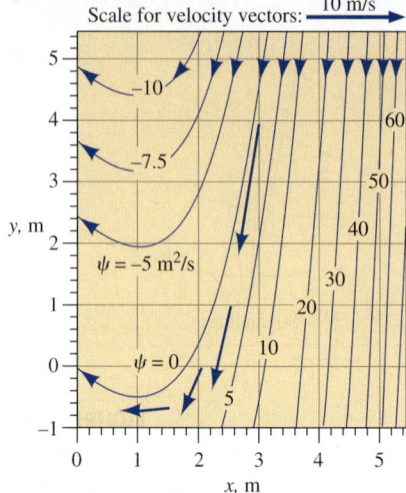

FIGURE 9–20
Streamlines for the velocity field of Example 9–8; the value of constant ψ is indicated for each streamline, and velocity vectors are shown at four locations.

incompressible continuity equation in the *xy*-plane is automatically satisfied by the very definition of ψ. We conclude that **the flow is indeed incompressible.** (*c*) To plot streamlines, we solve the given equation for either *y* as a function of *x* and ψ, or *x* as a function of *y* and ψ. In this case, the former is easier, and we have

Equation for a streamline: $$y = \frac{\psi - ax^3 - cx}{b}$$

This equation is plotted in Fig. 9–20 for several values of ψ, and for the provided values of *a*, *b*, and *c*. The flow is nearly straight down at large values of *x*, but veers upward for $x < 1$ m.

Discussion You can verify that $v = 0$ at $x = 1$ m. In fact, v is negative for $x > 1$ m and positive for $x < 1$ m. The direction of the flow can also be determined by picking an arbitrary point in the flow, say ($x = 3$ m, $y = 4$ m), and calculating the velocity there. We get $u = -2.0$ m/s and $v = -12.0$ m/s at this point, either of which shows that fluid flows to the lower left in this region of the flow field. For clarity, the velocity vector at this point is also plotted in Fig. 9–20; it is clearly parallel to the streamline near that point. Velocity vectors at three other locations are also plotted.

EXAMPLE 9–9   **Calculation of Stream Function for a Known Velocity Field**

Consider a steady, two-dimensional, incompressible velocity field with $u = ax + b$ and $v = -ay + cx$, where *a*, *b*, and *c* are constants: $a = 0.50$ s^{-1}, $b = 1.5$ m/s, and $c = 0.35$ s^{-1}. Generate an expression for the stream function and plot some streamlines of the flow in the upper-right quadrant.

SOLUTION For a given velocity field we are to generate an expression for ψ and plot several streamlines for given values of constants *a*, *b*, and *c*.
Assumptions **1** The flow is steady. **2** The flow is incompressible. **3** The flow is two-dimensional in the *xy*-plane, implying that $w = 0$ and neither u nor v depend on *z*.
Analysis We start by picking one of the two parts of Eq. 9–20 that define the stream function (it doesn't matter which part we choose—the solution will be identical).

$$\frac{\partial \psi}{\partial y} = u = ax + b$$

Next we integrate with respect to *y*, noting that this is a *partial* integration, so we add an arbitrary function of the other variable, *x*, rather than a constant of integration,

$$\psi = axy + by + g(x) \qquad (1)$$

Now we choose the other part of Eq. 9–20, differentiate Eq. 1, and rearrange as follows:

$$v = -\frac{\partial \psi}{\partial x} = -ay - g'(x) \qquad (2)$$

where $g'(x)$ denotes dg/dx since g is a function of only one variable, *x*. We now have two expressions for velocity component v, the equation given in the problem statement and Eq. 2. We equate these and integrate with respect to x to find $g(x)$,

$$v = -ay + cx = -ay - g'(x) \quad \rightarrow \quad g'(x) = -cx \quad \rightarrow \quad g(x) = -c\frac{x^2}{2} + C \quad (3)$$

Note that here we have added an arbitrary constant of integration C since g is a function of x only. Finally, substituting Eq. 3 into Eq. 1 yields the final expression for ψ,

Solution: $$\psi = axy + by - c\frac{x^2}{2} + C \quad (4)$$

To plot the streamlines, we note that Eq. 4 represents a *family* of curves, one unique curve for each value of the constant ($\psi - C$). Since C is arbitrary, it is common to set it equal to zero, although it can be set to any desired value. For simplicity we set $C = 0$ and solve Eq. 4 for y as a function of x, yielding

Equation for streamlines: $$y = \frac{\psi + cx^2/2}{ax + b} \quad (5)$$

For the given values of constants a, b, and c, we plot Eq. 5 for several values of ψ in Fig. 9–21; these curves of constant ψ are streamlines of the flow. From Fig. 9–21 we see that this is a smoothly converging flow in the upper-right quadrant.

Discussion It is always good to check your algebra. In this example, you should substitute Eq. 4 into Eq. 9–20 to verify that the correct velocity components are obtained.

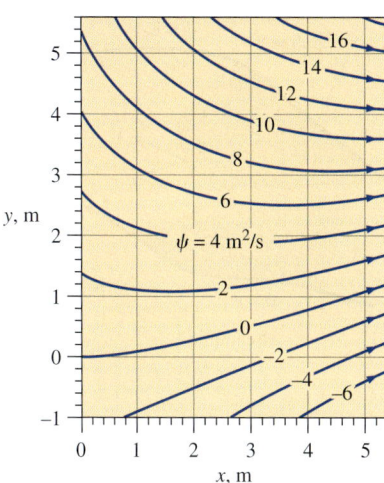

FIGURE 9–21
Streamlines for the velocity field of Example 9–9; the value of constant ψ is indicated for each streamline.

There is another physically significant fact about the stream function:

> The difference in the value of ψ from one streamline to another is equal to the volume flow rate per unit width between the two streamlines.

This statement is illustrated in Fig. 9–22. Consider two streamlines, ψ_1 and ψ_2, and imagine two-dimensional flow in the xy-plane, of unit width into the page (1 m in the $-z$-direction). By definition, *no flow can cross a streamline*. Thus, the fluid that happens to occupy the space between these two streamlines remains confined between the same two streamlines. It follows that the mass flow rate through any cross-sectional slice between the streamlines is the same at any instant in time. The cross-sectional slice can be any shape, provided that it starts at streamline 1 and ends at streamline 2. In Fig. 9–22, for example, slice A is a smooth arc from one streamline to the other while slice B is wavy. For steady, incompressible, two-dimensional flow in the xy-plane, the volume flow rate $\dot{V}$ between the two streamlines (per unit width) must therefore be a constant. If the two streamlines spread apart, as they do from cross-sectional slice A to cross-sectional slice B, the average velocity between the two streamlines decreases accordingly, such that the volume flow rate remains the same ($\dot{V}_A = \dot{V}_B$). In Fig. 9–20 of Example 9–8, velocity vectors at four locations in the flow field between streamlines $\psi = 0$ m^2/s and $\psi = 5$ m^2/s are plotted. You can clearly see that as the streamlines diverge from each other, the velocity vector decays in magnitude. Likewise, when streamlines *converge*, the average velocity between them must increase.

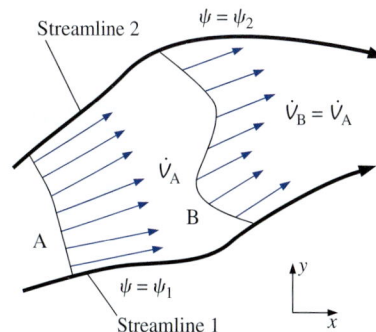

FIGURE 9–22
For two-dimensional streamlines in the xy-plane, the volume flow rate $\dot{V}$ per unit width between two streamlines is the same through any cross-sectional slice.

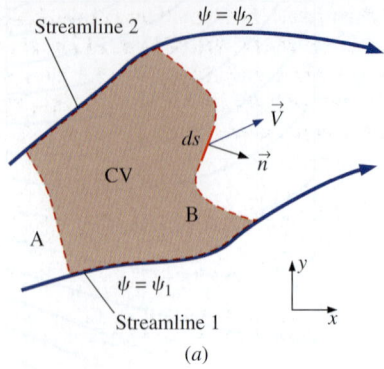

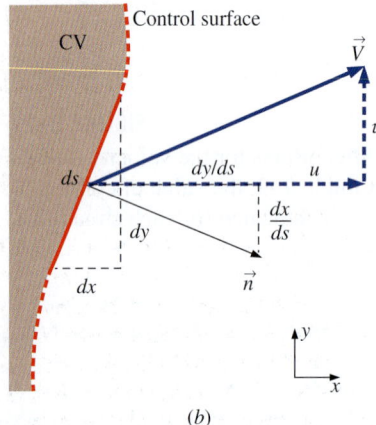

FIGURE 9–23
(a) Control volume bounded by streamlines ψ_1 and ψ_2 and slices A and B in the xy-plane; (b) magnified view of the region around infinitesimal length ds.

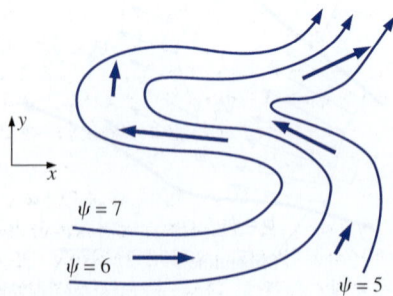

FIGURE 9–24
Illustration of the "left-side convention." In the xy-plane, the value of the stream function always increases to the left of the flow direction.

We prove the given statement mathematically by considering a control volume bounded by the two streamlines of Fig. 9–22 and by cross-sectional slice A and cross-sectional slice B (Fig. 9–23). An infinitesimal length ds along slice B is illustrated in Fig. 9–23a, along with its unit normal vector $\vec{n}$. A magnified view of this region is sketched in Fig. 9–23b for clarity. As shown, the two components of ds are dx and dy; thus the unit normal vector is

$$\vec{n} = \frac{dy}{ds}\vec{i} - \frac{dx}{ds}\vec{j}$$

The volume flow rate per unit width through segment ds of the control surface is

$$d\dot{V} = \vec{V} \cdot \vec{n}\, dA = (u\vec{i} + v\vec{j}) \cdot \left(\frac{dy}{ds}\vec{i} - \frac{dx}{ds}\vec{j}\right) \underbrace{ds}_{ds} \quad (9\text{–}23)$$

where $dA = ds$ times $1 = ds$, where the 1 indicates a unit width into the page, regardless of the unit system. When we expand the dot product of Eq. 9–23 and apply Eq. 9–20, we get

$$d\dot{V} = u\, dy - v\, dx = \frac{\partial \psi}{\partial y} dy + \frac{\partial \psi}{\partial x} dx = d\psi \quad (9\text{–}24)$$

We find the total volume flow rate through cross-sectional slice B by integrating Eq. 9–24 from streamline 1 to streamline 2,

$$\dot{V}_B = \int_B \vec{V} \cdot \vec{n}\, dA = \int_B d\dot{V} = \int_{\psi=\psi_1}^{\psi=\psi_2} d\psi = \psi_2 - \psi_1 \quad (9\text{–}25)$$

Thus, the volume flow rate per unit width through slice B is equal to the difference between the values of the two stream functions that bound slice B. Now consider the entire control volume of Fig. 9–23a. Since we know that no flow crosses the streamlines, conservation of mass demands that the volume flow rate into the control volume through slice A be identical to the volume flow rate out of the control volume through slice B. Finally, since we may choose a cross-sectional slice of any shape or location between the two streamlines, the statement is proven.

When dealing with stream functions, the direction of flow is obtained by what we might call the "left-side convention." Namely, if you are looking down the z-axis at the xy-plane (Fig. 9–24) and are moving in the direction of the flow, the stream function increases to your left.

The value of ψ increases to the left of the direction of flow in the xy-plane.

In Fig. 9–24, for example, the stream function increases to the left of the flow direction, regardless of how much the flow twists and turns. Notice also that when the streamlines are far apart (lower right of Fig. 9–24), the magnitude of velocity (the fluid speed) in that vicinity is small relative to the speed in locations where the streamlines are close together (middle region of Fig. 9–24). This is easily explained by conservation of mass. As the streamlines converge, the cross-sectional area between them decreases, and the velocity must increase to maintain the flow rate between the streamlines.

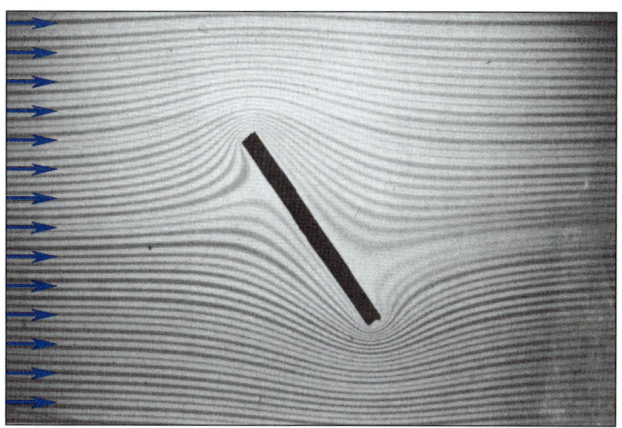

FIGURE 9–25
Streaklines produced by Hele–Shaw flow over an inclined plate. The streaklines model streamlines of potential flow (Chap. 10) over a two-dimensional inclined plate of the same cross-sectional shape.
Courtesy Howell Peregrine, School of Mathematics, University of Bristol. Used by permission.

■ **EXAMPLE 9–10**   **Relative Velocity Deduced from Streamlines**

Hele–Shaw flow is produced by forcing a liquid through a thin gap between parallel plates. An example of Hele–Shaw flow is provided in Fig. 9–25 for flow over an inclined plate. Streaklines are generated by introducing dye at evenly spaced points upstream of the field of view. Since the flow is steady, the streaklines are coincident with streamlines. The fluid is water and the glass plates are 1.0 mm apart. Discuss how you can tell from the streamline pattern whether the flow speed in a particular region of the flow field is (relatively) large or small.

SOLUTION   For the given set of streamlines, we are to discuss how we can tell the relative speed of the fluid.
Assumptions   **1** The flow is steady. **2** The flow is incompressible. **3** The flow models two-dimensional potential flow in the *xy*-plane.
Analysis   When equally spaced streamlines of a stream function spread away from each other, it indicates that the flow speed has decreased in that region. Likewise, if the streamlines come closer together, the flow speed has increased in that region. In Fig. 9–25 we infer that the flow far upstream of the plate is straight and uniform, since the streamlines are equally spaced. The fluid decelerates as it approaches the underside of the plate, especially near the stagnation point, as indicated by the wide gap between streamlines. The flow accelerates rapidly to very high speeds around the sharp corners of the plate, as indicated by the tightly spaced streamlines.
Discussion   The streaklines of Hele–Shaw flow turn out to be similar to those of potential flow, which is discussed in Chap. 10.

■ **EXAMPLE 9–11**   **Volume Flow Rate Deduced from Streamlines**

Water is sucked through a narrow slot on the bottom wall of a water channel. The water in the channel flows from left to right at uniform velocity $V = 1.0$ m/s. The slot is perpendicular to the *xy*-plane, and runs along the *z*-axis across the entire channel, which is $w = 2.0$ m wide. The flow is thus approximately two-dimensional in the *xy*-plane. Several streamlines of the flow are plotted and labeled in Fig. 9–26.

456
DIFFERENTIAL ANALYSIS OF FLUID FLOW

FIGURE 9–26
Streamlines for free-stream flow along a wall with a narrow suction slot; streamline values are shown in units of m²/s; the thick streamline is the dividing streamline. The direction of the velocity vector at point A is determined by the left-side convention.

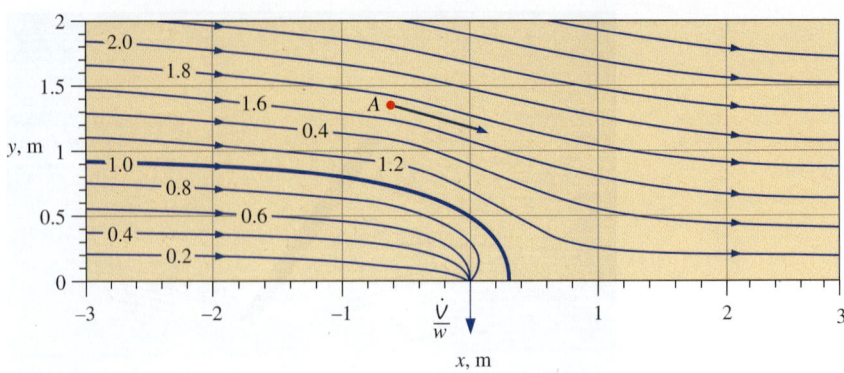

The thick streamline in Fig. 9–26 is called the **dividing streamline** because it divides the flow into two parts. Namely, all the water below this dividing streamline gets sucked into the slot, while all the water above the dividing streamline continues on its way downstream. What is the volume flow rate of water being sucked through the slot? Estimate the magnitude of the velocity at point A.

SOLUTION For the given set of streamlines, we are to determine the volume flow rate through the slot and estimate the fluid speed at a point.
Assumptions **1** The flow is steady. **2** The flow is incompressible. **3** The flow is two-dimensional in the xy-plane. **4** Friction along the bottom wall is neglected.
Analysis By Eq. 9–25, the volume flow rate per unit width between the bottom wall ($\psi_{wall} = 0$) and the dividing streamline ($\psi_{dividing} = 1.0$ m²/s) is

$$\frac{\dot{V}}{w} = \psi_{dividing} - \psi_{wall} = (1.0 - 0) \text{ m}^2/\text{s} = 1.0 \text{ m}^2/\text{s}$$

All of this flow must go through the slot. Since the channel is 2.0 m wide, the total volume flow rate through the slot is

$$\dot{V} = \frac{\dot{V}}{w} w = (1.0 \text{ m}^2/\text{s})(2.0 \text{ m}) = \mathbf{2.0 \text{ m}^3/\text{s}}$$

To estimate the speed at point A, we measure the distance δ between the two streamlines that enclose point A. We find that streamline 1.8 is about 0.21 m away from streamline 1.6 in the vicinity of point A. The volume flow rate per unit width (into the page) between these two streamlines is equal to the difference in value of the stream function. We thus estimate the speed at point A,

$$V_A \cong \frac{\dot{V}}{w\delta} = \frac{1}{\delta}\frac{\dot{V}}{w} = \frac{1}{\delta}(\psi_{1.8} - \psi_{1.6}) = \frac{1}{0.21 \text{ m}}(1.8 - 1.6) \text{ m}^2/\text{s} = \mathbf{0.95 \text{ m/s}}$$

Our estimate is close to the known free-stream speed (1.0 m/s), indicating that the fluid in the vicinity of point A flows at nearly the same speed as the free-stream flow, but points slightly downward.
Discussion The streamlines of Fig. 9–26 were generated by superposition of a uniform stream and a line sink, assuming irrotational (potential) flow. We discuss such superposition in Chap. 10.

The Stream Function in Cylindrical Coordinates

For two-dimensional flow in the *xy*-plane, we can also define the stream function in cylindrical coordinates, which is more convenient for many problems. Note that by *two-dimensional* we mean that there are only two relevant independent spatial coordinates—with no dependence on the third component. There are two possibilities. The first is **planar flow,** just like that of Eqs. 9–19 and 9–20, but in terms of (r, θ) and (u_r, u_θ) instead of (x, y) and (u, v) (see Fig. 9–10*a*). In this case, there is no dependence on coordinate z. We simplify the incompressible continuity equation, Eq. 9–18, for two-dimensional planar flow in the $r\theta$-plane,

$$\frac{\partial(ru_r)}{\partial r} + \frac{\partial(u_\theta)}{\partial \theta} = 0 \tag{9-26}$$

We define the stream function as follows:

Incompressible, planar stream function in cylindrical coordinates:

$$u_r = \frac{1}{r}\frac{\partial \psi}{\partial \theta} \quad \text{and} \quad u_\theta = -\frac{\partial \psi}{\partial r} \tag{9-27}$$

We note again that the signs are reversed in some textbooks. You can substitute Eq. 9–27 into Eq. 9–26 to convince yourself that Eq. 9–26 is identically satisfied for any smooth function $\psi(r, \theta)$, since the order of differentiation (r then θ versus θ then r) is irrelevant for a smooth function.

The second type of two-dimensional flow in cylindrical coordinates is **axisymmetric flow,** in which r and z are the relevant spatial variables, u_r and u_z are the nonzero velocity components, and there is no dependence on θ (Fig. 9–27). Examples of axisymmetric flow include flow around spheres, bullets, and the fronts of many objects like torpedoes and missiles, which would be axisymmetric everywhere if not for their fins. For incompressible axisymmetric flow, the continuity equation is

$$\frac{1}{r}\frac{\partial(ru_r)}{\partial r} + \frac{\partial(u_z)}{\partial z} = 0 \tag{9-28}$$

The stream function ψ is defined such that it satisfies Eq. 9–28 exactly, provided of course that ψ is a smooth function of r and z,

Incompressible, axisymmetric stream function in cylindrical coordinates:

$$u_r = -\frac{1}{r}\frac{\partial \psi}{\partial z} \quad \text{and} \quad u_z = \frac{1}{r}\frac{\partial \psi}{\partial r} \tag{9-29}$$

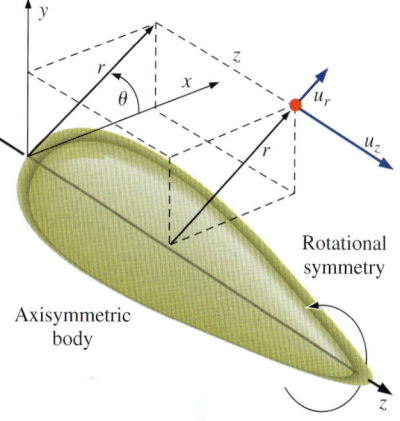

FIGURE 9–27
Flow over an axisymmetric body in cylindrical coordinates with rotational symmetry about the z-axis; neither the geometry nor the velocity field depend on θ, and $u_\theta = 0$.

We also note that there is another way to describe axisymmetric flows, namely, by using Cartesian coordinates (x, y) and (u, v), but forcing coordinate x to be the axis of symmetry. This can lead to confusion because the equations of motion must be modified accordingly to account for the axisymmetry. Nevertheless, this is often the approach used in CFD codes. The advantage is that after one sets up a grid in the *xy*-plane, the *same* grid can be used for both planar flow (flow in the *xy*-plane with no *z*-dependence) and axisymmetric flow (flow in the *xy*-plane with rotational symmetry about the *x*-axis). We do not discuss the equations for this alternative description of axisymmetric flows.

DIFFERENTIAL ANALYSIS OF FLUID FLOW

EXAMPLE 9–12 Stream Function in Cylindrical Coordinates

Consider a line vortex, defined as steady, planar, incompressible flow in which the velocity components are $u_r = 0$ and $u_\theta = K/r$, where K is a constant. This flow is represented in Fig. 9–15a. Derive an expression for the stream function $\psi(r, \theta)$, and prove that the streamlines are circles.

SOLUTION For a given velocity field in cylindrical coordinates, we are to derive an expression for the stream function and show that the streamlines are circular.
Assumptions 1 The flow is steady. 2 The flow is incompressible. 3 The flow is planar in the $r\theta$-plane.
Analysis We use the definition of stream function given by Eq. 9–27. We can choose either component to start with; we choose the tangential component,

$$\frac{\partial \psi}{\partial r} = -u_\theta = -\frac{K}{r} \quad \rightarrow \quad \psi = -K \ln r + f(\theta) \tag{1}$$

Now we use the other component of Eq. 9–27,

$$u_r = \frac{1}{r}\frac{\partial \psi}{\partial \theta} = \frac{1}{r} f'(\theta) \tag{2}$$

where the prime denotes a derivative with respect to θ. By equating u_r from the given information to Eq. 2, we see that

$$f'(\theta) = 0 \quad \rightarrow \quad f(\theta) = C$$

where C is an arbitrary constant of integration. Equation 1 is thus

Solution: $$\psi = -K \ln r + C \tag{3}$$

Finally, we see from Eq. 3 that curves of constant ψ are produced by setting r to a constant value. Since curves of constant r are circles by definition, **streamlines (curves of constant ψ) must therefore be circles about the origin, as in Fig. 9–15a.**
For given values of C and ψ, we solve Eq. 3 for r to plot the streamlines,

Equation for streamlines: $$r = e^{-(\psi - C)/K} \tag{4}$$

For $K = 10$ m²/s and $C = 0$, streamlines from $\psi = 0$ to 22 are plotted in Fig. 9–28.
Discussion Notice that for a uniform increment in the value of ψ, the streamlines get closer and closer together near the origin as the tangential velocity increases. This is a direct result of the statement that the difference in the value of ψ from one streamline to another is equal to the volume flow rate per unit width between the two streamlines.

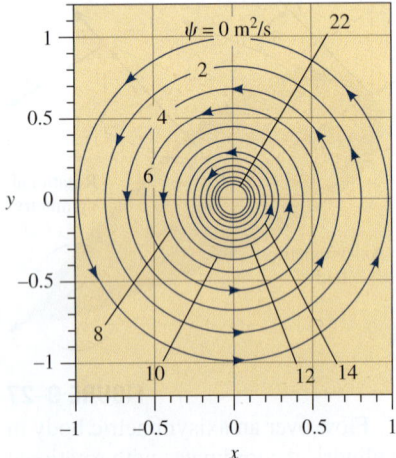

FIGURE 9–28
Streamlines for the velocity field of Example 9–12, with $K = 10$ m²/s and $C = 0$; the value of constant ψ is indicated for several streamlines.

The Compressible Stream Function*

We extend the stream function concept to steady, *compressible*, two-dimensional flow in the *xy*-plane. The compressible continuity equation (Eq. 9–14) in Cartesian coordinates reduces to the following for steady two-dimensional flow:

$$\frac{\partial (\rho u)}{\partial x} + \frac{\partial (\rho v)}{\partial y} = 0 \tag{9–30}$$

---

* This section can be skipped without loss of continuity (no pun intended).

We introduce a **compressible stream function**, which we denote as ψ_ρ,

Steady, compressible, two-dimensional stream function in Cartesian coordinates:

$$\rho u = \frac{\partial \psi_\rho}{\partial y} \quad \text{and} \quad \rho v = -\frac{\partial \psi_\rho}{\partial x} \qquad (9\text{--}31)$$

By definition, ψ_ρ of Eq. 9–31 satisfies Eq. 9–30 exactly, provided that ψ_ρ is a smooth function of x and y. Many of the features of the compressible stream function are the same as those of the incompressible ψ as discussed previously. For example, curves of constant ψ_ρ are still streamlines. However, the difference in ψ_ρ from one streamline to another is *mass* flow rate per unit width rather than volume flow rate per unit width. Although not as popular as its incompressible counterpart, the compressible stream function finds use in some commercial CFD codes.

9–4 ▪ THE DIFFERENTIAL LINEAR MOMENTUM EQUATION—CAUCHY'S EQUATION

Through application of the Reynolds transport theorem (Chap. 4), we have the general expression for the linear momentum equation as applied to a control volume,

$$\sum \vec{F} = \int_{CV} \rho \vec{g}\, dV + \int_{CS} \sigma_{ij} \cdot \vec{n}\, dA = \int_{CV} \frac{\partial}{\partial t}(\rho \vec{V})\, dV + \int_{CS} (\rho \vec{V}) \vec{V} \cdot \vec{n}\, dA \qquad (9\text{--}32)$$

where σ_{ij} is the **stress tensor** introduced in Chap. 6. Components of σ_{ij} on the positive faces of an infinitesimal rectangular control volume are shown in Fig. 9–29. Equation 9–32 applies to both fixed and moving control volumes, provided that $\vec{V}$ is the absolute velocity (as seen from a fixed observer). For the special case of flow with well defined inlets and outlets, Eq. 9–32 is simplified as follows:

$$\sum \vec{F} = \sum \vec{F}_{\text{body}} + \sum \vec{F}_{\text{surface}} = \int_{CV} \frac{\partial}{\partial t}(\rho \vec{V})\, dV + \sum_{\text{out}} \beta \dot{m} \vec{V} - \sum_{\text{in}} \beta \dot{m} \vec{V} \qquad (9\text{--}33)$$

where $\vec{V}$ in the last two terms is taken as the average velocity at an inlet or outlet, and β is the momentum flux correction factor (Chap. 6). In words, the total force acting on the control volume is equal to the rate at which momentum changes within the control volume plus the rate at which momentum flows out of the control volume minus the rate at which momentum flows into the control volume. Equation 9–33 applies to *any* control volume, regardless of its size. To generate a differential linear momentum equation, we imagine the control volume shrinking to infinitesimal size. In the limit, the entire control volume shrinks to a *point* in the flow (Fig. 9–2). We take the same approach here as we did for conservation of mass; namely, we show more than one way to derive the differential form of the linear momentum equation.

Derivation Using the Divergence Theorem

The most straightforward (and most elegant) way to derive the differential form of the momentum equation is to apply the divergence theorem of Eq. 9–3. A more general form of the divergence theorem applies not only to vectors, but to other quantities as well, such as tensors, as illustrated in

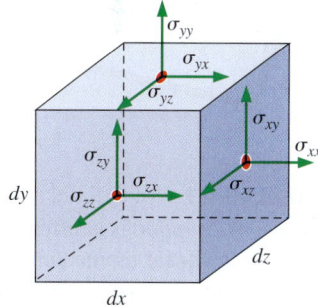

FIGURE 9–29
Positive components of the stress tensor in Cartesian coordinates on the positive (right, top, and front) faces of an infinitesimal rectangular control volume. The red dots indicate the center of each face. Positive components on the negative (left, bottom, and back) faces are in the opposite direction of those shown here.

DIFFERENTIAL ANALYSIS OF FLUID FLOW

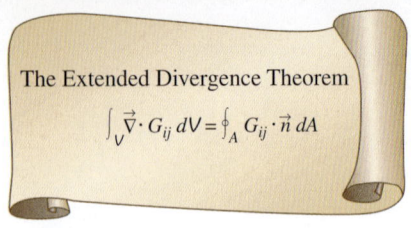

FIGURE 9–30
An extended form of the divergence theorem is useful not only for vectors, but also for tensors. In the equation, G_{ij} is a second-order tensor, V is a volume, and A is the surface area that encloses and defines the volume.

Fig. 9–30. Specifically, if we replace G_{ij} in the extended divergence theorem of Fig. 9–30 with the quantity $(\rho\vec{V})\vec{V}$, a second-order tensor, the last term in Eq. 9–32 becomes

$$\int_{CS} (\rho\vec{V})\vec{V}\cdot\vec{n}\, dA = \int_{CV} \vec{\nabla}\cdot(\rho\vec{V}\vec{V})\, dV \qquad (9\text{--}34)$$

where $\vec{V}\vec{V}$ is a vector product called the *outer product* of the velocity vector with itself. (The outer product of two vectors is *not* the same as the inner or dot product, nor is it the same as the cross product of the two vectors.) Similarly, if we replace G_{ij} in Fig. 9–30 by the stress tensor σ_{ij}, the second term on the left-hand side of Eq. 9–32 becomes

$$\int_{CS} \sigma_{ij}\cdot\vec{n}\, dA = \int_{CV} \vec{\nabla}\cdot\sigma_{ij}\, dV \qquad (9\text{--}35)$$

Thus, the two surface integrals of Eq. 9–32 become volume integrals by applying Eqs. 9–34 and 9–35. We combine and rearrange the terms, and rewrite Eq. 9–32 as

$$\int_{CV}\left[\frac{\partial}{\partial t}(\rho\vec{V}) + \vec{\nabla}\cdot(\rho\vec{V}\vec{V}) - \rho\vec{g} - \vec{\nabla}\cdot\sigma_{ij}\right] dV = 0 \qquad (9\text{--}36)$$

Finally, we argue that Eq. 9–36 must hold for *any* control volume regardless of its size or shape. This is possible only if the integrand (enclosed by square brackets) is identically zero. Hence, we have a general differential equation for linear momentum, known as **Cauchy's equation,**

Cauchy's equation: $\qquad \dfrac{\partial}{\partial t}(\rho\vec{V}) + \vec{\nabla}\cdot(\rho\vec{V}\vec{V}) = \rho\vec{g} + \vec{\nabla}\cdot\sigma_{ij} \qquad (9\text{--}37)$

Equation 9–37 is named in honor of the French engineer and mathematician Augustin Louis de Cauchy (1789–1857). It is valid for compressible as well as incompressible flow since we have not made any assumptions about incompressibility. It is valid at any point in the flow domain (Fig. 9–31). Note that Eq. 9–37 is a *vector* equation, and thus represents three scalar equations, one for each coordinate axis in three-dimensional problems.

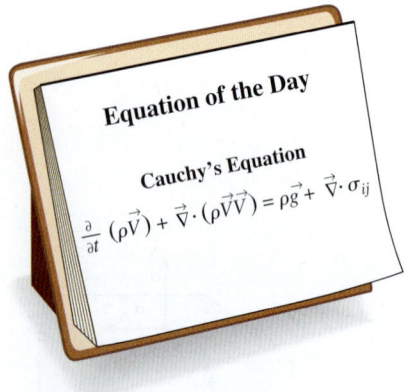

FIGURE 9–31
Cauchy's equation is a differential form of the linear momentum equation. It applies to any type of fluid.

Derivation Using an Infinitesimal Control Volume

We derive Cauchy's equation a second way, using an infinitesimal control volume on which we apply the linear momentum equation (Eq. 9–33). We consider the same box-shaped control volume we used to derive the continuity equation (Fig. 9–3). At the center of the box, as previously, we define the density as ρ and the velocity components as u, v, and w. We also define the stress tensor as σ_{ij} at the center of the box. For simplicity, we consider the x-component of Eq. 9–33, obtained by setting $\Sigma\vec{F}$ equal to its x-component, ΣF_x, and $\vec{V}$ equal to its x-component, u. This not only simplifies the diagrams, but enables us to work with a scalar equation, namely,

$$\Sigma F_x = \Sigma F_{x,\,body} + \Sigma F_{x,\,surface} = \int_{CV} \frac{\partial}{\partial t}(\rho u)\, dV + \sum_{out}\beta\dot{m}u - \sum_{in}\beta\dot{m}u \qquad (9\text{--}38)$$

As the control volume shrinks to a point, the first term on the right-hand side of Eq. 9–38 becomes

Rate of change of x-momentum within the control volume:

$$\int_{CV} \frac{\partial}{\partial t}(\rho u)\, dV \cong \frac{\partial}{\partial t}(\rho u)\, dx\, dy\, dz \qquad (9\text{--}39)$$

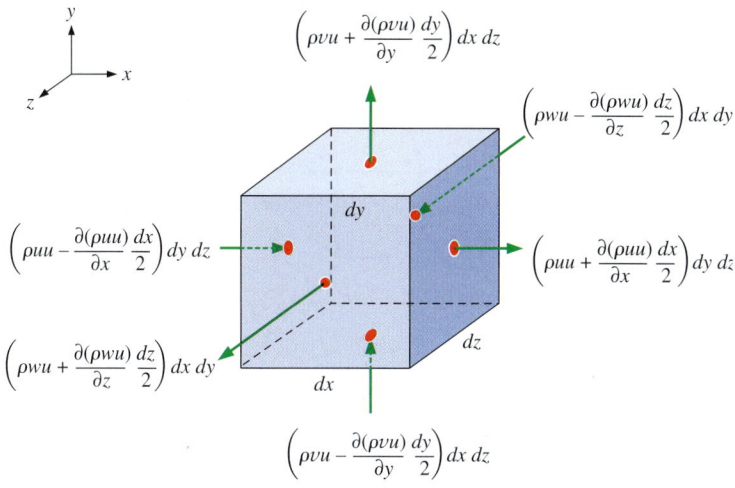

FIGURE 9–32
Inflow and outflow of the x-component of linear momentum through each face of an infinitesimal control volume; the red dots indicate the center of each face.

since the volume of the differential element is $dx\,dy\,dz$. We apply first-order truncated Taylor series expansions at locations away from the center of the control volume to approximate the inflow and outflow of momentum in the x-direction. Figure 9–32 shows these momentum fluxes at the center point of each of the six faces of the infinitesimal control volume. Only the *normal* velocity component at each face needs to be considered, since the tangential velocity components contribute no mass flow out of (or into) the face, and hence no momentum flow through the face either.

By summing all the outflows and subtracting all the inflows shown in Fig. 9–32, we obtain an approximation for the last two terms of Eq. 9–38,

Net outflow of x-momentum through the control surface:

$$\sum_{\text{out}} \beta \dot{m} u - \sum_{\text{in}} \beta \dot{m} u \cong \left(\frac{\partial}{\partial x}(\rho u u) + \frac{\partial}{\partial y}(\rho v u) + \frac{\partial}{\partial z}(\rho w u) \right) dx\,dy\,dz \quad \textbf{(9–40)}$$

where β is set equal to one at all faces, consistent with our first-order approximation.

Next, we sum all the forces acting on our infinitesimal control volume in the x-direction. As was done in Chap. 6, we need to consider both body forces and surface forces. Gravity force (weight) is the only body force we take into account. For the general case in which the coordinate system may not be aligned with the z-axis (or with any coordinate axis for that matter), as sketched in Fig. 9–33, the gravity vector is written as

$$\vec{g} = g_x \vec{i} + g_y \vec{j} + g_z \vec{k}$$

Thus, in the x-direction, the body force on the control volume is

$$\sum F_{x,\text{body}} = \sum F_{x,\text{gravity}} \cong \rho g_x\, dx\,dy\,dz \quad \textbf{(9–41)}$$

Next we consider the net surface force in the x-direction. Recall that stress tensor σ_{ij} has dimensions of force per unit area. Thus, to obtain a force, we must multiply each stress component by the surface area of the face on

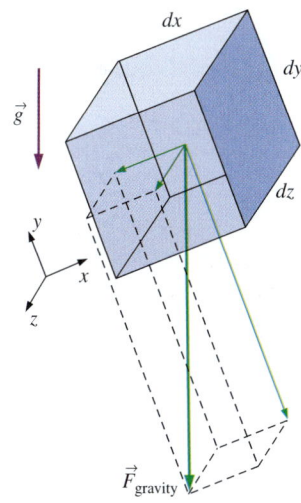

FIGURE 9–33
The gravity vector is not necessarily aligned with any particular axis, in general, and there are three components of the body force acting on an infinitesimal fluid element.

DIFFERENTIAL ANALYSIS OF FLUID FLOW

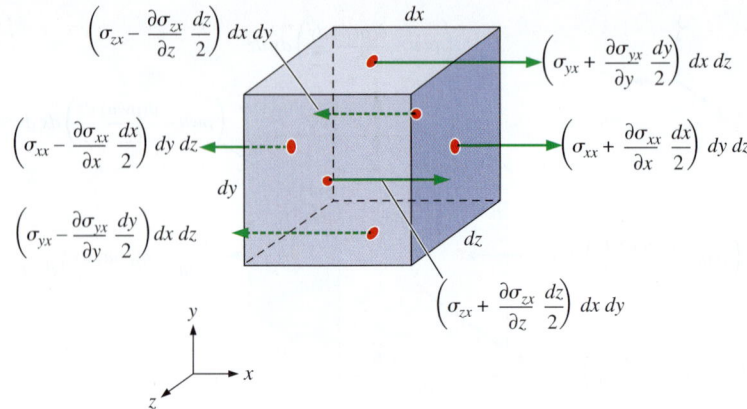

FIGURE 9–34
Sketch illustrating the surface forces acting in the x-direction due to the appropriate stress tensor component on each face of the differential control volume; the red dots indicate the center of each face.

which it acts. We need to consider only those components that point in the x- (or −x-) direction. (The other components of the stress tensor, although they may be nonzero, do not contribute to a net force in the x-direction.) Using truncated Taylor series expansions, we sketch all the surface forces that contribute to a net x-component of surface force acting on our differential fluid element (Fig. 9–34).

Summing all the surface forces illustrated in Fig. 9–34, we obtain an approximation for the net surface force acting on the differential fluid element in the x-direction,

$$\sum F_{x,\text{ surface}} \cong \left(\frac{\partial}{\partial x} \sigma_{xx} + \frac{\partial}{\partial y} \sigma_{yx} + \frac{\partial}{\partial z} \sigma_{zx} \right) dx\, dy\, dz \qquad (9\text{–}42)$$

We now substitute Eqs. 9–39 through 9–42 into Eq. 9–38, noting that the volume of the differential element of fluid, $dx\, dy\, dz$, appears in all terms and can be eliminated. After some rearrangement we obtain the differential form of the x-momentum equation,

$$\frac{\partial(\rho u)}{\partial t} + \frac{\partial(\rho u u)}{\partial x} + \frac{\partial(\rho v u)}{\partial y} + \frac{\partial(\rho w u)}{\partial z} = \rho g_x + \frac{\partial}{\partial x} \sigma_{xx} + \frac{\partial}{\partial y} \sigma_{yx} + \frac{\partial}{\partial z} \sigma_{zx} \qquad (9\text{–}43)$$

In similar fashion, we generate differential forms of the y- and z-momentum equations,

$$\frac{\partial(\rho v)}{\partial t} + \frac{\partial(\rho u v)}{\partial x} + \frac{\partial(\rho v v)}{\partial y} + \frac{\partial(\rho w v)}{\partial z} = \rho g_y + \frac{\partial}{\partial x} \sigma_{xy} + \frac{\partial}{\partial y} \sigma_{yy} + \frac{\partial}{\partial z} \sigma_{zy} \qquad (9\text{–}44)$$

and

$$\frac{\partial(\rho w)}{\partial t} + \frac{\partial(\rho u w)}{\partial x} + \frac{\partial(\rho v w)}{\partial y} + \frac{\partial(\rho w w)}{\partial z} = \rho g_z + \frac{\partial}{\partial x} \sigma_{xz} + \frac{\partial}{\partial y} \sigma_{yz} + \frac{\partial}{\partial z} \sigma_{zz} \qquad (9\text{–}45)$$

respectively. Finally, we combine Eqs. 9–43 through 9–45 into one vector equation,

Cauchy's equation: $\quad \dfrac{\partial}{\partial t}(\rho \vec{V}) + \vec{\nabla} \cdot (\rho \vec{V} \vec{V}) = \rho \vec{g} + \vec{\nabla} \cdot \sigma_{ij}$

This equation is identical to Cauchy's equation (Eq. 9–37); thus we confirm that our derivation using the differential fluid element yields the same result as our derivation using the divergence theorem. Note that the product $\vec{V}\vec{V}$ is a second-order tensor (Fig. 9–35).

Alternative Form of Cauchy's Equation

Applying the product rule to the first term on the left side of Eq. 9–37, we get

$$\frac{\partial}{\partial t}(\rho \vec{V}) = \rho \frac{\partial \vec{V}}{\partial t} + \vec{V}\frac{\partial \rho}{\partial t} \qquad (9\text{–}46)$$

The second term of Eq. 9–37 is written as

$$\vec{\nabla}\cdot(\rho \vec{V}\vec{V}) = \vec{V}\vec{\nabla}\cdot(\rho\vec{V}) + \rho(\vec{V}\cdot\vec{\nabla})\vec{V} \qquad (9\text{–}47)$$

Thus we have eliminated the second-order tensor represented by $\vec{V}\vec{V}$. After some rearrangement, substitution of Eqs. 9–46 and 9–47 into Eq. 9–37 yields

$$\rho\frac{\partial \vec{V}}{\partial t} + \vec{V}\left[\frac{\partial \rho}{\partial t} + \vec{\nabla}\cdot(\rho\vec{V})\right] + \rho(\vec{V}\cdot\vec{\nabla})\vec{V} = \rho\vec{g} + \vec{\nabla}\cdot\sigma_{ij}$$

But the expression in square brackets in this equation is identically zero by the continuity equation, Eq. 9–5. By combining the remaining two terms on the left side, we write

Alternative form of Cauchy's equation:

$$\rho\left[\frac{\partial \vec{V}}{\partial t} + (\vec{V}\cdot\vec{\nabla})\vec{V}\right] = \rho\frac{D\vec{V}}{Dt} = \rho\vec{g} + \vec{\nabla}\cdot\sigma_{ij} \qquad (9\text{–}48)$$

where we have recognized the expression in square brackets as the material acceleration—the acceleration following a fluid particle (see Chap. 4).

Derivation Using Newton's Second Law

We derive Cauchy's equation by yet a third method. Namely, we take the differential fluid element as a *material element* instead of a control volume. In other words, we think of the fluid within the differential element as a tiny *system* of fixed identity, moving with the flow (Fig. 9–36). The acceleration of this fluid element is $\vec{a} = D\vec{V}/Dt$ by definition of the material acceleration. By Newton's second law applied to a material element of fluid,

$$\sum \vec{F} = m\vec{a} = m\frac{D\vec{V}}{Dt} = \rho\,dx\,dy\,dz\,\frac{D\vec{V}}{Dt} \qquad (9\text{–}49)$$

At the instant in time represented in Fig. 9–36, the net force on the differential fluid element is found in the same way as that calculated earlier on the differential control volume. Thus the total force acting on the fluid element is the sum of Eqs. 9–41 and 9–42, extended to vector form. Substituting these into Eq. 9–49 and dividing by $dx\,dy\,dz$, we once again generate the alternative form of Cauchy's equation,

$$\rho\frac{D\vec{V}}{Dt} = \rho\vec{g} + \vec{\nabla}\cdot\sigma_{ij} \qquad (9\text{–}50)$$

Equation 9–50 is identical to Eq. 9–48. In hindsight, we could have started with Newton's second law from the beginning, avoiding some algebra. Nevertheless, derivation of Cauchy's equation by three methods certainly boosts our confidence in the validity of the equation!

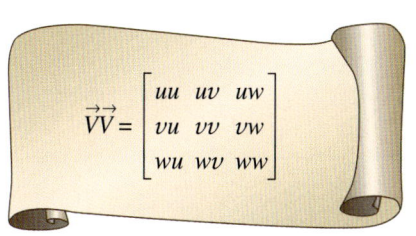

FIGURE 9–35
The outer product of vector $\vec{V} = (u, v, w)$ with itself is a second-order tensor. The product shown is in Cartesian coordinates and is illustrated as a nine-component matrix.

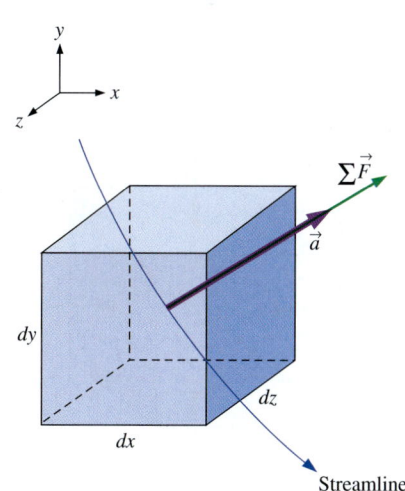

FIGURE 9–36
If the differential fluid element is a material element, it moves with the flow and Newton's second law applies directly.

We must be very careful when expanding the last term of Eq. 9–50, which is the divergence of a second-order tensor. In Cartesian coordinates, the three components of Cauchy's equation are

$$\text{x-component:} \quad \rho \frac{Du}{Dt} = \rho g_x + \frac{\partial \sigma_{xx}}{\partial x} + \frac{\partial \sigma_{yx}}{\partial y} + \frac{\partial \sigma_{zx}}{\partial z} \quad (9\text{–}51a)$$

$$\text{y-component:} \quad \rho \frac{Dv}{Dt} = \rho g_y + \frac{\partial \sigma_{xy}}{\partial x} + \frac{\partial \sigma_{yy}}{\partial y} + \frac{\partial \sigma_{zy}}{\partial z} \quad (9\text{–}51b)$$

$$\text{z-component:} \quad \rho \frac{Dw}{Dt} = \rho g_z + \frac{\partial \sigma_{xz}}{\partial x} + \frac{\partial \sigma_{yz}}{\partial y} + \frac{\partial \sigma_{zz}}{\partial z} \quad (9\text{–}51c)$$

We conclude this section by noting that we cannot solve any fluid mechanics problems using Cauchy's equation by itself (even when combined with continuity). The problem is that the stress tensor σ_{ij} needs to be expressed in terms of the primary unknowns in the problem, namely, density, pressure, and velocity. This is done for the most common type of fluid in Section 9–5.

9–5 ▪ THE NAVIER–STOKES EQUATION

Introduction

Cauchy's equation (Eq. 9–37 or its alternative form Eq. 9–48) is not very useful to us as is, because the stress tensor σ_{ij} contains nine components, six of which are independent (because of symmetry). Thus, in addition to density and the three velocity components, there are six additional unknowns, for a total of 10 unknowns. (In Cartesian coordinates the unknowns are ρ, u, v, w, σ_{xx}, σ_{xy}, σ_{xz}, σ_{yy}, σ_{yz}, and σ_{zz}). Meanwhile, we have discussed only four equations so far—continuity (one equation) and Cauchy's equation (three equations). Of course, to be mathematically solvable, the number of equations must equal the number of unknowns, and thus we need six more equations. These equations are called **constitutive equations,** and they enable us to write the components of the stress tensor in terms of the velocity field and pressure field.

The first thing we do is separate the pressure stresses and the viscous stresses. When a fluid is at rest, the only stress acting at *any* surface of *any* fluid element is the local hydrostatic pressure P, which always acts *inward* and *normal* to the surface (Fig. 9–37). Thus, regardless of the orientation of the coordinate axes, for a fluid at rest the stress tensor reduces to

$$\text{Fluid at rest:} \quad \sigma_{ij} = \begin{pmatrix} \sigma_{xx} & \sigma_{xy} & \sigma_{xz} \\ \sigma_{yx} & \sigma_{yy} & \sigma_{yz} \\ \sigma_{zx} & \sigma_{zy} & \sigma_{zz} \end{pmatrix} = \begin{pmatrix} -P & 0 & 0 \\ 0 & -P & 0 \\ 0 & 0 & -P \end{pmatrix} \quad (9\text{–}52)$$

Hydrostatic pressure P in Eq. 9–52 is the same as the **thermodynamic pressure** with which we are familiar from our study of thermodynamics. P is related to temperature and density through some type of **equation of state** (e.g., the ideal gas law). As a side note, this further complicates a compressible fluid flow analysis because we introduce yet another unknown, namely, temperature T. This new unknown requires another equation—the differential form of the energy equation—which is not discussed in this text.

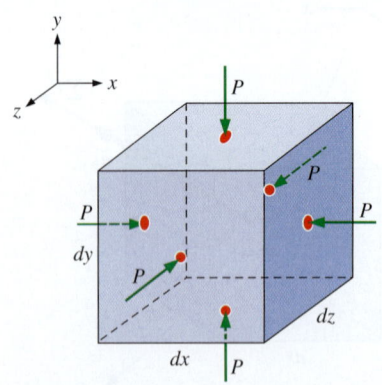

FIGURE 9–37
For fluids at rest, the only stress on a fluid element is the hydrostatic pressure, which always acts inward and normal to any surface.

When a fluid is *moving*, pressure still acts inwardly normal, but viscous stresses may also exist. We generalize Eq. 9–52 for moving fluids as

Moving fluids:

$$\sigma_{ij} = \begin{pmatrix} \sigma_{xx} & \sigma_{xy} & \sigma_{xz} \\ \sigma_{yx} & \sigma_{yy} & \sigma_{yz} \\ \sigma_{zx} & \sigma_{zy} & \sigma_{zz} \end{pmatrix} = \begin{pmatrix} -P & 0 & 0 \\ 0 & -P & 0 \\ 0 & 0 & -P \end{pmatrix} + \begin{pmatrix} \tau_{xx} & \tau_{xy} & \tau_{xz} \\ \tau_{yx} & \tau_{yy} & \tau_{yz} \\ \tau_{zx} & \tau_{zy} & \tau_{zz} \end{pmatrix} \quad (9\text{--}53)$$

where we have introduced a new tensor, τ_{ij}, called the **viscous stress tensor** or the **deviatoric stress tensor.** Mathematically, we have not helped the situation because we have replaced the six unknown components of σ_{ij} with six unknown components of τ_{ij}, and have added *another* unknown, pressure P. Fortunately, however, there are constitutive equations that express τ_{ij} in terms of the velocity field and measurable fluid properties such as viscosity. The actual form of the constitutive relations depends on the type of fluid, as discussed shortly.

As a side note, there are some subtleties associated with the pressure in Eq. 9–53. If the fluid is *incompressible*, we have no equation of state (it is replaced by the equation ρ = constant), and we can no longer define P as the thermodynamic pressure. Instead, we define P in Eq. 9–53 as the **mechanical pressure,**

Mechanical pressure: $\quad P_m = -\dfrac{1}{3}(\sigma_{xx} + \sigma_{yy} + \sigma_{zz}) \quad$ (9–54)

We see from Eq. 9–54 that *mechanical pressure is the mean normal stress acting inwardly on a fluid element*. It is therefore also called **mean pressure** by some authors. Thus, when dealing with incompressible fluid flows, pressure variable P is always interpreted as the mechanical pressure P_m. For *compressible* flow fields however, pressure P in Eq. 9–53 *is* the thermodynamic pressure, but the mean normal stress felt on the surfaces of a fluid element is not necessarily the same as P (pressure variable P and mechanical pressure P_m are not necessarily equivalent). You are referred to Panton (1996) or Kundu et al., (2011) for a more detailed discussion of mechanical pressure.

Newtonian versus Non-Newtonian Fluids

The study of the deformation of flowing fluids is called **rheology;** the rheological behavior of various fluids is sketched in Fig. 9–38. In this text, we concentrate on **Newtonian fluids,** defined as *fluids for which the shear stress is linearly proportional to the shear strain rate*. Newtonian fluids (stress proportional to strain rate) are analogous to elastic solids (Hooke's law: stress proportional to strain). Many common fluids, such as air and other gases, water, kerosene, gasoline, and other oil-based liquids, are Newtonian fluids. Fluids for which the shear stress is *not* linearly related to the shear strain rate are called **non-Newtonian fluids.** Examples include slurries and colloidal suspensions, polymer solutions, blood, paste, and cake batter. Some non-Newtonian fluids exhibit a "memory"—the shear stress depends not only on the local strain rate, but also on its *history*. A fluid that returns (either fully or partially) to its original shape after the applied stress is released is called **viscoelastic.**

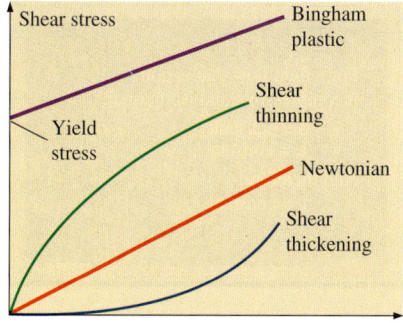

FIGURE 9–38
Rheological behavior of fluids—shear stress as a function of shear strain rate.

FIGURE 9–39
When an engineer falls into quicksand (a *dilatant fluid*), the faster he tries to move, the more viscous the fluid becomes.

Some non-Newtonian fluids are called **shear thinning fluids** or **pseudoplastic fluids,** because the more the fluid is sheared, the less viscous it becomes. A good example is paint. Paint is very viscous when poured from the can or when picked up by a paintbrush, since the shear rate is small. However, as we apply the paint to the wall, the thin layer of paint between the paintbrush and the wall is subjected to a large shear rate, and it becomes much less viscous. **Plastic fluids** are those in which the shear thinning effect is extreme. In some fluids a finite stress called the **yield stress** is required before the fluid begins to flow at all; such fluids are called **Bingham plastic fluids.** Certain pastes such as acne cream and toothpaste are examples of Bingham plastic fluids. If you hold the tube upside down, the paste does not flow, even though there is a nonzero stress due to gravity. However, if you squeeze the tube (greatly increasing the stress), the paste flows like a very viscous fluid. Other fluids show the opposite effect and are called **shear thickening fluids** or **dilatant fluids;** the more the fluid is sheared, the *more* viscous it becomes. The best example is quicksand, a thick mixture of sand and water. As we all know from Hollywood movies, it is easy to move *slowly* through quicksand, since the viscosity is low; but if you panic and try to move quickly, the viscous resistance increases considerably and you get "stuck" (Fig. 9–39). You can create your own quicksand by mixing two parts cornstarch with one part water—try it! Shear thickening fluids are used in some exercise equipment—the faster you pull, the more resistance you encounter.

Derivation of the Navier–Stokes Equation for Incompressible, Isothermal Flow

From this point on, we limit our discussion to Newtonian fluids, where by definition the stress tensor is linearly proportional to the strain rate tensor. The general result (for compressible flow) is rather involved and is not included here. Instead, we assume incompressible flow (ρ = constant). We also assume nearly isothermal flow—namely, that local changes in temperature are small or nonexistent; this eliminates the need for a differential energy equation. A further consequence of the latter assumption is that fluid properties, such as dynamic viscosity μ and kinematic viscosity ν, are constant as well (Fig. 9–40). With these assumptions, it can be shown (Kundu et al., 2011) that the viscous stress tensor reduces to

Viscous stress tensor for an incompressible Newtonian fluid with constant properties:

$$\tau_{ij} = 2\mu\varepsilon_{ij} \quad (9\text{–}55)$$

where ε_{ij} is the strain rate tensor defined in Chap. 4. Equation 9–55 shows that stress is linearly proportional to strain. In Cartesian coordinates, the nine components of the viscous stress tensor are listed, only six of which are independent due to symmetry:

For a fluid flow that is both incompressible and isothermal:
- ρ = constant
- μ = constant

And therefore:
- ν = constant

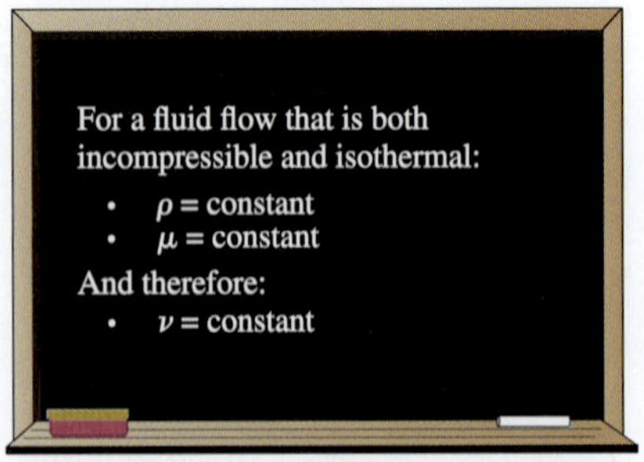

FIGURE 9–40
The incompressible flow approximation implies constant density, and the isothermal approximation implies constant viscosity.

$$\tau_{ij} = \begin{pmatrix} \tau_{xx} & \tau_{xy} & \tau_{xz} \\ \tau_{yx} & \tau_{yy} & \tau_{yz} \\ \tau_{zx} & \tau_{zy} & \tau_{zz} \end{pmatrix} = \begin{pmatrix} 2\mu\dfrac{\partial u}{\partial x} & \mu\left(\dfrac{\partial u}{\partial y}+\dfrac{\partial v}{\partial x}\right) & \mu\left(\dfrac{\partial u}{\partial z}+\dfrac{\partial w}{\partial x}\right) \\ \mu\left(\dfrac{\partial v}{\partial x}+\dfrac{\partial u}{\partial y}\right) & 2\mu\dfrac{\partial v}{\partial y} & \mu\left(\dfrac{\partial v}{\partial z}+\dfrac{\partial w}{\partial y}\right) \\ \mu\left(\dfrac{\partial w}{\partial x}+\dfrac{\partial u}{\partial z}\right) & \mu\left(\dfrac{\partial w}{\partial y}+\dfrac{\partial v}{\partial z}\right) & 2\mu\dfrac{\partial w}{\partial z} \end{pmatrix} \quad (9\text{–}56)$$

In Cartesian coordinates the stress tensor of Eq. 9–53 thus becomes

$$\sigma_{ij} = \begin{pmatrix} -P & 0 & 0 \\ 0 & -P & 0 \\ 0 & 0 & -P \end{pmatrix} + \begin{pmatrix} 2\mu \dfrac{\partial u}{\partial x} & \mu\left(\dfrac{\partial u}{\partial y} + \dfrac{\partial v}{\partial x}\right) & \mu\left(\dfrac{\partial u}{\partial z} + \dfrac{\partial w}{\partial x}\right) \\ \mu\left(\dfrac{\partial v}{\partial x} + \dfrac{\partial u}{\partial y}\right) & 2\mu \dfrac{\partial v}{\partial y} & \mu\left(\dfrac{\partial v}{\partial z} + \dfrac{\partial w}{\partial y}\right) \\ \mu\left(\dfrac{\partial w}{\partial x} + \dfrac{\partial u}{\partial z}\right) & \mu\left(\dfrac{\partial w}{\partial y} + \dfrac{\partial v}{\partial z}\right) & 2\mu \dfrac{\partial w}{\partial z} \end{pmatrix} \quad (9\text{–}57)$$

Now we substitute Eq. 9–57 into the three Cartesian components of Cauchy's equation. Let's consider the *x-component* first. Equation 9–51a becomes

$$\rho \frac{Du}{Dt} = -\frac{\partial P}{\partial x} + \rho g_x + 2\mu \frac{\partial^2 u}{\partial x^2} + \mu \frac{\partial}{\partial y}\left(\frac{\partial v}{\partial x} + \frac{\partial u}{\partial y}\right) + \mu \frac{\partial}{\partial z}\left(\frac{\partial w}{\partial x} + \frac{\partial u}{\partial z}\right) \quad (9\text{–}58)$$

Notice that since pressure consists of a normal stress only, it contributes only one term to Eq. 9–58. However, since the viscous stress tensor consists of both normal and shear stresses, it contributes *three* terms. (This is a direct result of taking the divergence of a second-order tensor, by the way.)

We note that as long as the velocity components are smooth functions of x, y, and z, the order of differentiation is irrelevant. For example, the first part of the last term in Eq. 9–58 can be rewritten as

$$\mu \frac{\partial}{\partial z}\left(\frac{\partial w}{\partial x}\right) = \mu \frac{\partial}{\partial x}\left(\frac{\partial w}{\partial z}\right)$$

After some clever rearrangement of the viscous terms in Eq. 9–58,

$$\rho \frac{Du}{Dt} = -\frac{\partial P}{\partial x} + \rho g_x + \mu \left[\frac{\partial^2 u}{\partial x^2} + \frac{\partial}{\partial x}\frac{\partial u}{\partial x} + \frac{\partial}{\partial x}\frac{\partial v}{\partial y} + \frac{\partial^2 u}{\partial y^2} + \frac{\partial}{\partial x}\frac{\partial w}{\partial z} + \frac{\partial^2 u}{\partial z^2}\right]$$

$$= -\frac{\partial P}{\partial x} + \rho g_x + \mu\left[\frac{\partial}{\partial x}\left(\frac{\partial u}{\partial x} + \frac{\partial v}{\partial y} + \frac{\partial w}{\partial z}\right) + \frac{\partial^2 u}{\partial x^2} + \frac{\partial^2 u}{\partial y^2} + \frac{\partial^2 u}{\partial z^2}\right]$$

The term in parentheses is zero because of the continuity equation for incompressible flow (Eq. 9–17). We also recognize the last three terms as the **Laplacian** of velocity component u in Cartesian coordinates (Fig. 9–41). Thus, we write the *x*-component of the momentum equation as

$$\rho \frac{Du}{Dt} = -\frac{\partial P}{\partial x} + \rho g_x + \mu \nabla^2 u \quad (9\text{–}59a)$$

Similarly, the *y*- and *z*-components of the momentum equation reduce to

$$\rho \frac{Dv}{Dt} = -\frac{\partial P}{\partial y} + \rho g_y + \mu \nabla^2 v \quad (9\text{–}59b)$$

and

$$\rho \frac{Dw}{Dt} = -\frac{\partial P}{\partial z} + \rho g_z + \mu \nabla^2 w \quad (9\text{–}59c)$$

respectively. Finally, we combine the three components into one vector equation; the result is the **Navier–Stokes equation** for incompressible flow with constant viscosity.

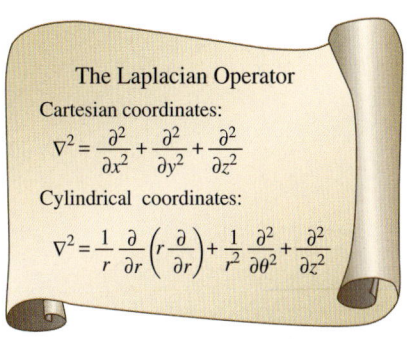

FIGURE 9–41
The Laplacian operator, shown here in both Cartesian and cylindrical coordinates, appears in the viscous term of the incompressible Navier–Stokes equation.

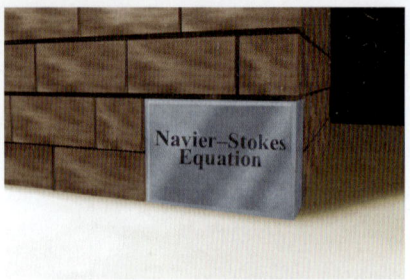

FIGURE 9–42
The Navier–Stokes equation is the cornerstone of fluid mechanics.

Incompressible Navier–Stokes equation:

$$\rho \frac{D\vec{V}}{Dt} = -\vec{\nabla}P + \rho\vec{g} + \mu\nabla^2\vec{V} \quad (9\text{–}60)$$

Although we derived the components of Eq. 9–60 in Cartesian coordinates, the vector form of Eq. 9–60 is valid in any orthogonal coordinate system. This famous equation is named in honor of the French engineer Louis Marie Henri Navier (1785–1836) and the English mathematician Sir George Gabriel Stokes (1819–1903), who both developed the viscous terms, although independently of each other.

The Navier–Stokes equation is the cornerstone of fluid mechanics (Fig. 9–42). It may look harmless enough, but it is an unsteady, nonlinear, second-order, partial differential equation. If we were able to solve this equation for flows of any geometry, this book would be about half as thick. Unfortunately, analytical solutions are unobtainable except for very simple flow fields. It is not too far from the truth to say that the rest of this book is devoted to solving Eq. 9–60! In fact, many researchers have spent their entire careers trying to solve the Navier–Stokes equation.

Equation 9–60 has four unknowns (three velocity components and pressure), yet it represents only three equations (three components since it is a vector equation). Obviously we need another equation to make the problem solvable. The fourth equation is the incompressible continuity equation (Eq. 9–16). Before we attempt to solve this set of differential equations, we need to choose a coordinate system and expand the equations in that coordinate system.

Continuity and Navier–Stokes Equations in Cartesian Coordinates

The continuity equation (Eq. 9–16) and the Navier–Stokes equation (Eq. 9–60) are expanded in Cartesian coordinates (x, y, z) and (u, v, w):

Incompressible continuity equation:

$$\frac{\partial u}{\partial x} + \frac{\partial v}{\partial y} + \frac{\partial w}{\partial z} = 0 \quad (9\text{–}61a)$$

x-component of the incompressible Navier–Stokes equation:

$$\rho\left(\frac{\partial u}{\partial t} + u\frac{\partial u}{\partial x} + v\frac{\partial u}{\partial y} + w\frac{\partial u}{\partial z}\right) = -\frac{\partial P}{\partial x} + \rho g_x + \mu\left(\frac{\partial^2 u}{\partial x^2} + \frac{\partial^2 u}{\partial y^2} + \frac{\partial^2 u}{\partial z^2}\right) \quad (9\text{–}61b)$$

y-component of the incompressible Navier–Stokes equation:

$$\rho\left(\frac{\partial v}{\partial t} + u\frac{\partial v}{\partial x} + v\frac{\partial v}{\partial y} + w\frac{\partial v}{\partial z}\right) = -\frac{\partial P}{\partial y} + \rho g_y + \mu\left(\frac{\partial^2 v}{\partial x^2} + \frac{\partial^2 v}{\partial y^2} + \frac{\partial^2 v}{\partial z^2}\right) \quad (9\text{–}61c)$$

z-component of the incompressible Navier–Stokes equation:

$$\rho\left(\frac{\partial w}{\partial t} + u\frac{\partial w}{\partial x} + v\frac{\partial w}{\partial y} + w\frac{\partial w}{\partial z}\right) = -\frac{\partial P}{\partial z} + \rho g_z + \mu\left(\frac{\partial^2 w}{\partial x^2} + \frac{\partial^2 w}{\partial y^2} + \frac{\partial^2 w}{\partial z^2}\right) \quad (9\text{–}61d)$$

Continuity and Navier–Stokes Equations in Cylindrical Coordinates

The continuity equation (Eq. 9–16) and the Navier–Stokes equation (Eq. 9–60) are expanded in cylindrical coordinates (r, θ, z) and (u_r, u_θ, u_z):

Incompressible continuity equation:
$$\frac{1}{r}\frac{\partial(ru_r)}{\partial r} + \frac{1}{r}\frac{\partial(u_\theta)}{\partial \theta} + \frac{\partial(u_z)}{\partial z} = 0 \quad (9\text{–}62a)$$

r-component of the incompressible Navier–Stokes equation:
$$\rho\left(\frac{\partial u_r}{\partial t} + u_r\frac{\partial u_r}{\partial r} + \frac{u_\theta}{r}\frac{\partial u_r}{\partial \theta} - \frac{u_\theta^2}{r} + u_z\frac{\partial u_r}{\partial z}\right)$$
$$= -\frac{\partial P}{\partial r} + \rho g_r + \mu\left[\frac{1}{r}\frac{\partial}{\partial r}\left(r\frac{\partial u_r}{\partial r}\right) - \frac{u_r}{r^2} + \frac{1}{r^2}\frac{\partial^2 u_r}{\partial \theta^2} - \frac{2}{r^2}\frac{\partial u_\theta}{\partial \theta} + \frac{\partial^2 u_r}{\partial z^2}\right] \quad (9\text{–}62b)$$

θ-component of the incompressible Navier–Stokes equation:
$$\rho\left(\frac{\partial u_\theta}{\partial t} + u_r\frac{\partial u_\theta}{\partial r} + \frac{u_\theta}{r}\frac{\partial u_\theta}{\partial \theta} + \frac{u_r u_\theta}{r} + u_z\frac{\partial u_\theta}{\partial z}\right)$$
$$= -\frac{1}{r}\frac{\partial P}{\partial \theta} + \rho g_\theta + \mu\left[\frac{1}{r}\frac{\partial}{\partial r}\left(r\frac{\partial u_\theta}{\partial r}\right) - \frac{u_\theta}{r^2} + \frac{1}{r^2}\frac{\partial^2 u_\theta}{\partial \theta^2} + \frac{2}{r^2}\frac{\partial u_r}{\partial \theta} + \frac{\partial^2 u_\theta}{\partial z^2}\right] \quad (9\text{–}62c)$$

z-component of the incompressible Navier–Stokes equation:
$$\rho\left(\frac{\partial u_z}{\partial t} + u_r\frac{\partial u_z}{\partial r} + \frac{u_\theta}{r}\frac{\partial u_z}{\partial \theta} + u_z\frac{\partial u_z}{\partial z}\right)$$
$$= -\frac{\partial P}{\partial z} + \rho g_z + \mu\left[\frac{1}{r}\frac{\partial}{\partial r}\left(r\frac{\partial u_z}{\partial r}\right) + \frac{1}{r^2}\frac{\partial^2 u_z}{\partial \theta^2} + \frac{\partial^2 u_z}{\partial z^2}\right] \quad (9\text{–}62d)$$

The first two viscous terms in Eqs. 9–62b and 9–62c can be manipulated to a different form that is often more useful when solving these equations (Fig. 9–43). The derivation is left as an exercise. The "extra" terms on both sides of the r- and θ-components of the Navier–Stokes equation (Eqs. 9–62b and 9–62c) arise because of the special nature of cylindrical coordinates. Namely, as we move in the θ-direction, the unit vector $\vec{e}_r$ also changes direction; thus the r- and θ-components are *coupled* (Fig. 9–44). (This coupling effect is not present in Cartesian coordinates, and thus there are no "extra" terms in Eqs. 9–61.)

For completeness, the six independent components of the viscous stress tensor are listed here in cylindrical coordinates,

$$\tau_{ij} = \begin{pmatrix} \tau_{rr} & \tau_{r\theta} & \tau_{rz} \\ \tau_{\theta r} & \tau_{\theta\theta} & \tau_{\theta z} \\ \tau_{zr} & \tau_{z\theta} & \tau_{zz} \end{pmatrix}$$

$$= \begin{pmatrix} 2\mu\dfrac{\partial u_r}{\partial r} & \mu\left[r\dfrac{\partial}{\partial r}\left(\dfrac{u_\theta}{r}\right) + \dfrac{1}{r}\dfrac{\partial u_r}{\partial \theta}\right] & \mu\left(\dfrac{\partial u_r}{\partial z} + \dfrac{\partial u_z}{\partial r}\right) \\ \mu\left[r\dfrac{\partial}{\partial r}\left(\dfrac{u_\theta}{r}\right) + \dfrac{1}{r}\dfrac{\partial u_r}{\partial \theta}\right] & 2\mu\left(\dfrac{1}{r}\dfrac{\partial u_\theta}{\partial \theta} + \dfrac{u_r}{r}\right) & \mu\left(\dfrac{\partial u_\theta}{\partial z} + \dfrac{1}{r}\dfrac{\partial u_z}{\partial \theta}\right) \\ \mu\left(\dfrac{\partial u_r}{\partial z} + \dfrac{\partial u_z}{\partial r}\right) & \mu\left(\dfrac{\partial u_\theta}{\partial z} + \dfrac{1}{r}\dfrac{\partial u_z}{\partial \theta}\right) & 2\mu\dfrac{\partial u_z}{\partial z} \end{pmatrix} \quad (9\text{–}63)$$

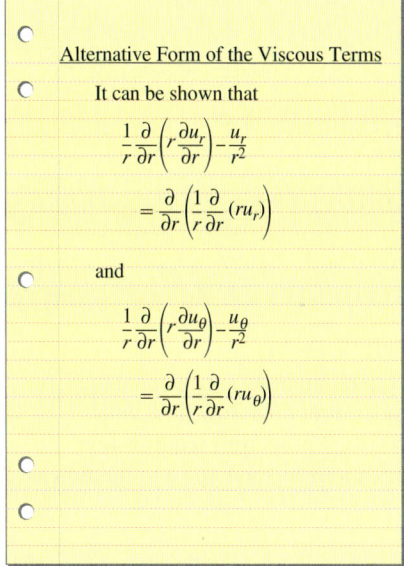

Alternative Form of the Viscous Terms

It can be shown that

$$\frac{1}{r}\frac{\partial}{\partial r}\left(r\frac{\partial u_r}{\partial r}\right) - \frac{u_r}{r^2}$$
$$= \frac{\partial}{\partial r}\left(\frac{1}{r}\frac{\partial}{\partial r}(ru_r)\right)$$

and

$$\frac{1}{r}\frac{\partial}{\partial r}\left(r\frac{\partial u_\theta}{\partial r}\right) - \frac{u_\theta}{r^2}$$
$$= \frac{\partial}{\partial r}\left(\frac{1}{r}\frac{\partial}{\partial r}(ru_\theta)\right)$$

FIGURE 9–43
An alternative form for the first two viscous terms in the r- and θ-components of the Navier–Stokes equation.

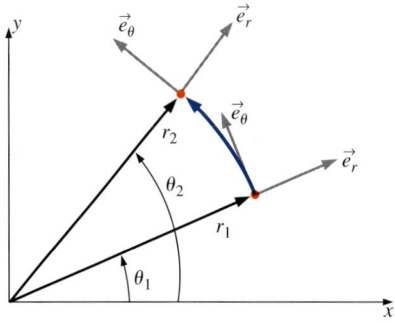

FIGURE 9–44
Unit vectors $\vec{e}_r$ and $\vec{e}_\theta$ in cylindrical coordinates are *coupled*: movement in the θ-direction causes $\vec{e}_r$ to change direction, and leads to extra terms in the r- and θ-components of the Navier–Stokes equation.

DIFFERENTIAL ANALYSIS OF FLUID FLOW

Three-Dimensional Incompressible Flow

Four variables or unknowns:
- Pressure P
- Three components of velocity $\vec{V}$

Four equations of motion:
- Continuity,
  $$\vec{\nabla} \cdot \vec{V} = 0$$
- Three components of Navier–Stokes,
  $$\rho \frac{D\vec{V}}{Dt} = -\vec{\nabla}P + \rho \vec{g} + \mu \nabla^2 \vec{V}$$

FIGURE 9–45
A general three-dimensional but incompressible flow field with constant properties requires four equations to solve for four unknowns.

9–6 ■ DIFFERENTIAL ANALYSIS OF FLUID FLOW PROBLEMS

In this section we show how to apply the differential equations of motion in both Cartesian and cylindrical coordinates. There are two types of problems for which the differential equations (continuity and Navier–Stokes) are useful:

- Calculating the pressure field for a known velocity field
- Calculating both the velocity and pressure fields for a flow of known geometry and known boundary conditions

For simplicity, we consider only incompressible flow, eliminating calculation of ρ as a variable. In addition, the form of the Navier–Stokes equation derived in Section 9–5 is valid only for Newtonian fluids with constant properties (viscosity, thermal conductivity, etc.). Finally, we assume negligible temperature variations, so that T is not a variable. We are left with four variables or unknowns (pressure plus three components of velocity), and we have four differential equations (Fig. 9–45).

Calculation of the Pressure Field for a Known Velocity Field

The first set of examples involves calculation of the pressure field for a known velocity field. Since pressure does not appear in the continuity equation, we can theoretically generate a velocity field based solely on conservation of mass. However, since velocity appears in both the continuity equation and the Navier–Stokes equation, these two equations are *coupled*. In addition, pressure appears in all three components of the Navier–Stokes equation, and thus the velocity and pressure fields are also coupled. This intimate coupling between velocity and pressure enables us to calculate the pressure field for a known velocity field.

EXAMPLE 9–13   Calculating the Pressure Field in Cartesian Coordinates

Consider the steady, two-dimensional, incompressible velocity field of Example 9–9, namely, $\vec{V} = (u, v) = (ax + b)\vec{i} + (-ay + cx)\vec{j}$. Calculate the pressure as a function of x and y.

SOLUTION For a given velocity field, we are to calculate the pressure field.
Assumptions **1** The flow is steady and incompressible. **2** The fluid has constant properties. **3** The flow is two-dimensional in the xy-plane. **4** Gravity does not act in either the x- or y-direction.
Analysis First we check whether the given velocity field satisfies the two-dimensional, incompressible continuity equation:

$$\underbrace{\frac{\partial u}{\partial x}}_{a} + \underbrace{\frac{\partial v}{\partial y}}_{-a} + \underbrace{\frac{\partial w}{\partial z}}_{0 \text{ (2-D)}} = a - a = 0 \qquad (1)$$

Thus, continuity is indeed satisfied by the given velocity field. If continuity were *not* satisfied, we would stop our analysis—the given velocity field would not be physically possible, and we could not calculate a pressure field.

Next, we consider the *y*-component of the Navier–Stokes equation:

$$\rho\left(\underbrace{\frac{\partial v}{\partial t}}_{0\text{ (steady)}} + u\underbrace{\frac{\partial v}{\partial x}}_{(ax+b)c} + v\underbrace{\frac{\partial v}{\partial y}}_{(-ay+cx)(-a)} + w\underbrace{\frac{\partial v}{\partial z}}_{0\text{ (2-D)}}\right) = -\frac{\partial P}{\partial y} + \underbrace{\rho g_y}_{0} + \mu\left(\underbrace{\frac{\partial^2 v}{\partial x^2}}_{0} + \underbrace{\frac{\partial^2 v}{\partial y^2}}_{0} + \underbrace{\frac{\partial^2 v}{\partial z^2}}_{0\text{ (2-D)}}\right)$$

The *y*-momentum equation reduces to

$$\frac{\partial P}{\partial y} = \rho(-acx - bc - a^2y + acx) = \rho(-bc - a^2y) \quad (2)$$

The *y*-momentum equation is satisfied if we can generate a pressure field that satisfies Eq. 2. In similar fashion, the *x*-momentum equation reduces to

$$\frac{\partial P}{\partial x} = \rho(-a^2x - ab) \quad (3)$$

The *x*-momentum equation is satisfied if we can generate a pressure field that satisfies Eq. 3.

In order for a steady flow solution to exist, *P* cannot be a function of time. Furthermore, a physically realistic steady, incompressible flow field requires a pressure field *P(x, y)* that is a smooth function of *x* and *y* (there can be no sudden discontinuities in either *P* or a derivative of *P*). Mathematically, this requires that the order of differentiation (*x* then *y* versus *y* then *x*) should not matter (Fig. 9–46). We check whether this is so by cross-differentiating Eqs. 2 and 3, respectively,

$$\frac{\partial^2 P}{\partial x\,\partial y} = \frac{\partial}{\partial x}\left(\frac{\partial P}{\partial y}\right) = 0 \quad \text{and} \quad \frac{\partial^2 P}{\partial y\,\partial x} = \frac{\partial}{\partial y}\left(\frac{\partial P}{\partial x}\right) = 0 \quad (4)$$

Equation 4 shows that *P* is indeed a smooth function of *x* and *y*. Thus, *the given velocity field satisfies the steady, two-dimensional, incompressible Navier–Stokes equation.*

If at this point in the analysis, the cross-differentiation of pressure were to yield two incompatible relationships (in other words if the equation in Fig. 9–46 were not satisfied) we would conclude that the given velocity field could not satisfy the steady, two-dimensional, incompressible Navier–Stokes equation, and we would abandon our attempt to calculate a steady pressure field.

To calculate *P(x, y)*, we partially integrate Eq. 2 (with respect to *y*)

Pressure field from y-momentum:

$$P(x, y) = \rho\left(-bcy - \frac{a^2y^2}{2}\right) + g(x) \quad (5)$$

Note that we add an arbitrary function of the other variable *x* rather than a constant of integration since this is a partial integration. We then take the partial derivative of Eq. 5 with respect to *x* to obtain

$$\frac{\partial P}{\partial x} = g'(x) = \rho(-a^2x - ab) \quad (6)$$

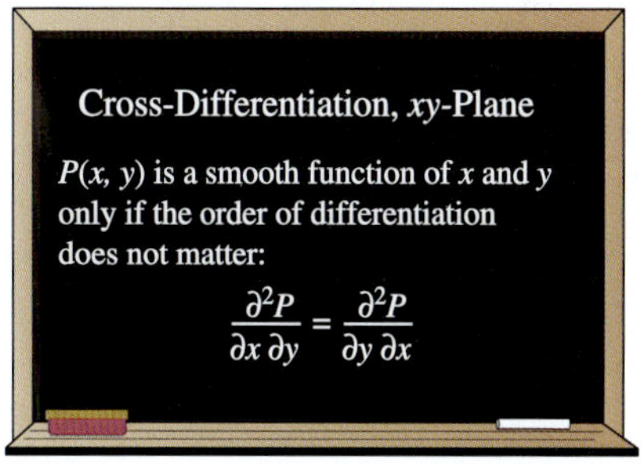

Cross-Differentiation, *xy*-Plane

P(x, y) is a smooth function of *x* and *y* only if the order of differentiation does not matter:

$$\frac{\partial^2 P}{\partial x\,\partial y} = \frac{\partial^2 P}{\partial y\,\partial x}$$

FIGURE 9–46
For a two-dimensional flow field in the *xy*-plane, cross-differentiation reveals whether pressure *P* is a smooth function.

where we have equated our result to Eq. 3 for consistency. We now integrate Eq. 6 to obtain the function $g(x)$:

$$g(x) = \rho\left(-\frac{a^2 x^2}{2} - abx\right) + C_1 \quad (7)$$

where C_1 is an arbitrary constant of integration. Finally, we substitute Eq. 7 into Eq. 5 to obtain our final expression for $P(x, y)$. The result is

$$P(x, y) = \rho\left(-\frac{a^2 x^2}{2} - \frac{a^2 y^2}{2} - abx - bcy\right) + C_1 \quad (8)$$

Discussion For practice, and as a check of our algebra, you should differentiate Eq. 8 with respect to both y and x, and compare to Eqs. 2 and 3. In addition, try to obtain Eq. 8 by starting with Eq. 3 rather than Eq. 2; you should get the same answer.

Notice that the final equation (Eq. 8) for pressure in Example 9–13 contains an arbitrary constant C_1. This illustrates an important point about the pressure field in an incompressible flow; namely,

The velocity field in an incompressible flow is not affected by the absolute magnitude of pressure, but only by pressure differences.

This should not be surprising if we look at the Navier–Stokes equation, where P appears only as a *gradient*, never by itself. Another way to explain this statement is that it is not the absolute magnitude of pressure that matters, but only pressure *differences* (Fig. 9–47). A direct result of the statement is that we can calculate the pressure field to within an arbitrary constant, but in order to determine that constant (C_1 in Example 9–13), we must measure (or otherwise obtain) P somewhere in the flow field. In other words, we require a pressure boundary condition.

We illustrate this point with an example generated using **computational fluid dynamics (CFD),** where the continuity and Navier–Stokes equations are solved numerically (Chap. 15). Consider downward flow of air through a channel in which there is a nonsymmetrical blockage (Fig. 9–48). (Note that the computational flow domain extends much further upstream and downstream than shown in Fig. 9–48.) We calculate two cases that are identical except for the pressure condition. In case 1 we set the gage pressure far downstream of the blockage to zero. In case 2 we set the pressure at the same location to 500 Pa gage pressure. The gage pressure at the top center of the field of view and at the bottom center of the field of view are shown in Fig. 9–48 for both cases, as generated by the two CFD solutions. You can see that the pressure field for case 2 is identical to that of case 1 except that the pressure is everywhere increased by 500 Pa. Also shown in Fig. 9–48 are a velocity vector plot and a streamline plot for each case. The results are identical, confirming our statement that the velocity field is not affected by the absolute magnitude of the pressure, but only by pressure *differences*. Subtracting the pressure at the bottom from that at the top, we see that $\Delta P = 12.784$ Pa for both cases.

The statement about pressure differences is *not* true for *compressible* flow fields, where P is the thermodynamic pressure rather than the mechanical

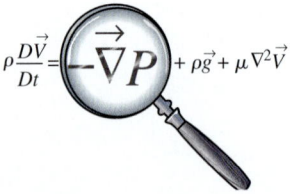

FIGURE 9–47
Since pressure appears only as a gradient in the incompressible Navier–Stokes equation, the absolute magnitude of pressure is not relevant—only pressure *differences* matter.

pressure. In such cases, P is coupled with density and temperature through an equation of state, and the absolute magnitude of pressure *is* important. A compressible flow solution requires not only mass and momentum equations, but also an energy equation and an equation of state.

We take this opportunity to comment further about the CFD results shown in Fig. 9–48. You can learn a lot about the physics of fluid flow by studying relatively simple flows like this. Notice that most of the pressure drop occurs across the throat of the channel where the flow is rapidly accelerated. There is also flow separation downstream of the blockage; rapidly moving air cannot turn around a sharp corner, and the flow separates off the walls as it exits the opening. The streamlines indicate large recirculating regions on both sides of the channel downstream of the blockage. Pressure is low in these recirculating regions. The velocity vectors indicate an inverse bell-shaped velocity profile exiting the opening—much like an exhaust jet. Because of the nonsymmetric nature of the geometry, the jet turns to the right, and the flow reattaches to the right wall much sooner than to the left wall. The pressure increases somewhat in the region where the jet impinges on the right wall, as you might expect. Finally, notice that as the air accelerates to squeeze through the orifice, the streamlines converge (as discussed in Section 9–3). As the jet of air fans out downstream, the streamlines diverge somewhat. Notice also that the streamlines in the recirculating zones are very far apart, indicating that the velocities are relatively small there; this is verified by the velocity vector plots.

Finally, we note that most CFD codes do *not* calculate pressure by integration of the Navier–Stokes equation as we have done in Example 9–13. Instead, some kind of **pressure correction algorithm** is used. Most of the commonly used algorithms work by combining the continuity and Navier–Stokes equations in such a way that pressure appears in the continuity equation. The most popular pressure correction algorithms result in a form of **Poisson's equation** for the change in pressure ΔP from one iteration (n) to the next $(n + 1)$,

Poisson's equation for ΔP: $\qquad \nabla^2(\Delta P) = \text{RHS}_{(n)}$ \hfill (9–64)

Then, as the computer iterates toward a solution, the modified continuity equation is used to "correct" the pressure field at iteration $(n + 1)$ from its values at iteration (n),

Correction for P: $\qquad P_{(n+1)} = P_{(n)} + \Delta P$

Details associated with the development of pressure correction algorithms is beyond the scope of the present text. An example for two-dimensional flows is developed in Gerhart, Gross, and Hochstein (1992).

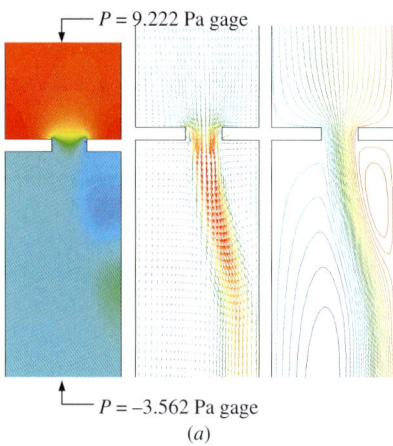

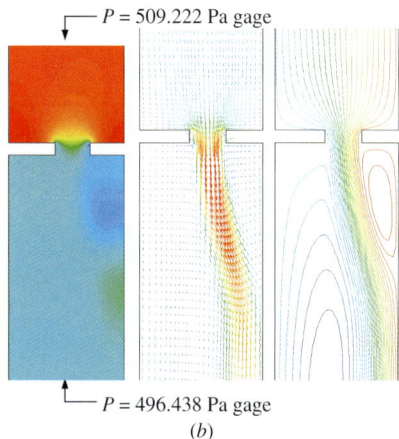

FIGURE 9–48
Filled pressure contour plot, velocity vector plot, and streamlines for downward flow of air through a channel with blockage: (*a*) case 1; (*b*) case 2—identical to case 1, except P is everywhere increased by 500 Pa. On the contour plots, blue is low pressure and red is high pressure.

■ **EXAMPLE 9–14**   **Calculating the Pressure Field in Cylindrical Coordinates**

Consider the steady, two-dimensional, incompressible velocity field of Example 9–5 with function $f(\theta, t)$ equal to 0. This represents a line vortex whose axis lies along the z-coordinate (Fig. 9–49). The velocity components are $u_r = 0$ and $u_\theta = K/r$, where K is a constant. Calculate the pressure as a function of r and θ.

474
DIFFERENTIAL ANALYSIS OF FLUID FLOW

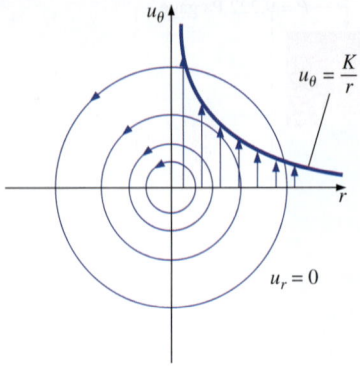

FIGURE 9–49
Streamlines and velocity profiles for a line vortex.

SOLUTION For a given velocity field, we are to calculate the pressure field.
Assumptions 1 The flow is steady. 2 The fluid is incompressible with constant properties. 3 The flow is two-dimensional in the $r\theta$-plane. 4 Gravity does not act in either the r- or the θ-direction.
Analysis The flow field must satisfy both the continuity and the momentum equations, Eqs. 9–62. For steady, two-dimensional, incompressible flow,

Incompressible continuity:
$$\underbrace{\frac{1}{r}\frac{\partial(ru_r)}{\partial r}}_{0} + \underbrace{\frac{1}{r}\frac{\partial(u_\theta)}{\partial \theta}}_{0} + \underbrace{\frac{\partial(u_z)}{\partial z}}_{0} = 0$$

Thus, the incompressible continuity equation is satisfied. Now we look at the θ component of the Navier–Stokes equation (Eq. 9–62c):

$$\rho\left(\underbrace{\frac{\partial u_\theta}{\partial t}}_{0 \text{ (steady)}} + \underbrace{u_r\frac{\partial u_\theta}{\partial r}}_{(0)\left(-\frac{K}{r^2}\right)} + \underbrace{\frac{u_\theta}{r}\frac{\partial u_\theta}{\partial \theta}}_{\left(\frac{K}{r^2}\right)(0)} + \underbrace{\frac{u_r u_\theta}{r}}_{0} + \underbrace{u_z\frac{\partial u_\theta}{\partial z}}_{0 \text{ (2-D)}}\right)$$

$$= -\frac{1}{r}\frac{\partial P}{\partial \theta} + \underbrace{\rho g_\theta}_{0} + \mu\left(\underbrace{\frac{1}{r}\frac{\partial}{\partial r}\left(r\frac{\partial u_\theta}{\partial r}\right)}_{\frac{K}{r^3}} - \underbrace{\frac{u_\theta}{r^2}}_{\frac{K}{r^3}} + \underbrace{\frac{1}{r^2}\frac{\partial^2 u_\theta}{\partial \theta^2}}_{0} + \underbrace{\frac{2}{r^2}\frac{\partial u_r}{\partial \theta}}_{0} + \underbrace{\frac{\partial^2 u_\theta}{\partial z^2}}_{0 \text{ (2-D)}}\right)$$

The θ-momentum equation therefore reduces to

θ-momentum:
$$\frac{\partial P}{\partial \theta} = 0 \qquad (1)$$

Thus, the θ-momentum equation is satisfied if we can generate an appropriate pressure field that satisfies Eq. 1. In similar fashion, the r-momentum equation (Eq. 9–62b) reduces to

r-momentum:
$$\frac{\partial P}{\partial r} = \rho\frac{K^2}{r^3} \qquad (2)$$

Thus, the r-momentum equation is satisfied if we can generate a pressure field that satisfies Eq. 2.

In order for a steady flow solution to exist, P cannot be a function of time. Furthermore, a physically realistic steady, incompressible flow field requires a pressure field $P(r, \theta)$ that is a smooth function of r and θ. Mathematically, this requires that the order of differentiation (r then θ versus θ then r) should not matter (Fig. 9–50). We check whether this is so by cross-differentiating the pressure:

$$\frac{\partial^2 P}{\partial r\,\partial \theta} = \frac{\partial}{\partial r}\left(\frac{\partial P}{\partial \theta}\right) = 0 \quad \text{and} \quad \frac{\partial^2 P}{\partial \theta\,\partial r} = \frac{\partial}{\partial \theta}\left(\frac{\partial P}{\partial r}\right) = 0 \qquad (3)$$

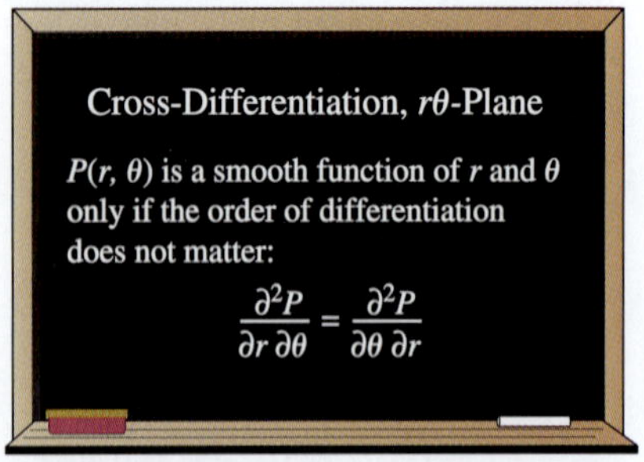

Equation 3 shows that P is indeed a smooth function of r and θ. Thus, the given velocity field satisfies the steady, two-dimensional, incompressible Navier–Stokes equation.

We integrate Eq. 1 with respect to θ to obtain an expression for $P(r, \theta)$,

Pressure field from θ-momentum:
$$P(r, \theta) = 0 + g(r) \qquad (4)$$

Cross-Differentiation, $r\theta$-Plane

$P(r, \theta)$ is a smooth function of r and θ only if the order of differentiation does not matter:

$$\frac{\partial^2 P}{\partial r\,\partial \theta} = \frac{\partial^2 P}{\partial \theta\,\partial r}$$

FIGURE 9–50
For a two-dimensional flow field in the $r\theta$-plane, cross-differentiation reveals whether pressure P is a smooth function.

Note that we added an arbitrary function of the other variable r, rather than a constant of integration, since this is a partial integration. We take the partial derivative of Eq. 4 with respect to r to obtain

$$\frac{\partial P}{\partial r} = g'(r) = \rho \frac{K^2}{r^3} \qquad (5)$$

where we have equated our result to Eq. 2 for consistency. We integrate Eq. 5 to obtain the function $g(r)$:

$$g(r) = -\frac{1}{2}\rho \frac{K^2}{r^2} + C \qquad (6)$$

where C is an arbitrary constant of integration. Finally, we substitute Eq. 6 into Eq. 4 to obtain our final expression for $P(r, \theta)$. The result is

$$P(r, \theta) = -\frac{1}{2}\rho \frac{K^2}{r^2} + C \qquad (7)$$

Thus the pressure field for a line vortex decreases like $1/r^2$ as we approach the origin. (The origin itself is a singular point.) This flow field is a simplistic model of a tornado or hurricane, and the low pressure at the center is the "eye of the storm" (Fig. 9–51). We note that this flow field is irrotational, and thus Bernoulli's equation can be used instead to calculate the pressure. If we call the pressure P_∞ far away from the origin ($r \to \infty$), where the local velocity approaches zero, Bernoulli's equation shows that at any distance r from the origin,

Bernoulli equation: $\quad P + \frac{1}{2}\rho V^2 = P_\infty \quad \to \quad P = P_\infty - \frac{1}{2}\rho \frac{K^2}{r^2} \qquad (8)$

Equation 8 agrees with our solution (Eq. 7) from the Navier–Stokes equation if we set constant C equal to P_∞. A region of rotational flow near the origin would avoid the singularity there and would yield a more physically realistic model of a tornado.

Discussion For practice, try to obtain Eq. 7 by starting with Eq. 2 rather than Eq. 1; you should get the same answer.

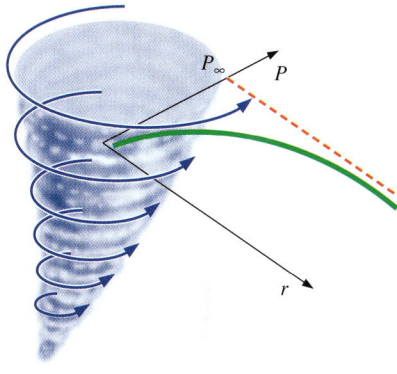

FIGURE 9–51
The two-dimensional line vortex is a simple approximation of a tornado; the lowest pressure is at the center of the vortex.

Exact Solutions of the Continuity and Navier–Stokes Equations

The remaining example problems in this section are exact solutions of the differential equation set consisting of the incompressible continuity and Navier–Stokes equations. As you will see, these problems are by necessity simple, so that they are solvable. Most of them assume infinite boundaries and fully developed conditions so that the advective terms on the left side of the Navier–Stokes equation disappear. In addition, they are laminar, two-dimensional, and either steady or dependent on time in a predefined manner. There are six basic steps in the procedure used to solve these problems, as listed in Fig. 9–52. Step 2 is especially critical, since the boundary conditions determine the uniqueness of the solution. Step 4 is not possible analytically except for simple problems. In step 5, enough boundary conditions must be available to solve for all the constants of integration produced in step 4. Step 6 involves verifying that all the differential equations and boundary conditions

Step 1: Set up the problem and geometry (sketches are helpful), identifying all relevant dimensions and parameters.

Step 2: List all appropriate assumptions, approximations, simplifications, and boundary conditions.

Step 3: Simplify the differential equations of motion (continuity and Navier–Stokes) as much as possible.

Step 4: Integrate the equations, leading to one or more constants of integration.

Step 5: Apply boundary conditions to solve for the constants of integration.

Step 6: Verify your results.

FIGURE 9–52
Procedure for solving the incompressible continuity and Navier–Stokes equations.

476
DIFFERENTIAL ANALYSIS OF FLUID FLOW

are satisfied. We advise you to follow these steps, even in cases where some of the steps seem trivial, in order to learn the procedure.

While the examples shown here are simple, they adequately illustrate the procedure used to solve these differential equations. In Chap. 15 we discuss how computers have enabled us to solve the Navier–Stokes equations *numerically* for much more complicated flows using computational fluid dynamics (CFD). You will see that the same procedure is used there—specification of geometry, application of boundary conditions, integration of the differential equations, etc., although the steps are not always followed in the same order.

Boundary Conditions

Since boundary conditions are so critical to a proper solution, we discuss the types of boundary conditions that are commonly encountered in fluid flow analyses. The most-used boundary condition is the **no-slip condition,** which states that for a fluid in contact with a solid wall, *the velocity of the fluid must equal that of the wall,*

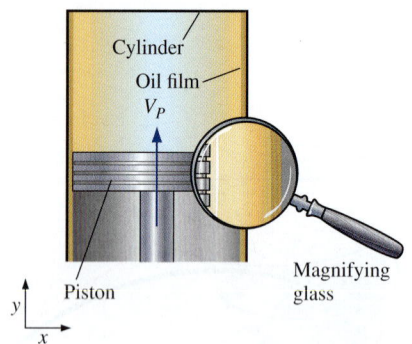

FIGURE 9–53
A piston moving at speed V_P in a cylinder. A thin film of oil is sheared between the piston and the cylinder; a magnified view of the oil film is shown. The *no-slip boundary condition* requires that the velocity of fluid adjacent to a wall equal that of the wall.

No-slip boundary condition: $\qquad \vec{V}_{\text{fluid}} = \vec{V}_{\text{wall}} \qquad$ (9–65)

In other words, as its name implies, there is no "slip" between the fluid and the wall. Fluid particles adjacent to the wall adhere to the surface of the wall and move at the same velocity as the wall. A special case of Eq. 9–65 is for a stationary wall with $\vec{V}_{\text{wall}} = 0$; *the fluid adjacent to a stationary wall has zero velocity.* For cases in which temperature effects are also considered, the temperature of the fluid must equal that of the wall, i.e., $T_{\text{fluid}} = T_{\text{wall}}$. You must be careful to assign the no-slip condition according to your chosen *frame of reference*. Consider, for example, the thin film of oil between a piston and its cylinder wall (Fig. 9–53). From a stationary frame of reference, the fluid adjacent to the cylinder is at rest, and the fluid adjacent to the moving piston has velocity $\vec{V}_{\text{fluid}} = \vec{V}_{\text{wall}} = V_P \vec{j}$. From a frame of reference *moving with the piston*, however, the fluid adjacent to the piston has zero velocity, but the fluid adjacent to the cylinder has velocity $\vec{V}_{\text{fluid}} = \vec{V}_{\text{wall}} = -V_P \vec{j}$. An exception to the no-slip condition occurs in rarefied gas flows, such as during reentry of a spaceship or in the study of motion of extremely small (submicron) particles. In such flows the air can actually slip along the wall, but these flows are beyond the scope of the present text.

When two fluids (fluid A and fluid B) meet at an interface, the **interface boundary conditions** are

Interface boundary conditions: $\qquad \vec{V}_A = \vec{V}_B \quad$ and $\quad \tau_{s,A} = \tau_{s,B} \qquad$ (9–66)

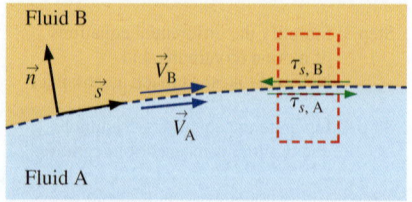

FIGURE 9–54
At an interface between two fluids, the velocity of the two fluids must be equal. In addition, the shear stress parallel to the interface must be the same in both fluids.

where, in addition to the condition that the velocities of the two fluids must be equal, the shear stress τ_s acting on a fluid particle adjacent to the interface in the direction parallel to the interface must also match between the two fluids (Fig. 9–54). Note that in the figure, $\tau_{s,A}$ is drawn on the *top* of the fluid particle in fluid A, while $\tau_{s,B}$ is drawn on the *bottom* of the fluid particle in fluid B, and we have considered the *direction* of shear stress carefully. Because of the sign convention on shear stress, the direction of the arrows in Fig. 9–54 is opposite (a consequence of Newton's third law). We note that although velocity is continuous across the interface, its slope is *not*. Also, if temperature effects are considered, $T_A = T_B$ at the interface, but there may be a discontinuity in the slope of temperature at the interface as well.

What about pressure at an interface? If surface tension effects are negligible or if the interface is nearly flat, $P_A = P_B$. If the interface is sharply curved, however, as in the meniscus of liquid rising in a capillary tube, the pressure on one side of the interface can be substantially different than that on the other side. You should recall from Chap. 2 that the pressure jump across an interface is inversely proportional to the radius of curvature of the interface, as a result of surface tension effects.

A degenerate form of the interface boundary condition occurs at the *free surface* of a liquid, meaning that fluid A is a liquid and fluid B is a gas (usually air). We illustrate a simple case in Fig. 9–55 where fluid A is liquid water and fluid B is air. The interface is flat and surface tension effects are negligible, but the water is moving horizontally (like water flowing in a calm river). In this case, the air and water velocities must match at the surface and the shear stress acting on a water particle on the surface of the water must equal that acting on an air particle just above the surface. According to Eq. 9–66,

Boundary conditions at water–air interface:

$$u_{\text{water}} = u_{\text{air}} \quad \text{and} \quad \tau_{s,\text{water}} = \mu_{\text{water}} \left(\frac{\partial u}{\partial y}\right)_{\text{water}} = \tau_{s,\text{air}} = \mu_{\text{air}} \left(\frac{\partial u}{\partial y}\right)_{\text{air}} \quad (9\text{–}67)$$

A quick glance at the fluid property tables reveals that μ_{water} is over 50 times greater than μ_{air}. In order for the shear stresses to be equal, Eq. 9–67 requires that slope $(\partial u/\partial y)_{\text{air}}$ be more than 50 times greater than $(\partial u/\partial y)_{\text{water}}$. Thus, it is reasonable to approximate the shear stress acting at the surface of the water as negligibly small compared to shear stresses elsewhere in the water. Another way to say this is that the moving water drags air along with it with little resistance from the air; in contrast, the air doesn't slow down the water by any significant amount. In summary, for the case of a liquid in contact with a gas, and with negligible surface tension effects, the **free-surface boundary conditions** are

Free-surface boundary conditions: $\quad P_{\text{liquid}} = P_{\text{gas}} \quad \text{and} \quad \tau_{s,\text{liquid}} \cong 0 \quad (9\text{–}68)$

Other boundary conditions arise depending on the problem setup. For example, we often need to define **inlet boundary conditions** at a boundary of a flow domain where fluid enters the domain. Likewise, we define **outlet boundary conditions** at an outflow. **Symmetry boundary conditions** are useful along an axis or plane of symmetry. For example, the appropriate symmetry boundary conditions along a horizontal plane of symmetry are illustrated in Fig. 9–56. For unsteady flow problems we also need to define **initial conditions** (at the starting time, usually $t = 0$).

In Examples 9–15 through 9–19, we apply boundary conditions from Eqs. 9–65 through 9–68 where appropriate. These and other boundary conditions are discussed in much greater detail in Chap. 15 where we apply them to CFD solutions.

■ EXAMPLE 9–15   Fully Developed Couette Flow

Consider steady, incompressible, laminar flow of a Newtonian fluid in the narrow gap between two infinite parallel plates (Fig. 9–57). The top plate is moving at speed V, and the bottom plate is stationary. The distance between these two plates is h, and gravity acts in the negative z-direction (into the page in Fig. 9–57). There is no applied pressure other than hydrostatic

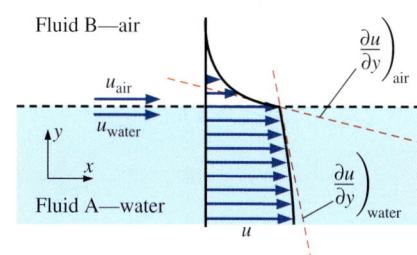

FIGURE 9–55
Along a horizontal *free surface* of water and air, the water and air velocities must be equal and the shear stresses must match. However, since $\mu_{\text{air}} \ll \mu_{\text{water}}$, a good approximation is that the shear stress at the water surface is negligibly small.

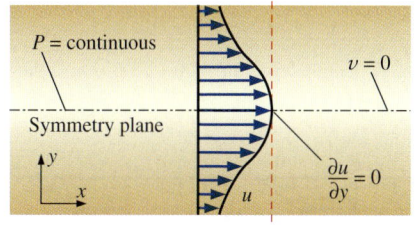

FIGURE 9–56
Boundary conditions along a plane of symmetry are defined so as to ensure that the flow field on one side of the symmetry plane is a *mirror image* of that on the other side, as shown here for a horizontal symmetry plane.

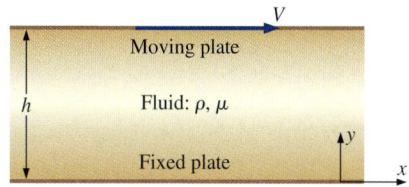

FIGURE 9–57
Geometry of Example 9–15: viscous flow between two infinite plates; upper plate moving and lower plate stationary.

478
DIFFERENTIAL ANALYSIS OF FLUID FLOW

pressure due to gravity. This flow is called **Couette flow.** Calculate the velocity and pressure fields, and estimate the shear force per unit area acting on the bottom plate.

SOLUTION For a given geometry and set of boundary conditions, we are to calculate the velocity and pressure fields, and then estimate the shear force per unit area acting on the bottom plate.

Assumptions **1** The plates are infinite in x and z. **2** The flow is steady, i.e., $\partial/\partial t$ of anything is zero. **3** This is a parallel flow (we assume that the y-component of velocity, v, is zero). **4** The fluid is incompressible and Newtonian with constant properties, and the flow is laminar. **5** Pressure $P =$ constant with respect to x. In other words, there is no applied pressure gradient pushing the flow in the x-direction; the flow establishes itself due to viscous stresses caused by the moving upper plate. **6** The velocity field is purely two-dimensional, meaning here that $w = 0$ and $\partial/\partial z$ of any velocity component is zero. **7** Gravity acts in the negative z-direction (into the page in Fig. 9–57). We express this mathematically as $\vec{g} = -g\vec{k}$, or $g_x = g_y = 0$ and $g_z = -g$.

Analysis To obtain the velocity and pressure fields, we follow the step-by-step procedure outlined in Fig. 9–52.

Step 1 *Set up the problem and the geometry.* See Fig. 9–57.

Step 2 *List assumptions and boundary conditions.* We have numbered and listed seven assumptions (above). The boundary conditions come from imposing the no-slip condition: (1) At the bottom plate ($y = 0$), $u = v = w = 0$. (2) At the top plate ($y = h$), $u = V$, $v = 0$, and $w = 0$.

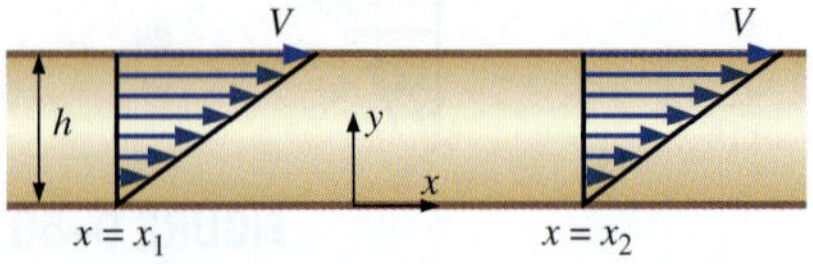

Step 3 *Simplify the differential equations.* We start with the incompressible continuity equation in Cartesian coordinates, Eq. 9–61a,

$$\frac{\partial u}{\partial x} + \underbrace{\frac{\partial v}{\partial y}}_{\text{assumption 3}} + \underbrace{\frac{\partial w}{\partial z}}_{\text{assumption 6}} = 0 \quad \rightarrow \quad \frac{\partial u}{\partial x} = 0 \quad (1)$$

Equation 1 tells us that u is not a function of x. In other words, it doesn't matter where we place our origin—the flow is the same at any x-location. The phrase **fully developed** is often used to describe this situation (Fig. 9–58). This can also be obtained directly from assumption 1, which tells us that there is nothing special about any x-location since the plates are infinite in length. Furthermore, since u is not a function of time (assumption 2) or z (assumption 6), we conclude that u is at most a function of y,

Result of continuity: $\quad\quad\quad u = u(y)$ only $\quad\quad\quad$ (2)

We now simplify the x-momentum equation (Eq. 9–61b) as far as possible. It is good practice to list the reason for crossing out a term, as we do here:

$$\rho\left(\underbrace{\frac{\partial u}{\partial t}}_{\text{assumption 2}} + \underbrace{u\frac{\partial u}{\partial x}}_{\text{continuity}} + \underbrace{v\frac{\partial u}{\partial y}}_{\text{assumption 3}} + \underbrace{w\frac{\partial u}{\partial z}}_{\text{assumption 6}}\right) = -\underbrace{\frac{\partial P}{\partial x}}_{\text{assumption 5}} + \underbrace{\rho g_x}_{\text{assumption 7}}$$

$$+ \mu\left(\underbrace{\frac{\partial^2 u}{\partial x^2}}_{\text{continuity}} + \frac{\partial^2 u}{\partial y^2} + \underbrace{\frac{\partial^2 u}{\partial z^2}}_{\text{assumption 6}}\right) \quad\rightarrow\quad \frac{d^2 u}{dy^2} = 0 \quad (3)$$

FIGURE 9–58
A *fully developed* region of a flow field is a region where the velocity profile does not change with downstream distance. Fully developed flows are encountered in long, straight channels and pipes. Fully developed Couette flow is shown here—the velocity profile at x_2 is identical to that at x_1.

Notice that the material acceleration (left-hand side of Eq. 3) is zero, implying that fluid particles are not accelerating in this flow field, neither by local (unsteady) acceleration, nor by advective acceleration. Since the advective acceleration terms make the Navier–Stokes equation nonlinear, this greatly simplifies the problem. In fact, all other terms in Eq. 3 have disappeared except for a lone viscous term, which must then itself equal zero. Also notice that we have changed from a partial derivative ($\partial/\partial y$) to a total derivative (d/dy) in Eq. 3 as a direct result of Eq. 2. We do not show the details here, but you can show in similar fashion that every term except the pressure term in the y-momentum equation (Eq. 9–61c) goes to zero, forcing that lone term to also be zero,

$$\frac{\partial P}{\partial y} = 0 \quad (4)$$

In other words, P is not a function of y. Since P is also not a function of time (assumption 2) or x (assumption 5), P is at most a function of z,

Result of y-momentum: $\qquad P = P(z)$ only $\qquad (5)$

Finally, by assumption 6 the z-component of the Navier–Stokes equation (Eq. 9–61d) simplifies to

$$\frac{\partial P}{\partial z} = -\rho g \quad \rightarrow \quad \frac{dP}{dz} = -\rho g \quad (6)$$

where we used Eq. 5 to convert from a partial derivative to a total derivative.

Step 4 *Solve the differential equations.* Continuity and y-momentum have already been "solved," resulting in Eqs. 2 and 5, respectively. Equation 3 (x-momentum) is integrated twice to get

$$u = C_1 y + C_2 \quad (7)$$

where C_1 and C_2 are constants of integration. Equation 6 (z-momentum) is integrated once, resulting in

$$P = -\rho g z + C_3 \quad (8)$$

Step 5 *Apply boundary conditions.* We begin with Eq. 8. Since we have not specified boundary conditions for pressure, C_3 remains an arbitrary constant. (Recall that for incompressible flow, the absolute pressure can be specified only if P is known somewhere in the flow.) For example, if we let $P = P_0$ at $z = 0$, then $C_3 = P_0$ and Eq. 8 becomes

Final solution for pressure field: $\qquad \boldsymbol{P = P_0 - \rho g z} \qquad (9)$

Alert readers will notice that Eq. 9 represents a simple **hydrostatic pressure distribution** (pressure decreasing linearly as z increases). We conclude that, at least for this problem, *hydrostatic pressure acts independently of the flow.* More generally, we make the following statement (see also Fig. 9–59):

For incompressible flow fields without free surfaces, hydrostatic pressure does not contribute to the dynamics of the flow field.

In fact, in Chap. 10 we show how hydrostatic pressure can actually be *removed* from the equations of motion through use of a modified pressure.

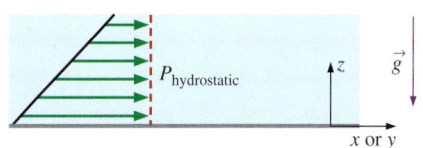

FIGURE 9–59

For incompressible flow fields *without free surfaces*, hydrostatic pressure does not contribute to the dynamics of the flow field.

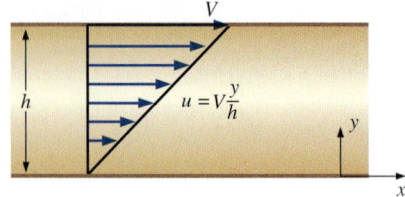

FIGURE 9–60
The linear velocity profile of Example 9–15: Couette flow between parallel plates.

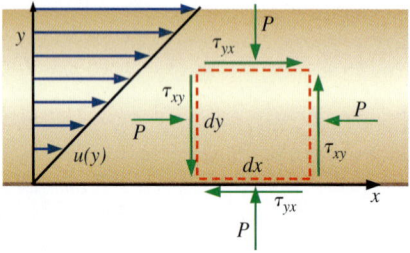

FIGURE 9–61
Stresses acting on a differential two-dimensional rectangular fluid element whose bottom face is in contact with the bottom plate of Example 9–15.

We next apply boundary conditions (1) and (2) from step 2 to obtain constants C_1 and C_2.

Boundary condition (1): $\quad u = C_1 \times 0 + C_2 = 0 \quad \rightarrow \quad C_2 = 0$

and

Boundary condition (2): $\quad u = C_1 \times h + 0 = V \quad \rightarrow \quad C_1 = V/h$

Finally, Eq. 7 becomes

Final result for velocity field: $\quad u = V\dfrac{y}{h} \quad$ (10)

The velocity field reveals a simple linear velocity profile from $u = 0$ at the bottom plate to $u = V$ at the top plate, as sketched in Fig. 9–60.

Step 6 *Verify the results.* Using Eqs. 9 and 10, you can verify that all the differential equations and boundary conditions are satisfied.

To calculate the shear force per unit area acting on the bottom plate, we consider a rectangular fluid element whose bottom face is in contact with the bottom plate (Fig 9–61). Mathematically positive viscous stresses are shown. In this case, these stresses are in the proper direction since fluid above the differential element pulls it to the right while the wall below the element pulls it to the left. From Eq. 9–56, we write out the components of the viscous stress tensor,

$$\tau_{ij} = \begin{pmatrix} 2\mu\dfrac{\partial u}{\partial x} & \mu\left(\dfrac{\partial u}{\partial y} + \dfrac{\partial v}{\partial x}\right) & \mu\left(\dfrac{\partial u}{\partial z} + \dfrac{\partial w}{\partial x}\right) \\ \mu\left(\dfrac{\partial v}{\partial x} + \dfrac{\partial u}{\partial y}\right) & 2\mu\dfrac{\partial v}{\partial y} & \mu\left(\dfrac{\partial v}{\partial z} + \dfrac{\partial w}{\partial y}\right) \\ \mu\left(\dfrac{\partial w}{\partial x} + \dfrac{\partial u}{\partial z}\right) & \mu\left(\dfrac{\partial w}{\partial y} + \dfrac{\partial v}{\partial z}\right) & 2\mu\dfrac{\partial w}{\partial z} \end{pmatrix} = \begin{pmatrix} 0 & \mu\dfrac{V}{h} & 0 \\ \mu\dfrac{V}{h} & 0 & 0 \\ 0 & 0 & 0 \end{pmatrix} \quad (11)$$

Since the dimensions of stress are force per unit area by definition, the force per unit area acting on the bottom face of the fluid element is equal to $\tau_{yx} = \mu V/h$ and acts in the negative x-direction, as sketched. The shear force per unit area on the *wall* is equal and opposite to this (Newton's third law); hence,

Shear force per unit area acting on the wall: $\quad \dfrac{\vec{F}}{A} = \mu\dfrac{V}{h}\vec{i} \quad$ (12)

The direction of this force agrees with our intuition; namely, the fluid tries to pull the bottom wall to the right, due to viscous effects (friction).

Discussion The z-component of the linear momentum equation is *uncoupled* from the rest of the equations; this explains why we get a hydrostatic pressure distribution in the z-direction, even though the fluid is not static, but moving. Equation 11 reveals that the viscous stress tensor is constant *everywhere* in the flow field, not just at the bottom wall (notice that none of the components of τ_{ij} is a function of location).

You may be questioning the usefulness of the final results of Example 9–15. After all, when do we encounter two infinite parallel plates, one of which is moving? Actually there *are* several practical flows for which the Couette flow solution is a very good approximation. One such flow occurs inside a **rotational viscometer** (Fig. 9–62), an instrument used to measure viscosity. It is constructed of two concentric circular cylinders of length L—a solid, rotating inner cylinder of radius R_i and a hollow, stationary outer cylinder of radius R_o. (L is into the page in Fig. 9–62; the z-axis is out of the page.) The gap between the two cylinders is very small and contains the fluid whose viscosity is to be measured. The magnified region of Fig. 9–62 is a nearly identical setup as that of Fig. 9–57 since the gap is small, i.e. $(R_o - R_i) \ll R_o$. In a viscosity measurement, the angular velocity of the inner cylinder, ω, is measured, as is the applied torque, $T_{applied}$, required to rotate the cylinder. From Example 9–15, we know that the viscous shear stress acting on a fluid element adjacent to the inner cylinder is approximately equal to

$$\tau = \tau_{yx} \cong \mu \frac{V}{R_o - R_i} = \mu \frac{\omega R_i}{R_o - R_i} \tag{9-69}$$

where the speed V of the moving upper plate in Fig. 9–57 is replaced by the counterclockwise speed ωR_i of the rotating wall of the inner cylinder. In the magnified region at the bottom of Fig. 9–62, τ acts to the right on the fluid element adjacent to the inner cylinder wall; hence, the force per unit area acting on the inner cylinder at this location acts to the left with magnitude given by Eq. 9–69. The total *clockwise* torque acting on the inner cylinder wall due to fluid viscosity is thus equal to this shear stress times the wall area times the moment arm,

$$T_{viscous} = \tau A R_i \cong \mu \frac{\omega R_i}{R_o - R_i} \left(2\pi R_i L \right) R_i \tag{9-70}$$

Under steady conditions, the clockwise torque $T_{viscous}$ is balanced by the applied counterclockwise torque $T_{applied}$. Equating these and solving Eq. 9–70 for the fluid viscosity yields

Viscosity of the fluid:
$$\mu = T_{applied} \frac{(R_o - R_i)}{2\pi \omega R_i^3 L}$$

A similar analysis can be performed on an unloaded journal bearing in which a viscous oil flows in the small gap between the inner rotating shaft and the stationary outer housing. (When the bearing is loaded, the inner and outer cylinders cease to be concentric and a more involved analysis is required.)

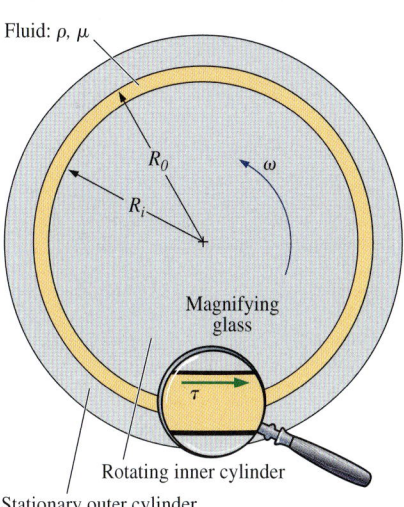

FIGURE 9–62
A rotational viscometer; the inner cylinder rotates at angular velocity ω, and a torque $T_{applied}$ is applied, from which the viscosity of the fluid is calculated.

■ **EXAMPLE 9–16** Couette Flow with an Applied Pressure Gradient

Consider the same geometry as in Example 9–15, but instead of pressure being constant with respect to x, let there be an applied pressure gradient in the x-direction (Fig. 9–63). Specifically, let the pressure gradient in the x-direction, $\partial P/\partial x$, be some constant value given by

Applied pressure gradient:
$$\frac{\partial P}{\partial x} = \frac{P_2 - P_1}{x_2 - x_1} = \text{constant} \tag{1}$$

where x_1 and x_2 are two arbitrary locations along the x-axis, and P_1 and P_2 are the pressures at those two locations. Everything else is the same as for Example 9–15. (*a*) Calculate the velocity and pressure field. (*b*) Plot a family of velocity profiles in dimensionless form.

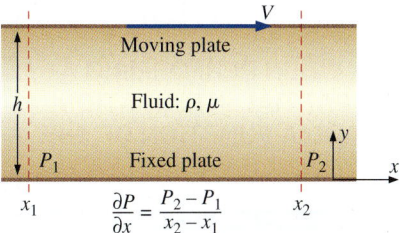

FIGURE 9–63
Geometry of Example 9–16: viscous flow between two infinite plates with a constant applied pressure gradient $\partial P/\partial x$; the upper plate is moving and the lower plate is stationary.

SOLUTION We are to calculate the velocity and pressure field for the flow sketched in Fig. 9–63 and plot a family of velocity profiles in dimensionless form.

Assumptions The assumptions are identical to those of Example 9–15, except assumption 5 is replaced by the following: A constant pressure gradient is applied in the x-direction such that pressure changes linearly with respect to x according to Eq. 1.

Analysis (a) We follow the same procedure as in Example 9–15. Much of the algebra is identical, so to save space we discuss only the differences.

Step 1 See Fig. 9–63.

Step 2 Same as Example 9–15 except for assumption 5.

Step 3 The continuity equation is simplified in the same way as in Example 9–15,

Result of continuity: $\quad u = u(y)$ only $\quad$ (2)

The x-momentum equation is simplified in the same manner as in Example 9–15 except that the pressure gradient term remains. The result is

Result of x-momentum: $\quad \dfrac{d^2 u}{dy^2} = \dfrac{1}{\mu} \dfrac{\partial P}{\partial x}$ $\quad$ (3)

Likewise, the y-momentum and z-momentum equations simplify to

Result of y-momentum: $\quad \dfrac{\partial P}{\partial y} = 0$ $\quad$ (4)

and

Result of z-momentum: $\quad \dfrac{\partial P}{\partial z} = -\rho g$ $\quad$ (5)

We cannot convert from a partial derivative to a total derivative in Eq. 5, because P is a function of both x and z in this problem, unlike in Example 9–15 where P was a function of z only.

Step 4 We integrate Eq. 3 (x-momentum) twice, noting that $\partial P/\partial x$ is a constant,

Integration of x-momentum: $\quad u = \dfrac{1}{2\mu} \dfrac{\partial P}{\partial x} y^2 + C_1 y + C_2$ $\quad$ (6)

where C_1 and C_2 are constants of integration. Equation 5 (z-momentum) is integrated once, resulting in

Integration of z-momentum: $\quad P = -\rho g z + f(x)$ $\quad$ (7)

Note that since P is now a function of both x and z, we add a function of x instead of a constant of integration in Eq. 7. This is a *partial* integration with respect to z, and we must be careful when performing partial integrations (Fig. 9–64).

Step 5 From Eq. 7, we see that the pressure varies hydrostatically in the z-direction, and we have specified a linear change in pressure in the x-direction. Thus the function f(x) must equal a constant plus $\partial P/\partial x$ times x. If we set $P = P_0$ along the line $x = 0$, $z = 0$ (the y-axis), Eq. 7 becomes

Final result for pressure field: $\quad P = P_0 + \dfrac{\partial P}{\partial x} x - \rho g z$ $\quad$ (8)

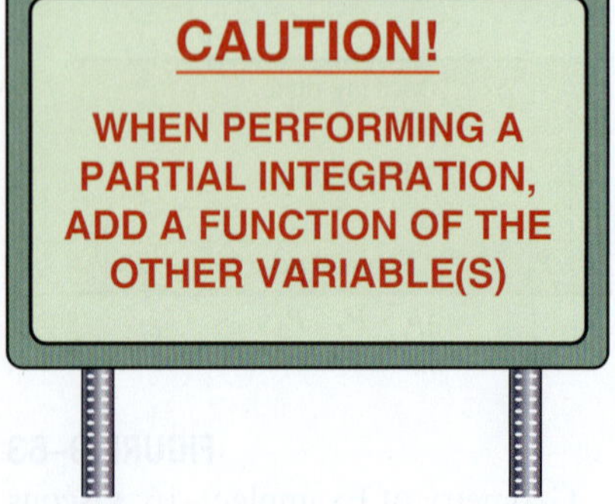

FIGURE 9–64
A caution about partial integration.

We next apply the velocity boundary conditions (1) and (2) from step 2 of Example 9–15 to obtain constants C_1 and C_2.

Boundary condition (1):

$$u = \frac{1}{2\mu}\frac{\partial P}{\partial x} \times 0 + C_1 \times 0 + C_2 = 0 \quad \rightarrow \quad C_2 = 0$$

and

Boundary condition (2):

$$u = \frac{1}{2\mu}\frac{\partial P}{\partial x}h^2 + C_1 \times h + 0 = V \quad \rightarrow \quad C_1 = \frac{V}{h} - \frac{1}{2\mu}\frac{\partial P}{\partial x}h$$

Finally, Eq. 6 becomes

$$u = \frac{Vy}{h} + \frac{1}{2\mu}\frac{\partial P}{\partial x}(y^2 - hy) \qquad (9)$$

Equation 9 indicates that the velocity field consists of the superposition of two parts: a linear velocity profile from $u = 0$ at the bottom plate to $u = V$ at the top plate, and a parabolic distribution that depends on the magnitude of the applied pressure gradient. If the pressure gradient is zero, the parabolic portion of Eq. 9 disappears and the profile is linear, just as in Example 9–15; this is sketched as the dashed red line in Fig. 9–65. If the pressure gradient is negative (pressure decreasing in the x-direction, causing flow to be pushed from left to right), $\partial P/\partial x < 0$ and the velocity profile looks like the one sketched in Fig. 9–65. A special case is when $V = 0$ (top plate stationary); the linear portion of Eq. 9 vanishes, and the velocity profile is parabolic and symmetric about the center of the channel ($y = h/2$); this is sketched as the dotted line in Fig. 9–65.

Step 6 You can use Eqs. 8 and 9 to verify that all the differential equations and boundary conditions are satisfied.

(*b*) We use dimensional analysis to generate the dimensionless groups (Π groups). We set up the problem in terms of velocity component u as a function of y, h, V, μ, and $\partial P/\partial x$. There are six variables (including the dependent variable u), and since there are three primary dimensions represented in the problem (mass, length, and time), we expect $6 - 3 = 3$ dimensionless groups. When we pick h, V, and μ as our repeating variables, we get the following result using the method of repeating variables (details are left for you to do on your own—this is a good review of Chap. 7 material):

Result of dimensional analysis: $\quad \dfrac{u}{V} = f\left(\dfrac{y}{h}, \dfrac{h^2}{\mu V}\dfrac{\partial P}{\partial x}\right) \qquad (10)$

Using these three dimensionless groups, we rewrite Eq. 9 as

Dimensionless form of velocity field: $\quad u^* = y^* + \dfrac{1}{2}P^*y^*(y^* - 1) \qquad (11)$

where the dimensionless parameters are

$$u^* = \frac{u}{V} \quad y^* = \frac{y}{h} \quad P^* = \frac{h^2}{\mu V}\frac{\partial P}{\partial x}$$

In Fig. 9–66, u^* is plotted as a function of y^* for several values of P^*, using Eq. 11.

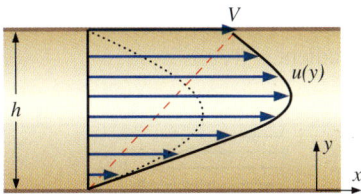

FIGURE 9–65
The velocity profile of Example 9–16: Couette flow between parallel plates with an applied negative pressure gradient; the dashed red line indicates the profile for a zero pressure gradient, and the dotted line indicates the profile for a negative pressure gradient with the upper plate stationary ($V = 0$).

484
DIFFERENTIAL ANALYSIS OF FLUID FLOW

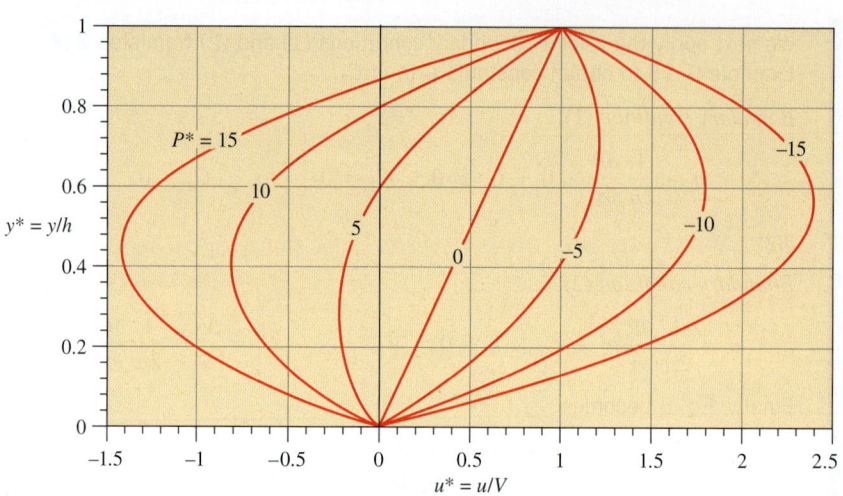

FIGURE 9–66
Nondimensional velocity profiles for Couette flow with an applied pressure gradient; profiles are shown for several values of nondimensional pressure gradient.

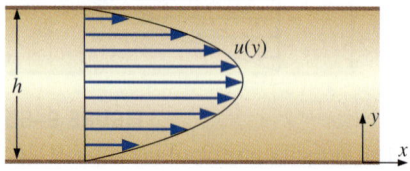

FIGURE 9–67
The velocity profile for fully developed two-dimensional channel flow (planar Poiseuille flow).

Discussion When the result is nondimensionalized, we see that Eq. 11 represents a *family* of velocity profiles. We also see that when the pressure gradient is *positive* (flow being pushed from right to left) and of sufficient magnitude, we can have *reverse flow* in the bottom portion of the channel. For all cases, the boundary conditions reduce to $u^* = 0$ at $y^* = 0$ and $u^* = 1$ at $y^* = 1$. If there is a pressure gradient but both walls are stationary, the flow is called two-dimensional channel flow, or **planar Poiseuille flow** (Fig. 9–67). We note, however, that most authors reserve the name *Poiseuille flow* for fully developed *pipe* flow—the axisymmetric analog of two-dimensional channel flow (see Example 9–18).

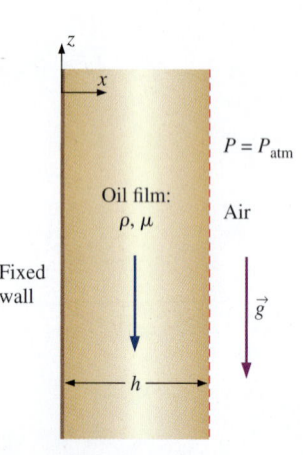

FIGURE 9–68
Geometry of Example 9–17: a viscous film of oil falling by gravity along a vertical wall.

EXAMPLE 9–17   Oil Film Flowing Down a Vertical Wall by Gravity

Consider steady, incompressible, parallel, laminar flow of a film of oil falling slowly down an infinite vertical wall (Fig. 9–68). The oil film thickness is h, and gravity acts in the negative z-direction (downward in Fig. 9–68). There is no applied (forced) pressure driving the flow—the oil falls by gravity alone. Calculate the velocity and pressure fields in the oil film and sketch the normalized velocity profile. You may neglect changes in the hydrostatic pressure of the surrounding air.

SOLUTION For a given geometry and set of boundary conditions, we are to calculate the velocity and pressure fields and plot the velocity profile.
Assumptions **1** The wall is infinite in the yz-plane (y is into the page for a right-handed coordinate system). **2** The flow is steady (all partial derivatives with respect to time are zero). **3** The flow is parallel (the x-component of velocity, u, is zero everywhere). **4** The fluid is incompressible and Newtonian with constant properties, and the flow is laminar. **5** Pressure $P = P_\text{atm} =$ constant at the free surface. In other words, there is no applied pressure gradient pushing the flow; the flow establishes itself due to a balance between gravitational forces and viscous forces. In addition, since there is no gravity force in the

horizontal direction, $P = P_{atm}$ everywhere. **6** The velocity field is purely two-dimensional, which implies that velocity component $v = 0$ and all partial derivatives with respect to y are zero. **7** Gravity acts in the negative z-direction. We express this mathematically as $\vec{g} = -g\vec{k}$, or $g_x = g_y = 0$ and $g_z = -g$.

Analysis We obtain the velocity and pressure fields by following the step-by-step procedure for differential fluid flow solutions. (Fig. 9–52).

Step 1 *Set up the problem and the geometry.* See Fig. 9–68.

Step 2 *List assumptions and boundary conditions.* We have listed seven assumptions. The boundary conditions are: (1) There is no slip at the wall; at $x = 0$, $u = v = w = 0$. (2) At the free surface ($x = h$), there is negligible shear (Eq. 9–68), which for a vertical free surface in this coordinate system means $\partial w/\partial x = 0$ at $x = h$.

Step 3 *Write out and simplify the differential equations.* We start with the incompressible continuity equation in Cartesian coordinates,

$$\underbrace{\frac{\partial u}{\partial x}}_{\text{assumption 3}} + \underbrace{\frac{\partial v}{\partial y}}_{\text{assumption 6}} + \frac{\partial w}{\partial z} = 0 \quad \rightarrow \quad \frac{\partial w}{\partial z} = 0 \qquad (1)$$

Equation 1 tells us that w is not a function of z; i.e., it doesn't matter where we place our origin—the flow is the same at *any* z-location. In other words, the flow is *fully developed*. Since w is not a function of time (assumption 2), z (Eq. 1), or y (assumption 6), we conclude that w is at most a function of x,

Result of continuity: $\qquad w = w(x)$ only $\qquad (2)$

We now simplify each component of the Navier–Stokes equation as far as possible. Since $u = v = 0$ everywhere, and gravity does not act in the x- or y-directions, the x- and y-momentum equations are satisfied exactly (in fact all terms are zero in both equations). The z-momentum equation reduces to

$$\rho\left(\underbrace{\frac{\partial w}{\partial t}}_{\text{assumption 2}} + \underbrace{u\frac{\partial w}{\partial x}}_{\text{assumption 3}} + \underbrace{v\frac{\partial w}{\partial y}}_{\text{assumption 6}} + \underbrace{w\frac{\partial w}{\partial z}}_{\text{continuity}}\right) = -\underbrace{\frac{\partial P}{\partial z}}_{\text{assumption 5}} + \underbrace{\rho g_z}_{-\rho g}$$

$$+ \mu\left(\frac{\partial^2 w}{\partial x^2} + \underbrace{\frac{\partial^2 w}{\partial y^2}}_{\text{assumption 6}} + \underbrace{\frac{\partial^2 w}{\partial z^2}}_{\text{continuity}}\right) \quad \rightarrow \quad \frac{d^2 w}{dx^2} = \frac{\rho g}{\mu} \qquad (3)$$

The material acceleration (left side of Eq. 3) is zero, implying that fluid particles are not accelerating in this flow field, neither by local nor advective acceleration. Since the advective acceleration terms make the Navier–Stokes equation nonlinear, this greatly simplifies the problem. We have changed from a partial derivative ($\partial/\partial x$) to a total derivative (d/dx) in Eq. 3 as a direct result of Eq. 2, reducing the partial differential equation (PDE) to an ordinary differential equation (ODE). ODEs are of course much easier than PDEs to solve (Fig. 9–69).

Step 4 *Solve the differential equations.* The continuity and x- and y-momentum equations have already been "solved." Equation 3 (z-momentum) is integrated twice to get

$$w = \frac{\rho g}{2\mu}x^2 + C_1 x + C_2 \qquad (4)$$

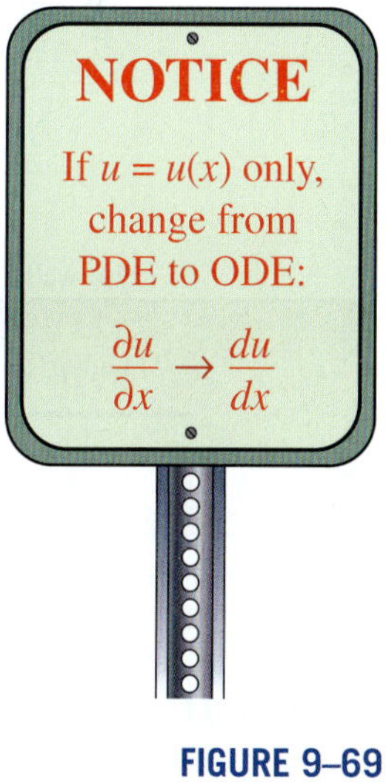

NOTICE

If $u = u(x)$ only, change from PDE to ODE:

$$\frac{\partial u}{\partial x} \rightarrow \frac{du}{dx}$$

FIGURE 9–69
In Examples 9–15 through 9–18, the equations of motion are reduced from *partial differential equations* to *ordinary differential equations*, making them much easier to solve.

486
DIFFERENTIAL ANALYSIS OF FLUID FLOW

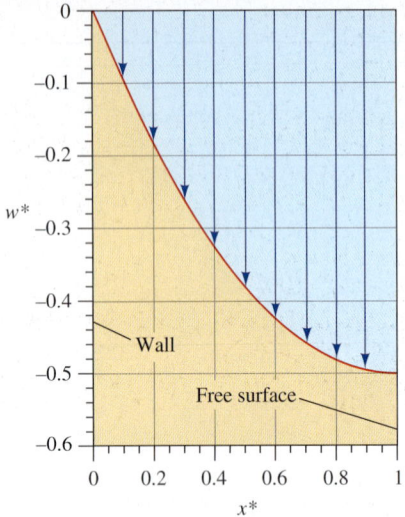

FIGURE 9–70
The normalized velocity profile of Example 9–17: an oil film falling down a vertical wall.

Step 5 *Apply boundary conditions.* We apply boundary conditions (1) and (2) from step 2 to obtain constants C_1 and C_2.

Boundary condition (1): $\quad w = 0 + 0 + C_2 = 0 \quad C_2 = 0$

and

Boundary condition (2): $\quad \left.\dfrac{dw}{dx}\right)_{x=h} = \dfrac{\rho g}{\mu} h + C_1 = 0 \;\rightarrow\; C_1 = -\dfrac{\rho g h}{\mu}$

Finally, Eq. 4 becomes

Velocity field: $\quad w = \dfrac{\rho g}{2\mu} x^2 - \dfrac{\rho g}{\mu} h x = \dfrac{\rho g x}{2\mu}(x - 2h) \quad$ (5)

Since $x < h$ in the film, w is negative everywhere, as expected (flow is downward). The pressure field is trivial; namely, $P = P_{\text{atm}}$ **everywhere**.

Step 6 *Verify the results.* You can verify that all the differential equations and boundary conditions are satisfied.

We normalize Eq. 5 by inspection: we let $x^* = x/h$ and $w^* = w\mu/(\rho g h^2)$. Equation 5 becomes

Normalized velocity profile: $\quad w^* = \dfrac{x^*}{2}(x^* - 2) \quad$ (6)

We plot the normalized velocity field in Fig. 9–70.

Discussion The velocity profile has a large slope near the wall due to the no-slip condition there ($w = 0$ at $x = 0$), but zero slope at the free surface, where the boundary condition is zero shear stress ($\partial w/\partial x = 0$ at $x = h$). We could have introduced a factor of -2 in the definition of w^* so that w^* would equal 1 instead of $-\tfrac{1}{2}$ at the free surface.

The solution procedure used in Examples 9–15 through 9–17 in Cartesian coordinates can also be used in any other coordinate system. In Example 9–18 we present the classic problem of fully developed flow in a round pipe, for which we use cylindrical coordinates.

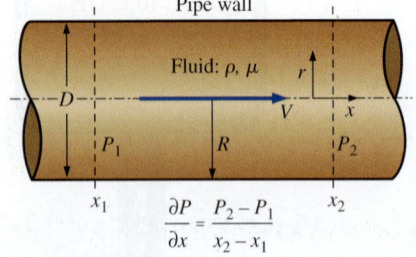

FIGURE 9–71
Geometry of Example 9–18: steady laminar flow in a long round pipe with an applied pressure gradient $\partial P/\partial x$ pushing fluid through the pipe. The pressure gradient is usually produced by a pump and/or gravity.

■ **EXAMPLE 9–18**   **Fully Developed Flow in a Round Pipe—Poiseuille Flow**

Consider steady, incompressible, laminar flow of a Newtonian fluid in an infinitely long round pipe of diameter D or radius $R = D/2$ (Fig. 9–71). We ignore the effects of gravity. A constant pressure gradient $\partial P/\partial x$ is applied in the x-direction,

Applied pressure gradient: $\quad \dfrac{\partial P}{\partial x} = \dfrac{P_2 - P_1}{x_2 - x_1} = \text{constant} \quad$ (1)

where x_1 and x_2 are two arbitrary locations along the x-axis, and P_1 and P_2 are the pressures at those two locations. Note that we adopt a modified cylindrical coordinate system here with x instead of z for the axial component, namely, (r, θ, x) and (u_r, u_θ, u). Derive an expression for the velocity field inside the pipe and estimate the viscous shear force per unit surface area acting on the pipe wall.

SOLUTION For flow inside a round pipe we are to calculate the velocity field, and then estimate the viscous shear stress acting on the pipe wall.

Assumptions **1** The pipe is infinitely long in the x-direction. **2** The flow is steady (all partial time derivatives are zero). **3** This is a parallel flow (the r-component of velocity, u_r, is zero). **4** The fluid is incompressible and Newtonian with constant properties, and the flow is laminar (Fig. 9–72). **5** A constant pressure gradient is applied in the x-direction such that pressure changes linearly with respect to x according to Eq. 1. **6** The velocity field is axisymmetric with no swirl, implying that $u_\theta = 0$ and all partial derivatives with respect to θ are zero. **7** We ignore the effects of gravity.

Analysis To obtain the velocity field, we follow the step-by-step procedure outlined in Fig. 9–52.

Step 1 *Lay out the problem and the geometry.* See Fig. 9–71.

Step 2 *List assumptions and boundary conditions.* We have listed seven assumptions. The first boundary condition comes from imposing the no-slip condition at the pipe wall: (1) at $r = R$, $\vec{V} = 0$. The second boundary condition comes from the fact that the centerline of the pipe is an axis of symmetry: (2) at $r = 0$, $\partial u/\partial r = 0$.

Step 3 *Write out and simplify the differential equations.* We start with the incompressible continuity equation in cylindrical coordinates, a modified version of Eq. 9–62a,

$$\underbrace{\frac{1}{r}\frac{\partial(ru_r)}{\partial r}}_{\text{assumption 3}} + \underbrace{\frac{1}{r}\frac{\partial(u_\theta)}{\partial \theta}}_{\text{assumption 6}} + \frac{\partial u}{\partial x} = 0 \quad \rightarrow \quad \frac{\partial u}{\partial x} = 0 \qquad (2)$$

Equation 2 tells us that u is not a function of x. In other words, it doesn't matter where we place our origin—the flow is the same at any x-location. This can also be inferred directly from assumption 1, which tells us that there is nothing special about any x-location since the pipe is infinite in length—the flow is fully developed. Furthermore, since u is not a function of time (assumption 2) or θ (assumption 6), we conclude that u is at most a function of r,

Result of continuity: $\qquad u = u(r)$ only $\qquad (3)$

We now simplify the axial momentum equation (a modified version of Eq. 9–62d) as far as possible:

$$\rho\left(\underbrace{\frac{\partial u}{\partial t}}_{\text{assumption 2}} + \underbrace{u_r\frac{\partial u}{\partial r}}_{\text{assumption 3}} + \underbrace{\frac{u_\theta}{r}\frac{\partial u}{\partial \theta}}_{\text{assumption 6}} + \underbrace{u\frac{\partial u}{\partial x}}_{\text{continuity}}\right)$$

$$= -\frac{\partial P}{\partial x} + \underbrace{\rho g_x}_{\text{assumption 7}} + \mu\left(\frac{1}{r}\frac{\partial}{\partial r}\left(r\frac{\partial u}{\partial r}\right) + \underbrace{\frac{1}{r^2}\frac{\partial^2 u}{\partial \theta^2}}_{\text{assumption 6}} + \underbrace{\frac{\partial^2 u}{\partial x^2}}_{\text{continuity}}\right)$$

or

$$\frac{1}{r}\frac{d}{dr}\left(r\frac{du}{dr}\right) = \frac{1}{\mu}\frac{\partial P}{\partial x} \qquad (4)$$

As in Examples 9–15 through 9–17, the material acceleration (entire left side of the x-momentum equation) is zero, implying that fluid particles are

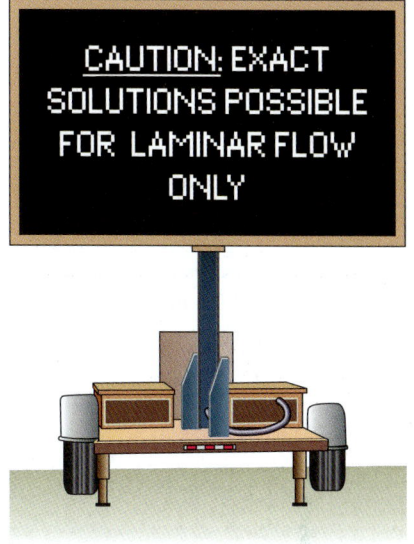

FIGURE 9–72
Exact analytical solutions of the Navier-Stokes equations, as in the examples provided here, are not possible if the flow is turbulent.

488
DIFFERENTIAL ANALYSIS OF FLUID FLOW

The Navier–Stokes Equation

$$\rho\left(\frac{\partial \vec{V}}{\partial t} + \boxed{(\vec{V}\cdot\vec{\nabla})\vec{V}}\right) = -\vec{\nabla}P + \rho\vec{g} + \mu\nabla^2\vec{V}$$

Nonlinear term

FIGURE 9–73
For incompressible flow solutions in which the advective terms in the Navier–Stokes equation are zero, the equation becomes *linear* since the advective term is the only nonlinear term in the equation.

not accelerating at all in this flow field, and linearizing the Navier–Stokes equation (Fig. 9–73). We have replaced the partial derivative operators for the *u*-derivatives with total derivative operators because of Eq. 3.

In similar fashion, every term in the r-momentum equation (Eq. 9–62b) except the pressure gradient term is zero, forcing that lone term to also be zero,

r-momentum:
$$\frac{\partial P}{\partial r} = 0 \quad (5)$$

In other words, *P* is not a function of *r*. Since *P* is also not a function of time (assumption 2) or θ (assumption 6), *P* can be at most a function of *x*,

Result of r-momentum: $\quad P = P(x)$ only $\quad (6)$

Therefore, we replace the partial derivative operator for the pressure gradient in Eq. 4 by the total derivative operator since *P* varies only with *x*. Finally, all terms of the θ-component of the Navier–Stokes equation (Eq. 9–62c) go to zero.

Step 4 *Solve the differential equations.* Continuity and *r*-momentum have already been "solved," resulting in Eqs. 3 and 6, respectively. The θ-momentum equation has vanished, and thus we are left with Eq. 4 (*x*-momentum). After multiplying both sides by *r*, we integrate once to obtain

$$r\frac{du}{dr} = \frac{r^2}{2\mu}\frac{dP}{dx} + C_1 \quad (7)$$

where C_1 is a constant of integration. Note that the pressure gradient dP/dx is a constant here. Dividing both sides of Eq. 7 by *r*, we integrate a second time to get

$$u = \frac{r^2}{4\mu}\frac{dP}{dx} + C_1 \ln r + C_2 \quad (8)$$

where C_2 is a second constant of integration.

Step 5 *Apply boundary conditions.* First, we apply boundary condition (2) to Eq. 7,

Boundary condition (2): $\quad 0 = 0 + C_1 \quad \rightarrow \quad C_1 = 0$

An alternative way to interpret this boundary condition is that *u* must remain finite at the centerline of the pipe. This is possible only if constant C_1 is equal to 0, since ln(0) is undefined in Eq. 8. Now we apply boundary condition (1),

Boundary condition (1): $\quad u = \frac{R^2}{4\mu}\frac{dP}{dx} + 0 + C_2 = 0 \quad \rightarrow \quad C_2 = -\frac{R^2}{4\mu}\frac{dP}{dx}$

Finally, Eq. 8 becomes

Axial velocity: $\quad u = \frac{1}{4\mu}\frac{dP}{dx}(r^2 - R^2) \quad (9)$

The axial velocity profile is thus in the shape of a paraboloid, as sketched in Fig. 9–74.

Step 6 *Verify the results.* You can verify that all the differential equations and boundary conditions are satisfied.

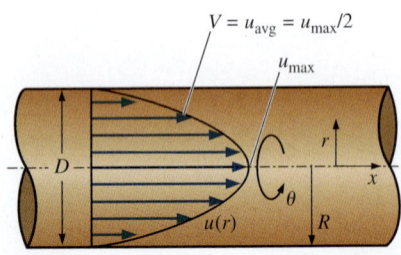

FIGURE 9–74
Axial velocity profile of Example 9–18: steady laminar flow in a long round pipe with an applied constant-pressure gradient dP/dx pushing fluid through the pipe.

We calculate some other properties of fully developed laminar pipe flow as well. For example, the maximum axial velocity obviously occurs at the centerline of the pipe (Fig. 9–74). Setting $r = 0$ in Eq. 9 yields

Maximum axial velocity: $$u_{max} = -\frac{R^2}{4\mu}\frac{dP}{dx} \quad (10)$$

The volume flow rate through the pipe is found by integrating Eq. 9 through a cross-section of the pipe,

$$\dot{V} = \int_{\theta=0}^{2\pi}\int_{r=0}^{R} u r\, dr\, d\theta = \frac{2\pi}{4\mu}\frac{dP}{dx}\int_{r=0}^{R}(r^2 - R^2)r\, dr = -\frac{\pi R^4}{8\mu}\frac{dP}{dx} \quad (11)$$

Since volume flow rate is also equal to the average axial velocity times cross-sectional area, we easily determine the average axial velocity V:

Average axial velocity: $$V = \frac{\dot{V}}{A} = \frac{(-\pi R^4/8\mu)(dP/dx)}{\pi R^2} = -\frac{R^2}{8\mu}\frac{dP}{dx} \quad (12)$$

Comparing Eqs. 10 and 12 we see that for fully developed laminar pipe flow, the average axial velocity is equal to exactly half of the maximum axial velocity.

To calculate the viscous shear force per unit surface area acting on the pipe wall, we consider a differential fluid element adjacent to the bottom portion of the pipe wall (Fig. 9–75). Pressure stresses and mathematically positive viscous stresses are shown. From Eq. 9–63 (modified for our coordinate system), we write the viscous stress tensor as

$$\tau_{ij} = \begin{pmatrix} \tau_{rr} & \tau_{r\theta} & \tau_{rx} \\ \tau_{\theta r} & \tau_{\theta\theta} & \tau_{\theta x} \\ \tau_{xr} & \tau_{x\theta} & \tau_{xx} \end{pmatrix} = \begin{pmatrix} 0 & 0 & \mu\frac{\partial u}{\partial r} \\ 0 & 0 & 0 \\ \mu\frac{\partial u}{\partial r} & 0 & 0 \end{pmatrix} \quad (13)$$

We use Eq. 9 for u, and set $r = R$ at the pipe wall; component τ_{rx} of Eq. 13 reduces to

Viscous shear stress at the pipe wall: $$\tau_{rx} = \mu\frac{du}{dr} = \frac{R}{2}\frac{dP}{dx} \quad (14)$$

For flow from left to right, dP/dx is negative, so the viscous shear stress on the bottom of the fluid element at the wall is in the direction opposite to that indicated in Fig. 9–75. (This agrees with our intuition since the pipe wall exerts a retarding force on the fluid.) The shear force per unit area on the *wall* is equal and opposite to this; hence,

Viscous shear force per unit area acting on the wall: $$\frac{\vec{F}}{A} = -\frac{R}{2}\frac{dP}{dx}\vec{i} \quad (15)$$

The direction of this force again agrees with our intuition; namely, the fluid tries to pull the bottom wall to the right, due to friction, when dP/dx is negative.

Discussion Since $du/dr = 0$ at the centerline of the pipe, $\tau_{rx} = 0$ there. You are encouraged to try to obtain Eq. 15 by using a control volume approach instead, taking your control volume as the fluid in the pipe between any two

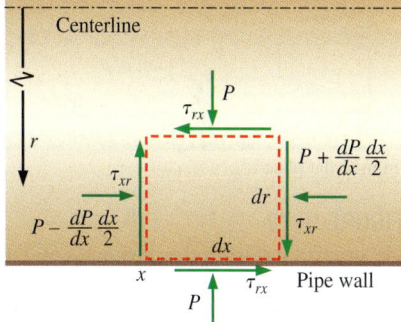

FIGURE 9–75
Pressure and viscous shear stresses acting on a differential fluid element whose bottom face is in contact with the pipe wall.

490
DIFFERENTIAL ANALYSIS OF FLUID FLOW

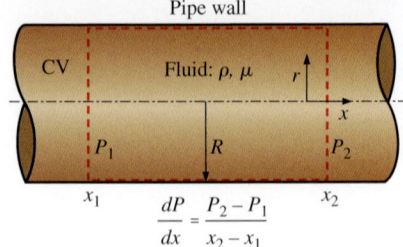

FIGURE 9–76
Control volume used to obtain Eq. 15 of Example 9–18 by an alternative method.

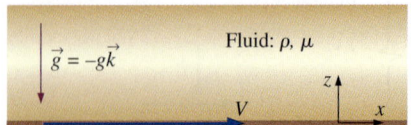

FIGURE 9–77
Geometry and setup for Example 9–19; the y-coordinate is into the page.

x-locations, x_1 and x_2 (Fig. 9–76). You should get the same answer. (*Hint*: Since the flow is fully developed, the axial velocity profile at location 1 is identical to that at location 2.) Note that when the volume flow rate through the pipe exceeds a critical value, instabilities in the flow occur, and the solution presented here is no longer valid. Specifically, flow in the pipe becomes turbulent rather than laminar; turbulent pipe flow is discussed in more detail in Chap. 8. This problem is also solved in Chap. 8 using an alternative approach.

So far, all our Navier–Stokes solutions have been for steady flow. You can imagine how much more complicated the solutions must get if the flow is allowed to be unsteady, and the time derivative term in the Navier–Stokes equation does not disappear. Nevertheless, there are some unsteady flow problems that can be solved analytically. We present one of these in Example 9–19.

EXAMPLE 9–19   Sudden Motion of an Infinite Flat Plate

Consider a viscous Newtonian fluid on top of an infinite flat plate lying in the xy-plane at $z = 0$ (Fig. 9–77). The fluid is at rest until time $t = 0$, when the plate suddenly starts moving at speed V in the x-direction. Gravity acts in the $-z$-direction. Determine the pressure and velocity fields.

SOLUTION The velocity and pressure fields are to be calculated for the case of fluid on top of an infinite flat plate that suddenly starts moving.
Assumptions 1 The wall is infinite in the x- and y-directions; thus, nothing is special about any particular x- or y-location. 2 The flow is *parallel* everywhere ($w = 0$). 3 Pressure P = constant with respect to x. In other words, there is no applied pressure gradient pushing the flow in the x-direction; flow occurs due to viscous stresses caused by the moving plate. 4 The fluid is incompressible and Newtonian with constant properties, and the flow is laminar. 5 The velocity field is two-dimensional in the xz-plane; therefore, $v = 0$, and all partial derivatives with respect to y are zero. 6 Gravity acts in the $-z$-direction.
Analysis To obtain the velocity and pressure fields, we follow the step-by-step procedure outlined in Fig. 9–52.

Step 1 *Lay out the problem and the geometry.* (See Fig. 9–77.)

Step 2 *List assumptions and boundary conditions.* We have listed six assumptions. The boundary conditions are: (1) At $t = 0$, $u = 0$ everywhere (no flow until the plate starts moving); (2) at $z = 0$, $u = V$ for all values of x and y (no-slip condition at the plate); (3) as $z \to \infty$, $u = 0$ (far from the plate, the effect of the moving plate is not felt); and (4) at $z = 0$, $P = P_{wall}$ (the pressure at the wall is constant at any x- or y-location along the plate).

Step 3 *Write out and simplify the differential equations.* We start with the incompressible continuity equation in Cartesian coordinates (Eq. 9–61a),

$$\frac{\partial u}{\partial x} + \underbrace{\frac{\partial \cancel{v}}{\cancel{\partial y}}}_{\text{assumption 5}} + \underbrace{\frac{\partial \cancel{w}}{\cancel{\partial z}}}_{\text{assumption 2}} = 0 \quad \to \quad \frac{\partial u}{\partial x} = 0 \qquad (1)$$

Equation 1 tells us that u is not a function of x. Furthermore, since u is not a function of y (assumption 5), we conclude that u is at most a function of z and t,

Result of continuity: $\quad u = u(z, t)$ only $\quad$ (2)

The *y*-momentum equation reduces to

$$\frac{\partial P}{\partial y} = 0 \quad (3)$$

by assumptions 5 and 6 (all terms with v, the *y-component* of velocity, vanish, and gravity does not act in the *y*-direction). Equation 3 simply tells us that pressure is not a function of y; hence,

Result of y-momentum: $\quad P = P(z, t)$ only $\quad$ (4)

Similarly the *z*-momentum equation reduces to

$$\frac{\partial P}{\partial z} = -\rho g \quad (5)$$

We now simplify the *x*-momentum equation (Eq. 9–61b) as far as possible.

$$\rho\left(\underbrace{\frac{\partial u}{\partial t}}_{} + \underbrace{u\frac{\partial u}{\partial x}}_{\text{continuity}} + \underbrace{v\frac{\partial u}{\partial y}}_{\text{assumption 5}} + \underbrace{w\frac{\partial u}{\partial z}}_{\text{assumption 2}}\right) = -\underbrace{\frac{\partial P}{\partial x}}_{\text{assumption 3}} + \underbrace{\rho g_x}_{\text{assumption 6}}$$

$$+ \mu\left(\underbrace{\frac{\partial^2 u}{\partial x^2}}_{\text{continuity}} + \underbrace{\frac{\partial^2 u}{\partial y^2}}_{\text{assumption 5}} + \frac{\partial^2 u}{\partial z^2}\right) \quad \rightarrow \quad \rho\frac{\partial u}{\partial t} = \mu\frac{\partial^2 u}{\partial z} \quad (6)$$

It is convenient to combine the viscosity and density into the kinematic viscosity, defined as $\nu = \mu/\rho$. Equation 6 reduces to the well-known **one-dimensional diffusion equation** (Fig. 9–78),

Result of x-momentum: $\quad \dfrac{\partial u}{\partial t} = \nu\dfrac{\partial^2 u}{\partial z^2} \quad (7)$

Step 4 Solve the differential equations. Continuity and *y*-momentum have already been "solved," resulting in Eqs. 2 and 4, respectively. Equation 5 (*z*-momentum) is integrated once, resulting in

$$P = -\rho g z + f(t) \quad (8)$$

where we have added a function of time instead of a constant of integration since *P* is a function of two variables, *z* and *t* (see Eq. 4). Equation 7 (*x*-momentum) is a linear partial differential equation whose solution is obtained by combining the two independent variables *z* and *t* into one independent variable. The result is called a **similarity solution,** the details of which are beyond the scope of this text. Note that the one-dimensional diffusion equation occurs in many other fields of engineering, such as diffusion of species (mass diffusion) and diffusion of heat (conduction); details about the solution can be found in books on these subjects. The solution of Eq. 7 is intimately tied to the boundary condition that the plate is impulsively started, and the result is

Integration of x-momentum: $\quad u = C_1\left[1 - \text{erf}\left(\dfrac{z}{2\sqrt{\nu t}}\right)\right] \quad (9)$

where **erf** in Eq. 9 is the **error function** (Çengel, 2010), defined as

Error function: $\quad \text{erf}(\xi) = \dfrac{2}{\sqrt{\pi}}\int_0^\xi e^{-\eta^2} d\eta \quad (10)$

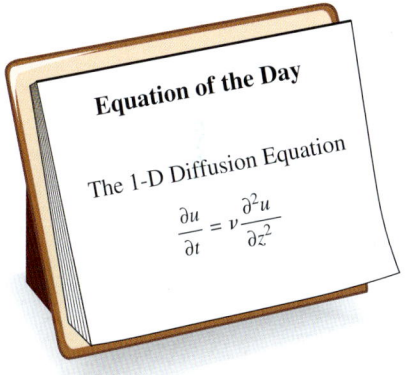

FIGURE 9–78
The one-dimensional diffusion equation is *linear*, but it is a *partial differential equation* (PDE). It occurs in many fields of science and engineering.

492
DIFFERENTIAL ANALYSIS OF FLUID FLOW

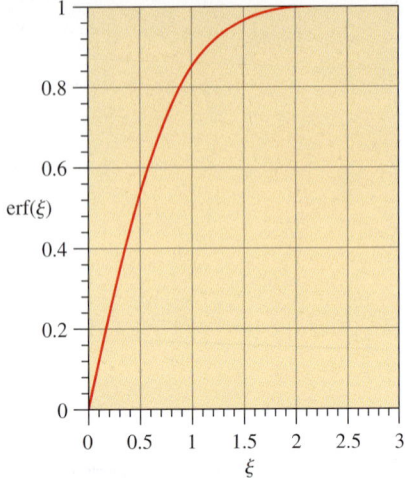

FIGURE 9–79
The error function ranges from 0 at $\xi = 0$ to 1 as $\xi \to \infty$.

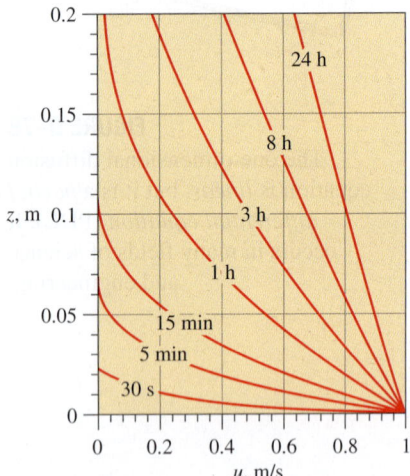

FIGURE 9–80
Velocity profiles of Example 9–19: flow of water above an impulsively started infinite plate; $\nu = 1.004 \times 10^{-6}$ m²/s and $V = 1.0$ m/s.

The error function is commonly used in probability theory and is plotted in Fig. 9–79. Tables of the error function can be found in many reference books, and some calculators and spreadsheets can calculate the error function directly. It is also provided as a function in the EES software that comes with this text.

Step 5 *Apply boundary conditions.* We begin with Eq. 8 for pressure. Boundary condition (4) requires that $P = P_{wall}$ at $z = 0$ for all times, and Eq. 8 becomes

Boundary condition (4): $\qquad P = 0 + f(t) = P_{wall} \quad \to \quad f(t) = P_{wall}$

In other words, the arbitrary function of time, $f(t)$, turns out not to be a function of time at all, but merely a constant. Thus,

Final result for pressure field: $\qquad P = P_{wall} - \rho g z \qquad$ (11)

which is simply hydrostatic pressure. We conclude that *hydrostatic pressure acts independently of the flow.* Boundary conditions (1) and (3) from step 2 have already been applied in order to obtain the solution of the x-momentum equation in step 4. Since erf(0) = 0, the second boundary condition yields

Boundary condition (2): $\qquad u = C_1(1 - 0) = V \quad \to \quad C_1 = V$

and Eq. 9 becomes

Final result for velocity field: $\qquad u = V\left[1 - \text{erf}\left(\dfrac{z}{2\sqrt{\nu t}}\right)\right] \qquad$ (12)

Several velocity profiles are plotted in Fig. 9–80 for the specific case of water at room temperature ($\nu = 1.004 \times 10^{-6}$ m²/s) with $V = 1.0$ m/s. At $t = 0$, there is no flow. As time goes on, the motion of the plate is felt farther and farther into the fluid, as expected. Notice how long it takes for viscous diffusion to penetrate into the fluid—after 15 min of flow, the effect of the moving plate is not felt beyond about 10 cm above the plate!

We define normalized variables u^* and z^* as

Normalized variables: $\qquad u^* = \dfrac{u}{V} \quad \text{and} \quad z^* = \dfrac{z}{2\sqrt{\nu t}}$

Then we rewrite Eq. 12 in terms of nondimensional parameters:

Normalized velocity field: $\qquad u^* = 1 - \text{erf}(z^*) \qquad$ (13)

The combination of unity minus the error function occurs often in engineering and is given the special name **complementary error function** and symbol **erfc**. Thus Eq. 13 can also be written as

Alternative form of the velocity field: $\qquad u^* = \text{erfc}(z^*) \qquad$ (14)

The beauty of the normalization is that this one equation for u^* as a function of z^* is valid for any fluid (with any kinematic viscosity ν) above a plate moving at any speed V and at any location z in the fluid at any time t! The normalized velocity profile of Eq. 13 is sketched in Fig. 9–81. All the profiles of Fig. 9–80 collapse into the single profile of Fig. 9–81; such a profile is called a **similarity profile.**

Step 6 *Verify the results.* You can verify that all the differential equations and boundary conditions are satisfied.

Discussion The time required for momentum to diffuse into the fluid seems much longer than we would expect based on our intuition. This is because the solution presented here is valid only for laminar flow. It turns out that if the plate's speed is large enough, or if there are significant vibrations in the plate or disturbances in the fluid, the flow will become turbulent. In a turbulent flow, large eddies mix rapidly moving fluid near the wall with slowly moving fluid away from the wall. This mixing process occurs rather quickly, so that turbulent diffusion is usually orders of magnitude faster than laminar diffusion.

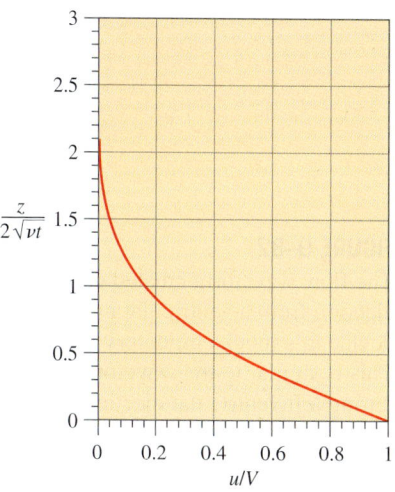

FIGURE 9–81
Normalized velocity profile of Example 9–19: laminar flow of a viscous fluid above an impulsively started infinite plate.

Examples 9–15 through 9–19 are for incompressible laminar flow. The same set of differential equations (incompressible continuity and Navier–Stokes) is valid for incompressible *turbulent* flow. However, turbulent flow solutions are much more complicated because the flow contains disordered, unsteady, three-dimensional eddies that mix the fluid. Furthermore, these eddies may range in size over several orders of magnitude. In a turbulent flow field, none of the terms in the equations can be ignored (with the exception of the gravity term in some cases), and thus solutions can be obtained only through numerical computations. Computational fluid dynamics (CFD) is discussed in Chap. 15.

Differential Analysis of Biofluid Mechanics Flows*

In Example 9–18 we derived fully developed flow in a round pipe, or what is commonly referred to as Poiseuille flow. The solution to the Navier-Stokes equation for this particular example is quite straightforward but is based on a number of assumptions and approximations. These approximations hold true for standard pipe flow with most water systems, for example. However, when applied to blood flow in the human body, the approximations must be closely monitored and evaluated for their applicability. Traditionally as a first-order attempt, cardiovascular fluid dynamists have used the Poiseuille flow derivation to understand blood flow in arteries. This can provide the engineer with a first-order approximation for the velocity and flow rate, but if the engineer were interested in a more sophisticated and, frankly realistic, understanding of blood flow, it is important to examine the main approximations used to arrive at Poiseuille flow.

Before delving in, let's retain the basic approximations about the fluid, or blood in this case. The fluid will remain incompressible, the flow will continue to be laminar, and gravity remains negligible. The approximation of fully developed flow will also remain, though in reality this is not applicable in the cardiovascular system. Based on only these approximations, this leaves the other main approximations of steady, parallel, axisymmetric Newtonian flow, and the pipe approximated as a rigid circular tube.

Recall that the heart pumps blood continuously at an average rate of 75 beats per minute for a healthy adult human at rest. As an example of

---

* This section was contributed by Professor Keefe Manning of Penn State University.

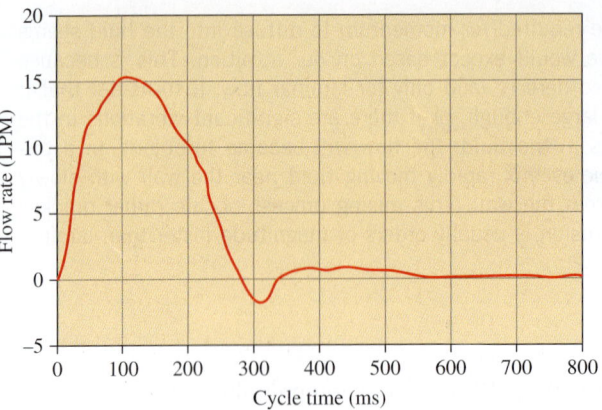

FIGURE 9–82
The flow waveform created during ejection from a ventricular assist device in a mock circulatory loop. This is similar to the waveform created during left ventricular ejection.

the flow waveform generated by the ventricular contraction simulated in a mock circulatory system (Figure 9–82), the flow rate changes temporally for this 800 ms cycle. Therefore, fundamentally to model blood flow through the arteries, the steady flow approximation is inappropriate, making modeling blood flow as Poiseuille flow unsuitable for just this one approximation alone. There is a rapid acceleration and deceleration of flow within a short time period (~300 ms). However, the wave propagation that is initiated at the heart diminishes with distance from it, and as the arteries become progressively smaller to the capillary level, the magnitude of pulsatility decreases. When focused on the venous side as blood returns to the heart, the steady flow approximation can be applied with more confidence, but it should be noted that there remains flow disruption, in particular, from the lower limbs as venous valves (similar to heart valves) help bring blood back to the heart.

The rigid, circular tube approximation is equally as inappropriate when applied to cardiovascular blood flow. As mentioned in Chapter 8, the blood vessels continually taper from the main vessel (the aorta) to smaller vessels (arteries, arterioles, and capillaries). There are no abrupt changes in diameter as might be seen in a commercial piping network. Therefore, one geometric consideration is the fact that a segment of blood vessel from one end to the other end will have a continual change in diameter. With respect to a circular tube cross-section, the vessels are not perfectly circular but rather more elliptical in their cross-section, so there is a major axis and minor axis. The most important approximation here that applies to Poiseuille flow is the fact that pipes are typically considered rigid. However, healthy vessels are *not* rigid; these structures are compliant and flexible. For example, the aorta emanating from the left ventricle can double in diameter to accommodate the sharp increase in blood volume during left ventricular ejection over a brief time period. One of the major exceptions to using this approximation is when studying pathologic states like atherosclerosis or studying blood flow in the elderly. The basic result of both is that the vessels will harden. In doing so, the rigidity approximation can be applied. There is also a secondary effect as the vessels harden, namely, the pulsatility of blood dampens more

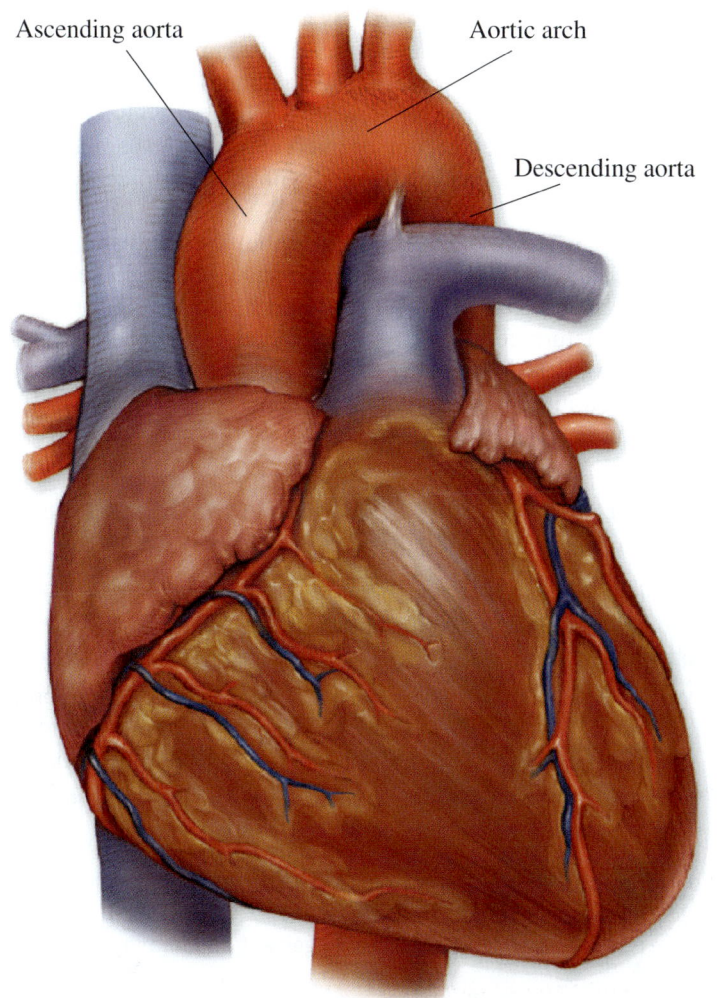

FIGURE 9–83
An anatomical figure illustrating the ascending aorta, aortic arch, and descending aorta coming from the left ventricle (on the backside of the heart in this view). The illustration demonstrates how the aorta moves toward the spinal cord.
McGraw-Hill Education

quickly, which can influence the steady flow approximation in the arterioles in these particular patient populations.

With respect to parallel flow and axisymmetric flow, these both can be invalidated as inappropriate approximations applied to blood flow, by focusing on one location of the cardiovascular system. Considering the aorta in Figure 9–83 (ascending from the left ventricle, the aortic arch, and descending from the arch), there are significant changes in geometry that influence the flow field. What is commonly not displayed in two-dimensional pictures of the cardiovascular system (like Figure 8–82) is the fact that the aorta does not remain in one plane as typically depicted. Actually, the aorta (as one looks at another person) will start from the left ventricle and move towards the spinal column (towards the back of the person) moving the flow into other planes due to pure anatomy. What this geometry does is create Dean flow in this region. As a result, the flow that is created moving around this bend and backwards, is a double helical swirling pattern (think about the DNA helix but the helixes are streamlines). With all this swirling, the approximations of parallel and axisymmetric flow are inappropriate. This is the most extreme case of flow in the human body (except for cases of pathology or with medical device intervention). The parallel and axisymmetric flow approximations can be used with more confidence in the rest of the circulatory system.

It should be mentioned that flow within the capillaries is *not* Poisueille flow since the red blood cells have to squeeze into these vessels and what results is a two-phase flow where a red blood cell is followed by plasma, which is in turn followed by a red blood cell; this continues, creating a unique flow field to facilitate oxygen and nutrient exchange. Finally, blood is not Newtonian, as illustrated in Example 9–20.

EXAMPLE 9–20 Fully Developed Flow in a Round Pipe with a Simple Blood Viscosity Model

Consider Example 9-18 and all the approximations to arrive at Poiseuille flow and the axial velocity profile shown in Fig. 9–74. In this example, we will change the basic assumption of a Newtonian fluid and instead use a non-Newtonian fluid viscosity model. Blood behaves as a viscoelastic fluid but for our purposes, we assume a shear thinning or pseudoplastic model and apply a generalized power law viscosity model. The power law model effectively comes from the viscous stress tensor and is $\tau_{rz} = -\mu \left(\dfrac{du}{dr} \right)^n$ where we introduce a negative sign for direction, and where $0 < n < 1$.

SOLUTION We take Example 9-18 up to Equation 4 in that example: $\dfrac{1}{r}\dfrac{d}{dr}\left(r\dfrac{du}{dr}\right) = \dfrac{1}{\mu}\dfrac{dP}{dx}$. Through rearrangement and one integration with respect to r, we arrive at $\dfrac{r}{2}\dfrac{dP}{dx} = \mu\dfrac{dP}{dx}$, which is also $\dfrac{r}{2}\dfrac{dP}{dx} = \mu\dfrac{dP}{dx} = \tau_{rz}$.

Then we can equate the power law model to this as well, and arrive at a new relationship, $\dfrac{r}{2}\dfrac{dP}{dx} = -\mu\left(\dfrac{du}{dr}\right)^n$. When we move the negative sign to the other side, multiple by $1/n$ on both sides, and solve for $\dfrac{du}{dr}$, we arrive at

$$\dfrac{du}{dr} = \left(-\dfrac{r}{2\mu}\dfrac{dP}{dx}\right)^{\frac{1}{n}}.$$

We integrate and then apply the second boundary condition from Example 9-18 (centerline of the pipe is an axis of symmetry). Our velocity then becomes

$$u = \dfrac{R^{\left(\frac{n+1}{n}\right)} - r^{\left(\frac{n+1}{n}\right)}}{\left(\dfrac{n+1}{n}\right)} \left(\dfrac{1}{2\mu}\dfrac{dP}{dx}\right)^{\frac{1}{n}}$$

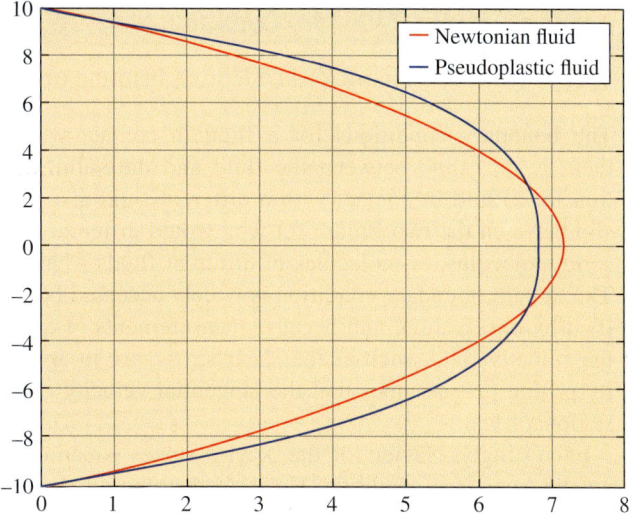

FIGURE 9–84

Assuming all values are the same in the velocity equations and the pipe is the same diameter, the pseudoplastic fluid causes the velocity profile to be more blunt compared to the parabolic profile generated for a Newtonian fluid.

We now have a generalized velocity profile for a power law fluid or a type of non-Newtonian fluid, which might be a rudimentary model for blood. As mentioned, we approximate blood as a pseudoplastic fluid; as such, we arbitrarily set $n = 0.5$. The actual velocity then becomes

$$u = \frac{R^3 - r^3}{3}\left(\frac{1}{2\mu}\frac{dP}{dx}\right)^2$$

Note that if we were to use $n = 1$ instead, we would get the following, $u = (R^2 - r^2)\left(\frac{1}{4\mu}\frac{dP}{dx}\right)$, which is the axial velocity for a Newtonian fluid.

We plot both the Newtonian and pseudoplastic velocity profiles in Fig. 9-84. Note how the viscosity alters the flow profile making it more blunt. To calculate the volume flow rate, we integrate over the cross section of the pipe using the equation $\dot{V} = \int_0^R 2\pi r u \, dr$ and using the generalized form for u. Once we integrate and do some algebraic manipulation, our flow rate becomes

$$\dot{V} = \frac{n\pi R^3}{3n + 1}\left(\frac{R}{2\mu}\frac{dP}{dx}\right)^{\frac{1}{n}}$$

For our example pseudoplastic fluid ($n = 0.5$), the flow rate simplifies to

$$\dot{V} = \frac{\pi R^5}{5}\left(\frac{1}{2\mu}\frac{dP}{dx}\right)^2$$

Discussion When $n = 1$, the general equation for volume flow rate reduces to that for Poiseuille flow, as it must.

DIFFERENTIAL ANALYSIS OF FLUID FLOW

APPLICATION SPOTLIGHT ■ The No-Slip Boundary Condition

Guest Author: *Minami Yoda, Georgia Institute of Technology*

The boundary conditions for a fluid in contact with a solid states that there is no "slip" between the fluid and the solid. The boundary condition for a fluid in contact with a different fluid also states that there is no slip between the two fluids. Yet why would different substances—fluid and solid molecules, or molecules of different fluids—have the same behavior? The no-slip boundary condition is widely accepted because it has been verified by observation, and because measurements of quantities derived from the velocity field, such as the shear stress, are in agreement with a velocity profile that assumes that the tangential velocity component is zero at a stationary wall.

Interestingly, Navier (of the Navier-Stokes equations) did not propose a no-slip boundary condition. He instead proposed the *partial-slip* boundary condition (Fig. 9–85) for a fluid in contact with a solid boundary: the fluid velocity component parallel to the wall at the wall, u_f, is proportional to the fluid shear stress at the wall, τ_s:

$$u_f = b\tau_s = b\mu_f \left(\frac{\partial u}{\partial y}\right)_f \tag{1}$$

FIGURE 9–85
Navier's partial-slip boundary condition.

where the constant of proportionality b, which has dimensions of length, is called the **slip length.** The no-slip condition is the special case of Eq. 1 where $b = 0$. Although some recent studies in very small (< 0.1 mm diameter) channels suggest that the no-slip condition may not hold within a few nanometers of the wall (recall that 1 nm = 10^{-9} m = 10 Ångstroms), the no-slip condition appears to be the correct boundary condition for a fluid in contact with a wall for a fluid that is a continuum.

Nevertheless, engineers also exploit the no-slip boundary condition to reduce friction (or viscous) drag. As discussed in this Chapter, the no-slip boundary condition at a free surface, or a water-air interface, makes the viscous stress τ_s, and thus the friction drag, very small in the liquid (Eq. 9-68). One way to create a free surface over a solid surface, like the hull of a ship, is to inject air to create a film of air that (at least partially) covers the hull surface (Fig. 9–86). In theory, the drag on the ship, and hence its fuel consumption, can be greatly reduced by creating a free-surface boundary condition over the ship hull. Maintaining a stable air film remains a major engineering challenge, however.

FIGURE 9–86
Proposed injection of air bubbles to form an air film over the bottom hull of a cargo ship [based on a picture courtesy of Y. Murai and Y. Oishi, Hokkaido University and the Monohakobi Technology Institute (MTI), Nippon Yusen Kaisha (NYK) and NYK-Hinode Lines].

References
Lauga, E., Brenner, M. and Stone, H., "Microfluidics: The No-Slip Boundary Condition," *Springer Handbook of Experimental Fluid Mechanics* (eds. C. Tropea, A. Yarin, J. F. Foss), Ch. 19, pp. 1219-1240, 2007.
http://www.nature.com/news/2008/080820/full/454924a.html

SUMMARY

In this chapter we derive the differential forms of conservation of mass (the *continuity equation*) and the linear momentum equation (the *Navier–Stokes equation*). For incompressible flow of a Newtonian fluid with constant properties, the continuity equation is

$$\vec{\nabla} \cdot \vec{V} = 0$$

and the Navier–Stokes equation is

$$\rho \frac{D\vec{V}}{Dt} = -\vec{\nabla} P + \rho \vec{g} + \mu \nabla^2 \vec{V}$$

For incompressible two-dimensional flow, we also define the stream function ψ. In Cartesian coordinates,

$$u = \frac{\partial \psi}{\partial y} \quad v = -\frac{\partial \psi}{\partial x}$$

We show that the difference in the value of ψ from one streamline to another is equal to the volume flow rate per unit width between the two streamlines and that curves of constant ψ are streamlines of the flow.

We provide several examples showing how the differential equations of fluid motion are used to generate an expression for the pressure field for a given velocity field and to generate expressions for both velocity and pressure fields for a flow with specified geometry and boundary conditions. The solution procedure learned here can be extended to much more complicated flows whose solutions require the aid of a computer.

The Navier–Stokes equation is the cornerstone of fluid mechanics. Although we know the necessary differential equations that describe fluid flow (continuity and Navier–Stokes), it is another matter to *solve* them. For some simple (usually infinite) geometries, the equations reduce to equations that we can solve analytically. For more complicated geometries, the equations are nonlinear, coupled, second-order, partial differential equations that cannot be solved with pencil and paper. We must then resort to either *approximate* solutions (Chap. 10) or *numerical* solutions (Chap. 15).

REFERENCES AND SUGGESTED READING

1. Y. A. Çengel. *Heat Transfer: A Practical Approach*, 4th ed. New York: McGraw-Hill, 2010.
2. R. W. Fox and A. T. McDonald. *Introduction to Fluid Mechanics*, 8th ed. New York: Wiley, 2011.
3. P. M. Gerhart, R. J. Gross, and J. I. Hochstein. *Fundamentals of Fluid Mechanics*, 2nd ed. Reading, MA: Addison-Wesley, 1992.
4. R. J. Heinsohn and J. M. Cimbala. *Indoor Air Quality Engineering*. New York: Marcel-Dekker, 2003.
5. P. K. Kundu, I. M. Cohen., and D. R. Dowling. *Fluid Mechanics*, ed. 5. San Diego, CA: Academic Press, 2011.
6. R. L. Panton. *Incompressible Flow*, 2nd ed. New York: Wiley, 2005.
7. M. R. Spiegel. *Vector Analysis, Schaum's Outline Series, Theory and Problems*. New York: McGraw-Hill Trade, 1968.
8. M. Van Dyke. *An Album of Fluid Motion*. Stanford, CA: The Parabolic Press, 1982.

PROBLEMS*

General and Mathematical Background Problems

9–1C Explain the fundamental differences between a *flow domain* and a *control volume*.

---

* Problems designated by a "C" are concept questions, and students are encouraged to answer them all. Problems with the icon are solved using EES, and complete solutions together with parametric studies are included on the text website. Problems with the icon are comprehensive in nature and are intended to be solved with an equation solver such as EES.

9–2C What does it mean when we say that two or more differential equations are *coupled*?

9–3C For a three-dimensional, unsteady, incompressible flow field in which temperature variations are insignificant, how many unknowns are there? List the equations required to solve for these unknowns.

9–4C For an unsteady, compressible flow field that is two-dimensional in the *x-y* plane and in which temperature and density variations *are* significant, how many unknowns are there? List the equations required to solve for these unknowns. (*Note*: Assume other flow properties like viscosity, thermal conductivity, etc., can be treated as constants.)

9–5C For an unsteady, incompressible flow field that is two-dimensional in the x-y plane and in which temperature variations are insignificant, how many unknowns are there? List the equations required to solve for these unknowns.

9–6 Transform the position $\vec{x} = (2, 4, -1)$ from Cartesian (x, y, z) coordinates to cylindrical (r, θ, z) coordinates, including units. The values of $\vec{x}$ are in units of meters.

9–7 Transform the position $x = (5 \text{ m}, \pi/3 \text{ radians}, 1.27 \text{ m})$ from cylindrical (r, θ, z) coordinates to Cartesian (x, y, z) coordinates, including units. Write all three components of $\vec{x}$ in units of meters.

9–8 A *Taylor series expansion* of function $f(x)$ about some x-location x_0 is given as

$$f(x_0 + dx) = f(x_0) + \left(\frac{df}{dx}\right)_{x=x_0} dx$$
$$+ \frac{1}{2!}\left(\frac{d^2f}{dx^2}\right)_{x=x_0} dx^2 + \frac{1}{3!}\left(\frac{d^3f}{dx^3}\right)_{x=x_0} dx^3 + \cdots$$

Consider the function $f(x) = \exp(x) = e^x$. Suppose we know the value of $f(x)$ at $x = x_0$, i.e., we know the value of $f(x_0)$, and we want to estimate the value of this function at some x location near x_0. Generate the first four terms of the Taylor series expansion for the given function (up to order dx^3 as in the above equation). For $x_0 = 0$ and $dx = -0.1$, use your truncated Taylor series expansion to estimate $f(x_0 + dx)$. Compare your result with the exact value of $e^{-0.1}$. How many digits of accuracy do you achieve with your truncated Taylor series?

9–9 Let vector $\vec{G}$ be given by $\vec{G} = 2xz\vec{i} - \frac{1}{2}x^2\vec{j} - z^2\vec{k}$. Calculate the divergence of $\vec{G}$, and simplify as much as possible. Is there anything special about your result? *Answer:* 0

9–10 The outer product of two vectors is a second-order tensor with nine components. In Cartesian coordinates, it is

$$\vec{F}\vec{G} = \begin{bmatrix} F_x G_x & F_x G_y & F_x G_z \\ F_y G_x & F_y G_y & F_y G_z \\ F_z G_x & F_z G_y & F_z G_z \end{bmatrix}$$

The *product rule* applied to the divergence of the product of two vectors $\vec{F}$ and $\vec{G}$ is written as $\vec{\nabla} \cdot (\vec{F}\vec{G}) = \vec{G}(\vec{\nabla} \cdot \vec{F}) + (\vec{F} \cdot \vec{\nabla})\vec{G}$. Expand both sides of this equation in Cartesian coordinates and verify that it is correct.

9–11 Use the product rule of Prob. 9–10 to show that $\vec{\nabla} \cdot (\rho \vec{V}\vec{V}) = \vec{V}\vec{\nabla} \cdot (\rho \vec{V}) + \rho(\vec{V} \cdot \vec{\nabla})\vec{V}$.

9–12 On many occasions we need to transform a velocity from Cartesian (x, y, z) coordinates to cylindrical (r, θ, z) coordinates (or vice versa). Using Fig. P9–12 as a guide, transform cylindrical velocity components (u_r, u_θ, u_z) into Cartesian velocity components (u, v, w). (*Hint*: Since the z-component of velocity remains the same in such a transformation, we need only to consider the xy-plane, as in Fig. P9–12.)

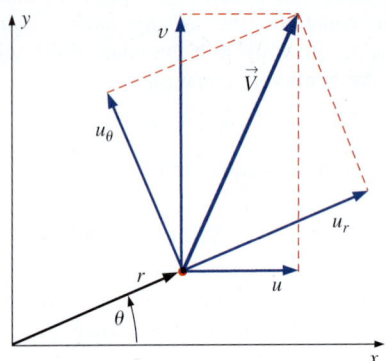

FIGURE P9–12

9–13 Using Fig. P9–12 as a guide, transform Cartesian velocity components (u, v, w) into cylindrical velocity components (u_r, u_θ, u_z). (*Hint*: Since the z-component of velocity remains the same in such a transformation, we need only to consider the xy-plane.)

9–14 Beth is studying a rotating flow in a wind tunnel. She measures the u and v components of velocity using a hot-wire anemometer. At $x = 0.40$ m and $y = 0.20$ m, $u = 10.3$ m/s and $v = -5.6$ m/s. Unfortunately, the data analysis program requires input in cylindrical coordinates (r, θ) and (u_r, u_θ). Help Beth transform her data into cylindrical coordinates. Specifically, calculate r, θ, u_r, and u_θ at the given data point.

9–15 A steady, two-dimensional, incompressible velocity field has Cartesian velocity components $u = Cy/(x^2 + y^2)$ and $v = -Cx/(x^2 + y^2)$, where C is a constant. Transform these Cartesian velocity components into cylindrical velocity components u_r and u_θ, simplifying as much as possible. You should recognize this flow. What kind of flow is this? *Answers:* 0, $-C/r$, line vortex

9–16 Consider a spiraling line vortex/sink flow in the xy- or $r\theta$-plane as sketched in Fig. P9–16. The two-dimensional cylindrical velocity components (u_r, u_θ) for this flow field are $u_r = C/2\pi r$ and $u_\theta = \Gamma/2\pi r$, where C and Γ are constants (m is negative and Γ is positive). Transform these two-dimensional cylindrical velocity components into two-dimensional Cartesian velocity components (u, v). Your final answer should contain no r or θ—only x and y. As a check of your algebra, calculate V^2 using Cartesian coordinates, and compare to V^2 obtained from the given velocity components in cylindrical components.

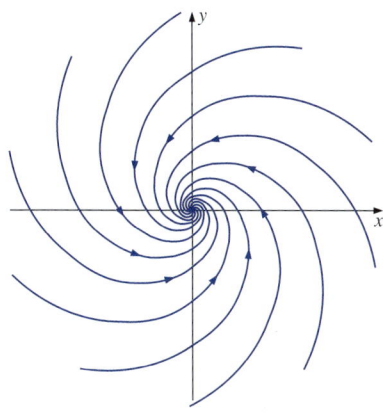

FIGURE P9–16

9–17 Let vector $\vec{G}$ be given by $\vec{G} = 4xz\vec{i} - y^2\vec{j} + yz\vec{k}$ and let V be the volume of a cube of unit length with its corner at the origin, bounded by $x = 0$ to 1, $y = 0$ to 1, and $z = 0$ to 1 (Fig. P9–17). Area A is the surface area of the cube. Perform both integrals of the divergence theorem and verify that they are equal. Show all your work.

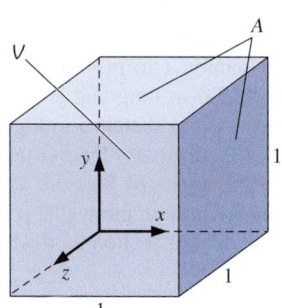

FIGURE P9–17

9–18 The *product rule* can be applied to the divergence of scalar f times vector $\vec{G}$ as: $\vec{\nabla} \cdot (f\vec{G}) = \vec{G} \cdot \vec{\nabla}f + f\vec{\nabla} \cdot \vec{G}$. Expand both sides of this equation in Cartesian coordinates and verify that it is correct.

Continuity Equation

9–19C In this chapter we derive the continuity equation in two ways: by using the divergence theorem and by summing mass flow rates through each face of an infinitesimal control volume. Explain why the former is so much less involved than the latter.

9–20C If a flow field is compressible, what can we say about the *material derivative* of density? What about if the flow field is incompressible?

9–21 Repeat Example 9–1 (gas compressed in a cylinder by a piston), but without using the continuity equation. Instead, consider the fundamental definition of density as mass div. by volume. Verify that Eq. 5 of Example 9–1 is correct.

9–22 The compressible form of the continuity equation is $(\partial \rho/\partial t) + \vec{\nabla} \cdot (\rho \vec{V}) = 0$. Expand this equation as far as possible in Cartesian coordinates (x, y, z) and (u, v, w).

9–23 In Example 9–6 we derive the equation for volumetric strain rate, $(1/V)(DV/Dt) = \vec{\nabla} \cdot \vec{V}$. Write this as a word equation and discuss what happens to the volume of a fluid element as it moves around in a compressible fluid flow field (Fig. P9–23).

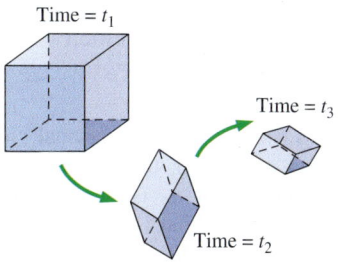

FIGURE P9–23

9–24 Verify that the spiraling line vortex/sink flow in the $r\theta$-plane of Prob. 9–16 satisfies the two-dimensional incompressible continuity equation. What happens to conservation of mass at the origin? Discuss.

9–25 Verify that the steady, two-dimensional, incompressible velocity field of Prob. 9–15 satisfies the continuity equation. Stay in Cartesian coordinates and show all your work.

9–26 Consider the steady, two-dimensional velocity field given by $\vec{V} = (u, v) = (1.6 + 1.8x)\vec{i} + (1.5 - 1.8y)\vec{j}$. Verify that this flow field is incompressible.

9–27 Consider steady flow of water through an axisymmetric garden hose nozzle (Fig. P9–27). The axial component of velocity increases linearly from $u_{z,\text{entrance}}$ to $u_{z,\text{exit}}$ as sketched. Between $z = 0$ and $z = L$, the axial velocity component is given by $u_z = u_{z,\text{entrance}} + [(u_{z,\text{exit}} - u_{z,\text{entrance}})/L]z$. Generate an expression for the *radial* velocity component u_r between $z = 0$ and $z = L$. You may ignore frictional effects on the walls.

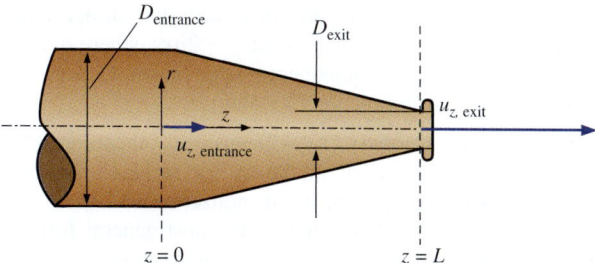

FIGURE P9–27

9–28 Consider the following steady, three-dimensional velocity field in Cartesian coordinates: $\vec{V} = (u, v, w) = (axy^2 - b)\vec{i} - 2cy^3\vec{j} + dxy\vec{k}$, where a, b, c, and d are constants. Under what conditions is this flow field incompressible? *Answer: $a = 6c$*

9–29 Consider the following steady, three-dimensional velocity field in Cartesian coordinates: $\vec{V} = (u, v, w) = (ax^2y + b)\vec{i} + cxy^2\vec{j} + dx^2y\vec{k}$ where a, b, c, and d are constants. Under what conditions is this flow field incompressible?

9–30 The u velocity component of a steady, two-dimensional, incompressible flow field is $u = ax + b$, where a and b are constants. Velocity component v is unknown. Generate an expression for v as a function of x and y.

9–31 Imagine a steady, two-dimensional, incompressible flow that is *purely circular* in the xy- or $r\theta$-plane. In other words, velocity component u_θ is nonzero, but u_r is zero everywhere (Fig. P9–31). What is the most general form of velocity component u_θ that does not violate conservation of mass?

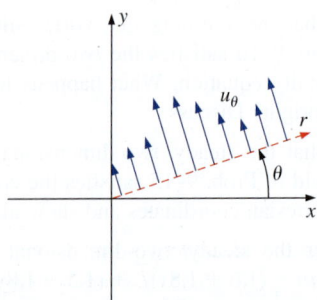

FIGURE P9–31

9–32 The u velocity component of a steady, two-dimensional, incompressible flow field is $u = ax + by$, where a and b are constants. Velocity component v is unknown. Generate an expression for v as a function of x and y. *Answer: $-ay + f(x)$*

9–33 The u velocity component of a steady, two-dimensional, incompressible flow field is $u = 3ax^2 - 2bxy$, where a and b are constants. Velocity component v is unknown. Generate an expression for v as a function of x and y.

9–34 Imagine a steady, two-dimensional, incompressible flow that is *purely radial* in the xy- or $r\theta$-plane. In other words, velocity component u_r is nonzero, but u_θ is zero everywhere (Fig. P9–34). What is the most general form of velocity component u_r that does not violate conservation of mass?

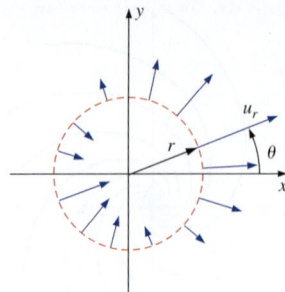

FIGURE P9–34

9–35 Two velocity components of a steady, incompressible flow field are known: $u = 2ax + bxy + cy^2$ and $v = axz - byz^2$, where a, b, and c are constants. Velocity component w is missing. Generate an expression for w as a function of x, y, and z.

9–36 A two-dimensional diverging duct is being designed to diffuse the high-speed air exiting a wind tunnel. The x-axis is the centerline of the duct (it is symmetric about the x-axis), and the top and bottom walls are to be curved in such a way that the axial wind speed u decreases approximately linearly from $u_1 = 300$ m/s at section 1 to $u_2 = 100$ m/s at section 2 (Fig. P9–36). Meanwhile, the air density ρ is to increase approximately linearly from $\rho_1 = 0.85$ kg/m³ at section 1 to $\rho_2 = 1.2$ kg/m³ at section 2. The diverging duct is 2.0 m long and is 1.60 m high at section 1 (only the upper half is sketched in Fig. P9–36; the half-height at section 1 is 0.80 m). (*a*) Predict the *y-component* of velocity, $v(x, y)$, in the duct. (*b*) Plot the approximate shape of the duct, ignoring friction on the walls. (*c*) What should be the half-height of the duct at section 2?

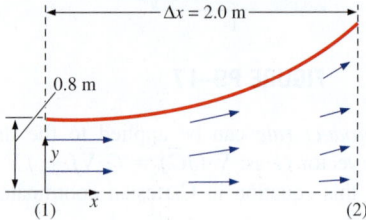

FIGURE P9–36

Stream Function

9–37C Consider two-dimensional flow in the xy-plane. What is the significance of the difference in value of stream function ψ from one streamline to another?

9–38C In CFD lingo, the stream function is often called a non-primitive variable, while velocity and pressure are called primitive variables. Why do you suppose this is the case?

9–39C What restrictions or conditions are imposed on stream function ψ so that it exactly satisfies the two-dimensional incompressible continuity equation by definition? Why are these restrictions necessary?

9–40C What is significant about curves of constant stream function? Explain why the stream function is useful in fluid mechanics.

9–41 Consider a steady, two-dimensional, incompressible flow field called a **uniform stream**. The fluid speed is V everywhere, and the flow is aligned with the x-axis (Fig. P9–41). The Cartesian velocity components are $u = V$ and $v = 0$. Generate an expression for the stream function for this flow. Suppose $V = 6.94$ m/s. If ψ_2 is a horizontal line at $y = 0.5$ m and the value of ψ along the x-axis is zero, calculate the volume flow rate per unit width (into the page of Fig. P9–41) between these two streamlines.

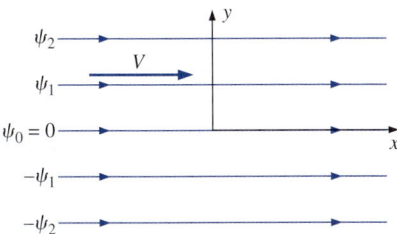

FIGURE P9–41

9–42 A common flow encountered in practice is the cross-flow of a fluid approaching a long cylinder of radius R at a free stream speed of U_∞. For incompressible inviscid flow, the velocity field of the flow is given as

$$u_r = U_\infty\left(1 - \frac{R^2}{r^2}\right)\cos\theta$$

$$u_\theta = -U_\infty\left(1 + \frac{R^2}{r^2}\right)\sin\theta$$

Show that the velocity field satisfies the continuity equation, and determine the stream function corresponding to this velocity field.

9–43 The stream function of an unsteady two-dimensional flow field is given by

$$\psi = \frac{4x}{y^2}t$$

Sketch a few streamlines for the given flow on the x-y plane, and derive expressions for the velocity components $u(x, y, t)$ and $v(x, y, t)$. Also determine the pathlines at $t = 0$.

9–44 Consider fully developed *Couette flow*—flow between two infinite parallel plates separated by distance h, with the top plate moving and the bottom plate stationary as illustrated in Fig. P9–44. The flow is steady, incompressible, and two-dimensional in the xy-plane. The velocity field is given by $\vec{V} = (u, v) = (Vy/h)\vec{i} + 0\vec{j}$. Generate an expression for stream function ψ along the vertical dashed line in Fig. P9–44. For convenience, let $\psi = 0$ along the bottom wall of the channel. What is the value of ψ along the top wall?
Answers: $Vy^2/2h$, $Vh/2$

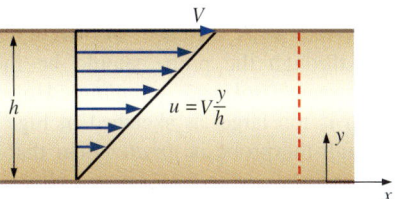

FIGURE P9–44

9–45 As a follow-up to Prob. 9–44, calculate the volume flow rate per unit width into the page of Fig. P9–44 from first principles (integration of the velocity field). Compare your result to that obtained directly from the stream function. Discuss.

9–46 Consider the Couette flow of Fig. P9–44. For the case in which $V = 3$ m/s and $h = 3$ cm, plot several streamlines using evenly spaced values of stream function. Are the streamlines themselves equally spaced? Discuss why or why not.

9–47 Consider fully developed, two-dimensional channel flow—flow between two infinite parallel plates separated by distance h, with both the top plate and bottom plate stationary, and a forced pressure gradient dP/dx driving the flow as illustrated in Fig. P9–47. (dP/dx is constant and negative.) The flow is steady, incompressible, and two-dimensional in the xy-plane. The velocity components are given by $u = (1/2\mu)(dP/dx)(y^2 - hy)$ and $v = 0$, where μ is the fluid's viscosity. Generate an expression for stream function ψ along the vertical dashed line in Fig. P9–47. For convenience, let $\psi = 0$ along the bottom wall of the channel. What is the value of ψ along the top wall?

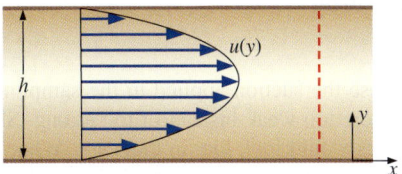

FIGURE P9–47

9–48 As a follow-up to Prob. 9–47, calculate the volume flow rate per unit width into the page of Fig. P9–47 from first principles (integration of the velocity field). Compare your result to that obtained directly from the stream function. Discuss.

9–49 Consider the channel flow of Fig. P9–47. The fluid is water at 20°C. For the case in which $dP/dx = -20,000$ N/m³ and $h = 1.20$ mm, plot several streamlines using evenly spaced values of stream function. Are the streamlines themselves equally spaced? Discuss why or why not.

9–50 In the field of air pollution control, one often needs to sample the quality of a moving airstream. In such measurements a sampling probe is aligned with the flow as sketched in Fig. P9–50. A suction pump draws air through the probe at volume flow rate $\dot{V}$ as sketched. For accurate sampling, the air speed through the probe should be the same as that of the airstream (*isokinetic sampling*). However, if the applied suction is too large, as sketched in Fig. P9–50, the air speed through the probe is *greater* than that of the airstream (*superisokinetic sampling*). For simplicity consider a two-dimensional case in which the sampling probe height is $h = 4.58$ mm and its width (into the page of Fig. P9–50) is $W = 39.5$ mm. The values of the stream function corresponding to the lower and upper dividing streamlines are $\psi_l = 0.093$ m²/s and $\psi_u = 0.150$ m²/s, respectively. Calculate the volume flow rate through the probe (in units of m³/s) and the average speed of the air sucked through the probe. *Answers:* 0.00225 m³/s, 12.4 m/s

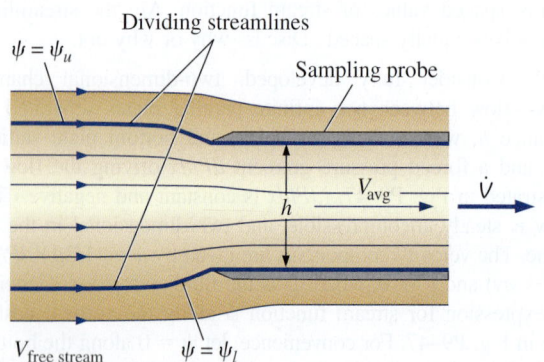

FIGURE P9–50

9–51 Suppose the suction applied to the sampling probe of Prob. 9–50 were too weak instead of too strong. Sketch what the streamlines would look like in that case. What would you call this kind of sampling? Label the lower and upper dividing streamlines.

9–52 Consider the air sampling probe of Prob. 9–50. If the upper and lower streamlines are 6.24 mm apart in the airstream far upstream of the probe, estimate the free stream speed $V_{\text{free stream}}$.

9–53 There are numerous occasions in which a fairly uniform free-stream flow of speed V in the x-direction encounters a long circular cylinder of radius a aligned normal to the flow (Fig. P9–53). Examples include air flowing around a car antenna, wind blowing against a flag pole or telephone pole, wind hitting electric wires, and ocean currents impinging on the submerged round beams that support oil platforms. In all these cases, the flow at the rear of the cylinder is separated and unsteady and usually turbulent. However, the flow in the front half of the cylinder is much more steady and predictable. In fact, except for a very thin boundary layer near the cylinder surface, the flow field can be approximated by the following steady, two-dimensional stream function in the xy- or $r\theta$-plane, with the cylinder centered at the origin: $\psi = V \sin \theta (r - a^2/r)$. Generate expressions for the radial and tangential velocity components.

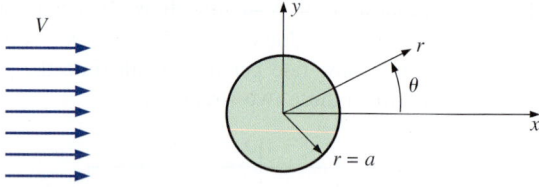

FIGURE P9–53

9–54 Consider steady, incompressible, *axisymmetric* flow (r, z) and (u_r, u_z) for which the stream function is defined as $u_r = -(1/r)(\partial\psi/\partial z)$ and $u_z = (1/r)(\partial\psi/\partial r)$. Verify that ψ so defined satisfies the continuity equation. What conditions or restrictions are required on ψ?

9–55 A uniform stream of speed V is inclined at angle α from the x-axis (Fig. P9–55). The flow is steady, two-dimensional, and incompressible. The Cartesian velocity components are $u = V \cos \alpha$ and $v = V \sin \alpha$. Generate an expression for the stream function for this flow.

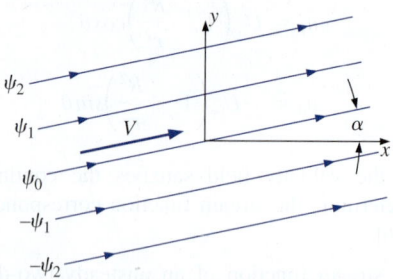

FIGURE P9–55

9–56 A steady, two-dimensional, incompressible flow field in the xy-plane has the following stream function: $\psi = ax^2 + bxy + cy^2$, where a, b, and c are constants. (*a*) Obtain expressions for velocity components u and v. (*b*) Verify that the flow field satisfies the incompressible continuity equation.

9–57 For the velocity field of Prob. 9–56, plot streamlines $\psi = 0, 1, 2, 3, 4, 5,$ and 6 m²/s. Let constants a, b, and c have the following values: $a = 0.50$ s⁻¹,

$b = -1.3 \text{ s}^{-1}$, and $c = 0.50 \text{ s}^{-1}$. For consistency, plot streamlines between $x = -2$ and 2 m, and $y = -4$ and 4 m. Indicate the direction of flow with arrows.

9–58 A steady, two-dimensional, incompressible flow field in the xy-plane has a stream function given by $\psi = ax^2 - by^2 + cx + dxy$, where a, b, c, and d are constants. (a) Obtain expressions for velocity components u and v. (b) Verify that the flow field satisfies the incompressible continuity equation.

9–59 Repeat Prob. 9–58, except make up your own stream function. You may create any function $\psi(x, y)$ that you desire, as long as it contains at least three terms and is not the same as an example or problem in this text. Discuss.

9–60 A steady, incompressible, two-dimensional CFD calculation of flow through an asymmetric two-dimensional branching duct reveals the streamline pattern sketched in Fig. P9–60, where the values of ψ are in units of m²/s, and W is the width of the duct into the page. The values of stream function ψ on the duct walls are shown. What percentage of the flow goes through the *upper* branch of the duct? *Answer:* 53.9%

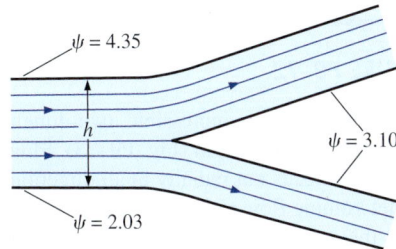

FIGURE P9–60

9–61 If the average velocity in the main branch of the duct of Prob. 9–60 is 13.4 m/s, calculate duct height h in units of cm. Obtain your result in two ways, showing all your work. You may use the results of Prob. 9–60 in only one of the methods.

9–62 Consider the garden hose nozzle of Prob. 9–27. Generate an expression for the stream function corresponding to this flow field.

9–63 Consider the garden hose nozzle of Probs. 9–27 and 9–62. Let the entrance and exit nozzle diameters be 1.25 and 0.35 cm, respectively, and let the nozzle length be 5 cm. The volume flow rate through the nozzle is 8 L/min. (a) Calculate the axial speeds (m/s) at the nozzle entrance and at the nozzle exit. (b) Plot several streamlines in the rz-plane inside the nozzle, and design the appropriate nozzle shape.

9–64 Flow separates at a sharp corner along a wall and forms a recirculating **separation bubble** as sketched in Fig. P9–64 (streamlines are shown). The value of the stream function at the wall is zero, and that of the uppermost streamline shown is some positive value ψ_{upper}. Discuss the value of the stream function inside the separation bubble. In particular, is it positive or negative? Why? Where in the flow is ψ a minimum?

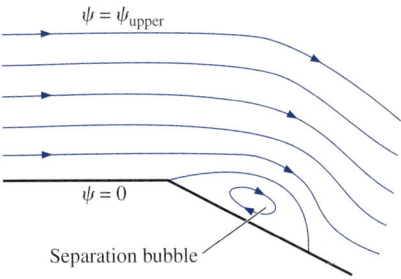

FIGURE P9–64

9–65 A graduate student is running a CFD code for his MS research project and generates a plot of flow streamlines (contours of constant stream function). The contours are of equally spaced values of stream function. Professor I. C. Flows looks at the plot and immediately points to a region of the flow and says, "Look how fast the flow is moving here!" What did Professor Flows notice about the streamlines in that region and how did she know that the flow was fast in that region?

9–66 Streaklines are shown in Fig. P9–66 for flow of water over the front portion of a blunt, axisymmetric cylinder aligned with the flow. Streaklines are generated by introducing air bubbles at evenly spaced points upstream of the field of view. Only the top half is shown since the flow is symmetric about the horizontal axis. Since the flow is steady, the streaklines are coincident with streamlines. Discuss how you can tell from the streamline pattern whether the flow speed in a particular region of the flow field is (relatively) large or small.

FIGURE P9–66

Courtesy ONERA. Photograph by Werlé.

9–67 A sketch of flow streamlines (contours of constant stream function) is shown in Fig. P9–67 for steady, incompressible, two-dimensional flow of air in a curved duct. (a) Draw arrows on the streamlines to indicate the direction of flow. (b) If $h = 4$ cm, what is the approximate speed of the air at point P? (c) Repeat part (b) if the fluid were water instead of air. Discuss. *Answers:* (b) 0.3 m/s, (c) 0.3 m/s

506
DIFFERENTIAL ANALYSIS OF FLUID FLOW

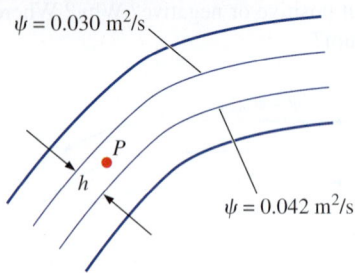

FIGURE P9–67

9–68 We briefly mention the *compressible stream function* ψ_ρ in this chapter, defined in Cartesian coordinates as $\rho u = (\partial \psi_\rho / \partial y)$ and $\rho v = -(\partial \psi_\rho / \partial x)$. What are the primary dimensions of ψ_ρ? Write the unit of ψ_ρ in primary SI units.

9–69 In Example 9–2, we provide expressions for u, v, and ρ for flow through a compressible converging duct. Generate an expression for the compressible stream function ψ_ρ that describes this flow field. For consistency, set $\psi_\rho = 0$ along the x-axis.

9–70 In Prob. 9–36 we developed expressions for u, v, and ρ for flow through the compressible, two-dimensional, diverging duct of a high-speed wind tunnel. Generate an expression for the compressible stream function ψ_ρ that describes this flow field. For consistency, set $\psi_\rho = 0$ along the x-axis. Plot several streamlines and verify that they agree with those you plotted in Prob. 9–36. What is the value of ψ_ρ at the top wall of the diverging duct?

9–71 Steady, incompressible, two-dimensional flow over a newly designed small hydrofoil of chord length $c = 9.0$ mm is modeled with a commercial computational fluid dynamics (CFD) code. A close-up view of flow streamlines (contours of constant stream function) is shown in Fig. P9–71. Values of the stream function are in units of m²/s. The fluid is water at room temperature. (*a*) Draw an arrow on the plot to indicate the direction and relative magnitude of the velocity at point *A*. Repeat for point *B*. Discuss how your results can be used to explain how such a body creates lift. (*b*) What is the approximate speed of the air at point *A*? (Point *A* is between streamlines 1.65 and 1.66 in Fig. P9–71.)

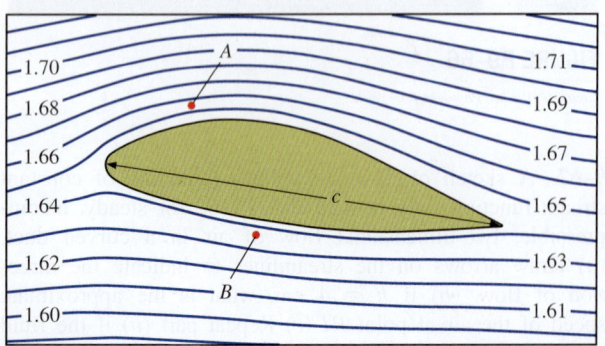

FIGURE P9–71

9–72 Time-averaged, turbulent, incompressible, two-dimensional flow over a square block of dimension $h = 1$ m sitting on the ground is modeled with a computational fluid dynamics (CFD) code. A close-up view of flow streamlines (contours of constant stream function) is shown in Fig. P9–72. The fluid is air at room temperature. Note that contours of constant *compressible stream function* are plotted in Fig. P9–72, even though the flow itself is approximated as incompressible. Values of ψ_ρ are in units of kg/m·s. (*a*) Draw an arrow on the plot to indicate the direction and relative magnitude of the velocity at point *A*. Repeat for point *B*. (*b*) What is the approximate speed of the air at point *B*? (Point *B* is between streamlines 5 and 6 in Fig. P9–72.)

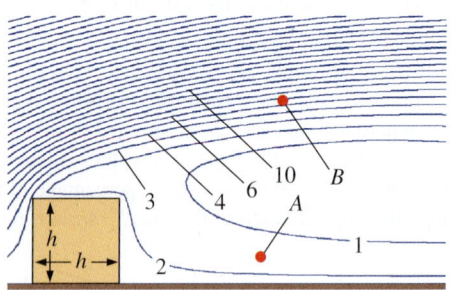

FIGURE P9–72

9–73 Consider steady, incompressible, two-dimensional flow due to a *line source* at the origin (Fig. P9–73). Fluid is created at the origin and spreads out radially in all directions in the xy-plane. The net volume flow rate of created fluid per unit width is $\dot{V}/L$ (into the page of Fig. P9–73), where L is the width of the line source into the page in Fig. P9–73. Since mass must be conserved everywhere except at the origin (a singular point), the volume flow rate per unit width through a circle of any radius r must also be $\dot{V}/L$. If we (arbitrarily) specify stream function ψ to be zero along the positive x-axis ($\theta = 0$), what is the value of ψ along the positive y-axis ($\theta = 90°$)? What is the value of ψ along the negative x-axis ($\theta = 180°$)?

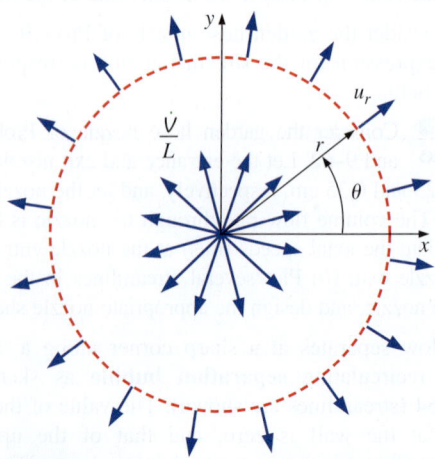

FIGURE P9–73

9–74 Repeat Prob. 9–73 for the case of a line sink instead of a line source. Let $\dot{V}/L$ be a positive value, but the flow is everywhere in the opposite direction.

Linear Momentum Equation, Boundary Conditions, and Applications

9–75C What is *mechanical pressure* P_m, and how is it used in an incompressible flow solution?

9–76C What are *constitutive equations*, and to which fluid mechanics equation are they applied?

9–77C An airplane flies at constant velocity $\vec{V}_{airplane}$ (Fig. P9–77C). Discuss the velocity boundary conditions on the air adjacent to the surface of the airplane from two frames of reference: (*a*) standing on the ground, and (*b*) moving with the airplane. Likewise, what are the far-field velocity boundary conditions of the air (far away from the airplane) in both frames of reference?

FIGURE P9–77C

9–78C What is the main distinction between a Newtonian fluid and a non-Newtonian fluid? Name at least three Newtonian fluids and three non-Newtonian fluids.

9–79C Define or describe each type of fluid: (*a*) viscoelastic fluid, (*b*) pseudoplastic fluid, (*c*) dilatant fluid, (*d*) Bingham plastic fluid.

9–80C The general control volume form of the linear momentum equation is

$$\underbrace{\int_{CV} \rho \vec{g}\, dV}_{I} + \underbrace{\int_{CS} \sigma_{ij} \cdot \vec{n}\, dA}_{II}$$

$$= \underbrace{\int_{CV} \frac{\partial}{\partial t}(\rho\vec{V})\, dV}_{III} + \underbrace{\int_{CS} (\rho\vec{V})\vec{V}\cdot\vec{n}\, dA}_{IV}$$

Discuss the meaning of each term in this equation. The terms are labeled for convenience. Write the equation as a word equation.

9–81 Consider liquid in a cylindrical tank. Both the tank and the liquid rotate as a rigid body (Fig. P9–81). The free surface of the liquid is exposed to room air. Surface tension effects are negligible. Discuss the boundary conditions required to solve this problem. Specifically, what are the velocity boundary conditions in terms of cylindrical coordinates (r, θ, z) and velocity components (u_r, u_θ, u_z) at all surfaces, including the tank walls and the free surface? What pressure boundary conditions are appropriate for this flow field? Write mathematical equations for each boundary condition and discuss.

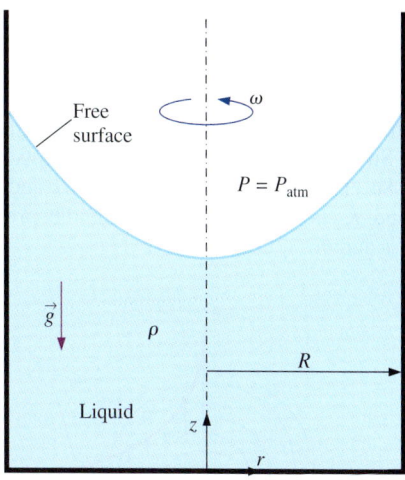

FIGURE P9–81

9–82 The $r\theta$-component of the viscous stress tensor in cylindrical coordinates is

$$\tau_{r\theta} = \tau_{\theta r} = \mu\left[r\frac{\partial}{\partial r}\left(\frac{u_\theta}{r}\right) + \frac{1}{r}\frac{\partial u_r}{\partial \theta}\right] \quad (1)$$

Some authors write this component instead as

$$\tau_{r\theta} = \tau_{\theta r} = \mu\left[\frac{1}{r}\left(\frac{\partial u_r}{\partial \theta} - u_\theta\right) + \frac{\partial u_\theta}{\partial r}\right] \quad (2)$$

Are these the same? In other words is Eq. 2 equivalent to Eq. 1, or do these other authors define their viscous stress tensor differently? Show all your work.

9–83 Engine oil at $T = 60°C$ is forced to flow between two very large, stationary, parallel flat plates separated by a thin gap height $h = 3.60$ mm (Fig. P9–83). The plate dimensions are $L = 1.25$ m and $W = 0.550$ m. The outlet pressure is atmospheric, and the inlet pressure is 1 atm gage pressure. Estimate the volume flow rate of oil. Also calculate the Reynolds number of the oil flow, based on gap height h and average velocity V. Is the flow laminar or turbulent? *Answers:* 2.39×10^{-3} m³/s, 51.8, laminar

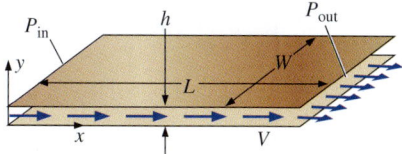

FIGURE P9–83

9–84 Consider the steady, two-dimensional, incompressible velocity field, $\vec{V} = (u, v) = (ax + b)\vec{i} + (-ay + c)\vec{j}$, where a, b, and c are constants. Calculate the pressure as a function of x and y.

9–85 Consider the following steady, two-dimensional, incompressible velocity field: $\vec{V} = (u, v) = (-ax^2)\vec{i} +$

$(2axy)\vec{j}$, where a is a constant. Calculate the pressure as a function of x and y.

9–86 Consider steady, two-dimensional, incompressible flow due to a spiraling line vortex/sink flow centered on the z-axis. Streamlines and velocity components are shown in Fig. P9–86. The velocity field is $u_r = C/r$ and $u_\theta = K/r$, where C and K are constants. Calculate the pressure as a function of r and θ.

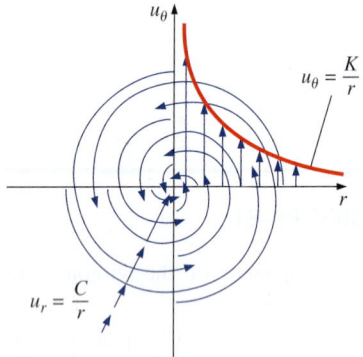

FIGURE P9–86

9–87 Consider the following steady, two-dimensional, incompressible velocity field: $\vec{V} = (u, v) = (ax + b)\vec{i} + (-ay + cx^2)\vec{j}$, where a, b, and c are constants. Calculate the pressure as a function of x and y. *Answer:* cannot be found

9–88 Consider steady, incompressible, parallel, laminar flow of a viscous fluid falling between two infinite vertical walls (Fig. P9–88). The distance between the walls is h, and gravity acts in the negative z-direction (downward in the figure). There is no applied (forced) pressure driving the flow—the fluid falls by gravity alone. The pressure is constant everywhere in the flow field. Calculate the velocity field and sketch the velocity profile using appropriate nondimensionalized variables.

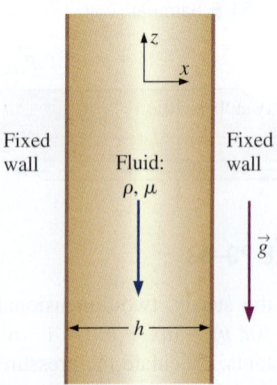

FIGURE P9–88

9–89 For the fluid falling between two parallel vertical walls (Prob. 9–88), generate an expression for the volume flow rate per unit width ($\dot{V}/L$) as a function of ρ, μ, h, and g. Compare your result to that of the same fluid falling along *one* vertical wall with a free surface replacing the second wall (Example 9–17), all else being equal. Discuss the differences and provide a physical explanation. *Answer:* $\rho g h^3 / 12 \mu$ downward

9–90 Repeat Example 9–17, except for the case in which the wall is inclined at angle α (Fig. P9–90). Generate expressions for both the pressure and velocity fields. As a check, make sure that your result agrees with that of Example 9–17 when $\alpha = 90°$. [*Hint:* It is most convenient to use the (s, y, n) coordinate system with velocity components (u_s, v, u_n), where y is into the page in Fig. P9–90. Plot the dimensionless velocity profile u_s^* versus n^* for the case in which $\alpha = 60°$.]

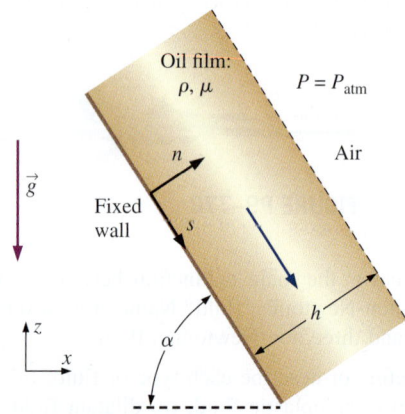

FIGURE P9–90

9–91 For the falling oil film of Prob. 9–90, generate an expression for the volume flow rate per unit width of oil falling down the wall ($\dot{V}/L$) as a function of ρ, μ, h, and g. Calculate ($\dot{V}/L$) for an oil film of thickness 5.0 mm with $\rho = 888$ kg/m^3 and $\mu = 0.80$ kg/m·s.

9–92 The first two viscous terms in the θ-component of the Navier–Stokes equation (Eq. 9–62c) are $\mu \left[\dfrac{1}{r} \dfrac{\partial}{\partial r} \left(r \dfrac{\partial u_\theta}{\partial r} \right) - \dfrac{u_\theta}{r^2} \right]$.

Expand this expression as far as possible using the product rule, yielding three terms. Now combine all three terms into one term. (*Hint:* Use the product rule in reverse—some trial and error may be required.)

9–93 An incompressible Newtonian liquid is confined between two concentric circular cylinders of infinite length—a solid inner cylinder of radius R_i and a hollow, stationary outer cylinder of radius R_o (Fig. P9–93; the z-axis is out of the page). The inner cylinder rotates at angular velocity ω_i. The flow is steady, laminar, and two-dimensional in the $r\theta$-plane. The flow is also *rotationally symmetric*, meaning that nothing is a function of coordinate θ (u_θ and P are functions of radius r only). The flow is also circular, meaning that velocity component $u_r = 0$ everywhere. Generate an exact expression for velocity component u_θ as a function of radius r and the other parameters in the problem. You may ignore gravity. (*Hint*: The result of Prob. 9–92 is useful.)

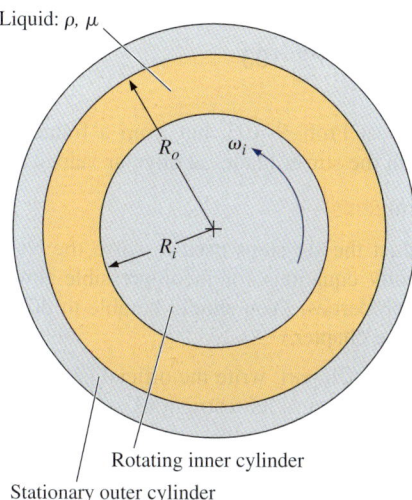

FIGURE P9–93

9–94 Repeat Prob. 9–93, but let the inner cylinder be stationary and the outer cylinder rotate at angular velocity ω_o. Generate an exact solution for $u_\theta(r)$ using the step-by-step procedure discussed in this chapter.

9–95 Analyze and discuss two limiting cases of Prob. 9–93: (*a*) The gap is very small. Show that the velocity profile approaches linear from the outer cylinder wall to the inner cylinder wall. In other words, for a very tiny gap the velocity profile reduces to that of simple two-dimensional Couette flow. (*Hint*: Define $y = R_o - r$, h = gap thickness = $R_o - R_i$, and V = speed of the "upper plate" = $R_i\omega_i$.) (*b*) The outer cylinder radius approaches infinity, while the inner cylinder radius is very small. What kind of flow does this approach?

9–96 Repeat Prob. 9–93 for the more general case. Namely, let the inner cylinder rotate at angular velocity ω_i and let the outer cylinder rotate at angular velocity ω_o. All else is the same as Prob. 9–93. Generate an exact expression for velocity component u_θ as a function of radius r and the other parameters in the problem. Verify that when $\omega_o = 0$ your result simplifies to that of Prob. 9–93.

9–97 Analyze and discuss a limiting case of Prob. 9–96 in which there is no inner cylinder ($R_i = \omega_i = 0$). Generate an expression for u_θ as a function of r. What kind of flow is this? Describe how this flow could be set up experimentally. *Answer:* $\omega_o r$

9–98 Consider steady, incompressible, laminar flow of a Newtonian fluid in an infinitely long round pipe annulus of inner radius R_i and outer radius R_o (Fig. P9–98). Ignore the effects of gravity. A constant negative pressure gradient $\partial P/\partial x$ is applied in the x-direction, $(\partial P/\partial x) = (P_2 - P_1)/(x_2 - x_1)$, where x_1 and x_2 are two arbitrary locations along the x-axis, and P_1 and P_2 are the pressures at those two locations. The pressure gradient may be caused by a pump and/or gravity. Note that we adopt a modified cylindrical coordinate system here with x instead of z for the axial component, namely, (r, θ, x) and (u_r, u_θ, u). Derive an expression for the velocity field in the annular space in the pipe.

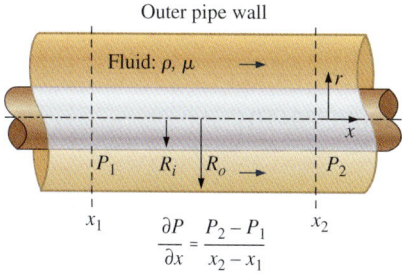

FIGURE P9–98

9–99 Consider again the pipe annulus sketched in Fig. P9–98. Assume that the pressure is constant everywhere (there is no forced pressure gradient driving the flow). However, let the inner cylinder be moving at steady velocity V to the right. The outer cylinder is stationary. (This is a kind of axisymmetric Couette flow.) Generate an expression for the x-component of velocity u as a function of r and the other parameters in the problem.

9–100 Repeat Prob. 9–99 except swap the stationary and moving cylinder. In particular, let the inner cylinder be stationary, and let the outer cylinder be moving at steady velocity V to the right, all else being equal. Generate an expression for the x-component of velocity u as a function of r and the other parameters in the problem.

9–101 Consider a modified form of Couette flow in which there are two immiscible fluids sandwiched between two infinitely long and wide, parallel flat plates (Fig. P9–101). The flow is steady, incompressible, parallel, and laminar. The top plate moves at velocity V to the right, and the bottom plate is stationary. Gravity acts in the $-z$-direction (downward in the figure). There is no forced pressure gradient pushing the fluids through the channel—the flow is set up solely by viscous effects created by the moving upper plate. You may ignore surface tension effects and assume that the interface is horizontal. The pressure at the bottom of the flow ($z = 0$) is equal to P_0. (*a*) List all the appropriate boundary conditions on both velocity and pressure. (*Hint*: There are six required boundary conditions.) (*b*) Solve for the velocity field. (*Hint*: Split up the solution into two portions, one for each fluid. Generate expressions for u_1 as a function of z and u_2 as a function of z.) (*c*) Solve for the pressure field. (*Hint*: Again split up the solution. Solve for P_1 and P_2.) (d) Let fluid 1 be water and let fluid 2 be unused engine oil, both at 80°C. Also let $h_1 = 5.0$ mm, $h_2 = 8.0$ mm, and $V = 10.0$ m/s. Plot u as a function of z across the entire channel. Discuss the results.

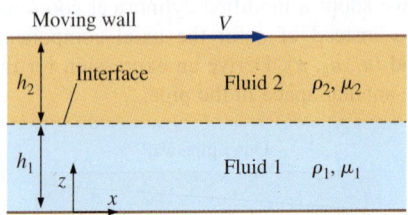

FIGURE P9–101

9–102 Consider steady, incompressible, laminar flow of a Newtonian fluid in an infinitely long round pipe of diameter D or radius $R = D/2$ inclined at angle α (Fig. P9–102). There is no applied pressure gradient ($\partial P/\partial x = 0$). Instead, the fluid flows down the pipe due to gravity alone. We adopt the coordinate system shown, with x down the axis of the pipe. Derive an expression for the x-component of velocity u as a function of radius r and the other parameters of the problem. Calculate the volume flow rate and average axial velocity through the pipe. *Answers:* $\rho g (\sin \alpha)(R^2 - r^2)/4\mu$, $\rho g (\sin \alpha)\pi R^4/8\mu$, $\rho g (\sin \alpha) R^2/8\mu$

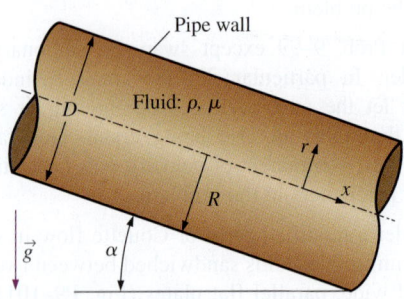

FIGURE P9–102

9–103 A stirrer mixes liquid chemicals in a large tank (Fig. P9–103). The free surface of the liquid is exposed to room air. Surface tension effects are negligible. Discuss the boundary conditions required to solve this problem. Specifically, what are the velocity boundary conditions in terms of cylindrical coordinates (r, θ, z) and velocity components (u_r, u_θ, u_z) at all surfaces, including the blades and the free surface? What pressure boundary conditions are appropriate for this flow field? Write mathematical equations for each boundary condition and discuss.

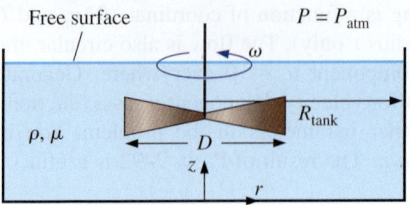

FIGURE P9–103

9–104 Repeat Prob. 9–103, but from a frame of reference rotating with the stirrer blades at angular velocity ω.

Review Problems

9–105C List the six steps used to solve the Navier–Stokes and continuity equations for incompressible flow with constant fluid properties. (You should be able to do this without peeking at the chapter.)

9–106C For each part, write the official name for the differential equation, discuss its restrictions, and describe what the equation represents physically.

(a) $\dfrac{\partial \rho}{\partial t} + \vec{\nabla} \cdot (\rho \vec{V}) = 0$

(b) $\dfrac{\partial}{\partial t}(\rho \vec{V}) + \vec{\nabla} \cdot (\rho \vec{V} \vec{V}) = \rho \vec{g} + \vec{\nabla} \cdot \sigma_{ij}$

(c) $\rho \dfrac{D\vec{V}}{Dt} = -\vec{\nabla} P + \rho \vec{g} + \mu \nabla^2 \vec{V}$

9–107C Explain why the incompressible flow approximation and the constant temperature approximation usually go hand in hand.

9–108C For each statement, choose whether the statement is true or false and discuss your answer briefly. For each statement it is assumed that the proper boundary conditions and fluid properties are known.

(a) A general incompressible flow problem with constant fluid properties has four unknowns.

(b) A general compressible flow problem has five unknowns.

(c) For an incompressible fluid mechanics problem, the continuity equation and Cauchy's equation provide enough equations to match the number of unknowns.

(d) For an incompressible fluid mechanics problem involving a Newtonian fluid with constant properties, the continuity equation and the Navier–Stokes equation provide enough equations to match the number of unknowns.

9–109C Discuss the relationship between volumetric strain rate and the continuity equation. Base your discussion on fundamental definitions.

9–110 Repeat Example 9–17, except for the case in which the wall is moving upward at speed V. As a check, make sure that your result agrees with that of Example 9–17 when $V = 0$. Nondimensionalize your velocity profile equation using the same normalization as in Example 9–17, and show that a Froude number and a Reynolds number emerge. Plot the profile w^* versus x^* for cases in which Fr = 0.5 and Re = 0.5, 1.0, and 5.0. Discuss.

9–111 For the falling oil film of Prob. 9–110, calculate the volume flow rate per unit width of oil falling down the wall ($\dot{V}/L$) as a function of wall speed V and the other parameters in the problem. Calculate the wall speed required such that there is no net volume flow of oil either up or down. Give your answer for V in terms of the other parameters in the problem, namely, ρ, μ, h, and g. Calculate V for zero volume flow rate for an oil film of thickness 4.12 mm with $\rho = 888$ kg/m^3 and $\mu = 0.801$ kg/m·s. *Answer:* 0.0615 m/s

9–112 Consider the following steady, three-dimensional velocity field in Cartesian coordinates: $\vec{V} = (u, v, w) = (axz^2 - by)\vec{i} + cxyz\vec{j} + (dz^3 + exz^2)\vec{k}$, where a, b, c, d, and e are constants. Under what conditions is this flow field incompressible? What are the primary dimensions of constants a, b, c, d, and e?

9–113 Simplify the Navier–Stokes equation as much as possible for the case of an incompressible liquid being *accelerated as a rigid body* in an arbitrary direction (Fig. P9–113). Gravity acts in the $-z$-direction. Begin with the incompressible vector form of the Navier–Stokes equation, explain how and why some terms can be simplified, and give your final result as a vector equation.

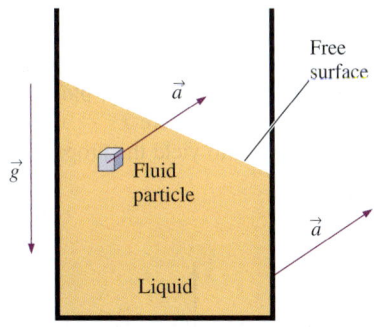

FIGURE P9–113

9–114 Simplify the Navier–Stokes equation as much as possible for the case of incompressible *hydrostatics*, with gravity acting in the negative z-direction. Begin with the incompressible vector form of the Navier–Stokes equation, explain how and why some terms can be simplified, and give your final result as a vector equation. *Answer:* $\vec{\nabla}P = -\rho g\vec{k}$

9–115 Bob uses a computational fluid dynamics code to model steady flow of an incompressible fluid through a two-dimensional sudden contraction as sketched in Fig. P9–115. Channel height changes from $H_1 = 12.0$ cm to $H_2 = 4.6$ cm. Uniform velocity $\vec{V}_1 = 18.5\vec{i}$ m/s is to be specified on the left boundary of the computational domain. The CFD code uses a numerical scheme in which the stream function must be specified along all boundaries of the computational domain. As shown in Fig. P9–115, ψ is specified as zero along the entire bottom wall of the channel. (*a*) What value of ψ should Bob specify on the top wall of the channel? (*b*) How should Bob specify ψ on the left side of the computational domain? (*c*) Discuss how Bob might specify ψ on the right side of the computational domain.

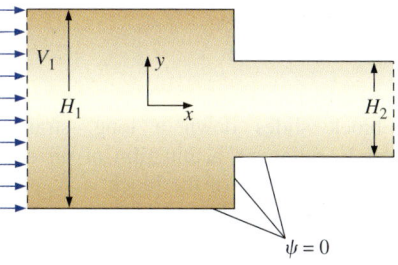

FIGURE P9–115

9–116 For each of the listed equations, write down the equation in vector form and decide if it is linear or nonlinear. If it is nonlinear, which term(s) make it so? (*a*) incompressible continuity equation, (*b*) compressible continuity equation, and (*c*) incompressible Navier–Stokes equation.

9–117 A **boundary layer** is a thin region near a wall in which viscous (frictional) forces are very important due to the no-slip boundary condition. The steady, incompressible, two-dimensional, boundary layer developing along a flat plate aligned with the free-stream flow is sketched in Fig. P9–117. The flow upstream of the plate is uniform, but boundary layer thickness δ grows with x along the plate due to viscous effects. Sketch some streamlines, both within the boundary layer and above the boundary layer. Is $\delta(x)$ a streamline? (*Hint*: Pay particular attention to the fact that for steady, incompressible, two-dimensional flow the volume flow rate per unit width between any two streamlines is constant.)

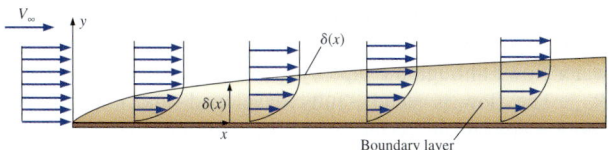

FIGURE P9–117

9–118 Consider steady, two-dimensional, incompressible flow in the xz-plane rather than in the xy-plane. Curves of constant stream function are shown in Fig. P9–118. The nonzero velocity components are (u, w). Define a stream function such that flow is from right to left in the xz-plane when ψ increases in the z-direction.

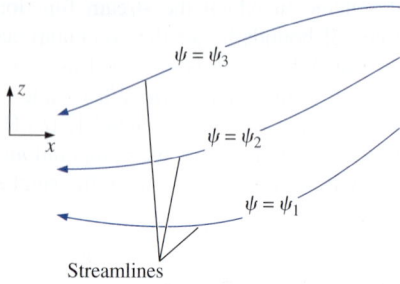

FIGURE P9–118

9–119 A block slides down a long, straight, inclined wall at speed V, riding on a thin film of oil of thickness h (Fig. P9–119). The weight of the block is W, and its surface area in contact with the oil film is A. Suppose V is measured, and W, A, angle α, and viscosity μ are also known. Oil film thickness h is not known. (a) Generate an exact analytical expression for h as a function of the known parameters V, A, W, α, and μ. (b) Use dimensional analysis to generate a dimensionless expression for h as a function of the given parameters. Construct a relationship between your Π's that matches the exact analytical expression of part (a).

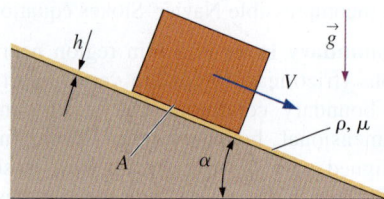

FIGURE P9–119

9–120 Look up the definition of *Poisson's equation* in one of your math textbooks or on the Internet. Write Poisson's equation in standard form. How is Poisson's equation similar to Laplace's equation? How do these two equations differ?

9–121 Water flows down a long, straight, inclined pipe of diameter D and length L (Fig. P9–121). There is no forced pressure gradient between points 1 and 2; in other words, the water flows through the pipe by gravity alone, and $P_1 = P_2 = P_{atm}$. The flow is steady, fully developed, and laminar. We adopt a coordinate system in which x follows the axis of the pipe. (a) Use the control volume technique of Chap. 8 to generate an expression for average velocity V as a function of the given parameters ρ, g, D, Δz, μ, and L. (b) Use differential analysis to generate an expression for V as a function of the given parameters. Compare with your result of part (a) and discuss. (c) Use dimensional analysis to generate a dimensionless expression for V as a function of the given parameters. Construct a relationship between your Π's that matches the exact analytical expression.

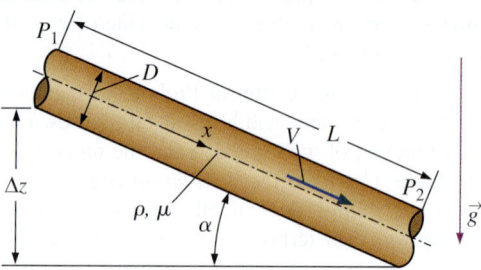

FIGURE P9–121

9–122 We approximate the flow of air into a vacuum cleaner's floor attachment by the stream function

$$\psi = \frac{-\dot{V}}{2\pi L} \arctan \frac{\sin 2\theta}{\cos 2\theta + b^2/r^2}$$

in the center plane (the xy-plane) in cylindrical coordinates, where L is the length of the attachment, b is the height of the attachment above the floor, and $\dot{V}$ is the volume flow rate of air being sucked into the hose. Shown in Fig. P9–122 is a three-dimensional view with the floor in the xz-plane; we model a two-dimensional slice of the flow in the xy-plane through the centerline of the attachment. Note that we have (arbitrarily) set $\psi = 0$ along the positive x-axis ($\theta = 0$). (a) What are the primary dimensions of the given stream function? (b) Nondimensionalize the stream function by defining $\psi^* = (2\pi L/\dot{V})\psi$ and $r^* = r/b$. (c) Solve your nondimensionalized equation for r^* as a function of ψ^* and θ. Use this equation to plot several nondimensional streamlines of the flow. For consistency, plot in the range $-2 < x^* < 2$ and $0 < y^* < 4$, where $x^* = x/b$ and $y^* = y/b$. (Hint: ψ^* must be *negative* to yield the proper flow direction.)

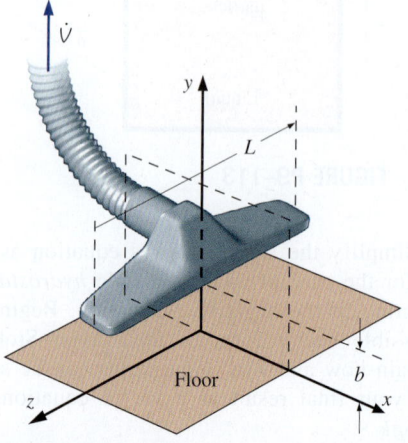

FIGURE P9–122

9–123 Taking all the Poiseuille flow approximations except that the fluid is Newtonian, determine the velocity profile and flow rate assuming blood is a Bingham plastic fluid based on the shear stress relationship below. Plot the velocity profile of a Newtonian fluid, a pseudoplastic fluid, and a Bingham plastic fluid. How do they differ? Determine the flow rate assuming a Bingham plastic fluid.

$$\tau_{rz} = -\mu \frac{du}{dr} + \tau_y$$

Fundamentals of Engineering (FE) Exam Problems

9–124 The continuity equation is also known as
(a) Conservation of mass (b) Conservation of energy
(c) Conservation of momentum (d) Newton's second law
(e) Cauchy's equation

9–125 The Navier-Stokes equation is also known as
(a) Newton's first law (b) Newton's second law
(c) Newton's third law (d) Continuity equation
(e) Energy equation

9–126 Which choice is the general differential equation form of the continuity equation for a control volume?

(a) $\int_{CS} \rho \vec{V} \cdot \vec{n} \, dA = 0$ (b) $\int_{CV} \frac{\partial \rho}{\partial t} dV + \int_{CS} \rho \vec{V} \cdot \vec{n} \, dA = 0$

(c) $\vec{\nabla} \cdot (\rho \vec{V}) = 0$ (d) $\frac{\partial \rho}{\partial t} + \vec{\nabla} \cdot (\rho \vec{V}) = 0$

(e) None of these

9–127 Which choice is the differential, incompressible, two-dimensional continuity equation in Cartesian coordinates?

(a) $\int_{CS} \rho \vec{V} \cdot \vec{n} \, dA = 0$ (b) $\frac{1}{r}\frac{\partial (r u_r)}{\partial r} + \frac{1}{r}\frac{\partial (u_\theta)}{\partial \theta} = 0$

(c) $\vec{\nabla} \cdot (\rho \vec{V}) = 0$ (d) $\vec{\nabla} \cdot \vec{V} = 0$

(e) $\frac{\partial u}{\partial x} + \frac{\partial v}{\partial y} = 0$

9–128 A steady velocity field is given by $\vec{V} = (u, v, w) = 2ax^2 y\vec{i} + 3bxy^2\vec{j} + cy\vec{k}$, where a, b, and c are constants. Under what conditions is this flow field incompressible?

(a) $a = b$ (b) $a = -b$ (c) $2a = -3b$
(d) $3a = 2b$ (e) $a = 2b$

9–129 A steady, two-dimensional, incompressible flow field in the xy-plane has a stream function given by $\psi = ax^2 + by^2 + cy$, where a, b, and c are constants. The expression for the velocity component u is

(a) $2ax$ (b) $2by + c$ (c) $-2ax$
(d) $-2by - c$ (e) $2ax + 2by + c$

9–130 A steady, two-dimensional, incompressible flow field in the xy-plane has a stream function given by $\psi = ax^2 + by^2 + cy$, where a, b, and c are constants. The expression for the velocity component v is

(a) $2ax$ (b) $2by + c$ (c) $-2ax$ (d) $-2by - c$
(e) $2ax + 2by + c$

9–131 If a fluid flow is both incompressible and isothermal, which property is not expected to be constant?
(a) Temperature (b) Density (c) Dynamic viscosity
(d) Kinematic viscosity (e) Specific heat

9–132 Which choice is the incompressible Navier-Stokes equation with constant viscosity?

(a) $\rho \frac{D\vec{V}}{Dt} + \vec{\nabla}P - \rho\vec{g} = 0$ (b) $-\vec{\nabla}P + \rho\vec{g} + \mu\vec{\nabla}^2\vec{V} = 0$

(c) $\rho \frac{D\vec{V}}{Dt} = -\vec{\nabla}P - \mu\vec{\nabla}^2\vec{V}$

(d) $\rho \frac{D\vec{V}}{Dt} = -\vec{\nabla}P + \rho\vec{g} + \mu\vec{\nabla}^2\vec{V}$

(e) $\rho \frac{D\vec{V}}{Dt} = -\vec{\nabla}P + \rho\vec{g} + \mu\vec{\nabla}^2\vec{V} + \vec{\nabla} \cdot \vec{V} = 0$

9–133 Which choice is not correct regarding the Navier-Stokes equation?
(a) Nonlinear equation (b) Unsteady equation
(c) Second-order equation (d) Partial differential equation
(e) None of these

9–134 In fluid flow analyses, which boundary condition can be expressed as $\vec{V}_{fluid} = \vec{V}_{wall}$
(a) No-slip (b) Interface (c) Free-surface
(d) Symmetry (e) Inlet

CHAPTER 10

APPROXIMATE SOLUTIONS OF THE NAVIER–STOKES EQUATION

In this chapter we look at several approximations that eliminate term(s), reducing the Navier–Stokes equation to a simplified form that is more easily solvable. Sometimes these approximations are appropriate in a whole flow field, but in most cases, they are appropriate only in certain *regions* of the flow field. We first consider *creeping flow*, where the Reynolds number is so low that the viscous terms dominate (and eliminate) the inertial terms. Following that, we look at two approximations that are appropriate in regions of flow away from walls and wakes: *inviscid flow* and *irrotational flow* (also called *potential flow*). In these regions, the opposite holds; i.e., inertial terms dominate viscous terms. Finally, we discuss the *boundary layer approximation*, in which both inertial and viscous terms remain, but some of the viscous terms are negligible. This last approximation is appropriate at *very high* Reynolds numbers (the opposite of creeping flow) and near walls, the opposite of potential flow.

OBJECTIVES
When you finish reading this chapter, you should be able to

- Appreciate why approximations are necessary to solve many fluid flow problems, and know when and where such approximations are appropriate
- Understand the effects of the lack of inertial terms in the creeping flow approximation, including the disappearance of density from the equations
- Understand superposition as a method of solving potential flow problems
- Predict boundary layer thickness and other boundary layer properties

In this chapter, we discuss several approximations that simplify the Navier-Stokes equation, including creeping flow, where viscous terms dominate inertial terms. The flow of lava from a volcano is an example of creeping flow—the viscosity of molten rock is so large that the Reynolds number is small even though the length scales are large.
StockTrek/Getty Images

APPROXIMATE SOLUTIONS OF THE N–S EQ

10–1 ■ INTRODUCTION

In Chap. 9, we derived the differential equation of linear momentum for an incompressible Newtonian fluid with constant properties—the *Navier–Stokes equation*. We showed some examples of analytical solutions to the continuity and Navier–Stockes equations for simple (usually infinite) geometries, in which most of the terms in the component equations are eliminated and the resulting differential equations are analytically solvable. Unfortunately, there aren't very many known analytical solutions available in the literature; in fact, we can count the number of such solutions on the fingers of a few students. The vast majority of practical fluid mechanics problems *cannot* be solved analytically and require either (1) further approximations or (2) computer assistance. We consider option 1 here; option 2 is discussed in Chap. 15. For simplicity, we consider only incompressible flow of Newtonian fluids in this chapter.

We emphasize first that the Navier–Stokes equation itself is not *exact*, but rather is a *model* of fluid flow that involves several inherent approximations (Newtonian fluid, constant thermodynamic and transport properties, etc.). Nevertheless, it is an *excellent* model and is the foundation of modern fluid mechanics. In this chapter we distinguish between "exact" solutions and approximate solutions (Fig. 10–1). The term *exact* is used when the solution starts with the *full* Navier–Stokes equation. The solutions discussed in Chap. 9 are exact solutions because we begin each of them with the full form of the equation. Some terms are eliminated in a specific problem due to the specified geometry or other simplifying assumptions in the problem. In a different solution, the terms that get eliminated may not be the same ones, but depend on the geometry and assumptions of that particular problem. We define an **approximate solution,** on the other hand, as one in which the Navier–Stokes equation is *simplified* in some region of the flow *before we even start the solution*. In other words, term(s) are eliminated *a priori* depending on the class of problem, which may differ from one region of the flow to another.

For example, we have already discussed one approximation, namely, *fluid statics* (Chap. 3). This can be considered to be an approximation of the Navier–Stokes equation in a region of the flow field where the fluid velocity is not necessarily zero, but the fluid is nearly stagnant, and we neglect all terms involving velocity. In this approximation, the Navier–Stokes equation reduces to just two terms, pressure and gravity, i.e., $\vec{\nabla}P = \rho\vec{g}$. The approximation is that the inertial and viscous terms in the Navier–Stokes equation are negligibly small compared to the pressure and gravity terms.

Although approximations render the problem more tractable, there is a danger associated with any approximate solution. Namely, if the approximation is not appropriate to begin with, the solution will be incorrect—even if we perform all the mathematics correctly. Why? Because we start with equations that do not apply to the problem at hand. For example, we may solve a problem using the creeping flow approximation and obtain a solution that satisfies all assumptions and boundary conditions. However, if the Reynolds number of the flow is too high, the creeping flow approximation is inappropriate from the start, and our solution (regardless of how proud of it we may be) is not physically correct. Another common mistake is to

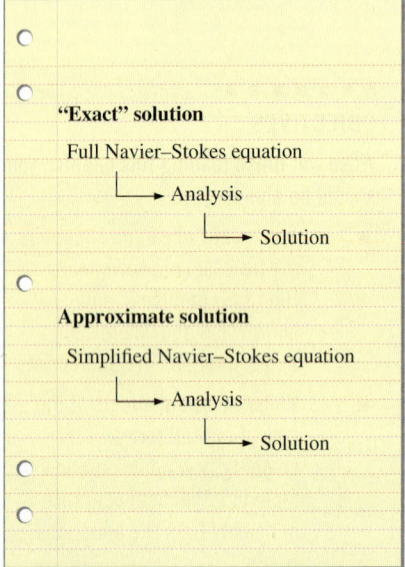

FIGURE 10–1
"Exact" solutions begin with the full Navier–Stokes equation, while approximate solutions begin with a simplified form of the Navier–Stokes equation right from the start.

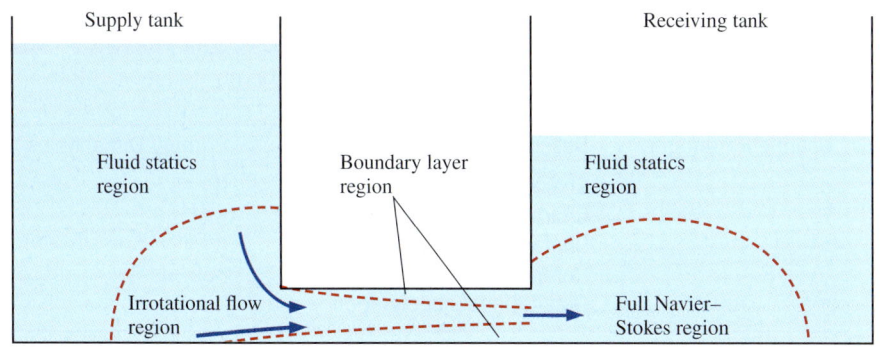

FIGURE 10–2
A particular approximation of the Navier–Stokes equation is appropriate only in certain *regions* of the flow field; other approximations may be appropriate in other regions of the flow field.

assume irrotational flow in regions of the flow where the assumption of irrotationality is not appropriate. *The bottom line is that we must be very careful of the approximations we apply, and we should always verify and justify our approximations wherever possible.*

Finally, we stress that in most practical fluid flow problems, a particular approximation may be appropriate in a certain *region* of the flow field, but not in other regions, where a different approximation may perhaps be more appropriate. Figure 10–2 illustrates this point qualitatively for flow of a liquid from one tank to another. The fluid statics approximation is appropriate in a region of the supply tank far away from the connecting pipe, and to a lesser extent in the receiving tank. The irrotational flow approximation is appropriate near the inlet to the connecting pipe and through the middle portion of the pipe where strong viscous effects are absent. Near the walls, the boundary layer approximation is appropriate. The flow in some regions does not meet the criteria for *any* approximations, and the full Navier–Stokes equation must be solved there (e.g., downstream of the pipe outlet in the receiving tank). How do we determine if an approximation is appropriate? We do this by comparing the orders of magnitude of the various terms in the equations of motion to see if any terms are negligibly small compared to other terms.

10–2 ■ NONDIMENSIONALIZED EQUATIONS OF MOTION

Our goal in this section is to nondimensionalize the equations of motion so that we can properly compare the orders of magnitude of the various terms in the equations. We begin with the incompressible continuity equation,

$$\vec{\nabla} \cdot \vec{V} = 0 \qquad (10\text{–}1)$$

and the vector form of the Navier–Stokes equation, valid for incompressible flow of a Newtonian fluid with constant properties,

$$\rho \frac{D\vec{V}}{Dt} = \rho \left[\frac{\partial \vec{V}}{\partial t} + (\vec{V} \cdot \vec{\nabla})\vec{V} \right] = -\vec{\nabla}P + \rho \vec{g} + \mu \nabla^2 \vec{V} \qquad (10\text{–}2)$$

We introduce in Table 10–1 some characteristic (reference) *scaling parameters* that are used to nondimensionalize the equations of motion.

APPROXIMATE SOLUTIONS OF THE N–S EQ

TABLE 10–1

Scaling parameters used to nondimensionalize the continuity and momentum equations, along with their primary dimensions

Scaling Parameter	Description	Primary Dimensions
L	Characteristic length	{L}
V	Characteristic speed	{Lt^{-1}}
f	Characteristic frequency	{t^{-1}}
$P_0 - P_\infty$	Reference pressure difference	{mL^{-1}t^{-2}}
g	Gravitational acceleration	{Lt^{-2}}

Cartesian coordinates

$$\vec{\nabla} = \left(\frac{\partial}{\partial x}, \frac{\partial}{\partial y}, \frac{\partial}{\partial z}\right)$$

$$= \left(\frac{\partial}{L\partial\left(\frac{x}{L}\right)}, \frac{\partial}{L\partial\left(\frac{y}{L}\right)}, \frac{\partial}{L\partial\left(\frac{z}{L}\right)}\right)$$

$$= \frac{1}{L}\left(\frac{\partial}{\partial x^*}, \frac{\partial}{\partial y^*}, \frac{\partial}{\partial z^*}\right) = \frac{1}{L}\vec{\nabla}^*$$

Cylindrical coordinates

$$\vec{\nabla} = \left(\frac{\partial}{\partial r}, \frac{1}{r}\frac{\partial}{\partial\theta}, \frac{\partial}{\partial z}\right)$$

$$= \left(\frac{\partial}{L\partial\left(\frac{r}{L}\right)}, \frac{1}{L\left(\frac{r}{L}\right)}\frac{\partial}{\partial\theta}, \frac{\partial}{L\partial\left(\frac{z}{L}\right)}\right)$$

$$= \frac{1}{L}\left(\frac{\partial}{\partial r^*}, \frac{1}{r^*}\frac{\partial}{\partial\theta}, \frac{\partial}{\partial z^*}\right) = \frac{1}{L}\vec{\nabla}^*$$

FIGURE 10–3

The gradient operator is nondimensionalized by Eq. 10–3, regardless of our choice of coordinate system.

We then define several *nondimensional variables* and one *nondimensional operator* based on the scaling parameters in Table 10–1,

$$t^* = ft \qquad \vec{x}^* = \frac{\vec{x}}{L} \qquad \vec{V}^* = \frac{\vec{V}}{V}$$

$$P^* = \frac{P - P_\infty}{P_0 - P_\infty} \qquad \vec{g}^* = \frac{\vec{g}}{g} \qquad \vec{\nabla}^* = L\vec{\nabla} \qquad (10\text{–}3)$$

Notice that we define the nondimensional pressure variable in terms of a pressure *difference*, based on our discussion about pressure versus pressure differences in Chap. 9. Each of the starred quantities in Eq. 10–3 is nondimensional. For example, although each component of the gradient operator $\vec{\nabla}$ has dimensions of {L^{-1}}, each component of $\vec{\nabla}^*$ has dimensions of {1} (Fig. 10–3). We substitute Eq. 10–3 into Eqs. 10–1 and 10–2, treating each term carefully. For example, $\vec{\nabla} = \vec{\nabla}^*/L$ and $\vec{V} = V\vec{V}^*$, so the advective acceleration term in Eq. 10–2 becomes

$$\rho(\vec{V} \cdot \vec{\nabla})\vec{V} = \rho\left(V\vec{V}^* \cdot \frac{\vec{\nabla}^*}{L}\right)V\vec{V}^* = \frac{\rho V^2}{L}\left(\vec{V}^* \cdot \vec{\nabla}^*\right)\vec{V}^*$$

We perform similar algebra on each term in Eqs. 10–1 and 10–2. Equation 10–1 is rewritten in terms of nondimensional variables as

$$\frac{V}{L}\vec{\nabla}^* \cdot \vec{V}^* = 0$$

After dividing both sides by V/L to make the equation dimensionless, we get

Nondimensionalized continuity: $\qquad \vec{\nabla}^* \cdot \vec{V}^* = 0 \qquad (10\text{–}4)$

Similarly, Eq. 10–2 is rewritten as

$$\rho Vf\frac{\partial\vec{V}^*}{\partial t^*} + \frac{\rho V^2}{L}\left(\vec{V}^* \cdot \vec{\nabla}^*\right)\vec{V}^* = -\frac{P_0 - P_\infty}{L}\vec{\nabla}^* P^* + \rho g\vec{g}^* + \frac{\mu V}{L^2}\nabla^{*2}\vec{V}^*$$

which, after multiplication by the collection of constants $L/(\rho V^2)$ to make all the terms dimensionless, becomes

$$\left[\frac{fL}{V}\right]\frac{\partial\vec{V}^*}{\partial t^*} + \left(\vec{V}^* \cdot \vec{\nabla}^*\right)\vec{V}^* = -\left[\frac{P_0 - P_\infty}{\rho V^2}\right]\vec{\nabla}^* P^* + \left[\frac{gL}{V^2}\right]\vec{g}^* + \left[\frac{\mu}{\rho VL}\right]\nabla^{*2}\vec{V}^* \quad (10\text{–}5)$$

Each of the terms in square brackets in Eq. 10–5 is a nondimensional grouping of parameters—a *Pi group* (Chap. 7). With the help of Table 7–5, we name each of these dimensionless parameters: The one on the left is the

Strouhal number, St = fL/V; the first one on the right is the *Euler number*, Eu = $(P_0 - P_\infty)/\rho V^2$; the second one on the right is the reciprocal of the square of the *Froude number*, $Fr^2 = V^2/gL$; and the last one is the reciprocal of the *Reynolds number*, Re = $\rho VL/\mu$. Equation 10–5 thus becomes

Nondimensionalized Navier–Stokes:

$$[St]\frac{\partial \vec{V}^*}{\partial t^*} + (\vec{V}^* \cdot \vec{\nabla}^*)\vec{V}^* = -[Eu]\vec{\nabla}^* P^* + \left[\frac{1}{Fr^2}\right]\vec{g}^* + \left[\frac{1}{Re}\right]\nabla^{*2}\vec{V}^* \quad (10\text{–}6)$$

Before we discuss specific approximations in detail, there is much to comment about the nondimensionalized equation set consisting of Eqs. 10–4 and 10–6:

- The nondimensionalized continuity equation contains *no* additional dimensionless parameters. Hence, Eq. 10–4 must be satisfied as is—we cannot simplify continuity further, because all the terms are of the same order of magnitude.

- The order of magnitude of the nondimensional variables is unity if they are nondimensionalized using a length, speed, frequency, etc., that are characteristic of the flow field. Thus, $t^* \sim 1$, $\vec{x}^* \sim 1$, $\vec{V}^* \sim 1$, etc., where we use the notation $\sim$ to denote order of magnitude. It follows that terms like $(\vec{V}^* \cdot \vec{\nabla}^*)\vec{V}^*$ and $\vec{\nabla}^* P^*$ in Eq. 10–6 are also order of magnitude unity and are the same order of magnitude as each other. Thus, *the relative importance of the terms in Eq. 10–6 depends only on the relative magnitudes of the dimensionless parameters* St, Eu, Fr, and Re. For example, if St and Eu are of order 1, but Fr and Re are very large, we may consider ignoring the gravitational and viscous terms in the Navier–Stokes equation.

- Since there are four dimensionless parameters in Eq. 10–6, *dynamic similarity* between a model and a prototype requires all four of these to be the same for the model and the prototype ($St_{model} = St_{prototype}$, $Eu_{model} = Eu_{prototype}$, $Fr_{model} = Fr_{prototype}$, and $Re_{model} = Re_{prototype}$), as illustrated in Fig. 10–4.

- If the flow is *steady*, then $f = 0$ and the Strouhal number drops out of the list of dimensionless parameters (St = 0). The first term on the left side of Eq. 10–6 then disappears, as does its corresponding unsteady term $\partial \vec{V}/\partial t$ in Eq. 10–2. If the characteristic frequency f is *very small* such that St $\ll$ 1, the flow is called **quasi-steady.** This means that at any instant in time (or at any phase of a slow periodic cycle), we can solve the problem as if the flow were steady, and the unsteady term in Eq. 10–6 again drops out.

- The effect of gravity is usually important only in flows with *free-surface effects* (e.g., waves, ship motion, spillways from hydroelectric dams, flow of rivers). For many engineering problems there is *no* free surface (pipe flow, fully submerged flow around a submarine or torpedo, automobile motion, flight of airplanes, birds, insects, etc.). In such cases, the only effect of gravity on the flow dynamics is a *hydrostatic pressure distribution* in the vertical direction superposed on the pressure field due to the fluid flow. In other words,

> For flows without free-surface effects, gravity does not affect the dynamics of the flow—its only effect is to superpose a hydrostatic pressure on the dynamic pressure field.

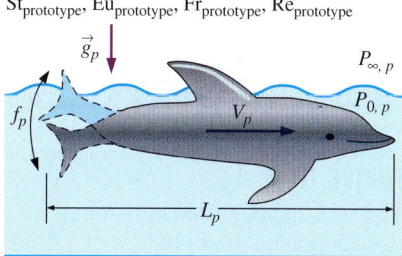

Prototype
$St_{prototype}$, $Eu_{prototype}$, $Fr_{prototype}$, $Re_{prototype}$

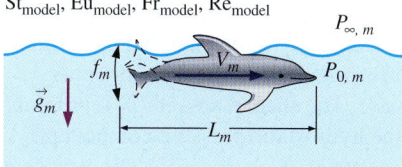

Model
St_{model}, Eu_{model}, Fr_{model}, Re_{model}

FIGURE 10–4
For complete dynamic similarity between prototype (subscript *p*) and model (subscript *m*), the model must be geometrically similar to the prototype, and (in general) all four dimensionless parameters, St, Eu, Fr, and Re, must match. As discussed in Chapter 7, however, this may not always be possible in a model test.
(Top) © James Gritz/Getty RF

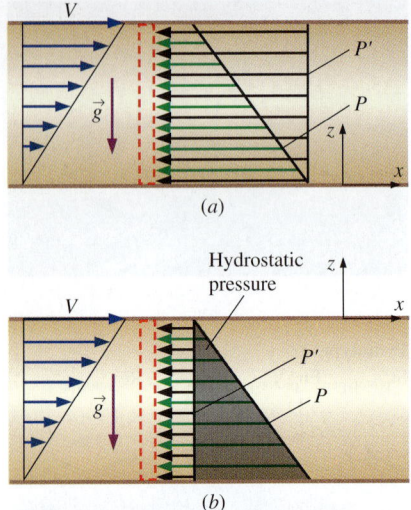

FIGURE 10–5
Pressure and modified pressure distribution on the right face of a fluid element in Couette flow between two infinite, parallel, horizontal plates: (a) $z = 0$ at the bottom plate, and (b) $z = 0$ at the top plate. The *modified pressure P'* is constant, but the *actual pressure P* is *not* constant in either case. The shaded area in (b) represents the hydrostatic pressure component.

- We define a **modified pressure** P' that absorbs the effect of hydrostatic pressure. For the case in which z is defined vertically upward (opposite to the direction of the gravity vector), and in which we define some arbitrary reference datum plane at $z = 0$,

Modified pressure: $$P' = P + \rho g z \quad (10\text{–}7)$$

The idea is to replace the two terms $-\vec{\nabla}P + \rho\vec{g}$ in Eq. 10–2 with *one* term $-\vec{\nabla}P'$ using the modified pressure of Eq. 10–7. The Navier–Stokes equation (Eq. 10–2) is written in modified form as

$$\rho\frac{D\vec{V}}{Dt} = \rho\left[\frac{\partial\vec{V}}{\partial t} + (\vec{V}\cdot\vec{\nabla})\vec{V}\right] = -\vec{\nabla}P' + \mu\nabla^2\vec{V} \quad (10\text{–}8)$$

With P replaced by P', and with the gravity term removed from Eq. 10–2, the Froude number drops out of the list of dimensionless parameters. The advantage is that we can solve a form of the Navier–Stokes equation that has *no gravity term*. After solving the Navier–Stokes equation in terms of modified pressure P', it is a simple matter to add back the hydrostatic pressure distribution using Eq. 10–7. An example is shown in Fig. 10–5 for the case of two-dimensional Couette flow. Modified pressure is often used in computational fluid dynamics (CFD) codes to separate gravitational effects (hydrostatic pressure in the vertical direction) from fluid flow (dynamic) effects. Note that modified pressure should *not* be used in flows with free-surface effects.

Now we are ready to make some approximations, in which we eliminate one or more of the terms in Eq. 10–2 by comparing the relative magnitudes of the dimensionless parameters associated with the corresponding terms in Eq. 10–6.

10–3 ■ THE CREEPING FLOW APPROXIMATION

Our first approximation is the class of fluid flow called **creeping flow.** Other names for this class of flow include **Stokes flow** and **low Reynolds number flow.** As the latter name implies, these are flows in which the Reynolds number is very small (Re $\ll$ 1). By inspection of the definition of the Reynolds number, Re $= \rho V L/\mu$, we see that creeping flow is encountered when either ρ, V, or L is very small or viscosity is very large (or some combination of these). You encounter creeping flow when you pour syrup (a very viscous liquid) on your pancakes or when you dip a spoon into a jar of honey (also very viscous) to add to your tea (Fig. 10–6).

Another example of creeping flow is all around us and inside us, although we can't see it, namely, flow around microscopic organisms. Microorganisms live their entire lives in the creeping flow regime since they are very small, their size being of order a few microns (1 μm $= 10^{-6}$ m), and they move very slowly, even though they may move in air or swim in water with a viscosity that can hardly be classified as "large" ($\mu_{air} \cong 1.8 \times 10^{-5}$ N·s/m^2 and $\mu_{water} \cong 1.0 \times 10^{-3}$ N·s/m^2 at room temperature). Figure 10–7 shows a *Salmonella* bacterium swimming through water. The bacterium's body is only about 1 μm long; its *flagella* (hairlike tails) extend several microns behind the body and serve as its propulsion mechanism. The Reynolds number associated with its motion is much smaller than 1.

FIGURE 10–6
The slow flow of a very viscous liquid like honey is classified as creeping flow.

Creeping flow also occurs in the flow of lubricating oil in the very small gaps and channels of a lubricated bearing. In this case, the speeds may not be small, but the gap size is very small (on the order of tens of microns), and the viscosity is relatively large ($\mu_{oil} \sim 1$ N·s/m² at room temperature).

For simplicity, we assume that gravitational effects are negligible, or that they contribute only to a hydrostatic pressure component, as discussed previously. We also assume either steady flow or oscillating flow, with a Strouhal number of order unity (St ~ 1) or smaller, so that the unsteady acceleration term $[St]\partial \vec{V}^*/\partial t^*$ is orders of magnitude smaller than the viscous term $[1/Re]\vec{\nabla}^{*2}\vec{V}^*$ (the Reynolds number is very small). The advective term in Eq. 10–6 is of order 1, $(\vec{V}^* \cdot \vec{\nabla}^*)\vec{V}^* \sim 1$, so this term drops out as well. Thus, we ignore the entire left side of Eq. 10–6, which reduces to

Creeping flow approximation: $\qquad [Eu]\vec{\nabla}^* P^* \cong \left[\dfrac{1}{Re}\right]\nabla^{*2}\vec{V}^*$ (10–9)

In words, pressure forces in the flow (left side) must be large enough to balance the (relatively) large viscous forces on the right side. However, since the nondimensional variables in Eq. 10–9 are of order 1, the only way for the two sides to balance is if Eu is of the same order of magnitude as 1/Re. Equating these,

$$[Eu] = \frac{P_0 - P_\infty}{\rho V^2} \sim \left[\frac{1}{Re}\right] = \frac{\mu}{\rho V L}$$

After some algebra,

Pressure scale for creeping flow: $\qquad P_0 - P_\infty \sim \dfrac{\mu V}{L}$ (10–10)

Equation 10–10 reveals two interesting properties of creeping flow. First, we are used to *inertially* dominated flows, in which pressure differences scale like ρV^2 (e.g., the Bernoulli equation). Here, however, pressure differences scale like $\mu V/L$ instead, since creeping flow is a *viscously* dominated flow. In fact, *all the inertial terms of the Navier–Stokes equation disappear in creeping flow*. Second, *density has completely dropped out as a parameter in the Navier–Stokes equation* (Fig. 10–8). We see this more clearly by writing the *dimensional* form of Eq. 10–9,

Approximate Navier–Stokes equation for creeping flow: $\quad \vec{\nabla}P \cong \mu \nabla^2 \vec{V}$ (10–11)

Alert readers may point out that density still has a *minor* role in creeping flow. Namely, it is needed in the calculation of the Reynolds number. However, once we have determined that Re is very small, density is no longer needed since it does not appear in Eq. 10–11. Density also pops up in the hydrostatic pressure term, but this effect is usually negligible in creeping flow, since the vertical distances involved are often measured in millimeters or micrometers. Besides, if there are no free-surface effects, we can use modified pressure instead of physical pressure in Eq. 10–11.

Let's discuss the lack of inertia terms in Eq. 10–11 in somewhat more detail. You rely on inertia when you swim (Fig. 10–9). For example, you take a stroke, and then you are able to glide for some distance before you need to take another stroke. When you swim, the inertial terms in the Navier–Stokes equation are much larger than the viscous terms, since the

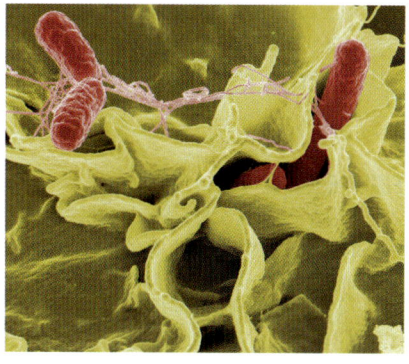

(a)

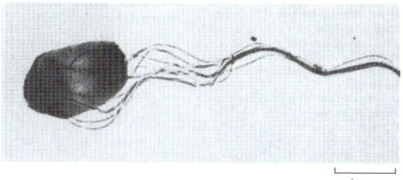

(b)

FIGURE 10–7
(a) *Salmonella typhimurium* invading cultured human cells.
(b) The bacterium *Salmonella abortusequi* swimming through water.

(a) *NIAID, NIH, Rocky Mantain Laboratories*
(b) *From* Comparative Physiology Functional Aspects of Structural Materials: Proceedings of the International Conference on Comparative Physiology, *Ascona, 1974, published by North-Holland Pub. Co., 1975.*

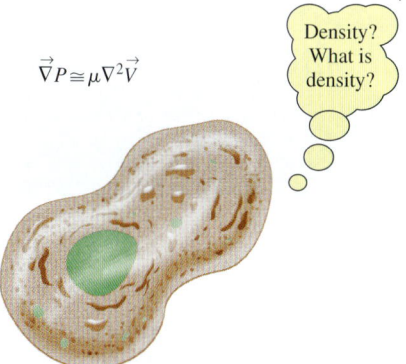

FIGURE 10–8
In the creeping flow approximation, density does not appear in the momentum equation.

522
APPROXIMATE SOLUTIONS OF THE N–S EQ

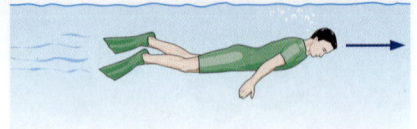

FIGURE 10–9
A person swims at a very high Reynolds number, and inertial terms are large; thus the person is able to glide long distances without moving.

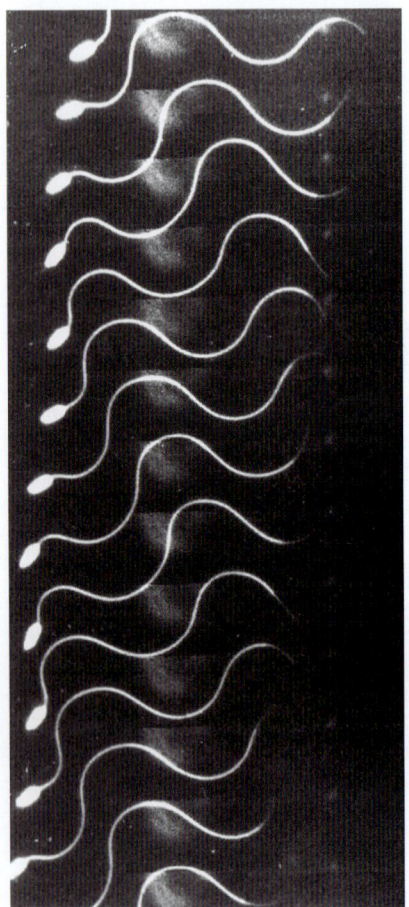

FIGURE 10–10
A sperm of the sea squirt *Ciona* swimming in seawater; flash photographs at 200 frames per second, with each image positioned directly below the one before it.

Courtesy of Professor Charlotte Omoto, Washington State University, School of Biological Sciences.

Reynolds number is very large. (Believe it or not, even extremely *slow* swimmers move at very *large* Reynolds numbers!)

For microorganisms swimming in the creeping flow regime, however, there is negligible inertia, and thus no gliding is possible. In fact, the lack of inertial terms in Eq. 10–11 has a substantial impact on how microorganisms are designed to swim. A flapping tail like that of a dolphin would get them nowhere. Instead, their long, narrow tails (*flagella*) undulate in a sinusoidal motion to propel them forward, as illustrated in Fig. 10–10 for the case of a sperm. Without any inertia, the sperm does not move unless his tail is moving. The instant his tail stops, the sperm stops moving. If you have ever seen a video clip of swimming sperm or other microorganisms, you may have noticed how hard they have to work just to move a short distance. That is the nature of creeping flow, and it is due to the lack of inertia. Careful study of Fig. 10–10 reveals that the sperm's tail has completed approximately two complete undulation cycles, yet the sperm's head has moved to the left by only about two head lengths.

It is very difficult for us humans to imagine moving in creeping flow conditions, since we are so used to the effects of inertia. Some authors have suggested that you imagine trying to swim in a vat of honey. We suggest instead that you go to a fast-food restaurant where they have a children's play area and watch a child play in a pool of plastic spheres (Fig. 10–11). When the child tries to "swim" among the balls (without touching the walls or the bottom), he or she can move forward only by certain snakelike wriggling body motions. The instant the child stops wriggling, all motion stops, since there is negligible inertia. The child must work very hard to move forward a short distance. There is a weak analogy between a child "swimming" in this kind of situation and a microorganism swimming in creeping flow conditions.

We next discuss the lack of density in Eq. 10–11. At high Reynolds numbers, the aerodynamic drag on an object increases proportionally with ρ. (Denser fluids exert more pressure force on the body as the fluid impacts the body.) However, this is actually an inertial effect, and inertia is negligible in creeping flow. In fact, aerodynamic drag cannot even be a *function* of density in a creeping flow, since density has disappeared from the Navier–Stokes equation. Example 10–1 illustrates this situation through the use of dimensional analysis.

EXAMPLE 10–1   Drag on an Object in Creeping Flow

Since density has vanished from the Navier–Stokes equation, aerodynamic drag on an object in creeping flow is a function only of its speed V, some characteristic length scale L of the object, and fluid viscosity μ (Fig. 10–12). Use dimensional analysis to generate a relationship for F_D as a function of these independent variables.

SOLUTION   We are to use dimensional analysis to generate a functional relationship between F_D and variables V, L, and μ.
Assumptions   **1** We assume Re $\ll$ 1 so that the creeping flow approximation applies. **2** Gravitational effects are irrelevant. **3** No parameters other than those listed in the problem statement are relevant to the problem.

Analysis We follow the step-by-step method of repeating variables discussed in Chap. 7; the details are left as an exercise. There are four parameters in this problem ($n = 4$). There are three primary dimensions: mass, length, and time, so we set $j = 3$ and use independent variables V, L, and μ as our repeating variables. We expect only one Pi since $k = n - j = 4 - 3 = 1$, and that Pi must equal a constant. The result is

$$F_D = \text{constant} \cdot \mu V L$$

Thus, we have shown that for creeping flow around any three-dimensional object, the aerodynamic drag force is simply a constant multiplied by μVL.
Discussion This result is significant, because all that is left to do is find the constant, which is a function only of the shape of the object.

FIGURE 10–11
A child trying to move in a pool of plastic balls is analogous to a microorganism trying to propel itself without the benefit of inertia.
Photo by Laura L. Pauley.

Drag on a Sphere in Creeping Flow

As shown in Example 10–1, the drag force F_D on a three-dimensional object of characteristic dimension L moving under creeping flow conditions at speed V through a fluid with viscosity μ is $F_D = \text{constant} \cdot \mu VL$. Dimensional analysis cannot predict the value of the constant, since it depends on the shape and orientation of the body in the flow field.

For the particular case of a *sphere*, Eq. 10–11 can be solved analytically. The details are beyond the scope of this text, but can be found in graduate-level fluid mechanics books (White, 2005; Panton, 2005). It turns out that the constant in the drag equation is equal to 3π if L is taken as the sphere's diameter D (Fig. 10–13).

Drag force on a sphere in creeping flow: $\quad F_D = 3\pi\mu VD \quad$ **(10–12)**

As a side note, two-thirds of this drag is due to viscous forces and the other one-third is due to pressure forces. This confirms that the viscous terms and the pressure terms in Eq. 10–11 are of the same order of magnitude, as mentioned previously.

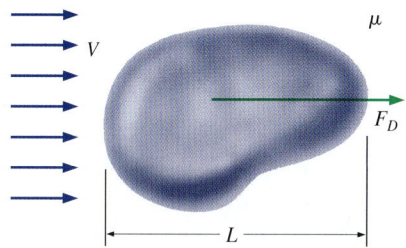

FIGURE 10–12
For creeping flow over a three-dimensional object, the aerodynamic drag on the object does not depend on density, but only on speed V, some characteristic size of the object L, and fluid viscosity μ.

■ **EXAMPLE 10–2**   **Terminal Velocity of a Particle from a Volcano**

A volcano has erupted, spewing stones, steam, and ash several thousand meters into the atmosphere (Fig. 10–14). After some time, the particles begin to settle to the ground. Consider a nearly spherical ash particle of diameter 50 μm, falling in air whose temperature is −50°C and whose pressure is 55 kPa. The density of the particle is 1240 kg/m³. Estimate the terminal velocity of this particle at this altitude.

SOLUTION We are to estimate the terminal velocity of a falling ash particle.
Assumptions **1** The Reynolds number is very small (we will need to verify this assumption after we obtain the solution). **2** The particle is spherical.
Properties At the given temperature and pressure, the ideal gas law gives $\rho = 0.8588$ kg/m³. Since viscosity is a very weak function of pressure, we use the value at −50°C and atmospheric pressure, $\mu = 1.474 \times 10^{-5}$ kg/m·s.
Analysis We treat the problem as quasi-steady. Once the falling particle has reached its terminal settling velocity, the net downward force (weight)

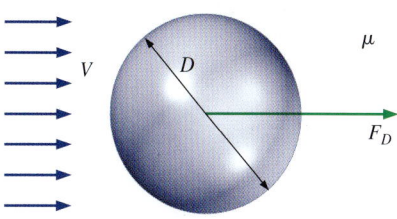

FIGURE 10–13
The aerodynamic drag on a sphere of diameter D in creeping flow is equal to $3\pi\mu VD$.

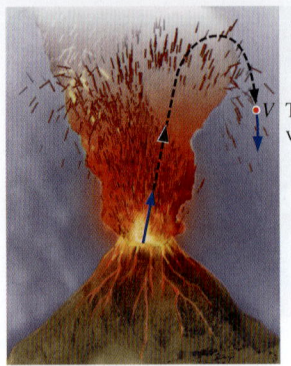

FIGURE 10–14
Small ash particles spewed from a volcanic eruption settle slowly to the ground; the creeping flow approximation is reasonable for this type of flow field.

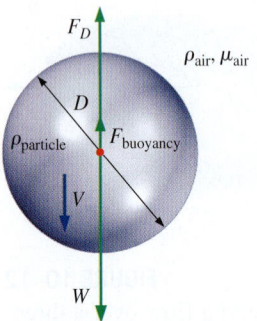

FIGURE 10–15
A particle falling at a steady terminal velocity has no acceleration; therefore, its weight is balanced by aerodynamic drag and the buoyancy force acting on the particle.

balances the net upward force (aerodynamic drag + buoyancy), as illustrated in Fig. 10–15.

Downward force:
$$F_{\text{down}} = W = \pi \frac{D^3}{6} \rho_{\text{particle}} g \qquad (1)$$

The aerodynamic drag force acting on the particle is obtained from Eq. 10–12, and the buoyancy force is the weight of the displaced air. Thus,

Upward force:
$$F_{\text{up}} = F_D + F_{\text{buoyancy}} = 3\pi\mu V D + \pi \frac{D^3}{6} \rho_{\text{air}} g \qquad (2)$$

We equate Eqs. 1 and 2, and solve for terminal velocity V,

$$V = \frac{D^2}{18\mu}(\rho_{\text{particle}} - \rho_{\text{air}})g$$

$$= \frac{(50 \times 10^{-6}\text{ m})^2}{18(1.474 \times 10^{-5}\text{ kg/m·s})}[(1240 - 0.8588)\text{ kg/m}^3](9.81\text{ m/s}^2)$$

$$= \mathbf{0.115\text{ m/s}}$$

Finally, we verify that the Reynolds number is small enough that creeping flow is an appropriate approximation,

$$\text{Re} = \frac{\rho_{\text{air}} V D}{\mu} = \frac{(0.8588\text{ kg/m}^3)(0.115\text{ m/s})(50 \times 10^{-6}\text{ m})}{1.474 \times 10^{-5}\text{ kg/m·s}} = 0.335$$

Thus the Reynolds number is less than 1, but certainly not *much* less than 1.
Discussion Although the equation for creeping flow drag on a sphere (Eq. 10–12) was derived for a case with Re $\ll$ 1, it turns out that the approximation is reasonable up to Re $\cong$ 1. A more involved calculation, including a Reynolds number correction and a correction based on the mean free path of air molecules, yields a terminal velocity of 0.110 m/s (Heinsohn and Cimbala, 2003); the error of the creeping flow approximation is less than 5 percent.

A consequence of the disappearance of density from the equations of motion for creeping flow is clearly seen in Example 10–2. Namely, air density is not important in any calculations except to verify that the Reynolds number is small. (Note that since ρ_{air} is so small compared to ρ_{particle}, the buoyancy force could have been ignored with negligible loss of accuracy.) Suppose instead that the air density were one-half of the actual density in Example 10–2, but all other properties were unchanged. The terminal velocity would be the same (to three significant digits), except that the Reynolds number would be smaller by a factor of 2. Thus,

> The terminal velocity of a dense, small particle in creeping flow conditions is nearly independent of fluid density, but highly dependent on fluid viscosity.

Since the viscosity of air varies with altitude by only about 25 percent, a small particle settles at nearly constant speed regardless of elevation, even though the air density increases by more than a factor of 10 as the particle falls from an altitude of 15,000 m to sea level.

For nonspherical three-dimensional objects, the creeping flow aerodynamic drag is still given by $F_D = \text{constant} \cdot \mu V L$; however, the constant is not 3π, but depends on both the shape and orientation of the body. The constant can be thought of as a kind of **drag coefficient** for creeping flow.

10–4 ▪ APPROXIMATION FOR INVISCID REGIONS OF FLOW

There is much confusion in the fluid mechanics literature about the word **inviscid** and the phrase **inviscid flow.** The apparent meaning of inviscid is *not viscous*. Inviscid flow would then seem to refer to flow of a fluid with no viscosity. However, that is *not* what is meant by the phrase *inviscid flow*! All fluids of engineering relevance have viscosity, regardless of the flow field. Authors who use the phrase inviscid flow actually mean flow of a *viscous fluid* in a *region* of the flow in which *net viscous forces are negligible compared to pressure and/or inertial forces* (Fig. 10–16). Some authors use the phrase "frictionless flow" as a synonym of inviscid flow. This causes more confusion, because even in regions of the flow where net viscous forces are negligible, *friction still acts on fluid elements*, and there may still be significant *viscous stresses*. It's just that these stresses cancel each other out, leaving no significant *net* viscous force on fluid elements. It can be shown that significant *viscous dissipation* may also be present in such regions. As is discussed in Section 10–5, fluid elements in an *irrotational* region of the flow also have negligible net viscous forces—not because there is no friction, but because the frictional (viscous) stresses cancel each other out. Because of the confusion caused by the terminology, the present authors discourage use of the phrases "inviscid flow" and "frictionless flow." Instead, we advocate use of the phrases *inviscid regions of flow* or *regions of flow with negligible net viscous forces*.

Regardless of the terminology used, if net viscous forces are very small compared to inertial and/or pressure forces, the last term on the right side of Eq. 10–6 is negligible. This is true only if 1/Re is small. Thus, inviscid regions of flow are regions of *high Reynolds number*—the opposite of creeping flow regions. In such regions, the Navier–Stokes equation (Eq. 10–2) loses its viscous term and reduces to the **Euler equation,**

Euler equation:
$$\rho\left[\frac{\partial \vec{V}}{\partial t} + (\vec{V}\cdot\vec{\nabla})\vec{V}\right] = -\vec{\nabla}P + \rho\vec{g} \qquad (10\text{–}13)$$

The Euler equation is simply the Navier–Stokes equation with the viscous term neglected; it is an *approximation* of the Navier–Stokes equation.

Because of the no-slip condition at solid walls, frictional forces are *not* negligible in a region of flow very near a solid wall. In such a region, called a **boundary layer,** the velocity gradients normal to the wall are large enough to offset the small value of 1/Re. An alternate explanation is that the characteristic length scale of the body (L) is no longer the most appropriate length scale inside a boundary layer and must be replaced by a much smaller length scale associated with the distance from the wall. When we define the Reynolds number with this smaller length scale, Re is no longer large, and the viscous term in the Navier–Stokes equation cannot be neglected.

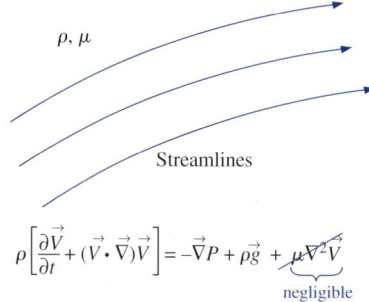

FIGURE 10–16
An inviscid region of flow is a region where net viscous forces are negligible compared to inertial and/or pressure forces because the Reynolds number is large; the fluid itself is still a viscous fluid.

526
APPROXIMATE SOLUTIONS OF THE N–S EQ

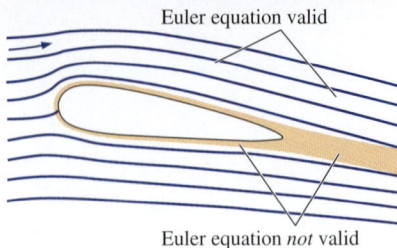

FIGURE 10–17
The Euler equation is an approximation of the Navier–Stokes equation, appropriate only in regions of the flow where the Reynolds number is large and where net viscous forces are negligible compared to inertial and/or pressure forces.

A similar argument can be made in the *wake* of a body, where velocity gradients are relatively large and the viscous terms are not negligible compared to inertial terms (Fig. 10–17). In practice, therefore, it turns out that

> The Euler equation approximation is appropriate in high Reynolds number regions of the flow, where net viscous forces are negligible, away from walls and wakes.

The term that is neglected in the Euler approximation of the Navier–Stokes equation ($\mu \nabla^2 \vec{V}$) is the term that contains the highest-order derivatives of velocity. Mathematically, loss of this term reduces the number of boundary conditions that we can specify. It turns out that when we use the Euler equation approximation, we *cannot* specify the no-slip boundary condition at solid walls, although we still specify that fluid cannot flow *through* the wall (the wall is *impermeable*). Solutions of the Euler equation are therefore not physically meaningful near solid walls, since flow is allowed to slip there. Nevertheless, as we show in Section 10–6, the Euler equation is often used as the *first step* in a boundary layer approximation. Namely, the Euler equation is applied over the whole flow field, including regions close to walls and wakes, where we know the approximation is not appropriate. Then, a thin boundary layer is inserted in these regions as a correction to account for viscous effects.

Finally, we point out that the Euler equation (Eq. 10–13) is sometimes used as a first approximation in CFD calculations in order to reduce CPU time (and cost).

Derivation of the Bernoulli Equation in Inviscid Regions of Flow

In Chap. 5, we derived the Bernoulli equation along a streamline. Here we show an alternative derivation based on the Euler equation. For simplicity, we assume steady incompressible flow. The advective term in Eq. 10–13 can be rewritten through use of a vector identity,

Vector identity: $(\vec{V} \cdot \vec{\nabla})\vec{V} = \vec{\nabla}\left(\dfrac{V^2}{2}\right) - \vec{V} \times (\vec{\nabla} \times \vec{V})$ (10–14)

where V is the magnitude of vector $\vec{V}$. We recognize the second term in parentheses on the right side as the *vorticity vector* $\vec{\zeta}$ (see Chap. 4); thus,

$$(\vec{V} \cdot \vec{\nabla})\vec{V} = \vec{\nabla}\left(\dfrac{V^2}{2}\right) - \vec{V} \times \vec{\zeta}$$

and an alternate form of the steady Euler equation is written as

$$\vec{\nabla}\left(\dfrac{V^2}{2}\right) - \vec{V} \times \vec{\zeta} = -\dfrac{\vec{\nabla}P}{\rho} + \vec{g} = \vec{\nabla}\left(-\dfrac{P}{\rho}\right) + \vec{g} \quad (10\text{–}15)$$

where we have divided each term by the density and moved ρ within the gradient operator, since density is constant in an incompressible flow.

We make the further assumption that gravity acts only in the $-z$-direction (Fig. 10–18), so that

$$\vec{g} = -g\vec{k} = -g\vec{\nabla}z = \vec{\nabla}(-gz) \quad (10\text{–}16)$$

where we have used the fact that the gradient of coordinate z is unit vector $\vec{k}$ in the z-direction. Note also that g is a constant, which allows us to move it

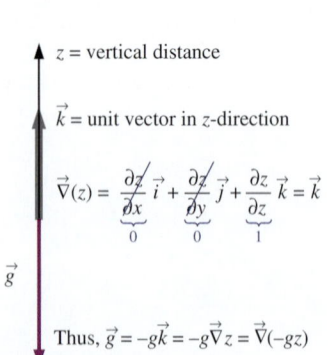

FIGURE 10–18
When gravity acts in the $-z$-direction, gravity vector $\vec{g}$ can be written as $\vec{\nabla}(-gz)$.

(and the negative sign) within the gradient operator. We substitute Eq. 10–16 into Eq. 10–15, and rearrange by combining three terms within one gradient operator,

$$\vec{\nabla}\left(\frac{P}{\rho} + \frac{V^2}{2} + gz\right) = \vec{V} \times \vec{\zeta} \qquad (10\text{–}17)$$

From the definition of the cross product of two vectors, $\vec{C} = \vec{A} \times \vec{B}$, the vector $\vec{C}$ is perpendicular to both $\vec{A}$ and $\vec{B}$. The left side of Eq. 10–17 must therefore be a vector everywhere perpendicular to the local velocity vector $\vec{V}$, since $\vec{V}$ appears in the cross product on the right side of Eq. 10–17. Now consider flow along a three-dimensional streamline (Fig. 10–19), which by definition is everywhere *parallel* to the local velocity vector. At every point along the streamline, $\vec{\nabla}(P/\rho + V^2/2 + gz)$ must be perpendicular to the streamline. Now dust off your vector algebra book and recall that the gradient of a scalar points in the direction of *maximum increase* of the scalar. Furthermore, the gradient of a scalar is a vector that points perpendicular to an imaginary surface on which the scalar is constant. Thus, we argue that the scalar $(P/\rho + V^2/2 + gz)$ must be *constant along a streamline*. This is true even if the flow is *rotational* ($\vec{\zeta} \neq 0$). Thus, we have derived a version of the steady incompressible Bernoulli equation, appropriate in regions of flow with negligible net viscous forces, i.e., in so-called inviscid regions of flow.

Steady incompressible Bernoulli equation in inviscid regions of flow:

$$\frac{P}{\rho} + \frac{V^2}{2} + gz = C = \text{constant along streamlines} \qquad (10\text{–}18)$$

Note that the Bernoulli "constant" C in Eq. 10–18 is constant only along a streamline; the constant may change from streamline to streamline.

You may be wondering if it is physically possible to have a rotational region of flow that is also inviscid, since rotationality is usually *caused* by viscosity. Yes, it *is* possible, and we give one simple example—*solid body rotation* (Fig. 10–20). Although the rotation may have been *generated* by viscous forces, a region of flow in solid body rotation has *no shear* and *no net viscous force*; it is an inviscid region of flow, even though it is also rotational. As a consequence of the rotational nature of this flow field, Eq. 10–18 applies to every streamline in the flow, but the Bernoulli constant C differs from streamline to streamline, as illustrated in Fig. 10–20.

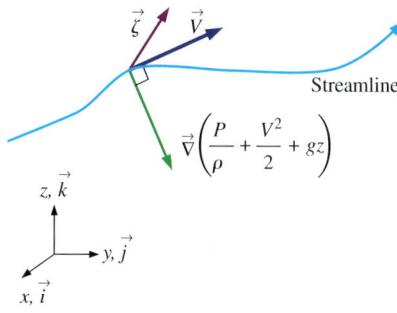

FIGURE 10–19
Along a streamline, $\vec{\nabla}(P/\rho + V^2/2 + gz)$ is a vector everywhere perpendicular to the streamline; hence, $P/\rho + V^2/2 + gz$ is constant along the streamline.

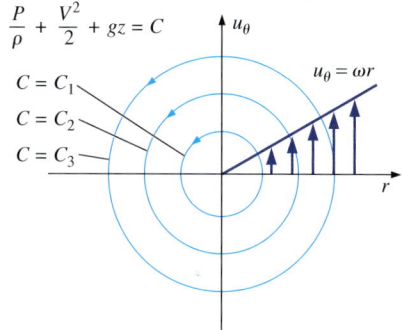

FIGURE 10–20
Solid body rotation is an example of an inviscid region of flow that is also rotational. The Bernoulli constant C differs from streamline to streamline but is constant along any particular streamline.

EXAMPLE 10–3   Pressure Field in Solid Body Rotation

A fluid is rotating as a rigid body (solid body rotation) around the z-axis as illustrated in Fig. 10–20. The steady incompressible velocity field is given by $u_r = 0$, $u_\theta = \omega r$, and $u_z = 0$. The pressure at the origin is equal to P_0. Calculate the pressure field everywhere in the flow, and determine the Bernoulli constant along each streamline.

SOLUTION   For a given velocity field, we are to calculate the pressure field and the Bernoulli constant along each streamline.

Assumptions 1 The flow is steady and incompressible. 2 Since there is no flow in the z- (vertical) direction, a hydrostatic pressure distribution exists in the vertical direction. 3 The entire flow field is approximated as an inviscid region of flow since viscous forces are zero. 4 There is no variation of any flow variable in the θ-direction.

Analysis Equation 10–18 can be applied directly because of assumption 3,

Bernoulli equation: $$P = \rho C - \frac{1}{2}\rho V^2 - \rho g z \quad (1)$$

where C is the Bernoulli constant that changes radially across streamlines as illustrated in Fig. 10–20. At any radial location r, $V^2 = \omega^2 r^2$, and Eq. 1 becomes

$$P = \rho C - \rho \frac{\omega^2 r^2}{2} - \rho g z \quad (2)$$

At the origin ($r = 0$, $z = 0$), the pressure is equal to P_0 (from the given boundary condition). Thus we calculate $C = C_0$ at the origin ($r = 0$),

Boundary condition at the origin: $\quad P_0 = \rho C_0 \rightarrow C_0 = \dfrac{P_0}{\rho}$

But how can we find C at an arbitrary radial location r? Equation 2 alone is insufficient since both C and P are unknowns. The answer is that we must use the Euler equation. Since there is no free surface, we employ the modified pressure of Eq. 10–7. The r-component of the Euler equation in cylindrical coordinates (see Eq. 9–62b without the viscous terms) reduces to

r-component of Euler equation: $$\frac{\partial P'}{\partial r} = \rho \frac{u_\theta^2}{r} = \rho \omega^2 r \quad (3)$$

where we have substituted the given value of u_θ. Since hydrostatic pressure is already included in the modified pressure, P' is not a function of z. By assumptions 1 and 4, respectively, P' is also not a function of t or θ. Thus P' is a function of r only, and we replace the partial derivative in Eq. 3 with a total derivative. Integration yields

Modified pressure field: $$P' = \rho \frac{\omega^2 r^2}{2} + B_1 \quad (4)$$

where B_1 is a constant of integration. At the origin, modified pressure P' is equal to actual pressure P, since $z = 0$ there. Thus, constant B_1 is found by applying the known pressure boundary condition at the origin. It turns out therefore that B_1 is equal to P_0. We now convert Eq. 4 back to actual pressure using Eq. 10–7, $P = P' - \rho g z$,

Actual pressure field: $$P = \rho \frac{\omega^2 r^2}{2} + P_0 - \rho g z \quad (5)$$

At the reference datum plane ($z = 0$), we plot nondimensional pressure as a function of nondimensional radius, where some arbitrary radial location $r = R$ is chosen as a characteristic length scale in the flow (Fig. 10–21). The pressure distribution is parabolic with respect to r.

Finally, we equate Eqs. 2 and 5 to solve for C,

Bernoulli constant as a function of r: $$C = \frac{P_0}{\rho} + \omega^2 r^2 \quad (6)$$

At the origin, $C = C_0 = P_0/\rho$, which agrees with our previous calculation.

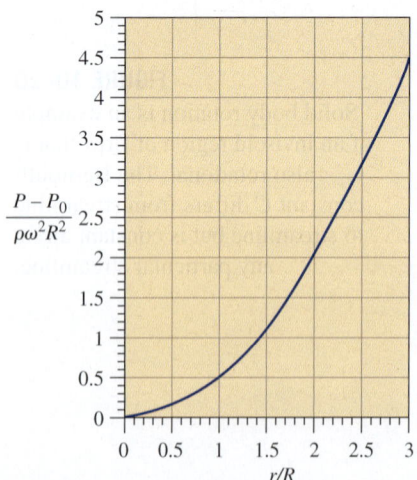

FIGURE 10–21
Nondimensional pressure as a function of nondimensional radial location at zero elevation for a fluid in solid body rotation.

Discussion For a fluid in solid body rotation, the Bernoulli constant increases as r^2. This is not surprising, since fluid particles move faster at larger values of r, and thus they possess more energy. In fact, Eq. 5 reveals that pressure itself increases as r^2. Physically, the pressure gradient in the (inward) radial direction provides the centripetal force necessary to keep fluid particles revolving about the origin.

10–5 · THE IRROTATIONAL FLOW APPROXIMATION

As was pointed out in Chap. 4, there are regions of flow in which fluid particles have *no net rotation*; these regions are called **irrotational.** You must keep in mind that the assumption of irrotationality is an *approximation*, which may be appropriate in some regions of a flow field, but not in other regions (Fig. 10–22). In general, inviscid regions of flow far away from solid walls and wakes of bodies are also irrotational, although as pointed out previously, there are situations in which an inviscid region of flow may *not* be irrotational (e.g., solid body rotation). Solutions obtained for the class of flow defined by irrotationality are thus *approximations* of full Navier–Stokes solutions. Mathematically, the approximation is that vorticity is negligibly small,

Irrotational approximation: $\quad \vec{\zeta} = \vec{\nabla} \times \vec{V} \cong 0 \quad$ (10-19)

We now examine the effect of this approximation on both the continuity and momentum equations.

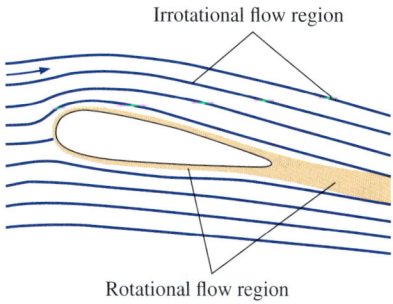

FIGURE 10–22
The irrotational flow approximation is appropriate only in certain regions of the flow where the vorticity is negligible.

Continuity Equation

If you shake some more dust off your vector algebra book, you will find a vector identity concerning the curl of the gradient of any scalar function ϕ, and hence the curl of any vector $\vec{V}$,

Vector identity: $\quad \vec{\nabla} \times \vec{\nabla}\phi = 0 \quad$ Thus, if $\vec{\nabla} \times \vec{V} = 0$, then $\vec{V} = \vec{\nabla}\phi$. (10-20)

This can easily be proven in Cartesian coordinates (Fig. 10–23), but applies to any orthogonal coordinate system as long as ϕ is a smooth function. In words, if the curl of a vector is zero, the vector can be expressed as the gradient of a scalar function ϕ, called the **potential function.** In fluid mechanics, vector $\vec{V}$ is the velocity vector, the curl of which is the vorticity vector $\vec{\zeta}$, and thus we call ϕ the **velocity potential function.** We write

For irrotational regions of flow: $\quad \vec{V} = \vec{\nabla}\phi \quad$ (10-21)

We should point out that the sign convention in Eq. 10–21 is not universal—in some fluid mechanics textbooks, a negative sign is inserted in the definition of the velocity potential function. We state Eq. 10–21 in words as follows:

> In an irrotational region of flow, the velocity vector can be expressed as the gradient of a scalar function called the *velocity potential function*.

Regions of irrotational flow are therefore also called **regions of potential flow.** Note that we have not restricted ourselves to two-dimensional flows;

Proof of the vector identity:
$\vec{\nabla} \times \vec{\nabla}\Phi = 0$

Expand in Cartesian coordinates,
$\vec{\nabla} \times \vec{\nabla}\phi =$

$\left(\dfrac{\partial^2\phi}{\partial y\,\partial z} - \dfrac{\partial^2\phi}{\partial z\,\partial y}\right)\vec{i} + \left(\dfrac{\partial^2\phi}{\partial z\,\partial x} - \dfrac{\partial^2\phi}{\partial x\,\partial z}\right)\vec{j}$

$+ \left(\dfrac{\partial^2\phi}{\partial x\,\partial y} - \dfrac{\partial^2\phi}{\partial y\,\partial x}\right)\vec{k} = 0$

The identity is proven if Φ is a smooth function of x, y, and z.

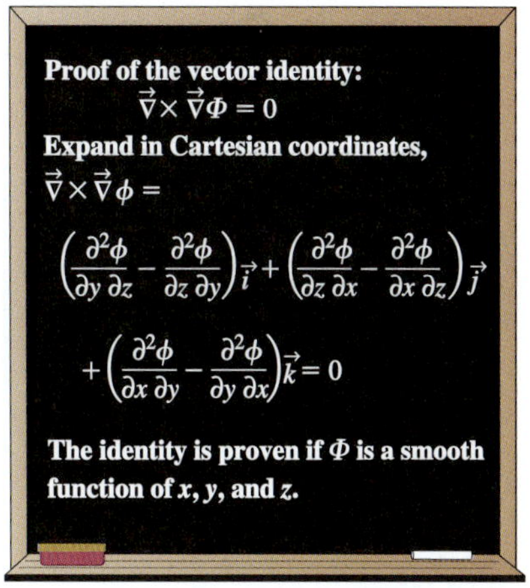

FIGURE 10–23
The vector identity of Eq. 10–20 is easily proven by expanding the terms in Cartesian coordinates.

530
APPROXIMATE SOLUTIONS OF THE N–S EQ

Eq. 10–21 is valid for three-dimensional flow fields, as long as the approximation of irrotationality is appropriate in the region of flow under study. In Cartesian coordinates,

$$u = \frac{\partial \phi}{\partial x} \qquad v = \frac{\partial \phi}{\partial y} \qquad w = \frac{\partial \phi}{\partial z} \qquad (10\text{–}22)$$

and in cylindrical coordinates,

$$u_r = \frac{\partial \phi}{\partial r} \qquad u_\theta = \frac{1}{r}\frac{\partial \phi}{\partial \theta} \qquad u_z = \frac{\partial \phi}{\partial z} \qquad (10\text{–}23)$$

The usefulness of Eq. 10–21 becomes apparent when it is substituted into Eq. 10–1, the incompressible continuity equation: $\vec{\nabla} \cdot \vec{V} = 0 \to \vec{\nabla} \cdot \vec{\nabla}\phi = 0$, or

For irrotational regions of flow: $\qquad \nabla^2 \phi = 0 \qquad (10\text{–}24)$

where the **Laplacian operator** ∇^2 is a scalar operator defined as $\vec{\nabla} \cdot \vec{\nabla}$, and Eq. 10–24 is called the **Laplace equation.** We stress that Eq. 10–24 is valid only in regions where the irrotational flow approximation is reasonable (Fig. 10–24). In Cartesian coordinates,

$$\nabla^2 \phi = \frac{\partial^2 \phi}{\partial x^2} + \frac{\partial^2 \phi}{\partial y^2} + \frac{\partial^2 \phi}{\partial z^2} = 0$$

and in cylindrical coordinates,

$$\nabla^2 \phi = \frac{1}{r}\frac{\partial}{\partial r}\left(r\frac{\partial \phi}{\partial r}\right) + \frac{1}{r^2}\frac{\partial^2 \phi}{\partial \theta^2} + \frac{\partial^2 \phi}{\partial z^2} = 0$$

The beauty of this approximation is that we have combined three unknown velocity components (u, v, and w, or u_r, u_θ, and u_z, depending on our choice of coordinate system) into *one* unknown scalar variable ϕ, eliminating two of the equations required for a solution (Fig. 10–25). Once we obtain a solution of Eq. 10–24 for ϕ, we can calculate all three components of the velocity field using Eq. 10–22 or 10–23.

The Laplace equation is well known since it shows up in several fields of physics, applied mathematics, and engineering. Various solution techniques, both analytical and numerical, are available in the literature. Solutions of the Laplace equation are dominated by the *geometry* (i.e., *boundary conditions*). Although Eq. 10–24 comes from conservation of mass, mass itself (or density, which is mass per unit volume) has dropped out of the equation altogether. With a given set of boundary conditions surrounding the entire irrotational region of the flow field, we can thus solve Eq. 10–24 for ϕ, regardless of the fluid properties. Once we have calculated ϕ, we can then calculate $\vec{V}$ everywhere in that region of the flow field (using Eq. 10–21), without ever having to solve the Navier–Stokes equation. The solution is valid for any incompressible fluid, regardless of its density or its viscosity, in regions of the flow in which the irrotational approximation is appropriate.

The solution is even valid instantaneously for an *unsteady* flow, since time does not appear in the incompressible continuity equation. In other words, at any instant in time, the incompressible flow field instantly adjusts itself so as to satisfy the Laplace equation and the boundary conditions that exist at that instant in time.

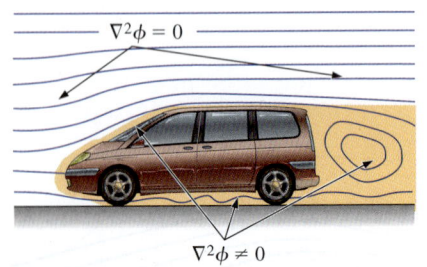

FIGURE 10–24
The Laplace equation for velocity potential function ϕ is valid in both two and three dimensions and in any coordinate system, but only in irrotational regions of flow (generally away from walls and wakes).

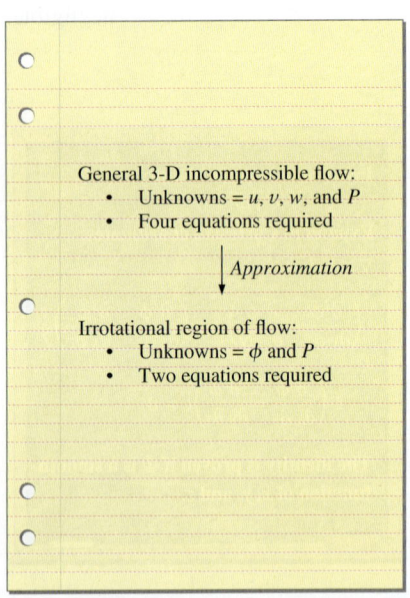

FIGURE 10–25
In irrotational regions of flow, three unknown scalar components of the velocity vector are combined into *one* unknown scalar function—the velocity potential function.

Momentum Equation

We now turn our attention to the differential linear momentum equation—the Navier–Stokes equation (Eq. 10–2). We have just shown that in an irrotational region of flow, we can obtain the velocity field without application of the Navier–Stokes equation. Why then do we need it at all? The answer is that once we have established the velocity field through use of the velocity potential function, *we use the Navier–Stokes equation to solve for the pressure field.* A simplified form of the Navier–Stokes equation is the second required equation mentioned in Fig. 10–25 for solution of two unknowns, ϕ and P, in an irrotational region of flow.

We begin our analysis by applying the irrotational flow approximation, (Eq. 10–21), to the viscous term of the Navier–Stokes equation (Eq. 10–2). Provided that ϕ is a smooth function, that term becomes

$$\mu \nabla^2 \vec{V} = \mu \nabla^2 (\vec{\nabla}\phi) = \mu \vec{\nabla}(\underbrace{\nabla^2 \phi}_{0}) = 0$$

where we have applied Eq. 10–24. Thus, the Navier–Stokes equation reduces to the *Euler equation* in irrotational regions of the flow,

For irrotational regions of flow: $\quad \rho\left[\dfrac{\partial \vec{V}}{\partial t} + (\vec{V}\cdot\vec{\nabla})\vec{V}\right] = -\vec{\nabla}P + \rho\vec{g}$ (10–25)

We emphasize that although we get the same Euler equation as we did for an inviscid region of flow (Eq. 10–13), the viscous term vanishes here for a *different reason*, namely, that the flow in this region is assumed to be irrotational rather than inviscid (Fig. 10–26).

Derivation of the Bernoulli Equation in Irrotational Regions of Flow

In Section 10–4 we derived the Bernoulli equation along a streamline for inviscid regions of flow, based on the Euler equation. We now do a similar derivation beginning with Eq. 10–25 for irrotational regions of flow. For simplicity, we again assume steady incompressible flow. We use the same vector identity used previously (Eq. 10–14), leading to the alternative form of the Euler equation of Eq. 10–15. Here, however, the vorticity vector $\vec{\zeta}$ is negligibly small since we are considering an irrotational region of flow (Eq. 10–19). Thus, for gravity acting in the negative z-direction, Eq. 10–17 reduces to

$$\vec{\nabla}\left(\dfrac{P}{\rho} + \dfrac{V^2}{2} + gz\right) = 0 \quad (10\text{–}26)$$

We now argue that if the gradient of some scalar quantity (the quantity in parentheses in Eq. 10–26) is zero everywhere, the scalar quantity itself must be a constant. Thus, we generate the Bernoulli equation for irrotational regions of flow,

Steady incompressible Bernoulli equation in irrotational regions of flow:

$$\dfrac{P}{\rho} + \dfrac{V^2}{2} + gz = C = \text{constant everywhere} \quad (10\text{–}27)$$

It is useful to compare Eqs. 10–18 and 10–27. In an inviscid region of flow, the Bernoulli equation holds along streamlines, and the Bernoulli constant

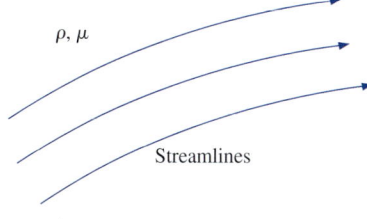

FIGURE 10–26
An irrotational region of flow is a region where net viscous forces are negligible compared to inertial and/or pressure forces because of the irrotational approximation. All irrotational regions of flow are therefore also inviscid, but not all inviscid regions of flow are irrotational. The fluid itself is still a viscous fluid in either case.

532
APPROXIMATE SOLUTIONS OF THE N–S EQ

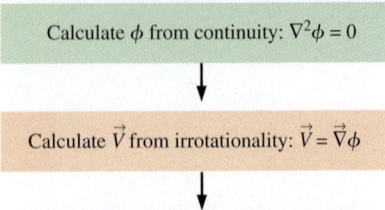

FIGURE 10–27
Flowchart for obtaining solutions in an irrotational region of flow. The velocity field is obtained from continuity and irrotationality, and then pressure is obtained from the Bernoulli equation.

may change from streamline to streamline. In an irrotational region of flow, the Bernoulli constant is the same everywhere, so the Bernoulli equation holds everywhere in the irrotational region of flow, even across streamlines. Thus, *the irrotational approximation is more restrictive than the inviscid approximation.*

A summary of the equations and solution procedure relevant to irrotational regions of flow is provided in Fig. 10–27. In a region of irrotational flow, the velocity field is obtained first by solution of the Laplace equation for velocity potential function ϕ (Eq. 10–24), followed by application of Eq. 10–21 to obtain the velocity field. To solve the Laplace equation, we must provide boundary conditions for ϕ everywhere along the boundary of the flow field of interest. Once the velocity field is known, we use the Bernoulli equation (Eq. 10–27) to obtain the pressure field, where the Bernoulli constant C is obtained from a boundary condition on P somewhere in the flow.

Example 10–4 illustrates a situation in which the flow field consists of two separate regions—an inviscid, rotational region and an inviscid, irrotational region.

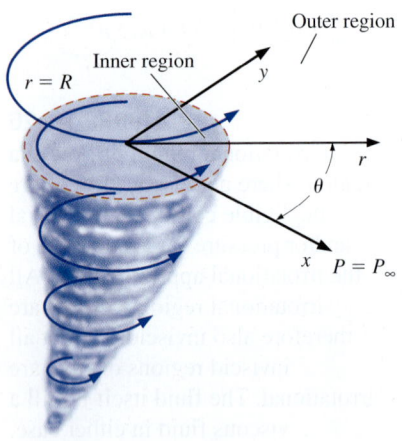

FIGURE 10–28
A horizontal slice through a tornado can be modeled by two regions—an inviscid but rotational inner region of flow ($r < R$) and an irrotational outer region of flow ($r > R$).

EXAMPLE 10–4 A Two-Region Model of a Tornado

A horizontal slice through a tornado (Fig. 10–28) is modeled by two distinct regions. The *inner or core region* ($0 < r < R$) is modeled by solid body rotation—a rotational but inviscid region of flow as discussed earlier. The *outer region* ($r > R$) is modeled as an irrotational region of flow. The flow is two-dimensional in the $r\theta$-plane, and the components of the velocity field $\vec{V} = (u_r, u_\theta)$ are given by

Velocity components:
$$u_r = 0 \qquad u_\theta = \begin{cases} \omega r & 0 < r < R \\ \dfrac{\omega R^2}{r} & r > R \end{cases} \qquad (1)$$

where ω is the magnitude of the angular velocity in the inner region. The ambient pressure (far away from the tornado) is equal to P_∞. Calculate the pressure field in a horizontal slice of the tornado for $0 < r < \infty$. What is the pressure at $r = 0$? Plot the pressure and velocity fields.

SOLUTION We are to calculate the pressure field $P(r)$ in a horizontal radial slice through a tornado for which the velocity components are approximated by Eq. 1. We are also to calculate the pressure in this horizontal slice at $r = 0$.
Assumptions 1 The flow is steady and incompressible. 2 Although R increases and ω decreases with increasing elevation z, R and ω are assumed to be constants when considering a particular horizontal slice. 3 The flow in the horizontal slice is two-dimensional in the $r\theta$-plane (no dependence on z and no w-component of velocity). 4 The effects of gravity are negligible within a particular horizontal slice (an additional hydrostatic pressure field exists in the z-direction, of course, but this does not affect the dynamics of the flow, as discussed previously).
Analysis In the inner region, the Euler equation is an appropriate approximation of the Navier–Stokes equation, and the pressure field is found by integration. In Example 10–3 we showed that for solid body rotation,

Pressure field in inner region ($r < R$): $\qquad P = \rho \dfrac{\omega^2 r^2}{2} + P_0 \qquad (2)$

where P_0 is the (unknown) pressure at $r = 0$ and we have neglected the gravity term. Since the outer region is a region of irrotational flow, the Bernoulli equation is appropriate and the Bernoulli constant is the same everywhere from $r = R$ outward to $r \to \infty$. The Bernoulli constant is found by applying the boundary condition far from the tornado, namely, as $r \to \infty$, $u_\theta \to 0$ and $P \to P_\infty$ (Fig. 10–29). Equation 10–27 yields

As $r \to \infty$:
$$\underbrace{\frac{P}{\rho}}_{P_\infty/\rho} + \underbrace{\frac{V^2}{2}}_{V \to 0 \text{ as } r \to \infty} + \underbrace{gz}_{\text{assumption 4}} = C \quad \to \quad C = \frac{P_\infty}{\rho} \qquad (3)$$

The pressure field anywhere in the outer region is obtained by substituting the value of constant C from Eq. 3 into the Bernoulli equation (Eq. 10–27). Neglecting gravity,

In outer region ($r > R$): $\quad P = \rho C - \frac{1}{2}\rho V^2 = P_\infty - \frac{1}{2}\rho V^2 \qquad (4)$

We note that $V^2 = u_\theta^2$. After substitution of Eq. 1 for u_θ, Eq. 4 reduces to

Pressure field in outer region ($r > R$): $\quad \boldsymbol{P = P_\infty - \frac{\rho}{2}\frac{\omega^2 R^4}{r^2}} \qquad (5)$

At $r = R$, the interface between the inner and outer regions, the pressure must be continuous (no sudden jumps in P), as illustrated in Fig. 10–30. Equating Eqs. 2 and 5 at this interface yields

Pressure at $r = R$: $\quad P_{r=R} = \rho\frac{\omega^2 R^2}{2} + P_0 = P_\infty - \frac{\rho}{2}\frac{\omega^2 R^4}{R^2} \qquad (6)$

from which the pressure P_0 at $r = 0$ is found,

Pressure at $r = 0$: $\quad \boldsymbol{P_0 = P_\infty - \rho\omega^2 R^2} \qquad (7)$

Equation 7 provides the value of pressure in the middle of the tornado—the eye of the storm. This is the lowest pressure in the flow field. Substitution of Eq. 7 into Eq. 2 enables us to rewrite Eq. 2 in terms of the given far-field ambient pressure P_∞,

In inner region ($r < R$): $\quad \boldsymbol{P = P_\infty - \rho\omega^2\left(R^2 - \frac{r^2}{2}\right)} \qquad (8)$

Instead of plotting P as a function of r in this horizontal slice, we plot a *nondimensional* pressure distribution instead, so that the plot is valid for *any* horizontal slice. In terms of nondimensional variables,

Inner region ($r < R$): $\quad \dfrac{u_\theta}{\omega R} = \dfrac{r}{R} \quad \dfrac{P - P_\infty}{\rho\omega^2 R^2} = \dfrac{1}{2}\left(\dfrac{r}{R}\right)^2 - 1$

Outer region ($r > R$): $\quad \dfrac{u_\theta}{\omega R} = \dfrac{R}{r} \quad \dfrac{P - P_\infty}{\rho\omega^2 R^2} = -\dfrac{1}{2}\left(\dfrac{R}{r}\right)^2 \qquad (9)$

Figure 10–31 shows both nondimensional tangential velocity and nondimensional pressure as functions of nondimensional radial location.

Discussion In the outer region, pressure increases as speed decreases—a direct result of the Bernoulli equation, which applies with the *same* Bernoulli constant everywhere in the outer region. You are encouraged to calculate P

FIGURE 10–29
A good place to obtain boundary conditions for this problem is the far field; this is true for many problems in fluid mechanics.

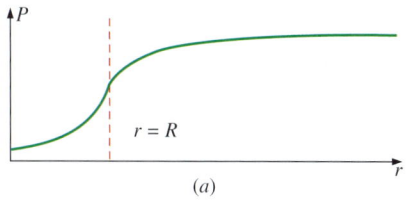

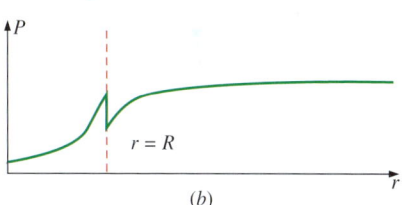

FIGURE 10–30
For our model of the tornado to be valid, the pressure can have a discontinuity in *slope* at $r = R$, but cannot have a sudden jump of value there; (a) is valid, but (b) is not.

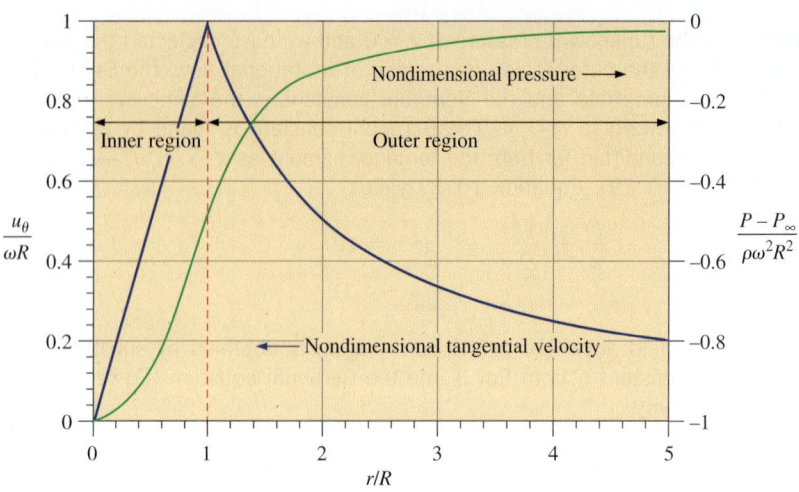

FIGURE 10–31
Nondimensional tangential velocity distribution (blue curve) and nondimensional pressure distribution (black curve) along a horizontal radial slice through a tornado. The inner and outer regions of flow are marked.

FIGURE 10–32
The lowest pressure occurs at the center of the tornado, and the flow in that region can be approximated by solid body rotation.

in the outer region by an alternate method—direct integration of the Euler equation without use of the Bernoulli equation; you should get the same result. In the inner region, P increases parabolically with r even though speed also increases; this is because the Bernoulli constant changes from streamline to streamline (as also pointed out in Example 10–3). Notice that even though there is a discontinuity in the slope of tangential velocity at $r/R = 1$, the pressure has a fairly smooth transition between the inner and outer regions. The pressure is lowest in the center of the tornado and rises to atmospheric pressure in the far field (Fig. 10–32). Finally, the flow in the inner region is *rotational* but *inviscid*, since viscosity plays no role in that region of the flow. The flow in the outer region is *irrotational* but *viscous*. Note, however, that viscosity still acts on fluid particles in the outer region. (Viscosity causes the fluid particles to shear and distort, even though the *net viscous force* on any fluid particle in the outer region is zero.)

Two-Dimensional Irrotational Regions of Flow

In irrotational regions of flow, Eqs. 10–24 and 10–21 apply for both two- and three-dimensional flow fields, and we solve for the velocity field in these regions by solving the Laplace equation for velocity potential function ϕ. If the flow is also *two-dimensional*, we are able to make use of the *stream function* as well (Fig. 10–33). The two-dimensional approximation is not limited to flow in the *xy*-plane, nor is it limited to Cartesian coordinates. In fact, we can assume two-dimensionality in any region of the flow where only *two* directions of motion are important and where there is no significant variation in the third direction. The two most common examples are **planar flow** (flow in a plane with negligible variation in the direction normal to the plane) and **axisymmetric flow** (flow in which there is rotational symmetry about some axis). We may also choose to work in Cartesian coordinates, cylindrical coordinates, or spherical polar coordinates, depending on the geometry of the problem at hand.

Planar Irrotational Regions of Flow

We consider planar flow first, since it is the simplest. For a steady, incompressible, planar, irrotational region of flow in the *xy*-plane in Cartesian coordinates (Fig. 10–34), the Laplace equation for ϕ is

$$\nabla^2 \phi = \frac{\partial^2 \phi}{\partial x^2} + \frac{\partial^2 \phi}{\partial y^2} = 0 \qquad (10\text{--}28)$$

For incompressible planar flow in the *xy*-plane, the stream function ψ is defined as (Chap. 9)

Stream function: $\qquad u = \dfrac{\partial \psi}{\partial y} \qquad v = -\dfrac{\partial \psi}{\partial x} \qquad (10\text{--}29)$

Note that Eq. 10–29 holds whether the region of flow is rotational or irrotational. In fact, the stream function is *defined* such that it always satisfies the continuity equation, regardless of rotationality. If we restrict our approximation to *irrotational* regions of flow, Eq. 10–19 must also hold; namely, the vorticity is zero or negligibly small. For general two-dimensional flow in the *xy*-plane, the *z*-component of vorticity is the only nonzero component. Thus, in an irrotational region of flow,

$$\zeta_z = \frac{\partial v}{\partial x} - \frac{\partial u}{\partial y} = 0$$

Substitution of Eq. 10–29 into this equation yields

$$\frac{\partial}{\partial x}\left(-\frac{\partial \psi}{\partial x}\right) - \frac{\partial}{\partial y}\left(\frac{\partial \psi}{\partial y}\right) = -\frac{\partial^2 \psi}{\partial x^2} - \frac{\partial^2 \psi}{\partial y^2} = 0$$

We recognize the Laplacian operator in this latter equation. Thus,

$$\nabla^2 \psi = \frac{\partial^2 \psi}{\partial x^2} + \frac{\partial^2 \psi}{\partial y^2} = 0 \qquad (10\text{--}30)$$

We conclude that the Laplace equation is applicable, not only for ϕ (Eq. 10–28), but also for ψ (Eq. 10–30) in steady, incompressible, irrotational, planar regions of flow.

Curves of constant values of ψ define *streamlines* of the flow, while curves of constant values of ϕ define **equipotential lines.** (Note that some authors use the phrase *equipotential lines* to refer to both streamlines *and* lines of constant ϕ rather than exclusively for lines of constant ϕ.) In planar irrotational regions of flow, it turns out that streamlines intersect equipotential lines at right angles, a condition known as **mutual orthogonality** (Fig. 10–35). In addition, the potential functions ψ and ϕ are intimately related to each other—both satisfy the Laplace equation, and from either ψ or ϕ we can determine the velocity field. Mathematicians call solutions of ψ and ϕ **harmonic functions,** and ψ and ϕ are called **harmonic conjugates** of each other. Although ψ and ϕ are related, their origins are somewhat opposite; it is perhaps best to say that ψ and ϕ are *complementary* to each other:

- *The stream function is defined by continuity; the Laplace equation for ψ results from irrotationality.*
- *The velocity potential is defined by irrotationality; the Laplace equation for ϕ results from continuity.*

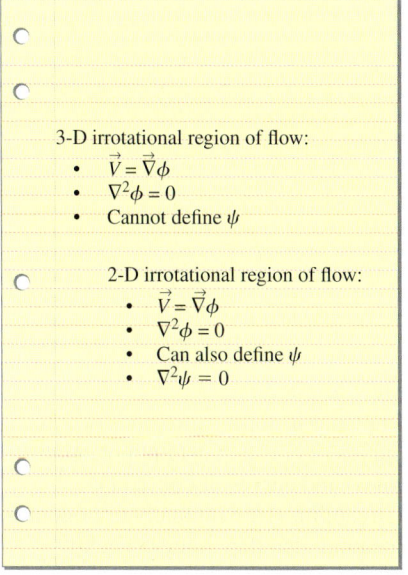

FIGURE 10–33

Two-dimensional flow is a *subset* of three-dimensional flow; in two-dimensional regions of flow we can define a stream function, but we cannot do so in three-dimensional flow. The velocity potential function, however, can be defined for any *irrotational* region of flow.

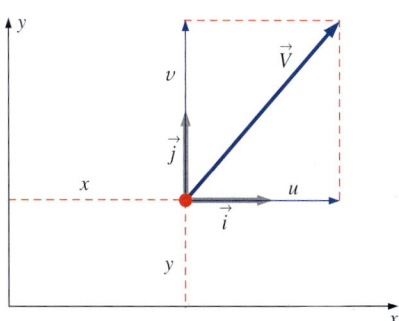

FIGURE 10–34

Velocity components and unit vectors in Cartesian coordinates for planar two-dimensional flow in the *xy*-plane. There is no variation normal to this plane.

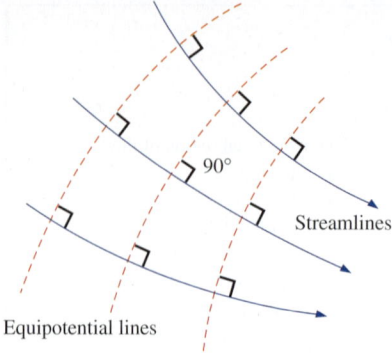

FIGURE 10–35
In planar irrotational regions of flow, curves of constant ϕ (equipotential lines) and curves of constant ψ (streamlines) are mutually orthogonal, meaning that they intersect at 90° angles everywhere.

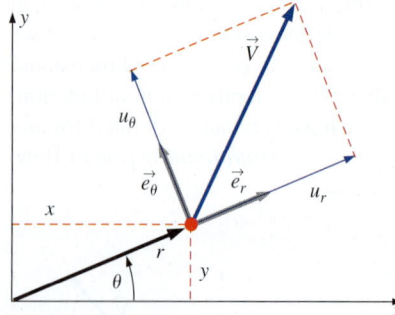

FIGURE 10–36
Velocity components and unit vectors in cylindrical coordinates for planar flow in the $r\theta$-plane. There is no variation normal to this plane.

In practice, we may perform a potential flow analysis using either ψ or ϕ, and we should achieve the same results either way. However, it is often more convenient to use ψ, since boundary conditions on ψ are usually easier to specify.

Planar flow in the xy-plane can also be described in cylindrical coordinates (r, θ) and (u_r, u_θ), as shown in Fig. 10–36. Again, there is no z-component of velocity, and velocity does not vary in the z-direction. In cylindrical coordinates,

Laplace equation, planar flow in (r, θ):
$$\frac{1}{r}\frac{\partial}{\partial r}\left(r\frac{\partial \phi}{\partial r}\right) + \frac{1}{r^2}\frac{\partial^2 \phi}{\partial \theta^2} = 0 \quad (10\text{–}31)$$

The stream function ψ for planar flow in Cartesian coordinates is defined by Eq. 10–29, and the irrotationality condition causes ψ to also satisfy the Laplace equation. In cylindrical coordinates we perform a similar analysis. Recall from Chap. 9,

Stream function:
$$u_r = \frac{1}{r}\frac{\partial \psi}{\partial \theta} \qquad u_\theta = -\frac{\partial \psi}{\partial r} \quad (10\text{–}32)$$

It is left as an exercise for you to show that the stream function defined by Eq. 10–32 also satisfies the Laplace equation in cylindrical coordinates for regions of two-dimensional planar irrotational flow. (Verify your results by replacing ϕ by ψ in Eq. 10–31 to obtain the Laplace equation for the stream function.)

Axisymmetric Irrotational Regions of Flow

Axisymmetric flow is a special case of two-dimensional flow that can be described in either cylindrical coordinates or spherical polar coordinates. In cylindrical coordinates, r and z are the relevant spatial variables, and u_r and u_z are the nonzero velocity components (Fig. 10–37). There is no dependence on angle θ since rotational symmetry is defined about the z-axis. This is a type of two-dimensional flow because there are only two independent spatial variables, r and z. (Imagine rotating the radial component r in Fig. 10–37 in the θ-direction about the z-axis without changing the magnitude of r.) Because of rotational symmetry about the z-axis, the magnitudes of velocity components u_r and u_z remain unchanged after such a rotation. The Laplace equation for velocity potential ϕ for the case of axisymmetric irrotational regions of flow in cylindrical coordinates is

$$\frac{1}{r}\frac{\partial}{\partial r}\left(r\frac{\partial \phi}{\partial r}\right) + \frac{\partial^2 \phi}{\partial z^2} = 0$$

In order to obtain expressions for the stream function for axisymmetric flow, we begin with the incompressible continuity equation in r- and z-coordinates,

$$\frac{1}{r}\frac{\partial}{\partial r}(ru_r) + \frac{\partial u_z}{\partial z} = 0 \quad (10\text{–}33)$$

After some algebra, we define a stream function that identically satisfies Eq. 10–33,

Stream function:
$$u_r = -\frac{1}{r}\frac{\partial \psi}{\partial z} \qquad u_z = \frac{1}{r}\frac{\partial \psi}{\partial r}$$

Following the same procedure as for planar flow, we generate an equation for ψ for axisymmetric irrotational regions of flow by forcing the vorticity to be zero. In this case, only the θ-component of vorticity is relevant since the velocity vector always lies in the rz-plane. Thus, in an irrotational region of flow,

$$\frac{\partial u_r}{\partial z} - \frac{\partial u_z}{\partial r} = \frac{\partial}{\partial z}\left(-\frac{1}{r}\frac{\partial \psi}{\partial z}\right) - \frac{\partial}{\partial r}\left(\frac{1}{r}\frac{\partial \psi}{\partial r}\right) = 0$$

After taking r outside the z-derivative (since r is not a function of z), we get

$$r\frac{\partial}{\partial r}\left(\frac{1}{r}\frac{\partial \psi}{\partial r}\right) + \frac{\partial^2 \psi}{\partial z^2} = 0 \qquad (10\text{–}34)$$

Note that Eq. 10–34 is *not* the same as the Laplace equation for ψ. You cannot use the Laplace equation for the stream function in axisymmetric irrotational regions of flow (Fig. 10–38).

> For planar irrotational regions of flow, the Laplace equation is valid for both ϕ and ψ; but for axisymmetric irrotational regions of flow, the Laplace equation is valid for ϕ but not for ψ.

A direct consequence of this statement is that curves of constant ψ and curves of constant ϕ in axisymmetric irrotational regions of flow are *not* mutually orthogonal. This is a fundamental difference between planar and axisymmetric flows. Finally, even though Eq. 10–34 is not the same as the Laplace equation, it is still a *linear* partial differential equation. This allows us to use the technique of superposition with either ψ or ϕ when solving for the flow field in axisymmetric irrotational regions of flow. Superposition is discussed shortly.

Summary of Two-Dimensional Irrotational Regions of Flow

Equations for the velocity components for both planar and axisymmetric irrotational regions of flow are summarized in Table 10–2.

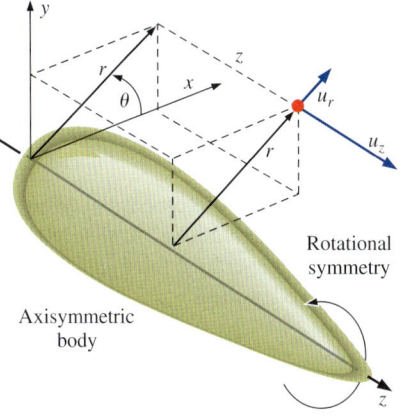

FIGURE 10–37
Flow over an axisymmetric body in cylindrical coordinates with rotational symmetry about the z-axis. Neither the geometry nor the velocity field depend on θ; and $u_\theta = 0$.

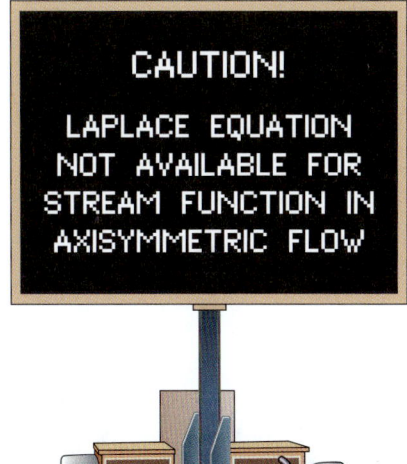

FIGURE 10–38
The equation for the stream function in axisymmetric irrotational flow (Eq. 10–34) is *not* the Laplace equation.

TABLE 10–2
Velocity components for steady, incompressible, irrotational, two-dimensional regions of flow in terms of velocity potential function and stream function in various coordinate systems

Description and Coordinate System	Velocity Component 1	Velocity Component 2
Planar; Cartesian coordinates	$u = \dfrac{\partial \phi}{\partial x} = \dfrac{\partial \psi}{\partial y}$	$v = \dfrac{\partial \phi}{\partial y} = -\dfrac{\partial \psi}{\partial x}$
Planar; cylindrical coordinates	$u_r = \dfrac{\partial \phi}{\partial r} = \dfrac{1}{r}\dfrac{\partial \psi}{\partial \theta}$	$u_\theta = \dfrac{1}{r}\dfrac{\partial \phi}{\partial \theta} = -\dfrac{\partial \psi}{\partial r}$
Axisymmetric; cylindrical coordinates	$u_r = \dfrac{\partial \phi}{\partial r} = -\dfrac{1}{r}\dfrac{\partial \psi}{\partial z}$	$u_z = \dfrac{\partial \phi}{\partial z} = \dfrac{1}{r}\dfrac{\partial \psi}{\partial r}$

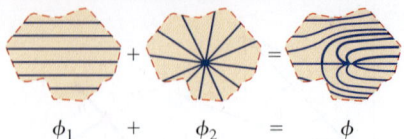

FIGURE 10–39
Superposition is the process of adding two or more irrotational flow solutions together to generate a third (more complicated) solution.

Superposition in Irrotational Regions of Flow

Since the Laplace equation is a *linear* homogeneous differential equation, the linear combination of two or more solutions of the equation must also be a solution. For example, if ϕ_1 and ϕ_2 are each solutions of the Laplace equation, then $A\phi_1$, $(A + \phi_1)$, $(\phi_1 + \phi_2)$, and $(A\phi_1 + B\phi_2)$ are also solutions, where A and B are arbitrary constants. By extension, you may combine *several* solutions of the Laplace equation, and the combination is guaranteed to also be a solution. If a region of irrotational flow is modeled by the sum of two or more separate irrotational flow fields, e.g., a source located in a free-stream flow, one can simply add the velocity potential functions for each individual flow to describe the combined flow field. This process of adding two or more known solutions to create a third, more complicated solution is known as **superposition** (Fig. 10–39).

For the case of two-dimensional irrotational flow regions, a similar analysis can be performed using the *stream function* rather than the velocity potential function. We stress that the concept of superposition is useful, but is valid only for *irrotational* flow fields for which the equations for ϕ and ψ are *linear*. You must be careful to ensure that the two flow fields you wish to add vectorially are both irrotational. For example, the flow field for a jet should never be added to the flow field for an inlet or for free-stream flow, because the velocity field associated with a jet is strongly affected by viscosity, is not irrotational, and cannot be described by potential functions.

It also turns out that since the potential function of the composite field is the sum of the potential functions of the individual flow fields, the velocity at any point in the composite field is the *vector sum* of the velocities of the individual flow fields. We prove this in Cartesian coordinates by considering a planar irrotational flow field that is the superposition of two independent planar irrotational flow fields denoted by subscripts 1 and 2. The composite velocity potential function is given by

Superposition of two irrotational flow fields:
$$\phi = \phi_1 + \phi_2$$

Using the equations for planar irrotational flow in Cartesian coordinates in Table 10–2, the x-component of velocity of the composite flow is

$$u = \frac{\partial \phi}{\partial x} = \frac{\partial (\phi_1 + \phi_2)}{\partial x} = \frac{\partial \phi_1}{\partial x} + \frac{\partial \phi_2}{\partial x} = u_1 + u_2$$

You can generate an analogous expression for v. Thus, superposition enables us to simply add the individual velocities vectorially at any location in the flow region to obtain the velocity of the composite flow field at that location (Fig. 10–40).

Composite velocity field from superposition:
$$\vec{V} = \vec{V}_1 + \vec{V}_2 \qquad (10\text{–}35)$$

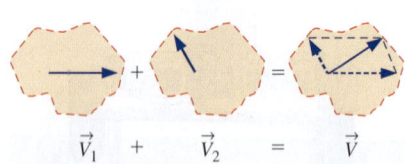

FIGURE 10–40
In the superposition of two irrotational flow solutions, the two velocity vectors at any point in the flow region add vectorially to produce the composite velocity at that point.

Elementary Planar Irrotational Flows

Superposition enables us to add two or more simple irrotational flow solutions to create a more complex (and hopefully more physically significant) flow field. It is therefore useful to establish a collection of elementary-*building block* irrotational flows, with which we can construct a variety of more practical flows (Fig. 10–41). Elementary planar irrotational flows are described in xy- and/or $r\theta$-coordinates, depending on which pair is more useful in a particular problem.

Building Block 1—Uniform Stream

The simplest building block flow we can think of is a **uniform stream** of flow moving at constant velocity V in the x-direction (left to right). In terms of the velocity potential and stream function (Table 10–2),

Uniform stream: $\quad u = \dfrac{\partial \phi}{\partial x} = \dfrac{\partial \psi}{\partial y} = V \quad v = \dfrac{\partial \phi}{\partial y} = -\dfrac{\partial \psi}{\partial x} = 0$

By integrating the first of these with respect to x, and then differentiating the result with respect to y, we generate an expression for the velocity potential function for a uniform stream,

$$\phi = Vx + f(y) \quad \rightarrow \quad v = \dfrac{\partial \phi}{\partial y} = f'(y) = 0 \quad \rightarrow \quad f(y) = \text{constant}$$

The constant is arbitrary since velocity components are always derivatives of ϕ. We set the constant equal to zero, knowing that we can always add an arbitrary constant later on if desired. Thus,

Velocity potential function for a uniform stream: $\quad \phi = Vx \quad$ **(10–36)**

In a similar manner we generate an expression for the stream function for this elementary planar irrotational flow,

Stream function for a uniform stream: $\quad \psi = Vy \quad$ **(10–37)**

Shown in Fig. 10–42 are several streamlines and equipotential lines for a uniform stream. Notice the mutual orthogonality.

It is often convenient to express the stream function and velocity potential function in cylindrical coordinates rather than rectangular coordinates, particularly when superposing a uniform stream with some other planar irrotational flow(s). The conversion relations are obtained from the geometry of Fig. 10–36,

$$x = r \cos \theta \quad y = r \sin \theta \quad r = \sqrt{x^2 + y^2} \quad \textbf{(10–38)}$$

From Eq. 10–38 and a bit of trigonometry, we derive relationships for u and v in terms of cylindrical coordinates,

Transformation: $\quad u = u_r \cos \theta - u_\theta \sin \theta \quad v = u_r \sin \theta + u_\theta \cos \theta \quad$ **(10–39)**

In cylindrical coordinates, Eqs. 10–36 and 10–37 for ϕ and ψ become

Uniform stream: $\quad \phi = Vr \cos \theta \quad \psi = Vr \sin \theta \quad$ **(10–40)**

We may modify the uniform stream so that the fluid flows uniformly at speed V at an angle of inclination α from the x-axis. For this situation, $u = V \cos \alpha$ and $v = V \sin \alpha$ as shown in Fig. 10–43. It is left as an exercise to show that the velocity potential function and stream function for a uniform stream inclined at angle α are

Uniform stream inclined at angle α:
$$\phi = V(x \cos \alpha + y \sin \alpha)$$
$$\psi = V(y \cos \alpha - x \sin \alpha)$$
(10–41)

When necessary, Eq. 10–41 can easily be converted to cylindrical coordinates through use of Eq. 10–38.

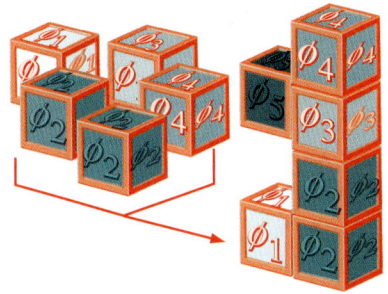

FIGURE 10–41
With superposition we build up a complicated irrotational flow field by adding together elementary "building block" irrotational flow fields.

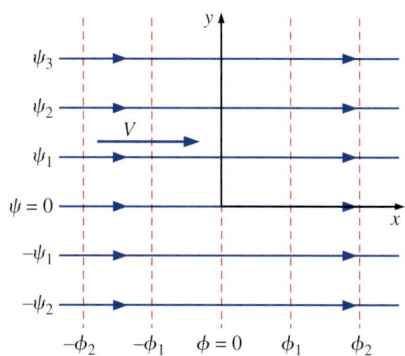

FIGURE 10–42
Streamlines (solid) and equipotential lines (dashed) for a uniform stream in the x-direction.

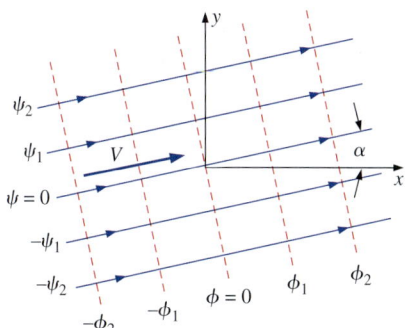

FIGURE 10–43
Streamlines (solid) and equipotential lines (dashed) for a uniform stream inclined at angle α.

540
APPROXIMATE SOLUTIONS OF THE N–S EQ

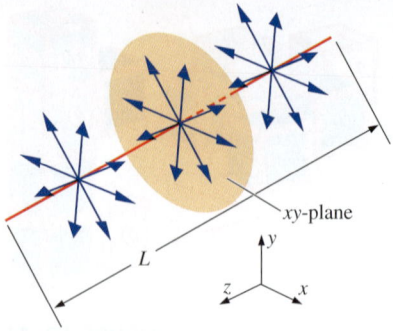

FIGURE 10–44
Fluid emerging uniformly from a finite line segment of length L. As L approaches infinity, the flow becomes a line source, and the xy-plane is taken as normal to the axis of the source.

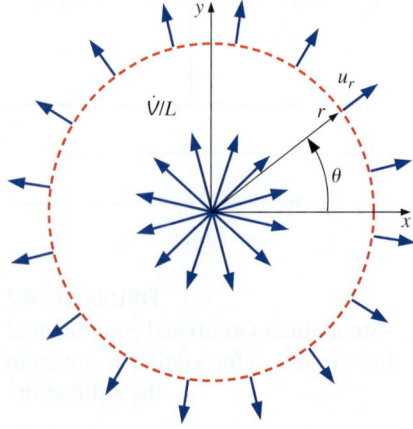

FIGURE 10–45
Line source of strength $\dot{V}/L$ located at the origin in the xy-plane; the total volume flow rate per unit depth through a circle of radius r must equal $\dot{V}/L$ regardless of the value of r.

Building Block 2—Line Source or Line Sink

Our second building block flow is a line source. Imagine a line segment of length L parallel to the z-axis, along which fluid emerges and flows uniformly outward in all directions normal to the line segment (Fig. 10–44). The total volume flow rate is equal to $\dot{V}$. As length L approaches infinity, the flow becomes two-dimensional in planes perpendicular to the line, and the line from which the fluid escapes is called a **line source**. For an infinite line, $\dot{V}$ also approaches infinity; thus, it is more convenient to consider the *volume flow rate per unit depth*, $\dot{V}/L$, called the **line source strength** (often given the symbol m).

A **line sink** is the opposite of a line source; fluid flows *into* the line from all directions in planes normal to the axis of the line sink. By convention, positive $\dot{V}/L$ signifies a line source and negative $\dot{V}/L$ signifies a line sink.

The simplest case occurs when the line source is located at the origin of the xy-plane, with the line itself lying along the z-axis. In the xy-plane, the line source looks like a point at the origin from which fluid is spewed outward in all directions in the plane (Fig. 10–45). At any radial distance r from the line source, the radial velocity component u_r is found by applying conservation of mass. Namely, the entire volume flow rate per unit depth from the line source must pass through the circle defined by radius r. Thus,

$$\frac{\dot{V}}{L} = 2\pi r u_r \qquad u_r = \frac{\dot{V}/L}{2\pi r} \qquad (10\text{–}42)$$

Clearly, u_r decreases with increasing r as we would expect. Notice also that u_r is infinite at the origin since r is zero in the denominator of Eq. 10–42. We call this a **singular point** or a **singularity**—it is certainly unphysical, but keep in mind that planar irrotational flow is merely an *approximation*, and the line source is still useful as a building block for superposition in irrotational flow. As long as we stay away from the immediate vicinity of the center of the line source, the rest of the flow field produced by superposition of a line source and other building block(s) may still be a good representation of a region of irrotational flow in a physically realistic flow field.

We now generate expressions for the velocity potential function and the stream function for a line source of strength $\dot{V}/L$. We use cylindrical coordinates, beginning with Eq. 10–42 for u_r and also recognize that u_θ is zero everywhere. Using Table 10–2, the velocity components are

Line source: $\quad u_r = \dfrac{\partial \phi}{\partial r} = \dfrac{1}{r}\dfrac{\partial \psi}{\partial \theta} = \dfrac{\dot{V}/L}{2\pi r} \qquad u_\theta = \dfrac{1}{r}\dfrac{\partial \phi}{\partial \theta} = -\dfrac{\partial \psi}{\partial r} = 0$

To generate the stream function, we (arbitrarily) choose one of these equations (we choose the second one), integrate with respect to r, and then differentiate with respect to the other variable θ,

$$\frac{\partial \psi}{\partial r} = -u_\theta = 0 \quad \rightarrow \quad \psi = f(\theta) \quad \rightarrow \quad \frac{\partial \psi}{\partial \theta} = f'(\theta) = r u_r = \frac{\dot{V}/L}{2\pi}$$

from which we integrate to obtain

$$f(\theta) = \frac{\dot{V}/L}{2\pi}\theta + \text{constant}$$

Again we set the arbitrary constant of integration equal to zero, since we can add back a constant as desired at any time without changing the flow.

After a similar analysis for ϕ, we obtain the following expressions for a line source at the origin:

Line source at the origin: $\quad \phi = \dfrac{\dot{V}/L}{2\pi} \ln r \quad \text{and} \quad \psi = \dfrac{\dot{V}/L}{2\pi} \theta \quad$ (10–43)

Several streamlines and equipotential lines are sketched for a line source in Fig. 10–46. As expected, the streamlines are *rays* (lines of constant θ), and the equipotential lines are *circles* (lines of constant r). The streamlines and equipotential lines are mutually orthogonal everywhere except at the origin, which is a singular point.

In situations where we would like to place a line source somewhere other than the origin, we must transform Eq. 10–43 carefully. Sketched in Fig. 10–47 is a source located at an arbitrary point (a, b) in the xy-plane. We define r_1 as the distance from the source to some point P in the flow, where P is located at (x, y) or (r, θ). Similarly, we define θ_1 as the angle from the source to point P, as measured from a line parallel to the x-axis. We analyze the flow as if the source were at a new origin at absolute location (a, b). Equations 10–43 for ϕ and ψ are thus still usable, but r and θ must be replaced by r_1 and θ_1. Some trigonometry is required to convert r_1 and θ_1 back to (x, y) or (r, θ). In Cartesian coordinates, for example,

Line source at point (a, b):
$$\phi = \dfrac{\dot{V}/L}{2\pi} \ln r_1 = \dfrac{\dot{V}/L}{2\pi} \ln \sqrt{(x-a)^2 + (y-b)^2}$$
$$\psi = \dfrac{\dot{V}/L}{2\pi} \theta_1 = \dfrac{\dot{V}/L}{2\pi} \arctan \dfrac{y-b}{x-a}$$
(10–44)

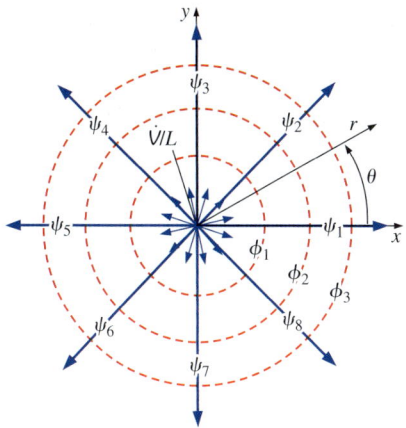

FIGURE 10–46
Streamlines (solid) and equipotential lines (dashed) for a line source of strength $\dot{V}/L$ located at the origin in the xy-plane.

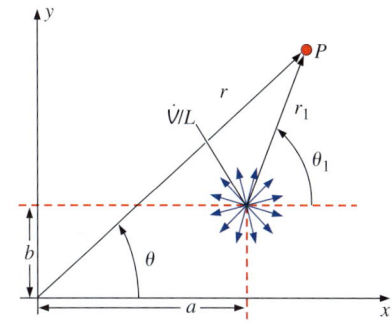

FIGURE 10–47
Line source of strength $\dot{V}/L$ located at some arbitrary point (a, b) in the xy-plane.

■ **EXAMPLE 10–5** **Superposition of a Source and Sink of Equal Strength**

Consider an irrotational region of flow composed of a line source of strength $\dot{V}/L$ at location $(-a, 0)$ and a line sink of the same strength (but opposite sign) at $(a, 0)$, as sketched in Fig. 10–48. Generate an expression for the stream function in both Cartesian and cylindrical coordinates.

SOLUTION We are to superpose a source and a sink, and generate an expression for ψ in both Cartesian and cylindrical coordinates.
Assumptions The region of flow under consideration is incompressible and irrotational.
Analysis We use Eq. 10–44 to obtain ψ for the source,

Line source at $(-a, 0)$: $\quad \psi_1 = \dfrac{\dot{V}/L}{2\pi} \theta_1 \quad$ where $\quad \theta_1 = \arctan \dfrac{y}{x+a} \quad$ (1)

Similarly for the sink,

Line sink at $(a, 0)$: $\quad \psi_2 = \dfrac{-\dot{V}/L}{2\pi} \theta_2 \quad$ where $\quad \theta_2 = \arctan \dfrac{y}{x-a} \quad$ (2)

Superposition enables us to simply add the two stream functions, Eqs. 1 and 2, to obtain the composite stream function,

Composite stream function: $\quad \psi = \psi_1 + \psi_2 = \dfrac{\dot{V}/L}{2\pi} (\theta_1 - \theta_2) \quad$ (3)

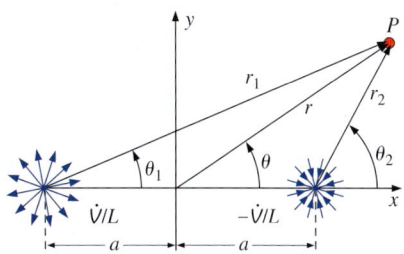

FIGURE 10–48
Superposition of a line source of strength $\dot{V}/L$ at $(-a, 0)$ and a line sink (source of strength $-\dot{V}/L$) at $(a, 0)$.

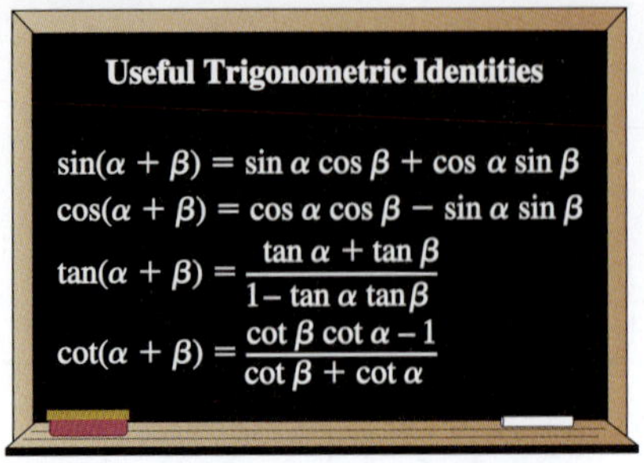

542
APPROXIMATE SOLUTIONS OF THE N–S EQ

FIGURE 10–49
Some useful trigonometric identities.

Useful Trigonometric Identities:
$$\sin(\alpha + \beta) = \sin \alpha \cos \beta + \cos \alpha \sin \beta$$
$$\cos(\alpha + \beta) = \cos \alpha \cos \beta - \sin \alpha \sin \beta$$
$$\tan(\alpha + \beta) = \frac{\tan \alpha + \tan \beta}{1 - \tan \alpha \tan \beta}$$
$$\cot(\alpha + \beta) = \frac{\cot \beta \cot \alpha - 1}{\cot \beta + \cot \alpha}$$

We rearrange Eq. 3 and take the tangent of both sides to get

$$\tan \frac{2\pi\psi}{\dot{V}/L} = \tan(\theta_1 - \theta_2) = \frac{\tan \theta_1 - \tan \theta_2}{1 + \tan \theta_1 \tan \theta_2} \quad (4)$$

where we have used a trigonometric identity (Fig. 10–49).

We substitute Eqs. 1 and 2 for θ_1 and θ_2 and perform some algebra to obtain an expression for the stream function,

$$\tan \frac{2\pi\psi}{\dot{V}/L} = \frac{\dfrac{y}{x+a} - \dfrac{y}{x-a}}{1 + \dfrac{y}{x+a}\dfrac{y}{x-a}} = \frac{-2ay}{x^2 + y^2 - a^2}$$

or, taking the arctangent of both sides,

Final result, Cartesian coordinates: $\quad \psi = \dfrac{-\dot{V}/L}{2\pi} \arctan \dfrac{2ay}{x^2 + y^2 - a^2} \quad (5)$

We translate to cylindrical coordinates by using Eqs. 10–38,

Final result, cylindrical coordinates: $\quad \psi = \dfrac{-\dot{V}/L}{2\pi} \arctan \dfrac{2ar \sin \theta}{r^2 - a^2} \quad (6)$

Discussion If the source and sink were to switch places, the result would be the same, except that the negative sign on source strength $\dot{V}/L$ would disappear.

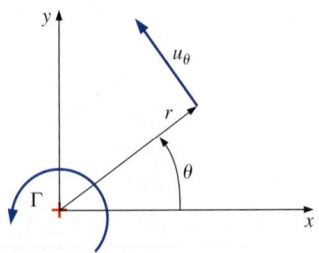

FIGURE 10–50
Line vortex of strength Γ located at the origin in the xy-plane.

Building Block 3—Line Vortex

Our third building block flow is a **line vortex** parallel to the z-axis. As with the previous building block, we start with the simple case in which the line vortex is located at the origin (Fig. 10–50). Again we use cylindrical coordinates for convenience. The velocity components are

Line vortex: $\quad u_r = \dfrac{\partial \phi}{\partial r} = \dfrac{1}{r}\dfrac{\partial \psi}{\partial \theta} = 0 \quad u_\theta = \dfrac{1}{r}\dfrac{\partial \phi}{\partial \theta} = -\dfrac{\partial \psi}{\partial r} = \dfrac{\Gamma}{2\pi r} \quad (10\text{–}45)$

where Γ is called the **circulation** or the **vortex strength.** Following the standard convention in mathematics, positive Γ represents a counterclockwise vortex, while negative Γ represents a clockwise vortex. It is left as an exercise to integrate Eq. 10–45 to obtain expressions for the stream function and the velocity potential function,

Line vortex at the origin: $\quad \phi = \dfrac{\Gamma}{2\pi}\theta \quad \psi = -\dfrac{\Gamma}{2\pi}\ln r \quad (10\text{–}46)$

Comparing Eqs. 10–43 and 10–46, we see that a line source and line vortex are somewhat complementary in the sense that the expressions for ϕ and ψ are reversed.

For situations in which we would like to place the vortex somewhere other than the origin, we must transform Eq. 10–46 as we did for a line source. Sketched in Fig. 10–51 is a line vortex located at an arbitrary point (a, b)

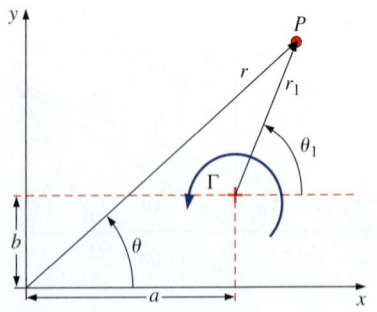

FIGURE 10–51
Line vortex of strength Γ located at some arbitrary point (a, b) in the xy-plane.

in the *xy*-plane. We define r_1 and θ_1 as previously (Fig. 10–47). To obtain expressions for ϕ and ψ, we replace r and θ by r_1 and θ_1 in Eqs. 10–46 and then transform to regular coordinates, either Cartesian or cylindrical. In Cartesian coordinates,

Line vortex at point (a, b):

$$\phi = \frac{\Gamma}{2\pi}\theta_1 = \frac{\Gamma}{2\pi}\arctan\frac{y-b}{x-a}$$

$$\psi = -\frac{\Gamma}{2\pi}\ln r_1 = -\frac{\Gamma}{2\pi}\ln\sqrt{(x-a)^2 + (y-b)^2}$$

(10–47)

EXAMPLE 10–6 Velocity in a Flow Composed of Three Components

An irrotational region of flow is formed by superposing a line source of strength $(\dot{V}/L)_1 = 2.00\ \text{m}^2/\text{s}$ at $(x, y) = (0, -1)$, a line source of strength $(\dot{V}/L)_2 = -1.00\ \text{m}^2/\text{s}$ at $(x, y) = (1, -1)$, and a line vortex of strength $\Gamma = 1.50\ \text{m}^2/\text{s}$ at $(x, y) = (1, 1)$, where all spatial coordinates are in meters. [Source number 2 is actually a sink, since $(\dot{V}/L)_2$ is negative.] The locations of the three building blocks are shown in Fig. 10–52. Calculate the fluid velocity at the point $(x, y) = (1, 0)$.

SOLUTION For the given superposition of two line sources and a vortex, we are to calculate the velocity at the point $(x, y) = (1, 0)$.
Assumptions **1** The region of flow being modeled is steady, incompressible, and irrotational. **2** The velocity at the location of each component is infinite (they are singularities), and the flow in the vicinity of each of these singularities is unphysical; however, these regions are ignored in the present analysis.
Analysis There are several ways to solve this problem. We could sum the three stream functions using Eqs. 10–44 and 10–47, and then take derivatives of the composite stream function to calculate the velocity components. Alternatively, we could do the same for velocity potential function. An easier approach is to recognize that velocity *itself* can be superposed; we simply add the velocity vectors induced by each of the three individual singularities to form the composite velocity at the given point. This is illustrated in Fig. 10–53. Since the vortex is located 1 m above the point (1, 0), the velocity induced by the vortex is to the right and has a magnitude of

$$V_{\text{vortex}} = \frac{\Gamma}{2\pi r_{\text{vortex}}} = \frac{1.50\ \text{m}^2/\text{s}}{2\pi(1.00\ \text{m})} = 0.239\ \text{m/s}$$ (1)

Similarly, the first source induces a velocity at point (1, 0) at a 45° angle from the *x*-axis as shown in Fig. 10–53. Its magnitude is

$$V_{\text{source 1}} = \frac{|(\dot{V}/L)_1|}{2\pi r_{\text{source 1}}} = \frac{2.00\ \text{m}^2/\text{s}}{2\pi(\sqrt{2}\ \text{m})} = 0.225\ \text{m/s}$$ (2)

Finally, the second source (the sink) induces a velocity straight down with magnitude

$$V_{\text{source 2}} = \frac{|(\dot{V}/L)_2|}{2\pi r_{\text{source 2}}} = \frac{|-1.00\ \text{m}^2/\text{s}|}{2\pi(1.00\ \text{m})} = 0.159\ \text{m/s}$$ (3)

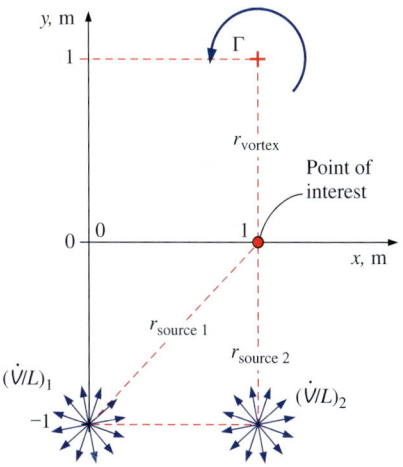

FIGURE 10–52
Superposition of two line sources and a line vortex in the *xy*-plane (Example 10–6).

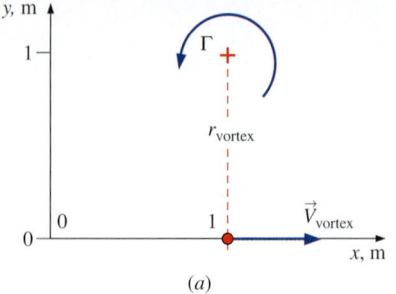

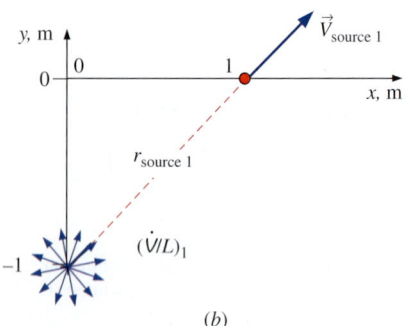

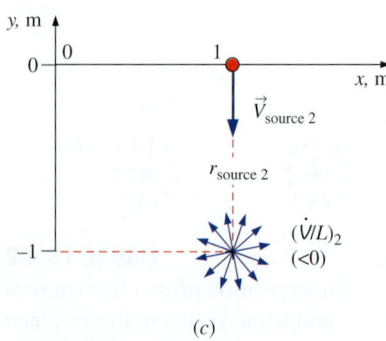

FIGURE 10–53
Induced velocity due to (*a*) the vortex, (*b*) source 1, and (*c*) source 2 (noting that source 2 is *negative*) (Example 10–6).

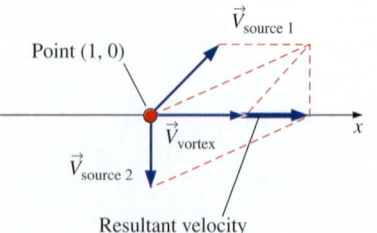

FIGURE 10–54
Vector summation of the three induced velocities of Example 10–6.

We sum these velocities vectorially by completing the parallelograms, as illustrated in Fig. 10–54. Using Eq. 10–35, the resultant velocity is

$$\vec{V} = \underbrace{\vec{V}_{vortex}}_{0.239\vec{i}\ m/s} + \underbrace{\vec{V}_{source\ 1}}_{\left(\frac{0.225}{\sqrt{2}}\vec{i}\ +\ \frac{0.225}{\sqrt{2}}\vec{j}\right)\ m/s} + \underbrace{\vec{V}_{source\ 2}}_{-0.159\vec{j}\ m/s} = (0.398\vec{i} + 0\vec{j})\ m/s \quad (4)$$

The superposed velocity at point (1, 0) is 0.398 m/s to the right.

Discussion This example demonstrates that velocity can be superposed just as stream function or velocity potential function can be superposed. Superposition of velocity is valid in irrotational regions of flow because the differential equations for ϕ and ψ are *linear*; the linearity extends to their derivatives as well.

Building Block 4—Doublet

Our fourth and final building block flow is called a **doublet**. Although we treat it as a building block for use with superposition, the doublet itself is generated by superposition of two earlier building blocks, namely, a line source and a line sink of equal magnitude, as discussed in Example 10–5. The composite stream function was obtained in that example problem and the result is repeated here:

Composite stream function: $\qquad \psi = \dfrac{-\dot{V}/L}{2\pi}\arctan\dfrac{2ar\sin\theta}{r^2 - a^2} \qquad$ **(10–48)**

Now imagine that the distance *a* from the origin to the source and from the origin to the sink approaches zero (Fig. 10–55). You should recall that arctan β approaches β for very small values of angle β in radians. Thus, as distance *a* approaches zero, Eq. 10–48 reduces to

Stream function as $a \to 0$: $\qquad \psi \to \dfrac{-a(\dot{V}/L)r\sin\theta}{\pi(r^2 - a^2)} \qquad$ **(10–49)**

If we shrink *a* while maintaining the same source and sink strengths ($\dot{V}/L$ and $-\dot{V}/L$), the source and sink cancel each other out when $a = 0$, leaving us with no flow at all. However, imagine that as the source and sink approach each other, their strength $\dot{V}/L$ increases inversely with distance *a* such that *the product $a(\dot{V}/L)$ remains constant*. In that case, $r \gg a$ at any point *P* except very close to the origin, and Eq. 10–49 reduces to

Doublet along the x-axis: $\qquad \psi = -\dfrac{a(\dot{V}/L)}{\pi}\dfrac{\sin\theta}{r} = -K\dfrac{\sin\theta}{r} \qquad$ **(10–50)**

where we have defined **doublet strength** $K = a(\dot{V}/L)/\pi$ for convenience. The velocity potential function is obtained in similar fashion,

Doublet along the x-axis: $\qquad \phi = K\dfrac{\cos\theta}{r} \qquad$ **(10–51)**

Several streamlines and equipotential lines for a doublet are plotted in Fig. 10–56. It turns out that the streamlines are circles tangent to the *x*-axis, and the equipotential lines are circles tangent to the *y*-axis. The circles intersect at 90° angles everywhere except at the origin, which is a singular point.

If K is negative, the doublet is "backwards," with the sink located at $x = 0^-$ (infinitesimally to the left of the origin) and the source located at $x = 0^+$ (infinitesimally to the right of the origin). In that case all the streamlines in Fig. 10–56 would be identical in shape, but the flow would be in the opposite direction. It is left as an exercise to construct expressions for a doublet that is aligned at some angle α from the x-axis.

Irrotational Flows Formed by Superposition

Now that we have a set of building block irrotational flows, we are ready to construct some more interesting irrotational flow fields by the superposition technique. We limit our examples to planar flows in the xy-plane; examples of superposition with axisymmetric flows can be found in more advanced textbooks (e.g., Kundu et al., 2011; Panton, 2005; Heinsohn and Cimbala, 2003). Note that even though ψ for axisymmetric irrotational flow does not satisfy the Laplace equation, the differential equation for ψ (Eq. 10–34) is still *linear*, and thus superposition is still valid.

Superposition of a Line Sink and a Line Vortex

Our first example is superposition of a line source of strength $\dot{V}/L$ ($\dot{V}/L$ is a negative quantity in this example) and a line vortex of strength Γ, both located at the origin (Fig. 10–57). This represents a region of flow above a drain in a sink or bathtub where fluid spirals in toward the drain. We can superpose either ψ or ϕ. We choose ψ and generate the composite stream function by adding ψ for a source (Eq. 10–43) and ψ for a line vortex (Eq. 10–46),

Superposition:
$$\psi = \frac{\dot{V}/L}{2\pi}\theta - \frac{\Gamma}{2\pi}\ln r \quad (10\text{–}52)$$

To plot streamlines of the flow, we pick a value of ψ and then solve for either r as a function of θ or θ as a function of r. We choose the former; after some algebra we get

Streamlines:
$$r = \exp\left(\frac{(\dot{V}/L)\theta - 2\pi\psi}{\Gamma}\right) \quad (10\text{–}53)$$

We pick some arbitrary values for $\dot{V}/L$ and Γ so that we can generate a plot; namely, we set $\dot{V}/L = -1.00$ m²/s and $\Gamma = 1.50$ m²/s. Note that $\dot{V}/L$ is negative for a sink. Also note that the units for $\dot{V}/L$ and Γ are obtained easily since we know that the dimensions of stream function in planar flow are {length²/time}. Streamlines are calculated for several values of ψ using Eq. 10–53 and are plotted in Fig. 10–58.

The velocity components at any point in this irrotational flow are obtained by differentiating Eq. 10–52,

Velocity components:
$$u_r = \frac{1}{r}\frac{\partial \psi}{\partial \theta} = \frac{\dot{V}/L}{2\pi r} \qquad u_\theta = -\frac{\partial \psi}{\partial r} = \frac{\Gamma}{2\pi r}$$

We notice that in this simple example, the radial velocity component is due entirely to the sink, since there is no contribution to radial velocity from the vortex. Similarly, the tangential velocity component is due entirely to the vortex. The composite velocity at any point in the flow is the vector sum of these two components, as sketched in Fig. 10–57.

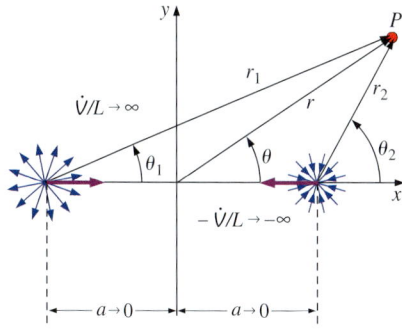

FIGURE 10–55

A doublet is formed by superposition of a line source at $(-a, 0)$ and a line sink at $(a, 0)$; a decreases to zero while $\dot{V}/L$ increases to infinity such that the product $a\dot{V}/L$ remains constant.

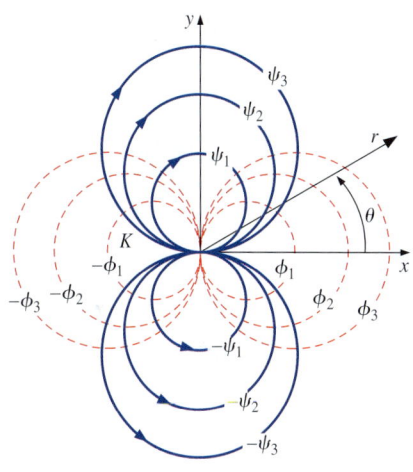

FIGURE 10–56

Streamlines (solid) and equipotential lines (dashed) for a doublet of strength K located at the origin in the xy-plane and aligned with the x-axis.

Superposition of a Uniform Stream and a Doublet—Flow over a Circular Cylinder

Our next example is a classic in the field of fluid mechanics, namely, the superposition of a uniform stream of speed V_∞ and a doublet of strength K located at the origin (Fig. 10–59). We superpose the stream function by adding Eq. 10–40 for a uniform stream and Eq. 10–50 for a doublet at the origin. The composite stream function is thus

Superposition:
$$\psi = V_\infty r \sin\theta - K\frac{\sin\theta}{r} \quad (10\text{–}54)$$

For convenience we set $\psi = 0$ when $r = a$ (the reason for this will soon become apparent). Equation 10–54 if then solved for doublet strength K,

Doublet strength:
$$K = V_\infty a^2$$

and Eq. 10–54 becomes

Alternate form of stream function:
$$\psi = V_\infty \sin\theta \left(r - \frac{a^2}{r}\right) \quad (10\text{–}55)$$

It is clear from Eq. 10–55 that one of the streamlines ($\psi = 0$) is a circle of radius a (Fig. 10–60). We can plot this and other streamlines by solving Eq. 10–55 for r as a function of θ or vice versa. However, as you should be aware by now, it is usually better to present results in terms of *nondimensional* parameters. By inspection, we define three nondimensional parameters,

$$\psi^* = \frac{\psi}{V_\infty a} \qquad r^* = \frac{r}{a} \qquad \theta$$

where angle θ is already dimensionless. In terms of these parameters, Eq. 10–55 is written as

$$\psi^* = \sin\theta \left(r^* - \frac{1}{r^*}\right) \quad (10\text{–}56)$$

We solve Eq. 10–56 for r^* as a function of θ through use of the quadratic rule,

Nondimensional streamlines:
$$r^* = \frac{\psi^* \pm \sqrt{(\psi^*)^2 + 4\sin^2\theta}}{2\sin\theta} \quad (10\text{–}57)$$

Using Eq. 10–57, we plot several nondimensional streamlines in Fig. 10–61. Now you see why we chose the circle $r = a$ (or $r^* = 1$) as the zero streamline—this streamline can be thought of as a solid wall, and this flow represents *potential flow over a circular cylinder*. Not shown are streamlines *inside* the circle—they exist, but are of no concern to us.

There are two stagnation points in this flow field, one at the nose of the cylinder and one at the tail. Streamlines near the stagnation points are far apart since the flow is very slow there. By contrast, streamlines near the top and bottom of the cylinder are close together, indicating regions of fast flow. Physically, fluid must accelerate around the cylinder since it is acting as an obstruction to the flow.

Notice also that the flow is symmetric about both the *x*- and *y*-axes. While top-to-bottom symmetry is not surprising, fore-to-aft symmetry is perhaps unexpected, since we know that real flow around a cylinder generates a

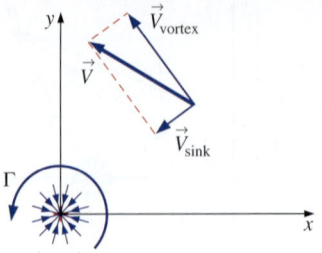

FIGURE 10–57
Superposition of a line source of strength $\dot{V}/L$ and a line vortex of strength Γ located at the origin. Vector velocity addition is shown at some arbitrary location in the *xy*-plane.

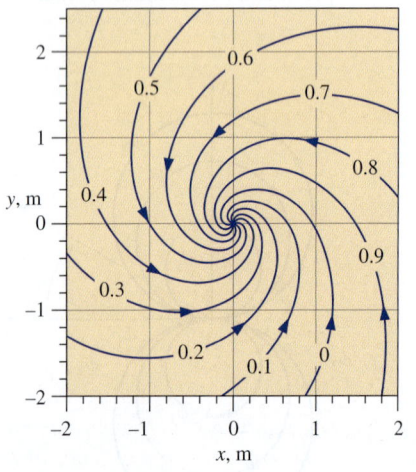

FIGURE 10–58
Streamlines created by superposition of a line sink and a line vortex at the origin. Values of ψ are in units of m²/s.

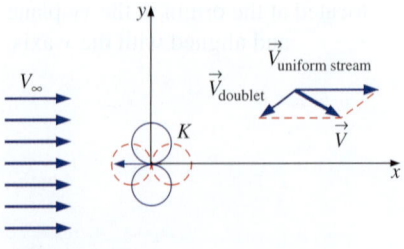

FIGURE 10–59
Superposition of a uniform stream and a doublet; vector velocity addition is shown at some arbitrary location in the *xy*-plane.

wake region behind the cylinder, and the streamlines are *not* symmetric. However, we must keep in mind that the results here are only *approximations* of a real flow. We have assumed irrotationality everywhere in the flow field, and we know that this approximation is not true near walls and in wake regions.

We calculate the velocity components everywhere in the flow field by differentiating Eq. 10–55,

$$u_r = \frac{1}{r}\frac{\partial \psi}{\partial \theta} = V_\infty \cos\theta\left(1 - \frac{a^2}{r^2}\right) \quad u_\theta = -\frac{\partial \psi}{\partial r} = -V_\infty \sin\theta\left(1 + \frac{a^2}{r^2}\right) \quad (10\text{–}58)$$

A special case is on the surface of the cylinder itself ($r = a$), where Eqs. 10–58 reduce to

On the surface of the cylinder: $\quad u_r = 0 \quad u_\theta = -2V_\infty \sin\theta \quad$ (10–59)

Since the no-slip condition at solid walls cannot be satisfied when making the irrotational approximation, there is slip at the cylinder wall. In fact, at the top of the cylinder ($\theta = 90°$), the fluid speed at the wall is *twice* that of the free stream.

EXAMPLE 10–7 Pressure Distribution on a Circular Cylinder

Using the irrotational flow approximation, calculate and plot the nondimensional static pressure distribution on the surface of a circular cylinder of radius a in a uniform stream of speed V_∞ (Fig. 10–62). Discuss the results. The pressure far away from the cylinder is P_∞.

SOLUTION We are to calculate and plot the nondimensional static pressure distribution along the surface of a circular cylinder in a free-stream flow.
Assumptions 1 The region of flow being modeled is steady, incompressible, and irrotational. 2 The flow field is two-dimensional in the *xy*-plane.
Analysis First of all, static pressure is the pressure that would be measured by a pressure probe moving with the fluid. Experimentally, we measure this pressure on a surface through use of a **static pressure tap**, which is basically a tiny hole drilled normal to the surface (Fig. 10–63). At the other end of the tap is a tube leading to a pressure measuring device. Experimental data of the static pressure distribution along the surface of a cylinder are available in the literature, and we compare our results to some of those experimental data.

From Chap. 7 we recognize that the appropriate nondimensional pressure is the **pressure coefficient**,

Pressure coefficient: $\quad C_p = \dfrac{P - P_\infty}{\frac{1}{2}\rho V_\infty^2} \quad$ (1)

Since the flow in the region of interest is irrotational, we use the Bernoulli equation (Eq. 10–27) to calculate the pressure anywhere in the flow field. Ignoring the effects of gravity,

Bernoulli equation: $\quad \dfrac{P}{\rho} + \dfrac{V^2}{2} = \text{constant} = \dfrac{P_\infty}{\rho} + \dfrac{V_\infty^2}{2} \quad$ (2)

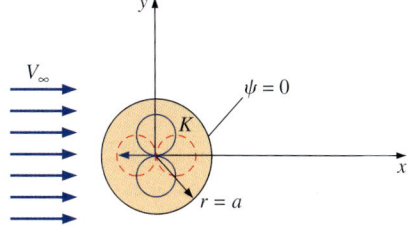

FIGURE 10–60
Superposition of a uniform stream and a doublet yields a streamline that is a circle.

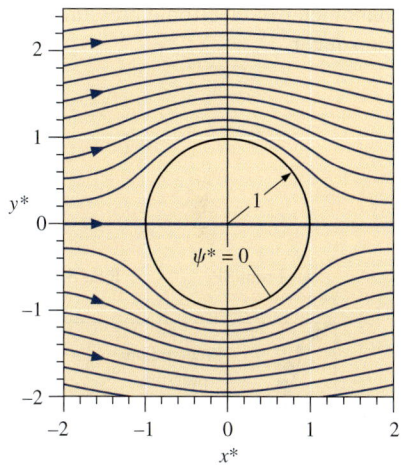

FIGURE 10–61
Nondimensional streamlines created by superposition of a uniform stream and a doublet at the origin; $\psi^* = \psi/(V_\infty a)$, $\Delta\psi^* = 0.2$, $x^* = x/a$, and $y^* = y/a$, where a is the cylinder radius.

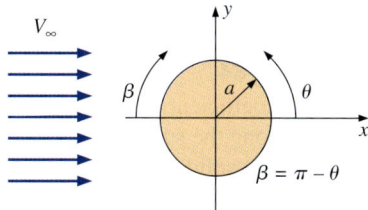

FIGURE 10–62
Planar flow over a circular cylinder of radius a immersed in a uniform stream of speed V_∞ in the *xy*-plane. Angle β is defined from the front of the cylinder by convention.

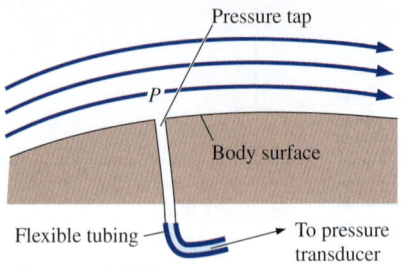

FIGURE 10–63
Static pressure on a surface is measured through use of a static pressure tap connected to a pressure manometer or electronic pressure transducer.

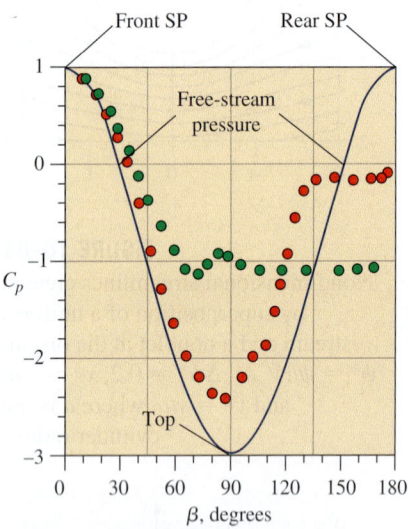

FIGURE 10–64
Pressure coefficient as a function of angle β along the surface of a circular cylinder; the solid blue curve is the irrotational flow approximation, green circles are from experimental data at Re = 2×10^5 – laminar boundary layer separation, and red circles are from typical experimental data at Re = 7×10^5 – turbulent boundary layer separation.
Data from Kundu et al., (2011).

Rearranging Eq. 2 into the form of Eq. 1, we get

$$C_p = \frac{P - P_\infty}{\frac{1}{2}\rho V_\infty^2} = 1 - \frac{V^2}{V_\infty^2} \quad (3)$$

We substitute our expression for tangential velocity on the cylinder surface, Eq. 10–59, since along the surface $V^2 = u_\theta^2$; Eq. 3 becomes

Surface pressure coefficient: $\quad C_p = 1 - \dfrac{(-2V_\infty \sin\theta)^2}{V_\infty^2} = 1 - 4\sin^2\theta$

In terms of angle β, defined from the front of the body (Fig. 10–62), we use the transformation $\beta = \pi - \theta$ to obtain

C_p in terms of angle β: $\quad C_p = 1 - 4\sin^2\beta \quad (4)$

We plot the pressure coefficient on the top half of the cylinder as a function of angle β in Fig. 10–64, solid blue curve. (Because of top–bottom symmetry, there is no need to also plot the pressure distribution on the bottom half of the cylinder.) The first thing we notice is that the pressure distribution is symmetric fore and aft. This is not surprising since we already know that the *streamlines* are also symmetric fore and aft (Fig. 10–61).

The front and rear stagnation points (at β = 0° and 180°, respectively) are labeled SP on Fig. 10–64. The pressure coefficient is unity there, and these two points have the highest pressure in the entire flow field. In physical variables, static pressure P at the stagnation points is equal to $P_\infty + \rho V_\infty^2/2$. In other words, the **full dynamic pressure** (also called **impact pressure**) of the oncoming fluid is felt as a static pressure on the nose of the body as the fluid is decelerated to zero speed at the stagnation point. At the very top of the cylinder (β = 90°), the speed along the surface is twice the free-stream velocity ($V = 2V_\infty$), and the pressure coefficient is lowest there ($C_p = -3$). Also marked on Fig. 10–64 are the two locations where $C_p = 0$, namely at β = 30° and 150°. At these locations, the static pressure along the surface is equal to that of the free stream ($P = P_\infty$).

Discussion Typical experimental data for laminar and turbulent flow over the surface of a circular cylinder are indicated by the green circles and red circles, respectively, in Fig. 10–64. It is clear that near the front of the cylinder, the irrotational flow approximation is excellent. However, for β greater than about 60°, and especially near the rear portion of the cylinder (right side of the plot), the irrotational flow results do not match well at all with experimental data. In fact, it turns out that for flow over bluff body shapes like this, the irrotational flow approximation usually does a fairly good job on the front half of the body, but a very poor job on the rear half of the body. The irrotational flow approximation agrees better with experimental turbulent data than with experimental laminar data; this is because flow separation occurs farther downstream for the case with a turbulent boundary layer, as discussed in more detail in Section 10–6.

One immediate consequence of the symmetry of the pressure distribution in Fig. 10–64 is that there is *no net pressure drag* on the cylinder (pressure forces in the front half of the body are exactly balanced by those on the rear half of the body). In this irrotational flow approximation, the pressure fully recovers at the rear stagnation point, so that the pressure there is the same as that at the front stagnation point. We also predict that there is no net viscous drag on the body, since we cannot specify the no-slip condition on the body surface when we make the irrotational approximation. Hence, the net aerodynamic drag on the cylinder in irrotational flow is identically zero. This is one example of a more general statement that applies to bodies of *any* shape (even unsymmetrical shapes) when the irrotational flow approximation is made, namely, the famous paradox first stated by Jean-le-Rond d'Alembert (1717–1783) in the year 1752:

> **D'Alembert's paradox**: With the irrotational flow approximation, the aerodynamic drag force on any nonlifting body of any shape immersed in a uniform stream is zero.

D'Alembert recognized the paradox of his statement, of course, knowing that there *is* aerodynamic drag on real bodies immersed in real fluids. In a real flow, the pressure on the back surface of the body is significantly *less* than that on the front surface, leading to a nonzero pressure drag on the body. This pressure difference is enhanced if the body is bluff and there is flow separation, as sketched in Fig. 10–65. Even for streamlined bodies, however (such as airplane wings at low angles of attack), the pressure near the back of the body never fully recovers. In addition, the no-slip condition on the body surface leads to a nonzero viscous drag as well. Thus, the irrotational flow approximation falls short in its prediction of aerodynamic drag for two reasons: it predicts no pressure drag and it predicts no viscous drag.

The pressure distribution at the front end of any rounded body shape is qualitatively similar to that plotted in Fig. 10–64. Namely, the pressure at the front stagnation point (SP) is the highest pressure on the body: $P_{SP} = P_\infty + \rho V^2/2$, where V is the free-stream velocity (we have dropped the subscript ∞), and $C_p = 1$ there. Moving downstream along the body surface, pressure drops to some minimum value for which P is less than P_∞ ($C_p < 0$). This point, where the velocity just above the body surface is largest and the pressure is smallest, is often called the **aerodynamic shoulder** of the body. Beyond the shoulder, the pressure slowly rises. With the irrotational flow approximation, the pressure always rises back to the dynamic pressure at the rear stagnation point, where $C_p = 1$. However, in a real flow, the pressure never fully recovers, leading to pressure drag as discussed previously.

Somewhere between the front stagnation point and the aerodynamic shoulder is a point on the body surface where the speed just above the body is equal to V, the pressure P is equal to P_∞, and $C_p = 0$. This point is called the **zero pressure point,** where the phrase is obviously based on *gage* pressure, not absolute pressure. At this point, the pressure acting normal to the body surface is the *same* ($P = P_\infty$), regardless of how fast the body moves through the fluid. This fact is a factor in the location of fish eyes (Fig. 10–66). If a fish's eye were located closer to its nose, the eye would experience an increase in water pressure as the fish swims—the faster it would swim, the higher the water pressure on its eye would be. This would cause the soft eyeball to distort, affecting the fish's vision. Likewise, if the eye were located farther back, near the aerodynamic shoulder, the eye would experience a relative *suction* pressure when the fish would swim, again distorting its eyeball and blurring its vision. Experiments have revealed that the fish's eye is instead located very close to the zero-pressure point where $P = P_\infty$, and the fish can swim at any speed without distorting its vision. Incidentally, the back of the

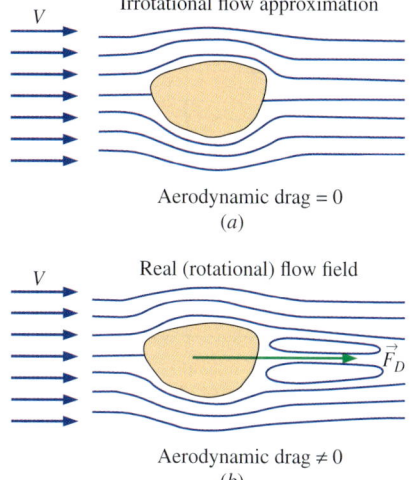

FIGURE 10–65

(*a*) D'Alembert's paradox is that the aerodynamic drag on *any* nonlifting body of *any* shape is predicted to be zero when the irrotational flow approximation is invoked; (*b*) in real flows there is a nonzero drag on bodies immersed in a uniform stream.

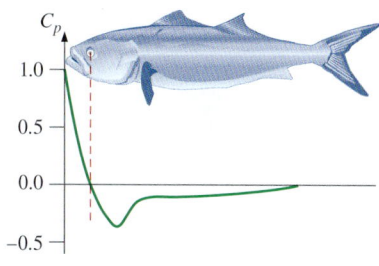

FIGURE 10–66

A fish's body is designed such that its eye is located near the zero-pressure point so that its vision is not distorted while it swims. Data shown are along the side of a bluefish.

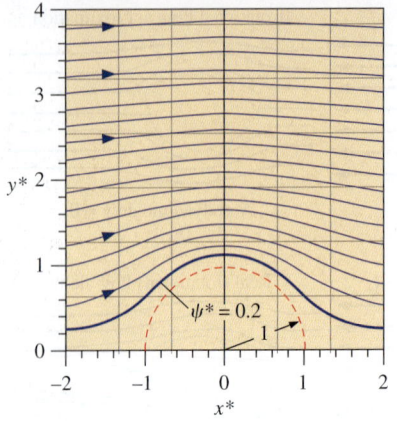

FIGURE 10–67
The same nondimensionalized streamlines as in Fig. 10–61, except streamline $\psi^* = 0.2$ is modeled as a solid wall. This flow represents flow of air over a symmetric hill.

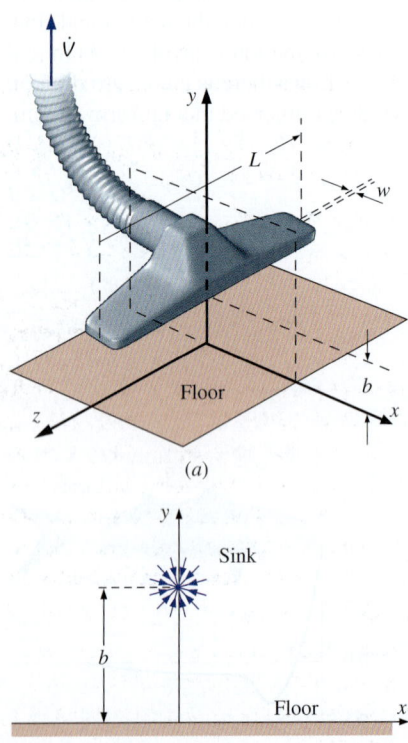

FIGURE 10–68
Vacuum cleaner hose with floor attachment; (*a*) three-dimensional view with floor in the *xz*-plane, and (*b*) view of a slice in the *xy*-plane with suction modeled by a line sink.

gills is located near the aerodynamic shoulder so that the suction pressure there helps the fish to "exhale." The heart is also located near this lowest-pressure point to increase the heart's stroke volume during rapid swimming.

If we think about the irrotational flow approximation a little more closely, we realize that the circle we modeled as a solid cylinder in Example 10–7 is not really a solid wall at all—it is just a streamline in the flow field that we are *modeling* as a solid wall. The particular streamline we model as a solid wall just happens to be a circle. We could have just as easily picked some *other* streamline in the flow to model as a solid wall. Since flow cannot cross a streamline by definition, and since we cannot satisfy the no-slip condition at a wall, we state the following:

> With the irrotational flow approximation, any streamline can be thought of as a solid wall.

For example, we can model *any* streamline in Fig. 10–61 as a solid wall. Let's take the first streamline above the circle, and model it as a wall. (This streamline has a nondimensional value of $\psi^* = 0.2$.) Several streamlines are plotted in Fig. 10–67; we have not shown any streamlines below the streamline $\psi^* = 0.2$—they are still there, it's just that we are no longer concerned with them. What kind of flow does this represent? Well, imagine wind flowing over a hill; the irrotational approximation shown in Fig. 10–67 is representative of this flow. We might expect inconsistencies very close to the ground, and perhaps on the downstream side of the hill, but the approximation is probably very good on the front side of the hill.

You may have noticed a problem with this kind of superposition. Namely, we perform the superposition *first*, and then try to define some physical problems that might be modeled by the flow we generate. While useful as a learning tool, this technique is not always practical in real-life engineering. For example, it is unlikely that we will encounter a hill shaped exactly like the one modeled in Fig. 10–67. Instead, we usually already *have* a geometry and wish to model flow over or through this geometry. There are more sophisticated superposition techniques available that are better suited to engineering design and analysis. Namely, there are techniques in which numerous sources and sinks are placed at appropriate locations so as to model flow over a predetermined geometry. These techniques can even be extended to fully three-dimensional irrotational flow fields, but require a computer because of the amount of calculations involved (Kundu et al., 2011). We do not discuss these techniques here.

EXAMPLE 10–8 Flow into a Vacuum Cleaner Attachment

Consider the flow of air into the floor attachment nozzle of a typical household vacuum cleaner (Fig. 10–68*a*). The width of the nozzle inlet slot is $w = 2.0$ mm, and its length is $L = 35.0$ cm. The slot is held a distance $b = 2.0$ cm above the floor, as shown. The total volume flow rate through the vacuum hose is $\dot{V} = 0.110$ m³/s. Predict the flow field in the center plane of the attachment (the *xy*-plane in Fig. 10–68*a*). Specifically, plot several streamlines and calculate the velocity and pressure distribution along the *x*-axis. What is the maximum speed along the floor, and where does it occur? Where along the floor is the vacuum cleaner most effective?

SOLUTION We are to predict the flow field in the center plane of a vacuum cleaner attachment, plot velocity and pressure along the floor (*x*-axis), predict the location and value of the maximum velocity along the floor, and predict where along the floor the vacuum cleaner is most effective.

Assumptions **1** The flow is steady and incompressible. **2** The flow in the *xy*-plane is two-dimensional (planar). **3** The majority of the flow field is irrotational. **4** The room is infinitely large and free of air currents that might influence the flow.

Analysis We approximate the slot on the vacuum cleaner attachment as a line sink (a line source with negative source strength), located at distance *b* above the *x*-axis, as sketched in Fig. 10–68*b*. With this approximation, we are ignoring the finite width of the slot (*w*); instead we model flow into the slot as flow into the line sink, which is simply a point in the *xy*-plane at (0, *b*). We are also ignoring any effects of the hose or the body of the attachment. The strength of the line source is obtained by dividing total volume flow rate by the length *L* of the slot,

Strength of line source: $$\frac{\dot{V}}{L} = \frac{-0.110 \text{ m}^3/\text{s}}{0.35 \text{ m}} = -0.314 \text{ m}^2/\text{s} \quad (1)$$

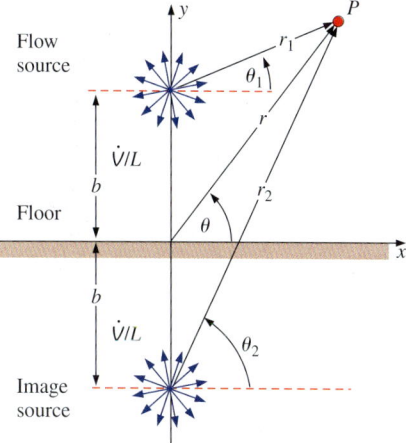

FIGURE 10–69
Superposition of a line source of strength $\dot{V}/L$ at (0, *b*) and a line source of the same strength at (0, −*b*). The bottom source is a mirror image of the top source, making the *x*-axis a streamline.

where we include a negative sign since this is a sink instead of a source.

Clearly this line sink by itself (Fig. 10–68*b*) is not sufficient to model the flow, since air would flow into the sink from all directions, including up through the floor. To avoid this problem, we add another elementary irrotational flow (building block) to model the effect of the floor. A clever way to do this is through the **method of images**. With this technique, we place a *second* identical sink *below* the floor at point (0, −*b*). We call this second sink the **image sink**. Since the *x*-axis is now a line of symmetry, the *x*-axis is itself a streamline of the flow, and hence can be thought of as the floor. The irrotational flow field to be analyzed is sketched in Fig. 10–69. Two sources of strength $\dot{V}/L$ are shown. The top one is called the flow source, and represents suction into the vacuum cleaner attachment. The bottom one is the image source. Keep in mind that source strength $\dot{V}/L$ is negative in this problem (Eq. 1), so that both sources are actually sinks.

We use superposition to generate the stream function for the irrotational approximation of this flow field. The algebra here is similar to that of Example 10–5; in that case we had a source and a sink on the *x*-axis, while here we have two sources on the *y*-axis. We use Eq. 10–44 to obtain ψ for the flow source,

Line source at (0, b): $$\psi_1 = \frac{\dot{V}/L}{2\pi} \theta_1 \quad \text{where} \quad \theta_1 = \arctan \frac{y - b}{x} \quad (2)$$

Similarly for the image source,

Line source at (0, −b): $$\psi_2 = \frac{\dot{V}/L}{2\pi} \theta_2 \quad \text{where} \quad \theta_2 = \arctan \frac{y + b}{x} \quad (3)$$

Superposition enables us to simply add the two stream functions, Eqs. 2 and 3, to obtain the composite stream function,

Composite stream function: $$\psi = \psi_1 + \psi_2 = \frac{\dot{V}/L}{2\pi} (\theta_1 + \theta_2) \quad (4)$$

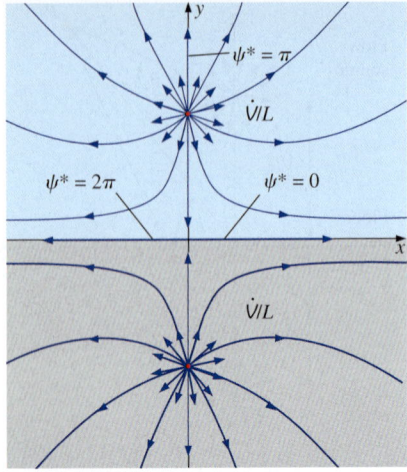

FIGURE 10–70
The x-axis is the dividing streamline that separates air produced by the top source (blue) from air produced by the bottom source (gray).

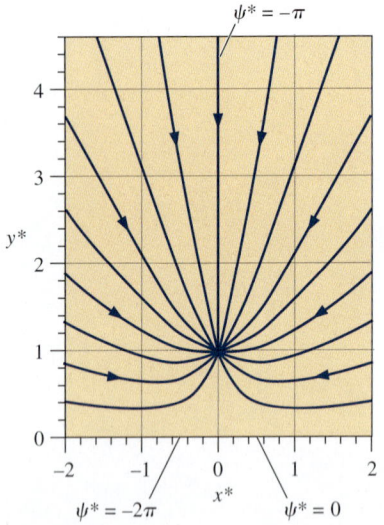

FIGURE 10–71
Nondimensional streamlines for the two sources of Fig. 10–69 for the case in which the source strengths are *negative* (they are *sinks*). ψ^* is incremented uniformly from -2π (negative x-axis) to 0 (positive x-axis), and only the upper half of the flow is shown. The flow is *toward* the sink at location (0, 1).

We rearrange Eq. 4 and take the tangent of both sides to get

$$\tan\frac{2\pi\psi}{\dot{V}/L} = \tan(\theta_1 + \theta_2) = \frac{\tan\theta_1 + \tan\theta_2}{1 - \tan\theta_1\tan\theta_2} \quad (5)$$

where we have again used a trigonometric identity (Fig. 10–49).

We substitute Eqs. 2 and 3 for θ_1 and θ_2 and perform some algebra to obtain our final expression for the stream function in Cartesian coordinates,

$$\psi = \frac{\dot{V}/L}{2\pi}\arctan\frac{2xy}{x^2 - y^2 + b^2} \quad (6)$$

We translate to cylindrical coordinates using Eq. 10–38 and nondimensionalize. After some algebra,

Nondimensional stream function: $\qquad \psi^* = \arctan\dfrac{\sin 2\theta}{\cos 2\theta + 1/r^{*2}} \quad (7)$

where $\psi^* = 2\pi\psi/(\dot{V}/L)$, $r^* = r/b$, and we used trigonometric identities from Fig. 10–49.

Because of symmetry about the x-axis, all the air that is produced by the upper line source must remain *above* the x-axis. Likewise, all the image air that is produced at the lower line source must remain *below* the x-axis. If we were to color air from the upper (north) source blue, and air from the lower (south) source gray (Fig. 10–70), all the blue air would stay above the x-axis, and all the gray air would stay below the x-axis. Thus, the x-axis acts as a **dividing streamline,** separating the blue from the gray. Furthermore, recall from Chap. 9 that the difference in value of ψ from one streamline to the next in planar flow is equal to the volume flow rate per unit width flowing between the two streamlines. We set ψ equal to zero along the positive x-axis. Following the *left-side convention*, introduced in Chap. 9, we know that ψ on the negative x-axis must equal the total volume flow rate per unit width produced by the upper line source, i.e., $\dot{V}/L$. Namely,

$$\psi_{-x\text{-axis}} - \underbrace{\psi_{+x\text{-axis}}}_{0} = \dot{V}/L \quad\rightarrow\quad \psi^*_{-x\text{-axis}} = 2\pi \quad (8)$$

These streamlines are labeled in Fig. 10–70. In addition, the nondimensional streamline $\psi^* = \pi$ is also labeled. It coincides with the y-axis since there is symmetry about that axis as well. The origin (0, 0) is a stagnation point, since the velocity induced by the lower source exactly cancels out that induced by the upper source.

For the case of the vacuum cleaner being modeled here, the source strengths are negative (they are sinks). Thus, the direction of flow is reversed, and the values of ψ^* are of opposite sign to those in Fig. 10–70. Using the left-side convention again, we plot the nondimensional stream function for $-2\pi < \psi^* < 0$ (Fig. 10–71). To do so, we solve Eq. 7 for r^* as a function of θ for various values of ψ^*,

Nondimensional streamlines: $\qquad r^* = \pm\sqrt{\dfrac{\tan\psi^*}{\sin 2\theta - \cos 2\theta \tan\psi^*}} \quad (9)$

Only the upper half is plotted, since the lower half is symmetric and is merely the mirror image of the upper half. For the case of negative $\dot{V}/L$, air gets sucked into the vacuum cleaner from all directions as indicated by the arrows on the streamlines.

To calculate the velocity distribution on the floor (the x-axis), we can either differentiate Eq. 6 and apply the definition of stream function for planar flow

(Eq. 10–29), or we can do a vector summation. The latter is simpler and is illustrated in Fig. 10–72 for an arbitrary location along the x-axis. The induced velocity from the upper source (or sink) has magnitude $(\dot{V}/L)/(2\pi r_1)$, and its direction is in line with r_1 as shown. Because of symmetry, the induced velocity from the image source has identical magnitude, but its direction is in line with r_2. The vector sum of these two induced velocities lies along the x-axis since the two horizontal components add together, but the two vertical components cancel each other out. After a bit of trigonometry, we conclude that

Axial velocity along the x-axis: $$u = V = \frac{(\dot{V}/L)x}{\pi(x^2 + b^2)} \qquad (10)$$

where V is the magnitude of the resultant velocity vector along the floor as sketched in Fig. 10–72. Since we have made the irrotational flow approximation, the Bernoulli equation can be used to generate the pressure field. Ignoring gravity,

Bernoulli equation: $$\frac{P}{\rho} + \frac{V^2}{2} = \text{constant} = \frac{P_\infty}{\rho} + \underbrace{\frac{V_\infty^2}{2}}_{0} \qquad (11)$$

To generate a pressure coefficient, we need a reference velocity for the denominator. Having none, we generate one from the known parameters, namely $V_{\text{ref}} = -(\dot{V}/L)/b$, where we insert the negative sign to make V_{ref} positive (since $\dot{V}/L$ is negative for our model of the vacuum cleaner). Then we define C_p as

Pressure coefficient: $$C_p = \frac{P - P_\infty}{\frac{1}{2}\rho V_{\text{ref}}^2} = -\frac{V^2}{V_{\text{ref}}^2} = -\frac{b^2 V^2}{(\dot{V}/L)^2} \qquad (12)$$

where we have also applied Eq. 11. Substituting Eq. 10 for V, we get

$$C_p = -\frac{b^2 x^2}{\pi^2 (x^2 + b^2)^2} \qquad (13)$$

We introduce nondimensional variables for axial velocity and distance,

Nondimensional variables: $$u^* = \frac{u}{V_{\text{ref}}} = -\frac{ub}{\dot{V}/L} \qquad x^* = \frac{x}{b} \qquad (14)$$

We note that C_p is already nondimensional. In dimensionless form, Eqs. 10 and 13 become

Along the floor: $$u^* = -\frac{1}{\pi}\frac{x^*}{1 + x^{*2}} \qquad C_p = -\left(\frac{1}{\pi}\frac{x^*}{1 + x^{*2}}\right)^2 = -u^{*2} \qquad (15)$$

Curves showing u^* and C_p as functions of x^* are plotted in Fig. 10–73.

We see from Fig. 10–73 that u^* increases slowly from 0 at $x^* = -\infty$ to a maximum value of about 0.159 at $x^* = -1$. The velocity is positive (to the right) for negative values of x^* as expected since air is being sucked into the vacuum cleaner. As speed increases, pressure decreases; C_p is 0 at $x = -\infty$ and decreases to its minimum value of about -0.0253 at $x^* = -1$. Between $x^* = -1$ and $x^* = 0$ the speed decreases to zero while the pressure increases to zero at the stagnation point directly below the vacuum cleaner nozzle. To the right of the nozzle (positive values of x^*), the velocity is antisymmetric, while the pressure is symmetric.

The maximum speed (minimum pressure) along the floor occurs at $x^* = \pm 1$, which is the same distance as the height of the nozzle above the

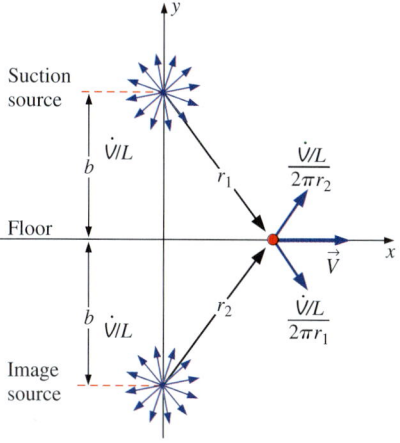

FIGURE 10–72
Vector sum of the velocities induced by the two sources; the resultant velocity is horizontal at any location on the x-axis due to symmetry.

554
APPROXIMATE SOLUTIONS OF THE N–S EQ

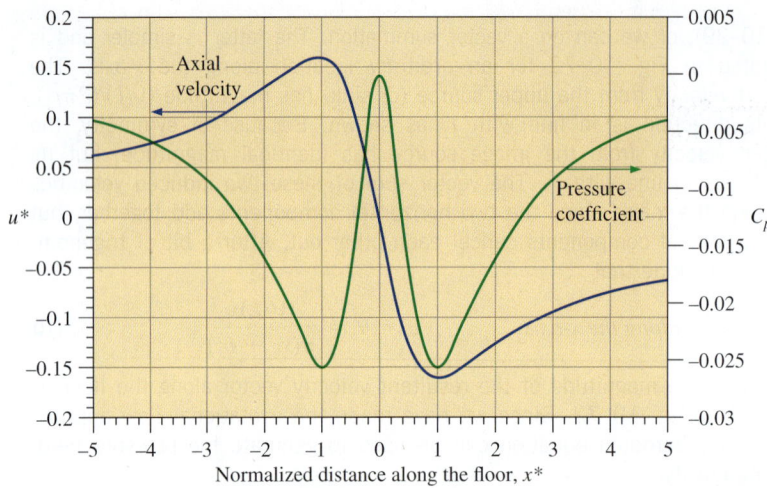

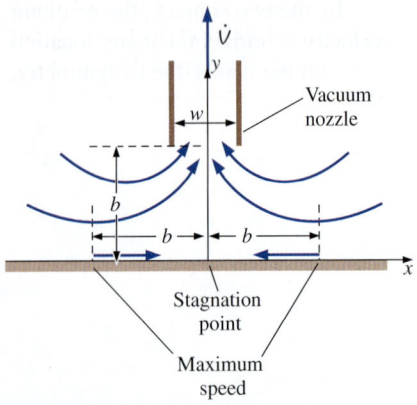

FIGURE 10–73
Nondimensional axial velocity (blue curve) and pressure coefficient (green curve) along the floor below a vacuum cleaner modeled as an irrotational region of flow.

FIGURE 10–74
Based on an irrotational flow approximation, the maximum speed along the floor beneath a vacuum cleaner nozzle occurs at $x = \pm b$. A stagnation point occurs directly below the nozzle.

floor (Fig. 10–74). In dimensional terms, **the maximum speed along the floor occurs at $x = \pm b$**, and the speed there is

Maximum speed along the floor:

$$|u|_{max} = -|u^*|_{max} \frac{\dot{V}/L}{b} = -0.159 \left(\frac{-0.314 \text{ m}^2/\text{s}}{0.020 \text{ m}} \right) = \mathbf{2.50 \text{ m/s}} \quad (16)$$

We expect that the vacuum cleaner is most effective at sucking up dirt from the floor when the speed along the floor is greatest and the pressure along the floor is lowest. Thus, contrary to what you may have thought, *the best performance is* not *directly below the suction inlet, but rather at $x = \pm b$*, as illustrated in Fig. 10–74.

Discussion Notice that we never used the width w of the vacuum nozzle in our analysis, since a line sink has no length scale. You can convince yourself that a vacuum cleaner works best at $x \cong \pm b$ by performing a simple experiment with a vacuum cleaner and some small granular material (like sugar or salt) on a hard floor. It turns out that the irrotational approximation is quite realistic for flow into the inlet of a vacuum cleaner everywhere except very close to the floor, because the flow is rotational there.

We conclude this section by emphasizing that although the irrotational flow approximation is mathematically simple, and velocity and pressure fields are easy to obtain, we must be very careful where we apply it. The irrotational flow approximation breaks down in regions of non-negligible vorticity, especially near solid walls, where fluid particles rotate because of viscous stresses caused by the no-slip condition at the wall. This leads us to the final section in this chapter (Section 10–6) in which we discuss the boundary layer approximation.

10–6 • THE BOUNDARY LAYER APPROXIMATION

As discussed in Sections 10–4 and 10–5, there are at least two flow situations in which the viscous term in the Navier–Stokes equation can be neglected. The first occurs in high Reynolds number regions of flow where net viscous forces are known to be negligible compared to inertial and/or pressure forces; we call these *inviscid regions of flow*. The second situation occurs when the vorticity is negligibly small; we call these *irrotational* or *potential regions of flow*. In either case, removal of the viscous terms from the Navier–Stokes equation yields the Euler equation (Eq. 10–13 and also Eq. 10–25). While the math is greatly simplified by dropping the viscous terms, there are some serious deficiencies associated with application of the Euler equation to practical engineering flow problems. High on the list of deficiencies is the inability to specify the no-slip condition at solid walls. This leads to unphysical results such as zero viscous shear forces on solid walls and zero aerodynamic drag on bodies immersed in a free stream. We can therefore think of the Euler equation and the Navier–Stokes equation as two mountains separated by a huge chasm (Fig. 10–75a). We make the following statement about the boundary layer approximation:

> The boundary layer approximation bridges the gap between the Euler equation and the Navier–Stokes equation, and between the slip condition and the no-slip condition at solid walls (Fig. 10–75b).

From a historical perspective, by the mid-1800s, the Navier–Stokes equation was known, but couldn't be solved except for flows of very simple geometries. Meanwhile, mathematicians were able to obtain beautiful analytical solutions of the Euler equation and of the potential flow equations for flows of complex geometry, but their results were often physically meaningless. Hence, the only reliable way to study fluid flows was empirically, i.e., with experiments. A major breakthrough in fluid mechanics occurred in 1904 when Ludwig Prandtl (1875–1953) introduced the **boundary layer approximation.** Prandtl's idea was to divide the flow into two regions: an **outer flow region** that is inviscid and/or irrotational, and an inner flow region called a **boundary layer**—a very thin region of flow near a solid wall where viscous forces and rotationality cannot be ignored (Fig. 10–76). In the outer flow region, we use the continuity and Euler equations to obtain the outer flow velocity field, and the Bernoulli equation to obtain the pressure field. Alternatively, if the outer flow region is irrotational, we may use the potential flow techniques discussed in Section 10–5 (e.g., superposition) to obtain the outer flow velocity field. In either case, we solve for the outer flow region *first*, and then fit in a thin boundary layer in regions where rotationality and viscous forces cannot be neglected. Within the boundary layer we solve the **boundary layer equations,** to be discussed shortly. (Note that the boundary layer equations are themselves approximations of the full Navier–Stokes equation, as we will see.)

The boundary layer approximation corrects some of the major deficiencies of the Euler equation by providing a way to enforce the no-slip condition at solid walls. Hence, viscous shear forces can exist along walls, bodies immersed in a free stream can experience aerodynamic drag, and flow separation in regions of adverse pressure gradient can be predicted more accurately. The boundary layer concept therefore became the workhorse of engineering fluid mechanics throughout most of the 1900s. However, the advent of fast, inexpensive computers and computational fluid dynamics (CFO) software in the latter part of the twentieth century enabled numerical solution of the Navier–Stokes equation for flows of complex geometry. Today, therefore, it is no longer necessary to split the flow into outer flow regions and boundary layer regions—we can use CFD to solve the

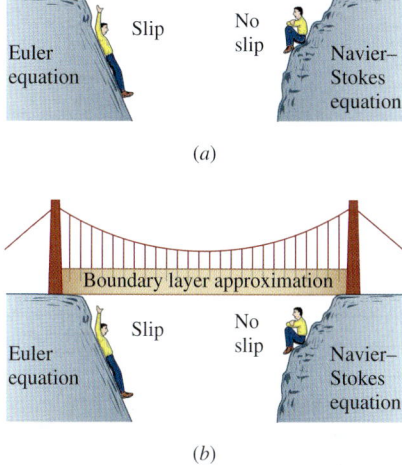

FIGURE 10–75
(*a*) A huge gap exists between the Euler equation (which allows slip at walls) and the Navier–Stokes equation (which supports the no-slip condition); (*b*) the boundary layer approximation bridges that gap.

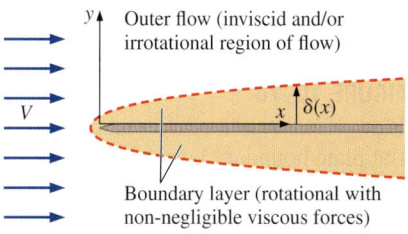

FIGURE 10–76
Prandtl's boundary layer concept splits the flow into an outer flow region and a thin boundary layer region (not to scale).

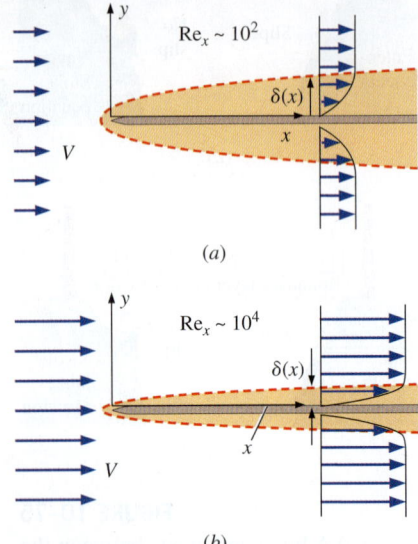

FIGURE 10–77
Flow of a uniform stream parallel to a flat plate (drawings not to scale): (a) $Re_x \sim 10^2$, (b) $Re_x \sim 10^4$. The larger the Reynolds number, the thinner the boundary layer along the plate at a given x-location.

full set of equations of motion (continuity plus Navier–Stokes) throughout the whole flow field. Nevertheless, boundary layer theory is still useful in some engineering applications, since it takes much less time to arrive at a solution. In addition, there is a lot we can learn about the behavior of flowing fluids by studying boundary layers. We stress again that boundary layer solutions are only *approximations* of full Navier–Stokes solutions, and we must be careful where we apply this or any approximation.

The key to successful application of the boundary layer approximation is the assumption that the boundary layer is very *thin*. The classic example is a uniform stream flowing parallel to a long flat plate aligned with the x-axis. **Boundary layer thickness** δ at some location x along the plate is sketched in Fig. 10–77. By convention, δ is usually defined as the distance away from the wall at which the velocity component parallel to the wall is 99 percent of the fluid speed outside the boundary layer. It turns out that for a given fluid and plate, the higher the free-stream speed V, the thinner the boundary layer (Fig. 10–77). In nondimensional terms, we define the Reynolds number based on distance x along the wall,

Reynolds number along a flat plate: $\qquad Re_x = \dfrac{\rho V x}{\mu} = \dfrac{V x}{\nu} \qquad$ (10–60)

Hence,

> At a given x-location, the higher the Reynolds number, the thinner the boundary layer.

In other words, the higher the Reynolds number, all else being equal, the more reliable the boundary layer approximation. We are confident that the boundary layer is thin when $\delta \ll x$ (or, expressed nondimensionally, $\delta/x \ll 1$).

The shape of the boundary layer profile can be obtained experimentally by flow visualization. An example is shown in Fig. 10–78 for a laminar boundary layer on a flat plate. Taken over 60 years ago by F. X. Wortmann, this is now considered a classic photograph of a laminar flat plate boundary layer profile.

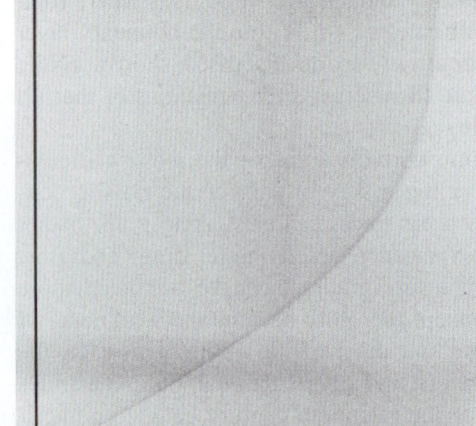

FIGURE 10–78
Flow visualization of a laminar flat plate boundary layer profile. Photograph taken by F. X. Wortmann in 1953 as visualized with the tellurium method. Flow is from left to right, and the leading edge of the flat plate is far to the left of the field of view.

Wortmann, F. X. 1977 AGARD Conf. Proc. no. 224, paper 12.

The no-slip condition is clearly verified at the wall, and the smooth increase in flow speed away from the wall verifies that the flow is indeed laminar.

Note that although we are discussing boundary layers in connection with the thin region near a solid wall, the boundary layer approximation is *not* limited to wall-bounded flow regions. The same equations may be applied to **free shear layers** such as jets, wakes, and mixing layers (Fig. 10–79), provided that the Reynolds number is sufficiently high that these regions are *thin*. The regions of these flow fields with non-negligible viscous forces and finite vorticity can also be considered to be boundary layers, even though a solid wall boundary may not even be present. Boundary layer thickness $\delta(x)$ is labeled in each of the sketches in Fig. 10–79. As you can see, by convention δ is usually defined based on *half* of the total thickness of the free shear layer. We define δ as the distance from the centerline to the edge of the boundary layer where the change in speed is 99 percent of the maximum change in speed from the centerline to the outer flow. Boundary layer thickness is not a constant, but varies with downstream distance x. In the examples discussed here (flat plate, jet, wake, and mixing layer), $\delta(x)$ *increases* with x. There are flow situations however, such as rapidly accelerating outer flow along a wall, in which $\delta(x)$ *decreases* with x.

A common misunderstanding among beginning students of fluid mechanics is that the curve representing δ as a function of x is a *streamline* of the flow—it is *not*! In Fig. 10–80 we sketch both streamlines and $\delta(x)$ for the boundary layer growing on a flat plate. As the boundary layer thickness grows downstream, streamlines passing through the boundary layer must diverge slightly upward in order to satisfy conservation of mass. The amount of this upward displacement is smaller than the growth of $\delta(x)$. Since streamlines *cross* the curve $\delta(x)$, $\delta(x)$ is clearly *not* a streamline (streamlines cannot cross each other or else mass would not be conserved).

For a laminar boundary layer growing on a flat plate, as in Fig. 10–80, boundary layer thickness δ is at most a function of V, x, and fluid properties ρ and μ. It is a simple exercise in dimensional analysis to show that δ/x is a function of Re_x. In fact, it turns out that δ is proportional to the *square root* of Re_x. You must note, however, that this result is valid only for a *laminar* boundary layer on a flat plate. As we move down the plate to larger and larger values of x, Re_x increases linearly with x. At some point, infinitesimal disturbances in the flow begin to grow, and the boundary layer cannot remain laminar—it begins a **transition** process toward turbulent flow. For a smooth flat plate with a uniform free stream, the transition process begins at a **critical Reynolds number**, $\text{Re}_{x,\,\text{critical}} \cong 1 \times 10^5$, and continues until the boundary layer is fully turbulent at the **transition Reynolds number**, $\text{Re}_{x,\,\text{transition}} \cong 3 \times 10^6$ (Fig. 10–81). The transition process is quite complicated, and details are beyond the scope of this text.

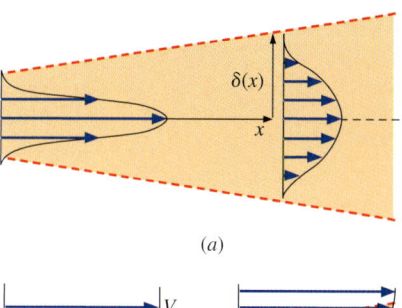
(a)

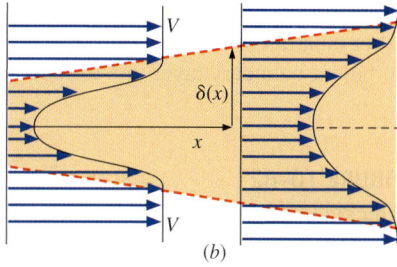
(b)

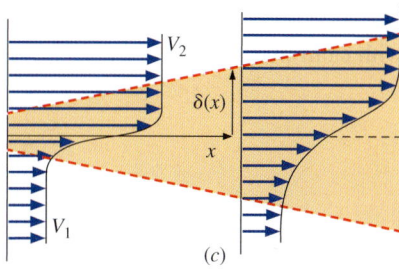
(c)

FIGURE 10–79
Three additional flow regions where the boundary layer approximation may be appropriate: (*a*) jets, (*b*) wakes, and (*c*) mixing layers.

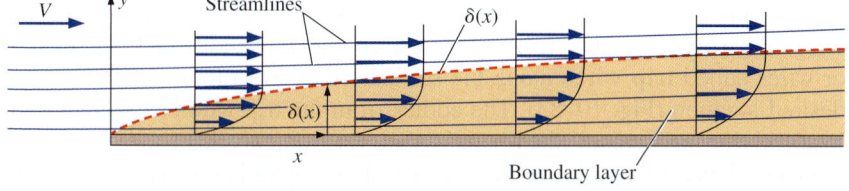

FIGURE 10–80
Comparison of streamlines and the curve representing δ as a function of x for a flat plate boundary layer. Since streamlines cross the curve $\delta(x)$, $\delta(x)$ cannot itself be a streamline of the flow.

558
APPROXIMATE SOLUTIONS OF THE N–S EQ

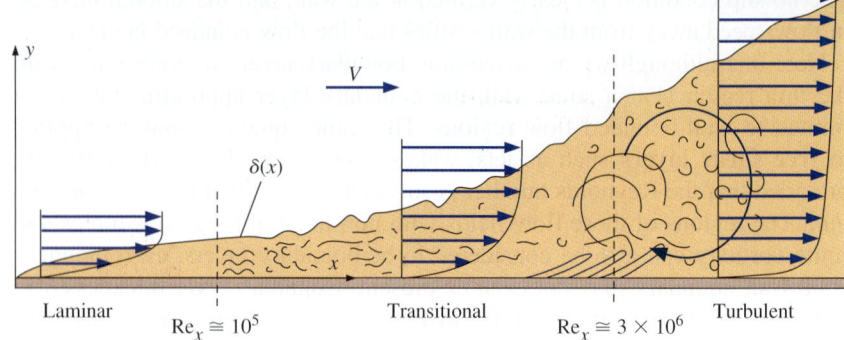

FIGURE 10–81
Transition of the laminar boundary layer on a flat plate into a fully turbulent boundary layer (not to scale).

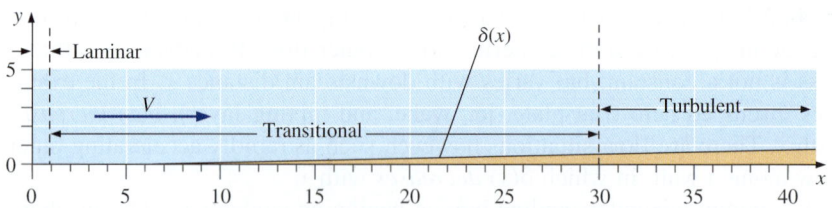

FIGURE 10–82
Thickness of the boundary layer on a flat plate, drawn to scale. Laminar, transitional, and turbulent regions are indicated for the case of a smooth wall with calm free-stream conditions.

Note that in Fig. 10–81 the vertical scale has been greatly exaggerated, and the horizontal scale has been shortened (in reality, since $\text{Re}_{x,\,\text{transition}} \cong$ 30 times $\text{Re}_{x,\,\text{critical}}$, the transitional region is much longer than indicated in the figure). To give you a better feel for how thin a boundary layer actually is, we have plotted δ as a function of *x to scale* in Fig. 10–82. To generate the plot, we carefully selected the parameters such that $\text{Re}_x = 100{,}000x$ regardless of the units of x. Thus, $\text{Re}_{x,\,\text{transition}}$ occurs at $x \cong 1$ and $\text{Re}_{x,\,\text{critical}}$ occurs at $x \cong 30$ in the plot. Notice how thin the boundary layer is and how long the transitional region is when plotted to scale.

In real-life engineering flows, transition to turbulent flow usually occurs more abruptly and much earlier (at a lower value of Re_x) than the values given for a smooth flat plate with a calm free stream. Factors such as roughness along the surface, free-stream disturbances, acoustic noise, flow unsteadiness, vibrations, and curvature of the wall contribute to an earlier transition location. Because of this, an *engineering critical Reynolds number* of $\text{Re}_{x,\,\text{cr}} = 5 \times 10^5$ is often used to determine whether a boundary layer is most likely laminar ($\text{Re}_x < \text{Re}_{x,\,\text{cr}}$) or most likely turbulent ($\text{Re}_x > \text{Re}_{x,\,\text{cr}}$). It is also common in heat transfer to use this value as the critical Re; in fact, relations for average friction and heat transfer coefficients are derived by assuming the flow to be laminar for Re_x lower than $\text{Re}_{x,\,\text{cr}}$, and turbulent otherwise. The logic here is to ignore transition by treating the first part of transition as laminar and the remaining part as turbulent. We follow this convention throughout the rest of the book unless noted otherwise.

The transition process is *unsteady* as well and is difficult to predict, even with modern CFD codes. In some cases, engineers install rough sandpaper or wires called **trip wires** along the surface, in order to force transition at a desired location (Fig. 10–83). The eddies from the trip wire cause enhanced local mixing and create disturbances that very quickly lead to a turbulent boundary layer. Again, the vertical scale in Fig. 10–83 is greatly exaggerated for illustrative purposes.

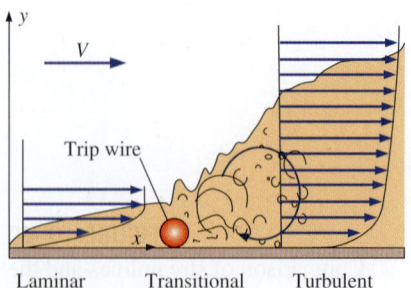

FIGURE 10–83
A trip wire is often used to initiate early transition to turbulence in a boundary layer (not to scale).

EXAMPLE 10–9 Laminar or Turbulent Boundary Layer?

Water flows over the fin of a small underwater vehicle at a speed of $V = 10$ km/h (Fig. 10–84). The temperature of the water is 5°C, and the chord length c of the fin is 0.5 m. Is the boundary layer on the surface of the fin laminar or turbulent or transitional?

SOLUTION We are to assess whether the boundary layer on the surface of a fin is laminar or turbulent or transitional.
Assumptions 1 The flow is steady and incompressible. 2 The fin surface is smooth.
Properties The density and viscosity of water at $T = 5$°C are 999.9 kg/m³ and 1.519×10^{-3} kg/m·s, respectively. The kinematic viscosity is thus $\nu = 1.519 \times 10^{-6}$ m²/s.
Analysis Although the fin is not a flat-plate, the flat plate boundary-layer values are useful as a reasonable first approximation to determine whether the boundary layer is laminar or turbulent. We calculate the Reynolds number at the trailing edge of the fin, using c as the approximate streamwise distance along the flat plate,

$$\text{Re}_x = \frac{Vx}{\nu} = \frac{(10 \text{ km/h})(0.5 \text{ m})}{1.519 \times 10^{-6} \text{ m}^2/\text{s}} \left(\frac{1000 \text{ m}}{\text{km}}\right)\left(\frac{\text{h}}{3600 \text{ s}}\right) = 9.14 \times 10^5 \quad (1)$$

The critical Reynolds number for transition to turbulence is 1×10^5 for the case of a smooth flat plate with very clean, low-noise free-stream conditions. Our Reynolds number is higher than this. The engineering value of the critical Reynolds number for real engineering flows is $\text{Re}_{x,cr} = 5 \times 10^5$. Since Re_x is greater than $\text{Re}_{x,cr}$, but less than $\text{Re}_{x,\text{transition}}$ (30×10^5), **the boundary layer is most likely transitional, but may be fully turbulent by the trailing edge of the fin.**
Discussion In a real-life situation, the free-stream flow is not very "clean"—there are eddies and other disturbances, the fin surface is not perfectly smooth, and the vehicle may be vibrating. Thus, transition and turbulence are likely to occur much earlier than predicted for a smooth flat plate.

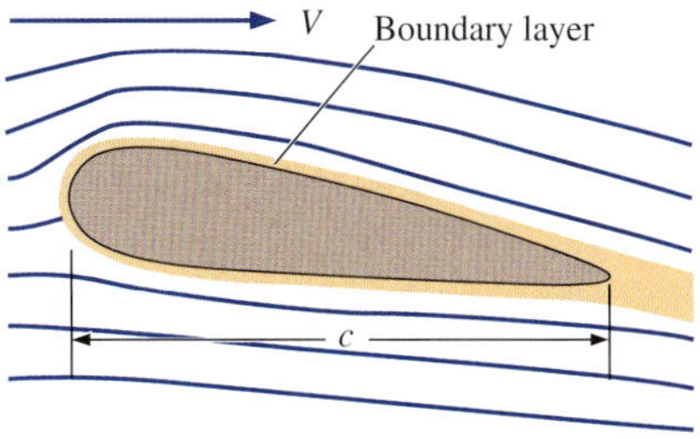

FIGURE 10–84
Boundary layer growing along the fin of an underwater vehicle. Boundary layer thickness is exaggerated for clarity.

The Boundary Layer Equations

Now that we have a physical feel for boundary layers, we need the equations of motion to be used in boundary layer calculations—the **boundary layer equations.** For simplicity we consider only steady, two-dimensional flow in the xy-plane in Cartesian coordinates. The methodology used here can be extended, however, to axisymmetric boundary layers or to three-dimensional boundary layers in any coordinate system. We neglect gravity since we are not dealing with free surfaces or with buoyancy-driven flows (free convection flows), where gravitational effects dominate. We consider only *laminar* boundary layers; turbulent boundary layer equations are beyond the scope of this text. For the case of a boundary layer along a solid wall, we adopt a coordinate system in which x is everywhere parallel to the wall and y is everywhere normal to the wall (Fig. 10–85). This coordinate system is called a **boundary layer coordinate system.** When we solve the boundary layer equations, we do so at one x-location at a time, using this coordinate system *locally*, and it is *locally orthogonal*. It is not critical where we define $x = 0$, but for flow over a body, as in Fig. 10–85, we typically set $x = 0$ at the front stagnation point.

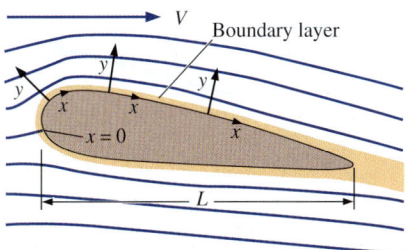

FIGURE 10–85
The boundary layer coordinate system for flow over a body; x follows the surface and is typically set to zero at the front stagnation point of the body, and y is everywhere normal to the surface locally.

We begin with the nondimensionalized Navier–Stokes equation derived at the beginning of this chapter. With the unsteady term and the gravity term neglected, Eq. 10–6 becomes

$$(\vec{V}^* \cdot \vec{\nabla}^*)\vec{V}^* = -[\text{Eu}]\vec{\nabla}^* P^* + \left[\frac{1}{\text{Re}}\right]\nabla^{*2}\vec{V}^* \quad (10\text{–}61)$$

The Euler number is of order unity, since pressure differences outside the boundary layer are determined by the Bernoulli equation and $\Delta P = P - P_\infty \sim \rho V^2$. We note that V is a characteristic velocity scale of the outer flow, typically equal to the free-stream velocity for bodies immersed in a uniform flow. The characteristic length scale used in this nondimensionalization is L, some characteristic size of the body. For boundary layers, x is of order of magnitude L, and the Reynolds number in Eq. 10–61 can be thought of as Re_x (Eq. 10–60). Re_x is very large in typical applications of the boundary layer approximation. It would seem then that we could neglect the last term in Eq. 10–61 in boundary layers. However, doing so would result in the Euler equation, along with all its deficiencies discussed previously. So, we must keep at least *some* of the viscous terms in Eq. 10–61.

How do we decide which terms to keep and which to neglect? To answer this question, we redo the nondimensionalization of the equations of motion based on appropriate length and velocity scales within the boundary layer. A magnified view of a portion of the boundary layer of Fig. 10–85 is sketched in Fig. 10–86. Since the order of magnitude of x is L, we use L as an appropriate length scale for distances in the streamwise direction and for derivatives of velocity and pressure with respect to x. However, this length scale is much too large for derivatives with respect to y. It makes more sense to use δ as the length scale for distances in the direction normal to the streamwise direction and for derivatives with respect to y. Similarly, while the characteristic velocity scale is V for the whole flow field, it is more appropriate to use U as the characteristic velocity scale for boundary layers, where *U is the magnitude of the velocity component parallel to the wall at a location just above the boundary layer* (Fig. 10–86). U is in general a function of x. Thus, within the boundary layer at some value of x, the orders of magnitude are

$$u \sim U \quad P - P_\infty \sim \rho U^2 \quad \frac{\partial}{\partial x} \sim \frac{1}{L} \quad \frac{\partial}{\partial y} \sim \frac{1}{\delta} \quad (10\text{–}62)$$

The order of magnitude of velocity component v is not specified in Eq. 10–62, but is instead obtained from the continuity equation. Applying the orders of magnitude in Eq. 10–62 to the incompressible continuity equation in two dimensions,

$$\underbrace{\frac{\partial u}{\partial x}}_{\sim U/L} + \underbrace{\frac{\partial v}{\partial y}}_{\sim v/\delta} = 0 \quad \rightarrow \quad \frac{U}{L} \sim \frac{v}{\delta}$$

Since the two terms have to balance each other, they must be of the same order of magnitude. Thus we obtain the order of magnitude of velocity component v,

$$v \sim \frac{U\delta}{L} \quad (10\text{–}63)$$

Since $\delta/L \ll 1$ in a boundary layer (the boundary layer is very thin), we conclude that $v \ll u$ in a boundary layer (Fig. 10–87). From Eqs. 10–62

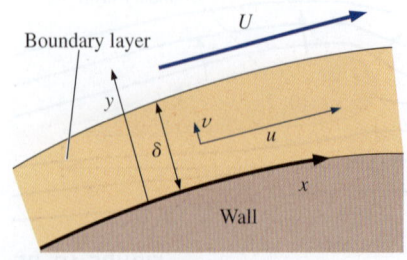

FIGURE 10–87
Highly magnified view of the boundary layer along the surface of a body, showing that velocity component v is much smaller than u.

and 10–63, we define the following nondimensional variables within the boundary layer:

$$x^* = \frac{x}{L} \quad y^* = \frac{y}{\delta} \quad u^* = \frac{u}{U} \quad v^* = \frac{vL}{U\delta} \quad P^* = \frac{P - P_\infty}{\rho U^2}$$

Since we used appropriate scales, all these nondimensional variables are of order unity—i.e., they are *normalized* variables (Chap. 7).

We now consider the *x*- and *y*-components of the Navier–Stokes equation. We substitute these nondimensional variables into the *y*-momentum equation, giving

$$\underbrace{u}_{u^*U}\underbrace{\frac{\partial v}{\partial x}}_{\frac{\partial}{\partial x^*}\frac{v^*U\delta}{L^2}} + \underbrace{v}_{v^*\frac{U\delta}{L}}\underbrace{\frac{\partial v}{\partial y}}_{\frac{\partial}{\partial y^*}\frac{v^*U\delta}{L\delta}} = \underbrace{-\frac{1}{\rho}\frac{\partial P}{\partial y}}_{\frac{1}{\rho}\frac{\partial}{\partial y^*}\frac{P^*\rho U^2}{\delta}} + \underbrace{\nu\frac{\partial^2 v}{\partial x^2}}_{\nu\frac{\partial^2}{\partial x^{*2}}\frac{v^*U\delta}{L^3}} + \underbrace{\nu\frac{\partial^2 v}{\partial y^2}}_{\nu\frac{\partial^2}{\partial y^{*2}}\frac{v^*U\delta}{L\delta^2}}$$

After some algebra and after multiplying each term by $L^2/(U^2\delta)$, we get

$$u^*\frac{\partial v^*}{\partial x^*} + v^*\frac{\partial v^*}{\partial y^*} = -\left(\frac{L}{\delta}\right)^2\frac{\partial P^*}{\partial y^*} + \left(\frac{\nu}{UL}\right)\frac{\partial^2 v^*}{\partial x^{*2}} + \left(\frac{\nu}{UL}\right)\left(\frac{L}{\delta}\right)^2\frac{\partial^2 v^*}{\partial y^{*2}} \quad \textbf{(10–64)}$$

Comparing terms in Eq. 10–64, the middle term on the right side is clearly orders of magnitude smaller than any other term since $Re_L = UL/\nu \gg 1$. For the same reason, the last term on the right is much smaller than the first term on the right. Neglecting these two terms leaves the two terms on the left and the first term on the right. However, since $L \gg \delta$, the pressure gradient term is orders of magnitude greater than the advective terms on the left side of the equation. Thus, the only term left in Eq. 10–64 is the pressure term. Since no other term in the equation can balance that term, we have no choice but to set it equal to zero. Thus, the nondimensional *y*-momentum equation reduces to

$$\frac{\partial P^*}{\partial y^*} \cong 0$$

or, in terms of the physical variables,

Normal pressure gradient through a boundary layer: $\quad \dfrac{\partial P}{\partial y} \cong 0 \quad$ **(10–65)**

In words, although pressure may vary *along* the wall (in the *x*-direction), there is negligible change in pressure in the direction *normal* to the wall. This is illustrated in Fig. 10–88. At $x = x_1$, $P = P_1$ at all values of *y* across the boundary layer from the wall to the outer flow. At some other *x*-location, $x = x_2$, the pressure may have changed, but $P = P_2$ at all values of *y* across that portion of the boundary layer.

The pressure across a boundary layer (*y*-direction) is nearly constant.

Physically, because the boundary layer is so thin, streamlines within the boundary layer have negligible *curvature* when observed at the scale of the boundary layer thickness. Curved streamlines require a *centripetal acceleration*, which comes from a pressure gradient along the radius of curvature. Since the streamlines are not significantly curved in a thin boundary layer, there is no significant pressure gradient across the boundary layer.

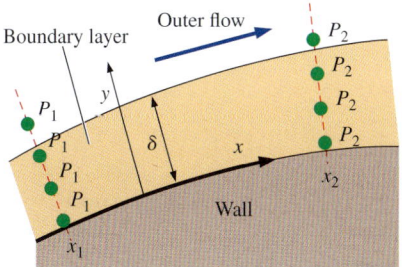

FIGURE 10–88
Pressure may change *along* a boundary layer (*x*-direction), but the change in pressure *across* a boundary layer (*y*-direction) is negligible.

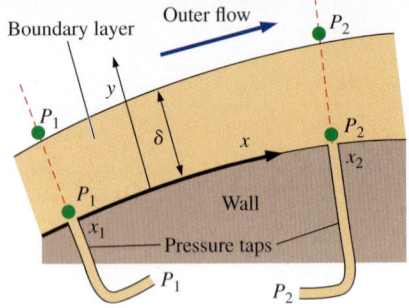

FIGURE 10–89
The pressure in the irrotational region of flow outside of a boundary layer can be measured by static pressure taps in the surface of the wall. Two such pressure taps are sketched.

One immediate consequence of Eq. 10–65 and the statement just presented is that at any x-location along the wall, the pressure at the outer edge of the boundary layer ($y \cong \delta$) is the *same* as that at the wall ($y = 0$). This leads to a tremendous practical application; namely, the pressure at the outer edge of a boundary layer can be measured experimentally by a static pressure tap *at the wall* directly beneath the boundary layer (Fig. 10–89). Experimentalists routinely take advantage of this fortunate situation, and countless airfoil shapes for airplane wings and turbomachinery blades were tested with such pressure taps over the past century.

The experimental pressure data shown in Fig. 10–64 for flow over a circular cylinder were measured with pressure taps at the cylinder's surface, yet they are used to compare with the pressure calculated by the irrotational outer flow approximation. Such a comparison is valid, because the pressure obtained *outside* of the boundary layer (from the Euler equation or potential flow analysis coupled with the Bernoulli equation) applies all the way through the boundary layer to the wall.

Returning to the development of the boundary layer equations, we use Eq. 10–65 to greatly simplify the x-component of the momentum equation. Specifically, since P is not a function of y, we replace $\partial P/\partial x$ by dP/dx, where P is the value of pressure calculated from our outer flow approximation (using either continuity plus Euler, or the potential flow equations plus Bernoulli). The x-component of the Navier–Stokes equation becomes

$$\underbrace{u}_{u^*U} \underbrace{\frac{\partial u}{\partial x}}_{\frac{\partial}{\partial x^*}\frac{u^*U}{L}} + \underbrace{v}_{v^*\frac{U\delta}{L}} \underbrace{\frac{\partial u}{\partial y}}_{\frac{\partial}{\partial y^*}\frac{u^*U}{\delta}} = \underbrace{-\frac{1}{\rho}\frac{dP}{dx}}_{\frac{1}{\rho}\frac{\partial}{\partial x^*}\frac{P^*\rho U^2}{L}} + \underbrace{\nu\frac{\partial^2 u}{\partial x^2}}_{\nu\frac{\partial^2}{\partial x^{*2}}\frac{u^*U}{L^2}} + \underbrace{\nu\frac{\partial^2 u}{\partial y^2}}_{\nu\frac{\partial^2}{\partial y^{*2}}\frac{u^*U}{\delta^2}}$$

After some algebra, and after multiplying each term by L/U^2, we get

$$u^*\frac{\partial u^*}{\partial x^*} + v^*\frac{\partial u^*}{\partial y^*} = -\frac{dP^*}{dx^*} + \left(\frac{\nu}{UL}\right)\frac{\partial^2 u^*}{\partial x^{*2}} + \left(\frac{\nu}{UL}\right)\left(\frac{L}{\delta}\right)^2 \frac{\partial^2 u^*}{\partial y^{*2}} \quad (10\text{–}66)$$

Comparing terms in Eq. 10–66, the middle term on the right side is clearly orders of magnitude smaller than the terms on the left side, since $\text{Re}_L = UL/\nu \gg 1$. What about the last term on the right? If we neglect this term, we throw out all the viscous terms and are back to the Euler equation. Clearly this term must remain. Furthermore, since all the remaining terms in Eq. 10–66 are of order unity, the combination of parameters in parentheses in the last term on the right side of Eq. 10–66 must also be of order unity,

$$\left(\frac{\nu}{UL}\right)\left(\frac{L}{\delta}\right)^2 \sim 1$$

Again recognizing that $\text{Re}_L = UL/\nu$, we see immediately that

$$\frac{\delta}{L} \sim \frac{1}{\sqrt{\text{Re}_L}} \quad (10\text{–}67)$$

This confirms our previous statement that at a given streamwise location along the wall, the larger the Reynolds number, the thinner the boundary layer. If we substitute x for L in Eq. 10–67, we also conclude that for a laminar boundary layer on a flat plate, where $U(x) = V = \text{constant}$, δ grows like the square root of x (Fig. 10–90).

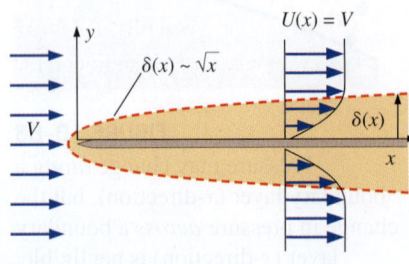

FIGURE 10–90
An order-of-magnitude analysis of the laminar boundary layer equations along a flat plate reveals that δ grows like $\sqrt{x}$ (not to scale).

In terms of the original (physical) variables, Eq. 10–66 is written as

x-momentum boundary layer equation: $u\dfrac{\partial u}{\partial x} + v\dfrac{\partial u}{\partial y} = -\dfrac{1}{\rho}\dfrac{dP}{dx} + \nu\dfrac{\partial^2 u}{\partial y^2}$ (10–68)

Note that the last term in Eq. 10–68 is not negligible in the boundary layer, since the y-derivative of velocity gradient $\partial u/\partial y$ is sufficiently large to offset the (typically small) value of kinematic viscosity ν. Finally, since we know from our y-momentum equation analysis that the pressure across the boundary layer is the same as that outside the boundary layer (Eq. 10–65), we apply the Bernoulli equation to the outer flow region. Differentiating with respect to x we get

$$\dfrac{P}{\rho} + \dfrac{1}{2}U^2 = \text{constant} \quad \rightarrow \quad \dfrac{1}{\rho}\dfrac{dP}{dx} = -U\dfrac{dU}{dx} \quad (10\text{–}69)$$

where we note that both P and U are functions of x only, as illustrated in Fig. 10–91. Substitution of Eq. 10–69 into Eq. 10–68 yields

$$u\dfrac{\partial u}{\partial x} + v\dfrac{\partial u}{\partial y} = U\dfrac{dU}{dx} + \nu\dfrac{\partial^2 u}{\partial y^2} \quad (10\text{–}70)$$

and we have eliminated pressure from the boundary layer equations.

We summarize the set of equations of motion for a steady, incompressible, laminar boundary layer in the *xy*-plane without significant gravitational effects,

Boundary layer equations:
$$\begin{aligned}\dfrac{\partial u}{\partial x} + \dfrac{\partial v}{\partial y} &= 0 \\ u\dfrac{\partial u}{\partial x} + v\dfrac{\partial u}{\partial y} &= U\dfrac{dU}{dx} + \nu\dfrac{\partial^2 u}{\partial y^2}\end{aligned} \quad (10\text{–}71)$$

Mathematically, the full Navier–Stokes equation is **elliptic** in space, which means that boundary conditions are required over the entire boundary of the flow domain. Physically, flow information is passed in all directions, both upstream and downstream. On the other hand, the x-momentum boundary layer equation (the second equation of Eq. 10–71) is **parabolic.** This means that we need to specify boundary conditions on only three sides of the (two-dimensional) flow domain. Physically, flow information is not passed in the direction opposite to the flow (from downstream). This fact greatly reduces the level of difficulty in solving the boundary layer equations. Specifically, we don't need to specify boundary conditions *downstream*, only upstream and on the top and bottom of the flow domain (Fig. 10–92). For a typical boundary layer problem along a wall, we specify the no-slip condition at the wall ($u = v = 0$ at $y = 0$), the outer flow condition at the edge of the boundary layer and beyond [$u = U(x)$ as $y \rightarrow \infty$], and a starting profile at some upstream location [$u = u_{\text{starting}}(y)$ at $x = x_{\text{starting}}$, where x_{starting} may or may not be zero]. With these boundary conditions, we simply march downstream in the x-direction, solving the boundary layer equations as we go. This is particularly attractive for numerical boundary layer computations, because once we know the profile at one x-location (x_i), we can march to the next x-location (x_{i+1}), and then use this newly calculated profile as the starting profile to march to the next x-location (x_{i+2}), etc.

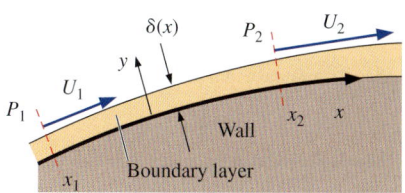

FIGURE 10–91

Outer flow speed parallel to the wall is $U(x)$ and is obtained from the outer flow pressure, $P(x)$. This speed appears in the x-component of the boundary layer momentum equation, Eq. 10–70.

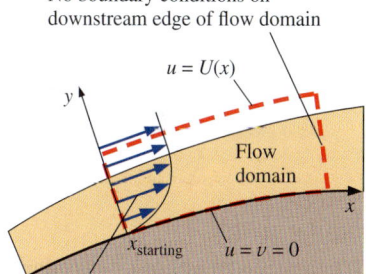

FIGURE 10–92

The boundary layer equation set is parabolic, so boundary conditions need to be specified on only three sides of the flow domain.

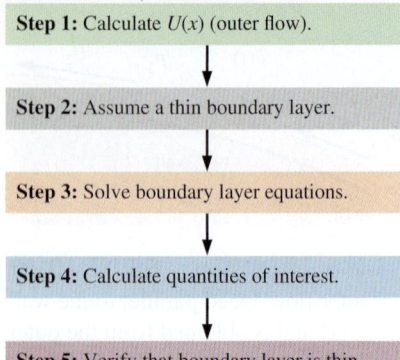

FIGURE 10–93
Summary of the boundary layer procedure for steady, incompressible, two-dimensional boundary layers in the *xy*-plane.

The Boundary Layer Procedure

When the boundary layer approximation is employed, we use a general step-by-step procedure. We outline the procedure here and in condensed form in Fig. 10–93.

Step 1 Solve for the outer flow, ignoring the boundary layer (assuming that the region of flow outside the boundary layer is approximately inviscid and/or irrotational). Transform coordinates as necessary to obtain $U(x)$.

Step 2 Assume a thin boundary layer—so thin, in fact, that it does not affect the outer flow solution of step 1.

Step 3 Solve the boundary layer equations (Eqs. 10–71), using appropriate boundary conditions: the no-slip boundary condition at the wall, $u = v = 0$ at $y = 0$; the known outer flow condition at the edge of the boundary layer, $u \to U(x)$ as $y \to \infty$; and some known starting profile, $u = u_{starting}(y)$ at $x = x_{starting}$.

Step 4 Calculate quantities of interest in the flow field. For example, once the boundary layer equations have been solved (step 3), we calculate $\delta(x)$, shear stress along the wall, total skin friction drag, etc.

Step 5 Verify that the boundary layer approximations are appropriate. In other words, verify that the boundary layer is *thin*—otherwise the approximation is not justified.

Before we do any examples, we list here some of the limitations of the boundary layer approximation. These are *red flags* to look for when performing boundary layer calculations:

- The boundary layer approximation breaks down if the Reynolds number is not large enough. How large is large enough? It depends on the desired accuracy of the approximation. Using Eq. 10–67 as a guideline, $\delta/L \sim 0.03$ (3 percent) for $Re_L = 1000$, and $\delta/L \sim 0.01$ (1 percent) for $Re_L = 10,000$.
- The assumption of zero pressure gradient in the *y*-direction (Eq. 10–65) breaks down if the wall curvature is of similar magnitude as δ (Fig. 10–94). In such cases, centripetal acceleration effects due to streamline curvature cannot be ignored. Physically, the boundary layer is not thin enough for the approximation to be appropriate when δ is not $\ll R$.
- When the Reynolds number is too *high*, the boundary layer does not remain laminar, as discussed previously. The boundary layer approximation itself may still be appropriate, but Eqs. 10–71 are *not* valid if the flow is transitional or fully turbulent. As noted before, the laminar boundary layer on a smooth flat plate under clean flow conditions begins to transition toward turbulence at $Re_x \cong 1 \times 10^5$. In practical engineering applications, walls may not be smooth and there may be vibrations, noise, and fluctuations in the free-stream flow above the wall, all of which contribute to an even earlier start of the transition process.
- If flow separation occurs, the boundary layer approximation is no longer appropriate in the separated flow region. The main reason for this is that a separated flow region contains *reverse flow*, and the parabolic nature of the boundary layer equations is lost.

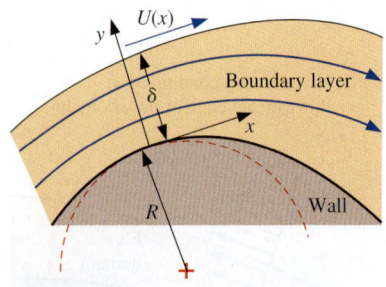

FIGURE 10–94
When the local radius of curvature of the wall (R) is small enough to be of the same magnitude as δ, centripetal acceleration effects cannot be ignored and $\partial P/\partial y \neq 0$. The thin boundary layer approximation is not appropriate in such regions.

EXAMPLE 10-10 Laminar Boundary Layer on a Flat Plate

A uniform free stream of speed V flows parallel to an infinitesimally thin semi-infinite flat plate as sketched in Fig. 10–95. The coordinate system is defined such that the plate begins at the origin. Since the flow is symmetric about the x-axis, only the upper half of the flow is considered. Calculate the boundary layer velocity profile along the plate and discuss.

SOLUTION We are to calculate the boundary layer velocity profile (u as a function of x and y) as the laminar boundary layer grows along the flat plate.
Assumptions **1** The flow is steady, incompressible, and two-dimensional in the xy-plane. **2** The Reynolds number is high enough that the boundary layer approximation is reasonable. **3** The boundary layer remains laminar over the range of interest.
Analysis We follow the step-by-step procedure outlined in Fig. 10–93.
 Step 1 The outer flow is obtained by ignoring the boundary layer altogether, since it is assumed to be very, very thin. Recall that any streamline in an irrotational flow can be thought of as a wall since there is no flow *through* a streamline. In this case, the x-axis can be thought of as a streamline of uniform free-stream flow, one of our building block flows in Section 10–5; this streamline can also be thought of as an infinitesimally thin plate (Fig. 10–96). Thus,

Outer flow: $\qquad U(x) = V = \text{constant} \qquad (1)$

For convenience, we use U instead of $U(x)$ from here on, since it is a constant.
 Step 2 We assume a very thin boundary layer along the wall (Fig. 10–97). The key here is that the boundary layer is so thin that it has negligible effect on the outer flow calculated in step 1.
 Step 3 We must now solve the boundary layer equations. We see from Eq. 1 that $dU/dx = 0$; in other words, no pressure gradient term remains in the x-momentum boundary layer equation. This is why the boundary layer on a flat plate is often called a **zero pressure gradient boundary layer.** The continuity and x-momentum equations for the boundary layer (Eqs. 10–71) become

$$\frac{\partial u}{\partial x} + \frac{\partial v}{\partial y} = 0 \qquad u\frac{\partial u}{\partial x} + v\frac{\partial u}{\partial y} = \nu\frac{\partial^2 u}{\partial y^2} \qquad (2)$$

There are four required boundary conditions,

$$u = 0 \text{ at } y = 0 \qquad u = U \text{ as } y \to \infty$$
$$v = 0 \text{ at } y = 0 \qquad u = U \text{ for all } y \text{ at } x = 0 \qquad (3)$$

The last of the boundary conditions in Eq. 3 is the starting profile; we assume that the plate has not yet influenced the flow at the starting location of the plate ($x = 0$).
 These equations and boundary conditions seem simple enough, but unfortunately *no convenient analytical solution is available*. However, a series solution of Eqs. 2 was obtained in 1908 by P. R. Heinrich Blasius (1883–1970). As a side note, Blasius was a Ph.D. student of Prandtl. In those days, of course, computers were not yet available, and all the calculations were performed *by hand*. Today we can solve these equations

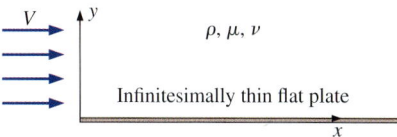

FIGURE 10–95
Setup for Example 10–10; flow of a uniform stream parallel to a semi-infinite flat plate along the x-axis.

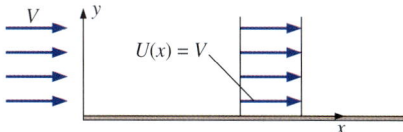

FIGURE 10–96
The outer flow of Example 10–10 is trivial since the x-axis is a streamline of the flow, and $U(x) = V = $ constant.

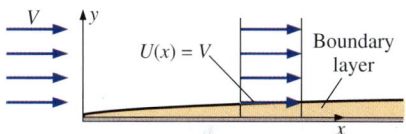

FIGURE 10–97
The boundary layer is so thin that it does not affect the outer flow; boundary layer thickness is exaggerated here for clarity.

APPROXIMATE SOLUTIONS OF THE N–S EQ

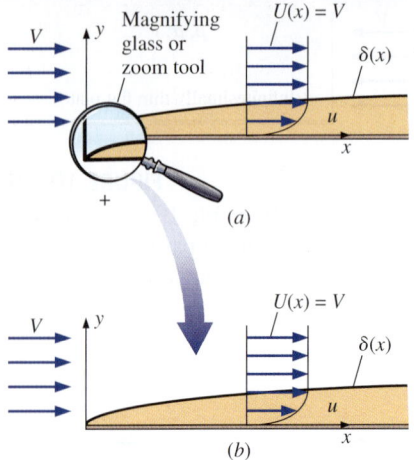

FIGURE 10–98
A useful result of the similarity assumption is that the flow looks the same (is *similar*) regardless of how far we zoom in or out; (*a*) view from a distance, as a person might see, (*b*) close-up view, as an ant might see.

on a computer in a few seconds. The key to the solution is the assumption of **similarity.** In simple terms, similarity can be assumed here because there is *no characteristic length scale* in the geometry of the problem. Physically, since the plate is infinitely long in the *x*-direction, we always see the *same* flow pattern no matter how much we zoom in or zoom out (Fig. 10–98).

Blasius introduced a **similarity variable** η that combines independent variables *x* and *y* into one nondimensional independent variable,

$$\eta = y\sqrt{\frac{U}{\nu x}} \qquad (4)$$

and he solved for a nondimensionalized form of the *x*-component of velocity,

$$f' = \frac{u}{U} = \text{function of } \eta \qquad (5)$$

When we substitute Eqs. 4 and 5 into Eqs. 2, subjected to the boundary conditions of Eq. 3, we get an ordinary differential equation for nondimensional speed $f'(\eta) = u/U$ as a function of similarity variable η. We use the popular Runge–Kutta numerical technique to obtain the results shown in Table 10–3 and in Fig. 10–99. Details of the numerical technique are beyond the scope of this text (see Heinsohn and Cimbala, 2003). There is also a small *y*-component of velocity v away from the wall, but $v \ll u$, and is not discussed here. The beauty of the similarity solution is that this one unique velocity profile shape applies to *any x*-location when plotted in similarity variables, as in Fig. 10–99. The agreement of the calculated profile shape in Fig. 10–99 to experimentally obtained data (circles in Fig. 10–99) and to the visualized profile shape of Fig. 10–78 is remarkable. The Blasius solution is a stunning success.

TABLE 10–3
Solution of the Blasius laminar flat plate boundary layer in similarity variables*

η	f''	f'	f	η	f''	f'	f
0.0	0.33206	0.00000	0.00000	2.4	0.22809	0.72898	0.92229
0.1	0.33205	0.03321	0.00166	2.6	0.20645	0.77245	1.07250
0.2	0.33198	0.06641	0.00664	2.8	0.18401	0.81151	1.23098
0.3	0.33181	0.09960	0.01494	3.0	0.16136	0.84604	1.39681
0.4	0.33147	0.13276	0.02656	3.5	0.10777	0.91304	1.83770
0.5	0.33091	0.16589	0.04149	4.0	0.06423	0.95552	2.30574
0.6	0.33008	0.19894	0.05973	4.5	0.03398	0.97951	2.79013
0.8	0.32739	0.26471	0.10611	5.0	0.01591	0.99154	3.28327
1.0	0.32301	0.32978	0.16557	5.5	0.00658	0.99688	3.78057
1.2	0.31659	0.39378	0.23795	6.0	0.00240	0.99897	4.27962
1.4	0.30787	0.45626	0.32298	6.5	0.00077	0.99970	4.77932
1.6	0.29666	0.51676	0.42032	7.0	0.00022	0.99992	5.27923
1.8	0.28293	0.57476	0.52952	8.0	0.00001	1.00000	6.27921
2.0	0.26675	0.62977	0.65002	9.0	0.00000	1.00000	7.27921
2.2	0.24835	0.68131	0.78119	10.0	0.00000	1.00000	8.27921

* η is the similarity variable defined in Eq. 4 above, and function $f(\eta)$ is solved using the Runge–Kutta numerical technique. Note that f'' is proportional to the shear stress τ, f' is proportional to the *x*-component of velocity in the boundary layer ($f' = u/U$), and f itself is proportional to the stream function. f' is plotted as a function of η in Fig. 10–99.

Step 4 We next calculate several quantities of interest in this boundary layer. First, based on a numerical solution with finer resolution than that shown in Table 10-3, we find that $u/U = 0.990$ at $\eta \cong 4.91$. This 99 percent boundary layer thickness is sketched in Fig. 10-99. Using Eq. 4 and the definition of δ, we conclude that $y = \delta$ when

$$\eta = 4.91 = \sqrt{\frac{U}{\nu x}} \delta \quad \rightarrow \quad \frac{\delta}{x} = \frac{4.91}{\sqrt{Re_x}} \quad (6)$$

This result agrees qualitatively with Eq. 10-67, obtained from a simple order-of-magnitude analysis. The constant 4.91 in Eq. 6 is rounded to 5.0 by many authors, but we prefer to express the result to three significant digits for consistency with other quantities obtained from the Blasius profile.

Another quantity of interest is the shear stress at the wall τ_w,

$$\tau_w = \mu \left. \frac{\partial u}{\partial y} \right)_{y=0} \quad (7)$$

Sketched in Fig. 10-99 is the slope of the nondimensional velocity profile at the wall ($y = 0$ and $\eta = 0$). From our similarity results (Table 10-3), the nondimensional slope at the wall is

$$\left. \frac{d(u/U)}{d\eta} \right)_{\eta=0} = f''(0) = 0.332 \quad (8)$$

After substitution of Eq. 8 into Eq. 7 and some algebra (transformation of similarity variables back to physical variables), we obtain

Shear stress in physical variables: $\quad \tau_w = 0.332 \dfrac{\rho U^2}{\sqrt{Re_x}} \quad (9)$

Thus, we see that the wall shear stress decays with x like $x^{-1/2}$, as sketched in Fig. 10-100. At $x = 0$, Eq. 9 predicts that τ_w is infinite, which is unphysical. The boundary layer approximation is not appropriate at the leading edge ($x = 0$), because the boundary layer thickness is not small compared to x. Furthermore, any real flat plate has finite thickness, and there is a stagnation point at the front of the plate, with the outer flow accelerating quickly to $U(x) = V$. We may ignore the region very close to $x = 0$ without loss of accuracy in the rest of the flow.

Equation 9 is nondimensionalized by defining a **skin friction coefficient** (also called a **local friction coefficient**),

Local friction coefficient, laminar flat plate: $\quad C_{f,x} = \dfrac{\tau_w}{\frac{1}{2}\rho U^2} = \dfrac{0.664}{\sqrt{Re_x}} \quad (10)$

Notice that Eq. 10 for $C_{f,x}$ has the same form as Eq. 6 for δ/x, but with a different constant—both decay like the inverse of the square root of Reynolds number. In Chap. 11, we integrate Eq. 10 to obtain the total friction drag on a flat plate of length L.

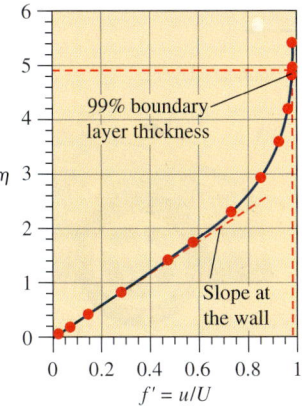

FIGURE 10-99
The Blasius profile in similarity variables for the boundary layer growing on a semi-infinite flat plate. Experimental data (circles) are at $Re_x = 3.64 \times 10^5$.

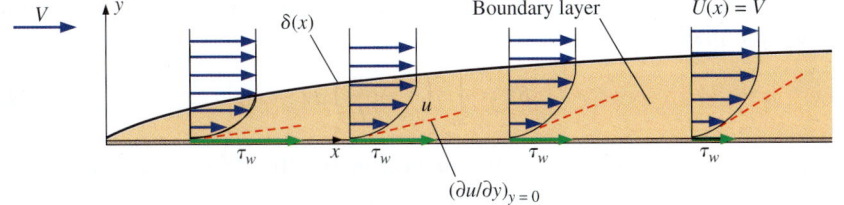

FIGURE 10-100
For a laminar flat plate boundary layer, wall shear stress decays like $x^{-1/2}$ as the slope $\partial u/\partial y$ at the wall decreases downstream. The front portion of the plate contributes more skin friction drag than does the rear portion.

568
APPROXIMATE SOLUTIONS OF THE N–S EQ

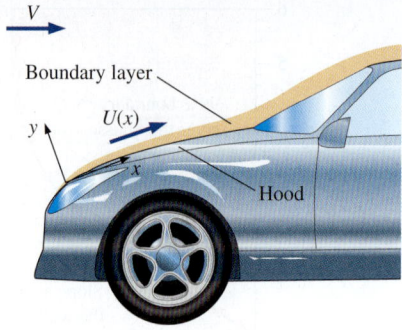

FIGURE 10–101
The boundary layer growing on the hood of a car. Boundary layer thickness is exaggerated for clarity.

Step 5 We need to verify that the boundary layer is thin. Consider the practical example of flow over the hood of your car (Fig. 10–101) while you are driving downtown at 30 km/h on a hot day. The kinematic viscosity of the air is $\nu = 1.7 \times 10^{-5}$ m²/s. We approximate the hood as a flat plate of length 1.2 m moving horizontally at a speed of $V = 30$ km/h. First, we approximate the Reynolds number at the end of the hood using Eq. 10–60,

$$\mathrm{Re}_x = \frac{Vx}{\nu} = \frac{(30 \text{ km/h})(1.2 \text{ m})}{1.7 \times 10^{-5} \text{ m}^2/\text{s}} \left(\frac{1000 \text{ m}}{\text{km}}\right)\left(\frac{\text{h}}{3600 \text{ s}}\right) = 5.9 \times 10^5$$

Since Re_x is very close to the ballpark critical Reynolds number, $\mathrm{Re}_{x,\,cr} = 5 \times 10^5$, the assumption of laminar flow may or may not be appropriate. Nevertheless, we use Eq. 6 to estimate the thickness of the boundary layer, assuming that the flow remains laminar,

$$\delta = \frac{4.91x}{\sqrt{\mathrm{Re}_x}} = \frac{4.91(1.2 \text{ m})}{\sqrt{5.9 \times 10^5}}\left(\frac{100 \text{ cm}}{\text{m}}\right) = 0.77 \text{ cm} \quad (11)$$

By the end of the hood the boundary layer is only about a quarter of an inch thick, and our assumption of a very thin boundary layer is verified.

Discussion The Blasius boundary layer solution is valid only for flow over a flat plate perfectly aligned with the flow. However, it is often used as a quick approximation for the boundary layer developing along solid walls that are not necessarily flat nor exactly parallel to the flow, as in the car hood. As illustrated in step 5, it is not difficult in practical engineering problems to achieve Reynolds numbers greater than the critical value for transition to turbulence. You must be careful not to apply the laminar boundary layer solution presented here when the boundary layer becomes turbulent.

Displacement Thickness

As was shown in Fig. 10–80, streamlines within and outside a boundary layer must bend slightly outward away from the wall in order to satisfy conservation of mass as the boundary layer thickness grows downstream. This is because the y-component of velocity, v, is small but finite and positive. Outside of the boundary layer, the outer flow is affected by this deflection of the streamlines. We define **displacement thickness** δ^* as the distance that a streamline just outside of the boundary layer is deflected, as sketched in Fig. 10–102.

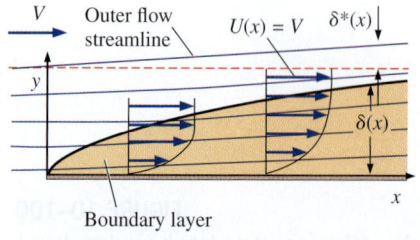

FIGURE 10–102
Displacement thickness defined by a streamline outside of the boundary layer. Boundary layer thickness is exaggerated.

> Displacement thickness is the distance that a streamline just outside of the boundary layer is deflected away from the wall due to the effect of the boundary layer.

We generate an expression for δ^* for the boundary layer along a flat plate by performing a control volume analysis using conservation of mass. The details are left as an exercise for the reader; the result at any x-location along the plate is

Displacement thickness:
$$\delta^* = \int_0^\infty \left(1 - \frac{u}{U}\right) dy \quad (10\text{–}72)$$

Note that the upper limit of the integral in Eq. 10–72 is shown as ∞, but since $u = U$ everywhere above the boundary layer, it is necessary to integrate only out to some finite distance above δ. Obviously δ* grows with x as the boundary layer grows (Fig. 10–103). For a laminar flat plate, we integrate the numerical (Blasius) solution of Example 10–10 to obtain

Displacement thickness, laminar flat plate: $\dfrac{\delta^*}{x} = \dfrac{1.72}{\sqrt{\mathrm{Re}_x}}$ (10–73)

The equation for δ* is the same as that for δ, but with a different constant. In fact, for laminar flow over a flat plate, δ* at any x-location turns out to be approximately three times smaller than δ at that same x-location (Fig. 10–103).

There is an alternative way to explain the physical meaning of δ* that turns out to be more useful for practical engineering applications. Namely, we can think of displacement thickness as an imaginary or apparent increase in thickness of the wall from the point of view of the inviscid and/or irrotational outer flow region. For our flat plate example, the outer flow no longer "sees" an infinitesimally thin flat plate; rather it sees a finite-thickness plate shaped like the displacement thickness of Eq. 10–73, as illustrated in Fig. 10–104.

> Displacement thickness is the imaginary increase in thickness of the wall, as seen by the outer flow, due to the effect of the growing boundary layer.

If we were to solve the Euler equation for the flow around this imaginary thicker plate, the outer flow velocity component U(x) would differ from the original calculation. We could then use this apparent U(x) to improve our boundary layer analysis. You can imagine a modification to the boundary layer procedure of Fig. 10–93 in which we go through the first four steps, calculate δ*(x), and then go back to step 1, this time using the imaginary (thicker) body shape to calculate an apparent U(x). Following this, we re-solve the boundary layer equations. We could repeat the loop as many times as necessary until convergence. In this way, the outer flow and the boundary layer would be more consistent with each other.

The usefulness of this interpretation of displacement thickness becomes obvious if we consider uniform flow entering a channel bounded by two parallel walls (Fig. 10–105). As the boundary layers grow on the upper and lower walls, the irrotational core flow must accelerate to satisfy conservation of mass (Fig. 10–105a). From the point of view of the core flow between the boundary layers, the boundary layers cause the channel walls to appear to converge—the apparent distance between the walls decreases as x increases. This imaginary increase in thickness of one of the walls is equal to δ*(x), and the *apparent U(x)* of the core flow must increase accordingly, as sketched, to satisfy conservation of mass.

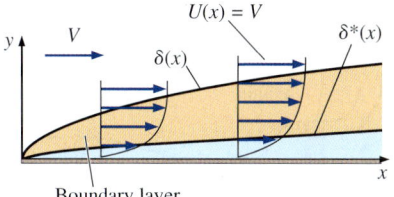

FIGURE 10–103
For a laminar flat plate boundary layer, the displacement thickness is roughly one-third of the 99 percent boundary layer thickness.

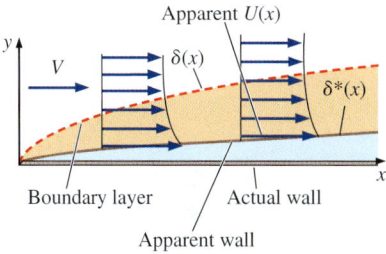

FIGURE 10–104
The boundary layer affects the irrotational outer flow in such a way that the wall appears to take the shape of the displacement thickness. The apparent U(x) differs from the original approximation because of the "thicker" wall.

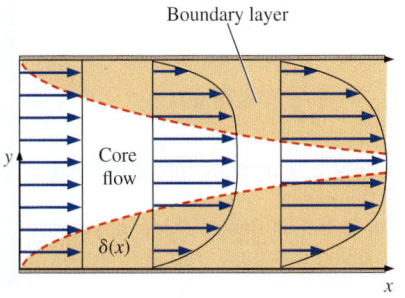
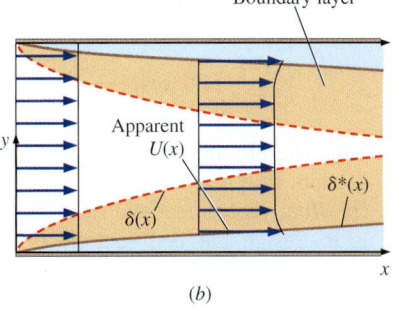

FIGURE 10–105
The effect of boundary layer growth on flow entering a two-dimensional channel: the irrotational flow between the top and bottom boundary layers accelerates as indicated by (a) actual velocity profiles, and (b) change in apparent core flow due to the displacement thickness of the boundary layer (boundary layers greatly exaggerated for clarity).

570
APPROXIMATE SOLUTIONS OF THE N–S EQ

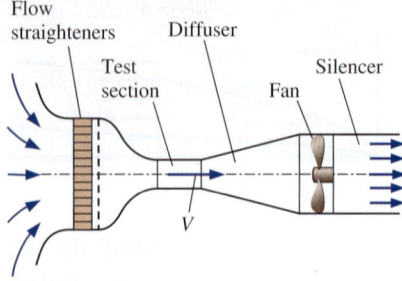

FIGURE 10–106
Schematic diagram of the wind tunnel of Example 10–11.

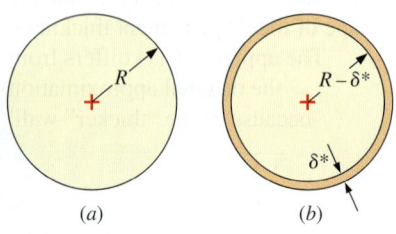

FIGURE 10–107
Cross-sectional views of the test section of the wind tunnel of Example 10–11: (*a*) beginning of test section and (*b*) end of test section.

EXAMPLE 10–11   Displacement Thickness in the Design of a Wind Tunnel

A small low-speed wind tunnel (Fig. 10–106) is being designed for calibration of hot wires. The air is at 19°C. The test section of the wind tunnel is 30 cm in diameter and 30 cm in length. The flow through the test section must be as uniform as possible. The wind tunnel speed ranges from 1 to 8 m/s, and the design is to be optimized for an air speed of $V = 4.0$ m/s through the test section. (*a*) For the case of nearly uniform flow at 4.0 m/s at the test section inlet, by how much will the centerline air speed accelerate by the end of the test section? (*b*) Recommend a design that will lead to a more uniform test section flow.

SOLUTION   The acceleration of air through the round test section of a wind tunnel is to be calculated, and a redesign of the test section is to be recommended.
Assumptions   **1** The flow is steady and incompressible. **2** The walls are smooth, and disturbances and vibrations are kept to a minimum. **3** The boundary layer is laminar.
Properties   The kinematic viscosity of air at 19°C is $\nu = 1.507 \times 10^{-5}$ m²/s.
Analysis   (*a*) The Reynolds number at the end of the test section is approximately

$$\mathrm{Re}_x = \frac{Vx}{\nu} = \frac{(4.0 \text{ m/s})(0.30 \text{ m})}{1.507 \times 10^{-5} \text{ m}^2/\text{s}} = 7.96 \times 10^4$$

Since Re_x is lower than the engineering critical Reynolds number, $\mathrm{Re}_{x,\,cr} = 5 \times 10^5$, and is even lower than $\mathrm{Re}_{x,\,critical} = 1 \times 10^5$, and since the walls are smooth and the flow is clean, we may assume that the boundary layer on the wall remains laminar throughout the length of the test section. As the boundary layer grows along the wall of the wind tunnel test section, air in the region of irrotational flow in the central portion of the test section accelerates as in Fig. 10–105 in order to satisfy conservation of mass. We use Eq. 10–73 to estimate the displacement thickness at the end of the test section,

$$\delta^* \cong \frac{1.72x}{\sqrt{\mathrm{Re}_x}} = \frac{1.72(0.30 \text{ m})}{\sqrt{7.96 \times 10^4}} = 1.83 \times 10^{-3} \text{ m} = 1.83 \text{ mm} \quad (1)$$

Two cross-sectional views of the test section are sketched in Fig. 10–107, one at the beginning and one at the end of the test section. The effective radius at the end of the test section is reduced by δ^* as calculated by Eq. 1. We apply conservation of mass to calculate the average air speed at the end of the test section,

$$V_{end} A_{end} = V_{beginning} A_{beginning} \rightarrow V_{end} = V_{beginning} \frac{\pi R^2}{\pi (R - \delta^*)^2} \quad (2)$$

which yields

$$V_{end} = (4.0 \text{ m/s}) \frac{(0.15 \text{ m})^2}{(0.15 \text{ m} - 1.83 \times 10^{-3} \text{ m})^2} = 4.10 \text{ m/s} \quad (3)$$

Thus the air speed increases by approximately 2.5 percent through the test section, due to the effect of displacement thickness.
(*b*) What recommendation can we make for a better design? One possibility is to design the test section as a slowly diverging duct, rather than as a

straight-walled cylinder (Fig. 10–108). If the radius were designed so as to increase like δ*(x) along the length of the test section, the displacement effect of the boundary layer would be eliminated, and the test section air speed would remain fairly constant. Note that there is still a boundary layer growing on the wall, as illustrated in Fig. 10–108. However, the core flow speed outside the boundary layer remains constant, unlike the situation of Fig. 10–105. The diverging wall recommendation would work well at the design operating condition of 4.0 m/s and would help somewhat at other flow speeds. Another option is to apply suction along the wall of the test section in order to remove some of the air along the wall. The advantage of this design is that the suction can be carefully adjusted as wind tunnel speed is varied so as to ensure constant air speed through the test section at any operating condition. This recommendation is the more complicated, and probably more expensive, option.

Discussion Wind tunnels have been constructed that use either the diverging wall option or the wall suction option to carefully control the uniformity of the air speed through the wind tunnel test section. The same displacement thickness technique is applied to larger wind tunnels, where the boundary layer is turbulent; however, a different equation for δ*(x) is required.

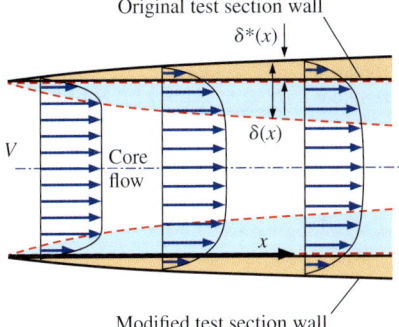

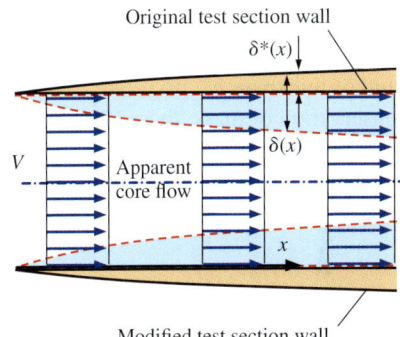

FIGURE 10–108
A diverging test section would eliminate flow acceleration due to the displacement effect of the boundary layer: (*a*) actual flow and (*b*) apparent irrotational core flow.

Momentum Thickness

Another measure of boundary layer thickness is **momentum thickness**, commonly given the symbol θ. Momentum thickness is best explained by analyzing the control volume of Fig. 10–109 for a flat plate boundary layer. Since the bottom of the control volume is the plate itself, no mass or momentum can cross that surface. The top of the control volume is taken as a streamline of the outer flow. Since no flow can cross a streamline, there can be no mass or momentum flux across the upper surface of the control volume. When we apply conservation of mass to this control volume, we find that the mass flow entering the control volume from the left (at $x = 0$) must equal the mass flow exiting from the right (at some arbitrary location x along the plate),

$$0 = \int_{CS} \rho \vec{V} \cdot \vec{n} \, dA = \underbrace{w\rho \int_0^{Y+\delta^*} u \, dy}_{\text{at location } x} - \underbrace{w\rho \int_0^Y U \, dy}_{\text{at } x = 0} \quad (10\text{–}74)$$

where w is the width into the page in Fig. 10–109, which we take arbitrarily as unit width, and Y is the distance from the plate to the outer streamline at $x = 0$, as indicated in Fig. 10–109. Since $u = U =$ constant everywhere along the left surface of the control volume, and since $u = U$ between $y = Y$ and $y = Y + \delta^*$ along the right surface of the control volume, Eq. 10–74 reduces to

$$\int_0^Y (U - u) \, dy = U\delta^* \quad (10\text{–}75)$$

Physically, the mass flow *deficit* within the boundary layer (the lower blue-shaded region in Fig. 10–109) is replaced by a chunk of free-stream flow of thickness δ* (the upper blue-shaded region in Fig. 10–109). Equation 10–75 verifies that these two shaded regions have the *same area*. We zoom in to show these areas more clearly in Fig. 10–110.

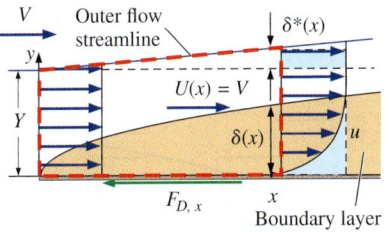

FIGURE 10–109
A control volume is defined by the thick dashed line, bounded above by a streamline outside of the boundary layer, and bounded below by the flat plate; $F_{D,x}$ is the viscous force of the plate acting on the control volume.

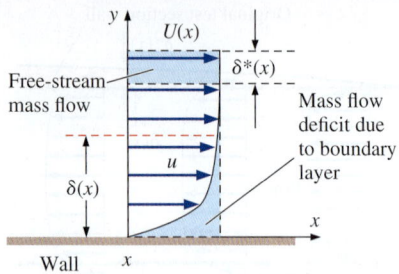

FIGURE 10–110
Comparison of the area under the boundary layer profile, representing the mass flow deficit, and the area generated by a chunk of free-stream fluid of thickness δ^*. To satisfy conservation of mass, these two areas must be identical.

Now consider the x-component of the control volume momentum equation. Since no momentum crosses the upper or lower control surfaces, the net force acting on the control volume must equal the momentum flux exiting the control volume minus that entering the control volume,

Conservation of x-momentum for the control volume:

$$\sum F_x = -F_{D,x} = \int_{CS} \rho u \vec{V} \cdot \vec{n} \, dA = \underbrace{\rho w \int_0^{Y+\delta^*} u^2 \, dy}_{\text{at location } x} - \underbrace{\rho w \int_0^{Y} U^2 \, dy}_{\text{at } x=0} \quad (10\text{–}76)$$

where $F_{D,x}$ is the drag force due to friction on the plate from $x = 0$ to location x. After some algebra, including substitution of Eq. 10–75, Eq. 10–76 reduces to

$$F_{D,x} = \rho w \int_0^Y u(U - u) \, dy \quad (10\text{–}77)$$

Finally, we define momentum thickness θ such that the viscous drag force on the plate per unit width into the page is equal to ρU^2 times θ, i.e.,

$$\frac{F_{D,x}}{w} = \rho \int_0^Y u(U - u) \, dy \equiv \rho U^2 \theta \quad (10\text{–}78)$$

In words,

Momentum thickness is defined as the loss of momentum flux per unit width divided by ρU^2 due to the presence of the growing boundary layer.

Equation 10–78 reduces to

$$\theta = \int_0^Y \frac{u}{U}\left(1 - \frac{u}{U}\right) dy \quad (10\text{–}79)$$

Streamline height Y can be any value, as long as the streamline taken as the upper surface of the control volume is above the boundary layer. Since $u = U$ for any y greater than Y, we may replace Y by infinity in Eq. 10–79 with no change in the value of θ,

Momentum thickness:
$$\theta = \int_0^\infty \frac{u}{U}\left(1 - \frac{u}{U}\right) dy \quad (10\text{–}80)$$

For the specific case of the Blasius solution for a laminar flat plate boundary layer (Example 10–10), we integrate Eq. 10–80 numerically to obtain

Momentum thickness, laminar flat plate:
$$\frac{\theta}{x} = \frac{0.664}{\sqrt{Re_x}} \quad (10\text{–}81)$$

We note that the equation for θ is the same as that for δ or for δ^* but with a different constant. In fact, for laminar flow over a flat plate, θ turns out to be approximately 13.5 percent of δ at any x-location, as indicated in Fig. 10–111. It is no coincidence that θ/x (Eq. 10–81) is identical to $C_{f,x}$ (Eq. 10 of Example 10–10)—both are derived from skin friction drag on the plate.

Turbulent Flat Plate Boundary Layer

It is beyond the scope of this text to derive or attempt to solve the turbulent flow boundary layer equations. Expressions for the boundary layer profile

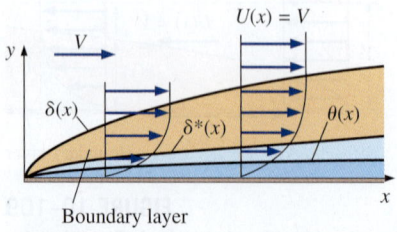

FIGURE 10–111
For a laminar flat plate boundary layer, displacement thickness is 35.0 percent of δ, and momentum thickness is 13.5 percent of δ.

shape and other properties of the turbulent boundary layer are obtained *empirically* (or at best *semi-empirically*), since we cannot solve the boundary layer equations for turbulent flow. Note also that turbulent flows are inherently *unsteady*, and the instantaneous velocity profile shape varies with time (Fig. 10–112). Thus, all turbulent expressions discussed here represent *time-averaged values*. One common empirical approximation for the time-averaged velocity profile of a turbulent flat plate boundary layer is the **one-seventh-power law**,

$$\frac{u}{U} \cong \left(\frac{y}{\delta}\right)^{1/7} \text{ for } y \leq \delta, \quad \rightarrow \quad \frac{u}{U} \cong 1 \text{ for } y > \delta \quad (10\text{–}82)$$

Note that in the approximation of Eq. 10–82, δ is *not* the 99 percent boundary layer thickness, but rather the actual edge of the boundary layer, unlike the definition of δ for laminar flow. Equation 10–82 is plotted in Fig. 10–113. For comparison, the laminar flat plate boundary layer profile (a numerical solution of the Blasius equations Fig. 10–99) is also plotted in Fig. 10–113, using y/δ for the vertical axis in place of similarity variable η. You can see that if the laminar and turbulent boundary layers were the same thickness, the turbulent one would be much *fuller* than the laminar one. In other words, the turbulent boundary layer would "hug" the wall more closely, filling the boundary layer with higher-speed flow close to the wall. This is due to the large turbulent eddies that transport high-speed fluid from the outer part of the boundary layer down to the lower parts of the boundary layer (and vice versa). In other words, a turbulent boundary layer has a much greater degree of mixing when compared to a laminar boundary layer. In the laminar case, fluid mixes slowly due to viscous diffusion. However, the large eddies in a turbulent flow promote much more rapid and thorough mixing.

The approximate turbulent boundary layer velocity profile shape of Eq. 10–82 is not physically meaningful very close to the wall ($y \rightarrow 0$) since it predicts that the slope ($\partial u/\partial y$) is infinite at $y = 0$. While the slope at the wall is very large for a turbulent boundary layer, it is nevertheless finite. This large slope at the wall leads to a very high wall shear stress, $\tau_w = \mu(\partial u/\partial y)_{y=0}$, and, therefore, correspondingly high skin friction along the surface of the plate (as compared to a laminar boundary layer of the same thickness). The skin friction drag produced by both laminar and turbulent boundary layers is discussed in greater detail in Chap. 11.

A nondimensionalized plot such as that of Fig. 10–113 is somewhat misleading, since the turbulent boundary layer would actually be much *thicker* than the corresponding laminar boundary layer at the same Reynolds number. This fact is illustrated in physical variables in Example 10–12.

We compare in Table 10–4 expressions for δ, δ^*, θ, and $C_{f,x}$ for laminar and turbulent boundary layers on a smooth flat plate. The turbulent expressions are based on the one-seventh-power law of Eq. 10–82. Note that the expressions in Table 10–4 for the turbulent flat plate boundary layer are valid only for a very *smooth* surface. Even a small amount of surface roughness greatly affects properties of the turbulent boundary layer, such as momentum thickness and local skin friction coefficient. The effect of surface roughness on a turbulent flat plate boundary layer is discussed in greater detail in Chap. 11.

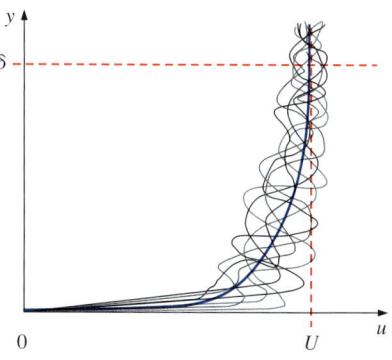

FIGURE 10–112
Illustration of the unsteadiness of a turbulent boundary layer; the thin, wavy black lines are instantaneous profiles, and the thick blue line is a long time-averaged profile.

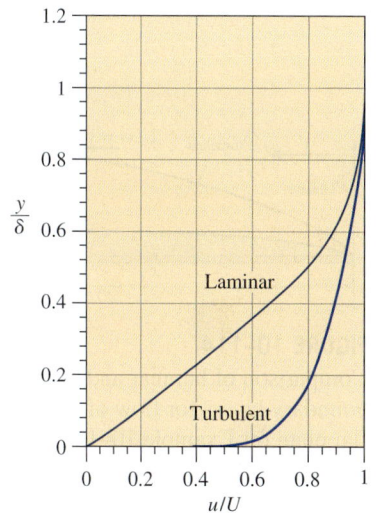

FIGURE 10–113
Comparison of laminar and turbulent flat plate boundary layer profiles, nondimensionalized by boundary layer thickness.

TABLE 10–4

Summary of expressions for laminar and turbulent boundary layers on a smooth flat plate aligned parallel to a uniform stream*

Property	Laminar	(a) Turbulent[†]	(b) Turbulent[‡]
Boundary layer thickness	$\dfrac{\delta}{x} = \dfrac{4.91}{\sqrt{Re_x}}$	$\dfrac{\delta}{x} \cong \dfrac{0.16}{(Re_x)^{1/7}}$	$\dfrac{\delta}{x} \cong \dfrac{0.38}{(Re_x)^{1/5}}$
Displacement thickness	$\dfrac{\delta^*}{x} = \dfrac{1.72}{\sqrt{Re_x}}$	$\dfrac{\delta^*}{x} \cong \dfrac{0.020}{(Re_x)^{1/7}}$	$\dfrac{\delta^*}{x} \cong \dfrac{0.048}{(Re_x)^{1/5}}$
Momentum thickness	$\dfrac{\theta}{x} = \dfrac{0.664}{\sqrt{Re_x}}$	$\dfrac{\theta}{x} \cong \dfrac{0.016}{(Re_x)^{1/7}}$	$\dfrac{\theta}{x} \cong \dfrac{0.037}{(Re_x)^{1/5}}$
Local skin friction coefficient	$C_{f,x} = \dfrac{0.664}{\sqrt{Re_x}}$	$C_{f,x} \cong \dfrac{0.027}{(Re_x)^{1/7}}$	$C_{f,x} \cong \dfrac{0.059}{(Re_x)^{1/5}}$

* Laminar values are exact and are listed to three significant digits, but turbulent values are listed to only two significant digits due to the large uncertainty affiliated with all turbulent flow fields.
† Obtained from one-seventh-power law.
‡ Obtained from one-seventh-power law combined with empirical data for turbulent flow through smooth pipes.

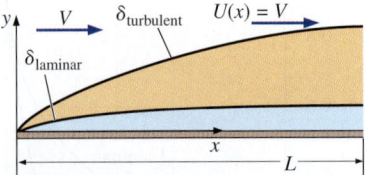

FIGURE 10–114
Comparison of laminar and turbulent boundary layers for flow of air over a flat plate for Example 10–12 (boundary layer thickness exaggerated).

EXAMPLE 10–12   Comparison of Laminar and Turbulent Boundary Layers

Air at 20°C flows at $V = 10.0$ m/s over a smooth flat plate of length $L = 1.52$ m (Fig. 10–114). (*a*) Plot and compare the laminar and turbulent boundary layer profiles in physical variables (*u* as a function of *y*) at $x = L$. (*b*) Compare the values of local skin friction coefficient for the two cases at $x = L$. (*c*) Plot and compare the growth of the laminar and turbulent boundary layers.

SOLUTION   We are to compare laminar versus turbulent boundary layer profiles, local skin friction coefficient, and boundary layer thickness at the end of a flat plate.
Assumptions   **1** The plate is smooth, and the free stream is calm and uniform. **2** The flow is steady in the mean. **3** The plate is infinitesimally thin and is aligned parallel to the free stream.
Properties   The kinematic viscosity of air at 20°C is $\nu = 1.516 \times 10^{-5}$ m²/s.
Analysis   (*a*) First we calculate the Reynolds number at $x = L$,

$$Re_x = \frac{Vx}{\nu} = \frac{(10.0 \text{ m/s})(1.52 \text{ m})}{1.516 \times 10^{-5} \text{ m}^2/\text{s}} = 1.00 \times 10^6$$

This value of Re_x is in the transitional region between laminar and turbulent, according to Fig. 10–81. Thus, a comparison between the laminar and turbulent velocity profiles is appropriate. For the laminar case, we multiply the y/δ values of Fig. 10–113 by δ_{laminar}, where

$$\delta_{\text{laminar}} = \frac{4.91x}{\sqrt{Re_x}} = \frac{4.91(1520 \text{ mm})}{\sqrt{1.00 \times 10^6}} = \mathbf{7.46 \text{ mm}} \quad (1)$$

This gives us y-values in units of mm. Similarly, we multiply the u/U values of Fig. 10–113 by U ($U = V = 10.0$ m/s) to obtain u in units of m/s. We plot the laminar boundary layer profile in physical variables in Fig. 10–115.

We calculate the turbulent boundary layer thickness at this same x-location using the equation provided in Table 10–4, column (a),

$$\delta_{turbulent} \cong \frac{0.16x}{(Re_x)^{1/7}} = \frac{0.16(1520 \text{ mm})}{(1.00 \times 10^6)^{1/7}} = \mathbf{34 \text{ mm}} \quad (2)$$

[The value of $\delta_{turbulent}$ based on column (b) of Table 10–4 is somewhat higher, namely 36 mm.] Comparing Eqs. 1 and 2, we see that the turbulent boundary layer is about 4.5 times thicker than the laminar boundary layer at a Reynolds number of 1.0×10^6. The turbulent boundary layer velocity profile of Eq. 10–82 is converted to physical variables and plotted in Fig. 10–115 for comparison with the laminar profile. The two most striking features of Fig. 10–115 are (1) the turbulent boundary layer is much thicker than the laminar one, and (2) the slope of u versus y near the wall is much steeper for the turbulent case. (Keep in mind, of course, that very close to the wall the one-seventh-power law does not adequately represent the actual turbulent boundary layer profile.)

(*b*) We use the expressions in Table 10–4 to compare the local skin friction coefficient for the two cases. For the laminar boundary layer,

$$C_{f, x, \text{laminar}} = \frac{0.664}{\sqrt{Re_x}} = \frac{0.664}{\sqrt{1.00 \times 10^6}} = \mathbf{6.64 \times 10^{-4}} \quad (3)$$

and for the turbulent boundary layer, column (a),

$$C_{f, x, \text{turbulent}} \cong \frac{0.027}{(Re_x)^{1/7}} = \frac{0.027}{(1.00 \times 10^6)^{1/7}} = \mathbf{3.8 \times 10^{-3}} \quad (4)$$

Comparing Eqs. 3 and 4, the turbulent skin friction value is more than five times larger than the laminar value. If we had used the other expression for turbulent skin friction coefficient, column (b) of Table 10–4, we would have obtained $C_{f, x, \text{turbulent}} = 3.7 \times 10^{-3}$, very close to the value calculated in Eq. 4.

(*c*) The turbulent calculation assumes that the boundary layer is turbulent from the beginning of the plate. In reality, there is a region of laminar flow, followed by a transition region, and then finally a turbulent region, as illustrated in Fig. 10–81. Nevertheless, it is interesting to compare how $\delta_{laminar}$ and $\delta_{turbulent}$ grow as functions of x for this flow, assuming either all laminar flow or all turbulent flow. Using the expressions in Table 10–4, both of these are plotted in Fig. 10–116 for comparison.

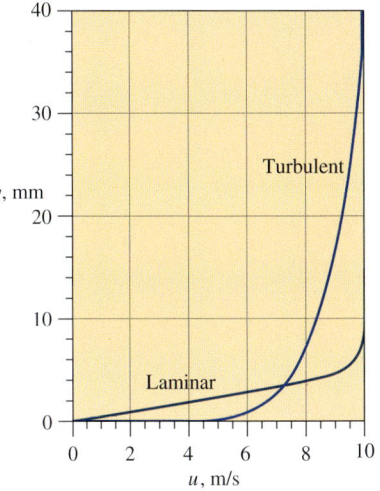

FIGURE 10–115
Comparison of laminar and turbulent flat plate boundary layer profiles in physical variables at the same x-location. The Reynolds number is $Re_x = 1.0 \times 10^6$.

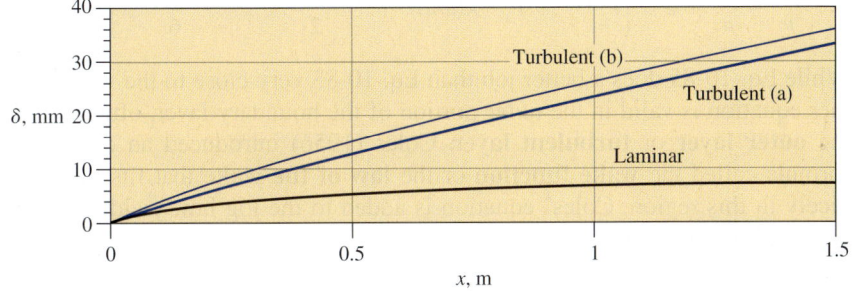

FIGURE 10–116
Comparison of the growth of a laminar boundary layer and a turbulent boundary layer for the flat plate of Example 10–12.

Discussion The ordinate in Fig. 10–116 is in mm, while the abscissa is in m for clarity—the boundary layer is incredibly thin, even for the turbulent case. The difference between the turbulent (a) and (b) cases (see Table 10–4) is explained by discrepancies between empirical curve fits and semi-empirical approximations used to obtain the expressions in Table 10–4. This reinforces our decision to report turbulent boundary layer values to at most two significant digits. The real value of δ will most likely lie somewhere between the laminar and turbulent values plotted in Fig. 10–116 since the Reynolds number by the end of the plate is within the transitional region.

The one-seventh-power law is not the only turbulent boundary layer approximation used by fluid mechanicians. Another common approximation is the **log law**, a semi-empirical expression that turns out to be valid not only for flat plate boundary layers but also for fully developed turbulent pipe flow velocity profiles (Chap. 8). In fact, the log law turns out to be applicable for nearly *all* wall-bounded turbulent boundary layers, not just flow over a flat plate. (This fortunate situation enables us to employ the log law approximation close to solid walls in computational fluid dynamics codes, as discussed in Chap. 15.) The log law is commonly expressed in variables nondimensionalized by a characteristic velocity called the **friction velocity** u_*. (Note that most authors use u^* instead of u_*. We use a subscript to distinguish u_*, a *dimensional* quantity, from u^*, which we use to indicate a nondimensional velocity.)

The log law:
$$\frac{u}{u_*} = \frac{1}{\kappa} \ln \frac{yu_*}{\nu} + B \tag{10-83}$$

where

Friction velocity:
$$u_* = \sqrt{\frac{\tau_w}{\rho}} \tag{10-84}$$

and κ and B are constants; their usual values are $\kappa = 0.40$ to 0.41 and $B = 5.0$ to 5.5. Unfortunately, the log law suffers from the fact that it does not work very close to the wall (ln 0 is undefined). It also deviates from experimental values close to the boundary layer edge. Nevertheless, Eq. 10–83 applies across a significant portion of the turbulent flat plate boundary layer and is useful because it relates the velocity profile shape to the local value of wall shear stress through Eq. 10–84.

A clever expression that is valid all the way to the wall was created by D. B. Spalding in 1961 and is called **Spalding's law of the wall**,

$$\frac{yu_*}{\nu} = \frac{u}{u_*} + e^{-\kappa B}\left[e^{\kappa(u/u_*)} - 1 - \kappa(u/u_*) - \frac{[\kappa(u/u_*)]^2}{2} - \frac{[\kappa(u/u_*)]^3}{6}\right] \tag{10-85}$$

While Eq. 10-85 does a better job than Eq. 10-83 very close to the wall, neither equation is valid in the *outer* portion of the boundary layer, often called the **outer layer** or **turbulent layer**. Coles (1956) introduced an empirical formula called the **wake function** or the **law of the wake** that fits the data nicely in this region. Coles' equation is added to the log law, yielding what some call the **wall-wake law**,

$$\frac{u}{u^*} = \frac{1}{\kappa} \ln \frac{yu^*}{\nu} + B + \frac{2\Pi}{\kappa} W\left(\frac{y}{\delta}\right) \tag{10-86}$$

Where $\Pi = 0.44$ for a flat plate boundary layer, and several expressions for W have been suggested, all of which smoothly change from 0 at the wall ($y/\delta = 0$) to 1 at the outer edge of the boundary layer ($y/\delta = 1$). One popular expression is

$$W\left(\frac{y}{\delta}\right) = \sin^2\left(\frac{\pi}{2}\left(\frac{y}{\delta}\right)\right) \quad \text{for} \quad \frac{y}{\delta} < 1 \quad (10\text{–}87)$$

■ **EXAMPLE 10–13**   **Comparison of Turbulent Boundary Layer Profile Equations**

Air at 20°C flows at $V = 10.0$ m/s over a smooth flat plate of length $L = 15.2$ m (Fig. 10–117). Plot the turbulent boundary layer profile in physical variables (u as a function of y) at $x = L$. Compare the profile generated by the one-seventh-power law, the log law, and Spalding's law of the wall, assuming that the boundary layer is fully turbulent from the beginning of the plate.

SOLUTION We are to plot the mean boundary layer profile $u(y)$ at the end of a flat plate using three different approximations.
Assumptions **1** The plate is smooth, but there are free-stream fluctuations that tend to cause the boundary layer to transition to turbulence sooner than usual—the boundary layer is turbulent from the beginning of the plate. **2** The flow is steady in the mean. **3** The plate is infinitesimally thin and is aligned parallel to the free stream.
Properties The kinematic viscosity of air at 20°C is $\nu = 1.516 \times 10^{-5}$ m²/s.
Analysis First we calculate the Reynolds number at $x = L$,

$$\text{Re}_x = \frac{Vx}{\nu} = \frac{(10.0 \text{ m/s})(15.2 \text{ m})}{1.516 \times 10^{-5} \text{ m}^2/\text{s}} = 1.00 \times 10^7$$

This value of Re_x is well above the transitional Reynolds number for a flat plate boundary layer (Fig. 10–81), so the assumption of turbulent flow from the beginning of the plate is reasonable.

Using the column (a) values of Table 10–4, we estimate the boundary layer thickness and the local skin friction coefficient at the end of the plate,

$$\delta \cong \frac{0.16x}{(\text{Re}_x)^{1/7}} = 0.240 \text{ m} \quad C_{f,x} \cong \frac{0.027}{(\text{Re}_x)^{1/7}} = 2.70 \times 10^{-3} \quad (1)$$

We calculate the friction velocity by using its definition (Eq. 10–84) and the definition of $C_{f,x}$ (left part of Eq. 10 of Example 10–10),

$$u_* = \sqrt{\frac{\tau_w}{\rho}} = U\sqrt{\frac{C_{f,x}}{2}} = (10.0 \text{ m/s})\sqrt{\frac{2.70 \times 10^{-3}}{2}} = 0.367 \text{ m/s} \quad (2)$$

where $U =$ constant $= V$ everywhere for a flat plate. It is trivial to generate a plot of the one-seventh-power law (Eq. 10–82), but the log law (Eq. 10–83) is implicit for u as a function of y. Instead, we solve Eq. 10–83 for y as a function of u,

$$y = \frac{\nu}{u_*} e^{\kappa(u/u_* - B)} \quad (3)$$

Since we know that u varies from 0 at the wall to U at the boundary layer edge, we are able to plot the log law velocity profile in physical variables using Eq. 3. Finally, Spalding's law of the wall (Eq. 10–85) is also written in terms of y as a function of u. We plot all three profiles on the same plot for comparison (Fig. 10–118). All three are close, and we cannot distinguish the log law from Spalding's law on this scale.

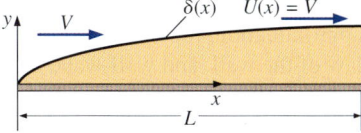

FIGURE 10–117
The turbulent boundary layer generated by flow of air over a flat plate for Example 10–13 (boundary layer thickness exaggerated).

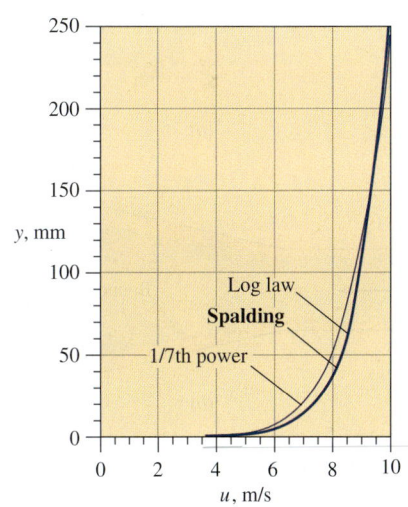

FIGURE 10–118
Comparison of turbulent flat plate boundary layer profile expressions in physical variables at $\text{Re}_x = 1.0 \times 10^7$: one-seventh-power approximation, log law, and Spalding's law of the wall.

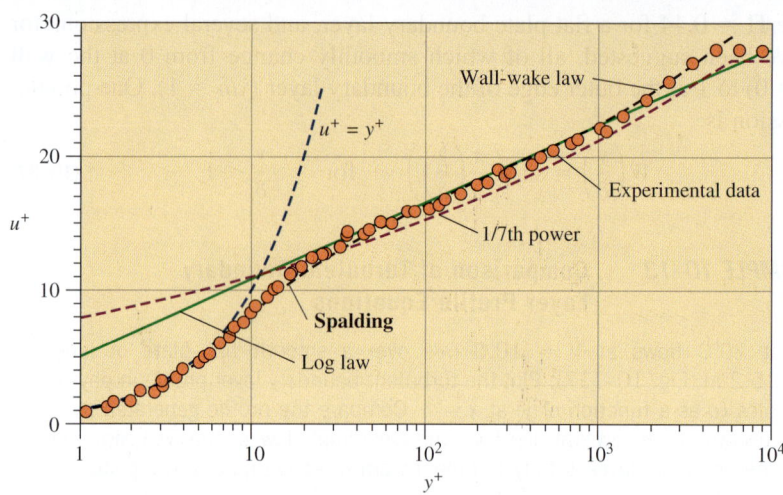

FIGURE 10–119
Comparison of turbulent flat plate boundary layer profile expressions in law of the wall variables at $Re_x = 1.0 \times 10^7$: one-seventh-power approximation, log law, Spalding's law of the wall, and wall-wake law. Typical experimental data and the viscous sublayer equation ($u^+ = y^+$) are also shown for comparison.

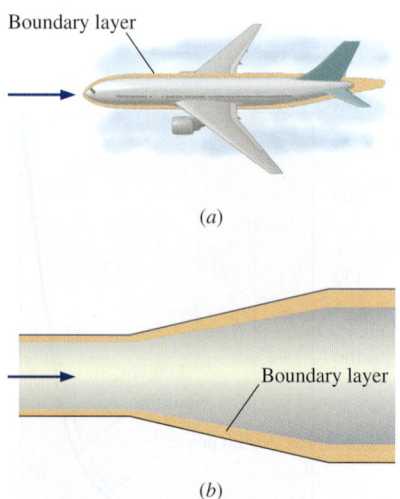

FIGURE 10–120
Boundary layers with nonzero pressure gradients occur in both external flows and internal flows: (a) boundary layer developing along the fuselage of an airplane and into the wake, and (b) boundary layer growing on the wall of a diffuser (boundary layer thickness exaggerated in both cases).

Instead of a physical variable plot with linear axes as in Fig. 10–118, a semi-log plot of nondimensional variables is often drawn to magnify the near-wall region. The most common notation in the boundary layer literature for the nondimensional variables is y^+ and u^+ (**inner variables** or **law of the wall variables**), where

Law of the wall variables:
$$y^+ = \frac{yu_*}{\nu} \qquad u^+ = \frac{u}{u_*} \qquad (4)$$

As you can see, y^+ is a type of Reynolds number, and friction velocity u_* is used to nondimensionalize both y and u. Figure 10–118 is redrawn in Fig. 10–119 using law of the wall variables. The differences between the three approximations, especially near the wall, are much clearer when plotted in this fashion. Typical experimental data are also plotted in Fig. 10–119 for comparison. Spalding's formula does the best job overall and is the only expression that follows experimental data near the wall. In the outer part of the boundary layer, the experimental values of u^+ level off beyond some value of y^+, as does the one-seventh-power law. However, both the log law and Spalding's formula continue indefinitely as a straight line on this semi-log plot.

Discussion Also plotted in Fig. 10–119 is the linear equation $u^+ = y^+$. The region *very* close to the wall ($0 < y^+ < 5$ or 6) is called the **viscous sublayer**. In this region, turbulent fluctuations are suppressed due to the close proximity of the wall, and the velocity profile is nearly *linear*. Other names for this region are **linear sublayer** and **laminar sublayer**. We see that Spalding's equation captures the viscous sublayer and blends smoothly into the log law. Neither the one-seventh-power law nor the log law are valid this close to the wall.

Boundary Layers with Pressure Gradients

So far we have spent most of our discussion on flat plate boundary layers. Of more practical concern for engineers are boundary layers on walls of arbitrary shape. These include external flows over bodies immersed in a free stream (Fig. 10–120a), as well as some internal flows like the walls of wind tunnels and other large ducts in which boundary layers develop along the walls (Fig. 10–120b). Just as with the zero pressure gradient flat plate boundary layer discussed earlier, boundary layers with nonzero pressure gradients may be laminar or turbulent. We often use the flat plate boundary layer results as ballpark estimates for such things as location of transition to turbulence, boundary layer thickness, skin friction, etc. However, when more accuracy is needed we must solve the boundary layer equations (Eqs. 10–71 for the steady, laminar, two-dimensional case) using the procedure outlined in Fig. 10–93. The analysis is harder than that for a flat plate since the pressure gradient term ($U\, dU/dx$) in the x-momentum equation is nonzero. Such an analysis can quickly get quite involved, especially for the case of three-dimensional flows. Therefore, we discuss only some *qualitative* features of boundary layers with pressure gradients, leaving detailed solutions of the boundary layer equations to higher-level fluid mechanics textbooks (e.g., Panton, 2005, and White, 2005).

First some terminology. When the flow in the inviscid and/or irrotational outer flow region (outside of the boundary layer) *accelerates*, $U(x)$ increases and $P(x)$ decreases. We refer to this as a **favorable pressure gradient.** It is favorable or desirable because the boundary layer in such an accelerating flow is usually thin, hugs closely to the wall, and therefore is not likely to separate from the wall. When the outer flow *decelerates*, $U(x)$ decreases, $P(x)$ increases, and we have an **unfavorable** or **adverse pressure gradient.** As its name implies, this condition is not desirable because the boundary layer is usually thicker, does not hug closely to the wall, and is much more likely to separate from the wall.

In a typical external flow, such as flow over an airplane wing (Fig. 10–121), the boundary layer in the front portion of the body is subjected to a favorable pressure gradient, while that in the rear portion is subjected to an adverse pressure gradient. If the adverse pressure gradient is strong enough ($dP/dx = -U\, dU/dx$ is large), the boundary layer is likely to **separate** off the wall. Examples of flow separation are shown in Fig. 10–122 for both external and internal flows. In Fig. 10–122a is sketched an airfoil at a moderate angle of attack. The boundary layer remains attached over the entire lower surface of the airfoil, but it separates somewhere near the rear of the upper surface as sketched. The closed streamline indicates a region of recirculating flow called a **separation bubble.** As pointed out previously, the boundary layer equations are parabolic, meaning that no information can be passed upstream from the downstream boundary. However, separation leads to **reverse flow** near the wall, destroying the parabolic nature of the flow field, and rendering the boundary layer equations inapplicable.

> The boundary layer equations are not valid downstream of a separation point because of reverse flow in the separation bubble.

In such cases, the full Navier–Stokes equations must be used in place of the boundary layer approximation. From the point of view of the boundary layer procedure of Fig. 10–93, the procedure breaks down because the outer flow calculated in step 1 is no longer valid when separation occurs, especially beyond the separation point (compare Fig. 10–121 to Fig. 10–122a).

Figure 10–122b shows the classic case of an airfoil at too high of an angle of attack, in which the separation point moves near the front of the airfoil;

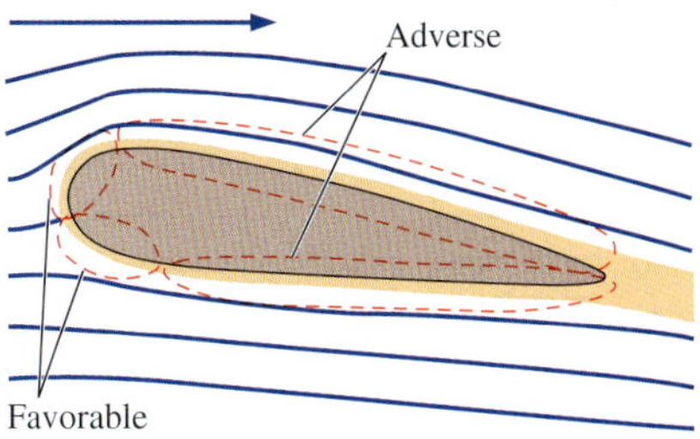

FIGURE 10–121
The boundary layer along a body immersed in a free stream is typically exposed to a favorable pressure gradient in the front portion of the body and an adverse pressure gradient in the rear portion of the body.

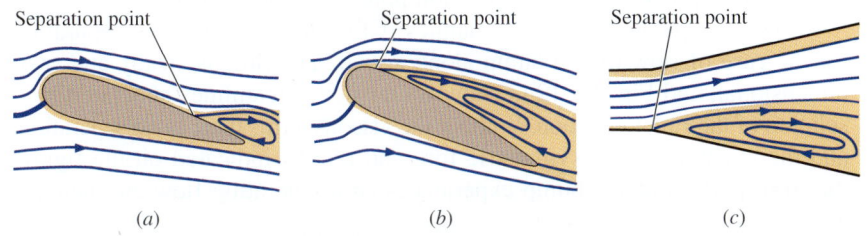

(a)   (b)   (c)

FIGURE 10–122
Examples of boundary layer separation in regions of adverse pressure gradient: (*a*) an airplane wing at a moderate angle of attack, (*b*) the same wing at a high angle of attack (a stalled wing), and (*c*) a wide-angle diffuser in which the boundary layer cannot remain attached and separates on one side.

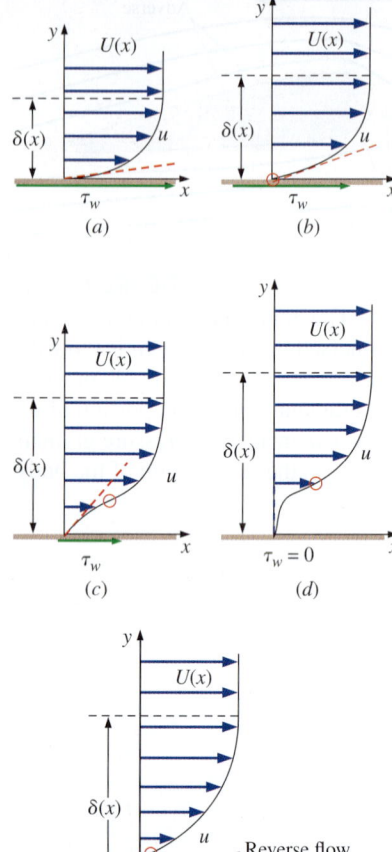

FIGURE 10–123
Comparison of boundary layer profile shape as a function of pressure gradient ($dP/dx = -U\, dU/dx$): (a) favorable, (b) zero, (c) mild adverse, (d) critical adverse (separation point), and (e) large adverse; inflection points are indicated by red circles, and wall shear stress $\tau_w = \mu\,(\partial u/\partial y)_{y=0}$ is sketched for each case.

the separation bubble covers nearly the entire upper surface of the airfoil—a condition known as **stall.** Stall is accompanied by a loss of lift and a marked increase in aerodynamic drag, as discussed in more detail in Chap. 11. Flow separation may also occur in internal flows, such as in the adverse pressure gradient region of a diffuser (Fig. 10–122c). As sketched, separation often occurs asymmetrically on one side of the diffuser only. As with an airfoil with flow separation, the outer flow calculation in the diffuser is no longer meaningful, and the boundary layer equations are not valid. Flow separation in a diffuser leads to a significant decrease of pressure recovery, and such conditions in a diffuser are also referred to as stall conditions.

We can learn a lot about the velocity profile shape under various pressure gradient conditions by examining the boundary layer momentum equation right at the wall. Since the velocity is zero at the wall (no-slip condition), the entire left side of Eq. 10–71b disappears, leaving only the pressure gradient term and the viscous term, which must balance,

At the wall: $$\nu\left(\frac{\partial^2 u}{\partial y^2}\right)_{y=0} = -U\frac{dU}{dx} = \frac{1}{\rho}\frac{dP}{dx} \qquad (10\text{–}88)$$

Under favorable pressure gradient conditions (accelerating outer flow), dU/dx is positive, and by Eq. 10–88, the second derivative of u at the wall is negative, i.e., $(\partial^2 u/\partial y^2)_{y=0} < 0$. We know that $\partial^2 u/\partial y^2$ must *remain* negative as u approaches $U(x)$ at the edge of the boundary layer. Thus, we expect the velocity profile across the boundary layer to be rounded, without any inflection point, as sketched in Fig. 10–123a. Under zero pressure gradient conditions, $(\partial^2 u/\partial y^2)_{y=0}$ is zero, implying a linear growth of u with respect to y near the wall, as sketched in Fig. 10–123b. (This is verified by the Blasius boundary layer profile for the zero pressure gradient boundary layer on a flat plate, as shown in Fig. 10–99.) For *adverse* pressure gradients, dU/dx is negative and Eq. 10–86 demands that $(\partial^2 u/\partial y^2)_{y=0}$ be positive. However, since $\partial^2 u/\partial y^2$ must be negative as u approaches $U(x)$ at the edge of the boundary layer, there has to be an *inflection point* ($\partial^2 u/\partial y^2 = 0$) somewhere in the boundary layer, as illustrated in Fig. 10–123c.

The *first* derivative of u with respect to y at the wall is directly proportional to τ_w, the wall shear stress [$\tau_w = \mu\,(\partial u/\partial y)_{y=0}$]. Comparison of $(\partial u/\partial y)_{y=0}$ in Fig. 10–123a through c reveals that τ_w is largest for favorable pressure gradients and smallest for adverse pressure gradients. Boundary layer thickness increases as the pressure gradient changes sign, as also illustrated in Fig. 10–123. If the adverse pressure gradient is large enough, $(\partial u/\partial y)_{y=0}$ becomes zero (Fig. 10–123d); this location along a wall is the *separation point*, beyond which there is reverse flow and a separation bubble (Fig. 10–123e). Notice that beyond the separation point τ_w is *negative* due to the negative value of $(\partial u/\partial y)_{y=0}$. As mentioned previously, the boundary layer equations break down in regions of reverse flow. Thus, the boundary layer approximation may be appropriate up to the separation point, but not beyond.

We use computational fluid dynamics (CFD) to illustrate flow separation for the case of flow over a bump along a wall. The flow is steady and two-dimensional, and Fig. 10–124a shows outer flow streamlines generated by a solution of the Euler equation. Without the viscous terms there is no separation, and the streamlines are symmetric fore and aft. As indicated on the figure, the front portion of the bump experiences an accelerating flow and hence a

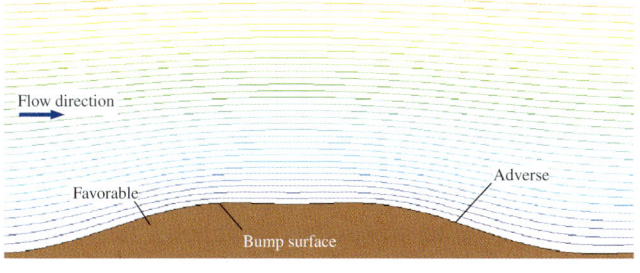
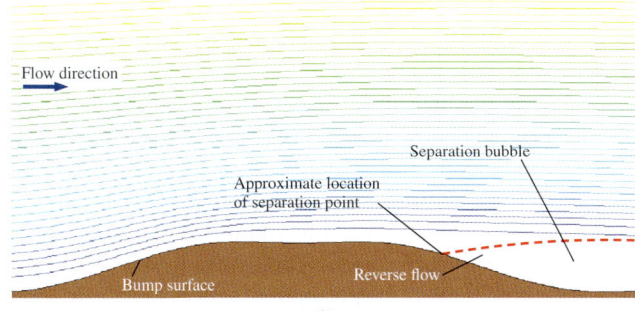
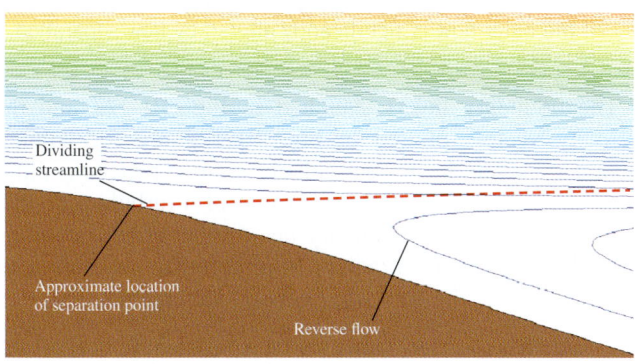
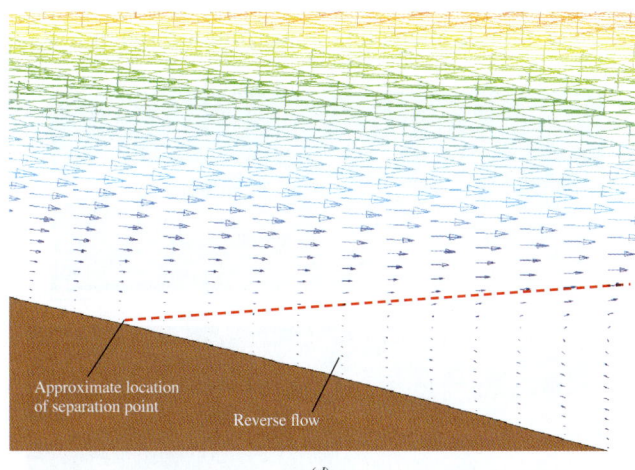

FIGURE 10–124
CFD calculations of flow over a two-dimensional bump: (*a*) solution of the Euler equation with outer flow streamlines plotted (no flow separation), (*b*) laminar flow solution showing flow separation on the downstream side of the bump, (*c*) close-up view of streamlines near the separation point, and (*d*) close-up view of velocity vectors, same view as (*c*). The dashed red line is a *dividing streamline* – fluid below this streamline is "trapped" in the recirculating separation bubble.

favorable pressure gradient. The rear portion experiences a decelerating flow and an adverse pressure gradient. When the full (laminar) Navier–Stokes equation is solved, the viscous terms lead to flow separation off the rear end of the bump, as seen in Fig. 10–124b. Keep in mind that this is a Navier–Stokes solution, not a boundary layer solution; nevertheless it illustrates the process of flow separation in the boundary layer. The approximate location of the separation point is indicated in Fig. 10–124b, and the dashed red line is a type of **dividing streamline.** Fluid below this streamline is caught in the separation bubble, while fluid above this streamline continues downstream. A close-up view of streamlines is shown in Fig. 10–124c, and velocity vectors are plotted in Fig. 10–124d using the same close-up view. Reverse flow in the lower portion of the separation bubble is clearly visible. Also, there is a strong y-component of velocity beyond the separation point, and the outer flow is no longer nearly parallel to the wall. In fact, the separated outer flow is nothing like the original outer flow of Fig. 10–124a. This is typical and represents a serious deficiency in the boundary layer approach. Namely, the boundary layer equations may be able to predict the location of the separation point fairly well, but cannot predict anything beyond the separation point. In some cases the outer flow changes significantly *upstream* of the separation point as well, and the boundary layer approximation gives erroneous results.

> The boundary layer approximation is only as good as the outer flow solution; if the outer flow is significantly altered by flow separation, the boundary layer approximation is erroneous.

The boundary layers sketched in Fig. 10–123 and the flow separation velocity vectors plotted in Fig. 10–124 are for laminar flow. Turbulent boundary layers have qualitatively similar behavior, although as discussed previously, the mean velocity profile of a turbulent boundary layer is much fuller than a laminar boundary layer under similar conditions. Thus a stronger adverse pressure gradient is required to separate a turbulent boundary layer. We make the following general statement:

> Turbulent boundary layers are more resistant to flow separation than are laminar boundary layers exposed to the same adverse pressure gradient.

Experimental evidence for this statement is shown in Fig. 10–125, in which the outer flow is attempting a sharp turn through a 20° angle. The laminar

(a)
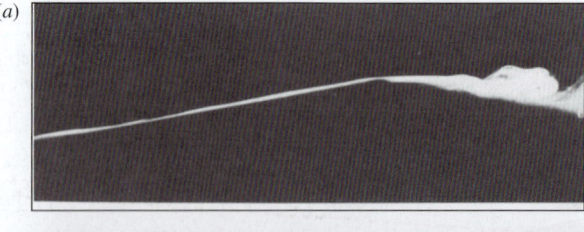

(b)

FIGURE 10–125

Flow visualization comparison of laminar and turbulent boundary layers in an adverse pressure gradient; flow is from left to right. (a) The laminar boundary layer separates at the corner, but (b) the turbulent one does not. Photographs taken by M. R. Head in 1982 as visualized with titanium tetrachloride.

Head, M. R. 1982 in Flow Visualization II, *W. Merzkirch, ed., pp. 399–403. Washington: Hemisphere.*

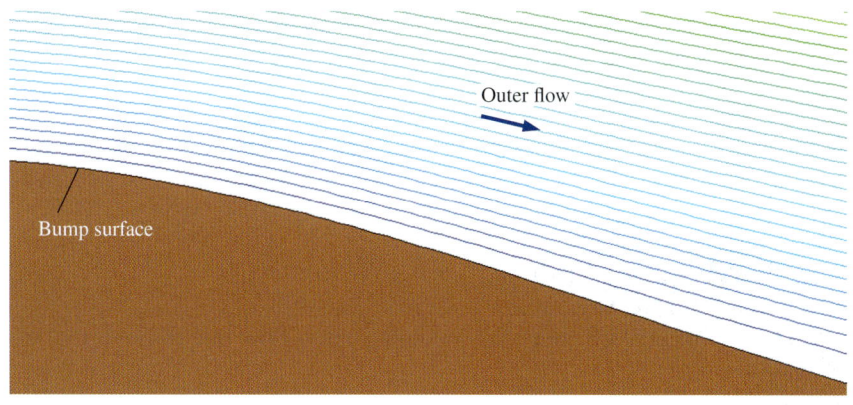

FIGURE 10–126
CFD calculation of turbulent flow over the same bump as that of Fig. 10–124. Compared to the laminar result of Fig. 10–124b, the turbulent boundary layer is more resistant to flow separation and does not separate in the adverse pressure gradient region in the rear portion of the bump.

boundary layer (Fig. 10–125a) cannot negotiate the sharp turn, and separates at the corner. The turbulent boundary layer on the other hand (Fig. 10–125b) manages to remain attached around the sharp corner.

As another example, flow over the same bump as that of Fig. 10–124 is recalculated, but with turbulence modeled in the simulation. Streamlines generated by the turbulent CFD calculation are shown in Fig. 10–126. Notice that the turbulent boundary layer remains attached (no flow separation), in contrast to the laminar boundary layer that separates off the rear portion of the bump. In the turbulent case, the outer flow Euler solution (Fig. 10–124a) is a reasonable approximation over the entire bump since there is no flow separation and since the boundary layer remains very thin.

A similar situation occurs for flow over bluff objects like spheres. A smooth golf ball, for example, would maintain a laminar boundary layer on its surface, and the boundary layer would separate fairly easily, leading to large aerodynamic drag. Golf balls have dimples (a type of surface roughness) in order to create an early transition to a turbulent boundary layer. Flow still separates from the golf ball surface, but much farther downstream in the boundary layer, resulting in significantly reduced aerodynamic drag. This is discussed in more detail in Chap. 11.

The Momentum Integral Technique for Boundary Layers

In many practical engineering applications, we do not need to know all the details inside the boundary layer; rather we seek reasonable estimates of gross features of the boundary layer such as boundary layer thickness and skin friction coefficient. The **momentum integral technique** utilizes a control volume approach to obtain such quantitative approximations of boundary layer properties along surfaces with zero or nonzero pressure gradients. The momentum integral technique is straightforward, and in some applications does not require use of a computer. It is valid for both laminar and turbulent boundary layers.

We begin with the control volume sketched in Fig. 10–127. The bottom of the control volume is the wall at $y = 0$, and the top is at $y = Y$, high enough to enclose the entire height of the boundary layer. The control volume is an infinitesimally thin slice of width dx in the x-direction. In accordance with

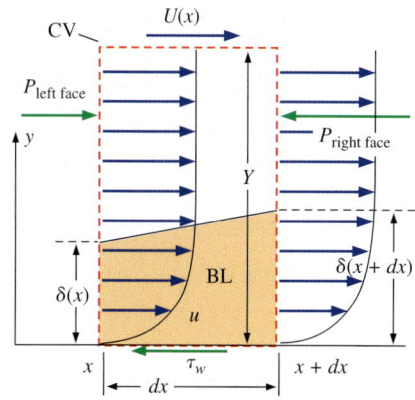

FIGURE 10–127
Control volume (dashed red line) used in derivation of the momentum integral equation.

the boundary layer approximation, $\partial P/\partial y = 0$, so we assume that pressure P acts along the entire left face of the control volume,

$$P_{\text{left face}} = P$$

In the general case with nonzero pressure gradient, the pressure on the right face of the control volume differs from that on the left face. Using a first-order truncated Taylor series approximation (Chap. 9), we set

$$P_{\text{right face}} = P + \frac{dP}{dx} dx$$

In a similar manner we write the incoming mass flow rate through the left face as

$$\dot{m}_{\text{left face}} = \rho w \int_0^Y u \, dy \qquad (10\text{–}89)$$

and the outgoing mass through the right face as

$$\dot{m}_{\text{right face}} = \rho w \left[\int_0^Y u \, dy + \frac{d}{dx}\left(\int_0^Y u \, dy\right) dx \right] \qquad (10\text{–}90)$$

where w is the width of the control volume into the page in Fig. 10–127. If you prefer, you can set w to unit width; it will cancel out later anyway.

Since Eq. 10–90 differs from Eq. 10–89, and since no flow crosses the bottom of the control volume (the wall), mass must flow into or out of the *top* face of the control volume. We illustrate this in Fig. 10–128 for the case of a growing boundary layer in which $\dot{m}_{\text{right face}} < \dot{m}_{\text{left face}}$, and $\dot{m}_{\text{top}}$ is positive (mass flows out). Conservation of mass over the control volume yields

$$\dot{m}_{\text{top}} = -\rho w \frac{d}{dx}\left(\int_0^Y u \, dy\right) dx \qquad (10\text{–}91)$$

We now apply conservation of x-momentum for the chosen control volume. The x-momentum is brought in through the left face and is removed through the right and top faces of the control volume. The net momentum flux out of the control volume must be balanced by the force due to the shear stress acting on the control volume by the wall and the net pressure force on the control surface, as shown in Fig. 10–127. The steady control volume x-momentum equation is thus

$$\underbrace{\sum F_{x,\text{body}}}_{\text{ignore gravity}} + \underbrace{\sum F_{x,\text{surface}}}_{YwP - Yw\left(P + \frac{dP}{dx}dx\right) - w\,dx\,\tau_w}$$

$$= \underbrace{\int_{\text{left face}} \rho u \vec{V} \cdot \vec{n} \, dA}_{-\rho w \int_0^Y u^2 \, dy} + \underbrace{\int_{\text{right face}} \rho u \vec{V} \cdot \vec{n} \, dA}_{\rho w \left[\int_0^Y u^2 \, dy + \frac{d}{dx}\left(\int_0^Y u^2 \, dy\right) dx\right]} + \underbrace{\int_{\text{top}} \rho u \vec{V} \cdot \vec{n} \, dA}_{\dot{m}_{\text{top}} U}$$

where the momentum flux through the top surface of the control volume is taken as the mass flow rate through that surface times U. Some of the terms cancel, and we rewrite the equation as

$$-Y\frac{dP}{dx} - \tau_w = \rho \frac{d}{dx}\left(\int_0^Y u^2 \, dy\right) - \rho U \frac{d}{dx}\left(\int_0^Y u \, dy\right) \qquad (10\text{–}92)$$

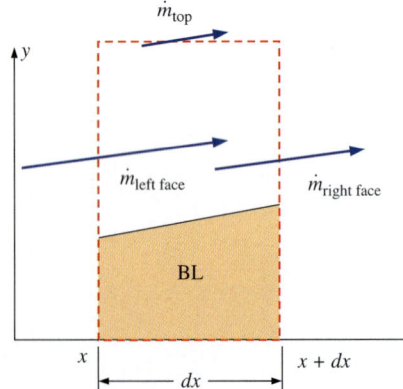

FIGURE 10–128
Mass flow balance on the control volume of Fig. 10–127.

where we have used Eq. 10–89 for $\dot{m}_{\text{top}}$, and w and dx cancel from each remaining term. For convenience we note that $Y = \int_0^Y dy$. From the outer flow (Euler equation), $dP/dx = -\rho U\, dU/dx$. After dividing each term in Eq. 10–90 by density ρ, we get

$$U\frac{dU}{dx}\int_0^Y dy - \frac{\tau_w}{\rho} = \frac{d}{dx}\left(\int_0^Y u^2\, dy\right) - U\frac{d}{dx}\left(\int_0^Y u\, dy\right) \quad (10\text{–}93)$$

We simplify Eq. 10–93 by utilizing the product rule of differentiation in reverse (Fig. 10–129). After some rearrangement, Eq. 10–91 becomes

$$\frac{d}{dx}\left(\int_0^Y u(U - u)\, dy\right) + \frac{dU}{dx}\int_0^Y (U - u)\, dy = \frac{\tau_w}{\rho}$$

where we are able to put U inside the integrals since at any given x-location, U is constant with respect to y (U is a function of x only).

We multiply and divide the first term by U^2 and the second term by U to get

$$\frac{d}{dx}\left(U^2\int_0^\infty \frac{u}{U}\left(1 - \frac{u}{U}\right) dy\right) + U\frac{dU}{dx}\int_0^\infty \left(1 - \frac{u}{U}\right) dy = \frac{\tau_w}{\rho} \quad (10\text{–}94)$$

where we have also substituted infinity in place of Y in the upper limit of each integral since $u = U$ for all y greater than Y, and thus the value of the integral does not change by this substitution.

We previously defined displacement thickness δ^* (Eq. 10–72) and momentum thickness θ (Eq. 10–80) for a flat plate boundary layer. In the general case with nonzero pressure gradient, we define δ^* and θ in the same way, except we use the *local* value of outer flow velocity, $U = U(x)$, at a given x-location in place of the constant U since U now varies with x. Equation 10–94 is thus written in more compact form as

Kármán integral equation: $\qquad \dfrac{d}{dx}(U^2\theta) + U\dfrac{dU}{dx}\delta^* = \dfrac{\tau_w}{\rho} \quad (10\text{–}95)$

Equation 10–95 is called the **Kármán integral equation** in honor of Theodor von Kármán (1881–1963), a student of Prandtl, who was the first to derive the equation in 1921.

An alternate form of Eq. 10–95 is obtained by performing the product rule on the first term, dividing by U^2, and rearranging,

Kármán integral equation, alternative form: $\qquad \dfrac{C_{f,x}}{2} = \dfrac{d\theta}{dx} + (2 + H)\dfrac{\theta}{U}\dfrac{dU}{dx} \quad (10\text{–}96)$

where we define **shape factor** H as

Shape factor: $\qquad H = \dfrac{\delta^*}{\theta} \quad (10\text{–}97)$

and **local skin friction coefficient** $C_{f,x}$ as

Local skin friction coefficient: $\qquad C_{f,x} = \dfrac{\tau_w}{\frac{1}{2}\rho U^2} \quad (10\text{–}98)$

Note that both H and $C_{f,x}$ are functions of x for the general case of a boundary layer with a nonzero pressure gradient developing along a surface.

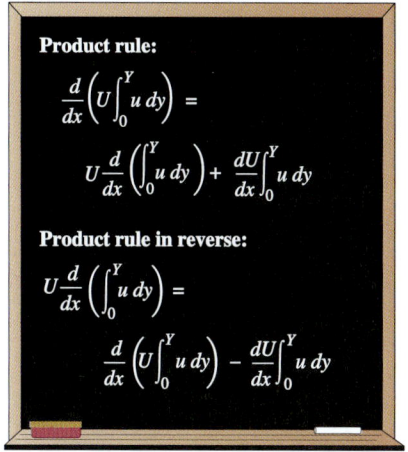

FIGURE 10–129
The product rule is utilized in reverse in the derivation of the momentum integral equation.

We emphasize again that the derivation of the Kármán integral equation and Eqs. 10–95 through 10–98 are valid for any steady incompressible boundary layer along a wall, regardless of whether the boundary layer is laminar, turbulent, or somewhere in between. For the special case of the boundary layer on a flat plate, $U(x) = U =$ constant, and Eq. 10–96 reduces to

Kármán integral equation, flat plat boundary layer: $\quad C_{f,x} = 2\dfrac{d\theta}{dx}$ (10–99)

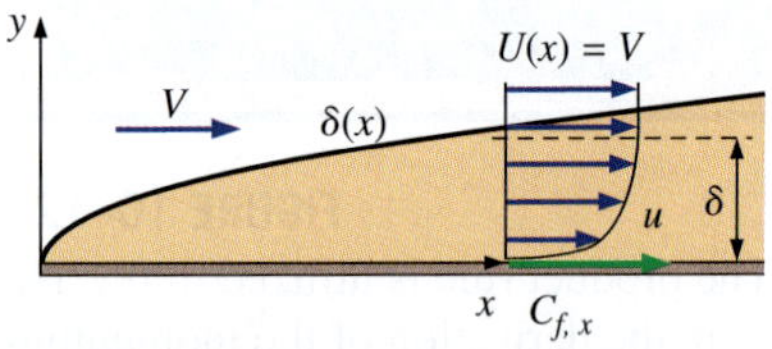

FIGURE 10–130
The turbulent boundary layer generated by flow over a flat plate for Example 10–14 (boundary layer thickness exaggerated).

EXAMPLE 10–14   Flat Plate Boundary Layer Analysis Using the Kármán Integral Equation

Suppose we know only two things about the turbulent boundary layer over a flat plate, namely, the local skin friction coefficient (Fig. 10–130),

$$C_{f,x} \cong \frac{0.027}{(\text{Re}_x)^{1/7}} \quad (1)$$

and the one-seventh-power law approximation for the boundary layer profile shape,

$$\frac{u}{U} \cong \left(\frac{y}{\delta}\right)^{1/7} \text{ for } y \leq \delta \quad \frac{u}{U} \cong 1 \text{ for } y > \delta \quad (2)$$

Using the definitions of displacement thickness and momentum thickness and employing the Kármán integral equation, estimate how δ, δ^*, and θ vary with x.

SOLUTION We are to estimate δ, δ^*, and θ based on Eqs 1 and 2.
Assumptions **1** The flow is turbulent, but steady in the mean. **2** The plate is thin and is aligned parallel to the free stream, so that $U(x) = V =$ constant.
Analysis First we substitute Eq. 2 into Eq. 10–80 and integrate to find momentum thickness,

$$\theta = \int_0^\infty \frac{u}{U}\left(1 - \frac{u}{U}\right) dy = \int_0^\delta \left(\frac{y}{\delta}\right)^{1/7}\left(1 - \left(\frac{y}{\delta}\right)^{1/7}\right) dy = \frac{7}{72}\delta \quad (3)$$

Similarly, we find displacement thickness by integrating Eq. 10–72,

$$\delta^* = \int_0^\infty \left(1 - \frac{u}{U}\right) dy = \int_0^\delta \left(1 - \left(\frac{y}{\delta}\right)^{1/7}\right) dy = \frac{1}{8}\delta \quad (4)$$

The Kármán integral equation reduces to Eq. 10–97 for a flat plate boundary layer. We substitute Eq. 3 into Eq. 10–97 and rearrange to get

$$C_{f,x} = 2\frac{d\theta}{dx} = \frac{14}{72}\frac{d\delta}{dx}$$

from which

$$\frac{d\delta}{dx} = \frac{72}{14} C_{f,x} = \frac{72}{14} 0.027(\text{Re}_x)^{-1/7} \quad (5)$$

where we have substituted Eq. 1 for the local skin friction coefficient. Equation 5 can be integrated directly, yielding

Boundary layer thickness: $\quad \dfrac{\delta}{x} \cong \dfrac{0.16}{(\text{Re}_x)^{1/7}}$ (6)

Finally, substitution of Eqs. 3 and 4 into Eq. 6 gives approximations for δ^* and θ,

Displacement thickness:
$$\frac{\delta^*}{x} \cong \frac{0.020}{(\text{Re}_x)^{1/7}} \quad (7)$$

and

Momentum thickness:
$$\frac{\theta}{x} \cong \frac{0.016}{(\text{Re}_x)^{1/7}} \quad (8)$$

Discussion The results agree with the expressions given in column (a) of Table 10–4 to two significant digits. Indeed, many of the expressions in Table 10–4 were *generated* with the help of the Kármán integral equation.

While fairly simple to use, the momentum integral technique suffers from a serious deficiency. Namely, we must know (or guess) the boundary layer profile shape in order to apply the Kármán integral equation (Fig. 10–131). For the case of boundary layers with pressure gradients, boundary layer shape changes with x (as illustrated in Fig. 10–123), further complicating the analysis. Fortunately, the shape of the velocity profile does not need to be known precisely, since integration is very forgiving. Several techniques have been developed that utilize the Kármán integral equation to predict gross features of the boundary layer. Some of these techniques, such as Thwaite's method, do a very good job for laminar boundary layers. Unfortunately, the techniques that have been proposed for turbulent boundary layers have not been as successful. Many of the techniques require the assistance of a computer and are beyond the scope of the present textbook.

FIGURE 10–131
Integration of a known (or assumed) velocity profile is required when using the Kármán integral equation.

■ **EXAMPLE 10–15**   **Drag on the Wall of a Wind Tunnel Test Section**

■ A boundary layer develops along the walls of a rectangular wind tunnel. The air is at 20°C and atmospheric pressure. The boundary layer starts upstream of the contraction and grows into the test section (Fig. 10–132). By the time it reaches the test section, the boundary layer is fully turbulent. The boundary layer profile and its thickness are measured at both the beginning ($x = x_1$) and the end ($x = x_2$) of the bottom wall of the wind tunnel test section. The test section is 1.8 m long and 0.50 m wide (into the page in Fig. 10–132). The following measurements are made:

$$\delta_1 = 4.2 \text{ cm} \quad \delta_2 = 7.7 \text{ cm} \quad V = 10.0 \text{ m/s} \quad (1)$$

At both locations the boundary layer profile fits better to a one-eighth-power law approximation than to the standard one-seventh-power law approximation,

$$\frac{u}{U} \cong \left(\frac{y}{\delta}\right)^{1/8} \text{ for } y \leq \delta \quad \frac{u}{U} \cong 1 \text{ for } y > \delta \quad (2)$$

Estimate the total skin friction drag force F_D acting on the bottom wall of the wind tunnel test section.

SOLUTION We are to estimate the skin friction drag force on the bottom wall of the test section of the wind tunnel (between $x = x_1$ and $x = x_2$).

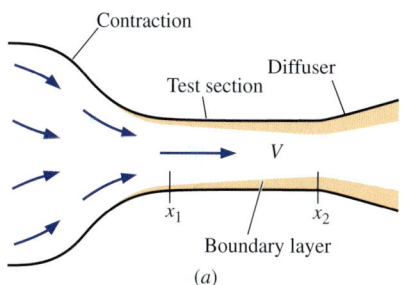

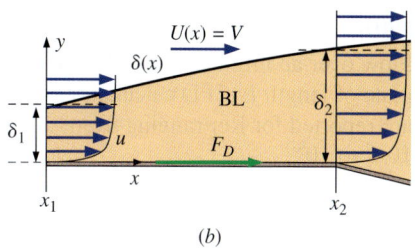

FIGURE 10–132
Boundary layer developing along the wind tunnel walls of Example 10–15: (*a*) overall view, and (*b*) close-up view of the bottom wall of the test section (boundary layer thickness exaggerated).

Properties For air at 20°C, $\nu = 1.516 \times 10^{-5}$ m²/s and $\rho = 1.204$ kg/m³.
Assumptions 1 The flow is steady in the mean. 2 The wind tunnel walls diverge slightly to ensure that $U(x) = V$ = constant.
Analysis First we substitute Eq. 2 into Eq. 10–80 and integrate to find momentum thickness θ,

$$\theta = \int_0^\infty \frac{u}{U}\left(1 - \frac{u}{U}\right) dy = \int_0^\delta \left(\frac{y}{\delta}\right)^{1/8}\left[1 - \left(\frac{y}{\delta}\right)^{1/8}\right] dy = \frac{4}{45}\delta \quad (3)$$

The Kármán integral equation reduces to Eq. 10–97 for a flat plate boundary layer. In terms of the shear stress along the wall, Eq. 10–97 is

$$\tau_w = \frac{1}{2}\rho U^2 C_{f,x} = \rho U^2 \frac{d\theta}{dx} \quad (4)$$

We integrate Eq. 4 from $x = x_1$ to $x = x_2$ to find the skin friction drag force,

$$F_D = w\int_{x_1}^{x_2} \tau_w \, dx = w\rho U^2 \int_{x_1}^{x_2} \frac{d\theta}{dx} dx = w\rho U^2(\theta_2 - \theta_1) \quad (5)$$

where w is the width of the wall into the page in Fig. 10–132. After substitution of Eq. 3 into Eq. 5 we obtain

$$F_D = w\rho U^2 \frac{4}{45}(\delta_2 - \delta_1) \quad (6)$$

Finally, substitution of the given numerical values into Eq. 6 yields the drag force,

$$F_D = (0.50 \text{ m})(1.204 \text{ kg/m}^3)(10.0 \text{ m/s})^2 \frac{4}{45}(0.077 - 0.042) \text{ m} \left(\frac{\text{s}^2\cdot\text{N}}{\text{kg}\cdot\text{m}}\right) = \mathbf{0.19 \text{ N}}$$

Discussion This is a very small force since the newton is itself a small unit of force. The Kármán integral equation would be more difficult to apply if the outer flow velocity $U(x)$ were not constant.

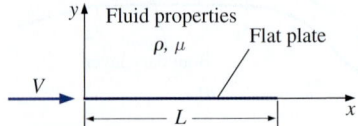

FIGURE 10–133
Flow over an infinitesimally thin flat plate of length L. CFD calculations are reported for Re_L ranging from 10^{-1} to 10^5.

We end this chapter with some illuminating results from CFD calculations of flow over a two-dimensional, infinitesimally thin flat plate aligned with the free stream (Fig. 10–133). In all cases the plate is 1 m long ($L = 1$ m), and the fluid is air with constant properties $\rho = 1.23$ kg/m³ and $\mu = 1.79 \times 10^{-5}$ kg/m·s. We vary free-stream velocity V so that the Reynolds number at the end of the plate ($Re_L = \rho VL/\mu$) ranges from 10^{-1} (creeping flow) to 10^5 (laminar but ready to start transitioning to turbulent). All cases are incompressible, steady, laminar Navier–Stokes solutions generated by a commercial CFD code. In Fig. 10–134, we plot velocity vectors for four Reynolds number cases at three x-locations: $x = 0$ (beginning of the plate), $x = 0.5$ m (middle of the plate), and $x = 1$ m (end of the plate). We also plot streamlines in the vicinity of the plate for each case.

In Fig. 10–134*a*, $Re_L = 0.1$, and the *creeping flow approximation* is reasonable. The flow field is nearly symmetric fore and aft—typical of creeping flow over symmetric bodies. Notice how the flow diverges around the plate as if it were of finite thickness. This is due to the large displacement effect caused by viscosity and the no-slip condition. In essence, the flow

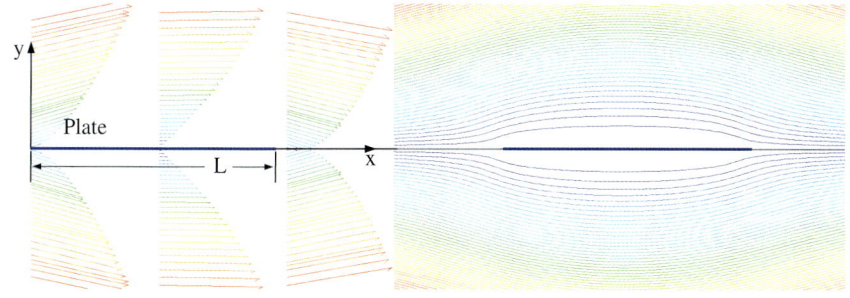

(a) $Re_L = 1\ 3\ 10^{-1}$

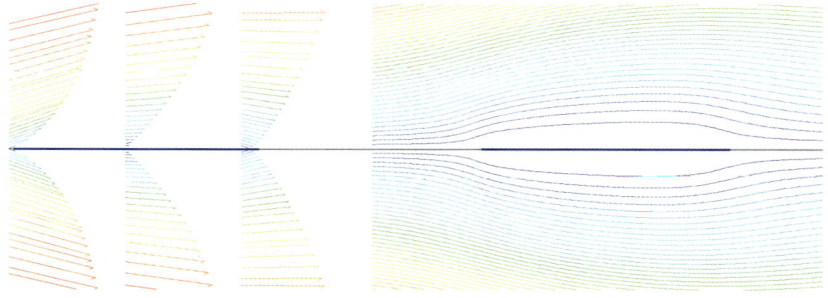

(b) $Re_L = 1\ 3\ 10^{1}$

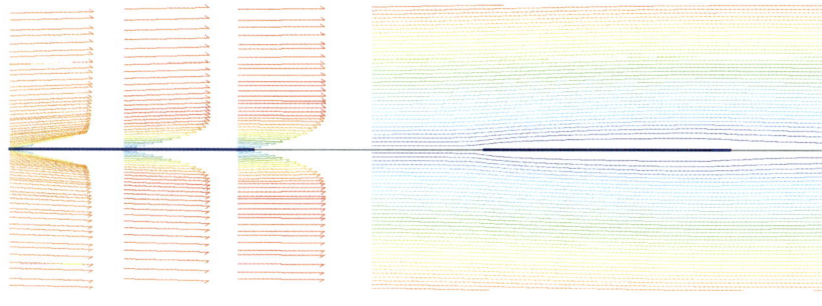

(c) $Re_L = 1\ 3\ 10^{3}$

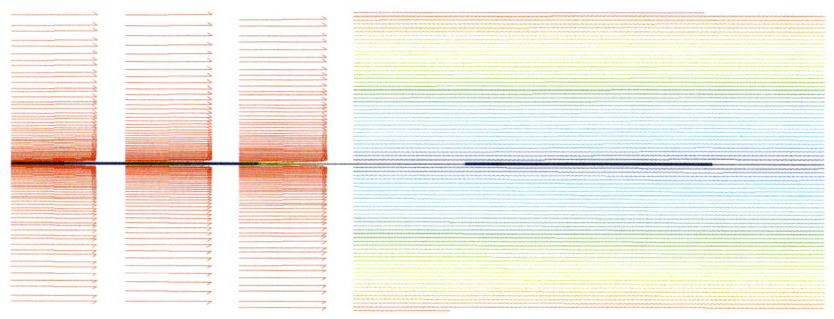

(d) $Re_L = 1\ 3\ 10^{5}$

FIGURE 10–134
CFD calculations of steady, incompressible, two-dimensional laminar flow from left to right over a 1-m-long flat plate of infinitesimal thickness; velocity vectors are shown in the left column at three locations along the plate, and streamlines near the plate are shown in the right column. $Re_L =$ (a) 0.1, (b) 10, (c) 1000, and (d) 100,000; only the upper half of the flow field is solved—the lower half is a mirror image. The computational domain extends hundreds of plate lengths beyond what is shown here in order to approximate "infinite" far-field conditions at the edges of the computational domain.

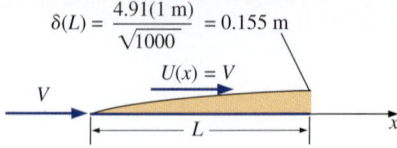

FIGURE 10–135

Calculation of boundary layer thickness for a laminar boundary layer on a flat plate at $Re_L = 1000$. This result is compared to the CFD-generated velocity profile at $x = L$ shown in Fig. 10–134c at this same Reynolds number.

velocity near the plate is so small that the rest of the flow "sees" it as a blockage around which the flow must be diverted. The y-component of velocity is significant near both the front and rear of the plate. Finally, the influence of the plate extends tens of plate lengths in all directions into the rest of the flow, which is also typical of creeping flows.

The Reynolds number is increased by two orders of magnitude to $Re_L = 10$ in the results shown in Fig. 10–134b. This Reynolds number is too high to be considered creeping flow, but too low for the boundary layer approximation to be appropriate. We notice some of the same features as those of the lower Reynolds number case, such as a large displacement of the streamlines and a significant y-component of velocity near the front and rear of the plate. The displacement effect is not as strong, however, and the flow is no longer symmetric fore and aft. We are seeing the effects of *inertia* as fluid leaves the end of the flat plate; inertia sweeps fluid into the developing wake behind the plate. The influence of the plate on the rest of the flow is still large, but much less so than for the flow at $Re_L = 0.1$.

In Fig. 10–134c are shown results of the CFD calculations at $Re_L = 1000$, another increase of two orders of magnitude. At this Reynolds number, inertial effects are starting to dominate over viscous effects throughout the majority of the flow field, and we can start calling this a *boundary layer* (albeit a fairly thick one). In Fig. 10–135 we calculate the boundary layer thickness using the laminar expression given in Table 10–4. The predicted value of $\delta(L)$ is about 15 percent of the plate length at $Re_L = 1000$, which is in reasonable agreement with the velocity vector plot at $x = L$ in Fig. 10–134c. Compared to the lower Reynolds number cases of Fig. 10–134a and b, the displacement effect is greatly reduced and any trace of fore–aft symmetry is gone.

Finally, the Reynolds number is once again increased by two orders of magnitude to $Re_L = 100{,}000$ in the results shown in Fig. 10–134d. There is no question about the appropriateness of the boundary layer approximation at this large Reynolds number. The CFD results show an extremely thin boundary layer with negligible effect on the outer flow. The streamlines of Fig. 10–134d are nearly parallel everywhere, and you must look closely to see the thin wake region behind the plate. The streamlines in the wake are slightly farther apart there than in the rest of the flow because in the wake region, the velocity is significantly less than the free-stream velocity. The y-component of velocity is negligible, as is expected in a very thin boundary layer, since the displacement thickness is so small.

Profiles of the x-component of velocity are plotted in Fig. 10–136 for each of the four Reynolds numbers of Fig. 10–134, plus some additional cases at other values of Re_L. We use a log scale for the vertical axis (y in units of m), since y spans several orders of magnitude. We nondimensionalize the abscissa as u/U so that the velocity profile shapes can be compared. All the profiles have a somewhat similar shape when plotted in this fashion. However, we notice that some of the profiles have a significant **velocity overshoot** ($u > U$) near the outer portion of the velocity profile. This is a direct result of the displacement effect and the effect of inertia as discussed before. At very *low* values of Re_L ($Re_L \leq 10^0$), where the displacement effect is most prominent, the velocity overshoot is almost nonexistent. This is explained by the lack of inertia at these low

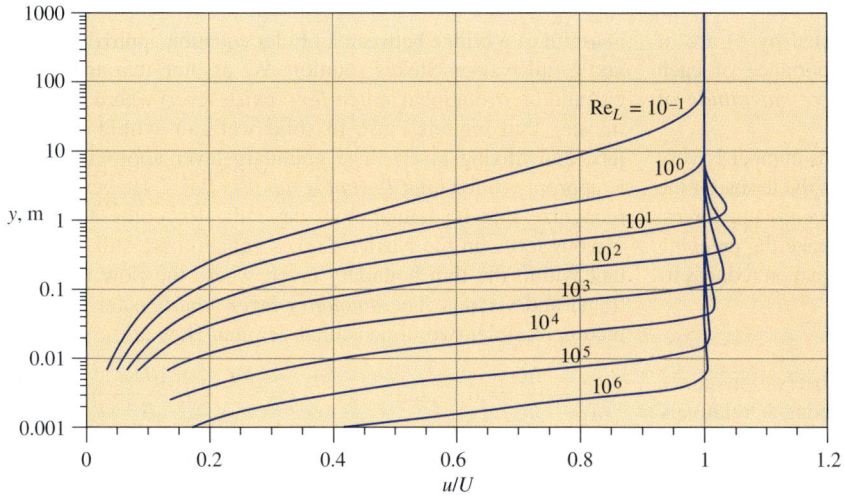

FIGURE 10–136
CFD calculations of steady, incompressible, two-dimensional laminar flow over a flat plate of infinitesimal thickness: nondimensional x velocity component u/U at the end of the plate ($x = L$) is plotted against vertical distance from the plate, y. Prominent velocity overshoot is observed at moderate Reynolds numbers, but disappears at very low and very high values of Re_L.

Reynolds numbers. Without inertia, there is no mechanism to accelerate the flow around the plate; rather, viscosity *retards* the flow everywhere in the vicinity of the plate, and the influence of the plate extends tens of plate lengths beyond the plate in all directions. For example, at $Re_L = 10^{-1}$, u does not reach 99 percent of U until $y \cong 320$ m—more than 300 plate lengths above the plate! At *moderate* values of the Reynolds number (Re_L between about 10^1 and 10^4), the displacement effect is significant, and inertial terms are no longer negligible. Hence, fluid is able to accelerate around the plate and the velocity overshoot is significant. For example, the maximum velocity overshoot is about 5 percent at $Re_L = 10^2$. At very *high* values of the Reynolds number ($Re_L \geq 10^5$), inertial terms dominate viscous terms, and the boundary layer is so thin that the displacement effect is almost negligible. The small displacement effect leads to very small velocity overshoot. For example, at $Re_L = 10^6$ the maximum velocity overshoot is only about 0.4 percent. Beyond $Re_L = 10^6$, laminar flow is no longer physically realistic, and the CFD calculations would need to include the effects of turbulence.

SUMMARY

Since the Navier–Stokes equation is difficult to solve, *approximations* are often used for practical engineering analyses. As with any approximation, however, we must be sure that the approximation is appropriate in the region of flow being analyzed. In this chapter we examine several approximations and show examples of flow situations in which they are useful. First we nondimensionalize the Navier–Stokes equation, yielding several nondimensional parameters: the *Strouhal number* (St), *Froude number* (Fr), *Euler number* (Eu), and *Reynolds number* (Re). Furthermore, for flows without free-surface effects, the hydrostatic pressure component due to gravity can be incorporated into a *modified pressure* P', effectively eliminating the gravity term (and the Froude number) from the Navier–Stokes equation. The nondimensionalized Navier–Stokes equation with modified pressure is

$$[\text{St}]\frac{\partial \vec{V}^*}{\partial t^*} + (\vec{V}^* \cdot \vec{\nabla}^*)\vec{V}^* = -[\text{Eu}]\vec{\nabla}^*P'^* + \left[\frac{1}{\text{Re}}\right]\vec{\nabla}^{*2}\vec{V}^*$$

When the nondimensional variables (indicated by *) are of order of magnitude unity, the relative importance of each term in the equation depends on the *relative magnitude* of the nondimensional parameters.

For regions of flow in which the Reynolds number is very small, the last term in the equation dominates the terms on the left side, and hence pressure forces must balance viscous forces. If we ignore inertial forces completely, we make the *creeping flow* approximation, and the Navier–Stokes equation reduces to

$$\vec{\nabla} P' \cong \mu \nabla^2 \vec{V}$$

Creeping flow is foreign to our everyday observations since our bodies, our automobiles, etc., move about at relatively high Reynolds numbers. The lack of inertia in the creeping flow approximation leads to some very interesting peculiarities, as discussed in this chapter.

We define *inviscid regions of flow* as regions where the viscous terms are negligible compared to the inertial terms (opposite of creeping flow). In such regions of flow the Navier–Stokes equation reduces to the *Euler equation*,

$$\rho \left(\frac{\partial \vec{V}}{\partial t} + (\vec{V} \cdot \vec{\nabla}) \vec{V} \right) = -\vec{\nabla} P'$$

In inviscid regions of flow, the Euler equation can be manipulated to derive the *Bernoulli equation*, valid along streamlines of the flow.

Regions of flow in which individual fluid particles do not rotate are called *irrotational regions of flow*. In such regions, the vorticity of fluid particles is negligibly small, and the viscous term in the Navier–Stokes equation can be neglected, leaving us again with the Euler equation. In addition, the Bernoulli equation becomes less restrictive, since the Bernoulli constant is the same everywhere, not just along streamlines. A nice feature of irrotational flow is that elementary flow solutions (*building block flows*) can be added together to generate more complicated flow solutions, a process known as *superposition*.

Since the Euler equation cannot support the no-slip boundary condition at solid walls, the *boundary layer approximation* is useful as a bridge between an Euler equation approximation and a full Navier–Stokes solution. We assume that an inviscid and/or irrotational *outer flow* exists everywhere except in very thin regions close to solid walls or within wakes, jets, and mixing layers. The boundary layer approximation is appropriate for *high Reynolds number flows*. However, we recognize that no matter how large the Reynolds number, viscous terms in the Navier–Stokes equation are still important within the thin boundary layer, where the flow is rotational and viscous. The *boundary layer equation* for steady, incompressible, two-dimensional, laminar flow are

$$\frac{\partial u}{\partial x} + \frac{\partial v}{\partial y} = 0 \quad \text{and} \quad u \frac{\partial u}{\partial x} + v \frac{\partial u}{\partial y} = U \frac{dU}{dx} + \nu \frac{\partial^2 u}{\partial y^2}$$

We define several measures of boundary layer thickness, including the *99 percent thickness* δ, the *displacement thickness* δ^*, and the *momentum thickness* θ. These quantities can be calculated exactly for a laminar boundary layer growing along a flat plate, under conditions of *zero pressure gradient*. As the Reynolds number increases down the plate, the boundary layer transitions to turbulence; semi-empirical expressions are given in this chapter for a turbulent flat plate boundary layer.

The *Kármán integral equation* is valid for both laminar and turbulent boundary layers exposed to arbitrary nonzero pressure gradients,

$$\frac{d}{dx}(U^2 \theta) + U \frac{dU}{dx} \delta^* = \frac{\tau_w}{\rho}$$

This equation is useful for "back of the envelope" estimations of gross boundary layer properties such as boundary layer thickness and skin friction.

The approximations presented in this chapter are applied to many practical problems in engineering. Potential flow analysis is useful for calculation of airfoil lift (Chap. 11). We utilize the inviscid approximation in the analysis of compressible flow (Chap. 12), open-channel flow (Chap. 13), and turbomachinery (Chap. 14). In cases where these approximations are not justified, or where more precise calculations are required, the continuity and Navier–Stokes equations are solved numerically using CFD (Chap. 15).

REFERENCES AND SUGGESTED READING

1. D. E. Coles. "The Law of the Wake in the Turbulent Boundary Layer," *J. Fluid Mechanics*, 1, pp. 191–226.
2. R. J. Heinsohn and J. M. Cimbala. *Indoor Air Quality Engineering*. New York: Marcel-Dekker, 2003.
3. P. K. Kundu, I. M. Cohen., and D. R. Dowling. *Fluid Mechanics*, ed. 5. San Diego, CA: Academic Press, 2011.
4. R. L. Panton. *Incompressible Flow*, 3rd ed. New York: Wiley, 2005.
5. M. Van Dyke. *An Album of Fluid Motion*. Stanford, CA: The Parabolic Press, 1982.
6. F. M. White. *Viscous Fluid Flow*, 3rd ed. New York: McGraw-Hill, 2005.
7. G. T. Yates. "How Microorganisms Move through Water," *American Scientist*, 74, pp. 358–365, July–August, 1986.

APPLICATION SPOTLIGHT ■ Droplet Formation

Guest Authors: James A. Liburdy and Brian Daniels, Oregon State University

Droplet formation is a complex interaction of inertial, surface tension, and viscous forces. The actual break-off of a drop from a stream of liquid, although studied for almost 200 years, has still not been fully explained. *Droplet-on Demand* (DoD) is used for such diverse applications as ink-jet printing and DNA analysis in microscale "lab-on-a-chip" devices. DoD requires very uniform droplet sizes, controlled velocities and trajectories, and a high rate of sequential droplet formation. For example, in ink-jet printing, the typical size of a droplet is 25 to 50 microns (barely visible with the naked eye), the velocities are on the order of 10 m/s, and the droplet formation rate can be higher than 20,000 per second.

The most common method for forming droplets involves accelerating a stream of liquid, and then allowing surface tension to induce an instability in the stream, which breaks up into individual droplets. In 1879, Lord Rayleigh developed a classical theory for the instability associated with this break-up; his theory is still widely used today to define droplet break-up conditions. A small perturbation to the surface of the liquid stream sets up an undulating pattern along the length of the stream, which causes the stream to break up into droplets whose size is determined by the radius of the stream and the surface tension of the liquid. However, most DoD systems rely on acceleration of the stream with time-dependent forcing functions in the form of a pressure wave exerted at the inlet of a nozzle. If the pressure wave is very rapid, viscous effects at the walls are negligible, and the potential flow approximation can be used to predict the flow.

Two important nondimensional parameters in DoD are the *Ohnesorge number* Oh = $\mu/(\rho\sigma_s a)^{1/2}$ and the *Weber number* We = $\rho V a/\sigma_s$, where a is the radius of the nozzle, σ_s is the surface tension, and V is the velocity. The Ohnesorge number determines when viscous forces are important relative to surface tension forces. In addition, the nondimensional pressure required to form an unstable fluid stream, $P_c = Pa/\sigma_s$, is called the *capillary pressure*, and the associated *capillary time scale* for droplets to form is $t_c = (\rho a/\sigma_s)^{1/2}$. When Oh is small, the potential flow approximation is applicable, and the surface shape is controlled by a balance between surface tension and fluid acceleration.

Example surfaces of flow emerging from a nozzle are shown in Fig. 10–137a and b. Surface shape depends on the pressure amplitude and the time scale of the perturbation, and is predicted well using the potential flow approximation. When the pressure is large enough and the pulse is fast enough, the surface ripples, and the center forms a jet stream that eventually breaks off into a droplet (Fig. 10–137c). An area of active research is how to control the size and velocity of these droplets, while producing thousands per second.

References

Rayleigh, Lord, "On the Instability of Jets," *Proc. London Math. Soc.*, 10, pp. 4–13, 1879.

Daniels, B. J., and Liburdy, J. A., "Oscillating Free-Surface Displacement in an Orifice Leading to Droplet Formation," *J. Fluids Engr.*, 10, pp. 7–8, 2004.

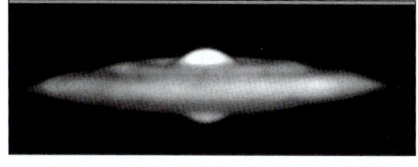

(a)

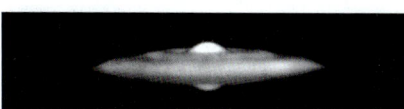

(b)

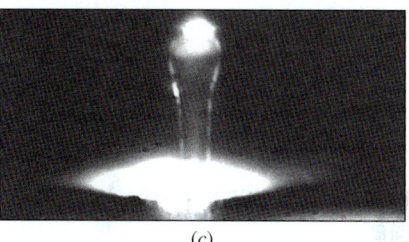

(c)

FIGURE 10–137

Droplet formation starts when a surface becomes unstable to a pressure pulse. Shown here are water surfaces in (a) an 800-micron orifice disturbed by a 5000-Hz pulse and (b) a 1200-micron orifice disturbed by an 8100-Hz pulse. Reflection from the surface causes the image to appear as if the surface wave is both up and down. The wave is axisymmetric, at least for small-amplitude pressure pulses. The higher the frequency, the shorter the wavelength and the smaller the central node. The size of the central node defines the diameter of the liquid jet, which then breaks up into a droplet. (c) Droplet formation from a high-frequency pressure pulse ejected from a 50-micron-diameter orifice. The center liquid stream produces the droplet and is only about 25 percent of the orifice diameter. Ideally, a single droplet forms, but unwanted, "satellite" droplets are often generated along with the main droplet.

Courtesy James A. Liburdy and Brian Daniels, Oregon State University. Used by permission.

594
APPROXIMATE SOLUTIONS OF THE N–S EQ

PROBLEMS*

Introductory Problems and Modified Pressure

10–1C A box fan sits on the floor of a very large room (Fig. P10–1C). Label regions of the flow field that may be approximated as static. Label regions in which the irrotational approximation is likely to be appropriate. Label regions where the boundary layer approximation may be appropriate. Finally, label regions in which the full Navier–Stokes equation most likely needs to be solved (i.e., regions where no approximation is appropriate).

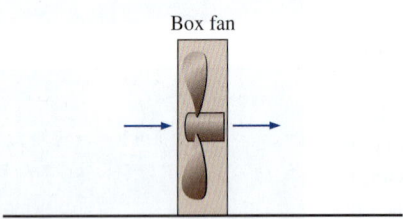

FIGURE P10–1C

10–2C Explain the difference between an "*exact*" solution of the Navier–Stokes equation (as discussed in Chap. 9) and an *approximate solution* (as discussed in this chapter).

10–3C Which nondimensional parameter in the nondimensionalized Navier–Stokes equation is eliminated by use of modified pressure instead of actual pressure? Explain.

10–4C What criteria can you use to determine whether an approximation of the Navier–Stokes equation is appropriate or not? Explain.

10–5C In the nondimensionalized incompressible Navier–Stokes equation (Eq. 10–5), there are four nondimensional parameters. Name each one, explain its physical significance (e.g., the ratio of pressure forces to viscous forces), and discuss what it means physically when the parameter is very small or very large.

10–6C What is the most important criterion for use of the *modified pressure P'* rather than the thermodynamic pressure P in a solution of the Navier–Stokes equation?

10–7C What is the most significant danger associated with an approximate solution of the Navier–Stokes equation? Give an example that is different than the ones given in this chapter.

---

* Problems designated by a "C" are concept questions, and students are encouraged to answer them all. Problems with the icon are solved using EES, and complete solutions together with parametric studies are included on the text website. Problems with the icon are comprehensive in nature and are intended to be solved with an equation solver such as EES.

10–8 Write out the three components of the Navier–Stokes equation in Cartesian coordinates in terms of modified pressure. Insert the definition of modified pressure and show that the x-, y-, and z-components are identical to those in terms of regular pressure. What is the advantage of using modified pressure?

10–9 Consider steady, incompressible, laminar, fully developed, planar Poiseuille flow between two parallel, horizontal plates (velocity and pressure profiles are shown in Fig. P10–9). At some horizontal location $x = x_1$, the pressure varies linearly with vertical distance z, as sketched. Choose an appropriate datum plane ($z = 0$), sketch the profile of *modified pressure* all along the vertical slice, and shade in the region representing the hydrostatic pressure component. Discuss.

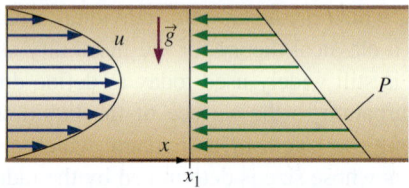

FIGURE P10–9

10–10 Consider the planar Poiseuille flow of Prob. 10–9. Discuss how modified pressure varies with downstream distance x. In other words, does modified pressure increase, stay the same, or decrease with x? If P' increases or decreases with x, how does it do so (e.g., linearly, quadratically, exponentially)? Use a sketch to illustrate your answer.

10–11 In Chap. 9 (Example 9–15), we generated an "exact" solution of the Navier–Stokes equation for fully developed Couette flow between two horizontal flat plates (Fig. P10–11), with gravity acting in the negative z-direction (into the page of Fig. P10–11). We used the actual pressure in that example. Repeat the solution for the x-component of velocity u and pressure P, but use the *modified pressure* in your equations. The pressure is P_0 at $z = 0$. Show that you get the same result as previously. Discuss.
Answers: $u = Vy/h$, $P = P_0 - \rho g z$

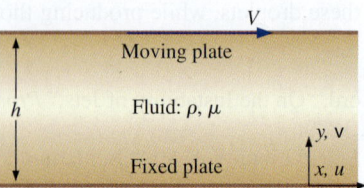

FIGURE P10–11

10–12 Consider flow of water through a small hole in the bottom of a large cylindrical tank (Fig. P10–12). The flow is laminar everywhere. Jet diameter d is much smaller than tank diameter D, but D is of the same order of magnitude as tank height H. Carrie reasons that she can use the fluid statics approximation everywhere in the tank except near the hole, but wants to validate this approximation mathematically. She lets the characteristic velocity scale in the tank be $V = V_{tank}$. The characteristic length scale is tank height H, the characteristic time is the time required to drain the tank t_{drain}, and the reference pressure difference is $\rho g H$ (pressure difference from the water surface to the bottom of the tank, assuming fluid statics). Substitute all these scales into the nondimensionalized incompressible Navier–Stokes equation (Eq. 10–6) and verify by order-of-magnitude analysis that for $d \ll D$, only the pressure and gravity terms remain. In particular, compare the order of magnitude of each term and each of the four nondimensional parameters St, Eu, Fr, and Re. (*Hint*: $V_{jet} \sim \sqrt{gH}$.) Under what criteria is Carrie's approximation appropriate?

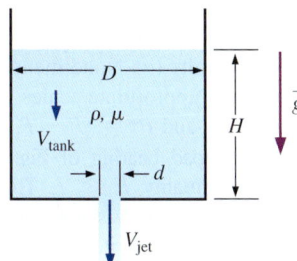

FIGURE P10–12

10–13 A flow field is simulated by a computational fluid dynamics code that uses the modified pressure in its calculations. A profile of modified pressure along a vertical slice through the flow is sketched in Fig. P10–13. The actual pressure at a point midway through the slice is known, as indicated on Fig. P10–13. Sketch the profile of actual pressure all along the vertical slice. Discuss.

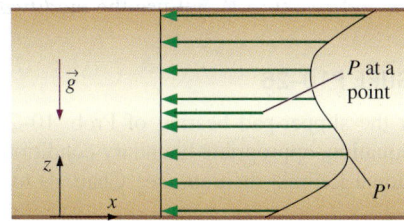

FIGURE P10–13

10–14 In Example 9–18 we solved the Navier–Stokes equation for steady, fully developed, laminar flow in a round pipe (Poiseuille flow), neglecting gravity. Now, add back the effect of gravity by re-solving that same problem, but use modified pressure P' instead of actual pressure P. Specifically, calculate the actual pressure field and the velocity field. Assume the pipe is horizontal, and let the datum plane $z = 0$ be at some arbitrary distance under the pipe. Is the actual pressure at the top of the pipe greater than, equal to, or less than that at the bottom of the pipe? Discuss.

Creeping Flow

10–15C Discuss why fluid density has negligible influence on the aerodynamic drag on a particle moving in the creeping flow regime.

10–16C Write a one-word description of each of the five terms in the incompressible Navier–Stokes equation,

$$\rho \frac{\partial \vec{V}}{\partial t} + \rho(\vec{V} \cdot \vec{\nabla})\vec{V} = -\vec{\nabla}P + \rho \vec{g} + \mu \nabla^2 \vec{V}$$
$$\quad\text{I} \qquad\quad \text{II} \qquad\qquad \text{III} \quad\;\; \text{IV} \quad\;\; \text{V}$$

When the creeping flow approximation is made, only two of the five terms remain. Which two terms remain, and why is this significant?

10–17 A person drops 3 aluminum balls of diameters 2 mm, 4 mm, and 10 mm into a tank filled with glycerin at 22°C ($\mu = 1$ kg·m/s), and measured the terminal velocities to be 3.2 mm/s, 12.8 mm/s, and 60.4 mm/s, respectively. The measurements are to be compared with theory using Stokes law for drag force acting on a spherical object of diameter D expressed as $F_D = 3\pi\mu DV$ for Re $\ll 1$. Compare experimental velocities values with those predicted theoretically.

10–18 Repeat Prob. 10–17 by considering the general form of the Stokes law expressed as $F_D = 3\pi\mu DV + (9\pi/16)\rho V^2 D^2$.

10–19 The viscosity of clover honey is listed as a function of temperature in Table P10–19. The specific gravity of the honey is about 1.42 and is not a strong function of temperature. The honey is squeezed through a small hole of diameter $D = 6.0$ mm in the lid of an inverted honey jar. The room and the honey are at $T = 20°C$. Estimate the maximum speed of the honey through the hole such that the flow can be approximated as creeping flow. (Assume that Re must be less than 0.1 for the creeping flow approximation to be appropriate.) Repeat your calculation if the temperature is 50°C. Discuss. *Answers*: 0.22 m/s, 0.012 m/s

TABLE P10–19
Viscosity of clover honey at 16 percent moisture content

T, °C	μ, poise*
14	600
20	190
30	65
40	20
50	10
70	3

* Poise = g/cm·s.

Data from Airborne Honey, Ltd., www.airborne.co.nz.

10–20 A good swimmer can swim 100 m in about a minute. If a swimmer's body is 1.85 m long, how many body lengths does he swim per second? Repeat the calculation for the sperm of Fig. 10–10. In other words, how many body lengths does the sperm swim per second? Use the sperm's whole body length, not just that of his head, for the calculation. Compare the two results and discuss.

10–21 A drop of water in a rain cloud has diameter $D = 42.5\ \mu m$ (Fig. P10–21). The air temperature is 25°C, and its pressure is standard atmospheric pressure. How fast does the air have to move vertically so that the drop will remain suspended in the air? *Answer: 0.0531 m/s*

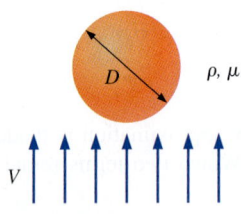

FIGURE P10–21

10–22 A *slipper-pad bearing* (Fig. P10–22) is often encountered in lubrication problems. Oil flows between two blocks; the upper one is stationary, and the lower one is moving in this case. The drawing is not to scale; in actuality, $h \ll L$. The thin gap between the blocks converges with increasing x. Specifically, gap height h decreases linearly from h_0 at $x = 0$ to h_L at $x = L$. Typically, the gap height length scale h_0 is much smaller than the axial length scale L. This problem is more complicated than simple Couette flow between parallel plates because of the changing gap height. In particular, axial velocity component u is a function of both x and y, and pressure P varies nonlinearly from $P = P_0$ at $x = 0$ to $P = P_L$ at $x = L$. ($\partial P/\partial x$ is not constant). Gravity forces are negligible in this flow field, which we approximate as two-dimensional, steady, and laminar. In fact, since h is so small and oil is so viscous, the creeping flow approximations are used in the analysis of such lubrication problems. Let the characteristic length scale associated with x be L, and let that associated with y be h_0 ($x \sim L$ and $y \sim h_0$). Let $u \sim V$. Assuming creeping flow, generate a characteristic scale for pressure difference $\Delta P = P - P_0$ in terms of L, h_0, μ, and V. *Answer: $\mu VL/h_0^2$*

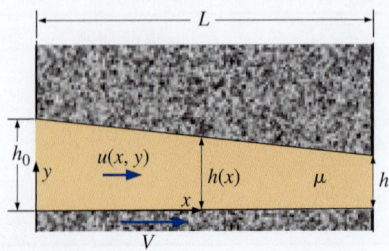

FIGURE P10–22

10–23 Consider the slipper-pad bearing of Prob. 10–22. (*a*) Generate a characteristic scale for v, the y-component of velocity. (*b*) Perform an order-of-magnitude analysis to compare the inertial terms to the pressure and viscous terms in the x-momentum equation. Show that when the gap is small ($h_0 \ll L$) and the Reynolds number is small (Re $= \rho V h_0/\mu \ll 1$), the creeping flow approximation is appropriate. (*c*) Show that when $h_0 \ll L$, the creeping flow equations may still be appropriate even if the Reynolds number (Re $= \rho V h_0/\mu$) is not less than 1. Explain. *Answer: (a) Vh_0/L*

10–24 Consider again the slipper-pad bearing of Prob. 10–22. Perform an order-of-magnitude analysis on the y-momentum equation, and write the final form of the y-momentum equation. (*Hint*: You will need the results of Probs. 10–22 and 10–23.) What can you say about pressure gradient $\partial P/\partial y$?

10–25 Consider again the slipper-pad bearing of Prob. 10–22. (*a*) List appropriate boundary conditions on u. (*b*) Solve the creeping flow approximation of the x-momentum equation to obtain an expression for u as a function of y (and indirectly as a function of x through h and dP/dx, which are functions of x). You may assume that P is *not* a function of y. Your final expression should be written as $u(x, y) = f(y, h, dP/dx, V,$ and $\mu)$. Name the two distinct components of the velocity profile in your result. (*c*) Nondimensionalize your expression for u using these appropriate scales: $x^* = x/L$, $y^* = y/h_0$, $h^* = h/h_0$, $u^* = u/V$, and $P^* = (P - P_0)h_0^2/\mu VL$.

10–26 Consider the slipper-pad bearing of Fig. P10–26. The drawing is not to scale; in actuality, $h \ll L$. This case differs from that of Prob. 10–22 in that $h(x)$ is not linear; rather h is some known, arbitrary function of x. Write an expression for axial velocity component u as a function of y, h, dP/dx, V, and μ. Discuss any differences between this result and that of Prob. 10–25.

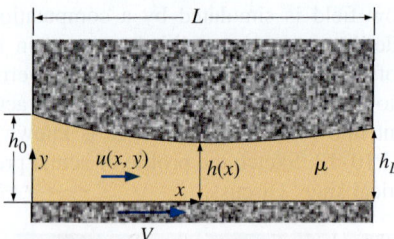

FIGURE P10–26

10–27 For the slipper-pad bearing of Prob. 10–22, use the continuity equation, appropriate boundary conditions, and the one-dimensional Leibniz theorem (see Chap. 4) to show that

$$\frac{d}{dx}\int_0^h u\,dy = 0.$$

10–28 Combine the results of Probs. 10–25 and 10–27 to show that for a two-dimensional slipper-pad bearing, pressure gradient dP/dx is related to gap height h by $\dfrac{d}{dx}\left(h^3\dfrac{dP}{dx}\right) = 6\mu U\dfrac{dh}{dx}$. This is the steady, two-dimensional form of the more general **Reynolds equation** for lubrication (Panton, 2005).

10–29 Consider flow through a two-dimensional slipper-pad bearing with linearly decreasing gap height from h_0 to h_L (Fig. P10–22), namely, $h = h_0 + \alpha x$, where α is the nondimensional convergence of the gap, $\alpha = (h_L - h_0)/L$. We note that $\tan \alpha \cong \alpha$ for very small values of α. Thus, α is approximately the angle of convergence of the upper plate in Fig. P10–22 (α is *negative* for this case). Assume that the oil is exposed to atmospheric pressure at both ends of the slipper-pad, so that $P = P_0 = P_{atm}$ at $x = 0$ and $P = P_L = P_{atm}$ at $x = L$. Integrate the Reynolds equation (Prob. 10–28) for this slipper-pad bearing to generate an expression for P as a function of x.

10–30 Estimate the speed at which you would need to swim in room temperature water to be in the creeping flow regime. (An order-of-magnitude estimate will suffice.) Discuss.

10–31 For each case, calculate an appropriate Reynolds number and indicate whether the flow can be approximated by the creeping flow equations. (*a*) A microorganism of diameter 5.0 μm swims in room temperature water at a speed of 0.25 mm/s. (*b*) Engine oil at 140°C flows in the small gap of a lubricated automobile bearing. The gap is 0.0012 mm thick, and the characteristic velocity is 15 m/s. (*c*) A fog droplet of diameter 10 μm falls through 30°C air at a speed of 2.5 mm/s.

10–32 Estimate the speed and Reynolds number of the sperm shown in Fig. 10–10. Is this microorganism swimming under creeping flow conditions? Assume it is swimming in room-temperature water.

Inviscid Flow

10–33C What is the main difference between the steady, incompressible Bernoulli equation for irrotational regions of flow, and the steady incompressible Bernoulli equation for rotational but inviscid regions of flow?

10–34C In what way is the Euler equation an approximation of the Navier–Stokes equation? Where in a flow field is the Euler equation an appropriate approximation?

10–35 In a certain region of steady, two-dimensional, incompressible flow, the velocity field is given by $\vec{V} = (u, v) = (ax + b)\vec{i} + (-ay + cx)\vec{j}$. Show that this region of flow can be considered inviscid.

10–36 In the derivation of the Bernoulli equation for regions of inviscid flow, we rewrite the steady, incompressible Euler equation into a form showing that the gradient of three scalar terms is equal to the velocity vector crossed with the vorticity vector, noting that z is vertically upward,

$$\vec{\nabla}\left(\frac{P}{\rho} + \frac{V^2}{2} + gz\right) = \vec{V} \times \vec{\zeta}$$

We then employ some arguments about the direction of the gradient vector and the direction of the cross product of two vectors to show that the sum of the three scalar terms must be constant along a streamline. In this problem you will use a different approach to achieve the same result. Namely, take the dot product of both sides of the Euler equation with velocity vector $\vec{V}$ and apply some fundamental rules about the dot product of two vectors. Sketches may be helpful.

10–37 Write out the components of the Euler equation as far as possible in Cartesian coordinates (x, y, z) and (u, v, w). Assume gravity acts in some arbitrary direction.

10–38 Write out the components of the Euler equation as far as possible in cylindrical coordinates (r, θ, z) and (u_r, u_θ, u_z). Assume gravity acts in some arbitrary direction.

10–39 Water at $T = 20$°C rotates as a rigid body about the z-axis in a spinning cylindrical container (Fig. P10–39). There are no viscous stresses since the water moves as a solid body; thus the Euler equation is appropriate. (We neglect viscous stresses caused by air acting on the water surface.) Integrate the Euler equation to generate an expression for pressure as a function of r and z everywhere in the water. Write an equation for the shape of the free surface ($z_{surface}$ as a function of r). (*Hint*: $P = P_{atm}$ everywhere on the free surface. The flow is rotationally symmetric about the z-axis.) *Answer*: $z_{surface} = \omega^2 r^2/2g$

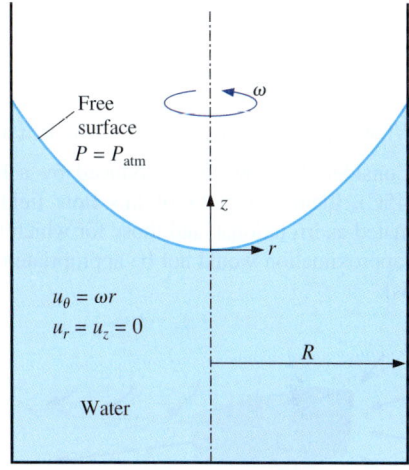

FIGURE P10–39

10–40 Repeat Prob. 10–39, except let the rotating fluid be engine oil at 60°C. Discuss.

10–41 Using the results of Prob. 10–39, calculate the Bernoulli constant as a function of radial coordinate r. *Answer*: $\frac{P_{atm}}{\rho} + \omega^2 r^2$

10–42 Consider steady, incompressible, two-dimensional flow of fluid into a converging duct with straight walls (Fig. P10–42). The volume flow rate is $\dot{V}$, and the velocity is in the radial direction only, with u_r a function of r only. Let b be the width into the page. At the inlet into the converging duct ($r = R$), u_r is known; $u_r = u_r(R)$. Assuming inviscid flow everywhere, generate an expression for u_r as a function of r, R, and $u_r(R)$ *only*. Sketch what the velocity profile at radius r would look like if friction were not neglected (i.e., a real flow) at the same volume flow rate.

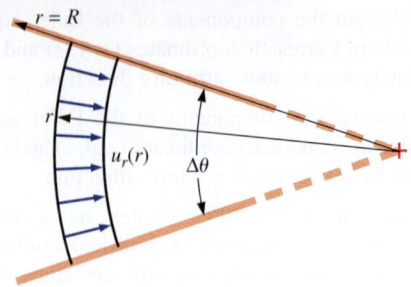

FIGURE P10–42

10–43 In the derivation of the Bernoulli equation for regions of inviscid flow, we use the vector identity

$$(\vec{V} \cdot \vec{\nabla})\vec{V} = \vec{\nabla}\left(\frac{V^2}{2}\right) - \vec{V} \times (\vec{\nabla} \times \vec{V})$$

Show that this vector identity is satisfied for the case of velocity vector $\vec{V}$ in Cartesian coordinates, i.e., $\vec{V} = u\vec{i} + v\vec{j} + w\vec{k}$. For full credit, expand each term as far as possible and show all your work.

Irrotational (Potential) Flow

10–44C What is D'Alembert's paradox? Why is it a paradox?

10–45C Consider the flow field produced by a hair dryer (Fig. P10–45C). Identify regions of this flow field that can be approximated as irrotational, and those for which the irrotational flow approximation would not be appropriate (rotational flow regions).

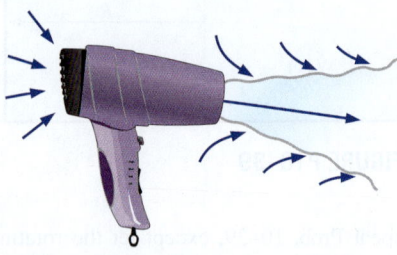

FIGURE P10–45C

10–46C In an irrotational region of flow, the velocity field can be calculated without need of the momentum equation by solving the Laplace equation for velocity potential function ϕ, and then solving for the components of $\vec{V}$ from the definition of ϕ, namely, $\vec{V} = \vec{\nabla}\phi$. Discuss the role of the momentum equation in an irrotational region of flow.

10–47C A subtle point, often missed by students of fluid mechanics (and even their professors!), is that an inviscid region of flow is *not* the same as an irrotational (potential) region of flow (Fig. P10–47C). Discuss the differences and similarities between these two approximations. Give an example of each.

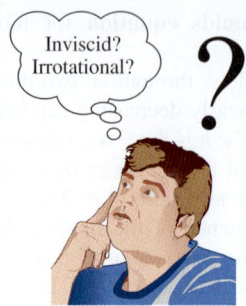

FIGURE P10–47C

10–48C What flow property determines whether a region of flow is rotational or irrotational? Discuss.

10–49 Write the Bernoulli equation, and discuss how it differs between an inviscid, rotational region of flow and a viscous, irrotational region of flow. Which case is more restrictive (in regards to the Bernoulli equation)?

10–50 Streamlines in a steady, two-dimensional, incompressible flow field are sketched in Fig. P10–50. The flow in the region shown is also approximated as irrotational. Sketch what a few equipotential curves (curves of constant potential function) might look like in this flow field. Explain how you arrive at the curves you sketch.

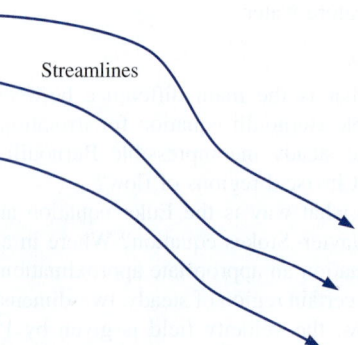

FIGURE P10–50

10–51 Consider the following steady, two-dimensional, incompressible velocity field: $\vec{V} = (u, v) = (ax + b)\vec{i} + (-ay + c)\vec{j}$. Is this flow field irrotational? If so, generate an expression for the velocity potential function. *Answers:* Yes, $a(x^2 - y^2)/2 + bx + cy + $ constant

10–52 Consider the following steady, two-dimensional, incompressible velocity field: $\vec{V} = (u, v) = (\frac{1}{2}ay^2 + b)\vec{i} + (axy + c)\vec{j}$. Is this flow field irrotational? If so, generate an expression for the velocity potential function.

10–53 Consider an irrotational line source of strength $\dot{V}/L$ in the xy- or $r\theta$-plane. The velocity components are $u_r = \dfrac{\partial \phi}{\partial r} =$

$\dfrac{1}{r}\dfrac{\partial \psi}{\partial \theta} = \dfrac{\dot{V}/L}{2\pi r}$ and $u_\theta = \dfrac{1}{r}\dfrac{\partial \phi}{\partial \theta} = -\dfrac{\partial \psi}{\partial r} = 0$. In this chapter, we started with the equation for u_θ to generate expressions for the velocity potential function and the stream function for the line source. Repeat the analysis, except start with the equation for u_r, showing all your work.

10–54 Consider a steady, two-dimensional, incompressible, irrotational velocity field specified by its velocity potential function, $\phi = 3(x^2 - y^2) + 4xy - 2x - 5y + 2$. (a) Calculate velocity components u and v. (b) Verify that the velocity field is irrotational in the region in which ϕ applies. (c) Generate an expression for the stream function in this region.

10–55 Consider a steady, two-dimensional, incompressible, irrotational velocity field specified by its velocity potential function, $\phi = 4(x^2 - y^2) + 6x - 4y$. (a) Calculate velocity components u and v. (b) Verify that the velocity field is irrotational in the region in which ϕ applies. (c) Generate an expression for the stream function in this region.

10–56 Consider a planar irrotational region of flow in the $r\theta$-plane. Show that stream function ψ satisfies the Laplace equation in cylindrical coordinates.

10–57 In this chapter, we describe axisymmetric irrotational flow in terms of cylindrical coordinates r and z and velocity components u_r and u_z. An alternative description of axisymmetric flow arises if we use *spherical polar coordinates* and set the x-axis as the axis of symmetry. The two relevant directional components are now r and θ, and their corresponding velocity components are u_r and u_θ. In this coordinate system, radial location r is the distance from the origin, and polar angle θ is the angle of inclination between the radial vector and the axis of rotational symmetry (the x-axis), as sketched in Fig. P10–57; a slice defining the $r\theta$-plane is shown. This is a type of two-dimensional flow because there are only two independent spatial variables, r and θ. In other words, a solution of the velocity and pressure fields in *any* $r\theta$-plane is sufficient to characterize the entire region of axisymmetric irrotational flow. Write the Laplace equation for ϕ in spherical polar coordinates, valid in regions of axisymmetric irrotational flow. (*Hint:* You may consult a textbook on vector analysis.)

10–58 Show that the incompressible continuity equation for axisymmetric flow in spherical polar coordinates, $\dfrac{1}{r}\dfrac{\partial}{\partial r}(r^2 u_r) + \dfrac{1}{\sin\theta}\dfrac{\partial}{\partial \theta}(u_\theta \sin\theta) = 0$, is identically satisfied by a stream function defined as $u_r = -\dfrac{1}{r^2 \sin\theta}\dfrac{\partial \psi}{\partial \theta}$ and $u_\theta = \dfrac{1}{r\sin\theta}\dfrac{\partial \psi}{\partial r}$, so long as ψ is a smooth function of r and θ.

10–59 Consider a uniform stream of magnitude V inclined at angle α (Fig. P10–59). Assuming incompressible planar irrotational flow, find the velocity potential function and the stream function. Show all your work. *Answers:* $\phi = Vx\cos\alpha + Vy\sin\alpha$, $\psi = Vy\cos\alpha - Vx\sin\alpha$

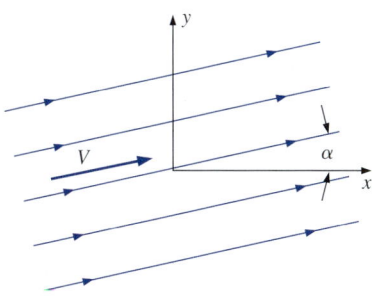

FIGURE P10–59

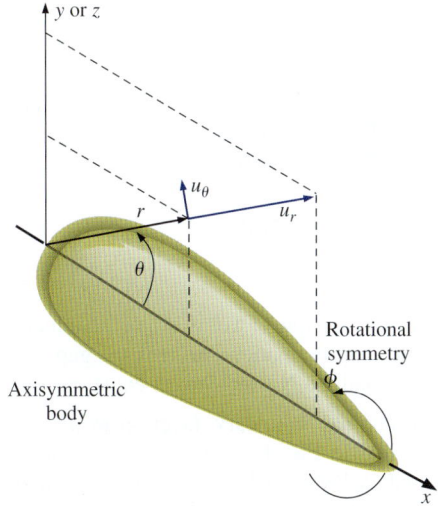

FIGURE P10–57

10–60 Consider the following steady, two-dimensional, incompressible velocity field: $\vec{V} = (u, v) = (\tfrac{1}{2}ay^2 + b)\vec{i} + (axy^2 + c)\vec{j}$. Is this flow field irrotational? If so, generate an expression for the velocity potential function.

10–61 In an irrotational region of flow, we write the velocity vector as the gradient of the scalar velocity potential function, $\vec{V} = \vec{\nabla}\phi$. The components of $\vec{V}$ in cylindrical coordinates, (r, θ, z) and (u_r, u_θ, u_z), are

$$u_r = \dfrac{\partial \phi}{\partial r} \quad u_\theta = \dfrac{1}{r}\dfrac{\partial \phi}{\partial \theta} \quad u_z = \dfrac{\partial \phi}{\partial z}$$

From Chap. 9, we also write the components of the vorticity vector in cylindrical coordinates as $\zeta_r = \dfrac{1}{r}\dfrac{\partial u_z}{\partial \theta} - \dfrac{\partial u_\theta}{\partial z}$, $\zeta_\theta = \dfrac{\partial u_r}{\partial z} - \dfrac{\partial u_z}{\partial r}$, and $\zeta_z = \dfrac{1}{r}\dfrac{\partial}{\partial r}(r u_\theta) - \dfrac{1}{r}\dfrac{\partial u_r}{\partial \theta}$. Substitute the velocity components into the vorticity components to show that all three components of the vorticity vector are indeed zero in an irrotational region of flow.

10–62 Substitute the components of the velocity vector given in Prob. 10–61 into the Laplace equation in cylindrical coordinates. Showing all your algebra, verify that the Laplace equation is valid in an irrotational region of flow.

10–63 Consider an irrotational line vortex of strength Γ in the xy- or $r\theta$-plane. The velocity components are $u_r = \dfrac{\partial \phi}{\partial r} = \dfrac{1}{r}\dfrac{\partial \psi}{\partial \theta} = 0$ and $u_\theta = \dfrac{1}{r}\dfrac{\partial \phi}{\partial \theta} = -\dfrac{\partial \psi}{\partial r} = \dfrac{\Gamma}{2\pi r}$. Generate expressions for the velocity potential function and the stream function for the line vortex, showing all your work.

10–64 Water at atmospheric pressure and temperature ($\rho = 998.2$ kg/m^3, and $\mu = 1.003 \times 10^{-3}$ kg/m·s) at free stream velocity $V = 0.100481$ m/s flows over a two-dimensional circular cylinder of diameter $d = 1.00$ m. Approximate the flow as potential flow. (a) Calculate the Reynolds number, based on cylinder diameter. Is Re large enough that potential flow should be a reasonable approximation? (b) Estimate the minimum and maximum speeds $|V|_{min}$ and $|V|_{max}$ (speed is the magnitude of velocity) and the maximum and minimum pressure difference $P - P_\infty$ in the flow, along with their respective locations.

10–65 The stream function for steady, incompressible, two-dimensional flow over a circular cylinder of radius a and free-stream velocity V_∞ is $\psi = V_\infty \sin\theta (r - a^2/r)$ for the case in which the flow field is approximated as irrotational (Fig. P10–65). Generate an expression for the velocity potential function ϕ for this flow as a function of r and θ, and parameters V_∞ and a.

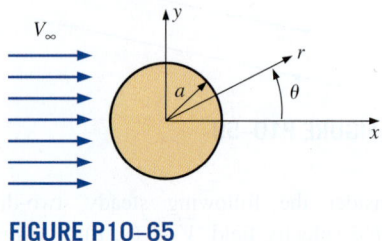

FIGURE P10–65

10–66 Superpose a uniform stream of velocity V_∞ and a line source of strength $\dot{V}/L$ at the origin. This generates potential flow over a two-dimensional half-body called the Rankine half-body (Fig. P10–66). One unique streamline is the **dividing streamline** that forms a dividing line between free-stream fluid coming from the left and fluid coming from the source. (a) Generate an equation for the dividing stream function $\psi_{dividing}$ as a function of $\dot{V}/L$. (Hint: The dividing streamline intersects the stagnation point at the nose of the body.) (b) Generate an expression for half-height b as a function of V_∞ and $\dot{V}/L$. (Hint: Consider the flow far downstream.) (c) Generate an equation for the dividing stream function in the form of r as a function of θ, V_∞, and $\dot{V}/L$. (d) Generate an expression for stagnation point distance a as a function of V_∞ and $\dot{V}/L$. (e) Generate an expression for $(V/V_\infty)^2$ (the squared nondimensional velocity magnitude) anywhere in the flow as a function of a, r, and θ.

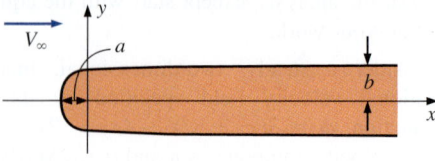

FIGURE P10–66

Boundary Layers

10–67C We usually think of boundary layers as occurring along solid walls. However, there are other flow situations in which the boundary layer approximation is also appropriate. Name three such flows, and explain why the boundary layer approximation is appropriate.

10–68C For each statement, choose whether the statement is true or false and discuss your answer briefly. These statements concern a laminar boundary layer on a flat plate (Fig. P10–68C). (a) At a given x-location, if the Reynolds number were to increase, the boundary layer thickness would also increase. (b) As outer flow velocity increases, so does the boundary layer thickness. (c) As the fluid viscosity increases, so does the boundary layer thickness. (d) As the fluid density increases, so does the boundary layer thickness.

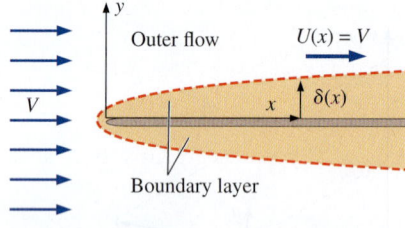

FIGURE P10–68C

10–69C In this chapter, we make a statement that the boundary layer approximation "bridges the gap" between the Euler equation and the Navier–Stokes equation. Explain.

10–70C A laminar boundary layer growing along a flat plate is sketched in Fig. P10–70C. Several velocity profiles and the boundary layer thickness $\delta(x)$ are also shown. Sketch several streamlines in this flow field. Is the curve representing $\delta(x)$ a streamline?

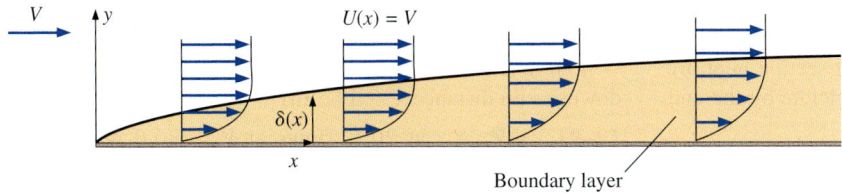

FIGURE P10–70C

10–71C What is a *trip wire*, and what is its purpose?

10–72C Discuss the implication of an inflection point in a boundary layer profile. Specifically, does the existence of an inflection point infer a favorable or adverse pressure gradient? Explain.

10–73C Compare flow separation for a laminar versus turbulent boundary layer. Specifically, which case is more resistant to flow separation? Why? Based on your answer, explain why golf balls have dimples.

10–74C In your own words, summarize the five steps of the boundary layer procedure.

10–75C In your own words, list at least three "red flags" to look out for when performing laminar boundary layer calculations.

10–76C Two definitions of displacement thickness are given in this chapter. Write both definitions in your own words. For the laminar boundary layer growing on a flat plate, which is larger—boundary layer thickness δ or displacement thickness δ^*? Discuss.

10–77C Explain the difference between a *favorable* and an *adverse* pressure gradient in a boundary layer. In which case does the pressure increase downstream? Why?

10–78 On a hot day ($T = 30°C$), a truck moves along the highway at 29.1 m/s. The flat side of the truck is treated as a simple, smooth flat–plate boundary layer, to first approximation. Estimate the *x*-location along the plate where the boundary layer begins to transition to turbulence. How far downstream from the beginning of the plate do you expect the boundary layer to become fully turbulent? Give both answers to one significant digit.

10–79 A boat moves through water ($T = 5°C$), at 42.0 km/h. A flat portion of the boat hull is 0.73 m long, and is treated as a simple smooth flat plate boundary layer, to first approximation. Is the boundary layer on this flat part of the hull laminar, transitional, or turbulent? Discuss.

10–80 Air flows parallel to a speed limit sign along the highway at speed $V = 8.5$ m/s. The temperature of the air is 25°C, and the width W of the sign parallel to the flow direction (i.e., its length) is 0.45 m. Is the boundary layer on the sign laminar or turbulent or transitional?

10–81 Static pressure P is measured at two locations along the wall of a laminar boundary layer (Fig. P10–81). The measured pressures are P_1 and P_2, and the distance between the taps is small compared to the characteristic body dimension ($\Delta x = x_2 - x_1 \ll L$). The outer flow velocity above the boundary layer at point 1 is U_1. The fluid density and viscosity are ρ and μ, respectively. Generate an approximate expression for U_2, the outer flow velocity above the boundary layer at point 2, in terms of P_1, P_2, Δx, U_1, ρ, and μ.

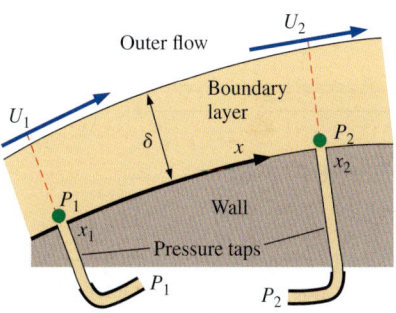

FIGURE P10–81

10–82 Consider two pressure taps along the wall of a laminar boundary layer as in Fig. P10–81. The fluid is air at 25°C, $U_1 = 10.3$ m/s, and the static pressure P_1 is 2.44 Pa greater than static pressure P_2, as measured by a very sensitive differential pressure transducer. Is outer flow velocity U_2 greater than, equal to, or less than outer flow velocity U_1? Explain. Estimate U_2. *Answers:* Less than, 10.1 m/s

10–83 Consider the Blasius solution for a laminar flat plate boundary layer. The nondimensional slope at the wall is given by Eq. 8 of Example 10–10. Transform this result to physical variables, and show that Eq. 9 of Example 10–10 is correct.

10–84 Calculate the value of shape factor H for the limiting case of a boundary layer that is infinitesimally thin (Fig. P10–84). This value of H is the minimum possible value.

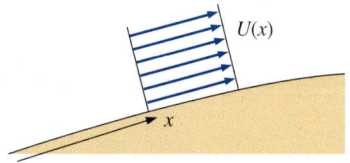

FIGURE P10–84

10–85 A laminar flow wind tunnel has a test section that is 30 cm in diameter and 80 cm in length. The air is at 20°C. At a uniform air speed of 2.0 m/s at the test section inlet, by how much will the centerline air speed accelerate by the end of the test section? *Answer:* Approx. 6%

10–86 Repeat the calculation of Prob. 10–85, except for a test section of square rather than round cross section, with a 30 cm × 30 cm cross section and a length of 80 cm. Compare the result to that of Prob. 10–85 and discuss.

10–87 Air at 20°C flows at $V = 8.5$ m/s parallel to a flat plate (Fig. P10–87). The front of the plate is well rounded, and the plate is 40 cm long. The plate thickness is $h = 0.75$ cm, but because of boundary layer displacement effects, the flow outside the boundary layer "sees" a plate that has larger apparent thickness. Calculate the apparent thickness of the plate (include both sides) at downstream distance $x = 10$ cm. *Answer:* 0.895 cm

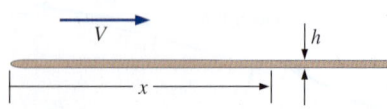

FIGURE P10–87

10–88 A small, axisymmetric, low-speed wind tunnel is built to calibrate hot wires. The diameter of the test section is 17.0 cm, and its length is 25.4 cm. The air is at 20°C. At a uniform air speed of 1.5 m/s at the test section inlet, by how much will the centerline air speed accelerate by the end of the test section? What should the engineers do to eliminate this acceleration?

10–89 Air at 20°C flows parallel to a smooth, thin, flat plate at 4.75 m/s. The plate is 3.23 m long. Determine whether the boundary layer on the plate is most likely laminar, turbulent, or somewhere in between (transitional). Compare the boundary layer thickness at the end of the plate for twocases: (a) the boundary layer is laminar everywhere, and (b) the boundary layer is turbulent everywhere. Discuss.

10–90 In order to avoid boundary layer interference, engineers design a "boundary layer scoop" to skim off the boundary

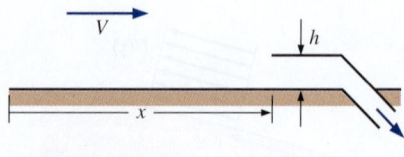

FIGURE P10–90

layer in a large wind tunnel (Fig. P10–90). The scoop is constructed of thin sheet metal. The air is at 20°C, and flows at $V = 45.0$ m/s. How high (dimension h) should the scoop be at downstream distance $x = 1.45$ m?

10–91 Air at 20°C flows at $V = 80.0$ m/s over a smooth flat plate of length $L = 17.5$ m. Plot the turbulent boundary layer profile in physical variables (u as a function of y) at $x = L$. Compare the profile generated by the one-seventh-power law, the log law, and Spalding's law of the wall, assuming that the boundary layer is fully turbulent from the beginning of the plate.

10–92 The streamwise velocity component of a steady, incompressible, laminar, flat plate boundary layer of boundary layer thickness δ is approximated by the simple linear expression, $u = Uy/\delta$ for $y < \delta$, and $u = U$ for $y > \delta$ (Fig. P10–92). Generate expressions for displacement thickness and momentum thickness as functions of δ, based on this linear approximation. Compare the approximate values of δ^*/δ and θ/δ to the values of δ^*/δ and θ/δ obtained from the Blasius solution. *Answers:* 0.500, 0.167

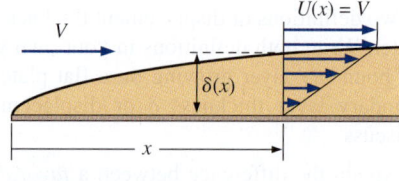

FIGURE P10–92

10–93 For the linear approximation of Prob. 10–92, use the definition of local skin friction coefficient and the Kármán integral equation to generate an expression for δ/x. Compare your result to the Blasius expression for δ/x. (Note: You will need the results of Prob. 10–92 to do this problem.)

10–94 Compare *shape factor H* (defined in Eq. 10–95) for a laminar versus a turbulent boundary layer on a flat plate, assuming that the turbulent boundary layer is turbulent from the beginning of the plate. Discuss. Specifically, why do you suppose H is called a "shape factor"? *Answers:* 2.59, 1.25 to 1.30

10–95 One dimension of a rectangular flat plate is twice the other. Air at uniform speed flows parallel to the plate, and a laminar boundary layer forms on both sides of the plate. Which orientation—long dimension parallel to the wind (Fig. P10–95a) or short dimension parallel to the wind (Fig. P10–95b)—has the higher drag? Explain.

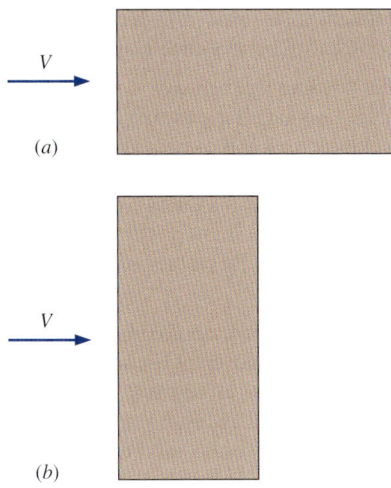

FIGURE P10–95

10–96 Integrate Eq. 5 to obtain Eq. 6 of Example 10–14, showing all your work.

10–97 Consider a turbulent boundary layer on a flat plate. Suppose only two things are known: $C_{f,x} \cong 0.059 \cdot (Re_x)^{-1/5}$ and $\theta \cong 0.097\delta$. Use the Kármán integral equation to generate an expression for δ/x, and compare your result to column (b) of Table 10–4.

10–98 Air at 30°C flows at a uniform speed of 35.0 m/s along a smooth flat plate. Calculate the approximate x-location along the plate where the boundary layer begins the transition process toward turbulence. At approximately what x-location along the plate is the boundary layer likely to be fully turbulent? *Answers:* 4 to 5 cm, 1 to 2 m

10–99 An aluminum canoe moves horizontally along the surface of a lake at 5.6 km/h (Fig. P10–99). The temperature of the lake water is 10°C. The bottom of the canoe is 6.1 m long and is flat. Is the boundary layer on the canoe bottom laminar or turbulent?

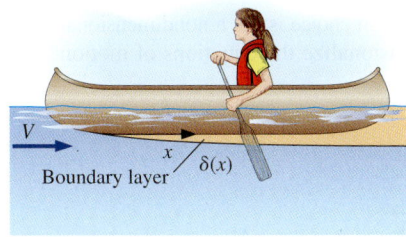

FIGURE P10–99

Review Problems

10–100C For each statement, choose whether the statement is true or false, and discuss your answer briefly.

(*a*) The velocity potential function can be defined for three-dimensional flows.

(*b*) The vorticity must be zero in order for the stream function to be defined.

(*c*) The vorticity must be zero in order for the velocity potential function to be defined.

(*d*) The stream function can be defined only for two-dimensional flow fields.

10–101 In this chapter, we discuss solid body rotation (Fig. P10–101) as an example of an inviscid flow that is also rotational. The velocity components are $u_r = 0$, $u_\theta = \omega r$, and $u_z = 0$. Compute the viscous term of the θ-component of the Navier–Stokes equation, and discuss. Verify that this velocity field is indeed rotational by computing the z-component of vorticity. *Answer:* $\zeta_z = 2\omega$

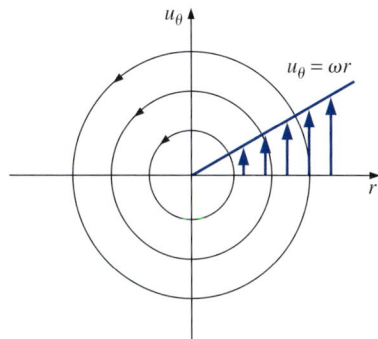

FIGURE P10–101

10–102 Calculate the nine components of the viscous stress tensor in cylindrical coordinates (see Chap. 9) for the velocity field of Prob. 10–101. Discuss your results.

10–103 In this chapter, we discuss the line vortex (Fig. P10–103) as an example of an irrotational flow field. The velocity components are $u_r = 0$, $u_\theta = \Gamma/(2\pi r)$, and $u_z = 0$. Compute the viscous term of the θ-component of the Navier–Stokes equation, and discuss. Verify that this velocity field is indeed irrotational by computing the z-component of vorticity.

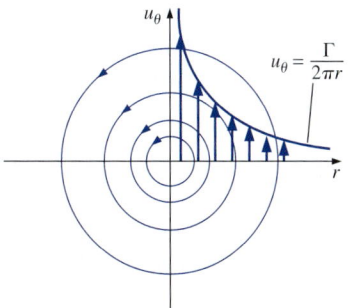

FIGURE P10–103

10–104 Calculate the nine components of the viscous stress tensor in cylindrical coordinates (see Chap. 9) for the velocity field of Prob. 10–103. Discuss.

10–105 Water falls down a vertical pipe *by gravity alone*. The flow between vertical locations z_1 and z_2 is fully developed, and velocity profiles at these two locations are sketched in Fig. P10–105. Since there is no forced pressure gradient, pressure P is constant everywhere in the flow ($P = P_{atm}$). Calculate the *modified pressure* at locations z_1 and z_2. Sketch profiles of modified pressure at locations z_1 and z_2. Discuss.

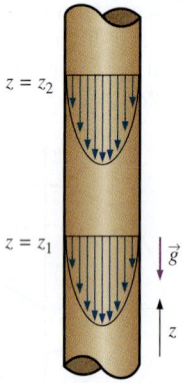

FIGURE P10–105

10–106 Suppose the vertical pipe of Prob. 10–105 is now *horizontal* instead. In order to achieve the same volume flow rate as that of Prob. 10–105, we must supply a forced pressure gradient. Calculate the required pressure drop between two axial locations in the pipe that are the same distance apart as z_2 and z_1 of Fig. P10–105. How does modified pressure P' change between the vertical and horizontal cases?

10–107 The Blasius boundary layer profile is an exact solution of the boundary layer equations for flow over a flat plate. However, the results are somewhat cumbersome to use, since the data appear in tabular form (the solution is numerical). Thus, a simple sine wave approximation (Fig. P10–107) is often used in place of the Blasius solution, namely, $u(y) \cong U \sin\left(\dfrac{\pi}{2}\dfrac{y}{\delta}\right)$ for $y < \delta$, and $u = U$ for $y \ll \delta$, where δ is the boundary layer thickness. Plot the Blasius profile and the sine wave approximation on the same plot, in nondimensional form (u/U versus y/δ), and compare. Is the sine wave profile a reasonable approximation?

10–108 The streamwise velocity component of a steady, incompressible, laminar, flat plate boundary layer of boundary layer thickness δ is approximated by the sine wave profile of Prob. 10–107. Generate expressions for displacement thickness and momentum thickness as functions of δ, based on this sine wave approximation. Compare the approximate values of δ^*/δ and θ/δ to the values of δ^*/δ and θ/δ obtained from the Blasius solution.

10–109 For the sine wave approximation of Prob. 10–107, use the definition of local skin friction coefficient and the Kármán integral equation to generate an expression for δ/x. Compare your result to the Blasius expression for δ/x. (Note: You will also need the results of Prob. 10–108 to do this problem.)

Fundamentals of Engineering (FE) Exam Problems

10–110 If the fluid velocity is zero in a flow field, the Navier-Stokes equation becomes
(a) $\vec{\nabla} P - \rho \vec{g} = 0$
(b) $-\vec{\nabla} P + \rho \vec{g} + \mu \vec{\nabla}^2 \vec{V} = 0$
(c) $\rho \dfrac{D\vec{V}}{Dt} = -\vec{\nabla} P + \mu \vec{\nabla}^2 \vec{V}$
(d) $\rho \dfrac{D\vec{V}}{Dt} = -\vec{\nabla} P + \rho \vec{g} + \mu \vec{\nabla}^2 \vec{V}$
(e) $\rho \dfrac{D\vec{V}}{Dt} + \vec{\nabla} P - \rho \vec{g} = 0$

10–111 Which choice is not a scaling parameter used to nondimensionalize the equations of motion?
(a) Characteristic length, L
(b) Characteristic speed, V
(c) Characteristic viscosity, μ
(d) Characteristic frequency, f
(e) Gravitational acceleration, g

10–112 Which choice is not a nondimensional variable defined to nondimensionalize the equations of motion?
(a) $t^* = ft$
(b) $\vec{x}^* = \dfrac{\vec{x}}{L}$
(c) $\vec{V}^* = \dfrac{\vec{V}}{V}$
(d) $\vec{g}^* = \dfrac{\vec{g}}{g}$
(e) $P^* = \dfrac{P}{P_0}$

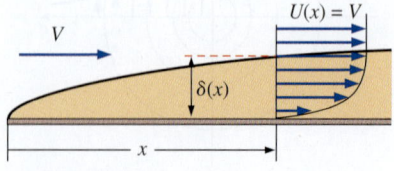

FIGURE P10–107

10–113 Which dimensionless parameter does not appear in the nondimensionalized Navier-Stokes equation?
(a) Reynolds number
(b) Prandtl number
(c) Strouhal number
(d) Euler number
(e) Froude number

10–114 Which dimensionless parameter is zero in the nondimensionalized Navier-Stokes equation when the flow is quasi-steady?
(a) Euler number
(b) Prandtl number
(c) Froude number
(d) Strouhal number
(e) Reynolds number

10–115 If pressure P is replaced by modified pressure $P' = P + \rho g z$ in the nondimensionalized Navier-Stokes equation, which dimensionless parameter drops out?
(a) Froude number
(b) Reynolds number
(c) Strouhal number
(d) Euler number
(e) Prandtl number

10–116 In creeping flow, the value of Reynolds number is typically
(a) Re < 1
(b) Re $<< 1$
(c) Re > 1
(d) Re $>> 1$
(e) Re $= 0$

10–117 Which equation is the proper approximate Navier-Stokes equation in dimensional form for creeping flow?
(a) $\vec{\nabla}P - \rho\vec{g} = 0$
(b) $-\vec{\nabla}P + \mu\vec{\nabla}^2\vec{V} = 0$
(c) $-\vec{\nabla}P + \rho\vec{g} + \mu\vec{\nabla}^2\vec{V} = 0$
(d) $\rho\dfrac{D\vec{V}}{Dt} = -\vec{\nabla}P + \rho\vec{g} + \mu\vec{\nabla}^2\vec{V}$
(e) $\rho\dfrac{D\vec{V}}{Dt} + \vec{\nabla}P - \rho\vec{g} = 0$

10–118 For creeping flow over a three-dimensional object, the aerodynamic drag on the object does not depend on
(a) Velocity, V
(b) Fluid viscosity, μ
(c) Characteristic length, L
(d) Fluid density, ρ
(e) None of these

10–119 Consider a spherical ash particle of diameter 65 μm, falling from a volcano at a high elevation in air whose temperature is $-50°$C and whose pressure is 55 kPa. The density of air is 0.8588 kg/m^3 and its viscosity is 1.474×10^{-5} kg/m·s. The density of the particle is 1240 kg/m^3. The drag force on a sphere in creeping flow is given by $F_D = 3\pi\mu VD$. The terminal velocity of this particle at this altitude is
(a) 0.096 m/s
(b) 0.123 m/s
(c) 0.194 m/s
(d) 0.225 m/s
(e) 0.276 m/s

10–120 Which statement is not correct regarding inviscid regions of flow?
(a) Inertial forces are not negligible.
(b) Pressure forces are not negligible.
(c) Reynolds number is large.
(d) Not valid in boundary layers and wakes.
(e) Solid body rotation of a fluid is an example.

10–121 For which regions of flow is the Laplace equation $\vec{\nabla}^2\phi = 0$ applicable?
(a) Irrotational
(b) Inviscid
(c) Boundary layer
(d) Wake
(e) Creeping

10–122 A very thin region of flow near a solid wall where viscous forces and rotationality cannot be ignored is called
(a) Inviscid region of flow
(b) Irrotational flow
(c) Boundary layer
(d) Outer flow region
(e) Creeping flow

10–123 Which one of the following is not a flow region where the boundary layer approximation may be appropriate?
(a) Jet
(b) Inviscid region
(c) Wake
(d) Mixing layer
(e) Thin region near a solid wall

10–124 Which statement is not correct regarding the boundary layer approximation?
(a) The higher the Reynolds number, the thinner the boundary layer.
(b) The boundary layer approximation may be appropriate for free shear layers.
(c) The boundary layer equations are approximations of the Navier-Stokes equation.
(d) The curve representing boundary layer thickness δ as a function of x is a streamline.
(e) The boundary layer approximation bridges the gap between the Euler equation and the Navier-Stokes equation.

10–125 For a laminar boundary layer growing on a horizontal flat plate, the boundary layer thickness δ is not a function of
(a) Velocity, V
(b) Distance from the leading edge, x
(c) Fluid density, ρ
(d) Fluid viscosity, μ
(e) Gravitational acceleration, g

10–126 For flow along a flat plate with x being the distance from the leading edge, the boundary layer thickness grows like
(a) x
(b) $\sqrt{x}$
(c) x^2
(d) $1/x$
(e) $1/x^2$

10–127 Air flows at 25°C with a velocity of 3 m/s in a wind tunnel whose test section is 25 cm long. The displacement thickness at the end of the test section is (the kinematic viscosity of air is 1.562×10^{-5} m^2/s).
(a) 0.955 mm
(b) 1.18 mm
(c) 1.33 mm
(d) 1.70 mm
(e) 1.96 mm

10–128 Air flows at 25°C with a velocity of 6 m/s over a flat plate whose length is 40 cm. The momentum thickness at the center of the plate is (the kinematic viscosity of air is 1.562×10^{-5} m^2/s).
(a) 0.479 mm
(b) 0.678 mm
(c) 0.832 mm
(d) 1.08 mm
(e) 1.34 mm

10–129 Water flows at 20°C with a velocity of 1.1 m/s over a flat plate whose length is 15 cm. The boundary layer thickness at the end of the plate is (the density and viscosity of water are 998 kg/m^3 and 1.002×10^3 kg/m·s, respectively).
(a) 1.14 mm
(b) 1.35 mm
(c) 1.56 mm
(d) 1.82 mm
(e) 2.09 mm

10–130 Air flows at 15°C with a velocity of 12 m/s over a flat plate whose length is 80 cm. Using one-seventh power law of the turbulent flow, what is the boundary layer thickness at the end of the plate? (The kinematic viscosity of air is 1.470×10^{-5} m^2/s.)
(a) 1.54 cm
(b) 1.89 cm
(c) 2.16 cm
(d) 2.45 cm
(e) 2.82 cm

10–131 Air at 15°C flows at 10 m/s over a flat plate of length 2 m. Using one-seventh power law of the turbulent flow, what is the ratio of local skin friction coefficient for the turbulent and laminar flow cases? (The kinematic viscosity of air is 1.470×10^{-5} m^2/s.)
(a) 1.25
(b) 3.72
(c) 6.31
(d) 8.64
(e) 12.0

Design and Essay Problem

10–132 Explain why there is a significant velocity overshoot for the midrange values of the Reynolds number in the velocity profiles of Fig. 10–136, but not for the very small values of Re or for the very large values of Re.

EXTERNAL FLOW: DRAG AND LIFT

CHAPTER 11

In this chapter we consider *external flow*—flow over bodies that are immersed in a fluid, with emphasis on the resulting lift and drag forces. In external flow, the viscous effects are confined to a portion of the flow field such as the boundary layers and wakes, which are surrounded by an outer flow region that involves small velocity and temperature gradients.

When a fluid moves over a solid body, it exerts pressure forces normal to the surface and shear forces parallel to the surface of the body. We are usually interested in the *resultant* of the pressure and shear forces acting on the body rather than the details of the distributions of these forces along the entire surface of the body. The component of the resultant pressure and shear forces that acts in the flow direction is called the *drag force* (or just *drag*), and the component that acts normal to the flow direction is called the *lift force* (or just *lift*).

We start this chapter with a discussion of drag and lift, and explore the concepts of pressure drag, friction drag, and flow separation. We continue with the drag coefficients of various two- and three-dimensional geometries encountered in practice and determine the drag force using experimentally determined drag coefficients. We then examine the development of the velocity boundary layer during parallel flow over a flat surface, and develop relations for the skin friction and drag coefficients for flow over flat plates, cylinders, and spheres. Finally, we discuss the lift developed by airfoils and the factors that affect the lift characteristics of bodies.

OBJECTIVES

When you finish reading this chapter, you should be able to

- Have an intuitive understanding of the various physical phenomena associated with external flow such as drag, friction and pressure drag, drag reduction, and lift
- Calculate the drag force associated with flow over common geometries
- Understand the effects of flow regime on the drag coefficients associated with flow over cylinders and spheres
- Understand the fundamentals of flow over airfoils, and calculate the drag and lift forces acting on airfoils

The wake of a Boeing 767 disrupts the top of a cumulus cloud and clearly shows the counter-rotating trailing vortices.

Photo by Steve Morris, used by permission.

11–1 · INTRODUCTION

Fluid flow over solid bodies frequently occurs in practice, and it is responsible for numerous physical phenomena such as the *drag force* acting on automobiles, power lines, trees, and underwater pipelines; the *lift* developed by bird or airplane wings; *upward draft* of rain, snow, hail, and dust particles in high winds; the transportation of red blood cells by blood flow; the entrainment and disbursement of liquid droplets by sprays; the vibration and noise generated by bodies moving in a fluid; and the power generated by wind turbines (Fig. 11–1). Therefore, developing a good understanding of external flow is important in the design of many engineering systems such as aircraft, automobiles, buildings, ships, submarines, and all kinds of turbines. Late-model cars, for example, have been designed with particular emphasis on aerodynamics. This has resulted in significant reductions in fuel consumption and noise, and considerable improvement in handling.

Sometimes a fluid moves over a stationary body (such as the wind blowing over a building), and other times a body moves through a quiescent fluid (such as a car moving through air). These two seemingly different processes are equivalent to each other; what matters is the relative motion between the fluid and the body. Such motions are conveniently analyzed by fixing the coordinate system on the body and are referred to as **flow over bodies** or **external flow.** The aerodynamic aspects of different airplane wing designs, for example, are studied conveniently in a lab by placing the wings in a wind tunnel and blowing air over them by large fans. Also, a flow can be classified as being steady or unsteady, depending on the reference frame selected. Flow around an airplane, for example, is always unsteady with respect to the ground, but it is steady with respect to a frame of reference moving with the airplane at cruise conditions.

The flow fields and geometries for most external flow problems are too complicated to be solved analytically, and thus we have to rely on correlations based on experimental data. The availability of high-speed computers has made it possible to conduct a series of "numerical experiments" quickly by solving the governing equations numerically (Chap. 15), and to resort to the expensive and time-consuming testing and experimentation only in the final stages of design. Such testing is done in wind tunnels. H. F. Phillips (1845–1912) built the first wind tunnel in 1894 and measured lift and drag. In this chapter we mostly rely on relations developed experimentally.

The velocity of the fluid approaching a body is called the **free-stream velocity** and is denoted by V. It is also denoted by u_∞ or U_∞ when the flow is aligned with the x-axis since u is used to denote the x-component of velocity. The fluid velocity ranges from zero at the body surface (the no-slip condition) to the free-stream value away from the body surface, and the subscript "infinity" serves as a reminder that this is the value at a distance where the presence of the body is not felt. The free-stream velocity may vary with location and time (e.g., the wind blowing past a building). But in the design and analysis, the free-stream velocity is usually assumed to be *uniform* and *steady* for convenience, and this is what we do in this chapter.

The shape of a body has a profound influence on the flow over the body and the velocity field. The flow over a body is said to be **two-dimensional** when the body is very long and of constant cross section and the flow is

609
CHAPTER 11

FIGURE 11–1
Flow over bodies is commonly encountered in practice.
(a) Royalty-Free/CORBIS; (b) Imagestate Media/John Foxx RF; (c) © IT Stock/age fotostock RF; (d) Royalty-Free/CORBIS; (e) © StockTrek/Superstock RF; (f) Royalty-Free/CORBIS; (g) © Roy H. Photography/Getty RF

EXTERNAL FLOW: DRAG AND LIFT

Long cylinder (2-D)
(a)

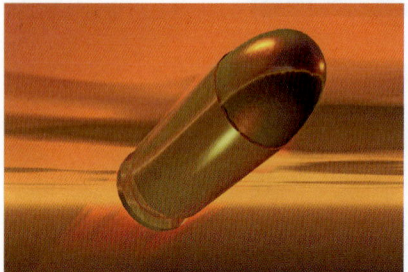

Bullet (axisymmetric)
(b)

Car (3-D)
(c)

FIGURE 11–2
Two-dimensional, axisymmetric, and three-dimensional flows.
(a) Photo by John M. Cimbala; (b) © CorbisRF
(c) Hannu Liivaar/Alamy.

normal to the body. The wind blowing over a long pipe perpendicular to its axis is an example of two-dimensional flow. Note that the velocity component in the axial direction is zero in this case, and thus the velocity is two-dimensional.

The two-dimensional idealization is appropriate when the body is sufficiently long so that the end effects are negligible and the approach flow is uniform. Another simplification occurs when the body possesses rotational symmetry about an axis in the flow direction. The flow in this case is also two-dimensional and is said to be **axisymmetric.** A bullet piercing through air is an example of axisymmetric flow. The velocity in this case varies with the axial distance x and the radial distance r. Flow over a body that cannot be modeled as two-dimensional or axisymmetric, such as flow over a car, is **three-dimensional** (Fig. 11–2).

Flow over bodies can also be classified as **incompressible flows** (e.g., flows over automobiles, submarines, and buildings) and **compressible flows** (e.g., flows over high-speed aircraft, rockets, and missiles). Compressibility effects are negligible at low velocities (flows with $Ma \lesssim 0.3$), and such flows can be treated as incompressible with little loss in accuracy. Compressible flow is discussed in Chap. 12, and flows that involve partially immersed bodies with a free surface (such as a ship cruising in water) are beyond the scope of this introductory text.

Bodies subjected to fluid flow are classified as being streamlined or bluff, depending on their overall shape. A body is said to be **streamlined** if a conscious effort is made to align its shape with the anticipated streamlines in the flow. Streamlined bodies such as race cars and airplanes appear to be contoured and sleek. Otherwise, a body (such as a building) tends to block the flow and is said to be **bluff** or blunt. Usually it is much easier to force a streamlined body through a fluid, and thus streamlining has been of great importance in the design of vehicles and airplanes (Fig. 11–3).

11–2 ■ DRAG AND LIFT

It is a common experience that a body meets some resistance when it is forced to move through a fluid, especially a liquid. As you may have noticed, it is very difficult to walk in water because of the much greater resistance it offers to motion compared to air. Also, you may have seen high winds knocking down trees, power lines, and even trailers and felt the strong "push" the wind exerts on your body (Fig. 11–4). You experience the same feeling when you extend your arm out of the window of a moving car. A fluid may exert forces and moments on a body in and about various directions. The force a flowing fluid exerts on a body in the flow direction is called **drag.** The drag force can be measured directly by simply attaching the body subjected to fluid flow to a calibrated spring and measuring the displacement in the flow direction (just like measuring weight with a spring scale). More sophisticated drag-measuring devices, called drag balances, use flexible beams fitted with strain gages to measure the drag electronically.

Drag is usually an undesirable effect, like friction, and we do our best to minimize it. Reduction of drag is closely associated with the reduction of fuel consumption in automobiles, submarines, and aircraft; improved safety and durability of structures subjected to high winds; and reduction of noise and vibration. But in some cases drag produces a beneficial effect and we try to maximize it. Friction, for example, is a "life saver" in the brakes of automobiles. Likewise, it is the drag that makes it possible for people to parachute, for pollens to fly to distant locations, and for us all to enjoy the waves of the oceans and the relaxing movements of the leaves of trees.

A stationary fluid exerts only normal pressure forces on the surface of a body immersed in it. A moving fluid, however, also exerts tangential shear forces on the surface because of the no-slip condition caused by viscous effects. Both of these forces, in general, have components in the direction of flow, and thus the drag force is due to the combined effects of pressure and wall shear forces in the flow direction. The components of the pressure and wall shear forces in the direction *normal* to the flow tend to move the body in that direction, and their sum is called **lift.**

For two-dimensional flows, the resultant of the pressure and shear forces can be split into two components: one in the direction of flow, which is the drag force, and another in the direction normal to flow, which is the lift, as shown in Fig. 11–5. For three-dimensional flows, there is also a side force component in the direction normal to the page that tends to move the body in that direction.

The fluid forces may also generate moments and cause the body to rotate. The moment about the flow direction is called the *rolling moment,* the moment about the lift direction is called the *yawing moment,* and the moment about the side force direction is called the *pitching moment.* For bodies that possess symmetry about the lift–drag plane such as cars, airplanes, and ships, the time-averaged side force, yawing moment, and rolling moment are zero when the wind and wave forces are aligned with the body. What remain for such bodies are the drag and lift forces and the pitching moment. For axisymmetric bodies aligned with the flow, such as a bullet, the only time-averaged force exerted by the fluid on the body is the drag force.

The pressure and shear forces acting on a differential area dA on the surface are $P\,dA$ and $\tau_w\,dA$, respectively. The differential drag force and the lift force acting on dA in two-dimensional flow are (Fig. 11–5)

$$dF_D = -P\,dA\cos\theta + \tau_w\,dA\sin\theta \tag{11-1}$$

and

$$dF_L = -P\,dA\sin\theta - \tau_w\,dA\cos\theta \tag{11-2}$$

where θ is the angle that the outer normal of dA makes with the positive flow direction. The total drag and lift forces acting on the body are determined by integrating Eqs. 11–1 and 11–2 over the entire surface of the body,

Drag force: $$F_D = \int_A dF_D = \int_A (-P\cos\theta + \tau_w\sin\theta)\,dA \tag{11-3}$$

and

Lift force: $$F_L = \int_A dF_L = -\int_A (P\sin\theta + \tau_w\cos\theta)\,dA \tag{11-4}$$

These are the equations used to predict the net drag and lift forces on bodies when the flow is simulated on a computer (Chap. 15). However, when we perform experimental analyses, Eqs. 11–3 and 11–4 are not practical since the detailed distributions of pressure and shear forces are difficult to obtain

FIGURE 11–3
It is much easier to force a streamlined body than a blunt body through a fluid.

FIGURE 11–4
High winds knock down trees, power lines, and even people as a result of the drag force.

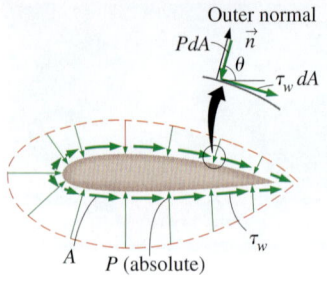

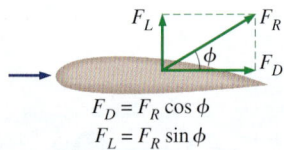

FIGURE 11–5
The pressure and viscous forces acting on a two-dimensional body and the resultant lift and drag forces.

by measurements. Fortunately, this information is often not needed. Usually all we need to know is the resultant drag force and lift acting on the entire body, which can be measured directly and easily in a wind tunnel.

Equations 11–1 and 11–2 show that both the skin friction (wall shear) and pressure, in general, contribute to the drag and the lift. In the special case of a thin *flat plate* aligned parallel to the flow direction, the drag force depends on the wall shear only and is independent of pressure since $\theta = 90°$. When the flat plate is placed normal to the flow direction, however, the drag force depends on the pressure only and is independent of wall shear since the shear stress in this case acts in the direction normal to flow and $\theta = 0°$ (Fig. 11–6). If the flat plate is tilted at an angle relative to the flow direction, then the drag force depends on both the pressure and the shear stress.

The wings of airplanes are shaped and positioned specifically to generate lift with minimal drag. This is done by maintaining an angle of attack during cruising, as shown in Fig. 11–7. Both lift and drag are strong functions of the angle of attack, as we discuss later in this chapter. The pressure difference between the top and bottom surfaces of the wing generates an upward force that tends to lift the wing and thus the airplane to which it is connected. For slender bodies such as wings, the shear force acts nearly parallel to the flow direction, and thus its contribution to the lift is small. The drag force for such slender bodies is mostly due to shear forces (the skin friction).

The drag and lift forces depend on the density ρ of the fluid, the upstream velocity V, and the size, shape, and orientation of the body, among other things, and it is not practical to list these forces for a variety of situations. Instead, it is more convenient to work with appropriate dimensionless numbers that represent the drag and lift characteristics of the body. These numbers are the **drag coefficient** C_D, and the **lift coefficient** C_L, and they are defined as

Drag coefficient:
$$C_D = \frac{F_D}{\frac{1}{2}\rho V^2 A} \quad (11\text{–}5)$$

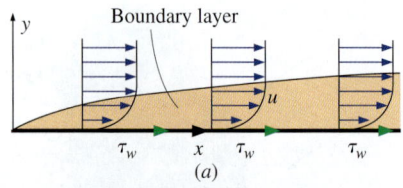

Lift coefficient:
$$C_L = \frac{F_L}{\frac{1}{2}\rho V^2 A} \quad (11\text{–}6)$$

where A is ordinarily the **frontal area** (the area projected on a plane normal to the direction of flow) of the body. In other words, A is the area seen by a person looking at the body from the direction of the approaching fluid. The frontal area of a cylinder of diameter D and length L, for example, is $A = LD$. In lift and drag calculations of some thin bodies, such as airfoils, A is taken to be the **planform area**, which is the area seen by a person looking at the body from above in a direction normal to the body. The drag and lift coefficients are primarily functions of the shape of the body, but in some cases they also depend on the Reynolds number and the surface roughness. The term $\frac{1}{2}\rho V^2$ in Eqs. 11–5 and 11–6 is the **dynamic pressure.**

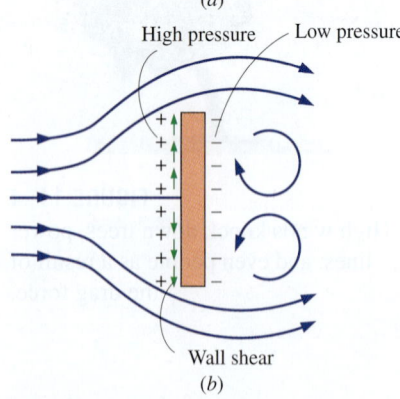

FIGURE 11–6
(*a*) Drag force acting on a flat plate parallel to the flow depends on wall shear only. (*b*) Drag force acting on a flat plate normal to the flow depends on the pressure only and is independent of the wall shear, which acts normal to the free-stream flow.

The local drag and lift coefficients vary along the surface as a result of changes in the velocity boundary layer in the flow direction. We are usually interested in the drag and lift forces for the *entire* surface, which can be determined using the *average* drag and lift coefficients. Therefore, we present correlations for both local (identified with the subscript x) and average drag and lift coefficients. When relations for local drag and lift coefficients for a surface of length L are available, the *average* drag and lift coefficients for the entire surface are determined by integration from

$$C_D = \frac{1}{L}\int_0^L C_{D,x}\,dx \qquad (11\text{-}7)$$

and

$$C_L = \frac{1}{L}\int_0^L C_{L,x}\,dx \qquad (11\text{-}8)$$

The forces acting on a falling body are usually the drag force, the buoyant force, and the weight of the body. When a body is dropped into the atmosphere or a lake, it first accelerates under the influence of its weight. The motion of the body is resisted by the drag force, which acts in the direction opposite to motion. As the velocity of the body increases, so does the drag force. This continues until all the forces balance each other and the net force acting on the body (and thus its acceleration) is zero. Then the velocity of the body remains constant during the rest of its fall if the properties of the fluid in the path of the body remain essentially constant. This is the maximum velocity a falling body can attain and is called the **terminal velocity** (Fig. 11–8).

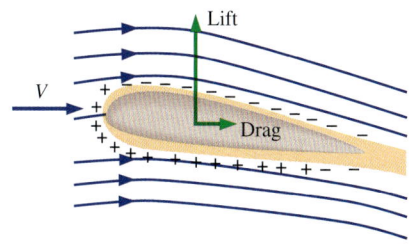

FIGURE 11–7
Airplane wings are shaped and positioned to generate sufficient lift during flight while keeping drag at a minimum. Pressures above and below atmospheric pressure are indicated by plus and minus signs, respectively.

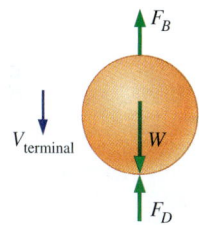

$F_D = W - F_B$
(No acceleration)

FIGURE 11–8
During a free fall, a body reaches its *terminal velocity* when the drag force equals the weight of the body minus the buoyant force.

■ **EXAMPLE 11–1**   **Measuring the Drag Coefficient of a Car**

The drag coefficient of a car at the design conditions of 1 atm, 20°C, and 95 km/h is to be determined experimentally in a large wind tunnel in a full-scale test (Fig. 11–9). The frontal area of the car is 2.07 m². If the force acting on the car in the flow direction is measured to be 300 N, determine the drag coefficient of this car.

SOLUTION The drag force acting on a car is measured in a wind tunnel. The drag coefficient of the car at test conditions is to be determined.
Assumptions **1** The flow of air is steady and incompressible. **2** The cross section of the tunnel is large enough to simulate free flow over the car. **3** The bottom of the tunnel is also moving at the speed of air to approximate actual driving conditions or this effect is negligible.
Properties The density of air at 1 atm and 20°C is $\rho = 1.204$ kg/m³.
Analysis The drag force acting on a body and the drag coefficient are given by

$$F_D = C_D A \frac{\rho V^2}{2} \quad \text{and} \quad C_D = \frac{2 F_D}{\rho A V^2}$$

where A is the frontal area. Substituting and noting that 1 m/s = 3.6 km/h, the drag coefficient of the car is determined to be

$$C_D = \frac{2\times(300\text{ N})}{(1.204\text{ kg/m}^3)(2.07\text{ m}^2)(95/3.6\text{ m/s})^2}\left(\frac{1\text{ kg}\cdot\text{m/s}^2}{1\text{ N}}\right) = \mathbf{0.35}$$

Discussion Note that the drag coefficient depends on the design conditions, and its value may be different at different conditions such as the Reynolds number. Therefore, the published drag coefficients of different vehicles can be compared meaningfully only if they are determined under dynamically similar conditions or if Reynolds number independence is demonstrated (Chap. 7). This shows the importance of developing standard testing procedures.

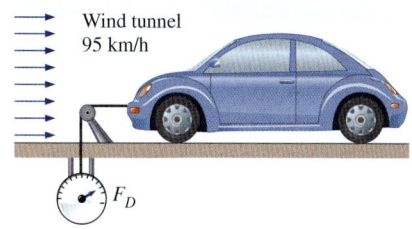

FIGURE 11–9
Schematic for Example 11–1.

11–3 · FRICTION AND PRESSURE DRAG

As mentioned in Section 11–2, the drag force is the net force exerted by a fluid on a body in the direction of flow due to the combined effects of wall shear and pressure forces. It is often instructive to separate the two effects, and study them separately.

The part of drag that is due directly to wall shear stress τ_w is called the **skin friction drag** (or just *friction drag* $F_{D,\,\text{friction}}$) since it is caused by frictional effects, and the part that is due directly to pressure P is called the **pressure drag** (also called the *form drag* because of its strong dependence on the form or shape of the body). The friction and pressure drag coefficients are defined as

$$C_{D,\,\text{friction}} = \frac{F_{D,\,\text{friction}}}{\frac{1}{2}\rho V^2 A} \quad \text{and} \quad C_{D,\,\text{pressure}} = \frac{F_{D,\,\text{pressure}}}{\frac{1}{2}\rho V^2 A} \quad (11\text{–}9)$$

When the friction and pressure drag coefficients (based on the same area A) or forces are available, the total drag coefficient or drag force is determined by simply adding them,

$$C_D = C_{D,\,\text{friction}} + C_{D,\,\text{pressure}} \quad \text{and} \quad F_D = F_{D,\,\text{friction}} + F_{D,\,\text{pressure}} \quad (11\text{–}10)$$

The *friction drag* is the component of the wall shear force in the direction of flow, and thus it depends on the orientation of the body as well as the magnitude of the wall shear stress τ_w. The friction drag is *zero* for a flat surface normal to the flow, and *maximum* for a flat surface parallel to the flow since the friction drag in this case equals the total shear force on the surface. Therefore, for parallel flow over a flat surface, the drag coefficient is equal to the *friction drag coefficient*, or simply the *friction coefficient*. Friction drag is a strong function of viscosity, and increases with increasing viscosity.

The Reynolds number is inversely proportional to the viscosity of the fluid. Therefore, the contribution of friction drag to total drag for blunt bodies is less at higher Reynolds numbers and may be negligible at very high Reynolds numbers. The drag in such cases is mostly due to pressure drag. At low Reynolds numbers, most drag is due to friction drag. This is especially the case for highly streamlined bodies such as airfoils. The friction drag is also proportional to the surface area. Therefore, bodies with a larger surface area experience a larger friction drag. Large commercial airplanes, for example, reduce their total surface area and thus their drag by retracting their wing extensions when they reach cruising altitudes to save fuel. The friction drag coefficient is independent of *surface roughness* in laminar flow, but is a strong function of surface roughness in turbulent flow due to surface roughness elements protruding further into the boundary layer. The *friction drag coefficient* is analogous to the *friction factor* in pipe flow discussed in Chap. 8, and its value depends on the flow regime.

The pressure drag is proportional to the frontal area and to the *difference* between the pressures acting on the front and back of the immersed body. Therefore, the pressure drag is usually dominant for blunt bodies, small for streamlined bodies such as airfoils, and zero for thin flat plates parallel to the flow (Fig. 11–10). The pressure drag becomes most significant

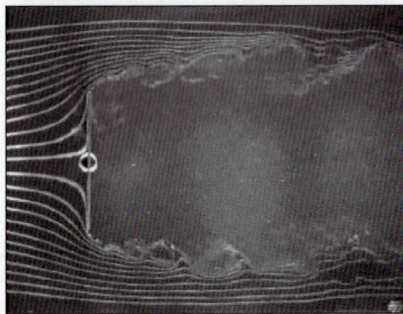

FIGURE 11–10
Drag is due entirely to *friction drag* for a flat plate parallel to the flow; it is due entirely to pressure drag for a flat plate normal to the flow; and it is due to *both* (but mostly *pressure drag*) for a cylinder normal to the flow. The total drag coefficient C_D is lowest for a parallel flat plate, highest for a vertical flat plate, and in between (but close to that of a vertical flat plate) for a cylinder.

From G. M. Homsy, et al. (2004).

when the velocity of the fluid is too high for the fluid to be able to follow the curvature of the body, and thus the fluid *separates* from the body at some point and creates a very low pressure region in the back. The pressure drag in this case is due to the large pressure difference between the front and back sides of the body.

Reducing Drag by Streamlining

The first thought that comes to mind to reduce drag is to streamline a body in order to reduce flow separation and thus to reduce pressure drag. Even car salespeople are quick to point out the low drag coefficients of their cars, owing to streamlining. But streamlining has opposite effects on pressure and friction drag forces. It decreases pressure drag by delaying boundary layer separation and thus reducing the pressure difference between the front and back of the body and increases the friction drag by increasing the surface area. The end result depends on which effect dominates. Therefore, any optimization study to reduce the drag of a body must consider both effects and must attempt to minimize the *sum* of the two, as shown in Fig. 11–11. The minimum total drag occurs at $D/L = 0.25$ for the case shown in Fig. 11–11. For the case of a circular cylinder with the same thickness as the streamlined shape of Fig. 11–11, the drag coefficient would be about five times as much. Therefore, it is possible to reduce the drag of a cylindrical component to nearly one-fifth by the use of proper fairings.

The effect of streamlining on the drag coefficient is described best by considering long elliptical cylinders with different aspect (or length-to-thickness) ratios L/D, where L is the length in the flow direction and D is the thickness, as shown in Fig. 11–12. Note that the drag coefficient decreases drastically as the ellipse becomes slimmer. For the special case of $L/D = 1$ (a circular cylinder), the drag coefficient is $C_D \cong 1$ at this Reynolds number. As the aspect ratio is decreased and the cylinder resembles a flat plate, the drag coefficient increases to 1.9, the value for a flat plate normal to flow. Note that the curve becomes nearly flat for aspect ratios greater than about 4. Therefore, for a given diameter D, elliptical shapes with an aspect ratio of about $L/D \cong 4$ usually offer a good compromise between the total drag coefficient and length L. The reduction in the drag coefficient at high aspect ratios is primarily due to the boundary layer staying attached to the surface longer and the resulting pressure recovery. The pressure drag on an elliptical cylinder with an aspect ratio of 4 or greater is negligible (less than 2 percent of total drag at this Reynolds number).

As the aspect ratio of an elliptical cylinder is increased by flattening it (i.e., decreasing D while holding L constant), the drag coefficient starts increasing and tends to infinity as $L/D \to \infty$ (i.e., as the ellipse resembles a flat plate parallel to flow). This is due to the frontal area, which appears in the denominator in the definition of C_D, approaching zero. It does not mean that the drag force increases drastically (actually, the drag force decreases) as the body becomes flat. This shows that the frontal area is inappropriate for use in the drag force relations for slim bodies such as thin airfoils and flat plates. In such cases, the drag coefficient is defined

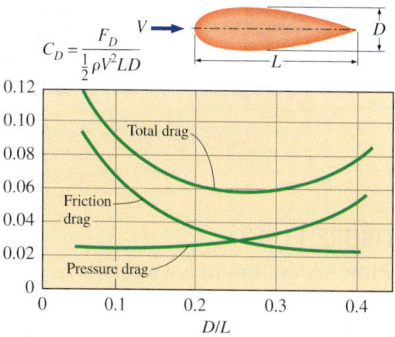

FIGURE 11–11
The variation of friction, pressure, and total drag coefficients of a two-dimensional streamlined strut with thickness-to-chord length ratio for Re = 4×10^4. Note that C_D for airfoils and other thin bodies is based on *planform* area rather than frontal area.
Data from Abbott and von Doenhoff (1959).

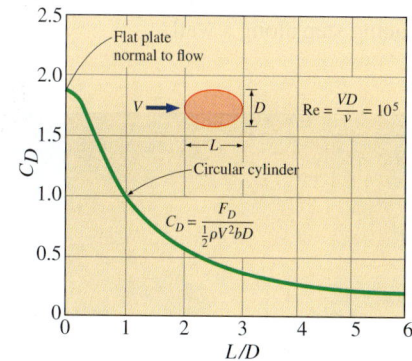

FIGURE 11–12
The variation of the drag coefficient of a long elliptical cylinder with aspect ratio. Here C_D is based on the frontal area bD where b is the width of the body.
Data from Blevins (1984).

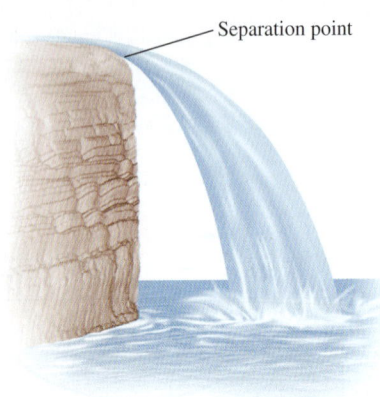

FIGURE 11–13
Flow separation in a waterfall.

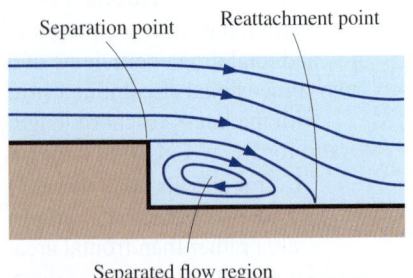

FIGURE 11–14
Flow separation over a backward-facing step along a wall.

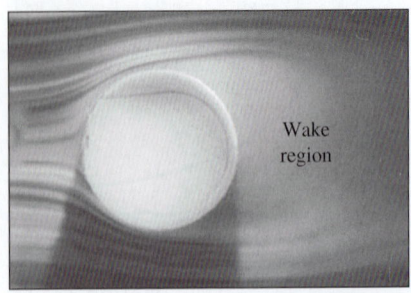

FIGURE 11–15
Flow separation and the wake region for flow over a tennis ball.
Courtesy NASA and Cislunar Aerospace, Inc.

on the basis of the *planform area*, which is simply the surface area of one side (top or bottom) of a flat plate parallel to the flow. This is quite appropriate since for slim bodies the drag is almost entirely due to friction drag, which is proportional to the surface area.

Streamlining has the added benefit of *reducing vibration and noise*. Streamlining should be considered only for bluff bodies that are subjected to high-velocity fluid flow (and thus high Reynolds numbers) for which flow separation is a real possibility. It is not necessary for bodies that typically involve low Reynolds number flows (e.g., creeping flows in which Re < 1) as discussed in Chap. 10, since the drag in those cases is almost entirely due to friction drag, and streamlining would only increase the surface area and thus the total drag. Therefore, careless streamlining may actually increase drag instead of decreasing it.

Flow Separation

When driving on country roads, it is a common safety measure to slow down at sharp turns in order to avoid being thrown off the road. Many drivers have learned the hard way that a car refuses to comply when forced to turn curves at excessive speeds. We can view this phenomenon as "the separation of cars" from roads. This phenomenon is also observed when fast vehicles jump off hills. At low velocities, the wheels of the vehicle always remain in contact with the road surface. But at high velocities, the vehicle is too fast to follow the curvature of the road and takes off at the hill, losing contact with the road.

A fluid acts much the same way when forced to flow over a curved surface at high velocities. A fluid follows the front portion of the curved surface with no problem, but it has difficulty remaining attached to the surface on the back side. At sufficiently high velocities, the fluid stream detaches itself from the surface of the body. This is called **flow separation** (Fig. 11–13). Flow can separate from a surface even if it is fully submerged in a liquid or immersed in a gas (Fig. 11–14). The location of the separation point depends on several factors such as the Reynolds number, the surface roughness, and the level of fluctuations in the free stream, and it is usually difficult to predict exactly where separation will occur unless there are sharp corners or abrupt changes in the shape of the solid surface.

When a fluid separates from a body, it forms a separated region between the body and the fluid stream. This low-pressure region behind the body where recirculating and backflows occur is called the **separated region.** The larger the separated region, the larger the pressure drag. The effects of flow separation are felt far downstream in the form of reduced velocity (relative to the upstream velocity). The region of flow trailing the body where the effects of the body on velocity are felt is called the **wake** (Fig. 11–15). The separated region comes to an end when the two flow streams reattach. Therefore, the separated region is an enclosed volume, whereas the wake keeps growing behind the body until the fluid in the wake region regains its velocity and the velocity profile becomes nearly flat again. Viscous and rotational effects are the most significant in the boundary layer, the separated region, and the wake.

The occurrence of separation is not limited to bluff bodies. Complete separation over the entire back surface may also occur on a streamlined body such as an airplane wing at a sufficiently large **angle of attack** (larger than about 15° for most airfoils), which is the angle the incoming fluid stream makes with the **chord** (the line that connects the nose and the trailing edge) of the wing. Flow separation on the top surface of a wing reduces lift drastically and may cause the airplane to **stall.** Stalling has been blamed for many airplane accidents and loss of efficiencies in turbomachinery (Fig. 11–16).

Note that drag and lift are strongly dependent on the shape of the body, and any effect that causes the shape to change has a profound effect on the drag and lift. For example, snow accumulation and ice formation on airplane wings may change the shape of the wings sufficiently to cause significant loss of lift. This phenomenon has caused many airplanes to lose altitude and crash and many others to abort takeoff. Therefore, it has become a routine safety measure to check for ice or snow buildup on critical components of airplanes before takeoff in bad weather. This is especially important for airplanes that have waited a long time on the runway before takeoff because of heavy traffic.

An important consequence of flow separation is the formation and shedding of circulating fluid structures, called **vortices,** in the wake region. The periodic generation of these vortices downstream is referred to as **vortex shedding.** This phenomenon usually occurs during normal flow over long cylinders or spheres for Re ≳ 90. The vibrations generated by vortices near the body may cause the body to resonate to dangerous levels if the frequency of the vortices is close to the natural frequency of the body—a situation that must be avoided in the design of equipment that is subjected to high-velocity fluid flow such as the wings of airplanes and suspended bridges subjected to steady high winds.

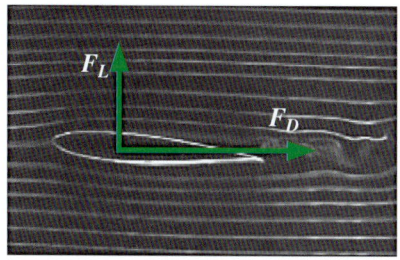

(*a*) 5°

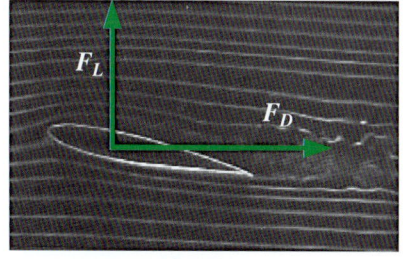

(*b*) 15°

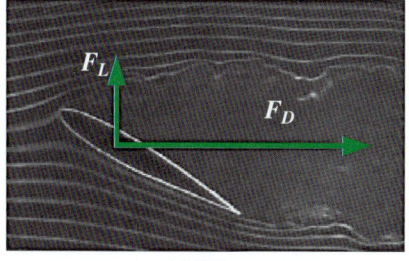

(*c*) 30°

FIGURE 11–16
At large angles of attack (usually larger than 15°), flow may separate completely from the top surface of an airfoil, reducing lift drastically and causing the airfoil to stall.
From G. M. Homsy, et al. (2004).

11–4 ■ DRAG COEFFICIENTS OF COMMON GEOMETRIES

The concept of drag has important consequences in daily life, and the drag behavior of various natural and human-made bodies is characterized by their drag coefficients measured under typical operating conditions. Although drag is caused by two different effects (friction and pressure), it is usually difficult to determine them separately. Besides, in most cases, we are interested in the *total* drag rather than the individual drag components, and thus usually the *total* drag coefficient is reported. The determination of drag coefficients has been the topic of numerous studies (mostly experimental), and there is a huge amount of drag coefficient data in the literature for just about any geometry of practical interest.

The drag coefficient, in general, depends on the *Reynolds number*, especially for Reynolds numbers below about 10^4. At higher Reynolds numbers, the drag coefficients for most geometries remain essentially constant (Fig. 11–17). This is due to the flow at high Reynolds numbers becoming fully turbulent. However, this is not the case for rounded bodies such as

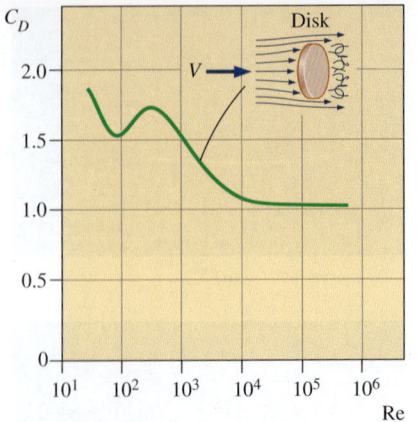

FIGURE 11–17
The drag coefficients for most geometries (but not all) remain essentially constant at Reynolds numbers above about 10^4.

circular cylinders and spheres, as we discuss later in this section. The reported drag coefficients are usually applicable only to flows at high Reynolds numbers.

The drag coefficient exhibits different behavior in the low (creeping), moderate (laminar), and high (turbulent) regions of the Reynolds number. The inertia effects are negligible in low Reynolds number flows (Re ≲ 1), called *creeping flows* (Chap. 10), and the fluid wraps around the body smoothly. The drag coefficient in this case is inversely proportional to the Reynolds number, and for a sphere it is determined to be

Sphere: $$C_D = \frac{24}{\text{Re}} \quad (\text{Re} \lesssim 1) \quad (11\text{–}11)$$

Then the drag force acting on a spherical object at low Reynolds numbers becomes

$$F_D = C_D A \frac{\rho V^2}{2} = \frac{24}{\text{Re}} A \frac{\rho V^2}{2} = \frac{24}{\rho V D / \mu} \frac{\pi D^2}{4} \frac{\rho V^2}{2} = 3\pi\mu V D \quad (11\text{–}12)$$

which is known as **Stokes law,** after British mathematician and physicist G. G. Stokes (1819–1903). This relation shows that at very low Reynolds numbers, the drag force acting on spherical objects is proportional to the diameter, the velocity, and the viscosity of the fluid. This relation is often applicable to dust particles in the air and suspended solid particles in water.

The drag coefficients for low Reynolds number flows past some other geometries are given in Fig. 11–18. Note that at low Reynolds numbers, the shape of the body does not have a major influence on the drag coefficient.

The drag coefficients for various two- and three-dimensional bodies are given in Tables 11–1 and 11–2 for large Reynolds numbers. We make several observations from these tables about the drag coefficient at high Reynolds numbers. First of all, the *orientation* of the body relative to the direction of flow has a major influence on the drag coefficient. For example, the drag coefficient for flow over a hemisphere is 0.4 when the spherical side faces the flow, but it increases threefold to 1.2 when the flat side faces the flow (Fig. 11–19).

For blunt bodies with sharp corners, such as flow over a rectangular block or a flat plate normal to the flow, separation occurs at the edges of the front and back surfaces, with no significant change in the character of flow. Therefore, the drag coefficient of such bodies is nearly independent of the Reynolds number. Note that the drag coefficient of a long rectangular rod is reduced almost by half from 2.2 to 1.2 by rounding the corners.

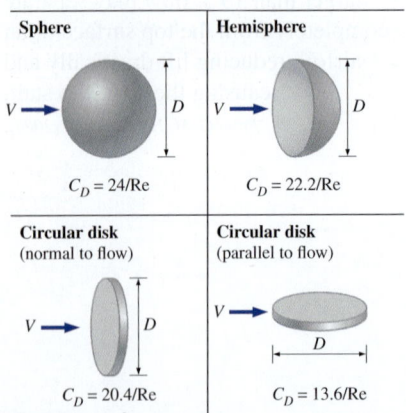

FIGURE 11–18
Drag coefficients C_D at low Reynolds numbers (Re ≲ 1 where Re = VD/ν and $A = \pi D^2/4$).

Biological Systems and Drag

The concept of drag also has important consequences for biological systems. For example, the bodies of *fish*, especially the ones that swim fast for long distances (such as dolphins), are highly streamlined to minimize

CHAPTER 11

TABLE 11–1

Drag coefficients C_D of various two-dimensional bodies for $Re > 10^4$ based on the frontal area $A = bD$, where b is the length in direction normal to the page (for use in the drag force relation $F_D = C_D A \rho V^2/2$ where V is the upstream velocity)

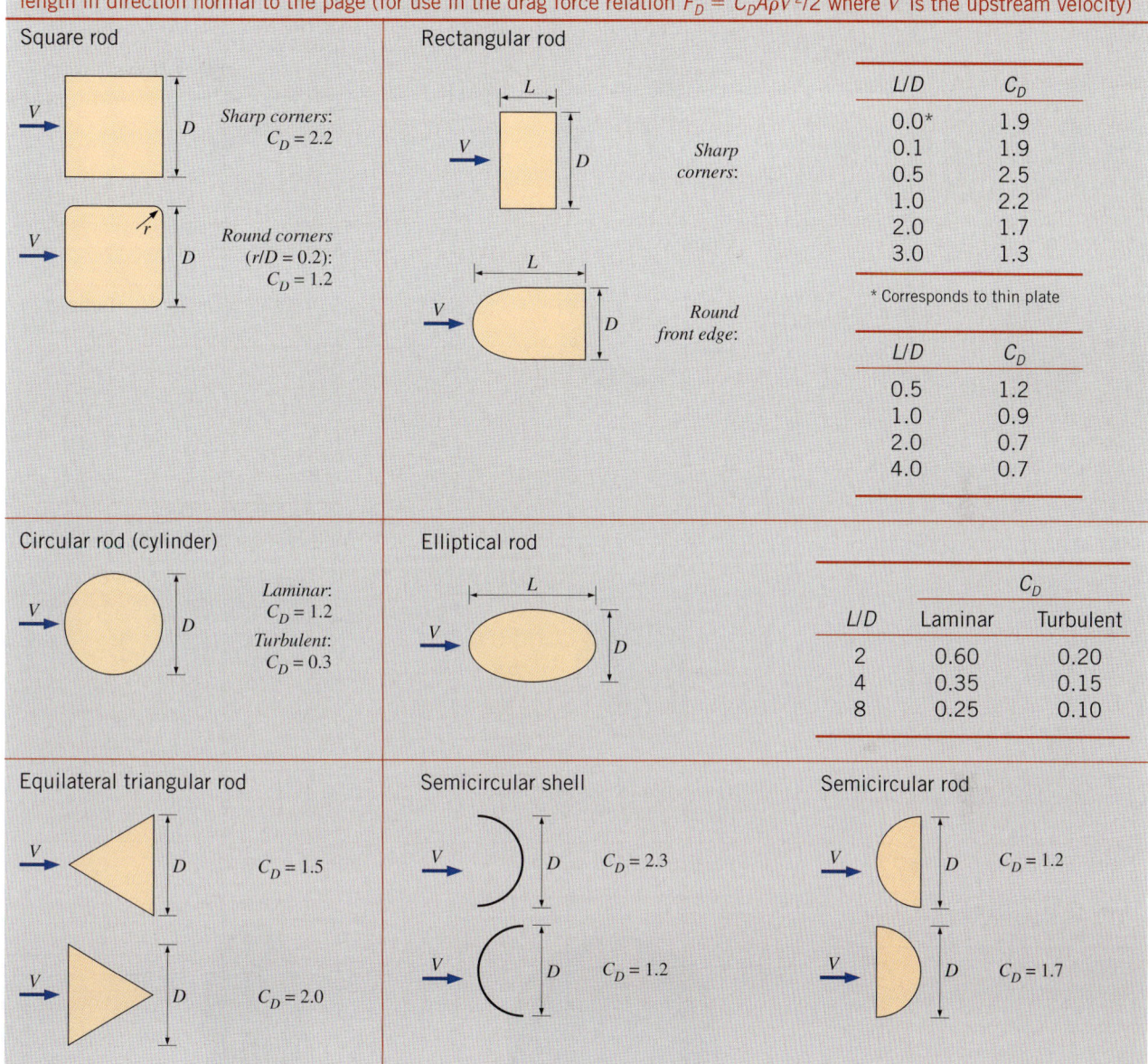

drag (the drag coefficient of dolphins based on the wetted skin area is about 0.0035, comparable to the value for a flat plate in turbulent flow). So it is no surprise that we build submarines that mimic large fish. Tropical fish with fascinating beauty and elegance, on the other hand, swim short distances only. Obviously grace, not high speed and drag, was the primary consideration in their design. Birds teach us a lesson on drag reduction by extending their beak forward and folding their feet backward

TABLE 11–2

Representative drag coefficients C_D for various three-dimensional bodies based on the frontal area for $Re > 10^4$ unless stated otherwise (for use in the drag force relation $F_D = C_D A \rho V^2 / 2$ where V is the upstream velocity)

Cube, $A = D^2$	Thin circular disk, $A = \pi D^2/4$	Cone (for $\theta = 30°$), $A = \pi D^2/4$
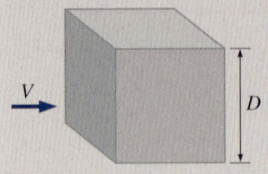 $C_D = 1.05$	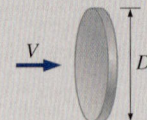 $C_D = 1.1$	 $C_D = 0.5$

Sphere, $A = \pi D^2/4$

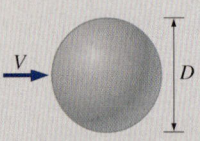

Laminar: $Re \leq 2 \times 10^5$, $C_D = 0.5$
Turbulent: $Re \geq 2 \times 10^6$, $C_D = 0.2$

See Fig. 11–36 for C_D vs. Re for smooth and rough spheres.

Ellipsoid, $A = \pi D^2/4$

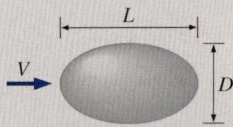

	C_D	
L/D	Laminar $Re \leq 2 \times 10^5$	Turbulent $Re \geq 2 \times 10^6$
0.75	0.5	0.2
1	0.5	0.2
2	0.3	0.1
4	0.3	0.1
8	0.2	0.1

Hemisphere, $A = \pi D^2/4$

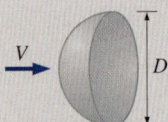 $C_D = 0.4$

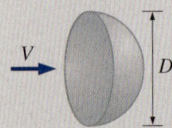 $C_D = 1.2$

Finite cylinder, vertical, $A = LD$

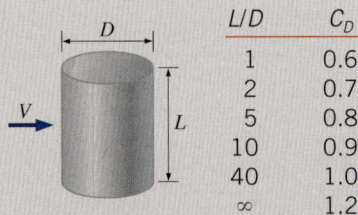

L/D	C_D
1	0.6
2	0.7
5	0.8
10	0.9
40	1.0
∞	1.2

Values are for laminar flow ($Re \leq 2 \times 10^5$)

Finite cylinder, horizontal, $A = \pi D^2/4$

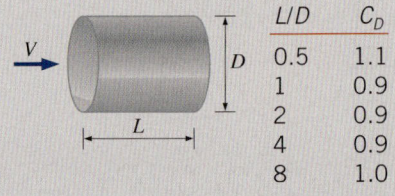

L/D	C_D
0.5	1.1
1	0.9
2	0.9
4	0.9
8	1.0

Streamlined body, $A = \pi D^2/4$

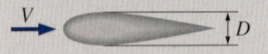 $C_D = 0.04$

Rectangular plate, $A = LD$

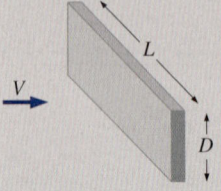

$C_D = 1.10 + 0.02 (L/D + D/L)$
for $1/30 < (L/D) < 30$

Parachute, $A = \pi D^2/4$

$C_D = 1.3$

Tree, A = frontal area

A = frontal area

V, m/s	C_D
10	0.4–1.2
20	0.3–1.0
30	0.2–0.7

(continues)

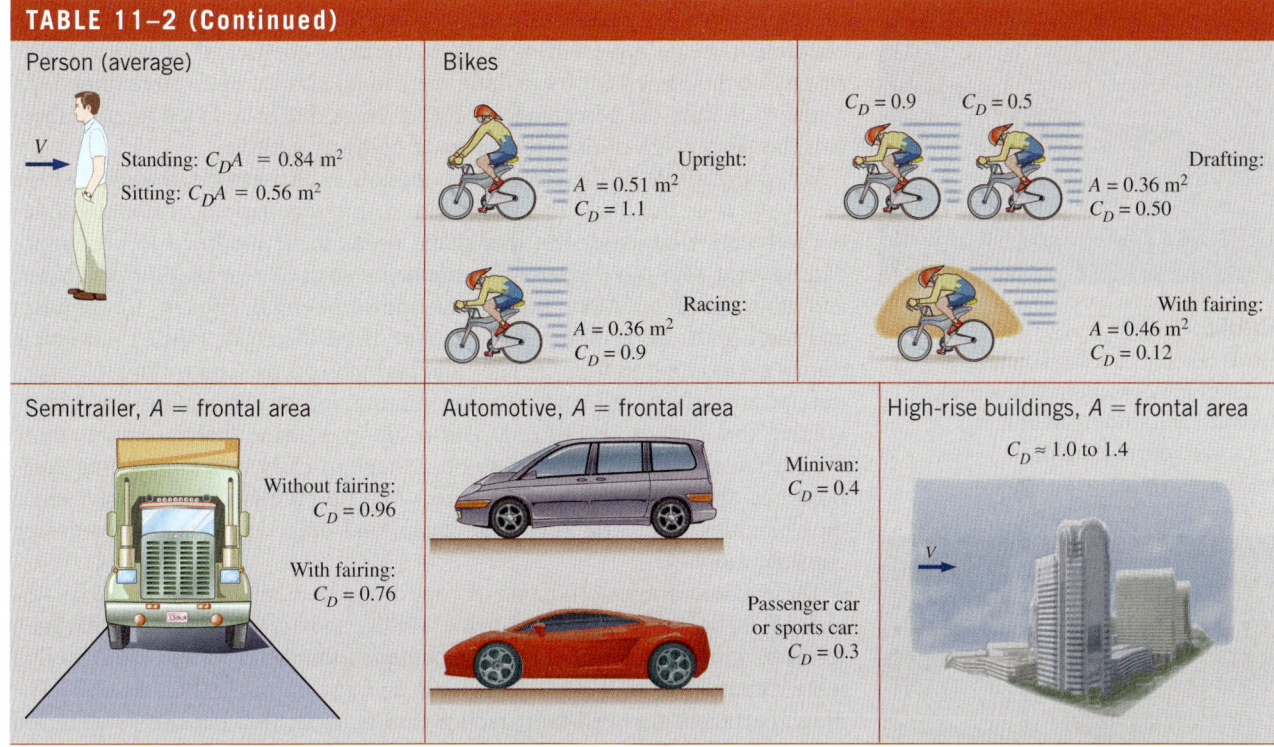

during flight (Fig. 11–20). Airplanes, which look somewhat like large birds, retract their wheels after takeoff in order to reduce drag and thus fuel consumption.

The flexible structure of plants enables them to reduce drag at high winds by changing their shapes. Large flat leaves, for example, curl into a low-drag conical shape at high wind speeds, while tree branches cluster to reduce drag. Flexible trunks bend under the influence of the wind to reduce drag, and the bending moment is lowered by reducing frontal area.

If you watch the Olympic games, you have probably observed many instances of conscious effort by the competitors to reduce drag. Some examples: During 100-m running, the runners hold their fingers together and straight and move their hands parallel to the direction of motion to reduce the drag on their hands. Swimmers with long hair cover their head with a tight and smooth cover to reduce head drag. They also wear well-fitting one-piece swimming suits. Horse and bicycle riders lean forward as much as they can to reduce drag (by reducing both the drag coefficient and frontal area). Speed skiers do the same thing.

Drag Coefficients of Vehicles

The term *drag coefficient* is commonly used in various areas of daily life. Car manufacturers try to attract consumers by pointing out the *low drag coefficients* of their cars (Fig. 11–21). The drag coefficients of vehicles range from about 1.0 for large semitrailers to 0.4 for minivans and 0.3 for

A hemisphere at two different orientations for Re > 10^4

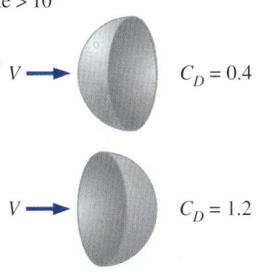

FIGURE 11–19
The drag coefficient of a body may change drastically by changing the body's orientation (and thus shape) relative to the direction of flow.

622
EXTERNAL FLOW: DRAG AND LIFT

FIGURE 11–20
Birds teach us a lesson on drag reduction by extending their beak forward and folding their feet backward during flight.
Photodisc/Getty Images

FIGURE 11–21
This sleek-looking Toyota Prius has a drag coefficient of 0.26—one of the lowest for a passenger car.
Courtesy Toyota.

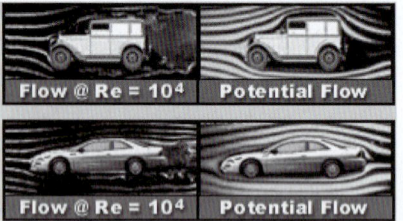

FIGURE 11–22
Streamlines around an aerodynamically designed modern car closely resemble the streamlines around the car in the ideal potential flow (assumes negligible friction), except near the rear end, resulting in a low drag coefficient.
From G. M. Homsy, et al. (2004).

passenger cars. In general, the more blunt the vehicle, the higher the drag coefficient. Installing a fairing reduces the drag coefficient of tractor-trailer rigs by about 20 percent by making the frontal surface more streamlined. As a rule of thumb, the percentage of fuel savings due to reduced drag is about half the percentage of drag reduction at highway speeds.

When the effect of the road on air motion is disregarded, the ideal shape of a *vehicle* is the basic *teardrop*, with a drag coefficient of about 0.1 for the turbulent flow case. But this shape needs to be modified to accommodate several necessary external components such as wheels, mirrors, axles, and door handles. Also, the vehicle must be high enough for comfort and there must be a minimum clearance from the road. Further, a vehicle cannot be too long to fit in garages and parking spaces. Controlling the material and manufacturing costs requires minimizing or eliminating any "dead" volume that cannot be utilized. The result is a shape that resembles more a box than a teardrop, and this was the shape of early cars with a drag coefficient of about 0.8 in the 1920s. This wasn't a problem in those days since the velocities were low, fuel was cheap, and drag was not a major design consideration.

The average drag coefficients of cars dropped to about 0.70 in the 1940s, to 0.55 in the 1970s, to 0.45 in the 1980s, and to 0.30 in the 1990s as a result of improved manufacturing techniques for metal forming and paying more attention to the shape of the car and streamlining (Fig. 11–22). The drag coefficient for well-built racing cars is about 0.2, but this is achieved after making the comfort of drivers a secondary consideration. Noting that the theoretical lower limit of C_D is about 0.1 and the value for racing cars is 0.2, it appears that there is only a little room for further improvement in the drag coefficient of passenger cars from the current value of about 0.3. The drag coefficient of a Mazda 3, for example, is 0.29. For trucks and buses, the drag coefficient can be reduced further by optimizing the front and rear contours (by rounding, for example) to the extent it is practical while keeping the overall length of the vehicle the same.

When traveling as a group, a sneaky way of reducing drag is **drafting**, a phenomenon well known by bicycle riders and car racers. It involves approaching a moving body from behind and being *drafted* into the low-pressure region in the rear of the body. The drag coefficient of a racing bicyclist, for example, is reduced from 0.9 to 0.5 (Table 11–2) by drafting, as also shown in Fig. 11–23.

We also can help reduce the overall drag of a vehicle and thus fuel consumption by being more conscientious drivers. For example, drag force is proportional to the square of velocity. Therefore, driving over the speed limit on the highways not only increases the chances of getting speeding tickets or getting into an accident, but it also increases the amount of fuel consumption per mile. Therefore, driving at moderate speeds is safe and economical. Also, anything that extends from the car, even an arm, increases the drag coefficient. Driving with the windows rolled down also increases the drag and fuel consumption. At highway speeds, a driver can often save fuel in hot weather by running the air conditioner instead of driving with the windows rolled down. For many low-drag automobiles, the turbulence and additional drag generated by open windows consume more fuel than does the air conditioner, but this is not the case for high-drag vehicles.

Superposition

The shapes of many bodies encountered in practice are not simple. But such bodies can be treated conveniently in drag force calculations by considering them to be composed of two or more simple bodies. A satellite dish mounted on a roof with a cylindrical bar, for example, can be considered to be a combination of a hemispherical body and a cylinder. Then the drag coefficient of the body can be determined approximately by using **superposition.** Such a simplistic approach does not account for the effects of components on each other, and thus the results obtained should be interpreted accordingly.

FIGURE 11–23
The drag coefficients of bodies following other moving bodies closely is reduced considerably due to drafting (i.e., entering into the low pressure region created by the body in front).
Getty Images

EXAMPLE 11–2 Effect of Frontal Area on Fuel Efficiency of a Car

Two common methods of improving fuel efficiency of a vehicle are to reduce the drag coefficient and the frontal area of the vehicle. Consider a car (Fig. 11–24) whose width (*W*) and height (*H*) are 1.85 m and 1.70 m, respectively, with a drag coefficient of 0.30. Determine the amount of fuel and money saved per year as a result of reducing the car height to 1.55 m while keeping its width the same. Assume the car is driven 18,000 km a year at an average speed of 95 km/h. Take the density and price of gasoline to be 0.74 kg/L and $0.95/L, respectively. Also take the density of air to be 1.20 kg/m³, the heating value of gasoline to be 44,000 kJ/kg, and the overall efficiency of the car's drive train to be 30 percent.

SOLUTION The frontal area of a car is reduced by redesigning it. The resulting fuel and money savings per year are to be determined.
Assumptions **1** The car is driven 18,000 km a year at an average speed of 95 km/h. **2** The effect of reduction of the frontal area on the drag coefficient is negligible.
Properties The densities of air and gasoline are given to be 1.20 kg/m³ and 0.74 kg/L, respectively. The heating value of gasoline is given to be 44,000 kJ/kg.
Analysis The drag force acting on a body is

$$F_D = C_D A \frac{\rho V^2}{2}$$

where *A* is the frontal area of the body. The drag force acting on the car before redesigning is

$$F_D = 0.3(1.85 \times 1.70 \text{ m}^2) \frac{(1.20 \text{ kg/m}^3)(95 \text{ km/h})^2}{2} \left(\frac{1 \text{ m/s}}{3.6 \text{ km/h}}\right)^2 \left(\frac{1 \text{ N}}{1 \text{ kg} \cdot \text{m/s}^2}\right)$$

$$= 394 \text{ N}$$

FIGURE 11–24
Schematic for Example 11–2.

Noting that work is force times distance, the amount of work done to overcome this drag force and the required energy input for a distance of 18,000 km are

$$W_{\text{drag}} = F_D L = (394 \text{ N})(18{,}000 \text{ km/year})\left(\frac{1000 \text{ m}}{1 \text{ km}}\right)\left(\frac{1 \text{ kJ}}{1000 \text{ N·m}}\right)$$
$$= 7.092 \times 10^6 \text{ kJ/year}$$

$$E_{\text{in}} = \frac{W_{\text{drag}}}{\eta_{\text{car}}} = \frac{7.092 \times 10^6 \text{ kJ/year}}{0.30} = 2.364 \times 10^7 \text{ kJ/year}$$

The amount and the cost of the fuel that supplies this much energy are

$$\text{Amount of fuel} = \frac{m_{\text{fuel}}}{\rho_{\text{fuel}}} = \frac{E_{\text{in}}/\text{HV}}{\rho_{\text{fuel}}} = \frac{(2.364 \times 10^7 \text{ kJ/year})/(44{,}000 \text{ kJ/kg})}{0.74 \text{ kg/L}}$$
$$= 726 \text{ L/year}$$

Cost = (Amount of fuel)(Unit cost) = (726 L/year)($0.95/L) = $690/year

That is, the car uses about 730 liters of gasoline at a total cost of about $690 per year to overcome the drag.

The drag force and the work done to overcome it are directly proportional to the frontal area. Then the percent reduction in the fuel consumption due to reducing the frontal area is equal to the percent reduction in the frontal area:

$$\text{Reduction ratio} = \frac{A - A_{\text{new}}}{A} = \frac{H - H_{\text{new}}}{H} = \frac{1.70 - 1.55}{1.70} = 0.0882$$

Amount reduction = (Reduction ratio)(Amount)

Fuel reduction = 0.0882(726 L/year) = **64 L/year**

Cost reduction = (Reduction ratio)(Cost) = 0.0882($690/year) = **$61/year**

Therefore, reducing the car's height reduces the fuel consumption due to drag by nearly 9 percent.

Discussion Answers are given to 2 significant digits. This example demonstrates that significant reductions in drag and fuel consumption can be achieved by reducing the frontal area of a vehicle as well as its drag coefficient.

Example 11–2 is indicative of the tremendous amount of effort put into redesigning various parts of cars such as the window moldings, the door handles, the windshield, and the front and rear ends in order to reduce aerodynamic drag. For a car moving on a level road at constant speed, the power developed by the engine is used to overcome rolling resistance, friction between moving components, aerodynamic drag, and driving the auxiliary equipment. The aerodynamic drag is negligible at low speeds, but becomes significant at speeds above about 50 km/h. Reduction of the frontal area of the cars (to the dislike of tall drivers) has also contributed greatly to the reduction of drag and fuel consumption.

11–5 • PARALLEL FLOW OVER FLAT PLATES

Consider the flow of a fluid over a *flat plate,* as shown in Fig. 11–25. Surfaces that are slightly contoured (such as turbine blades) also can be approximated as flat plates with reasonable accuracy. The *x*-coordinate is measured along the plate surface from the *leading edge* of the plate in the direction of the flow, and *y* is measured from the surface in the normal direction. The fluid approaches the plate in the *x*-direction with a uniform velocity V, which is equivalent to the velocity over the plate away from the surface.

For the sake of discussion, we consider the fluid to consist of adjacent layers piled on top of each other. The velocity of the particles in the first fluid layer adjacent to the plate is zero because of the no-slip condition. This motionless layer slows down the particles of the neighboring fluid layer as a result of friction between the particles of these two adjoining fluid layers at different velocities. This fluid layer then slows down the molecules of the next layer, and so on. Thus, the presence of the plate is felt up to some normal distance δ from the plate beyond which the free-stream velocity remains virtually unchanged. As a result, the *x*-component of the fluid velocity, *u*, varies from 0 at $y = 0$ to nearly V (typically $0.99V$) at $y = \delta$ (Fig. 11–26).

The region of the flow above the plate bounded by δ in which the effects of the viscous shearing forces caused by fluid viscosity are felt is called the **velocity boundary layer.** The *boundary layer thickness* δ is typically defined as the distance *y* from the surface at which $u = 0.99V$.

The hypothetical line of $u = 0.99V$ divides the flow over a plate into two regions: the **boundary layer region,** in which the viscous effects and the velocity changes are significant, and the **irrotational flow region,** in which the frictional effects are negligible and the velocity remains essentially constant.

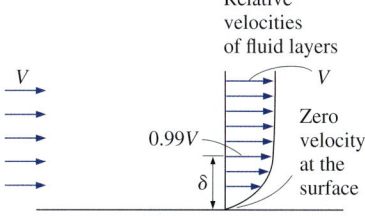

FIGURE 11–26
The development of a boundary layer on a surface is due to the no-slip condition and friction.

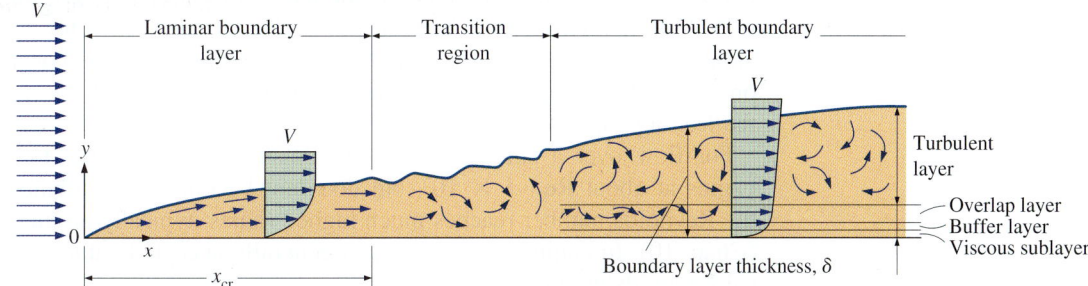

FIGURE 11–25
The development of the boundary layer for flow over a flat plate, and the different flow regimes. Not to scale.

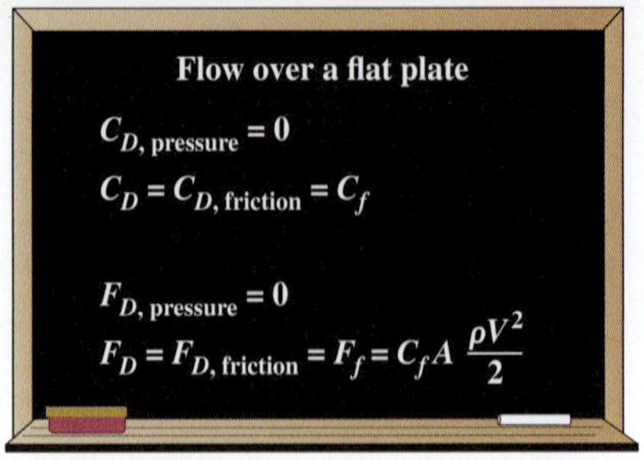

FIGURE 11–27
For parallel flow over a flat plate, the pressure drag is zero, and thus the drag coefficient is equal to the friction coefficient and the drag force is equal to the friction force.

Blackboard:
Flow over a flat plate
$C_{D,\,pressure} = 0$
$C_D = C_{D,\,friction} = C_f$

$F_{D,\,pressure} = 0$
$F_D = F_{D,\,friction} = F_f = C_f A \dfrac{\rho V^2}{2}$

For parallel flow over a flat plate, the pressure drag is zero, and thus the drag coefficient is equal to the *friction drag coefficient*, or simply the *friction coefficient* (Fig. 11–27). That is,

Flat plate: $\qquad C_D = C_{D,\,friction} = C_f \qquad$ (11–13)

Once the average friction coefficient C_f is available, the drag (or friction) force over the surface is determined from

Friction force on a flat plate: $\qquad F_D = F_f = \tfrac{1}{2} C_f A \rho V^2 \qquad$ (11–14)

where A is the surface area of the plate exposed to fluid flow. When both sides of a thin plate are subjected to flow, A becomes the total area of the top and bottom surfaces. Note that both the average friction coefficient C_f and the local friction coefficient $C_{f,\,x}$, in general, vary with location along the surface.

Typical average velocity profiles in laminar and turbulent flow are sketched in Fig. 11–25. Note that the velocity profile in turbulent flow is much fuller than that in laminar flow, with a sharp drop near the surface. The turbulent boundary layer can be considered to consist of four regions, characterized by the distance from the wall. The very thin layer next to the wall where viscous effects are dominant is the **viscous sublayer.** The velocity profile in this layer is very nearly *linear*, and the flow is nearly parallel. Next to the viscous sublayer is the **buffer layer,** in which turbulent effects are becoming significant, but the flow is still dominated by viscous effects. Above the buffer layer is the **overlap layer,** in which the turbulent effects are much more significant, but still not dominant. Above that is the **turbulent** (or **outer**) **layer** in which turbulent effects dominate over viscous effects. Note that the turbulent boundary layer profile on a flat plate closely resembles the boundary layer profile in fully developed turbulent pipe flow (Chap. 8).

The transition from laminar to turbulent flow depends on the *surface geometry, surface roughness, upstream velocity, surface temperature,* and the *type of fluid,* among other things, and is best characterized by the Reynolds number. The Reynolds number at a distance x from the leading edge of a flat plate is expressed as

$$\text{Re}_x = \dfrac{\rho V x}{\mu} = \dfrac{V x}{\nu} \qquad (11\text{–}15)$$

where V is the upstream velocity and x is the characteristic length of the geometry, which, for a flat plate, is the length of the plate in the flow direction. Note that unlike pipe flow, the Reynolds number varies for a flat plate along the flow, reaching $\text{Re}_L = VL/\nu$ at the end of the plate. For any point on a flat plate, the characteristic length is the distance x of the point from the leading edge in the flow direction.

For flow over a smooth flat plate, transition from laminar to turbulent begins at about Re 1×10^5, but does not become fully turbulent before the Reynolds number reaches much higher values, typically around 3×10^6 (Chap. 10). In engineering analysis, a generally accepted value for the critical Reynolds number is

$$\text{Re}_{x,\,cr} = \dfrac{\rho V x_{cr}}{\mu} = 5 \times 10^5$$

The actual value of the engineering critical Reynolds number for a flat plate may vary somewhat from about 10^5 to 3×10^6 depending on the surface roughness, the turbulence level, and the variation of pressure along the surface, as discussed in more detail in Chap. 10.

Friction Coefficient

The friction coefficient for laminar flow over a flat plate can be determined theoretically by solving the conservation of mass and linear momentum equations numerically (Chap. 10). For turbulent flow, however, it must be determined experimentally and expressed by empirical correlations.

The local friction coefficient *varies* along the surface of the flat plate as a result of the changes in the velocity boundary layer in the flow direction. We are usually interested in the drag force on the *entire* surface, which can be determined using the *average* friction coefficient. But sometimes we are also interested in the drag force at a certain location, and in such cases, we need to know the *local* value of the friction coefficient. With this in mind, we present correlations for both local (identified with the subscript x) and average friction coefficients over a flat plate for *laminar, turbulent,* and *combined laminar and turbulent* flow conditions. Once the local values are available, the *average* friction coefficient for the entire plate is determined by integration as

$$C_f = \frac{1}{L}\int_0^L C_{f,x}\, dx \quad (11\text{-}16)$$

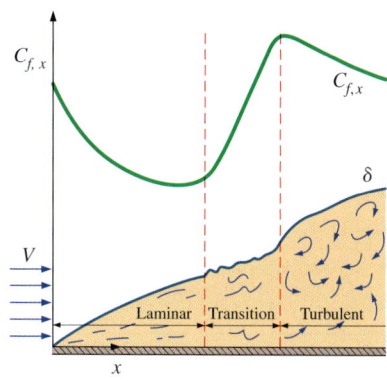

FIGURE 11–28
The variation of the local friction coefficient for flow over a flat plate. Note that the vertical scale of the boundary layer is greatly exaggerated in this sketch.

Based on analysis, the boundary layer thickness and the local friction coefficient at location x for laminar flow over a flat plate were determined in Chap. 10 to be

Laminar: $\quad \delta = \dfrac{4.91x}{\text{Re}_x^{1/2}} \quad \text{and} \quad C_{f,x} = \dfrac{0.664}{\text{Re}_x^{1/2}}, \quad \text{Re}_x \lesssim 5 \times 10^5 \quad (11\text{-}17)$

The corresponding relations for turbulent flow are

Turbulent: $\quad \delta = \dfrac{0.38x}{\text{Re}_x^{1/5}} \quad \text{and} \quad C_{f,x} = \dfrac{0.059}{\text{Re}_x^{1/5}}, \quad 5 \times 10^5 \lesssim \text{Re}_x \lesssim 10^7 \quad (11\text{-}18)$

where x is the distance from the leading edge of the plate and $\text{Re}_x = Vx/\nu$ is the Reynolds number at location x. Note that $C_{f,x}$ is proportional to $1/\text{Re}_x^{1/2}$ and thus to $x^{-1/2}$ for laminar flow and it is proportional to $x^{-1/5}$ for turbulent flow. In either case, $C_{f,x}$ is infinite at the leading edge ($x = 0$), and therefore Eqs. 11–17 and 11–18 are not valid close to the leading edge. The variation of the boundary layer thickness δ and the friction coefficient $C_{f,x}$ along a flat plate is sketched in Fig. 11–28. The local friction coefficients are higher in turbulent flow than they are in laminar flow because of the intense mixing that occurs in the turbulent boundary layer. Note that $C_{f,x}$ reaches its highest values when the flow becomes fully turbulent, and then decreases by a factor of $x^{-1/5}$ in the flow direction, as shown in the figure.

The *average* friction coefficient over the entire plate is determined by substituting Eqs. 11–17 and 11–18 into Eq. 11–16 and performing the integrations (Fig. 11–29). We get

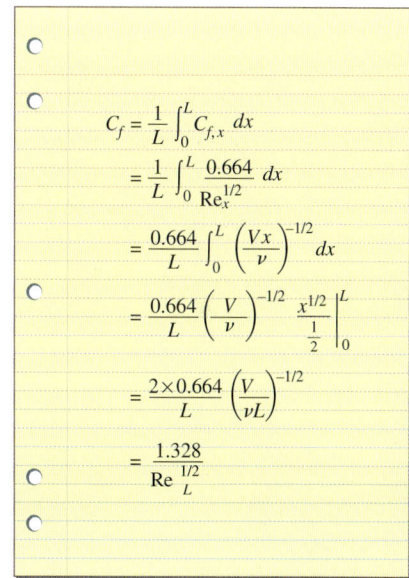

FIGURE 11–29
The average friction coefficient over a surface is determined by integrating the local friction coefficient over the entire surface. The values shown here are for a laminar flat plate boundary layer.

Laminar: $\quad C_f = \dfrac{1.33}{\text{Re}_L^{1/2}} \quad \text{Re}_L \lesssim 5 \times 10^5 \quad (11\text{-}19)$

Turbulent: $\quad C_f = \dfrac{0.074}{\text{Re}_L^{1/5}} \quad 5 \times 10^5 \lesssim \text{Re}_L \lesssim 10^7 \quad (11\text{-}20)$

Relative Roughness, ε/L	Friction Coefficient, C_f
0.0*	0.0029
1×10^{-5}	0.0032
1×10^{-4}	0.0049
1×10^{-3}	0.0084

*Smooth surface for Re = 10^7. Others calculated from Eq. 11–23 for fully rough flow.

FIGURE 11–30

For turbulent flow, surface roughness may cause the friction coefficient to increase severalfold.

The first of these relations gives the average friction coefficient for the entire plate when the flow is *laminar* over the *entire* plate. The second relation gives the average friction coefficient for the entire plate only when the flow is *turbulent* over the *entire* plate, or when the laminar flow region of the plate is negligibly small relative to the turbulent flow region (that is, $x_{cr} \ll L$ where the length of the plate x_{cr} over which the flow is laminar is determined from $Re_{cr} = 5 \times 10^5 = Vx_{cr}/\nu$).

In some cases, a flat plate is sufficiently long for the flow to become turbulent, but not long enough to disregard the laminar flow region. In such cases, the *average* friction coefficient over the entire plate is determined by performing the integration in Eq. 11–16 over two parts: the laminar region $0 \le x \le x_{cr}$ and the turbulent region $x_{cr} < x \le L$ as

$$C_f = \frac{1}{L}\left(\int_0^{x_{cr}} C_{f,x,\text{laminar}}\, dx + \int_{x_{cr}}^{L} C_{f,x,\text{turbulent}}\, dx\right) \quad (11\text{–}21)$$

Note that we included the transition region with the turbulent region. Again taking the critical Reynolds number to be $Re_{cr} = 5 \times 10^5$ and performing these integrations after substituting the indicated expressions, the *average* friction coefficient over the *entire* plate is determined to be

$$C_f = \frac{0.074}{Re_L^{1/5}} - \frac{1742}{Re_L} \qquad 5 \times 10^5 \le Re_L \le 10^7 \quad (11\text{–}22)$$

The constants in this relation would be different for different critical Reynolds numbers. Also, the surfaces are assumed to be *smooth,* and the free stream to be of very low turbulence intensity. For laminar flow, the friction coefficient depends on only the Reynolds number, and the surface roughness has no effect. For turbulent flow, however, surface roughness causes the friction coefficient to increase severalfold, to the point that in the fully rough turbulent regime the friction coefficient is a function of surface roughness alone and is independent of the Reynolds number (Fig. 11–30). This is analogous to flow in pipes.

A curve fit of experimental data for the average friction coefficient in this regime is given by Schlichting (1979) as

Fully rough turbulent regime: $\qquad C_f = \left(1.89 - 1.62 \log \frac{\varepsilon}{L}\right)^{-2.5} \quad (11\text{–}23)$

where ε is the surface roughness and L is the length of the plate in the flow direction. In the absence of a better one, this relation can be used for turbulent flow on rough surfaces for Re > 10^6, especially when $\varepsilon/L > 10^{-4}$.

Friction coefficients C_f for parallel flow over smooth and rough flat plates are plotted in Fig. 11–31 for both laminar and turbulent flows. Note that C_f increases severalfold with roughness in turbulent flow. Also note that C_f is independent of the Reynolds number in the fully rough region. This chart is the flat-plate analog of the Moody chart for pipe flows.

EXAMPLE 11–3   Flow of Hot Oil over a Flat Plate

Engine oil at 40°C flows over a 5-m-long flat plate with a free-stream velocity of 2 m/s (Fig. 11–32). Determine the drag force acting on the top side of the plate per unit width.

SOLUTION Engine oil flows over a flat plate. The drag force per unit width of the plate is to be determined.
Assumptions 1 The flow is steady and incompressible. 2 The critical Reynolds number is $\text{Re}_{cr} = 5 \times 10^5$.
Properties The density and kinematic viscosity of engine oil at 40°C are $\rho = 876$ kg/m³ and $\nu = 2.485 \times 10^{-4}$ m²/s.
Analysis Noting that $L = 5$ m, the Reynolds number at the end of the plate is

$$\text{Re}_L = \frac{VL}{\nu} = \frac{(2 \text{ m/s})(5 \text{ m})}{2.485 \times 10^{-4} \text{ m}^2/\text{s}} = 4.024 \times 10^4$$

which is less than the critical Reynolds number. Thus we have *laminar flow* over the entire plate, and the average friction coefficient is (Fig. 11–29)

$$C_f = 1.328 \text{Re}_L^{-0.5} = 1.328 \times (4.024 \times 10^4)^{-0.5} = 0.00662$$

Noting that the pressure drag is zero and thus $C_D = C_f$ for parallel flow over a flat plate, the drag force acting on the plate per unit width becomes

$$F_D = C_f A \frac{\rho V^2}{2} = 0.00662(5 \times 1 \text{ m}^2) \frac{(876 \text{ kg/m}^3)(2 \text{ m/s})^2}{2} \left(\frac{1 \text{ N}}{1 \text{ kg·m/s}^2} \right) = \mathbf{58.0 \text{ N}}$$

The total drag force acting on the entire plate can be determined by multiplying the value just obtained by the width of the plate.
Discussion The force per unit width corresponds to the weight of a mass of about 6 kg. Therefore, a person who applies an equal and opposite force to the plate to keep it from moving will feel like he or she is using as much force as is necessary to hold a 6-kg mass from dropping.

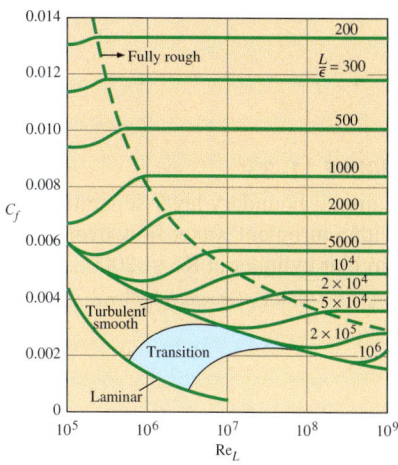

FIGURE 11–31
Friction coefficient for parallel flow over smooth and rough flat plates.
Data from White (2010).

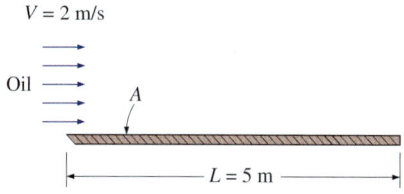

FIGURE 11–32
Schematic for Example 11–3.

11–6 ■ FLOW OVER CYLINDERS AND SPHERES

Flow over cylinders and spheres is frequently encountered in practice. For example, the tubes in a shell-and-tube heat exchanger involve both *internal flow* through the tubes and *external flow* over the tubes, and both flows must be considered in the analysis of the heat exchanger. Also, many sports such as soccer, tennis, and golf involve flow over spherical balls.

The characteristic length for a circular cylinder or sphere is taken to be the *external diameter D*. Thus, the Reynolds number is defined as $\text{Re} = VD/\nu$ where V is the uniform velocity of the fluid as it approaches the cylinder or sphere. The critical Reynolds number for flow across a circular cylinder or sphere is about $\text{Re}_{cr} \cong 2 \times 10^5$. That is, the boundary layer remains laminar for about $\text{Re} \lesssim 2 \times 10^5$, is transitional for $2 \times 10^5 \lesssim \text{Re} \lesssim 2 \times 10^6$, and becomes fully turbulent for $\text{Re} \gtrsim 2 \times 10^6$.

Cross-flow over a cylinder exhibits complex flow patterns, as shown in Fig. 11–33. The fluid approaching the cylinder branches out and encircles the cylinder, forming a boundary layer that wraps around the cylinder. The fluid particles on the midplane strike the cylinder at the stagnation

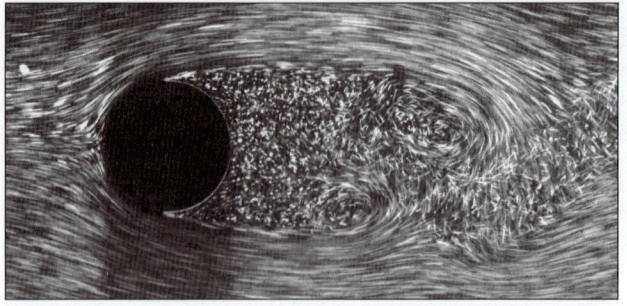

FIGURE 11–33
Laminar boundary layer separation with a turbulent wake; flow over a circular cylinder at Re = 2000.

Courtesy ONERA, photograph by Werlé.

point, bringing the fluid to a complete stop and thus raising the pressure at that point. The pressure decreases in the flow direction while the fluid velocity increases.

At very low upstream velocities (Re ≲ 1), the fluid completely wraps around the cylinder and the two arms of the fluid meet on the rear side of the cylinder in an orderly manner. Thus, the fluid follows the curvature of the cylinder. At higher velocities, the fluid still hugs the cylinder on the frontal side, but it is too fast to remain attached to the surface as it approaches the top (or bottom) of the cylinder. As a result, the boundary layer detaches from the surface, forming a separation region behind the cylinder. Flow in the wake region is characterized by periodic vortex formation and pressures much lower than the stagnation point pressure.

The nature of the flow across a cylinder or sphere strongly affects the total drag coefficient C_D. Both the *friction drag* and the *pressure drag* can be significant. The high pressure in the vicinity of the stagnation point and the low pressure on the opposite side in the wake produce a net force on the body in the direction of flow. The drag force is primarily due to friction drag at low Reynolds numbers (Re ≲ 10) and to pressure drag at high Reynolds numbers (Re ≳ 5000). Both effects are significant at intermediate Reynolds numbers.

The average drag coefficients C_D for cross-flow over a smooth single circular cylinder and a sphere are given in Fig. 11–34. The curves exhibit different behaviors in different ranges of Reynolds numbers:

- For Re ≲ 1, we have creeping flow (Chap. 10), and the drag coefficient decreases with increasing Reynolds number. For a sphere, it is $C_D = 24/Re$. There is no flow separation in this regime.

- At about Re ≅ 10, separation starts occurring on the rear of the body with vortex shedding starting at about Re ≅ 90. The region of separation increases with increasing Reynolds number up to about Re ≅ 10^3. At this point, the drag is mostly (about 95 percent) due to pressure drag. The drag coefficient continues to decrease with increasing Reynolds number in this range of 10 ≲ Re ≲ 10^3. (A decrease in the drag coefficient does not necessarily indicate a decrease in drag. The drag force is proportional to the square of the velocity, and the increase in velocity at higher Reynolds numbers usually more than offsets the decrease in the drag coefficient.)

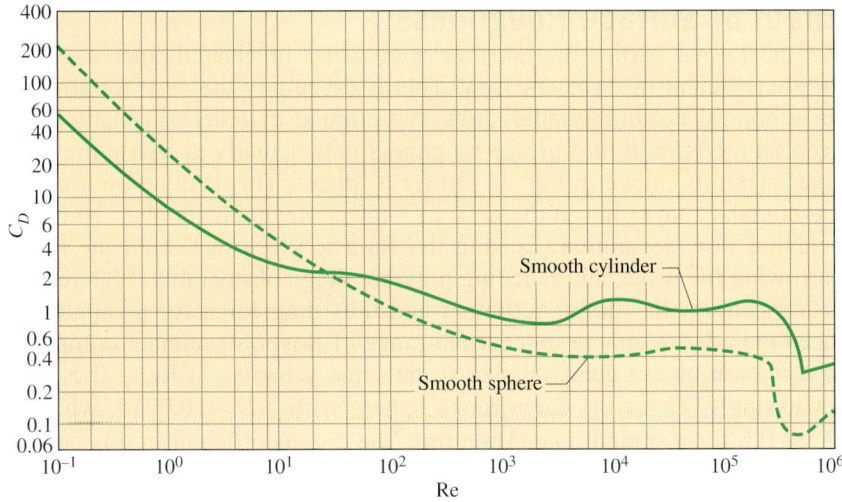

FIGURE 11–34
Average drag coefficient for cross-flow over a smooth circular cylinder and a smooth sphere.
Data from H. Schlichting.

- In the moderate range of $10^3 \lesssim \text{Re} \lesssim 10^5$, the drag coefficient remains relatively constant. This behavior is characteristic of bluff bodies. The flow in the boundary layer is laminar in this range, but the flow in the separated region past the cylinder or sphere is highly turbulent with a wide turbulent wake.
- There is a sudden drop in the drag coefficient somewhere in the range of $10^5 \lesssim \text{Re} \lesssim 10^6$ (usually, at about 2×10^5). This large reduction in C_D is due to the flow in the boundary layer becoming *turbulent*, which moves the separation point further on the rear of the body, reducing the size of the wake and thus the magnitude of the pressure drag. This is in contrast to streamlined bodies, which experience an increase in the drag coefficient (mostly due to friction drag) when the boundary layer becomes turbulent.
- There is a "transitional" regime for $2 \times 10^5 \lesssim \text{Re} \lesssim 2 \times 10^6$, in which C_D dips to a minimum value and then slowly rises to its final turbulent value.

Flow separation occurs at about $\theta \cong 80°$ (measured from the front stagnation point of a cylinder) when the boundary layer is *laminar* and at about $\theta \cong 140°$ when it is *turbulent* (Fig. 11–35). The delay of separation in turbulent flow is caused by the rapid fluctuations of the fluid in the transverse direction, which enables the turbulent boundary layer to travel farther along the surface before separation occurs, resulting in a narrower wake and a smaller pressure drag. Keep in mind that turbulent flow has a fuller velocity profile as compared to the laminar case, and thus it requires a stronger adverse pressure gradient to overcome the additional momentum close to the wall. In the range of Reynolds numbers where the flow changes from laminar to turbulent, even the drag force F_D decreases as the velocity (and thus the Reynolds number) increases. This results in a sudden decrease in drag of a flying body (sometimes called the *drag crisis*) and instabilities in flight.

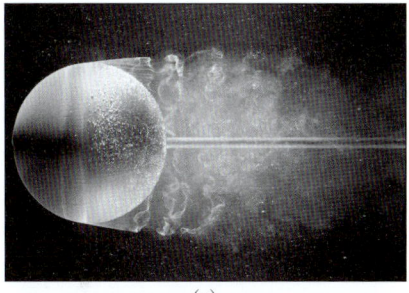
(a)

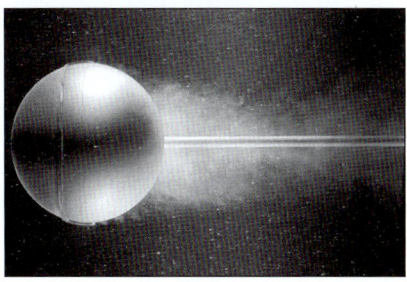
(b)

FIGURE 11–35
Flow visualization of flow over (a) a smooth sphere at Re = 15,000, and (b) a sphere at Re = 30,000 with a trip wire. The delay of boundary layer separation is clearly seen by comparing the two photographs.
Courtesy ONERA, photograph by Werlé.

Effect of Surface Roughness

We mentioned earlier that *surface roughness*, in general, increases the drag coefficient in turbulent flow. This is especially the case for streamlined bodies. For blunt bodies such as a circular cylinder or sphere, however, an increase in the surface roughness may actually *decrease* the drag coefficient, as shown in Fig. 11–36 for a sphere. This is done by tripping the boundary layer into turbulence at a lower Reynolds number, and thus delaying flow separation, causing the fluid to close in behind the body, narrowing the wake, and reducing pressure drag considerably. This results in a much smaller drag coefficient and thus drag force for a rough-surfaced cylinder or sphere in a certain range of Reynolds number compared to a smooth one of identical size at the same velocity. At Re = 2×10^5, for example, $C_D \cong 0.1$ for a rough sphere with $\varepsilon/D = 0.0015$, whereas $C_D \cong 0.5$ for a smooth one. Therefore, the drag coefficient in this case is reduced by a factor of 5 by simply roughening the surface. Note, however, that at Re = 10^6, $C_D \cong 0.4$ for a very rough sphere while $C_D \cong 0.1$ for the smooth one. Obviously, roughening the sphere in this case increases the drag by a factor of 4 (Fig. 11–37).

The preceding discussion shows that roughening the surface can be used to great advantage in reducing drag, but it can also backfire on us if we are not careful—specifically, if we do not operate in the right range of the Reynolds number. With this consideration, golf balls are intentionally roughened to induce *turbulence* at a lower Reynolds number to take advantage of the sharp *drop* in the drag coefficient at the onset of turbulence in the boundary layer (the typical velocity range of golf balls is 15 to 150 m/s, and the Reynolds number is less than 4×10^5). The critical Reynolds number of dimpled golf balls is about 4×10^4. The occurrence of turbulent flow at this Reynolds number reduces the drag coefficient of a golf ball by about half, as shown in Fig. 11–36. For a given hit, this means a longer distance for the ball. Experienced golfers also give the ball a spin during the hit, which helps the rough

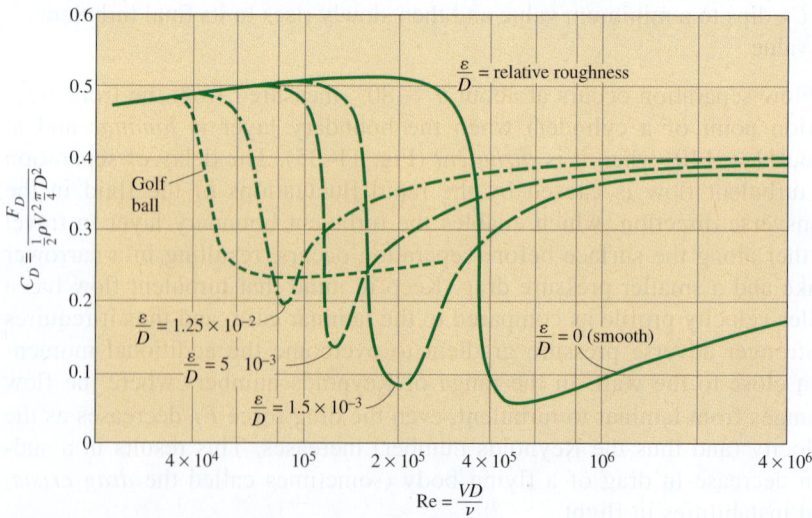

FIGURE 11–36
The effect of surface roughness on the drag coefficient of a sphere.
Data from Blevins (1984).

ball develop a lift and thus travel higher and farther. A similar argument can be given for a tennis ball. For a table tennis ball, however, the speeds are slower and the ball is smaller—it never reaches the turbulent range. Therefore, the surfaces of table tennis balls are smooth.

Once the drag coefficient is available, the drag force acting on a body in cross-flow is determined from Eq. 11–5 where A is the *frontal area* ($A = LD$ for a cylinder of length L and $A = \pi D^2/4$ for a sphere). It should be kept in mind that free-stream turbulence and disturbances by other bodies in the flow (such as flow over tube bundles) may affect the drag coefficient significantly.

	C_D	
Re	Smooth Surface	Rough Surface, $\varepsilon/D = 0.0015$
2×10^5	0.5	0.1
10^6	0.1	0.4

FIGURE 11–37
Surface roughness may increase or decrease the drag coefficient of a spherical object, depending on the value of the Reynolds number.

EXAMPLE 11–4   Drag Force Acting on a Pipe in a River

A 2.2-cm-outer-diameter pipe is to span across a river at a 30-m-wide section while being completely immersed in water (Fig. 11–38). The average flow velocity of water is 4 m/s and the water temperature is 15°C. Determine the drag force exerted on the pipe by the river.

SOLUTION A pipe is submerged in a river. The drag force that acts on the pipe is to be determined.
Assumptions **1** The outer surface of the pipe is smooth so that Fig. 11–34 can be used to determine the drag coefficient. **2** Water flow in the river is steady. **3** The direction of water flow is normal to the pipe. **4** Turbulence in river flow is not considered.
Properties The density and dynamic viscosity of water at 15°C are $\rho = 999.1$ kg/m³ and $\mu = 1.138 \times 10^{-3}$ kg/m·s.
Analysis Noting that $D = 0.022$ m, the Reynolds number is

$$\mathrm{Re} = \frac{VD}{\nu} = \frac{\rho VD}{\mu} = \frac{(999.1 \text{ kg/m}^3)(4 \text{ m/s})(0.022 \text{ m})}{1.138 \times 10^{-3} \text{ kg/m·s}} = 7.73 \times 10^4$$

The drag coefficient corresponding to this value is, from Fig. 11–34, $C_D = 1.0$. Also, the frontal area for flow past a cylinder is $A = LD$. Then the drag force acting on the pipe becomes

$$F_D = C_D A \frac{\rho V^2}{2} = 1.0(30 \times 0.022 \text{ m}^2)\frac{(999.1 \text{ kg/m}^3)(4 \text{ m/s})^2}{2}\left(\frac{1 \text{ N}}{1 \text{ kg·m/s}^2}\right)$$

$$= 5275 \text{ N} \cong \mathbf{5300 \text{ N}}$$

Discussion Note that this force is equivalent to the weight of a mass over 500 kg. Therefore, the drag force the river exerts on the pipe is equivalent to hanging a total of over 500 kg in mass on the pipe supported at its ends 30 m apart. The necessary precautions should be taken if the pipe cannot support this force. If the river were to flow at a faster speed or if turbulent fluctuations in the river were more significant, the drag force would be even larger. *Unsteady* forces on the pipe might then be significant.

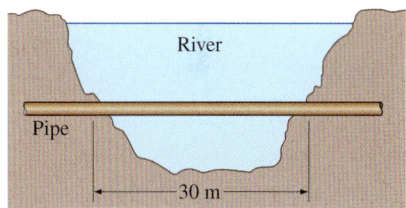

FIGURE 11–38
Schematic for Example 11–4.

634
EXTERNAL FLOW: DRAG AND LIFT

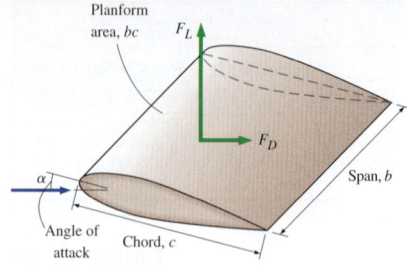

FIGURE 11–39
Definition of various terms associated with an airfoil.

11–7 · LIFT

Lift was defined earlier as the component of the net force (due to viscous and pressure forces) that is perpendicular to the flow direction, and the lift coefficient was expressed in Eq. 11–6 as

$$C_L = \frac{F_L}{\frac{1}{2}\rho V^2 A}$$

where A in this case is normally the *planform area*, which is the area that would be seen by a person looking at the body from above in a direction normal to the body, and V is the upstream velocity of the fluid (or, equivalently, the velocity of a flying body in a quiescent fluid). For an airfoil of width (or span) b and chord length c (the length between the leading and trailing edges), the planform area is $A = bc$. The distance between the two ends of a wing or airfoil is called the **wingspan** or just the **span.** For an aircraft, the wingspan is taken to be the total distance between the tips of the two wings, which includes the width of the fuselage between the wings (Fig. 11–39). The average lift per unit planform area F_L/A is called the **wing loading,** which is simply the ratio of the weight of the aircraft to the planform area of the wings (since lift equals weight when flying at constant altitude).

Airplane flight is based on lift, and thus developing a better understanding of lift as well as improving the lift characteristics of bodies have been the focus of numerous studies. Our emphasis in this section is on devices such as *airfoils* that are specifically designed to generate lift while keeping the drag at a minimum. But it should be kept in mind that some devices such as *spoilers* and *inverted airfoils* on racing cars are designed for the opposite purpose of avoiding lift or even generating negative lift to improve traction and control (some early race cars actually "took off" at high speeds as a result of the lift produced, which alerted the engineers to come up with ways to reduce lift in their design).

For devices that are intended to generate lift such as airfoils, the contribution of *viscous effects* to lift is usually negligible since the bodies are streamlined, and wall shear is parallel to the surfaces of such devices and thus nearly normal to the direction of lift (Fig. 11–40). Therefore, lift in practice can be approximated as due entirely to the pressure distribution on the surfaces of the body, and thus the shape of the body has the primary influence on lift. Then the primary consideration in the design of airfoils is minimizing the average pressure at the upper surface while maximizing it at the lower surface. The Bernoulli equation can be used as a guide in identifying the high- and low-pressure regions: *Pressure is low at locations where the flow velocity is high, and pressure is high at locations where the flow velocity is low.* Also, at moderate angles of attack, lift is practically independent of the surface roughness since roughness affects the wall shear, not the pressure. The contribution of shear to lift is significant only for very small (lightweight) bodies that fly at low velocities (and thus low Reynolds numbers).

Noting that the contribution of viscous effects to lift is negligible, we should be able to determine the lift acting on an airfoil by simply integrating the pressure distribution around the airfoil. The pressure changes in the flow

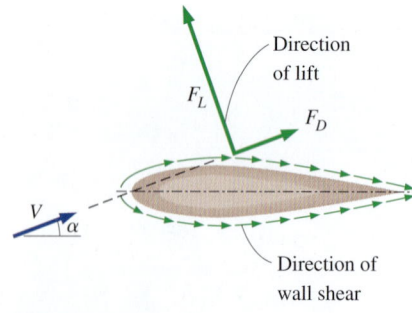

FIGURE 11–40
For airfoils, the contribution of viscous effects to lift is usually negligible since wall shear is parallel to the surfaces and thus nearly normal to the direction of lift.

direction along the surface, but it remains essentially constant through the boundary layer in a direction normal to the surface (Chap. 10). Therefore, it seems reasonable to ignore the very thin boundary layer on the airfoil and calculate the pressure distribution around the airfoil from the relatively simple potential flow theory (zero vorticity, irrotational flow) for which net viscous forces are zero for flow past an airfoil.

The flow fields obtained from such calculations are sketched in Fig. 11–41 for both symmetrical and nonsymmetrical airfoils by ignoring the thin boundary layer. At zero angle of attack, the lift produced by the symmetrical airfoil is zero, as expected because of symmetry, and the stagnation points are at the leading and trailing edges. For the nonsymmetrical airfoil, which is at a small angle of attack, the front stagnation point has moved down below the leading edge, and the rear stagnation point has moved up to the upper surface close to the trailing edge. To our surprise, the lift produced is calculated again to be zero—a clear contradiction of experimental observations and measurements. Obviously, the theory needs to be modified to bring it in line with the observed phenomenon.

The source of inconsistency is the rear stagnation point being at the upper surface instead of the trailing edge. This requires the lower side fluid to make a nearly U-turn and flow around the sharp trailing edge toward the stagnation point while remaining attached to the surface, which is a physical impossibility since the observed phenomenon is the separation of flow at sharp turns (imagine a car attempting to make this turn at high speed). Therefore, the lower side fluid separates smoothly off the trailing edge, and the upper side fluid responds by pushing the rear stagnation point downstream. In fact, the stagnation point at the upper surface moves all the way to the trailing edge. This way the two flow streams from the top and the bottom sides of the airfoil meet at the trailing edge, yielding a smooth flow downstream parallel to the sharp trailing edge. Lift is generated because the flow velocity at the top surface is higher, and thus the pressure on that surface is lower due to the Bernoulli effect.

The potential flow theory and the observed phenomenon can be reconciled as follows: Flow starts out as predicted by theory, with no lift, but the lower fluid stream separates at the trailing edge when the velocity reaches a certain value. This forces the separated upper fluid stream to close in at the trailing edge, initiating clockwise circulation around the airfoil. This clockwise circulation increases the velocity of the upper stream while decreasing that of the lower stream, causing lift. A **starting vortex** of opposite sign (counterclockwise circulation) is then shed downstream (Fig. 11–42), and smooth streamlined flow is established over the airfoil. When the potential flow theory is modified by the addition of an appropriate amount of circulation to move the stagnation point down to the trailing edge, excellent agreement is obtained between theory and experiment for both the flow field and the lift.

It is desirable for airfoils to generate the most lift while producing the least drag. Therefore, a measure of performance for airfoils is the **lift-to-drag ratio,** which is equivalent to the ratio of the lift-to-drag coefficients C_L/C_D. This information is provided either by plotting C_L versus C_D for different values of the angle of attack (a lift–drag polar) or by plotting the ratio C_L/C_D versus the angle of attack. The latter is done for a particular airfoil design in Fig. 11–43. Note that the C_L/C_D ratio increases with the angle of

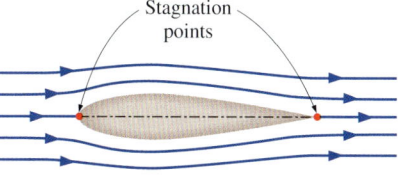

(*a*) Irrotational flow past a symmetrical airfoil (zero lift)

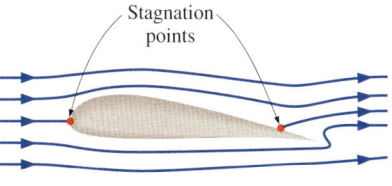

(*b*) Irrotational flow past a nonsymmetrical airfoil (zero lift)

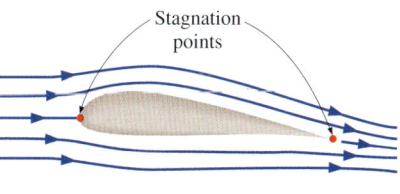

(*c*) Actual flow past a nonsymmetrical airfoil (positive lift)

FIGURE 11–41
Irrotational and actual flow past symmetrical and nonsymmetrical two-dimensional airfoils.

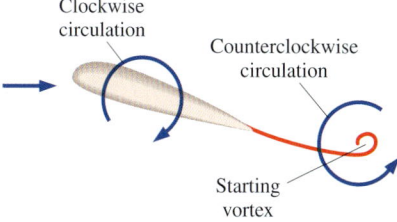

FIGURE 11–42
Shortly after a sudden increase in angle of attack, a counterclockwise starting vortex is shed from the airfoil, while clockwise circulation appears around the airfoil, causing lift to be generated.

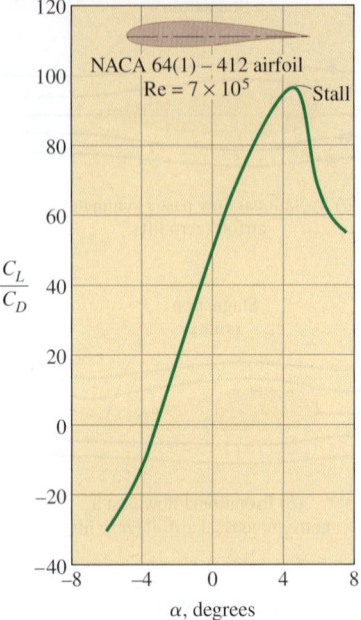

FIGURE 11–43

The variation of the lift-to-drag ratio with angle of attack for a two-dimensional airfoil.

Data from Abbott, von Doenhoff, and Stivers (1945).

attack until the airfoil stalls, and the value of the lift-to-drag ratio can be of the order of 100 for a two-dimensional airfoil.

One obvious way to change the lift and drag characteristics of an airfoil is to change the angle of attack. On an airplane, for example, the entire plane is pitched up to increase lift, since the wings are fixed relative to the fuselage. Another approach is to change the shape of the airfoil by the use of movable *leading edge* and *trailing edge flaps,* as is commonly done in modern large aircraft (Fig. 11–44). The flaps are used to alter the shape of the wings during takeoff and landing to maximize lift and to enable the aircraft to land or take off at low speeds. The increase in drag during this takeoff and landing is not much of a concern because of the relatively short time periods involved. Once at cruising altitude, the flaps are retracted, and the wing is returned to its "normal" shape with minimal drag coefficient and adequate lift coefficient to minimize fuel consumption while cruising at a constant altitude. Note that even a small lift coefficient can generate a large lift force during normal operation because of the large cruising velocities of aircraft and the proportionality of lift to the square of flow velocity.

The effects of flaps on the lift and drag coefficients are shown in Fig. 11–45 for an airfoil. Note that the maximum lift coefficient increases from about 1.5 for the airfoil with no flaps to 3.5 for the double-slotted flap case. But also note that the maximum drag coefficient increases from about 0.06 for the airfoil with no flaps to about 0.3 for the double-slotted flap case. This is a fivefold increase in the drag coefficient, and the engines must work much harder to provide the necessary thrust to overcome this drag. The angle of attack of the flaps can be increased to maximize the lift coefficient. Also, the flaps extend the chord length, and thus enlarge the wing area A. The Boeing 727 uses a triple-slotted flap at the trailing edge and a slot at the leading edge.

The minimum flight velocity is determined from the requirement that the total weight W of the aircraft be equal to lift and $C_L = C_{L,\,max}$. That is,

$$W = F_L = \tfrac{1}{2} C_{L,\,max} \rho V_{min}^2 A \quad \rightarrow \quad V_{min} = \sqrt{\frac{2W}{\rho C_{L,\,max} A}} \qquad (11\text{–}24)$$

For a given weight, the landing or takeoff speed can be minimized by maximizing the product of the lift coefficient and the wing area, $C_{L,\,max} A$. One way of doing that is to use flaps, as already discussed. Another way is to control the boundary layer, which can be accomplished simply by leaving flow sections (slots) between the flaps, as shown in Fig. 11–46. Slots are used to prevent the separation of the boundary layer from the upper surface of the wings and the flaps. This is done by allowing air to move from the high-pressure region under the wing into the low-pressure region at the top surface. Note that the lift

FIGURE 11–44

The lift and drag characteristics of an airfoil during takeoff and landing are changed by changing the shape of the airfoil by the use of movable flaps.

Photos by Yunus Çengel.

(*a*) Flaps extended (landing)

(*b*) Flaps retracted (cruising)

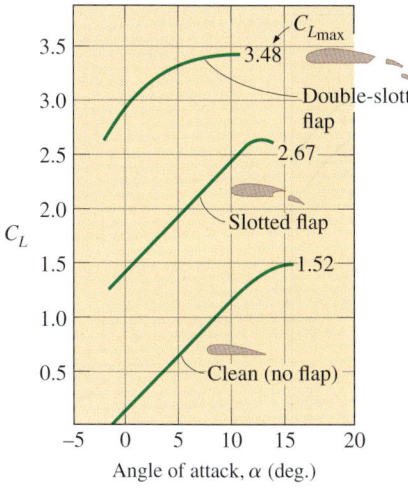
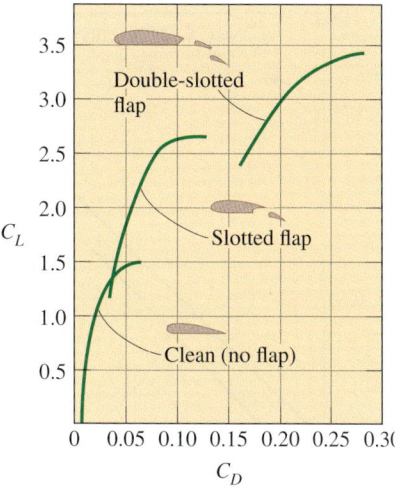

FIGURE 11–45
Effect of flaps on the lift and drag coefficients of an airfoil.
Data from Abbott and von Doenhoff, for NACA 23012 (1959).

coefficient reaches its maximum value $C_L = C_{L,\,max}$, and thus the flight velocity reaches its minimum, at stall conditions, which is a region of unstable operation and must be avoided. The Federal Aviation Administration (FAA) does not allow operation below 1.2 times the stall speed for safety.

Another thing we notice from this equation is that the minimum velocity for takeoff or landing is inversely proportional to the square root of density. Noting that air density decreases with altitude (by about 15 percent at 1500 m), longer runways are required at airports at higher altitudes such as Denver to accommodate higher minimum takeoff and landing velocities. The situation becomes even more critical on hot summer days since the density of air is inversely proportional to temperature.

The development of efficient (low-drag) airfoils was the subject of intense experimental investigations in the 1930s. These airfoils were standardized by the National Advisory Committee for Aeronautics (NACA, which is now NASA), and extensive lists of data on lift coefficients were reported. The variation of the lift coefficient C_L with angle of attack for two 2-D (infinite span) airfoils (NACA 0012 and NACA 2412) is given in Fig. 11–47. We make the following observations from this figure:

- The lift coefficient increases almost linearly with angle of attack α, reaches a maximum at about $\alpha = 16°$, and then starts to decrease sharply. This decrease of lift with further increase in the angle of attack is called *stall*, and it is caused by flow separation and the formation of a wide wake region over the top surface of the airfoil. Stall is highly undesirable since it also increases drag.
- At zero angle of attack ($\alpha = 0°$), the lift coefficient is zero for symmetrical airfoils but nonzero for nonsymmetrical ones with greater curvature at the top surface. Therefore, planes with symmetrical wing sections must fly with their wings at higher angles of attack in order to produce the same lift.
- The lift coefficient is increased by severalfold by adjusting the angle of attack (from 0.25 at $\alpha = 0°$ for the nonsymmetrical airfoil to 1.25 at $\alpha = 10°$).
- The drag coefficient also increases with angle of attack, often exponentially (Fig. 11–48). Therefore, large angles of attack should be used sparingly for short periods of time for fuel efficiency.

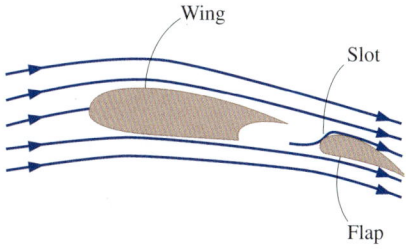

FIGURE 11–46
A flapped airfoil with a slot to prevent the separation of the boundary layer from the upper surface and to increase the lift coefficient.

638
EXTERNAL FLOW: DRAG AND LIFT

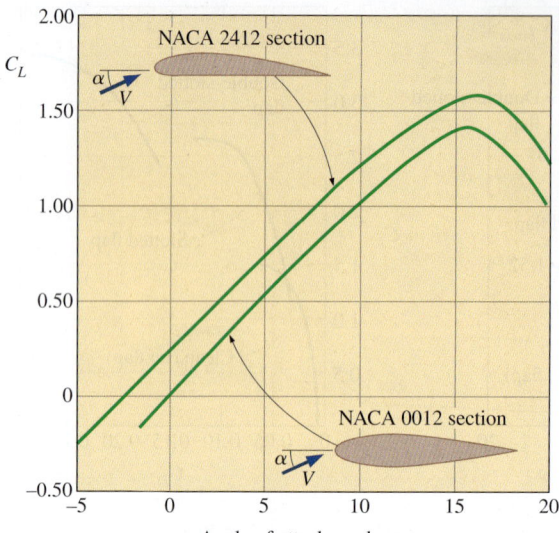

FIGURE 11–47
The variation of the lift coefficient with angle of attack for a symmetrical and a nonsymmetrical airfoil.
Data from Abbott (1945, 1959).

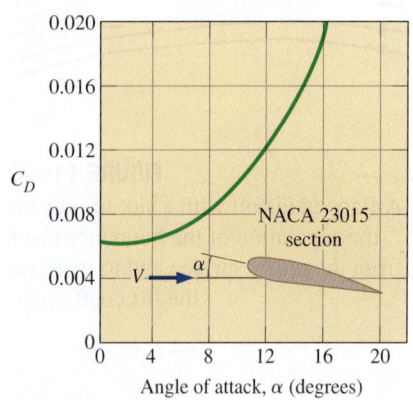

FIGURE 11–48
The variation of the drag coefficient of an airfoil with angle of attack.
Data from Abbott and von Doenhoff (1959).

Finite-Span Wings and Induced Drag

For airplane wings and other airfoils of finite span, the end effects at the tips become important because of the fluid leakage between the lower and upper surfaces. The pressure difference between the lower surface (high-pressure region) and the upper surface (low-pressure region) drives the fluid at the tips upward while the fluid is swept toward the back because of the relative motion between the fluid and the wing. This results in a swirling motion that spirals along the flow, called the **tip vortex,** at the tips of both wings. Vortices are also formed along the airfoil between the tips of the wings. These distributed vortices collect toward the edges after being shed from the trailing edges of the wings and combine with the tip vortices to form two streaks of powerful **trailing vortices** along the tips of the wings (Fig. 11–49). Trailing vortices generated by large aircraft persist for a long time for long distances (over 10 km) before they gradually disappear due to viscous dissipation. Such vortices and the accompanying downdraft are strong enough to cause a small aircraft to lose control and flip over if it flies through the wake of a larger aircraft. Therefore, following a large aircraft closely (within 10 km) poses a real danger for smaller aircraft. This issue is the controlling factor that governs the spacing of aircraft at take-off, which limits the flight capacity at airports. In nature, this effect is used to advantage by birds that migrate in V-formation by utilizing the updraft generated by the bird in front. It has been determined that the birds in a typical flock can fly to their destination in V-formation with one-third less energy. Military jets also occasionally fly in V-formation for the same reason (Fig. 11–50).

Tip vortices that interact with the free stream impose forces on the wing tips in all directions, including the flow direction. The component of the force in the flow direction adds to drag and is called **induced drag.** The total drag of a wing is then the sum of the induced drag (3-D effects) and the drag of the airfoil section (2-D effects).

The ratio of the square of the average span of an airfoil to the planform area is called the **aspect ratio.** For an airfoil with a rectangular planform of chord c and span b, it is expressed as

$$\text{AR} = \frac{b^2}{A} = \frac{b^2}{bc} = \frac{b}{c} \qquad (11-25)$$

Therefore, the aspect ratio is a measure of how (relatively) narrow an airfoil is in the flow direction. The lift coefficient of wings, in general, increases while the drag coefficient decreases with increasing aspect ratio. This is because a long narrow wing (large aspect ratio) has a shorter tip length and thus smaller tip losses and smaller induced drag than a short and wide wing of the same planform area. Therefore, bodies with large aspect ratios fly more efficiently, but they are less maneuverable because of their larger moment of inertia (owing to the greater distance from the center). Bodies with smaller aspect ratios maneuver better since the wings are closer to the central part. So it is no surprise that *fighter planes* (and fighter birds like falcons) have short and wide wings while *large commercial planes* (and soaring birds like albatrosses) have long and narrow wings.

The end effects can be minimized by attaching **endplates** or **winglets** at the tips of the wings perpendicular to the top surface. The endplates function by blocking some of the leakage around the wing tips, which results in a considerable reduction in the strength of the tip vortices and the induced drag. Wing tip feathers on birds fan out for the same purpose (Fig. 11–51).

Lift Generated by Spinning

You have probably experienced giving a spin to a tennis ball or making a drop shot on a tennis or ping-pong ball by giving a fore spin in order to alter the lift characteristics and cause the ball to produce a more desirable trajectory and bounce of the shot. Golf, soccer, and baseball players also utilize spin in their games. The phenomenon of producing lift by the rotation of a solid body is called the **Magnus effect** after the German scientist Heinrich Magnus (1802–1870), who was the first to study the lift of rotating bodies, which is illustrated in Fig. 11–52 for the simplified case of irrotational (potential) flow. When the ball is not spinning, the lift is zero because of top–bottom symmetry. But when the cylinder is rotated about its axis, the cylinder drags some fluid around because of the no-slip condition and the flow field reflects the superposition of the spinning and nonspinning flows. The stagnation points shift down, and the flow is no longer symmetric about the horizontal plane that passes through the center of the cylinder. The average pressure on the upper half is less than the average pressure on the lower half because of the Bernoulli effect, and thus there is a *net upward force* (lift) acting on the cylinder. A similar argument can be given for the lift generated on a spinning ball.

The effect of the rate of rotation on the lift and drag coefficients of a smooth sphere is shown in Fig. 11–53. Note that the lift coefficient strongly depends on the rate of rotation, especially at low angular velocities. The effect of the rate of rotation on the drag coefficient is small. Roughness also affects the drag and lift coefficients. In a certain range of Reynolds number, roughness produces the desirable effect of increasing the lift coefficient while decreasing

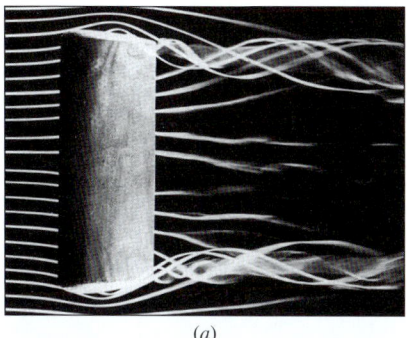

(a)

(b)

(c)

FIGURE 11–49
Trailing vortices visualized in various ways: (*a*) Smoke streaklines in a wind tunnel show vortex cores leaving the trailing edge of a rectangular wing; (*b*) Four contrails initially formed by condensation of water vapor in the low pressure region behind the jet engines eventually merge into the two counter-rotating trailing vortices that persist very far downstream; (*c*) A crop duster flies through smoky air which swirls around in one of the tip vortices from the aircraft's wing.

(a) Courtesy of the Parabolic Press, Stanford, California; (b) Geostock/Getty Images; (c) NASA Langley Research Center

the drag coefficient. Therefore, golf balls with the right amount of roughness travel higher and farther than smooth balls for the same hit.

FIGURE 11–50
(a) Geese flying in their characteristic V-formation to save energy.
(b) Military jets imitating nature.
(a) © Royalty-Free/CORBIS
(b) © Charles Smith/Corbis RF

(a) A bearded vulture with its wing feathers fanned out during flight.

(b) Winglets are used on this sailplane to reduce induced drag.

FIGURE 11–51
Induced drag is reduced by (a) wing tip feathers on bird wings and (b) endplates or other disruptions on airplane wings.
(a) © Jeremy Woodhouse/Getty RF; (b) Courtesy of Jacques Noel, Schempp-Hirth. Used by permission.

EXAMPLE 11–5 Lift and Drag of a Commercial Airplane

A commercial airplane has a total mass of 70,000 kg and a wing planform area of 150 m² (Fig. 11–54). The plane has a cruising speed of 558 km/h and a cruising altitude of 12,000 m, where the air density is 0.312 kg/m³. The plane has double-slotted flaps for use during takeoff and landing, but it cruises with all flaps retracted. Assuming the lift and the drag characteristics of the wings can be approximated by NACA 23012 (Fig. 11–45), determine (a) the minimum safe speed for takeoff and landing with and without extending the flaps, (b) the angle of attack to cruise steadily at the cruising altitude, and (c) the power that needs to be supplied to provide enough thrust to overcome wing drag.

SOLUTION The cruising conditions of a passenger plane and its wing characteristics are given. The minimum safe landing and takeoff speeds, the angle of attack during cruising, and the power required are to be determined.
Assumptions 1 The drag and lift produced by parts of the plane other than the wings, such as the fuselage are not considered. 2 The wings are assumed to be two-dimensional airfoil sections, and the tip effects of the wings are not considered. 3 The lift and the drag characteristics of the wings are approximated by NACA 23012 so that Fig. 11–45 is applicable. 4 The average density of air on the ground is 1.20 kg/m³.
Properties The density of air is 1.20 kg/m³ on the ground and 0.312 kg/m³ at cruising altitude. The maximum lift coefficient $C_{L,\,max}$ of the wing is 3.48 and 1.52 with and without flaps, respectively (Fig. 11–45).
Analysis (a) The weight and cruising speed of the airplane are

$$W = mg = (70{,}000 \text{ kg})(9.81 \text{ m/s}^2)\left(\frac{1 \text{ N}}{1 \text{ kg} \cdot \text{m/s}^2}\right) = 686{,}700 \text{ N}$$

$$V = (558 \text{ km/h})\left(\frac{1 \text{ m/s}}{3.6 \text{ km/h}}\right) = 155 \text{ m/s}$$

The minimum velocities corresponding to the stall conditions without and with flaps, respectively, are obtained from Eq. 11–24,

$$V_{\min 1} = \sqrt{\frac{2W}{\rho C_{L,\,max\,1} A}} = \sqrt{\frac{2(686{,}700 \text{ N})}{(1.2 \text{ kg/m}^3)(1.52)(150 \text{ m}^2)}}\left(\frac{1 \text{ kg} \cdot \text{m/s}^2}{1 \text{ N}}\right) = 70.9 \text{ m/s}$$

$$V_{\min 2} = \sqrt{\frac{2W}{\rho C_{L,\,max\,2} A}} = \sqrt{\frac{2(686{,}700 \text{ N})}{(1.2 \text{ kg/m}^3)(3.48)(150 \text{ m}^2)}}\left(\frac{1 \text{ kg} \cdot \text{m/s}^2}{1 \text{ N}}\right) = 46.8 \text{ m/s}$$

Then the "safe" minimum velocities to avoid the stall region are obtained by multiplying the values above by 1.2:

Without flaps: $V_{\min 1,\,safe} = 1.2 V_{\min 1} = 1.2(70.9 \text{ m/s}) = 85.1 \text{ m/s} = $ **306 km/h**

With flaps: $V_{\min 2,\,safe} = 1.2 V_{\min 2} = 1.2(46.8 \text{ m/s}) = 56.2 \text{ m/s} = $ **202 km/h**

since 1 m/s = 3.6 km/h. Note that the use of flaps allows the plane to take off and land at considerably lower velocities, and thus on a shorter runway.

(b) When an aircraft is cruising steadily at a constant altitude, the lift must be equal to the weight of the aircraft, $F_L = W$. Then the lift coefficient is

$$C_L = \frac{F_L}{\frac{1}{2}\rho V^2 A} = \frac{686{,}700 \text{ N}}{\frac{1}{2}(0.312 \text{ kg/m}^3)(155 \text{ m/s})^2(150 \text{ m}^2)}\left(\frac{1 \text{ kg·m/s}^2}{1 \text{ N}}\right) = 1.22$$

For the case with no flaps, the angle of attack corresponding to this value of C_L is determined from Fig. 11–45 to be $\alpha \cong \mathbf{10°}$.

(c) When the aircraft is cruising steadily at a constant altitude, the net force acting on the aircraft is zero, and thus thrust provided by the engines must be equal to the drag force. The drag coefficient corresponding to the cruising lift coefficient of 1.22 is determined from Fig. 11–45 to be $C_D \cong 0.03$ for the case with no flaps. Then the drag force acting on the wings becomes

$$F_D = C_D A \frac{\rho V^2}{2} = (0.03)(150 \text{ m}^2)\frac{(0.312 \text{ kg/m}^3)(155 \text{ m/s})^2}{2}\left(\frac{1 \text{ kN}}{1000 \text{ kg·m/s}^2}\right)$$

$$= 16.9 \text{ kN}$$

Noting that power is force times velocity (distance per unit time), the power required to overcome this drag is equal to the thrust times the cruising velocity:

$$\text{Power} = \text{Thrust} \times \text{Velocity} = F_D V = (16.9 \text{ kN})(155 \text{ m/s})\left(\frac{1 \text{ kW}}{1 \text{ kN·m/s}}\right)$$

$$= \mathbf{2620 \text{ kW}}$$

Therefore, the engines must supply 2620 kW of power to overcome the drag on the wings during cruising. For a propulsion efficiency of 30 percent (i.e., 30 percent of the energy of the fuel is utilized to propel the aircraft), the plane requires energy input at a rate of 8730 kJ/s.

Discussion The power determined is the power to overcome the drag that acts on the wings only and does not include the drag that acts on the remaining parts of the aircraft (the fuselage, the tail, etc.). Therefore, the total power required during cruising will be much greater. Also, it does not consider induced drag, which can be dominant during takeoff when the angle of attack is high (Fig. 11–45 is for a 2-D airfoil, and does not include 3-D effects).

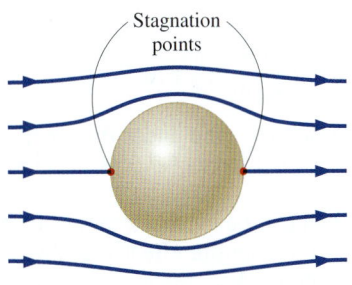

(a) Potential flow over a stationary cylinder

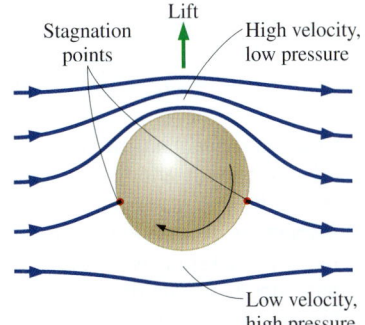

(b) Potential flow over a rotating cylinder

FIGURE 11–52
Generation of lift on a rotating circular cylinder for the case of "idealized" potential flow (the actual flow involves flow separation in the wake region).

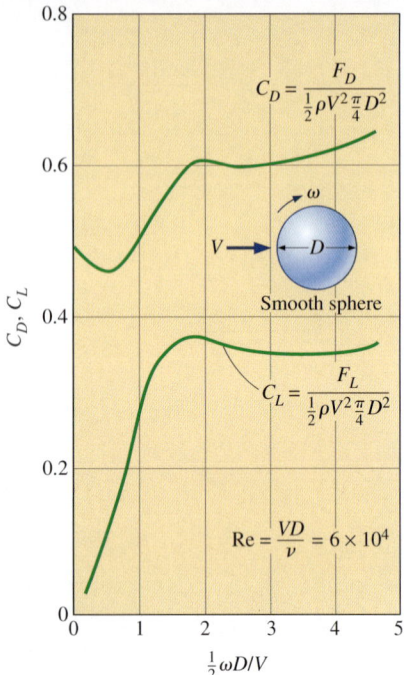

FIGURE 11–53
The variation of lift and drag coefficients of a smooth sphere with the nondimensional rate of rotation for Re = $VD/\nu = 6 \times 10^4$.
Data from Goldstein (1938).

FIGURE 11–54
Schematic for Example 11–5.

EXAMPLE 11–6   Effect of Spin on a Tennis Ball

A tennis ball with a mass of 0.0570 kg and a diameter of 6.37 cm is hit at 72 km/h with a backspin of 4800 rpm (Fig. 11–55). Determine if the ball will fall or rise under the combined effect of gravity and lift due to spinning shortly after being hit in air at 1 atm and 25°C.

SOLUTION A tennis ball is hit with a backspin. It is to be determined whether the ball will fall or rise after being hit.
Assumptions 1 The surface of the ball is smooth enough for Fig. 11–53 to be applicable (this is a stretch for a tennis ball). 2 The ball is hit horizontally so that it starts its motion horizontally.
Properties The density and kinematic viscosity of air at 1 atm and 25°C are $\rho = 1.184$ kg/m³ and $\nu = 1.562 \times 10^{-5}$ m²/s.
Analysis The ball is hit horizontally, and thus it would normally fall under the effect of gravity without the spin. The backspin generates a lift, and the ball will rise if the lift is greater than the weight of the ball. The lift is determined from

$$F_L = C_L A \frac{\rho V^2}{2}$$

where A is the frontal area of the ball, which is $A = \pi D^2/4$. The translational and angular velocities of the ball are

$$V = (72 \text{ km/h})\left(\frac{1000 \text{ m}}{1 \text{ km}}\right)\left(\frac{1 \text{ h}}{3600 \text{ s}}\right) = 20 \text{ m/s}$$

$$\omega = (4800 \text{ rev/min})\left(\frac{2\pi \text{ rad}}{1 \text{ rev}}\right)\left(\frac{1 \text{ min}}{60 \text{ s}}\right) = 502 \text{ rad/s}$$

Then, the nondimensional rate of rotation is

$$\frac{\omega D}{2V} = \frac{(502 \text{ rad/s})(0.0637 \text{ m})}{2(20 \text{ m/s})} = 0.80 \text{ rad}$$

From Fig. 11–53, the lift coefficient corresponding to this value is $C_L = 0.21$. Then the lift force acting on the ball is

$$F_L = (0.21)\frac{\pi(0.0637 \text{ m})^2}{4}\frac{(1.184 \text{ kg/m}^3)(20 \text{ m/s})^2}{2}\left(\frac{1 \text{ N}}{1 \text{ kg·m/s}^2}\right)$$

$$= 0.158 \text{ N}$$

The weight of the ball is

$$W = mg = (0.0570 \text{ kg})(9.81 \text{ m/s}^2)\left(\frac{1 \text{ N}}{1 \text{ kg·m/s}^2}\right) = 0.559 \text{ N}$$

which is more than the lift. Therefore, the ball will **drop** under the combined effect of gravity and lift due to spinning with a net force of 0.559 − 0.158 = 0.401 N.
Discussion This example shows that the ball can be hit much farther by giving it a backspin. Note that a topspin has the opposite effect (negative lift) and speeds up the drop of the ball to the ground. Also, the Reynolds number for this problem is 8×10^4, which is sufficiently close to the 6×10^4 for which Fig. 11–53 is prepared.

Also keep in mind that although some spin may increase the distance traveled by a ball, there is an optimal spin that is a function of launch angle, as most golfers are now more aware. Too much spin decreases distance by introducing more induced drag.

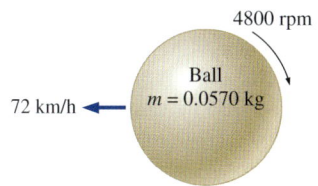

FIGURE 11–55
Schematic for Example 11–6.

No discussion on lift and drag would be complete without mentioning the contributions of Wilbur (1867–1912) and Orville (1871–1948) Wright. The Wright Brothers are truly the most impressive engineering team of all time. Self-taught, they were well informed of the contemporary theory and practice in aeronautics. They both corresponded with other leaders in the field and published in technical journals. While they cannot be credited with developing the concepts of lift and drag, they used them to achieve the first powered, manned, heavier-than-air, controlled flight (Fig. 11–56). They succeeded, while so many before them failed, because they evaluated and designed parts separately. Before the Wrights, experimenters were building and testing whole airplanes. While intuitively appealing, the approach did not allow the determination of how to make the craft better. When a flight lasts only a moment, you can only guess at the weakness in the design. Thus, a new craft did not necessarily perform any better than its predecessor. Testing was simply one belly flop followed by another. The Wrights changed all that. They studied each part using scale and full-size models in wind tunnels and in the field. Well before the first powered flyer was assembled, they knew the area required for their best wing shape to support a plane carrying a man and the engine horsepower required to provide adequate thrust with their improved impeller. The Wright Brothers not only showed the world how to fly, they showed engineers how to use the equations presented here to design even better aircraft.

FIGURE 11–56
The Wright Brothers take flight at Kitty Hawk.
Library of Congress Prints & Photographs Division [LC-DIG-ppprs-00626].

SUMMARY

In this chapter, we study flow of fluids over immersed bodies with emphasis on the resulting lift and drag forces. A fluid may exert forces and moments on a body in and about various directions. The force a flowing fluid exerts on a body in the flow direction is called *drag* while that in the direction normal to the flow is called *lift*. The part of drag that is due directly to wall shear stress τ_w is called the *skin friction drag* since it is caused by frictional effects, and the part that is due directly to pressure P is called the *pressure drag* or *form drag* because of its strong dependence on the form or shape of the body.

The *drag coefficient* C_D and the *lift coefficient* C_L are dimensionless numbers that represent the drag and the lift characteristics of a body and are defined as

$$C_D = \frac{F_D}{\frac{1}{2}\rho V^2 A} \quad \text{and} \quad C_L = \frac{F_L}{\frac{1}{2}\rho V^2 A}$$

where A is usually the *frontal area* (the area projected on a plane normal to the direction of flow) of the body. For plates and airfoils, A is taken to be the *planform area*, which is the area that would be seen by a person looking at the body from directly above. The drag coefficient, in general, depends on the Reynolds number, especially for Reynolds numbers below 10^4. At higher Reynolds numbers, the drag coefficients for many geometries remain essentially constant.

A body is said to be *streamlined* if a conscious effort is made to align its shape with the anticipated streamlines in the flow in order to reduce drag. Otherwise, a body (such as a building) tends to block the flow and is said to be *bluff*. At sufficiently high velocities, the fluid stream detaches itself from the surface of the body. This is called *flow separation*. When a fluid stream separates from the body, it forms a *separated region* between the body and the fluid stream. Separation may also occur on a streamlined body such as an airplane wing at a sufficiently large *angle of attack,* which is the angle the incoming fluid stream makes with the *chord* (the line that connects the nose and the end) of the body. Flow separation on the top surface of a wing reduces lift drastically and may cause the airplane to *stall*.

The region of flow above a surface in which the effects of the viscous shearing forces caused by fluid viscosity are felt is called the *velocity boundary layer* or just the *boundary layer.* The *thickness* of the boundary layer, δ, is defined as the distance from the surface at which the velocity is $0.99V$. The hypothetical line of velocity $0.99V$ divides the flow over a plate into two regions: the *boundary layer region,* in which the viscous effects and the velocity changes are significant, and the *irrotational outer flow region,* in which the frictional effects are negligible and the velocity remains essentially constant.

For external flow, the Reynolds number is expressed as

$$\mathrm{Re}_L = \frac{\rho V L}{\mu} = \frac{V L}{\nu}$$

where V is the upstream velocity and L is the characteristic length of the geometry, which is the length of the plate in the flow direction for a flat plate and the diameter D for a cylinder or sphere. The *average* friction coefficients over an entire flat plate are

Laminar flow: $\quad C_f = \dfrac{1.33}{\mathrm{Re}_L^{1/2}} \quad \mathrm{Re}_L \lesssim 5 \times 10^5$

Turbulent flow: $\quad C_f = \dfrac{0.074}{\mathrm{Re}_L^{1/5}} \quad 5 \times 10^5 \lesssim \mathrm{Re}_L \lesssim 10^7$

If the flow is approximated as laminar up to the engineering critical number of $\mathrm{Re}_{cr} = 5 \times 10^5$, and then turbulent beyond, the average friction coefficient over the entire flat plate becomes

$$C_f = \frac{0.074}{\mathrm{Re}_L^{1/5}} - \frac{1742}{\mathrm{Re}_L} \quad 5 \times 10^5 \lesssim \mathrm{Re}_L \lesssim 10^7$$

A curve fit of experimental data for the average friction coefficient in the fully rough turbulent regime is

Rough surface: $\quad C_f = \left(1.89 - 1.62 \log \dfrac{\varepsilon}{L}\right)^{-2.5}$

where ε is the surface roughness and L is the length of the plate in the flow direction. In the absence of a better one, this relation can be used for turbulent flow on rough surfaces for $\mathrm{Re} > 10^6$, especially when $\varepsilon/L > 10^{-4}$.

Surface roughness, in general, increases the drag coefficient in turbulent flow. For bluff bodies such as a circular cylinder or sphere, however, an increase in the surface roughness may *decrease* the drag coefficient. This is done by tripping the flow into turbulence at a lower Reynolds number, and thus causing the fluid to close in behind the body, narrowing the wake and reducing pressure drag considerably.

It is desirable for airfoils to generate the most lift while producing the least drag. Therefore, a measure of performance for airfoils is the *lift-to-drag ratio,* C_L/C_D.

The minimum safe flight velocity of an aircraft is determined from

$$V_{\min} = \sqrt{\frac{2W}{\rho C_{L,\max} A}}$$

For a given weight, the landing or takeoff speed can be minimized by maximizing the product of the lift coefficient and the wing area, $C_{L,\max} A$.

For airplane wings and other airfoils of finite span, the pressure difference between the lower and the upper surfaces drives the fluid at the tips upward. This results in swirling eddies, called *tip vortices.* Tip vortices that interact with the free stream impose forces on the wing tips in all directions, including the flow direction. The component of the force in the flow direction adds to drag and is called *induced drag.* The total drag of a wing is then the sum of the induced drag (3-D effects) and the drag of the airfoil section (2-D effects).

It is observed that lift develops when a cylinder or sphere in flow is rotated at a sufficiently high rate. The phenomenon of producing lift by the rotation of a solid body is called the *Magnus effect.*

Some external flows, complete with flow details including plots of velocity fields, are solved using computational fluid dynamics, and presented in Chap. 15.

REFERENCES AND SUGGESTED READING

1. I. H. Abbott. "The Drag of Two Streamline Bodies as Affected by Protuberances and Appendages," *NACA Report* 451, 1932.

2. I. H. Abbott and A. E. von Doenhoff. *Theory of Wing Sections, Including a Summary of Airfoil Data.* New York: Dover, 1959.

3. I. H. Abbott, A. E. von Doenhoff, and L. S. Stivers. "Summary of Airfoil Data," *NACA Report* 824, Langley Field, VA, 1945.

4. J. D. Anderson. *Fundamentals of Aerodynamics,* 5th ed. New York: McGraw-Hill, 2010.

(continues on page 646)

APPLICATION SPOTLIGHT ■ Drag Reduction

Guest Author: Werner J. A. Dahm, The University of Michigan

A reduction of just a few percent in the drag that acts on an air vehicle, a naval surface vehicle, or an undersea vehicle can translate into large reductions in fuel weight and operating costs, or increases in vehicle range and payload. One approach to achieve such drag reduction is to actively control naturally occurring streamwise vortices in the viscous sublayer of the turbulent boundary layer at the vehicle surface. The thin viscous sublayer at the base of any turbulent boundary layer is a powerful nonlinear system, capable of amplifying small microactuator-induced perturbations into large reductions in the vehicle drag. Numerous experimental, computational, and theoretical studies have shown that reductions of 15 to 25 percent in the wall shear stress are possible by properly controlling these sublayer structures. The challenge has been to develop large, dense arrays of microactuators that can manipulate these structures to achieve drag reduction on practical aeronautical and hydronautical vehicles (Fig. 11–57). The sublayer structures are typically a few hundred microns, and thus well matched to the scale of *microelectromechanical systems* (MEMS).

Figure 11–58 shows an example of one type of such microscale actuator array based on the electrokinetic principle that is potentially suitable for active sublayer control on real vehicles. Electrokinetic flow provides a way to move small amounts of fluid on very fast time scales in very small devices. The actuators impulsively displace a fixed volume of fluid between the wall and the viscous sublayer in a manner that counteracts the effect of the sublayer vortices. A system architecture based on independent unit cells, appropriate for large arrays of such microactuators, provides greatly reduced control processing requirements within individual unit cells, which consist of a relatively small number of individual sensors and actuators. Fundamental consideration of the scaling principles governing electrokinetic flow, as well as the sublayer structure and dynamics and microfabrication technologies, have been used to develop and produce full-scale electrokinetic microactuator arrays that can meet many of the requirements for active sublayer control of turbulent boundary layers under real-vehicle conditions.

Such microelectrokinetic actuator (MEKA) arrays, when fabricated with wall shear stress sensors also based on microelectromechanical systems fabrication, may in the future allow engineers to achieve dramatic reductions in the drag acting on practical aeronautical and hydronautical vehicles.

References

Diez-Garias, F. J., Dahm, W. J. A., and Paul, P. H., "Microactuator Arrays for Sublayer Control in Turbulent Boundary Layers Using the Electrokinetic Principle," *AIAA Paper No. 2000-0548*, AIAA, Washington, DC, 2000.

Diez, F. J., and Dahm, W. J. A., "Electrokinetic Microactuator Arrays and System Architecture for Active Sublayer Control of Turbulent Boundary Layers," *AIAA Journal*, Vol. 41, pp. 1906–1915, 2003.

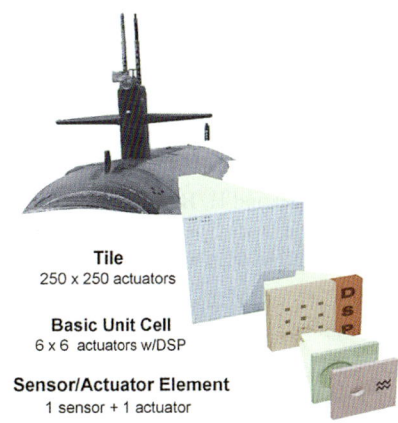

FIGURE 11–57

Drag-reducing microactuator arrays on the hull of a submarine. Shown is the system architecture with tiles composed of unit cells containing sensors and actuators.

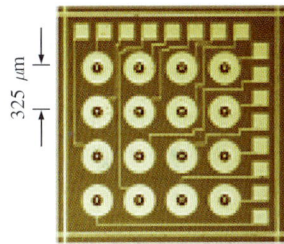

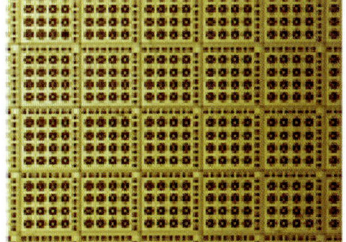

FIGURE 11–58

Microelectrokinetic actuator array (MEKA-5) with 25,600 individual actuators at 325-μm spacing for full-scale hydronautical drag reduction. Close-up of a single unit cell (*top*) and partial view of the full array (*bottom*).

5. R. D. Blevins. *Applied Fluid Dynamics Handbook.* New York: Van Nostrand Reinhold, 1984.

6. S. W. Churchill and M. Bernstein. "A Correlating Equation for Forced Convection from Gases and Liquids to a Circular Cylinder in Cross Flow," *Journal of Heat Transfer* 99, pp. 300–306, 1977.

7. S. Goldstein. *Modern Developments in Fluid Dynamics.* London: Oxford Press, 1938.

8. J. Happel and H. Brenner. *Low Reynolds Number Hydrodynamics with Special Applications to Particulate Media.* Norwell, MA: Kluwer Academic Publishers, 2003.

9. S. F. Hoerner. *Fluid-Dynamic Drag.* [Published by the author.] Library of Congress No. 64, 1966.

10. J. D. Holmes. *Wind Loading of Structures* 2nd ed. London: Spon Press (Taylor and Francis), 2007.

11. G. M. Homsy, H. Aref, K. S. Breuer, S. Hochgreb, J. R. Koseff, B. R. Munson, K. G. Powell, C. R. Roberston, S. T. Thoroddsen. *Multi-Media Fluid Mechanics* (CD) 2nd ed. Cambridge University Press, 2004.

12. W. H. Hucho. *Aerodynamics of Road Vehicles* 4th ed. London: Butterworth-Heinemann, 1998.

13. H. Schlichting. *Boundary Layer Theory,* 7th ed. New York: McGraw-Hill, 1979.

14. M. Van Dyke. *An Album of Fluid Motion.* Stanford, CA: The Parabolic Press, 1982.

15. J. Vogel. *Life in Moving Fluids,* 2nd ed. Boston: Willard Grand Press, 1994.

16. F. M. White. *Fluid Mechanics,* 7th ed. New York: McGraw-Hill, 2010.

PROBLEMS*

Drag, Lift, and Drag Coefficients

11–1C Consider laminar flow over a flat plate. How does the local friction coefficient change with position?

11–2C Define the frontal area of a body subjected to external flow. When is it appropriate to use the frontal area in drag and lift calculations?

11–3C Define the planform area of a body subjected to external flow. When is it appropriate to use the planform area in drag and lift calculations?

11–4C Explain when an external flow is two-dimensional, three-dimensional, and axisymmetric. What type of flow is the flow of air over a car?

11–5C What is the difference between the upstream velocity and the free-stream velocity? For what types of flow are these two velocities equal to each other?

11–6C What is the difference between streamlined and bluff bodies? Is a tennis ball a streamlined or bluff body?

11–7C Name some applications in which a large drag is desired.

11–8C What is drag? What causes it? Why do we usually try to minimize it?

11–9C What is lift? What causes it? Does wall shear contribute to the lift?

11–10C During flow over a given body, the drag force, the upstream velocity, and the fluid density are measured. Explain how you would determine the drag coefficient. What area would you use in the calculations?

11–11C During flow over a given slender body such as a wing, the lift force, the upstream velocity, and the fluid density are measured. Explain how you would determine the lift coefficient. What area would you use in the calculations?

11–12C What is terminal velocity? How is it determined?

11–13C What is the difference between skin friction drag and pressure drag? Which is usually more significant for slender bodies such as airfoils?

11–14C What is the effect of surface roughness on the friction drag coefficient in laminar and turbulent flows?

11–15C What is the effect of streamlining on (*a*) friction drag and (*b*) pressure drag? Does the total drag acting on a body necessarily decrease as a result of streamlining? Explain.

11–16C What is flow separation? What causes it? What is the effect of flow separation on the drag coefficient?

11–17C What is drafting? How does it affect the drag coefficient of the drafted body?

---

* Problems designated by a "C" are concept questions, and students are encouraged to answer them all. Problems with the icon are solved using EES, and complete solutions together with parametric studies are included on the text website. Problems with the icon are comprehensive in nature and are intended to be solved with an equation solver such as EES.

11–18C In general, how does the drag coefficient vary with the Reynolds number at (*a*) low and moderate Reynolds numbers and (*b*) at high Reynolds numbers (Re > 10^4)?

11–19C Fairings are attached to the front and back of a cylindrical body to make it look more streamlined. What is the effect of this modification on the (*a*) friction drag, (*b*) pressure drag, and (*c*) total drag? Assume the Reynolds number is high enough so that the flow is turbulent for both cases.

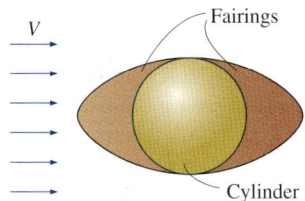

FIGURE P11–19C

11–20 The drag coefficient of a car at the design conditions of 1 atm, 25°C, and 90 km/h is to be determined experimentally in a large wind tunnel in a full-scale test. The height and width of the car are 1.25 m and 1.65 m, respectively. If the horizontal force acting on the car is measured to be 220 N, determine the total drag coefficient of this car. *Answer:* 0.29

11–21 The resultant of the pressure and wall shear forces acting on a body is measured to be 580 N, making 35° with the direction of flow. Determine the drag and the lift forces acting on the body.

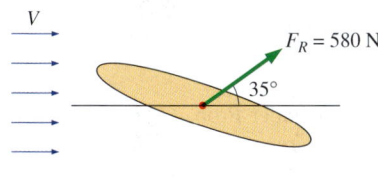

FIGURE P11–21

11–22 During a high Reynolds number experiment, the total drag force acting on a spherical body of diameter $D = 12$ cm subjected to airflow at 1 atm and 5°C is measured to be 5.2 N. The pressure drag acting on the body is calculated by integrating the pressure distribution (measured by the use of pressure sensors throughout the surface) to be 4.9 N. Determine the friction drag coefficient of the sphere. *Answer:* 0.0115

11–23 A car is moving at a constant velocity of 110 km/h. Determine the upstream velocity to be used in fluid flow analysis if (*a*) the air is calm, (*b*) wind is blowing against the direction of motion of the car at 30 km/h, and (*c*) wind is blowing in the same direction of motion of the car at 30 km/h.

11–24 A circular sign has a diameter of 50 cm and is subjected to normal winds up to 150 km/h at 10°C and 100 kPa. Determine the drag force acting on the sign. Also determine the bending moment at the bottom of its pole whose height from the ground to the bottom of the sign is 1.5 m. Disregard the drag on the pole.

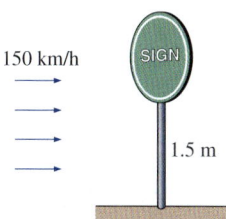

FIGURE P11–24

11–25 Bill gets a job delivering pizzas. The pizza company makes him mount a sign on the roof of his car. The frontal area of the sign is $A = 0.0569$ m², and he estimates the drag coefficient to be $C_D = 0.94$ at nearly all air speeds. Estimate how much additional money it costs Bill per year in fuel to drive with the sign on his roof compared to without the sign. Use the following additional information: He drives about 16,000 km per year at an average speed of 72 km/h. The overall car efficiency is 0.332, $\rho_{fuel} = 804$ kg/m³, and the heating value of the fuel is 45,700 kJ/kg. The fuel costs $0.925 per liter. Use standard air properties. Be careful with unit conversions.

11–26 Advertisement signs are commonly carried by taxicabs for additional income, but they also increase the fuel cost. Consider a sign that consists of a 0.30-m-high, 0.9-m-wide, and 0.9-m-long rectangular block mounted on top of a taxicab such that the sign has a frontal area of 0.3 m by 0.9 m from all four sides. Determine the increase in the annual fuel cost of this taxicab due to this sign. Assume the taxicab is driven 60,000 km a year at an average speed of 50 km/h and the overall efficiency of the engine is 28 percent. Take the density, unit price, and heating value of gasoline to be 0.72 kg/L, $1.10/L, and 42,000 kJ/kg, respectively, and the density of air to be 1.25 kg/m³.

FIGURE P11–26

11–27 At highway speeds, about half of the power generated by the car's engine is used to overcome aerodynamic drag, and thus the fuel consumption is nearly proportional to the drag force on a level road. Determine the percentage increase in fuel consumption of a car per unit time when a person who normally drives at 90 km/h now starts driving at 120 km/h.

11–28 A submarine can be treated as an ellipsoid with a diameter of 5 m and a length of 25 m. Determine the power required for this submarine to cruise horizontally and steadily at 40 km/h in seawater whose density is 1025 kg/m^3. Also determine the power required to tow this submarine in air whose density is 1.30 kg/m^3. Assume the flow is turbulent in both cases.

FIGURE P11–28

11–29 Wind loading is a primary consideration in the design of the supporting mechanisms of billboards, as evidenced by many billboards being knocked down during high winds. Determine the wind force acting on a 3.7-m-high, 6-m-wide billboard due to 90-km/h winds in the normal direction when the atmospheric conditions are 98 kPa and 5°C. *Answer:* 17,000 N

11–30 During major windstorms, high vehicles such as RVs and semis may be thrown off the road and boxcars off their tracks, especially when they are empty and in open areas. Consider a 5000-kg semi that is 9 m long, 2.5 m high, and 2 m wide. The distance between the bottom of the truck and the road is 0.75 m. Now the truck is exposed to winds from its side surface. Determine the wind velocity that will tip the truck over to its side. Take the air density to be 1.1 kg/m^3 and assume the weight to be uniformly distributed.

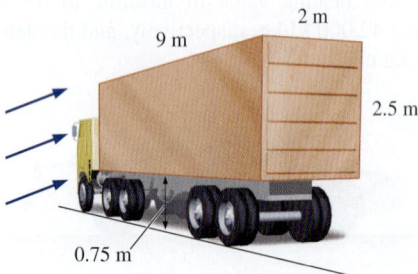

FIGURE P11–30

11–31 A 70-kg bicyclist is riding her 15-kg bicycle downhill on a road with a slope of 8° without pedaling or braking. The bicyclist has a frontal area of 0.45 m^2 and a drag coefficient of 1.1 in the upright position, and a frontal area of 0.4 m^2 and a drag coefficient of 0.9 in the racing position. Disregarding the rolling resistance and friction at the bearings, determine the terminal velocity of the bicyclist for both positions. Take the air density to be 1.25 kg/m^3. *Answers:* 70 km/h, 82 km/h

11–32 A wind turbine with two or four hollow hemispherical cups connected to a pivot is commonly used to measure wind speed. Consider a wind turbine with four 8-cm-diameter cups with a center-to-center distance of 40 cm, as shown in Fig. P11–32. The pivot is stuck as a result of some malfunction, and the cups stop rotating. For a wind speed of 15 m/s and air density of 1.25 kg/m^3, determine the maximum torque this turbine applies on the pivot.

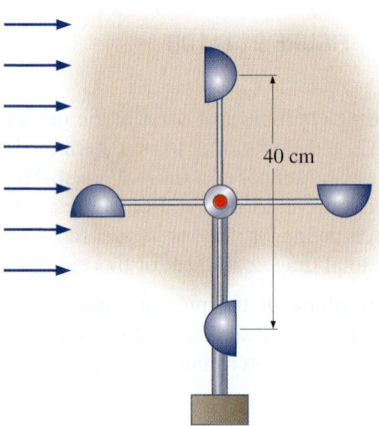

FIGURE P11–32

11–33 Reconsider Prob. 11–32. Using EES (or other) software, investigate the effect of wind speed on the torque applied on the pivot. Let the wind speed vary from 0 to 50 m/s in increments of 5 m/s. Tabulate and plot the results.

11–34 During steady motion of a vehicle on a level road, the power delivered to the wheels is used to overcome aerodynamic drag and rolling resistance (the product of the rolling resistance coefficient and the weight of the vehicle), assuming the friction at the bearings of the wheels is negligible. Consider a car that has a total mass of 950 kg, a drag coefficient of 0.32, a frontal area of 1.8 m^2, and a rolling resistance coefficient of 0.04. The maximum power the engine can deliver to the wheels is 80 kW. Determine (*a*) the speed at which the rolling resistance is equal to the aerodynamic drag force and (*b*) the maximum speed of this car. Take the air density to be 1.20 kg/m^3.

11–35 Reconsider Prob. 11–34. Using EES (or other) software, investigate the effect of car speed on the required power to overcome (*a*) rolling resistance, (*b*) the aerodynamic drag, and (*c*) their combined effect. Let the car speed vary from 0 to 150 km/h in increments of 15 km/h. Tabulate and plot the results.

11–36 Suzy likes to drive with a silly sun ball on her car antenna. The frontal area of the ball is $A = 2.08 \times 10^{-3}$ m². As gas prices rise, her husband is concerned that she is wasting fuel because of the additional drag on the ball. He runs a quick test in the wind tunnel at his university and measures the drag coefficient to be $C_D = 0.87$ at nearly all air speeds. Estimate how many liters of fuel she wastes per year by having this ball on her antenna. Use the following additional information: She drives about 15,000 km per year at an average speed of 20.8 m/s. The overall car efficiency is 0.312, $\rho_{\text{fuel}} = 0.802$ kg/L, and the heating value of the fuel is 44,020 kJ/kg. Use standard air properties. Is the amount of wasted fuel significant?

FIGURE P11–36

Photo by Suzanne Cimbala.

11–37 An 0.90-m-diameter, 1.1-m-high garbage can is found in the morning tipped over due to high winds during the night. Assuming the average density of the garbage inside to be 150 kg/m³ and taking the air density to be 1.25 kg/m³, estimate the wind velocity during the night when the can was tipped over. Take the drag coefficient of the can to be 0.7. *Answer:* 159 km/h

11–38 A 6-mm-diameter plastic sphere whose density is 1150 kg/m³ is dropped into water at 20°C. Determine the terminal velocity of the sphere in water.

11–39 A 7-m-diameter hot air balloon that has a total mass of 350 kg is standing still in air on a windless day. The balloon is suddenly subjected to 40 km/h winds. Determine the initial acceleration of the balloon in the horizontal direction.

11–40 The drag coefficient of a vehicle increases when its windows are rolled down or its sunroof is opened. A sports car has a frontal area of 1.7 m² and a drag coefficient of 0.32 when the windows and sunroof are closed. The drag coefficient increases to 0.41 when the sunroof is open. Determine the additional power consumption of the car when the sunroof is opened at (*a*) 55 km/h and (*b*) 110 km/h. Take the density of air to be 1.2 kg/m³.

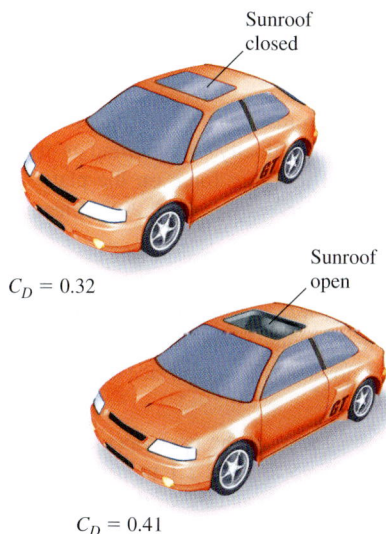

FIGURE P11–40

11–41 To reduce the drag coefficient and thus to improve the fuel efficiency of cars, the design of side rearview mirrors has changed drastically in recent decades from a simple circular plate to a streamlined shape. Determine the amount of fuel and money saved per year as a result of replacing a 13-cm-diameter flat mirror by one with a hemispherical back, as shown in the figure. Assume the car is driven 24,000 km a year at an average speed of 95 km/h. Take the density and price of gasoline to be 0.75 kg/L and $0.90/L, respectively; the heating value of gasoline to be 44,000 kJ/kg; and the overall efficiency of the engine to be 30 percent.

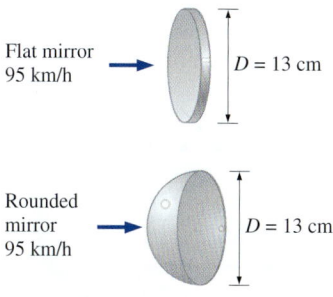

FIGURE P11–41

Flow over Flat Plates

11–42C How is the average friction coefficient determined in flow over a flat plate?

11–43C What fluid property is responsible for the development of the velocity boundary layer? What is the effect of the velocity on the thickness of the boundary layer?

11–44C What does the friction coefficient represent in flow over a flat plate? How is it related to the drag force acting on the plate?

11–45 Consider laminar flow of a fluid over a flat plate. Now the free-stream velocity of the fluid is tripled. Determine the change in the drag force on the plate. Assume the flow to remain laminar. *Answer:* A 5.20-fold increase

11–46 The local atmospheric pressure in Denver, Colorado (elevation 1610 m) is 83.4 kPa. Air at this pressure and at 25°C flows with a velocity of 9 m/s over a 2.5-m × 5-m flat plate. Determine the drag force acting on the top surface of the plate if the air flows parallel to the (*a*) 5-m-long side and (*b*) the 2.5-m-long side.

11–47 The top surface of the passenger car of a train moving at a velocity of 95 km/h is 2.1 m wide and 8 m long. If the outdoor air is at 1 atm and 25°C, determine the drag force acting on the top surface of the car.

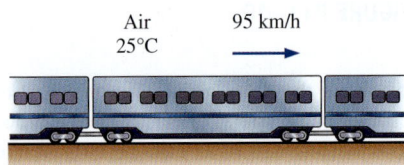

FIGURE P11–47

11–48 The forming section of a plastics plant puts out a continuous sheet of plastic that is 1.2 m wide and 2 mm thick at a rate of 18 m/min. The sheet is subjected to airflow at a velocity of 4 m/s on both top and bottom surfaces normal to the direction of motion of the sheet. The width of the air cooling section is such that a fixed point on the plastic sheet passes through that section in 2 s. Using properties of air at 1 atm and 60°C, determine the drag force the air exerts on the plastic sheet in the direction of airflow.

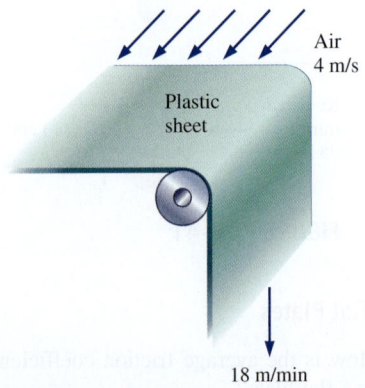

FIGURE P11–48

11–49 Light oil at 20°C flows over a 4.5-m-long flat plate with a free-stream velocity of 2 m/s. Determine the total drag force per unit width of the plate.

11–50 Consider a refrigeration truck traveling at 105 km/h at a location where the air is at 1 atm and 25°C. The refrigerated compartment of the truck can be considered to be a 2.7-m-wide, 2.4-m-high, and 6-m-long rectangular box. Assuming the airflow over the entire outer surface to be turbulent and attached (no flow separation), determine the drag force acting on the top and side surfaces and the power required to overcome this drag.

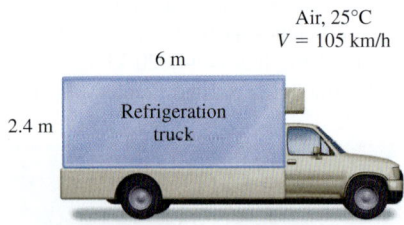

FIGURE P11–50

11–51 Reconsider Prob. 11–50. Using EES (or other) software, investigate the effect of truck speed on the total drag force acting on the top and side surfaces, and the power required to overcome it. Let the truck speed vary from 0 to 150 km/h in increments of 10 km/h. Tabulate and plot the results.

11–52 Air at 25°C and 1 atm is flowing over a long flat plate with a velocity of 8 m/s. Determine the distance from the leading edge of the plate where the flow becomes turbulent, and the thickness of the boundary layer at that location.

11–53 Repeat Prob. 11–52 for water.

11–54 During a winter day, wind at 55 km/h, 5°C, and 1 atm is blowing parallel to a 4-m-high and 10-m-long wall of a house. Approximating the wall surfaces as smooth, determine the friction drag acting on the wall. What would your answer be if the wind velocity has doubled? How realistic is it to treat the flow over side wall surfaces as flow over a flat plate? *Answers:* 16 N, 58 N

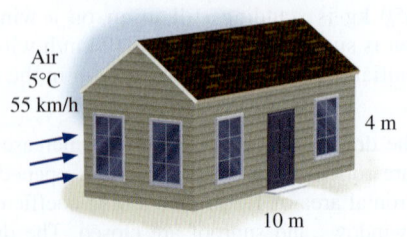

FIGURE P11–54

11–55 The weight of a thin flat plate 50 cm × 50 cm in size is balanced by a counterweight that has a mass of 2 kg, as shown in Fig. P11–55. Now a fan is turned on, and air at 1 atm and 25°C flows downward over both surfaces of the plate (front and back in the sketch) with a free-stream velocity of 10 m/s. Determine the mass of the counterweight that needs to be added in order to balance the plate in this case.

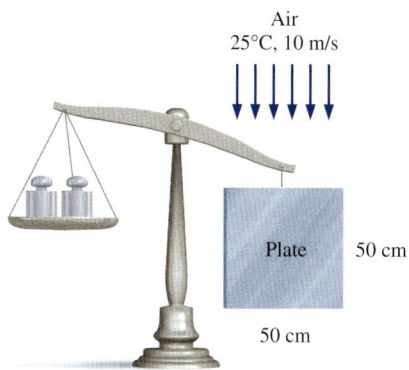

FIGURE P11–55

Flow across Cylinders and Spheres

11–56C Why is flow separation in flow over cylinders delayed when the boundary layer is turbulent?

11–57C In flow over bluff bodies such as a cylinder, how does the pressure drag differ from the friction drag?

11–58C In flow over cylinders, why does the drag coefficient suddenly drop when the boundary layer becomes turbulent? Isn't turbulence supposed to increase the drag coefficient instead of decreasing it?

11–59 A 0.1-mm-diameter dust particle whose density is 2.1 g/cm³ is observed to be suspended in the air at 1 atm and 25°C at a fixed point. Estimate the updraft velocity of air motion at that location. Assume Stokes law to be applicable. Is this a valid assumption? *Answer:* 0.62 m/s

11–60 A long 5-cm-diameter steam pipe passes through some area open to the wind. Determine the drag force acting on the pipe per unit of its length when the air is at 1 atm and 10°C and the wind is blowing across the pipe at a speed of 50 km/h.

11–61 Consider 0.8-cm-diameter hail that is falling freely in atmospheric air at 1 atm and 5°C. Determine the terminal velocity of the hail. Take the density of hail to be 910 kg/m³.

11–62 A 3-cm-outer-diameter pipe is to span across a river at a 30-m-wide section while being completely immersed in water. The average flow velocity of the water is 3 m/s, and its temperature is 20°C. Determine the drag force exerted on the pipe by the river. *Answer:* 4450 N

11–63 Dust particles of diameter 0.06 mm and density 1.6 g/cm³ are unsettled during high winds and rise to a height of 200 m by the time things calm down. Estimate how long it takes for the dust particles to fall back to the ground in still air at 1 atm and 30°C, and their velocity. Disregard the initial transient period during which the dust particles accelerate to their terminal velocity, and assume Stokes law to be applicable.

11–64 A 2-m-long, 0.2-m-diameter cylindrical pine log (density = 513 kg/m³) is suspended by a crane in the horizontal position. The log is subjected to normal winds of 40 km/h at 5°C and 88 kPa. Disregarding the weight of the cable and its drag, determine the angle θ the cable will make with the horizontal and the tension on the cable.

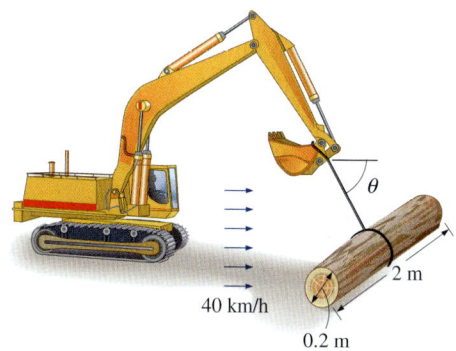

FIGURE P11–64

11–65 A 6-mm-diameter electrical transmission line is exposed to windy air. Determine the drag force exerted on a 160-m-long section of the wire during a windy day when the air is at 1 atm and 15°C and the wind is blowing across the transmission line at 65 km/h.

11–66 One of the popular demonstrations in science museums involves the suspension of a ping-pong ball by an upward air jet. Children are amused by the ball always coming back to the center when it is pushed by a finger to the side of the jet. Explain this phenomenon using the Bernoulli equation. Also determine the velocity of air if the ball has a mass of 3.1 g and a diameter of 4.2 cm. Assume the air is at 1 atm and 25°C.

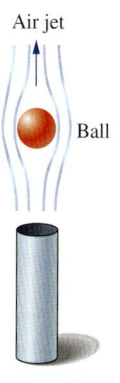

FIGURE P11–66

Lift

11–67C Why is the contribution of viscous effects to lift usually negligible for airfoils?

11–68C Air is flowing past a symmetrical airfoil at an angle of attack of 5°. Is the (*a*) lift and (*b*) drag acting on the airfoil zero or nonzero?

11–69C What is stall? What causes an airfoil to stall? Why are commercial aircraft not allowed to fly at conditions near stall?

11–70C Air is flowing past a nonsymmetrical airfoil at zero angle of attack. Is the (*a*) lift and (*b*) drag acting on the airfoil zero or nonzero?

11–71C Air is flowing past a symmetrical airfoil at zero angle of attack. Is the (*a*) lift and (*b*) drag acting on the airfoil zero or nonzero?

11–72C Both the lift and the drag of an airfoil increase with an increase in the angle of attack. In general, which increases at a higher rate, the lift or the drag?

11–73C Why are flaps used at the leading and trailing edges of the wings of large aircraft during takeoff and landing? Can an aircraft take off or land without them?

11–74C Air is flowing past a spherical ball. Is the lift exerted on the ball zero or nonzero? Answer the same question if the ball is spinning.

11–75C What is the effect of wing tip vortices (the air circulation from the lower part of the wings to the upper part) on the drag and the lift?

11–76C What is induced drag on wings? Can induced drag be minimized by using long and narrow wings or short and wide wings?

11–77C Explain why endplates or winglets are added to some airplane wings.

11–78C How do flaps affect the lift and the drag of wings?

11–79 A small aircraft has a wing area of 35 m², a lift coefficient of 0.45 at takeoff settings, and a total mass of 4000 kg. Determine (*a*) the takeoff speed of this aircraft at sea level at standard atmospheric conditions, (*b*) the wing loading, and (*c*) the required power to maintain a constant cruising speed of 300 km/h for a cruising drag coefficient of 0.035.

11–80 Consider an aircraft that takes off at 260 km/h when it is fully loaded. If the weight of the aircraft is increased by 10 percent as a result of overloading, determine the speed at which the overloaded aircraft will take off. *Answer:* 273 km/h

11–81 Consider an airplane whose takeoff speed is 220 km/h and that takes 15 s to take off at sea level. For an airport at an elevation of 1600 m (such as Denver), determine (*a*) the takeoff speed, (*b*) the takeoff time, and (*c*) the additional runway length required for this airplane. Assume constant acceleration for both cases.

FIGURE P11–81

11–82 An airplane is consuming fuel at a rate of 20 L/min when cruising at a constant altitude of 3,000 m at constant speed. Assuming the drag coefficient and the engine efficiency to remain the same, determine the rate of fuel consumption at an altitude of 9,000 m at the same speed.

11–83 A jumbo jet airplane has a mass of about 400,000 kg when fully loaded with over 400 passengers and takes off at a speed of 250 km/h. Determine the takeoff speed when the airplane has 100 empty seats. Assume each passenger with luggage is 140 kg and the wing and flap settings are maintained the same. *Answer:* 246 km/h

11–84 Reconsider Prob. 11–83. Using EES (or other) software, investigate the effect of passenger count on the takeoff speed of the aircraft. Let the number of passengers vary from 0 to 500 in increments of 50. Tabulate and plot the results.

11–85 A tennis ball with a mass of 57 g and a diameter of 6.4 cm is hit with an initial velocity of 105 km/h and a backspin of 4200 rpm. Determine if the ball falls or rises under the combined effect of gravity and lift due to spinning shortly after hitting. Assume air is at 1 atm and 25°C.

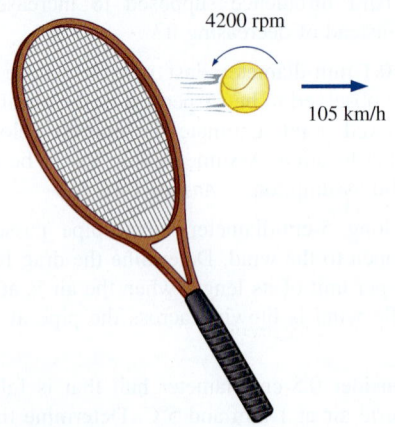

FIGURE P11–85

11–86 A 6.1-cm-diameter smooth ball rotating at 500 rpm is dropped in a water stream at 15°C flowing at 1.2 m/s. Determine the lift and the drag force acting on the ball when it is first dropped in the water.

11–87 The NACA 64(1)–412 airfoil has a lift-to-drag ratio of 50 at 0° angle of attack, as shown in Fig. 11–43. At what angle of attack does this ratio increase to 80?

11–88 Consider a light plane that has a total weight of 11,000 N and a wing area of 39 m² and whose wings resemble the NACA 23012 airfoil with no flaps. Using data from Fig. 11–45, determine the takeoff speed at an angle of attack of 5° at sea level. Also determine the stall speed. *Answers:* 99.7 km/h, 62.7 km/h

11–89 A small airplane has a total mass of 1800 kg and a wing area of 42 m². Determine the lift and drag coefficients of this airplane while cruising at an altitude of 4000 m at a constant speed of 280 km/h and generating 190 kW of power.

11–90 An airplane has a mass of 50,000 kg, a wing area of 300 m², a maximum lift coefficient of 3.2, and a cruising drag coefficient of 0.03 at an altitude of 12,000 m. Determine (*a*) the takeoff speed at sea level, assuming it is 20 percent over the stall speed, and (*b*) the thrust that the engines must deliver for a cruising speed of 700 km/h.

Review Problems

11–91 Consider a blimp that can be approximated as a 3-m diameter, 8-m long ellipsoid and is connected to the ground. On a windless day, the rope tension due to the net buoyancy effect is measured to be 120 N. Determine the rope tension when there are 50 km/h winds blowing along the blimp (parallel to the blimp axis).

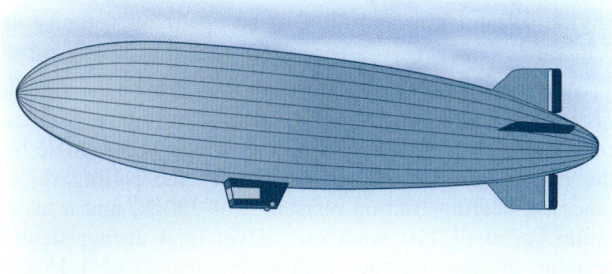

FIGURE P11–91

11–92 A 1.2-m-external-diameter spherical tank is located outdoors at 1 atm and 25°C and is subjected to winds at 48 km/h. Determine the drag force exerted on it by the wind. *Answer:* 16.7 N

11–93 A 2-m-high, 4-m-wide rectangular advertisement panel is attached to a 4-m-wide, 0.15-m-high rectangular concrete block (density = 2300 kg/m³) by two 5-cm-diameter, 4-m-high (exposed part) poles, as shown in Fig. P11–93. If the sign is to withstand 150 km/h winds from any direction, determine (*a*) the maximum drag force on the panel, (*b*) the drag force acting on the poles, and (*c*) the minimum length *L* of the concrete block for the panel to resist the winds. Take the density of air to be 1.30 kg/m³.

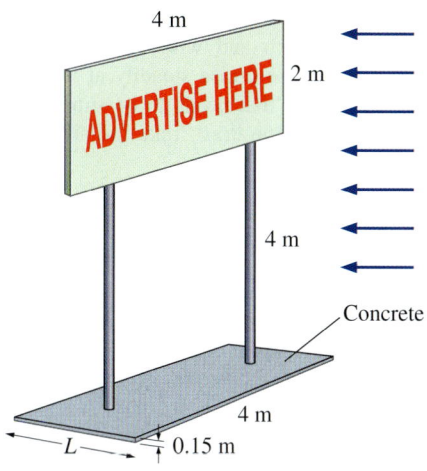

FIGURE P11–93

11–94 A plastic boat whose bottom surface can be approximated as a 1.5-m-wide, 2-m-long flat surface is to move through water at 15°C at speeds up to 45 km/h. Determine the friction drag exerted on the boat by the water and the power needed to overcome it.

FIGURE P11–94

11–95 Reconsider Prob. 11–94. Using EES (or other) software, investigate the effect of boat speed on the drag force acting on the bottom surface of the boat, and the power needed to overcome it. Let the boat speed vary from 0 to 100 km/h in increments of 10 km/h. Tabulate and plot the results.

11–96 The cylindrical chimney of a factory has an external diameter of 1.1 m and is 20 m high. Determine the bending moment at the base of the chimney when winds at 110 km/h are blowing across it. Take the atmospheric conditions to be 20°C and 1 atm.

11–97 A commercial airplane has a total mass of 70,000 kg and a wing planform area of 170 m². The plane has a cruising speed of 900 km/h and a cruising altitude of 11,500 m where the air density is 0.333 kg/m³. The plane has double-slotted flaps for use during takeoff and landing, but it cruises with all flaps retracted. Assuming the lift and drag characteristics of the wings can be approximated by NACA 23012, determine (*a*) the minimum safe speed for takeoff and landing with and without extending the flaps, (*b*) the angle of attack to cruise steadily at the cruising altitude, and (*c*) the power that needs to be supplied to provide enough thrust to overcome drag. Take the air density on the ground to be 1.2 kg/m³.

11–98 An automotive engine can be approximated as a 0.4-m-high, 0.60-m-wide, and 0.7-m-long rectangular block. The ambient air is at 1 atm and 15°C. Determine the drag force acting on the bottom surface of the engine block as the car travels at a velocity of 120 km/h. Assume the flow to be turbulent over the entire surface because of the constant agitation of the engine block. *Answer:* 1.22 N

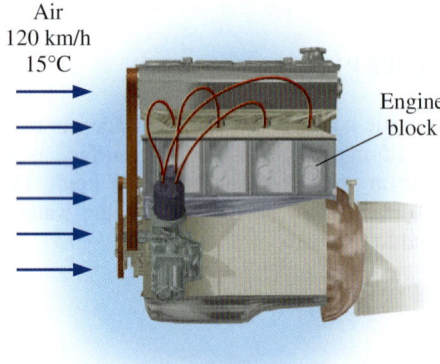

FIGURE P11–98

11–99 A paratrooper and his 8-m-diameter parachute weigh 950 N. Taking the average air density to be 1.2 kg/m³, determine the terminal velocity of the paratrooper. *Answer:* 4.9 m/s

FIGURE P11–99

11–100 It is proposed to meet the water needs of a recreational vehicle (RV) by installing a 3-m-long, 0.5-m-diameter cylindrical tank on top of the vehicle. Determine the additional power requirement of the RV at a speed of 80 km/h when the tank is installed such that its circular surfaces face (*a*) the front and back (as sketched) and (*b*) the sides of the RV. Assume atmospheric conditions are 87 kPa and 20°C. *Answers:* (a) 1.05 kW, (b) 6.77 kW

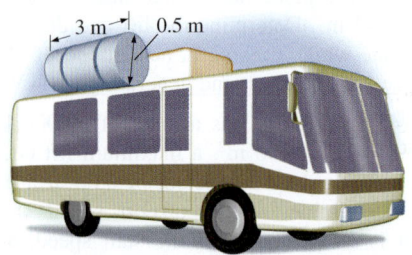

FIGURE P11–100

11–101 A 9-cm-diameter smooth sports ball has a velocity of 36 km/h during a typical hit. Determine the percent increase in the drag coefficient if the ball is given a spin of 3500 rpm in air at 1 atm and 25°C.

11–102 Calculate the thickness of the boundary layer during flow over a 2.5-m-long flat plate at intervals of 25 cm and plot the boundary layer over the plate for the flow of (*a*) air, (*b*) water, and (*c*) engine oil at 1 atm and 20°C at an upstream velocity of 3 m/s.

11–103 A 17,000-kg tractor-trailer rig has a frontal area of 9.2 m², a drag coefficient of 0.96, a rolling resistance coefficient of 0.05 (multiplying the weight of a vehicle by the rolling resistance coefficient gives the rolling resistance), a bearing friction resistance of 350 N, and a maximum speed of 110 km/h on a level road during steady cruising in calm weather with an air density of 1.25 kg/m³. Now a fairing is installed to the front of the rig to suppress separation and to streamline the flow to the top surface, and the drag coefficient is reduced to 0.76. Determine the maximum speed of the rig with the fairing. *Answer:* 133 km/h

11–104 Janie likes to drive with a tennis ball on her car antenna. The ball diameter is $D = 6.65$ cm and its equivalent roughness factor is $\varepsilon/D = 1.5 \times 10^{-3}$. Her friends tell her she is wasting gas because of the additional drag on the ball. Estimate how much money (in dollars) she wastes per year by driving around with this tennis ball on her antenna. Use the following additional information: She drives mostly on the highway, about 25,000 km per year at an average speed of 90 km/h. The overall car efficiency is 0.308, $\rho_{\text{fuel}} = 804$ kg/m³, and the heating value of the fuel is 43,900 kJ/kg. The fuel costs $1.06 per liter. Use standard air properties. Be careful with unit conversions. Should Janie remove the tennis ball?

FIGURE P11–1104
Photo by John M. Cimbala.

11–105 During an experiment, three aluminum balls ($\rho_s = 2600$ kg/m^3) having diameters 2, 4, and 10 mm, respectively, are dropped into a tank filled with glycerin at 22°C ($\rho_f = 1274$ kg/m^3 and $\mu = 1$ kg/m·s). The terminal settling velocities of the balls are measured to be 3.2, 12.8, and 60.4 mm/s, respectively. Compare these values with the velocities predicted by Stokes law for drag force $F_D = 3\pi\mu DV$, which is valid for very low Reynolds numbers (Re $\ll$ 1). Determine the error involved for each case and assess the accuracy of Stokes law.

11–106 Repeat Prob. 11–105 by considering the more general form of Stokes law expressed as $F_{V_D} = 3\pi\mu DV + (9\pi/16)\rho V^2 D^2$ where ρ is the fluid density.

11–107 A small aluminum ball with $D = 2$ mm and $\rho_s = 2700$ kg/m^3 is dropped into a large container filled with oil at 40°C ($\rho_f = 876$ kg/m^3 and $\mu = 0.2177$ kg/m·s). The Reynolds number is expected to be low and thus Stokes law for drag force $F_D = 3\pi\mu DV$ to be applicable. Show that the variation of velocity with time can be expressed as $V = (a/b)(1 - e^{-bt})$ where $a = g(1 - \rho_f/\rho_s)$ and $b = 18\mu/(\rho_s D^2)$. Plot the variation of velocity with time, and calculate the time it takes for the ball to reach 99 percent of its terminal velocity.

11–108 Engine oil at 40°C is flowing over a long flat plate with a velocity of 6 m/s. Determine the distance x_{cr} from the leading edge of the plate where the flow becomes turbulent, and calculate and plot the thickness of the boundary layer over a length of $2x_{cr}$.

11–109 Stokes law can be used to determine the viscosity of a fluid by dropping a spherical object in it and measuring the terminal velocity of the object in that fluid. This can be done by plotting the distance traveled against time and observing when the curve becomes linear. During such an experiment a 3-mm-diameter glass ball ($\rho = 2500$ kg/m^3) is dropped into a fluid whose density is 875 kg/m^3, and the terminal velocity is measured to be 0.12 m/s. Disregarding the wall effects, determine the viscosity of the fluid.

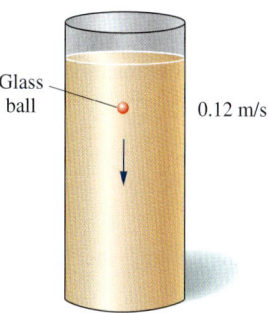

FIGURE P11–109

Fundamentals of Engineering (FE) Exam Problems

11–110 Which quantities are physical phenomena associated with fluid flow over bodies?
I. Drag force acting on automobiles
II. The lift developed by airplane wings
III. Upward draft of rain and snow
IV. Power generated by wind turbines
(a) I and II
(b) I and III
(c) II and III
(d) I, II, and III
(e) I, II, III, and IV

11–111 The sum of the components of the pressure and wall shear forces in the direction normal to the flow is called
(a) Drag
(b) Friction
(c) Lift
(d) Bluff
(e) Blunt

11–112 A car is moving at a speed of 70 km/h in air at 20°C. The frontal area of the car is 2.4 m^2. If the drag force acting on the car in the flow direction is 205 N, the drag coefficient of the car is
(a) 0.312
(b) 0.337
(c) 0.354
(d) 0.375
(e) 0.391

11–113 A person is driving his motorcycle at a speed of 110 km/h in air at 20°C. The frontal area of the motorcycle and driver is 0.75 m^2. If the drag coefficient under these conditions is estimated to be 0.90, the drag force acting on the car in the flow direction is
(a) 379 N
(b) 220 N
(c) 283 N
(d) 308 N
(e) 450 N

11–114 The manufacturer of a car reduces the drag coefficient of the car from 0.38 to 0.33 as a result of some modifications in its shape and design. If, on average, the aerodynamic drag accounts for 20 percent of the fuel consumption, the percent reduction in the fuel consumption of the car due to reducing the drag coefficient is
(a) 15%
(b) 13%
(c) 6.6%
(d) 2.6%
(e) 1.3%

11–115 The region of flow trailing the body where the effects of the body are felt is called
(a) Wake
(b) Separated region
(c) Stall
(d) Vortice
(e) Irrotational

11–116 The turbulent boundary layer can be considered to consist of four regions. Which choice is not one of them?
(a) Buffer layer
(b) Overlap layer
(c) Transition layer
(d) Viscous layer
(e) Turbulent layer

11–117 Water at 10°C flows over a 1.1-m-long flat plate with a velocity of 0.55 m/s. If the width of the plate is 2.5 m, calculate the drag force acting on the top side of the plate. (Water properties at 10°C are: $\rho = 999.7$ kg/m^3, $\mu = 1.307 \times 10^{-3}$ kg/m·s.)
(a) 0.46 N
(b) 0.81 N
(c) 2.75 N
(d) 4.16 N
(e) 6.32 N

11–118 Water at 10°C flows over a 3.75-m-long flat plate with a velocity of 1.15 m/s. If the width of the plate is 6.5 m, calculate the average friction coefficient over the entire plate. (Water properties at 10°C are: $\rho = 999.7$ kg/m^3, $\mu = 1.307 \times 10^{-3}$ kg/m·s.)
(a) 0.00508
(b) 0.00447
(c) 0.00302
(d) 0.00367
(e) 0.00315

11–119 Air at 30°C flows over a 3.0-cm-outer-diameter, 45-m-long pipe with a velocity of 6 m/s. Calculate the drag force exerted on the pipe by the air. (Air properties at 30°C are: $\rho = 1.164$ kg/m^3, $\nu = 1.608 \times 10^{-5}$ m^2/s.)
(a) 19.3 N
(b) 36.8 N
(c) 49.3 N
(d) 53.9 N
(e) 60.1 N

11–120 A 0.8-m-outer-diameter spherical tank is completely submerged in a flowing water stream at a velocity of 2.5 m/s. Calculate the drag force acting on the tank. (Water properties are: $\rho = 998.0$ kg/m^3, $\mu = 1.002 \times 10^{-3}$ kg/m·s.)
(a) 878 N
(b) 627 N
(c) 545 N
(d) 356 N
(e) 220 N

11–121 An airplane has a total mass of 18,000 kg and a wing planform area of 35 m^2. The density of air at the ground is 1.2 kg/m^3. The maximum lift coefficient is 3.48. The minimum safe speed for takeoff and landing while extending the flaps is
(a) 305 km/h
(b) 173 km/h
(c) 194 km/h
(d) 212 km/h
(e) 246 km/h

11–122 An airplane has a total mass of 35,000 kg and a wing planform area of 65 m^2. The airplane is cruising at 10,000 m altitude with a velocity of 1100 km/h. The density of air on cruising altitude is 0.414 kg/m^3. The lift coefficient of this airplane at the cruising altitude is
(a) 0.273
(b) 0.290
(c) 0.456
(d) 0.874
(e) 1.22

11–123 An airplane is cruising at a velocity of 800 km/h in air whose density is 0.526 kg/m^3. The airplane has a wing planform area of 90 m^2. The lift and drag coefficients on cruising conditions are estimated to be 2.0 and 0.06, respectively. The power that needs to be supplied to provide enough trust to overcome wing drag is
(a) 9760 kW
(b) 11,300 kW
(c) 15,600 kW
(d) 18,200 kW
(e) 22,600 kW

Design and Essay Problems

11–124 Write a report on the history of the reduction of the drag coefficients of cars and obtain the drag coefficient data for some recent car models from the catalogs of car manufacturers or from the Internet.

11–125 Write a report on the flaps used at the leading and trailing edges of the wings of large commercial aircraft.

Discuss how the flaps affect the drag and lift coefficients during takeoff and landing.

11–126 Large commercial airplanes cruise at high altitudes (up to about 12,000 m) to save fuel. Discuss how flying at high altitudes reduces drag and saves fuel. Also discuss why small planes fly at relatively low altitudes.

11–127 Many drivers turn off their air conditioners and roll down the car windows in hopes of saving fuel. But it is claimed that this apparent "free cooling" actually increases the fuel consumption of some cars. Investigate this matter and write a report on which practice saves gasoline under what conditions.

CHAPTER 12

COMPRESSIBLE FLOW

For the most part, we have limited our consideration so far to flows for which density variations and thus compressibility effects are negligible. In this chapter, we lift this limitation and consider flows that involve significant changes in density. Such flows are called *compressible flows,* and they are frequently encountered in devices that involve the flow of gases at very high speeds. Compressible flow combines fluid dynamics and thermodynamics in that both are necessary to the development of the required theoretical background. In this chapter, we develop the general relations associated with compressible flows for an ideal gas with constant specific heats.

We start this chapter by reviewing the concepts of *stagnation state, speed of sound,* and *Mach number* for compressible flows. The relationships between the static and stagnation fluid properties are developed for isentropic flows of ideal gases, and they are expressed as functions of specific heat ratios and the Mach number. The effects of area changes for one-dimensional isentropic subsonic and supersonic flows are discussed. These effects are illustrated by considering the isentropic flow through *converging* and *converging–diverging nozzles.* The concept of *shock waves* and the variation of flow properties across normal and oblique shock waves are discussed. Finally, we consider the effects of friction and heat transfer on compressible flows and develop relations for property changes.

OBJECTIVES

When you finish reading this chapter, you should be able to

- Appreciate the consequences of compressibility in gas flow
- Understand why a nozzle must have a diverging section to accelerate a gas to supersonic speeds
- Predict the occurrence of shocks and calculate property changes across a shock wave
- Understand the effects of friction and heat transfer on compressible flows

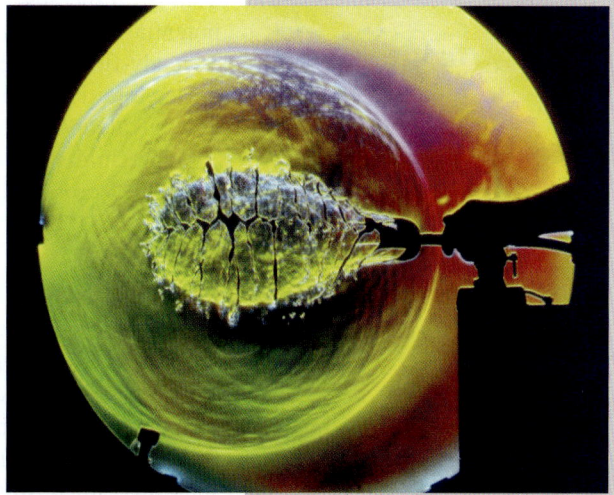

High-speed color schlieren image of the bursting of a toy balloon overfilled with compressed air. This 1-microsecond exposure captures the shattered balloon skin and reveals the bubble of compressed air inside beginning to expand. The balloon burst also drives a weak spherical shock wave, visible here as a circle surrounding the balloon. The silhouette of the photographer's hand on the air valve can be seen at center right.

Photo by G. S. Settles, Penn State University. Used by permission.

660
COMPRESSIBLE FLOW

(a)

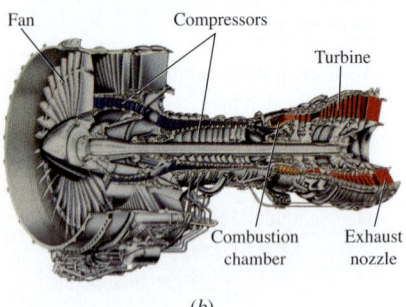

(b)

FIGURE 12–1
Aircraft and jet engines involve high speeds, and thus the kinetic energy term should always be considered when analyzing them.

(a) © Corbis RF; (b) Photo courtesy of United Technologies Corporation/Pratt & Whitney. Used by permission. All rights reserved.

12–1 ■ STAGNATION PROPERTIES

When analyzing control volumes, we find it very convenient to combine the *internal energy* and the *flow energy* of a fluid into a single term, *enthalpy,* defined per unit mass as $h = u + P/\rho$. Whenever the kinetic and potential energies of the fluid are negligible, as is often the case, the enthalpy represents the *total energy* of a fluid. For high-speed flows, such as those encountered in jet engines (Fig. 12–1), the potential energy of the fluid is still negligible, but the kinetic energy is not. In such cases, it is convenient to combine the enthalpy and the kinetic energy of the fluid into a single term called **stagnation** (or **total**) **enthalpy** h_0, defined per unit mass as

$$h_0 = h + \frac{V^2}{2} \quad \text{(kJ/kg)} \tag{12–1}$$

When the potential energy of the fluid is negligible, the stagnation enthalpy represents the *total energy of a flowing fluid stream* per unit mass. Thus it simplifies the thermodynamic analysis of high-speed flows.

Throughout this chapter the ordinary enthalpy h is referred to as the **static enthalpy,** whenever necessary, to distinguish it from the stagnation enthalpy. Notice that the stagnation enthalpy is a combination property of a fluid, just like the static enthalpy, and these two enthalpies are identical when the kinetic energy of the fluid is negligible.

Consider the steady flow of a fluid through a duct such as a nozzle, diffuser, or some other flow passage where the flow takes place adiabatically and with no shaft or electrical work, as shown in Fig. 12–2. Assuming the fluid experiences little or no change in its elevation and its potential energy, the energy balance relation ($\dot{E}_{in} = \dot{E}_{out}$) for this single-stream steady-flow device reduces to

$$h_1 + \frac{V_1^2}{2} = h_2 + \frac{V_2^2}{2} \tag{12–2}$$

or

$$h_{01} = h_{02} \tag{12–3}$$

That is, in the absence of any heat and work interactions and any changes in potential energy, the stagnation enthalpy of a fluid remains constant during a steady-flow process. Flows through nozzles and diffusers usually satisfy these conditions, and any increase in fluid velocity in these devices creates an equivalent decrease in the static enthalpy of the fluid.

If the fluid were brought to a complete stop, then the velocity at state 2 would be zero and Eq. 12–2 would become

$$h_1 + \frac{V_1^2}{2} = h_2 = h_{02}$$

Thus the *stagnation enthalpy* represents the *enthalpy of a fluid when it is brought to rest adiabatically.*

During a stagnation process, the kinetic energy of a fluid is converted to enthalpy (internal energy + flow energy), which results in an increase in the fluid temperature and pressure. The properties of a fluid at the stagnation state are called **stagnation properties** (stagnation temperature, stagnation

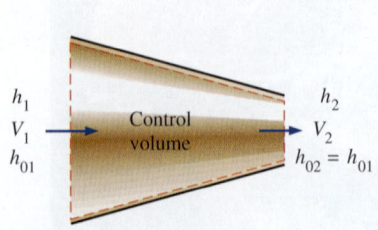

FIGURE 12–2
Steady flow of a fluid through an adiabatic duct.

pressure, stagnation density, etc.). The stagnation state and the stagnation properties are indicated by the subscript 0.

The stagnation state is called the **isentropic stagnation state** when the stagnation process is reversible as well as adiabatic (i.e., isentropic). The entropy of a fluid remains constant during an isentropic stagnation process. The actual (irreversible) and isentropic stagnation processes are shown on an *h-s* diagram in Fig. 12–3. Notice that the stagnation enthalpy of the fluid (and the stagnation temperature if the fluid is an ideal gas) is the same for both cases. However, the actual stagnation pressure is lower than the isentropic stagnation pressure since entropy increases during the actual stagnation process as a result of fluid friction. Many stagnation processes are approximated to be isentropic, and isentropic stagnation properties are simply referred to as stagnation properties.

When the fluid is approximated as an *ideal gas* with constant specific heats, its enthalpy can be replaced by $c_p T$ and Eq. 12–1 is expressed as

$$c_p T_0 = c_p T + \frac{V^2}{2}$$

or

$$T_0 = T + \frac{V^2}{2c_p} \quad (12\text{–}4)$$

Here, T_0 is called the **stagnation** (or **total**) **temperature,** and it represents *the temperature an ideal gas attains when it is brought to rest adiabatically.* The term $V^2/2c_p$ corresponds to the temperature rise during such a process and is called the **dynamic temperature.** For example, the dynamic temperature of air flowing at 100 m/s is $(100 \text{ m/s})^2/(2 \times 1.005 \text{ kJ/kg·K}) = 5.0$ K. Therefore, when air at 300 K and 100 m/s is brought to rest adiabatically (at the tip of a temperature probe, for example), its temperature rises to the stagnation value of 305 K (Fig. 12–4). Note that for low-speed flows, the stagnation and static (or ordinary) temperatures are practically the same. But for high-speed flows, the temperature measured by a stationary probe placed in the fluid (the stagnation temperature) may be significantly higher than the static temperature of the fluid.

The pressure a fluid attains when brought to rest isentropically is called the **stagnation pressure** P_0. For ideal gases with constant specific heats, P_0 is related to the static pressure of the fluid by

$$\frac{P_0}{P} = \left(\frac{T_0}{T}\right)^{k/(k-1)} \quad (12\text{–}5)$$

By noting that $\rho = 1/v$ and using the isentropic relation $Pv^k = P_0 v_0^k$, the ratio of the stagnation density to static density is expressed as

$$\frac{\rho_0}{\rho} = \left(\frac{T_0}{T}\right)^{1/(k-1)} \quad (12\text{–}6)$$

When stagnation enthalpies are used, there is no need to refer explicitly to kinetic energy. Then the energy balance $\dot{E}_{in} = \dot{E}_{out}$ for a single-stream, steady-flow device can be expressed as

$$q_{in} + w_{in} + (h_{01} + gz_1) = q_{out} + w_{out} + (h_{02} + gz_2) \quad (12\text{–}7)$$

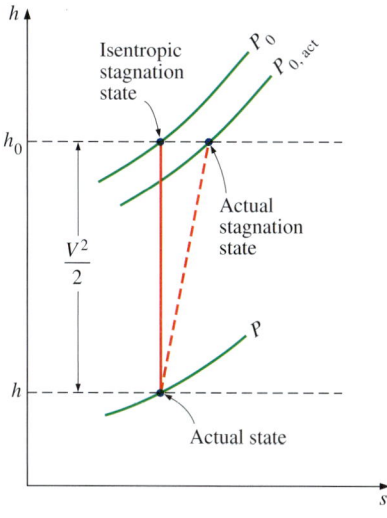

FIGURE 12–3
The actual state, actual stagnation state, and isentropic stagnation state of a fluid on an *h-s* diagram.

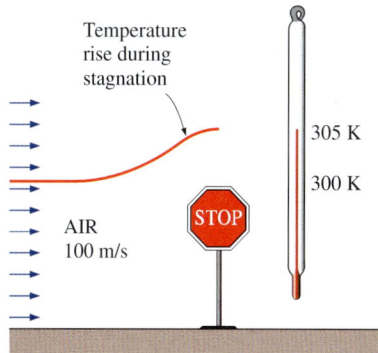

FIGURE 12–4
The temperature of an ideal gas flowing at a velocity V rises by $V^2/2c_p$ when it is brought to a complete stop.

where h_{01} and h_{02} are the stagnation enthalpies at states 1 and 2, respectively. When the fluid is an ideal gas with constant specific heats, Eq. 12–7 becomes

$$(q_{in} - q_{out}) + (w_{in} - w_{out}) = c_p(T_{02} - T_{01}) + g(z_2 - z_1) \quad \text{(12–8)}$$

where T_{01} and T_{02} are the stagnation temperatures.

Notice that kinetic energy terms do not explicitly appear in Eqs. 12–7 and 12–8, but the stagnation enthalpy terms account for their contribution.

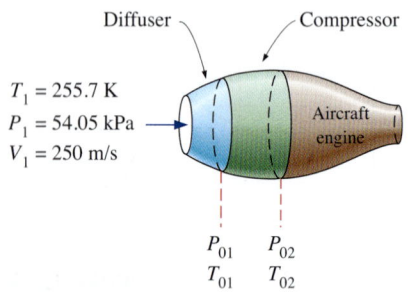

FIGURE 12–5
Schematic for Example 12–1.

EXAMPLE 12–1 Compression of High-Speed Air in an Aircraft

An aircraft is flying at a cruising speed of 250 m/s at an altitude of 5000 m where the atmospheric pressure is 54.05 kPa and the ambient air temperature is 255.7 K. The ambient air is first decelerated in a diffuser before it enters the compressor (Fig. 12–5). Approximating both the diffuser and the compressor to be isentropic, determine (a) the stagnation pressure at the compressor inlet and (b) the required compressor work per unit mass if the stagnation pressure ratio of the compressor is 8.

SOLUTION High-speed air enters the diffuser and the compressor of an aircraft. The stagnation pressure of the air and the compressor work input are to be determined.
Assumptions 1 Both the diffuser and the compressor are isentropic. 2 Air is an ideal gas with constant specific heats at room temperature.
Properties The constant-pressure specific heat c_p and the specific heat ratio k of air at room temperature are

$$c_p = 1.005 \text{ kJ/kg·K} \quad \text{and} \quad k = 1.4$$

Analysis (a) Under isentropic conditions, the stagnation pressure at the compressor inlet (diffuser exit) can be determined from Eq. 12–5. However, first we need to find the stagnation temperature T_{01} at the compressor inlet. Under the stated assumptions, T_{01} is determined from Eq. 12–4 to be

$$T_{01} = T_1 + \frac{V_1^2}{2c_p} = 255.7 \text{ K} + \frac{(250 \text{ m/s})^2}{(2)(1.005 \text{ kJ/kg·K})}\left(\frac{1 \text{ kJ/kg}}{1000 \text{ m}^2/\text{s}^2}\right)$$

$$= 286.8 \text{ K}$$

Then from Eq. 12–5,

$$P_{01} = P_1\left(\frac{T_{01}}{T_1}\right)^{k/(k-1)} = (54.05 \text{ kPa})\left(\frac{286.8 \text{ K}}{255.7 \text{ K}}\right)^{1.4/(1.4-1)}$$

$$= 80.77 \text{ kPa}$$

That is, the temperature of air would increase by 31.1°C and the pressure by 26.72 kPa as air is decelerated from 250 m/s to zero velocity. These increases in the temperature and pressure of air are due to the conversion of the kinetic energy into enthalpy.

(b) To determine the compressor work, we need to know the stagnation temperature of air at the compressor exit T_{02}. The stagnation pressure ratio across the compressor P_{02}/P_{01} is specified to be 8. Since the compression process is approximated as isentropic, T_{02} can be determined from the ideal-gas isentropic relation (Eq. 12–5):

$$T_{02} = T_{01}\left(\frac{P_{02}}{P_{01}}\right)^{(k-1)/k} = (286.8 \text{ K})(8)^{(1.4-1)/1.4} = 519.5 \text{ K}$$

Disregarding potential energy changes and heat transfer, the compressor work per unit mass of air is determined from Eq. 12–8:

$$w_{in} = c_p(T_{02} - T_{01})$$
$$= (1.005 \text{ kJ/kg·K})(519.5 \text{ K} - 286.8 \text{ K})$$
$$= \mathbf{233.9 \text{ kJ/kg}}$$

Thus the work supplied to the compressor is 233.9 kJ/kg.

Discussion Notice that using stagnation properties automatically accounts for any changes in the kinetic energy of a fluid stream.

12–2 ■ ONE-DIMENSIONAL ISENTROPIC FLOW

An important parameter in the study of compressible flow is the **speed of sound** c, which was shown in Chap. 2 to be related to other fluid properties as

$$c = \sqrt{(\partial P/\partial \rho)_s} \qquad (12\text{--}9)$$

or

$$c = \sqrt{k(\partial P/\partial \rho)_T} \qquad (12\text{--}10)$$

For an ideal gas it simplifies to

$$c = \sqrt{kRT} \qquad (12\text{--}11)$$

where k is the specific heat ratio of the gas and R is the specific gas constant. The ratio of the speed of the flow to the speed of sound is the dimensionless Mach number Ma,

$$\text{Ma} = \frac{V}{c} \qquad (12\text{--}12)$$

During fluid flow through many devices such as nozzles, diffusers, and turbine blade passages, flow quantities vary primarily in the flow direction only, and the flow can be approximated as one-dimensional isentropic flow with good accuracy. Therefore, it merits special consideration. Before presenting a formal discussion of one-dimensional isentropic flow, we illustrate some important aspects of it with an example.

EXAMPLE 12–2 Gas Flow through a Converging–Diverging Duct

Carbon dioxide flows steadily through a varying cross-sectional area duct such as a nozzle shown in Fig. 12–6 at a mass flow rate of 3.00 kg/s. The carbon dioxide enters the duct at a pressure of 1400 kPa and 200°C with a low velocity, and it expands in the nozzle to an exit pressure of 200 kPa. The duct is designed so that the flow can be approximated as isentropic. Determine the density, velocity, flow area, and Mach number at each location along the duct that corresponds to an overall pressure drop of 200 kPa.

SOLUTION Carbon dioxide enters a varying cross-sectional area duct at specified conditions. The flow properties are to be determined along the duct.

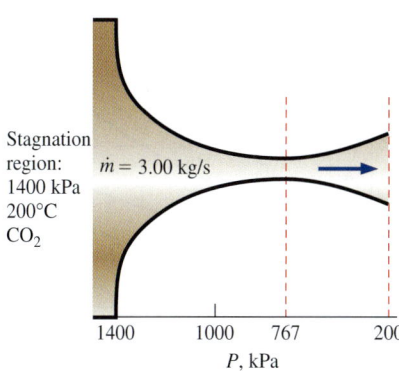

FIGURE 12–6
Schematic for Example 12–2.

Assumptions 1 Carbon dioxide is an ideal gas with constant specific heats at room temperature. 2 Flow through the duct is steady, one-dimensional, and isentropic.
Properties For simplicity we use $c_p = 0.846$ kJ/kg·K and $k = 1.289$ throughout the calculations, which are the constant-pressure specific heat and specific heat ratio values of carbon dioxide at room temperature. The gas constant of carbon dioxide is $R = 0.1889$ kJ/kg·K.
Analysis We note that the inlet temperature is nearly equal to the stagnation temperature since the inlet velocity is small. The flow is isentropic, and thus the stagnation temperature and pressure throughout the duct remain constant. Therefore,

$$T_0 \cong T_1 = 200°C = 473 \text{ K}$$

and

$$P_0 \cong P_1 = 1400 \text{ kPa}$$

To illustrate the solution procedure, we calculate the desired properties at the location where the pressure is 1200 kPa, the first location that corresponds to a pressure drop of 200 kPa.
From Eq. 12–5,

$$T = T_0 \left(\frac{P}{P_0}\right)^{(k-1)/k} = (473 \text{ K})\left(\frac{1200 \text{ kPa}}{1400 \text{ kPa}}\right)^{(1.289-1)/1.289} = 457 \text{ K}$$

From Eq. 12–4,
$$V = \sqrt{2c_p(T_0 - T)}$$
$$= \sqrt{2(0.846 \text{ kJ/kg·K})(473 \text{ K} - 457 \text{ K})\left(\frac{1000 \text{ m}^2/\text{s}^2}{1 \text{ kJ/kg}}\right)}$$
$$= 164.5 \text{ m/s} \cong \mathbf{164 \text{ m/s}}$$

From the ideal-gas relation,

$$\rho = \frac{P}{RT} = \frac{1200 \text{ kPa}}{(0.1889 \text{ kPa·m}^3/\text{kg·K})(457 \text{ K})} = \mathbf{13.9 \text{ kg/m}^3}$$

From the mass flow rate relation,

$$A = \frac{\dot{m}}{\rho V} = \frac{3.00 \text{ kg/s}}{(13.9 \text{ kg/m}^3)(164.5 \text{ m/s})} = 13.1 \times 10^{-4} \text{ m}^2 = \mathbf{13.1 \text{ cm}^2}$$

From Eqs. 12–11 and 12–12,

$$c = \sqrt{kRT} = \sqrt{(1.289)(0.1889 \text{ kJ/kg·K})(457 \text{ K})\left(\frac{1000 \text{ m}^2/\text{s}^2}{1 \text{ kJ/kg}}\right)} = 333.6 \text{ m/s}$$

$$\text{Ma} = \frac{V}{c} = \frac{164.5 \text{ m/s}}{333.6 \text{ m/s}} = \mathbf{0.493}$$

The results for the other pressure steps are summarized in Table 12–1 and are plotted in Fig. 12–7.
Discussion Note that as the pressure decreases, the temperature and speed of sound decrease while the fluid velocity and Mach number increase in the flow direction. The density decreases slowly at first and rapidly later as the fluid velocity increases.

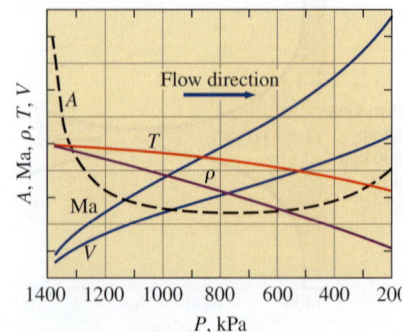

FIGURE 12–7
Variation of normalized fluid properties and cross-sectional area along a duct as the pressure drops from 1400 to 200 kPa.

TABLE 12–1

Variation of fluid properties in flow direction in the duct described in Example 12–2 for $\dot{m} = 3$ kg/s = constant

P, kPa	T, K	V, m/s	ρ, kg/m^3	c, m/s	A, cm^2	Ma
1400	473	0	15.7	339.4	∞	0
1200	457	164.5	13.9	333.6	13.1	0.493
1000	439	240.7	12.1	326.9	10.3	0.736
800	417	306.6	10.1	318.8	9.64	0.962
767*	413	317.2	9.82	317.2	9.63	1.000
600	391	371.4	8.12	308.7	10.0	1.203
400	357	441.9	5.93	295.0	11.5	1.498
200	306	530.9	3.46	272.9	16.3	1.946

* 767 kPa is the critical pressure where the local Mach number is unity.

We note from Example 12–2 that the flow area decreases with decreasing pressure down to a critical-pressure value where the Mach number is unity, and then it begins to increase with further reductions in pressure. The Mach number is unity at the location of smallest flow area, called the **throat** (Fig. 12–8). Note that the velocity of the fluid keeps increasing after passing the throat although the flow area increases rapidly in that region. This increase in velocity past the throat is due to the rapid decrease in the fluid density. The flow area of the duct considered in this example first decreases and then increases. Such ducts are called **converging–diverging nozzles.** These nozzles are used to accelerate gases to supersonic speeds and should not be confused with *Venturi nozzles*, which are used strictly for incompressible flow. The first use of such a nozzle occurred in 1893 in a steam turbine designed by a Swedish engineer, Carl G. B. de Laval (1845–1913), and therefore converging–diverging nozzles are often called *Laval nozzles*.

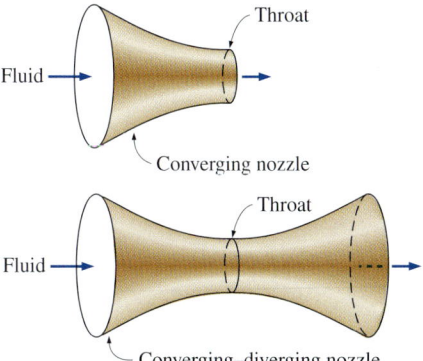

FIGURE 12–8
The cross section of a nozzle at the smallest flow area is called the *throat*.

Variation of Fluid Velocity with Flow Area

It is clear from Example 12–2 that the couplings among the velocity, density, and flow areas for isentropic duct flow are rather complex. In the remainder of this section we investigate these couplings more thoroughly, and we develop relations for the variation of static-to-stagnation property ratios with the Mach number for pressure, temperature, and density.

We begin our investigation by seeking relationships among the pressure, temperature, density, velocity, flow area, and Mach number for one-dimensional isentropic flow. Consider the mass balance for a steady-flow process:

$$\dot{m} = \rho A V = \text{constant}$$

Differentiating and dividing the resultant equation by the mass flow rate, we obtain

$$\frac{d\rho}{\rho} + \frac{dA}{A} + \frac{dV}{V} = 0 \quad (12\text{–}13)$$

Neglecting the potential energy, the energy balance for an isentropic flow with no work interactions is expressed in differential form as (Fig. 12–9)

$$\frac{dP}{\rho} + V\, dV = 0 \quad (12\text{–}14)$$

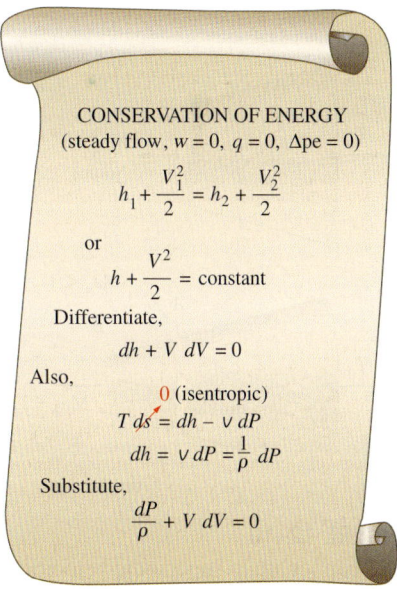

FIGURE 12–9
Derivation of the differential form of the energy equation for steady isentropic flow.

This relation is also the differential form of Bernoulli's equation when changes in potential energy are negligible, which is a form of Newton's second law of motion for steady-flow control volumes. Combining Eqs. 12–13 and 12–14 gives

$$\frac{dA}{A} = \frac{dP}{\rho}\left(\frac{1}{V^2} - \frac{d\rho}{dP}\right) \qquad (12\text{–}15)$$

Rearranging Eq. 12–9 as $(\partial \rho / \partial P)_s = 1/c^2$ and substituting into Eq. 12–15 yield

$$\frac{dA}{A} = \frac{dP}{\rho V^2}(1 - \text{Ma}^2) \qquad (12\text{–}16)$$

This is an important relation for isentropic flow in ducts since it describes the variation of pressure with flow area. We note that A, ρ, and V are positive quantities. For *subsonic* flow (Ma < 1), the term $1 - \text{Ma}^2$ is positive; and thus dA and dP must have the same sign. That is, the pressure of the fluid must increase as the flow area of the duct increases and must decrease as the flow area of the duct decreases. Thus, at subsonic velocities, the pressure decreases in converging ducts (subsonic nozzles) and increases in diverging ducts (subsonic diffusers).

In *supersonic* flow (Ma > 1), the term $1 - \text{Ma}^2$ is negative, and thus dA and dP must have opposite signs. That is, the pressure of the fluid must increase as the flow area of the duct decreases and must decrease as the flow area of the duct increases. Thus, at supersonic velocities, the pressure decreases in diverging ducts (supersonic nozzles) and increases in converging ducts (supersonic diffusers).

Another important relation for the isentropic flow of a fluid is obtained by substituting $\rho V = -dP/dV$ from Eq. 12–14 into Eq. 12–16:

$$\frac{dA}{A} = -\frac{dV}{V}(1 - \text{Ma}^2) \qquad (12\text{–}17)$$

This equation governs the shape of a nozzle or a diffuser in subsonic or supersonic isentropic flow. Noting that A and V are positive quantities, we conclude the following:

$$\text{For subsonic flow (Ma < 1)}, \quad \frac{dA}{dV} < 0$$

$$\text{For supersonic flow (Ma > 1)}, \quad \frac{dA}{dV} > 0$$

$$\text{For sonic flow (Ma = 1)}, \quad \frac{dA}{dV} = 0$$

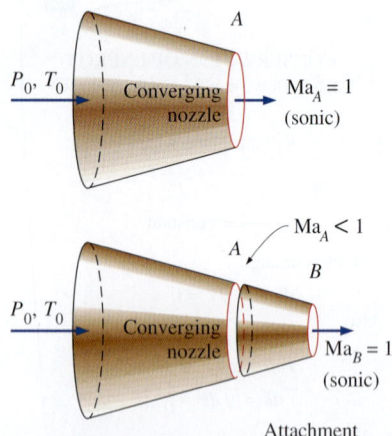

FIGURE 12–10
We cannot attain supersonic velocities by extending the converging section of a converging nozzle. Doing so will only move the sonic cross section farther downstream and decrease the mass flow rate.

Thus the proper shape of a nozzle depends on the highest velocity desired relative to the sonic velocity. To accelerate a fluid, we must use a converging nozzle at subsonic velocities and a diverging nozzle at supersonic velocities. The velocities encountered in most familiar applications are well below the sonic velocity, and thus it is natural that we visualize a nozzle as a converging duct. However, the highest velocity we can achieve by a converging nozzle is the sonic velocity, which occurs at the exit of the nozzle. If we extend the converging nozzle by further decreasing the flow area, in hopes of accelerating the fluid to supersonic velocities, as shown in Fig. 12–10, we are up for disappointment. Now the sonic velocity will occur at the exit of the converging extension, instead of the exit of the original nozzle, and the mass flow rate through the nozzle will decrease because of the reduced exit area.

Based on Eq. 12–16, which is an expression of the conservation of mass and energy principles, we must add a diverging section to a converging nozzle to accelerate a fluid to supersonic velocities. The result is a converging–diverging nozzle. The fluid first passes through a subsonic (converging) section, where the Mach number increases as the flow area of the nozzle decreases, and then reaches the value of unity at the nozzle throat. The fluid continues to accelerate as it passes through a supersonic (diverging) section. Noting that $\dot{m} = \rho AV$ for steady flow, we see that the large decrease in density makes acceleration in the diverging section possible. An example of this type of flow is the flow of hot combustion gases through a nozzle in a gas turbine.

The opposite process occurs in the engine inlet of a supersonic aircraft. The fluid is decelerated by passing it first through a supersonic diffuser, which has a flow area that decreases in the flow direction. Ideally, the flow reaches a Mach number of unity at the diffuser throat. The fluid is further decelerated in a subsonic diffuser, which has a flow area that increases in the flow direction, as shown in Fig. 12–11.

Property Relations for Isentropic Flow of Ideal Gases

Next we develop relations between the static properties and stagnation properties of an ideal gas in terms of the specific heat ratio k and the Mach number Ma. We assume the flow is isentropic and the gas has constant specific heats.

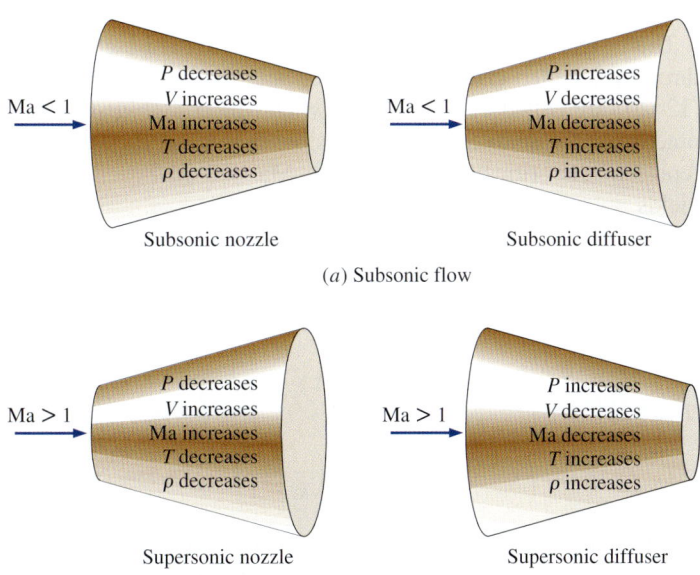

FIGURE 12–11
Variation of flow properties in subsonic and supersonic nozzles and diffusers.

The temperature T of an ideal gas anywhere in the flow is related to the stagnation temperature T_0 through Eq. 12–4:

$$T_0 = T + \frac{V^2}{2c_p}$$

or

$$\frac{T_0}{T} = 1 + \frac{V^2}{2c_p T}$$

Noting that $c_p = kR/(k - 1)$, $c^2 = kRT$, and $Ma = V/c$, we see that

$$\frac{V^2}{2c_p T} = \frac{V^2}{2[kR/(k-1)]T} = \left(\frac{k-1}{2}\right)\frac{V^2}{c^2} = \left(\frac{k-1}{2}\right)Ma^2$$

Substitution yields

$$\frac{T_0}{T} = 1 + \left(\frac{k-1}{2}\right)Ma^2 \tag{12–18}$$

which is the desired relation between T_0 and T.

The ratio of the stagnation to static pressure is obtained by substituting Eq. 12–18 into Eq. 12–5:

$$\frac{P_0}{P} = \left[1 + \left(\frac{k-1}{2}\right)Ma^2\right]^{k/(k-1)} \tag{12–19}$$

The ratio of the stagnation to static density is obtained by substituting Eq. 12–18 into Eq. 12–6:

$$\frac{\rho_0}{\rho} = \left[1 + \left(\frac{k-1}{2}\right)Ma^2\right]^{1/(k-1)} \tag{12–20}$$

Numerical values of T/T_0, P/P_0, and ρ/ρ_0 are listed versus the Mach number in Table A–13 for $k = 1.4$, which are very useful for practical compressible flow calculations involving air.

The properties of a fluid at a location where the Mach number is unity (the throat) are called **critical properties,** and the ratios in Eqs. (12–18) through (12–20) are called **critical ratios** when $Ma = 1$ (Fig. 12–12). It is standard practice in the analysis of compressible flow to let the superscript asterisk (*) represent the critical values. Setting $Ma = 1$ in Eqs. 12–18 through 12–20 yields

$$\frac{T^*}{T_0} = \frac{2}{k+1} \tag{12–21}$$

$$\frac{P^*}{P_0} = \left(\frac{2}{k+1}\right)^{k/(k-1)} \tag{12–22}$$

$$\frac{\rho^*}{\rho_0} = \left(\frac{2}{k+1}\right)^{1/(k-1)} \tag{12–23}$$

These ratios are evaluated for various values of k and are listed in Table 12–2. The critical properties of compressible flow should not be confused with the thermodynamic properties of substances at the *critical point* (such as the critical temperature T_c and critical pressure P_c).

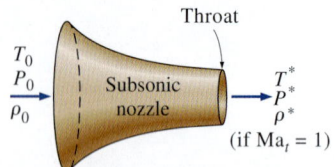

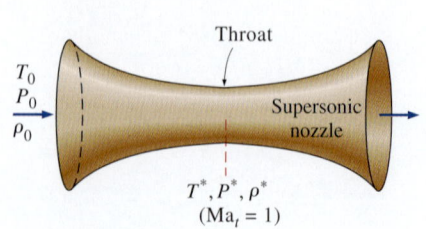

FIGURE 12–12
When $Ma_t = 1$, the properties at the nozzle throat are the critical properties.

TABLE 12–2

The critical-pressure, critical-temperature, and critical-density ratios for isentropic flow of some ideal gases

	Superheated steam, $k = 1.3$	Hot products of combustion, $k = 1.33$	Air, $k = 1.4$	Monatomic gases, $k = 1.667$
$\dfrac{P^*}{P_0}$	0.5457	0.5404	0.5283	0.4871
$\dfrac{T^*}{T_0}$	0.8696	0.8584	0.8333	0.7499
$\dfrac{\rho^*}{\rho_0}$	0.6276	0.6295	0.6340	0.6495

EXAMPLE 12–3   Critical Temperature and Pressure in Gas Flow

Calculate the critical pressure and temperature of carbon dioxide for the flow conditions described in Example 12–2 (Fig. 12–13).

SOLUTION For the flow discussed in Example 12–2, the critical pressure and temperature are to be calculated.
Assumptions 1 The flow is steady, adiabatic, and one-dimensional. 2 Carbon dioxide is an ideal gas with constant specific heats.
Properties The specific heat ratio of carbon dioxide at room temperature is $k = 1.289$.
Analysis The ratios of critical to stagnation temperature and pressure are determined to be

$$\frac{T^*}{T_0} = \frac{2}{k+1} = \frac{2}{1.289 + 1} = 0.8737$$

$$\frac{P^*}{P_0} = \left(\frac{2}{k+1}\right)^{k/(k-1)} = \left(\frac{2}{1.289+1}\right)^{1.289/(1.289-1)} = 0.5477$$

Noting that the stagnation temperature and pressure are, from Example 12–2, $T_0 = 473$ K and $P_0 = 1400$ kPa, we see that the critical temperature and pressure in this case are

$$T^* = 0.8737 T_0 = (0.8737)(473 \text{ K}) = \mathbf{413 \text{ K}}$$

$$P^* = 0.5477 P_0 = (0.5477)(1400 \text{ kPa}) = \mathbf{767 \text{ kPa}}$$

Discussion Note that these values agree with those listed in the 5th row of Table 12–1, as expected. Also, property values other than these at the throat would indicate that the flow is not critical, and the Mach number is not unity.

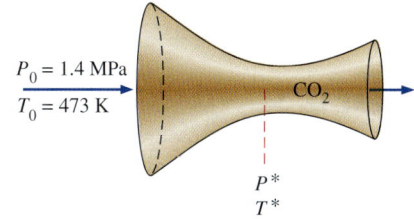

FIGURE 12–13
Schematic for Example 12–3.

12–3 · ISENTROPIC FLOW THROUGH NOZZLES

Converging or converging–diverging nozzles are found in many engineering applications including steam and gas turbines, aircraft and spacecraft propulsion systems, and even industrial blasting nozzles and torch nozzles. In this section we consider the effects of **back pressure** (i.e., the pressure applied

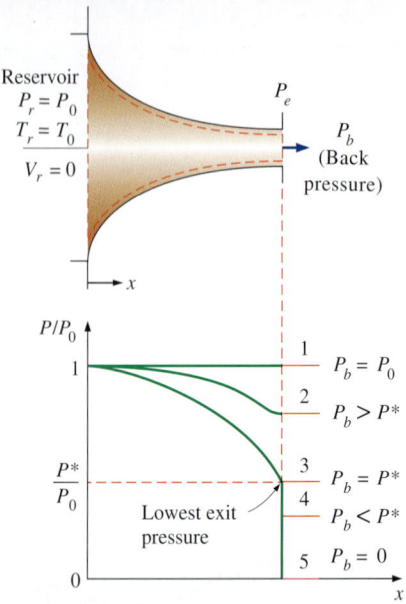

FIGURE 12–14
The effect of back pressure on the pressure distribution along a converging nozzle.

at the nozzle discharge region) on the exit velocity, the mass flow rate, and the pressure distribution along the nozzle.

Converging Nozzles

Consider the subsonic flow through a converging nozzle as shown in Fig. 12–14. The nozzle inlet is attached to a reservoir at pressure P_r and temperature T_r. The reservoir is sufficiently large so that the nozzle inlet velocity is negligible. Since the fluid velocity in the reservoir is zero and the flow through the nozzle is approximated as isentropic, the stagnation pressure and stagnation temperature of the fluid at any cross section through the nozzle are equal to the reservoir pressure and temperature, respectively.

Now we begin to reduce the back pressure and observe the resulting effects on the pressure distribution along the length of the nozzle, as shown in Fig. 12–14. If the back pressure P_b is equal to P_1, which is equal to P_r, there is no flow and the pressure distribution is uniform along the nozzle. When the back pressure is reduced to P_2, the exit plane pressure P_e also drops to P_2. This causes the pressure along the nozzle to decrease in the flow direction.

When the back pressure is reduced to P_3 (= P^*, which is the pressure required to increase the fluid velocity to the speed of sound at the exit plane or throat), the mass flow reaches a maximum value and the flow is said to be **choked**. Further reduction of the back pressure to level P_4 or below does not result in additional changes in the pressure distribution, or anything else along the nozzle length.

Under steady-flow conditions, the mass flow rate through the nozzle is constant and is expressed as

$$\dot{m} = \rho A V = \left(\frac{P}{RT}\right) A (\text{Ma}\sqrt{kRT}) = PA\text{Ma}\sqrt{\frac{k}{RT}}$$

Solving for T from Eq. 12–18 and for P from Eq. 12–19 and substituting,

$$\dot{m} = \frac{A\text{Ma}P_0\sqrt{k/(RT_0)}}{[1 + (k-1)\text{Ma}^2/2]^{(k+1)/[2(k-1)]}} \qquad (12\text{–}24)$$

Thus the mass flow rate of a particular fluid through a nozzle is a function of the stagnation properties of the fluid, the flow area, and the Mach number. Equation 12–24 is valid at any cross section, and thus $\dot{m}$ can be evaluated at any location along the length of the nozzle.

For a specified flow area A and stagnation properties T_0 and P_0, the maximum mass flow rate can be determined by differentiating Eq. 12–24 with respect to Ma and setting the result equal to zero. It yields Ma = 1. Since the only location in a nozzle where the Mach number can be unity is the location of minimum flow area (the throat), the mass flow rate through a nozzle is a maximum when Ma = 1 at the throat. Denoting this area by A^*, we obtain an expression for the maximum mass flow rate by substituting Ma = 1 in Eq. 12–24:

$$\dot{m}_{\max} = A^* P_0 \sqrt{\frac{k}{RT_0}} \left(\frac{2}{k+1}\right)^{(k+1)/[2(k-1)]} \qquad (12\text{–}25)$$

Thus, for a particular ideal gas, the maximum mass flow rate through a nozzle with a given throat area is fixed by the stagnation pressure and temperature of the inlet flow. The flow rate can be controlled by changing the stagnation pressure or temperature, and thus a converging nozzle can be used as a flowmeter. The flow rate can also be controlled, of course, by varying the throat area. This principle is very important for chemical processes, medical devices, flowmeters, and anywhere the mass flux of a gas must be known and controlled.

A plot of $\dot{m}$ versus P_b/P_0 for a converging nozzle is shown in Fig. 12–15. Notice that the mass flow rate increases with decreasing P_b/P_0, reaches a maximum at $P_b = P^*$, and remains constant for P_b/P_0 values less than this critical ratio. Also illustrated on this figure is the effect of back pressure on the nozzle exit pressure P_e. We observe that

$$P_e = \begin{cases} P_b & \text{for } P_b \geq P^* \\ P^* & \text{for } P_b < P^* \end{cases}$$

To summarize, for all back pressures lower than the critical pressure P^*, the pressure at the exit plane of the converging nozzle P_e is equal to P^*, the Mach number at the exit plane is unity, and the mass flow rate is the maximum (or choked) flow rate. Because the velocity of the flow is sonic at the throat for the maximum flow rate, a back pressure lower than the critical pressure cannot be sensed in the nozzle upstream flow and does not affect the flow rate.

The effects of the stagnation temperature T_0 and stagnation pressure P_0 on the mass flow rate through a converging nozzle are illustrated in Fig. 12–16 where the mass flow rate is plotted against the static-to-stagnation pressure ratio at the throat P_t/P_0. An increase in P_0 (or a decrease of T_0) will increase the mass flow rate through the converging nozzle; a decrease in P_0 (or an increase in T_0) will decrease it. We could also conclude this by carefully observing Eqs. 12–24 and 12–25.

A relation for the variation of flow area A through the nozzle relative to throat area A^* can be obtained by combining Eqs. 12–24 and 12–25 for the same mass flow rate and stagnation properties of a particular fluid. This yields

$$\frac{A}{A^*} = \frac{1}{\text{Ma}}\left[\left(\frac{2}{k+1}\right)\left(1 + \frac{k-1}{2}\text{Ma}^2\right)\right]^{(k+1)/[2(k-1)]} \quad (12\text{–}26)$$

Table A–13 gives values of A/A^* as a function of the Mach number for air ($k = 1.4$). There is one value of A/A^* for each value of the Mach number, but there are two possible values of the Mach number for each value of A/A^*—one for subsonic flow and another for supersonic flow.

Another parameter sometimes used in the analysis of one-dimensional isentropic flow of ideal gases is Ma*, which is the ratio of the local velocity to the speed of sound at the throat:

$$\text{Ma}^* = \frac{V}{c^*} \quad (12\text{–}27)$$

Equation 12–27 can also be expressed as

$$\text{Ma}^* = \frac{V}{c}\frac{c}{c^*} = \frac{\text{Ma}\, c}{c^*} = \frac{\text{Ma}\sqrt{kRT}}{\sqrt{kRT^*}} = \text{Ma}\sqrt{\frac{T}{T^*}}$$

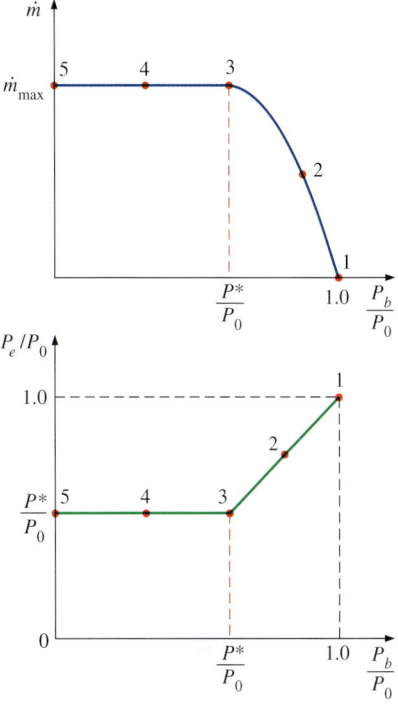

FIGURE 12–15
The effect of back pressure P_b on the mass flow rate $\dot{m}$ and the exit pressure P_e of a converging nozzle.

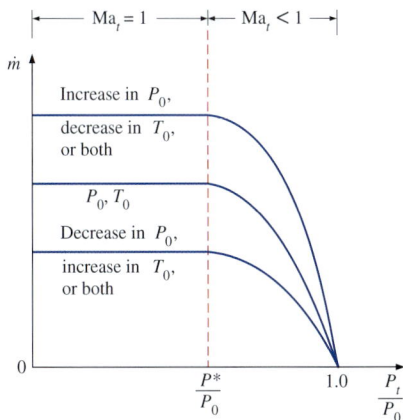

FIGURE 12–16
The variation of the mass flow rate through a nozzle with inlet stagnation properties.

Ma	Ma*	$\frac{A}{A^*}$	$\frac{P}{P_0}$	$\frac{\rho}{\rho_0}$	$\frac{T}{T_0}$
⋮	⋮	⋮	⋮	⋮	⋮
0.90	0.9146	1.0089	0.5913		
1.00	1.0000	1.0000	0.5283		
1.10	1.0812	1.0079	0.4684		
⋮	⋮	⋮	⋮	⋮	⋮

FIGURE 12–17
Various property ratios for isentropic flow through nozzles and diffusers are listed in Table A–13 for $k = 1.4$ (air) for convenience.

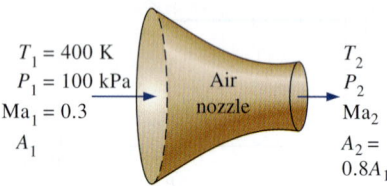

FIGURE 12–18
Schematic for Example 12–4.

where Ma is the local Mach number, T is the local temperature, and T^* is the critical temperature. Solving for T from Eq. 12–18 and for T^* from Eq. 12–21 and substituting, we get

$$\text{Ma}^* = \text{Ma}\sqrt{\frac{k+1}{2+(k-1)\text{Ma}^2}} \qquad (12\text{–}28)$$

Values of Ma* are also listed in Table A–13 versus the Mach number for $k = 1.4$ (Fig. 12–17). Note that the parameter Ma* differs from the Mach number Ma in that Ma* is the local velocity nondimensionalized with respect to the sonic velocity at the *throat*, whereas Ma is the local velocity nondimensionalized with respect to the *local* sonic velocity. (Recall that the sonic velocity in a nozzle varies with temperature and thus with location.)

EXAMPLE 12–4   Effect of Back Pressure on Mass Flow Rate

Air at 1 MPa and 600°C enters a converging nozzle, shown in Fig. 12–18, with a velocity of 150 m/s. Determine the mass flow rate through the nozzle for a nozzle throat area of 50 cm² when the back pressure is (*a*) 0.7 MPa and (*b*) 0.4 MPa.

SOLUTION   Air enters a converging nozzle. The mass flow rate of air through the nozzle is to be determined for different back pressures.
Assumptions   **1** Air is an ideal gas with constant specific heats at room temperature. **2** Flow through the nozzle is steady, one-dimensional, and isentropic.
Properties   The constant pressure specific heat and the specific heat ratio of air are $c_p = 1.005$ kJ/kg·K and $k = 1.4$.
Analysis   We use the subscripts i and t to represent the properties at the nozzle inlet and the throat, respectively. The stagnation temperature and pressure at the nozzle inlet are determined from Eqs. 12–4 and 12–5:

$$T_{0i} = T_i + \frac{V_i^2}{2c_p} = 873 \text{ K} + \frac{(150 \text{ m/s})^2}{2(1.005 \text{ kJ/kg·K})}\left(\frac{1 \text{ kJ/kg}}{1000 \text{ m}^2/\text{s}^2}\right) = 884 \text{ K}$$

$$P_{0i} = P_i\left(\frac{T_{0i}}{T_i}\right)^{k/(k-1)} = (1 \text{ MPa})\left(\frac{884 \text{ K}}{873 \text{ K}}\right)^{1.4/(1.4-1)} = 1.045 \text{ MPa}$$

These stagnation temperature and pressure values remain constant throughout the nozzle since the flow is assumed to be isentropic. That is,

$$T_0 = T_{0i} = 884 \text{ K} \quad \text{and} \quad P_0 = P_{0i} = 1.045 \text{ MPa}$$

The critical-pressure ratio is determined from Table 12–2 (or Eq. 12–22) to be $P^*/P_0 = 0.5283$.

(*a*) The back pressure ratio for this case is

$$\frac{P_b}{P_0} = \frac{0.7 \text{ MPa}}{1.045 \text{ MPa}} = 0.670$$

which is greater than the critical-pressure ratio, 0.5283. Thus the exit plane pressure (or throat pressure P_t) is equal to the back pressure in this case. That is, $P_t = P_b = 0.7$ MPa, and $P_t/P_0 = 0.670$. Therefore, the flow is not choked. From Table A–13 at $P_t/P_0 = 0.670$, we read $\text{Ma}_t = 0.778$ and $T_t/T_0 = 0.892$.

The mass flow rate through the nozzle can be calculated from Eq. 12–24. But it can also be determined in a step-by-step manner as follows:

$$T_t = 0.892 T_0 = 0.892(884 \text{ K}) = 788.5 \text{ K}$$

$$\rho_t = \frac{P_t}{RT_t} = \frac{700 \text{ kPa}}{(0.287 \text{ kPa·m}^3/\text{kg·K})(788.5 \text{ K})} = 3.093 \text{ kg/m}^3$$

$$V_t = \text{Ma}_t c_t = \text{Ma}_t \sqrt{kRT_t}$$

$$= (0.778)\sqrt{(1.4)(0.287 \text{ kJ/kg·K})(788.5 \text{ K})\left(\frac{1000 \text{ m}^2/\text{s}^2}{1 \text{ kJ/kg}}\right)}$$

$$= 437.9 \text{ m/s}$$

Thus,

$$\dot{m} = \rho_t A_t V_t = (3.093 \text{ kg/m}^3)(50 \times 10^{-4} \text{ m}^2)(437.9 \text{ m/s}) = \mathbf{6.77 \text{ kg/s}}$$

(*b*) The back pressure ratio for this case is

$$\frac{P_b}{P_0} = \frac{0.4 \text{ MPa}}{1.045 \text{ MPa}} = 0.383$$

which is less than the critical-pressure ratio, 0.5283. Therefore, sonic conditions exist at the exit plane (throat) of the nozzle, and Ma = 1. The flow is choked in this case, and the mass flow rate through the nozzle is calculated from Eq. 12–25:

$$\dot{m} = A^* P_0 \sqrt{\frac{k}{RT_0}} \left(\frac{2}{k+1}\right)^{(k+1)/[2(k-1)]}$$

$$= (50 \times 10^{-4} \text{ m}^2)(1045 \text{ kPa})\sqrt{\frac{1.4}{(0.287 \text{ kJ/kg·K})(884 \text{ K})}} \left(\frac{2}{1.4+1}\right)^{2.4/0.8}$$

$$= \mathbf{7.10 \text{ kg/s}}$$

since $\text{kPa·m}^2 \sqrt{\text{kJ/kg}} = \sqrt{1000} \text{ kg/s}$.

Discussion This is the maximum mass flow rate through the nozzle for the specified inlet conditions and nozzle throat area.

EXAMPLE 12–5   Air Loss from a Flat Tire

Air in an automobile tire is maintained at a pressure of 220 kPa (gage) in an environment where the atmospheric pressure is 94 kPa. The air in the tire is at the ambient temperature of 25°C. A 4-mm-diameter leak develops in the tire as a result of an accident (Fig. 12–19). Approximating the flow as isentropic determine the initial mass flow rate of air through the leak.

SOLUTION A leak develops in an automobile tire as a result of an accident. The initial mass flow rate of air through the leak is to be determined.
Assumptions 1 Air is an ideal gas with constant specific heats. 2 Flow of air through the hole is isentropic.

FIGURE 12–19
Schematic for Example 12–5.

Properties The specific gas constant of air is $R = 0.287$ kPa·m³/kg·K. The specific heat ratio of air at room temperature is $k = 1.4$.

Analysis The absolute pressure in the tire is

$$P = P_{\text{gage}} + P_{\text{atm}} = 220 + 94 = 314 \text{ kPa}$$

The critical pressure is (from Table 12–2)

$$P^* = 0.5283 P_o = (0.5283)(314 \text{ kPa}) = 166 \text{ kPa} > 94 \text{ kPa}$$

Therefore, the flow is choked, and the velocity at the exit of the hole is the sonic speed. Then the flow properties at the exit become

$$\rho_0 = \frac{P_0}{RT_0} = \frac{314 \text{ kPa}}{(0.287 \text{ kPa·m}^3/\text{kg·K})(298 \text{ K})} = 3.671 \text{ kg/m}^3$$

$$\rho^* = \rho\left(\frac{2}{k+1}\right)^{1/(k-1)} = (3.671 \text{ kg/m}^3)\left(\frac{2}{1.4+1}\right)^{1/(1.4-1)} = 2.327 \text{ kg/m}^3$$

$$T^* = \frac{2}{k+1} T_0 = \frac{2}{1.4+1}(298 \text{ K}) = 248.3 \text{ K}$$

$$V = c = \sqrt{kRT^*} = \sqrt{(1.4)(0.287 \text{ kJ/kg·K})\left(\frac{1000 \text{ m}^2/\text{s}^2}{1 \text{ kJ/kg}}\right)(248.3 \text{ K})}$$

$$= 315.9 \text{ m/s}$$

Then the initial mass flow rate through the hole is

$$\dot{m} = \rho A V = (2.327 \text{ kg/m}^3)[\pi(0.004 \text{ m})^2/4](315.9 \text{ m/s}) = 0.00924 \text{ kg/s}$$

$$= \textbf{0.554 kg/min}$$

Discussion The mass flow rate decreases with time as the pressure inside the tire drops.

Converging–Diverging Nozzles

When we think of nozzles, we ordinarily think of flow passages whose cross-sectional area decreases in the flow direction. However, the highest velocity to which a fluid can be accelerated in a converging nozzle is limited to the sonic velocity (Ma = 1), which occurs at the exit plane (throat) of the nozzle. Accelerating a fluid to supersonic velocities (Ma > 1) can be accomplished only by attaching a diverging flow section to the subsonic nozzle at the throat. The resulting combined flow section is a converging–diverging nozzle, which is standard equipment in supersonic aircraft and rocket propulsion (Fig. 12–20).

Forcing a fluid through a converging–diverging nozzle is no guarantee that the fluid will be accelerated to a supersonic velocity. In fact, the fluid may find itself decelerating in the diverging section instead of accelerating if the back pressure is not in the right range. The state of the nozzle flow is determined by the overall pressure ratio P_b/P_0. Therefore, for given inlet conditions, the flow through a converging–diverging nozzle is governed by the back pressure P_b, as will be explained.

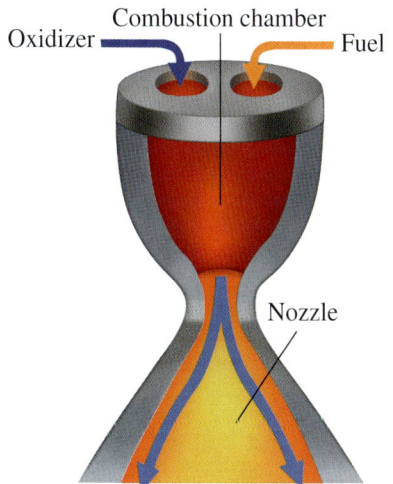

FIGURE 12–20
Converging–diverging nozzles are commonly used in rocket engines to provide high thrust.
(Right) NASA

Consider the converging–diverging nozzle shown in Fig. 12–21. A fluid enters the nozzle with a low velocity at stagnation pressure P_0. When $P_b = P_0$ (case A), there is no flow through the nozzle. This is expected since the flow in a nozzle is driven by the pressure difference between the nozzle inlet and the exit. Now let us examine what happens as the back pressure is lowered.

1. When $P_0 > P_b > P_C$, the flow remains subsonic throughout the nozzle, and the mass flow is less than that for choked flow. The fluid velocity increases in the first (converging) section and reaches a maximum at the throat (but Ma < 1). However, most of the gain in velocity is lost in the second (diverging) section of the nozzle, which acts as a diffuser. The pressure decreases in the converging section, reaches a minimum at the throat, and increases at the expense of velocity in the diverging section.
2. When $P_b = P_C$, the throat pressure becomes P^* and the fluid achieves sonic velocity at the throat. But the diverging section of the nozzle still acts as a diffuser, slowing the fluid to subsonic velocities. The mass flow rate that was increasing with decreasing P_b also reaches its maximum value. Recall that P^* is the lowest pressure that can be obtained at the throat, and the sonic velocity is the highest velocity that can be achieved with a converging nozzle. Thus, lowering P_b further has no influence on the fluid flow in the converging part of the nozzle or the mass flow rate through the nozzle. However, it does influence the character of the flow in the diverging section.
3. When $P_C > P_b > P_E$, the fluid that achieved a sonic velocity at the throat continues accelerating to supersonic velocities in the diverging section as the pressure decreases. This acceleration comes to a sudden stop, however, as a **normal shock** develops at a section between the throat and the exit plane, which causes a sudden drop in velocity to subsonic levels and a sudden increase in pressure. The fluid then continues to decelerate further

676
COMPRESSIBLE FLOW

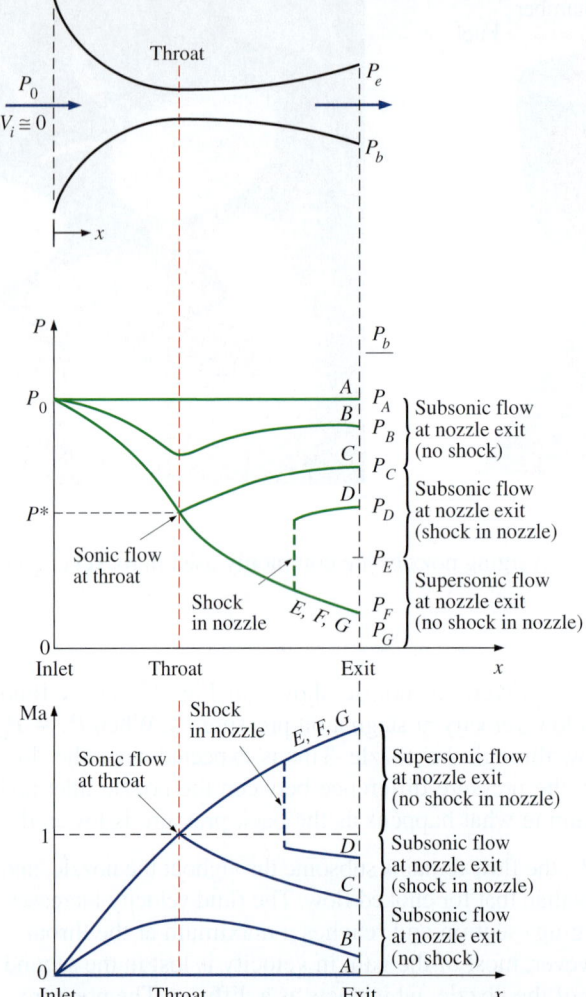

FIGURE 12–21
The effects of back pressure on the flow through a converging–diverging nozzle.

in the remaining part of the converging–diverging nozzle. Flow through the shock is highly irreversible, and thus it cannot be approximated as isentropic. The normal shock moves downstream away from the throat as P_b is decreased, and it approaches the nozzle exit plane as P_b approaches P_E.

When $P_b = P_E$, the normal shock forms at the exit plane of the nozzle. The flow is supersonic through the entire diverging section in this case, and it can be approximated as isentropic. However, the fluid velocity drops to subsonic levels just before leaving the nozzle as it crosses the normal shock. Normal shock waves are discussed in Section 12–4.

4. When $P_E > P_b > 0$, the flow in the diverging section is supersonic, and the fluid expands to P_F at the nozzle exit with no normal shock forming within the nozzle. Thus, the flow through the nozzle can be approximated as isentropic. When $P_b = P_F$, no shocks occur within or outside the nozzle. When $P_b < P_F$, irreversible mixing and expansion waves occur downstream of the exit plane of the nozzle. When $P_b > P_F$, however, the pressure of the fluid increases from P_F to P_b irreversibly in the wake of the nozzle exit, creating what are called *oblique shocks*.

EXAMPLE 12–6   Airflow through a Converging–Diverging Nozzle

Air enters a converging–diverging nozzle, shown in Fig. 12–22, at 1.0 MPa and 800 K with negligible velocity. The flow is steady, one-dimensional, and isentropic with $k = 1.4$. For an exit Mach number of Ma = 2 and a throat area of 20 cm², determine (a) the throat conditions, (b) the exit plane conditions, including the exit area, and (c) the mass flow rate through the nozzle.

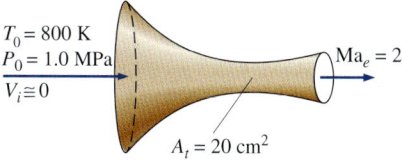

FIGURE 12–22
Schematic for Example 12–6.

SOLUTION Air flows through a converging–diverging nozzle. The throat and the exit conditions and the mass flow rate are to be determined.
Assumptions 1 Air is an ideal gas with constant specific heats at room temperature. 2 Flow through the nozzle is steady, one-dimensional, and isentropic.
Properties The specific heat ratio of air is given to be $k = 1.4$. The gas constant of air is 0.287 kJ/kg·K.
Analysis The exit Mach number is given to be 2. Therefore, the flow must be sonic at the throat and supersonic in the diverging section of the nozzle. Since the inlet velocity is negligible, the stagnation pressure and stagnation temperature are the same as the inlet temperature and pressure, $P_0 = 1.0$ MPa and $T_0 = 800$ K. Assuming ideal-gas behavior, the stagnation density is

$$\rho_0 = \frac{P_0}{RT_0} = \frac{1000 \text{ kPa}}{(0.287 \text{ kPa·m}^3/\text{kg·K})(800 \text{ K})} = 4.355 \text{ kg/m}^3$$

(a) At the throat of the nozzle Ma = 1, and from Table A–13 we read

$$\frac{P^*}{P_0} = 0.5283 \qquad \frac{T^*}{T_0} = 0.8333 \qquad \frac{\rho^*}{\rho_0} = 0.6339$$

Thus,

$$P^* = 0.5283 P_0 = (0.5283)(1.0 \text{ MPa}) = \mathbf{0.5283 \text{ MPa}}$$
$$T^* = 0.8333 T_0 = (0.8333)(800 \text{ K}) = \mathbf{666.6 \text{ K}}$$
$$\rho^* = 0.6339 \rho_0 = (0.6339)(4.355 \text{ kg/m}^3) = \mathbf{2.761 \text{ kg/m}^3}$$

Also,

$$V^* = c^* = \sqrt{kRT^*} = \sqrt{(1.4)(0.287 \text{ kJ/kg·K})(666.6 \text{ K})\left(\frac{1000 \text{ m}^2/\text{s}^2}{1 \text{ kJ/kg}}\right)}$$
$$= \mathbf{517.5 \text{ m/s}}$$

(b) Since the flow is isentropic, the properties at the exit plane can also be calculated by using data from Table A–13. For Ma = 2 we read

$$\frac{P_e}{P_0} = 0.1278 \quad \frac{T_e}{T_0} = 0.5556 \quad \frac{\rho_e}{\rho_0} = 0.2300 \quad \text{Ma}_e^* = 1.6330 \quad \frac{A_e}{A^*} = 1.6875$$

Thus,

$$P_e = 0.1278 P_0 = (0.1278)(1.0 \text{ MPa}) = \mathbf{0.1278 \text{ MPa}}$$
$$T_e = 0.5556 T_0 = (0.5556)(800 \text{ K}) = \mathbf{444.5 \text{ K}}$$
$$\rho_e = 0.2300 \rho_0 = (0.2300)(4.355 \text{ kg/m}^3) = \mathbf{1.002 \text{ kg/m}^3}$$
$$A_e = 1.6875 A^* = (1.6875)(20 \text{ cm}^2) = \mathbf{33.75 \text{ cm}^2}$$

and

$$V_e = \text{Ma}_e^* c^* = (1.6330)(517.5 \text{ m/s}) = \mathbf{845.1 \text{ m/s}}$$

The nozzle exit velocity could also be determined from $V_e = \text{Ma}_e c_e$, where c_e is the speed of sound at the exit conditions:

$$V_e = \text{Ma}_e c_e = \text{Ma}_e \sqrt{kRT_e} = 2\sqrt{(1.4)(0.287 \text{ kJ/kg·K})(444.5 \text{ K})\left(\frac{1000 \text{ m}^2/\text{s}^2}{1 \text{ kJ/kg}}\right)}$$

$$= 845.2 \text{ m/s}$$

(c) Since the flow is steady, the mass flow rate of the fluid is the same at all sections of the nozzle. Thus it may be calculated by using properties at any cross section of the nozzle. Using the properties at the throat, we find that the mass flow rate is

$$\dot{m} = \rho^* A^* V^* = (2.761 \text{ kg/m}^3)(20 \times 10^{-4} \text{ m}^2)(517.5 \text{ m/s}) = \mathbf{2.86 \text{ kg/s}}$$

Discussion Note that this is the highest possible mass flow rate that can flow through this nozzle for the specified inlet conditions.

12–4 ▪ SHOCK WAVES AND EXPANSION WAVES

We discussed in Chap. 2 that sound waves are caused by infinitesimally small pressure disturbances, and they travel through a medium at the speed of sound. We have also seen in the present chapter that for some back pressure values, abrupt changes in fluid properties occur in a very thin section of a converging–diverging nozzle under supersonic flow conditions, creating a **shock wave**. It is of interest to study the conditions under which shock waves develop and how they affect the flow.

Normal Shocks

First we consider shock waves that occur in a plane normal to the direction of flow, called **normal shock waves**. The flow process through the shock wave is highly irreversible and *cannot* be approximated as being isentropic.

Next we follow the footsteps of Pierre Laplace (1749–1827), G. F. Bernhard Riemann (1826–1866), William Rankine (1820–1872), Pierre Henry Hugoniot (1851–1887), Lord Rayleigh (1842–1919), and G. I. Taylor (1886–1975) and develop relationships for the flow properties before and after the shock. We do this by applying the conservation of mass, momentum, and energy relations as well as some property relations to a stationary control volume that contains the shock, as shown in Fig. 12–23. The normal shock waves are extremely thin, so the entrance and exit flow areas for the control volume are approximately equal (Fig 12–24).

We assume steady flow with no heat and work interactions and no potential energy changes. Denoting the properties upstream of the shock by the subscript 1 and those downstream of the shock by 2, we have the following:

Conservation of mass: $\qquad \rho_1 A V_1 = \rho_2 A V_2 \qquad$ (12–29)

or

$$\rho_1 V_1 = \rho_2 V_2$$

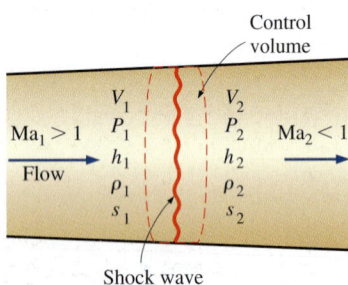

FIGURE 12–23
Control volume for flow across a normal shock wave.

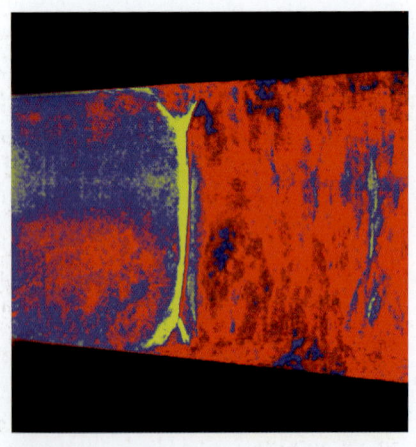

FIGURE 12–24
Schlieren image of a normal shock in a Laval nozzle. The Mach number in the nozzle just upstream (to the left) of the shock wave is about 1.3. Boundary layers distort the shape of the normal shock near the walls and lead to flow separation beneath the shock.
Photo by G. S. Settles, Penn State University. Used by permission.

Conservation of energy:
$$h_1 + \frac{V_1^2}{2} = h_2 + \frac{V_2^2}{2} \quad (12\text{–}30)$$

or

$$h_{01} = h_{02} \quad (12\text{–}31)$$

Linear momentum equation: Rearranging Eq. 12–14 and integrating yield

$$A(P_1 - P_2) = \dot{m}(V_2 - V_1) \quad (12\text{–}32)$$

Increase of entropy:
$$s_2 - s_1 \geq 0 \quad (12\text{–}33)$$

We can combine the conservation of mass and energy relations into a single equation and plot it on an *h-s* diagram, using property relations. The resultant curve is called the **Fanno line**, and it is the locus of states that have the same value of stagnation enthalpy and mass flux (mass flow per unit flow area). Likewise, combining the conservation of mass and momentum equations into a single equation and plotting it on the *h-s* diagram yield a curve called the **Rayleigh line.** Both these lines are shown on the *h-s* diagram in Fig. 12–25. As proved later in Example 12–7, the points of maximum entropy on these lines (points *a* and *b*) correspond to Ma = 1. The state on the upper part of each curve is subsonic and on the lower part supersonic.

The Fanno and Rayleigh lines intersect at two points (points 1 and 2), which represent the two states at which all three conservation equations are satisfied. One of these (state 1) corresponds to the state before the shock, and the other (state 2) corresponds to the state after the shock. Note that the flow is supersonic before the shock and subsonic afterward. Therefore the flow must change from supersonic to subsonic if a shock is to occur. The larger the Mach number before the shock, the stronger the shock will be. In the limiting case of Ma = 1, the shock wave simply becomes a sound wave. Notice from Fig. 12–25 that entropy increases, $s_2 > s_1$. This is expected since the flow through the shock is adiabatic but irreversible.

The conservation of energy principle (Eq. 12–31) requires that the stagnation enthalpy remain constant across the shock; $h_{01} = h_{02}$. For ideal gases $h = h(T)$, and thus

$$T_{01} = T_{02} \quad (12\text{–}34)$$

That is, the stagnation temperature of an ideal gas also remains constant across the shock. Note, however, that the stagnation pressure decreases across the shock because of the irreversibilities, while the ordinary (static) temperature rises drastically because of the conversion of kinetic energy into enthalpy due to a large drop in fluid velocity (see Fig. 12–26).

We now develop relations between various properties before and after the shock for an ideal gas with constant specific heats. A relation for the ratio of the static temperatures T_2/T_1 is obtained by applying Eq. 12–18 twice:

$$\frac{T_{01}}{T_1} = 1 + \left(\frac{k-1}{2}\right)\text{Ma}_1^2 \quad \text{and} \quad \frac{T_{02}}{T_2} = 1 + \left(\frac{k-1}{2}\right)\text{Ma}_2^2$$

Dividing the first equation by the second one and noting that $T_{01} = T_{02}$, we have

$$\frac{T_2}{T_1} = \frac{1 + \text{Ma}_1^2(k-1)/2}{1 + \text{Ma}_2^2(k-1)/2} \quad (12\text{–}35)$$

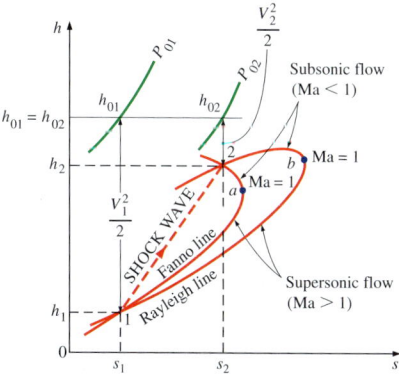

FIGURE 12–25
The *h-s* diagram for flow across a normal shock.

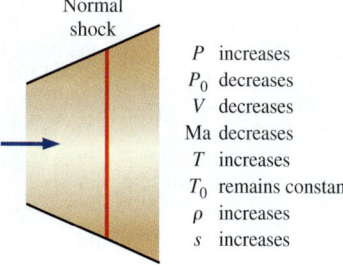

FIGURE 12–26
Variation of flow properties across a normal shock in an ideal gas.

From the ideal-gas equation of state,

$$\rho_1 = \frac{P_1}{RT_1} \quad \text{and} \quad \rho_2 = \frac{P_2}{RT_2}$$

Substituting these into the conservation of mass relation $\rho_1 V_1 = \rho_2 V_2$ and noting that $\text{Ma} = V/c$ and $c = \sqrt{kRT}$, we have

$$\frac{T_2}{T_1} = \frac{P_2 V_2}{P_1 V_1} = \frac{P_2 \text{Ma}_2 c_2}{P_1 \text{Ma}_1 c_1} = \frac{P_2 \text{Ma}_2 \sqrt{T_2}}{P_1 \text{Ma}_1 \sqrt{T_1}} = \left(\frac{P_2}{P_1}\right)^2 \left(\frac{\text{Ma}_2}{\text{Ma}_1}\right)^2 \quad (12\text{-}36)$$

Combining Eqs. 12–35 and 12–36 gives the pressure ratio across the shock:

Fanno line:
$$\frac{P_2}{P_1} = \frac{\text{Ma}_1 \sqrt{1 + \text{Ma}_1^2(k-1)/2}}{\text{Ma}_2 \sqrt{1 + \text{Ma}_2^2(k-1)/2}} \quad (12\text{-}37)$$

Equation 12–37 is a combination of the conservation of mass and energy equations; thus, it is also the equation of the Fanno line for an ideal gas with constant specific heats. A similar relation for the Rayleigh line is obtained by combining the conservation of mass and momentum equations. From Eq. 12–32,

$$P_1 - P_2 = \frac{\dot{m}}{A}(V_2 - V_1) = \rho_2 V_2^2 - \rho_1 V_1^2$$

However,

$$\rho V^2 = \left(\frac{P}{RT}\right)(\text{Ma } c)^2 = \left(\frac{P}{RT}\right)(\text{Ma}\sqrt{kRT})^2 = Pk\,\text{Ma}^2$$

Thus,

$$P_1(1 + k\text{Ma}_1^2) = P_2(1 + k\text{Ma}_2^2)$$

or

Rayleigh line:
$$\frac{P_2}{P_1} = \frac{1 + k\text{Ma}_1^2}{1 + k\text{Ma}_2^2} \quad (12\text{-}38)$$

Combining Eqs. 12–37 and 12–38 yields

$$\text{Ma}_2^2 = \frac{\text{Ma}_1^2 + 2/(k-1)}{2\text{Ma}_1^2 k/(k-1) - 1} \quad (12\text{-}39)$$

This represents the intersections of the Fanno and Rayleigh lines and relates the Mach number upstream of the shock to that downstream of the shock.

The occurrence of shock waves is not limited to supersonic nozzles only. This phenomenon is also observed at the engine inlet of supersonic aircraft, where the air passes through a shock and decelerates to subsonic velocities before entering the diffuser of the engine (Fig. 12–27). Explosions also produce powerful expanding spherical normal shocks, which can be very destructive (Fig. 12–28).

Various flow property ratios across the shock are listed in Table A–14 for an ideal gas with $k = 1.4$. Inspection of this table reveals that Ma_2 (the Mach

FIGURE 12–27
The air inlet of a supersonic fighter jet is designed such that a shock wave at the inlet decelerates the air to subsonic velocities, increasing the pressure and temperature of the air before it enters the engine.
© StockTrek/Getty RF

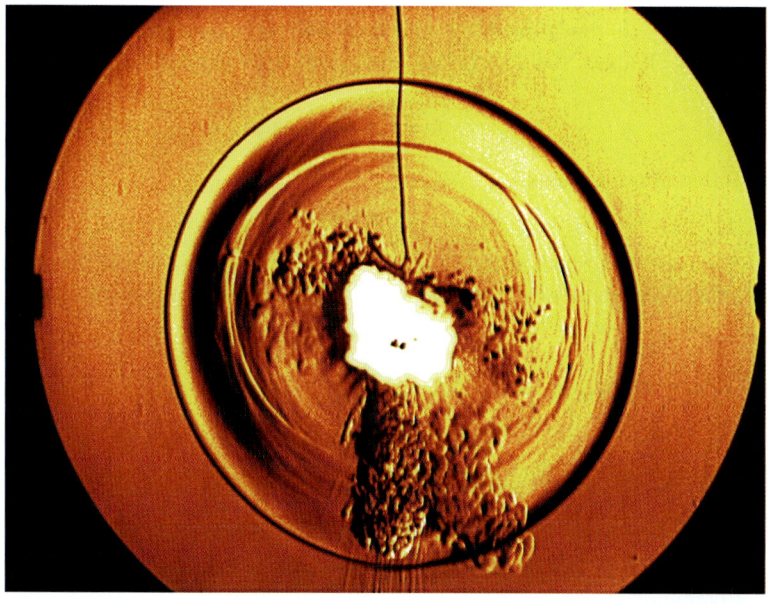

FIGURE 12–28
Schlieren image of the blast wave (expanding spherical normal shock) produced by the explosion of a firecracker. The shock expanded radially outward in all directions at a supersonic speed that decreased with radius from the center of the explosion.
A microphone sensed the sudden change in pressure of the passing shock wave and triggered the microsecond flashlamp that exposed the photograph.
StockTrek/Getty Images

number after the shock) is always less than 1 and that the larger the supersonic Mach number before the shock, the smaller the subsonic Mach number after the shock. Also, we see that the static pressure, temperature, and density all increase after the shock while the stagnation pressure decreases.

The entropy change across the shock is obtained by applying the entropy-change equation for an ideal gas across the shock:

$$s_2 - s_1 = c_p \ln \frac{T_2}{T_1} - R \ln \frac{P_2}{P_1} \quad (12\text{--}40)$$

which can be expressed in terms of k, R, and Ma_1 by using the relations developed earlier in this section. A plot of nondimensional entropy change across the normal shock $(s_2 - s_1)/R$ versus Ma_1 is shown in Fig. 12–29. Since the flow across the shock is adiabatic and irreversible, the second law of thermodynamics requires that the entropy increase across the shock wave. Thus, a shock wave cannot exist for values of Ma_1 less than unity where the entropy change would be negative. For adiabatic flows, shock waves can exist only for supersonic flows, $Ma_1 > 1$.

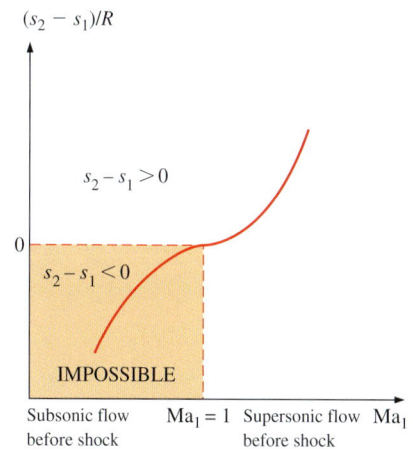

FIGURE 12–29
Entropy change across a normal shock.

EXAMPLE 12–7 The Point of Maximum Entropy on the Fanno Line

Show that the point of maximum entropy on the Fanno line (point a of Fig. 12–25) for the adiabatic steady flow of a fluid in a duct corresponds to the sonic velocity, $Ma = 1$.

SOLUTION It is to be shown that the point of maximum entropy on the Fanno line for steady adiabatic flow corresponds to sonic velocity.
Assumption The flow is steady, adiabatic, and one-dimensional.

Analysis In the absence of any heat and work interactions and potential energy changes, the steady-flow energy equation reduces to

$$h + \frac{V^2}{2} = \text{constant}$$

Differentiating yields

$$dh + V\,dV = 0$$

For a very thin shock with negligible change of duct area across the shock, the steady-flow continuity (conservation of mass) equation is expressed as

$$\rho V = \text{constant}$$

Differentiating, we have

$$\rho\,dV + V\,d\rho = 0$$

Solving for dV gives

$$dV = -V\frac{d\rho}{\rho}$$

Combining this with the energy equation, we have

$$dh - V^2 \frac{d\rho}{\rho} = 0$$

which is the equation for the Fanno line in differential form. At point a (the point of maximum entropy) $ds = 0$. Then from the second $T\,ds$ relation ($T\,ds = dh - v\,dP$) we have $dh = v\,dP = dP/\rho$. Substituting yields

$$\frac{dP}{\rho} - V^2 \frac{d\rho}{\rho} = 0 \quad \text{at } s = \text{constant}$$

Solving for V, we have

$$V = \left(\frac{\partial P}{\partial \rho}\right)_s^{1/2}$$

which is the relation for the speed of sound, Eq. 12–9. Thus $V = c$ and the proof is complete.

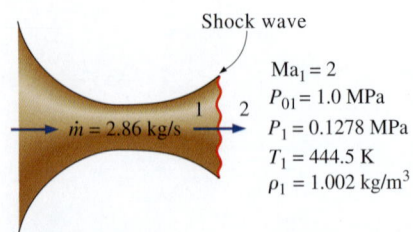

FIGURE 12–30
Schematic for Example 12–8.

EXAMPLE 12–8 Shock Wave in a Converging–Diverging Nozzle

If the air flowing through the converging–diverging nozzle of Example 12–6 experiences a normal shock wave at the nozzle exit plane (Fig. 12–30), determine the following after the shock: (*a*) the stagnation pressure, static pressure, static temperature, and static density; (*b*) the entropy change across the shock; (*c*) the exit velocity; and (*d*) the mass flow rate through the nozzle. Approximate the flow as steady, one-dimensional, and isentropic with $k = 1.4$ from the nozzle inlet to the shock location.

SOLUTION Air flowing through a converging–diverging nozzle experiences a normal shock at the exit. The effect of the shock wave on various properties is to be determined.

Assumptions **1** Air is an ideal gas with constant specific heats at room temperature. **2** Flow through the nozzle is steady, one-dimensional, and isentropic before the shock occurs. **3** The shock wave occurs at the exit plane.
Properties The constant-pressure specific heat and the specific heat ratio of air are $c_p = 1.005$ kJ/kg·K and $k = 1.4$. The gas constant of air is 0.287 kJ/kg·K.
Analysis (a) The fluid properties at the exit of the nozzle just before the shock (denoted by subscript 1) are those evaluated in Example 12–6 at the nozzle exit to be

$$P_{01} = 1.0 \text{ MPa} \quad P_1 = 0.1278 \text{ MPa} \quad T_1 = 444.5 \text{ K} \quad \rho_1 = 1.002 \text{ kg/m}^3$$

The fluid properties after the shock (denoted by subscript 2) are related to those before the shock through the functions listed in Table A–14. For $Ma_1 = 2.0$, we read

$$Ma_2 = 0.5774 \quad \frac{P_{02}}{P_{01}} = 0.7209 \quad \frac{P_2}{P_1} = 4.5000 \quad \frac{T_2}{T_1} = 1.6875 \quad \frac{\rho_2}{\rho_1} = 2.6667$$

Then the stagnation pressure P_{02}, static pressure P_2, static temperature T_2, and static density ρ_2 after the shock are

$$P_{02} = 0.7209 P_{01} = (0.7209)(1.0 \text{ MPa}) = \mathbf{0.721 \text{ MPa}}$$
$$P_2 = 4.5000 P_1 = (4.5000)(0.1278 \text{ MPa}) = \mathbf{0.575 \text{ MPa}}$$
$$T_2 = 1.6875 T_1 = (1.6875)(444.5 \text{ K}) = \mathbf{750 \text{ K}}$$
$$\rho_2 = 2.6667 \rho_1 = (2.6667)(1.002 \text{ kg/m}^3) = \mathbf{2.67 \text{ kg/m}^3}$$

(b) The entropy change across the shock is

$$s_2 - s_1 = c_p \ln \frac{T_2}{T_1} - R \ln \frac{P_2}{P_1}$$
$$= (1.005 \text{ kJ/kg·K}) \ln (1.6875) - (0.287 \text{ kJ/kg·K}) \ln (4.5000)$$
$$= \mathbf{0.0942 \text{ kJ/kg·K}}$$

Thus, the entropy of the air increases as it passes through a normal shock, which is highly irreversible.
(c) The air velocity after the shock is determined from $V_2 = Ma_2 c_2$, where c_2 is the speed of sound at the exit conditions after the shock:

$$V_2 = Ma_2 c_2 = Ma_2 \sqrt{kRT_2}$$
$$= (0.5774)\sqrt{(1.4)(0.287 \text{ kJ/kg·K})(750.1 \text{ K})\left(\frac{1000 \text{ m}^2/\text{s}^2}{1 \text{ kJ/kg}}\right)}$$
$$= \mathbf{317 \text{ m/s}}$$

(d) The mass flow rate through a converging–diverging nozzle with sonic conditions at the throat is not affected by the presence of shock waves in the nozzle. Therefore, the mass flow rate in this case is the same as that determined in Example 12–6:

$$\dot{m} = \mathbf{2.86 \text{ kg/s}}$$

Discussion This result can easily be verified by using property values at the nozzle exit after the shock at all Mach numbers significantly greater than unity.

FIGURE 12–31
When a lion tamer cracks his whip, a weak spherical shock wave forms near the tip and spreads out radially; the pressure inside the expanding shock wave is higher than ambient air pressure, and this is what causes the crack when the shock wave reaches the lion's ear.

© Joshua Ets-Hokin/Getty RF

Example 12–8 illustrates that the stagnation pressure and velocity decrease while the static pressure, temperature, density, and entropy increase across the shock (Fig. 12–31). The rise in the temperature of the fluid downstream of a shock wave is of major concern to the aerospace engineer because it creates heat transfer problems on the leading edges of wings and nose cones of space reentry vehicles and the recently proposed hypersonic space planes. Overheating, in fact, led to the tragic loss of the space shuttle *Columbia* in February of 2003 as it was reentering earth's atmosphere.

Oblique Shocks

Not all shock waves are normal shocks (perpendicular to the flow direction). For example, when the space shuttle travels at supersonic speeds through the atmosphere, it produces a complicated shock pattern consisting of inclined shock waves called **oblique shocks** (Fig. 12–32). As you can see, some portions of an oblique shock are curved, while other portions are straight.

First, we consider straight oblique shocks, like that produced when a uniform supersonic flow ($Ma_1 > 1$) impinges on a slender, two-dimensional wedge of half-angle δ (Fig. 12–33). Since information about the wedge cannot travel upstream in a supersonic flow, the fluid "knows" nothing about the wedge until it hits the nose. At that point, since the fluid cannot flow *through* the wedge, it turns suddenly through an angle called the **turning angle** or **deflection angle** θ. The result is a straight oblique shock wave, aligned at **shock angle** or **wave angle** β, measured relative to the oncoming flow (Fig. 12–34). To conserve mass, β must obviously be greater than δ. Since the Reynolds number of supersonic flows is typically large, the boundary layer growing along the wedge is very thin, and we ignore its effects. The flow therefore turns by the same angle as the wedge; namely, deflection angle θ is equal to wedge half-angle δ. If we take into account the displacement thickness effect of the boundary layer (Chap. 10), the deflection angle θ of the oblique shock turns out to be slightly greater than wedge half-angle δ.

FIGURE 12–32
Schlieren image of a small model of the space shuttle orbiter being tested at Mach 3 in the supersonic wind tunnel of the Penn State Gas Dynamics Lab. Several *oblique shocks* are seen in the air surrounding the spacecraft.

© Joshua Ets-Hokin/Getty Images

Like normal shocks, the Mach number decreases across an oblique shock, and oblique shocks are possible only if the upstream flow is supersonic. However, unlike normal shocks, in which the downstream Mach number is always subsonic, Ma_2 downstream of an oblique shock can be subsonic, sonic, or supersonic, depending on the upstream Mach number Ma_1 and the turning angle.

We analyze a straight oblique shock in Fig. 12–34 by decomposing the velocity vectors upstream and downstream of the shock into normal and tangential components, and considering a small control volume around the shock. Upstream of the shock, all fluid properties (velocity, density, pressure, etc.) along the lower left face of the control volume are identical to those along the upper right face. The same is true downstream of the shock. Therefore, the mass flow rates entering and leaving those two faces cancel each other out, and conservation of mass reduces to

$$\rho_1 V_{1,n} A = \rho_2 V_{2,n} A \rightarrow \rho_1 V_{1,n} = \rho_2 V_{2,n} \quad (12\text{–}41)$$

where A is the area of the control surface that is parallel to the shock. Since A is the same on either side of the shock, it has dropped out of Eq. 12–41.

As you might expect, the tangential component of velocity (parallel to the oblique shock) does not change across the shock, i.e., $V_{1,t} = V_{2,t}$. This is easily proven by applying the tangential momentum equation to the control volume.

When we apply conservation of momentum in the direction *normal* to the oblique shock, the only forces are pressure forces, and we get

$$P_1 A - P_2 A = \rho V_{2,n} A V_{2,n} - \rho V_{1,n} A V_{1,n} \rightarrow P_1 - P_2 = \rho_2 V_{2,n}^2 - \rho_1 V_{1,n}^2 \quad (12\text{–}42)$$

Finally, since there is no work done by the control volume and no heat transfer into or out of the control volume, stagnation enthalpy does *not* change across an oblique shock, and conservation of energy yields

$$h_{01} = h_{02} = h_0 \rightarrow h_1 + \tfrac{1}{2} V_{1,n}^2 + \tfrac{1}{2} V_{1,t}^2 = h_2 + \tfrac{1}{2} V_{2,n}^2 + \tfrac{1}{2} V_{2,t}^2$$

But since $V_{1,t} = V_{2,t}$, this equation reduces to

$$h_1 + \tfrac{1}{2} V_{1,n}^2 = h_2 + \tfrac{1}{2} V_{2,n}^2 \quad (12\text{–}43)$$

Careful comparison reveals that the equations for conservation of mass, momentum, and energy (Eqs. 12–41 through 12–43) across an oblique shock are identical to those across a normal shock, except that they are written in terms of the *normal* velocity component only. Therefore, the normal shock relations derived previously apply to oblique shocks as well, but must be written in terms of Mach numbers $Ma_{1,n}$ and $Ma_{2,n}$ normal to the oblique shock. This is most easily visualized by rotating the velocity vectors in Fig. 12–34 by angle $\pi/2 - \beta$, so that the oblique shock appears to be vertical (Fig. 12–35). Trigonometry yields

$$Ma_{1,n} = Ma_1 \sin \beta \quad \text{and} \quad Ma_{2,n} = Ma_2 \sin(\beta - \theta) \quad (12\text{–}44)$$

where $Ma_{1,n} = V_{1,n}/c_1$ and $Ma_{2,n} = V_{2,n}/c_2$. From the point of view shown in Fig. 12–35, we see what looks like a normal shock, but with some superposed tangential flow "coming along for the ride." Thus,

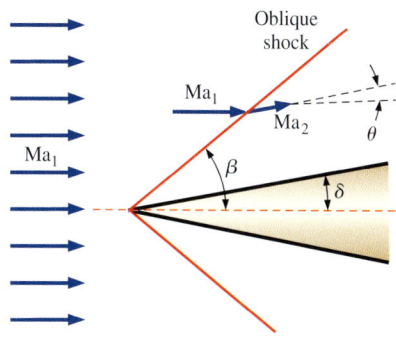

FIGURE 12–33
An oblique shock of *shock angle* β formed by a slender, two-dimensional wedge of half-angle δ. The flow is turned by *deflection angle* θ downstream of the shock, and the Mach number decreases.

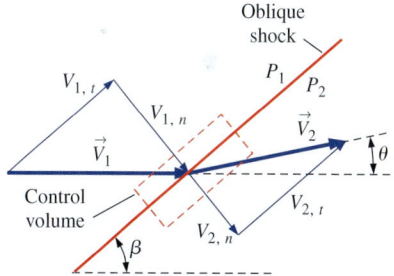

FIGURE 12–34
Velocity vectors through an oblique shock of shock angle β and deflection angle θ.

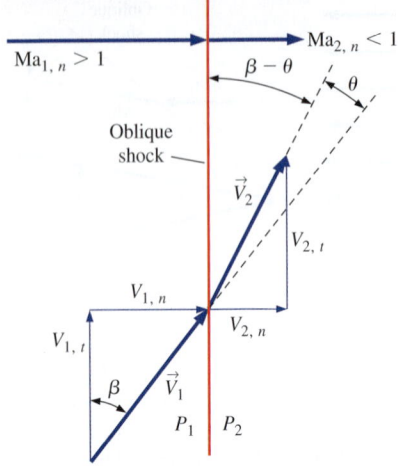

FIGURE 12–35
The same velocity vectors of Fig. 12–34, but rotated by angle $\pi/2 - \beta$, so that the oblique shock is vertical. Normal Mach numbers $Ma_{1,n}$ and $Ma_{2,n}$ are also defined.

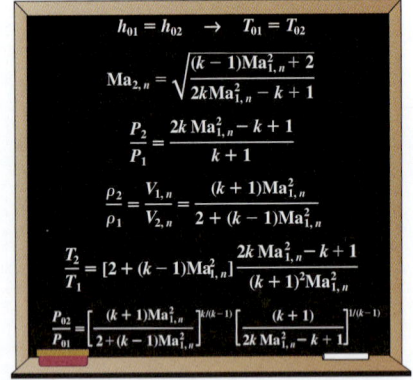

FIGURE 12–36
Relationships across an oblique shock for an ideal gas in terms of the normal component of upstream Mach number $Ma_{1,n}$.

All the equations, shock tables, etc., for normal shocks apply to oblique shocks as well, provided that we use only the **normal** components of the Mach number.

In fact, you may think of normal shocks as special oblique shocks in which shock angle $\beta = \pi/2$, or 90°. We recognize immediately that an oblique shock can exist only if $Ma_{1,n} > 1$ and $Ma_{2,n} < 1$. The normal shock equations appropriate for oblique shocks in an ideal gas are summarized in Fig. 12–36 in terms of $Ma_{1,n}$.

For known shock angle β and known upstream Mach number Ma_1, we use the first part of Eq. 12–44 to calculate $Ma_{1,n}$, and then use the normal shock tables (or their corresponding equations) to obtain $Ma_{2,n}$. If we also knew the deflection angle θ, we could calculate Ma_2 from the second part of Eq. 12–44. But, in a typical application, we know either β or θ, but not both. Fortunately, a bit more algebra provides us with a relationship between θ, β, and Ma_1. We begin by noting that $\tan \beta = V_{1,n}/V_{1,t}$ and $\tan(\beta - \theta) = V_{2,n}/V_{2,t}$ (Fig. 12–35). But since $V_{1,t} = V_{2,t}$, we combine these two expressions to yield

$$\frac{V_{2,n}}{V_{1,n}} = \frac{\tan(\beta - \theta)}{\tan \beta} = \frac{2 + (k-1)Ma_{1,n}^2}{(k+1)Ma_{1,n}^2} = \frac{2 + (k-1)Ma_1^2 \sin^2 \beta}{(k+1)Ma_1^2 \sin^2 \beta} \quad (12\text{–}45)$$

where we have also used Eq. 12–44 and the fourth equation of Fig. 12–36. We apply trigonometric identities for $\cos 2\beta$ and $\tan(\beta - \theta)$, namely,

$$\cos 2\beta = \cos^2 \beta - \sin^2 \beta \quad \text{and} \quad \tan(\beta - \theta) = \frac{\tan \beta - \tan \theta}{1 + \tan \beta \tan \theta}$$

After some algebra, Eq. 12–45 reduces to

The θ-β-Ma relationship: $\quad \tan \theta = \dfrac{2 \cot \beta (Ma_1^2 \sin^2 \beta - 1)}{Ma_1^2(k + \cos 2\beta) + 2} \quad$ (12–46)

Equation 12–46 provides deflection angle θ as a unique function of shock angle β, specific heat ratio k, and upstream Mach number Ma_1. For air ($k = 1.4$), we plot θ versus β for several values of Ma_1 in Fig. 12–37. We note that this plot is often presented with the axes reversed (β versus θ) in compressible flow textbooks, since, physically, shock angle β is determined by deflection angle θ.

Much can be learned by studying Fig. 12–37, and we list some observations here:

- Figure 12–37 displays the full range of possible shock waves at a given free-stream Mach number, from the weakest to the strongest. For any value of Mach number Ma_1 greater than 1, the possible values of θ range from $\theta = 0°$ at some value of β between 0 and 90°, to a maximum value $\theta = \theta_{max}$ at an intermediate value of β, and then back to $\theta = 0°$ at $\beta = 90°$. Straight oblique shocks for θ or β outside of this range *cannot* and *do not* exist. At $Ma_1 = 1.5$, for example, straight oblique shocks cannot exist in air with shock angle β less than about 42°, nor with deflection angle θ greater than about 12°. If the wedge half-angle is greater than θ_{max}, the shock becomes curved and detaches from the nose of the wedge, forming what is called a **detached oblique shock** or a **bow wave** (Fig. 12–38). The shock angle β of the detached shock is 90° at the nose, but β decreases

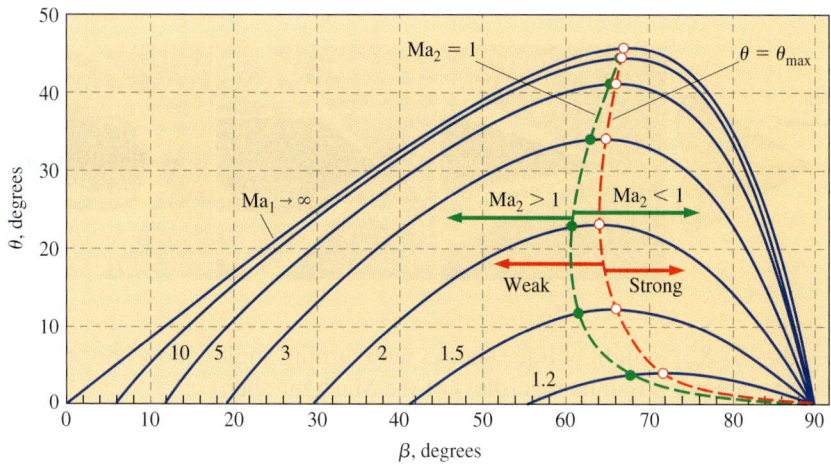

FIGURE 12–37
The dependence of straight oblique shock deflection angle θ on shock angle β for several values of upstream Mach number Ma_1. Calculations are for an ideal gas with $k = 1.4$. The dashed red line connects points of maximum deflection angle ($\theta = \theta_{max}$). *Weak oblique shocks* are to the left of this line, while *strong oblique shocks* are to the right of this line. The dashed green line connects points where the downstream Mach number is *sonic* ($Ma_2 = 1$). *Supersonic downstream flow* ($Ma_2 > 1$) is to the left of this line, while *subsonic downstream flow* ($Ma_2 < 1$) is to the right of this line.

as the shock curves downstream. Detached shocks are much more complicated than simple straight oblique shocks to analyze. In fact, no simple solutions exist, and prediction of detached shocks requires computational methods (Chap. 15).

- Similar oblique shock behavior is observed in *axisymmetric flow* over cones, as in Fig. 12–39, although the θ-β-Ma relationship for axisymmetric flows differs from that of Eq. 12–46.

- When supersonic flow impinges on a blunt (or bluff) body—a body *without* a sharply pointed nose, the wedge half-angle δ at the nose is 90°, and an attached oblique shock cannot exist, regardless of Mach number. In fact, a detached oblique shock occurs in front of *all* such blunt-nosed bodies, whether two-dimensional, axisymmetric, or fully three-dimensional. For example, a detached oblique shock is seen in front of the space shuttle model in Fig. 12–32 and in front of a sphere in Fig. 12–40.

- While θ is a unique function of Ma_1 and β for a given value of k, there are *two* possible values of β for $\theta < \theta_{max}$. The dashed red line in Fig. 12–37 passes through the locus of θ_{max} values, dividing the shocks into **weak oblique shocks** (the smaller value of β) and **strong oblique shocks** (the larger value of β). At a given value of θ, the weak shock is more common and is "preferred" by the flow unless the downstream pressure conditions are high enough for the formation of a strong shock.

- For a given upstream Mach number Ma_1, there is a unique value of θ for which the downstream Mach number Ma_2 is exactly 1. The dashed green line in Fig. 12–37 passes through the locus of values where $Ma_2 = 1$. To the left of this line, $Ma_2 > 1$, and to the right of this line, $Ma_2 < 1$. Downstream sonic conditions occur on the weak shock side of the plot, with θ very close to θ_{max}. Thus, the flow downstream of a strong oblique shock is *always subsonic* ($Ma_2 < 1$). The flow downstream of a weak oblique shock remains *supersonic*, except for a narrow range of θ just below θ_{max}, where it is subsonic, although it is still called a weak oblique shock.

- As the upstream Mach number approaches infinity, straight oblique shocks become possible for any β between 0 and 90°, but the maximum possible turning angle for $k = 1.4$ (air) is $\theta_{max} \cong 45.6°$, which occurs at

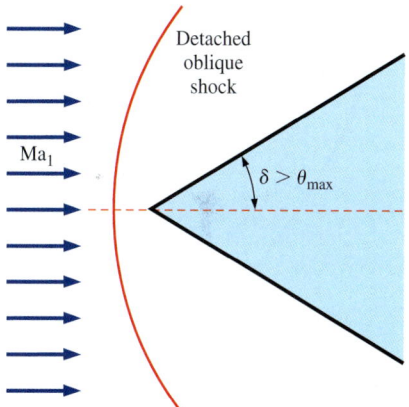

FIGURE 12–38
A *detached oblique shock* occurs upstream of a two-dimensional wedge of half-angle δ when δ is greater than the maximum possible deflection angle θ. A shock of this kind is called a *bow wave* because of its resemblance to the water wave that forms at the bow of a ship.

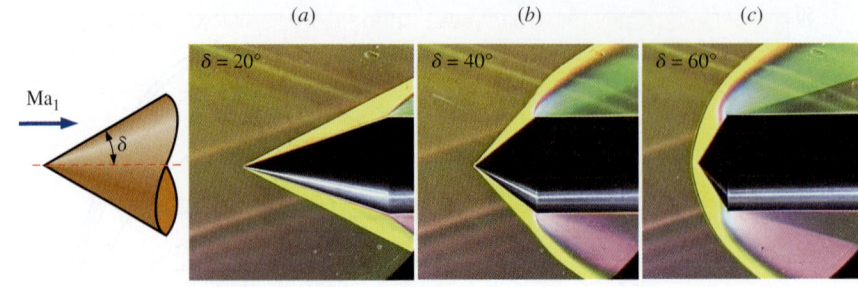

FIGURE 12–39
Still frames from schlieren videography illustrating the detachment of an oblique shock from a cone with increasing cone half-angle δ in air at Mach 3. At (a) δ = 20° and (b) δ = 40°, the oblique shock remains attached, but by (c) δ = 60°, the oblique shock has detached, forming a bow wave.

Photos by G. S. Settles, Penn State University. Used by permission.

$\beta = 67.8°$. Straight oblique shocks with turning angles above this value of θ_{max} are not possible, regardless of the Mach number.

- For a given value of upstream Mach number, there are two shock angles where there is *no turning of the flow* ($\theta = 0°$): the strong case, $\beta = 90°$, corresponds to a *normal shock*, and the weak case, $\beta = \beta_{min}$, represents the weakest possible oblique shock at that Mach number, which is called a **Mach wave**. Mach waves are caused, for example, by very small non-uniformities on the walls of a supersonic wind tunnel (several can be seen in Figs. 12–32 and 12–39). Mach waves have no effect on the flow, since the shock is vanishingly weak. In fact, in the limit, Mach waves are *isentropic*. The shock angle for Mach waves is a unique function of the Mach number and is given the symbol μ, not to be confused with the coefficient of viscosity. Angle μ is called the **Mach angle** and is found by setting θ equal to zero in Eq. 12–46, solving for $\beta = \mu$, and taking the smaller root. We get

Mach angle: $$\mu = \sin^{-1}(1/Ma_1) \tag{12-47}$$

Since the specific heat ratio appears only in the denominator of Eq. 12–46, μ is independent of k. Thus, we can estimate the Mach number of any supersonic flow simply by measuring the Mach angle and applying Eq. 12–47.

Prandtl–Meyer Expansion Waves

We now address situations where supersonic flow is turned in the *opposite* direction, such as in the upper portion of a two-dimensional wedge at an angle of attack greater than its half-angle δ (Fig. 12–41). We refer to this type of flow as an **expanding flow**, whereas a flow that produces an oblique shock may be called a **compressing flow**. As previously, the flow changes direction to conserve mass. However, unlike a compressing flow, an expanding flow does *not* result in a shock wave. Rather, a continuous expanding region called an **expansion fan** appears, composed of an infinite number of Mach waves called **Prandtl–Meyer expansion waves**. In other words, the flow does not turn suddenly, as through a shock, but *gradually*—each successive Mach wave turns the flow by an infinitesimal amount. Since each individual expansion wave is nearly isentropic, the flow across the entire expansion fan is also nearly isentropic. The Mach number downstream of the expansion *increases* ($Ma_2 > Ma_1$), while pressure, density, and temperature *decrease*, just as they do in the supersonic (expanding) portion of a converging–diverging nozzle.

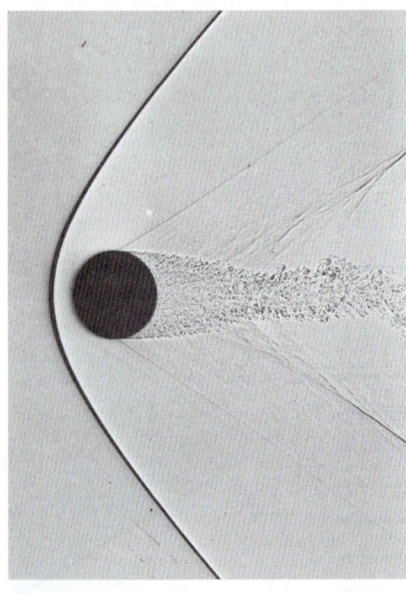

FIGURE 12–40
Shadowgram of a 12.7-mm-diameter sphere in free flight through air at Ma = 1.53. The flow is subsonic behind the part of the bow wave that is ahead of the sphere and over its surface back to about 45°. At about 90° the laminar boundary layer separates through an oblique shock wave and quickly becomes turbulent. The fluctuating wake generates a system of weak disturbances that merge into the second "recompression" shock wave.

Photo by A. C. Charters, as found in Van Dyke (1982).

Prandtl–Meyer expansion waves are inclined at the local Mach angle μ, as sketched in Fig. 12–41. The Mach angle of the first expansion wave is easily determined as $\mu_1 = \sin^{-1}(1/\text{Ma}_1)$. Similarly, $\mu_2 = \sin^{-1}(1/\text{Ma}_2)$, where we must be careful to measure the angle relative to the *new* direction of flow downstream of the expansion, namely, parallel to the upper wall of the wedge in Fig. 12–41 if we neglect the influence of the boundary layer along the wall. But how do we determine Ma_2? It turns out that the turning angle θ across the expansion fan can be calculated by integration, making use of the isentropic flow relationships. For an ideal gas, the result is (Anderson, 2003),

Turning angle across an expansion fan: $\quad \theta = \nu(\text{Ma}_2) - \nu(\text{Ma}_1) \quad$ (12–48)

where $\nu(\text{Ma})$ is an angle called the **Prandtl–Meyer function** (not to be confused with the kinematic viscosity),

$$\nu(\text{Ma}) = \sqrt{\frac{k+1}{k-1}} \tan^{-1}\left(\sqrt{\frac{k-1}{k+1}(\text{Ma}^2-1)}\right) - \tan^{-1}\left(\sqrt{\text{Ma}^2-1}\right) \quad (12\text{–}49)$$

Note that $\nu(\text{Ma})$ is an angle, and can be calculated in either degrees or radians. Physically, $\nu(\text{Ma})$ is the angle through which the flow must expand, starting with $\nu = 0$ at $\text{Ma} = 1$, in order to reach a supersonic Mach number, $\text{Ma} > 1$.

To find Ma_2 for known values of Ma_1, k, and θ, we calculate $\nu(\text{Ma}_1)$ from Eq. 12–49, $\nu(\text{Ma}_2)$ from Eq. 12–48, and then Ma_2 from Eq. 12–49, noting that the last step involves solving an implicit equation for Ma_2. Since there is no heat transfer or work, and the flow can be approximated as isentropic through the expansion, T_0 and P_0 remain constant, and we use the isentropic flow relations derived previously to calculate other flow properties downstream of the expansion, such as T_2, ρ_2, and P_2.

Prandtl–Meyer expansion fans also occur in axisymmetric supersonic flows, as in the corners and trailing edges of a cone-cylinder (Fig. 12–42). Some very complex and, to some of us, beautiful interactions involving both

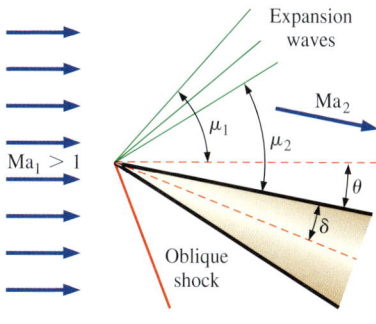

FIGURE 12–41

An expansion fan in the upper portion of the flow formed by a two-dimensional wedge at an angle of attack in a supersonic flow. The flow is turned by angle θ, and the Mach number increases across the expansion fan. Mach angles upstream and downstream of the expansion fan are indicated. Only three expansion waves are shown for simplicity, but in fact, there are an infinite number of them. (An oblique shock is also present in the bottom portion of this flow.)

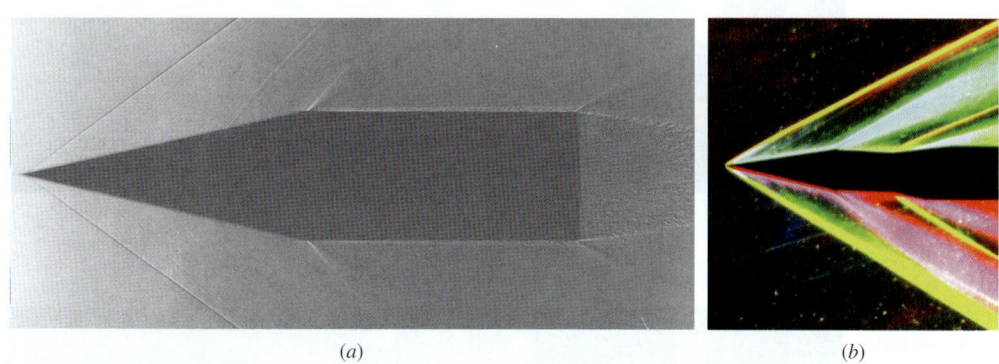

(a) (b)

FIGURE 12–42

(a) A cone-cylinder of 12.5° half-angle in a Mach number 1.84 flow. The boundary layer becomes turbulent shortly downstream of the nose, generating Mach waves that are visible in this shadowgraph. Expansion waves are seen at the corners and at the trailing edge of the cone. (b) A similar pattern for Mach 3 flow over an 11° 2-D wedge.

(a) Photo by A. C. Charters, as found in Van Dyke (1982). (b) Photo by G. S. Settles, Penn State University. Used by permission.

shock waves and expansion waves occur in the supersonic jet produced by an "overexpanded" nozzle, as in Fig. 12–43. When such patterns are visible in the exhaust of a jet engine, pilots refer to it as a "tiger tail." Analysis of such flows is beyond the scope of the present text; interested readers are referred to compressible flow textbooks such as Thompson (1972), Leipmann and Roshko (2001), and Anderson (2003).

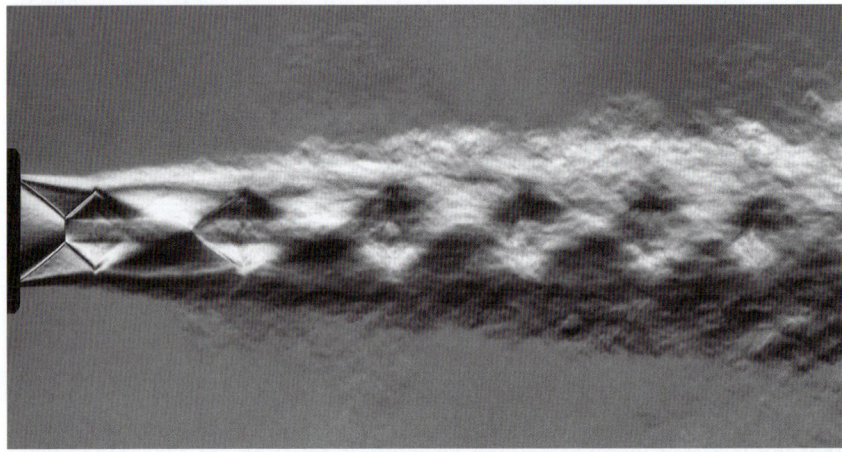

(a)

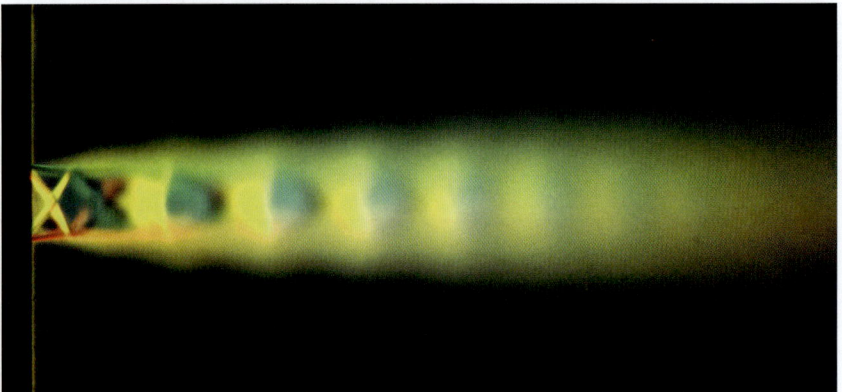

(b)

(c)

FIGURE 12–43

The complex interactions between shock waves and expansion waves in an "overexpanded" supersonic jet. (a) The flow is visualized by a schlieren-like differential interferogram. (b) Color shlieren image. (c) Tiger tail shock pattern.

(a) Photo by H. Oertel sen. Reproduced by courtesy of the French-German Research Institute of Saint-Louis, ISL. Used with permission. (b) Photo by G. S. Settles, Penn State University. Used by permission. (c) Photo courtesy of Joint Strike Fighter Program, Department of Defense.

EXAMPLE 12–9   Estimation of the Mach Number from Mach Lines

Estimate the Mach number of the free-stream flow upstream of the space shuttle in Fig. 12–32 from the figure alone. Compare with the known value of Mach number provided in the figure caption.

SOLUTION We are to estimate the Mach number from a figure and compare it to the known value.
Analysis Using a protractor, we measure the angle of the Mach lines in the free-stream flow: $\mu \cong 19°$. The Mach number is obtained from Eq. 12–47,

$$\mu = \sin^{-1}\left(\frac{1}{Ma_1}\right) \rightarrow Ma_1 = \frac{1}{\sin 19°} \rightarrow Ma_1 = 3.07$$

Our estimated Mach number agrees with the experimental value of 3.0 ± 0.1.
Discussion The result is independent of the fluid properties.

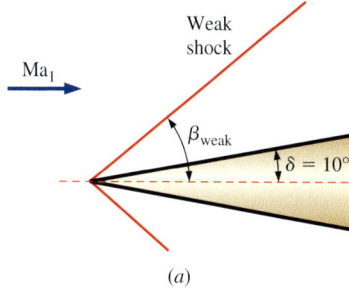

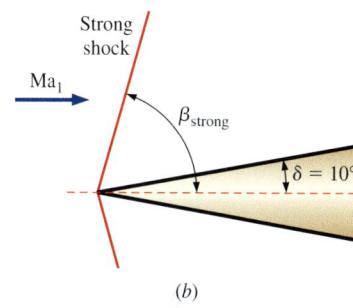

FIGURE 12–44
Two possible oblique shock angles, (a) β_{weak} and (b) β_{strong}, formed by a two-dimensional wedge of half-angle $\delta = 10°$.

EXAMPLE 12–10   Oblique Shock Calculations

Supersonic air at $Ma_1 = 2.0$ and 75.0 kPa impinges on a two-dimensional wedge of half-angle $\delta = 10°$ (Fig. 12–44). Calculate the two possible oblique shock angles, β_{weak} and β_{strong}, that could be formed by this wedge. For each case, calculate the pressure and Mach number downstream of the oblique shock, compare, and discuss.

SOLUTION We are to calculate the shock angle, Mach number, and pressure downstream of the weak and strong oblique shock formed by a two-dimensional wedge.
Assumptions **1** The flow is steady. **2** The boundary layer on the wedge is very thin.
Properties The fluid is air with $k = 1.4$.
Analysis Because of assumption 2, we approximate the oblique shock deflection angle as equal to the wedge half-angle, i.e., $\theta \cong \delta = 10°$. With $Ma_1 = 2.0$ and $\theta = 10°$, we solve Eq. 12–46 for the two possible values of oblique shock angle β: $\beta_{weak} = \mathbf{39.3°}$ and $\beta_{strong} = \mathbf{83.7°}$. From these values, we use the first part of Eq. 12–44 to calculate upstream normal Mach number $Ma_{1,n}$,

Weak shock: $\quad Ma_{1,n} = Ma_1 \sin \beta \rightarrow Ma_{1,n} = 2.0 \sin 39.3° = 1.267$

and

Strong shock: $\quad Ma_{1,n} = Ma_1 \sin \beta \rightarrow Ma_{1,n} = 2.0 \sin 83.7° = 1.988$

We substitute these values of $Ma_{1,n}$ into the second equation of Fig. 12–36 to calculate the downstream normal Mach number $Ma_{2,n}$. For the weak shock, $Ma_{2,n} = 0.8032$, and for the strong shock, $Ma_{2,n} = 0.5794$. We also calculate the downstream pressure for each case, using the third equation of Fig. 12–36, which gives

Weak shock:

$$\frac{P_2}{P_1} = \frac{2k\,\text{Ma}_{1,n}^2 - k + 1}{k + 1} \rightarrow P_2 = (75.0\text{ kPa})\frac{2(1.4)(1.267)^2 - 1.4 + 1}{1.4 + 1} = \mathbf{128\text{ kPa}}$$

and

Strong shock:

$$\frac{P_2}{P_1} = \frac{2k\,\text{Ma}_{1,n}^2 - k + 1}{k + 1} \rightarrow P_2 = (75.0\text{ kPa})\frac{2(1.4)(1.988)^2 - 1.4 + 1}{1.4 + 1} = \mathbf{333\text{ kPa}}$$

Finally, we use the second part of Eq. 12–44 to calculate the downstream Mach number,

Weak shock: $$\text{Ma}_2 = \frac{\text{Ma}_{2,n}}{\sin(\beta - \theta)} = \frac{0.8032}{\sin(39.3° - 10°)} = \mathbf{1.64}$$

and

Strong shock: $$\text{Ma}_2 = \frac{\text{Ma}_{2,n}}{\sin(\beta - \theta)} = \frac{0.5794}{\sin(83.7° - 10°)} = \mathbf{0.604}$$

The changes in Mach number and pressure across the strong shock are much greater than the changes across the weak shock, as expected.

Discussion Since Eq. 12–46 is implicit in β, we solve it by an iterative approach or with an equation solver such as EES. For both the weak and strong oblique shock cases, $\text{Ma}_{1,n}$ is supersonic and $\text{Ma}_{2,n}$ is subsonic. However, Ma_2 is *supersonic* across the weak oblique shock, but *subsonic* across the strong oblique shock. We could also use the normal shock tables in place of the equations, but with loss of precision.

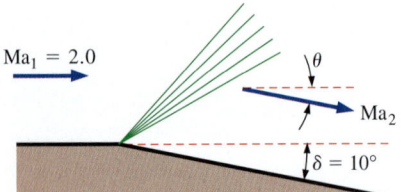

FIGURE 12–45
An expansion fan caused by the sudden expansion of a wall with $\delta = 10°$.

EXAMPLE 12–11   Prandtl–Meyer Expansion Wave Calculations

Supersonic air at $\text{Ma}_1 = 2.0$ and 230 kPa flows parallel to a flat wall that suddenly expands by $\delta = 10°$ (Fig. 12–45). Ignoring any effects caused by the boundary layer along the wall, calculate downstream Mach number Ma_2 and pressure P_2.

SOLUTION We are to calculate the Mach number and pressure downstream of a sudden expansion along a wall.
Assumptions 1 The flow is steady. 2 The boundary layer on the wall is very thin.
Properties The fluid is air with $k = 1.4$.
Analysis Because of assumption 2, we approximate the total deflection angle as equal to the wall expansion angle, i.e., $\theta \cong \delta = 10°$. With $\text{Ma}_1 = 2.0$, we solve Eq. 12–49 for the upstream Prandtl–Meyer function,

$$\nu(\text{Ma}) = \sqrt{\frac{k+1}{k-1}}\,\tan^{-1}\!\left(\sqrt{\frac{k-1}{k+1}(\text{Ma}^2 - 1)}\right) - \tan^{-1}\!\left(\sqrt{\text{Ma}^2 - 1}\right)$$

$$= \sqrt{\frac{1.4+1}{1.4-1}}\,\tan^{-1}\!\left(\sqrt{\frac{1.4-1}{1.4+1}(2.0^2 - 1)}\right) - \tan^{-1}\!\left(\sqrt{2.0^2 - 1}\right) = 26.38°$$

Next, we use Eq. 12–48 to calculate the downstream Prandtl–Meyer function,

$$\theta = \nu(Ma_2) - \nu(Ma_1) \rightarrow \nu(Ma_2) = \theta + \nu(Ma_1) = 10° + 26.38° = 36.38°$$

Ma_2 is found by solving Eq. 12–49, which is implicit—an equation solver is helpful. We get Ma_2 = **2.38.** There are also compressible flow calculators on the Internet that solve these implicit equations, along with both normal and oblique shock equations; e.g., see www.aoe.vt.edu/~devenpor/aoe3114/calc.html.

We use the isentropic relations to calculate the downstream pressure,

$$P_2 = \frac{P_2/P_0}{P_1/P_0} P_1 = \frac{\left[1 + \left(\frac{k-1}{2}\right)Ma_2^2\right]^{-k/(k-1)}}{\left[1 + \left(\frac{k-1}{2}\right)Ma_1^2\right]^{-k/(k-1)}} (230 \text{ kPa}) = \textbf{126 kPa}$$

Since this is an expansion, Mach number increases and pressure decreases, as expected.

Discussion We could also solve for downstream temperature, density, etc., using the appropriate isentropic relations.

12–5 ▪ DUCT FLOW WITH HEAT TRANSFER AND NEGLIGIBLE FRICTION (RAYLEIGH FLOW)

So far we have limited our consideration mostly to *isentropic flow*, also called *reversible adiabatic flow* since it involves no heat transfer and no irreversibilities such as friction. Many compressible flow problems encountered in practice involve chemical reactions such as combustion, nuclear reactions, evaporation, and condensation as well as heat gain or heat loss through the duct wall. Such problems are difficult to analyze exactly since they may involve significant changes in chemical composition during flow, and the conversion of latent, chemical, and nuclear energies to thermal energy (Fig. 12–46).

The essential features of such complex flows can still be captured by a simple analysis by modeling the generation or absorption of thermal energy as heat transfer through the duct wall at the same rate and disregarding any changes in chemical composition. This simplified problem is still too complicated for an elementary treatment of the topic since the flow may involve friction, variations in duct area, and multidimensional effects. In this section, we limit our consideration to one-dimensional flow in a duct of constant cross-sectional area with negligible frictional effects.

Consider steady one-dimensional flow of an ideal gas with constant specific heats through a constant-area duct with heat transfer, but with negligible friction. Such flows are referred to as **Rayleigh flows** after Lord Rayleigh (1842–1919). The conservation of mass, momentum, and

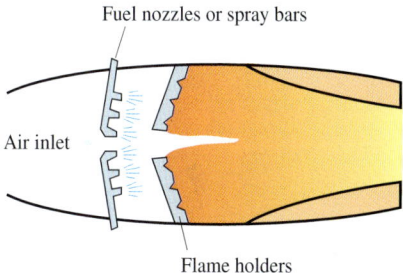

FIGURE 12–46
Many practical compressible flow problems involve combustion, which may be modeled as heat gain through the duct wall.

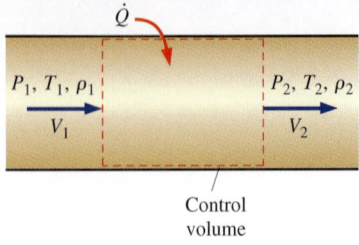

FIGURE 12–47
Control volume for flow in a constant-area duct with heat transfer and negligible friction.

energy equations for the control volume shown in Fig. 12–47 are written as follows:

Continuity equation Noting that the duct cross-sectional area A is constant, the relation $\dot{m}_1 = \dot{m}_2$ or $\rho_1 A_1 V_1 = \rho_2 A_2 V_2$ reduces to

$$\rho_1 V_1 = \rho_2 V_2 \quad (12\text{--}50)$$

***x*-Momentum equation** Noting that the frictional effects are negligible and thus there are no shear forces, and assuming there are no external and body forces, the momentum equation $\sum \vec{F} = \sum_{\text{out}} \beta \dot{m} \vec{V} - \sum_{\text{in}} \beta \dot{m} \vec{V}$ in the flow (or *x*-) direction becomes a balance between static pressure forces and momentum transfer. Noting that the flows are high speed and turbulent and we are ignoring friction, the momentum flux correction factor is approximately 1 ($\beta \cong 1$) and thus can be neglected. Then,

$$P_1 A_1 - P_2 A_2 = \dot{m} V_2 - \dot{m} V_1 \rightarrow P_1 - P_2 = (\rho_2 V_2) V_2 - (\rho_1 V_1) V_1$$

or

$$P_1 + \rho_1 V_1^2 = P_2 + \rho_2 V_2^2 \quad (12\text{--}51)$$

Energy equation The control volume involves no shear, shaft, or other forms of work, and the potential energy change is negligible. If the rate of heat transfer is $\dot{Q}$ and the heat transfer per unit mass of fluid is $q = \dot{Q}/\dot{m}$, the steady-flow energy balance $\dot{E}_{\text{in}} = \dot{E}_{\text{out}}$ becomes

$$\dot{Q} + \dot{m}\left(h_1 + \frac{V_1^2}{2}\right) = \dot{m}\left(h_2 + \frac{V_2^2}{2}\right) \rightarrow q + h_1 + \frac{V_1^2}{2} = h_2 + \frac{V_2^2}{2} \quad (12\text{--}52)$$

For an ideal gas with constant specific heats, $\Delta h = c_p \Delta T$, and thus

$$q = c_p(T_2 - T_1) + \frac{V_2^2 - V_1^2}{2} \quad (12\text{--}53)$$

or

$$q = h_{02} - h_{01} = c_p(T_{02} - T_{01}) \quad (12\text{--}54)$$

Therefore, the stagnation enthalpy h_0 and stagnation temperature T_0 change during Rayleigh flow (both increase when heat is transferred to the fluid and thus q is positive, and both decrease when heat is transferred from the fluid and thus q is negative).

Entropy change In the absence of any irreversibilities such as friction, the entropy of a system changes by heat transfer only: it increases with heat gain, and decreases with heat loss. Entropy is a property and thus a state function, and the entropy change of an ideal gas with constant specific heats during a change of state from 1 to 2 is given by

$$s_2 - s_1 = c_p \ln \frac{T_2}{T_1} - R \ln \frac{P_2}{P_1} \quad (12\text{--}55)$$

The entropy of a fluid may increase or decrease during Rayleigh flow, depending on the direction of heat transfer.

Equation of state Noting that $P = \rho RT$, the properties P, ρ, and T of an ideal gas at states 1 and 2 are related to each other by

$$\frac{P_1}{\rho_1 T_1} = \frac{P_2}{\rho_2 T_2} \qquad (12\text{-}56)$$

Consider a gas with known properties R, k, and c_p. For a specified inlet state 1, the inlet properties P_1, T_1, ρ_1, V_1, and s_1 are known. The five exit properties P_2, T_2, ρ_2, V_2, and s_2 can be determined from Equations 12–50, 12–51, 12–53, 12–55, and 12–56 for any specified value of heat transfer q. When the velocity and temperature are known, the Mach number can be determined from $\text{Ma} = V/c = V/\sqrt{kRT}$.

Obviously there is an infinite number of possible downstream states 2 corresponding to a given upstream state 1. A practical way of determining these downstream states is to assume various values of T_2, and calculate all other properties as well as the heat transfer q for each assumed T_2 from Eqs. 12–50 through 12–56. Plotting the results on a T-s diagram gives a curve passing through the specified inlet state, as shown in Fig. 12–48. The plot of Rayleigh flow on a T-s diagram is called the **Rayleigh line,** and several important observations can be made from this plot and the results of the calculations:

1. All the states that satisfy the conservation of mass, momentum, and energy equations as well as the property relations are on the Rayleigh line. Therefore, for a given initial state, the fluid cannot exist at any downstream state outside the Rayleigh line on a T-s diagram. In fact, the Rayleigh line is the locus of all physically attainable downstream states corresponding to an initial state.
2. Entropy increases with heat gain, and thus we proceed to the right on the Rayleigh line as heat is transferred to the fluid. The Mach number is $\text{Ma} = 1$ at point a, which is the point of maximum entropy (see Example 12–12 for proof). The states on the upper arm of the Rayleigh line above point a are subsonic, and the states on the lower arm below point a are supersonic. Therefore, a process proceeds to the right on the Rayleigh line with heat addition and to the left with heat rejection regardless of the initial value of the Mach number.
3. Heating increases the Mach number for subsonic flow, but decreases it for supersonic flow. The flow Mach number approaches $\text{Ma} = 1$ in both cases (from 0 in subsonic flow and from ∞ in supersonic flow) during heating.
4. It is clear from the energy balance $q = c_p(T_{02} - T_{01})$ that heating increases the stagnation temperature T_0 for both subsonic and supersonic flows, and cooling decreases it. (The maximum value of T_0 occurs at $\text{Ma} = 1$.) This is also the case for the static temperature T except for the narrow Mach number range of $1/\sqrt{k} < \text{Ma} < 1$ in subsonic flow (see Example 12–12). Both temperature and the Mach

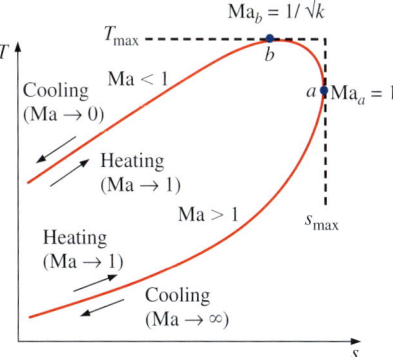

FIGURE 12–48
T-s diagram for flow in a constant-area duct with heat transfer and negligible friction (Rayleigh flow).

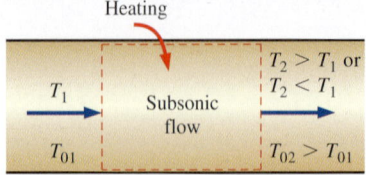

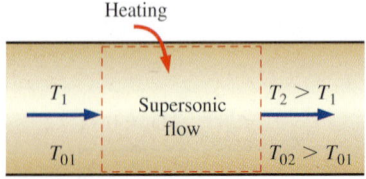

FIGURE 12–49
During heating, fluid temperature always increases if the Rayleigh flow is supersonic, but the temperature may actually drop if the flow is subsonic.

number increase with heating in subsonic flow, but T reaches a maximum T_{max} at Ma = $1/\sqrt{k}$ (which is 0.845 for air), and then decreases. It may seem peculiar that the temperature of a fluid drops as heat is transferred to it. But this is no more peculiar than the fluid velocity increasing in the diverging section of a converging–diverging nozzle. The cooling effect in this region is due to the large increase in the fluid velocity and the accompanying drop in temperature in accordance with the relation $T_0 = T + V^2/2c_p$. Note also that heat rejection in the region $1/\sqrt{k} <$ Ma < 1 causes the fluid temperature to increase (Fig. 12–49).

5. The momentum equation $P + KV =$ constant, where $K = \rho V =$ constant (from the continuity equation), reveals that velocity and static pressure have opposite trends. Therefore, static pressure decreases with heat gain in subsonic flow (since velocity and the Mach number increase), but increases with heat gain in supersonic flow (since velocity and the Mach number decrease).

6. The continuity equation $\rho V =$ constant indicates that density and velocity are inversely proportional. Therefore, density decreases with heat transfer to the fluid in subsonic flow (since velocity and the Mach number increase), but increases with heat gain in supersonic flow (since velocity and the Mach number decrease).

7. On the left half of Fig. 12–48, the lower arm of the Rayleigh line is steeper than the upper arm (in terms of s as a function of T), which indicates that the entropy change corresponding to a specified temperature change (and thus a given amount of heat transfer) is larger in supersonic flow.

The effects of heating and cooling on the properties of Rayleigh flow are listed in Table 12–3. Note that heating or cooling has opposite effects on most properties. Also, the stagnation pressure decreases during heating and increases during cooling regardless of whether the flow is subsonic or supersonic.

TABLE 12–3

The effects of heating and cooling on the properties of Rayleigh flow

Property	Heating Subsonic	Heating Supersonic	Cooling Subsonic	Cooling Supersonic
Velocity, V	Increase	Decrease	Decrease	Increase
Mach number, Ma	Increase	Decrease	Decrease	Increase
Stagnation temperature, T_0	Increase	Increase	Decrease	Decrease
Temperature, T	Increase for Ma $< 1/k^{1/2}$ Decrease for Ma $> 1/k^{1/2}$	Increase	Decrease for Ma $< 1/k^{1/2}$ Increase for Ma $> 1/k^{1/2}$	Decrease
Density, ρ	Decrease	Increase	Increase	Decrease
Stagnation pressure, P_0	Decrease	Decrease	Increase	Increase
Pressure, P	Decrease	Increase	Increase	Decrease
Entropy, s	Increase	Increase	Decrease	Decrease

EXAMPLE 12–12 Extrema of Rayleigh Line

Consider the *T-s* diagram of Rayleigh flow, as shown in Fig. 12–50. Using the differential forms of the conservation equations and property relations, show that the Mach number is $Ma_a = 1$ at the point of maximum entropy (point *a*), and $Ma_b = 1/\sqrt{k}$ at the point of maximum temperature (point *b*).

SOLUTION It is to be shown that $Ma_a = 1$ at the point of maximum entropy and $Ma_b = 1/\sqrt{k}$ at the point of maximum temperature on the Rayleigh line.
Assumptions The assumptions associated with Rayleigh flow (i.e., steady one-dimensional flow of an ideal gas with constant properties through a constant cross-sectional area duct with negligible frictional effects) are valid.
Analysis The differential forms of the continuity (ρV = constant), momentum [rearranged as $P + (\rho V)V$ = constant], ideal gas ($P = \rho R T$), and enthalpy change ($\Delta h = c_p \Delta T$) equations are expressed as

$$\rho V = \text{constant} \;\rightarrow\; \rho\, dV + V\, d\rho = 0 \;\rightarrow\; \frac{d\rho}{\rho} = -\frac{dV}{V} \quad (1)$$

$$P + (\rho V)V = \text{constant} \;\rightarrow\; dP + (\rho V)\, dV = 0 \;\rightarrow\; \frac{dP}{dV} = -\rho V \quad (2)$$

$$P = \rho R T \;\rightarrow\; dP = \rho R\, dT + RT\, d\rho \;\rightarrow\; \frac{dP}{P} = \frac{dT}{T} + \frac{d\rho}{\rho} \quad (3)$$

The differential form of the entropy change relation (Eq. 12–40) of an ideal gas with constant specific heats is

$$ds = c_p \frac{dT}{T} - R \frac{dP}{P} \quad (4)$$

Substituting Eq. 3 into Eq. 4 gives

$$ds = c_p \frac{dT}{T} - R\left(\frac{dT}{T} + \frac{d\rho}{\rho}\right) = (c_p - R)\frac{dT}{T} - R\frac{d\rho}{\rho} = \frac{R}{k-1}\frac{dT}{T} - R\frac{d\rho}{\rho} \quad (5)$$

since

$$c_p - R = c_v \;\rightarrow\; kc_v - R = c_v \;\rightarrow\; c_v = R/(k-1)$$

Dividing both sides of Eq. 5 by dT and combining with Eq. 1,

$$\frac{ds}{dT} = \frac{R}{T(k-1)} + \frac{R}{V}\frac{dV}{dT} \quad (6)$$

Dividing Eq. 3 by dV and combining it with Eqs. 1 and 2 give, after rearranging,

$$\frac{dT}{dV} = \frac{T}{V} - \frac{V}{R} \quad (7)$$

Substituting Eq. 7 into Eq. 6 and rearranging,

$$\frac{ds}{dT} = \frac{R}{T(k-1)} + \frac{R}{T - V^2/R} = \frac{R(kRT - V^2)}{T(k-1)(RT - V^2)} \quad (8)$$

Setting $ds/dT = 0$ and solving the resulting equation $R(kRT - V^2) = 0$ for V give the velocity at point *a* to be

$$V_a = \sqrt{kRT_a} \quad \text{and} \quad Ma_a = \frac{V_a}{c_a} = \frac{\sqrt{kRT_a}}{\sqrt{kRT_a}} = 1 \quad (9)$$

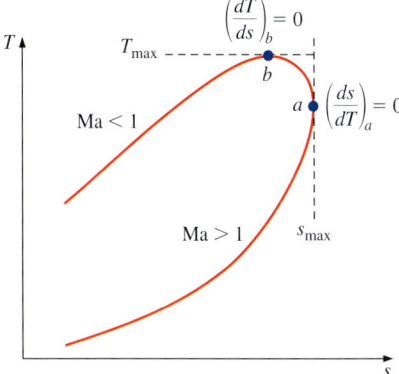

FIGURE 12–50
The *T-s* diagram of Rayleigh flow considered in Example 12–12.

Therefore, sonic conditions exist at point a, and thus the Mach number is 1.
Setting $dT/ds = (ds/dT)^{-1} = 0$ and solving the resulting equation $T(k-1) \times (RT - V^2) = 0$ for velocity at point b give

$$V_b = \sqrt{RT_b} \quad \text{and} \quad \text{Ma}_b = \frac{V_b}{c_b} = \frac{\sqrt{RT_b}}{\sqrt{kRT_b}} = \frac{1}{\sqrt{k}} \qquad (10)$$

Therefore, the Mach number at point b is $\text{Ma}_b = 1\sqrt{k}$. For air, $k = 1.4$ and thus $\text{Ma}_b = 0.845$.

Discussion Note that in Rayleigh flow, sonic conditions are reached as the entropy reaches its maximum value, and maximum temperature occurs during subsonic flow.

EXAMPLE 12–13 Effect of Heat Transfer on Flow Velocity

Starting with the differential form of the energy equation, show that the flow velocity increases with heat addition in subsonic Rayleigh flow, but decreases in supersonic Rayleigh flow.

SOLUTION It is to be shown that flow velocity increases with heat addition in subsonic Rayleigh flow and that the opposite occurs in supersonic flow.
Assumptions **1** The assumptions associated with Rayleigh flow are valid. **2** There are no work interactions and potential energy changes are negligible.
Analysis Consider heat transfer to the fluid in the differential amount of δq. The differential forms of the energy equations are expressed as

$$\delta q = dh_0 = d\left(h + \frac{V^2}{2}\right) = c_p \, dT + V \, dV \qquad (1)$$

Dividing by $c_p T$ and factoring out dV/V give

$$\frac{\delta q}{c_p T} = \frac{dT}{T} + \frac{V \, dV}{c_p T} = \frac{dV}{V}\left(\frac{V}{dV}\frac{dT}{T} + \frac{(k-1)V^2}{kRT}\right) \qquad (2)$$

where we also used $c_p = kR/(k-1)$. Noting that $\text{Ma}^2 = V^2/c^2 = V^2/kRT$ and using Eq. 7 for dT/dV from Example 12–12 give

$$\frac{\delta q}{c_p T} = \frac{dV}{V}\left(\frac{V}{T}\left(\frac{T}{V} - \frac{V}{R}\right) + (k-1)\text{Ma}^2\right) = \frac{dV}{V}\left(1 - \frac{V^2}{TR} + k\,\text{Ma}^2 - \text{Ma}^2\right) \qquad (3)$$

Canceling the two middle terms in Eq. 3 since $V^2/TR = k\,\text{Ma}^2$ and rearranging give the desired relation,

$$\frac{dV}{V} = \frac{\delta q}{c_p T}\frac{1}{(1-\text{Ma}^2)} \qquad (4)$$

In subsonic flow, $1 - \text{Ma}^2 > 0$ and thus heat transfer and velocity change have the same sign. As a result, heating the fluid ($\delta q > 0$) increases the flow velocity while cooling decreases it. In supersonic flow, however, $1 - \text{Ma}^2 < 0$ and heat transfer and velocity change have opposite signs. **As a result, heating the fluid ($\delta q > 0$) decreases the flow velocity while cooling increases it** (Fig. 12–51).
Discussion Note that heating the fluid has the opposite effect on flow velocity in subsonic and supersonic Rayleigh flows.

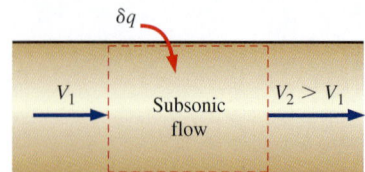
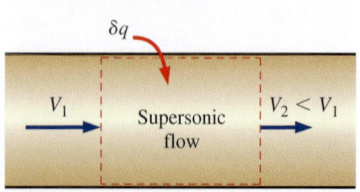

FIGURE 12–51
Heating increases the flow velocity in subsonic flow, but decreases it in supersonic flow.

Property Relations for Rayleigh Flow

It is often desirable to express the variations in properties in terms of the Mach number Ma. Noting that $Ma = V/c = V/\sqrt{kRT}$ and thus $V = Ma\sqrt{kRT}$,

$$\rho V^2 = \rho kRT Ma^2 = kP Ma^2 \qquad (12\text{-}57)$$

since $P = \rho RT$. Substituting into the momentum equation (Eq. 12–51) gives $P_1 + kP_1 Ma_1^2 = P_2 + kP_2 Ma_2^2$, which can be rearranged as

$$\frac{P_2}{P_1} = \frac{1 + kMa_1^2}{1 + kMa_2^2} \qquad (12\text{-}58)$$

Again utilizing $V = Ma\sqrt{kRT}$, the continuity equation $\rho_1 V_1 = \rho_2 V_2$ is expressed as

$$\frac{\rho_1}{\rho_2} = \frac{V_2}{V_1} = \frac{Ma_2 \sqrt{kRT_2}}{Ma_1 \sqrt{kRT_1}} = \frac{Ma_2 \sqrt{T_2}}{Ma_1 \sqrt{T_1}} \qquad (12\text{-}59)$$

Then the ideal-gas relation (Eq. 12–56) becomes

$$\frac{T_2}{T_1} = \frac{P_2 \rho_1}{P_1 \rho_2} = \left(\frac{1 + kMa_1^2}{1 + kMa_2^2}\right)\left(\frac{Ma_2 \sqrt{T_2}}{Ma_1 \sqrt{T_1}}\right) \qquad (12\text{-}60)$$

Solving Eq. 12–60 for the temperature ratio T_2/T_1 gives

$$\frac{T_2}{T_1} = \left(\frac{Ma_2(1 + kMa_1^2)}{Ma_1(1 + kMa_2^2)}\right)^2 \qquad (12\text{-}61)$$

Substituting this relation into Eq. 12–59 gives the density or velocity ratio as

$$\frac{\rho_2}{\rho_1} = \frac{V_1}{V_2} = \frac{Ma_1^2(1 + kMa_2^2)}{Ma_2^2(1 + kMa_1^2)} \qquad (12\text{-}62)$$

Flow properties at sonic conditions are usually easy to determine, and thus the critical state corresponding to Ma = 1 serves as a convenient reference point in compressible flow. Taking state 2 to be the sonic state ($Ma_2 = 1$, and superscript * is used) and state 1 to be any state (no subscript), the property relations in Eqs. 12–58, 12–61, and 12–62 reduce to (Fig. 12–52)

$$\frac{P}{P^*} = \frac{1 + k}{1 + kMa^2} \quad \frac{T}{T^*} = \left(\frac{Ma(1 + k)}{1 + kMa^2}\right)^2 \quad \text{and} \quad \frac{V}{V^*} = \frac{\rho^*}{\rho} = \frac{(1 + k)Ma^2}{1 + kMa^2} \qquad (12\text{-}63)$$

Similar relations can be obtained for dimensionless stagnation temperature and stagnation pressure as follows:

$$\frac{T_0}{T_0^*} = \frac{T_0}{T}\frac{T}{T^*}\frac{T^*}{T_0^*} = \left(1 + \frac{k-1}{2}Ma^2\right)\left(\frac{Ma(1+k)}{1+kMa^2}\right)^2\left(1 + \frac{k-1}{2}\right)^{-1} \qquad (12\text{-}64)$$

which simplifies to

$$\frac{T_0}{T_0^*} = \frac{(k+1)Ma^2[2 + (k-1)Ma^2]}{(1 + kMa^2)^2} \qquad (12\text{-}65)$$

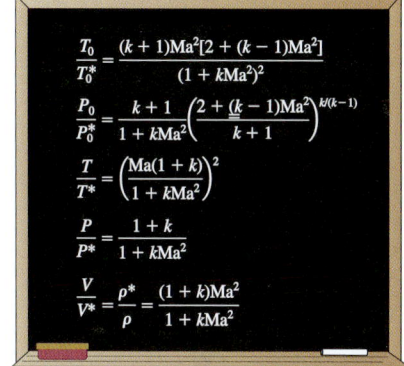

FIGURE 12–52
Summary of relations for Rayleigh flow.

Also,

$$\frac{P_0}{P_0^*} = \frac{P_0}{P}\frac{P}{P^*}\frac{P^*}{P_0^*} = \left(1 + \frac{k-1}{2}\text{Ma}^2\right)^{k/(k-1)}\left(\frac{1+k}{1+k\text{Ma}^2}\right)\left(1 + \frac{k-1}{2}\right)^{-k/(k-1)} \quad (12\text{–}66)$$

which simplifies to

$$\frac{P_0}{P_0^*} = \frac{k+1}{1+k\text{Ma}^2}\left(\frac{2+(k-1)\text{Ma}^2}{k+1}\right)^{k/(k-1)} \quad (12\text{–}67)$$

The five relations in Eqs. 12–63, 12–65, and 12–67 enable us to calculate the dimensionless pressure, temperature, density, velocity, stagnation temperature, and stagnation pressure for Rayleigh flow of an ideal gas with a specified k for any given Mach number. Representative results are given in tabular and graphical form in Table A–15 for $k = 1.4$.

Choked Rayleigh Flow

It is clear from the earlier discussions that subsonic Rayleigh flow in a duct may accelerate to sonic velocity (Ma = 1) with heating. What happens if we continue to heat the fluid? Does the fluid continue to accelerate to supersonic velocities? An examination of the Rayleigh line indicates that the fluid at the critical state of Ma = 1 cannot be accelerated to supersonic velocities by heating. Therefore, the flow is *choked*. This is analogous to not being able to accelerate a fluid to supersonic velocities in a converging nozzle by simply extending the converging flow section. If we keep heating the fluid, we will simply move the critical state further downstream and reduce the flow rate since fluid density at the critical state will now be lower. Therefore, for a given inlet state, the corresponding critical state fixes the maximum possible heat transfer for steady flow (Fig. 12–53). That is,

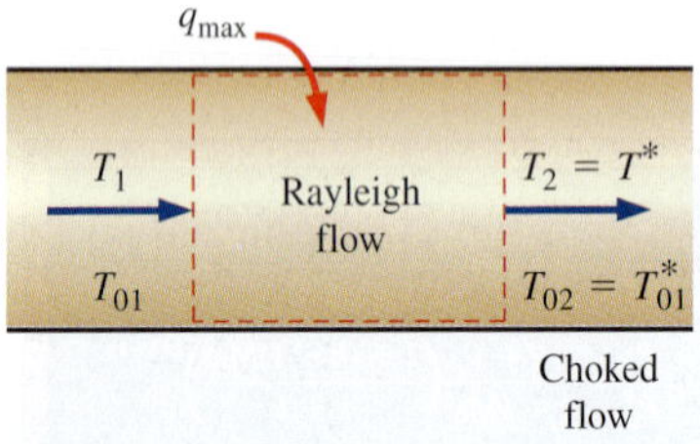

$$q_{\text{max}} = h_0^* - h_{01} = c_p(T_0^* - T_{01}) \quad (12\text{–}68)$$

Further heat transfer causes choking and thus the inlet state to change (e.g., inlet velocity will decrease), and the flow no longer follows the same Rayleigh line. Cooling the subsonic Rayleigh flow reduces the velocity, and the Mach number approaches zero as the temperature approaches absolute zero. Note that the stagnation temperature T_0 is maximum at the critical state of Ma = 1.

In supersonic Rayleigh flow, heating decreases the flow velocity. Further heating simply increases the temperature and moves the critical state farther downstream, resulting in a reduction in the mass flow rate of the fluid. It may seem like supersonic Rayleigh flow can be cooled indefinitely, but it turns out that there is a limit. Taking the limit of Eq. 12–65 as the Mach number approaches infinity gives

$$\lim_{\text{Ma}\to\infty} \frac{T_0}{T_0^*} = 1 - \frac{1}{k^2} \quad (12\text{–}69)$$

which yields $T_0/T_0^* = 0.49$ for $k = 1.4$. Therefore, if the critical stagnation temperature is 1000 K, air cannot be cooled below 490 K in Rayleigh flow. Physically this means that the flow velocity reaches infinity by the time the temperature reaches 490 K—a physical impossibility. When supersonic flow cannot be sustained, the flow undergoes a normal shock wave and becomes subsonic.

FIGURE 12–53
For a given inlet state, the maximum possible heat transfer occurs when sonic conditions are reached at the exit state.

EXAMPLE 12–14 Rayleigh Flow in a Tubular Combustor

A combustion chamber consists of tubular combustors of 15-cm diameter. Compressed air enters the tubes at 550 K, 480 kPa, and 80 m/s (Fig. 12–54). Fuel with a heating value of 42,000 kJ/kg is injected into the air and is burned with an air–fuel mass ratio of 40. Approximating combustion as a heat transfer process to air, determine the temperature, pressure, velocity, and Mach number at the exit of the combustion chamber.

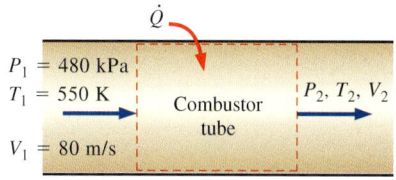

FIGURE 12–54
Schematic of the combustor tube analyzed in Example 12–14.

SOLUTION Fuel is burned in a tubular combustion chamber with compressed air. The exit temperature, pressure, velocity, and Mach number are to be determined.
Assumptions 1 The assumptions associated with Rayleigh flow (i.e., steady one-dimensional flow of an ideal gas with constant properties through a constant cross-sectional area duct with negligible frictional effects) are valid. 2 Combustion is complete, and it is treated as a heat addition process, with no change in the chemical composition of the flow. 3 The increase in mass flow rate due to fuel injection is disregarded.
Properties We take the properties of air to be $k = 1.4$, $c_p = 1.005$ kJ/kg·K, and $R = 0.287$ kJ/kg·K.
Analysis The inlet density and mass flow rate of air are

$$\rho_1 = \frac{P_1}{RT_1} = \frac{480 \text{ kPa}}{(0.287 \text{ kJ/kg·K})(550 \text{ K})} = 3.041 \text{ kg/m}^3$$

$$\dot{m}_{air} = \rho_1 A_1 V_1 = (3.041 \text{ kg/m}^3)[\pi(0.15 \text{ m})^2/4](80 \text{ m/s}) = 4.299 \text{ kg/s}$$

The mass flow rate of fuel and the rate of heat transfer are

$$\dot{m}_{fuel} = \frac{\dot{m}_{air}}{AF} = \frac{4.299 \text{ kg/s}}{40} = 0.1075 \text{ kg/s}$$

$$\dot{Q} = \dot{m}_{fuel} \text{ HV} = (0.1075 \text{ kg/s})(42{,}000 \text{ kJ/kg}) = 4514 \text{ kW}$$

$$q = \frac{\dot{Q}}{\dot{m}_{air}} = \frac{4514 \text{ kJ/s}}{4.299 \text{ kg/s}} = 1050 \text{ kJ/kg}$$

The stagnation temperature and Mach number at the inlet are

$$T_{01} = T_1 + \frac{V_1^2}{2c_p} = 550 \text{ K} + \frac{(80 \text{ m/s})^2}{2(1.005 \text{ kJ/kg·K})}\left(\frac{1 \text{ kJ/kg}}{1000 \text{ m}^2/\text{s}^2}\right) = 553.2 \text{ K}$$

$$c_1 = \sqrt{kRT_1} = \sqrt{(1.4)(0.287 \text{ kJ/kg·K})(550 \text{ K})\left(\frac{1000 \text{ m}^2/\text{s}^2}{1 \text{ kJ/kg}}\right)} = 470.1 \text{ m/s}$$

$$\text{Ma}_1 = \frac{V_1}{c_1} = \frac{80 \text{ m/s}}{470.1 \text{ m/s}} = 0.1702$$

The exit stagnation temperature is, from the energy equation $q = c_p(T_{02} - T_{01})$,

$$T_{02} = T_{01} + \frac{q}{c_p} = 553.2 \text{ K} + \frac{1050 \text{ kJ/kg}}{1.005 \text{ kJ/kg·K}} = 1598 \text{ K}$$

The maximum value of stagnation temperature T_0^* occurs at Ma = 1, and its value can be determined from Table A–15 or from Eq. 12–65. At $Ma_1 = 0.1702$ we read $T_0/T_0^* = 0.1291$. Therefore,

$$T_0^* = \frac{T_{01}}{0.1291} = \frac{553.2 \text{ K}}{0.1291} = 4284 \text{ K}$$

The stagnation temperature ratio at the exit state and the Mach number corresponding to it are, from Table A–15,

$$\frac{T_{02}}{T_0^*} = \frac{1598 \text{ K}}{4284 \text{ K}} = 0.3730 \rightarrow Ma_2 = 0.3142 \cong \mathbf{0.314}$$

The Rayleigh flow functions corresponding to the inlet and exit Mach numbers are (Table A–15):

$$Ma_1 = 0.1702: \quad \frac{T_1}{T^*} = 0.1541 \quad \frac{P_1}{P^*} = 2.3065 \quad \frac{V_1}{V^*} = 0.0668$$

$$Ma_2 = 0.3142: \quad \frac{T_2}{T^*} = 0.4389 \quad \frac{P_2}{P^*} = 2.1086 \quad \frac{V_2}{V^*} = 0.2082$$

Then the exit temperature, pressure, and velocity are determined to be

$$\frac{T_2}{T_1} = \frac{T_2/T^*}{T_1/T^*} = \frac{0.4389}{0.1541} = 2.848 \rightarrow T_2 = 2.848 T_1 = 2.848(550 \text{ K}) = \mathbf{1570 \text{ K}}$$

$$\frac{P_2}{P_1} = \frac{P_2/P^*}{P_1/P^*} = \frac{2.1086}{2.3065} = 0.9142 \rightarrow P_2 = 0.9142 P_1 = 0.9142(480 \text{ kPa}) = \mathbf{439 \text{ kPa}}$$

$$\frac{V_2}{V_1} = \frac{V_2/V^*}{V_1/V^*} = \frac{0.2082}{0.0668} = 3.117 \rightarrow V_2 = 3.117 V_1 = 3.117(80 \text{ m/s}) = \mathbf{249 \text{ m/s}}$$

Discussion Note that the temperature and velocity increase and pressure decreases during this subsonic Rayleigh flow with heating, as expected. This problem can also be solved using appropriate relations instead of tabulated values, which can likewise be coded for convenient computer solutions.

12–6 · ADIABATIC DUCT FLOW WITH FRICTION (FANNO FLOW)

Wall friction associated with high-speed flow through short devices with large cross-sectional areas such as large nozzles is often negligible, and flow through such devices can be approximated as being frictionless. But wall friction is significant and should be considered when studying flows through long flow sections, such as long ducts, especially when the cross-sectional area is small. In this section we consider compressible flow with significant wall friction but negligible heat transfer in ducts of constant cross-sectional area.

Consider steady, one-dimensional, adiabatic flow of an ideal gas with constant specific heats through a constant-area duct with significant frictional effects. Such flows are referred to as **Fanno flows.** The conservation of mass, momentum, and energy equations for the control volume shown in Fig. 12–55 are written as follows:

Continuity equation Noting that the duct cross-sectional area A is constant (and thus $A_1 = A_2 = A_c$), the relation $\dot{m}_1 = \dot{m}_2$ or $\rho_1 A_1 V_1 = \rho_2 A_2 V_2$ reduces to

$$\rho_1 V_1 = \rho_2 V_2 \rightarrow \rho V = \text{constant} \quad (12\text{–}70)$$

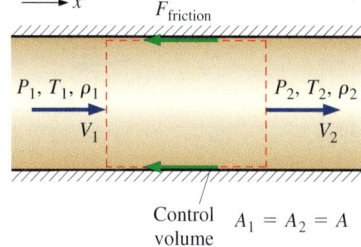

FIGURE 12–55
Control volume for adiabatic flow in a constant-area duct with friction.

x-Momentum equation Denoting the friction force exerted on the fluid by the inner surface of the duct by F_{friction} and assuming there are no other external and body forces, the momentum equation $\sum \vec{F} = \sum_{\text{out}} \beta \dot{m} \vec{V} - \sum_{\text{in}} \beta \dot{m} \vec{V}$ in the flow direction can be expressed as

$$P_1 A - P_2 A - F_{\text{friction}} = \dot{m} V_2 - \dot{m} V_1 \rightarrow P_1 - P_2 - \frac{F_{\text{friction}}}{A}$$
$$= (\rho_2 V_2) V_2 - (\rho_1 V_1) V_1$$

where even though there is friction at the walls, and the velocity profiles are not uniform, we approximate the momentum flux correction factor β as 1 for simplicity since the flow is usually fully developed and turbulent. The equation is rewritten as

$$P_1 + \rho_1 V_1^2 = P_2 + \rho_2 V_2^2 + \frac{F_{\text{friction}}}{A} \quad (12\text{–}71)$$

Energy equation The control volume involves no heat or work interactions and the potential energy change is negligible. Then the steady-flow energy balance $\dot{E}_{\text{in}} = \dot{E}_{\text{out}}$ becomes

$$h_1 + \frac{V_1^2}{2} = h_2 + \frac{V_2^2}{2} \rightarrow h_{01} = h_{02} \rightarrow h_0 = h + \frac{V^2}{2} = \text{constant} \quad (12\text{–}72)$$

For an ideal gas with constant specific heats, $\Delta h = c_p \Delta T$ and thus

$$T_1 + \frac{V_1^2}{2c_p} = T_2 + \frac{V_2^2}{2c_p} \rightarrow T_{01} = T_{02} \rightarrow T_0 = T + \frac{V^2}{2c_p} = \text{constant} \quad (12\text{–}73)$$

Therefore, the stagnation enthalpy h_0 and stagnation temperature T_0 remain constant during Fanno flow.

Entropy change In the absence of any heat transfer, the entropy of a system can be changed only by irreversibilities such as friction, whose effect is always to increase entropy. Therefore, the entropy of the fluid must increase during Fanno flow. The entropy change in this case is equivalent to entropy increase or entropy generation, and for an ideal gas with constant specific heats it is expressed as

$$s_2 - s_1 = c_p \ln \frac{T_2}{T_1} - R \ln \frac{P_2}{P_1} > 0 \quad (12\text{–}74)$$

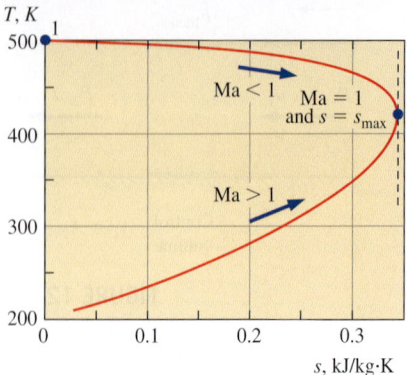

FIGURE 12–56

T-s diagram for adiabatic frictional flow in a constant-area duct (Fanno flow). Numerical values are for air with $k = 1.4$ and inlet conditions of $T_1 = 500$ K, $P_1 = 600$ kPa, $V_1 = 80$ m/s, and an assigned value of $s_1 = 0$.

Equation of state Noting that $P = \rho RT$, the properties P, ρ, and T of an ideal gas at states 1 and 2 are related to each other by

$$\frac{P_1}{\rho_1 T_1} = \frac{P_2}{\rho_2 T_2} \qquad (12\text{–}75)$$

Consider a gas with known properties R, k, and c_p flowing in a duct of constant cross-sectional area A. For a specified inlet state 1, the inlet properties P_1, T_1, ρ_1, V_1, and s_1 are known. The five exit properties P_2, T_2, ρ_2, V_2, and s_2 can be determined from Eqs. 12–70 through 12–75 for any specified value of the friction force F_{friction}. Knowing the velocity and temperature, we can also determine the Mach number at the inlet and the exit from the relation $\text{Ma} = V/c = V\sqrt{kRT}$.

Obviously there is an infinite number of possible downstream states 2 corresponding to a given upstream state 1. A practical way of determining these downstream states is to assume various values of T_2, and calculate all other properties as well as the friction force for each assumed T_2 from Eqs. 12–70 through 12–75. Plotting the results on a *T-s* diagram gives a curve passing through the specified inlet state, as shown in Fig. 12–56. The plot of Fanno flow on a *T-s* diagram is called the **Fanno line,** and several important observations can be made from this plot and the results of calculations:

1. All states that satisfy the conservation of mass, momentum, and energy equations as well as the property relations are on the Fanno line. Therefore, for a given inlet state, the fluid cannot exist at any downstream state outside the Fanno line on a *T-s* diagram. In fact, the Fanno line is the locus of all possible downstream states corresponding to an initial state. Note that if there were no friction, the flow properties would have remained constant along the duct during Fanno flow.
2. Friction causes entropy to increase, and thus a process always proceeds to the right along the Fanno line. At the point of maximum entropy, the Mach number is $\text{Ma} = 1$. All states on the upper part of the Fanno line are subsonic, and all states on the lower part are supersonic.
3. Friction increases the Mach number for subsonic Fanno flow, but decreases it for supersonic Fanno flow. The Mach number approaches unity ($\text{Ma} = 1$) in both cases.
4. The energy balance requires that stagnation temperature $T_0 = T + V^2/2c_p$ remain constant during Fanno flow. But the actual temperature may change. Velocity increases and thus temperature decreases during subsonic flow, but the opposite occurs during supersonic flow (Fig. 12–57).
5. The continuity equation $\rho V = $ constant indicates that density and velocity are inversely proportional. Therefore, the effect of friction is to decrease density in subsonic flow (since velocity and Mach number increase), but to increase it in supersonic flow (since velocity and Mach number decrease).

The effects of friction on the properties of Fanno flow are listed in Table 12–4. Note that frictional effects on most properties in subsonic flow are opposite to those in supersonic flow. However, the effect of friction is to always decrease stagnation pressure, regardless of whether the flow is

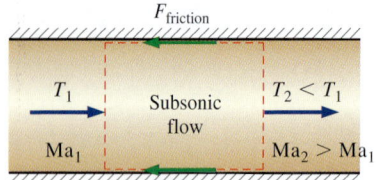

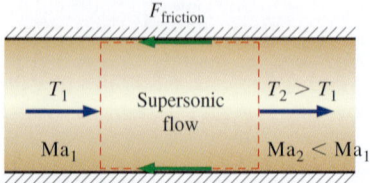

FIGURE 12–57

Friction causes the Mach number to increase and the temperature to decrease in subsonic Fanno flow, but it does the opposite in supersonic Fanno flow.

TABLE 12–4
The effects of friction on the properties of Fanno flow

Property	Subsonic	Supersonic
Velocity, V	Increase	Decrease
Mach number, Ma	Increase	Decrease
Stagnation temperature, T_0	Constant	Constant
Temperature, T	Decrease	Increase
Density, ρ	Decrease	Increase
Stagnation pressure, P_0	Decrease	Decrease
Pressure, P	Decrease	Increase
Entropy, s	Increase	Increase

subsonic or supersonic. But friction has no effect on stagnation temperature since friction simply causes the mechanical energy to be converted to an equivalent amount of thermal energy.

Property Relations for Fanno Flow

In compressible flow, it is convenient to express the variation of properties in terms of Mach number, and Fanno flow is no exception. However, Fanno flow involves the friction force, which is proportional to the square of the velocity even when the friction factor is constant. But in compressible flow, velocity varies significantly along the flow, and thus it is necessary to perform a differential analysis to account for the variation of the friction force properly. We begin by obtaining the differential forms of the conservation equations and property relations.

Continuity equation The differential form of the continuity equation is obtained by differentiating the continuity relation $\rho V =$ constant and rearranging,

$$\rho \, dV + V \, d\rho = 0 \quad \rightarrow \quad \frac{d\rho}{\rho} = -\frac{dV}{V} \quad (12\text{–}76)$$

x-Momentum equation Noting that $\dot{m}_1 = \dot{m}_2 = \dot{m} = \rho A V$ and $A_1 = A_2 = A$, applying the momentum equation

$$\sum \vec{F} = \sum_{\text{out}} \beta \dot{m} \vec{V} - \sum_{\text{in}} \beta \dot{m} \vec{V}$$

to the differential control volume in Fig. 12–58 gives

$$PA_c - (P + dP)A - \delta F_{\text{friction}} = \dot{m}(V + dV) - \dot{m}V$$

where we have again approximated the momentum flux correction factor β as 1. This equation simplifies to

$$-dPA - \delta F_{\text{friction}} = \rho A V \, dV \quad \text{or} \quad dP + \frac{\delta F_{\text{friction}}}{A} + \rho V \, dV = 0 \quad (12\text{–}77)$$

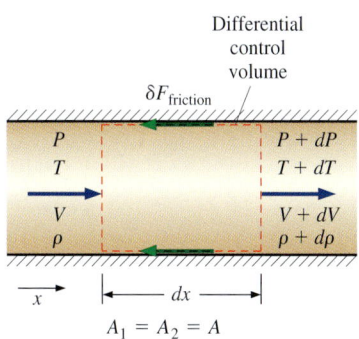

FIGURE 12–58
Differential control volume for adiabatic flow in a constant-area duct with friction.

The friction force is related to the wall shear stress τ_w and the local friction factor f_x by

$$\delta F_{\text{friction}} = \tau_w \, dA_s = \tau_w p \, dx = \left(\frac{f_x}{8}\rho V^2\right)\frac{4A}{D_h} dx = \frac{f_x}{2}\frac{A\,dx}{D_h}\rho V^2 \quad (12\text{--}78)$$

where dx is the length of the flow section, p is the perimeter, and $D_h = 4A/p$ is the hydraulic diameter of the duct (note that D_h reduces to ordinary diameter D for a duct of circular cross section). Substituting,

$$dP + \frac{\rho V^2 f_x}{2 D_h} dx + \rho V \, dV = 0 \quad (12\text{--}79)$$

Noting that $V = \text{Ma}\sqrt{kRT}$ and $P = \rho RT$, we have $\rho V^2 = \rho k R T \text{Ma}^2 = kP\text{Ma}^2$ and $\rho V = kP\text{Ma}^2/V$. Substituting into Eq. 12–79,

$$\frac{1}{k\text{Ma}^2}\frac{dP}{P} + \frac{f_x}{2 D_h} dx + \frac{dV}{V} = 0 \quad (12\text{--}80)$$

Energy equation Noting that $c_p = kR/(k-1)$ and $V^2 = \text{Ma}^2 kRT$, the energy equation $T_0 = \text{constant}$ or $T + V^2/2c_p = \text{constant}$ is expressed as

$$T_0 = T\left(1 + \frac{k-1}{2}\text{Ma}^2\right) = \text{constant} \quad (12\text{--}81)$$

Differentiating and rearranging give

$$\frac{dT}{T} = -\frac{2(k-1)\text{Ma}^2}{2+(k-1)\text{Ma}^2}\frac{d\text{Ma}}{\text{Ma}} \quad (12\text{--}82)$$

which is an expression for the differential change in temperature in terms of a differential change in Mach number.

Mach number The Mach number relation for ideal gases can be expressed as $V^2 = \text{Ma}^2 kRT$. Differentiating and rearranging give

$$2V\,dV = 2\text{Ma}kRT\,d\text{Ma} + kR\text{Ma}^2\,dT \rightarrow \quad (12\text{--}83)$$

$$2V\,dV = 2\frac{V^2}{\text{Ma}}d\text{Ma} + \frac{V^2}{T}dT$$

Dividing each term by $2V^2$ and rearranging,

$$\frac{dV}{V} = \frac{d\text{Ma}}{\text{Ma}} + \frac{1}{2}\frac{dT}{T} \quad (12\text{--}84)$$

Combining Eq. 12–84 with Eq. 12–82 gives the velocity change in terms of the Mach number as

$$\frac{dV}{V} = \frac{d\text{Ma}}{\text{Ma}} - \frac{(k-1)\text{Ma}^2}{2+(k-1)\text{Ma}^2}\frac{d\text{Ma}}{\text{Ma}} \quad \text{or} \quad \frac{dV}{V} = \frac{2}{2+(k-1)\text{Ma}^2}\frac{d\text{Ma}}{\text{Ma}} \quad (12\text{--}85)$$

Ideal gas The differential form of the ideal-gas equation is obtained by differentiating the equation $P = \rho RT$,

$$dP = \rho R\,dT + RT\,d\rho \rightarrow \frac{dP}{P} = \frac{dT}{T} + \frac{d\rho}{\rho} \quad (12\text{--}86)$$

Combining with the continuity equation (Eq. 12–76) gives

$$\frac{dP}{P} = \frac{dT}{T} - \frac{dV}{V} \quad (12\text{–}87)$$

Now combining with Eqs. 12–82 and 12–84 gives

$$\frac{dP}{P} = -\frac{2 + 2(k-1)\text{Ma}^2}{2 + (k-1)\text{Ma}^2}\frac{d\text{Ma}}{\text{Ma}} \quad (12\text{–}88)$$

which is an expression for differential changes in P with Ma.

Substituting Eqs. 12–85 and 12–88 into 12–80 and simplifying give the differential equation for the variation of the Mach number with x as

$$\frac{f_x}{D_h}dx = \frac{4(1-\text{Ma}^2)}{k\text{Ma}^3[2+(k-1)\text{Ma}^2]}d\text{Ma} \quad (12\text{–}89)$$

Considering that all Fanno flows tend to Ma = 1, it is again convenient to use the critical point (i.e., the sonic state) as the reference point and to express flow properties relative to the critical point properties, even if the actual flow never reaches the critical point. Integrating Eq. 12–89 from any state (Ma = Ma and $x = x$) to the critical state (Ma = 1 and $x = x_{cr}$) gives

$$\frac{fL^*}{D_h} = \frac{1-\text{Ma}^2}{k\text{Ma}^2} + \frac{k+1}{2k}\ln\frac{(k+1)\text{Ma}^2}{2+(k-1)\text{Ma}^2} \quad (12\text{–}90)$$

where f is the average friction factor between x and x_{cr}, which is assumed to be constant, and $L^* = x_{cr} - x$ is the channel length required for the Mach number to reach unity under the influence of wall friction. Therefore, L^* represents the distance between a given section where the Mach number is Ma and a section (an imaginary section if the duct is not long enough to reach Ma = 1) where sonic conditions occur (Fig. 12–59).

Note that the value of fL^*/D_h is fixed for a given Mach number, and thus values of fL^*/D_h can be tabulated versus Ma for a specified k. Also, the value of duct length L^* needed to reach sonic conditions (or the "sonic length") is inversely proportional to the friction factor. Therefore, for a given Mach number, L^* is large for ducts with smooth surfaces and small for ducts with rough surfaces.

The actual duct length L between two sections where the Mach numbers are Ma_1 and Ma_2 can be determined from

$$\frac{fL}{D_h} = \left(\frac{fL^*}{D_h}\right)_1 - \left(\frac{fL^*}{D_h}\right)_2 \quad (12\text{–}91)$$

The average friction factor f, in general, is different in different parts of the duct. If f is approximated as constant for the entire duct (including the hypothetical extension part to the sonic state), then Eq. 12–91 simplifies to

$$L = L_1^* - L_2^* \quad (f = \text{constant}) \quad (12\text{–}92)$$

Therefore, Eq. 12–90 can be used for short ducts that never reach Ma = 1 as well as long ones with Ma = 1 at the exit.

The friction factor depends on the Reynolds number Re = $\rho V D_h/\mu$, which varies along the duct, and the roughness ratio ε/D_h of the surface. The variation of Re is mild, however, since ρV = constant (from continuity), and any change in Re is due to the variation of viscosity with temperature.

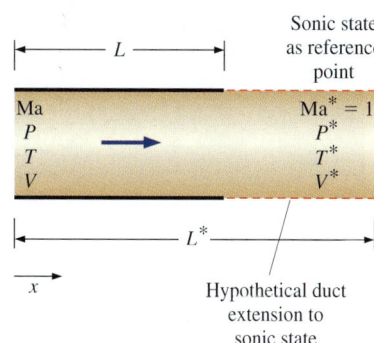

FIGURE 12–59
The length L^* represents the distance between a given section where the Mach number is Ma and a real or imaginary section where Ma* = 1.

Therefore, it is a reasonable approximation to evaluate f from the Moody chart or Colebrook equation discussed in Chap. 8 at the average Reynolds number and to treat it as a constant. This is the case for subsonic flow since the temperature changes involved are relatively small. The treatment of the friction factor for supersonic flow is beyond the scope of this text. The Colebrook equation is implicit in f, and thus it is more convenient to use the explicit Haaland relation expressed as

$$\frac{1}{\sqrt{f}} \cong -1.8 \log\left[\frac{6.9}{\text{Re}} + \left(\frac{\varepsilon/D}{3.7}\right)^{1.11}\right] \quad (12\text{-}93)$$

The Reynolds numbers encountered in compressible flow are typically high, and at very high Reynolds numbers (fully rough turbulent flow) the friction factor is independent of the Reynolds number. For Re $\to \infty$, the Colebrook equation reduces to $1\sqrt{f} = -2.0 \log[(\varepsilon/D_h)/3.7]$.

Relations for other flow properties can be determined similarly by integrating the dP/P, dT/T, and dV/V relations from Eqs. 12–79, 12–82, and 12–85, respectively, from any state (no subscript and Mach number Ma) to the sonic state (with a superscript asterisk and Ma = 1) with the following results (Fig. 12–60):

$$\frac{P}{P^*} = \frac{1}{\text{Ma}}\left(\frac{k+1}{2+(k-1)\text{Ma}^2}\right)^{1/2} \quad (12\text{-}94)$$

$$\frac{T}{T^*} = \frac{k+1}{2+(k-1)\text{Ma}^2} \quad (12\text{-}95)$$

$$\frac{V}{V^*} = \frac{\rho^*}{\rho} = \text{Ma}\left(\frac{k+1}{2+(k-1)\text{Ma}^2}\right)^{1/2} \quad (12\text{-}96)$$

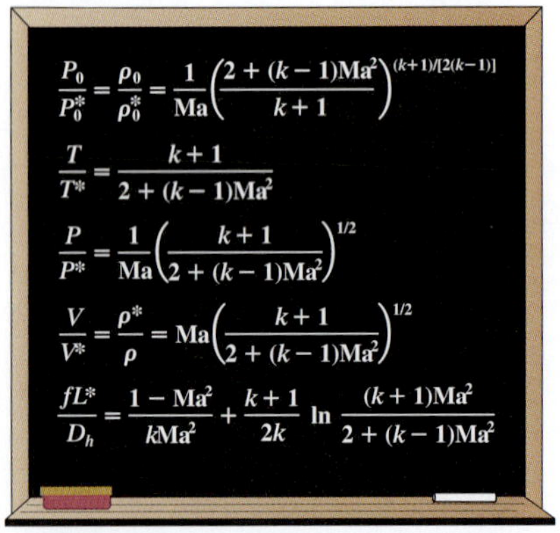

A similar relation can be obtained for the dimensionless stagnation pressure as follows:

$$\frac{P_0}{P_0^*} = \frac{P_0}{P}\frac{P}{P^*}\frac{P^*}{P_0^*} = \left(1 + \frac{k-1}{2}\text{Ma}^2\right)^{k/(k-1)}\frac{1}{\text{Ma}}\left(\frac{k+1}{2+(k-1)\text{Ma}^2}\right)^{1/2}\left(1 + \frac{k-1}{2}\right)^{-k/(k-1)}$$

which simplifies to

$$\frac{P_0}{P_0^*} = \frac{\rho_0}{\rho_0^*} = \frac{1}{\text{Ma}}\left(\frac{2+(k-1)\text{Ma}^2}{k+1}\right)^{(k+1)/[2(k-1)]} \quad (12\text{-}97)$$

Note that the stagnation temperature T_0 is constant for Fanno flow, and thus $T_0/T_0^* = 1$ everywhere along the duct.

Eqs. 12–90 through 12–97 enable us to calculate the dimensionless pressure, temperature, density, velocity, stagnation pressure, and fL^*/D_h for Fanno flow of an ideal gas with a specified k for any given Mach number. Representative results are given in tabular and graphical form in Table A–16 for $k = 1.4$.

Choked Fanno Flow

It is clear from the previous discussions that friction causes subsonic Fanno flow in a constant-area duct to accelerate toward sonic velocity, and the Mach number becomes exactly unity at the exit for a certain duct length. This duct length is referred to as the **maximum length,** the **sonic length,** or the **critical length,** and is denoted by L^*. You may be curious to know what happens if we extend the duct length beyond L^*. In particular, does the flow accelerate to

FIGURE 12–60
Summary of relations for Fanno flow.

supersonic velocities? The answer to this question is a definite *no* since at Ma = 1 the flow is at the point of maximum entropy, and proceeding along the Fanno line to the supersonic region would require the entropy of the fluid to decrease—a violation of the second law of thermodynamics. (Note that the exit state must remain on the Fanno line to satisfy all conservation requirements.) Therefore, the flow is choked. This again is analogous to not being able to accelerate a gas to supersonic velocities in a converging nozzle by simply extending the converging flow section. If we extend the duct length beyond L^* anyway, we simply move the critical state further downstream and reduce the flow rate. This causes the inlet state to change (e.g., inlet velocity decreases), and the flow shifts to a different Fanno line. Further increase in duct length further decreases the inlet velocity and thus the mass flow rate.

Friction causes supersonic Fanno flow in a constant-area duct to decelerate and the Mach number to decrease toward unity. Therefore, the exit Mach number again becomes Ma = 1 if the duct length is L^*, as in subsonic flow. But unlike subsonic flow, increasing the duct length beyond L^* cannot choke the flow since it is already choked. Instead, it causes a normal shock to occur at such a location that the continuing subsonic flow becomes sonic again exactly at the duct exit (Fig. 12–61). As the duct length increases, the location of the normal shock moves further upstream. Eventually, the shock occurs at the duct inlet. Further increase in duct length moves the shock to the diverging section of the converging–diverging nozzle that originally generates the supersonic flow, but the mass flow rate still remains unaffected since the mass flow rate is fixed by the sonic conditions at the throat of the nozzle, and it does not change unless the conditions at the throat change.

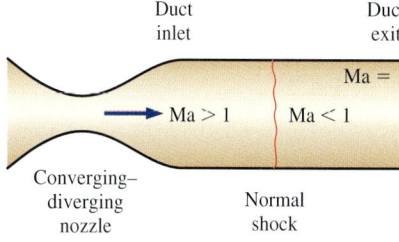

FIGURE 12–61
If duct length L is greater than L^*, supersonic Fanno flow is always sonic at the duct exit. Extending the duct will only move the location of the normal shock further upstream.

■ EXAMPLE 12–15 Choked Fanno Flow in a Duct

Air enters a 3-cm-diameter smooth adiabatic duct at $Ma_1 = 0.4$, $T_1 = 300$ K, and $P_1 = 150$ kPa (Fig. 12–62). If the Mach number at the duct exit is 1, determine the duct length and temperature, pressure, and velocity at the duct exit. Also determine the percentage of stagnation pressure lost in the duct.

SOLUTION Air enters a constant-area adiabatic duct at a specified state and leaves at the sonic state. The duct length, exit temperature, pressure, velocity, and the percentage of stagnation pressure lost in the duct are to be determined.
Assumptions **1** The assumptions associated with Fanno flow (i.e., steady, frictional flow of an ideal gas with constant properties through a constant cross-sectional area adiabatic duct) are valid. **2** The friction factor is constant along the duct.
Properties We take the properties of air to be $k = 1.4$, $c_p = 1.005$ kJ/kg·K, $R = 0.287$ kJ/kg·K, and $\nu = 1.58 \times 10^{-5}$ m²/s.
Analysis We first determine the inlet velocity and the inlet Reynolds number,

$$c_1 = \sqrt{kRT_1} = \sqrt{(1.4)(0.287 \text{ kJ/kg·K})(300 \text{ K})\left(\frac{1000 \text{ m}^2/\text{s}^2}{1 \text{ kJ/kg}}\right)} = 347 \text{ m/s}$$

$$V_1 = Ma_1 c_1 = 0.4(347 \text{ m/s}) = 139 \text{ m/s}$$

$$Re_1 = \frac{V_1 D}{\nu} = \frac{(139 \text{ m/s})(0.03 \text{ m})}{1.58 \times 10^{-5} \text{ m}^2/\text{s}} = 2.637 \times 10^5$$

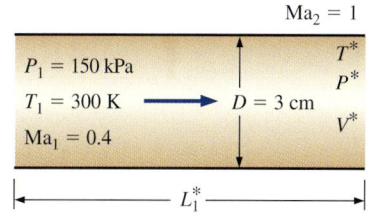

FIGURE 12–62
Schematic for Example 12–15.

The friction factor is determined from the Colebrook equation,

$$\frac{1}{\sqrt{f}} = -2.0 \log\left(\frac{\varepsilon/D}{3.7} + \frac{2.51}{\mathrm{Re}\sqrt{f}}\right) \rightarrow \frac{1}{\sqrt{f}} = -2.0 \log\left(\frac{0}{3.7} + \frac{2.51}{2.637 \times 10^5 \sqrt{f}}\right)$$

Its solution is

$$f = 0.0148$$

The Fanno flow functions corresponding to the inlet Mach number of 0.4 are (Table A–16):

$$\frac{P_{01}}{P_0^*} = 1.5901 \quad \frac{T_1}{T^*} = 1.1628 \quad \frac{P_1}{P^*} = 2.6958 \quad \frac{V_1}{V^*} = 0.4313 \quad \frac{fL_1^*}{D} = 2.3085$$

Noting that * denotes sonic conditions, which exist at the exit state, the duct length and the exit temperature, pressure, and velocity are determined to be

$$L_1^* = \frac{2.3085 D}{f} = \frac{2.3085(0.03 \text{ m})}{0.0148} = \mathbf{4.68 \text{ m}}$$

$$T^* = \frac{T_1}{1.1628} = \frac{300 \text{ K}}{1.1628} = \mathbf{258 \text{ K}}$$

$$P^* = \frac{P_1}{2.6958} = \frac{150 \text{ kPa}}{2.6958} = \mathbf{55.6 \text{ kPa}}$$

$$V^* = \frac{V_1}{0.4313} = \frac{139 \text{ m/s}}{0.4313} = \mathbf{322 \text{ m/s}}$$

Thus, for the given friction factor, the duct length must be 4.68 m for the Mach number to reach Ma = 1 at the duct exit. The fraction of inlet stagnation pressure P_{01} lost in the duct due to friction is

$$\frac{P_{01} - P_0^*}{P_{01}} = 1 - \frac{P_0^*}{P_{01}} = 1 - \frac{1}{1.5901} = 0.371 \text{ or } \mathbf{37.1\%}$$

Discussion This problem can also be solved using appropriate relations instead of tabulated values for the Fanno functions. Also, we determined the friction factor at the inlet conditions and assumed it to remain constant along the duct. To check the validity of this assumption, we calculate the friction factor at the outlet conditions. It can be shown that the friction factor at the duct outlet is 0.0121—a drop of 18 percent, which is large. Therefore, we should repeat the calculations using the average value of the friction factor (0.0148 + 0.0121)/2 = 0.0135. This would give the duct length to be $L_1^* = 2.3085(0.03\text{m})/0.0135 = \mathbf{5.13 \text{ m}}$, and we take this to be the required duct length.

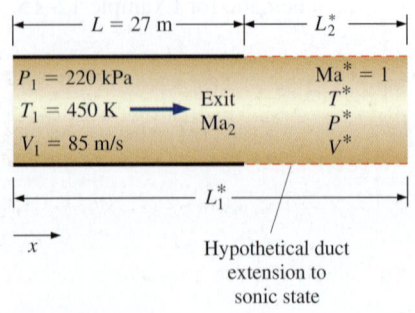

FIGURE 12–63
Schematic for Example 12–16.

EXAMPLE 12–16   Exit Conditions of Fanno Flow in a Duct

Air enters a 27-m-long 5-cm-diameter adiabatic duct at V_1 = 85 m/s, T_1 = 450 K, and P_1 = 220 kPa (Fig. 12–63). The average friction factor for the duct is estimated to be 0.023. Determine the Mach number at the duct exit and the mass flow rate of air.

SOLUTION Air enters a constant-area adiabatic duct of given length at a specified state. The exit Mach number and the mass flow rate are to be determined.

Assumptions **1** The assumptions associated with Fanno flow (i.e., steady, frictional flow of an ideal gas with constant properties through a constant cross-sectional area adiabatic duct) are valid. **2** The friction factor is constant along the duct.

Properties We take the properties of air to be $k = 1.4$, $c_p = 1.005$ kJ/kg·K, and $R = 0.287$ kJ/kg·K.

Analysis The first thing we need to know is whether the flow is choked at the exit or not. Therefore, we first determine the inlet Mach number and the corresponding value of the function fL^*/D_h,

$$c_1 = \sqrt{kRT_1} = \sqrt{(1.4)(0.287 \text{ kJ/kg·K})(450 \text{ K})\left(\frac{1000 \text{ m}^2/\text{s}^2}{1 \text{ kJ/kg}}\right)} = 425 \text{ m/s}$$

$$\text{Ma}_1 = \frac{V_1}{c_1} = \frac{85 \text{ m/s}}{425 \text{ m/s}} = 0.200$$

Corresponding to this Mach number we read, from Table A–16, $(fL^*/D_h)_1 = 14.5333$. Also, using the actual duct length L, we have

$$\frac{fL}{D_h} = \frac{(0.023)(27 \text{ m})}{0.05 \text{ m}} = 12.42 < 14.5333$$

Therefore, flow is *not* choked and the exit Mach number is less than 1. The function fL^*/D_h at the exit state is calculated from Eq. 12–91,

$$\left(\frac{fL^*}{D_h}\right)_2 = \left(\frac{fL^*}{D_h}\right)_1 - \frac{fL}{D_h} = 14.5333 - 12.42 = 2.1133$$

The Mach number corresponding to this value of fL^*/D is 0.42, obtained from Table A–16. Therefore, the Mach number at the duct exit is

$$\text{Ma}_2 = \mathbf{0.420}$$

The mass flow rate of air is determined from the inlet conditions to be

$$\rho_1 = \frac{P_1}{RT_1} = \frac{220 \text{ kPa}}{(0.287 \text{ kJ/kg·K})(450 \text{ K})}\left(\frac{1 \text{ kJ}}{1 \text{ kPa·m}^3}\right) = 1.703 \text{ kg/m}^3$$

$$\dot{m}_{\text{air}} = \rho_1 A_1 V_1 = (1.703 \text{ kg/m}^3)\,[\pi(0.05 \text{ m})^2/4]\,(85 \text{ m/s}) = \mathbf{0.284 \text{ kg/s}}$$

Discussion Note that it takes a duct length of 27 m for the Mach number to increase from 0.20 to 0.42, but only 4.6 m to increase from 0.42 to 1. Therefore, the Mach number rises at a much higher rate as sonic conditions are approached.

To gain some insight, let's determine the lengths corresponding to fL^*/D_h values at the inlet and the exit states. Noting that f is assumed to be constant for the entire duct, the maximum (or sonic) duct lengths at the inlet and exit states are

$$L_{\text{max},1} = L_1^* = 14.5333 \frac{D_h}{f} = 14.5333 \frac{0.05 \text{ m}}{0.023} = 31.6 \text{ m}$$

$$L_{\text{max},2} = L_2^* = 2.1133 \frac{D_h}{f} = 2.1133 \frac{0.05 \text{ m}}{0.023} = 4.59 \text{ m}$$

(or, $L_{\text{max},2} = L_{\text{max},1} - L = 31.6 - 27 = 4.6$ m). Therefore, the flow would reach sonic conditions if a 4.6-m-long section were added to the existing duct.

COMPRESSIBLE FLOW

APPLICATION SPOTLIGHT ■ Shock-Wave/Boundary-Layer Interactions

Guest Author: Gary S. Settles, The Pennsylvania State University

Shock waves and boundary layers are among nature's most incompatible phenomena. Boundary layers, as described in Chap. 10, are susceptible to separation from aerodynamic surfaces wherever strong adverse pressure gradients occur. Shock waves, on the other hand, produce very strong adverse pressure gradients, since a finite rise in static pressure occurs across a shock wave over a negligibly short streamwise distance. Thus, when a boundary layer encounters a shock wave, a complicated flow pattern develops and the boundary layer often separates from the surface to which it was attached.

There are important cases in high-speed flight and wind tunnel testing where such a clash is unavoidable. For example, commercial jet transport aircraft cruise in the bottom edge of the transonic flow regime, where the airflow over their wings actually goes supersonic and then returns to subsonic flow through a normal shock wave (Fig. 12–64). If such an aircraft flies significantly faster than its design cruise Mach number, serious aerodynamic disturbances arise due to shock-wave/boundary-layer interactions causing flow separation on the wings. This phenomenon thus limits the speed of passenger aircraft around the world. Some military aircraft are designed to avoid this limit and fly supersonically, but shock-wave/boundary-layer interactions are still limiting factors in their engine air inlets.

FIGURE 12–64
Normal shock wave above the wing of an L-1011 commercial jet aircraft in transonic flight, made visible by background distortion of low clouds over the Pacific Ocean.

U.S. Govt. photo by Carla Thomas, NASA Dryden Research Center.

The interaction of a shock wave and a boundary layer is a type of *viscous–inviscid interaction* in which the viscous flow in the boundary layer encounters the essentially inviscid shock wave generated in the free stream. The boundary layer is slowed and thickened by the shock and may separate. The shock, on the other hand, bifurcates when flow separation occurs (Fig. 12–65). Mutual changes in both the shock and the boundary layer continue until an equilibrium condition is reached. Depending upon boundary conditions, the interaction can vary in either two or three dimensions and may be steady or unsteady.

Such a strongly interacting flow is difficult to analyze, and no simple solutions exist. Moreover, in most of the problems of practical interest, the boundary layer in question is turbulent. Modern computational methods are able to predict many features of these flows by supercomputer solutions of the Reynolds-averaged Navier–Stokes equations. Wind tunnel experiments play a key role in guiding and validating such computations. Overall, the shock-wave/boundary-layer interaction has become one of the pacing problems of modern fluid dynamics research.

FIGURE 12–65
Shadowgram of the swept interaction generated by a fin mounted on a flat plate at Mach 3.5. The oblique shock wave generated by the fin (at top of image) bifurcates into a "λ-foot" beneath which the boundary layer separates and rolls up. The airflow through the λ-foot above the separation zone forms a supersonic "jet" that curves downward and impinges upon the wall. This three-dimensional interaction required a special optical technique known as conical shadowgraphy to visualize the flow.

Photo by F. S. Alvi and G. S. Settles.

References

Knight, D. D., et al., "Advances in CFD Prediction of Shock Wave Turbulent Boundary Layer Interactions," *Progress in Aerospace Sciences* 39(2-3), pp. 121–184, 2003.

Alvi, F. S., and Settles, G. S., "Physical Model of the Swept Shock Wave/Boundary-Layer Interaction Flowfield," *AIAA Journal* 30, pp. 2252–2258, Sept. 1992.

SUMMARY

In this chapter the effects of compressibility on gas flow are examined. When dealing with compressible flow, it is convenient to combine the enthalpy and the kinetic energy of the fluid into a single term called *stagnation* (or *total*) *enthalpy* h_0, defined as

$$h_0 = h + \frac{V^2}{2}$$

The properties of a fluid at the stagnation state are called *stagnation properties* and are indicated by the subscript zero. The *stagnation temperature* of an ideal gas with constant specific heats is

$$T_0 = T + \frac{V^2}{2c_p}$$

which represents the temperature an ideal gas would attain if it is brought to rest adiabatically. The stagnation properties of an ideal gas are related to the static properties of the fluid by

$$\frac{P_0}{P} = \left(\frac{T_0}{T}\right)^{k/(k-1)} \quad \text{and} \quad \frac{\rho_0}{\rho} = \left(\frac{T_0}{T}\right)^{1/(k-1)}$$

The velocity at which an infinitesimally small pressure wave travels through a medium is the *speed of sound*. For an ideal gas it is expressed as

$$c = \sqrt{\left(\frac{\partial P}{\partial \rho}\right)_s} = \sqrt{kRT}$$

The *Mach number* is the ratio of the actual velocity of the fluid to the speed of sound at the same state:

$$\text{Ma} = \frac{V}{c}$$

The flow is called *sonic* when Ma = 1, *subsonic* when Ma < 1, *supersonic* when Ma > 1, *hypersonic* when Ma $\gg$ 1, and *transonic* when Ma $\cong$ 1.

Nozzles whose flow area decreases in the flow direction are called *converging* nozzles. Nozzles whose flow area first decreases and then increases are called *converging–diverging nozzles*. The location of the smallest flow area of a nozzle is called the *throat*. The highest velocity to which a fluid can be accelerated in a converging nozzle is the sonic velocity. Accelerating a fluid to supersonic velocities is possible only in converging–diverging nozzles. In all supersonic converging–diverging nozzles, the flow velocity at the throat is the velocity of sound.

The ratios of the stagnation to static properties for ideal gases with constant specific heats can be expressed in terms of the Mach number as

$$\frac{T_0}{T} = 1 + \left(\frac{k-1}{2}\right)\text{Ma}^2$$

$$\frac{P_0}{P} = \left[1 + \left(\frac{k-1}{2}\right)\text{Ma}^2\right]^{k/(k-1)}$$

and

$$\frac{\rho_0}{\rho} = \left[1 + \left(\frac{k-1}{2}\right)\text{Ma}^2\right]^{1/(k-1)}$$

When Ma = 1, the resulting static-to-stagnation property ratios for the temperature, pressure, and density are called *critical ratios* and are denoted by the superscript asterisk:

$$\frac{T^*}{T_0} = \frac{2}{k+1} \quad \frac{P^*}{P_0} = \left(\frac{2}{k+1}\right)^{k/(k-1)}$$

and

$$\frac{\rho^*}{\rho_0} = \left(\frac{2}{k+1}\right)^{1/(k-1)}$$

The pressure outside the exit plane of a nozzle is called the *back pressure*. For all back pressures lower than P^*, the pressure at the exit plane of the converging nozzle is equal to P^*, the Mach number at the exit plane is unity, and the mass flow rate is the maximum (or choked) flow rate.

In some range of back pressure, the fluid that achieved a sonic velocity at the throat of a converging–diverging nozzle and is accelerating to supersonic velocities in the diverging section experiences a *normal shock*, which causes a sudden rise in pressure and temperature and a sudden drop in velocity to subsonic levels. Flow through the shock is highly irreversible, and thus it cannot be approximated as isentropic. The properties of an ideal gas with constant specific heats before (subscript 1) and after (subscript 2) a shock are related by

$$T_{01} = T_{02} \quad \text{Ma}_2 = \sqrt{\frac{(k-1)\text{Ma}_1^2 + 2}{2k\text{Ma}_1^2 - k + 1}}$$

$$\frac{T_2}{T_1} = \frac{2 + \text{Ma}_1^2(k-1)}{2 + \text{Ma}_2^2(k-1)}$$

and

$$\frac{P_2}{P_1} = \frac{1 + k\text{Ma}_1^2}{1 + k\text{Ma}_2^2} = \frac{2k\text{Ma}_1^2 - k + 1}{k+1}$$

These equations also hold across an oblique shock, provided that the component of the Mach number *normal* to the oblique shock is used in place of the Mach number.

Steady one-dimensional flow of an ideal gas with constant specific heats through a constant-area duct with heat transfer and negligible friction is referred to as *Rayleigh flow*. The property relations and curves for Rayleigh flow are given in Table A–15. Heat transfer during Rayleigh flow are determined from

$$q = c_p(T_{02} - T_{01}) = c_p(T_2 - T_1) + \frac{V_2^2 - V_1^2}{2}$$

Steady, frictional, and adiabatic flow of an ideal gas with constant specific heats through a constant-area duct is referred to as *Fanno flow*. The channel length required for the Mach number to reach unity under the influence of wall friction is denoted by L^* and is expressed as

$$\frac{fL^*}{D_h} = \frac{1 - \mathrm{Ma}^2}{k\mathrm{Ma}^2} + \frac{k+1}{2k}\ln\frac{(k+1)\mathrm{Ma}^2}{2 + (k-1)\mathrm{Ma}^2}$$

where f is the average friction factor. The duct length between two sections where the Mach numbers are Ma_1 and Ma_2 is determined from

$$\frac{fL}{D_h} = \left(\frac{fL^*}{D_h}\right)_1 - \left(\frac{fL^*}{D_h}\right)_2$$

During Fanno flow, the stagnation temperature T_0 remains constant. Other property relations and curves for Fanno flow are given in Table A–16.

This chapter provides an overview of compressible flow and is intended to motivate the interested student to undertake a more in-depth study of this exciting subject. Some compressible flows are analyzed in Chap. 15 using computational fluid dynamics.

REFERENCES AND SUGGESTED READING

1. J. D. Anderson. *Modern Compressible Flow with Historical Perspective,* 3rd ed. New York: McGraw-Hill, 2003.
2. Y. A. Çengel and M. A. Boles. *Thermodynamics: An Engineering Approach,* 7th ed. New York: McGraw-Hill, 2011.
3. H. Cohen, G. F. C. Rogers, and H. I. H. Saravanamuttoo. *Gas Turbine Theory,* 3rd ed. New York: Wiley, 1987.
4. W. J. Devenport. Compressible Aerodynamic Calculator, http://www.aoe.vt.edu/~devenpor/aoe3114/calc.html.
5. R. W. Fox and A. T. McDonald. *Introduction to Fluid Mechanics,* 8th ed. New York: Wiley, 2011.
6. H. Liepmann and A. Roshko. *Elements of Gas Dynamics,* Dover Publications, Mineola, NY, 2001.
7. C. E. Mackey, responsible NACA officer and curator. *Equations, Tables, and Charts for Compressible Flow.* NACA Report 1135.
8. A. H. Shapiro. *The Dynamics and Thermodynamics of Compressible Fluid Flow,* vol. 1. New York: Ronald Press Company, 1953.
9. P. A. Thompson. *Compressible-Fluid Dynamics,* New York: McGraw-Hill, 1972.
10. United Technologies Corporation. *The Aircraft Gas Turbine and Its Operation,* 1982.
11. M. Van Dyke, *An Album of Fluid Motion.* Stanford, CA: The Parabolic Press, 1982.
12. F. M. White. *Fluid Mechanics,* 7th ed. New York: McGraw-Hill, 2010.

PROBLEMS*

Stagnation Properties

12–1C What is dynamic temperature?

12–2C In air-conditioning applications, the temperature of air is measured by inserting a probe into the flow stream.

---

* Problems designated by a "C" are concept questions, and students are encouraged to answer them all. Problems with the [EES] icon are solved using EES, and complete solutions together with parametric studies are included on the text website. Problems with the [EES] icon are comprehensive in nature and are intended to be solved with an equation solver such as EES.

Thus, the probe actually measures the stagnation temperature. Does this cause any significant error?

12–3 Air flows through a device such that the stagnation pressure is 0.6 MPa, the stagnation temperature is 400°C, and the velocity is 570 m/s. Determine the static pressure and temperature of the air at this state. *Answers:* 519 K, 0.231 MPa

12–4 Air at 320 K is flowing in a duct at a velocity of (*a*) 1, (*b*) 10, (*c*) 100, and (*d*) 1000 m/s. Determine the temperature that a stationary probe inserted into the duct will read for each case.

12–5 Calculate the stagnation temperature and pressure for the following substances flowing through a duct: (*a*) helium at 0.25 MPa, 50°C, and 240 m/s; (*b*) nitrogen at 0.15 MPa, 50°C, and 300 m/s; and (*c*) steam at 0.1 MPa, 350°C, and 480 m/s.

12–6 Determine the stagnation temperature and stagnation pressure of air that is flowing at 36 kPa, 238 K, and 325 m/s. *Answers:* 291 K, 72.4 kPa

12–7 Steam flows through a device with a stagnation pressure of 800 kPa, a stagnation temperature of 400°C, and a velocity of 300 m/s. Assuming ideal-gas behavior, determine the static pressure and temperature of the steam at this state.

12–8 Air enters a compressor with a stagnation pressure of 100 kPa and a stagnation temperature of 35°C, and it is compressed to a stagnation pressure of 900 kPa. Assuming the compression process to be isentropic, determine the power input to the compressor for a mass flow rate of 0.04 kg/s. *Answer:* 10.8 kW

12–9 Products of combustion enter a gas turbine with a stagnation pressure of 0.75 MPa and a stagnation temperature of 690°C, and they expand to a stagnation pressure of 100 kPa. Taking $k = 1.33$ and $R = 0.287$ kJ/kg·K for the products of combustion, and assuming the expansion process to be isentropic, determine the power output of the turbine per unit mass flow.

One-Dimensional Isentropic Flow

12–10C Is it possible to accelerate a gas to a supersonic velocity in a converging nozzle? Explain.

12–11C A gas initially at a subsonic velocity enters an adiabatic diverging duct. Discuss how this affects (*a*) the velocity, (*b*) the temperature, (*c*) the pressure, and (*d*) the density of the fluid.

12–12C A gas at a specified stagnation temperature and pressure is accelerated to Ma = 2 in a converging–diverging nozzle and to Ma = 3 in another nozzle. What can you say about the pressures at the throats of these two nozzles?

12–13C A gas initially at a supersonic velocity enters an adiabatic converging duct. Discuss how this affects (*a*) the velocity, (*b*) the temperature, (*c*) the pressure, and (*d*) the density of the fluid.

12–14C A gas initially at a supersonic velocity enters an adiabatic diverging duct. Discuss how this affects (*a*) the velocity, (*b*) the temperature, (*c*) the pressure, and (*d*) the density of the fluid.

12–15C Consider a converging nozzle with sonic speed at the exit plane. Now the nozzle exit area is reduced while the nozzle inlet conditions are maintained constant. What will happen to (*a*) the exit velocity and (*b*) the mass flow rate through the nozzle?

12–16C A gas initially at a subsonic velocity enters an adiabatic converging duct. Discuss how this affects (*a*) the velocity, (*b*) the temperature, (*c*) the pressure, and (*d*) the density of the fluid.

12–17 Helium enters a converging–diverging nozzle at 0.7 MPa, 800 K, and 100 m/s. What are the lowest temperature and pressure that can be obtained at the throat of the nozzle?

12–18 Consider a large commercial airplane cruising at a speed of 1050 km/h in air at an altitude of 10 km where the standard air temperature is −50°C. Determine if the speed of this airplane is subsonic or supersonic.

12–19 Calculate the critical temperature, pressure, and density of (*a*) air at 200 kPa, 100°C, and 250 m/s, and (*b*) helium at 200 kPa, 40°C, and 300 m/s.

12–20 Air enters a converging–diverging nozzle at a pressure of 1200 kPa with negligible velocity. What is the lowest pressure that can be obtained at the throat of the nozzle? *Answer:* 634 kPa

12–21 In March 2004, NASA successfully launched an experimental supersonic-combustion ramjet engine (called a *scramjet*) that reached a record-setting Mach number of 7. Taking the air temperature to be −20°C, determine the speed of this engine. *Answer:* 8040 km/h

12–22 Reconsider the scram jet engine discussed in Prob. 12–21. Determine the speed of this engine in km/h corresponding to a Mach number of 7 in air at a temperature of −18°C.

12–23 Air at 200 kPa, 100°C, and Mach number Ma = 0.8 flows through a duct. Calculate the velocity and the stagnation pressure, temperature, and density of the air.

12–24 Reconsider Prob. 12–23. Using EES (or other) software, study the effect of Mach numbers in the range 0.1 to 2 on the velocity, stagnation pressure, temperature, and density of air. Plot each parameter as a function of the Mach number.

12–25 An aircraft is designed to cruise at Mach number Ma = 1.1 at 12,000 m where the atmospheric temperature is 236.15 K. Determine the stagnation temperature on the leading edge of the wing.

12–26 Quiescent carbon dioxide at 1200 kPa and 600 K is accelerated isentropically to a Mach number of 0.6. Determine the temperature and pressure of the carbon dioxide after acceleration. *Answers:* 570 K, 957 kPa

Isentropic Flow through Nozzles

12–27C Is it possible to accelerate a fluid to supersonic velocities with a velocity other than the sonic velocity at the throat? Explain

12–28C What would happen if we tried to further accelerate a supersonic fluid with a diverging diffuser?

12–29C How does the parameter Ma* differ from the Mach number Ma?

12–30C Consider subsonic flow in a converging nozzle with specified conditions at the nozzle inlet and critical pressure at the nozzle exit. What is the effect of dropping the back pressure well below the critical pressure on (a) the exit velocity, (b) the exit pressure, and (c) the mass flow rate through the nozzle?

12–31C Consider a converging nozzle and a converging–diverging nozzle having the same throat areas. For the same inlet conditions, how would you compare the mass flow rates through these two nozzles?

12–32C Consider gas flow through a converging nozzle with specified inlet conditions. We know that the highest velocity the fluid can have at the nozzle exit is the sonic velocity, at which point the mass flow rate through the nozzle is a maximum. If it were possible to achieve hypersonic velocities at the nozzle exit, how would it affect the mass flow rate through the nozzle?

12–33C Consider subsonic flow in a converging nozzle with fixed inlet conditions. What is the effect of dropping the back pressure to the critical pressure on (a) the exit velocity, (b) the exit pressure, and (c) the mass flow rate through the nozzle?

12–34C Consider the isentropic flow of a fluid through a converging–diverging nozzle with a subsonic velocity at the throat. How does the diverging section affect (a) the velocity, (b) the pressure, and (c) the mass flow rate of the fluid?

12–35C What would happen if we attempted to decelerate a supersonic fluid with a diverging diffuser?

12–36 Nitrogen enters a converging–diverging nozzle at 700 kPa and 400 K with a negligible velocity. Determine the critical velocity, pressure, temperature, and density in the nozzle.

12–37 For an ideal gas obtain an expression for the ratio of the speed of sound where Ma = 1 to the speed of sound based on the stagnation temperature, c^*/c_0.

12–38 Air enters a converging–diverging nozzle at 1.2 MPa with a negligible velocity. Approximating the flow as isentropic, determine the back pressure that would result in an exit Mach number of 1.8. *Answer:* 209 kPa

12–39 An ideal gas flows through a passage that first converges and then diverges during an adiabatic, reversible, steady-flow process. For subsonic flow at the inlet, sketch the variation of pressure, velocity, and Mach number along the length of the nozzle when the Mach number at the minimum flow area is equal to unity.

12–40 Repeat Prob. 12–39 for supersonic flow at the inlet.

12–41 Explain why the maximum flow rate per unit area for a given ideal gas depends only on $P_0/\sqrt{T_0}$. For an ideal gas with $k = 1.4$ and $R = 0.287$ kJ/kg·K, find the constant a such that $\dot{m}/A^* = aP_0/\sqrt{T_0}$.

12–42 An ideal gas with $k = 1.4$ is flowing through a nozzle such that the Mach number is 1.8 where the flow area is 36 cm². Approximating the flow as isentropic, determine the flow area at the location where the Mach number is 0.9.

12–43 Repeat Prob. 12–42 for an ideal gas with $k = 1.33$.

12–44 Air enters a converging–diverging nozzle of a supersonic wind tunnel at 1 MPa and 37°C with a low velocity. The flow area of the test section is equal to the exit area of the nozzle, which is 0.5 m². Calculate the pressure, temperature, velocity, and mass flow rate in the test section for a Mach number Ma = 2. Explain why the air must be very dry for this application. *Answers:* 128 kPa, 172 K, 526 m/s, 680 kg/s

12–45 Air enters a nozzle at 0.5 MPa, 420 K, and a velocity of 110 m/s. Approximating the flow as isentropic, determine the pressure and temperature of air at a location where the air velocity equals the speed of sound. What is the ratio of the area at this location to the entrance area? *Answers:* 355 K, 278 kPa, 0.428

12–46 Repeat Prob. 12–45 assuming the entrance velocity is negligible.

12–47 Air at 900 kPa and 400 K enters a converging nozzle with a negligible velocity. The throat area of the nozzle is 10 cm². Approximating the flow as isentropic, calculate and plot the exit pressure, the exit velocity, and the mass flow rate versus the back pressure P_b for $0.9 \geq P_b \geq 0.1$ MPa.

12–48 Reconsider Prob. 12–47. Using EES (or other) software, solve the problem for the inlet conditions of 0.8 MPa and 1200 K.

Shock Waves and Expansion Waves

12–49C Are the isentropic relations of ideal gases applicable for flows across (a) normal shock waves, (b) oblique shock waves, and (c) Prandtl–Meyer expansion waves?

12–50C What do the states on the Fanno line and the Rayleigh line represent? What do the intersection points of these two curves represent?

12–51C It is claimed that an oblique shock can be analyzed like a normal shock provided that the normal component of velocity (normal to the shock surface) is used in the analysis. Do you agree with this claim?

12–52C How does the normal shock affect (a) the fluid velocity, (b) the static temperature, (c) the stagnation temperature, (d) the static pressure, and (e) the stagnation pressure?

12–53C How do oblique shocks occur? How do oblique shocks differ from normal shocks?

12–54C For an oblique shock to occur, does the upstream flow have to be supersonic? Does the flow downstream of an oblique shock have to be subsonic?

12–55C Can the Mach number of a fluid be greater than 1 after a normal shock wave? Explain.

12–56C Consider supersonic airflow approaching the nose of a two-dimensional wedge and experiencing an oblique shock. Under what conditions does an oblique shock detach from the nose of the wedge and form a bow wave? What is the numerical value of the shock angle of the detached shock at the nose?

12–57C Consider supersonic flow impinging on the rounded nose of an aircraft. Is the oblique shock that forms in front of the nose an attached or a detached shock? Explain.

12–58C Can a shock wave develop in the converging section of a converging–diverging nozzle? Explain.

12–59 Air enters a normal shock at 26 kPa, 230 K, and 815 m/s. Calculate the stagnation pressure and Mach number upstream of the shock, as well as pressure, temperature, velocity, Mach number, and stagnation pressure downstream of the shock.

12–60 Calculate the entropy change of air across the normal shock wave in Problem 12–59.   *Answer:* 0.242 kJ/kg·K

12–61 For an ideal gas flowing through a normal shock, develop a relation for V_2/V_1 in terms of k, Ma_1, and Ma_2.

12–62 Air enters a converging–diverging nozzle with low velocity at 2.0 MPa and 100°C. If the exit area of the nozzle is 3.5 times the throat area, what must the back pressure be to produce a normal shock at the exit plane of the nozzle?   *Answer:* 0.661 MPa

12–63 What must the back pressure be in Prob. 12–62 for a normal shock to occur at a location where the cross-sectional area is twice the throat area?

12–64 Air enters a converging–diverging nozzle of a supersonic wind tunnel at 1 MPa and 300 K with a low velocity. If a normal shock wave occurs at the exit plane of the nozzle at Ma = 2.4, determine the pressure, temperature, Mach number, velocity, and stagnation pressure after the shock wave.   *Answers:* 448 kPa, 284 K, 0.523, 177 m/s, 540 kPa

12–65 Using EES (or other) software, calculate and plot the entropy change of air across the normal shock for upstream Mach numbers between 0.5 and 1.5 in increments of 0.1. Explain why normal shock waves can occur only for upstream Mach numbers greater than Ma = 1.

12–66 Consider supersonic airflow approaching the nose of a two-dimensional wedge at a Mach number of 5. Using Fig. 12–37, determine the minimum shock angle and the maximum deflection angle a straight oblique shock can have.

12–67 Air flowing at 32 kPa, 240 K, and Ma_1 = 3.6 is forced to undergo an expansion turn of 15°. Determine the Mach number, pressure, and temperature of air after the expansion.   *Answers:* 4.81, 6.65 kPa, 153 K

12–68 Consider the supersonic flow of air at upstream conditions of 70 kPa and 260 K and a Mach number of 2.4 over a two-dimensional wedge of half-angle 10°. If the axis of the wedge is tilted 25° with respect to the upstream air flow, determine the downstream Mach number, pressure, and temperature above the wedge.   *Answers:* 3.105, 23.8 kPa, 191 K

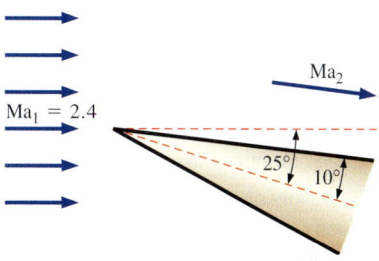

FIGURE P12–68

12–69 Reconsider Prob. 12–68. Determine the downstream Mach number, pressure, and temperature below the wedge for a strong oblique shock for an upstream Mach number of 5.

12–70 Air at 55 kPa, −7°C, and a Mach number of 2.0 is forced to turn upward by a ramp that makes an 8° angle off the flow direction. As a result, a weak oblique shock forms. Determine the wave angle, Mach number, pressure, and temperature after the shock.

12–71 Air flowing at 55 kPa, 265 K, and Ma_1 = 2.0 is forced to undergo a compression turn of 15°. Determine the Mach number, pressure, and temperature of air after the compression.

12–72 Air flowing at 60 kPa, 240 K, and a Mach number of 3.4 impinges on a two-dimensional wedge of half-angle 8°. Determine the two possible oblique shock angles, β_{weak} and β_{strong}, that could be formed by this wedge. For each case, calculate the pressure, temperature, and Mach number downstream of the oblique shock.

12–73 Air flowing steadily in a nozzle experiences a normal shock at a Mach number of Ma = 2.6. If the pressure and temperature of air are 58 kPa and 270 K, respectively, upstream of the shock, calculate the pressure, temperature, velocity, Mach number, and stagnation pressure downstream of the shock. Compare these results to those for helium undergoing a normal shock under the same conditions.

12–74 Calculate the entropy changes of air and helium across the normal shock wave in Prob. 12–73.

Duct Flow with Heat Transfer and Negligible Friction (Rayleigh Flow)

12–75C What is the effect of heating the fluid on the flow velocity in subsonic Rayleigh flow? Answer the same questions for supersonic Rayleigh flow.

12–76C On a T-s diagram of Rayleigh flow, what do the points on the Rayleigh line represent?

12–77C What is the effect of heat gain and heat loss on the entropy of the fluid during Rayleigh flow?

12–78C Consider subsonic Rayleigh flow of air with a Mach number of 0.92. Heat is now transferred to the fluid and the Mach number increases to 0.95. Does the temperature T of the fluid increase, decrease, or remain constant during this process? How about the stagnation temperature T_0?

12–79C What is the characteristic aspect of Rayleigh flow? What are the main assumptions associated with Rayleigh flow?

12–80C Consider subsonic Rayleigh flow that is accelerated to sonic velocity (Ma = 1) at the duct exit by heating. If the fluid continues to be heated, will the flow at duct exit be supersonic, subsonic, or remain sonic?

12–81 Argon gas enters a constant cross-sectional area duct at $Ma_1 = 0.2$, $P_1 = 320$ kPa, and $T_1 = 400$ K at a rate of 1.2 kg/s. Disregarding frictional losses, determine the highest rate of heat transfer to the argon without reducing the mass flow rate.

12–82 Air is heated as it flows subsonically through a duct. When the amount of heat transfer reaches 67 kJ/kg, the flow is observed to be choked, and the velocity and the static pressure are measured to be 680 m/s and 270 kPa. Disregarding frictional losses, determine the velocity, static temperature, and static pressure at the duct inlet.

12–83 Compressed air from the compressor of a gas turbine enters the combustion chamber at $T_1 = 700$ K, $P_1 = 600$ kPa, and $Ma_1 = 0.2$ at a rate of 0.3 kg/s. Via combustion, heat is transferred to the air at a rate of 150 kJ/s as it flows through the duct with negligible friction. Determine the Mach number at the duct exit, and the drop in stagnation pressure $P_{01} - P_{02}$ during this process. *Answers:* 0.271, 12.7 kPa

12–84 Repeat Prob. 12–83 for a heat transfer rate of 300 kJ/s.

12–85 Air flows with negligible friction through a 10-cm-diameter duct at a rate of 2.3 kg/s. The temperature and pressure at the inlet are $T_1 = 450$ K and $P_1 = 200$ KPa, and the Mach number at the exit is $Ma_2 = 1$. Determine the rate of heat transfer and the pressure drop for this section of the duct.

12–86 Air enters an approximately frictionless duct with $V_1 = 70$ m/s, $T_1 = 600$ K, and $P_1 = 350$ kPa. Letting the exit temperature T_2 vary from 600 to 5000 K, evaluate the entropy change at intervals of 200 K, and plot the Rayleigh line on a T-s diagram.

12–87 Air is heated as it flows through a 10 cm × 10 cm square duct with negligible friction. At the inlet, air is at $T_1 = 400$ K, $P_1 = 550$ kPa, and $V_1 = 80$ m/s. Determine the rate at which heat must be transferred to the air to choke the flow at the duct exit, and the entropy change of air during this process.

12–88 Air enters a rectangular duct at $T_1 = 300$ K, $P_1 = 420$ kPa, and $Ma_1 = 2$. Heat is transferred to the air in the amount of 55 kJ/kg as it flows through the duct. Disregarding frictional losses, determine the temperature and Mach number at the duct exit. *Answers:* 386 K, 1.64

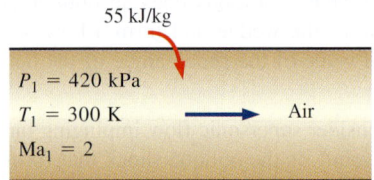

FIGURE P12–88

12–89 Repeat Prob. 12–88 assuming air is cooled in the amount of 55 kJ/kg.

12–90 Consider a 16-cm-diameter tubular combustion chamber. Air enters the tube at 450 K, 380 kPa, and 55 m/s. Fuel with a heating value of 39,000 kJ/kg is burned by spraying it into the air. If the exit Mach number is 0.8, determine the rate at which the fuel is burned and the exit temperature. Assume complete combustion and disregard the increase in the mass flow rate due to the fuel mass.

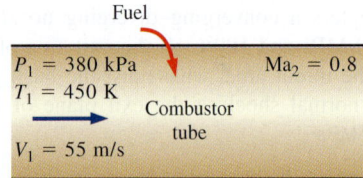

FIGURE P12–90

12–91 Consider supersonic flow of air through a 7-cm-diameter duct with negligible friction. Air enters the duct at $Ma_1 = 1.8$, $P_{01} = 140$ kPa, and $T_{01} = 600$ K, and it is decelerated by heating. Determine the highest temperature that air can be heated by heat addition while the mass flow rate remains constant.

Adiabatic Duct Flow with Friction (Fanno Flow)

12–92C What is the effect of friction on flow velocity in subsonic Fanno flow? Answer the same question for supersonic Fanno flow.

12–93C On a T-s diagram of Fanno flow, what do the points on the Fanno line represent?

12–94C What is the effect of friction on the entropy of the fluid during Fanno flow?

12–95C Consider supersonic Fanno flow that is decelerated to sonic velocity (Ma = 1) at the duct exit as a result of frictional effects. If the duct length is increased further, will the flow at the duct exit be supersonic, subsonic, or remain sonic? Will the mass flow rate of the fluid increase, decrease, or remain constant as a result of increasing the duct length?

12–96C Consider supersonic Fanno flow of air with an inlet Mach number of 1.8. If the Mach number decreases to 1.2 at the duct exit as a result of friction, does the (a) stagnation temperature T_0, (b) stagnation pressure P_0, and (c) entropy s of the fluid increase, decrease, or remain constant during this process?

12–97C What is the characteristic aspect of Fanno flow? What are the main approximations associated with Fanno flow?

12–98C Consider subsonic Fanno flow accelerated to sonic velocity (Ma = 1) at the duct exit as a result of frictional effects. If the duct length is increased further, will the flow at the duct exit be supersonic, subsonic, or remain sonic? Will the mass flow rate of the fluid increase, decrease, or remain constant as a result of increasing the duct length?

12–99C Consider subsonic Fanno flow of air with an inlet Mach number of 0.70. If the Mach number increases to 0.90 at the duct exit as a result of friction, will the (a) stagnation temperature T_0, (b) stagnation pressure P_0, and (c) entropy s of the fluid increase, decrease, or remain constant during this process?

12–100 Air enters a 12-cm-diameter adiabatic duct at $Ma_1 = 0.4$, $T_1 = 550$ K, and $P_1 = 200$ kPa. The average friction factor for the duct is estimated to be 0.021. If the Mach number at the duct exit is 0.8, determine the duct length, temperature, pressure, and velocity at the duct exit.

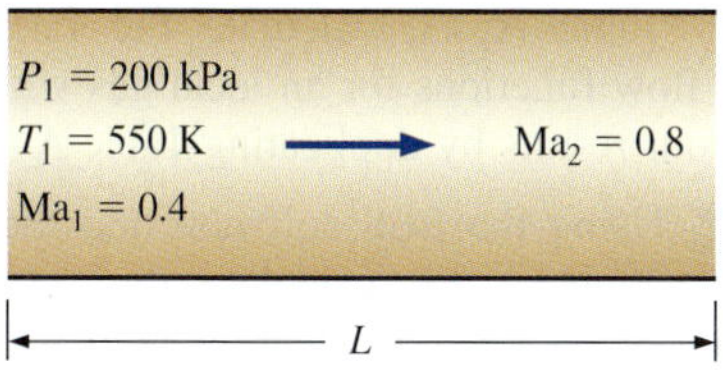

FIGURE P12–100

12–101 Air enters a 15-m-long, 4-cm-diameter adiabatic duct at $V_1 = 70$ m/s, $T_1 = 500$ K, and $P_1 = 300$ kPa. The average friction factor for the duct is estimated to be 0.023. Determine the Mach number at the duct exit, the exit velocity, and the mass flow rate of air.

12–102 Air enters a 5-cm-diameter, 4-m-long adiabatic duct with inlet conditions of $Ma_1 = 2.8$, $T_1 = 380$ K, and $P_1 = 80$ kPa. It is observed that a normal shock occurs at a location 3 m from the inlet. Taking the average friction factor to be 0.007, determine the velocity, temperature, and pressure at the duct exit. *Answers:* 572 m/s, 813 K, 328 kPa

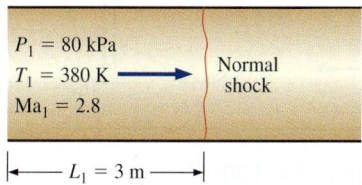

FIGURE P12–102

12–103 Helium gas with $k = 1.667$ enters a 15-cm-diameter duct at $Ma_1 = 0.2$, $P_1 = 400$ kPa, and $T_1 = 325$ K. For an average friction factor of 0.025, determine the maximum duct length that will not cause the mass flow rate of helium to be reduced. *Answer:* 87.2 m

12–104 Air enters a 15-cm-diameter adiabatic duct with inlet conditions of $V_1 = 150$ m/s, $T_1 = 500$ K, and $P_1 = 200$ kPa. For an average friction factor of 0.014, determine the duct length from the inlet where the inlet velocity doubles. Also determine the pressure drop along that section of the duct.

12–105 Consider subsonic airflow through a 20-cm-diameter adiabatic duct with inlet conditions of $T_1 = 330$ K, $P_1 = 180$ kPa, and $Ma_1 = 0.1$. Taking the average friction factor to be 0.02, determine the duct length required to accelerate the flow to a Mach number of unity. Also, calculate the duct length at Mach number intervals of 0.1, and plot the duct length against the Mach number for $0.1 \leq Ma \leq 1$. Discuss the results.

12–106 Repeat Prob. 12–105 for helium gas.

12–107 Argon gas with $k = 1.667$, $c_p = 0.5203$ kJ/kg·K, and $R = 0.2081$ kJ/kg·K enters an 8-cm-diameter adiabatic duct with $V_1 = 70$ m/s, $T_1 = 520$ K, and $P_1 = 350$ kPa. Taking the average friction factor to be 0.005 and letting the exit temperature T_2 vary from 540 K to 400 K, evaluate the entropy change at intervals of 10 K, and plot the Fanno line on a T-s diagram.

12–108 Air in a room at $T_0 = 300$ K and $P_0 = 100$ kPa is drawn steadily by a vacuum pump through a 1.4-cm-diameter, 35-cm-long adiabatic tube equipped with a converging nozzle at the inlet. The flow in the nozzle section can be approximated as isentropic, and the average friction factor for the duct can be taken to be 0.018. Determine the maximum mass flow rate of air that can be sucked through this tube and the Mach number at the tube inlet. *Answers:* 0.0305 kg/s, 0.611

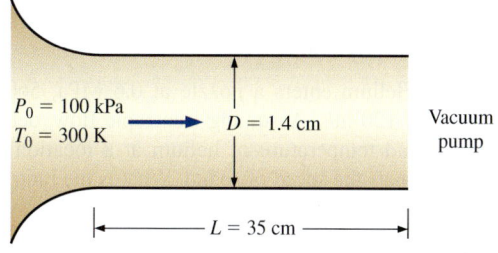

FIGURE P12–108

12–109 Repeat Prob. 12–108 for a friction factor of 0.025 and a tube length of 1 m.

Review Problems

12–110 The thrust developed by the engine of a Boeing 777 is about 380 kN. Assuming choked flow in the nozzles, determine the mass flow rate of air through the nozzle. Take the ambient conditions to be 220 K and 40 kPa.

12–111 A stationary temperature probe inserted into a duct where air is flowing at 190 m/s reads 85°C. What is the actual temperature of the air?   *Answer:* 67.0°C

12–112 Nitrogen enters a steady-flow heat exchanger at 150 kPa, 10°C, and 100 m/s, and it receives heat in the amount of 150 kJ/kg as it flows through it. The nitrogen leaves the heat exchanger at 100 kPa with a velocity of 200 m/s. Determine the stagnation pressure and temperature of the nitrogen at the inlet and exit states.

12–113 Plot the mass flow parameter $\dot{m}\sqrt{RT_0}/(AP_0)$ versus the Mach number for $k = 1.2$, 1.4, and 1.6 in the range of $0 \leq \mathrm{Ma} \leq 1$.

12–114 Obtain Eq. 12–10 by starting with Eq. 12–9 and using the cyclic rule and the thermodynamic property relations
$$\frac{c_p}{T} = \left(\frac{\partial s}{\partial T}\right)_P \text{ and } \frac{c_v}{T} = \left(\frac{\partial s}{\partial T}\right)_v.$$

12–115 For ideal gases undergoing isentropic flows, obtain expressions for P/P^*, T/T^*, and ρ/ρ^* as functions of k and Ma.

12–116 Using Eqs. 12–4, 12–13, and 12–14, verify that for the steady flow of ideal gases $dT_0/T = dA/A + (1 - \mathrm{Ma}^2)\, dV/V$. Explain the effect of heating and area changes on the velocity of an ideal gas in steady flow for (a) subsonic flow and (b) supersonic flow.

12–117 A subsonic airplane is flying at a 5000-m altitude where the atmospheric conditions are 54 kPa and 256 K. A Pitot static probe measures the difference between the static and stagnation pressures to be 16 kPa. Calculate the speed of the airplane and the flight Mach number.   *Answers:* 199 m/s, 0.620

12–118 Derive an expression for the speed of sound based on van der Waals' equation of state $P = RT/(v - b) - a/v^2$. Using this relation, determine the speed of sound in carbon dioxide at 80°C and 320 kPa, and compare your result to that obtained by assuming ideal-gas behavior. The van der Waals constants for carbon dioxide are $a = 364.3$ kPa·m^6/kmol^2 and $b = 0.0427$ m^3/kmol.

12–119 Helium enters a nozzle at 0.6 MPa, 560 K, and a velocity of 120 m/s. Assuming isentropic flow, determine the pressure and temperature of helium at a location where the velocity equals the speed of sound. What is the ratio of the area at this location to the entrance area?

12–120 Repeat Problem 12–119 assuming the entrance velocity is negligible.

12–121 Air at 0.9 MPa and 400 K enters a converging nozzle with a velocity of 180 m/s. The throat area is 10 cm^2. Assuming isentropic flow, calculate and plot the mass flow rate through the nozzle, the exit velocity, the exit Mach number, and the exit pressure–stagnation pressure ratio versus the back pressure–stagnation pressure ratio for a back pressure range of $0.9 \geq P_b \geq 0.1$ MPa.

12–122 Nitrogen enters a duct with varying flow area at 400 K, 100 kPa, and a Mach number of 0.3. Assuming a steady, isentropic flow, determine the temperature, pressure, and Mach number at a location where the flow area has been reduced by 20 percent.

12–123 Repeat Prob. 12–122 for an inlet Mach number of 0.5.

12–124 Nitrogen enters a converging–diverging nozzle at 620 kPa and 310 K with a negligible velocity, and it experiences a normal shock at a location where the Mach number is Ma = 3.0. Calculate the pressure, temperature, velocity, Mach number, and stagnation pressure downstream of the shock. Compare these results to those of air undergoing a normal shock at the same conditions.

12–125 An aircraft flies with a Mach number Ma$_1$ = 0.9 at an altitude of 7000 m where the pressure is 41.1 kPa and the temperature is 242.7 K. The diffuser at the engine inlet has an exit Mach number of Ma$_2$ = 0.3. For a mass flow rate of 38 kg/s, determine the static pressure rise across the diffuser and the exit area.

12–126 Consider an equimolar mixture of oxygen and nitrogen. Determine the critical temperature, pressure, and density for stagnation temperature and pressure of 550 K and 350 kPa.

12–127 Using the EES software and the relations in Table A–13, calculate the one-dimensional compressible flow functions for an ideal gas with $k = 1.667$, and present your results by duplicating Table A–13.

12–128 Using the EES software and the relations in Table A–14, calculate the one-dimensional normal shock functions for an ideal gas with $k = 1.667$, and present your results by duplicating Table A–14.

12–129 Helium expands in a nozzle from 1 MPa, 500 K, and negligible velocity to 0.1 MPa. Calculate the throat and exit areas for a mass flow rate of 0.46 kg/s, assuming the nozzle is isentropic. Why must this nozzle be converging–diverging?   *Answers:* 6.46 cm^2, 10.8 cm^2

12–130 In compressible flow, velocity measurements with a Pitot probe can be grossly in error if relations developed for incompressible flow are used. Therefore, it is essential that compressible flow relations be used when evaluating flow velocity from Pitot probe measurements. Consider supersonic flow of air through a channel. A probe inserted

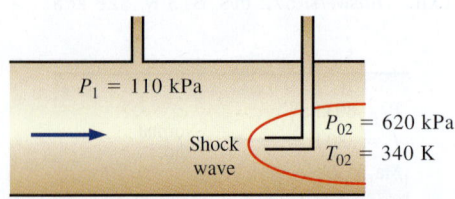

FIGURE P12–130

into the flow causes a shock wave to occur upstream of the probe, and it measures the stagnation pressure and temperature to be 620 kPa and 340 K, respectively. If the static pressure upstream is 110 kPa, determine the flow velocity.

12–131 Using EES (or other) software and the relations given in Table A–14, generate the one-dimensional normal shock functions by varying the upstream Mach number from 1 to 10 in increments of 0.5 for air with $k = 1.4$.

12–132 Repeat Prob. 12–131 for methane with $k = 1.3$.

12–133 Air in a room at $T_0 = 290$ K and $P_0 = 90$ kPa is to be drawn by a vacuum pump through a 3-cm-diameter, 2-m-long adiabatic tube equipped with a converging nozzle at the inlet. The flow in the nozzle section can be approximated as isentropic. The static pressure is measured to be 87 kPa at the tube inlet and 55 kPa at the tube exit. Determine the mass flow rate of air through the duct, the air velocity at the duct exit, and the average friction factor for the duct.

12–134 Air enters a 5.5-cm-diameter adiabatic duct with inlet conditions of $Ma_1 = 2.2$, $T_1 = 250$ K, and $P_1 = 70$ kPa, and exits at a Mach number of $Ma_2 = 1.8$. Taking the average friction factor to be 0.03, determine the velocity, temperature, and pressure at the exit.

12–135 Consider supersonic airflow through a 12-cm-diameter adiabatic duct with inlet conditions of $T_1 = 500$ K, $P_1 = 80$ kPa, and $Ma_1 = 3$. Taking the average friction factor to be 0.03, determine the duct length required to decelerate the flow to a Mach number of unity. Also, calculate the duct length at Mach number intervals of 0.25, and plot the duct length against the Mach number for $1 \leq Ma \leq 3$. Discuss the results.

12–136 Air is heated as it flows subsonically through a 10 cm × 10 cm square duct. The properties of air at the inlet are maintained at $Ma_1 = 0.6$, $P_1 = 350$ kPa, and $T_1 = 420$ K at all times. Disregarding frictional losses, determine the highest rate of heat transfer to the air in the duct without affecting the inlet conditions. *Answer:* 716 kW

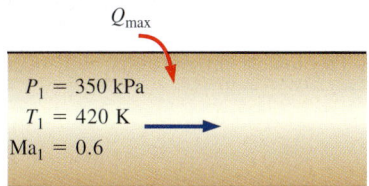

FIGURE P12–136

12–137 Repeat Prob. 12–136 for helium.

12–138 Air is accelerated as it is heated in a duct with negligible friction. Air enters at $V_1 = 100$ m/s, $T_1 = 400$ K, and $P_1 = 35$ kPa and the exits at a Mach number of $Ma_2 = 0.8$. Determine the heat transfer to the air, in kJ/kg. Also determine the maximum amount of heat transfer without reducing the mass flow rate of air.

12–139 Air at sonic conditions and at static temperature and pressure of 340 K and 250 kPa, respectively, is to be accelerated to a Mach number of 1.6 by cooling it as it flows through a channel with constant cross-sectional area. Disregarding frictional effects, determine the required heat transfer from the air, in kJ/kg. *Answer:* 47.5 kJ/kg

12–140 Combustion gases with an average specific heat ratio of $k = 1.33$ and a gas constant of $R = 0.280$ kJ/kg·K enter a 10-cm-diameter adiabatic duct with inlet conditions of $Ma_1 = 2$, $T_1 = 510$ K, and $P_1 = 180$ kPa. If a normal shock occurs at a location 2 m from the inlet, determine the velocity, temperature, and pressure at the duct exit. Take the average friction factor of the duct to be 0.010.

12–141 Air is cooled as it flows through a 20-cm-diameter duct. The inlet conditions are $Ma_1 = 1.2$, $T_{01} = 350$ K, and $P_{01} = 240$ kPa and the exit Mach number is $Ma_2 = 2.0$. Disregarding frictional effects, determine the rate of cooling of air.

12–142 Air is flowing through a 6-cm-diameter adiabatic duct with inlet conditions of $V_1 = 120$ m/s, $T_1 = 400$ K, and $P_1 = 100$ kPa and an exit Mach number of $Ma_2 = 1$. To study the effect of duct length on the mass flow rate and the inlet velocity, the duct is now extended until its length is doubled while P_1 and T_1 are held constant. Taking the average friction factor to be 0.02, calculate the mass flow rate, and the inlet velocity, for various extension lengths, and plot them against the extension length. Discuss the results.

12–143 Using EES (or other) software, determine the shape of a converging–diverging nozzle for air for a mass flow rate of 3 kg/s and inlet stagnation conditions of 1400 kPa and 200°C. Approximate the flow as isentropic. Repeat the calculations for 50-kPa increments of pressure drop to an exit pressure of 100 kPa. Plot the nozzle to scale. Also, calculate and plot the Mach number along the nozzle.

12–144 Steam at 6.0 MPa and 700 K enters a converging nozzle with a negligible velocity. The nozzle throat area is 8 cm². Approximating the flow as isentropic, plot the exit pressure, the exit velocity, and the mass flow rate through the nozzle versus the back pressure P_b for $6.0 \geq P_b \geq 3.0$ MPa. Treat the steam as an ideal gas with $k = 1.3$, $c_p = 1.872$ kJ/kg·K, and $R = 0.462$ kJ/kg·K.

12–145 Find the expression for the ratio of the stagnation pressure after a shock wave to the static pressure before the shock wave as a function of k and the Mach number upstream of the shock wave Ma_1.

12–146 Using EES (or other) software and the relations given in Table A–13, calculate the one-dimensional isentropic compressible-flow functions by varying the upstream Mach number from 1 to 10 in increments of 0.5 for air with $k = 1.4$.

12–147 Repeat Prob. 12–146 for methane with $k = 1.3$.

Fundamentals of Engineering (FE) Exam Problems

12–148 An aircraft is cruising in still air at 5°C at a velocity of 400 m/s. The air temperature at the nose of the aircraft where stagnation occurs is
(a) 5°C
(b) 25°C
(c) 55°C
(d) 80°C
(e) 85°C

12–149 Air is flowing in a wind tunnel at 25°C, 80 kPa, and 250 m/s. The stagnation pressure at the location of a probe inserted into the flow section is
(a) 87 kPa
(b) 93 kPa
(c) 113 kPa
(d) 119 kPa
(e) 125 kPa

12–150 An aircraft is reported to be cruising in still air at −20°C and 40 kPa at a Mach number of 0.86. The velocity of the aircraft is
(a) 91 m/s
(b) 220 m/s
(c) 186 m/s
(d) 280 m/s
(e) 378 m/s

12–151 Air is flowing in a wind tunnel at 12°C and 66 kPa at a velocity of 230 m/s. The Mach number of the flow is
(a) 0.54 m/s
(b) 0.87 m/s
(c) 3.3 m/s
(d) 0.36 m/s
(e) 0.68 m/s

12–152 Consider a converging nozzle with a low velocity at the inlet and sonic velocity at the exit plane. Now the nozzle exit diameter is reduced by half while the nozzle inlet temperature and pressure are maintained the same. The nozzle exit velocity will
(a) remain the same
(b) double
(c) quadruple
(d) go down by half
(e) go down by one-fourth

12–153 Air is approaching a converging–diverging nozzle with a low velocity at 12°C and 200 kPa, and it leaves the nozzle at a supersonic velocity. The velocity of air at the throat of the nozzle is
(a) 338 m/s
(b) 309 m/s
(c) 280 m/s
(d) 256 m/s
(e) 95 m/s

12–154 Argon gas is approaching a converging–diverging nozzle with a low velocity at 20°C and 120 kPa, and it leaves the nozzle at a supersonic velocity. If the cross-sectional area of the throat is 0.015 m², the mass flow rate of argon through the nozzle is
(a) 0.41 kg/s
(b) 3.4 kg/s
(c) 5.3 kg/s
(d) 17 kg/s
(e) 22 kg/s

12–155 Carbon dioxide enters a converging–diverging nozzle at 60 m/s, 310°C, and 300 kPa, and it leaves the nozzle at a supersonic velocity. The velocity of carbon dioxide at the throat of the nozzle is
(a) 125 m/s
(b) 225 m/s
(c) 312 m/s
(d) 353 m/s
(e) 377 m/s

12–156 Consider gas flow through a converging–diverging nozzle. Of the five following statements, select the one that is incorrect:
(a) The fluid velocity at the throat can never exceed the speed of sound.
(b) If the fluid velocity at the throat is below the speed of sound, the diversion section will act like a diffuser.
(c) If the fluid enters the diverging section with a Mach number greater than one, the flow at the nozzle exit will be supersonic.
(d) There will be no flow through the nozzle if the back pressure equals the stagnation pressure.
(e) The fluid velocity decreases, the entropy increases, and stagnation enthalpy remains constant during flow through a normal shock.

12–157 Combustion gases with $k = 1.33$ enter a converging nozzle at stagnation temperature and pressure of 350°C and 400 kPa, and are discharged into the atmospheric air at 20°C and 100 kPa. The lowest pressure that will occur within the nozzle is
(a) 13 kPa
(b) 100 kPa
(c) 216 kPa
(d) 290 kPa
(e) 315 kPa

Design and Essay Problems

12–158 Find out if there is a supersonic wind tunnel on your campus. If there is, obtain the dimensions of the wind tunnel and the temperatures and pressures as well as the Mach number at several locations during operation. For what typical experiments is the wind tunnel used?

12–159 Assuming you have a thermometer and a device to measure the speed of sound in a gas, explain how you can determine the mole fraction of helium in a mixture of helium gas and air.

12–160 Design a 1-m-long cylindrical wind tunnel whose diameter is 25 cm operating at a Mach number of 1.8. Atmospheric air enters the wind tunnel through a converging–diverging nozzle where it is accelerated to supersonic velocities. Air leaves the tunnel through a converging–diverging diffuser where it is decelerated to a very low velocity before entering the fan section. Disregard any irreversibilities. Specify the temperatures and pressures at several locations as well as the mass flow rate of air at steady-flow conditions. Why is it often necessary to dehumidify the air before it enters the wind tunnel?

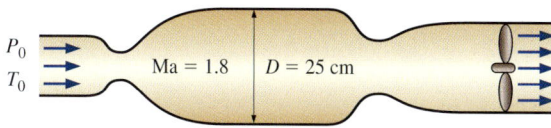

FIGURE P12–160

CHAPTER 13

OPEN-CHANNEL FLOW

Open-channel flow implies flow in a channel open to the atmosphere, but flow in a conduit is also open-channel flow if the liquid does not fill the conduit completely, and thus there is a free surface. An open-channel flow involves liquids only (typically water or wastewater) exposed to a gas (usually air, which is at atmospheric pressure).

Flow in pipes is driven by gravity and/or a pressure difference, whereas flow in a channel is driven naturally by gravity. Water flow in a river, for example, is driven by the upstream and downstream elevation difference. The flow rate in an open channel is established by the dynamic balance between gravity and friction. Inertia of the flowing liquid also becomes important in unsteady flow. The free surface coincides with the hydraulic grade line (HGL) and the pressure is constant along the free surface. But the height of the free surface from the channel bottom and thus all dimensions of the flow cross-section along the channel are not known *a priori*—they change along with average flow velocity.

In this chapter we present the basic principles of open-channel flows and the associated correlations for steady one-dimensional flow in channels of common cross sections. Detailed information can be obtained from several books written on the topic, some of which are listed in the references.

OBJECTIVES

When you finish reading this chapter, you should be able to

- Understand how flow in open channels differs from pressurized flow in pipes
- Learn the different flow regimes in open channels and their characteristics
- Predict if hydraulic jumps are to occur during flow, and calculate the fraction of energy dissipated during hydraulic jumps
- Understand how flow rates in open channels are measured using sluice gates and weirs

Any flow of a liquid with a free surface is a type of open-channel flow. In this photograph, the Nicholson River meanders through northern Australia.
© Digital Vision/Getty RF

(a)

(b)

FIGURE 13–1
Natural and human-made open-channel flows are characterized by a free surface open to the atmosphere.
(a) © Doug Sherman/Geofile RF;
(b) Royalty-Free/CORBIS

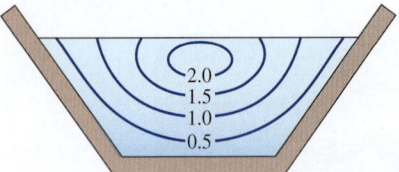

FIGURE 13–2
Typical constant axial velocity contours in an open channel of trapezoidal cross section; values are relative to the average velocity.

13–1 ▪ CLASSIFICATION OF OPEN-CHANNEL FLOWS

Open-channel flow refers to the flow of liquids in channels open to the atmosphere or in partially filled conduits and is characterized by the presence of a liquid–gas interface called the *free surface* (Fig. 13–1). Most natural flows encountered in practice, such as the flow of water in creeks, rivers, and floods, as well as the draining of rainwater off highways, parking lots, and roofs are open-channel flows. Human-made open-channel flow systems include irrigation systems, sewer lines, drainage ditches, and gutters, and the design of such systems is an important application area of engineering.

In an open channel, the flow velocity is zero at the side and bottom surfaces because of the no-slip condition, and maximum at the midplane for symmetric geometries, typically somewhat below the free surface, as shown in Fig. 13–2. (Because of secondary flows that occur even in straight channels when they are narrow, the maximum axial velocity occurs *below* the free surface, typically within the top 25 percent of depth.) Furthermore, flow velocity also varies in the flow direction in most cases. Therefore, the velocity distribution (and thus flow) in open channels is, in general, three-dimensional. In engineering practice, however, the equations are written in terms of the average velocity at a cross section of the channel. Since the average velocity varies only with streamwise distance x, V is a **one-dimensional** variable. The one-dimensionality makes it possible to solve significant real-world problems in a simple manner by hand calculations, and we restrict our consideration in this chapter to flows with one-dimensional average velocity. Despite its simplicity, the one-dimensional equations provide remarkably accurate results and are commonly used in practice.

The no-slip condition on the channel walls gives rise to velocity gradients, and wall shear stress τ_w develops along the wetted surfaces. The wall shear stress varies along the wetted perimeter at a given cross section and offers resistance to flow. The magnitude of this resistance depends on the viscosity of the fluid as well as the velocity gradients at the wall surface, which in turn depend on wall roughness.

Open-channel flows are also classified as being steady or unsteady. A flow is said to be **steady** if there is no change with time at a given location. The representative quantity in open-channel flows is the **flow depth** (or alternately, the average velocity), which may vary along the channel. The flow is said to be *steady* if the flow depth does not vary with time at any given location along the channel (although it may vary from one location to another). Otherwise, the flow is *unsteady*. In this chapter we deal with steady flow only.

Uniform and Varied Flows

Flow in open channels is also classified as being *uniform* or *nonuniform* (also called *varied*), depending on how the flow depth y (the distance of the free surface from the bottom of the channel measured in the vertical direction) varies along the channel. The flow in a channel is said to be **uniform** if the flow depth (and thus the average velocity) remains constant. Otherwise, the flow is said to be **nonuniform** or **varied,** indicating that the flow depth varies with distance in the flow direction. Uniform flow conditions are commonly encountered in practice in long straight sections of channels with constant slope, constant roughness, and constant cross section.

In open channels of constant slope and constant cross section, the liquid accelerates until the head loss due to frictional effects equals the elevation drop. The liquid at this point reaches its terminal velocity, and uniform flow is established. The flow remains uniform as long as the slope, cross section, and surface roughness of the channel remain unchanged. The flow depth in uniform flow is called the **normal depth** y_n, which is an important characteristic parameter for open-channel flows (Fig. 13–3).

The presence of an obstruction in the channel, such as a gate or a change in slope or cross section, causes the flow depth to vary, and thus the flow to become **varied** or **nonuniform.** Such varied flows are common in both natural and human-made open channels such as rivers, irrigation systems, and sewer lines. The varied flow is called **rapidly varied flow (RVF)** if the flow depth changes markedly over a relatively short distance in the flow direction (such as the flow of water past a partially open gate or over a falls), and **gradually varied flow (GVF)** if the flow depth changes gradually over a long distance along the channel. A gradually varied flow region typically occurs between rapidly varied and uniform flow regions, as shown in Fig. 13–4.

In gradually varied flows, we can work with the one-dimensional average velocity just as we can with uniform flows. However, average velocity is not always the most useful or most appropriate parameter for rapidly varying flows. Therefore, the analysis of rapidly varied flows is rather complicated, especially when the flow is unsteady (such as the breaking of waves on the shore). For a known discharge rate, the flow height in a gradually varied flow region (i.e., the profile of the free surface) in a specified open channel can be determined in a step-by-step manner by starting the analysis at a cross section where the flow conditions are known, and evaluating head loss, elevation drop, and then the average velocity for each step.

Laminar and Turbulent Flows in Channels

Like pipe flow, open-channel flow can be laminar, transitional, or turbulent, depending on the value of the **Reynolds number** expressed as

$$\text{Re} = \frac{\rho V R_h}{\mu} = \frac{V R_h}{\nu} \quad (13\text{–}1)$$

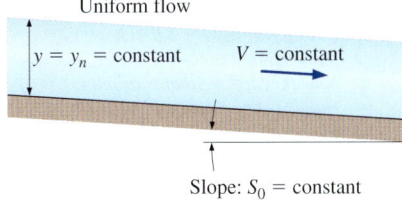

FIGURE 13–3
For uniform flow in an open channel, the flow depth y and the average flow velocity V remain constant.

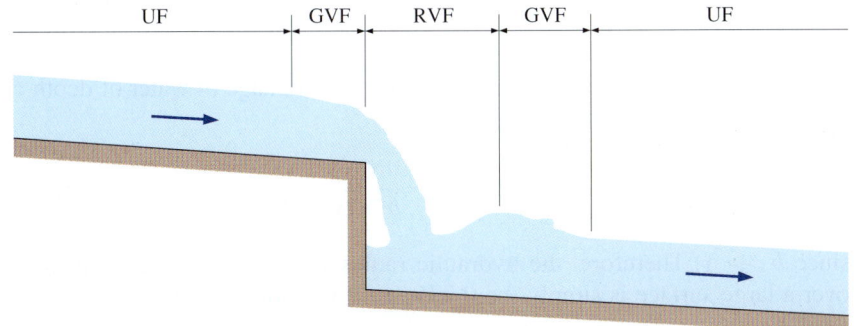

FIGURE 13–4
Uniform flow (UF), gradually varied flow (GVF), and rapidly varied flow (RVF) in an open channel.

Here V is the average liquid velocity, ν is the kinematic viscosity, and R_h is the **hydraulic radius** defined as the ratio of the cross-sectional flow area A_c and the wetted perimeter p,

Hydraulic radius: $$R_h = \frac{A_c}{p} \quad (m) \tag{13-2}$$

Considering that open channels come with rather irregular cross sections, the hydraulic radius serves as the characteristic dimension and brings uniformity to the treatment of open channels. Also, the Reynolds number is constant for the entire uniform flow section of an open channel.

You might expect that the hydraulic radius would be defined as half the hydraulic diameter, but this is unfortunately not the case. Recall that the hydraulic diameter D_h for pipe flow is defined as $D_h = 4A_c/p$ so that the hydraulic diameter reduces to the pipe diameter for circular pipes. The relation between hydraulic radius and hydraulic diameter is

Hydraulic diameter: $$D_h = \frac{4A_c}{p} = 4R_h \tag{13-3}$$

So, we see that the hydraulic radius is in fact *one-fourth*, rather than one-half, of the hydraulic diameter (Fig. 13–5).

Therefore, a Reynolds number based on the hydraulic radius is one-fourth of the Reynolds number based on hydraulic diameter as the characteristic dimension. So it will come as no surprise that the flow is laminar for Re ≲ 2000 in pipe flow, but for Re ≲ 500 in open-channel flow. Also, open-channel flow is usually turbulent for Re ≳ 2500 and transitional for 500 ≲ Re ≲ 2500. Laminar flow is encountered when a thin layer of water (such as the rainwater draining off a road or parking lot) flows at a low velocity.

The kinematic viscosity of water at 20°C is 1.00×10^{-6} m²/s, and the average flow velocity in open channels is usually above 0.5 m/s. Also, the hydraulic radius is usually greater than 0.1 m. Therefore, the Reynolds number associated with water flow in open channels is typically above 50,000, and thus the flow is almost always turbulent.

Note that the wetted perimeter includes the sides and the bottom of the channel in contact with the liquid—it does not include the free surface and the parts of the sides exposed to air. For example, the wetted perimeter and the cross-sectional flow area for a rectangular channel of height h and width b containing water of depth y are $p = b + 2y$ and $A_c = yb$, respectively. Then,

Rectangular channel: $$R_h = \frac{A_c}{p} = \frac{yb}{b + 2y} = \frac{y}{1 + 2y/b} \tag{13-4}$$

As another example, the hydraulic radius for the drainage of water of depth y off a parking lot of width b is (Fig. 13–6)

Liquid layer of thickness y: $$R_h = \frac{A_c}{p} = \frac{yb}{b + 2y} \cong \frac{yb}{b} \cong y \tag{13-5}$$

since $b \gg y$. Therefore, the hydraulic radius for the flow of a liquid film over a large surface is simply the thickness of the liquid layer.

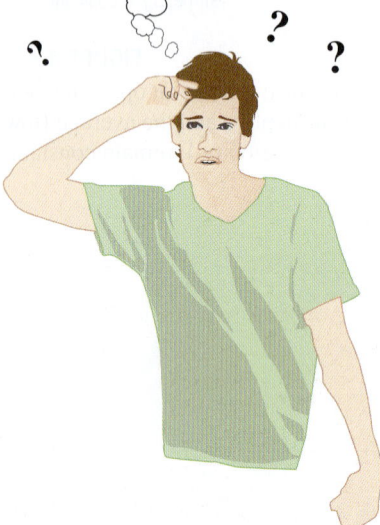

FIGURE 13–5
The relationship between the hydraulic radius and hydraulic diameter is not what you might expect.

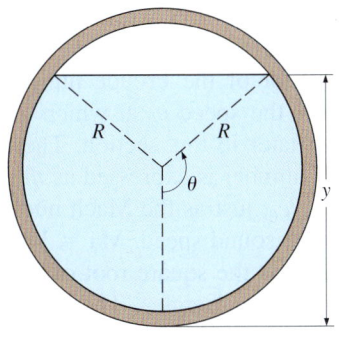

$A_c = R^2(\theta - \sin\theta \cos\theta)$
$p = 2R\theta$
$R_h = \dfrac{A_c}{p} = \dfrac{\theta - \sin\theta \cos\theta}{2\theta} R$

(a) Circular channel (θ in rad)

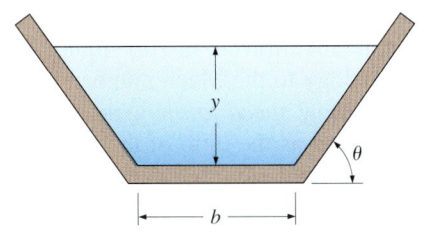

$R_h = \dfrac{A_c}{p} = \dfrac{y(b + y/\tan\theta)}{b + 2y/\sin\theta}$

(b) Trapezoidal channel

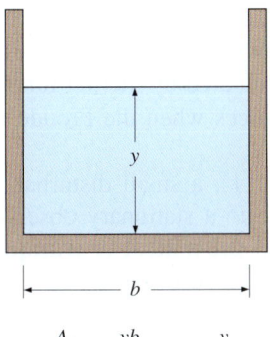

$R_h = \dfrac{A_c}{p} = \dfrac{yb}{b + 2y} = \dfrac{y}{1 + 2y/b}$

(c) Rectangular channel

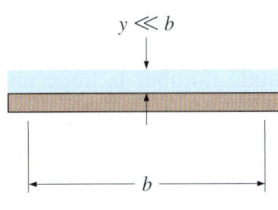

$R_h = \dfrac{A_c}{p} = \dfrac{yb}{b + 2y} \cong \dfrac{yb}{b} \cong y$

(d) Liquid film of thickness y

FIGURE 13–6
Hydraulic radius relations for various open-channel geometries.

13–2 · FROUDE NUMBER AND WAVE SPEED

Open-channel flow is also classified as *subcritical*, *critical*, or *supercritical*, depending on the value of the dimensionless Froude number mentioned in Chap. 7 and defined as

Froude number:
$$\text{Fr} = \dfrac{V}{\sqrt{gL_c}} \quad (13\text{-}6)$$

where g is the gravitational acceleration, V is the average liquid velocity at a cross section, and L_c is the characteristic length. L_c is taken to be the flow depth y for wide rectangular channels, and $\text{Fr} = V/\sqrt{gy}$. The Froude number is an important parameter that governs the character of flow in open channels. The flow is classified as

$$\begin{array}{ll} \text{Fr} < 1 & \text{Subcritical or tranquil flow} \\ \text{Fr} = 1 & \text{Critical flow} \\ \text{Fr} > 1 & \text{Supercritical or rapid flow} \end{array} \quad (13\text{-}7)$$

Compressible Flow	Open-Channel Flow
Ma = V/c	Fr = V/c₀
Ma < 1 Subsonic	Fr < 1 Subcritical
Ma = 1 Sonic	Fr = 1 Critical
Ma > 1 Supersonic	Fr > 1 Supercritical

V = speed of flow
$c = \sqrt{kRT}$ = speed of sound (ideal gas)
$c_0 = \sqrt{gy}$ = speed of wave (liquid)

FIGURE 13–7
Analogy between the Mach number for compressible flow and the Froude number for open-channel flow.

This resembles the classification of compressible flow with respect to the Mach number: subsonic for Ma < 1, sonic for Ma = 1, and supersonic for Ma > 1 (Fig. 13–7). Indeed, the denominator of the Froude number has the dimensions of velocity, and it represents the speed c_0 at which a small disturbance travels in still liquid, as shown later in this section. Therefore, in analogy to the Mach number, the Froude number is expressed as *the ratio of the flow speed to the wave speed*, Fr = V/c_0, just as the Mach number is expressed as the ratio of the flow speed to the sound speed, Ma = V/c.

The Froude number can also be thought of as the square root of the ratio of inertia (or dynamic) force to gravity force (or weight). This is demonstrated by multiplying both the numerator and the denominator of the square of the Froude number V^2/gL_c by ρA, where ρ is density and A is a representative area, which gives

$$\text{Fr}^2 = \frac{V^2}{gL_c}\frac{\rho A}{\rho A} = \frac{2(\frac{1}{2}\rho V^2 A)}{mg} \propto \frac{\text{Inertia force}}{\text{Gravity force}} \quad (13\text{–}8)$$

Here $L_c A$ represents volume, $\rho L_c A$ is the mass of this fluid volume, and mg is the weight. The numerator is twice the inertial force $\frac{1}{2}\rho V^2 A$, which can be thought of as the dynamic pressure $\frac{1}{2}\rho V^2$ times the cross-sectional area, A. Therefore, the flow in an open channel is dominated by inertial forces when the Froude number is large and by gravity forces when the Froude number is small.

It follows that at *low flow velocities* (Fr < 1), a small disturbance travels upstream (with a velocity $c_0 - V$ relative to a stationary observer) and affects the upstream conditions. This is called **subcritical** or **tranquil** flow. But at *high flow velocities* (Fr > 1), a small disturbance cannot travel upstream (in fact, the wave is washed downstream at a velocity of $V - c_0$ relative to a stationary observer) and thus the upstream conditions cannot be influenced by the downstream conditions. This is called **supercritical** or **rapid** flow, and the flow in this case is controlled by the upstream conditions. Therefore, a surface wave travels upstream when Fr < 1, is swept downstream when Fr > 1, and appears frozen on the surface when Fr = 1. Also, when the water is shallow compared to the wavelength of the disturbance, the surface wave speed increases with flow depth y, and thus a surface disturbance propagates much faster in deep channels than it does in shallow ones.

Consider the flow of a liquid in an open rectangular channel of cross-sectional area A_c with volume flow rate $\dot{V}$. When the flow is critical, Fr = 1 and the average flow velocity is $V = \sqrt{gy_c}$, where y_c is the **critical depth**. Noting that $\dot{V} = A_c V = A_c \sqrt{gy_c}$, the critical depth is expressed as

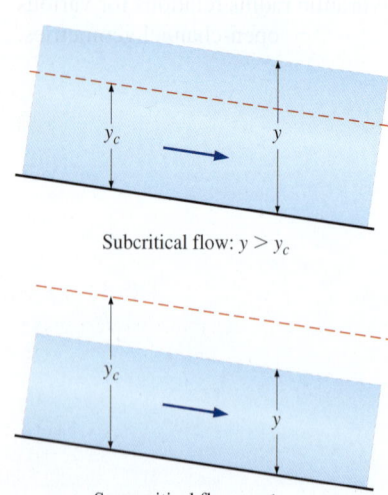

Subcritical flow: $y > y_c$

Supercritical flow: $y < y_c$

FIGURE 13–8
Definitions of subcritical flow and supercritical flow in terms of critical depth.

Critical depth (general): $\quad y_c = \dfrac{\dot{V}^2}{gA_c^2} \quad (13\text{–}9)$

For a rectangular channel of width b we have $A_c = by_c$, and the critical depth relation reduces to

Critical depth (rectangular): $\quad y_c = \left(\dfrac{\dot{V}^2}{gb^2}\right)^{1/3} \quad (13\text{–}10)$

The liquid depth is $y > y_c$ for subcritical flow and $y < y_c$ for supercritical flow (Fig. 13–8).

As in compressible flow, a liquid can accelerate from subcritical to supercritical flow. Of course, it can also decelerate from supercritical to subcritical flow, and it can do so by undergoing a shock. The shock in this case is called a **hydraulic jump,** which corresponds to a *normal shock* in compressible flow. Therefore, the analogy between open-channel flow and compressible flow is remarkable.

Speed of Surface Waves

We are all familiar with the waves forming on the free surfaces of oceans, lakes, rivers, and even swimming pools. The surface waves can be very high, like the ones we see on the oceans, or barely noticeable. Some are smooth; some break on the surface. A basic understanding of wave motion is necessary for the study of certain aspects of open-channel flow, and here we present a brief description. A detailed treatment of wave motion can be found in numerous books written on the subject.

An important parameter in the study of open-channel flow is the **wave speed** c_0, which is the speed at which a surface disturbance travels through a liquid. Consider a long, wide channel that initially contains a still liquid of height y. One end of the channel is moved with speed δV, generating a surface wave of height δy propagating at a speed of c_0 into the still liquid, as shown in Fig. 13–9a.

Now consider a control volume that encloses the wave front and moves with it, as shown in Fig. 13–9b. To an observer traveling with the wave front, the liquid to the right appears to be moving toward the wave front with speed c_0 and the liquid to the left appears to be moving away from the wave front with speed $c_0 - \delta V$. Of course the observer would think the control volume that encloses the wave front (and herself or himself) is stationary, and he or she would be witnessing a steady-flow process.

The steady-flow mass balance $\dot{m}_1 = \dot{m}_2$ (or the continuity relation) for this control volume of width b is expressed as

$$\rho c_0 y b = \rho (c_0 - \delta V)(y + \delta y) b \quad \rightarrow \quad \delta V = c_0 \frac{\delta y}{y + \delta y} \quad (13\text{--}11)$$

We make the following approximations: (1) the velocity is nearly constant across the channel and thus the momentum flux correction factors (β_1 and β_2) are one, (2) the distance across the wave is short and thus friction at the bottom surface and air drag at the top are negligible, (3) the dynamic effects are negligible and thus the pressure in the liquid varies hydrostatically; in terms of gage pressure, $P_{1,\text{avg}} = \rho g h_{1,\text{avg}} = \rho g (y/2)$ and $P_{2,\text{avg}} = \rho g h_{2,\text{avg}} = \rho g (y + \delta y)/2$, (4) the mass flow rate is constant with $\dot{m}_1 = \dot{m}_2 = \rho c_0 y b$, and (5) there are no external forces or body forces and thus the only forces acting on the control volume in the horizontal *x*-direction are the pressure forces. Then, the momentum equation $\sum \vec{F} = \sum_{\text{out}} \beta \dot{m} \vec{V} - \sum_{\text{in}} \beta \dot{m} \vec{V}$ in the *x*-direction becomes a balance between hydrostatic pressure forces and momentum transfer,

$$P_{2,\text{avg}} A_2 - P_{1,\text{avg}} A_1 = \dot{m}(-V_2) - \dot{m}(-V_1) \quad (13\text{--}12)$$

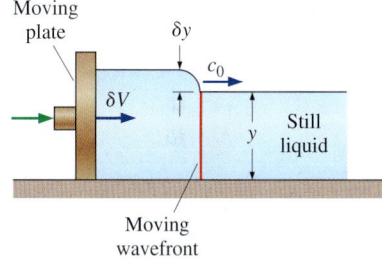

(*a*) Generation and propagation of a wave

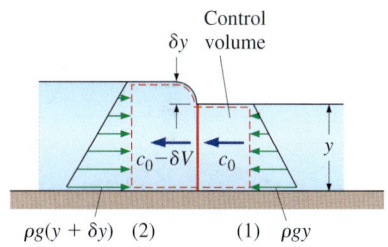

(*b*) Control volume relative to an observer traveling with the wave, with gage pressure distributions shown

FIGURE 13–9
The generation and analysis of a wave in an open channel.

732
OPEN-CHANNEL FLOW

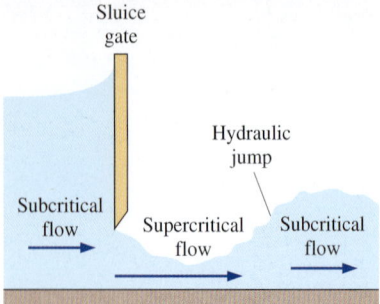

FIGURE 13–10
Supercritical flow through a sluice gate.

FIGURE 13–11
A hydraulic jump can be observed on a dinner plate when (*a*) it is right-side-up, but not when (*b*) it is upside down.
Photos by Abel Po-Ya Chuang. Used by permission.

Note that both the inlet and the outlet average velocities are negative since they are in the negative *x*-direction. Substituting,

$$\frac{\rho g(y + \delta y)^2 b}{2} - \frac{\rho g y^2 b}{2} = \rho c_0 y b(-c_0 + \delta V) - \rho c_0 y b(-c_0) \quad (13\text{–}13)$$

or,

$$g\left(1 + \frac{\delta y}{2y}\right)\delta y = c_0 \, \delta V \quad (13\text{–}14)$$

Combining the momentum and continuity relations and rearranging give

$$c_0^2 = gy\left(1 + \frac{\delta y}{y}\right)\left(1 + \frac{\delta y}{2y}\right) \quad (13\text{–}15)$$

Therefore, the wave speed c_0 is proportional to the wave height δy. For infinitesimal surface waves, $\delta y \ll y$ and thus

Infinitesimal surface waves: $\quad c_0 = \sqrt{gy} \quad (13\text{–}16)$

Therefore, the speed of infinitesimal surface waves is proportional to the square root of liquid depth. Again note that this analysis is valid only for shallow liquid bodies, such as those encountered in open channels. Otherwise, the wave speed is independent of liquid depth for deep bodies of liquid, such as the oceans. The wave speed can also be determined by using the energy balance relation instead of the momentum equation together with the continuity relation. Note that the waves eventually die out because of the viscous effects that are neglected in the analysis. Also, for flow in channels of non-rectangular cross-section, the **hydraulic depth** defined as $y_h = A_c/L_t$ where L_t is the *top width* of the flow section should be used in the calculation of Froude number in place of the flow depth y. For a half-full circular channel, for example, the hydraulic depth is $y_h = (\pi R^2/2)/2R = \pi R/4$.

We know from experience that when a rock is thrown into a lake, the concentric waves that form propagate evenly in all directions and vanish after some distance. But when the rock is thrown into a river, the upstream side of the wave moves upstream if the flow is tranquil or subcritical ($V < c_0$), moves downstream if the flow is rapid or supercritical ($V > c_0$), and remains stationary at the location where it is formed if the flow is critical ($V = c_0$).

You may be wondering why we pay so much attention to flow being subcritical or supercritical. The reason is that the character of the flow is strongly influenced by this phenomenon. For example, a rock at the riverbed may cause the water level at that location to rise or to drop, depending on whether the flow is subcritical or supercritical. Also, the liquid level drops gradually in the flow direction in subcritical flow, but a sudden rise in liquid level, called a hydraulic jump, may occur in supercritical flow (Fr > 1) as the flow decelerates to subcritical (Fr < 1) velocities.

This phenomenon can occur downstream of a sluice gate as shown in Fig. 13–10. The liquid approaches the gate with a subcritical velocity, but the upstream liquid level is sufficiently high to accelerate the liquid to a supercritical level as it passes through the gate (just like a gas flowing in a converging–diverging nozzle). But if the downstream section of the channel is not sufficiently sloped down, it cannot maintain this supercritical velocity, and the liquid jumps up to a higher level with a larger cross-sectional area, and thus to a lower subcritical velocity. Finally, the flow in rivers, canals, and irrigation systems is typically subcritical. But the flow past sluice gates and spillways is typically supercritical.

You can create a beautiful hydraulic jump the next time you wash dishes (Fig. 13–11). Let the water from the faucet hit the middle of a dinner plate. As the water spreads out radially, its depth decreases and the flow is supercritical. Eventually, a hydraulic jump occurs, which you can see as a sudden increase in water depth. Try it!

13–3 ■ SPECIFIC ENERGY

Consider the flow of a liquid in a channel at a cross section where the flow depth is y, the average flow velocity is V, and the elevation of the bottom of the channel at that location relative to some reference datum is z. For simplicity, we ignore the variation of liquid speed over the cross section and assume the speed to be V everywhere. The total mechanical energy of this liquid in the channel in terms of heads is expressed as (Fig. 13–12)

$$H = z + \frac{P}{\rho g} + \frac{V^2}{2g} = z + y + \frac{V^2}{2g} \tag{13-17}$$

where z is the *elevation head*, $P/\rho g = y$ is the *gage pressure head*, and $V^2/2g$ is the *velocity* or *dynamic head*. The total energy as expressed in Eq. 13–17 is not a realistic representation of the true energy of a flowing fluid since the choice of the reference datum and thus the value of the elevation head z is rather arbitrary. The intrinsic energy of a fluid at a cross section is represented more realistically if the reference datum is taken to be the bottom of the channel so that $z = 0$ there. Then the total mechanical energy of a fluid in terms of heads becomes the sum of the pressure and dynamic heads. The sum of the pressure and dynamic heads of a liquid in an open channel is called the **specific energy** E_s and is expressed as (Bakhmeteff, 1932)

$$E_s = y + \frac{V^2}{2g} \tag{13-18}$$

as shown in Fig. 13–12.

Consider flow in an open channel of rectangular cross section and of constant width b. Noting that the volume flow rate is $\dot{V} = A_c V = ybV$, the average flow velocity is

$$V = \frac{\dot{V}}{yb} \tag{13-19}$$

Substituting into Eq. 13–18, the specific energy becomes

$$E_s = y + \frac{\dot{V}^2}{2gb^2y^2} \tag{13-20}$$

This equation is very instructive as it shows the variation of the specific energy with flow depth. During steady flow in an open channel the flow rate is constant, and a plot of E_s versus y for constant $\dot{V}$ and b is given in Fig. 13–13. We observe the following from this figure:

- The distance from a point on the vertical y-axis to the curve represents the specific energy at that y-value. The part between the $E_s = y$ line and the curve corresponds to dynamic head (or kinetic energy head) of the liquid, and the remaining part to pressure head (or potential energy head).

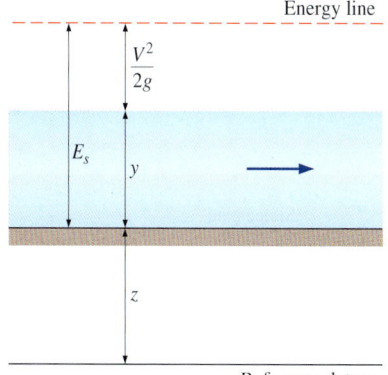

FIGURE 13–12
The specific energy E_s of a liquid in an open channel is the total mechanical energy (expressed as a head) relative to the bottom of the channel.

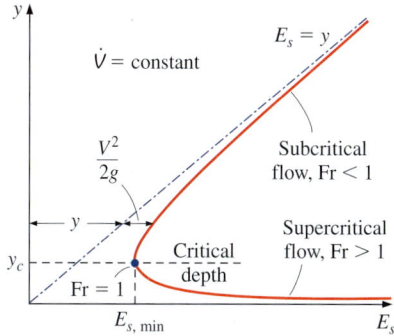

FIGURE 13–13
Variation of specific energy E_s with depth y for a specified flow rate.

- The specific energy tends to infinity as $y \to 0$ (due to the velocity approaching infinity), and it becomes equal to flow depth y for large values of y (due to the velocity and thus the kinetic energy becoming very small). The specific energy reaches a minimum value $E_{s,\,min}$ at some intermediate point, called the **critical point,** characterized by the **critical depth** y_c and **critical velocity** V_c. The minimum specific energy is also called the **critical energy.**

- There is a minimum specific energy $E_{s,\,min}$ required to support the specified flow rate $\dot{V}$. Therefore, E_s cannot be below $E_{s,\,min}$ for a given $\dot{V}$.

- A horizontal line intersects the specific energy curve at one point only, and thus a fixed value of flow depth corresponds to a fixed value of specific energy. This is expected since the velocity has a fixed value when $\dot{V}$, b, and y are specified. However, for $E_s > E_{s,\,min}$, a vertical line intersects the curve at *two* points, indicating that a flow can have two different depths (and thus two different velocities) corresponding to a fixed value of specific energy. These two depths are called **alternate depths.** For flow through a sluice gate with negligible frictional losses (and thus E_s = constant), the upper depth corresponds to the upstream flow, and the lower depth to the downstream flow (Fig. 13–14).

- A small change in specific energy near the critical point causes a large difference between alternate depths and may cause violent fluctuations in flow level. Therefore, operation near the critical point should be avoided in the design of open channels.

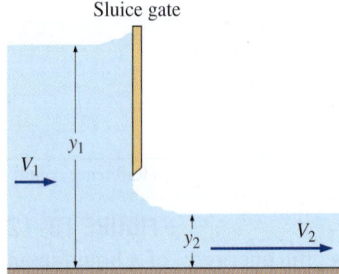

FIGURE 13–14
A sluice gate illustrates alternate depths—the deep liquid upstream of the sluice gate and the shallow liquid downstream of the sluice gate.

The value of the minimum specific energy and the critical depth at which it occurs is determined by differentiating E_s from Eq. 13–20 with respect to y for constant b and $\dot{V}$, and setting the derivative equal to zero:

$$\frac{dE_s}{dy} = \frac{d}{dy}\left(y + \frac{\dot{V}^2}{2gb^2y^2}\right) = 1 - \frac{\dot{V}^2}{gb^2y^3} = 0 \qquad (13\text{–}21)$$

Solving for y, which is the critical flow depth y_c, gives

$$y_c = \left(\frac{\dot{V}^2}{gb^2}\right)^{1/3} \qquad (13\text{–}22)$$

The flow rate at the critical point can be expressed as $\dot{V} = y_c b V_c$. Substituting, the critical velocity is determined to be

$$V_c = \sqrt{gy_c} \qquad (13\text{–}23)$$

which is the wave speed. The Froude number at this point is

$$\text{Fr} = \frac{V}{\sqrt{gy}} = \frac{V_c}{\sqrt{gy_c}} = 1 \qquad (13\text{–}24)$$

indicating that *the point of minimum specific energy is indeed the critical point,* and *the flow becomes critical when the specific energy reaches its minimum value.*

It follows that the flow is subcritical at lower flow velocities and thus higher flow depths (the upper arm of the curve in Fig. 13–13), supercritical at higher velocities and thus lower flow depths (the lower arm of the curve), and critical at the critical point (the point of minimum specific energy).

Noting that $V_c = \sqrt{gy_c}$, the minimum (or critical) specific energy can be expressed in terms of the critical depth alone as

$$E_{s,\,min} = y_c + \frac{V_c^2}{2g} = y_c + \frac{gy_c}{2g} = \frac{3}{2}y_c \qquad (13\text{--}25)$$

In uniform flow, the flow depth and the flow velocity, and thus the specific energy, remain constant since $E_s = y + V^2/2g$. The head loss is made up by the decline in elevation (the channel is sloped downward in the flow direction). In nonuniform flow, however, the specific energy may increase or decrease, depending on the slope of the channel and the frictional losses. If the decline in elevation across a flow section is more than the head loss in that section, for example, the specific energy increases by an amount equal to the difference between elevation drop and head loss. The specific energy concept is a particularly useful tool when studying varied flows.

■ **EXAMPLE 13–1**   **Character of Flow and Alternate Depth**

■ Water is flowing steadily in a 0.4-m-wide rectangular open channel at a rate
■ of 0.2 m³/s (Fig. 13–15). If the flow depth is 0.15 m, determine the flow
velocity and if the flow is subcritical or supercritical. Also determine the
alternate flow depth if the character of flow were to change.

SOLUTION   Water flow in a rectangular open channel is considered. The character of flow, the flow velocity, and the alternate depth are to be determined.
Assumptions   The specific energy is constant.
Analysis   The average flow velocity is determined from

$$V = \frac{\dot{V}}{A_c} = \frac{\dot{V}}{yb} = \frac{0.2\text{ m}^3/\text{s}}{(0.15\text{ m})(0.4\text{ m})} = \mathbf{3.33\text{ m/s}}$$

The critical depth for this flow is

$$y_c = \left(\frac{\dot{V}^2}{gb^2}\right)^{1/3} = \left(\frac{(0.2\text{ m}^3/\text{s})^2}{(9.81\text{ m/s}^2)(0.4\text{ m})^2}\right)^{1/3} = 0.294\text{ m}$$

Therefore, the flow is **supercritical** since the actual flow depth is $y = 0.15$ m, and $y < y_c$. Another way to determine the character of flow is to calculate the Froude number,

$$\text{Fr} = \frac{V}{\sqrt{gy}} = \frac{3.33\text{ m/s}}{\sqrt{(9.81\text{ m/s}^2)(0.15\text{ m})}} = 2.75$$

Again the flow is supercritical since Fr > 1. The specific energy for the given conditions is

$$E_{s1} = y_1 + \frac{\dot{V}^2}{2gb^2 y_1^2} = (0.15\text{ m}) + \frac{(0.2\text{ m}^3/\text{s})^2}{2(9.81\text{ m/s}^2)(0.4\text{ m})^2(0.15\text{ m})^2} = 0.7163\text{ m}$$

Then the alternate depth is determined from $E_{s1} = E_{s2}$ to be

$$E_{s2} = y_2 + \frac{\dot{V}^2}{2gb^2 y_2^2} \quad \rightarrow \quad 0.7163\text{ m} = y_2 + \frac{(0.2\text{ m}^3/\text{s})^2}{2(9.81\text{ m/s}^2)(0.4\text{ m})^2 y_2^2}$$

Solving for y_2 gives the alternate depth to be $y_2 = \mathbf{0.69\text{ m}}$. Therefore, if the character of flow were to change from supercritical to subcritical while holding the specific energy constant, the flow depth would rise from 0.15 to 0.69 m.

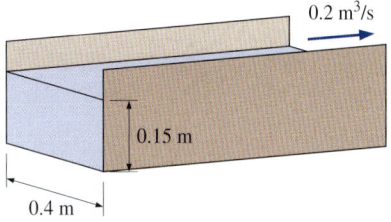

FIGURE 13–15
Schematic for Example 13–1.

> **Discussion** Note that if the water underwent a hydraulic jump at constant specific energy (the frictional losses being equal to the drop in elevation), the flow depth would rise to 0.69 m, assuming of course that the side walls of the channel are high enough.

13–4 · CONSERVATION OF MASS AND ENERGY EQUATIONS

Open-channel flows involve liquids whose densities are nearly constant, and thus the one-dimensional steady-flow conservation of mass equation is expressed as

$$\dot{V} = A_c V = \text{constant} \tag{13–26}$$

That is, the product of the flow cross section and the average flow velocity remains constant throughout the channel. Equation 13–26 between two sections along the channel is expressed as

Continuity equation: $$A_{c1}V_1 = A_{c2}V_2 \tag{13–27}$$

which is identical to the steady-flow conservation of mass equation for liquid flow in a pipe. Note that both the flow cross section and the average flow velocity may vary during flow, but, as stated, their product remains constant.

To determine the total energy of a liquid flowing in an open channel relative to a reference datum, as shown in Fig. 13–16, consider a point A in the liquid at a distance a from the free surface (and thus a distance $y - a$ from the channel bottom). Noting that the elevation, pressure (hydrostatic pressure relative to the free surface), and velocity at point A are $z_A = z + (y - a)$, $P_A = \rho g a$, and $V_A = V$, respectively, the total energy of the liquid in terms of heads is

$$H_A = z_A + \frac{P_A}{\rho g} + \frac{V_A^2}{2g} = z + (y - a) + \frac{\rho g a}{\rho g} + \frac{V^2}{2g} = z + y + \frac{V^2}{2g} \tag{13–28}$$

which is independent of the location of the point A at a cross section. Therefore, the total mechanical energy of a liquid at any cross section of an open channel can be expressed in terms of heads as

$$H = z + y + \frac{V^2}{2g} \tag{13–29}$$

where y is the flow depth, z is the elevation of the channel bottom, and V is the average flow velocity. Then the one-dimensional energy equation for open-channel flow between an upstream section 1 and a downstream section 2 is written as

Energy equation: $$z_1 + y_1 + \frac{V_1^2}{2g} = z_2 + y_2 + \frac{V_2^2}{2g} + h_L \tag{13–30}$$

The head loss h_L due to frictional effects is expressed as in pipe flow as

$$h_L = f \frac{L}{D_h} \frac{V^2}{2g} = f \frac{L}{R_h} \frac{V^2}{8g} \tag{13–31}$$

where f is the average friction factor and L is the length of channel between sections 1 and 2. The relation $D_h = 4R_h$ should be observed when using the hydraulic radius instead of the hydraulic diameter.

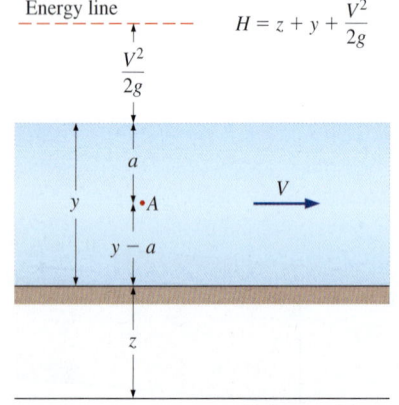

FIGURE 13–16
The total energy of a liquid flowing in an open channel.

Flow in open channels is gravity driven, and thus a typical channel is slightly sloped down. The slope of the bottom of the channel is expressed as

$$S_0 = \tan \alpha = \frac{z_1 - z_2}{x_2 - x_1} \cong \frac{z_1 - z_2}{L} \quad (13\text{-}32)$$

where α is the angle the channel bottom makes with the horizontal. In general, the bottom slope S_0 is very small, and thus the channel bottom is nearly horizontal. Therefore, $L \cong x_2 - x_1$, where x is the distance in the horizontal direction. Also, the flow depth y, which is measured in the vertical direction, can be taken to be the depth normal to the channel bottom with negligible error.

If the channel bottom is straight so that the bottom slope is constant, the vertical drop between sections 1 and 2 can be expressed as $z_1 - z_2 = S_0 L$. Then the energy equation (Eq. 13–30) becomes

Energy equation:
$$y_1 + \frac{V_1^2}{2g} + S_0 L = y_2 + \frac{V_2^2}{2g} + h_L \quad (13\text{-}33)$$

This equation has the advantage that it is independent of a reference datum for elevation.

In the design of open-channel systems, the bottom slope is selected such that it provides adequate elevation drop to overcome the frictional head loss and thus to maintain flow at the desired rate. Therefore, there is a close connection between the head loss and the bottom slope, and it makes sense to express the head loss as a slope (or the tangent of an angle). This is done by defining a **friction slope** as

Friction slope:
$$S_f = \frac{h_L}{L} \quad (13\text{-}34)$$

Then the energy equation is written as

Energy equation:
$$y_1 + \frac{V_1^2}{2g} = y_2 + \frac{V_2^2}{2g} + (S_f - S_0)L \quad (13\text{-}35)$$

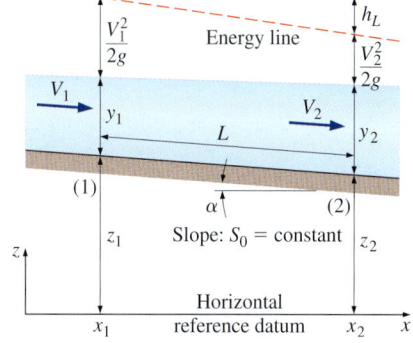

FIGURE 13–17
The total energy of a liquid at two sections of an open channel.

Note that the friction slope is equal to the bottom slope when the head loss is equal to the elevation drop. That is, $S_f = S_0$ when $h_L = z_1 - z_2$.

Figure 13–17 also shows the energy line, which is a distance $z + y + V^2/2g$ (total mechanical energy of the liquid expressed as a head) above the horizontal reference datum. The energy line is typically sloped down like the channel itself as a result of frictional losses, the vertical drop being equal to the head loss h_L and thus the slope being the same as the friction slope. Note that if there were *no* head loss, the energy line would be horizontal even when the channel is not. The elevation and velocity heads ($z + y$ and $V^2/2g$) would then be able to convert to each other during flow in this case, but their sum would remain constant.

13–5 • UNIFORM FLOW IN CHANNELS

We mentioned in Sec. 13–1 that flow in a channel is called *uniform flow* if the flow depth (and thus the average flow velocity since $\dot{V} = A_c V =$ constant in steady flow) remains constant. Uniform flow conditions are commonly encountered in practice in long straight runs of channels with constant slope,

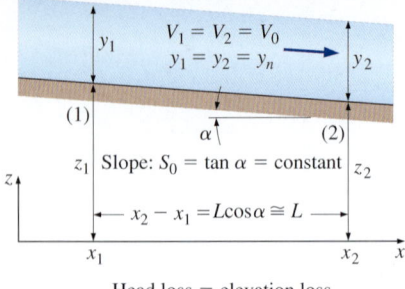

FIGURE 13–18
In uniform flow, the flow depth y, the average flow velocity V, and the bottom slope S_0 remain constant, and the head loss equals the elevation loss, $h_L = z_1 - z_2 = S_f L = S_0 L$.

constant cross section, and constant surface lining. In the design of open channels, it is very desirable to have uniform flow in the majority of the system since this means having a channel of constant wall height, which is easier to design and build.

The flow depth in uniform flow is called the **normal depth** y_n, and the average flow velocity is called the **uniform-flow velocity** V_0. The flow remains uniform as long as the slope, cross section, and surface roughness of the channel remain unchanged (Fig. 13–18). When the bottom slope is increased, the flow velocity increases and the flow depth decreases. Therefore, a new uniform flow is established with a new (lower) flow depth. The opposite occurs if the bottom slope is decreased.

During flow in open channels of constant slope S_0, constant cross section A_c, and constant surface friction factor f, the terminal velocity is reached and thus uniform flow is established when the head loss equals the elevation drop. Therefore,

$$h_L = f \frac{L}{D_h} \frac{V^2}{2g} \quad \text{or} \quad S_0 L = f \frac{L}{R_h} \frac{V_0^2}{8g} \quad (13\text{–}36)$$

since $h_L = S_0 L$ in uniform flow and $D_h = 4R_h$. Solving the second relation for V_0, the uniform-flow velocity and the flow rate are determined to be

$$V_0 = C\sqrt{S_0 R_h} \quad \text{and} \quad \dot{V} = CA_c\sqrt{S_0 R_h} \quad (13\text{–}37)$$

where

$$C = \sqrt{8g/f} \quad (13\text{–}38)$$

is called the **Chezy coefficient.** The Eqs. 13–37 and the coefficient C are named in honor of the French engineer Antoine Chezy (1718–1798), who first proposed a similar relationship in about 1769. The Chezy coefficient is a dimensional quantity, and its value ranges from about 30 m$^{1/2}$/s for small channels with rough surfaces to 90 m$^{1/2}$/s for large channels with smooth surfaces.

The Chezy coefficient can be determined in a straightforward manner from Eq. 13–38 by first determining the friction factor f as done for pipe flow in Chap. 8 from the Moody chart or the Colebrook equation for the fully rough turbulent limit (Re $\to \infty$),

$$f = [2.0 \log(14.8 R_h/\varepsilon)]^{-2} \quad (13\text{–}39)$$

Here, ε is the mean surface roughness. Note that open-channel flow is typically turbulent, and the flow is *fully developed* by the time uniform flow is established. Therefore, it is reasonable to use the friction factor relation for fully developed turbulent flow. Also, at large Reynolds numbers, the friction factor curves corresponding to specified relative roughness are nearly horizontal, and thus the friction factor is independent of the Reynolds number. The flow in that region is called *fully rough turbulent flow* (Chap. 8).

Since the introduction of the Chezy equations, considerable effort has been devoted by numerous investigators to the development of simpler empirical relations for the average velocity and flow rate. The most widely used equation was developed independently by the Frenchman Philippe-Gaspard Gauckler (1826–1905) in 1868 and the Irishman Robert Manning (1816–1897) in 1889.

Both Gauckler and Manning made recommendations that the constant in the Chezy equation be expressed as

$$C = \frac{a}{n} R_h^{1/6} \quad (13\text{-}40)$$

where n is called the **Manning coefficient,** whose value depends on the roughness of the channel surfaces. Substituting into Eqs. 13–37 gives the following empirical relations known as the **Manning equations** (also referred to as **Gauckler–Manning equations** since they were first proposed by Gauckler) for the uniform-flow velocity and the flow rate,

Uniform flow: $\quad V_0 = \dfrac{a}{n} R_h^{2/3} S_0^{1/2} \quad \text{and} \quad \dot{V} = \dfrac{a}{n} A_c R_h^{2/3} S_0^{1/2} \quad (13\text{-}41)$

The factor a is a dimensional constant whose value in SI units is $a = 1 \text{ m}^{1/3}/\text{s}$. Noting that 1 m = 3.2808 ft, its value in English units is

$$a = 1 \text{ m}^{1/3}/\text{s} = (3.2808 \text{ ft})^{1/3}/\text{s} = 1.486 \text{ ft}^{1/3}/\text{s} \quad (13\text{-}42)$$

Note that the bottom slope S_0 and the Manning coefficient n are dimensionless quantities, and Eqs. 13–41 give the velocity in m/s and the flow rate in m³/s in SI units when R_h is expressed in m.

Experimentally determined values of n are given in Table 13–1 for numerous natural and artificial channels. More extensive tables are available in the literature. Note that the value of n varies from 0.010 for a glass channel to 0.150 for a floodplain laden with trees (15 times that of a glass channel). There is considerable uncertainty in the value of n, especially in natural channels, as you would expect, since no two channels are exactly alike. The scatter can be 20 percent or more. Nevertheless, coefficient n is approximated as being independent of the size and shape of the channel—it varies only with the surface roughness.

Critical Uniform Flow

Flow through an open channel becomes critical flow when the Froude number Fr = 1 and thus the flow speed equals the wave speed $V_c = \sqrt{gy_c}$, where y_c is the critical flow depth, defined previously (Eq. 13–9). When the volume flow rate $\dot{V}$, the channel slope S_0, and the Manning coefficient n are known, the normal flow depth y_n can be determined from the Manning equation (Eq. 13–41). However, since A_c and R_h are both functions of y_n, the equation often ends up being implicit in y_n and requires a numerical (or trial and error) approach to solve. If $y_n = y_c$, the flow is *uniform critical flow*, and bottom slope S_0 equals the critical slope S_c in this case. When flow depth y_n is known instead of the flow rate $\dot{V}$, the flow rate can be determined from the Manning equation and the critical flow depth from Eq. 13–9. Again the flow is critical only if $y_n = y_c$.

During uniform critical flow, $S_0 = S_c$ and $y_n = y_c$. Replacing $\dot{V}$ and S_0 in the Manning equation by $\dot{V} = A_c\sqrt{gy_c}$ and S_c, respectively, and solving for S_c gives the following general relation for the critical slope,

Critical slope (general): $\quad S_c = \dfrac{gn^2 y_c}{a^2 R_h^{4/3}} \quad (13\text{-}43)$

TABLE 13–1

Mean values of the Manning coefficient n for water flow in open channels*

From Chow (1959).

Wall Material	n
A. Artificially lined channels	
Glass	0.010
Brass	0.011
Steel, smooth	0.012
Steel, painted	0.014
Steel, riveted	0.015
Cast iron	0.013
Concrete, finished	0.012
Concrete, unfinished	0.014
Wood, planed	0.012
Wood, unplaned	0.013
Clay tile	0.014
Brickwork	0.015
Asphalt	0.016
Corrugated metal	0.022
Rubble masonry	0.025
B. Excavated earth channels	
Clean	0.022
Gravelly	0.025
Weedy	0.030
Stony, cobbles	0.035
C. Natural channels	
Clean and straight	0.030
Sluggish with deep pools	0.040
Major rivers	0.035
Mountain streams	0.050
D. Floodplains	
Pasture, farmland	0.035
Light brush	0.050
Heavy brush	0.075
Trees	0.150

* The uncertainty in n can be ± 20 percent or more.

For film flow or flow in a wide rectangular channel with $b \gg y_c$, Eq. 13–43 simplifies to

Critical slope ($b \gg y_c$): $$S_c = \frac{gn^2}{a^2 y_c^{1/3}}$$ (13–44)

This equation gives the slope necessary to maintain a critical flow of depth y_c in a wide rectangular channel having a Manning coefficient of n.

Superposition Method for Nonuniform Perimeters

The surface roughness and thus the Manning coefficient for most natural and some human-made channels vary along the wetted perimeter and even along the channel. A river, for example, may have a stony bottom for its regular bed but a surface covered with bushes for its extended floodplain. There are several methods for solving such problems, either by finding an effective Manning coefficient n for the entire channel cross section, or by considering the channel in subsections and applying the superposition principle. For example, a channel cross section can be divided into N subsections, each with its own uniform Manning coefficient and flow rate. When determining the perimeter of a section, only the wetted portion of the boundary for that section is considered, and the imaginary boundaries are ignored. The flow rate through the channel is the sum of the flow rates through all the sections, as illustrated in Example 13–4.

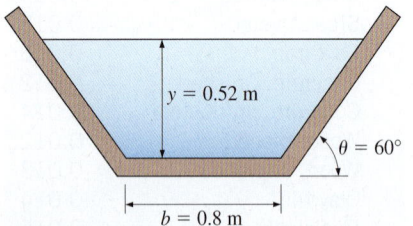

FIGURE 13–19
Schematic for Example 13–2.

EXAMPLE 13–2 Flow Rate in an Open Channel in Uniform Flow

Water is flowing in a weedy excavated earth channel of trapezoidal cross section with a bottom width of 0.8 m, trapezoid angle of 60°, and a bottom slope angle of 0.3°, as shown in Fig. 13–19. If the flow depth is measured to be 0.52 m, determine the flow rate of water through the channel. What would your answer be if the bottom angle were 1°?

SOLUTION Water is flowing in a weedy trapezoidal channel of given dimensions. The flow rate corresponding to a measured value of flow depth is to be determined.
Assumptions **1** The flow is steady and uniform. **2** The bottom slope is constant. **3** The roughness of the wetted surface of the channel and thus the friction coefficient are constant.
Properties The Manning coefficient for an open channel with weedy surfaces is $n = 0.030$.
Analysis The cross-sectional area, perimeter, and hydraulic radius of the channel are

$$A_c = y\left(b + \frac{y}{\tan\theta}\right) = (0.52 \text{ m})\left(0.8 \text{ m} + \frac{0.52 \text{ m}}{\tan 60°}\right) = 0.5721 \text{ m}^2$$

$$p = b + \frac{2y}{\sin\theta} = 0.8 \text{ m} + \frac{2 \times 0.52 \text{ m}}{\sin 60°} = 2.001 \text{ m}$$

$$R_h = \frac{A_c}{p} = \frac{0.5721 \text{ m}^2}{2.991 \text{ m}} = 0.2859 \text{ m}$$

The bottom slope of the channel is

$$S_0 = \tan\alpha = \tan 0.3° = 0.005236$$

Then the flow rate through the channel is determined from the Manning equation to be

$$\dot{V} = \frac{a}{n} A_c R_h^{2/3} S_0^{1/2} = \frac{1 \text{ m}^{1/3}/\text{s}}{0.030}(0.5721 \text{ m}^2)(0.2859 \text{ m})^{2/3}(0.005236)^{1/2} = \mathbf{0.60 \text{ m}^3/\text{s}}$$

The flow rate for a bottom angle of 1° is determined by using $S_0 = \tan \alpha = \tan 1° = 0.01746$ in the last relation. It gives $\dot{V} = \mathbf{1.1 \text{ m}^3/\text{s}}$.

Discussion Note that the flow rate is a strong function of the bottom angle. Also, there is considerable uncertainty in the value of the Manning coefficient, and thus in the flow rate calculated. A 10 percent uncertainty in *n* results in a 10 percent uncertainty in the flow rate. Final answers are therefore given to only two significant digits.

■ **EXAMPLE 13–3**   **The Height of a Rectangular Channel**

Water is to be transported in an unfinished-concrete rectangular channel with a bottom width of 1.2 m at a rate of 1.5 m³/s. The terrain is such that the channel bottom drops 0.6 m per 300 m length. Determine the minimum height of the channel under uniform-flow conditions (Fig. 13–20). What would your answer be if the bottom drop is just 0.3 m per 300 m length?

SOLUTION Water is flowing in an unfinished-concrete rectangular channel with a specified bottom width. The minimum channel height corresponding to a specified flow rate is to be determined.
Assumptions **1** The flow is steady and uniform. **2** The bottom slope is constant. **3** The roughness of the wetted surface of the channel and thus the friction coefficient are constant.
Properties The Manning coefficient for an open channel with unfinished-concrete surfaces is $n = 0.014$.
Analysis The cross-sectional area, perimeter, and hydraulic radius of the channel are

$$A_c = by = (1.2 \text{ m})y \quad p = b + 2y = (1.2 \text{ m}) + 2y \quad R_h = \frac{A_c}{p} = \frac{1.2y}{1.2 + 2y}$$

The bottom slope of the channel is $S_0 = 0.6/300 = 0.002$. Using the Manning equation, the flow rate through the channel is expressed as

$$\dot{V} = \frac{a}{n} A_c R_h^{2/3} S_0^{1/2}$$

$$1.5 \text{ m}^3/\text{s} = \frac{1 \text{ m}^{1/3}/\text{s}}{0.014}(1.2y \text{ m}^2)\left(\frac{1.2y}{1.2 + 2y} \text{ m}\right)^{2/3}(0.002)^{1/2}$$

which is a nonlinear equation in *y*. Using an equation solver such as EES or an itirative approach, the flow depth is determined to be

$$y = \mathbf{0.799 \text{ m}}$$

If the bottom drop were just 0.3 m per 300 m length, the bottom slope would be $S_0 = 0.001$, and the flow depth would be $y = \mathbf{1.05 \text{ m}}$.
Discussion Note that *y* is the flow depth, and thus this is the minimum value for the channel height. Also, there is considerable uncertainty in the value of the Manning coefficient *n*, and this should be considered when deciding the height of the channel to be built.

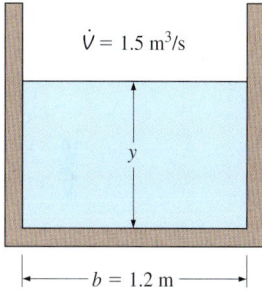

FIGURE 13–20
Schematic for Example 13–3.

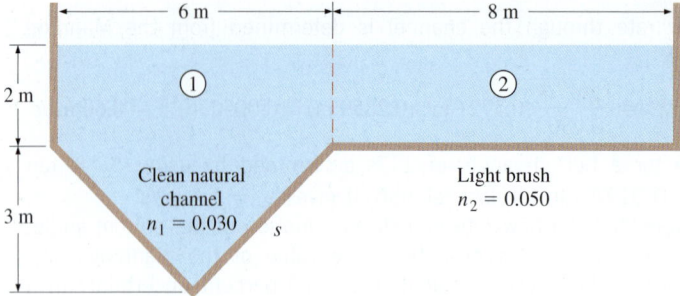

FIGURE 13–21
Schematic for Example 13–4.

EXAMPLE 13–4   Channels with Nonuniform Roughness

Water flows in a channel whose bottom slope is 0.003 and whose cross section is shown in Fig. 13–21. The dimensions and the Manning coefficients for the surfaces of different subsections are also given on the figure. Determine the flow rate through the channel and the effective Manning coefficient for the channel.

SOLUTION Water is flowing through a channel with nonuniform surface properties. The flow rate and the effective Manning coefficient are to be determined.
Assumptions 1 The flow is steady and uniform. 2 The bottom slope is constant. 3 The Manning coefficients do not vary along the channel.
Analysis The channel involves two parts with different roughnesses, and thus it is appropriate to divide the channel into two subsections as indicated in Fig. 13–21. The flow rate for each subsection is determined from the Manning equation, and the total flow rate is determined by adding them up.

The side length of the triangular channel is $s = \sqrt{3^2 + 3^2} = 4.243$ m. Then the flow area, perimeter, and hydraulic radius for each subsection and the entire channel become

Subsection 1:
$$A_{c1} = 21 \text{ m}^2 \quad p_1 = 10.486 \text{ m} \quad R_{h1} = \frac{A_{c1}}{p_1} = \frac{21 \text{ m}^2}{10.486 \text{ m}} = 2.00 \text{ m}$$

Subsection 2:
$$A_{c2} = 16 \text{ m}^2 \quad p_2 = 10 \text{ m} \quad R_{h2} = \frac{A_{c2}}{p_2} = \frac{16 \text{ m}^2}{10 \text{ m}} = 1.60 \text{ m}$$

Entire channel:
$$A_c = 37 \text{ m}^2 \quad p = 20.486 \text{ m} \quad R_h = \frac{A_c}{p} = \frac{37 \text{ m}^2}{20.486 \text{ m}} = 1.806 \text{ m}$$

Using the Manning equation for each subsection, the total flow rate through the channel is determined to be

$$\dot{V} = \dot{V}_1 + \dot{V}_2 = \frac{a}{n_1} A_{c1} R_{h1}^{2/3} S_0^{1/2} + \frac{a}{n_2} A_{c2} R_{h2}^{2/3} S_0^{1/2}$$

$$= (1 \text{ m}^{1/3}/\text{s}) \left[\frac{(21 \text{ m}^2)(2 \text{ m})^{2/3}}{0.030} + \frac{(16 \text{ m}^2)(1.60 \text{ m})^{2/3}}{0.050} \right] (0.003)^{1/2}$$

$$= 84.8 \text{ m}^3/\text{s} \cong \mathbf{85 \text{ m}^3/\text{s}}$$

Knowing the total flow rate, the effective Manning coefficient for the entire channel is determined from the Manning equation,

$$n_{\text{eff}} = \frac{aA_c R_h^{2/3} S_0^{1/2}}{\dot{V}} = \frac{(1\ \text{m}^{1/3}/\text{s})(37\ \text{m}^2)(1.806\ \text{m})^{2/3}(0.003)^{1/2}}{84.8\ \text{m}^3/\text{s}} = 0.035$$

Discussion The effective Manning coefficient n_{eff} of the channel turns out to lie between the two n values, as expected. The weighted average of the Manning coefficient of the channel is $n_{\text{avg}} = (n_1 p_1 + n_2 p_2)/p = 0.040$, which is quite different than n_{eff}. Therefore, using a weighted average Manning coefficient for the entire channel may be tempting, but it would not be so accurate.

13–6 ▪ BEST HYDRAULIC CROSS SECTIONS

Open-channel systems are usually designed to transport a liquid to a location at a lower elevation at a specified rate under the influence of gravity at the lowest possible cost. Noting that no energy input is required, the cost of an open-channel system consists primarily of the initial construction cost, which is proportional to the physical size of the system. Therefore, for a given channel length, the perimeter of the channel is representative of the system cost, and it should be kept to a minimum in order to minimize the size and thus the cost of the system.

From another perspective, resistance to flow is due to wall shear stress τ_w and the wall area, which is equivalent to the wetted perimeter per unit channel length. Therefore, for a given flow cross-sectional area A_c, the smaller the wetted perimeter p, the smaller the resistance force, and thus the larger the average velocity and the flow rate.

From yet another perspective, for a specified channel geometry with a specified bottom slope S_0 and surface lining (and thus the roughness coefficient n), the flow velocity is given by the Manning formula as $V = aR_h^{2/3}S_0^{1/2}/n$. Therefore, the flow velocity increases with the hydraulic radius, and the hydraulic radius must be maximized (and thus the perimeter must be minimized since $R_h = A_c/p$) in order to maximize the average flow velocity or the flow rate per unit cross-sectional area. Thus we conclude the following:

> The best hydraulic cross section for an open channel is the one with the maximum hydraulic radius or, equivalently, the one with the minimum wetted perimeter for a specified cross-sectional area.

The shape with the minimal perimeter per unit area is a circle. Therefore, on the basis of minimum flow resistance, the best cross section for an open channel is a semicircular one (Fig. 13–22). However, it is usually cheaper to construct an open channel with straight sides (such as channels with trapezoidal or rectangular cross sections) instead of semicircular ones, and the general shape of the channel may be specified a priori. Thus it makes sense to analyze each geometric shape separately for the best cross section.

As a motivational example, consider a rectangular channel of finished concrete ($n = 0.012$) of width b and flow depth y with a bottom slope of 1° (Fig. 13–23). To determine the effects of the aspect ratio y/b on the hydraulic radius R_h and the flow rate $\dot{V}$ for a cross-sectional area of 1 m², R_h and $\dot{V}$ are

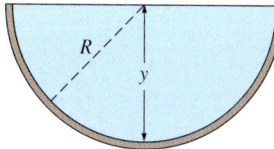

FIGURE 13–22
The best hydraulic cross section for an open channel is a semicircular one since it has the minimum wetted perimeter for a specified cross-sectional area, and thus the minimum flow resistance.

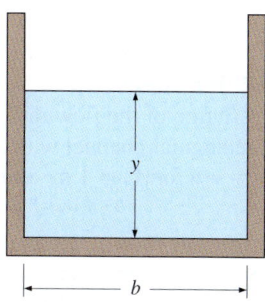

FIGURE 13–23
A rectangular open channel of width b and flow depth y. For a given cross-sectional area, the highest flow rate occurs when $y = b/2$.

TABLE 13-2

Variation of the hydraulic radius R_h and the flow rate $\dot{V}$ with aspect ratio y/b for a rectangular channel with $A_c = 1 \text{ m}^2$, $S_0 = \tan 1°$, and $n = 0.012$

Aspect Ratio y/b	Channel Width b, m	Flow Depth y, m	Perimeter p, m	Hydraulic Radius R_h, m	Flow Rate $\dot{V}$, m³/s
0.1	3.162	0.316	3.795	0.264	4.53
0.2	2.236	0.447	3.130	0.319	5.14
0.3	1.826	0.548	2.921	0.342	5.39
0.4	1.581	0.632	2.846	0.351	5.48
0.5	1.414	0.707	2.828	0.354	5.50
0.6	1.291	0.775	2.840	0.352	5.49
0.7	1.195	0.837	2.869	0.349	5.45
0.8	1.118	0.894	2.907	0.344	5.41
0.9	1.054	0.949	2.951	0.339	5.35
1.0	1.000	1.000	3.000	0.333	5.29
1.5	0.816	1.225	3.266	0.306	5.00
2.0	0.707	1.414	3.536	0.283	4.74
3.0	0.577	1.732	4.041	0.247	4.34
4.0	0.500	2.000	4.500	0.222	4.04
5.0	0.447	2.236	4.919	0.203	3.81

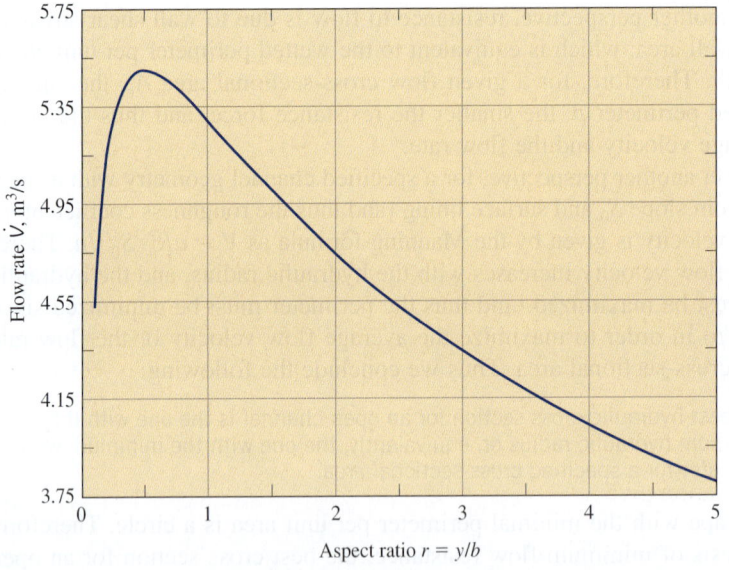

FIGURE 13-24
Variation of the flow rate in a rectangular channel with aspect ratio $r = y/b$ for $A_c = 1 \text{ m}^2$ and $S_0 = \tan 1°$.

evaluated from the Manning formula. The results are tabulated in Table 13–2 and plotted in Fig. 13–24 for aspect ratios from 0.1 to 5. We observe from this table and the plot that the flow rate $\dot{V}$ increases as the flow aspect ratio y/b is increased, reaches a maximum at $y/b = 0.5$, and then starts to decrease (the numerical values for $\dot{V}$ can also be interpreted as the flow velocities in m/s since $A_c = 1 \text{ m}^2$). We see the same trend for the hydraulic radius, but the opposite trend for the wetted perimeter p. These results confirm that the best cross section for a given shape is the one with the maximum hydraulic radius, or equivalently, the one with the minimum perimeter.

Rectangular Channels

Consider liquid flow in an open channel of rectangular cross section of width b and flow depth y. The cross-sectional area and the wetted perimeter at a flow section are

$$A_c = yb \quad \text{and} \quad p = b + 2y \quad (13\text{–}45)$$

Solving the first relation of Eq. 13–45 for b and substituting it into the second relation give

$$p = \frac{A_c}{y} + 2y \quad (13\text{–}46)$$

Now we apply the criterion that the best hydraulic cross section for an open channel is the one with the minimum wetted perimeter for a given cross-sectional area. Taking the derivative of p with respect to y while holding A_c constant gives

$$\frac{dp}{dy} = -\frac{A_c}{y^2} + 2 = -\frac{by}{y^2} + 2 = -\frac{b}{y} + 2 \quad (13\text{–}47)$$

Setting $dp/dy = 0$ and solving for y, the criterion for the best hydraulic cross section is determined to be

Best hydraulic cross section (rectangular channel): $\quad y = \dfrac{b}{2} \quad (13\text{–}48)$

Therefore, a rectangular open channel should be designed such that the liquid height is half the channel width to minimize flow resistance or to maximize the flow rate for a given cross-sectional area. This also minimizes the perimeter and thus the construction costs. This result confirms the finding from Table 13–2 that $y = b/2$ gives the best cross section.

Trapezoidal Channels

Now consider liquid flow in an open channel of trapezoidal cross section of bottom width b, flow depth y, and trapezoid angle θ measured from the horizontal, as shown in Fig. 13–25. The cross-sectional area and the wetted perimeter at a flow section are

$$A_c = \left(b + \frac{y}{\tan \theta}\right)y \quad \text{and} \quad p = b + \frac{2y}{\sin \theta} \quad (13\text{–}49)$$

Solving the first relation of Eq. 13–49 for b and substituting it into the second relation give

$$p = \frac{A_c}{y} - \frac{y}{\tan \theta} + \frac{2y}{\sin \theta} \quad (13\text{–}50)$$

Taking the derivative of p with respect to y while holding A_c and θ constant gives

$$\frac{dp}{dy} = -\frac{A_c}{y^2} - \frac{1}{\tan \theta} + \frac{2}{\sin \theta} = -\frac{b + y/\tan \theta}{y} - \frac{1}{\tan \theta} + \frac{2}{\sin \theta} \quad (13\text{–}51)$$

Setting $dp/dy = 0$ and solving for y, the criterion for the best hydraulic cross section for any specified trapezoid angle θ is determined to be

Best hydraulic cross section (trapezoidal channel): $\quad y = \dfrac{b \sin \theta}{2(1 - \cos \theta)} \quad (13\text{–}52)$

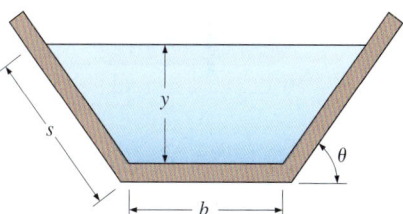

$$R_h = \frac{A_c}{p} = \frac{y(b + y/\tan \theta)}{b + 2y/\sin \theta}$$

FIGURE 13–25
Parameters for a trapezoidal channel.

For the special case of $\theta = 90°$ (a rectangular channel), this relation reduces to $y = b/2$, as expected.

The hydraulic radius R_h for a trapezoidal channel can be expressed as

$$R_h = \frac{A_c}{p} = \frac{y(b + y/\tan\theta)}{b + 2y/\sin\theta} = \frac{y(b\sin\theta + y\cos\theta)}{b\sin\theta + 2y} \quad (13\text{-}53)$$

Rearranging Eq. 13–52 as $b\sin\theta = 2y(1 - \cos\theta)$, substituting into Eq. 13–53 and simplifying, the hydraulic radius for a trapezoidal channel with the best cross section becomes

Hydraulic radius for the best cross section: $\quad R_h = \dfrac{y}{2} \quad (13\text{-}54)$

Therefore, the hydraulic radius is half the flow depth for trapezoidal channels with the best cross section regardless of the trapezoid angle θ.

Similarly, the trapezoid angle for the best hydraulic cross section is determined by taking the derivative of p (Eq. 13–50) with respect to θ while holding A_c and y constant, setting $dp/d\theta = 0$, and solving the resulting equation for θ. This gives

Best trapezoid angle: $\quad \theta = 60° \quad (13\text{-}55)$

Substituting the best trapezoid angle $\theta = 60°$ into the best hydraulic cross section relation $y = b\sin\theta/(2 - 2\cos\theta)$ gives

Best flow depth for $\theta = 60°$: $\quad y = \dfrac{\sqrt{3}}{2}b \quad (13\text{-}56)$

Then the length of the side edge of the flow section and the flow area become

$$s = \frac{y}{\sin 60°} = \frac{b\sqrt{3}/2}{\sqrt{3}/2} = b \quad (13\text{-}57)$$

$$p = 3b \quad (13\text{-}58)$$

$$A_c = \left(b + \frac{y}{\tan\theta}\right)y = \left(b + \frac{b\sqrt{3}/2}{\tan 60°}\right)(b\sqrt{3}/2) = \frac{3\sqrt{3}}{4}b^2 \quad (13\text{-}59)$$

since $\tan 60° = \sqrt{3}$. Therefore, the best cross section for trapezoidal channels is *half of a hexagon* (Fig. 13–26). This is not surprising since a hexagon closely approximates a circle, and a half-hexagon has the least perimeter per unit cross-sectional area of all trapezoidal channels.

Best hydraulic cross sections for other channel shapes can be determined in a similar manner. For example, the best hydraulic cross section for a circular channel of diameter D can be shown to be $y = D/2$.

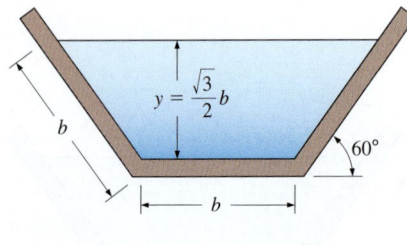

FIGURE 13–26
The best cross section for trapezoidal channels is *half of a hexagon*.

> **EXAMPLE 13–5**   **Best Cross Section of an Open Channel**
>
> Water is to be transported at a rate of 2 m³/s in uniform flow in an open channel whose surfaces are asphalt lined. The bottom slope is 0.001. Determine the dimensions of the best cross section if the shape of the channel is (a) rectangular and (b) trapezoidal (Fig. 13–27).

SOLUTION Water is to be transported in an open channel at a specified rate. The best channel dimensions are to be determined for rectangular and trapezoidal shapes.

Assumptions **1** The flow is steady and uniform. **2** The bottom slope is constant. **3** The roughness of the wetted surface of the channel and thus the friction coefficient are constant.

Properties The Manning coefficient for an open channel with asphalt lining is $n = 0.016$.

Analysis (*a*) The best cross section for a rectangular channel occurs when the flow height is half the channel width, $y = b/2$. Then the cross-sectional area, perimeter, and hydraulic radius of the channel are

$$A_c = by = \frac{b^2}{2} \qquad p = b + 2y = 2b \qquad R_h = \frac{A_c}{p} = \frac{b}{4}$$

Substituting into the Manning equation,

$$\dot{V} = \frac{a}{n} A_c R_h^{2/3} S_0^{1/2} \quad \rightarrow \quad b = \left(\frac{2n\dot{V}4^{2/3}}{a\sqrt{S_0}}\right)^{3/8} = \left(\frac{2(0.016)(2 \text{ m}^3/\text{s})4^{2/3}}{(1 \text{ m}^{1/3}/\text{s})\sqrt{0.001}}\right)^{3/8}$$

which gives $b = 1.84$ m. Therefore, $A_c = 1.70$ m², $p = 3.68$ m, and the dimensions of the best rectangular channel are

$$b = \mathbf{1.84 \text{ m}} \quad \text{and} \quad y = \mathbf{0.92 \text{ m}}$$

(*b*) The best cross section for a trapezoidal channel occurs when the trapezoid angle is 60° and flow height is $y = b\sqrt{3}/2$. Then,

$$A_c = y(b + b\cos\theta) = 0.5\sqrt{3}b^2(1 + \cos 60°) = 0.75\sqrt{3}b^2$$

$$p = 3b \qquad R_h = \frac{y}{2} = \frac{\sqrt{3}}{4}b$$

Substituting into the Manning equation,

$$\dot{V} = \frac{a}{n} A_c R_h^{2/3} S_0^{1/2} \quad \rightarrow \quad b = \left(\frac{(0.016)(2 \text{ m}^3/\text{s})}{0.75\sqrt{3}(\sqrt{3}/4)^{2/3}(1 \text{ m}^{1/3}/\text{s})\sqrt{0.001}}\right)^{3/8}$$

which yields $b = 1.12$ m. Therefore, $A_c = 1.64$ m², $p = 3.37$ m, and the dimensions of the best trapezoidal channel are

$$b = \mathbf{1.12 \text{ m}} \quad y = \mathbf{0.973 \text{ m}} \quad \text{and} \quad \theta = \mathbf{60°}$$

Discussion Note that the trapezoidal cross section is better since it has a smaller perimeter (3.37 m versus 3.68 m) and thus lower construction cost. This is why many man-made waterways are trapezoidal in shape (Fig. 13–28). However, the average velocity through the trapezoidal channel is larger since A_c is smaller.

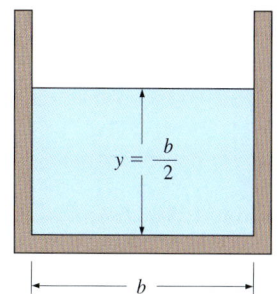
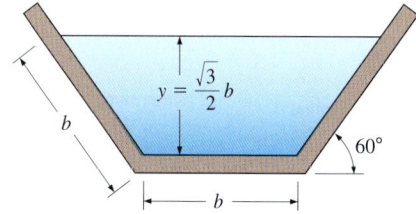

FIGURE 13–27
Schematic for Example 13–5.

(*a*)

(*b*)

FIGURE 13–28
Many man-made water channels are trapezoidal in shape because of low construction cost and good performance.
(*a*) © Pixtal/AGE Fotostock RF;
(*b*) Photo by Bryan Lewis.

13–7 ▪ GRADUALLY VARIED FLOW

To this point we considered *uniform flow* during which the flow depth y and the flow velocity V remain constant. In this section we consider *gradually varied flow* (GVF), which is a form of steady nonuniform flow characterized by gradual variations in flow depth and velocity (small slopes and no abrupt changes) and a free surface that always remains smooth (no discontinuities or zigzags). Flows that involve rapid changes in flow depth and velocity, called *rapidly varied flows* (RVF), are considered in Section 13–8. A change in the bottom slope or cross section of a channel or an obstruction in the path of flow may cause the uniform flow in a channel to become gradually or rapidly varied flow.

Rapidly varied flows occur over a short section of the channel with relatively small surface area, and thus frictional losses associated with wall shear are negligible. Head losses in RVF are highly localized and are due to intense agitation and turbulence. Losses in GVF, on the other hand, are primarily due to frictional effects along the channel and can be determined from the Manning formula.

In gradually varied flow, the flow depth and velocity vary slowly, and the free surface is stable. This makes it possible to formulate the variation of flow depth along the channel on the basis of the conservation of mass and energy principles and to obtain relations for the profile of the free surface.

In uniform flow, the slope of the energy line is equal to the slope of the bottom surface. Therefore, the friction slope equals the bottom slope, $S_f = S_0$. In gradually varied flow, however, these slopes are different (Fig. 13–29).

Consider steady flow in a rectangular open channel of width b, and assume any variation in the bottom slope and water depth to be rather gradual. We again write the equations in terms of average velocity V and approximate the pressure distribution as hydrostatic. From Eq. 13–17, the total head of the liquid at any cross section is $H = z_b + y + V^2/2g$, where z_b is the vertical distance of the bottom surface from the reference datum. Differentiating H with respect to x gives

$$\frac{dH}{dx} = \frac{d}{dx}\left(z_b + y + \frac{V^2}{2g}\right) = \frac{dz_b}{dx} + \frac{dy}{dx} + \frac{V}{g}\frac{dV}{dx} \qquad (13\text{–}60)$$

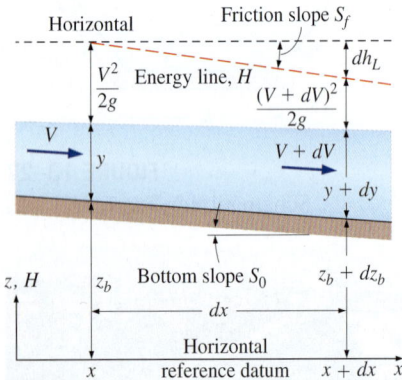

FIGURE 13–29
Variation of properties over a differential flow section in an open channel under conditions of gradually varied flow (GVF).

But H is the total energy of the liquid and thus dH/dx is the slope of the energy line (a negative quantity), which is equal to the negative of the friction slope, as shown in Fig. 13–29. Also, dz_b/dx is the negative of the bottom slope. Therefore,

$$\frac{dH}{dx} = -\frac{dh_L}{dx} = -S_f \quad \text{and} \quad \frac{dz_b}{dx} = -S_0 \qquad (13\text{–}61)$$

Substituting Eqs. 13–61 into Eq. 13–60 gives

$$S_0 - S_f = \frac{dy}{dx} + \frac{V}{g}\frac{dV}{dx} \qquad (13\text{–}62)$$

The conservation of mass equation for steady flow in a rectangular channel is $\dot{V} = ybV = $ constant. Differentiating with respect to x gives

$$0 = bV\frac{dy}{dx} + yb\frac{dV}{dx} \rightarrow \frac{dV}{dx} = -\frac{V}{y}\frac{dy}{dx} \qquad (13\text{–}63)$$

Substituting Eq. 13–63 into Eq. 13–62 and noting that $V/\sqrt{gy}$ is the Froude number,

$$S_0 - S_f = \frac{dy}{dx} - \frac{V^2}{gy}\frac{dy}{dx} = \frac{dy}{dx} - \text{Fr}^2\frac{dy}{dx} \qquad (13\text{–}64)$$

Solving for dy/dx gives the desired relation for the rate of change of flow depth (or the surface profile) in gradually varied flow in an open channel,

The GVF equation: $$\frac{dy}{dx} = \frac{S_0 - S_f}{1 - \text{Fr}^2} \qquad (13\text{–}65)$$

which is analogous to the variation of flow area as a function of the Mach number in compressible flow. This relation is derived for a rectangular channel, but it is also valid for channels of other constant cross sections provided that the Froude number is expressed accordingly. An analytical or numerical solution of this differential equation gives the flow depth y as a function of x for a given set of parameters, and the function $y(x)$ is the *surface profile*.

The general trend of flow depth—whether it increases, decreases, or remains constant along the channel—depends on the sign of dy/dx, which depends on the signs of the numerator and the denominator of Eq. 13–65. The Froude number is always positive and so is the friction slope S_f (except for the idealized case of flow with negligible frictional effects for which both h_L and S_f are zero). The bottom slope S_0 is positive for downward-sloping sections (typically the case), zero for horizontal sections, and negative for upward-sloping sections of a channel (adverse flow). The flow depth increases when $dy/dx > 0$, decreases when $dy/dx < 0$, and remains constant (and thus the free surface is parallel to the channel bottom, as in uniform flow) when $dy/dx = 0$ and thus $S_0 = S_f$ (Fig. 13–30). For specified values of S_0 and S_f, the term dy/dx may be positive or negative, depending on whether the Froude number is less than or greater than 1. Therefore, the flow behavior is opposite in subcritical and supercritical flows. For $S_0 - S_f > 0$, for example, the flow depth increases in the flow direction in subcritical flow, but it decreases in supercritical flow.

The determination of the sign of the denominator $1 - \text{Fr}^2$ is easy: it is positive for subcritical flow ($\text{Fr} < 1$), and negative for supercritical flow ($\text{Fr} > 1$). But the sign of the numerator depends on the relative magnitudes of S_0 and S_f. Note that the friction slope S_f is always positive, and its value is equal to the channel slope S_0 in uniform flow, $y = y_n$. The friction slope is a quantity that varies with streamwise distance, and is calculated from the Manning equation, based upon the depth at each streamwise location, as demonstrated in Example 13–6. Noting that head loss increases with increasing velocity, and that the velocity is inversely proportional to flow depth for a given flow rate, $S_f > S_0$ and thus $S_0 - S_f < 0$ when $y < y_n$, and $S_f < S_0$ and thus $S_0 - S_f > 0$ when $y > y_n$. The numerator $S_0 - S_f$ is always negative for horizontal ($S_0 = 0$) and upward-sloping ($S_0 < 0$) channels, and thus the flow depth decreases in the flow direction during subcritical flows in such channels.

FIGURE 13–30
A slow-moving river of approximately constant depth and cross section, such as the Chicago River shown here, is an example of uniform flow with $S_0 \approx S_f$ and $dy/dx \approx 0$.
© *Hisham F. Ibrahim/Getty RF*

Liquid Surface Profiles in Open Channels, *y*(*x*)

Open-channel systems are designed and built on the basis of the projected flow depths along the channel. Therefore, it is important to be able to predict the flow depth for a specified flow rate and specified channel geometry. A plot of flow depth versus downstream distance is the **surface profile** $y(x)$ of the flow. The general characteristics of surface profiles for gradually varied flow depend on the bottom slope and flow depth relative to the critical and normal depths.

A typical open channel involves various sections of different bottom slopes S_0 and different flow regimes, and thus various sections of different surface profiles. For example, the general shape of the surface profile in a downward-sloping section of a channel is different than that in an upward-sloping section. Likewise, the profile in subcritical flow is different than the

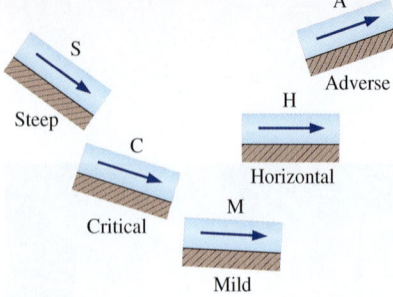

FIGURE 13–31
Designation of the letters S, C, M, H, and A for liquid surface profiles for different types of slopes.

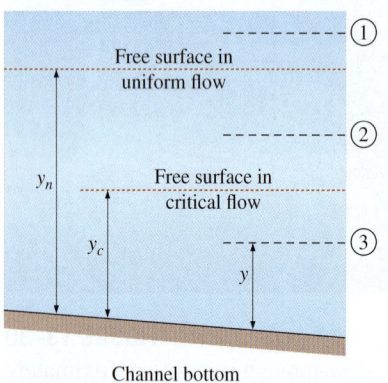

FIGURE 13–32
Designation of the numbers 1, 2, and 3 for liquid surface profiles based on the value of the flow depth relative to the normal and critical depths.

profile in supercritical flow. Unlike uniform flow that does not involve inertial forces, gradually varied flow involves acceleration and deceleration of liquid, and the surface profile reflects the dynamic balance between liquid weight, shear force, and inertial effects.

Each surface profile is identified by a letter that indicates the slope of the channel and by a number that indicates flow depth relative to the critical depth y_c and normal depth y_n. The slope of the channel can be steep (S), critical (C), mild (M), horizontal (H), or adverse (A) (Fig. 13–31). The channel slope is said to be mild if $y_n > y_c$, steep if $y_n < y_c$, critical if $y_n = y_c$, horizontal if $S_0 = 0$ (zero bottom slope), and adverse if $S_0 < 0$ (negative slope). Note that a liquid flows uphill in an open channel that has an adverse slope.

The classification of a channel section depends on the flow rate and the channel cross section as well as the slope of the channel bottom. A channel section that is classified to have a mild slope for one flow may have a steep slope for another flow, and even a critical slope for a third flow. Therefore, we need to calculate the critical depth y_c and the normal depth y_n before we can assess the slope.

The number designation indicates the initial position of the liquid surface for a given channel slope relative to the surface levels in critical and uniform flows, as shown in Fig. 13–32. A surface profile is designated by 1 if the flow depth is above both critical and normal depths ($y > y_c$ and $y > y_n$), by 2 if the flow depth is between the two ($y_n > y > y_c$ or $y_n < y < y_c$), and by 3 if the flow depth is below both the critical and normal depths ($y < y_c$ and $y < y_n$). Therefore, three different profiles are possible for a specified type of channel slope. But for channels with zero or adverse slopes, type 1 flow cannot exist since the flow can never be uniform in horizontal and upward channels, and thus normal depth is not defined. Also, type 2 flow does not exist for channels with critical slope since normal and critical depths are identical in this case.

The five classes of slopes and the three types of initial positions discussed give a total of 12 distinct configurations for surface profiles in GVF, all tabulated and sketched in Table 13–3. The Froude number is also given for each case, with Fr > 1 for $y < y_c$, as well as the sign of the slope dy/dx of the surface profile determined from Eq. 13–65, $dy/dx = (S_0 - S_f)/(1 - \text{Fr}^2)$. Note that $dy/dx > 0$, and thus the flow depth increases in the flow direction when both $S_0 - S_f$ and $1 - \text{Fr}^2$ are positive or negative. Otherwise $dy/dx < 0$ and the flow depth decreases. In type 1 flows, the flow depth increases in the flow direction and the surface profile approaches the horizontal plane asymptotically. In type 2 flows, the flow depth decreases and the surface profile approaches the lower of y_c or y_n. In type 3 flows, the flow depth increases and the surface profile approaches the lower of y_c or y_n. These trends in surface profiles continue as long as there is no change in bottom slope or roughness.

Consider the case in Table 13–3 designated M1 (mild channel slope and $y > y_n > y_c$). The flow is subcritical since $y > y_c$ and thus Fr < 1 and $1 - \text{Fr}^2 > 0$. Also, $S_f < S_0$ and thus $S_0 - S_f > 0$ since $y > y_n$, and thus the flow velocity is less than the velocity in normal flow. Therefore, the slope of the surface profile $dy/dx = (S_0 - S_f)/(1 - \text{Fr}^2) > 0$, and the flow depth y increases in the flow direction. But as y increases, the flow velocity decreases, and thus S_f and Fr approach zero. Consequently, dy/dx approaches S_0 and the rate of increase in flow depth becomes equal to the channel slope. This requires the surface profile to become horizontal at

TABLE 13-3
Classification of surface profiles in gradually varied flow. The vertical scale is greatly exaggerated.

Channel Slope	Profile Notation	Flow Depth	Froude Number	Profile Slope	Surface Profile
Steep (S) $y_c > y_n$ $S_0 < S_c$	S1	$y > y_c$	Fr < 1	$\frac{dy}{dx} > 0$	
	S2	$y_n < y < y_c$	Fr > 1	$\frac{dy}{dx} < 0$	
	S3	$y < y_n$	Fr > 1	$\frac{dy}{dx} > 0$	Channel bottom, $S_0 > S_c$
Critical (C) $y_c = y_n$ $S_0 < S_c$	C1	$y > y_c$	Fr < 1	$\frac{dy}{dx} > 0$	
	C3	$y < y_c$	Fr > 1	$\frac{dy}{dx} > 0$	Channel bottom, $S_0 = S_c$
Mild (M) $y_c < y_n$ $S_0 < S_c$	M1	$y > y_n$	Fr < 1	$\frac{dy}{dx} > 0$	
	M2	$y_c < y < y_n$	Fr < 1	$\frac{dy}{dx} < 0$	
	M3	$y < y_c$	Fr > 1	$\frac{dy}{dx} > 0$	Channel bottom, $S_0 < S_c$
Horizontal (H) $y_n \to \infty$ $S_0 = 0$	H2	$y > y_c$	Fr < 1	$\frac{dy}{dx} < 0$	
	H3	$y < y_c$	Fr > 1	$\frac{dy}{dx} > 0$	Channel bottom, $S_0 = 0$
Adverse (A) $S_0 < 0$ y_n: does not exist	A2	$y > y_c$	Fr < 1	$\frac{dy}{dx} < 0$	
	A3	$y < y_c$	Fr > 1	$\frac{dy}{dx} > 0$	Channel bottom, $S_0 < 0$

large y. Then we conclude that the M1 surface profile first rises in the flow direction and then tends to a horizontal asymptote.

As $y \to y_c$ in subcritical flow (such as M2, H2, and A2), we have Fr $\to$ 1 and $1 - \text{Fr}^2 \to 0$, and thus the slope dy/dx tends to negative infinity. But as $y \to y_c$ in supercritical flow (such as M3, H3, and A3), we have Fr $\to$ 1 and $1 - \text{Fr}^2 \to 0$, and thus the slope dy/dx, which is a positive quantity, tends to infinity. That is, the free surface rises almost vertically and the flow depth increases very rapidly. This cannot be sustained physically, and the free surface breaks down. The result is a hydraulic jump. The one-dimensional approximation is no longer applicable when this happens.

Some Representative Surface Profiles

A typical open-channel system involves several sections of different slopes, with connections called *transitions*, and thus the overall surface profile of the flow is a continuous profile made up of the individual profiles described earlier. Some representative surface profiles commonly encountered in open channels, including some composite profiles, are given in Fig. 13–33. For each case, the change in surface profile is caused by a change in channel geometry such as an abrupt change in slope or an obstruction in the flow such as a sluice gate. More composite profiles can be found in specialized books listed in the references. A point on a surface profile represents the flow height at that point that satisfies the mass, momentum, and energy conservation relations. Note that $dy/dx \ll 1$ and $S_0 \ll 1$ in gradually varied flow, and the slopes of both the channels and the surface profiles in these sketches are highly exaggerated for better visualization. Many channels and surface profiles would appear nearly horizontal if drawn to scale.

Figure 13–33a shows the surface profile for gradually varied flow in a channel with mild slope and a sluice gate. The subcritical upstream flow (note that the flow is subcritical since the slope is mild) slows down as it approaches the gate (such as a river approaching a dam) and the liquid level rises. The flow past the gate is supercritical (since the height of the opening is less than the critical depth). Therefore, the surface profile is M1 before the gate and M3 after the gate prior to the hydraulic jump.

A section of an open channel may have a negative slope and involve uphill flow, as shown in Fig. 13–33b. Flow with an adverse slope cannot be maintained unless the inertia forces overcome the gravity and viscous forces that oppose the fluid motion. Therefore, an uphill channel section must be followed by a downhill section or a free outfall. For subcritical flow with an adverse slope approaching a sluice gate, the flow depth decreases as the gate is approached, yielding an A2 profile. Flow past the gate is typically supercritical, yielding an A3 profile prior to the hydraulic jump.

The open-channel section in Fig. 13–33c involves a slope change from steep to less steep. The flow velocity in the less steep part is lower (a smaller elevation drop to drive the flow), and thus the flow depth is higher when uniform flow is established again. Noting that uniform flow with steep slope must be supercritical ($y < y_c$), the flow depth increases from the initial to the new uniform level smoothly through an S3 profile.

Figure 13–33d shows a composite surface profile for an open channel that involves various flow sections. Initially the slope is mild, and the flow is uniform and subcritical. Then the slope changes to steep, and the flow

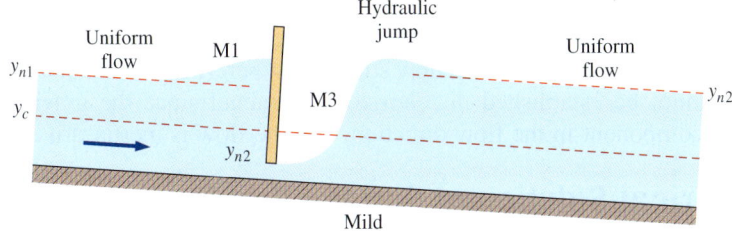

(a) Flow through a sluice gate in an open channel with mild slope

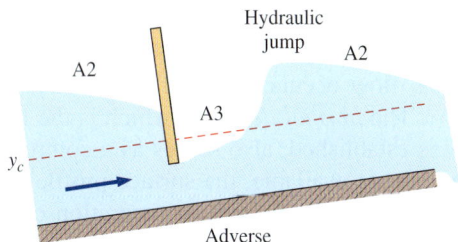

(b) Flow through a sluice gate in an open channel with adverse slope and free outfall

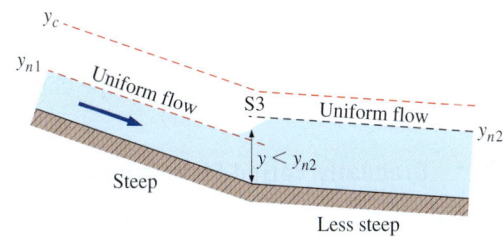

(c) Uniform supercritical flow changing from steep to less steep slope

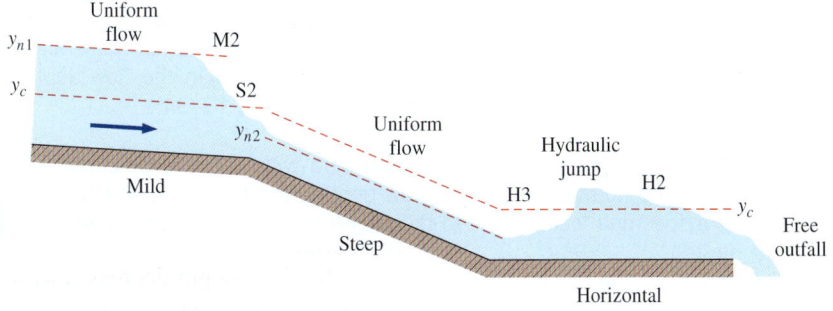

(d) Uniform subcritical flow changing from mild to steep to horizontal slope with free outfall

FIGURE 13–33
Some common surface profiles encountered in open-channel flow. All flows are from left to right.

becomes supercritical when uniform flow is established. The critical depth occurs at the break in grade. The change of slope is accompanied by a smooth decrease in flow depth through an M2 profile at the end of the mild section, and through an S2 profile at the beginning of the steep section. In the horizontal section, the flow depth increases first smoothly through an H3 profile, and then rapidly during a hydraulic jump. The flow depth then decreases through an H2 profile as the liquid accelerates toward the end of the channel to a free outfall. The flow becomes critical before reaching the end of the channel, and the outfall controls the upstream flow past the hydraulic jump. The outfalling flow stream is supercritical. Note that uniform flow cannot be established in a horizontal channel since the gravity force has no component in the flow direction, and the flow is inertia-driven.

Numerical Solution of Surface Profile

The prediction of the surface profile $y(x)$ is an important part of the design of open-channel systems. A good starting point for the determination of the surface profile is the identification of the points along the channel, called the **control points,** at which the flow depth can be calculated from a knowledge of flow rate. For example, the flow depth at a section of a rectangular channel where critical flow occurs, called the *critical point*, is determined from $y_c = (\dot{V}^2/gb^2)^{1/3}$. The *normal depth* y_n, which is the flow depth reached when uniform flow is established, also serves as a control point. Once flow depths at control points are available, the surface profile upstream or downstream is determined usually by numerical integration of the nonlinear differential equation (Eq. 13–65, repeated here)

$$\frac{dy}{dx} = \frac{S_0 - S_f}{1 - \mathrm{Fr}^2} \tag{13-66}$$

The friction slope S_f is determined from the uniform-flow conditions, and the Froude number from a relation appropriate for the channel cross section.

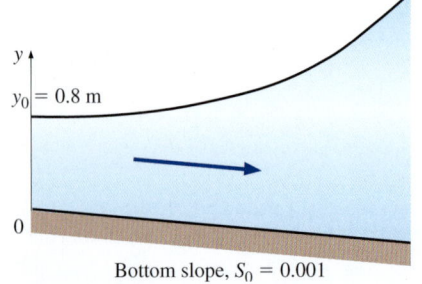

FIGURE 13–34
Schematic for Example 13–6.

EXAMPLE 13–6 Gradually Varied Flow with M1 Surface Profile

Gradually varied flow of water in a wide rectangular channel with a per-unit-width flow rate of 1 m³/s·m and a Manning coefficient of $n = 0.02$ is considered. The slope of the channel is 0.001, and at the location $x = 0$, the flow depth is measured to be 0.8 m. (*a*) Determine the normal and critical depths of the flow and classify the water surface profile, and (*b*) calculate the flow depth y at $x = 1000$ m by integrating the GVF equation numerically over the range $0 \le x \le 1000$ m. Repeat part (*b*) to obtain the flow depths for different x values, and plot the surface profile (Fig. 13–34).

SOLUTION Gradually varied flow of water in a wide rectangular channel is considered. The normal and critical flow depths, the flow type, and the flow depth at a specified location are to be determined, and the surface profile is to be plotted.
Assumptions 1 The channel is wide, and the flow is gradually varied. 2 The bottom slope is constant. 3 The roughness of the wetted surface of the channel and thus the friction coefficient are constant.
Properties The Manning coefficient of the channel is given to be $n = 0.02$.
Analysis (*a*) The channel is said to be wide, and thus the hydraulic radius is equal to the flow depth, $R_h \cong y$. Knowing the flow rate per unit width ($b = 1$ m), the normal depth is determined from the Manning equation to be

$$\dot{V} = \frac{a}{n} A_c R_h^{2/3} S_0^{1/2} = \frac{a}{n}(yb) y^{2/3} S_0^{1/2} = \frac{a}{n} b y^{5/3} S_0^{1/2}$$

$$y_n = \left(\frac{(\dot{V}/b)n}{aS_0^{1/2}}\right)^{3/5} = \left(\frac{(1\,\mathrm{m^2/s})(0.02)}{(1\,\mathrm{m^{1/3}/s})(0.001)^{1/2}}\right)^{3/5} = \mathbf{0.76\ m}$$

The critical depth for this flow is

$$y_c = \frac{\dot{V}^2}{gA_c^2} = \frac{\dot{V}^2}{g(by)^2} \rightarrow y_c = \left(\frac{(\dot{V}/b)^2}{g}\right)^{1/3} = \left(\frac{(1\,\text{m}^2/\text{s})^2}{(9.81\,\text{m/s}^2)}\right)^{1/3} = \mathbf{0.47\,m}$$

Noting that $y_c < y_n < y$ at $x = 0$, we see from Table 13-3 that the water surface profile during this GVF is classified as **M1**.

(b) Knowing the initial condition $y(0) = 0.8$ m, the flow depth y at any x location is determined by numerical integration of the GVF equation

$$\frac{dy}{dx} = \frac{S_0 - S_f}{1 - \text{Fr}^2}$$

where the Froude number for a wide rectangular channel is

$$\text{Fr} = \frac{V}{\sqrt{gy}} = \frac{\dot{V}/by}{\sqrt{gy}} = \frac{\dot{V}/b}{\sqrt{gy^3}}$$

and the friction slop is determined from the uniform-flow equation by setting $S_0 = S_f$,

$$\dot{V} = \frac{a}{n} b y^{5/3} S_f^{1/2} \rightarrow S_f = \left(\frac{(\dot{V}/b)n}{ay^{5/3}}\right)^2 = \frac{(\dot{V}/b)^2 n^2}{a^2 y^{10/3}}$$

Substituting, the GVF equation for a wide rectangular channel becomes

$$\frac{dy}{dx} = \frac{S_0 - (\dot{V}/b)^2 n^2/(a^2 y^{10/3})}{1 - (\dot{V}/b)^2/(gy^3)}$$

which is highly nonlinear, and thus it is difficult (if not impossible) to integrate analytically. Fortunately, nowadays solving nonlinear differential equations by integrating such nonlinear equations numerically using a program like EES or Matlab is easy. With this mind, the solution of the nonlinear first order differential equation subject to the initial condition $y(x_1) = y_1$ is expressed as

$$y = y_1 + \int_{x_1}^{x_2} f(x,y)dx \quad \text{where} \quad f(x,y) = \frac{S_0 - (\dot{V}/b)^2 n^2/(a^2 y^{10/3})}{1 - (\dot{V}/b)^2/(gy^3)}$$

and where $y = y(x)$ is the water depth at the specified location x. For given numerical values, this problem can be solved using EES as follows:

Vol = 1 "m^3/s, volume flow rate per unit width, b = 1 m"
b = 1 "m, width of channel"
n = 0.02 "Manning coefficient"
S_0 = 0.001 "slope of channel"
g = 9.81 "gravitational acceleration, m/s^2"

x1 = 0; y1=0.8 "m, initial condition"
x2 = 1000 "m, length of channel"

f_xy = (S_0-((Vol/b)^2*n^2/y(10/3)))/(1-(Vol/b)^2/(g*y^3)) "the GVF equation to be integrated"
y = y1+integral(f_xy, x, x1, x2) "integral equation with automatic step size."

Copying the mini program above into a blank EES screen and calculating gives the water depth at a location of 1000 m,

$$y(x_2) = y(1000\,\text{m}) = \mathbf{1.44\,m}$$

Distance along the channel, m	Water depth, m
0	0.80
100	0.82
200	0.86
300	0.90
400	0.96
500	1.03
600	1.10
700	1.18
800	1.26
900	1.35
1000	1.44

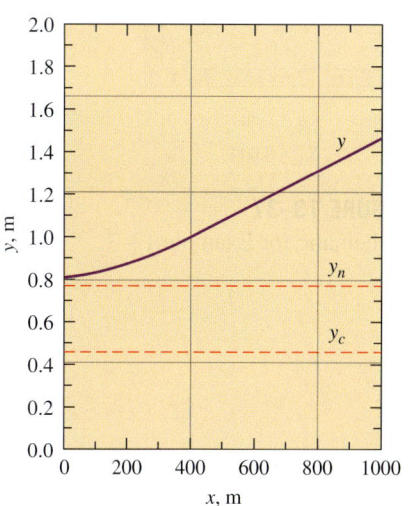

FIGURE 13–35

Flow depth and surface profile for the GVF problem discussed in Example 13–6.

```
clear all
domain=[0 1000]; % limits on integral
s0=.001; % channel slope
n=.02; % Manning roughness
q=1; % per-unit-width flowrate
g=9.81; % gravity (SI)
y0=.8; % initial condition on depth
[X,Y]=ode45('simple_flow_derivative',
[domain(1) domain (end)],y0,
[],s0,n,q,g,domain);

plot (X, Y, 'k')
axis([0 1000 0 max(Y)])
xlabel('x (m)');ylabel('y (m)');
**************

function
yprime=simple_flow_
derivative(x,y,flag,s0, n,q,g, (domain)
yprime=(s0-n.^2*q.^2./y.^(10/3))./(1-
q.^2/g./y.^3);
```

FIGURE 13–36
A Matlab program for solving the GVF problem of Example 13–6.

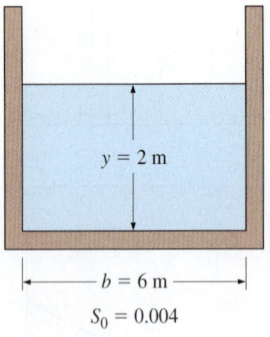

FIGURE 13–37
Schematic for Example 13–7.

Note that the built-in function "integral" performs integrations numerically between specified limits using an automatically adjusted step size. Water depths at different locations along the channel are obtained by repeating the calculations at different x_2 values. Plotting the results gives the surface profile, as shown in Fig. 13–34. Using the curve-fit feature of EES, we can even curve-fit the flow depth data into the following second-order polynomial,

$$y_{approx}(x) = 0.7930 + 0.0002789x + 3.7727 \times 10^{-7}x^2$$

It can be shown that the flow depth results obtained from this curve-fit formula do not differ from tabulated data by more than 1 percent.

Discussion The graphical result confirms the quantitative prediction from Table 13–3 that an M1 profile should yield increasing water depth in the downstream direction. This problem can also be solved using other programs, like Matlab, using the code given in Fig. 13–36.

EXAMPLE 13–7 Classification of Channel Slope

Water is flowing uniformly in a rectangular open channel with unfinished-concrete surfaces. The channel width is 6 m, the flow depth is 2 m, and the bottom slope is 0.004. Determine if the channel should be classified as mild, critical, or steep for this flow (Fig. 13–37).

SOLUTION Water is flowing uniformly in an open channel. It is to be determined whether the channel slope is mild, critical, or steep for this flow.
Assumptions 1 The flow is steady and uniform. 2 The bottom slope is constant. 3 The roughness of the wetted surface of the channel and thus the friction coefficient are constant.
Properties The Manning coefficient for an open channel with unfinished-concrete surfaces is $n = 0.014$.
Analysis The cross-sectional area, perimeter, and hydraulic radius are

$$A_c = yb = (2 \text{ m})(6 \text{ m}) = 12 \text{ m}^2$$
$$p = b + 2y = 6 \text{ m} + 2(2 \text{ m}) = 10 \text{ m}$$
$$R_h = \frac{A_c}{p} = \frac{12 \text{ m}^2}{10 \text{ m}} = 1.2 \text{ m}$$

The flow rate is determined from the Manning equation to be

$$\dot{V} = \frac{a}{n} A_c R_h^{2/3} S_0^{1/2} = \frac{1 \text{ m}^{1/3}/\text{s}}{0.014} (12 \text{ m}^2)(1.2 \text{ m})^{2/3}(0.004)^{1/2} = \mathbf{61.2 \text{ m}^3/\text{s}}$$

Noting that the flow is uniform, the specified flow rate is the normal depth and thus $y = y_n = 2$ m. The critical depth for this flow is

$$y_c = \frac{\dot{V}^2}{gA_c^2} = \frac{(61.2 \text{ m}^3/\text{s})^2}{(9.81 \text{ m/s}^2)(12 \text{ m}^2)^2} = 2.65 \text{ m}$$

This channel at these flow conditions is classified as **steep** since $y_n < y_c$, and the flow is supercritical.
Discussion If the flow depth were greater than 2.65 m, the channel slope would be said to be *mild*. Therefore, the bottom slope alone is not sufficient to classify a downhill channel as being mild, critical, or steep.

13–8 ▪ RAPIDLY VARIED FLOW AND THE HYDRAULIC JUMP

Recall that flow in open channels is called **rapidly varied flow (RVF)** if the flow depth changes markedly over a relatively short distance in the flow direction (Fig. 13–38). Such flows occur in sluice gates, broad- or sharp-crested weirs, waterfalls, and the transition sections of channels for expansion and contraction. A change in the cross section of the channel is one cause of rapidly varied flow. But some rapidly varied flows, such as flow through a sluice gate, occur even in regions where the channel cross section is constant.

Rapidly varied flows are typically complicated by the fact that they may involve significant multidimensional and transient effects, backflows, and flow separation (Fig. 13–39). Therefore, rapidly varied flows are usually studied experimentally or numerically. But despite these complexities, it is still possible to analyze some rapidly varied flows using the one-dimensional flow approximation with reasonable accuracy.

The flow in steep channels may be supercritical, and the flow must change to subcritical if the channel can no longer sustain supercritical flow due to a reduced slope of the channel or increased frictional effects. Any such change from supercritical to subcritical flow occurs through a *hydraulic jump*. A hydraulic jump involves considerable mixing and agitation, and thus a significant amount of mechanical energy dissipation.

Consider steady flow through a control volume that encloses the hydraulic jump, as shown in Fig. 13–39. To make a simple analysis possible, we make the following approximations:

1. The velocity is nearly constant across the channel at sections 1 and 2, and therefore the momentum-flux correction factors are $\beta_1 = \beta_2 \cong 1$.
2. The pressure in the liquid varies hydrostatically, and we consider gage pressure only since atmospheric pressure acts on all surfaces and its effect cancels out.
3. The wall shear stress and its associated losses are negligible relative to the losses that occur during the hydraulic jump due to the intense agitation.
4. The channel is wide and horizontal.
5. There are no external or body forces other than gravity.

For a channel of width b, the conservation of mass relation $\dot{m}_2 = \dot{m}_1$ is expressed as $\rho y_1 b V_1 = \rho y_2 b V_2$ or

$$y_1 V_1 = y_2 V_2 \tag{13-67}$$

Noting that the only forces acting on the control volume in the horizontal x-direction are the pressure forces, the momentum equation $\sum \vec{F} = \sum_{\text{out}} \beta \dot{m} \vec{V} - \sum_{\text{in}} \beta \dot{m} \vec{V}$ in the x-direction becomes a balance between hydrostatic pressure forces and momentum transfer,

$$P_{1,\text{avg}} A_1 - P_{2,\text{avg}} A_2 = \dot{m} V_2 - \dot{m} V_1 \tag{13-68}$$

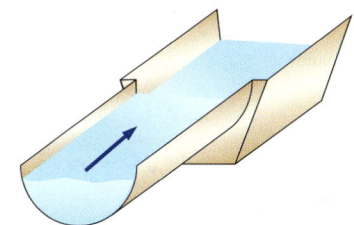

FIGURE 13–38
Rapidly varied flow occurs when there is a sudden change in flow, such as an abrupt change in cross section.

FIGURE 13–39
When riding the rapids, a kayaker encounters several features of both gradually varied flow (GVF) and rapidly varied flow (RVF), with the latter being more exciting.
© Karl Weatherly/Getty RF

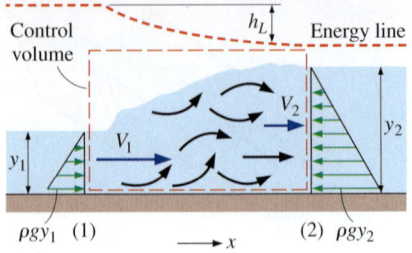

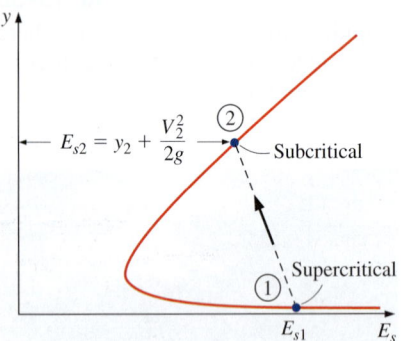

FIGURE 13–40
Schematic and flow depth-specific energy diagram for a hydraulic jump (specific energy decreases).

where $P_{1,\text{avg}} = \rho g y_1 / 2$ and $P_{2,\text{avg}} = \rho g y_2 / 2$. For a channel width of b, we have $A_1 = y_1 b$, $A_2 = y_2 b$, and $\dot{m} = \dot{m}_2 = \dot{m}_1 = \rho A_1 V_1 = \rho y_1 b V_1$. Substituting and simplifying, the momentum equation reduces to

$$y_1^2 - y_2^2 = \frac{2 y_1 V_1}{g}(V_2 - V_1) \qquad (13\text{–}69)$$

Eliminating V_2 by using $V_2 = (y_1/y_2)V_1$ from Eq. 13–67 gives

$$y_1^2 - y_2^2 = \frac{2 y_1 V_1^2}{g y_2}(y_1 - y_2) \qquad (13\text{–}70)$$

Canceling the common factor $y_1 - y_2$ from both sides and rearranging give

$$\left(\frac{y_2}{y_1}\right)^2 + \frac{y_2}{y_1} - 2\text{Fr}_1^2 = 0 \qquad (13\text{–}71)$$

where $\text{Fr}_1 = V_1/\sqrt{g y_1}$. This is a quadratic equation for y_2/y_1, and it has two roots—one negative and one positive. Noting that y_2/y_1 cannot be negative since both y_2 and y_1 are positive quantities, the depth ratio y_2/y_1 is determined to be

Depth ratio: $\qquad \dfrac{y_2}{y_1} = 0.5\left(-1 + \sqrt{1 + 8\text{Fr}_1^2}\right) \qquad (13\text{–}72)$

The energy equation (Eq. 13–30) for this horizontal flow section is

$$y_1 + \frac{V_1^2}{2g} = y_2 + \frac{V_2^2}{2g} + h_L \qquad (13\text{–}73)$$

Noting that $V_2 = (y_1/y_2)V_1$ and $\text{Fr}_1 = V_1/\sqrt{g y_1}$, the head loss associated with a hydraulic jump is expressed as

$$h_L = y_1 - y_2 + \frac{V_1^2 - V_2^2}{2g} = y_1 - y_2 + \frac{y_1 \text{Fr}_1^2}{2}\left(1 - \frac{y_1^2}{y_2^2}\right) \qquad (13\text{–}74)$$

The energy line for a hydraulic jump is shown in Fig. 13–40. The drop in the energy line across the jump represents the head loss h_L associated with the jump.

For given values of Fr_1 and y_1, the downstream flow depth y_2 and the head loss h_L can be calculated from Eqs. 13–72 and 13–74, respectively. Plotting h_L against Fr_1 would reveal that h_L becomes negative for $\text{Fr}_1 < 1$, which is impossible (it would correspond to negative entropy generation, which would be a violation of the second law of thermodynamics). Thus we conclude that the upstream flow must be supercritical ($\text{Fr}_1 > 1$) for a hydraulic jump to occur. In other words, it is impossible for subcritical flow to undergo a hydraulic jump. This is analogous to gas flow having to be supersonic (Mach number greater than 1) to undergo a shock wave.

Head loss is a measure of the mechanical energy dissipated via internal fluid friction, and head loss is usually undesirable as it represents the mechanical energy wasted. But sometimes hydraulic jumps are designed in conjunction with stilling basins and spillways of dams, and it is desirable to waste as much of the mechanical energy as possible to minimize the mechanical energy of the water and thus its potential to cause damage. This is done by first producing supercritical flow by converting high pressure to high linear velocity, and then allowing the flow to agitate and dissipate part of its kinetic energy as it breaks down and decelerates to a subcritical velocity. Therefore, a measure of performance of a hydraulic jump is its fraction of energy dissipation.

The specific energy of the liquid before the hydraulic jump is $E_{s1} = y_1 + V_1^2/2g$. Then the **energy dissipation ratio** (Fig. 13–41) is defined as

$$\text{Energy dissipation ratio} = \frac{h_L}{E_{s1}} = \frac{h_L}{y_1 + V_1^2/2g} = \frac{h_L}{y_1(1 + \text{Fr}_1^2/2)} \quad (13\text{–}75)$$

The fraction of energy dissipation ranges from just a few percent for weak hydraulic jumps ($\text{Fr}_1 < 2$) to 85 percent for strong jumps ($\text{Fr}_1 > 9$).

Unlike a normal shock in gas flow, which occurs at a cross section and thus has negligible thickness, the hydraulic jump occurs over a considerable channel length. In the Froude number range of practical interest, the length of the hydraulic jump is observed to be 4 to 7 times the downstream flow depth y_2.

Experimental studies indicate that hydraulic jumps can be classified into five categories as shown in Table 13–4, depending primarily on the value of the upstream Froude number Fr_1. For Fr_1 somewhat higher than 1, the liquid rises slightly during the hydraulic jump, producing standing waves. At larger Fr_1, highly damaging oscillating waves occur. The desirable range of Froude numbers is $4.5 < \text{Fr}_1 < 9$, which produces stable and well-balanced steady waves with high levels of energy dissipation within the jump. Hydraulic jumps with $\text{Fr}_1 > 9$ produce very rough waves. The depth ratio y_2/y_1 ranges from slightly over 1 for *undular jumps* that are mild and involve small rises in surface level to over 12 for *strong jumps* that are rough and involve high rises in surface level.

In this section we limit our consideration to wide horizontal rectangular channels so that edge and gravity effects are negligible. Hydraulic jumps in nonrectangular and sloped channels behave similarly, but the flow characteristics and thus the relations for depth ratio, head loss, jump length, and dissipation ratio are different.

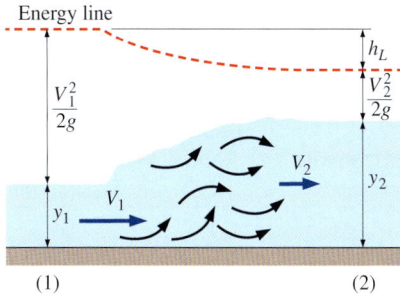

$$\text{Dissipation ratio} = \frac{h_L}{E_{s1}} = \frac{h_L}{y_1 + V_1^2/2g}$$

FIGURE 13–41
The energy dissipation ratio represents the fraction of mechanical energy dissipated during a hydraulic jump.

■ **EXAMPLE 13–8** Hydraulic Jump

■ Water discharging into a 10-m-wide rectangular horizontal channel from a sluice gate is observed to have undergone a hydraulic jump. The flow depth and velocity before the jump are 0.8 m and 7 m/s, respectively. Determine (a) the flow depth and the Froude number after the jump, (b) the head loss and the energy dissipation ratio, and (c) the wasted power production potential due to the hydraulic jump (Fig. 13–42).

SOLUTION Water at a specified depth and velocity undergoes a hydraulic jump in a horizontal channel. The depth and Froude number after the jump, the head loss and the dissipation ratio, and the wasted power potential are to be determined.
Assumptions 1 The flow is steady or quasi-steady. 2 The channel is sufficiently wide so that the end effects are negligible.
Properties The density of water is 1000 kg/m³.

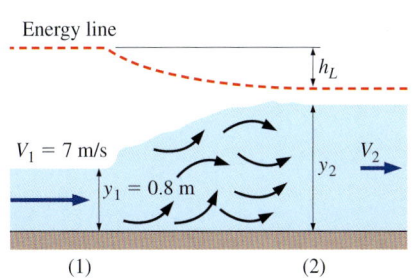

FIGURE 13–42
Schematic for Example 13–8.

TABLE 13–4
Classification of hydraulic jumps
Source: U.S. Bureau of Reclamation (1955).

Upstream Fr$_1$	Depth Ratio y_2/y_1	Fraction of Energy Dissipation	Description	Surface Profile
<1	1	0	*Impossible jump.* Would violate the second law of thermodynamics.	
1–1.7	1–2	<5%	*Undular jump* (or *standing wave*). Small rise in surface level. Low energy dissipation. Surface rollers develop near Fr = 1.7.	
1.7–2.5	2–3.1	5–15%	*Weak jump.* Surface rising smoothly, with small rollers. Low energy dissipation.	
2.5–4.5	3.1–5.9	15–45%	*Oscillating jump.* Pulsations caused by jets entering at the bottom generate large waves that can travel for miles and damage earth banks. Should be avoided in the design of stilling basins.	
4.5–9	5.9–12	45–70%	*Steady jump.* Stable, well-balanced, and insensitive to downstream conditions. Intense eddy motion and high level of energy dissipation within the jump. Recommended range for design.	
>9	>12	70–85%	*Strong jump.* Rough and intermittent. Very effective energy dissipation, but may be uneconomical compared to other designs because of the larger water heights involved.	

Analysis (a) The Froude number before the hydraulic jump is

$$\text{Fr}_1 = \frac{V_1}{\sqrt{gy_1}} = \frac{7 \text{ m/s}}{\sqrt{(9.81 \text{ m/s}^2)(0.8 \text{ m})}} = 2.50$$

which is greater than 1. Therefore, the flow is indeed supercritical before the jump. The flow depth, velocity, and Froude number after the jump are

$$y_2 = 0.5y_1(-1 + \sqrt{1 + 8\text{Fr}_1^2}) = 0.5(0.8 \text{ m})(-1 + \sqrt{1 + 8 \times 2.50^2}) = \mathbf{2.46 \text{ m}}$$

$$V_2 = \frac{y_1}{y_2}V_1 = \frac{0.8 \text{ m}}{2.46 \text{ m}}(7 \text{ m/s}) = 2.28 \text{ m/s}$$

$$\text{Fr}_2 = \frac{V_2}{\sqrt{gy_2}} = \frac{2.28 \text{ m/s}}{\sqrt{(9.81 \text{ m/s}^2)(2.46 \text{ m})}} = \mathbf{0.464}$$

Note that the flow depth triples and the Froude number reduces to about one-fifth after the jump.

(b) The head loss is determined from the energy equation to be

$$h_L = y_1 - y_2 + \frac{V_1^2 - V_2^2}{2g} = (0.8 \text{ m}) - (2.46 \text{ m}) + \frac{(7 \text{ m/s})^2 - (2.28 \text{ m/s})^2}{2(9.81 \text{ m/s}^2)}$$

$$= \mathbf{0.572 \text{ m}}$$

The specific energy of water before the jump and the dissipation ratio are

$$E_{s1} = y_1 + \frac{V_1^2}{2g} = (0.8 \text{ m}) + \frac{(7 \text{ m/s})^2}{2(9.81 \text{ m/s}^2)} = 3.30 \text{ m}$$

$$\text{Dissipation ratio} = \frac{h_L}{E_{s1}} = \frac{0.572 \text{ m}}{3.30 \text{ m}} = \mathbf{0.173}$$

Therefore, 17.3 percent of the available head (or mechanical energy) of the liquid is wasted (converted to thermal energy) as a result of frictional effects during this hydraulic jump.

(c) The mass flow rate of water is

$$\dot{m} = \rho \dot{V} = \rho b y_1 V_1 = (1000 \text{ kg/m}^3)(0.8 \text{ m})(10 \text{ m})(7 \text{ m/s}) = 56{,}000 \text{ kg/s}$$

Then the power dissipation corresponding to a head loss of 0.572 m becomes

$$\dot{E}_{\text{dissipated}} = \dot{m}gh_L = (56{,}000 \text{ kg/s})(9.81 \text{ m/s}^2)(0.572 \text{ m})\left(\frac{1 \text{ N}}{1 \text{ kg·m/s}^2}\right)$$

$$= 314{,}000 \text{ N·m/s} = \mathbf{314 \text{ kW}}$$

Discussion The results show that the hydraulic jump is a highly dissipative process, wasting 314 kW of power production potential in this case. That is, if the water were routed to a hydraulic turbine instead of being released from the sluice gate, up to 314 kW of power could be generated. But this potential is converted to useless thermal energy instead of useful power, causing a water temperature rise of

$$\Delta T = \frac{\dot{E}_{\text{dissipated}}}{\dot{m}c_p} = \frac{314 \text{ kJ/s}}{(56{,}000 \text{ kg/s})(4.18 \text{ kJ/kg·°C})} = 0.0013°\text{C}$$

Note that a 314-kW resistance heater would cause the same temperature rise for water flowing at a rate of 56,000 kg/s.

13–9 · FLOW CONTROL AND MEASUREMENT

The flow rate in pipes and ducts is controlled by various kinds of valves. Liquid flow in open channels, however, is not confined, and thus the flow rate is controlled by partially blocking the channel. This is done by either allowing the liquid to flow *over* the obstruction or *under* it. An obstruction that allows the liquid to flow over it is called a **weir** (Fig. 13–43), and an obstruction with an adjustable opening at the bottom that allows the liquid to flow underneath it is called an **underflow gate.** Such devices can be used to control the flow rate through the channel as well as to measure it.

(a)

(b)

FIGURE 13–43
A weir is a flow control device in which the water flows *over* the obstruction.

(a) © Design Pics RF/The Irish Image Collection/Getty RF; (b) Photo courtesy of Bryan Lewis.

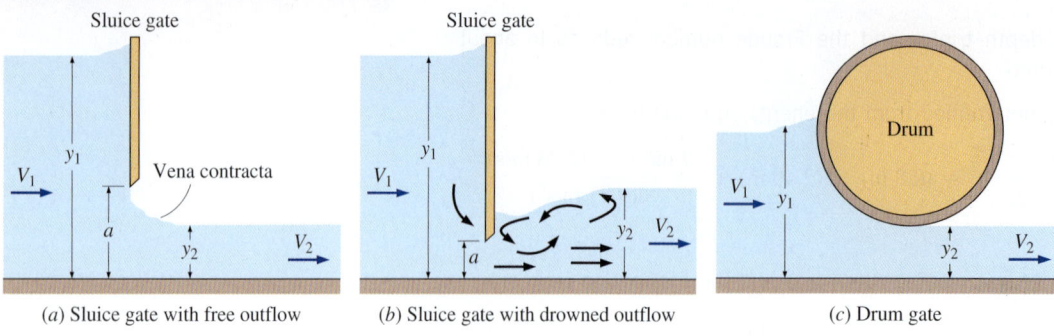

FIGURE 13–44
Common types of underflow gates to control flow rate.

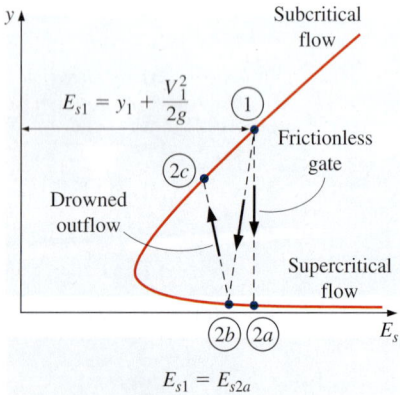

FIGURE 13–45
Schematic and flow depth-specific energy diagram for flow through underflow gates.

Underflow Gates

There are numerous types of underflow gates to control the flow rate, each with certain advantages and disadvantages. Underflow gates are located at the bottom of a wall, dam, or an open channel. Two common types of such gates, the **sluice gate** and the **drum gate,** are shown in Fig. 13–44. A sluice gate is typically vertical and has a plane surface, whereas a drum gate has a circular cross section with a streamlined surface.

When the gate is partially opened, the upstream liquid accelerates as it approaches the gate, reaches critical speed at the gate, and accelerates further to supercritical speeds past the gate. Therefore, an underflow gate is analogous to a converging–diverging nozzle in gas dynamics. The discharge from an underflow gate is called a *free outflow* if the liquid jet streaming out of the gate is open to the atmosphere (Fig. 13–44a), and it is called a *drowned* (or *submerged*) *outflow* if the discharged liquid flashes back and submerges the jet (Fig. 13–44b). In drowned flow, the liquid jet undergoes a hydraulic jump, and thus the downstream flow is subcritical. Also, drowned outflow involves a high level of turbulence and backflow, and thus a large head loss h_L.

The flow depth-specific energy diagram for flow through underflow gates with free and drowned outflow is given in Fig. 13–45. Note that the specific energy remains constant for idealized gates with negligible frictional effects (from point 1 to point 2a), but decreases for actual gates. The downstream is supercritical for a gate with free outflow (point 2b), but subcritical for one with drowned outflow (point 2c) since a drowned outflow also involves a hydraulic jump to subcritical flow, which involves considerable mixing and energy dissipation.

Approximating the frictional effects as negligible and the upstream (or reservoir) velocity to be low, it can be shown by using the Bernoulli equation that the discharge velocity of a free jet is (see Chap. 5 for details)

$$V = \sqrt{2gy_1} \qquad (13\text{–}76)$$

The frictional effects can be accounted for by modifying this relation with a **discharge coefficient** C_d. Then the discharge velocity at the gate and the flow rate become

$$V = C_d\sqrt{2gy_1} \quad \text{and} \quad \dot{V} = C_d b a \sqrt{2gy_1} \qquad (13\text{–}77)$$

where b and a are the width and the height of the gate opening, respectively.

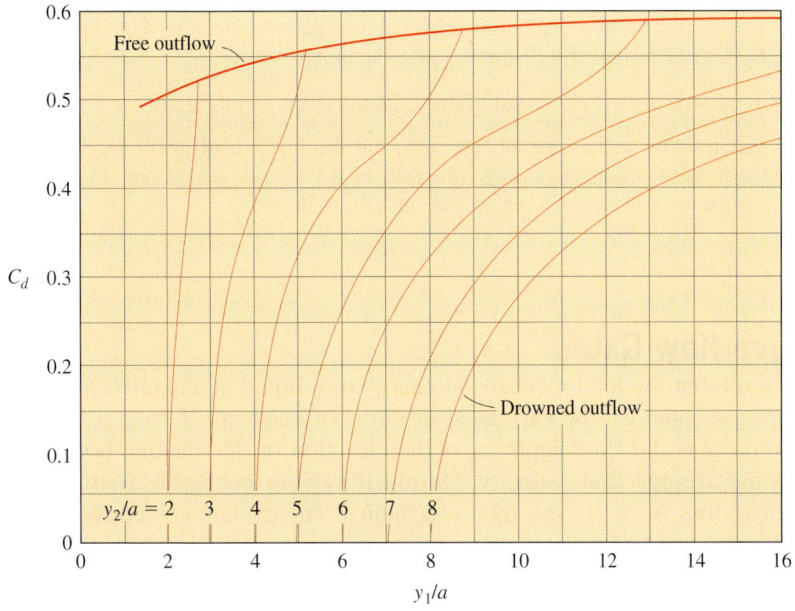

FIGURE 13–46
Discharge coefficients for drowned and free discharge from underflow gates.

Data from Henderson, Open Channel Flow, 1st Edition, © 1966. Reprinted by permission of Pearson Education, Inc., Upper Saddle River, NJ.

The discharge coefficient $C_d = 1$ for idealized flow, but $C_d < 1$ for actual flow through the gates. Experimentally determined values of C_d for underflow gates are plotted in Fig. 13–46 as functions of the contraction coefficient y_2/a and the depth ratio y_1/a. Note that most values of C_d for free outflow from a vertical sluice gate range between 0.5 and 0.6. The C_d values drop sharply for drowned outflow, as expected, and the flow rate decreases for the same upstream conditions. For a given value of y_1/a, the value of C_d decreases with increasing y_2/a.

■ **EXAMPLE 13–9   Sluice Gate with Drowned Outflow**

Water is released from a 3-m-deep reservoir into a 6-m-wide open channel through a sluice gate with a 0.25-m-high opening at the channel bottom. The flow depth after all turbulence subsides is measured to be 1.5 m. Determine the rate of discharge (Fig. 13–47).

SOLUTION Water is released from a reservoir through a sluice gate into an open channel. For specified flow depths, the rate of discharge is to be determined.
Assumptions 1 The flow is steady in the mean. 2 The channel is sufficiently wide so that the end effects are negligible.
Analysis The depth ratio y_1/a and the contraction coefficient y_2/a are

$$\frac{y_1}{a} = \frac{3 \text{ m}}{0.25 \text{ m}} = 12 \quad \text{and} \quad \frac{y_2}{a} = \frac{1.5 \text{ m}}{0.25 \text{ m}} = 6$$

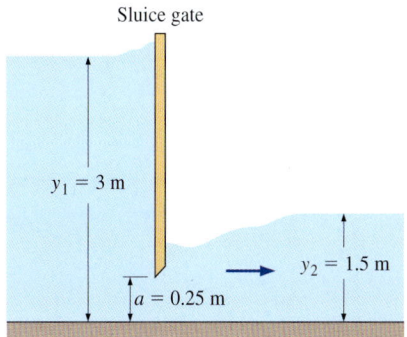

FIGURE 13–47
Schematic for Example 13–9.

The corresponding discharge coefficient is determined from Fig. 13–46 to be $C_d = 0.47$. Then the discharge rate becomes

$$\dot{V} = C_d b a \sqrt{2gy_1} = 0.47(6 \text{ m})(0.25 \text{ m})\sqrt{2(9.81 \text{ m/s}^2)(3 \text{ m})} = \mathbf{5.41 \text{ m}^3/\text{s}}$$

Discussion In the case of free flow, the discharge coefficient would be $C_d = 0.59$, with a corresponding flow rate of 6.78 m³/s. Therefore, the flow rate decreases considerably when the outflow is drowned.

Overflow Gates

Recall that the total mechanical energy of a liquid at any cross section of an open channel can be expressed in terms of heads as $H = z_b + y + V^2/2g$, where y is the flow depth, z_b is the elevation of the channel bottom, and V is the average flow velocity. During flow with negligible frictional effects (head loss $h_L = 0$), the total mechanical energy remains constant, and the one-dimensional energy equation for open-channel flow between upstream section 1 and downstream section 2 is written as

$$z_{b1} + y_1 + \frac{V_1^2}{2g} = z_{b2} + y_2 + \frac{V_2^2}{2g} \quad \text{or} \quad E_{s1} = \Delta z_b + E_{s2} \quad (13\text{–}78)$$

where $E_s = y + V^2/2g$ is the specific energy and $\Delta z_b = z_{b2} - z_{b1}$ is the elevation of the bottom point of flow at section 2 relative to that at section 1. Therefore, the specific energy of a liquid stream increases by $|\Delta z_b|$ during downhill flow (note that Δz_b is negative for channels inclined down), decreases by Δz_b during uphill flow, and remains constant during horizontal flow. (The specific energy also decreases by h_L for all cases if the frictional effects are not negligible.)

For a channel of constant width b, $\dot{V} = A_c V = byV = $ constant in steady flow and $V = \dot{V}/A_c$. Then the specific energy becomes

$$E_s = y + \frac{\dot{V}^2}{2gb^2y^2} \quad (13\text{–}79)$$

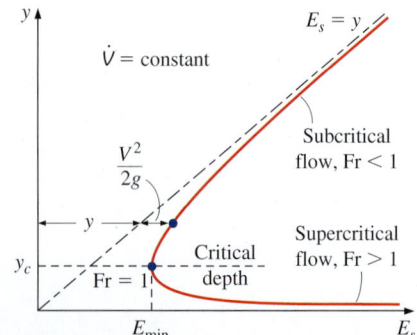

FIGURE 13–48
Variation of specific energy E_s with depth y for a specified flow rate in a channel of constant width.

The variation of the specific energy E_s with flow depth y for steady flow in a channel of constant width b is replotted in Fig. 13–48. This diagram is extremely valuable as it shows the allowable states during flow. Once the upstream conditions at a flow section 1 are specified, the state of the liquid at any section 2 on an E_s–y diagram must fall on a point on the specific energy curve that passes through point 1.

Flow over a Bump with Negligible Friction

Now consider steady flow with negligible friction over a bump of height Δz_b in a horizontal channel of constant width b, as shown in Fig. 13–47. The energy equation in this case is, from Eq. 13–78,

$$E_{s2} = E_{s1} - \Delta z_b \quad (13\text{–}80)$$

Therefore, the specific energy of the liquid decreases by Δz_b as it flows over the bump, and the state of the liquid on the E_s–y diagram shifts to the left by

Δz_b, as shown in Fig. 13–49. The conservation of mass equation for a channel of large width is $y_2 V_2 = y_1 V_1$ and thus $V_2 = (y_1/y_2)V_1$. Then the specific energy of the liquid over the bump can be expressed as

$$E_{s2} = y_2 + \frac{V_2^2}{2g} \quad \rightarrow \quad E_{s1} - \Delta z_b = y_2 + \frac{V_1^2}{2g} \frac{y_1^2}{y_2^2} \tag{13–81}$$

Rearranging,

$$y_2^3 - (E_{s1} - \Delta z_b)y_2^2 + \frac{V_1^2}{2g} y_1^2 = 0 \tag{13–82}$$

which is a third-degree polynomial equation in y_2 and thus has three solutions. Disregarding the negative solution, it appears that the flow depth over the bump can have two values.

Now the curious question is, does the liquid level rise or drop over the bump? Our intuition says the entire liquid body will follow the bump and thus the liquid surface will rise over the bump, but this is not necessarily so. Noting that specific energy is the sum of the flow depth and dynamic head, either scenario is possible, depending on how the velocity changes. The E_s–y diagram in Fig. 13–49 gives us the definite answer: If the flow before the bump is *subcritical* (state 1*a*), the flow depth y_2 decreases (state 2*a*). If the decrease in flow depth is greater than the bump height (i.e., $y_1 - y_2 > \Delta z_b$), the free surface is suppressed. But if the flow is *supercritical* as it approaches the bump (state 1*b*), the flow depth rises over the bump (state 2*b*), creating a bump along the free surface.

The situation is reversed if the channel has a depression of depth Δz_b instead of a bump: The specific energy in this case increases (so that state 2 is to the right of state 1 on the E_s–y diagram) since Δz_b is negative. Therefore, the flow depth increases if the approach flow is subcritical and decreases if it is supercritical.

Now let's reconsider flow over a bump with negligible friction, as discussed earlier. As the height of the bump Δz_b is increased, point 2 (either 2*a* or 2*b* for sub- or supercritical flow) continues shifting to the left on the E_s–y diagram, until finally reaching the critical point. That is, the flow over the bump is *critical* when the bump height is $\Delta z_c = E_{s1} - E_{sc} = E_{s1} - E_{\min}$, and the specific energy of the liquid reaches its minimum level.

The question that comes to mind is, what happens if the bump height is increased further? Does the specific energy of the liquid continue decreasing? The answer to this question is a resounding *no* since the liquid is already at its minimum energy level, and its energy cannot decrease any further. In other words, the liquid is already at the furthest left point on the E_s–y diagram, and no point further left can satisfy conservation of mass and energy and the momentum equation. Therefore, the flow must remain critical. The flow at this state is said to be **choked.** In gas dynamics, this is analogous to the flow in a converging nozzle accelerating as the back pressure is lowered, and reaching the speed of sound at the nozzle exit when the back pressure reaches the critical pressure. But the nozzle exit velocity remains at the sonic level no matter how much the back pressure is lowered. Here again, the flow is choked.

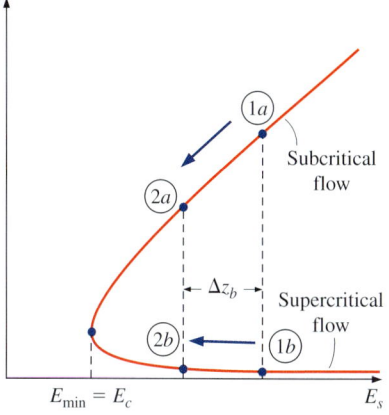

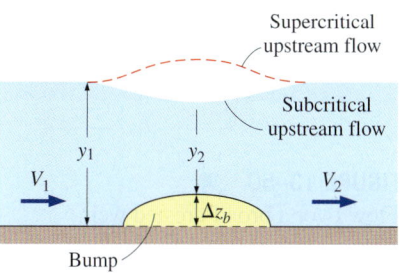

FIGURE 13–49
Schematic and flow depth-specific energy diagram for flow over a bump for subcritical and supercritical upstream flows.

Broad-Crested Weir

The discussions on flow over a high bump can be summarized as follows: *The flow over a sufficiently high obstruction in an open channel is always critical.* Such obstructions placed intentionally in an open channel to measure the flow rate are called *weirs*. Therefore, the flow velocity over a sufficiently broad weir is the critical velocity, which is expressed as $V = \sqrt{gy_c}$, where y_c is the critical depth. Then the flow rate over a weir of width b is expressed as

$$\dot{V} = A_c V = y_c b \sqrt{gy_c} = bg^{1/2} y_c^{3/2} \qquad (13\text{–}83)$$

A **broad-crested weir** is a rectangular block of height P_w and length L_w that has a horizontal crest over which critical flow occurs (Fig. 13–50). The upstream head above the top surface of the weir is called the **weir head** and is denoted by H. To obtain a relation for the critical depth y_c in terms of weir head H, we write the energy equation between a section upstream and a section over the weir for flow with negligible friction as

$$H + P_w + \frac{V_1^2}{2g} = y_c + P_w + \frac{V_c^2}{2g} \qquad (13\text{–}84)$$

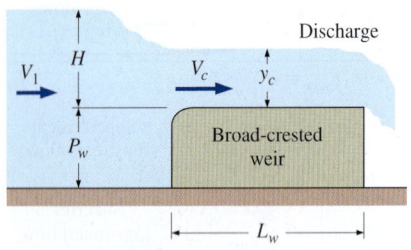

FIGURE 13–50
Flow over a broad-crested weir.

Cancelling P_w from both sides and substituting $V_c = \sqrt{gy_c}$ give

$$y_c = \frac{2}{3}\left(H + \frac{V_1^2}{2g}\right) \qquad (13\text{–}85)$$

Substituting into Eq. 13–83, the flow rate for this idealized flow case with negligible friction is determined to be

$$\dot{V}_{\text{ideal}} = b\sqrt{g}\left(\frac{2}{3}\right)^{3/2}\left(H + \frac{V_1^2}{2g}\right)^{3/2} \qquad (13\text{–}86)$$

This relation shows the functional dependence of the flow rate on the flow parameters, but it overpredicts the flow rate by several percent because it does not consider the frictional effects. These effects are typically accounted for by modifying the theoretical relation (Eq. 13–86) with an experimentally determined *weir discharge coefficient* C_{wd} as

Broad-crested weir: $\qquad \dot{V} = C_{wd,\,\text{broad}}\, b\sqrt{g}\left(\frac{2}{3}\right)^{3/2}\left(H + \frac{V_1^2}{2g}\right)^{3/2} \qquad (13\text{–}87)$

where reasonably accurate values of discharge coefficients for broad-crested weirs can be obtained from (Chow, 1959)

$$C_{wd,\,\text{broad}} = \frac{0.65}{\sqrt{1 + H/P_w}} \qquad (13\text{–}88)$$

More accurate but complicated relations for $C_{wd,\,\text{broad}}$ are also available in the literature (e.g., Ackers, 1978). Also, the upstream velocity V_1 is usually very low, and it can be disregarded. This is especially the case for high weirs. Then the flow rate is approximated as

Broad-crested weir with low V_1: $\qquad \dot{V} \cong C_{wd,\,\text{broad}}\, b\sqrt{g}\left(\frac{2}{3}\right)^{3/2} H^{3/2} \qquad (13\text{–}89)$

It should always be kept in mind that the basic requirement for the use of Eqs. 13–87 to 13–89 is the establishment of critical flow above the weir, and this puts some limitations on the weir length L_w. If the weir is too long ($L_w > 12H$), wall shear effects dominate and cause the flow over the weir to be subcritical. If the weir is too short ($L_w < 2H$), the liquid may not be able to accelerate to critical velocity. Based on observations, the proper length of the broad-crested weir is $2H < L_w < 12H$. Note that a weir that is too long for one flow may be too short for another flow, depending on the value of the weir head H. Therefore, the range of flow rates should be known before a weir can be selected.

Sharp-Crested Weirs

A sharp-crested weir is a vertical plate placed in a channel that forces the liquid to flow through an opening to measure the flow rate. The type of the weir is characterized by the shape of the opening. A vertical thin plate with a straight top edge is referred to as rectangular weir since the cross section of the flow over it is rectangular; a weir with a triangular opening is referred to as a triangular weir; etc.

Upstream flow is subcritical and becomes critical as it approaches the weir. The liquid continues to accelerate and discharges as a supercritical flow stream that resembles a free jet. The reason for acceleration is the steady decline in the elevation of the free surface, and the conversion of this elevation head into velocity head. The flow-rate correlations given below are based on the free overfall of liquid discharge past the weir, called a **nappe**, being clear from the weir. It may be necessary to ventilate the space under the nappe to assure atmospheric pressure underneath. Empirical relations for drowned weirs are also available.

Consider the flow of a liquid over a sharp-crested weir placed in a horizontal channel, as shown in Fig. 13–51. For simplicity, the velocity upstream of the weir is approximated as being nearly constant through vertical cross section 1. The total energy of the upstream liquid expressed as a head relative to the channel bottom is the specific energy, which is the sum of the flow depth and the velocity head. That is, $y_1 + V_1^2/2g$, where $y_1 = H + P_w$. The flow over the weir is not one-dimensional since the liquid undergoes large changes in velocity and direction over the weir. But the pressure within the nappe is atmospheric.

A simple relation for the variation of liquid velocity over the weir is obtained by assuming negligible friction and writing the Bernoulli equation between a point in upstream flow (point 1) and a point over the weir at a distance h from the upstream liquid level as

$$H + P_w + \frac{V_1^2}{2g} = (H + P_w - h) + \frac{u_2^2}{2g} \quad (13\text{–}90)$$

Cancelling the common terms and solving for u_2, the idealized velocity distribution over the weir is determined to be

$$u_2 = \sqrt{2gh + V_1^2} \quad (13\text{–}91)$$

In reality, the liquid surface level drops somewhat over the weir as the liquid starts its free overfall (the drawdown effect at the top) and the flow separation at the top edge of the weir further narrows the nappe (the contraction effect

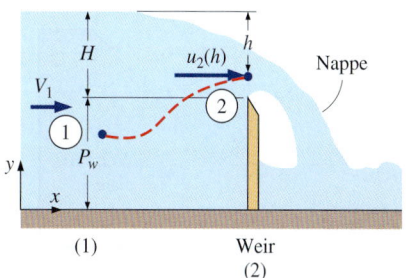

FIGURE 13–51
Flow over a sharp-crested weir.

at the bottom). As a result, the flow height over the weir is considerably smaller than H. When the drawdown and contraction effects are disregarded for simplicity, the flow rate is obtained by integrating the product of the flow velocity and the differential flow area over the entire flow area,

$$\dot{V} = \int_{A_c} u_2 \, dA_{c2} = \int_{h=0}^{H} \sqrt{2gh + V_1^2} \, w \, dh \tag{13-92}$$

where w is the width of the flow area at distance h from the upstream free surface.

In general, w is a function of h. But for a rectangular weir, $w = b$, which is constant. Then the integration can be performed easily, and the flow rate for a rectangular weir for idealized flow with negligible friction and negligible drawdown and contraction effects is determined to be

$$\dot{V}_{ideal} = \frac{2}{3} b \sqrt{2g} \left[\left(H + \frac{V_1^2}{2g} \right)^{3/2} - \left(\frac{V_1^2}{2g} \right)^{3/2} \right] \tag{13-93}$$

When the weir height is large relative to the weir head ($P_w \gg H$), the upstream velocity V_1 is low and the upstream velocity head can be neglected. That is, $V_1^2/2g \ll H$. Then,

$$\dot{V}_{ideal, \, rec} \cong \frac{2}{3} b \sqrt{2g} H^{3/2} \tag{13-94}$$

Therefore, the flow rate can be determined from knowledge of two geometric quantities: the crest width b and the weir head H, which is the vertical distance between the weir crest and the upstream free surface.

This simplified analysis gives the general form of the flow-rate relation, but it needs to be modified to account for the frictional and surface tension effects, which play a secondary role, as well as the drawdown and contraction effects. Again this is done by multiplying the ideal flow-rate relations by an experimentally determined weir discharge coefficient C_{wd}. Then the flow rate for a sharp-crested rectangular weir is expressed as

Sharp-crested rectangular weir: $\quad \dot{V}_{rec} = C_{wd, \, rec} \dfrac{2}{3} b \sqrt{2g} H^{3/2} \tag{13-95}$

where, from Ref. 1 (Ackers, 1978),

$$C_{wd, \, rec} = 0.598 + 0.0897 \frac{H}{P_w} \quad \text{for} \quad \frac{H}{P_w} \leq 2 \tag{13-96}$$

This formula is applicable over a wide range of upstream Reynolds number defined as Re = $V_1 H/\nu$. More precise but also more complex correlations are also available in the literature. Note that Eq. 13–95 is valid for *full-width* rectangular weirs. If the width of the weir is less than the channel width so that the flow is forced to contract, an additional coefficient for contraction correction should be incorporated to properly account for this effect.

Another type of sharp-crested weir commonly used for flow measurement is the *triangular weir* (also called the *V-notch weir*) shown in Fig. 13–52. The triangular weir has the advantage that it maintains a high weir head H even for small flow rates because of the decreasing flow area with decreasing H, and thus it can be used to measure a wide range of flow rates accurately.

From geometric consideration, the notch width can be expressed as $w = 2(H - h)\tan(\theta/2)$, where θ is the V-notch angle. Substituting into Eq. 13–92 and performing the integration give the ideal flow rate for a triangular weir to be

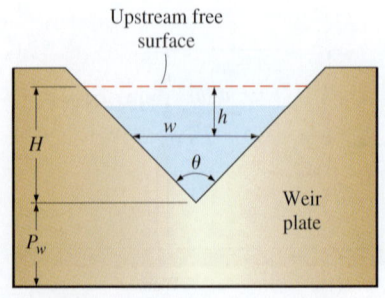

FIGURE 13–52
A triangular (or V-notch) sharp-crested weir plate geometry. The view is from downstream looking upstream.

$$\dot{V}_{ideal,\,tri} = \frac{8}{15}\tan\left(\frac{\theta}{2}\right)\sqrt{2g}H^{5/2} \qquad (13\text{–}97)$$

where we again neglected the upstream velocity head. The frictional and other dissipative effects are again accounted for conveniently by multiplying the ideal flow rate by a weir discharge coefficient. Then the flow rate for a sharp-crested triangular weir becomes

Sharp-crested triangular weir: $\qquad \dot{V} = C_{wd,\,tri}\dfrac{8}{15}\tan\left(\dfrac{\theta}{2}\right)\sqrt{2g}H^{5/2} \qquad (13\text{–}98)$

where the values of $C_{wd,\,tri}$ typically range between 0.58 and 0.62. Therefore, the fluid friction, the constriction of flow area, and other dissipative effects cause the flow rate through the V-notch to decrease by about 40 percent compared to the ideal case. For most practical cases ($H > 0.2$ m and $45° < \theta < 120°$), the value of the weir discharge coefficient is about $C_{wd,\,tri} = 0.58$. More precise values are available in the literature.

■ **EXAMPLE 13–10**   **Subcritical Flow over a Bump**

Water flowing in a wide horizontal open channel encounters a 15-cm-high bump at the bottom of the channel. If the flow depth is 0.80 m and the velocity is 1.2 m/s before the bump, determine if the water surface is depressed over the bump (Fig. 13–53) and if so, by how much.

SOLUTION Water flowing in a horizontal open channel encounters a bump. It will be determined if the water surface is depressed over the bump.
Assumptions **1** The flow is steady. **2** Frictional effects are negligible so that there is no dissipation of mechanical energy. **3** The channel is sufficiently wide so that the end effects are negligible.
Analysis The upstream Froude number and the critical depth are

$$\text{Fr}_1 = \frac{V_1}{\sqrt{gy_1}} = \frac{1.2 \text{ m/s}}{\sqrt{(9.81 \text{ m}^2/\text{s})(0.80 \text{ m})}} = 0.428$$

$$y_c = \left(\frac{\dot{V}^2}{gb^2}\right)^{1/3} = \left(\frac{(by_1V_1)^2}{gb^2}\right)^{1/3} = \left(\frac{y_1^2 V_1^2}{g}\right)^{1/3} = \left(\frac{(0.8 \text{ m})^2(1.2 \text{ m/s})^2}{9.81 \text{ m/s}^2}\right)^{1/3} = 0.455 \text{ m}$$

The flow is subcritical since Fr < 1 and therefore the flow depth decreases over the bump. The upstream specific energy is

$$E_{s1} = y_1 + \frac{V_1^2}{2g} = (0.80 \text{ m}) + \frac{(1.2 \text{ m/s})^2}{2(9.81 \text{ m/s}^2)} = 0.873 \text{ m}$$

The flow depth over the bump is determined from

$$y_2^3 - (E_{s1} - \Delta z_b)y_2^2 + \frac{V_1^2}{2g}y_1^2 = 0$$

Substituting,

$$y_2^3 - (0.873 - 0.15 \text{ m})y_2^2 + \frac{(1.2 \text{ m/s})^2}{2(9.81 \text{ m/s}^2)}(0.80 \text{ m})^2 = 0$$

or

$$y_2^3 - 0.723y_2^2 + 0.0470 = 0$$

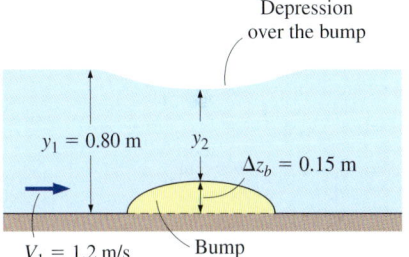

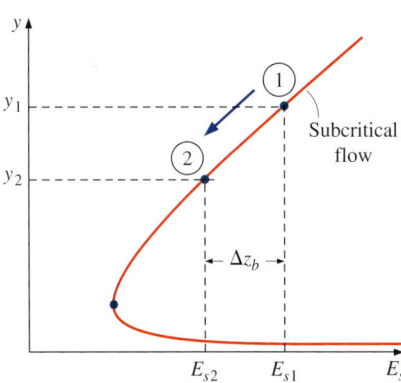

FIGURE 13–53
Schematic and flow depth-specific energy diagram for Example 13–10.

Using an equation solver, the three roots of this equation are determined to be 0.59 m, 0.36 m, and −0.22 m. We discard the negative solution as physically impossible. We also eliminate the solution 0.36 m since it is less than the critical depth, and it can occur only in supercritical flow. Thus the only meaningful solution for flow depth over the bump is $y_2 = 0.59$ m. Then the distance of the water surface over the bump from the channel bottom is $\Delta z_b + y_2 = 0.15 + 0.59 = 0.74$ m, which is less than $y_1 = 0.80$ m. Therefore, the water surface is **depressed** over the bump in the amount of

$$\text{Depression} = y_1 - (y_2 + \Delta z_b) = 0.80 - (0.59 + 0.15) = \mathbf{0.06 \text{ m}}$$

Discussion Note that having $y_2 < y_1$ does not necessarily indicate that the water surface is depressed (it may still rise over the bump). The surface is depressed over the bump only when the difference $y_1 - y_2$ is larger than the bump height Δz_b. Also, the actual value of depression may be different than 0.06 m because of the frictional effects that are neglected in the analysis.

EXAMPLE 13–11   Measuring Flow Rate by a Weir

The flow rate of water in a 5-m-wide horizontal open channel is being measured with a 0.60-m-high sharp-crested rectangular weir of equal width. If the water depth upstream is 1.5 m, determine the flow rate of water (Fig. 13–54).

SOLUTION The water depth upstream of a horizontal open channel equipped with a sharp-crested rectangular weir is measured. The flow rate is to be determined.
Assumptions **1** The flow is steady. **2** The upstream velocity head is negligible. **3** The channel is sufficiently wide so that the end effects are negligible.
Analysis The weir head is

$$H = y_1 - P_w = 1.5 - 0.60 = 0.90 \text{ m}$$

The discharge coefficient of the weir is

$$C_{wd,\,rec} = 0.598 + 0.0897 \frac{H}{P_w} = 0.598 + 0.0897 \frac{0.90}{0.60} = 0.733$$

The condition $H/P_w < 2$ is satisfied since $0.9/0.6 = 1.5$. Then the water flow rate through the channel becomes

$$\dot{V}_{rec} = C_{wd,\,rec} \frac{2}{3} b \sqrt{2g} H^{3/2}$$

$$= (0.733)\frac{2}{3}(5 \text{ m})\sqrt{2(9.81 \text{ m/s}^2)}(0.90 \text{ m})^{3/2}$$

$$= \mathbf{9.24 \text{ m}^3/\text{s}}$$

Discussion The upstream velocity and the upstream velocity head are

$$V_1 = \frac{\dot{V}}{by_1} = \frac{9.24 \text{ m}^3/\text{s}}{(5 \text{ m})(1.5 \text{ m})} = 1.23 \text{ m/s} \quad \text{and} \quad \frac{V_1^2}{2g} = \frac{(1.23 \text{ m/s})^2}{2(9.81 \text{ m/s}^2)} = 0.077 \text{ m}$$

This is 8.6 percent of the weir head, which is significant. When the upstream velocity head is considered, the flow rate becomes 10.2 m³/s, which is about 10 percent higher than the value determined. Therefore, it is good practice to consider the upstream velocity head unless the weir height P_w is very large relative to the weir head H.

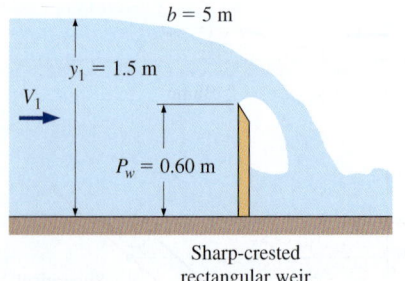

FIGURE 13–54
Schematic for Example 13–11.

APPLICATION SPOTLIGHT ■ Bridge Scour

Guest Author: Peggy A. Johnson, Penn State University

Bridge scour is the most common cause of bridge failure in the United States (Wardhana and Hadipriono, 2003). Bridge scour is the erosion of a stream or river channel bed in the vicinity of a bridge, including erosion around the bridge piers and abutments as well as the erosion and lowering of the entire channel bed. Scour around bridge foundations has been a leading cause of bridge failure for the nearly 400,000 bridges over waterways in the United States. A few recent examples of the damage that can be caused by high flows in rivers at bridges illustrate the magnitude of the problem. During the 1993 flood in the upper Mississippi and lower Missouri river basins, at least 22 of the 28 bridge failures were due to scour, at an estimated cost of more than $8 million (Kamojjala et al., 1994). During the "Super Flood" in Tennessee in 2010 in which more than 30 counties were declared major disaster areas, flooding in Tennessee's rivers caused scour and embankment erosion at 587 bridges and resulted in the closure of more than 50 bridges. In the fall of 2011, Hurricane Irene and Tropical Storm Lee in the mid-Atlantic and northeast U.S. caused flooding in rivers that resulted in numerous bridge failures and damage to bridges due to scour.

The mechanics of scour at bridge piers has been studied in laboratories and computer models. The primary mechanism is thought to be due to a "horseshoe" vortex that forms during floods as an adverse pressure gradient caused by the pier drives a portion of the approach flow downward just ahead of the pier (Arneson et al, 2012). The rate of erosion of the scour hole is directly associated with the magnitude of the downflow, which is directly related to the velocity of the approaching river flow. The strong vortex lifts the sediment out of the hole and deposits it downstream in the wake vortex. The result is a deep hole upstream of the bridge pier that can cause the bridge foundation to become unstable.

Protecting bridge piers over rivers and streams against the damaging floodwaters remains a major challenge for states across the country. Flood flows in channels have enormous capacity to move sediment and rock; thus, traditional protection, such as riprap, is often not sufficient. There has been considerable research on the use of vanes and similar structures in the river channel to help direct the flow around the bridge piers and abutments and provide a smoother transition of the flow through the bridge opening (Johnson et al, 2010).

FIGURE 13–55
A scour hole developed around this bridge pier near San Diego during high flows in the river channel.
Photo by Peggy Johnson, Penn State, used by permission.

FIGURE 13–56
Scour that developed around the bridge foundation during a 50 year flood in 1996 caused this bridge to fail in central PA. A temporary metal bridge was placed across the opening while a new bridge was being designed.
Photo by Peggy Johnson, Penn State, used by permission.

References

Arneson, L. A., L. W. Zevenbergen, P. F. Lagasse, P. E. Clopper (2012). Hydraulic Engineering Circular 18, Evaluating Scour at Bridges. Federal Highway Administration Report FHWA-HIF-12-003, HEC-18, Washington, D.C.

Johnson, P. A., Sheeder, S. A., Newlin, J. T. (2010). Waterway transitions at US bridges. Water and Environment Journal, 24 (2010), 274–281.

Kamojjala, S., Gattu, N. P. Parola. A. C., Hagerty, D. J. (1994), "Analysis of 1993 Upper Mississippi flood highway infrastructure damage," in ASCE Proceedings of the First International Conference of Water Resources Engineering, San Antonio, TX, pp. 1061–1065.

Wardhana, K., and Hadipriono, F. C., (2003). 17(3). ASCE Journal of Performance of Constructed Facilities, 144–150.

SUMMARY

Open-channel flow refers to the flow of liquids in channels open to the atmosphere or in partially filled conduits. The flow in a channel is said to be *uniform* if the flow depth (and thus the average velocity) remains constant. Otherwise, the flow is said to be *nonuniform* or *varied*. The *hydraulic radius* is defined as $R_h = A_c/p$. The dimensionless Froude number is defined as

$$\text{Fr} = \frac{V}{\sqrt{gL_c}} = \frac{V}{\sqrt{gy}}$$

The flow is classified as subcritical for $\text{Fr} < 1$, critical for $\text{Fr} = 1$, and supercritical for $\text{Fr} > 1$. Flow depth in critical flow is called the *critical depth* and is expressed as

$$y_c = \frac{\dot{V}^2}{gA_c^2} \quad \text{or} \quad y_c = \left(\frac{\dot{V}^2}{gb^2}\right)^{1/3}$$

where b is the channel width for wide channels.

The speed at which a surface disturbance travels through a liquid of depth y is the *wave speed* c_0, which is expressed as $c_0 = \sqrt{gy}$. The total mechanical energy of a liquid in a channel is expressed in terms of heads as

$$H = z_b + y + \frac{V^2}{2g}$$

where z_b is the elevation head, $P/\rho g = y$ is the pressure head, and $V^2/2g$ is the velocity head. The sum of the pressure and dynamic heads is called the *specific energy* E_s,

$$E_s = y + \frac{V^2}{2g}$$

The conservation of mass equation is $A_{c1}V_1 = A_{c2}V_2$. The energy equation is expressed as

$$y_1 + \frac{V_1^2}{2g} + S_0 L = y_2 + \frac{V_2^2}{2g} + h_L$$

Here h_L is the head loss and $S_0 = \tan\theta$ is the bottom slope of a channel. The *friction slope* is defined as $S_f = h_L/L$.

The flow depth in uniform flow is called the *normal depth* y_n, and the average flow velocity is called the *uniform-flow velocity* V_0. The velocity and flow rate in uniform flow are given by

$$V_0 = \frac{a}{n}R_h^{2/3}S_0^{1/2} \quad \text{and} \quad \dot{V} = \frac{a}{n}A_c R_h^{2/3}S_0^{1/2}$$

where n is the *Manning coefficient* whose value depends on the roughness of the channel surfaces, and $a = 1 \text{ m}^{1/3}/\text{s} = (3.2808 \text{ ft})^{1/3}/\text{s} = 1.486 \text{ ft}^{1/3}/\text{s}$. If $y_n = y_c$, the flow is uniform critical flow, and the bottom slope S_0 equals the critical slope S_c expressed as

$$S_c = \frac{gn^2 y_c}{a^2 R_h^{4/3}} \quad \text{which simplifies to} \quad S_c = \frac{gn^2}{a^2 y_c^{1/3}}$$

for film flow or flow in a wide rectangular channel with $b \gg y_c$.

The best hydraulic cross section for an open channel is the one with the maximum hydraulic radius, or equivalently, the one with the minimum wetted perimeter for a specified cross-sectional area. The criteria for best hydraulic cross section for a rectangular channel is $y = b/2$. The best cross section for trapezoidal channels is *half of a hexagon*.

In *gradually varied flow* (GVF), the flow depth changes gradually and smoothly with downstream distance. The *surface profile* $y(x)$ is calculated by integrating the GVF equation,

$$\frac{dy}{dx} = \frac{S_0 - S_f}{1 - \text{Fr}^2}$$

In *rapidly varied flow* (RVF), the flow depth changes markedly over a relatively short distance in the flow direction. Any change from supercritical to subcritical flow occurs through a *hydraulic jump*, which is a highly dissipative process. The depth ratio y_2/y_1, head loss, and energy dissipation ratio during hydraulic jump are expressed as

$$\frac{y_2}{y_1} = 0.5\left(-1 + \sqrt{1 + 8\text{Fr}_1^2}\right)$$

$$h_L = y_1 - y_2 + \frac{V_1^2 - V_2^2}{2g}$$

$$= y_1 - y_2 + \frac{y_1 \text{Fr}_1^2}{2}\left(1 - \frac{y_1^2}{y_2^2}\right)$$

$$\text{Dissipation ratio} = \frac{h_L}{E_{s1}} = \frac{h_L}{y_1 + V_1^2/2g}$$

$$= \frac{h_L}{y_1(1 + \text{Fr}_1^2/2)}$$

An obstruction that allows the liquid to flow over it is called a *weir*, and an obstruction with an adjustable opening at the bottom that allows the liquid to flow underneath it is called an *underflow gate*. The flow rate through a *sluice gate* is given by

$$\dot{V} = C_d ba\sqrt{2gy_1}$$

where b and a are the width and the height of the gate opening, respectively, and C_d is the *discharge coefficient*, which accounts for the frictional effects.

A *broad-crested weir* is a rectangular block that has a horizontal crest over which critical flow occurs. The upstream head above the top surface of the weir is called the *weir head*, H. The flow rate is expressed as

$$\dot{V} = C_{\text{wd, broad}} b\sqrt{g}\left(\frac{2}{3}\right)^{3/2}\left(H + \frac{V_1^2}{2g}\right)^{3/2}$$

where the discharge coefficient is

$$C_{wd,\,broad} = \frac{0.65}{\sqrt{1 + H/P_w}}$$

The flow rate for a sharp-crested rectangular weir is expressed as

$$\dot{V}_{rec} = C_{wd,\,rec} \frac{2}{3} b \sqrt{2g} H^{3/2}$$

where

$$C_{wd,\,rec} = 0.598 + 0.0897 \frac{H}{P_w} \quad \text{for} \quad \frac{H}{P_w} \leq 2$$

For a sharp-crested triangular weir, the flow rate is given as

$$\dot{V} = C_{wd,\,tri} \frac{8}{15} \tan\left(\frac{\theta}{2}\right) \sqrt{2g} H^{5/2}$$

where the values of $C_{wd,\,tri}$ typically range between 0.58 and 0.62.

Open-channel analysis is commonly used in the design of sewer systems, irrigation systems, floodways, and dams. Some open-channel flows are analyzed in Chap. 15 using computational fluid dynamics (CFD).

REFERENCES AND SUGGESTED READING

1. P. Ackers et al. *Weirs and Flumes for Flow Measurement.* New York: Wiley, 1978.
2. B. A. Bakhmeteff. *Hydraulics of Open Channels.* New York: McGraw-Hill, 1932.
3. M. H. Chaudhry. *Open Channel Flow.* Upper Saddle River, NJ: Prentice Hall, 1993.
4. V. T. Chow. *Open Channel Hydraulics.* New York: McGraw-Hill, 1959.
5. R. H. French. *Open Channel Hydraulics.* New York: McGraw-Hill, 1985.
6. F. M. Henderson. *Open Channel Flow.* New York: Macmillan, 1966.
7. C. C. Mei. *The Applied Dynamics of Ocean Surface Waves.* New York: Wiley, 1983.
8. U. S. Bureau of Reclamation. "Research Studies on Stilling Basins, Energy Dissipaters, and Associated Appurtenances," Hydraulic Lab Report Hyd.-399, June 1, 1955.

PROBLEMS*

Classification, Froude Number, and Wave Speed

13–1C How does the pressure change along the free surface in an open-channel flow?

13–2C Consider steady fully developed flow in an open channel of rectangular cross section with a constant slope of 5° for the bottom surface. Will the slope of the free surface also be 5°? Explain.

13–3C What causes the flow in an open channel to be varied (or nonuniform)? How does rapidly varied flow differ from gradually varied flow?

13–4C What is the driving force for flow in an open channel? How is the flow rate in an open channel established?

13–5C How does uniform flow differ from nonuniform flow in open channels? In what kind of channels is uniform flow observed?

13–6C Given the average flow velocity and the flow depth, explain how you would determine if the flow in open channels is tranquil, critical, or rapid.

13–7C The flow in an open channel is observed to have undergone a hydraulic jump. Is the flow upstream from the jump necessarily supercritical? Is the flow downstream from the jump necessarily subcritical?

13–8C What is critical depth in open-channel flow? For a given average flow velocity, how is it determined?

13–9C What is the Froude number? How is it defined? What is its physical significance?

13–10 A single wave is initiated in a sea by a strong jolt during an earthquake. Taking the average water depth to be 2 km and the density of seawater to be 1.030 kg/m³, determine the speed of propagation of this wave.

13–11 Consider the flow of water in a wide channel. Determine the speed of a small disturbance in the flow if the flow

---

* Problems designated by a "C" are concept questions, and students are encouraged to answer them all. Problems with the icon are solved using EES, and complete solutions together with parametric studies are included on the text website. Problems with the icon are comprehensive in nature and are intended to be solved with an equation solver such as EES.

depth is (*a*) 25 cm and (*b*) 80 cm. What would your answer be if the fluid were oil?

13–12 Water at 15°C is flowing uniformly in a 2-m-wide rectangular channel at an average velocity of 1.5 m/s. If the water depth is 24 cm, determine whether the flow is subcritical or supercritical. *Answer:* subcritical

13–13 After heavy rain, water flows on a concrete surface at an average velocity of 1.3 m/s. If the water depth is 2 cm, determine whether the flow is subcritical or supercritical.

13–14 Water at 20°C is flowing uniformly in a wide rectangular channel at an average velocity of 1.5 m/s. If the water depth is 0.16 m, determine (*a*) whether the flow is laminar or turbulent and (*b*) whether the flow is subcritical or supercritical.

13–15 Water at 10°C flows in a 3-m-diameter circular channel half-full at an average velocity of 2.5 m/s. Determine the hydraulic radius, the Reynolds number, and the flow regime (laminar or turbulent).

13–16 Repeat Prob. 13–15 for a channel diameter of 2 m.

13–17 Water at 20°C flows in a partially full 3-m-diameter circular channel at an average velocity of 2 m/s. If the maximum water depth is 0.75 m, determine the hydraulic radius, the Reynolds number, and the flow regime.

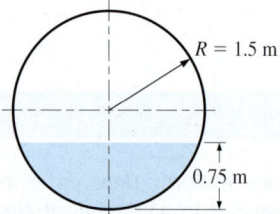

FIGURE P13–17

Specific Energy and the Energy Equation

13–18C Consider steady flow of water through two identical open rectangular channels at identical flow rates. If the flow in one channel is subcritical and in the other supercritical, can the specific energies of the water in these two channels be identical? Explain.

13–19C How is the specific energy of a fluid flowing in an open channel defined in terms of heads?

13–20C Consider steady flow of a liquid through a wide rectangular channel. It is claimed that the energy line of flow is parallel to the channel bottom when the frictional losses are negligible. Do you agree?

13–21C Consider steady one-dimensional flow through a wide rectangular channel. Someone claims that the total mechanical energy of the fluid at the free surface of a cross section is equal to that of the fluid at the channel bottom of the same cross section. Do you agree? Explain.

13–22C How is the total mechanical energy of a fluid during steady one-dimensional flow through a wide rectangular channel expressed in terms of heads? How is it related to the specific energy of the fluid?

13–23C Express the one-dimensional energy equation for open-channel flow between an upstream section 1 and downstream section 2, and explain how the head loss can be determined.

13–24C For a given flow rate through an open channel, the variation of specific energy with flow depth is studied. One person claims that the specific energy of the fluid will be minimum when the flow is critical, but another person claims that the specific energy will be minimum when the flow is subcritical. What is your opinion?

13–25C Consider steady supercritical flow of water through an open rectangular channel at a constant flow rate. Someone claims that the larger is the flow depth, the larger the specific energy of water. Do you agree? Explain.

13–26C During steady and uniform flow through an open channel of rectangular cross section, a person claims that the specific energy of the fluid remains constant. A second person claims that the specific energy decreases along the flow because of the frictional effects and thus head loss. With which person do you agree? Explain.

13–27C How is the friction slope defined? Under what conditions is it equal to the bottom slope of an open channel?

13–28 Water at 15°C flows at a depth of 0.4 m with an average velocity of 6 m/s in a rectangular channel. Determine (*a*) the critical depth, (*b*) the alternate depth, and (*c*) the minimum specific energy.

13–29 Water at 10°C flows in a 6-m-wide rectangular channel at a depth of 0.55 m and a flow rate of 12 m³/s. Determine (*a*) the critical depth, (*b*) whether the flow is subcritical or supercritical, and (*c*) the alternate depth. *Answers:* (*a*) 0.742 m, (*b*) supercritical, (*c*) 1.03 m

13–30 Water at 18°C flows at a depth of 42 cm with an average velocity of 6 m/s in a wide rectangular channel. Determine (*a*) the Froude number, (*b*) the critical depth, and (*c*) whether the flow is subcritical or supercritical. What would your response be if the flow depth were 6 cm?

13–31 Repeat Prob. 13–30 for an average velocity of 3 m/s.

13–32 Water flows steadily in a 1.4-m-wide rectangular channel at a rate of 0.7 m³/s. If the flow depth is 0.40 m, determine the flow velocity and if the flow is subcritical or supercritical. Also determine the alternate flow depth if the character of flow were to change.

13–33 Water at 20°C flows at a depth of 0.4 m with an average velocity of 4 m/s in a rectangular channel. Determine the specific energy of the water and whether the flow is subcritical or supercritical.

13–34 Water flows half-full through a hexagonal channel of bottom width 2 m at a rate of 60 m³/s. Determine (a) the average velocity and (b) whether the flow is subcritical and supercritical.

13–35 Repeat Prob. 13–34 for a flow rate of 30 m³/s.

13–36 Water flows half-full through a 50-cm-diameter steel channel at an average velocity of 2.8 m/s. Determine the volume flow rate and whether the flow is subcritical or supercritical.

13–37 Water flows through a 2-m-wide rectangular channel with an average velocity of 5 m/s. If the flow is critical, determine the flow rate of water. *Answer:* 25.5 m³/s

Uniform Flow and Best Hydraulic Cross Sections

13–38C When is the flow in an open channel said to be uniform? Under what conditions will the flow in an open channel remain uniform?

13–39C Which is a better hydraulic cross section for an open channel: one with a small or a large hydraulic radius?

13–40C Which is the best hydraulic cross section for an open channel: (a) circular, (b) rectangular, (c) trapezoidal, or (d) triangular?

13–41C The best hydraulic cross section for a rectangular open channel is one whose fluid height is (a) half, (b) twice, (c) equal to, or (d) one-third the channel width.

13–42C The best hydraulic cross section for a trapezoidal channel of base width b is one for which the length of the side edge of the flow section is (a) b, (b) b/2, (c) 2b, or (d) $\sqrt{3}b$.

13–43C During uniform flow in an open channel, someone claims that the head loss can be determined by simply multiplying the bottom slope by the channel length. Can it be this simple? Explain.

13–44C Consider uniform flow through a wide rectangular channel. If the bottom slope is increased, the flow depth will (a) increase, (b) decrease, or (c) remain constant.

13–45 Consider uniform flow through an open channel lined with bricks with a Manning coefficient of n = 0.015. If the Manning coefficient doubles (n = 0.030) as a result of some algae growth on surfaces while the flow cross section remains constant, the flow rate will (a) double, (b) decrease by a factor of $\sqrt{2}$, (c) remain unchanged, (d) decrease by half, or (e) decrease by a factor of $2^{1/3}$.

13–46 Water flows uniformly half-full in a 2-m-diameter circular channel that is laid on a grade of 1.5 m/km. If the channel is made of finished concrete, determine the flow rate of the water.

13–47 Water is flowing uniformly in a finished-concrete channel of trapezoidal cross section with a bottom width of 0.8 m, trapezoid angle of 50°, and a bottom angle of 0.4°. If the flow depth is measured to be 0.52 m, determine the flow rate of water through the channel.

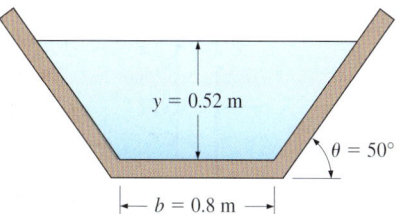

FIGURE P13–47

13–48 A 1-m-diameter semicircular channel made of unfinished concrete is to transport water to a distance of 1.5 km uniformly. If the flow rate is to reach 4 m³/s when the channel is full, determine the minimum elevation difference across the channel.

13–49 During uniform flow in open channels, the flow velocity and the flow rate can be determined from the Manning equations expressed as $V_0 = (a/n)R_h^{2/3}S_0^{1/2}$ and $\dot{V} = (a/n)A_c R_h^{2/3}S_0^{1/2}$. What is the value and dimension of the constant a in these equations in SI units? Also, explain how the Manning coefficient n can be determined when the friction factor f is known.

13–50 Show that for uniform critical flow, the general critical slope relation $S_c = \dfrac{gn^2 y_c}{a^2 R_h^{4/3}}$ reduces to $S_c = \dfrac{gn^2}{a^2 y_c^{1/3}}$ for film flow with $b \gg y_c$.

13–51 A trapezoidal channel with a bottom width of 6 m, free surface width of 12 m, and flow depth of 2.2 m discharges water at a rate of 120 m³/s. If the surfaces of the channel are lined with asphalt (n = 0.016), determine the elevation drop of the channel per km. *Answer:* 5.61 m

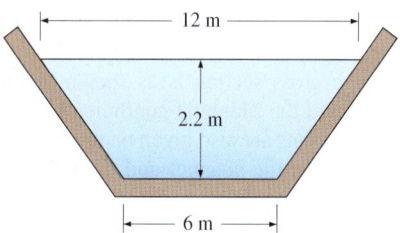

FIGURE P13–51

13–52 Reconsider Prob. 13–51. If the maximum flow height the channel can accommodate is 3.2 m, determine the maximum flow rate through the channel.

13–53 Consider water flow through two identical channels with square flow sections of 4 m × 4 m. Now the two channels are combined, forming a 8-m-wide channel. The flow rate is adjusted so that the flow depth remains constant at 4 m. Determine the percent increase in flow rate as a result of combining the channels.

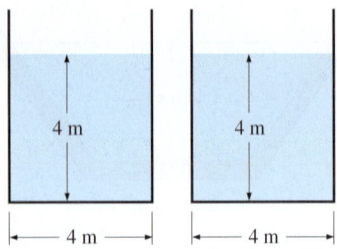

FIGURE P13–53

13–54 A cast iron V-shaped water channel shown in Fig. P13–54 has a bottom slope of 0.5°. For a flow depth of 0.75 m at the center, determine the discharge rate in uniform flow. *Answer:* 1.03 m³/s

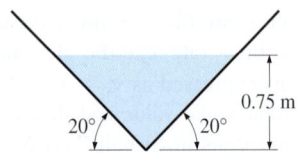

FIGURE P13–54

13–55 A clean-earth trapezoidal channel with a bottom width of 1.8 m and a side surface slope of 1:1 is to drain water uniformly at a rate of 8 m³/s to a distance of 1 km. If the flow depth is not to exceed 1.2 m, determine the required elevation drop. *Answer:* 3.90 m

13–56 A water draining system with a constant slope of 0.0025 is to be built of three circular channels made of finished concrete. Two of the channels have a diameter of 1.8 m and drain into the third channel. If all channels are to run half-full and the losses at the junction are negligible, determine the diameter of the third channel. *Answer:* 2.33 m

13–57 Water flows in a channel whose bottom slope is 0.002 and whose cross section is as shown in Fig. P13–57. The dimensions and the Manning coefficients for the surfaces of different subsections are also given on the figure. Determine the flow rate through the channel and the effective Manning coefficient for the channel.

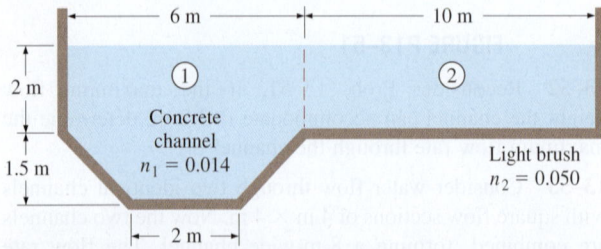

FIGURE P13–57

13–58 A 2-m-internal-diameter circular steel storm drain ($n = 0.012$) is to discharge water uniformly at a rate of 12 m³/s to a distance of 1 km. If the maximum depth is to be 1.5 m, determine the required elevation drop.

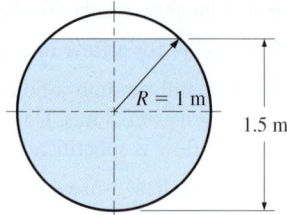

FIGURE P13–58

13–59 Water is to be transported at a rate of 10 m³/s in uniform flow in an open channel whose surfaces are asphalt lined. The bottom slope is 0.0015. Determine the dimensions of the best cross section if the shape of the channel is (*a*) circular of diameter *D*, (*b*) rectangular of bottom width *b*, and (*c*) trapezoidal of bottom width *b*.

13–60 Consider uniform flow in an asphalt-lined rectangular channel with a flow area of 2 m² and a bottom slope of 0.0003. By varying the depth-to-width ratio *y/b* from 0.1 to 2.0, calculate and plot the flow rate, and confirm that the best flow cross section occurs when the flow depth-to-width ratio is 0.5.

13–61 A rectangular channel with a bottom slope of 0.0004 is to be built to transport water at a rate of 20 m³/s. Determine the best dimensions of the channel if it is to be made of (*a*) unfinished concrete and (*b*) finished concrete. *Answer:* (*a*) 4.93 m × 2.47 m, (*b*) 4.66 m × 2.33 m

13–62 Repeat Prob. 13–61 for a flow rate of 17 m³/s.

13–63 A trapezoidal channel made of unfinished concrete has a bottom slope of 1°, base width of 5 m, and a side surface slope of 1:1, as shown in Fig. P13–63. For a flow rate of 25 m³/s, determine the normal depth *h*.

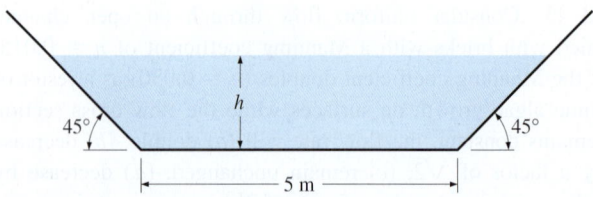

FIGURE P13–63

13–64 Repeat Prob. 13–63 for a weedy excavated earth channel with $n = 0.030$.

Gradually and Rapidly Varied Flows and Hydraulic Jump

13–65C How does gradually varied flow (GVF) differ from rapidly varied flow (RVF)?

13–66C How does nonuniform or varied flow differ from uniform flow?

13–67C Someone claims that frictional losses associated with wall shear on surfaces can be neglected in the analysis of rapidly varied flow, but should be considered in the analysis of gradually varied flow. Do you agree with this claim? Justify your answer.

13–68C Consider steady flow of water in an upward-sloped channel of rectangular cross section. If the flow is supercritical, the flow depth will (*a*) increase, (*b*) remain constant, or (*c*) decrease in the flow direction.

13–69C Is it possible for subcritical flow to undergo a hydraulic jump? Explain.

13–70C Why is the hydraulic jump sometimes used to dissipate mechanical energy? How is the energy dissipation ratio for a hydraulic jump defined?

13–71C Consider steady flow of water in a horizontal channel of rectangular cross section. If the flow is subcritical, the flow depth will (*a*) increase, (*b*) remain constant, or (*c*) decrease in the flow direction.

13–72C Consider steady flow of water in a downward-sloped channel of rectangular cross section. If the flow is subcritical and the flow depth is greater than the normal depth ($y > y_n$), the flow depth will (*a*) increase, (*b*) remain constant, or (*c*) decrease in the flow direction.

13–73C Consider steady flow of water in a horizontal channel of rectangular cross section. If the flow is supercritical, the flow depth will (*a*) increase, (*b*) remain constant, or (*c*) decrease in the flow direction.

13–74C Consider steady flow of water in a downward-sloped channel of rectangular cross section. If the flow is subcritical and the flow depth is less than the normal depth ($y < y_n$), the flow depth will (*a*) increase, (*b*) remain constant, or (*c*) decrease in the flow direction.

13–75 Water is flowing in a 90° V-shaped cast iron channel with a bottom slope of 0.002 at a rate of 3 m³/s. Determine if the slope of this channel should be classified as mild, critical, or steep for this flow. *Answer:* mild

13–76 Consider uniform water flow in a wide brick channel of slope 0.4°. Determine the range of flow depth for which the channel is classified as being steep.

13–77 Consider the flow of water through a 3.5-m-wide unfinished-concrete rectangular channel with a bottom slope of 0.5°. If the flow rate is 8.5 m³/s, determine if the slope of this channel is mild, critical, or steep. Also, for a flow depth of 0.9 m, classify the surface profile while the flow develops.

13–78 Water flows uniformly in a rectangular channel with finished-concrete surfaces. The channel width is 3 m, the flow depth is 1.2 m, and the bottom slope is 0.002. Determine if the channel should be classified as mild, critical, or steep for this flow.

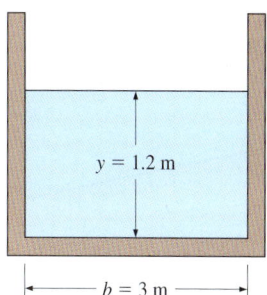

FIGURE P13–78

13–79 Water discharging into an 8-m-wide rectangular horizontal channel from a sluice gate is observed to have undergone a hydraulic jump. The flow depth and velocity before the jump are 1.2 m and 9 m/s, respectively. Determine (*a*) the flow depth and the Froude number after the jump, (*b*) the head loss and the dissipation ratio, and (*c*) the mechanical energy dissipated by the hydraulic jump.

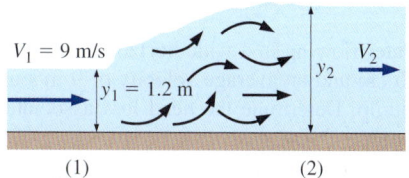

FIGURE P13–79

13–80 Consider the flow of water in a 10-m-wide channel at a rate of 70 m³/s and a flow depth of 0.50 m. The water now undergoes a hydraulic jump, and the flow depth after the jump is measured to be 4 m. Determine the mechanical power wasted during this jump. *Answer:* 4.35 MW

13–81 The flow depth and velocity of water after undergoing a hydraulic jump are measured to be 1.1 m and 1.75 m/s, respectively. Determine the flow depth and velocity before the jump, and the fraction of mechanical energy dissipated.

13–82 Consider uniform flow of water in a wide rectangular channel with a per-unit-width flow rate of 1.5 m³/s·m and a Manning coefficient of 0.03. The slope of the channel is 0.0005. (*a*) Calculate the normal and critical depths of the flow and determine if the uniform flow is subcritical or supercritical. (*b*) Next, a dam is installed (at $x = 0$) in order to impound a reservoir of water upstream. This raises the water surface profile upstream, creating a "backwater" curve (Fig. P13–82). The new water depth just upstream of the dam is 2.5 m. Determine how far upstream of the dam the "reservoir" extends. You may consider the reservoir boundary to be the point at which the water depth is within 5% of the original uniform water depth. *Answer:* 3500 m

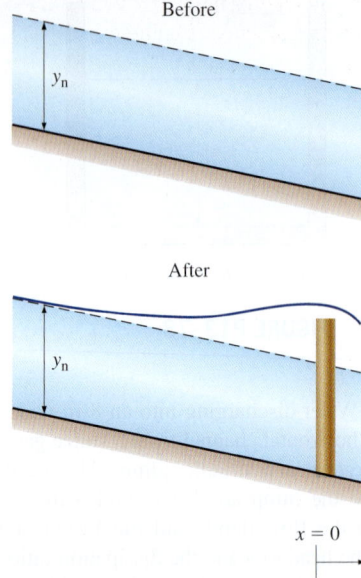

FIGURE P13–82

13–83 Water flowing in a wide horizontal channel at a flow depth of 56 cm and an average velocity of 9 m/s undergoes a hydraulic jump. Determine the head loss associated with the hydraulic jump.

13–84 During a hydraulic jump in a wide channel, the flow depth increases from 0.6 to 3 m. Determine the velocities and Froude numbers before and after the jump, and the energy dissipation ratio.

13–85 Consider gradually varied flow over a bump in a wide channel, as shown in Fig. P13–85. The initial flow velocity is 0.75 m/s, the initial flow depth is 1 m, the Manning parameter is 0.02, and the elevation of the channel bottom is prescribed to be

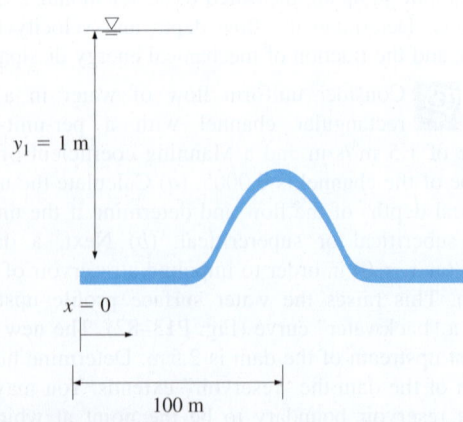

FIGURE P13–85

$$z_b = \Delta z_b \exp[-0.001(x-100)^2]$$

where the maximum bump height Δz_b is equal to 0.15 m and the crest of the bump is located at $x = 100$ m. (*a*) Calculate and plot the critical depth of the flow and (where it exists) the normal depth of the flow. (*b*) Integrate the GVF equation over the range $0 \leq x \leq 200$ m, and comment on the observed behavior of the free surface in light of the classification scheme presented in Table 13–3.

13–86 Consider a wide rectangular water channel with a per-unit-width flow rate of 5 m³/s·m and a Manning coefficient of $n = 0.02$. The channel is comprised of a 100 m length having a slope of $S_{01} = 0.01$ followed by a 100 m length having a slope of $S_{02} = 0.02$. (*a*) Calculate the normal and critical depths for the two channel segments. (*b*) Given an initial water depth of 1.25 m, calculate and graph the water surface profile over the full 200 m extent of the channel. Also classify the two channel segments (M1, A2, etc.).

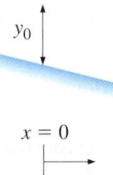

FIGURE P13–86

13–87 Repeat Problem 13–86 for the case of an initial water depth of 0.75 m instead of 1.25 m.

13–88 While the GVF equation cannot be used to predict a hydraulic jump directly, it can be coupled with the ideal hydraulic jump depth ratio equation in order to help locate the position at which a jump will occur in a channel. Consider a jump created in a wide ($R_h \approx y$) horizontal ($S_0 = 0$) laboratory flume having a length of 3 m and a Manning

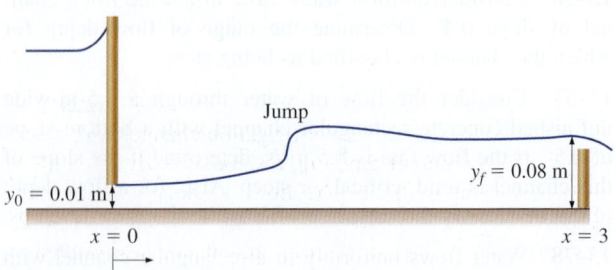

FIGURE P13–88

coefficient of 0.009. The supercritical flow under the head gate has an initial depth of 0.01 m at $x = 0$. The tailgate results in an overflow depth of 0.08 m at $x = 3$ m. The per-unit-width flow rate is 0.025 m³/s·m. (a) Calculate the critical depth of the flow and verify that the initial and final flows are supercritical and subcritical, respectively. (b) Determine the location of the hydraulic jump. *Hint*: integrate the GVF equation from $x = 0$ to a "guessed" location of the jump, apply the jump depth-ratio equation, and integrate the GVF equation using this new initial condition from the jump location to $x = 3$ m. If you do not obtain the desired overflow depth, try a new jump location. *Answer:* 1.80 m

13–89 Consider the gradually varied flow equation,

$$\frac{dy}{dx} = \frac{S_0 - S_f}{1 - \text{Fr}^2}$$

For the case of a wide rectangular channel, show that this can be reduced to the following form, which explicitly shows the importance of the relationship between y, y_n, and y_c:

$$\frac{dy}{dx} = \frac{S_0[1 - (y_n/y)^{10/3}]}{1 - (y_c/y)^3}$$

13–90 Consider gradually varied flow of water in a 6-m wide rectangular channel with a flow rate of 8.5 m³/s and a Manning coefficient of 0.008. The slope of the channel is 0.01, and at the location $x = 0$, the mean flow speed is measured to be 1.6 m/s. Determine the classification of the water surface profile, and, by integrating the GVF equation numerically, calculate the flow depth y at (a) $x = 150$ m, (b) 300 m, and (c) 600 m.

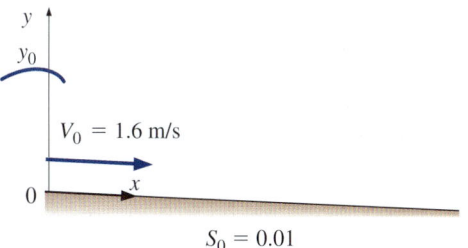

FIGURE P13–90

13–91 Consider gradually varied flow of water in a wide rectangular irrigation channel with a per-unit-width flow rate of 5m³/s·m, a slope of 0.01, and a Manning coefficient of 0.02. The flow is initially at uniform depth. At a given location, $x = 0$, the flow enters a 200m length of channel where lack of maintenance has resulted in a channel roughnness of 0.03. Following this stretch of channel, the roughness returns to the initial (maintained) value. (a) Calculate the normal and critical depths of the flow for the two distinct segments. (b) Numerically solve the gradually varied flow equation over the range $0 \leq x \leq 400$ m. Plot your solution (i.e., y vs. x) and comment about the behavior of the water surface.

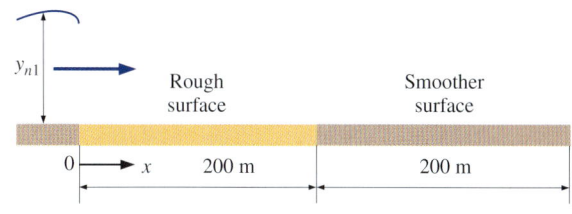

FIGURE P13–91

Flow Control and Measurement in Channels

13–92C What is a sharp-crested weir? On what basis are the sharp-crested weirs classified?

13–93C What is the basic principle of operation of a broad-crested weir used to measure flow rate through an open channel?

13–94C For sluice gates, how is the discharge coefficient C_d defined? What are typical values of C_d for sluice gates with free outflow? What is the value of C_d for the idealized frictionless flow through the gate?

13–95C Consider steady frictionless flow over a bump of height Δz in a horizontal channel of constant width b. Will the flow depth y increase, decrease, or remain constant as the fluid flows over the bump? Assume the flow to be subcritical.

13–96C Consider the flow of a liquid over a bump during subcritical flow in an open channel. The specific energy and the flow depth decrease over the bump as the bump height is increased. What will the character of flow be when the specific energy reaches its minimum value? Will the flow become supercritical if the bump height is increased even further?

13–97C Draw a flow depth-specific energy diagram for flow through underwater gates, and indicate the flow through the gate for cases of (a) frictionless gate, (b) sluice gate with free outflow, and (c) sluice gate with drowned outflow (including the hydraulic jump back to subcritical flow).

13–98 Consider uniform water flow in a wide rectangular channel with a depth of 2 m made of unfinished concrete laid on a slope of 0.0022. Determine the flow rate of water per m width of channel. Now water flows over a 15-cm-high bump. If the water surface over the bump remains flat (no rise or drop), determine the change in discharge rate of water per meter width of the channel. (*Hint*: Investigate if a flat surface over the bump is physically possible.)

13–99 Water flowing in a wide channel encounters a 22-cm-high bump at the bottom of the channel. If the flow depth is 1.2 m and the velocity is 2.5 m/s before the bump, determine if the flow is choked over the bump, and discuss.

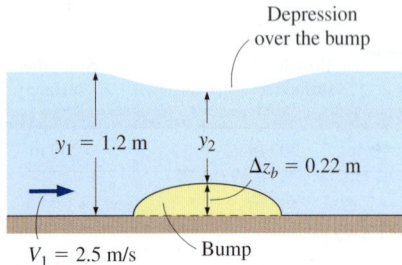

FIGURE P13–99

13–100 Consider the uniform flow of water in a wide channel with a velocity of 8 m/s and flow depth of 0.8 m. Now water flows over a 30-cm-high bump. Determine the change (increase or decrease) in the water surface level over the bump. Also determine if the flow over the bump is sub- or supercritical.

13–101 Water is released from a 12-m-deep reservoir into a 6-m-wide open channel through a sluice gate with a 1-m-high opening at the channel bottom. If the flow depth downstream from the gate is measured to be 3 m, determine the rate of discharge through the gate.

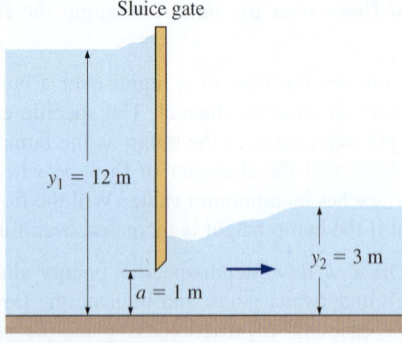

FIGURE P13–101

13–102 A full-width sharp-crested weir is to be used to measure the flow rate of water in a 2-m-wide rectangular channel. The maximum flow rate through the channel is 5 m³/s, and the flow depth upstream from the weir is not to exceed 1.5 m. Determine the appropriate height of the weir.

13–103 The flow rate of water in a 10-m-wide horizontal channel is being measured using a 1.3-m-high sharp-crested rectangular weir that spans across the channel. If the water depth upstream is 3.4 m, determine the flow rate of water. *Answer:* 66.8 m³/s

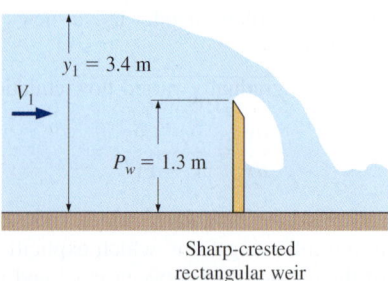

FIGURE P13–103

13–104 Repeat Prob. 13–103 for the case of a weir height of 1.6 m.

13–105 Water flows over a 2-m-high sharp-crested rectangular weir. The flow depth upstream of the weir is 3 m, and water is discharged from the weir into an unfinished-concrete channel of equal width where uniform-flow conditions are established. If no hydraulic jump is to occur in the downstream flow, determine the maximum slope of the downstream channel.

13–106 Water is to be discharged from an 8-m-deep lake into a channel through a sluice gate with a 5-m wide and 0.6-m-high opening at the bottom. If the flow depth downstream from the gate is measured to be 4 m, determine the rate of discharge through the gate.

13–107 Consider water flow through a wide channel at a flow depth of 2.5 m. Now water flows through a sluice gate with a 0.3-m-high opening, and the freely discharged outflow subsequently undergoes a hydraulic jump. Disregarding any losses associated with the sluice gate itself, determine the flow depth and velocities before and after the jump, and the fraction of mechanical energy dissipated during the jump.

13–108 The flow rate of water flowing in a 5-m-wide channel is to be measured with a sharp-crested triangular weir 0.5 m above the channel bottom with a notch angle of 80°. If the flow depth upstream from the weir is 1.5 m, determine the flow rate of water through the channel. Take the weir discharge coefficient to be 0.60. *Answer:* 1.19 m³/s

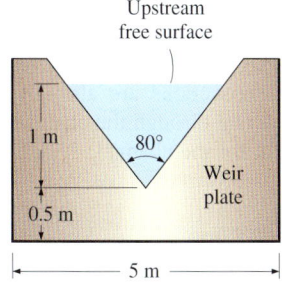

FIGURE P13–108

13–109 Repeat Prob. 13–108 for an upstream flow depth of 0.90 m.

13–110 A sharp-crested triangular weir with a notch angle of 100° is used to measure the discharge rate of water from a large lake into a spillway. If a weir with half the notch angle ($\theta = 50°$) is used instead, determine the percent reduction in the flow rate. Assume the water depth in the lake and the weir discharge coefficient remain unchanged.

13–111 A 0.80-m-high broad-crested weir is used to measure the flow rate of water in a 5-m-wide rectangular channel. The flow depth well upstream from the weir is 1.8 m. Determine the flow rate through the channel and the minimum flow depth above the weir.

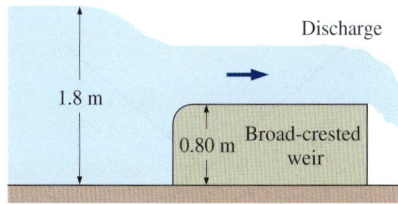

FIGURE P13–111

13–112 Repeat Prob. 13–111 for an upstream flow depth of 1.4 m.

13–113 Consider uniform water flow in a wide channel made of unfinished concrete laid on a slope of 0.0022. Now water

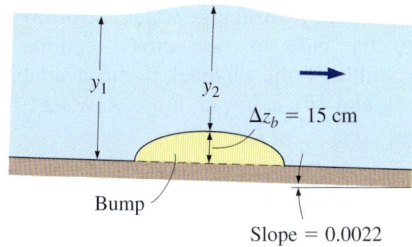

FIGURE P13–113

flows over a 15-cm-high bump. If the flow over the bump is exactly critical (Fr = 1), determine the flow rate and the flow depth over the bump per m width. *Answers:* 20.3 m³/s, 3.48 m

13–114 Consider water flow over a 0.80-m-high sufficiently long broad-crested weir. If the minimum flow depth above the weir is measured to be 0.50 m, determine the flow rate per meter width of channel and the flow depth upstream of the weir.

13–115 The flow rate of water through a 8-m-wide (into the paper) channel is controlled by a sluice gate. If the flow depths are measured to be 0.9 and 0.25 m upstream and downstream from the gates, respectively, determine the flow rate and the Froude number downstream from the gate.

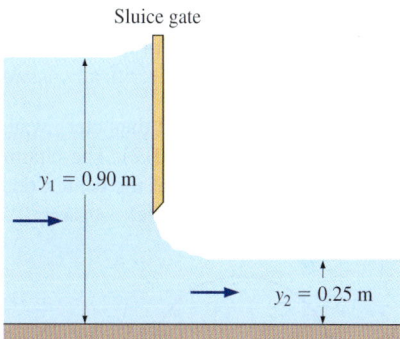

FIGURE P13–115

Review Problems

13–116 Water flows in a canal at an average velocity of 4 m/s. Determine if the flow is subcritical or supercritical for flow depths of (*a*) 0.2 m, (*b*) 2 m, and (*c*) 1.63 m.

13–117 A trapezoidal channel with a bottom width of 4 m and a side slope of 45° discharges water at a rate of 18 m³/s. If the flow depth is 0.6 m, determine if the flow is subcritical or supercritical.

13–118 A 5-m-wide rectangular channel lined with finished concrete is to be designed to transport water to a distance of 1 km at a rate of 12 m³/s. Using EES (or other) software, investigate the effect of bottom slope on flow depth (and thus on the required channel height). Let the bottom angle vary from 0.5 to 10° in increments of 0.5°. Tabulate and plot the flow depth against the bottom angle, and discuss the results.

13–119 Repeat Prob. 13–118 for a trapezoidal channel that has a base width of 5 m and a side surface angle of 45°.

13–120 A trapezoidal channel with brick lining has a bottom slope of 0.001 and a base width of 4 m, and the side surfaces are angled 25° from the horizontal, as shown in Fig. P13–120. If the normal depth is measured to be 1.5 m, estimate the flow rate of water through the channel. *Answer:* 22.5 m³/s

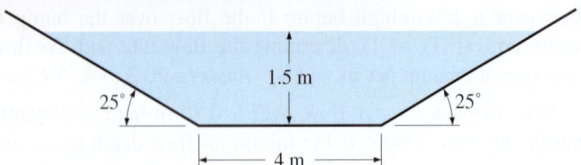

FIGURE P13–120

13–121 Water flows through a 2.2-m-wide rectangular channel with a Manning coefficient of $n = 0.012$. If the water is 0.9 m deep and the bottom slope of the channel is 0.6°, determine the rate of discharge of the channel in uniform flow.

13–122 A rectangular channel with a bottom width of 7 m discharges water at a rate of 45 m³/s. Determine the flow depth below which the flow is supercritical. *Answer:* 1.62 m

13–123 Consider a 1-m-internal-diameter water channel made of finished concrete ($n = 0.012$). The channel slope is 0.002. For a flow depth of 0.32 m at the center, determine the flow rate of water through the channel. *Answer:* 0.258 m³/s

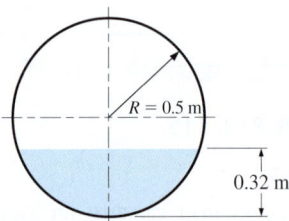

FIGURE P13–123

13–124 Reconsider Prob. 13–123. By varying the flow depth-to-radius ratio y/R from 0.1 to 1.9 while holding the flow area constant and evaluating the flow rate, show that the best cross section for flow through a circular channel occurs when the channel is half-full. Tabulate and plot your results.

13–125 Consider the flow of water through a parabolic notch shown in Fig. P13–125. Develop a relation for the flow rate, and calculate its numerical value for the ideal case in which the flow velocity is given by Torricelli's equation $V = \sqrt{2g(H - y)}$. *Answer:* 0.123 m³/s

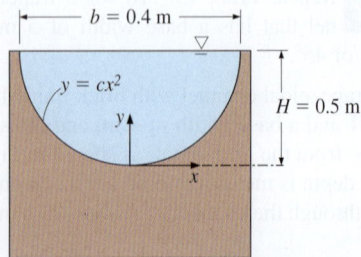

FIGURE P13–125

13–126 Water flows in a channel whose bottom slope is 0.5° and whose cross section is as shown in Fig. P13–126. The dimensions and the Manning coefficients for the surfaces of different subsections are also given on the figure. Determine the flow rate through the channel and the effective Manning coefficient for the channel.

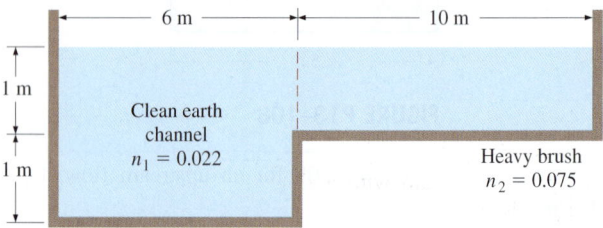

FIGURE P13–126

13–127 Consider two identical channels, one rectangular of bottom width b and one circular of diameter D, with identical flow rates, bottom slopes, and surface linings. If the flow height in the rectangular channel is also b and the circular channel is flowing half-full, determine the relation between b and D.

13–128 Consider water flow through a V-shaped channel. Determine the angle θ the channel makes from the horizontal for which the flow is most efficient.

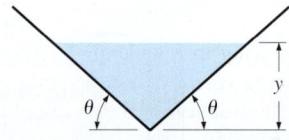

FIGURE P13–128

13–129 The flow rate of water in a 6-m-wide rectangular channel is to be measured using a 1.1-m-high sharp-crested rectangular weir that spans across the channel. If the head above the weir crest is 0.60 m upstream from the weir, determine the flow rate of water.

13–130 A rectangular channel with unfinished concrete surfaces is to be built to discharge water uniformly at a rate of 6 m³/s. For the case of best cross section, determine the bottom width of the channel if the available vertical drop is (*a*) 1 and (*b*) 2 m per km. *Answers:* (*a*) 2.65 m, (*b*) 2.32 m

13–131 Repeat Prob. 13–130 for the case of a trapezoidal channel of best cross section.

13–132 In practice, the V-notch is commonly used to measure flow rate in open channels. Using the idealized Torricelli's equation $V = \sqrt{2g(H - y)}$ for velocity, develop a relation for the flow rate through the V-notch in terms of the angle θ. Also, show the variation of the flow rate with θ by evaluating the flow rate for θ = 25, 40, 60, and 75°, and plotting the results.

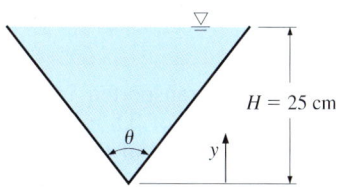

FIGURE P13–132

13–133 Water flows uniformly half-full in a 3.2-m-diameter circular channel laid with a slope of 0.004. If the flow rate of water is measured to be 4.5 m³/s, determine the Manning coefficient of the channel and the Froude number. *Answers:* 0.0487, 0.319

13–134 Consider water flow through a wide rectangular channel undergoing a hydraulic jump. Show that the ratio of the Froude numbers before and after the jump can be expressed in terms of flow depths y_1 and y_2 before and after the jump, respectively, as

$$Fr_1/Fr_2 = \sqrt{(y_2/y_1)^3}.$$

13–135 A sluice gate with free outflow is used to control the discharge rate of water through a channel. Determine the flow rate per unit width when the gate is raised to yield a gap of 50 cm and the upstream flow depth is measured to be 2.8 m. Also determine the flow depth and the velocity downstream.

13–136 Water flowing in a wide channel at a flow depth of 45 cm and an average velocity of 8 m/s undergoes a hydraulic jump. Determine the fraction of the mechanical energy of the fluid dissipated during this jump. *Answer:* 36.9 percent

13–137 Water flowing through a sluice gate undergoes a hydraulic jump, as shown in Fig. P13–137. The velocity of the water is 1.25 m/s before reaching the gate and 4 m/s after the jump. Determine the flow rate of water through the gate per meter of width, the flow depths y_1 and y_2, and the energy dissipation ratio of the jump.

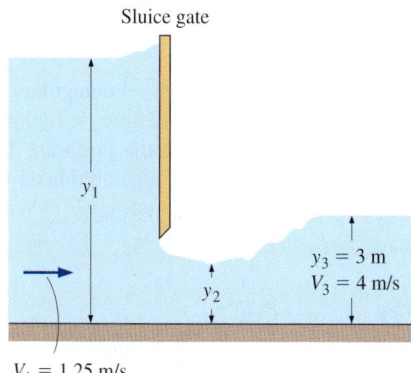

FIGURE P13–137

13–138 Repeat Prob. 13–137 for a velocity of 3.2 m/s after the hydraulic jump.

13–139 Water is discharged from a 5-m-deep lake into a finished concrete channel with a bottom slope of 0.004 through a sluice gate with a 0.5-m-high opening at the bottom. Shortly after supercritical uniform-flow conditions are established, the water undergoes a hydraulic jump. Determine the flow depth, velocity, and Froude number after the jump. Disregard the bottom slope when analyzing the hydraulic jump.

13–140 Water is discharged from a dam into a wide spillway to avoid overflow and to reduce the risk of flooding. A large fraction of the destructive power of the water is dissipated by a hydraulic jump during which the water depth rises from 0.70 to 5.0 m. Determine the velocities of water before and after the jump, and the mechanical power dissipated per meter width of the spillway.

13–141 Water flowing in a wide horizontal channel approaches a 20-cm-high bump with a velocity of 1.25 m/s and a flow depth of 1.8 m. Determine the velocity, flow depth, and Froude number over the bump.

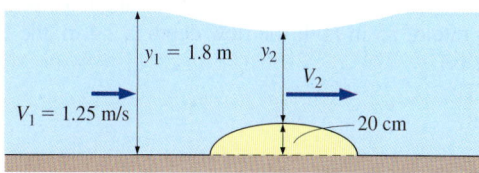

FIGURE P13–141

13–142 Reconsider Prob. 13–141. Determine the bump height for which the flow over the bump is critical (Fr = 1).

Fundamentals of Engineering (FE) Exam Problems

13–143 Which choices are examples of open-channel flow?
I. Flow of water in rivers
II. Draining of rainwater off highways
III. Upward draft of rain and snow
IV. Sewer lines
(*a*) I and II
(*b*) I and III
(*c*) II and III
(*d*) I, II, and IV
(*e*) I, II, III, and IV

13–144 If the flow depth remains constant in an open-channel flow, the flow is called
(*a*) Uniform flow
(*b*) Steady flow
(*c*) Varied flow
(*d*) Unsteady flow
(*e*) Laminar flow

13–145 Consider water flow in a rectangular open channel of height 2 m and width 5 m containing water of depth 1.5 m. The hydraulic radius for this flow is
(a) 0.47 m
(b) 0.94 m
(c) 1.5 m
(d) 3.8 m
(e) 5 m

13–146 Water flows in a rectangular open channel of width 5 m at a rate of 7.5 m³/s. The critical depth for this flow is
(a) 5 m
(b) 2.5 m
(c) 1.5 m
(d) 0.96 m
(e) 0.61 m

13–147 Water flows in a rectangular open channel of width 0.6 m at a rate of 0.25 m³/s. If the flow depth is 0.2 m, what is the alternate flow depth if the character of flow were to change?
(a) 0.2 m
(b) 0.26 m
(c) 0.35 m
(d) 0.6 m
(e) 0.8 m

13–148 Water flows in a 6-m-wide rectangular open channel at a rate of 55 m³/s. If the flow depth is 2.4 m, the Froude number is
(a) 0.531
(b) 0.787
(c) 1.0
(d) 1.72
(e) 2.65

13–149 Water flows in a clean and straight natural channel of rectangular cross section with a bottom width of 0.75 m and a bottom slope angle of 0.6°. If the flow depth is 0.15 m, the flow rate of water through the channel is
(a) 0.0317 m³/s
(b) 0.05 m³/s
(c) 0.0674 m³/s
(d) 0.0866 m³/s
(e) 1.14 m³/s

13–150 Water is to be transported in a finished-concrete rectangular channel with a bottom width of 1.2 m at a rate of 5 m³/s. The channel bottom drops 1 m per 500 m length. The minimum height of the channel under uniform-flow conditions is
(a) 1.9 m
(b) 1.5 m
(c) 1.2 m
(d) 0.92 m
(e) 0.60 m

13–151 Water is to be transported in a 4-m-wide rectangular open channel. The flow depth to maximize the flow rate is
(a) 1 m
(b) 2 m
(c) 4 m
(d) 6 m
(e) 8 m

13–152 Water is to be transported in a clay tile lined rectangular channel at a rate of 0.8 m³/s. The channel bottom slope is 0.0015. The width of the channel for the best cross section is
(a) 0.68 m
(b) 1.33 m
(c) 1.63 m
(d) 0.98 m
(e) 1.15 m

13–153 Water is to be transported in a clay tile lined trapezoidal channel at a rate of 0.8 m³/s. The channel bottom slope is 0.0015. The width of the channel for the best cross section is
(a) 0.48 m
(b) 0.70 m
(c) 0.84 m
(d) 0.95 m
(e) 1.22 m

13–154 Water flows uniformly in a finished-concrete rectangular channel with a bottom width of 0.85 m. The flow depth is 0.4 m and the bottom slope is 0.003. The channel should be classified as
(a) Steep
(b) Critical
(c) Mild
(d) Horizontal
(e) Adverse

13–155 Water discharges into a rectangular horizontal channel from a sluice gate and undergoes a hydraulic jump. The channel is 25-m-wide and the flow depth and velocity before the jump are 2 m and 9 m/s, respectively. The flow depth after the jump is
(a) 1.26 m
(b) 2 m
(c) 3.61 m
(d) 4.83 m
(e) 6.55 m

13–156 Water discharges into a rectangular horizontal channel from a sluice gate and undergoes a hydraulic jump. The flow depth and velocity before the jump are 1.25 m and 6 m/s, respectively. The percentage available head loss due to the hydraulic jump is
(a) 4.7%
(b) 6.2%
(c) 8.5%
(d) 13.9%
(e) 17.4%

13–157 Water discharges into a 7-m-wide rectangular horizontal channel from a sluice gate and undergoes a hydraulic jump. The flow depth and velocity before the jump are 0.65 m and 5 m/s, respectively. The wasted power potential due to the hydraulic jump is
(a) 158 kW
(b) 112 kW
(c) 67.3 kW
(d) 50.4 kW
(e) 37.6 kW

13–158 Water is released from a 0.8-m-deep reservoir into a 4-m-wide open channel through a sluice gate with a 0.1-m-high opening at the channel bottom. The flow depth after all turbulence subsides is 0.5 m. The rate of discharge is
(a) 0.92 m^3/s
(b) 0.79 m^3/s
(c) 0.66 m^3/s
(d) 0.47 m^3/s
(e) 0.34 m^3/s

13–159 The flow rate of water in a 3-m-wide horizontal open channel is being measured with a 0.4-m-high sharp-crested rectangular weir of equal width. If the water depth upstream is 0.9 m, the flow rate of water is
(a) 1.37 m^3/s
(b) 2.22 m^3/s
(c) 3.06 m^3/s
(d) 4.68 m^3/s
(e) 5.11 m^3/s

Design and Essay Problems

13–160 Using catalogs or websites, obtain information from three different weir manufacturers. Compare the different weir designs, and discuss the advantages and disadvantages of each design. Indicate the applications for which each design is best suited.

13–161 Consider water flow in the range of 10 to 15 m^3/s through a horizontal section of a 5-m-wide rectangular channel. A rectangular or triangular thin-plate weir is to be installed to measure the flow rate. If the water depth is to remain under 2 m at all times, specify the type and dimensions of an appropriate weir. What would your response be if the flow range were 0 to 15 m^3/s?

CHAPTER 14

TURBOMACHINERY

In this chapter we discuss the basic principles of a common and important application of fluid mechanics, *turbomachinery*. First we classify turbomachines into two broad categories, *pumps* and *turbines*. Then we discuss both of these turbomachines in more detail, mostly qualitatively, explaining the basic principles of their operation. We emphasize preliminary design and overall performance of turbomachines rather than detailed design. In addition, we discuss how to properly match the requirements of a fluid flow system to the performance characteristics of a turbomachine. A significant portion of this chapter is devoted to *turbomachinery scaling laws*—a practical application of dimensional analysis. We show how the scaling laws are used in the design of new turbomachines that are geometrically similar to existing ones.

OBJECTIVES

When you finish reading this chapter, you should be able to

- Identify various types of pumps and turbines, and understand how they work
- Apply dimensional analysis to design new pumps or turbines that are geometrically similar to existing pumps or turbines
- Perform basic vector analysis of the flow into and out of pumps and turbines
- Use specific speed for preliminary design and selection of pumps and turbines

The jet engines on modern commercial airplanes are highly complex turbomachines that include both pump (compressor) and turbine sections.
© Stockbyte/PunchStock RF

14–1 ■ CLASSIFICATIONS AND TERMINOLOGY

There are two broad categories of turbomachinery, **pumps** and **turbines.** The word *pump* is a general term for any fluid machine that *adds* energy to a fluid. Some authors call pumps **energy absorbing devices** since energy is supplied *to* them, and they transfer most of that energy to the fluid, usually via a rotating shaft (Fig. 14–1a). The increase in fluid energy is usually felt as an increase in the pressure of the fluid. Turbines, on the other hand, are **energy producing devices**—they extract energy *from* the fluid and transfer most of that energy to some form of mechanical energy output, typically in the form of a rotating shaft (Fig. 14–1b). The fluid at the outlet of a turbine suffers an energy loss, typically in the form of a loss of pressure.

An ordinary person may think that the energy supplied to a pump increases the speed of fluid passing through the pump and that a turbine extracts energy from the fluid by slowing it down. This is not necessarily the case. Consider a control volume surrounding a pump (Fig. 14–2). We assume steady conditions. By this we mean that neither the mass flow rate nor the rotational speed of the rotating blades changes with time. (The detailed flow field near the rotating blades inside the pump is *not* steady of course, but control volume analysis is not concerned with details inside the control volume.) By conservation of mass, we know that the mass flow rate into the pump must equal the mass flow rate out of the pump. If the flow is incompressible, the volume flow rates at the inlet and outlet must be equal as well. Furthermore, if the diameter of the outlet is the same as that of the inlet, conservation of mass requires that the average speed across the outlet must be identical to the average speed across the inlet. In other words, the pump does not necessarily increase the *speed* of the fluid passing through it; rather, it increases the *pressure* of the fluid. Of course, if the pump were turned off, there might be no flow at all. So, the pump *does* increase fluid speed compared to the case of no pump in the system. However, in terms of changes from the inlet to the outlet *across* the pump, fluid speed is not necessarily increased. (The output speed may even be *lower* than the input speed if the outlet diameter is larger than that of the inlet.)

> The purpose of a pump is to add energy to a fluid, resulting in an increase in fluid pressure, not necessarily an increase of fluid speed across the pump.

An analogous statement is made about the purpose of a turbine:

> The purpose of a turbine is to extract energy from a fluid, resulting in a decrease of fluid pressure, not necessarily a decrease of fluid speed across the turbine.

Fluid machines that move liquids are called **pumps,** but there are several other names for machines that move gases (Fig. 14–3). A **fan** is a gas pump with relatively low pressure rise and high flow rate. Examples include ceiling fans, house fans, and propellers. A **blower** is a gas pump with relatively moderate to high pressure rise and moderate to high flow rate. Examples include centrifugal blowers and squirrel cage blowers in automobile ventilation systems, furnaces, and leaf blowers. A **compressor** is a gas pump designed to deliver a very high pressure rise, typically at low to moderate flow rates. Examples include air compressors that run pneumatic tools and

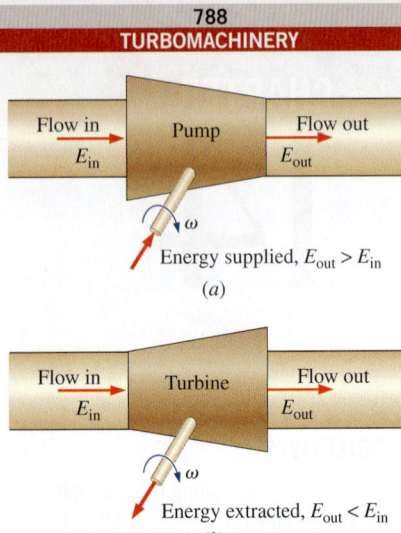

FIGURE 14–1
(a) A pump supplies energy to a fluid, while (b) a turbine extracts energy from a fluid.

FIGURE 14–2
For the case of steady flow, conservation of mass requires that the mass flow rate out of a pump must equal the mass flow rate into the pump; for incompressible flow with equal inlet and outlet cross-sectional areas ($D_{out} = D_{in}$), we conclude that $V_{out} = V_{in}$, but $P_{out} > P_{in}$.

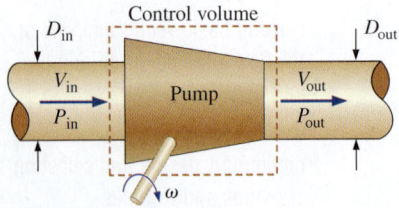

FIGURE 14–3
When used with gases, pumps are called *fans*, *blowers*, or *compressors*, depending on the relative values of pressure rise and volume flow rate.

inflate tires at automobile service stations, and refrigerant compressors used in heat pumps, refrigerators, and air conditioners.

Pumps and turbines in which energy is supplied or extracted by a rotating shaft are properly called **turbomachines,** since the Latin prefix *turbo* means "spin." Not all pumps or turbines utilize a rotating shaft, however. The hand-operated air pump you use to inflate the tires of your bicycle is a prime example (Fig. 14–4*a*). The up and down reciprocating motion of a plunger or piston replaces the rotating shaft in this type of pump, and it is more proper to call it simply a **fluid machine** instead of a turbomachine. An old-fashioned well pump operates in a similar manner to pump water instead of air (Fig. 14–4*b*). Nevertheless, the words *turbomachine* and *turbomachinery* are often used in the literature to refer to *all* types of pumps and turbines regardless of whether they utilize a rotating shaft or not.

Fluid machines may also be broadly classified as either *positive-displacement* machines or *dynamic* machines, based on the manner in which energy transfer occurs. In **positive-displacement machines,** fluid is directed into a closed volume. Energy transfer to the fluid is accomplished by movement of the boundary of the closed volume, causing the volume to expand or contract, thereby sucking fluid in or squeezing fluid out, respectively. Your heart is a good example of a **positive-displacement pump** (Fig. 14–5*a*). It is designed with one-way valves that open to let blood in as heart chambers expand, and other one-way valves that open as blood is pushed out of those chambers when they contract. An example of a **positive-displacement turbine** is the common water meter in your house (Fig. 14–5*b*), in which water forces itself into a closed chamber of expanding volume connected to an output shaft that turns as water enters the chamber. The boundary of the volume then collapses, turning the output shaft some more, and letting the water continue on its way to your sink, shower, etc. The water meter records each 360° rotation of the output shaft, and the meter is precisely calibrated to the known volume of fluid in the chamber.

FIGURE 14–4
Not all pumps have a rotating shaft; (*a*) energy is supplied to this manual tire pump by the up and down motion of a person's arm to pump air; (*b*) a similar mechanism is used to pump water with an old-fashioned well pump.

(a) Photo by Andrew Cimbala, with permission.
(b) © Bear Dancer Studios/Mark Dierker.

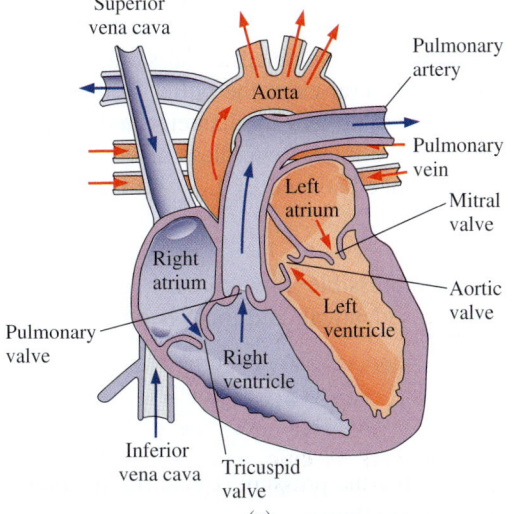

FIGURE 14–5
(*a*) The human heart is an example of a *positive-displacement pump*; blood is pumped by expansion and contraction of heart chambers called *ventricles*. (*b*) The common water meter in your house is an example of a *positive-displacement turbine*; water fills and exits a chamber of known volume for each revolution of the output shaft.

(b) Courtesy of Badger Meter, Inc. Used by permission.

FIGURE 14–6
A wind turbine is a good example of a dynamic machine of the open type; air turns the blades, and the output shaft drives an electric generator.

The Wind Turbine Company. Used by permission.

In **dynamic machines,** there is no closed volume; instead, rotating blades supply or extract energy to or from the fluid. For pumps, these rotating blades are called **impeller blades,** while for turbines, the rotating blades are called **runner blades** or **buckets.** Examples of **dynamic pumps** include **enclosed pumps** and **ducted pumps** (those with casings around the blades such as the water pump in your car's engine), and **open pumps** (those without casings such as the ceiling fan in your house, the propeller on an airplane, or the rotor on a helicopter). Examples of **dynamic turbines** include **enclosed turbines,** such as the hydroturbine that extracts energy from water in a hydroelectric dam, and **open turbines** such as the wind turbine that extracts energy from the wind (Fig. 14–6).

14–2 ▪ PUMPS

Some fundamental parameters are used to analyze the performance of a pump. The **mass flow rate** $\dot{m}$ of fluid through the pump is an obvious primary pump performance parameter. For incompressible flow, it is more common to use **volume flow rate** rather than mass flow rate. In the turbomachinery industry, volume flow rate is called **capacity** and is simply mass flow rate divided by fluid density,

Volume flow rate (capacity): $$\dot{V} = \frac{\dot{m}}{\rho}$$ (14–1)

The performance of a pump is characterized additionally by its **net head** H, defined as the change in **Bernoulli head** between the inlet and outlet of the pump,

Net head: $$H = \left(\frac{P}{\rho g} + \frac{V^2}{2g} + z\right)_{out} - \left(\frac{P}{\rho g} + \frac{V^2}{2g} + z\right)_{in}$$ (14–2)

The dimension of net head is length, and it is often listed as an equivalent column height of water, even for a pump that is not pumping water.

For the case in which a *liquid* is being pumped, the Bernoulli head at the inlet is equivalent to the **energy grade line** at the inlet, EGL_{in}, obtained by aligning a Pitot probe in the center of the flow as illustrated in Fig. 14–7. The energy grade line at the outlet EGL_{out} is obtained in the same manner, as also illustrated in the figure. In the general case, the outlet of the pump may be at a different elevation than the inlet, and its diameter and average speed may not be the same as those at the inlet. Regardless of these differences, net head H is equal to the difference between EGL_{out} and EGL_{in},

Net head for a liquid pump: $$H = EGL_{out} - EGL_{in}$$

Consider the special case of incompressible flow through a pump in which the inlet and outlet diameters are identical, and there is no change in elevation. Equation 14–2 reduces to

Special case with $D_{out} = D_{in}$ and $z_{out} = z_{in}$: $$H = \frac{P_{out} - P_{in}}{\rho g}$$

For this simplified case, net head is simply the pressure rise across the pump expressed as a head (column height of the fluid).

Net head is proportional to the useful power actually delivered to the fluid. It is traditional to call this power the **water horsepower,** even if the fluid being pumped is not water, and even if the power is not measured in units of horsepower. By dimensional reasoning, we must multiply the net head of Eq. 14–2 by mass flow rate and gravitational acceleration to obtain dimensions of power. Thus,

Water horsepower: $$\dot{W}_{\text{water horsepower}} = \dot{m}gH = \rho g \dot{V} H \quad (14\text{-}3)$$

All pumps suffer from irreversible losses due to friction, internal leakage, flow separation on blade surfaces, turbulent dissipation, etc. Therefore, the mechanical energy supplied to the pump must be *larger* than $\dot{W}_{\text{water horsepower}}$. In pump terminology, the external power supplied to the pump is called the **brake horsepower,** which we abbreviate as bhp. For the typical case of a rotating shaft supplying the brake horsepower,

Brake horsepower: $$\text{bhp} = \dot{W}_{\text{shaft}} = \omega T_{\text{shaft}} \quad (14\text{-}4)$$

where ω is the rotational speed of the shaft (rad/s) and T_{shaft} is the torque supplied to the shaft. We define **pump efficiency** η_{pump} as the ratio of useful power to supplied power,

Pump efficiency: $$\eta_{\text{pump}} = \frac{\dot{W}_{\text{water horsepower}}}{\dot{W}_{\text{shaft}}} = \frac{\dot{W}_{\text{water horsepower}}}{\text{bhp}} = \frac{\rho g \dot{V} H}{\omega T_{\text{shaft}}} \quad (14\text{-}5)$$

Pump Performance Curves and Matching a Pump to a Piping System

The maximum volume flow rate through a pump occurs when its net head is zero, $H = 0$; this flow rate is called the pump's **free delivery.** The free delivery condition is achieved when there is no flow restriction at the pump inlet or outlet—in other words when there is no **load** on the pump. At this operating point, $\dot{V}$ is large, but H is zero; the pump's efficiency is zero because the pump is doing no useful work, as is clear from Eq. 14–5. At the other extreme, the **shutoff head** is the net head that occurs when the volume flow rate is zero, $\dot{V} = 0$, and is achieved when the outlet port of the pump is blocked off. Under these conditions, H is large but $\dot{V}$ is zero; the pump's efficiency (Eq. 14–5) is again zero, because the pump is doing no useful work. Between these two extremes, from shutoff to free delivery, the pump's net head may increase from its shutoff value somewhat as the flow rate increases, but H must eventually decrease to zero as the volume flow rate increases to its free delivery value. The pump's efficiency reaches its maximum value somewhere between the shutoff condition and the free delivery condition; this operating point of maximum efficiency is appropriately called the **best efficiency point** (BEP), and is notated by an asterisk (H^*, $\dot{V}^*$, bhp*). Curves of H, η_{pump}, and bhp as functions of $\dot{V}$ are called **pump performance curves** (or *characteristic curves,* Chap. 8); typical curves at one rotational speed are plotted in Fig. 14–8. The pump performance curves change with rotational speed.

It is important to realize that *for steady conditions, a pump can operate only along its performance curve.* Thus, the operating point of a piping

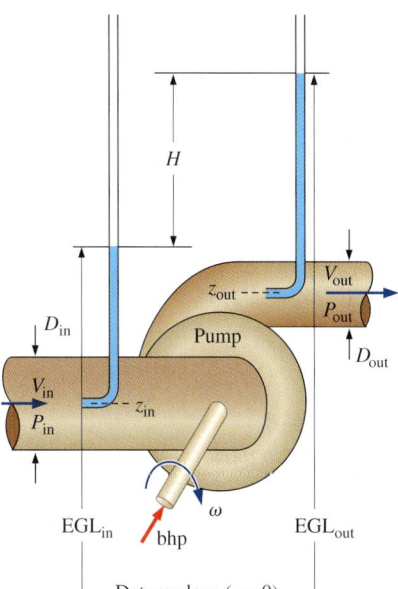

FIGURE 14–7

The *net head* of a pump, H, is defined as the change in Bernoulli head from inlet to outlet; for a liquid, this is equivalent to the change in the energy grade line, $H = \text{EGL}_{\text{out}} - \text{EGL}_{\text{in}}$, relative to some arbitrary datum plane; bhp is the *brake horsepower,* the external power supplied to the pump.

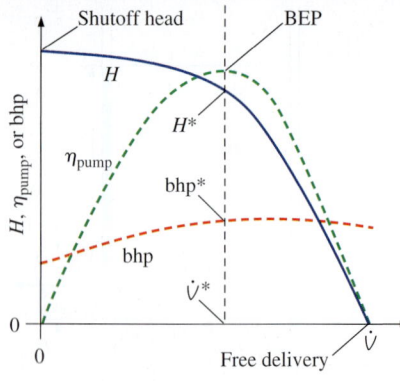

FIGURE 14–8

Typical *pump performance curves* for a centrifugal pump with backward-inclined blades; the curve shapes for other types of pumps may differ, and the curves change as shaft rotation speed is changed.

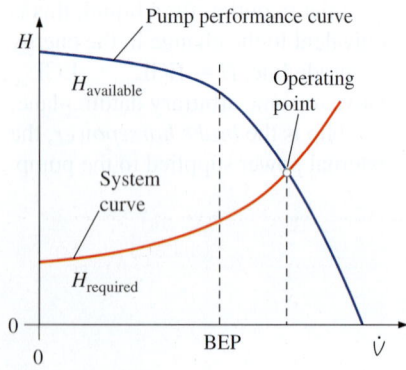

FIGURE 14–9

The *operating point* of a piping system is established as the volume flow rate where the system curve and the pump performance curve intersect.

system is determined by matching system requirements (*required* net head) to pump performance (*available* net head). In a typical application, $H_{required}$ and $H_{available}$ match at one unique value of flow rate—this is the **operating point** or **duty point** of the system.

> The steady operating point of a piping system is established at the volume flow rate where $H_{required} = H_{available}$.

For a given piping system with its major and minor losses, elevation changes, etc., the required net head *increases* with volume flow rate. On the other hand, the available net head of most pumps *decreases* with flow rate, as in Fig. 14–8, at least over the majority of its recommended operating range. Hence, the system curve and the pump performance curve intersect as sketched in Fig. 14–9, and this establishes the operating point. If we are lucky, the operating point is at or near the best efficiency point of the pump. In most cases, however, as illustrated in Fig. 14–9, the pump does not run at its optimum efficiency. If efficiency is of major concern, the pump should be carefully selected (or a new pump should be designed) such that the operating point is as close to the best efficiency point as possible. In some cases it may be possible to change the shaft rotation speed so that an existing pump can operate much closer to its design point (best efficiency point).

There are unfortunate situations where the system curve and the pump performance curve intersect at more than one operating point. This can occur when a pump that has a dip in its net head performance curve is mated to a system that has a fairly flat system curve, as illustrated in Fig. 14–10. Although rare, such situations are possible and should be avoided, because the system may "hunt" for an operating point, leading to an unsteady-flow situation.

It is fairly straightforward to match a piping system to a pump, once we realize that the term for **useful pump head** ($h_{pump,\,u}$) that we used in the head form of the energy equation (Chap. 5) is the same as the *net head* (H) used in the present chapter. Consider, for example, a general piping system with elevation change, major and minor losses, and fluid acceleration (Fig. 14–11). We begin by solving the energy equation for the **required net head** $H_{required}$,

$$H_{required} = h_{pump,\,u} = \frac{P_2 - P_1}{\rho g} + \frac{\alpha_2 V_2^2 - \alpha_1 V_1^2}{2g} + (z_2 - z_1) + h_{L,\,total} \quad (14\text{–}6)$$

where we assume that there is no turbine in the system, although that term can be added back in, if necessary. We have also included the kinetic energy correction factors in Eq. 14–6 for greater accuracy, even though it is common practice in the turbomachinery industry to ignore them (α_1 and α_2 are often assumed to be unity since the flow is turbulent).

Equation 14–6 is evaluated from the inlet of the piping system (point 1, upstream of the pump) to the outlet of the piping system (point 2, downstream of the pump). Equation 14–6 agrees with our intuition, because it tells us that the useful pump head delivered to the fluid does four things:

- It increases the *static pressure* of the fluid from point 1 to point 2 (first term on the right).
- It increases the *dynamic pressure* (kinetic energy) of the fluid from point 1 to point 2 (second term on the right).

- It raises the *elevation* (potential energy) of the fluid from point 1 to point 2 (third term on the right).
- It overcomes *irreversible head losses* in the piping system (last term on the right).

In a general system, the change in static pressure, dynamic pressure, and elevation may be either positive or negative, while irreversible head losses are *always positive*. In many mechanical and civil engineering problems in which the fluid is a liquid, the elevation term is important, but when the fluid is a gas, such as in ventilation and air pollution control problems, the elevation term is almost always negligible.

To match a pump to a system, and to determine the operating point, we equate $H_{required}$ of Eq. 14–6 to $H_{available}$, which is the (typically known) net head of the pump as a function of volume flow rate.

Operating point: $\qquad H_{required} = H_{available} \qquad$ (14–7)

The most common situation is that an engineer selects a pump that is somewhat heftier than actually required. The volume flow rate through the piping system is then a bit larger than needed, and a valve or damper is installed in the line so that the flow rate can be decreased as necessary.

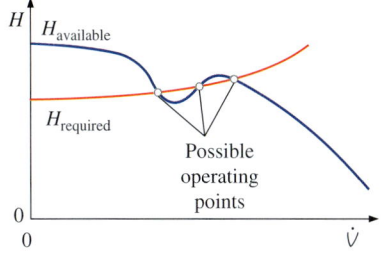

FIGURE 14–10
Situations in which there can be more than one unique operating point should be avoided. In such cases a different pump should be used.

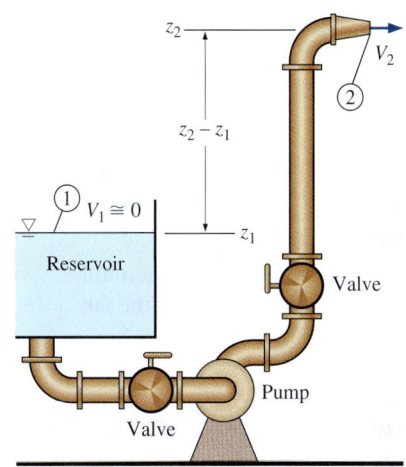

FIGURE 14–11
Equation 14–6 emphasizes the role of a pump in a piping system; namely, it increases (or decreases) the static pressure, dynamic pressure, and elevation of the fluid, and it overcomes irreversible losses.

EXAMPLE 14–1 Operating Point of a Fan in a Ventilation System

A *local ventilation system* (hood and exhaust duct) is used to remove air and contaminants produced by a dry-cleaning operation (Fig. 14–12). The duct is round and is constructed of galvanized steel with longitudinal seams and with joints every 0.76 m. The inner diameter (ID) of the duct is $D = 0.230$ m, and its total length is $L = 13.4$ m. There are five CD3-9 elbows along the duct. The equivalent roughness height of this duct is 0.15 mm, and each elbow has a minor (local) loss coefficient of $K_L = C_o = 0.21$. Note the notation C_o for minor loss coefficient, commonly used in the ventilation industry (ASHRAE, 2001). To ensure adequate ventilation, the minimum required volume flow rate through the duct is $\dot{V} = 600$ cfm (cubic feet per minute), or 0.283 m³/s at 25°C. Literature from the hood manufacturer lists the hood entry loss coefficient as 1.3 based on duct velocity. When the damper is fully open, its loss coefficient is 1.8. A centrifugal fan with 22.9-cm inlet and outlet diameters is available. Its performance data are shown in Table 14–1, as listed by the manufacturer. Predict the operating point of this local ventilation system, and draw a plot of required and available fan pressure rise as functions of volume flow rate. Is the chosen fan adequate?

SOLUTION We are to estimate the operating point for a given fan and duct system and to plot required and available fan pressure rise as functions of volume flow rate. We are then to determine if the selected fan is adequate.
Assumptions 1 The flow is steady. 2 The concentration of contaminants in the air is low; the fluid properties are those of air alone. 3 The flow at the outlet is fully developed turbulent pipe flow with $\alpha = 1.05$.
Properties For air at 25°C, $\nu = 1.562 \times 10^{-5}$ m²/s and $\rho = 1.184$ kg/m³. Standard atmospheric pressure is $P_{atm} = 101.3$ kPa.

794
TURBOMACHINERY

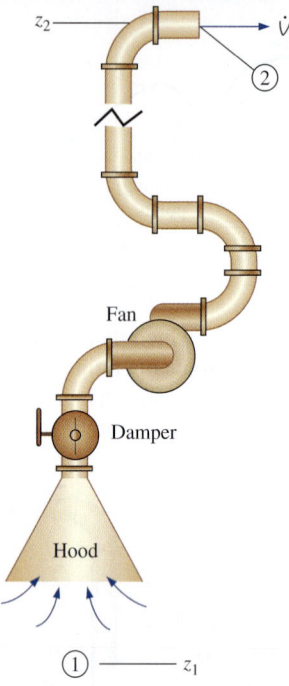

FIGURE 14–12
The local ventilation system for Example 14–1, showing the fan and all minor losses.

TABLE 14–1
Manufacturer's performance data for the fan of Example 14–1*

$\dot{V}$, cfm	$H_{available}$, inches H$_2$O
0	0.90
250	0.95
500	0.90
750	0.75
1000	0.40
1200	0.0

* Note that the head data are listed as inches of *water*, even though *air* is the fluid. This is common practice in the ventilation industry.

Analysis We apply the steady energy equation in head form (Eq. 14–6) from point 1 in the stagnant air region in the room to point 2 at the duct outlet,

$$H_{required} = \frac{P_2 - P_1}{\rho g} + \frac{\alpha_2 V_2^2 - \alpha_1 V_1^2}{2g} + (z_2 - z_1) + h_{L,\,total} \quad (1)$$

In Eq. 1 we may ignore the air speed at point 1 since it was chosen (wisely) far enough away from the hood inlet so that the air is nearly stagnant. At point 1, we let $P_1 = P_{atm}$. At point 2, P_2 is then equal to $P_{atm} - \rho g(z_2 - z_1)$ since the jet discharges into stagnant outside air at higher elevation z_2 on the roof of the building. Thus, the pressure terms cancel with the elevation terms, and Eq. 1 reduces to

Required net head:
$$H_{required} = \frac{\alpha_2 V_2^2}{2g} + h_{L,\,total} \quad (2)$$

The total head loss in Eq. 2 is a combination of major and minor losses and depends on volume flow rate. Since the duct diameter is constant,

Total irreversible head loss:
$$h_{L,\,total} = \left(f\frac{L}{D} + \sum K_L\right)\frac{V^2}{2g} \quad (3)$$

The dimensionless roughness factor is ε/D = (0.15 mm)/(230 mm) = 6.52×10^{-4}. The Reynolds number of the air flowing through the duct is

Reynolds number:
$$\text{Re} = \frac{DV}{\nu} = \frac{D}{\nu}\frac{4\dot{V}}{\pi D^2} = \frac{4\dot{V}}{\nu \pi D} \quad (4)$$

The Reynolds number varies with volume flow rate. At the minimum required flow rate, the air speed through the duct is $V = V_2 = 6.81$ m/s, and the Reynolds number is

$$\text{Re} = \frac{4(0.283 \text{ m}^3/\text{s})}{(1.562 \times 10^{-5} \text{ m}^2/\text{s})\pi(0.230 \text{ m})} = 1.00 \times 10^5$$

From the Moody chart (or the Colebrook equation) at this Reynolds number and roughness factor, the friction factor is $f = 0.0209$. The sum of all the minor loss coefficients is

Minor losses:
$$\sum K_L = 1.3 + 5(0.21) + 1.8 = 4.15 \quad (5)$$

Substituting these values at the minimum required flow rate into Eq. 2, the required net head of the fan at the minimum flow rate is

$$H_{required} = \left(\alpha_2 + f\frac{L}{D} + \sum K_L\right)\frac{V^2}{2g}$$

$$= \left(1.05 + 0.0209\frac{13.4 \text{ m}}{0.230 \text{ m}} + 4.15\right)\frac{(6.81 \text{ m/s})^2}{2(9.81 \text{ m/s}^2)} = 15.2 \text{ m of air} \quad (6)$$

Note that the head is expressed naturally in units of equivalent column height of the pumped fluid, which is air in this case. We convert to an equivalent column height of *water* by multiplying by the ratio of air density to water density,

$$H_{\text{required, inches of water}} = H_{\text{required, air}} \frac{\rho_{\text{air}}}{\rho_{\text{water}}}$$

$$= (15.2 \text{ m}) \frac{1.184 \text{ kg/m}^3}{998.0 \text{ kg/m}^3} \left(\frac{1 \text{ in}}{0.0254 \text{ m}}\right)$$

$$= 0.709 \text{ inches of water} \quad (7)$$

We repeat the calculations at several values of volume flow rate, and compare to the available net head of the fan in Fig. 14–13. The operating point is at a volume flow rate of about **650 cfm,** at which both the required and available net head equal about **0.83 inches of water**. We conclude that **the chosen fan is more than adequate for the job.**

Discussion The purchased fan is somewhat more powerful than required, yielding a higher flow rate than necessary. The difference is small and is acceptable; the butterfly damper valve could be partially closed to cut back the flow rate to 600 cfm if necessary. For safety reasons, it is clearly better to oversize than undersize a fan when used with an air pollution control system.

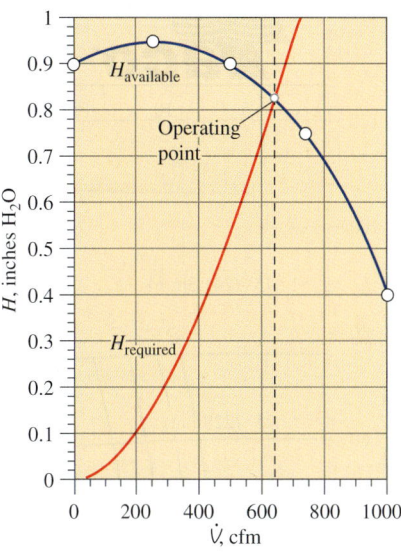

FIGURE 14–13
Net head as a function of volume flow rate for the ventilation system of Example 14–1. The point where the available and required values of H intersect is the operating point.

It is common practice in the pump industry to offer several choices of impeller diameter for a single pump casing. There are several reasons for this: (1) to save manufacturing costs, (2) to enable capacity increase by simple impeller replacement, (3) to standardize installation mountings, and (4) to enable reuse of equipment for a different application. When plotting the performance of such a "family" of pumps, pump manufacturers do not plot separate curves of H, η_{pump}, and bhp for each impeller diameter in the form sketched in Fig. 14–8. Instead, they prefer to combine the performance curves of an entire family of pumps of different impeller diameters onto a single plot (Fig. 14–14). Specifically, they plot a curve of H as a function of $\dot{V}$ for each impeller diameter in the same way as in Fig. 14–8, but create *contour lines* of constant efficiency, by drawing smooth curves through points that have the same value of η_{pump} for the various choices of impeller diameter. Contour lines of constant bhp are often drawn on the same plot in similar fashion. An example is provided in Fig. 14–15 for a family of centrifugal pumps manufactured by Taco, Inc. In this case, five impeller diameters are available, but the identical pump casing is used for all five options. As seen in Fig. 14–15, pump manufacturers do not always plot their pumps' performance curves all the way to free delivery. This is because the pumps are usually not operated there due to the low values of net head and efficiency. If higher values of flow rate and/or net head are required, the customer should step up to the next larger casing size, or consider using additional pumps in series or parallel.

It is clear from the performance plot of Fig. 14–15 that for a given pump casing, the larger the impeller, the higher the maximum achievable efficiency. Why then would anyone buy the smaller impeller pump? To answer this question, we must recognize that the customer's application requires a certain combination of flow rate and net head. If the requirements match a particular impeller diameter, it may be more cost effective to sacrifice pump efficiency in order to satisfy these requirements.

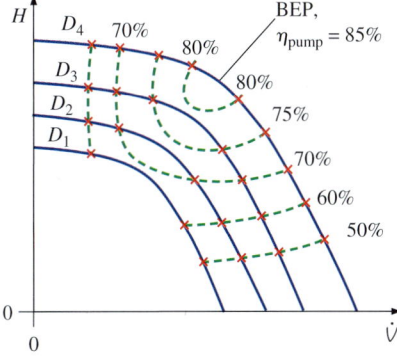

FIGURE 14–14
Typical pump performance curves for a *family* of centrifugal pumps of the same casing diameter but different impeller diameters.

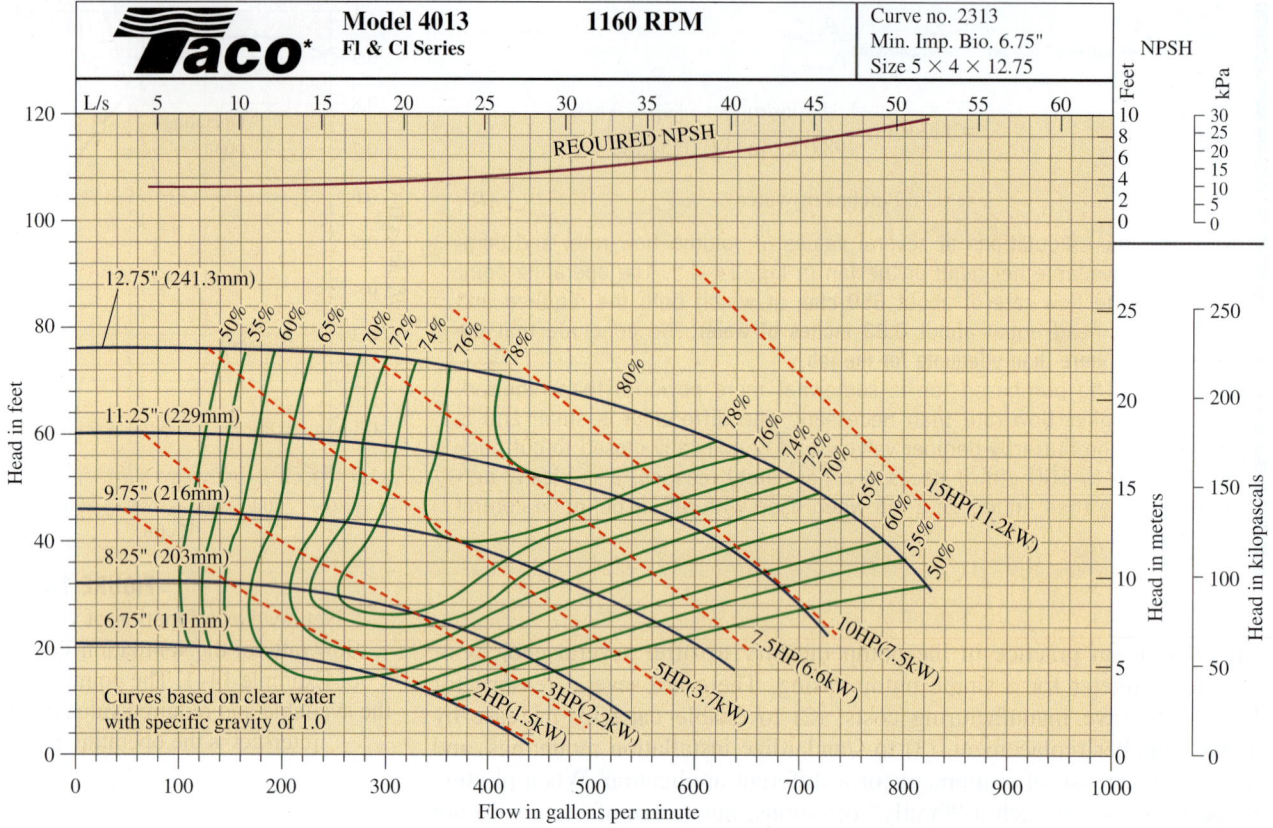

FIGURE 14–15

Example of a manufacturer's performance plot for a family of centrifugal pumps. Each pump has the same casing, but a different impeller diameter.

Courtesy of Taco, Inc., Cranston, RI. Used by permission.

EXAMPLE 14–2 Selection of Pump Impeller Size

A washing operation at a power plant requires 370 gallons per minute (gpm) or 0.0233 m^3/s of water. The required net head is about 24 ft (7.3 m) at this flow rate. A newly hired engineer looks through some catalogs and decides to purchase the 8.25-in (203-mm) impeller option of the Taco Model 4013 Fl Series centrifugal pump of Fig. 14–15. If the pump operates at 1160 rpm, as specified in the performance plot, she reasons, its performance curve intersects 370 gpm (0.0233 m^3/s) at $H = 24$ ft (7.3 m). The chief engineer, who is very concerned about efficiency, glances at the performance curves and notes that the efficiency of this pump at this operating point is only 70 percent. He sees that the 12.75-in (241.3-mm) impeller option achieves a higher efficiency (about 76.5 percent) at the same flow rate. He notes that a throttle valve can be installed downstream of the pump to increase the required net head so that the pump operates at this higher efficiency. He asks the junior engineer to justify her choice of impeller diameter. Namely, he asks her to calculate which impeller option (8.25-in or 12.75-in) would need the least amount of electricity to operate (Fig. 14–16). Perform the comparison and discuss.

SOLUTION For a given flow rate and net head, we are to calculate which impeller size uses the least amount of power, and we are to discuss our results.

Assumptions **1** The water is at 20°C. **2** The flow requirements (volume flow rate and head) are constant.

Properties For water at 20°C, $\rho = 998$ kg/m³.

Analysis From the contours of brake horsepower that are shown on the performance plot of Fig. 14–15, the junior engineer estimates that the pump with the smaller impeller requires about 3.2 hp (2.4 kW) from the motor. She verifies this estimate by using Eq. 14–5,

Required bhp for the 8.25-in impeller option:

$$\text{bhp} = \frac{\rho g \dot{V} H}{\eta_{\text{pump}}} = \frac{(998 \text{ kg/m}^3)(9.81 \text{ m/s}^2)(0.0233 \text{ m}^3/\text{s})(7.3 \text{ m})}{0.70}$$

$$\times \left(\frac{1 \text{ N}}{1 \text{ kg·m/s}^2}\right)\left(\frac{1 \text{ kW}}{1000 \text{ N·m/s}}\right) = 2.38 \text{ kW}$$

Similarly, the larger-diameter impeller option requires

Required bhp for the 12.75-in impeller option: $\quad$ bhp $= 6.56$ kW

using the operating point of that pump, namely, $\dot{V} = 370$ gpm (0.0233 m³/s), $H = 72.0$ ft (22.0 m) and $\eta_{\text{pump}} = 76.5$ percent (Fig. 14–15). Clearly, **the smaller-diameter impeller option is the better choice in spite of its lower efficiency, because it uses less than half the power.**

Discussion Although the larger impeller pump would operate at a somewhat higher value of efficiency, it would deliver about 72 ft (22 m) of net head at the required flow rate. This is overkill, and the throttle valve would be required to make up the difference between this net head and the required flow head of 24 ft (7.3 m) of water. A throttle valve does nothing more than waste mechanical energy, however; so the gain in efficiency of the pump is more than offset by losses through the throttle valve. If the flow head or capacity requirements increase at some time in the future, a larger impeller can be purchased for the same casing.

FIGURE 14–16
In some applications, a less efficient pump from the same family of pumps may require less energy to operate. An even better choice, however, would be a pump whose best efficiency point occurs at the required operating point of the pump, but such a pump is not always commercially available.

Pump Cavitation and Net Positive Suction Head

When pumping liquids, it is possible for the local pressure inside the pump to fall below the **vapor pressure** of the liquid, P_v. (P_v is also called the **saturation pressure** P_{sat} and is listed in thermodynamics tables as a function of saturation temperature.) When $P < P_v$, vapor-filled bubbles called **cavitation bubbles** appear. In other words, the liquid *boils* locally, typically on the suction side of the rotating impeller blades where the pressure is lowest (Fig. 14–17). After the cavitation bubbles are formed, they are transported through the pump to regions where the pressure is higher, causing rapid collapse of the bubbles. It is this *collapse* of the bubbles that is undesirable, since it causes noise, vibration, reduced efficiency, and most importantly, damage to the impeller blades. Repeated bubble collapse near a blade surface leads to pitting or erosion of the blade and eventually catastrophic blade failure.

To avoid cavitation, we must ensure that the local pressure everywhere inside the pump stays *above* the vapor pressure. Since pressure is most easily

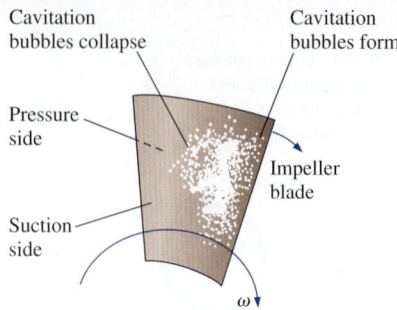

FIGURE 14–17
Cavitation bubbles forming and collapsing on the suction side of an impeller blade.

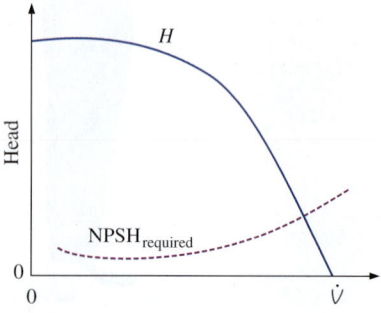

FIGURE 14–18
Typical pump performance curve in which net head and required net positive suction head are plotted versus volume flow rate.

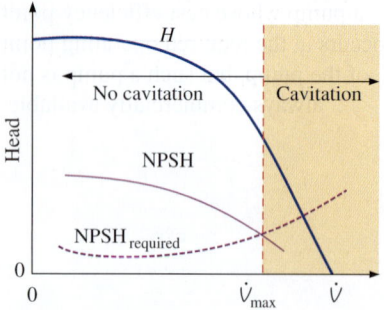

FIGURE 14–19
The volume flow rate at which the actual NPSH and the required NPSH intersect represents the maximum flow rate that can be delivered by the pump without the occurrence of cavitation.

measured (or estimated) at the inlet of the pump, cavitation criteria are typically specified *at the pump inlet*. It is useful to employ a flow parameter called **net positive suction head (NPSH),** defined as *the difference between the pump's inlet stagnation pressure head and the vapor pressure head,*

Net positive suction head: $\text{NPSH} = \left(\dfrac{P}{\rho g} + \dfrac{V^2}{2g}\right)_{\text{pump inlet}} - \dfrac{P_v}{\rho g}$ (14–8)

Pump manufacturers test their pumps for cavitation in a pump test facility by varying the volume flow rate and inlet pressure in a controlled manner. Specifically, at a given flow rate and liquid temperature, the pressure at the pump inlet is slowly lowered until cavitation occurs somewhere inside the pump. The value of NPSH is calculated using Eq. 14–8 and is recorded at this operating condition. The process is repeated at several other flow rates, and the pump manufacturer then publishes a performance parameter called the **required net positive suction head (NPSH$_{\text{required}}$)**, defined as the *minimum NPSH necessary to avoid cavitation in the pump*. The measured value of NPSH$_{\text{required}}$ varies with volume flow rate, and therefore NPSH$_{\text{required}}$ is often plotted on the same pump performance curve as net head (Fig. 14–18). When expressed properly in units of head of the liquid being pumped, NPSH$_{\text{required}}$ is independent of the type of liquid. However, if the required net positive suction head is expressed for a particular liquid in pressure units such as pascals or psi, the engineer must be careful to convert this pressure to the equivalent column height of the actual liquid being pumped. Note that since NPSH$_{\text{required}}$ is usually much smaller than H over the majority of the performance curve, it is often plotted on a separate expanded vertical axis for clarity (see Fig. 14–15) or as contour lines when being shown for a family of pumps. NPSH$_{\text{required}}$ typically increases with volume flow rate, although for some pumps it decreases with $\dot{V}$ at low flow rates where the pump is not operating very efficiently, as sketched in Fig. 14–18.

In order to ensure that a pump does not cavitate, the actual or available NPSH must be greater than NPSH$_{\text{required}}$. It is important to note that the value of NPSH varies not only with flow rate, but also with liquid temperature, since P_v is a function of temperature. NPSH also depends on the type of liquid being pumped, since there is a unique P_v versus T curve for each liquid. Since irreversible head losses through the piping system upstream of the inlet *increase* with flow rate, the pump inlet stagnation pressure head *decreases* with flow rate. Therefore, the value of NPSH *decreases* with $\dot{V}$, as sketched in Fig. 14–19. By identifying the volume flow rate at which the curves of actual NPSH and NPSH$_{\text{required}}$ intersect, we estimate the maximum volume flow rate that can be delivered by the pump without cavitation (Fig. 14–19).

EXAMPLE 14–3   Maximum Flow Rate to Avoid Pump Cavitation

The 11.25-in (229 mm) impeller option of the Taco Model 4013 FI Series centrifugal pump of Fig. 14–15 is used to pump water at 25°C from a reservoir whose surface is 4.0 ft (1.2 m) above the centerline of the pump inlet (Fig. 14–20). The piping system from the reservoir to the pump consists of 10.5 ft (3.2 m) of cast iron pipe with an ID of 4.0 in (10.2 cm) and an average inner roughness height of 0.02 in (0.51 mm). There are several minor losses: a sharp-edged inlet ($K_L = 0.5$), three flanged smooth 90° regular elbows ($K_L = 0.3$ each), and a fully open flanged globe valve ($K_L = 6.0$). Estimate the maximum volume flow rate (in units of gpm) that can be pumped without cavitation. If the water were warmer, would this maximum flow rate increase or decrease? Why? Discuss how you might increase the maximum flow rate while still avoiding cavitation.

SOLUTION For a given pump and piping system we are to estimate the maximum volume flow rate that can be pumped without cavitation. We are also to discuss the effect of water temperature and how we might increase the maximum flow rate.

Assumptions 1 The flow is steady. 2 The liquid is incompressible. 3 The flow at the pump inlet is turbulent and fully developed, with $\alpha = 1.05$.

Properties For water at $T = 25°C$, $\rho = 997.0$ kg/m³, $\mu = 8.91 \times 10^{-4}$ kg/m·s, and $P_v = 3.169$ kPa. Standard atmospheric pressure is $P_{atm} = 101.3$ kPa.

Analysis We apply the steady energy equation in head form along a streamline from point 1 at the reservoir surface to point 2 at the pump inlet,

$$\frac{P_1}{\rho g} + \frac{\alpha_1 V_1^2}{2g} + z_1 + \cancel{h_{pump,u}} = \frac{P_2}{\rho g} + \frac{\alpha_2 V_2^2}{2g} + z_2 + \cancel{h_{turbine,e}} + h_{L,total} \quad (1)$$

In Eq. 1 we have ignored the water speed at the reservoir surface ($V_1 \cong 0$). There is no turbine in the piping system. Also, although there is a pump in the system, there is no pump between points 1 and 2; hence the pump head term also drops out. We solve Eq. 1 for $P_2/\rho g$, which is the pump inlet pressure expressed as a head,

Pump inlet pressure head:
$$\frac{P_2}{\rho g} = \frac{P_{atm}}{\rho g} + (z_1 - z_2) - \frac{\alpha_2 V_2^2}{2g} - h_{L,total} \quad (2)$$

Note that in Eq. 2, we have recognized that $P_1 = P_{atm}$ since the reservoir surface is exposed to atmospheric pressure.

The available net positive suction head at the pump inlet is obtained from Eq. 14–8. After substitution of Eq. 2, we get

Available NPSH:
$$\text{NPSH} = \frac{P_{atm} - P_v}{\rho g} + (z_1 - z_2) - h_{L,total} - \frac{(\alpha_2 - 1)V_2^2}{2g} \quad (3)$$

Since we know P_{atm}, P_v, and the elevation difference, all that remains is to estimate the total irreversible head loss through the piping system, which depends on volume flow rate. Since the pipe diameter is constant,

Irreversible head loss:
$$h_{L,total} = \left(f\frac{L}{D} + \sum K_L\right)\frac{V^2}{2g} \quad (4)$$

The rest of the problem is most easily solved on a computer. For a given volume flow rate, we calculate speed V and Reynolds number Re. From Re and the known pipe roughness, we use the Moody chart (or the Colebrook equation) to obtain friction factor f. The sum of all the minor loss coefficients is

Minor losses:
$$\sum K_L = 0.5 + 3 \times 0.3 + 6.0 = 7.4 \quad (5)$$

We make one calculation by hand for illustrative purposes. At $\dot{V} = 400$ gpm (0.02523 m³/s), the average speed of water through the pipe is

$$V = \frac{\dot{V}}{A} = \frac{4\dot{V}}{\pi D^2} = \frac{4(0.02523 \text{ m}^3/\text{s})}{\pi(4.0 \text{ in})^2}\left(\frac{1 \text{ in}}{0.0254 \text{ m}}\right)^2 = 3.112 \text{ m/s} \quad (6)$$

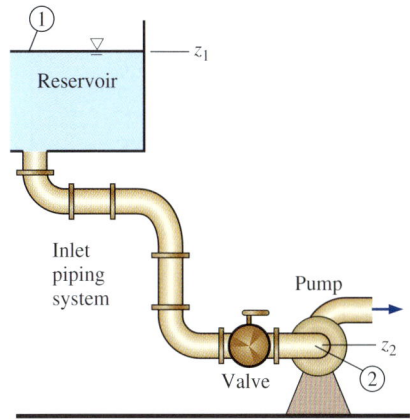

FIGURE 14–20
Inlet piping system from the reservoir (1) to the pump inlet (2) for Example 14–3.

800
TURBOMACHINERY

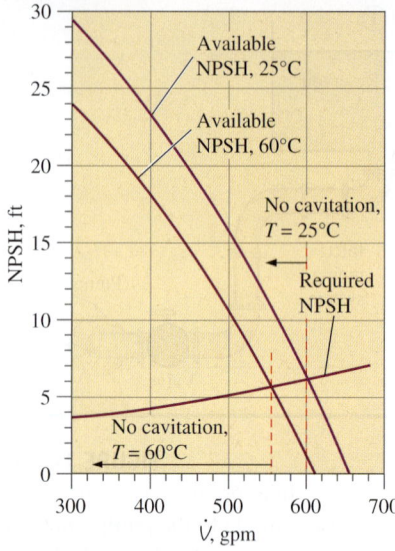

FIGURE 14–21
Net positive suction head as a function of volume flow rate for the pump of Example 14–3 at two temperatures. Cavitation is predicted to occur at flow rates greater than the point where the available and required values of NPSH intersect.

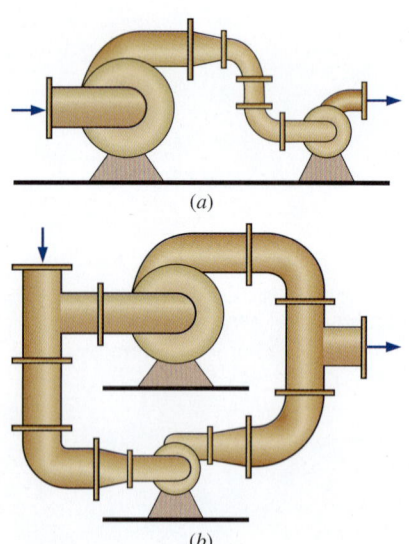

FIGURE 14–22
Arranging two very dissimilar pumps in (a) series or (b) parallel can sometimes lead to problems.

which produces a Reynolds number of Re = $\rho VD/\mu$ = 3.538 × 10^5. At this Reynolds number, and with roughness factor ϵ/D = 0.005, the Colebrook equation yields f = 0.0306. Substituting the given properties, along with f, D, L, and Eqs. 4, 5, and 6, into Eq. 3, we calculate the available net positive suction head at this flow rate,

$$\text{NPSH} = \frac{(10{,}300 - 3169) \text{ N/m}^2}{(997.0 \text{ kg/m}^3)(9.81 \text{ m/s}^2)} \left(\frac{\text{kg·m/s}^2}{\text{N}}\right) + 1.219 \text{ m}$$

$$- \left(0.0306 \frac{10.5 \text{ ft}}{0.3333 \text{ ft}} + 7.4 - (1.05 - 1)\right) \frac{(3.112 \text{ m/s})^2}{2(9.81 \text{ m/s}^2)}$$

$$= 7.148 \text{ m} = 23.5 \text{ ft} \quad (7)$$

The required net positive suction head is obtained from Fig. 14–15. At our example flow rate of 400 gpm, NPSH$_{required}$ is just above 4.0 ft. Since the actual NPSH is much higher than this, we need not worry about cavitation at this flow rate. We use EES (or a spreadsheet) to calculate NPSH as a function of volume flow rate, and the results are plotted in Fig. 14–21. It is clear from this plot that at 25°C, **cavitation occurs at flow rates above approximately 600 gpm**—close to the free delivery.

If the water were warmer than 25°C, the vapor pressure would increase, the viscosity would decrease, and the density would decrease slightly. The calculations are repeated at T = 60°C, at which ρ = 983.3 kg/m^3, μ = 4.67 × 10^{-4} kg/m·s, and P_v = 19.94 kPa. The results are also plotted in Fig. 14–21, where we see that **the maximum volume flow rate without cavitation decreases with temperature** (to about 555 gpm at 60°C). This decrease agrees with our intuition, since warmer water is already closer to its boiling point from the start.

Finally, how can we increase the maximum flow rate? *Any modification that increases the available NPSH helps.* We can raise the height of the reservoir surface (to increase the hydrostatic head). We can reroute the piping so that only one elbow is necessary and replace the globe valve with a ball valve (to decrease the minor losses). We can increase the diameter of the pipe and decrease the surface roughness (to decrease the major losses). In this particular problem, the minor losses have the greatest influence, but in many problems, the major losses are more significant, and increasing the pipe diameter is most effective. That is one reason why many centrifugal pumps have a larger inlet diameter than outlet diameter.

Discussion Note that NPSH$_{required}$ does not depend on water temperature, but the actual or available NPSH decreases with temperature (Fig. 14–21).

Pumps in Series and Parallel

When faced with the need to increase volume flow rate or pressure rise by a small amount, you might consider adding an additional smaller pump in series or in parallel with the original pump. While series or parallel arrangement is acceptable for some applications, arranging *dissimilar* pumps in series or in parallel may lead to problems, especially if one pump is much larger than the other (Fig. 14–22). A better course of action is to increase the original pump's speed and/or input power (larger electric motor), replace the impeller with a larger one, or replace the entire pump with a larger one. The logic for this decision can be seen from the pump performance curves, realizing that *pressure rise and volume flow rate are related.* Arranging dissimilar pumps in series may create problems because the volume flow rate through each pump must be the same, but the overall pressure rise is equal to the pressure rise of one pump plus that of the other. If the pumps have widely different performance curves, the smaller pump may be forced to operate beyond its free delivery flow rate, whereupon it acts like a head *loss,* reducing the total volume flow rate. Arranging dissimilar pumps in parallel may create problems because the overall pressure rise must be the same, but the net volume flow rate is the sum of that through each branch. If the pumps are not sized properly, the smaller pump may not be able to handle the large head imposed on it, and the flow in its branch could actually be *reversed;* this would inadvertently reduce the overall pressure rise. In either case, the power supplied to the smaller pump would be wasted.

Keeping these cautions in mind, there are many applications where two or more similar (usually identical) pumps are operated in series or in parallel. When operated in *series,* the combined net head is simply the sum of the net heads of each pump (at a given volume flow rate),

Combined net head for n pumps in series: $\quad H_{\text{combined}} = \sum_{i=1}^{n} H_i \quad$ (14–9)

Equation 14–9 is illustrated in Fig. 14–23 for three pumps in series. In this example, pump 3 is the strongest and pump 1 is the weakest. The shutoff head of the three pumps combined in series is equal to the sum of the shutoff head of each individual pump. For low values of volume flow rate, the net head of the three pumps in series is equal to $H_1 + H_2 + H_3$. Beyond the free delivery of pump 1 (to the right of the first vertical dashed red line in Fig. 14–23), *pump 1 should be shut off and bypassed.* Otherwise it would be running beyond its maximum designed operating point, and the pump or its motor could be damaged. Furthermore, the net head across this pump would be *negative* as previously discussed, contributing to a net loss in the system. With pump 1 bypassed, the combined net head becomes $H_2 + H_3$. Similarly, beyond the free delivery of pump 2, that pump should also be shut off and bypassed, and the combined net head is then equal to H_3 alone, as indicated to the right of the second vertical dashed gray line in Fig. 14–23.

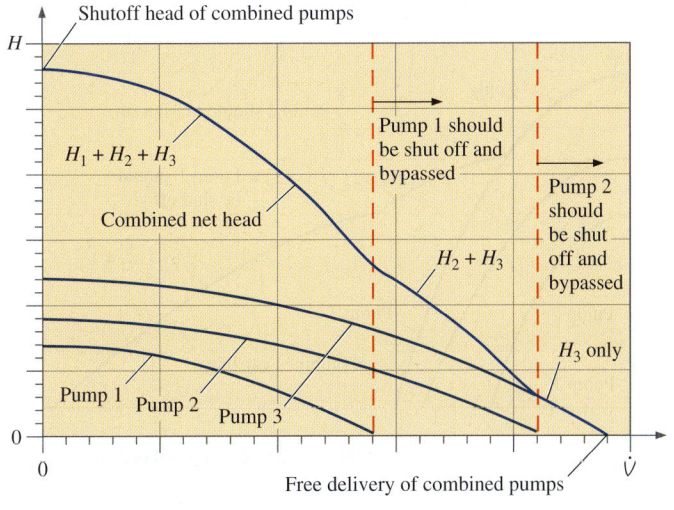

FIGURE 14–23
Pump performance curve (dark blue) for three dissimilar pumps in *series.* At low values of volume flow rate, the combined net head is equal to the sum of the net head of each pump by itself. However, to avoid pump damage and loss of combined net head, any individual pump should be shut off and bypassed at flow rates larger than that pump's free delivery, as indicated by the vertical dashed red lines. If the three pumps were identical, it would not be necessary to turn off any of the pumps, since the free delivery of each pump would occur at the same volume flow rate.

In this case, the combined free delivery is the same as that of pump 3 alone, assuming that the other two pumps are bypassed.

When two or more identical (or similar) pumps are operated in *parallel*, their individual volume flow rates (rather than net heads) are summed,

Combined capacity for n pumps in parallel: $\quad \dot{V}_{\text{combined}} = \sum_{i=1}^{n} \dot{V}_i \quad$ (14–10)

As an example, consider the *same* three pumps, but arranged in parallel rather than in series. The combined pump performance curve is shown in Fig. 14–24. The free delivery of the three combined pumps is equal to the sum of the free delivery of each individual pump. For low values of net head, the capacity of the three pumps in parallel is equal to $\dot{V}_1 + \dot{V}_2 + \dot{V}_3$. Above the shutoff head of pump 1 (above the first horizontal dashed red line in Fig. 14–24), *pump 1 should be shut off and its branch should be blocked* (with a valve). Otherwise it would be running beyond its maximum designed operating point, and the pump or its motor could be damaged. Furthermore, the volume flow rate through this pump would be *negative* as previously discussed, contributing to a net loss in the system. With pump 1 shut off and blocked, the combined capacity becomes $\dot{V}_2 + \dot{V}_3$. Similarly, above the shutoff head of pump 2, that pump should also be shut off and blocked. The combined capacity is then equal to $\dot{V}_3$ alone, as indicated above, the second horizontal dashed gray line in Fig. 14–24. In this case, the combined shutoff head is the same as that of pump 3 alone, assuming that the other two pumps are shut off and their branches are blocked.

In practice, several pumps may be combined in parallel to deliver a large volume flow rate (Fig. 14–25). Examples include banks of pumps used to circulate water in cooling towers and chilled water loops (Wright, 1999). Ideally all the pumps should be identical so that we don't need to worry about shutting any of them off (Fig. 14–24). Also, it is wise to install check valves in each branch so that when a pump needs to be shut down

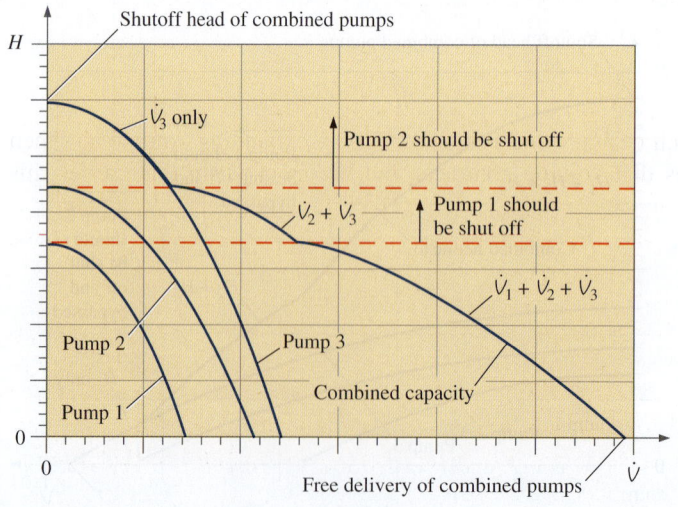

FIGURE 14–24
Pump performance curve (dark blue) for three pumps in *parallel*. At a low value of net head, the combined capacity is equal to the sum of the capacity of each pump by itself. However, to avoid pump damage and loss of combined capacity, any individual pump should be shut off at net heads larger than that pump's shutoff head, as indicated by the horizontal dashed gray lines. That pump's branch should also be blocked with a valve to avoid reverse flow. If the three pumps were identical, it would not be necessary to turn off any of the pumps, since the shutoff head of each pump would occur at the same net head.

FIGURE 14–25
Several identical pumps are often run in a parallel configuration so that a large volume flow rate can be achieved when necessary. Three parallel pumps are shown.
Courtesy of Goulds Pumps, ITT Corporation. Used by permission.

(for maintenance or when the required flow rate is low), backflow through the pump is avoided. Note that the extra valves and piping required for a parallel pump network add additional head losses to the system; thus the overall performance of the combined pumps suffers somewhat.

Positive-Displacement Pumps

People have designed numerous positive-displacement pumps throughout the centuries. In each design, fluid is sucked into an expanding volume and then pushed along as that volume contracts, but the mechanism that causes this change in volume differs greatly among the various designs. Some designs are very simple, like the flexible-tube *peristaltic pump* (Fig. 14–26a) that compresses a tube by small wheels, pushing the fluid along. (This mechanism is somewhat similar to peristalsis in your esophagus or intestines, where muscles rather than wheels compress the tube.) Others are more complex, using rotating cams with synchronized lobes (Fig. 14–26b), interlocking gears (Fig. 14–26c), or screws (Fig. 14–26d). Positive-displacement pumps are ideal for high-pressure applications like pumping viscous liquids or thick slurries, and for applications where precise amounts of liquid are to be dispensed or metered, as in medical applications.

TURBOMACHINERY

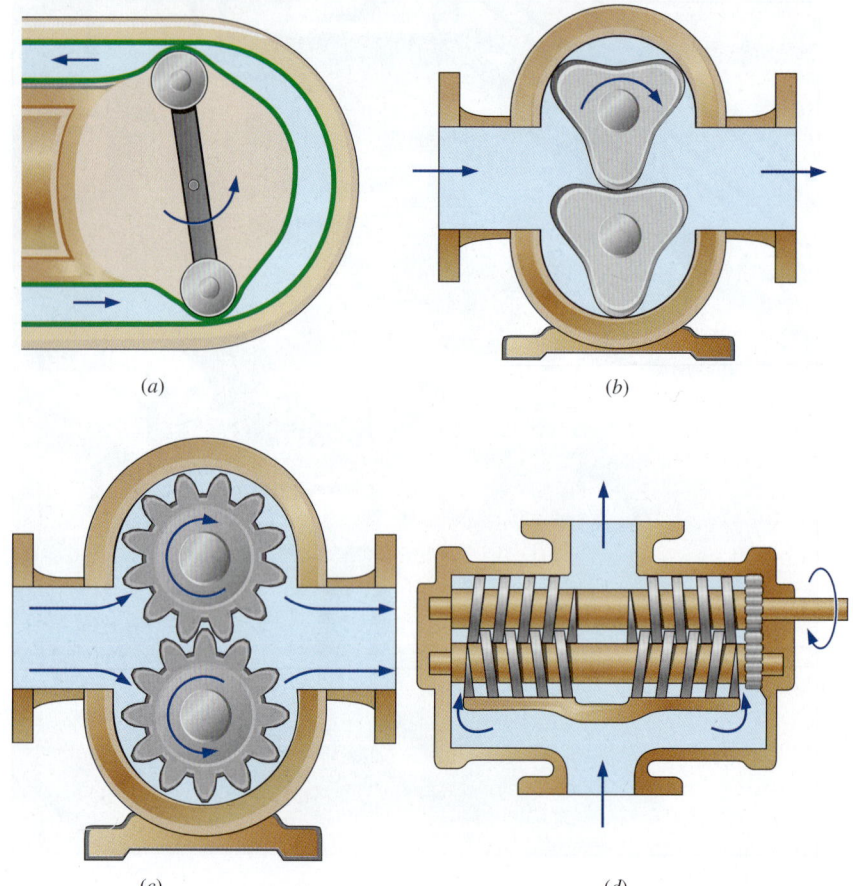

FIGURE 14–26
Examples of positive-displacement pumps: (*a*) flexible-tube peristaltic pump, (*b*) three-lobe rotary pump, (*c*) gear pump, and (*d*) double screw pump.

Adapted from F. M. White, Fluid Mechanics *4/e. Copyright © 1999. The McGraw-Hill Companies, Inc. With permission.*

FIGURE 14–27
Four phases (one-eighth of a turn apart) in the operation of a two-lobe rotary pump, a type of positive-displacement pump. The blue region represents a chunk of fluid pushed through the top rotor, while the red region represents a chunk of fluid pushed through the bottom rotor, which rotates in the opposite direction. Flow is from left to right.

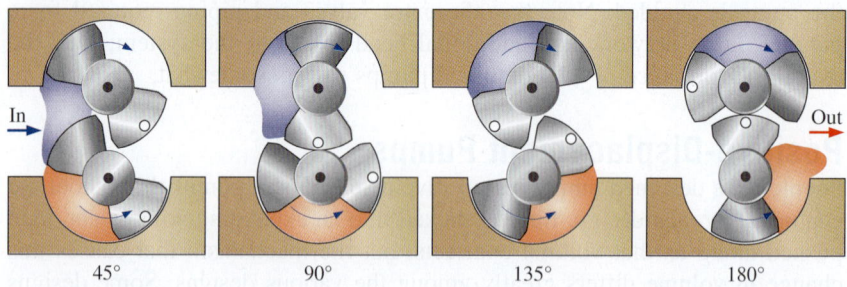

To illustrate the operation of a positive-displacement pump, we sketch four phases of half of a cycle of a simple **rotary pump** with two lobes on each rotor (Fig. 14–27). The two rotors are synchronized by an external gear box so as to rotate at the same angular speed, but in opposite directions. In the diagram, the top rotor turns clockwise and the bottom rotor turns counterclockwise, sucking in fluid from the left and discharging it to the right. A white dot is drawn on one lobe of each rotor to help you visualize the rotation.

Gaps exist between the rotors and the housing and between the lobes of the rotors themselves, as illustrated (and exaggerated) in Fig. 14–27. Fluid can leak through these gaps, reducing the pump's efficiency. High-viscosity fluids cannot penetrate the gaps as easily; hence the net head (and efficiency) of a rotary pump generally *increases* with fluid viscosity, as shown in Fig. 14–28. This is one reason why rotary pumps (and other types of positive-displacement pumps) are a good choice for pumping highly viscous fluids and slurries. They are used, for example, as automobile engine oil pumps and in the foods industry to pump heavy liquids like syrup, tomato paste, and chocolate, and slurries like soups.

The pump performance curve (net head versus capacity) of a rotary pump is nearly vertical throughout its recommended operating range, since the capacity is fairly constant regardless of load at a given rotational speed (Fig. 14–28). However, as indicated by the dashed blue curve in Fig. 14–28, at very high values of net head, corresponding to very high pump outlet pressure, leaks become more severe, even for high-viscosity fluids. In addition, the motor driving the pump cannot overcome the large torque caused by this high outlet pressure, and the motor begins to suffer stall or overload, which may burn out the motor. Therefore, rotary pump manufacturers do not recommend operation of the pump above a certain maximum net head, which is typically well below the shutoff head. The pump performance curves supplied by the manufacturer often do not even show the pump's performance outside of its recommended operating range.

Positive-displacement pumps have many advantages over dynamic pumps. For example, a positive-displacement pump is better able to handle shear sensitive liquids since the induced shear is much less than that of a dynamic pump operating at similar pressure and flow rate. Blood is a shear sensitive liquid, and this is one reason why positive-displacement pumps are used for artificial hearts. A well-sealed positive-displacement pump can create a significant vacuum pressure at its inlet, even when dry, and is thus able to lift a liquid from several meters below the pump. We refer to this kind of pump as a **self-priming pump** (Fig. 14–29). Finally, the rotor(s) of a positive-displacement pump run at lower speeds than the rotor (impeller) of a dynamic pump at similar loads, extending the useful lifetime of seals, etc.

There are some disadvantages of positive-displacement pumps as well. Their volume flow rate cannot be changed unless the rotation rate is changed. (This is not as simple as it sounds, since most AC electric motors are designed to operate at one or more *fixed* rotational speeds.) They create very high pressure at the outlet side, and if the outlet becomes blocked, ruptures may occur or electric motors may overheat, as previously discussed. Overpressure protection (e.g., a pressure-relief valve) is often required for this reason. Because of their design, positive-displacement pumps sometimes deliver a pulsating flow, which may be unacceptable for some applications.

Analysis of positive-displacement pumps is fairly straightforward. From the geometry of the pump, we calculate the **closed volume** (V_{closed}) that is filled (and expelled) for every n rotations of the shaft. Volume flow rate is then equal to rotation rate $\dot{n}$ times V_{closed} divided by n,

Volume flow rate, positive-displacement pump: $\qquad \dot{V} = \dot{n} \dfrac{V_{closed}}{n} \qquad$ (14–11)

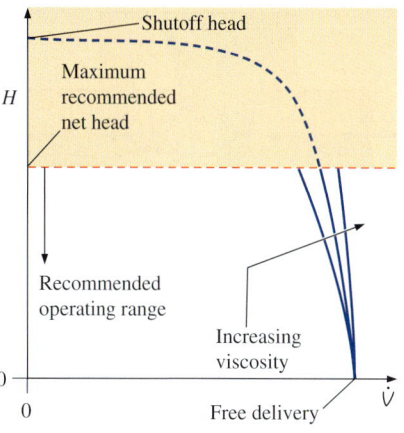

FIGURE 14–28
Comparison of the pump performance curves of a rotary pump operating at the same speed, but with fluids of various viscosities. To avoid motor overload the pump should not be operated in the shaded region.

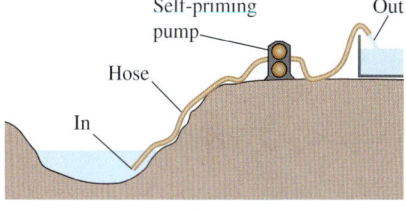

FIGURE 14–29
A pump that can lift a liquid even when the pump itself is "empty" is called a self-priming pump.

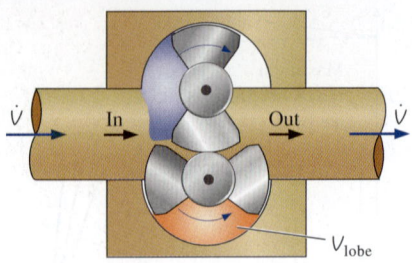

FIGURE 14–30
The two-lobe rotary pump of Example 14–4. Flow is from left to right.

EXAMPLE 14–4   Volume Flow Rate through a Positive-Displacement Pump

A two-lobe rotary positive-displacement pump, similar to that of Fig. 14–27, moves 0.45 cm³ of SAE 30 motor oil in each lobe volume V_{lobe}, as sketched in Fig. 14–30. Calculate the volume flow rate of oil for the case where $\dot{n} = 900$ rpm.

SOLUTION   We are to calculate the volume flow rate of oil through a positive-displacement pump for given values of lobe volume and rotation rate.
Assumptions   **1** The flow is steady in the mean. **2** There are no leaks in the gaps between lobes or between lobes and the casing. **3** The oil is incompressible.
Analysis   By studying Fig. 14–27, we see that for half of a rotation (180° for $n = 0.5$ rotations) of the two counter-rotating shafts, the total volume of oil pumped is $V_{closed} = 2V_{lobe}$. The volume flow rate is then calculated from Eq. 14–11,

$$\dot{V} = \dot{n}\frac{V_{closed}}{n} = (900 \text{ rot/min})\frac{2(0.45 \text{ cm}^3)}{0.5 \text{ rot}} = \mathbf{1620 \text{ cm}^3/\text{min}}$$

Discussion   If there were leaks in the pump, the volume flow rate would be lower. The oil's density is not needed for calculation of the volume flow rate. However, the higher the fluid density, the higher the required shaft torque and brake horsepower.

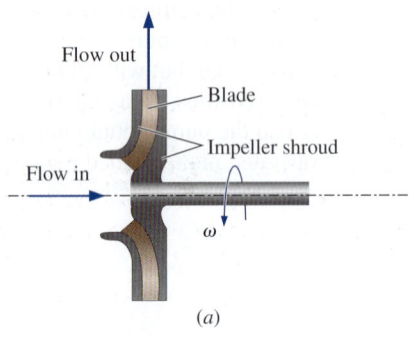

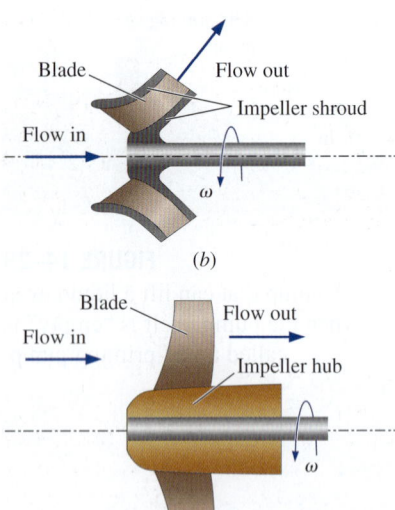

FIGURE 14–31
The *impeller* (rotating portion) of the three main categories of dynamic pumps: (*a*) *centrifugal flow*, (*b*) *mixed flow*, and (*c*) *axial flow*.

Dynamic Pumps

There are three main types of *dynamic pumps* that involve rotating blades called **impeller blades** or **rotor blades,** which impart momentum to the fluid. For this reason they are sometimes called **rotodynamic pumps** or simply **rotary pumps** (not to be confused with rotary positive-displacement pumps, which use the same name). There are also some nonrotary dynamic pumps, such as jet pumps and electromagnetic pumps; these are not discussed in this text. Rotary pumps are classified by the manner in which flow exits the pump: *centrifugal flow*, *axial flow*, and *mixed flow* (Fig. 14–31). In a **centrifugal-flow pump,** fluid enters axially (in the same direction as the axis of the rotating shaft) in the center of the pump, but is discharged radially (or tangentially) along the outer radius of the pump casing. For this reason centrifugal pumps are also called **radial-flow pumps.** In an **axial-flow pump,** fluid enters and leaves axially, typically along the outer portion of the pump because of blockage by the shaft, motor, hub, etc. A **mixed-flow pump** is intermediate between centrifugal and axial, with the flow entering axially, not necessarily in the center, but leaving at some angle between radially and axially.

Centrifugal Pumps

Centrifugal pumps and blowers can be easily identified by their snail-shaped casing, called the **scroll** (Fig. 14–32). They are found all around your home—in dishwashers, hot tubs, clothes washers and dryers, hairdryers, vacuum cleaners, kitchen exhaust hoods, bathroom exhaust fans, leaf blowers, furnaces, etc. They are used in cars—the water pump in the engine, the air blower in the heater/air conditioner unit, etc. Centrifugal pumps are ubiquitous in industry as well; they are used in building ventilation systems, washing operations, cooling ponds and cooling towers, and in numerous other industrial operations in which fluids are pumped.

A schematic diagram of a centrifugal pump is shown in Fig. 14–33. Note that a **shroud** often surrounds the impeller blades to increase blade stiffness. In pump terminology, the rotating assembly that consists of the shaft, the hub, the impeller blades, and the impeller shroud is called the **impeller** or **rotor**. Fluid enters axially through the hollow middle portion of the pump (the **eye**), after which it encounters the rotating blades. It acquires tangential and radial velocity by momentum transfer with the impeller blades, and acquires additional radial velocity by so-called centrifugal forces, which are actually a lack of sufficient *centripetal* forces to sustain circular motion. The flow leaves the impeller after gaining both speed and pressure as it is flung radially outward into the scroll (also called the **volute**). As sketched in Fig. 14–33, the scroll is a snail-shaped **diffuser** whose purpose is to decelerate the fast-moving fluid leaving the trailing edges of the impeller blades, thereby further increasing the fluid's pressure, and to combine and direct the flow from all the blade passages toward a common outlet. As mentioned previously, if the flow is steady in the mean, if the fluid is incompressible, and if the inlet and outlet diameters are the same, the average flow speed at the outlet is identical to that at the inlet. Thus, it is not necessarily the speed, but the *pressure* that increases from inlet to outlet through a centrifugal pump.

There are three types of centrifugal pump that warrant discussion, based on impeller blade geometry, as sketched in Fig. 14–34: *backward-inclined blades*, *radial blades*, and *forward-inclined blades*. Centrifugal pumps with **backward-inclined blades** (Fig. 14–34a) are the most common. These yield the highest efficiency of the three because fluid flows into and out of the blade passages with the least amount of turning. Sometimes the blades are airfoil shaped, yielding similar performance but even higher efficiency. The pressure rise is intermediate between the other two types of centrifugal

FIGURE 14–32
A typical centrifugal blower with its characteristic snail-shaped scroll.
Courtesy of The New York Blower Company, Willowbrook, IL. Used by permission.

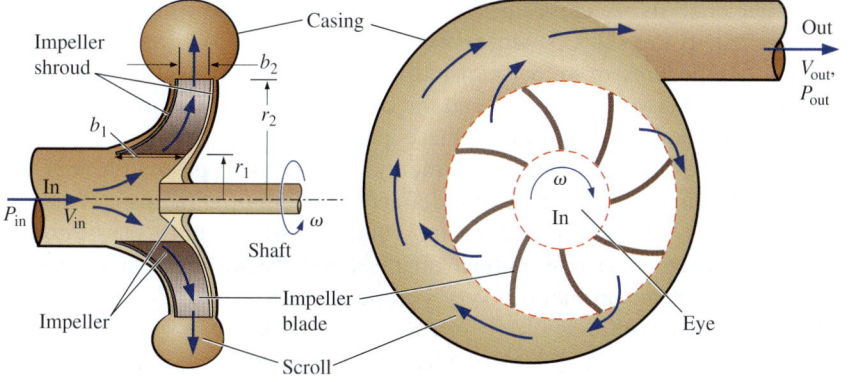

FIGURE 14–33
Side view and frontal view of a typical *centrifugal pump*. Fluid enters axially in the middle of the pump (the *eye*), is flung around to the outside by the rotating blade assembly (*impeller*), is diffused in the expanding diffuser (*scroll*), and is discharged out the side of the pump. We define r_1 and r_2 as the radial locations of the impeller blade inlet and outlet, respectively; b_1 and b_2 are the axial blade widths at the impeller blade inlet and outlet, respectively.

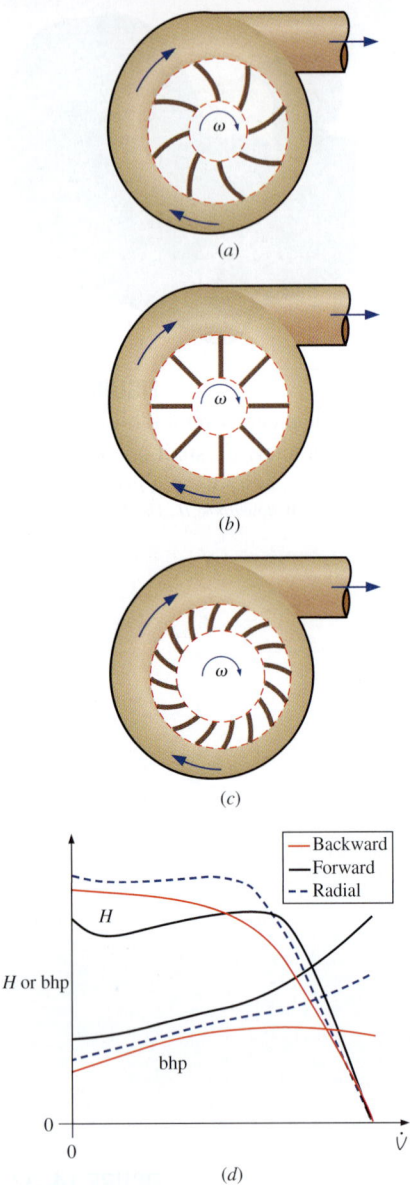

FIGURE 14–34
The three main types of centrifugal pumps are those with (a) *backward-inclined blades*, (b) *radial blades*, and (c) *forward-inclined blades*; (d) comparison of net head and brake horsepower performance curves for the three types of centrifugal pumps.

pumps. Centrifugal pumps with **radial blades** (also called **straight blades,** Fig. 14–34b) have the simplest geometry and produce the largest pressure rise of the three for a wide range of volume flow rates, but the pressure rise decreases rapidly after the point of maximum efficiency. Centrifugal pumps with **forward-inclined blades** (Fig. 14–34c) produce a pressure rise that is nearly constant, albeit lower than that of radial or backward-inclined blades, over a wide range of volume flow rates. Forward-inclined centrifugal pumps generally have more blades, but the blades are smaller, as sketched in Fig. 14–34c. Centrifugal pumps with forward-inclined blades generally have a lower maximum efficiency than do straight-bladed pumps. Radial and backward-inclined centrifugal pumps are preferred for applications where one needs to provide volume flow rate and pressure rise within a narrow range of values. If a wider range of volume flow rates and/or pressure rises are desired, the performance of radial pumps and backward-inclined pumps may not be able to satisfy the new requirements; these types of pumps are less forgiving (less robust). The performance of forward-inclined pumps is more forgiving and accommodates a wider variation, at the cost of lower efficiency and less pressure rise per unit of input power. If a pump is needed to produce large pressure rise over a wide range of volume flow rates, the forward-inclined centrifugal pump is attractive.

Net head and brake horsepower performance curves for these three types of centrifugal pump are compared in Fig. 14–34d. The curves have been adjusted such that each pump achieves the same free delivery (maximum volume flow rate at zero net head). Note that these are qualitative sketches for comparison purposes only—actual measured performance curves may differ significantly in shape, depending on details of the pump design.

For any inclination of the impeller blades (backward, radial, or forward), we can analyze the velocity vectors through the blades. The actual flow field is unsteady, fully three-dimensional, and perhaps compressible. For simplicity in our analysis we consider steady flow in both the absolute reference frame and in the relative frame of reference rotating with the impeller. We consider only incompressible flow, and we consider only the radial or normal velocity component (subscript n) and the circumferential or tangential velocity component (subscript t) from blade inlet to blade outlet. We do not consider the axial velocity component (to the right in Fig. 14–35 and into the page in the frontal view of Fig. 14–33). In other words, although there is a nonzero axial component of velocity through the impeller, it does not enter our analysis. A close-up side view of a simplified centrifugal pump is sketched in Fig. 14–35, where we define $V_{1,n}$ and $V_{2,n}$ as the average normal components of velocity at radii r_1 and r_2, respectively. Although a gap is shown between the blade and the casing, we assume in our simplified analysis that no leakage occurs in these gaps.

The volume flow rate $\dot{V}$ entering the eye of the pump passes through the circumferential cross-sectional area defined by width b_1 at radius r_1. Conservation of mass requires that this same volume flow rate must pass through the circumferential cross-sectional area defined by width b_2 at radius r_2. Using the average normal velocity components $V_{1,n}$ and $V_{2,n}$ defined in Fig. 14–35, we write

Volume flow rate: $$\dot{V} = 2\pi r_1 b_1 V_{1,n} = 2\pi r_2 b_2 V_{2,n} \qquad (14\text{–}12)$$

from which we obtain

$$V_{2,n} = V_{1,n} \frac{r_1 b_1}{r_2 b_2} \quad (14\text{--}13)$$

It is clear from Eq. 14–13 that $V_{2,n}$ may be less than, equal to, or greater than $V_{1,n}$, depending on the values of b and r at the two radii.

We sketch a close-up frontal view of one impeller blade in Fig. 14–36, where we show both radial and tangential velocity components. We have drawn a backward-inclined blade, but the same analysis holds for blades of *any* inclination. The inlet of the blade (at radius r_1) moves at tangential velocity ωr_1. Likewise, the outlet of the blade moves at tangential velocity ωr_2. It is clear from Fig. 14–36 that these two tangential velocities differ not only in magnitude, but also in direction, because of the inclination of the blade. We define **leading edge angle** β_1 as the blade angle relative to the reverse tangential direction at radius r_1. In like manner we define **trailing edge angle** β_2 as the blade angle relative to the reverse tangential direction at radius r_2.

We now make a significant simplifying approximation. We assume that the flow impinges on the blade *parallel to the blade's leading edge* and exits the blade *parallel to the blade's trailing edge*. In other words,

> We assume that the flow is everywhere tangent to the blade surface when viewed from a reference frame rotating with the blade.

At the inlet, this approximation is sometimes called the **shockless entry condition,** not to be confused with shock waves (Chap. 12). Rather, the terminology implies smooth flow into the impeller blade without a sudden turning "shock." Inherent in this approximation is the assumption that there is *no flow separation* anywhere along the blade surface. If the centrifugal pump operates at or near its design conditions, this assumption is valid. However, when the pump operates far off design conditions, the flow may separate off the blade surface (typically on the suction side where there are adverse pressure gradients), and our simplified analysis breaks down.

Velocity vectors $\vec{V}_{1,\text{relative}}$ and $\vec{V}_{2,\text{relative}}$ are drawn in Fig. 14–36 parallel to the blade surface, in accordance with our simplifying assumption. These are the velocity vectors seen from the relative reference frame of an observer moving with the rotating blade. When we vectorially add tangential velocity ωr_1 (the velocity of the blade at radius r_1) to $\vec{V}_{1,\text{relative}}$ by completing the parallelogram as sketched in Fig. 14–36, the resultant vector is the *absolute* fluid velocity $\vec{V}_1$ at the blade inlet. In exactly similar fashion, we obtain $\vec{V}_2$, the absolute fluid velocity at the blade outlet (also sketched in Fig. 14–36). For completeness, normal velocity components $V_{1,n}$ and $V_{2,n}$ are also shown in Fig. 14–36. Notice that these normal velocity components are independent of which frame of reference we use, absolute or relative.

To evaluate the torque on the rotating shaft, we apply the angular momentum relation for a control volume, as discussed in Chap. 6. We choose a control volume surrounding the impeller blades, from radius r_1 to radius r_2, as sketched in Fig. 14–37. We also introduce in Fig. 14–37 angles α_1 and α_2, defined as the angle of departure of the absolute velocity vector from the normal direction at radii r_1 and r_2, respectively. In keeping with the concept of treating a control volume like a "black box," we ignore details of individual impeller blades. Instead we make the approximation that flow

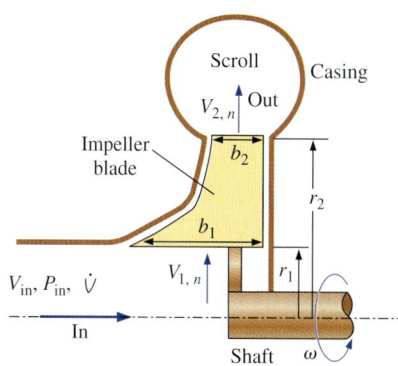

FIGURE 14–35
Close-up side view of the simplified centrifugal flow pump used for elementary analysis of the velocity vectors; $V_{1,n}$ and $V_{2,n}$ are defined as the average normal (radial) components of velocity at radii r_1 and r_2, respectively.

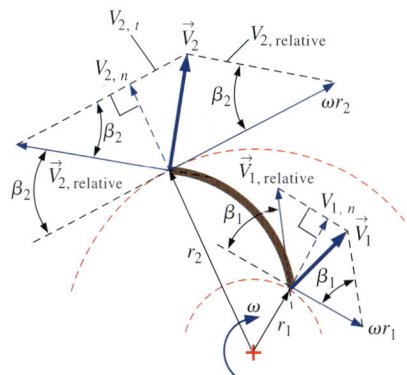

FIGURE 14–36
Close-up frontal view of the simplified centrifugal flow pump used for elementary analysis of the velocity vectors. Absolute velocity vectors of the fluid are shown as bold arrows. It is assumed that the flow is everywhere tangent to the blade surface when viewed from a reference frame rotating with the blade, as indicated by the relative velocity vectors.

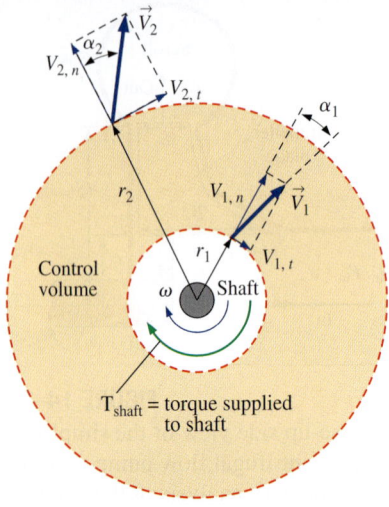

FIGURE 14–37
Control volume (shaded) used for angular momentum analysis of a centrifugal pump; absolute tangential velocity components $V_{1,t}$ and $V_{2,t}$ are labeled.

enters the control volume with uniform absolute velocity $\vec{V}_1$ around the entire circumference at radius r_1 and exits with uniform absolute velocity $\vec{V}_2$ around the entire circumference at radius r_2.

Since moment of momentum is defined as the cross product $\vec{r} \times \vec{V}$, only the *tangential* components of $\vec{V}_1$ and $\vec{V}_2$ are relevant to the shaft torque. These are shown as $V_{1,t}$ and $V_{2,t}$ in Fig. 14–37. It turns out that shaft torque is equal to the change in moment of momentum from inlet to outlet, as given by the **Euler turbomachine equation** (also called **Euler's turbine formula**), derived in Chap. 6,

Euler turbomachine equation: $\qquad T_{shaft} = \rho \dot{V}(r_2 V_{2,t} - r_1 V_{1,t}) \qquad$ (14–14)

Or, in terms of angles α_1 and α_2 and the magnitudes of the absolute velocity vectors,

Alternative form, Euler turbomachine equation:
$$T_{shaft} = \rho \dot{V}(r_2 V_2 \sin \alpha_2 - r_1 V_1 \sin \alpha_1) \qquad (14\text{–}15)$$

In our simplified analysis there are no irreversible losses. Hence, pump efficiency $\eta_{pump} = 1$, implying that water horsepower $\dot{W}_{water\ horsepower}$ and brake horsepower bhp are the same. Using Eqs. 14–3 and 14–4,

$$bhp = \omega T_{shaft} = \rho \omega \dot{V}(r_2 V_{2,t} - r_1 V_{1,t}) = \dot{W}_{water\ horsepower} = \rho g \dot{V} H \qquad (14\text{–}16)$$

which is solved for net head H,

Net head: $\qquad H = \dfrac{1}{g}(\omega r_2 V_{2,t} - \omega r_1 V_{1,t}) \qquad$ (14–17)

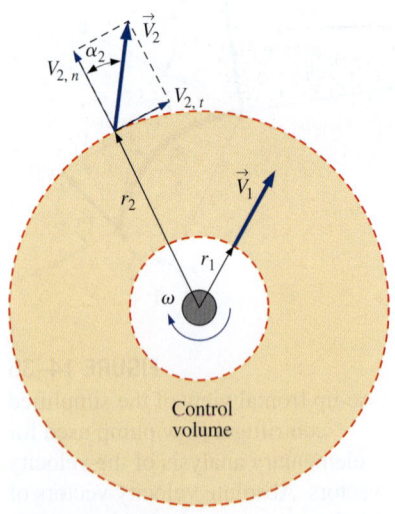

FIGURE 14–38
Control volume and absolute velocity vectors for the centrifugal blower of Example 14–5. The view is along the blower axis.

> **EXAMPLE 14–5** Idealized Blower Performance
>
> A centrifugal blower rotates at $\dot{n} = 1750$ rpm (183.3 rad/s). Air enters the impeller normal to the blades ($\alpha_1 = 0°$) and exits at an angle of 40° from radial ($\alpha_2 = 40°$) as sketched in Fig. 14–38. The inlet radius is $r_1 = 4.0$ cm, and the inlet blade width $b_1 = 5.2$ cm. The outlet radius is $r_2 = 8.0$ cm, and the outlet blade width $b_2 = 2.3$ cm. The volume flow rate is 0.13 m³/s. For the idealized case, i.e., 100 percent efficiency, calculate the net head produced by this blower in equivalent millimeters of water column height. Also calculate the required brake horsepower in watts.
>
> **SOLUTION** We are to calculate the brake horsepower and net head of an idealized blower at a given volume flow rate and rotation rate.
> **Assumptions** 1 The flow is steady in the mean. 2 There are no leaks in the gaps between rotor blades and blower casing. 3 The air flow is incompressible. 4 The efficiency of the blower is 100 percent (no irreversible losses).
> **Properties** We take the density of air to be $\rho_{air} = 1.20$ kg/m³.
> **Analysis** Since the volume flow rate (capacity) is given, we calculate the normal velocity components at the inlet using Eq. 14–12,
>
> $$V_{1,n} = \dfrac{\dot{V}}{2\pi r_1 b_1} = \dfrac{0.13 \text{ m}^3/\text{s}}{2\pi(0.040 \text{ m})(0.052 \text{ m})} = 9.947 \text{ m/s} \qquad (1)$$
>
> $V_1 = V_{1,n}$, and $V_{1,t} = 0$, since $\alpha_1 = 0°$. Similarly, $V_{2,n} = 11.24$ m/s, and
>
> $$V_{2,t} = V_{2,n} \tan \alpha_2 = (11.24 \text{ m/s}) \tan(40°) = 9.435 \text{ m/s} \qquad (2)$$

Now we use Eq. 14–17 to predict the net head,

$$H = \frac{\omega}{g}(r_2 V_{2,t} - r_1 \underbrace{V_{1,t}}_{0}) = \frac{183.3 \text{ rad/s}}{9.81 \text{ m/s}^2}(0.080 \text{ m})(9.435 \text{ m/s}) = 14.1 \text{ m} \quad (3)$$

Note that the net head of Eq. 3 is in meters of *air*, the pumped fluid. To convert to pressure in units of equivalent millimeters of water column, we multiply by the ratio of air density to water density,

$$H_{\text{water column}} = H \frac{\rho_{\text{air}}}{\rho_{\text{water}}}$$

$$= (14.1 \text{ m}) \frac{1.20 \text{ kg/m}^3}{998 \text{ kg/m}^3}\left(\frac{1000 \text{ mm}}{1 \text{ m}}\right) = \mathbf{17.0 \text{ mm of water}} \quad (4)$$

Finally, we use Eq. 14–16 to predict the required brake horsepower,

$$\text{bhp} = \rho g \dot{V} H = (1.20 \text{ kg/m}^3)(9.81 \text{ m/s}^2)(0.13 \text{ m}^3/\text{s})(14.1 \text{ m})\left(\frac{\text{W} \cdot \text{s}}{\text{kg} \cdot \text{m/s}^2}\right)$$

$$= \mathbf{21.6 \text{ W}} \quad (5)$$

Discussion Note the unit conversion in Eq. 5 from kilograms, meters, and seconds to watts; this conversion turns out to be useful in many turbomachinery calculations. The actual net head delivered to the air will be lower than that predicted by Eq. 3 due to inefficiencies. Similarly, actual brake horsepower will be higher than that predicted by Eq. 5 due to inefficiencies in the blower, friction on the shaft, etc.

In order to design the shape of the impeller blades, we must use trigonometry to obtain expressions for $V_{1,t}$ and $V_{2,t}$ in terms of blade angles β_1 and β_2. Applying the *law of cosines* (Fig. 14–39) to the triangle in Fig. 14–36 formed by absolute velocity vector $\vec{V}_2$, relative velocity vector $\vec{V}_{2,\text{relative}}$, and the tangential velocity of the blade at radius r_2 (of magnitude ωr_2) we get

$$V_2^2 = V_{2,\text{relative}}^2 + \omega^2 r_2^2 - 2\omega r_2 V_{2,\text{relative}} \cos \beta_2 \quad (14\text{–}18)$$

But we also see from Fig. 14–36 that

$$V_{2,\text{relative}} \cos \beta_2 = \omega r_2 - V_{2,t}$$

Substitution of this equation into Eq. 14–18 yields

$$\omega r_2 V_{2,t} = \frac{1}{2}(V_2^2 - V_{2,\text{relative}}^2 + \omega^2 r_2^2) \quad (14\text{–}19)$$

A similar equation results for the blade inlet (change all subscripts 2 in Eq. 14–19 to subscript 1). Substitution of these into Eq. 14–17 yields

Net head: $\quad H = \frac{1}{2g}[(V_2^2 - V_1^2) + (\omega^2 r_2^2 - \omega^2 r_1^2) - (V_{2,\text{relative}}^2 - V_{1,\text{relative}}^2)] \quad (14\text{–}20)$

In words, Eq. 14–20 states that in the ideal case (no irreversible losses), the net head is proportional to the change in absolute kinetic energy plus the rotor-tip kinetic energy change minus the change in relative kinetic energy from inlet to outlet of the impeller. Finally, equating Eq. 14–20 and Eq. 14–2,

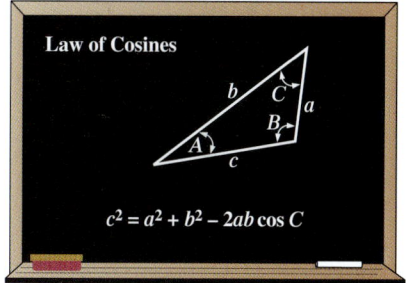

FIGURE 14–39
The law of cosines is utilized in the analysis of a centrifugal pump.

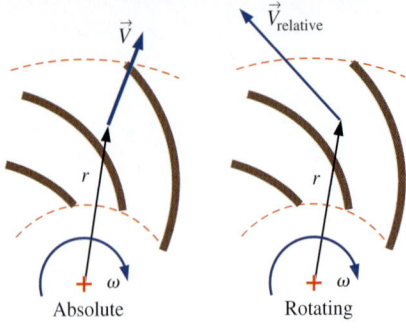

FIGURE 14–40
For the approximation of flow through an impeller with no irreversible losses, it is often more convenient to work with a relative frame of reference rotating with the impeller; in that case, the Bernoulli equation gets an additional term, as indicated in Eq. 14–22.

where we set subscript 2 as the outflow and subscript 1 as the inflow, we see that

$$\left(\frac{P}{\rho g} + \frac{V_{relative}^2}{2g} - \frac{\omega^2 r^2}{2g} + z\right)_{out} = \left(\frac{P}{\rho g} + \frac{V_{relative}^2}{2g} - \frac{\omega^2 r^2}{2g} + z\right)_{in} \quad (14\text{–}21)$$

Note that we are not limited to analysis of only the inlet and outlet. In fact, we may apply Eq. 14–21 to *any* two radii along the impeller. In general then, we write an equation that is commonly called the **Bernoulli equation in a rotating reference frame**:

$$\frac{P}{\rho g} + \frac{V_{relative}^2}{2g} - \frac{\omega^2 r^2}{2g} + z = \text{constant} \quad (14\text{–}22)$$

We see that Eq. 14–22 is the same as the usual Bernoulli equation, except that since the speed used is the *relative* speed (in the rotating reference frame), an "extra" term (the third term on the left of Eq. 14–22) appears in the equation to account for rotational effects (Fig. 14–40). We emphasize that Eq. 14–22 is an approximation, valid only for the ideal case in which there are no irreversible losses through the impeller. Nevertheless, it is valuable as a first-order approximation for flow through the impeller of a centrifugal pump.

We now examine Eq. 14–17, the equation for net head, more closely. Since the term containing $V_{1,t}$ carries a negative sign, we obtain the maximum H by setting $V_{1,t}$ to zero. (We are assuming that there is no mechanism in the eye of the pump that can generate a *negative* value of $V_{1,t}$.) Thus, a first-order approximation for the **design condition** of the pump is to set $V_{1,t} = 0$. In other words, we select the blade inlet angle β_1 such that the flow into the impeller blade is purely radial from an absolute reference frame, and $V_{1,n} = V_1$. The velocity vectors at $r = r_1$ in Fig. 14–36 are magnified and redrawn in Fig. 14–41. Using some trigonometry we see that

$$V_{1,t} = \omega r_1 - \frac{V_{1,n}}{\tan \beta_1} \quad (14\text{–}23)$$

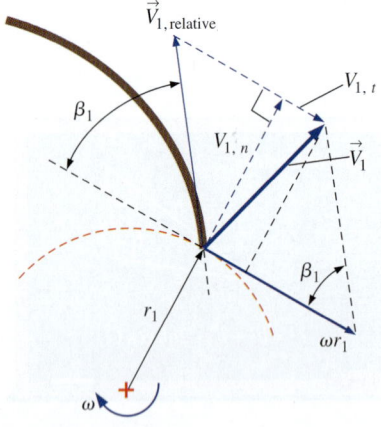

FIGURE 14–41
Close-up frontal view of the velocity vectors at the impeller blade inlet. The absolute velocity vector is shown as a bold arrow.

A similar expression is obtained for $V_{2,t}$ (replace subscript 1 by 2), or in fact for any radius between r_1 and r_2. When $V_{1,t} = 0$ and $V_{1,n} = V_1$,

$$\omega r_1 = \frac{V_{1,n}}{\tan \beta_1} \quad (14\text{–}24)$$

Finally, combining Eq. 14–24 with Eq. 14–12, we have an expression for volume flow rate as a function of inlet blade angle β_1 and rotational speed,

$$\dot{V} = 2\pi b_1 \omega r_1^2 \tan \beta_1 \quad (14\text{–}25)$$

Equation 14–25 can be used for preliminary design of the impeller blade shape as illustrated by Example 14–6.

EXAMPLE 14–6   Preliminary Design of a Centrifugal Pump

A centrifugal pump is being designed to pump liquid refrigerant R-134a at room temperature and atmospheric pressure. The impeller inlet and outlet radii are $r_1 = 100$ and $r_2 = 180$ mm, respectively (Fig. 14–42). The impeller inlet and outlet widths are $b_1 = 50$ and $b_2 = 30$ mm (into the page of Fig. 14–42). The pump is to deliver 0.25 m³/s of the liquid at a net head of 14.5 m when the impeller rotates at 1720 rpm. Design the blade shape for the case in which these operating conditions are the *design conditions* of the pump ($V_{1,t} = 0$, as sketched in the figure); specifically, calculate angles β_1 and β_2, and discuss the shape of the blade. Also predict the horsepower required by the pump.

SOLUTION For a given flow rate, net head, and dimensions of a centrifugal pump, we are to design the blade shape (leading and trailing edge angles). We are also to estimate the horsepower required by the pump.
Assumptions 1 The flow is steady. 2 The liquid is incompressible. 3 There are no irreversible losses through the impeller. 4 This is only a preliminary design.
Properties For refrigerant R-134a at $T = 20°C$, $v_f = 0.0008157$ m³/kg. Thus $\rho = 1/v_f = 1226$ kg/m³.
Analysis We calculate the required water horsepower from Eq. 14–3,

$$\dot{W}_{\text{water horsepower}} = \rho g \dot{V} H$$

$$= (1226 \text{ kg/m}^3)(9.81 \text{ m/s}^2)(0.25 \text{ m}^3/\text{s})(14.5 \text{ m})\left(\frac{\text{W}\cdot\text{s}}{\text{kg}\cdot\text{m/s}^2}\right)$$

$$= 43{,}600 \text{ W}$$

The required brake horsepower will be greater than this in a real pump. However, in keeping with the approximations for this preliminary design, we assume 100 percent efficiency such that bhp is approximately equal to $\dot{W}_{\text{water horsepower}}$,

$$\text{bhp} \cong \dot{W}_{\text{water horsepower}} = 43{,}600 \text{ W}\left(\frac{\text{hp}}{745.7 \text{ W}}\right) = 58.5 \text{ hp}$$

We report the final result to two significant digits in keeping with the precision of the given quantities; thus, bhp ≈ **59 horsepower.**

In all calculations with rotation, we need to convert the rotational speed from $\dot{n}$ (rpm) to ω (rad/s), as illustrated in Fig. 14–43,

$$\omega = 1720 \frac{\text{rot}}{\text{min}}\left(\frac{2\pi \text{ rad}}{\text{rot}}\right)\left(\frac{1 \text{ min}}{60 \text{ s}}\right) = 180.1 \text{ rad/s} \qquad (1)$$

We calculate the blade inlet angle using Eq. 14–25,

$$\beta_1 = \arctan\left(\frac{\dot{V}}{2\pi b_1 \omega r_1^2}\right) = \arctan\left(\frac{0.25 \text{ m}^3/\text{s}}{2\pi(0.050 \text{ m})(180.1 \text{ rad/s})(0.10 \text{ m})^2}\right) = 23.8°$$

We find β_2 by utilizing the equations derived earlier for our elementary analysis. First, for the design condition in which $V_{1,t} = 0$, Eq. 14–17 reduces to

Net head: $$H = \frac{1}{g}(\omega r_2 V_{2,t} - \omega r_1 \underbrace{V_{1,t}}_{0}) = \frac{\omega r_2 V_{2,t}}{g}$$

from which we calculate the tangential velocity component,

$$V_{2,t} = \frac{gH}{\omega r_2} \qquad (2)$$

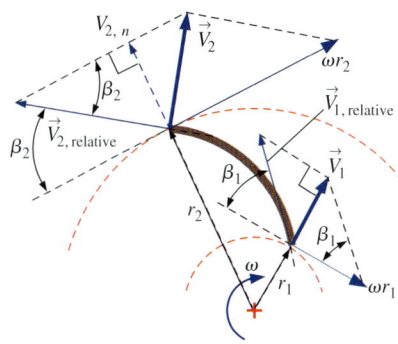

FIGURE 14–42
Relative and absolute velocity vectors and geometry for the centrifugal pump impeller design of Example 14–6.

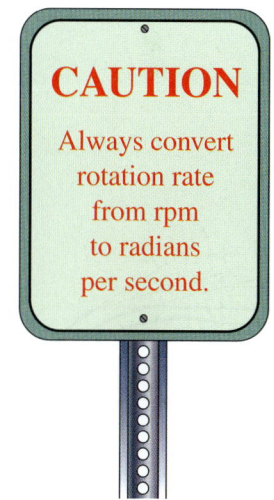

FIGURE 14–43
Proper unit conversion requires the units of rotation rate to be rad/s.

Using Eq. 14–12, we calculate the normal velocity component,

$$V_{2,n} = \frac{\dot{V}}{2\pi r_2 b_2} \quad (3)$$

Next, we perform the same trigonometry used to derive Eq. 14–23, but on the *trailing edge* of the blade rather than the leading edge. The result is

$$V_{2,t} = \omega r_2 - \frac{V_{2,n}}{\tan \beta_2}$$

from which we finally solve for β_2,

$$\beta_2 = \arctan\left(\frac{V_{2,n}}{\omega r_2 - V_{2,t}}\right) \quad (4)$$

After substitution of Eqs. 2 and 3 into Eq. 4, and insertion of the numerical values, we obtain

$$\beta_2 = 14.7°$$

We report the final results to only two significant digits. Thus our preliminary design requires *backward-inclined* impeller blades with $\beta_1 \cong \mathbf{24°}$ and $\beta_2 \cong \mathbf{15°}$.

Once we know the leading and trailing edge blade angles, we design the detailed *shape* of the impeller blade by smoothly varying blade angle β from β_1 to β_2 as radius increases from r_1 to r_2. As sketched in Fig. 14–44, the blade can be of various shapes while still keeping $\beta_1 \cong 24°$ and $\beta_2 \cong 15°$, depending on how we vary β with the radius. In the figure, all three blades begin at the same location (zero absolute angle) at radius r_1; the leading edge angle for all three blades is $\beta_1 = 24°$. The medium length blade (the brown one in Fig. 14–44) is constructed by varying β *linearly* with r. Its trailing edge intercepts radius r_2 at an absolute angle of approximately 93°. The longer blade (the black one in the figure) is constructed by varying β more rapidly near r_1 than near r_2. In other words, the blade curvature is more pronounced near its leading edge than near its trailing edge. It intercepts the outer radius at an absolute angle of about 114°. Finally, the shortest blade (the blue blade in Fig. 14–44) has less blade curvature near its leading edge, but more pronounced curvature near its trailing edge. It intercepts r_2 at an absolute angle of approximately 77°. *It is not immediately obvious which blade shape is best.*

Discussion Keep in mind that this is a preliminary design in which irreversible losses are ignored. A real pump would have losses, and the required brake horsepower would be higher (perhaps 20 to 30 percent higher) than the value estimated here. In a real pump with losses, a shorter blade has less skin friction drag, but the normal stresses on the blade are larger because the flow is turned more sharply near the trailing edge where the velocities are largest; this may lead to structural problems if the blades are not very thick, especially when pumping dense liquids. A longer blade has higher skin friction drag, but lower normal stresses. In addition, you can see from a simple blade volume estimate in Fig. 14–44 that for the same number of blades, the longer the blades, the more flow blockage, since the blades are of finite thickness. In addition, the displacement thickness effect of boundary layers growing along the blade surfaces (Chap. 10) leads to even more pronounced blockage for the long blades. Obviously some engineering optimization is required to determine the exact shape of the blade.

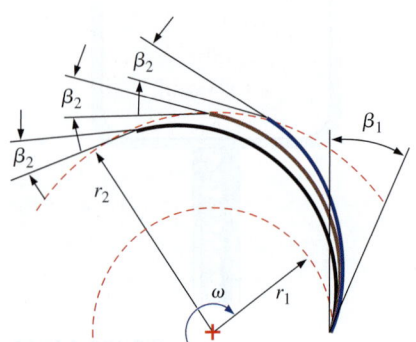

FIGURE 14–44
Three possible blade shapes for the centrifugal pump impeller design of Example 14–6. All three blades have leading edge angle $\beta_1 = 24°$ and trailing edge angle $\beta_2 = 15°$, but differ in how β is varied with the radius. The drawing is to scale.

How *many* blades should we use in an impeller? If we use too few blades, **circulatory flow loss** will be high. Circulatory flow loss occurs because there is a finite number of blades. Recall that in our preliminary analysis, we assume a uniform tangential velocity $V_{2,t}$ around the entire circumference of the outlet of the control volume (Fig. 14–37). This is strictly correct only if we have an infinite number of infinitesimally thin blades. In a real pump, of course, the number of blades is finite, and the blades are not infinitesimally thin. As a result, the tangential component of the absolute velocity vector is not uniform, but drops off in the spaces between blades as illustrated in Fig. 14–45a. The net result is an effectively smaller value of $V_{2,t}$, which in turn decreases the actual net head. This loss of net head (and pump efficiency) is called *circulatory flow loss*. On the other hand, if we have too many blades (as in Fig. 14–45b) there will be excessive flow blockage losses and losses due to the growing boundary layers, again leading to nonuniform flow speeds at the outer radius of the pump and lower net head and efficiency. These losses are called **passage losses.** The bottom line is that some engineering optimization is necessary in order to choose both the blade shape and number of blades. Such analysis is beyond the scope of the present text. A quick perusal through the turbomachinery literature shows that 11, 14, and 16 are common numbers of rotor blades for medium-sized centrifugal pumps.

Once we have designed the pump for specified net head and flow rate (design conditions), we can estimate its net head at conditions *away* from design conditions. In other words, keeping b_1, b_2, r_1, r_2, β_1, β_2, and ω fixed, we vary the volume flow rate above and below the design flow rate. We have all the equations: Eq. 14–17 for net head H in terms of absolute tangential velocity components $V_{1,t}$ and $V_{2,t}$, Eq. 14–23 for $V_{1,t}$ and $V_{2,t}$ as functions of absolute normal velocity components $V_{1,n}$ and $V_{2,n}$, and Eq. 14–12 for $V_{1,n}$ and $V_{2,n}$ as functions of volume flow rate $\dot{V}$. In Fig. 14–46 we combine these equations to generate a plot of H versus $\dot{V}$ for the pump designed in Example 14–6. The solid blue line is the predicted performance, based on our preliminary analysis. The predicted performance curve is nearly linear with $\dot{V}$ both above and below design conditions since the term $\omega r_1 V_{1,t}$ in Eq. 14–17 is small compared to the term $\omega r_2 V_{2,t}$. Recall that at the predicted design conditions, we had set $V_{1,t} = 0$. For volume flow rates higher than this, $V_{1,t}$ is predicted by Eq. 14–23 to be *negative*. In keeping with our previous assumptions, however, it is not possible to have negative values of $V_{1,t}$. Thus, the slope of the predicted performance curve changes suddenly beyond the design conditions.

Also sketched in Fig. 14–46 is the *actual* performance of this centrifugal pump. While the predicted performance is close to the actual performance at design conditions, the two curves deviate substantially away from design conditions. At all volume flow rates, the actual net head is *lower* than the predicted net head. This is due to irreversible effects such as friction along blade surfaces, leakage of fluid between the blades and the casing, prerotation (swirl) of fluid in the region of the eye, flow separation on the leading edges of the blades (shock losses) or in the expanding portions of the flow passages, circulatory flow loss, passage loss, and irreversible dissipation of swirling eddies in the volute, among other things.

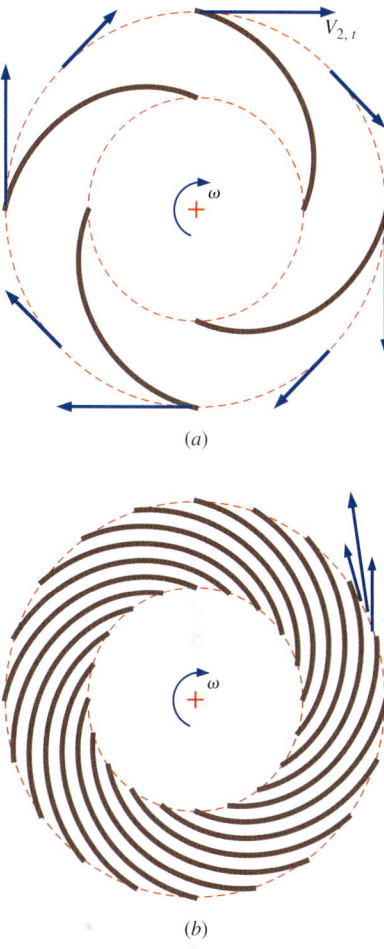

FIGURE 14–45
(*a*) A centrifugal pump impeller with too few blades leads to excessive *circulatory flow loss*—the tangential velocity at outer radius r_2 is smaller in the gaps between blades than at the trailing edges of the blades (absolute tangential velocity vectors are shown). (*b*) On the other hand, since real impeller blades have finite thickness, an impeller with too many blades leads to *passage losses* due to excessive flow blockage and large skin friction drag (velocity vectors in a frame of reference rotating with the impeller are shown exiting one blade row). The bottom line is that pump engineers must optimize both blade shape *and* number of blades.

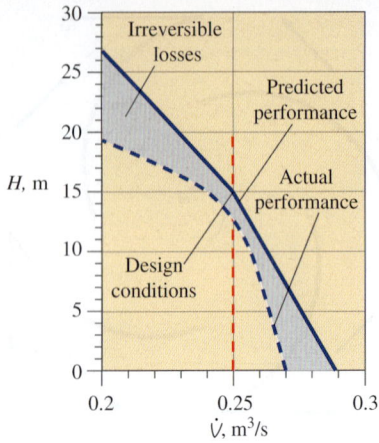

FIGURE 14–46
Net head as a function of volume flow rate for the pump of Example 14-6. The difference between predicted and actual performance is due to unaccounted irreversibilities in the prediction.

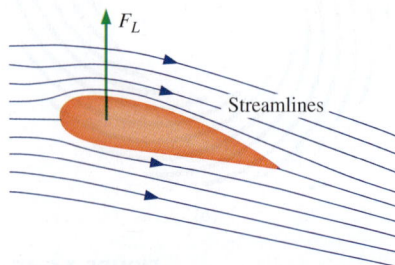

FIGURE 14–47
The blades of an axial-flow pump behave like the wing of an airplane. The air is turned downward by the wing as it generates lift force F_L.

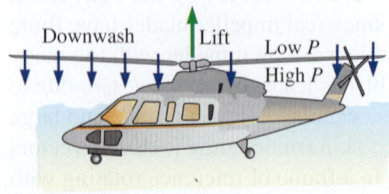

FIGURE 14–48
Downwash and pressure rise across the rotor plane of a helicopter, which is a type of axial-flow pump.

Axial Pumps

Axial pumps do not utilize so-called centrifugal forces. Instead, the impeller blades behave more like the wing of an airplane (Fig. 14–47), producing lift by changing the momentum of the fluid as they rotate. The rotor of a helicopter, for example, is a type of axial-flow pump (Fig. 14–48). The lift force on the blade is caused by pressure differences between the top and bottom surfaces of the blade, and the change in flow direction leads to **downwash** (a column of descending air) through the rotor plane. From a time-averaged perspective, there is a pressure jump across the rotor plane that induces a downward airflow (Fig. 14–48).

Imagine turning the rotor plane vertically; we now have a **propeller** (Fig. 14–49a). Both the helicopter rotor and the airplane propeller are examples of **open axial-flow fans**, since there is no duct or casing around the tips of the blades. The common window fan you install in your bedroom window in the summer operates under the same principles, but the goal is to blow air rather than to provide a force. Be assured, however, that there *is* a net force acting on the fan housing. If air is blown from left to right, the force on the fan acts to the left, and the fan is held down by the window sash. The casing around the house fan also acts as a short duct, which helps to direct the flow and eliminate some losses at the blade tips. The small cooling fan inside your computer is typically an axial-flow fan; it looks like a miniature window fan (Fig. 14–49b) and is an example of a **ducted axial-flow fan**.

If you look closely at the airplane propeller blade in Fig. 14–49a, the rotor blade of a helicopter, the propeller blade of a radio-controlled model airplane, or even the blade of a well-designed window fan, you will notice some **twist** in the blade. Specifically, the airfoil at a cross section near the hub or root of the blade is at a higher **pitch angle** (θ) than the airfoil at a cross section near the tip, $\theta_{root} > \theta_{tip}$ (Fig. 14–50). This is because the tangential speed of the blade increases linearly with radius,

$$u_\theta = \omega r \tag{14-26}$$

At a given radius then, the velocity $\vec{V}_{relative}$ of the air *relative to the blade* is estimated to first order as the vector sum of inlet velocity $\vec{V}_{in}$ and the negative of blade velocity $\vec{V}_{blade}$,

$$\vec{V}_{relative} \cong \vec{V}_{in} - \vec{V}_{blade} \tag{14-27}$$

where the magnitude of $\vec{V}_{blade}$ is equal to the tangential blade speed u_θ, as given by Eq. 14–26. The direction of $\vec{V}_{blade}$ is tangential to the rotational path of the blade. At the blade position sketched in Fig. 14–50, $\vec{V}_{blade}$ is to the left.

In Fig. 14–51 we compute $\vec{V}_{relative}$ graphically using Eq. 14–27 at two radii—the root radius and the tip radius of the rotor blade sketched in Fig. 14–50. As you can see, the relative angle of attack α is the same in either case. In fact, the amount of twist is determined by setting pitch angle θ such that α is the same at any radius.

Note also that the magnitude of the relative velocity $\vec{V}_{relative}$ increases from the root to the tip. It follows that the dynamic pressure encountered by

CHAPTER 14

(a)

(b)

FIGURE 14–49
Axial-flow fans may be open or ducted: (*a*) a propeller is an open fan, and (*b*) a computer cooling fan is a ducted fan.
Photos by John M. Cimbala.

cross sections of the blade increases with radius, and the lift force per unit width into the page in Fig. 14–51 also increases with radius. Propellers tend to be narrower at the root and wider toward the tip in order to take advantage of the larger lift contribution available toward the tip. At the very tip, however, the blade is usually rounded off to avoid excessive *induced drag* (Chap. 11) that would exist if the blade were simply chopped off abruptly as in Fig. 14–50.

Equation 14–27 is not exact for several reasons. First, the rotating motion of the rotor introduces some **swirl** to the airflow (Fig. 14–52). This reduces the effective tangential speed of the blade relative to the incoming air. Second, since the hub of the rotor is of finite size, the air accelerates around it, causing the air speed to increase locally at cross sections of the blade close to the root. Third, the axis of the rotor or propeller may not be aligned exactly parallel to the incoming air. Finally, the air speed itself is not easily determined because it turns out that the air accelerates as it approaches the

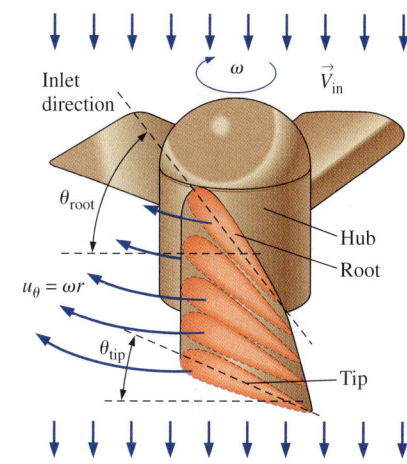

FIGURE 14–50
A well-designed rotor blade or propeller blade has *twist*, as shown by the blue cross-sectional slices through one of the three blades; blade pitch angle θ is higher at the root than at the tip because the tangential speed of the blade increases with radius.

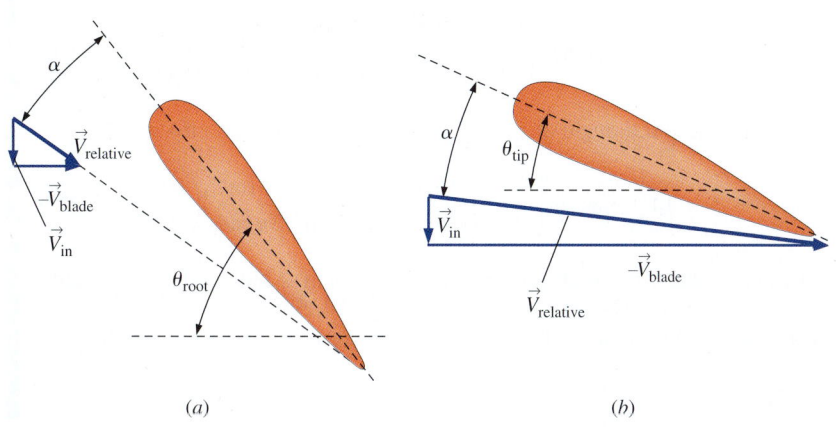

(a)   (b)

FIGURE 14–51
Graphical computation of vector $\vec{V}_{\text{relative}}$ at two radii: (*a*) root, and (*b*) tip of the rotor blade sketched in Fig. 14–50.

818
TURBOMACHINERY

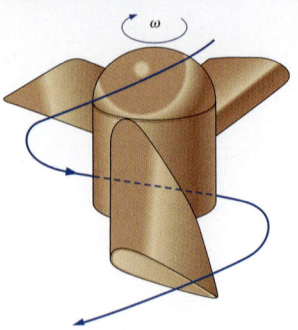

FIGURE 14–52
The rotating blades of a rotor or propeller induce swirl in the surrounding fluid.

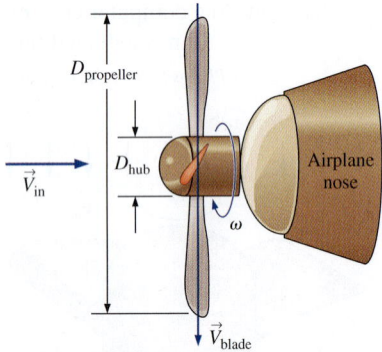

FIGURE 14–53
Setup for the design of the model airplane propeller of Example 14–7, not to scale.

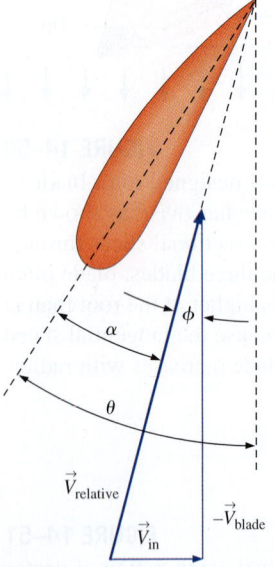

FIGURE 14–54
Velocity vectors at some arbitrary radius r of the propeller of Example 14–7.

whirling rotor. There are methods available to approximate these and other secondary effects, but they are beyond the scope of the present text. The first-order approximation given by Eq. 14–27 is adequate for preliminary rotor and propeller design, as illustrated in Example 14–7.

EXAMPLE 14–7   Calculation of Twist in an Airplane Propeller

Suppose you are designing the propeller of a radio-controlled model airplane. The overall diameter of the propeller is 34.0 cm, and the hub assembly diameter is 5.5 cm (Fig. 14–53). The propeller rotates at 1700 rpm, and the airfoil chosen for the propeller cross section achieves its maximum efficiency at an angle of attack of 14°. When the airplane flies at 30 mi/h (13.4 m/s), calculate the blade pitch angle from the root to the tip of the blade such that $\alpha = 14°$ everywhere along the propeller blade.

SOLUTION   We are to calculate blade pitch angle θ from the root to the tip of the propeller such that the angle of attack is $\alpha = 14°$ at every radius along the propeller blade.
Assumptions   **1** The air at these low speeds is incompressible. **2** We neglect the secondary effects of swirl and acceleration of the air as it approaches the propeller; i.e., the magnitude of $\vec{V}_{in}$ is approximated to be equal to the speed of the aircraft. **3** The airplane flies level, such that the propeller axis is parallel to the incoming air velocity.
Analysis   The velocity of the air relative to the blade is approximated to first order at any radius by using Eq. 14–27. A sketch of the velocity vectors at some arbitrary radius r is shown in Fig. 14–54. From the geometry we see that

Pitch angle at arbitrary radius r: $\quad \theta = \alpha + \phi \quad$ (1)

and

$$\phi = \arctan \frac{|\vec{V}_{in}|}{|\vec{V}_{blade}|} = \arctan \frac{|\vec{V}_{in}|}{\omega r} \quad (2)$$

where we have also used Eq. 14–26 for the blade speed at radius r. At the root ($r = D_{hub}/2 = 2.75$ cm), Eq. 2 becomes

$$\theta = \alpha + \phi = 14° + \arctan\left[\frac{13.4 \text{ m/s}}{(1700 \text{ rot/min})(0.0275 \text{ m})}\left(\frac{1 \text{ rot}}{2\pi \text{ rad}}\right)\left(\frac{60 \text{ s}}{\text{min}}\right)\right] = \mathbf{83.9°}$$

Similarly, the pitch angle at the tip ($r = D_{propeller}/2 = 17.0$ cm) is

$$\theta = \alpha + \phi = 14° + \arctan\left[\frac{13.4 \text{ m/s}}{(1700 \text{ rot/min})(0.17 \text{ m})}\left(\frac{1 \text{ rot}}{2\pi \text{ rad}}\right)\left(\frac{60 \text{ s}}{\text{min}}\right)\right] = \mathbf{37.9°}$$

At radii between the root and the tip, Eqs. 1 and 2 are used to calculate θ as a function of r. Results are plotted in Fig. 14–55.
Discussion   The pitch angle is not linear because of the arctangent function in Eq. 2.

Airplane propellers have **variable pitch,** meaning that the pitch of the entire blade can be adjusted by rotating the blades through mechanical linkages in the hub. For example, when a propeller-driven airplane is sitting at the airport, warming up its engines at high rpm, why does it not start moving? Well, for one thing, the brakes are being applied. But more importantly, propeller pitch is adjusted so that the average angle of attack of the airfoil cross sections is nearly zero—little or no net thrust is provided. While the airplane taxies to the runway, the pitch is adjusted so as to produce a small amount of thrust. As the plane takes off, the engine rpm is high, and the blade pitch is adjusted such that the propeller delivers maximum thrust. In many cases the pitch can even be adjusted "backward" (negative angle of attack) to provide **reverse thrust** to slow down the airplane after landing.

We plot qualitative performance curves for a typical propeller fan in Fig. 14–56. Unlike centrifugal fans, brake horsepower tends to *decrease* with flow rate. In addition, the efficiency curve leans more to the right compared to that of centrifugal fans (see Fig. 14–8). The result is that efficiency drops off rapidly for volume flow rates higher than that at the best efficiency point. The net head curve also decreases continuously with flow rate (although there are some wiggles), and its shape is much different than that of a centrifugal flow fan. If the head requirements are not severe, propeller fans can be operated beyond the point of maximum efficiency to achieve higher volume flow rates. Since bhp decreases at high values of $\dot{V}$, there is not a power penalty when the fan is run at high flow rates. For this reason it is tempting to install a slightly *undersized* fan and push it beyond its best efficiency point. At the other extreme, if operated *below* its maximum efficiency point, the flow may be noisy and unstable, which indicates that the fan may be *oversized* (larger than necessary). For these reasons, it is usually best to run a propeller fan at, or slightly above, its maximum efficiency point.

When used to move flow in a duct, a single-impeller axial-flow fan is called a **tube-axial fan** (Fig. 14–57a). In many practical engineering applications of axial-flow fans, such as exhaust fans in kitchens, building ventilation duct fans, fume hood fans, and automotive radiator cooling fans, the swirling flow produced by the rotating blades (Fig. 14–57a) is of no concern. But the swirling motion and increased turbulence intensity can continue for quite some distance downstream, and there are applications where swirl (or its affiliated noise and turbulence) is highly undesirable. Examples include wind tunnel fans, torpedo fans, and some specialized mine shaft ventilation fans. There are two basic designs that largely eliminate swirl: A second rotor that rotates in the *opposite direction* can be added in series with the existing rotor to form a pair of counter-rotating rotor blades; such a fan is called a **counter-rotating axial-flow fan** (Fig. 14–57b). The swirl caused by the upstream rotor is cancelled by an opposite swirl caused by the downstream rotor. Alternatively, a set of **stator blades** can be added either upstream or downstream of the rotating impeller. As implied by their name, stator blades are *stationary* (nonrotating) guide vanes that simply redirect the fluid. An axial-flow fan with a set of rotor blades (the **impeller** or the **rotor**) *and* a set of stator blades called **vanes** (the **stator**) is called a **vane-axial fan** (Fig. 14–57c). The stator blade design of the vane-axial fan is

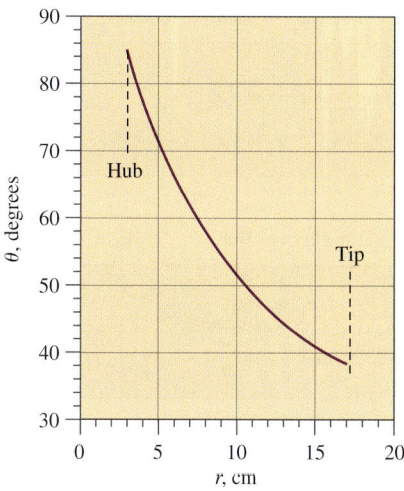

FIGURE 14–55
Blade pitch angle as a function of radius for the propeller of Example 14–7.

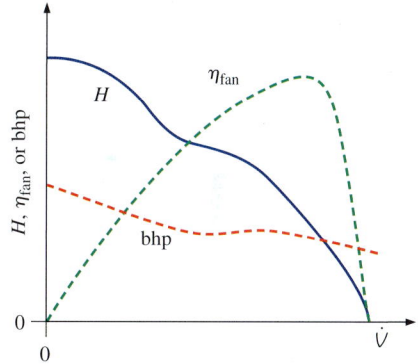

FIGURE 14–56
Typical *fan performance curves* for a propeller (axial-flow) fan.

820
TURBOMACHINERY

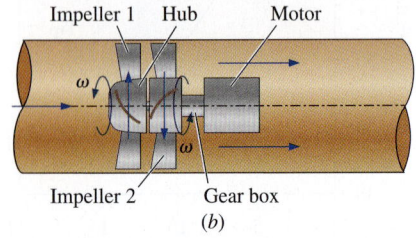

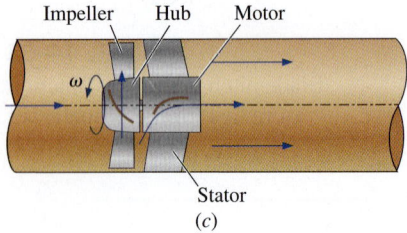

FIGURE 14–57
A *tube-axial fan* (*a*) imparts swirl to the exiting fluid, while (*b*) a *counter-rotating axial-flow fan* and (*c*) a *vane-axial fan* are designed to remove the swirl.

much simpler and less expensive to implement than is the counter-rotating axial-flow fan design.

The swirling fluid downstream of a tube-axial fan wastes kinetic energy and has a high level of turbulence; the vane-axial fan partially recovers this wasted kinetic energy and reduces the level of turbulence. Vane-axial fans are thus both quieter and more energy efficient than tube-axial fans. A properly designed counter-rotating axial-flow fan may be even quieter and more energy efficient. Furthermore, since there are two sets of rotating blades, a higher pressure rise can be obtained with the counter-rotating design. The construction of a counter-rotating axial-flow fan is more complex, of course, requiring either two synchronized motors or a gear box.

Axial-flow fans can be either belt driven or direct drive. The motor of a direct-drive vane-axial fan is mounted in the middle of the duct. It is common practice (and good design) to use the *stator blades* to provide physical support for the motor. Photographs of a belt-driven tube-axial fan and a direct-drive vane-axial fan are provided in Fig. 14–58. The stator blades of the vane-axial fan can be seen behind (downstream of) the rotor blades in Fig. 14–58*b*. An alternative design is to place the stator blades *upstream* of the impeller, imparting **preswirl** to the fluid. The swirl caused by the rotating impeller blades then removes this preswirl.

It is fairly straightforward to design the shape of the blades in all these axial-flow fan designs, at least to first order. For simplicity, we assume thin blades (e.g., blades made out of sheet metal) rather than airfoil-shaped blades. Consider, for example, a vane-axial flow fan with rotor blades upstream of stator blades (Fig. 14–59). The distance between the rotor and stator has been exaggerated in this figure to enable velocity vectors to be drawn between the blades. The hub radius of the stator is assumed to be the same as the hub radius of the rotor so that the cross-sectional area of flow remains constant. As we did previously with the propeller, we consider the cross section of one impeller blade as it passes vertically in front of us. Since there are multiple blades, the next blade passes by shortly thereafter. At a chosen radius r, we make the two-dimensional approximation that the

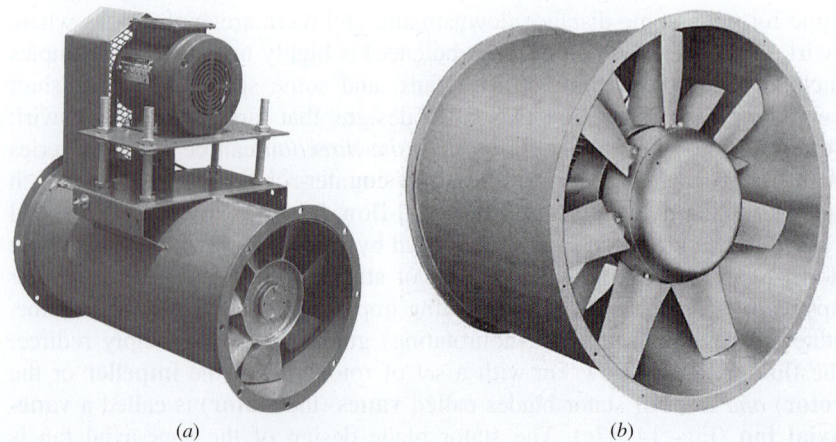

FIGURE 14–58
Axial-flow fans: (*a*) a belt-driven tube-axial fan without stator blades, and (*b*) a direct-drive vane-axial fan with stator blades to reduce swirl and improve efficiency.
(*a*) © PennBarry 2012. Used by permission.
(*b*) Photo courtesy of Howden. Used by permission.

blades pass by as *an infinite series* of two-dimensional blades called a **blade row** or **cascade**. A similar assumption is made for the stator blades, even though they are stationary. Both blade rows are sketched in Fig. 14–59.

In Fig. 14–59b, the velocity vectors are seen from an absolute reference frame, i.e., that of a fixed observer looking horizontally at the vane-axial flow fan. Flow enters from the left at speed V_{in} in the horizontal (axial) direction. The rotor blade row moves at constant speed ωr vertically upward in this reference frame, as indicated. Flow is turned by these moving blades and leaves the trailing edge upward and to the right as indicated in Fig. 14–59b as vector $\vec{V}_{rt}$. (The subscript notation indicates rotor trailing edge.) To find the magnitude and direction of $\vec{V}_{rt}$, we redraw the blade rows and vectors in a *relative* reference frame (the frame of reference of the rotating rotor blade) in Fig. 14–59c. This reference frame is obtained by subtracting the rotor blade velocity (adding a vector of magnitude ωr pointing vertically downward) from all velocity vectors. As shown in Fig. 14–59c, the velocity vector relative to the leading edge of the rotor blade is $\vec{V}_{in,\,relative}$, calculated as the vector sum of $\vec{V}_{in}$ and the downward vector of magnitude ωr. We adjust the pitch of the rotor blade such that $\vec{V}_{in,\,relative}$ is parallel (tangential) to the leading edge of the rotor blade at this cross section.

Flow is turned by the rotor blade. We assume that the flow leaving the rotor blade is parallel to the blade's trailing edge (from the relative reference frame), as sketched in Fig. 14–59c as vector $\vec{V}_{rt,\,relative}$. We also know that the horizontal (axial) component of $\vec{V}_{rt,\,relative}$ must equal $\vec{V}_{in}$ in order to conserve mass. Note that we are assuming incompressible flow and constant flow area normal to the page in Fig. 14–59. Thus, the axial component of velocity must be everywhere equal to V_{in}. This piece of information establishes the magnitude of vector $\vec{V}_{rt,\,relative}$, which is not the same as the magnitude of $\vec{V}_{in,\,relative}$. Returning to the absolute reference frame of Fig. 14–59b, absolute velocity $\vec{V}_{rt}$ is calculated as the vector sum of $\vec{V}_{rt,\,relative}$ and the vertically upward vector of magnitude ωr.

Finally, the stator blade is designed such that $\vec{V}_{rt}$ is parallel to the leading edge of the stator blade. The flow is once again turned, this time by the stator blade. Its trailing edge is horizontal so that the flow leaves axially (without any swirl). The final outflow velocity must be identical to the inflow velocity by conservation of mass if we assume incompressible flow and constant flow area normal to the page. In other words, $\vec{V}_{out} = \vec{V}_{in}$. For completeness, the outflow velocity in the relative reference frame is sketched in Fig. 14–59c. We also see that $\vec{V}_{out,\,relative} = \vec{V}_{in,\,relative}$.

Now imagine repeating this analysis for *all* radii from the hub to the tip. As with the propeller, we would design our blades with some *twist* since the value of ωr increases with radius. A modest improvement in efficiency can be gained at design conditions by using airfoils instead of sheet metal blades; the improvement is more significant at off-design conditions.

If there are, say, seven rotor blades in a vane-axial fan, how many stator blades should there be? You might at first say seven so that the stator matches the rotor—but this would be a very poor design! Why? Because at the instant in time when one blade of the rotor passes directly in front of a stator blade, all six of its brothers would do the same. Each stator blade would simultaneously encounter the disturbed flow in the wake of a rotor blade. The resulting flow would be both pulsating and noisy, and the entire

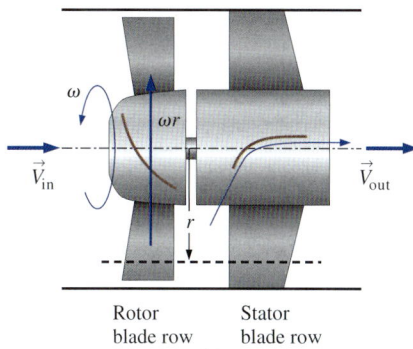
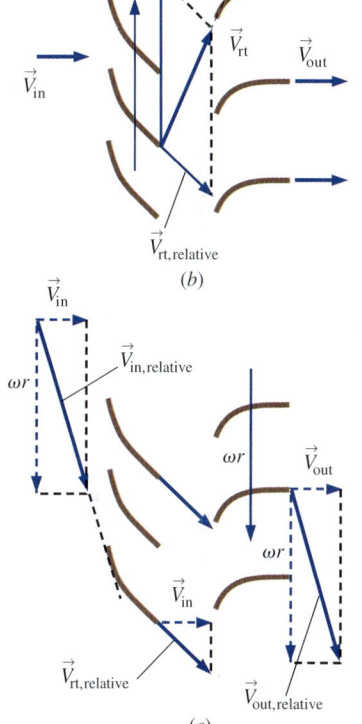

FIGURE 14–59
Analysis of a vane-axial flow fan at radius r using the two-dimensional blade row approximation; (a) overall view, (b) absolute reference frame, and (c) reference frame relative to the rotating rotor blades (impeller).

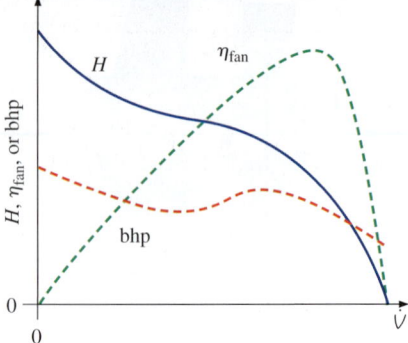

FIGURE 14–60

Typical *fan performance curves* for a vane-axial flow fan.

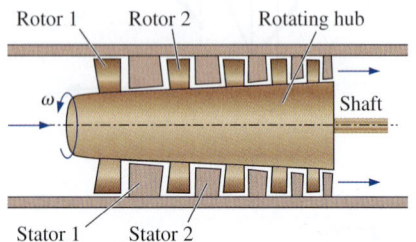

FIGURE 14–61

A multistage axial-flow pump consists of two or more rotor–stator pairs.

unit would vibrate severely. Instead, it is good design practice to choose the number of stator blades such that it has *no common denominator* with the number of rotor blades. Combinations like seven and eight, seven and nine, six and seven, or nine and eleven are good choices. Combinations like eight and ten (common denominator of two) or nine and twelve (common denominator of three) are *not* good choices.

We plot the performance curves of a typical vane-axial flow fan in Fig. 14–60. The general shapes are very similar to those of a propeller fan (Fig. 14–56), and you are referred to the discussion there. After all, a vane-axial flow fan is really the same as a propeller fan or tube-axial flow fan except for the additional stator blades that straighten the flow and tend to smooth out the performance curves.

As discussed previously, an axial-flow fan delivers high volume flow rate, but fairly low pressure rise. Some applications require both high flow rate *and* high pressure rise. In such cases, several stator–rotor pairs can be combined *in series*, typically with a common shaft and common hub (Fig. 14–61). When two or more rotor–stator pairs are combined like this we call it a **multistage axial-flow pump.** A blade row analysis similar to the one of Fig. 14–59 is applied to each successive stage. The details of the analysis can get complicated, however, because of compressibility effects and because the flow area from the hub to the tip may not remain constant. In a **multistage axial-flow compressor,** for example, the flow area decreases downstream. The blades of each successive stage get smaller as the air gets further compressed. In a **multistage axial-flow turbine,** the flow area typically *grows* downstream as pressure is lost in each successive stage of the turbine.

One well-known example of a turbomachine that utilizes both multistage axial-flow compressors and multistage axial-flow turbines is the **turbofan engine** used to power modern commercial airplanes. A cutaway schematic

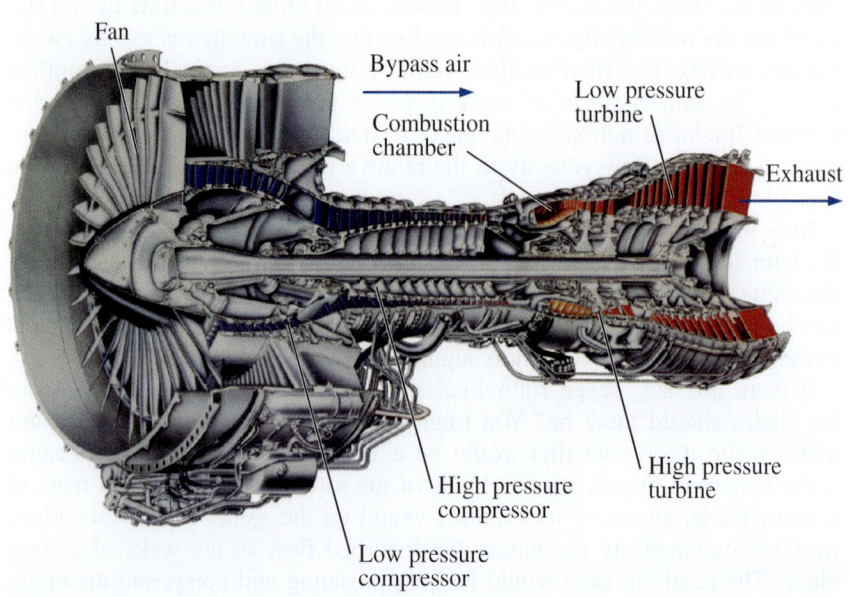

FIGURE 14–62

Pratt & Whitney PW4000 turbofan engine; an example of a multistage axial-flow turbomachine.

Photo courtesy of United Technologies Corporation/Pratt & Whitney. Used by permission. All rights reserved.

diagram of a turbofan engine is shown in Fig. 14–62. Some of the air passes through the fan, which delivers thrust much like a propeller. The rest of the air passes through a low-pressure compressor, a high-pressure compressor, a combustion chamber, a high-pressure turbine, and then finally a low-pressure turbine. The air and products of combustion are then exhausted at high speed to provide even more thrust. Computational fluid dynamics (CFD) codes are obviously quite useful in the design of such complex turbomachines (Chap. 15).

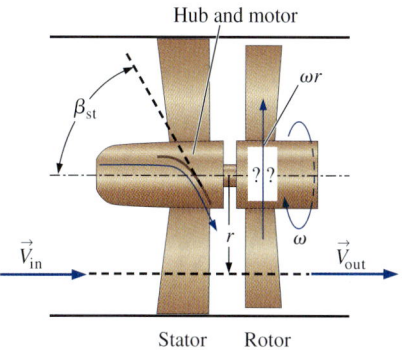

FIGURE 14–63
Schematic diagram of the vane-axial flow fan of Example 14–8. The stator precedes the rotor, and the shape of the rotor blade is unknown—it is to be designed.

■ **EXAMPLE 14–8   Design of a Vane-Axial Flow Fan for a Wind Tunnel**

A vane-axial flow fan is being designed to power a wind tunnel. There must not be any swirl in the flow downstream of the fan. It is decided that the stator blades should be *upstream* of the rotor blades (Fig. 14–63) to protect the impeller blades from damage by objects that might accidentally get blown into the fan. To reduce expenses, both the stator and rotor blades are to be constructed of sheet metal. The leading edge of each stator blade is aligned axially ($\beta_{sl} = 0.0°$) and its trailing edge is at angle $\beta_{st} = 60.0°$ from the axis as shown in the sketch. (The subscript notation "sl" indicates stator leading edge and "st" indicates stator trailing edge.) There are 16 stator blades. At design conditions, the axial-flow speed through the blades is 47.1 m/s, and the impeller rotates at 1750 rpm. At radius $r = 0.40$ m, calculate the leading and trailing edge angles of the rotor blade, and sketch the shape of the blade. How many rotor blades should there be?

SOLUTION For given flow conditions and stator blade shape at a given radius, we are to design the rotor blade. Specifically, we are to calculate the leading and trailing edge angles of the rotor blade and sketch its shape. We are also to decide how many rotor blades to construct.
Assumptions 1 The air is nearly incompressible. 2 The flow area between the hub and tip is constant. 3 Two-dimensional blade row analysis is appropriate.
Analysis First we analyze flow through the stator from an absolute reference frame, using the two-dimensional approximation of a cascade (blade row) of stator blades (Fig. 14–64). Flow enters axially (horizontally) and is turned 60.0° downward. Since the axial component of velocity must remain constant to conserve mass, the magnitude of the velocity leaving the trailing edge of the stator, $\vec{V}_{st}$, is calculated as

$$V_{st} = \frac{V_{in}}{\cos \beta_{st}} = \frac{47.1 \text{ m/s}}{\cos(60.0°)} = 94.2 \text{ m/s} \quad (1)$$

The direction of $\vec{V}_{st}$ is assumed to be that of the stator trailing edge. In other words, we assume that the flow turns nicely through the blade row and exits parallel to the trailing edge of the blade, as shown in Fig. 14–64.
   We convert $\vec{V}_{st}$ to the *relative* reference frame moving with the rotor blades. At a radius of 0.40 m, the tangential velocity of the rotor blades is

$$u_\theta = \omega r = (1750 \text{ rot/min})\left(\frac{2\pi \text{ rad}}{\text{rot}}\right)\left(\frac{1 \text{ min}}{60 \text{ s}}\right)(0.40 \text{ m}) = 73.30 \text{ m/s} \quad (2)$$

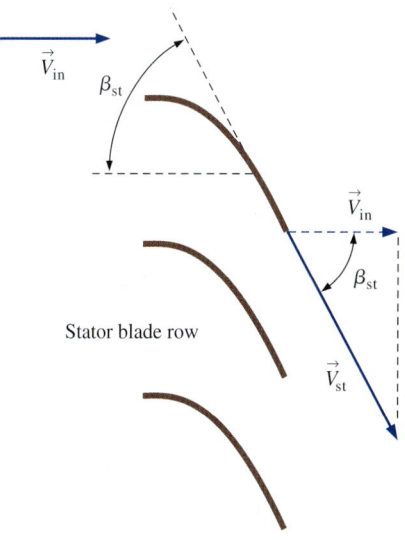

FIGURE 14–64
Velocity vector analysis of the stator blade row of the vane-axial flow fan of Example 14–8; absolute reference frame.

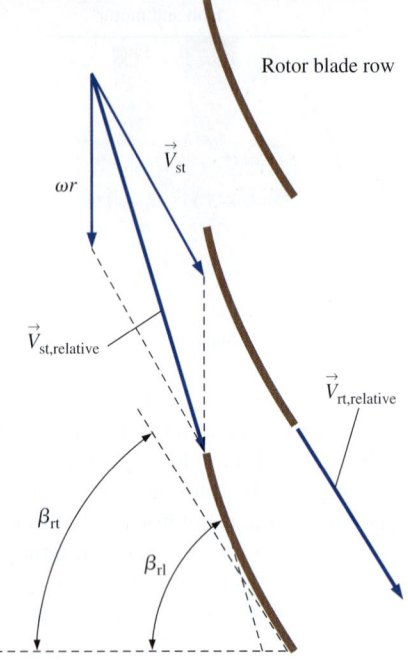

FIGURE 14–65
Analysis of the stator trailing edge velocity of Example 14-8 as it impinges on the rotor leading edge; relative reference frame.

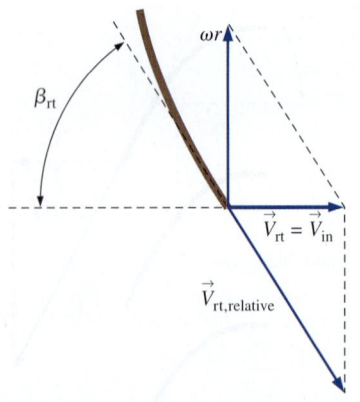

FIGURE 14–66
Analysis of the rotor trailing edge velocity of Example 14-8; absolute reference frame.

Since the rotor blade row moves upward in Fig. 14–63, we add a *downward* velocity with magnitude given by Eq. 2 to translate $\vec{V}_{st}$ into the rotating reference frame sketched in Fig. 14–65. The angle of the leading edge of the rotor, β_{rl}, is calculated by using trigonometry,

$$\beta_{rl} = \arctan \frac{\omega r + V_{in} \tan \beta_{st}}{V_{in}}$$

$$= \arctan \frac{(73.30 \text{ m/s}) + (47.1 \text{ m/s}) \tan(60.0°)}{47.1 \text{ m/s}} = 73.09° \quad (3)$$

The air must now be turned by the rotor blade row in such a way that it leaves the trailing edge of the rotor blade at a zero angle (axially, no swirl) from an absolute reference frame. This determines the rotor's trailing edge angle, β_{rt}. Specifically, when we add an *upward* velocity of magnitude ωr (Eq. 2) to the relative velocity exiting the trailing edge of the rotor, $\vec{V}_{rt,\,relative}$, we convert back to the absolute reference frame, and obtain $\vec{V}_{rt}$, the velocity leaving the rotor trailing edge. It is this velocity, $\vec{V}_{rt}$, that must be axial (horizontal). Furthermore, to conserve mass, $\vec{V}_{rt}$ must equal $\vec{V}_{in}$ since we are assuming incompressible flow. Working backwards, we construct $\vec{V}_{rt,\,relative}$ in Fig. 14–66. Trigonometry reveals that

$$\beta_{rt} = \arctan \frac{\omega r}{V_{in}} = \arctan \frac{73.30 \text{ m/s}}{47.1 \text{ m/s}} = 57.28° \quad (4)$$

We conclude that the rotor blade at this radius has a leading edge angle of about **73.1°** (Eq. 3) and a trailing edge angle of about **57.3°** (Eq. 4). A sketch of the rotor blade at this radius is provided in Fig. 14–65; the total curvature is small, being less than 16° from leading to trailing edge.

Finally, to avoid interaction of the stator blade wakes with the rotor blade leading edges, we choose the number of rotor blades such that it has no common denominator with the number of stator blades. Since there are 16 stator blades, we pick a number like **13, 15,** or **17** rotor blades. Choosing 14 would not be appropriate since it shares a common denominator of 2 with the number 16. Choosing 12 would be worse since it shares both 2 and 4 as common denominators.

Discussion We can repeat the calculation for all radii from hub to tip, completing the design of the entire rotor. There would be twist, as discussed previously.

14–3 · PUMP SCALING LAWS

Dimensional Analysis

Turbomachinery provides a very practical example of the power and usefulness of *dimensional analysis* (Chap. 7). We apply the *method of repeating variables* to the relationship between gravity times net head (gH) and pump properties such as volume flow rate ($\dot{V}$); some characteristic length, typically the diameter of the impeller blades (D); blade surface roughness height (ϵ); and impeller rotational speed (ω), along with fluid properties density (ρ) and viscosity (μ). Note that we treat the group gH as one variable.

The dimensionless Pi groups are shown in Fig. 14–67; the result is the following relationship involving dimensionless parameters:

$$\frac{gH}{\omega^2 D^2} = \text{function of}\left(\frac{\dot{V}}{\omega D^3}, \frac{\rho \omega D^2}{\mu}, \frac{\varepsilon}{D}\right) \quad (14\text{–}28)$$

A similar analysis with input brake horsepower as a function of the same variables results in

$$\frac{\text{bhp}}{\rho \omega^3 D^5} = \text{function of}\left(\frac{\dot{V}}{\omega D^3}, \frac{\rho \omega D^2}{\mu}, \frac{\varepsilon}{D}\right) \quad (14\text{–}29)$$

The second dimensionless parameter (or Π group) on the right side of both Eqs. 14–28 and 14–29 is obviously a *Reynolds number* since ωD is a characteristic velocity,

$$\text{Re} = \frac{\rho \omega D^2}{\mu}$$

The third Π on the right is the *nondimensional roughness parameter*. The three new dimensionless groups in these two equations are given symbols and named as follows:

Dimensionless pump parameters:

$$C_H = \text{Head coefficient} = \frac{gH}{\omega^2 D^2}$$

$$C_Q = \text{Capacity coefficient} = \frac{\dot{V}}{\omega D^3} \quad (14\text{–}30)$$

$$C_P = \text{Power coefficient} = \frac{\text{bhp}}{\rho \omega^3 D^5}$$

Note the subscript Q in the symbol for capacity coefficient. This comes from the nomenclature found in many fluid mechanics and turbomachinery textbooks that Q rather than $\dot{V}$ is the volume flow rate through the pump. We use the notation C_Q for consistency with turbomachinery convention, even though we use $\dot{V}$ for volume flow rate to avoid confusion with heat transfer.

When pumping liquids, cavitation may be of concern, and we need another dimensionless parameter related to the required net positive suction head. Fortunately, we can simply substitute NPSH$_\text{required}$ in place of H in the dimensional analysis, since they have identical dimensions (length). The result is

$$C_{\text{NPSH}} = \text{Suction head coefficient} = \frac{g\text{NPSH}_\text{required}}{\omega^2 D^2} \quad (14\text{–}31)$$

Other variables, such as gap thickness between blade tips and pump housing and blade thickness, can be added to the dimensional analysis if necessary. Fortunately, these variables typically are of only minor importance and are not considered here. In fact, you may argue that two pumps are not even strictly *geometrically similar* unless gap thickness, blade thickness, and surface roughness scale geometrically.

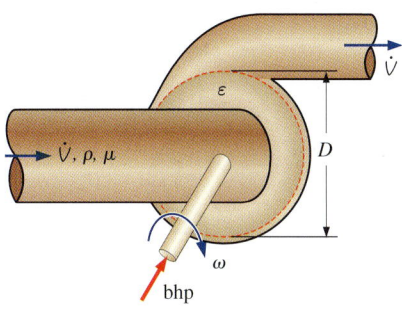

$gH = f(\dot{V}, D, \varepsilon, \omega, \rho, \mu)$
$k = n - j = 7 - 3 = 4$ Π's expected.

$$\Pi_1 = \frac{gH}{\omega^2 D^2} \quad \Pi_2 = \frac{\dot{V}}{\omega D^3}$$

$$\Pi_3 = \frac{\rho \omega D^2}{\mu} \quad \Pi_4 = \frac{\varepsilon}{D}$$

FIGURE 14–67
Dimensional analysis of a pump.

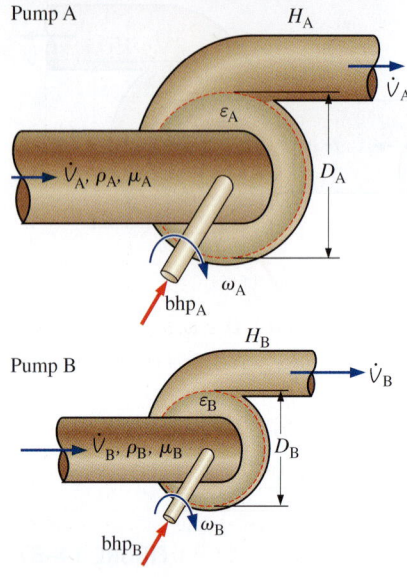

FIGURE 14–68
Dimensional analysis is useful for scaling two *geometrically similar* pumps. If all the dimensionless pump parameters of pump A are equivalent to those of pump B, the two pumps are *dynamically similar*.

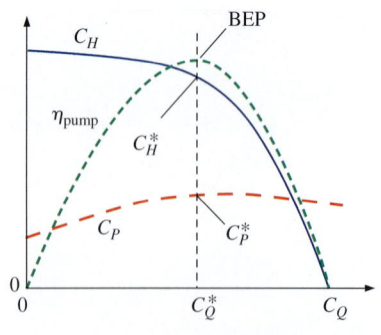

FIGURE 14–69
When plotted in terms of dimensionless pump parameters, the performance curves of all pumps in a family of geometrically similar pumps collapse onto one set of *nondimensional pump performance curves.* Values at the best efficiency point are indicated by asterisks.

Relationships derived by dimensional analysis, such as Eqs. 14–28 and 14–29, are interpreted as follows: If two pumps, A and B, are *geometrically similar* (pump A is geometrically proportional to pump B, although they may be of different sizes), and if the *independent* Π's are equal to each other (in this case if $C_{Q,A} = C_{Q,B}$, $Re_A = Re_B$, and $\varepsilon_A/D_A = \varepsilon_B/D_B$), then the *dependent* Π's are guaranteed to also be equal to each other as well. In particular, $C_{H,A} = C_{H,B}$ from Eq. 14–28 and $C_{P,A} = C_{P,B}$ from Eq. 14–29. If such conditions are established, the two pumps are said to be *dynamically similar* (Fig. 14–68). When dynamic similarity is achieved, the operating point on the pump performance curve of pump A and the corresponding operating point on the pump performance curve of pump B are said to be **homologous.**

The requirement of equality of all three of the independent dimensionless parameters can be relaxed somewhat. If the Reynolds numbers of both pump A and pump B exceed several thousand, turbulent flow conditions exist inside the pump. It turns out that for turbulent flow, if the values of Re_A and Re_B are not equal, but not too far apart, dynamic similarity between the two pumps is still a reasonable approximation. This fortunate condition is due to **Reynolds number independence** (Chap. 7). (Note that if the pumps operate in the *laminar* regime, or at low Re, the Reynolds number must usually remain as a scaling parameter.) In most cases of practical turbomachinery engineering analysis, the effect of differences in the roughness parameter is also small, unless the roughness differences are large, as when one is scaling from a very small pump to a very large pump (or vice versa). Thus, for many practical problems, we may neglect the effect of both Re and ε/D. Equations 14–28 and 14–29 then reduce to

$$C_H \cong \text{function of } C_Q \qquad C_P \cong \text{function of } C_Q \qquad (14\text{--}32)$$

As always, dimensional analysis cannot predict the *shape* of the functional relationships of Eq. 14–32, but once these relationships are obtained for a particular pump, they can be generalized for geometrically similar pumps that are of different diameters, operate at different rotational speeds and flow rates, and operate even with fluids of different density and viscosity.

We transform Eq. 14–5 for pump efficiency into a function of the dimensionless parameters of Eq. 14–30,

$$\eta_{\text{pump}} = \frac{\rho(\dot{V})(gH)}{\text{bhp}} = \frac{\rho(\omega D^3 C_Q)(\omega^2 D^2 C_H)}{\rho \omega^3 D^5 C_P} = \frac{C_Q C_H}{C_P} \cong \text{function of } C_Q \qquad (14\text{--}33)$$

Since η_{pump} is already dimensionless, it is another dimensionless pump parameter all by itself. Note that since Eq. 14–33 reveals that η_{pump} can be formed by the combination of three other Π's, η_{pump} is not *necessary* for pump scaling. It is, however, certainly a *useful* parameter. Since C_H, C_P, and η_{pump} are approximated as functions only of C_Q, we often plot these three parameters as functions of C_Q on the same plot, generating a set of **nondimensional pump performance curves.** An example is provided in Fig. 14–69 for the case of a typical centrifugal pump. The curve shapes for other types of pumps would, of course, be different.

The simplified similarity laws of Eqs. 14–32 and 14–33 break down when the full-scale prototype is significantly larger than its model (Fig. 14–70);

the prototype's performance is generally *better*. There are several reasons for this: The prototype pump often operates at high Reynolds numbers that are not achievable in the laboratory. We know from the Moody chart that the friction factor decreases with Re, as does boundary layer thickness. Hence, the influence of viscous boundary layers is less significant as pump size increases, since the boundary layers occupy a less significant percentage of the flow path through the impeller. In addition, the relative roughness (ϵ/D) on the surfaces of the prototype impeller blades may be significantly smaller than that on the model pump blades unless the model surfaces are micropolished. Finally, large full-scale pumps have smaller tip clearances relative to the blade diameter; therefore, tip losses and leakage are less significant. Some empirical equations have been developed to account for the increase in efficiency between a small model and a full-scale prototype. One such equation was suggested by Moody (1926) for turbines, but it can be used as a first-order correction for pumps as well,

Moody efficiency correction equation for pumps:

$$\eta_{pump,\,prototype} \cong 1 - (1 - \eta_{pump,\,model})\left(\frac{D_{model}}{D_{prototype}}\right)^{1/5} \quad (14\text{-}34)$$

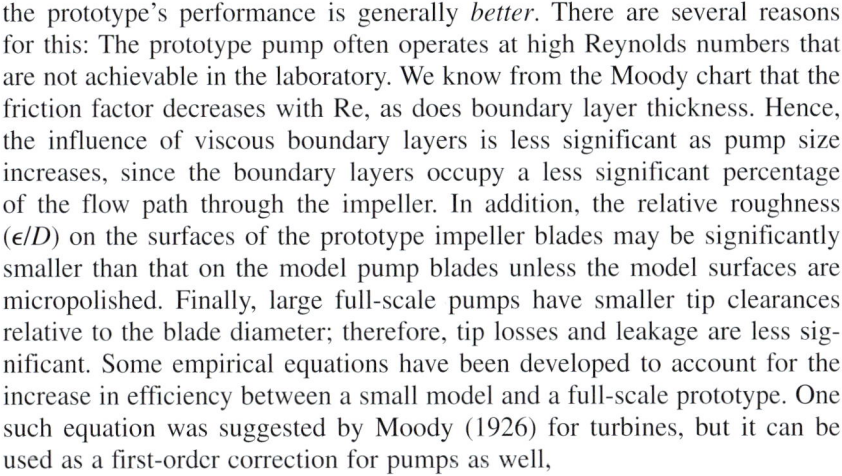

FIGURE 14–70
When a small-scale model is tested to predict the performance of a full-scale prototype pump, the measured efficiency of the model is typically somewhat *lower* than that of the prototype. Empirical correction equations such as Eq. 14–34 have been developed to account for the improvement of pump efficiency with pump size.

Pump Specific Speed

Another useful dimensionless parameter called **pump specific speed** (N_{Sp}) is formed by a combination of parameters C_Q and C_H:

Pump specific speed:
$$N_{Sp} = \frac{C_Q^{1/2}}{C_H^{3/4}} = \frac{(\dot{V}/\omega D^3)^{1/2}}{(gH/\omega^2 D^2)^{3/4}} = \frac{\omega \dot{V}^{1/2}}{(gH)^{3/4}} \quad (14\text{-}35)$$

If all engineers watched their units carefully, N_{Sp} would always be listed as a dimensionless parameter. Unfortunately, practicing engineers have grown accustomed to using inconsistent units in Eq. 14–35, which renders the perfectly fine dimensionless parameter N_{Sp} into a cumbersome dimensional quantity (Fig. 14–71). Further confusion results because some engineers prefer units of rotations per minute (rpm) for rotational speed, while others use rotations per second (Hz), the latter being more common in Europe. In addition, practicing engineers in the United States typically ignore the gravitational constant in the definition of N_{Sp}. In this book, we add subscripts "Eur" or "US" to N_{Sp} in order to distinguish the dimensional forms of pump specific speed from the nondimensional form. In the United States, it is customary to write H in units of feet (net head expressed as an equivalent column height of the fluid being pumped), $\dot{V}$ in units of gallons per minute (gpm), and rotation rate in terms of $\dot{n}$ (rpm) instead of ω (rad/s). Using Eq. 14–35 we define

Pump specific speed, customary U.S. units:
$$N_{Sp,\,US} = \frac{(\dot{n},\,\text{rpm})(\dot{V},\,\text{gpm})^{1/2}}{(H,\,\text{ft})^{3/4}} \quad (14\text{-}36)$$

In Europe it is customary to write H in units of meters (and to include $g = 9.81$ m/s² in the equation), $\dot{V}$ in units of m³/s, and rotation rate $\dot{n}$ in

FIGURE 14–71
Even though pump specific speed is a dimensionless parameter, it is common practice to write it as a dimensional quantity using an inconsistent set of units.

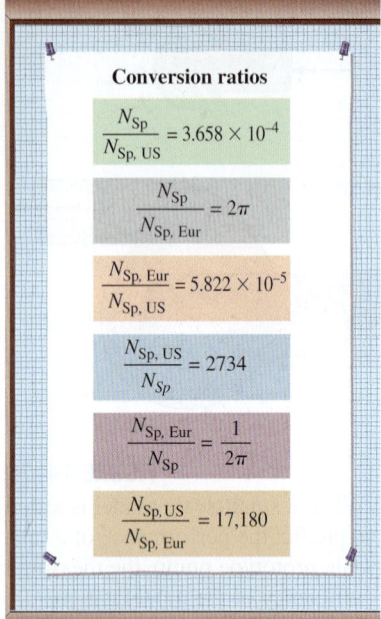

FIGURE 14–72
Conversions between the dimensionless, conventional U.S., and conventional European definitions of pump specific speed. Numerical values are given to four significant digits. The conversions for $N_{Sp, US}$ assume standard earth gravity.

units of rotations per *second* (Hz) instead of ω (rad/s) or $\dot{n}$ (rpm). Using Eq. 14–35 we define

Pump specific speed, customary European units:

$$N_{Sp, Eur} = \frac{(\dot{n}, \text{Hz})(\dot{V}, \text{m}^3/\text{s})^{1/2}}{(gH, \text{m}^2/\text{s}^2)^{3/4}} \quad (14-37)$$

The conversions between these three forms of pump specific speed are provided as ratios for your convenience in Fig. 14–72. When you become a practicing engineer, you will need to be very careful that you know which form of pump specific speed is being used, although it may not always be obvious.

Technically, pump specific speed could be applied at any operating condition and would just be another function of C_Q. That is not how it is typically used, however. Instead, it is common to define pump specific speed at *only one operating point*, namely, the best efficiency point (BEP) of the pump. The result is a single number that characterizes the pump.

> Pump specific speed is used to characterize the operation of a pump at its optimum conditions (best efficiency point) and is useful for preliminary pump selection and/or design.

As plotted in Fig. 14–73, centrifugal pumps perform optimally for N_{Sp} near 1, while mixed-flow and axial pumps perform best at N_{Sp} near 2 and 5, respectively. It turns out that if N_{Sp} is less than about 1.5, a centrifugal pump is the best choice. If N_{Sp} is between about 1.5 and 3.5, a mixed-flow pump is a better choice. When N_{Sp} is greater than about 3.5, an axial pump should be used. These ranges are indicated in Fig. 14–73 in terms of N_{Sp}, $N_{Sp, US}$, and $N_{Sp, Eur}$. Sketches of the blade types are also provided on the plot for reference.

EXAMPLE 14–9   Using Pump Specific Speed for Preliminary Pump Design

A pump is being designed to deliver 320 gpm of gasoline at room temperature. The required net head is 23.5 ft (of gasoline). It has already been determined that the pump shaft is to rotate at 1170 rpm. Calculate the pump specific speed in both nondimensional form and customary U.S. form. Based on your result, decide which kind of dynamic pump would be most suitable for this application.

SOLUTION We are to calculate pump specific speed and then determine whether a centrifugal, mixed-flow, or axial pump would be the best choice for this particular application.
Assumptions 1 The pump operates near its best efficiency point. 2 The maximum efficiency versus pump specific speed curve follows Fig. 14–73 reasonably well.
Analysis First, we calculate pump specific speed in customary U.S. units,

$$N_{Sp, US} = \frac{(1170 \text{ rpm})(320 \text{ gpm})^{1/2}}{(23.5 \text{ ft})^{3/4}} = \textbf{1960} \quad (1)$$

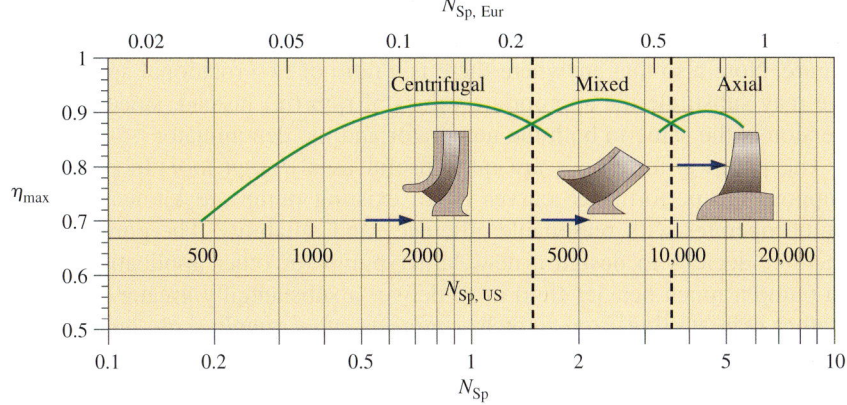

FIGURE 14–73
Maximum efficiency as a function of pump specific speed for the three main types of dynamic pump. The horizontal scales show nondimensional pump specific speed (N_{Sp}), pump specific speed in customary U.S. units ($N_{Sp,\,US}$), and pump specific speed in customary European units ($N_{Sp,\,Eur}$).

We convert to normalized pump specific speed using the conversion factor given in Fig. 14–72,

$$N_{Sp} = N_{Sp,\,US}\left(\frac{N_{Sp}}{N_{Sp,\,US}}\right) = 1960(3.658 \times 10^{-4}) = \mathbf{0.717} \quad (2)$$

Using either Eq. 1 or 2, Fig. 14–73 shows that **a centrifugal flow pump is the most suitable choice.**

Discussion Notice that the properties of the fluid never entered our calculations. The fact that we are pumping gasoline rather than some other liquid like water is irrelevant. However, the brake horsepower required to run the pump *does* depend on the fluid density.

Affinity Laws

We have developed dimensionless groups that are useful for relating any two pumps that are both geometrically similar and dynamically similar. It is convenient to summarize the similarity relationships as *ratios*. Some authors call these relationships **similarity rules,** while others call them **affinity laws.** For any two homologous states A and B,

Affinity laws:

$$\frac{\dot{V}_B}{\dot{V}_A} = \frac{\omega_B}{\omega_A}\left(\frac{D_B}{D_A}\right)^3 \quad (14\text{–}38a)$$

$$\frac{H_B}{H_A} = \left(\frac{\omega_B}{\omega_A}\right)^2\left(\frac{D_B}{D_A}\right)^2 \quad (14\text{–}38b)$$

$$\frac{bhp_B}{bhp_A} = \frac{\rho_B}{\rho_A}\left(\frac{\omega_B}{\omega_A}\right)^3\left(\frac{D_B}{D_A}\right)^5 \quad (14\text{–}38c)$$

Equations 14–38 apply to both pumps and turbines. States A and B can be *any* two homologous states between *any* two geometrically similar turbomachines, or even between two homologous states of the *same* machine. Examples include changing rotational speed or pumping a different fluid with the same pump. For the simple case of a given pump in which ω is varied, but the same fluid is pumped, $D_A = D_B$, and $\rho_A = \rho_B$. In such a

830 TURBOMACHINERY

V: Volume flow rate	$\dfrac{\dot{V}_B}{\dot{V}_A} = \left(\dfrac{\omega_B}{\omega_A}\right)^1 = \left(\dfrac{\dot{n}_B}{\dot{n}_A}\right)^1$
H: Head	$\dfrac{H_B}{H_A} = \left(\dfrac{\omega_B}{\omega_A}\right)^2 = \left(\dfrac{\dot{n}_B}{\dot{n}_A}\right)^2$
P: Power	$\dfrac{bhp_B}{bhp_A} = \left(\dfrac{\omega_B}{\omega_A}\right)^3 = \left(\dfrac{\dot{n}_B}{\dot{n}_A}\right)^3$

FIGURE 14–74
When the affinity laws are applied to a single pump in which the only thing that is varied is shaft rotational speed ω, or shaft rpm, $\dot{n}$, Eqs. 14–38 reduce to those shown above, for which a mnemonic can be used to help us remember the exponent on ω (or on $\dot{n}$):

Very Hard Problems are as easy as 1, 2, 3.

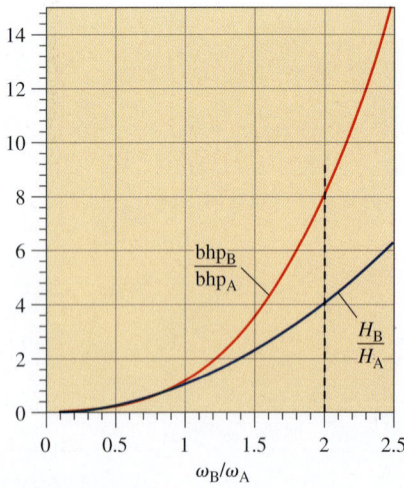

FIGURE 14–75
When the speed of a pump is increased, net head increases rapidly; brake horsepower increases even more rapidly.

case, Eqs. 14–38 reduce to the forms shown in Fig. 14–74. A mnemonic has been developed to help us remember the exponent on ω, as indicated in the figure. Note also that anywhere there is a ratio of two rotational speeds (ω), we may substitute the appropriate values of rpm ($\dot{n}$) instead, since the conversion is the same in both the numerator and the denominator.

The pump affinity laws are quite useful as a *design tool*. In particular, suppose the performance curves of an existing pump are known, and the pump operates with reasonable efficiency and reliability. The pump manufacturer decides to design a new, larger pump for other applications, e.g., to pump a much heavier fluid or to deliver a substantially greater net head. Rather than starting from scratch, *engineers often simply scale up an existing design*. The pump affinity laws enable such scaling to be accomplished with a minimal amount of effort.

EXAMPLE 14–10   The Effects of Doubling Pump Speed

Professor Seymour Fluids uses a small closed-loop water tunnel to perform flow visualization research. He would like to double the water speed in the test section of the tunnel and realizes that the least expensive way to do this is to double the rotational speed of the flow pump. What he doesn't realize is how much more powerful the new electric motor will need to be! If Professor Fluids doubles the flow speed, by approximately what factor will the motor power need to be increased?

SOLUTION For a doubling of ω, we are to calculate by what factor the power to the pump motor must increase.
Assumptions **1** The water remains at the same temperature. **2** After doubling pump speed, the pump runs at conditions homologous to the original conditions.
Analysis Since neither diameter nor density has changed, Eq. 14–38c reduces to

Ratio of required shaft power: $\dfrac{bhp_B}{bhp_A} = \left(\dfrac{\omega_B}{\omega_A}\right)^3$ (1)

Setting $\omega_B = 2\omega_A$ in Eq. 1 gives $bhp_B = 8\,bhp_A$. Thus **the power to the pump motor must be increased by a factor of 8.** A similar analysis using Eq. 14–38b shows that the pump's net head increases by a factor of 4. As seen in Fig. 14–75, both net head and power increase rapidly as pump speed is increased.
Discussion The result is only approximate since we have not included any analysis of the piping system. While doubling the flow speed through the pump increases available head by a factor of 4, doubling the flow speed through the water tunnel does not necessarily increase the *required* head of the system by the same factor of 4 (e.g., the friction factor decreases with the Reynolds number except at very high values of Re). In other words, our assumption 2 is not necessarily correct. The system will, of course, adjust to an operating point at which required and available heads match, but this point will not necessarily be homologous with the original operating point. Nevertheless, the approximation is useful as a first-order result. Professor Fluids may also need to be concerned with the possibility of cavitation at the higher speed.

EXAMPLE 14–11 Design of a New Geometrically Similar Pump

After graduation, you work for a pump manufacturing company. One of your company's best-selling products is a water pump, which we shall call pump A. Its impeller diameter is $D_A = 6.0$ cm, and its performance data when operating at $\dot{n}_A = 1725$ rpm ($\omega_A = 180.6$ rad/s) are shown in Table 14–2. The marketing research department is recommending that the company design a new product, namely, a larger pump (which we shall call pump B) that will be used to pump liquid refrigerant R-134a at room temperature. The pump is to be designed such that its best efficiency point occurs as close as possible to a volume flow rate of $\dot{V}_B = 2400$ cm^3/s and at a net head of $H_B = 450$ cm (of R-134a). The chief engineer (your boss) tells you to perform some preliminary analyses using pump scaling laws to determine if a geometrically scaled-up pump could be designed and built to meet the given requirements. (a) Plot the performance curves of pump A in both dimensional and dimensionless form, and identify the best efficiency point. (b) Calculate the required pump diameter D_B, rotational speed $\dot{n}_B$, and brake horsepower bhp$_B$ for the new product.

SOLUTION (a) For a given table of pump performance data for a water pump, we are to plot both dimensional and dimensionless performance curves and identify the BEP. (b) We are to design a new geometrically similar pump for refrigerant R-134a that operates at its BEP at given design conditions.
Assumptions 1 The new pump can be manufactured so as to be geometrically similar to the existing pump. 2 Both liquids (water and refrigerant R-134a) are incompressible. 3 Both pumps operate under steady conditions.
Properties At room temperature (20°C), the density of water is $\rho_{water} = 998.0$ kg/m^3 and that of refrigerant R-134a is $\rho_{R\text{-}134a} = 1226$ kg/m^3.
Analysis (a) First, we apply a second-order least-squares polynomial curve fit to the data of Table 14–2 to obtain smooth pump performance curves. These are plotted in Fig. 14–76, along with a curve for brake horsepower, which is obtained from Eq. 14–5. A sample calculation, including unit conversions, is shown in Eq. 1 for the data at $\dot{V}_A = 500$ cm^3/s, which is approximately the best efficiency point:

$$\text{bhp}_A = \frac{\rho_{water} g \dot{V}_A H_A}{\eta_{pump,A}}$$

$$= \frac{(998.0 \text{ kg/m}^3)(9.81 \text{ m/s}^2)(500 \text{ cm}^3/\text{s})(150 \text{ cm})}{0.81} \left(\frac{1 \text{ m}}{100 \text{ cm}}\right)^4 \left(\frac{\text{W} \cdot \text{s}}{\text{kg} \cdot \text{m}/\text{s}^2}\right)$$

$$= 9.07 \text{ W} \qquad (1)$$

Note that the actual value of bhp$_A$ plotted in Fig. 14–76 at $\dot{V}_A = 500$ cm^3/s differs slightly from that of Eq. 1 due to the fact that the least-squares curve fit smoothes out scatter in the original tabulated data.

Next we use Eqs. 14–30 to convert the dimensional data of Table 14–2 into nondimensional pump similarity parameters. Sample calculations are shown in Eqs. 2 through 4 at the same operating point as before (at the approximate location of the BEP). At $\dot{V}_A = 500$ cm^3/s the capacity coefficient is approximately

$$C_Q = \frac{\dot{V}}{\omega D^3} = \frac{500 \text{ cm}^3/\text{s}}{(180.6 \text{ rad/s})(6.0 \text{ cm})^3} = 0.0128 \qquad (2)$$

TABLE 14–2

Manufacturer's performance data for a water pump operating at 1725 rpm and room temperature (Example 14–11)*

$\dot{V}$, cm^3/s	H, cm	η_{pump}, %
100	180	32
200	185	54
300	175	70
400	170	79
500	150	81
600	95	66
700	54	38

* Net head is in centimeters of water.

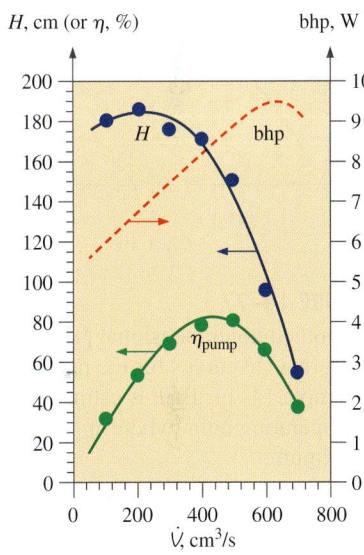

FIGURE 14–76
Data points and smoothed dimensional pump performance curves for the water pump of Example 14–11.

The head coefficient at this flow rate is approximately

$$C_H = \frac{gH}{\omega^2 D^2} = \frac{(9.81 \text{ m/s}^2)(1.50 \text{ m})}{(180.6 \text{ rad/s})^2(0.060 \text{ m})^2} = 0.125 \qquad (3)$$

Finally, the power coefficient at $\dot{V}_A = 500$ cm³/s is approximately

$$C_P = \frac{\text{bhp}}{\rho\omega^3 D^5} = \frac{9.07 \text{ W}}{(998 \text{ kg/m}^3)(180.6 \text{ rad/s})^3(0.060 \text{ m})^5}\left(\frac{\text{kg·m/s}^2}{\text{W·s}}\right) = 0.00198 \qquad (4)$$

These calculations are repeated (with the aid of a spreadsheet) at values of $\dot{V}_A$ between 100 and 700 cm³/s. The curve-fitted data are used so that the normalized pump performance curves are smooth; they are plotted in Fig. 14–77. Note that η_pump is plotted as a fraction rather than as a percentage. In addition, in order to fit all three curves on one plot with a single ordinate, and with the abscissa centered nearly around unity, we have multiplied C_Q by 100, C_H by 10, and C_P by 100. You will find that these scaling factors work well for a wide range of pumps, from very small to very large. A vertical line at the BEP is also sketched in Fig. 14–77 from the smoothed data. The curve-fitted data yield the following nondimensional pump performance parameters at the BEP:

$$C_Q^* = 0.0112 \quad C_H^* = 0.133 \quad C_P^* = 0.00184 \quad \eta_\text{pump}^* = 0.812 \qquad (5)$$

(b) We design the new pump such that its best efficiency point is homologous with the BEP of the original pump, but with a different fluid, a different pump diameter, and a different rotational speed. Using the values identified in Eq. 5, we use Eqs. 14–30 to obtain the operating conditions of the new pump. Namely, since both $\dot{V}_B$ and H_B are known (design conditions), we solve simultaneously for D_B and ω_B. After some algebra in which we eliminate ω_B, we calculate the design diameter for pump B,

$$D_B = \left(\frac{\dot{V}_B^2 C_H^*}{(C_Q^*)^2 g H_B}\right)^{1/4} = \left(\frac{(0.0024 \text{ m}^3/\text{s})^2(0.133)}{(0.0112)^2(9.81 \text{ m/s}^2)(4.50 \text{ m})}\right)^{1/4} = \mathbf{0.108 \text{ m}} \qquad (6)$$

In other words, pump A needs to be scaled up by a factor of $D_B/D_A = 10.8$ cm/6.0 cm = 1.80. With the value of D_B known, we return to Eqs. 14–30 to solve for ω_B, the design rotational speed for pump B,

$$\omega_B = \frac{\dot{V}_B}{(C_Q^*)D_B^3} = \frac{0.0024 \text{ m}^3/\text{s}}{(0.0112)(0.108 \text{ m})^3} = 168 \text{ rad/s} \rightarrow \dot{n}_B = \mathbf{1610 \text{ rpm}} \qquad (7)$$

Finally, the required brake horsepower for pump B is calculated from Eqs. 14–30,

$$\text{bhp}_B = (C_P^*)\rho_B \omega_B^3 D_B^5$$

$$= (0.00184)(1226 \text{ kg/m}^3)(168 \text{ rad/s})^3(0.108 \text{ m})^5\left(\frac{\text{W·s}}{\text{kg·m}^2/\text{s}}\right) = \mathbf{160 \text{ W}} \qquad (8)$$

An alternative approach is to use the affinity laws directly, eliminating some intermediate steps. We solve Eqs. 14–38a and b for D_B by eliminating

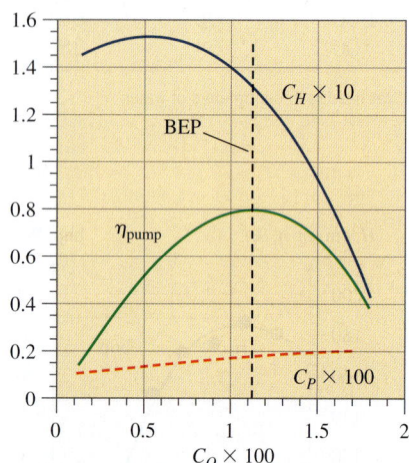

FIGURE 14–77
Smoothed nondimensional pump performance curves for the pumps of Example 14–11; BEP is estimated as the operating point where η_pump is a maximum.

the ratio ω_B/ω_A. We then plug in the known value of D_A and the curve-fitted values of $\dot{V}_A$ and H_A at the BEP (Fig. 14–78). The result agrees with those calculated before. In a similar manner we can calculate ω_B and bhp$_B$.

Discussion Although the desired value of ω_B has been calculated precisely, a practical issue is that it is difficult (if not impossible) to find an electric motor that rotates at exactly the desired rpm. Standard single-phase, 60-Hz, 120-V AC electric motors typically run at 1725 or 3450 rpm. Thus, we may not be able to meet the rpm requirement with a direct-drive pump. Of course, if the pump is belt-driven or if there is a gear box or a frequency controller, we can easily adjust the configuration to yield the desired rotation rate. Another option is that since ω_B is only slightly smaller than ω_A, we drive the new pump at standard motor speed (1725 rpm), providing a somewhat stronger pump than necessary. The disadvantage of this option is that the new pump would then operate at a point not exactly at the BEP.

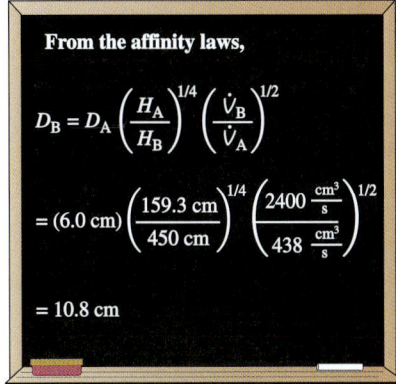

FIGURE 14–78
The affinity laws are manipulated to obtain an expression for the new pump diameter D_B. ω_B and bhp$_B$ can be obtained in similar fashion (not shown).

14–4 ■ TURBINES

Turbines have been used for centuries to convert freely available mechanical energy from rivers and wind into useful mechanical work, usually through a rotating shaft. Whereas the rotating part of a pump is called the impeller, the rotating part of a hydroturbine is called the **runner**. When the working fluid is water, the turbomachines are called **hydraulic turbines** or **hydroturbines**. When the working fluid is air, and energy is extracted from the wind, the machine is properly called a **wind turbine**. The word **windmill** should technically be applied only when the mechanical energy output is used to grind grain, as in ancient times (Fig. 14–79). However, most people use the word *windmill* to describe any wind turbine, whether used to grind grain, pump water, or generate electricity. In coal or nuclear power plants, the working fluid is usually steam; hence, the turbomachines that convert energy from the steam into mechanical energy of a rotating shaft are called **steam turbines**. A more generic name for turbines that employ a compressible gas as the working fluid is **gas turbine**. (The turbine in a modern commercial jet engine is a type of gas turbine.)

In general, energy-producing turbines have somewhat higher overall efficiencies than do energy-absorbing pumps. Large hydroturbines, for example, achieve overall efficiencies above 95 percent, while the best efficiency of large pumps is a little more than 90 percent. There are several reasons for this. First, pumps normally operate at higher rotational speeds than do turbines; therefore, shear stresses and frictional losses are higher. Second, conversion of kinetic energy into flow energy (pumps) has inherently higher losses than does the reverse (turbines). You can think of it this way: Since pressure *rises* across a pump (adverse pressure gradient), but *drops* across a turbine (favorable pressure gradient), boundary layers are less likely to separate in a turbine than in a pump. Third, turbines (especially hydroturbines) are often much larger than pumps, and viscous losses become less important as size increases. Finally, while pumps often operate over a wide range of flow rates, most electricity-generating turbines run within a narrower operating range and at a controlled constant speed; they can therefore be designed to operate most

FIGURE 14–79
A restored windmill in Brewster, MA, that was used in the 1800s to grind grain. (Note that the blades must be covered to function.) Modern "windmills" that generate electricity are more properly called *wind turbines*.

© Visions of America/Joe Sohm/Photodisc/Getty Images

efficiently at those conditions. In the United States, the standard AC electrical supply is 60 Hz (3600 cycles per minute); thus most wind, water, and steam turbines operate at speeds that are natural fractions of this, namely, 7200 rpm divided by the number of poles on the generator, usually an even number. Large hydroturbines usually operate at low speeds like 7200/60 = 120 rpm or 7200/48 = 150 rpm. Gas turbines used for power generation run at much higher speeds, some up to 7200/2 = 3600 rpm!

As with pumps, we classify turbines into two broad categories, *positive displacement* and *dynamic*. For the most part, positive-displacement turbines are small devices used for volume flow rate measurement, while dynamic turbines range from tiny to huge and are used for both flow measurement and power production. We provide details about both of these categories.

Positive-Displacement Turbines

A **positive-displacement turbine** may be thought of as a positive-displacement pump running backward—as fluid pushes into a closed volume, it turns a shaft or displaces a reciprocating rod. The closed volume of fluid is then pushed out as more fluid enters the device. There is a net head loss through the positive-displacement turbine; in other words, energy is extracted from the flowing fluid and is turned into mechanical energy. However, positive-displacement turbines are generally *not* used for power production, but rather for flow rate or flow volume measurement.

The most common example is the water meter in your house (Fig. 14–80). Many commercial water meters use a **nutating disc** that wobbles and spins as water flows through the meter. The disc has a sphere in its center with appropriate linkages that transfer the eccentric spinning motion of the nutating disc into rotation of a shaft. The volume of fluid that passes through the device per 360° rotation of the shaft is known precisely, and thus the total volume of water used is recorded by the device. When water is flowing at moderate speed from a spigot in your house, you can sometimes hear a bubbly sound coming from the water meter—this is the sound of the nutating disc wobbling inside the meter. There are, of course, other positive-displacement turbine designs, just as there are various designs of positive-displacement pumps.

Dynamic Turbines

Dynamic turbines are used both as flow measuring devices and as power generators. For example, meteorologists use a three-cup anemometer to measure wind speed (Fig. 14–81a). Experimental fluid mechanics researchers use small turbines of various shapes (most of which look like small propellers) to measure air speed or water speed (Chap. 8). In these applications, the shaft power output and the efficiency of the turbine are of little concern. Rather, these instruments are designed such that their rotational speed can be accurately calibrated to the speed of the fluid. Then, by electronically counting the number of blade rotations per second, the speed of the fluid is calculated and displayed by the device.

A novel application of a dynamic turbine is shown in Fig. 14–81b. NASA researchers mounted turbines at the wing tips of a Piper PA28 research aircraft to extract energy from wing tip vortices (Chap. 11); the extracted energy was converted to electricity to be used for on-board power requirements.

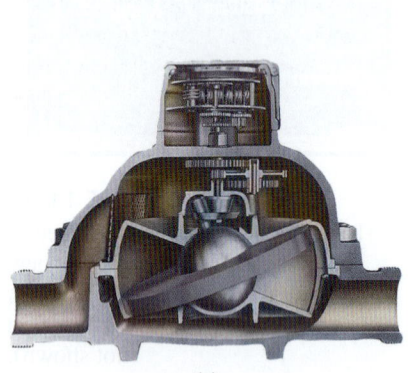

(a)

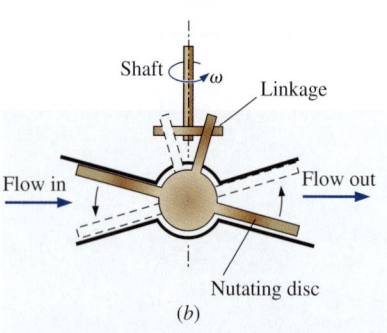

(b)

FIGURE 14–80

The *nutating disc fluid flowmeter* is a type of *positive-displacement turbine* used to measure volume flow rate: (a) cutaway view and (b) diagram showing motion of the nutating disc. This type of flowmeter is commonly used as a water meter in homes.

Photo courtesy of Niagara Meters, Spartanburg, SC.

(a)   (b)

FIGURE 14–81
Examples of dynamic turbines: (a) a typical three-cup anemometer used to measure wind speed, and (b) a Piper PA28 research airplane with turbines designed to extract energy from the wing tip vortices.

(a) © matthias engelien/Alamy. (b) NASA Langley Research Center.

In this chapter, we emphasize large dynamic turbines that are designed to produce electricity. Most of our discussion concerns hydroturbines that utilize the large elevation change across a dam to generate electricity, and wind turbines that generate electricity from blades rotated by the wind. There are two basic types of dynamic turbine—*impulse* and *reaction*, each of which are discussed in some detail. Comparing the two power-producing dynamic turbines, impulse turbines require a higher head, but can operate with a smaller volume flow rate. Reaction turbines can operate with much less head, but require a higher volume flow rate.

Impulse Turbines

In an **impulse turbine,** the fluid is sent through a nozzle so that most of its available mechanical energy is converted into kinetic energy. The high-speed jet then impinges on bucket-shaped vanes that transfer energy to the turbine shaft, as sketched in Fig. 14–82. The modern and most efficient type of impulse turbine was invented by Lester A. Pelton (1829–1908) in 1878, and the rotating wheel is now called a **Pelton wheel** in his honor. The buckets of a Pelton wheel are designed so as to split the flow in half, and turn the flow nearly 180° around (with respect to a frame of reference moving with the bucket), as illustrated in Fig. 14–82b. According to legend, Pelton modeled the splitter ridge shape after the nostrils of a cow's nose. A portion of the outermost part of each bucket is cut out so that the majority of the jet can pass through the bucket that is not aligned with the jet (bucket $n + 1$ in Fig. 14–82a) to reach the most aligned bucket (bucket n in Fig. 14–82a). In this way, the maximum amount of momentum from the jet is utilized. These details are seen in a photograph of a Pelton wheel (Fig. 14–83). Figure 14–84 shows a Pelton wheel in operation; the splitting and turning of the water jet is clearly seen.

We analyze the power output of a Pelton wheel turbine by using the Euler turbomachine equation. The power output of the shaft is equal to ωT_{shaft}, where T_{shaft} is given by Eq. 14–14,

Euler turbomachine equation for a turbine:

$$\dot{W}_{shaft} = \omega T_{shaft} = \rho \omega \dot{V}(r_2 V_{2,t} - r_1 V_{1,t}) \quad (14\text{–}39)$$

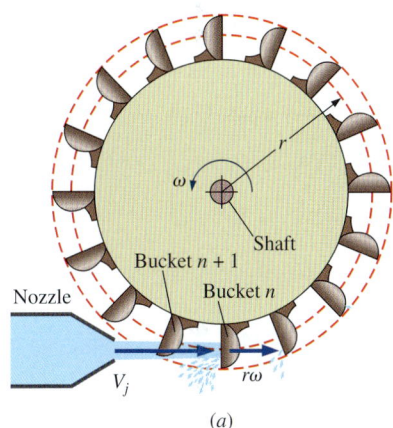

(a)

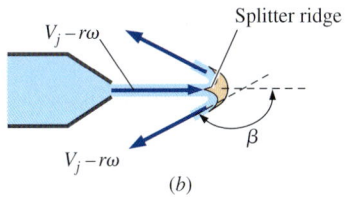

(b)

FIGURE 14–82
Schematic diagram of a Pelton-type *impulse turbine*; the turbine shaft is turned when high-speed fluid from one or more jets impinges on buckets mounted to the turbine shaft. (a) Side view, absolute reference frame, and (b) bottom view of a cross section of bucket n, rotating reference frame.

FIGURE 14–83
A close-up view of a Pelton wheel showing the detailed design of the buckets; the electrical generator is on the right. This Pelton wheel is on display at the Waddamana Power Station Museum near Bothwell, Tasmania.

Courtesy of Hydro Tasmania, www.hydro.com.au. Used by permission.

FIGURE 14–84
A view from the bottom of an operating Pelton wheel illustrating the splitting and turning of the water jet in the bucket. The water jet enters from the left, and the Pelton wheel is turning to the right.

Courtesy of VA TECH HYDRO. Used by permission.

We must be careful of negative signs since this is an energy-*producing* rather than an energy-*absorbing* device. For turbines, it is conventional to define point 2 as the inlet and point 1 as the outlet. The center of the bucket moves at tangential velocity $r\omega$, as illustrated in Fig. 14–82. We simplify the analysis by assuming that since there is an opening in the outermost part of each bucket, the entire jet strikes the bucket that happens to be at the direct bottom of the wheel at the instant of time under consideration (bucket n in Fig. 14–82a). Furthermore, since both the size of the bucket and the diameter of the water jet are small compared to the wheel radius, we approximate r_1

and r_2 as equal to r. Finally, we make the approximation that the water is turned through angle β without losing any speed; in the relative frame of reference moving with the bucket, the relative exit speed is thus $V_j - r\omega$ (the same as the relative inlet speed) as sketched in Fig. 14–82b. Returning to the absolute reference frame, which is necessary for the application of Eq. 14–39, the tangential component of velocity at the inlet, $V_{2,t}$, is simply the jet speed itself, V_j. We construct a velocity diagram in Fig. 14–85 as an aid in calculating the tangential component of absolute velocity at the outlet, $V_{1,t}$. After some trigonometry, which you can verify after noting that $\sin(\beta - 90°) = -\cos\beta$,

$$V_{1,t} = r\omega + (V_j - r\omega)\cos\beta$$

Upon substitution of this equation, Eq. 14–39 yields

$$\dot{W}_{shaft} = \rho r\omega \dot{V}\{V_j - [r\omega + (V_j - r\omega)\cos\beta]\}$$

which simplifies to

Output shaft power: $$\dot{W}_{shaft} = \rho r\omega \dot{V}(V_j - r\omega)(1 - \cos\beta) \quad (14\text{–}40)$$

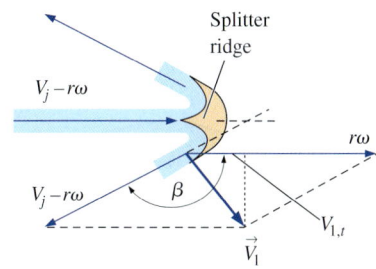

FIGURE 14–85
Velocity diagram of flow into and out of a Pelton wheel bucket. We translate outflow velocity from the moving reference frame to the absolute reference frame by adding the speed of the bucket ($r\omega$) to the right.

Obviously, the maximum power is achieved theoretically if $\beta = 180°$. However, if that were the case, the water exiting one bucket would strike the back side of its neighbor coming along behind it, reducing the generated torque and power. It turns out that in practice, the maximum power is achieved by reducing β to around 160° to 165°. The efficiency factor due to β being less than 180° is

Efficiency factor due to β: $$\eta_\beta = \frac{\dot{W}_{shaft,\,actual}}{\dot{W}_{shaft,\,ideal}} = \frac{1 - \cos\beta}{1 - \cos(180°)} \quad (14\text{–}41)$$

When $\beta = 160°$, for example, $\eta_\beta = 0.97$—a loss of only about 3 percent.

Finally, we see from Eq. 14–40 that the shaft power output $\dot{W}_{shaft}$ is zero if $r\omega = 0$ (wheel not turning at all). $\dot{W}_{shaft}$ is also zero if $r\omega = V_j$ (bucket moving at the jet speed). Somewhere in between these two extremes lies the optimum wheel speed. By setting the derivative of Eq. 14–40 with respect to $r\omega$ to zero, we find that this occurs when $r\omega = V_j/2$ (bucket moving at half the jet speed, as shown in Fig. 14–86).

For an actual Pelton wheel turbine, there are other losses besides that reflected in Eq. 14–41: mechanical friction, aerodynamic drag on the buckets, friction along the inside walls of the buckets, nonalignment of the jet and bucket as the bucket turns, backsplashing, and nozzle losses. Even so, the efficiency of a well-designed Pelton wheel turbine can approach 90 percent. In other words, up to 90 percent of the available mechanical energy of the water is converted to rotating shaft energy.

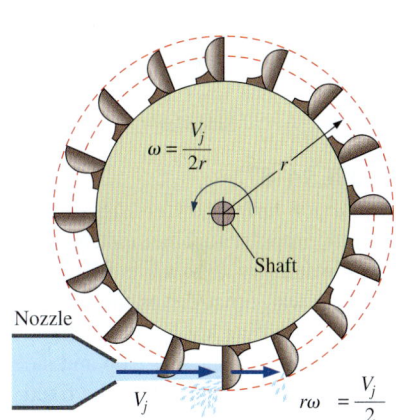

FIGURE 14–86
The theoretical maximum power achievable by a Pelton turbine occurs when the wheel rotates at $\omega = V_j/(2r)$, i.e., when the bucket moves at half the speed of the water jet.

Reaction Turbines

The other main type of energy-producing hydroturbine is the **reaction turbine**, which consists of fixed guide vanes called **stay vanes**, adjustable guide vanes called **wicket gates**, and rotating blades called **runner blades** (Fig. 14–87). Flow enters tangentially at high pressure, is turned toward

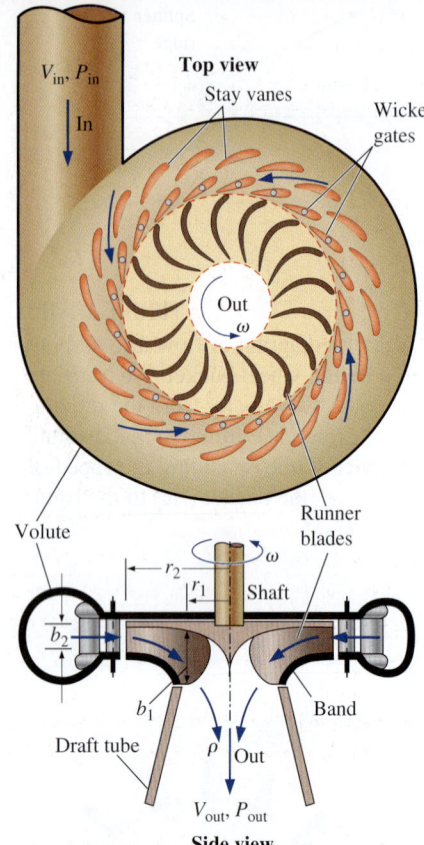

FIGURE 14–87
A *reaction turbine* differs significantly from an impulse turbine; instead of using water jets, a *volute* is filled with swirling water that drives the runner. For hydroturbine applications, the axis is typically vertical. Top and side views are shown, including the fixed *stay vanes* and adjustable *wicket gates*.

the runner by the stay vanes as it moves along the spiral casing or **volute**, and then passes through the wicket gates with a large tangential velocity component. Momentum is exchanged between the fluid and the runner as the runner rotates, and there is a large pressure drop. Unlike the impulse turbine, the water completely fills the casing of a reaction turbine. For this reason, a reaction turbine generally produces more power than an impulse turbine of the same diameter, net head, and volume flow rate. The angle of the wicket gates is adjustable so as to control the volume flow rate through the runner. (In most designs the wicket gates can close on each other, cutting off the flow of water into the runner.) At design conditions the flow leaving the wicket gates impinges parallel to the runner blade leading edge (from a rotating frame of reference) to avoid shock losses. Note that in a good design, the number of wicket gates does not share a common denominator with the number of runner blades. Otherwise there would be severe vibration caused by simultaneous impingement of two or more wicket gate wakes onto the leading edges of the runner blades. For example, in Fig. 14–87 there are 17 runner blades and 20 wicket gates. These are typical numbers for many large reaction hydroturbines, as shown in the photographs in Figs. 14–89 and 14–90. The number of stay vanes and wicket gates is usually the same (there are 20 stay vanes in Fig. 14–87). This is not a problem since neither of them rotate, and unsteady wake interaction is not an issue.

There are two main types of reaction turbine—*Francis* and *Kaplan*. The **Francis turbine** is somewhat similar in geometry to a centrifugal or mixed-flow pump, but with the flow in the opposite direction. Note, however, that a typical pump running backward would *not* be a very efficient turbine. The Francis turbine is named in honor of James B. Francis (1815–1892), who developed the design in the 1840s. In contrast, the **Kaplan turbine** is somewhat like an *axial-flow* fan running backward. If you have ever seen a window fan start spinning in the wrong direction when a gust of wind blows through the window, you can visualize the basic operating principle of a Kaplan turbine. The Kaplan turbine is named in honor of its inventor, Viktor Kaplan (1876–1934). There are actually several subcategories of both Francis and Kaplan turbines, and the terminology used in the hydroturbine field is not always standard.

Recall that we classify dynamic pumps according to the angle at which the flow exits the impeller blade—centrifugal (radial), mixed flow, or axial (see Fig. 14–31). In a similar but reversed manner, we classify reaction turbines according to the angle that the flow *enters* the runner (Fig. 14–88). If the flow enters the runner radially as in Fig. 14–88*a*, the turbine is called a **Francis radial-flow turbine** (see also Fig. 14–87). If the flow enters the runner at some angle between radial and axial (Fig. 14–88*b*), the turbine is called a **Francis mixed-flow turbine.** The latter design is more common. Some hydroturbine engineers use the term "Francis turbine" only when there is a **band** on the runner as in Fig. 14–88*b*. Francis turbines are most suited for heads that lie between the high heads of Pelton wheel turbines and the low heads of Kaplan turbines. A typical large Francis turbine may have 16 or more runner blades and can achieve a turbine efficiency of 90 to 95 percent. If the runner has no band, and flow enters the runner partially turned, it is called a **propeller mixed-flow turbine** or simply a **mixed-flow**

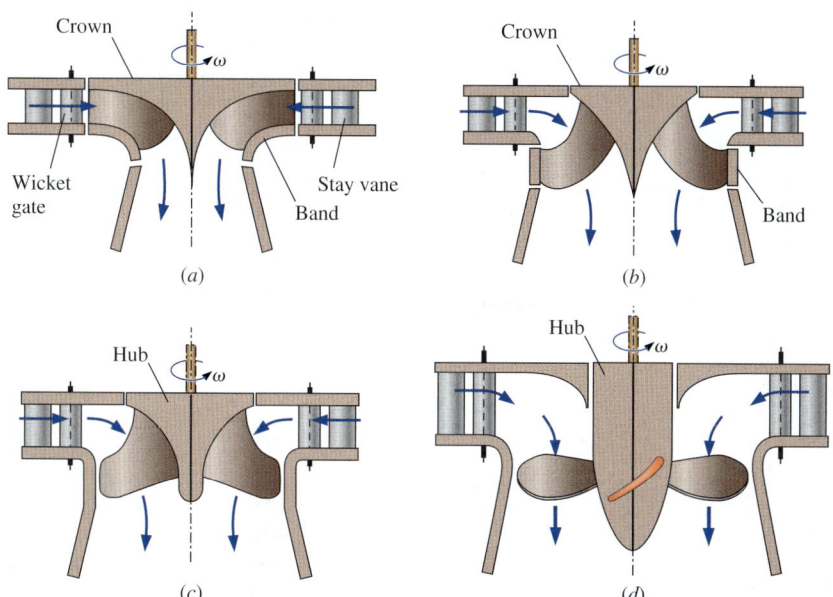

FIGURE 14–88
The distinguishing characteristics of the four subcategories of reaction turbines: (*a*) *Francis radial flow*, (*b*) *Francis mixed flow*, (*c*) *propeller mixed flow*, and (*d*) *propeller axial flow*. The main difference between (*b*) and (*c*) is that Francis mixed-flow runners have a *band* that rotates with the runner, while propeller mixed-flow runners do not. There are two types of propeller mixed-flow turbines: *Kaplan turbines* have adjustable pitch blades, while *propeller turbines* do not. Note that the terminology used here is not universal among turbomachinery textbooks nor among hydroturbine manufacturers.

turbine (Fig. 14–88*c*). Finally, if the flow is turned completely axially *before* entering the runner (Fig. 14–88*d*), the turbine is called an **axial-flow turbine.** The runners of an axial-flow turbine typically have only three to eight blades, a lot fewer than Francis turbines. Of these there are two types: Kaplan turbines and propeller turbines. Kaplan turbines are called **double regulated** because the flow rate is controlled in two ways—by turning the wicket gates and by adjusting the pitch on the runner blades. **Propeller turbines** are nearly identical to Kaplan turbines except that the blades are fixed (pitch is not adjustable), and the flow rate is regulated only by the wicket gates (**single regulated**). Compared to the Pelton and Francis turbines, Kaplan turbines and propeller turbines are most suited for low head, high volume flow rate conditions. Their efficiencies rival those of Francis turbines and may be as high as 94 percent.

Figure 14–89 is a photograph of the radial-flow runner of a Francis radial-flow turbine. The workers are shown to give you an idea of how large the runners are in a hydroelectric power plant. Figure 14–90 is a photograph of the mixed-flow runner of a Francis turbine, and Fig. 14–91 is a photograph of an axial-flow propeller turbine. The view is from the inlet (top).

We sketch in Fig. 14–92 a typical hydroelectric dam that utilizes Francis reaction turbines to generate electricity. The overall or **gross head** H_{gross} is defined as the elevation difference between the reservoir surface upstream of the dam and the surface of the water exiting the dam, $H_{\text{gross}} = z_A - z_E$. If there were no irreversible losses *anywhere* in the system, the maximum amount of power that could be generated per turbine would be

Ideal power production: $$\dot{W}_{\text{ideal}} = \rho g \dot{V} H_{\text{gross}} \qquad (14\text{–}42)$$

Of course, there are irreversible losses throughout the system, so the power actually produced is lower than the ideal power given by Eq. 14–42.

FIGURE 14–89
The runner of a Francis radial-flow turbine used at the Round Butte hydroelectric power station in Madras, OR. There are 17 runner blades of outer diameter 3.60 m. The turbine rotates at 180 rpm and produces 119 MW of power at a volume flow rate of 127 m³/s from a net head of 105 m.
Courtesy of American Hydro Corporation, York, PA. Used by permission.

840
TURBOMACHINERY

FIGURE 14–90

The runner of a Francis mixed-flow turbine used at the Smith Mountain hydroelectric power station in Roanoke, VA. There are 17 runner blades of outer diameter 6.19 m. The turbine rotates at 100 rpm and produces 194 MW of power at a volume flow rate of 375 m^3/s from a net head of 54.9 m.

Courtesy of American Hydro Corporation, York, PA. Used by permission.

FIGURE 14–91

The five-bladed propeller turbine used at the Warwick hydroelectric power station in Cordele, GA. There are five runner blades of outer diameter 3.87 m. The turbine rotates at 100 rpm and produces 5.37 MW of power at a volume flow rate of 63.7 m^3/s from a net head of 9.75 m.

Photo courtesy of Weir American Hydro Corporation, York, PA. Used by permission.

We follow the flow of water through the whole system of Fig. 14–92, defining terms and discussing losses along the way. We start at point *A* upstream of the dam where the water is still, at atmospheric pressure, and at its highest elevation, z_A. Water flows at volume flow rate $\dot{V}$ through a large tube through the dam called the **penstock.** Flow to the penstock can be cut off by closing a large gate valve called a **head gate** at the penstock inlet. If we were to insert a Pitot probe at point *B* at the end of the penstock just before the turbine, as illustrated in Fig. 14–92, the water in the tube would rise to a column height equal to the energy grade line EGL$_{in}$ at the inlet of the turbine. This column height is lower than the water level at point *A*, due to irreversible losses in the penstock and its inlet. The flow then passes through the turbine, which is connected by a shaft to the electric generator. Note that the electric generator itself has irreversible losses. From a fluid mechanics perspective, however, we are interested only in the losses through the turbine and downstream of the turbine.

After passing through the turbine runner, the exiting fluid (point *C*) still has appreciable kinetic energy, and perhaps swirl. To recover some of this kinetic energy (which would otherwise be wasted), the flow enters an expanding area diffuser called a **draft tube,** which turns the flow horizontally and slows down the flow speed, while increasing the pressure prior to discharge into the downstream water, called the **tailrace.** If we were to imagine another Pitot probe at point *D* (the exit of the draft tube), the water in the tube would rise to a column height equal to the energy grade line labeled EGL$_{out}$ in Fig. 14–92. Since the draft tube is considered to be an

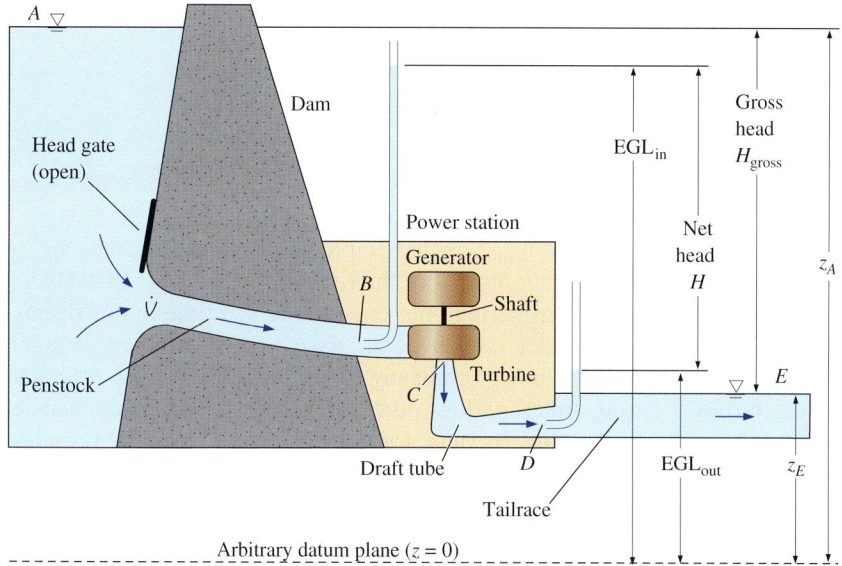

FIGURE 14–92
Typical setup and terminology for a hydroelectric plant that utilizes a Francis turbine to generate electricity; drawing not to scale. The Pitot probes are shown for illustrative purposes only.

integral part of the turbine assembly, the net head across the turbine is specified as the difference between EGL_{in} and EGL_{out},

Net head for a hydraulic turbine: $\qquad H = EGL_{in} - EGL_{out} \qquad$ (14-43)

In words,

> The net head of a turbine is defined as the difference between the energy grade line just upstream of the turbine and the energy grade line at the exit of the draft tube.

At the draft tube exit (point D) the flow speed is significantly slower than that at point C upstream of the draft tube; however, it is *finite*. All the kinetic energy leaving the draft tube is dissipated in the tailrace. This represents an irreversible head loss and is the reason why EGL_{out} is higher than the elevation of the tailrace surface, z_E. Nevertheless, significant pressure recovery occurs in a well-designed draft tube. The draft tube causes the pressure at the outlet of the runner (point C) to decrease *below* atmospheric pressure, thereby enabling the turbine to utilize the available head most efficiently. In other words, the draft tube causes the pressure at the runner outlet to be lower than it would have been without the draft tube—increasing the change in pressure from the inlet to the outlet of the turbine. Designers must be careful, however, because subatmospheric pressures may lead to cavitation, which is undesirable for many reasons, as discussed previously.

If we were interested in the net efficiency of the entire hydroelectric plant, we would define this efficiency as the ratio of actual electric power produced to ideal power (Eq. 14–42), based on gross head. Of more concern in this chapter is the efficiency of the turbine itself. By convention, **turbine efficiency** is based on net head H rather than gross head H_{gross}. Specifically, $\eta_{turbine}$ is defined as the ratio of brake horsepower output (actual turbine

842
TURBOMACHINERY

output shaft power) to water horsepower (power extracted from the water flowing through the turbine),

$$\text{Turbine efficiency:} \quad \eta_{\text{turbine}} = \frac{\dot{W}_{\text{shaft}}}{\dot{W}_{\text{water horsepower}}} = \frac{\text{bhp}}{\rho g H \dot{V}} \quad (14\text{--}44)$$

Note that turbine efficiency η_{turbine} is the reciprocal of pump efficiency η_{pump}, since bhp is the *actual output* instead of the *required input* (Fig. 14–93).

Note also that we are considering only one turbine at a time in this discussion. Most large hydroelectric power plants have *several* turbines arranged in parallel. This offers the power company the opportunity to turn off some of the turbines during times of low power demand and for maintenance. Hoover Dam in Boulder City, Nevada, for example, has 17 parallel turbines, 15 of which are identical large Francis turbines that can produce approximately 130 MW of electricity each (Fig. 14–94). The maximum gross head is 180 m. The total peak power production of the power plant exceeds 2 GW (2000 MW).

We perform preliminary design and analysis of turbines in the same way we did previously for pumps, using the Euler turbomachine equation and velocity diagrams. In fact, we keep the same notation, namely r_1 for the inner radius and r_2 for the outer radius of the rotating blades. For a turbine, however, the flow direction is opposite to that of a pump, so the inlet is at radius r_2 and the outlet is at radius r_1. For a first-order analysis we approximate the blades as being infinitesimally thin. We also assume that the blades are aligned such that the flow is always tangent to the blade surface, and we ignore viscous effects (boundary layers) at the surfaces. Higher-order corrections are best obtained with a computational fluid dynamics code.

Consider for example the top view of the Francis turbine of Fig. 14–87. Velocity vectors are drawn in Fig. 14–95 for both the absolute reference

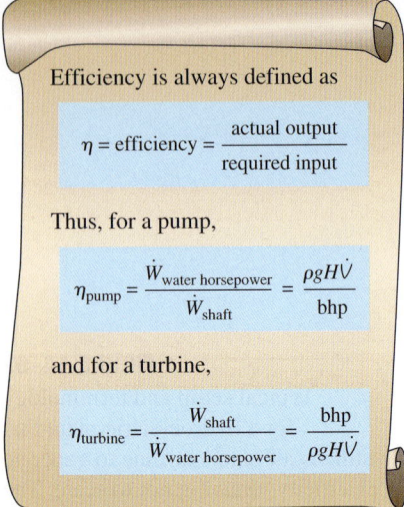

FIGURE 14–93
By definition, efficiency must always be less than unity. The efficiency of a turbine is the reciprocal of the efficiency of a pump.

FIGURE 14–94
(*a*) An aerial view of Hoover Dam and (*b*) the top (visible) portion of several of the parallel electric generators driven by hydraulic turbines at Hoover Dam.
(*a*) © Corbis RF (*b*) © Brand X Pictures RF

(*a*)   (*b*)

frame and the relative reference frame rotating with the runner. Beginning with the stationary guide vane (thick black line in Fig. 14–95), the flow is turned so that it strikes the runner blade (thick brown line) at absolute velocity $\vec{V}_2$. But the runner blade is rotating counterclockwise, and at radius r_2 it moves tangentially to the lower left at speed ωr_2. To translate into the rotating reference frame, we form the vector sum of $\vec{V}_2$ and the *negative* of ωr_2, as shown in the sketch. The resultant is vector $\vec{V}_{2,\,\text{relative}}$, which is parallel to the runner blade leading edge (angle β_2 from the tangent line of circle r_2). The tangential component $V_{2,\,t}$ of the absolute velocity vector $\vec{V}_2$ is required for the Euler turbomachine equation (Eq. 14–39). After some trigonometry,

Runner leading edge:
$$V_{2,\,t} = \omega r_2 - \frac{V_{2,\,n}}{\tan \beta_2} \quad (14\text{–}45)$$

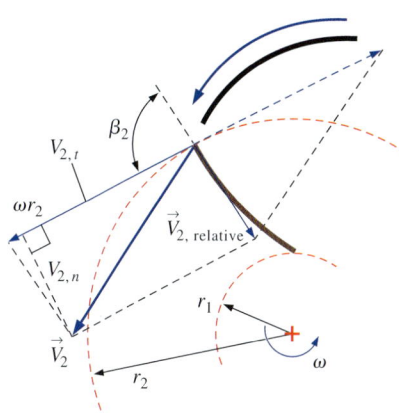

FIGURE 14–95
Relative and absolute velocity vectors and geometry for the outer radius of the runner of a Francis turbine. Absolute velocity vectors are bold.

Following the flow along the runner blade in the relative (rotating) reference frame, we see that the flow is turned such that it exits parallel to the trailing edge of the runner blade (angle β_1 from the tangent line of circle r_1). Finally, to translate back to the absolute reference frame we vectorially add $\vec{V}_{1,\,\text{relative}}$ and blade speed ωr_1, which acts to the left as sketched in Fig. 14–96. The resultant is absolute vector $\vec{V}_1$. Since mass must be conserved, the normal components of the absolute velocity vectors $V_{1,\,n}$ and $V_{2,\,n}$ are related through Eq. 14–12, where axial blade widths b_1 and b_2 are defined in Fig. 14–87. After some trigonometry (which turns out to be identical to that at the leading edge), we generate an expression for the tangential component $V_{1,\,t}$ of absolute velocity vector $\vec{V}_1$ for use in the Euler turbomachine equation,

Runner trailing edge:
$$V_{1,\,t} = \omega r_1 - \frac{V_{1,\,n}}{\tan \beta_1} \quad (14\text{–}46)$$

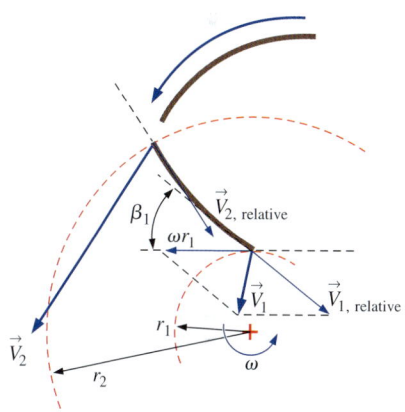

FIGURE 14–96
Relative and absolute velocity vectors and geometry for the inner radius of the runner of a Francis turbine. Absolute velocity vectors are bold.

Alert readers will notice that Eq. 14–46 for a turbine is identical to Eq. 14–23 for a pump. This is not just fortuitous, but results from the fact that the velocity vectors, angles, etc., are defined in the same way for a turbine as for a pump except that everything is flowing in the opposite direction.

For some hydroturbine runner applications, high power/high flow operation can result in $V_{1,\,t} < 0$. Here the runner blade turns the flow so much that the flow at the runner outlet rotates in the direction opposite to runner rotation, a situation called **reverse swirl** (Fig. 14–97). The Euler turbomachine equation predicts that maximum power is obtained when $V_{1,\,t} < 0$, so we suspect that reverse swirl should be part of a good turbine design. In practice, however, it has been found that the best efficiency operation of most hydroturbines occurs when the runner imparts a small amount of **with-rotation swirl** to the flow exiting the runner (swirl in the same direction as runner rotation). This improves draft tube performance. A large amount of swirl (either reverse or with-rotation) is not desirable, because it leads to much higher losses in the draft tube. (High swirl velocities result in "wasted" kinetic energy.) Obviously, much fine tuning needs to be done in order to design the most efficient hydroturbine system (including the draft tube as an integral component) within imposed design constraints. Also keep in mind that the flow is three-dimensional; there is an *axial* component of the velocity as the flow is turned into the draft tube, and there are differences in velocity in the *circumferential* direction as well. It doesn't take long before you realize that computer simulation tools are enormously useful to turbine designers. In fact, with the help of modern CFD codes, the efficiency of hydroturbines has increased to the point where retrofits of old turbines in hydroelectric plants are economically wise and common. An example CFD output is shown in Fig. 14–98 for a Francis mixed-flow turbine.

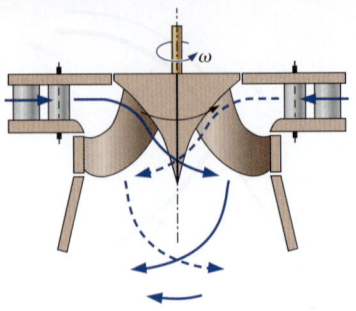

Reverse swirl

FIGURE 14–97
In some Francis mixed-flow turbines, high-power, high-volume flow rate conditions sometimes lead to *reverse swirl*, in which the flow exiting the runner swirls in the direction opposite to that of the runner itself, as sketched here.

EXAMPLE 14–12   Effect of Component Efficiencies on Plant Efficiency

A hydroelectric power plant is being designed. The gross head from the reservoir to the tailrace is 325 m, and the volume flow rate of water through each turbine is 12.8 m³/s at 20°C. There are 12 identical parallel turbines, each with an efficiency of 95.2 percent, and all other mechanical energy losses (through the penstock, etc.) are estimated to reduce the output by 3.5 percent. The generator itself has an efficiency of 94.5 percent. Estimate the electric power production from the plant in MW.

SOLUTION   We are to estimate the power production from a hydroelectric plant.
Properties   The density of water at $T = 20°C$ is 998 kg/m³.

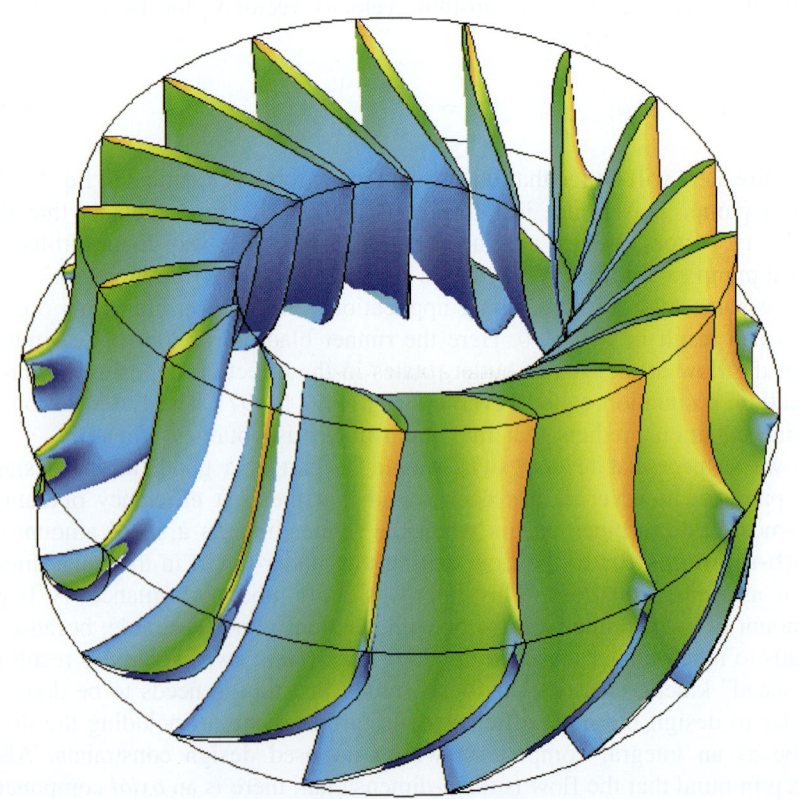

FIGURE 14–98
Contour plot of the static pressure distribution on runner blade surfaces as calculated by CFD; pressure is in pascals. Shown is a 17-blade Francis mixed-flow turbine runner that rotates counterclockwise about the *z*-axis. Only one blade passage is modeled, but the image is reproduced 16 times due to the symmetry. The highest pressures (red regions) are encountered near the leading edges of the pressure surfaces of the runner, while the lowest pressures (blue regions) occur on the suction surface of the runner near the trailing edge.

Photo courtesy of Weir American Hydro Corporation, York, PA. Used by permission.

Analysis The ideal power produced by one hydroturbine is

$$\dot{W}_{ideal} = \rho g \dot{V} H_{gross}$$

$$= (998 \text{ kg/m}^3)(9.81 \text{ m/s}^2)(12.8 \text{ m}^3/\text{s})(325 \text{ m})$$

$$\times \left(\frac{1 \text{ N}}{1 \text{ kg} \cdot \text{m/s}^2}\right)\left(\frac{1 \text{ W}}{1 \text{ N} \cdot \text{m/s}}\right)\left(\frac{1 \text{ MW}}{10^6 \text{ W}}\right)$$

$$= 40.73 \text{ MW}$$

But inefficiencies in the turbine, the generator, and the rest of the system reduce the actual electrical power output. For each turbine,

$$\dot{W}_{electrical} = \dot{W}_{ideal} \eta_{turbine} \eta_{generator} \eta_{other} = (40.73 \text{ MW})(0.952)(0.945)(1 - 0.035)$$

$$= 35.4 \text{ MW}$$

Finally, since there are 12 turbines in parallel, the total power produced is

$$\dot{W}_{\text{total electrical}} = 12 \, \dot{W}_{electrical} = 12(35.4 \text{ MW}) = \mathbf{425 \text{ MW}}$$

Discussion A small improvement in any of the efficiencies ends up increasing the power output and it thus increases the power company's profitability.

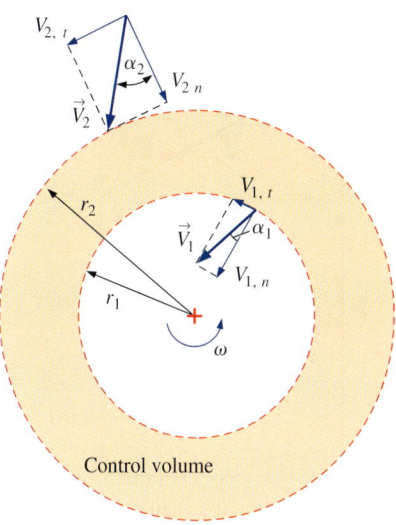

FIGURE 14–99
Top view of the absolute velocities and flow angles associated with the runner of a Francis turbine being designed for a hydroelectric dam (Example 14-13). The control volume is from the inlet to the outlet of the runner.

EXAMPLE 14–13 Hydroturbine Design

A retrofit Francis radial-flow hydroturbine is being designed to replace an old turbine in a hydroelectric dam. The new turbine must meet the following design restrictions in order to properly couple with the existing setup: The runner inlet radius is $r_2 = 2.50$ m and its outlet radius is $r_1 = 1.77$ m. The runner blade widths are $b_2 = 0.914$ m and $b_1 = 2.62$ m at the inlet and outlet, respectively. The runner must rotate at $\dot{n} = 120$ rpm ($\omega = 12.57$ rad/s) to turn the 60-Hz electric generator. The wicket gates turn the flow by angle $\alpha_2 = 33°$ from radial at the runner inlet, and the flow at the runner outlet is to have angle α_1 between $-10°$ and $10°$ from radial (Fig. 14–99) for proper flow through the draft tube. The volume flow rate at design conditions is 599 m³/s, and the gross head provided by the dam is $H_{gross} = 92.4$ m. (a) Calculate the inlet and outlet runner blade angles β_2 and β_1, respectively, and predict the power output and required net head if irreversible losses are neglected for the case with $\alpha_1 = 10°$ from radial (with-rotation swirl). (b) Repeat the calculations for the case with $\alpha_1 = 0°$ from radial (no swirl). (c) Repeat the calculations for the case with $\alpha_1 = -10°$ from radial (reverse swirl).

SOLUTION For a given set of hydroturbine design criteria we are to calculate runner blade angles, required net head, and power output for three cases—two with swirl and one without swirl at the runner outlet.
Assumptions 1 The flow is steady. 2 The fluid is water at 20°C. 3 The blades are infinitesimally thin. 4 The flow is everywhere tangent to the runner blades. 5 We neglect irreversible losses through the turbine.
Properties For water at 20°C, $\rho = 998.0$ kg/m³.

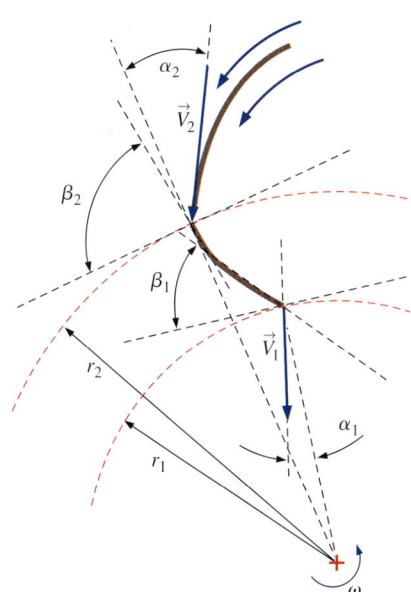

FIGURE 14–100
Sketch of the runner blade design of Example 14-13, top view. A guide vane and absolute velocity vectors are also shown.

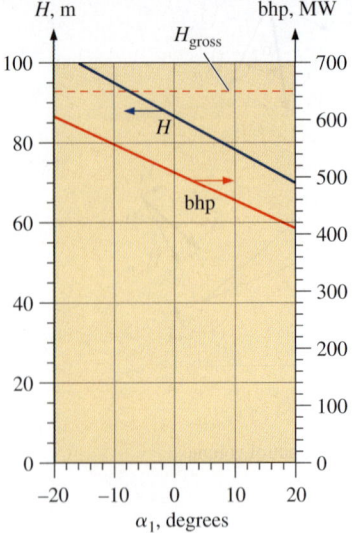

FIGURE 14–101
Ideal required net head and brake horsepower output as functions of runner outlet flow angle for the turbine of Example 14–13.

Analysis (*a*) We solve for the normal component of velocity at the inlet using Eq. 14–12,

$$V_{2,n} = \frac{\dot{V}}{2\pi r_2 b_2} = \frac{599 \text{ m}^3/\text{s}}{2\pi(2.50 \text{ m})(0.914 \text{ m})} = 41.7 \text{ m/s} \quad (1)$$

Using Fig. 14–99 as a guide, the tangential velocity component at the inlet is

$$V_{2,t} = V_{2,n} \tan \alpha_2 = (41.7 \text{ m/s}) \tan 33° = 27.1 \text{ m/s} \quad (2)$$

We now solve Eq. 14–45 for the runner leading edge angle β_2,

$$\beta_2 = \arctan\left(\frac{V_{2,n}}{\omega r_2 - V_{2,t}}\right)$$

$$= \arctan\left(\frac{41.7 \text{ m/s}}{(12.57 \text{ rad/s})(2.50 \text{ m}) - 27.1 \text{ m/s}}\right) = \mathbf{84.1°} \quad (3)$$

Equations 1 through 3 are repeated for the runner outlet, with the following results:

Runner outlet: $V_{1,n} = 20.6$ m/s, $V_{1,t} = 3.63$ m/s, $\beta_1 = \mathbf{47.9°}$ (4)

The top view of this runner blade is sketched (to scale) in Fig. 14–100.
Using Eqs. 2 and 4, the shaft output power is estimated from the Euler turbomachine equation, Eq. 14–39,

$$\dot{W}_{\text{shaft}} = \rho \omega \dot{V}(r_2 V_{2,t} - r_1 V_{1,t}) = (998.0 \text{ kg/m}^3)(12.57 \text{ rad/s})(599 \text{ m}^3/\text{s})$$

$$\times [(2.50 \text{ m})(27.2 \text{ m/s}) - (1.77 \text{ m})(3.63 \text{ m/s})]\left(\frac{\text{MW}\cdot\text{s}}{10^6 \text{ kg}\cdot\text{m}^2/\text{s}^2}\right)$$

$$= 461 \text{ MW} = \mathbf{6.18 \times 10^5 \text{ hp}} \quad (5)$$

Finally, we calculate the required net head using Eq. 14–44, assuming that $\eta_{\text{turbine}} = 100$ percent since we are ignoring irreversibilities,

$$H = \frac{\text{bhp}}{\rho g \dot{V}} = \frac{461 \text{ MW}}{(998.0 \text{ kg/m}^3)(9.81 \text{ m/s}^2)(599 \text{ m}^3/\text{s})}\left(\frac{10^6 \text{ kg}\cdot\text{m}^2/\text{s}^2}{\text{MW}\cdot\text{s}}\right) = \mathbf{78.6 \text{ m}} \quad (6)$$

(*b*) When we repeat the calculations with no swirl at the runner outlet ($\alpha_1 = 0°$), the runner blade trailing edge angle reduces to **42.8°**, and the output power increases to 509 MW (**6.83 × 10^5 hp**). The required net head increases to **86.8 m**.
(*c*) When we repeat the calculations with *reverse* swirl at the runner outlet ($\alpha_1 = -10°$), the runner blade trailing edge angle reduces to **38.5°**, and the output power increases to 557 MW (**7.47 × 10^5 hp**). The required net head increases to **95.0 m**. A plot of power and net head as a function of runner outlet flow angle α_1 is shown in Fig. 14–101. You can see that both bhp and H increase with decreasing α_1.

Discussion The theoretical output power increases by about 10 percent by eliminating swirl from the runner outlet and by nearly another 10 percent when there is 10° of reverse swirl. However, the gross head available from the dam is only 92.4 m. Thus, the reverse swirl case of part (*c*) is clearly impossible, since the predicted net head is required to be greater than H_{gross}. Keep in mind that this is a preliminary design in which we are neglecting irreversibilities. The actual output power will be lower and the actual required net head will be higher than the values predicted here.

Gas and Steam Turbines

Most of our discussion so far has concerned hydroturbines. We now discuss turbines that are designed for use with *gases*, like combustion products or steam. In a coal or nuclear power plant, high-pressure steam is produced by a boiler and then sent to a steam turbine to produce electricity. Because of reheat, regeneration, and other efforts to increase overall efficiency, these steam turbines typically have two stages (high pressure and low pressure). Most power plant steam turbines are multistage axial-flow devices like that shown in Fig. 14–102. Not shown are the stator vanes (called **nozzles**) that direct the flow between each set of turbine blades (called *buckets*). Analysis of axial-flow turbines is very similar to that of axial-flow fans, as discussed in Section 14–2, and is not repeated here.

Similar axial-flow turbines are used in jet aircraft engines (Fig. 14–62) and gas turbine generators (Fig. 14–103). A gas turbine generator is similar to a jet engine except that instead of providing thrust, the turbomachine is designed to transfer as much of the fuel's energy as possible into the rotating shaft, which is connected to an electric generator. Gas turbines used for power generation are typically much larger than jet engines, of course, since they are ground-based. As with hydroturbines, a significant gain in efficiency is realized as overall turbine size increases.

Wind Turbines*

As global demand for energy increases, the supply of fossil fuels diminishes and the price of energy continues to rise. To keep up with global energy demand, renewable sources of energy such as solar, wind, wave, tidal, hydroelectric, and geothermal must be tapped more extensively. In this section we concentrate on wind turbines used to generate electricity. We note the distinction between the terms *windmill* used for *mechanical* power generation (grinding grain, pumping water, etc.) and *wind turbine* used for electrical power generation, although technically both devices are turbines since they extract energy from the fluid. Although the wind is "free" and renewable, modern wind turbines are expensive and suffer from one obvious disadvantage compared to most other power generation devices – they produce power only when the wind is blowing, and the power output of a wind turbine is thus inherently unsteady. Furthermore and equally obvious is the fact that wind turbines need to be located where the wind blows, which is often far from traditional power grids, requiring construction of new high-voltage power lines. Nevertheless, wind turbines are expected to play an ever-increasing role in the global supply of energy for the foreseeable future.

Numerous innovative wind turbine designs have been proposed and tested over the centuries as sketched in Fig. 14–104. We generally categorize wind turbines by the orientation of their axis of rotation: **horizontal axis wind turbines** (**HAWTs**) and **vertical axis wind turbines** (**VAWTs**). An alternative way to categorize them is by the mechanism that provides torque to the rotating shaft: lift or drag. So far, none of the VAWT designs or drag-type

---

* Much of the material for this section is condensed from Manwell et al. (2010), and the authors acknowledge Professors J. F. Manwell, J. G. McGowan, and A. L. Rogers for their help in reviewing this section.

FIGURE 14–102
The turbine blades (called *buckets*) of a typical two-stage steam turbine used in a coal or nuclear power plant. The flow is from left to right, with the high-pressure stage on the left and the low-pressure stage on the right.

© Brand X Pictures/PunchStock

FIGURE 14–103
The rotor assembly of the MS7001F gas turbine being lowered into the bottom half of the gas turbine casing. Flow is from right to left, with the upstream set of rotor blades (called *blades*) comprising the multistage compressor and the downstream set of rotor blades (called *buckets*) comprising the multistage turbine. Compressor stator blades (called *vanes*) and turbine stator blades (called *nozzles*) can be seen in the bottom half of the gas turbine casing. This gas turbine spins at 3600 rpm and produces over 135 MW of power.

Courtesy of GE Energy.

Horizontal axis turbines
Single bladed
Double bladed
Three bladed
U.S. farm windmill
multi-bladed
Bicycle
multi-bladed
Up-wind
Down-wind
Enfield-Andreau
Sail wing
Multi-rotor
Counter - rotating blades
Cross-wind
Savonius
Cross-wind
paddles
Diffuser
Concentrator
Unconfined vortex

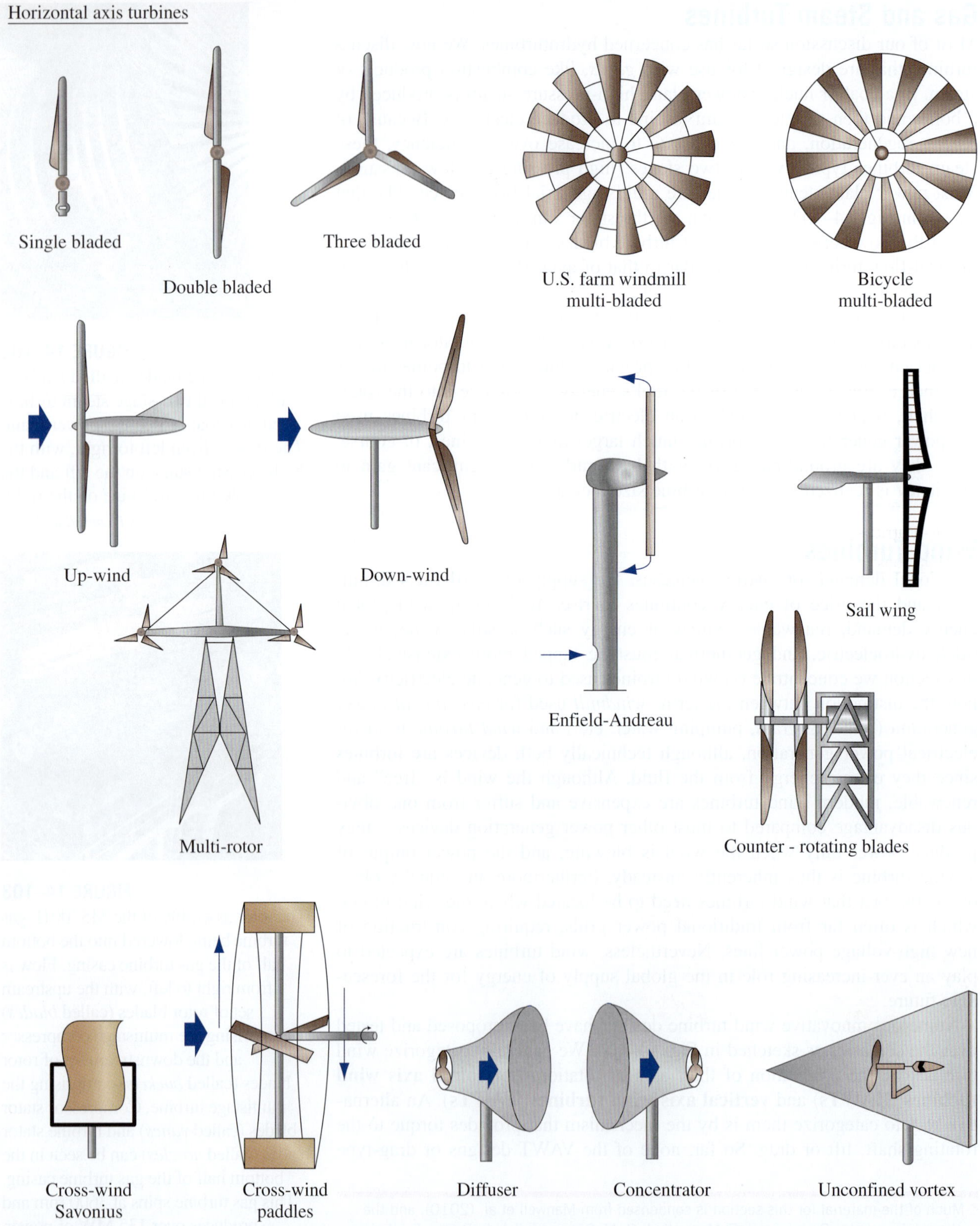

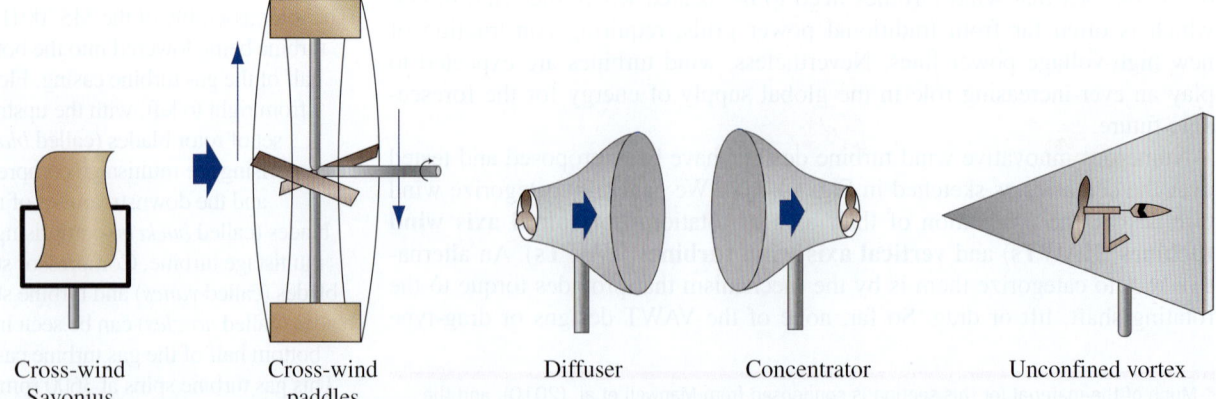

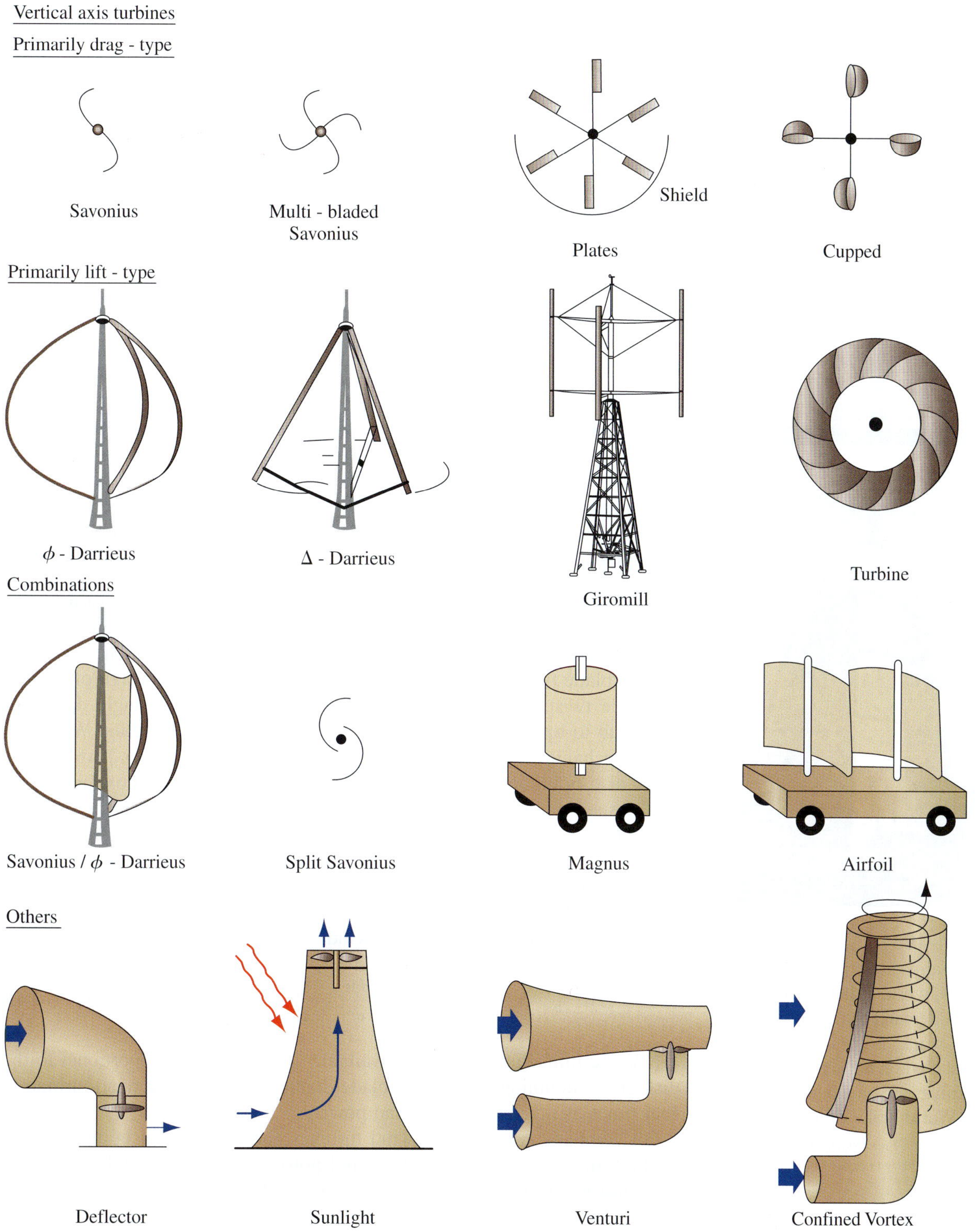

CHAPTER 14

Vertical axis turbines

Primarily drag - type

Savonius

Multi - bladed Savonius

Shield

Plates

Cupped

Primarily lift - type

ϕ - Darrieus

Δ - Darrieus

Giromill

Turbine

Combinations

Savonius / ϕ - Darrieus

Split Savonius

Magnus

Airfoil

Others

Deflector

Sunlight

Venturi

Confined Vortex

FIGURE 14–104
Various wind turbine designs and their categorization. Adapted from Manwell et al. (2010).

FIGURE 14–105

(a) Wind farms are popping up all over the world to help reduce the global demand for fossil fuels. (b) Some wind turbines are even being installed on buildings! (These three turbines are on a building at the Bahrain World Trade Center.)

(a) © Digital Vision/Punchstock RF
(b) © Adam Jam/Getty Images

designs has achieved the efficiency or success of the lift-type HAWT. This is why the vast majority of wind turbines being built around the world are of this type, often in clusters affectionately called *wind farms* (Fig. 14–105). For this reason, the lift-type HAWT is the only type of wind turbine discussed in any detail in this section. [See Manwell et al. (2010) for a detailed discussion as to why drag-type devices have inherently lower efficiency than lift-type devices.]

Every wind turbine has a characteristic power performance curve; a typical one is sketched in Fig. 14–106, in which electrical power output is plotted as a function of wind speed V at the height of the turbine's axis. We identify three key locations on the wind-speed scale:

- **Cut-in speed** is the minimum wind speed at which useful power can be generated.
- **Rated speed** is the wind speed that delivers the rated power, usually the maximum power.
- **Cut-out speed** is the maximum wind speed at which the wind turbine is designed to produce power. At wind speeds greater than the cut-out speed, the turbine blades are stopped by some type of braking mechanism to avoid damage and for safety issues. The short section of dashed blue line indicates the power that *would* be produced if cut-out were not implemented.

The design of HAWT turbine blades includes tapering and twist to maximize performance and is similar to the design of axial flow fans (propellers), as discussed in Section 14–2 and is not repeated here. The design of turbine blade twist, for example, is nearly identical to the design of propeller blade twist, as in Example 14–7, and the blade pitch angle decreases from hub to tip in much the same manner as that of a propeller. While the fluid mechanics of wind turbine design is critical, the power performance curve also is influenced by the electrical generator, the gearbox, and structural issues. Inefficiencies appear in every component of course, as in all machines.

We define the **disk area** A of a wind turbine as the area normal to the wind direction swept out by the turbine blades as they rotate (Fig. 14–107). The **available wind power** $\dot{W}_{\text{available}}$ in the disk area is calculated as the rate of change of kinetic energy of the wind,

$$\dot{W}_{\text{available}} = \frac{d(\tfrac{1}{2}mV^2)}{dt} = \frac{1}{2}V^2\frac{dm}{dt} = \frac{1}{2}V^2\dot{m} = \frac{1}{2}V^2\rho VA = \frac{1}{2}\rho V^3 A \quad (14\text{–}47)$$

We notice immediately that the available wind power is proportional to the disk area—doubling the turbine blade diameter exposes the wind turbine to four times as much available wind power.

For comparison of various wind turbines and locations, it is more useful to think in terms of the available wind power *per unit area*, which we call the **wind power density,** typically in units of W/m²,

Wind power density: $$\frac{\dot{W}_{\text{available}}}{A} = \frac{1}{2}\rho V^3 \quad (14\text{–}48)$$

Thus,

- The wind power density is directly proportional to air density—cold air has a larger wind power density than warm air blowing at the same speed, although this effect is not as significant as wind speed.
- The wind power density is proportional to the cube of the wind speed—doubling the wind speed increases the wind power density by a factor of 8. It should be obvious then why wind farms are located where the wind speed is high!

Equation 14–48 is an instantaneous equation. As we all know, however, wind speed varies greatly throughout the day and throughout the year. For this reason, it is useful to define the **average wind power density** in terms of annual average wind speed $\overline{V}$, based on hourly averages as

Average wind power density:
$$\frac{\overline{\dot{W}_{available}}}{A} = \frac{1}{2}\rho_{avg}\overline{V}^3 K_e \quad (14\text{–}49)$$

where K_e is a correction factor called the **energy pattern factor.** In principle, it is analogous to the kinetic energy factor α that we use in control volume analyses (Chap. 5). K_e is defined as

$$K_e = \frac{1}{N\overline{V}^3}\sum_{i=1}^{N}V_i^3 \quad (14\text{–}50)$$

where $N = 8760$, which is the number of hours in a year. As a general rule of thumb, a location is considered poor for construction of wind turbines if the average wind power density is less than about 100 W/m², good if it is around 400 W/m², and great if it is greater than about 700 W/m². Other factors affect the choice of a wind turbine site, such as atmospheric turbulence intensity, terrain, obstacles (buildings, trees, etc.), environmental impact, etc. See Manwell, et al. (2010) for further details.

For analysis purposes, we consider a given wind speed V and define the aerodynamic efficiency of a wind turbine as the fraction of available wind power that is extracted by the turbine blades. This efficiency is commonly called the **power coefficient,** C_P,

Power coefficient:
$$C_p = \frac{\dot{W}_{\text{rotor shaft output}}}{\dot{W}_{\text{available}}} = \frac{\dot{W}_{\text{rotor shaft output}}}{\frac{1}{2}\rho V^3 A} \quad (14\text{–}51)$$

It is fairly simple to calculate the maximum possible power coefficient for a wind turbine, and this was first done by Albert Betz (1885–1968) in the mid 1920s. We consider two control volumes surrounding the disk area—a large control volume and a small control volume—as sketched in Fig. 14–108, with upstream wind speed V taken as V_1.

The axisymmetric stream tube (enclosed by streamlines as drawn on the top and bottom of Fig. 14–108) can be thought of as forming an imaginary "duct" for the flow of air through the turbine. The control volume momentum equation for the large control volume for steady flow is

$$\sum \vec{F} = \sum_{\text{out}}\beta\dot{m}\vec{V} - \sum_{\text{in}}\beta\dot{m}\vec{V}$$

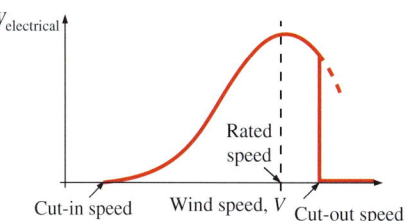

FIGURE 14–106
Typical qualitative wind-turbine power performance curve with definitions of cut-in, rated, and cut-out speeds.

FIGURE 14–107
The disk area of a wind turbine is defined as the swept area or frontal area of the turbine as "seen" by the oncoming wind, as sketched here in red. The disk area is (*a*) circular for a horizontal axis turbine and (*b*) rectangular for a vertical axis turbine.

(*a*) © Construction Photography/Corbis RF
(*b*) © VisionofAmerica/Joe Sohm/Photodisc/Getty RF

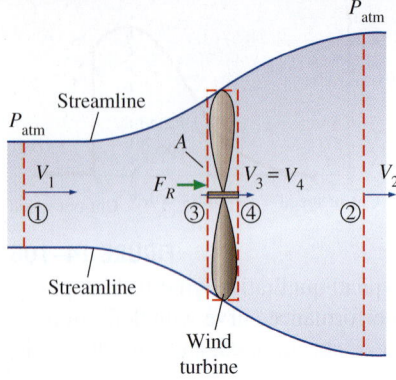

FIGURE 14–108
The large and small control volumes for analysis of ideal wind turbine performance bounded by an axisymmetric diverging stream tube.

and is analyzed in the streamwise (x) direction. Since locations 1 and 2 are sufficiently far from the turbine, we take $P_1 = P_2 = P_{atm}$, yielding no net pressure force on the control volume. We approximate the velocities at the inlet (1) and outlet (2) to be uniform at V_1 and V_2, respectively; and the momentum flux correction factors are thus $\beta_1 = \beta_2 = 1$. The momentum equation reduces to

$$F_R = \dot{m}V_2 - \dot{m}V_1 = \dot{m}(V_2 - V_1) \quad (14\text{-}52)$$

The smaller control volume in Fig. 14–108 encloses the turbine, but $A_3 = A_4 = A$, since this control volume is infinitesimally thin in the limit (we approximate the turbine as a disk). Since the air is considered to be incompressible, $V_3 = V_4$. However, the wind turbine extracts energy from the air, causing a pressure drop. Thus, $P_3 \neq P_4$. When we apply the streamwise component of the control volume momentum equation on the small control volume, we get

$$F_R + P_3 A - P_4 A = 0 \quad \rightarrow \quad F_R = (P_4 - P_3)A \quad (14\text{-}53)$$

The Bernoulli equation is certainly *not* applicable across the turbine, since it is extracting energy from the air. However, it is a reasonable approximation between locations 1 and 3 and between locations 4 and 2:

$$\frac{P_1}{\rho g} + \frac{V_1^2}{2g} + z_1 = \frac{P_3}{\rho g} + \frac{V_3^2}{2g} + z_3 \quad \text{and} \quad \frac{P_4}{\rho g} + \frac{V_4^2}{2g} + z_4 = \frac{P_2}{\rho g} + \frac{V_2^2}{2g} + z_2$$

In this ideal analysis, the pressure starts at atmospheric pressure far upstream ($P_1 = P_{atm}$), rises smoothly from P_1 to P_3, drops suddenly from P_3 to P_4 across the turbine disk, and then rises smoothly from P_4 to P_2, ending at atmospheric pressure far downstream ($P_2 = P_{atm}$) (Fig. 14–109). We add Eqs. 14–52 and 14–53, setting $P_1 = P_2 = P_{atm}$ and $V_3 = V_4$. In addition, since the wind turbine is horizontally inclined, $z_1 = z_2 = z_3 = z_4$ (gravitational effects are negligible in air anyway). After some algebra, this yields

$$\frac{V_1^2 - V_2^2}{2} = \frac{P_3 - P_4}{\rho} \quad (14\text{-}54)$$

Substituting $\dot{m} = \rho V_3 A$ into Eq. 14–52 and then combining the result with Eqs. 14–53 and 14–54 yields

$$V_3 = \frac{V_1 + V_2}{2} \quad (14\text{-}55)$$

Thus, we conclude that *the average velocity of the air through an ideal wind turbine is the arithmetic average of the far upstream and far downstream velocities.* Of course, the validity of this result is limited by the applicability of the Bernoulli equation.

For convenience, we define a new variable a as the fractional loss of velocity from far upstream to the turbine disk as

$$a = \frac{V_1 - V_3}{V_1} \quad (14\text{-}56)$$

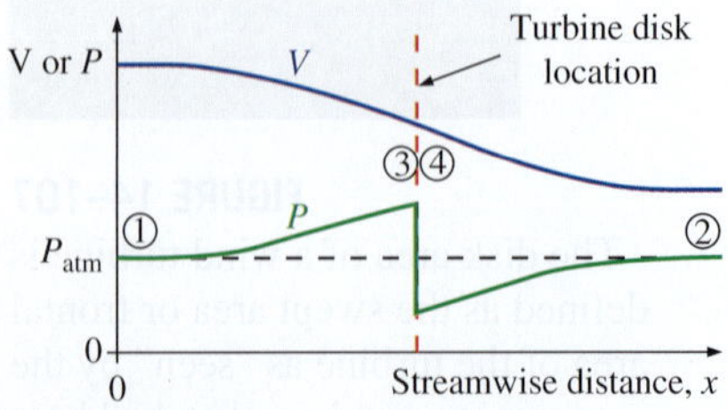

FIGURE 14–109
Qualitative sketch of average streamwise velocity and pressure profiles through a wind turbine.

The velocity through the turbine thus becomes $V_3 = V_1(1 - a)$, and the mass flow rate through the turbine becomes $\dot{m} = \rho A V_3 = \rho A V_1(1 - a)$. Combining this expression for V_3 with Eq. 14–55 yields

$$V_2 = V_1(1 - 2a) \quad (14\text{–}57)$$

For an ideal wind turbine without irreversible losses such as friction, the power generated by the turbine is simply the difference between the incoming and outgoing kinetic energies. Performing some algebra, we get

$$\dot{W}_{ideal} = \dot{m}\frac{V_1^2 - V_2^2}{2} = \rho A V_1(1-a)\frac{V_1^2 - V_1^2(1-2a)^2}{2} = 2\rho A V_1^3 a(1-a)^2 \quad (14\text{–}58)$$

Again assuming no irreversible losses in transferring power from the turbine to the turbine shaft, the efficiency of the wind turbine is expressed as the power coefficient defined in Eq. 14–51 as

$$C_P = \frac{\dot{W}_{rotor\ shaft\ output}}{\frac{1}{2}\rho V_1^3 A} = \frac{\dot{W}_{ideal}}{\frac{1}{2}\rho V_1^3 A} = \frac{2\rho A V_1^3 a(1-a)^2}{\frac{1}{2}\rho V_1^3 A} = 4a(1-a)^2 \quad (14\text{–}59)$$

Finally, as any good engineer knows, we calculate the maximum possible value of C_P by setting $dC_P/da = 0$ and solving for a (Fig. 14–110). This yields $a = 1$ or $1/3$, and the details are left as an exercise. Since $a = 1$ is the trivial case (no power generated), we conclude that a must equal $1/3$ for maximum possible power coefficient. Substituting $a = 1/3$ into Eq. 14–59 gives

$$C_{P,\ max} = 4\frac{1}{3}\left(1 - \frac{1}{3}\right)^2 = \frac{16}{27} \cong 0.5926 \quad (14\text{–}60)$$

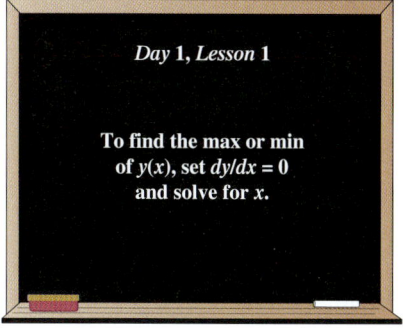

FIGURE 14–110
The use of derivatives to calculate minima or maxima is one of the first things that engineers learn.

This value of $C_{P,\ max}$ represents the *maximum possible power coefficient of any wind turbine* and is known as the **Betz limit.** All real wind turbines have a maximum achievable power coefficient less than this due to irreversible losses which have been ignored in this ideal analysis.

Figure 14–111 shows power coefficient C_P as a function of the ratio of turbine blade tip speed ωR to wind speed V for several types of wind turbines, where ω is the angular velocity of the wind turbine blades and R is their radius. From this plot, we see that an ideal propeller-type wind turbine approaches the Betz limit as $\omega R/V$ approaches infinity. However, the power coefficient of real wind turbines reaches a maximum at some *finite* value of $\omega R/V$ and then drops beyond that. In practice, three primary effects lead to a maximum achievable power coefficient that is lower than the Betz limit:

- Rotation of the wake behind the rotor (swirl)
- Finite number of rotor blades and their associated tip losses (tip vortices are generated in the wake of rotor blades for the same reason they are generated on finite airplane wings since both produce "lift") (see Chap. 11)
- Non-zero aerodynamic drag on the rotor blades (frictional drag as well as induced drag–see Chap. 11)

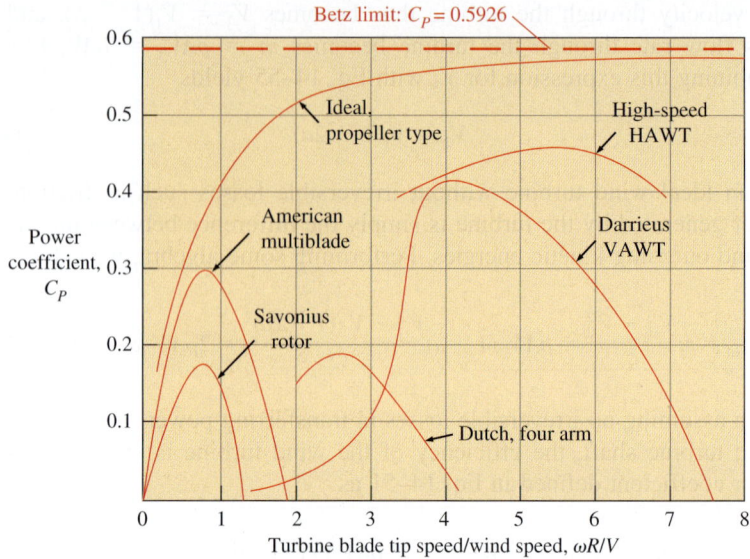

FIGURE 14–111
Performance (power coefficient) of various types of wind turbines as a function of the ratio of turbine blade tip speed to wind speed. So far, no design has achieved better performance than the horizontal axis wind turbine (HAWT). Adapted from Robinson (1981, Ref. 10).

See Manwell, et al. (2010) for further discussion about how to account for these losses.

In addition, mechanical losses due to shaft friction lead to even lower maximum achievable power coefficients. Other mechanical and electrical losses in the gearbox, generator, etc., also reduce the overall wind turbine efficiency, as previously mentioned. As seen in Fig. 14–111, the "best" wind turbine is the high-speed HAWT, and that is why you see this type of wind turbine being installed throughout the world. In summary, wind turbines provide a "green" alternative to fossil fuels, and as the price of fossil fuels rises, wind turbines will become more commonplace.

EXAMPLE 14–14   Power Generated by a Wind Turbine

To save money, a school plans to generate some of their own electricity using a HAWT wind turbine on top of a hill where it is fairly windy. As a conservative estimate based on the data of Fig. 14–111, they hope to achieve a power coefficient of 40 percent. The combined efficiency of the gearbox and generator is estimated to be 85 percent. If the diameter of the wind turbine disk is 12.5 m, estimate the electrical power production when the wind blows at 10.0 m/s.

SOLUTION   We are to estimate the power generated by a wind turbine.
Assumptions   **1** The power coefficient is 0.40 and the combined efficiency of the gearbox and generator is 0.85. **2** The air is at 20°C.
Properties   At 20°C, the air density is 1.204 kg/m³.
Analysis   From the definition of power coefficient,

$$\dot{W}_{\text{rotor shaft output}} = C_P \frac{1}{2} \rho V^3 A = C_P \frac{1}{2} \rho V^3 (\pi D^2/4)$$

But the actual electrical power produced is lower than this because of gearbox and generator inefficiencies,

$$\dot{W}_{\text{electrical output}} = \eta_{\text{gearbox/generator}} \frac{C_P \pi \rho V^3 D^2}{8}$$

$$= (0.85) \frac{(0.40)\pi \left(1.204 \frac{\text{kg}}{\text{m}^3}\right)\left(10.0 \frac{\text{m}}{\text{s}}\right)^3 (12.5 \text{ m})^2}{8} \left(\frac{\text{N}}{\text{kg} \cdot \text{m/s}^2}\right)\left(\frac{\text{W}}{\text{N} \cdot \text{m/s}}\right)$$

$$= 25118 \text{ W} \cong \mathbf{25 \text{ kW}}$$

Discussion We give the final answer to two significant digits since we cannot expect any better than that, based on the given information and approximations. To give you a feel for how much electrical power this is, consider that a typical hair dryer draws around 1500 W, so this is enough power to run more than 16 hair dryers simultaneously. The school would need to do a cost analysis to calculate how long it would take for the wind turbine to pay for itself considering the reduction in electricity purchased from the power company.

14–5 · TURBINE SCALING LAWS

Dimensionless Turbine Parameters

We define dimensionless groups (Pi groups) for turbines in much the same way as we did in Section 14–3 for pumps. Neglecting Reynolds number and roughness effects, we deal with the same dimensional variables: gravity times net head (gH), volume flow rate ($\dot{V}$), some characteristic diameter of the turbine (D), runner rotational speed (ω), output brake horsepower (bhp), and fluid density (ρ), as illustrated in Fig. 14–112. In fact, the dimensional analysis is identical whether analyzing a pump or a turbine, except for the fact that for turbines, we take bhp instead of $\dot{V}$ as the independent variable. In addition, η_{turbine} (Eq. 14–44) is used in place of η_{pump} as the nondimensional efficiency. A summary of the dimensionless parameters is provided here:

Dimensionless turbine parameters:

$$C_H = \text{Head coefficient} = \frac{gH}{\omega^2 D^2} \qquad C_Q = \text{Capacity coefficient} = \frac{\dot{V}}{\omega D^3}$$

$$C_P = \text{Power coefficient} = \frac{\text{bhp}}{\rho \omega^3 D^5} \qquad \eta_{\text{turbine}} = \text{Turbine efficiency} = \frac{\text{bhp}}{\rho g H \dot{V}}$$

(14–61)

When plotting turbine performance curves, we use C_P instead of C_Q as the independent parameter. In other words, C_H and C_Q are functions of C_P, and η_{turbine} is thus also a function of C_P, since

$$\eta_{\text{turbine}} = \frac{C_P}{C_Q C_H} = \text{function of } C_P \qquad (14\text{–}62)$$

The affinity laws (Eqs. 14–38) can be applied to turbines as well as to pumps, allowing us to scale turbines up or down in size (Fig. 14–113).

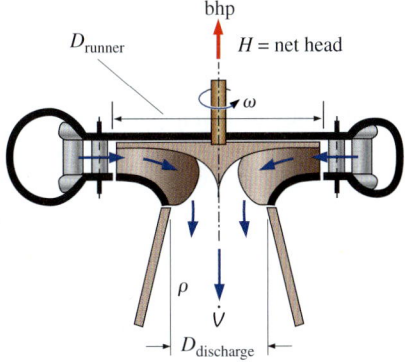

FIGURE 14–112
The main variables used for dimensional analysis of a turbine. The characteristic turbine diameter D is typically either the runner diameter D_{runner} or the discharge diameter $D_{\text{discharge}}$.

856
TURBOMACHINERY

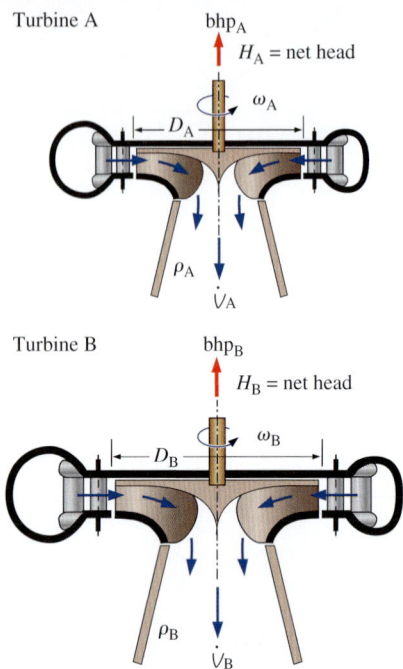

FIGURE 14–113
Dimensional analysis is useful for scaling two *geometrically similar* turbines. If all the dimensionless turbine parameters of turbine A are equivalent to those of turbine B, the two turbines are *dynamically similar*.

We also use the affinity laws to predict the performance of a given turbine operating at different speeds and flow rates in the same way as we did previously for pumps.

The simple similarity laws are strictly valid only if the model and the prototype operate at identical Reynolds numbers and are exactly geometrically similar (including relative surface roughness and tip clearance). Unfortunately, it is not always possible to satisfy all these criteria when performing model tests, because the Reynolds number achievable in the model tests is generally much smaller than that of the prototype, and the model surfaces have larger relative roughness and tip clearances. When the full-scale prototype is significantly larger than its model, the prototype's performance is generally *better*, for the same reasons discussed previously for pumps. Some empirical equations have been developed to account for the increase in efficiency between a small model and a full-scale prototype. One such equation was suggested by Moody (1926), and can be used as a first-order correction,

Moody efficiency correction equation for turbines:

$$\eta_{\text{turbine, prototype}} \cong 1 - (1 - \eta_{\text{turbine, model}})\left(\frac{D_{\text{model}}}{D_{\text{prototype}}}\right)^{1/5} \quad (14\text{–}63)$$

Note that Eq. 14–63 is also used as a first-order correction when scaling model *pumps* to full scale (Eq. 14–34).

In practice, hydroturbine engineers generally find that the actual increase in efficiency from model to prototype is only about two-thirds of the increase given by Eq. 14–63. For example, suppose the efficiency of a one-tenth scale model is 93.2 percent. Equation 14–63 predicts a full-scale efficiency of 95.7 percent, or an increase of 2.5 percent. In practice, we expect only about two-thirds of this increase, or 93.2 + 2.5(2/3) = 94.9 percent. Some more advanced correction equations are available from the **International Electrotechnical Commission (IEC),** a worldwide organization for standardization.

EXAMPLE 14–15   Application of Turbine Affinity Laws

A Francis turbine is being designed for a hydroelectric dam. Instead of starting from scratch, the engineers decide to geometrically scale up a previously designed hydroturbine that has an excellent performance history. The existing turbine (turbine A) has diameter $D_A = 2.05$ m, and spins at $\dot{n}_A = 120$ rpm ($\omega_A = 12.57$ rad/s). At its best efficiency point, $\dot{V}_A = 350$ m³/s, $H_A = 75.0$ m of water, and $bhp_A = 242$ MW. The new turbine (turbine B) is for a larger facility. Its generator will spin at the same speed (120 rpm), but its net head will be higher ($H_B = 104$ m). Calculate the diameter of the new turbine such that it operates most efficiently, and calculate $\dot{V}_B$, bhp_B, and $\eta_{\text{turbine, B}}$.

SOLUTION We are to design a new hydroturbine by scaling up an existing hydroturbine. Specifically we are to calculate the new turbine diameter, volume flow rate, and brake horsepower.
Assumptions **1** The new turbine is geometrically similar to the existing turbine. **2** Reynolds number effects and roughness effects are negligible. **3** The new penstock is also geometrically similar to the existing penstock so that the flow entering the new turbine (velocity profile, turbulence intensity, etc.) is similar to that of the existing turbine.

Properties The density of water at 20°C is $\rho = 998.0$ kg/m³.

Analysis Since the new turbine (B) is dynamically similar to the existing turbine (A), we are concerned with only one particular homologous operating point of both turbines, namely, the best efficiency point. We solve Eq. 14–38b for D_B,

$$D_B = D_A \sqrt{\frac{H_B}{H_A} \frac{\dot{n}_A}{\dot{n}_B}} = (2.05 \text{ m}) \sqrt{\frac{104 \text{ m}}{75.0 \text{ m}} \frac{120 \text{ rpm}}{120 \text{ rpm}}} = \mathbf{2.41 \text{ m}}$$

We then solve Eq. 14–38a for $\dot{V}_B$,

$$\dot{V}_B = \dot{V}_A \left(\frac{\dot{n}_B}{\dot{n}_A}\right)\left(\frac{D_B}{D_A}\right)^3 = (350 \text{ m}^3/\text{s})\left(\frac{120 \text{ rpm}}{120 \text{ rpm}}\right)\left(\frac{2.41 \text{ m}}{2.05 \text{ m}}\right)^3 = \mathbf{572 \text{ m}^3/\text{s}}$$

Finally, we solve Eq. 14–38c for bhp$_B$,

$$\text{bhp}_B = \text{bhp}_A \left(\frac{\rho_B}{\rho_A}\right)\left(\frac{\dot{n}_B}{\dot{n}_A}\right)^3 \left(\frac{D_B}{D_A}\right)^5$$

$$= (242 \text{ MW})\left(\frac{998.0 \text{ kg/m}^3}{998.0 \text{ kg/m}^3}\right)\left(\frac{120 \text{ rpm}}{120 \text{ rpm}}\right)^3 \left(\frac{2.41 \text{ m}}{2.05 \text{ m}}\right)^5 = \mathbf{548 \text{ MW}}$$

As a check, we calculate the dimensionless turbine parameters of Eq. 14–61 for both turbines to show that these two operating points are indeed homologous, and the turbine efficiency is calculated to be 0.942 for both turbines (Fig. 14–114). As discussed previously, however, total dynamic similarity may not actually be achieved between the two turbines because of scale effects (larger turbines generally have higher efficiency). The diameter of the new turbine is about 18 percent greater than that of the existing turbine, so the increase in efficiency due to turbine size should not be significant. We verify this by using the Moody efficiency correction equation (Eq. 14–63), considering turbine A as the "model" and B as the "prototype,"

Efficiency correction:

$$\eta_{\text{turbine, B}} \cong 1 - (1 - \eta_{\text{turbine, A}})\left(\frac{D_A}{D_B}\right)^{1/5} = 1 - (1 - 0.942)\left(\frac{2.05 \text{ m}}{2.41 \text{ m}}\right)^{1/5} = \mathbf{0.944}$$

or 94.4 percent. Indeed, the first-order correction yields a predicted efficiency for the larger turbine that is only a fraction of a percent greater than that of the smaller turbine.

Discussion If the flow entering the new turbine from the penstock were not similar to that of the existing turbine (e.g., velocity profile and turbulence intensity), we could not expect exact dynamic similarity.

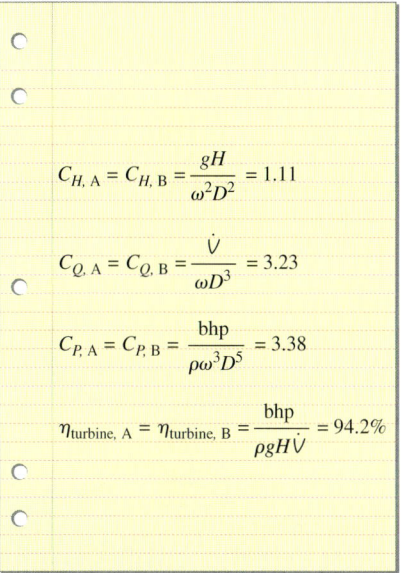

FIGURE 14–114
Dimensionless turbine parameters for both turbines of Example 14–15. Since the two turbines operate at homologous points, their dimensionless parameters must match.

Turbine Specific Speed

In our discussion of pump scaling laws (Sec. 14–3), we defined another useful dimensionless parameter, pump specific speed (N_{Sp}), based on C_Q and C_H. We could use the same definition of specific speed for turbines, but since C_P rather than C_Q is the independent dimensionless parameter for turbines, we define **turbine specific speed** (N_{St}) differently, namely, in terms of C_P and C_H,

Turbine specific speed:

$$N_{St} = \frac{C_P^{1/2}}{C_H^{5/4}} = \frac{(\text{bhp}/\rho\omega^3 D^5)^{1/2}}{(gH/\omega^2 D^2)^{5/4}} = \frac{\omega(\text{bhp})^{1/2}}{\rho^{1/2}(gH)^{5/4}} \qquad (14\text{–}64)$$

858
TURBOMACHINERY

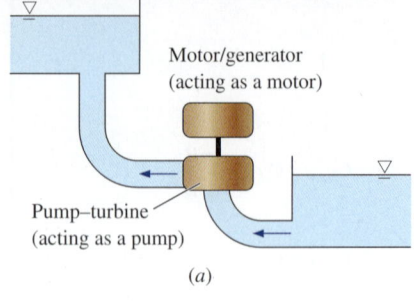

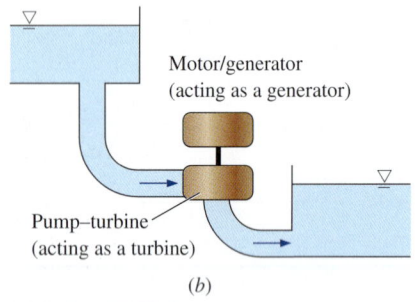

FIGURE 14–115
A *pump–turbine* is used by some power plants for energy storage: (*a*) water is pumped by the pump–turbine during periods of low demand for power, and (*b*) electricity is generated by the pump–turbine during periods of high demand for power.

Turbine specific speed is also called **power specific speed** in some textbooks. It is left as an exercise to compare the definitions of pump specific speed (Eq. 14–35) and turbine specific speed (Eq. 14–64) in order to show that

Relationship between N_{St} and N_{Sp}: $\qquad N_{St} = N_{Sp}\sqrt{\eta_{turbine}}$ (14–65)

Note that Eq. 14–51 does *not* apply to a pump running backward as a turbine or vice versa. There *are* applications in which the *same* turbomachine is used as both a pump *and* a turbine; these devices are appropriately called **pump–turbines.** For example, a coal or nuclear power plant may pump water to a higher elevation during times of low power demand, and then run that water through the same turbomachine (operating as a turbine) during times of high power demand (Fig. 14–115). Such facilities often take advantage of natural elevation differences at mountainous sites and can achieve significant gross heads (upward of 300 m) without construction of a dam. A photograph of a pump–turbine is shown in Fig. 14–116.

Note that there are inefficiencies in the pump–turbine when operating as a pump and also when operating as a turbine. Moreover, since one turbomachine must be designed to operate as both a pump *and* a turbine, neither η_{pump} nor $\eta_{turbine}$ are as high as they would be for a dedicated pump or turbine. Nevertheless, the overall efficiency of this type of energy storage is around 80 percent for a well-designed pump–turbine unit.

In practice, the pump–turbine may operate at a different flow rate and rpm when it is acting as a turbine compared to when it is acting as a pump, since the best efficiency point of the turbine is not necessarily the same as that of the pump. However, for the simple case in which the flow rate and rpm are the same for both the pump and turbine operations, we use Eqs. 14–35 and 14–64 to compare pump specific speed and turbine specific speed. After some algebra,

Pump–turbine specific speed relationship at same flow rate and rpm:

$$N_{St} = N_{Sp}\sqrt{\eta_{turbine}}\left(\frac{H_{pump}}{H_{turbine}}\right)^{3/4} = N_{Sp}(\eta_{turbine})^{5/4}(\eta_{pump})^{3/4}\left(\frac{bhp_{pump}}{bhp_{turbine}}\right)^{3/4} \quad (14\text{–}66)$$

FIGURE 14–116
The runner of a pump–turbine used at the Yards Creek pumped storage station in Blairstown, NJ. There are seven runner blades of outer diameter 5.27 m. The turbine rotates at 240 rpm and produces 112 MW of power at a volume flow rate of 56.6 m³/s from a net head of 221 m.
Courtesy of American Hydro Corporation, York, PA. Used by permission.

We previously discussed some problems with the units of pump specific speed. Unfortunately, these same problems also occur with turbine specific speed. Namely, although N_{St} is by definition a dimensionless parameter, practicing engineers have grown accustomed to using inconsistent units that transform N_{St} into a cumbersome dimensional quantity. In the United States, most turbine engineers write the rotational speed in units of rotations per minute (rpm), bhp in units of horsepower, and H in units of feet. Furthermore, they ignore gravitational constant g and density ρ in the definition of N_{St}. (The turbine is assumed to operate on earth and the working fluid is assumed to be water.) We define

Turbine specific speed, customary U.S. units:

$$N_{St, US} = \frac{(\dot{n}, \text{rpm})(bhp, \text{hp})^{1/2}}{(H, \text{ft})^{5/4}} \quad (14\text{--}67)$$

There is some discrepancy in the turbomachinery literature over the conversions between the two forms of turbine specific speed. To convert $N_{St, US}$ to N_{St} we divide by $g^{5/4}$ and $\rho^{1/2}$, and then use conversion ratios to cancel all units. We set $g = 32.174$ ft/s^2 and assume water at density $\rho = 62.40$ lbm/ft^3. When done properly by converting ω to rad/s, the conversion is $N_{St, US} = 0.02301 N_{St}$ or $N_{St} = 43.46 N_{St, US}$. However, some authors convert ω to *rotations* per second, introducing a factor of 2π in the conversion, i.e., $N_{St, US} = 0.003662 N_{St}$ or $N_{St} = 273.1 N_{St, US}$. The former conversion is more common and is summarized in Fig. 14–117.

There is also a metric or SI version of turbine specific speed that is becoming more popular these days and is preferred by many hydroturbine designers. It is defined in the same way as the customary U.S. pump specific speed (Eq. 14–36), except that SI units are used (m^3/s instead of gpm and m instead of ft),

$$N_{St, SI} = \frac{(\dot{n}, \text{rpm})(\dot{V}, \text{m}^3/\text{s})^{1/2}}{(H, \text{m})^{3/4}} \quad (14\text{--}68)$$

We may call this **capacity specific speed** to distinguish it from power specific speed (Eq. 14–64). One advantage is that $N_{St, SI}$ can be compared more directly to pump specific speed and is thus useful for analyzing pump-turbines. It is less useful, however, to compare $N_{St, SI}$ to previously published values of N_{St} or $N_{St, US}$ because of the fundamental difference between their definitions.

Technically, turbine specific speed could be applied at any operating condition and would just be another function of C_P. That is not how it is typically used, however. Instead, it is common to define turbine specific speed only at the best efficiency point (BEP) of the turbine. The result is a single number that characterizes the turbine.

> Turbine specific speed is used to characterize the operation of a turbine at its optimum conditions (best efficiency point) and is useful for preliminary turbine selection.

As plotted in Fig. 14–118, impulse turbines perform optimally for N_{St} near 0.15, while Francis turbines and Kaplan or propeller turbines perform best at N_{St} near 1 and 2.5, respectively. It turns out that if N_{St} is less than about 0.3, an impulse turbine is the best choice. If N_{St} is between about 0.3 and 2, a Francis turbine is a better choice. When N_{St} is greater than about 2, a

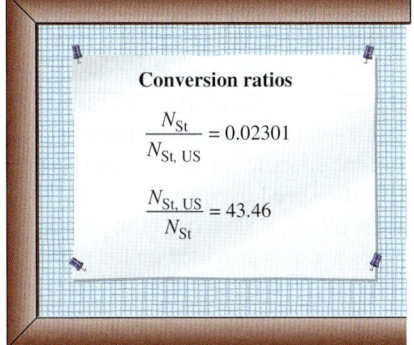

FIGURE 14–117
Conversions between the dimensionless and the conventional U.S. definitions of turbine specific speed. Numerical values are given to four significant digits. The conversions assume earth gravity and water as the working fluid.

860
TURBOMACHINERY

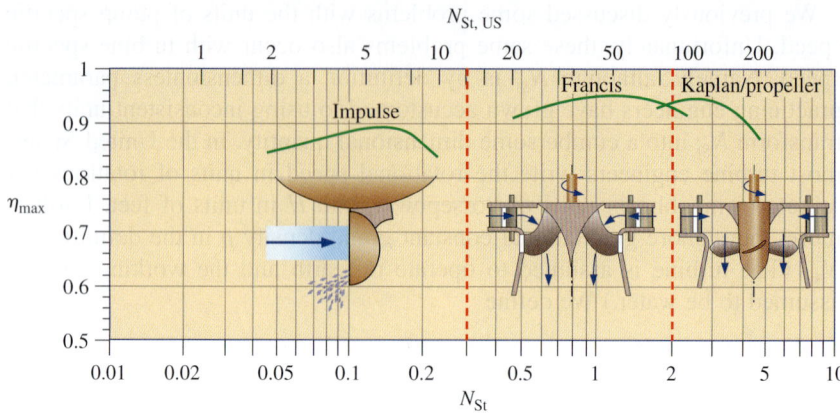

FIGURE 14–118
Maximum efficiency as a function of turbine specific speed for the three main types of dynamic turbine. Horizontal scales show nondimensional turbine specific speed (N_{St}) and turbine specific speed in customary U.S. units ($N_{St, US}$). Sketches of the blade types are also provided on the plot for reference.

Kaplan or propeller turbine should be used. These ranges are indicated in Fig. 14–118 in terms of N_{St} and $N_{St, US}$.

EXAMPLE 14–16   Turbine Specific Speed

Calculate and compare the turbine specific speed for both the small (A) and large (B) turbines of Example 14–15.

SOLUTION The turbine specific speed of two dynamically similar turbines is to be compared.
Properties The density of water at $T = 20°C$ is $\rho = 998.0$ kg/m³.
Analysis We calculate the dimensionless turbine specific speed for turbine A,

$$N_{St, A} = \frac{\omega_A (\text{bhp}_A)^{1/2}}{\rho_A^{1/2} (gH_A)^{5/4}}$$

$$= \frac{(12.57 \text{ rad/s})(242 \times 10^6 \text{ W})^{1/2}}{(998.0 \text{ kg/m}^3)^{1/2}[(9.81 \text{ m/s}^2)(75.0 \text{ m})]^{5/4}} \left(\frac{\text{kg} \cdot \text{m/s}^2}{\text{W} \cdot \text{s}}\right)^{1/2} = 1.615 \cong \mathbf{1.62}$$

and for turbine B,

$$N_{St, B} = \frac{\omega_B (\text{bhp}_B)^{1/2}}{\rho_B^{1/2} (gH_B)^{5/4}}$$

$$= \frac{(12.57 \text{ rad/s})(548 \times 10^6 \text{ W})^{1/2}}{(998.0 \text{ kg/m}^3)^{1/2}[(9.81 \text{ m/s}^2)(104 \text{ m})]^{5/4}} \left(\frac{\text{kg} \cdot \text{m/s}^2}{\text{W} \cdot \text{s}}\right)^{1/2} = 1.615 \cong \mathbf{1.62}$$

We see that the turbine specific speeds of the two turbines are the same. As a check of our algebra we calculate N_{St} in Fig. 14–119 a different way using its definition in terms of C_P and C_H (Eq. 14–64). The result is the same (except for roundoff error). Finally, we calculate the turbine specific speed in customary U.S. units from the conversions of Fig. 14–117,

$$N_{St, US, A} = N_{St, US, B} = 43.46 N_{St} = (43.46)(1.615) = \mathbf{70.2}$$

Discussion Since turbines A and B operate at homologous points, it is no surprise that their turbine specific speeds are the same. In fact, if they weren't the same, it would be a sure sign of an algebraic or calculation error. From Fig. 14–118, a Francis turbine is indeed the appropriate choice for a turbine specific speed of 1.6.

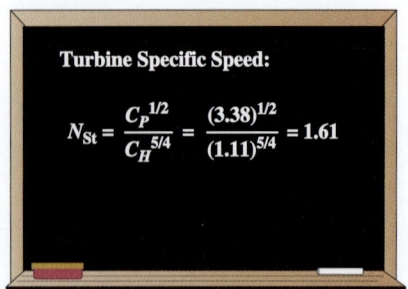

FIGURE 14–119
Calculation of turbine specific speed using the dimensionless parameters C_P and C_H for Example 14–16. (See Fig. 14–114 for values of C_P and C_H for turbine A and turbine B.)

Turbine Specific Speed:

$$N_{St} = \frac{C_P^{1/2}}{C_H^{5/4}} = \frac{(3.38)^{1/2}}{(1.11)^{5/4}} = 1.61$$

CHAPTER 14

APPLICATION SPOTLIGHT ■ Rotary Fuel Atomizers

Guest Author: Werner J. A. Dahm, The University of Michigan

The very high rotation rates at which small gas turbine engines operate, often approaching 100,000 rpm, allow rotary centrifugal atomizers to create the liquid fuel spray that is burned in the combustor. Note that a 10-cm-diameter atomizer rotating at 30,000 rpm imparts 490,000 m/s^2 of acceleration (50,000 g) to the liquid fuel, which allows such fuel atomizers to potentially produce very small drop sizes.

The actual drop sizes depend on the fluid properties, including the liquid and gas densities ρ_L and ρ_G, the viscosities μ_L and μ_G, and the liquid–gas surface tension σ_s. Figure 14–120 shows such a rotary atomizer rotating at rate ω, with radial channels in the rim at nominal radius $R \equiv (R_1 + R_2)/2$. Fuel flows into the channels due to the acceleration $R\omega^2$ and forms a liquid film on the channel walls. The large acceleration leads to a typical film thickness t of only about 10 μm. The channel shape is chosen to produce desirable atomization performance. For a given shape, the resulting drop sizes depend on the cross-flow velocity $V_c \equiv R\omega$ into which the film issues at the channel exit, together with the liquid and gas properties. From these, there are four dimensionless groups that determine the atomization performance: the liquid–gas density and viscosity ratios $r \equiv [\rho_L/\rho_G]$ and $m \equiv [\mu_L/\mu_G]$, the film *Weber number* $We_t \equiv [\rho_G V_c^2 t/\sigma_s]$, and the *Ohnesorge number* $Oh_t \equiv [\mu_L/(\rho_L \sigma_s t)^{1/2}]$.

Note that We_t gives the characteristic ratio of the aerodynamic forces that the gas exerts on the liquid film to the surface tension forces that act on the liquid surface, while Oh_t gives the ratio of the viscous forces in the liquid film to the surface tension forces that act on the film. Together these express the relative importance of the three main physical effects involved in the atomization process: *inertia*, *viscous diffusion*, and *surface tension*.

Figure 14–121 shows examples of the resulting liquid breakup process for several channel shapes and rotation rates, visualized using 10-ns pulsed-laser photography. The drop sizes turn out to be relatively insensitive to changes in the Ohnesorge number, since the values for practical fuel atomizers are in the limit $Oh_t \ll 1$ and thus viscous effects are relatively unimportant. The Weber number, however, remains crucial since surface tension and inertia effects dominate the atomization process. At small We_t, the liquid undergoes *subcritical* breakup in which surface tension pulls the thin liquid film into a single column that subsequently breaks up to form relatively large drops. At *supercritical* values of We_t, the thin liquid film breaks up aerodynamically into fine drop sizes on the order of the film thickness t. From results such as these, engineers can successfully develop rotary fuel atomizers for practical applications.

Reference

Dahm, W. J. A., Patel, P. R., and Lerg, B. H., "Visualization and Fundamental Analysis of Liquid Atomization by Fuel Slingers in Small Gas Turbines," *AIAA Paper No. 2002-3183*, AIAA, Washington, DC, 2002.

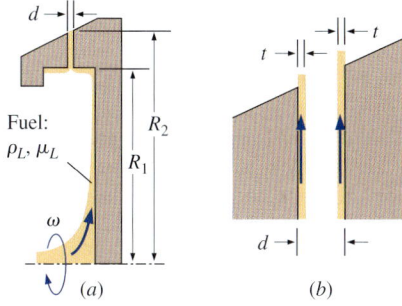

FIGURE 14–120
Schematic diagram of *(a)* a rotary fuel atomizer, and *(b)* a close-up of the liquid fuel film along the channel walls.

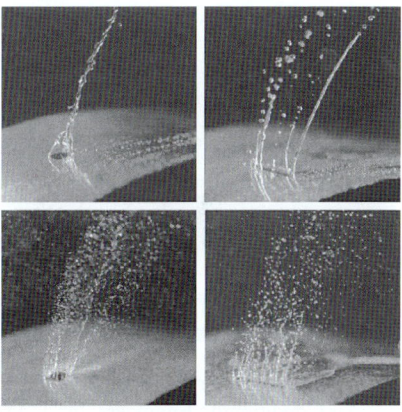

FIGURE 14–121
Visualizations of liquid breakup by rotary fuel atomizers, showing subcritical breakup at relatively low values of We_t *(top)*, for which surface tension effects are sufficiently strong relative to inertia to pull the thin liquid film into large columns; and supercritical breakup at higher values of We_t *(bottom)*, for which inertia dominates over surface tension and the thin film breaks into fine droplets.

Reprinted by permission of Werner J. A. Dahm, University of Michigan.

SUMMARY

We classify turbomachinery into two broad categories, *pumps* and *turbines*. The word *pump* is a general term for any fluid machine that *adds* energy *to* a fluid. We explain how this energy transfer occurs for several types of pump designs—both *positive-displacement pumps* and *dynamic pumps*. The word *turbine* refers to a fluid machine that *extracts* energy *from* a fluid. There are also *positive-displacement turbines* and *dynamic turbines* of several varieties.

The most useful equation for preliminary turbomachinery design is the *Euler turbomachine equation*,

$$T_{shaft} = \rho \dot{V}(r_2 V_{2,t} - r_1 V_{1,t})$$

Note that for pumps, the inlet and outlet are at radii r_1 and r_2, respectively, while for turbines, the inlet is at radius r_2 and the outlet is at radius r_1. We show several examples where blade shapes for both pumps and turbines are designed based on desired flow velocities. Then, using the Euler turbomachine equation, the performance of the turbomachine is predicted.

The *turbomachinery scaling laws* illustrate a practical application of dimensional analysis. The scaling laws are used in the design of new turbomachines that are geometrically similar to existing turbomachines. For both pumps and turbines, the main dimensionless parameters are head coefficient, capacity coefficient, and power coefficient, defined respectively as

$$C_H = \frac{gH}{\omega^2 D^2} \qquad C_Q = \frac{\dot{V}}{\omega D^3} \qquad C_P = \frac{bhp}{\rho \omega^3 D^5}$$

In addition to these, we define *pump efficiency* and *turbine efficiency* as reciprocals of each other,

$$\eta_{pump} = \frac{\dot{W}_{water\ horsepower}}{\dot{W}_{shaft}} = \frac{\rho g \dot{V} H}{bhp}$$

$$\eta_{turbine} = \frac{\dot{W}_{shaft}}{\dot{W}_{water\ horsepower}} = \frac{bhp}{\rho g \dot{V} H}$$

Finally, two other useful dimensionless parameters called *pump specific speed* and *turbine specific speed* are defined, respectively, as

$$N_{Sp} = \frac{C_Q^{1/2}}{C_H^{3/4}} = \frac{\omega \dot{V}^{1/2}}{(gH)^{3/4}} \qquad N_{St} = \frac{C_P^{1/2}}{C_H^{5/4}} = \frac{\omega (bhp)^{1/2}}{\rho^{1/2}(gH)^{5/4}}$$

These parameters are useful for preliminary design and for selection of the type of pump or turbine that is most appropriate for a given application.

We discuss the basic design features of both *hydroturbines* and *wind turbines*. For the latter we derive an upper limit to the power coefficient, namely the *Betz limit*,

$$C_{P,\ max} = 4\frac{1}{3}\left(1 - \frac{1}{3}\right)^2 = \frac{16}{27} \cong 0.5926$$

Turbomachinery design assimilates knowledge from several key areas of fluid mechanics, including mass, energy, and momentum analysis (Chaps. 5 and 6); dimensional analysis and modeling (Chap. 7); flow in pipes (Chap. 8); differential analysis (Chaps. 9 and 10); and aerodynamics (Chap. 11). In addition, for gas turbines and other types of turbomachines that involve gases, compressible flow analysis (Chap. 12) is required. Finally, computational fluid dynamics (Chap. 15) plays an ever-increasing role in the design of highly efficient turbomachines.

REFERENCES AND SUGGESTED READING

1. ASHRAE (American Society of Heating, Refrigerating and Air Conditioning Engineers, Inc.). *ASHRAE Fundamentals Handbook*, ASHRAE, 1791 Tullie Circle, NE, Atlanta, GA, 30329; editions every four years: 1993, 1997, 2001, etc.

2. L. F. Moody. "The Propeller Type Turbine," *ASCE Trans.*, 89, p. 628, 1926.

3. Earl Logan, Jr., ed. *Handbook of Turbomachinery*. New York: Marcel Dekker, Inc., 1995.

4. A. J. Glassman, ed. *Turbine Design and Application*. NASA Sp-290, NASA Scientific and Technical Information Program. Washington, DC, 1994.

5. D. Japikse and N. C. Baines. *Introduction to Turbomachinery*. Norwich, VT: Concepts ETI, Inc., and Oxford: Oxford University Press, 1994.

6. Earl Logan, Jr. *Turbomachinery: Basic Theory and Applications*, 2nd ed. New York: Marcel Dekker, Inc., 1993.

7. R. K. Turton. *Principles of Turbomachinery*, 2nd ed. London: Chapman & Hall, 1995.

8. Terry Wright. *Fluid Machinery: Application, Selection, and Design*. Boca Raton, FL: CRC Press, 2009.

9. J. F. Manwell, J. G. McGowan, and A. L. Rogers. *Wind Energy Explained – Theory, Design, and Application*, 2nd ed. West Sussex, England: John Wiley & Sons, LTC, 2010.

10. M. L. Robinson. "The Darrieus Wind Turbine for Electrical Power Generation," *J. Royal Aeronautical Society*, Vol. 85, pp. 244–255, June 1981.

PROBLEMS*

General Problems

14–1C What are the primary differences between fans, blowers, and compressors? Discuss in terms of pressure rise and volume flow rate.

14–2C What is the more common term for an *energy-producing turbomachine*? How about an *energy-absorbing turbomachine*? Explain this terminology. In particular, from which frame of reference are these terms defined—that of the fluid or that of the surroundings?

14–3C Discuss the primary difference between a *positive-displacement turbomachine* and a *dynamic turbomachine*. Give an example of each for both pumps and turbines.

14–4C Explain why there is an "extra" term in the Bernoulli equation in a rotating reference frame.

14–5C For a turbine, discuss the difference between *brake horsepower* and *water horsepower*, and also define turbine efficiency in terms of these quantities.

14–6C For a pump, discuss the difference between *brake horsepower* and *water horsepower*, and also define pump efficiency in terms of these quantities.

14–7 An air compressor increases the pressure ($P_{out} > P_{in}$) and the density ($\rho_{out} > \rho_{in}$) of the air passing through it (Fig. P14–7). For the case in which the outlet and inlet diameters are equal ($D_{out} = D_{in}$), how does average air speed change across the compressor? In particular, is V_{out} less than, equal to, or greater than V_{in}? Explain.   *Answer:* less than

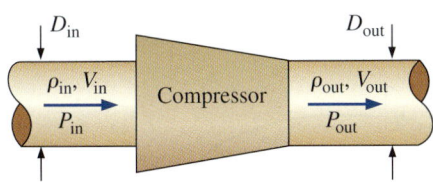

FIGURE P14–7

* Problems designated by a "C" are concept questions, and students are encouraged to answer them all. Problems with the [icon] icon are solved using EES, and complete solutions together with parametric studies are included on the text website. Problems with the [EES] icon are comprehensive in nature and are intended to be solved with an equation solver such as EES.

14–8 A water pump increases the pressure of the water passing through it (Fig. P14–8). The flow is assumed to be incompressible. For each of the three cases listed below, how does average water speed change across the pump? In particular, is V_{out} less than, equal to, or greater than V_{in}? Show your equations, and explain.

(*a*) Outlet diameter is less than inlet diameter ($D_{out} < D_{in}$)
(*b*) Outlet and inlet diameters are equal ($D_{out} = D_{in}$)
(*c*) Outlet diameter is greater than inlet diameter ($D_{out} > D_{in}$)

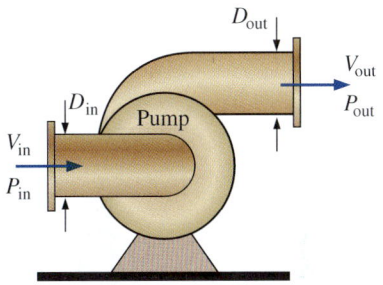

FIGURE P14–8

Pumps

14–9C Define *net positive suction head* and *required net positive suction head*, and explain how these two quantities are used to ensure that cavitation does not occur in a pump.

14–10C For each statement about centrifugal pumps, choose whether the statement is true or false, and discuss your answer briefly:

(*a*) A centrifugal pump with radial blades has higher efficiency than the same pump with backward-inclined blades.

(*b*) A centrifugal pump with radial blades produces a larger pressure rise than the same pump with backward- or forward-inclined blades over a wide range of $\dot{V}$.

(*c*) A centrifugal pump with forward-inclined blades is a good choice when one needs to provide a large pressure rise over a wide range of volume flow rates.

(*d*) A centrifugal pump with forward-inclined blades would most likely have less blades than a pump of the same size with backward-inclined or radial blades.

14–11C Figure P14–11C shows two possible locations for a water pump in a piping system that pumps water from the lower tank to the upper tank. Which location is better? Why?

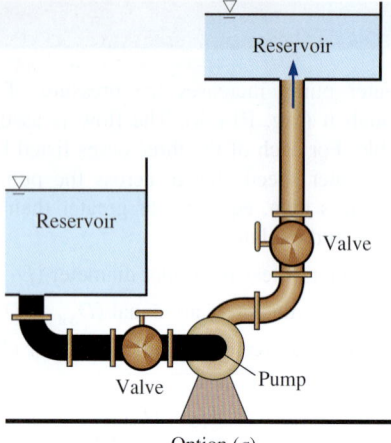

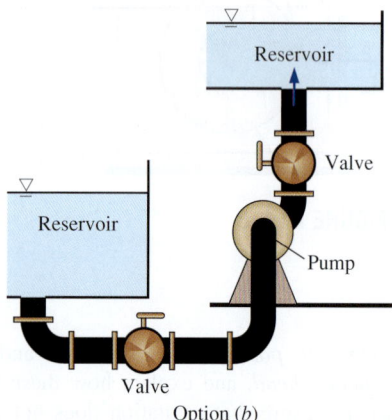

FIGURE P14–11

14–12C There are three main categories of dynamic pumps. List and define them.

14–13C Consider flow through a water pump. For each statement, choose whether the statement is true or false, and discuss your answer briefly:

(a) The faster the flow through the pump, the more likely that cavitation will occur.
(b) As water temperature increases, NPSH$_{required}$ also increases.
(c) As water temperature increases, the available NPSH also increases.
(d) As water temperature increases, cavitation is less likely to occur.

14–14C Write the equation that defines actual (available) net positive suction head NPSH. From this definition, discuss at least five ways you can decrease the likelihood of cavitation in the pump, for the same liquid, temperature, and volume flow rate.

14–15C Consider a typical centrifugal liquid pump. For each statement, choose whether the statement is true or false, and discuss your answer briefly:

(a) $\dot{V}$ at the pump's *free delivery* is greater than $\dot{V}$ at its *best efficiency point*.
(b) At the pump's *shutoff head*, the *pump efficiency* is zero.
(c) At the pump's *best efficiency point*, its *net head* is at its maximum value.
(d) At the pump's *free delivery*, the *pump efficiency* is zero.

14–16C Explain why it is usually not wise to arrange two (or more) dissimilar pumps in series or in parallel.

14–17C Consider steady, incompressible flow through two identical pumps (pumps 1 and 2), either in series or in parallel. For each statement, choose whether the statement is true or false, and discuss your answer briefly:

(a) The volume flow rate through the two pumps in series is equal to $\dot{V}_1 + \dot{V}_2$.
(b) The overall net head across the two pumps in series is equal to $H_1 + H_2$.
(c) The volume flow rate through the two pumps in parallel is equal to $\dot{V}_1 + \dot{V}_2$.
(d) The overall net head across the two pumps in parallel is equal to $H_1 + H_2$.

14–18C In Fig. P14–18C is shown a plot of pump net head as a function of pump volume flow rate, or capacity. On the figure, label the shutoff head, the free delivery, the pump performance curve, the system curve, and the operating point.

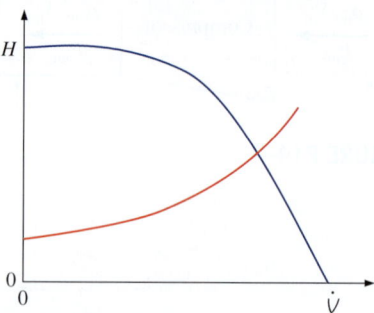

FIGURE P14–18C

14–19 Suppose the pump of Fig. P14–18C is situated between two water tanks with their free surfaces open to the atmosphere. Which free surface is at a higher elevation—the one corresponding to the tank supplying water to the pump inlet, or the one corresponding to the tank connected to the pump outlet? Justify your answer through use of the energy equation between the two free surfaces.

14–20 Suppose the pump of Fig. P14–18C is situated between two large water tanks with their free surfaces open to the atmosphere. Explain qualitatively what would happen to the pump performance curve if the free surface of the outlet tank were raised in elevation, all else being equal. Repeat for the system curve. What would happen to the operating point—would the volume flow rate at the operating point decrease, increase, or remain the same? Indicate the change on a qualitative plot of H versus $\dot{V}$, and discuss. (*Hint*: Use the energy equation between the free surface of the tank upstream of the pump and the free surface of the tank downstream of the pump.)

14–21 Suppose the pump of Fig. P14–18C is situated between two large water tanks with their free surfaces open to the atmosphere. Explain qualitatively what would happen to the pump performance curve if a valve in the piping system were changed from 100 percent open to 50 percent open, all else being equal. Repeat for the system curve. What would happen to the operating point—would the volume flow rate at the operating point decrease, increase, or remain the same? Indicate the change on a qualitative plot of H versus $\dot{V}$, and discuss. (*Hint*: Use the energy equation between the free surface of the upstream tank and the free surface of the downstream tank.) *Answer:* decrease

14–22 Consider the flow system sketched in Fig. P14–22. The fluid is water, and the pump is a centrifugal pump. Generate a *qualitative* plot of the pump net head as a function of the pump capacity. On the figure, label the shutoff head, the free delivery, the pump performance curve, the system curve, and the operating point. (*Hint*: Carefully consider the required net head at conditions of zero flow rate.)

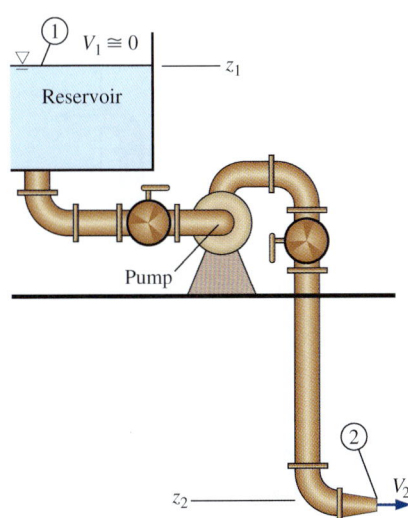

FIGURE P14–22

14–23 Suppose the pump of Fig. P14–22 is operating at *free delivery conditions*. The pipe, both upstream and downstream of the pump, has an inner diameter of 2.0 cm and nearly zero roughness. The minor loss coefficient associated with the sharp inlet is 0.50, each valve has a minor loss coefficient of 2.4, and each of the three elbows has a minor loss coefficient of 0.90. The contraction at the exit reduces the diameter by a factor of 0.60 (60% of the pipe diameter), and the minor loss coefficient of the contraction is 0.15. Note that this minor loss coefficient is based on the average *exit* velocity, not the average velocity through the pipe itself. The total length of pipe is 8.75 m, and the elevation difference is $(z_1 - z_2) = 4.6$ m. Estimate the volume flow rate through this piping system. *Answer:* 34.4 Lpm

14–24 Repeat Prob. 14–23, but with a rough pipe—pipe roughness $\epsilon = 0.12$ mm. Assume that a modified pump is used, such that the new pump operates at its free delivery conditions, just as in Prob. 14–23. Assume all other dimensions and parameters are the same as in that problem. Do your results agree with intuition? Explain.

14–25 Consider the piping system of Fig. P14–22, with all the dimensions, parameters, minor loss coefficients, etc., of Prob. 14–23. The pump's performance follows a parabolic curve fit, $H_{available} = H_0 - a\dot{V}^2$, where $H_0 = 19.8$ m is the pump's shutoff head, and $a = 0.00426$ m/(Lpm)^2 is a coefficient of the curve fit. Estimate the operating volume flow rate $\dot{V}$ in Lpm (liters per minute), and compare with that of Prob. 14–23. Discuss.

14–26 Repeat Prob. 14–25, but instead of a smooth pipe, let the pipe roughness = 0.12 mm. Compare to the smooth pipe case and discuss—does the result agree with your intuition?

14–27 The performance data for a centrifugal water pump are shown in Table P14–27 for water at 20°C (Lpm = liters per minute). (*a*) For each row of data, calculate the pump efficiency (percent). *Show all units and unit conversions for full credit.* (*b*) Estimate the volume flow rate (Lpm) and net head (m) at the BEP of the pump.

TABLE P14–27

$\dot{V}$, Lpm	H, m	bhp, W
0.0	47.5	133
6.0	46.2	142
12.0	42.5	153
18.0	36.2	164
24.0	26.2	172
30.0	15.0	174
36.0	0.0	174

14–28 For the centrifugal water pump of Prob. 14–27, plot the pump's performance data: H (m), bhp (W), and η_{pump} (percent) as functions of $\dot{V}$ (Lpm), using symbols only (no lines). Perform linear least-squares polynomial curve fits for all three parameters, and plot the fitted curves as lines (no symbols) on the same plot. For consistency, use a first-order curve fit for H as a function of $\dot{V}^2$, use a second-order curve fit for bhp as a function of both $\dot{V}$ and $\dot{V}^2$, and use a third-order curve fit for η_{pump} as a function of $\dot{V}$, $\dot{V}^2$, and $\dot{V}^3$. List all curve-fitted equations and coefficients (with units) for full credit. Calculate the BEP of the pump based on the curve-fitted expressions.

14–29 Suppose the pump of Probs. 14–27 and 14–28 is used in a piping system that has the system requirement $H_{required} = (z_2 - z_1) + b\dot{V}^2$, where the elevation difference $z_2 - z_1 = 21.7$ m, and coefficient $b = 0.0185$ m/(Lpm)^2. Estimate the operating point of the system, namely, $\dot{V}_{operating}$ (Lpm) and $H_{operating}$ (m).

14–30 Suppose you are looking into purchasing a water pump with the performance data shown in Table P14–30. Your supervisor asks for some more information about the pump. (a) Estimate the shutoff head H_0 and the free delivery $\dot{V}_{max}$ of the pump. [*Hint:* Perform a least-squares curve fit (regression analysis) of $H_{available}$ versus $\dot{V}^2$, and calculate the best-fit values of coefficients H_0 and a that translate the tabulated data of Table P14–30 into the parabolic expression, $H_{available} = H_0 - a\dot{V}^2$. From these coefficients, estimate the free delivery of the pump.] (b) The application requires 57.0 Lpm of flow at a pressure rise across the pump of 40 kPa. Is this pump capable of meeting the requirements? Explain.

TABLE P14–30

$\dot{V}$, Lpm	H, m
20	21
30	18.4
40	14
50	7.6

14–31 The performance data of a water pump follow the curve fit $H_{available} = H_0 - a\dot{V}^2$, where the pump's shutoff head $H_0 = 7.46$ m, coefficient $a = 0.0453$ m/(Lpm)^2, the units of pump head H are meters, and the units of $\dot{V}$ are liters per minute (Lpm). The pump is used to pump water from one large reservoir to another large reservoir at a higher elevation. The free surfaces of both reservoirs are exposed to atmospheric pressure. The system curve simplifies to $H_{required} = (z_2 - z_1) + b\dot{V}^2$, where elevation difference $z_2 - z_1 = 3.52$ m, and coefficient $b = 0.0261$ m/(Lpm)^2. Calculate the operating point of the pump ($\dot{V}_{operating}$ and $H_{operating}$) in appropriate units (Lpm and meters, respectively). *Answers:* 7.43 Lpm, 4.96 m

14–32 For the application at hand, the flow rate of Prob. 14–31 is not adequate. At least 9 Lpm is required. Repeat Prob. 14–31 for a more powerful pump with $H_0 = 8.13$ m and $a = 0.0297$ m/(Lpm)^2. Calculate the percentage improvement in flow rate compared to the original pump. Is this pump able to deliver the required flow rate?

14–33 A manufacturer of small water pumps lists the performance data for a family of its pumps as a parabolic curve fit, $H_{available} = H_0 - a\dot{V}^2$, where H_0 is the pump's shutoff head and a is a coefficient. Both H_0 and a are listed in a table for the pump family, along with the pump's free delivery. The pump head is given in units of meter of water column, and capacity is given in units of liters per minute. (a) What are the units of coefficient a? (b) Generate an expression for the pump's free delivery $\dot{V}_{max}$ in terms of H_0 and a. (c) Suppose one of the manufacturer's pumps is used to pump water from one large reservoir to another at a higher elevation. The free surfaces of both reservoirs are exposed to atmospheric pressure. The system curve simplifies to $H_{required} = (z_2 - z_1) + b\dot{V}^2$. Calculate the operating point of the pump ($\dot{V}_{operating}$ and $H_{operating}$) in terms of H_0, a, b, and elevation difference $z_2 - z_1$.

14–34 A water pump is used to pump water from one large reservoir to another large reservoir that is at a higher elevation. The free surfaces of both reservoirs are exposed to atmospheric pressure, as sketched in Fig. P14–34. The dimensions and minor loss coefficients are provided in the figure. The pump's performance is approximated by the expression $H_{available} = H_0 - a\dot{V}^2$, where the shutoff head $H_0 = 40$ m of water column, coefficient $a = 0.053$ m/Lpm^2, available pump head $H_{available}$ is in units of meter of water column, and capacity $\dot{V}$ is in units of liters per minute (Lpm). Estimate the capacity delivered by the pump. *Answer:* 24.7 Lpm

$z_2 - z_1 = 6.7$ m (elevation difference)
$D = 3.0$ cm (pipe diameter)
$K_{L, entrance} = 0.50$ (pipe entrance)
$K_{L, valve\ 1} = 2.0$ (valve 1)
$K_{L, valve\ 2} = 6.8$ (valve 2)
$K_{L, elbow} = 0.34$ (each elbow—there are 3)
$K_{L, exit} = 1.05$ (pipe exit)
$L = 40$ m (total pipe length)
$\varepsilon = 0.0028$ cm (pipe roughness)

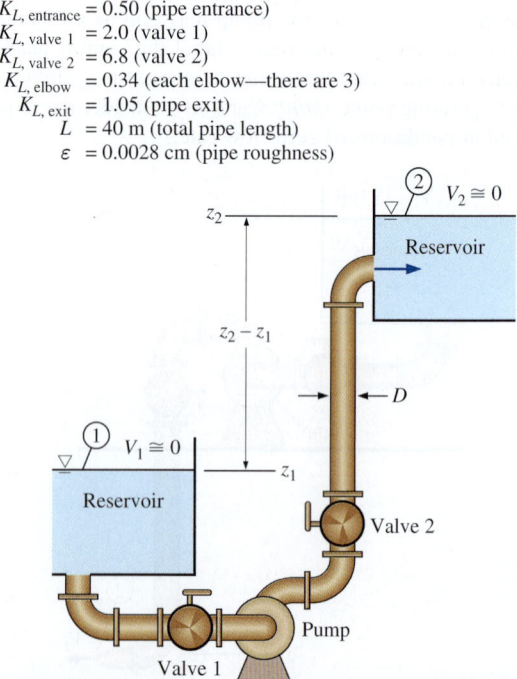

FIGURE P14–34

14–35 For the pump and piping system of Prob. 14–34, plot the required pump head $H_{required}$ (m of water column) as a function of volume flow rate $\dot{V}$ (Lpm). On the same plot, compare the available pump head $H_{available}$ versus $\dot{V}$, and mark the operating point. Discuss.

14–36 Suppose that the two reservoirs in Prob. 14–34 are 300 m farther apart horizontally, but at the same elevations. All the constants and parameters are identical to those of Prob. 14–34 except that the total pipe length is 340 m instead of 40 m. Calculate the volume flow rate for this case and compare with the result of Prob. 14–34. Discuss.

14–37 Paul realizes that the pump being used in Prob. 14–34 is not well-matched for this application, since its shutoff head (40 m) is much larger than its required net head (less than 10 m), and its capacity is fairly low. In other words, this pump is designed for high-head, low-capacity applications, whereas the application at hand is fairly low-head, and a higher capacity is desired. Paul tries to convince his supervisor that a less expensive pump, with lower shutoff head but higher free delivery, would result in a significantly increased flow rate between the two reservoirs. Paul looks through some online brochures, and finds a pump with the performance data shown in Table P14–37. His supervisor asks him to predict the volume flow rate between the two reservoirs if the existing pump were replaced with the new pump. (a) Perform a least-squares curve fit (regression analysis) of $H_{available}$ versus $\dot{V}^2$, and calculate the best-fit values of coefficients H_0 and a that translate the tabulated data of Table P14–37 into the parabolic expression $H_{available} = H_0 - a\dot{V}^2$. Plot the data points as symbols and the curve fit as a line for comparison. (b) Estimate the operating volume flow rate of the new pump if it were to replace the existing pump, all else being equal. Compare to the result of Prob. 14–34 and discuss. Is Paul correct? (c) Generate a plot of required net head and available net head as functions of volume flow rate and indicate the operating point on the plot.

TABLE P14–37

$\dot{V}$, Lpm	H, m
0	11.4
15	11.1
30	10.2
45	8.7
60	6.3
75	3.6
90	0

14–38 A water pump is used to pump water from one large reservoir to another large reservoir that is at a higher elevation. The free surfaces of both reservoirs are exposed to atmospheric pressure, as sketched in Fig. P14–38. The dimensions and minor loss coefficients are provided in the figure. The pump's performance is approximated by the expression $H_{available} = H_0 - a\dot{V}^2$, where shutoff head $H_0 = 24.4$ m of water column, coefficient $a = 0.0678$ m/Lpm^2, available pump head $H_{available}$ is in units of meters of water column, and capacity $\dot{V}$ is in units of liters per minute (Lpm). Estimate the capacity delivered by the pump. *Answer:* 11.6 Lpm

$z_2 - z_1$ = 7.85 m (elevation difference)
D = 2.03 cm (pipe diameter)
$K_{L, entrance}$ = 0.50 (pipe entrance)
$K_{L, valve}$ = 17.5 (valve)
$K_{L, elbow}$ = 0.92 (each elbow—there are 5)
$K_{L, exit}$ = 1.05 (pipe exit)
L = 176.5 m (total pipe length)
ε = 0.25 mm (pipe roughness)

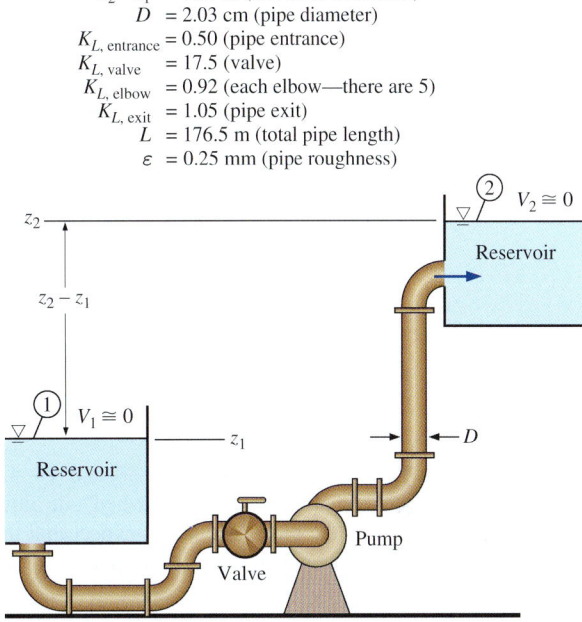

FIGURE P14–38

14–39 For the pump and piping system of Prob. 14–38, plot required pump head $H_{required}$ (m of water column) as a function of volume flow rate $\dot{V}$ (Lpm). On the same plot, compare available pump head $H_{available}$ versus $\dot{V}$, and mark the operating point. Discuss.

14–40 Suppose that the free surface of the inlet reservoir in Prob. 14–38 is 3.0 m lower in elevation, such that $z_2 - z_1 = 10.85$ m. All the constants and parameters are identical to those of Prob. 14–38 except for the elevation difference. Calculate the volume flow rate for this case and compare with the result of Prob. 14–38. Discuss.

14–41 April's supervisor asks her to find a replacement pump that will increase the flow rate through the piping system of Prob. 14–38 by a factor of 2 or greater. April looks through some online brochures, and finds a pump with the performance data shown in Table P14–41. All dimensions and parameters remain the same as in Prob. 14–38—only the pump is changed. (a) Perform a least-squares curve fit (regression analysis) of $H_{available}$ versus $\dot{V}^2$, and calculate the best-fit values of coefficients H_0 and a that translate the tabulated data of Table P14–41 into the parabolic expression $H_{available} = H_0 - a\dot{V}^2$. Plot the data points as symbols and the curve fit as a line for comparison. (b) Use the expression obtained in part (a) to estimate the operating volume flow rate of the new pump if it were to replace the existing pump, all else being equal. Compare to the result of Prob. 14–38 and discuss. Has April achieved her goal? (c) Generate a plot of required net head and available net head as functions of volume flow rate, and indicate the operating point on the plot.

TABLE P14–41

$\dot{V}$, Lpm	H, m
0	46.5
5	46
10	42
15	37
20	29
25	16.5
30	0

14–42 Calculate the volume flow rate between the reservoirs of Prob. 14–38 for the case in which the pipe diameter is doubled, all else remaining the same. Discuss.

14–43 Comparing the results of Probs. 14–38 and 14–42, the volume flow rate increases as expected when one doubles the inner diameter of the pipe. One might expect that the Reynolds number increases as well. Does it? Explain.

14–44 Repeat Prob. 14–38, but neglect all minor losses. Compare the volume flow rate with that of Prob. 14–38. Are minor losses important in this problem? Discuss.

14–45 Consider the pump and piping system of Prob. 14–38. Suppose that the lower reservoir is huge, and its surface does not change elevation, but the upper reservoir is not so big, and its surface rises slowly as the reservoir fills. Generate a curve of volume flow rate $\dot{V}$ (Lpm) as a function of $z_2 - z_1$ in the range 0 to the value of $z_2 - z_1$ at which the pump ceases to pump any more water. At what value of $z_2 - z_1$ does this occur? Is the curve linear? Why or why not? What would happen if $z_2 - z_1$ were greater than this value? Explain.

14–46 A local ventilation system (a hood and duct system) is used to remove air and contaminants from a pharmaceutical lab (Fig. P14–46). The inner diameter (ID) of the duct is $D = 150$ mm, its average roughness is 0.15 mm, and its total length is $L = 24.5$ m. There are three elbows along the duct, each with a minor loss coefficient of 0.21. Literature from the hood manufacturer lists the hood entry loss coefficient as 3.3 based on duct velocity. When the damper is fully open, its loss coefficient is 1.8. The minor loss coefficient through the 90° tee is 0.36. Finally, a one-way valve is installed to prevent contaminants from a second hood from flowing "backward" into the room. The minor loss coefficient of the (open) one-way valve is 6.6. The performance data of the fan fit a parabolic curve of the form $H_{available} = H_0 - a\dot{V}^2$, where shutoff head $H_0 = 60.0$ mm of water column, coefficient $a =$ 2.50×10^{-7} mm of water column per (Lpm)², available head $H_{available}$ is in units of mm of water column, and capacity $\dot{V}$ is in units of Lpm of air. Estimate the volume flow rate in Lpm through this ventilation system. *Answer:* 7090 Lpm

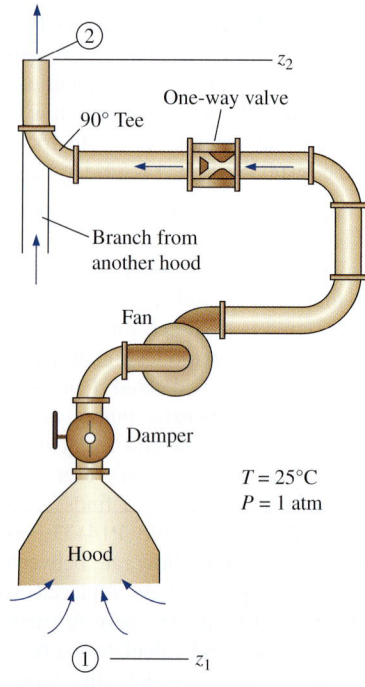

FIGURE P14–46

14–47 For the duct system of Prob. 14–46, plot required fan head $H_{required}$ (mm of water column) as a function of volume flow rate $\dot{V}$ (Lpm). On the same plot, compare available fan head $H_{available}$ versus $\dot{V}$, and mark the operating point. Discuss.

14–48 Repeat Prob. 14–46, ignoring all minor losses. How important are the minor losses in this problem? Discuss.

14–49 Suppose the one-way valve of Fig. P14–46 malfunctions due to corrosion and is stuck in its fully closed position (no air can get through). The fan is on, and all other conditions are identical to those of Prob. 14–46. Calculate the gage pressure (in pascals and in mm of water column) at a point just downstream of the fan. Repeat for a point just upstream of the one-way valve.

14–50 A local ventilation system (a hood and duct system) is used to remove air and contaminants produced by a welding operation (Fig. P14–50). The inner diameter (ID) of the duct is $D = 23$ cm, its average roughness is 0.015 cm, and its total length is $L = 10.4$ m. There are three elbows along the duct, each with a minor loss coefficient of 0.21. Literature from the hood manufacturer lists the hood entry loss coefficient as 4.6 based on duct velocity. When the damper is fully open, its loss coefficient is 1.8. A squirrel cage centrifugal fan with a 22.9 cm inlet is available. Its performance data fit a parabolic curve of the form $H_{available} = H_0 - a\dot{V}^2$, where shutoff head $H_0 = 5.8$ cm of water column, coefficient $a = 96.9$ cm of water column per $(m^3/s)^2$, available head $H_{available}$ is in units of cm of water column, and capacity $\dot{V}$ is in units of m^3/s (at 25°C). Estimate the volume flow rate in m^3/s through this ventilation system. *Answer:* 0.212 m³/s

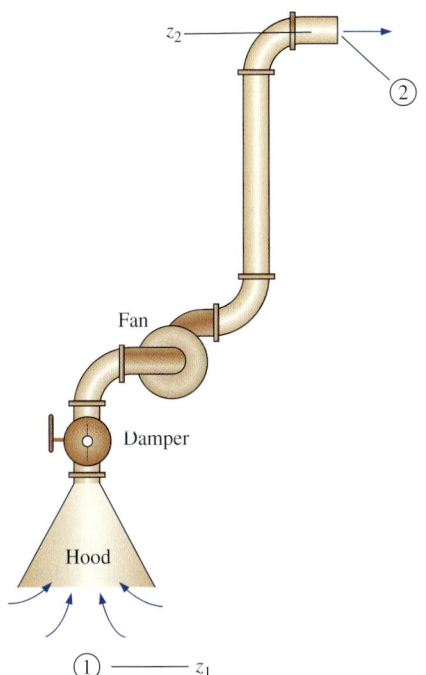

FIGURE P14–50

14–51 For the duct system and fan of Prob. 14–50, partially closing the damper would decrease the flow rate. All else being unchanged, estimate the minor loss coefficient of the damper required to decrease the volume flow rate by a factor of 2.

14–52 Repeat Prob. 14–50, ignoring all minor losses. How important are the minor losses in this problem? Discuss.

14–53 A self-priming centrifugal pump is used to pump water at 25°C from a reservoir whose surface is 2.2 m above the centerline of the pump inlet (Fig. P14–53). The pipe is PVC pipe with an ID of 24.0 mm and negligible average inner roughness height. The pipe length from the submerged pipe inlet to the pump inlet is 2.8 m. There are only two minor losses in the piping system from the pipe inlet to the pump inlet: a sharp-edged reentrant inlet ($K_L = 0.85$), and a flanged smooth 90° regular elbow ($K_L = 0.3$). The pump's required net positive suction head is provided by the manufacturer as a curve fit: NPSH$_{required}$ = 2.2 m + (0.0013 m/Lpm²)$\dot{V}^2$,

where volume flow rate is in Lpm. Estimate the maximum volume flow rate (in units of Lpm) that can be pumped without cavitation.

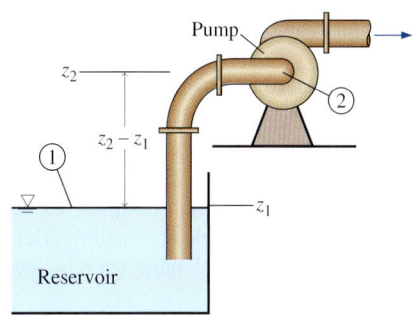

FIGURE P14–53

14–54 Repeat Prob. 14–53, but at a water temperature of 80°C. Repeat for 90°C. Discuss.

14–55 Repeat Prob. 14–53, but with the pipe diameter increased by a factor of 2 (all else being equal). Does the volume flow rate at which cavitation occurs in the pump increase or decrease with the larger pipe? Discuss.

14–56 The two-lobe rotary pump of Fig. P14–56 moves 0.42 L of a coal slurry in each lobe volume $\dot{V}_{lobe}$. Calculate the volume flow rate of the slurry (in Lpm) for the case where $\dot{n} = 175$ rpm.   *Answer:* 294 Lpm

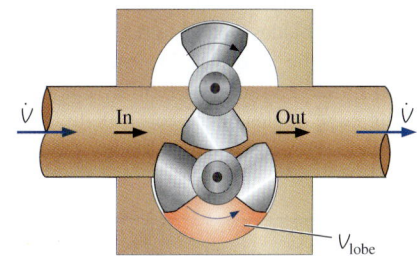

FIGURE P14–56

14–57 Repeat Prob. 14–56 for the case in which the pump has *three* lobes on each rotor instead of two, and $\dot{V}_{lobe} = 0.312$ L.

14–58 A two-lobe rotary positive-displacement pump, similar to that of Fig. 14–30, moves 3.64 cm³ of tomato paste in each lobe volume $\dot{V}_{lobe}$. Calculate the volume flow rate of tomato paste for the case where $\dot{n} = 336$ rpm.

14–59 Consider the gear pump of Fig. 14–26c. Suppose the volume of fluid confined between two gear teeth is 0.350 cm³. How much fluid volume is pumped per rotation? *Answer:* 9.80 cm³

14–60 A centrifugal pump rotates at $\dot{n}$ = 750 rpm. Water enters the impeller normal to the blades (α_1 = 0°) and exits at an angle of 35° from radial (α_2 = 35°). The inlet radius is r_1 = 12.0 cm, at which the blade width b_1 = 18.0 cm. The outlet radius is r_2 = 24.0 cm, at which the blade width b_2 = 16.2 cm. The volume flow rate is 0.573 m³/s. Assuming 100 percent efficiency, calculate the net head produced by this pump in cm of water column height. Also calculate the required brake horsepower in W.

14–61 Suppose the pump of Prob. 14–60 has some swirl at the inlet such that α_1 = 7° instead of 0°. Calculate the net head and required horsepower and compare to Prob. 14–60. Discuss. In particular, is the angle at which the fluid impinges on the impeller blade a critical parameter in the design of centrifugal pumps?

14–62 Suppose the pump of Prob. 14–60 has some reverse swirl at the inlet such that α_1 = −10° instead of 0°. Calculate the net head and required horsepower and compare to Prob. 14–60. Discuss. In particular, is the angle at which the fluid impinges on the impeller blade a critical parameter in the design of centrifugal pumps? Does a small amount of reverse swirl increase or decrease the net head of the pump—in other words, is it desirable? *Note:* Keep in mind that we are neglecting losses here.

14–63 A vane-axial flow fan is being designed with the stator blades upstream of the rotor blades (Fig. P14–63). To reduce expenses, both the stator and rotor blades are to be constructed of sheet metal. The stator blade is a simple circular arc with its leading edge aligned axially and its trailing edge at angle β_{st} = 26.6° from the axis as shown in the sketch. (The subscript notation indicates stator trailing edge.) There are 18 stator blades. At design conditions, the axial-flow speed through the blades is 31.4 m/s, and the impeller rotates at 1800 rpm. At a radius of 0.50 m, calculate the leading and trailing edge angles of the rotor blade, and sketch the shape of the blade. How many rotor blades should there be?

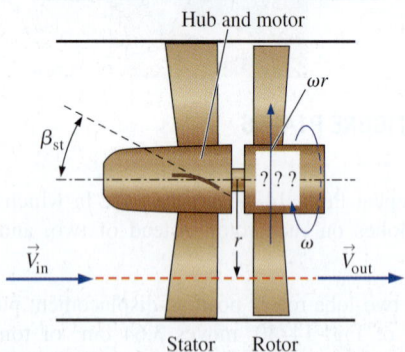

FIGURE P14–63

14–64 Two water pumps are arranged in *series*. The performance data for both pumps follow the parabolic curve fit $H_{available} = H_0 - a\dot{V}^2$. For pump 1, H_0 = 6.33 m and coefficient a = 0.0633 m/Lpm²; for pump 2, H_0 = 9.25 m and coefficient a = 0.0472 m/Lpm². In either case, the units of net pump head H are m, and the units of capacity $\dot{V}$ are Lpm. Calculate the combined shutoff head and free delivery of the two pumps working together in series. At what volume flow rate should pump 1 be shut off and bypassed? Explain. *Answers:* 15.6 m, 14.0 Lpm, 10.0 Lpm

14–65 The same two water pumps of Prob. 14–64 are arranged in *parallel*. Calculate the shutoff head and free delivery of the two pumps working together in parallel. At what combined net head should pump 1 be shut off and bypassed? Explain.

Turbines

14–66C What is a *draft tube*, and what is its purpose? Describe what would happen if turbomachinery designers did not pay attention to the design of the draft tube.

14–67C Name and briefly describe the differences between the two basic types of dynamic turbine.

14–68C Discuss the meaning of *reverse swirl* in reaction hydroturbines, and explain why some reverse swirl may be desirable. Use an equation to support your answer. Why is it *not* wise to have too much reverse swirl?

14–69C Give at least two reasons why turbines often have greater efficiencies than do pumps.

14–70C Briefly discuss the main difference in the way that dynamic pumps and reaction turbines are classified as centrifugal (radial), mixed flow, or axial.

14–71 A hydroelectric plant has 14 identical Francis turbines, a gross head of 284 m, and a volume flow rate of 13.6 m³/s through each turbine. The water is at 25°C. The efficiencies are $\eta_{turbine}$ = 95.9%, $\eta_{generator}$ = 94.2%, and η_{other} = 95.6%, where η_{other} accounts for all other mechanical energy losses. Estimate the electrical power production from this plant in MW.

14–72 A Pelton wheel is used to produce hydroelectric power. The average radius of the wheel is 1.83 m, and the jet velocity is 102 m/s from a nozzle of exit diameter equal to 10.0 cm. The turning angle of the buckets is β = 165°. (a) Calculate the volume flow rate through the turbine in m³/s. (b) What is the optimum rotation rate (in rpm) of the wheel (for maximum power)? (c) Calculate the output shaft power in MW if the efficiency of the turbine is 82 percent. *Answers:* (a) 0.801 m³/s, (b) 266 rpm, (c) 3.35 MW

14–73 Some engineers are evaluating potential sites for a small hydroelectric dam. At one such site, the gross head is 340 m, and they estimate that the volume flow rate of water through each turbine would be 0.95 m³/s. Estimate the ideal power production per turbine in MW.

14–74 Prove that for a given jet speed, volume flow rate, turning angle, and wheel radius, the maximum shaft power produced by a Pelton wheel occurs when the turbine bucket moves at half the jet speed.

14–75 Wind ($\rho = 1.204$ kg/m^3) blows through a HAWT wind turbine. The turbine diameter is 45.0 m. The combined efficiency of the gearbox and generator is 88 percent. (*a*) For a realistic power coefficient of 0.42, estimate the electrical power production when the wind blows at 7.8 m/s. (*b*) Repeat and compare using the Betz limit, assuming the same gearbox and generator.

14–76 A Francis radial-flow hydroturbine is being designed with the following dimensions: $r_2 = 2.00$ m, $r_1 = 1.42$ m, $b_2 = 0.731$ m, and $b_1 = 2.20$ m. The runner rotates at $\dot{n} = 180$ rpm. The wicket gates turn the flow by angle $\alpha_2 = 30°$ from radial at the runner inlet, and the flow at the runner outlet is at angle $\alpha_1 = 10°$ from radial (Fig. P14–76). The volume flow rate at design conditions is 340 m^3/s, and the gross head provided by the dam is $H_{gross} = 90.0$ m. For the preliminary design, irreversible losses are neglected. Calculate the inlet and outlet runner blade angles β_2 and β_1, respectively, and predict the power output (MW) and required net head (m). Is the design feasible?

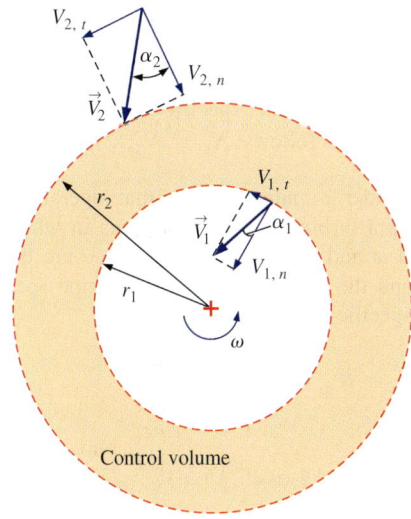

FIGURE P14–76

14–77 Reconsider Prob. 14–76. Using EES (or other) software, investigate the effect of the runner outlet angle α_1 on the required net head and the output power. Let the outlet angle vary from $-20°$ to $20°$ in increments of 1°, and plot your results. Determine the minimum possible value of α_1 such that the flow does not violate the laws of thermodynamics.

14–78 A Francis radial-flow hydroturbine has the following dimensions, where location 2 is the inlet and location 1 is the outlet: $r_2 = 2.00$ m, $r_1 = 1.30$ m, $b_2 = 0.85$ m, and $b_1 = 2.10$ m. The runner blade angles are $\beta_2 = 71.4°$ and $\beta_1 = 15.3°$ at the turbine inlet and outlet, respectively. The runner rotates at $\dot{n} = 160$ rpm. The volume flow rate at design conditions is 80.0 m^3/s. Irreversible losses are neglected in this preliminary analysis. Calculate the angle α_2 through which the wicket gates should turn the flow, where α_2 is measured from the radial direction at the runner inlet (Fig. P14–76). Calculate the swirl angle α_1, where α_1 is measured from the radial direction at the runner outlet (Fig. P14–76). Does this turbine have forward or reverse swirl? Predict the power output (MW) and required net head (m).

14–79 A simple single-stage axial turbine is being designed to produce power from water flowing through a tube as in Fig. P14–79. We approximate both the stator and rotor as thin (bent sheet metal). The 16 stator (upstream) blades have $\beta_{sl} = 0°$ and $\beta_{st} = 50.3°$, where subscripts "sl" and "st" mean stator leading edge and stator trailing edge, respectively. At design conditions, the axial flow speed is 8.31 m/s, the rotor turns at 360 rpm, and it is desired that there be no swirl downstream of the turbine. At a radius of 0.324 m, calculate angles β_{rl} and β_{rt} (rotor leading and trailing edge angles), sketch what the rotor vanes should look like, and specify how many rotor vanes there should be.

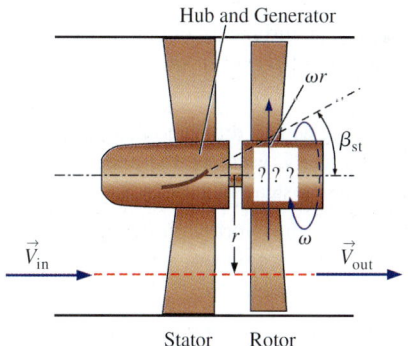

FIGURE P14–79

14–80 In the section on wind turbines, an expression was derived for the ideal power coefficient of a wind turbine, $C_P = 4a(1-a)^2$. Prove that the maximum possible power coefficient occurs when $a = 1/3$.

14–81 A hydroelectric power plant is being designed. The gross head from the reservoir to the tailrace is 262 m, and the volume flow rate of water through each turbine is 717 m^3/min at 10°C. There are 10 identical parallel turbines, each with an efficiency of 96.3 percent, and all other mechanical energy losses (through the penstock, etc.) are estimated to reduce the output by 3.6 percent. The generator itself has an efficiency of 93.9 percent. Estimate the electric power production from the plant in MW.

14–82 The average wind speed at a proposed HAWT wind farm site is 12.5 m/s. The power coefficient of each wind turbine is predicted to be 0.41, and the combined efficiency of the gearbox and generator is 92 percent. Each wind turbine must produce 2.5 MW of electrical power when the wind blows at 12.5 m/s. (*a*) Calculate the required diameter of each turbine disk. Take the average air density to be $\rho = 1.2$ kg/m^3. (b) If 30 such turbines are built on the site and an average home in the area consumes approximately 1.5 kW of electrical power, estimate how many homes can be powered by this wind farm, assuming an additional efficiency of 96 percent to account for the powerline losses.

Pump and Turbine Scaling Laws

14–83C Pump specific speed and turbine specific speed are "extra" parameters that are not necessary in the scaling laws for pumps and turbines. Explain, then, their purpose.

14–84C For each statement, choose whether the statement is true or false, and discuss your answer briefly:

(a) If the rpm of a pump is doubled, all else staying the same, the capacity of the pump goes up by a factor of about 2.

(b) If the rpm of a pump is doubled, all else staying the same, the net head of the pump goes up by a factor of about 2.

(c) If the rpm of a pump is doubled, all else staying the same, the required shaft power goes up by a factor of about 4.

(d) If the rpm of a turbine is doubled, all else staying the same, the output shaft power of the turbine goes up by a factor of about 8.

14–85C Discuss which dimensionless pump performance parameter is typically used as the independent parameter. Repeat for turbines instead of pumps. Explain.

14–86C Look up the word *affinity* in a dictionary. Why do you suppose some engineers refer to the turbomachinery scaling laws as *affinity laws*?

14–87 Consider the fan of Prob. 14–46. The fan diameter is 30.0 cm, and it operates at $\dot{n} = 600$ rpm. Nondimensionalize the fan performance curve, i.e., plot C_H versus C_Q. Show sample calculations of C_H and C_Q at $\dot{V} = 13{,}600$ Lpm.

14–88 Calculate the fan specific speed of the fan of Probs. 14–46 and 14–87 at the best efficiency point for the case in which the BEP occurs at 13,600 Lpm. Provide answers in both dimensionless form and in customary U.S. units. What kind of fan is it?

14–89 Calculate the pump specific speed of the pump of Example 14–11 at its best efficiency point. Provide answers in both dimensionless form and in customary U.S. units. What kind of pump is it?

14–90 Len is asked to design a small water pump for an aquarium. The pump should deliver 14.0 Lpm of water at a net head of 1.5 m at its best efficiency point. A motor that spins at 1200 rpm is available. What kind of pump (centrifugal, mixed, or axial) should Len design? Show all your calculations and justify your choice. Estimate the maximum pump efficiency Len can hope for with this pump. *Answers:* centrifugal, 81.0%

14–91 Consider the pump of Prob. 14–90. Suppose the pump is modified by attaching a different motor, for which the rpm is 1800 rpm. If the pumps operate at homologous points (namely, at the BEP) for both cases, predict the volume flow rate and net head of the modified pump. Calculate the pump specific speed of the modified pump, and compare to that of the original pump. Discuss.

14–92 A large water pump is being designed for a nuclear reactor. The pump should deliver 9500 Lpm of water at a net head of 14 m at its best efficiency point. A motor that spins at 300 rpm is available. What kind of pump (centrifugal, mixed, or axial) should be designed? Show all your calculations and justify your choice. Estimate the maximum pump efficiency that can be hoped for with this pump. Estimate the power (brake horsepower) required to run the pump.

14–93 Consider the pump of Prob. 14–38. The pump diameter is 1.80 cm, and it operates at $\dot{n} = 4200$ rpm. Nondimensionalize the pump performance curve, i.e., plot C_H versus C_Q. Show sample calculations of C_H and C_Q at $\dot{V} = 14.0$ Lpm.

14–94 Calculate the pump specific speed of the pump of Prob. 14–93 at the best efficiency point for the case in which the BEP occurs at 14.0 Lpm. Provide answers in both dimensionless form and in customary U.S. units. What kind of pump is it? *Answers:* 0.199, 545, centrifugal

14–95 Verify that turbine specific speed and pump specific speed are related as follows: $N_{St} = N_{Sp}\sqrt{\eta_{turbine}}$.

14–96 Consider a pump–turbine that operates both as a pump and as a turbine. Under conditions in which the rotational speed ω and the volume flow rate $\dot{V}$ are the same for the pump and the turbine, verify that turbine specific speed and pump specific speed are related as

$$N_{St} = N_{Sp}\sqrt{\eta_{turbine}}\left(\frac{H_{pump}}{H_{turbine}}\right)^{3/4}$$

$$= N_{Sp}(\eta_{turbine})^{5/4}(\eta_{pump})^{3/4}\left(\frac{bhp_{pump}}{bhp_{turbine}}\right)^{3/4}$$

14–97 Apply the necessary conversion factors to prove the relationship between dimensionless turbine specific speed and conventional U.S. turbine specific speed, $N_{St} = 43.46 N_{St, US}$. Note that we assume water as the fluid and standard earth gravity.

14–98 Calculate the turbine specific speed of the turbine in Prob. 14–76. Provide answers in both dimensionless form and in customary U.S. units. Is it in the normal range for a Francis turbine? If not, what type of turbine would be more appropriate?

14–99 Calculate the turbine specific speed of the Smith Mountain hydroturbine of Fig 14–90. Does it fall within the range of N_{St} appropriate for that type of turbine?

14–100 Calculate the turbine specific speed of the Warwick hydroturbine of Fig 14–91. Does it fall within the range of N_{St} appropriate for that type of turbine?

14–101 Calculate the turbine specific speed of the turbine of Example 14–13 for the case where $\alpha_1 = 10°$. Provide answers in both dimensionless form and in customary U.S. units. Is it in the normal range for a Francis turbine? If not, what type of turbine would be more appropriate?

14–102 Calculate the turbine specific speed of the turbine in Prob. 14–78. Provide answers in both dimensionless form and in customary U.S. units. Is it in the normal range for a Francis turbine? If not, what type of turbine would be more appropriate?

14–103 Calculate the turbine specific speed of the Round Butte hydroturbine of Fig 14–89. Does it fall within the range of N_{St} appropriate for that type of turbine?

14–104 A one-fifth scale model of a water turbine is tested in a laboratory at $T = 20°C$. The diameter of the model is 8.0 cm, its volume flow rate is 25.5 m³/h, it spins at 1500 rpm, and it operates with a net head of 15.0 m. At its best efficiency point, it delivers 720 W of shaft power. Calculate the efficiency of the model turbine. What is the most likely kind of turbine being tested? *Answers:* 69.2%, impulse

14–105 The prototype turbine corresponding to the one-fifth scale model turbine discussed in Prob. 14–104 is to operate across a net head of 50 m. Determine the appropriate rpm and volume flow rate for best efficiency. Predict the brake horsepower output of the prototype turbine, assuming exact geometric similarity.

14–106 Prove that the model turbine (Prob. 14–104) and the prototype turbine (Prob. 14–105) operate at homologous points by comparing turbine efficiency and turbine specific speed for both cases.

14–107 In Prob. 14–106, we scaled up the model turbine test results to the full-scale prototype assuming exact dynamic similarity. However, as discussed in the text, a large prototype typically yields higher efficiency than does the model. Estimate the actual efficiency of the prototype turbine. Briefly explain the higher efficiency.

Review Problems

14–108C What is a *pump–turbine*? Discuss an application where a pump–turbine is useful.

14–109C The common water meter found in most homes can be thought of as a type of turbine, since it extracts energy from the flowing water to rotate the shaft connected to the volume-counting mechanism (Fig. P14–109C). From the point of view of a piping system, however (Chap. 8), what kind of device is a water meter? Explain.

FIGURE P14–109C

14–110C For each statement, choose whether the statement is true or false, and discuss your answer briefly:

(*a*) A gear pump is a type of positive-displacement pump.
(*b*) A rotary pump is a type of positive-displacement pump.
(*c*) The pump performance curve (net head versus capacity) of a positive-displacement pump is nearly vertical through-out its recommended operating range at a given rotational speed.
(*d*) At a given rotational speed, the net head of a positive-displacement pump *decreases* with fluid viscosity.

14–111 For two dynamically similar pumps, manipulate the dimensionless pump parameters to show that $D_B = D_A(H_A/H_B)^{1/4}(\dot{V}_B/\dot{V}_A)^{1/2}$. Does the same relationship apply to two dynamically similar *turbines*?

14–112 For two dynamically similar turbines, manipulate the dimensionless turbine parameters to show that $D_B = D_A(H_A/H_B)^{3/4}(\rho_A/\rho_B)^{1/2}(bhp_B/bhp_A)^{1/2}$. Does the same relationship apply to two dynamically similar *pumps*?

14–113 A group of engineers is designing a new hydro-turbine by scaling up an existing one. The existing turbine (turbine A) has diameter $D_A = 1.50$ m, and spins at $\dot{n}_A = 150$ rpm. At its best efficiency point, $\dot{V}_A = 162$ m³/s, $H_A = 90.0$ m of water, and $bhp_A = 132$ MW. The new turbine (turbine B) will spin at 105 rpm, and its net head will be $H_B = 95$ m. Calculate the diameter of the new turbine such that it operates most efficiently, and calculate $\dot{V}_B$ and bhp_B. *Answers:* 2.20 m, 359 m³/h, 308 MW

14–114 Calculate and compare the efficiency of the two turbines of Prob. 14–113. They should be the same since we are assuming dynamic similarity. However, the larger turbine will actually be slightly more efficient than the smaller turbine. Use the Moody efficiency correction equation to predict the actual expected efficiency of the new turbine. Discuss.

14–115 Calculate and compare the turbine specific speed for both the small (A) and large (B) turbines of Prob. 14–113. What kind of turbine are these most likely to be?

Fundamentals of Engineering (FE) Exam Problems

14–116 Which turbomachine is designed to deliver a very high pressure rise, typically at low to moderate flow rates?
(*a*) Compressor
(*b*) Blower
(*c*) Turbine
(*d*) Pump
(*e*) Fan

14–117 In the turbomachinery industry, capacity refers to
(*a*) Power
(*b*) Mass flow rate
(*c*) Volume flow rate
(*d*) Net head
(*e*) Energy grade line

14–118 A pump increases the pressure of water from 100 kPa to 3 MPa at a rate of 0.5 m³/min. The inlet and outlet diameters are identical and there is no change in elevation

across the pump. If the efficiency of the pump is 77 percent, the power supplied to the pump is
(a) 18.5 kW
(b) 21.8 kW
(c) 24.2 kW
(d) 27.6 kW
(e) 31.4 kW

14–119 A pump increases the pressure of water from 100 kPa to 900 kPa to an elevation of 35 m. The inlet and outlet diameters are identical. The net head of the pump is
(a) 143 m
(b) 117 m
(c) 91 m
(d) 70 m
(e) 35 m

14–120 The brake horsepower and water horsepower of a pump are determined to be 15 kW and 12 kW, respectively. If the flow rate of water to the pump under these conditions is 0.05 m^3/s, the total head loss of the pump is
(a) 11.5 m
(b) 9.3 m
(c) 7.7 m
(d) 6.1 m
(e) 4.9 m

14–121 In the pump performance curve, the point at which the net head is zero is called
(a) Best efficiency point
(b) Free delivery
(c) Shutoff head
(d) Operating point
(e) Duty point

14–122 A power plant requires 940 L/min of water. The required net head is 5 m at this flow rate. An examination of pump performance curves indicates that two centrifugal pumps with different impeller diameters can deliver this flow rate. The pump with an impeller diameter of 203 mm has a pump efficiency of 73 percent and delivers 10 m of net head. The pump with an impeller diameter of 111 mm has a lower pump efficiency of 67 percent and delivers 5 m of net head. What is the ratio of the required brake horse power (bhp) of the pump with 203-mm-diameter impeller to that of the pump with 111-mm-diameter impeller?
(a) 0.45
(b) 0.68
(c) 0.86
(d) 1.84
(e) 2.11

14–123 Water enters the pump of a steam power plant at 20 kPa and 50°C at a rate of 0.15 m^3/s. The diameter of the pipe at the pump inlet is 0.25 m. What is the net positive suction head (NPSH) at the pump inlet?

(a) 2.14 m
(b) 1.89 m
(c) 1.66 m
(d) 1.42 m
(e) 1.26 m

14–124 Which quantities are added when two pumps are connected in series and parallel?
(a) Series: Pressure change. Parallel: Net head
(b) Series: Net head. Parallel: Pressure change
(c) Series: Net head. Parallel: Flow rate
(d) Series: Flow rate. Parallel: Net head
(e) Series: Flow rate. Parallel: Pressure change

14–125 Three pumps are connected in series. According to pump performance curves, the free delivery of each pump is as follows:

Pump 1: 1600 L/min   Pump 2: 2200 L/min
Pump 3: 2800 L/min

If the flow rate for this pump system is 2500 L/min, which pump(s) should be shut off?
(a) Pump 1
(b) Pump 2
(c) Pump 3
(d) Pumps 1 and 2
(e) Pumps 2 and 3

14–126 Three pumps are connected in parallel. According to pump performance curves, the shutoff head of each pump is as follows:

Pump 1: 7 m   Pump 2: 10 m   Pump 3: 15 m

If the net head for this pump system is 9 m, which pump(s) should be shut off?
(a) Pump 1
(b) Pump 2
(c) Pump 3
(d) Pumps 1 and 2
(e) Pumps 2 and 3

14–127 A two-lobe rotary positive-displacement pump moves 0.60 cm^3 of motor oil in each lobe volume. For every 90° of rotation of the shaft, one lobe volume is pumped. If the rotation rate is 550 rpm, the volume flow rate of oil is
(a) 330 cm^3/min
(b) 660 cm^3/min
(c) 1320 cm^3/min
(d) 2640 cm^3/min
(e) 3550 cm^3/min

14–128 The snail-shaped casing of centrifugal pumps is called
(a) Rotor
(b) Scroll
(c) Volute
(d) Impeller
(e) Shroud

14–129 A centrifugal blower rotates at 1400 rpm. Air enters the impeller normal to the blades ($\alpha_1 = 0°$) and exits at an angle of 25° ($\alpha_2 = 25°$). The inlet radius is $r_1 = 6.5$ cm, and the inlet blade width $b_1 = 8.5$ cm. The outlet radius and blade width are $r_2 = 12$ cm and $b_2 = 4.5$ cm, respectively. The volume flow rate is 0.22 m³/s. What is the net head produced by this blower in meters of air?
(a) 12.3 m
(b) 3.9 m
(c) 8.8 m
(d) 5.4 m
(e) 16.4 m

14–130 A pump is designed to deliver 9500 L/min of water at a required head of 8 m. The pump shaft rotates at 1500 rpm. The pump specific speed in nondimensional form is
(a) 0.377
(b) 0.540
(c) 1.13
(d) 1.48
(e) 1.84

14–131 The net head delivered by a pump at a rotational speed of 1000 rpm is 10 m. If the rotational speed is doubled, the net head delivered will be
(a) 5 m
(b) 10 m
(c) 20 m
(d) 40 m
(e) 80 m

14–132 The rotating part of a turbine is called
(a) Propeller
(b) Scroll
(c) Blade ro
(d) Impeller
(e) Runner

14–133 Which choice is correct for the comparison of the operation of impulse and reaction turbines?
(a) Impulse: Higher flow rate
(b) Impulse: Higher head
(c) Reaction: Higher head
(d) Reaction: Smaller flow rate
(e) None of these

14–134 Which turbine type is an impulse turbine?
(a) Kaplan
(b) Francis
(c) Pelton
(d) Propeller
(e) Centrifugal

14–135 A turbine is placed at the bottom of a 20-m-high water body. Water flows through the turbine at a rate of 30 m³/s. If the shaft power delivered by the turbine is 5 MW, the turbine efficiency is

(a) 85%
(b) 79%
(c) 88%
(d) 74%
(e) 82%

14–136 A hydroelectric power plant is to be built at a dam with a gross head of 200 m. The head losses in the head gate and penstock are estimated to be 6 m. The flow rate through the turbine is 18,000 L/min. The efficiencies of the turbine and the generator are 88 percent and 96 percent, respectively. The electricity production from this turbine is
(a) 6910 kW
(b) 6750 kW
(c) 6430 kW
(d) 6170 kW
(e) 5890 kW

14–137 In a hydroelectric power plant, water flows through a large tube through the dam. This tube is called a
(a) Tailrace
(b) Draft tube
(c) Runner
(d) Penstock
(e) Propeller

14–138 In wind turbines, the minimum wind speed at which useful power can be generated is called the
(a) Rated speed
(b) Cut-in speed
(c) Cut-out speed
(d) Available speed
(e) Betz speed

14–139 A wind turbine is installed in a location where the wind blows at 8 m/s. The air temperature is 10°C and the diameter of turbine blade is 30 m. If the overall turbine-generator efficiency is 35 percent, the electrical power production is
(a) 79 kW
(b) 109 kW
(c) 142 kW
(d) 154 kW
(e) 225 kW

14–140 The available power from a wind turbine is calculated to be 50 kW when the wind speed is 5 m/s. If the wind velocity is doubled, the available wind power becomes
(a) 50 kW
(b) 100 kW
(c) 200 kW
(d) 400 kW
(e) 800 kW

14–141 A new hydraulic turbine is to be designed to be similar to an existing turbine with following parameters at its best efficiency point: $D_A = 3$ m, $\dot{n}_A = 90$ rpm, $\dot{V}_A = 200$ m³/s, $H_A = 55$ m, bhp$_A$ = 100 MW. The new turbine

will have a speed of 110 rpm and the net head will be 40 m. What is the bhp of the new turbine such that it operates most efficiently?
(a) 17.6 MW
(b) 23.5 MW
(c) 30.2 MW
(d) 40.0 MW
(e) 53.7 MW

14–142 A hydraulic turbine operates at the following parameters at its best efficiency point: $\dot{n} = 90$ rpm, $\dot{V} = 200$ m^3/s, $H = 55$ m, bhp = 100 MW. The turbine specific speed of this turbine is
(a) 0.71
(b) 0.18
(c) 1.57
(d) 2.32
(e) 1.15

Design and Essay Problem

14–143 Develop a general-purpose computer application (using EES or other software) that employs the affinity laws to design a new pump (B) that is dynamically similar to a given pump (A). The inputs for pump A are diameter, net head, capacity, density, rotational speed, and pump efficiency. The inputs for pump B are density (ρ_B may differ from ρ_A), desired net head, and desired capacity. The outputs for pump B are diameter, rotational speed, and required shaft power. Test your program using the following inputs: $D_A = 5.0$ cm, $H_A = 120$ cm, $\dot{V}_A = 400$ cm³/s, $\rho_A = 998.0$ kg/m³, $\dot{n}_A = 1725$ rpm, $\eta_{pump, A} = 81$ percent, $\rho_B = 1226$ kg/m³, $H_B = 450$ cm, and $\dot{V}_B = 2400$ cm³/s. Verify your results manually. *Answers:* $D_B = 8.80$ cm, $\dot{n}_B = 1898$ rpm, and $bhp_B = 160$ MW

14–144 Experiments on an existing pump (A) yield the following BEP data: $D_A = 10.0$ cm, $H_A = 210$ cm, $\dot{V}_A = 1350$ cm³/s, $\rho_A = 998.0$ kg/m³, $\dot{n}_A = 1500$ rpm, $\eta_{pump, A} = 87$ percent. You are to design a new pump (B) that has the following requirements: $\rho_B = 998.0$ kg/m³, $H_B = 570$ cm, and $\dot{V}_B = 3670$ cm³/s. Apply the computer program you developed in Prob. 14–143 to calculate D_B (cm), $\dot{n}_B$ (rpm), and bhp_B (W). Also calculate the pump specific speed. What kind of pump is this (most likely)?

14–145 Develop a general-purpose computer application (using EES or other software) that employs the affinity laws to design a new turbine (B) that is dynamically similar to a given turbine (A). The inputs for turbine A are diameter, net head, capacity, density, rotational speed, and brake horsepower. The inputs for turbine B are density (ρ_B may differ from ρ_A), available net head, and rotational speed. The outputs for turbine B are diameter, capacity, and brake horsepower. Test your program using the following inputs: $D_A = 1.40$ m, $H_A = 80.0$ m, $\dot{V}_A = 162$ m³/s, $\rho_A = 998.0$ kg/m³, $\dot{n}_A = 150$ rpm, $bhp_A = 118$ MW, $\rho_B = 998.0$ kg/m³, $H_B = 95.0$ m, and $\dot{n}_B = 120$ rpm. Verify your results manually. *Answers:* $D_B = 1.91$ m, $\dot{V}_B = 328$ m³/s, and $bhp_B = 283$ MW

14–146 Experiments on an existing turbine (A) yield the following data: $D_A = 86.0$ cm, $H_A = 22.0$ m, $\dot{V}_A = 69.5$ m³/s, $\rho_A = 998.0$ kg/m³, $\dot{n}_A = 240$ rpm, $bhp_A = 11.4$ MW. You are to design a new turbine (B) that has the following requirements: $\rho_B = 998.0$ kg/m³, $H_B = 95.0$ m, and $\dot{n}_B = 210$ rpm. Apply the computer program you developed in Prob. 14–145 to calculate D_B (m), $\dot{V}_B$ (m³/s), and bhp_B (MW). Also calculate the turbine specific speed. What kind of turbine is this (most likely)?

14–147 Calculate and compare the efficiency of the two turbines of Prob. 14–146. They should be the same since we are assuming dynamic similarity. However, the larger turbine will actually be slightly more efficient than the smaller turbine. Use the Moody efficiency correction equation to predict the actual expected efficiency of the new turbine. Discuss.

CHAPTER 15

INTRODUCTION TO COMPUTATIONAL FLUID DYNAMICS

A brief introduction to computational fluid dynamics (CFD) is presented in this chapter. While any intelligent, computer-literate person can run a CFD code, the results he or she obtains may not be physically correct. In fact, if the grid is not properly generated, or if the boundary conditions or flow parameters are improperly applied, the results may even be completely erroneous. Therefore, the goal of this chapter is to present *guidelines* about how to generate a grid, how to specify boundary conditions, and how to determine if the computer output is meaningful. We stress the *application* of CFD to engineering problems, rather than details about grid generation techniques, discretization schemes, CFD algorithms, or numerical stability.

The examples presented here have been obtained with the commercial computational fluid dynamics code **ANSYS-FLUENT.** Other CFD codes would yield similar, but not identical results. Sample CFD solutions are shown for incompressible and compressible laminar and turbulent flows, flows with heat transfer, and flows with free surfaces. As always, one learns best by hands-on practice. For this reason, we provide several homework problems that utilize many additional CFD problems are provided or the books website at www.mheducation.asia/olc/cengel.

Objectives

When you finish reading this chapter, you should be able to

- Understand the importance of a high-quality, good resolution mesh
- Apply appropriate boundary conditions to computational domains
- Understand how to apply CFD to basic engineering problems and how to determine whether the output is physically meaningful
- Realize that you need much further study and practice to use CFD successfully

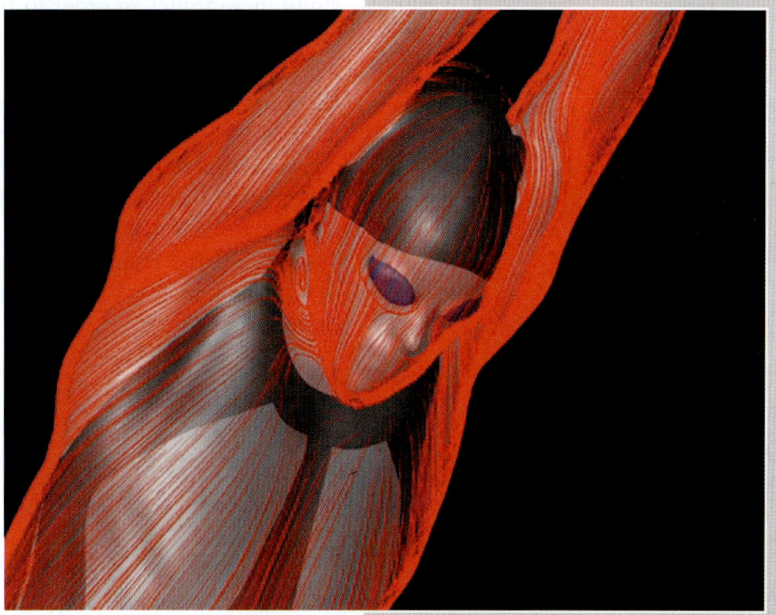

Flow over a male swimmer simulated using the ANSYS-FLUENT CFD code. The image shows simulated oil flow lines along the surface of the body. Flow separation in the region of the neck is visible.

Photo used with the permission of the owner, Speedo International Limited.

COMPUTATIONAL FLUID DYNAMICS

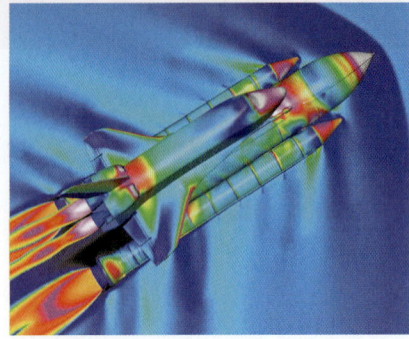

FIGURE 15–1
CFD calculations of the ascent of the space shuttle launch vehicle (SSLV). The grid consists of more than 16 million points, and filled pressure contours are shown. Free-stream conditions are Ma = 1.25, and the angle of attack is −3.3°.

NASA/Photo by Ray J. Gomez. Used by permission.

15–1 · INTRODUCTION AND FUNDAMENTALS

Motivation

There are two fundamental approaches to design and analysis of engineering systems that involve fluid flow: experimentation and calculation. The former typically involves construction of models that are tested in wind tunnels or other facilities (Chap. 7), while the latter involves solution of differential equations, either analytically (Chaps. 9 and 10) or computationally. In the present chapter, we provide a brief introduction to **computational fluid dynamics** (**CFD**), the field of study devoted to solution of the equations of fluid flow through use of a computer (or, more recently, *several* computers working in parallel). Modern engineers apply both experimental *and* CFD analyses, and the two complement each other. For example, engineers may obtain *global properties*, such as lift, drag, pressure drop, or power, experimentally, but use CFD to obtain *details* about the flow field, such as shear stresses, velocity and pressure profiles (Fig. 15–1), and flow streamlines. In addition, experimental data are often used to *validate* CFD solutions by matching the computationally and experimentally determined global quantities. CFD is then employed to shorten the design cycle through carefully controlled parametric studies, thereby reducing the required amount of experimental testing.

The current state of computational fluid dynamics is that CFD can handle laminar flows with ease, but turbulent flows of practical engineering interest are impossible to solve without invoking *turbulence models*. Unfortunately, no turbulence model is *universal*, and a turbulent CFD solution is only as good as the appropriateness of the turbulence model. In spite of this limitation, the standard turbulence models yield reasonable results for many practical engineering problems.

There are several aspects of CFD that are not covered in this chapter—grid generation techniques, numerical algorithms, finite difference and finite volume schemes, stability issues, turbulence modeling, etc. You need to study these topics in order to fully understand both the capabilities and limitations of computational fluid dynamics. In this chapter, we merely scratch the surface of this exciting field. Our goal is to present the fundamentals of CFD from a *user's* point of view, providing guidelines about how to generate a grid, how to specify boundary conditions, and how to determine if the computer output is physically meaningful.

We begin this section by presenting the differential equations of fluid flow that are to be solved, and then we outline a solution procedure. Subsequent sections of this chapter are devoted to example CFD solutions for laminar flow, turbulent flow, flows with heat transfer, compressible flow, and open-channel flow.

Equations of Motion

For steady laminar flow of a viscous, incompressible, Newtonian fluid without free-surface effects, the equations of motion are the *continuity equation*

$$\vec{\nabla} \cdot \vec{V} = 0 \qquad (15\text{–}1)$$

and the *Navier–Stokes equation*

$$(\vec{V} \cdot \vec{\nabla})\vec{V} = -\frac{1}{\rho}\vec{\nabla}P' + \nu \nabla^2 \vec{V} \qquad (15\text{–}2)$$

Strictly speaking, Eq. 15–1 is a **conservation equation,** while Eq. 15–2 is a **transport equation** that represents transport of linear momentum throughout the computational domain. In Eqs. 15–1 and 15–2, $\vec{V}$ is the velocity of the fluid, ρ is its density, and ν is its kinematic viscosity ($\nu = \mu/\rho$). The lack of free-surface effects enables us to use the *modified pressure P'*, thereby eliminating the gravity term from Eq. 15–2 (see Chap. 10). Note that Eq. 15–1 is a *scalar* equation, while Eq. 15–2 is a *vector* equation. Equations 15–1 and 15–2 apply only to incompressible flows in which we also assume that both ρ and ν are constants. Thus, for three-dimensional flow in Cartesian coordinates, there are *four* coupled differential equations for *four* unknowns, u, v, w, and P' (Fig. 15–2). If the flow were compressible, Eqs. 15–1 and 15–2 would need to be modified appropriately, as discussed in Section 15–5. Liquid flows can almost always be treated as incompressible, and for many gas flows, the gas is at a low enough Mach number that it behaves as a nearly incompressible fluid.

Solution Procedure

To solve Eqs. 15–1 and 15–2 numerically, the following steps are performed. Note that the order of some of the steps (particularly steps 2 through 5) is interchangeable.

1. A **computational domain** is chosen, and a **grid** (also called a **mesh**) is generated; the domain is divided into many small elements called **cells.** For two-dimensional (2-D) domains, the cells are *areas*, while for three-dimensional (3-D) domains the cells are *volumes* (Fig. 15–3). You can think of each cell as a tiny control volume in which discretized versions of the conservation equations are solved. Note that we limit our discussion here to cell-centered finite volume CFD codes. The quality of a CFD solution is highly dependent on the quality of the grid. Therefore, you are advised to make sure that your grid is of high quality before proceeding to the next step (Fig. 15–4).
2. **Boundary conditions** are specified on each **edge** of the computational domain (2-D flows) or on each **face** of the domain (3-D flows).
3. The type of fluid (water, air, gasoline, etc.) is specified, along with fluid properties (temperature, density, viscosity, etc.). Many CFD codes

Continuity:
$$\frac{\partial u}{\partial x} + \frac{\partial v}{\partial y} + \frac{\partial w}{\partial z} = 0$$

x-momentum:
$$u\frac{\partial u}{\partial x} + v\frac{\partial u}{\partial y} + w\frac{\partial u}{\partial z} =$$
$$-\frac{1}{\rho}\frac{\partial P'}{\partial x} + \nu\left(\frac{\partial^2 u}{\partial x^2} + \frac{\partial^2 u}{\partial y^2} + \frac{\partial^2 u}{\partial z^2}\right)$$

y-momentum:
$$u\frac{\partial v}{\partial x} + v\frac{\partial v}{\partial y} + w\frac{\partial v}{\partial z} =$$
$$-\frac{1}{\rho}\frac{\partial P'}{\partial y} + \nu\left(\frac{\partial^2 v}{\partial x^2} + \frac{\partial^2 v}{\partial y^2} + \frac{\partial^2 v}{\partial z^2}\right)$$

z-momentum:
$$u\frac{\partial w}{\partial x} + v\frac{\partial w}{\partial y} + w\frac{\partial w}{\partial z} =$$
$$-\frac{1}{\rho}\frac{\partial P'}{\partial z} + \nu\left(\frac{\partial^2 w}{\partial x^2} + \frac{\partial^2 w}{\partial y^2} + \frac{\partial^2 w}{\partial z^2}\right)$$

FIGURE 15–2
The equations of motion to be solved by CFD for the case of steady, incompressible, laminar flow of a Newtonian fluid with constant properties and without free-surface effects. A Cartesian coordinate system is used. There are four equations and four unknowns: u, v, w, and P'.

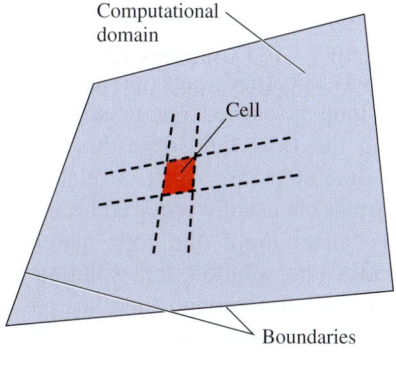
(a)

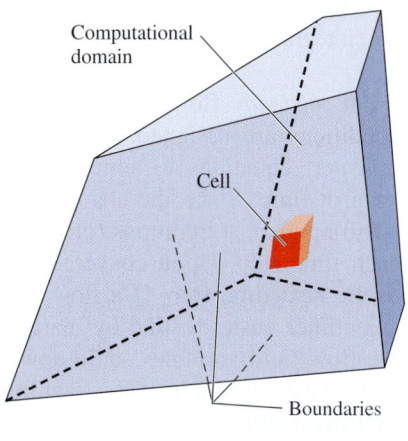
(b)

FIGURE 15–3
A *computational domain* is the region in space in which the equations of motion are solved by CFD. A *cell* is a small subset of the computational domain. Shown are (*a*) a two-dimensional domain and quadrilateral cell, and (*b*) a three-dimensional domain and hexahedral cell. The boundaries of a 2-D domain are called *edges*, while those of a 3-D domain are called *faces*.

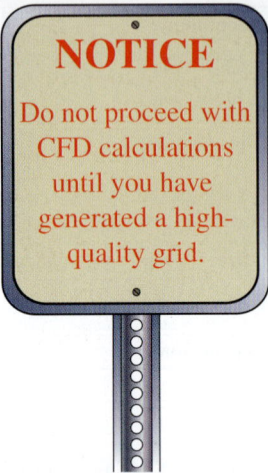

FIGURE 15–4
A quality grid is essential to a quality CFD simulation.

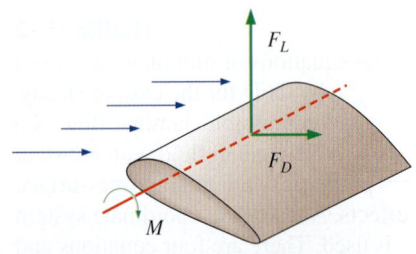

FIGURE 15–5
Global and integral properties of a flow, such as forces and moments on an object, are calculated after a CFD solution has converged. They can also be calculated during the iteration process to monitor convergence.

have built-in property databases for common fluids, making this step relatively painless.

4. Numerical parameters and solution algorithms are selected. These are specific to each CFD code and are not discussed here. The default settings of most modern CFD codes are appropriate for the simple problems discussed in this chapter.

5. Starting values for all flow field variables are specified for each cell. These are *initial conditions,* which may or may not be correct, but are necessary as a starting point, so that the iteration process may proceed (step 6). We note that for proper *unsteady*-flow calculations, the initial conditions *must* be correct.

6. Beginning with the initial guesses, discretized forms of Eqs. 15–1 and 15–2 are solved iteratively, usually at the center of each cell. If one were to put all the terms of Eq. 15–2 on one side of the equation, the solution would be "exact" when the sum of these terms, defined as the **residual,** is zero for every cell in the domain. In a CFD solution, however, the sum is *never* identically zero, but (hopefully) decreases with progressive iterations. A residual can be thought of as a measure of how much the solution to a given transport equation deviates from exact, and you monitor the average residual associated with each transport equation to help determine when the solution has converged. Sometimes hundreds or even thousands of iterations are required to converge on a final solution, and the residuals may decrease by several orders of magnitude.

7. Once the solution has converged, flow field variables such as velocity and pressure are plotted and analyzed graphically. You can also define and analyze additional custom functions that are formed by algebraic combinations of flow field variables. Most commercial CFD codes have built in **postprocessors,** designed for quick graphical analysis of the flow field. There are also stand-alone postprocessor software packages available for this purpose. Since the graphics output is often displayed in vivid colors, CFD has earned the nickname *colorful fluid dynamics*.

8. *Global properties* of the flow field, such as pressure drop, and *integral properties,* such as forces (lift and drag) and moments acting on a body, are calculated from the converged solution (Fig. 15–5). With most CFD codes, this can also be done "on the fly" as the iterations proceed. In many cases, in fact, it is wise to monitor these quantities along with the residuals during the iteration process; when a solution has converged, the global and integral properties should settle down to constant values as well.

For *unsteady* flow, a physical time step is specified, appropriate initial conditions are assigned, and an iteration loop is carried out to solve the transport equations to simulate changes in the flow field over this small span of time. Since the changes between time steps are small, a relatively small number of iterations (on the order of tens) is usually required between each time step. Upon convergence of this "inner loop," the code marches to the next time step. If a flow has a steady-state solution, that solution is sometimes easier to find by marching in time—after enough time has past, the flow field variables settle down to their steady-state values. Most CFD codes take advantage of this fact by internally specifying a pseudo-time step (**artificial time**) and marching toward a steady-state solution. In such cases, the pseudo-time step can even be different for different cells in the computational domain and can be tuned appropriately to decrease convergence time.

Other "tricks" are often used to reduce computation time, such as **multigridding**, in which the flow field variables are updated first on a coarse grid so that gross features of the flow are quickly established. That solution is then interpolated to finer and finer grids, the final grid being the one specified by the user (Fig. 15–6). In some commercial CFD codes, several layers of multigridding may occur "behind the scenes" during the iteration process, without user input (or awareness). You can learn more about computational algorithms and other numerical techniques that improve convergence by reading books devoted to computational methods, such as Tannehill, Anderson, and Pletcher (2012).

Additional Equations of Motion

If energy conversion or heat transfer is important in the problem, another transport equation, the *energy equation*, must also be solved. If temperature differences lead to significant changes in density, an *equation of state* (such as the ideal-gas law) is used. If buoyancy is important, the effect of temperature on density is reflected in the gravity term (which must then be separated from the modified pressure term in Eq. 15–2).

For a given set of boundary conditions, a laminar flow CFD solution approaches an "exact" solution, limited only by the accuracy of the discretization scheme used for the equations of motion, the level of convergence, and the degree to which the grid is resolved. The same would be true of a turbulent flow simulation if the grid could be fine enough to resolve all the unsteady, three-dimensional turbulent eddies. Unfortunately, this kind of direct simulation of turbulent flow is usually not possible for practical engineering applications due to computer limitations. Instead, additional approximations are made in the form of turbulence models so that turbulent flow solutions are possible. The turbulence models generate additional transport equations that model the enhanced mixing and diffusion of turbulence; these additional transport equations must be solved along with those of mass and momentum. Turbulence modeling is discussed in more detail in Section 15–3.

Modern CFD codes include options for calculation of particle trajectories, species transport, heat transfer, and turbulence. The codes are easy to use, and solutions can be obtained without knowledge about the equations or their limitations. Herein lies the danger of CFD: When in the hands of someone without knowledge of fluid mechanics, erroneous results are likely to occur (Fig. 15–7). It is critical that users of CFD possess some fundamental knowledge of fluid mechanics so that they can discern whether a CFD solution makes physical sense or not.

Grid Generation and Grid Independence

The first step (and arguably the most important step) in a CFD solution is generation of a grid that defines the cells on which flow variables (velocity, pressure, etc.) are calculated throughout the computational domain. Modern commercial CFD codes come with their own grid generators, and third-party grid generation programs are also available. The grids used in this chapter are generated with ANSYS-FLUENT's grid generation package.

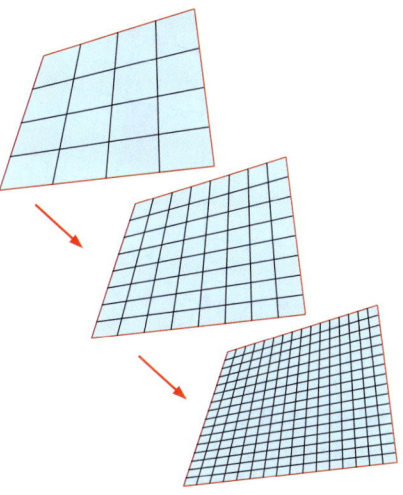

FIGURE 15–6
With *multigridding*, solutions of the equations of motion are obtained on a coarse grid first, followed by successively finer grids. This speeds up convergence.

FIGURE 15–7
CFD solutions are easy to obtain, and the graphical outputs can be beautiful; but correct answers depend on correct inputs and knowledge about the flow field.

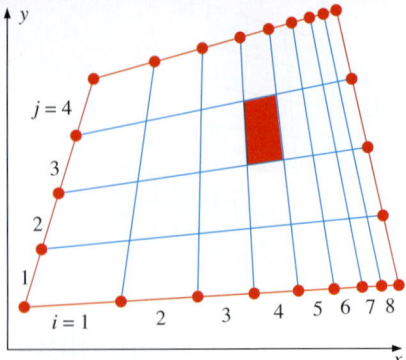

FIGURE 15–8
Sample structured 2-D grid with nine nodes and eight intervals on the top and bottom edges, and five nodes and four intervals on the left and right edges. Indices i and j are shown. The red cell is at $(i = 4, j = 3)$.

Many CFD codes can run with either structured or unstructured grids. A **structured grid** consists of planar cells with four edges (2-D) or volumetric cells with six faces (3-D). Although the cells may be distorted from rectangular, each cell is numbered according to indices (i, j, k) that do not necessarily correspond to coordinates x, y, and z. An illustration of a 2-D structured grid is shown in Fig. 15–8. To construct this grid, nine **nodes** are specified on the top and bottom edges; these nodes correspond to eight **intervals** along these edges. Similarly, five nodes are specified on the left and right edges, corresponding to four intervals along these edges. The intervals correspond to $i = 1$ through 8 and $j = 1$ through 4, and are numbered and marked in Fig. 15–8. An internal grid is then generated by connecting nodes one-for-one across the domain such that rows (j = constant) and columns (i = constant) are clearly defined, even though the cells themselves may be distorted (not necessarily rectangular). In a 2-D structured grid, each cell is uniquely specified by an index pair (i, j). For example, the shaded cell in Fig. 15–8 is at $(i = 4, j = 3)$. You should be aware that some CFD codes number *nodes* rather than intervals.

An **unstructured grid** consists of cells of various shapes, but typically triangles or quadrilaterals (2-D) and tetrahedrons or hexahedrons (3-D) are used. Two unstructured grids for the same domain as that of Fig. 15–8 are generated, using the *same* interval distribution on the edges; these grids are shown in Fig. 15–9. Unlike the structured grid, one cannot uniquely identify cells in the unstructured grid by indices i and j; instead, cells are numbered in some other fashion internally in the CFD code.

For complex geometries, an unstructured grid is usually much easier for the user of the grid generation code to create. However, there are some advantages to structured grids. For example, some (usually older) CFD codes are written specifically for structured grids; these codes converge more rapidly, and often more accurately, by utilizing the index feature of structured grids. For modern general-purpose CFD codes that can handle both structured and unstructured grids, however, this is no longer an issue. More importantly, *fewer cells are usually generated with a structured grid than with an unstructured grid*. In Fig. 15–8, for example, the structured grid has $8 \times 4 = 32$ cells, while the unstructured triangular grid of Fig. 15–9a has 76 cells, and the unstructured quadrilateral grid has 38 cells, even though the identical node distribution is applied at the edges in all three cases. In boundary

FIGURE 15–9
Sample 2-D unstructured grids with nine nodes and eight intervals on the top and bottom edges, and five nodes and four intervals on the left and right edges. These grids use the same node distribution as that of Fig. 15–8: (*a*) unstructured triangular grid, and (*b*) unstructured quadrilateral grid. The red cell in the upper right corner of (*a*) is moderately skewed.

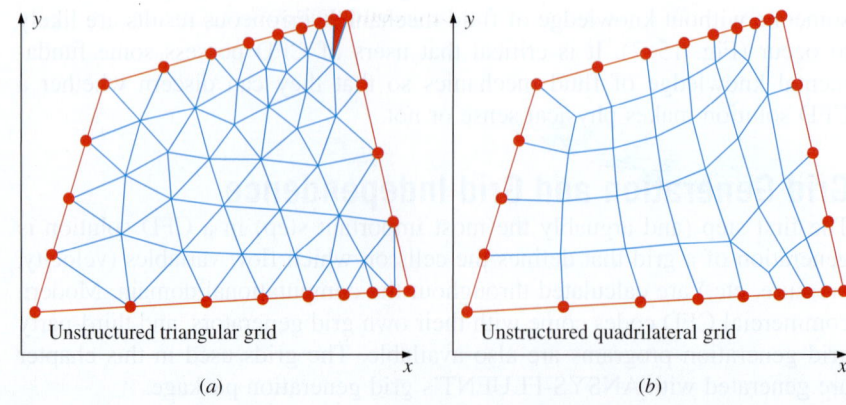

layers, where flow variables change rapidly normal to the wall and highly resolved grids are required close to the wall, structured grids enable much finer resolution than do unstructured grids for the same number of cells. This can be seen by comparing the grids of Figs. 15–8 and 15–9 near the far right edge. The cells of the structured grid are thin and tightly packed near the right edge, while those of the unstructured grids are not.

We must emphasize that regardless of the type of grid you choose (structured or unstructured, quadrilateral or triangular, etc.), it is the *quality* of the grid that is most critical for reliable CFD solutions. In particular, you must always be careful that individual cells are not highly skewed, as this can lead to convergence difficulties and inaccuracies in the numerical solution. The shaded cell in Fig. 15–9a is an example of a cell with moderately high **skewness,** defined as the departure from symmetry. There are various kinds of skewness, for both two- and three-dimensional cells. Three-dimensional cell skewness is beyond the scope of the present textbook—the type of skewness most appropriate for *two-dimensional* cells is **equiangle skewness,** defined as

Equiangle skewness: $$Q_{EAS} = \text{MAX}\left(\frac{\theta_{max} - \theta_{equal}}{180° - \theta_{equal}}, \frac{\theta_{equal} - \theta_{min}}{\theta_{equal}}\right) \quad (15\text{-}3)$$

where θ_{min} and θ_{max} are the minimum and maximum angles (in degrees) between any two edges of the cell, and θ_{equal} is the angle between any two edges of an ideal equilateral cell with the same number of edges. For triangular cells $\theta_{equal} = 60°$ and for quadrilateral cells $\theta_{equal} = 90°$. You can show by Eq. 15–3 that $0 < Q_{EAS} < 1$ for any 2-D cell. By definition, an equilateral triangle has zero skewness. In the same way, a square or rectangle has zero skewness. A grossly distorted triangular or quadrilateral element may have unacceptably high skewness (Fig. 15–10). Some grid generation codes use numerical schemes to smooth the grid so as to minimize skewness.

Other factors affect the quality of the grid as well. For example, abrupt changes in cell size can lead to numerical or convergence difficulties in the CFD code. Also, cells with a very large aspect ratio can sometimes cause problems. While you can often minimize the cell count by using a structured grid instead of an unstructured grid, a structured grid is not always the best choice, depending on the shape of the computational domain. You must always be cognizant of grid quality. Keep in mind that *a high-quality unstructured grid is better than a poor-quality structured grid.* An example is shown in Fig. 15–11 for the case of a computational domain with a small

(a) Triangular cells

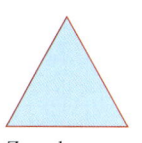

Zero skewness   High skewness

(b) Quadrilateral cells

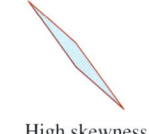

Zero skewness   High skewness

FIGURE 15–10

Skewness is shown in two dimensions: (*a*) an equilateral triangle has zero skewness, but a highly distorted triangle has high skewness. (*b*) Similarly, a rectangle has zero skewness, but a highly distorted quadrilateral cell has high skewness.

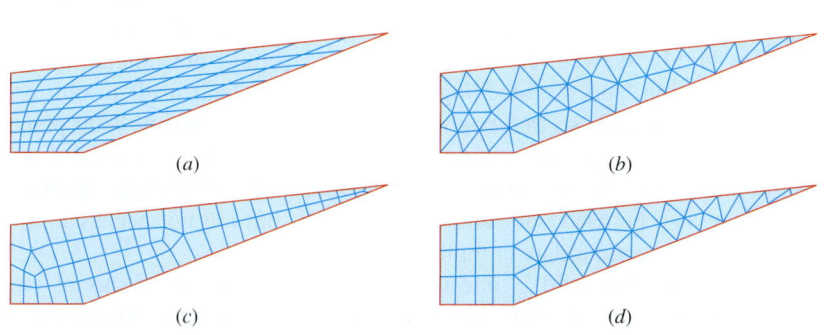

FIGURE 15–11

Comparison of four 2-D grids for a highly distorted computational domain: (*a*) structured 8 × 8 grid with 64 cells and $(Q_{EAS})_{max} = 0.83$, (*b*) unstructured triangular grid with 70 cells and $(Q_{EAS})_{max} = 0.76$, (*c*) unstructured quadrilateral grid with 67 cells and $(Q_{EAS})_{max} = 0.87$, and (*d*) hybrid grid with 62 cells and $(Q_{EAS})_{max} = 0.76$.

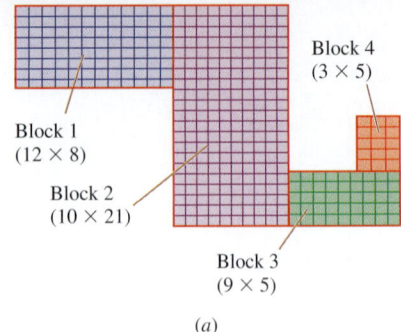

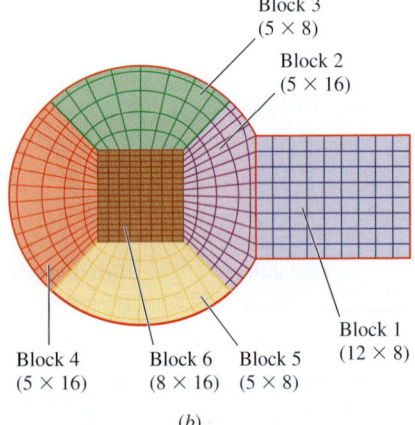

FIGURE 15–12

Examples of structured grids generated for multiblock CFD analysis: (*a*) a simple 2-D computational domain composed of rectangular four-sided blocks, and (*b*) a more complicated 2-D domain with curved surfaces, but again composed of four-sided blocks and quadrilateral cells. The number of *i*- and *j*-intervals is shown in parentheses for each block. There are, of course, acceptable alternative ways to divide these computational domains into blocks.

acute angle at the upper-right corner. For this example we have adjusted the node distribution so that the grid in any case contains between 60 and 70 cells for direct comparison. The structured grid (Fig. 15–11*a*) has $8 \times 8 = 64$ cells; but even after smoothing, the maximum equiangle skewness is 0.83—cells near the upper right corner are highly skewed. The unstructured triangular grid (Fig. 15–11*b*) has 70 cells, but the maximum skewness is reduced to 0.76. More importantly, the overall skewness is lower throughout the entire computational domain. The unstructured quad grid (Fig. 15–11*c*) has 67 cells. Although the overall skewness is better than that of the structured mesh, the maximum skewness is 0.87—higher than that of the structured mesh. The hybrid grid shown in Fig. 15–11*d* is discussed shortly.

Situations arise in which a structured grid is preferred (e.g., the CFD code requires structured grids, boundary layer zones need high resolution, or the simulation is pushing the limits of available computer memory). Generation of a structured grid is straightforward for geometries with straight edges. All we need to do is divide the computational domain into four-sided (2-D) or six-sided (3-D) **blocks** or **zones.** Inside each block, we generate a structured grid (Fig. 15–12*a*). Such an analysis is called **multiblock** analysis. For more complicated geometries with curved surfaces, we need to determine how the computational domain can be divided into individual blocks that may or may not have flat edges (2-D) or faces (3-D). A two-dimensional example involving circular arcs is shown in Fig. 15–12*b*. Most CFD codes require that the nodes match on the common edges and faces between blocks.

Many commercial CFD codes allow you to split the edges or faces of a block and assign different boundary conditions to each segment of the edge or face. In Fig. 15–12*a* for example, the left edge of block 2 is split about two-thirds of the way up to accommodate the junction with block 1. The lower segment of this edge is a wall, and the upper segment of this edge is an interior edge. (These and other boundary conditions are discussed shortly.) A similar situation occurs on the right edge of block 2 and on the top edge of block 3. Some CFD codes accept only **elementary blocks,** namely, *blocks whose edges or faces cannot be split.* For example, the four-block grid of Fig. 15–12*a* requires seven elementary blocks under this limitation (Fig. 15–13). The total number of cells is the same, which you can verify. Finally, for CFD codes that allow blocks with split edges or faces, we can sometimes combine two or more blocks into one. For example, it is left as an exercise to show how the structured grid of Fig. 5-11*b* can be simplified to just *three* nonelementary blocks.

When developing the block topology with complicated geometries as in Fig. 15–12*b*, the goal is to create blocks in such a way that no cells in the grid are highly skewed. In addition, cell size should not change abruptly in any direction, and the blocking topology should lend itself to clustering cells near solid walls so that boundary layers can be resolved. With practice you can master the art of creating sophisticated multiblock structured grids. Multiblock grids are *necessary* for structured grids of complex geometry. They may also be used with unstructured grids, but are not necessary since the cells can accommodate complex geometries.

Finally, a **hybrid grid** is one that combines regions or blocks of structured and unstructured grids. For example, you can mate a structured grid block close to a wall with an unstructured grid block outside of the region of influence of the boundary layer. A hybrid grid is often used to enable high resolution near a wall without requiring high resolution away from the wall (Fig. 15–14). When generating any type of grid (structured, unstructured, or hybrid), you must always be careful that individual cells are not highly skewed. For example, none of the cells in Fig. 15–14 has any significant skewness. Another example of a hybrid grid is shown in Fig. 15–11*d*. Here we have split the computational domain into two blocks. The four-sided block on the left is meshed with a structured grid, while the three-sided block on the right is meshed with an unstructured triangular grid. The maximum skewness is 0.76, the same as that of the unstructured triangular grid of Fig. 15–11*b*, but the total number of cells is reduced from 70 to 62.

Computational domains with very small angles like the one shown in Fig. 15–11 are difficult to mesh at the sharp corner, regardless of the type of cells used. One way to avoid large values of skewness at a sharp corner is to simply chop off or round off the sharp corner. This can be done very close to the corner so that the geometric modification is imperceptible from an overall view and has little if any effect on the flow, yet greatly improves the performance of the CFD code by reducing the skewness. For example, the troublesome sharp corner of the computational domain of Fig. 15–11 is chopped off and replotted in Fig. 15–15. Through use of multiblocking and hybrid grids, the grid shown in Fig. 15–15 has 62 cells and a maximum skewness of only 0.53—a vast improvement over any of the grids in Fig. 15–11.

The examples shown here are for two dimensions. In three dimensions, you can still choose between structured, unstructured, and hybrid grids. If a four-sided 2-D face with structured cells is swept in the third dimension, a fully structured 3-D mesh is produced, consisting of **hexahedral** cells ($n = 6$ faces per cell). When a 2-D face with unstructured triangular cells is swept in the third direction, the 3-D mesh can consist of **prism** cells ($n = 5$ faces per cell) or **tetrahedral** cells ($n = 4$ faces per cell—like a pyramid). These are illustrated in Fig. 15–16. When a hexahedral mesh is impractical to apply (e.g., complex geometry), a tetrahedral mesh (also called a tet mesh) is a common alternative approach. Automatic grid generation codes often generate a tet mesh by default. However, just as in the 2-D case, a 3-D unstructured tet mesh results in greater overall cell count than a structured hexahedral mesh with the same resolution along boundaries.

The most recent enhancement in grid generation is the use of **polyhedral meshes**. As the name implies, such a mesh consists of cells of many faces, called polyhedral cells. Some modern grid generators can create unstructured three-dimensional meshes with a mixture of *n*-sided cells, where *n*

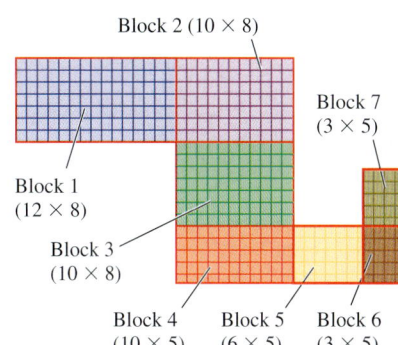

FIGURE 15–13
The multiblock grid of Fig. 15–12*a* modified for a CFD code that can handle only *elementary blocks*.

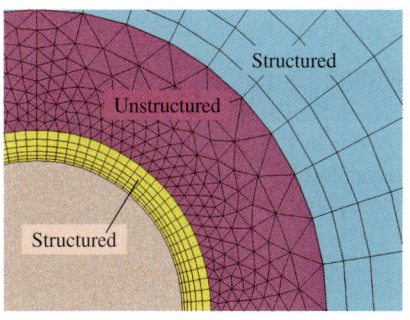

FIGURE 15–14
Sample two-dimensional hybrid grid near a curved surface; two structured regions and one unstructured region are labeled.

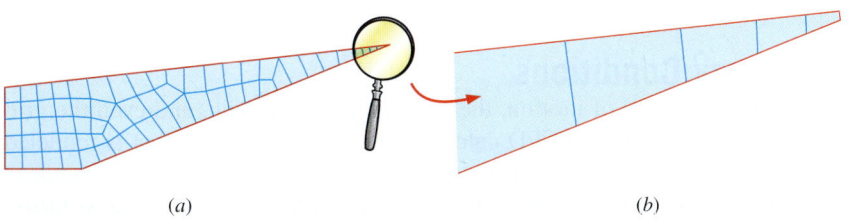

FIGURE 15–15
Hybrid grid for the computational domain of Fig. 15–11 with the sharp corner chopped off: (*a*) overall view—the grid contains 62 cells with $(Q_{EAS})_{max} = 0.53$, (*b*) magnified view of the chopped off corner.

COMPUTATIONAL FLUID DYNAMICS

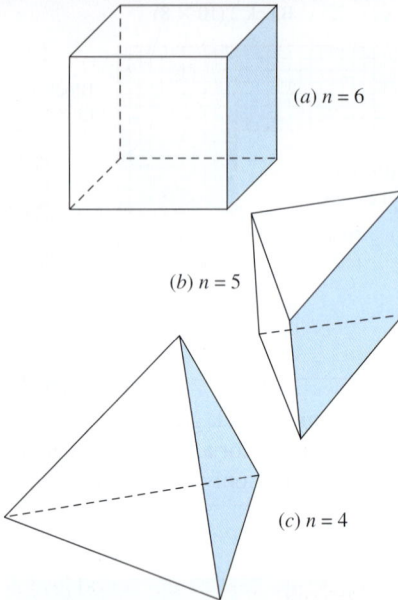

FIGURE 15–16
Examples of three-dimensional cells: (*a*) hexahedral, (*b*) prism, and (*c*) tetrahedral, along with the number of faces *n* for each case.

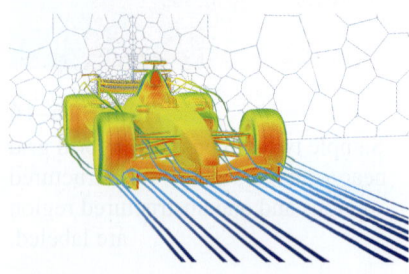

FIGURE 15–17
This Formula 1 car is modeled using a polyhedral mesh to reduce cell count and simulation time and is simulated using the ANSYS-FLUENT CFD code. The image depicts shaded pressure contours on the car body (red color indicating higher pressure) and pathlines over the body (shaded by time). Because of the symmetry between the right and left sides of the car, the analysis is performed on only half of the car; the results depict a mirror image (about the center plane) of the solution domain.
Photo courtesy of ANSYS.

can be any integer greater than 3. An example polyhedral mesh is shown in Fig. 15–17. In some codes, the polyhedral cells are formed by merging tetrahedral cells, reducing total cell count. This saves a significant amount of computer memory and speeds up the CFD calculations. Overall cell-count reductions (and corresponding CPU time savings) by a factor of as much as 5 have been reported without compromising solution accuracy. Another advantage of polyhedral meshes is that cell skewness can be reduced, improving the overall mesh quality and also speeding up convergence. Finally, polyhedral cells with large *n* have many more neighbor cells than do simple tetrahedral or prism cells. This is advantageous for tasks such as calculating gradients (derivatives) of flow parameters—details are beyond the level of the present text.

Generation of a good grid is often tedious and time consuming; engineers who use CFD on a regular basis will agree that grid generation usually takes more of their time than does the CFD solution itself (engineer's time, not CPU time). However, *time spent generating a good grid is time well spent*, since the CFD results will be more reliable and may converge more rapidly (Fig. 15–18). A high-quality grid is critical to an accurate CFD solution; a poorly resolved or low-quality grid may even lead to an *incorrect* solution. It is important, therefore, for users of CFD to test if their solution is **grid independent.** The standard method to test for grid independence is to increase the resolution (by a factor of 2 in all directions if feasible) and repeat the simulation. If the results do not change appreciably, the original grid is probably adequate. If, on the other hand, there are significant differences between the two solutions, the original grid is likely of inadequate resolution. In such a case, an even finer grid should be tried until the grid is adequately resolved. This method of testing for grid independence is time consuming, and unfortunately, not always feasible, especially for large engineering problems in which the solution pushes computer resources to their limits. In a 2-D simulation, if one doubles the number of intervals on each edge, the number of cells increases by a factor of $2^2 = 4$; the required computation time for the CFD solution also increases by approximately a factor of 4. For three-dimensional flows, doubling the number of intervals in each direction increases the cell count by a factor of $2^3 = 8$. You can see how grid independence studies can easily get beyond the range of a computer's memory capacity and/or CPU availability. If you cannot double the number of intervals because of computer limitations, a good rule of thumb is that you should increase the number of intervals by at least 20 percent in all directions to test for grid independence.

On a final note about grid generation, the trend in CFD today is automated grid generation, coupled with automated grid refinement based on error estimates. Yet even in the face of these emerging trends, it is critical that you understand how the grid impacts the CFD solution.

Boundary Conditions

While the equations of motion, the computational domain, and even the grid may be the same for two CFD calculations, the type of flow that is modeled is determined by the imposed boundary conditions. *Appropriate boundary conditions are required in order to obtain an accurate CFD solution*

(Fig. 15–19). There are several types of boundary conditions available; the most relevant ones are listed and briefly described in the following. The names are those used by ANSYS-FLUENT; other CFD codes may use somewhat different terminology, and the details of their boundary conditions may differ. In the descriptions given, the words *face* or *plane* are used, implying three-dimensional flow. For a two-dimensional flow, the word *edge* or *line* should be substituted for *face* or *plane*.

Wall Boundary Conditions

The simplest boundary condition is that of a **wall.** Since fluid cannot pass through a wall, the normal component of velocity is set to zero relative to the wall along a face on which the wall boundary condition is prescribed. In addition, because of the no-slip condition, we usually set the tangential component of velocity at a stationary wall to zero as well. In Fig. 15–19, for example, the upper and lower edges of this simple domain are specified as wall boundary conditions with no slip. If the energy equation is being solved, either wall temperature or wall heat flux must also be specified (but not both; see Section 15–4). If a turbulence model is being used, turbulence transport equations are solved, and wall roughness may need to be specified, since turbulent boundary layers are influenced greatly by the roughness of the wall. In addition, you must choose among various kinds of turbulence wall treatments (**wall functions,** etc.). These turbulence options are beyond the scope of the present text (see Wilcox, 2006); fortunately the default options of most modern CFD codes are sufficient for many applications involving turbulent flow.

Moving walls and walls with specified shear stresses can also be simulated in many CFD codes. There are situations where we desire to let the fluid slip along the wall (we call this an "inviscid wall"). For example, we can specify a zero-shear-stress wall boundary condition along the free surface of a swimming pool or hot tub when simulating such a flow (Fig. 15–20). Note that with this simplification, the fluid is allowed to "slip" along the surface, since the viscous shear stress caused by the air above it is negligibly small (Chap. 9). When making this approximation, however, surface waves and their associated pressure fluctuations cannot be taken into account.

Inflow/Outflow Boundary Conditions

There are several options at the boundaries through which fluid enters the computational domain (inflow) or leaves the domain (outflow). They are generally categorized as either *velocity-specified conditions* or *pressure-specified conditions*. At a **velocity inlet,** we specify the velocity of the incoming flow along the inlet face. If energy and/or turbulence equations are being solved, the temperature and/or turbulence properties of the incoming flow need to be specified as well.

At a **pressure inlet,** we specify the total pressure along the inlet face (for example, flow coming into the computational domain from a pressurized tank of known pressure or from the far field where the ambient pressure is known). At a **pressure outlet,** fluid flows *out of* the computational domain. We specify the static pressure along the outlet face; in many cases this is atmospheric pressure (zero gage pressure). For example, the pressure is

FIGURE 15–18
Time spent generating a good grid is time well spent.

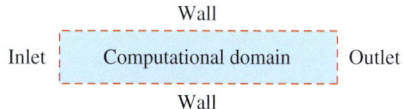

FIGURE 15–19
Boundary conditions must be carefully applied at *all* boundaries of the computational domain. Appropriate boundary conditions are required in order to obtain an accurate CFD solution.

atmospheric at the outlet of a subsonic exhaust pipe open to ambient air (Fig. 15–21). Flow properties, such as temperature, and turbulence properties are also specified at pressure inlets and pressure outlets. For outlets, however, these properties are not used unless the solution demands **reverse flow** across the outlet. *Reverse flow at a pressure outlet is usually an indication that the computational domain is not large enough.* If reverse flow warnings persist as the CFD solution iterates, the computational domain should be extended.

Pressure is *not* specified at a velocity inlet, as this would lead to mathematical overspecification, since pressure and velocity are *coupled* in the equations of motion. Rather, pressure at a velocity inlet adjusts itself to match the rest of the flow field. In similar fashion, velocity is not specified at a pressure inlet or outlet, as this would also lead to mathematical over-specification. Rather, velocity at a pressure-specified boundary condition adjusts itself to match the rest of the flow field (Fig. 15–22).

Another option at an outlet of the computational domain is the **outflow** boundary condition. At an outflow boundary, no flow properties are specified; instead, flow properties such as velocity, turbulence quantities, and temperature are forced to have *zero gradients normal to the outflow face* (Fig. 15–23). For example, if a duct is sufficiently long so that the flow is *fully developed* at the outlet, the outflow boundary condition would be appropriate, since velocity does not change in the direction normal to the outlet face. Note that the flow direction is not constrained to be perpendicular to the outflow boundary, as also illustrated in Fig. 15–23. If the flow is still developing, but the pressure at the outlet is known, a pressure outlet boundary condition would be more appropriate than an outflow boundary condition. The outflow boundary condition is often preferred over the pressure outlet in rotating flows since the swirling motion leads to radial pressure gradients that are not easily handled by a pressure outlet.

A common situation in a simple CFD application is to specify one or more velocity inlets along portions of the boundary of the computational domain, and one or more pressure outlets or outflows at other portions of

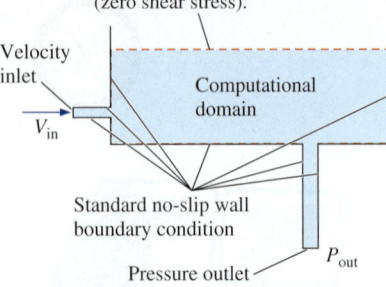

FIGURE 15–20
The standard *wall* boundary condition is imposed on stationary solid boundaries, where we also impose either a wall temperature or a wall heat flux. The shear stress along the wall can be set to zero to simulate the free surface of a liquid, as shown here for the case of a swimming pool. There is slip along this "wall" that simulates the free surface (in contact with air).

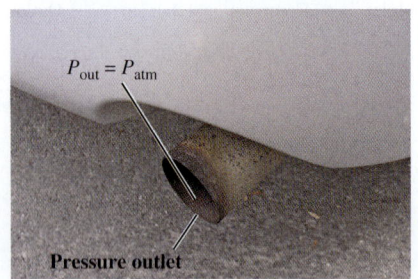

FIGURE 15–21
When modeling an incompressible flow field, with the outlet of a pipe or duct exposed to ambient air, the proper boundary condition is a pressure outlet with $P_{out} = P_{atm}$. Shown here is the tail pipe of an automobile.

Photo by Po-Ya Abel Chuang. Used by permission.

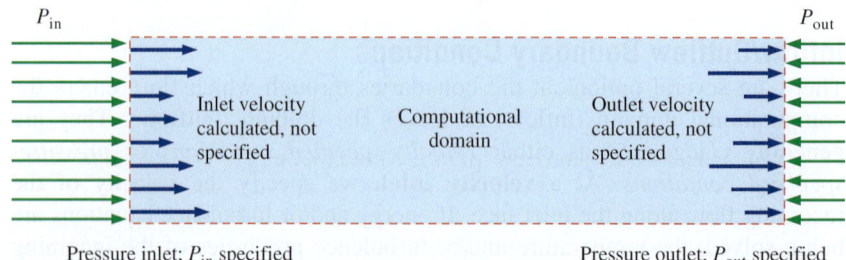

FIGURE 15–22
At a *pressure inlet* or a *pressure outlet*, we specify the pressure on the face, but we cannot specify the velocity through the face. As the CFD solution converges, the velocity adjusts itself such that the prescribed pressure boundary conditions are satisfied.

the boundary, with walls defining the geometry of the rest of the computational domain. For example, in our swimming pool (Fig. 15–20), we set the left-most face of the computational domain as a velocity inlet and the bottom-most face as a pressure outlet. The rest of the faces are walls, with the free surface modeled as a wall with zero shear stress.

Finally, for compressible flow simulations, the inlet and outlet boundary conditions are further complicated by introduction of Riemann invariants and characteristic variables related to incoming and outgoing waves, discussion of which is beyond the scope of the present text. Fortunately, many CFD codes have a **pressure far field** boundary condition for compressible flows. This boundary condition is used to specify the Mach number, pressure, and temperature at an inlet. The same boundary condition can be applied at an outlet; when flow exits the computational domain, flow variables at the outlet are extrapolated from the interior of the domain. Again, you must ensure that there is no reverse flow at an outlet.

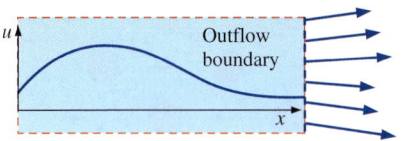

FIGURE 15–23
At an *outflow* boundary condition, the gradient or slope of velocity normal to the outflow face is zero, as illustrated here for u as a function of x along a horizontal line. Note that neither pressure nor velocity are specified at an outflow boundary.

Miscellaneous Boundary Conditions

Some boundaries of a computational domain are neither walls nor inlets or outlets, but rather enforce some kind of symmetry or periodicity. For example, the **periodic** boundary condition is useful when the geometry involves repetition. Flow field variables along one face of a periodic boundary are numerically **linked** to a second face of identical shape (and in most CFD codes, also identical face *mesh*). Thus, flow leaving (crossing) the first periodic boundary can be imagined as entering (crossing) the second periodic boundary with identical properties (velocity, pressure, temperature, etc.). Periodic boundary conditions always occur in *pairs* and are useful for flows with repetitive geometries, such as flow between the blades of a turbomachine or through an array of heat exchanger tubes (Fig. 15–24). The periodic boundary condition enables us to work with a computational domain that is much smaller than the full flow field, thereby conserving computer resources. In Fig. 15–24, you can imagine an infinite number of repeated domains (dashed lines) above and below the actual computational domain (the light blue shaded region). Periodic boundary conditions must be specified as either **translational** (periodicity applied to two parallel faces, as in Fig. 15–24) or **rotational** (periodicity applied to two radially oriented faces). The region of flow between two neighboring blades of a fan (a **flow passage**) is an example of a rotationally periodic domain (see Fig. 15–58).

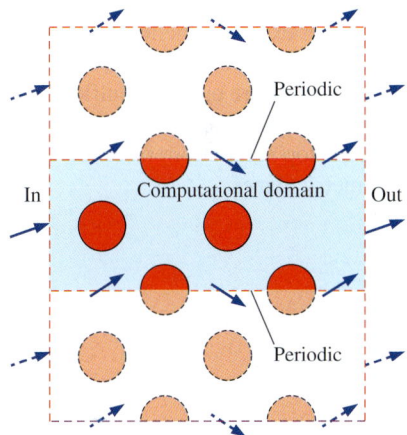

FIGURE 15–24
The *periodic* boundary condition is imposed on two identical faces. Whatever happens at one of the faces must also happen at its periodic partner face, as illustrated by the velocity vectors crossing the periodic faces.

The **symmetry** boundary condition forces flow field variables to be *mirror-imaged* across a symmetry plane. Mathematically, *gradients* of most flow field variables in the direction normal to the symmetry plane are set to zero across the plane of symmetry, although some variables are specified as even functions and some as odd functions across a symmetry boundary condition. For physical flows with one or more symmetry planes, this boundary condition enables us to model a *portion* of the physical flow domain, thereby conserving computer resources. The symmetry boundary differs from the periodic boundary in that no "partner" boundary is required for the symmetry case. In addition, fluid can flow *parallel* to a symmetry boundary, but not *through* a symmetry boundary, whereas flow can *cross* a periodic boundary. Consider, for example, flow across an array of heat exchanger tubes (Fig. 15–24). If we assume that no flow crosses the periodic boundaries of that computational domain, we can use symmetry boundary conditions instead. Alert readers will notice that we can even cut the size of the computational domain in half by wise choice of symmetry planes (Fig. 15–25).

For *axisymmetric* flows, the **axis** boundary condition is applied to a straight edge that represents the axis of symmetry (Fig. 15–26a). Fluid can flow *parallel* to the axis, but cannot flow *through* the axis. The axisymmetric option enables us to solve the flow in only two dimensions, as sketched in Fig. 15–26b. The computational domain is simply a rectangle in the *xy*-plane; you can imagine rotating this plane about the *x*-axis to generate the axisymmetry. In the case of swirling axisymmetric flows, fluid may also flow *tangentially* in a circular path around the axis of symmetry. Swirling axisymmetric flows are sometimes called **rotationally symmetric.**

Internal Boundary Conditions

The final classification of boundary conditions is imposed on faces or edges that do not define a boundary of the computational domain, but rather exist *inside* the domain. When an **interior** boundary condition is specified on a face, flow crosses through the face without any user-forced changes, just as it would cross from one interior cell to another (Fig. 15–27). This boundary condition is necessary for situations in which the computational domain is divided into separate blocks or zones, and enables communication between blocks. We have found this boundary condition to be useful for postprocessing as well, since a predefined face is present in the flow field, on whose surface we can plot velocity vectors, pressure contours, etc. In more sophisticated CFD applications in which there is a sliding or rotating mesh, the interface between the two blocks is called upon to smoothly transfer information from one block to another.

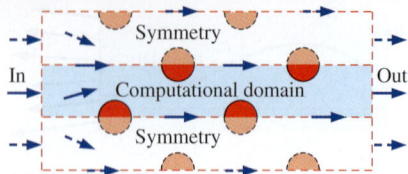

FIGURE 15–25
The *symmetry* boundary condition is imposed on a face so that the flow across that face is a mirror image of the calculated flow. We sketch imaginary domains (dashed lines) above and below the computational domain (the light blue shaded region) in which the velocity vectors are mirror images of those in the computational domain. In this heat exchanger example, the left face of the domain is a velocity inlet, the right face is a pressure outlet or outflow outlet, the cylinders are walls, and both the top and bottom faces are symmetry planes.

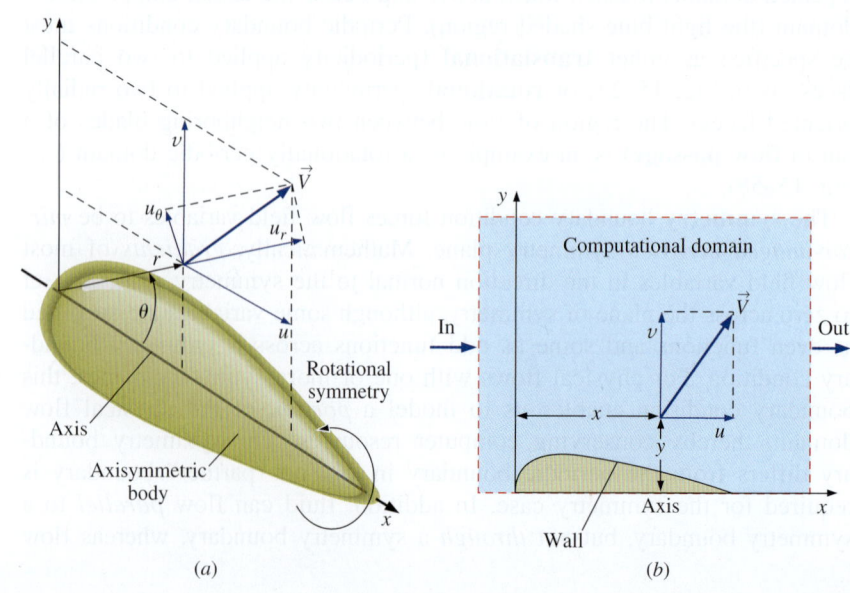

FIGURE 15–26
The *axis* boundary condition is applied to the axis of symmetry (here the *x*-axis) in an axisymmetric flow, since there is rotational symmetry about that axis. (*a*) A slice defining the *xy*- or *rθ*-plane is shown, and the velocity components can be either (u, v) or (u_r, u_θ). (*b*) The computational domain (light blue shaded region) for this problem is reduced to a plane in two dimensions (*x* and *y*). In many CFD codes, *x* and *y* are used as axisymmetric coordinates, with *y* being understood as the distance from the *x*-axis.

The **fan** boundary condition is specified on a plane across which a sudden pressure increase (or decrease) is to be assigned. This boundary condition is similar to an interior boundary condition except for the forced pressure rise. The CFD code does not solve the detailed, unsteady flow field through individual fan blades, but simply models the plane as an infinitesimally thin fan that changes the pressure across the plane. The fan boundary condition is useful, for example, as a simple model of a fan inside a duct (Fig. 15–27), a ceiling fan in a room, or the propeller or jet engine that provides thrust to an airplane. If the pressure rise across the fan is specified as zero, this boundary condition behaves the same as an interior boundary condition.

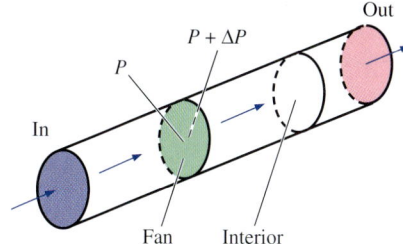

FIGURE 15–27
The *fan* boundary condition imposes an abrupt change in pressure across the fan face to simulate an axial-flow fan in a duct. When the specified pressure rise is zero, the fan boundary condition degenerates to an *interior* boundary condition.

Practice Makes Perfect

The best way to learn computational fluid dynamics is through examples and *practice*. You are encouraged to experiment with various grids, boundary conditions, numerical parameters, etc., in order to get a feel for CFD. Before tackling a complicated problem, it is best to solve simpler problems, especially ones for which analytical or empirical solutions are known (for comparison and verification). For this reason, dozens of practice problems are provided on the book's website.

In the following sections, we solve several example problems of general engineering interest to illustrate many of the capabilities and limitations of CFD. We start with laminar flows, and then provide some introductory turbulent flow examples. Finally we provide examples of flows with heat transfer, compressible flows, and liquid flows with free surfaces.

15–2 ■ LAMINAR CFD CALCULATIONS

Computational fluid dynamics does an excellent job at computing incompressible, steady or unsteady, laminar flow, provided that the grid is well resolved and the boundary conditions are properly specified. We show several simple examples of laminar flow solutions, paying particular attention to grid resolution and appropriate application of boundary conditions. In all examples in this section, the flows are incompressible and two-dimensional (or axisymmetric).

Pipe Flow Entrance Region at Re = 500

Consider flow of room-temperature water inside a smooth round pipe of length $L = 40.0$ cm and diameter $D = 1.00$ cm. We assume that the water enters at a uniform speed equal to $V = 0.05024$ m/s. The kinematic viscosity of the water is $\nu = 1.005 \times 10^{-6}$ m²/s, producing a Reynolds number of $Re = VD/\nu = 500$. We assume incompressible, steady, laminar flow. We are interested in the entrance region in which the flow gradually becomes fully developed. Because of the axisymmetry, we set up a computational domain that is a two-dimensional slice from the axis to the wall, rather than a three-dimensional cylindrical volume (Fig. 15–28). We generate six structured grids for this computational domain: *very coarse* (40 intervals in the axial direction × 8 intervals in the radial direction), *coarse* (80 × 16), *medium* (160 × 32), *fine* (320 × 64), *very fine* (640 × 128), and *ultrafine* (1280 × 256). (Note that the number of intervals is doubled in both directions for

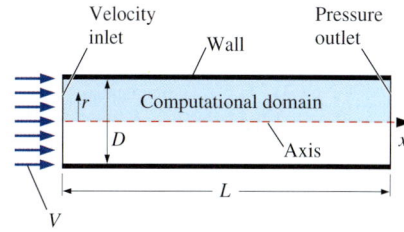

FIGURE 15–28
Because of axisymmetry about the *x*-axis, flow through a round pipe can be solved computationally with a two-dimensional slice through the pipe from $r = 0$ to $D/2$. The computational domain is the light blue shaded region, and the drawing is not to scale. Boundary conditions are indicated.

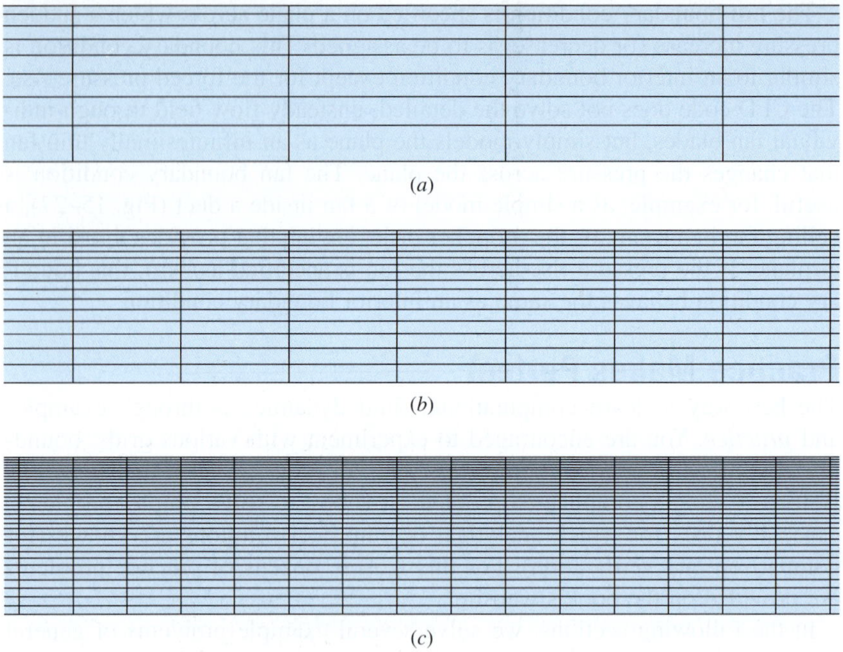

FIGURE 15–29
Portions of the three coarsest structured grids generated for the laminar pipe flow example: (*a*) very coarse (40 × 8), (*b*) coarse (80 × 16), and (*c*) medium (160 × 32). The number of computational cells is 320, 1280, and 5120, respectively. In each view, the pipe wall is at the top and the pipe axis is at the bottom, as in Fig. 15–28.

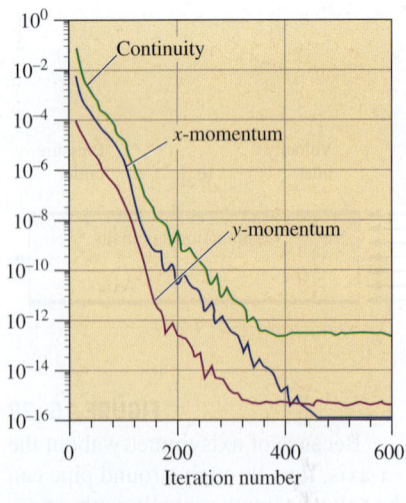

FIGURE 15–30
Decay of the residuals with iteration number for the very coarse grid laminar pipe flow solution (double precision arithmetic).

each successive grid; the number of computational cells increases by a factor of 4 for each grid.) In all cases the nodes are evenly distributed axially, but are concentrated near the wall radially, since we expect larger velocity gradients near the pipe wall. Close-up views of the first three of these grids are shown in Fig. 15–29.

We run the CFD program ANSYS-FLUENT in double precision for all six cases. (Double precision arithmetic is not always necessary for engineering calculations—we use it here to obtain the best possible precision in our comparisons.) Since the flow is laminar, incompressible, and axisymmetric, only three transport equations are solved—continuity, *x*-momentum, and *y*-momentum. Note that coordinate *y* is used in the CFD code instead of *r* as the distance from the axis of rotation (Fig. 15–26). The CFD code is run until convergence (all the residuals level off). Recall that a residual is a measure of how much the solution to a given transport equation deviates from exact; the lower the residual, the better the convergence. For the very coarse grid case, this occurs in about 500 iterations, and the residuals level off to less than 10^{-12} (relative to their initial values). The decay of the residuals is plotted in Fig. 15–30 for the very coarse case. Note that for more complicated flow problems with finer grids, you cannot always expect such low residuals; in some CFD solutions, the residuals level off at much higher values, like 10^{-3}.

We define P_1 as the average pressure at an axial location one pipe diameter downstream of the inlet. Similarly we define P_{20} at 20 pipe diameters. The average axial pressure drop from 1 to 20 diameters is thus $\Delta P = P_1 - P_{20}$, and is equal to 4.404 Pa (to four significant digits of precision) for the very coarse grid case. Centerline pressure and axial velocity are plotted in Fig. 15–31*a* as functions of downstream distance. The solution

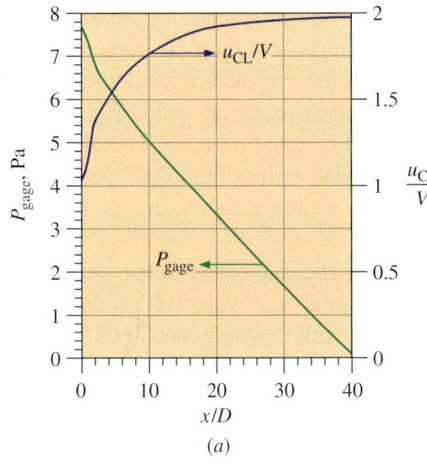
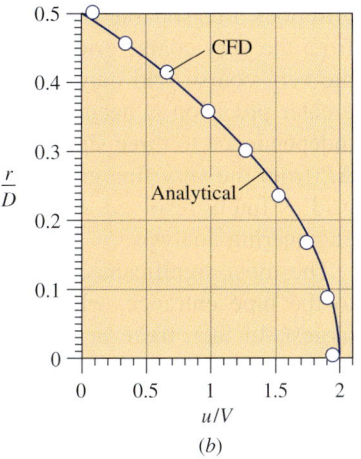

FIGURE 15–31
CFD results for the very coarse grid laminar pipe flow simulation: (a) development of centerline pressure and centerline axial velocity with downstream distance, and (b) axial velocity profile at pipe outlet compared to analytical prediction.

appears to be physically reasonable. We see the increase of centerline axial velocity to conserve mass as the boundary layer on the pipe wall grows downstream. We see a sharp drop in pressure near the pipe entrance where viscous shear stresses on the pipe wall are highest. The pressure drop approaches linear toward the end of the entrance region where the flow is nearly fully developed, as expected. Finally, we compare in Fig. 15–31b the axial velocity profile at the end of the pipe to the known analytical solution for fully developed laminar pipe flow (see Chap. 8). The agreement is excellent, especially considering that there are only eight intervals in the radial direction.

Is this CFD solution grid independent? To find out, we repeat the calculations using the coarse, medium, fine, very fine, and ultrafine grids. The convergence of the residuals is qualitatively similar to that of Fig. 15–30 for all cases, but CPU time increases significantly as grid resolution improves, and the levels of the final residuals are not as low as those of the coarse resolution case. The number of iterations required until convergence also increases with improved grid resolution. The pressure drop from $x/D = 1$ to 20 is listed in Table 15–1 for all six cases. ΔP is also plotted as a function of number of cells in Fig. 15–32. We see that even the very coarse

TABLE 15–1

Pressure drop from $x/D = 1$ to 20 for the various grid resolution cases in the entrance flow region of axisymmetric pipe flow

Case	Number of Cells	ΔP, Pa
Very coarse	320	4.404
Coarse	1280	3.983
Medium	5120	3.998
Fine	20,480	4.016
Very fine	81,920	4.033
Ultrafine	327,680	4.035

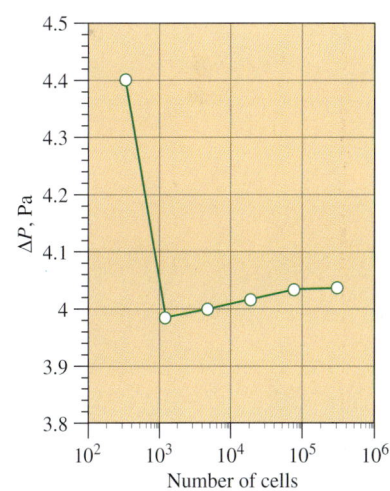

FIGURE 15–32
Calculated pressure drop from $x/D = 1$ to 20 in the entrance flow region of axisymmetric pipe flow as a function of number of cells.

grid does a reasonable job at predicting ΔP. The difference in pressure drop from the very coarse grid to the ultrafine grid is less than 10 percent. Thus, the very coarse grid may be adequate for some engineering calculations. If greater precision is needed, however, we must use a finer grid. We see grid independence to three significant digits by the very fine case. The change in ΔP from the very fine grid to the ultrafine grid is less than 0.07 percent—a grid as finely resolved as the ultrafine grid is unnecessary in any practical engineering analysis.

The most significant differences between the six cases occur very close to the pipe entrance, where pressure gradients and velocity gradients are largest. In fact, there is a *singularity* at the inlet, where the axial velocity changes suddenly from V to zero at the wall because of the no-slip condition. We plot in Fig. 15–33 contour plots of normalized axial velocity, u/V near the pipe entrance. We see that although global properties of the flow field (like overall pressure drop) vary by only a few percent as the grid is refined, *details* of the flow field (like the velocity contours shown here) change considerably with grid resolution. You can see that as the grid is continually refined, the axial velocity contour shapes become smoother and more well defined. The greatest differences in the contour shapes occur near the pipe wall.

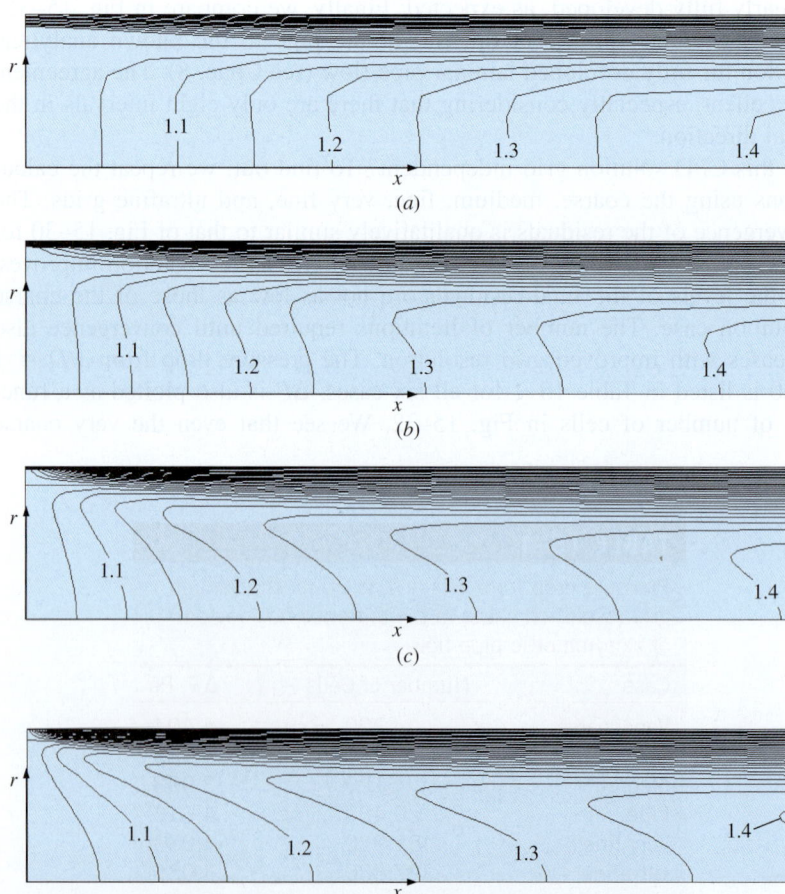

FIGURE 15–33
Normalized axial velocity contours (u/V) for the laminar pipe flow example. Shown is a close-up view of the entrance region of the pipe for each of the first four grids: (*a*) very coarse (40 × 8), (*b*) coarse (80 × 16), (*c*) medium (160 × 32), and (*d*) fine (320 × 64).

Flow around a Circular Cylinder at Re = 150

To illustrate that reliable CFD results require correct problem formulation, consider the seemingly simple problem of steady, incompressible, two-dimensional flow of air over a circular cylinder of diameter $D = 2.0$ cm (Fig. 15–34). The two-dimensional computational domain used for this simulation is sketched in Fig. 15–35. Only the upper half of the flow field is solved, due to symmetry along the bottom edge of the computational domain; a symmetry boundary condition is specified along this edge to ensure that no flow crosses the plane of symmetry. With this boundary condition imposed, the required computational domain size is reduced by a factor of 2. A stationary, no-slip wall boundary condition is applied at the cylinder surface. The left half of the far field outer edge of the domain has a velocity inlet boundary condition, on which is specified the velocity components $u = V$ and $v = 0$. A pressure outlet boundary condition is specified along the right half of the outer edge of the domain. (The gage pressure there is set to zero, but since the velocity field in an incompressible CFD code depends only on pressure *differences*, not absolute value of pressure, the value of pressure specified for the pressure outlet boundary condition is irrelevant.)

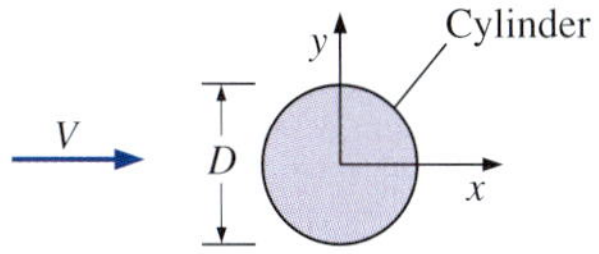

Three two-dimensional structured grids are generated for comparison: *coarse* (30 radial intervals × 60 intervals along the cylinder surface = 1800 cells), *medium* (60 × 120 = 7200 cells), and *fine* (120 × 240 = 28,800 cells), as seen in Fig. 15–36. Note that only a small portion of the computational domain is shown here; the full domain extends 15 cylinder diameters outward from the origin, and the cells get progressively larger further away from the cylinder.

We apply a free-stream flow of air at a temperature of 25°C, at standard atmospheric pressure, and at velocity $V = 0.1096$ m/s from left to right around this circular cylinder. The Reynolds number of the flow, based on cylinder diameter ($D = 2.0$ cm), is thus Re $= \rho VD/\mu = 150$. Experiments at this Reynolds number reveal that the boundary layer is laminar and separates almost 10° *before* the top of the cylinder, at $\alpha \cong 82°$ from the front stagnation point. The wake also remains laminar. Experimentally measured values of drag coefficient at this Reynolds number show much discrepancy in the literature; the range is from $C_D \cong 1.1$ to 1.4, and the differences are most likely due to the quality of the free-stream and three-dimensional effects (oblique vortex shedding, etc.). (Recall that $C_D = 2F_D/\rho V^2 A$, where A is the frontal area of the cylinder, and $A = D$ times the span of the cylinder, taken as unit length in a two-dimensional CFD calculation.)

CFD solutions are obtained for each of the three grids shown in Fig. 15–36, assuming steady laminar flow. All three cases converge without problems, but

FIGURE 15–34
Flow of fluid at free-stream speed V over a two-dimensional circular cylinder of diameter D.

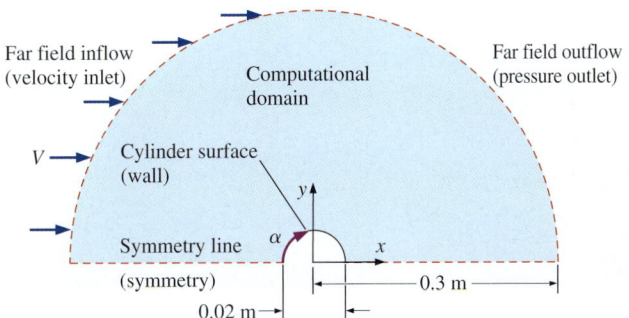

FIGURE 15–35
Computational domain (light blue shaded region) used to simulate steady two-dimensional flow over a circular cylinder (not to scale). It is assumed that the flow is symmetric about the x-axis. Applied boundary conditions are shown for each edge in parentheses. We also define α, the angle measured along the cylinder surface from the front stagnation point.

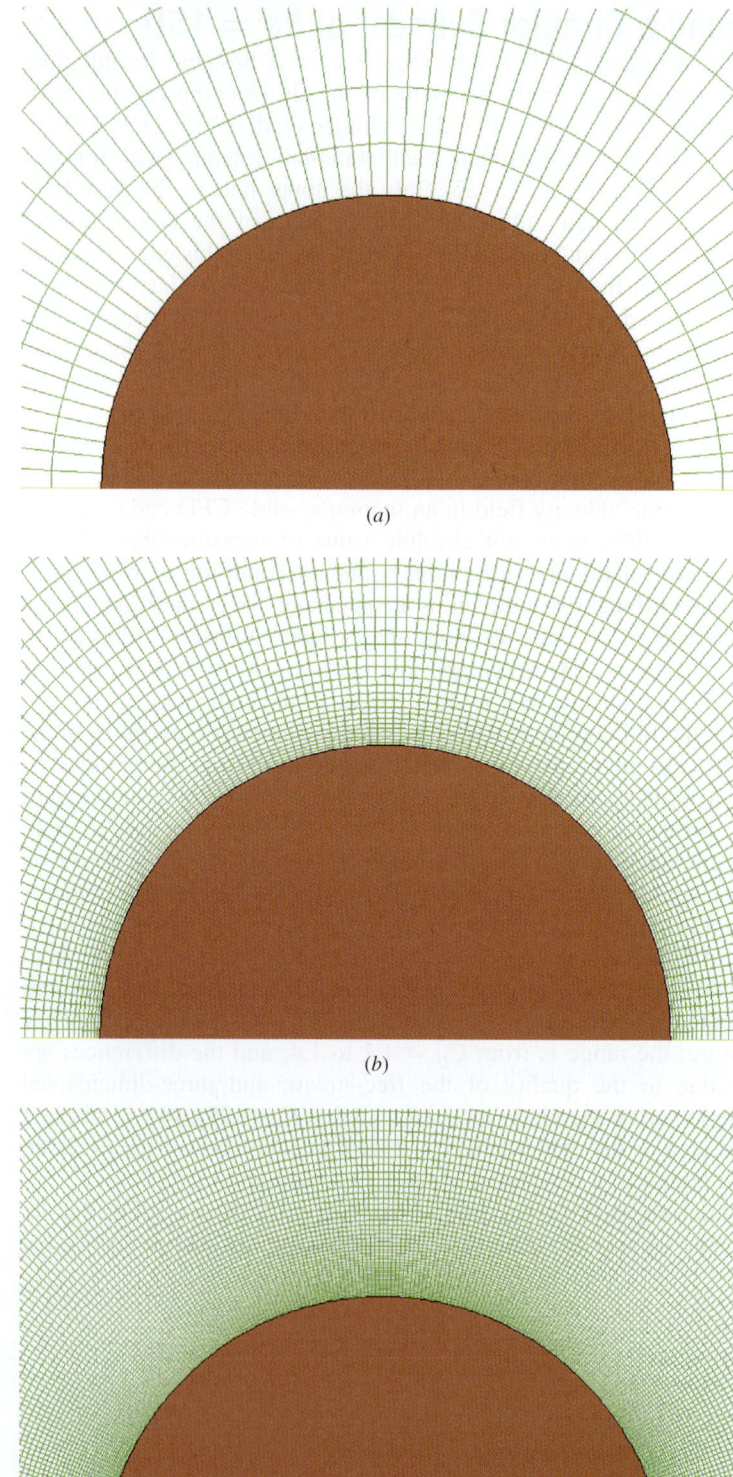

FIGURE 15–36
Structured two-dimensional grids around the upper half of a circular cylinder: (a) coarse (30 × 60), (b) medium (60 × 120), and (c) fine (120 × 240). The bottom edge is a line of symmetry. Only a portion of each computational domain is shown—the domain extends well beyond the portion shown here.

the results do not necessarily agree with physical intuition or with experimental data. Streamlines are shown in Fig. 15–37 for the three grid resolutions. In all cases, the image is mirrored about the symmetry line so that even though only the top half of the flow field is solved, the full flow field is displayed.

For the coarse resolution case (Fig. 15–37a), the boundary layer separates at $\alpha = 120°$, well past the top of the cylinder, and C_D is 1.00. The boundary layer is not well enough resolved to yield the proper boundary layer separation point, and the drag is somewhat smaller than it should be. Two large counter-rotating separation bubbles are seen in the wake; they stretch several cylinder diameters downstream. For the medium resolution case (Fig. 15–37b), the flow field is significantly different. The boundary layer separates a little further upstream at $\alpha = 110°$, which is more in line with the experimental results, but C_D has decreased to about 0.982—further away from the experimental value. The separation bubbles in the cylinder's wake have grown much longer than those of the coarse grid case. Does refining the grid even further improve the numerical results? Figure 15–37c shows streamlines for the fine resolution case. The results look qualitatively similar to those of the medium resolution case, with $\alpha = 109°$, but the drag coefficient is even smaller ($C_D = 0.977$), and the separation bubbles are even longer. A fourth calculation (not shown) at even finer grid resolution shows the same trend—the separation bubbles stretch downstream and the drag coefficient decreases somewhat.

Shown in Fig. 15–38 is a contour plot of tangential velocity component (u_θ) for the medium resolution case. We plot values of u_θ over a very small range around zero, so that we can clearly see where along the cylinder the flow changes direction. This is thus a clever way to locate the separation point along a cylinder wall. Note that this works only for a circular cylinder because of its unique geometry. A more general way to determine the separation point is to identify the point along the wall where the wall shear stress τ_w is zero; this technique works for bodies of any shape. From Fig. 15–38, we see that the boundary layer separates at an angle of $\alpha = 110°$ from the front stagnation point, much further downstream than the experimentally obtained value of 82°. In fact, all our CFD results predict boundary layer separation on the *rear* side rather than the front side of the cylinder.

These CFD results are unphysical—such elongated separation bubbles could not remain stable in a real flow situation, the separation point is too far downstream, and the drag coefficient is too low compared to experimental data. Furthermore, repeated grid refinement does *not* lead to better results as we would hope; on the contrary, *the results get worse with grid refinement*. Why do these CFD simulations yield such poor agreement with experiment? The answer is twofold:

1. We have forced the CFD solution to be steady, when in fact flow over a circular cylinder at this Reynolds number is *not* steady. Experiments show that a periodic **Kármán vortex street** forms behind the cylinder (Tritton, 1977; see also Fig. 4–25 of this text).
2. All three cases in Fig. 15–37 are solved for the upper half-plane only, and symmetry is enforced about the *x*-axis. In reality, flow over a circular cylinder is highly nonsymmetric; vortices are shed alternately from the top and the bottom of the cylinder, forming the Kármán vortex street.

To correct both of these problems, we need to run an *unsteady* CFD simulation with a *full* grid (top and bottom)—without imposing the symmetry

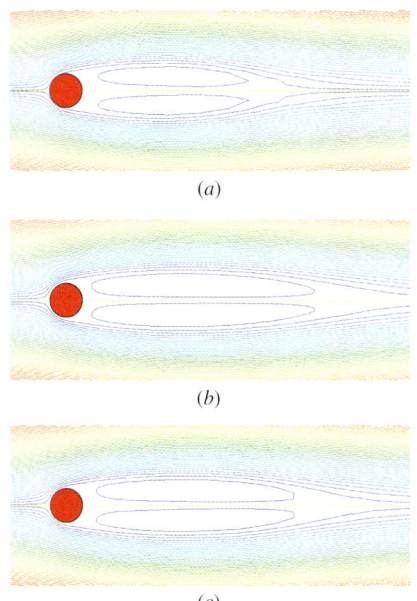

FIGURE 15–37
Streamlines produced by steady-state CFD calculations of flow over a circular cylinder at Re = 150: (a) coarse grid (30 × 60), (b) medium grid (60 × 120), and (c) fine grid (120 × 240). Note that only the top half of the flow is calculated—the bottom half is displayed as a mirror image of the top.

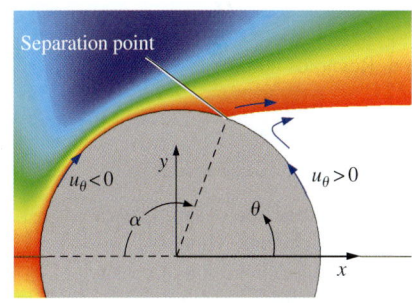

FIGURE 15–38
Contour plot of tangential velocity component u_θ for flow over a circular cylinder at Re = 150 and for the medium grid resolution case (60 × 120). Values in the range $-10^{-4} < u_\theta < 10^{-4}$ m/s are plotted, so as to reveal the precise location of boundary layer separation, i.e., where u_θ changes sign just outside the cylinder wall, as sketched. For this case, the flow separates at $\alpha = 110°$.

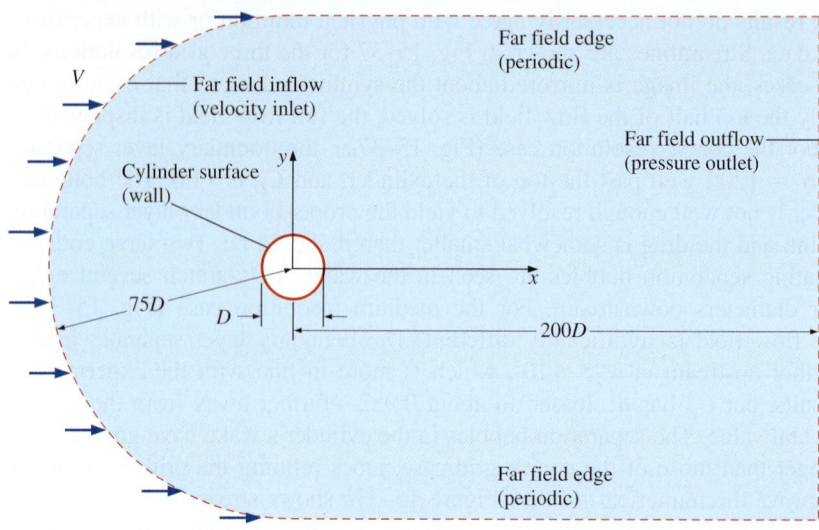

FIGURE 15–39
Computational domain (light blue shaded region) used to simulate unsteady, two-dimensional, laminar flow over a circular cylinder (not to scale). Applied boundary conditions are in parentheses.

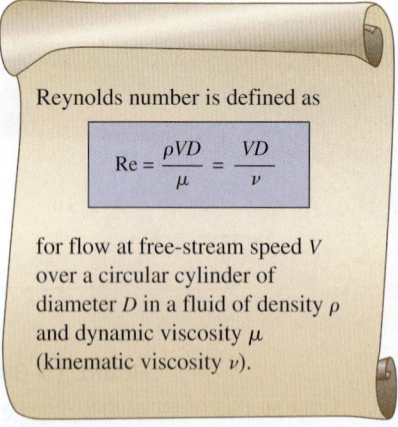

FIGURE 15–40
In an incompressible CFD simulation of flow around a cylinder, the choice of free-stream speed, cylinder diameter, or even type of fluid is not critical, provided that the desired Reynolds number is achieved.

condition. We run the simulation as an unsteady two-dimensional laminar flow, using the computational domain sketched in Fig. 15–39. The top and bottom (far field) edges are specified as a periodic boundary condition pair so that nonsymmetric oscillations in the wake are not suppressed (flow can cross these boundaries as necessary). The far field boundaries are also very far away (75 to 200 cylinder diameters), so that their effect on the calculations is insignificant.

The mesh is very fine near the cylinder to resolve the boundary layer. The grid is also fine in the wake region to resolve the shed vortices as they travel downstream. For this particular simulation, we use a hybrid grid somewhat like that shown in Fig. 15–14. The fluid is air, the cylinder diameter is 1.0 m, and the free-stream air speed is set to 0.00219 m/s. These values produce a Reynolds number of 150 based on cylinder diameter. Note that the Reynolds number is the important parameter in this problem—the choices of D, V, and type of fluid are not critical, so long as they produce the desired Reynolds number (Fig. 15–40).

As we march in time, small nonuniformities in the flow field amplify, and the flow becomes unsteady and antisymmetric with respect to the x-axis. A Kármán vortex street forms naturally. After sufficient CPU time, the simulated flow settles into a periodic vortex shedding pattern, much like the real flow. A contour plot of vorticity at one instant in time is shown in Fig. 15–41, along with a photograph showing streaklines of the same flow obtained experimentally in a wind tunnel. It is clear from the CFD simulation that the Kármán vortices decay downstream, since the magnitude of vorticity in the vortices decreases with downstream distance. This decay is partly physical (viscous), and partly artificial (numerical dissipation). Nevertheless, physical experiments verify the decay of the Kármán vortices. The decay is not so obvious in the streakline photograph (Fig. 15–41b); this is due to the time-integrating property of streaklines, as was pointed out in Chap. 4. A close-up view of vortices shedding from the cylinder at a particular instant in time is shown in Fig. 15–42, again with a comparison between CFD results

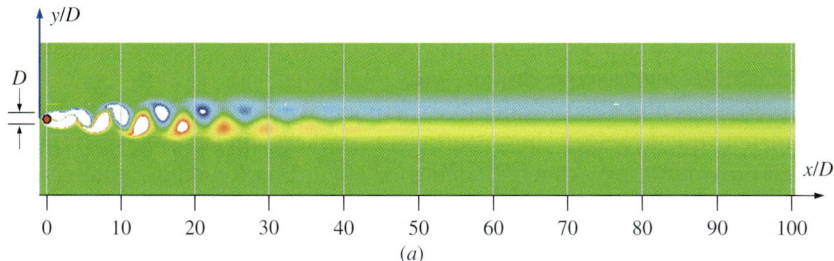
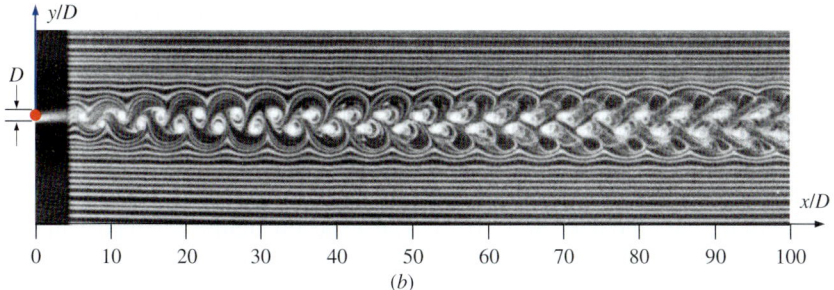

FIGURE 15–41
Laminar flow in the wake of a circular cylinder at Re ≅ 150: (*a*) an instantaneous snapshot of vorticity contours produced by CFD, and (*b*) time-integrated streaklines produced by a smoke wire located at $x/D = 5$. The vorticity contours show that Kármán vortices decay rapidly in the wake, whereas the streaklines retain a "memory" of their history from upstream, making it appear that the vortices continue for a great distance downstream.

Photo from Cimbala et al., 1988.

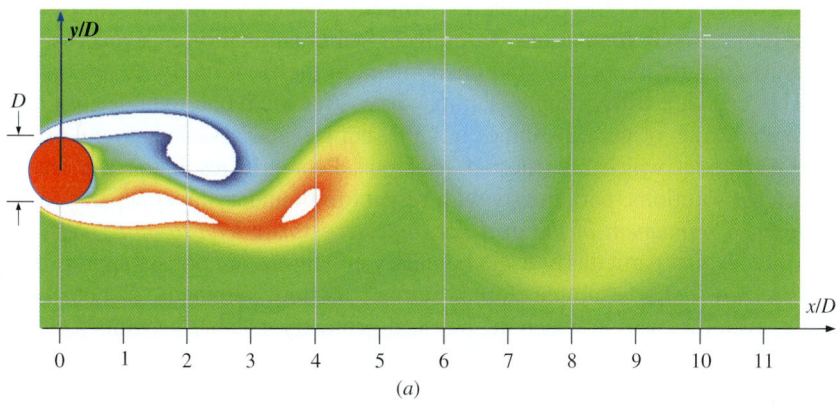
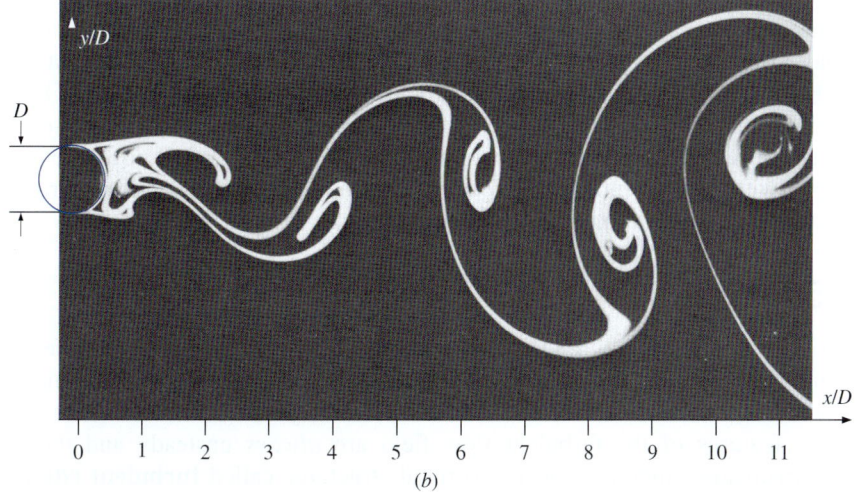

FIGURE 15–42
Close-up view of vortices shedding from a circular cylinder: (*a*) instantaneous vorticity contour plot produced by CFD at Re = 150, and (*b*) dye streaklines produced by dye introduced at the cylinder surface at Re = 140. An animated version of this CFD picture is available on the book's website.

Photo (b) reprinted by permission of Sadatoshi Taneda.

TABLE 15-2

Comparison of CFD results and experimental results for unsteady laminar flow over a circular cylinder at Re = 150*

	C_D	St
Experiment	1.1 to 1.4	0.18
CFD	1.14	0.16

* The main cause of the disagreement is most likely due to three-dimensional effects rather than grid resolution or numerical issues.

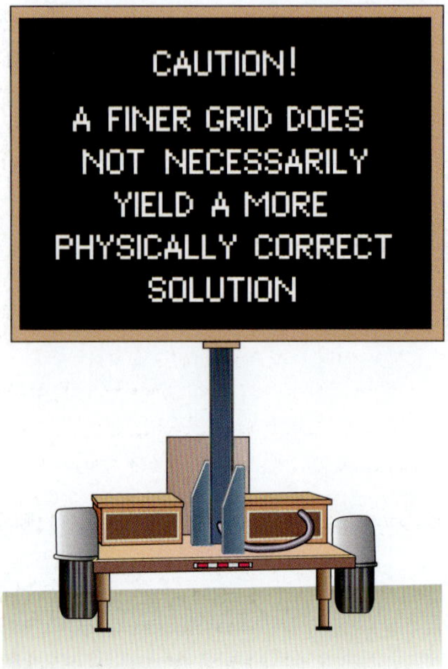

FIGURE 15-43

Poor grid resolution can lead to incorrect CFD results, but a finer grid does not guarantee a more physically correct solution. If the boundary conditions are not specified properly, the results may be unphysical, regardless of how fine the grid.

---

and experimental results—this time from experiments in a water channel. An animated version of Fig. 15–42 is provided on the book's website so that you can watch the dynamic process of vortex shedding.

We compare the CFD results to experimental results in Table 15–2. The calculated time-averaged drag coefficient on the cylinder is 1.14. As mentioned previously, experimental values of C_D at this Reynolds number vary from about 1.1 to 1.4, so the agreement is within the experimental scatter. Note that the present simulation is two-dimensional, inhibiting any kind of oblique vortex shedding or other three-dimensional nonuniformities. This may be why our calculated drag coefficient is on the lower end of the reported experimental range. The Strouhal number of the Kármán vortex street is defined as

Strouhal number: $$\text{St} = \frac{f_{\text{shedding}} D}{V} \qquad (15\text{-}4)$$

where f_{shedding} is the shedding frequency of the vortex street. From our CFD simulation, we calculate St = 0.16. The experimentally obtained value of Strouhal number at this Reynolds number is about 0.18 (Williamson, 1989), so again the agreement is reasonable, although the CFD results are a bit low compared to experiment. Perhaps a finer grid would help somewhat, but the major reason for the discrepancy is more likely due to unavoidable three-dimensional effects in the experiments, which are not present in these two-dimensional simulations. Overall this CFD simulation is a success, as it captures all the major physical phenomena in the flow field.

This exercise with "simple" laminar flow over a circular cylinder has demonstrated some of the capabilities of CFD, but has also revealed several aspects of CFD about which one must be cautious. Poor grid resolution can lead to incorrect solutions, particularly with respect to boundary layer separation, but *continued refinement of the grid does not necessarily lead to more physically correct results,* particularly if the boundary conditions are not set appropriately (Fig. 15–43). For example, forced numerical flow symmetry is not always wise, even for cases in which the physical geometry is entirely symmetric.

Symmetric geometry does not guarantee symmetric flow.

In addition, forced steady flow may yield incorrect results when the flow is inherently unstable and/or oscillatory. Likewise, forced two-dimensionality may yield incorrect results when the flow is inherently three-dimensional.

How then can we ensure that a laminar CFD calculation is correct? Only by systematic study of the effects of computational domain size, grid resolution, boundary conditions, flow regime (steady or unsteady, 2-D or 3-D, etc.), along with experimental validation. As with most other areas of engineering, *experience* is of paramount importance.

15–3 ▪ TURBULENT CFD CALCULATIONS

CFD simulations of turbulent flow are much more difficult than those of laminar flow, even for cases in which the flow field is steady in the mean (statisticians refer to this condition as **stationary**). The reason is that the finer features of the turbulent flow field are *always* unsteady and three-dimensional—random, swirling, vortical structures called **turbulent eddies** of all orientations arise in a turbulent flow (Fig. 15–44). Some CFD calculations use a technique called **direct numerical simulation** (**DNS**), in which an attempt is made to resolve the unsteady motion of *all* the scales of the turbulent flow. However, the size difference and the time scale difference between the largest and smallest eddies can be several orders of magnitude ($L \gg \eta$ in Fig. 15–44). Furthermore, these differences increase with the Reynolds number (Tennekes and Lumley, 1972), making DNS calculations of turbulent flows even more difficult as the Reynolds number increases. DNS solutions require extremely fine, fully three-dimensional grids, large computers, and an enormous amount of CPU time. With today's computers, DNS results are not yet feasible for practical high Reynolds number turbulent flows of engineering interest such as flow over a full-scale airplane. This situation is not expected to change for several more decades, even if the fantastic rate of computer improvement continues at today's pace.

Thus, we find it necessary to make some simplifying assumptions in order to simulate complex, high Reynolds number, turbulent flow fields. The next level below DNS is **large eddy simulation** (**LES**). With this technique, large unsteady features of the turbulent eddies are resolved, while small-scale dissipative turbulent eddies are *modeled* (Fig. 15–45). The basic assumption is that the smaller turbulent eddies are **isotropic**; i.e., it is assumed that the small eddies are independent of coordinate system orientation and always behave in a statistically similar and predictable way, regardless of the turbulent flow field. LES requires significantly less computer resources than does DNS, because we eliminate the need to resolve the smallest eddies in the flow field. In spite of this, the computer requirements for practical engineering analysis and design are nevertheless still formidable using today's technology. Further discussion about DNS and LES is beyond the scope of the present text, but these are areas of much current research.

The next lower level of sophistication is to model *all* the unsteady turbulent eddies with some kind of **turbulence model**. No attempt is made to resolve the unsteady features of *any* of the turbulent eddies, not even the largest ones (Fig. 15–46). Instead, mathematical models are employed to take into account the enhanced mixing and diffusion caused by turbulent eddies. For simplicity, we consider only steady (that is, *stationary*), incompressible flow. When using a turbulence model, the steady Navier–Stokes equation (Eq. 15–2) is replaced by what is called the **Reynolds-averaged Navier–Stokes** (**RANS**) equation, shown here for steady (stationary), incompressible, turbulent flow,

Steady RANS equation: $\quad (\vec{V} \cdot \vec{\nabla})\vec{V} = -\frac{1}{\rho}\vec{\nabla}P' + \nu\nabla^2\vec{V} + \vec{\nabla} \cdot (\tau_{ij,\,\text{turbulent}})$ (15–5)

Compared to Eq. 15–2, there is an additional term on the right side of Eq. 15–5 that accounts for the turbulent fluctuations. $\tau_{ij,\,\text{turbulent}}$ is a tensor known as the **specific Reynolds stress tensor**, so named because it acts in a similar fashion as the viscous stress tensor τ_{ij} (Chap. 9). In Cartesian coordinates, $\tau_{ij,\,\text{turbulent}}$ is

$$\tau_{ij,\,\text{turbulent}} = -\begin{pmatrix} \overline{u'^2} & \overline{u'v'} & \overline{u'w'} \\ \overline{u'v'} & \overline{v'^2} & \overline{v'w'} \\ \overline{u'w'} & \overline{v'w'} & \overline{w'^2} \end{pmatrix} \quad (15\text{–}6)$$

FIGURE 15–44
All turbulent flows, even those that are steady in the mean (*stationary*), contain unsteady, three-dimensional *turbulent eddies* of various sizes. Shown is the average velocity profile and some of the eddies; the smallest turbulent eddies (size η) are orders of magnitude smaller than the largest turbulent eddies (size L). *Direct numerical simulation* (DNS) is a CFD technique that simulates *all* relevant turbulent eddies in the flow.

FIGURE 15–45
Large eddy simulation (LES) is a simplification of direct numerical simulation in which only the *large* turbulent eddies are resolved—the small eddies are *modeled*, significantly reducing computer requirements. Shown is the average velocity profile and the resolved eddies.

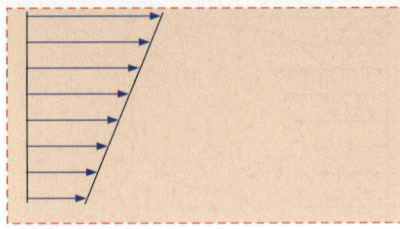

FIGURE 15–46

When a *turbulence model* is used in a CFD calculation, *all* the turbulent eddies are modeled, and only Reynolds-averaged flow properties are calculated. Shown is the average velocity profile. There are *no* resolved turbulent eddies.

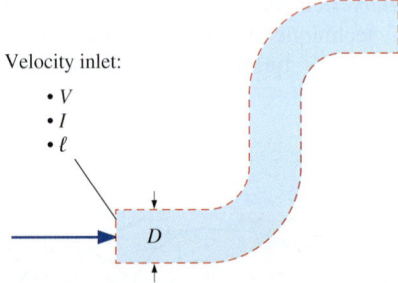

FIGURE 15–47

A useful rule of thumb for turbulence properties at a pressure inlet or velocity inlet boundary condition is to specify a turbulence intensity of 10 percent and a turbulent length scale of one-half of some characteristic length scale in the problem ($\ell = D/2$).

where the overbar indicates the time average of the product of two fluctuating velocity components and primes denote fluctuating velocity components. Since the Reynolds stress is symmetric, six (rather than nine) additional unknowns are introduced into the problem. These new unknowns are modeled in various ways by turbulence models. A detailed description of turbulence models is beyond the scope of this text; you are referred to Wilcox (2006) or Chen and Jaw (1998) for further discussion.

There are many turbulence models in use today, including algebraic, one-equation, two-equation, and Reynolds stress models. Three of the most popular turbulence models are the k-ε model, the k-ω model, and the q-ω model. These so-called **two-equation turbulence models** add two more transport equations, which must be solved simultaneously with the equations of mass and linear momentum (and also energy if this equation is being used). Along with the two additional transport equations that must be solved when using a two-equation turbulence model, two additional *boundary conditions* must be specified for the turbulence properties at inlets and at outlets. (Note that the properties specified at outlets are not used unless reverse flow is encountered at the outlet.) For example, in the k-ε model you may specify both k (**turbulent kinetic energy**) and ε (**turbulent dissipation rate**). However, appropriate values of these variables are not always known. A more useful option is to specify **turbulence intensity** I (ratio of characteristic turbulent eddy velocity to free-stream velocity or some other characteristic or average velocity) and **turbulent length scale** ℓ (characteristic length scale of the energy-containing turbulent eddies). If detailed turbulence data are not available, a good rule of thumb at inlets is to set I to 10 percent and to set ℓ to one-half of some characteristic length scale in the flow field (Fig. 15–47).

We emphasize that turbulence models are *approximations* that rely heavily on empirical constants for mathematical closure of the equations. The models are calibrated with the aid of direct numerical simulation and experimental data obtained from simple flow fields like flat plate boundary layers, shear layers, and isotropic decaying turbulence downstream of screens. Unfortunately, no turbulence model is **universal,** meaning that although the model works well for flows similar to those used for calibration, it is not guaranteed to yield a physically correct solution when applied to general turbulent flow fields, especially those involving flow separation and reattachment and/or large-scale unsteadiness.

> Turbulent flow CFD solutions are only as good as the appropriateness and validity of the turbulence model used in the calculations.

We emphasize also that this statement remains true regardless of how fine we make the computational grid. When applying CFD to laminar flows, we can usually improve the physical accuracy of the simulation by refining the grid (provided that the boundary conditions are properly specified, of course). This is *not* always the case for turbulent flow CFD analyses using turbulence models, even when the boundary conditions are correct. While a refined grid produces better *numerical accuracy*, the *physical accuracy* of the solution is always limited by the physical accuracy of the turbulence model itself.

With these cautions in mind, we now present some practical examples of CFD calculations of turbulent flow fields. In all the turbulent flow examples discussed in this chapter, we employ the k-ε turbulence model with wall functions. This model is the default turbulence model in many commercial CFD codes such as ANSYS-FLUENT. In all cases we assume stationary flow; no attempt is made to model unsteady features of the flow, such as vortex shedding in the wake of a bluff body. *It is assumed that the turbulence model accounts for all the inherent unsteadiness due to turbulent eddies in the flow field.* Note that *unsteady* (nonstationary) turbulent flows are also solvable with turbulence models, through the use of time-marching schemes (unsteady RANS calculations), but only when the time scale of the unsteadiness is much longer than that of individual turbulent eddies. For example, suppose you are calculating the forces and moments on a blimp during a gust of wind (Fig. 15–48). At the inlet boundary, you would impose the time-varying wind velocity and turbulence levels, and an unsteady turbulent flow solution could then be calculated using turbulence models. The large-scale, overall features of the flow (flow separation, forces and moments on the body, etc.) would be unsteady, but the fine-scale features of the turbulent boundary layer, for example, would be modeled by the quasi-steady turbulence model.

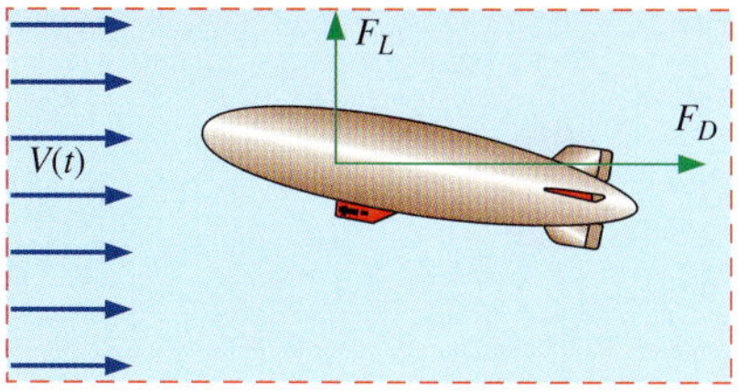

FIGURE 15–48
While most CFD calculations with turbulence models are *stationary* (steady in the mean), it is also possible to calculate *unsteady* turbulent flow fields using turbulence models. In the case of flow over a body, we may impose unsteady boundary conditions and march in time to predict gross features of the unsteady flow field.

Flow around a Circular Cylinder at Re = 10,000

As our first example of a turbulent flow CFD solution, we calculate flow over a circular cylinder at Re = 10,000. For illustration, we use the same two-dimensional computational domain that was used for the laminar cylinder flow calculations, as sketched in Fig. 15–35. As with the laminar flow calculation, only the upper half of the flow field is solved here, due to symmetry along the bottom edge of the computational domain. We use the same three grids used for the laminar flow case as well—coarse, medium, and fine resolution (Fig. 15–36). We point out, however, that grids designed for turbulent flow calculations (especially those employing turbulence models with wall functions) are generally not the same as those designed for laminar flow of the same geometry, particularly near walls.

We apply a free-stream flow of air at 25°C and at velocity $V = 7.304$ m/s from left to right around this circular cylinder. The Reynolds number of the flow, based on cylinder diameter ($D = 2.0$ cm), is approximately 10,000. Experiments at this Reynolds number reveal that the boundary layer is laminar and separates several degrees upstream of the top of the cylinder (at $\alpha \cong 82°$). The wake, however, is turbulent; such a mixture of laminar and turbulent flow is particularly difficult for CFD codes. The measured drag coefficient at this Reynolds number is $C_D \cong 1.15$ (Tritton, 1977). CFD solutions are obtained for each of the three grids, assuming stationary (steady in the mean) turbulent flow. We employ the k-ε turbulence model with wall functions. The inlet turbulence level is set to 10 percent with a length scale of 0.01 m (half of the cylinder diameter). All three cases converge nicely. Streamlines are plotted in Fig. 15–49 for the three grid resolution cases. In each plot, the image is mirrored about the symmetry line so that even though only the top half of the flow field is solved, the full flow field is visualized.

For the coarse resolution case (Fig. 15–49a), the boundary layer separates well past the top of the cylinder, at $\alpha \approx 140°$. Furthermore, the drag coefficient C_D is only 0.647, almost a factor of 2 smaller than it should be. Let's

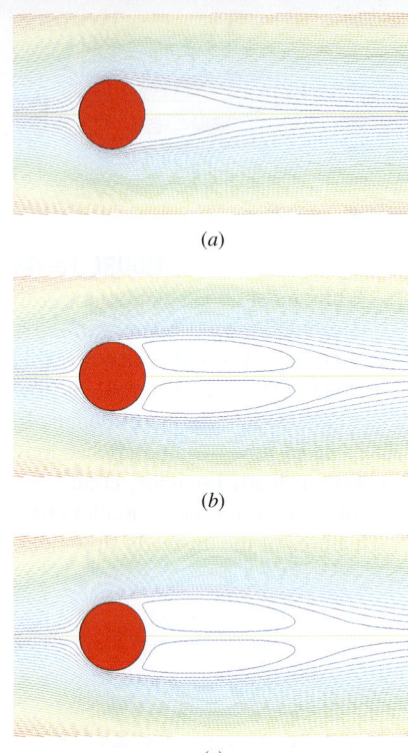

FIGURE 15–49
Streamlines produced by CFD calculations of stationary turbulent flow over a circular cylinder at Re = 10,000: (a) coarse grid (30 × 60), (b) medium grid (60 × 120), and (c) fine grid (120 × 240). Note that only the top half of the flow is calculated—the bottom half is displayed as a mirror image of the top.

see if a finer mesh improves the agreement with experimental data. For the medium resolution case (Fig. 15–49b), the flow field is significantly different. The boundary layer separates nearer to the top of the cylinder, at $\alpha = 104°$, and C_D has increased to about 0.742—closer, but still significantly less than the experimental value. We also notice that the recirculating eddies in the cylinder's wake have grown in length by nearly a factor of 2 compared to those of the coarse grid case. Figure 15–49c shows streamlines for the fine resolution case. The results look very similar to those of the medium resolution case, and the drag coefficient has increased only slightly ($C_D = 0.753$). The boundary layer separation point for this case is at $\alpha = 102°$.

Further grid refinement (not shown) does not change the results significantly from those of the fine grid case. In other words, the fine grid appears to be sufficiently resolved, yet the results do not agree with experiment. Why? There are several problems with our calculations: we are modeling a steady flow, even though the actual physical flow is unsteady; we are enforcing symmetry about the *x*-axis, even though the physical flow is unsymmetric (a Kármán vortex street can be observed in experiments at this Reynolds number); and we are using a turbulence model instead of resolving all the small eddies of the turbulent flow. Another significant source of error in our calculations is that the CFD code is run with turbulence turned on in order to reasonably model the wake region, which is turbulent; however, the boundary layer on the cylinder surface is actually still *laminar*. The predicted location of the separation point downstream of the top of the cylinder is more in line with *turbulent* boundary layer separation, which does not occur until much higher values of Reynolds number (after the "drag crisis" at Re greater than 2×10^5).

The bottom line is that CFD codes have a hard time in the transitional regime between laminar and turbulent flow, and when there is a mixture of laminar and turbulent flow in the same computational domain. In fact, most commercial CFD codes give the user a choice between laminar and turbulent—there is no "middle ground." In the present calculations, we model the boundary layer as turbulent, even though the physical boundary layer is laminar; it is not surprising, then, that the results of our calculations do not agree well with experiment. If we would have instead specified laminar flow over the entire computational domain, the CFD results would have been even worse (less physical).

Is there any way around this problem of poor physical accuracy for the case of mixed laminar and turbulent flow? Perhaps. In some CFD codes you can specify the flow to be laminar or turbulent in different regions of the flow. But even then, the transitional process from laminar to turbulent flow is somewhat abrupt, again not physically correct. Furthermore, you would need to know where the transition takes place in advance—this defeats the purpose of a stand-alone CFD calculation for fluid flow prediction. Advanced wall treatment models are being generated that may some day do a better job in the transitional region. In addition, some new turbulence models are being developed that are better tuned to low Reynolds number turbulence. Transition is an area of active research in CFD.

In summary, we cannot accurately model the mixed laminar/turbulent flow problem of flow over a circular cylinder at Re ~ 10,000 using standard turbulence models and the steady Reynolds-averaged Navier–Stokes (RANS) equation. It appears that accurate results can be obtained only through use of time-accurate (unsteady RANS), LES, or DNS solutions that are orders of magnitude more computationally demanding.

Flow around a Circular Cylinder at Re = 10^7

As a final cylinder example, we use CFD to calculate flow over a circular cylinder at Re = 10^7—well beyond the drag crisis. The cylinder for this case is of 1.0 m diameter, and the fluid is water. The free-stream velocity is 10.05 m/s. At this value of Reynolds number the experimentally measured value of drag coefficient is around 0.7 (Tritton, 1977). The boundary layer is turbulent at the separation point, which occurs at around 120°. Thus we do not have the mixed laminar/turbulent boundary layer problem as in the lower Reynolds number example—the boundary layer is turbulent everywhere except near the nose of the cylinder, and we should expect better results from the CFD predictions. We use a two-dimensional half-grid similar to that of the fine resolution case of the previous examples, but the mesh near the cylinder wall is adapted appropriately for this high Reynolds number. As previously, we use the k-ε turbulence model with wall functions. The inlet turbulence level is set to 10 percent with a length scale of 0.5 m. Unfortunately, the drag coefficient is calculated to be 0.262—less than half of the experimental value at this Reynolds number. Streamlines are shown in Fig. 15–50. The boundary layer separates a bit too far downstream, at $\alpha = 129°$. There are several possible reasons for the discrepancy. We are forcing the simulated flow to be steady and symmetric, whereas the actual flow is neither, due to vortex shedding. (Vortices are shed even at high Reynolds numbers.) In addition, the turbulence model and its near wall treatment (wall functions) may not be capturing the proper physics of the flow field. Again we must conclude that accurate results for flow over a circular cylinder can be obtained only through use of a full grid rather than a half grid, and with time-accurate (unsteady RANS), LES, or DNS solutions that are orders of magnitude more computationally demanding.

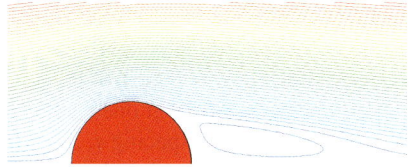

FIGURE 15–50
Streamlines produced by CFD calculations of stationary turbulent flow over a circular cylinder at Re = 10^7. Unfortunately, the predicted drag coefficient is still not accurate for this case.

Design of the Stator for a Vane-Axial Flow Fan

The next turbulent flow CFD example involves design of the stator for a vane-axial flow fan that is to be used to drive a wind tunnel. The overall fan diameter is $D = 1.0$ m, and the design point of the fan is at an axial-flow speed of $V = 50$ m/s. The stator vanes span from radius $r = r_{hub} = 0.25$ m at the hub to $r = r_{tip} = 0.50$ m at the tip. The stator vanes are upstream of the rotor blades in this design (Fig. 15–51). A preliminary stator vane shape is chosen that has a trailing edge angle of $\beta_{st} = 63°$ and a chord length of 20 cm. At any value of radius r, the actual amount of turning depends on the number of stator vanes—we expect that the fewer the number of vanes, the smaller the *average* angle at which the flow is turned by the stator vanes because of the greater spacing between vanes. It is our goal to determine the minimum number of stator vanes required so that the flow impinging on the leading edges of the rotor blades (located one chord length downstream of the stator vane trailing edges) is turned at an average angle of at least 45°. We also require there to be no significant flow separation from the stator vane surface.

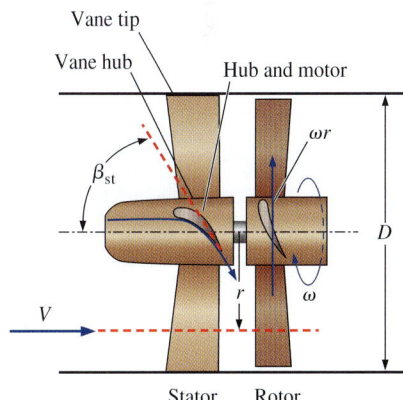

FIGURE 15–51
Schematic diagram of the vane-axial flow fan being designed. The stator precedes the rotor, and the flow through the stator vanes is to be modeled with CFD.

FIGURE 15–52

Definition of *blade spacing s*: (*a*) frontal view of the stator, and (*b*) the stator modeled as a two-dimensional cascade in edge view. Twelve radial stator vanes are shown in the frontal view, but the actual number of vanes is to be determined. Three stator vanes are shown in the cascade, but the actual cascade consists of an infinite number of vanes, each displaced by blade spacing *s*, which increases with radius *r*. The two-dimensional cascade is an approximation of the three-dimensional flow at one value of radius *r* and blade spacing *s*. Chord length *c* is defined as the horizontal length of the stator vane.

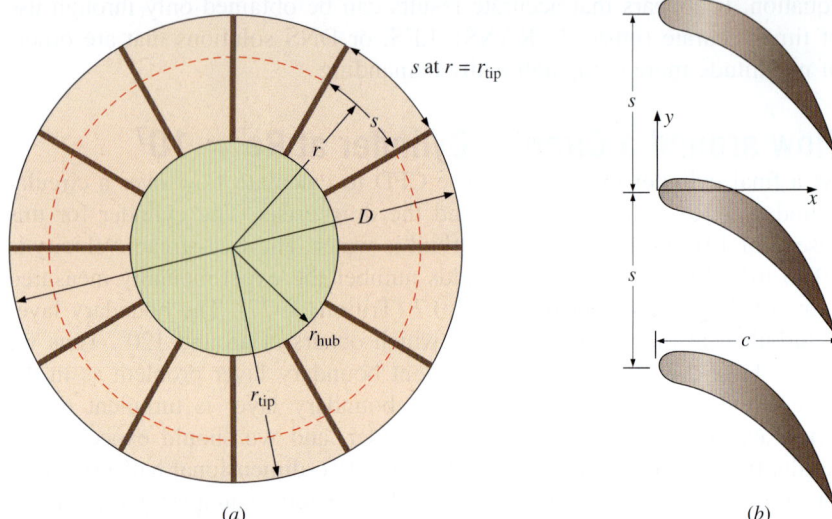

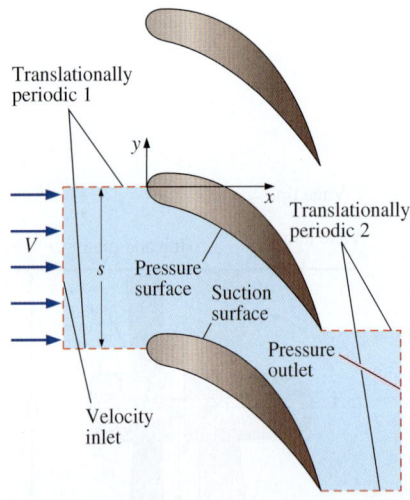

FIGURE 15–53

Computational domain (light blue shaded region) defined by one flow passage through two stator vanes. The top wall of the passage is the pressure surface, and the bottom wall is the suction surface. Two translationally periodic pairs are defined: periodic 1 upstream and periodic 2 downstream.

As a first approximation, we model the stator vanes at any desired value of *r* as a two-dimensional *cascade* of vanes (see Chap. 14). Each vane is separated by **blade spacing** *s* at this radius, as defined in Fig. 15–52. We use CFD to predict the maximum allowable value of *s*, from which we estimate the minimum number of stator vanes that meet the given requirements of the design.

Since the flow through the two-dimensional cascade of stator vanes is infinitely periodic in the *y*-direction, we need to model only *one* flow passage through the vanes, specifying two pairs of periodic boundary conditions on the top and bottom edges of the computational domain (Fig. 15–53). We run six cases, each with a different value of blade spacing. We choose $s = 10$, 20, 30, 40, 50, and 60 cm, and generate a structured grid for each of these values of blade spacing. The grid for the case with $s = 20$ cm is shown in Fig. 15–54; the other grids are similar, but more intervals are specified in the *y*-direction as *s* increases. Notice how we have made the grid spacing fine near the pressure and suction surfaces so that the boundary layer on these surfaces can be better resolved. We specify $V = 50$ m/s at the velocity inlet, zero gage pressure at the pressure outlet, and a smooth wall boundary condition with no slip at both the pressure and suction surfaces. Since we are modeling the flow with a turbulence model (k-ε with wall functions), we must also specify turbulence properties at the velocity inlet. For these simulations we specify a turbulence intensity of 10 percent and a turbulence length scale of 0.01 m (1.0 cm).

We run the CFD calculations long enough to converge as far as possible for all six cases, and we plot streamlines in Fig. 15–55 for six blade spacings: $s = 10$, 20, 30, 40, 50, and 60 cm. Although we solve for flow through only one flow passage, we plot several *duplicate* flow passages, stacked one on top of the other, in order to visualize the flow field as a periodic cascade. The streamlines for the first three cases look very similar at first glance, but closer inspection reveals that the average angle of flow downstream of the trailing edge of the stator vane *decreases* with *s*. (We define

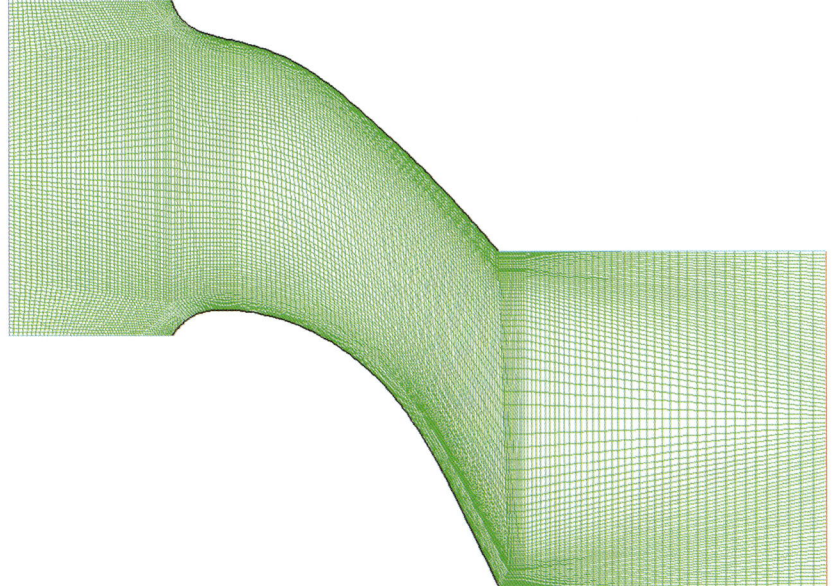

FIGURE 15–54
Structured grid for the two-dimensional stator vane cascade at blade spacing $s = 20$ cm. The outflow region in the wake of the vanes is intentionally longer than that at the inlet to avoid backflow at the pressure outlet in case of flow separation on the suction surface of the stator vane. The outlet is one chord length downstream of the stator vane trailing edges; the outlet is also the location of the leading edges of the rotor blades (not shown).

flow angle β relative to horizontal as sketched in Fig. 15–55a.) Also, the gap (white space) between the wall and the closest streamline to the suction surface increases in size as s increases, indicating that the flow speed in that region decreases. In fact, it turns out that the boundary layer on the suction surface of the stator vane must resist an ever-increasingly adverse pressure gradient (decelerating flow speed and positive pressure gradient) as blade spacing is increased. At large enough s, the boundary layer on the suction surface cannot withstand the severely adverse pressure gradient and separates off the wall. For $s = 40$, 50, and 60 cm (Fig. 15–55d through f), flow separation off the suction surface is clearly seen in these streamline plots. Furthermore, the severity of the flow separation increases with s. This is not unexpected if we imagine the limit as $s \rightarrow \infty$. In that case, the stator vane is isolated from its neighbors, and we surely expect massive flow separation since the vane has such a high degree of camber.

We list average outlet flow angle β_{avg}, average outlet flow speed V_{avg}, and predicted drag force on a stator vane per unit depth F_D/b in Table 15–3 as functions of blade spacing s. (Depth b is into the page of Fig. 15–55 and is assumed to be 1 m in two-dimensional calculations such as these.) While β_{avg} and V_{avg} decrease continuously with s, F_D/b first rises to a maximum for the $s = 20$ cm case, and then decreases from there on.

You may recall from the previously stated design criteria for this example that the average outlet flow angle must be greater than 45°, and there must be no significant flow separation. From our CFD results, it appears that both of these criteria break down somewhere between $s = 30$ and 40 cm. We obtain a better picture of flow separation by plotting vorticity contours (Fig. 15–56). In these color contour plots, blue represents large negative vorticity (clockwise rotation), red represents large positive vorticity (counterclockwise rotation), and green is zero vorticity. When the boundary layer remains attached, we expect the vorticity to be concentrated within thin

TABLE 15–3

Variation of average outlet flow angle β_{avg}, average outlet flow speed V_{avg}, and predicted drag force per unit depth F_D/b as functions of blade spacing s*

s, cm	β_{avg}, degrees	V_{avg}, m/s	F_D/b, N/m
10	60.8	103	554
20	56.1	89.6	722
30	49.7	77.4	694
40	43.2	68.6	612
50	37.2	62.7	538
60	32.3	59.1	489

* All calculated values are reported to three significant digits. The CFD calculations are performed using the k-ϵ turbulence model with wall functions.

910
COMPUTATIONAL FLUID DYNAMICS

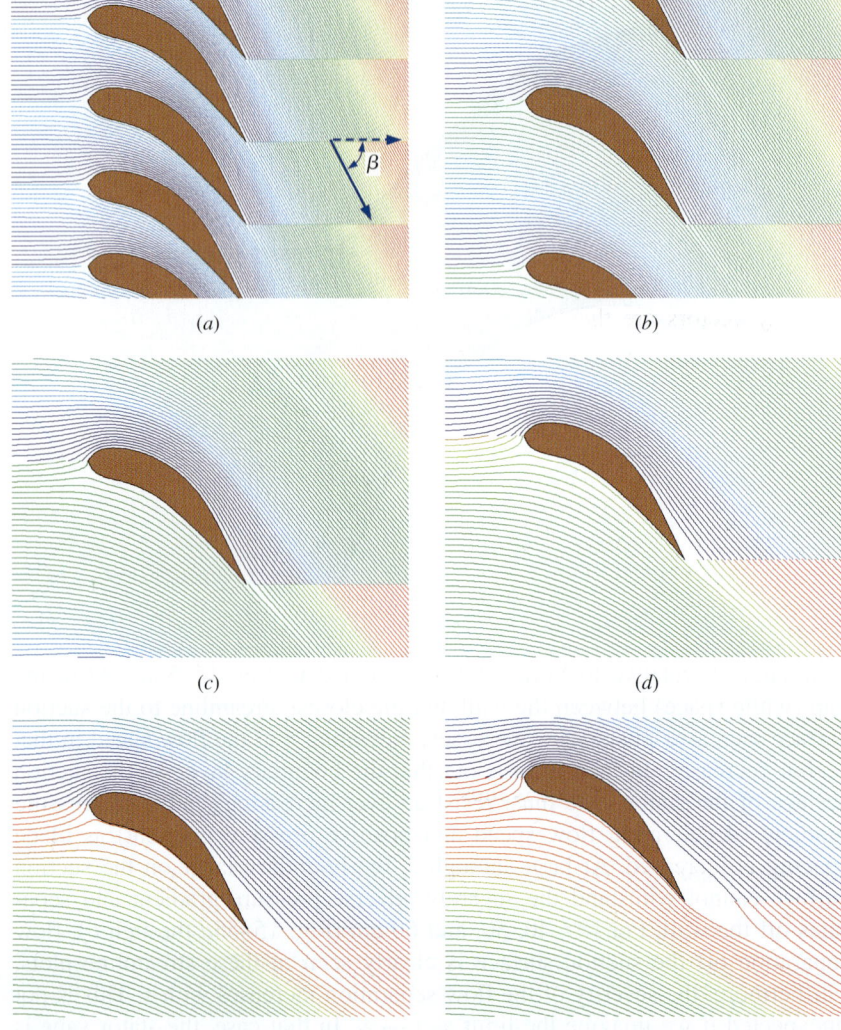

FIGURE 15–55
Streamlines produced by CFD calculations of stationary turbulent flow through a stator vane flow passage: (*a*) blade spacing $s = 10$, (*b*) 20, (*c*) 30, (*d*) 40, (*e*) 50, and (*f*) 60 cm. The CFD calculations are performed using the k-ε turbulence model with wall functions. Flow angle β is defined in image (*a*) as the average angle of flow, relative to horizontal, just downstream of the trailing edge of the stator vane.

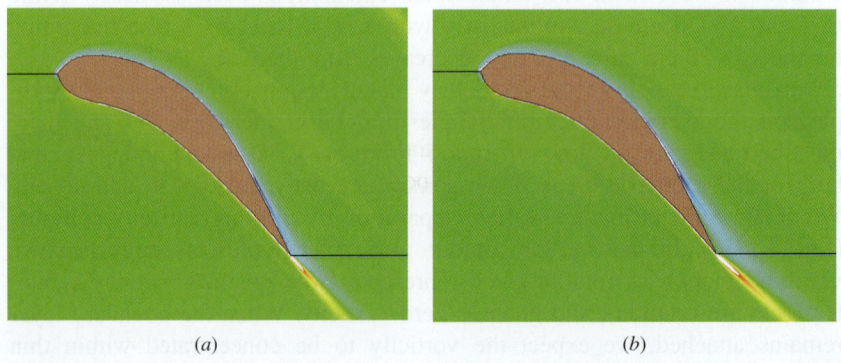

FIGURE 15–56
Vorticity contour plots produced by CFD calculations of stationary turbulent flow through a stator vane flow passage: blade spacing (*a*) $s = 30$ cm and (*b*) $s = 40$ cm. The flow field is largely irrotational (zero vorticity) except in the thin boundary layer along the walls and in the wake region. However, when the boundary layer separates, as in case (*b*), the vorticity spreads throughout the separated flow region.

boundary layers along the stator vane surfaces, as is the case in Fig. 15–56a for $s = 30$ cm. When the boundary layer separates, however, the vorticity suddenly spreads out away from the suction surface, as seen in Fig. 15–56b for $s = 40$ cm. These results verify that significant flow separation occurs somewhere between $s = 30$ and 40 cm. As a side note, notice how the vorticity is concentrated not only in the boundary layer, but also in the *wake* for both cases shown in Fig. 15–56.

Finally, we compare velocity vector plots in Fig. 15–57 for three cases: $s = 20$, 40, and 60 cm. We generate several equally spaced parallel lines in the computational domain; each line is tilted at 45° from the horizontal. Velocity vectors are then plotted along each of these parallel lines. When $s = 20$ cm (Fig. 15–57a), the boundary layer remains attached on both the suction and pressure surfaces of the stator vane all the way to its trailing edge. When $s = 40$ cm (Fig. 15–57b), flow separation and reverse flow along the suction surface appears. When $s = 60$ cm (Fig. 15–57c), the separation bubble and the reverse flow region have grown – this is a "dead" flow region, in which the air speeds are very small. In all cases, the flow on the pressure surface (lower left side) of the stator vane remains attached.

How many vanes (N) does a blade spacing of $s = 30$ cm represent? We can easily calculate N by noting that at the vane tip ($r = r_{\text{tip}} = D/2 = 50$ cm), where s is largest, the total available circumference (C) is

Available circumference: $C = 2\pi r_{\text{tip}} = \pi D$ (15–7)

The number of vanes that can be placed within this circumference with a blade spacing of $s = 30$ cm is thus

Maximum number of vanes: $N = \dfrac{C}{s} = \dfrac{\pi D}{s} = \dfrac{\pi(100 \text{ cm})}{30 \text{ cm}} = 10.5$ (15–8)

Obviously we can have only an integer value of N, so we conclude from our preliminary analysis that we should have at least 10 or 11 stator vanes.

How good is our approximation of the stator as a two-dimensional cascade of vanes? To answer this question, we perform a full three-dimensional CFD analysis of the stator. Again, we take advantage of the periodicity by modeling only one flow passage—a three-dimensional passage between two radial stator vanes (Fig. 15–58). We choose $N = 10$ stator vanes by specifying an angle of periodicity of 360/10 = 36°. From Eq. 15–8, this represents $s = 31.4$ at the vane tips and $s = 15.7$ at the hub, for an average value of $s_{\text{avg}} = 23.6$. We generate a hexagonal structured grid in a computational domain bounded by a velocity inlet, an outflow outlet, a section of cylindrical wall at the hub and another at the tip, the pressure surface of the vane, the suction surface of the vane, and two pairs of periodic boundary conditions. In this three-dimensional case, the periodic boundaries are *rotationally* periodic instead of translationally periodic. Note that we use an outflow boundary condition rather than a pressure outlet boundary condition, because we expect the swirling motion to produce a radial pressure distribution on the outlet face. The grid is finer near the walls than elsewhere (as usual), to better resolve the boundary layer. The incoming velocity, turbulence level, turbulence model, etc., are all the same as those used for the

(a)

(b)

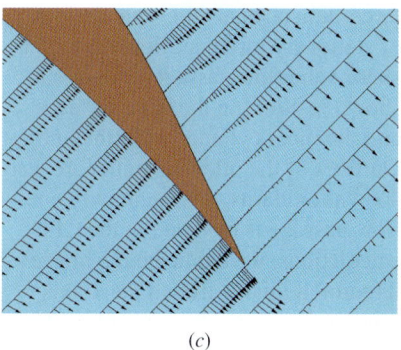
(c)

FIGURE 15–57
Velocity vectors produced by CFD calculations of stationary turbulent flow through a stator vane flow passage: blade spacing $s =$ (a) 20 cm, (b) 40 cm, and (c) 60 cm.

912
COMPUTATIONAL FLUID DYNAMICS

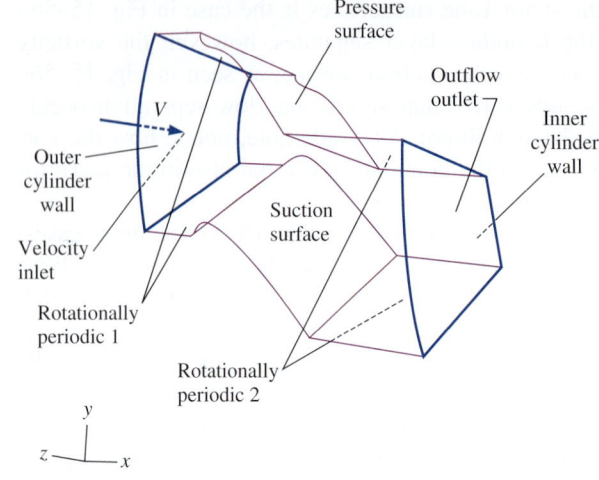

FIGURE 15–58
Three-dimensional computational domain defined by one flow passage through two stator vanes for $N = 10$ (angle between vanes = 36°). The computational domain volume is defined between the pressure and suction surfaces of the stator vanes, between the inner and outer cylinder walls, and from the inlet to the outlet. Two pairs of rotationally periodic boundary conditions are defined as shown.

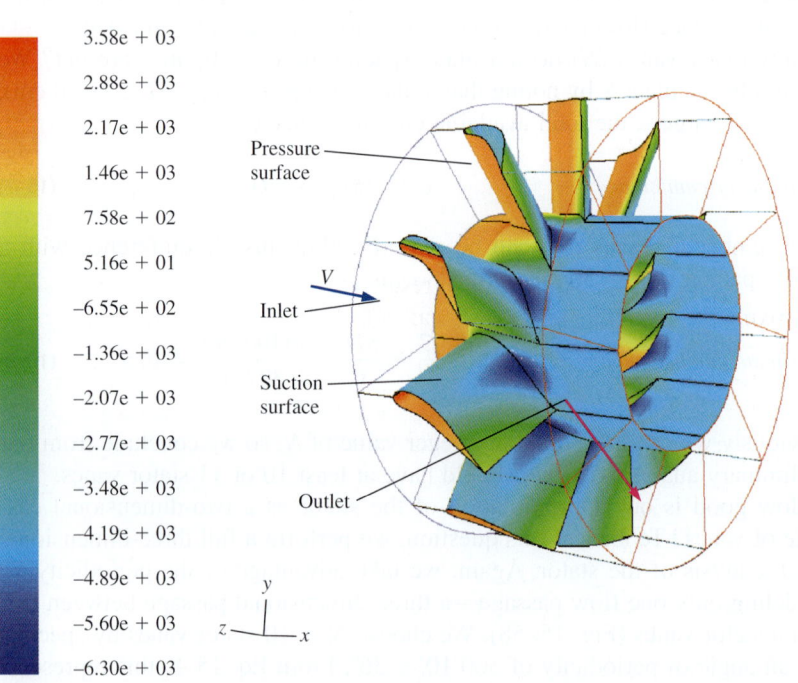

FIGURE 15–59
Pressure contour plot produced by three-dimensional CFD calculations of stationary turbulent flow through a stator vane flow passage. Pressure is shown in N/m^2 on the vane surfaces and the inner cylinder wall (the hub). Outlines of the inlet and outlet are also shown for clarity. Although only one flow passage is modeled in the CFD calculations, we duplicate the image circumferentially around the *x*-axis nine times to visualize the entire stator flow field. In this image, high pressures (as on the pressure surfaces of the vanes) are red, while low pressures (as on the suction surfaces of the vanes, especially near the hub) are blue.

two-dimensional approximation. The total number of computational cells is almost 800,000.

Pressure contours on the stator vane surfaces and on the inner cylindrical wall are plotted in Fig. 15–59. This view is from the same angle as that of Fig. 15–60, but we have zoomed out and duplicated the computational domain nine times circumferentially about the axis of rotation (the *x*-axis) for a total of 10 flow passages to aid in visualization of the flow field. You can see that the pressure is higher (red) on the pressure surface than on the suction surface (blue). You can also see an overall drop in pressure along the hub surface from upstream to downstream of the stator. The change in average pressure from the inlet to the outlet is calculated to be 3.29 kPa.

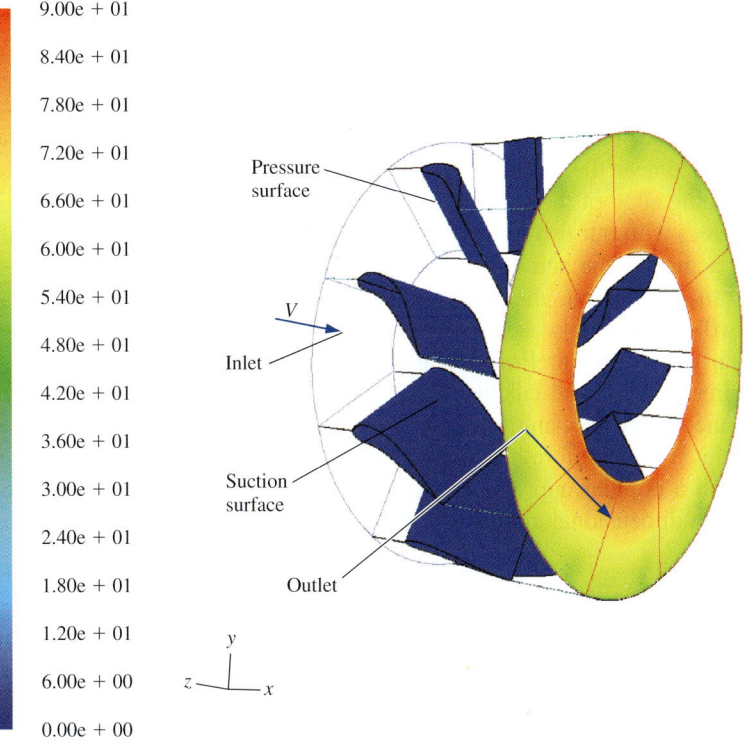

FIGURE 15–60
Tangential velocity contour plot produced by three-dimensional CFD calculations of stationary turbulent flow through a stator vane flow passage. The tangential velocity component is shown in m/s at the outlet of the computational domain (and also on the vane surfaces, where the velocity is zero). An outline of the inlet to the computational domain is also shown for clarity. Although only one flow passage is modeled, we duplicate the image circumferentially around the x-axis nine times to visualize the entire stator flow field. In this image, the tangential velocity ranges from 0 (blue) to 90 m/s (red).

To compare our three-dimensional results directly with the two-dimensional approximation, we run one additional two-dimensional case at the average blade spacing, $s = s_{avg} = 23.6$ cm. A comparison between the two- and three-dimensional cases is shown in Table 15–4. From the three-dimensional calculation, the net axial force on one stator vane is $F_D = 183$ N. We compare this to the two-dimensional value by converting to force per unit depth (force per unit span of the stator vane). Since the stator vane spans 0.25 m, $F_D/b = $ (183 N)/(0.25 m) = 732 N/m. The corresponding two-dimensional value from Table 15–4 is $F_D/b = 724$ N/m, so the agreement is very good ($\cong 1$ percent difference). The average speed at the outlet of the three-dimensional domain is $V_{avg} = 84.7$ m/s, almost identical to the two-dimensional value of 84.8 m/s in Table 15–4. The two-dimensional approximation differs by less than 1 percent. Finally, the average outlet flow angle β_{avg} obtained from our full three-dimensional calculation is 53.3°, which easily meets the design criterion of 45°. We compare this to the two-dimensional approximation of 53.9° in Table 15–4; the agreement is again around 1 percent.

Contours of tangential velocity component at the outlet of the computational domain are plotted in Fig. 15–60. We see that the tangential velocity distribution is not uniform; it decreases as we move radially outward from hub to tip as we should expect, since blade spacing s increases from hub to tip. We also find (not shown here) that the outflow pressure increases radially from hub to tip. This also agrees with our intuition, since we know that a radial pressure gradient is required to sustain a tangential flow—the pressure rise with increasing radius provides the centripetal acceleration necessary to turn the flow about the x-axis.

Another comparison can be made between the three-dimensional and two-dimensional calculations by plotting vorticity contours in a slice through the

TABLE 15–4

CFD results for flow through a stator vane flow passage: the two-dimensional cascade approximation at the average blade spacing, ($s = s_{avg} = 23.6$ cm) is compared to the full three-dimensional calculation*

	2-D, $s = 23.6$ cm	Full 3-D
β_{avg}	53.9°	53.3°
V_{avg}, m/s	84.8	84.7
F_D/b, N/m	724	732

* Values are shown to three significant digits.

computational domain within the flow passage between vanes. Two such slices are created—a slice close to the hub and a slice close to the tip, and vorticity contours are plotted in Fig. 15–61. In both slices, the vorticity is confined to the thin boundary layer and wake. There is no flow separation near the hub, but we see that near the tip, the flow has just begun to separate on the suction surface near the trailing edge of the stator vane. Notice that the air leaves the trailing edge of the vane at a steeper angle at the hub than at the tip. This also agrees with our two-dimensional approximation (and our intuition), since blade spacing s at the hub (15.7 cm) is smaller than s at the tip (31.4 cm).

In conclusion, the approximation of this three-dimensional stator as a two-dimensional cascade of stator vanes turns out to be quite good overall, particularly for preliminary analysis. The discrepancy between the two- and three-dimensional calculations for gross flow features, such as force on the vane, outlet flow angle, etc., is around 1 percent or less for all reported quantities. It is therefore no wonder that the two-dimensional cascade approach is such a popular approximation in turbomachinery design. The more detailed three-dimensional analysis gives us confidence that a stator with 10 vanes is

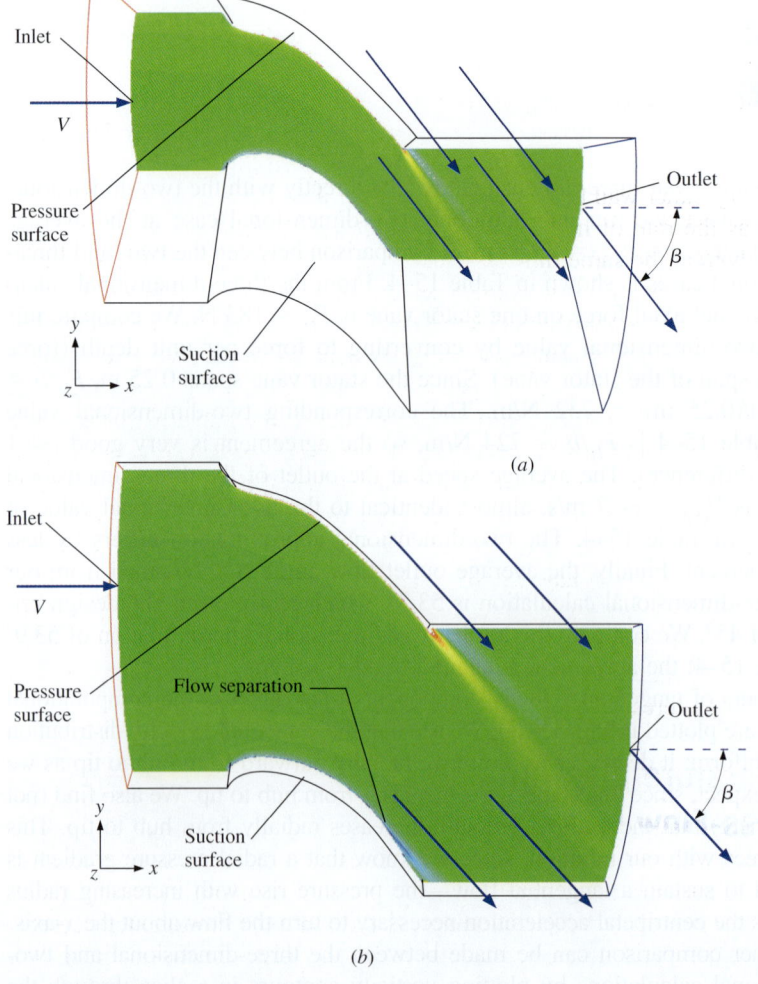

FIGURE 15–61
Vorticity contour plots produced by three-dimensional stationary turbulent CFD calculations of flow through a stator vane flow passage: (*a*) a slice near the hub or root of the vanes and (*b*) a slice near the tip of the vanes. Contours of *z*-vorticity are plotted, since the faces are nearly perpendicular to the *z*-axis. In these images, blue regions (as in the upper half of the wake and in the flow separation zone) represent negative (clockwise) *z*-vorticity, while red regions (as in the lower half of the wake) represent positive (counterclockwise) *z*-vorticity. Near the hub, there is no sign of flow separation, but near the tip, there is some indication of flow separation near the trailing edge of the suction side of the vane. Also shown are arrows indicating how the periodic boundary condition works. Flow leaving the *bottom* of the periodic boundary enters at the same speed and direction into the *top* of the periodic boundary. Outflow angle β is larger near the hub than near the tip of the stator vanes, because blade spacing s is smaller at the hub than at the tip, and also because of the mild flow separation near the tip.

sufficient to meet the imposed design criteria for this axial-flow fan. However, our three-dimensional calculations have revealed a small separated region near the tip of the stator vane. It may be wise to apply some **twist** to the stator vanes (reduce the pitch angle or angle of attack toward the tip) in order to avoid this separation. (Twist is discussed in more detail in Chap. 14.) Alternatively, we can increase the number of stator vanes to 11 or 12 to hopefully eliminate flow separation at the vane tips.

As a final comment on this example flow field, all the calculations were performed in a fixed coordinate system. Modern CFD codes contain options for modeling zones in the flow field with *rotating coordinate systems* so that similar analyses can be performed on *rotor* blades as well as on stator vanes.

15–4 ■ CFD WITH HEAT TRANSFER

By coupling the differential form of the energy equation with the equations of fluid motion, we can use a computational fluid dynamics code to calculate properties associated with **heat transfer** (e.g., temperature distributions or rate of heat transfer from a solid surface to a fluid). Since the energy equation is a scalar equation, only *one* extra transport equation (typically for either temperature or enthalpy) is required, and the computational expense (CPU time and RAM requirements) is not increased significantly. Heat transfer capability is built into most commercially available CFD codes, since many practical problems in engineering involve both fluid flow *and* heat transfer. As mentioned previously, additional boundary conditions related to heat transfer need to be specified. At solid wall boundaries, we may specify either wall temperature T_{wall} (K) or the *wall heat flux* $\dot{q}_{wall}$ (W/m²), defined as the rate of heat transfer per unit area from the wall to the fluid (but not *both* at the same time, as illustrated in Fig. 15–62). When we model a zone in a computational domain as a solid body that involves the generation of thermal energy via electric heating (as in electronic components) or chemical or nuclear reactions (as in nuclear fuel rods), we may instead specify the heat generation rate per unit volume $\dot{g}$ (W/m³) within the solid since the ratio of the total heat generation rate to the exposed surface area must equal the average wall heat flux. In that case, neither T_{wall} nor $\dot{q}_{wall}$ is specified; both converge to values that match the specified heat generation rate. In addition, the temperature distribution inside the solid object itself can then be calculated. Other boundary conditions (such as those associated with radiation heat transfer) may also be applied in CFD codes.

In this section we do not go into details about the equations of motion or the numerical techniques used to solve them. Rather, we show some basic examples that illustrate the capability of CFD to calculate practical flows of engineering interest that involve heat transfer.

Temperature Rise through a Cross-Flow Heat Exchanger

Consider flow of cool air through a series of hot tubes as sketched in Fig. 15–63. In heat exchanger terminology, this geometrical configuration is called a **cross-flow heat exchanger.** If the airflow were to enter horizontally ($\alpha = 0$) at all times, we could cut the computational domain in half

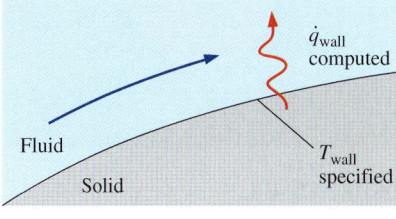
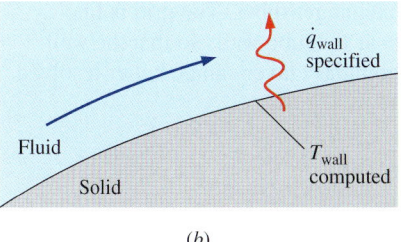

FIGURE 15–62
At a wall boundary, we may specify either (*a*) the wall temperature or (*b*) the wall heat flux, but not both, as this would be mathematically overspecified.

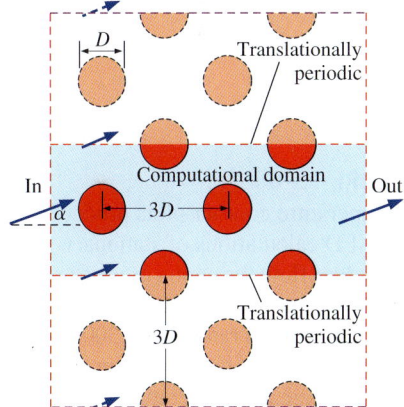

FIGURE 15–63
The computational domain (light blue shaded region) used to model turbulent flow through a cross-flow heat exchanger. Flow enters from the left at angle α from the horizontal.

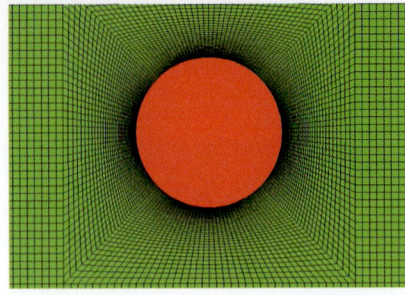

FIGURE 15–64
Close-up view of the structured grid near one of the cross-flow heat exchanger tubes. The grid is fine near the tube walls so that the wall boundary layer can be better resolved.

and apply symmetry boundary conditions on the top and bottom edges of the domain (see Fig. 15–25). In the case under consideration, however, we allow the airflow to enter the computational domain at some angle ($\alpha \neq 0$). Thus, we impose *translationally periodic* boundary conditions on the top and bottom edges of the domain as sketched in Fig. 15–63. We set the inlet air temperature to 300 K and the surface temperature of each tube to 500 K. The diameter of the tubes and the speed of the air are chosen such that the Reynolds number is approximately 1×10^5 based on tube diameter. The tube surfaces are assumed to be hydrodynamically smooth (zero roughness) in this first set of calculations. The hot tubes are staggered as sketched in Fig. 15–63 and are spaced three diameters apart both horizontally and vertically. We assume two-dimensional stationary turbulent flow without gravity effects and set the turbulence intensity of the inlet air to 10 percent. We run two cases for comparison: $\alpha = 0$ and $10°$. Our goal is to see whether the heat transfer to the air is enhanced or inhibited by a nonzero value of α. Which case do you think will provide greater heat transfer?

We generate a two-dimensional, multiblock, structured grid with very fine resolution near the tube walls as shown in Fig. 15–64, and we run the CFD code to convergence for both cases. Temperature contours are shown for the $\alpha = 0°$ case in Fig. 15–65, and for the $\alpha = 10°$ case in Fig. 15–66. The average rise of air temperature leaving the outlet of the control volume for the case with $\alpha = 0°$ is 5.51 K, while that for $\alpha = 10°$ is 5.65 K. Thus we conclude that the off-axis inlet flow leads to more effective heating of the air, although the improvement is only about 2.5 percent. We compute a third case (not shown) in which $\alpha = 0°$ but the turbulence intensity of the incoming air is increased to 25 percent. This leads to improved mixing, and the average air temperature rise from inlet to outlet increases by about 6.5 percent to 5.87 K.

Finally, we study the effect of rough tube surfaces. We model the tube walls as rough surfaces with a characteristic roughness height of 0.01 m (1 percent of cylinder diameter). Note that we had to coarsen the grid somewhat near each tube so that the distance from the center of the closest computational cell to the wall is greater than the roughness height; otherwise the roughness model in the CFD code is unphysical. The flow inlet angle is set to $\alpha = 0°$ for this case, and flow conditions are identical to those of Fig. 15–65.

FIGURE 15–65
Temperature contour plots produced by CFD calculations of stationary turbulent flow through a cross-flow heat exchanger at $\alpha = 0°$ with smooth tubes. Contours range from 300 K (blue) to 315 K (red) or higher (white). The average air temperature at the outlet increases by 5.51 K compared to the inlet air temperature. Note that although the calculations are performed in the computational domain of Fig. 15–63, the image is duplicated here three times for purposes of illustration.

CHAPTER 15

FIGURE 15–66
Temperature contour plots produced by CFD calculations of stationary turbulent flow through a cross-flow heat exchanger at $\alpha = 10°$ with smooth tubes. Contours range from 300 K (blue) to 315 K (red) or higher (white). The average air temperature at the outlet increases by 5.65 K compared to the inlet air temperature. Thus, off-axis inlet flow ($\alpha = 10°$) yields a ΔT that is 2.5 percent higher than that for the on-axis inlet flow ($\alpha = 0°$).

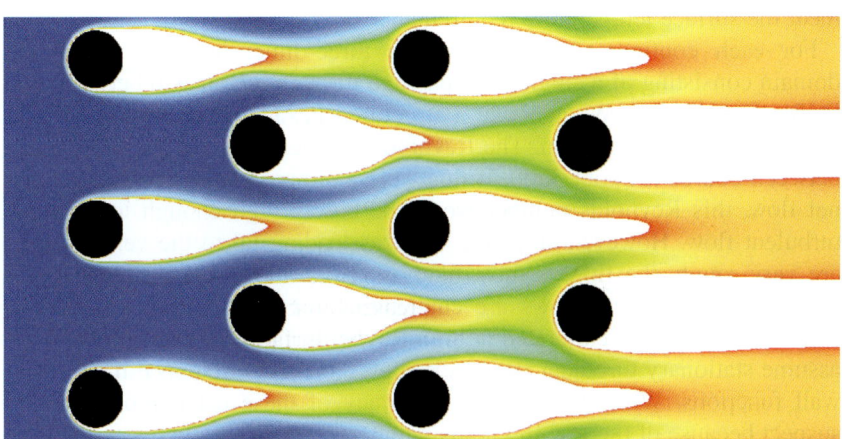

FIGURE 15–67
Temperature contour plots produced by CFD calculations of stationary turbulent flow through a cross-flow heat exchanger at $\alpha = 0°$ with rough tubes (average wall roughness equal to 1 percent of tube diameter; wall functions utilized in the CFD calculations). Contours range from 300 K (blue) to 315 K (red) or higher (white). The average air temperature at the outlet increases by 14.48 K compared to the inlet air temperature. Thus, even this small amount of surface roughness yields a ΔT that is 163 percent higher than that for the case with smooth tubes.

Temperature contours are plotted in Fig. 15–67. Pure white regions in the contour plot represent locations where the air temperature is greater than 315 K. The average air temperature rise from inlet to outlet is 14.48 K, a 163 percent increase over the smooth wall case at $\alpha = 0°$. Thus we see that wall roughness is a critical parameter in turbulent flows. This example provides some insight as to why the tubes in heat exchangers are often purposely roughened.

Cooling of an Array of Integrated Circuit Chips

In electronics equipment, instrumentation, and computers, electronic components, such as **integrated circuits (ICs** or "chips"), resistors, transistors, diodes, and capacitors, are soldered onto **printed circuit boards (PCBs).** The PCBs are often stacked in rows as sketched in Fig. 15–68. Because many of these electronic components must dissipate heat, cooling air is often blown through the air gap between each pair of PCBs to keep the components from getting too hot. Consider the design of a PCB for an outer space application. Several identical PCBs are to be stacked as in Fig. 15–68. Each PCB is 10 cm high and 30 cm long, and the spacing between boards is

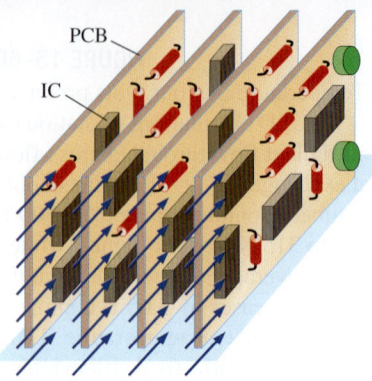

FIGURE 15–68
Four printed circuit boards (PCBs) stacked in rows, with air blown in between each PCB to provide cooling.

2.0 cm. Cooling air enters the gap between the PCBs at a speed of 2.60 m/s and a temperature of 30°C. The electrical engineers must fit eight identical ICs on a 10 cm × 15 cm portion of each board. Each of the ICs dissipates 6.24 W of heat: 5.40 W from its top surface and 0.84 W from its sides. (There is assumed to be no heat transfer from the bottom of the chip to the PCB.) The rest of the components on the board have negligible heat transfer compared to that from the eight ICs. To ensure adequate performance, the average temperature on the chip surface should not exceed 150°C, and the maximum temperature anywhere on the surface of the chip should not exceed 180°C. Each chip is 2.5 cm wide, 4.5 cm long, and 0.50 cm thick. The electrical engineers come up with two possible configurations of the eight chips on the PCB as sketched in Fig. 15–69: in the long configuration, the chips are aligned with their *long* dimension parallel to the flow, and in the short configuration, the chips are aligned with their *short* dimension parallel to the flow. The chips are staggered in both cases to enhance cooling. We are to determine which arrangement leads to the lower maximum surface temperature on the chips, and whether the electrical engineers will meet the surface temperature requirements.

For each configuration, we define a three-dimensional computational domain consisting of a single flow passage through the air gap between two PCBs (Fig. 15–70). We generate a structured hexagonal grid with 267,520 cells for each configuration. The Reynolds number based on the 2.0 cm gap between boards is about 3600. If this were a simple two-dimensional channel flow, this Reynolds number would be barely high enough to establish turbulent flow. However, since the surfaces leading up to the velocity inlet are very rough, the flow is most likely turbulent. We note that low Reynolds number turbulent flows are challenging for most turbulence models, since the models are calibrated at high Reynolds numbers. Nevertheless, we assume stationary turbulent flow and employ the k-ε turbulence model with wall functions. While the absolute accuracy of these calculations may be suspect because of the low Reynolds number, comparisons between the long and short configurations should be reasonable. We ignore buoyancy effects in the calculations since this is a space application. The inlet is specified as

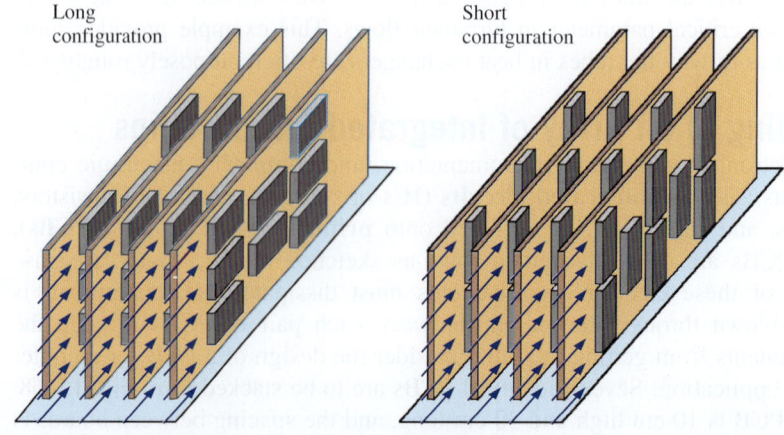

FIGURE 15–69
Two possible configurations of the eight ICs on the PCB: long configuration and short configuration. Without peeking ahead, which configuration do you think will offer the best cooling to the chips?

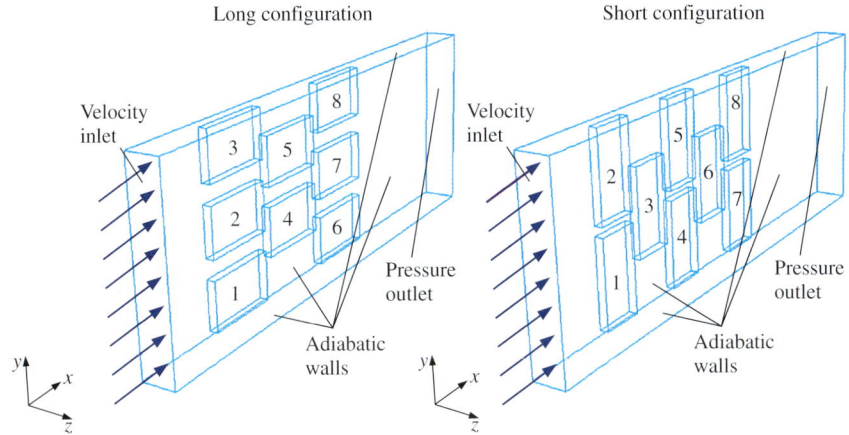

FIGURE 15–70
Computational domains for the chip cooling example. Air flowing through the gap between two PCBs is modeled. Two separate grids are generated, one for the long configuration and one for the short configuration. Chips 1 through 8 are labeled for reference. The surfaces of these chips transfer heat to the air; all other walls are adiabatic.

a velocity inlet (air) with $V = 2.60$ m/s and $T_\infty = 30°C$; we set the inlet turbulence intensity to 20 percent and the turbulent length scale to 1.0 mm. The outlet is a pressure outlet at zero gage pressure. The PCB is modeled as a smooth adiabatic wall (zero heat transfer from the wall to the air). The top and sides of the computational domain are also approximated as smooth adiabatic walls.

Based on the given chip dimensions, the surface area of the top of a chip is 4.5 cm × 2.5 cm = 11.25 cm². The total surface area of the four sides of the chip is 7.0 cm². From the given heat transfer rates, we calculate the rate of heat transfer per unit area from the top surface of each chip,

$$\dot{q}_{top} = \frac{5.4 \text{ W}}{11.25 \text{ cm}^2} = 0.48 \text{ W/cm}^2$$

So, we model the top surface of each chip as a smooth wall with a surface heat flux of 4800 W/m² from the wall to the air. Similarly, the rate of heat transfer per unit area from the sides of each chip is

$$\dot{q}_{sides} = \frac{0.84 \text{ W}}{7.0 \text{ cm}^2} = 0.12 \text{ W/cm}^2$$

Since the sides of the chip have electrical leads, we model each side surface of each chip as a rough wall with an equivalent roughness height of 0.50 mm and a surface heat flux of 1200 W/m² from the wall to the air.

The CFD code ANSYS-FLUENT is run for each case to convergence. Results are summarized in Table 15–5, and temperature contours are plotted in Figs. 15–71 and 15–72. The average temperature on the top surfaces of the chips is about the same for either configuration (144.4°C for the long case and 144.7°C for the short case) and is below the recommended limit of 150°C. There is more of a difference in average temperature on the *side* surfaces of the chips, however (84.2°C for the long case and 91.4°C for the short case), although these values are well below the limit. Of greatest concern are the maximum temperatures. For the long configuration, $T_{max} = 187.5°C$ and occurs on the top surface of chip 7 (the middle chip of the last row). For the

COMPUTATIONAL FLUID DYNAMICS

TABLE 15–5

Comparison of CFD results for the chip cooling example, long and short configurations

	Long	Short
T_{max}, top surfaces of chips	187.5°C	182.1°C
T_{avg}, top surfaces of chips	144.5°C	144.7°C
T_{max}, side surfaces of chips	154.0°C	170.6°C
T_{avg}, side surfaces of chips	84.2°C	91.4°C
Average ΔT, inlet to outlet	7.83°C	7.83°C
Average ΔP, inlet to outlet	−5.14 Pa	−5.58 Pa

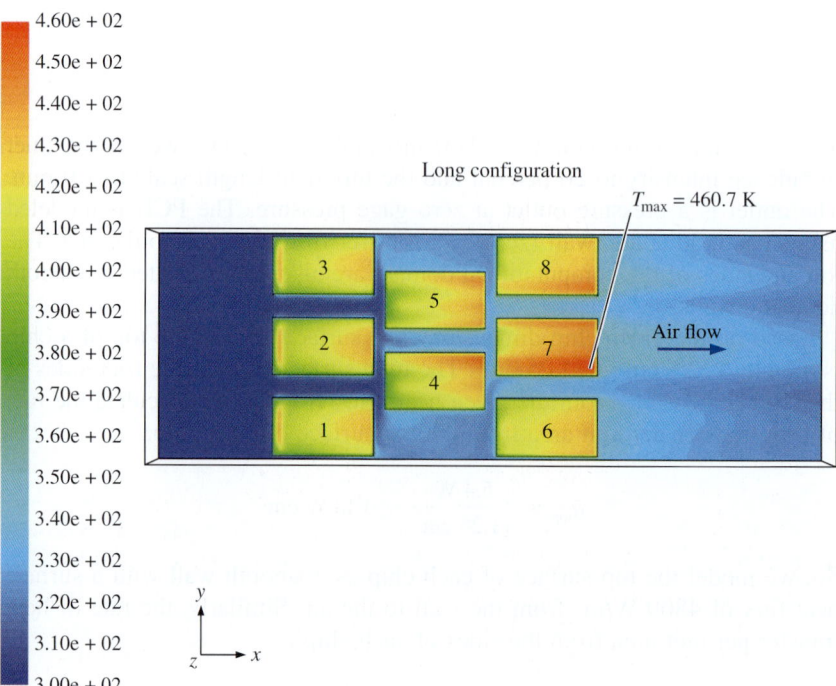

FIGURE 15–71

CFD results for the chip cooling example, long configuration: temperature contours as viewed from directly above the chip surfaces, with T values in K on the legend. The location of maximum surface temperature is indicated, it occurs near the end of chip 7. Red regions near the leading edges of chips 1, 2, and 3 are also seen, indicating high surface temperatures at those locations.

short configuration, $T_{max} = 182.1$°C and occurs close to midboard on the top surfaces of chips 7 and 8 (the two chips in the last row). For both configurations these values exceed the recommended limit of 180°C, although not by much. The short configuration does a better job at cooling the top surfaces of the chips, but at the expense of a slightly larger pressure drop and poorer cooling along the side surfaces of the chips.

Notice from Table 15–5 that the average change in air temperature from inlet to outlet is identical for both configurations (7.83°C). This should not be surprising, because the total rate of heat transferred from the chips to the air is the same regardless of chip configuration. In fact, in a CFD analysis it is wise to check values like this—if average ΔT were *not* the same between the two configurations, we would suspect some kind of error in our calculations.

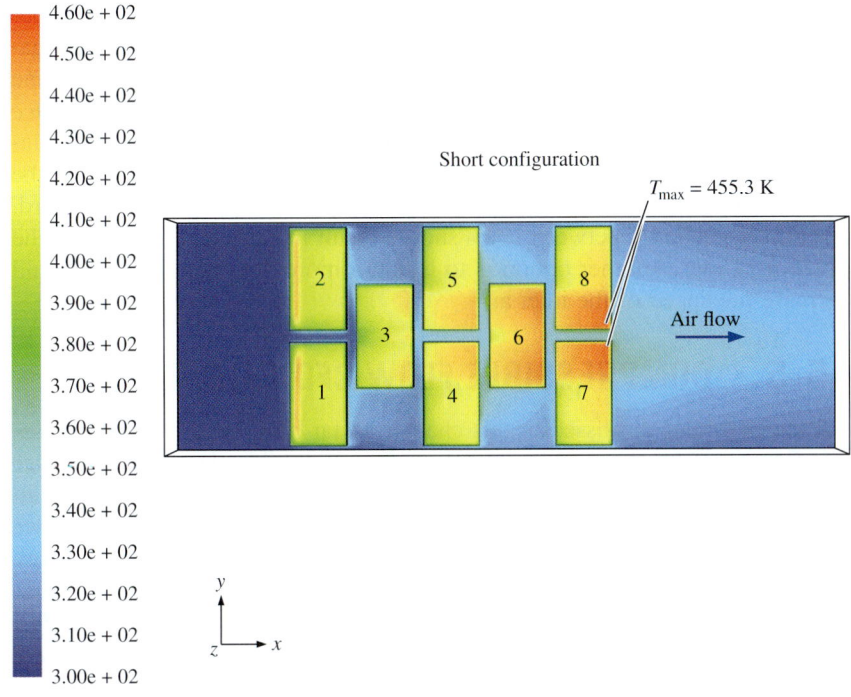

FIGURE 15–72
CFD results for the chip cooling example, short configuration: temperature contours as viewed from directly above the chip surfaces, with T values in K on the legend. The same temperature scale is used here as in Fig. 15–71. The locations of maximum surface temperature are indicated; they occur near the end of chips 7 and 8 near the center of the PCB. Red regions near the leading edges of chips 1 and 2 are also seen, indicating high surface temperatures at those locations.

We point out many other interesting features of these flow fields. For either configuration, the average surface temperature on the downstream chips is greater than that on the upstream chips. This makes sense physically, since the first chips receive the coolest air, while those downstream are cooled by air that has already been warmed up somewhat. We notice that the front chips (1, 2, and 3 in the long configuration and 1 and 2 in the short configuration) have regions of high temperature just downstream of their leading edges. A close-up view of the temperature distribution on one of these chips is shown in Fig. 15–73a. Why is the temperature so high there? It turns out that the flow separates off the sharp corner at the front of the chip and forms a recirculating eddy called a **separation bubble** on the top of the chip (Fig. 15–73b). The air speed is slow in that region, especially along the **reattachment line** where the flow reattaches to the surface. The slow air speed leads to a local "hot spot" in that region of the chip surface since convective cooling is minimal there. Finally, we notice in Fig. 15–73a that downstream of the separation bubble, T increases down the chip surface. There are two reasons for this: (1) the air warms up as it travels down the chip, and (2) the boundary layer on the chip surface grows downstream. The larger the boundary layer thickness, the lower the air speed near the surface, and thus the lower the amount of convective cooling at the surface.

In summary, our CFD calculations have predicted that the short configuration leads to a lower value of maximum temperature on the chip surfaces and appears at first glance to be the preferred configuration for heat transfer. However, the short configuration demands a higher pressure drop at the same volume flow rate (Table 15–5). For a given cooling fan, this additional pressure drop would shift the operating point of the fan to a lower volume

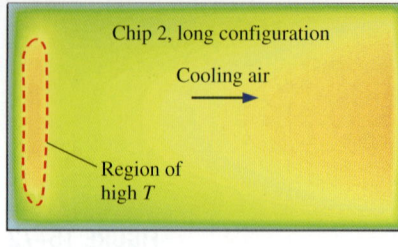

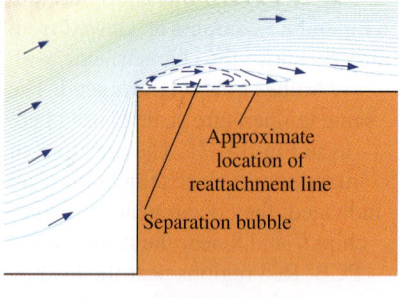

FIGURE 15–73
(*a*) Close-up top view of temperature contours on the surface of chip 2 of the long configuration. The region of high temperature is outlined. Temperature contour levels are the same as in Fig. 15–71. (*b*) An even closer view (an edge view) of streamlines outlining the separation bubble in that region. The approximate location of the reattachment line on the chip surface is also shown.

flow rate (Chap. 14), decreasing the cooling effect. It is not known whether this shift would be enough to favor the long configuration—more information about the fan and more analysis would be required. The bottom line in either case is that *there is not sufficient cooling to keep the chip surface temperature below 180°C everywhere on every chip*. To rectify the situation, we recommend that the designers spread the eight hot chips over the entire PCB rather than in the limited 10 cm × 15 cm area of the board. The increased space between chips should result in sufficient cooling for the given flow rate. Another option is to install a more powerful fan that would increase the speed of the inlet air.

15–5 ■ COMPRESSIBLE FLOW CFD CALCULATIONS

All the examples discussed in this chapter so far have been for incompressible flow (ρ = constant). When the flow is *compressible*, density is no longer a constant, but becomes an additional variable in the equation set. We limit our discussion here to *ideal gases*. When we apply the ideal-gas law, we introduce yet *another* unknown, namely, temperature T. Hence, the energy equation must be solved along with the compressible forms of the equations of conservation of mass and conservation of momentum (Fig. 15–74). In addition, fluid properties, such as viscosity and thermal conductivity, are no longer necessarily treated as constants, since they are functions of temperature; thus, they appear inside the derivative operators in the differential equations of Fig. 15–74. While the equation set looks ominous, many commercially available CFD codes are able to handle compressible flow problems, including shock waves.

When solving compressible flow problems with CFD, the boundary conditions are somewhat different than those of incompressible flow. For example, at a pressure inlet we need to specify both stagnation pressure *and* static pressure, along with stagnation temperature. A special boundary condition (called *pressure far field* in ANSYS-FLUENT) is also available for compressible flows. With this boundary condition, we specify the Mach number, the static pressure, and the temperature; it can be applied to both inlets *and* outlets and is well-suited for supersonic external flows.

The equations of Fig. 15–74 are for laminar flow, whereas many compressible flow problems occur at high flow speeds in which the flow is *turbulent*. Therefore, the equations of Fig. 15–74 must be modified accordingly (into the RANS equation set) to include a turbulence model, and more transport equations must be added, as discussed previously. The equations then get quite long and complicated and are not included here. Fortunately, in many situations, we can approximate the flow as *inviscid*, eliminating the viscous terms from the equations of Fig. 15–74 (the Navier–Stokes equation reduces to the Euler equation). As we shall see, the inviscid flow approximation turns out to be quite good for many practical high-speed flows, since the boundary layers along walls are very thin at high Reynolds numbers. In fact, compressible CFD calculations can predict flow features that are often quite difficult to obtain experimentally. For example, many experimental measurement techniques require optical access, which is limited in three-dimensional flows, and even in some axisymmetric flows. CFD is not limited in this way.

Continuity: $\dfrac{\partial(\rho u)}{\partial x} + \dfrac{\partial(\rho v)}{\partial y} + \dfrac{\partial(\rho w)}{\partial z} = 0$ **Ideal gas law:** $P = \rho RT$

x-momentum: $\rho\left(u\dfrac{\partial u}{\partial x} + v\dfrac{\partial u}{\partial y} + w\dfrac{\partial u}{\partial z}\right) = \rho g_x - \dfrac{\partial P}{\partial x} + \dfrac{\partial}{\partial x}\left(2\mu\dfrac{\partial u}{\partial x} + \lambda\vec{\nabla}\cdot\vec{V}\right) + \dfrac{\partial}{\partial y}\left[\mu\left(\dfrac{\partial u}{\partial y} + \dfrac{\partial v}{\partial x}\right)\right] + \dfrac{\partial}{\partial z}\left[\mu\left(\dfrac{\partial w}{\partial x} + \dfrac{\partial u}{\partial z}\right)\right]$

y-momentum: $\rho\left(u\dfrac{\partial v}{\partial x} + v\dfrac{\partial v}{\partial y} + w\dfrac{\partial v}{\partial z}\right) = \rho g_y - \dfrac{\partial P}{\partial y} + \dfrac{\partial}{\partial x}\left[\mu\left(\dfrac{\partial v}{\partial x} + \dfrac{\partial u}{\partial y}\right)\right] + \dfrac{\partial}{\partial y}\left(2\mu\dfrac{\partial v}{\partial y} + \lambda\vec{\nabla}\cdot\vec{V}\right) + \dfrac{\partial}{\partial z}\left[\mu\left(\dfrac{\partial v}{\partial z} + \dfrac{\partial w}{\partial y}\right)\right]$

z-momentum: $\rho\left(u\dfrac{\partial w}{\partial x} + v\dfrac{\partial w}{\partial y} + w\dfrac{\partial w}{\partial z}\right) = \rho g_z - \dfrac{\partial P}{\partial z} + \dfrac{\partial}{\partial x}\left[\mu\left(\dfrac{\partial w}{\partial x} + \dfrac{\partial u}{\partial z}\right)\right] + \dfrac{\partial}{\partial y}\left[\mu\left(\dfrac{\partial v}{\partial z} + \dfrac{\partial w}{\partial y}\right)\right] + \dfrac{\partial}{\partial z}\left(2\mu\dfrac{\partial w}{\partial z} + \lambda\vec{\nabla}\cdot\vec{V}\right)$

Energy: $\rho c_p\left(u\dfrac{\partial T}{\partial x} + v\dfrac{\partial T}{\partial y} + w\dfrac{\partial T}{\partial z}\right) = \beta T\left(u\dfrac{\partial P}{\partial x} + v\dfrac{\partial P}{\partial y} + w\dfrac{\partial P}{\partial z}\right) + \vec{\nabla}\cdot(k\vec{\nabla}T) + \Phi$

FIGURE 15–74

The equations of motion for the case of steady, compressible, laminar flow of a Newtonian fluid in Cartesian coordinates. There are six equations and six unknowns: ρ, u, v, w, T, and P. Five of the equations are nonlinear partial differential equations, while the ideal-gas law is an algebraic equation. R is the specific ideal-gas constant, λ is the second coefficient of viscosity, often set equal to $-2\mu/3$; c_p is the specific heat at constant pressure; k is the thermal conductivity; β is the coefficient of thermal expansion, and Φ is the dissipation function, given by White (2005) as

$$\Phi = 2\mu\left(\dfrac{\partial u}{\partial x}\right)^2 + 2\mu\left(\dfrac{\partial v}{\partial y}\right)^2 + 2\mu\left(\dfrac{\partial w}{\partial z}\right)^2 + \mu\left(\dfrac{\partial v}{\partial x} + \dfrac{\partial u}{\partial y}\right)^2 + \mu\left(\dfrac{\partial w}{\partial y} + \dfrac{\partial v}{\partial z}\right)^2 + \mu\left(\dfrac{\partial u}{\partial z} + \dfrac{\partial w}{\partial x}\right)^2 + \lambda\left(\dfrac{\partial u}{\partial x} + \dfrac{\partial v}{\partial y} + \dfrac{\partial w}{\partial z}\right)^2$$

Compressible Flow through a Converging–Diverging Nozzle

For our first example, we consider compressible flow of air through an axisymmetric converging–diverging nozzle. The computational domain is shown in Fig. 15–75. The inlet radius is 0.10 m, the throat radius is 0.075 m, and the outlet radius is 0.12 m. The axial distance from the inlet to the throat is 0.30 m—the same as the axial distance from the throat to the outlet. A structured grid with approximately 12,000 quadrilateral cells is used in the calculations. At the pressure inlet boundary, the stagnation pressure $P_{0,\,\text{inlet}}$ is set to 220 kPa (absolute), the static pressure P_{inlet} is set to 210 kPa, and the stagnation temperature $T_{0,\,\text{inlet}}$ is set to 300 K. For the first case, we set the static pressure Pb at the pressure outlet boundary to 50.0 kPa ($Pb/P_{0,\,\text{inlet}} = 0.227$)—low enough that the flow is supersonic through the entire diverging section of the nozzle, without any normal shocks in the nozzle. This back pressure ratio corresponds to a value between cases E and F in Fig. 12–22, in which a complex shock pattern occurs downstream of the nozzle exit; these shock waves do not influence the flow in the nozzle itself, since the flow exiting the nozzle is supersonic. We do not attempt to model the flow downstream of the nozzle exit.

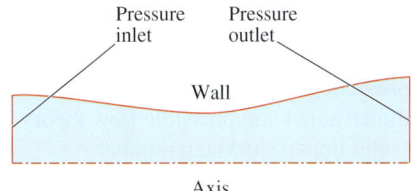

FIGURE 15–75

Computational domain for compressible flow through a converging–diverging nozzle. Since the flow is axisymmetric, only one two-dimensional slice is needed for the CFD solution.

The CFD code is run to convergence in its steady, inviscid, compressible flow mode. The average values of the Mach number Ma and pressure ratio $P/P_{0,\,inlet}$ are calculated at 25 axial locations along the converging–diverging nozzle (every 0.025 m) and are plotted in Fig. 15–76a. The results match almost perfectly with the predictions of one-dimensional isentropic flow (Chap. 12). At the throat ($x = 0.30$ m), the average Mach number is 0.997, and the average value of $P/P_{0,\,inlet}$ is 0.530. One-dimensional isentropic flow theory predicts Ma = 1 and $P/P_{0,\,inlet} = 0.528$ at the throat. The small discrepancies between CFD and theory are due to the fact that the computed flow is *not* one-dimensional, since there is a radial velocity component and, therefore, a radial variation of the Mach number and static pressure. Careful examination of the Mach number contour lines of Fig. 15–76b reveal that they are curved, not straight as would be predicted by one-dimensional isentropic theory. The sonic line (Ma = 1) is identified for clarity in the figure. Although Ma = 1 right at the wall of the throat, sonic conditions along the axis of the nozzle are not reached until somewhat downstream of the throat.

Next, we run a series of cases in which back pressure P_b is varied, while keeping all other boundary conditions fixed. Results for three cases are

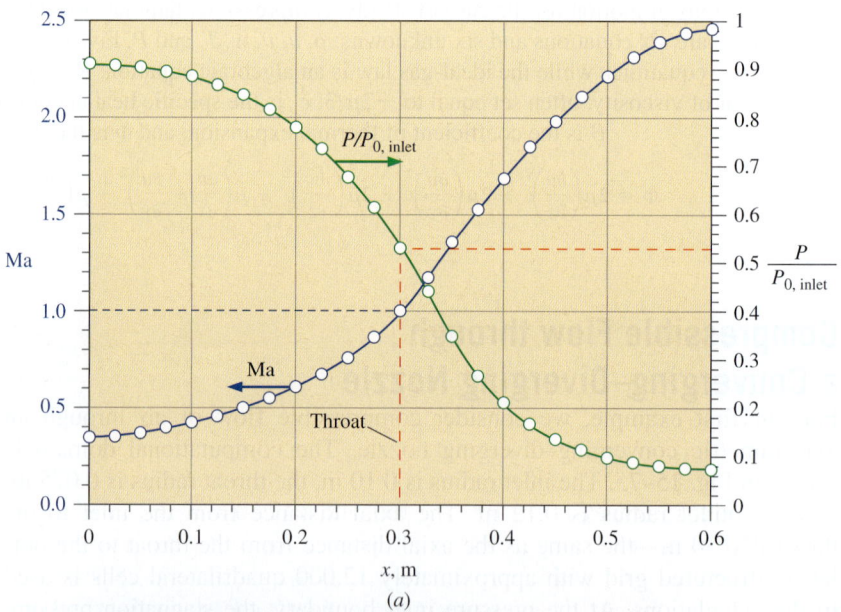

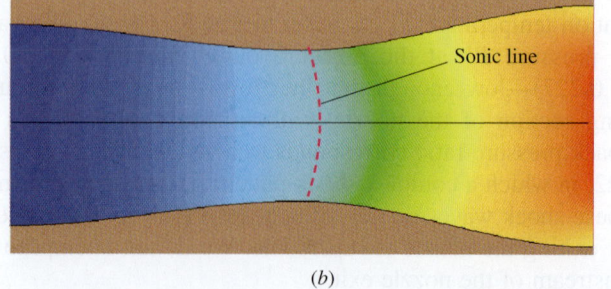

FIGURE 15–76
CFD results for steady, adiabatic, inviscid compressible flow through an axisymmetric converging–diverging nozzle: (*a*) calculated average Mach number and pressure ratio at 25 axial locations (circles), compared to predictions from isentropic, one-dimensional compressible flow theory (solid lines); (*b*) Mach number contours, ranging from Ma = 0.3 (blue) to 2.7 (red). Although only the top half is calculated, a mirror image about the *x*-axis is shown for clarity. The sonic line (Ma = 1) is also highlighted. It is parabolic rather than straight in this axisymmetric flow due to the radial component of velocity, as discussed in Schreier (1982).

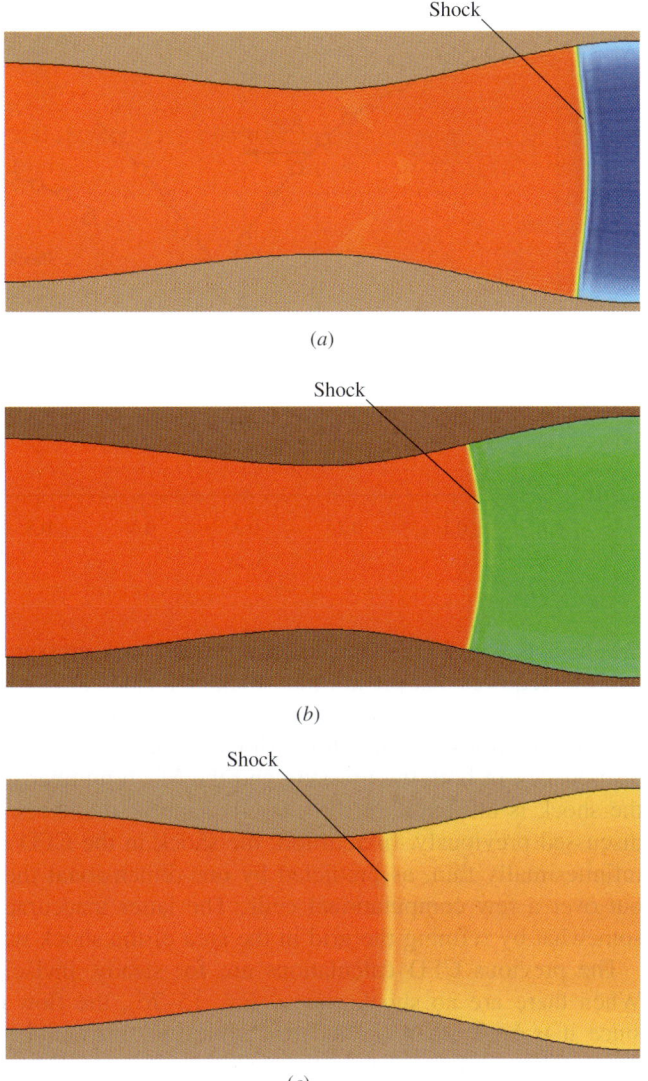

FIGURE 15–77

CFD results for steady, adiabatic, inviscid, compressible flow through a converging-diverging nozzle: contours of stagnation pressure ratio $P_0/P_{0,\,\text{inlet}}$ are shown for $P_b/P_{0,\,\text{inlet}} =$ (a) 0.455; (b) 0.682; and (c) 0.909. Since stagnation pressure is constant upstream of the shock and decreases suddenly across the shock, it serves as a convenient indicator of the location and strength of the normal shock in the nozzle. In these contour plots, $P_0/P_{0,\,\text{inlet}}$ ranges from 0.5 (blue) to 1.01 (red). It is clear from the colors downstream of the shock that the farther downstream the shock, the stronger the shock (larger magnitude of stagnation pressure drop across the shock). We also note the shape of the shocks—curved rather than straight, because of the radial component of velocity.

shown in Fig. 15–77: P_b = (a) 100, (b) 150, and (c) 200 kPa, i.e., $P_b/P_{0,\,\text{inlet}} =$ (a) 0.455, (b) 0.682, and (c) 0.909, respectively. For all three cases, a normal shock occurs in the diverging portion of the nozzle. Furthermore, as back pressure increases, the shock moves upstream toward the throat, and decreases in strength. Since the flow is choked at the throat, the mass flow rate is identical in all three cases (and also in the previous case shown in Fig. 15–76). We notice that the normal shock is not straight, but rather is curved due to the radial component of velocity, as previously mentioned.

For case (b), in which $P_b/P_{0,\,\text{inlet}} = 0.682$, the average values of the Mach number and pressure ratio $P/P_{0,\,\text{inlet}}$ are calculated at 25 axial locations along the converging–diverging nozzle (every 0.025 m), and are plotted in Fig. 15–78. For comparison with theory, the one-dimensional isentropic flow relations are used upstream and downstream of the shock, and the normal shock relations are used to calculate the pressure jump *across*

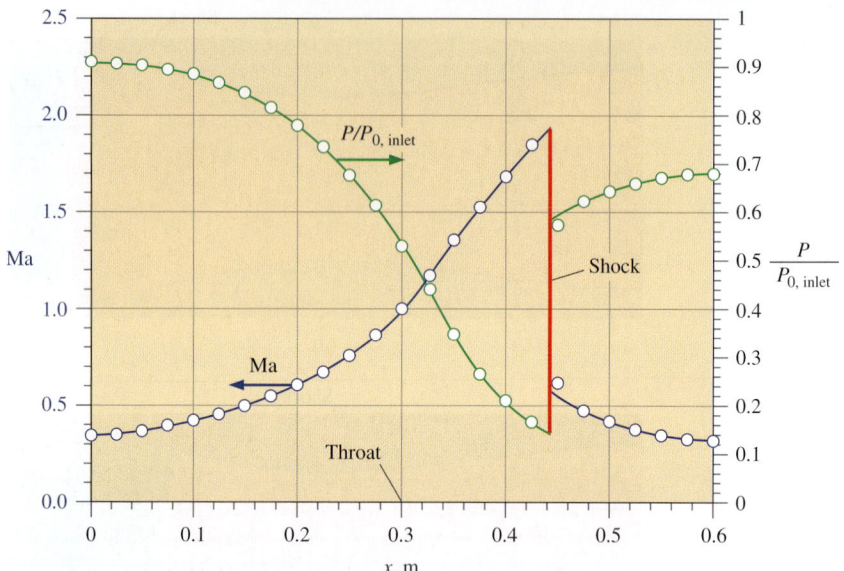

FIGURE 15–78

Mach number and pressure ratio as functions of axial distance along a converging–diverging nozzle for the case in which $P_b/P_{0,\,inlet} = 0.682$. Averaged CFD results at 25 axial locations (circles) for steady, inviscid, adiabatic, compressible flow are compared to predictions from one-dimensional compressible flow theory (solid lines).

the shock (Chap. 12). To match the specified back pressure, one-dimensional analysis requires that the normal shock be located at $x = 0.4436$ m, accounting for the change in both P_0 and A^* across the shock. The agreement between CFD calculations and one-dimensional theory is again excellent. The small discrepancy in both the pressure and the Mach number just downstream of the shock is attributed to the curved shape of the shock (Fig. 15–77b), as discussed previously. In addition, the shock in the CFD calculations is not infinitesimally thin, as predicted by one-dimensional theory, but is spread out over a few computational cells. The latter inaccuracy can be reduced somewhat by refining the grid in the area of the shock wave (not shown).

The previous CFD calculations are for steady, inviscid, adiabatic flow. When there are no shock waves (Fig. 15–76), the flow is also *isentropic*, since it is both adiabatic and reversible (no irreversible losses). However, when a shock wave exists in the flow field (Fig. 15–77), the flow is no longer isentropic since there are irreversible losses across the shock, although it is still adiabatic.

One final CFD case is run in which two additional irreversibilities are included, namely, *friction* and *turbulence*. We modify case (b) of Fig. 15–77 by running a steady, adiabatic, turbulent case using the k-ε turbulence model with wall functions. The turbulence intensity at the inlet is set to 10 percent with a turbulence length scale of 0.050 m. A contour plot of $P/P_{0,\,inlet}$ is shown in Fig. 15–79, using the same color contour scale as in Fig. 15–77. Comparison of Figs. 15–77b and 15–79 reveals that the shock wave for the turbulent case occurs further upstream and is therefore somewhat weaker. In addition, the stagnation pressure is small in a very thin region along the channel walls. This is due to frictional losses in the thin boundary layer. Turbulent and viscous irreversibilities in the boundary layer region are responsible for this decrease in stagnation pressure. Furthermore, the boundary layer separates just downstream of the shock, leading to more irreversibilities. A close-up view of velocity vectors in the vicinity of the

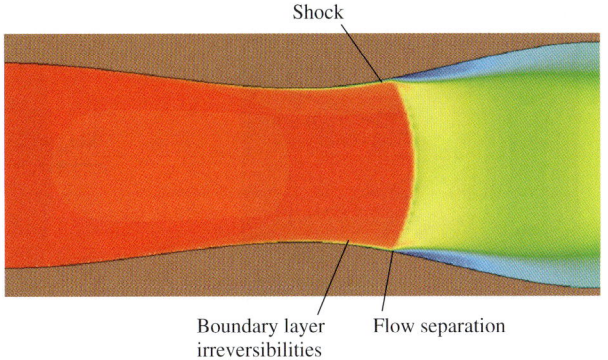

FIGURE 15–79

CFD results for stationary, adiabatic, turbulent, compressible flow through a converging–diverging nozzle. Contours of stagnation pressure ratio $P_0/P_{0,\text{inlet}}$ are shown for the case with $P_b/P_{0,\text{inlet}} = 0.682$, the same back pressure and color scale as that of Fig. 15–77b. Flow separation and irreversibilities in the boundary layer are identified.

separation point along the wall is shown in Fig. 15–80. We note that this case does not converge well and is inherently unsteady; the interaction between shock waves and boundary layers is a very difficult task for CFD. Because we use wall functions, flow details within the turbulent boundary layer are not resolved in this CFD calculation. Experiments reveal, however, that the shock wave interacts much more significantly with the boundary layer, producing "λ-feet," as discussed in the Application Spotlight of Chap. 12.

Finally, we compare the mass flow rate for this viscous, turbulent case to that of the inviscid case, and find that $\dot{m}$ has decreased by about 0.7 percent. Why? As discussed in Chap. 10, a boundary layer along a wall impacts the outer flow such that the wall appears to be thicker by an amount equal to the displacement thickness δ^*. *The effective throat area is thus reduced somewhat by the presence of the boundary layer*, leading to a reduction in mass flow rate through the converging–diverging nozzle. The effect is small in this example since the boundary layers are so thin relative to the dimensions of the nozzle, and it turns out that the inviscid approximation is quite good (less than one percent error).

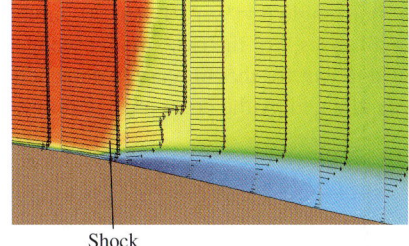

FIGURE 15–80

Close-up view of velocity vectors and stagnation pressure contours in the vicinity of the separated flow region of Fig. 15–79. The sudden decrease in velocity magnitude across the shock is seen, as is the reverse flow region downstream of the shock.

Oblique Shocks over a Wedge

For our final compressible flow example, we model steady, adiabatic, two-dimensional, inviscid, compressible flow of air over a wedge of half-angle θ (Fig. 15–81). Since the flow has top–bottom symmetry, we model only the upper half of the flow and use a symmetry boundary condition along the bottom edge. We run three cases: $\theta = 10$, 20, and 30°, at an inlet Mach number of 2.0. CFD results are shown in Fig. 15–82 for all three cases. In the CFD plots, a mirror image of the computational domain is projected across the line of symmetry for clarity.

For the 10° case (Fig. 15–82a), a straight oblique shock originating at the apex of the wedge is observed, as also predicted by inviscid theory. The flow turns across the oblique shock by 10° so that it is parallel to the wedge wall. The shock angle β predicted by inviscid theory is 39.31°, and the predicted Mach number downstream of the shock is 1.64. Measurements with a protractor on Fig. 15–82a yield $\beta \cong 40°$, and the CFD calculation of the Mach number downstream of the shock is 1.64; the agreement with theory is thus excellent.

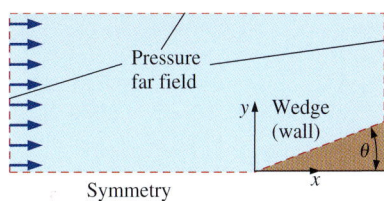

FIGURE 15–81

Computational domain and boundary conditions for compressible flow over a wedge of half-angle θ. Since the flow is symmetric about the x-axis, only the upper half is modeled in the CFD analysis.

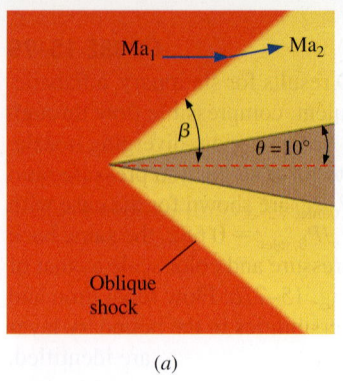 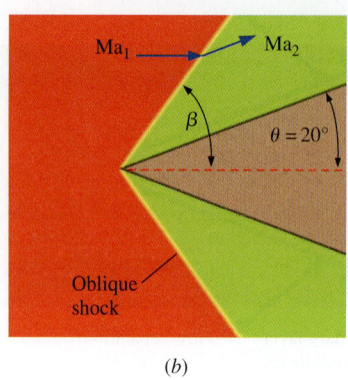 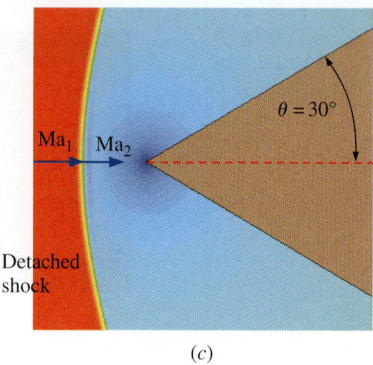

FIGURE 15–82
CFD results (Mach number contours) for steady, adiabatic, inviscid, compressible flow at $Ma_1 = 2.0$ over a wedge of half-angle $\theta = $ (a) 10°, (b) 20°, and (c) 30°. The Mach number contours range from $Ma = 0.2$ (blue) to 2.0 (red) in all cases. For the two smaller wedge half-angles, an attached weak oblique shock forms at the leading edge of the wedge, but for the 30° case, a detached shock (bow wave) forms ahead of the wedge. Shock strength increases with θ, as indicated by the color change downstream of the shock as θ increases.

For the 20° case (Fig. 15–82b), the CFD calculations yield a Mach number of 1.21 downstream of the shock. The shock angle measured from the CFD calculations is about 54°. Inviscid theory predicts a Mach number of 1.21 and a shock angle of 53.4°, so again the agreement between theory and CFD is excellent. Since the shock for the 20° case is at a steeper angle (closer to a normal shock), it is stronger than the shock for the 10° case, as indicated by the redder coloring in the Mach contours downstream of the shock for the 20° case.

At Mach number 2.0 in air, inviscid theory predicts that a straight oblique shock can form up to a maximum wedge half-angle of about 23° (Chap. 12). At wedge half-angles greater than this, the shock must move upstream of the wedge (become detached), forming a **detached shock,** which takes the shape of a **bow wave** (Chap. 12). The CFD results at $\theta = 30°$ (Fig. 15–82c) show that this is indeed the case. The portion of the detached shock just upstream of the leading edge is a normal shock, and thus the flow downstream of that portion of the shock is subsonic. As the shock curves backward, it becomes progressively weaker, and the Mach number downstream of the shock increases, as indicated by the coloring.

15–6 · OPEN-CHANNEL FLOW CFD CALCULATIONS

So far, all our examples have been for one single-phase fluid (air or water). However, many commercially available CFD codes can handle flow of a mixture of gases (e.g., carbon monoxide in air), flow with two phases of the same fluid (e.g., steam and liquid water), and even flow of two fluids of different phase (e.g., liquid water and gaseous air). The latter case is of interest in this section, namely, the flow of water with a free surface, above which is gaseous air, i.e., open-channel flow. We present here some simple examples of CFD solutions of open-channel flows.

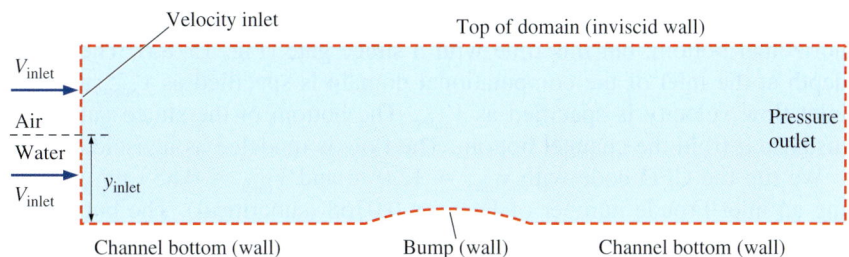

FIGURE 15–83
Computational domain for steady, incompressible, two-dimensional flow of water over a bump along the bottom of a channel, with boundary conditions identified. Two fluids are modeled in the flow field—liquid water and air above the free surface of the water. Liquid depth y_{inlet} and inlet speed V_{inlet} are specified.

Flow over a Bump on the Bottom of a Channel

Consider a two-dimensional channel with a flat, horizontal bottom. At a certain location along the bottom of the channel, there is a smooth bump, 1.0 m long and 0.10 m high at its center (Fig. 15–83). The velocity inlet is split into two parts—the lower part for liquid water and the upper part for air. In the CFD calculations, the inlet velocity of both the air and the water is specified as V_{inlet}. The water depth at the inlet of the computational domain is specified as y_{inlet}, but the location of the water surface in the rest of the domain is calculated. The flow is modeled as inviscid.

We consider cases with both subcritical and supercritical inlets (Chap. 13). Results from the CFD calculations are shown in Fig. 15–84 for three cases for comparison. For the first case (Fig. 15–84a), y_{inlet} is specified as 0.30 m, and V_{inlet} is specified as 0.50 m/s. The corresponding Froude number is calculated to be

Froude number: $$\text{Fr} = \frac{V_{inlet}}{\sqrt{gy_{inlet}}} = \frac{0.50 \text{ m/s}}{\sqrt{(9.81 \text{ m/s}^2)(0.30 \text{ m})}} = 0.291$$

Since Fr < 1, the flow at the inlet is *subcritical*, and the liquid surface *dips* slightly above the bump (Fig. 15–84a). The flow remains subcritical downstream of the bump, and the liquid surface height slowly rises back to its prebump level. The flow is thus subcritical everywhere.

For the second case (Fig. 15–84b), y_{inlet} is specified as 0.50 m, and V_{inlet} is specified as 4.0 m/s. The corresponding Froude number is calculated to be 1.81. Since Fr > 1, the flow at the inlet is *supercritical*, and the liquid surface *rises* above the bump (Fig. 15–84b). Far downstream, the liquid depth returns to 0.50 m, and the average velocity returns to 4.0 m/s, yielding Fr = 1.81—the same as at the inlet. Thus, this flow is supercritical everywhere.

Finally, we show results for a third case (Fig. 15–84c) in which the flow entering the channel is subcritical (y_{inlet} = 0.50 m, V_{inlet} = 1.0 m/s, and Fr = 0.452). In this case, the water surface dips downward over the bump, as expected for subcritical flow. However, on the downstream side of the bump, y_{outlet} = 0.25 m, V_{outlet} = 2.0 m/s, and Fr = 1.28. Thus, this flow starts subcritical, but changes to supercritical downstream of the bump. If the domain had extended further downstream, we would likely see a *hydraulic jump* that would bring the Froude number back below unity (subcritical).

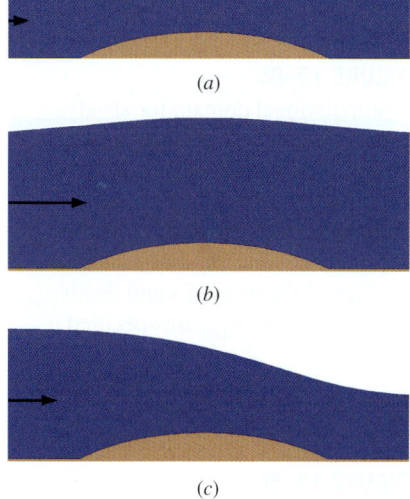

FIGURE 15–84
CFD results for incompressible, two-dimensional flow of water over a bump along the channel bottom. Phase contours are plotted, where blue indicates liquid water and white indicates gaseous air: (a) subcritical-to-subcritical, (b) supercritical-to-supercritical, and (c) subcritical-to-supercritical.

Flow through a Sluice Gate (Hydraulic Jump)

As a final example, we consider a two-dimensional channel with a flat, horizontal bottom, but this time with a sluice gate (Fig. 15–85). The water depth at the inlet of the computational domain is specified as y_{inlet}, and the inlet flow velocity is specified as V_{inlet}. The bottom of the sluice gate is at distance a from the channel bottom. The flow is modeled as inviscid.

We run the CFD code with y_{inlet} = 12.0 m and V_{inlet} = 0.833 m/s, yielding an inlet Froude number of Fr_{inlet} = 0.0768 (subcritical). The bottom of the sluice gate is at a = 0.125 m from the channel bottom. Results from the CFD calculations are shown in Fig. 15–86. After the water passes under the sluice gate, its average velocity increases to 12.8 m/s, and its depth decreases to y = 0.78 m. Thus, Fr = 4.63 (supercritical) downstream of the sluice gate and upstream of the hydraulic jump. Some distance downstream, we see a hydraulic jump in which the average water depth increases to y = 3.54 m, and the average water velocity decreases to 2.82 m/s. The Froude number downstream of the hydraulic jump is thus Fr = 0.478 (subcritical). We notice that the downstream water depth is significantly lower than that upstream of the sluice gate, indicating relatively large dissipation through the hydraulic jump and a corresponding decrease in the specific energy of the flow (Chap. 13). The analogy between specific energy loss through a hydraulic jump in open-channel flow and stagnation pressure loss through a shock wave in compressible flow is reinforced.

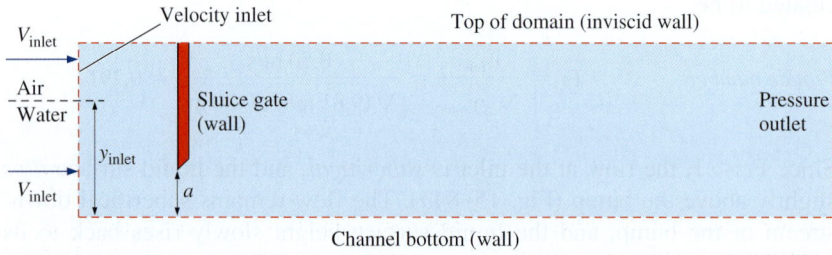

FIGURE 15–85
Computational domain for steady, incompressible, two-dimensional flow of water through a sluice gate, with boundary conditions identified. Two fluids are modeled in the flow field—liquid water, and air above the free surface of the water. Liquid depth y_{inlet} and inlet speed V_{inlet} are specified.

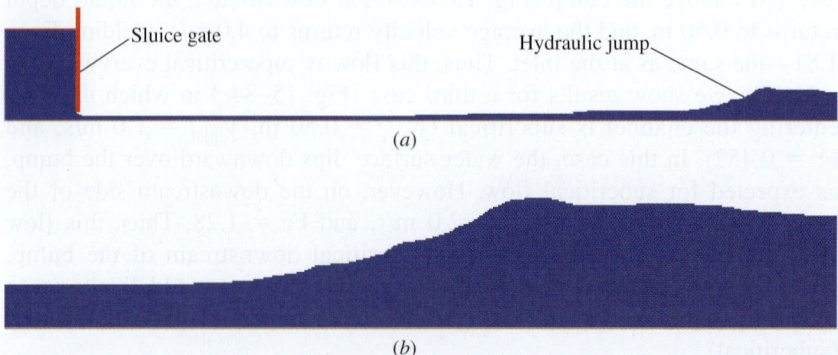

FIGURE 15–86
CFD results for incompressible, two-dimensional flow of water through a sluice gate in an open channel. Phase contours are plotted, where blue indicates liquid water and white indicates gaseous air: (*a*) overall view of the sluice gate and hydraulic jump, and (*b*) close-up view of the hydraulic jump. The flow is highly unsteady, and these are instantaneous snapshots at an arbitrary time.

CHAPTER 15

APPLICATION SPOTLIGHT ■ A Virtual Stomach

Guest Authors: James G. Brasseur and Anupam Pal,
The Pennsylvania State University

The mechanical function of the stomach (called gastric "motility") is central to proper nutrition, reliable drug delivery, and many gastric dysfunctions such as gastroparesis. Figure 15–87 shows a magnetic resonance image (MRI) of the stomach. The stomach is a mixer, a grinder, a storage chamber, and a sophisticated pump that controls the release of liquid and solid gastric content into the small intestines where nutrient uptake occurs. Nutrient release is controlled by the opening and closing of a valve at the end of the stomach (the pylorus) and the time variations in pressure difference between the stomach and duodenum. Gastric pressure is controlled by muscle tension over the stomach wall and peristaltic contraction waves that pass through the antrum (Fig. 15–87). These antral peristaltic contraction waves also break down food particles and mix material within the stomach, both food and drugs. It is currently impossible, however, to measure the mixing fluid motions in the human stomach. The MRI, for example, gives only an *outline* of special magnetized fluid within the stomach. In order to study these invisible fluid motions and their effects, we have developed a computer model of the stomach using computational fluid dynamics.

The mathematics underlying our computational model is derived from the laws of fluid mechanics. The model is a way of extending MRI measurements of time-evolving stomach geometry to the fluid motions within. Whereas computer models cannot describe the full complexity of gastric physiology, they have the great advantage of allowing controlled systematic variation of parameters, so sensitivities that cannot be measured experimentally can be studied computationally. Our virtual stomach applies a numerical method called the "lattice Boltzmann" algorithm that is well suited to fluid flows in complex geometries, and the boundary conditions are obtained from MRI data. In Fig. 15–88 we predict the motions, breakdown, and mixing of 1-cm-size extended-release drug tablets in the stomach. In this numerical experiment the drug tablet is denser than the surrounding highly viscous meal. We predict that the antral peristaltic waves generate recirculating eddies and retropulsive "jets" within the stomach, which in turn generate high shear stresses that wear away the tablet surface and release the drug. The drug then mixes by the same fluid motions that release the drug. We find that gastric fluid motions and mixing depend on the details of the time variations in stomach geometry and pylorus.

References

Indireshkumar, K., Brasseur, J. G., Faas, H., Hebbard, G. S., Kunz, P., Dent, J., Boesinger, P., Feinle, C., Fried, M., Li, M., and Schwizer, W., "Relative Contribution of 'Pressure Pump' and 'Peristaltic Pump' to Slowed Gastric Emptying," *Amer J Physiol*, 278, pp. G604–616, 2000.

Pal, A., Indireshkumar, K., Schwizer, W., Abrahamsson, B., Fried, M., Brasseur, J. G., "2004 Gastric Flow and Mixing Studied Using Computer Simulation," *Proc. Royal Soc. London, Biological Sciences*, October 2004.

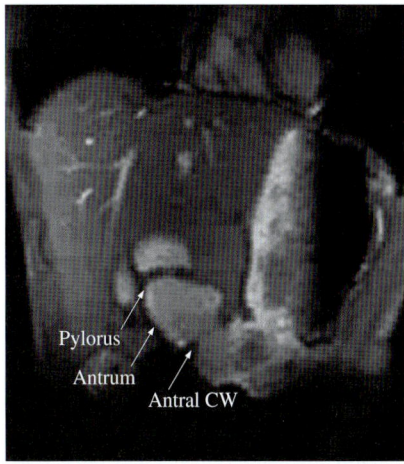

FIGURE 15–87
Magnetic resonance image of the human stomach in vivo at one instant in time showing peristaltic (i.e., propagating) contraction waves (CW) in the end region of the stomach (the antrum). The pylorus is a sphincter, or valve, that allows nutrients into the duodenum (small intestines).

Developed by Anupam Pal and James Brasseur. Used by permission.

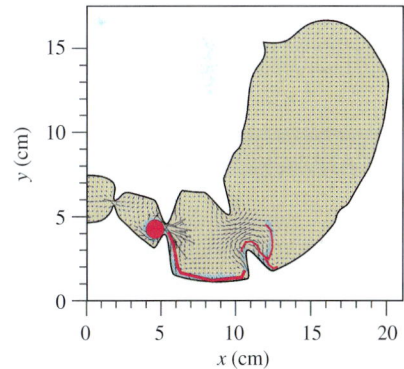

FIGURE 15–88
Computer simulation of fluid motions within the stomach (velocity vectors) from peristaltic antral contraction waves (Fig. 15–87), and the release of a drug (red trail) from an extended release tablet (red circle).

Developed by Anupam Pal and James Brasseur. Used by permission.

SUMMARY

Although neither as ubiquitous as spreadsheets, nor as easy to use as mathematical solvers, computational fluid dynamics codes are continually improving and are becoming more commonplace. Once the realm of specialized scientists who wrote their own codes and used supercomputers, commercial CFD codes with numerous features and user-friendly interfaces can now be obtained for personal computers at a reasonable cost and are available to engineers of all disciplines. As shown in this chapter, however, a poor grid, improper choice of laminar versus turbulent flow, inappropriate boundary conditions, and/or any of a number of other miscues can lead to CFD solutions that are physically incorrect, even though the colorful graphical output always looks pretty. Therefore, it is imperative that CFD users be well grounded in the fundamentals of fluid mechanics in order to avoid erroneous answers from a CFD simulation. In addition, appropriate comparisons should be made to experimental data whenever possible to validate CFD predictions. Bearing these cautions in mind, CFD has enormous potential for diverse applications involving fluid flows.

We show examples of both laminar and turbulent CFD solutions. For incompressible laminar flow, computational fluid dynamics does an excellent job, even for unsteady flows with separation. In fact, laminar CFD solutions are "exact" to the extent that they are limited by grid resolution and boundary conditions. Unfortunately, many flows of practical engineering interest are *turbulent*, not laminar. *Direct numerical simulation* (DNS) has great potential for simulation of complex turbulent flow fields, and algorithms for solving the equations of motion (the three-dimensional continuity and Navier–Stokes equations) are well established. However, resolution of all the fine scales of a high Reynolds number complex turbulent flow requires computers that are orders of magnitude faster than today's fastest machines. It will be decades before computers advance to the point where DNS is useful for practical engineering problems. In the meantime, the best we can do is employ *turbulence models*, which are semi-empirical transport equations that model (rather than solve) the increased mixing and diffusion caused by turbulent eddies. When running CFD codes that utilize turbulence models, we must be careful that we have a fine-enough mesh and that all boundary conditions are properly applied. In the end, however, regardless of how fine the mesh, or how valid the boundary conditions, *turbulent CFD results are only as good as the turbulence model used.* Nevertheless, while no turbulence model is *universal* (applicable to *all* turbulent flows), we obtain reasonable performance for many practical flow simulations.

We also demonstrate in this chapter that CFD can yield useful results for flows with heat transfer, compressible flows, and open-channel flows. In all cases, however, users of CFD must be careful that they choose an appropriate computational domain, apply proper boundary conditions, generate a good grid, and use the proper models and approximations. As computers continue to become faster and more powerful, CFD will take on an ever-increasing role in design and analysis of complex engineering systems.

We have only scratched the surface of computational fluid dynamics in this brief chapter. In order to become proficient and competent at CFD, you must take advanced courses of study in numerical methods, fluid mechanics, turbulence, and heat transfer. We hope that, if nothing else, this chapter has spurred you on to further study of this exciting topic.

REFERENCES AND SUGGESTED READING

1. C-J. Chen and S-Y. Jaw. *Fundamentals of Turbulence Modeling*. Washington, DC: Taylor & Francis, 1998.
2. J. M. Cimbala, H. Nagib, and A. Roshko. "Large Structure in the Far Wakes of Two-Dimensional Bluff Bodies," *Fluid Mech.*, 190, pp. 265–298, 1988.
3. S. Schreier. *Compressible Flow*. New York: Wiley-Interscience, Chap. 6 (Transonic Flow), pp. 285–293, 1982.
4. J. C. Tannehill, D. A. Anderson, and R. H. Pletcher. *Computational Fluid Mechanics and Heat Transfer*, 3rd ed. Washington, DC: Taylor & Francis, 2012.
5. H. Tennekes and J. L. Lumley. *A First Course in Turbulence*. Cambridge, MA: The MIT Press, 1972.
6. D. J. Tritton. *Physical Fluid Dynamics*. New York: Van Nostrand Reinhold Co., 1977.
7. M. Van Dyke. *An Album of Fluid Motion*. Stanford, CA: The Parabolic Press, 1982.
8. F. M. White. *Viscous Fluid Flow*, 3rd ed. New York: McGraw-Hill, 2005.
9. D. C. Wilcox. *Turbulence Modeling for CFD*, 3rd ed. La Cañada, CA: DCW Industries, Inc., 2006.
10. C. H. K. Williamson. "Oblique and Parallel Modes of Vortex Shedding in the Wake of a Circular Cylinder at Low Reynolds Numbers," *J. Fluid Mech.*, 206, pp. 579–627, 1989.
11. Tu, J., Yeoh, G.H., and Liu, C. *Computational Fluid Dynamics: A Practical Approach*. Burlington, MA: Elsevier, 2008.

PROBLEMS*

Fundamentals, Grid Generation, and Boundary Conditions

15–1C A CFD code is used to solve a two-dimensional (x and y), incompressible, laminar flow without free surfaces. The fluid is Newtonian. Appropriate boundary conditions are used. List the variables (unknowns) in the problem, and list the corresponding equations to be solved by the computer.

15–2C Write a brief (a few sentences) definition and description of each of the following, and provide example(s) if helpful: (*a*) computational domain, (*b*) mesh, (*c*) transport equation, (*d*) coupled equations.

15–3C What is the difference between a *node* and an *interval* and how are they related to *cells*? In Fig. P15–3C, how many nodes and how many intervals are on each edge?

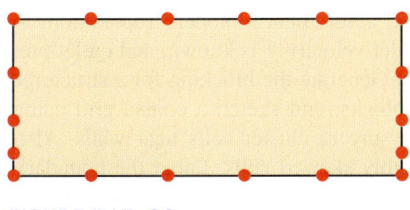

FIGURE P15–3C

15–4C For the two-dimensional computational domain of Fig. P15–3C, with the given node distribution, sketch a simple structured grid using four-sided cells and sketch a simple unstructured grid using three-sided cells. How many cells are in each? Discuss.

15–5C For the two-dimensional computational domain of Fig. P15–3C, with the given node distribution, sketch a simple structured grid using four-sided cells and sketch a simple unstructured polyhedral grid using at least one of each: 3-sided, 4-sided, and 5-sided cells. Try to avoid large skewness. Compare the cell count for each case and discuss your results.

15–6C Summarize the eight steps involved in a typical CFD analysis of a steady, laminar flow field.

15–7C Suppose you are using CFD to simulate flow through a duct in which there is a circular cylinder as in Fig. P15–7C. The duct is long, but to save computer resources you choose a computational domain in the vicinity of the cylinder only. Explain why the downstream edge of the computational domain should be further from the cylinder than the upstream edge.

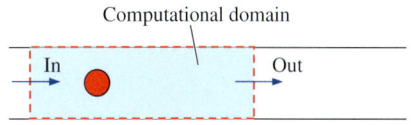

FIGURE P15–7C

15–8C Write a brief (a few sentences) discussion about the significance of each of the following in regards to an iterative CFD solution: (*a*) initial conditions, (*b*) residual, (*c*) iteration, (*d*) postprocessing.

15–9C Briefly discuss how each of the following is used by CFD codes to speed up the iteration process: (*a*) multigridding and (*b*) artificial time.

15–10C Of the boundary conditions discussed in this chapter, list all the boundary conditions that may be applied to the right edge of the two-dimensional computational domain sketched in Fig. P15–10C. Why can't the *other* boundary conditions be applied to this edge?

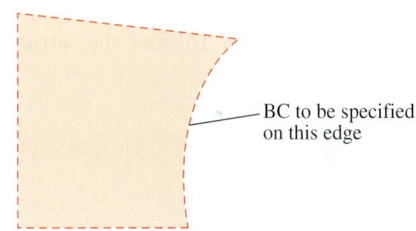

FIGURE P15–10C

15–11C What is the standard method to test for adequate grid resolution when using CFD?

15–12C What is the difference between a pressure inlet and a velocity inlet boundary condition? Explain why you cannot specify both pressure and velocity at a velocity inlet boundary condition or at a pressure inlet boundary condition.

15–13C An incompressible CFD code is used to simulate the flow of air through a two-dimensional rectangular channel (Fig. P15–13C). The computational domain consists of four blocks, as indicated. Flow enters block 4 from the upper right and exits block 1 to the left as shown. Inlet velocity V is known and outlet pressure P_{out} is also known. Label the boundary conditions that should be applied to every edge of every block of this computational domain.

---

* Problems designated by a "C" are concept questions, and students are encouraged to answer them all. Additional CFD problems are posted on the text website.

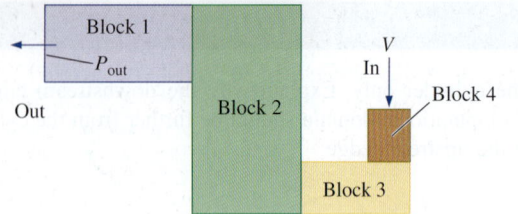

FIGURE P15–13C

15–14C Consider Prob. 15–13C again, except let the boundary condition on the common edge between blocks 1 and 2 be a *fan* with a specified pressure rise from right to left across the fan. Suppose an incompressible CFD code is run for both cases (with and without the fan). All else being equal, will the pressure at the inlet increase or decrease? Why? What will happen to the velocity at the outlet? Explain.

15–15C List six boundary conditions that are used with CFD to solve incompressible fluid flow problems. For each one, provide a brief description and give an example of how that boundary condition is used.

15–16 A CFD code is used to simulate flow over a two-dimensional airfoil at an angle of attack. A portion of the computational domain near the airfoil is outlined in Fig. P15–16 (the computational domain extends well beyond the region outlined by the dashed line). Sketch a coarse structured grid using four-sided cells and sketch a coarse unstructured grid using three-sided cells in the region shown. Be sure to cluster the cells where appropriate. Discuss the advantages and disadvantages of each grid type.

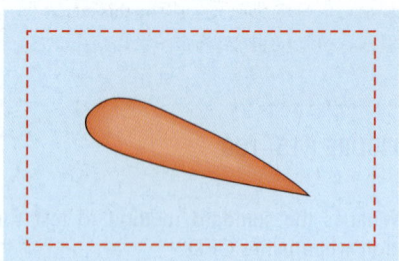

FIGURE P15–16

15–17 For the airfoil of Prob. 15–16, sketch a coarse hybrid grid and explain the advantages of such a grid.

15–18 An incompressible CFD code is used to simulate the flow of water through a two-dimensional rectangular channel in which there is a circular cylinder (Fig. P15–18). A time-averaged turbulent flow solution is generated using a turbulence model. Top–bottom symmetry about the cylinder is assumed. Flow enters from the left and exits to the right as shown. Inlet velocity V is known, and outlet pressure P_{out} is also known. Generate the blocking for a structured grid using four-sided blocks, and sketch a coarse grid using four-sided cells, being sure to cluster cells near walls. Also be careful to avoid highly skewed cells. Label the boundary conditions that should be applied to every edge of every block of your computational domain. (*Hint:* Six to seven blocks are sufficient.)

FIGURE P15–18

15–19 An incompressible CFD code is used to simulate the flow of gasoline through a two-dimensional rectangular channel in which there is a large circular settling chamber (Fig. P15–19). Flow enters from the left and exits to the right as shown. A time-averaged turbulent flow solution is generated using a turbulence model. Top–bottom symmetry is assumed. Inlet velocity V is known, and outlet pressure P_{out} is also known. Generate the blocking for a structured grid using four-sided blocks, and sketch a coarse grid using four-sided cells, being sure to cluster cells near walls. Also be careful to avoid highly skewed cells. Label the boundary conditions that should be applied to every edge of every block of your computational domain.

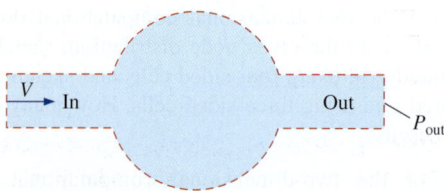

FIGURE P15–19

15–20 Redraw the structured multiblock grid of Fig. 15–12*b* for the case in which your CFD code can handle only *elementary blocks*. Renumber all the blocks and indicate how many *i*- and *j*-intervals are contained in each block. How many elementary blocks do you end up with? Add up all the cells, and verify that the total number of cells does not change.

15–21 Suppose your CFD code can handle *nonelementary blocks*. Combine as many blocks of Fig. 15–12*b* as you can. The only restriction is that in any one block, the number of *i*-intervals and the number of *j*-intervals must be constants. Show that you can create a structured grid with only *three* nonelementary blocks. Renumber all the blocks and indicate how many *i*- and *j*-intervals are contained in each block. Add up all the cells and verify that the total number of cells does not change.

15–22 A new heat exchanger is being designed with the goal of mixing the fluid downstream of each stage as thoroughly as possible. Anita comes up with a design whose cross section for one stage is sketched in Fig. P15–22. The geometry extends periodically up and down beyond the region shown here. She uses several dozen rectangular tubes inclined at a high angle of attack to ensure that the flow separates and mixes in the wakes. The performance of this geometry is to be tested using two-dimensional time-averaged CFD simulations with a turbulence model, and the results will be compared to those of competing geometries. Sketch the simplest possible computational domain that can be used to simulate this flow. Label and indicate all boundary conditions on your diagram. Discuss.

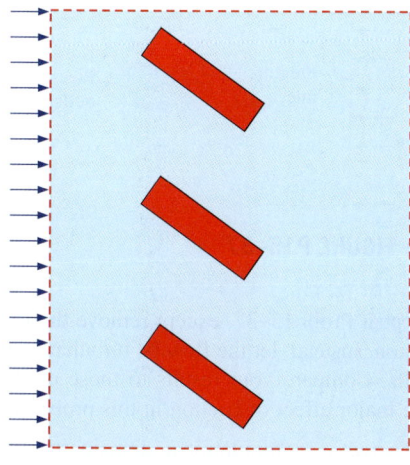

FIGURE P15–22

15–23 Sketch a coarse structured multiblock grid with four-sided elementary blocks and four-sided cells for the computational domain of Prob. 15–22.

15–24 Anita runs a CFD code using the computational domain and grid developed in Probs. 15–22 and 15–23. Unfortunately, the CFD code has a difficult time converging and Anita realizes that there is *reverse flow* at the outlet (far right edge of the computational domain). Explain why there is reverse flow, and discuss what Anita should do to correct the problem.

15–25 As a follow-up to the heat exchanger design of Prob. 15–22, suppose Anita's design is chosen based on the results of a preliminary single-stage CFD analysis. Now she is asked to simulate *two* stages of the heat exchanger. The second row of rectangular tubes is staggered and inclined oppositely to that of the first row to promote mixing (Fig. P15–25). The geometry extends periodically up and down beyond the region shown here. Sketch a computational domain that can be used to simulate this flow. Label and indicate all boundary conditions on your diagram. Discuss.

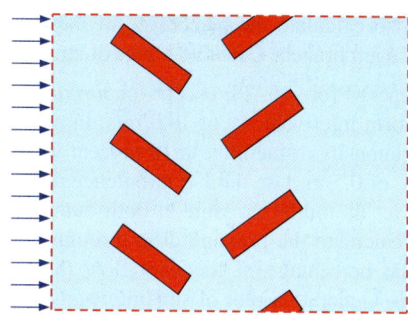

FIGURE P15–25

15–26 Sketch a structured multiblock grid with four-sided elementary blocks for the computational domain of Prob. 15–25. Each block is to have four-sided structured cells, but you do not have to sketch the grid, just the block topology. Try to make all the blocks as rectangular as possible to avoid highly skewed cells in the corners. Assume that the CFD code requires that the node distribution on periodic pairs of edges be identical (the two edges of a periodic pair are "linked" in the grid generation process). Also assume that the CFD code does not allow a block's edges to be split for application of boundary conditions.

General CFD Problems*

15–27 Consider the two-dimensional wye of Fig. P15–27. Dimensions are in meters, and the drawing is not to scale. Incompressible flow enters from the left, and splits into two parts. Generate three coarse grids, with identical node distributions on all edges of the computational domain: (*a*) structured multiblock grid, (*b*) unstructured triangular grid, and (*c*) unstructured quadrilateral grid. Compare the number of cells in each case and comment about the quality of the grid in each case.

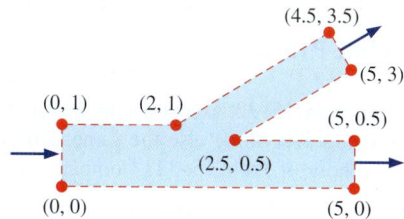

FIGURE P15–27

15–28 Choose one of the grids generated in Prob. 15–27, and run a CFD solution for laminar flow of air with a uniform inlet velocity of 0.02 m/s. Set the outlet pressure at both outlets to the same value, and calculate the pressure drop through

---

* These problems require CFD software, although not any particular brand. Students must do the following problems "from scratch," including generation of an appropriate mesh.

the wye. Also calculate the percentage of the inlet flow that goes out of each branch. Generate a plot of streamlines.

15–29 Repeat Prob. 15–28, except for *turbulent* flow of air with a uniform inlet velocity of 10.0 m/s. In addition, set the turbulence intensity at the inlet to 10 percent with a turbulent length scale of 0.5 m. Use the k-ε turbulence model with wall functions. Set the outlet pressure at both outlets to the same value, and calculate the pressure drop through the wye. Also calculate the percentage of the inlet flow that goes out of each branch. Generate a plot of streamlines. Compare results with those of laminar flow (Prob. 15–28).

15–30 Generate a computational domain to study the laminar boundary layer growing on a flat plate at Re = 10,000. Generate a very coarse mesh, and then continually refine the mesh until the solution becomes grid independent. Discuss.

15–31 Repeat Prob. 15–30, except for a *turbulent* boundary layer at Re = 10^6. Discuss.

15–32 Generate a computational domain to study ventilation in a room (Fig. P15–32). Specifically, generate a rectangular room with a velocity inlet in the ceiling to model the supply air, and a pressure outlet in the ceiling to model the return air. You may make a two-dimensional approximation for simplicity (the room is infinitely long in the direction normal to the page in Fig. P15–32). Use a structured rectangular grid. Plot streamlines and velocity vectors. Discuss.

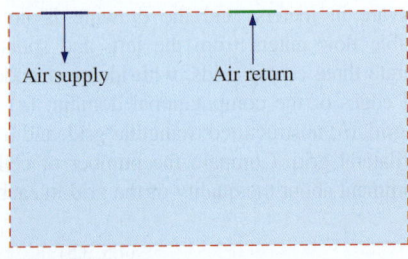

FIGURE P15–32

15–33 Repeat Prob. 15–32, except use an unstructured triangular grid, keeping everything else the same. Do you get the same results as those of Prob. 15–32? Compare and discuss.

15–34 Repeat Prob. 15–32, except move the supply and/or return vents to various locations in the ceiling. Compare and discuss.

15–35 Choose one of the room geometries of Probs. 15–32 and 15–34, and add the energy equation to the calculations. In particular, model a room with *air-conditioning*, by specifying the supply air as cool (T = 18°C), while the walls, floor, and ceiling are warm (T = 26°C). Adjust the supply air speed until the average temperature in the room is as close as possible to 22°C. How much ventilation (in terms of number of room air volume changes per hour) is required to cool this room to an average temperature of 22°C? Discuss.

15–36 Repeat Prob. 15–35, except create a *three-dimensional* room, with an air supply and an air return in the ceiling. Compare the two-dimensional results of Prob. 15–35 with the more realistic three-dimensional results of this problem. Discuss.

15–37 Generate a computational domain to study compressible flow of air through a converging nozzle with atmospheric pressure at the nozzle exit (Fig. P15–37). The nozzle walls may be approximated as inviscid (zero shear stress). Run several cases with various values of inlet pressure. How much inlet pressure is required to choke the flow? What happens if the inlet pressure is higher than this value? Discuss.

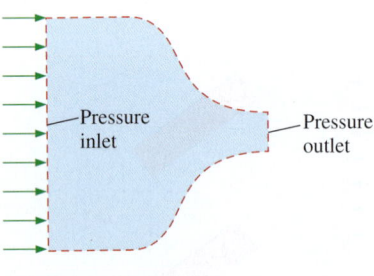

FIGURE P15–37

15–38 Repeat Prob. 15–37, except remove the inviscid flow approximation. Instead, let the flow be turbulent, with smooth, no-slip walls. Compare your results to those of Prob. 15–37. What is the major effect of friction in this problem? Discuss.

15–39 Generate a computational domain to study incompressible, laminar flow over a two-dimensional streamlined body (Fig. P15–39). Generate various body shapes, and calculate the drag coefficient for each shape. What is the smallest value of C_D that you can achieve? (*Note*: For fun, this problem can be turned into a contest between students. Who can generate the lowest-drag body shape?)

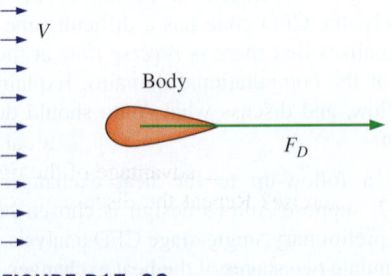

FIGURE P15–39

15–40 Repeat Prob. 15–39, except for an *axisymmetric*, rather than a two-dimensional, body. Compare to the two-dimensional case. For the same sectional slice shape, which has the lower drag coefficient? Discuss.

15–41 Repeat Prob. 15–40, except for *turbulent*, rather than laminar, flow. Compare to the laminar case. Which has the lower drag coefficient? Discuss.

15–42 Generate a computational domain to study Mach waves in a two-dimensional supersonic channel (Fig. P15–42). Specifically, the domain should consist of a simple rectangular channel with a supersonic inlet (Ma = 2.0), and with a very small bump on the lower wall. Using air with the inviscid flow approximation, generate a Mach wave, as sketched. Measure the Mach angle, and compare with theory (Chap. 12). Also discuss what happens when the Mach wave hits the opposite wall. Does it disappear, or does it reflect, and if so, what is the reflection angle? Discuss.

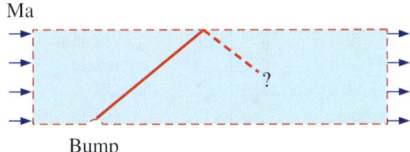

FIGURE P15–42

15–43 Repeat Prob. 15–42, except for several values of the Mach number, ranging from 1.10 to 3.0. Plot the calculated Mach angle as a function of Mach number and compare to the theoretical Mach angle (Chap. 12). Discuss.

Review Problems

15–44C For each statement, choose whether the statement is true or false, and discuss your answer briefly:
(*a*) The physical validity of a CFD solution always improves as the grid is refined.
(*b*) The *x*-component of the Navier–Stokes equation is an example of a transport equation.
(*c*) For the same number of nodes in a two-dimensional mesh, a structured grid typically has fewer cells than an unstructured triangular grid.
(*d*) A time-averaged turbulent flow CFD solution is only as good as the turbulence model used in the calculations.

15–45C In Prob. 15–19 we take advantage of top–bottom symmetry when constructing our computational domain and grid. Why can't we also take advantage of the right–left symmetry in this exercise? Repeat the discussion for the case of potential flow.

15–46C Gerry creates the computational domain sketched in Fig. P15–46C to simulate flow through a sudden contraction in a two-dimensional duct. He is interested in the time-averaged pressure drop and the minor loss coefficient created by the sudden contraction. Gerry generates a grid and calculates the flow with a CFD code, assuming steady, turbulent, incompressible flow (with a turbulence model).

(*a*) Discuss one way that Gerry could improve his computational domain and grid so that he would get the *same* results in approximately half the computer time.
(*b*) There may be a fundamental flaw in how Gerry has set up his computational domain. What is it? Discuss what should be different about Gerry's setup.

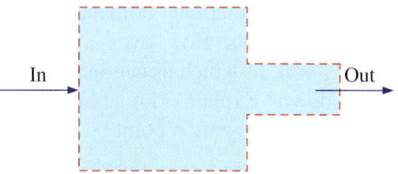

FIGURE P15–46C

15–47C Think about modern high-speed, large-memory computer systems. What feature of such computers lends itself nicely to the solution of CFD problems using a multiblock grid with approximately equal numbers of cells in each individual block? Discuss.

15–48C What is the difference between *multigridding* and *multiblocking*? Discuss how each may be used to speed up a CFD calculation. Can these two be applied together?

15–49C Suppose you have a fairly complex geometry and a CFD code that can handle unstructured grids with triangular cells. Your grid generation code can create an unstructured grid very quickly. Give some reasons why it might be wiser to take the time to create a multiblock structured grid instead. In other words, is it worth the effort? Discuss.

15–50 Generate a computational domain and grid, and calculate flow through the single-stage heat exchanger of Prob. 15–22, with the heating elements set at a 45° angle of attack with respect to horizontal. Set the inlet air temperature to 20°C, and the wall temperature of the heating elements to 120°C. Calculate the average air temperature at the outlet.

15–51 Repeat the calculations of Prob. 15–50 for several angles of attack of the heating elements, from 0 (horizontal) to 90° (vertical). Use identical inlet conditions and wall conditions for each case. Which angle of attack provides the most heat transfer to the air? Specifically, which angle of attack yields the highest average outlet temperature?

15–52 Generate a computational domain and grid, and calculate flow through the two-stage heat exchanger of Prob. 15–25, with the heating elements of the first stage set at a 45° angle of attack with respect to horizontal, and those of the second stage set to an angle of attack of −45°. Set the inlet air temperature to 20°C, and the wall temperature of the heating elements to 120°C. Calculate the average air temperature at the outlet.

15–53 Repeat the calculations of Prob. 15–52 for several angles of attack of the heating elements, from 0 (horizontal) to 90° (vertical). Use identical inlet conditions and wall conditions for each case. Note that the second stage of heating elements should always be set to an angle of attack that is the negative of that of the first stage. Which angle of attack provides the most heat transfer to the air? Specifically, which angle of attack yields the highest average outlet temperature? Is this the same angle as calculated for the single-stage heat exchanger of Prob. 15–51? Discuss.

15–54 Generate a computational domain and grid, and calculate stationary turbulent flow over a spinning circular cylinder (Fig. P15–54). In which direction is the side force on the body—up or down? Explain. Plot streamlines in the flow. Where is the upstream stagnation point?

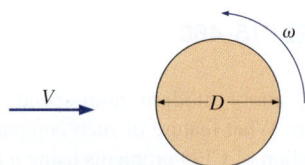

FIGURE P15–54

15–55 For the spinning cylinder of Fig. P15–54, generate a dimensionless parameter for rotational speed relative to free-stream speed (combine variables ω, D, and V into a nondimensional Pi group). Repeat the calculations of Prob. 15–54 for several values of angular velocity ω. Use identical inlet conditions for each case. Plot lift and drag coefficients as functions of your dimensionless parameter. Discuss.

15–56 Consider the flow of air into a two-dimensional slot along the floor of a large room, where the floor is coincident with the x-axis (Fig. P15–56). Generate an appropriate computational domain and grid. Using the inviscid flow approximation in the CFD code, calculate vertical velocity component v as a function of distance away from the slot along the y-axis. Compare with the potential flow results of Chap. 10 for flow into a line sink. Discuss.

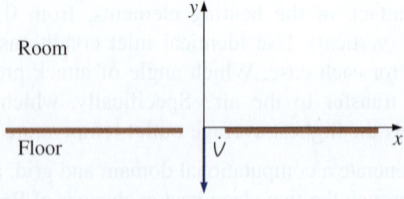

FIGURE P15–56

15–57 For the slot flow of Prob. 15–56, change to laminar flow instead of inviscid flow, and recompute the flow field. Compare your results to the inviscid flow case and to the potential flow case of Chap. 10. Plot contours of vorticity. Where is the irrotational flow approximation appropriate? Discuss.

15–58 Generate a computational domain and grid, and calculate the flow of air into a two-dimensional vacuum cleaner inlet (Fig. P15–58), using the inviscid flow approximation in the CFD code. Compare your results with those predicted in Chap. 10 for potential flow. Discuss.

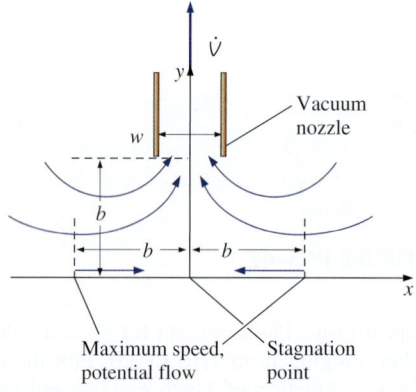

FIGURE P15–58

15–59 For the vacuum cleaner of Prob. 15–58, change to laminar flow instead of inviscid flow, and recompute the flow field. Compare your results to the inviscid flow case and to the potential flow case of Chap. 10. Discuss.

APPENDIX

PROPERTY TABLES AND CHARTS*

TABLE A–1   Molar Mass, Gas Constant, and Ideal-Gas Specfic Heats of Some Substances   940
TABLE A–2   Boiling and Freezing Point Properties   941
TABLE A–3   Properties of Saturated Water   942
TABLE A–4   Properties of Saturated Refrigerant-134a   943
TABLE A–5   Properties of Saturated Ammonia   944
TABLE A–6   Properties of Saturated Propane   945
TABLE A–7   Properties of Liquids   946
TABLE A–8   Properties of Liquid Metals   947
TABLE A–9   Properties of Air at 1 atm Pressure   948
TABLE A–10   Properties of Gases at 1 atm Pressure   949
TABLE A–11   Properties of the Atmosphere at High Altitude   951
FIGURE A–12   The Moody Chart for the Friction Factor for Fully Developed Flow in Circular Pipes   952
TABLE A–13   One-Dimensional Isentropic Compressible Flow Functions for an Ideal Gas with $k = 1.4$ 953
TABLE A–14   One-Dimensional Normal Shock Functions for an Ideal Gas with $k = 1.4$ 954
TABLE A–15   Rayleigh Flow Functions for an Ideal Gas with $k = 1.4$ 955
TABLE A–16   Fanno Flow Functions for an Ideal Gas with $k = 1.4$ 956

---

* Most properties in the tables are obtained from the property database of EES, and the original sources are listed under the tables. Properties are often listed to more significant digits than the claimed accuracy for the purpose of minimizing accumulated round-off error in hand calculations and ensuring a close match with the results obtained with EES.

TABLE A–1

Molar mass, gas constant, and ideal-gas specfic heats of some substances

Substance	Molar Mass M, kg/kmol	Gas Constant R, kJ/kg·K*	c_p, kJ/kg·K	c_v, kJ/kg·K	$k = c_p/c_v$
Air	28.97	0.2870	1.005	0.7180	1.400
Ammonia, NH_3	17.03	0.4882	2.093	1.605	1.304
Argon, Ar	39.95	0.2081	0.5203	0.3122	1.667
Bromine, Br_2	159.81	0.05202	0.2253	0.1732	1.300
Isobutane, C_4H_{10}	58.12	0.1430	1.663	1.520	1.094
n-Butane, C_4H_{10}	58.12	0.1430	1.694	1.551	1.092
Carbon dioxide, CO_2	44.01	0.1889	0.8439	0.6550	1.288
Carbon monoxide, CO	28.01	0.2968	1.039	0.7417	1.400
Chlorine, Cl_2	70.905	0.1173	0.4781	0.3608	1.325
Chlorodifluoromethane (R-22), $CHClF_2$	86.47	0.09615	0.6496	0.5535	1.174
Ethane, C_2H_6	30.070	0.2765	1.744	1.468	1.188
Ethylene, C_2H_4	28.054	0.2964	1.527	1.231	1.241
Fluorine, F_2	38.00	0.2187	0.8237	0.6050	1.362
Helium, He	4.003	2.077	5.193	3.116	1.667
n-Heptane, C_7H_{16}	100.20	0.08297	1.649	1.566	1.053
n-Hexane, C_6H_{14}	86.18	0.09647	1.654	1.558	1.062
Hydrogen, H_2	2.016	4.124	14.30	10.18	1.405
Krypton, Kr	83.80	0.09921	0.2480	0.1488	1.667
Methane, CH_4	16.04	0.5182	2.226	1.708	1.303
Neon, Ne	20.183	0.4119	1.030	0.6180	1.667
Nitrogen, N_2	28.01	0.2968	1.040	0.7429	1.400
Nitric oxide, NO	30.006	0.2771	0.9992	0.7221	1.384
Nitrogen dioxide, NO_2	46.006	0.1889	0.8060	0.6171	1.306
Oxygen, O_2	32.00	0.2598	0.9180	0.6582	1.395
n-Pentane, C_5H_{12}	72.15	0.1152	1.664	1.549	1.074
Propane, C_3H_8	44.097	0.1885	1.669	1.480	1.127
Propylene, C_3H_6	42.08	0.1976	1.531	1.333	1.148
Steam, H_2O	18.015	0.4615	1.865	1.403	1.329
Sulfur dioxide, SO_2	64.06	0.1298	0.6228	0.4930	1.263
Tetrachloromethane, CCl_4	153.82	0.05405	0.5415	0.4875	1.111
Tetrafluoroethane (R-134a), $C_2H_2F_4$	102.03	0.08149	0.8334	0.7519	1.108
Trifluoroethane (R-143a), $C_2H_3F_3$	84.04	0.09893	0.9291	0.8302	1.119
Xenon, Xe	131.30	0.06332	0.1583	0.09499	1.667

Specific Heat Data at 25°C

* The unit kJ/kg·K is equivalent to kPa·m³/kg·K. The gas constant is calculated from $R = R_u/M$, where R_u = 8.31447 kJ/kmol·K is the universal gas constant and M is the molar mass.

Source: Specific heat values are obtained primarily from the property routines prepared by The National Institute of Standards and Technology (NIST), Gaithersburg, MD.

TABLE A–2
Boiling and freezing point properties

Substance	Normal Boiling Point, °C	Latent Heat of Vaporization h_{fg}, kJ/kg	Freezing Point, °C	Latent Heat of Fusion h_{if}, kJ/kg	Temperature, °C	Density ρ, kg/m^3	Specific Heat c_p, kJ/kg·K
Ammonia	−33.3	1357	−77.7	322.4	−33.3	682	4.43
					−20	665	4.52
					0	639	4.60
					25	602	4.80
Argon	−185.9	161.6	−189.3	28	−185.6	1394	1.14
Benzene	80.2	394	5.5	126	20	879	1.72
Brine (20% sodium chloride by mass)	103.9	—	−17.4	—	20	1150	3.11
n-Butane	−0.5	385.2	−138.5	80.3	−0.5	601	2.31
Carbon dioxide	−78.4*	230.5 (at 0°C)	−56.6		0	298	0.59
Ethanol	78.2	838.3	−114.2	109	25	783	2.46
Ethyl alcohol	78.6	855	−156	108	20	789	2.84
Ethylene glycol	198.1	800.1	−10.8	181.1	20	1109	2.84
Glycerine	179.9	974	18.9	200.6	20	1261	2.32
Helium	−268.9	22.8	—	—	−268.9	146.2	22.8
Hydrogen	−252.8	445.7	−259.2	59.5	−252.8	70.7	10.0
Isobutane	−11.7	367.1	−160	105.7	−11.7	593.8	2.28
Kerosene	204–293	251	−24.9	—	20	820	2.00
Mercury	356.7	294.7	−38.9	11.4	25	13,560	0.139
Methane	−161.5	510.4	−182.2	58.4	−161.5	423	3.49
					−100	301	5.79
Methanol	64.5	1100	−97.7	99.2	25	787	2.55
Nitrogen	−195.8	198.6	−210	25.3	−195.8	809	2.06
					−160	596	2.97
Octane	124.8	306.3	−57.5	180.7	20	703	2.10
Oil (light)					25	910	1.80
Oxygen	−183	212.7	−218.8	13.7	−183	1141	1.71
Petroleum	—	230–384			20	640	2.0
Propane	−42.1	427.8	−187.7	80.0	−42.1	581	2.25
					0	529	2.53
					50	449	3.13
Refrigerant-134a	−26.1	216.8	−96.6	—	−50	1443	1.23
					−26.1	1374	1.27
					0	1295	1.34
					25	1207	1.43
Water	100	2257	0.0	333.7	0	1000	4.22
					25	997	4.18
					50	988	4.18
					75	975	4.19
					100	958	4.22

*Sublimation temperature. (At pressures below the triple-point pressure of 518 kPa, carbon dioxide exists as a solid or gas. Also, the freezing-point temperature of carbon dioxide is the triple-point temperature of −56.5°C.)

TABLE A-3

Properties of saturated water

Temp. T, °C	Saturation Pressure P_{sat}, kPa	Density ρ, kg/m³ Liquid	Density ρ, kg/m³ Vapor	Enthalpy of Vaporization h_{fg}, kJ/kg	Specific Heat c_p, J/kg·K Liquid	Specific Heat c_p, J/kg·K Vapor	Thermal Conductivity k, W/m·K Liquid	Thermal Conductivity k, W/m·K Vapor	Dynamic Viscosity μ, kg/m·s Liquid	Dynamic Viscosity μ, kg/m·s Vapor	Prandtl Number Pr Liquid	Prandtl Number Pr Vapor	Volume Expansion Coefficient β, 1/K Liquid	Surface Tension, N/m Liquid
0.01	0.6113	999.8	0.0048	2501	4217	1854	0.561	0.0171	1.792×10^{-3}	0.922×10^{-5}	13.5	1.00	-0.068×10^{-3}	0.0756
5	0.8721	999.9	0.0068	2490	4205	1857	0.571	0.0173	1.519×10^{-3}	0.934×10^{-5}	11.2	1.00	0.015×10^{-3}	0.0749
10	1.2276	999.7	0.0094	2478	4194	1862	0.580	0.0176	1.307×10^{-3}	0.946×10^{-5}	9.45	1.00	0.733×10^{-3}	0.0742
15	1.7051	999.1	0.0128	2466	4186	1863	0.589	0.0179	1.138×10^{-3}	0.959×10^{-5}	8.09	1.00	0.138×10^{-3}	0.0735
20	2.339	998.0	0.0173	2454	4182	1867	0.598	0.0182	1.002×10^{-3}	0.973×10^{-5}	7.01	1.00	0.195×10^{-3}	0.0727
25	3.169	997.0	0.0231	2442	4180	1870	0.607	0.0186	0.891×10^{-3}	0.987×10^{-5}	6.14	1.00	0.247×10^{-3}	0.0720
30	4.246	996.0	0.0304	2431	4178	1875	0.615	0.0189	0.798×10^{-3}	1.001×10^{-5}	5.42	1.00	0.294×10^{-3}	0.0712
35	5.628	994.0	0.0397	2419	4178	1880	0.623	0.0192	0.720×10^{-3}	1.016×10^{-5}	4.83	1.00	0.337×10^{-3}	0.0704
40	7.384	992.1	0.0512	2407	4179	1885	0.631	0.0196	0.653×10^{-3}	1.031×10^{-5}	4.32	1.00	0.377×10^{-3}	0.0696
45	9.593	990.1	0.0655	2395	4180	1892	0.637	0.0200	0.596×10^{-3}	1.046×10^{-5}	3.91	1.00	0.415×10^{-3}	0.0688
50	12.35	988.1	0.0831	2383	4181	1900	0.644	0.0204	0.547×10^{-3}	1.062×10^{-5}	3.55	1.00	0.451×10^{-3}	0.0679
55	15.76	985.2	0.1045	2371	4183	1908	0.649	0.0208	0.504×10^{-3}	1.077×10^{-5}	3.25	1.00	0.484×10^{-3}	0.0671
60	19.94	983.3	0.1304	2359	4185	1916	0.654	0.0212	0.467×10^{-3}	1.093×10^{-5}	2.99	1.00	0.517×10^{-3}	0.0662
65	25.03	980.4	0.1614	2346	4187	1926	0.659	0.0216	0.433×10^{-3}	1.110×10^{-5}	2.75	1.00	0.548×10^{-3}	0.0654
70	31.19	977.5	0.1983	2334	4190	1936	0.663	0.0221	0.404×10^{-3}	1.126×10^{-5}	2.55	1.00	0.578×10^{-3}	0.0645
75	38.58	974.7	0.2421	2321	4193	1948	0.667	0.0225	0.378×10^{-3}	1.142×10^{-5}	2.38	1.00	0.607×10^{-3}	0.0636
80	47.39	971.8	0.2935	2309	4197	1962	0.670	0.0230	0.355×10^{-3}	1.159×10^{-5}	2.22	1.00	0.653×10^{-3}	0.0627
85	57.83	968.1	0.3536	2296	4201	1977	0.673	0.0235	0.333×10^{-3}	1.176×10^{-5}	2.08	1.00	0.670×10^{-3}	0.0617
90	70.14	965.3	0.4235	2283	4206	1993	0.675	0.0240	0.315×10^{-3}	1.193×10^{-5}	1.96	1.00	0.702×10^{-3}	0.0608
95	84.55	961.5	0.5045	2270	4212	2010	0.677	0.0246	0.297×10^{-3}	1.210×10^{-5}	1.85	1.00	0.716×10^{-3}	0.0599
100	101.33	957.9	0.5978	2257	4217	2029	0.679	0.0251	0.282×10^{-3}	1.227×10^{-5}	1.75	1.00	0.750×10^{-3}	0.0589
110	143.27	950.6	0.8263	2230	4229	2071	0.682	0.0262	0.255×10^{-3}	1.261×10^{-5}	1.58	1.00	0.798×10^{-3}	0.0570
120	198.53	943.4	1.121	2203	4244	2120	0.683	0.0275	0.232×10^{-3}	1.296×10^{-5}	1.44	1.00	0.858×10^{-3}	0.0550
130	270.1	934.6	1.496	2174	4263	2177	0.684	0.0288	0.213×10^{-3}	1.330×10^{-5}	1.33	1.01	0.913×10^{-3}	0.0529
140	361.3	921.7	1.965	2145	4286	2244	0.683	0.0301	0.197×10^{-3}	1.365×10^{-5}	1.24	1.02	0.970×10^{-3}	0.0509
150	475.8	916.6	2.546	2114	4311	2314	0.682	0.0316	0.183×10^{-3}	1.399×10^{-5}	1.16	1.02	1.025×10^{-3}	0.0487
160	617.8	907.4	3.256	2083	4340	2420	0.680	0.0331	0.170×10^{-3}	1.434×10^{-5}	1.09	1.05	1.145×10^{-3}	0.0466
170	791.7	897.7	4.119	2050	4370	2490	0.677	0.0347	0.160×10^{-3}	1.468×10^{-5}	1.03	1.05	1.178×10^{-3}	0.0444
180	1,002.1	887.3	5.153	2015	4410	2590	0.673	0.0364	0.150×10^{-3}	1.502×10^{-5}	0.983	1.07	1.210×10^{-3}	0.0422
190	1,254.4	876.4	6.388	1979	4460	2710	0.669	0.0382	0.142×10^{-3}	1.537×10^{-5}	0.947	1.09	1.280×10^{-3}	0.0399
200	1,553.8	864.3	7.852	1941	4500	2840	0.663	0.0401	0.134×10^{-3}	1.571×10^{-5}	0.910	1.11	1.350×10^{-3}	0.0377
220	2,318	840.3	11.60	1859	4610	3110	0.650	0.0442	0.122×10^{-3}	1.641×10^{-5}	0.865	1.15	1.520×10^{-3}	0.0331
240	3,344	813.7	16.73	1767	4760	3520	0.632	0.0487	0.111×10^{-3}	1.712×10^{-5}	0.836	1.24	1.720×10^{-3}	0.0284
260	4,688	783.7	23.69	1663	4970	4070	0.609	0.0540	0.102×10^{-3}	1.788×10^{-5}	0.832	1.35	2.000×10^{-3}	0.0237
280	6,412	750.8	33.15	1544	5280	4835	0.581	0.0605	0.094×10^{-3}	1.870×10^{-5}	0.854	1.49	2.380×10^{-3}	0.0190
300	8,581	713.8	46.15	1405	5750	5980	0.548	0.0695	0.086×10^{-3}	1.965×10^{-5}	0.902	1.69	2.950×10^{-3}	0.0144
320	11,274	667.1	64.57	1239	6540	7900	0.509	0.0836	0.078×10^{-3}	2.084×10^{-5}	1.00	1.97		0.0099
340	14,586	610.5	92.62	1028	8240	11,870	0.469	0.110	0.070×10^{-3}	2.255×10^{-5}	1.23	2.43		0.0056
360	18,651	528.3	144.0	720	14,690	25,800	0.427	0.178	0.060×10^{-3}	2.571×10^{-5}	2.06	3.73		0.0019
374.14	22,090	317.0	317.0	0	—	—	—	—	0.043×10^{-3}	4.313×10^{-5}	—	—		0

Note 1: Kinematic viscosity ν and thermal diffusivity α can be calculated from their definitions, $\nu = \mu/\rho$ and $\alpha = k/\rho c_p = \nu/\text{Pr}$. The temperatures 0.01°C, 100°C, and 374.14°C are the triple-, boiling-, and critical-point temperatures of water, respectively. The properties listed above (except the vapor density) can be used at any pressure with negligible error except at temperatures near the critical-point value.

Note 2: The unit kJ/kg·°C for specific heat is equivalent to kJ/kg·K, and the unit W/m·°C for thermal conductivity is equivalent to W/m·K.

Source: Viscosity and thermal conductivity data are from J. V. Sengers and J. T. R. Watson, *Journal of Physical and Chemical Reference Data* 15 (1986), pp. 1291–1322. Other data are obtained from various sources or calculated.

TABLE A–4

Properties of saturated refrigerant-134a

Temp. T, °C	Saturation Pressure P, kPa	Density ρ, kg/m³ Liquid	Density ρ, kg/m³ Vapor	Enthalpy of Vaporization h_{fg}, kJ/kg	Specific Heat c_p, J/kg·K Liquid	Specific Heat c_p, J/kg·K Vapor	Thermal Conductivity k, W/m·K Liquid	Thermal Conductivity k, W/m·K Vapor	Dynamic Viscosity μ, kg/m·s Liquid	Dynamic Viscosity μ, kg/m·s Vapor	Prandtl Number Pr Liquid	Prandtl Number Pr Vapor	Volume Expansion Coefficient β, 1/K Liquid	Surface Tension, N/m Liquid
−40	51.2	1418	2.773	225.9	1254	748.6	0.1101	0.00811	4.878×10^{-4}	2.550×10^{-6}	5.558	0.235	0.00205	0.01760
−35	66.2	1403	3.524	222.7	1264	764.1	0.1084	0.00862	4.509×10^{-4}	3.003×10^{-6}	5.257	0.266	0.00209	0.01682
−30	84.4	1389	4.429	219.5	1273	780.2	0.1066	0.00913	4.178×10^{-4}	3.504×10^{-6}	4.992	0.299	0.00215	0.01604
−25	106.5	1374	5.509	216.3	1283	797.2	0.1047	0.00963	3.882×10^{-4}	4.054×10^{-6}	4.757	0.335	0.00220	0.01527
−20	132.8	1359	6.787	213.0	1294	814.9	0.1028	0.01013	3.614×10^{-4}	4.651×10^{-6}	4.548	0.374	0.00227	0.01451
−15	164.0	1343	8.288	209.5	1306	833.5	0.1009	0.01063	3.371×10^{-4}	5.295×10^{-6}	4.363	0.415	0.00233	0.01376
−10	200.7	1327	10.04	206.0	1318	853.1	0.0989	0.01112	3.150×10^{-4}	5.982×10^{-6}	4.198	0.459	0.00241	0.01302
−5	243.5	1311	12.07	202.4	1330	873.8	0.0968	0.01161	2.947×10^{-4}	6.709×10^{-6}	4.051	0.505	0.00249	0.01229
0	293.0	1295	14.42	198.7	1344	895.6	0.0947	0.01210	2.761×10^{-4}	7.471×10^{-6}	3.919	0.553	0.00258	0.01156
5	349.9	1278	17.12	194.8	1358	918.7	0.0925	0.01259	2.589×10^{-4}	8.264×10^{-6}	3.802	0.603	0.00269	0.01084
10	414.9	1261	20.22	190.8	1374	943.2	0.0903	0.01308	2.430×10^{-4}	9.081×10^{-6}	3.697	0.655	0.00280	0.01014
15	488.7	1244	23.75	186.6	1390	969.4	0.0880	0.01357	2.281×10^{-4}	9.915×10^{-6}	3.604	0.708	0.00293	0.00944
20	572.1	1226	27.77	182.3	1408	997.6	0.0856	0.01406	2.142×10^{-4}	1.075×10^{-5}	3.521	0.763	0.00307	0.00876
25	665.8	1207	32.34	177.8	1427	1028	0.0833	0.01456	2.012×10^{-4}	1.160×10^{-5}	3.448	0.819	0.00324	0.00808
30	770.6	1188	37.53	173.1	1448	1061	0.0808	0.01507	1.888×10^{-4}	1.244×10^{-5}	3.383	0.877	0.00342	0.00742
35	887.5	1168	43.41	168.2	1471	1098	0.0783	0.01558	1.772×10^{-4}	1.327×10^{-5}	3.328	0.935	0.00364	0.00677
40	1017.1	1147	50.08	163.0	1498	1138	0.0757	0.01610	1.660×10^{-4}	1.408×10^{-5}	3.285	0.995	0.00390	0.00613
45	1160.5	1125	57.66	157.6	1529	1184	0.0731	0.01664	1.554×10^{-4}	1.486×10^{-5}	3.253	1.058	0.00420	0.00550
50	1318.6	1102	66.27	151.8	1566	1237	0.0704	0.01720	1.453×10^{-4}	1.562×10^{-5}	3.231	1.123	0.00456	0.00489
55	1492.3	1078	76.11	145.7	1608	1298	0.0676	0.01777	1.355×10^{-4}	1.634×10^{-5}	3.223	1.193	0.00500	0.00429
60	1682.8	1053	87.38	139.1	1659	1372	0.0647	0.01838	1.260×10^{-4}	1.704×10^{-5}	3.229	1.272	0.00554	0.00372
65	1891.0	1026	100.4	132.1	1722	1462	0.0618	0.01902	1.167×10^{-4}	1.771×10^{-5}	3.255	1.362	0.00624	0.00315
70	2118.2	996.2	115.6	124.4	1801	1577	0.0587	0.01972	1.077×10^{-4}	1.839×10^{-5}	3.307	1.471	0.00716	0.00261
75	2365.8	964	133.6	115.9	1907	1731	0.0555	0.02048	9.891×10^{-5}	1.908×10^{-5}	3.400	1.612	0.00843	0.00209
80	2635.2	928.2	155.3	106.4	2056	1948	0.0521	0.02133	9.011×10^{-5}	1.982×10^{-5}	3.558	1.810	0.01031	0.00160
85	2928.2	887.1	182.3	95.4	2287	2281	0.0484	0.02233	8.124×10^{-5}	2.071×10^{-5}	3.837	2.116	0.01336	0.00114
90	3246.9	837.7	217.8	82.2	2701	2865	0.0444	0.02357	7.203×10^{-5}	2.187×10^{-5}	4.385	2.658	0.01911	0.00071
95	3594.1	772.5	269.3	64.9	3675	4144	0.0396	0.02544	6.190×10^{-5}	2.370×10^{-5}	5.746	3.862	0.03343	0.00033
100	3975.1	651.7	376.3	33.9	7959	8785	0.0322	0.02989	4.765×10^{-5}	2.833×10^{-5}	11.77	8.326	0.10047	0.00004

Note 1: Kinematic viscosity ν and thermal diffusivity α can be calculated from their definitions, $\nu = \mu/\rho$ and $\alpha = k/\rho c_p = \nu/\text{Pr}$. The properties listed here (except the vapor density) can be used at any pressures with negligible error except at temperatures near the critical-point value.

Note 2: The unit kJ/kg·°C for specific heat is equivalent to kJ/kg·K, and the unit W/m·°C for thermal conductivity is equivalent to W/m·K.

Source: Data generated from the EES software developed by S. A. Klein and F. L. Alvarado. Original sources: R. Tillner-Roth and H. D. Baehr, "An International Standard Formulation for the Thermodynamic Properties of 1,1,1,2-Tetrafluoroethane (HFC-134a) for Temperatures from 170 K to 455 K and Pressures up to 70 MPa," *J. Phys. Chem, Ref. Data*, Vol. 23, No. 5, 1994; M. J. Assael, N. K. Dalaouti, A. A. Griva, and J. H. Dymond, "Viscosity and Thermal Conductivity of Halogenated Methane and Ethane Refrigerants," *IJR*, Vol. 22, pp. 525–535, 1999; NIST REFPROP 6 program (M. O. McLinden, S. A. Klein, E. W. Lemmon, and A. P. Peskin, Physical and Chemical Properties Division, National Institute of Standards and Technology, Boulder, CO 80303, 1995).

TABLE A-5

Properties of saturated ammonia

Temp. T, °C	Saturation Pressure P, kPa	Density ρ, kg/m³ Liquid	Density ρ, kg/m³ Vapor	Enthalpy of Vaporization h_{fg}, kJ/kg	Specific Heat c_p, J/kg·K Liquid	Specific Heat c_p, J/kg·K Vapor	Thermal Conductivity k, W/m·K Liquid	Thermal Conductivity k, W/m·K Vapor	Dynamic Viscosity μ, kg/m·s Liquid	Dynamic Viscosity μ, kg/m·s Vapor	Prandtl Number Pr Liquid	Prandtl Number Pr Vapor	Volume Expansion Coefficient β, 1/K Liquid	Surface Tension, N/m Liquid
−40	71.66	690.2	0.6435	1389	4414	2242	—	0.01792	2.926×10^{-4}	7.957×10^{-6}	—	0.9955	0.00176	0.03565
−30	119.4	677.8	1.037	1360	4465	2322	—	0.01898	2.630×10^{-4}	8.311×10^{-6}	—	1.017	0.00185	0.03341
−25	151.5	671.5	1.296	1345	4489	2369	0.5968	0.01957	2.492×10^{-4}	8.490×10^{-6}	1.875	1.028	0.00190	0.03229
−20	190.1	665.1	1.603	1329	4514	2420	0.5853	0.02015	2.361×10^{-4}	8.669×10^{-6}	1.821	1.041	0.00194	0.03118
−15	236.2	658.6	1.966	1313	4538	2476	0.5737	0.02075	2.236×10^{-4}	8.851×10^{-6}	1.769	1.056	0.00199	0.03007
−10	290.8	652.1	2.391	1297	4564	2536	0.5621	0.02138	2.117×10^{-4}	9.034×10^{-6}	1.718	1.072	0.00205	0.02896
−5	354.9	645.4	2.886	1280	4589	2601	0.5505	0.02203	2.003×10^{-4}	9.218×10^{-6}	1.670	1.089	0.00210	0.02786
0	429.6	638.6	3.458	1262	4617	2672	0.5390	0.02270	1.896×10^{-4}	9.405×10^{-6}	1.624	1.107	0.00216	0.02676
5	516	631.7	4.116	1244	4645	2749	0.5274	0.02341	1.794×10^{-4}	9.593×10^{-6}	1.580	1.126	0.00223	0.02566
10	615.3	624.6	4.870	1226	4676	2831	0.5158	0.02415	1.697×10^{-4}	9.784×10^{-6}	1.539	1.147	0.00230	0.02457
15	728.8	617.5	5.729	1206	4709	2920	0.5042	0.02492	1.606×10^{-4}	9.978×10^{-6}	1.500	1.169	0.00237	0.02348
20	857.8	610.2	6.705	1186	4745	3016	0.4927	0.02573	1.519×10^{-4}	1.017×10^{-5}	1.463	1.193	0.00245	0.02240
25	1003	602.8	7.809	1166	4784	3120	0.4811	0.02658	1.438×10^{-4}	1.037×10^{-5}	1.430	1.218	0.00254	0.02132
30	1167	595.2	9.055	1144	4828	3232	0.4695	0.02748	1.361×10^{-4}	1.057×10^{-5}	1.399	1.244	0.00264	0.02024
35	1351	587.4	10.46	1122	4877	3354	0.4579	0.02843	1.288×10^{-4}	1.078×10^{-5}	1.372	1.272	0.00275	0.01917
40	1555	579.4	12.03	1099	4932	3486	0.4464	0.02943	1.219×10^{-4}	1.099×10^{-5}	1.347	1.303	0.00287	0.01810
45	1782	571.3	13.8	1075	4993	3631	0.4348	0.03049	1.155×10^{-4}	1.121×10^{-5}	1.327	1.335	0.00301	0.01704
50	2033	562.9	15.78	1051	5063	3790	0.4232	0.03162	1.094×10^{-4}	1.143×10^{-5}	1.310	1.371	0.00316	0.01598
55	2310	554.2	18.00	1025	5143	3967	0.4116	0.03283	1.037×10^{-4}	1.166×10^{-5}	1.297	1.409	0.00334	0.01493
60	2614	545.2	20.48	997.4	5234	4163	0.4001	0.03412	9.846×10^{-5}	1.189×10^{-5}	1.288	1.452	0.00354	0.01389
65	2948	536.0	23.26	968.9	5340	4384	0.3885	0.03550	9.347×10^{-5}	1.213×10^{-5}	1.285	1.499	0.00377	0.01285
70	3312	526.3	26.39	939.0	5463	4634	0.3769	0.03700	8.879×10^{-5}	1.238×10^{-5}	1.287	1.551	0.00404	0.01181
75	3709	516.2	29.90	907.5	5608	4923	0.3653	0.03862	8.440×10^{-5}	1.264×10^{-5}	1.296	1.612	0.00436	0.01079
80	4141	505.7	33.87	874.1	5780	5260	0.3538	0.04038	8.030×10^{-5}	1.292×10^{-5}	1.312	1.683	0.00474	0.00977
85	4609	494.5	38.36	838.6	5988	5659	0.3422	0.04232	7.645×10^{-5}	1.322×10^{-5}	1.338	1.768	0.00521	0.00876
90	5116	482.8	43.48	800.6	6242	6142	0.3306	0.04447	7.284×10^{-5}	1.354×10^{-5}	1.375	1.871	0.00579	0.00776
95	5665	470.2	49.35	759.8	6561	6740	0.3190	0.04687	6.946×10^{-5}	1.389×10^{-5}	1.429	1.999	0.00652	0.00677
100	6257	456.6	56.15	715.5	6972	7503	0.3075	0.04958	6.628×10^{-5}	1.429×10^{-5}	1.503	2.163	0.00749	0.00579

Note 1: Kinematic viscosity ν and thermal diffusivity α can be calculated from their definitions, $\nu = \mu/\rho$ and $\alpha = k/\rho c_p = \nu/\text{Pr}$. The properties listed here (except the vapor density) can be used at any pressures with negligible error except at temperatures near the critical-point value.

Note 2: The unit kJ/kg·°C for specific heat is equivalent to kJ/kg·K, and the unit W/m·°C for thermal conductivity is equivalent to W/m·K.

Source: Data generated from the EES software developed by S. A. Klein and F. L. Alvarado. Original sources: Tillner-Roth, Harms-Watzenberg, and Baehr, "Eine neue Fundamentalgleichung fur Ammoniak," DKV-Tagungsbericht 20:167–181, 1993; Liley and Desai, "Thermophysical Properties of Refrigerants," *ASHRAE*, 1993, ISBN 1-1883413-10-9.

TABLE A-9
Properties of air at 1 atm pressure

Temp. T, °C	Density ρ, kg/m³	Specific Heat c_p J/kg·K	Thermal Conductivity k, W/m·K	Thermal Diffusivity α, m²/s	Dynamic Viscosity μ, kg/m·s	Kinematic Viscosity ν, m²/s	Prandtl Number Pr
−150	2.866	983	0.01171	4.158×10^{-6}	8.636×10^{-6}	3.013×10^{-6}	0.7246
−100	2.038	966	0.01582	8.036×10^{-6}	1.189×10^{-6}	5.837×10^{-6}	0.7263
−50	1.582	999	0.01979	1.252×10^{-5}	1.474×10^{-5}	9.319×10^{-6}	0.7440
−40	1.514	1002	0.02057	1.356×10^{-5}	1.527×10^{-5}	1.008×10^{-5}	0.7436
−30	1.451	1004	0.02134	1.465×10^{-5}	1.579×10^{-5}	1.087×10^{-5}	0.7425
−20	1.394	1005	0.02211	1.578×10^{-5}	1.630×10^{-5}	1.169×10^{-5}	0.7408
−10	1.341	1006	0.02288	1.696×10^{-5}	1.680×10^{-5}	1.252×10^{-5}	0.7387
0	1.292	1006	0.02364	1.818×10^{-5}	1.729×10^{-5}	1.338×10^{-5}	0.7362
5	1.269	1006	0.02401	1.880×10^{-5}	1.754×10^{-5}	1.382×10^{-5}	0.7350
10	1.246	1006	0.02439	1.944×10^{-5}	1.778×10^{-5}	1.426×10^{-5}	0.7336
15	1.225	1007	0.02476	2.009×10^{-5}	1.802×10^{-5}	1.470×10^{-5}	0.7323
20	1.204	1007	0.02514	2.074×10^{-5}	1.825×10^{-5}	1.516×10^{-5}	0.7309
25	1.184	1007	0.02551	2.141×10^{-5}	1.849×10^{-5}	1.562×10^{-5}	0.7296
30	1.164	1007	0.02588	2.208×10^{-5}	1.872×10^{-5}	1.608×10^{-5}	0.7282
35	1.145	1007	0.02625	2.277×10^{-5}	1.895×10^{-5}	1.655×10^{-5}	0.7268
40	1.127	1007	0.02662	2.346×10^{-5}	1.918×10^{-5}	1.702×10^{-5}	0.7255
45	1.109	1007	0.02699	2.416×10^{-5}	1.941×10^{-5}	1.750×10^{-5}	0.7241
50	1.092	1007	0.02735	2.487×10^{-5}	1.963×10^{-5}	1.798×10^{-5}	0.7228
60	1.059	1007	0.02808	2.632×10^{-5}	2.008×10^{-5}	1.896×10^{-5}	0.7202
70	1.028	1007	0.02881	2.780×10^{-5}	2.052×10^{-5}	1.995×10^{-5}	0.7177
80	0.9994	1008	0.02953	2.931×10^{-5}	2.096×10^{-5}	2.097×10^{-5}	0.7154
90	0.9718	1008	0.03024	3.086×10^{-5}	2.139×10^{-5}	2.201×10^{-5}	0.7132
100	0.9458	1009	0.03095	3.243×10^{-5}	2.181×10^{-5}	2.306×10^{-5}	0.7111
120	0.8977	1011	0.03235	3.565×10^{-5}	2.264×10^{-5}	2.522×10^{-5}	0.7073
140	0.8542	1013	0.03374	3.898×10^{-5}	2.345×10^{-5}	2.745×10^{-5}	0.7041
160	0.8148	1016	0.03511	4.241×10^{-5}	2.420×10^{-5}	2.975×10^{-5}	0.7014
180	0.7788	1019	0.03646	4.593×10^{-5}	2.504×10^{-5}	3.212×10^{-5}	0.6992
200	0.7459	1023	0.03779	4.954×10^{-5}	2.577×10^{-5}	3.455×10^{-5}	0.6974
250	0.6746	1033	0.04104	5.890×10^{-5}	2.760×10^{-5}	4.091×10^{-5}	0.6946
300	0.6158	1044	0.04418	6.871×10^{-5}	2.934×10^{-5}	4.765×10^{-5}	0.6935
350	0.5664	1056	0.04721	7.892×10^{-5}	3.101×10^{-5}	5.475×10^{-5}	0.6937
400	0.5243	1069	0.05015	8.951×10^{-5}	3.261×10^{-5}	6.219×10^{-5}	0.6948
450	0.4880	1081	0.05298	1.004×10^{-4}	3.415×10^{-5}	6.997×10^{-5}	0.6965
500	0.4565	1093	0.05572	1.117×10^{-4}	3.563×10^{-5}	7.806×10^{-5}	0.6986
600	0.4042	1115	0.06093	1.352×10^{-4}	3.846×10^{-5}	9.515×10^{-5}	0.7037
700	0.3627	1135	0.06581	1.598×10^{-4}	4.111×10^{-5}	1.133×10^{-4}	0.7092
800	0.3289	1153	0.07037	1.855×10^{-4}	4.362×10^{-5}	1.326×10^{-4}	0.7149
900	0.3008	1169	0.07465	2.122×10^{-4}	4.600×10^{-5}	1.529×10^{-4}	0.7206
1000	0.2772	1184	0.07868	2.398×10^{-4}	4.826×10^{-5}	1.741×10^{-4}	0.7260
1500	0.1990	1234	0.09599	3.908×10^{-4}	5.817×10^{-5}	2.922×10^{-4}	0.7478
2000	0.1553	1264	0.11113	5.664×10^{-4}	6.630×10^{-5}	4.270×10^{-4}	0.7539

Note: For ideal gases, the properties c_p, k, μ, and Pr are independent of pressure. The properties ρ, ν, and α at a pressure P (in atm) other than 1 atm are determined by multiplying the values of ρ at the given temperature by P and by dividing ν and α by P.

Source: Data generated from the EES software developed by S. A. Klein and F. L. Alvarado. Original sources: Keenan, Chao, Keyes, Gas Tables, Wiley, 198; and Thermophysical Properties of Matter, Vol. 3: Thermal Conductivity, Y. S. Touloukian, P. E. Liley, S. C. Saxena, Vol. 11: Viscosity, Y. S. Touloukian, S. C. Saxena, and P. Hestermans, IFI/Plenun, NY, 1970, ISBN 0-306067020-8.

TABLE A-6
Properties of saturated propane

Temp. T, °C	Saturation Pressure P, kPa	Density ρ, kg/m³ Liquid	Density ρ, kg/m³ Vapor	Enthalpy of Vaporization h_{fg}, kJ/kg	Specific Heat c_p J/kg·K Liquid	Specific Heat c_p J/kg·K Vapor	Thermal Conductivity k, W/m·K Liquid	Thermal Conductivity k, W/m·K Vapor	Dynamic Viscosity μ, kg/m·s Liquid	Dynamic Viscosity μ, kg/m·s Vapor	Prandtl Number Pr Liquid	Prandtl Number Pr Vapor	Volume Expansion Coefficient β, 1/K Liquid	Surface Tension, N/m Liquid
−120	0.4053	664.7	0.01408	498.3	2003	1115	0.1802	0.00589	6.136×10^{-4}	4.372×10^{-6}	6.820	0.827	0.00153	0.02630
−110	1.157	654.5	0.03776	489.3	2021	1148	0.1738	0.00645	5.054×10^{-4}	4.625×10^{-6}	5.878	0.822	0.00157	0.02486
−100	2.881	644.2	0.08872	480.4	2044	1183	0.1672	0.00705	4.252×10^{-4}	4.881×10^{-6}	5.195	0.819	0.00161	0.02344
−90	6.406	633.8	0.1870	471.5	2070	1221	0.1606	0.00769	3.635×10^{-4}	5.143×10^{-6}	4.686	0.817	0.00166	0.02202
−80	12.97	623.2	0.3602	462.4	2100	1263	0.1539	0.00836	3.149×10^{-4}	5.409×10^{-6}	4.297	0.817	0.00171	0.02062
−70	24.26	612.5	0.6439	453.1	2134	1308	0.1472	0.00908	2.755×10^{-4}	5.680×10^{-6}	3.994	0.818	0.00177	0.01923
−60	42.46	601.5	1.081	443.5	2173	1358	0.1407	0.00985	2.430×10^{-4}	5.956×10^{-6}	3.755	0.821	0.00184	0.01785
−50	70.24	590.3	1.724	433.6	2217	1412	0.1343	0.01067	2.158×10^{-4}	6.239×10^{-6}	3.563	0.825	0.00192	0.01649
−40	110.7	578.8	2.629	423.1	2258	1471	0.1281	0.01155	1.926×10^{-4}	6.529×10^{-6}	3.395	0.831	0.00201	0.01515
−30	167.3	567.0	3.864	412.1	2310	1535	0.1221	0.01250	1.726×10^{-4}	6.827×10^{-6}	3.266	0.839	0.00213	0.01382
−20	243.8	554.7	5.503	400.3	2368	1605	0.1163	0.01351	1.551×10^{-4}	7.136×10^{-6}	3.158	0.848	0.00226	0.01251
−10	344.4	542.0	7.635	387.8	2433	1682	0.1107	0.01459	1.397×10^{-4}	7.457×10^{-6}	3.069	0.860	0.00242	0.01122
0	473.3	528.7	10.36	374.2	2507	1768	0.1054	0.01576	1.259×10^{-4}	7.794×10^{-6}	2.996	0.875	0.00262	0.00996
5	549.8	521.8	11.99	367.0	2547	1814	0.1028	0.01637	1.195×10^{-4}	7.970×10^{-6}	2.964	0.883	0.00273	0.00934
10	635.1	514.7	13.81	359.5	2590	1864	0.1002	0.01701	1.135×10^{-4}	8.151×10^{-6}	2.935	0.893	0.00286	0.00872
15	729.8	507.5	15.85	351.7	2637	1917	0.0977	0.01767	1.077×10^{-4}	8.339×10^{-6}	2.909	0.905	0.00301	0.00811
20	834.4	500.0	18.13	343.4	2688	1974	0.0952	0.01836	1.022×10^{-4}	8.534×10^{-6}	2.886	0.918	0.00318	0.00751
25	949.7	492.2	20.68	334.8	2742	2036	0.0928	0.01908	9.702×10^{-5}	8.738×10^{-6}	2.866	0.933	0.00337	0.00691
30	1076	484.2	23.53	325.8	2802	2104	0.0904	0.01982	9.197×10^{-5}	8.952×10^{-6}	2.850	0.950	0.00358	0.00633
35	1215	475.8	26.72	316.2	2869	2179	0.0881	0.02061	8.710×10^{-5}	9.178×10^{-6}	2.837	0.971	0.00384	0.00575
40	1366	467.1	30.29	306.1	2943	2264	0.0857	0.02142	8.240×10^{-5}	9.417×10^{-6}	2.828	0.995	0.00413	0.00518
45	1530	458.0	34.29	295.3	3026	2361	0.0834	0.02228	7.785×10^{-5}	9.674×10^{-6}	2.824	1.025	0.00448	0.00463
50	1708	448.5	38.79	283.9	3122	2473	0.0811	0.02319	7.343×10^{-5}	9.950×10^{-6}	2.826	1.061	0.00491	0.00408
60	2110	427.5	49.66	258.4	3283	2769	0.0765	0.02517	6.487×10^{-5}	1.058×10^{-5}	2.784	1.164	0.00609	0.00303
70	2580	403.2	64.02	228.0	3595	3241	0.0717	0.02746	5.649×10^{-5}	1.138×10^{-5}	2.834	1.343	0.00811	0.00204
80	3127	373.0	84.28	189.7	4501	4173	0.0663	0.03029	4.790×10^{-5}	1.249×10^{-5}	3.251	1.722	0.01248	0.00114
90	3769	329.1	118.6	133.2	6977	7239	0.0595	0.03441	3.807×10^{-5}	1.448×10^{-5}	4.465	3.047	0.02847	0.00037

Note 1: Kinematic viscosity ν and thermal diffusivity α can be calculated from their definitions, $\nu = \mu/\rho$ and $\alpha = k/\rho c_p = \nu/\text{Pr}$. The properties listed here (except the vapor density) can be used at any pressures with negligible error except at temperatures near the critical-point value.

Note 2: The unit kJ/kg·°C for specific heat is equivalent to kJ/kg·K, and the unit W/m·°C for thermal conductivity is equivalent to W/m·K.

Source: Data generated from the EES software developed by S. A. Klein and F. L. Alvarado. Original sources: Reiner Tillner-Roth, "Fundamental Equations of State," Shaker, Verlag, Aachan, 1998; B. A. Younglove and J. F. Ely, "Thermophysical Properties of Fluids. II Methane, Ethane, Propane, Isobutane, and Normal Butane," J. Phys. Chem. Ref. Data, Vol. 16, No. 4, 1987; G.R. Somayajulu, "A Generalized Equation for Surface Tension from the Triple-Point to the Critical-Point," International Journal of Thermophysics, Vol. 9, No. 4, 1988.

TABLE A–7
Properties of liquids

Temp. T, °C	Density ρ, kg/m³	Specific Heat c_p, J/kg·K	Thermal Conductivity k, W/m·K	Thermal Diffusivity α, m²/s	Dynamic Viscosity μ, kg/m·s	Kinematic Viscosity ν, m²/s	Prandtl Number Pr	Volume Expansion Coeff. β, 1/K
\multicolumn{9}{c}{Methane (CH₄)}
−160	420.2	3492	0.1863	1.270×10^{-7}	1.133×10^{-4}	2.699×10^{-7}	2.126	0.00352
−150	405.0	3580	0.1703	1.174×10^{-7}	9.169×10^{-5}	2.264×10^{-7}	1.927	0.00391
−140	388.8	3700	0.1550	1.077×10^{-7}	7.551×10^{-5}	1.942×10^{-7}	1.803	0.00444
−130	371.1	3875	0.1402	9.749×10^{-8}	6.288×10^{-5}	1.694×10^{-7}	1.738	0.00520
−120	351.4	4146	0.1258	8.634×10^{-8}	5.257×10^{-5}	1.496×10^{-7}	1.732	0.00637
−110	328.8	4611	0.1115	7.356×10^{-8}	4.377×10^{-5}	1.331×10^{-7}	1.810	0.00841
−100	301.0	5578	0.0967	5.761×10^{-8}	3.577×10^{-5}	1.188×10^{-7}	2.063	0.01282
−90	261.7	8902	0.0797	3.423×10^{-8}	2.761×10^{-5}	1.055×10^{-7}	3.082	0.02922
\multicolumn{9}{c}{Methanol [CH₃(OH)]}
20	788.4	2515	0.1987	1.002×10^{-7}	5.857×10^{-4}	7.429×10^{-7}	7.414	0.00118
30	779.1	2577	0.1980	9.862×10^{-8}	5.088×10^{-4}	6.531×10^{-7}	6.622	0.00120
40	769.6	2644	0.1972	9.690×10^{-8}	4.460×10^{-4}	5.795×10^{-7}	5.980	0.00123
50	760.1	2718	0.1965	9.509×10^{-8}	3.942×10^{-4}	5.185×10^{-7}	5.453	0.00127
60	750.4	2798	0.1957	9.320×10^{-8}	3.510×10^{-4}	4.677×10^{-7}	5.018	0.00132
70	740.4	2885	0.1950	9.128×10^{-8}	3.146×10^{-4}	4.250×10^{-7}	4.655	0.00137
\multicolumn{9}{c}{Isobutane (R600a)}
−100	683.8	1881	0.1383	1.075×10^{-7}	9.305×10^{-4}	1.360×10^{-6}	12.65	0.00142
−75	659.3	1970	0.1357	1.044×10^{-7}	5.624×10^{-4}	8.531×10^{-7}	8.167	0.00150
−50	634.3	2069	0.1283	9.773×10^{-8}	3.769×10^{-4}	5.942×10^{-7}	6.079	0.00161
−25	608.2	2180	0.1181	8.906×10^{-8}	2.688×10^{-4}	4.420×10^{-7}	4.963	0.00177
0	580.6	2306	0.1068	7.974×10^{-8}	1.993×10^{-4}	3.432×10^{-7}	4.304	0.00199
25	550.7	2455	0.0956	7.069×10^{-8}	1.510×10^{-4}	2.743×10^{-7}	3.880	0.00232
50	517.3	2640	0.0851	6.233×10^{-8}	1.155×10^{-4}	2.233×10^{-7}	3.582	0.00286
75	478.5	2896	0.0757	5.460×10^{-8}	8.785×10^{-5}	1.836×10^{-7}	3.363	0.00385
100	429.6	3361	0.0669	4.634×10^{-8}	6.483×10^{-5}	1.509×10^{-7}	3.256	0.00628
\multicolumn{9}{c}{Glycerin}
0	1276	2262	0.2820	9.773×10^{-8}	10.49	8.219×10^{-3}	84,101	
5	1273	2288	0.2835	9.732×10^{-8}	6.730	5.287×10^{-3}	54,327	
10	1270	2320	0.2846	9.662×10^{-8}	4.241	3.339×10^{-3}	34,561	
15	1267	2354	0.2856	9.576×10^{-8}	2.496	1.970×10^{-3}	20,570	
20	1264	2386	0.2860	9.484×10^{-8}	1.519	1.201×10^{-3}	12,671	
25	1261	2416	0.2860	9.388×10^{-8}	0.9934	7.878×10^{-4}	8,392	
30	1258	2447	0.2860	9.291×10^{-8}	0.6582	5.232×10^{-4}	5,631	
35	1255	2478	0.2860	9.195×10^{-8}	0.4347	3.464×10^{-4}	3,767	
40	1252	2513	0.2863	9.101×10^{-8}	0.3073	2.455×10^{-4}	2,697	
\multicolumn{9}{c}{Engine Oil (unused)}
0	899.0	1797	0.1469	9.097×10^{-8}	3.814	4.242×10^{-3}	46,636	0.00070
20	888.1	1881	0.1450	8.680×10^{-8}	0.8374	9.429×10^{-4}	10,863	0.00070
40	876.0	1964	0.1444	8.391×10^{-8}	0.2177	2.485×10^{-4}	2,962	0.00070
60	863.9	2048	0.1404	7.934×10^{-8}	0.07399	8.565×10^{-5}	1,080	0.00070
80	852.0	2132	0.1380	7.599×10^{-8}	0.03232	3.794×10^{-5}	499.3	0.00070
100	840.0	2220	0.1367	7.330×10^{-8}	0.01718	2.046×10^{-5}	279.1	0.00070
120	828.9	2308	0.1347	7.042×10^{-8}	0.01029	1.241×10^{-5}	176.3	0.00070
140	816.8	2395	0.1330	6.798×10^{-8}	0.006558	8.029×10^{-6}	118.1	0.00070
150	810.3	2441	0.1327	6.708×10^{-8}	0.005344	6.595×10^{-6}	98.31	0.00070

Source: Data generated from the EES software developed by S. A. Klein and F. L. Alvarado. Originally based on various sources.

TABLE A–8
Properties of liquid metals

Temp. T, °C	Density ρ, kg/m³	Specific Heat c_p, J/kg·K	Thermal Conductivity k, W/m·K	Thermal Diffusivity α, m²/s	Dynamic Viscosity μ, kg/m·s	Kinematic Viscosity ν, m²/s	Prandtl Number Pr	Volume Expansion Coeff. β, 1/K
\multicolumn{9}{c}{Mercury (Hg) Melting Point: −39°C}
0	13595	140.4	8.18200	4.287×10^{-6}	1.687×10^{-3}	1.241×10^{-7}	0.0289	1.810×10^{-4}
25	13534	139.4	8.51533	4.514×10^{-6}	1.534×10^{-3}	1.133×10^{-7}	0.0251	1.810×10^{-4}
50	13473	138.6	8.83632	4.734×10^{-6}	1.423×10^{-3}	1.056×10^{-7}	0.0223	1.810×10^{-4}
75	13412	137.8	9.15632	4.956×10^{-6}	1.316×10^{-3}	9.819×10^{-8}	0.0198	1.810×10^{-4}
100	13351	137.1	9.46706	5.170×10^{-6}	1.245×10^{-3}	9.326×10^{-8}	0.0180	1.810×10^{-4}
150	13231	136.1	10.07780	5.595×10^{-6}	1.126×10^{-3}	8.514×10^{-8}	0.0152	1.810×10^{-4}
200	13112	135.5	10.65465	5.996×10^{-6}	1.043×10^{-3}	7.959×10^{-8}	0.0133	1.815×10^{-4}
250	12993	135.3	11.18150	6.363×10^{-6}	9.820×10^{-4}	7.558×10^{-8}	0.0119	1.829×10^{-4}
300	12873	135.3	11.68150	6.705×10^{-6}	9.336×10^{-4}	7.252×10^{-8}	0.0108	1.854×10^{-4}
\multicolumn{9}{c}{Bismuth (Bi) Melting Point: 271°C}
350	9969	146.0	16.28	1.118×10^{-5}	1.540×10^{-3}	1.545×10^{-7}	0.01381	
400	9908	148.2	16.10	1.096×10^{-5}	1.422×10^{-3}	1.436×10^{-7}	0.01310	
500	9785	152.8	15.74	1.052×10^{-5}	1.188×10^{-3}	1.215×10^{-7}	0.01154	
600	9663	157.3	15.60	1.026×10^{-5}	1.013×10^{-3}	1.048×10^{-7}	0.01022	
700	9540	161.8	15.60	1.010×10^{-5}	8.736×10^{-4}	9.157×10^{-8}	0.00906	
\multicolumn{9}{c}{Lead (Pb) Melting Point: 327°C}
400	10506	158	15.97	9.623×10^{-6}	2.277×10^{-3}	2.167×10^{-7}	0.02252	
450	10449	156	15.74	9.649×10^{-6}	2.065×10^{-3}	1.976×10^{-7}	0.02048	
500	10390	155	15.54	9.651×10^{-6}	1.884×10^{-3}	1.814×10^{-7}	0.01879	
550	10329	155	15.39	9.610×10^{-6}	1.758×10^{-3}	1.702×10^{-7}	0.01771	
600	10267	155	15.23	9.568×10^{-6}	1.632×10^{-3}	1.589×10^{-7}	0.01661	
650	10206	155	15.07	9.526×10^{-6}	1.505×10^{-3}	1.475×10^{-7}	0.01549	
700	10145	155	14.91	9.483×10^{-6}	1.379×10^{-3}	1.360×10^{-7}	0.01434	
\multicolumn{9}{c}{Sodium (Na) Melting Point: 98°C}
100	927.3	1378	85.84	6.718×10^{-5}	6.892×10^{-4}	7.432×10^{-7}	0.01106	
200	902.5	1349	80.84	6.639×10^{-5}	5.385×10^{-4}	5.967×10^{-7}	0.008987	
300	877.8	1320	75.84	6.544×10^{-5}	3.878×10^{-4}	4.418×10^{-7}	0.006751	
400	853.0	1296	71.20	6.437×10^{-5}	2.720×10^{-4}	3.188×10^{-7}	0.004953	
500	828.5	1284	67.41	6.335×10^{-5}	2.411×10^{-4}	2.909×10^{-7}	0.004593	
600	804.0	1272	63.63	6.220×10^{-5}	2.101×10^{-4}	2.614×10^{-7}	0.004202	
\multicolumn{9}{c}{Potassium (K) Melting Point: 64°C}
200	795.2	790.8	43.99	6.995×10^{-5}	3.350×10^{-4}	4.213×10^{-7}	0.006023	
300	771.6	772.8	42.01	7.045×10^{-5}	2.667×10^{-4}	3.456×10^{-7}	0.004906	
400	748.0	754.8	40.03	7.090×10^{-5}	1.984×10^{-4}	2.652×10^{-7}	0.00374	
500	723.9	750.0	37.81	6.964×10^{-5}	1.668×10^{-4}	2.304×10^{-7}	0.003309	
600	699.6	750.0	35.50	6.765×10^{-5}	1.487×10^{-4}	2.126×10^{-7}	0.003143	
\multicolumn{9}{c}{Sodium–Potassium (%22Na-%78K) Melting Point: −11°C}
100	847.3	944.4	25.64	3.205×10^{-5}	5.707×10^{-4}	6.736×10^{-7}	0.02102	
200	823.2	922.5	26.27	3.459×10^{-5}	4.587×10^{-4}	5.572×10^{-7}	0.01611	
300	799.1	900.6	26.89	3.736×10^{-5}	3.467×10^{-4}	4.339×10^{-7}	0.01161	
400	775.0	879.0	27.50	4.037×10^{-5}	2.357×10^{-4}	3.041×10^{-7}	0.00753	
500	751.5	880.1	27.89	4.217×10^{-5}	2.108×10^{-4}	2.805×10^{-7}	0.00665	
600	728.0	881.2	28.28	4.408×10^{-5}	1.859×10^{-4}	2.553×10^{-7}	0.00579	

Source: Data generated from the EES software developed by S. A. Klein and F. L. Alvarado. Originally based on various sources.

TABLE A–10

Properties of gases at 1 atm pressure

Temp. T, °C	Density ρ, kg/m³	Specific Heat c_p J/kg·K	Thermal Conductivity k, W/m·K	Thermal Diffusivity α, m²/s	Dynamic Viscosity μ, kg/m·s	Kinematic Viscosity ν, m²/s	Prandtl Number Pr
Carbon Dioxide, CO_2							
−50	2.4035	746	0.01051	5.860×10^{-6}	1.129×10^{-5}	4.699×10^{-6}	0.8019
0	1.9635	811	0.01456	9.141×10^{-6}	1.375×10^{-5}	7.003×10^{-6}	0.7661
50	1.6597	866.6	0.01858	1.291×10^{-5}	1.612×10^{-5}	9.714×10^{-6}	0.7520
100	1.4373	914.8	0.02257	1.716×10^{-5}	1.841×10^{-5}	1.281×10^{-5}	0.7464
150	1.2675	957.4	0.02652	2.186×10^{-5}	2.063×10^{-5}	1.627×10^{-5}	0.7445
200	1.1336	995.2	0.03044	2.698×10^{-5}	2.276×10^{-5}	2.008×10^{-5}	0.7442
300	0.9358	1060	0.03814	3.847×10^{-5}	2.682×10^{-5}	2.866×10^{-5}	0.7450
400	0.7968	1112	0.04565	5.151×10^{-5}	3.061×10^{-5}	3.842×10^{-5}	0.7458
500	0.6937	1156	0.05293	6.600×10^{-5}	3.416×10^{-5}	4.924×10^{-5}	0.7460
1000	0.4213	1292	0.08491	1.560×10^{-4}	4.898×10^{-5}	1.162×10^{-4}	0.7455
1500	0.3025	1356	0.10688	2.606×10^{-4}	6.106×10^{-5}	2.019×10^{-4}	0.7745
2000	0.2359	1387	0.11522	3.521×10^{-4}	7.322×10^{-5}	3.103×10^{-4}	0.8815
Carbon Monoxide, CO							
−50	1.5297	1081	0.01901	1.149×10^{-5}	1.378×10^{-5}	9.012×10^{-6}	0.7840
0	1.2497	1048	0.02278	1.739×10^{-5}	1.629×10^{-5}	1.303×10^{-5}	0.7499
50	1.0563	1039	0.02641	2.407×10^{-5}	1.863×10^{-5}	1.764×10^{-5}	0.7328
100	0.9148	1041	0.02992	3.142×10^{-5}	2.080×10^{-5}	2.274×10^{-5}	0.7239
150	0.8067	1049	0.03330	3.936×10^{-5}	2.283×10^{-5}	2.830×10^{-5}	0.7191
200	0.7214	1060	0.03656	4.782×10^{-5}	2.472×10^{-5}	3.426×10^{-5}	0.7164
300	0.5956	1085	0.04277	6.619×10^{-5}	2.812×10^{-5}	4.722×10^{-5}	0.7134
400	0.5071	1111	0.04860	8.628×10^{-5}	3.111×10^{-5}	6.136×10^{-5}	0.7111
500	0.4415	1135	0.05412	1.079×10^{-4}	3.379×10^{-5}	7.653×10^{-5}	0.7087
1000	0.2681	1226	0.07894	2.401×10^{-4}	4.557×10^{-5}	1.700×10^{-4}	0.7080
1500	0.1925	1279	0.10458	4.246×10^{-4}	6.321×10^{-5}	3.284×10^{-4}	0.7733
2000	0.1502	1309	0.13833	7.034×10^{-4}	9.826×10^{-5}	6.543×10^{-4}	0.9302
Methane, CH_4							
−50	0.8761	2243	0.02367	1.204×10^{-5}	8.564×10^{-6}	9.774×10^{-6}	0.8116
0	0.7158	2217	0.03042	1.917×10^{-5}	1.028×10^{-5}	1.436×10^{-5}	0.7494
50	0.6050	2302	0.03766	2.704×10^{-5}	1.191×10^{-5}	1.969×10^{-5}	0.7282
100	0.5240	2443	0.04534	3.543×10^{-5}	1.345×10^{-5}	2.567×10^{-5}	0.7247
150	0.4620	2611	0.05344	4.431×10^{-5}	1.491×10^{-5}	3.227×10^{-5}	0.7284
200	0.4132	2791	0.06194	5.370×10^{-5}	1.630×10^{-5}	3.944×10^{-5}	0.7344
300	0.3411	3158	0.07996	7.422×10^{-5}	1.886×10^{-5}	5.529×10^{-5}	0.7450
400	0.2904	3510	0.09918	9.727×10^{-5}	2.119×10^{-5}	7.297×10^{-5}	0.7501
500	0.2529	3836	0.11933	1.230×10^{-4}	2.334×10^{-5}	9.228×10^{-5}	0.7502
1000	0.1536	5042	0.22562	2.914×10^{-4}	3.281×10^{-5}	2.136×10^{-4}	0.7331
1500	0.1103	5701	0.31857	5.068×10^{-4}	4.434×10^{-5}	4.022×10^{-4}	0.7936
2000	0.0860	6001	0.36750	7.120×10^{-4}	6.360×10^{-5}	7.395×10^{-4}	1.0386
Hydrogen, H_2							
−50	0.11010	12635	0.1404	1.009×10^{-4}	7.293×10^{-6}	6.624×10^{-5}	0.6562
0	0.08995	13920	0.1652	1.319×10^{-4}	8.391×10^{-6}	9.329×10^{-5}	0.7071
50	0.07603	14349	0.1881	1.724×10^{-4}	9.427×10^{-6}	1.240×10^{-4}	0.7191
100	0.06584	14473	0.2095	2.199×10^{-4}	1.041×10^{-5}	1.582×10^{-4}	0.7196
150	0.05806	14492	0.2296	2.729×10^{-4}	1.136×10^{-5}	1.957×10^{-4}	0.7174
200	0.05193	14482	0.2486	3.306×10^{-4}	1.228×10^{-5}	2.365×10^{-4}	0.7155
300	0.04287	14481	0.2843	4.580×10^{-4}	1.403×10^{-5}	3.274×10^{-4}	0.7149
400	0.03650	14540	0.3180	5.992×10^{-4}	1.570×10^{-5}	4.302×10^{-4}	0.7179
500	0.03178	14653	0.3509	7.535×10^{-4}	1.730×10^{-5}	5.443×10^{-4}	0.7224
1000	0.01930	15577	0.5206	1.732×10^{-3}	2.455×10^{-5}	1.272×10^{-3}	0.7345
1500	0.01386	16553	0.6581	2.869×10^{-3}	3.099×10^{-5}	2.237×10^{-3}	0.7795
2000	0.01081	17400	0.5480	2.914×10^{-3}	3.690×10^{-5}	3.414×10^{-3}	1.1717

(Continued)

TABLE A–10

Properties of gases at 1 atm pressure (*Continued*)

Temp. T, °C	Density ρ, kg/m³	Specific Heat c_p J/kg·K	Thermal Conductivity k, W/m·K	Thermal Diffusivity α, m²/s	Dynamic Viscosity μ, kg/m·s	Kinematic Viscosity ν, m²/s	Prandtl Number Pr
\multicolumn{8}{c}{*Nitrogen, N_2*}
−50	1.5299	957.3	0.02001	1.366×10^{-5}	1.390×10^{-5}	9.091×10^{-6}	0.6655
0	1.2498	1035	0.02384	1.843×10^{-5}	1.640×10^{-5}	1.312×10^{-5}	0.7121
50	1.0564	1042	0.02746	2.494×10^{-5}	1.874×10^{-5}	1.774×10^{-5}	0.7114
100	0.9149	1041	0.03090	3.244×10^{-5}	2.094×10^{-5}	2.289×10^{-5}	0.7056
150	0.8068	1043	0.03416	4.058×10^{-5}	2.300×10^{-5}	2.851×10^{-5}	0.7025
200	0.7215	1050	0.03727	4.921×10^{-5}	2.494×10^{-5}	3.457×10^{-5}	0.7025
300	0.5956	1070	0.04309	6.758×10^{-5}	2.849×10^{-5}	4.783×10^{-5}	0.7078
400	0.5072	1095	0.04848	8.727×10^{-5}	3.166×10^{-5}	6.242×10^{-5}	0.7153
500	0.4416	1120	0.05358	1.083×10^{-4}	3.451×10^{-5}	7.816×10^{-5}	0.7215
1000	0.2681	1213	0.07938	2.440×10^{-4}	4.594×10^{-5}	1.713×10^{-4}	0.7022
1500	0.1925	1266	0.11793	4.839×10^{-4}	5.562×10^{-5}	2.889×10^{-4}	0.5969
2000	0.1502	1297	0.18590	9.543×10^{-4}	6.426×10^{-5}	4.278×10^{-4}	0.4483
\multicolumn{8}{c}{*Oxygen, O_2*}
−50	1.7475	984.4	0.02067	1.201×10^{-5}	1.616×10^{-5}	9.246×10^{-6}	0.7694
0	1.4277	928.7	0.02472	1.865×10^{-5}	1.916×10^{-5}	1.342×10^{-5}	0.7198
50	1.2068	921.7	0.02867	2.577×10^{-5}	2.194×10^{-5}	1.818×10^{-5}	0.7053
100	1.0451	931.8	0.03254	3.342×10^{-5}	2.451×10^{-5}	2.346×10^{-5}	0.7019
150	0.9216	947.6	0.03637	4.164×10^{-5}	2.694×10^{-5}	2.923×10^{-5}	0.7019
200	0.8242	964.7	0.04014	5.048×10^{-5}	2.923×10^{-5}	3.546×10^{-5}	0.7025
300	0.6804	997.1	0.04751	7.003×10^{-5}	3.350×10^{-5}	4.923×10^{-5}	0.7030
400	0.5793	1025	0.05463	9.204×10^{-5}	3.744×10^{-5}	6.463×10^{-5}	0.7023
500	0.5044	1048	0.06148	1.163×10^{-4}	4.114×10^{-5}	8.156×10^{-5}	0.7010
1000	0.3063	1121	0.09198	2.678×10^{-4}	5.732×10^{-5}	1.871×10^{-4}	0.6986
1500	0.2199	1165	0.11901	4.643×10^{-4}	7.133×10^{-5}	3.243×10^{-4}	0.6985
2000	0.1716	1201	0.14705	7.139×10^{-4}	8.417×10^{-5}	4.907×10^{-4}	0.6873
\multicolumn{8}{c}{*Water Vapor, H_2O*}
−50	0.9839	1892	0.01353	7.271×10^{-6}	7.187×10^{-6}	7.305×10^{-6}	1.0047
0	0.8038	1874	0.01673	1.110×10^{-5}	8.956×10^{-6}	1.114×10^{-5}	1.0033
50	0.6794	1874	0.02032	1.596×10^{-5}	1.078×10^{-5}	1.587×10^{-5}	0.9944
100	0.5884	1887	0.02429	2.187×10^{-5}	1.265×10^{-5}	2.150×10^{-5}	0.9830
150	0.5189	1908	0.02861	2.890×10^{-5}	1.456×10^{-5}	2.806×10^{-5}	0.9712
200	0.4640	1935	0.03326	3.705×10^{-5}	1.650×10^{-5}	3.556×10^{-5}	0.9599
300	0.3831	1997	0.04345	5.680×10^{-5}	2.045×10^{-5}	5.340×10^{-5}	0.9401
400	0.3262	2066	0.05467	8.114×10^{-5}	2.446×10^{-5}	7.498×10^{-5}	0.9240
500	0.2840	2137	0.06677	1.100×10^{-4}	2.847×10^{-5}	1.002×10^{-4}	0.9108
1000	0.1725	2471	0.13623	3.196×10^{-4}	4.762×10^{-5}	2.761×10^{-4}	0.8639
1500	0.1238	2736	0.21301	6.288×10^{-4}	6.411×10^{-5}	5.177×10^{-4}	0.8233
2000	0.0966	2928	0.29183	1.032×10^{-3}	7.808×10^{-5}	8.084×10^{-4}	0.7833

Note: For ideal gases, the properties c_p, k, μ, and Pr are independent of pressure. The properties ρ, ν, and α at a pressure P (in atm) other than 1 atm are determined by multiplying the values of ρ at the given temperature by P and by dividing ν and α by P.

Source: Data generated from the EES software developed by S. A. Klein and F. L. Alvarado. Originally based on various sources.

TABLE A-11
Properties of the atmosphere at high altitude

Altitude, m	Temperature, °C	Pressure, kPa	Gravity g, m/s^2	Speed of Sound, m/s	Density, kg/m^3	Viscosity μ, kg/m·s	Thermal Conductivity, W/m·K
0	15.00	101.33	9.807	340.3	1.225	1.789×10^{-5}	0.0253
200	13.70	98.95	9.806	339.5	1.202	1.783×10^{-5}	0.0252
400	12.40	96.61	9.805	338.8	1.179	1.777×10^{-5}	0.0252
600	11.10	94.32	9.805	338.0	1.156	1.771×10^{-5}	0.0251
800	9.80	92.08	9.804	337.2	1.134	1.764×10^{-5}	0.0250
1000	8.50	89.88	9.804	336.4	1.112	1.758×10^{-5}	0.0249
1200	7.20	87.72	9.803	335.7	1.090	1.752×10^{-5}	0.0248
1400	5.90	85.60	9.802	334.9	1.069	1.745×10^{-5}	0.0247
1600	4.60	83.53	9.802	334.1	1.048	1.739×10^{-5}	0.0245
1800	3.30	81.49	9.801	333.3	1.027	1.732×10^{-5}	0.0244
2000	2.00	79.50	9.800	332.5	1.007	1.726×10^{-5}	0.0243
2200	0.70	77.55	9.800	331.7	0.987	1.720×10^{-5}	0.0242
2400	−0.59	75.63	9.799	331.0	0.967	1.713×10^{-5}	0.0241
2600	−1.89	73.76	9.799	330.2	0.947	1.707×10^{-5}	0.0240
2800	−3.19	71.92	9.798	329.4	0.928	1.700×10^{-5}	0.0239
3000	−4.49	70.12	9.797	328.6	0.909	1.694×10^{-5}	0.0238
3200	−5.79	68.36	9.797	327.8	0.891	1.687×10^{-5}	0.0237
3400	−7.09	66.63	9.796	327.0	0.872	1.681×10^{-5}	0.0236
3600	−8.39	64.94	9.796	326.2	0.854	1.674×10^{-5}	0.0235
3800	−9.69	63.28	9.795	325.4	0.837	1.668×10^{-5}	0.0234
4000	−10.98	61.66	9.794	324.6	0.819	1.661×10^{-5}	0.0233
4200	−12.3	60.07	9.794	323.8	0.802	1.655×10^{-5}	0.0232
4400	−13.6	58.52	9.793	323.0	0.785	1.648×10^{-5}	0.0231
4600	−14.9	57.00	9.793	322.2	0.769	1.642×10^{-5}	0.0230
4800	−16.2	55.51	9.792	321.4	0.752	1.635×10^{-5}	0.0229
5000	−17.5	54.05	9.791	320.5	0.736	1.628×10^{-5}	0.0228
5200	−18.8	52.62	9.791	319.7	0.721	1.622×10^{-5}	0.0227
5400	−20.1	51.23	9.790	318.9	0.705	1.615×10^{-5}	0.0226
5600	−21.4	49.86	9.789	318.1	0.690	1.608×10^{-5}	0.0224
5800	−22.7	48.52	9.785	317.3	0.675	1.602×10^{-5}	0.0223
6000	−24.0	47.22	9.788	316.5	0.660	1.595×10^{-5}	0.0222
6200	−25.3	45.94	9.788	315.6	0.646	1.588×10^{-5}	0.0221
6400	−26.6	44.69	9.787	314.8	0.631	1.582×10^{-5}	0.0220
6600	−27.9	43.47	9.786	314.0	0.617	1.575×10^{-5}	0.0219
6800	−29.2	42.27	9.785	313.1	0.604	1.568×10^{-5}	0.0218
7000	−30.5	41.11	9.785	312.3	0.590	1.561×10^{-5}	0.0217
8000	−36.9	35.65	9.782	308.1	0.526	1.527×10^{-5}	0.0212
9000	−43.4	30.80	9.779	303.8	0.467	1.493×10^{-5}	0.0206
10,000	−49.9	26.50	9.776	299.5	0.414	1.458×10^{-5}	0.0201
12,000	−56.5	19.40	9.770	295.1	0.312	1.422×10^{-5}	0.0195
14,000	−56.5	14.17	9.764	295.1	0.228	1.422×10^{-5}	0.0195
16,000	−56.5	10.53	9.758	295.1	0.166	1.422×10^{-5}	0.0195
18,000	−56.5	7.57	9.751	295.1	0.122	1.422×10^{-5}	0.0195

Source: U.S. Standard Atmosphere Supplements, U.S. Government Printing Office, 1966. Based on year-round mean conditions at 45° latitude and varies with the time of the year and the weather patterns. The conditions at sea level ($z = 0$) are taken to be $P = 101.325$ kPa, $T = 15$°C, $\rho = 1.2250$ kg/m^3, $g = 9.80665$ m^2/s.

952
PROPERTY TABLES AND CHARTS

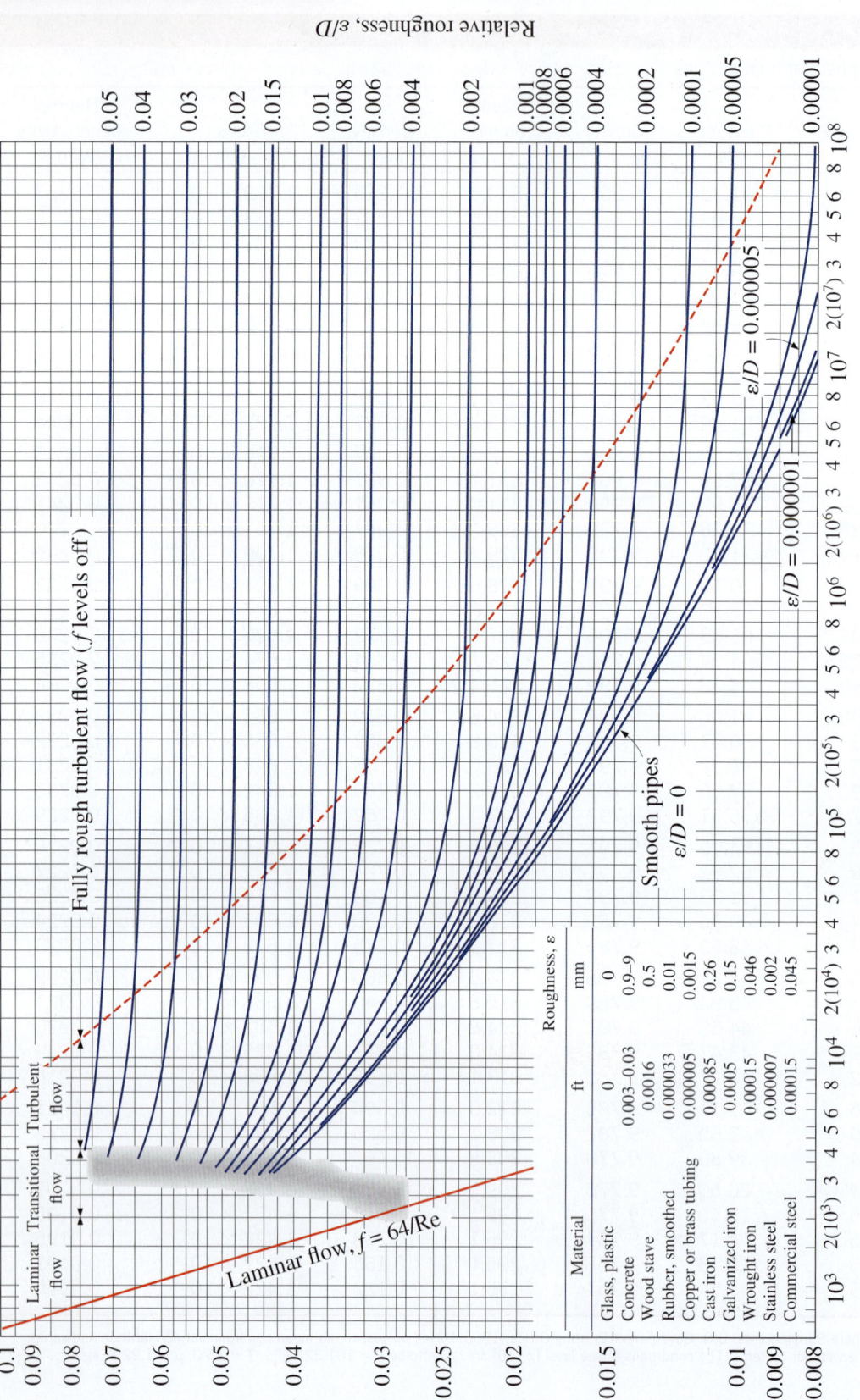

FIGURE A–12
The Moody chart for the friction factor for fully developed flow in circular pipes for use in the head loss relation $h_L = f \dfrac{L}{D} \dfrac{V^2}{2g}$. Friction factors in the turbulent flow are evaluated from the Colebrook equation $\dfrac{1}{\sqrt{f}} = -2 \log_{10}\left(\dfrac{\varepsilon/D}{3.7} + \dfrac{2.51}{\mathrm{Re}\sqrt{f}}\right)$.

APPENDIX

$$\text{Ma}^* = \text{Ma}\sqrt{\frac{k+1}{2+(k-1)\text{Ma}^2}}$$

$$\frac{A}{A^*} = \frac{1}{\text{Ma}}\left[\left(\frac{2}{k+1}\right)\left(1+\frac{k-1}{2}\text{Ma}^2\right)\right]^{0.5(k+1)/(k-1)}$$

$$\frac{P}{P_0} = \left(1+\frac{k-1}{2}\text{Ma}^2\right)^{-k/(k-1)}$$

$$\frac{\rho}{\rho_0} = \left(1+\frac{k-1}{2}\text{Ma}^2\right)^{-1/(k-1)}$$

$$\frac{T}{T_0} = \left(1+\frac{k-1}{2}\text{Ma}^2\right)^{-1}$$

TABLE A–13

One-dimensional isentropic compressible flow functions for an ideal gas with $k = 1.4$

Ma	Ma*	A/A*	P/P₀	ρ/ρ₀	T/T₀
0	0	∞	1.0000	1.0000	1.0000
0.1	0.1094	5.8218	0.9930	0.9950	0.9980
0.2	0.2182	2.9635	0.9725	0.9803	0.9921
0.3	0.3257	2.0351	0.9395	0.9564	0.9823
0.4	0.4313	1.5901	0.8956	0.9243	0.9690
0.5	0.5345	1.3398	0.8430	0.8852	0.9524
0.6	0.6348	1.1882	0.7840	0.8405	0.9328
0.7	0.7318	1.0944	0.7209	0.7916	0.9107
0.8	0.8251	1.0382	0.6560	0.7400	0.8865
0.9	0.9146	1.0089	0.5913	0.6870	0.8606
1.0	1.0000	1.0000	0.5283	0.6339	0.8333
1.2	1.1583	1.0304	0.4124	0.5311	0.7764
1.4	1.2999	1.1149	0.3142	0.4374	0.7184
1.6	1.4254	1.2502	0.2353	0.3557	0.6614
1.8	1.5360	1.4390	0.1740	0.2868	0.6068
2.0	1.6330	1.6875	0.1278	0.2300	0.5556
2.2	1.7179	2.0050	0.0935	0.1841	0.5081
2.4	1.7922	2.4031	0.0684	0.1472	0.4647
2.6	1.8571	2.8960	0.0501	0.1179	0.4252
2.8	1.9140	3.5001	0.0368	0.0946	0.3894
3.0	1.9640	4.2346	0.0272	0.0760	0.3571
5.0	2.2361	25.000	0.0019	0.0113	0.1667
∞	2.2495	∞	0	0	0

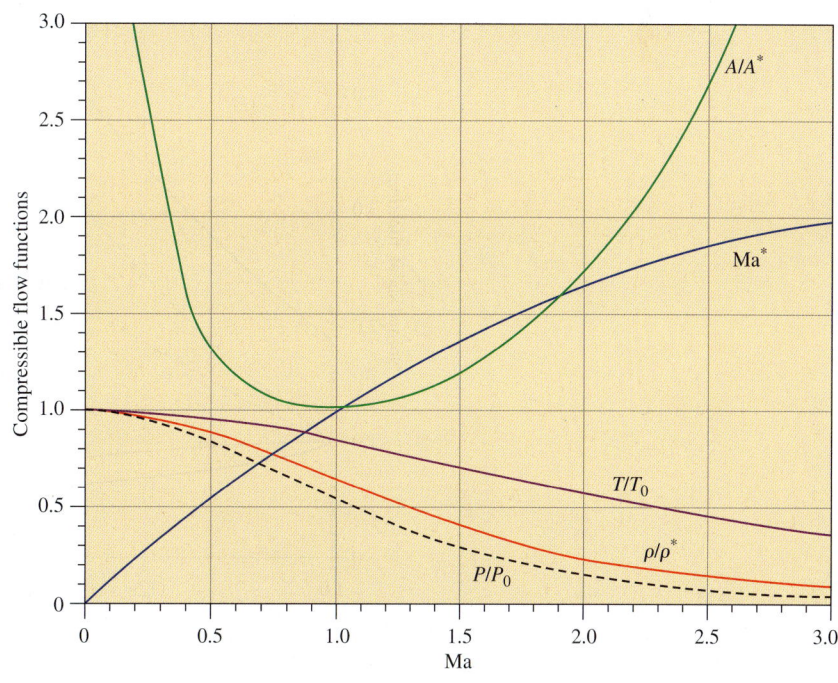

954
PROPERTY TABLES AND CHARTS

$T_{01} = T_{02}$

$\mathrm{Ma}_2 = \sqrt{\dfrac{(k-1)\mathrm{Ma}_1^2 + 2}{2k\mathrm{Ma}_1^2 - k + 1}}$

$\dfrac{P_2}{P_1} = \dfrac{1 + k\mathrm{Ma}_1^2}{1 + k\mathrm{Ma}_2^2} = \dfrac{2k\mathrm{Ma}_1^2 - k + 1}{k + 1}$

$\dfrac{\rho_2}{\rho_1} = \dfrac{P_2/P_1}{T_2/T_1} = \dfrac{(k+1)\mathrm{Ma}_1^2}{2 + (k-1)\mathrm{Ma}_1^2} = \dfrac{V_1}{V_2}$

$\dfrac{T_2}{T_1} = \dfrac{2 + \mathrm{Ma}_1^2(k-1)}{2 + \mathrm{Ma}_2^2(k-1)}$

$\dfrac{P_{02}}{P_{01}} = \dfrac{\mathrm{Ma}_1}{\mathrm{Ma}_2}\left[\dfrac{1 + \mathrm{Ma}_2^2(k-1)/2}{1 + \mathrm{Ma}_1^2(k-1)/2}\right]^{(k+1)/[2(k-1)]}$

$\dfrac{P_{02}}{P_1} = \dfrac{(1 + k\mathrm{Ma}_1^2)[1 + \mathrm{Ma}_2^2(k-1)/2]^{k/(k-1)}}{1 + k\mathrm{Ma}_2^2}$

TABLE A–14
One-dimensional normal shock functions for an ideal gas with $k = 1.4$

Ma_1	Ma_2	P_2/P_1	ρ_2/ρ_1	T_2/T_1	P_{02}/P_{01}	P_{02}/P_1
1.0	1.0000	1.0000	1.0000	1.0000	1.0000	1.8929
1.1	0.9118	1.2450	1.1691	1.0649	0.9989	2.1328
1.2	0.8422	1.5133	1.3416	1.1280	0.9928	2.4075
1.3	0.7860	1.8050	1.5157	1.1909	0.9794	2.7136
1.4	0.7397	2.1200	1.6897	1.2547	0.9582	3.0492
1.5	0.7011	2.4583	1.8621	1.3202	0.9298	3.4133
1.6	0.6684	2.8200	2.0317	1.3880	0.8952	3.8050
1.7	0.6405	3.2050	2.1977	1.4583	0.8557	4.2238
1.8	0.6165	3.6133	2.3592	1.5316	0.8127	4.6695
1.9	0.5956	4.0450	2.5157	1.6079	0.7674	5.1418
2.0	0.5774	4.5000	2.6667	1.6875	0.7209	5.6404
2.1	0.5613	4.9783	2.8119	1.7705	0.6742	6.1654
2.2	0.5471	5.4800	2.9512	1.8569	0.6281	6.7165
2.3	0.5344	6.0050	3.0845	1.9468	0.5833	7.2937
2.4	0.5231	6.5533	3.2119	2.0403	0.5401	7.8969
2.5	0.5130	7.1250	3.3333	2.1375	0.4990	8.5261
2.6	0.5039	7.7200	3.4490	2.2383	0.4601	9.1813
2.7	0.4956	8.3383	3.5590	2.3429	0.4236	9.8624
2.8	0.4882	8.9800	3.6636	2.4512	0.3895	10.5694
2.9	0.4814	9.6450	3.7629	2.5632	0.3577	11.3022
3.0	0.4752	10.3333	3.8571	2.6790	0.3283	12.0610
4.0	0.4350	18.5000	4.5714	4.0469	0.1388	21.0681
5.0	0.4152	29.000	5.0000	5.8000	0.0617	32.6335
∞	0.3780	∞	6.0000	∞	0	∞

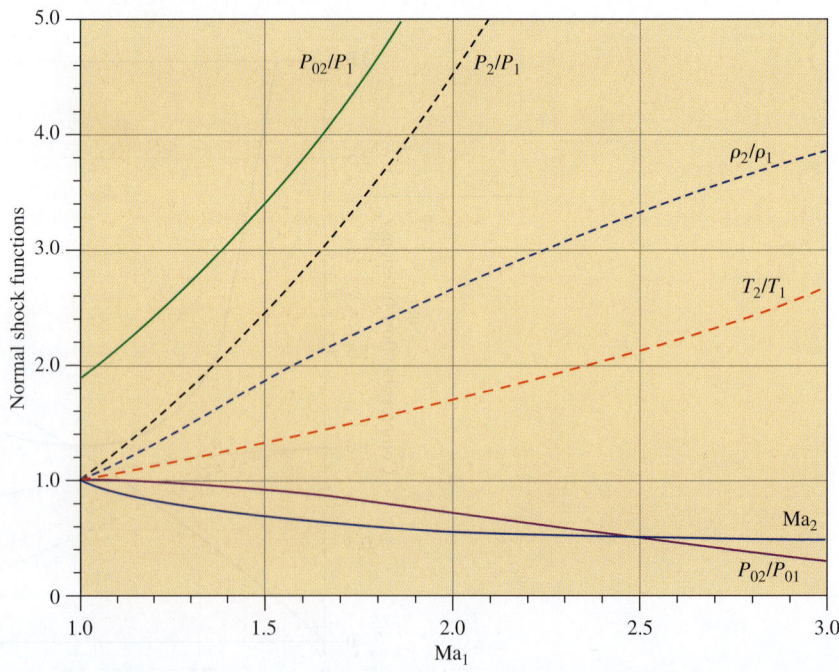

$$\frac{T_0}{T_0^*} = \frac{(k+1)\text{Ma}^2[2+(k-1)\text{Ma}^2]}{(1+k\text{Ma}^2)^2}$$

$$\frac{P_0}{P_0^*} = \frac{k+1}{1+k\text{Ma}^2}\left(\frac{2+(k-1)\text{Ma}^2}{k+1}\right)^{k/(k-1)}$$

$$\frac{T}{T^*} = \left(\frac{\text{Ma}(1+k)}{1+k\text{Ma}^2}\right)^2$$

$$\frac{P}{P^*} = \frac{1+k}{1+k\text{Ma}^2}$$

$$\frac{V}{V^*} = \frac{\rho^*}{\rho} = \frac{(1+k)\text{Ma}^2}{1+k\text{Ma}^2}$$

TABLE A–15
Rayleigh flow functions for an ideal gas with $k = 1.4$

Ma	T_0/T_0^*	P_0/P_0^*	T/T^*	P/P^*	V/V^*
0.0	0.0000	1.2679	0.0000	2.4000	0.0000
0.1	0.0468	1.2591	0.0560	2.3669	0.0237
0.2	0.1736	1.2346	0.2066	2.2727	0.0909
0.3	0.3469	1.1985	0.4089	2.1314	0.1918
0.4	0.5290	1.1566	0.6151	1.9608	0.3137
0.5	0.6914	1.1141	0.7901	1.7778	0.4444
0.6	0.8189	1.0753	0.9167	1.5957	0.5745
0.7	0.9085	1.0431	0.9929	1.4235	0.6975
0.8	0.9639	1.0193	1.0255	1.2658	0.8101
0.9	0.9921	1.0049	1.0245	1.1246	0.9110
1.0	1.0000	1.0000	1.0000	1.0000	1.0000
1.2	0.9787	1.0194	0.9118	0.7958	1.1459
1.4	0.9343	1.0777	0.8054	0.6410	1.2564
1.6	0.8842	1.1756	0.7017	0.5236	1.3403
1.8	0.8363	1.3159	0.6089	0.4335	1.4046
2.0	0.7934	1.5031	0.5289	0.3636	1.4545
2.2	0.7561	1.7434	0.4611	0.3086	1.4938
2.4	0.7242	2.0451	0.4038	0.2648	1.5252
2.6	0.6970	2.4177	0.3556	0.2294	1.5505
2.8	0.6738	2.8731	0.3149	0.2004	1.5711
3.0	0.6540	3.4245	0.2803	0.1765	1.5882

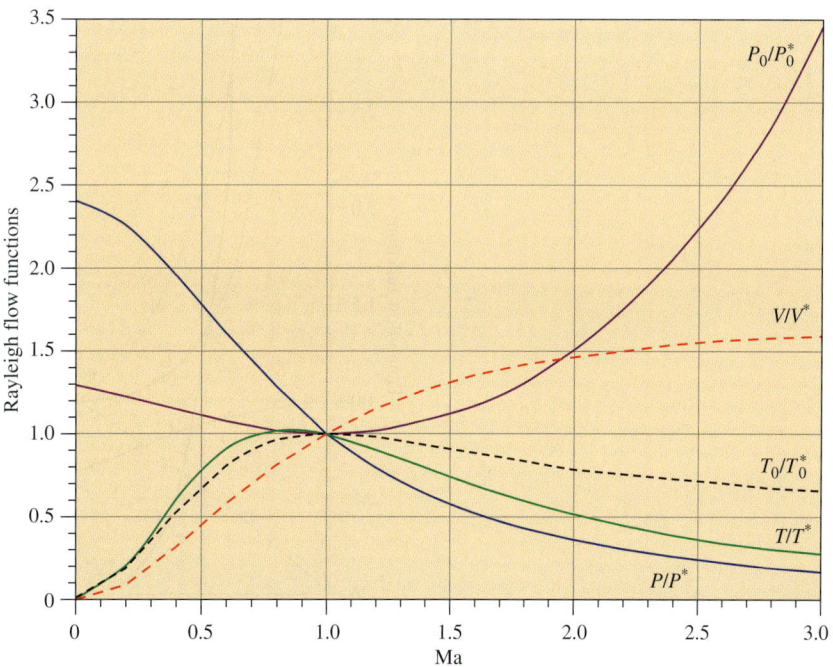

$$T_0 = T_0^*$$
$$\frac{P_0}{P_0^*} = \frac{\rho_0}{\rho_0^*} = \frac{1}{\text{Ma}}\left(\frac{2 + (k-1)\text{Ma}^2}{k+1}\right)^{(k+1)/2(k-1)}$$
$$\frac{T}{T^*} = \frac{k+1}{2 + (k-1)\text{Ma}^2}$$
$$\frac{P}{P^*} = \frac{1}{\text{Ma}}\left(\frac{k+1}{2 + (k-1)\text{Ma}^2}\right)^{1/2}$$
$$\frac{V}{V^*} = \frac{\rho^*}{\rho} = \text{Ma}\left(\frac{k+1}{2 + (k-1)\text{Ma}^2}\right)^{1/2}$$
$$\frac{fL^*}{D} = \frac{1 - \text{Ma}^2}{k\text{Ma}^2} + \frac{k+1}{2k}\ln\frac{(k+1)\text{Ma}^2}{2 + (k-1)\text{Ma}^2}$$

TABLE A–16

Fanno flow functions for an ideal gas with $k = 1.4$

Ma	P_0/P_0^*	T/T^*	P/P^*	V/V^*	fL^*/D
0.0	∞	1.2000	∞	0.0000	∞
0.1	5.8218	1.1976	10.9435	0.1094	66.9216
0.2	2.9635	1.1905	5.4554	0.2182	14.5333
0.3	2.0351	1.1788	3.6191	0.3257	5.2993
0.4	1.5901	1.1628	2.6958	0.4313	2.3085
0.5	1.3398	1.1429	2.1381	0.5345	1.0691
0.6	1.1882	1.1194	1.7634	0.6348	0.4908
0.7	1.0944	1.0929	1.4935	0.7318	0.2081
0.8	1.0382	1.0638	1.2893	0.8251	0.0723
0.9	1.0089	1.0327	1.1291	0.9146	0.0145
1.0	1.0000	1.0000	1.0000	1.0000	0.0000
1.2	1.0304	0.9317	0.8044	1.1583	0.0336
1.4	1.1149	0.8621	0.6632	1.2999	0.0997
1.6	1.2502	0.7937	0.5568	1.4254	0.1724
1.8	1.4390	0.7282	0.4741	1.5360	0.2419
2.0	1.6875	0.6667	0.4082	1.6330	0.3050
2.2	2.0050	0.6098	0.3549	1.7179	0.3609
2.4	2.4031	0.5576	0.3111	1.7922	0.4099
2.6	2.8960	0.5102	0.2747	1.8571	0.4526
2.8	3.5001	0.4673	0.2441	1.9140	0.4898
3.0	4.2346	0.4286	0.2182	1.9640	0.5222

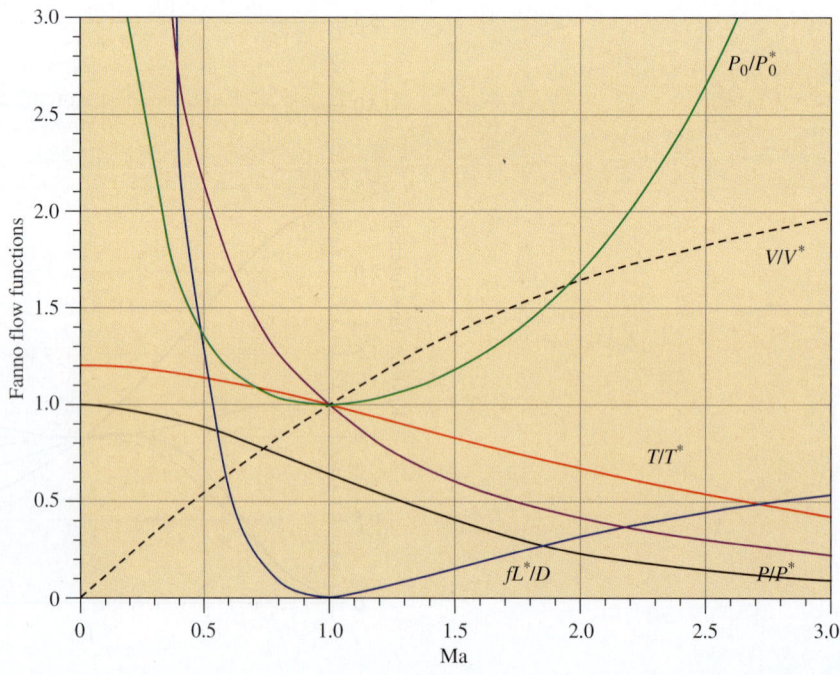

GLOSSARY

Guest Author: James G. Brasseur, The Pennsylvania State University

Note: **Boldface color** glossary terms correspond to **boldface color** terms in the text. *Italics* indicates a term defined elsewhere in the glossary.

Boldface terms without page numbers are concepts that are not defined in the text but are defined or cross-referenced in the glossary for students to review.

absolute pressure: See *stress, pressure stress*. Contrast with *gage pressure*.

absolute viscosity: See *viscosity*.

acceleration field: See *field*.

adiabatic process: A process with no heat transfer.

advective acceleration: In order to reduce confusion of terminology in flows where *buoyancy forces* generate convective fluid motions, the term "convective acceleration" is often replaced with the term "advective acceleration."

aerodynamics: The application of *fluid dynamics* to air, land, and water-going vehicles. Often the term is specifically applied to the flow surrounding, and forces and moments on, flight vehicles in air, as opposed to vehicles in water or other liquids (*hydrodynamics*).

angle of attack: The angle between an airfoil or wing and the free-stream flow velocity vector.

average: An area/volume/time average of a fluid property is the integral of the property over an area/volume/time period divided by the corresponding area/volume/time period. Also called *mean*.

axisymmetric flow: A flow that when specified appropriately using cylindrical coordinates (r, θ, x) does not vary in the azimuthal (θ) direction. Thus, all partial derivatives in θ are zero. The flow is therefore either one-dimensional or two-dimensional (see also *dimensionality* and *planar flow*).

barometer: A device that measures atmospheric pressure.

basic dimensions: See *dimensions*.

Bernoulli equation: A useful reduction of *conservation of momentum* (and *conservation of energy*) that describes a balance between pressure (*flow work*), velocity (*kinetic energy*), and position of *fluid particles* relative to the gravity vector (potential energy) in regions of a fluid flow where frictional force on fluid particles is negligible compared to pressure force in that region of the flow (see *inviscid flow*).

Note: This glossary covers **boldface color terms** found in Chapters 1 to 11.

There are multiple forms of the Bernoulli equation for incompressible vs. compressible, steady vs. nonsteady, and derivations through *Newton's law* vs. the *first law of thermodynamics*. The most commonly used forms are for steady incompressible fluid flow derived through conservation of momentum.

bluff (or blunt) body: A moving object with a blunt rear portion. Bluff bodies have *wakes* resulting from massive *flow separation* over the rear of the body.

boundary condition: In solving for flow field variables (velocity, temperature) from governing equations, it is necessary to mathematically specify a function of the variable at the surface. These mathematical statements are called boundary conditions. The no-slip condition that the flow velocity must equal the surface velocity at the surface is an example of a boundary condition that is used with the Navier–Stokes equation to solve for the velocity field.

boundary layer: At high *Reynolds numbers* relatively thin "boundary layers" exist in the flow adjacent to surfaces where the flow is brought to rest (see *no-slip condition*). Boundary layers are characterized by high *shear* with the highest velocities away from the surface. *Frictional force, viscous stress,* and *vorticity* are significant in boundary layers. The approximate form of the two components of the Navier– Stokes equation, simplified by neglecting the terms that are small within the boundary layer, are called the *boundary layer equations*. The associated approximation based on the existence of thin boundary layers surrounded by *irrotational* or *inviscid* flow is called the *boundary layer approximation*.

boundary layer approximation: See *boundary layer*.

boundary layer equations: See *boundary layer*.

boundary layer thickness measures: Different measures of the thickness of a boundary layer as a function of downstream distance are used in fluid flow analyses. These are:

boundary layer thickness: The full thickness of the viscous layer that defines the boundary layer, from the surface to the edge. Defining the edge is difficult to do precisely, so the "edge" of the boundary layer is often defined as the point where the boundary layer velocity is a large fraction of the free-stream velocity (e.g., δ_{99} is the distance from the surface to the point where the streamwise velocity component is 99 percent of the free-stream velocity).

displacement thickness: A boundary layer thickness measure that quantifies the deflection of fluid streamlines in the direction away from the surface as a result of friction-induced reduction in mass flow adjacent to the surface. Displacement thickness (δ^*) is a measure of the thickness of this mass flow rate deficit layer. In all boundary layers, $\delta^* < \delta$.

momentum thickness: A measure of the layer of highest deficit in momentum flow rate adjacent to the surface as a result of frictional resisting force (shear stress). Because Newton's second law states that force equals time rate of momentum change, momentum thickness θ is proportional to surface shear stress. In all boundary layers, $\theta < \delta^*$.

Buckingham Pi theorem: A mathematical theorem used in *dimensional analysis* that predicts the number of nondimensional groups that must be functionally related from a set of dimensional parameters that are thought to be functionally related.

buffer layer: The part of a turbulent boundary layer, close to the wall, lying between the *viscous* and *inertial sublayers*. This thin layer is a transition from the friction-dominated layer adjacent to the wall where *viscous stresses* are large, to the inertial layer where *turbulent stresses* are large compared to viscous stresses.

bulk modulus of elasticity: See *compressibility*.

buoyant force: The net upward hydrostatic pressure force acting on an object submerged, or partially submerged, in a fluid.

cavitation: The formation of vapor bubbles in a liquid as a result of pressure going below the *vapor pressure*.

center of pressure: The effective point of application of pressure distributed over a surface. This is the point where a counteracting force (equal to integrated pressure) must be placed for the net moment from pressure about that point to be zero.

centripetal acceleration: Acceleration associated with the change in the direction of the velocity (vector) of a material particle.

closed system: See *system*.

coefficient of compressibility: See *compressibility*.

compressibility: The extent to which a *fluid particle* changes volume when subjected to either a change in pressure or a change in temperature.

bulk modulus of elasticity: Synonymous with *coefficient of compressibility*.

coefficient of compressibility: The ratio of pressure change to relative change in volume of a *fluid particle*. This coefficient quantifies compressibility in response to pressure change, an important effect in high Mach number flows.

coefficient of volume expansion: The ratio of relative density change to change in temperature of a *fluid particle*. This coefficient quantifies compressibility in response to temperature change.

computational fluid dynamics (CFD): The application of the conservation laws with boundary and initial conditions in mathematical discretized form to estimate field variables quantitatively on a discretized grid (or mesh) spanning part of the flow field.

conservation laws: The fundamental principles upon which all engineering analysis is based, whereby the material properties of mass, momentum, energy, and entropy can change only in balance with other physical properties involving forces, work, and *heat transfer*. These laws are predictive when written in mathematical form and appropriately combined with boundary conditions, initial conditions, and constitutive relationships.

conservation of energy principle: This is the *first law of thermodynamics*, a fundamental law of physics stating that the time rate of change of total *energy* of a fixed mass (*system*) is balanced by the net rate at which *work* is done on the mass and *heat energy* is transferred to the mass.

Note: To mathematically convert the time derivative of mass, momentum, and energy of fluid mass in a system to that in a *control volume*, one applies the *Reynolds transport theorem*.

conservation of mass principle: A fundamental law of physics stating that a volume always containing the same atoms and molecules (*system*) must always contain the same mass. Thus the time rate of change of mass of a system is zero. This law of physics must be revised when matter moves at speeds approaching the speed of light so that mass and energy can be exchanged as per Einstein's laws of relativity.

conservation of momentum: This is *Newton's second law* of motion, a fundamental law of physics stating that the time rate of change of momentum of a fixed mass (*system*) is balanced by the net sum of all forces applied to the mass.

constitutive equations: An empirical relationship between a physical variable in a *conservation law of physics* and other physical variables in the equation that are to be predicted. For example, the energy equation written for temperature includes the *heat flux* vector. It is known from experiments that heat flux for most common materials is accurately approximated as proportional to the gradient in temperature (this is called Fourier's law). In *Newton's law* written for a *fluid particle*, the *viscous stress tensor* (see *stress*) must be written as a function of velocity to solve the equation. The most common constitutive relationship for viscous stress is that for a *Newtonian fluid*. See also *rheology*.

continuity equation: Mathematical form of *conservation of mass* applied to a *fluid particle* in a flow.

continuum: Treatment of matter as a continuous (without holes) distribution of finite mass *differential volume* elements. Each volume element must contain huge numbers of molecules so that the macroscopic effect of the molecules can be modeled without considering individual molecules.

GLOSSARY

contour plot: Also called an *isocontour plot,* this is a way of plotting data as lines of constant variable through a flow *field. Streamlines,* for example, may be identified as lines of constant *stream function* in *two-dimensional* incompressible steady flows.

control mass: See *system.*

control volume: A volume specified for analysis where flow enters and/or exits through some portion(s) of the volume surface. Also called an *open system* (see *system*).

convective acceleration: Synonymous with *advective acceleration,* this term must be added to the partial time derivative of velocity to properly quantify the acceleration of a *fluid particle* within an *Eulerian* frame of reference. For example, a fluid particle moving through a contraction in a *steady flow* speeds up as it moves, yet the time derivative is zero. The additional convective acceleration term required to quantify fluid acceleration (e.g., in *Newton's second law*) is called the *convective derivative.* See also *Eulerian description, Lagrangian description, material derivative,* and *steady flow.*

convective derivative: See *material derivative* and *convective acceleration.*

creeping flow: Fluid flow in which frictional forces dominate fluid accelerations to the point that the flow can be well modeled with the acceleration term in Newton's second law set to zero. Such flows are characterized by Reynolds numbers that are small compared to 1 (Re $\ll$ 1). Since Reynolds number typically can be written as characteristic velocity times characteristic length divided by kinematic viscosity (VL/ν), creeping flows are often slow-moving flows around very small objects (e.g., sedimentation of dust particles in air or motion of spermatozoa in water), or with very viscous fluids (e.g., glacier and tar flows). Also called Stokes flow.

deformation rate: See *strain rate.*

derived dimensions: See *dimensions.*

deviatoric stress tensor: Another term for *viscous stress tensor.* See *stress.*

differential analysis: Analysis at a point in the flow (as opposed to over a *control volume*).

differential volume/area/length: A small volume δV, area δA, or length δx in the limit of the volume/area/length shrinking to a point. Derivatives are often produced in this limit. (Note that δ is sometimes written as Δ or d.)

dimensional analysis: A process of analysis based solely on the variables of relevance to the flow system under study, the dimensions of the variables, and dimensional homogeneity. After determining the other variables on which a variable of interest depends (e.g., drag on a car depends on the speed and size of the car, fluid viscosity, fluid density, and surface roughness), one applies the principle of dimensional homogeneity with the *Buckingham Pi theorem* to relate an appropriately nondimensionalized variable of interest (e.g., drag) with the other variables appropriately nondimensionalized (e.g., Reynolds numbers, roughness ratio, and Mach number).

dimensional homogeneity: The requirement that summed terms must have the same *dimensions* (e.g., ρV^2, pressure P, and shear stress τ_{xy} are dimensionally homogeneous while *power,* specific enthalpy h, and $P\dot{m}$ are not). Dimensional homogeneity is the basis of *dimensional analysis.*

dimensionality: The number of spatial coordinates in whose direction velocity components and/or other variables vary for a specified coordinate system. For example, *fully developed flow* in a tube is one-dimensional (1-D) in the radial direction r since the only nonzero velocity component (the axial, or x-, component) is constant in the x- and θ-directions, but varies in the r-direction. *Planar flows* are two-dimensional (2-D). Flows over *bluff bodies* such as cars, airplanes, and buildings are three-dimensional (3-D). Spatial derivatives are nonzero only in the directions of dimensionality.

dimensions: The required specification of a physical quantity beyond its numerical value. See also *units.*

derived (or secondary) dimensions: Combinations of fundamental dimensions. Examples of derived dimensions are: velocity (L/t), stress or pressure (F/L^2 = m/(Lt^2), energy or work (mL^2/t^2 = FL), density (m/L^3), specific weight (F/L^3), and specific gravity (unitless).

fundamental (primary, basic) dimensions: Mass (m), length (L), time (t), temperature (T), electrical current (I), amount of light (C), and amount of matter (N) without reference to a specific system of units. Note that the force dimension is obtained through Newton's law as F = mL/t^2 (thus, the mass dimension can be replaced with a force dimension by replacing m with Ft^2/L).

drag coefficient: Nondimensional drag given by the *drag force* on an object nondimensionalized by *dynamic pressure* of the free-stream flow times frontal area of the object:

$$C_D \equiv \frac{F_D}{\frac{1}{2}\rho V^2 A}$$

Note that at high Reynolds numbers (Re $\gg$ 1), C_D is a normalized variable, whereas at Re $\ll$ 1, C_D is nondimensional but is not normalized (see *normalization*). See also *lift coefficient.*

drag force: The force on an object opposing the motion of the object. In a frame of reference moving with the object, this is the force on the object in the direction of flow. There are multiple components to drag force:

friction drag: The part of the drag on an object resulting from integrated surface *shear stress* in the direction of flow relative to the object.

induced drag: The component of the drag force on a finite-span wing that is "induced" by lift and associated with the *tip vortices* that form at the tips of the wing and "downwash" behind the wing.

pressure (or form) drag: The part of the drag on an object resulting from integrated surface *pressure* in the direction of flow relative to the object. Larger pressure on the front of a moving *bluff body* (such as a car) relative to the rear results from massive *flow separation* and *wake* formation at the rear.

dynamic pressure: When the *Bernoulli equation* in *incompressible steady flow* and/or the *conservation of energy* equation along a streamline are written in forms where each term in the equations has the *dimensions* force/area, dynamic pressure is the *kinetic energy* (per unit volume) term (i.e., $\frac{1}{2}\rho V^2$).

dynamic similarity: See *similarity*.

dynamic viscosity: See *viscosity*.

dynamics: When contrasted with *statics* the term refers to the application of Newton's second law of motion to moving matter. When contrasted with *kinematics* the term refers to forces or accelerations through Newton's law force balances.

eddy viscosity: See *turbulence models*.

efficiency: A ratio that describes levels of losses of useful power obtained from a device. Efficiency of 1 implies no losses in the particular function of the device for which a particular definition of efficiency is designed. For example, mechanical efficiency of a pump is defined as the ratio of useful mechanical power transferred to the flow by the pump to the mechanical energy, or shaft work, required to drive the pump. Pump-motor efficiency of a pump is defined as the ratio of useful mechanical power transferred to the flow over the electrical power required to drive the pump. Pump-motor efficiency, therefore, includes additional losses and is thus lower than mechanical pump efficiency.

energy: A state of matter described by the first law of thermodynamics that can be altered at the macroscopic level by work, and at the microscopic level through adjustments in thermal energy.

    flow energy: Synonymous with *flow work*. The work associated with *pressure* acting on a flowing *fluid*.

    heat (transfer): The term "heat" is generally used synonymously with *thermal energy*. Heat transfer is the transfer of thermal energy from one physical location to another.

    internal energy: Forms of energy arising from the microscopic motions of molecules and atoms, and from the structure and motions of the subatomic particles comprising the atoms and molecules, within matter.

    kinetic energy: Macroscopic (or mechanical) form of energy arising from the speed of matter relative to an inertial frame of reference.

    mechanical energy: The nonthermal components of energy; examples include kinetic and potential energy.

    potential energy: A mechanical form of energy that changes as a result of macroscopic displacement of matter relative to the gravitational vector.

    thermal energy: Internal energy associated with microscopic motions of molecules and atoms. For single-phase systems, it is the energy represented by temperature.

    total energy: Sum of all forms of energy. Total energy is the sum of kinetic, potential, and internal energies. Equivalently, total energy is the sum of mechanical and thermal energies.

    work energy: The integral of force over the distance in which a mass is moved by the force. Work is energy associated with the movement of matter by a force.

energy grade line: See *grade lines*.

English system: See *units*.

entry length: The entry flow region in a pipe or duct flow where the wall boundary layers are thickening toward the center with axial distance x of the duct, so that axial derivatives are nonzero. As with the *fully developed* region, the *hydrodynamic entry length* involves growth of a velocity boundary layer, and the *thermal entry length* involves growth of a temperature boundary layer.

Eulerian derivative: See *material derivative*.

Eulerian description: In contrast with a *Lagrangian description*, an Eulerian analysis of fluid flow is developed from a frame of reference through which the *fluid particles* move. In this frame the acceleration of fluid particles is not simply the time derivative of fluid velocity, and must include another term, called *convective acceleration*, to describe the change in velocity of fluid particles as they move through a *velocity field*. Note that velocity fields are always defined in an Eulerian frame of reference.

extensional strain rate: See *strain rate*.

extensive property: A fluid property that depends on total volume or total mass (e.g., total internal energy). See *intensive property*.

field: The representation of a flow variable as a function of Eulerian coordinates (x, y, z). For example, the *velocity* and *acceleration fields* are the fluid velocity and acceleration vectors ($\vec{V}, \vec{a}$) as functions of position (x, y, z) in the *Eulerian description* at a specified time t.

    flow field: The field of flow variables. Generally, this term refers to the velocity field, but it may also mean all field variables in a fluid flow.

first law of thermodynamics: See *conservation laws, conservation of energy*.

flow separation: A phenomenon where a *boundary layer* adjacent to a surface is forced to leave, or "separate" from, the surface due to "adverse" pressure forces (i.e., increasing pressure) in the flow direction. Flow separation occurs in regions of high surface curvature, for example, at the rear of an automobile and other bluff bodies.

flow work: The work term in *first law of thermodynamics* applied to fluid flow associated with pressure forces on the flow. See *energy, flow energy*.

GLOSSARY

fluid: A material that when sheared deforms continuously in time during the period that shear forces are applied. By contrast, shear forces applied to a *solid* cause the material either to deform to a fixed static position (after which deformation stops), or cause the material to fracture. Consequently, whereas solid deformations are generally analyzed using strain and shear, fluid flows are analyzed using rates of strain and shear (see *strain rate*).

fluid mechanics/dynamics: The study and analysis of fluids through the macroscopic conservation laws of physics, i.e., conservation of mass, momentum (*Newton's second law*), and energy (first law of thermodynamics), and the second law of thermodynamics.

fluid particle/element: A *differential* particle, or element, embedded in a fluid flow containing always the same atoms and molecules. Thus a fluid particle has fixed mass δm and moves with the flow with local flow velocity $\vec{V}$, acceleration $\vec{a}_{particle} = D\vec{V}/Dt$ and trajectory $(x_{particle}(t), y_{particle}(t), t_{particle}(t))$. See also *material derivative, material particle, material position vector,* and *pathline*.

forced flow: Flow resulting from an externally applied force. Examples include liquid flow through tubes driven by a pump and fan-driven airflow for cooling computer components. *Natural flows,* in contrast, result from internal buoyancy forces driven by temperature (i.e., density) variations within a fluid in the presence of a gravitational field. Examples include buoyant plumes around a human body or in the atmosphere.

friction/frictional: See *Newtonian fluid, viscosity,* and *viscous force*.

friction factor: It can be shown from *dimensional analysis* and *conservation of momentum* applied to a *steady fully developed* pipe flow that the frictional contribution to the pressure drop along the pipe, nondimensionalized by flow *dynamic pressure* ($\frac{1}{2}\rho V_{avg}^2$), is proportional to the length-to-diameter ratio (L/D) of the pipe. The proportionality factor f is called the friction factor. The friction factor is quantified from experiment (turbulent flow) and theory (laminar flow) in empirical relationships, and in the *Moody chart,* as a function of the Reynolds number and nondimensional roughness. Conservation of momentum shows that the friction factor is proportional to the nondimensional wall shear stress (i.e., the *skin friction*).

frictionless flow: Mathematical treatments of fluid flows sometimes use conservation of momentum and energy equations without the frictional terms. Such mathematical treatments "assume" that the flow is "frictionless," implying no *viscous force* (*Newton's second law*), nor *viscous dissipation* (*first law of thermodynamics*). However, no real fluid flow of engineering interest can exist without viscous forces, dissipation, and/or head losses in regions of practical importance. The engineer should always identify the flow regions where frictional effects are concentrated. When developing models for prediction, the engineer should consider the role of these viscous regions in the prediction of variables of interest and should estimate levels of error in simplified treatments of the viscous regions. In high *Reynolds number* flows, frictional regions include boundary layers, *wakes, jets, shear layers,* and flow regions surrounding *vortices*.

Froude number: An order-of-magnitude estimate of the ratio of the inertial term in Newton's law of motion to the gravity force term. The Froude number is an important nondimensional group in free-surface flows, as is generally the case in channels, rivers, surface flows, etc.

fully developed: Used by itself, the term is generally understood to imply hydrodynamically fully developed, a flow region where the velocity field is constant along a specified direction in the flow. In the fully developed region of pipe or duct flow, the velocity field is constant in the axial direction, x (i.e., it is independent of x), so that x-derivatives of velocity are zero in the fully developed region. There also exists the concept of "thermally fully developed" for the temperature field; however, unlike hydrodynamically fully developed regions where both the magnitude and shape of the velocity profile are constant in x, in thermally fully developed regions only the shape of the temperature profile is constant in x. See also *entry length*.

fundamental dimensions: See *dimensions*.

gage pressure: Pressure (P) relative to atmospheric pressure (P_{atm}). That is, $P_{gage} = P - P_{atm}$. See also *stress, pressure stress*. Thus $P_{gage} > 0$ or $P_{gage} < 0$ is simply the pressure above or below atmospheric pressure.

gas dynamics: The study and analysis of gases and vapors through the macroscopic conservation laws of physics (see *fluid mechanics/dynamics*).

geometric similarity: See *similarity*.

grade lines: Lines of *head* summations.

    energy grade line: Line describing the sum of *pressure head, velocity head,* and *elevation head*. See *head*.

    hydraulic grade line: Line describing the sum of *pressure head* and *elevation head*. See *head*.

Hagen–Poiseuille flow: See *Poiseuille flow*.

head: A quantity (pressure, kinetic energy, etc.) expressed as an equivalent column height of a fluid. *Conservation of energy* for *steady flow* written for a *control volume* surrounding a central *streamline* with one inlet and one outlet, or shrunk to a streamline, can be written such that each term has the *dimensions* of length. Each of these terms is called a head term:

    elevation head: The term in the head form of *conservation of energy* (see *head*) involving distance in the direction opposite to the gravitational vector relative to a predefined datum (z).

head loss: The term in the head form of *conservation of energy* (see *head*) that contains frictional losses and other irreversibilities. Without this term, the energy equation for streamlines becomes the *Bernoulli equation* in head form.

pressure head: The term in the head form of *conservation of energy* (see *head*) involving pressure ($P/\rho g$).

velocity head: The (*kinetic energy*) term in the head form of *conservation of energy* (see *head*) involving velocity ($V^2/2g$).

heat: See *energy*.

hot-film anemometer: Similar to a *hot-wire anemometer* except using a metallic film rather than a wire; used primarily for liquid flows. The measurement portion of a hot-film probe is generally larger and more rugged than that of a hot-wire probe.

hot-wire anemometer: A device used to measure a velocity component locally in a gas flow based on the relationship between the flow around a thin heated wire (the hot wire), temperature of the wire, and heating of the wire resulting from a current. See also *hot-film anemometer*.

hydraulic grade line: See *grade lines*.

hydraulics: The *hydrodynamics* of liquid and vapor flow in pipes, ducts, and open channels. Examples include water piping systems and ventilation systems.

hydrodynamic entry length: See *entry length*.

hydrodynamically fully developed: See *fully developed*.

hydrodynamics: The study and analysis of liquids through the macroscopic conservation laws of physics (see *fluid mechanics/dynamics*). The term is sometimes applied to *incompressible* vapor and gas flows, but when the fluid is air, the term *aerodynamics* is generally used instead.

hydrostatic pressure: The component of *pressure* variation in a fluid flow that would exist in the absence of flow as a result of gravitational body force. This term appears in the hydrostatic equation and in the *Bernoulli equation*. See also *dynamic* and *static pressure*.

hypersonic: An order of magnitude or more above the speed of sound (Mach number $\gg 1$).

ideal fluid: See *perfect fluid*.

ideal gas: A gas at low enough density and/or high enough temperature that (a) density, pressure, and temperature are related by the ideal-gas equation of state, $P = \rho RT$, and (b) specific internal energy and enthalpy are functions only of temperature.

incompressible flow: A fluid flow where variations in density are sufficiently small to be negligible. Flows are generally incompressible either because the fluid is incompressible (liquids) or because the Mach number is low (roughly < 0.3).

induced drag: See *drag force*.

inertia/inertial: The acceleration term in Newton's second law, or effects related to this term. Thus, a flow with higher inertia requires larger deceleration to be brought to rest.

inertial sublayer: A highly turbulent part of a turbulent boundary layer, close to the wall but just outside the *viscous sublayer* and *buffer layer*, where *turbulent stresses* are large compared to *viscous stresses*.

intensive property: A fluid property that is independent of total volume or total mass (i.e., an *extensive property* per unit mass or sometimes per unit volume).

internal energy: See *energy*.

inviscid (region of) flow: Region of a fluid flow where viscous forces are sufficiently small relative to other forces (typically, pressure force) on *fluid particles* in that region of the flow to be neglected in *Newton's second law* of motion to a good level of approximation (compare with *viscous flow*). See also *frictionless flow*. An inviscid region of flow is not necessarily *irrotational*.

irrotational (region of) flow: A region of a flow with negligible *vorticity* (i.e., *fluid particle* rotation). Also called *potential flow*. An irrotational region of flow is also inviscid.

isocontour plot: See *contour plot*.

jet: A friction-dominated region issuing from a tube or orifice and formed by surface boundary layers that have been swept behind by the mean velocity. Jets are characterized by high *shear* with the highest velocities in the center of the jet and lowest velocities at the edges. *Frictional force, viscous stress,* and *vorticity* are significant in jets.

Kármán vortex street: The *two-dimensional* alternating unsteady pattern of *vortices* that is commonly observed behind circular cylinders in a flow (e.g., the vortex street behind wires in the wind is responsible for the distinct tone sometimes heard).

kinematic similarity: See *similarity*.

kinematic viscosity: Fluid *viscosity* divided by density.

kinematics: In contrast with *dynamics*, the kinematic aspects of a fluid flow are those that do not directly involve Newton's second law force balance. Kinematics refers to descriptions and mathematical derivations based only on conservation of mass (continuity) and definitions related to flow and deformation.

kinetic energy: See *energy*.

kinetic energy correction factor: *Control volume* analysis of the *conservation of energy* equation applied to tubes contains area integrals of kinetic energy flux. The integrals are often approximated as proportional to kinetic energy formed with area-averaged velocity, V_{avg}. The inaccuracy in this approximation can be significant, so a kinetic energy correction factor, α, multiplies the term to improve the approximation. The correction α depends on the shape of the *velocity profile*, is largest for *laminar* profiles (*Poiseuille flow*), and is closest to 1 in *turbulent* pipe flows at very high *Reynolds numbers*.

Lagrangian derivative: See *material derivative*.

Lagrangian description: In contrast with the *Eulerian description*, a Lagrangian analysis is developed from a frame of reference attached to moving material particles. For example, solid particle acceleration in the standard Newton's second law form, $\vec{F} = m\vec{a}$, is in a coordinate system that moves with the particle so that acceleration $\vec{a}$ is given by the time derivative of particle velocity. This is the typical analytical approach used for analysis of the motion of solid objects.

laminar flow: A stable well-ordered state of fluid flow in which all pairs of adjacent *fluid particles* move alongside one another forming laminates. A flow that is not laminar is either *turbulent* or *transitional* to turbulence, which occurs above a critical *Reynolds number*.

laser Doppler velocimetry (LDV): Also called laser Doppler anemometry (LDA). A technique for measuring a velocity component locally in a flow based on the Doppler shift associated with the passage of small particles in the flow through the small target volume formed by the crossing of two laser beams. Unlike *hot-wire* and *hot-film anemometry* and like *particle image velocimetry*, there is no interference to the flow.

lift coefficient: Nondimensional lift given by the lift force on a lifting object (such as an airfoil or wing) nondimensionalized by dynamic pressure of the free-stream flow times planform area of the object:

$$C_L \equiv \frac{F_L}{\frac{1}{2}\rho V^2 A}$$

Note that at high Reynolds numbers (Re $\gg$ 1), C_L is a normalized variable, whereas at Re $\ll$ 1, C_L is nondimensional but is not normalized (see *normalization*). See also *drag coefficient*.

lift force: The net aerodynamic force on an object perpendicular to the motion of the object.

linear strain rate: Synonymous with *extensional strain rate*. See *strain rate*.

losses: Frictional *head losses* in pipe flows are separated into those losses in the fully developed pipe flow regions of a piping network, the *major losses*, plus head losses in other flow regions of the network, the *minor losses*. Minor loss regions include *entry lengths*, pipe couplings, bends, valves, etc. It is not unusual for minor losses to be larger than major losses.

Mach number: *Nondimensional* ratio of the characteristic speed of the flow to the speed of sound. Mach number characterizes the level of *compressibility* in response to pressure variations in the flow.

major losses: See *losses*.

manometer: A device that measures pressure based on hydrostatic pressure principles in liquids.

material acceleration: The acceleration of a *fluid particle* at the point (x, y, z) in a flow at time t. This is given by the *material derivative* of fluid velocity: $D\vec{V}(x, y, z, t)/Dt$.

material derivative: Synonymous terms are *total derivative*, *substantial derivative*, and *particle derivative*. These terms mean the time rate of change of fluid variables (temperature, velocity, etc.) moving with a *fluid particle*. Thus, the material derivative of temperature at a point (x, y, z) at time t is the time derivative of temperature attached to a moving *fluid particle* at the point (x, y, z) in the flow at the time t. In a *Lagrangian* frame of reference (i.e., a frame attached to the moving particle), particle temperature $T_{particle}$ depends only on time, so a time derivative is a total derivative $dT_{particle}(t)/dt$. In an Eulerian frame, the temperature *field* $T(x, y, z, t)$ depends on both position (x, y, z) and time t, so the *material derivative* must include both a partial derivative in time and a *convective derivative*: $dT_{particle}(t)/dt \equiv DT(x, y, z, t)/Dt = \partial T/\partial t + \vec{V}\cdot\vec{\nabla}T$. See also *field*.

material particle: A *differential* particle, or element, that contains always the same atoms and molecules. Thus a material particle has fixed mass δm. In a fluid flow, this is the same as a *fluid particle*.

material position vector: A vector $[x_{particle}(t), y_{particle}(t), z_{particle}(t)]$ that defines the location of a *material particle* as a function of time. Thus the material position vector in a fluid flow defines the trajectory of a *fluid particle* in time.

mean: Synonymous with *average*.

mechanical energy: See *energy*.

mechanics: The study and analysis of matter through the macroscopic conservation laws of physics (mass, momentum, energy, second law).

minor losses: See *losses*.

mixing length: See *turbulence models*.

momentum: The momentum of a *material particle* (or *fluid particle*) is the mass of the material particle times its velocity. The momentum of a macroscopic volume of material particles is the integrated momentum per unit volume over the volume, where momentum per unit volume is the density of the material particle times its velocity. Note that momentum is a vector.

momentum flux correction factor: A correction factor added to correct for approximations made in the simplification of the area integrals for the momentum flux terms in the control volume form of *conservation of momentum*.

Moody chart: A commonly used plot of the *friction factor* as a function of the Reynolds number and roughness parameter for fully developed pipe flow. The chart is a combination of flow theory for laminar flow with a graphical representation of an empirical formula by Colebrook to a large set of experimental data for turbulent pipe flow of various values of "sandpaper" roughness.

natural flow: Contrast with *forced flow*.

Navier–Stokes equation: *Newton's second law* of fluid motion (or *conservation of momentum*) written for a fluid particle (the *differential* form) with the *viscous stress tensor* replaced by the *constitutive relationship* between *stress* and *strain rate* for Newtonian fluids. Thus the Navier–Stokes equation is simply Newton's law written for Newtonian fluids.

Newtonian fluid: When a fluid is subjected to a *shear stress*, the fluid continuously changes shape (deformation). If the fluid is Newtonian, the rate of deformation (i.e., strain rate) is proportional to the applied shear stress and the constant of proportionality is called *viscosity*. In general flows, the rate of deformation of a *fluid particle* is described mathematically by a *strain rate* tensor and the *stress* by a *stress tensor*. In flows of Newtonian fluids, the stress tensor is proportional to the strain rate tensor, and the constant of proportionality is called *viscosity*. Most common fluids (water, oil, gasoline, air, most gases and vapors) without particles or large molecules in suspension are Newtonian.

Newton's second law: See *conservation of momentum*.

nondimensionalization: The process of making a dimensional variable dimensionless by dividing the variable by a *scaling parameter* (a single variable or a combination of variables) that has the same dimensions. For example, the surface pressure on a moving ball might be nondimensionalized by dividing it by ρV^2, where ρ is fluid density and V is free-stream velocity. See also *normalization*.

non-Newtonian fluid: A non-Newtonian fluid is one that deforms at a rate that is not linearly proportional to the stress causing the deformation. Depending on the manner in which *viscosity* varies with *strain rate*, non-Newtonian fluids can be labeled shear thinning (viscosity decreases with increasing strain rate), shear thickening (viscosity increases with increasing strain rate), and viscoelastic (when the shearing forces are removed, the fluid particles return partially to an earlier shape). Suspensions and liquids with long-chain molecules are generally non-Newtonian. See also *Newtonian fluid* and *viscosity*.

normal stress: See *stress*.

normalization: A particular *nondimensionalization* where the *scaling parameter* is chosen so that the nondimensionalized variable attains a maximum value that is of order 1 (say, within roughly 0.5 to 2). Normalization is more restrictive (and more difficult to do properly) than nondimensionalization. For example, $P/(\rho V^2)$ discussed under *nondimensionalization* is also normalized pressure on a flying baseball (where Reynolds number Re $\gg$ 1), but is simply nondimensionalization of surface pressure on a small glass bead dropping slowly through honey (where Re $\ll$ 1).

no-slip condition: The requirement that at the interface between a fluid and a solid surface, the fluid velocity and surface velocity are equal. Thus if the surface is fixed, the fluid must obey the *boundary condition* that fluid velocity $= 0$ at the surface.

one-dimensional: See *dimensionality*.

open system: Same as *control volume*.

particle derivative: See *material derivative*.

particle image velocimetry (PIV): A technique for measuring a velocity component locally in a flow based on tracking the movement of small particles in the flow over a short time using pulsed lasers. Unlike *hot-wire* and *hot-film anemometry* and like laser Doppler velocimetry, there is no interference to the flow.

pathline: A curve mapping the trajectory of a *fluid particle* as it travels through a flow over a period of time. Mathematically, this is the curve through the points mapped out by the *material position vector* $[x_{particle}(t), y_{particle}(t), z_{particle}(t)]$ over a defined period of time. Thus, pathlines are formed over time, and each fluid particle has its own pathline. In a steady flow, fluid particles move along streamlines, so pathlines and streamlines coincide. In a nonsteady flow, however, pathlines and streamlines are generally very different. Contrast with *streamline*.

perfect fluid: Also called an *ideal fluid*, the concept of a fictitious fluid that can flow in the absence of all frictional effects. There is no such thing as a perfect fluid, even as an approximation, so the engineer need not consider the concept further.

periodic: An unsteady flow in which the flow oscillates about a steady mean.

Pitot-static probe: A device used to measure fluid velocity through the application of the Bernoulli equation with simultaneous measurement of *static* and *stagnation pressures*. Also called a Pitot-Darcy probe.

planar flow: A *two-dimensional* flow with two nonzero components of velocity in Cartesian coordinates that vary only in the two coordinate directions of the flow. Thus, all partial derivatives perpendicular to the plane of the flow are zero. See also *axisymmetric flow* and *dimensionality*.

Poiseuille flow: *Fully developed laminar flow* in a pipe or duct. Also called *Hagen–Poiseuille flow*. The mathematical model relationships for Poiseuille flow relating the flow rate and/or velocity profile to the pressure drop along the pipe/duct, fluid viscosity and geometry are sometimes referred to as *Poiseuille's law* (although strictly not a "law" of mechanics). The velocity profile of all Poiseuille flows is parabolic, and the rate of axial pressure drop is constant.

Poiseuille's law: See *Poiseuille flow*.

potential energy: See *energy*.

potential flow: Synonymous with *irrotational flow*. This is a region of a flow with negligible *vorticity* (i.e., *fluid particle* rotation). In such regions, a velocity *potential function* exists (thus the name).

potential function: If a region of a flow has zero *vorticity* (*fluid particle* spin), the velocity vector in that region can be written as the gradient of a scalar function called the velocity potential function, or simply the potential function. In practice, potential functions are often used to model flow regions where vorticity levels are small but not necessarily zero.

power: *Work* per unit time; time rate at which work is done.

pressure: See *stress*.

pressure force: As applicable to Newton's second law, this is the force acting on a *fluid particle* that arises from spatial gradients in pressure within the flow. See also *stress, pressure stress*.

pressure work: See *flow work*.

primary dimension: See *dimensions*.

profile plot: A graphical representation of the spatial variation of a fluid property (temperature, pressure, strain rate, etc.) through a region of a fluid flow. A profile plot defines property variations in part of a *field* (e.g., a temperature profile might define the variation of temperature along a line within the temperature field).

> **velocity profile:** The spatial variation in a velocity component or vector through a region of a fluid flow. For example, in a pipe flow the velocity profile generally defines the variation in axial velocity with radius across the pipe cross section, while a *boundary layer* velocity profile generally defines variation in axial velocity normal to the surface. The velocity profile is part of a velocity *field*.

quasi-steady flow: See *steady flow*.

Reynolds number: An order-of-magnitude estimate of the ratio of the following two terms in Newton's second law of motion over a region of the flow: the *inertial* (or acceleration) term over the viscous force term. Most but not all Reynolds numbers can be written as an appropriate characteristic velocity V times a characteristic length scale L consistent with the velocity V, divided by the kinematic viscosity ν of the fluid: $Re = VL/\nu$. The Reynolds number is arguably the most important nondimensional *similarity* parameter in fluid flow analysis since it gives a rough estimate of the importance of frictional force in the overall flow.

Reynolds stress: Velocity components (and other variables) in turbulent flows are separated into mean plus fluctuating components. When the equation for mean streamwise velocity component is derived from the *Navier–Stokes equation*, six new terms appear given by fluid density times the averaged product of two velocity components. Because these terms have the same units as *stress* (force/area), they are called turbulent stresses or Reynolds stresses (in memory of Osborne Reynolds who first quantified turbulent variables as mean + fluctuation). Just as *viscous stresses* can be written as a tensor (or matrix), we define a Reynolds stress tensor with Reynolds normal stress components and Reynolds shear stress components. Although Reynolds stresses are not true stresses, they have qualitatively similar effects as do viscous stresses, but as a result of the large chaotic *vortical* motions of turbulence rather than the microscopic molecular motions that underlie viscous stresses.

Reynolds transport theorem: The mathematical relationship between the time rate of change of a fluid property in a *system* (volume of fixed mass moving with the flow) and the time rate of change of a fluid property in a *control volume* (volume, usually fixed in space, with fluid mass moving across its surface). This finite volume expression is closely related to the *material (time) derivative* of a fluid property attached to a moving *fluid particle*. See also *conservation laws*.

rheology: The study and mathematical representation of the deformation of different fluids in response to surface forces, or *stress*. The mathematical relationships between stress and deformation rate (or strain rate) are called *constitutive equations*. The Newtonian relationship between *stress* and *strain rate* is the simplest example of a rheological constitutive equation. See also *Newtonian* and *non-Newtonian fluid*.

rotation rate: The angular velocity, or rate of spin, of a *fluid particle* (a vector, with units rad/s, given by 1/2 the curl of the velocity vector). See also *vorticity*.

rotational flow: Synonymous with *vortical flow*, this term describes a flow field, or a region of a flow field, with significant levels of *vorticity*.

saturation pressure: The pressure at which the phase of a simple compressible substance changes between liquid and vapor at fixed temperature.

saturation temperature: The temperature at which the phase of a simple compressible substance changes between liquid and vapor at fixed pressure.

scaling parameter: A single variable, or a combination of variables, that is chosen to nondimensionalize a variable of interest. See also *nondimensionalization* and *normalization*.

schlieren technique: An experimental technique to visualize flows based on the refraction of light from varying fluid density. The illuminance level in a schlieren image responds to the first spatial derivative of density.

secondary dimensions: See *dimensions*.

shadowgraph technique: An experimental technique to visualize flows based on the refraction of light from varying fluid density. The illuminance level in a shadowgraph image responds to the second spatial derivative of density.

shear: Refers to gradients (derivatives) in velocity components in directions normal to the velocity component.

> **shear force:** See *stress, shear stress*.
>
> **shear layer:** A quasi two-dimensional flow region with a high gradient in streamwise velocity component in the transverse flow direction. Shear layers are inherently *viscous* and *vortical* in nature.
>
> **shear rate:** The gradient in streamwise velocity in the direction perpendicular to the velocity. Thus, if streamwise (x) velocity u varies in y, the shear rate is du/dy. The term is applied to *shear flows*, where shear rate is twice the *shear strain rate*. See also *strain rate*.
>
> **shear strain:** See *strain rate*.
>
> **shear stress:** See *stress, shear stress*.

shear thickening fluid: See *non-Newtonian fluid*.

shear thinning fluid: See *non-Newtonian fluid*.

SI system: See *units*.

similarity: The principle that allows one to quantitatively relate one flow to another when certain conditions are met. *Geometric similarity*, for example, must be true before one can hope for *kinematic* or *dynamic similarity*. The quantitative relationship that relates one flow to another is developed using a combination of dimensional analysis and data (generally, experimental, but also numerical or theoretical).

> **dynamic similarity:** If two objects are *geometrically* and *kinematically similar*, then if the ratios of all forces (pressure, viscous stress, gravity force, etc.) between a point in the flow surrounding one object, and the same point scaled appropriately in the flow surrounding the other object, are all the same at all corresponding pairs of points, the flow is *dynamically similar*.
>
> **geometric similarity:** Two objects of different size are geometrically similar if they have the same geometrical shape (i.e., if all dimensions of one are a constant multiple of the corresponding dimensions of the other).
>
> **kinematic similarity:** If two objects are *geometrically similar*, then if the ratios of all velocity components between a point in the flow surrounding one object, and the same point scaled appropriately in the flow surrounding the other object, are all the same at all corresponding pairs of points, the flow is *kinematically similar*.

skin friction: Surface shear stress τ_w nondimensionalized by an appropriate *dynamic pressure* $\frac{1}{2}\rho V^2$. Also called the skin friction coefficient, C_f.

solid: A material that when sheared either deforms to a fixed static position (after which deformation stops) or fractures. See also *fluid*.

sonic: At the speed of sound (*Mach number* = 1).

specific gravity: Fluid density nondimensionalized by the density of liquid water at 4°C and atmospheric pressure (1 g/cm^3 or 1000 kg/m^3). Thus, specific gravity, SG = ρ/ρ_{water}.

specific weight: The weight of a fluid per unit volume, i.e., fluid density times acceleration due to gravity (specific weight, $\gamma \equiv \rho g$).

spin: See *rotation rate* and *vorticity*.

stability: A general term that refers to the tendency of a material particle or object (fluid or solid) to move away from or return when displaced slightly from its original position.

> **neutrally stable:** See *stability*. When displaced slightly, the particle or object will remain in its displaced position.
>
> **stable:** See *stability*. When displaced slightly, the particle or object will return to its original position.
>
> **unstable:** See *stability*. When displaced slightly, the particle or object will continue to move from its original position.

stagnation point: A point in a fluid flow where the velocity goes to zero. For example, the point on the *streamline* that intersects the nose of a moving projectile is a stagnation point.

stall: The phenomenon of massive *flow separation* from the surface of a wing when *angle of attack* exceeds a critical value, and consequent dramatic loss of lift and increase in drag. A plane in stall drops rapidly and must have its nose brought down to reestablish attached boundary layer flow and regenerate lift and reduce drag.

static pressure: Another term for *pressure*, used in context with the *Bernoulli equation* to distinguish it from *dynamic pressure*.

statics: The mechanical study and analysis of material that is fully at rest in a specific frame of reference.

steady flow: A flow in which all fluid variables (velocity, pressure, density, temperature, etc.) at all fixed points in the flow are constant in time (but generally vary from place to place). Thus, in steady flows all partial derivatives in time are zero. Flows that are not precisely steady but that change sufficiently slowly in time to neglect time derivative terms with relatively little error are called *quasi-steady*.

Stokes flow: See *creeping flow*.

strain: See *strain rate*.

strain rate: Strain rate can also be called *deformation rate*. This is the rate at which a *fluid particle* deforms (i.e., changes shape) at a given position and time in a fluid flow. To fully quantify all possible changes in shape of a *three-dimensional* fluid particle require six numbers. Mathematically, these are the six independent components of a second-rank symmetric strain rate tensor, generally written as a symmetric 3 × 3 matrix. Strain is time-integrated strain rate and describes deformation of a fluid particle after a period of time. See *stress*.

GLOSSARY

extensional strain rate: The components of strain rate that describe elongation or compression of a *fluid particle* in one of the three coordinate directions. These are the three diagonal elements of the strain rate tensor. The definition of extensional strain depends on one's choice of coordinate axes. Also called *linear strain rate*.

shear strain rate: The components of strain rate that describe deformation of a *fluid particle* in response to shear changing an angle between planes mutually perpendicular to the three coordinate axes. These are the off-diagonal elements of the strain rate tensor. The definition of shear strain depends on one's choice of coordinate axes.

volumetric strain rate: Rate of change of volume of a *fluid particle* per unit volume. Also called bulk strain rate and rate of volumetric dilatation.

streakline: Used in flow visualization of fluid flows, this is a curve defined over time by the release of a marker (dye or smoke) from a fixed point in the flow. Contrast with *pathline* and *streamline*. In a steady flow, streamlines, *pathlines,* and *streaklines* all coincide. In a nonsteady flow, however, these sets of curves are each different from one another.

stream function: The two velocity components in a *two-dimensional* steady incompressible flow can be defined in terms of a single two-dimensional function ψ that automatically satisfies conservation of mass (the continuity equation), reducing the solution of the two-component velocity field to the solution of this single stream function. This is done by writing the two velocity components as spatial derivatives of the stream function. A wonderful property of the stream function is that (iso)contours of constant ψ define *streamlines* in the flow.

streamline: A curve that is everywhere tangent to a velocity vector in a fluid velocity *field* at a fixed instant in time. Thus, the streamlines indicate the direction of the fluid motions at each point. In a *steady flow,* streamlines are constant in time and *fluid particles* move along streamlines. In a *nonsteady flow* the streamlines change with time and fluid particles do not move along streamlines. Contrast with *pathline*.

streamtube: A bundle of streamlines. A streamtube is usually envisioned as a surface formed by an infinite number of streamlines initiated within the flow on a circular circuit and tending to form a tubelike surface in some region of the flow.

stress: A component of a force distributed over an area is written as the integral of a stress over that area. Thus, stress is the force component dF_i on a differential area element divided by the area of the element dA_j (in the limit $dA_j \to 0$), where i and j indicate a coordinate direction $x, y,$ or z. Stress $\sigma_{ij} = dF_i/dA_j$ is therefore a force component per unit area in the i-direction on surface j. To obtain the surface force from stress, one integrates stress over the corresponding surface area. Mathematically, there are six independent components of a second-rank symmetric *stress tensor,* generally written as a symmetric 3×3 matrix.

normal stress: A stress (force component per unit area) that acts perpendicular to the area. Therefore σ_{xx}, σ_{yy}, and σ_{zz} are normal stresses. The normal force over a surface is the net force from shear stress, given by integrating the shear stress over the surface area. The normal stresses are the diagonal elements of the *stress tensor*.

pressure stress: In a fluid at rest all stresses are normal stresses and all stresses act inward on a surface. At a fixed point, the three normal stresses are equal and the magnitude of these equal normal stresses is called pressure. Thus, in a static fluid $\sigma_{xx} = \sigma_{yy} = \sigma_{zz} = -P$, where P is pressure. In a moving fluid, stresses in addition to pressure are *viscous stresses*. A pressure force on a surface is the pressure stress integrated over the surface. The pressure force per unit volume on a *fluid particle* for Newton's second law, however, is the negative of the gradient (spatial derivatives) of pressure at that point.

Reynolds stress: See *Reynolds stress*.

shear stress: A stress (force component per unit area) that acts tangent to the area. Therefore, σ_{xy}, σ_{yx}, σ_{xz}, σ_{zx}, σ_{yz}, and σ_{zy} are shear stresses. The shear force over a surface is the net force from shear stress, given by integrating the shear stress over the surface area. The shear stresses are the off-diagonal elements of the *stress tensor*.

turbulent stress: See *Reynolds stress*.

viscous stress: Flow creates stresses in the fluid that are in addition to hydrostatic pressure stresses. These additional stresses are viscous since they arise from friction-induced fluid deformations within the flow. For example, $\sigma_{xx} = -P + \tau_{xx}$, $\sigma_{yy} = -P + \tau_{yy}$, and $\sigma_{zz} = -P + \tau_{zz}$, where τ_{xx}, τ_{yy}, and τ_{zz} are viscous normal stresses. All shear stresses result from friction in a flow and are therefore viscous stresses. A viscous force on a surface is a viscous stress integrated over the surface. The viscous force per unit volume on a *fluid particle* for Newton's second law, however, is the divergence (spatial derivatives) of the viscous stress tensor at that point.

stress tensor: See *stress*.

subsonic: Below the speed of sound (*Mach number* < 1).

substantial derivative: See *material derivative*.

supersonic: Above the speed of sound (*Mach number* > 1).

surface tension: The force per unit length at a liquid–vapor or liquid–liquid interface resulting from the imbalance in attractive forces among like liquid molecules at the interface.

system: Usually when the word *system* is used by itself, *closed system* is implied, in contrast with a *control volume* or *open system*.

closed system: A volume specified for analysis that encloses always the same *fluid particles*. Therefore, no flow crosses any part of the volume's surface and a closed system must move with the flow. Note that Newton's law analysis of solid particles is generally a *closed system* analysis, sometimes referred to as a free body.

open system: A volume specified for analysis where flow crosses at least part of the volume's surface. Also called a *control volume*.

thermal energy: See *energy*.

three-dimensional: See *dimensionality*.

timeline: Used for visualization of fluid flows, this is a curve defined at some instant in time by the release of a marker from a line in the flow at some earlier instant in time. The timeline, often used to approximate a *velocity profile* in a laboratory flow, is very different from *streaklines, pathlines,* and *streamlines*.

tip vortex: *Vortex* formed off each tip of an airplane wing as a byproduct of lift. Synonymous with *trailing vortex*. See also *induced drag*.

total derivative: See *material derivative*.

total energy: See *energy*.

trailing vortex: See *tip vortex*.

trajectory: See *pathline*.

transient period: A time-dependent period of flow evolution leading to a new equilibrium period that is generally, but not necessarily, steady. An example is the start-up period after a jet engine is switched on, leading to a steady (equilibrium) jet flow.

transitional flow: An unstable *vortical* fluid flow at a Reynolds number higher than a critical value that is large relative to 1, but is not sufficiently high that the flow has reached a fully *turbulent flow* state. Transitional flows often oscillate randomly between *laminar* and turbulent states.

turbulence models: Constitutive model relationships between *Reynolds stresses* and the mean velocity field in turbulent flows. Such model equations are necessary to solve the equation for mean velocity. A simple and widely used modeled form for the Reynolds stresses is to write them like the Newtonian relationship for viscous stresses, as proportional to the mean strain rate, with the proportionality being a *turbulent viscosity* or *eddy viscosity*. However, unlike Newtonian fluids, the eddy viscosity is a strong function of the flow itself, and the different ways in which eddy viscosity is modeled as a function of other calculated flow field variables constitute different eddy viscosity models. One traditional approach to modeling eddy viscosity is in terms of a *mixing length*, which is made proportional to a length set by the flow.

turbulent flow: An unstable disordered state of *vortical* fluid flow that is inherently *unsteady* and that contains eddying motions over a wide range of sizes (or scales). Turbulent flows are always at *Reynolds numbers* above a critical value that is large relative to 1. Mixing is hugely enhanced, surface shear stresses are much higher, and head loss is greatly increased in turbulent flows as compared to corresponding *laminar flows*.

turbulent stress: See *Reynolds stress*.

turbulent viscosity: See *turbulence models*.

two-dimensional: See *dimensionality*.

units: A specific system to quantify numerically the dimensions of a physical quantity. The most common systems of units are SI (kg, N, m, s), English (lbm, lbf, ft, s), BGS (slug, lb, ft, s), and cgs (g, dyne, cm, s). See also *dimensions*.

unsteady flow: A flow in which at least one variable at a fixed point in the flow changes with time. Thus, in unsteady flows a partial derivative in time is nonzero for at least one point in the flow.

vapor pressure: The pressure below which a fluid, at a given temperature, will exist in the vapor state. See also *cavitation* and *saturation pressure*.

velocity: A vector that quantifies the rate of change in position and the direction of motion of a material particle.

velocity field: See *field*.

velocity profile: See *profile plots*.

viscoelastic fluid: See *non-Newtonian fluid*.

viscosity: See *Newtonian fluid*. Viscosity is a property of a fluid that quantifies the ratio of shear stress to rate of deformation (strain rate) of a *fluid particle*. (Therefore viscosity has the dimensions of stress/strain rate, or $Ft/L^2 = m/Lt$.) Qualitatively, viscosity quantifies the level by which a particular fluid resists deformation when subjected to shear stress (frictional resistance or *friction*). Viscosity is a measured property of a fluid and is a function of temperature. For Newtonian fluids, viscosity is independent of the rate of applied stress and strain rate. The viscous nature of *non-Newtonian fluids* is more difficult to quantify in part because viscosity varies with strain rate. The terms *absolute viscosity, dynamic viscosity,* and *viscosity* are synonymous. See also *kinematic viscosity*.

viscous (regions of) flow: Regions of a fluid flow where *viscous forces* are significant relative to other forces (typically, pressure force) on *fluid particles* in that region of the flow, and therefore cannot be neglected in *Newton's second law* of motion (compare with *inviscid flow*).

viscous (or frictional) force: As applicable to Newton's second law, this is the force acting on a *fluid particle* that arises from spatial gradients in viscous (or frictional) stresses within the flow. The viscous force on a surface is the viscous stress integrated over the surface. See also *stress, viscous stress*.

viscous stress tensor: See *stress*. Also called the *deviatoric stress tensor*.

viscous sublayer: The part of a turbulent boundary layer adjacent to the surface that contains the highest *viscous stresses*. The velocity gradient in this layer adjacent to the wall is exceptionally high. See also *inertial layer* and *buffer layer*.

vortex: A local structure in a fluid flow characterized by a concentration of vorticity (i.e., *fluid particle* spin or rotation) in a tubular core with circular streamlines around the core axis. A tornado, hurricane, and bathtub vortex are common examples of vortices. Turbulent flow is filled with small vortices of various sizes, strengths, and orientations.

vortical flow: Synonymous with *rotational flow,* this term describes a flow field, or a region of a flow field, with significant levels of *vorticity*.

vorticity: Twice the angular velocity, or rate of spin, of a *fluid particle* (a vector, with units rad/s, given by the curl of the velocity vector). See also *rotation rate*.

wake: The friction-dominated region behind a body formed by surface boundary layers that are swept to the rear by the free-stream velocity. Wakes are characterized by high *shear* with the lowest velocities in the center of the wake and highest velocities at the edges. *Frictional force, viscous stress,* and *vorticity* are significant in wakes.

work: See *energy*.

INDEX

Abscissa, 149
Absolute (dynamic) viscosity, 52, 54
Absolute pressure, 76–77, 90
Absolute temperature (T), 46–47
Absolute velocity, 164, 191, 245–246
Acceleration (a), 106–107, 135, 136–140, 199–200, 265, 561
  advective (convective), 137–138
  Bernoulli equation and, 199–200
  centripetal, 265, 561
  convective (advective), 137–138
  Euclidean flow description, 135, 136–139
  field (vector), 135, 136–139
  first-order difference approximation, 139
  fluid (material) flow, 135–140, 199–200
  fluid rigid-body motion, 106–107
  gradient (del) operator, 137
  Lagrangian description, 136–139
  local, 137
  material (derivative of), 139–140
  material (fluid) particles, 136–138
  material position vectors, 136–137
  Newton's second law for, 136
  normal (a_n), 199
  partial derivative operator (δ), 137
  particle streamlines (paths) of, 199
  point function, as a, 139
  residence time, 138–139
  rotation and, 265
  straight path, 106–107
  streamwise (a_s), 199
  total derivative operator (d), 137
Accuracy error, 28–30
Adhesive forces, 58
Adiabatic duct flow, 702–711.
    See also Fanno flow
Adiabatic process, 215
Advective (convective) acceleration, 137–138
Aerodynamic drag, 302–303, 548–550
Aerodynamic drag coefficient, 322–323
Aerodynamic shoulder, 549–550
Aerodynamics, study of, 2
Affinity laws, 829–830
Air, properties of at 1 atm pressure, 948
Air flow, 14, 40–41. *See also* Ideal gases
Aircraft, 578–583, 612–613, 616–617, 634–643, 662–663, 674–688, 688–693, 816–819
  airfoils, 634–638
  angle of attack (α), 612, 617
  back pressure (P_b) effects, 674–678
  blade twist for, 816–818
  boundary layer approximation for, 578–583
  compression of air in, 662–663
  converging–diverging nozzles, 675–678
  drag force on, 612–613
  efficiency of, 637–638
  finite-span wings, 638–639
  flaps, 636–637
  flow fields, 637
  flow separation, 578–583, 616–617
  induced drag and, 638–639
  lift force, 612–613, 634–643
  lift-to-drag ratio, 635–636

National Advisory Committee for Aeronautics (NACA) standards for, 637–638
  normal shockwaves, 675–676
  oblique shockwaves, 676
  open axial-flow fans, 816, 816–819
  pitch angle (θ), 816–819
  Prandtl–Meyer expansion waves, 688–693
  pressure forces acting on, 612–613, 634–635
  pressure gradient effects, 578–583
  propellers (rotor), 816–819
  propulsion, 674–678
  reverse thrust of, 819
  rotor airflow swirl, 817–818
  separation bubble, 579–580
  separation point, 580–581
  shock (wave) angle (δ), 684–685
  stall conditions, 580, 616, 637
  starting vortex, 635
  takeoff and landing speeds, 636–637
  turning (deflective) angle (θ), 684–685
  variable pitch, 819
  viscous forces acting on, 612–613, 634–635
  vortex shedding, 617
  wings, 612–613, 617, 634–639
  wingspan (span), 634
  Wright Brothers' impact on, 643
Airfoils, 634–638. *See also* Aircraft
Alternate depth, 734
Ammonia saturation properties, 944
Ampere (A), unit of, 16
Analytical problem approaches, 21–22
Anemometers, 402–404. *See also* Flow rate
Angle of attack (α), 315, 612, 617, 634–637
Angle of deformation (α), 2
Angle valves, 375, 380
Angular displacement (α), 2
Angular momentum, 244–245, 263–273
  conservation of, 245
  equations, 244–245, 264–267
  external forces and, 265–268
  Euler's turbine formula for, 269–270
  fixed control volume (CV), 267
  moment, 266
  moment forces (F) of, 264–265
  momentum analysis and, 244–245, 263–265
  Newton's second law and, 244–245
  no external moments, 268
  radial-flow devices, 269–270
  Reynolds transport theorem (RTM) and, 266
  rotation (ω) and, 244–245, 263–265
  steady flow, 267–268
Angular velocity (rate of rotation), 151–152, 264–265
Apparent viscosity, 52
Approximate solutions, 515–616. *See also* Navier–Stokes equation
Archimedes number (Ar), 309
Area, moments of, 90–91
Aspect ratio (AR), 309, 639
Atmosphere properties at high altitudes, 951
Atmospheric pressure (P_{atm}), 81–83, 89–90, 249
Available wind power ($W_{available}$), 850–851
Average velocity (V_{avg}), 188, 348–349

Axial-flow turbine, 839
Axial pumps, 806, 816–824
Axisymmetric flow, 14, 457, 534, 536–537, 610, 687–688

Back pressure (P_b), 669–678
Backward-inclined blades, 807–808
Ball valve, 380
Barometers, 81–84
Barometric pressure (P_{atm}), 81–83
Beam splitter, 405
Bends, pipe flow losses at, 377–380
Bernoulli equation, 199–214, 221, 294, 392–393, 526–527, 531–534, 812
  acceleration of fluid particles and, 199–200
  applications of, 207–214
  approximation solutions using, 526–527, 531–534
  boundary layers and wake regions for, 199
  compressible flow, 200–202
  dimensional homogeneity of, 294
  energy grade line (EGL), 205–207
  fluid particle acceleration and path, 199–200
  frictionless flow and, 199, 204
  hydraulic grade line (HGL), 205–207
  impeller rotation, 812
  incompressible flow, 201, 205, 221
  internal flow rate from, 392–393
  inviscid flow regions and, 199, 526–527
  irrotational flow regions, 531–534
  law of thermodynamics for, 202, 221
  limitations on use of, 204–205
  linear momentum for, 199–201
  mechanical energy balance, 201–202, 207
  Navier–Stokes equation and, 526–527, 531–534
  negligible viscous effects and, 204
  Newton's second law for, 200–202
  negligible heat transfer and, 205
  no shaft work and, 204–205
  pressure representation, 202–204
  rotating reference frame, 812
  stagnation pressure, 203–204
  steady flow, 199–202, 204
  streamlines and, 199–200, 202–205
  unsteady, compressible flow, 202
  vector identity for, 526–527
  velocity measurement using, 392–393
Bernoulli head, 790
Best efficiency point (BEP), 791
Betz limit, 853–854
Bias error, 28
Bingham plastic fluids, 466
Biofluid mechanics, 408–416, 493–497
  blood flow studies, 410–415
  cardiovascular system, 408–410
  differential analysis of flow, 493–497
  flow measurement and, 408–416
  particle image velocimetry (PIV) for, 410, 416
  Poiseuille flow comparisons, 493–496
  pulsatile pediatric ventricular device (PVAD) for, 410–411

INDEX

Biological systems, drag coefficient (C_D), 618–621
Biot number (Bi), 309
Blades, 806–809, 815–823, 907–915.
   See also Runners
Blasius similarity variable (η), 565–567
Blockage, 321
Blood flow studies, 410–415
Blower, 788
Bluff (blunt) bodies, 610
Body cavitation, 62
Body forces, 104–105, 246–249
Boiling properties, 941
Bond number, 309
Boundary conditions, 14–15, 161, 214–215, 438, 475–477, 888–893
   axis, 892
   closed system, 14–15
   computational fluid dynamics (CFD) use of, 888–893
   continuity equations, 475–477
   control surface, 161
   control volume (CV), 15, 161
   differential analysis use of, 438, 475–477
   energy transfer and, 214–215
   exact solution for equations using, 475–477
   fan, 892
   flow passage (fan), 891
   fluid flow systems and, 14–15
   free-surface, 477
   inflow/outflow, 889–890
   initial, 477
   inlet, 477
   interface, 476
   interior, 892–893
   Navier–Stokes equations, 475–477
   no-slip, 476
   open systems, 15
   outlet, 477
   periodic, 891
   pressure far field, 891
   pressure inlet/outlet, 889–890
   reverse flow, 890
   rotational, 891
   symmetry, 477, 891–892
   translational, 891
   velocity inlet, 889
   wall functions, 889
   water–air interface, 477
Boundary layers, 7, 9, 199, 351–352, 364–367, 525–526, 554–591, 625–627, 712
   approximation, 525–526, 554–591
   arbitrary shapes, 578–583
   Bernoulli equation for, 199,
   Blasius similarity variable (η) for, 565–567
   buffer, 364–365, 626
   computational fluid dynamics (CFD) calculations for, 580–583, 588–591
   coordinate system, 559
   curvature and, 561–562
   displacement thickness (δ^*), 568–571, 574
   equations, 555, 559–563
   external parallel flow, 625–627
   flat plates, 556–558, 572–578, 583–591
   flow region and, 9, 526–527
   free shear layers and, 557
   historical significance of, 7
   inertial sublayer, 364
   internal flow, 351–352
   inviscid regions of flow, 525–526, 554–555
   irrotational (core) regions of flow, 351, 625
   laminar flow, 557–572

logarithmic (log) law for, 366, 576
momentum integral technique for, 583–591
momentum thickness (θ), 571–572, 574
Navier–Stokes equation for, 525–526, 560–563
no slip condition and, 525–526, 555
one-seventh-power law for, 366, 573–574
outer (turbulent), 364–366, 626
overlap (transition), 364–366, 626
pressure gradient for, 561–562, 564–565, 578–583
procedure for approximation, 564–568
profile comparisons, 573–578
regions, 351–352, 554–591, 625–627
Reynolds number (Re) for, 557–559
shock wave interactions, 712
Spaulding's law of the wall for, 365–366, 576, 578
thickness (δ), 556, 562–564, 574, 625
transitional flow, 557–559
turbulent flow, 364–367, 557–558, 572–578
velocity profile for, 351–352, 364–367
viscous sublayer, 364–367, 626
wall-wake law for, 576–577
zero pressure gradient, 561, 564–565
Bourdon tube, 88
Bow wave, 686–687, 928
Brake horsepower (bhp), 791
Bridge scour, 771
British thermal unit (btu), unit of, 18, 43
Broad-crested weir, 766–767
Buckets, 790
Buckingham Pi theorem, 303–319
Buffer layer, 364–365, 626
Bulk modulus of elasticity (κ), 44–46
Bumps, open-channel flow, 764–765, 929
Buoyancy, 32, 47, 98–103

Calorie (cal), unit of, 18, 43
Candela (cd), unit of, 16
Capacity (volume flow rate), 790, 805–806
Capillary effect, 58–60
Cardiovascular system, 408–410
Cartesian coordinates, 13–14, 157, 247–249, 440–442, 445, 450–456, 468, 470–472
   continuity equation in, 440–442, 445, 468
   control volume (CV) and, 247–249
   differential analysis applications, 470–472
   fluid flow dimensions, 13–14
   gravitational forces in, 247
   Navier–Stokes equation in, 468
   rotational flow, 157
   stream functions in, 450–456
   vorticity in, 157
Cauchy's equation, 459–464
Cavitation, 41–43, 62, 207, 797–800
   avoidance of, 797–800
   bubbles, 42, 797–798
   net positive suction head (NPSH) for, 798–800
   pumps and, 797–800
   saturation pressure (P_{sat}) and, 41–43, 797–798
   saturation temperature (T_{sat}) and, 41–43
   sonar dome study of, 63
   sonoluminescence, 63
   vapor pressure (P_v) and, 41–43, 797
   vaporous (gaseous), 63
Cavitation number (Ca), 309
Centrifugal pumps, 806–815

Centripetal acceleration and force, 265, 561
Centroid, 91–92
Channels, 727–729, 737–759, 771
   hydraulic cross sections for, 728–729
   hydraulic diameter (D_h), 728
   hydraulic radius (R_h) for, 728–729
   open-channel flow and, 727–729, 737–759, 771
   rectangular, 745
   trapezoidal, 745–746
Characteristic (performance) curves, 383–384, 791–797
Chezy coefficient (C), 738–739
Choked flow, 670, 700, 708–711, 765
Circular fluid flows, comparison of, 159–160
Circular pipe flow, 348, 353–357, 952
Circulatory flow loss, 815
Closed system, 14–15
Closed volume, 805–806
Cohesive forces, 58
Colebrook equation, 367–368
Compressible flow, 10–11, 44–50, 200–202, 439–445, 610, 659–723, 953–956
   adiabatic, 702–711
   aircraft and, 662–663, 674–688
   Bernoulli equation and, 200–202
   bulk modulus of elasticity (κ), 44–46
   CFD calculations for, 922–928
   choked, 670, 700, 708–711
   compressing flow, 678–688
   computational fluid mechanics (CFD) for, 922–928
   continuity equations for, 439–445
   converging nozzles, 665, 670–674
   density and, 10–11, 46–47, 667–669
   external, 610
   Fanno flow, 702–711, 956
   friction and, 702–711
   heat transfer (Q) and, 693–702
   ideal gases, 45–46, 663, 667–669, 693–702, 953–956
   isentropic, 663–678, 953
   Mach number (Ma) for, 11, 50, 663–669, 670–672
   nozzles for, 665–678, 923–927
   one-dimensional, 663–669, 953–954
   Prandtl–Meyer expansion waves, 688–693
   property tables for, 953–956
   Rayleigh flow, 693–702, 955
   shock waves, 675–688, 712, 927–928
   sonic, 11, 50, 666–667
   speed of sound (c) and, 11, 48–50, 663–665
   stagnation properties of, 660–663, 704–705
   steady, 200–201
   subsonic, 11, 50, 666–667, 687
   supersonic, 11, 50, 666–667, 687
   three-dimensional, 610
   transonic, 50
   viscous-inviscid interactions, 712
Compressible stream function ($\psi\rho$), 458–459
Compressing flow, 678–688. *See also* Shock waves
Compressive force, 77–78
Compressors, 788–789
Computational fluid dynamics (CFD), 27, 32, 141, 149–151, 318–319, 406, 472–473, 580–583, 588–591, 879–938
   boundary conditions for, 888–893
   boundary layer approximation, 580–583, 588–591
   bumps, calculations for flow over, 929

INDEX

cells, 881–882, 887–888
compressible flow calculations, 922–928
computational domain for, 881
continuity equation solution, 472–473
contour plots, 150–151, 916–917, 924–928
converging–diverging nozzle flow calculations, 923–927
creeping flow approximation, 588–591
cross-flow heat exchanger, 915–917
cylinders, calculations of flow around, 897–902, 905–907
differential analysis using, 472–473
direct numerical simulations (DHS), 903
engineering software use, 27, 32
equations of motion solutions, 880–883
Euler equation solution, 580–581
flow separation, 580–583, 921–922
flow visualization and, 141, 582
grids, 883–888
heat transfer calculations, 915–922
hydraulic jump calculations, 930
integrated circuit (IC) chips, 917–922
laminar flow calculations, 893–902
magnetic resonance image (MRI) simulations, 929
multigridding for, 883
oblique shock wave calculations, 927–928
open-channel flow calculations, 928–930
particle image velocimetry (PIV) for, 406
pipe flow calculations, 893–896
plots for flow data, 149–151, 916–917, 924–928
postprocessors for, 882
pressure correction algorithms for, 473
pressure gradients, 580–583
pressure contour plots, 924–928
residual of terms, 882
stator blade (vane) design, 907–915
turbulence models, 903–905
turbulent flow calculations, 902–915
vector plots, 149–150
velocity overshoot and, 590–591
Conservation of energy, see Energy (E)
Conservation of mass, see Continuity equations; Mass (m)
Constant error, 28
Constitutive equations, 464–465. See also Navier–Stokes equation
Contact angle (ϕ), 58
Continuity equation, 186, 438–450, 468–469, 475–493, 517–518, 529–530, 694, 703, 880
boundary conditions for, 475–477
Cartesian coordinates, 440–442, 468
compressible flow, 439–445
computational fluid dynamics (CFD) solution to, 472–473, 880
conservation of mass and, 186, 438–450
cylindrical coordinates, 442–443, 444–445, 469
differential analysis using, 470–493
divergence (Gauss') theorem for, 439–440, 443–444
exact solutions of, 475–493
Fanno flow
incompressible flow, 443–445
infinitesimal control volume and, 440–443
irrotational flow, 529–530
Laplacian operator (∇) for, 530
continuity equation for, 529–530
material element (derivative) for, 443–444, 463–464
Navier–Stokes equation coupled with, 475–493
Rayleigh flow, 694
steady, compressible flow, 444–445
Taylor series expansion for, 440–443
velocity potential function (ϕ) and, 529–530
Continuum, 38–39, 134
Contour plots, 150–151, 916–917, 924–928
Contracted/inner (dot) product, 248
Control mass, 15
Control points, 754
Control volume (CV), 15, 32, 134–134, 160–168, 186–187, 189–191, 217–219, 245–251, 267, 440–443, 460–463, 583–591, 809–810, 851–854
analysis, 32
atmospheric pressure acting on, 249
body forces, 246–248
boundary (fixed position) of, 15, 161
Cauchy's equation derived using, 460–463
closed system relationship to, 160–162
conservation of energy and, 186–187, 217–219
conservation of mass (m_{CV}) and, 186, 189–191, 440–443
continuity equation derived using, 440–443
deforming, 191, 245–246
energy transfer and, 217–219
extensive property of, 162
fixed, 163, 217–219, 245, 250, 267
forces acting on, 246–249
gravity acting on, 247
infinitesimal, 440–443, 460–463
Leibniz theorem for, 165–167
material derivative and, 167–168
momentum analysis of, 245–249, 267
relative velocity of, 163–165, 191, 245–246
Reynolds transport theorem (RTT) for, 160–168
selection of, 245–246
steady flow, 251
stress tensor for, 247–249
surface forces, 246–248
Taylor series expansion for, 440–441, 462
unit outer normal of, 162–163
velocity (moving) of, 245–246
vortex structures, 32
wind turbine power, 851–854
Convective (advective) acceleration, 137–138
Converging–diverging nozzles, 665, 674–678, 923–927
back pressure (P_b) effects, 675–678
isentropic flow though, 674–678
supersonic flow and, 674, 678–681
Converging nozzles, 665, 670–673
back pressure (P_b) effects, 670–674
choked flow, 670
isentropic flow though, 670–674
Mach number (Ma) and, 665, 670–672
Couette flow, 477–484
Counter-rotating axial-flow fans, 819–820
Coupled equations, 438, 470–475
Creeping flow, 520–525, 588–591
approximation, 520–525, 588–591
CFD calculations for, 588–591
drag force on a sphere in, 523–525
Navier–Stokes equation for, 520–525
Reynolds number (Re) for, 520–522
terminal velocity of, 523–524
Critical depth (y_c), 730, 734
Critical energy, 734
Critical flow, 729–732, 739–737
Critical property values (*), 668–669
Critical ratios, 668
Critical Reynolds number (Re_{cr}), 350, 557–559
Critical uniform flow, 736–740
Critical velocity (V_c), 734
Cross-flow heat exchanger, 915–917
Curvature of boundary layer, 561–562
Curved surfaces, hydrostatic forces on, 95–97
Cut-in and cut-out speeds, 850
Cylinders, 629–633, 897–902, 905–907
CFD calculations for flow around, 897–902, 905–907
drag force on, 629–633
external diameter (D), 629
external flow over, 629–633
flow separation, 631
Kármán vortex street formation, 899–902
laminar flow around, 897–902
Reynolds number (Re) for, 629–633
stagnation points, 629–631
surface roughness effects, 632–633
turbulent flow around, 905–907
Cylindrical coordinates, 13–14, 107–110, 158–159, 442–445, 457–458, 469, 473–475
continuity equation in, 442–445, 469
differential analysis in, 442–445, 457–458, 469, 473–475
fluid flow dimensions, 13–14
Navier–Stokes equation in, 469
stream functions in, 457–458
vorticity in, 158–159

d'Alembert's paradox, 548–549
Darcy friction factor (f), 309, 317, 355–356, 358, 367–369, 952
dimensional analysis use of, 317
laminar flow analysis, 355–356
Moody charts for, 367–369, 952
pipe cross sections and, 358
pipe flow and, 952
ratio of significance, 309
relative roughness (ϵ/D) and, 367–369
turbulent flow analysis, 367–369
Deadweight testers, 88–89
Deformation, 2, 151–156
angle of (α), 2
fluid kinematic properties of, 151–156
linear strain rate, 151–153
rates of flow, 2, 151–156
rotation, rate of, 151–152
shear strain rate, 153–154
volumetric (bulk) strain rate, 153
Density (ρ), 10–11, 32, 39–41, 46–47, 667–669, 850–851
compressible flow and, 10–11, 46–47, 667–669
critical, 667–669
fluid properties of, 39–41
ideal gases, 40–41, 667–669
isentropic flow, 667–669
Mach number (Ma) and, 11, 668
specific gravity (SG) and, 39, 41
volume expansion (β) and, 46–47
vortex structure and, 32
wind power, 850–851
Dependent Π (Pi), 300–301
Depth, pressure variation with, 78–81
Derivatives, study of, 22
Derived (secondary) dimensions, 15
Detached oblique shock wave, 686–687, 928

INDEX

Deviatoric (viscous) stress tensor, 465–467
Differential analysis, 32, 437–513, 515–606, 705–708
  applications of, 470–497
  approximate solutions of, 515–606
  biofluid mechanics flows, 493–497
  boundary conditions for, 438, 475–477
  Cartesian coordinates, 440–442, 450–456, 468, 470–472
  Cauchy's equation for, 459–464
  compressible flow, 439–445
  computational fluid dynamics (CFD) for, 472–473
  conservation of mass and, 438–450
  constitutive equations for, 464–465
  continuity equation for, 438–450, 468–469, 475–493
  Couette flow, 477–484
  coupled equations, 438, 470–475
  cylindrical coordinates, 442–445, 457–458, 469, 473–475
  differential linear momentum equation, 459–464
  divergence (Gauss') theorem for, 439–440, 443–444, 459–460
  error function (erf), 491–492
  exact solutions for, 475–493
  Fanno flow, friction effects and, 705–708
  flow domain, 438
  incompressible flow, 443–445, 466–468
  infinitesimal control volume for, 440–443, 460–463
  Navier–Stokes equation for, 464–469, 475–493, 515–606
  Poiseuille flow, 484–490, 493–496
  pressure field calculation, 470–475
  similarity solution for, 491
  steady flow, 470–490
  stream functions (ψ) for, 450–459
  stress tensors (σ) for, 459–460, 465
  Taylor series expansion for, 440–441
  unsteady flow, 490–493
  viscosity (μ) of fluids, 480–481
Differential equations, use of, 22
Differential manometer, 85–86
Diffusers for minor loses, 379
Dilatant (shear thickening) fluids, 52, 466
Dimensional analysis, 291–345, 824–827, 855–857
  Froude number (Fr), 296–299, 323–325
  geometric similarity, 299–300
  incomplete similarity, 320–323
  insect flight and, 326
  inspectional analysis, 294–299
  kinematic similarity, 299–300
  method of repeating variables, 303–319
  models and prototypes for, 299–303, 320–325
  nondimensional equations for, 294–299
  nondimensional parameters, 294–319
  scaling laws, 824–827, 855–857
  pumps, 824–827
  repeating variables, method of, 303–319
  Reynolds number (Re), 320–326
  similarity of models and prototypes, 299–303
  turbines, 855–857
  units and, 292
Dimensional homogeneity, 19–20, 293–299
  Bernoulli equation example of, 294
  Froude number (Fr) for, 296–299
  inspectional analysis and, 294–299

  law of, 293
  nondimensionalization of equations, 294–299
  nondimensional parameters, 294–296
  normalized equations, 294
  pure constants, 295
  scaling parameters, 295–296
  units and, 19–20
Dimensional variables, 294
Dimensionless (nondimensional) variables, 295
Dimensions, 15–16, 292
Direct numerical simulations (DNS), 903
Discharge coefficient (C_d), 393–394, 762–764
Disk area (A), 850–851
Displacement thickness (δ^*), 568–571
Distorted flow models, 323–325
Dividing streamline, 552, 581–582
Divergence (Gauss') theorem, 439–440, 443–444, 459–460
Dividing streamline, 552, 581–582
Doppler-effect ultrasonic flowmeters, 399–401
Double-regulated turbine, 839
Doublet, 544–545, 546–547
Downwash, 816
Draft tube, 840
Drafting for drag reduction, 622–623
Drag coefficient (C_D), 309, 320–323, 525, 612–625, 630–631
  aerodynamic, 322–323
  average, 612–613
  biological systems and, 618–621
  creeping flow, 525
  drafting, 622–623
  drag reduction and, 615–616
  external flow and, 612–614, 617–625, 630–631
  frontal area, 612
  planform area, 612
  dynamic pressure, 612
  ratio of significance, 309
  Reynolds number for, 320–323, 617–618
  Stokes law, 618
  streamlining effects on, 615–616
  superposition of, 623–625
  surface roughness effects, 612, 614–615, 626–628, 632–633
  three-dimensional bodies, 611, 620–621
  total, 617–618
  two-dimensional bodies, 611–612, 619
  vehicles, 621–623
Drag force (F_D), 51, 302–303, 523–525, 548–550, 587–588, 610–617, 629–633, 638–639, 645
  aerodynamic, 302–303, 548–550
  angle of attack (α), 612, 617
  balance and, 302–303
  creeping flow approximation and, 523–525
  cylinders with external flow over, 629–633
  d'Alembert's paradox, 548–549
  differential analysis of, 302–303
  external flow and, 610–617, 638–639
  flat plates with external flow over, 612, 625–629
  flow separation, 616–617
  friction and, 51, 612, 614–617
  incompressible flow and, 610–617, 638–639
  induced, 638–639
  irrotational flow and, 548–550
  lift force and, 610–613, 638–639
  lift-to-drag ratio, 638–639
  pressure, 612–613, 614–617
  skin friction (wall shear), 567, 612, 614

  spheres and, 523–525, 629–633
  streamlining, 615–616
  surface roughness effects, 612, 614–615, 626–628, 632–633
  vehicles, 621–623
  wing design and, 612–613
Droplet-on Demand (DoD), 593
Drum gate, 762
Ducted axial-flow fan, 816–817, 819–824
Ducted pumps, 790
Ducts, 348, 358, 663–665, 693–711. *See* Converging-diverging nozzles; Pipe flow
Dynamic machines, 790, 806–824, 834–855
Dynamic pressure, 202–203, 612
Dynamic similarity, 300–301, 315–316, 318–319
Dynamic temperature, 667
Dynamic (absolute) viscosity, 52, 54
Dynamics, study of, 2

Eckert number (Ec), 309
Eddies, 361–364, 902–903
Eddy (turbulent) viscosity, 363–364
Edge of computational domain (2-D flow), 881
Efficiency (η), 195–197, 381–390, 637–638, 791–797, 815, 841–842
  aircraft, 637–638
  best efficiency point (BEP), 791
  centrifugal pumps, 815
  characteristic (performance) curves for, 383–384
  circulatory flow loss, 815
  combined (overall), 196–197
  generator, 196
  impeller blades and, 815
  mechanical, 195–197
  motor, 196
  operating point, 384, 792–796
  passage losses, 815
  performance (characteristic) curves, 383–384, 791–797
  piping systems and pump selection, 381–390, 791–797
  pump, 196, 383–390, 791–797
  pump–motor, 383–384
  reaction turbines, 841–842
  turbine, 196, 383, 815, 841–842
Elbows, pipe flow losses at, 377–380
Electromagnetic flowmeters, 401–402
Elevation head, 205
Enclosed pumps and turbines, 790
Energy (E), 38, 43–44, 186–187, 194–199, 214–228, 403–404, 693–702, 703, 733–737. *See also* Heat transfer; Power
  conservation of, 186–187, 198–199, 214–228, 736–737
  control volume (CV) and, 186, 217–219
  enthalpy (h), 43–44
  Fanno flow, 703
  first law of thermodynamics, 214
  flow work (P/ρ), 43–44, 194–195, 218–219
  fluid properties of, 43–44
  friction slope (S_f) and, 737
  heat (Q), transfer by, 215, 693–702
  incompressible flow, analysis of, 221–222
  internal (U), 43
  loss, 216, 219–221
  kinetic (ke), 43, 195
  kinetic energy correction factor (α), 221–222
  macroscopic, 43

INDEX

mechanical (E_{mech}), 194–199, 215–228
microscopic, 43
open-channel flow, 733–737
potential (pe), 43, 195
Rayleigh flow, 694–702
single-stream devices, 219
specific, 733–736
specific heats and, 43–44
specific total (e), 38, 44
steady flows, analysis of, 219–221
thermal, 43
total, 43–44
units of, 43
work, transfer by, 215–219
Energy absorbing devices, 788. *See also* Pumps
Energy dissipation ratio, 759
Energy grade line (EGL), 205–207, 790, 840–841
Energy pattern factor (K_e), 851
Energy producing devices, 788. *See also* Turbines
Engineering Equation Solver (EES), 26–27
Engineering software packages, 25–27
English system of units, 16–19, 292
Enthalpy (h), 43–44, 660–662
  compressed flow and, 660–662
  energy and, 43–44
  ideal gas, 661
  stagnation and, 660–662
  total, 660–662
Entrance region, 351–352, 893–896
Entropy change, 694–695, 703
Entry length, 352–353
Equations, 40, 105–106, 108, 185–242, 244–245, 249–250, 264–267, 269–270, 293–299, 437–513, 517–520, 555, 559–563, 880–883
  angular momentum, 244–245, 264–267
  approximate solutions for, 517–520
  Bernoulli, 199–214
  boundary layer, 555, 559–563
  Cauchy's, 459–464
  computational fluid dynamic (CFD) solution, 880–883
  conservation of momentum principle, 186
  continuity, 438–450, 468–469, 475–493, 880
  coupled, 438, 470–475
  differential analysis, 437–513
  dimensional homogeneity and, 293–299
  efficiency, 194–199
  energy, 186–187, 194–199, 214–228
  Euler's turbine formula, 269–270
  ideal gases, 40
  inspectional analysis of, 294–299
  linear momentum, 186, 244–245, 249–250
  mass, 186, 187–194
  motion, 105–106, 108, 517–520, 880–883
  Navier–Stokes, 464–469, 475–493, 880
  nondimensionalization of, 294–299, 517–520
  normalized, 294
  of state, 40
  Taylor series expansion, 440–441
Equipotential lines, 535–536
Equivalent length (L_{equiv}), 375
Equivalent roughness (ϵ), 368
Errors, 28–31, 88–89, 221–222, 251–253, 491–492
Euler equation, 525–526, 531, 580–581
Euler number (Eu), 309, 519
Eulerian description, 134–140, 167–168
  acceleration field (vector), 135, 136–139
  field variables, 134–136
  flow domain (control volume), 134

flow field, 135
  gradient (del) operator, 137
  material (substantial) derivative, 139–140, 167–168
  material position vectors, 136–137
  partial derivative operator (δ), 137
  pressure field (scalar), 134
  Reynolds transport theorem (RTM) and, 167–168
  total derivative operator (d), 137
  vector variables, 134–135
  velocity field (vector), 134
Euler's turbine formula, 269–270
Euler's turbomachine equation, 810
Expanding flow, 688–693
Expansion fan, 688–689
Experimental problem approaches, 21–22
Extensional (linear) strain, 151
Extensive property of fluid flow, 38, 162
External diameter (D), 629
External flow, 10, 607–657, 678–693
  aircraft, 612–613, 616–617, 634–643
  angle of attack (α), 612, 617
  axisymmetric, 610
  bluff (blunt) bodies, 610
  compressible, 610, 678–693
  cylinders, over, 629–633
  drag coefficient (C_D), 612–614, 617–625, 630–631
  drag force (F_D), 610–617, 638–639
  drag reduction, 610–611, 615–616, 645
  flat plates, over, 612, 625–629
  flow fields for, 608–610
  flow separation, 616–617, 630–631
  free-stream velocity, 608
  friction and, 612, 614–617, 625–629
  incompressible, 610–657
  internal flow compared to, 10
  lift force, 610–613, 634–643
  parallel, 625–629
  Prandtl–Meyer expansion waves, 688–690
  resultant forces of, 607
  Reynolds number (Re) for, 612, 617–618, 629–631
  shock waves, 678–693, 712
  spheres, over, 629–633
  stagnation points, 629–631, 635, 639
  streamlined bodies, 610, 614–616
  surface roughness effects, 612, 614–615, 626–628, 632–633
  three-dimensional bodies, 610, 611, 620–621
  two-dimensional bodies, 608, 610, 611–612, 619
External forces, 249–250, 254–255, 265–268

Face of computational domain (3-D flow), 881
Fanning friction factor (C_f), 309, 317, 355–356
Fanno flow, 702–711, 956
  choked, 708–711
  continuity equation for, 703, 705
  differential analysis of, 705–708
  energy equation for, 703, 705
  entropy change of, 703
  equation of state for, 704
  friction effects on, 704–705
  ideal-gases, 702–711, 956
  Mach number (Ma) for, 704–705
  maximum duct (sonic or critical) length, 707–708

momentum equation for, 703–705
property functions, 956
property relations of, 705–708
Reynolds number (Re) for, 707–708
T-s diagrams for, 704–705
Fanno line (curve), 679–682
Fans, 788, 816–817, 819–824, 907–915
  axial pumps, 788, 816, 819–824
  blade row (cascade), 821
  CFD model for, 907–915
  counter-rotating, 819–820
  ducted-axial flow, 816–817, 819–824
  open-axial flow, 816
  stator blades (vanes), 819–822, 907–915
  tube-axial, 819–824
  vane-axial, 819–820, 907–915
Field variables, 134–136
Finite-span wings, 638–639
First-order difference approximation, 139
Flaps, lift effects from, 636–637
Flat plate analysis, 89–92, 98, 556–558, 572–578, 583–591, 612, 625–629
  boundary layer approximation of, 556–558, 572–578, 583–591
  boundary layer regions, 625–627
  buffer layer, 626
  buoyant force on, 98
  drag force and, 612, 625–629
  external flow over, 612, 625–629
  friction coefficient for, 627–629
  hydrostatic forces on, 89–92
  irrotational flow region, 625
  Kármán integral equation, 585–588
  laminar flow and, 556–558
  momentum integral technique for, 583–591
  overlap layer, 626
  parallel flow over, 625–629
  Reynolds number (Re) for, 626–628
  skin friction and, 612
  surface, 89–92
  surface roughness and, 628
  turbulent flow and, 572–578
  turbulent layer, 626
  velocity boundary layer, 625
  viscous sublayer, 626
Flow depth, 726–727, 751
Flow domain (control volume), 134, 438
Flow field regions, 9–10, 156–157, 199, 273–274, 350–352, 517, 525–591, 608–610
Flow rate, 187–189, 191, 367–370, 391–416, 670–674, 805–806, 808–809
  discharge coefficient (C_d) for, 393–394
  Doppler-effect ultrasonic flowmeters, 399–401
  electromagnetic flowmeters, 401–402
  flowmeters, 392–402
  internal flow, 367–370, 391–408
  isentropic nozzle flow, 670–674
  King's law for energy balance, 403–404
  laser Doppler velocimetry (LDV), 404–406
  mass (m), 187–188, 191, 670–674
  obstruction flowmeters, 392–396
  paddlewheel flowmeters, 397–398
  particle image velocimetry (PIV), 406–408, 410, 416
  pipe length and, 369–370
  Pitot formula for, 392
  Pitot probes, 391–392
  positive displacement flowmeters, 396

INDEX

Flow rate (*continued*)
  pump capacity, 805–806, 808–809
  Strouhal number (St) for, 402
  thermal anemometers, 402–404
  turbine (propeller) flowmeters, 397
  ultrasonic flowmeters, 399–401
  variable-area flowmeters (rotameters), 398
  velocimetry, 404–408
  velocity measurement, 391–408
  volume, 188–189, 391, 805–806, 808–809
  vortex flowmeters, 402
Flow separation, 9, 578–583, 616–617, 630–631, 921–922
  angle of attack (α), 612, 617
  boundary layer approximation for, 578–583
  CFD calculations for, 921–922
  friction drag and, 616–617
  pressure drag and, 616–617
  pressure gradient effects, 578–583
Flow visualization, 32, 141–148, 582, 631
Flow work (P/ρ), 43–44, 194–195, 218–219
Flowmeters, 392–402. *See also* Flow rate
Fluctuating components of turbulent flow, 361–362
Fluid, defined, 2
Fluid flow, 1–35, 52–54, 133–183, 191–195, 199–214, 250–273, 299–303, 312–325, 347–435, 437–513, 663–669, 726, 880, 882
  acceleration of, 135–140, 199–200
  accuracy of measurements, 28–30
  analysis, 1–35
  Bernoulli equation applications of, 199–214
  biofluid mechanics, 408–416
  boundary layers, 8–9, 351–352, 364–367
  circular, 159–160
  classification of, 9–14
  compressible, 10–11, 200–202
  control volume, 15, 160–168
  differential analysis of, 32, 437–513
  engineering software packages for, 25–27
  Eulerian description, 134–140
  external, 10
  field variables, 134–136
  forced, 11
  forces (F) and, 2–3
  frictionless, 199, 204
  fully developed (one-dimensional), 13–14
  global properties of, CFD, 880, 882
  incompressible, 10–11, 192–194, 201, 205, 221
  internal, 10, 347–435
  inviscid, 10, 199
  irrotational, 156–157
  Lagrangian description, 134–140
  laminar, 11, 349–361
  modeling (mathematical), 21–23
  molecular structure and, 3–4
  natural (unforced), 11
  no-slip condition, 8–9
  no-temperature-jump condition, 9
  one-dimensional (fully developed), 13–14, 663–669
  precision of measurements, 28–29
  problem-solving technique, 23–25
  Reynolds transport theorem (RTM) for, 160–168
  rotational (ω), 151–152, 156–159
  significant digits and, 28–31
  steady, 12–13, 191, 199–202, 251, 253–254
  strain rate and, 2–3
  streaklines, 144–146
  streamlines, 141–142
  systems, 14–15
  three-dimensional, 13–14
  turbulent, 11, 349–353, 361–374
  two-dimensional, 13–14
  units of measurement, 15–21
  unsteady, 12–13, 202, 726
  viscous, 10
Fluid kinematics, 133–183
  acceleration field (vector), 135, 136–139
  angular velocity (rate of rotation), 151–152
  circular flows, comparison of, 159–160
  deformation of flow, 151–156
  Eulerian description, 134–140, 167–168
  Lagrangian description, 134–140, 167–168
  linear strain rate (ϵ), 151–153
  material derivative, 139–140, 167–168
  motion of flow, 151–160
  pathline, 142–144
  plots, 148–151
  refractive flow visualization, 147–148
  Reynolds transport theorem (RTM), 160–168
  rotation, 151–152, 156–160
  streaklines, 144–146
  streamlines, 141–142
  surface flow visualization, 148
  timelines, 146–147
  translation, 151
  velocity vector (rate of translation), 134, 151
  vorticity, 156–160
Fluid machines, *see* Turbomachines
Fluid mechanics, 2, 4–5, 6–8, 14–15, 21–31, 185–242, 243–289, 291–345
  applications of, 4–5
  Bernoulli equation, 199–214, 221
  categories of, 2
  dimensional analysis, 291–345
  efficiency (η), 195–197
  energy (E), conservation of, 186–187, 198–199, 214–228
  energy grade line, 205–207
  engineering and, 21–31
  equations for, 185–242
  flow systems, 14–15, 243–289, 293
  history of, 6–8
  hydraulic grade line, 205–207
  linear momentum, 186
  mass (M) conservation of, 186, 187–194
  mechanical energy, 194–199, 201–202, 207, 215–228
  momentum analysis, 243–289
Fluid properties, 37–74
  capillary effect, 58–60
  cavitation, 41–43, 62
  compressibility (κ), 44–50
  continuum, 38–39
  density, 39–41
  energy, 43–44
  equations of state, 40
  ideal gases, 40–41
  saturation and, 41–43
  specific gravity (SG), 39, 41
  specific heats, 43–44
  speed of sound, 48–50
  state postulate for, 38
  surface tension, 55–60
  vapor pressure, 41–43
  viscosity, 50–55
  volume expansion, 46–48
Fluid statics, 75–131. *See also* Pressure
  buoyancy, 98–103
Force (F), 2–4, 10, 17–19, 50–52, 58–59, 76–81, 89–103, 104–105, 216–219, 244–250, 254–255, 264–267, 607–617, 634–643
  adhesive, 58
  angular momentum and, 244–245, 264–266
  body, 104–105, 246–248
  buoyant, 98–103
  capillary effect and, 58–59
  centripetal, 265
  cohesive, 58
  compressive, 77–78
  control volume, acting on, 246–250
  drag, 51, 610–617, 638–639
  external, 249–250, 254–255, 265–267
  flow and, 2–4, 249–250, 254–255
  friction, 10, 50–51
  gravity as, 18, 247
  hydrostatic, 89–97
  intermolecular (pressure), 3–4
  lift, 51, 610–613, 634–643
  linear momentum and, 249–250, 254–255
  moment of, 265–266
  momentum analysis and, 244–250, 254–255, 264–267
  net, 244–245
  pressure and, 76–81, 216–219
  resultant, 90–91, 95–96, 607
  rigid-body motion, 104–105
  shear, 52
  stresses and, 2–3
  surface, 104–105, 246–248
  thrust, 254–255
  units of, 17–19
  viscosity and, 10, 50–52
  weight as, 17–18
  work as, 18–19, 216–219
Forced flow, 11
Forward-inclined blades, 808
Fourier number (Fo), 309
Fractional factorial test matrix, 319
Francis turbine, 838–864
Free delivery, 384, 791
Free shear layers, approximation of, 557
Free-stream velocity, 608
Free-surface boundary conditions, 477
Free-surface flow, 323–325
Freezing point properties, 941
Friction, 10, 50–51, 199, 204, 348–349, 355–356, 358, 365, 367–369, 567, 612, 614–617, 625–629, 702–711, 952
  absence of (frictionless) in flow, 199, 204
  coefficient, 355–356, 614–615, 627–629
  Darcy friction factor (f), 355–356, 367–369
  drag, 612, 614–617, 625–629
  equivalent roughness and, 368
  external flow, effects on, 612, 614–617, 625–629
  factor (f), 355–356, 358, 614, 952
  Fanning friction factor, 309, 317, 355–356
  Fanno flow, 702–711
  flow separation, 616–617
  force, 10, 50–51, 199, 204
  internal flow and, 348–349
  laminar flow and, 355–356, 358
  local coefficient, 567
  Moody chart for pipe flow, 367–369, 952
  parallel flow on flat plates, 625–629
  pipe flow and, 348–349, 355–356, 358

INDEX

relative roughness (ϵ/D) and, 367–369
skin (wall shear), 567, 612, 614
streamlining, 615–616
surface roughness and, 612, 614–615, 628
turbulent flow and, 365, 367–368, 576
velocity, 365, 576
viscosity (μ) and, 10, 50–51
Friction lines, 148
Friction slope (S_f), 737, 751
Frictionless flow, 199, 204
Fringe lines, 405
Frontal area, 612
Froude number (Fr), 296–299, 309, 323–325, 519, 729–730, 750–751, 929–930
  CFD calculations using, 929–930
  critical depth of, 730
  critical flow, 729–730
  dimensional analysis using, 323–325
  free-surface flow, 323–325
  Navier–Stokes nondimensionalization using, 519
  nondimensionalization using parameter of, 296–299
  open-channel flow, 729–730, 929–930
  ratio of significance, 309
  subcritical (tranquil) flow and, 729–730
  supercritical (rapid) flow and, 729–730
  surface profiles, 750–751
  wave speed (c_0) and, 729–730
Full dynamic pressure, 548
Full factorial test matrix, 319
Fully developed flow, 13–14. *See also* One-dimensional flow
Fundamental (primary) dimensions, 15, 292

Gage pressure (P_{gage}), 76–77, 82–83
Gas dynamics, 2
Gas turbines, 833, 847
Gaseous cavitation, 62
Gases, 3–4, 53–54, 99, 940, 948
  See also Air; Ideal gases
Gate valves, 374, 380
Gates, 761–770, 930. *See also* Flow control
Gaukler-Manning equations, 738–739
Gauss' (divergence) theorem, 439–440, 443–444, 459–460
General Conference of Weights and Measures (CGMP), 16
Generator efficiency ($\eta_{generator}$), 196
Geometric similarity, 299–300
Geometric similarity, 32
Global properties of flow, 880, 882
Globe valve, 380
Gradient (del) operator, 137
Gradually varied flow (GVF), 727, 747–756
Grashof number (Gr), 309
Gravity, 18, 39, 41, 102–103, 247, 357–358, 461, 519, 526–527
  buoyancy and, 102–103
  Cauchy's equation and, 461
  center of, 102
  control volume, acting on, 247
  force of, 18, 247
  hydrostatic pressure and, 519
  inviscid flow regions and, 526–527
  laminar flow, effects of on, 357–358
  metacentric height, 102–103
  specific (SG), 39, 41
  stability and, 102–103

Grids, 881–883, 883–888
  equiangle skewness, 885
  face (3-D flow), 881
  generation, 883–888
  hexahedral cells, 887–888
  hybrid, 886–887
  independence, 888
  intervals, 884
  multiblock analysis using, 886–887
  nodes, 884
  polyhedral meshes, 887–888
  prism cells, 887–888
  skewness, 885
  structured, 884, 886
  tetrahedral cells, 887–888
  three-dimensional (3-D), 881, 886–887
  two-dimensional (2-D), 881, 885–887
  unstructured, 884–885
Gross head (H_{gross}), 839

Harmonic functions, 535–537
Head gate, 840
Head loss (h_L), 220–221, 348, 356–357, 369–370, 375–376, 736–737, 758–759
  head loss (h_L), 736–737, 758–759
  friction slope (S_f) for, 737
  internal flow, 348, 356–357, 369–370, 375–376
  irreversible, 220–221
  laminar flow, 356–357
  minor losses, 375–376
  open-channel flow, 736–737, 758–759
  steady flow analysis and, 220–221
  total, 375–376
  turbulent flow, 369–370
Heads (h), 79, 205–207, 221, 383–384, 790, 791–800, 839–842
  Bernoulli, 790
  cavitation and, 797–800
  elevation, 205
  energy grade line and, 205–207, 790, 840–841
  gross for, 839, 841
  hydraulic grade line, 205–207
  net (total), 205, 790–795, 841–842
  net positive suction, 798–800
  piping system efficiency and, 383–384, 791–797
  pressure, 79, 205
  pump, 221, 384, 792–793
  pump performance and, 790–797
  required net, 792–793
  shutoff, 384, 791–792
  turbine, 221, 383–384
  turbine performance and, 839–842
  useful pump, 792
  velocity, 205
Heat transfer (Q), 43, 205, 215, 693–702, 915–922
  CFD calculations for, 915–922
  compressible duct flow, 693–702
  cross-flow heat exchanger, 915–917
  energy (E) transfer and, 43, 215
  integrated circuit (IC) chips, cooling of, 917–922
  negligible effects of, 205
  printed circuit boards (PCB) and, 917–922
  rate, 215, 919
  Rayleigh flow, 693–702

reattachment line, 921–922
separation bubbles, 921–922
temperature contour plots, 916–917
thermal energy compared to, 43, 215
Homologous operating points, 826
Horizontal axis wind turbines (HAWTs), 847–850
Horsepower (hp), unit of, 18–19
Hot film anemometers, 403
Hot wire anemometers, 403–404
Hydraulic cross sections, 728–729
Hydraulic depth (y_h), 732–733
Hydraulic diameter (D_h), 350, 728
Hydraulic grade line (HGL), 205–207
Hydraulic jump, 731–733, 757–761, 930
Hydraulic radius (R_h), 728–729
Hydraulic turbines, 833
Hydraulics, study of, 2
Hydrodynamic entrance region, 351–352
Hydrodynamic entry length (L_h), 351–353
Hydrodynamics, 2
Hydrogen bubble wire, 147
Hydrostatics, 83–84, 89–97, 464, 479, 519–520
  absolute pressure, 76–77, 90
  area, moments of, 90–91
  atmospheric pressure, 89–90
  center of pressure (point of application), 89–91
  centroid, 91–92
  curved surfaces, 95–97
  inertia, moments of, 91–92
  forces, 89–97
  gravity effects on, 519
  magnitude, 90–91, 96
  modified pressure and, 520
  multilayered fluid and, 96
  nondimensionalized equations and, 519–520
  rectangular plates, 92–94
  plane surfaces, 89–92
  pressure, 83–84, 464
  pressure distribution, 479, 519
  pressure prism, 91–92
  resultant forces (F_R), 90–91, 95–96
  submerged surfaces, 89–97
Hypersonic flow, 11, 50

Ideal gases, 40–41, 44–46, 661, 663–665, 667–669, 693–711, 940, 953–956
  compressibility of (κ), 45–46
  compressible flow of, 45–46, 661, 663–665, 667–669, 693–702, 953–956
  converging-diverging duct flow of, 663–665
  critical property values (*), 668–669
  density of, 40–41
  duct flow, 693–702
  energy of, 44
  energy equation for, 694, 703
  enthalpy of, 661
  equation of state, 40, 695, 704
  Fanno flow, 702–711, 956
  fluid properties of, 44–46
  isentropic flow, 663–665, 667–669, 953
  Mach number (Ma) for, 663–665, 667–669
  property relations, 699–700, 705–708
  property tables for, 940, 953–956
  flow property variations, 667–669
  Rayleigh flow, 693–702, 955
  shock functions for, 954
  specific heat of, 44, 667–669, 940
  speed of sound (c) for, 663
  volume expansion (β) of, 46

INDEX

Image sink, 551
Immersed bodies, buoyancy of, 101–103
Impact pressure, 548
Impellers (rotor), 806–808, 811–812, 819–822
  axial pumps, 819–822
  backward-inclined blades, 807–808
  centrifugal pumps, 806–807, 811–812
  forward-inclined blades, 808
  radial (straight) blades, 808
  shroud, 807
  stator blades (vanes) for, 819–822
Impulse turbines, 835–837
Impurities, 57
In series pipes, 381–382
Inclined manometer, 85
Inclined pipe flow, 357–358
Incomplete similarity, 320–323
Incompressible flow, 10–11, 44, 192–194, 201, 205, 221–222, 443–445, 466–468, 610–613, 634–643. *See also* External flow
  Bernoulli equation for, 201–202, 205, 221
  compressible flow compared to, 10–11
  conservation of mass and, 192–194, 443–445
  continuity equations for, 443–445
  determination of, 44
  differential analysis of, 443–445, 466–468
  drag force and, 610–617, 638–639
  energy analysis of, 221–222
  external, 610–657
  isothermal, 466–468
  kinetic energy correction factor (α), 221–222
  lift force and, 610–613, 634–643
  mass balance for, 192–194
  material elements and, 443–444
  Navier–Stokes equation for, 466–468
  separation, 630–631
  specific heat of, 44
  steady, 201–202, 221
  streamlined bodies, 610, 614–616
  surface roughness effects, 626–628, 632–633
  three-dimensional, 610, 611
  two-dimensional, 608, 610, 611–612
Independent Π (Pi), 300–301
Induced drag, 638–639
Inertia, moments of, 91–92
Inertial sublayer, 364
Infinitesimal control volume, 440–443, 460–463
Initial boundary conditions, 477
Inlet and outlet flow, 164–165, 251, 254
Inlet boundary conditions, 477
Inner flow region, 555. *See also* Boundary layers
Insect flight, dimensional analysis and, 326
Insertion electromagnetic flowmeter, 401
Inspectional analysis, 294–299
Integral properties of flow, 882
Integrated circuit chips (ICs), 917–922
Intensive property of fluid flow, 38, 162
Interface boundary conditions, 476
Interferometry, 147
Intermolecular bonds (pressure), 3–4
Internal energy (U), 43
Internal flow, 10, 347–435
  Bernoulli equation for, 392–393
  biofluid mechanics, 408–416
  Colebrook equation for, 367–374
  ducts, 348, 358
  entrance region, 351–352
  entry length, 352–353
  external flow compared to, 10
  flow rate, 369–370
  friction effects on, 348–349

  gravity effects on, 357–358
  head loss (h_L), 348, 356–357, 369–370, 375–376
  laminar, 349–361
  measurement of velocity, 391–408
  minor losses, 374–381
  Moody chart for, 367–374
  pipes, 348–349, 351–390
  Hagen-Poiseuille flow, 356
  pressure drop (ΔP), 348, 355–357, 369
  pumps, 381–390
  Reynolds number (Re) for, 350–351
  transitional, 349–351
  tubes, 348
  turbulent, 349–353, 361–374
  valves and, 364–365, 367
  velocity of, 348–349, 357–358, 364–367, 391–408
International System (SI) of units, 16–19, 292
Interrogation regions, PIV, 407
Inviscid flow, 10, 199, 525–529
  approximation, 525–529
  Bernoulli equation for, 199, 526–527
  boundary layers and, 525–526
  Euler equation for, 525–526
  gravity effects on, 526–527
  Navier–Stokes equation and, 525–529
Irrotational flow, 156–157, 351, 529–554, 625
  aerodynamic drag and, 548–550
  approximation, 529–554
  axisymmetric regions, 534, 536–537
  Bernoulli equation for regions of, 531–534
  boundary layers and, 351, 625
  circular, 156–157
  circulation (vortex strength) of, 542–543, 545–546
  continuity equation for, 529–530
  core region, 351
  d'Alembert's paradox, 548–549
  doublet of, 544–545, 546–547
  doublet strength of, 544–545
  equipotential lines for, 535–536
  Euler equation for, 531
  external parallel flow, 625
  harmonic functions for, 535–537
  Laplace equation for, 530, 535–537
  Laplacian operator for, 530
  line sink of, 540–542, 545
  line source of, 540–542
  line vortex of, 542–543, 545
  momentum equation for, 531
  Navier–Stokes equations for, 529–554
  planar regions, 534–537, 538–554
  regions of potential flow, 351, 529–530
  stream functions for, 535–537, 545–547
  superposition of, 538, 545–554
  two-dimensional regions of, 534–537
  uniform stream of, 539, 546–549
  velocity components for, 537
  velocity potential function for, 529–530, 535–537
  vortex strength (circulation) of, 542–543, 545–546
  vorticity (ζ) of, 156–157
  zero pressure point of, 549–550
Isentropic flow, 663–678, 953
Isobars, 106
Isolated system, 15
Isothermal compressibility (α), 46

Jakob number (Ja), 309
Joule (J), unit of, 18, 43

Kaplan turbine, 838–839
Kármán integral equation, 585–588
Kármán vortex street, 145, 900–902
Kelvin (°K) temperature scale, 16, 40
Kilogram (kg), unit of, 16–17
Kilojoule (kj), unit of, 43
Kilopascal (kPa), unit of, 76
Kilowatt-hour (kWh), unit of, 19
Kinematic eddy (turbulent) viscosity, 363–364
Kinematic similarity, 32, 299–300
Kinematic viscosity (ν), 53
Kinematics, laws of, 32. *See also* Fluid kinematics
Kinetic energy (ke), 43, 194–195, 265
Kinetic energy correction factor (α), 221–222, 357
Knudsen number (Kn), 309

Lagrangian description, 134–140, 167–168
Laminar flow, 11, 349–361, 556–572, 578–583, 727–729, 893–902
  Blasius similarity variable (η) for, 565–567
  boundary layer approximation, 556–572, 578–583
  circular pipe, 353–357
  computational fluid dynamic (CFD) calculations, 580–583, 893–902
  cylinders, CFD calculations for flow around, 897–902
  Darcy friction factor (f), 355–356
  dividing streamline, 581–582
  entrance region, 351–352, 893–896
  entry length, 352–353
  flat plate boundary layer, 556–558
  flow separation, 580–583
  fluid behavior (flow regime) of, 349–351
  friction factor (f), 355–356, 358
  fully developed, 335–355
  head loss (h_L), 356–357
  hydraulic diameter and radius for, 728–729
  hydrodynamic entry length (L_h), 352–353
  inclined pipes, 357–358
  internal, 349–361
  kinetic energy correction factor (α) for, 357
  Navier–Stokes equation for, 559–563
  noncircular pipes, 358
  open-channel flow, 727–729
  pipe flow, CFD calculations for, 893–896
  Pouseuille's law for, 356
  pressure drop (ΔP), 355–357
  pressure gradient effects, 578–583
  pressure loss (ΔP_L), 355
  Reynolds number (Re) for, 350–351, 556, 727–729
  skin (local) friction coefficient for, 567
  transition to turbulent, 349–351, 557–558
  turbulent flow compared to, 557–559, 580–583
  velocity profile for, 351–352, 353–355
  zero pressure gradient, 561, 564–565
Laplace equation for irrotational flow, 530, 535–537
Laplacian operator (∇), 530
Large eddy simulation (LES), 903
Laser Doppler anemometry (LDA), 404
Laser Doppler velocimetry (LDV), 404–406, 410
Laval nozzles, 665, 678

INDEX

Law of the wall, 365–366, 576, 578
Leading edge angle, 809, 843
Leading edge flaps, 636
Leibniz theorem, 165–167
Lewis number (Le), 309
Lift coefficient, 309, 314–315, 612–613, 634–637
Lift force, 313–316, 610–613, 634–643
   airfoils, 634–639
   angle of attack, 315, 612, 634–637
   aspect ratio (AR) for, 639
   dimensional analysis of, 313–316
   drag and, 610–613, 638–639
   finite-span wings, 638–639
   flaps, effects from, 636–637
   frontal area, 612
   incompressible flow and, 610–613, 634–643
   induced drag and, 638–639
   lift-to-drag ratio, 635–636
   Mach number (Ma) and, 315–316
   Magnus effect, 639–643
   planform area, 313, 612, 639
   pressure effects on, 634–635
   spinning, generation of by, 639–643
   stagnation points, 635, 639
   stall conditions, 637
   starting vortex, 635
   takeoff and landing speeds, 636–637
   viscous effects on, 634–635
   wing prototype for, 313–316
Lift-to-drag ratio, 635–636
Line of action, 91
Line (contour) plot, 150–151
Line sink, 160, 540–542, 545
Line source, 540–542
Line source strength, 540–541
Line vortex, 542–543, 545
Linear momentum, 186, 199–201, 244, 249–263, 459–464
   Cauchy's equation for, 459–464
   conservation of, 199, 244
   control volume (CV), 250–251, 459–460
   differential analysis and, 459–464
   divergence theorem for, 459–460
   energy balance and, 199–200
   equation, 186, 244, 249–250
   external forces and, 249–250, 254–255
   momentum analysis and, 244, 249–263
   momentum-flux correction factor (b), 251–253
   Newton's second law as, 200, 244, 249–250
   no external forces and, 254–255
   Reynolds transport theorem for, 250
   steady flow, 251, 253–254
   thrust, 254–255
   uniform flow, 251
Linear strain rate (ϵ), 151–153
Liquids, 3–4, 53–60, 946
   properties of, 946
   viscosity of, 53–54
Local acceleration, 137
Local friction coefficient, 567
Logarithmic (log) law, 366, 576
Loss (resistance) coefficient (K_L), 374–379
Loss of energy in transfer, 216, 219–221

Mach angle (μ), 688
Mach number (Ma), 11, 50, 310, 315–316, 663–669, 670–672, 679–681, 685–688, 688–689, 691, 695–696, 704
   back pressure (P_b) and, 669–678
   compressible flow, 663–669, 670–672, 679–681, 685–688, 688–689
   converging nozzles, 665, 670–672
   converging–diverging nozzles, 665
   critical property values (*), 668–669
   dimensional analysis use of, 315–316
   estimation of, 691
   Fanno flow, 704
   hypersonic flow, 50
   ideal gases, 663–665, 667–669, 695–696, 704
   isentropic flow, 663–669, 672
   lift force analysis and, 315–316
   nozzle shapes and, 665–669
   Rayleigh flow, 695–696
   shock waves and, 679–681, 685–688
   sonic flow, 50, 666–667
   subsonic flow, 50, 666–667, 687
   supersonic flow, 50, 666–667, 687
   transonic flow, 50
   unity, 665–665, 668
Mach wave, 688
Macroscopic energy, 43
Magnetic resonance image (MRI) simulations, 929
Magnitude, 90–91, 96
Magnus effect, 639–643
Manning coefficient (n), 739
Mass (m), 21, 165, 186, 187–194, 349, 382, 393, 438–450, 670–674, 736, 790, 940
   absolute velocity for, 191
   average velocity for, 188, 349
   balance, 189, 191–194, 393
   conservation of, 186, 187–194, 349, 382, 438–450, 736
   continuity equations for, 186, 438–450
   control volume, 186, 189–191, 440–443
   differential analysis and, 438–450
   divergence (Gauss') theorem for, 439–440
   incompressible flow, 192–194
   molar (M), 940
   open-channel flow, 736
   pipe flow analysis and, 382
   principle of conservation, 189–191
   pumps, 790
   relative velocity for, 191
   steady-flow processes, 191
   Taylor series expansion for, 440–441
   time rate of change, 187
   volume flow rate and, 188–189, 790
   weight and, 21
Material acceleration, 139–140
Material derivative, 139–140, 167–168, 443–444, 463–464
   acceleration, 139–140
   Cauchy's equation and, 463–464
   continuity equation and, 443–444
   control volume and, 167–168
   differential analysis using, 443–444, 463–464
   fluid particle flow and, 139–140
   pressure, 139–140
   Reynolds transport theorem (RTM) and, 167–168
Material element, 443–444, 463–464
Material (fluid) particles, 134, 136–138
Material position vectors, 136–137
Material volume, 165–167
Mean pressure (P_m), 465
Mechanical energy (E_{mech}), 194–199, 201–202, 207, 215–228
   conversion of, 195–197
   efficiency, 195–197
   energy grade line (EGL), 207
   flow energy, 195
   flow work, 194–195, 218–219
   head loss, 220–221
   hydraulic grade line (HGL), 207
   irreversible losses, 220–221
   kinetic energy and, 194–195
   loss, 216, 219–221
   potential energy and, 194–195
   power transfer from, 215–217
   pressure forces ($W_{pressure}$) and, 216–219
   shaft work (W_{shaft}) as, 195–196, 199, 216
   steady flows, analysis of, 219–228
   transfer of, 195–197, 205–228
Mechanical pressure (P_m), 465
Mechanics, study of, 2
Megapascal (MPa), unit of, 76
Meniscus, 58
Mesh, see Grids
Metacentric height (GM), 102–103
Metals, properties of liquid, 947
Meter (m), unit of, 16–17
Microelectrokinetic actuator arrays (MEKA), 645
Micromechanical systems (MEMS), 645
Microscopic energy, 43
Minor losses, 374–381
Mixed-flow pumps, 790
Mixed-flow turbines, 838–840
Mixing length (l_m), 364
Models, 299–303. *See also* Dimensional analysis; Similarity
Modified pressure (P'), 520
Molar mass (M), 940
Mole (mol), unit of, 16
Moment forces (F), 264–265
Moments, 90–92, 102
   area, 90–91
   inertia, 91–92
   restoring, 102
Momentum, 186, 244, 531, 571–572, 583–591, 694, 703–705
   approximation solution using, 531
   conservation of, 186, 244
   equation, 531, 694, 703–705
   Fanno flow, 703–705
   integral technique, 583–591
   irrotational flow, 531
   Rayleigh flow, 694
   thickness (θ), 571–572
Momentum analysis, 243–289
   angular momentum and, 244–245, 263–273
   atmospheric pressure (P_{atm}) and, 249
   body forces, 246–248
   conservation of momentum, 244–245
   control volume (CV), forces acting on, 246–249
   gravity and, 247
   linear momentum and, 244, 249–263
   net forces and, 244–245
   Newton's laws and, 244–245
   Reynolds transport theorem (RTM) for, 250
   right-hand rule for, 266
   rotational motion and, 244–245, 247–248, 263–265
   surface forces, 246–248
   tensors for, 247–248
   thrust, 254–255
   torque, 263–264
   vortex shedding, 273–274
Momentum-flux correction factor (β), 251–253
Moody chart, 367–374, 952

INDEX

Motion, 103–110, 134–140, 151–168, 199–200, 243–289, 464–465, 880–883. *See also* Continuity equation; Navier–Stokes equation
  acceleration (*a*), 106–107, 135, 136–139, 199–200, 265
  angular momentum, 244–245, 263–273
  angular velocity, 151–152
  Bernoulli's equation and, 199–200
  circular, 159–160
  computational fluid dynamics (CFD) solutions, 880–883
  constitutive equations for, 464–465
  deformation rates of flow, 151–156
  equations of, 105–106, 108
  Eulerian description, 134–140, 167–168
  field variables, 134–136
  fluid flow, 149–160
  Lagrangian description, 134–140, 167–168
  linear momentum, 199–200, 244, 249–263
  material (substantial) derivative, 139–140
  momentum analysis, 243–289
  Newton's laws of, 103, 136, 244–245
  Reynolds transport theorem (RTM), 160–168
  rigid bodies, fluids in, 103–110
  rotation (ω), 107–110, 151–152, 156–159, 244–245, 247–248, 263–265
  thrust, 254–255
  torque, 263–264
  translation (velocity vector), 134, 151
Motor efficiency (η_{motor}), 196
Moving belt, 302
Multilayered fluid, hydrostatics and, 96
Multistage axial-flow turbomachines, 822–823, 847
Mutual orthogonality, 535–536

National Advisory Committee for Aeronautics (NACA) standards, 637–638
Natural (unforced) flow, 11
Navier–Stokes equation, 464–469, 475–493, 515–606, 903
  approximate solutions of, 515–606
  boundary conditions for, 475–477
  boundary layer approximation, 525–526, 560–563
  Cartesian coordinates of, 468
  exact and approximate solution comparison, 516–517
  computational fluid dynamics (CFD) solution to, 472–473, 580–583
  constitutive equations for, 464–465
  continuity equation coupled with, 475–493
  creeping flow, 520–525, 588–591
  cylindrical coordinates of, 469
  differential analysis using, 475–493
  Euler equation for, 525–526, 531, 580–581
  Euler number for, 519
  exact solutions of, 475–493
  Froude number for, 519
  incompressible, isothermal flow, 466–468
  inviscid flow regions, 525–529
  irrotational flow, 529–554
  Laplacian operator for, 530
  modified pressure and, 520
  Newtonian versus non-Newtonian fluids, 465–466
  nondimensionalization of equations for, 517–520

Reynolds-averaged Navier-Stokes (RANS) equation, 903
Reynolds number (Re) for, 519, 520–522, 524
  scaling parameters for, 517–518
  Strouhal number (St) for, 519
  velocity components for, 467–468
  viscous (deviatoric) stress tensor for, 465–467
Net (total) head (*H*), 205, 790–795, 841–842
Net positive suction head (NPSH), 798–800
Neutrally stable, 101–102
Newton (N), unit of, 16–17
Newtonian fluids, 52, 465–466
Newton's laws of motion, 17, 78, 103, 136, 200–202, 244–245, 249–250, 463–464
Non-Newtonian fluids, 465–466
Noncircular pipe flow, 348
Nondimensional equations, 294–299, 517–520
  approximate solution using, 517–520
  Buckingham Pi theorem, 303–319
  computational fluid dynamics (CFD) prediction for, 318–319
  continuity, 517–518
  Euler number (Eu) for, 519
  Froude number (Fr), 296–299, 519
  generation of, 303–319
  independent Π (Pi), 300–301
  motion, 517–520
  Navier–Stokes, 519–520
  parameters for, 294–319, 518–519
  persons honored by, 311
  Pi (Π) grouping parameters, 300–319, 516–517
  process for (nondimensionalization), 294–299, 517–520
  ratios of significance, 309–310
  repeating parameters, 304–306, 312, 314
  Reynolds number (Re), 301, 315–318
  scaling parameters, 295–296, 517–518
  similarity of model and prototype using, 300–303
  Strouhal number (St) for, 519
Nonfixed control volume, 164
Normal acceleration (a_n), 199
Normal depth (y_n), 727, 738
Normal shock waves, 675–676, 678–684
Normal stress (σ), 3, 247
Normalized equations, 294
No-slip boundary condition, 8–9, 51, 476, 498–500, 525–526, 555
No-temperature-jump condition, 9
Nozzle meter, 393–394
Nozzles, 379, 665–678, 837–838, 923–927
Nusselt number (Nu), 310
Nutating discs, 834
Nutating disk flowmeters, 396

Oblique shock waves, 676, 684–688, 691–692, 927–928
  back pressure (P_b) and, 676
  bow wave, 686–687, 928
  calculations for, 691–692
  CFD calculations for, 927–928
  detached, 686–687, 928
  isentropic flow and, 676, 688
  Mach angle, 688
  Mach number for, 685–688
  Mach wave, 688
  strong, 687
  turning (deflective) angle, 684–685
  wedges, over a, 927–928

Obstruction flowmeters, 392–396
Ohnesorge number (Oh), 593
One-dimensional (fully developed) flow, 13–14, 351–352, 353–355, 663–669, 953–954
  compressible, 663–669, 953–954
  entrance region for, 13, 351–352
  flow functions, 953
  fluid velocity variation, 665–667
  hydrodynamically fully developed region, 351–352
  ideal gases, 663–665, 667–669, 953–954
  isentropic, 663–669, 953
  laminar, 353–355
  property functions for, 953–954
  radial direction of, 13–14
  shock functions, 954
  velocity profile for, 13–14, 351–352, 353–355
One-dimensional variables, 726
One-seventh-power law velocity profile, 366, 573–574
Open pumps and turbines, 790
Open system, *see* Control volume (CV)
Open-channel flow, 10, 725–785, 928–930
  bridge scour, 771
  bumps, 764–765, 929
  CFD calculations for, 928–930
  channels, 727–729, 737–747, 771
  choked, 765
  classification of, 726–729
  conservation of mass, 736
  critical depth (y_c) of, 730, 734
  critical, 729–732, 739–740
  discharge coefficient (C_d) for, 762–764
  energy equations for, 736–737
  flow control and measurement, 761–770
  flow depth, 726–727
  friction slope (S_f) of, 737
  Froude number (F_r) for, 729–730, 929–930
  gates, 761–770, 930
  gradually varied flow (GVF), 727, 747–756
  head loss (h_L), 736–737, 758–759
  hydraulic cross sections, 743–747
  hydraulic depth (y_h), 732–733
  hydraulic diameter (D_h), 728
  hydraulic jump, 731–733, 757–761, 930
  hydraulic radius (R_h) for, 728–729
  laminar, 727–728
  one-dimensional variables for, 726
  rapidly varied flow (RVF), 727, 757–761
  Reynolds number (Re) for, 727–728
  slope (*S*) of, 737, 748–752
  specific energy (E_s) of, 733–736
  steady/unsteady determination, 726
  subcritical (tranquil), 729–732
  supercritical (rapid), 729–732
  surface profiles, 749–756
  surface waves, 731–733
  turbulent, 727–728
  uniform (UF), 726–727, 737–743
  varied (nonuniform), 726–727, 740–743
  wave speed (c_0), 729–733
  weirs, 761, 766–770
Operating (duty) point, 384, 792–793, 826
Ordinary differential equations (ODE), 485
Ordinate, 149
Orifice meter, 393–394
Outer (turbulent) layer, 364–366, 626
Outer flow region, 555. *See also* Inviscid flow; Irrotational flow

INDEX

Outlet boundary conditions, 477
Overflow gates, 764–770
Overlap (transition) layer, 364–366, 626

Paddlewheel flowmeters, 397–398
Paraboloids of revolution, 108–109
Parabolic velocity profile, 351–352, 354–355
Parallel axis theorem, 91
Parallel pipe flow, 382
Parameters, *see* Nondimensional equations
Partial derivative operator (δ), 137
Partial differential equations (PDE), 485
Partial pressure, 41
Particle image velocimetry (PIV), 143, 406–408, 410, 416
Pascal (Pa), unit of, 52, 76
Pascal's law, 80–81, 85–86
Passage losses, 815
Path functions, 187–188
Pathlines, 142–144
Peclet number (Pe), 310
Pelton wheel, 835–837
Penstock, 840
Performance (characteristic) curves, 383–384, 791–797
Pi (Π) grouping parameters, 300–319, 518–519
Piezoelectric transducers, 88
Piezometer (tube), 203
Pipe flow, 348–349, 351–390, 791–797, 893–896, 952
   analysis of networks for, 381–382
   bend sections, 377–380
   CFD calculations for, 893–896
   circular, 348, 353–357, 952
   conservation of mass for, 382
   Darcy friction factor for, 355–356, 367–368
   diameter (D) problems, 369–370
   efficiency (η) of, 381–390, 791–797
   elbow sections, 377–380
   entrance region, 351–352, 893–896
   entry length, 352–353
   equivalent length, 375
   equivalent roughness and, 368
   friction effects on, 348–349, 355–356, 367–369
   friction factor for, 355–356, 358, 952
   in series, 381–382
   inclined, 357–358
   internal, 348–349, 351–390
   kinetic energy correction factor for, 357
   laminar, 351–361, 893–896
   loss (resistance) coefficient (K_L), 374–379
   minor losses, 374–381
   Moody chart for, 367–374, 952
   noncircular shapes, 348
   parallel, 382
   pump efficiency and, 383–390, 791–797
   relative roughness of, 367–369
   sizes of pipe for, 369
   sudden expansion, 379
   systems (networks), 381–390, 791–797
   turbulent, 361–374
   valves for, 374–375, 378, 380
   velocity in, 384–385
   velocity profile for, 351–352, 353–355, 364–367
   vena contracta region, 376
Pitch angle (θ), 816–819
Pitching moment, 611
Pitot formula for flow rate, 392
Pitot probes, 391–392
Pitot tube, 203
Pitot-static probes, 391–392
Planar flow, 457, 484, 534–537, 538–554
   doublet of, 544–545, 546–547
   harmonic functions for, 535–537
   irrotational regions of, 534–537, 538–554
   line sink of, 540–542, 545
   line source of, 540–542
   line vortex of, 542–543, 545
   Poiseuille flow, 484
   singular point (singularity) of, 540–541
   stream functions for, 457, 535–537, 545–547
   superposition of, 538, 545–554
   two-dimensional, 534–537
   uniform stream of, 539, 546–549
   velocity potential function for, 535–537
Plane surfaces, 89–94, 98
Planform area, 313, 612, 639
Plastic fluids, 466
Plots, 148–151, 916–917, 924–928
   abscissa, 149
   computational fluid dynamics (CFD) use of, 149–151, 916–917, 924–928
   contour, 150–151, 916–917, 924–928
   fluid flow data using, 148–151
   ordinate, 149
   pressure contour, CFD, 924–928
   profile, 149
   temperature, 916–917
   vector, 149–150
Poise, unit of, 52
Poiseuille flow, 484–490, 493–496
Poisson's equation, 473
Polar coordinates, *see* Cylindrical coordinates
Position vector, 134
Positive-displacement machines, 396, 789, 803–806, 834, 834
   closed volume for, 805–806
   flowmeters, 396
   fluid flow in, 803–806
   nutating discs, 834
   peristaltic pumps, 803–804
   pumps, 789, 803–806
   rotary pumps, 804–805
   self-priming pumps, 805
   turbines, 789, 834
   water meters, 834
Potential energy, 43, 194–195
Potential flow, *see* Irrotational flow
Potential function (ϕ), 529–530, 535–537
Pound-force (lbf), unit of, 17–18, 76
Pound-mass (lbm), unit of, 17–18
Pouseuille's law, 356
Power, 18–19, 215–217, 847–855
   available wind power ($W_{available}$) for, 850–851
   Betz limit, 853–854
   coefficient, 851, 853–854
   energy pattern factor, 851
   wind power density, 850–851
   wind turbines, 847–855
Power-law velocity profile, 366–367, 496–497
Power number (N_P), 310
Prandtl equation, 368
Prandtl number (Pr), 310
Prandtl–Meyer expansion waves, 688–693. *See also* Expanding flow
Prandtl–Meyer function, 689
Precision error, 28–29
Pressure (P), 3–4, 41–44, 75–131, 139–140, 202–204, 216–219, 249, 464–465, 470–475, 519–520, 547–550, 593, 612–617, 634–635, 661–662, 667–678, 705, 924–928
   absolute, 76–77, 90
   aerodynamic drag, 548–550
   aerostatics, 89
   aircraft wing design and, 612–613
   atmospheric, 81–83, 89–90, 249
   back, 669–678
   barometric, 81–83
   Bernoulli equation representation, 202–204
   buoyant force, 98–103
   capillary, 593
   calibration for, 88–89
   center of pressure (point of application), 89–91
   choked flow from, 670
   compressible flow, 661–662, 669–678, 705, 924–928
   compressive force as, 77–78
   constitutive equations for, 464–465
   contour plots, CFD, 924–928
   critical, 667–669
   depth, variation of with, 78–81
   differential analysis of fields, 470–475
   drag affected by, 612–617
   dynamic, 202–203, 612
   external flow and, 612–617, 634–634
   flow separation and, 616–617
   flow work (P/ρ), 43–44, 218–219
   force as, 76–81, 216–219
   fluid flow and, 3–4, 202–204
   fluid statics and, 75–131
   full dynamic, 548
   gage, 76–77, 82–83
   hydrostatic, 83–84, 464, 519–520
   hydrostatic forces and, 89–97
   ideal gases, 644–646
   irrotational flow, 547–550
   isentropic nozzle flow and, 669–678
   lift affected by, 612–613, 634–635
   Mach number (Ma) and, 669–678
   material derivative of, 139–140
   measurement of, 4, 81–89, 203–204
   mechanical, 465
   modified, 520
   Navier–Stokes equation and, 464–465, 470–475
   nondimensionalized equations and, 519–520
   normal force as, 76–77
   partial, 41
   Pascal's law for, 80–81, 85–86
   power from work of, 216–217
   prism, 91–92
   rigid-body motion of fluids and, 103–110
   saturation, 41–43
   scalar quantity of, 77–78
   shock waves from, 675–678
   stability and, 98–103
   stagnation (P_0), 203–204, 661–662, 705
   static, 202–203
   streamlining, 615–616
   superposition and, 547–550
   thermodynamic (hydrostatic), 464
   total, 203
   units of, 76–77

INDEX

Pressure (*continued*)
  vacuum, 76–77
  vapor, 41–43
  work by, 216–219
Pressure coefficient (C_P), 310, 523, 547–549
Pressure correction algorithms, CFD, 473
Pressure drop (ΔP), 348, 355–357, 369–370, 382
Pressure field (scalar), 134
Pressure gradient, 104, 561–562, 564–565, 578–583
  boundary layer approximation with, 561–562, 564–565, 578–583
  computational fluid dynamics (CFD) calculations for, 580–583
  curvature and, 561–562
  dividing streamline, 581–582
  favorable, 579
  flow separation, 578–583
  reverse flow and, 579–580, 582
  rigid-body motion, 104
  separation bubble, 579–580
  stall condition, 580
  unfavorable (adverse), 579
  velocity profile of, 580
  zero, 561, 564–565
Pressure head, 79, 205
Pressure loss (ΔP_L), 355
Pressure transducers, 88
Preswirl, 820
Primary (fundamental) dimensions, 15–16, 292
Problem-solving technique, 23–25
Profile plots, 149
Propane saturation properties, 945
Propellers (rotor), 816–819, 839–840
Property, defined, 38
Prototypes, 299–303. *See also* Dimensional analysis; Similarity
Pseudoplastic (shear thinning) fluids, 52, 466
Pulsatile pediatric ventricular device (PVAD), 410–411
Pump–motor efficiency, 383–384
Pump–turbines, 858
Pumps, 196, 221, 381–390, 788–790, 790–833
  affinity laws, 829–830
  axial, 806, 816–824
  best efficiency point (BEP), 791
  blower, 788
  brake horsepower (bhp), 791
  capacity (volume flow rate) for, 790, 805–806
  cavitation and, 797–800
  centrifugal, 806–815
  compressors, 788–789
  dimensional analysis for, 824–827
  ducted, 790
  dynamic, 790, 806–824
  efficiency, 196, 383–390, 791–797
  enclosed, 790
  energy absorbing devices, as, 788
  energy grade line (EGL), 790
  fans, 788, 816, 819–824
  flow analysis for selection of, 381–382
  flow rate curves for, 383–384
  free delivery, 384, 791
  head, 221, 383–384, 792–793
  impellers, 790, 806–808, 811–812, 819–822
  in-series systems, 381–382, 800–803
  internal flow and, 381–390
  mass flow rate for, 790
  net head for, 790–791
  net positive suction head (NPSH), 798–800
  open, 790

operating (duty) point, 384, 792–793, 826
parallel systems, 382, 800–803
performance (characteristic) curves, 383–384, 791–797
piping networks and, 383–390791–797
positive-displacement, 789, 803–806
Reynolds number (Re) for, 825–826
rotary (rotodynamic), 806
scaling laws, 824–833
selection of, 381–390, 791–797
shutoff head, 384, 791–792
specific speed, 827–829
system (demand) curve, 383–384
volume flow rate (capacity), 790
Pure constants, 295

Quasi-steady flow, 519

Radial (straight) blades, 808
Radial-flow devices, 269–270, 806–815, 838–839. *See also* Centrifugal pumps
Rake of streaklines, 145
Rankine temperature scale, 40
Rapidly varied flow, 727, 757–761
Rarified gas flow theory, 39
Rate of rotation (angular velocity), 151–152
Rated speed, 850
Rayleigh flow, 693–702, 955
  choked, 700
  continuity equation for, 694
  control volume energy equations for, 694–695
  effects of heating and cooling on, 696
  energy balance equations, 694
  entropy change of, 694–695
  equation of state for, 695
  heat transfer and, 693–702
  heating and cooling effects on, 696
  ideal gases, 693–702
  Mach number (Ma) and properties of, 695–696
  momentum equation for, 694
  property functions for, 955
  property relations for, 699–700
Rayleigh line (curve), 679–680, 695
Rayleigh number, 310
Reaction turbines, 837–846
  axial-flow, 839
  double-regulated, 839
  draft tube, 840
  efficiency of, 841–842
  Francis, 838–864
  efficiency of, 841–842
  energy grade line (EGL), 840–841
  gross head for, 839
  head gate, 840
  Kaplan, 838–839
  mixed-flow, 838–840
  net head, 841–842
  propeller, 839–840
  radial-flow, 838–839
  runner blades (rotor), 837–839, 843–846
  single-regulated, 839
  stay vanes, 837–838
  tailrace, 840
  volute, 838
  wicket gates, 837–838
Reattachment line, 921–922
Rectangular hydraulic cross sections, 745
Refractive flow visualization, 147–148
Refrigerant-134a saturation properties, 943

Regression analysis, 320
Relative density, 39
Relative roughness (ϵ/D), 367–369
Relative velocity, 163–165, 191, 245–246
Repeating variables, method of, 303–319
Required net head, ($H_{required}$), 792–793
Residence time, 138–139
Resistance (loss) coefficient (K_L), 374–379
Restoring moment, 102
Resultant forces (F_R), 90–91, 95–96, 607
Reverse flow, 579–580, 582
Reverse swirl, 843–844
Reverse thrust, 819
Reversible adiabatic flow, *see* Isentropic flow
Revolution, *see* Rotation
Revolutions per minute, 264–165
Reynolds-averaged Navier-Stokes (RANS) equation, 903
Reynolds number (Re), 11, 301, 310, 315–318, 320–326, 350–351, 519, 520–522, 524, 557–559, 617–618, 626–633, 707–708, 727–729
  aerodynamic drag coefficient of, 322–323
  boundary layer approximation, 557–559
  creeping flow approximation and, 520–522, 524
  critical, 350, 557–559
  cylinders, 629–633
  dimensional analysis use of, 301, 315–318, 320–326
  drag coefficient (C_D) and, 320–303, 617–618
  external flow, 612, 617–618, 629–631
  Fanno flow, 707–708
  flat plate analysis, 626–628
  free-surface flow, 323–325
  hydraulic diameter for, 350, 728
  hydraulic radius for, 728–729
  incomplete similarity, 320–321
  independent flow, 321–322
  insect flight and, 326
  internal flow, 350–351
  Mach number (Ma) and, 315–316
  Navier–Stokes nondimensionalization using, 519
  open-channel flow, 727–729
  ratio of significance, 310
  spheres, 629–633
  Stokes law, 618
  surface roughness and, 626–628, 632–633
  transition, 557–559
  transitional flow and, 350–351, 557–559
Reynolds (turbulent) stresses, 363
Reynolds transport theorem (RTM), 160–168, 250, 266
  angular momentum, 266
  closed/open system relationships, 160–162
  control volume (CV) approach, 160–168
  extensive flow properties, 162
  fixed control volume, 163
  inlet and outlet crossings, 164–165
  intensive flow properties, 162
  Leibniz theorem and, 165–167
  linear momentum, 250
  mass flow rate and, 165
  material derivative and, 167–168
  material volume applications, 165–167
  momentum analysis and, 250, 266
  relative velocity for, 163–165
  streamlines of flow and, 161–162
  system-to-control volume transformation, 163
  nonfixed control volume, 164
  unit outer normal for, 162–163

INDEX

Richardson number (Ri), 310
Right-hand rule, 266
Rigid-body motion, 103–110
    acceleration (a) on a straight path, 106–107
    body forces for, 104–105
    equations of motion for, 105–106, 108
    fluids at rest, 105
    fluids in, 103–110
    forced vortex motion, 107–108
    free-fall of a fluid body, 105
    isobars, 106
    Newton's second law of motion for, 103
    paraboloids of revolution, 108–109
    pressure gradient, 104
    rotation in cylindrical containers, 107–110
    surface forces for, 104–105
Rocket propulsion, 675–688. *See also* Aircraft; Shock waves
Rolling moment, 611
Rotameters (variable-area flowmeters), 398
Rotary fuel atomizers, 861
Rotary (rotodynamic) pumps, 806. *See also* Dynamic machines
Rotation (ω), 107–110, 151–152, 156–159, 244–245, 247–248, 263–265
    angular momentum and, 244–245, 263–265
    angular velocity, 151–152, 264–265
    Cartesian coordinates for, 157, 247–248
    centripetal acceleration and force, 265
    control volume (CV) surface forces, 247–248
    cylindrical containers, fluid in, 107–110
    cylindrical coordinates for, 158–159
    forced vortex motion, 107–108
    kinetic energy and, 265
    momentum analysis and, 244–245, 247–248
    paraboloids of revolution, 108–109
    rate of, 151–152
    rigid-body motion of fluids, 107–110
    shaft power and, 264–265
    tensor notation for, 247–248
    torque, 263–264
    vorticity (ζ) and, 156–159
Rotation swirl, 843–844
Rotational stability, 102
Rotational viscometer, 480–481
Runners, 790, 833, 837–839, 843–844, 847
    band, 838–839
    blades, 790, 837–839, 843–846
    buckets, 790, 847
    gas and steam turbines, 847
    leading edge (angle), 843
    reaction turbines, 837–839, 843–844
    reverse swirl, 843–844
    rotation swirl, 843–844
    trailing edge (angle), 843

Saturation, 41–43, 797–798, 942–945
    ammonia properties, 944
    cavitation and, 41–43, 797–798
    fluid properties and, 41–43
    pressure, 41–43, 797–798
    propane properties, 945
    refrigerant-134a properties, 943
    temperature, 41–42
    vapor pressure and, 41–43
    water properties, 942
Scaling laws, 824–833, 855–860
    affinity laws, 829–830
    dimensional analysis for, 824–827, 855–857
    homologous operating points, 826
    performance (characteristic) curves for, 826
    pumps, 824–833
    Reynolds number (Re) and, 825–826
    similarity ratios and, 825–826, 829–830
    specific speed (N_{Sp} or N_{St}), 827–829, 857–860
    turbines, 855–860
    turbomachinery, 824–833, 855–860
Scaling parameters, 295–296, 517–518
Schlieren images, 678, 684
Schlieren technique, 147–148
Schmidt number (Sc), 310
Scroll (diffuser), 806–807
Second (s), unit of, 16–17
Secondary (derived) dimensions, 15
Seeding particles (seeds), 4016
Separation bubble, 579–580, 921–922
Shadowgraph technique, 147–148
Shaft work, 195–196, 199, 204–205, 216, 264–265. *See also* Mechanical energy
    efficiency of, 195–196
    energy transfer (W_{shaft}) by, 216
    mechanical energy as, 195–196, 199
Shape factor (H), 585
Sharp-crested weir, 767–770
Shear force (F), 52
Shear strain (ϵ), 2, 151, 153–154
Shear stress (τ), 2–3, 52, 247, 363–364
Shear thinning and thickening fluids, 466
Sherwood number (Sh), 310
Shock (wave) angle (δ), 684–685
Shock waves, 32, 445, 675–688, 712, 927–928, 954
    aircraft and rocket propulsion, 675–688
    axisymmetric flow, 687–688
    back pressure (P_b) and, 675–678
    boundary layer interactions, 712
    bow wave, 686–687, 928
    CFD calculations for, 927–928
    compressible flow and, 675–678, 712, 927–928
    continuity equation for, 445
    converging–diverging nozzles and, 675–678
    detached oblique, 686–687, 928
    Fanno line (curve), 679–682
    ideal gas, 954
    incompressible flow and, 445
    isentropic flow, 675–678
    Mach angle (μ), 688
    Mach number (Ma) for, 679–681, 685–688
    normal, 675–676, 678–684
    oblique, 676, 684–688, 927–928
    property functions for, 954
    Rayleigh line (curve), 679–680
    Schlieren images, 678, 684
    shock (wave) angle (δ), 684–685
    turning (deflective) angle, 684–685
    viscous-inviscid interactions, 712
    vorticity, 32
Shockless entry condition, 809
Shutoff head, 384, 791–792
Significant digits, 28–31
Similarity, 299–303, 315–316, 320–323, 491, 565–567, 825–826, 829–830
    characteristics of, 299–303
    dimensional analysis for, 299–303, 315–316, 318–319
    dynamic, 300–301, 315–316, 318–319
    geometric, 299–300
    incomplete, 320–323
    kinematic, 299–300
    models and prototypes, 299–303, 320–323
    pumps, 825–826, 829–830
    scaling laws, 825–826, 829–830
    ratios and, 825–826, 829–830
    solution, 491
    variable (η), 565–567
Single-regulated turbines, 839
Single-stream devices, 219
Single-stream systems, 254
Singular point (singularity), 540–541
Skin friction coefficient, 567, 585
Skin friction (surface) drag, 9
Slope (S), 498, 727, 737–740, 748–754
    adverse, 750–751, 753
    classification of, 750–751, 756
    critical (S_c), 739–740
    friction (S_f), 737, 751
    gradually varied flow (GVF), 748–749
    length, 498
    mild, 750–751, 753
    open-channel flow, 727, 737–740, 748–752
    steep, 750–751, 753
    surface profiles and, 749–752
    transition connections, 752–754
    uniform flow, 727, 738–739, 753
Sluice gate, 762, 930
Smoke wire, 144–145
Software packages for engineering, 25–27
Solid phase, 3–4
Sonar dome, 62
Sonic flow, 11, 50, 666–667
Sonic speed, *see* Speed of sound
Sonoluminescence, 62
Sound navigation and ranging (sonar), 62
Specific energy (E), 733–735
Specific gravity (SG), 39, 41
Specific heat, 43–44, 310, 667–669, 940
    energy and, 43–44
    ideal gases, 940
    isentropic gas flow and, 667–669
    Mach number (Ma) relationships, 667–669
Specific properties, 38
Specific Reynolds stress tensor, 903–904
Specific speed (N_{Sp} or N_{St}), 827–829, 857–860
Specific total energy (e), 38, 44
Specific volume (v), 39
Specific weight (γ_s), 17, 39
Speed of sound (c), 11, 48–50, 315–316, 663–665
    analysis of, 48–50
    compressible flow and, 11, 48–50, 663–665
    flow direction variations, 665
    fluid flow regimes, 11, 48–50
    ideal gas, 663
    isentropic flow, 663–665
    lift on a wing and, 315–316
    Mach number (Ma) for, 11, 50, 663–665
Speeds for takeoff and landing, 636–637
Spheres, 523–525, 629–633
    creeping flow approximation, 525–525
    drag force on, 523–525, 629–633
    external diameter (D), 629
    external flow over, 629–633
    flow separation, 631
    Reynolds number (Re) for, 524, 629–633
    stagnation points, 629–631
    surface roughness effects, 632–633
Spinning, generation of by lift, 639–643
Stability, 101–103
    buoyancy and, 101–103
    center of gravity and, 102
    metacentric height, 102–103
    rotational, 102

INDEX

Stagnation, 203–204, 629–631, 635, 639, 660–663, 667–669, 704–705
  enthalpy, 660–662
  external flow and, 629–631, 635, 639
  friction effects on, 704–705
  isentropic ideal gas flow properties, 667–669
  isentropic state, 661
  lift force and, 635, 639
  Mach number (Ma) relationships, 667–669
  points, 204, 629–631, 635, 639
  pressure, 203–204, 661–662, 705
  properties of, 660–663, 705
  specific heat, 667–669
  spheres, flow over and, 629–631
  streamline, 204
  temperature, 661–662, 705
Stall conditions, 580, 617, 637
Stanton number (St), 310
Starting vortex, 635
State postulate, 38
Static enthalpy (h), 660
Static pressure, 202–203
Static pressure tap, 203, 547
Statics, study of, 2
Stator blades (vanes), 819–822, 907–915
Stay vanes, 837–838
Steady flow, 12–13, 191, 199–202, 204, 219–221, 251, 253–254, 267–268, 444–445, 470–490, 726
  angular momentum of, 267–268
  Bernoulli equation for, 199–202, 204
  compressible, 200–202, 219–221, 444–445
  continuity equation for, 444–445
  devices, 12
  differential analysis of, 470–490
  energy analysis of, 219–221
  energy losses in, 219–221
  ideal (no mechanical loss), 219
  incompressible, 201–202, 221
  inlet/outlet crossings, 251, 254
  irreversible head loss, 220–221
  linear momentum of, 251, 253–254
  mass balance for, 191
  mechanical energy balance, 220–221
  open-channel flow, 726
  real (mechanical loss), 219
  single-stream devices, 219
  unsteady flow compared to, 12–13
Steam turbines, 833, 847
Stoke, unit of, 53
Stokes flow, 520. *See also* Creeping flow
Stokes law, 618
Stokes number (Stk or St), 310
Straight path, fluid acceleration on, 106–107
Strain (ϵ), 2, 151–154
Strain-gage pressure transducers, 88
Streaklines, 144–146
Stream functions (ψ), 450–459, 535–537, 545–547
  axisymmetric flow, 457, 536–537
  Cartesian coordinates, 450–456
  compressible ($\psi\rho$), 458–459
  cylindrical coordinates, 457–458
  differential analysis and, 450–459
  irrotational flow, 535–537, 545–547
  mutual orthogonality of, 535–536
  planar flow, 457, 535–537, 545–547
  streamlines and, 451–456
  velocity potential function and, 535–536
Streamlined bodies, 610, 614–616

Streamlines, 141–142, 161–162, 199–200, 202–205, 451–456, 552, 615–616
  Bernoulli's equation and, 199–200, 202–205
  dividing, 552
  drag reduction by, 615–616
  flow visualization using, 141–142
  force balance across, 202
  nondimensional, 552
  particle acceleration and, 199–200
  Reynolds transport theorem (RTT) and, 161–162
  stagnation, 204
  stream functions and, 451–456
  superposition of irrotational flow and, 552
Streamtubes, 142
Streamwise acceleration (a_s), 199
Stress (σ), 2–3, 52, 247–249, 363–364, 459–460, 465–467
  control volume (CV) and, 247–249
  differential analysis and, 459–460, 465–467
  flow and, 2–3
  normal (σ), 3, 247
  Reynolds (turbulent), 363–364
  shear (τ), 2–3, 52, 247, 363–364
  tensor (σ), 247–249, 459–460, 465
  turbulent shear (τ_{turb}), 363–364
  viscous (deviatoric) tensor, 465–467
  yield, 466
Strouhal number (St or Sr), 273, 310, 402, 519, 902
  Kármán vortex street, 902
  Navier–Stokes nondimensionalization using, 519
  ratio of significance, 310
  vortex flowmeters use of, 402
  vortex shedding use of, 273
Subcritical (tranquil) flow, 729–732
Submerged bodies, 89–101
  buoyancy of, 98–101
  curved surfaces, 95–97
  hydrostatics of, 89–97
  plane surfaces, 89–94, 98
Subsonic flow, 11, 50, 666–667, 687
  downstream, 687
  Mach number (Ma) for, 11, 50, 666–667
Substantial derivative, *see* Material derivative
Sudden expansion, 379
Supercritical (rapid) flow, 729–732
Superposition, 538, 545–554, 623–625, 740
  aerodynamic drag and, 548–550
  approximation solutions using, 538, 545–554
  d'Alembert's paradox, 548–549
  dividing streamline for, 552
  drag coefficient (C_D), 623–625
  flow over a circular cylinder, 546–554
  irrotational flow, 538, 545–554
  line sink and line vortex, 545
  method of images for, 551
  open-channel flow, 740
  planar flow, 538, 545–554
  pressure coefficient (C_p) for, 547–549
  two-dimensional flow, 538
  uniform flow with nonuniform parameters, 740
  uniform stream and doublet, 546–554
  velocity of composite flow field from, 538
Supersonic flow, 11, 50, 666–667, 674, 678–681, 687

  downstream, 687
  Mach number (Ma) for, 11, 50, 666–667
  shock waves from, 674, 678–681
Surface (skin friction) drag, 9, 567, 612, 614
Surface forces, 104–105, 246–248
Surface oil visualization, 148
Surface profiles, 749–756
Surface roughness, 612, 614–615, 626–628, 632–633
Surface tension (σ_s), 55–60
  capillary effect and, 58–60
  contact angle (ϕ), 58
  fluid properties of, 55–58
  impurities and, 57
  meniscus, 58
  work by expansion ($W_{expansion}$), 57–58
  visualization of, 56
Surface waves, open-channel flow, 731–733
Surfactants, 57
Surge tower, 45
Swirl, 817–818, 820, 843–844
Symmetry boundary conditions, 477, 891–892
System (demand) curve 383–384
Systematic error, 28
Systems, 14–21, 160–168, 243–289, 293. *See also* Pipe flow
  boundary conditions for, 14–15
  closed, 14–15
  closed and open relationships, 160–162
  control mass, 15
  control volume (CV), 15, 160–165
  dimensions of, 15, 19–20
  isolated, 15
  Leibniz theorem for, 165–167
  material derivative and, 167–168
  material volume, 166–168
  momentum analysis of, 243–289
  open, 15
  relative velocity of, 163–165
  Reynolds transport theorem (RTT) for, 160–168
  single-stream, 254
  surroundings of, 14
  total energy of, 293
  units of, 15–21
System-to-control volume transformation, 163

T-s diagrams, 695–696, 704–705
Tailrace, 840
Taylor series expansion, 440–441
Temperature (T), 16, 41–42, 46–47, 667, 661–662, 668–669, 705, 915–922
  absolute, 46–47
  cavitation and, 41–43
  computational fluid mechanics (CFD) for, 915–922
  contour plots, 916–917
  cooling integrated circuit chips (ICs), 917–922
  critical (T^*), 668–669
  dynamic, 667
  ideal gases, 661–662
  Kelvin (°K) scale, 16, 40
  Mach number (Ma) and, 668–669
  Rankine (R) scale, 40
  rise in cross-flow heat exchanger, 915–917
  saturation, 41–42
  stagnation, 661–662, 705
  volume expansion (β) and, 46–47

INDEX

Tensors, 154, 247–249
    contracted (inner) product of, 248
    control volume and, 247–249
    fluid element deformation and, 154
    notation, 247
    strain rate, 154
    stress, 247–249
Terminal velocity, 523–524, 613
Thermal anemometers, 402–404
Thermal energy, 43, 215
Thermodynamics, first law of, 202, 221
Thickness (δ), boundary layers, 556, 562–564, 574, 625
Three-dimensional bodies, drag coefficients (C_D) for, 611, 620–621
Three-dimensional (3-D) CFD grids, 881, 886–887
Three-dimensional flow, 13–14
Throat size of nozzles, 665, 668, 672
Thrust (force), 254–255, 819
Time rate of change, 187
Timelines, 146–147
Tip vortex, 638
Torque, 263–264. See also Angular momentum
Total derivative operator (d), 137
Total energy (E), 43–44
Total enthalpy, (h_0), 660–662
Total head (H) of flow, 205–206
Total head loss (h_L), 375–376
Total pressure, 203
Tracer particles, 143–144
Trailing edge angle, 809, 843
Trailing edge flaps, 636
Trailing vortices, 638–639
Transducers, 88
Transient flow, 12
Transit-time ultrasonic flowmeter, 399
Transition connections, 752–754
Transition Reynolds number, 557–559
Transitional flow, 11, 349–351, 557–559
Translation, rate of, 151
Transonic flow, 50
Trapezoidal hydraulic cross sections, 745–746
Trip wires, 558
Tube-axial fans, 819–824
Tubes, 348. See also Pipe flow
Tufts, 148
Turbine (propeller) flowmeters, 397
Turbines, 196, 221, 380, 788–790, 833–860
    dimensional analysis for, 855–857
    dynamic, 834–855
    efficiency ($\eta_{turbine}$), 196, 380, 815, 841–842
    enclosed, 790
    energy grade line (EGL), 840–841
    energy producing devices, as, 788
    gas, 833, 847
    gross head for, 839
    head ($h_{turbine}$), 221, 383–384
    hydraulic, 833
    impulse, 835–837
    net head for, 841–842
    open, 790
    positive-displacement, 790, 834
    power coefficient, 851, 853–854
    reaction, 837–846
    runners, 790, 833, 837–839
    scaling laws, 855–860
    specific speed, 857–860
    steam, 833, 847
    turbomachinery, as, 788–790, 843–855
    wind, 833, 847–855

Turbofan engine, 822–823
Turbomachinery, 265, 787–877
    angular momentum equation for, 265
    classifications of, 788–790
    dynamic machines, 790, 806–824, 834–855
    energy grade line (EGL), 790, 840–841
    multistage, 822–823, 847
    positive-displacement machines, 789, 803–806, 834
    pump–turbines, 858
    pumps, 788–790, 790–833
    rotary fuel atomizers, 861
    scaling laws, 824–833, 855–860
    turbines, 788–790, 833–860
Turbulence dissipation rate (ϵ), 904
Turbulence intensity (I), 904
Turbulence kinetic energy (k), 904
Turbulent flow, 11, 253, 349–353, 361–374, 557–558, 572–583, 727–728, 902–915
    boundary layer approximation, 557–558, 572–583
    buffer layer, 364–365
    Colebrook equation for, 367–368
    computational fluid dynamics (CFD) calculations, 580–583, 902–915
    cylinders, CFD calculations for flow around, 905–907
    Darcy friction factor for, 367–368
    direct numerical simulations (DNS) for, 903
    eddies, 361–364, 902–903
    eddy viscosity of, 363–364
    entrance region, 351–352
    entry length, 352–353
    flat-plate boundary layer, 572–578
    flow separation, 580–583
    flow visualization, 582
    fluctuating components of, 361–362
    fluid behavior (flow regime) of, 349–351
    free shear layers, 557
    friction velocity of, 365
    head loss (h_L), 369–370
    hydraulic diameter and radius for, 728–729
    hydrodynamic entry length (L_h), 353
    internal, 349–353, 361–374
    kinematic eddy viscosity of, 363–364
    laminar flow compared to, 557–559, 580–583
    large eddy simulation (LES) for, 903
    law of the wall, 365–366, 576, 578
    logarithmic (log) law, 366, 576
    mixing length, 364
    momentum-flux correction factor for, 253
    Moody chart for, 367–374
    Navier–Stokes equation for, 557–558, 572–578
    one-seventh-power law for, 366, 573–574
    open-channel flow, 727–728
    outer (turbulent) layer, 364–366
    overlap (transition) layer, 364–366
    pipes, 361–374
    power-law velocity profile for, 366–367
    Prandtl equation for, 368
    pressure drop (ΔP), 369
    pressure gradient effects, 578–583
    Reynolds number (Re) for, 350–351, 727–728
    Reynolds (turbulent) stresses, 363–364
    shear stress (τ_{turb}) of, 363–364
    Spaulding's law of the wall, 576, 578
    stator design, CFD model for, 907–915
    transitional flow to, 349–351, 557–558
    trip wires for transition, 558
    turbulence models for, 903–905

    vane-axial flow fan, CFD model for, 907–915
    velocity defect law, 366
    velocity profile for, 351–352, 364–367
    viscous length, 365
    viscous sublayer, 364–367
    von Kármán equation for, 368–369
Turbulent length scale (l), 904
Turning (deflective) angle (θ), 684–685
Twist on propeller blades, 816–818
Two-dimensional bodies, drag coefficients (C_D) for, 612–613, 619
Two-dimensional (2-D) CFD grids, 881, 885–887
Two-dimensional flow, 13–14, 534–537, 608, 610–613

Ultrasonic flowmeters, 399–401
Underflow gates, 761–764
Uniform flow, 12, 251, 539, 546–549, 726–727, 737–743
    Chezy coefficient for, 738–739
    critical, 736–740
    Gaukler-Manning equations for, 738–739
    irrotational flow, 539, 546–549
    linear momentum of, 251
    Manning coefficient for, 739
    normal depth for, 727, 738
    open-channel flow, 726–727, 737–743
    stream of, 539, 546–549
    superposition for nonuniform parameters of, 740
    varied (nonuniform) flow compared to, 726–727
    velocity (V_0), 738
Unit outer normal, 162–163
United States Customary System (USCS) of units, 16–19
Units of measurement, 15–21, 43, 52, 53, 76–77, 292, 827–828, 859–860
    dimensional homogeneity and, 19–20
    dimensions and, 15, 292
    energy, 43
    English system, 16–19, 292
    General Conference of Weights and Measures (CGMP), 16
    importance of, 15–19
    International System (SI), 16–19, 292
    pressure, 76–77
    specific speed, 827–828, 859–860
    United States Customary System (USCS), 16–19
    unity conversion ratios, 20–21
    viscosity (μ), 52, 53
Unity conversion ratios, 20–21
Universal (turbulent) velocity profile, 366
Universal gas constant (R_u), 40
Unstable situations, 101–102
Unsteady flow, 12–13, 202, 490–493, 726
Useful pump head, ($h_{pump, u}$), 792

Vacuum pressure, 76–77
Valves for pipe flow, 374–375, 378, 380
Vane-axial fan, 819–820, 907–915
Vanes, 819–822, 837–838
Vapor phase, 4
Vapor pressure (P_v), 41–43, 797
Vaporous cavitation, 62
Variable-area flowmeters (rotameters), 398

INDEX

Variable pitch, 819
Varied (nonuniform) flow, 726–727, 740–743
Vector identity, 526–527
Vector plots, 149–150
Vector variables, 134–135
Vehicles, drag coefficient (C_D) for, 621–623
Velocimetry, 404–408. *See also* Flow rate
Velocity (V), 51–52, 151–152, 163–165, 188, 245–246, 264–265, 348–349, 357–358, 364–367, 391–408, 523–524, 537, 576, 608, 610, 612–613, 665–667, 734
   absolute, 164, 191, 245–246
   angular (rate of rotation), 151–152, 264–265
   average, 188, 348–349
   conservation of mass and, 188, 191, 349
   control volume, 163–165, 245–246
   critical, 734
   external flow, 608, 610, 612–613
   flow rate and, 391–408
   fluid motion and, 151–152
   free-stream, 608
   friction, 365, 576
   internal flow, 348–349, 357–358, 364–367, 391–408
   isentropic flow, variations in, 665–667
   Mach number (Ma) and, 665–667
   measurement of, 391–408
   nozzle shapes and, 665–667
   relative, 163–165, 191, 245–246
   Reynolds transport theorem (RTT), 163–165
   rotation and, 151–152, 264–265
   terminal, 523–524, 613
   two-dimensional irrotational flow, components for, 537
   viscosity and, 51–52
Velocity boundary layer, 351
Velocity defect law, 366
Velocity field (vector), 134
Velocity gradient, 51–52
Velocity head, 205
Velocity overshoot, 590–591
Velocity potential function (ϕ), 529–530, 535–537
Velocity profile, 13–14, 51, 351–352, 353–355, 364–367, 573–578, 738
   boundary layer approximation, 573–578
   boundary layer development, 351–352, 364–367
   boundary layer profile comparisons, 573–578
   entrance region and, 351–352
   flat (full), 351–352, 365
   internal (pipe) flow, 351–352, 353–355, 364–367
   laminar flow, 351–352, 353–355
   logarithmic (log) law, 366, 576
   one-dimensional (fully developed) flow, 13–14, 351–352
   one-seventh-power law, 366, 573–574
   parabolic, 351–352, 354
   power-law velocity profile for, 366–367
   Spaulding's law of the wall, 365–366, 576, 578
   turbulent flow, 351–352, 364–367, 573–578
   uniform-flow, 738
   universal, 366
   velocity defect law, 366
   viscosity and, 51, 364–367

viscous length, 365
viscous sublayer, 364–367
wall-wake law, 576–577
Velocity vector (rate of translation), 134, 151
Vena contracta region, 376
Venturi meter, 393–394
Venturi nozzles, 665
Vertical axis wind turbines (VAWTs), 847, 849
Viscoelastic fluid, 465
Viscometer, 55
Viscosity (μ), 9, 11, 50–55, 204, 363–364, 480–481, 612
   apparent, 52
   coefficient of, 52–54
   differential analysis for, 480–481
   dilatant (shear thickening) fluids, 52
   drag force and, 51, 612
   dynamic (absolute), 52, 54
   eddy, 363–364
   fluid flow and, 9, 11, 52–54
   fluid properties of, 50–55
   friction force and, 10, 50–51
   gases, 53–54
   kinematic, 53
   kinematic eddy, 363–364
   liquids, 53–54
   Newtonian fluids and, 52
   negligible effects of, 204
   no-slip condition and, 9, 51
   pseudoplastic (shear thinning) fluids, 52
   Reynolds number (Re) and, 11
   rotational viscometer for, 480–481
   shear force and, 52
   turbulent pipe flow and, 363–364
   units for, 52, 53
   velocity gradient of, 51–52
   velocity profile for, 51
Viscous flow, 10, 612–613, 634–635
Viscous-inviscid interactions, 712
Viscous length, 365
Viscous (deviatoric) stress tensor, 465–467
Viscous sublayer, 364–367, 626
Volume (V), 39, 46–47, 165–167, 186, 188–189, 790, 805–806. *See also* Control volume (CV)
   capacity, 790, 805–806
   closed, 805–806
   conservation of mass and, 188–189
   expansion (β), coefficient of, 46–47
   flow rate, 188–189, 790, 805–806, 808–809
   Leibniz theorem for, 165–167
   linear momentum and, 186
   material, 166–167
   pumps, 790, 805–806, 808–809
   specific volume, 39
Volumetric (bulk) strain rate, 153
Volute, 838
von Kármán equation, 368–369
Vortex flood formation and damages, 771
Vortex flowmeters, 402
Vortex motion, forced, 107–108
Vortex shedding, 273–274, 617
Vortex strength (circulation) (Γ) of, 542–543, 545–546
Vortices, 32, 144–145, 900–902
Vorticity (ζ), 32, 156–159

Wake, flow region of, 199, 273–274
Wake function, 576–577
Wall-wake law, 576–577
Water–air interface boundary conditions, 477
Water hammer, 45
Water meters, 835
Water saturation properties, 942
Watt (W), unit of, 18–19
Wave (shock) angle (δ), 684–685
Wave speed (c_0), 729–733
Weber number (We), 310, 313, 593
Wedges, oblique shock waves over, 927–928
Weight (W), 17–18, 21, 39
Weirs, 761, 766–770. *See also* Flow control
Wicket gates, 837–838
Wind power density, 850–851
Wind tunnels, 313–316, 320–323, 823–824
   lift determination using, 313–316
   testing, 320–323
   vane-axial flow fan design for, 823–824
Wind turbines, 833, 847–855
   available wind power for, 850–851
   Betz limit, 853–854
   disk area for, 850–851
   energy pattern factor, 851
   horizontal axis, 847–850
   power coefficient, 851, 853–854
   rated speed, 850
   vertical axis, 847, 849
   wind power density, 850–851
Wings, 612–613, 617, 634–639
   endplates, 639–640
   external flow efficiency, 612–613
   finite-span, 638–639
   flaps, 636–637
   flow fields, 635
   ice and snow loads, 617
   lift force, 612–613, 634–639
   loading, 634
   shape, 612–613
   tip vortex, 638
   trailing vortices, 638–639
   winglets, 639–640
   wingspan (span), 634
Wingspan (span), 634
Work (W), 18–19, 43–44, 57–58, 194–195, 215–219
   by expansion, 57–58
   energy transfer by, 215–219
   flow, 43–44, 194–195, 218–219
   mechanical energy and, 194–195
   power as, 18–19, 215–216
   pressure force, 216–219
   shaft, 216
   surface tension and, 57–58
   units of, 18–19

Yawing moment, 611
Yield stress, 466

Zero pressure gradient, 561, 564–565
Zero pressure point, 549–550

NOMENCLATURE

a	Manning constant, $m^{1/3}$/s; height from channel bottom to bottom of sluice gate, m
$\vec{a}, a$	Acceleration and its magnitude, m/s^2
A, A_c	Area, m^2; cross-sectional area, m^2
Ar	Archimedes number
AR	Aspect ratio
b	Width or other distance, m; intensive property in RTT analysis; turbomachinery blade width, m
bhp	Brake horsepower, hp or kW
B	Center of buoyancy; extensive property in RTT analysis
Bi	Biot number
Bo	Bond number
c	Specific heat for incompressible substance, $kJ/kg \cdot K$; speed of sound, m/s; speed of light in a vacuum, m/s; chord length of an airfoil, m
c_0	Wave speed, m/s
c_p	Constant-pressure specific heat, $kJ/kg \cdot K$
c_v	Constant-volume specific heat, $kJ/kg \cdot K$
C	Dimension of the amount of light
C	Bernoulli constant, m^2/s^2 or $m/t^2 \cdot L$, depending on the form of Bernoulli equation; Chezy coefficient, $m^{1/2}$/s; circumference, m
Ca	Cavitation number
$C_D, C_{D,x}$	Drag coefficient; local drag coefficient
C_d	Discharge coefficient
$C_f, C_{f,x}$	Fanning friction factor or skin friction coefficient; local skin friction coefficient
C_H	Head coefficient
$C_L, C_{L,x}$	Lift coefficient; local lift coefficient
C_{NPSH}	Suction head coefficient
CP	Center of pressure
C_p	Pressure coefficient
C_P	Power coefficient
C_Q	Capacity coefficient
CS	Control surface
CV	Control volume
C_{wd}	Weir discharge coefficient
D or d	Diameter, m (d typically for a smaller diameter than D)
D_{AB}	Species diffusion coefficient, m^2/s
D_h	Hydraulic diameter, m
D_p	Particle diameter, m
e	Specific total energy, kJ/kg
$\vec{e}_r, \vec{e}_\theta$	Unit vector in r- and θ-direction, respectively
E	Voltage, V
$E, \dot{E}$	Total energy, kJ; and rate of energy, kJ/s
Ec	Eckert number
EGL	Energy grade line, m
E_s	Specific energy in open-channel flows, m
Eu	Euler number
f	Frequency, cycles/s; Blasius boundary layer dependent similarity variable
f, f_x	Darcy friction factor; and local Darcy friction factor
$\vec{F}, F$	Force and its magnitude, N
F_B	Magnitude of buoyancy force, N
F_D	Magnitude of drag force, N
F_f	Magnitude of drag force due to friction, N
F_L	Magnitude of lift force, N
Fo	Fourier number
Fr	Froude number
F_T	Magnitude of tension force, N
$\vec{g}, g$	Gravitational acceleration and its magnitude, m/s^2
$\dot{g}$	Heat generation rate per unit volume, W/m^3
G	Center of gravity
GM	Metacentric height, m
Gr	Grashof number
h	Specific enthalpy, kJ/kg; height, m; head, m; convective heat transfer coefficient, $W/m^2 \cdot K$
h_{fg}	Latent heat of vaporization, kJ/kg
h_L	Head loss, m
H	Boundary layer shape factor; height, m; net head of a pump or turbine, m; total energy of a liquid in open-channel flow, expressed as a head, m; weir head, m
$\vec{H}, H$	Moment of momentum and its magnitude, $N \cdot m \cdot s$
HGL	Hydraulic grade line, m
H_{gross}	Gross head acting on a turbine, m
i	Index of intervals in a CFD grid (typically in x-direction)
$\vec{i}$	Unit vector in x-direction
I	Dimension of electric current
I	Moment of inertia, $N \cdot m \cdot s^2$; current, A; turbulence intensity
I_{xx}	Second moment of inertia, m^4
j	Reduction in Buckingham Pi theorem; index of intervals in a CFD grid (typically in y-direction)

Symbol	Description
$\vec{j}$	Unit vector in y-direction
Ja	Jakob number
k	Specific heat ratio; expected number of Πs in Buckingham Pi theorem; thermal conductivity, W/m·K; turbulent kinetic energy per unit mass, m²/s²; index of intervals in a CFD grid (typically in z-direction)
$\vec{k}$	Unit vector in z-direction
ke	Specific kinetic energy, kJ/kg
K	Doublet strength, m³/s
KE	Kinetic energy, kJ
K_L	Minor loss coefficient
Kn	Knudsen number
ℓ	Length or distance, m; turbulent length scale, m
L	Dimension of length
L	Length or distance, m
Le	Lewis number
L_c	Chord length of an airfoil, m; characteristic length, m
L_h	Hydrodynamic entry length, m
L_w	Weir length, m
m	Dimension of mass
$m, \dot{m}$	Mass, kg; and mass flow rate, kg/s
M	Molar mass, kg/kmol
$\vec{M}, M$	Moment of force and its magnitude, N·m
Ma	Mach number
n	Number of parameters in Buckingham Pi theorem; Manning coefficient
$n, \dot{n}$	Number of rotations; and rate of rotation, rpm
$\vec{n}$	Unit normal vector
N	Dimension of the amount of matter
N	Number of moles, mol or kmol; number of blades in a turbomachine
N_P	Power number
NPSH	Net positive suction head, m
N_{Sp}	Pump specific speed
N_{St}	Turbine specific speed
Nu	Nusselt number
p	Wetted perimeter, m
pe	Specific potential energy, kJ/kg
P, P'	Pressure and modified pressure, N/m² or Pa
PE	Potential energy, kJ
Pe	Peclet number
P_{gage}	Gage pressure, N/m² or Pa
P_m	Mechanical pressure, N/m² or Pa
Pr	Prandtl number
P_{sat} or P_v	Saturation pressure or vapor pressure, kPa
P_{vac}	Vacuum pressure, N/m² or Pa
P_w	Weir height, m
q	Heat transfer per unit mass, kJ/kg
$\dot{q}$	Heat flux (rate of heat transfer per unit area), W/m²
$Q, \dot{Q}$	Total heat transfer, kJ; and rate of heat transfer, W or kW
Q_{EAS}	Equiangle skewness in a CFD grid
$\vec{r}; r$	Moment arm and its magnitude, m; radial coordinate, m; radius, m
R	Gas constant, kJ/kg·K; radius, m; electrical resistance, Ω
Ra	Rayleigh number
Re	Reynolds number
R_h	Hydraulic radius, m
Ri	Richardson number
R_u	Universal gas constant, kJ/kmol·K
s	Submerged distance along the plane of a plate, m; distance along a surface or streamline, m; specific entropy, kJ/kg·K; fringe spacing in LDV, m; turbomachinery blade spacing, m
S_0	Slope of the bottom of a channel in open-channel flow
Sc	Schmidt number
S_c	Critical slope in open-channel flow
S_f	Friction slope in open-channel flow
SG	Specific gravity
Sh	Sherwood number
SP	Property at a stagnation point
St	Stanton number; Strouhal number
Stk	Stokes number
t	Dimension of time
t	Time, s
T	Dimension of temperature
T	Temperature, °C or K
$\vec{T}, T$	Torque and its magnitude, N·m
u	Specific internal energy, kJ/kg; Cartesian velocity component in x-direction, m/s
u_*	Friction velocity in turbulent boundary layer, m/s
u_r	Cylindrical velocity component in r-direction, m/s
u_θ	Cylindrical velocity component in θ-direction, m/s
u_z	Cylindrical velocity component in z-direction, m/s
U	Internal energy, kJ; x-component of velocity outside a boundary layer (parallel to the wall), m/s
v	Cartesian velocity component in y-direction, m/s
$\mathcal{v}$	Specific volume, m³/kg
$V, \dot{V}$	Volume, m³; and volume flow rate, m³/s
$\vec{V}, V$	Velocity and its magnitude (speed), m/s; average velocity, m/s
V_0	Uniform-flow velocity in open-channel flow, m/s
w	Work per unit mass, kJ/kg; Cartesian velocity component in z-direction, m/s; width, m
W	Weight, N; width, m
$W, \dot{W}$	Work transfer, kJ; and rate of work (power), W or kW
We	Weber number
x	Cartesian coordinate (usually to the right), m
$\vec{x}$	Position vector, m

y	Cartesian coordinate (usually up or into the page), m; depth of liquid in open-channel flow, m	θ	Angle or angular coordinate; boundary layer momentum thickness, m; pitch angle of a turbomachinery blade; turning or deflection angle of oblique shock
y_n	Normal depth in open-channel flow, m	ρ	Density, kg/m³
z	Cartesian coordinate (usually up), m	σ	Normal stress, N/m²

Greek Letters

α	Angle; angle of attack; kinetic energy correction factor; thermal diffusivity, m²/s; isothermal compressibility, kPa⁻¹ or atm⁻¹	σ_{ij}	Stress tensor, N/m²
		σ_s	Surface tension, N/m
		τ	Shear stress, N/m²
		τ_{ij}	Viscous stress tensor (also called shear stress tensor), N/m²
$\vec{\alpha}, \alpha$	Angular acceleration and its magnitude, s⁻²	$\tau_{ij,\,turbulent}$	Specific Reynolds stress tensor, m²/s²
β	Coefficient of volume expansion, K⁻¹; momentum-flux correction factor; angle; diameter ratio in obstruction flowmeters; oblique shock angle; turbomachinery blade angle	$\vec{\omega}, \omega$	Angular velocity vector and its magnitude, rad/s; angular frequency, rad/s
		ψ	Stream function, m²/s
		$\vec{\zeta}, \zeta$	Vorticity vector and its magnitude, s⁻¹
δ	Boundary layer thickness, m; distance between streamlines, m; angle; small change in a quantity		

Subscripts

∞	Property of the far field
δ^*	Boundary layer displacement thickness, m
0	Stagnation property; property at the origin or at a reference point
ε	Mean surface roughness, m; turbulent dissipation rate, m²/s³
abs	Absolute
atm	Atmospheric
ε_{ij}	Strain rate tensor, s⁻¹
avg	Average quantity
Φ	Dissipation function, kg/m·s³
b	Property of the back or exit of a nozzle, e.g., back pressure P_b
ϕ	Angle; velocity potential function, m²/s
γ_s	Specific weight, N/m³
C	Acting at the centroid
Γ	Circulation or vortex strength, m²/s
c	Pertaining to a cross section
η	Efficiency; Blasius boundary layer independent similarity variable
cr	Critical property
		CL	Pertaining to the centerline
κ	Bulk modulus of compressibility, kPa or atm; log law constant in turbulent boundary layer	CS	Pertaining to a control surface
		CV	Pertaining to a control volume
λ	Mean free path length, m; wavelength, m; second coefficient of viscosity, kg/m·s	e	Property at an exit; extracted portion
		eff	Effective property
μ	Viscosity, kg/m·s; Mach angle	f	Property of a fluid, usually of a liquid
ν	Kinematic viscosity m²/s	H	Acting horizontally
ν(Ma)	Prandtl–Meyer function for expansion waves, degrees or rad	lam	Property of a laminar flow
		L	Portion lost by irreversibilities
Π	Nondimensional parameter in dimensional analysis	m	Property of a model

max	Maximum value	P	Acting at the center of pressure
mech	Mechanical property	p	Property of a prototype; property of a particle; property of a piston
min	Minimum value	R	Resultant
n	Normal component	r	Relative (moving frame of reference)
		rec	Rectangular property
		rl	Property of the rotor leading edge
		rt	Property of the rotor trailing edge
		S	Acting on a surface
		s	Property of a solid
		sat	Saturation property; property of a satellite
		sl	Property of the stator leading edge
		st	Property of the stator trailing edge
		sub	Submerged portion
		sys	Pertaining to a system
		t	Tangential component
		tri	Triangular property
		turb	Property of a turbulent flow
		u	Useful portion
		V	Acting vertically
		v	Property of a vapor
		vac	Vacuum
		w	Property at a wall

Superscripts

¯ (overbar)	Averaged quantity
· (overdot)	Quantity per unit time; time derivative
′ (prime)	Fluctuating quantity; derivative of a variable; modified variable
*	Nondimensional property; sonic property
+	Law of the wall variable in turbulent boundary layer
→ (over arrow)	Vector quantity